22nd International Slovak-Polls Conference on Machine Modelling and Simulations (MMS 2017)

MATEC Web of Conferences Volume 157 (2018)

Sklene Teplice, Slovakia
5 – 8 September 2017

Part 1 of 2

Editors:

M. Vasko
M. Handrik
L. Jakubovicova
P. Kopas
M. Blatnicka
V. Baniari
O. Stalmach
A. Sapietova
M. Saga

ISBN: 978-1-5108-5977-7

Printed from e-media with permission by:

Curran Associates, Inc.
57 Morehouse Lane
Red Hook, NY 12571

proceedings.com

Some format issues inherent in the e-media version may also appear in this print version.

Printed by Curran Associates, Inc. (2018)

For additional information, please contact EDP Sciences – Web of Conferences at the address below.

EDP Sciences – Web of Conferences
17, Avenue du Hoggar
Parc d'Activité de Courtabœuf
BP 112
F-91944 Les Ulis Cedex A
France

Phone: +33 (0) 1 69 18 75 75
Fax: +33 (0) 1 69 28 84 91

contact-edps@webofconferences.org

Additional copies of this publication are available from:

Curran Associates, Inc.
57 Morehouse Lane
Red Hook, NY 12571 USA
Phone: 845-758-0400
Fax: 845-758-2633
Email: curran@proceedings.com
Web: www.proceedings.com

TABLE OF CONTENTS

PART 1

PART 2

Ladies and Gentlemen, dear Readers,

theoretical and experimental solutions of technical problems using advanced approaches and modern computational and measuring engineering is nowadays only way to obtain new knowledge in all field of science and technique. Machine modelling and simulations using advanced computer and experimental technologies play an important role in many areas of industrial applications.

Our scientific meetings began on a regular footing in 1996 on the basis of an agreement between Poznan University of Technology and the University of Žilina back in 1994. Under its auspices, 21 international scientific conferences have been organized alternately in Slovakia or Poland and in one case in the Czech Republic since 1996. Today we are starting the 22nd chapter of these international meetings.

Main conference organizer is Faculty of mechanical Engineering, Department of Applied Mechanics, University of Žilina, Slovakia. The University of Žilina has more than 65 years of tradition and currently has seven faculties and over 10,000 students. The University of Žilina is a modern university providing a full range of technological, economics, management and a limited range of humanistic and natural science education at under graduate, graduate and post-graduate levels.

This scientific publication presents results of the chosen international research centres and universities, mainly from Slovakia and Poland. The publication contains eight chapters and summarized scientific activities of more than 150 authors in field of structural modelling, design, simulation and machine optimization, experimental evaluation and modelling of materials.

The scientific committee determined the following conference topics:
- Methods and systems in machine design,
- Modelling and simulation, structural optimization,
- Machine dynamics and multibody systems simulations,
- Advanced industrial, automotive and green energy applications,
- Experimental mechanics, identification and validation,
- Modelling of structural materials, composites and nanomaterials,
- Physical and chemical properties of materials,
- Theoretical and applied mathematics in engineering.

XXII. Slovak–Polish Scientific Conference

on

Machine Modelling and Simulations 2017

MMS 2017

September 5–8, 2017
Sklené Teplice, Slovak Republic

Under the auspices of the rector of the University of Žilina
Dr.h.c. Prof. Ing. Tatiana Čorejová, PhD.

Organizing Committee

Assoc. Prof. Ing. Alžbeta Sapietová, PhD. (chairman)
Assoc. Prof. Ing. Milan Vaško, PhD.
Ing. Marián Handrik, PhD.
Ing. Lenka Jakubovičová, PhD.
Ing. Peter Kopas, PhD.

Editors

Assoc. Prof. Ing. Milan Vaško, PhD.
Ing. Marián Handrik, PhD.
Ing. Lenka Jakubovičová, PhD.
Ing. Peter Kopas, PhD.
Ing. Mária Blatnická
Ing. Vladislav Baniari
Ing. Ondrej Štalmach
Assoc. Prof. Ing. Alžbeta Sapietová, PhD.
Prof. Dr. Ing. Milan Sága

Conference Secretariat

Ing. Lenka Jakubovičová, PhD.
Department of Applied Mechanics
Faculty of Mechanical Engineering
University of Žilina
Univerzitná 1
010 26 Žilina
E-mail: mms2017@mmsconf.eu
https://www.mmsconf.eu

Scientific Committee and Reviewers

Additional Scientific Reviewers

Assoc Prof. Ladislav Écsi
Prof. Jerzy Winczek
Assoc. Prof. Michał Śledziński
Assoc. Prof. Richard Pastirčák
Assoc. Prof. František Nový
Assoc. Prof. Zuzana Murčinková
Assoc. Prof. Bohuš Leitner
Assoc. Prof. Tomáš Lack
Assoc. Prof. Marta Kianicová
Asocc. Prof. Jan Krmela
Assoc. Prof. Piotr Krawiec
Assoc. Prof. Kateryna Kravchenko
Prof. Jerzy Jaroszewicz
Assoc. Prof. Juraj Grenčík
Prof. Juraj Gerlici
Assoc. Prof. Tadeusz Andrzej Złoto
MSc., PhD. Branislav Ftorek
MSc., PhD. Urszula Siedlecka

Presentation of the MMS 2017

Conference opening ceremony, speaker Prof. Dr. Ing. Milan Sága, dean of the Faculty of mechanical engineering, University of Žilina (Photo: Tomasz Geisler)

Conference opening ceremony, speaker Dr.h.c. Prof. Ing. Tatiana Čorejová, PhD., rector of the University of Žilina (Photo: Tomasz Geisler)

Conference participants (Photo: Tomasz Geisler)

Dear Colleagues,

The 22nd International Slovak – Polish Conference on **Machine modelling and Simulations 2017** continues in Slovakia and will be held from **5th to 8th** of **September 2017** in Sklené Teplice.

This international conference was established in 1996 and is held annually since then every year. Papers from last two conferences was published in Elsevier journal **Procedia engineering** registered on **WoS** and **SCOPUS**.

Scientific Committee

Prof. Milan Sága – Dean of the Faculty
Prof. Jerzy Bajkowski
Prof. Jozef Bocko
PhD. Eng. Maciej Bodnicki
Prof. Andrej Czan
Assoc. Prof. Vladimír Dekýš
PhD. Eng. Tomasz Domański
Prof. Grzegorz Domek
Prof. Marian Dudziak
Prof. Pavel Élesztös
Assoc. Prof. Robert Grega
Assoc. Prof. Radim Halama
Dr.h.c. Prof. Petr Horyl
Prof. Danuta Jasińska-Choromańska
Prof. Małgorzata Klimek
Assoc. Prof. Andrzej Kołodziej
PhD. Eng. Marcin Kubiak
Prof. Ivan Kuric
Prof. Ireneusz Malujda
Prof. Janusz Mielniczuk
Prof. Justín Murín
Assoc. Prof. Milan Naď
Prof. Darina Ondrušová
Prof. Mariana Pajtášová
Prof. Františka Pešlová
Assoc. Prof. Wiesława Piekarska
Prof. Bogdan Posiadała
Assoc. Prof. Jerzy Pojek
Assoc. Prof. Alžbeta Sapietová
Prof. Norbert Sczygiol
Assoc. Prof. Jozef Soukup
PhD. Eng. Ksawery Szykiedans
PhD. Eng. Krzysztof Talaśka
Prof. Eva Tillova
Assoc. Prof. Stefan Tomaszek
Dr.h.c. mult. Prof. František Trebuňa
Assoc. Prof. Milan Vaško
Prof. Ján Vavro
Assoc. Prof. Ján Vavro, jr.
Assoc. Prof. Jan Vimmr
Prof. Robert Zalewski
Prof. Jaroslav Zapoměl
Prof. Milan Žmindák

Organizing Committee

Assoc. Prof. Alžbeta Sapietová, PhD. - chairman
Assoc. Prof. Milan Vaško, PhD.
Ing. Lenka Jakubovičová, PhD.
Ing. Peter Kopas, PhD.
Ing. Marián Handrik, PhD.

Main Topics

- Methods and systems in machine design.
- Modelling and simulation, structural optimization.
- Machine dynamics and multibody systems simulations.
- Advanced industrial, automotive and green energy applications.
- Experimental mechanics, identification and validation.
- Modelling of structural materials, composites and nanomaterials.
- Physical and chemical properties of materials.
- Theoretical and applied mathematics in engineering.

Conference Secretariat

Ing. Lenka Jakubovičová, PhD.
Department of Applied Mechanics
Faculty of Mechanical Engineering
University of Žilina
Univerzitná 1
Žilina 010 26
E-mail: mms@uniza.sk
https://mms.uniza.sk

Conference fees

Participant **260 €** (VAT included)
PhD. Student **210 €** (VAT included)
The conference fee for participant and PhD. Students include:

- Book of abstracts and journal publications.
- Participation in all scientific sessions.
- Coffee breaks and social evening.
- Organizing expenses.

Accompany person **60 €** (VAT included)
The conference fee for accompany person include:

- Coffee breaks and social evening.

Accommodation and food are payed by cash or by credit card at venue of a conference (Visa Card or Master Card) 141 € (VAT included)

Payment

The conference fee payment should be made before 19th of June, 2017 via the bank transfer to the account.
Payment information:
Bank name: Štátna pokladnica
BIC-SWIFT: SPSRSKBA
IBAN: SK34 8180 0000 0070 0026 9861
Details of Payment:
Name_Surname_MMS2017/102240

Conference Contributions

Conference contributions will include invited lectures (30 min) and contributed papers (10 min).

Publication of the papers

Full papers in English will be published in "**MATEC Web of Conferences**", www.matec-conferences.org (indexed in **WoS**, **SCOPUS** ...). Only paper with active presentation will be published!

Main Topics

The conference languages are English, Slovak and Polish with no simultaneous translation service.

Statement of Peer review

In submitting conference proceedings to MATEC Web of Conferences, I certify to the Publisher that I adhere to the **Policy on Publishing Integrity** of the journal in order to safeguard good scientific practice in publishing.

1. All articles have been subjected to peer review administered by the proceedings editors.
2. Reviews have been conducted by expert referees, who have been requested to provide unbiased and constructive comments aimed, whenever possible, at improving the work.
3. Proceedings editors have taken all reasonable steps to ensure the quality of the materials they publish and their decision to accept or reject a paper for publication has been based only on the merits of the work and the relevance to the journal.

Title, date and place of the conference

Machine Modelling and Simulations 2017 (MMS 2017)
September 5–8, 2017
Sklené Teplice, Slovak Republic

Proceedings editor(s):

Assoc. Prof. Ing. Milan Vaško, PhD.
Ing. Marián Handrik, PhD.
Ing. Lenka Jakubovičová, PhD.
Ing. Peter Kopas, PhD.
Ing. Mária Blatnická
Ing. Vladislav Baniari
Ing. Ondrej Štalmach
Assoc. Prof. Ing. Alžbeta Sapietová, PhD.
Prof. Dr. Ing. Milan Sága

Date and editor's signature

23.2.2018

Design of pallet, frame and chain of soldering station conveyor

Mária Blatnická[1,*], *Milan* Sága[1], *Miroslav* Blatnický[2]

[1]Department of Applied Mechanics, Faculty of Mechanical Engineering, University of Žilina, Univerzitná 1, 010 26 Žilina, Slovak Republic
[2]Department of Transport and Handling Machines, Faculty of Mechanical Engineering, University of Žilina, Univerzitná 1, 010 26 Žilina, Slovak Republic

Abstract. The paper deals with constructional procedure and design of the soldering station for compressor outlet piping. The design of the mechanism is based on the need to create a device that will be able to move the pallet with the compressor between individual operational positions of the entire conveyor line.

Keywords: Chain conveyor, constructional design, 3D modelling, soldering station

1 Introduction

The successful development of mankind to the state we see today has its origin in distant past when man had to fight for survival in the original living conditions. Repeating successful activities that have brought about improved living conditions has also triggered brain activity, i.e. from simple logical operations to memorization and analysis. The result of this process was the creations of man, whose improvement led to progress consisting of the construction of machines and heavy machinery. This process of developing technical systems that relieve man from repetitive physical work is called mechanization, and there is no need to talk too much about the importance of machinery for manipulating material. Mechanisation is an important means of increasing productivity, quality and competitiveness of production. The effort is to make the individual working operations as short and simple as possible and to make them easy to learn while demanding a minimum of human effort. Mechanisation greatly relieves man from heavy physical work, for example, in dangerous or harmful environments [1-3].

Today's advanced production is characterized by automation. Automation is a process in which the managerial function of a person is replaced by the activity of various machines and devices [4, 5]. Automation is a very complex process including very simple control operations performed on a relatively simple devices, as well as very complex management of large production units. Controlling is a purposeful activity that evaluates and processes information about a controlled object or process, activities in that process (which may be data of measuring devices, statuses of signalising equipment and devices), and controls the devices in question to achieve the same prescribed purpose [6, 7, 8] - in this case, it is the

* Corresponding author: maria.blatnicka@fstroj.uniza.sk

Reviewers: *Robert* Grega, *Kateryna* Kravchenko

aim of designing a chain conveyor (Fig. 1) for transporting a handling unit between individual working positions, the design must meet the requirements of the customer's technical standards. According to these standards and customer requirements, a flame soldering station for compressor outlet piping will be designed [9].

Fig. 1. 3D model of the chain conveyor design

2 Device requirements

The device should be formed by a chain conveyor with compressor pallets attached to it (Fig. 2). Pallets are formed to shape a tube and at the same time cover the compressor so that it is not damaged by the temperature field from the flame when welding. Technical description of the device positions:

Position 1: Manual loading the compressor into the pallet, loading the soldering ring and pipes.

Position 2: Flame welding. The pallet with compressor stops at the welding station, a burner (the entire Messer assembly) comes to the welding point and welding takes place. The compressor is mounted on the pallet in a slanting manner so that melt of the soldering ring flows towards the pipe joint.

Position 3: Is a manual station where an operator tests tightness manually by inserting the compressed air hose into the compressor tube and checks tightness with spray.

Fig. 2. 2D view of the compressor pallet design assembly

The design of a chain conveyor is based on a need to create device that is able to move the pallet with the component (SECOP compressor) from the first workstation (manual position - loading the compressor into the pallet, loading the soldering ring and pipe) successively through the second workplace containing flame welding (automatic position - pressure pipe soldering) to the third position (checking and unloading the component into the pallet). The process itself consists of several steps. The choice of a suitable chain and drive is important, with consequent further calculations or selection of other conveyor components. Subsequently, a load-bearing structure will be created, in the creation of which it is desirable to adhere to the correct principles of construction methodology. From the given requirements it was necessary to choose basic dimensional parameters:

- Pallet length: $Ldp = 250$ mm,
- Pallet width: $L\check{s}p = 250$ mm,
- Conveyor manipulator length: $LDM = 3, 200$ mm,
- Total number of pallets: $iC = 12$ (-).

3 Pallet design

Firstly, a design of the pallet with a bed for the outlet pipe soldered in position no. 2 was created. The design of the pallet began by selecting basic dimensions of a base plate (Fig. 3) and the subsequent two-pin bedding for centering the compressor to the desired position.

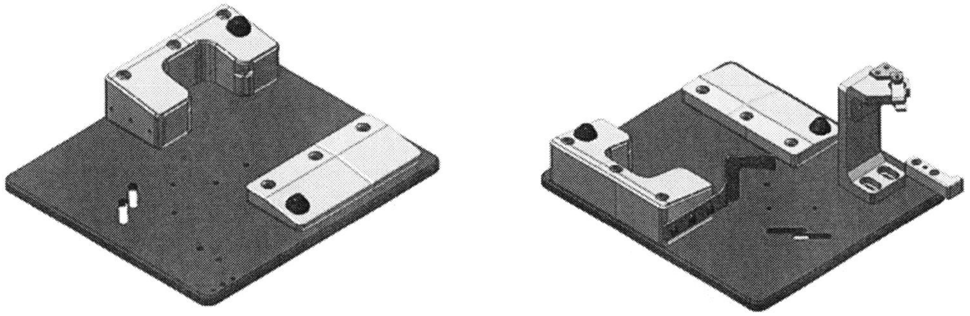

Fig. 3. Method of pallet design as a compressor attachment

As the component itself (the compressor) already has a considerable weight (6.5 kg), it was necessary to design the pallet so as to make it as lightweight as possible. This was the reason for choosing a base plate and 2 beds (Fig. 3 left) of AlMgSi07 material. After the compressor bed was placed, the design of the bed for the soldered pipe was created with securing against vertical displacement (Fig. 3 right) so that it was height adjustable. To ensure correct loading of the pipe in the horizontal position, it was necessary to design a flip-over locking mechanism for the soldering tube (red tube Fig. 5), which serves to accurately load the pipe and secure it in the desired soldering position.

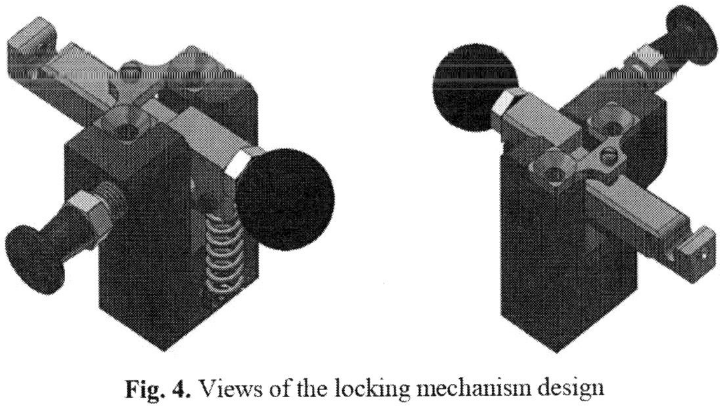

Fig. 4. Views of the locking mechanism design

The proposed mechanism (Fig. 4) is designed so that it can be automatically locked off in position no. 2 for welding in a given location. Thus, the proposed pallet assembly includes a pallet base plate, a compressor bed, an anti-fall pipe locking device, and an outlet pipe housing.

Fig. 5. 3D view of the compressor pallet design assembly

4 Frame design

When designing a frame, it was necessary to first choose the basic dimensions of the conveyor. The design was based primarily on the number of pallets on the device (12 pieces). Subsequently, the necessary spacing between the pallets and overall length of the frame was chosen. The support frame is designed from AL Bosch profile 80x120 mm (Fig. 6). Probably it was also possible to choose a profile with smaller dimensions but this type of beam was suitable for construction from a structural point of view [10-12].

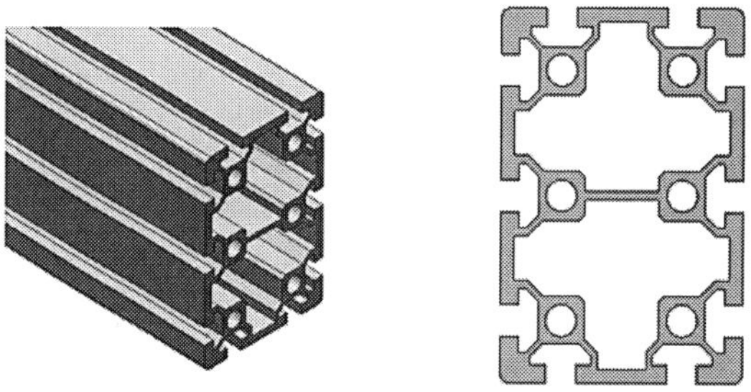

Fig. 6. Bosch Rexroth profile 80x120 mm

Struts and crossbeams are designed from the same profiles but with dimensions 45x45 mm (Fig. 7). Supporting profiles are joined by 90x90 mm angle bars and other profiles using 45 ° 45x45 mm angle bar and Bosch Rexroth pin joint.

Fig. 7. Design of conveyor frame

5 Conveyor chain design

After the design of the frame and subsequent placement of the pallets, the chain and the gearwheels could be made. Chains are dimensioned according to the maximum force in the chain resulting from all power losses and tension. Static and dynamic forces are created in the chain at motion. The main objective is to predict the power (and thus the size) of the engine and also to pre-determine the type of roller chain serving as a traction element. Determination of these two components is very important for the design because, for example, according to the chain, a suitable sprocket or chain guide must be selected. A choice of the engine and the gearbox affects a size of the drive shaft, bearings and spatial arrangement of the entire drive. Depending on the purpose of the conveyor, various types of chains can be used [13-15]. In this case, "standard roller chain with attachment 1 – DIN 8187 with bend attachment K1/01 double sided" for better pallet attachment (Fig. 8) was selected. The basic parameters required for the design of the chain drive were chosen according to Tab. 1 from HABERKORN.

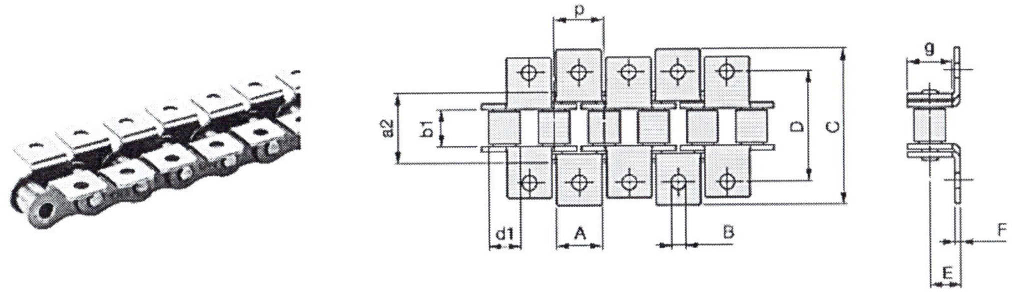

Fig. 8. Geometry of the selected chain

We select a chain with the label of link 10 B-1 5/8 x 3/8
Chain length / (number of links): LR = 5,820.6 mm,
Pitch spacing of pallets: $R1$ = 485 mm,
Pitch of first and last sprocket: LRK = 2,635.4 mm.

Table 1. Chains from Haberkorn

DIN ISO labeling	store labeling p x b1	Pitch p	Width b1 min.	Roller d1 max.	a2 max.	g max.	A	B	C max.	D	E	F	Force [N]	Mass in kg/m	Order code
06 B-1	3/8" x 7/32"	9.525	5.72	6.35	13.5	8.26	8.0	3.5	27.0	19.0	6.5	1.03	10,700	0.61	552666
08 B-1	1/2" x 5/16"	12.700	7.75	8.51	17.0	11.81	9.5	4.3	37.2	28.2	8.4	1.40	21,100	1.01	552692
10 B-1	5/8" x 3/8"	15.875	9.65	10.16	19.6	14.73	14.1	5.2	47.6	31.8	10.4	1.63	27,400	1.32	552726
12 B-1	3/4" x 7/16"	19.050	11.68	12.07	22.7	16.13	15.8	5.6	51.6	35.0	12.0	1.80	32,300	1.66	552689
16 B-1	1" x 17.02 mm	25.400	17.02	15.88	36.1	21.08	19.0	6.8	74.8	50.8	15.9	3.00	70,600	4.27	552716

6 Design of chain drive and tensioner

For most applications, drive and tensioner are two separate units. According to the design methodology, it is appropriate to place the drive at the beginning of the conveyor and the tensioning unit at its end [16, 17]. A plate is rigidly screwed by four screws onto the frame of the conveyor beam. To this plate is mounted a gearbox with an electric motor. There is a secondary drive shaft with a sprocket (Fig. 9 left) located together with the chain transmission from the electric motor to the main drive shaft. Chain transmission is made by a pair of wheels with the dimensions shown in Tab. 2.

Table 2. Sprocket wheels dimensions

Usage	Number of teeth	d_e [mm]	d_p [mm]	*D1* [mm]	*B1* [mm]	Profile radius of curvature *r3* [mm]	Tooth curvature *C* [mm]
Sprockets of the conveyor	18	78	73.14	35	7	13	1.3
Drive wheel of the shaft and transmission	15	65.9	61.09	35	7	13	1.3

On the main drive shaft there is a bearing housing, clamping sleeves with a torque limit and a corresponding sprocket. For safety reasons, the limiter is used on all drives. In the case of undesirable collision, the load increases and the critical torque is exceeded. At this point, the limiter ensures slipping of the shaft and the sprocket. The main drive gear on the shaft is secured by two 250.0120 Halder clamping sleeves (Fig. 9 right) on both sides.

Fig. 9. Drive sprocket (left) and clamping case 250.0120 Halder (right)

The shaft is stored in bearings that are pressed in the housing in the base plate. The shaft is secured with a securing ring against axial movement. The complete proposed drive device is shown in Fig. 10 left. The reversing tensioning part consists of a gear that is mounted in bearings fixed in solid housings in the same way as on the driving part.

Fig. 10. Designed drive unit (left) and tensioning device (right)

If the wheel fastening is firm, the chain is tensioned by its own weight [18, 19]. In the case of using sliding housings on the driven side, a chain is tensioned by a pair of tensioning screws. The tensioning force used to be between 1,000 and 4,000 N. The conveyor is driven by an electric motor through the spur gearbox. This power station is located below the conveyor. For safety reasons, the drive chain will be covered.

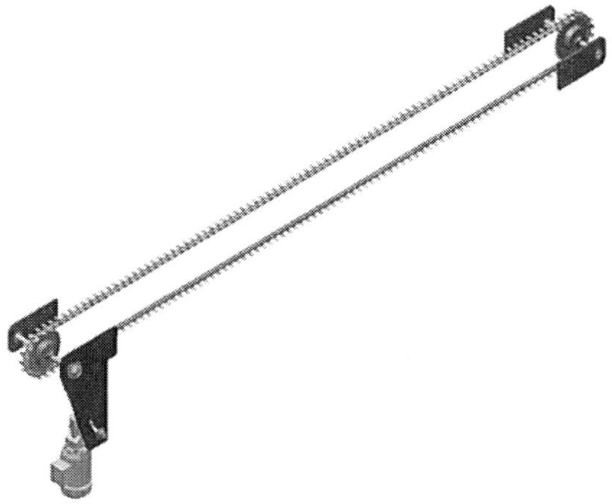

Fig. 11. Overview of the drive, chain and tensioning device of the proposed chain conveyor

Conclusion

The aim of the article was to partially design a chain conveyor of an automatic soldering station consisting of the bearing pallet design, which is also a jig (tool) for ensuring the correct position of the compressor and the pipe; the conveyor frame design from standardized profiles from Bosch and the design of the chain with its drive and tensioning device. We can say that this goal has been met. In the next step of solving this problem, it will be necessary to design support and guide of the chain links, to propose design solutions of the individual work positions of the station and to make a dimensional calculation of the driving electric motor. After this step, the prototype of the flame soldering station for compressor outlet piping can be manufactured and brought into operation.

This work has been supported by KEGA project No. 015ŽU-4/2017.

References

1. M. Macko, J. Flizikowski, Z. Szczepański, K. Tyszczuk, G. Śmigielski, A. Mroziński, J. Czerniak, A. Tomporowski, *CAD/CAE applications in mill's design and investigation*. Proceedings of the 13th International Scientific Conference on Computer Aided Engineering, Springer International Publishing AG, Lecture Notes in Mechanical Engineering, Chapter **35**, 343-351 (2017)
2. R. Halama, M. Fusek, Z. Poruba, *Influence of mean stress and stress amplitude on uniaxial and biaxial ratcheting of ST52 steel and its prediction by the AbdelKarim–Ohno model*. International Journal of Fatigue **91**, 313-321 (2016)
3. J. Vavro, J. Vavro jr., P. Kováčiková, P. Kopas, M. Handrik, *Simulation and analysis of defect distribution in passenger car tire under dynamic loading*. Applied Mechanics and Materials 611, 544 – 547 (2014)
4. J. Gerlici, T. Lack, J. Harušinec, *Realistic simulation of railway operation on the RAILBCOT test stand*. Applied mechanics and materials **486**, 387-395 (2014)

5. A. Sapietová, V. Dekýš, M. Sapieta, P. Pecháč, *Application of computational and design approaches to improve carrier stability*. Procedia Eng. **96**, 410-418 (2014)
6. J. Czerniak, D. Ewald, M. Macko, G. Śmigielski, K. Tyszczuk, *Approach to the Monitoring of Energy Consumption in Eco-grinder Based on ABC optimization*, in: S. Kozielski et al. (Eds.): BDAS 2015, Communications in Computer and Information Science **521**, 516-529, (2015)
7. P. Šťastniak, M. Moravčík, *Freight bogie prototype properties analysis by means of simulation computations*. Manufacturing technology **17** (3), 381-388 (2017)
8. G. Domek, A. Kołodziej: *Design of the tendon structure in timing belts* Procedia Engineering **136**, 365-369 (2016)
9. J. Galliková, R. Poprocký, P. Volna, *Implementation of FMEA method in maintenance of semi-trailer combination*. Diagnostyka **17** (4), 85-92 (2016)
10. V. Bulej, J. Uricek, V. Poppeova, R. Zahoransky, M. Rupikova, *Study of the Workspace of Hybrid Mechanism Trivariant*. Appl. Mech. and Mat., Mech. Eng. and Manufacturing **436**, 366-373 (2013)
11. J. Stancek, V. Bulej, *Design of driving system for scissor lifting mechanism*. Acad. J. of Manufacturing Eng. **13**, 38-43 (2015)
12. M. Cisar, I. Kuric, N. Cubonova, *Proposal of device for monitoring of educational machine tools*. Advances in Sci. and Technology-Res. J. **31**, 80-86 (2016)
13. M. Dudziak, G. Domek, A. Kołodziej, K. Talaśka, *Contact Problems Between the Hub and the Shaft with a Four-Angular Shape of Cross-Section for Different Angular Positions*. Applied Mechanics and Materials **816**, 54-62 (2015)
14. J. Vavro jr., J. Vavro, P. Kováčiková, P. Kopas, M. Handrik, R. Bezdedová, *Numerical analysis of stress states for graphitic cast iron structures*. Applied Mechanics and Materials **611,** 252-255 (2014)
15. J. Gerlici, T. Lack, *Railway wheel and rail head profiles development based on the geometric characteristics shapes*. Wear: an international journal on the science and technology of friction, lubrication and wear **271** (1-2), 246-258 (2011)
16. G. Śmigielski, W. Toczek, R. Dygdała, K. Stefański, *Metrological analysis of precision of the system of delivering a water capsule for explosive production of water aerosol*. Metrology and Measurement Systems, **23** (1), 47-58 (2016)
17. M. Sapieta, A. Sapietová, L. Gajdoš, *Determine the fatigue life of flange of bearings test station*. Procedia Engineering **177**, 548-553 (2017)
18. R. Halama, M. Fusek, F. Fojtík, *An industrial application of a method for spatial motion measurement*. EAN 2010, 48th International Scientific Conference on Experimental Stress Analysis, 75-80 (2010)
19. M. Handrik, M. Vaško, P. Kopas, V. Mozer, *The linear and nonlinear stability loss of structures due to thermal load*. Procedia Engineering **136**, 359-364 (2016)

Structural design of soldering station chain conveyor working positions

Miroslav Blatnický[1], *Ján* Dižo[1,*], *Mária* Blatnická[2]

[1]Faculty of Mechanical Engineering, University of Žilina, Department of Transport and Handling Machines, Univerzitná 8215/1, 010 26 Žilina, Slovak Republic
[2]Faculty of Mechanical engineering, University of Žilina, Department of Applied Mechanics, Univerzitná 8215/1, 010 26 Žilina, Slovak Republic

Abstract. The paper deals with design of the support and guide of the conveyor chain links, the individual working positions of the handling mechanism for the conveyor line, i.e. positions of manual loading of components on the line and automatic soldering of the component. The entire design will serve as an input for the dimensional calculation of the propulsion engine of the proposed conveyor line in order to enable the production of a real functional prototype.

Keywords: chain conveyor, structural design, 3D modelling, soldering station

1 Introduction

An important role of engineers is to analyse used and newly designed workflows to find the optimal way to perform the operation.

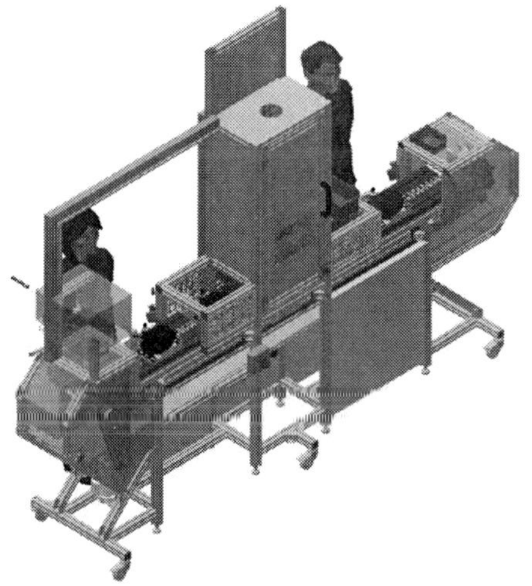

Fig. 1. General view of the proposed soldering station

* Corresponding author: jan.dizo@fstroj.uniza.sk

Reviewers: *Jan* Krmela, *Ján* Vavro *Jr*.

The best workflow is generally considered to be the one that minimizes costs for performance, which can be achieved by mechanization. Mechanization is an important means of increasing productivity, quality and competitiveness of production. It is necessary to know and understand the physical dependence of the performed operations for the successful implementation of mechanization. The operations are carried out by transferring mechanical, electrical, pneumatic or hydraulic power [1]. The effort is to make the individual workings as short and simple as possible to make them easy to learn while demanding a minimum of human effort. Mechanism greatly relieves man from heavy physical work, for example, in a hazardous or harmful environment. Such a mechanism is also the proposed chain conveyor of the soldering station (Fig. 1), which greatly facilitates the handling process with great operability, and with the condition of minimal production costs we achieve the most cost-effective ratio.

2 Supporting and guiding chain links

Chain conveyors, as one of the types of transport and handling equipment, interfere with their activities in all areas of the national economy. By doing so, they make work easier and increase the efficiency of production. It belongs to a group of conveyors with a towing element and we can define them as transport equipment that fulfils the function of the towing element by one or more closed chains circulating around the end sprockets. These conveyors are used in various engineering branches. There are two types of the conveyor support and guide, namely sliding or rolling, and its task is to absorb the gravitational forces from the belt and the conveyed material. These forces used to be large and have an impact on the conveyor movement power losses and therefore the sliding guide is only recommended for short, low-capacity conveyors. Since the conveyor will have a total length of about 3 m, the sliding guide of the belt (Fig. 2) has been selected during construction.

Fig. 2. Detail of the chain guide design

Chains can be led in either chain metal guides or plastic guides that are sufficiently resistant to wear and also reduce chain wear [2]. In view of this, plastic chains made of POM have been chosen that are designed so that the chain is guided in plastic.

3 Design of the position no. 1

At position no. 1 (Fig. 3) operator manually loads the compressor into the pallet bed and bends the compressor hose to the position according to the jig.

Fig. 3. Operator's view of work position no. 1

They put on the solder and the outlet pipe, then spread colophony on the joint and press the green OK button (start). The palette moves automatically by one position to position no. 2, where automatic soldering is carried out by burner.

4 Design of the position no. 2

Design of the soldering position (position 2) was one of the more complex designs of this device. It was necessary to design a vertical movement of the soldering burner with adjustable stops at both positions and the mechanism of the lock off / lock in of the outlet pipe in the pallet, subsequent fixation of the burner on the sliding mechanism, the positioning of the auxiliary burner for ignition of the main (welding) burner and the locking mechanism for stopping the pallet in the desired position.

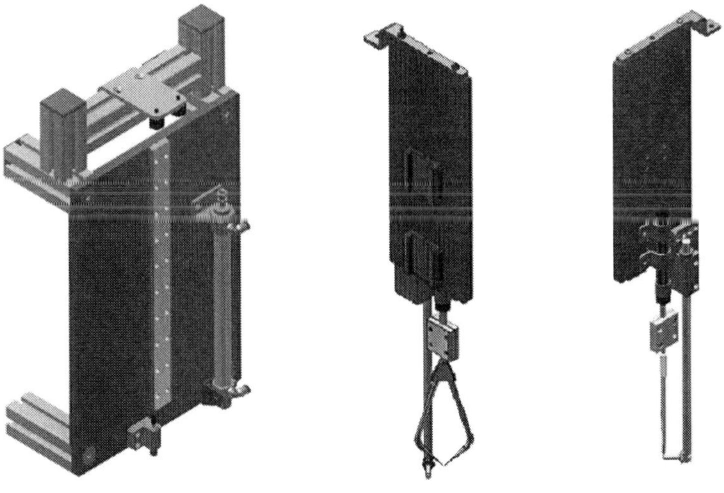

Fig. 4. Linear line mechanism design (left) and proposed sliding mechanism (right)

When designing the sliding mechanism, it was necessary to choose such a pneumatic cylinder [3, 4, 5] to be able to lift the mechanism that is being moved in the linear guide (two carts on the rail). A cylinder from Festo (www.festo.com) with the designation 0822033208 was configured. The cylinder configuration was dependent on its desired lift. For this mechanism, the rail with designation of WEH17CA has been designed to provide accurate guidance in the vertical direction. The rail is mounted on a rigid board (orange plate Fig. 4 left) along with the designed 45x45 mm (Bosch Rexroth) profile [5, 6, 7] and they both are mounted on the profile frame of the conveyor. On the sliding mechanism, two carts (HIWIN WET17CA) are bolted from the bottom by 4 screws on the sliding plate (blue plate in Fig. 4 right), which assures movement towards and away from of the soldering (welding) device on the rail [8, 9, 10]. At the end positions, a stop and a pneumatic damper is used to cushion the impact when moving the pneumatic cylinder.

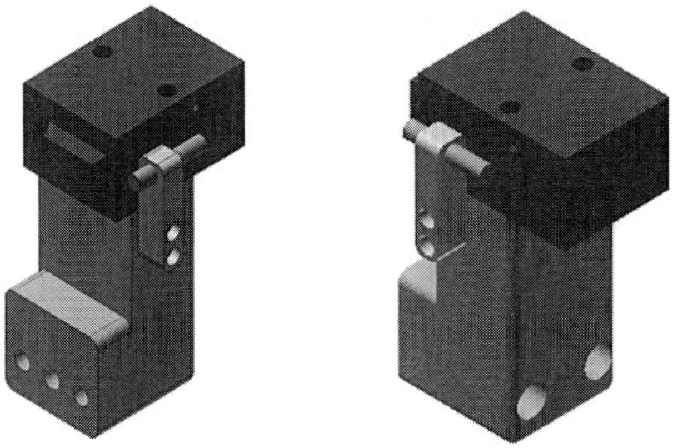

Fig. 5. Pallet stopping mechanism (position no. 2)

At position no. 2 it was also necessary to solve the pallet stop in the desired position. A stop mechanism has been designed for this purpose (Fig. 5). It consists of a pneumatic stopper, a holder to which the stopper is screwed by two screws and two pins and an inductive sensor which is screwed from the side of the holder. The sensor serves to indicate the presence of the pallet at position no. 2.

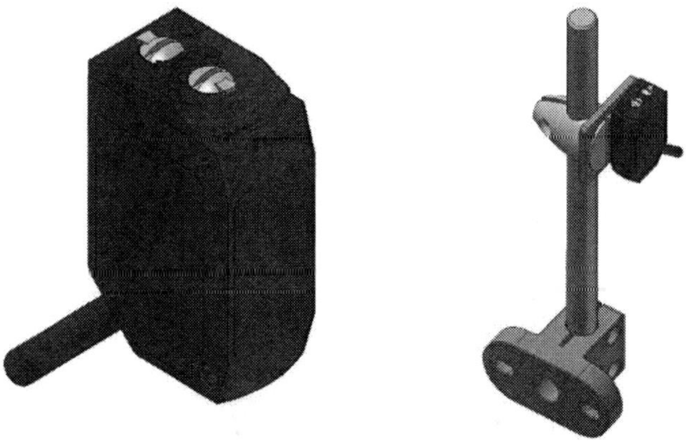

Fig. 6. Optical sensor (left) with designed holder (right)

Furthermore, it was necessary to sense the presence of the compressor on the pallet at this position. The reason was to avoid soldering (welding) in case an empty pallet without a compressor (component) would arrive at the workstation. The optical sensor 06H-300 SICK (Fig. 6) was used for this purpose. It was necessary to design a holder that would allow the sensor to be set in any direction and at the same time make it easy to produce [11].

The proposed holder consists of two purchased BAM parts, a stainless steel rod of 12 mm diameter and a holder with sensor screwed on an adjustable joint (Fig. 6). The holder is bolted by two screws on the support profile of the conveyor (Fig. 6). Since there is a door in the workspace at position no. 2, it was necessary to prevent any intrusion (hand, various hand gripped objects) into the working space during the operation (soldering). The SICK I200-EO323 automatic electric lock was used for security (Fig. 7).

Fig. 7. Optical sensor (left) with designed holder (right)

To bolt the lock it was designed a holder for mounting the main part to the conveyor frame. The second part (so-called fork) was mounted to the door frame. This lock serves to automatically switch off the pneumatic and electrical supply to the designed device for safety reasons.

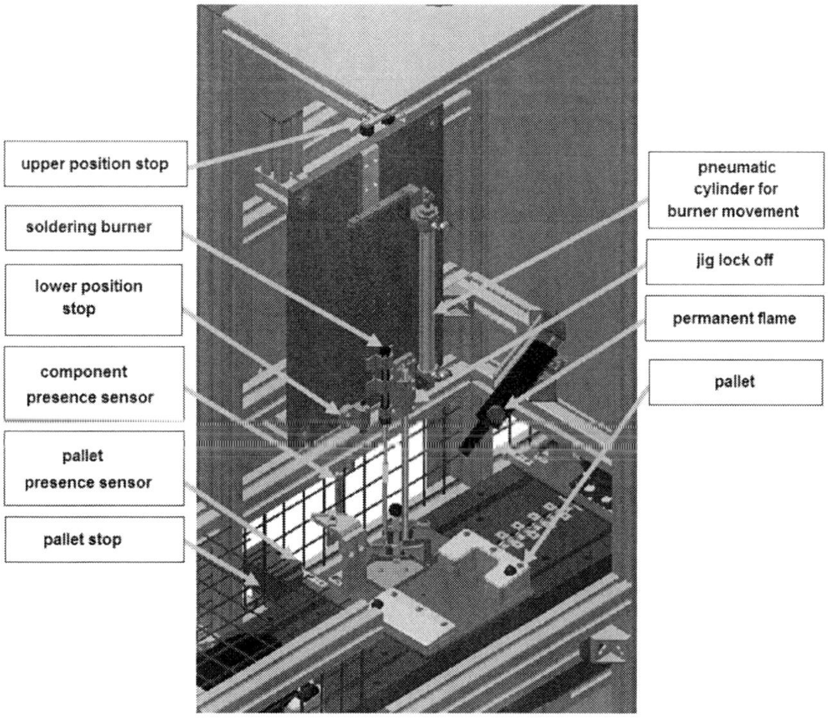

Fig. 8. View of working position no. 2

Working position no. 2 (Fig. 8) is automatic. In this position, the outlet pipe is soldered. The pallet comes to the stop and the presence of the piece on the pallet is sensed. The main burner activates and using the pneumatic cylinder, the burner is moved to the working position where the outlet pipe is soldered. After soldering, the burner switches off and returns to the base position. Another task in designing the device was to design a placement for an electrical switchboard [12-14]. It is located under the conveyor structure at position no. 1. The operating panel, which serves as a software for managing and controlling the entire operating line, was placed in the frame of the transport line at position no. 2 from the operator's side, where the buttons for starting and stopping the machine are located. Another problem was the location of a pneumatic block, which serves to control the pneumatic components used on the device.

Conclusion

The aim of the article was to create a design of the first two of the three working positions of the automatic soldering station conveyor (Fig. 9), namely the first manual position where the components are loaded into the jigs and the second fully automated welding position of the components.

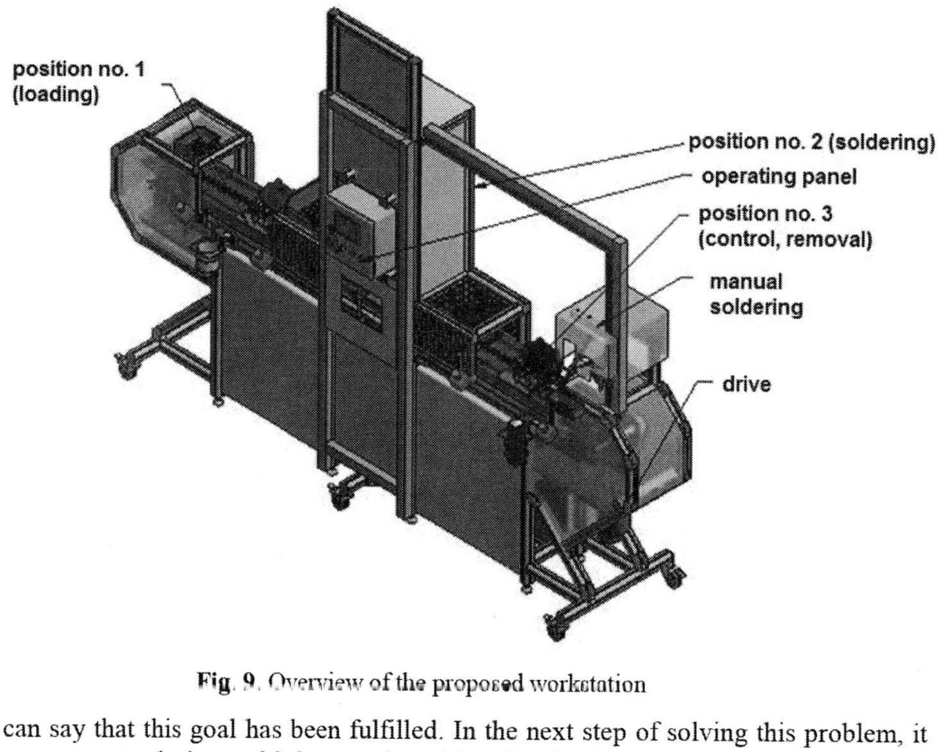

Fig. 9. Overview of the proposed workstation

We can say that this goal has been fulfilled. In the next step of solving this problem, it will be necessary to design a third manual position for checking the tightness of the joint and to perform the dimensional calculation of the drive electric motor. After this step, the prototype of the compressor discharge line soldering station of the compressor can be used to bring the functional device into operation.

This article was supported by the project VEGA 1/0927/15 "Research of the use of alternative fuels and hybrid drives on traction vehicles with aim to reduce fuel consumption and air pollutants production".
The work was supported by the Cultural and Educational Grant Agency of the Ministry of Education of the Slovak Republic in project No. KEGA 077ŽU-4/2017: Modernization of the Vehicles and engines study program.
This publication is the result of the project implementation: Modern methods of teaching control and diagnostic systems of engine vehicles ITMS 26110230107 supported by the Operational Programme Education funded by the ESF.

VÝSKUMNÁ AGENTÚRA Európska únia Štrukturálne fondy

References

1. L. Kašiar, P. Zvolenský, D. Barta, L. Bavlna, M. Mikolajčík, P. Drożdziel, *Diagnostics of electric motor of locomotive series 757*. Diagnostyka **3**, 95-101 (2016)
2. J. Gerlici, T. Lack, *Railway wheel and rail head profiles development based on the geometric characteristics shapes*. Wear **1-2**, 246-258 (2011)
3. J. Harušinec, M. Maňurová, A. Suchánek, *Optimization of a brake unit in terms of control range*. Manufacturing Technology **5**, 917-923 (2016)
4. J. Gerlici, T. Lack, *Rail vehicles brake components test bench utilisation*. Applied mechanics and materials **486**, 379-386 (2014)
5. J. Gerlici, T. Lack, *Contact geometry influence on the rail/wheel surface stress distribution*. Procedia Engineering **1**, 2249-2257 (2010)
6. J. Gerlici, T. Lack, J. Harušinec, *Realistic simulation of railway operation on the RAILBCOT test stand*. Applied mechanics and materials **486**, 387-395 (2014)
7. J. Gerlici, T. Lack, J. Harušinec, *The test stand load modulus implementation for the realistic railway operation in the laboratory conditions*. Manufacturing Technology **4**, 444-4495 (2013)
8. J. Gerlici, T. Lack, *Modified HHT method for vehicle vibration analysis in time domain utilisation*. Applied mechanics and materials **486**, 396-405 (2014)
9. T. Lack, J. Gerlici, *A modified strip method to speed up the calculation of tangential stress between wheel and rail*. Applied mechanics and materials **486**, 371-378 (2014)
10. T. Lack, J. Gerlici, *A modified strip method to speed up the calculation of normal stress between wheel and rail*. Applied mechanics and materials **486**, 359-370 (2014)
11. J. Skočilas, B. Skočilasová, J. Soukup, *Determination of the rheological properties of thin plate under transient vibration*. Latin American Journal Of Solids And Structures **1**, 189-195 (2013)
12. J. Zapoměl, V. Dekýš, P. Ferfecki, A. Sapietová, M. Sága, M. Žmindák, *Identification of material damping of a carbon composite bar and study of its effect on attenuation of its transient lateral vibrations*. International journal of applied mechanics **7/6**, 18 p. (2015)
13. M. Sága, M. Vaško, P. Kopas, L. Jakubovičová, *Application of Karray-Bouc hysteretic model for cumulative damage calculation using energy fatigue curve*. Applied mechanics and materials **611**, 32-39 (2014)

14. J. Gerlici, I. A. Shedchikova, I. V. Nikitchenko, J. A. Romanchenko, *Investigation of influence of separator magnetic system configuration with permanent magnets on magnetic field distribution in working area*. Electrical engineering & electromechanics **2**, 13-17 (2017)

The welded CNC machine tool frame

Piotr Boral[1,*], *Tadeusz* Nieszporek[1], *Rafał* Gołębski[1]

[1]Czestochowa University of Technology, Institute of Mechanical Technologies, Al. Armii Krajowej 21, 42-201 Czestochowa, Poland

Abstract. Machine tools are built based on cast-iron bodies. Cast iron well dumps down vibrations and is dimensionally stable. Therefore, many conventional machine tools are modernized and modified into CNC machine tools based on the seasoned old bodies of those machine tools. Nevertheless, CNC machine tools are also manufactured based on polymer concrete, which vibration damping ability is even better. On the other hand, attempts made in the past by leading machine tool manufacturers to use welded frames in CNC machine tools failed. The revival of interest in welded frames was spurred by economic considerations, as steel systems are much thinner compared to cast-iron ones. They are able to carry the same loads. The construction of CNC machine tools with a welded steel frame undertaken at the Institute of Mechanical Technologies (ITM) of the Czestochowa University of Technology is an attempt to solve this constructional problem. The machine tool frame was designed with use of CAD solid works system. The ultimate success of using the welded frame is determined not only by the frame itself, but also generally the design of the machine tool and its operation parameters. The welded frame was used for a high-accuracy three-axis milling machine. In this respect, it is equal of professional machine tools manufactured by reputed companies. It can be either a production machine tool or a test stand.

Keywords: machine tools, the welded frame

1 Introduction

Conventional machine tools and, generally, also CNC (Computer Numerical Control) machine tools are manufactured based on cast frames of either grey engineering cast iron or modified (malleable or nodular) cast iron [1]. Cast iron has a good vibration damping capability and is dimensionally stable, and the machine tool is rigid and accurate. For these reasons, many conventional machine tools are modernized and modified into CNC machine tools just based on the well-seasoned old frames of such machine tools.

Nevertheless, CNC machine tools are also manufactured based on polymer concrete frames, whose vibration damping capability is even better [2] – Table I. A machine tool of this type must be very accurately levelled and statically stable, since should it breaks, the frame would be irreparable. Investigations are also being undertaken into hybrid frames that combine cast iron with a mineral casting [3]. Polymer concrete may also provide a filling of cast iron or steel machine tool frames as a vibration damping mass.

* Corresponding author: piotrek@itm.pcz.czest.pl

Reviewers: *Andrej* Czán, *Bogdan* Posiadala

On the other hand, attempts made in the past, even by leading machine tool manufacturers, to use welded steel frames in CNC machine tools failed on a wider scale have so far ended in failure.

Table 1. Comparison of the properties of steel, cast iron and polymer concrete [2]

Properties	Units	Steel	Grey cast iron	Polymer concrete
Density	g/cm^3	7.85	7.2	2.1-2.4
Compressive strength	N/mm^2	250-1200	350-450	140-170
Bending strength	N/mm^2	400-1600	150-400	25-40
Young's modulus	kN/mm^2	210	80-120	30-40
Heat-transfer coefficient	W/(m*K)	50	50	1.3-20
Thermal expansion coefficient	μm/(m*K)	12	10	12-20
Specific heat	kJ/(kg*K)	0.45-0.5	0.5	0.9-1.3
Logarithmic decrement		0.002	0.003	0.02-0.03
Cost	EUR/dm^3	7.67	5.11	2.56
Manufacturing energy demand	MJ/dm^3	160	120	25

Trials on manufacturing such machine tools are under way, since in comparison with cast frames, welded frames allow a twofold reduction of their weight with the same rigidity of the frame (the module of elasticity of steel is 1.8–2.4 times greater than that of cast iron) [4]. For example, a special welded-frame machine tool, Millturn M100, intended for the production of worms is manufactured by Austria's WFL company. Extrusion machines designed for the processing of large-molecular plastics require worms of a length of up to 6500 mm to be used. The company PROMUS of Czestochowa has made a welded grinder frame for the company JOTES of Lodz. This is a machine tool frame with the required high accuracy and dimensional stability. PROMUS of Czestochowa has also made about eighty welded machine frames for a German customer [5].

The construction of a CNC machine tool with a welded steel frame, undertaken at the Institute of Mechanical Technologies (ITM) of the Czestochowa University of Technology, is another attempt to solve this problem. The ultimate constructional success is determined not only by the building of the frame itself, but, in general, also by the design of the machine tool and its operation parameters. On the assumption, this is a three-axis CNC milling machine with high accuracy, in this respect being the equal of professional machine tools manufactured by reputed companies. It will be a production machine tool and, at the same time, a test stand. The machine tool is to be designed for manufacturing geometrically complex parts by the Step-by-Step (SbS) method using a small-diameter cylindrical mill. The machine tool is now under construction. The reason for the manufacturer's interest in frames of this type are economic considerations.

For their casting, cast frames require wooden patterns, which are used for manufacturing several dozen castings and are very expensive (the price of a pattern is comparable to that of the frame). Therefore, in the case of building a single machine tool and making one frame, this technology was considered economically unjustified. For this reason, it was decided on a welded machine tool frame. All machine tools with a cast frame based on the same pattern belong to the same type series. By contrast, welded frame machine tools may be different in terms of the table size and the magnitude of displacements along individual axes, as needed, which can be an asset of this technology. This is important especially in the automation of production and building of production lines composed of several specially designed machine tools.

2 Machine tool development trends

Contemporary CNC machine tools differ significantly from conventional machine tools – Fig. 1.

Fig. 1. A conventional machine tool

Above all, conventional machine tools have one main drive, which is cinematically and mechanically linked with all axes [6], and have no tool stores [7]. In conventional machine tools there is a clear division into turning lathes, milling machines, grinding machines, gear machining machine tools, slotting machines, etc. For CNC machine tools, this division is increasingly difficult, especially in the case of multi-axis machine tools. For example, the QTN (Quick Turn Nexus) MSY lathe by MAZAK is equipped with a turret with driven tools situated in the axial and radial directions, thus featuring milling functions [8] – Fig. 2.

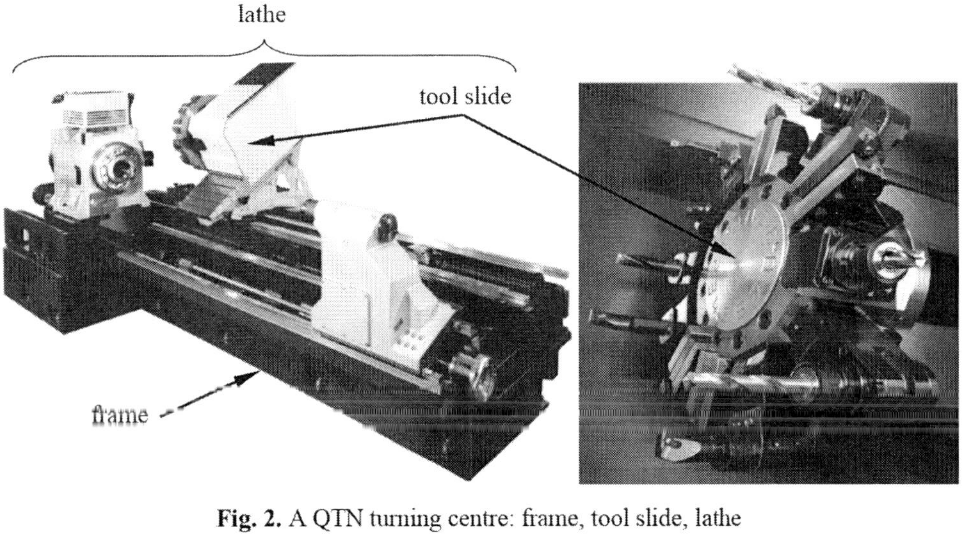

Fig. 2. A QTN turning centre: frame, tool slide, lathe

In addition, it has the capability to control and displace the tool slide in the Y axis. The C axis of the workpiece spindle is also controlled and an intercepting spindle can be situated on the opposite side, while the rotations of both spindles can be synchronized, which allows the workpiece to be fixed in the other spindle and being machined on the other side without having to switch off the rotations. This increases the productivity and accuracy of machining, which corresponds to the Do-In-One idea – machining a workpiece in a single fixture on one machine – which has been promoting by MAZAK for years.

During machine tool operation, chips are transferred with a conveyor to the outside at the side of the machine tool. In turn, a special arm intercepts machined small workpieces and throws them into a container in front of the machine tool. The main workpiece spindle has normally a clearance, which enables the use of a feeder and machining of a bar in an automatic cycle. Thus, the machine tool can operate, for instance, through the whole shift without the operator's intervention. The machine tool can operate with a robot, and is furnished with special functions for this purpose. The machine tool's working space is closed, which enables intensive cooling with cooling and lubricating liquid or with oil mist. For cleaning the machine tool, the spindles and the slide, compressed air is used. The control systems enables the programming of sub-micrometric 0.0001 [mm] feeds and the workpiece spindle rotation angle with an accuracy of 0.0001 [deg]. This machine tool is classified as a turning centre.

In CNC machine tools, some manufacturers employ their own control systems. For example, the MAZAK company uses the universal MAZATROL MATRIX CAM control system [9]. There is a very large number of machine tool manufacturing companies worldwide, therefore control systems such as SINUMERIC [10], FANUC [11] and HaidenHein are commonly used [12]. These are universal control systems with great capabilities and a wide range of applications in different machine tool types. Generally, machine tools are sold with the basic version of control systems, and their possible expansion by additional modules and functions involves added, often high costs. The capability of machine tools to operate in a local ETHERNET network and to be programmed off-line enables their programming using CAD/CAM systems. All contemporary control systems practically operate in the WINDOWS environment, thanks to which the interface is user-friendly, which makes the operation and programming of the machine tool easy.

With the development of CNC machine tools and tooling, an intensive development of CAD/CAM (Computer Aided Design/Computer Aided Manufacturing) and control systems follows. Programs, such as CATIA [13], EdgeCAM [14], SolidWorks [15], MasterCAM [16] and others enable the 3D solid design of parts, generation of 2D working drawings, simulation of the machine operation kinematics, MES strength analysis, as well as 3D machining simulation and the generation of the CNC machine tool control code.

The advantage of these solutions lies in the fact that, although the writing of a machine tool control program for machining a simple part on a three-axis machine tool in either a dialogue system or manually in G codes is possible, the writing of a program for machining a geometrically complex part requiring a five-axis machine tool for its machining is very difficult, if not impossible, without special software.

3 The three-axis CNC milling machine

The design of the milling machine (see Fig. 3) was developed by using the SolidWorks program. This is a three-axis milling machine of a welded construction from S355J2 steel plates. The core of the solution lies, among other things, in the use of a high-speed tool spindle with a rotational speed of 18000 rpm. In conventional machine tools, the spindle rotational speed is low; for example, for the Heckert FSS 400 VI/2 vertical milling machine, it is 1400 rpm [17]. At a low spindle rotational speed in machining with a multi-blade mill, especially a large-diameter one, an impact character of tool blade operation occurs, which results in considerable vibrations. In the adopted milling machine design, at a low spindle rotational speed and a small mill diameter, this phenomenon will not occur.

The welded frame is made from 20 mm-thick plates and is densely ribbed, which imparts it high rigidity – Fig. 4.

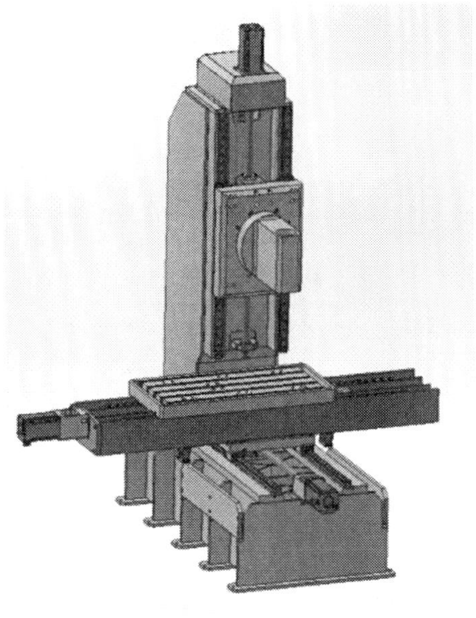

Fig. 3. The design of the XYZ CNC milling machine

The frame was constructed of enhanced-strength steel S355J2. It was made by the MAG method employing the following welding parameters: current intensity – welding currents, 280 - 290 A; voltage, 26V; 1.2 mm-diameter wire in grade G4Si1. V and X-type grooves were designed for making welds, depending on the access to them. In the future, for similar solutions, robotized welding with off-line programming assisting can be used [18]. If further design and technological changes are made, it will be possible to reduce the mass by employing other welded materials [19] and more accurate weld dimensioning [20]. The analysis of the above-mentioned activities and the industrial implementation of the frame under discussion may lead to further engineering & design changes, similarly as for the nodes described in work [21].

Fig. 4. Machine tool pillars and bases during seasoning

The welds were examined for the correctness of execution by an ultrasonic method using a KRAUTKRAMER USM 35XS no. 14374a flaw detector (with wallpaper glue being used as a coupling agent).

The frame was subjected to vibration stress relieving. The vibration was carried out using a WIBROPOL VM120 set with a VC220 vibrator.

The deformation of the frame base after stress relieving over a length of 2 m amounted to 5 mm (0.25%) and was removed after machining. This result indicates a possible and effective reduction of welding stresses and generation of deformations that can be subsequently removed by further machining.

A similar character of operation is exhibited by welds in truck chassis frames, which are subjected to cyclic stresses. They may cause their damage in the form of fatigue cracking [22].

Using stress relieving vibration results in an advantageous reduction of peak stresses in the welds that are decisive to their fatigue strength – Fig. 5.

Fig. 5. The CNC milling machine frame after seasoning, vibrating and machining

The power transmission systems were based on ball screws (by Steinmeyer [23]) and prismatic guides (by Schneeberger [24]) of the highest accuracy. Three servo-drives manufactured by Beckhoff, each with a moment of 29 Nm, were employed. The control was based on the Beckhoff system with the PLC CX5010 controller, which provides the capability to control up to 100 axes and is very efficient [25]. This is a control system that is very rarely found in machine tools, which provides a great potential for being further expanded.

The system includes a CNC interface that can be modified, which is impossible in the case of other common control systems [26].

The controller has the in-built WINDOWS CE and the TwinCAT program installed [27]. To program this controller, it is sufficient to connect a monitor, a keyboard and a mouse to it (there is no need for connecting a PC), which distinguishes it from other controllers. The programs can be written in the TwinCAT [28] or the CoDeSys system [29], practically in 6 languages [30, 31]. Textual languages: IL – Instruction List, ST – Structured Text. Graphical languages: SFC – Sequential Function Chart, FBD – Function Block Diagram, LD – Ladder Diagram, CFC – Continuous Function Chart. It is a fully professional CNC machine tool.

The constructed machine tool will be the subject of further investigation. It will also be both a test stand and a production machine tool.

Conclusions

The stresses and deformations of a frame after welding can be effectively removed by seasoning and vibrating followed by machining. The above-mentioned operations are especially important for new-generation machine tools, such as CNC machine tools with three and more axes.

The linear welding deformations, amounting to 0.25%, which form due to vibration are contained within machining allowances and can be foreseen by the designer.

The welded frame construction discussed in the present study can be used for similar applications in numerically controlled machine tools, as well as for other engineering purposes.

References

1. A. Łukomski, *The frames of machines and equipment, Part 1*. Des. Eng. Cons. **93**, (2015)
2. E. Baumeister, *Hohlkugelkomposit - Characterization of thermal and mechanical properties of a new lightweight construction material*. (Doctoral dissertation. Magdeburg 2004)
3. N. Kępczak, W. Pawłowski, *Theoretical research of dynamic properties of machine tool bodies made of cast iron and hybrid connection of cast iron and mineral cast*. Mech. **8/9**, 177-203F (2015)
4. A. Łukomski, *The frames of machines and equipment, Part 2*. Design. Eng. Constr. **94/95**, (2015)
5. http://www.promus.pl/index.html
6. J. Honczarenko, *Numerically controlled machine tools*. (WNT, Warsaw, 2008)
7. P. Cichosz, *Cutting tools*. (WNT, Warsaw, 2009)
8. Mazak, *Quick Turn Nexus 250-II MSY*. https://www.mazakusa.com/machines
9. Mazak, *MazaCAM - Mazak Mazatrol Programming System*. http://www.mazacam.com
10. Siemens, *Numerical control SiNUMERIC* – Siemens. https://www.automatyka.siemens.pl/solutionandproducts dt/845.htm (2011)
11. FANUC, *Programming of machine tools Fanuc, FANUC seria 16i-TB FANUC seria 18i-TB FANUC seria 160i-TB FANUC seria 180i-TB, User's Guide*. (2002)
12. Heidenhain, *User's Manulal. Conversational Programming TNC 406 TNC 416 NC Software*. (2001)
13. J. Pobożniak, *Programming numerically controlled machine tools in the CAD/CAM CATIA V5 system*. (Helion, Gliwice, 2014)
14. P. Kochan, *Edgecam. Multi-axial CNC milling*. (Helion, Gliwice, 2014)

15. J. Domański, *SolidWorks 2014. Designs of machines and structures. Practical examples*. (Helion, Gliwice, 2015)
16. P. Niesłony, *Basics of programming CNC machine in CAD/CAM system - Mastercam*. (Helion, Gliwice, 2012)
17. Fritz Heckert, *Heckert FSS 400 VI/2 vertical milling machine*. Technical documentation
18. M. Nowak, J. Buchowski, D. Wiśniewski, *Robot montioned large structures welding with off-line support programing*. Wel. Tech. Rev. **85**, 19-28 (2013)
19. Z. Mirski, Z. Fałek, *Welded containers for scrap and debris transport of 450 Hardox steel*. Wel. Tech. Rev. **87**, 5-9 (2015)
20. K. Kudła, K. Wojsyk, *Method for analytical computation of stresses in fillet joints for complex load states – according to eurocode 3*. Wel. Tech. Rev. **86**, 8-14 (2014)
21. T. Urbański, *Hybrid node – predictions of transverse shrinkage in butt join on the basis of twodimensional designed experiment*. Wel. Tech. Rev. **87**, 22-26 (2015)
22. J. Słania, Ł Kuk, *Process of joining materials to build vehicles and motor-car bodies in the automotive industry*. Wel. Tech. Rev. **86**, 40-46 (2014)
23. Steinmeyer, *Ball scrwe drives. Welcome to whete precision is*. www.steinmeyer.com
24. Schneeberger, *MONORAIL and AMS. Profiled linear guideways and integrated measuring systems*. (2015)
25. Beckhoff, *Beckhoff TwinCAT The Windows Control and Automation Technology PC Control Introduction Beckhoff*. http:www.beckhoff.com/twincat, (2001)
26. Beckhoff, *Beckoff New Automation Technology TwinCAT 3*. Beckhoff, Brian McClure, http:www.beckhoff.com/twincat, (2013)
27. Beckhoff, *Basics of program handling TwinCAT System Manager Part 1*. docplayer.pl, (Warszawa, 2009)
28. Beckhoff, *Basics of program handling TwinCAT PLC Control Part 2*. docplayer.pl, (Warszawa, 2009)
29. Moeller, *User's Guide, Programming PLCs in CoDeSys 2.3*. www.moeller.pl/ Documentation/Literatura/h1437pl.pdf, 3S - Smart Software Solutions GMBH, (2006)
30. Moeller, *Programming PLCs in easy Soft CoDeSys*. (2010)
31. Moeller, *Easy Soft environment – CoDeSys*. www.moeller.pl (2015)

Utilization of educational machine tool for training of technical diagnostics based on positioning performance

Miroslav Cisar[1,*], *Ivan* Kuric[1], *Nadezda* Cubonova[1], *Ivan* Zajacko[1], *Tomas* Dodok[1], *Anna* Rudawska[2]

[1]University of Zilina, Faculty of Mechanical Engineering, Slovakia
[2]Lublin University of Technology, Faculty of Mechanical Engineering, Lublin, Poland

Abstract. The technical diagnostics as essential part of machine tool maintenance provides information about condition of examined machine tool but it requires not only special equipment but also certain level of skill of operators in order to achieve optimal performance. The article is aimed at the set of methods and tools capable of providing realistic simulation of various machine tool conditions in order to allow hone skills necessary for effective diagnostician.

Keywords: Machine tool diagnostics, training, error simulation

1 Introduction

In order to meet customers' demands for high quality and low prices, manufacturers have to pay close attention to production resources and quality of production. The quality of machined parts is closely connected to its accuracy and by extension to quality and condition of used machine tool and its individual parts. Technical diagnostic provides various methods and tools to obtain not only current condition of machine tool but also its future progress and allows to estimate probability of fault occurrence. This allows to avoid unnecessary costs due machinery fault, unexpected downtime and production of insufficiently accurate components [1, 2].

Technical diagnostics is discipline that includes wide range of methods and physical phenomena in order to detect various types of fails and changes in performance of examined device. In terms of machine tool diagnostics, there are two common elements for each of method. The first one is the machine tool on which diagnostic is performed and second one is the diagnostician. It is necessary for such person to have necessary amount of knowledge and certain level of skills in order to provide diagnostic procedures effectively and reliably. In order to keep such skills fresh or to teach new diagnostician is necessary to train certain routines. Proper training of machine tool diagnostics requires performing tasks on machines in various condition in order to gain experiences with real situations. Training of diagnostic routines in various condition requires access to machine tool, or machine

* Corresponding author: tomas.dodok@fstroj.uniza.sk

Reviewers: *Andrej* Czán, *Alžbeta* Sapietová

tools, in various technical conditions. This article is aimed to simulating machine tool faults by manipulating its positioning performance [1, 3].

One way to achieve changes in positioning precision of machine tool is using same tools that are normally meant to reducing geometrical and positioning errors such as loosened fit in feed mechanism or using compensation tables in control system. Another approach is to change programmed tool path same way as desired error. The approach that uses modification of NC program can also be easily used with different devices such as industrial robots and manipulators [4, 5].

2 Devices used for experiment

Emco Concept Mill 105 (CM 105) (Fig. 1) is compact machine tool with interchangeable control system designed for industrial training and educational purposes. Such machine tools are commonly used at technical high schools and universities. The machine tool is controlled via PC with WinNC software and interchangeable control panels. In terms of training and education it is good to have one machine that can be controlled with multiple control systems but implementation of such thing reduces some functions of machine tool. For example, it is not possible to use compensation table directly in control system what reduces possibilities of training machine tool diagnostics [6, 7].

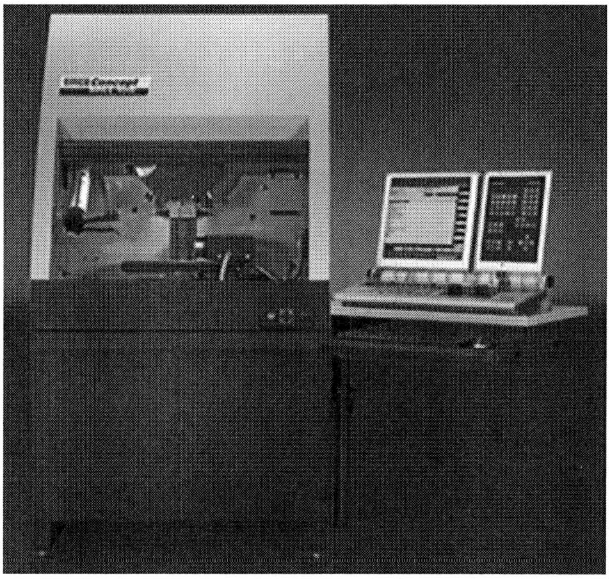

Fig. 1. EMCO Concept MILL 105 [6]

In order to confirm supposition about the possibility of simulating machine tool error by editing NC code, the Renishaw Ballbar QC20-W (Fig. 2) was selected as measuring device. Renishaw Ballbar consists of precise linear sensor, terminated by polished steel balls that fits exactly into magnetic mounts. The first ball fits to the mount held as tool in spindle; second one on pivot that is usually mounted on worktable. Unlocked mount of pivot have three degrees of freedom. That allows fast and simple aligning of mounts using setting ball in order to set center of measuring circle [7, 8, 9].

Essentially, the Renishaw Ballbar is a linear displacement sensor designed to perform a simple, rapid check of a machine tool's positioning performance according to recognized internationals standards based on measuring deviations in the radius of circular movement measured by a transducer and captured by the computer software. The software provided with this device is able to detect and measure wide range of faults and reveal its possible sources. For machine tools or devices with perpendicular feeds it is possible to identify following errors and faults: backlash, cyclic error, lateral play, master-slave changeover, offset change, plot rotation, positional tolerance, radius change, reversal spikes, scaling error, servo mismatch, spiral error, squareness, slick slip, straightness, and vibration of machine. Measured data are stored in *.b5r file that is formatted similarly to XML standard which allows reading, processing, and further analysis of measured values with custom software [9, 10, 11].

Fig. 2. Renishaw Ballbar QC20-W

3 Simulation of machine tool geometrical errors

As mentioned above, the CM 105 have no compensation tables in simulated control system; hence the only direct way of simulating inaccuracies and faults is actually creating them by adjusting mechanical components that are meant to correct such faults. For example it is possible to adjust position of spacer wedges in guiding (Fig. 3) and cause changes in backlash, straightness, squareness and so on.

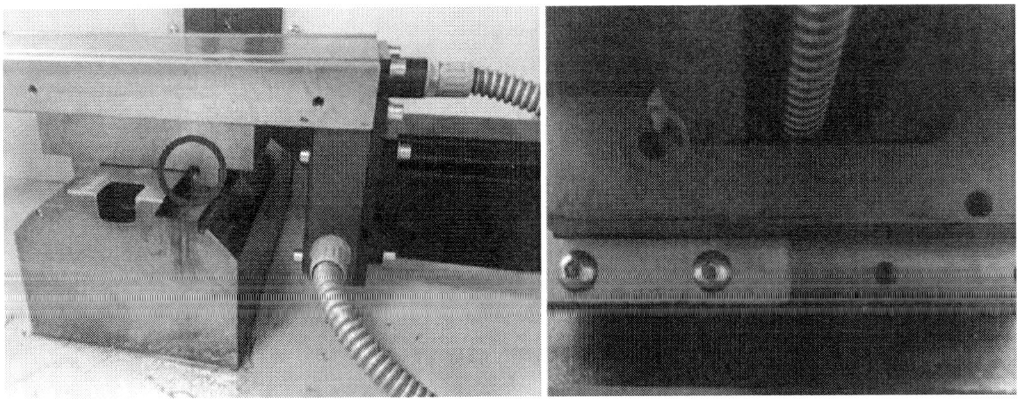

Fig. 3. Location of screws for setting position of spacer wedges on Emco Concept Mill 105

The result of measurement with loosened wedges can be additionally affected by changing of weight distribution on worktable using weights such as magnetic stands. However, this approach is relatively impractical as it requires removing casing and covers of guiding and feeds and correcting settings after measurement.

It is not possible to easily alternate actual shape of tool path programmed thru circular interpolation and therefore circular interpolation (G02 and G03) was replaced by approximated path that consists of multiple short lines (G01). Custom software "circgen" (Fig. 4) was created at Department of Automation and Production Systems, Faculty of Mechanical Engineering, University of Zilina, in order to generate such approximated tool path. The software allows to set number of points on circle, start angle of measurement, angle of run-in and run-out path for measurement, and order of directions of measuring circles. The machine tool uses three decimal places to show coordinates, but its actual precision of positioning is far worse. Calculation of coordinates for apexes of polygon, that replaces circular path, uses radius of circumscription circle and it also includes factor of rounding coordinates to three decimal places in order to minimize deviation from radius of measured circle.

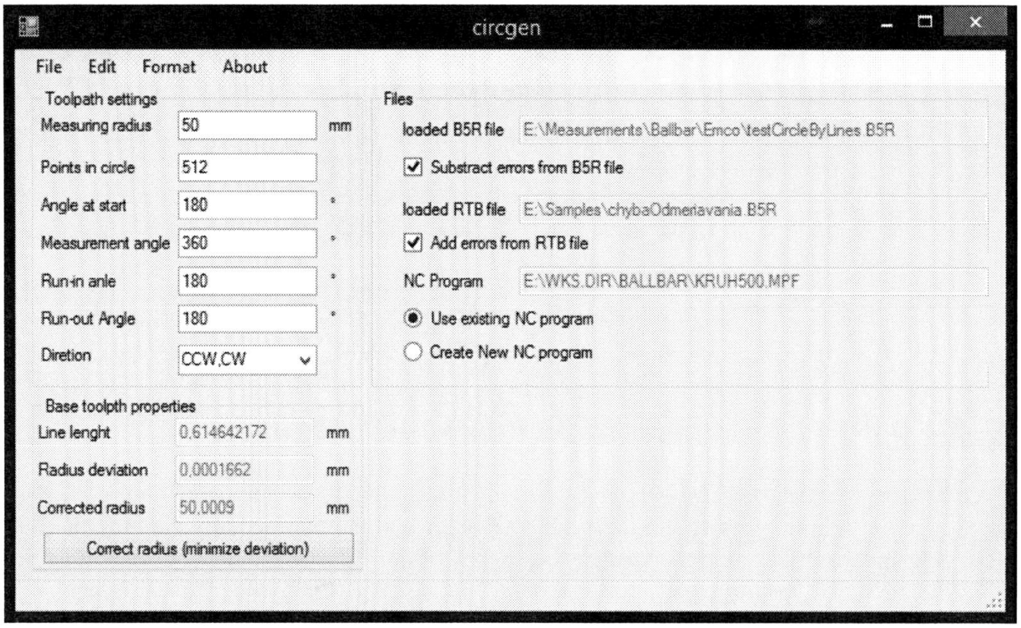

Fig. 4. User interface of the circgen software

First tests with NC program generated with circgen for control system Sinumerik 840D revealed that tool stops for moment at the end of each line that rendered measurement to be unusable. This problem was solved by adding instruction G64 (Continuous path mode). However measurement with ballbar still revealed drop in velocity. For measurement with 512 points on circle with 50mm radius, the maximum velocity reached was 460 mm.min-1 regardless feedrate set in NC program. However, measurement with actual velocity of 460 mm.min-1 is still acceptable even for measurement with velocity set to 500 mm.min-1. The results of measurement with standard NC program that uses circular interpolation was essentially the same as with the NC program with polygonal path in terms of the occurrence and size of measured errors.

Next step necessary for simulation of errors is to eliminate errors that are currently occurring and can be measured. Therefore the capability of reading ballbar measurement was added to circgen software. The measured profile contains much more values than number of points used in NC program. The software uses harmonic analysis to get actual components of errors and to filter out measurement error. Main measurement error is eccentricity of measured circle and real center, it can be eliminated by setting first harmonic to zero. Then the corrected set of points is generated by calculation of deviations for each point using fast Fourier transform and recalculating X and Y coordinates for each point.

The NC program generated with subtracted errors from previous measurement should bring measured polar diagraph closer to ideal circle but it is not possible to compensate errors in full range. Process of measuring and subtracting errors can be repeated but further subtraction have minimal to no effect as shown in Fig. 5, where the 1st iteration is the measurement that uses the uncorrected NC program and following iterations was measurement with path corrected according previous measurements. For comparison 0th iteration shows measurement with standard NC program that uses normal circular interpolation.

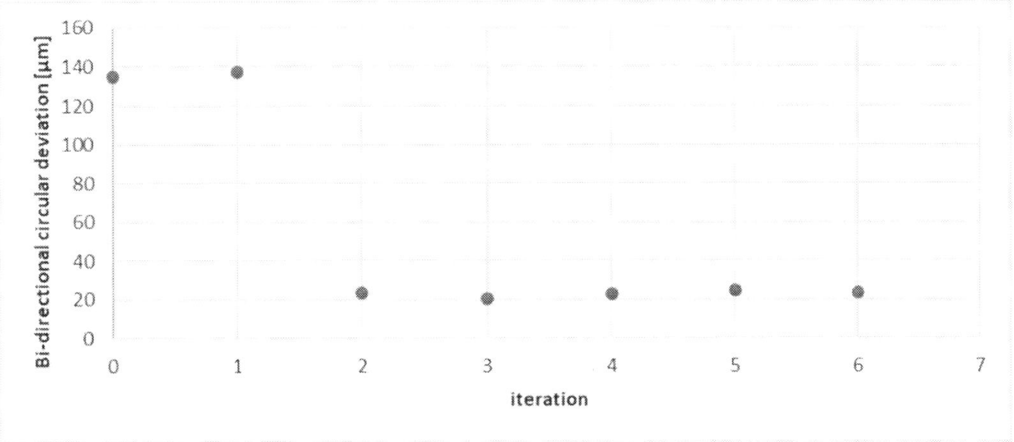

Fig. 5. Bi-directional circular deviation development after subtraction of errors measured from previous iteration

Last step for generating NC program that simulates path corresponding with some machine tool errors is adding deviations to coordinates of apexes of polygon. The source of these deviations can be the file generated in Ballbar plot simulator (Fig. 6).

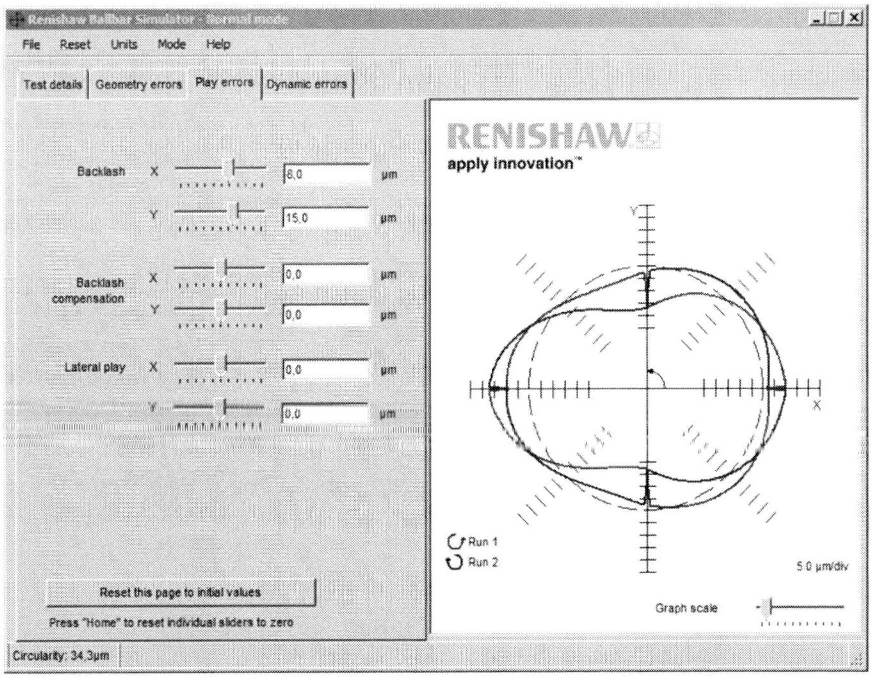

Fig. 6. Ballbar plot simulator

This software can be used to simulate effect of various errors to shape of polar diagraph measured on ballbar. This diagraph can be saved to RTB file which contain 512 records for deviation. The count of records in RTB file was real reason to select 512 points for polygon replacing measuring circle in order to keep calculations as simple as possible.

The size of errors can be simply set by slider or by filling textboxes and saved to RTB file. Another approach to generate errors would require mathematically describe changes in circle shape caused by specific errors and adding it ideal circle to create corrected one. [9]

Conclusion

The article shows two approaches to simulating positioning errors on training machine tool Emco Concept Mill 105 and describes software tool "circgen" developed at Department of Automation and Production Systems, Faculty of Mechanical Engineering, University of Zilina. Experiments revealed that it is impossible to completely eliminate existing errors which was reduced close to 15% of original value. Such reduction is sufficient for simulating other machine tool faults without necessity of change machine tool setup mechanically. Results of measurements allows us to assume that similar technique can be applied for other methods of machine tool diagnostics involving positioning precision measurement such as laser interferometry.

This article was made under support projects KEGA 037ZU-4/2014 and KEGA 024ZU-4/2016.

References

1. M. Cisar, *Diagnosis of Equipments*. Mechanization and Automation Equipment for Processing, Cluj-Napoca : Publishing House Alma Mater, 209-240 (2015)
2. I. Kuric, et al., *Analytical Intelligence Tools for Multicriterial Diagnostics of CNC Machines*. Advances in Sci. and Tech. Res. J. **10** (32), 59-64 (2016)
3. N. Cubonova, M. Cisar, *Design of Camera Mount and its Application for Monitoring Machining Process*. Advances in Sci. and Tech. Res. J. **9** (26), 34-40 (2015)
4. A. Rengevic, D. Kumicakova, *New Possibilities of Robot Arm Motion Simulation*. Communications – Sci. Let. of the Uni. of Zilina **18** (1), 81-86 (2016)
5. D. Kumicakova et al., *Specialised Robotic Hand Designing and Object Grasping Simulation*. Applied Mechanics and Materials **282**, 90-98 (2013)
6. Didactic Martin, s.r.o., *Technical Description of Concept MILL 105* (in Slovak), Online: <www.didactic.sk/Data/802/UserFiles/emco_tech_popisy/spec_cm105-sk.pdf> (2007)
7. N. Cubonova, I. Kuric, *Data Structures Implementation of the Protocol STEP-NC at CNC Machines Programming*. Communications - Sci. Let. of the Uni. of Zilina **16** (3A), 176-183 (2014)
8. N. Cubonova, *Computer Support Programming of Control System Sinumerik 840D for e-learning Education*. Proc. of 12th intern. sci. conf. Engineering for Rural Development **12**, 563-568 (2013)
9. Renishaw Plc., *Manual: QC20-W ballbar* (Czech). <http://resources.renishaw.com/details/Manual%3a+QC20-W+ballbar+(English)(138102)(33885)> (2011)
10. M. Saga, M. Vasko, N. Cubonova, W. Piekarska, *Optimisation Algorithms in Mechanical Engineering Applications*. (Pearson Education Limited, 230 p., 2015)

11. P. Pechac, M. Saga, et al., *Implementation of Memetic Algorithms into Structural Optimization*. Communications Sci. Let. of the Uni. of Zilina **18** (1A), 64-69 (2016)

Analysis of residual stress in subsurface layers after precision hard machining of forging tools

Andrej Czan[1,*], *Jozef* Holubjak[1], *Igor* Daniš[1], *Juraj* Martinček[1], *Matej* Mikloš[1], *Robert* Čep[2], *Michal* Hatala[3], *Šarka* Malotova[2]

[1]Faculty of Mechanical Engineering, University of Zilina, Univerzitna 8215/1, 010 26 Zilina, Slovakia
[2]Faculty of Mechanical Engineering, VŠB – Technical University of Ostrava, 17. Listopadu 15/2172 708 33, Ostrava, Czech Republic
[3]Faculty of Manufacturing Technologies Technical University of Košice with a seat in Presov, Bayerova 1, 080 01 Presov, Slovakia

Abstract. This paper is focused on analysis of residual stress of functional surfaces and subsurface layers created by precision technologies of hard machining for progressive constructional materials of forging tools. Methods of experiments are oriented on monitoring of residual stress in surface which is created by hard turning (roughing and finishing operations). Subsequently these surfaces were etched in thin layers by electro-chemical polishing. The residual stress was monitored in each etched layer. The measuring was executed by portable X-ray diffractometer for detection of residual stress and structural phases. The results significantly indicate rise and distribution of residual stress in surface and subsurface layers and their impact on functional properties of surface integrity.

Keywords: x-ray diffraction, residual stress, triaxial stress tensor, high feed machining

1 Residual stress in subsurface layers

Residual stresses are an integral part of produced workpieces, whether they are introduced deliberately, as a part of the design, as a by-product of a process carried out during the manufacturing process, or are present as the product of the component's service history. Residual stresses are additive with the stresses existing in the workpieces as a result of service loads. Clearly, they may be considered beneficial to the workpieces and therefore desirable, they may be irrelevant and can be ignored, or they are a potential detriment to the workpieces and their continued service life [1, 2, 3]. In practice, residual stress is stress which affects the entire volume of part or the majority of itself, i.e. macroscopic character. For full classification, it should be noted that residual stresses are called sometimes as technological stresses, because they arise from the action of technological processes during the producing of parts. Direction of residual stress (tension or compression) depends on the

* Corresponding author: andrej.czan@fstroj.uniza.sk

Reviewers: *Ivan* Kuric, *Alžbeta* Sapietová

kind of deformation. Permanent residual stresses have the largest share on the functionality of part, and they cannot be detected by conventional methods [4, 5, 6].

Stress conditions are ones of demonstrations of used machining technologies. After manufacturing process, they remain in parts and constructions and they operate continuously even without load. Their impact significantly affects the functionality of machined surfaces [2, 7].

Normal stress is defined as the stress acting normal to the surface of a plane; the plane on which these stresses are acting is usually denoted by subscripts. For example, consider the general case as shown in Figure 1, where stresses acting normal to the faces of an elemental cube are identified by the subscripts that also identify the direction in which the stress acts, e.g. σ_x is the normal stress acting in the x direction. Since σ_x is a normal stress it must act on the plane perpendicular to the x direction. The convention used is that positive values of normal stress de-note tensile stress, and negative values denote a compressive stress [8, 9, 10].

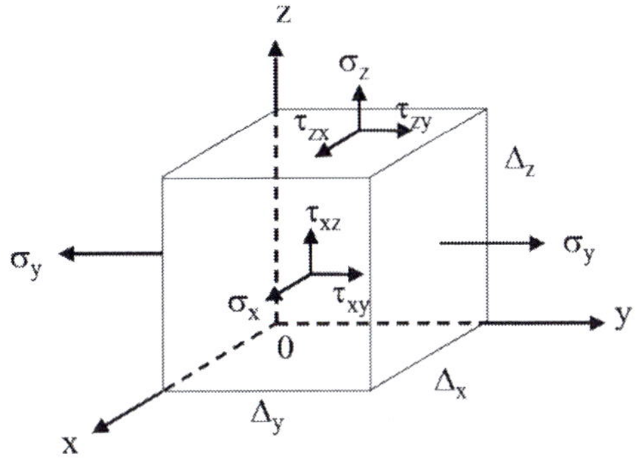

Fig. 1. Scheme of triaxial stress acting on an elemental unit cube [10]

Macroscopic stresses, which extend over distances that are large relative to the grain size of the material, are of general interest in design and failure analysis. Macroscopic stresses are tensor quantities, with magnitudes varying with direction at a single point in a component. The macroscopic stress for a given location and direction is determined by measuring the strain in that direction at a single point. When macroscopic stresses are determined in at least three known directions, and a condition of plane stress is assumed, the three stresses can be combined using Mohr's circle for stress to determine the maximum and minimum residual stresses, the maximum shear stress, and their orientation relative to a reference direction. Macroscopic stresses strain many crystals periodically in the surface. This periodical distortion of the crystal lattice shifts the angular position of the diffraction peak selected for residual stress measurement [1, 11, 12].

Microscopic stresses are scalar properties of the sample, such as percent of cold work or hardness, which are without direction and result from imperfections in the crystal lattice. Microscopic stresses are associated with strains within the crystal lattice that traverse distances on the order or less than the dimensions of the crystals. Microscopic stresses change within the crystal lattice, altering the lattice spacing and broadening the diffraction peak. Macroscopic stresses and microscopic stresses can be determined individually from the diffraction peak position and breadth [5, 13, 14].

2 Principles of X-ray diffraction stress measurement

X-ray diffraction is specific method that can measure residual stress quantitatively in crystalline and semi-crystalline materials, which include virtually all metals and their alloys, and most ceramic materials [7, 8, 15].

The residual stress determined using X-ray diffraction is the arithmetic average stress in a volume of material defined by the irradiated area, which may vary from square centimeters to square millimeters, and the depth of penetration of the X-ray beam. The linear absorption coefficient of the material for the radiation used governs the depth of penetration, which can vary considerably. However, in iron, nickel, and aluminium-based alloys, 50% of the radiation is diffracted from a layer approximately 0.005 mm deep for the radiations generally used for stress measurement. This shallow depth of penetration allows determination of macro and microscopic residual stresses as functions of depth, with depth resolution approximately 10 to 100 times that possible using other methods. Although in principle virtually any interplanar spacing may be used to measure strain in the crystal lattice, the availability of the wavelengths produced by commercial X-ray tubes limits the choice to a few possible planes. The choice of a diffraction peak selected for residual stress measurement impacts significantly on the precision of the method. The higher the diffraction angle, the greater the precision. Practical techniques generally require diffraction angles, 2θ, greater than 120° [4, 16].

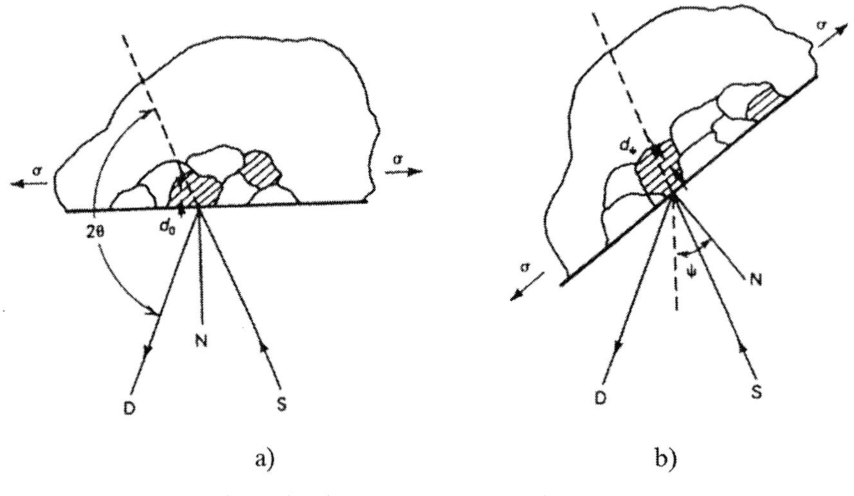

Fig. 2. (a) out rotation ψ = 0. (b) angle of rotation ψ = ψ sample rotated through some known angle ψ, D- detector: S- source; N - normal to the surface.

Figure 2 describes the diffraction of a monochromatic beam of X-rays at a high diffraction angle 2θ from the surface of a stressed sample for two orientations of the sample relative to the X-ray beam. The angle ψ, defining the orientation of the sample surface, is the angle between the normal of the surface and the incident and diffracted beam bisector, which is also the angle between the normal to the diffracting lattice planes and the sample surface [2].

Diffraction occurs at an angle 2θ, which is defined by Bragg's Law: nλ = 2d sin θ, where n is a number denoting the order of diffraction, λ is the X-ray wavelength, d is dimension of the lattice spacing of crystal planes, and θ is the diffraction angle. For the monochromatic X-rays produced by the metallic target of an X-ray tube, the wavelength is known to 1 part in 105. Any change in the lattice spacing d results in a corresponding shift in the diffraction angle 2θ. Figure 1a shows the sample in the ψ = 0 orientation. The presence of a tensile stress in the sample results in a Poisson's ratio contraction, reducing the lattice spacing and slightly increasing the diffraction angle, 2θ. If the sample is then rotated through some known angle ψ (Fig. 2b), the tensile stress present in the surface increases the lattice spacing over the stress-free state and decreases 2θ. Measuring the change in the angular position of the diffraction peak for at least two orientations of the sample defined by the angle ψ enables calculation of the stress present in the sample surface lying in the plane of diffraction, which contains the incident and diffracted X-ray beams. To measure the stress in different directions at the same point, the sample is rotated about its surface normal so that the direction of interest coincides with the diffraction plane. Because only the elastic strain changes the mean lattice spacing, only elastic strains are measured using X-ray diffraction for the determination of macroscopic stresses. When the elastic limit is exceeded, further strain results in dislocation motion, disruption of the crystal lattice, and the formation of microscopic stresses, but no additional increase in macroscopic stress. Although residual stresses result from non-uniform plastic deformation, all residual macrostresses remaining after deformation are necessarily elastic [2, 9].

3 Experiment conditions

Experiment was performed on constructional non-doped steel C56E2, which is used for production of bearing rings and other rolling parts of bearings. This steel is also suitable for production of shafts, gear wheels, pivots etc. It is characterized by worse weldability, and it is used mostly for production of parts with higher wear resistance.

Sample for the experiment was machined by turning technology on CNC turning centre. We monitored the residual stress with constant cutting speed (v_c= 50 m.s^{-1} for roughing and v_c=75m.s^{-1} for finishing) and depth (a_p=2mm for roughing and a_p=0.5mm for finishing) and increasing values of feed (f=0.17÷0.55mm for roughing and f=0.05÷0.1mm for finishing). The measuring of residual normal and shear stress was performed with Proto XRD diffractometer (Fig. 3) and electro-chemical polisher for removing of minimal layers of surface up to 200μm.

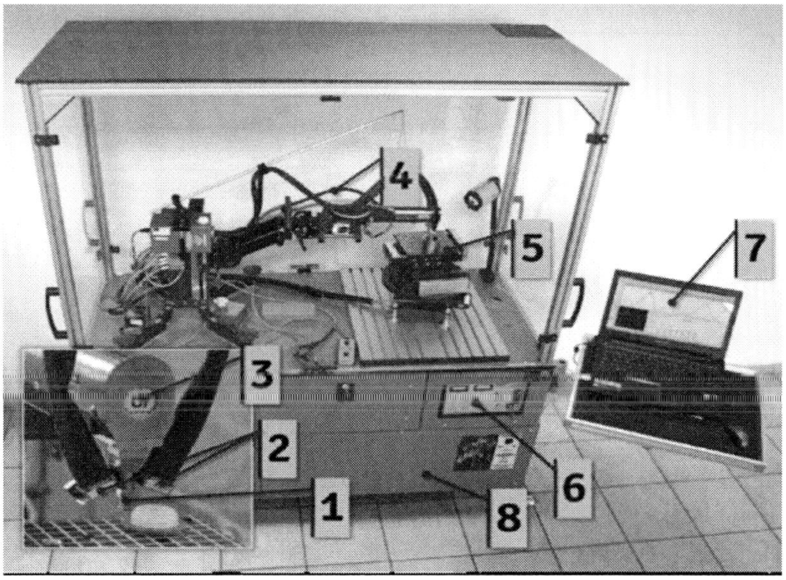

Fig. 3. X-ray diffraction residual stress measurement system (stationary: 1 - collimator; 2 – two detectors for diffraction cone; 3 - X-ray tube; 4 - Cobralink® flexible arm; 5 - positionable and rotary table; 6 – control unit; 7 – PC with software; 8 - laboratory stand

From the theory of elasticity, the relationship between residual stress (σ) and strain (ε) on the sample surface under plane stress is given by the Bragg equation, $\lambda=2d\sin\theta$, relating incident X-ray wavelength (λ), lattice inter-planar spacing (d) and diffraction angle (θ) (Fig. 4). The direction of maximum residual stress, that can be tensile or compressive, is assumed to occur in the cutting direction during most machining operations.

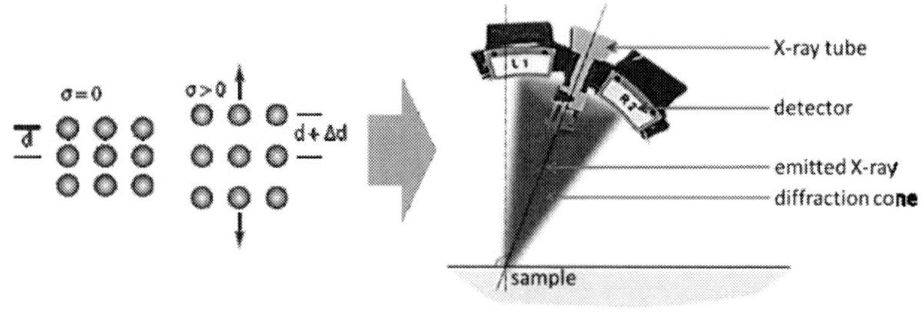

Fig. 4. Principle of measuring of residual stress by X-ray diffractometry based on Brag's Law

4 Results of experiments

Input data for simulation were measured as cloud of points or square map (10×10mm) consisting from point regularly disposed within its. Measurements were made in surface layer (depth of 0mm) and regular subsurface layers up to 200μm. On the following figures, it can be seen the selected courses of normal residual stress through the measured depth in one place (at one point).

Selected figures (Fig. 5 and 6) present the courses of residual stress through the measured depth. Almost all cases, with the given cutting conditions, has similar course. In the surface layer and higher subsurface layers, the residual stress had compressive character. With the increasing depth, these residual stress decreased, or changed character to the tensile.

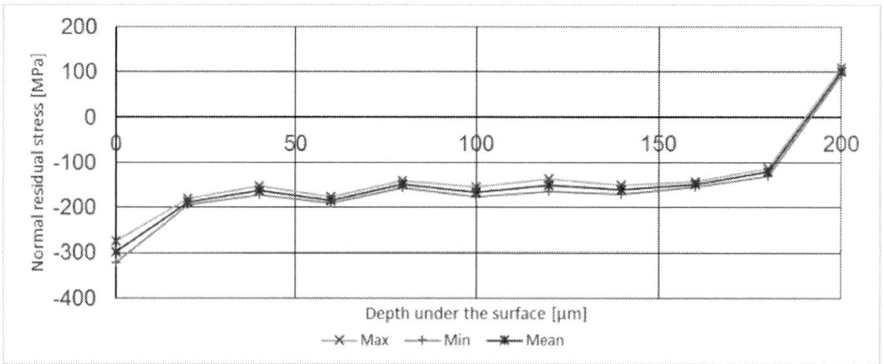

Fig. 5. Measured values of normal residual stress into the depth after roughing (cutting conditions: $v_c = 50m.s^{-1}$, $f = 0.17mm$, $a_p = 2mm$)

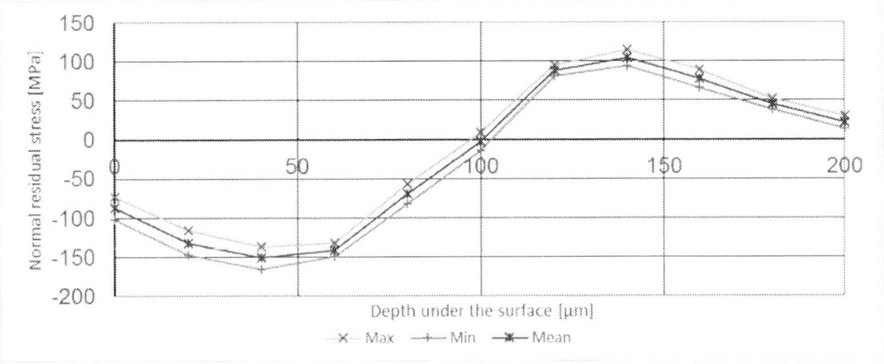

Fig. 6. Measured values of normal residual stress into the depth after finishing (cutting conditions: v_c = 75m.s^{-1}, f = 0.05mm, a_p = 0.5mm)

From measured data, there was computed via statistical software simulation of distribution of normal residual stress in three directions: direction of cutting speed, direction of cutting feed (or speed of cutting feed) and direction of depth. Subsequently, there were generated 3D graphs of these distributions.

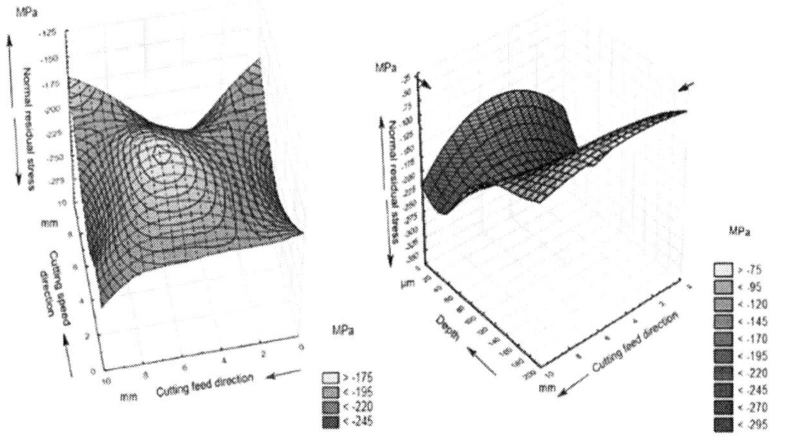

a) b)

Fig. 7. Simulation of distribution normal residual stress in various direction after roughing operations (cutting conditions: v_c = 50m.s^{-1}, f = 0.37mm, a_p = 2mm)

3D simulation of distribution of normal residual stress after roughing operations shows changes of normal residual stress in surface layer in direction of cutting speed and cutting feed (Fig. 7a). It can be seen no step changes: normal residual stress is in range between 175MPa and 245MPa with compressive character. Roughing operations with cutting depth 2mm are relatively suitable for next finishing operations without inheritance of visible signs of possible rise of errors. Simulation into the depth (Fig. 7b) of part material shows in deeper layers, residual stress with tensile character and range of values of very large (from 292MPa of compressive residual stress up to 74MPa of tensile residual stress). The risk of rise of errors is increased, but in many causes it can be improved by finishing operations or other treatments of surface.

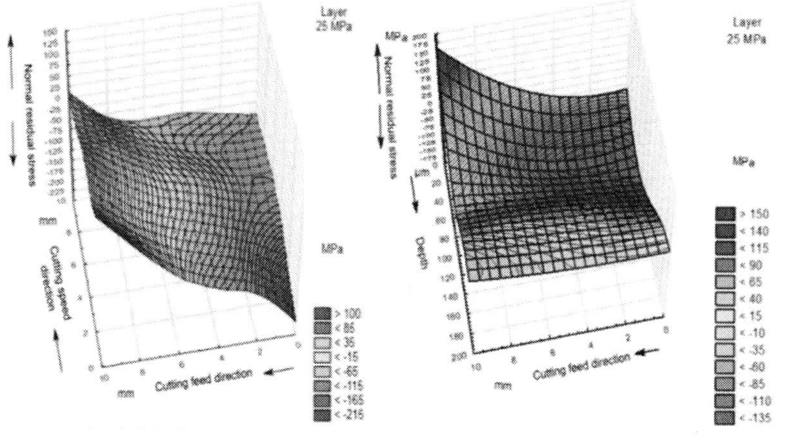

a) b)

Fig. 8. Simulation of distribution normal residual stress in various direction after finishing operations (cutting conditions: $v_c = 75 m.s^{-1}$, $f = 0.05mm$, $a_p = 0.5mm$)

Simulation of distribution normal residual stress, for finishing operations, shows rise of tensile residual stress with direction of cutting feed (Fig. 8a). It can be caused by gradual increase in the volume of extruded material by the cutting tool due to relatively low cutting depth (a_p=0.5mm). Simulation into the depth (Fig. 8b) shows increased appearance of residual stress with tensile character. Range of normal residual stress on surface layer and deepest measured layer (200μm) is very large again. It is not suitable for final state of part. This problem can be solved via improvements of surface and subsurface layers by applying other finishing methods such as shot peening.

Conclusions

Measuring of residual stress is actual problem by which we can predict the functional properties of parts. Some destructive methods allow to measure residual stress in subsurface layers, but due to its principles, they are not so detailed and accurate. Principles of X-ray diffraction can measure the residual stress at microstructure levels and so push the boundaries of measuring in this area. Automation of this device allows to measure not only individual points or small lines, but we can obtain cloud of points or maps of residual stress distribution. Through the mathematical-statistical procedures, we can minimalize the number of required measured points and so create sufficiently precise simulations of residual stress distribution. And finally, from this data we can see detailed course of residual stress depending on cutting conditions into the depth of machined material. In showed experiments, we can see, that course of residual stress can be changed dramatically in range of several micrometres. Applying by this simulations, we can design suitable cutting conditions and so to minimalize rise of potential errors, which can cause formation of cracks and subsequent failures.

This article was funded by the University of Žilina project APVV 15-0405 – "Complex use of X-ray diffractometry for identification and quantification of functional properties of dynamically loaded structural elements from important technical materials" and project of University of Zilina: OPVav-2009/2.2/04-SORO number (26220220101) – "Intelligent system for nondestructive technologies on evaluation for the functional properties of components of X-ray diffraction" and KEGA project 025ŽU-4/2017 Integration of innovative detectional-visaulization multi-disciplinary technologies as on-line tool for learning of progressive CNC technologies.

References

1. C. A. Griffiths, *Closing the Loop on Product Integrity on Bearings with Engineered Source Approval*. (Split Ballbearing Features, Company Newsletter, 1989)
2. N. Ganev, I. Kraus, *X-ray diffraction measurement of residual stresses*. Material Structure **9** (2), (2002)
3. K. Tiitto, et al., *Testing Shot Peening Stresses in the Field*. The Shot Peener **4**, ISSN 1069-2010 (1991)
4. R. Cep, J. Petru, T. Zlamal, J. Valicek, M. Harnicarova, A. Janasek, L. Cepova, *Influence of feed speed on machined surface quality*. 22nd International Conference on Metallurgy and Materials, Conference proceedings, ISBN 978-808729441-3, 1033-1038 (2013)
5. Z. Pala, K. Kolarik, N. Ganev, J. Capek, *Study of residual stress surface distribution on laser welded steel sheets*. on 51st Annual of the International Scientific Conference on Experimental Stress Analysis, EAN (2013)
6. M. Handrik, P. Kopas, V. Baniari, M. Vaško, M. Sága, *Analysis of stress and strain of fatigue specimens localised in the cross-sectional area of the gauge section testing on bi-axial fatigue machine loaded in the high-cycle fatigue region*. Procedia Engineering **177**, 516-519 (2017)
7. P. S. Prevéy, *X-ray Diffraction Residual Stress Techniques*. Metals Handbook **10**, Metals Park: American Society for Metals, p. 380-392 (1986)
8. M. E. Fitzpatrick, et.al., *Determination of Residual Stresses by X-ray Diffraction*. Measurement Good Practice **52**, National Physical Laboratory, Teddlingtoon
9. A. Czan, M. Sajgalik, A. Martikan, V. Kuzdak, *Identification of residual stress and structure properties in surface and subsurface layers of austenitic steel designed for nuclear energetics*. EAN 2015 – 53rd Conference on Experimental Stress Analysis, 36-40, ISBN 978-800105734-6 (2015)
10. M. Brezina, L. Kupca, *Stress corrosion cracking of austenitic stainless steel in a nuclear power plant enviroment*. Enviroment inducted cracking of materials. Department of Structural Analysis, VUJE Inc, Trnava, 431-436 (2008)
11. A. Gorog, D. Stancekova, M. Kapustova, I. Gorogova, I. Miturska, *Influence of Selected Cutting Conditions on Cutting Forces*. Advances In Science And Technology-Research Journal **11** (1), 179-185, ISSN 2299-8624
12. A. Czan, E. Tillova, J. Semcer, J. Pilc, *Surface and subsurface residual stresses after machining and their analysis by X-ray diffraction*. Comunications **15** (2), 69-76 (2013)
13. D. Mital, J. Zajac, Z. Mitalova, *Monitoring influence of selected parameters on surface quality after polymer powder coating*. MM Science Journal **2016** (October), 1077-1081 (2016)
14. N. Náprstková, F. Holešovský, *Admeasurement of Grinding Wheel Loss at EPTM*. 24th International Colloquium (Advanced manufacturing and repair technologies), 159-164, ISBN 978-80-7194-962-6 (2007)
15. P. Martikan, M. Drbul, J. Holubjak, J. Mrazik, R. Joch, *The Issue of Determining the Geometric Position Deviation of the Threaded Holes*. Advances in Science and Technology-Research J. **10** (32), 47-52 (2016)
16. R. Cep, A. Janasek, J. Petru, L. Cepova, A. Czan, J. Valicek, *Hard Machinable Machining of Cobalt-based Superalloy*. Manufacturing Technology **XIII** (13), UJEP: Ústi n. Labem, 226-231, ISSN 1213-2489

Diagnosis of machine tool with using Renishaw ball-bar system

Robert Čep[1,*], *Šárka* Malotová[1], *Jiří* Kratochvíl[1], *Dana* Stančeková[2], *Andrej* Czán[2], *Tomáš* Jakab[1]

[1]Faculty of Mechanical Engineering, VŠB – Technical University of Ostrava, 17. Listopadu 15/2172 708 33, Ostrava, Czech Republic
[2]Faculty of Mechanical Engineering, University of Zilina, Univerzitna 8215/1, 010 26 Zilina, Slovakia

Abstract. The article deals with accuracy of machine tools. The aim was to determine the accuracy with which already operates used machine – HAAS CNC Milling machine MiniMill. For evaluation of accuracy of the machine tool was used Renishaw QC20-W Ballbar system, which has the Department of Machining, Assembly and Engineering Metrology of VŠB – TUO. The theoretical part describes the measurement principle using a telescopic Ballbar system. The principle of measurement is the comparison of the actual deviation from the ideal circle which is programmed. Experimental part is focused on the diagnosis of the whole process by the direct method of measurement during changing feed rate. Specifically, the determination of circularity deviation at machine tool was in accordance with international standards, e.g. ISO 230-4.

Keywords: ball-bar system, milling machine, circularity, circularity deviation

1 Introduction

3-5 axis machine tools are increasingly applied in all areas of modern production. The main criterion of these machine tools is production of accurate components. This may be achieved under controlled process. Accuracy and reliability of the machine tools are in the first place. Many users think that buying expensive machine tool achieves a high quality production. Complex accuracy of machine tool is dependent on the geometric accuracy of the individual components of machine tool and on the toughness all the system. Identification, description and elimination of sources of error can improve the final working accuracy of machine tool. For these sources of error can be considered kinetic and thermo mechanical errors loads, dynamic forces, imperfection and deviation from individual functional parts of machine tools and their wear. Last but not least, there is an emphasis on the correct choice of machine strategy, cutting conditions, cutting tools and material of workpieces [1, 2].

Accuracy of machining can be characterized as degree of compliance of machined parts with its geometrical pattern on drawing. The agreement is bigger, so the accuracy of

---

* Corresponding author: rober.cep@vsb.cz

Reviewers: *Marcin* Kubiak, *Ivan* Kuric

machining is higher. On the other hand, relevant deviation (error) represents degree of compliance of part with geometrical pattern – inaccuracy. So that component could fulfil its function for which is specified, its real shape and surface have to deviate from ideal shape in definite acceptable limits. These limits are defined as deviations shape, position, dimensions from nominal values and surface roughness [3, 4].

2 Accuracy of machine tools

Diagnostic accuracy of machine tools is very important, and therefore is even bigger effort of service technician to approach the desired tolerances every thousandth of a millimetre. The machine tools are before putting on production process, submit tests geometric and working accuracy on the basis of standards ČSN ISO 230 – Test code for machine tools. Machine tools are under certain loading conditions and thanks these tests can be detected deficiencies, which have effect on functioning of the machine tools [5].

2.1 Test code for machine tools - part 4: circular tests for numerically controlled machine tools

Circular test is described by standards ISO 230-4. To determine the circular interpolation follow two methods.

Direct method – determination of accuracy of cylindrical surface milled at the workpiece with minimal removal of material. The principle of the method is the measurement of accuracy of a circle of the certain radius on the workpiece using the CMM. The advantage of this method is high accuracy of acquired results. The main disadvantage is duration causes by machining and measurement. Direct method of circular test gives to us fast results. For this test is chosen special device connected with control system of machine tool.

The main disadvantage of indirect method is that cutting forces which can cause inaccuracies of the machine tool during machining are not reflected in the measurement. Between other standards dealing with circular interpolation, can be classified ASME B5.54, ASME B5.57 (GB) or JIS B6194 (Japan) [5, 6].

Circularity deviation is one of deviation of shape and position. Real surfaces show deviation given nominal values. In this tolerance field must be the actual element. Circularity deviation is the biggest perpendicular distance measured to envelope circle (see Fig. 1).

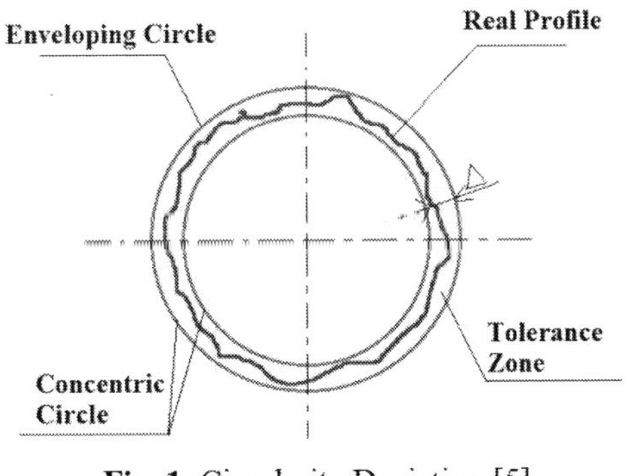

Fig. 1. Circularity Deviation [5]

Circularity deviations are measured with special touch device, which rotates around fixed component, respectively conversely or used CMM. Circularity deviation (Δ) of points of actual profile of enveloping circle is the biggest distance. These deviations together with surface roughness significantly influence the level of vibration of rotating parts, hence they need to be properly diagnosed and improving the whole process of production. The experiment on diagnosis of accuracy of machine tool was realized with using Ball-Bar system from Renishaw Company [3].

3 Measurement of accuracy

The aim of the article was to evaluate the accuracy with which operates the machine tool, HAAS CNC MiniMill, which is commonly used for the production of the medium-sized components. This vertical milling center operates with power 5.6 kW. Maximum speed is 6 000 rpm, working range of axis X/Y/Z is (406/305/254) mm. Diagnosis of accuracy of machine tool was evaluated on the basis Ball-Bar system QC20 – W. The device operates on the principle of circular interpolation using the direct method. The measuring of accuracy of machine tool was realized in central position, which is common used for machining of parts. Given small range of working table was not possible to apply the system in more position. In our case the circularity deviation was evaluated on unloaded machine in the XY plane at 5 different feed rates (see below). The Ball-Bar system was purchased by the Department of Machining, Assembly and Engineering Metrology, Faculty of Mechanical Engineering, VŠB – TUO.

Table 1. Measurement parameters

v_{f1}	v_{f2}	v_{f3}	v_{f4}	v_{f5}[1]
1 000 mm·min^{-1}	2 000 mm·min^{-1}	3 000 mm·min^{-1}	4 000 mm·min^{-1}	5 000 mm·min^{-1}

3.1 Diagnosis by system ballbar QC20 – W

System Ball-Bar QC20 – W falls into a group of telescopic systems from Renishaw using for analysis accuracy CNC machine tools. The ball bar is a very high accuracy, telescoping linear sensor with precision balls at each end. The machine performs two consecutive circular arcs (clockwise - CW and counter clockwise - CCW) in any one of the machines test planes (XY, YZ, ZX) and very accurately measures any variations in the test circle radius traced by the machine during the test. Software then analyses the measured data to give results in accordance with various international standards [1, 6].

Fig. 2. Sensor Ball-Bar [7]

Figure 2 shows the sensor Ball-Bar. The set also includes extension with lengths of 50, 150, and 300 mm. Ball holder allows testing in free orthogonal planes without moving the central pivot. Table 2 shows technical parameters of system Ball-Bar.

Table 2. Technical parameters of system Ball-Bar [7]

Technical parameters of system Ball-Bar	
Sensor resolution	0.1 μm
Ball-Bar measurement accuracy	± (0.7 + 0.3% L*) μm
Ball-Bar software measurement range	± 1.0 mm
Stroke sensor	-1.25 mm to +1.75 mm
Max. scanning rate	1 000 Hz
Transmission range (Bluetooth)	typically 10 m
Dimensions	(395 x 300 x 105) mm
Weight system	3.75 kg

*L – length of measuring errors

For evaluation of deviation was used standard ISO 230 – 4. All analysis was in the form of a graphical report, which specifically states of size of deviation. For connection of the system with machine tool we used Bluetooth connectivity. We created program for acquisition of data. The machine tool had to draw the circle 2times in clockwise (CW) and 2times counter clockwise (CCW). The resulting deviations were drawn in report. As shown in figure 3a, scanning data was taken place in second phase – after covering 180°, scanning arc (360°). Figure 3b shows setting o system Ball-Bar.

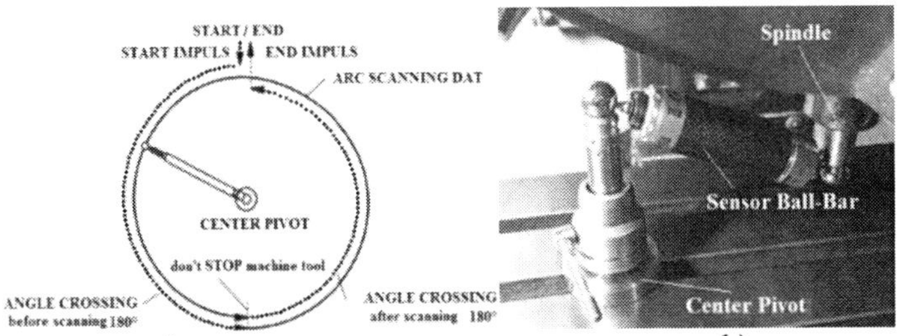

a) b)

Fig. 3. (a) Arc scanning data; (b) ball-bar sensor clamped in machine tool

4 Results of measurement

The values measured using a system Ball-Bar were evaluated based on ISO 230 – 4. Table 2 shows deviations in clockwise. For each feed rate measurement was performed four times.

Table 3. Measured deviation ISO 230 – 4 (CW) [3]

FEED RATE v_f **[mm·min^{-1}]**					
	1000	**2000**	**3000**	**4000**	**5000**
no.	**Δ [μm]**				
1	10.50	10.50	10.40	11.40	19.80
2	10.40	10.40	10.40	11.20	19.50
3	10.20	9.80	10.20	10.50	18.70
4	10.60	9.30	10.30	10.10	19.30
$\bar{\Delta}$ **[μm]**	(10.43 ± 0.17)	(10.00 ± 0.56)	(10.33 ± 0.10)	(10.55±0.47)	(19.33 ± 0.47)

The resulting circularity deviation for feed rate (1 000 – 4 000) mm·min^{-1} achieves almost constant values – about 10.5 μm. For feed rate 2 000 mm·min^{-1} – deviation for third and fourth measurement descends by 1 μm. Overall, the greatest measured value was during the highest testing feed rate – 5 000 mm·min^{-1}. The deviation was double – Δ = (19.33 ± 0.47) μm.

Figure 4 shows four shapes for measuring of accuracy of machine tool in clockwise for feed rate 2 000 mm·min^{-1}. In third and fourth case you can see these changes of deviation size.

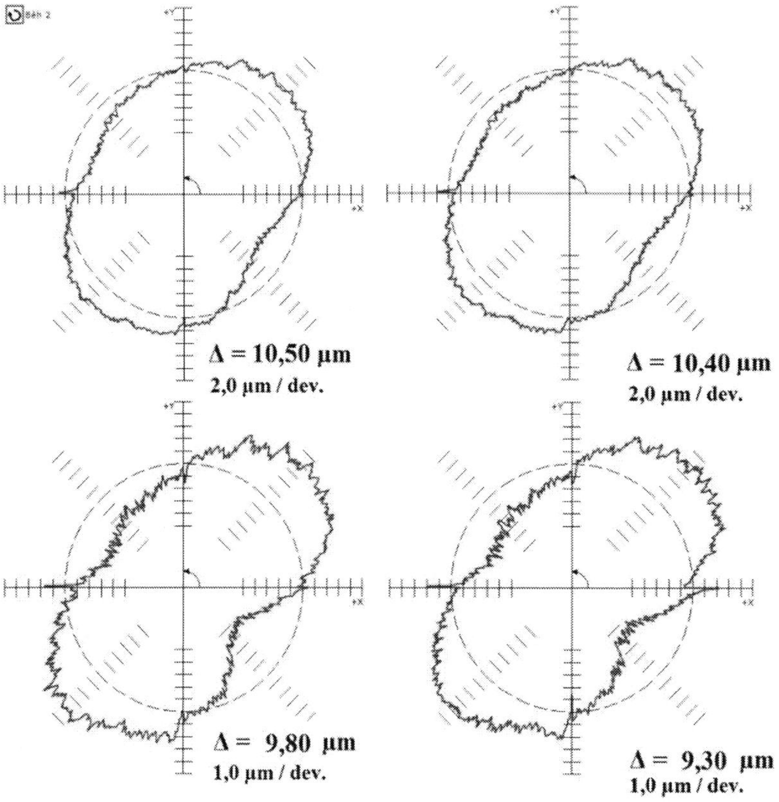

Fig. 4. Circularity Deviation – Clock Wise – v_f= 2000 mm·min^{-1}

Table 4. Measured deviation ISO 230 – 4 (CCW)[3]

CCW – COUNTER CLOCK WISE					
FEED RATE v_f [mm·min^{-1}]					
no.	1000 **Δ [μm]**	2000	3000	4000	5000
1	10.00	10.30	10.20	11.40	18.40
2	10.10	10.10	9.90	10.80	19.80
3	9.90	10.10	10.10	10.40	18.80
4	10.00	9.90	10.10	10.60	19.00
$\bar{\Delta}$ [μm]	(10.00 ± 0.08)	(10.10 ± 0.16)	(10.08 ± 0.13)	(10.80±0.43)	(19.03 ± 0.58)

Table 3 shows values of the deviations in the same conditions, but in counter clockwise. Measured values for feed rate (1000 – 3000) mm·min^{-1} have almost constant progress again. The average deviation with increasing feed rate increased linearly. The big change was measured for feed rate 5 000 mm·min^{-1}– Δ = (19.03 ± 0.58) μm. Figure 5 shows again shapes of deviations for the same feed rate, but in counterclockwise (CCW) for comparison with the previous case.

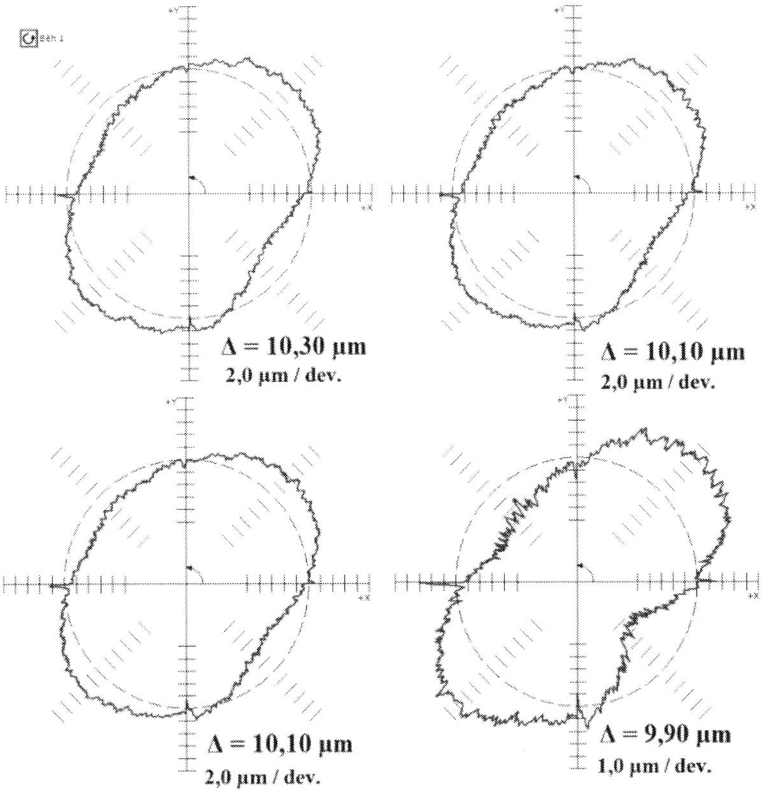

Fig. 5. Circularity Deviation – Counter Clock Wise - v_f= 2000 mm·min^{-1}

Conclusion

At companies are nowadays high demands on the accuracy of manufactured parts. One of the ways to achieve prescribed parameters is to keep machine tools in good condition and continuously evaluate his situation. Only machine tools, which comply with predetermined amount of deviation, may be used for final machining and ensures of production in tolerances. A suitable solution would be to use measuring system Renishaw Ball-Bar QC20 – W, when the machine tool is measured in the unloaded condition and monitored circularity deviation from ideal circle. The advantage is the evaluation according to variety of standards.

The aim of article was to determine accuracy of the selected milling machine HAAS MiniMill. Circularity deviations were evaluated in plane XY during unloaded phase of machine tool. The measuring equipment was put in the central of machine tool, because the most of workpieces are situated here for machining. It was set five different feed rates from 1 to 5 m·min^{-1}. For individual feed rate was detected circularity deviation in clockwise and counter clockwise. All measurements were measured always 4times at the same parameters and calculated arithmetic mean and standard deviation. In the both cases (CW and CCW) were evaluated, that with increasing feed rate, to value 4 000 mm.min^{-1}, circularity deviations were on the same level. With feed rate 5 000 mm.min^{-1}, circularity deviation increased sharply up to twice the amount (see in fig. 6). The accuracy of machine tool and machined surface decreased with this value of feed rate. This machine tool is suitable to use for finishing operations for feed rates up to 4 m.min^{-1}. At higher feed rates only use on roughing or preparatory operations.

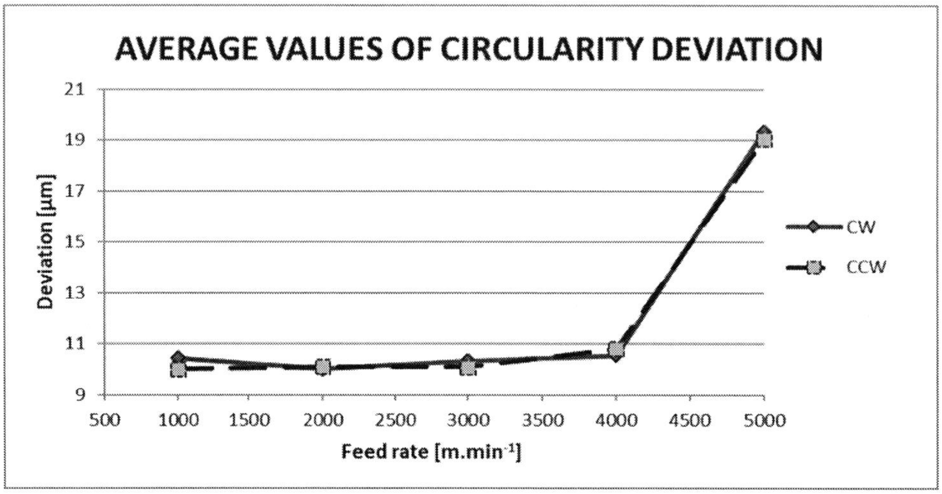

Fig. 6. Scheme of the average CW and CCW deviations

Article has been done in connection with projects Education system for personal resource of development and research in field of modern trend of surface engineering - surface integrity, reg. no. CZ.1.07/2.3.00/20.0037 financed by Structural Founds of Europe Union and from the means of state budget of the Czech Republic and by project Students Grant Competition SP2016/172 and SP2016/174 financed by the Ministry of Education, Youth and Sports and Faculty of Mechanical Engineering VŠB-TUO.

References

1. J. Dekan, *Diagnostic equipment for monitoring of manufacturing machines*. Strojárstvo /Strojírenství: Stroje a technológie [online] [cit. 2016-04-08] http://www.engineering.sk/index.php/clanky2/stroje-a-technologie/546-diagnosticke-zariadenia-monitorujuce-stav-obrabacich-strojov (2012)
2. H. Schwenke, et al., *Geometric error measurement and compensation of machines-An update*. ELSEVIER - Manufacturing Technology **16** (57), (2008)
3. T. Jakab, *Diagnostics of machining equipment by using of Renishaw system QC20-W BALLBAR*. Ostrava, Diploma thesis. VŠB - TUO (2015)
4. P. Demeč, *Accuracy of manufacturing machines and their mathematical modelling*. (Edícia vedeckej a odbornej literatúry – SjF TU v Košiciach, Vienala Košice, 15-28, ISBN 80-7099-620-X, 2001)
5. ČSN ISO 230- 4. *Zásady zkoušek obráběcích strojů: – část 4: zkoušky kruhové interpolace u číslicově řízených obráběcích strojů*. (2010)
6. M. Beňo, *Circular interpolation and positioning accuracy deviation measurement on five axis machine tools with different structures*. Technical Gazette **20** (3), 6, ISSN 1330-3651 (2013)
7. *Ball bar testing explained*. [online] Renishaw plc. [cit. 2016-04-08] Available at: http://www.renishaw.com/en/ballbar-testing-explained--6818 (2001)

The stability analysis of two-wheeled vehicle model

Juraj Gerlici[1], *Vladimir* Sakhno[2], *Alla* Yefymenko[2], *Vladimir* Verbitskii[3], *Alexandr* Kravchenko[4], *Kateryna* Kravchenko[1,*]

[1]University of Žilina, Department of Transport and Handling Machines, Univerzitná 1, 010 26 Žilina, Slovak Republic
[2]National Transport University, Department of Automobiles, Suvorova 1, 01010, Kiev, Ukraine
[3]Zaporizhzhia State Engineering Academy, Department of Automation System Maintenance, Soborny Ave. 226, 69006, Zaporizhzhia, Ukraine
[4]Zhytomyr State Technological University, Department of Automobiles and Motor Industry, Chernyakhovsky 103, 10005, Zhytomyr, Ukraine

Abstract. This research investigates the stability of a two-wheeled vehicle model on the basis of numerical determination of full range of eigenvalues of a linear approximation matrix in the vicinity of the rectilinear driving mode. The received result was checked by numerical integration of the initial equations system of the disturbed motion of the model. The discrepancy of two research techniques is explained by the specialty of the considered mathematical model in which two pairs of complex conjugate eigenvalues close to each other are realized, that explains the emergence of standard derivations at calculating their numerical determination. The model is asymptotically stable in the range much wider than an operational interval (up to 100 m/s). In order to provide more intensive dampening of initial disturbances, it is possible to introduce additional resilient and damping elements between the trucks and the body in the design of the wheeled vehicle that will counteract the yaw mode of trucks.

Keywords: wheeled vehicle, truck, guiding wheeled module, mathematical model

1 Introduction

Monorail vehicles are used in many countries. That provides safety work of transport systems in big cities. The increase of the population in metropolises of Ukraine was a cause of new problems in public transport loading level. To solve these problems it is possible to implement monorail systems that will have several benefits, such as compactness, the cost of implementation and the safety of passengers' transit. Based on the fact that monorail vehicle is rapid-transit, the crucial task of the investigation is the stability of its running. The practical interest is aroused to the monorail vehicle with means of support «Alweg», discussed in Ryan R. Kennedy's works and online www.STC-IN.com & www.skytraincorp.com (2004) [1, 2]. In Sakhno's and Yefymenko's recent researches

* Corresponding author: Kateryna.Kravchenko@fstroj.uniza.sk

Reviewers: *Milan* Sága, *Milan* Naď

(2014, 2015, 2016) [3, 4, 5] the structural features of a monorail vehicle (hereinafter referred to as a wheeled vehicle) were investigated, the base unit of which is a truck that moves on four carrying pneumatic tires.

Traffic safety from a convergence or tipover, relative to the overpass, is provided by the guiding wheels. The mathematical model of the wheeled vehicle motion is considered in Korotenko's (1984) [6] and Yefymenko's (2015) [7] works where the authors analyse the rectilinear motion of the car and the movement along the transitional curves of the overpass, taking into account the influence of the forces and withdrawal moments arising in the contact spot of the bearing wheel with the road surface, as well as the elastic forces and moments generated by the guiding wheels.

The self-sustained vibrations in above mentioned works were not analysed; the task of the research was to establish the fact of stability or instability by the linear approximation (Korotenko, 1984) [6].

The key feature of the wheeled vehicle construction considered in this work is the fact that each guiding truck has one traction wheel (Fig. 1). This design solution is proved by weight reduction and simplification of the wheeled vehicle operation. However, it can lead to undesirable self-sustained vibrations. Various approaches to the analysis of rail and wheel vehicles self-sustained vibrations are presented in the works by Domin (1984) and Lobas (1990) [8, 9]. Sakhno (2016) [10] used the numerical and analytical method and obtained an assessment of self-sustained vibrations amplitudes of a truck model of the wheeled vehicle in the straight lines of the overpass. Based on the methods of determining the stationary modes of the model motion, supplemented by an algorithm for constructing the bifurcation set in conjunction with the Poincare Index method, the work (2013) [11] analysed the number of fixed modes, and the stability boundaries were defined in the plane of the controlled parameters.

The purpose of the article. The purpose of this article is to analyse the stability of the two-wheeled vehicle model in the range of operating speed and the rational choice of its design values.

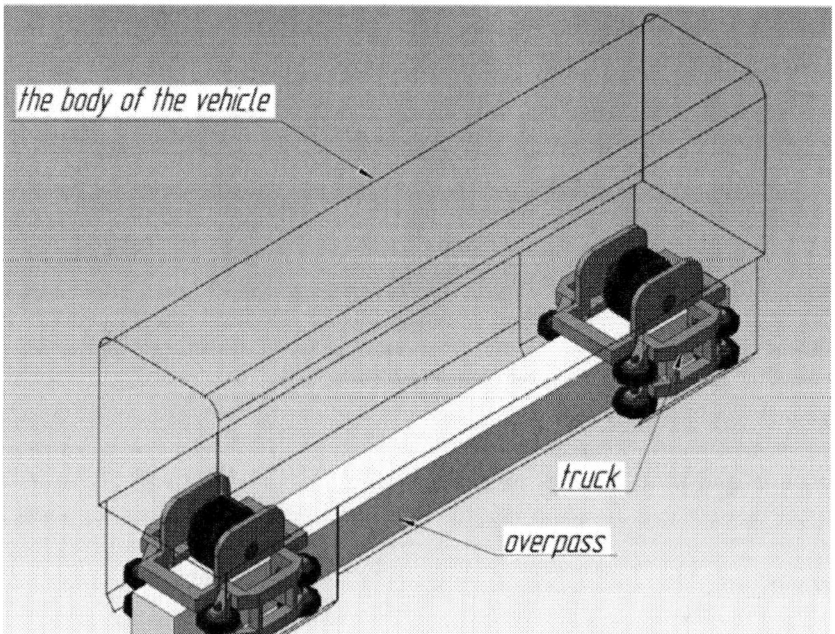

Fig. 1. Two-wheeled vehicle model

2 The stability analysis of straight-line motion of the two-wheeled vehicle model

The wheeled vehicle consists of two trucks moving on pneumatical tires and performing the functions of traction wheels 1 and guiding wheels 2 (Fig. 2). The initial set of equations of the "two-wheeled" vehicle motion (1) on a straightway overpass is given relative to the inertial coordinate system moving with uniform velocity along the overpass centerline (Fig. 3).

Fig. 2. The scheme of the wheeled vehicle: 1 – traction wheels, 2 – guiding wheels

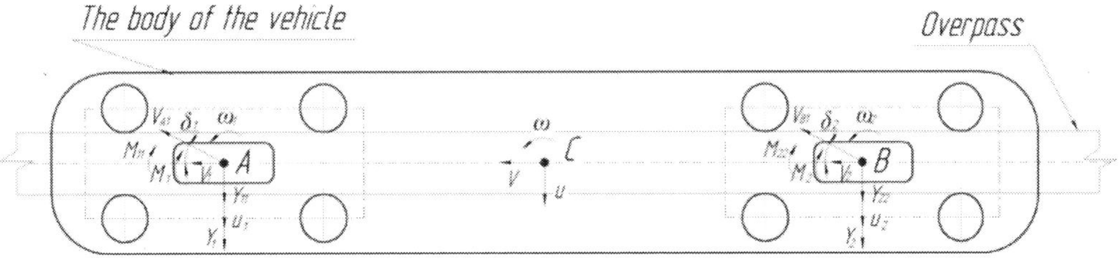

Fig. 3. Calculated bicycle scheme of the wheeled vehicle with two trucks

It is accepted in the work that the body position of the wheeled vehicle is calculated by the position of its mass center (x_C, y_C) and by the heading angle ψ, the position of the trucks is calculated by the rotation angles ψ_1 and ψ_2, relative to the axis of the wheeled vehicle body.

The dynamic equations of the body plane-parallel motion are described by two half degrees of freedom: the transverse component of the mass center velocity u and angular velocity relative to the vertical axis ω.

The external forces and moments that have effect on the wheeled vehicle:

- the reduced elastic force Y_i and the elastic moment M_i caused by the guiding module from the overpass correspond to the ideas accepted before (Yefymenko, 2015 and Korotenko, 1984);
- forces Y_{ii} and moments M_{ii} of the traction wheel slip (i - truck number);
- X_{ii} – longitudinal forces in a spot of the traction wheels contact.

In rotational articulation (cylindrical joints) of the truck with the wheeled vehicle body the internal forces appear, which are formally withdrawn at the derivation of the equations of the whole system motion.

Thus, the equations of plane-parallel motion of the system, assuming that the longitudinal component of the vehicle mass center (parameter v) is kept constant, are as follows:

$$-m_1(V_1-\omega_1 u_1)-XA\cdot\cos\psi_1+X_{11}-YA\cdot\sin\psi_1=0;$$
$$-m_1(U_1+\omega_1 v_1)+XA\cdot\sin\psi_1-YA\cdot\cos\psi_1+Y_{11}+Y_1=0; \quad (1)$$

$$-m_2\left(V_2-\omega_2 u_2\right)-\mathrm{XB}\cdot\cos\psi_2+X_{22}-\mathrm{YB}\cdot\sin\psi_2=0;$$

$$-m_2\left(U_2+\omega_2 v_2\right)+\mathrm{XB}\cdot\sin\psi_2-\mathrm{YB}\cdot\cos\psi_2+Y_{22}+Y_2=0.$$

The unknown reactions in rotational articulations with the wheeled vehicle body YA, XA, YB, XB are calculated by the set of equations (1) of the plane-parallel motion of the guiding trucks.

$$\mathrm{XA}=\frac{-\cos\psi_1\cdot m_1\omega_1\mathrm{u}_1\text{-}\sin\psi_1 m_1\omega_1\mathrm{v}_1+\cos\psi_1\mathrm{V}_1 m_1\text{-}\sin\psi_1\mathrm{U}_1 m_1\text{-}\cos\psi_1 X_{11}+\sin\psi_1 Y_1+\sin\psi_1 Y_{11}}{\cos\psi_1{}^2+\sin\psi_1{}^2};$$

$$\mathrm{XB}=-\frac{-\cos\psi_2 m_2\omega_2\mathrm{u}_2-\sin\psi_2 m_2\omega_2\mathrm{v}_2+\cos\psi_2\mathrm{V}_2 m_2-\sin\psi_2\mathrm{U}_2 m_2-\cos\psi_2 X_{22}+\sin\psi_2 Y_2+\sin\psi_2 Y_{22}}{\cos\psi_1{}^2+\sin\psi_1{}^2};$$

$$\mathrm{YA}=\frac{-m_1\omega_1\mathrm{v}_1\cos\psi_1+\sin\psi_1 m_1\omega_1\mathrm{u}_1+\mathrm{U}_1 m_1\cos\psi_1+\sin\psi_1\mathrm{V}_1 m_1-Y_1\cos\psi_1-Y_{11}\cos\psi_1-\sin\psi_1 X_{11}}{\cos\psi_1{}^2+\sin\psi_1{}^2};$$

$$\mathrm{YB}=-\frac{m_2\omega_2\mathrm{v}_2\cos\psi_2-\sin\psi_2 m_2\omega_2\mathrm{u}_2+\mathrm{U}_2 m_2\cos\psi_2+\sin\psi_2\mathrm{V}_2 m_2-Y_2\cdot\cos\psi_2-Y_{22}\cdot\cos\psi_2-\sin\psi_2\cdot X_{22}}{\cos\psi_1{}^2+\sin\psi_1{}^2}.$$

where ω - the vehicle body angular velocity with according to the vertical axis, ψ_1 - relative angular velocity of the *i-th* truck, m - mass of the vehicle body, m_i - mass of the *i*-th truck.

The main kinematic relations used in the research are given below. The angular velocity of the guiding wheeled module is:

$$\omega_1=\omega+\psi_1,\ \ \omega_2=\omega+\psi_2 \tag{2}$$

The angular acceleration of the guiding wheeled module is:

$$\Omega_1=\Omega+\psi_1,\ \ \Omega_2=\Omega+\psi_2 \tag{3}$$

where Ω - the angular acceleration of the vehicle body.

The determination of lateral and longitudinal projections of the mass center velocity of the guiding wheeled module "A" was calculated using the following equations:

$$u_1=-v\cdot\sin\psi_1+\left(u+\mathrm{LA}\cdot\omega\right)\cos\psi_1;$$
$$v_1=v\cdot\cos\psi_1+\left(u+\mathrm{LA}\cdot\omega\right)\sin\psi_1. \tag{4}$$

The determination of lateral and longitudinal projections of the mass center velocity of the guiding wheeled module "B" was calculated by (5):

$$u_2=-v\cdot\sin\psi_2+\left(u-\mathrm{LB}\cdot\omega\right)\cos\psi_2;$$
$$v_2=v\cdot\cos\psi_2+\left(u-\mathrm{LB}\cdot\omega\right)\sin\psi_2. \tag{5}$$

LA, LB is the distance from the car mass center to the hinges "A" and "B".

The determination of lateral and longitudinal projections of the mass center acceleration of the guiding wheeled modules:

$$U_1=-V\cdot\sin\ \psi_1-v\cdot\cos\ \psi_1\cdot\psi_1+(U+\mathrm{LA}\cdot\Omega)\cdot\cos\ \psi_1+(u+\mathrm{LA}\cdot\omega)\cdot\sin\ \psi_1\cdot\psi_1;$$
$$V_1=V\cdot\cos\ \psi_1-v\cdot\sin\ \psi_1\cdot\psi_1+(U+\mathrm{LA}\cdot\Omega)\cdot\sin\ \psi_1+(u+\mathrm{LA}\cdot\omega)\cdot\cos\ \psi_1\cdot\psi_1; \tag{6}$$

$$U_2 = -V\cdot\sin\psi_2 - v\cdot\cos\psi_2\cdot\psi_2 + (U - \mathrm{LB}\cdot\Omega)\cdot\cos\psi_2 + (u - \mathrm{LB}\cdot\omega)\cdot\sin\psi_2\cdot\psi_2;$$
$$V_2 = V\cdot\cos\psi_2 - v\cdot\sin\psi_2\cdot\psi_2 + (U - \mathrm{LB}\cdot\Omega)\cdot\sin\psi_2 + (u - \mathrm{LB}\cdot\omega)\cdot\cos\psi_2\cdot\psi_2. \quad (7)$$

After the substitution of certain reactions (rotational articulations) in the equations of the plane-parallel wheeled vehicle motion, we have the closed equation system of motion (8-9):

$$\begin{aligned}
m(V-\omega u) &= \mathrm{XA}+\mathrm{XB} = -\mathrm{V}m_1 - \mathrm{V}m_2 + \cos\psi_2 X_{22} - \sin\psi_2 Y_2 - \sin\psi_2 Y_{22} + \cos\psi_1 X_{11} - \\
&- \sin\psi_1 Y_1 - \sin\psi_1 Y_{11} + \mathrm{LA}\cdot m_1\omega^2 - \mathrm{LB}\cdot m_2\omega^2 + m_1\omega\mathrm{u} + m_2\omega\mathrm{u};\\
m(\omega V+U) &= \mathrm{YA}+\mathrm{YB} = Y_1\cos\psi_1 + Y_{11}\cos\psi_1 + \sin\psi_1 X_{11} + Y_2\cos\psi_2 + Y_{22}\cos\psi_2 + \\
&+ \sin\psi_2 X_{22} + \mathrm{LB}\cdot\Omega m_2 - \mathrm{LA}\cdot\Omega m_1 - m_1\omega\mathrm{v} - m_2\omega\mathrm{v} - \mathrm{U}m_2 - \mathrm{U}m_1;\\
\mathrm{J}\Omega &= -\mathrm{LA}\cdot m_1\omega\mathrm{v} + \mathrm{LB}\cdot m_2\omega\mathrm{v} + \mathrm{LB}\cdot\mathrm{U}m_2 - \mathrm{LA}\cdot\mathrm{U}m_1 - \mathrm{LB}^2\omega m_2 - \mathrm{LA}^2\omega m_1 - \\
&- \mathrm{LB}\cdot\cos\psi_2 Y_2 - \mathrm{LB}\cdot\cos\psi_2 Y_{22} - \mathrm{LB}\cdot\sin\psi_2 X_{22} + \mathrm{LA}\cdot\cos\psi_1 Y_1 + \\
&+ \mathrm{LA}\cdot\cos\psi_1 Y_{11} + \mathrm{LA}\cdot\sin\psi_1 X_{11},
\end{aligned} \quad (8)$$

The equations of the truck guiding wheeled modules rotary motion:

$$\mathrm{J}_1\Omega_1 = M_1 - M_{11};\ (\Omega_1 = \Omega + \psi_1);$$
$$\mathrm{J}_2\Omega_2 = M_2 - M_{22};\ (\Omega_2 = \Omega + \psi_2). \quad (9)$$

M_1 - the total moment of elastic lateral reactions on the i-th truck;
M_{11} - the moment of deviation of the carrier wheel on the i-th truck;
J_1 - the central moments of inertia of the vehicle body with respect to the vertical axis.
J_2 - the central moments of inertia of the truck with respect to the vertical axis.
Forces in a coupling point "A" (projections to longitudinal axes and lateral axes):

$$\mathrm{XA} = \mathrm{LA}\cdot m_1\cdot\omega^2 + m_1\cdot\omega\cdot\mathrm{u} - \mathrm{V}\cdot m_1 + \cos\psi_1\cdot X_{11} - \sin\psi_1\cdot Y_1 - \sin\psi_1 Y_{11};$$
$$\mathrm{YA} = -\mathrm{LA}\ \Omega\ m_1 - m_1\omega\mathrm{v} - \mathrm{U}m_1 + Y_1\cdot\cos\psi_1 + Y_{11}\cos\psi_1 + \sin\psi_1\cdot X_{11}. \quad (10)$$

The equations of the dynamics of the plane-parallel motion of a wheeled vehicle assuming that the system maintains a constant value of the longitudinal component of the body mass center velocity (parameter v) and there are no longitudinal forces X_{ii} (LA, LB is the distance from the car mass center to the hinges "A" and "B") are written in the form:

$$\begin{aligned}
m(\omega\mathrm{v}+U) &= Y_1\cdot\cos\psi_1 + Y_{11}\cdot\cos\psi_1 + \sin\psi_1\cdot X_{11} + Y_2\cdot\cos\psi_2 + Y_{22}\cdot\cos\psi_2 + \\
&+ \sin\psi_2\cdot X_{22} + \mathrm{LB}\cdot\Omega\cdot\mathrm{m}_2 - \mathrm{LA}\cdot\Omega\cdot m_1 - m_1\omega\mathrm{v} - m_2\omega\mathrm{v} - \mathrm{U}m_2 - \mathrm{U}m_1;
\end{aligned}$$

$$\begin{aligned}
\mathrm{J}\Omega &= -\mathrm{LA}\cdot m_1\omega\mathrm{v} - \mathrm{LB}^2\cdot\Omega m_2 - \mathrm{LA}^2\cdot\Omega m_1 + \mathrm{LB}\cdot\mathrm{U}m_2 + \mathrm{LB}\cdot m_2\omega\mathrm{v} - \mathrm{LB}\cdot\cos\psi_2\cdot Y_2 - \\
&- \mathrm{LB}\cdot\cos\psi_2\cdot Y_{22} - \mathrm{LB}\cdot\sin\psi_2\cdot X_{22} + \mathrm{LA}\cdot\cos\psi_1\cdot Y_1 + \mathrm{LA}\cdot\cos\psi_1\cdot Y_{11} + \mathrm{LA}\cdot\sin\psi_1\cdot X_{11} - \\
&- \mathrm{LA}\cdot\mathrm{U}m_1;
\end{aligned} \quad (11)$$

$$J_1(\Omega + \psi_1) = -M_1 - M_{11};\ J_2(\Omega + \psi_2) = -M_2 - M_{22}.$$

Approximation of the nonlinear dependences of forces and withdrawal moments can be presented in the form:

$$Y_{ii} = k_i \cdot \delta_i / \sqrt{1 + \left(k_i\delta_i / (\varphi_i N_i)\right)^2}; \tag{12}$$

$$M_{ii} = A_i \cdot \delta_i / (B_i\delta_i^4 + C_i\delta_i^2 + 1); \tag{13}$$

k_i - slip drag coefficient;
δ_i - slip angle of the traction wheel of the truck i;
φ - coefficient of friction of a wheel in the transverse direction;
N_i - load on the traction wheel of the truck i;
A_i, B_i, C_i - coefficients that are determined empirically.
The slip angles on the traction wheels of the truck i.

$$\delta_1 = -\,arctg\frac{-v\sin\psi_1 + (\mathrm{LA}\omega + u)\cos\psi_1}{v\cos\psi_1 + (\mathrm{LA}\omega + u)\sin\psi_1};\; \delta_2 = arctg\frac{v\sin\psi_2 - (-\mathrm{LB}\omega + u)\cos\psi_2}{v\cos\psi_2 + (-\mathrm{LB}\omega + u)\sin\psi_2}. \tag{14}$$

The elastic forces Y_i (15) and the elastic moment M_i (16) linearly depend on the transverse displacement of the truck mass center relative to the program trajectory and the difference of the truck heading angles and the program curve respectively.

$$\begin{gathered} Y_1 = -C_y \cdot y_A;\; Y_2 = -C_y \cdot y_B; \\ x_A = x_C + \mathrm{LA}\cdot\cos\psi;\; y_A = y_C + \mathrm{LA}\cdot\sin\psi; \end{gathered} \tag{15}$$

$$\begin{gathered} x_B = x_C - \mathrm{LB}\cdot\cos\psi;\; y_B = y_C - \mathrm{LB}\cdot\sin\psi; \\ M_1 = Cy\cdot ll^2(\psi + \psi_1);\; M_2 = Cy\cdot ll^2(\psi + \psi_2); \end{gathered} \tag{16}$$

C_y - lateral stiffness of pneumatic wheels;
x_A, y_A, x_B, y_B - coordinates of the wheeled vehicle truck mass center;
ll - base regulating wheeled module;
ψ_1, ψ_2 - heading angles of the trucks.

3 Research results

The paper analyzes the stability of the rectilinear motion regime of a two-wheeled vehicle model based on the numerical determination of the complete set of eigenvalues of the system of linear approximation equations for the numerical values of the design parameters of the system listed below:
$m_1 = m_2 = 1320$ kg; $m = 16000$ kg; $J_1 = J_2 = 4224$ kg·m^2; $J = 274400$ kg·m^2; LA=LB = 7 m; $k_1 = k_2 = 360000$ N; $C_y = 250000$ N/m; $g = 9{,}8$ m/s^2; $N_1 = N_2 = 45715$; $ll = 1{,}5$ m; $\varphi = 0{,}8$.

The speed parameter was discretely varied to determine the flutter (oscillatory instability) boundary when one control parameter was changed. In the general case, a pair of complex conjugate eigenvalues goes over into the right half-plane. However, the results of numerical integration of the initial equations of the disturbed motion indicated the damping of the initial perturbations in the velocity range, which is much broader than the boundary of the oscillatory instability thus found.

The analysis of this discrepancy is based on the numerical integration of the wheeled vehicle system at the speed of 12.3 m/s and 22.3 m/s with a fixed initial disturbance at the heading angle $\psi_2 = 0.001$ rad (second truck).

The position of the eigenvalues of the system of linear approximation in the complex plane at a velocity of 12.3 m/s and 22.3 m /s is shown in Fig. 3. a), b). The first value of the velocity parameter belongs to the stability interval, the second supposedly should belong to the area of the oscillatory instability.

Both in the first and in the second case there are two pairs of complex-conjugate roots, which differ little from each other. As it turned out, this fact is a characteristic feature of this system and remains valid in a fairly wide range of speeds, since the subsystems corresponding to the heading angles have practically identical partial circular frequencies, the latter are generated by the elastic moments of the guiding wheeled modules, and a small discrepancy between the partial frequencies is entered by the stabilizing moments M_{11} and M_{22}. Thus, the presence of sufficiently close two pairs of eigenvalues (located near the imaginary axis) can lead to the emergence of an error at their numerical determination.

This conclusion is fully confirmed by the results of numerical integration of the initial system, presented in Figures 5-7. Indeed, they illustrate the well-known phenomenon of synchronization of oscillations of two subsystems with close eigenfrequencies, which is observed in a wide range of the velocity parameter; on a time interval of the order of several tens of seconds, energy is "pumped" from one subsystem to another against the background of dissipative processes that in the final analysis provides asymptotic stability to undisturbed rectilinear motion.

To demonstrate close vibration frequencies with respect to the variables ψ_1, ψ_2, the fragments of the combined integral curves $\psi_1(t)$, $\psi_2(t)$ in a small time interval are represented in Fig. 7, and the integral curve with the phase variable ψ_1 corresponds to a harmonic with a smaller value of the amplitude.

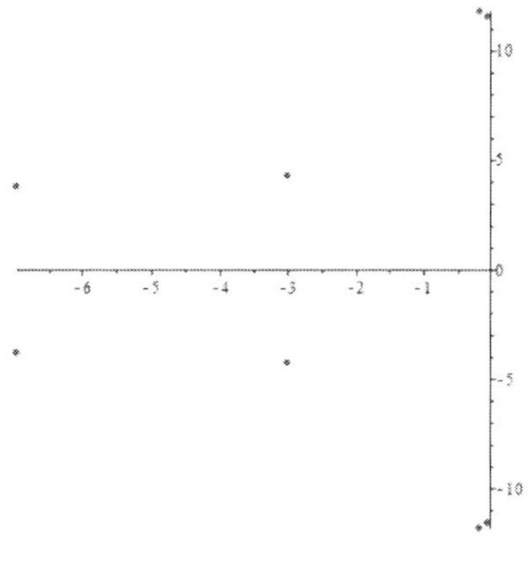

a)

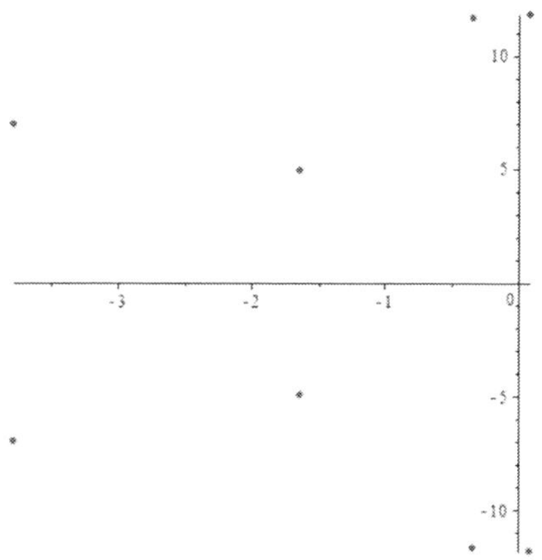

b)

Fig. 4. The position of eigenvalues of the linear approximation system near the rectilinear motion: a) - v = 12.3 m/s; b) - v = 22.3 m/s

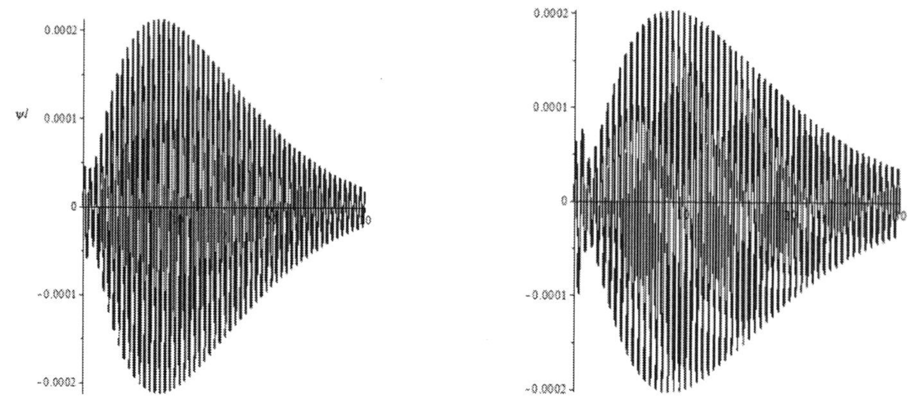

a) b)

Fig. 5. The heading angle of the first truck of the wheeled vehicle:
a) - v =12.3 m/s; b) - v =22.3 m/s

At a speed of v = 22.3 m/s, based on the analysis of integral curves, a practical coincidence of the oscillation frequencies with respect to the variables ψ_1, ψ_2 is observed, which explains, in the final analysis, the cause of the error in the numerical determination of the set of eigenvalues.

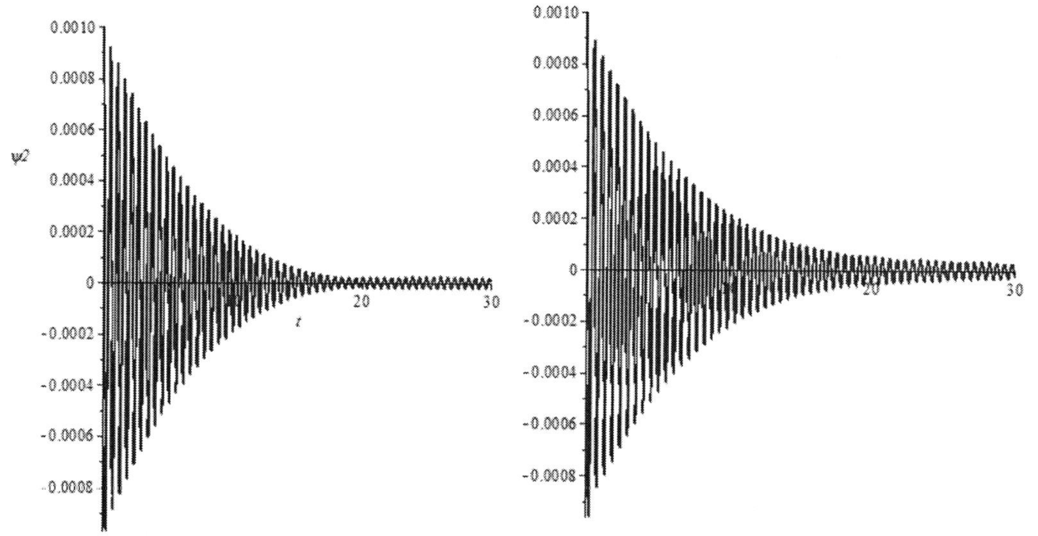

a) b)

Fig. 6. The heading angle of the second truck of the wheeled vehicle:
a) - v =12.3 m/s; b) - v =22.3 m/s

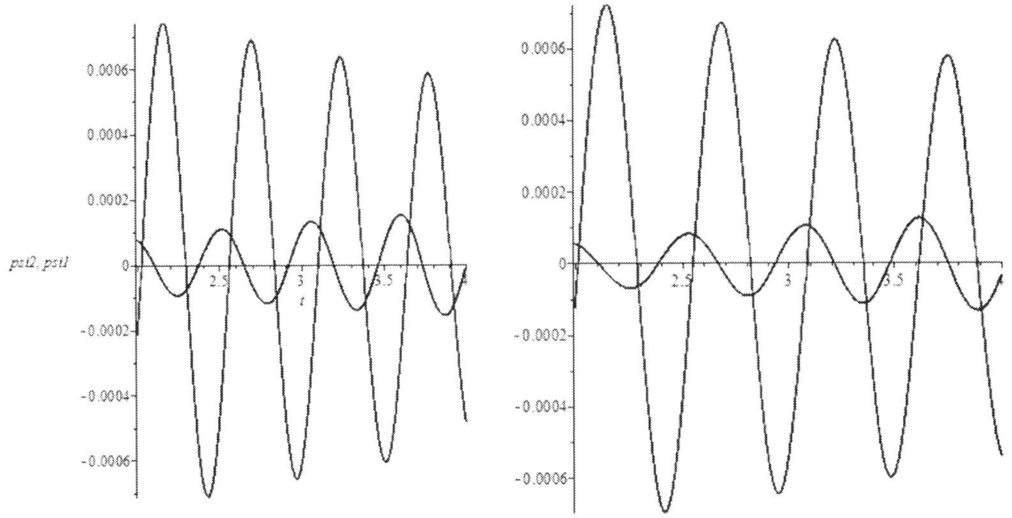

a) b)

Fig. 7. The heading angles of both trucks of the wheeled vehicle:
a) - v =12.3 m/s; b) - v =22.3 m/s

Conclusions

The presented plane (limited) model of a two-wheeled vehicle is asymptotically stable in a fairly wide range of speed (up to 100 m/s), however, as it can be seen from the results of numerical integration, the subsystems corresponding to the yawing angles of the trucks turned out to be synchronized. As a consequence, unwanted dynamic transient processes occur on the time intervals of the order of three to four tens of seconds, which can lead to some discomfort and fatigue of passengers. To ensure a more intensive attenuation of the initial disturbances, it is possible to introduce additional elastic and damping elements into the car's construction between the trucks and the body, which will parry the yawing of the trucks.

References

1. R. R. Kennedy, *Considering Monorail Rapid Transit for North American Cities.* available at: http://www.monorails.org/webpix%202/RyanRKennedy.pdf (2004)
2. www.STC-IN.com & www.skytraincorp.com, *Monorail History and Technology of Successful Technology.* available at: http://www.skytraincorp.com/pp/stc_mht.pdf (2004)
3. V. P. Sakhno, A. M. Yefymenko, *About choosing the rational values of transverse stiffness parameters of the trolley guiding wheel module on pneumatic tires of the wheeled vehicle (analog of the system Alweg).* The National Transport University Bulletin. Series: Technical Sciences, A Scientific and Technical Journal **2** (35), 204-210 (2016)
4. A. M. Yefymenko, A. V. Kostenko, *Analysis of approximation methods of torque allocation hypotheses.* SevNTU Bulletin: collection of scientific works, Series: Mechanical and instrument engineering **122/2011**, 107-110 (2011)
5. A. M. Yefymenko, *Analysis of the dynamics of a monorail van.* Works of III International Scientific and Practical Internet Conference "Modern Technologies and

Prospects of Motor Transport Development", April, 14-16, 2015: collection of scientific works / Vinnytsia National Technical University [and others], 62-66 (2015)

6. M. L. Korotenko, N.V. Dontsova *Stability of the movement of a monorail elevated road van* [Text]. Interuniversity collection of scientific works «Problems of the dynamics and strength of the railway rolling stock» DIIT **232**, 53-58 (1984)
7. A. M. Yefymenko, V. H. Verbitsky, V. A. Demchenko, *About the construction of mathematical model of a system Alweg monorail van*. XXI Scientific and Technical Conference with International Participation "Transport, Ecology – Stable Development", (May 14-16, 2015, Varna), collection of reports, 341-346 (2015)
8. Y. V. Domin, L. A. Dluhach, M. L. Korotenko, O. M. Markova, *Autooscillations of rail-tracked vehicles*. 160 (1984)
9. L. H. Lobas, V. H. Verbitsky. *Qualitative and analytical methods in the dynamics of wheeled machines*. 232 (1990)
10. V. P. Sakhno, A. M. Yefymenko, *About the oscillations analysis of a wheeled vehicle model at its moving along the straight overhead road*. ZSTU Bulletin **2** (77), 248-256 (2016)
11. V. H. Verbitsky, V. P. Sakhno, A. P. Kravchenko, A. V. Kostenko, A. E. Danilenko *Automobiles. Stability*. Monograph, 176 (2013)

Analysis of motion of the three wheeled mobile platform

Anna Jaskot[1], *Bogdan* Posiadała[1,*]

[1]Institute of Mechanics and Machine Design Fundamentals, Czestochowa University of Technology, ul. Dąbrowskiego 73, 42-201, Częstochowa, Poland

Abstract. The work is dedicated to the designing motion of the three wheeled mobile platform under the unsteady conditions. In this paper the results of the analysis based on the dynamics model of the three wheeled mobile robot, with two rear wheels and one front wheel has been included The prototype has been developed by the author's construction assumptions that is useful to realize the motion of the platform in a various configurations of wheel drives, including control of the active forces and the direction of their settings while driving. Friction forces, in longitudinal and in the transverse directions, are considered in the proposed model. Relation between friction and active forces are also included. The motion parameters of the mobile platform has been determined by adopting classical approach of mechanics. The formulated initial problem of platform motion has been solved numerically using the Runge-Kutta method of the fourth order. Results of motion analysis with motion parameters values are determined and sample results are presented.

1 Introduction

Mobile platforms are designed as nonholonomic dynamical systems with its wheels assumed to be rolling without slipping or skidding. Research of rolling without slippage of both the kinematics and dynamics of motion have been performed and the results have been presented among others in [1]. Thus, in order to properly mapped the motion of the wheeled mobile platform the slippage need to be taken into account. The farm vehicles designed to accomplish agricultural tasks in order to detect sliding which is different from the one occurring in cars have been considered in [2]. The internal dynamics of the vehicle with one front steering wheel has been described in [3]. In this work the model and some sample results of motion simulations of the mobile platform with three steering wheels are presented. Approach of considering the slip during motion and the model of single wheel dynamics for three-wheeled mobile robot with friction coefficient as a function of wheel slip have been discussed in [4]. The three kinematical aspects of mobile robots, such as mobility, control, and positioning, have been detailed reviewed in [5]. The dynamical model is important. The control strategy with kinematics of the system and with the

* Corresponding author: bogdan.p@imipkm.pcz.pl

Reviewers: *Marek* Macko, *Ján* Vavro *Jr*.

velocity controller based on the robust sliding mode control technique have been discussed and developed in [6].

The novel method of tracking the reference trajectory and linearizing the nonlinear behaviour of robot under the influence of uncertainty and disturbance by using the hybrid controller has been introduced in [7]. In the same paper authors have presented the kinematic and dynamics models, and the results have been presented by using Langrange formulation.

In this work, besides the model of dynamics of motion of the mobile platform with three wheel drive, the simulation results have been included. In the model the linear velocity and acceleration as well as the angular velocity and acceleration of the center of mass, and the displacement of the center of mass of the platform have been performed by using numerical method. The aim of this research was to present an analysis of the behaviour of platform's motion.

The analysis of motion of the three wheeled mobile platform has been conducted. The conclusions have been formulated on the basis of comparison of the two cases of motion. First case concerns motion without skidding or slipping and the second when platform have been working under both mentioned: slipping and skidding. This comparison is useful in correct understanding and violating the method, presented in this work, and its results. The model of dynamics of motion has been formulated by adopting classical approach of the mechanics. The idea to design the three wheeled robot itself isn't new, but to design three wheeled mobile platform with three steering wheels equipped in three drive units is an author's model. By using such a system the analysis of motion in unsteady conditions have been accomplished.

In the next section the model of the platform has been introduced. In the third section the theoretical and mathematical models have been presented. In the fifth section the simulation results with analysis of the graphical representations have been included. The conclusions and vision of further research have been gathered and described in the last section of this paper.

2 Model of the three wheeled mobile platform

The model has been made on the basis of the prototype model of the four wheeled mobile platform described in [8]. A modular design is useful in analysing every possible position of the platform and its wheels. It was assumed, that each drive wheel will be steering wheel (possibility of rotate around the *Z*-axis), and the drive units could be subjected to different values of drive torques resulting from the construction and ability of the adopted electric motors.

Model of the analysed system, which is the mobile platform with three wheel drive has been presented in Fig. 1. The model of dynamics has been determined in the reference frame *OXYZ*. Point *S* has been denoted as a center of mass of the platform. The *Sxyz* coordinates represent the local coordinate system of the center of mass. Rotation angle between the *X* and *x* axes (and the same between the *Y* and *y* axes) has been denoted as β. It has been assumed that the platform will work on a flat surface (permissible inclination angle should not exceed 5°). The motion has been analyzed as a planar motion, and, since during motion has not occurred any changes between parts of the system, the platform has been considered as a rigid body. Each of the three steering wheels have been considered in order to determine forces occurring during platform work.

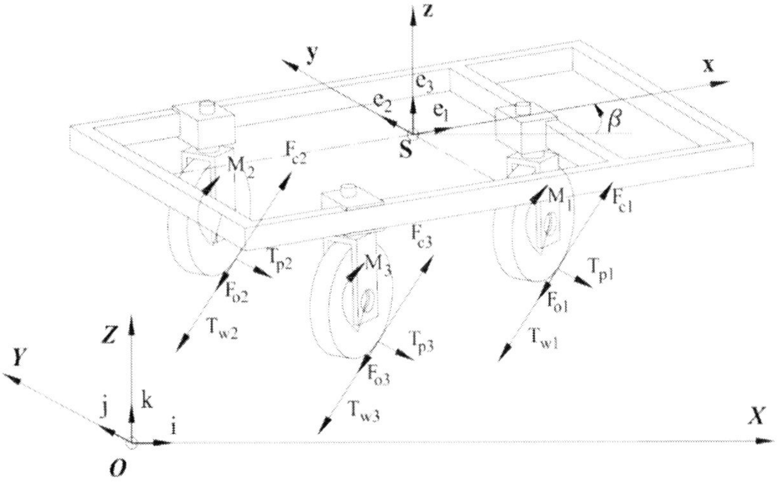

Fig. 1. The prototype model of the mobile platform in the reference frame

The $O_ix_iy_iz_i$ (Fig. 2) system is connected to every single wheel. The forces characterizing the dynamics of motion are schematically presented in Fig. 2. Based on the total mass of the platform calculated to a single wheel, the N_i force represents a reaction from the ground. The M_i is a drive torque and it is the initial data in this research. From the drive torque comes the active force $\mathbf{F}_i$. The friction forces have been designated as $\mathbf{T}_{wi}$ in longitudinal and $\mathbf{T}_{pi}$ in transverse directions. The $\mathbf{F}_{oi}$ force represents the resistance force, which prevented from starting movement of the platform and decelerated when the drive torque has been switched off.

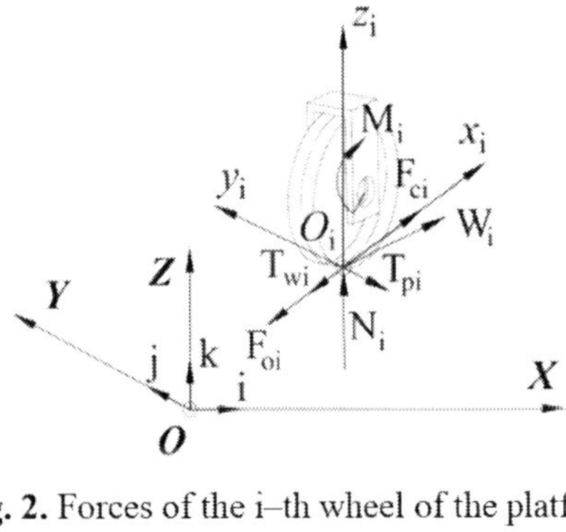

Fig. 2. Forces of the i–th wheel of the platform

3 Model of dynamics of motion

In this section the theoretical and mathematical models have been included. Model of dynamics of motion of the wheeled robots has been determined when considering the system as a rigid structure and each of the three wheels and the steering assembly. To complete obtaining the mathematical model of the dynamics of the mobile platform it is a very important to look into the dependencies between the ground, wheels and the system with taking into account all forces applied upon the mobile platform system. Based on the initial value of the drive torque the active force Fci can be determined from the Eq. 1.

$$\mathbf{F}_{ci} = \frac{M_i}{r} \cdot \mathbf{e}_1^i \tag{1}$$

where: r – radius of an i-th drive wheel.

By adopting the Coulomb friction model the friction forces in both longitudinal and the transverse directions can be obtained from the formulas in Eqs. 2 and 3, where the μ_w, and μ_p are the coefficients of friction (in the longitudinal and transverse direction), and the v_{wi}, and v_{pi} are the velocity components (in the longitudinal and transverse direction, as well).

$$\mathbf{T}_{wi} = -\mu_w \cdot N_i \cdot sign(v_{wi}) \cdot \mathbf{e}_1^i \tag{2}$$

$$\mathbf{T}_{pi} = -\mu_p \cdot N_i \cdot sign(v_{pi}) \cdot \mathbf{e}_2^i \tag{3}$$

where N_i is a reaction deriving from the wheel load on the ground.

The dependence between the active and passive forces cause the slippage or lack thereof. When the friction forces overcome the active forces, the platforms wheel will slip on the ground, otherwise - it won't. In this work both possibilities have been considered.

The motion description can be explained as the superposition of two motions compound of translation and rotation. The translational motion is when all wheels slide in the same direction (a lateral or side motion). The rotational motion is when the wheels does not slide exactly in the same way, this generates move around center of mass. The equation of a translational motion (the progressive motion) equation is presented in Eq. 4, and the rotational motion around Z-axis is presented in Eq. 5.

$$m\mathbf{a} = \sum_{i=1}^{3} \mathbf{W}_i \tag{4}$$

$$\frac{d\mathbf{K}}{dt} = \sum_{i=1}^{3} \mathbf{s}_i \times \mathbf{W}_i + \sum_{i=1}^{3} \mathbf{M}_{mi} \tag{5}$$

where: m is the total mass of the platform (with an equipment), $\mathbf{a}$ is the acceleration of the center of mass of the platform, $\mathbf{W}_i$ is the resultant force containing the described active and passive forces, $\mathbf{K}$ is the angular momentum vector of the whole platform, $\mathbf{s}_i$ is the location vectors of each of the drive wheels, $\mathbf{M}_{mi}$ is moment deriving from the rotational motion around center of mass during the slippage.

After neglecting the $\mathbf{M}_{mi}$, to obtain the motion parameters the transformation the above equations in form of the differential equations have been made Eqs. 6-8.

$$\ddot{X} = \frac{1}{m} \sum_{i=1}^{3} W_{ix} \tag{6}$$

$$\ddot{Y} = \frac{1}{m} \sum_{i=1}^{3} W_{iy} \tag{7}$$

$$\ddot{\beta} = \frac{1}{I_z} \sum_{i=1}^{3} (s_{ix} \cdot W_{iy} - s_{iy} \cdot W_{ix}) \tag{8}$$

where: $\ddot{X}$ is designated as the acceleration of center of mass on the X–axis in reference frame, $\ddot{Y}$ is the acceleration of center of mass on the Y–axis in reference frame, β is an angular acceleration around center of mass of the platform.

By using the Runge-Kutta method of the fourth order, the motion parameters have been determined and the sample results are presented.

4 Sample results of motion simulation

This chapter has been dedicated to the solution results of the dynamics of motion of the platform with three wheel drive. The results including positioning of the platforms wheel with its trajectories of motion, velocities and accelerations in two cases. First case, when none of the platform's wheels has slipped, and the second – when platform has been under slippage conditions. The second option has been made as a result of modification the friction coefficient in the second wheel and representing situation with moved center of mass of the platform from the central position. The size of the platform is equal one meter along the X axis and 0.7 m along the Y axis. The front wheel has been placed in local coordinates $x = 0.4$ m and $y = 0$. The two back wheels have been situated in $x = -0.2$ m (both), and the $y = 0.35$ m (second wheel) and $y = -0.35$ m (third wheel). Initially it was assumed, that by placed the load in the center of the platform will be convenient, because the weight distribution will be similar. Modification of placement load in different way will bring the new position of the center of mass of the platform, and those forces will be different.

The initial parameters, assumed on the basis of the engines from the catalogue [9] and of prototype design, have been gathered in the Tab. 1.

Table 1. Initial parameters in the first case

Initial values	**Symbol**	**Value**
Acceleration due to gravity [m/s^2]	g	9.81
Radius of wheel [m]	r	0.2
Total worktime [s]	t_c	20
Drive torque [Nm]	M_i	15
Coefficient of friction for the longitudinal direction	μ_w	0.2
Coefficient of friction for the transverse direction	μ_p	0.1
Mass of the platform [kg]	m	100
Starting position of the center of mass of the platform [m]	S	X=0.0, Y=0.0
Angle of inclination the platform [rad]	$\mathbf{b}_p$	Pi/6 (30deg)

The value of the drive torque means the maximum value of the drive torque, which is given in a trapezoidal function in Fig. 3.

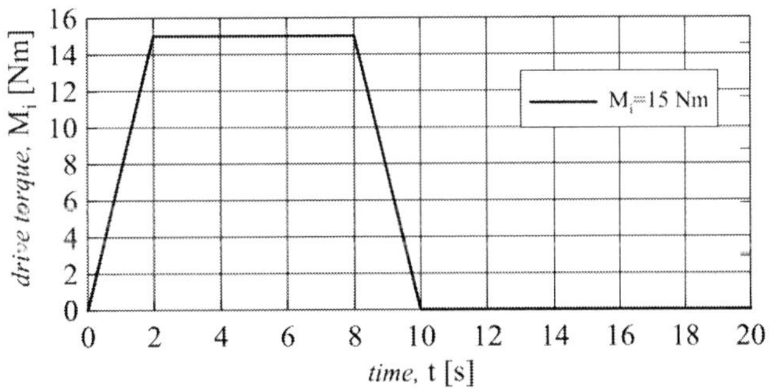

Fig. 3. Drive torque representation

From the initial parameter the active forces have been obtained. The passive forces, based on the total mass of the platform (constructional and external load) and the friction coefficients equal in every wheel, have been calculated. A graphical comparison of the relation between active and friction forces in every wheel has been shown in Fig. 4.

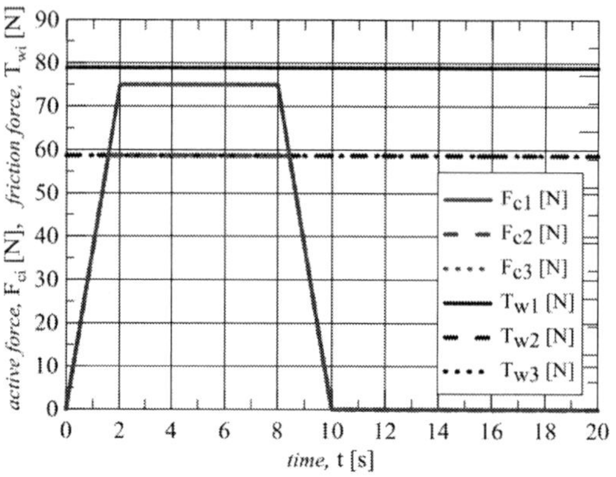

Fig. 4. Relation between active and friction forces during motion

The trajectory of motion of the center of mass of the platform without slippage has been presented in Fig. 6b. The representation of motion parameters in translational and rotational motions in this case are presented in Fig. 5a-b and 6a.

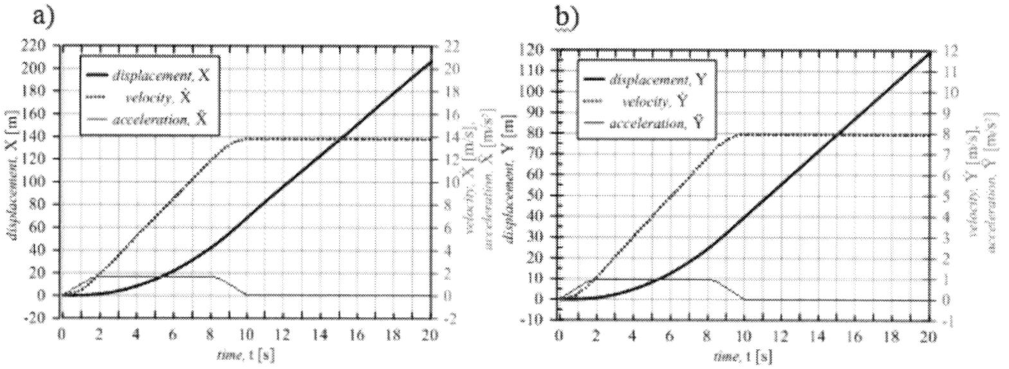

Fig. 5. Motion parameters a) in X-axis b) in Y-axis in the reference frame in translational motion

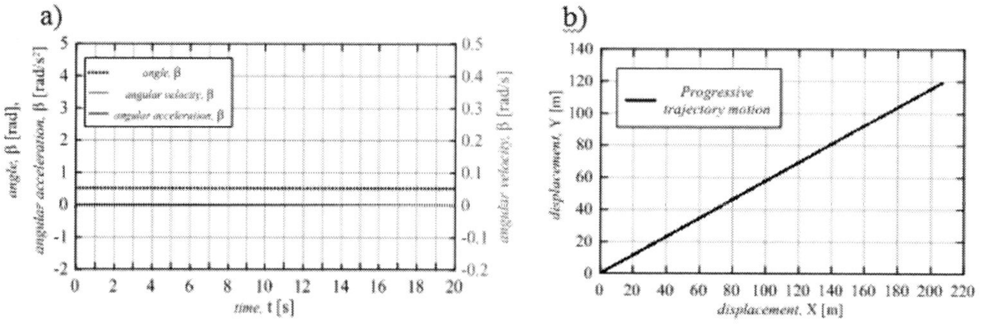

Fig. 6. a) Motion parameters in rotational motion, b) trajectory of motion of the center of mass

The second case of this analysis has been made on the basis of the description from the beginning of this section. The initial parameters in this case have been gathered and presented in Tab 2. The coefficients of friction have been reduced by a half of their initial values. This can represent an unpredictable situation on the surface at which the platform could work. The center of mass, as a result of different load localization on the platform, have been changed along the Y-axis at half distance.

Table 2. Initial parameters in the second case

Initial values	**Symbol**	**Value**
Acceleration due to gravity [m/s^2]	g	9.81
Radius of wheel [m]	r	0.2
Total worktime [s]	t_c	20
Drive torque [Nm]	M_i	15
Coeff. of friction in longitudinal direction (wheels: 1, 3)	μ_w	0.2
Coeff. of friction in transverse direction (wheels: 1, 3)	μ_p	0.1
Coeff. of friction in longitudinal direction (wheel: 2)	μ_w	0.1
Coeff. of friction in transverse direction (wheel: 2)	μ_p	0.05
Mass of the platform [kg]	m	100
Starting position of the center of mass of the platform [m]	S	X=0, Y=0.175m
Angle of inclination the platform [rad]	$\mathbf{b}_p$	Pi/6 (30deg)

In this case the same course of the drive torque from Fig. 3 have been used. Because of modification of the coefficients of friction the passive forces have changed and the relation between active and friction forces has been presented in Fig. 7.

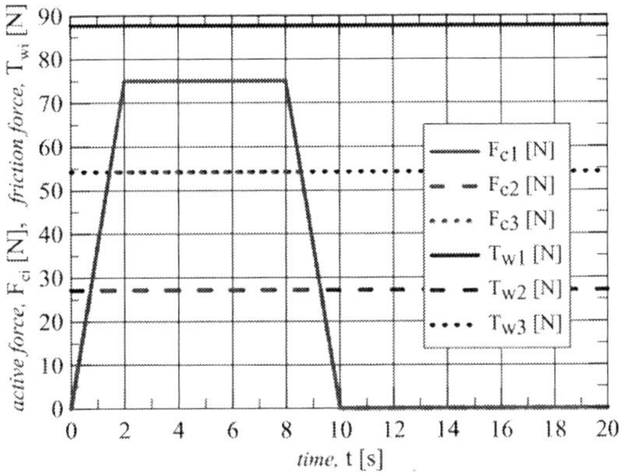

Fig. 7. Relation between active and friction forces during motion

The motion parameters have been presented in Fig. 8a-b. Those concerns progressive motion and have been shown in X and Y axes of the reference frame.

a) b)

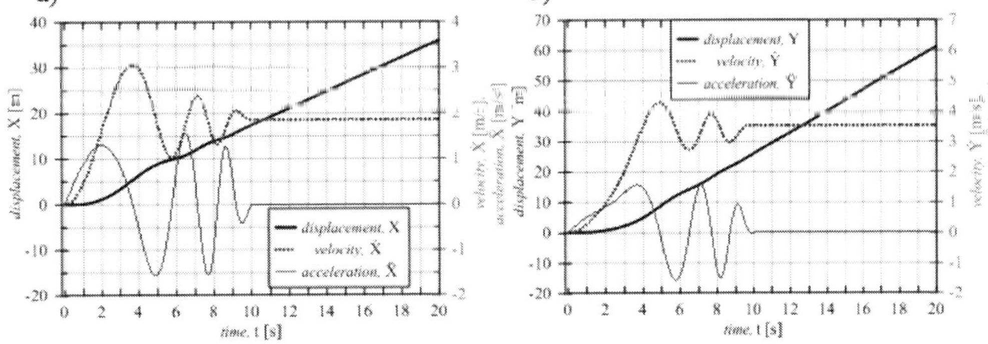

Fig. 8. Motion parameters a) in X-axis b) in Y-axis in the reference frame in translational motion in the second analyzed case

 https://doi.org/10.1051/matecconf/201815701008

The results deriving from the slippage and concerning the rotational motion of the platform have been presented in Fig. 9. As it is presented, the velocities after turning off the drive torque haven't changed. This is a result of assumption, that the resistance forces were equal zero.

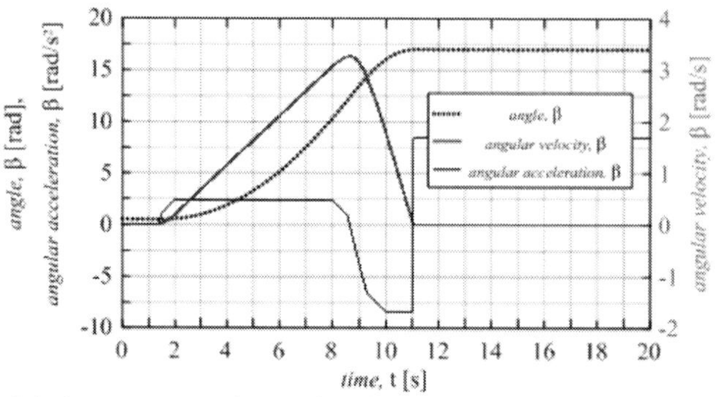

Fig. 9. Motion parameters in rotational motion when the slippage occurred

The trajectory of the superposition of the two motions is presented in Fig. 10.

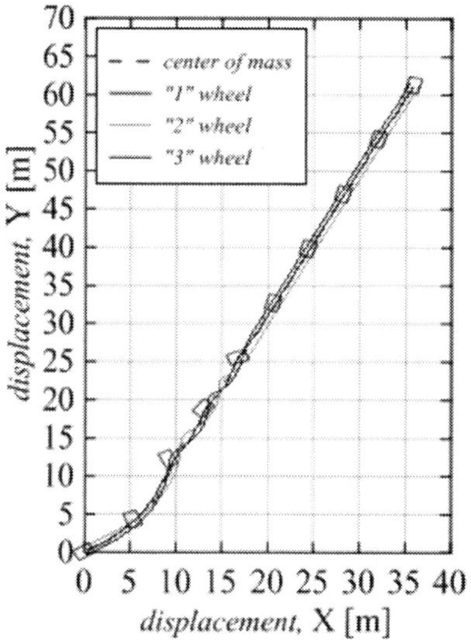

Fig. 10. Trajectory of motion of the center of mass in progressive motion and rotational motion around the Z-axis

The platforms rotations have been presented by the rectangles attached to the trajectory lines in the figure.

Conclusions

The analysis of motion under and without slipping has been performed. The model of dynamics has been determined and the sample results of platform motion analysis have been presented. The model is useful in examining wheeled vehicles designed to work under unsteady conditions. The solution presented in the paper is a base of further studies of the dynamics of motion with taking into account the slippage conditions. Presented in the paper the mileages of changes the friction forces can cause slippage with respect to the wheels, where the determined values of friction forces reach its limit values (exceed the developed friction).

The solution presented in the work allows to study the behaviour of the platform also while slippage and in the circumstances to refrain the platform from falling into the skid. Direction of further studies concerns the solution of refraining to fall into the skid.

References

1. A. Jaskot, B. Posiadała, S. Śpiewak, *Dynamics model of the mobile platform for its various configurations*. XXI International Polish-Slovak Scientific Conference Machine Modeling and Simulations, Hucisko, Poland (06-08 september 2016), Proceedia Engineering **177**, 162-167 (2016)
2. R. Lenain, B. Thuilot, C. Cariou, P. Marinet, *High accuracy path tracking for vehicles in presence of sliding: Application to farm vehicle automatic guidance for agricultural tasks*. Autonomous Robotics **21**, 79-97 (2006)
3. M. Eghtesad & D. S. Necsulescu, *Study of the internal dynamics of an autonomous mobile robot*. Robotics and Autonomous Systems **54**, 342-349 (2006)
4. R. Balakrishna, A. Ghosal, *Modeling of slip for wheeled mobile robots*. IEEE Transactions on Robotics and Automation **11** (1), 126-132 (1995)
5. N. Sidek, N. Sarkar, *Dynamic Modeling and Control of Nonholonomic Wheeled Mobile Robot when subjected to wheel slip*. (VDM Verlag Dr. Müller Aktiengesellschaft & Co. KG, Saarbrücken, 2009)
6. A. El-Sayed, I. Bayoumi, *Wheeled mobile robot trajectory tracking using sliding mode control*. Journal of Computer Science, 48-55 (2016)
7. Z. A. Ali, D. Wang, M. Safwan, W. Jiang, M. Shafiq, *Trajectory Tracking of a Nonholonomic Wheeled Mobile Robot Using Hybrid Controller*. International Journal of Modeling and Optimization **6** (3), 136-141 (2016)
8. A. Jaskot, B. Posiadała, S. Śpiewak, *Motion modeling of the four-wheeled mobile platform under slippage conditions*. Machine Dynamics Research **40** (2), 43-53 (2016)
9. http://pl.kwapil.com/downloads/maxon-ec-motor.pdf (24 july 2017)

Wear evaluation of elements of V-belt transmission with the application of optical microscope

Piotr Krawiec[1,*], *Konrad* Waluś[1] *Łukasz* Warguła[1], *Jarosław* Adamiec[1]

[1]Poznań University of Technology, Chair of Basics of MachineDesign, Piotrowo 3 60-965 Poznań, Poland

Abstract. The paper presents the analysis of reasons, form and wear of elements of V-belt transmission. Syndromes of loss of driving properties for chosen wheels and V-belts are specified. Methodology of failure evaluation of wheels and V-belts with the application of optical microscope is proposed. The applied measuring set allows for universal analysis of the measurement results. Operation principle of this device and its measuring and analysing abilities are shown. Failures of chosen V-belts and the influence of these failures on accuracy of kinematic features of chosen driving units are evaluated.

Keywords: V- belts, belt wear evaluation, optical measurements

1 Introduction

One of the first described applications of a rod was constructing a universal machine [1]. It was a basis to construct many other machines in Egyptian times i.e. first versions of turning lathe, transporting machines, etc. The beginning of using V-belts dates back to 1917 when John Gates used them in a car powertrain (fig. 1).

Fig. 1. The car in which V-belt was used for the first time [2]

* Corresponding author: piotr.krawiec@put.poznan.pl

Reviewers: *Andrej* Czán, *Juraj* Gerlici

Due to their properties and low production costs V-belts and multi-groove belts are widely used in many branches of industry. Their properties were also a starting point for development research conducted by research centers and belts producers. There are a few areas of research of V-belts and multi-groove belts such as: material, construction, production technology [3]. The issues concerning the designing of integrated V-belts are presented in articles [4-5]. The issues concerning the evaluation of properties of V-belts and multi-groove belts are presented in articles [6-9]. This article tries to present the evaluation of belt transmission used in alternator drive of engine of VW Transporter T4 1.9D. Multi-groove belts, poly-V belt pulleys and belt tensioners. The evaluation of wear of belts, tensioners' pulleys was done with the use of optical microscope Bresser BIOLUX ICD BINO 80x LED with camera CMOS 2 MP (fig. 2).

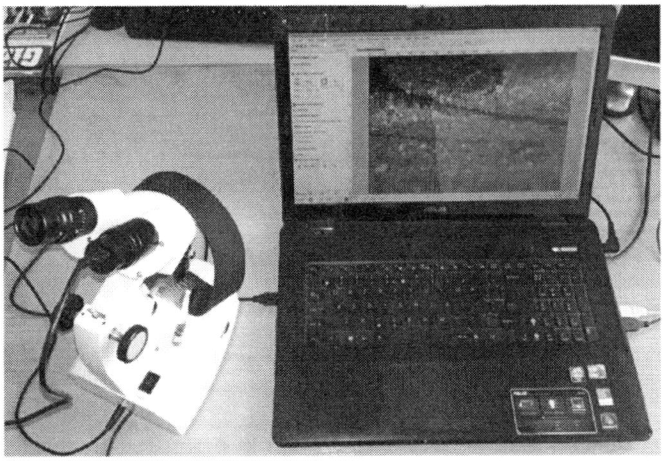

Fig. 2. Optical microscope with the measurement set

2 Typical causes of V-belts and Multi-groove belts failure

The most common failures of V-belts and multi-groove belts are:
– Cracking of ribs. On the working surface of belt transversal cracks are visible, alongside one or more ribs (fig. 3). This damage is probably caused by the influence of too high or too low temperatures. Other reason of such failure may be the interaction between the belt and pulley. It is especially important when the pulley diameter is small. If the interaction is incorrect it may cause tensions which lead to cracks. The cracks appear on the top of the rib and then deepen reaching down to the cord.

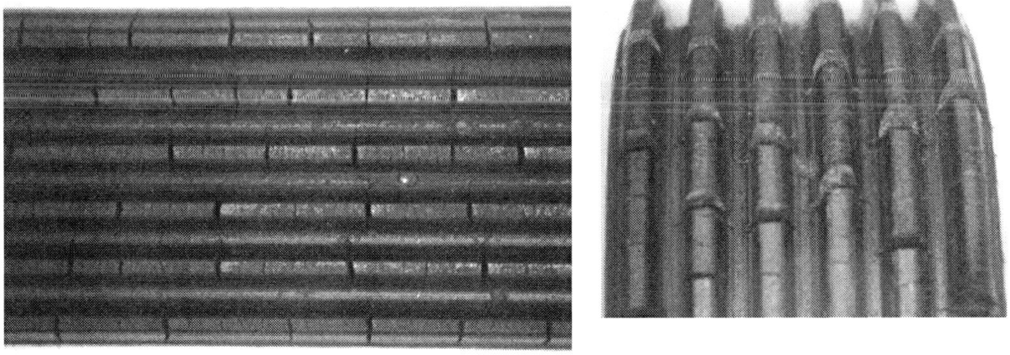

Fig. 3. Multi-groove belt transversal cracks

– Belt rubber flaking. When rubber flakes (fig. 4) then the belt may break immediately. Such state of the belt may be the result of many cracks cumulating in one place which are parallel to the cord strand. Such state of belt is caused by high temperature, the duration of using the belt and belt tensions.

Fig. 4. Rubber flaking in multi-grove belt

– The phenomenon of peeling. Peeling occurs (fig. 5) when the material of the belt is rubbed off the tops of the ribs and accumulates in its grooves. One of the reasons of such situation may be the improper tension of the belt, assembly mistakes or the wear of pulleys. This phenomenon is common for the belt drives of engines propelled by petrol or diesel oil.

Fig. 5. Peeling phenomenon on multi-groove Belt

– Friction wear. The symptom of friction wear (fig. 6) is a shinny outer surface of the belt. If there is a severe wear of the belt then on its surface a piece of fabric may be visible. This state of wear may be caused by a direct interaction of the belt with an object on its way such as a flange or a screw. Other reason of appearing the friction wear may be the incorrect tension of the belt.

Fig. 6. The friction wear of multi-groove belt

– The damage of the external belt ribs. The symptoms of such state of belt (fig. 7) appear on its sidewalls. The flat surface of the tie becomes shinny or in the most extreme cases the cord on the sidewalls is ragged. The outermost belt ribs may be rubbed off or torn off. The symptom of such failure may be the noise generated by the transmission. Such a result may be caused by the misalignment of belt pulleys.
The other symptom of such failure is bending or twisting of the belt Chile transmission works.

Fig. 7. The damage of the external belt ribs

– The uneven wear of the belt ribs. This kind of wear (fig. 8) is indicated by the damage of the sides of the tie, usually in a form of broken cord or ragged belt ribs. Other indicator of this state of belt is a rumbling or scraping sound in the transmission. The cause of this form of damage may be a foreign matter i.e. a small stone on the pulley. If nothing is done when these symptoms occur then it may result in the uneven wear of the belt, cutting the material of the belt ribs to the cord and eventually lead to its breaking.

Fig. 8. The uneven wear of the belt ribs

– The influence of contaminations. When on the outer (smooth) surface appear small punctures or swellings (fig. 9) and what is more, the fabric around the holes is ragged it may suggests that some contaminations penetrate into the belt. The cause of this form of damage is most often the situation when small stones, sand or gravel penetrate into the belt grooves and pulley grooves. Poly – V belts are produced with different shapes of belt ribs, some of them have sharp tops, others rounded and another ones cut off, as a result the interaction with pulleys differs so the specks of metal are the effect of the pulley rubbing off.

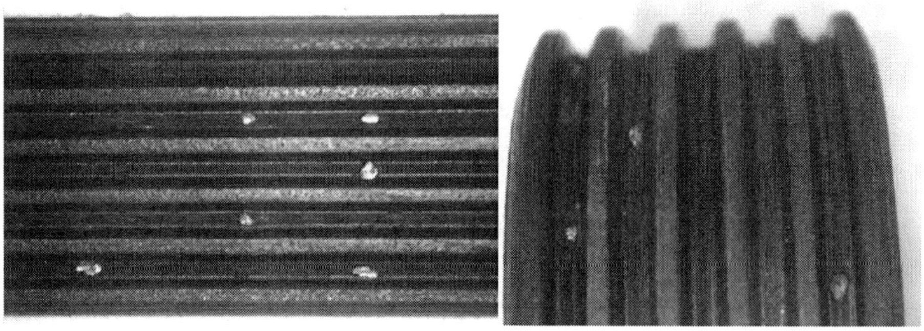

Fig. 9. The influence of contaminations on poly-v belt

– The band separation. This problem occurs when the outmost belt ribs start separating from the rest of the material (fig. 10). When this process lasts longer then the whole rib separates and causes the damage to the fabric and another belt ribs. The reason of such state of the belt is the improper assembly of the transmission. Due to the improper assembly one or more belt ribs is beyond the groove on the pulley what causes lack of corelation of belt sidewalls that lowers significantly the duration of proper transmission work.

Fig. 10. The band separation

– Contamination by fluids. The surface of the belt is flaking, sticky or swollen (fig. 11). This state of belt may be caused by oil, grease or chemical substances. These substances weaken significantly the internal structure of the belt, influencing its stiffness. The belt starts slipping on the surfaces of the pulleys. Inseparable symptom of this state is the increase of the belt temperature.

Fig. 11. Fluid contamination in poly-v belts

– Transversal cracking of the belt. This state of belt may be caused by a few reasons. One of them is thrusting of a foreign matter and as a result cutting a piece of belt including the cord. The second reason may be using too much strength when assembling the belt. Another reason for cracking the belt may be a sudden change of the transmission load caused by a blocked pulley.

– Excessive noise. If the assembly is improper and the pulleys are misaligned then the excess ive noise may occur. The noisiness of belt transmission with poly-v belt may also occur when the belt is very wide. In order to lower the noise level of transmission the belt may be cut into two pieces decreasing its width or use two poly-v belts that work parallel.

3 The damage of poly-v belt pulleys

Groove pulleys are widely use in the production of the means of transport. Cars used nowadays have belt transmissions used i.e. to drive the alternator, power steering and air conditioning compressor. As the cars have better performances (speed, torque) it causes vibrations and increases the noise level in the transmission. The natural solutions to this problem becomes applying in the car construction both silencers and transmission which working principle limits the negative effects. The reasons for belt pulley failure (fig. 12) may be: the assembly of the new pulley with the use of old or inappropriate screws as well as assembly mistakes such as incorrect torque while tightening the screws.

The symptoms of wear of multi-groove belt pulleys may be:

- loud work of the poly-v belt,
- slipping of the poly-v belt,
- rust on the outer part of the pulley,
- the increase in the level of vibrations,
- the deformation of the pulley grooves, cracks,
- rubbing of the parts of belt.

Fig. 12. New pulley and pulley exploited for about 250000 km

4 The failures of multi-groove pulley tensioners

The task of multi-groove pulley tensioners is to avoid excessive slippage, decrease the level of vibrations and noise, as well as proper tension of the belt in the drive.

Fig. 13. The sample failure of multi-groove pulley tensioners

The most common reasons of multi-groove pulley tensioner failures (fig. 13) are:
– rusty dripstones and cracks. Rusty liquid flows between tensioner arm and base or it drips from the tensioner. Rust leaks are the symptoms of internal wear of a part. Most of these types of failure occur where the limiters and tensioner mounting screws are.
– the wear of pulley bearing. The symptom of this failure is noise, resistance while turning the tensioner manually.
– the wear of the pulley. Correctly exploited pulleys of tensioners should be smooth and without any chirps, cracks or contaminations in the grooves. The spokes on the side surfaces of (plastic) pulleys should not be cracked or broken.
– assembly misalignment of the tensioner. The reason of this failure is lack of alignment while assembling the tensioner. The direct reason of such failures may be misaligned base of the tensioner, incorrect assembly of the tensioner or corrosion.
– noise coming from the tensioner. The noise from the tensioner is most of them caused by the failure of the bearings.
– misalignment of the tensioner arm. The main reason of this failure is improper belt leading on the tensioner's pulley. The visible symptoms of this situation are shiny, smooth strands or carves in the shield or tensioner arm. This situation may be caused by an interaction between metal arm and metal shield of the spring.
– excessive vibrations of tensioner arm. Vibrations of tensioner arm are caused by its tendency to fall into constant vibrations. Such state is caused by an impropriate suppression system or lowered spring tension.
– the loss of spring properties. If there is no resistance detected while turning the tensioner, it may mean that probably the spring lost its properties. Other symptoms of this failure are belt noise resulting from slipping or torque speed drop of the propelled part.

Conclusions and suggestions for designers

In order to assure proper time of failure- less work of belt pulley it is essential to:
– while exchanging or assembling the belt to change the distance between the pulleys in order to make it easier to fit the belt and then pull it to obtain a proper slippage (assuming 1,5 % tolerance of the belt total permissible lengthiness while work and 1% tolerance of the belt length),
– use only the pulleys that were manufactured in accordance with standards and catalogues of prestigious producers. In case of poly-V belt pulleys only belt sheaves should be used that consist of belts compiled especially in accordance with length tolerance,

– check if pulley and belt grooves are clean, damage less or without any anti-corrosion substances,
– pay an extra attention to a proper assembly of a tie. The belt should be set in grooves in such way that it adjoins its all sidewall with the walls of the groove,
– not use any substances that increase the adhesiveness of belt to the pulley,
– remember that the acceptable misalignment of transmission pulley axis should not exceed1 mm per 100 mm of axis length and the tolerance of mutual displacement of pulley grooves should not exceed 0,25% of distance between the axis, during the first hours of its work the belt stretches and then its tension should be corrected. As it results from experimental research after one hour of work under full load, the belt lengthens by about 60 % of the total acceptable lengthiness, which is 1, 5 % of the belt length,
– if one belt is broken in the belt transmission then all the belts should be exchanged,
– secure the transmission with a protective shield,
– use proper assembly of belt tensioner and monitor its work regularly.

References

1. F. M. Feldhaus, *Maszyny w dziejach ludzkości*. (PWN 1958)
2. www2.gates.com
3. M. Dudziak, *Przekładnie cięgnowe*. (PWN, Warszawa, 1997)
4. M. Dudziak, P. Krawiec, R. Mostowski, *Pasy zespolone Poli-V*. Zeszyt Specjalny Politechniki Lubelskiej, 229-234 (1995)
5. M. Dudziak, P. Krawiec, R. Mostowski, *Aspekty doboru i zastosowań przekładni pasowych z pasami zespolonymi*. Zeszyt Specjalny Politechniki Warszawskiej, 121-126 (1995)
6. K. Kubas, *A research stand for measuring friction parameters in a belt transmission*. The Archives of Automotive Engineering **75**, 69-82 (2017)
7. G. Sheng, H. Lee, V. Narravula, *Experimental characterization and analysis of wet belt friction and the vibro-acoustic behavior*. Tribology International **44**, 258-265 (2011)
8. L. Kátai, I. Szabó, *Identification of V-belt power losses with temperature measurement*. Journal of Mechanical Science and Technology **29,** 3195-3203 (2015)
9. www.machinedesign.com/technologies/pushing-belt-drive-efficiency

The application of the optical system ATOS II for rapid prototyping methods of non-classical models of cogbelt pulleys

Piotr Krawiec[1,*], *Grzegorz* Domek[2], *Łukasz* Warguła[1], *Konrad* Waluś[1], *Jarosław* Adamiec[1]

[1]Poznań University of Technology, Katedra PKM, Piotrowo 3 60-965 Poznań, Poland
[2]Kazimierz Wielki Uniwersity, Chodkiewicza 30, 85-064 Bydgoszcz, Poland

Abstract. The widespread application of both gear wheels and cogbelt pulleys with noncircular generating line in technique stimulates the development of manufacturing and measuring methods of these wheels. The paper presents the rapid prototyping methods of models of cogbelt pulleys with nocircular evelope. Evaluation method of manufacturing accuracy of cogbelt pulleys, which are applied in unevenrunning belt transmissions, are presented. These transmissions are widely applied in steering techniques and drives of machines and devices. Verification of mapping accuracy of shape of geometrical model of cogbelt pulley was done with the application of noncontact optical system i.e. coordinate optical scanner GOM Atos Compact Scan 5M GOM company.

Keywords: cogbelt pulleys, belt transmissions, ATOS Compact Scan

1 Introduction

One of the key factors of economic growth of enterprises is inter alia ensuring proper quality and competitiveness of produced goods. Competitiveness may be defined as the ability of panning, creating and selling a product which is more attractive than other products available on the market. It is vital to launch new products quickly and at the same time take into account individual needs of future customers.

In order to comply with these conditions it is necessary to use modern modelling tools, computer simulations and prototyping. One of the new methods of increasing the competitiveness of companies that deal with machines and devices production is using the designing and manufacturing technique called Rapid Prototyping RP.

This notion refers to all techniques of allow for rapid manufacturing physical models of parts and units. The task of Rapid Prototyping techniques is to obtain a full-value product with the same durability, visual and aesthetic features as can be obtained using traditional manufacturing methods. Modern process of designing and manufacturing should be restricted to feedback loop in a form of geometric accuracy analysis of a prototype using a control - measurement method named Rapid Inspection – RI (Fig. 1). This method means a quick checking of parameters' compliances of the product with the use of a chosen RP

* Corresponding author: piotr.krawiec@put.poznan.pl

Reviewers: *Andrej* Czán, *Milan* Sága

technique (mainly its shape and surface) and comparing it with a corresponding CAD model or standard.

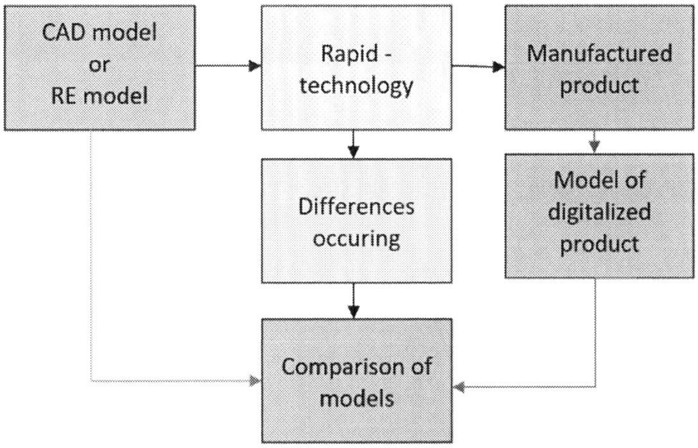

Fig. 1. The principle of Rapid Inspection process [1]

The Rapid Prototyping techniques were used to create an example of construction solution such as uneven-running strand transmissions [2] or cog belt transmissions with periodically variable transmission ratios [3]. Such transmissions allow to obtain periodically variable kinematic characteristics during one cycle of transmission by using elliptical cogbelts or pulleys (Fig. 2).

Fig. 2. Two-wheeled uneven-running strand transmission with a cog belt

The issues concerning the geometric characteristics of such transmissions are accessible in papers [3, 4], whereas the detailed description of possible manufacturing methods was characterised and exemplified in papers [5-7].

2 The proposed method of manufacturing the pulleys of uneven-running strand transmission

In search for precise and cheap methods of manufacturing elliptical pulleys the following universal technological machines were used: CNC milling machines, wire cutters, water jet cutters etc., or advanced systems like CAD and CAD/CAM. Also attention was paid to the need of using standard tools and ensuring the manufacturing precision in the area of geometric and stereometric characteristics of the surfaces. The method that allows fulfilling these requirements in case of elliptical pulleys is teeth processing with the use of CNC milling machine. Using standard and universal technological machines as well as classic tools is connected with simplifying the technological process as well as low production costs. In order to conduct this process it was necessary to prepare a software to steer the work of devices/tools. It was made with the use of commonly known ISO code and parts geometry in CAD system. In the case described here a 2D model was made in AutoCAD system which allows for designing conic sections and B-spline. This programme makes it is first to appoint and then exemplify with the use of any word processor the data necessary to create a software of processing the outline being shaped. The vital advantage of a universal CNC milling machine is the possibility of simulating the process (Fig. 3) directly from the level of steering panel. It enables to verify (without any mistakes) the steering programs and allows to detect possible interferences in the configuration: the processed object – machine tool – tool before starting the shaper.

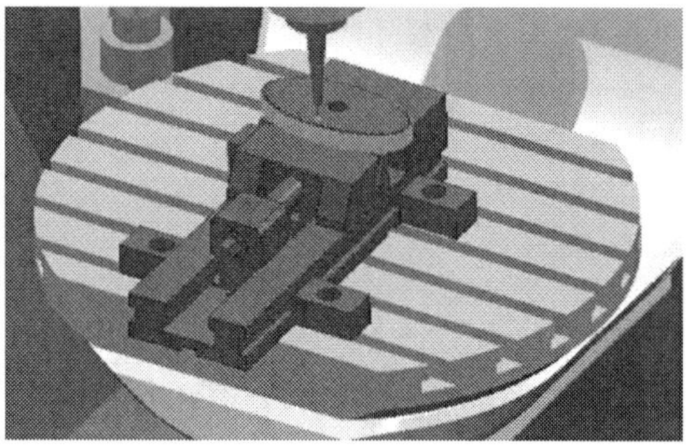

Fig. 3. The simulation of shaping the elliptical pulley

Fig. 4. Shaping the elliptical pulley with flutes

On the basis of preliminary analysis of construction documentation of the pulleys it could be stated that is necessary to involve three technological steps. The first of them was to remove excessive material from the rectangular- shaped semi-finished product with the use of flute milling cutter of ø6 mm (Fig. 4). In the second step the finishing milling cutter of ø4 mm was used and the final shape of teeth was obtained with the use of cutter of ø2 mm. The shaping effect of exemplary elliptical pulleys is shown in Figure 5.

Fig. 5. The exemplary elliptical pulleys made with the use of universal CNC milling machine

3 The proposal of the measurement methods of geometrical characteristics of nonstandard pulleys

The authors of this article have not so far come across literature that would inform about the methods or measuring procedures of nonstandard pulleys. One of the possible options to be chosen is identification and verification of accuracy of geometry projection of such pulleys using noncontact measurement methods. The noncontact measurement methods of 3D models geometry are: handheld laser scanning, laser scanning with head installed on CMMs, optical scanning, computer tomography. Laser scanning enables to measure objects with very complex shapes. The measurement may be done in two ways. The first one is connected with using scanning heads fixed directly on CMM or cantilever arm. The second one is based on using a self-positioning handheld laser scanning. These devices allow in the digitalisation process to convert geometrical objects into 3D digital form. The working principle of self-positioning handheld laser scanning is usually connected with two combined processes: laser triangulation and marker detection. Triangulation means measuring an object in two perpendicular directions in coordinate system. The data obtained from triangular measurements are linked into one object with the use of markers detection (Fig. 6)

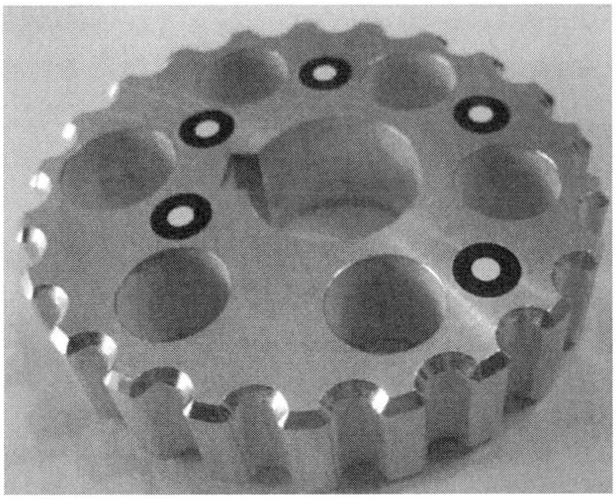

Fig. 6. Cogbelt pulley with marked reference points (markers)

Markers should firstly be placed, irregularly, onto the model or special mat that enables the scanner to recognize its position against the object. Handheld scanner during the measurement is moved freely by the operator around the measured object. The great advantage of such solution is the randomness of relative position of scanner and the object; it also eliminates the necessity of preparing a few point clouds from different positioning of the scanner and later linking them. Optical noncontact laser enables the measurement and to capture accurately the geometry of complex-shaped objects. The basis for scanning process is the analysis of line diffractions in a series of lined images projected on the object. When the surface is flat, the lines are parallel. If the surface of the object is curvilinear then the lines are diffracted in accordance with the shape. The diffraction is registered by a camera, processed by algorithms that take into account the parameters of optical system. The obtained set of points of x, y coordinates is a computer mapping of the surface of a scanned object. The software of scanning systems allows to: combine the scanned images in order to obtain the whole measured surface, remove the measurement inaccuracy, compete the missing parts of the surface and prepare a ready CAD model. The accuracy of results obtained in his way of measurement depends on the kind of probe head used and the measurement area of the surface of the object being measured as well as kind of ray emitted.

4 The measurements of geometrical characteristics of pulleys

The analysis of accuracy of mapping the geometrical shape of manufactured cogbelt pulleys was conducted with the use of noncontact optical system based on coordinate optical scanner GOM Atos Compact Scan 5M (Fig. 7). In order to conduct the measurement the measurement area SO 125 was used with the measurement range of 25x90x90 mm. The scanning accuracy for this measurement area is at the level of 0.010 mm.

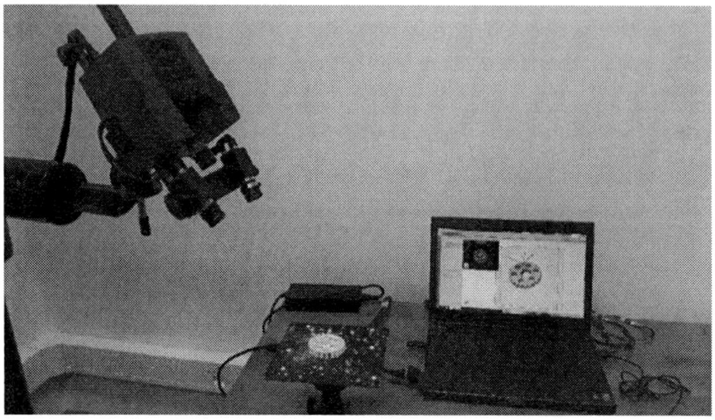

Fig . 7. The measurement of geometric characteristic of cogbelt pulleys with the use of ATOS II system

Before Canning the pulley to be measured was covered with a thin layer of chalk to obtain a matt surface. After scanning the model was put on CAD model with the use of Best fit method. The measurement analysis was made with the use of software GOM Inspect Professional V7 SR2. The exemplary measurement results are shown in Figs. 8-10.

The inaccuracy on the edges of the pulley (dark blue colour) is a result of an inaccurate deflection of the pulley edge in reference to CAD model. The same refers to the inaccuracies of holes lowering the weight of pulley. These inaccuracies are irrelevant for the proper functioning of the transmission.

Summary

The analysis of accuracy of mapping the geometrical shape of cogbelt pulleys with the use of optic scanning system ATOS II may be applied at all stages of the technological process of the production. The advantage of noncontact method is that the measurements may be carried out quickly and the results obtained and their analyses do not differ from the accuracy of contact measurements.

The analysis of accuracy of manufacturing the cogbelt pulleys may be visualized with the use of software GOM Inspect V7 SR2 showing the detailed deflections of dimensions specified in chosen points of pulley's surface and specified profiles of rim sections. As the detailed report is very vast, this article presents only some chosen figures. On the basis of the comparison of the researched pulleys with 3D-CAD models designed in software AutoCAD and Inventor it can be stated that there are visible divergences of dimensions within the range from -0,019 to +0,014 mm. The divergences occuring in the dimensions of the upper edge of the hub and teeth of the nominal model and the researched pulleys may result from the inaccurate mechanical working, but its value does not exclude this universal manufacturing method to be used for a mass production of pulleys.

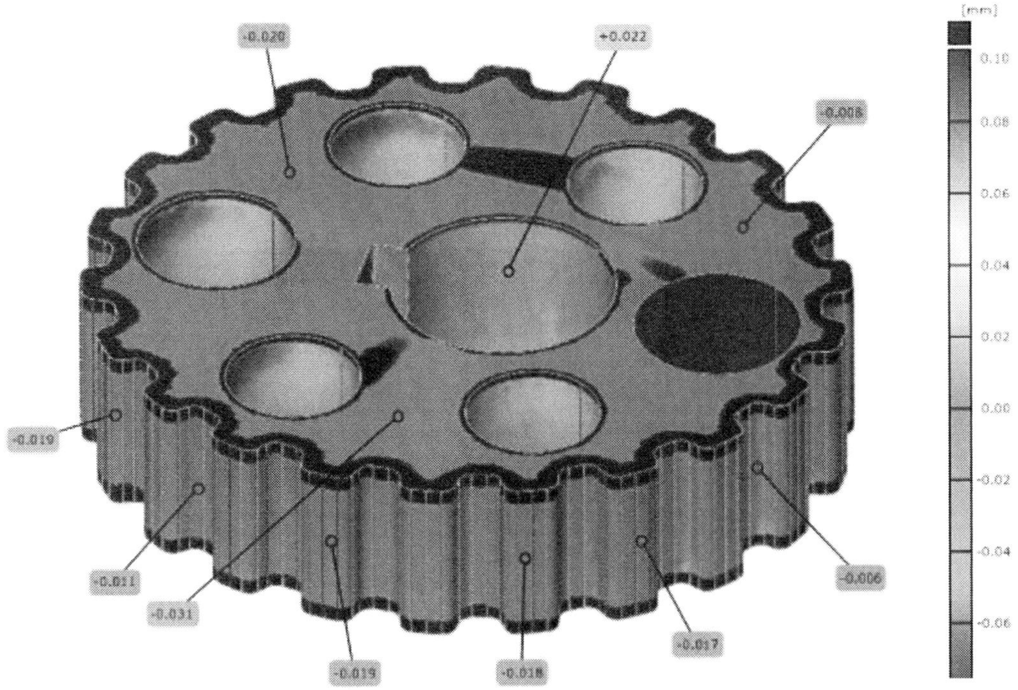

Fig. 8. The accuracy of mapping of geometrical shape of the cogbelt pulley

Fig. 9. The examplary section of a pulley

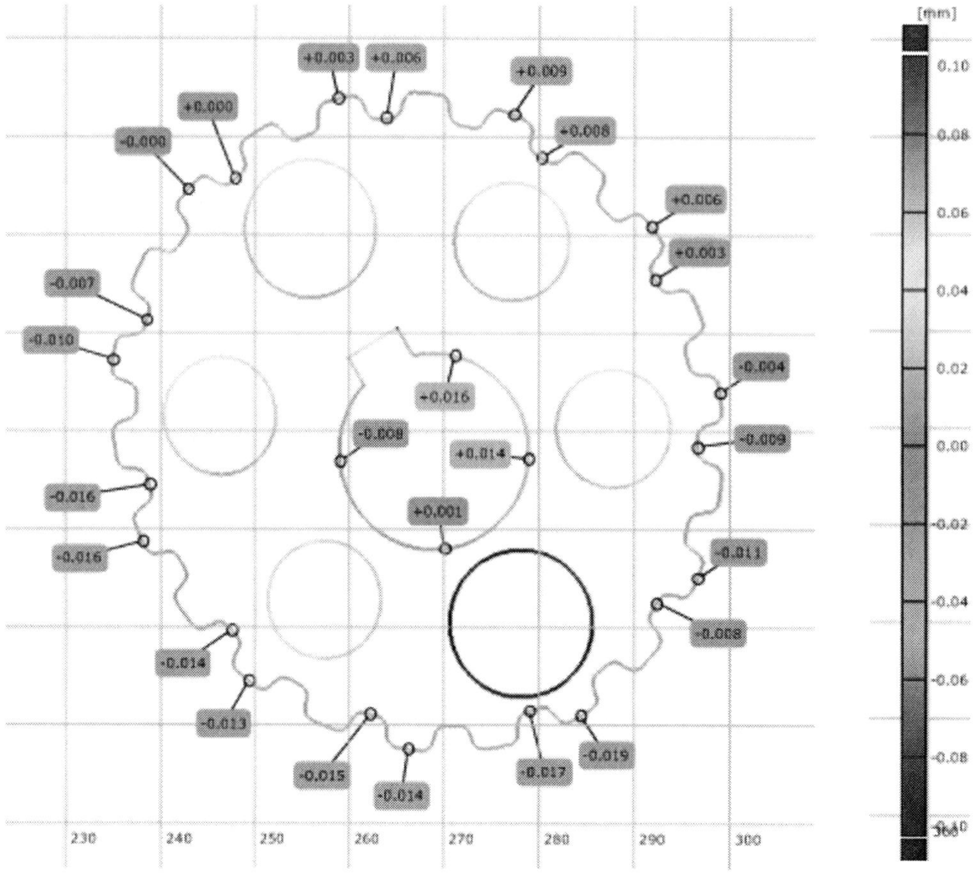

Fig. 10. The detailed values of the deflections of shape of cogbelt pulley

References

1. K. Oczoś, Cena I. *Rapid Inspecton – metody pomiarowo-kontrolne adekwatne do rapid technologii.* Mechanik **3**, 165-178 (2008)
2. S. Medvecká-Beňová, *Meshing Condition and Kinematic Properties of non-circular GearActa.* Mechanica Slovaca **20** (2), 16–21 (2016)
3. P. Krawiec, A. Marlewski, *Profile design of noncircular belt pulleys.* Journal of Theoretical and Applied Mechanics **54** (2), 561–570 (2016)
4. P. Krawiec, *Projektowanie i kształtowanie uzębień nieokrągłych kół pasowych*, Acta Mechanica et Automatica **4**, 52–55 (2010)
5. M. Kujawski, P. Krawiec, *Analysis of Generation Capabilities of Noncircular Cog belt Pulleys on the Example of a Gear with an Elliptical Pitch Line.* Journal of Manufacturing Science and Engineering. Transactions of the ASME **133** (5), 0510061–0510067 (2011)
6. M. Sobolak, G. Budzik, *Experimental method of tooth contact analysis (TCA) with rapid prototyping (RP) use.* Rapid Prototyping Journal **14** (4), 197–201 (2008)
7. J. Butler, D. Aitchison, *Using selective laser sintering for manufacturing.* Rapid Prototyping Journal **31** (3), 212–219 (2011)

The material parameters for computational modeling of long-fibre composites with textile

Jan Krmela[1,*], *Vladimíra* Krmelová[1]

[1]Alexander Dubček University of Trenčín, Faculty of Industrial Technologies, I. Krasku 491/30, 020 01 Púchov, Slovakia

Abstract. In this contribution, the composites with textile fibre (cord) and an elastomer matrix are studied based on computational modeling of car tires in combination with experiments. These composite are applied in tire casings for cars, conveyor belts etc. The aim of this research work of authors is creation of computational models namely for stress-strain analyses of selected parts of radial tire casing. The typically passenger car radial tire casing consists of one or two polyester plies in tire carcass and two steel-cord belts and one polyamide 66 cap ply below tread. The finite element method using the program system ANSYS is applied to the computational modeling. For the determination of the material parameters of elastomer and textile cords as input data to the computational models, it was necessary to perform tests as statically tensile tests. Also the experiments of composites as tests of low cyclic loading of composites are needed for verification analyses between computational results and experimental data. For computational modeling, the modulus of elasticity and Poisson ratio are used as material input parameters of textile reinforcements. The results from the computational modeling and selected results from the tests are presented in this contribution.

Keywords: tire, ANSYS, material parameters, tensile test, low cycle loading, PA66

1 Introduction

The paper deals with the long-fiber composites with textile fibre. A tire casing can be classified as a long-fiber composite with polymer matrixes – an elastomer and long-filament reinforcements – textile and steel cords.

It is necessary to have knowledge about geometry, material parameters, cross-section and structure of a tire casing (number of layers of a belt and carcass, information about a bead and cap ply) and other parameters for creations of computational models of a tire casing for the strain-stress analysis of a tire under the vertical load, modal analysis etc.

A standard automobile radial tire casing consists of elastomer parts and parts with textile-cords (Tab. 1) and steel-cords in a tire tread as reinforcements. The structure parts applied into passenger car radial tire casings are: textile carcass plies, a textile cap ply (called an overlap belt) and steel-cord belts [1, 2]. These structures of a tire have a different cord material, cord angle (e.g. for a steel belt there is applied an angle of 21–27° in a radial

* Corresponding author: jan2.krmela@post.cz

Reviewers: *Jozef* Bocko, *Justin* Murín

tire casing for a passenger car), the construction of cord and number of layers (single-layer or multi-layer). Therefore, tires have specific deformation properties. The textile carcass density (it is a number of ends per one meter of width, marked as EPM) is from 700 to 1 150 m^{-1}. The EPM of a textile cap ply is 1 100–1 200 m^{-1}. The PA66 and PES textile fibers are used for passenger tires, especially for common purposes.

Table 1. Materials of textile-cords for tires [2]

Material	**Labeling of construction [tex]**	**Application**
Polyester	PES 144x1x2 PES 167x1x3	Carcass plies
Polyamide 6	PA6 94x1x2 PA6 94x1	Cap ply
Polyamide 66	PA66 140x1x2	Cyp ply
Rayon (viscose)	VS 184x1x2 VS 244x1x2	Carcass plies
Aramid	Aramid 110x1x2	Carcass plies for sport tire casings

The input data for computational modeling are divided to geometric and materials parametres obtained in an experimental way. For computational modeling the elastic modulus of elasticity and Poisson ratio are used as material input parameters of textile reinforcements. The static tensile modulus of elasticity of PA66 is from 900 to 3 450 MPa or from 9 to 50 cN/dtex [2, 3]. It is a broad range. In case of rayon, the modulus is from 600 MPa to 11 GPa or 8–120 cN/dtex. The Poisson ration of PA66 is approx. 0.4. Typically, values of modulus of elasticity are in Tab. 2. The samples of cross-sections of tire casings in the area of tire-crown are in Fig. 1 for the tire Dunlop 215/40 R17 87V (low-profile, extra load) for sport cars.

Table 2. Moduli of elasticity of textile cords [3]

Material	**Modulus of elasticity [GPa]**
Polyester	4
Polyamide 66	3.4
Rayon (viscose)	11
Aramid	25

The EPM of the textile cap ply of 215/40 R17 is approx. 1 400 m^{-1}. The diameter of cords is 0.44 mm and the thickness of the cap ply is approx. 1.1 mm. The textile carcass has 0.5 mm diameter of cords and the thickness is approx. 1.5 mm. The steel cord belts consist of two plies with a density of 650–680 m^{-1} and the cord construction is 2+2x0.30 mm. The thickness of the steel cord belt (two plies altogether) is 1.4 mm. Further information about the structure is in the paper [4].

The geometric parameters, which were obtained by an image analysis of the cross-section, will be used as necessary geometric input data in order to create a computational model with a real configuration of cords. The results can be used for an optimization of the deposition angle of textile reinforcement in composites.

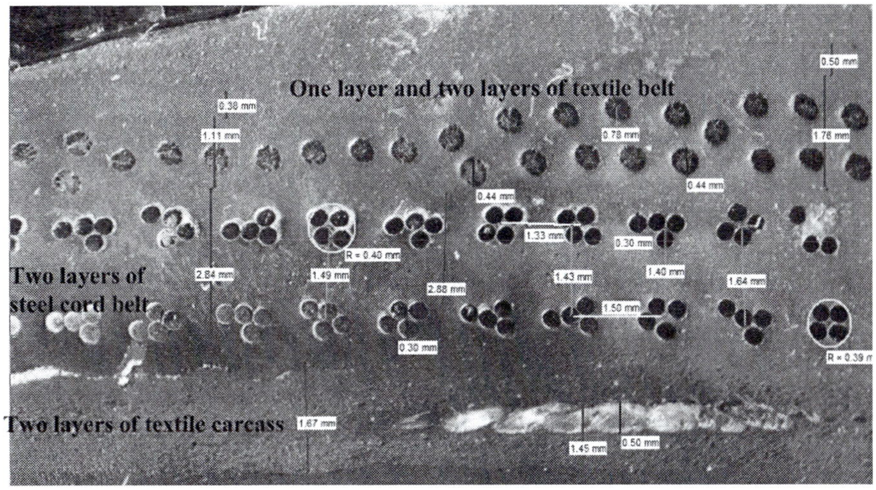

Fig. 1. The detail of structure of the tire-crown 215/40 R17

2 Experimental part

Tests of specific long-fiber composite materials with an elastomer are not standardized. The tire casing 215/40 R17 were cut with a water jet in the longitudinal and transverse direction in order to obtain the specimens for statical tensile tests The specimens were prepared with different width 15 and 20 mm. The test machine HOUNSFIELD with capacity 20 kN load was used.

The conditions of the tensile tests are: The initial length between the clamps of the test machine is 100 mm and the loading speed is 10 mm/min. The modulus of elasticity (Tab. 3) is obtained by two elongations by 4 and 8 %.

Table 3. Modulus of elasticity

Modulus of elasticity for elongations 4 / 8 % [MPa]	**Specimen width [mm]**	
	15	20
Longitudinal	125 / 134	155 / 183
Transverse	120 / 158	125 / 173

The conditions of the tensile tests were: initial length between the clamps of test machine is 100 mm, loading speed is 10 mm/min. The dependences from tensile tests for the specimens with different width from the tire casing 215/40 R17 are in Fig. 2. The moduli of elasticity were obtained by two elongations 4 and 8 %. The test data are used for verification analyses between calculated values and experimental data.

Next, the tests of low cyclic loading are realized for textile cap ply with PA66 on the test machine Autograph AG-X plus 5kN – Schimadzu with a video-extensometer. The specimens geometry is other than it was for tensile tests of structure parts of tire 215/40 R17. The geometric parameters of specimens are a length of 195 mm, a width of 35 mm, initial length between clamps of a test machine 100 mm and a thickness of the specimen of 1.05 mm. The cord diameter is 0.5 mm. The EPM is 870 m-1. The initial length between the points for a video-extensometer is 50 mm. The composite specimens with an angle of the textile reinforcement 0° (textile reinforcement is perpendicular to the direction of loading) and 45° are used for the cyclic tests.

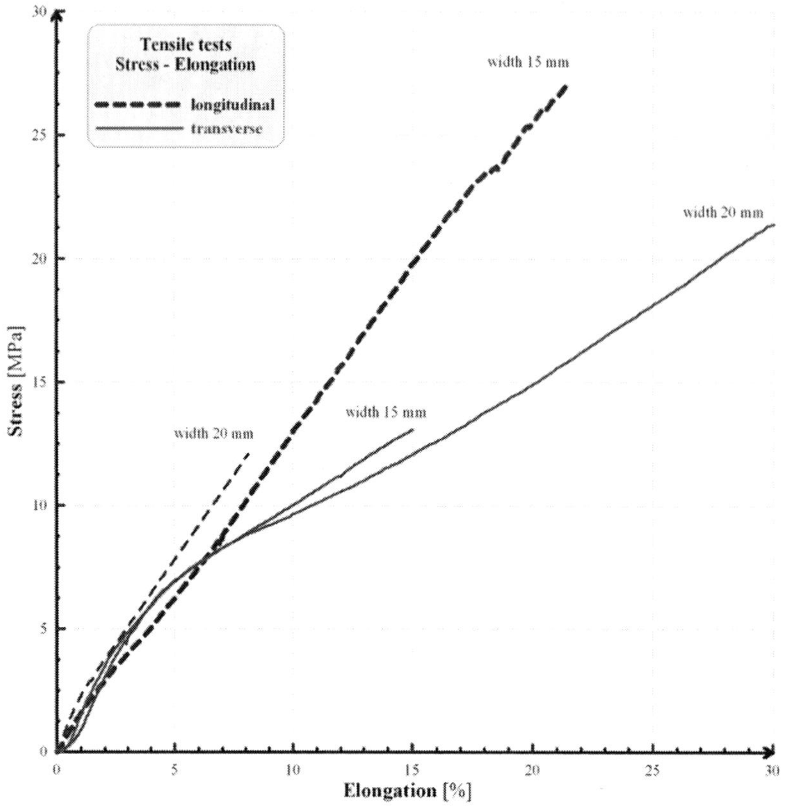

Fig. 2. Stress-strain dependencies of samples from tire casing 215/40 R17

The design of method for cycle loading tests of composites with a textile fiber and an elastomer matrix on a test machine with a video-extensometer has these conditions: the loading speed 250 mm/min for cycles and 50 mm/min for a pre-test by force value 2 N by standard [5]. Five cycle loops are applied. Every cycle loop consists of five cycles. Every cycle is defined as loading to a certain percentage of elongation between clamps of a test machine and unloading to a certain percentage of elongation between clamps of a test machine the first cycle loop consists of cycles with loading to 30 % and unloading to 3 % of elongation (not 0 % because the negative force is not possible during tensile testing for certain composite specimens with the textile reinforcement). The second loop consists of cycles with elongation higher by 10 %: loading to 40 % and unloading to 10 % of elongation. The third loop consists of five cycles that have loading to 50 % and unloading to 20 % of elongation, the fourth loop with cycles has loading to 60 % and unloading to 30 % of elongation. The fifth loop has cycles with different elongations, the loading to 60 % is the same as fourth loop and unloading is 5 % of elongation. The comparisons of the fifth cycles are given in Fig. 3.

3 Computational model

The aim of this research work of authors is creation of computational models namely for stress-strain analyses of selected parts of radial tire casing [6]. The finite element method using the program system ANSYS is applied to the computational modeling.

For the description of the behavior of elastomers in a tire casing, there is the most commonly used the two-parameter Mooney-Rivlin material model. The Mooney-Rivlin model (M-R) is the most commonly used for computational modeling of tires. The M-R model can be well applied for the range of elongation up to 150 %. To determine the basic M-R parameters, it is necessary to carry out the tensile test for elastomer samples. These parameters describe the strain-stress behavior of elastomer parts of a tire casing. Next tests of elastomer parts of a tire casing are needed for the determination of parameters C_{10} and C_{01} of the Mooney-Rivlin model.

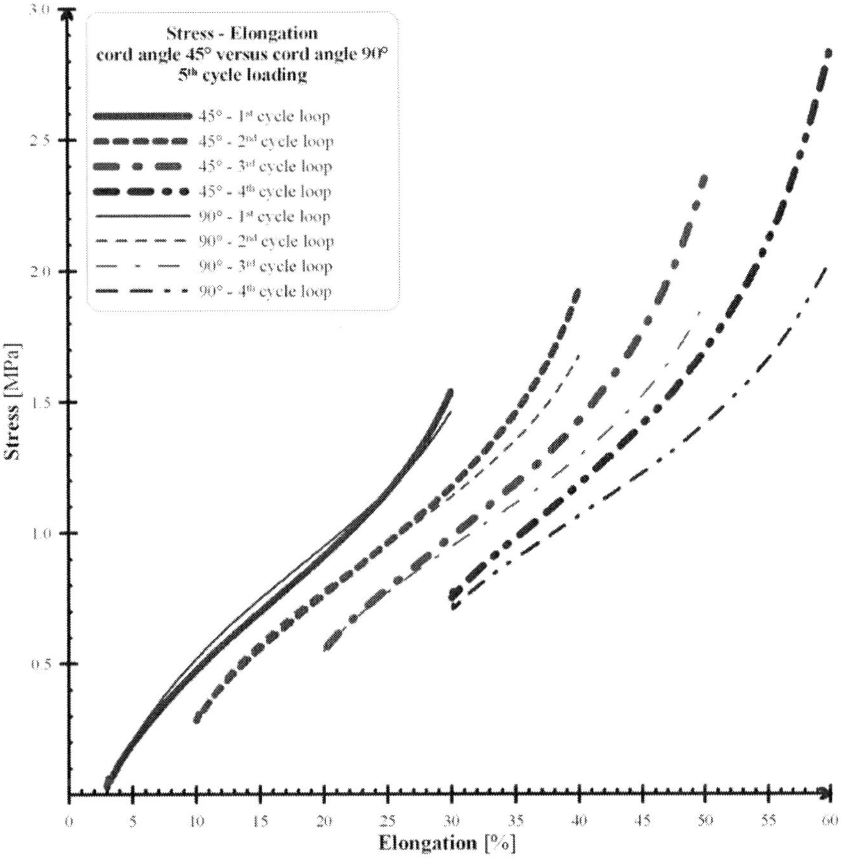

Fig. 3. Dependences of stress on elongation for fifth cycles

The M-R parameters are 0.638 MPa (as C_{10} in ANSYS) and 0.284 MPa (as C_{01} in ANSYS) and incompressibility parameter is 0.151 MPa^{-1}.

The computational model is presented in Fig. 4. The Solid 186 element type is used. The model has a length of 120 mm and the initial length between the clamps of the test machine 100 mm is defined by boundary conditions – by removing of degrees of freedom in two axes and applying displacements on the selected nodes in areas of clamps). The model is reverse loaded, the displacement in *z*-axis is defined and reaction force at the area of a sliding clamp of the test machine is searched. It is better for quick convergence and speed solution. The PCG solver is used. The increment of every substep is 0.2 mm in z-axis.

On the longitudinal edges of the model, there is characteristic "dentation" because the computational model is designed by APDL (ANSYS Graphical User Interface) procedures for the automatic creation of models from geometric parameters such as a cord diameter, cord distance and thickness of layer.

https://doi.org/10.1051/matecconf/201815701011

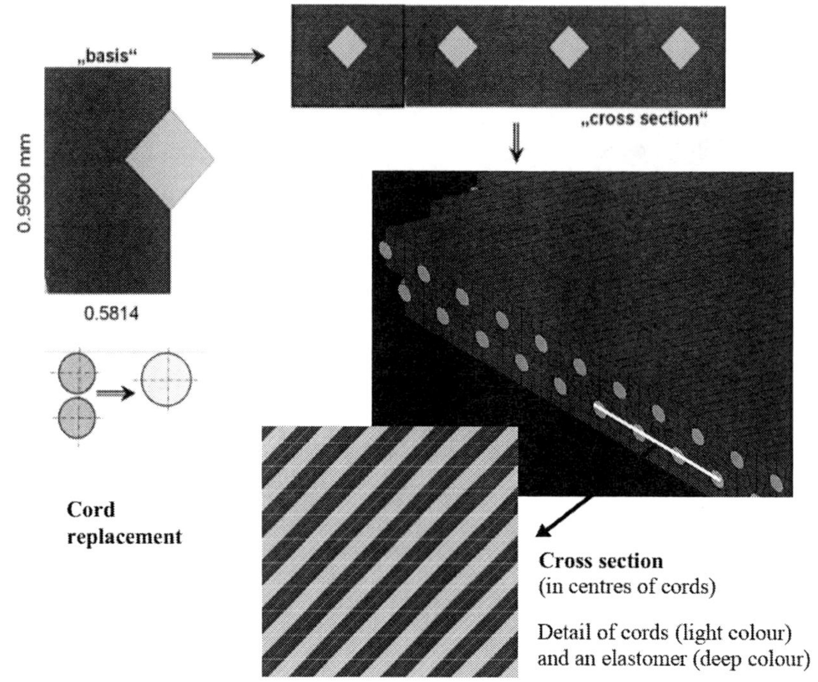

Fig. 4. Computational model of a specific layer of a steel-cord belt with the cross-sections of the model

4 Discussion

The approximate the dependences for a cord angle of 90° (Fig. 3) can be used as input material parameters of an elastomer matrix. The fifth cycle in every cycle loop can be considered as a stable cycle. The values of true stress for a cord angle of 45° are approximately 20–25 % higher than values for a cord angle of 90°.

The results from computational modeling of one layer of specific composite parts of a tire casing as length displacement (z-axis) and width displacement (x-axis) for elongation of 5 mm are in Figs. 5 and 6 and for elongation 10 mm in Figs. 7 and 8. The values of reaction forces are 14.8 N for elongation of 5 mm and 28.5 N for elongation of 10 mm. From test data, the tensile force 16.3 N causes the deformation of 5 mm and force 27.4 N causes the deformation of 10 mm. The computational model has tensile stiffness and the forces very similar. The force value differences between experiment and computation are 9.2 %.

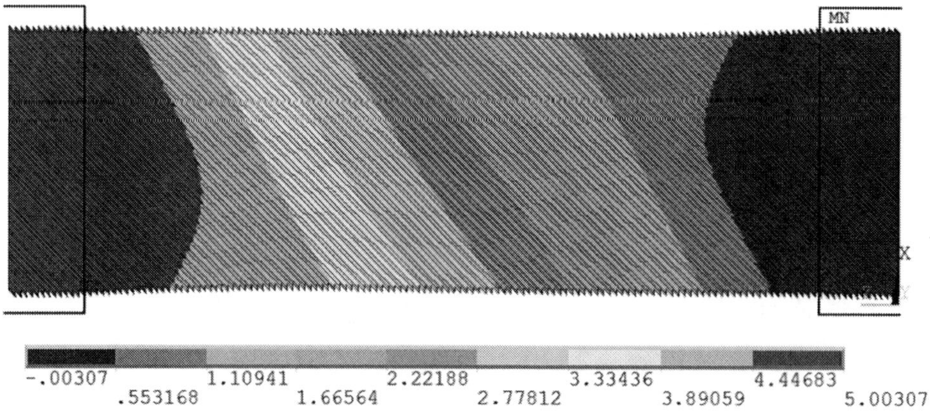

Fig. 5. Displacement in z-axis (length) for elongation 5 mm

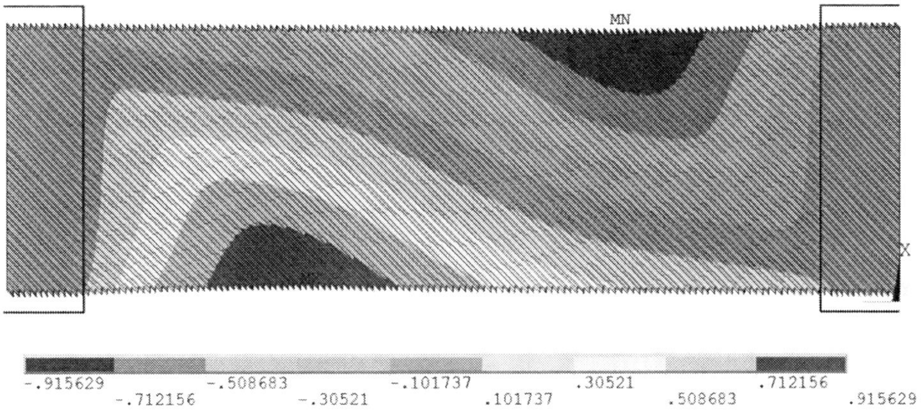

Fig. 6. Displacement in *x*-axis (width) for elongation 5 mm

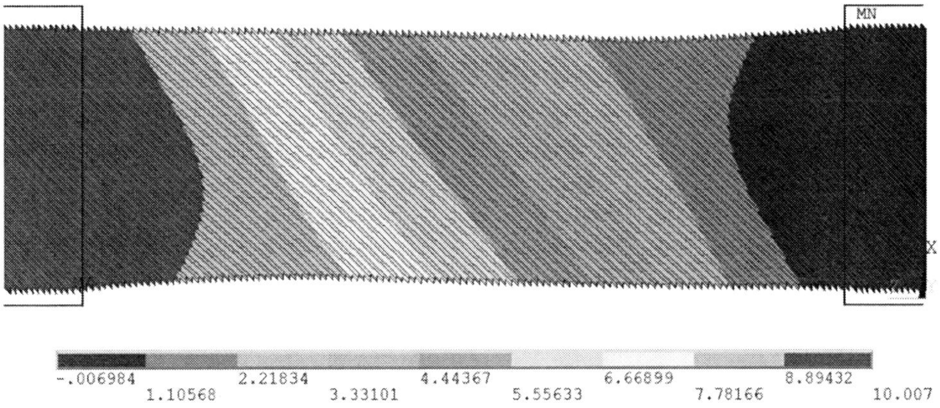

Fig. 7. Displacement in *z*-axis (length) for elongation 10 mm

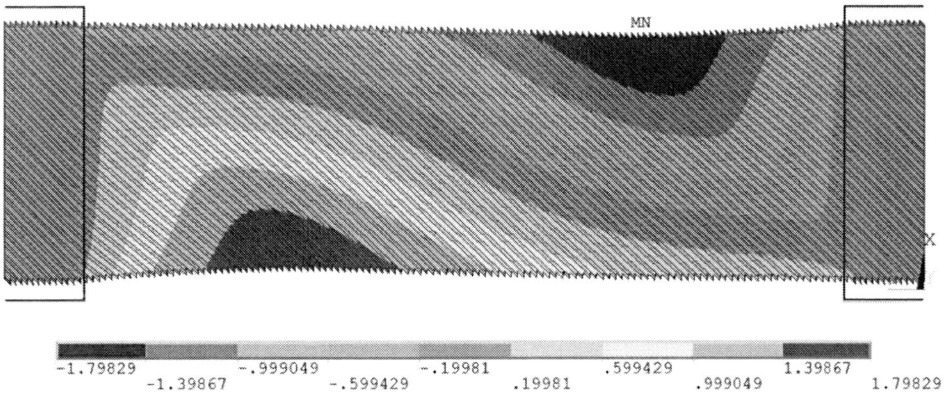

Fig. 8. Displacement in *x*-axis (width) for elongation 10 mm

Conclusions

Methods for testing of textile composites with five cycle loops for different elongation with five cycles in every cycle loop have been designed. The fifth cycle in every cycle loop can be considered as a stable cycle because difference between fourth and fifth cycle is small (about to 5 %).

It is quite difficult to compare the results from the calculations with the data or outputs in literatures because each one author deals either with the different dimensions of specimens or applies any other material parameters and material models for the computational modeling of parts of tire casing.

The test machine has temperature chamber allowing the tests from -70 to 180 °C and from 20 to 80 °C, it is possible to change the humidity from 30 to 95 %. Future research is needed for the tests of cyclic loading of composites, which will be realized at different temperatures. These results can be used in such conditions where the mentioned composites are exposed to the cyclic loading at an unsteady temperature: tire casing, conveyor belts, etc. The experiment results can be used for the optimization of the deposition angle of textile reinforcement in composites.

The contribution was supported by the Cultural and Educational Grant Agency of the Slovak Republic (KEGA), grant No. KEGA 005TnUAD-4/2016 and the project „Centrum pre testovanie kvality a diagnostiku materiálov" ITMS code 26210120046 of the Operational Program Research and Development funded from European Fund of Regional Development.

References

1. P. Koštial, J. Krmela, K. Frydrýšek, I. Ružiak, *The Chosen Aspects of Materials and Construction Influence on the Tire Safety*. Composites and Their Properties, InTech, 265-298 (2012), DOI: 10.5772/48181
2. Composite authors, *Rubber Technology*. (GC Tech Peter Gerši, ISBN 80-969189-5-8, 2006)
3. *The Pneumatic Tire*. Eds.: A. N. Gent, J.D. Walter, (2006)
4. J. Krmela, V. Krmelová, *Tire casing and their material characteristics for computational modeling of tires*. 16th International Scientific Conference Engineering for Rural Development, 206-211 (2011), DOI: 10.22616/ERDev2017.16.N043
5. DIN 53835-13: *Testing of textiles; determination of the elastic behaviour of textile fabrics by a single application of tensile load between constant extension limits*. (1993)
6. J. Krmela, V. Krmelová, *Replacement of Belt Structure for FEA of Tire*. Procedia Engineering **136**, 132-136 (2016)

Computer simulation of the load test of an extrusion blow moulded product

Michał Modławski[*,1], *Jacek* Nabiałek[1], *Tomasz* Jaruga[1]

[1]Department of Polymer Processing, Czestochowa University of Technology, Armii Krajowej 19c, 42-201 Częstochowa, Poland

Abstract. The results of compression "top load" test of an extrusion blow moulded container, performed in *ANSYS Mechanical* software, were presented. The non-uniform wall thickness of the blown product was considered because the thickness distribution was the result of the first step of the simulation in *ANSYS Polyflow* - the simulation of the product manufacturing by extrusion blow moulding. The hydrostatic pressure factor was taken into account during the "top load" simulation, because such a test is usually done for filled products. It was found that the biggest stress values occur around the inlet as well as in the side walls of the container.

Keywords: extrusion blow moulding simulation, top load test

1 Introduction

Extrusion blow moulding technology is used for many years, but because of the growing demand on plastics packagings in the last years this process is developing fastly, which is remarkable not only by increasing number of machines working in the industry but also by increasing the technological innovations [1].

Although the simulation of injection moulding process is widespread today, the simulation of blow moulding is still not used so much. The results of this simulation could contribute to the improvement in new products development thanks to the possibility of design modification at the product design stage. The computer simulation gives also an useful information about the processing parameters efficiency which is important while designing a product and a mould, that means before taking the costs of building a tool.

The research works regarding the computer simulation of extrusion blow moulding done so far were mainly about wall thickness distribution optimization [2-10], by controlling the extruded parison wall thickness distribution, for example. It is not only possible to optimize the wall thickness distribution, but also to reduce the mass of the product, using a computer simulation as a tool [7]. The processing conditions like blowing pressure, parison temperature or its wall thickness have a big impact on products' properties and on their quality. However, not all quality factors can be simulated nowadays – for example the surface roughness of a product and the degree of cavity mapping. These factors are an important quality indicators and they highly depend on processing parameters [11].

[*] Corresponding author: modlawski@ipp.pcz.pl

Reviewers: *Kateryna* Kravchenko, *Ján* Vavro

The modeling of the product strength [8-10] with the reference to the results of blow moulding computer simulation is especially important because it allows modifying product design and processing conditions to obtain the required strength. The wall thickness distribution result should be used to simulate the strain and stress of the loaded container or to simulate the stress in the case of free fall of the container which occurs during the typical "drop test" of such products [10].

A canister-type container is one of typical extrusion blow moulded products. This product is specific due to its cubicoidal shape which makes the risk to thin the wall significantly in the corners during the moulding process as well as due to the handle used for carrying, but also being problematic during moulding [12-13].

2 Simulation research

The aim of this article was to determine the stress and strain distribution in a canister-type product, during the compression test, using a computer simulation. This product should be manufactured from high-density polyethylene, in an extrusion blow moulding process. The canister was loaded with the concentrated force of 100N, coming from the palletization of the filled products during their storage and transport, and loaded with the hydrostatic pressure. The computer simulation of the extrusion blow moulding process was done using *ANSYS Polyflow* software. The stress and strain distribution was simulated using *ANSYS Mechanical* software which was convenient to use in this case because of the possibility to transfer the data of manufacturing simulation from *ANSYS Polyflow*.

2.1 The project of the canister

The canister was designed in *Siemens NX 8.5* CAD software. The following assumptions regarding the design and dimensions of this container were made:

- the canister volume is no less than 5 l. The level of the free surface of the filling liquid should be below the neck,
- easy demoulding of the product is possible because the surfaces paralell to the mould opening direction are rounded, except of the upper surface of the handle, which is flat, and except of the bottom surface of this container,
- the side walls of the canister are flat; in these places usually labels are placed,
- there are no geometrical elements making canister more stiff (increasing its strength) except of the bottom,
- the bottom of the canister is of the shape which enables the canister to be put on a flat surface and increasing the strength of this product,
- the inlet was extended and evened with the top surface of the handle (as the cap height). The thread was neglected to simplify the shape used in the computer simulation.

The designed shape of the canister is presented in Fig. 1 with its main dimensions. The shape of this container is symmetrical so only a half of the canister was assumed in the blow moulding process and mechanical strength simulation.

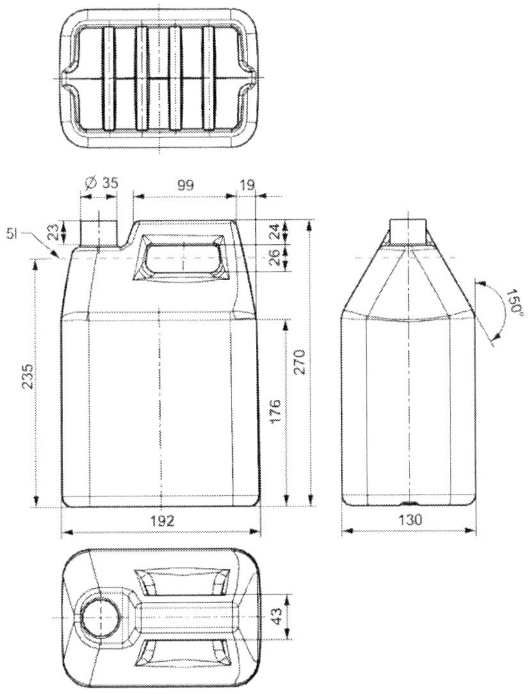

Fig. 1. The shape and main dimensions of the canister

2.2 Blow moulding computer simulation

During the blow moulding process simulation 1 second time of this process was considered. In this time the mould is closed with the velocity of 0.2 m/s (from 0 to 0.4 s) and after that the parison is blown by the gas overpressure in the time interval from 0.4 to 1.0 s. The assumed value of the gas overpressure increases between 0.4 and 0.5 s time and reaches the value of 0.4 MPa (Fig. 2).

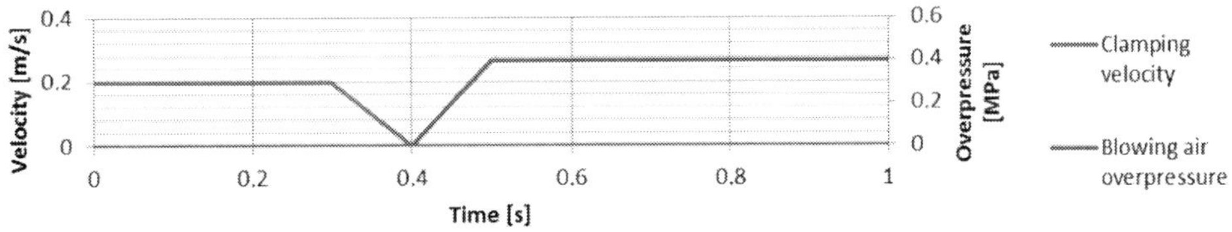

Fig. 2. The function of mould closing speed and blowing gas overpressure versus time

This type of container is manufactured very often from high-density polyethylene (PE-HD). In this simulation PE-HD Tipelin 6000B, produced by MOL Petrochemicals was selected. The rheological viscosity curve was determined for this material in the temperature of 160°C and in the shear rate range of 0.005 ÷ 0.5 s^{-1} using Rheolyst AR 1000N rotational rheometer, manufactured by TA Instruments. The average value of viscosity in this shear rate range is $56.4 \cdot 10^3$ Pa·s. The density of this material in the liquid state is 0.770 g/cm^3 in the temperature of 160°C. The density was determined using pvT-100 device.

The diameter of the extruded parison was 130 mm. Its initial thickness was computed by *Parison Programming* module of *ANSYS Polyflow* software. Some initial simulations were conducted to obtain this thickness. It was assumed that the minimal thickness of the canister should be no less than 0.6 mm. The thickness distribution of the parison along its height is shown in Fig. 3.

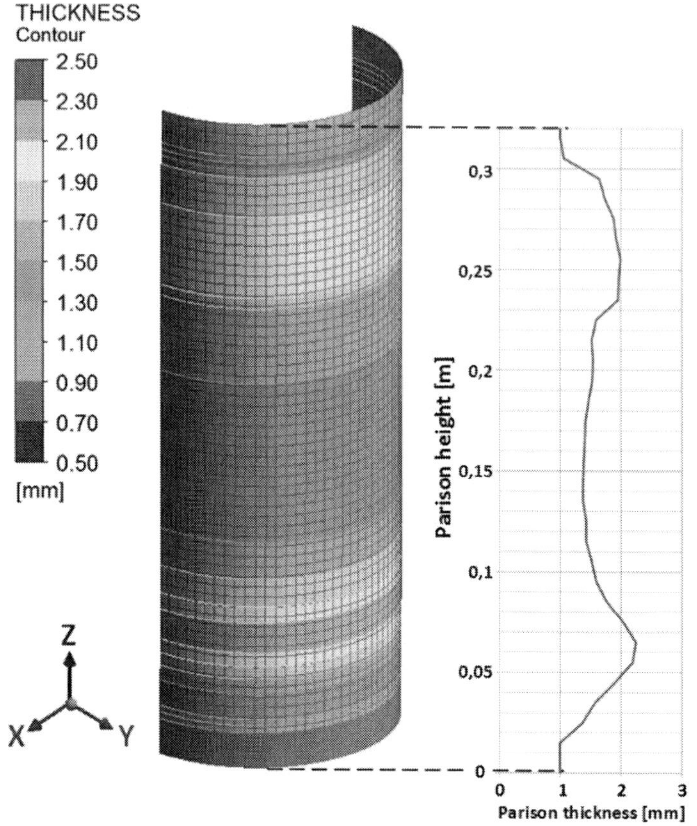

Fig. 3. Parison dimensions – wall thickness distribution

The boundary conditions were imposed on the horizontal and vertical edges of the parison. The condition on vertical edges involves plane symmetry with a normal vector directed along Y axis. The direction of Y axis corresponds to the direction of the mould closing and opening. The boundary condition on the horizontal edges is defined by zero value of: the normal velocity and the force tangent to the plane containing the horizontal parison edge. Between the finite elements of the parison and mould mesh a contact condition was imposed. The contact is applied when the distance between the nodes of these meshes is smaller than 0.4 mm.

The size of the parison mesh finite elements is 5 mm, which means the distance between the nodes. The mesh modifier *Mapped Face Meshing* was used to obtain even distribution of nodes, without mesh deformation. *Adaptive Meshing* function was also applied to achieve good mesh quality in the handle area. This function is used to increase the number of finite elements (mesh density) when:

- the distance between a node of the parison mesh and a node of the mould mesh – expression $1+\left(\frac{d}{B}\right)^2$ from the equation (1) - is smaller than the defined value,
- the radius of the parison mesh curvature – expression $A \cdot r + C$ is of a small value.

The maximum number of mesh divisions equal 2 was assumed. The mesh elements are divided when:

$$new\_size = max\left\{size\_min,\quad (A \cdot r + C)\cdot\left[1+\left(\frac{d}{B}\right)^2\right]\right\} \tag{1}$$

where:

- size_min – minimum size of an element in the parison –1.6 mm was assumed for computations,
- A – fraction of the radius of curvature – 0.2 was assumed,
- B – typical distance when mesh is begun to refine – value of 5 mm,
- C – coefficient of proportionality – the assumed value was 0,
- d – distance between node E of a mesh parison and the nearest node N of a mesh of the mould,
- r – local radius of curvature at node N [14].

The simulation results of blow moulding – the distribution of the canister wall thickness is shown in Fig. 4. The thickness distribution of the product - canister is homogeneous, except of some small thin areas in the handle and some thicker wall below the neck by the parting plane. This proves that distribution of the initial parison thickness was well determined.

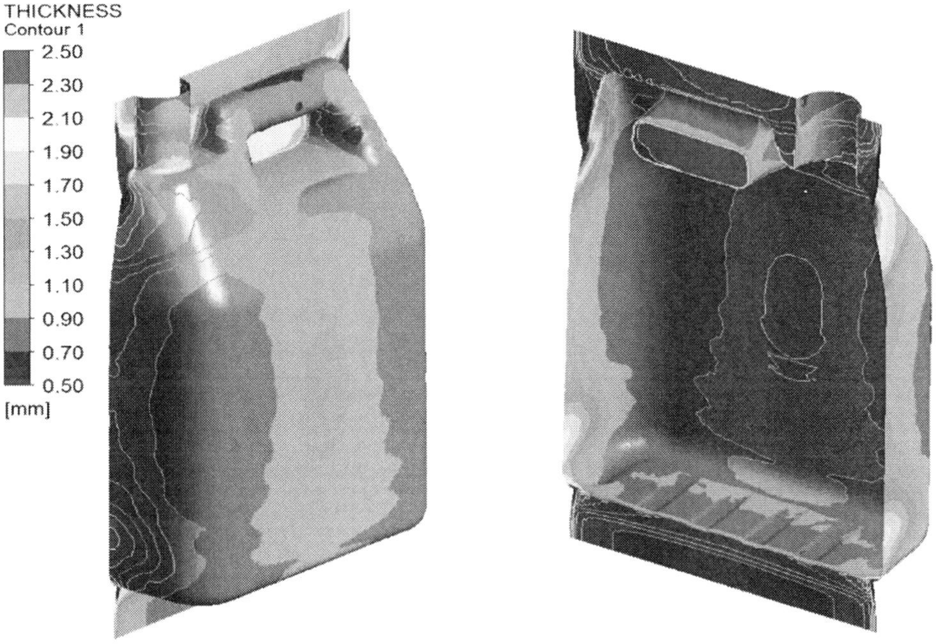

Fig. 4. Blow moulding simulation result

2.3 Computer simulation of the loaded product

The model was supplemented with two plates to simulate the load. The plate on the bottom of the canister is imposed with *fixed support* condition and the plate on the top is loaded with 100 N concentrated force, directed down along the axis Z-. The value of 100 N acts on one canister when being transported in the bottom layer, below two other layers of these products filled with water. This is an usual condition of the transport of such products. The numerical calculations were conducted for two cases. In the first case a load caused by hydrostatic pressure was not taken into account, but in the second case it was.

https://doi.org/10.1051/matecconf/201815701012

The contact condition was defined between the bottom surface of the canister and the upper surface of the bottom plate as well as between the upper surface of the handle together with the upper edge of an inlet and a bottom surface of the upper plate.

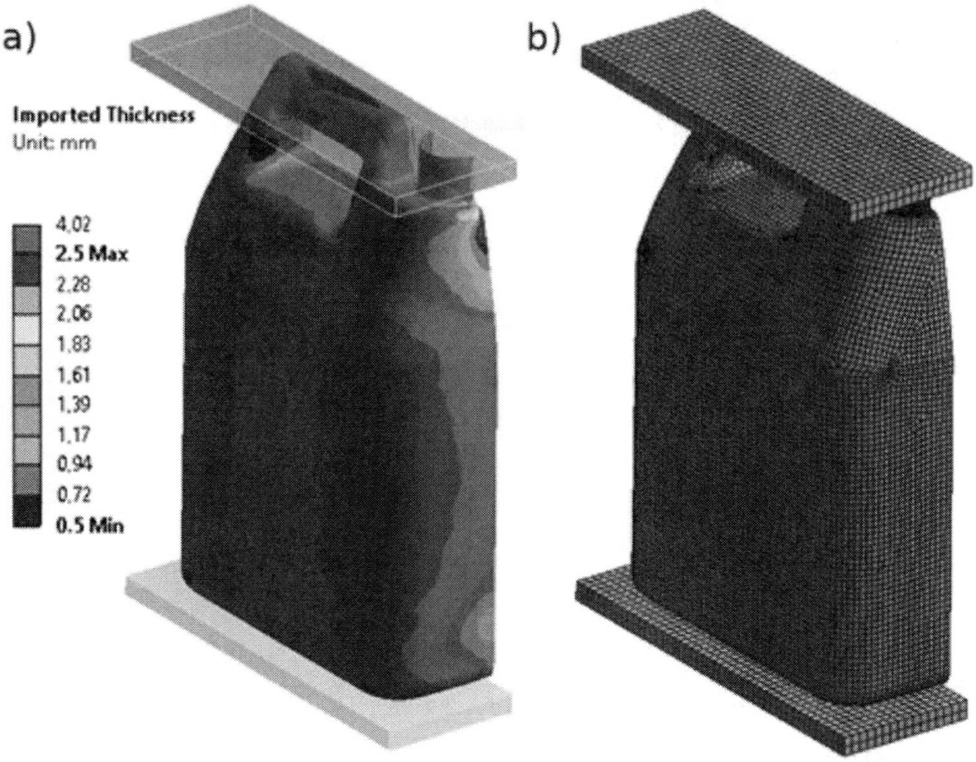

Fig. 5. Preparing load simulation a) mesh used for the load simulation, b) canister wall thickness distribution imported from ANSYS *Polyflow* solver

For the numerical computations of the load test the surface model with the additional thickness parameter was involved. The thickness values of the container were imported from a text file, generated during the parison blow moulding simulation in *Polyflow solver*. The thickness distribution after the blow moulding process computer simulation is presented in Fig. 5a. The thickness value was generated inside the canister so the outer surface of the container was unchanged. After mapping of the thickness 3D finite element mesh can be generated. The average size of the container mesh edges was equal 3 mm, after adding a mesh modifier (Body Sizing) to the canister model. The mesh is shown in Fig. 5b.

2.4 Results of computer simulation of the load

The numerical simulation results of the load test were: stress distribution and ready product deformation, computed for two cases:

- canister loaded by a vertical force equal 100 N,
- canister loaded by a vertical force equal 100 N and the internal walls loaded by hydrostatic pressure resulting from filling the container by a liquid.

Loading the designed canister with the vertical force equal 100 N causes that the stress is created in the product. The stress distribution is shown in Fig. 6. The maximum value equals 9.13 MPa and occurs in the area below the inlet. A big stress values are also in the zone of the vertical walls of the canister handle.

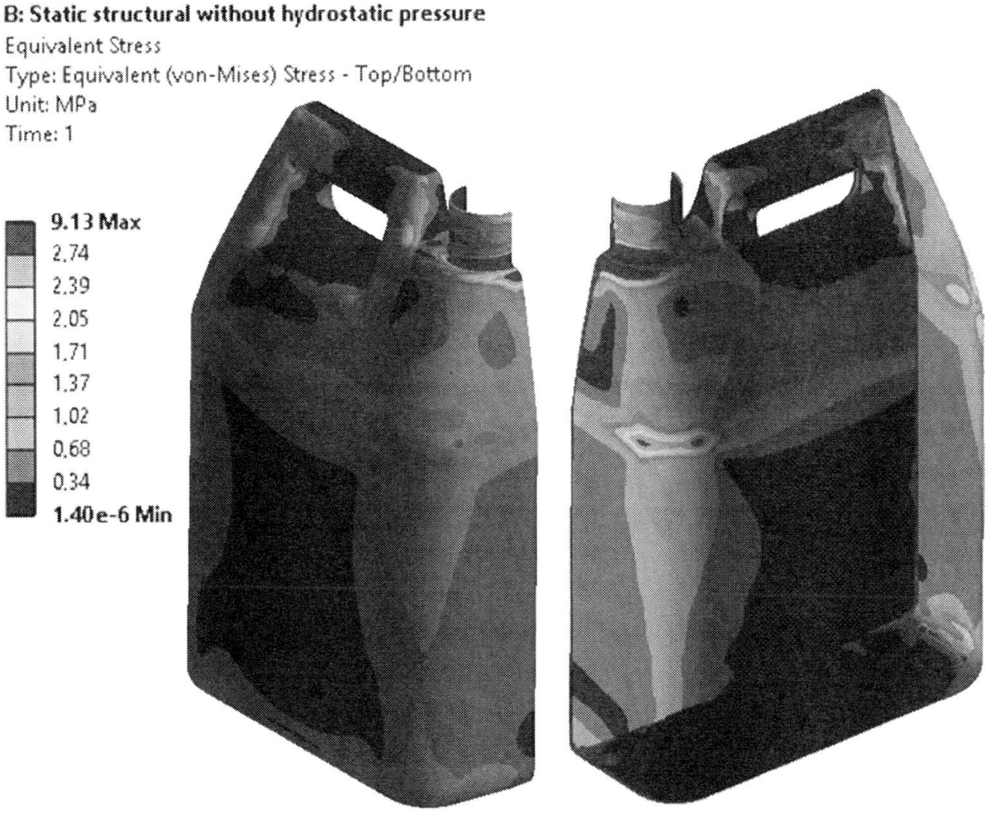

Fig. 6. Distribution of the stress without hydrostatic pressure (Equivalent Stress)

The canister deformation caused by the vertical force of 100 N is presented in Fig. 7. This force causes the biggest displacement of the rear wall in a place where the vertical surface changes into the handle wall at 11° angle. A significant deformation occurs also in the analogical zone on a front wall.

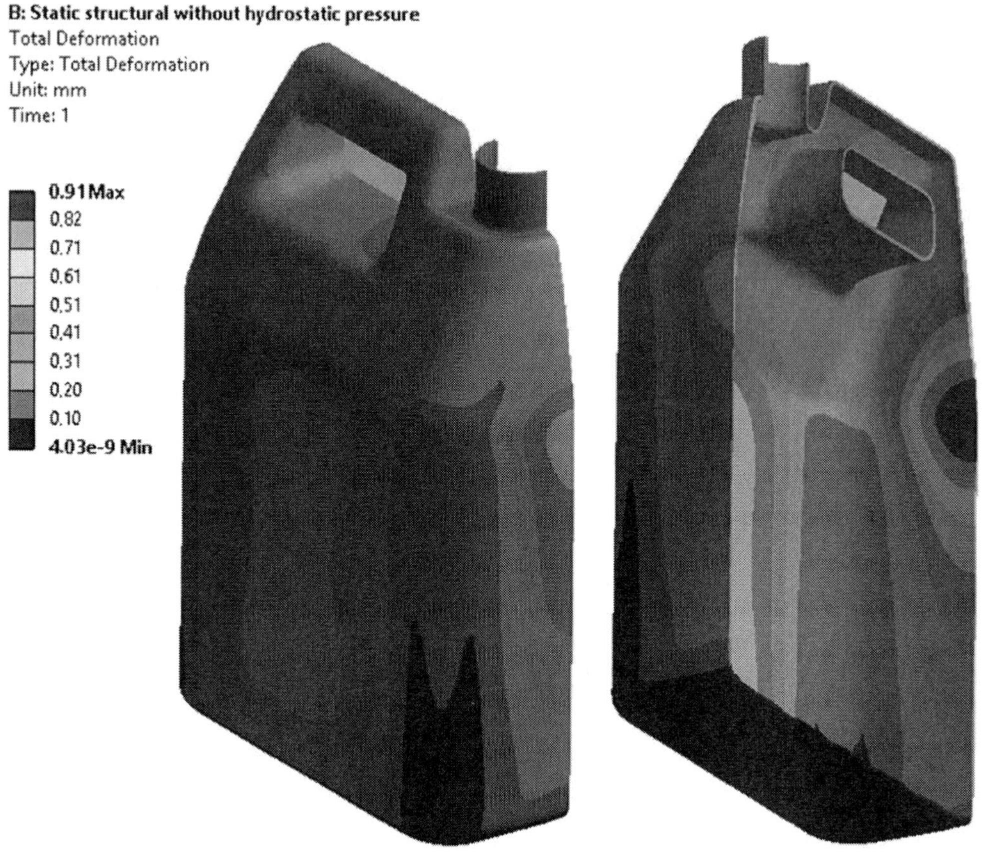

Fig. 7. Strain distribution – real scale (Total Deformation)

After adding a load resulting from hydrostatic pressure the maximum value of the stress increased by 19%. In this case the side walls of the canister are much more loaded – see Fig. 8 compared with Fig. 6. The maximum stress value calculated occurs in the bottom zone of the side walls.

The maximum value of stress equal 10.9 MPa does not exceed the maximum tensile strength at yield, declared by the manufacturer of this plastic: 32 MPa [15]. In order to predict whether the canister wall can be thinner than this obtained from the current simulation some numerical simulation of dynamical loads should be done.

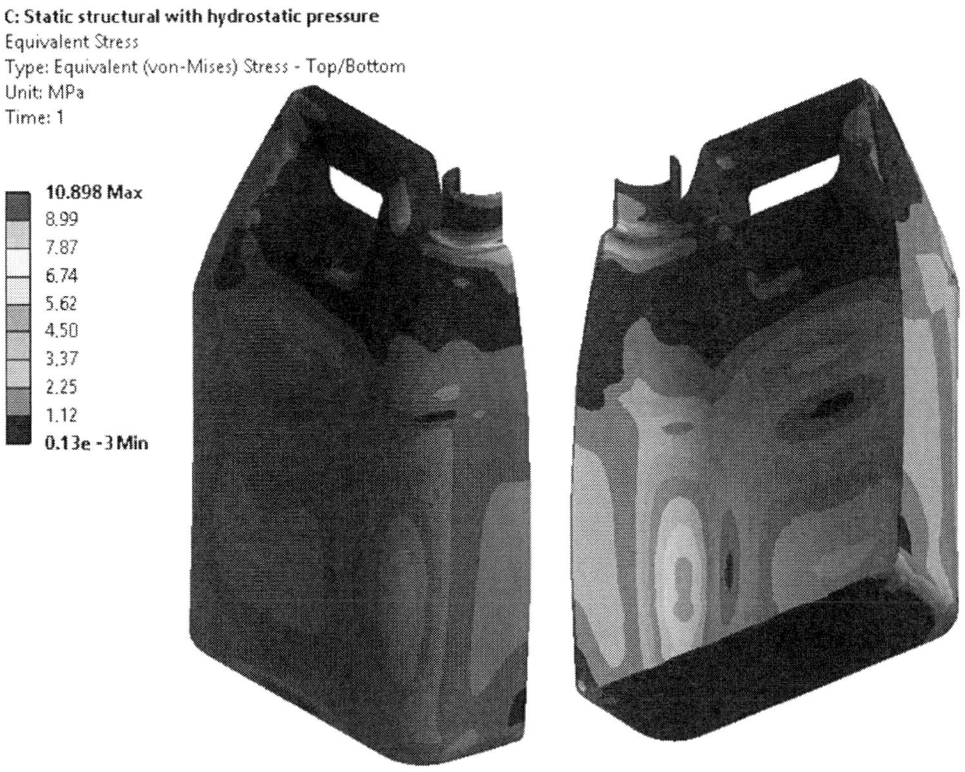

Fig. 8. Stress calculated considering hydrostatic pressure (Equivalent Stress)

Filling the canister with a liquid causes a large deformation of this product in the side wall zone and also in a zone of front and back vertical walls (Fig. 9). The displacement values caused by the vertical force are relatively small in comparison to the displacement values caused by the hydrostatic pressure.

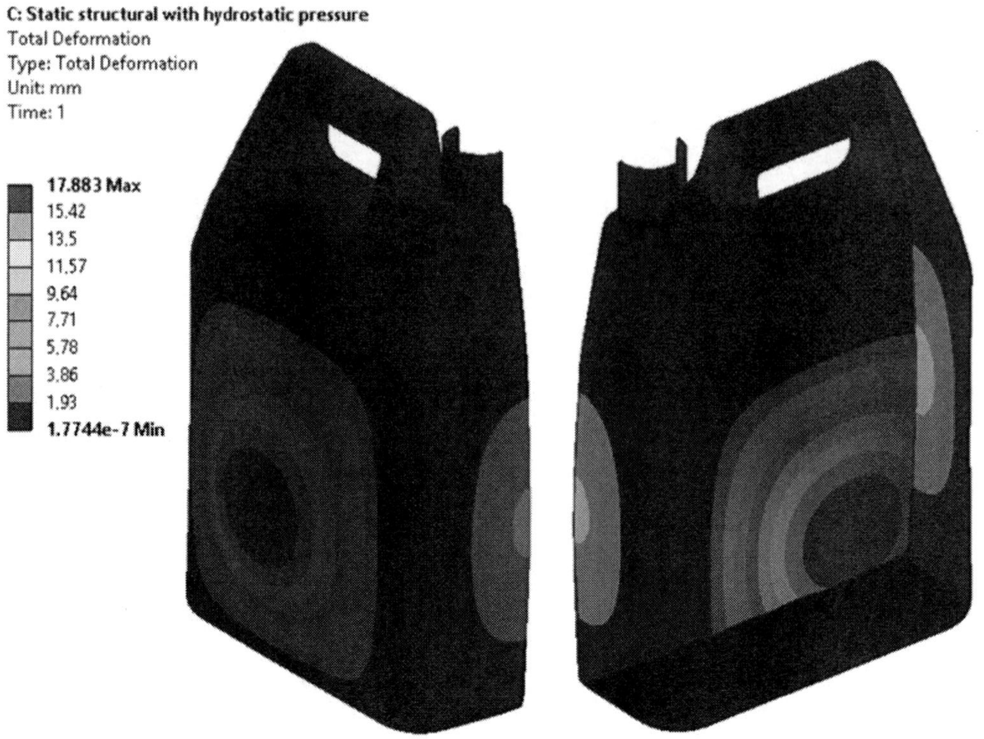

Fig. 9. Total deformation – in three axes directions

Conclusions

A study of a combined computer simulation was presented in this article. The mechanical strength simulation of a canister-type plastic container loaded with a compression force was made, but based on the results of the simulation of manufacturing this product by blow moulding – on the base of the wall thickness distribution result, since the wall thickness impacts the local product strength.

Results of numerical calculations of strength made using the data coming from he simulation of manufacturing process allow obtaining more accurate data, important regarding the predicted quality of a new product. Using the results of polymer processing simulation for strength calculations improves the quality of forecasting the properties of plastics products.

On the base on the computer simulation study the following conclusions can be drawn:

- simulation of the extrusion blow moulding process allows predicting many properties of a designed product which is going to be manufactured. A key result of this simulation is the distribution of the product wall thickness,
- using results of the polymer processing simulation further in mechanical strength computer simulation enables stress and strain analysis of a designed product without making real tests, and eliminating the problems with incorrect product design,
- contemporary computer simulation programs allow making virtual analyses concerning each aspect of the plastic product lifetime - starting from a shape design, through usage properties prediction up to the product utilization.

The authors wish to thank Department of Polymer Engineering and Technology from Wrocław University of Science and Technology for the rheological measurements and Department of Polymer Processing from Lublin University of Technology for p-v-T measurements.

References

1. H. Meetwall, *Virtual prototyping for the blow molding industry*. (ANSYS U.S.A. 2012)
2. S. Tanoue, T. Kajiwara, K. Funatsu, K. Terada, M. Yamabe, *Numerical Simulation of Blow Molding – Prediction of Parison Diameter and Thickness Distributions in the Parison Formation Process*. Polym. Eng. Sci. **36**, 2008-2017 (1996)
3. S. I. Tanifuji, *Overall Numerical Simulation of Extrusion Blow Molding Process*. Polym. Eng. Sci. **40**, 1878-1893 (2000)
4. K. Pepliński, M. Bieliński, *Polyflow software use to optimize the parison thickness in blowing extrusion*. Journal of POLISH CIMAC. R. **4** (3), (2009)
5. C. Gauvin, F. Thibauld, D. Laroche, *Optimization of Blow Molded Part Performance through Process Simulation*. Polym. Eng. Sci. **43**, 1407-1414 (2003)
6. Y. Fukuzawa, S. Tanoue, Y. Iemoto, R. Kawachi, H. Tomiyama, *Three-dimensional simulation on multi-layer parison shape at pinch-off stage in extrusion blow molding*. Polym. Eng. Sci. **50**, 1476-1484 (2010)
7. J. C. Yu, X. X. Chen, T. R. Hung, F. Thibauld, *Optimization of extrusion blow molding processes using soft computing and Taguchi's method*. J. Intell. Manuf. **15**, 625-634 (2004)
8. M. Modławski, J. Nabiałek, T. Jaruga, *Symulacja odkształcenia wyrobu rozdmuchowego o stałej i zmiennej grubości ścianki*. Tworzywa Sztuczne w Przemyśle **34**, 94-97 (2016)
9. D. Kwiatkowski, M. Modławski, T. Jaruga, *Symulacje komputerowe grubości ścianki butelki uzyskiwanej w procesie wytłaczania z rozdmuchiwaniem*. Przetwórstwo Tworzyw **21**, 256-261 (2015)
10. P. Klein, F. Fradet, H. Meetway, T. Marchal, *Virtual Prototyping Applied to a Blow-Molded Container*. (Proceedings of SPE ANTEC, Orlando, Florida, U.S.A., 2012)
11. K. Pepliński, M. Bieliński, *Wpływ warunków przetwórstwa na wybrane wskaźniki użytkowe wytworów wytłaczanych z rozdmuchiwaniem*. (Postęp w przetwórstwie materiałów polimerowych, Ed. J. Koszkul, E. Bociąga, 176-183, Częstochowa 2006)
12. D. V. Rosato, A. V. Rosato, D. P. DiMattia, *Blow Molding Handbook*. (Hanser Publishers, Munich, 2004)
13. K. Pepliński, *Problem rączki. Jak wyeliminować problemy w procesie wytłaczania z rozdmuchiwaniem podczas formowania rączek pojemników?*. Plastics Review **71-72**, 16-17 (2007)
14. ANSYS Polyflow help documents (2017)
15. TIPELIN 6000B Technical Data Sheet - MOL Petrochemicals: https://mol.hu/en/polymers/products/hdpe-medium-and-high-density-polyethylene, (Access: 07.2017)

The determination of wormwheel toothing surface

Tadeusz Nieszporek[1], *Rafał* Gołębski[1,*], *Piotr* Boral[1]

[1]Częstochowa University of Technology, Institute of Mechanical Technologies, Al. Armii Krajowej 21, 42-201 Częstochowa, Poland

Abstract. In heavy industry (metallurgy, mining), large-size worm gears designed to carry large loads are often used. However, their technology is very difficult and their manufacturing costs are very high. In practice, cone-derivative worm gears are most often used, which are machined by the envelope method using a rotary tool. The literature has given much coverage to the determination of the worm helical surface. The surface of wormwheel teeth is much less commonly described. Therefore, this paper presents an analytical and a numerical methods for generating the wormwheel toothing by the tangential and radial methods with a special cutter and with a modular hob.

Keywords: envelope method, worm gear, worm wheel

1 Introduction

Heavy industry (such as metallurgy or mining) often uses large-size worm gears designed for carrying large loads. However, their manufacturing costs are very high and their accuracy, durability and reliability should meet appropriate operation conditions [1] – Figure 1.

Fig. 1. A worm reduction gear with a rated power of 200 kW: a) a general view, b) a wormwheel machined with a modular hob; axial module, 20 mm; pitch diameter, 872 mm; number of teeth, 42

* Corresponding author: rafal@itm.pcz.pl

Reviewers: *Juraj* Gerlici, *Ireneusz* Malujda

2 Worm surface

The helical surface of a worm can be shaped using a rotary tool [2-8]. The axial profile of the tool action surface can be described the following equation – Fig. 2.

$$\underset{p}{\mathbf{x}}(u)=\left[0,\ \ \underset{p}{x}^{2}(u),\ \ \underset{p}{x}^{3}(u)\right]^{T},\quad m_p=tg\alpha_p \tag{1a-b}$$

where: α_p – angle of direction of the tangent to the tool axial profile at the point under consideration; u – profile parameter, m_p – slope of the tangent; p – subscript identifying the profile's coordinate system.

Considering the relative helical motion of the tool and the worm, the family of tool action surfaces in the worm's coordinate system can be notated with the following equation:

$$\underset{1}{x}(u,\ \varphi,\ v)=[3,\ -v]\underset{0}{x}+[0,\ 0,\ \pm pv]^{T} \tag{2}$$

where: v - relative helical motion parameter (tool action surface family parameter), φ - tool action surface parameter, $\underset{0}{\mathbf{x}}$ - vector of tool position in the worm coordinate system, p - worm helical surface parameter.

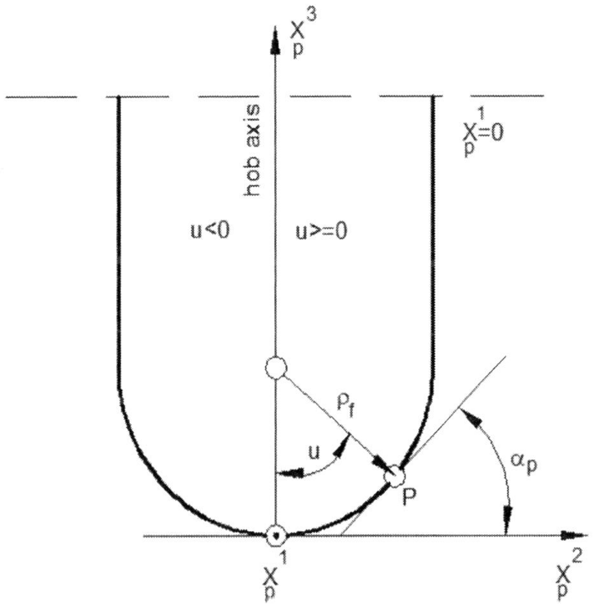

Fig. 2. The form of definition of the tool action surface axial profile

The envelope condition [2-6] should be added to Equation (2).

$$f_1=\frac{\partial \underset{1}{\mathbf{x}}}{\partial u}\frac{\partial \underset{1}{\mathbf{x}}}{\partial \varphi}\frac{\partial \underset{1}{\mathbf{x}}}{\partial v}=0 \tag{3}$$

In Equation (3), no relative tool and worm helical motion parameter occurs, so the worm convolution surfaces will be described by the system of equations:

$$x_1 = x_1(u, \quad \varphi, \quad v), \tag{4}$$

$$f_1(u, \quad \varphi) = 0 \tag{5}$$

The surface of the worm (cutter – a wormwheel machining tool) may be known, for example from measurements.

3 The worm – wormwheel technological gear

After the cutter has been cut in to the full depth by the tangential or radial method, the tool and the wormwheel being machined will form a worm–wormwheel technological gear [3, 4].

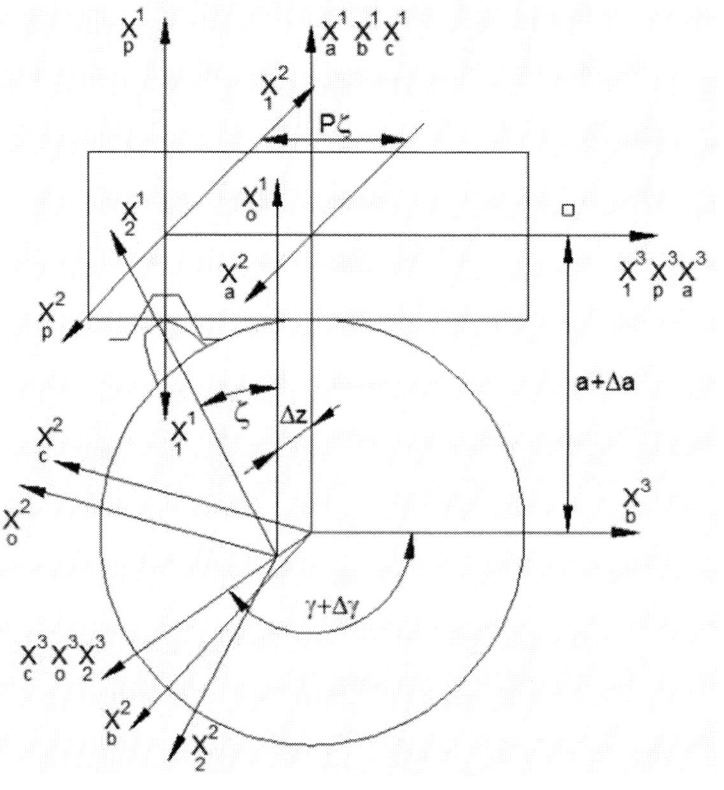

Fig. 3. Kinematic diagram of the worm (tool) – wormwheel technological gear with the axis-wise worm motion

The technological gear, in which the rotary motion of the both gear elements occurs can be substituted with an equivalent gear, in which axis-wise worm motion occurs with no rotation [3, 4] – Figure 3. To determine the wormwheel toothing surfaces, we must change over from the tool's system to the wormwheel's system.

$$\underset{2}{\mathbf{x}} = [3, \ i\xi][1, \ -(\gamma + \Delta\gamma)] \begin{pmatrix} [3, \ \pi]\underset{1}{\mathbf{x}} + \\ + [a + \Delta a, \ -\Delta z, \ -p\xi]^T \end{pmatrix} \tag{6}$$

where: ξ - relative tool and wormwheel turning motion parameter; γ - angle between the tool and wormwheel rotation axes; a - tool and wormwheel axis distance for machining by the tangential method; Δa - axis distance error with the tangential method or the quantity allowing for the tool cutting into the machined wormwheel with machining by the radial method; $\Delta\gamma, \Delta z$ - tool positioning errors; i - worm gear transmission ratio.

Every single wormwheel rotation, the wormwheel tooth surface needs to be determined and, by juxtaposing (comparing) thus obtained surfaces respectively with one another, the tooth surface shaped by the radial method should be determined.

If subsequent tool indentations are omitted and only tool and machined wormwheel turning for the full indentation depth is considered ($\Delta a = 0$), then the wormwheel tooth surface shaped by the tangential method will be obtained.

4 Machining with a special hob

The tool action surface can be described with the system of equations (4) and (5). From the envelope condition (5), the parameter φ can be determined as a function of the parameter u, and the worm surface equation will take on the following form:

$$\underset{1}{\mathbf{x}} = \underset{1}{\mathbf{x}}(u, \ \varphi(u), \ v) = \underset{1}{\mathbf{x}}(u, \ v) \tag{7}$$

The tool action surface family (6) in the wormwheel's system can be generally written using the following equation:

$$\underset{2}{\mathbf{x}} = \underset{2}{\mathbf{x}}(u, \ v, \ \xi) \tag{8}$$

In order to determine the wormwheel tooth surface, the envelope condition needs to be added:

$$f_2 = \frac{\partial \underset{2}{\mathbf{x}}}{\partial u} \frac{\partial \underset{2}{\mathbf{x}}}{\partial v} \frac{\partial \underset{2}{\mathbf{x}}}{\partial \xi} = \mathbf{abc} = 0 \tag{9}$$

where

$$\mathbf{a} = \frac{\partial \underset{1}{\mathbf{x}}}{\partial u} + \frac{\partial \underset{1}{\mathbf{x}}}{\partial \varphi} \frac{\partial \varphi}{\partial u}, \quad \mathbf{b} = \frac{\partial \underset{1}{\mathbf{x}}}{\partial v} \tag{10a,b}$$

$$\mathbf{c} = [3, \ \pi]^T \left\{ \begin{matrix} i[M] \cdot \\ \cdot \begin{pmatrix} [3, \ \pi]\underset{1}{\mathbf{x}} + \\ + [a + \Delta a, \ -\Delta z, \ -p\xi]^T \end{pmatrix} + \\ + [0, \ 0, \ -p]^T \end{matrix} \right\} \tag{10c}$$

$$[M]=\begin{bmatrix} 0 & c\alpha & -s\alpha \\ -c\alpha & 0 & 0 \\ s\alpha & 0 & 0 \end{bmatrix} \tag{10d}$$

As follows from Equations (10), the condition (9) is a linear function of the parameter ξ [3].

$$\xi=\frac{(\mathbf{a}\times\mathbf{b})\left\{[3,\ \ \pi]^T\left\{\begin{matrix} i[M] \\ \left(\begin{matrix}[3,\ \ \pi]\underset{1}{\mathbf{x}}+ \\ +[a+\Delta a,\ \ -\Delta z,\ \ 0]^T\end{matrix}\right)+ \\ +[0,\ \ 0,\ \ -p]^T \end{matrix}\right\}\right\}}{(\mathbf{a}\times\mathbf{b})[3,\ \ \pi]^T i[M][0,\ \ 0,\ \ p]^T} \tag{11}$$

It is accepted to determine the wormwheel tooth profile in the frontal section from condition below:

$$f_3=\underset{2}{x}^3-s=0 \tag{12}$$

where $\underset{2}{x}^3$ is the third component of the vector (6), while the parameter s defines the position of the cutting plane relative to the frontal symmetry plane (situated in the mid-width of the rim and perpendicular to the wormwheel axis).
For a given value of Δa, in a given cutting plane s, for successive values of the parameter u, the values of the parameters φ are determined from Equation (5). From Equation (12), the values of the parameter v are determined, whereas at each approximation from Equation (11), the value of the parameter ξ is determined. In each computation cycle, the parameters (u, φ, v, ξ), after being substituted in Equation (6), will define the successive points of the wormwheel tooth profile. After changing the cutting plane position, the computation cycle is repeated, yielding, as a result, the wormwheel tooth surface in a discrete form of a set of points for a specified tool "indentation" into the wormwheel being machined. After the change of Δa, the computation cycle is repeated for the same cutting planes. The obtained profiles should be compared with one another in these planes, while leaving the points on the machined material side. For $\Delta a=0$, the wormwheel toothing surface is obtained.

5 Normalizing the profile points

In individual cutting planes after successive tool passes (for varying cutting depths), the two curves given in a discrete form of sets of points should be each time compared with one another. Normalizing the points of the curve involves their approximation with the set of points lying on the abscissa grid lines introduced in the cutting plane (successive curves are substituted with the sets of points with the same abscissae, but different ordinates) [3, 4]. After considering the successive wormwheel rotations, until achieving the full cutting depth ($\Delta a=0$), the wormwheel tooth surface will be obtained in a discrete form of a set of points.

The helical surface can be described with the helical motion of an arbitrary curve lying on this surface, so it is possible to take into consideration also the characteristics (the curve of worm machining tool and worm contact when shaping the worm helical surface) [3, 4, 9, 10]. With the fixed value of the relative worm machining tool and worm helical motion parameter ν, Equations (4) and (5) are the equations of the characteristics. From Equation (5), for successive values of the parameter u, the values of the parameter φ can be determined, while from Equation (4), the characteristics sought for. So, at the fixed value of the parameter ξ, for successive values of the parameter u, after determining the value of the parameter φ from Equation (5), it is possible to determine the values of the parameter ν from Equation (12), while from Equation (6), the worm profile in a given cutting plane in the wormwheel's system. For successive values of the parameter ξ, successive curves can be obtained, which, after normalizing and comparing, enable us to determine the tooth flank profile in a given section for a given value of Δa. The computation cycle is similar as before. In this case, there is no need for determining the envelope condition (9) (this is a significant simplification).

6 Machining with a modular hob

In this case, the radial wormwheel machining method is used. Normally, the diameter of the hob does not correspond to the diameter of the worm (the module and the pitch must agree), which will mate the machined wormwheel. The hob thread lead angle on the pitch diameter will be different from the worm thread lead angle, and the $\Delta\gamma$ angle will be interpreted as a correction of hob positioning [11]. It is assumed that the action surface of the module hob is known (is the result of measurements or calculations) [11]. For the determination of the wormwheel tooth surface, the algorithm described above can be used, with a slight modification.

7 Computation results and conclusions

The radial method enables the illustration of tooth formation as the tool cuts into the wormwheel being machined – Figure 4. The developed program allows the wormwheel tooth tip relief to be eliminated. The analytical and numerical method of generating the wormwheel toothing, in contrast to the analytical method, enables the determination of the undercut or the transition curve of the tooth profile – Figure 5.

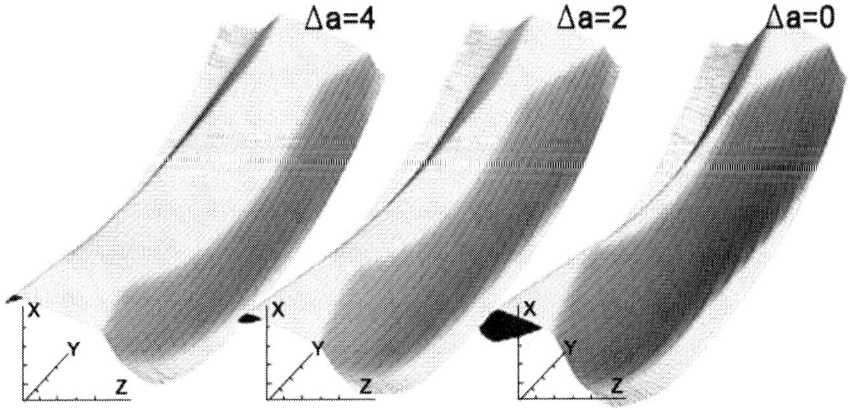

Fig. 4. The radial toothing cutting method – the parameter Δa defines the indentation; number of teeth, 31; module, 3 mm; inner diameter, 105 mm

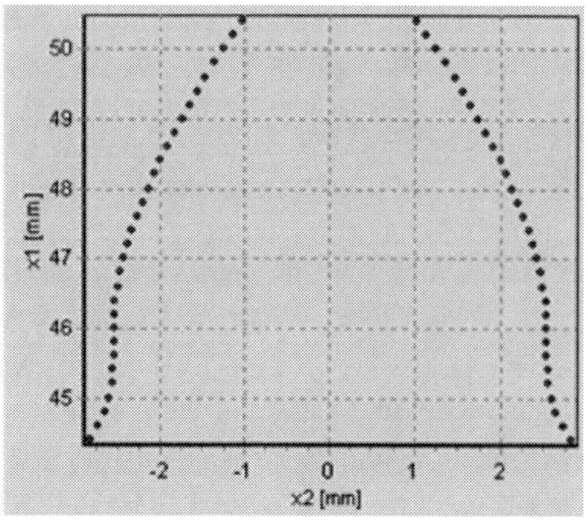

Fig. 5. The wormwheel tooth profile: number of teeth, 31; module, 3 mm; inner diameter, 105 mm

The wormwheel surface was determined for the case of machining by the tangential and radial method, respectively. As the tool, a special wormwheel machining hob and a modular hob were used. The relative tool and wormwheel positioning errors were allowed for in the toothing cutting process. In the second algorithm, there is no need for solving the envelope condition. The developed programs are analytical and numerical in character, and can be used in the process of the design, identification and analysis of the worm gear.

References

1. R. Chudzikiewicz, *The mechanization and automation of foundries*. (PWT, Warsaw, 1985)
2. I. Dudas, *The Theory and Practice of Worm Gear Drives*. (Penton Press, London, 2000)
3. R. Gołębski, *Geometrical analysis of the cylindrical worm gear*. (Doctoral dissertation, Czestochowa, 2006)
4. T. Nieszporek, *The design of cutting tools and the technology of outer cylindrical toothings*. (University Press of the Czestochowa University of Technology, Series Monografie **265**, Czestochowa, 2013)
5. F. L. Litvin, *Gear geometry and applied theory*. (Prentice Hall, Englewood, New Jersey, 1994)
6. F. L. Litvin, *Development of Gear Technology and Theory of Gearing*. (NASA RP-1406, 1998)
7. F L. Litvin, V. Kin, *Computerized Simulation of Meshing and Bearing Contact for Single Enveloping Worm Gear Drives*. J. of Mech. Des. **114** (1992)
8. I. Dudas, *Production of helicoids surfaces in inteligent systems*. (Annals of MTeM Cluj-Napoca, Romania, 2001)
9. T. Markowski, M. Sobolak, *Numerical method of analysis of geometry of tool and workpiece mating area*. (International Scientific Colloquium CASE TECHNIQUES, Bielfeld, 1995)
10. F. L. Litvin, C. L. Hasiao, *Computerized simulation of meshing and contact of eneveloping gear tooth surfaces*. Comp. Met. in App. Mech. and Eng. J. **102**, Holland (1993)

11. A. Piotrowski, *Enhancing the accuracy of modular hobs*. (Doctoral dissertation, Czestochowa, 2002)

Modeling and strength analysis of the prototype of the multi-tasking car trailer

Bogdan Posiadała[1,*], *Piotr* Ladra[1]

[1]Institute of Mechanics and Machine Design Fundamentals, Czestochowa University of Technology, ul. Dąbrowskiego 73, 42-201, Częstochowa, Poland

Abstract. The subject of the paper is modeling and strength analysis of the prototype of the multi-tasking car trailer. The innovativeness of the solutions applied consists in the combination of several single-purpose trailers into one, allowing for multitasking and the combination of several specialized trailers. Using the author's solution, the prototype of a multi-purpose trailer has been developed, which has the characteristics of several specialized trailers, such as for the transportation of general cargo, kayaks, motorcycles, quads etc. The model has been created by using SolidWorks package and it has been used for full strength analysis by simulation module. The sample analysis results cover various load configurations with respect to selected trailer construction components.

1 Introduction

The issues about the road transportation have been considered in this paper. This work is a result of progress of considerations of the studies described in [1-7]. By presenting in the paper an object of research, the proposal solution, answering the questions with difficulties occurring during the work of the car trailers, have been described.

An usual car trailers meet with problems of carrying the load intendent to a specialist trailers. In this context, the design of construction the special multi-tasking car trailer, which in its construction combines few one-tasking trailer, has been included in [2, 3]. This paper is a continuation of the proposed solution. The proposed construction solution is based on a traditional car trailer, which due to its modularity is able to transform into a specific application. What distinguishes it from other trailers is multitasking achieved thanks to innovative solutions. The composite trailer does not differ much from a standard trailer of this type, it is 1300 mm wide and 2200 mm long. But thanks to modifications, it can get dimensions of 2000 mm in width and 4000 mm in length (as platforms). GVW (gross vehicle weight) this construction is 1350 kg, what with the weight of the trailer not exceeding 300 kg, allows to load 1000 kg of cargo. The trailer frame is made of structural steel S355 with the following properties: 220 HB, Rm = 490-630 MPa, Re = 335 MPa.

* Corresponding author: bogdan.p@imipkm.pcz.pl

Reviewers: *Jozef* Bocko, *Grzegorz* Domek

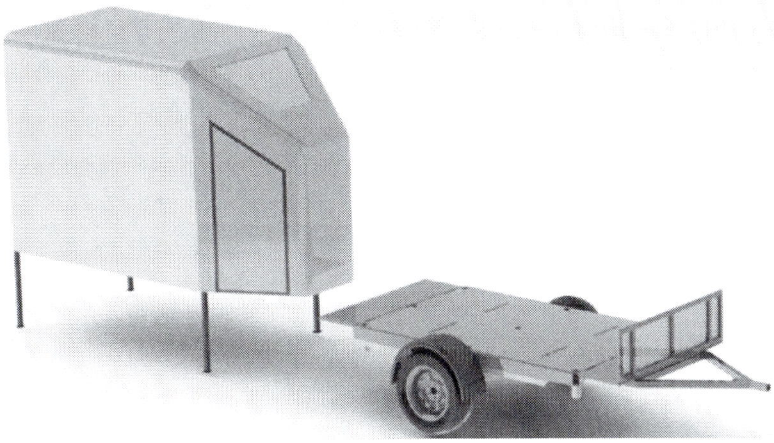

Fig. 1. Model of the car trailer – one of the applications

The trailer is adapted for various tasks, including specialist tasks, such as transporting large loads, motorcycles or kayaks. It has a hinged platform enabling entry of small vehicles, e.g. mowers or quads. One construction replaces several other trailers.

The strength analysis of the prototype of a multi-tasking car trailer has been performed in scope of this work.

This trailer in its construction contains some innovative solutions which haven't been used in designing such machines yet. The model of the prototype have been made in SolidWorks 2015 environment and it has been used as a basis to the strength studies in module: Simulation. The innovative constructional solutions have been reserved at the Patent Office of the Republic of Poland.

The proposed solution have been made as a result of the needs to design one universal trailer combining purposes of transporting in example: kayaks, gates, windows, or the self-propelled vehicles (lawn mower or quads), and motorcycles. The applied in the prototype solutions significantly expands the functionality of the system. Furthermore, by designing this construction, the adaptation of the different type od the superstructures is possible. There is a need to transform the car trailer to be used of the other purposes, among others to modify a car trailer into a panel van (wagon) or carriage of horses.

2 Strength analysis of the selected construction nodes of the prototype

The designed prototype has been made in SolidWorks 2015. The static analysis has been made on the basis of the frame model, and it has been performed in the Simulation Module. The frame model has been designed with all its components and all the system has been subjected to the analysis. Since the prototype has been composed of the two main subsections: external (see Fig. 2) and internal (see Fig. 3) frames, the two cases of the analysis has been performed. Cases of the analysis:

- Test no. 1- the external frame,
- Test no. 2- the internal frame.

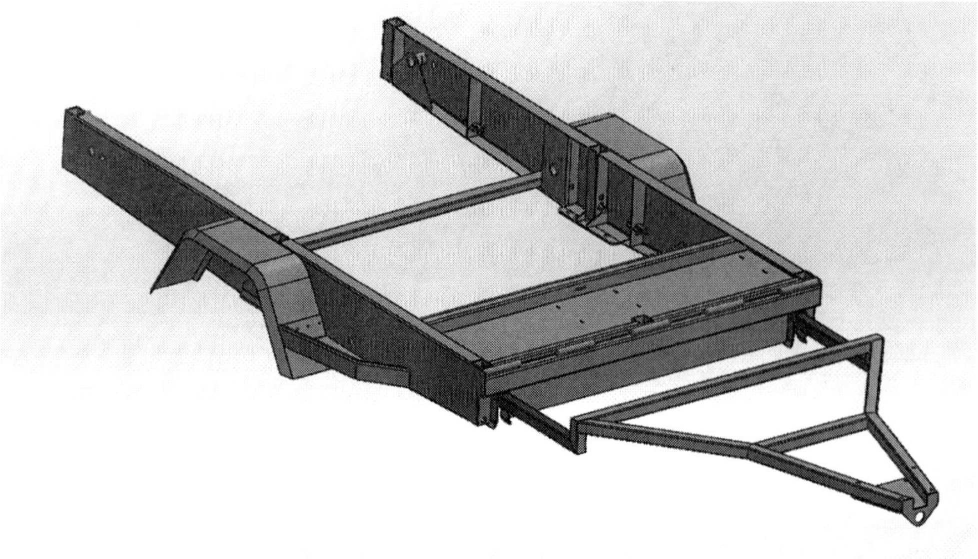

Fig. 2. Model of the external frame of the car trailer

By entering the description of boundary conditions of the model, the trailer's working conditions were reflected. These conditions are shown in Figure 4. The boundary conditions include the area for attaching the trailer axis and towing hook. The calculation model reflects the full nature of work while the trailer is in operation.

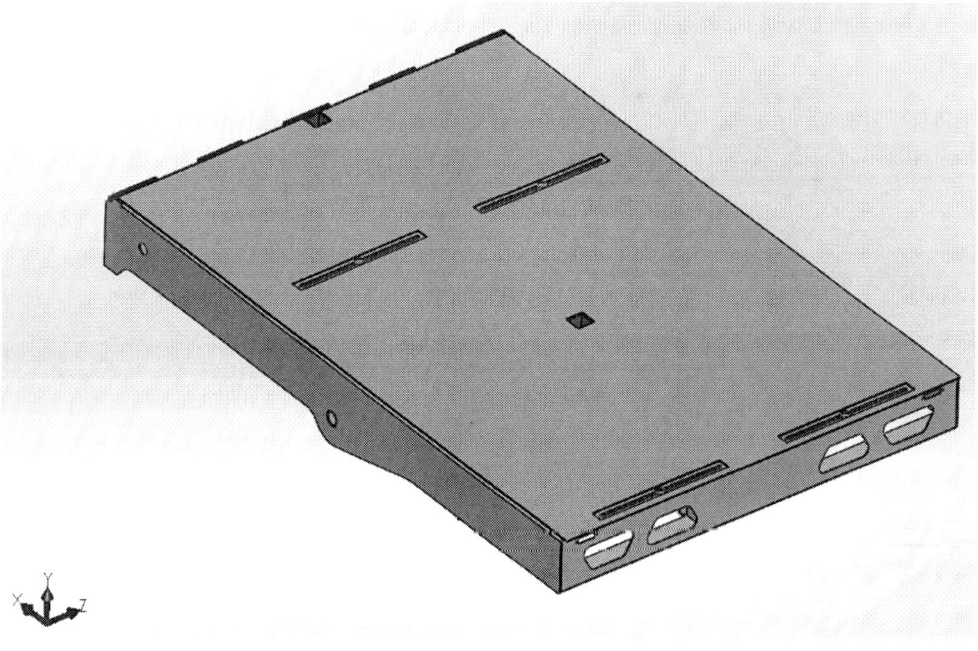

Fig. 3. Model of the internal frame of the car trailer

The static analysis, which included the distributions of displacement, reduced stresses and strains, have been considered. The two tasks have been described below.

2.1 Task no. 1 – the external frame

The external frame has been subjected to load on the external stringers with force equals 1500 N. Such a loading represents the situation, when the superstructure will be mounted to the trailer's frame. The boundary conditions as well as the load have been presented in the Fig. 4.

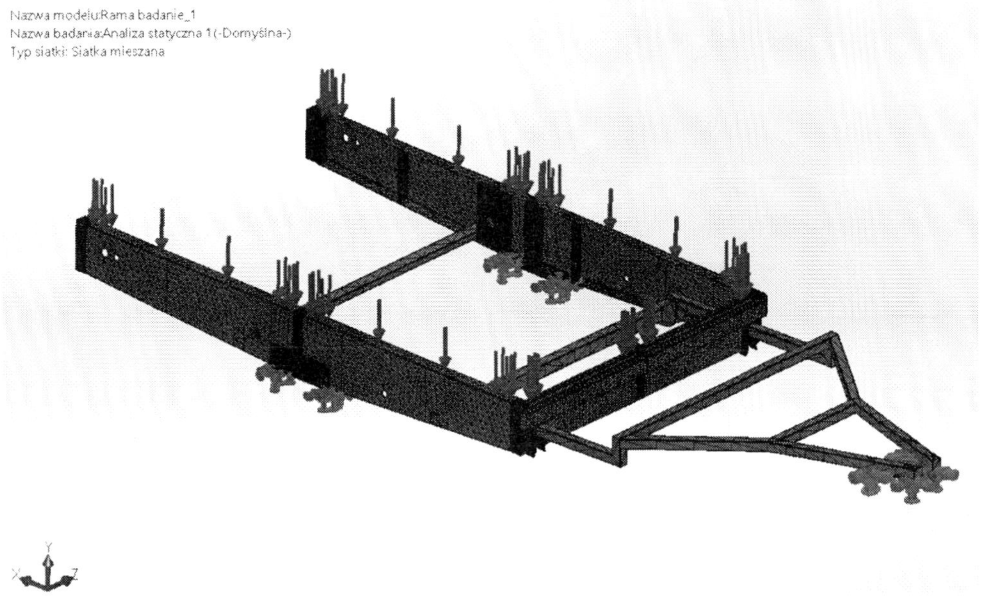

Fig. 4. Boundary conditions and the load of the external frame

The static analysis performed for study 1 provides a full picture of the changes that have occurred in the construction of this prototype. The results show the distribution of displacements, stresses and strains that occurred under the influence of a given load.

The study provides analysis results in any areas of the framework, including the maximum values that have been achieved and their location. During the simulation very good results of displacements, stresses according to the Huber-von Mises hypothesis and deformations were achieved. Figures 5 to 7 show graphic results from study 1.

The maximum displacement occurred on the stringers but these are very small values (1,049mm) and thus confirm the reliability of the prototype.

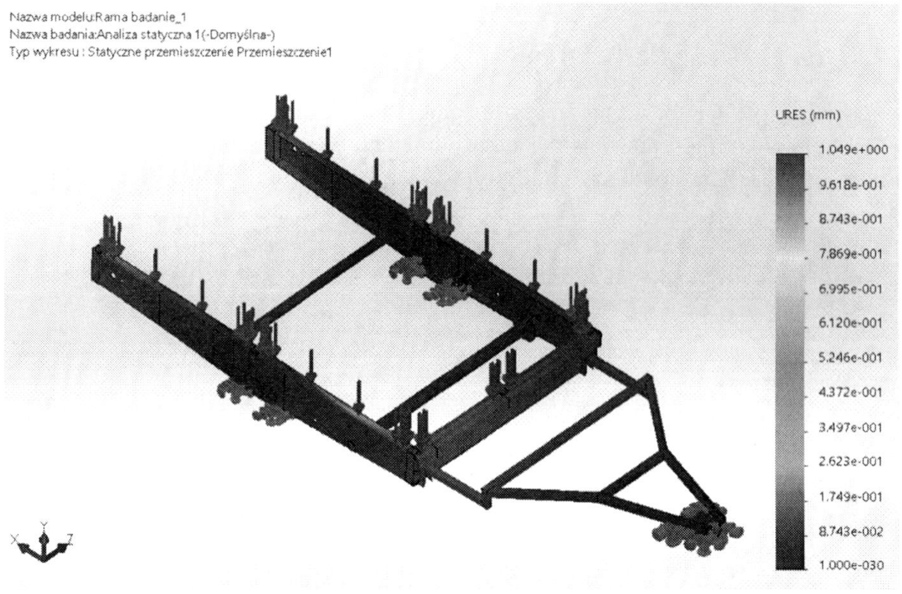

Fig. 5. Distribution of resulting displacement – first task

The highest values of the reduced stresses have been noticed on the extreme parts of the frame. This has been illustrated in Fig. 6. The results of the strain have been included in Fig. 7.

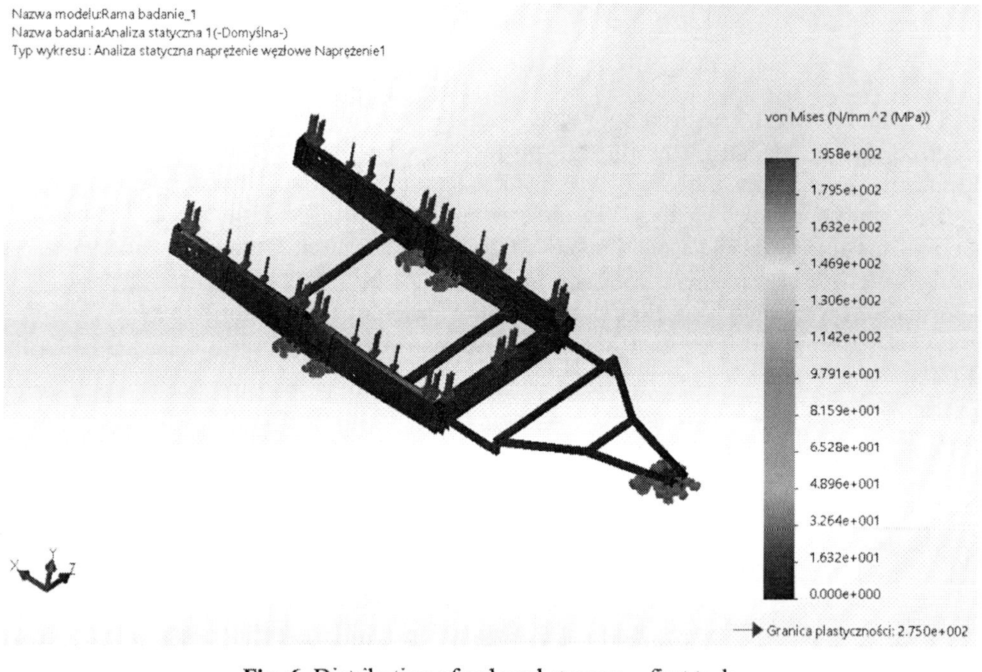

Fig. 6. Distribution of reduced stresses – first task

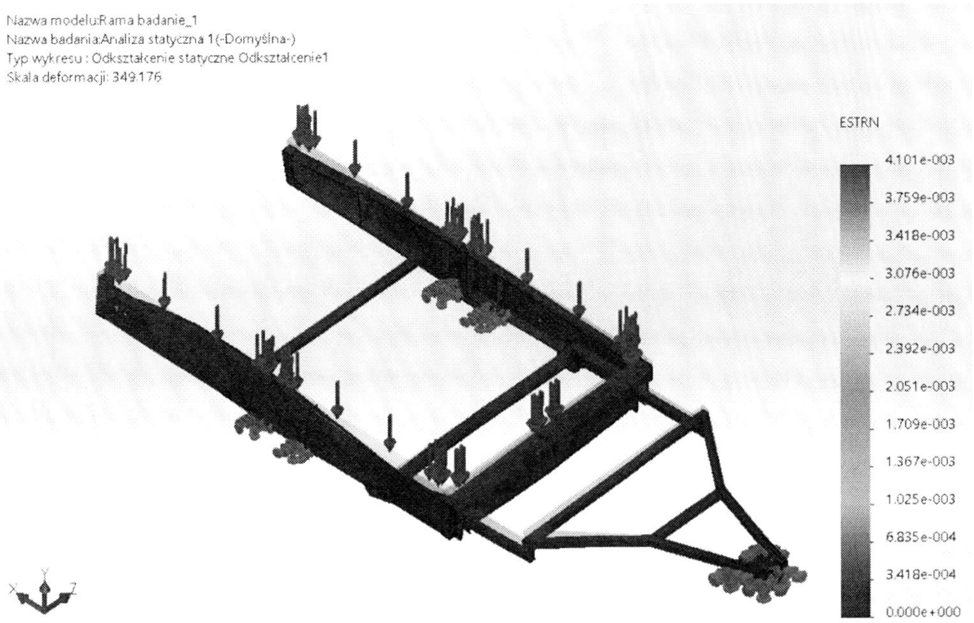

Fig. 7. Distribution of strains – first task

2.2 Task no. 2 – internal frame

The internal frame has been subjected to the load distributed in all surface of the floor. The value of the increased admissible load is equal 15000 N. The illustration of this is included in Fig. 8.

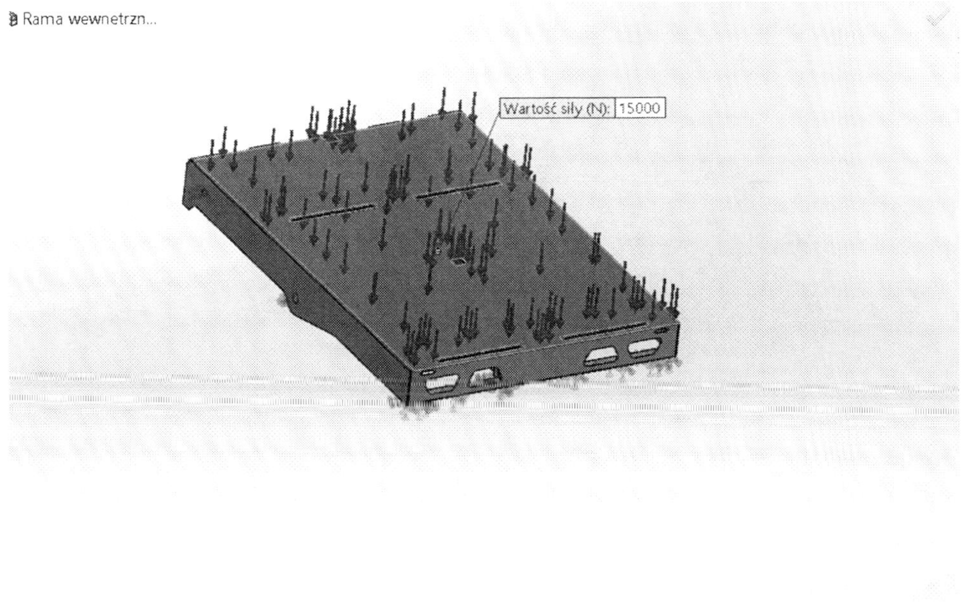

Fig. 8. Boundary conditions and the load of the internal frame

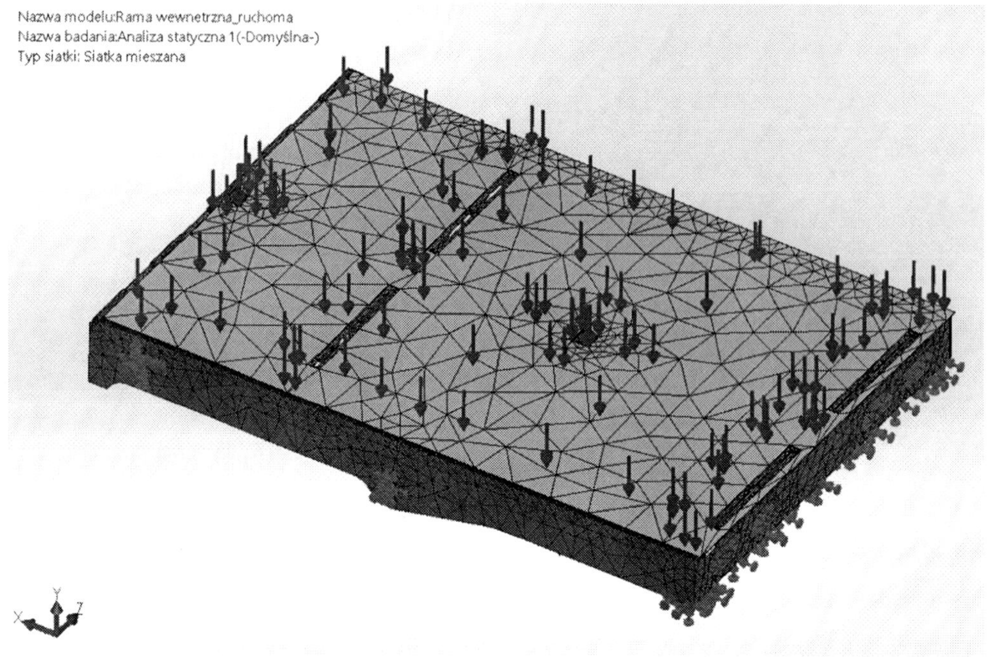

Fig. 9. Model of the analysis with mesh of the internal frame

Also for study 2 very good results were obtained, which are shown graphically in Figures 10-12. The obtained results confirm a solid construction (theoretical assumptions) which for displacements amounted to only 0.392 mm at small stresses 47.53 [MPa].

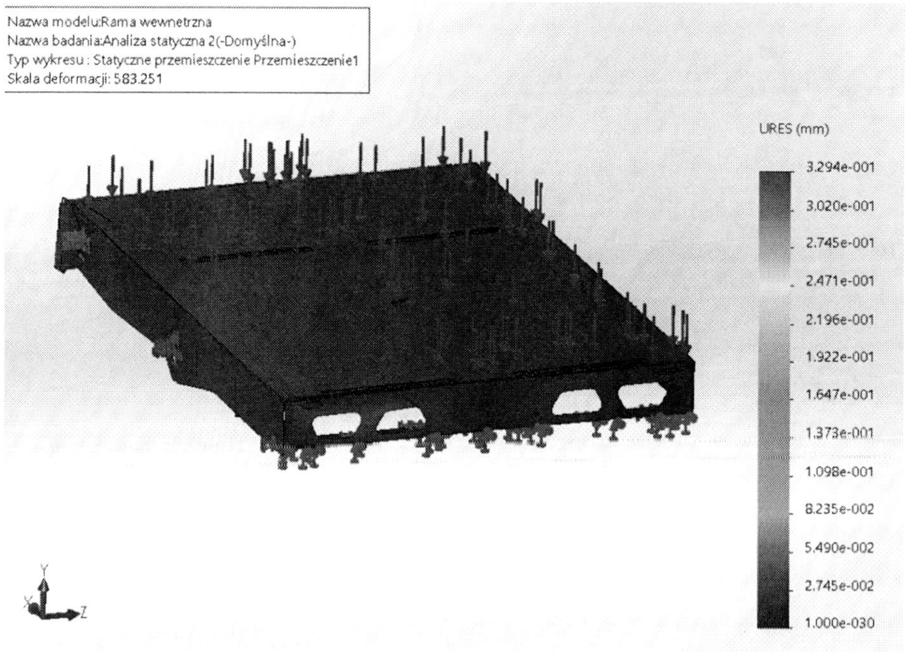

Fig. 10. Distribution of resulting displacement – second task

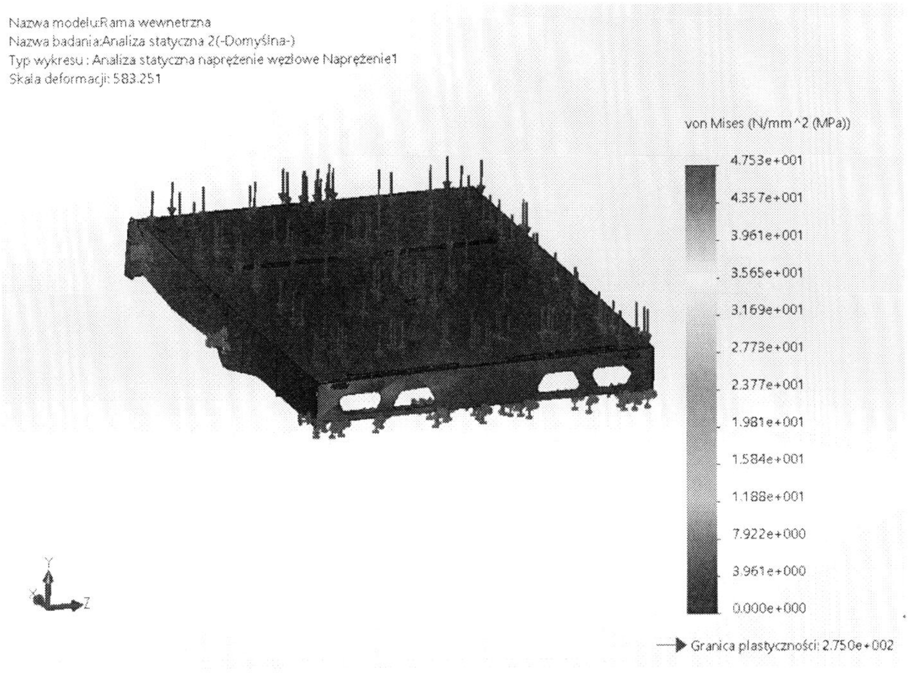

Fig. 11. Distribution of reduced stresses – second task

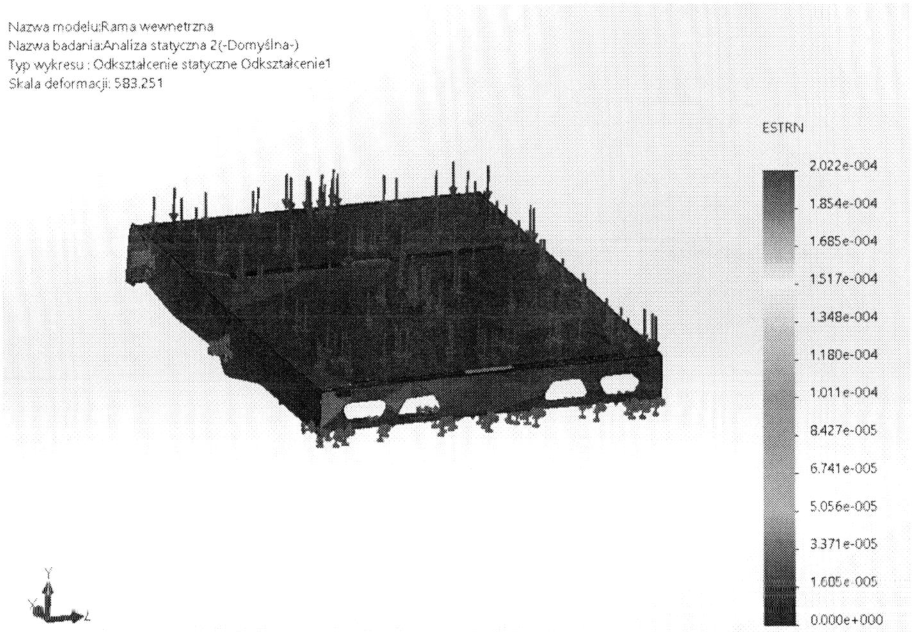

Fig. 12. Distribution of strains – second task

Conclusions

The studies of the prototype of the multi-tasking car trailer, which maintained the standards of the usual trailer and combine within construction the functionality of few specialized one-tasking trailers into a one multi-tasking trailer, have been performed. The proposed prototype of the system is useful in many specialized tasks, which have been mentioned in the introduction. The most satisfying of the innovative solutions are:

- the way of spreading the back side, which, after unfolding, prolongs the loading space of the trailer,
- multifunctional folding frame, which, after proper setting and fixing, creates the various configurations necessary to transport of the specified loading,
- a special backlight of the sliding back panel, in which the innovative solution is a way of its sliding without a need of unscrewing bolts or removing pins, which is dedicated to carrying long loads such as kayaks.

In this work the analysis of the strength analysis, in reference to specified construction nodes, have been performed and included. The analysis has been made on the SolidWorks 2015 environment, by using the finite element method. The external and internal frames have been examined in the analysis in order to ensure the constructional solution in the using the trailer i.e. during entry of quad type cargo. The construction of the trailer has been designed according to the constructional assumptions and satisfies the strength requirements.

The maximum analyzed values and areas where they occurred can be seen in the drawings placed under a particular study. All values obtained during the tests are consistent with the assumptions for the project and guarantee a solid construction and long-term use.

References

1. K. Bentkowska-Senator, Z. Kordel, J. Waśkiewicz, *Transport samochodowy ładunków*. (Wyd. ITS, 2009)
2. P. Ladra, B. Posiadała, *Modeling and Strength Analysis of the Specialized Car Trailer*. Rusiński E., Pietrusiak D.: Proceedings of the 13th International Scientific Conference, RESRB 2016, Lecture Notes in Mechanical Engineering, Springer, Cham, (2016)
3. P. Ladra, *Projekt komputerowy przyczepki samochodowej specjalizowanej*. (Praca magisterska Politechniki Częstochowskiej, Częstochowa, 2015)
4. J. Domański, *SolidWorks 2014 Projektowanie maszyn i konstrukcji. Praktyczne przykłady*. (Wydawnictwo Helion, 2014)
5. P. Kęska, *SolidWorks 2013. Konstrukcje spawane. Arkusze blach. Projektowanie w kontekście złożenia*. (Wydawnictwo CADvantage, 2013)
6. B. Posiadała, *Modelowanie i analiza zjawisk dynamicznych maszyn roboczych i ich elementów jako dyskretno-ciągłych układów mechanicznych*. (Wydawnictwo Politechniki Częstochowskiej, Częstochowa, 1999)
7. *Problematyka transportu drogowego*. Praca zbiorowa pod red. G. Nowackiego (Wyd. ITS, 2008)

Localization of delamination in composite test specimen

Milan Sapieta[1,*], *Peter* Šulka[1], *Martin* Svoboda[2]

[1]Faculty of Mechanical Engineering, Univerzity of Žilina. Unierzitná 1, 010 26 Žilina. Slovak Republic
[2]Faculty of production technology and management of Jan Evangelista Purkyně University in Ústí nad Labem, Na Okraji 1001, Czech Republic

Abstract. The main purpose of paper is presented an innovative approach in the field of infrared thermal imaging and its integration into the area of engineering applications. As the test specimen was selected composite rod with lengthwise composite fibers to each bonded. They have been excitated with ultrasound. For obtain the results was used Lock-in method. Last part of the paper is a description of the results and most appropriate settings of measurement technique for applications with chosen material.

Keywords: infrared radiation, non-destructive testing, lock-in, ultrasound

1 Introduction

Infrared thermography has developed from a rarely used method to the method growing in popularity among users in recent years. This progress contributed largely by the availability and constantly falling prices of infrared cameras. For this reason, some methods have been developed which make use of infrared thermography in industry [1].

Pulse thermography was invented for the purpose of non-destructive testing (NDT) of materials and components, hence the "looking under the surface 'bodies. In pulsed thermography is pulsed heat flux generated on the surface typically by means of irradiation with light (e.g. lamp) and then is surface temperature observed in time depending. If inhomogeneity for example: cracks, holes or cavities with different temperature parameters, lies below the surface of the test object, it will have an impact on local dynamic heat flux that passes through the specimen.

As an alternative technique is offered to lock-in thermography, which is also known as heat waves imaging technique, it can be described by the theory of oscillating heat waves. Excitation of the heat occurs periodically with a certain frequency in the body [2]. The advantage of lock-in thermography is that with applying processing of the registered image it may considerably increase its sensitivity compared with the nominal sensitivity of the thermal camera.

* Corresponding author: milan.sapieta@fstroj.uniza.sk

Reviewers: *Robert* Grega, *Wiesława* Piekarska

2 LOCK-IN thermography

A prerequisite for the use of this technology is that the excitation signal may be periodically pulsate or in any other way modulated by certain amplitude frequencies, called "lock-in frequency" $f_{lock-in}$. Lock-in thermography is a type of active lock-in method. It means that heat to excite the sample is generated periodically and correlation using the lock-in process is applied to all heat signatures of every single pixel on the image of the observed object.

Digital lock-in process correlation consists of averaging the results of the measured values F_k and system of weighting factors K_k to the total number of measured values M:

$$S = \frac{1}{M}\sum_{k=1}^{M} F_k\, K_k, \tag{1}$$

where S is the output signal.

If the excitation is harmonic then the most advantageous correlation function is also harmonic (sine, cosine) function. This kind of lock-in correlation is called sin/cos or narrowband correlation. It can be realized either by narrowing the bandwidth of the detected signal or using the values of harmonic functions for K_k in the equation (1).

The main advantage of the sin/cos correlation is that it enables to user to take into consideration the phase of the signal after the measurement (off-line) when it is used two-channel correlation. Two-channel correlation means that there were used two types of weighting factors, one approximates a sin-function and the second approximates cos-function. Correlation is transferred twice in parallel with both types of weighting factors:

$$K^0(t) = 2\,sin(2\pi f_{lock-in} t), \tag{2}$$

$$K^{\pi/2}(t) = 2\cos(2\pi f_{lock-in} t). \tag{3}$$

Then the first channel measures the component in-phase with the sin-function, and the other channel measures the component in-phase with the cos-function, which is $\pi/2$ phase-shifted to the sin-function [3].

If equations (2, 3) are inserted into equation (1) than the result of the two correlations over a complete number periods is:

$$S^0 = \frac{1}{n}\sum_{i=1}^{n} F_i\, K_i^0 \tag{4}$$

$$S^{\frac{\pi}{2}} = \frac{1}{n}\sum_{i=1}^{n} F_i\, K_i^{\frac{\pi}{2}} \tag{5}$$

Where S^0 is called the in-phase signal and $S^{\frac{\pi}{2}}$ is usually called the quasrature signal. Both signals may be either positive or negative [4].

3 Ultrasound excitation

Typical set of the lock-in analysis consists of a PC, an infrared camera, an amplifier and generator lock-in frequency. It is precisely the type of generator lock-in frequency will vary different sets, so one can say that it will divided according to the type of excitation.

We decided to use ultrasonic excitation type in other measurements, which is compared with the optical excitation more suitable for steel materials, which have the largest representation in engineering practice yet. Ultrasound lock-in thermography uses the interaction between mechanical and thermal waves to detect material defects. If some damage in components absorbs the excitation of high energy ultrasound waves than it is locally heated. The resulting temperature gradient is captured by an infrared camera on the sample surface and it is subsequently dissipative energy visualized.

This method is suitable for applications in the detection of cracks (as well as open and closed, regardless of its orientation), adhesion testing, rivets and welded joints. As with optical methods as well as ultrasound method, it is possible to detect delamination and impact damage in materials made of composite fibres.

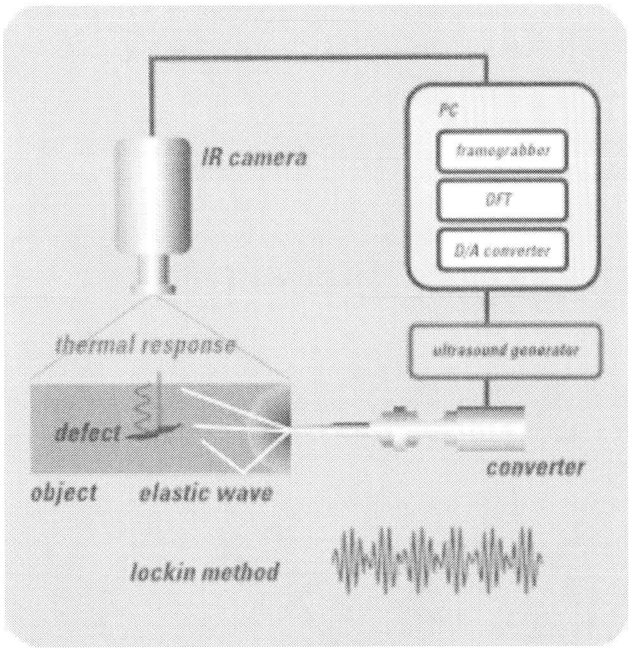

Fig. 1. Schematic representation of the lock-in principle with ultrasonic excitation Utvis.

4 Measurement

As the specimen was selected composite rod with longitudinally at each bonded composite fibres. The specimen was used to measure damping in the past where it was fixed in the jaws of a vice. This makes assumption of damage formation. The specimen was placed under ultrasound exciter and is subsoiled by a holder suitable shape Fig. 2.

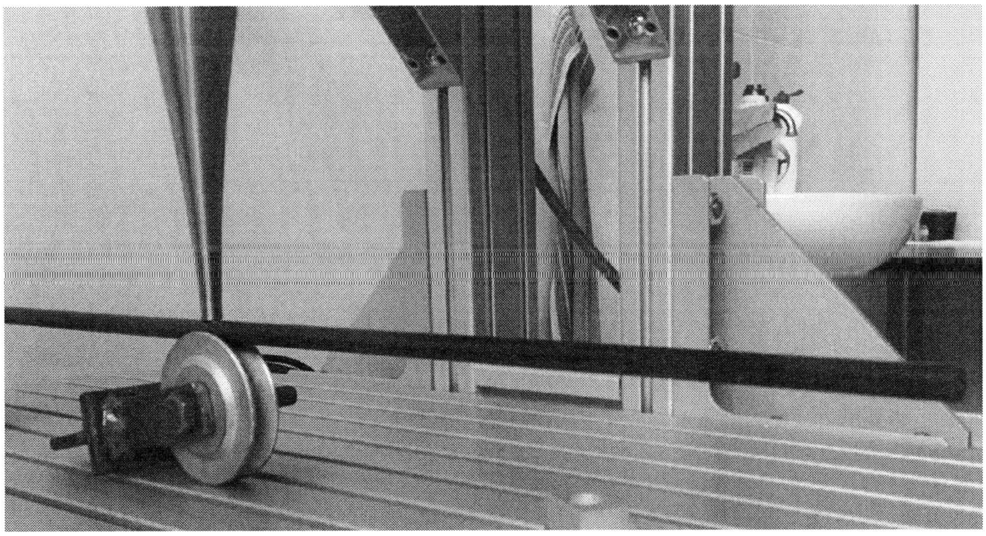

Fig. 2. Specimen placed under ultrasonic sonotrode.

Then it was excited by selected modulated frequency until the damage was displayed. The whole process of excitation was captured by an infrared camera. Setting of IR camera was identical for each single modulation of frequency. Frame rate of IR camera was selected 50 Hz. Value of emissivity does not affect the measurements of this type therefore it was set value 1 to the software interface. In this case the surface of the specimen is dark, so surface treatment was not necessary, as on the specimens with shiny metal surfaces, respectively coating.

The software, which was used to evaluate individual images, is called DisplayImg and it came with the excitation device. This software uses the theoretical basis of the lock-in method, as were described in the first chapter of this paper.

As a first type of excitation has been selected sweep excitation. This is excitation which passes through whole range of modulated frequencies. It is possible to observe the whole course of excitation and then choose the most appropriate modulated frequency at which is displayed defects in the measured object.

After finding the ideal modulated excitation frequency which was 0.1 Hz, layout of radiation energy has been succeeded drawn in the software like in phase and amplitude display Fig. 4 Fig. 5. On both it is possible observed the damage in about one third of the point of excitation. This damage is more pronounced on display in-phase where it is in the form of longitudinal light lines, marked with a red ellipse Fig. 4.

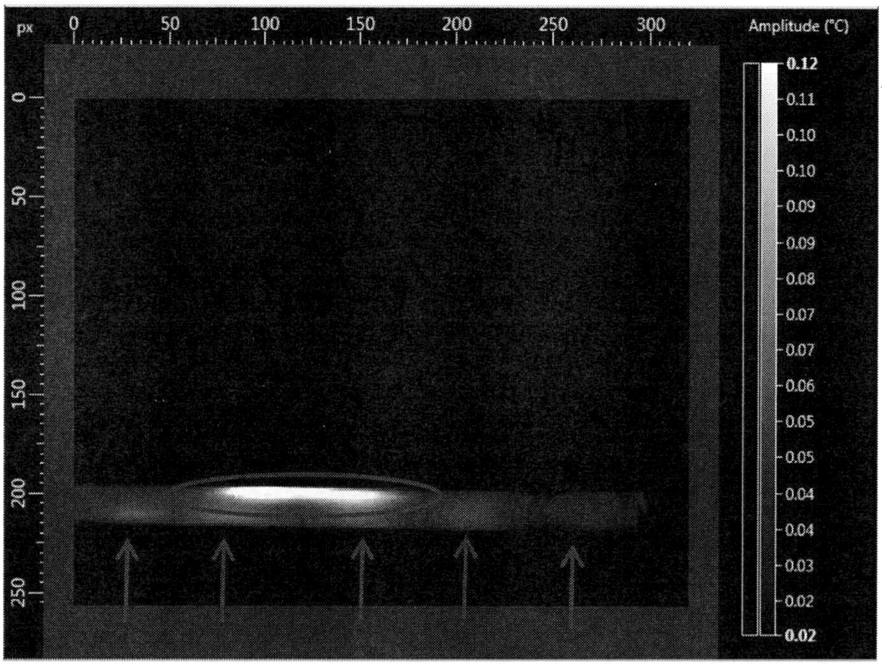

Fig. 3. Detection of damage on specimen, modulated frequency 0.1 Hz, amplitude view, side view and damage marked with red ellipse

It can be assumed that the most likely case of delamination, separate each fiber from each other due to either the previous loading in experiments or some manufacturing defect. In Amplitude display can be seen light areas perpendicular to the fibers, marked by red arrows Fig. 3. It can be assumed that these light areas are caused by individual excitation ultrasonic waves. Therefore, it is better to evaluate the damage on the image in the displayed in-phase.

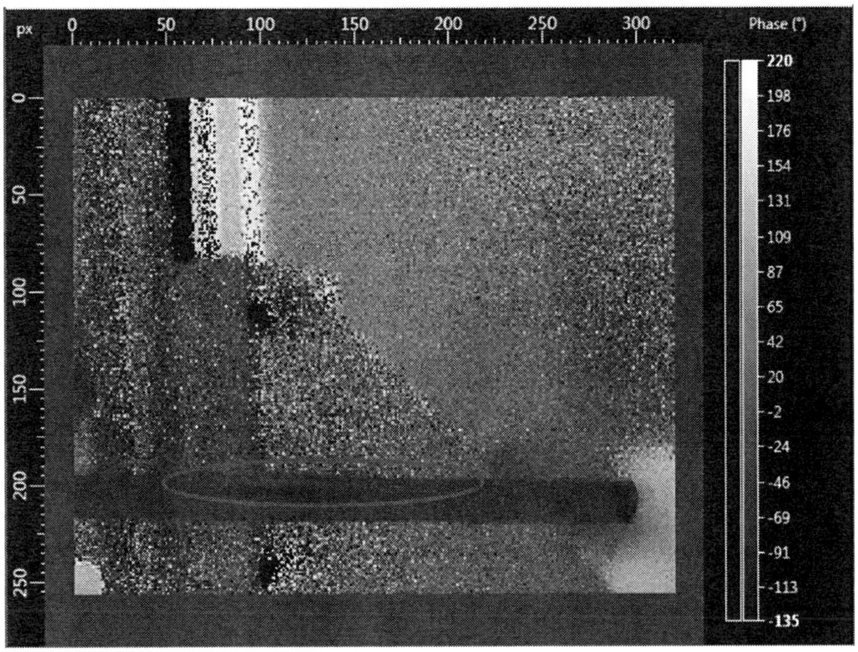

Fig. 4. Detection of damage on specimen, modulated frequency 0.1 Hz, in-phase view, side view and damage marked with red ellipse

Conclusion

The measurement was carried out for the purpose of non-destructive testing of material. As the specimen was selected composite rod with longitudinal fibres bonded to each other with damage, which was made by either the manufactured or previous used. Using a measuring system UTvis, a device with an ultrasonic excitation used by the lock-in method for NDT has been made to test of this specimen. First it was performed ultrasonic excitation across the whole range of modulated frequency, by this excitation was chosen modulated frequency at which was able detect cracks on the side of the specimen.

For significant knowledge can be considered the fact that, as of oscillation (when changing the excitation field frequency, then if the exciting frequency coincides with its own than is a registered resonances) and in this type of excitation (ultrasound) is a response in the IR area dependents on immediate parameters of the excitation signal [5-7].

Crack was formulated by the light areas by changing the radiation layout on interpreted images. This method has proven to be a suitable solution for NDT testing to detect delamination and damage to composite fibres.

The part of the results of this work has been supported by VEGA grant No. 1/0795/16 a KEGA No. 017ŽU-4/2017.

References

1. M. Sapieta, A. Sapietová, V. Dekýš, *A detection of deformation mechanisms using infrared thermography and acoustic emission*. Applied Mechanics and Materials **474**, 315-320 (2014)
2. V. Vavilov, *Thermal/Infrared testing*. Nondestructive testing, Publishing house Spektr., 732 p. (2009)

3. X. Xian, W Zheng, W. HU, *Evaluation of the Damage of Carbon/Epoxy Composite Laminates under Static and Fatigue Loading by Thermographic Analyses*.Proceedings of the International Symposium on Composite Materials and Structures, 710-716 (1986)
4. M. Handrik, P. Kopas, V. Baniari, M. Vaško, M. Sága, *Analysis of stress and strain of fatigue specimens localised in the cross-sectional area of the gauge section testing on bi-axial fatigue machine loaded in the high-cycle fatigue region*. Procedia Engineering **177**, 516-519 (2017)
5. P. Pecháč, M. Sága, *Controlling of local search methods' parameters in memetic algorithms using the principles of simulated annealing*. Proc. Eng. **136**, 70-76 (2016)
6. L. Jakubovičová, M. Sága, *Computational analysis of contact stress distribution in the case of mutual stewing of roller bearing rings*. Novel Trends in Production Devices and Systems, Applied Mechanics and Materials **474**, 363-368 (2014)
7. J. Zapoměl, V. Dekýš, P. Ferfecki, A. Sapietová, M. Sága, M. Žmindák, *Identification of material damping of a carbon composite bar and study of its effect on attenuation of its transient lateral vibrations*. Int. Journal of Applied Mechanics **7** (6), (2015)

Monitoring the fatigue crack on the test specimen during the cyclic loading

Milan Sapieta[1,*], *Peter* Šulka[1], *Martin* Svoboda[2]

[1]Faculty of Mechanical Engineering, Univerzity of Žilina. Unierzitná 1, 010 26 Žilina. Slovak Republic
[2]Faculty of production technology and management of Jan Evangelista Purkyně University in Ústí nad Labem, Na Okraji 1001, Czech Republic

Abstract. The main purpose of this paper is monitoring propagation of plastic area of fatigue crack. There were made notch on specimens and they were cyclically loaded with three-point bending. The record was evaluated after a certain number of loading cycles. The loading force is still constant throughout the measurement time for all specimens. The whole process was recorded using an infrared camera.

Keywords: crack propagation, cyclic load, infrared, radiation.

1 Introduction

Nowadays, a combination of numerous infrared detectors or thermal sensors with advanced signal processing generates a practical tool for investigating thermal-stress characteristics of materials and structures from the respective surface or part of the space.

Thermoelastic stress analysis has been used by engineers and scientists for more than 50 years to address practical problems. It utilises the principle of sensing the energy released during loading in the elastic domain. Once loading is removed, the body returns to its original position (elasticity) and its original temperature (thermoelasticity) [1].

2 Thermoelasticity and temperature variation

Thermoelastic stress analysis (TSA) is an experimental contactless method based on measuring the infrared radiation emitted from the component surface exposed to dynamically linear elastic strain (deformation). Kelvin was the first scientist to study the thermoelastic effect, and the basic equations to describe the thermoelastic [2] effect were formulated by Darken and Curry.

The general form of the heat conduction equation for elastic body is derived from the energy conservation equation, and it can be written as follows:

$$\rho c_\varepsilon \frac{dT}{dt} - \frac{\partial}{\partial x_j}\left(k\frac{\partial T}{\partial x_j}\right) = \rho r + \sigma_{ij}\frac{\partial \varepsilon_{ij}}{\partial t} - \rho\frac{\partial \Psi}{\partial V_k}\frac{dV_k}{dt} + \rho T\frac{\partial^2 \Psi}{\partial T\,\partial V_k}\frac{dV_k}{dt}. \quad (1)$$

* Corresponding author: milan.sapieta@fstroj.uniza.sk

Reviewers: *Juraj* Gerlici, *Milan* Naď

Equation (1) uses the Einstein summation convention, ρ is the density, c_ε is the specific heat capacity at constant deformation, T is the absolute temperature, k is the thermal conductivity tensor, σ_{ij} and ε_{ij} are the tensors of stress and strain, ε_{ij} is the inner heat source per unit of volume and Ψ is Helmholtz free energy that is dependent on k and independent of the internal state variables V_k.

Provided that the material elastic properties and constant material coefficients are temperature independent, then it can be derived a 3-dimensional heat conduction equation written as follows:

$$\rho C_\varepsilon \dot{T} - k\Delta^2 T = T_0\left(-\frac{E\alpha}{1-2\upsilon}\right)\dot{\varepsilon}_I^e + \alpha_p \sigma_{ij}\dot{\varepsilon}_{ij}^p \,. \tag{2}$$

The equation for heat conduction will include the creation of the thermoelastic and the thermoplastic heat. Parameter ε_I^e denotes the first invariant of elastic deformation tensor ε_{ij}^p is the plastic part of the deformation tensor, α is the coefficient of thermal expansion, E Young's modulus, υ Poisson's ratio and T_0 is the initial temperature [3]. Non-dimensional coefficient $\propto_p$ is the ratio of the total plastic work to plastic work, which is converted to heat. The value $\alpha_p \cong 1$ because of only a small part of the plastic work (obtained energy at cold forming) is used to change the inner properties of the material [4]. Therefor the thermoplastic area will be neglected and relationship will be formulated only for thermo-elastic area, which has the following form:

$$\Delta T = \frac{\alpha T_0}{\rho c_p}\Delta\sigma_{ii} \,, \tag{3}$$

where c_p is the specific heat capacity under constant pressure and its relationship with c_ε is as follows:

$$c_\varepsilon = c_p - \frac{2E\alpha^2 T}{\rho(1-v)} \,. \tag{4}$$

3 Measurement setup

Within this work it was carried out measurement of three-point bending of stainless steel beams. The specimen was 50 mm long and had a 10x10 mm square cross section. For this measurement it was chosen three types of material, namely: AISI 304, AISI 316L, and AISI 316TI.

The specimen was put on two supports and the loading forces acted from above onto the specimen centre between the supports Fig. 1. The supports were placed apart at a distance of 29 mm.

Fig. 1. Specimen position during the three-point bending loading

The notch width of 0.8 mm and a depth of 0.5 mm were created on the specimen. This notch was placed at the center of the bottom side of the specimen.
Again, it was chosen the infra-red camera maximum resolution 320x256 pixels, and corresponding frame rate 383 Hz. Loading frequencies differed for each material and each loading state – they fluctuated in the range of 100-105 Hz. Once again there was a slight subsampling. Since it was found from the previous measurements that this type of subsampling does not affect the quality of results, it chose that as the most suitable one in view of the resolution – sampling frequency ratio [5-7].

The intermediate ring was inserted between the camera and its lens. The intermediate ring extends the focal length of the lens to increase the view of the subject. Basically, the resolution of the camera remains 320x256 pixels, it only changes the overall frame size (decrease), the FOV and IFOV parameters. Taking advantage of the intermediate ring also brings its hardship and is not always possible to use it. As a result of its use, the depth of field also changes, making it necessary to place the camera closer to the subject, which is not always possible in terms of safety. It is very easy to damage the lens, it is not economically advantageous due to lens cost.

The load was chosen constant for all 3 types of materials, the static load component was 6.5 kN and the dynamic 6 kN. Specimens were cyclically loaded with frequency 105 Hz and were attempting to record after a certain number of cycles. Recording was done after 5000 cycles.

4 Evaluation of measured results

As a result of the AISI 304 and AISI 316L specimens not having suffered enough cycles (AISI 304 13 780 cycles and AISI 316L 6923 cycles), it was not possible to carry out a sufficient number of records to track fatigue crack propagation. It will further be evaluated formation and propagation of fatigue cracks only sample material AISI 316Ti. With this specimen, the fatigue fracture was up to 24,149 cycles, so up to 6 records were made.

The records were evaluated in unmodified DL (digital level) units. For maximum sensitivity, the lock-in method and the thermoelastic analysis were used. In this case, it is possible to determine the stress after calculating the temperature data, but only at the present time relationships are formulated for the elastic area (near the crack root is the expected plastic area). Therefore, the following observations were evaluated only quantitatively.

The first record was performed at the beginning of the load after a certain number of cycles, essentially as soon as the machine was able to stabilize the load frequency. The first record shows the distribution of radiated radiation in DL units Fig. 2.

Figure 2 shows the distribution of radiation energy where negative values can be observed at the top of the sample, which is the stress of the longitudinal fibers to the pressure and the positive values at the lower part representing the tensile of the longitudinal threads.

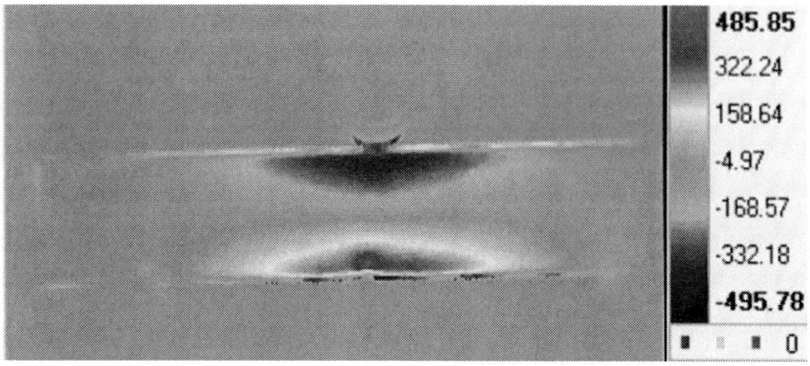

Fig. 2. Distribution of radiation energy recorded by the IR camera at the beginning of the fatigue test

The second record was performed approximately after 5000 cycles. As a result of the increase of the maximum value of the radiation energy, there was a scaling (this explains the background colour change, where the value should be zero). There is an increase in the upper and lower areas, but what is more important in the lower part above the notch are visible, the first signs of crack formation (purple) Fig. 3.

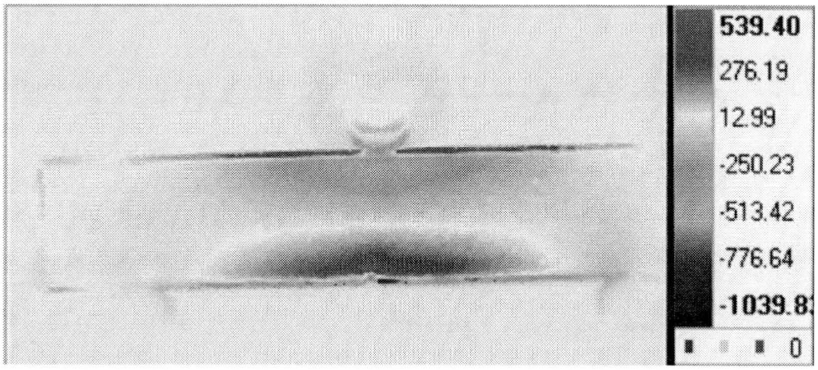

Fig. 3. Distribution of radiation energy recorded by the IR camera after 5000 cycles, the first signs of fatigue cracking

After 10,000 cycles, a third record was taken. There is a clearly visible crack in the image just above the notch Fig. 4. The crack is purple and this colour will be characteristic of this crack throughout the measurement time. Again, assuming an increase in the radiation values, depending on the number of cycles.

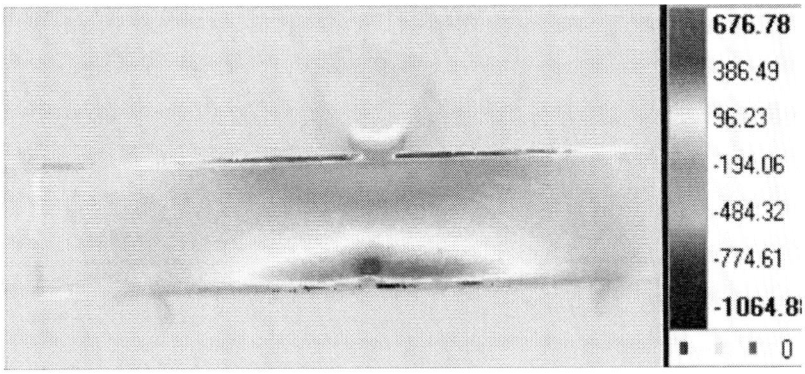

Fig. 4. Distribution of radiation energy recorded by the IR camera after 10 000 cycles

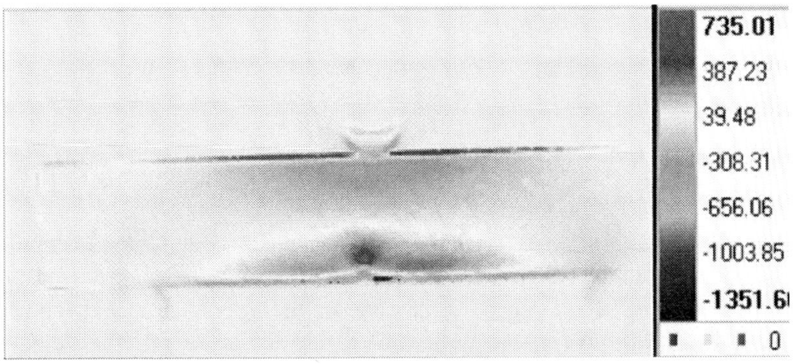

Fig. 5. Distribution of radiation energy recorded by the IR camera after 15 000 cycles

The crack retains its size, it can be observed its gradual shift towards the centre of the specimen Fig. 5, 6.

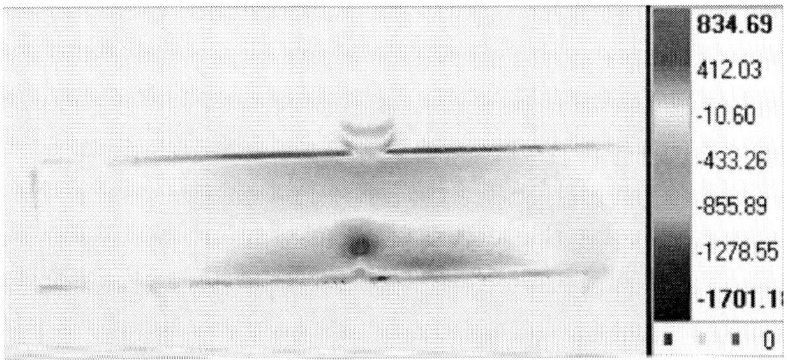

Fig. 6. Distribution of radiation energy recorded by the IR camera after 20 000 cycles

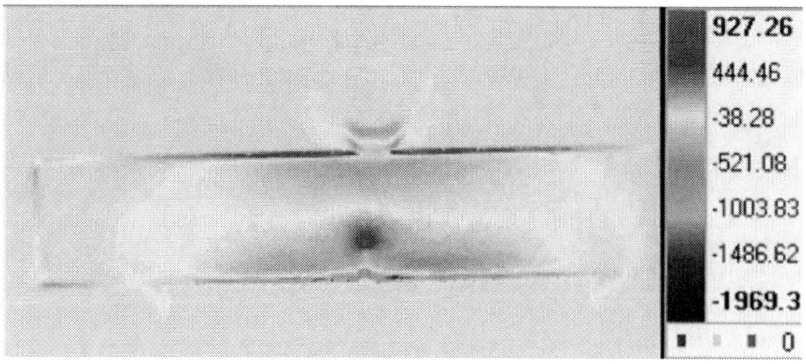

Fig. 7. Distribution of radiation energy recorded after 21 500 cycles, recording made just before the expected fatigue fracture

Fatigue fracture of a sample of this material occurred after 24,149 cycles Fig. 7. These images demonstrated the possibility of analysing the area around the crack root, based on the different radiated energy in the IR area. These differences are attributed to different energy accumulation in areas around the crack due to cyclic loading.

Conclusion

Interpretation of thermograms in this article is based on the lock-in signal processing theory. Areas are characterized by corresponding radiation, which is not evenly distributed in them. This, however, allows mapping the size of the plastic zone, respectively to study this area in more detail, since it can be assumed that the values of the detected radiation correspond to the degree of plastic area creation.

The part of the results of this work has been supported by VEGA grant No. 1/0795/16 a KEGA No. 017ŽU-4/2017.

References

1. M. Sapieta, A. Sapietová, V. Dekýš, *Comparison of the thermoelastic phenomenon expressions in stainless steels during cyclic loading*. Metalurgia **56**, 203-206 (2017)
2. V. Vavilov, *Thermal / Infrared testing*. Nondestructive testing, Publishing house Spektr., 732 p. (2009)
3. X. P. V. Maldague, *Theory and Practice of Infrared Technology for Nondestructive Testing*. Wiley, New York (2001)
4. M. Handrik, P. Kopas, V. Baniari, M. Vaško, M. Sága, *Analysis of stress and strain of fatigue specimens localised in the cross-sectional area of the gauge section testing on bi-axial fatigue machine loaded in the high-cycle fatigue region*. Procedia Engineering **177**, 516-519 (2017)
5. P. Pecháč, M. Sága, *Controlling of local search methods' parameters in memetic algorithms using the principles of simulated annealing*. Proc. Eng. **136**, 70-76 (2016)
6. L. Jakubovičová, M. Sága, *Computational analysis of contact stress distribution in the case of mutual stewing of roller bearing rings*. Novel Trends in Production Devices and Systems, Applied Mechanics and Materials **474**, 363-368 (2014)
7. J. Zapoměl, V. Dekýš, P. Ferfecki, A. Sapietová, M. Sága, M. Žmindák, *Identification of material damping of a carbon composite bar and study of its effect on attenuation of its transient lateral vibrations*. Int. Journal of Applied Mechanics **7** (6), (2015)

Contribution to fatigue damage prediction of thin shell finite element models

Milan Sága[1,*], *Milan* Vaško[1], *Peter* Kopas[1], *Wiesława* Piekarska[2], *Tomasz* Domański[2], *Marcin* Kubiak[2]

[1]University of Žilina, Faculty of Mechanical Engineering, Department of Applied Mechanics, Univerzitná 8215/1, Žilina, Slovakia
[2]Częstochowa University of Technology, Faculty of Mechanical Engineering and Computer Science, Institute of Mechanics and Machine Design Foundations, Dąbrowskiego 73, Częstochowa, Poland

Abstract. The paper deals with the implementation of the chosen fatigue damage criterions into FE analysis. Our study considers a shell finite element structural analysis in conjunction with the multiaxial rainflow counting and the fatigue damage prediction. The computational analysis considers stochastic response under random excitation in time domain and damage calculation based on so called critical plane approach (CPA).

Keywords: finite element analysis, multiaxial rainflow counting, fatigue damage

1 Introduction

Fatigue damage of structural features is a complex physical process which is governed by a great number of parameters related to, for example, local geometry and material properties of the structural region surrounding the crack growth path. There are different approaches and methods which can be used in fatigue life predictions [1].

It is commonly recognized that it is impossible for a physical model to account for all fatigue influencing parameters, thus a lot of approximate models have been conceived for practical fatigue assessments. In every stadium of fatigue damage cumulation dominates a definite mechanism controlled by more or less known and verified rules. There exists the stage of microplastic process in total volume of material with following stage of fatigue crack nucleation and stage of their growing with more or less detailed zoning. Despite of this research no results have been achieved, which could be considered as successful ones. This applies mainly to the cases of random and combined stress, where today's procedures used in one axis stress analysis fails [2].

There are plenty of hypotheses used for evaluating a degree of damage caused by variable load [3-5]. Life prediction methods which presume homogeneous material (free from cracks, inclusions or defects) at the outset of the investigation can be divided into strain-based (low-cycle fatigue) and stress-based (high-cycle fatigue) methods. This study assumes stress-based approaches use the elastic stress range (or amplitude) as the governing

* Corresponding author: milan.saga@fstroj.uniza.sk

Reviewers: *Juraj* Gerlici, *Marek* Macko

load parameter. At a sufficient load level, which may result in a fatigue life of approximately 10^7 cycles, a threshold referred to as the fatigue or endurance limit can be seen for many materials.

Research of mechanism and processes of fatigue failure of materials achieves great advance, however there still doesn't exist general failure model, which should be applicable for different conditions of activity. There is needed an integration for such a procedures into a modern systems of computing aided design (CAD) [6-8] in relationship to methods of strength computing transferred by finite element method (FEM) [9-11].

2 Thin shell stress calculation

We will consider well-known shell finite elements (Kirchhoff's or Mindlin's formulation) [1, 12]. The stiffness parameters depend on material constants and element geometry, mainly on its thickness. At first we have to prepare the stress calculation process. This process is based on the expression of the j-th element membrane forces and bending moments (without shear forces) [13, 14], i.e.

$$\begin{bmatrix} F_{xx} & F_{yy} & F_{xy} \end{bmatrix}_j^T = \mathbf{F}_m^{\,j} = \int_S \mathbf{E}_m^j \cdot \boldsymbol{\varepsilon}_m^j dS_j = \mathbf{E}_m^j \cdot \int_S \mathbf{B}_m^j dS_j \cdot \mathbf{u}_L^j = t_j \cdot \mathbf{D}_j \cdot \mathbf{I}_m^j \cdot \mathbf{u}_L^j \tag{1}$$

and

$$\begin{bmatrix} M_{xx} & M_{yy} & M_{xy} \end{bmatrix}_j^T = \mathbf{M}_b^{\,j} = \int_S \mathbf{E}_b^j \cdot \boldsymbol{\varepsilon}_b^j \cdot dS_j = \mathbf{E}_b^j \cdot \int_S \mathbf{B}_b^j \cdot dS_j \cdot \mathbf{u}_L^j = \frac{t_j^3}{12} \cdot \mathbf{D}_j \cdot \mathbf{I}_b^j \cdot \mathbf{u}_L^j \ . \tag{2}$$

The auxiliary matrices $\mathbf{I}_m$ and $\mathbf{I}_b$ can be calculated only using the numerical approach. Further details about $\mathbf{E}_m$, $\mathbf{E}_b$, $\mathbf{D}$, $\mathbf{B}_m$, $\mathbf{B}_b$, $\mathbf{u}_{el}$ and t are presented in [10, 15]. The extreme stress values can be expected at the top or at the bottom surface. Generally, it means

$$\begin{bmatrix} \boldsymbol{\sigma}_{mb}|_{top} \\ \boldsymbol{\sigma}_{mb}|_{bot} \end{bmatrix}^j = \begin{Bmatrix} \sigma_{xx,top} \\ \sigma_{yy,top} \\ \sigma_{xy,top} \\ \sigma_{xx,bot} \\ \sigma_{yy,bot} \\ \sigma_{xy,bot} \end{Bmatrix}^j = \begin{bmatrix} 1/t_j & 0 & 0 & 6/t_j^2 & 0 & 0 \\ 0 & 1/t_j & 0 & 0 & 6/t_j^2 & 0 \\ 0 & 0 & 1/t_j & 0 & 0 & 6/t_j^2 \\ 1/t_j & 0 & 0 & -6/t_j^2 & 0 & 0 \\ 0 & 1/t_j & 0 & 0 & -6/t_j^2 & 0 \\ 0 & 0 & 1/t_j & 0 & 0 & -6/t_j^2 \end{bmatrix} \cdot \begin{Bmatrix} F_{xx} \\ F_{yy} \\ F_{xy} \\ M_{xx} \\ M_{yy} \\ M_{xy} \end{Bmatrix}^j = \begin{bmatrix} \mathbf{A}_{t,top} \\ \mathbf{A}_{t,bot} \end{bmatrix}^j \cdot \begin{Bmatrix} \mathbf{F}_m \\ \mathbf{M}_b \end{Bmatrix}^j , \tag{3}$$

or

$$\boldsymbol{\sigma}_{mb\_L}^j = \mathbf{C}_L^j \cdot \mathbf{f}_L^j \ . \tag{4}$$

Let's build new material and auxiliary matrices

$$\mathbf{E}_{mb} = \begin{bmatrix} t_j \cdot \mathbf{I}_3 & \mathbf{0}_3 \\ \mathbf{0}_3 & \dfrac{t_j^3}{12} \cdot \mathbf{I}_3 \end{bmatrix} \cdot \begin{Bmatrix} \mathbf{D} \\ \mathbf{D} \end{Bmatrix}_j = \mathbf{D}_t \cdot \mathbf{D}_{mb} \ , \quad \mathbf{I}_{mb} = \begin{Bmatrix} \mathbf{I}_m^j \\ \mathbf{I}_b^j \end{Bmatrix} , \tag{5}$$

where the matrix $\mathbf{I}_3$ is the classical unit matrix. Then (3) can be rewritten as follows

$$\boldsymbol{\sigma}_{j\_mb}\Big|_{top} = \mathbf{A}_{t,top} \cdot \mathbf{E}_{mb} \cdot \mathbf{I}_{mb} \cdot \mathbf{u}_L^j = \mathbf{A}_{t,top} \cdot \mathbf{D}_t \cdot \mathbf{D}_{mb} \cdot \mathbf{I}_{mb} \cdot \mathbf{u}_L^j, \tag{6a}$$

$$\boldsymbol{\sigma}_{j\_mb}\Big|_{bot} = \mathbf{A}_{t,bot} \cdot \mathbf{E}_{mb} \cdot \mathbf{I}_{mb} \cdot \mathbf{u}_L^j = \mathbf{A}_{t,bot} \cdot \mathbf{D}_t \cdot \mathbf{D}_{mb} \cdot \mathbf{I}_{mb} \cdot \mathbf{u}_L^j. \tag{6b}$$

3 Chosen approaches for multiaxial fatigue damage calculation

Let's now focus on the cumulative damage counting by using multiaxial rainflow decomposition of the stress response. It should be noted that the fatigue damage calculation of the machine parts is generally problematic because the results are considerable changed in the principal stresses [2, 4, 5]. Using FE analysis we can get six components of the stress-time function (multiaxial stress) but it is very difficult to obtain an equivalent - uniaxial load spectrum by reason of comparison with applied computational fatigue curve.

In our case the rainflow analysis for random stresses known in classic uniaxial form as von Mises or Tresca hypotheses is impossible. It means that the important goal of this part will be to propose some approaches to estimate the high-cycle fatigue damage for multiaxial stresses caused by random vibration analysed structure [10, 16, 17]. Generally we can apply two fundamental approaches for multiaxial rainflow counting:

- Critical Plane Approach (CPA) [13], and
- Integral Approach (IA) [10].

It is well-known that the Wöhler curve (Fig. 1, sometimes called S-2N curve) is basic source of getting information of the material fatigue life. Generally the S-2N curve is statistically evaluating by experimental fatigue curve. This is a graph of the magnitude of a cyclical nominal stress σ_A against the logarithmic scale of cycles to failure $2N_f$. It is advantage to show it in logarithmical or semi logarithmical coordinates.

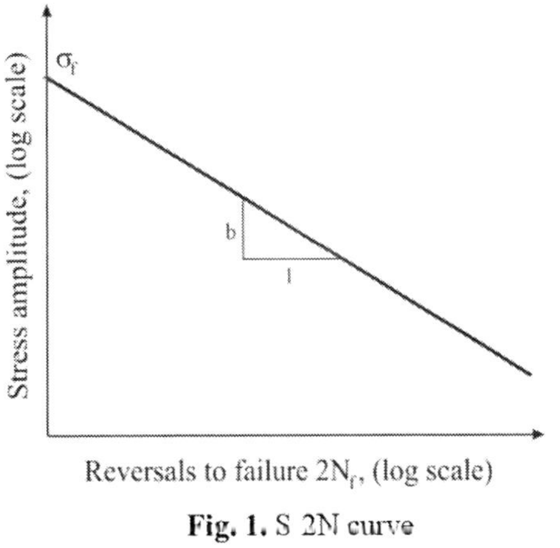

Fig. 1. S 2N curve

The σ_A–$2N_f$ relation can be written as follows

$$\sigma_A = \sigma_f \cdot \left(2N_f\right)^b, \tag{7}$$

where σ_f is the fatigue stress coefficient, $2N_f$ is number of cycles to failure, b is fatigue strength exponent and σ_A is stress amplitude to failure.

Some researches the relationship (7) rewrite into following form

$$\sigma_A^{\;m} \cdot (2N_f) = K\,, \tag{8}$$

where $m = -(1/b)$ and $K = \sigma_f^{(-1/b)} = \sigma_f^{\;m}$.

Considering mean stress modified version of the stress amplitude (using Goodman, Soderberg, Geber), Eq. (8) can be rewritten as follows

$$\left\{\sigma_A \cdot \left[1 - \left(\frac{\sigma_M}{R_F}\right)^k\right]^{-1}\right\}^m \cdot (2N_f) = K\,. \tag{9}$$

If $k = 1$ and $R_F = R_E$ (yield stress) the Soderberg's model is used, if $k = 1$ and $R_F = R_M$ (strength limit) the Goodman's model is used and if $k = 2$ and $R_F = R_M$ the Geber's model is used. Using the linear Palmgren-Miner law we can calculate fatigue damage for stress amplitude σ_{Ai} as follows

$$d_i = \frac{1}{2N_{f\_i}} = \left\{\frac{\sigma_{Ai}}{\sigma_f} \cdot \left[1 - \left(\frac{\sigma_M}{R_F}\right)^k\right]^{-1}\right\}^m\,. \tag{10}$$

Findley hypothesis

Findley has assumed the critical plane as a plane with maximum shear stress, i.e. the fatigue equivalent shear stress can be written as follows [10, 18]

$$\tau_{A\_Fin} = \tau_{max} + k \cdot \sigma_m\,, \tag{11}$$

where k is Findley's factor which value for tough metal can be about *0.3*. Using von Mises relationship between normal and shear stresses and classic plane stress analysis for top or bottom element surface it is possible rewrite (9) into following form

$$\sigma^j_{A\_Fin}\Big|_{top} = \\ = \sqrt{3} \cdot \left\{ sign\left(\mathbf{T}_2 \cdot \boldsymbol{\sigma}_{j\_mb}\big|_{top}\right) \cdot \left[\left(\boldsymbol{\sigma}_{j\_mb}\big|_{top}\right)^T \cdot \mathbf{T}_1 \cdot \left(\boldsymbol{\sigma}_{j\_mb}\big|_{top}\right)\right]^{-1/2} + k \cdot \mathbf{T}_2 \cdot \boldsymbol{\sigma}_{j\_mb}\big|_{top} \right\}, \tag{12a}$$

$$\sigma^j_{A\_Fin}\Big|_{bot} = \\ = \sqrt{3} \cdot \left\{ sign\left(\mathbf{T}_2 \cdot \boldsymbol{\sigma}_{j\_mb}\big|_{bot}\right) \cdot \left[\left(\boldsymbol{\sigma}_{j\_mb}\big|_{bot}\right)^T \cdot \mathbf{T}_1 \cdot \left(\boldsymbol{\sigma}_{j\_mb}\big|_{bot}\right)\right]^{-1/2} + k \cdot \mathbf{T}_2 \cdot \boldsymbol{\sigma}_{j\_mb}\big|_{bot} \right\}. \tag{12b}$$

where

$$\mathbf{T}_1 = \begin{bmatrix} 0.25 & -0.25 & 0 \\ -0.25 & 0.25 & 0 \\ 0 & 0 & 1 \end{bmatrix} \quad \text{and} \quad \mathbf{T}_2 = \begin{bmatrix} 0.5 & 0.5 & 0 \end{bmatrix}. \tag{13}$$

The damage calculation can be realised using Eq. (10). Presented relationships have been applicable for FE analyses. Numerical and experimental tests have confirmed that the factor $k = 0.3$ was overstated [4, 13] by author.

Dang Van hypothesis

Dang Van again has assumed the critical plane with shear stress but with difference in factor k, which can be calculated from normal and shear fatigue limit, i.e.

$$\tau_{A\_DV} = \tau_{max} + k \cdot p_m = \frac{\sigma_1 - \sigma_3}{2} + \frac{\tau_C - \frac{\sigma_C}{2}}{\frac{\sigma_C}{\sqrt{3}}} \cdot \frac{\sigma_1 + \sigma_2 + \sigma_3}{3}, \tag{14}$$

where τ_C is shear (torsional) fatigue limit, σ_C is normal (axial) fatigue limit, σ_1, σ_2, σ_3 are principal stresses. Relationship (11) is possible to use like that

$$\sigma_{A\_DV} = \sqrt{3} \cdot \left[sign\left(\frac{\sigma_1 + \sigma_3}{2}\right) \cdot \frac{\sigma_1 - \sigma_3}{2} + \frac{\tau_C - \frac{\sigma_C}{2}}{\frac{\sigma_C}{\sqrt{3}}} \cdot \frac{\sigma_1 + \sigma_2 + \sigma_3}{3} \right]. \tag{15}$$

Using von Mises hypothesis we can get [4]

$$\frac{\tau_C - \frac{\sigma_C}{2}}{\frac{\sigma_C}{\sqrt{3}}} \approx \frac{\tau_C - \frac{\sqrt{3} \cdot \tau_C}{2}}{\frac{\sqrt{3} \cdot \tau_C}{3}} \approx 0{,}232 \cdot \tag{16}$$

Using by application of the shell stress theory and Eqs. 6a,b we can obtain

$$\sigma^j_{A\_DV}\big|_{top} = \\ = \sqrt{3} \cdot \left\{ sign\left(\mathbf{T}_3 \cdot \boldsymbol{\sigma}_{j\_mb}\big|_{top}\right) \left[\left(\boldsymbol{\sigma}_{j\_mb}\big|_{top}\right)^T \cdot \mathbf{T}_1 \cdot \left(\boldsymbol{\sigma}_{j\_mb}\big|_{top}\right)\right]^{-1/2} + 0{,}232 \cdot \mathbf{T}_3 \cdot \boldsymbol{\sigma}_{j\_mb}\big|_{top} \right\}, \tag{17a}$$

$$\sigma^j_{A\_DV}\big|_{bot} = \\ = \sqrt{3} \cdot \left\{ sign\left(\mathbf{T}_3 \cdot \boldsymbol{\sigma}_{j\_mb}\big|_{bot}\right) \cdot \left[\left(\boldsymbol{\sigma}_{j\_mb}\big|_{bot}\right)^T \cdot \mathbf{T}_1 \cdot \left(\boldsymbol{\sigma}_{j\_mb}\big|_{bot}\right)\right]^{1/2} + 0{,}232 \cdot \mathbf{T}_3 \cdot \boldsymbol{\sigma}_{j\_mb}\big|_{bot} \right\}. \tag{17b}$$

Relations (12) and (17) present equivalent stresses applicable for rainflow decomposition for both proportional and non-proportional loading. The cumulative damage calculation can be realised using Eq. (10) again.

HMH modified hypothesis

Applying von Mises equivalent stress for CPA we can obtain following relationship

$$\sigma_{A\_HMH} = sign\left[\sigma(\mathbf{n}_{CPA})\right] \cdot \sqrt{\sigma(\mathbf{n}_{CPA})^2 + 3 \cdot \tau(\mathbf{n}_{CPA})^2}, \tag{18}$$

or in detail

$$\sigma^{j}_{A\_HMH}\Big|_{top}(\mathbf{n}_{CPA})= = sign\left[\boldsymbol{\sigma}^{T}_{j\_mb}\Big|_{top}(\mathbf{n}_{CPA})\cdot\mathbf{T}_{\sigma}\cdot\boldsymbol{\sigma}_{j\_mb}\Big|_{top}(\mathbf{n}_{CPA})\right]\cdot\left[\boldsymbol{\sigma}^{T}_{j\_mb}\Big|_{top}(\mathbf{n}_{CPA})\cdot\mathbf{T}_{4}\cdot\boldsymbol{\sigma}_{j\_mb}\Big|_{top}(\mathbf{n}_{CPA})\right]^{1/2}, \tag{19a}$$

$$\sigma^{j}_{A\_HMH}\Big|_{bot}(\mathbf{n}_{CPA})= = sign\left[\boldsymbol{\sigma}^{T}_{j\_mb}\Big|_{bot}(\mathbf{n}_{CPA})\cdot\mathbf{T}_{\sigma}\cdot\boldsymbol{\sigma}_{j\_mb}\Big|_{bot}(\mathbf{n}_{CPA})\right]\cdot\left[\boldsymbol{\sigma}^{T}_{j\_mb}\Big|_{bot}(\mathbf{n}_{CPA})\cdot\mathbf{T}_{4}\cdot\boldsymbol{\sigma}_{j\_mb}\Big|_{bot}(\mathbf{n}_{CPA})\right]^{1/2}, \tag{19b}$$

where

$$\mathbf{T}_4=\begin{bmatrix}1 & -0.5 & 0\\ -0.5 & 1 & 0\\ 0 & 0 & 3\end{bmatrix} \quad \text{and} \quad \mathbf{T}_\sigma=\begin{bmatrix}\sigma_x & \tau_{xy} & 0\\ \tau_{xy} & \sigma_y & 0\\ 0 & 0 & 0\end{bmatrix}. \tag{20}$$

In this case it should be noted that computational approach depends on a searching process of a critical plane normal vector $\mathbf{n}_{CPA}$. By reason rainflow analysis it is very important to know the sign of the calculated equivalent stress therefore the sign of this stress will be defined by sign of normal component. For searching process was used optimizing tools in Matlab [7] and optimizing problem for cumulative damage function is usually formulated as follows

$$D_{i\_max}\Big|_{HMH}=max\left[\sum_{j=1}^{mc}d_j(\mathbf{n}_{CPA})\right]= =max\left\{\left\{\frac{\sigma^{i}_{A\_HMH}\Big|_{bot}(\mathbf{n}_{CPA})_j}{\sigma_f}\cdot\left[1-\left(\frac{\sigma^{i}_{M\_HMH}\Big|_{bot}(\mathbf{n}_{CPA})_j}{R_F}\right)^{k}\right]^{-1}\right\}^{m}\right\} \tag{21}$$

for unknown vector $\mathbf{n}_{CPA}$ and stresses on bottom surface. The same computational process we can have to realise on top element surface.

Damage calculation using IA

The fundamental idea is to count rainflow cycles on all linear combinations of the stress random vector components [10], i.e.

$$\sigma_{A\_MRF}(t)= =c_1\cdot\sigma_x(t)+c_2\cdot\sigma_y(t)+c_3\cdot\sigma_z(t)+c_4\cdot\tau_{xy}(t)+c_5\cdot\tau_{yz}(t)+c_6\cdot\tau_{zx}(t) \tag{22}$$

on the assumption that the parameters c_i belong to a hypersphere $\sum_{i=1}^{n}c_i^2=1$. Practically if the stress state is biaxial (e.g. thin shell finite element) the stress components can be written under the form of three dimension vector $\boldsymbol{\sigma}=[\sigma_x, \sigma_y, \tau_{xy}]^T$ and the equivalent stress will be calculated as follows

$$\sigma_{A\_MRF}(t)=c_1\cdot\sigma_x(t)+c_2\cdot\sigma_y(t)+c_3\cdot\tau_{xy}(t)=\mathbf{c}\cdot\boldsymbol{\sigma} \tag{23}$$

on condition that $c_1^2 + c_2^2 + c_3^2 = 1$. In the case of shell element we can obtain again following relationships

$$\sigma^{j}_{A\_MRF}\Big|_{top}(\mathbf{c}) = \mathbf{c}\cdot\boldsymbol{\sigma}_{j\_mb}\Big|_{top}(\mathbf{c}) \quad \text{and} \quad \sigma^{j}_{A\_MRF}\Big|_{bot}(\mathbf{c}) = \mathbf{c}\cdot\boldsymbol{\sigma}_{j\_mb}\Big|_{bot}(\mathbf{c}). \tag{24}$$

Hence the next goal will be to find extreme value of the estimated damage for vector **c** and i-th element, i.e.

$$D_{i\_max}\Big|_{MRF} = max\left[\sum_{j=1}^{mc} d_j(\mathbf{c})\right] = max\left\{\left\{\frac{\sigma^{i}_{A\_MRF}\big|_{bot}(\mathbf{c})_j}{\sigma_f}\cdot\left[1-\left(\frac{\sigma^{i}_{M\_MRF}\big|_{bot}(\mathbf{c})_j}{R_F}\right)^{k}\right]^{-1}\right\}^{m}\right\}, \tag{25}$$

where $D_{i\text{-}max}|_{MRF}$ is the maximum value of the cumulative damage for i-th element, $2N_i$ is the number of cycles to failure, mc is the number of cycles after rain-flow decomposition of the stress. Naturally we have to observe the normality condition for **c** using following transformation

$$\mathbf{c} = \frac{\mathbf{c}'}{\sqrt{\mathbf{c}'^{T}\cdot\mathbf{c}'}}. \tag{26}$$

4 Testing example

Let's solve fatigue damage of the model of ribbed beam as a shell construction created in FEA software ADINA (Fig. 3). The real geometry of this component is shown in Fig. 2. On this model there was analyzed rate of cumulative fatigue damage with help of presented CPA methods. We considered calculation with the chosen multiaxial criterions which were applied for 16 identical random loading processes (2000 random values – Figs. 4 and 5): *Dang Van* criterion; *Findley* criterion and *Von Mises* criterion.

Fig. 2. Proportions of ribbed beam with wall thickness of 6 mm

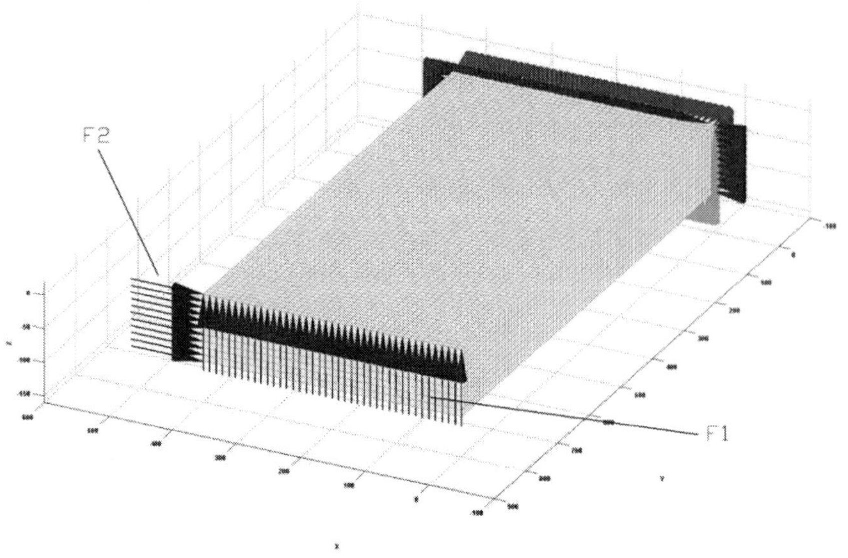

Fig. 3. Three-Dimensional model of ribbed beam

The Fig. 3 represents 3D model of ribbed beam and its mesh with detail of loading. We used the following parameters in finite element model: used material was typical construction steel with Young's modulus $E = 2.1 \cdot 10^{11}$ Pa and Poisson number $\mu = 0.3$. During the FE meshing it was used thin shell finite elements with four nodes. The model was excited under two stochastic forces with Gauss distribution. Stochastic character of this forces we can describe as follows: mean ($F1$) = 12000 N, standard deviation ($F1$) = 8000 N at direction of z axis and mean ($F2$) = 5000 N, standard deviation ($F2$) = 10000 N at direction of x axis. Time functionality of the forces $F1$ and $F2$ are shown in Figs. 4 and 5.

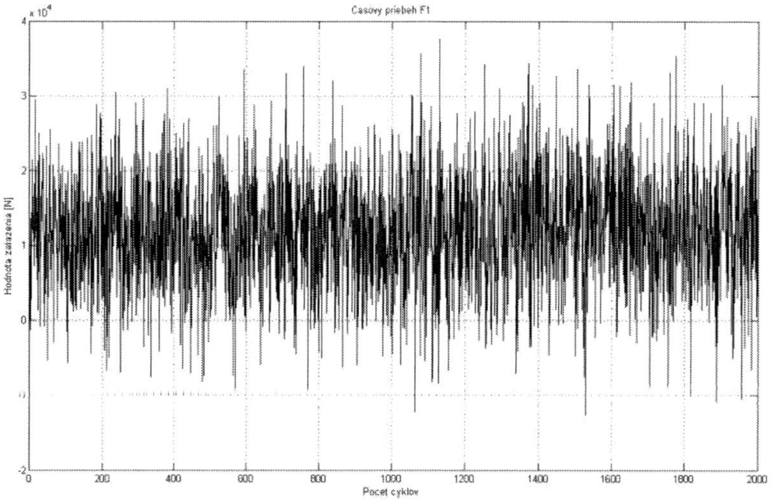

Fig. 4. Time course function of loading force $F1$

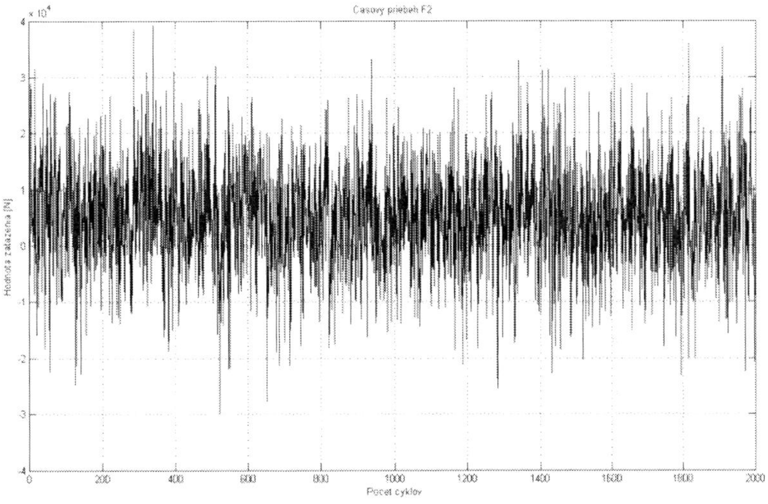

Fig. 5. Time course function of loading force $F2$

Cumulative Damage Calculations

In Table 1 is shown estimate of cumulative fatigue damage for ribbed beam. Graphical representations of cumulative fatigue damage of analyzed shell structure for three chosen hypothesis are shown in Fig. 6, Fig. 7 and Fig. 8.

Table 1. Cumulative damage for different multiaxial criterions

Used method	Node with maximal damage value [number]	Cumulative fatigue damage [cycles]
Dang Van	1	$5.2883 \cdot 10^{-4}$
Findley	1	$5.303 \cdot 10^{-4}$
Von Mises	1	$1.665 \cdot 10^{-2}$

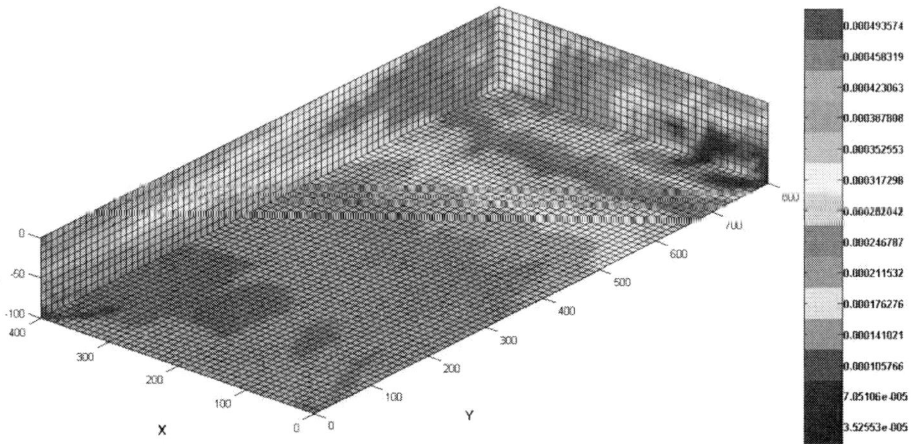

Fig. 6. Distribution of the cumulative fatigue damage using Dang Van criterion

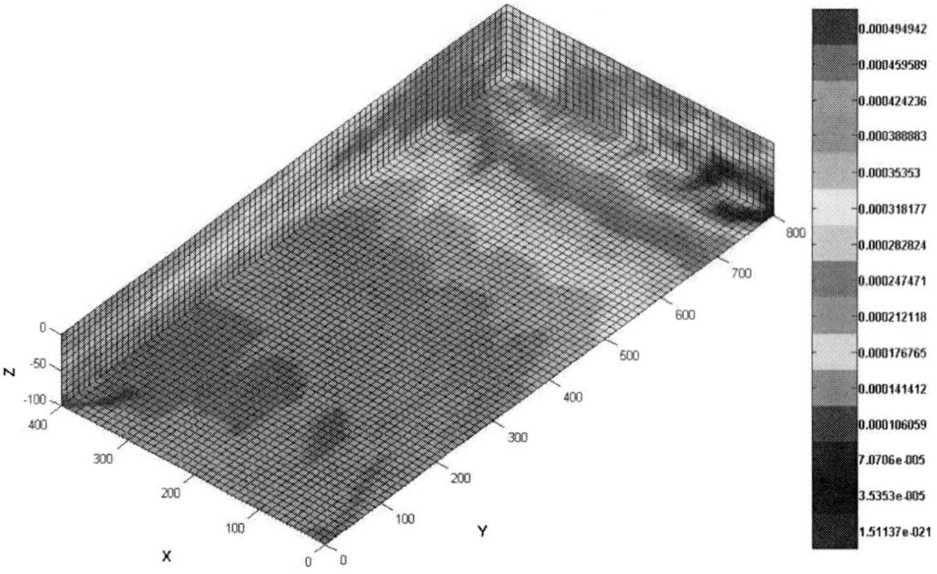

Fig. 7. Distribution of the cumulative fatigue damage using Findley criterion

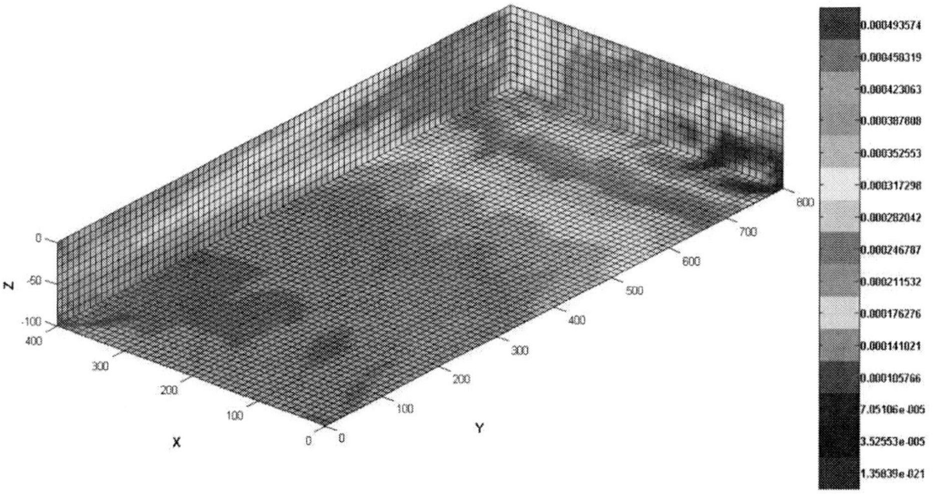

Fig. 8. Distribution of the cumulative fatigue damage using von Mises criterion

Our computational analysis confirms fact that the results of cumulative fatigue damage of ribbed beam is for Dang Van and Findley criterion very similar. The most difference was observed in von Mises criterion which also gave the worst results. Average value of obtained computational results was 0.0059 and standard deviation has value 0.0093.

Conclusion

The goal of this study has been to realized an overview and propose of possibilities of computational fatigue damage (life) prediction of modeled shell structure with help of finite element method. There was used a software facility interface between MATLAB and ADINA software.

They were also presented and compared different methodology about prediction of multiaxial fatigue criterions. Our benefit has been in application of the algorithmization process of the fatigue failure computational prediction in shell finite element models.

This work has been supported by the Slovak Research and Development Agency under the contract No. APVV-14-0096.

References

1. M. Balda, J. Svoboda, V. Fröhlich, *Using hypotheses for calculating fatigue lives of parts exposed to combined random loads*. Engineering Mechanics **10** (5), 12-15 (2003)
2. C. Chu, *Multiaxial fatigue life prediction methods in the ground vehicle industry*. Int. J. Fatigue **19** (1), 325-330 (1997)
3. D. F. Socie, *Multiaxial fatigue damage models*. J. Eng. Mater. Technol. **1069**, 293-298 (1987)
4. X. Pitoiset, A. Preumont, A. Kernilis, *Tools for a Multiaxial Fatigue Analysis of Structures Submitted to Random Vibrations*. Proceedings European Conference on Spacecraft Structures Materials and Mechanical Testing, Germany, 1-6 (1998)
5. A. Carpinteri, A. Spagnoli, S. Vantadori, *A multiaxial fatigue criterion for random loading*. Fatigue and Fracture of Engineering Materials and Structures **26** (6 - Special Issue), 515-522 (2003)
6. J. Flizikowski, M. Macko, J. Czerniak, A. Mroziński, *Implementation of genetic algorithms into development of mechatronic multi-edge's grinder design*. ASME 2011, Int. Mech. Eng. Congress and Exposition, IMECE, (2011)
7. M. Macko, J. Flizikowski, Z. Szczepański, K. Tyszczuk, G. Śmigielski, A. Mroziński, J. Czerniak, A. Tomporowski, *CAD/CAE applications in mill's design and investigation*. Proc. of the 13th Int. Scientific Conf. on Computer Aided Engineering: Lecture Notes in Mech. Eng., Chapt. **35**, Springer Intern. Publ. AG, 343-351 (2017)
8. G. Domek, A. Kołodziej, *Design of the tendon structure in timing belts*. Procedia Engineering **136**, 365-369 (2016)
9. M. Sága, Š. Medvecký, M. Kopecký, *The effective algorithm for discrete structural mass minimising subjected to fatigue life*. 6th World Congress on Structural and Multidisciplinary Optimization, Rio de Janeiro, Brasil, (full paper on CD)
10. M. Sága, M. Vaško, P. Pecháč, *Chosen numerical algorithms for interval finite element analysis*. Procedia Engineering **96**, 400-409 (2014)
11. J. Gerlici, T. Lack, *Modified HHT method for vehicle vibration analysis in time domain utilisation*. Applied mechanics and materials **486**, 396-405 (2014)
12. T. Lack, J. Gerlici, *Rail vehicles brake components test bench utilisation*. Applied mechanics and materials **486**, 379-386 (2014)
13. M. Dudziak, G. Domek, A. Kołodziej, K. Talaśka, *Contact Problems Between the Hub and the Shaft with a Four-Angular Shape of Cross-Section for Different Angular Positions*. Applied Mechanics and Materials **816**, 54-62 (2015)
14. R. Halama, J. Sedlák, M. Fusek, Z. Poruba, *Uniaxial and biaxial ratcheting of ST52 steel under variable amplitude loading-experiments and modeling*. Procedia Engineering **101** (C), 185-193 (2015)

15. V. Dekýš, A. Sapietová, O. Števka, *Understanding of the dynamical properties of machines based on the interpretation of spectral measurements and FRF*. Experimental stress analysis, Applied Mechanics and Materials **486**, 106-112 (2014)
16. J. Vavro, J. Vavro Jr., P. Kováčiková, R. Bezdedová, *The optimisation of the materials properties for the passenger cars in dependence on defect distribution at the dynamic loading*. Procedia Engineering **136**, 114-119 (2016)
17. J. Vavro, J. Vavro Jr., P. Kováčiková, P. Kopas, M. Handrik, *Simulation and analysis of defect distribution in passenger car tire under dynamic loading*. Applied Mechanics and Materials **611**, 544-547 (2014)
18. R. Halama, M. Fusek, Z. Poruba, *Influence of mean stress and stress amplitude on uniaxial and biaxial ratcheting of ST52 steel and its prediction by the Abdel Karim–Ohno model*. International Journal of Fatigue **91**, 313-321 (2016)

Shifted axis angular positioning mechanisms – unconventional use of exact constraint design

Ksawery Szykiedans[1,*], *Maciej* Bujwan[2]

[1]Warsaw University of Technology, Faculty of Mechatronics, 02-525 Warszawa, Boboli 8, Poland
[2]Alumnus of Warsaw University of Technology, Faculty of Mechatronics, 02-525 Warszawa, Boboli 8, Poland

Abstract. Paper presents some remarks about designing angular positioners having significant angular displacement and shifted axis of rotation. Those driving mechanism are useful in specific applications where mechanism wraps around other object or there is a need of rotation around virtual axis. Examples of those applications like orthotic robots or gimbals for panoramic motions where developed at Faculty of Mechatronics WUT. Usage of exact constraint method was described including its influence to a kinematic structure of designed mechanism.

Keywords: Exact constraint, Camera Head, 3DOF axial positioning.

1 Introduction

The engineers designing mechanisms use the principles of kinematics because under-constrained or over-constrained devices will not function. What the precision designer/engineer must be aware that at some scale, everything is a mechanism. Significantly deformable structure will not provide stability of nanometres to precise component. This is often the most important motivation for exact-constraint or kinematic design in precision machines, a need of a separation for a sensitive sub-system or a sensor from the influence of dimensional changing elements. I will present some influence of using exact constraint method to designing process of two kind of 3DOF manipulators. Both cases presents mechanisms of restricted range of angular motion having different aspects of possible shift of its axis of revolution.

Presented case describes very specific example of such conditions. Camera head - a 3 DOF positioning system – requires proper kinematics to ensure not a safety of a camera sensor. The aim is to get a proper motion picture especially during taking panoramic shots. Rotation point of a head and a cameras one has to be the same as no – parallax point of a lens (centre of an entrance pupil). This provides panoramic shot similar to natural viewing by moving only your eyeballs. To achieve panoramic similar to a one's head turning a few centimetres shift between pan axis of rotation and no-parallax point is needed.

---

* Corresponding author: k.szykiedans@mchtr.pw.edu.pl

Reviewers: *Wiesława* Piekarska, *Alžbeta* Sapietová

2 Exact constraint - method of design

Basics of exact constraint method are taken form the works of J.C. Maxwell [1] but still many engineers do not use this method or not understand it. A quotation from one of founders of exact constraint methodology describes why the method is not so widely or fully used:
«By "dusting off" the principles of kinematics and applying them to machine design, we arrive at the method of Exact Constraint. The method of Exact Constraint has been developed to the point where it comprises a body of knowledge which can be used to routinely create new machine designs which are both high in performance and low in cost. The results are so excellent, yet so obvious; so elegant, yet so simple; that at once they seem both profound and trivial! Perhaps it is this duality which has kept these principles so well hidden. One may ask: "Of what value could anything so trivial be?" And so these principles have been overlooked. They have become disused.» [Blanding, 1992] [2]

The term exact constraint is very overt and expressive when the idea is understood [3-6]. An unconstrained rigid body has six degrees of freedom (6DOF) - three translations and three rotations. A non-stiff object may have one or more degrees of flexibility that behaves as extra degrees of freedom, relatively speaking. The correct application of constraints would disregard degrees of freedom, one constrain to one DOF. The aim of exact-constraint design is to get desired freedom of motion or full bond by usage of the minimum number of constraints required. For most of the designing engineers previous statements are obvious and are known as true ones. But are they truly used in designing. Man conceptualizes in terms of an ideal constraint, which is absolutely rigid against motion in one or more degrees of freedom and is absolutely free in the remaining degrees of freedom.

The reference by Blanding [6] is an excellent introduction to exact-constraint design and rigid structures. Some other references provide accounts of more specific exact constraint designs such [5].

Most important statements of exact constraint method are quoted below [6, 2].

"Statement 1: Points on the object along the constraint line can move only at right angles to the constraint line, not along it.

Statement 2: Any constraint along a given constraint line is functionally equivalent to any other constraint along the same constraint line (for small motions).

Statement 3: Any pair of constraints whose constraint lines intersect at a given point, is functionally equivalent to any other pair in the same plane whose constraint lines intersect at the same point. This is true for small motions and where the two constraints lie on distinctly different constraint lines.

Statement 4: The axes of a body's rotational degrees of freedom will each intersect all constraints applied to the body

Statement 5: A constraint applied to a body removes that rotational degree of freedom about which it exerts a moment.

Statement 6: Any set of constraints whose constraint lines intersect a complete and independent set of rotational axes, is functionally equivalent to any other set of constraints whose constraint lines intersect the same or equivalent set of rotational axes. This is true for small motions and when each set contains the same number of independent constraints.

Statement 7: An Ideal Sheet Flexure imposes absolutely rigid constraint in its own plane (X, Y, and θz), but it allows three degrees of freedom: Z, θx, and θy.

Statement 8: An Ideal Wire Flexure imposes absolutely rigid constraint along its axis (X), but it allows five degrees of freedom: Y, Z, θx, θy, θz.

Statement 9: A constraint (C) properly applied to a body (i.e., without over-constraint) has the effect of removing one of the body's rotational degrees of freedom (R 's). The R removed is the one about which the constraint exerts a moment. A body constrained by n constraints will have 6 - n rotational degrees of freedom, each positioned such that no constraint exerts a moment about it. In other words, each R will intersect all C 's
Statement 10:Any pair of intersecting rotational degrees of freedom (R 's) is equivalent to any other pair intersecting at the same point and lying in the same plane. This holds true for small motions.
Statement 11:Two parallel R 's are equivalent to any two parallel R 's, parallel to the first pair and lying in the same plane. They are also equivalent to a single R parallel to the first pair and lying in the same plane; and a T perpendicular to that plane.
Statement 12:When parts are connected in series (cascaded), add the degrees of freedom. When the connections occur in parallel, add constraints."

Introductory statements of presented methodology, especially those about planar movements, are clear and so obvious that are threated piffling. But 3-dimensional analysis of constrains that are not so obvious and can led man to important consideration. I.e. *Statement 4* is true if each axis provides uncoupled rotation. An example of such a coupling is the rotation of a threaded part with its related translation. The consequence of infringement of this statement is a complex movement. Similarly statement 9 has the same problem – a single degree of freedom may consist of a coupled rotation and translation. Again it is a third form of kinematic couple of fifth class – lead screw. This behaviour happens when a constraint does not intersect the axis of rotation, thereby introducing translation along the same axis. What lead us to a mechanism with a shifted axis of rotation. Another thought that can be abstract for Blanding works is that over-constrain causes a need of a "perfect" dimensions manufacturing or adding a precise adjusting systems.

Following examples of understanding using statement from Blading works are about shifting axis of rotation in two similar mechanism. But first case will present device with intentionally added adjusting mechanism, second one will prove that general look on the whole kinematics and spotted elasticity can give simpler design.

3 Camera head for a film making crane

Camera heads some time called gimbals are 2 or 3-DOF rotary manipulators or positioning devices. It is a quite popular subject for a diploma, connected with media, quite complex thus challenging. Yet not so to stop a student form building one by himself, or with a small support from student's internet TV, as it took place twice, while author was a supervisor of such diplomas at Faculty of Mechatronics WUT. First camera head were built by Maciej Bujwan [7] in first half of 2014, now after being heavily worn will be replaced with a new design of Piotr Grygo.

Let's consider 3-DOF camera head kinematics (**Fig. 1**) – all possible options have 3 axis rotations, they vary in design of supporting yokes or in a ability of multiple rotation in Z axis.

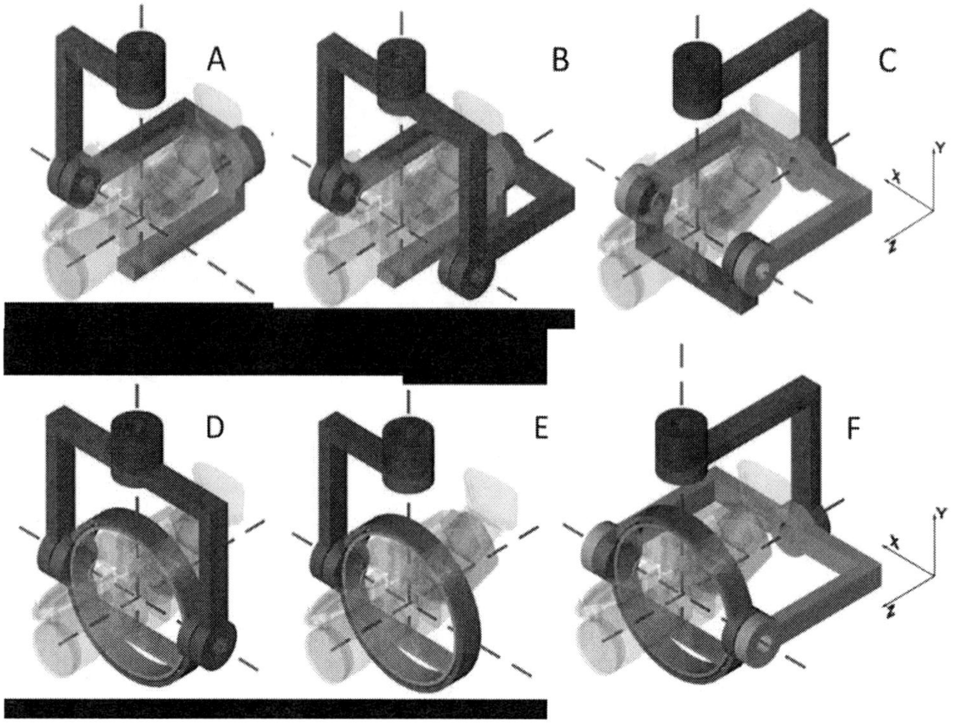

Fig. 1. Kinematics of 3-DOF camera heads

Considering camera head with a connection to a camera we must remember that camera movements not the goal but a way to get proper motion picture. As it was already stated in Introduction shiftage of a Pan axis is desired and should be made very precisely to set a camera to desired panoramic mode (eyeball or head movement). Following pictures (**Chyba! Nenašiel sa žiaden zdroj odkazov., Chyba! Nenašiel sa žiaden zdroj odkazov.**) explains difference between those two settings.

Fig. 2. Camera panoramic setting with paralaxy

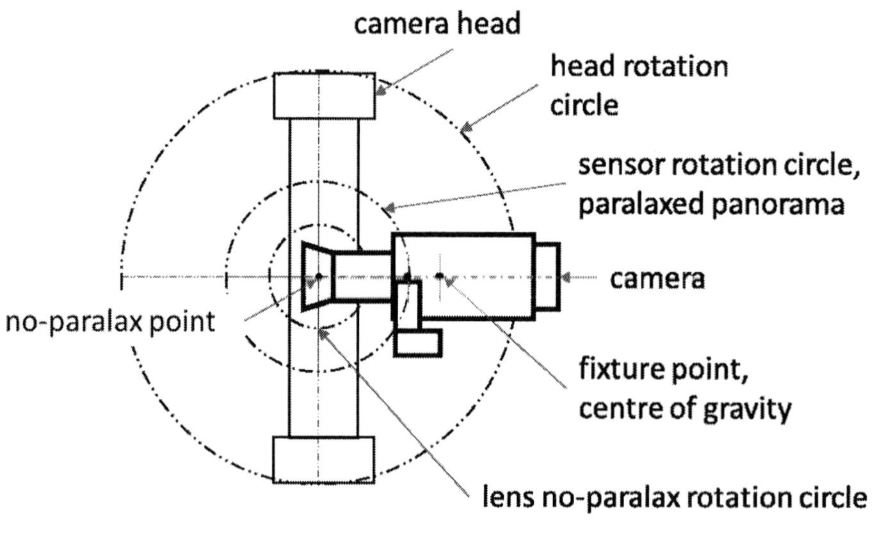

Fig. 3. No -paralax camera panoramic setting

According to statements of exact constraint method, especially 4 and 9, shifted axes of rotation will cause helical movement. Above case of parallax of camera made panorama is a specific occurrence of those statements because is it planar, that's why it cause only circle not a helix. But analysis of possible shifting of roll or tilt axis will also generate additional movement of a camera sensor (**Fig. 4**).

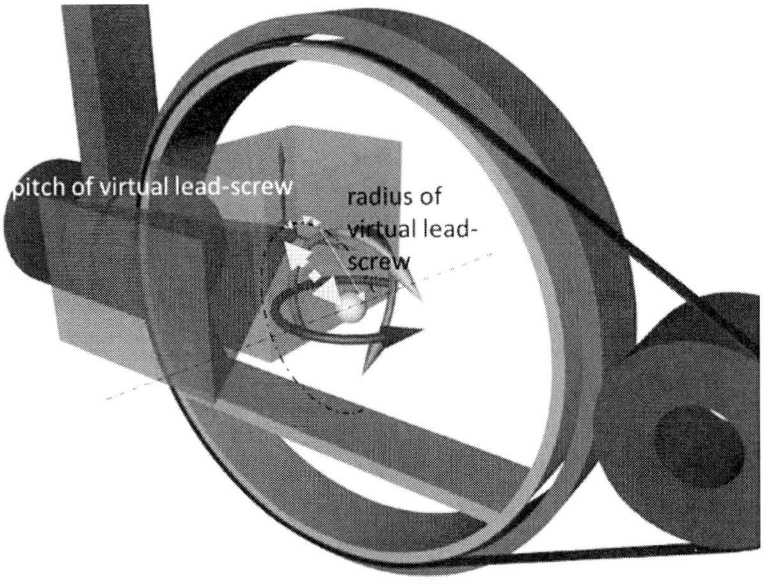

Fig. 4. Helical movement of camera sensor caused by shift between axis of rotation and axis of recording

Such a additional, inattedned movement of a camera is unwanted for at least one reason it wil cause change of image plane and can defocus recorded image or create unpleasent visual effect. To avoid metioned any camera head has to have adjusting mechanism, as well decriben one had (**Fig. 5**, **Chyba! Nenašiel sa žiaden zdroj odkazov.**).

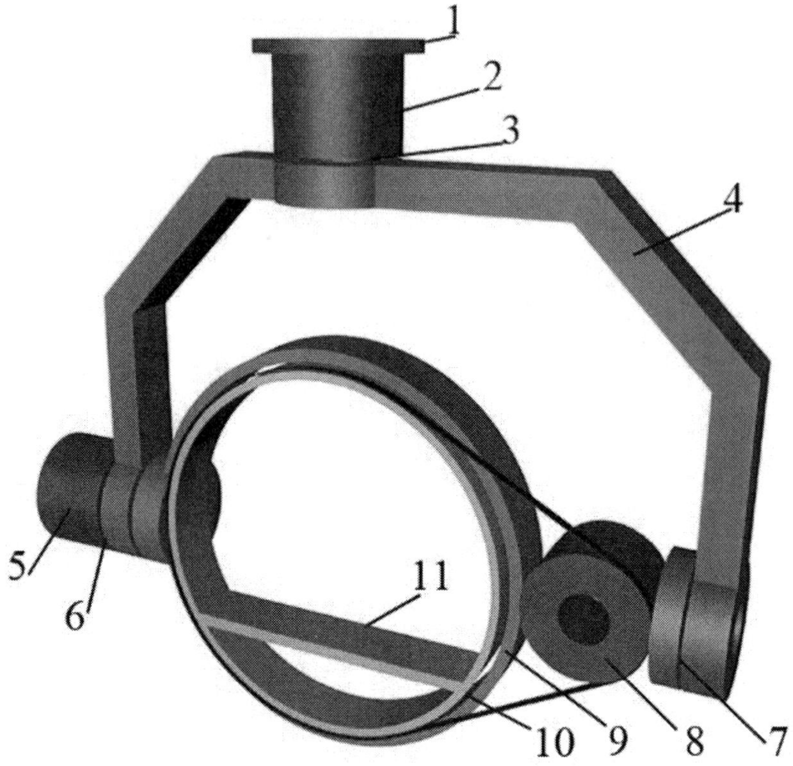

Fig. 5. Camera head designed by M.Bujwan as eengineering thesis at Faculty of Mechatronics WUT ; 1 – head main fixture, 2 – PAN drive, 3 – PAN axis bearing, 4 – frame, 5 – TILT drive, 6 – TILT axis bearing , 7 – ILT support bearing, 8 – ROLL belt drive, 9 – rim, 10 – ROLL ring, 11 – adjustable fixture

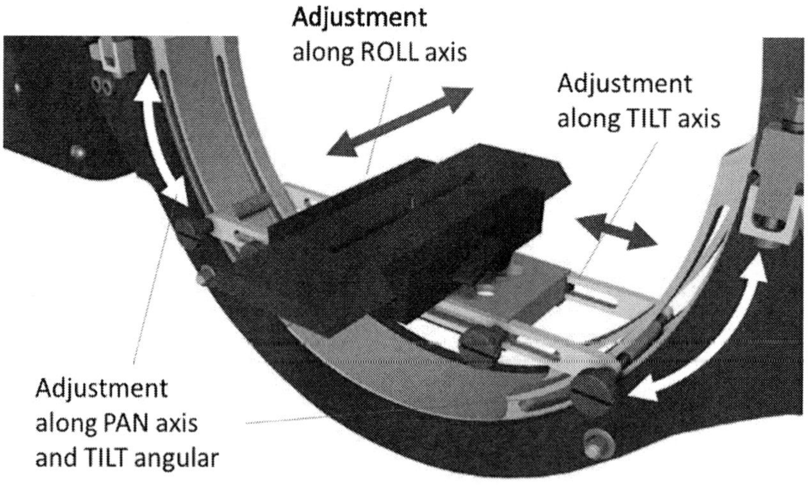

Fig. 6. Adjusting mechanism used in presented camera head

Another subsystem where exact constraint method was used is ROLL unit bearing system (**Fig. 7**). To provide space for different cameras and to make easy acces to it unit recieved form of a ring or a large bearing. To provide as low drag as possible rolling manner was choosen. Usage of such a big rolling bearing was superfluous, it was decided to design a system of same idea but simpler and cheaper. Rolling bearing is over-constrained system – only three rollers, balls or other supporting elements are required to constrain it and define rotation axis. Due to manufacturing and assembling ease frame that holds and supports ROLL unit where split in to symmetrical halves. Somehow it was natural to equip them with the same number of rolling elements.

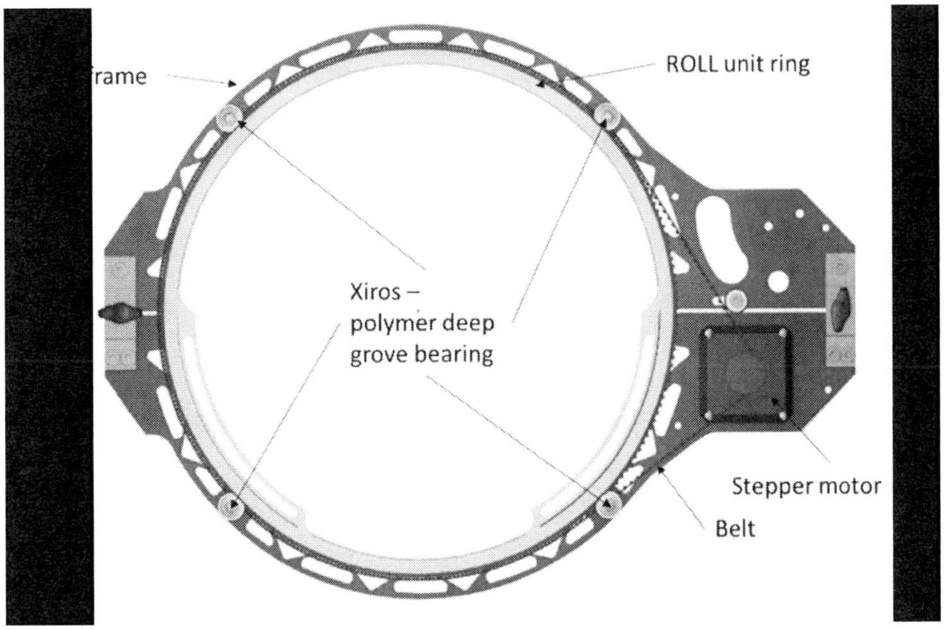

Fig. 7. ROLL unit view, note split frame and symetrical positions of bearings

To provide proper stiffness against twist in TILT axis ring where made wide so at least double row support was required. Such a design will cause single over-constraint, to avoid it extra degree of flexibility had to be added. Frame made as a sheet metal was the first choice but its flexing could affect operationality of whole head. Other solution where chosen. As supporting rolls polymer deep groove ball bearings where used (**Fig. 8**).

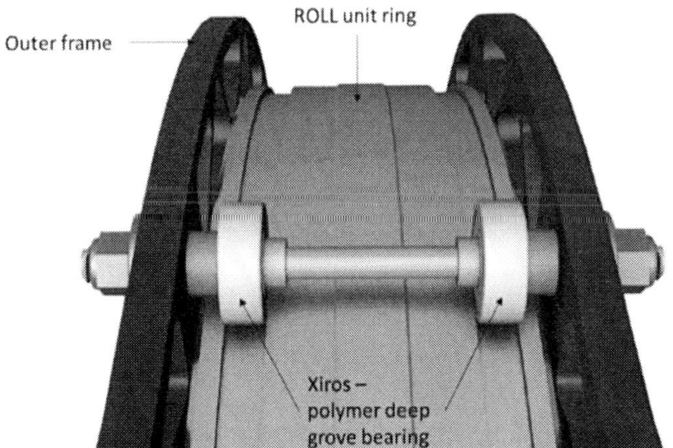

Fig. 8. Side view of support subsystem of ROLL unit

Those bearing have extra-large axial clearance and are made of flexible (comparing to steel metal) material. These two features provide to design extra degree of freedom to neglect influence of extra constraint.

Conclusion

Exact Constraint Method as it was published it was connected with designing precise machinery, sometimes heavy, sometimes very sophisticated but always intended to achieve fine/precision movements, processes or products. In presented paper authors gave on example of use exact constraint rules to slightly *ethereal* product as motion picture. Proper understanding of work of Ch.Evans, D.L. Blanding or A.Slocum does not restrict its usage to machines designing we believe that can be also transferred to virtual world of cinematography.

Prepared with a support of Statutory Found of Faculty of Mechatronics, Warsaw University of Technology.

References

1. J. C. Maxwell, *The Scientific Paper of James Clerk Maxwell* (*Vol. 2*). (Cambridge University Press, London, 1890)
2. L. C. Hale, *Principles and Techniques for Desiging Precision Machines*. MIT PhD Thesis, 67-82 and 174-204 (1999)
3. T. N. Whitehead, *The Design and Use of Instruments and Accurate Mechanism*. (Macmillan, New York, 1934)
4. C. J. Evans, *Precision Engineering : An Evolutionary View*. (Cranfield Press, Bedford, 1989)
5. A. H. Slocum, *Precision Machine Design*. (SME, Dearborn, Michigan, 1992)
6. D. L. Blanding, *Exact Constraint : Machines Design Using Kinematic Principles*. (ASME Press, New York, 1999)
7. M. Bujwan, *Mechanical design of camera crane head*. (WUT, Eng Thesis, 2014)

Investigation of the effects of the equivalent circuit parameters on induction motor torque using three different equivalent circuit models

Mehmet Murat Tezcan[1], *Asim Gökhan* Yetgin[1,*], *Ali Ihsan* Canakoglu[1], *Barış* Cevher[2], *Mustafa* Turan[2], *Murat* Ayaz[3]

[1]Dumlupınar University, Electrical and Electronics Engineering, Kutahya, 43100, Turkey
[2]Sakarya University, Electrical and Electronics Engineering, Sakarya, 54000, Turkey
[3]Kocaeli University, Technical Education Faculty, Kocaeli, 41000, Turkey

Abstract. In this study, the most vital characteristic properties of induction motor, motor starting, maximun and nominal torque are analyzed for what are affected. For the analysis of torque, L type, T type and IEEE 112 equivalent circuit model is employed. Torque change is investigated with the help of codes developed in MATLAB to be used in electrical machines course. First, the slip-moment characteristic curve of the induction motor is divided into three working zones and it was determined which equivalent circuit model gave the best results in these study zones. Later, such parameters as stator and rotor resistance, stator and rotor reactance, magnetization reactance, core resistance and slip are researched for how torque values are affected from. The obtained values for each of the three equivalent circuit models are given and compared with experimental results. 50 HP squirrel cage induction motor parameters are used in analysis. In the analysis, the most influential parameters on the starting torque are found as rotor resistance and stator and rotor reactance. Change of other parameters' impact ratio depending upon right proportion and inverse proportion is observed. When nominal torque variations are investigated, again rotor resistance is the most influential parameter and by selecting proper factor effecting rotor resistance during design, the result of acquiring desired rotor resistance is obtained. When the maximum torque variations are examined, it appears that the stator and rotor reactance values are effective. Starting, maximum and nominal torque expressions change with square of voltage alters the effect on them a lot. Therefore, triangle or star connection importance arises during machine design.

Keywords: Induction motor, Nominal torque, Starting torque, Maximum torque, Equivalent circuit models, Motor parameters.

1 Introduction

Induction machines is widely used in many applications in terms of specifications such as simple and robust structure, low cost, wide speed adjustment, low maintenance requirement, wide power range, reliable operation, mass production technology and etc. In

* Corresponding author: gokhan.yetgin@dpu.edu.tr

Reviewers: *Maciej* Bodnicki, *Josef* Soukup

addition, induction machines are easier and cheaper to produce compared with other electric machines because they do not contain permanent magnets, brushes, or collectors [1-6]. Especially, induction motors are the preferred motors in industrial applications as they can be operated directly on line voltage and their speeds remain constant under load.

In addition, speed and position controls of induction machines can be made easily and this specification makes them advantageous over other industrial machines [5]. However, the low efficiency compared to the permanent magnet synchronous motors and the need high current values for starting operation are the major disadvantages of induction machines. In order to eliminate this drawback, designer should make a preliminary modeling and determine the physical size via parametric optimization with modern software in accordance with the requirements of the application [4]. Machine designers improve performance of electric machines with optimization some parameters such as winding structure, geometry of rotor slots, size of the short circuit ring, geometry of rotor bars, number of stator slots, number of rotor bars, amount and type of magnetic steel and etc. Most of standard T-frame induction motors with squirrel cage are attempted to be designed to have "design B features" of the National Electrical Manufacturers Association (NEMA). Some applications may require torque-speed characteristics outside the "design B feature", and a higher starting torque may require (e.g., for a loaded conveyor) than the capacity of "design B feature". In this case, a machine produced having "design C features" of the NEMA can be selected or machine having higher output power with "design B features" must be selected [7]. Moment-speed characteristics of NEMA standards are collected in 4 different machine groups [8].

Harson et al. and Song et al. have investigated the effects of some design parameters such as air gap length, slot opening, skewed rotor bars, and slot combination on torque-speed characteristics of induction machines [1]. Kul has shown in his study that by choosing a large number of rotor slots, starting torque of the induction machines can be increased [9]. Usudum has stated in his study that the change of the air gap value causes two kinds of effects both as advantages and disadvantages in terms of machine efficiency and performance. As the air gap value decreases, the magnetic saturations in the teeth increase; so that torque ripples and harmonics of the phase current occur and the iron losses increase. On the other hand, as the air gap value decreases, both the starting current value and the nominal operation current value decrease and the copper losses decrease due to decreasing of the current value [10].

In this study, torque expressions of an induction machine with three-phase squirrel cage have been analyzed for using in electrical machinery lessons using three different equivalent circuit models. The slip-torque diagram of the induction machine has been divided into three parts, and it is determined which equivalent circuit model converges better to the experimental results in these parts. Then, it has been investigated how the starting, nominal and maximum torque values vary with changing the equivalent circuit parameters of the induction machine using three different circuit models. For this purpose, the parameters such as slip, stator and rotor resistances, stator and rotor leakage reactance, magnetization reactance and iron resistance have been changed at certain intervals and the change of torque values has been investigated.

2 Torque of induction motors

The torque, also referred to as the turning moment, is a force applied to rotate an object around an axis. In an electric machine, the torque is expressed as the ratio of the motor power to the angular velocity. It is clear that the torque value can be calculated with this basic expression, by using some known parameters of the machine. However, it is important to note that the moment-speed curve of each induction motor is unique. As can be seen from the expression of torque, when the speed is increased, a decrease in torque value will be seen. However, it should be taken consideration that the speed at which the maximum torque produced by the three-phase induction motor with a squirrel cage is generated and the speed at which the maximum power is generated are different from each other [11].

The torque-speed characteristic of induction motors is usually determined using equivalent circuit parameters. These parameters are derived from measurements of the open circuit and short circuit tests [12]. In induction motors, the torque values calculated by using equivalent circuit models and of L type, T type and IEEE type are given in Table 1. As can be seen, moment expression varies depending on many parameters. L type, T type and IEEE type equivalent circuit models of induction motor are shown in Fig. 1 [13-15].

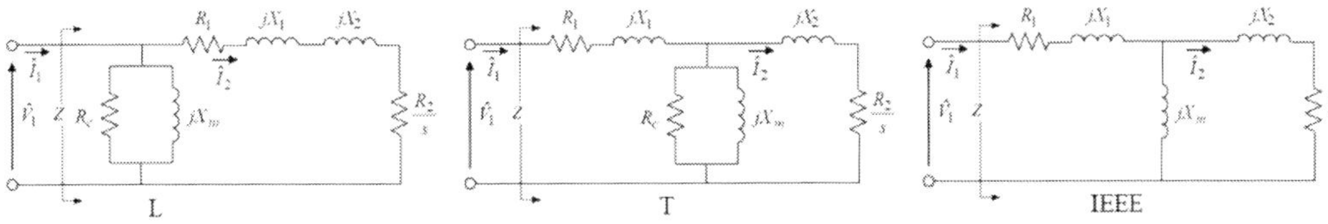

Fig. 1. L, T and IEEE equivalent circuit models of induction motor

In Fig. 1, V_1 is grid voltage, I_1 is the stator current, I_2 is rotor current, R_1 and R_2 are respectively stator winding resistance and rotor resistance with referred to stator winding, R_c is iron resistance, X_m is magnetization reactance, X_1 and X_2 are respectively the stator reactance and rotor reactance with referred to stator, Z is equivalent impedance and s refers slip value. As seen from Fig. 1, the equivalent impedance values obtained from all three models are different from each other. The variation of impedance value causes to change of current value drawn from the grid. Thus, the output power and shaft torque values to be obtained vary.

The moment calculations are given in Table 1 using L type, T type and IEEE type equivalent circuit models and parameters of induction motors. As can be seen, moment expressions vary depending on many parameters.

Table 1. Torque formulations of L, T and IEEE equivalent circuit models [6, 12, 13, 16-18]

	L	T	IEEE
Torque	$T=\frac{m.p.R_2}{2.\pi.f.s}\cdot\frac{V_1^2}{\left[\left(R_1+\frac{R_2}{s}\right)^2+(X_1+X_2)^2\right]}$	$T=\frac{m.p.R_2}{2.\pi.f.s}\cdot\frac{V_{th}^2}{\left[\left(R_{th}+\frac{R_2}{s}\right)^2+(X_{th}+X_2)^2\right]}$	$T=\frac{m.p.R_2}{2.\pi.f.s}\cdot\frac{V_{th}^2}{\left[\left(R_{th}+\frac{R_2}{s}\right)^2+(X_{th}+X_2)^2\right]}$
Voltage	V_1	$V_{th}=\frac{R_cX_m}{R_1+jX_1+R_cX_m}\cdot V_1$	$V_{th}=\frac{jX_m}{R_1+jX_1+jX_m}\cdot V_1$
Impedance	$Z_{th}=R_1+R2+jX_1+jX_2$	$Z_{th}=\frac{R_cX_m\cdot(R_1+jX_1)}{R_cX_m+R_1+jX_1}$	$Z_{th}=\frac{jX_m\cdot(R_1+jX_1)}{jX_m+R_1+jX_1}$
Maximum slip	$s_{\max}=\pm\frac{R_2}{\sqrt{R_1^2+(X_1+X_2)^2}}$	$s_{\max}=\pm\frac{R_2}{\sqrt{R_{th}^2+(X_{th}+X_2)^2}}$	$s_{\max}=\pm\frac{R_2}{\sqrt{R_{th}^2+(X_{th}+X_2)^2}}$

In Table 1, p is the number of poles, f is the frequency, R_{th} is the thevenin equivalent resistance, X_{th} is the thevenin reactance, V_{th} is the thevenin voltage, s_{max} is the maximum slip, T is the torque, m refers the number of phases. In order to find the nominal torque, the starting torque and the maximum torque in the calculations, the results is obtained by writing $s = s_n$, $s = 1$ and $s = s_{max}$ in place of the slip value in the torque expression in all three models.

As shown in Table 1, the torque expressions of equivalent circuit models of T type and IEEE type are identical to each other. Calculations of the torque value are made using thevenin method in both models. However, since the iron resistance is neglected in the IEEE model, the difference in torque expressions arises due to the change in the thevenin equivalent impedance and the thevenin voltage.

The slip-torque characteristic curve of the induction motor can be analyzed with divided into 3 operation regions. These operation regions are shown in Fig. 2 [19].

- The first operation region is the low slip (high speed) region.
- The second operation region is the unsteady region. The maximum torque is obtained in this region.
- The third operation region is the high slip (low speed) region.

The manufacturer's information and experimental data for the three-phase squirrel cage induction motor used in the calculations are given in Table 2.

Table 2. Manufacture and experimental data of the induction motor [20]

Parameters	Values
Stator resistance (ohm)	0.087
Rotor resistance (ohm)	0.228
Stator leakage reactance (ohm)	0.489
Rotor leakage reactance (ohm)	0.115
Core resistance (ohm)	23
Magnetization reactance (ohm)	13.08
Number of pole (2p)	4
Number of phase	3
Frequency (Hz)	60
Voltage (V)	460
Power (HP)	50
Winding connection	Star
Rotor speed (rpm)	1705
Stator speed (rpm)	1800
Nominal Torque (N·m)	234
Starting Torque (N·m)	529
Maximum Torque (N·m)	773

3 Obtained Results

3.1 Effect of slip

Most important performance parameter for induction motor is slip. Slip value for an induction motor is calculated with $s = \frac{n_S - n_r}{n_S} \cdot 100$. As known, n_r is the shaft speed, n_s is the synchronous speed or rotating magnetic field speed. For starting conditions, rotor speed is near zero and the slip value is near to 1. With the increasing of the shaft speed, slip value decreases and gets close to zero [19]. Fig. 2 shows the slip-output torque curves of the induction machine with the obtained results from three equivalent steady-state model.

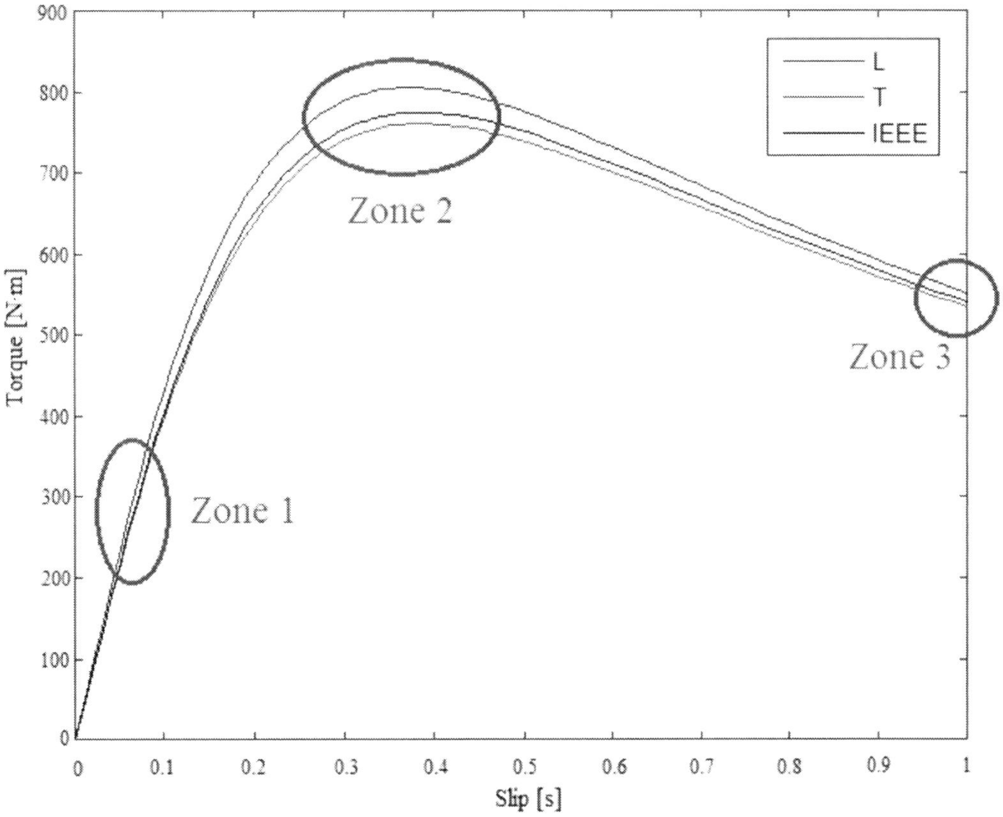

Fig. 2. Graph of slip-torque

When the starting torque values are analyzed, L type model's starting torque value is 551 N·m, IEEE type model's starting torque value is 540 N·m, T type model's starting torque value is 534 N·m respectively. Rated torque values are calculated for 1705 rpm shaft speed and s_n=0.0527 slip value. Obtained values from the calculations, 244 N·m for L type, 228 N·m for IEEE type, 225 N·m for T type equivalent circuit respectively. Finally, when the maximum torque values are analyzed, L type model's maximum torque value is 805 N·m, IEEE type model's maximum torque value is 774 N·m, T type model's maximum torque value is 760 N·m respectively.

When the obtained results and experimental results are investigated from the three model, table 3 shows the errors between real and calculated values of three models respectively.

Table 3. % Error values of torques for three equivalent circuit models

Models	Starting torque error	Maximum torque error	Nominal torque error
L Type	4.13	4.01	4.30
T Type	0.92	-1.71	-3.82
IEEE Type	2.08	0.08	-2.71

When the obtained error values from three models are examined, L type equivalent circuit has highest error value for all moment types. At starting conditions, the T type equivalent circuit gives the closest value with the experimental result. At maximum torque conditions, IEEE type equivalent circuit gives the almost same value with experimental result. At rated operation conditions, IEEE equivalent circuit also gives the closest value with the experimental result.

3.2 Effect of rotor resistance

One of the important parameters that influence the induction machine's performance is rotor winding resistance. Rotor winding resistance can be changed with different techniques. These techniques are, adding external resistor, changing the widths of the rotor slot openings, using double cage rotors, using long caged rotors, using different rotor slot numbers, using different materials for rotor cage structures etc. Besides, with the increasing of rotor winding resistance, starting the motor on maximum torque conditions can be provided. This situation can be seen in Fig. 3 below. Effect of the rotor resistance on starting torque, maximum torque and rated torque has been given in Fig. 3.

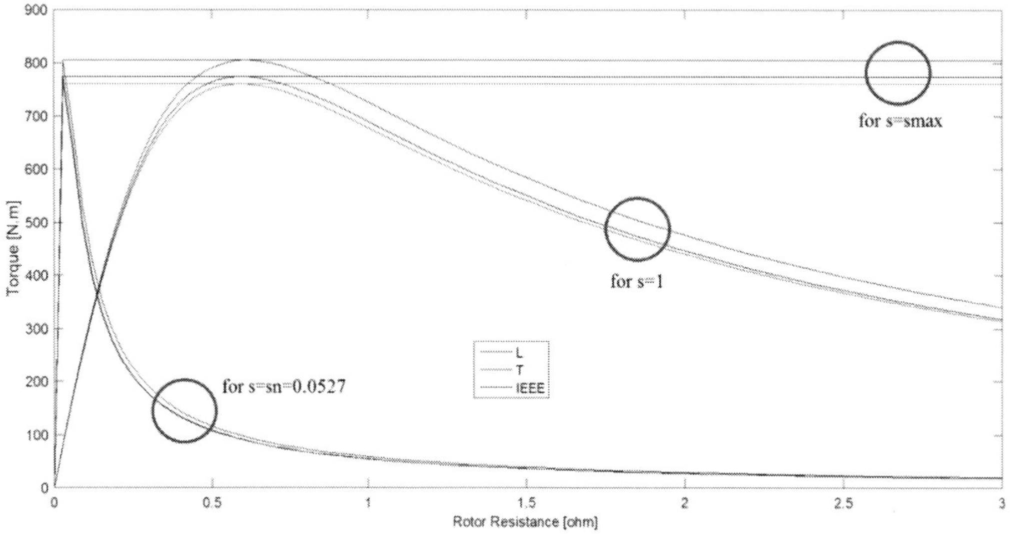

Fig. 3. Graph of rotor resistance-torque

In Fig. 3, increasing the rotor resistance does not affect the starting torque linearly. Rotor resistance increases at a certain value and starting torque increases too. After that value increasing of the rotor resistance does not affect the torque value. For the parameters such that starting torque and rated torque in the desired boundaries, an optimization process should be implemented.

3.3 Effect of stator resistance

In induction machines, stator winding resistance varies depended with winding conductor length, winding conductor cross-section, winding conductor number and winding conductor self-resistance. In general, stator winding resistance too smaller than rotor winding resistance. According to this feature, when calculating of the rated torque of the motor, stator winding resistance is neglected. Effect of the stator winding resistance on three torque items is given in Fig. 4.

According to Fig. 4, when stator winding resistance increases, total impedance of the equivalent circuit increases too. This effects the current and rated torque values negatively. Maximum torque and starting torque curves are quite similar but effect of the stator winding resistance on maximum torque variation is a bit bigger.

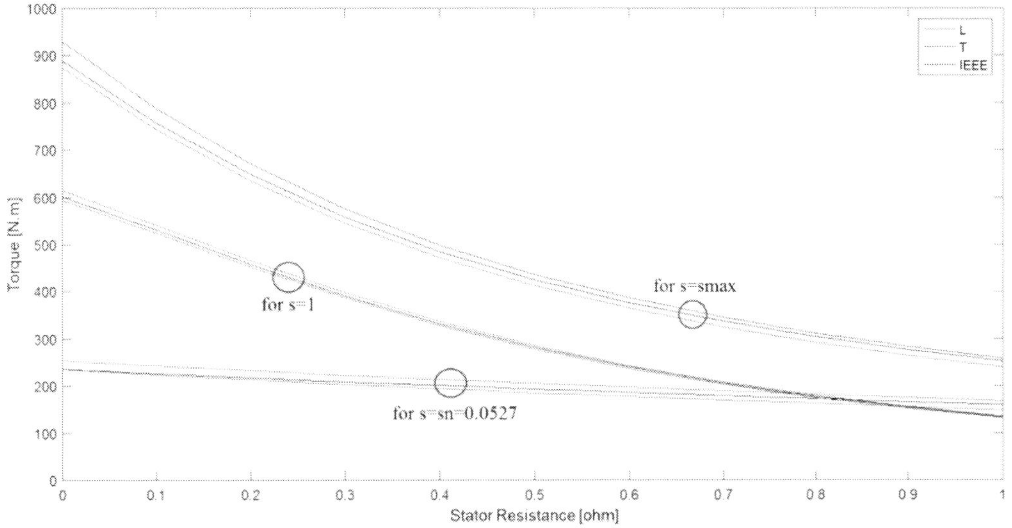

Fig. 4. Graph of stator resistance-torque

3.4 Effect of stator and rotor reactance

According to Biot-Savart law, in induction machines, magnetic field directly proportional with magnetic field (Φ_m) and current (I). Φ_m is produced by magnetizing current (I_μ) but, stator leakage reactance (X_1) blocks the magnetizing current. Therefore, increasing of the X_1 affects the magnetizing current and magnetizing flux, that multiplier of torque negatively.

Increasing of the rotor leakage reactance (X_2) affects the maximum torque value negatively. Conversely, increasing of the X_2 affects the slip negatively too. Therefore, increasing of the X_2 affects the motor's performance badly.

To provide a better motor performance, X_1 and X_2 must be chosen as smaller as possible but magnetization reactance (X_m) must be chosen as bigger as possible. Because of X_1+X_2 value important parameter for maximum torque, designers define the machine dimensions for a certain value of X_1+X_2 and they try to arrange the X_m value as bigger as they can [21]. Effects of the stator and rotor leakage reactances on motor's starting torque and maximum torque values is shown in Fig. 5 and Fig. 6 respectively.

As seen in Fig. 5, increasing of the stator leakage reactance has an important effect on starting torque and maximum torque values. As the increasing of the stator leakage reactance, starting torque value decreases from 2500 N·m to 150 N·m. Maximum torque curve changes quite similar with starting torque curve. Not like the starting torque or maximum torque, increasing of the stator leakage reactance has not an important effect on motor's rated torque. It decreases a little with the increasing of the stator leakage reactance.

When Fig. 6 is investigated, increasing of the rotor leakage reactance value causes the similar effects on both maximum torque and starting torque values. As the increasing of the rotor leakage reactance, starting torque value decreases. Maximum torque curve changes quite similar with starting torque curve. Not like the starting torque or maximum torque, increasing of the rotor leakage reactance has not an important effect on motor's rated torque. It decreases a little with the increasing of the stator leakage reactance.

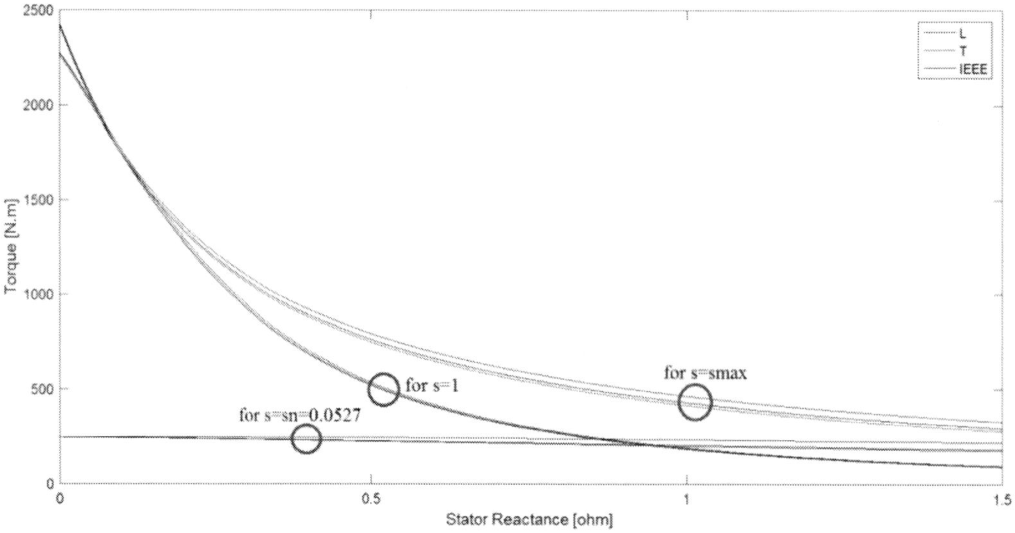

Fig. 5. Graph of stator reactance-torque

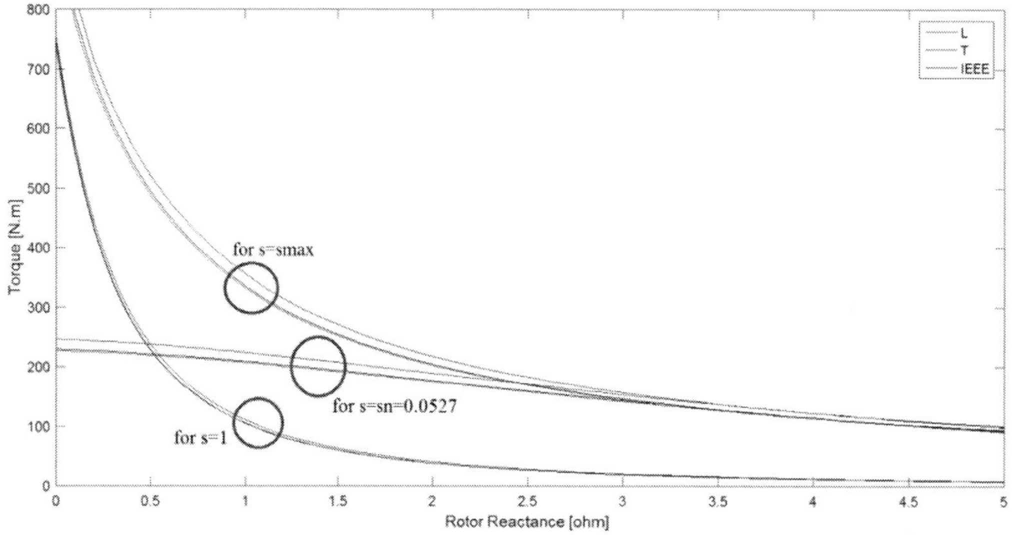

Fig. 6. Graph of rotor reactance-torque

3.5 Effect of core resistance and magnetization reactance

Iron resistance and magnetizing reactance are the parameters that obtained from induction motor no load test. These parameters have not any effect on starting torque, rated torque or maximum torque for L type equivalent circuit. On T type and IEEE type equivalent circuits, the magnetizing reactance has an important effect on these parameters. Iron resistance has an effect only T type equivalent circuit calculations. Fig. 7 and Fig. 8 shows the effects of the magnetizing reactance and iron resistance on three torque type respectively.

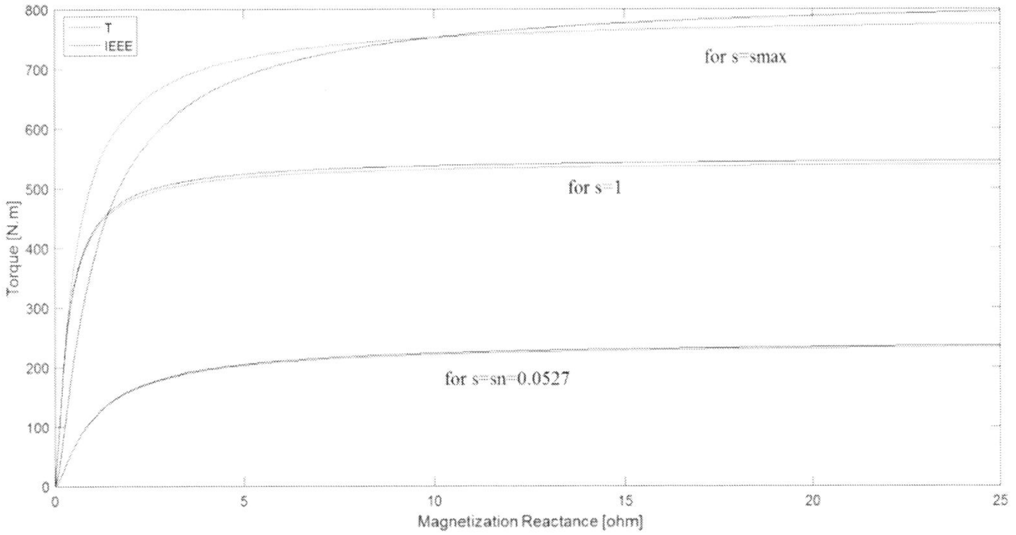

Fig. 7. Graph of magnetization reactance-torque

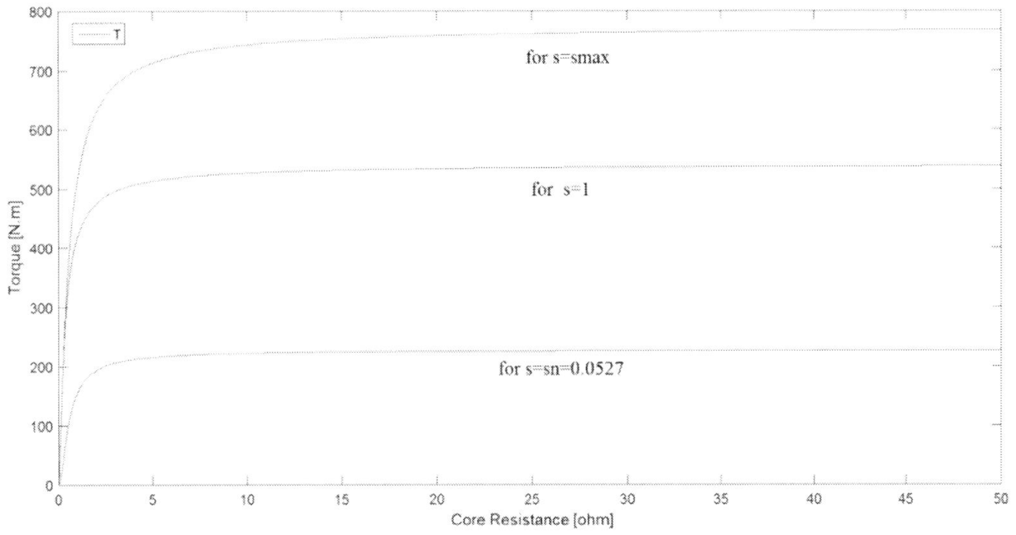

Fig. 8. Graph of core resistance-torque

As seen on Fig. 7 and Fig. 8, parameter-torque curves look like quite similar with the motor's B-H characteristic. Little changes on these two parameters have a significant influence on torque characteristics.

Conclusion

To be used in electric machinery courses, three different equivalent circuits of three-phase squirrel cage induction motor have been examined to observe the changes at starting torque, nominal torque and maximum torque. In the study, the effects of parameters such as resistance and leakage reactance of stator and rotor that affect the torque values have been investigated. The slip-torque diagram of the induction motor have been examined in three study zones and it has been determined which equivalent circuit model approximated best to the experimental results. The study is shown that important parameters for starting torque are rotor resistance and stator and rotor leakage reactances; for nominal torque, that is rotor resistance; and for maximum torque, those are stator and rotor reactances. The study is also shown that T-type equivalent circuit give the best approximation to experimental results in the starting torque region (Zone 3) while IEEE equivalent circuit is better in maximum torque region (Zone 2) and nominal torque region (Zone 1). However the results obtained from the L type equivalent circuit model are found to be far from the experimental results.

In this study, it is aimed to show the effect of the change of the equivalent circuit parameters on the torque effect of the induction motor in three important study zones according to three different equivalent circuit models which are widely used in the description of induction motors.

References

1. Y. Guan, Z.Q. Zhu, I.A.A. Afinowi, J.C. Mipo, P. Farah, *Difference in maximum torque-speed characteristics of induction machine between motor and generator operation modes for electric vehicle application.* Electr. Power Syst. Res. **136**, 406-414 (2016)
2. *Three phase induction motor.* Chapter **8**, 1-63.
3. M. Aydın, Y. Demir, M. Aydın, *Flux skew SEA ile üç fazlı bir asenkron motorun moment-kayma analizi.* MDS Bulletin **11**, 1-2 (2014)
4. M. Aydın, Y. Demir, M. Aydın, *SPEED yazılımı ile üç fazlı asenkron motor analizi.* MDS Bulletin **15**, 1-2 (2015)
5. B. Saraçoğlu, U. Güvenç, M. Dursun, G. Poyraz, S. Duman, *Estimation of the induction motor parameters using biogeography based optimization method.* J. Adv. Technol. Sci. **2**, 46-54 (2013)
6. D. Pejovski, B. Velkovski, *Calculation of induction motor starting parameters using MATLAB.* Infoteh Jahoriina **15**, 879-884 (2016)
7. *Torque characteristics of nema design a,b,c,d & e motors.* 1-2
8. S. Şal, M. Imeryüz, L.T. Ergene, *The analysis of the squirrel cage induction motor rotor bars under the cost constraint.* Chamber Electr. Eng. Sci. J. **2**, 23-28 (2012)
9. S. Kül, *Effect of rotor parameters to traction induction motor performance*, Power Systems Conference, 1-4 (2016)
10. A. Uşüdüm, *Standart bir sincap kafesli asenkron motorun, ansys rmxprt ve maxwell yazılımları ile analizi asenkron motor analizi.* Arge Dergisi, 28-32 (2013)
11. M. Fidan, M. Sersöz, *Motor torque effect on energy efficiency of CER motors.* 3rd International Symposium on Railway Systems Engineering **1**, 245-249 (2016)
12. P. Aree, *Analytical approach to determine speed-torque curve of induction motor from manufacturer data.* Procedia Comput. Sci. **86**, 293–296 (2016)
13. V.P. Sakthivel, S. Subramanian, *Bio-inspired optimization algorithms for parameter determination of three-phase induction motor. COMPEL*: Int. J. Comput. Math. Electr. Electron. Eng. **31,** 528-551 (2012)
14. H.K. Jafari, L.M.F. Corcoles, J. Pedra, *Using the instantaneous power of a free acceleration test for squirrel-cage motor parameters estimation.* IEEE Trans. Energy Convers. **30**, 974–982 (2015)
15. T. Phumiphak, C. Chat-uthai, *Estimation of induction motor parameters based on field test coupled with genetic algorithm.* International Conference on Power System Technology. Proceedings PowerCon 2002, 1199-1203 (2002)

16. M.A.A. Morad, E.F. El-Saadany, *Estimation of induction motor single-cage model parameters from manufacturer data*. Power and Energy Society General Meeting (PES), 2013 IEEE, Vancouver, BC, Canada, 1-5 (2013)
17. A.F. Mergen, S. Zorlu, *Elektrik Makinaları 2, Asenkron Makinalar*. Birsen Yayınevi, İstanbul, 1-256, (2005)
18. S. Sehra, K.K.Gautam, V. Bhuria, *Performance evaluation of three phase induction motor based on no load and blocked rotor test using matlab*. Int. J. Sci. Environ. Technol. **1**, 541–547 (2012)
19. *Asenkron motorlarda moment*. 1-28
20. P. Pillay, R. Nolan, T. Haque, *Application of genetic algorithms to motor parameter determination for transient torque calculations*. IEEE Trans. Ind. Appl. **33**, (1997)
21. M. Turan, *The analysis of the harmonic reactances and torques on the expanded equivalent circuit of the induction machine*. Master Thesis, Sakarya University, Institute of Science and Technology, Sakarya, 1-97 (1996)

Analysis and modeling of variables of the precise shredder's pneumatic material transport's system

Krzysztof Tyszczuk[1*], *Kazimierz* Peszyński[2], *Adam* Mroziński[2], *Grzegorz* Śmigielski[1]

[1] Kazimierz Wielki University in Bydgoszcz, Poland
[2] University of Science and Technology in Bydgoszcz, Poland

Abstract. The technological process of materials shredding, especially biological material, with a precise RPW shredder, requires control of its many elements. One of them, not methodologically recognized so far, is the flow of air conditioning an appropriate product's quality and shredding capacity. In the article, there are presented the first results of studies of the process of pneumatic transport in a shredder, its analysis and simulation attempts, which shall make it possible to reach the optimum product's capacity and quality, constituting the basis for biological diet supplements.

Keywords: pneumatic transport, flow of air, Prandtl tube, precise shredder

1 Precise shredding of oil seeds

High temperature and oxidation accompanying the shredding process, are damaging to vitamins, enzymes, volatile substances and many other active substances occurring in seeds being the base of diet supplements. In the standard shredding method, the process temperature may reach even 100 °C. In circumstances like that, all the volatile and thermo-sensitive active plant's ingredients are destroyed. In case of the discussed precise shredding with the RPW shredder, the increase of the material's temperature, as compared to the input, is little and it neither destructs nor lowers chemical and biological properties of active plant's ingredients. So, it makes it possible to obtain them in an intact state, in the form of powder of optimum parameters. The essence of the precise RPW shredder's operation consists in quick flow of material through the system of active and passive discs, and the degree of shredding and capacity are conditioned by the variables of the technological process and the design form of the working tool.

The basis of good performance of the oil seeds' shredding process (obtaining of the required degree of shredding) is, among the others, the selection of an appropriate flow of air of the required negative pressure and velocity, and the constant volume of feed material at the inlet to the shredder.

* Corresponding author: krzytysz@ukw.edu.pl

Reviewers: *Milan* Sága, *Alžbeta* Sapietová

So, here the process of shredding is based on the operating needs defined as to the value and quality, among the others:

1. for the type of material – arrangement and type of working discs in the shredder,
2. for the type of material and degree of shredding – rotational velocity of the shredder's knife (within the scope of 1420 – 3600 min^{-1}),
3. to maintain the capacity (efficiency of the process) – the volume of the negative pressure in the tank for the shredding product and on the material's inlet to the shredder – suction apparatus setting (pneumatic transport),
4. for high process's efficiency (no blocking by batch) – the volume of the material fed to the charging hopper – set by controlling of the motor of the feeding screw's storage bin.

The above presented operations and their state, must be controlled in the course of the process and have to be verified on a current basis (closed control system). Activities like these are based on the mechatronic control systems of the selected process' stages, are verified via computer-aided decisive processes, including the supported AI ones [1, 2, 3, 4, 5, 6, 7].

1.1 Construction of a test stand

A test stands (Fig. 1) consists of the following, main working sub-systems: material storage bin with outlet's adjustment, a precise shredder RPW-11TN, Z_1- tank for shredded material, with monitoring – a camera via the door viewer, E_4- cyclone, E_3- suction apparatus.

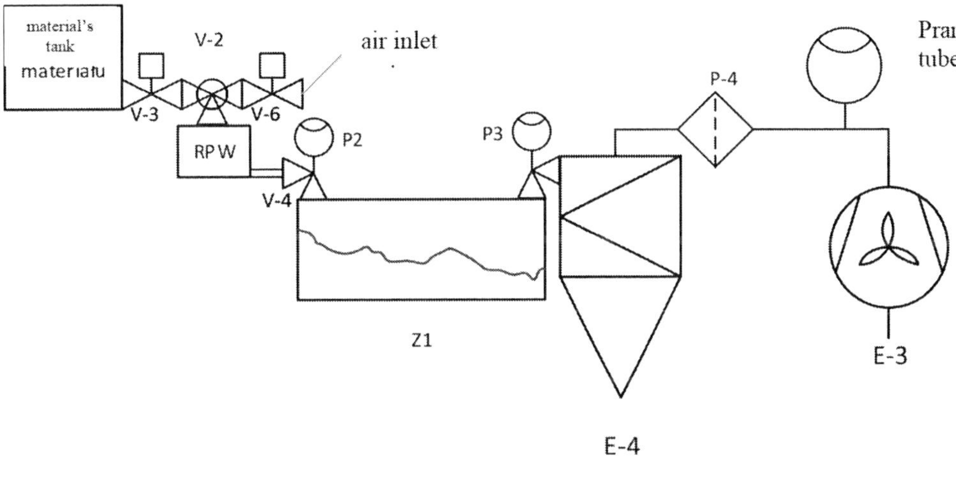

Fig. 1. A block scheme of a test stand of a precise shredder RPW with the pneumatic transport measurement

Material is shredded by thin knives shields and remains in the shredding area for only 2-3 seconds and then leaves the shredder not increasing its temperature. In this case, the pneumatic transport shall be the necessary procedural condition. It comes down to the flow of air forced by the E_3 suction apparatus, and then to the flow of the mixture *air-product*, conditioning an appropriate process's efficiency and the shredding degree. In case of excessive velocity of the flow of air, there may occur the loss of the smallest reduct fractions (shredded material), what shall be visible in the control cyclone. In order to recognize the pneumatic transport's phenomenon, in the stand there have been applied the control elements – a flow meter, vacuum meters and a Prandtl tube, all with digital measurement converters. A computer application making it possible to control the actuators and to visualise the process variables, has been developed in the LabVIEW environment [7] – view of the control panel presented in Fig. 2.

The conducted tests aim at determining of the allowable (for a given material) negative pressure and the velocity of flow.

1.2 The pneumatic transport's arrangement in the shredder

Establishment of the model shredding process with identification of the subsystem of the material and reduct's pneumatic transport on the experimental way, what occurs in case of the described activity, is based on identification of the conducted measurements [8]. On the basis of the experiment, there took place determination of the characteristic parameters' values, and the earlier analysis and knowledge of the *inlet-outlet* relationship in the process of precise shredding of seeds of high fat content was a considerable problem's simplification. By the analysis of the signal and the process, one comes to the model of the system, what constitutes the basis for the process' steering [8, 9], what means the project of interfering in the process that means determining of the adjustment values for the purposes of possibly easy reaching of the required process's functionality, in particular at constant design shredder's parameters being the test facility (Fig. 3) [10, 11, 12, 13, 14, 15, 16].

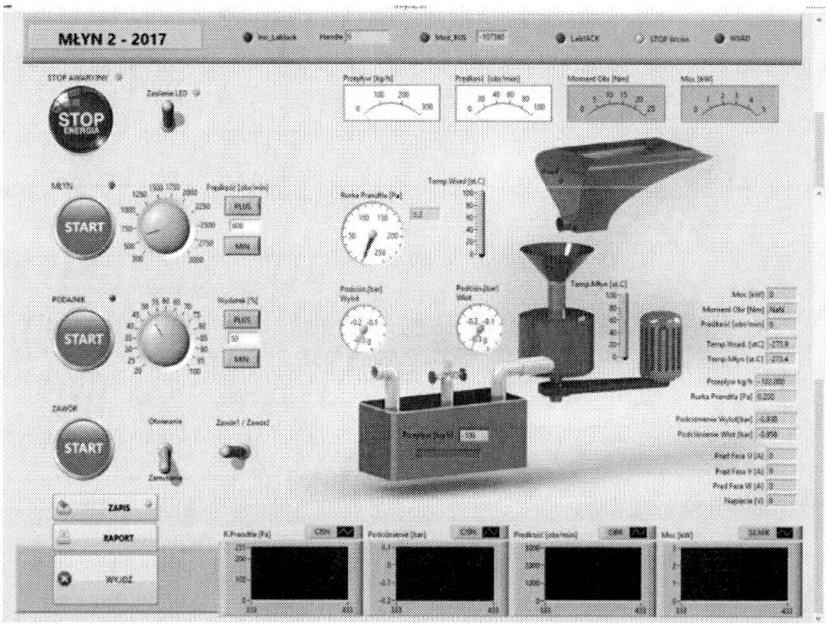

Fig. 2. View of the screen of the control desk of the precise RPW shredder's test stand with measurement of the selected pneumatic transport's parameters

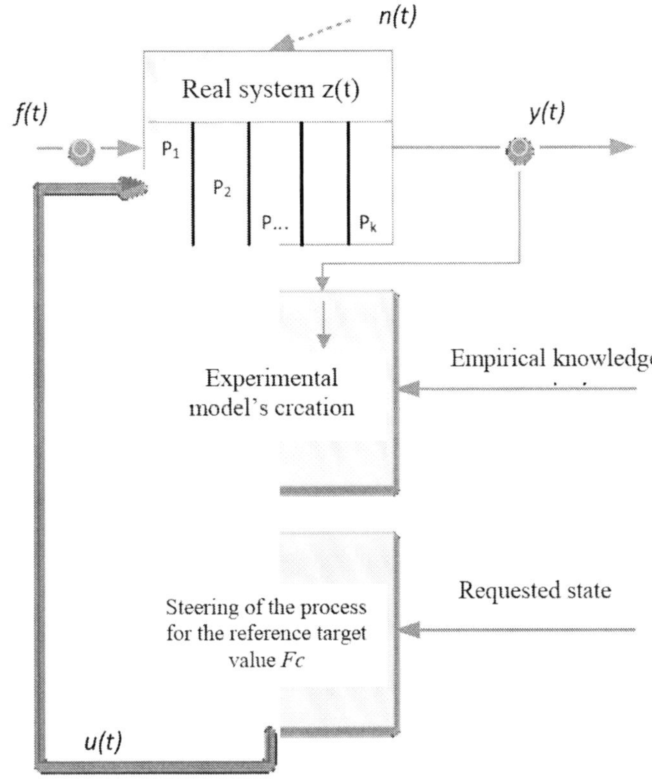

Fig. 3. The principle of the shredding process's control is based on its experimental model, f(t) – input vector (including the process and design factors), *z(t)* – the system's state vector depending on the state of subsystems P_1 - P_k , *u(t)* – control vector, *n(t)* – disturbances vector

2 Measurements of the pneumatic transport parameters

2.1 Introduction to parameters

The preliminary measurements used to determine the pneumatic (flow) shredder's properties, have been conducted with the use of the atmospheric air flow. The authors are aware of the fact, that the mixture of air and shredded seed flowing in the real system, shall have slightly different flow properties. However, the preliminary tests may already give many valuable hints making it possible to improve the system's design and its final modelling. In Fig. 4 there is presented the schematic diagram of the shredding system, referring to the arrangement presented in Fig. 1.

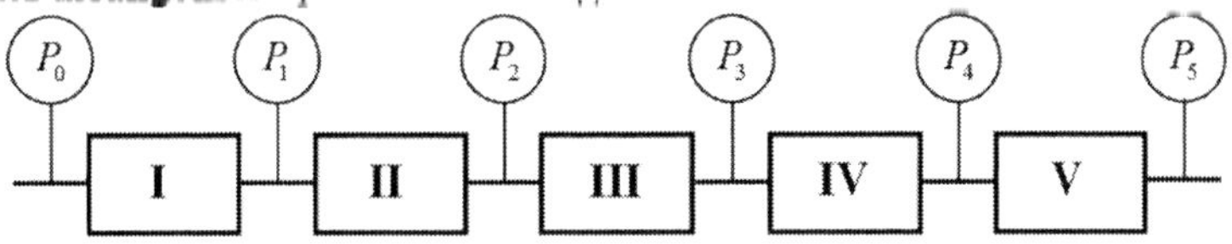

Fig. 4. The schematic diagram of the elements of the shredder's arrangement assumed at the time of experimental tests

Pursuant to the Fig. 4, the whole system has been divided into sections according to the direction of the air flow. In such a system, the suction apparatus E_4, which is the element generating the flow, is in the section **V**. The tests have been commenced from determining of its flow properties. The velocity of the flow of air in the pipeline exhausting air from the cyclone, was measured with the Prandtl tube, Ø6mm, the pipeline has the internal diameter of $d_p = 45 \cdot 10^{-3}$ m. The pneumatic resistances of devices preceding the suction apparatus on the air flow way (mixer, shredder, tank, cyclone) were simulated by assembly of a throttle diaphragm on the inlet pipeline (Fig. 5).

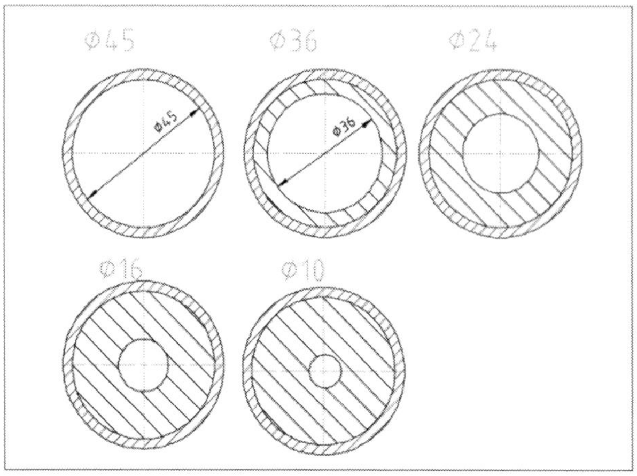

Fig. 5. Set of control holes for the for the measurement of the flow of air on the test stand of the RPW shredder

These diaphragms constitute local resistance deciding on the intensity of air flow through the examined system. The kinematic pressure P_k measured with the use of the Prandtl tube, is the measure of the flow velocity in the tube's section axis (maximum velocity v_C), according to the formula

$$v_C = \sqrt{\frac{2P_k}{\rho_{air}}} \tag{1}$$

where: ρ_{air} is the density of the flowing air.

It has been initially assumed, that at the time of operation of the shredding system, the flow is turbulent. For the turbulent flow, the velocity distribution curve cannot be determined analytically.

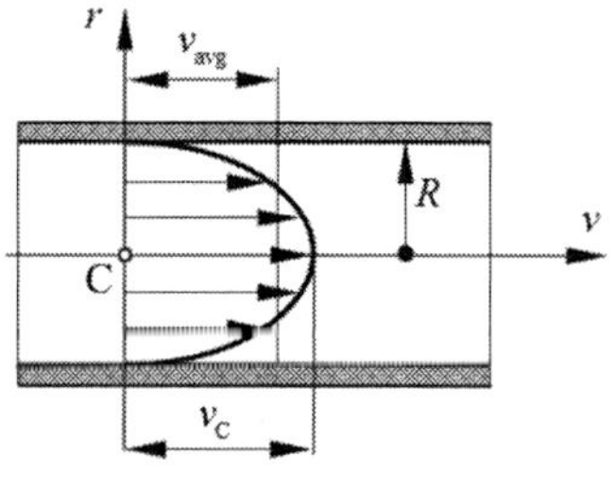

Fig. 6. Flow in the pipeline of a column section: a) laminar, b) turbulent

Due to that, the semi-empirical relations are used. The Prandtl power formula is one of the best-known dependencies describing the velocity distribution

$$v = v_C \left(1 - \frac{r}{R}\right)^{\frac{1}{n}} \tag{2}$$

where: v, v_c, r, R are the values presented in Fig. 6, while $1/n = 1/6 \div 1/10$ is the exponent for the range of Reynolds numbers $Re = 4 \cdot 10^3 \div 3,2 \cdot 10^6$, these values refer to smooth pipes [8, 9]. The equation 2 loses its importance by the same wall. The mean velocity v_{avg} in the turbulent flow may be determined by averaging of the velocity distribution. From the practical point of view, it would be very advantageous to determine the mean velocity depending on the maximum velocity v_C on the section axis and the coefficient n, at the assumed distribution curve (assumed coefficient n).

In order to do that, one should set the intensity of flow in a given pipeline.

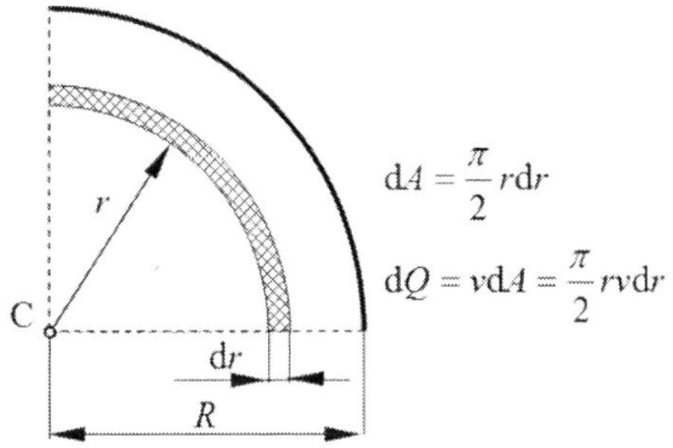

Fig. 7. Integration area in circular section

In case of assuming that the velocity distribution is determined by the power Prandtl formula and the elementary section is determined by the formula $\mathrm{d}A = 0,5\pi r\mathrm{d}r$, the elementary intensity of flow of ¼ of elementary section is determined by the formula $\mathrm{d}Q = 0,5\pi v(r) r\mathrm{d}r$ (Fig. 7) we may determine the mean velocity following substitution of $v = v(r)$ from the formula (2) on the basis of dependency:

$$v_{avg} = \frac{Q}{A} = \frac{4\frac{\pi}{2} v_C \int_0^R \left(1 - \frac{r}{R}\right)^{\frac{1}{n}} r\mathrm{d}r}{\pi R^2} = \frac{4 v_C \mathrm{I}}{\pi R^2} \tag{3}$$

where the integral $\mathrm{I} = \frac{\pi}{2}\int_0^R \left(1 - \frac{r}{R}\right)^{\frac{1}{n}} r\mathrm{d}r$ in the numerator of the formula (3) is the intensity of flow through ¼ of section, at the velocity $v_C = 1\,\mathrm{m \cdot s^{-1}}$. After its multiplication by $4v_C$ we shall receive the flow intensity in the section. The mentioned integral causes certain analytical problems, however it may be calculated with the use of a special function Γ.

Substituting $\frac{r}{R} = t$, where $r = Rt$ and $\mathrm{d}r = R\mathrm{d}t$ we receive:

$$\mathrm{I} = \frac{\pi}{2}\int_0^R (1-t)^{\frac{1}{n}} t R^2 \mathrm{d}t = \frac{\pi R^2}{2}\int_0^R (1-t)^{\frac{1}{n}+1-1} t^{2-1} \mathrm{d}t = \frac{\pi R^2}{2} \frac{\Gamma(2)\cdot\Gamma\left(1+\frac{1}{n}\right)}{\Gamma\left(3+\frac{1}{n}\right)} \tag{4}$$

Following subsequent transformations, we get:

$$\mathrm{I} = \frac{\pi R^2 n^2}{2(2n+1)(n+1)} \tag{5}$$

Due to what, the finally searched dependency $v_{avg} = f(v_c, n)$ is determined by formula

$$v_{avg} = \frac{Q}{A} = \frac{4v_C \mathrm{I}}{\pi R^2} = \frac{2\pi v_C \dfrac{R^2 n^2}{(2n+1)(n+1)}}{\pi R^2} = \frac{2n^2}{(2n+1)(n+1)} v_C \tag{6}$$

For the commonly assumed, at the time of turbulent flows' analysis, the value $n = 7$ was received

$$v_{avg} = 0{,}8167 v_C \tag{7}$$

The formula (7) was used to determine the mean velocity in the circular section and at the same time the intensity of flow during experimental tests. The basic assumption making it possible to use the formula (7) is the turbulent flow of air in the pipeline. The character of the flow in determined with the Reynolds number expressing the relations of the inertial force to the force of internal friction, following the use of the formula (7), determined with the formula

$$Re = \frac{\rho v_{avg} D}{\mu} = 0.8166 \frac{\rho v_C D}{\mu} \tag{8}$$

where: ρ is the mass density of the flowing air, $\mathrm{kg \cdot m^{-3}}$; μ – dynamic viscosity of the flowing air, $\mathrm{kg \cdot m^{-1} \cdot s^{-1}}$; D – internal diameter of the pipeline, m; v_C – velocity of air on the pipeline's axis, $\mathrm{m \cdot s^{-1}}$.

At the time of the examinations with the suction apparatus, the maximum velocity on the pipeline's axis v_C (measured with the Prandtl tube) was within the range $v_C = (5.77 \div 20{,}52)$ m/s. For the mean values $\rho = 1.2\,\mathrm{kg \cdot m^{-3}}$, $\mu = 1.7 \cdot 10^{-5}\,\mathrm{kg \cdot m^{-1} \cdot s^{-1}}$ and $D = 45 \cdot 10^{-3}$ m there were received the values $Re \approx 15000 \div 55000$, what proves the turbulent flow of air in the suction apparatus.

Conclusions

The basic achievement of the theoretical analysis, which is the introduction to the measurements of air flow through the suction apparatus with local resistances on the pipeline supplying air is determination of the formula (7), which makes it possible to determine the mean air velocity in the pipeline leading to the suction apparatus.

In the future studies on the pneumatic properties of the shredding system, the throttle holes shall be replaced with real elements (I ÷ IV, Fig. 4), what shall make it possible to determine the pneumatic local holes of the shredding system.

References

1. J. Flizikowski, M. Macko, J. Czerniak, A. Mroziński, *Implementation of genetic algorithms into development of mechatronic multi-edge's grinder design*. ASME 2011 Int. Mech. Eng. Congress and Exposition, IMECE, (2011)
2. M. Macko, J. Flizikowski, *The method of the selection of comminution design for non-brittle materials*. AIChE Annual Meeting, Conference Proceedings, (2010)
3. M. Macko, J. Flizikowski, K. Tyszczuk, G Śmigielski, I. Rojek, A. Mroziński, J. Czerniak, A. Tomporowski, *Mechatronic implementations used for the effective monitoring and data acquisition of the shredding process*. ASME 2016 International Design Engineering Technical Conferences and Computers and Information in Engineering Conference. IDETC2016, Charlotte, North Carolina, USA. August 21-24, (2016)
4. M. Macko, J. Flizikowski, Z. Szczepański, K. Tyszczuk, G. Śmigielski, A. Mroziński, J. Czerniak, A. Tomporowski, *CAD/CAE applications in mill's design and investigation*. Proceedings of the 13th International Scientific Conference: Computer Aided Engineering, Lecture Notes in Mechanical Engineering, Springer International Publishing AG, 343-351 (2017)
5. J. Czerniak, D. Ewald, M. Macko, G. Śmigielski, K. Tyszczuk, *Approach to the monitoring of energy consumption in eco-grinder based on ABC optimization*. Communications in Computer and Information Science **521**, Springer International Publishing Switzerland, 516-529 (2015)
6. P. Pecháč, M. Sága, *Controlling of local search methods' parameters in memetic algorithms using the principles of simulated annealing*. Proc. Eng. **136**, 70-76 (2016)
7. G. Śmigielski, R. Dygdała, H. Zarzycki, D. Lewandowski, *Real-time system of delivering water-capsule for firefighting*. Hard and Soft Computing for Artificial Intelligence, Multimedia and Security. Advances in Intelligent Systems and Computing, **534**, Springer International Publishing Switzerland, 102-111 (2017)
8. M. Sága, P. Kopas, M. Uhríčik, *Modeling and experimental analysis of the aluminium alloy fatigue damage in the case of bending-torsion loading*. Proc. Eng. **48**, 599-606 (2012)
9. M. Sága, R. Bednár, M. Vaško, *Contribution to modal and spectral interval finite element analysis*. Vibration Problems ICOVP 2011, Springer Proceedings in Physics **139**, 269-274 (2011)
10. J. Flizikowski, A. Mroziński, A. Tomporowski, *Active monitoring as cognitive control of grinders design*. AIP Conference Proceedings, **1822** 020006, doi: http://dx.doi.org/10.1063/1.4977680, (2017)
11. J. Flizikowski, A. Tomporowski, M. Macko, A. Mroziński, *Inżynieria rozdrabniania biomasy*. Monografia pod redakcją A. Mrozińskiego. ISBN: 978-83-64423-35-2. Wydawnictwo Grafpol Agnieszka Blicharz-Krupińska, Bydgoszcz (2016)
12. J. Flizikowski, T. Topoliński, M. Opielak, A. Tomporowski, A. Mroziński, *Research and analysis of operating characteristics of energetic biomass micronizer*. Eksploatacja i Niezawodność/Maintenance and Reliability **17**, No. 1, 19-26 (2015)
13. Y. A. Cengel, J. M. Cimbala: *Fluid Mechanics. Fundamentals and Applications*. McGraw-Hill, (2006)
14. K. Peszyński, D. Perczyński, E. Smyk, P. Kolber, *Experimental verification of velocity distribution in different cross-sectional ventilation ducts*. Proceedings of 23rd International Conference Engineering Mechanics 2017, 770 – 783 (2017)

15. M. Macko, *Economic-energetic analysis of multi-edge comminution of polymer recyclates*. Przemysł Chemiczny **92** (8), (2013)
16. J. Flizikowski, M. Macko, *Method of estimation of efficiency of quasi-cutting of recycled opto-telecommunication pipes*. Polimery/Polymers **46** (1), (2001)

The characteristics analysis of torque and rotation speed of working unit of branch grinder - introductory research

Łukasz Warguła[1,*], *Jarosław M.* Adamiec[1], *Konrad J.* Waluś[1], *Piotr* Krawiec[1]

[1]Poznan University of Technology, Chair of Basic of Machine Design, Piotrowo Street 3, 60-965 Poznan, Poland

Abstract. Carrying out a realistic research on working units of machines that grinder waste coming from tree and bush trimming requires designing machines that would be more efficient. It is neccesary both to introduce the analysis of already existing solutions as well as searching for innovative ones that would result in the increase of efficiency, lowering the costs and time needed as well as lower exhaust emission into the atmosphere. The results presented below will be used for the following purposes: the analysis of grinding process that supplies data to cunstruct simulating models that reflect the real working conditions of grinding machines. The results of the experiments will allow to estimate the influence of construction featrures, including the ones connected with automatisation. The importance of the carried out research is also applicable to the analisys of influence of grinding process on the form of flakes that result from the process, energy consumption. It also allows to make conclusions leading to the improvement of the processes that use grinded mass.The construction specifity of ths kind of machines and their working principles results from the fact, that the cutting process they perform concerns orthotropic plant material that has inhomogenic structure and physical-chemical properties requires further analysis. The article presents the construcion of a research stand designed to record the characteristisc of changes of the torque and rotation speed of working unit of mill chopper type MQS2800 Macalister. The introductory research results presented below point out that depending on the position of the sample (along the grain) against the mill chopper significantly influences on the form of generated changes of the torque. The research stand that has been designed and constructed allows the analysis of both construction features of the chopper as well via the assessment of grinding process also separte paremeters of processed wood.

Keywords: wood chipping process; actual research findings; wood chips; wood shavings; kinematics of the wood chipper; dynamics of wood chopping processes; processing of green waste

* Corresponding author: lukasz.wargula@put.poznan.pl

Reviewers: *Grzegorz* Domek, *Janusz* Mielniczuk

1 Introduction

The forestry, orchard, horticulture, energy and urban greenfield or roadside industries are constantly looking for new, more effective methods of tree care [1-4]. Many of these tasks can be improved by introducing automated systems or modifying existing ones [5, 6]. One of the important aspects of the design and modeling process of new efficient constructions is knowledge of the actual operating conditions of the equipment. Literature review indicates that there is a lack of real results of dynamic research of chipping process. The dynamic characteristics of the periodically-variable rotational speed and torque that are generated in these processes are necessary for describing the input conditions of simulation models that reproduce actual working conditions. The research has been done to help model the construction of the chipping machines, both the working unit and the drives, clutches and gear units. Testing the existing construction allows for a better understanding of the processes involved and supports their automation, which is the main reason for the increased efficiency of these devices [7- 9].

One of the most important issues in the abovementioned industries, but also in households, is the processing of waste generated by them [10]. Branches as waste from the trees and shrubs' care or their cultivation as sources of energy such as willow (biofuel), require proper treatment prior to application. Depending on the application i.e. directly as fuel or biomass; Energy is obtained by: pyrolysis, estimation or fermentation [11, 12]. Chips are also used as compost, mulch [13] and chipboard raw material [14]. Additionally chipping branches facilitate their transport [15], storage and exploitation in the aforementioned processes.

Chipping machines for branches, leaves or similar material use differentiated systems and grinding systems. Machines such as breaker row, crushers, disintegrators, cutters or mills are used for mechanical shredding of solids,. Woodchips are mainly used for the processing of green waste, due to way of use also referred to as recyclers. These are machines for chipping branches, and in case of large machines, even for chipping whole trees. Another division based on application is the division into household chippers and industrially used. Among the first group there are self-propelled wood-chippers or combustion-driven wood chippers, and chippers driven by farm tractors. The second group is professional woodchips, which are divided into: self-propelled wheeled machinery, self-propelled machinery and built-in machines. Due to the kinematics of the cutting mechanism, a number of chipper constructions can be distinguished: discs, beaters, screws, cylinders, drums, otherwise known as mills, when electrically driven for home use. In 2006, Naimi, Sokhansanj, Mani, Hoque, Bi, Womac, Narayan [16] conducted a review of the available chipper constructions together with their energy consumption studies.

The specificity of the construction of such devices and operating principles is due, among other things, to the fact that their cutting process involves orthotropic plant materials whose structure is heterogeneous, and the physicochemical properties are not fully identified [17-21].

An important aspect in the construction of machines is the construction of devices characterized by low energy consumption of work processes. It contributes to the development of a variety of cutting tool designs that are still being tested for optimization. Very often already used and well-know constructions due to technological development undergo structural changes, which lead to improvement of the characteristics of these structures in many aspects at the same time. Krajnc and Dolsak [22] conducted real and simulated drum chipper experiments in 2013, they controlled the material, construction and technological parameters influencing the crushing process. In addition, the thermal characteristics of the generated whirls were investigated, indicating that the chipping process that generates more dust reduces the energy properties of the raw material. In 2012,

Merhar and Bučar [23] presented the actual results of the wood cutting force survey indicating the variable value of this parameter. The cause of this phenomenon is the fracture of the workpiece right in front of the cutting edge. In the articles of 2010, Nati and Spinelli [24] and Nati, Spinelli and Fabbri [25] examined the actual working conditions of shredders by examining fuel consumption, analyzing the impact of blade wear and the type of shredded wood on this parameter. A year later (2011) Spinelli, Magagnotti, Paletto and Preti [26] presented results of studies on the impact of wood characteristics such as species, moisture and parts of the processed tree on the quality of the chipped product and fuel consumption. Conducting research on actual work units helps to design more efficient constructions, and exploration of obtained wood chips results conclusions leading to improved processes utilizing comminuted mass.

The subject of consideration is the scientific and technical importance of the problem of determining the torque characteristics and the rotational speed of the working unit in the process of shredding. It will allow to determine the nature of the load generated and the correct choice of powertrain and executive system of shredding process. The article presents a modern research stand together with experimental results.

2 Research stand

The research was carried out at a research station that allows research on the processes of chipping wood. The station was constructed on the basis of the MQS2800 Macalister electric wood chipper. A hinge assembly consisting of an electric motor, a belt transmission, a planetary gear and a working member was excluded. Research on the kinematic and dynamic characteristics of the chipping process required the development of a separate unit with a measuring device (Figure 1). The torque measurement was performed using a universal torque gauge designed by Chair of Basic of Machine Design, Poznan University of Technology. A MEGATRON MOB 2500/5/BZ/N encoder was used to record instantaneous changes in the angular speed of the working shaft of the shredding member. The results of torque and angular velocity measurements were recorded in the author's computer program, which is an integral part of the research stand.

Fig. 1. Research stand that records torque and rotational speeds characteristics during chipping processes: 1- encoder; 2-working unit; 3- planetary gear unit; 4-solid coupling; 5-torquemeter; 6-belt transmission; 7 -drive engine

3 Results of preliminary tests

Preliminary tests of the kinematic and dynamic characteristics of the working unit of the chipper has been carried out, determining its torque and rotational speed. In the first attempt, these values were measured without load on the working unit (Figure 2).

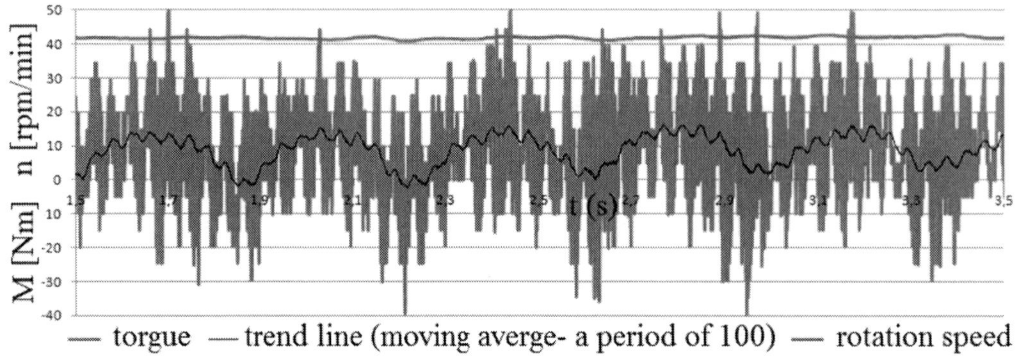

Fig. 2. Torque and rotation speed characteristics as a function of time, without load on the working unit

In the second and third set of tests, the torque value and the rotational speed generated during the pine beam shredding with a cross section of 18 x 18, with a moisture content of 21.5% MC were measured. The sample was chipped in two positions, due to the position of the tree grains. The cutting characteristics of the sample of the tree grains arranged perpendicularly to the cutting edge are shown in Figure 3, and parallel in Figure 4. In addition, the characteristics of the chipper workings phase are as follows: work without load a, transition b and chipping c.

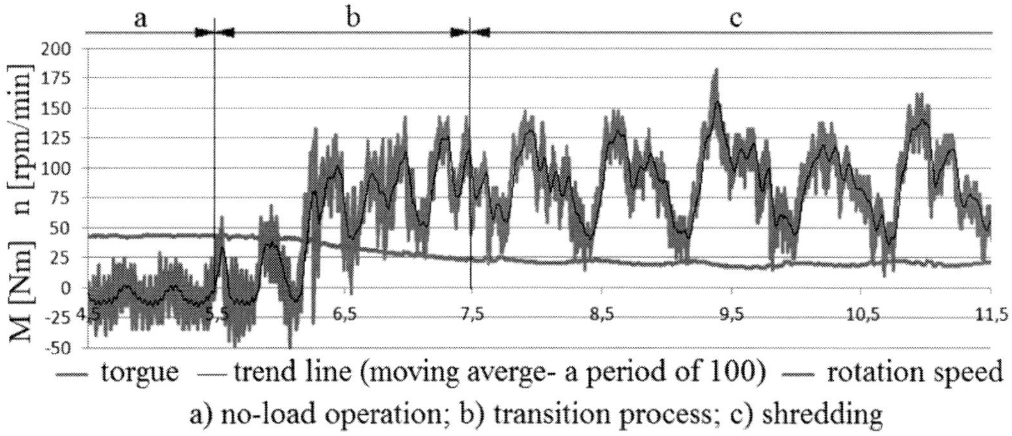

Fig. 3. Torque and rotation speed characteristics as a function of time, chipper working unit in different phases of work, test 2

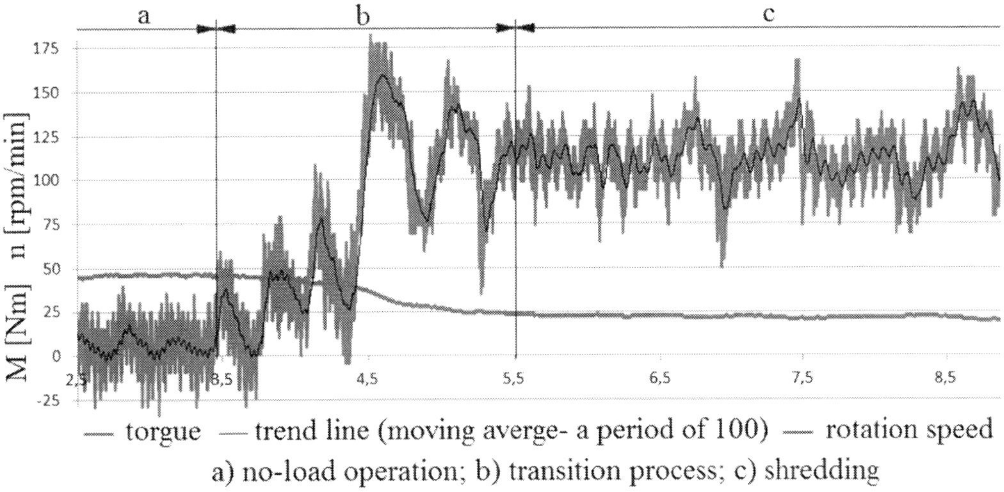

Fig. 4. Torque and rotation speed characteristics as a function of time, chipper working unit in different phases of work, test 3

4 Analysis of results

Actual dynamic characteristics of the rotary speed and torque of the MQS2800 Macalister chipper working unit are characterized by variability and signal distortion. However, the main character of the change is possible to be defined.

An unloaded working unit generates a constant speed of about 45 rpm. The generated torque oscillates between 0 and 10 Nm, due to the structure which is unbalanced as shown in Figure 2, and is additionally exposed to imbalances associated with blade defects created during exploitation.

The process of chipping is characterized by unique characteristics that result from the process of cutting orthotropic plant materials whose structure is an anisotropic. During the chipping of the aforementioned sample, the rotational speed decreases almost twice and the torque in the extremes reaches more than 150 Nm (Figures 5 and 6). However, due to the position of the tree grains during the chipping process, differences in the main character of the changes can be observed. The process of chipping when the tree grains of the workpiece is positioned as shown in Figure 5 generates a significantly different nature than the position of the sample in Figure 6. The first position of the sample generates a dynamic increase in the load which, after exceeding the extremity, significantly decreases the generated resistance up to 40 Nm. The reason for such a characteristic is the cracking of the wood which in this position is prone to layering, while generating evenly shredded chips. The second position of the sample to be processed results in a milder increase in the maximum torque and offload. However, in this position the minimum torque does not drop below 60 Nm and the chip is not cut The collected characteristics indicate that there is no significant cracking of the material at this position in front of the cutting edge. In addition, during this process the characteristics may also be affected by the fact that some of the fragmented parts may return to the work area and be re-fragmented. For the principal character of the changes (Figs. 5 and 6), the most common characteristics of the cut material were chosen for the two successive blades of the cutter.

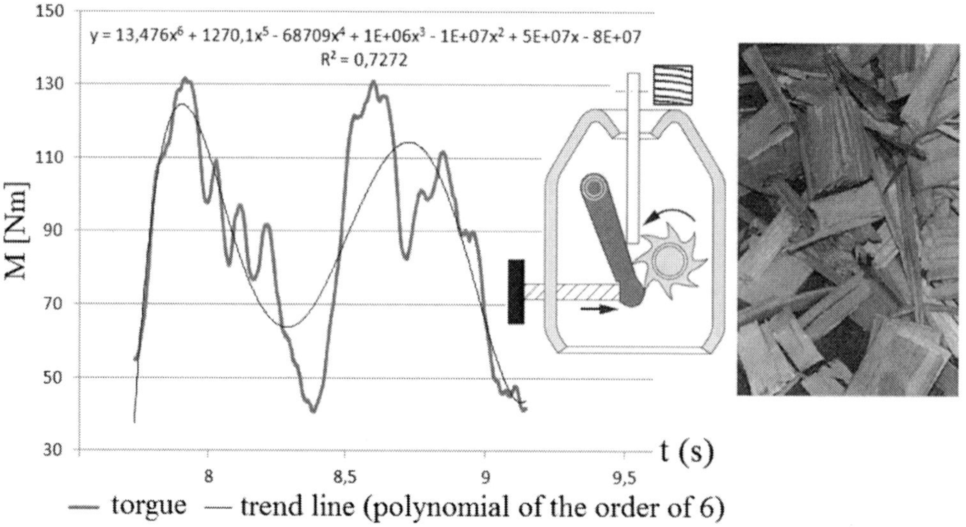

Fig. 5. Characteristics of the main character of torque changes, due to the location of the sample (I) and generated chips

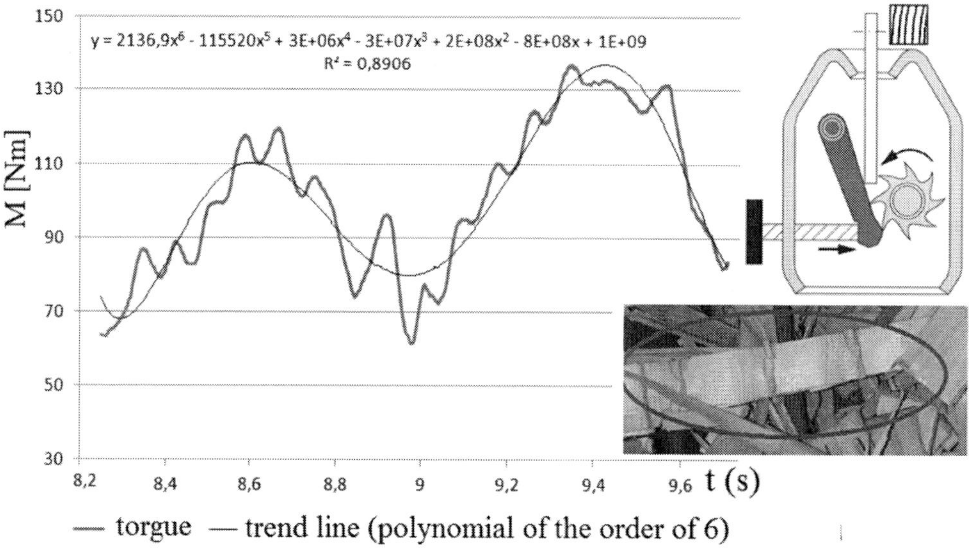

Fig. 6. Characteristics of the main character of torque changes, due to the location of the sample (II) and generated chips

Conclusions

Experimental preliminary research have shown that the constructed research station enables the study of the characteristics of the MQS2800 Macalister chipper working unit, a representative of electric chippers for use in households. The collected characteristics also allow the analysis of the chipping samples not only within their maximum strength values, but also in the processes taking place during cutting. The actual characteristics of the generated torque and the rotational speed can be used in the analysis of the strength of the structure. Modeling simulations of shear component life, i.e., working unit, transmission or drive loaded periodically-variable. These characteristics also allow us to evaluate the efficiency and energy consumption of this type of construction, together with the possibility to evaluate the generated raw material.

References

1. B. Fortuna-Antoszkiewicz, K. Gadomska, K. Gadomski, *Decoration and maintenance of green areas* (original title in Polish: *Urządzanie i pielęgnacja terenów zieleni*), part III, edit II, Hortpress Sp. z o.o., Warszawa (2010)
2. Municipal Office in Gliwice, *Nursing of trees and shrubs in the light of the Nature Conservation Act* (original title in Polish: *Pielęgnowanie drzew i krzewów w świetle ustawy o ochronie przyrody*), https://bip.gliwice.eu/strona=10612, March (2017)
3. H.B. Szczepanowska, *Good practices in the development of urban space with green infrastructure, especially with trees* (original title in Polish: *Dobre praktyki w zakresie zagospodarowania przestrzeni miejskiej zieloną infrastrukturą, zwłaszcza drzewami*) Miasto idealne - miasto zrównoważone, Planowanie przestrzenne terenów zurbanizowanych i jego wpływ na ograniczenie skutków zmian klimatu, edited by Kalinowskiej A., part II, B. Podejście ekosystemowi do miast- usługi na rzecz adaptacji do zmian klimatu, 201-209, Warszawa, ISBN 978-83-917679-6-2, (2015)
4. S. Tombesi, D. Farinelli, M. Molfese, M. Cipolletti, T. Visco, *Pruning technique in young high density hedgerow olive orchards*. Conference: 7th International Symposium on Olive Growing, Location: San Juan, ARGENTINA, Date: SEP 25-29, 2012. VII INTERNATIONAL SYMPOSIUM ON OLIVE GROWING, Book Series: Acta Horticulturae **1057**, 385-390 (2014)
5. V. Adamchuk, V. Bulgakov, N. Skorikov, T. Yezekyan, J. Olt, *Developing a new design of wood chopper for grape vine and fruit tree pruning and the results of field testing*. Agronomy Research **14** (5), 1519-1529 (2016)
6. A.B. Dias, S. Patrocinio, S. Pereira, T. Brites, V. Pita, J.M.M. Barroso, *Evaluation of the use of a disc-saw machine in winter pruning 'Rocha' pears orchards - an account of five years*. Conference: 12th International Pear Symposium, Location: Leuven, BELGIUM, Date: JUL 14-18, 2014, XII INTERNATIONAL PEAR SYMPOSIUM, Book Series: Acta Horticulturae **1094**, 281-288 (2015)
7. J. Przybylko, *The applications of autonomous systems to forestry management*. Massachusetts Institute of Technology, http://hdl.handle.net/1721.1/81014 (2013)
8. P.J.C. Stassen, *Development of pruning and training systems for higher density peach orchards in South Africa*. Conference: 8th International Peach Symposium, Location: Matera, ITALY, Date: JUN 17-20, 2013-2016, VIII INTERNATIONAL PEACH SYMPOSIUM Book Series: Acta Horticulturae **1084**, 343-350 (2015)
9. B. Velazquez-Marti, E. Fernandez-Gonzalez, *The influence of mechanical pruning in cost reduction, production of fruit, and biomass waste in citrus orchards*. APPLIED ENGINEERING IN AGRICULTURE **26** (4), 531-540 (2010)
10. M. Carlini, S. Castellucci, S. Cocchi, A. Manzo, *Waste wood biomass arising from pruning of urban green in Viterbo town: energy characterization and potential uses*. International Conference on Computational Science and Its Applications: ICCSA 2013, Part II, LNCS 7972, 242-255, DOI: 10.1007/978-3-642-39643-4_19 (2013)
11. G.Q. Luo, D.S. Chandler, L.C.A. Anjos, R.J. Eng, P., Jia, F.L.P. Resende, *Pyrolysis of whole wood chips and rods in a novel ablative reactor*. FUEL **194**, 229-238, DOI: 10.1016/j.fuel.2017.01.010 (2017)

12. N. Striugas, K. Zakarauskas, A. Džiugys, R. Navakas, R. Paulauskas, *An evaluation of performance of automatically operated multi-fuel downdraft gasifier for energy production*. Applied Thermal Engineering **73** (1), 1151-1159 DOI: 10.1016/j.applthermaleng.2014.09.007 (2014)
13. J. Calatrava, J.A. Franco, *Using pruning residues as mulch: Analysis of its adoption and process of diffusion in Southern Spain olive orchards*. Journal of Environmental Managment **92**, (3), 620-629, DOI:10.1016/j.jenvman.2010.09.023 (2011)
14. H.T. Sahin, M.B. Arslan, *Properties of orchard pruning and suitability for composite production*. Science and Engineering of Composite Materials **20** (4), 337-342 DOI: 10.1515/secm-2012-0033 (2013)
15. D. Timmons, C.V. Mejía, *Biomass energy from wood chips: Diesel fuel dependence?* Biomass and Bioenergy **34** (9), 1419-1425, DOI: 10.1016/j.biombioe.2010.04.010 (2010)
16. L.J. Naimi, S. Sokhansanj, S. Mani, M. Hoque, T. Bi, A.R. Womac, S. Narayan, *Cost and performance of woody biomass size reduction for energy production*. CSBE/SCGAB Annual Conference, Edmonton Alberta, July 16 - 19, 2006, paper **06-107**, 1-13 (2006)
17. R. Abdallah, S. Auchet, P.J. Méausoone, *Experimental study about the effects of disc chipper settings on the distribution of wood chip size*. Biomass and Bioenergy **35** (2), 843-852, DOI: 10.1016/j.biombioe.2010.11.009 (2011)
18. L.M. Hellström, P.A. Gradin, T. Carlberg, *A method for experimental investigation of the wood chipping process*. Nordic Pulp & Paper Research Journal **23** (3), 339-342 (2008)
19. R. Kuryjański, *Machining and machine tools* (original title in Polish: *Obróbka skrawaniem i obrabiarki*), Copyright Politechnika Warszawska, ISBN 83-89703-72-6 (2011)
20. K. Orlowski, T. Ochrymiuk, *A newly-developed model for predicting cutting power during wood sawing with circular saw blades*. Maderas. Cienc. tecnol. **19** (2), 149-162 (2017)
21. W. Spychała, Ł. Łowiński, T. Pawłowski, Z. Zbytek, *Simulation research of impast of cutting resistance on machine frame*. Journal of Research and Applications in Agricultural Engineering **55** (2), 134-137 (2010)
22. M. Krajnc, B. Dolsak, *Computer and experimental simulation of biomass production using drum chipper*. International Journal of Simulation Modelling **12** (1), 39-49, ISSN 1726-4529, DOI:10.2507/IJSIMM12(1)4.223 (2013)
23. M. Merhar, B. Bučar, *Cutting force variability as a consequence of exchangeable cleavage fracture and compressive breakdown of wood tissue*. Wood Science and Technology **46** (5), 965-977, doi:10.1007/s00226-011-0457-4 (2012)
24. C. Nati, R. Spinelli, *How blade wear of chippers can affect fuel consumption and wood chip size distribution*, Forest engeneering: Meeting the Needs of the Society and the Environment, FORMEC 2010, July 11–14, 2010, Padova – Italy, 1-7 (2010)
25. C. Nati, R. Spinelli, P. Fabbri, *Wood chips size distribution in relation to blade wear and screen use.* Biomass and Bioenergy **34** (5), 583-587, DOI: 10.1016/j.biombioe.2010.01.005 (2010)
26. R. Spinelli, N. Magagnotti, G. Paletto, C. Preti, *Determining the impact of some wood characteristics on the performance of a mobile chipper*. Silva Fennica **45** (1), 85-95 ISSN 0037-5330 (2011)

Analysis of the influence of the cutting edge geometry on parameters of the perforation process for conveyor and transmission belts

Dominik Wojtkowiak[1,*], *Krzysztof* Talaśka[1], *Ireneusz* Malujda[1], *Grzegorz* Domek[2]

[1]Chair of Basics of Machine Design, Poznan University of Technology, Piotrowo 3, 60-965 Poznań, Poland
[2]Faculty of Mathematics, Physics & Technical Sciences, Kazimierz Wielki University in Bydgoszcz, 85-092 Bydgoszcz, Poland

Abstract. Perforated belts, which are used in vacuum conveyor belts, can have significantly different mechanical properties like strength and elasticity due to a variety of used materials and can have different thickness from very thin (0,7 mm) to thick belts (6 mm). In order to design a complex machine for mechanical perforation, which can perforate whole range of belts, it is necessary to research the influence of the cutting edge geometry on the parameters of the perforation process. Three most important parameters, which describe the perforation process are the cutting force, the velocity and the temperature of the piercing punch. The results presented in this paper consider two different types of punching (a piercing punch with the punching die or with the reducer plate) and different cutting edge directions, angles, diameters and material properties. Test were made for different groups of composites belts – with polyurethane and polyester fabric, polyamide core or aramid-fibre reinforced polymers. The main goal of this research is to specify effective tools and parameters of the perforation process for each group of composites belts.

Keywords: conveyor belts, timing belts, punching, polymer composites, perforation.

1 Introduction

Conveyor belts are used to transport goods from one point to another and most commonly are made as multilayer composite flat belts. Basically there are two main types of conveyor belts: strong rigid belts with polyamide or aramid cord traction layer (Fig. 1a) and light elastic belts, which are made of polyurethane, polyester fabric or PVC (Fig. 1b) [1 3].

Timing belts (Fig. 1c) are toothed belts, which are used to synchronize a rotation of two toothed wheels or to convert a rotary motion of the shaft into a reciprocating motion of the carriage. Timing belts are made of polyurethane or rubber as a base material with tension member, which can be made of either aramid or glass fibers or even steel cord for applications with higher loads. In order to gain specific properties (e.g. lower friction, silent

* Corresponding author: dominik.wojtkowiak@put.poznan.pl

Reviewers: *Milan* Vaško, *Ján* Vavro

operation) both sides of the timing belt can be coated with various materials and structures [4-6].

Vacuum belts (Fig. 1d) are a highly elegant solution to transport light products, such as paper, foil, carton etc., as individual items due to the fact that their low mass makes them unsuitable for conveying by friction. That is why vacuum belt conveyors are commonly used in paper manufacturing and converting, printing machines, food industry, pharmaceutical industry and as conveyors for material handling. This type of belts can be made from both flat and toothed belts by using proper modifications, which are the perforation and the milling. During the perforation process cylindrical holes, which are used to suck the air through the belt, are machined. Milling can be used to create slots in order to increase suction surface area or to create flat suction track by machining part of teeth [4-6].

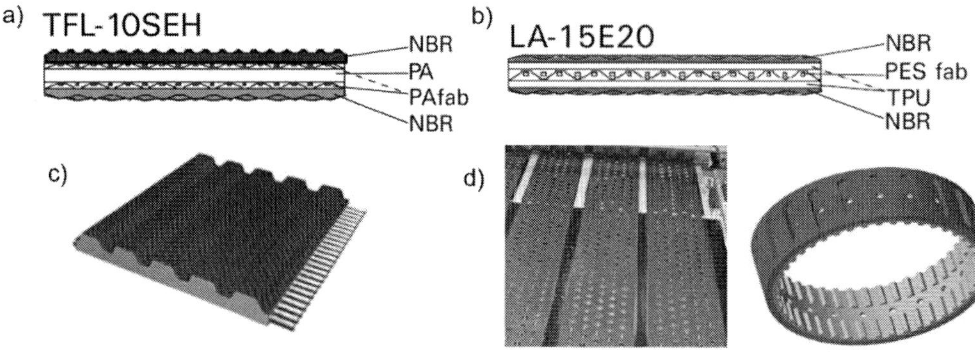

Fig. 1. Conveyor, timing and vacuum belts: a) Nitta PolyBelt with polyamide traction layer b) Nitta PolySprint with polyester fabric and polyurethane, c) BRECO polyurethane timing belt with steel tension member and polyamide fabric coating, d) vacuum belts (Norditec and NK Technics)

The most important stage of a vacuum belt manufacturing process is perforation. This paper shows the analysis of the influence of the cutting edge geometry, the temperature and the velocity of a piercing punch on the cutting force necessary to perforate a composite belt and on the quality of holes.

2 Methods of belt perforations

Basically we can distinguish four different perforation methods, which are suitable for polymer composites: laser or abrasive waterjet cutting (nonmechanical methods) and drilling or punching holes (mechanical methods). Nonmechanical methods have an advantage over the mechanical ones that there is no contact forces and abrasive wear of the tool, however its initial and exploitation cost are much higher [7].

Laser cutting allows to cut precisely very small holes with smooth edges and high speed, but due to the high temperature it may change the properties and lower the strength of a material in heat affected zone (HAZ). Moreover laser cutting may cause thermal deformation such as folding of the coating and delamination, changing belt colour or emitting dangerous substances – during the machining of aramid-fibre reinforced materials large amount of hydrogen cyanide are produced [7-11].

With abrasive waterjet (AWJ) technology it is possible to cut almost all materials, and it gives similar quality of the hole to laser cutting without any thermal deformation, but the velocity of cutting is low and it is necessary to drain water after cutting and dry a belt [7, 10-12].

Drilling has an ease of positioning, but increasing temperature during the process may cause thermal deformation and the characteristic load distribution during entering and exiting the material leads to delamination of the composite belt and creating uneven edges of a hole. Most of the disadvantages can be reduced or even eliminated by properly choosing the drilling tool and parameters of the drilling (feed rate, cutting speed), but it must be made for each case individually [7, 13, 14].

The most suitable method for mass production of perforated belt will be mechanical working using a punching die with a piercing punch, which can cooperate with the piercing die or the reducer plate. Punching is a cheap and simple mechanical machining process, which can be performed with high velocity and low thermal influence on a perforated belt. Holes created by punching can have different contours depending on the shape of a piercing punch. When considering multilayer and fibre-reinforced composites in a punching process, there is a chance that delamination or fibre pull-out may occur, but by choosing proper parameters and tool it can be prevented. There are three basic types of piercing punches: cylindrical with the punching die (Fig. 2a), hollow with the cutting edge towards inside (Fig. 2b) and hollow with the cutting edge towards outside (Fig. 2c) [15-18].

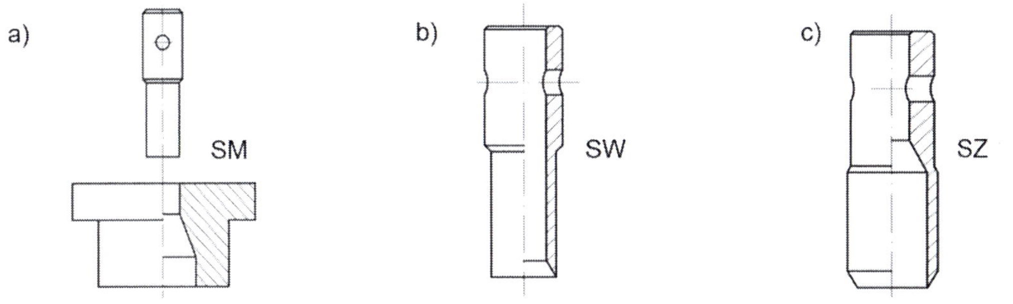

Fig. 2. Basic types of piercing punches: a) piercing punch with the piercing die, b, c) hollow piercing punch which cooperate with the reducer plate with the cutting edge towards inside (b) or outside (c)

3 Methodology of research

Perforation research was made by using the designed punching die (Fig. 3) and strength testing machine MTS. The presented construction allows to use different types of piercing punches and heat them up to 200°C. However preliminary research indicated, that the temperature of the piercing punch higher than 150°C causes permanent thermal deformations of the perforated belt. That is why all tests were performed for 4 different temperatures of the working tool: 25°C (room temperature), 50°C, 100°C and 150°C. Default value of the velocity of the piercing punch in all tests, except the one where the influence of velocity on the cutting force was checked, was 0,5 mm/s. During the research, by changing the setting of testing machine, it was possible to measure the cutting force for different (Nitta) belts, cutting edge geometry (outer diameter, angle of the cutting edge and type of a piercing punch), temperature and velocity. The results are presented following the key presented in Fig. 4.

a) b)

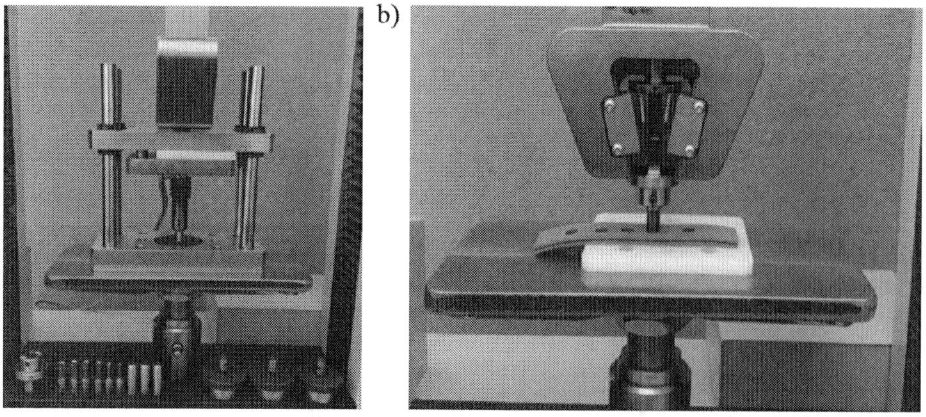

Fig. 3. Punching diees with all used tools during performed tests: a) piercing punch with the piercing die, b) hollow piercing punch which cooperates with the reducer plate

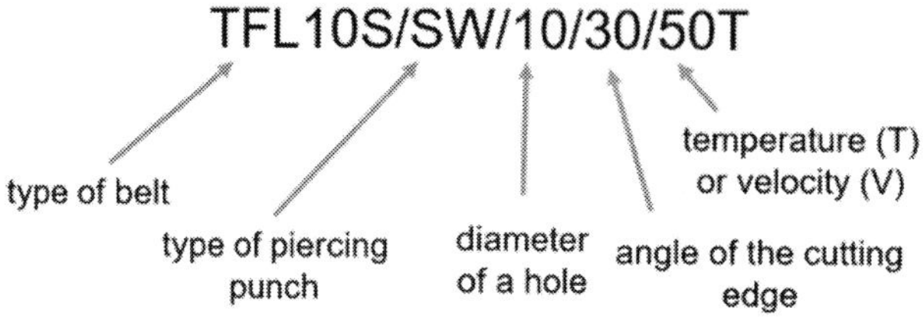

Fig. 4. Key used in signification of performed tests results

4 Results and discussion

First series of tests were made in order to determine the influence of the tool geometry on the value and characteristic of the cutting force. Tests were performed with default velocity and room temperature for 7 different piercing punches with diameters of 10 mm and for two belts, which represent both groups of conveyor belts – strong rigid belt with polyamide core TFL10S and light elastic belt with polyester fabric, as a traction layer LAB12E. Obtained results are presented as a cutting force in a function of a displacement on the charts in Fig. 5 and Fig. 6.

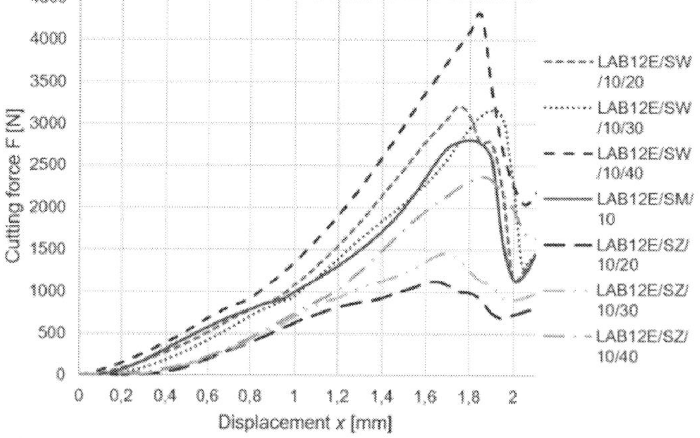

Fig. 5. Cutting force during perforation for different types of tools with diameter of 10 mm and various tool angles for PolySprint Nitta LAB12E

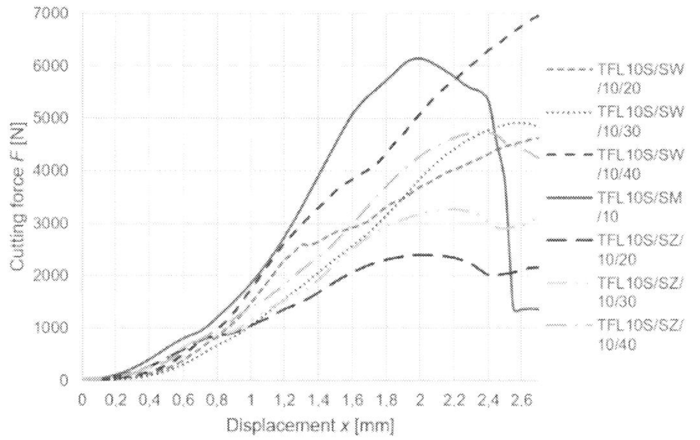

Fig. 6. Cutting force during perforation for different types of tools with diameter of 10 mm and various tool angles for PolyBelt Nitta TFL10S

By analysing shown characteristics, it is visible that in both cases the lowest amount of the cutting force was used, while perforating with the hollow piercing punch with the cutting edge towards outside (SZ). This phenomenon can be explained by describing the changes of the cutting force during the punching process for each type of the piercing punch. In case of the LAB12E belt (Fig. 5) the main strength depends on the polyester fabric, and this is the reason why there is a sudden drop in the force value at approximately 1,8 mm of thickness. By comparing SZ and SW types of the piercing punch it is clear that cutting through the fabric is much smoother and needs about 2 times lower force when the cutting edge is towards outside. This is the consequence of the type of stress, which occurs for different types of tools. If we use the SZ piercing punch, scrap must be compressed with the decrease of the inner diameter in a tool, while in the SW piercing punch belt it must be extended with the increase the outer diameter of a tool – inner diameter remains constant. Polyester fabric has an orthogonal properties, which depends on the configuration of the woven in the structures. It is much more elastic in the direction along belt axis, because in normal work conditions the conveyor belt is meant to carry tensile stress.

The piercing punch, which cooperates with the piercing die (SM) is placed in the middle, because in that case the main stress taking place in the perforation process, is the shearing stress. In that case the stage of the perforation process occurring just before the sudden drop of the cutting force, is connected with friction resistance during pushing the material through the channel in the piercing die. In all types of punching additional compression occurs due to the thrust force, which is the effect of the feed of the piercing punch. That is why there is a bending point, characteristically at 0,8 – 1 mm of displacement, which corresponds to the elastic-plastic compression model with reinforcement of the material, which was verified during compression tests made on this belt in preliminary tests.

On the other hand, if we observe the results for the strong rigid belt with polyamide film core TFL10S (Fig. 6), we can see a clear fall of the cutting force at the end of the process only for the piercing punch with the piercing die. The main reason for this may be the fact, that the cut through the belt occurs after the rupture of the hard polyamide film structure, which is not as sudden as in polyester fabric. Due to higher resistance and greater compression stress than for the elastic belt, the piercing punch must be plunged deeper into the reducer plate. Due to the similar properties of the core of TFL10S and the reducer plate made of PA6 or PP, the cutting force is reduced slightly or is not reduced at all after finishing the perforation process. The main difference between the belt LAB12E and TFL10S is that the highest value of force is necessary for the piercing punch with the piercing die due to higher Young modulus for compression in strong rigid belts. Other relations described for elastic belts correspond with the results for the belt TFL10S.

Another conclusion which can be observed on both charts is that decreasing the tool angle causes the fall of the cutting force, but it is not proportional. That is why the tool angle 30° can be taken as an effective one, because it is a compromise between the cutting force and the durability of the cutting edge. The obtained results also provide information that proper stress of the belt during the perforation process and the wear of the piercing punch have an effect on the belt perforation, especially when machining elastic conveyor belts. We can see, that regardless of the type of the piercing punch or type of the belt, the perforation process is a really complex issue and many factors have an influence on the cutting force.

The second part of the research was made to determine, if velocity has influence on the cutting force or not. The obtained results are presented in Fig. 7 and Fig. 8. For elastic belts the influence of velocity is insignificant, while for polyamide belts higher speed may reduce the cutting force, but not in a linear way. It is necessary to extend the tests in future research in order to draw more specific conclusions.

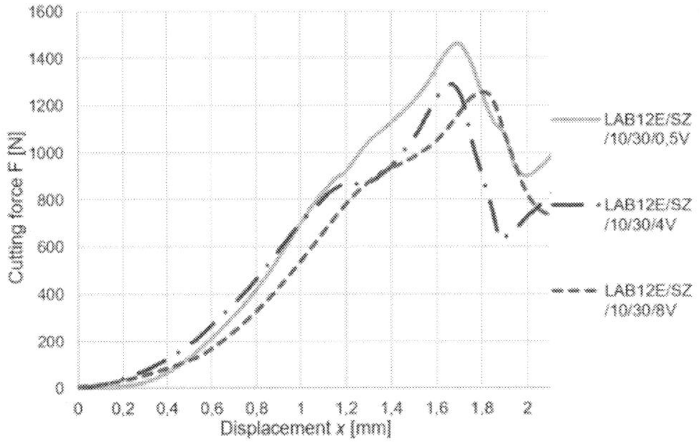

Fig. 7. Comparison of the cutting force during perforation by using the piercing punch with diameter of 10 mm, the tool angle 30° and the cutting edge towards outside with various perforation velocity for PolySprint Nitta LAB12E

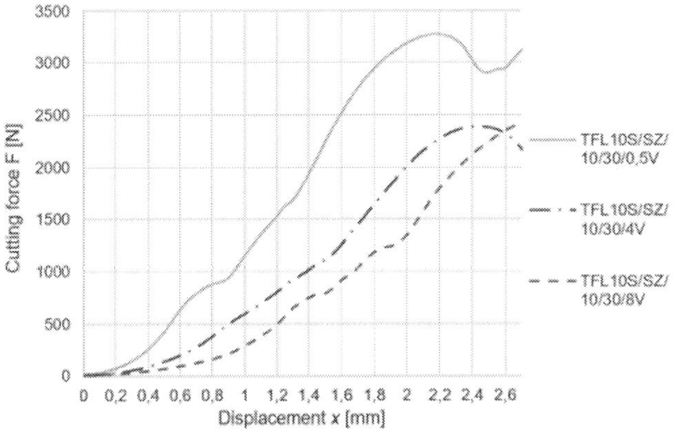

Fig. 8. Comparison of the cutting force during perforation by using the piercing punch with diameter of 10 mm, the tool angle 30° and the cutting edge towards outside with various perforation velocity for PolyBelt Nitta TFL10S

During the machining of the polymer composites increased temperature is almost always treated as a disadvantage of the method due some thermal deformation and mechanical properties changes. On the other hand properly controlled heat can be very helpful, as it softens the material, which makes it easier to machine. That is why, it is crucial to determine the influence of the increased temperature on the cutting force in punching of conveyor belts. The obtained results are presented in Fig. 9, 10, 11 and 12.

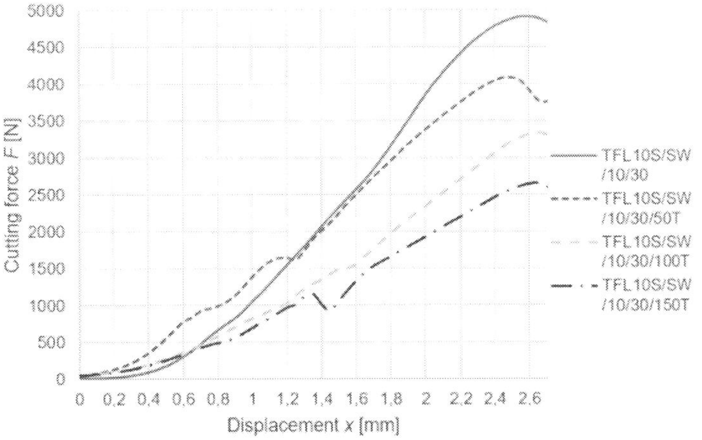

Fig. 9. Comparison of the cutting force during the perforation by using piercing punch with diameter of 10 mm, the tool angle 30° and the cutting edge towards inside, with various perforation temperature for PolyBelt Nitta TFL10S

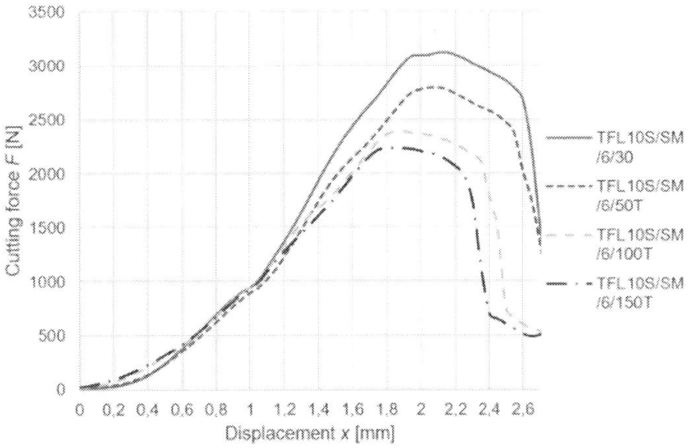

Fig. 10. Comparison of the cutting force during perforation by using the piercing punch with diameter 6 mm with the punching die, with various perforation temperature for PolyBelt Nitta TFL10S

If we take into account the strong rigid belt TFL10S (Fig 9, 10 and 11), it is visible that increased temperature of the piercing punch can decrease the amount of force necessary to perform perforation of the composite belt. To properly analyse the results punching must be divided into two stages: pure cutting with thrust compression of the belt, and cutting with additional resistance due to axial and transverse compression of the scrap, extension of the belt or friction in the channel of the piercing die. It is visible, that by using the piercing punch with the reducer plate, cutting force is lower already in the first stage, which is connected with penetration into the softened material – it is easier to compress the scrap or extend the belt. On the contrary to that, while using the piercing die differences are visible only in the final stage – it is easier to push softened scrap through the die channel. To overall maximum cutting force is reduced in all three cases and the decrease is approximately linear. A further temperature increase causes thermal deformation of the belt, which is why it is established that temperature 100°C - 150°C can be taken as an effective value for belt perforation.

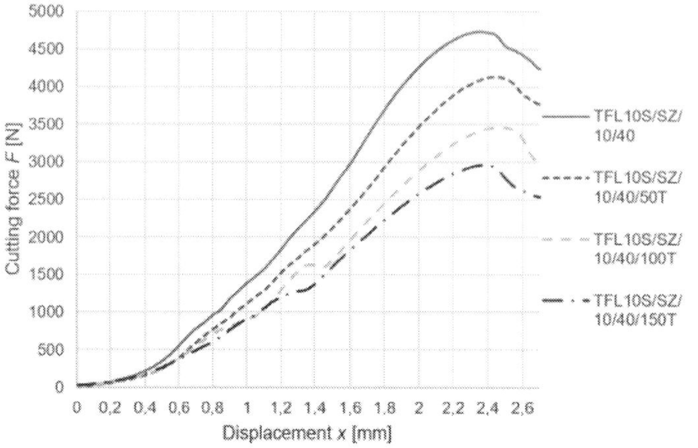

Fig. 11. Comparison of the cutting force during perforation by using the piercing punch with diameter of 10 mm, the tool angle 40° and the cutting edge towards outside, with various perforation temperature for PolyBelt Nitta TFL10S

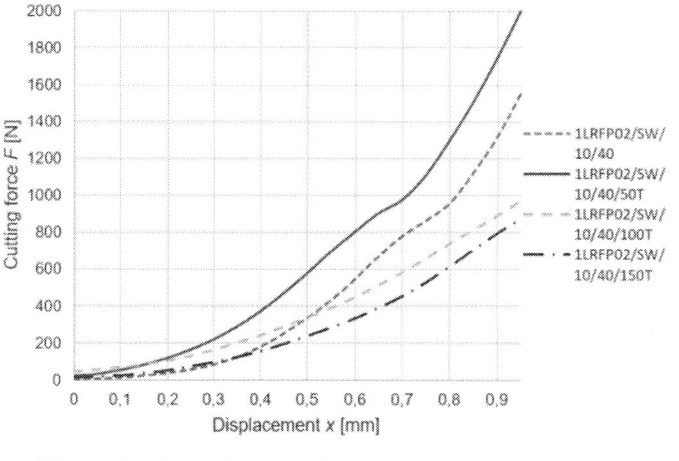

Fig. 12. Comparison of the cutting force during perforation by using the piercing punch with diameter of 10 mm, the tool angle 40° and the cutting edge towards inside, with various perforation temperature PolySprint Nitta 1LRFP02/W2 IM LSQ

On the other hand not all the material will behave the same way and in some cases increasing the temperature may even cause the increase of the force value. The example is the very thin and elastic conveyor belt made of thermosetting polyurethane 1LRFP02/W2 IM LSQ (Fig. 12). In that case temperature 50°C caused the hardening of the belt and as a result an increase of the cutting force. Using higher temperatures gave positive results, but the differences were very small due to high elasticity and almost pure shearing of the belt with compression resistance close to zero. It can be stated, that heating up the working tool has a positive effect in strong rigid polyamide belt perforation, while for light elastic conveyor belts the effect is close to neutral.

While analysing the perforation process it is very important to consider the quality of the holes along with the cutting force measurement. In Fig. 13. typical defects, which occurred during the perforation process of the conveyor belts are presented.

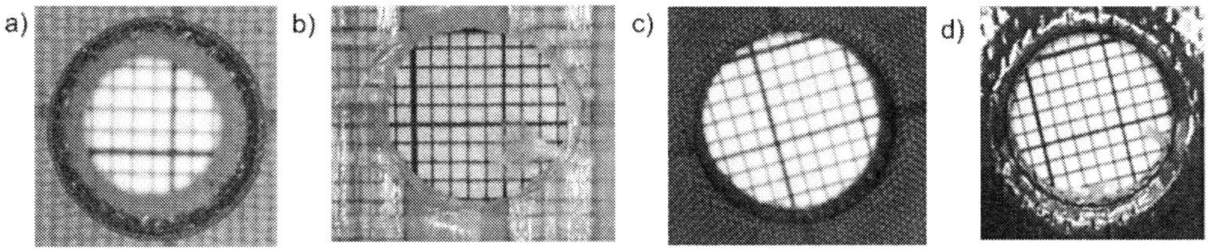

Fig. 13. Defects of perforated belts: a) conicity of the hole, which occurs in thick polyamide belts, b) ovality of the hole for belts with different axial and lateral elasticity and fibre pull-out for belts with fabrics and cords, c) folding of the coating for higher temperatures when cutting edge tilt towards outside, d) deformation of the belt for elastic PVC belts

While perforating thick belts with a polyamide core, the visible difference between entry and exit diameter occurs due the conicity of the hole along with fuzzy edges and uneven side surface of a hole (Fig. 13a and 13c). If the belt was reinforced with a fabric or a cord the pull-out of the fibres occurred (Fig. 13b and 13d). Too high temperature of the piercing punch caused permanent thermal deformation for elastic PCV belt (Fig. 13d) or folding of the coating along the edge of a hole (Fig 13c). Significant differences in the elasticity in longitudinal and traverse direction for light belts caused high roundness deviation whose result was an oval hole (Fig. 13b and 13d). The above-mentioned defects can be minimalized or eliminated by properly chosen perforation process parameters. For example, heating up the piercing punch, as well as using the hollow piercing punch with the cutting edge towards outside, almost eliminated the conicity of the hole. That is why extended research of these phenomena is necessary in order to develop the punching of the vacuum conveyor belts and specify effective process parameters (velocity, temperature, tool geometry) for different types of belts.

Conclusions

Based on the results, regardless of the belt type, the lowest value of the cutting force is necessary, when we use the piercing punch with the reducer plate and the cutting edge towards outside. By decreasing the tool angle, the fall of the cutting force is visible, but it is not proportional. The tool angle 30° is the effective one, because it is the compromise between the cutting force and the durability of the cutting edge. It was very visible that elastic belts are much more sensitive to the wear of the blade than strong polyamide belts, which is caused by the different cutting – compressing ratio of the perforation process. We can also conclude that for elastic belts the influence of velocity is insignificant, while for polyamide belts higher speed may reduce the cutting force. By increasing the temperature of the piercing punch it is possible to lower the cutting force and improve the quality of the holes (cylindrical holes) for polyamide belts, but for elastic thermosetting polyurethane belts 50°C may even harden the material and increase the cutting force or may cause the coating deformation. Increasing the temperature will affect mainly the compression phase, which is why the differences are visible mainly in the second stage of perforation (compressing scrap material inside the hollow piercing punch or pushing it through the piercing die channel). The defects discussed in this paper can be removed by properly choosing the tools and parameters of the perforation process, and this is why the research is so significant for the industry.

The presented results have been obtained in the course of realising the project "*Opracowanie innowacyjnej technologii systemu precyzyjnej mechanicznej perforacji pasów do transportu podciśnieniowego i sterowania optycznego – systemu automatyzującego proces produkcyjny pasów transportowych*", which is funded by Wilhelm Herm. Müller Polska Sp. z.o.o and Polska Agencja Rozwoju Przedsiębiorczości (PARP).

References

1. M. Hardygóra. *Taśmy przenośnikowe*. Wydawnictwo Naukowo-Techniczne, Chapter **1** and **2**, Warszawa (1999)
2. L. Gładyszewicz. *Przenośniki Taśmowe – teoria i obliczenia*. Wydawnictwo Politechniki Wrocławskiej (2003)
3. Wilhelm Herm. Müller, *Nitta Industries* – polish catalog. In: www.whm.pl (Access date: 06.07.2017)
4. R. Perneder, I. Osborne. *Handbook Timing Belts – Principles, Calculations, Applications*. Springer (2012)
5. Optibelt. *Technical manual polyurethane timing belts* – catalog. In: www.optibelt-usa.com (Access date: 06.07.2017)
6. Habasit. *HabaSYNC Timing Belts* – product media No. 4107. In: www.habasit.com (Access date: 06.07.2017)
7. J. Sheikh-Ahmad. *Machining of polymer composites. Springer (2009)*
8. G. Chryssolouris. K. Salonitis. *Fundamentals of laser machining of composites*, In: *Machining Technology for Composite Materials*, Chapter **10**, 266-287 (2011)
9. R. Negarestani, L. Li. *Laser machining of fibre-reinforced polymeric composite materials*. In: *Machining Technology for Composite Materials*, Elsevier, Chapter **11**, 288-308 (2011)
10. D. Krajcarz. *Comparison Metal Water Jet Cutting with Laser and Plasma Cutting*. Procedia Engineering **69**, 838-843 (2014)
11. A. Skoczylas, *Analiza porównawcza procesu cięcia wiązką laserową i strumieniem wodno- ściernym*. Postępy Nauki i Techniki **8**, 121-128 (2011)
12. D. Krajcarz, S. Spadło. *Wpływ wybranych parametrów procesu cięcia strugą wodno-ścierną na odchyłkę okrągłości otworów cylindrycznych*. Mechanik **01/2017**, 64-65 (2017)
13. F. Lachaud, R. Piquet, F. Collombet, L. Surcin. *Drilling of composite structures*. Composite Structures **52**, 511-516 (2001)
14. T. Leppert, T. Paczkowski, R. Polasik, D. Serwacki. *Delaminacja materiału kompozytowego włóknistego podczas wykonywania otworów*. Mechanik **10/2016**, 1422-1423 (2016)
15. I. Suchy. *Handbook of Die Design*. McGraw-Hill Companies (2006)
16. H.Y. Chan, A.B. Abdullah, Z. Samad. *Precision punching of hole on composite panels*. Indian Journal of Engineering & Material Sciences **22**, 641-651 (2015)
17. F. Lambiase, M. Durante. *Mechanical behavior of punched holes produced on thin glass fiber reinforced plastic laminates*. Composite Structures **173**, 25-34 (2017)
18. M.S.M. Zain, A.B. Abdullah, Z. Samad. *Effect of puncher profile on the precision of punched holes on composite panels*. International Journal Of Advanced Manufacturing Technology **89**, 3331-3336 (2017)

Inelastic damage using continuum damage mechanics in composite plate reinforced by unidirectional fibers

Milan Žmindák[1,*], *Martin* Dudinský[1]

[1]University of Žilina, Faculty of Mechanical Engineering, Department of Applied Mechanics, 010 26 Žilina, Slovak Republic

Abstract. It is well that a finite element method is very popular simulation method to predict the physical behavior of systems and structures. In the last years an increase of interest in a new type of numerical methods known as meshless methods was observed. The paper deals with application of radial basis functions on modelling of inelastic damage using continuum damage mechanics of layered plate composite structures reinforced with long unidirectional fibers. For numerical simulations of elastic-plastic damage of layered composite plates own computational programs were implemented in MATLAB programming language. We will use the Newton-Raphson method to solve nonlinear systems of equations. Evaluation damage during plasticity has been solved using return mapping algorithm. The results of elastic-plastic damage analysis of composite plate with unsymmetrical laminate stacking sequence are presented.

Keywords: Radial point interpolation method, continuum damage mechanics, layered composite plates, elastic-plastic damage

1 Introduction

Carbon fiber-reinforced composite materials are finding increased application in many areas [1]. The finite element method (FEM) is one of the most widely used and most popular numerical methods for damage of plate structures [2, 3]. In the last years an increase of interest in a new type of numerical methods known as meshless methods or meshfree was observed [4]. Meshless approaches for problems of continuum mechanics have attracted much attention during the past decade especially owing to their high adaptivity and low costs to prepare input data for numerical analyses. One of the areas where meshless methods are convenient to use is the analysis of plate and shell structures [5].

This paper deals with the application of radial basis function (RBF) on modelling of inelastic damage using continuum damage mechanics (CDM) of layered plate composite structures reinforced with long unidirectional fibers. For numerical simulations of elasto-plastic damage of layered composite plates own computational algorithms were

* Corresponding author: milan.zmindak@fstroj.uniza.sk

Reviewers: *Radim* Halama, *Jerzy* Winczek

implemented into the MATLAB programming language. For solving a system of nonlinear system equations the Newton-Raphson method was used. The evaluation of damage during plasticity has been solved using return mapping algorithm described in [6]. The results of elasto-plastic damage of a composite plate with unsymmetrical laminate stacking sequence are also presented.

2 Point interpolation methods

One of the common characteristics of all mesh-free methods is their ability to construct functional approximation or interpolation entirely from information at a set of scattered nodes or particles, among which there is no pre-specified connectivity or relationships. The point interpolation method (PIM) was proposed by Liu [7] to replace moving least square (MLS) approximation for creating shape functions in meshfree settings. In contrast to mesh-based methods, algorithms based on collocation methods are frequently called meshless or meshfree methods. The terminology "meshless" is not always clear, and it is sometimes also used for Galerkin or Petrov-Galerkin methods [8]. In both cases, a variational formulation of a physical problem is solved in an average sense around node points. However, these types of methods still require geometrical information about the connectivity of the nodes, i.e. the volume surrounding a node, and are therefore not true meshless methods.

The goal of this section is to describe mathematical formulation of computational models based on (RBF) for linear analysis of layered composite structures.

2.1 Radial point interpolation method

The RBF was originally introduced in the 1970s to multivariate scattered data approximation and function interpolation. Many types of meshless RBF methods have been proposed in the literature [9, 10]. Radial point interpolation method (RPIM) can be included among the methods that represent the functions of unknown field variables by series. This method, in addition to the least squares (MLS) method, belongs to the most commonly used interpolation / approximation techniques in meshless methods.

The function $u(\boldsymbol{x})$ defined on the local sub-area Ω_S is interpolated at point $\boldsymbol{x}[x\ y]$ as follows

$$u^h(\boldsymbol{x}) = \sum_{i=1}^{n} R_i(\boldsymbol{x}) a_i = \boldsymbol{R}^T(\boldsymbol{x})\boldsymbol{a}\,, \tag{1}$$

where n denotes the number of nodes supporting interpolation at a given point. These points can be selected using the concept of a local support domain or concept of the domain of influence [8]. Vector $\boldsymbol{a}$ is is vector of unknown functions and R_i is radial basis function with distance r_i defined as

$$r_i = \sqrt{(x - x_i)^2 + (y - y_i)^2}\,. \tag{2}$$

The most widely used RBF in the RPIM are the functions listed in Table 1 [7], where d_N is the average distance between nodes in the local sub/area. In these functions, there are several shape parameters that must be determined prior to performing the analysis. One of the advantages of the shape functions assembled using the RPIM is the property of the Kronecker delta function, which implies that no special techniques are required to introduce the essential boundary conditions. The RPIM which uses only RBF is not polynomial consistent because cannot reconstruct the polynomial fields accurately. For this reason, polynomial members are added to the base functions. The function will be interpolated as follows

Table 1. Typical RBF with dimensionless shape parameter used in RPIM

Full name	Expression	Shape parameters
Multi-quadratic radial base function	$R_i(x,y) = \left(r_i^2 + (\alpha_c d_N)^2\right)^q$	$\alpha_c \geq 0, q$
Gaussian radial base function	$R_i(x,y) = \exp\left[-\alpha_c \left(\frac{r_i}{d_N}\right)^2\right]$	α_c
Thin plate splines	$R_i(x,y) = r_i^\eta$	η
Logarithmic radial base function	$R_i(x,y) = r_i^\eta \log r_i$	η

$$u^h(\boldsymbol{x}) = \sum_{i=1}^{n} R_i(\boldsymbol{x})a_i + \sum_{j=1}^{m} p_j(\boldsymbol{x})b_j = \boldsymbol{R}^T(\boldsymbol{x})\boldsymbol{a} + \boldsymbol{p}^T(\boldsymbol{x})\boldsymbol{b} \tag{3}$$

where a_i is coefficient for radial base and b_j is coefficient for polynomial base $p_j(\boldsymbol{x})$. The number of RBF n is determined by the number of nodes in the support region and the number of polynomial bases m can be selected based on the reproducibility requirement [10].

The coefficients a_i and b_j can be determined from the condition that interpolation passes through all n nodes, which can be written in the matrix form as

$$\boldsymbol{d}_s = \boldsymbol{R}_Q\boldsymbol{a} + \boldsymbol{P}_m\boldsymbol{b}\ , \tag{4}$$

where $\boldsymbol{d}_s$ is vector of function values of the field variable in n nodes. The moment matrix $\boldsymbol{R}_Q$ corresponding to RBF is given as

$$\boldsymbol{R}_Q = \begin{bmatrix} R_1(r_1) & R_2(r_1) & \cdots & R_n(r_1) \\ R_1(r_2) & R_2(r_2) & \cdots & R_n(r_2) \\ \vdots & \vdots & \ddots & \vdots \\ R_1(r_n) & R_2(r_n) & \cdots & R_n(r_n) \end{bmatrix}, \tag{5}$$

where

$$r_k = \sqrt{(x_k - x_i)^2 + (y_k - y_i)^2} \tag{6}$$

is the distance independent of the direction, so it must be meet that $R_i(r_j) = R_j(r_i)$. Then it is valid that matrix $\boldsymbol{R}_Q$ and also matrix $\boldsymbol{G}$ (defined below) are symmetric. Note that matrix $\boldsymbol{R}_Q$ is invertible for arbitrary scattered nodes [11]. The moment matrix $\boldsymbol{P}_m$ is given as:

$$\boldsymbol{P}_m = \begin{bmatrix} P_1(x_1,y_1) & P_2(x_1,y_1) & \cdots & P_m(x_1,y_1) \\ P_1(x_2,y_2) & P_2(x_2,y_2) & \cdots & P_m(x_2,y_2) \\ \vdots & \vdots & \ddots & \vdots \\ P_1(x_n,y_n) & P_2(x_n,y_n) & \cdots & P_m(x_n,y_n) \end{bmatrix}. \tag{7}$$

Polynomial terms must meet extra requirements to guarantee a unique approximation of function. The following restrictions are often used [4]

$$\sum_{i=1}^{n} p_j(x_i,y_i)a_i = 0 \quad j = 1,2,\dots,m\ . \tag{8}$$

Combining Eq. (4) and Eq. (8) yields the following set of equations in the matrix form

$$\begin{bmatrix} \boldsymbol{R}_Q & \boldsymbol{P}_m \\ \boldsymbol{P}_m^T & 0 \end{bmatrix} \begin{Bmatrix} \boldsymbol{a} \\ \boldsymbol{b} \end{Bmatrix} = \begin{Bmatrix} \boldsymbol{d}_s \\ \boldsymbol{0} \end{Bmatrix} \tag{9}$$

or

$$\boldsymbol{G} \begin{Bmatrix} \boldsymbol{a} \\ \boldsymbol{b} \end{Bmatrix} = \begin{Bmatrix} \boldsymbol{d}_s \\ \boldsymbol{0} \end{Bmatrix} . \tag{10}$$

By solving the above system of equations we get the following relation for interpolation in point $\boldsymbol{x}$

$$u(\boldsymbol{x}) = [\boldsymbol{R}^T(\boldsymbol{x})\boldsymbol{S}_a + \boldsymbol{p}^T(\boldsymbol{x})\boldsymbol{S}_b]\boldsymbol{d}_s = \boldsymbol{\varphi}(\boldsymbol{x})\boldsymbol{d}_s , \tag{11}$$

where matrices $\boldsymbol{S}_b$ and $\boldsymbol{S}_a$ are defined as follows

$$\boldsymbol{S}_b = \left[\boldsymbol{P}_m^T \boldsymbol{R}_Q^{-1} \boldsymbol{P}_m\right]^{-1} \boldsymbol{P}_m^T \boldsymbol{R}_Q^{-1} , \tag{12}$$

$$\boldsymbol{S}_a = \boldsymbol{R}_Q^{-1}[1 - \boldsymbol{P}_m \boldsymbol{S}_b] = \boldsymbol{R}_Q^{-1} - \boldsymbol{R}_Q^{-1} \boldsymbol{P}_m \boldsymbol{S}_b . \tag{13}$$

From Eq. 11 the matrix of shape functions $\boldsymbol{\varphi}(\boldsymbol{x})$ is

$$\boldsymbol{\varphi}(\boldsymbol{x}) = [\boldsymbol{R}^T(\boldsymbol{x})\boldsymbol{S}_a + \boldsymbol{p}^T(\boldsymbol{x})\boldsymbol{S}_b] = [\phi_1(\boldsymbol{x}) \;\; \phi_2(\boldsymbol{x}) \;\; \dots \;\; \phi_n(\boldsymbol{x})] . \tag{14}$$

Shape function for k-th node is given as

$$\phi_k(\boldsymbol{x}) = \sum_{i=1}^{n} R_i(\boldsymbol{x}) S_{ik}^a + \sum_{j=1}^{m} p_j(\boldsymbol{x}) S_{jk}^b . \tag{15}$$

The first and the second derivative of shape function we get differentiation (15)

$$\phi_{k,i} = \sum_{c=1}^{n} R_{c,i} S_{ck}^a + \sum_{d=1}^{m} p_{d,i} S_{dk}^b , \tag{16}$$

$$\phi_{k,ij} = \sum_{c=1}^{n} R_{c,ij} S_{ck}^a + \sum_{d=1}^{m} p_{d,ij} S_{dk}^b . \tag{17}$$

In Fig. 1 to Fig.3 is described the shape function as well as its first and second derivations. These functions are assembled for a central node in a square area represented by 9x9 nodes, The domain of influence concept was used with a circular shape for construction of the shape function. A multi-quadratic RBF was used to which linear polynomial members were added, i.e. $m = 3$. The average node spacing was calculated as the average of the distances to the four closest nodes. The following parameters were used to construct shape functions: α_i=4,0, α_c=4.0, q=1.03.

2.2 Governing equations and discretization

Reddy's theory of shear deformations is one of many higher order theories. Higher-order theory can better describe deformation kinematics, do not require shear correction coefficients, and more accurately describe the distribution of transverse shear stresses. These theories, however, include internal higher order power quantities that are difficult to interpret physically and require higher calculation costs. In principle, it is possible to extend the displacement field expressed by thickness dimensions to any degree. Due to algebraic complexity and higher computational costs, theories higher than the third order are not used. The reason for which the relations for displacement were extended to the cubic member is to achieve the quadratic distribution of the transverse shear deformations and the stresses across the thickness of each layer. Therefore, shear correction coefficients are not required in these theories.

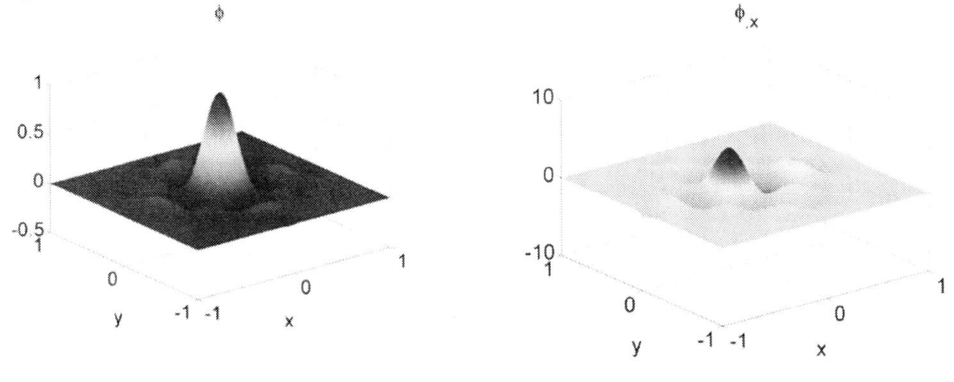

Fig. 1. Graph of shape function for central node (left) and its derivation according x (right), α_i=4.0, α_c=4.0, q=1.03

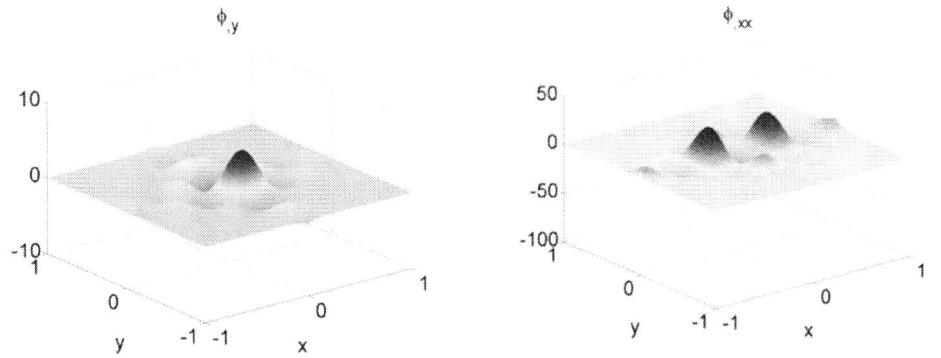

Fig. 2. Graph of derivation of shape function for central node (left) according *y* (left) and second derivation according *xx* (right)

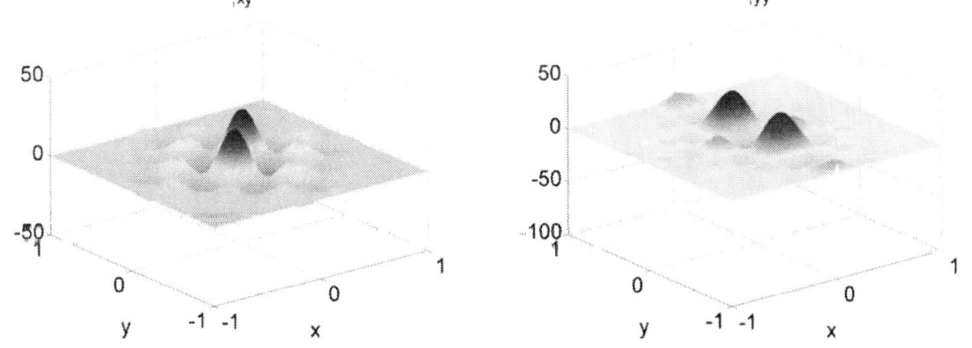

Fig. 3. Graph of second derivation of shape function for central node (left) according *yy* (left) and second derivation according *xy* (right)

In Reddy's theory, the displacement field is expressed by the unknown u_0, v_0, w_0, θ_x and θ_y as follows

$$u(x,y,z) = u_0(x,y) + z\theta_x(x,y) - \frac{4}{3h^2}z^3\left(\theta_x(x,y) + \frac{\partial w_0(x,y)}{\partial x}\right), \tag{18}$$

$$v(x,y,z) = v_0(x,y) + z\theta_y(x,y) - \frac{4}{3h^2} z^3 \left(\theta_y(x,y) + \frac{\partial w_0(x,y)}{\partial y}\right), \tag{19}$$

$$w(x,y,z) = w_0(x,y) . \tag{20}$$

The strain components is obtained by the equations of linear elasticity

$$\begin{bmatrix} \varepsilon_x \\ \varepsilon_y \\ \gamma_{xy} \end{bmatrix} = \begin{bmatrix} \varepsilon_x^{(0)} \\ \varepsilon_y^{(0)} \\ \gamma_{xy}^{(0)} \end{bmatrix} + z \begin{bmatrix} \varepsilon_x^{(1)} \\ \varepsilon_y^{(1)} \\ \gamma_{xy}^{(1)} \end{bmatrix} + z^3 \begin{bmatrix} \varepsilon_x^{(3)} \\ \varepsilon_y^{(3)} \\ \gamma_{xy}^{(3)} \end{bmatrix} \tag{21}$$

$$\begin{bmatrix} \gamma_{yz} \\ \gamma_{xz} \end{bmatrix} = \begin{bmatrix} \gamma_{yz}^{(0)} \\ \gamma_{xz}^{(0)} \end{bmatrix} + z^2 \begin{bmatrix} \gamma_{yz}^{(2)} \\ \gamma_{xz}^{(2)} \end{bmatrix} \tag{22}$$

where

$$\boldsymbol{\varepsilon}^{(0)} = \begin{bmatrix} \varepsilon_x^{(0)} \\ \varepsilon_y^{(0)} \\ \gamma_{xy}^{(0)} \end{bmatrix} = \begin{bmatrix} \frac{\partial u_0}{\partial x} \\ \frac{\partial v_0}{\partial y} \\ \frac{\partial u_0}{\partial y} + \frac{\partial v_0}{\partial x} \end{bmatrix}, \quad \boldsymbol{\gamma}^{(0)} = \begin{bmatrix} \gamma_{yz}^{(0)} \\ \gamma_{xz}^{(0)} \end{bmatrix} = \begin{bmatrix} \theta_y + \frac{\partial w_0}{\partial y} \\ \theta_x + \frac{\partial w_0}{\partial x} \end{bmatrix}, \tag{23}$$

$$\boldsymbol{\varepsilon}^{(1)} = \begin{bmatrix} \varepsilon_x^{(1)} \\ \varepsilon_y^{(1)} \\ \gamma_{xy}^{(1)} \end{bmatrix} = \begin{bmatrix} \frac{\partial \theta_x}{\partial x} \\ \frac{\partial \theta_y}{\partial y} \\ \frac{\partial \theta_x}{\partial y} + \frac{\partial \theta_y}{\partial x} \end{bmatrix}, \quad \boldsymbol{\gamma}^{(2)} = \begin{bmatrix} \gamma_{yz}^{(2)} \\ \gamma_{xz}^{(2)} \end{bmatrix} = -\frac{4}{h^2} \begin{bmatrix} \theta_y + \frac{\partial w_0}{\partial y} \\ \theta_x + \frac{\partial w_0}{\partial x} \end{bmatrix}, \tag{24}$$

$$\boldsymbol{\varepsilon}^{(3)} = \begin{bmatrix} \varepsilon_x^{(3)} \\ \varepsilon_y^{(3)} \\ \gamma_{xy}^{(3)} \end{bmatrix} = -\frac{4}{3h^2} \begin{bmatrix} \frac{\partial \theta_x}{\partial x} + \frac{\partial^2 w_0}{\partial x^2} \\ \frac{\partial \theta_y}{\partial y} + \frac{\partial^2 w_0}{\partial y^2} \\ \frac{\partial \theta_x}{\partial y} + \frac{\partial \theta_y}{\partial x} + 2\frac{\partial^2 w_0}{\partial x \partial y} \end{bmatrix}. \tag{25}$$

Integrating the stress over the thickness of the plate we can determine the stress resultants

$$\begin{bmatrix} N_{\alpha\beta} \\ M_{\alpha\beta} \\ P_{\alpha\beta} \end{bmatrix} = \sum_{k=1}^{N} \int_{z_{k-1}}^{z_k} \sigma_{\alpha\beta}^{(k)} \begin{bmatrix} 1 \\ z \\ z^3 \end{bmatrix} dz \ , \quad \begin{bmatrix} Q_\alpha \\ R_\alpha \end{bmatrix} = \sum_{k=1}^{N} \int_{z_{k-1}}^{z_k} \sigma_{\alpha\beta}^{(k)} \begin{bmatrix} 1 \\ z^2 \end{bmatrix} dz \ , \tag{26}$$

where the symbols α and β can take x and y, N denotes the normal and in-plane shear forces, M means the bending and twisting moments, Q is the shear force, P and S are higher-order stress resultants, respectively. If we use interpolation, then we can discretize equations (26)

$$u_i(\boldsymbol{x},t) = \boldsymbol{\varphi}^T(\boldsymbol{x})\hat{\boldsymbol{u}}(t) = \sum_{i=1}^{n} \varphi_i(\boldsymbol{x})\hat{u}_i(t) . \tag{27}$$

By placing them in the Reddy governing equations [12] we obtain for all nodes a system of linear equations

$$\mathbf{K}\,\hat{\mathbf{u}} = \mathbf{F} . \tag{28}$$

The shape functions (14) assembled using RBF possess the delta function property, so the imposition of essential boundary conditions can be applied similarly as in FEM.

3 Implementation and verification of material model

The material model was verified for T300/914 Carbon/Epoxy [13]. The elastic properties of T300/914 lamina arte given in Table 2. The strength and critical integrity values are shown in Table 3. The indices t and c at critical values of the damage variable denote tensile stress, and compression stress. The damaged in-plane shear moduli at imminent failure is $G_{12}^{*} =$ 3.410 GPa. The plastic strain and damage thresholds are $F_6^p = F_6^d = 17.8$ MPa.

The parameters required to define the damage surface and the plasticity surface as well as the parameters needed to define the hardening functions were calculated according to the procedure described in [6]. Only two terms were taken in Prony's series in calculation of the parameters. These parameters are listed in Tab. 4 -6.

Table 2. Elastic properties for unidirectional lamina 300/914

E_1	$E_2 = E_3$	$G_{12} = G_{13}$	G_{23}	ν_{12}
142 GPa	10.30 GPa	6.42 GPa	3.71 GPa	0.21

Table 3. Strength and critical integrity properties for a T300/914 unidirectional lamina

F_{1t}	F_{1c}	$F_{2t} = F_{3t}$	$F_{2c} = F_{3c}$	$F_5 = F_6$	F_4
1830 MPa	1096 MPa	57,0 MPa	57,0 MPa	89,1 MPa	78,0 MPa
	D_{1t}^{cr}	D_{1c}^{cr}	$D_{2t}^{cr} = D_{3t}^{cr}$	$D_{2c}^{cr} = D_{3c}^{cr}$	
	0.1161	0.1109	0.5	0.5	

Table 4. Damage parameters for defining surface damage for a T300/914 unidirectional lamina

Damage surface	J_1	$J_2 = J_3$	B_1	$B_2 = B_3$	A_1	$A_2 = A_3$
	$8.05156.10^{-3}$	0.21035	0.0	0.20821	1.12300	1.0

Table 5. Plasticity parameters for defining plasticity surface for a T300/914 unidirectional lamina

Plasticity surface	f_1	f_2	f_{11}	f_{22}
	$-3.28217.10^{-4}$	0	$3.91825.10^{-7}$	$7.69468.10^{-5}$
	f_{12}	f_{44}	$f_{55} = f_{66}$	
	$-2.74544.10^{-6}$	$4.63713.10^{-5}$	$6.69058.10^{-5}$	

Table. 6. The parameters for defining the hardening function for a T300/914 lamina

Damage hardening	γ_0	α_1^d	β_1^d	α_2^d	β_2^d
	$1.12596.10^{-2}$	$2.25337.10^{-3}$	-0.18214	0.42967	-2.35997
Plastic-strain hardening	R_0	α_1^p	β_1^p	α_2^p	β_2^p
	0.14560	-0.37184	0.07433	-0.55523	0.91852

Subsequently, the material model with these input parameters was used to determine the material response under a shear load in plane 1-2. This response was tested at one point with the prescribed ε_6 values. The results obtained using the programmed material model were compared with the results obtained from experimental data published in the literature [13].

4 Damage and material model for elasto-plastic damage

For polymer matrix composites reinforced by strong and stiff fibers, damage and its conjugate thermodynamic force was used. The formulation of the damage and material model is based on linking CDM with the classical plasticity theory by using thermodynamic principles of irreversible processes. This model is suitable for damage prediction and plastic deformations in composite materials reinforced by long uniaxial fibers with a polymer matrix. To represent the damage, the diagonal damage tensor **D** was used. Experimental data for T300/914 Carbon Epoxy are used to identify model.

The development of damage together with the development of plastic deformation is solved incrementally and is evaluated at each collocation point. The input and output data for the material model are described in Fig. 4, where input variables are: $\bar{\boldsymbol{C}}$ is the constitutive matrix in the effective configuration, A_i, B_i, J_i, f_i a f_{ij} are parameters for damage and plasticity surface, α_i^d, β_i^d, α_i^p, β_i^p are parameters for hardening functions, γ_0, R_0 are damage threshold and yield threshold, D_i are damage components, ε_i^p are components of plastic strain, p – is yield hardening variable and δ is damage hardening variable. The output variables are: D_i are damage variables, stress components and tangent constitutive matrix $\boldsymbol{C}^{epd}$ at the end of the last iteration.

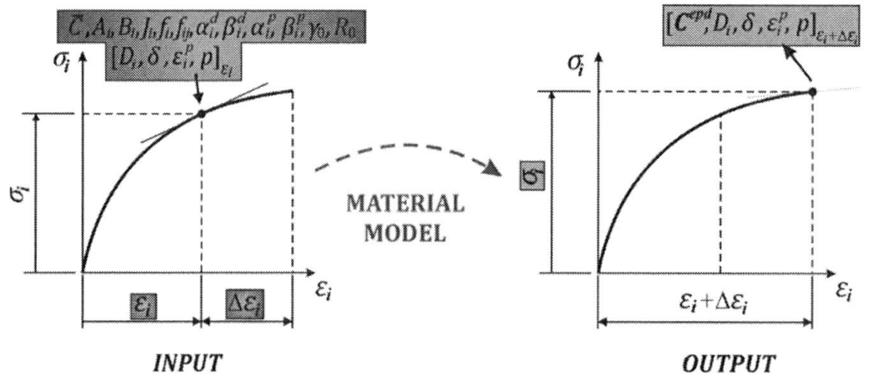

Fig. 4. Description of input and output data for elasto-plastic material model

5 Numerical results and discussion

As a verification example, a composite plate with the dimensions 100x100mm with a thickness of 2.5mm and ply stack orientation [0/45/-45/90] is selected. The plate is clamped on the edges and subjected to pressure load $q^* = 1$MPa. The material data for each ply are: $E_1 = 142$ GPa, $E_2 = 10.3$ GPa, $G_{12} = G_{13} = 6.42$ GPa, $G_{13} = 3.71$ GPa, $\nu_{12} = 0.21$. The implemented computational models were verified by comparing the results with results obtained from the FEM program ANSYS with a very fine mesh (40 000 quadratic shell91-type shell elements).

Unlike metal structures, when evaluating the results of the analysis of the composite structure it is necessary to evaluate quantities such as stresses or deformations in the coordinate system of the lamina in the individual locations of structure. This is because a single lamina of fiber reinforced composite behaves as an orthotropic material. The degree of stiffness reduction can be calculated from the values of the individual variables of damage and integrities. From the evaluation of the values of the damage variable $D_1$the degree of stiffness reduction in the direction of the fibers can be deduced (Fig.5). The degree of stiffness reduction in the direction perpendicular to the fibers in plane 1-2 can be deduced from the values of the damage variable D_2 (Fig.6). To evaluate the degree of stiffness reduction caused by the shear load in plane 1-2, the value of the integrity component $\Omega_1\Omega_2$ must be evaluated (Fig.7). In the case of the analysed composite plate with an asymmetrical layout, the most critical places from the point of view of the macro cracks caused by reduction of stiffness in the fiber direction are in the vicinity of the center of the edges are parallel to the x-axis, especially in the [45 °] layer. The most critical macro-crack spots due to the reduction of stiffness in the direction perpendicular to the fibers in the laminate plane are locations in the center of the edges parallel to the x-axis in the [0 °] layer.

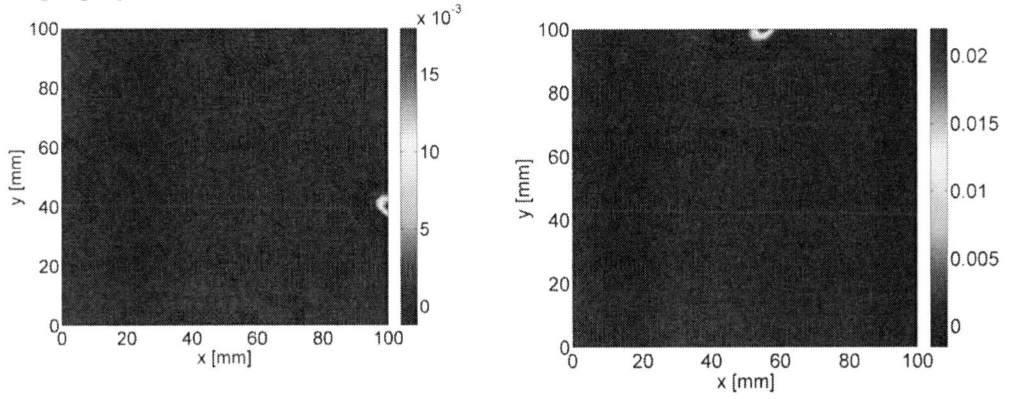

Fig. 5. Distribution of damage variable D_1 at medium plane of the layer [0°] (left) and at medium plane layer [45°] (right)

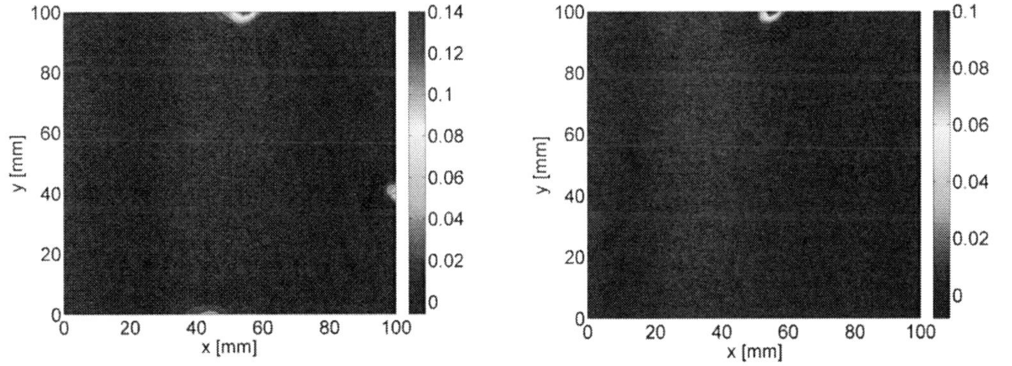

Fig. 6. Distribution of damage variable D_2 at medium plane of the layer [0°] (left) and at medium plate of the layer [45°] (right)

From the viewpoint of the macro cracks caused by the reduction of stiffness due to shear stress in plane 1-2, there are also places in the region of the center of the edges parallel to the x-axis in the layer [0 °]. As we can see from the Fig.8 to Fig.10, in the analyzed structure only very small plastic deformations occur at the given load. Stress distribution is described in Fig.11 to Fig.13.

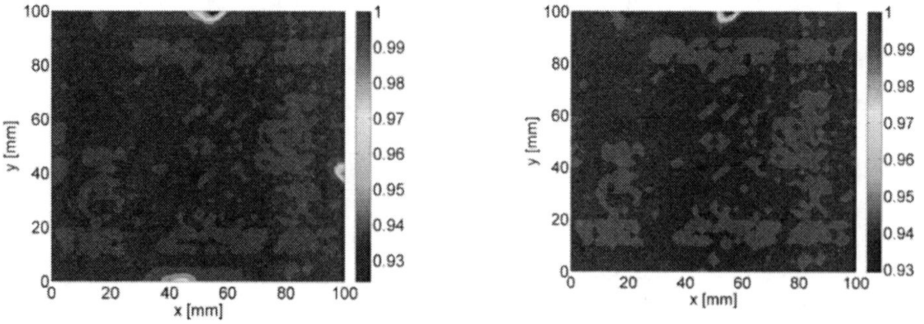

Fig. 7. Distribution of product of integrities $\Omega_1\Omega_2$ at medium plane of the layer [0°] (left) and at medium plate of the layer [45°] (right)

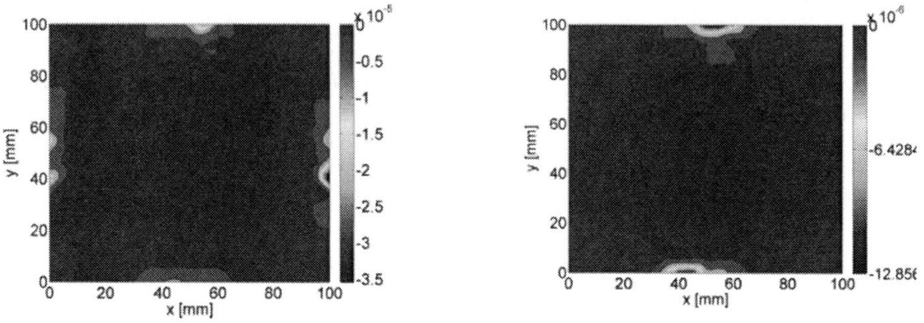

Fig. 8. Distribution of plastic deformation ε_1^p at [0°] (left) and at medium plate of the layer [45°] (right)

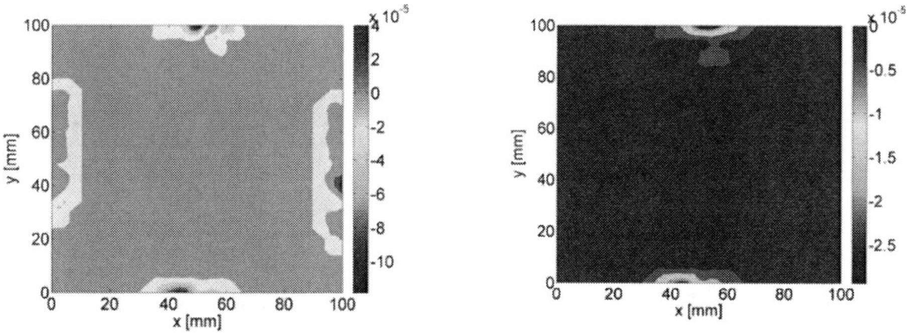

Fig. 9. Distribution of plastic deformation ε_2^p at [0°] (left) and at medium plate of the layer [45°] (right)

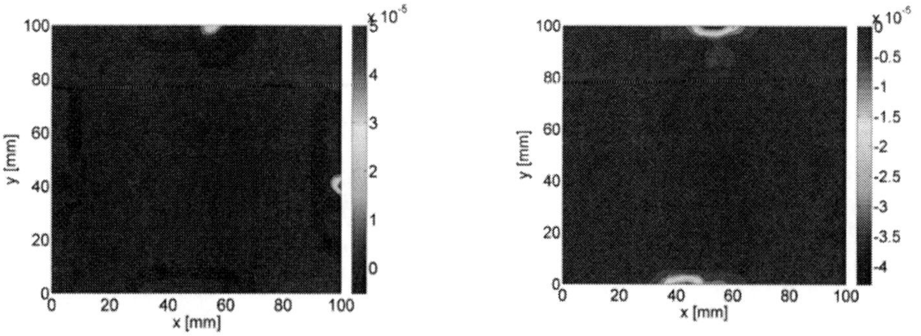

Fig. 10. Distribution of plastic deformation γ_6^p at [0°] (left) and at medium plate of the layer [45°] (right)

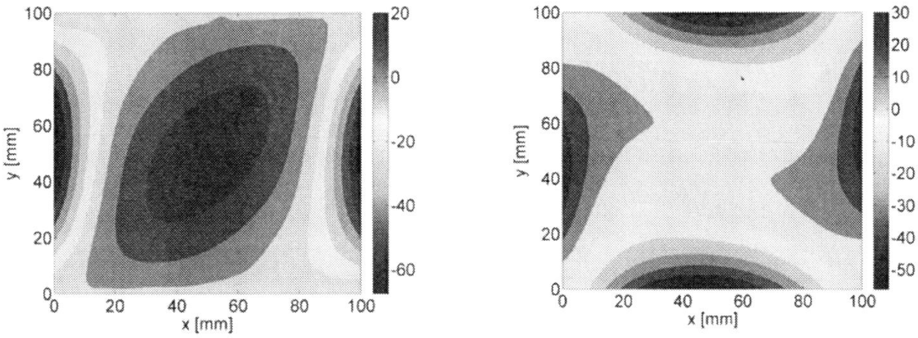

Fig. 11. Distribution of stress $\sigma_1$at [0°] (left) and at medium plate of the layer [45°] (right)

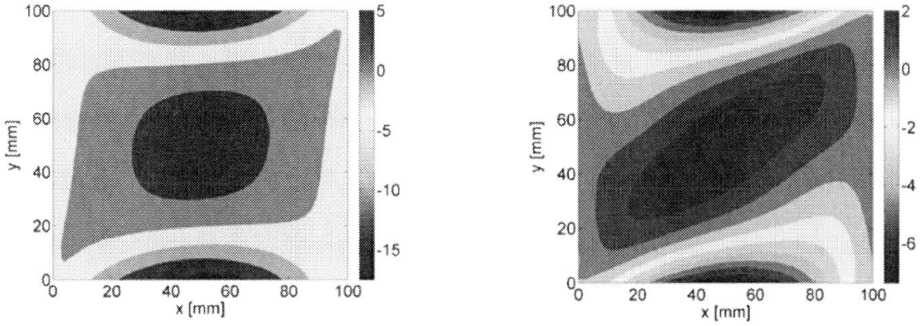

Fig. 12. Distribution of stress $\sigma_2$at [0°] (left) and at medium plate of the layer [45°] (right)

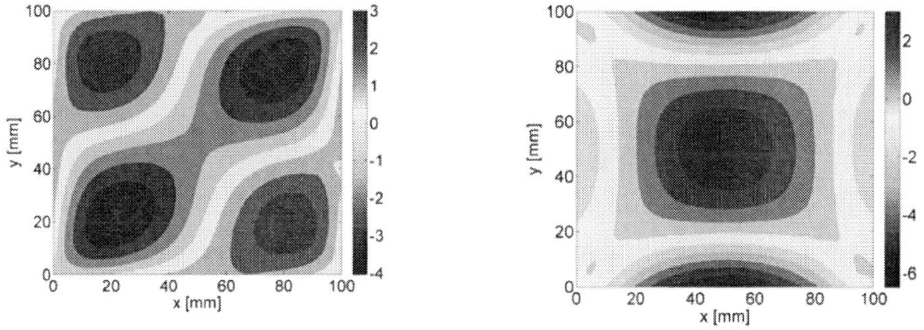

Fig. 13. Distribution of stress τ_6 at [0°] (left) and at medium plate of the layer [45°] (right)

Conclusion

In the presented paper a developed algorithm for the analysis of elasto-plastic damage by using RPIM based on strong form of governing equations based on Reddy's plate theories is presented. The shape functions constructed by RPIM have the Kronecker delta function property, thus there is no need to apply special techniques to apply essential boundary conditions. From the RPIM results we can see that implemented computational model be able to calculate the nodal displacements with very good accuracy using a relatively small number of nodes. The stresses σ_1, σ_2 and τ_6 are also calculated with good accuracy. However, when the transverse shear stress is compared with FEM calculations the errors are larger. In the case of the analysed composite plate with non-symmetrical layer arrangement, the maximum damage is in the fiber direction around the middle point parallel to the x-axis, especially in the layer [45 °]. In the direction perpendicular to the fiber, the maximum damage is in the area of the middle point of the x-axis parallel to the layer [0 °]. The maximum damage in shear plane 1-2 is in the central area of the edges parallel to the x-axis in the layer [0 °]. For a given load of the analysed composite plate only very small plastic deformations were observed.

This work was supported by the Slovak Research and Development Agency under the contract No. APVV-0736-12 and Slovak Grant Agency VEGA 1/0983/15.

References

1. P. Pastorek, P. Novak, P. Kopas, M. Močilan, *Finite element analysis of bond behaviour in steel reinforced concrete structure strengthened carbon fiber reinforced polymer (CFRP) strips*. Metalurgija, **56** (3-4), 405-408 (2017)
2. M. Žmindák, M. Dudinský, *Finite Element Implementation of Failure and Damage Simulation in Composite Plates*. Composites and their properties, N. Hu (eds.). (Intech, 131-152, 2012)
3. M. Žmindák, M. Dudinský, P. Frnka, *Modeling of damage evaluation in thin composite plate loaded by pressure loading*. Applied and Computational Mechanics, **6** (2), 151-162 (2012)
4. G. R. Liu, *Meshfree methods: Moving beyond the finite element method, Second edition* (CRP Press, 2010)
5. J. Sládek et al., *Analysis of orthotropic thick plates by meshless local Petrov- Galerkin (MLPG) method. Int. J. for Num. Meth. in Engng.,* **67/13,** 1830-1850, (2006)
6. E. J. Barbero, *Finite Element Analysis of Composite Materials Using ANSYS - Second Edition* (CRC Press, 2013)
7. G. R. Liu, Y. T. Gu, *An Introduction to meshfree methods and their programming* (Springer, 2005)
8. S. N. Atluri, S. P. Shen, *The Meshless Local Petrov-Galerkin (MLPG) method* (Tech Science Press, 2002)
9. P. Pechac, M. Saga, *Memetic algorithm with normalized RBF ANN for approximation of objective function and secondary RBF ANN for error mapping*, Procedia Engineering, **177,** 540-547 (2017)
10. W. Chen, Z. J. Fu, C. S. Chen, *Recent advances in radial basis function collocation methods* (Springer 2014)
11. R. Schaback, *Approximation of polynomials by radial basis functions*. Wavelets, Images and Surface Fitting, P. Laurent,at al. eds. Wellesley: A. K. Peters Ltd., 459-466 (1994)
12. A. Szekrényes, *Application of Reddy's third-order theory to delaminated orthotropic composite plates.* European J. of Mech. A/Solids **43**, 9-24, (2014)
13. E. J. Barbero, *Finite element analysis of composite materials*. (CRC Press, 2008)

Sensitivity analysis overlaps of friction elements in cartridge seals

Milan Žmindák[1,*], *Miloš* Falat[2]

[1]Faculty of Mechanical Engineering, Department of Applied Mechanics, University of Žilina, Univerzitná 1, 01026 Žilina, Slovak Republic
[2]Schaeffler Slovakia, s.r.o, Ul. Dr. G. Schaefflera 1, 024 01 Kysucké Nové Mesto, Slovak Republic

Abstract. Cartridge seals are self-contained units consisting of a shaft sleeve, seals, and gland plate. The applications of mechanical seals are numerous. The most common example of application is in bearing production for automobile industry. This paper deals with the sensitivity analysis of overlaps friction elements in cartridge seal and their influence on the friction torque sealing and compressive force. Furthermore, it describes materials for the manufacture of sealings, approaches usually used to solution of hyperelastic materials by FEM and short introduction into the topic wheel bearings. The practical part contains one of the approach for measurement friction torque, which results were used to specifying the methodology and precision of FEM calculation realized by software ANSYS WORKBENCH. This part also contains the sensitivity analysis of overlaps friction elements.

Keywords: sensitivity analysis, sealing, hyperelasticity, elastomer, friction torque

1 Introduction

Engineering modeling and simulations plays a role in designing the cars components [1]. Today engineers apply simulation tools and methods in the design phases of product development, replacing costly physical prototyping and testing with advanced numerical analyses. Mechanical face seals are used for sealing fluid at places were a rotating shafts enters an enclosure. The primary function of seals is to prevent the leakage of gaseous and liquid substances between the contact surfaces of the construction. They prevent lubricant from escaping and also protect the internal structure from the ingress of dirt. Well-sealed joint has a crucial impact on the performance and reliability of the construction. It also prevents the leakage of chemical substances from construction into the surrounding environment [2]. The degree of sealing refers to the flow of lubricant through the sealed joint per time unit.

Currently there are a lot of seals variations for wheel car bearings. Fig. 1 illustrates the development of mass-produced sealing systems in SKF [3] over the time. This diversity is historically determined by the development of increasing requirements imposed on the sealing systems as well as increasing experience of the development departments with the operation of the individual variants in practice.

* Corresponding author: milan.zmindak@fstroj.uniza.sk

Reviewers: *Juraj* Gerlici, *Ján* Vavro

Back in the 60's so-called "onelip" flat gaskets were used (Fig.1, the first left). It consists of a metal part that functions as a reinforcement for the rubber part. This part includes a flexibly deformable lip (friction element) that permanently touches the bearing ring. Its role is to prevent the permeation of impurities from the external environment to the internal construction of the bearing as well as the leakage of the lubricant.

Given the increasing demands on the sealing performance, the construction of the seal has been supplemented with another friction element (Fig. 1, second from the left) whose role was to catch those impurities that might possibly get through the first friction element. Progress was made in the 1970s when a flinger ring was introduced to be used as pre-sealing for the flat gasket. Unlike the flat gasket, it is attached to the internal ring of the bearing. Thanks to this, the contaminated water can easily flow out under the seallings lip.

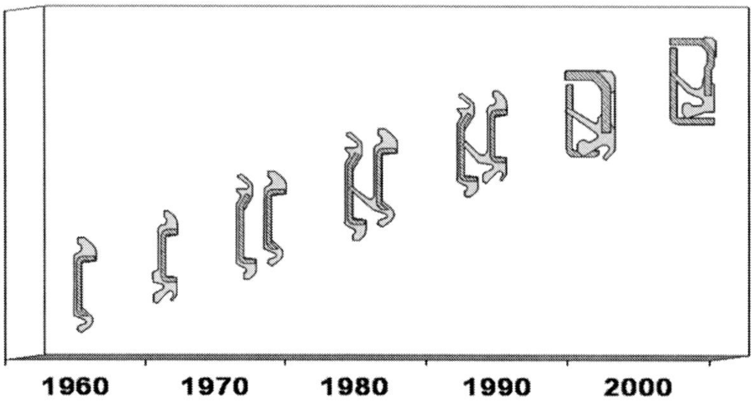

Fig. 1. Historical development of shaft seal designs used for wheel bearings

In the next decade, the concept of the flat gasket has been added with an axial friction element (Fig. 1, third from the left). Thanks to this, the gasket could function longer before the bearing got contaminated. The space created for the lubricant reduced the deterioration of sealing friction elements of the gasket; having at the same time tightened the impact of fine surface defects. Other customer requirements, such as the inclinability of the bearing for some applications, could no more be ensured through flat gaskets. Therefore, a cassette seal (Fig.2) was developed, consisting of two assembled parts.

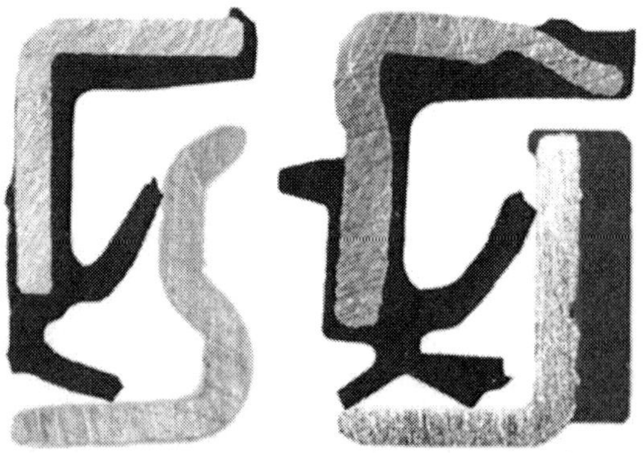

Fig. 2. Cutting through sealing gaskets. Left is cassette seal with two sealing lips, right is cassette seal with three lips and the Anti-lock Braking System (ABS) ring

One part consists of a molded sheetmetal with vulcanized sealing friction elements, while the other consists of a sheetmetal or, alternatively, of a sheetmetal with vulcanized ABS ring (Fig.2, right). The friction elements in the first part mostly move around the second part, and not around the ground ring of the bearing. Its advantage is a better quality of the touching surfaces for the friction elements [4] as well as a higher accuracy. The main difference lies in the fact that the seal's surfaces touching both the external and internal rings are flat. This did away with the inaccuracy that was a result of the compression pressing of flat gaskets into recess. Although this process is financially more demanding, it provides a better sealing performance compared to its predecessor as well as a simpler integrability of the ABS signal generator. In 2000, the demands on the price started increasing in addition to the tightness requirements. The original design of the cassette seal was modified by removing one radial friction element.

The main aim of this article is to apply an finite element method (FEM) in the process of the designing of sealing system with respect to the decrease of the friction torque in a bearing joint. The goal is to create a simple tool to calculate the final friction torque for the sealing system of the first-generation wheel bearing (Fig.3) using the commercial software ANSYS WORKBENCH. By way of an experiment we will verify the total friction torque of the first-generation bearing and identify the friction torque of the individual components of this bearing. This experiment will include the verification of the friction torque of the partial friction torques of the seal. Using the modified seals, we will next analyse the course of the friction torques of its friction elements. For the experiment, we will use the same seals with a different friction element removed in each of them. Finally, we will analyse the impact of the overhang of the friction elements of the sealing system on the magnitude of the contact force with respect to the change of the dimensions of the sealing elements in the admissible tolerance zone.

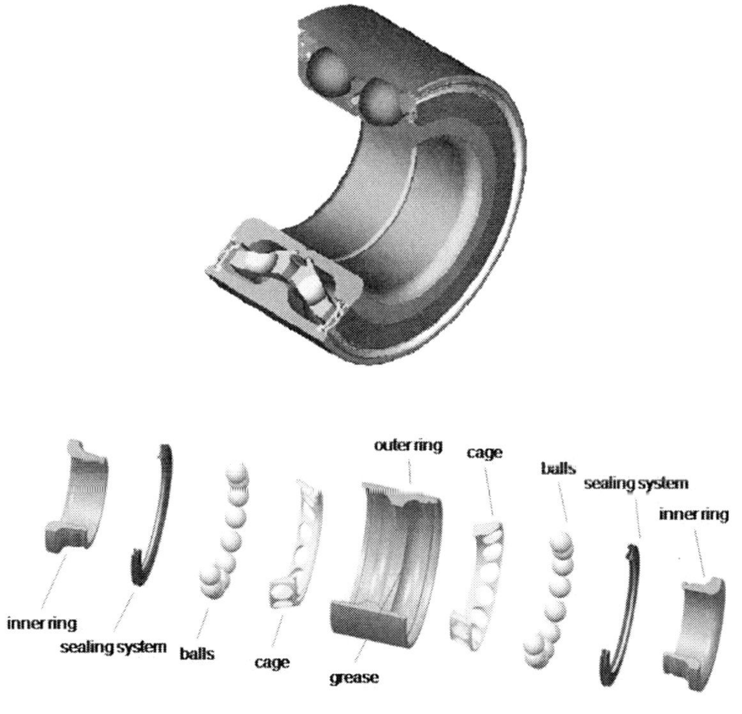

Fig. 3. First-generation wheel bearing and its components

2 Finite element model

The geometrical model of the seal was created in the Creo 2 program and imported into the ANSYS Workbench program. Fig. 4 depicts a cross-section of the bearing; Fig. 5 depicts its basic parts. The final elementary model was created in the ANSYS Workbench software. The task was pursued as a linear static with 2D-axially symmetrical final elements.

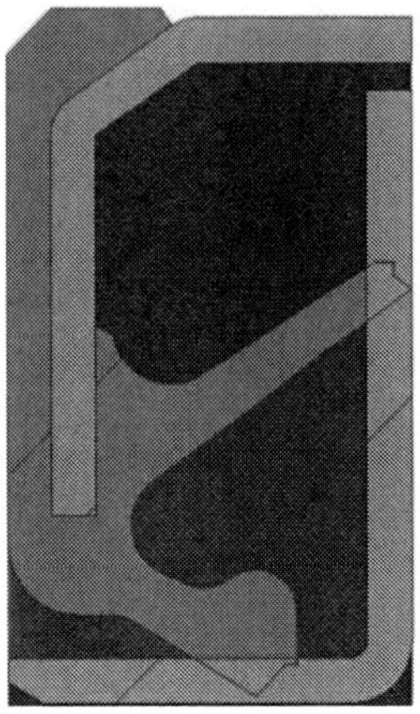

Fig. 4. Cutting through sealing

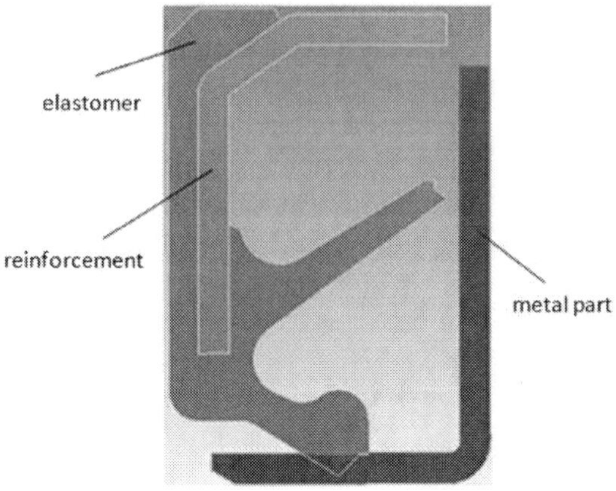

Fig. 5. 2D Seal model with the names of its components

The metal part was constructed from regular construction steel with the stress modulus of E = 2.1 e05 MPa and Poisson's ratio of v = 0.3. The material model for the rubber part of the seal is the first-order parametric Ogden model [5] with the material constants of MU1= 2.34 MPa, A1= 2 and D1= 0. Having created the geometrical model and defined the material properties, we specified the boundary conditions. Since this is an axially symmetrical task, it is necessary to take into consideration the impact of forces effective around the whole circumference. Next, the contact conditions among the individual parts of the model are defined (Fig.6). A perfect connection between the reinforcement and the rubber is envisioned (Fig.7). A contact is defined between the rubber and the sheetmetal in both the axial and the radial directions. In comparison to its original position, the sheetmetal has been shifted by 0.7 mm. The cover sheetmetal will be inserted to its original place during the calculation. It is therefore necessary to set up the actual contact conditions for the contact between the friction element and the cover sheetmetal so that the friction element is in the correct position. In both the axial and radial contact points we consider the friction between the rotating sheetmetal and the stationary rubber part. For the simulation purposes, we set up the detection of the contact surfaces by using the "Pinball Region" function [6]. To create an FEM web, an automatic generator of quadrangular elements is used (Fig.9).

3 Numerical results

The FEM analysis is divided into two separate analyses. In the first analysis, we calculate the friction torque of the axial sealing element (Fig. 10), while in the second part of the analysis and the friction torque of the radial sealing element (Fig.11). The FEM analysis was done for two time steps with the growth figure of 1 second. In the first step, the contact is identified between the radial friction element and the cover sheetmetal with the actual position of the radial friction element. In the second step, the cover sheetmetal is inserted by 0.7 mm into the original position so that the axial friction element gets into the correct position in the seal. Finally, we will calculate the reactions because the axial friction element moves in the direction of the X-axis so that it has an impact on the value of the radius r.

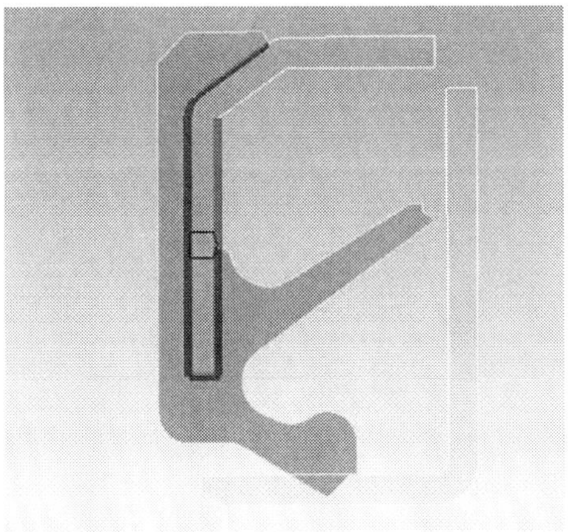

Fig. 6. Contact definition between rubber and reinforcement

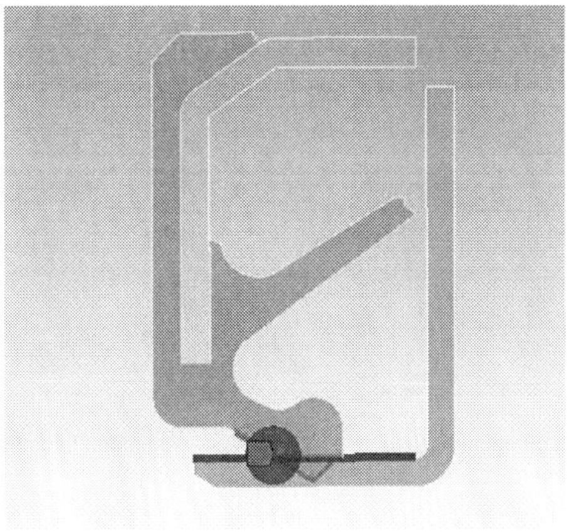

Fig. 7. Contact definition between rubber and metal part in radial direction

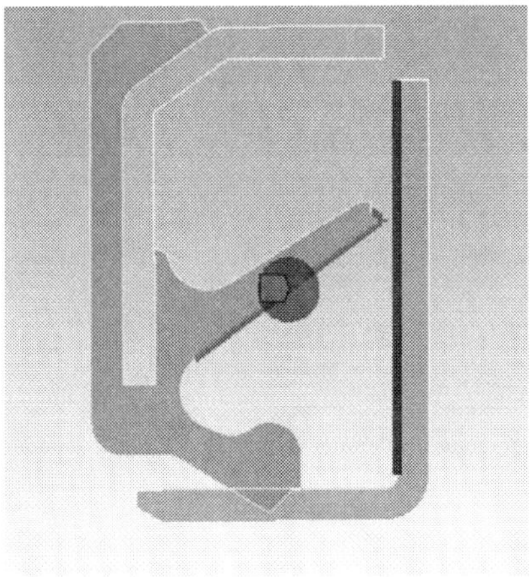

Fig. 8. Contact definition between rubber and metal part in axial direction

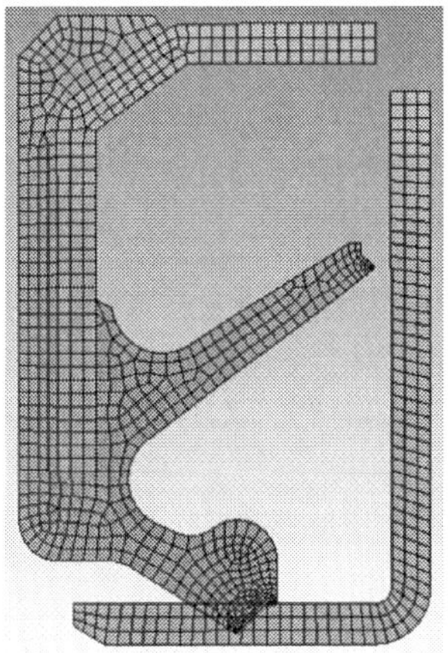

Fig. 9. Finite element mesh

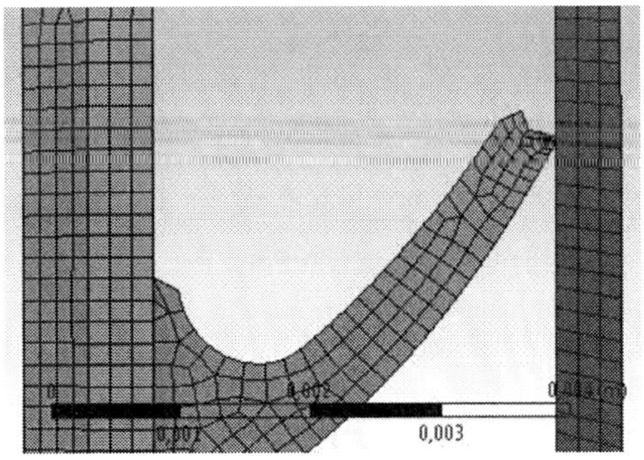

Fig. 10. Shape of axial friction element in assambled shape

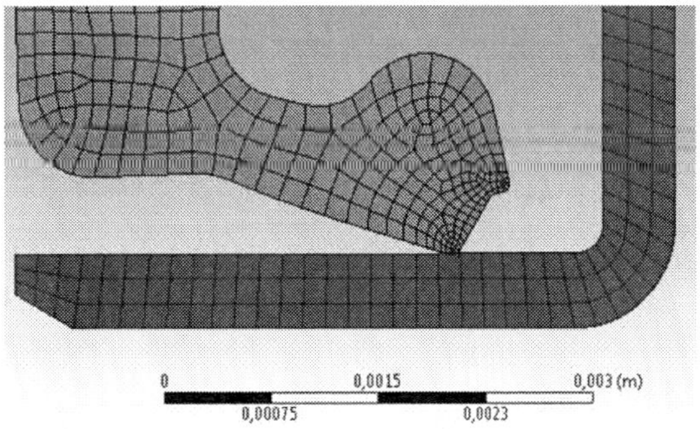

Fig. 11. Shape of radial friction element in assembled form

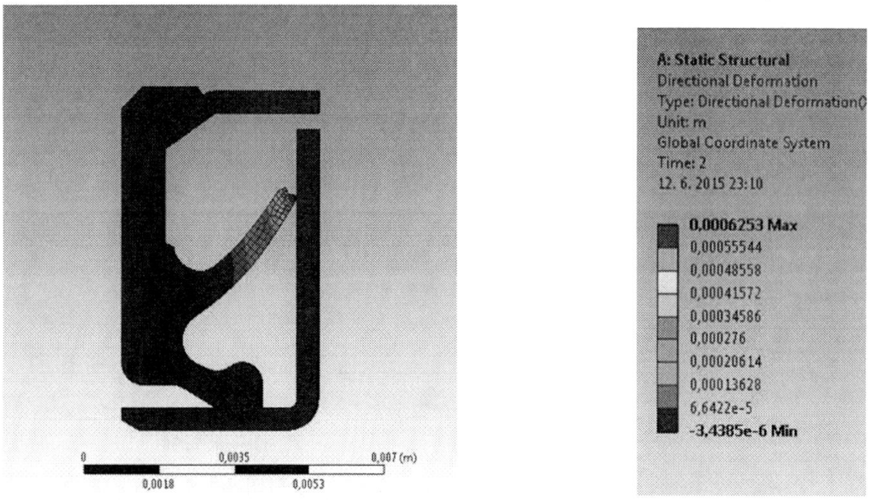

Fig. 12. Axial displacement of the friction element in an assembled and unassembled state

The calculated value of the shift of the end of the axial friction element (no-load condition) is depicted by Fig.12. It means that the change in the contact diameter of the friction element is 0.62 mm and it corresponds to the radius. Therefore, we can conclude that the seal touches down on the diameter of Ø61.24 mm, whereas the value on the drawing in a no-load condition is Ø60 mm. A reaction in the direction of the Y-axis (Fig. 13) has an impact on the other part of the seal and the direct effect has the magnitude of the friction torque.

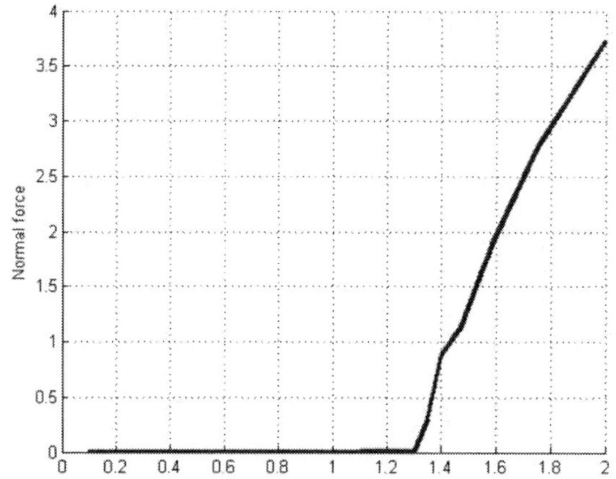

Fig. 13. Axial friction element compression force values in the axial direction in each calculation step

The value of the final compressive force of the axial friction element is 3.73 N, and therefore the value of the friction torque of the axial sealing element is $M_a = 11.42$ Ncm.

In the second step, we calculated the radial friction element (Fig.11). The calculation procedure was similar to that done for the axial friction element. Analogically to the calculation of the axial friction element, the values of the compressive force were achieved. The value of this force in the second 2 of the calculation is 5.1994 N. In case of the radial friction element it is not necessary to calculate the diameter on which the friction element will be positioned as here this value is given structurally and it is Ø51.85 mm. Then the value of the friction torque of the radial friction element is $M_r = 13.48$ Ncm and the friction torque of the whole seal is $M = M_a + M_r = 11.42 + 13.48 = 24.9$ Ncm.

4 Experimental verification of numerical results

This part includes a procedure for detecting the friction torque of the bearing, its components, and the friction elements of the seal on the experimental stand S110 (Fig.14). This experimental stand enables calculating the tested bearing up to the external diameter of 350 mm and exercising the load of up to 400N (radial direction, using an overhang in arrangement). The maximum revolutions of the experimental stand are 10,000 rpm. Fig. 15 depicts the experimental stand S110 with a description of its basic parts.

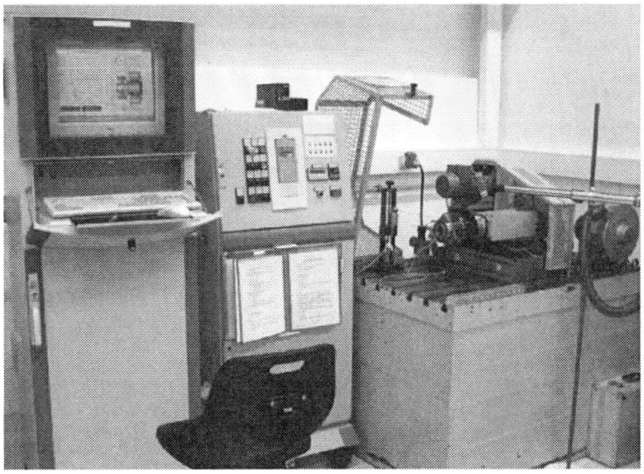

Fig. 14. Experimental stand S110

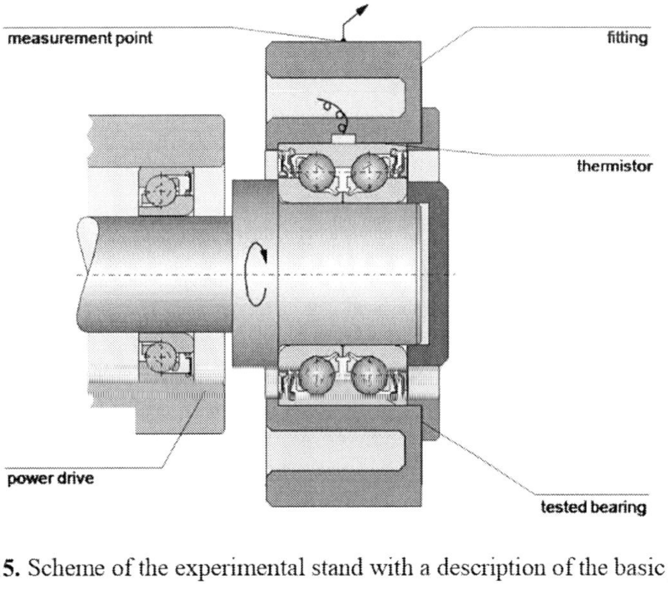

Fig. 15. Scheme of the experimental stand with a description of the basic parts

The calculation principle is based on the fact that the external construction of the bearing is connected to a sensor for measuring the force by way of a wire strand. The internal bearing ring revolves together with a spindle; this movement is in turn transmitted to the external ring due to the impact of the friction torque. The magnitude of the force, as measured on the sensor, is then converted using the radius to which the wire strand is attached. The experimental measurement of the friction torque was done on a first-generation wheel bearing with the external diameter of 90 mm and the internal diameter of 56 mm. Four measurements were done for the revolutions of 900 rpm; the measurement results are given in Table 1.

Table 1. Results of the friction torques for the tested bearing

Measurement	Bearing arrangement	Friction torque [Ncm]	Percentage share [%]
1st	Complete bearing	98.7	100
2nd	Without one seal	72.4	73.4
3rd	Without seals	48.6	49.2
4th	Without lubricant and oil	45.3	45.9

Table shows that the seal on the right side represented 26.6 % of the total friction torque. This corresponds to the value of 26.3 Ncm. The seal on the left side represented 24.2 % of the total friction torque of 23.8 Ncm. Note that for this experiment we used modified seals that were gradually integrated into the bearing whose friction torque we already know (approximately 45.9 Ncm). The surfaces of the friction elements that are in contact with the sheetmetal were lubricated with grease. Friction elements were removed from the seals so that one of them only was in contact at a time. The design of the modified seals is shown on Fig.16.

Fig. 16. Updated seals for friction torque measurement

The experiment was done in three steps. Into the bearing we first compressed a seal with a friction element removed in the radial direction, so that the friction torque was produced on the axial friction element only (blue seal; Fig.16). After the measurements were finished, we compressed this seal out of the bearing, instead compressing in a seal with friction elements removed in the axial direction, so that there was friction on the axial friction element (green seal; Fig.16). After the measurements were finished, we finally compressed the seal out, instead compressing in its place a seal with all friction elements (red seal; Fig.16). The evaluation of this experiment is depicted on Fig.17. It shows the influence of the friction elements of the bearing on the friction torque.

The friction torque of the axial friction element (blue seal) was approximately 12 Ncm, while the friction torque of the radial element (green seal) reached the value of approximately 14.5 Ncm. The value of the friction torque of an unmodified seal (red seal) was approximately 24.5 Ncm. However, this value does not have to be necessarily identical to the sum of the measured values of the friction elements because we did not measure the identical seals but rather three modified seals.

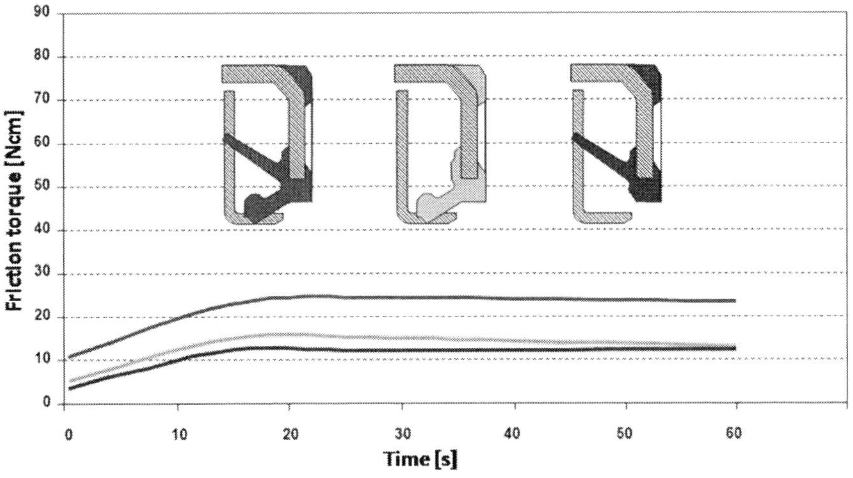

Fig. 17. Course of friction moment of friction lip of the tested seal

Table 2 compares the values of the friction torques achieved in both the experiment and the FEM analysis. The conformity of the results is very good. Deviations may be due to the inaccurate geometry of the used seals, manufacturing inaccuracies, and FEM discretization errors. By comparing the results was verified the accuracy of our calculation and the input data used in the FEM analysis.

Table 2. Results of the experiment and FEM calculation

	Experiment [Ncm]	FEM calculation [Ncm]
Axial lip	12	11.42
Radial lip	14.5	13.48
Axial + Radial lip	24.5	24.9

5 Sensitivity analysis

The sensitivity analysis is an important indicator for the designing of machinery components and their material properties, dimensions, etc. It is an analysis of the impact of the input construction parameters on the final characteristics of a model. If we define the input values of the calculation model as $\mathbf{X} = (X_1, X_2, ..., X_M)$ and assume that these are non-zero values, then we can observe the influence of these values on the output value Y

$$Y(\mathbf{X}) = f(X_1, X_2, ..., X_M). \quad (1)$$

Generally, the sensitivity analysis explores the output indefiniteness of the model caused by various reasons of the indefiniteness of model inputs [7].

This chapter brings a sensitivity analysis of the influence of the deviation of the seal from the nominal dimensions of both the axial and the radial friction elements (Fig. 18). Like the previous analyses, we did an analysis for both the axial and the radial directions. We analysed 60 models created in the Creo 2 program in the environment of the ANSYS WORKBENCH program. The course of the compressive force was assessed because the change in the contact diameter ranges in tenths of a millimetre only, and would have a negligible influence on the calculation itself. Moreover, the calculation would be equally influenced by a round-off error.

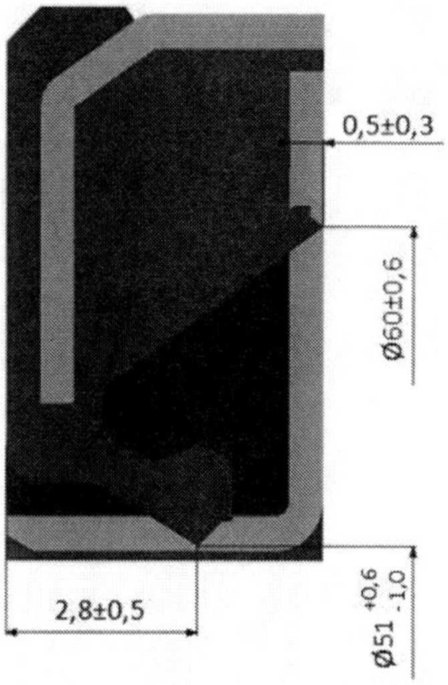

Fig. 18. Geometry of the seal with marked dimensions and tolerances

5.1 Axial friction torque

The analysis of an axial friction torque consists of an analysis of an axial friction element in both the axial and the radial directions. Fig. 18 shows the deviations of the geometrical model of the seal: for the axial direction, the deviation is ±0.3 mm, while the growth figure to the nominal value of 0.5 is 0.05 mm; for the radial direction the deviation is ±0.6 mm, while the growth figure is 0.1 mm, corresponding to the nominal value of Ø60 mm, and the magnitude of the tolerance zone ±0.4 mm. The results of the course of the compressive force are given in the diagrams in Fig. 19 and Fig. 20. Fig.19 shows the course of the compressive force of the axial friction element as well as the values depending on the overhang magnitude with respect to the cover sheetmetal. It is evident that the compressive force increases together with the increasing overhang. Fig. 20 confirms that the compressive force will increase as long as the diameter deviation in the radial direction is negative, and vice versa.

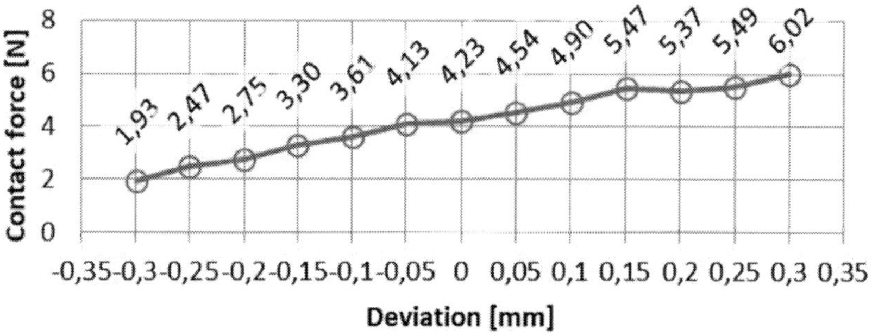

Fig. 19. The course of the axial friction element pressure force when changing its overlap (change of dimension in axial direction)

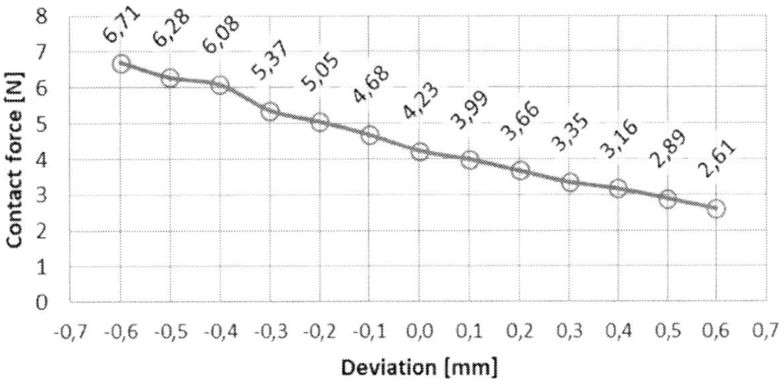

Fig. 20. The course of the axial friction element pressure force when changing its diameter (change of diameter in radial direction)

5.2 Radial friction torque

The calculation procedure for the radial friction element is analogical to the axial friction element. The modifications of the geometrical model used in this analysis are given in Fig. 18. For the radial direction ranging from +0.6 to -1 mm with the growth figure of 0.1 mm for the nominal value of Ø50.9 mm with the tolerance of ±0.2 mm. The analysis results are given in Fig. 21 and Fig. 22. Fig. 21 depicts the course of the compressive force for the change in the position of the radial friction element in the axial direction. We can see that

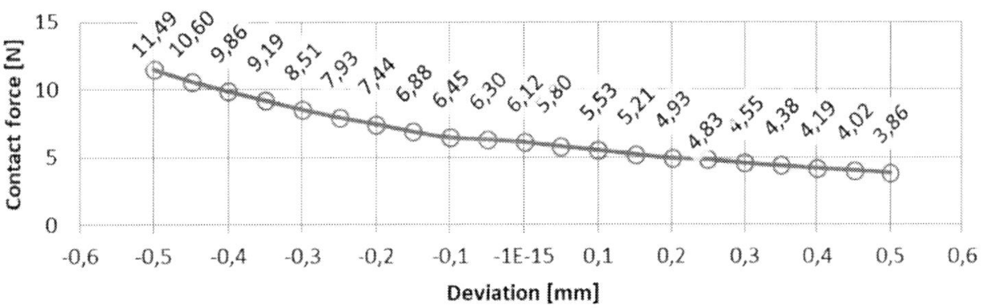

Fig. 21. The course of the radial friction element pressure force when changing its location (change of dimension in axial direction)

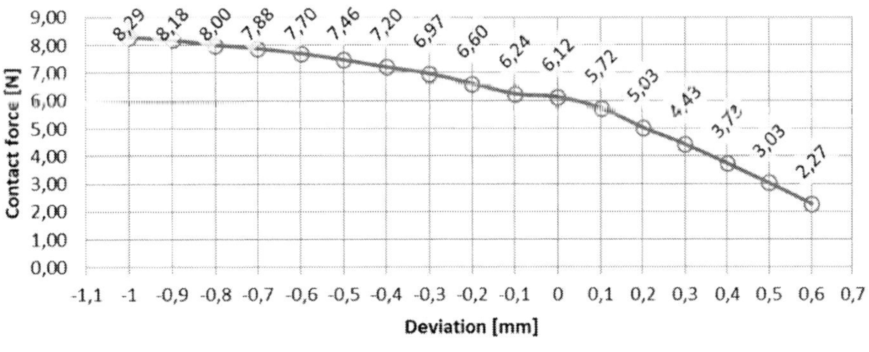

Fig. 22. The course of the radial friction element presure force when changing its overlap (change of dimension in radial direction)

the compressive force decreases with the increasing deviation. Fig. 22 shows the course of the compressive force, taking into consideration the change of the overhang of the radial friction element with respect to the cover sheetmetal. We can see that the change in the compressive force decreases with the increasing overhang. This is the development opposite to the case of the decreasing overhang when the change in the compressive force increases. The differences in the courses are probably a result of the FEM discretization errors.

Conclusion

The article brings a procedure for the analysis of sealing systems in the wheel bearing of an automobile. It mainly focused on analysing the friction torques of friction elements when their main dimensions are changed. The ANSYS WORKBECH program was used in the analysis. The geometry was developed in the Creo 2 program. The numerical results clearly show that the method used for the calculation of the friction torque does not require too sophisticated hardware equipment and that the analysis calculation time does not exceed tens of minutes when standard hardware equipment is available. The acquired results give us a good notion of the course of the contact force as well as the friction torque depending on the main dimensions of the friction elements of the sealing system.

In the experimental part, we established the actual values of the friction torques and the compressive forces of the friction elements of the sealing system of the wheel bearing on the experimental stand S110. Having compared the experimental results and the numerical results we can see that the agreement is very good. Minor differences are caused by the tolerance deviations from the drawing values of the sealing geometry as well as the FEM discretization errors.

The research work is partially supported by the Project of the Structural Funds of the EU, Operational Program Research and Development, Measure 2.2 Transfer of knowledge and technology from research and development into practice: Title of the project: Development of optimal technology for the analysis of the limit states of structural elements in contact, ITMS code: 26220220118.
We are support research activities in Slovakia / Project is cofounded from sources of ES.
This work was supported by the Slovak Grant Agency VEGA 1/0983/15.

References

1. M. Kekez, L. Radziszewski, *Modelling of pressure in the injection pipe of a diesel engine by computational intelligence.* Proc. of the Institution of Mechanical Engineering, part D-Journal of Automobile Engineering **225**, 1660-1670 (2011)
2. W. Tietze, *Handbuch Dichtungspraxis*. (Vulkan-Verlag Essen, 2003).
3. *SKF*, Available on internet [http://www.skf.com], (2015)
4. T. Kunstfeld, *Einfluss der Wellenoberfläche auf das Dichtverhalten von Radial-Wellendichtungen*. (der Universität Stuttgart, 2005).
5. Y. B. Fu, R. W. Ogden, eds., *Nonlinear Elasticity: Theory and Applications* (Cambridge University Press, 2001)
6. *ANSYS Mechanical User's Guide*, ANSYS, Inc., 2015.
7. A. Sapietová, V. Dekýš, *Dynamic Analysis of Rotating Machines MSC.ADAMS*. Procedia Engineering **136**, 143-149 (2016)

Modelling of road traffic noise

Andrzej Bąkowski[1*], *Leszek* Radziszewski[2], *Vladimir* Dekys[3]

[1]Kielce University of Technology, Aleja Tysiąclecia Państwa Polskiego 7, 25314 Kielce
[2]Kielce University of Technology, Aleja Tysiąclecia Państwa Polskiego 7, 25314 Kielce
[3]University of Žilina, Univerzitná 8215/1, 010 26 Žilina, Slovakia

Abstract. Effective implementation of the EC Communication relating to urban mobility is dependent on adopting relevant indicators that will allow comprehensive assessment of the measures undertaken under the urban mobility plan. One of the fundamental parameters for this assessment is the equivalent sound level indicator, measured over the 24-hour period, used to evaluate the annoyance of the road traffic noise. Not only is its value important but also its variability. This paper reports the results of the analysis of traffic noise along a major road, performed using various descriptors. The data recorded continuously throughout the year by the traffic rate and noise monitoring terminal were used to construct the measurement database. Analysis of the results was performed for the period between March and June in 2011 and 2016. The road was renovated between 2013 and 2015. The noise variability assessment was conducted using classical and positional indicators. Measurement uncertainties were evaluated.

Keywords: traffic noise, traffic noise variability, noise measurement uncertainty

1 Introduction

The Communication of the European Commission on urban mobility [1] promotes "...cooperation across different policy areas and sectors (transport, land use and spatial planning, environment, economic development, social policy, health, road safety, etc.); across different levels of government and administration; as well as with authorities in neighbouring areas – both urban and rural." For these guidelines to be effectively and broadly deployed, it is necessary to adopt the solutions available within the so-called urban intelligent transport system regulations, concerning urban access, road safety issues, citizen and stakeholder engagement as well as fostering changes in mobility behaviour. The EC recommends solutions that will encourage inhabitants to use multimodal transport, choose transit over car, shift towards cycling and walking, and combine different modes within one travel chain. As for the urban road traffic management, public transport should be given priority over private transport and the speed of the vehicles should be controlled. Enhancing the traffic flow and discouraging drivers from speeding, thus reducing brake/acceleration cycles, will contribute to increased road safety for all citizens and to reduced air pollution and noise [2]. The mobility plans will be effective provided that subsequent stages of the

* Corresponding author: abakowski@tu.kielce.pl
Reviewers: *Marcin* Kubiak, *Bogdan* Posiadala

plan implementation are monitored and assessed, with the assessment results adapted to the changing conditions. The measures undertaken under the mobility plans are evaluated using common descriptors. There are fixed indicators, on which the evaluation is based, and supplementary indicators determined when the need arises. The fixed indicators include:
- travel speed of cars and collective transport vehicles,
- fine particulate matter concentration, PM2.5, PM10, BaP at each assessment point,
- the number of inhabitants and dwellings exposed to road traffic noise, assessed using the L_{DWN} and L_N descriptors. Average values of some of these descriptors are determined periodically, once in five years, during noise mapping [3], but this is insufficient for the purpose of mobility change analysis. Several cities, including Kielce, have stationary monitoring systems constructed to register noise and traffic rates for 24 hours throughout the year [4]. These systems, e.g., Enviro 151, are able to remotely configure and program measurement sessions. Measurement results, available on the internet in real time, illustrate the variations in the structure and rate of traffic in the city and allow the implementation of the digital acoustic mapping project. Kielce has more than ten such stations, both in the centre and on the outskirts of the city. This paper analyses the results from the equivalent sound level measurements recorded by one of these stations in 2011 and 2016.

2 Measuring station

The data under analysis were recorded by automatic sound and traffic volume continuous monitoring station located in Popiełuszki Avenue in Kielce. Kielce is a medium-size town located in south-central Poland (the Świętokrzyskie Mountains) within the moderate climate zone. Temperatures within a year range approximately from -5°C in January to +17°C in July. Average monthly precipitation is from 34 mm in October to 96 mm in July. The wind, predominantly from the south and west, reaches an average speed of about 3 m/s over a year. Kielce gets on average 70 days a year of snow on the ground.

Popiełuszki Avenue is a road with four lanes of traffic separated by a 3m grass median. One side of the road is comprised of compact residential development, about 200m from the measuring station. The other side is the edge of a woods area. The road is part of the eastern bypass around Kielce and part of the national road No. 73 (Warszawa/Łódź – Kielce - Tarnów - Krosno), which is directly connected with the Trans-European Transport Networks (TEN-T). It is also the major part of the Tarnów-bound thoroughfare functioning as both the transit route and the city street.

The results of acoustic measurements performed during the period from 25/03/2011 to 25/06/2011 and from 25/03/2016 to 25/06/2016 were split into three sub-intervals of a 24-hour interval: day time, evening time and night time. In the years 2013 - 2015, Popiełuszki Avenue was thoroughly renovated. The acoustic measurements were carried out with the SVAN 958A, a four-channel digital vibration analyser and a class 1 sound level meter, operating within the measuring frequency range 0.5 Hz to 20 kHz, depending on a microphone used. The frequency range is 3.5 Hz to 20 kHz when a Microtech Gefell MK250 free-field, prepolarised 1/2" condenser microphone with a sensitivity of 50 mV/Pa, SV 12L preamplifier is used. The temperature range within which the device is operable is from -10° C to 50° C. The resolution of the signal *RMS* detector is 0.1 dB. The measurements were carried out 24 hours a day. The *RMS* values of the A sound level were registered in the buffer every 1 s and the results were recorded every 1 minute. The data collected formed the basis for equivalent sound level calculation for three time intervals, i.e., from 6:00 to 18:00, from 18:00 to 22:00 and from 22:00 to 6:00. The microphone for the sound pressure measurements was mounted at a distance of 4 m from the edge of the road at a height of 4 m. Traffic volume was measured with a digital radar 245 MHz by

WAVETRONIX. Weather data were recorded by the VAISALA WTX 510 automatic meteorological station.

3 Analysis of measurement results

The most common noise indicator used to assess annoyance is the equivalent sound level ($L_{Aeq,T}$), expressed in (dB), defined as [5]:

$$L_{Aeq,T} = 10 \cdot log\left[\frac{1}{T}\int_0^T \left(\frac{p_A(t)}{p_0}\right)^2 dt\right] = 10 \cdot log\left[\left(\frac{p_{A_{RMS}}}{p_0}\right)^2\right] \tag{1}$$

where:
T - represents the overall measurement time, s
$p_A(t)$ - *A*-weighted sound pressure level, Pa
p_0 - is the standardized reference sound pressure of 20 μPa
$p_{A_{RMS}}$ - represents the effective sound pressure.

Figure 1a compiles the measurement results for the equivalent sound level expressed as dB for T= 86400 s, recorded in 2011 and 2016. Figure 1b shows quantile plots and histograms with the function of probability density distribution for the standardized values of the data under analysis. The graphical representation of the spread of variables is shown on the box plots in Fig. 1c.

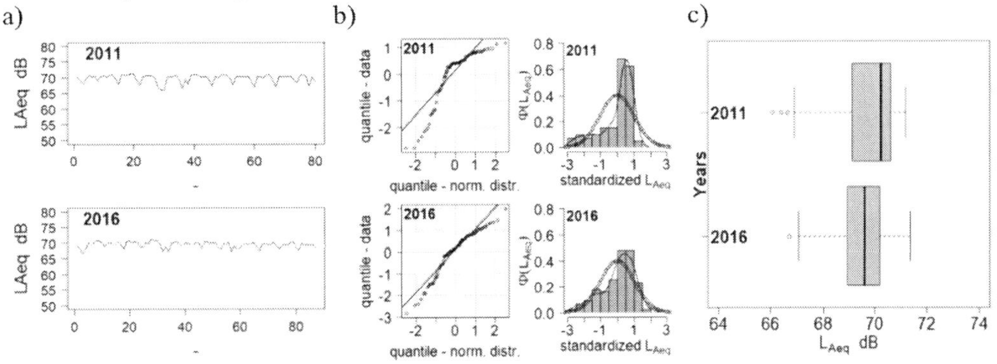

Fig. 1. Equivalent sound level results from the terminal in Popiełuszki Avenue recorded in 2011 and 2016 for all 24-hour measurement periods; a) equivalent sound level variations and the corresponding: b) quantile plots and histograms with the function of probability density distribution for the standardized data, c) box plots.

The graphs in Fig. 1a indicate that despite the thorough renovation of the road, the values of $L_{Aeq,T}$ underwent only slight changes. The range between the minimal values (Saturdays and Sundays) and the values recorded on the working days decreased. The quantile plots and histograms in Fig. 1b confirm that the data do not follow a normal distribution [6].

The non-linear character of logarithmic function changes is a limitation to the use of the $L_{Aeq,T}$ parameter. This impedes both the determination of standard deviation or measurement uncertainty and the performance of a comparative analysis. It may also affect the results of statistical tests for normal distribution of the data [5]. Therefore, the authors of this paper decided to determine the *RMS* sound pressure ($p_{A_{RMS}}$) in the *T* period from equation (1) and use this parameter in further analysis.

$$p_{A_{RMS}} = p_A = \sqrt{10^{(0.1 * L_{Aeq,T})} * p_0^2} \tag{2}$$

Standard uncertainty of the parameter, determined in the Type A evaluation, can be calculated from the following relationship:

$$u_A = \sqrt{\frac{1}{n(n-1)}\sum_{i=1}^{n}\left(p_{A_{RMS_i}} - \overline{p_{A_{RMS}}}\right)^2} \quad (3)$$

In this study, the authors analysed $p_{A_{RMS}}$ from (2) expressed in terms of Pa to be able to easily compare the fixed components (the expected value and the median) and variable components (standard deviation, coefficient of variation – classical and positional) of the sound pressure signals recorded. The tests for the variable components contained in the signals were based on the classical and positional measures: standard deviation ($\sigma_{p_{A_{RMS}}}$), coefficient of variation (COV), quartile deviation Q_{31}, quartile variation coefficient ($V_{Q_{31}}$), and quartile coefficient of dispersion (V_{Q1Q3}). Standard deviation is an absolute measure commonly used for the analysis of sound pressure variable component. Standard deviation is estimated from (4), where *n* is the amount of data:

$$\sigma_{p_{A_{RMS}}} = \sqrt{\frac{1}{n-1}\sum_{i=1}^{i=n}\left(p_{A_{RMS_i}} - \overline{p_{A_{RMS}}}\right)^2} \quad (4)$$

This parameter defines the average variation in individual sound pressure values from the arithmetic mean. Standard deviation can be related to the expected value of the signal being analysed to obtain the coefficient of variation (*COV*). The *COV* is a dimensionless relative measure that can be used to directly compare the variable components in its several realisations. For the sound pressure tested, the *COV* can be expressed as [7]:

$$COV_{p_{A_{RMS}}} = COV = \frac{\sigma_{p_{A_{RMS}}}}{\overline{p_{A_{RMS}}}} \cdot 100\% \quad (5)$$

The value of the *COV* is greatly influenced by atypical data, taken into account in the analyses. This influence is less significant when positional measures are used. The measure of dispersion of the variable is the average quartile deviation:

$$Q_{31} = 0.5 \cdot \left[Q_3\left(p_{A_{RMS}}\right) - Q_1\left(p_{A_{RMS}}\right)\right] \quad (6)$$

Quartile deviation is an absolute measure that defines the average variance of half of the measurement data around the median (after rejecting 25% data with the lowest values and 25% data of the highest values of sound pressure), expressed in terms of pascal. By relating it to the median, the positional coefficient of variation is calculated from (7):

$$V_{Q_{31}} = \frac{Q_{31}}{Med} \cdot 100\% \quad (7)$$

The quartile coefficient of dispersion is a relative measure of variance, calculated from (8):

$$V_{Q1Q3} = \frac{Q_3 - Q_1}{Q_1 + Q_3} \cdot 100\% \quad (8)$$

The positional coefficient of variation and the quartile coefficient of dispersion are positional measures of the data between the first and third quartiles. Thus, atypical data exert less influence on these coefficients. It has to be noted, however, that the data under analysis represent the measurements collected within 24-hour periods, thereby atypical data cannot be regarded as erroneous measurements.

Table 1 summarizes the results of the data analysis for 2011, the *RMS* sound pressure values calculated according to relationship (2). The calculations show that the pressure increases on working days and decreases by 10 to 20 mPa on Saturdays and Sundays. Note large changes in the standard deviation values that range from 2 mPa to 8 mPa. The *COV* values range from 3% to 13%, and the highest value is obtained on Mondays. By contrast, the values of V_{Q31} and V_{Q1Q3} are in the range from 2% to 5%, with the highest values recorded on Sundays. The measurement uncertainty is the lowest on Wednesdays (0.56 mPa), with 1.20 mPa on Sundays which is a noticeably lower value than that recorded for Mondays (2.45 mPa).

Table 1. Values of basic statistical measures of $p_{A_{RMS}}$, determined for all 24h periods of the week in 2011

	Med. mPa	$\overline{p_{A_{RMS}}}$ mPa	$\sigma_{p_{A_{RMS}}}$ mPa	*COV* %	Q_{31} mPa	$V_{Q_{31}}$ %	V_{Q1Q3} %	u_A mPa
Monday	65.41	63.18	8.13	12.86	1.11	1.70	1.72	2.45
Tuesday	65.31	64.60	5.45	8.44	1.54	2.35	2.33	1.72
Wednesday	65.89	66.55	1.95	2.94	1.72	2.61	2.58	0.56
Thursday	67.08	65.72	7.04	10.71	2.28	3.40	3.40	2.03
Friday	67.29	67.29	3.14	4.66	2.42	3.60	3.58	0.87
Saturday	56.25	56.05	4.58	8.18	1.34	2.39	2.36	1.32
Sunday	47.30	46.53	3.79	8.14	2.43	5.13	5.19	1.20

Table 2 compiles the analysis results for *RMS* sound pressure data collected in 2016. The calculations show that this value varies from about 60 mPa to 65 mPa on working days and decreases by 10 to 15 mPa on Saturdays and Sundays. Standard deviation values range from 4 mPa to 6 mPa. The *COV* is in the range 6% to 10%, with the highest value achieved on Tuesday. By contrast, V_{Q31} and V_{Q1Q3}values range from 3% to 6%, with the highest value recorded on Saturdays. The measurement uncertainty is 1.35 mPa on Saturdays and 1.04 mPa (the lowest) on Fridays.

Table 2. Values of basic statistical measures of $p_{A_{RMS}}$, determined for all 24h periods of the week in 2016

	Med. mPa	$\overline{p_{A_{RMS}}}$ mPa	$\sigma_{p_{A_{RMS}}}$ mPa	*COV* %	Q_{31} mPa	$V_{Q_{31}}$ %	V_{Q1Q3} %	u_A mPa
Monday	61.90	61.01	4.66	7.63	2.59	4.18	4.21	1.34
Tuesday	61.73	61.34	5.87	9.57	2.41	3.91	3.87	1.69
Wednesday	60.61	62.56	4.26	6.80	2.28	3.76	3.68	1.23
Thursday	63.05	61.52	5.86	9.53	2.64	4.19	4.24	1.69
Friday	65.33	64.84	3.90	6.01	1.81	2.78	2.77	1.04
Saturday	55.60	56.26	4.86	8.64	3.38	6.08	6.04	1.35
Sunday	50.68	50.80	4.12	8.11	2.38	4.70	4.72	1.19

Analysis of the results in Tables 1 and 2 indicates that the renovation of Popiełuszki Avenue reduced the *RMS* sound pressure on working days. This value changed slightly on Saturdays, but on Sundays it increased by about 10%. The average value of coefficient V_{Q31}increased from about 3% in 2011 to around 4% in 2016 while the *COV* did not change and remained at the level of about 8%. The lowest measurement uncertainty was 0.6 mPa in 2011 and 1.0 mPa in 2016. Figure 2 shows the box plots of the *RMS* sound pressures in 2011 and 2016

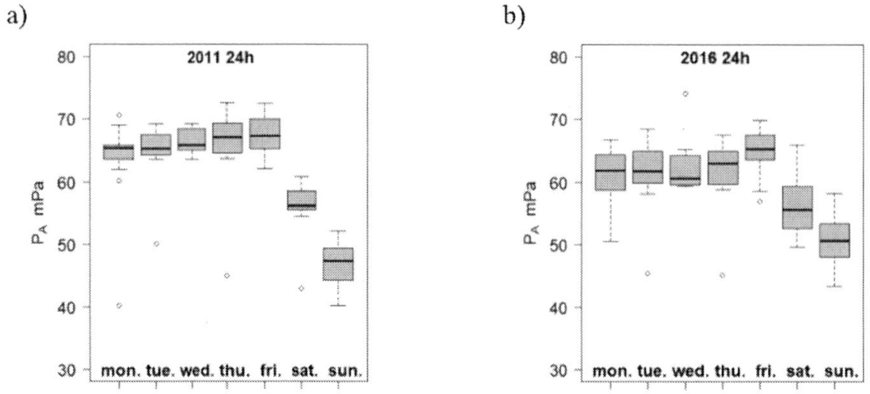

Fig. 2. Box plots of $p_{A_{RMS}}$ for each 24-hour period of the week: a) in 2011, b) in 2016.

The outliers visible in the graphs should be accounted for in the analysis because, in the authors' view, they do not represent measurement errors. Analysis of the measurement results database revealed the outliers on the following days of public holidays in 2011: Easter Monday – Monday, Constitution Day – Tuesday, Corpus Christi – Thursday, Easter Saturday – Saturday, and in 2016; Constitution Day – Tuesday, Corpus Christi – Thursday. The data confirmed a regularity between the box plots in Fig. 2a and in Fig. 2b, showing that the highest *RMS* sound pressure values occurred on Fridays. Figure 3 depicts the $V_{Q_{31}}$ values split into days of the week in 2011 and 2016. The nature of these changes in each year is different, but the coefficients are in the range of about 2% to 6%. The character of the V_{Q1Q3} changes is very similar to that shown in Fig. 3.

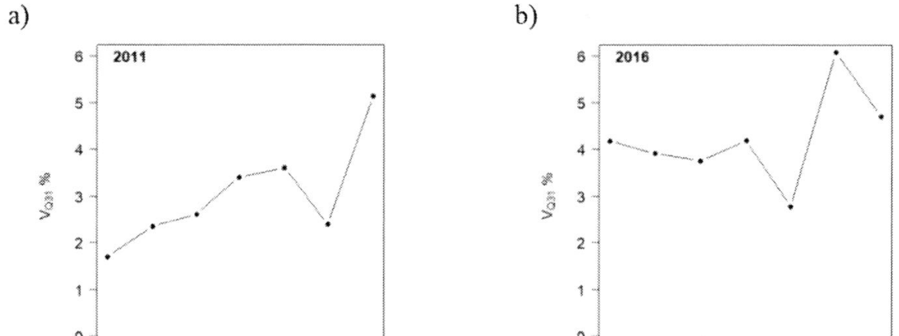

Fig. 3. Relationship between the value of $V_{Q_{31}}$ and particular days of the week a) in 2011, b) in 2016.

Figure 4 shows the changes in measurement uncertainty values of sound pressure u_A and *COV* on particular days of the week. The measurement uncertainty values of sound pressure vary with the year but remain lower than 2.5 mPa, and the differences on Saturdays and Sundays are minor. As for the *COV*, its values in the two years are different, but remain lower than 12.5%, with minor differences on Saturdays and Sundays.

a)   b)

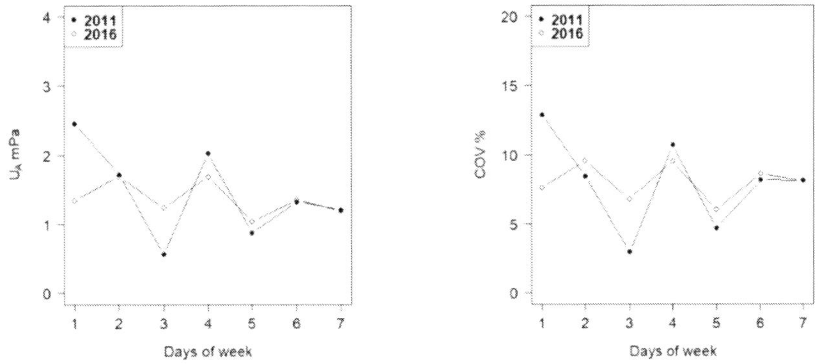

Fig. 4. Comparison of the values of a) sound pressure measurement uncertainty – Type *A* evaluation, b) the *COV*, recorded in 2011 and 2016.

The values of u_A and *COV* against the average value of *RMS* sound pressure are shown in Fig. 5a and 5b.

a)   b)

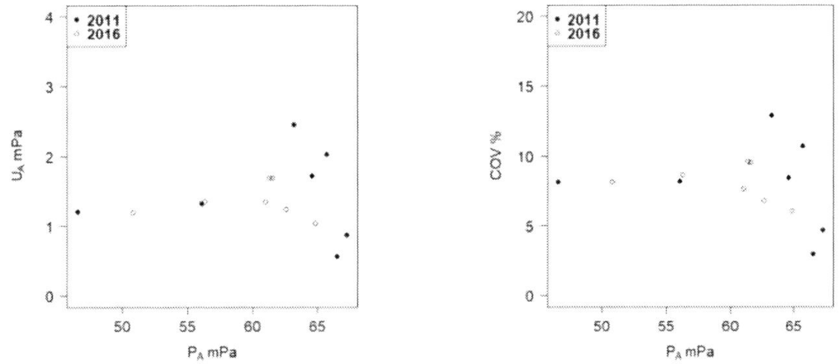

Fig. 5. Average value of the *RMS* sound pressure versus a) sound pressure measurement uncertainty – Type A evaluation, b) the *COV*, for the measurement data of 2011 and 2016.

Both drawings show some similarity despite the fact that they represent different physical quantities. For pressures up to about 60 mPa, the dependencies are close to linear. For higher pressure values, the dependency is difficult to indicate. Note that both parameters i.e., u_A and *COV* depend on the $\sigma_{p_{A_{RMS}}}$ standard deviation. The variability interval of standard deviation in 2011 is significantly higher than in 2016, which results in a large spread in Fig. 5. The $\sigma_{p_{A_{RMS}}}$ parameter values depend on the technical condition of the road and motor vehicles, which in 2011 was far worse than in 2016. As the use of roads causes their wear, the values of u_A, *COV*, V_{Q31} and V_{Q1Q3} will increase over time. This problem requires further analysis.

Conclusions

The data collected for the equivalent sound level, measured over the 24-hour period and generated by vehicles passing along Popiełuszki Avenue, Kielce between March and June in 2011, before road renovation and in the corresponding period of 2016 after the renovation, showed that the $L_{Aeq,T}$ values fluctuated slightly. Analysis of *RMS* sound pressure values revealed that compared with 2011, the *RMS* sound pressure decreased on working days in 2016. There was a minor change on Saturdays, while on Sundays the values increased by about 10%. The average value of the $V_{Q_{31}}$ coefficient of variation increased from about 3% in 2011 to about 4% in 2016 while the *COV* remained at the level of about 8%. The lowest measurement uncertainty was 0.6 mPa in 2011 and 1.0 mPa in 2016. The highest *RMS* sound pressure values were recorded on Fridays. The nature of changes in the sound pressure coefficient varies from day of the week, but is different in each year, but the values of the coefficients range from about 2% to 6%. The sound pressure measurement uncertainties in 2011 and in 2016 were different, but never exceeded 2.5 mPa, and on Saturdays and Sundays differed only slightly. Likewise, the *COV* values in 2011 and in 2016 were different, but never exceeded 12.5% and on Saturdays and Sundays only minor differences were recorded.

References

1. COM (2013) 913 final: *Together towards competitive and resource-efficient urban mobility*, Brussels, 17 December 2013
2. S, Shaho, A. Dehrashid, P. Nassiri, *Traffic Noise Assessment in the Main Roads of Sananda*j, Iran, Journal of low frequency noise, vibration and active control **34** (1), (2015)
3. Directive 2002/49/Ec of the European Parliament and of the Council of 25 June 2002 relating to the assessment and management of environmental Noise, Official Journal of the European Communities – 18.07.2002
4. A. Bąkowski, L. Radziszewski, Z. Skrobacki, *Assessment of uncertainty in urban traffic noise measurement*, Procedia Engineering **177**, (2017)
5. W. Batko, B. Stępień, Type *A Standard Uncertainty of Long-Term Noise Indicators*, ARCH. ACOUST. **39** (1), 25-36 (2014)
6. T.Wszołek, M. Kłaczyński, *Effect of traffic noise statistical distribution on Laeq,T measurement uncertainty* Arch. Acoust. **31**, 4 (Supplement), 311-318 (2006)
7. A. Bąkowski, L. Radziszewski, M. Žmindak, *Analysis of the coefficient of variation for injection pressure in a compression ignition engine*, Procedia Engineering **177** (2017)

Determination the maximum load capacity of the welded structure of the transport carriage in operation

Vladislav Baniari[1,*], *Marián* Handrik[1], *Milan* Vaško[1], *Ján* Kortiš[2], *Ľuboš* Daniel[2]

[1]University of Žilina, Faculty of Mechanical Engineering, Department of Applied Mechanics, Univerzitná 8215/1, Žilina, Slovakia

[2]University of Žilina, Faculty of Civil Engineering, Department of Structural Mechanics, Univerzitná 8215/1, Žilina, Slovakia

Abstract. The article presents a calculation procedure for determining the maximum permissible load on the welded structure of the transport carriage. During the transport carriages operation inside the production hall there was no breakage of the welding joints. The use of carriages outside the hall and with an uneven surface began to create cracks and fractures in the weld joints. The values of vertical accelerations that were generated by running on uneven surfaces were determined by experimental measurement. To avoid cracks in welded joints, the maximum permissible load value of the transport carriage was subsequently determined by calculation using FEM.

Keywords: weld, weld joint, crack, deformation, FEM, FE analysis

1 Introduction

The purpose of the carriage is transport some various kinds of costs. During use of the carriage in the exterior on the construction were started some deformations which are looked like a rift of welding joints. The carriage is pulling by the same standard electric tractor in the interior and in the exterior as well.

Based on the request there were made some experimental measurements by using the carriage in the exterior [1, 2]. Subsequently, their results were evaluated and they were used as an input data for the calculation by the Finite Element Method (FEM) [3].

2 Experimental measurement and evaluation of results

During the process of measurement there were used piezoelectric accelerometers which were fitted on unsuspended axle near the wheels (Fig. 1) [4, 5]. Accelerometers were marked ACC 1 till ACC 4 according Fig. 2.

* Corresponding author: vladislav.baniari@fstroj.uniza.sk

Reviewers: *Marcin* Kubiak, *Bohuš* Leitner

Fig. 1. Accelerometers on axles

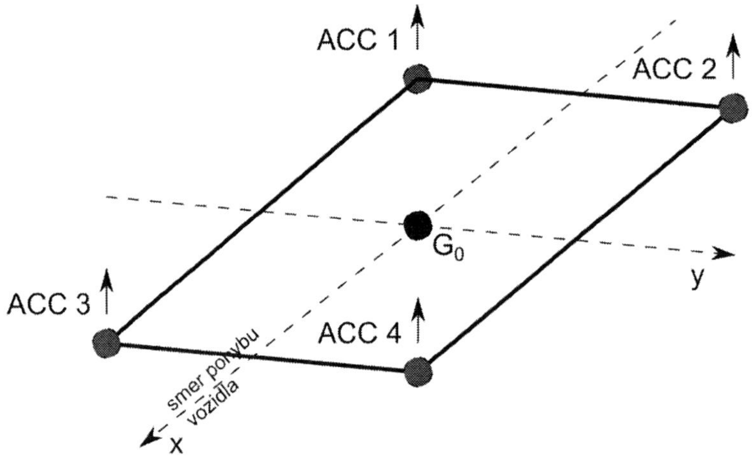

Fig. 2. Marking of accelerometer versus move direction

Steel frames were used as the burden. They were typed as the hardest possible way of the burden which can be transported by this carriage. Four types of measurements there were noted down by using the accelerometers [6, 7].

Several values with different tops were measured by the process of measurement. The highest tops were chosen and written into the Table 1.

Table 1. Measured accelerations

Motion	**max ACC1 [ms^{-2}]**		**max ACC2 [ms^{-2}]**		**max ACC3 [ms^{-2}]**		**max ACC4 [ms^{-2}]**	
	plus	**minus**	**plus**	**minus**	**plus**	**minus**	**plus**	**minus**
Slow motion on a paneled road (empty carriage)	5.114	-5.113	4.797	-5.12	5.116	-5.114	5.112	-5.117
Fast motion on a paneled road (empty carriage)	14.405	-8.046	14.699	-9.09	10.859	-10.14	11.428	-7.863
Slow motion on a railway crossing (empty carriage)	13.296	-7.82	10.314	-9.69	14.735	-10.5	11.085	-7.86
Slow motion on a railway crossing (loaded carriage)	12.927	-8.07	12.113	-8.64	12.674	-8.94	10.684	-8.21

3 FEM analyses

The basic request for the stress analysis of the transport carriage is to analyze the condition of the state of stress in the carriage parts of the construction by the transition of the carriage through the obstacle. We can see from the results of some experimental measurements that when the empty carriage went through the obstacle by the speed of 4–6 km·h^{-1} appeals on the construction additional acceleration in vertical way in the size $a_p = 14.735$ m·s^{-2}. This value is the highest of the measured values and it represents the biggest load which effects the construction of the carriage in the exterior [5, 8, 9].

On the model of the carriage parts was applied additional acceleration in the value of $a_p = 14.735$ m·s^{-2}. The complete acceleration was created by the gravitation and the additional acceleration of $a_v = 24.545$ m·s^{-2}.

The weight of superstructure included the door was substituted by the force effect in the places of connections of the superstructure into the carriage part of the construction of the transport carriage. The weight of the superstructure was determined as the conjunction of the result acceleration and the weight of the construction of the superstructure. The weight of the load represents the pressure on the floor which can be given as the weight on one square meter multiplied by the value of the result acceleration [1, 10, 11].

The construction will be suspended with the aim of the permission of deformations of the construction in the places of the wheels. The elastic constrain of the construction is chosen from the reason of small spring loading of the wheels and friction constrain of the wheel with the roadway.

The construction of the carriage is created by the rolled profiles. The thickness of the rolled profiles is much smaller than other dimensions of the profiles in these construction elements. This type of the construction in the working out of deformation-stress analysis can be substituted by shell elements [4, 12].

The linear elastic material model was used during the preparation of the mathematical model. The usage of the linear material model is adequate and it is not needed to use nonlinear material models, whereas there is the basic expectation that the construction has to be operated from the point of view of the secure use in the allowed stresses area. This ensures that the stress is lower than the Yield strength [1, 13].

3.1 Mesh of the geometric model

The geometrical model was created in CAD system and imported into the CAE software with divided parts. According to standard progress there were extracted middle planes parts which will be working out through the shell elements.

The remaining parts are made by the solid elements as we can see in the Fig. 3 and 4.

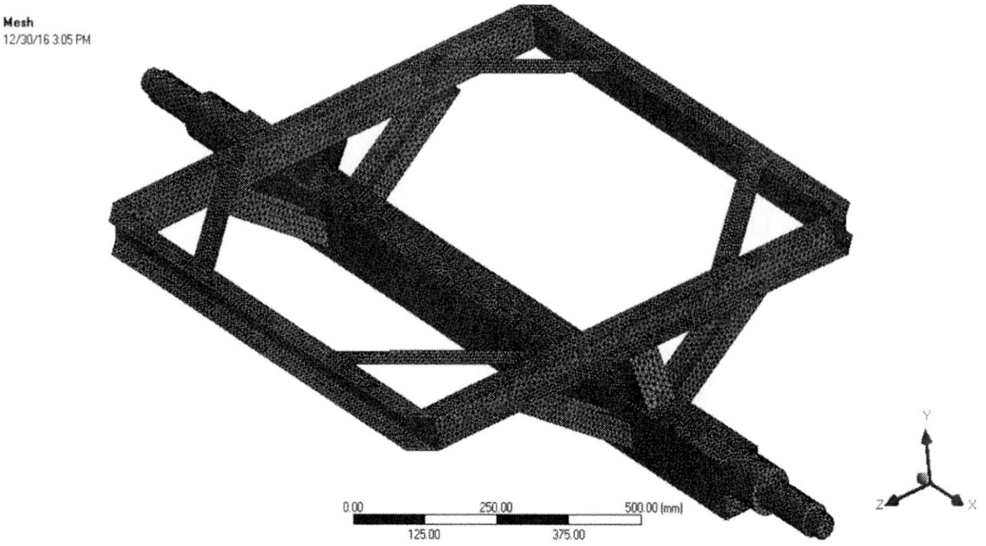

Fig. 3. Undercarriage with solid elements

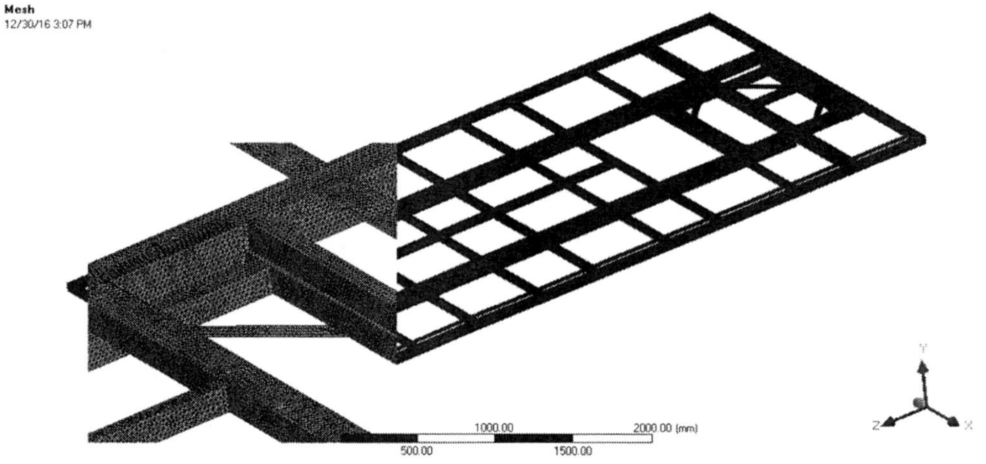

Fig. 4. Frame with shell elements

3.2 Boundary conditions

Springs were used in terms of deformation boundary conditions. Generally, there were used 12 springs on the model according to the Fig. 5. The forces boundary conditions were depended on the parts of the construction and their joints. The weight of the construction and substruction were applied as a force. The weight of load was applied by the way of pressure on the floor [14, 15].

The created finite element model of the transport carriage is made from 10 independent bodies and 10 independent meshes of the finite elements as well [16, 17]. The meshes are connected through the contact which is defined between following parts of the construction:

- the contact between the wheel pins and the undercarriage with the gluemesh in 4 places,
- the contact between the undercarriage and the turning platform with the gluemesh in 2 places,

- the contact between the frame and the turning platform with the gluemesh in the 2 places,
- the contact between the floor and the frame without friction in 1 place. The contact between the floor and the frame has to be carried by normal load; tangential components making the membrane load in the floor cannot be carried. The floor is laid on the frame - it is not connected hard.

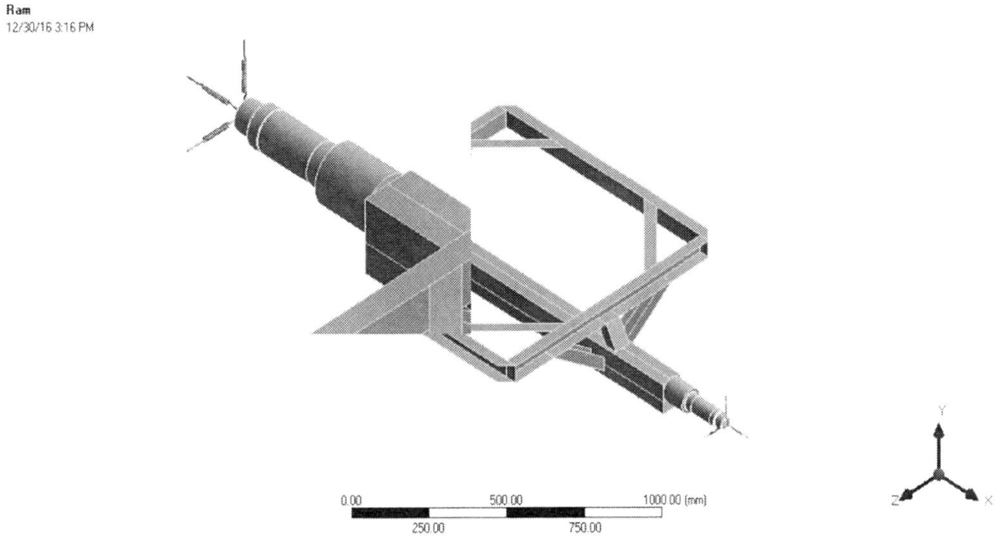

Fig. 5. Deformation boundary condition on the wheel shaft

4 FEA results

According to Figures 6 to 9 we can see two conditions which were solved by von Mises. The first condition is marked as Time 1 and its load is in the static way, i.e. without accelerations. The second condition is marked as Time 19 and it is dynamic, i.e. with accelerations.

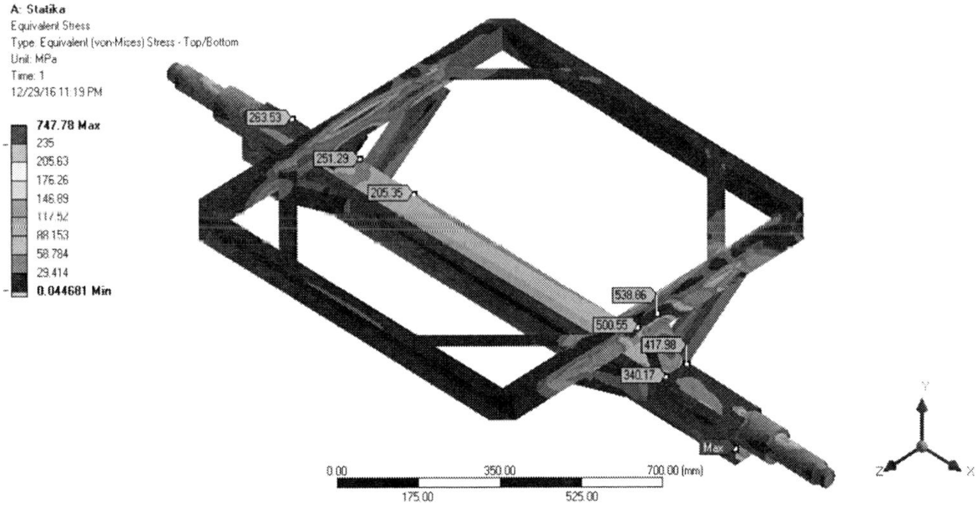

Fig. 6. Von Mises stress on undercarriage in Time 1

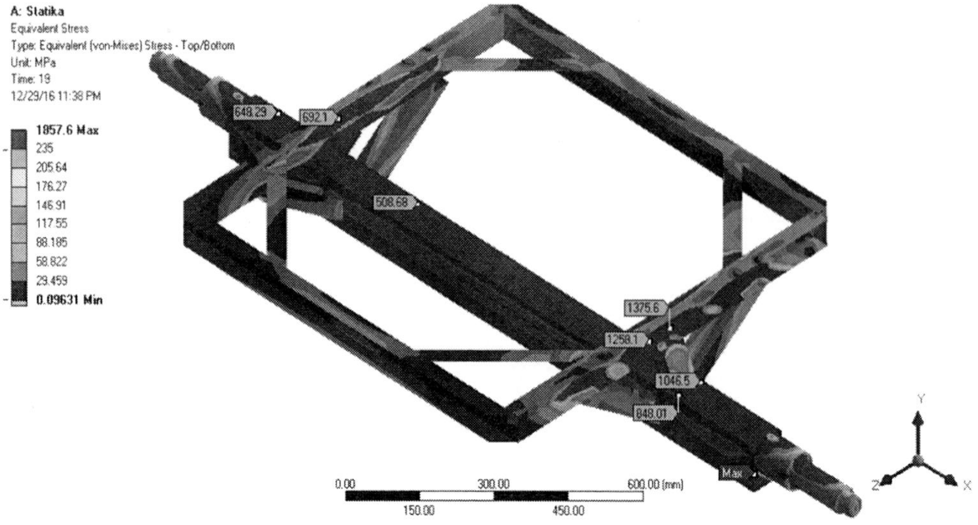

Fig. 7. Von Mises stress on undercarriage in Time 19

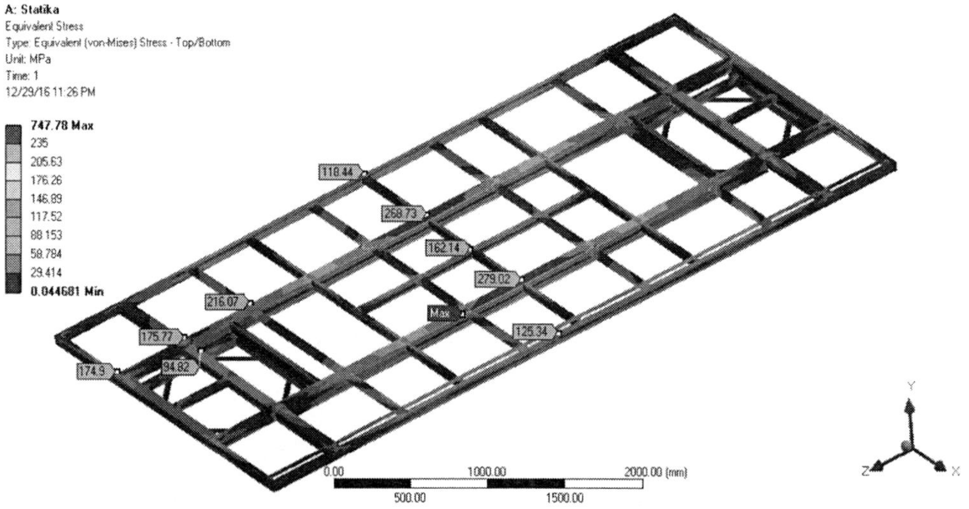

Fig. 8. Von Mises stress on a frame in Time 1

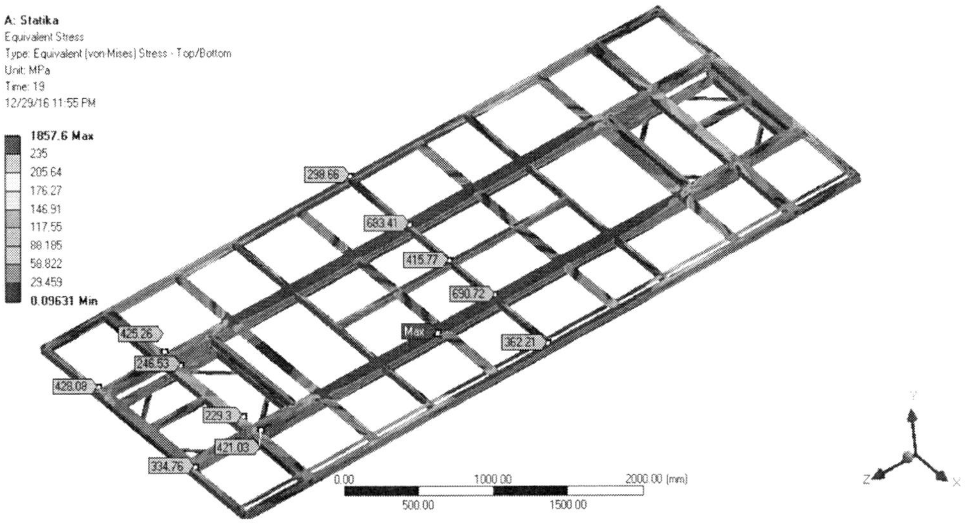

Fig. 9. Von Mises stress on a frame in Time 19

Solution for both problem versions clearly show that the construction cannot hold in the exterior and the caused stress in tolerable transcending the allowing Yield stress of the S 235 material. Welded joints are loaded excessively and dynamic effects may cause cracking of the joints and subsequent destruction of the frame.

Conclusion

The transport carriage has been designed to operate in the interior of the production hall. Usage the transport carriage during transport of the material in outdoor environments with significantly uneven surface causes additional loads from dynamic effects.

Local points where the stress state exceeds the yield strength were determined by analyzing the stress state of the welded structure of the carriage under static load mode. The stress state analysis was done also for dynamic load mode. Results of the analysis have shown that there is significantly exceeded yield strength and maximum allowable load of the carriage needs to be reduced to approximately 1/3 of the load.

The proposed reductions in the carrying capacity of the transport carriage allow safe operation and significantly reduce the probability of fatigue cracks at the weld joints. However, the carriage failure due to fatigue cracks cannot be completely ruled out due to the unknown history of the previous operation. Existing cracks that arose in the previous operation may continue to grow even at reduced load.

References

1. M. Handrik, P. Kopas, V. Baniari, M. Vaško, M. Sága, *Analysis of stress and strain of fatigue specimens localised in the cross-sectional area of the gauge section testing on bi-axial fatigue machine loaded in the high-cycle fatigue region*. Procedia Engineering **177**, 516-519 (2017)
2. R. Halama, P. Macura, L. Pečenka, F. Fojtík, M. Šofer, *Experimental Analysis of Residual Stresses in Backup Roll and FE Solution*. Experimental Stress Analysis - 49th Int. Scient. Conf. on Experimental Stress Analysis, 85-90 (2011)
3. K. J. Bathe, *Finite Element Procedures*. (Prentice Hall, New Jersey, 1982)

4. L. Jakubovičová, M. Sága, *Computational analysis of contact stress distribution in the case of mutual stewing of roller bearing rings*. Novel Trends in Production Devices and Systems, Applied Mechanics and Materials **474**, 363-368 (2014)
5. J. Gerlici, T. Lack, *Modified HHT method for vehicle vibration analysis in time domain utilisation*. Applied Mechanics and Materials **486**, 396-405 (2014)
6. J. Ondrouch, P. Ferfecki, Z. Poruba, *Active vibration reduction of rigid rotor by kinematic excitation of bushes of journal bearings*. Metalurgija **49** (2), 107-110 (2010)
7. R. Grega, J. Homišin, J. Krajňák, M. Urbanský, *Analysis of the impact of flexible couplings on gearbox vibrations*. Scientific Journal of Silesian University of Technology-Series Transport **91**, 43-50 (2016)
8. A. Sapietová, V. Dekýš, *Dynamic analysis of rotating machines in MSC.ADAMS*. Procedia Engineering **136**, 143-149 (2016)
9. R. Grega, J. Krajňák, L. Žuľová, G. Fedorko, V. Molnár, *Failure analysis of driveshaft of truck body caused by vibrations*. Engineering Failure Analysis **79**, 208-215 (2017)
10. J. Vavro, J. Vavro Jr., P. Kováčiková, R. Bezdedová, *The optimization of the materials properties for the passenger cars in dependence on defect distribution at the dynamic loading*. Procedia Engineering **136**, 114-119 (2016)
11. A. Vaško, J. Belan, L. Kuchariková, E. Tillová, *Low and high frequency fatigue tests of nodular cast irons*. Metalurgija **56** (1-2), 25-28 (2017)
12. J. Rojek, M. Hyrcza-Michalska, A. Bokota, W. Piekarska, *Determination of mechanical properties of the weld zone in tailor-welded blanks*. Archives of Civil and Mechanical Engineering **12** (2), 156-162 (2012)
13. P. Kopas, M. Blatnický, M. Sága, M. Vaško, *Identification of mechanical properties of weld joints of AlMgSi07.F25 aluminium alloy*. Metalurgija **56** (1-2), 99-102 (2017)
14. V. Bulej, J. Uricek, V. Poppeova, R. Zahoransky, M. Rupikova, *Study of the Workspace of Hybrid Mechanism Trivariant*. Appl. Mech. and Mat., Mech. Eng. and Manufacturing **436,** 366-373 (2013)
15. M. Bruna, D. Bolibruchova, R. Pastircak, *Numerical simulation of porosity for Al based alloys*. Procedia Engineering **177**, 488-495 (2017)
16. M. Sága, P. Kopas, M. Uhríčik, *Modeling and experimental analysis of the aluminium alloy fatigue damage in the case of bending-torsion loading*. Procedia Engineering **48**, 599-606 (2012)
17. B. Strnadel, P. Ferfecki, P. Židlík, *Statistical characteristics of fracture surfaces in X70 steel DWTT samples*. Engineering Fracture Mechanics **112**, 1-13 (2013)

Investigation of kinematic parameters of two nonconventional piston machines with wobble board

Peter Baran[1,*], *Pavol* Šťastniak[1], *Pavol* Kukuča[1], *Miloš* Brezáni[1]

[1]University of Žilina, Faculty of Mechanical Engineering, Department of Transport and Handling Machines, Univerzitná 8215/1, 010 26 Žilina, Slovak Republic

Abstract. Article deals with description of non-conventional mechanism FIK and mechanism with ring, which may be suitable for special applications in the engineering industry. It is possible to use not only classical crank mechanisms, but also non-conventional mechanisms in design of piston machines. Investigation of kinematic parameters of the machines is important step to know its basic movement behaviour, and from this, it is then possible to create the alternate dynamic model or force analysis. For kinematic comparison will be designed the same basic geometric conditions in both types of models. In terms of comparisons will be monitored mainly points on wobble board and the main part will be determination of piston stroke.

Keywords: Non-conventional mechanism, piston machine, kinematic analysis

1 Mechanism with ring and FIK

The principle of changing reciprocating motion of the pistons into a rotary motion of output shaft has both mechanisms through a swinging plate, but in a case of construction, they are not identical mechanisms.

1.1 Mechanism with ring

Mechanism with ring (Fig. 1) is a swinging system, which is characterized, that the centre of the swinging plate (4) acting circular motion around *z* axis during the rotation of the shaft (1). Ring (3) is used against parallel rotation of swinging plate with shaft. Reciprocating movement of the piston (6) is transformed to the rotational movement of shaft through ball joint segments (5) [1]. This schematic and virtual representation was designed for application in type of a Stirling engine, especially for connection of ball joint segment with swinging plate. Figure 1 shows simple CAD model with basic dimensions and position of piston in default case.

* Corresponding author: peter.baran@fstroj.uniza.sk

Reviewers: *Robert* Grega, *Krzysztof* Talaśka

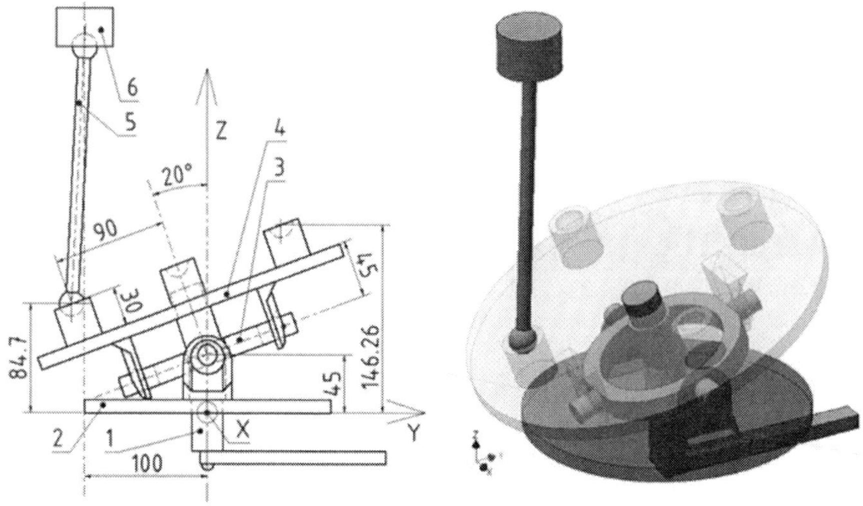

Fig. 1. Mechanism with ring - schematic representation (left side): 1 - output shaft, 2 - part of crankcase (fixed part), 3 - ring, 4 - swinging plate, 5 - ball joint segment, 6 - piston, CAD model (right side)

1.2 Mechanism FIK

FIK mechanism (Fig. 2) is also a swinging system. Centre (centre of gravity) of the swinging plate (4) makes circular motion during the rotation of the shaft (1). Bevel gear (3) is used against parallel rotation of swinging plate with shaft. Reciprocating movement of the pistons (6) is transformed to the rotational movement of shaft through ball joint segments (5). This schematic and virtual representation was designed for application in type of a Stirling engine, specifically for alpha configuration [1].

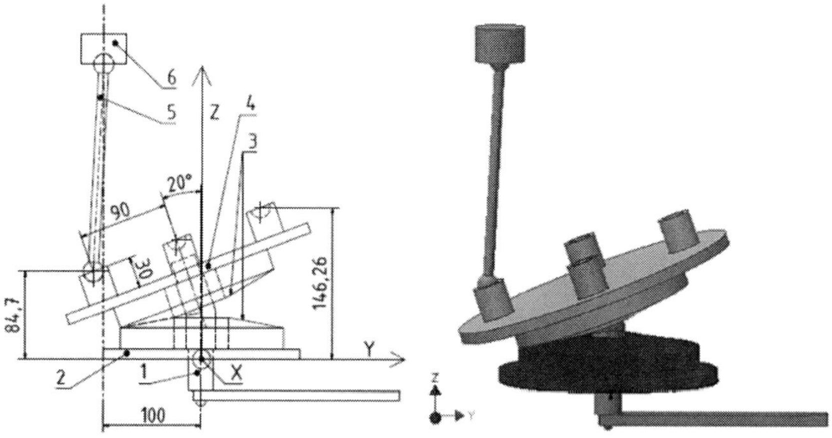

Fig. 2. Mechanism FIK: 1 - shaft, 2 – crankcase (fixed part), 3 – bevel wheels, 4 – swinging plate, 5 - ball joint segment, 6 – piston, CAD model (right side)

2 Initial conditions and dimensions

The models of both mechanisms are identical in terms of dimensions (basic geometry model in Fig. 3). The kinematic analysis of the ring mechanism was created in Autodesk Inventor dynamic simulation environment and also due to verification in MSC.Adams. The FIK mechanism was created through a mathematical model.

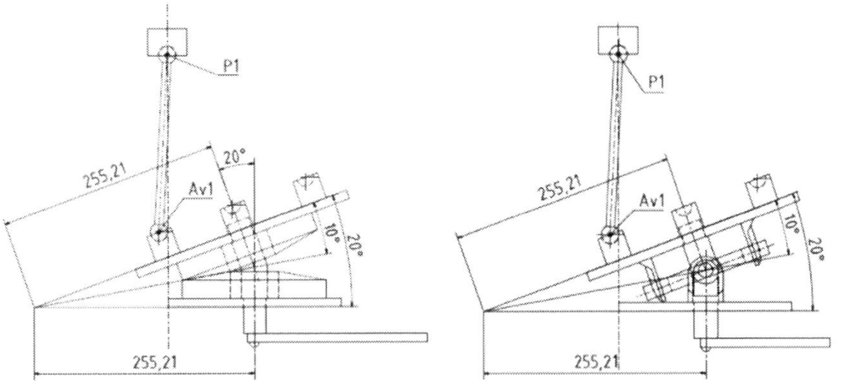

Fig. 3. Basic geometrical parameters of mechanism FIK and mechanism with ring

In a case of mechanism with a ring it was investigated five points on the board in angular range of 0 - 90° (see Fig. 4 right side). This analysis was made because of different trajectories of these points. Mechanism FIK was investigated only as one-point model design, because of immutability with regard to the position of the given point in the same pitch diameter on swinging plate.

2.1 Determination of points on the swinging plate

Basic dimensions of schematic model are the most important parameters for creating a CAD model [2]. Kinematic analysis refers to the observation of the movement (in all three axes x, y, z) of five points in the swinging plate (see Fig. 4). For each of these five positions will be determined waveforms and trajectories for points Av1 and P1. During the analysis of the piston stroke was used always the same ball joint segment (distance between spherical centres is 200 mm). The position of cylinder axis is also set by angular adjustment as points Av1.

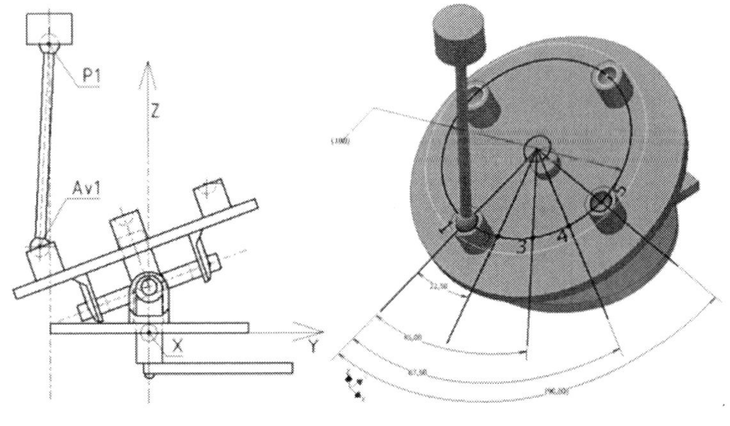

Fig. 4. Determination of five positions on swinging plate: schematic representation of examined points (left side), position of points (right side)

Point marking (as shown in Fig. 4):

- Av1 means point on a swinging plate, point P1 is point of piston connecting part with ball joint segment.
- Av1_1 means point on a swinging plate, first position, P1_1 means point of piston first position (it is only in case of mechanism with ring, marking range from Av1_1 to Av1_5 and from P1_1 to P1_5).
- Av1_FIK and P1_FIK means marking of points in mechanism FIK.

3 Kinematic analysis

3.1 Analysis of trajectories (mechanism with ring)

The first preview of the trajectories projections in the individual planes are shown in the Fig. 5. Point markings means Av1_1 is Av1 and Av1_5 means last fifth point. From figure 5 it is possible to see differences in the courses of individual points on swinging plate.

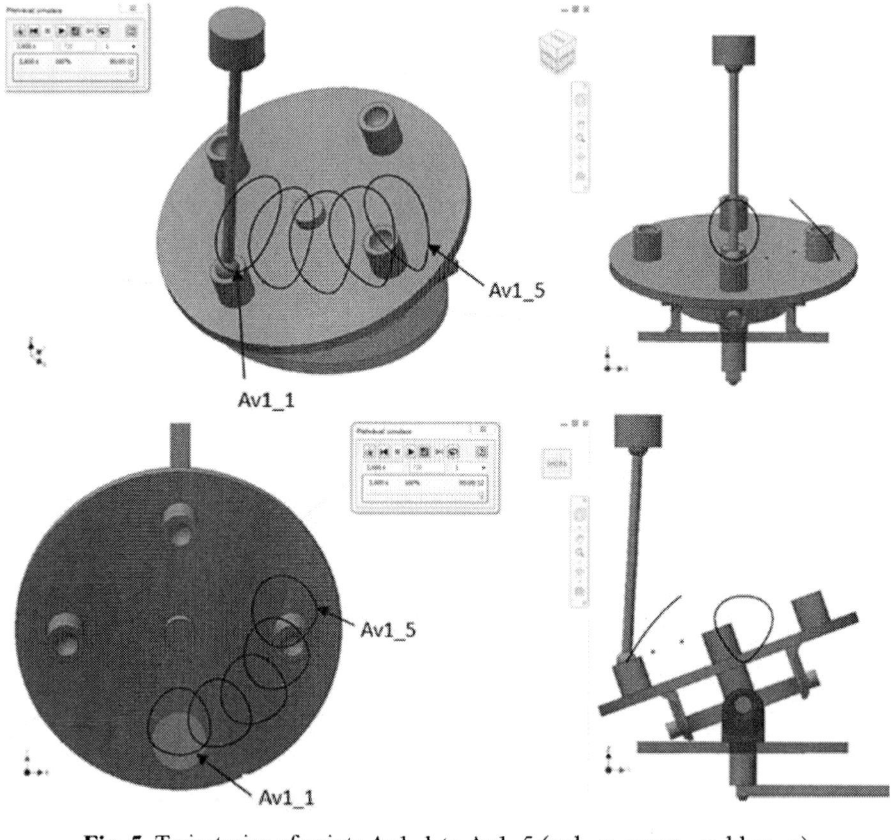

Fig. 5. Trajectories of points Av1_1 to Av1_5 (red - x, green - y, blue - z)

Setting conditions for changing the coordinate system position with grounded part of a model. In this case, the *x*, *y* axes will be rotated around the *z* axis, according to the positions of the individual points Av1_1 to Av_5, because it is easier for comparison of all waveforms. CAD model, kinematic analysis, and the courses were created in Autodesk Inventor. The simulation was set at a constant angular velocity of the output shaft and its duration was calculated to perform exactly one revolution. For the points on the swinging plate, were plotted the following projections in the individual planes (Fig. 6).

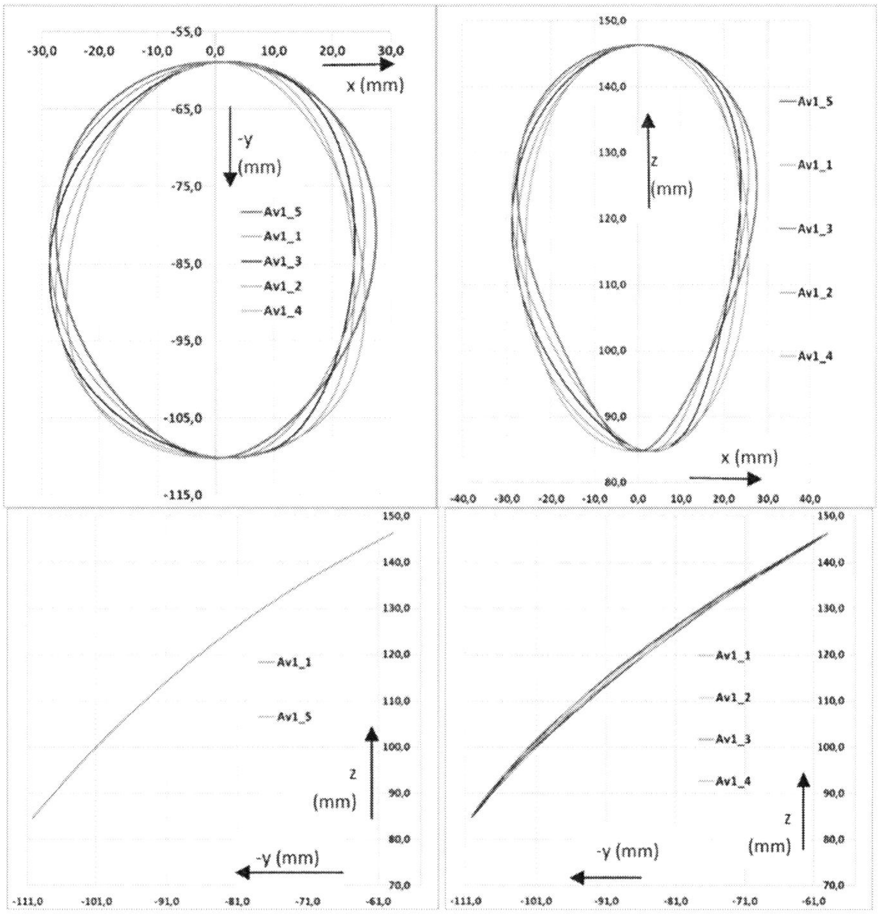

Fig. 6. Projections of trajectories of points (in millimeters) Av1_1 to Av1_5 in (at the top, from left side of picture) x-(-y), x-z and z-y plane (at the bottom)

3.2 Analysis of trajectories (mechanism FIK)

In this case, the model was described by mathematical equations (detailed description is in literature [1], pages 122-127). Movement of point Av1 is described by equations 1, 2, 3 (Cartesian coordinate system).

$$1\ x=\sin\alpha\cdot\left(R_{ko}\cdot\cos\alpha-R\right)\cdot\left(1-\cos\varphi\right) \tag{1}$$

$$y=\left(R\cdot\cos\alpha+R_{ko}\cdot\sin^2\alpha\right)\cdot\left(1-\cos\varphi\right)+R_{ko}\cdot\cos\varphi \tag{2}$$

$$z=\left(R-R_{ko}\cdot\cos\alpha\right)\cdot\sin\varphi \tag{3}$$

The resulting equation for calculating the stroke of point P1 is [1]:

$$z_p=z+\sqrt{l_0^2-(x_{ov}-x)^2-(y_{ov}-y)^2} \tag{4}$$

Where:
- α - rotation angle of shaft,
- R - radius of base rolling cone (see Fig. 3, distance 255,21 mm),

- R_{ko} - distance between angle cranked axis of shaft and centre of lower ball joint centre (Fig. 2, distance 90 mm),
- φ - angle between main axis and axis of cranked part of the shaft (by Fig. 2 model, 20°),
- l_0 - length of ball joint segment (length of the connecting rod),
- x_{ov}, y_{ov} - position in coordinates of the cylinder axis in the *x*-*y* plane.

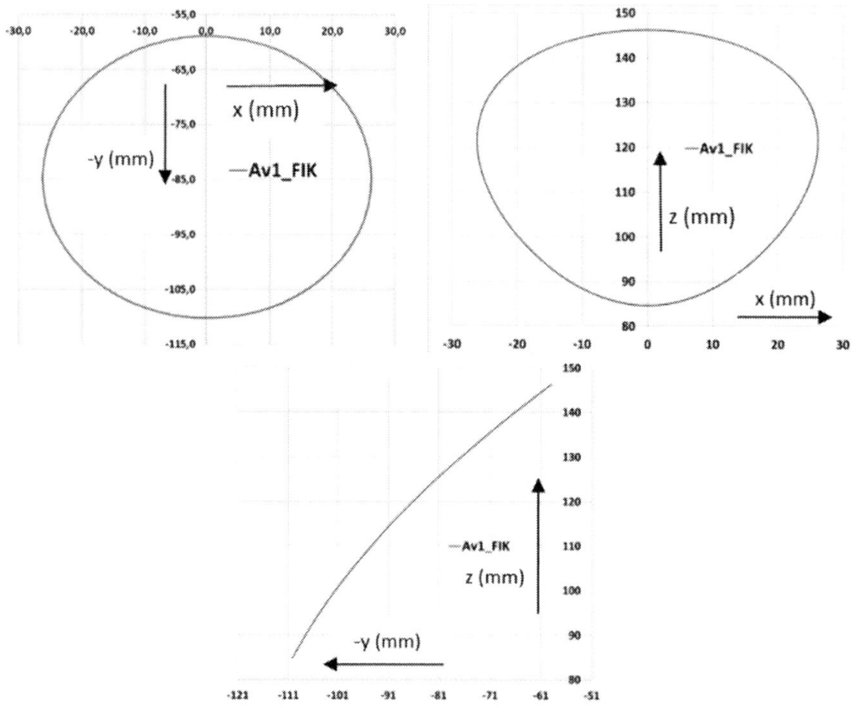

Fig. 7. Projections of trajectories of point (in millimeters) Av1_1 in (at the top, from left side of picture) x-(-y), x-z and z-y plane (at the bottom)

3.3 Piston stroke

Courses of the piston strokes will depend on the courses of Av1 points in the vertical (*z*) direction. The Fig. 8 shows the differences of courses of piston strokes (mechanism with ring) and mechanism FIK (according equation 4).

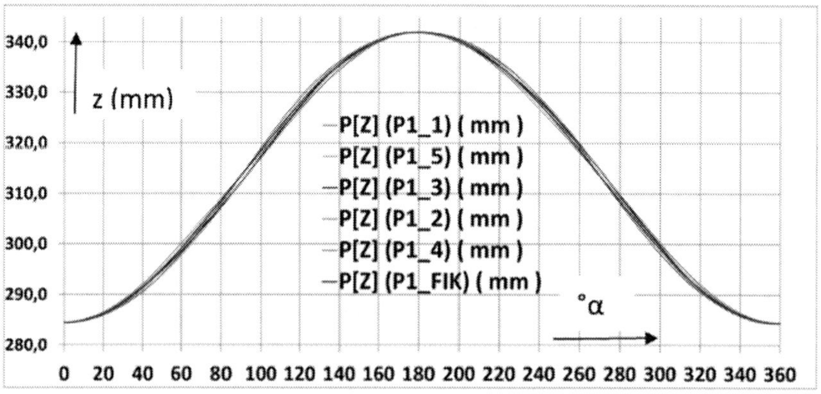

Fig. 8. The sequence of movements of the points P1_1 to P1_5 in z direction according to the angle of rotation of the output shaft

3.4 Investigation of the differences in strokes P1 and Av1

In this case it will be pointed to the differences in the plotted of the piston stroke and points Av1. The curves in Fig. 9 were modified so that the P1 stroke curve was approximately in the Av1 point curve region. This analysis is to show how are these curves different.

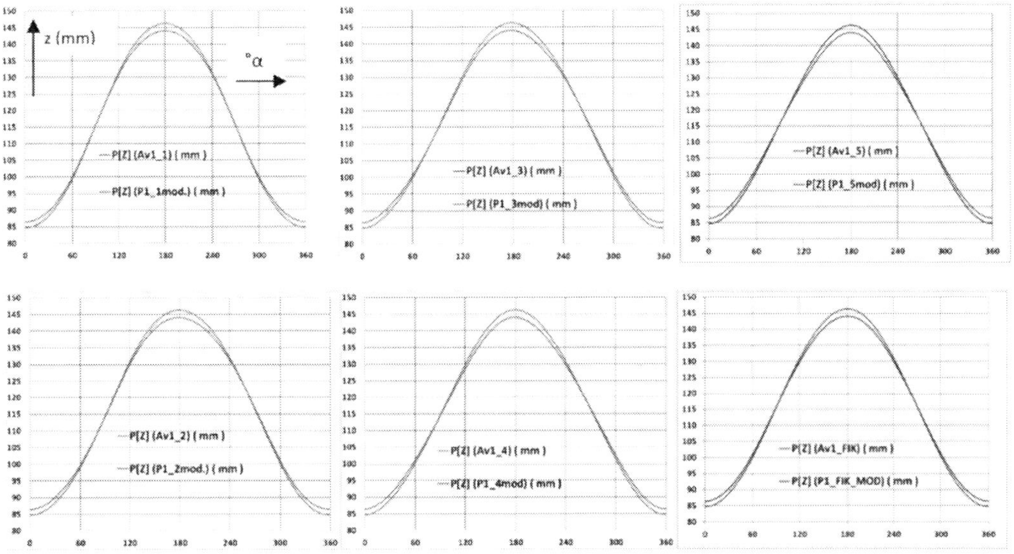

Fig. 9. Courses for P1 and Av1 (horizontal axis °α, vertical axis *z* for all diagram)

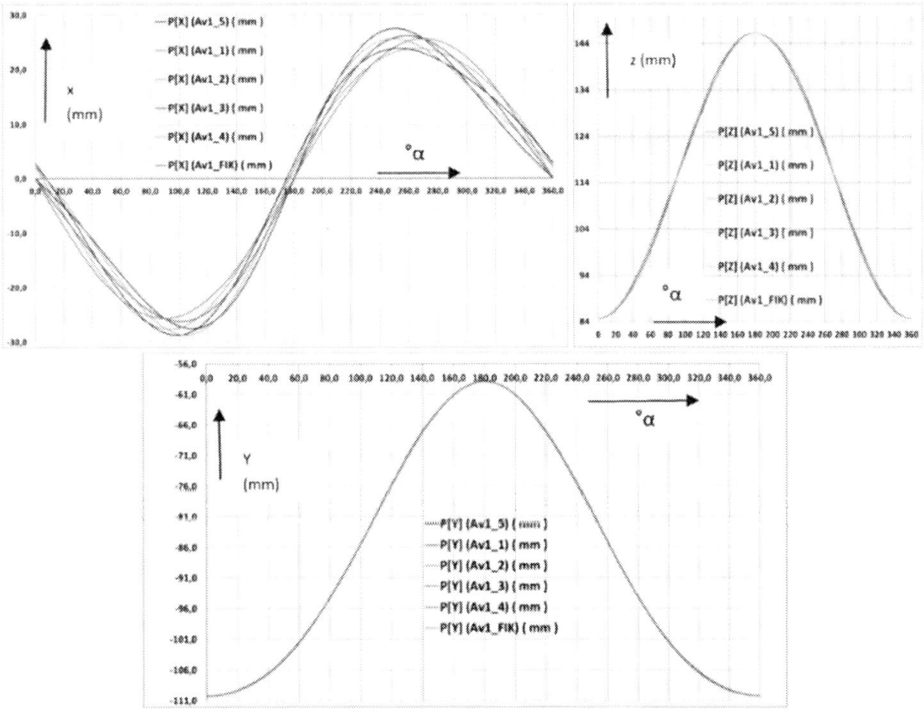

Fig. 10. The sequence of movements of the point Av1_1 to Av1_5 in all three planes according to the angle of rotation of the output shaft

3.5 Overall view of the swinging plate points

In Autodesk Inventor were finally generated trajectories of points in other quadrants of the swinging plate. The aim was an overview of the movement behaviour of the ring mechanism (Fig. 11). From Fig. 11 it can be seen that the trajectory Av1_1 (marked 1) and Av1_9 (marked 9) are similar, mirror-reversed. According to the same procedure as on figure 11, the mechanism FIK would have identical trajectories on the designated circle of the swinging plate.

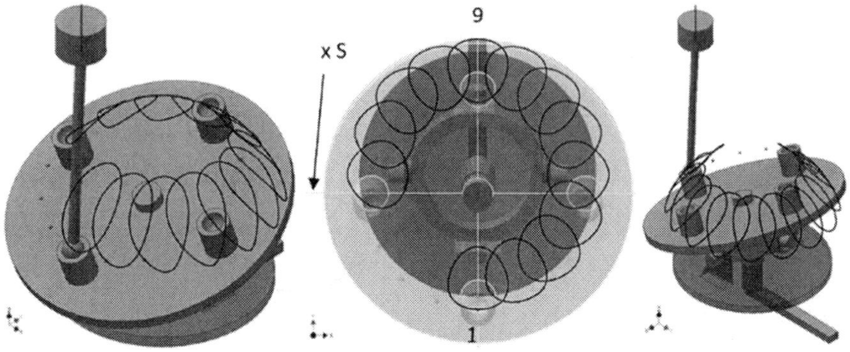

Fig. 11. Overall view of Av1 points in swinging plate (*x S* - means axis *x* of swinging plate)

Conclusion

The resulting graphs were imported from Autodesk Inventor to Microsoft Excel in case of mechanism with ring and in mechanism FIK were calculated according mathematical model in Microsoft Excel. This analysis shows the differences in kinematic waveforms at the set positions of the points on the plate and the position of the cylinder. From Fig. 6 and Fig. 7, there are visible differences in all planes, but most interesting is projection in y-z, between points Av1_1, Av1_5 with Av1_2 to Av1_3. These points also perform the curve (some type of a closed loop) path in this plane. From Fig. 10 it is possible to see again the differences in the projections of points in the direction x, z, but in the direction y, the deviations are minimal, resp. none. Fig. 9 compares Av1 points and piston strokes (P1) in z direction. From this analysis, it is possible to see only minimal differences in the individual progresses, mainly the Av1_4, P1_4 and Av1_5, P1_5, P1_5 have a slightly different curvature around the angle of rotation of the output shaft by 180°. Fig. 11 shows the waveforms tracks in several quadrants of swinging plate, which can be seen as mirrored projections of these trajectories (points 1 and 9 around the axis *x S*). In the difference of the trajectories of the examined points and their courses, depending on the angle of the output shaft, will be speeds and accelerations courses different as well. From this basic analysis it is possible to predict the degree of difficulty in creating a dynamic model and for example in order to creating balancing system on this mechanism. These parameters would have an effect on the torque curve of output shaft depending on its angle of rotation (α), for example in the case of an internal combustion engine. The resulting values generated by Autodesk Inventor were compared with models created with MSC Adams. On reaching the same results, we consider Autodesk Inventor's overall analysis to be reliable. Movement of points and the overall progress of courses will be different from other factors such as the angle of the output shaft breaks (Fig. 2 - 20° in our examined model), in positions of points Av1 for a distance of pitch and diameter of the wobble board (Fig. 1, 2 - 90mm), or positions of cylinders. Therefore, there is necessary to make another analysis of this mechanism. Both of these piston machines have a similar kinematic behaviour, but from the construction point of view they are two different mechanisms. Any other analysis, such as design proposal of balancing systems on these machines will be also different.

With this research and development subject deals other publication too [3, 4, 5, 6].

The work was supported by the Cultural and Educational Grant Agency of the Ministry of Education of the Slovak Republic in project No. KEGA 077ŽU-4/2017: Modernization of the Vehicles and engines study program.
This article was supported by the project VEGA 1/0927/15 "Research of the use of alternative fuels and hybrid drives on traction vehicles with aim to reduce fuel consumption and air pollutants production".

References

1. R. Isteník, *Distribution and nonconventional mechanisms of combustion engines*. Žilina, (2008)
2. Q. Ling, Yong He, Yuan He, C. Pang, *Dynamic response of multibody structure subjected to blast loading*. European journal of mechanics a-solids **64** (July–August), 46-57 (2017)
3. D. Barta, J. Saniga, M. Mruzek, M. Kendra, *Design of Selected Parts of Non-Conventional Stirling Engine with FIK Mechanism*. The Online J. of Science and Technology **3** (2), 129-139, ISSN 2146-7390 (2013)
4. R. Makaras, J. Sapragonas, A. Keršys, S. Pukalskas, *Dynamic model of a vehicle moving in the urban area*. Transport, Vilnius Technika **26** (1), 35-42 (2011)
5. R. Isteník, D. Barta, W. Mucha, *Influence of the Wheels on the Automobile Dynamics*. Communications - Scientific Letters of the University of Zilina **6** (1), 26-28 (2004)
6. J. Dižo, M. Blatnický, B. Skočilasová, *Computational Modelling of the Rail Vehicle Multibody System Including Flexible Bodies*. Communications - Scientific Letters of the University of Zilina **17** (3), 31-36 (2015)

Effect of the pressed ceramic filters on the reoxidation process

Marek Brůna[1,*], *Dana* Bolibruchová[1], *Richard* Pastirčák[1]

[1]University of Žilina, Faculty of Engineering, Department of technological engineering, Univerzitná 1, 01026 Žilina, Slovakia

Abstract. Article is focused on reoxidation processes during filtration of aluminium alloys. Many of our experiments proven, that when using filtration media in gating system, reoxidation occurs. Main goal of this article is to point out reoxidation problems occurring during filtration of aluminium alloys. Reason to conduct this work was anomalies discovered by experimental casts, which pointed out that after filtration of primary alloys, small oxide films was detected. Oxides was absent, when we poured without filtration media. Simulation software ProCAST was used during our research.

Keywords: Filtration, Aluminium alloys, simulation software

1 Introduction

Oxygen solubility in aluminum is very low, less than one oxygen atom in 10^{35} to 10^{40} aluminum atoms, yet it is generally known that aluminum and its alloys are full of oxides. However, oxides cannot penetrate the inner layers of the melt by chemical reaction with oxygen. Oxygen reacts only with the surface layer of the melt and the penetration of the oxides into the inner layers can occur only through the mechanical processes during which the surface oxides are entrained into the internal volume of the melt. Oxide films can be formed in aluminum alloys in 0.01 seconds and already have a thickness of 1 μm after 1 second, after 10 seconds they are thick about 10 μm. If the melt surface is exposed longer time to an atmosphere in the mold cavity, the resulting oxide film will be thickening during all filling process. However, if turbulence occurs, this thick oxide film will furl and entrain into the inner volume of the melt (future solidified casting), or it may cause melt-in-entry problems from wider cross-sections to narrower.

Oxide films can be considered relatively harmless when found on the surface of the melt. Solid oxide film grows from the melt surface, as each metal atom is combined with the incoming gas surrounding atoms. [1]

As a result of turbulence, the surface oxide film is entrained (fig. 1) and furl in a compact shape, resulting in a change in size. The oxide film changes the shape from a 1 mm to 10 mm sheet on a substantially smaller, furled area of 0.1 to 1 mm. In this compact

* Corresponding author: rapture.marko@gmail.com

Reviewers: *Eva* Tillová, *Mariana* Pajtášová

form, most of the oxide films are able even to penetrate through the pores of the filter media placed in the gating system.

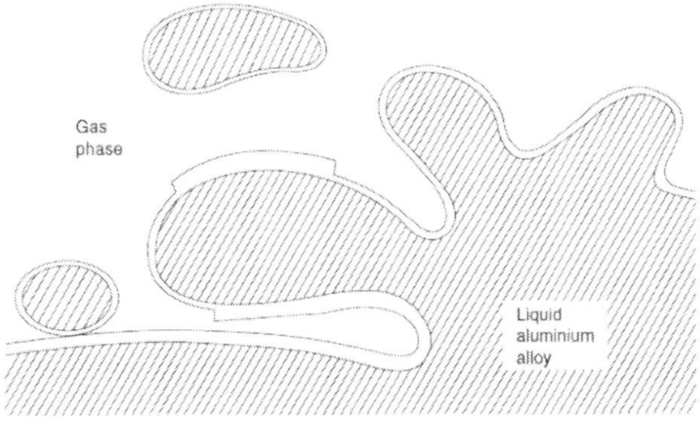

Fig. 1. Scheme of a surface entrainment event [1]

At the moment, when it comes to filling the mold, oxide films find themselves in a new, relatively quiet environment. Movement occurs only due to the heat transfer and slight motions of the liquid metal where the Reynolds number reaches only low values. The new environment around the oxide films creates the conditions for its gradual development, opening and rapid growth up to ten times its original size. In the developed form, the oxide film has the ability to substantially weaken the cross section of the casting and thereby affect its resulting mechanical properties. The driving force for redeveloping of the oxide film may be a number of mechanisms (most often by precipitation of hydrogen between the layers of the oxide film) [1, 2].

This does not mean, however, that the use of filters in the casting of aluminum alloys is not justified. The positive effect of the filters lies mainly on the fact that the narrow channels in the filters support the laminar flow pattern.

The aim of the present paper is to investigate the effect of the pressed ceramic filter on the reoxidation of aluminum alloy AlSi7Mg0.3 by simulating the flow in the area of the filtration medium [3].

2 Experimental part

Aluminum alloy AlSi7Mg0.3 casted into a metal mold with ceramic pressed filters of three different thicknesses (fig. 2. 50 x 50 x 10 - 15 - 22 mm, 314 channels, channels diameter 2, $2^{\pm 0.15}$ mm) in gating system was used for experimental work. Pouring temperature was 730 ± 5 °C. Placement of the filtration media and scheme of the whole experimental apparatus is shown in fig. 3.

Fig. 2. Ceramics pressed filters (22 mm – 15 mm – 10 mm)

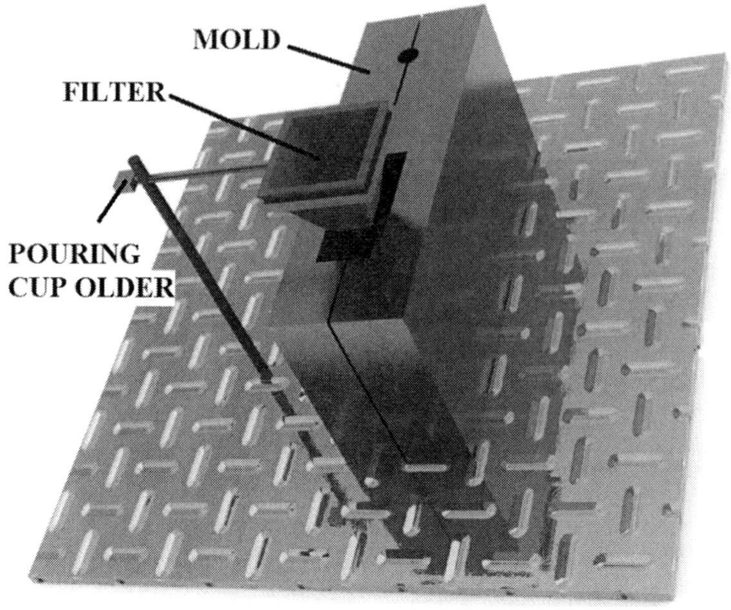

Fig. 3. Measurement apparatus

In order to observe the nature of the flow in the near vicinity of the ceramic filter, ProCAST software was used to simulate the filling of the mold. Instead of the standard function for the analysis of the melt flow in the filter area offered by the software, an innovative approach to the problem was proposed. Individual filter media were imported into the simulation software with precise geometry (created in the Solidworks CAD program). This fact has a considerable impact on the complexity of the simulation process, especially on the surface and subsequently the volume mesh creation, which eventually resulted in the time required for simulation (in the case of a pressed filter up to approximately 350 hours of calculations). However, this mode of simulation has great advantages over the conventional method offered by the software. By using normal method, the software does not take into account the geometry of the individual channels. The main role of the simulation method used is the possibility of more accurate observation of the melt flow in the area of the filtration medium. During the experiments, we mainly focused on the nature of the melt flow before the filter and behind the filter, in terms of the formation of the reoxidation products. The assumption is that the filter itself presents an obstacle in the gating system which may cause a turbulent flow by its presence on the inflow side of the filter. Another important and observed factor is the phenomenon behind the filter, where the continuous melt flow is divided on the sum of small flows from the individual filter channels which may result in a considerable increase of the free area exposed to the atmosphere.

Figure 4a shows a mold cavity and filter with thickness of 10 mm before the pouring, Fig. 4b represents the start of the mold filling and the first contact of the melt with the filter. Purple color of the melt is the liquid metal with the lowest probability of reoxidation occurrence (created during mold filling, because simulation software does not provide information about oxides incurred in previous metallurgical stages) and shades of blue, green, yellow and red, the possibility of oxide occurrence increases. Whole color spectrum in all figures represent interval from 0 to 1 cm^2/sec, that means area of free surface possibly covered in oxide in one second.

At the inlet side of the filter in the initial stage of filling, the melt does not accumulate and all its volume flows through the filter filling the mold cavity in a smooth and continuous flow of liquid metal, which is to counteract the formation of oxides optimal conditions. During the accumulation of the melt at the inlet side there is a prolonged contact of the melt free surface with the atmosphere, which would lead to increased formation of oxide films. Continuous stream of liquid metal on the downstream side may also be seen as a positive phenomenon, the separation of the stream to a smaller streams could increase the melt surface exposed to the atmosphere [4, 5].

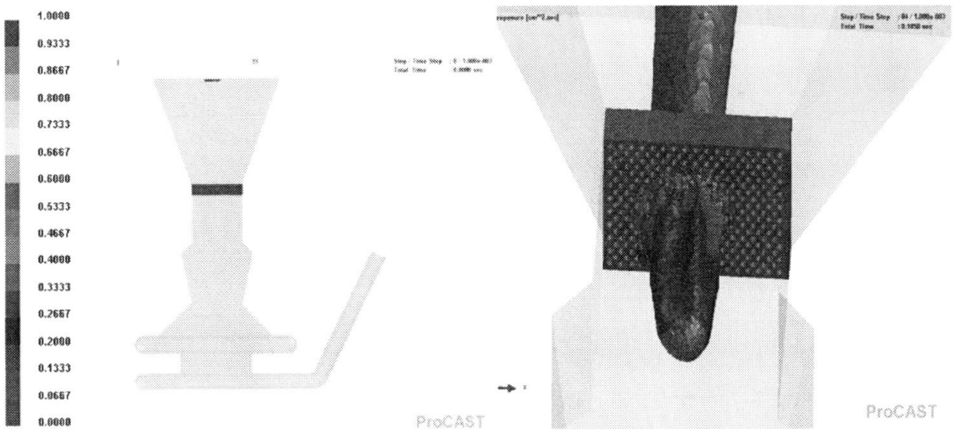

a) b)

Fig. 4. Simulation of starting stage of filling –filter thickness 10 mm

Comparing Figures 5a and 5b, which represent the later stages of filling, it can be stated that most oxide contamination occurs when the melt is on the bottom of the mold (red melt color on the mold bottom). The critical free-fall velocity of aluminum melt according to [3] is 0.5 m / s. The author claims that, once the given value has been exceeded, turbulence cause to submerge surface oxides into the internal volume of the melt and new oxide films will form on the surface of the melt. The simulation software has determined that the melt will have velocity up to 2.2 m / s when it reaches the bottom of the mold. Negative phenomena is also the reflection waves originating in the central area of the mold, which are also shown in red.

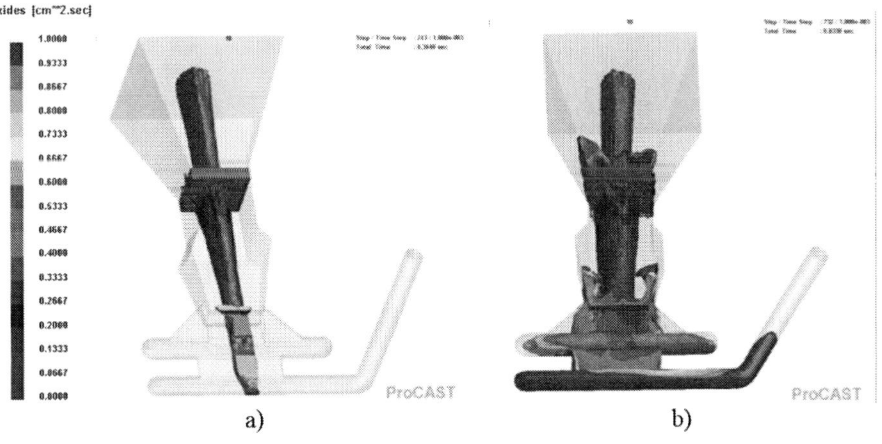

a) b)

Fig. 5. Simulation of later stage of filling –filter thickness 10 mm

Fig. 6a shows a mold cavity and filter with thickness of 15 mm before the pouring. In fig. 6b can be seen that melt passage through a thickened filter press produces the formation of separate melt streams emerging from the individual channels of the filter. This negative effect was reflected in an increase in the area of the melt exposed to the atmosphere in the inlet system, as can be seen in fig. 7 (melt color yellow to orange on the filter side) [6].

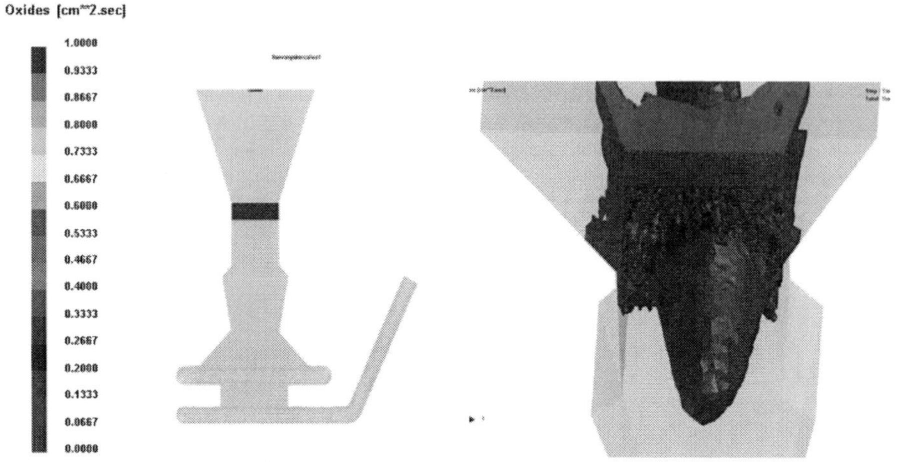

a) b)

Fig. 6. Simulation of starting stage of filling –filter thickness 15 mm

In Figure 7, redundant amount of melt at the inlet side of the filter can be seen; the red color at the free surface indicates the formation of oxides in the region, consequently they will be pulled deeper into the mold cavity. The excessive amount of melt at the inlet side of the filter and the turbulent flow pattern in this zone allows the oxide films to furl into compact form and thereby penetrate through the filter despite all filtration mechanisms. Due to the greater thickness of the filter used, the impact velocity of the melt was reduced to the bottom of the mold - 1.7 m / s (the color of the melt on the bottom of the mold blue and green).

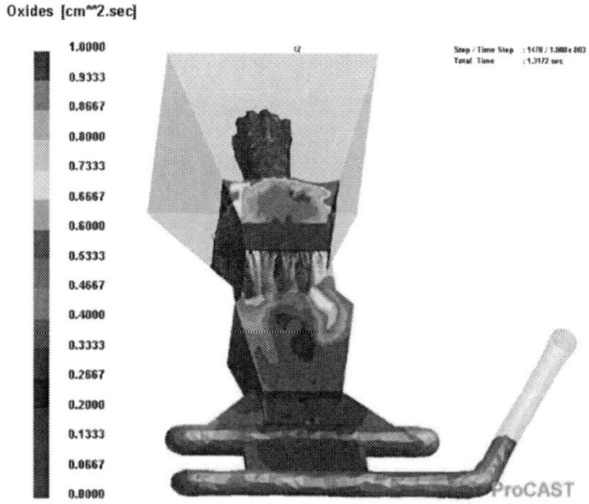

Fig. 7. Simulation of later stage of filling –filter thickness 15 mm

Fig. 8a shows a mold cavity and filter with thickness of 22 mm before the pouring. On fig. 8b, it is possible to notice the formation of an oxide film layer lining the entire volume of the melt at the inlet side of the filter. The impact velocity of the molten metal on the mold bottom at a 22 mm filter thickness was 1.3 m / s (the color of the blue and purple mold bottom - Figure 9).

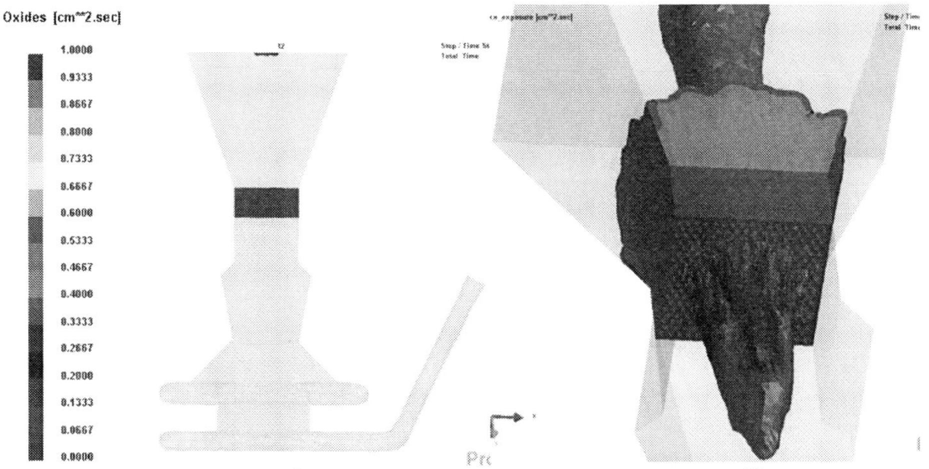

a) b)

Fig. 8. Simulation of starting stage of filling –filter thickness 22 mm

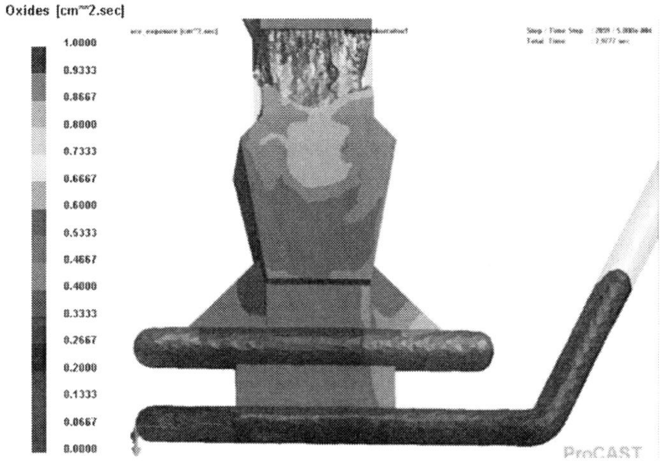

Fig. 9. Simulation of later stage of filling –filter thickness 22 mm

Figure 10 shows a comparison of the filling of the metal mold in the area of the flow of the filter medium (Figure 10a - Thickness filter 10 mm, Figure 10b - 22 mm thickness filter). In order to have a comparison value, in both cases the simulation was stopped at the same timeout (0.22 s). Seeing that with a different thickness of the filter, there is also a different way of filling the mold. In the case of a thinner filter, filling occurs faster, the melt does not stick to the inlet side of the filter, there is no excessive turbulence in the area, but the melt reaches higher speeds.

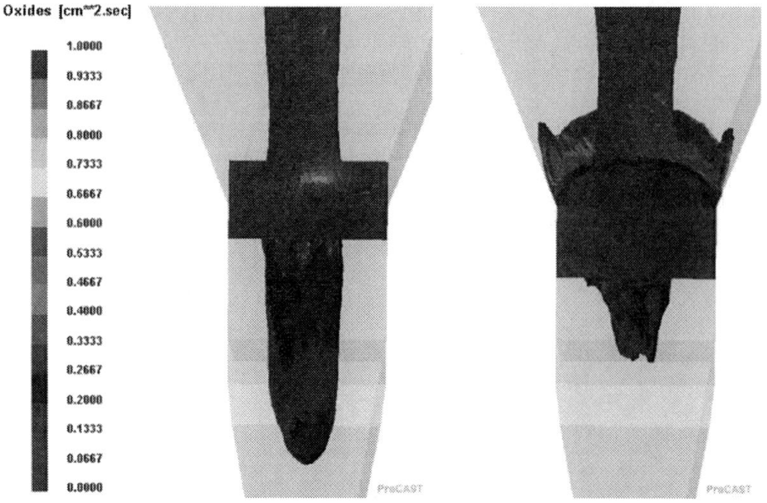

Fig. 10. Flow comparison –filter thickness 15 mm (left), 22 mm (right)

Conclusion

The aim of this work was to point out an atypical view of the mechanism of ceramic pressed filter effect during the pouring of aluminum alloys. The results of the work paradoxically point to increased reoxidation processes compared to the refining effect during filtration of aluminum alloys and extend the set of works that deal with the simulation of the liquid aluminium flow. The aim of the thesis was not to determine the optimum location of the filter in the inlet system, but to visualize the changes in the filling process, taking into account the change of the used filter thickness. Using a detailed analysis of the simulations of all three variants, it can be concluded that the thickness of the used filter has a significant effect on the flow of melt flow, melt velocity and the amount of oxide generated in the gating system. When using a 10 mm thick filter, the excessive flow of the melt is not accumulated at the inlet side, but the flow rate on the drain side is considerably larger than the thicker filters. The simulation software estimated that the optimum filter thickness of 22 mm is optimized for the purpose of suppressing the excessive reoxidation since the melt flow rate at the bottom of the mold approximates to a subcritical value of 0.5 m / s, and the long filter channels more strongly contribute to the laminarization of the melt stream.

This work was created within the framework of the VEGA project grant no. 1/0494/17 with the title: Vplyv pretavovania recyklovaných hliníkových zliatin na úžitkové vlastnosti náročných odliatkov automobilového priemyslu.

References

1. J. Campbell, *Complete casting handbook second edition*, Butterwirth-Heinemann (2015)
2. L. Zhang, L. Damoah, S. Li, W. Abebe, *Mechanisms of Inclusion Removal from Aluminum through Filtration*. Light Metals, TMS, 649-655 (2008)
3. J. Campbell, *Casting practice – the 10 rules of castings*. (Oxford, Butterworth-Heinemann, 224 p., ISBN 07506-4791-4, 2010)

4. M. Tropp, M. Handrik, P. Kopas, M. Sága, *Computer simulation of induction bending process*. Pipeline & Gas Journal, ISSN 32-0188, (2012)
5. R. Raiszadeh and W. D. Griffiths, *The Effect of Holding Liquid Aluminum Alloys on Oxide Film Content*. Metallurgical and Materials Transaction **42B**, 133-143 (2011)
6. D. Bolibruchová, *Filtrácia hliníkových zliatin*. (Žilina, Žilinská univerzita, 203 p., ISBN 978–80–55–0342–7, 1st ed., 2011)

Flap casting of turbine engine optimization by numerical simulation

Marek Brůna[1,*], *Anna* Remišová[1]

[1]University of Žilina, Faculty of Engineering, Department of technological engineering, Univerzitná 1, 01026 Žilina, Slovakia

Abstract. Submitted article focuses on possibilities of increasing internal quality of flap casting of turbine engine produced by investment casting technology. Material porosity will be eliminated by altering the gate system. ProCast software has been used to evaluate and analyse the design of the flap. The aim of the research was to evaluate the porosity of the cast after its solidification, the flow velocity of the melted material and visualise the turbulence of the flow during casting. Main aim will be to observe and analyse velocity, flow, turbulences, solidification and internal porosity for nickel based alloy.

Keywords: Numerical simulation, investment casting, porosity

1 Introduction

The ever-increasing demands of customers, together with the industrial advancement of production, force the founders to improve the quality of the castings while reducing the price. Casting production by investment casting method is one of the most advanced production technologies currently used for production of castings with very high demands on quality, especially used in the aircraft industry or precision mechanics.
This technology makes it possible to create components whose production by other technology would be too expensive or impossible. [1, 2]

Despite many positive attributes, defects can occur on castings, leading to economic losses and losses in quality. One solution to effectively avoid casting errors during the design process is to analyze the cause of the problem and to solve its removal. This work is focused on such a case. The working environment of the ProCAST simulation software was used to solve the porosity occurrence during the casting solidification, with subsequent application of the right solution.

Shrinkage formation involves heat flow, mass flow and some other complicated phenomena, a generally accepted practice to simulate shrinkage porosity is purely from thermal calculations. Through this simplest way, shrinkage can be recognized in terms of solidification patterns. It is defined that shrinkage appears in a spot within the closed loops formed by the isochronal contours of solidification time. This means a complete separation

* Corresponding author: rapture.marko@gmail.com

Reviewers: *Eva* Tillová, *Ladislav* Écsi

of the casting from the riser, simultaneously, accompanied by the disappearance of the feeding path

2 Experimental part

The main objective of the work was to design solutions to improve the internal quality of the turbine engine flap. The casting is manufactured by the investment casting technology by the company CPP s.r.o., whose initially designed casting exhibited unacceptable internal defects. For this reason, the company asked for a solution to this problem. Analysis and design of possible solutions were carried out in the ProCAST simulation program.

2.1 Characteristics of turbine engine flap

The flap is a part of the so-called double flapper check valve (Figure 1) of the Trent 7000 turbine engine, designed by Rolls-Royce, developed for the Airbus A330neo.

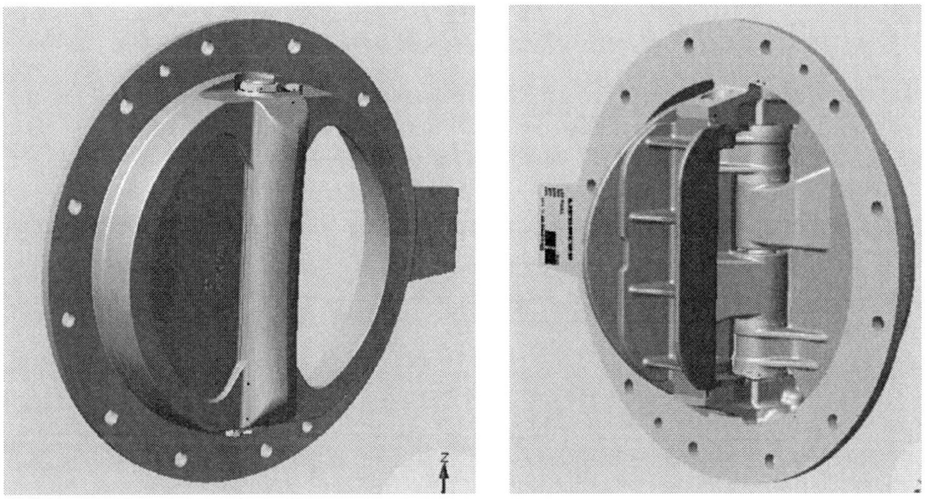

Fig. 1. Scheme of complete double flapper check valve with flap

Pumpage is a phenomenon that occurs when the air flow is disturbed by the engine, resulting in a reduction in the absolute flow rate of the airflow and the change in the rise angle of the flow on the blades. This causes the overpressure before the compressor and not beyond it, then gases from the combustion chamber expands to the front of the engine and loses the thrust. Rapid accelerating, respectively. Deceleration of the engine will result in an imbalance which has a negative effect on its life. Therefore, this phenomenon must be avoided by the check valves, which are opened when the pressure before the compressor reaches a critical value, after the pressure equalization, the flaps return to the original position. Placement of the flap as part of the valve is shown in Fig. 2. [3, 4]

https://doi.org/10.1051/matecconf/201815702005

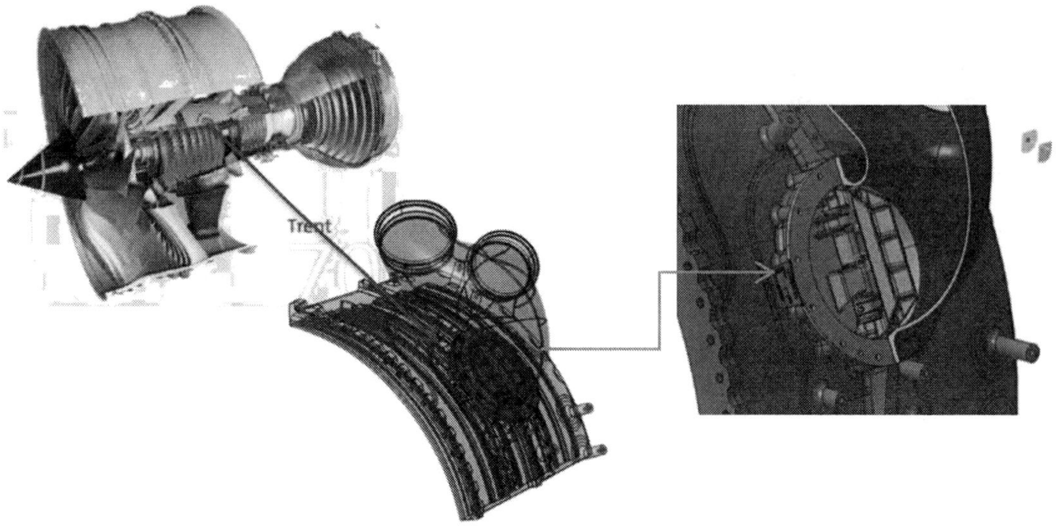

Fig. 2. Placement in the engine

The flap is made of INCONEL718. This material is classified among nickel super alloys and can be used up to temperatures of about 1050 ° C. The microstructure of the material comprises an austenitic matrix (c) in which further secondary phases are dissolved. The Inconel 718 is highly heat-treated with the ability to obtain high mechanical properties, in particular hardness and strength, with a high degree of resistance to corrosion and is suitable for working in aggressive environments.

2.2 Flap casting

The first design of castings exhibited internal porosity at the points of maximum flap force loading. Areas with excessive porosity are located at the sections A-A and B-B shown in Fig. 3. The first design of tree with a castings prepared in software Solidworks is shown in Fig. 4.

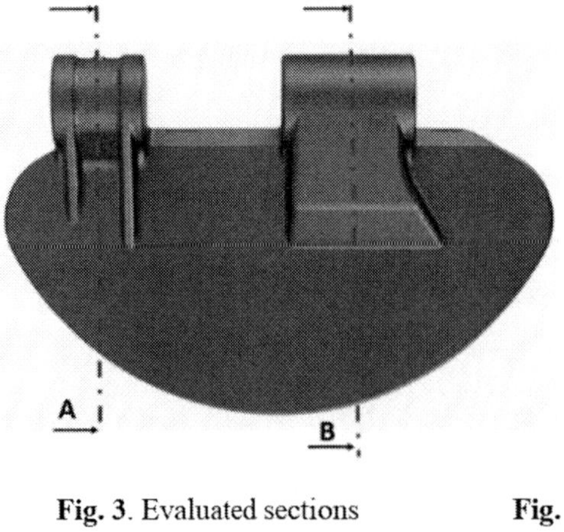

Fig. 3. Evaluated sections

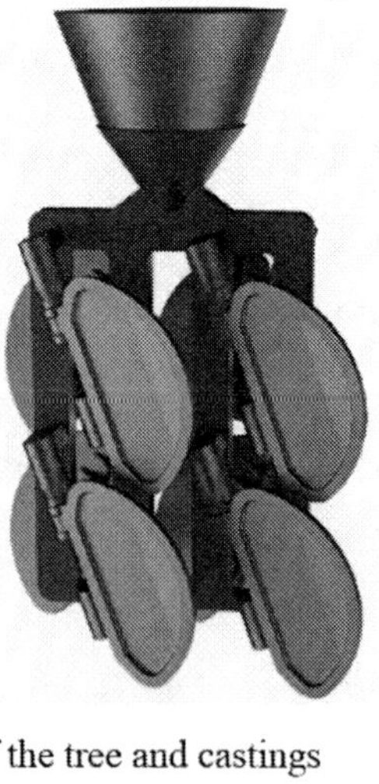

Fig. 4. First design of the tree and castings

Evaluation of section A-A

In section A-A, the porosity was evaluated using the brushed cuts (Figure 5 - left). The pore area represented 1.23% of the given area, which is supercritical for the customer. The rating scale set in ProCAST was 1%. The simulation result confirmed the porosity at critical point and evaluated it approximately in the same range (more than 1%). The critical area is shown in red in the simulation result (Figure 5 - right).

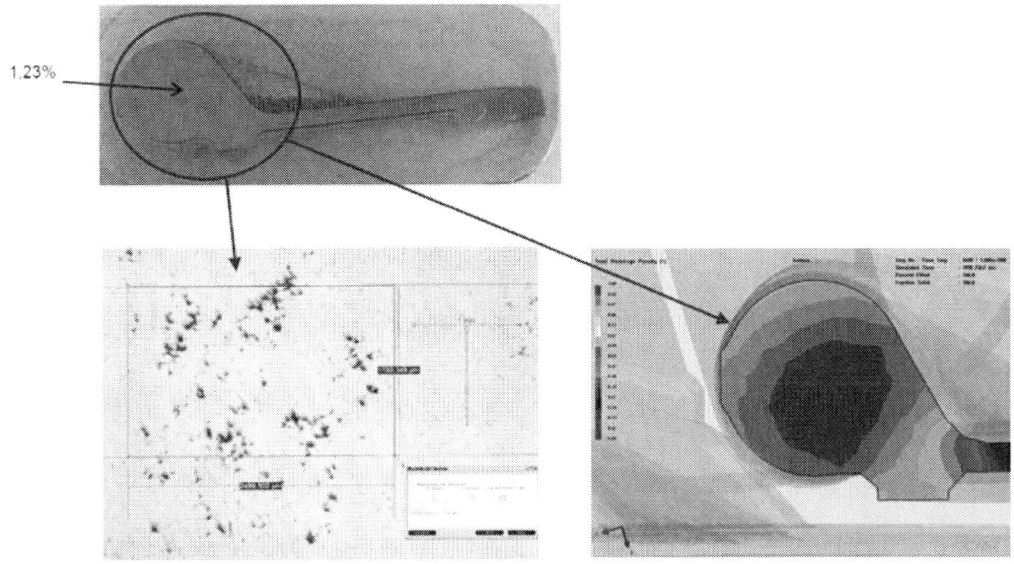

Fig. 5. A-A section cut with microstructure image – left; calculated porosity by ProCAST - right

Evaluation of section B-B

The same analysis was performed at B-B cut. The porosity detected at the cut was 3.04% (Figure 6 - left). The simulation again confirmed an identical critical area (Figure 6 - right). The resulting simulation image also shows a part of the gating system.

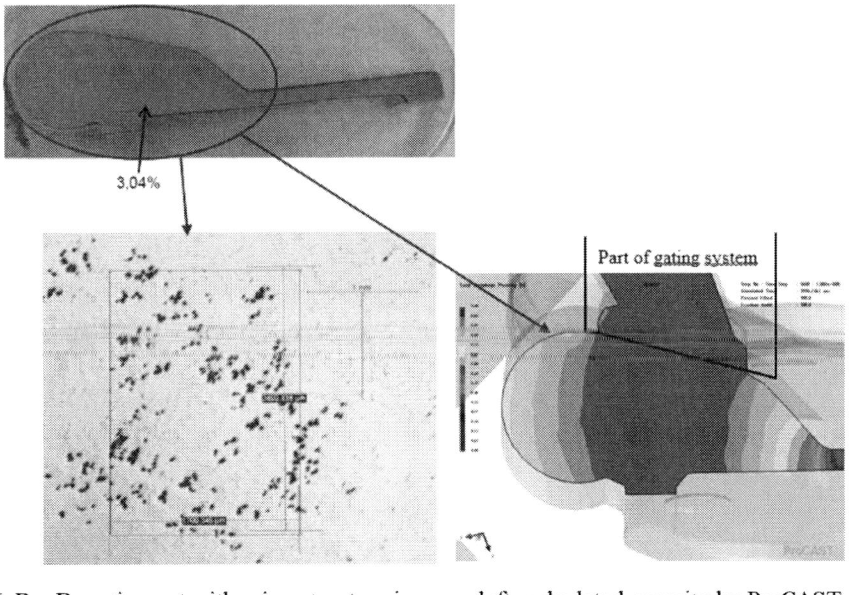

Fig. 6. B – B section cut with microstructure image – left; calculated porosity by ProCAST - right

2.2.1 Analysis and suggested solutions

Due to the requirements of the customer, it was not possible to interfere with changes in the technological parameters of the casting (shown in Table 1). Elimination of the problem was therefore focused on the design changes of the tree (gating system and casting orientation). The analysis of the problem was focused on monitoring the solidification, the occurrence of hot spots and the monitoring of turbulence in the system. The input parameters entered into the ProCAST system are identical to the real parameters.

Table 1. Technological parameters

Casting	*Vacuum furnace:* - pouring temperature: 1440 ± 5°C - Filling time: 3 sec. - Pouring height = 500 mm - crucible type: silica 5009 *Material:* - Inconel 718 - batch weight: 13.4 ± 0.2kg - Cooling on air
Shell mold	- number of layers: 8 - shell thickness: 8-11 mm *Shell composition:* - first layer: slurry based on zircon-kobalt-alumina - 2. layer: slurry based on zircon, covered by zircon - 3. layer: slurry based on zircon, covered by mullit - 4. – 8. layer: slurry based on mullit, covered by mulit - soaked in mullit *Annealing of the shell:* - temperature: 1050°C ±10°C - time: 2 hrs. 30min.

Solidification analysis

During the solidification, the heat transfer occurs when the mold dissipates heat from the liquid metal. The overall nature of the casting crystallization affects the rate of heat removal from the metal in a mold that is directly dependent on the physical and geometrical properties of the casting and mold. [5]

With the simulation, it was found that the condition of the directional solidification was respected in each design (Figure 7), but it is also necessary to keep the width of the two-phase interval during the solidification extended for the purpose of filling the melt into the places where is the risk of shrinkage porosity. For this reason, a modification was proposed in the form of insulation (Figure 8). The proposed insulation was located on the front part of the flap in the area of the thinnest wall. The insulator also covered the critical parts of the flap. One insulation layer with a thickness of 13 mm was proposed.

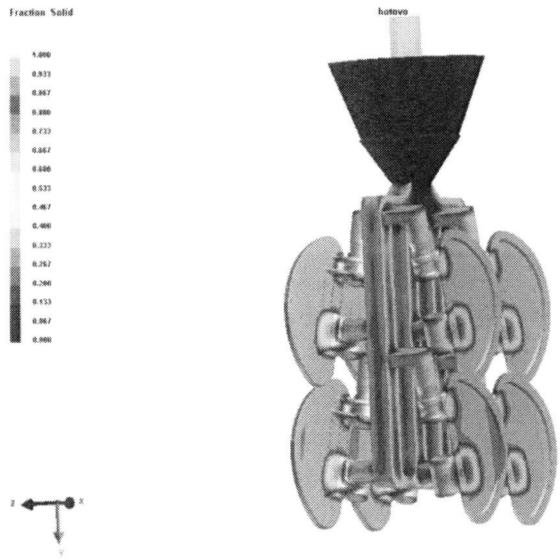

Fig. 7. Solidification result for first design

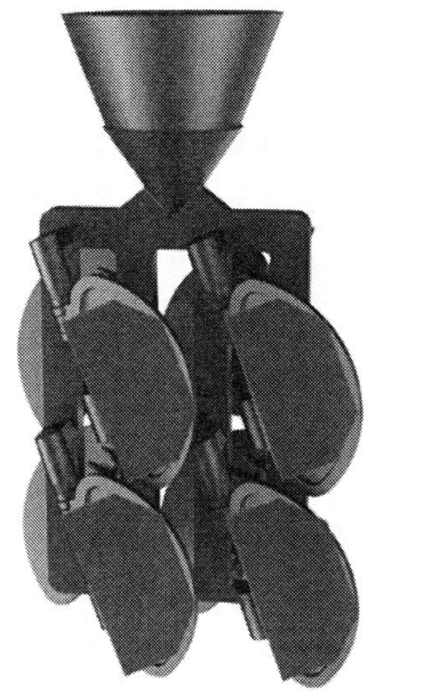

Fig. 8. Placement of the insulations

Solidification with an insulating wrap showed a decrease in porosity concentration in critical areas, confirming the simulation results of the porosity (Figure 9). However, solving the problem in this way was not sufficient as the porosity in the area B-B was still supercritical.

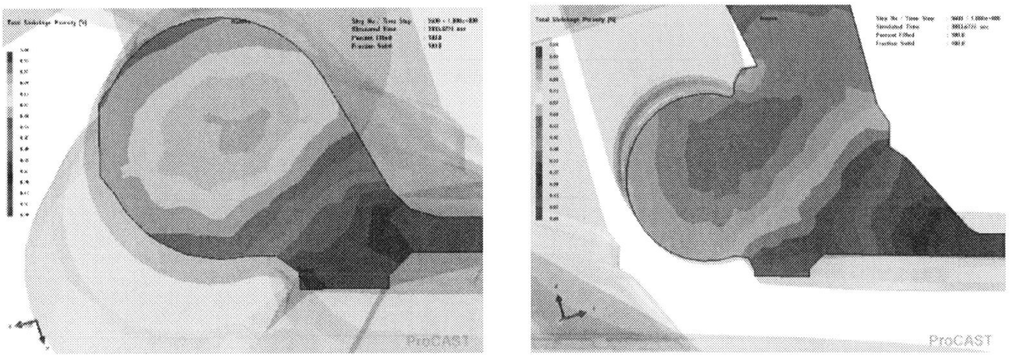

Fig. 9. Results for porosity occurence after insulation application

Metal turbulence analysis

The simulation result (Figure 10) analyzed the effect of metal turbulence. The filling process significantly influences the quality of the resulting casting, even in the case of casting in the vacuum. Although the rate of reoxidation is substantially reduced, the presence of oxides cannot be completely eliminated. For this reason, the nature of the flow is a very important factor because it influences the presence of oxides (bifilms). The individual flow paths determine the directional vector and, according to the indicated scale, are shown with color time spacing to analyze the flow pattern. Within the color spectrum, it can be concluded that the melt in the mold cavity is circulating undesirably and promotes the turbulent nature of the filling. Based on the assumption that the turbulence has an impact on the internal quality of the casts, a change of the gating system has been proposed. This change was centered on the creation of a single inlet channel, also with the change of the castings position.

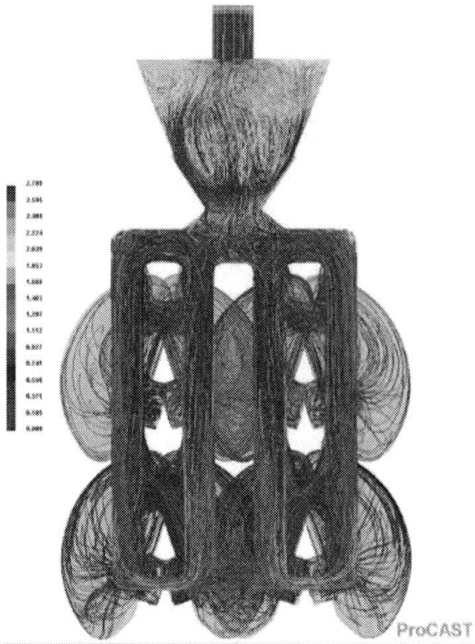

Fig. 10. Turbulence evaluation by ProCAST for the first design

Due to the positive effect of the tree's insulation, it was placed also on new design (Figure 11). Simulations for a new design focused on turbulences are shown in Figure 12.

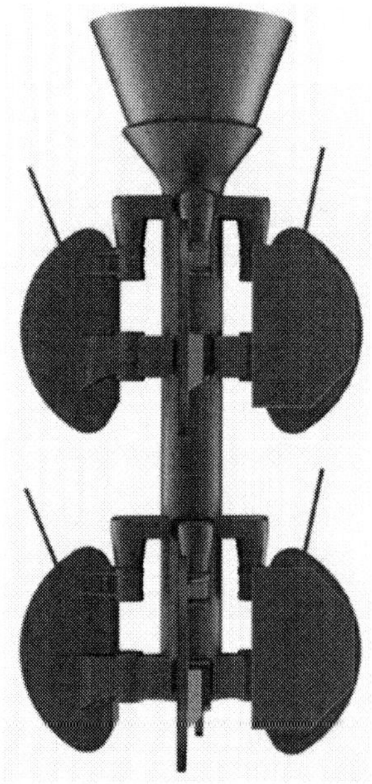

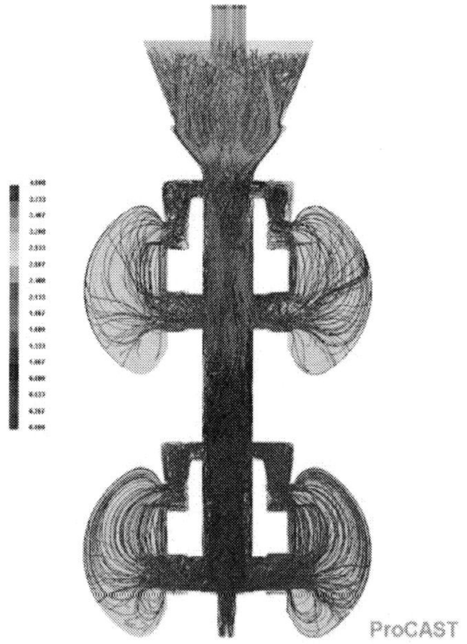

Fig. 11. Proposition of a new design **Fig. 12.** Turbulence evaluation by ProCAST for new design

The result of the turbulence evaluations shows a more direct flow of metal, which corresponds to the color spectrum of the individual vectors. The melt tends to fill the casting cavity and does not circulate in the inlet system. Subsequently, also result for solidification analysis was performed (Fig. 13). This variant also complied with the requirement for direct solidification. Based on the porosity analysis, the system has satisfactory results that do not show critical values in the monitored sections (Figure 14).

https://doi.org/10.1051/matecconf/201815702005

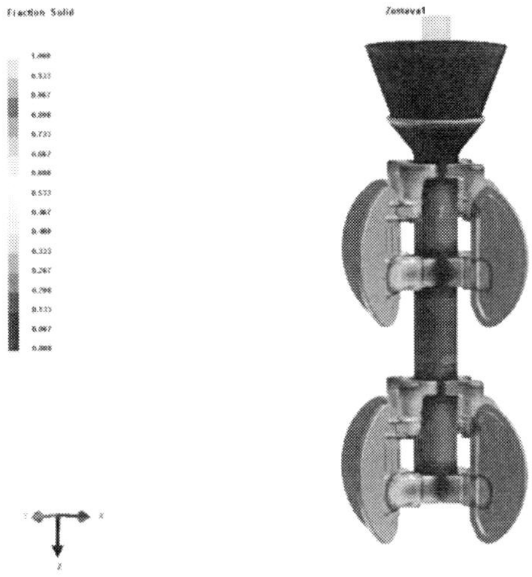

Fig. 13. Solidification evaluation by ProCAST for a new design

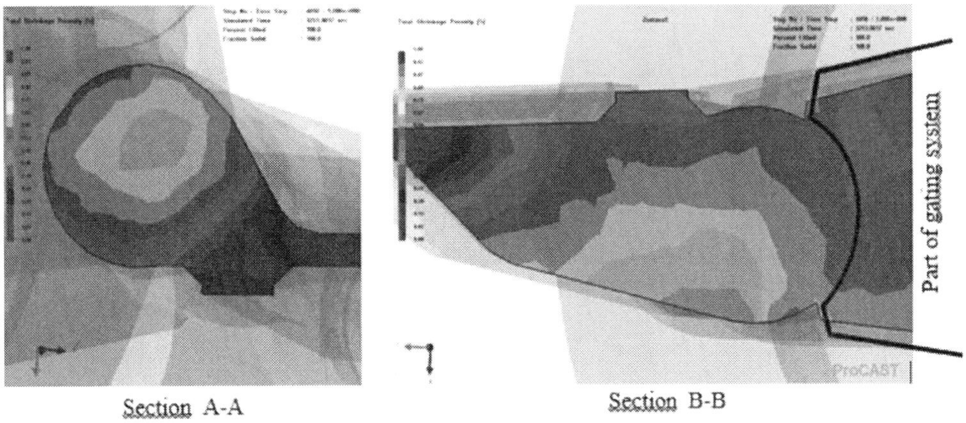

Fig. 14. Porosity evaluation by ProCAST for a new design

Based on the positive results of the ProCAST simulations evaluating the new design, along with the approval of this design by CPP s.r.o., it was processed and casted at the premises of the company (Figure 15).

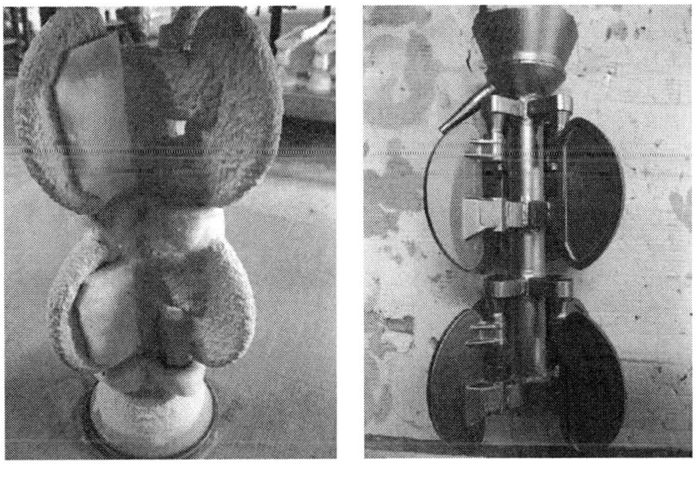

Fig. 15. Casting of a new design in company CPP s.r.o , shell mold – left, casting – right

Conclusion

In this paper optimization of a flap casting process using the ProCast commercial code for casting numerical simulation has been carried out. Flow velocity and turbulence during casting and shrinkage porosity of the cast after its solidification have been evaluated. The weak parts of the original design causing the shrinkage porosity have been identified and the gate system and casting orientation of the double flapper check valve have eventually been altered to improve the design of the casting process. Finally, the calculated porosity has been compared with the experimentally determined porosity which confirmed the calculated results.

This work was created within the framework of the VEGA project grant no. 1/0494/17 with the title: The effect of remelting of recycled aluminum alloys on the properties of advanced castings in the automotive industry.

References

1. H. Fredriksson, U. Akerlind. *Solidification and crystallization processing in metals and alloys*. (Wiley, 1st edition, 2012)
2. K. D. Carlson, C. Beckermann, *Prediction of shrinkage pore volume fraction using dimensionless niyama criterion*. Metallurgical and Materials Transactions A (40A), 163-175 (2009)
3. D. Bolibruchová, J. Macko, M. Bruna, *Elimination of negative effect of fe in secondary alloys alsi6cu4 (en ac 45 000, a 319) by nickel.* Archives of metallurgy and materials, ISSN 1733-3490, 717-721 (2014)
4. P. R. Beeley, R. F. Smart. *Investment Casting*. (The Institute of Materials, 2nd edition, 1995)
5. D. Z. Li, J. Campbell, Y. Y. Li, *Filling system for investment cast ni-base turbine blades*. Journal of Material Processing Technology **148**, 310-316 (2004)

The space distribution and transfer of positioning errors from actuators to the TCP point of parallel mechanism

Vladimír Bulej[1,*], *Ján* Stanček[1], *Ivan* Kuric[1], *Ivan* Zajacko[1]

[1]University of Zilina, Faculty of Mechanical Engineering, Department of Automation and Production Systems, Univerzitna 1, 010 26 Zilina, Slovakia

Abstract. The article deals with the analysis of the transfer and distribution of inaccuracies from actuators to the position of the point TCP on one of these machines. The real functional model of the mechanism was developed at the authors' workplace. The analyses of the workspace were carried out on CAD model. The results should be used for simulation of mechanism's behaviour, structural optimization as well as for generating of control algorithms for real functional prototype.

Keywords: parallel kinematic structure, hybrid kinematic structure, simulation and modelling, inaccuracies distribution in workspace

1 Introduction

High speed cutting (HSC) has its own special place in field of machining. High-speed manipulation is a visible similar trend in robotics. It appears that for both areas the machines with parallel and hybrid kinematic structure fit better than machines with conventional serial kinematics. These mechanisms are characterized above all by higher stiffness and higher dynamic parameters (mainly due to reduced moving mass and inertia) [1-4]. One of the well-known fully parallel mechanisms is called Hexapod or Gough-Stewart platform. Hexapod is multi-axis mechanism capable of full six degrees of freedom (DOFs) motion. Few years ago a research group at the University of Zilina started to deal with area of parallel mechanisms as well. During this period there were designed some mechanism concepts and different kind of simulation software for these mechanisms [5].

Main aim of this article is the analysis of space distribution and transfer of positioning errors from actuators to the TCP point of mechanism with parallel kinematic structure, concretely machine tool prototype called *Hexapod ZU*. When the end effector of hexapod is positioning in any axis, resulting motion is composed of several simple extensions of each separate actuator. It seems to be important to study more detailed the properties of this type of mechanisms to understand their behaviour. This article contains the first analysis results, which was carried out on 3D CAD model of the mechanism in software Pro Engineer Wildfire 4.

* Corresponding author: vladimir.bulej@fstroj.uniza.sk

Reviewers: *Andrej* Czán, *Ksawery* Szykiedans

2 Evaluation of the positioning error space distribution

The evaluation of the positioning error of a manipulator is a fundamental step in the design process if accuracy and stiffness are the guidelines for the mechanical project, as is the case for machining or assembly manipulators. In most of these works of these works, the analytic model is exploited to evaluate the stiffness of the manipulator in a series of configurations inside the workspace [6].

We created a three-dimensional grid of points (figure 1) for the distribution analysis of positioning errors by positioning in three axes [7]. This grid represents selected set of positions of moving platform where we take measurements as well as where we apply the inaccuracies into the arms. The inaccuracy was defined as an extension of 1mm implemented on one or more arm / arms. The result is an output value of final deviation between the required position of TCP point and the position of TCP affected by an error. The final deviation can be divided into each axis separately.

The grid size was chosen according to maximize workpiece size. The workpiece centre point is aligned with the centre of working table. Within the machine tool workspace was chosen a main coordinate system whose location is given by the axis of the machine tool and the top horizontal surface of a workpiece. Then we defined two horizontal planes, which are labelled as level 1 and level 2 and in each were defined 9 points. Both of them, level 1 as well as level 2 are parallel with the plane XY.

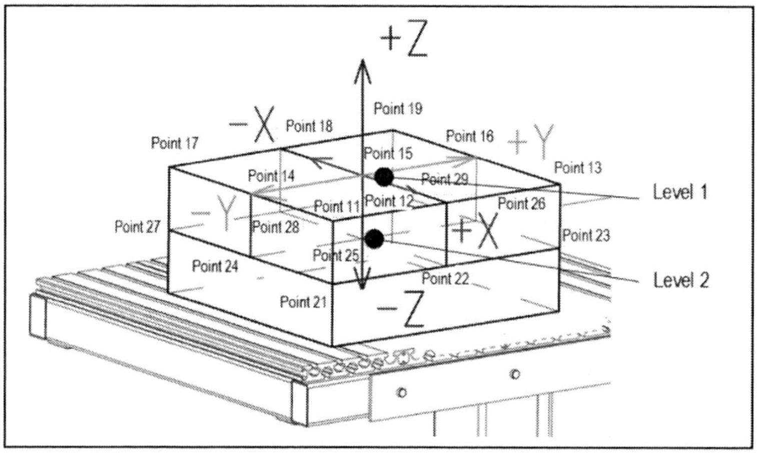

Fig. 1. The workspace divided into the 3D grid of points placed in 2 levels [7]

2.1 Reference acting values in all points

At the beginning it was necessary to determine the initialization values of extension for each linear actuator for all defined points (figure 1 and figure 2). These values are considered as reference positions. To this reference values will be continuously added the specified increment which will represent the source of positioning errors (input positioning error). Consequently will be measured the deviation from the expected position of TCP point (output positioning error) in axis X, Y and Z as well as the total space error.

The value of input positioning error was determined as +2 μm unit displacement (positive extension of actuator) with respect to the predicted accuracy of final mechanism. We decided for application of electromechanical linear actuators *Exlar IX30-1602* with a preloaded planetary roller screw system as arms in design of tested mechanism. These actuators offer the high load carrying capacity (nominal more than 4000 N per actuator) as well as up to 15 times the life of an equivalent ball screw.

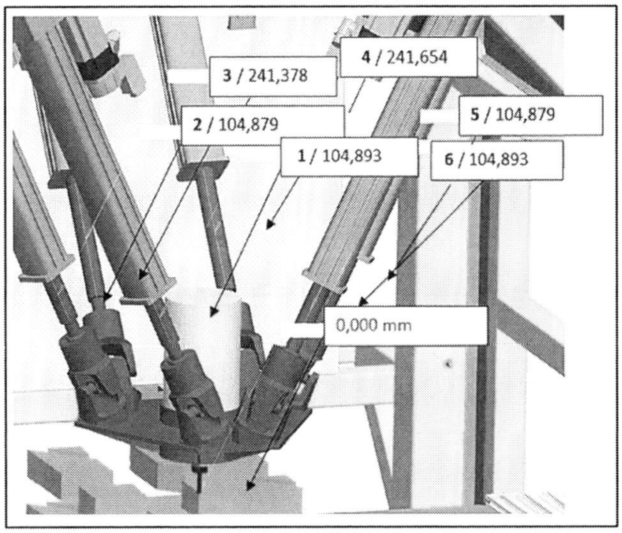

Fig. 2. Example how to reach the acting values (linear extensions) with respect to the position of moving platform (here is shown the case when moving platform is located in point 12)

The defined errors will be gradually applied on actuators according to following schema: error is applied on arm number 1; error is applied on arms number 1 and 2; then 1, 2 and 3 up to case when error is applied on all of the arms. There were done 2 other simulations (will be not mentioned in article) with different values of applied input positioning error: +1 mm and +1 µm. First of them were used only to illustrate the principle how are positional errors distributed in the 3D space and their transfer from actuators to the TCP point.

Table 1. Extension of each arms (from 1 to 6) to achieve separate grid points

Level	**Point**	**Extension of arm number: (mm)**					
		1	**2**	**3**	**4**	**5**	**6**
1	11	39,595	39,588	260,196	260,544	206,071	206,092
1	12	104,893	104,879	241,378	241,654	104,879	104,893
1	13	206,920	206,710	260,480	260,558	39,588	39,595
1	14	68,271	68,279	152,490	152,635	230,169	230,176
1	15	131,577	131,577	131,577	131,577	131,577	131,577
1	16	230,176	230,169	152,492	152,651	68,279	68,271
1	17	139,283	139,305	77,419	77,380	290,772	290,767
1	18	198,128	198,141	54,545	54,486	198,141	198,128
1	19	290,767	290,772	77,392	77,398	139,305	139,283
2	21	137,345	137,339	339,403	339,615	289,078	289,097
2	22	196,304	196,290	321,912	322,049	196,890	196,304
2	23	289,097	289,078	339,389	339,628	137,339	137,345
2	24	163,137	163,145	239,825	239,886	311,389	311,396
2	25	220,618	220,618	220,618	220,618	220,618	220,618
2	26	311,396	311,389	239,805	239,901	163,145	163,137
2	27	227,660	227,681	171,438	171,365	367,811	367,806
2	28	281,740	281,752	150,769	150,719	281,752	281,740
2	29	367,806	367,811	171,414	171,380	227,681	227,660

Whole process consists of following steps:

- dividing the subspace enveloped by maximal workpiece and creating of auxiliary planes to facilitate the definition of individual points,
- blocking the motion of the moving platform for translation in Z axis,
- measuring the actual length of each arm after they reach the desired point (table 1)

2.2 Positioning errors on arms and their influence on TCP point

We used the module ProEngineer / Mechanism for this experiment to carry out the dynamic simulations of the mechanism motion.
Process steps (mentioned by Zabensky [7]):

- deactivation the General constrain for moving platform;
- creation of the tables with input displacements of each arm at the time. Whole process started at point 11 (figure 2). The moving platform was placed into the selected point in every sequence step. Consequently the error of 1 mm was applied per defined arm / arms gradually from one arm up to all of them. So at the end the error was applied to all six arms. Finally, the arms return to the home position;
- creation the tables of output values from the simulation and preparation of the output text file;
- The same procedure was applied to all other points within the workspace. The results are represented by the length deviations in X, Y, Z axis as well as the absolute space deviation of the point TCP.

3 The measuring of the positioning errors

The measurement is carried out for each of the 18 points separately. At the point 11 we first define a coordinate system with the same sense of marking the axes as in figure 3. Consequently we are using the tool Measure for the measuring of the deviation between the TCP (Tool Centre Point) and defined CCS (Centre of Coordinate System) in all three axes and the overall spatial displacement.

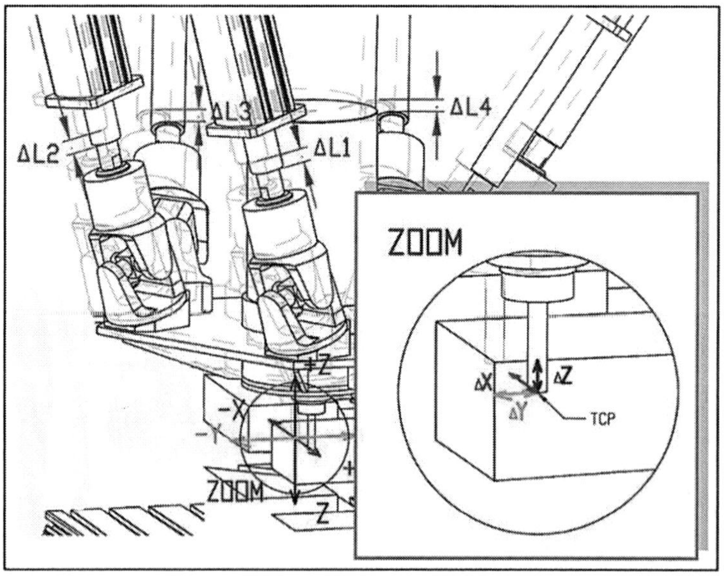

Fig. 3. Transfer of positioning errors from arms 1, 2, 3, 4 (point 11) into the position of TCP point (dashed lines shown the reference position)

3.1 Simulation results

The analysis results (Table 2) can be shown in a set of graphs of output positioning errors with respect to the reference coordinate system - for each axis separately and also for total deviation. We decided for processing of obtained data in Microsoft Excel. There can be observed mainly two different influences:

- position of moving platform within the workpiece volume
- and the number of error affected arm / arms.

Table 2. Output TCP positioning errors in individual points (level 1)

	Output deviation (mm)	Input positioning error 2 μm applied on arm / arms number:					
		1	**1,2**	**1,2,3**	**1,2,3,4**	**1,2,3,4,5**	**1,2,3,4,5,6**
Point:	in axis X	-0,302	-1,045	0,232	2,01	0,672	0,628
11	in axis Y	0,931	2,066	1,282	2,082	0,715	-0,543
	in axis Z	-0,615	-1,51	-1,082	-1,589	-2,454	-2,15
	TOTAL	1,156	2,764	1,694	3,301	2,643	2,305
Point:	in axis X	-0,252	-1,292	0,26	1,833	0,792	0,452
12	in axis Y	1,076	2,333	1,522	2,324	1,071	-0,042
	in axis Z	-0,184	-1,068	-1,201	-1,152	-2,022	-2,127
	TOTAL	1,120	2,873	1,956	3,176	2,421	2,175
Point:	in axis X	-0,422	-1,904	-0,112	1,174	0,458	-0,142
13	in axis Y	2,108	3,213	2,897	2,998	2,113	1,641
	in axis Z	-0,116	-0,123	-1,464	-1,054	-1,842	-2,221
	TOTAL	2,153	3,737	3,248	3,388	2,840	2,765
Point:	in axis X	-0,58	-1,324	-0,176	1,54	0,051	-0,01
14	in axis Y	1,35	2,226	1,817	2,043	1,156	-0,413
	in axis Z	-0,871	-1,245	-1,226	-2,005	-2,224	-2,157
	TOTAL	1,708	2,874	2,199	3,250	2,507	2,196
Point:	in axis X	-0,132	-1,175	0,203	1,601	0,534	0,209
15	in axis Y	1,342	2,312	1,849	2,354	1,368	-0,082
	in axis Z	-0,411	-0,822	-1,17	-1,567	-1,927	-2,325
	TOTAL	1,410	2,721	2,197	3,250	2,423	2,336
Point:	in axis X	-0,054	-1,512	0,173	1,322	0,575	0,018
16	in axis Y	1,538	2,232	2,125	2,461	1,725	0,462
	in axis Z	0,115	-0,201	-0,967	-0,944	-1,337	-2,201
	TOTAL	1,543	2,703	2,341	2,949	2,257	2,249
Point:	in axis X	-0,622	-1,428	-0,381	1,154	-0,362	-0,425
17	in axis Y	1,617	2,245	2,134	2,352	1,601	-0,353
	in axis Z	-1,009	-0,944	-1,275	-2,334	-2,101	-2,159
	TOTAL	2,005	2,823	2,515	3,509	2,666	2,229
Point:	in axis X	-0,348	-1,49	-0,219	1,066	-0,076	-0,431
18	in axis Y	1,818	2,589	2,466	2,553	1,821	0,005
	in axis Z	-0,548	-0,442	-1,107	-1,671	-1,545	-2,222
	TOTAL	1,930	3,020	2,712	3,232	2,389	2,263
Point:	in axis X	-0,145	-1,708	-0,144	0,908	0,125	-0,471
19	in axis Y	2,007	2,792	2,699	2,845	2,111	0,398
	in axis Z	-0,024	0,249	-0,803	-1,102	-1,087	-2,245
	TOTAL	2,012	3,282	2,820	3,183	2,378	2,328

All positioning errors in each point of level 1 were compared with the positioning errors in reference point 15 (middle of the worskspace). On this basis were created the graphs (figure 4, 5, 6 and 7) represented the relation between the point TCP from the vertical axis of the machine tool (workspace). The positioning errors are divided according to the axes X, Y, Z and total space distance.

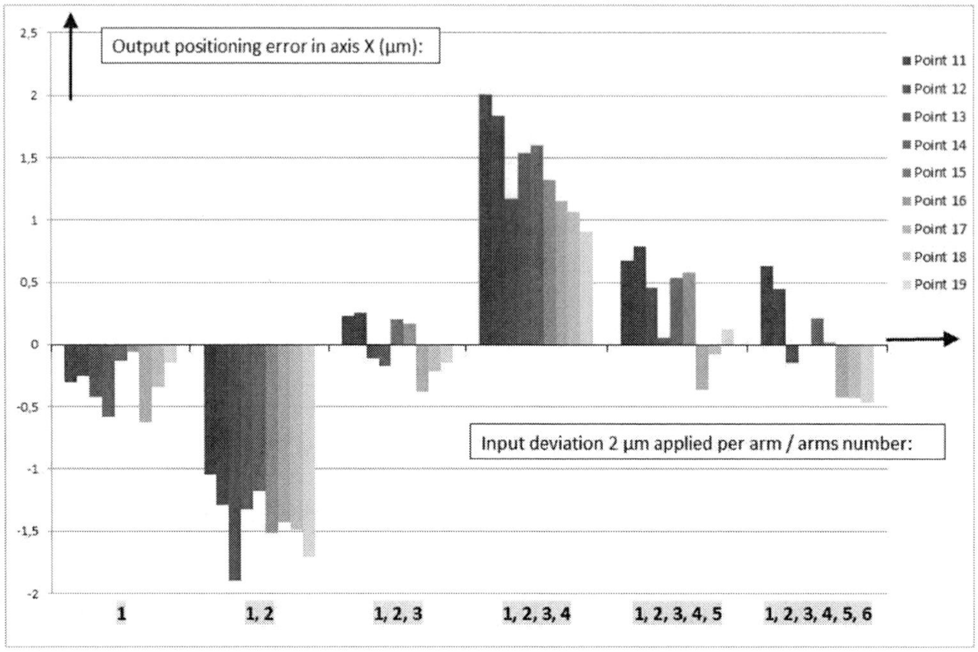

Fig. 4. Graph of output positioning error of point TCP in axis X (level 1)

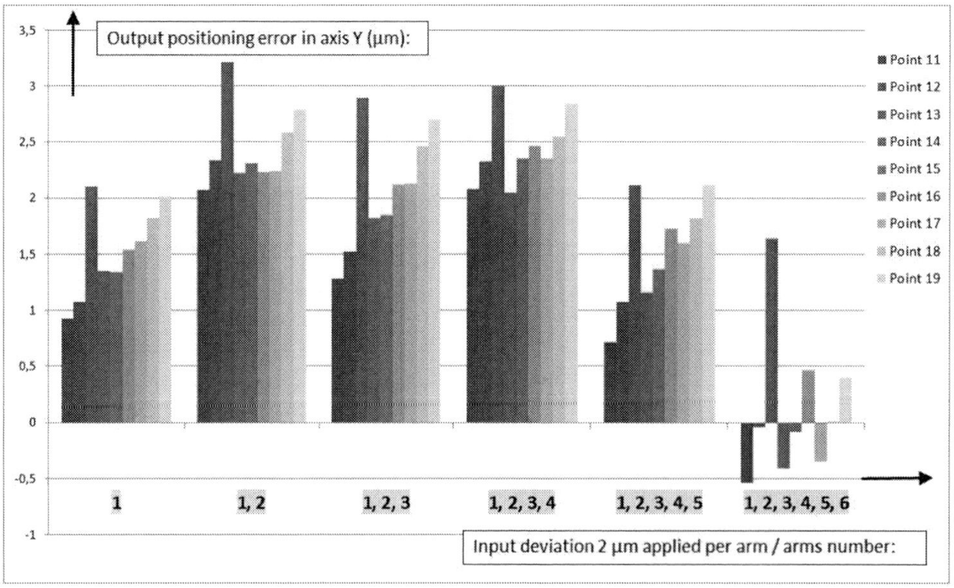

Fig. 5. Graph of output positioning error of point TCP in axis Y (level 1)

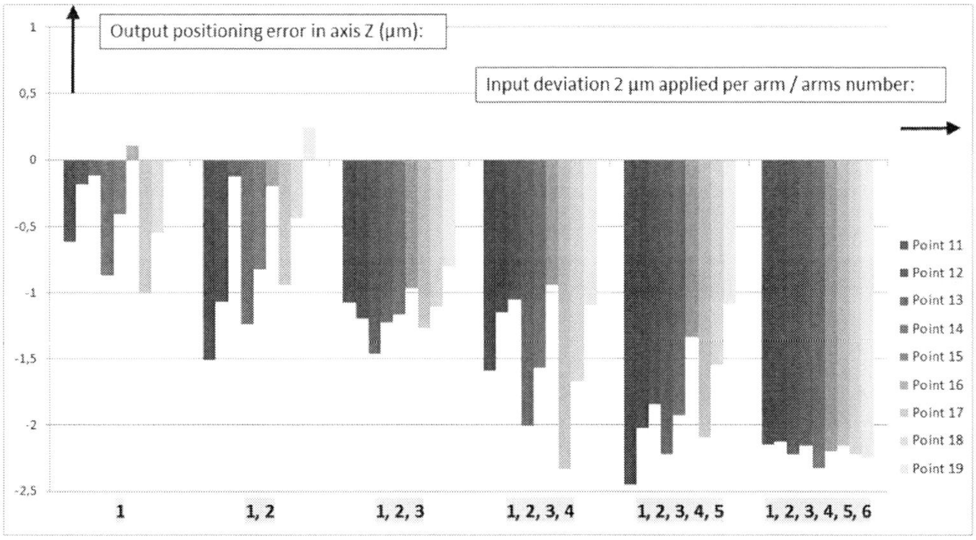

Fig. 6. Graph of output positioning error of point TCP in axis Z (level 1)

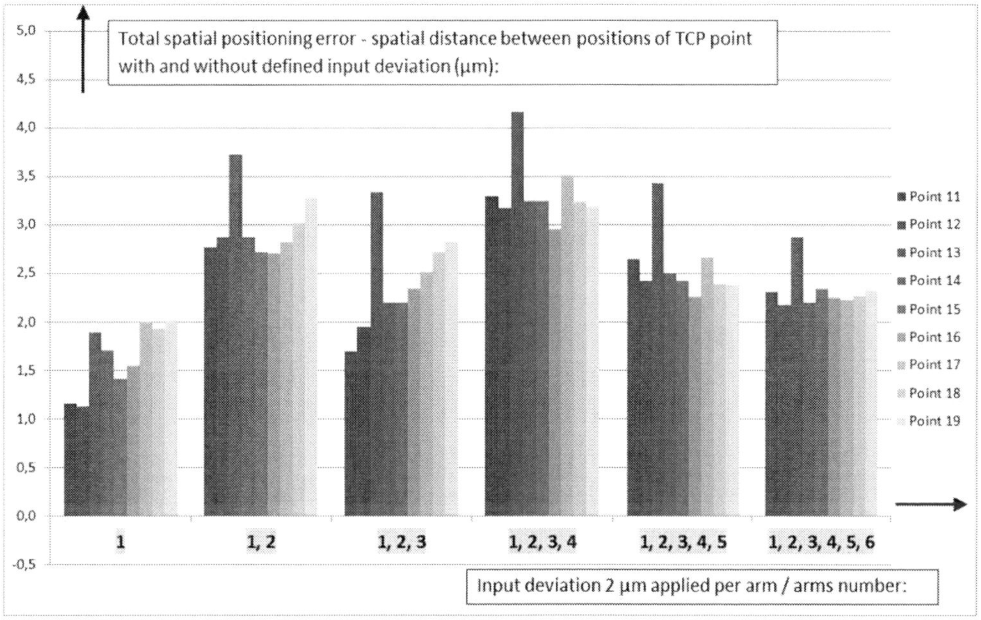

Fig. 7. Graph of total spatial output positioning error of point TCP (level 1)

Conclusions

In recent years, due to fewer driving elements, lower cost, and more compact structure, lower-mobility parallel mechanisms have shown great potential for application, and have has gradually attracted the attention of scholars both at home and abroad [8].

The main aim of this article was to demonstrate the uncertainties influence of telescopic arm control in case of mechanism with parallel kinematic structure called Hexapod ZU. There can be visible also the difference between measurement of accuracy on conventional machines and machines with parallel kinematics. The accuracy in case of parallel machines is much harder to achieve and to define. This was also the main reason why we decided to analyse the relationship between the accuracy in actuators and in moving platform.

On this analysis we can conclude that:

- the maximum deviation in X and Y axis as well as the final deviation in both levels happened by errors applied on arms 1, 2 or 1, 2, 3, 4 (together);
- the deviation in X and Y axis is close to zero in both levels when the errors is applied on all arms;
- the value of deviation in the Z-axis in both levels is directly proportional to the number of "non-precision" controlled arms;
- the impact of the distance from the vertical axis of the machine is in level 1 greater than in level 2;
- it's important to know the deviation in the Z-axis when we measure the deviation of circular interpolation in the XY plane for proper assessment of the results.

Therefore we approve that the size of the final TCP point deviation is not affected only by the error on each arm but by the current position of the platform as well. In level 1 were observed in all axes higher deviations from the reference point. From this we can conclude, that the size of deviations is changing with respect to the position of moving platform within the workspace.

The authors disclosed receipt of the following financial support for the publication of this article: This work is partly supported by the project VEGA 1/0504/17: Research and development of methods for multi-criteria diagnosis of CNC machine tools' accuracy.

References

1. V. Bulej, *Mechanization and automation equipment for processing: Innovative and non-standard applications of mechanisms with parallel kinematic structure*. (Publ. House Alma Mater, Cluj-Napoca, 2015)
2. J. Uricek, et al., *Design of Machine Tool with Parallel Kinematic Structure*. Sci. Reports - J. of Univ. of Appl. Sci. Mittweida **21,** 36-39 (2011)
3. V. Bulej et al., *The modelling of mechanism with parallel kinematic structure in software Matlab/Simulink.* Transactions of the VSB **62**, 1-8 (2016)
4. B. Leitner, M. Vaško, *Design and Modelling of Tank Wagon Assembly Operations in CAM Environment*. Transport Means 2015 **I & II**, 87-90 (2015)
5. M. Krajcovic, et al., *Intelligent manufacturing systems in concept of digital factory*. Communications : Sci. Letters of the Uni. of Zilina, **15**, 77-87 (2013)
6. G. Palmieri, *On the positioning error of a 2-DOF spherical parallel wrist with flexible links and joints – an FEM approach*. Mechanical Sci. **6**, 9-14 (2015)
7. M. Zabensky, *Evaluation of selected operational properties of machine tool prototype with parallel kinematic structure*. Diploma thesis, Zilina (2013)
8. X. Chen et al., *Rigid Dynamic Model and Analysis of 5-DOF Parallel Mechanism*. Int J Adv Robot Syst, 1-9 (2015)

Numerical simulations of progressive hardening by using ABAQUS FEA software

Tomasz Domański[1,*], *Wiesława* Piekarska[1], *Marcin* Kubiak[1], *Zbigniew* Saternus[1], *Dorota* Goszczyńska-Króliszewska[1]

[1]Institute of Mechanics and Machine Design Foundations, Czestochowa University of Technology, Dabrowskiego 73, 42-200 Częstochowa, Poland, EU

Abstract. The paper concerns numerical simulations of progressive hardening include phase transformations in solid state of steel. Abaqus FEA software is used for numerical analysis of temperature field and phase transformations. Numerical subroutines, written in fortran programming language are used in computer simulations where models of the distribution of movable heat source, kinetics of phase transformations in solid state as well as thermal and structural strain are implemented. Model for evaluation of fractions of phases and their kinetics is based on continuous heating diagram and continuous cooling diagram. The numerical analysis of thermal fields, phase fractions and strain associated progressive hardening of elements made of steel were done.

Keywords: phase transformations, temperature fields, hardening

1 Introduction

The anticipation of final properties of the element that undergoes progressive hardening is possible after defining the type of the formed microstructure accompany thermal treatment. In order to achieve this aim it is essential to consider such a treatment of thermal phenomena, phase transformations and mechanical phenomena in numerical modeling. Phenomena that accompany thermal treatment are complex, and so far they are incompletely described. Findings of the numerical simulation of the above-mentioned phenomena depend on, inter alia, the accuracy of determination of phase transformation kinetics in the solid state. Surface hardening technology is very often used. There is no comprehensive numerical models which allow provide reliable assessment of phenomena that accompany to such a process of hardening. Progressive hardening is performed for concentrated heat sources with high power. Near to this sources occurs a high temperature and its considerable gradients [1, 2]. Temperature and phase transformations in hardening steel elements are reasons to generated significant thermal and structural strains and in consequently to generate residual and temporal stresses.

* Corresponding author: domanski@imipkm.pcz.pl

Reviewers: *Ján* Vavro, *Milan* Žmindák

2 Model of thermal and mechanical phenomena

Heat transfer equation with convective unit was used to modeling of thermal phenomena. The arguments of searched temperature fields are spatial coordinates (Lagrange's coordinates) and time [3, 4]:

$$a\nabla^2 T(\mathbf{x},t) - \frac{\partial T(\mathbf{x},t)}{\partial t} + \nabla T(\mathbf{x},t)\cdot \mathbf{v} = -\frac{Q}{C} \tag{1}$$

where: $T=T(\mathbf{x},t)$ is temperature [K], a is thermal diffusivity, C is specific heat capacity, Q is element of volumetric heat sources that takes into account the heat generation from the source, $\mathbf{v}$ is the vector of velocity of the beam, $\mathbf{x}$ is the vector of position of considered particle (point), t is time [s].

To have the possibility to carry out a full analysis of thermal treatment it is necessary to have proper mathematical and numerical models that can provide information about instantaneous temperature fields, the change in time of fractions of particular phase proportions of the material (Fig.1).

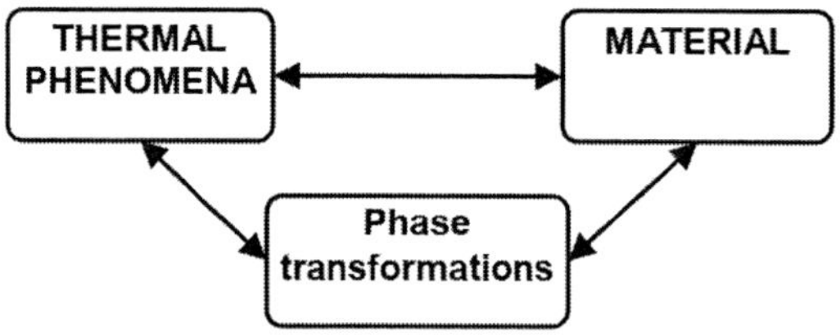

Fig. 1. Scheme of correlation of the hardening phenomena

Equation (1) is solved by methods proposed in [3]. Temperature distributions, obtained from the simulation after the determination of the heating process, are presented in Fig. 2.

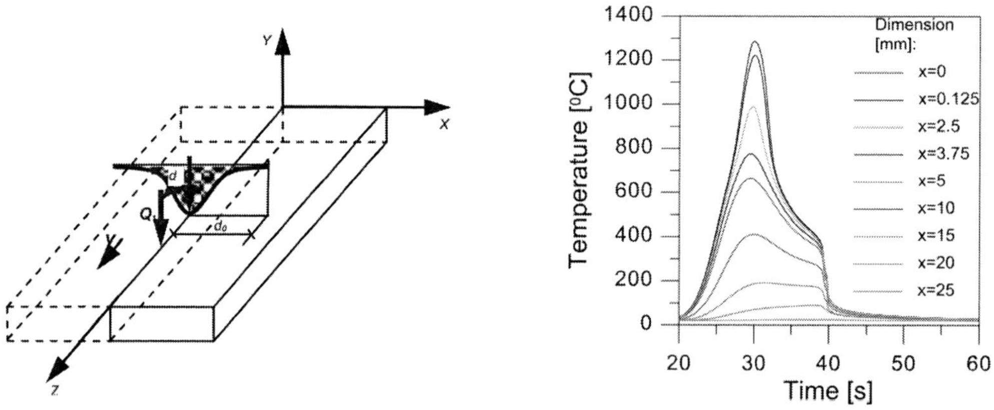

Fig. 2. Scheme of considered system, temperature distributions in z=0 the plane of operation of heat source

Two sources: first with a Gaussian distribution was heating source, and second at the distance $z = -d$ from the sprayed water stream. The distribution of second source has selected linear penetration ($x=x+d_0$). Water was sprayed at a distance of 25 mm for the source of a heating source. Width at the sprayed water was 20 mm. In an surface that was sprayed water cooling was carried out by the difference of temperature of the surrounding medium and the surface of the sample (condition Newton).

$$Q_1(\mathbf{x}) = \frac{P(1-R)}{2\pi r^2}\exp\left(-\frac{(x)^2+(z)^2}{2r^2} - \beta|y|\right) \tag{2}$$

where: P is power derived from the heating source, R is the reflectance, β is absorption coefficient. $z=z_0+v\cdot t$, z_0 is a central of heat source, v is speed of movement source, t is time. Fig. 1 shows the temperature distributions in time (thermal cycles) on the surface of hardened object along the x coordinate. While Fig. 3 has been represented the temperature distribution on the x-axis perpendicular to the direction of the heating source.

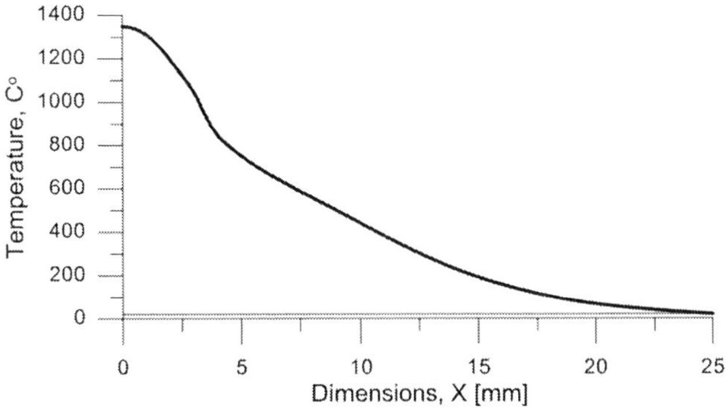

Fig. 3. Temperature distributions in x axis

The information about the stresses associated with hardening are obtained by solving the rate equilibrium equations, complemented by constitutive compounds and appropriate initial and boundary conditions. Young's modulus was depended on temperature. There are thus the opportunity to reflect changes of material constants of temperature and phase composition in the following increments of load [3, 5]:

$$\mathrm{div}(\dot{\boldsymbol{\sigma}}(\mathbf{x},t)) = 0,\ \boldsymbol{\sigma}^T = \boldsymbol{\sigma},\ \ \dot{\boldsymbol{\sigma}} = \mathbf{E}\circ\dot{\boldsymbol{\varepsilon}}^{e} + \dot{\mathbf{E}}\circ\boldsymbol{\varepsilon}^{e},$$

$$\boldsymbol{\sigma}(\mathbf{x},t_0) = \boldsymbol{\sigma}(\mathbf{x},T_{kr}) = \mathbf{0},\ \boldsymbol{\varepsilon}(\mathbf{x},t_0) = \boldsymbol{\varepsilon}(\mathbf{x},T_{kr}) = \mathbf{0} \tag{3}$$

where: $\boldsymbol{\sigma}=\boldsymbol{\sigma}(\sigma_{\alpha\beta})$ the stress tensor, $\mathbf{E}$ is the tensor of material constants, $\boldsymbol{\varepsilon}^e$ is the tensor of elastic strains ($\boldsymbol{\varepsilon}^e=\boldsymbol{\varepsilon}-\boldsymbol{\varepsilon}^p-\boldsymbol{\varepsilon}^{TPh}$), $\boldsymbol{\varepsilon}$ is a total strains tensor, $\boldsymbol{\varepsilon}^p$ plastic strains, $\boldsymbol{\varepsilon}^{TPh}$ is isotropic tensor of thermal strains and phase transformations ($\boldsymbol{\varepsilon}^{TPh}=\boldsymbol{\varepsilon}^T+\boldsymbol{\varepsilon}^{Ph}$).

3 Phase transformation is the solid state, structural and thermal strains

Model the kinetics of phase transformations in the solid state on the base on continuous cooling diagram (CCT) for C60 carbon steel [6,7]. For determined heating and cooling rates austenitization temperatures (A_{c1}, A_{c3}) during heating, start (F_s, P_s, B_s, M_s) and final temperatures (F_f, P_f, B_f, M_f) of each phase transformation and final fractions of structure components during cooling are determined in UEXPAN subroutine. In this subroutine interpolated CHT and CCT diagrams are implemented with diagram of fractions of structural components (Fig. 4).

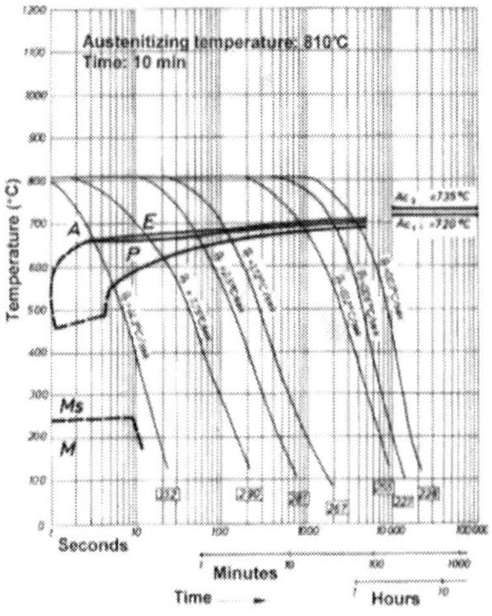

Fig. 4. Scheme of considered CCT diagrams

Phase fraction of austenite formed during heating is determined using a Machnienko model [7]. Phase fractions of the austenite during cooling process are determinated by temperature and cooling rate. Fraction of created phases such as bainite, ferrite and pearlite is also determinated by the model used for transformation of diffusion (Machnienko formula) taking account of the existing fractions of previously formed phases, ie.:

$$\tilde{\eta}_A(T,t)=1-\exp(-k\frac{T_{sA}-T}{T_{sA}-T_{fA}}), \quad \eta_{(\cdot)}(T,t)=\eta_{(\cdot)}^{\%}\tilde{\eta}_A(1-\exp)(-k(\frac{T_s^i-T}{T_s^i-T_f^i})) \tag{4}$$

Transformation of austenite into martensite is defined as follows [2,7]:

$$\tilde{\eta}_M(T)=\eta_{(\cdot)}^{\%}\tilde{\eta}_A(1-\exp(-k(\frac{M_S-T}{M_s-M_f})^m)), T\in[M_s,M_f] \tag{5}$$

where: M_s and M_f are receptively start and final temperatures of martensite transformation determined for specified cooling time in CCT diagram, while k and m factors are determined experimentally [2].

Isotropic strains from the temperature and phase transformations (structural strains) are determined by formulas [8, 10]:

$$d\varepsilon^{TPh}=\sum_{i=1}^{i=5}\alpha_i\eta_i dT-\varepsilon_A^{Ph}d\eta_A, \quad d\varepsilon^{TPh}=\sum_{i=1}^{i=5}\alpha_i\eta_i dT+\sum_{j=2}^{j=5}\varepsilon_j^{Ph}d\eta_j \tag{6}$$

where: $i,j = A, B, F, M.$ and P, $\alpha_i=\alpha_i(T)$ are thermal expansion coefficients of austenite, bainite, ferrite, martensite and pearlite, $\varepsilon_i^{Ph}=\varepsilon_i^{Ph}(T)$ are isotropic deformations of phase transformation: the initial structure in austenite, austenite in bainite, ferrite, martensite or in pearlite.

4 Example

Computer simulation of temperature field, phase fractions of progressive hardening process was performed in Abaqus FEA for dimensions 120x25x10 mm thin plate (Fig. 2) made of C60 steel with assumed base material structure consist of ferritic–pearlitic structure (60% ferrite and 40% pearlite). Assumed in calculations hardening process parameters were set to: heat source power Q_1=1.45 kW, beam radius r=10 mm, and rate the heating source v=2,5 mm/s. In order to reduce computational time, symmetry of the joint was assumed in calculations. Calculated temperature field in the cross section (plane YZ) of the plate at the time 30[s], presented in Fig. 5, and in the cross section (plane XY) of the plate at the same time, presented in Fig. 6 allows the determination of heat affected zone geometry.

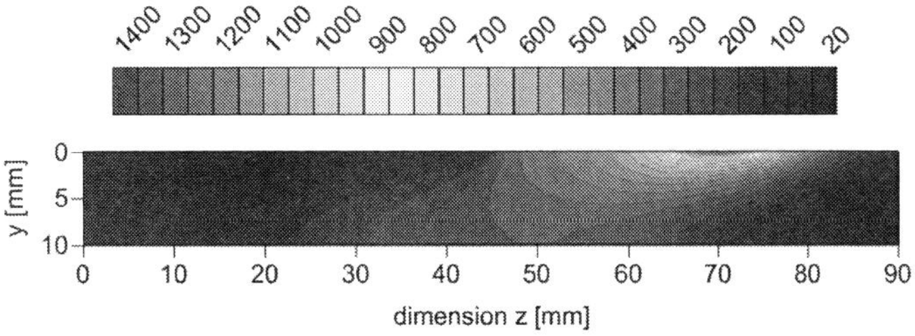

Fig. 5. Distributions (maps) of temperature (cross section YZ)

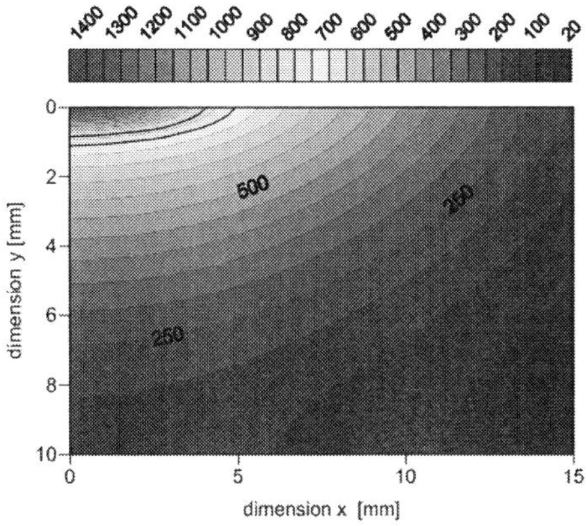

Fig. 6. Distributions (maps) of temperature (cross section YX)

Obtained distributions of phase fractions and kinetics of phase transformations from simulations (in cross section) are presented in Figure 7, and kinetics of phase transformations at point x = 0.

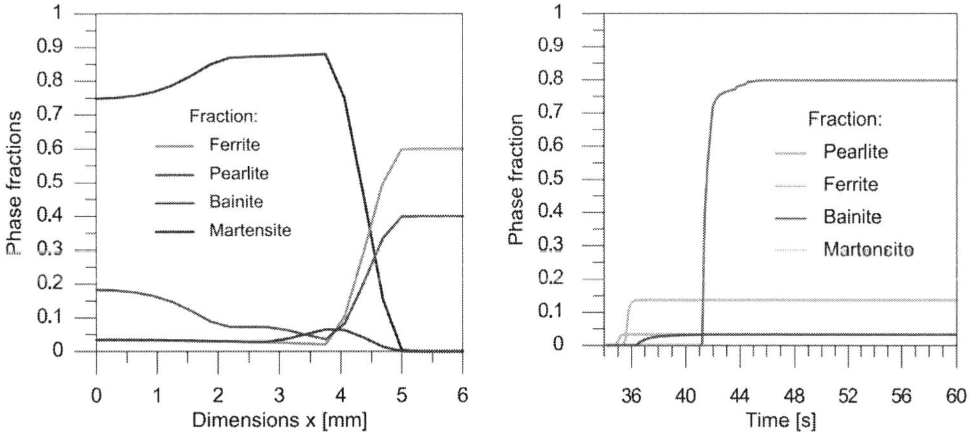

Fig. 7. Distributions of phase fractions in cross section, kinetics of phase transformations

Conclusions

During the analysis of findings obtained from the simulation of progressive hardening of the steel object, which was heated by means of the movable heat source, one may observe that after the adoption of the heating and cooling method with respect to the hardened flat, the retained martensite, bainite, pearlite and ferrite, occur only in surface layers. Presented model for determination of movable heat source power distribution, kinetics of phase transformation, CHT and CCT diagrams, implemented in additional subroutines used in ABAQUS FEA allowed numerical simulation of progressive hardening process.

In the cross-section YZ and YX obtained temperature field (Fig.5,6) allows the determination of fusion zone and heat affected zone geometry. With so selected parameters heating source depth of hardened zone reaches about 2 mm. In this zone behind the martensitic structure which has good mechanical properties. Developed in Abaqus FEA the 3D numerical model of hardening allows the prediction of joint structural composition. Estimated properties are helpful in determining a proper set of process parameters necessary to obtain desired geometry of object.

References

1. S. H. Kang, Y. T. Im, *Three-dimensional thermo-elastic-plastic finite element modeling of quenching process of plain carbon steel in coulee with phase transformation*. Journal of Materials Processing Technology **192-193**, 381-390 (2007)
2. T. Domanski, A. Bokota, *The numerical model prediction of phase components and stresses distributions in hardened tool steel for cold work*. International Journal of Mechanical Sciences **96-97**, 47-57 (2015)
3. O. C. Zienkiewicz, R. L. Taylor, *The finite element method*. (Butterworth-Heinemann, Fifth edition, vol. 1, 2, 3, 2000)
4. L. Huiping, Z. Guoqun, N. Shanting, H. Chuanzhen, *FEM simulation of quenching process and experimental verification of simulation results*. Material Science and Engineering A **452-453,** 705-714 (2007)
5. W. Piekarska, M. Kubiak, M. Żmindak, *Issues in numerical modeling of phase transformations in welded joint*. Procedia Engineering **177**, 141- 148 (2017)
6. T. Skrzypczak, E. Węgrzyn-Skrzypczak, *Modeling of thermal contact through gap with the use of finite element method*. J. Appl. Math. Comput. Mech. **14** (4), 145-152 (2015)
7. K. J. Lee, *Characteristics of heat generation during transformation in carbon steel*. Scripta Materialia **40**, 735-742 (1999)
8. L. Sowa, P. Kwiaton, *Numerical analysis of stress fields generated in the gantry crane beam*. Procedia Engineering **177**, 218-224 (2017)
9. D. Y. Ju, W. M. Zhang, Y. Zhang, *Modeling and experimental verification of martensitic transformation plastic behavior in carbon steel for quenching process*. Material Science and Engineering A **438-440**, 246-250 (2006)
10. Z. Saternus, *Computer Methods for Determination of Deformations in Welded Closed Profile*. Procedia Engineering **177**, 188 – 195 (2017)

A computer simulation of solidification taking into account the movement of the liquid phase

Robert Dyja[1,*], *Elżbieta* Gawrońska[1], *Andrzej* Grosser[1]

[1]Czestochowa University of Technology, ul. J.H. Dąbrowskiego 69, 42-201 Częstochowa, Poland

Abstract. In the paper, we present the results of solidification simulation taking into account the movement of the liquid phase. The results are obtained from an author software which is implemented on the base of a stabilized finite elements method (Petrov-Galerkin formulation). Using that formulation the Navier-Stokes equation is solved together with the convection term (Boussinesq approximation). The Finite Element Method (FEM) formulation is responsible for solidification, approximating the solution of the heat conduction equation (with the internal heat source term responsible for the heat released during the phase transition). The movement of the liquid phase in a solidifying cast that is caused by convection can significantly affect the process of heat transfer from the casting to the mold, which in turn has an influence on the temperature distribution in the cast and may cause a change in the location of the defects. The presented results allow to assess under what conditions the effect of convection on the solidification process is significant.

Keywords: solidification, Navier-Stokes, Streamline Upwind Petrov-Galerkin, Pressure Stabilized Petrov-Galerkin

1 Introduction

Solidification is still one of the most difficult process for modelling. Many physical processes that takes place from micro scale up to macro adds present significant difficulties, when someone tries to prepare suitable computational model. During years, many researchers came with models, that could give reasonable results [1, 2].

However, many models had some simplifications and limits. Example of such limit was omitting forces of convection in liquid metal. This can lead to significant differences between temperature map from simulation and temperatures observed in real casting. Additional problem is that without convection it is impossible to model such things like distribution of species in casting or solidification shrinkage, which are important factors of castings quality [3].

However, increasing complexity of models, present problem in their implementation. Nowadays, complex models must be implemented with use of technologies like parallel computers [4], accelerated architectures like GPU [5], or using special organization of calculations [6].

* Corresponding author: robert.dyja@icis.pcz.pl

Reviewers: *Juraj* Grenčík, *Kateryna* Kravchenko

2 Mathematical model

The governing equation for modelling solidification process is based on heat transfer equation with source term:

$$\rho c\dot{T} + \rho c(\mathbf{u}\cdot\nabla)T = \lambda\nabla^2 T + \dot{q}, \tag{1}$$

where ρ is density [kg/m^3], c is specific heat [J/(kg K)], T is temperature [K], $\mathbf{u}$ is velocity from convection force [m/s], λ is thermal conductivity [W/(m K)] and q is heat source connected with heat of solidification [J/m^3].

Boundary conditions, that are used in model with equation (1) are the Newton boundary condition on external sides of mold and contact condition for heat exchange between casting and mold.

With use of the apparent heat capacity formulation it is possible to obtain following form of equation (1):

$$c^*\dot{T} + \rho c(\mathbf{u}\cdot\nabla)T = \lambda\nabla^2 T, \tag{2}$$

where c^* is the approximation of the effective heat capacity. There are various methods for obtaining this approximation. Here, the Morgan method is used:

$$c^* = \frac{H^t - H^{t-1}}{T^t - T^{t-1}}, \tag{3}$$

where H is enthalpy [J] and t in upper script is time level.

Liquid metal in this model is assumed to be Newtonian fluid. This allows to write Navier-Stokes set of equations as:

$$\rho\dot{\mathbf{u}} + \rho\big((\mathbf{u}\cdot\nabla)\mathbf{u}\big) - \nabla p + \rho\mu\big((\nabla\mathbf{u}) + (\nabla\mathbf{u})^T\big) + \rho\mu\frac{f_l}{K_\epsilon}\mathbf{u} = \rho\mathbf{f} \tag{4}$$

$$\nabla\cdot\mathbf{u} = 0, \tag{5}$$

where p is pressure [Pa], μ is viscosity [kg/(m s)], f_l is liquid fraction [-] and K_ε is the permeability of the mushy zone [m].

This equation needs to be accompanied by appropriate set of initial and boundary conditions. Initial $\boldsymbol{u}$ is set as initial condition and between mold and casting the no-slip condition is used.

The right hand side part of equation (4) describes body forces that arose in liquid. The first part is connected with buoyance force that is approximated by Boussinesq formula:

$$\mathbf{f} = -\beta\mathbf{g}(T - T_0), \tag{6}$$

where β is expansion coefficient [1/K], $\boldsymbol{g}$ is gravitational acceleration [m/s^2], T_0 is reference temperature [K], which in this case was temperature from initial conditions.

The last part of left hand side of equation (4) is drag force that appears in mushy zone from interaction between liquid and already solidified metal. This model assumes that solidified part is immovable [7]. The permeability of the mushy zone is approximated by the Kozeny-Carman equation:

$$K_\varepsilon = K_0\frac{f_l^3}{(1-f_l)^2}, \tag{7}$$

where K_0 is secondary dendrite arm spacing [m] and f_l is liquid fraction [-].

Liquid fraction and solid fraction are connected with simple relation:

$$f_l = 1 - f_s, \tag{8}$$

where value of f_s is taken from phase equilibrium graph relationship:

$$f_s = \frac{1}{1-k}\frac{T_L - T}{T_M - T_L}, \tag{9}$$

where T_L is liquidus temperature [K], T_M is solidification temperature of pure component [K] and k is solute partition coefficient [-].

Equations (1), (4) and (5) are discretized using Finite Element Method, which leads to following set of equations:

$$\mathbf{M}^S\mathbf{T} + (\mathbf{N}^S\mathbf{T} + \mathbf{K}^S)\mathbf{T} = 0 \tag{10}$$

$$\mathbf{Mu} + (\mathbf{N} + \mathbf{K})\mathbf{u} - \mathbf{Gp} + \mathbf{Du} = \mathbf{F} \tag{11}$$

$$\mathbf{G}^T\mathbf{u} = 0, \tag{12}$$

where superscript S is used to set apart matrices used in solidification equation and **T**, **u**, **p** are vectors of unknown temperature, velocity and pressure. Elements of matrices $\mathbf{M}^S$, **M**, $\mathbf{N}^S$, **N**, $\mathbf{K}^S$, **K**, $\mathbf{G}^T$, **G** and **D** can be calculated from the following formulas:

$$M^S_{\alpha\beta} = \int_\Omega c^* S_a S_b \, \mathrm{d}\Omega \tag{13}$$

$$N^S_{\alpha\beta} = \int_\Omega \rho c S_a u_k S_{b,k} \, \mathrm{d}\Omega \tag{14}$$

$$K^S_{\alpha\beta} = \int_\Omega \lambda S_{a,j} S_{b,j} \, \mathrm{d}\Omega \tag{15}$$

$$M_{\alpha\beta} = \int_\Omega \rho c S_a S_b \, \mathrm{d}\Omega \tag{16}$$

$$K_{\alpha\beta} = \int_\Omega \mu S_{a,k} S_{b,k} \delta_{ij} \, \mathrm{d}\Omega + \int_\Omega \mu S_{a,j} S_{b,i} \, \mathrm{d}\Omega \tag{17}$$

$$N_{\alpha\beta} = \int_\Omega \rho S_a u_k S_{b,k} \delta_{ij} \, \mathrm{d}\Omega \tag{18}$$

$$G_\alpha = \int_\Omega S_{a,i} S_b \, \mathrm{d}\Omega \tag{19}$$

$$D_{\alpha\beta} = \int_\Omega \rho\mu \frac{f_l}{K_\epsilon} S_a S_b \delta_{ij} \, \mathrm{d}\Omega \tag{20}$$

$$G^T_\beta = \int_\Omega S_{b,j} S_a \, \mathrm{d}\Omega \tag{22}$$

$$F_\alpha = \int_\Omega \rho S_a f_i \, \mathrm{d}\Omega, \tag{23}$$

where α, β are indices of single element matrix, a and b have range up to number of nodes in element, i, j, k have range up to number of dimensions, S are shape functions of finite elements and δ is Kronecker delta.

Equations in this form can be solved with carefully selected finite elements [8]. In this work, approach that uses stabilized Finite Element is used [9], which allows to avoid limits imposed by the Ladyzhenskaya-Babuska-Breezi condition.

Special care is needed for treating drag force part in stabilization [10]. This work uses approach, that determines values of stabilization coefficient not only by velocity of liquid, but also limits them proportionally to volume of liquid fraction.

In calculations, it is assumed that small time step is used, hence in determination of actual values of material properties, when temperature was needed, temperature from previous time step was used. This strategy allows to treat solidification equation as linear and even solve it independently from Navier-Stokes equations, which gives better overall performance [11]. Moreover, such approach makes possible to use lumped mass matrix in solidification equation [11, 12].

With those assumptions in mind and using theta scheme for time integration, the final form of equations solved in the model is:

$$\begin{aligned}[\mathbf{M}^s + \mathbf{M}^s_{SUPG} + \Delta t\theta(\mathbf{N}^s + \mathbf{N}^s_{SUPG} + \mathbf{K}^s)]\mathbf{T}^{t+1} = \\ = [\mathbf{M}^s + \mathbf{M}^s_{SUPG} + \Delta t(1-\theta)(\mathbf{N}^s + \mathbf{N}^s_{SUPG} + \mathbf{K}^s)]\mathbf{T}^t\end{aligned} \quad (24)$$

$$\begin{aligned}\left[\mathbf{M} + \mathbf{M}_{SUPG} + \Delta t\theta\left(\mathbf{N} + \mathbf{N}_{SUPG} + \mathbf{K} + \mathbf{D} + \mathbf{D}_{SUPG}\right)\right]\mathbf{u}^{t+1} - \Delta t\mathbf{G}\mathbf{p}^{t+1} + \\ +\left[\mathbf{M} + \mathbf{M}_{SUPG} + \Delta t(1-\theta)\left(\mathbf{N} + \mathbf{N}_{SUPG} + \mathbf{K} + \mathbf{D} + \mathbf{D}_{SUPG}\right)\right]\mathbf{u}^t = \\ = \Delta t\left[\theta\left(\mathbf{F} + \mathbf{F}_{SUPG}\right) + (1-\theta)\left(\mathbf{F} + \mathbf{F}_{SUPG}\right)\right]\end{aligned} \quad (25)$$

$$\begin{aligned}\left[\mathbf{M}_{PSPG} + \Delta t\theta\left(\mathbf{G}^T + \mathbf{N}_{PSPG} + \mathbf{D}_{PSPG}\right)\right]\mathbf{u}^{t+1} - \Delta t\mathbf{G}_{PSPG}\mathbf{p}^{t+1} + \\ +\left[\mathbf{M}_{PSPG} + \Delta t(1-\theta)\left(\mathbf{G}^T + \mathbf{N}_{PSPG} + \mathbf{D}_{PSPG}\right)\right]\mathbf{u}^t = \\ = \Delta t[\theta\mathbf{F}_{PSPG} + (1-\theta)\mathbf{F}_{PSPG}],\end{aligned} \quad (26)$$

where matrices with *SUPG* and *PSPG* are terms supplied by stabilization, Δt is time step and θ is parameter determining type of time integration scheme (0 for Euler Backward and 0.5 for Cranck-Nicolson).

This model uses Newton-Raphson approach for linearization of Navier-Stokes equations.

3 Simulations

Model described in section 2 was implemented as an in-house C++ program. This program uses TalyFEM [13] and PETSc [14] libraries. TalyFEM library implements finite element formulas like calculations of shape functions or Gauss quadrature. PETSc library is used for linear algebra operations.

The results for two cases are presented. All simulations use linear triangle finite elements. First is small casting (square shape with 0.05 [m] side length), the second is larger one (0.5 [m] side length), see Fig. 1. Those two cases allow to asses importance of convection in solidification.

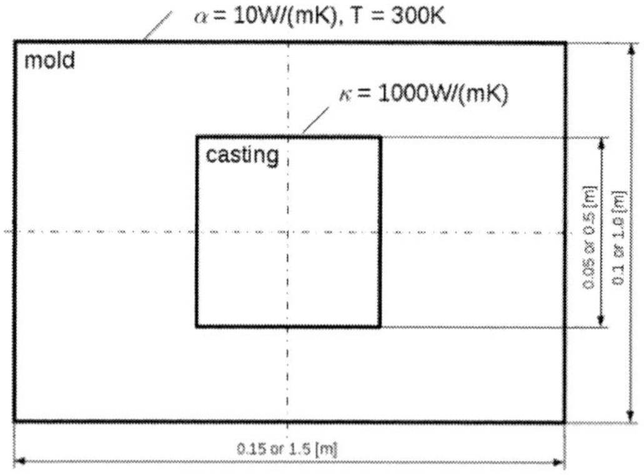

Fig. 1. View of domain shape used in simulations

The boundary conditions utilized following parameters: Newton boundary condition on all sides of mold had the environment temperature equal to 300 [K] and heat exchange coefficient was equal to 10 [W/(m K)]. Continuity condition assumed value of 1000 [W/(m K)] for heat transfer through separation layer.

The material properties are summarised in Table 1 (for casting) and Table 2 (for mold).

Table 1. Material properties for casting

Quantity name	Unit	Value
Density ρ_s	kg/m^3	2824
Density ρ_l	kg/m^3	2498
Specific heat c_s	J/(kg K)	1077
Specific heat c_l	J/(kg K)	1275
Thermal conductivity λ_s	W/(m K)	262
Thermal conductivity λ_l	W/(m K)	104
Solidus temperature T_s	K	853
Liquidus temperature T_l	K	926
Solidification temperature of pure component T_M	K	933
Eutectic temperature T_E	K	821
Heat of solidification L	J/kg	390000
Solute partition coefficient k	-	0.125
Viscosity μ	kg/(m s)	0.004
Expansion coefficient β	1/K	0.0001
Secondary dendrite arm spacing K_0	m	1.4*10^{-11}

Table 2. Material properties for mold

Quantity name	Unit	Value
Density ρ_s	kg/m^3	7500
Specific heat c_s	J/(kg K)	620
Thermal conductivity λ_s	W/(m K)	40

The first series of results present temperature map in four time moments (12.5 [s], 25.0 [s], 37.5 [s], 50.0 [s]) for small square. It is presented in Fig. 2. The effect of convection is only slightly visible in the beginning of simulation. With increasing time, effects of convection are less visible. However, in time moment 50 [s], location of hot spot can be viewed as minimally moved toward top. In this case symmetry of results is easily observable.

Corresponding pressure and velocity map is presented in Fig 3. The pressure is presented as colour map, while velocity is presented as black vectors. Their size and density corresponds to the magnitude of velocity. The main observation here is that even in places with mushy zone there are no visible fluctuations of pressure, which shows effectiveness of used stabilization technique. After 25 [s] there is almost no liquid phase movement.

In larger case the effect of convection is better visible. It can be observed as fluctuations of temperatures presented in Fig. 4. In Fig. 5, we can see that during whole simulation time, liquid metal had nonzero velocity.

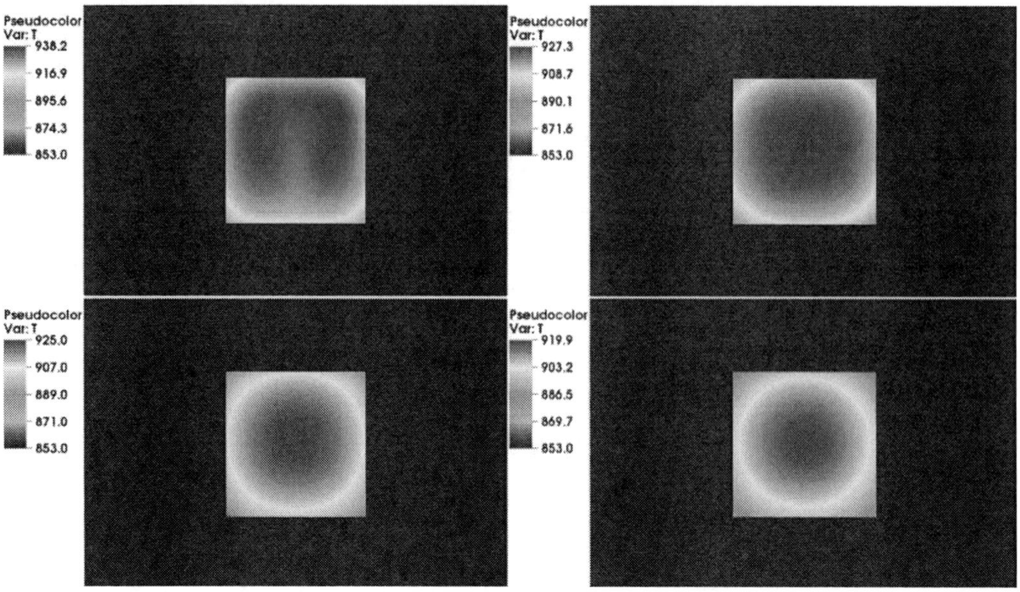

Fig. 2. Temperature in different time moments (12.5 [s], 25.0 [s], 37.5 [s], 50.0 [s] - starting from upper left corner). Size of single side of casting is 0.05 [m]

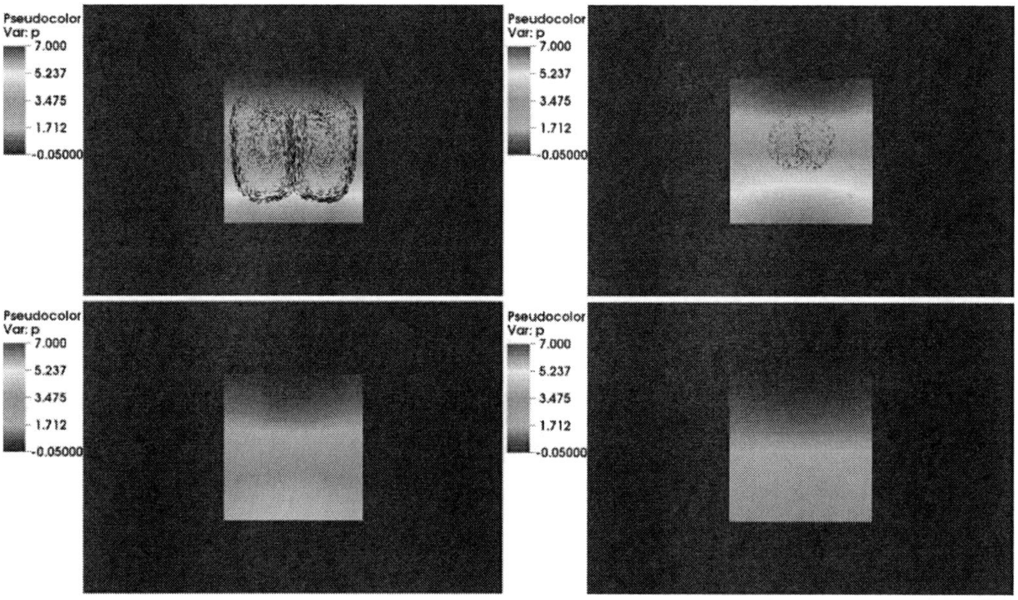

Fig. 3. Pressure with velocity vectors in different time moments (12.5 [s], 25.0 [s], 37.5 [s], 50.0 [s] - starting from upper left corner). Size of single side of casting is 0.05 [m]

More detailed comparison can be done with help of plots presented in Fig. 6 and Fig. 7, where also results from case without convection are used. Fig. 6 presents cooling curves taken in different places of casting (in centre, top, bottom and close to right wall). When there is no convection, curves from bottom, top and right spot are the same. However, with convection, it can be easily seen, that bottom cools faster then top spot.

Fig. 7 shows results of the same analysis, but for the large case. Here differences are even more visible. Without convection, the results from places close to walls are the same.

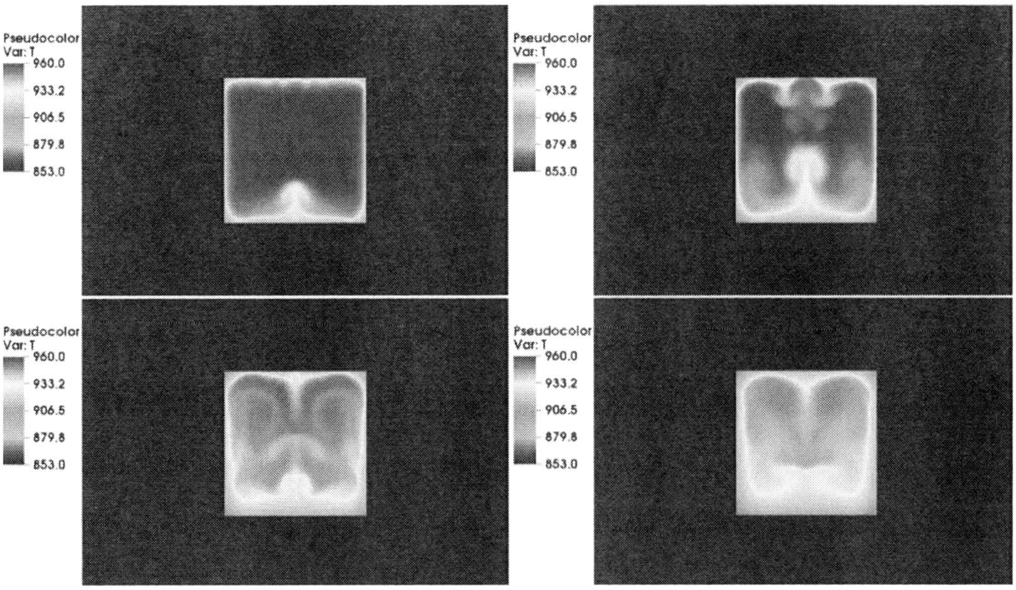

Fig. 4. Temperature in different time moments (12.5 [s], 25.0 [s], 37.5 [s], 50.0 [s] - starting from upper left corner). Size of single side of casting is 0.5 [m]

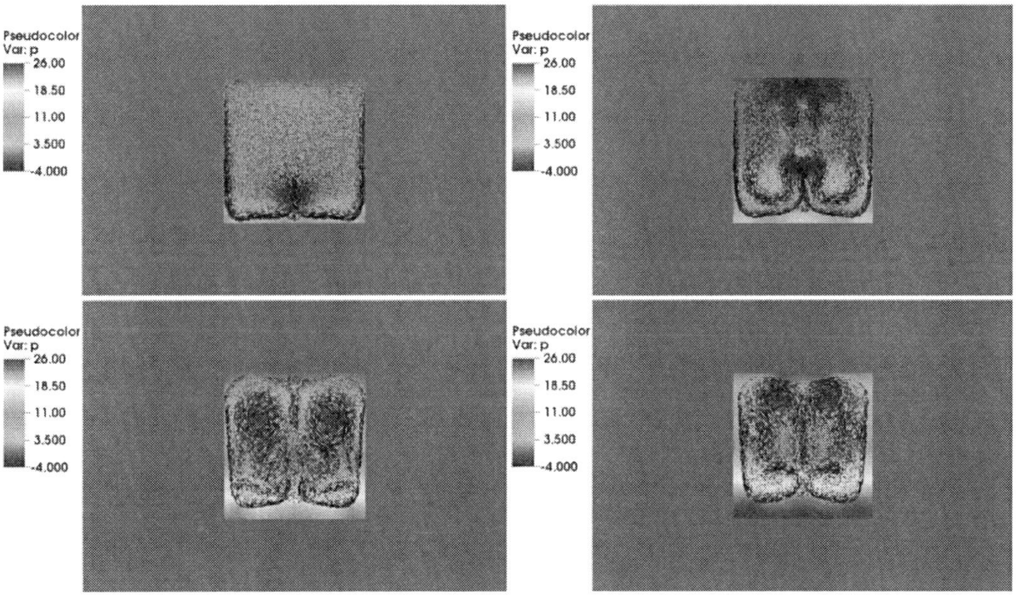

Fig. 5. Pressure with velocity vectors in different time moments (12.5 [s], 25.0 [s], 37.5 [s], 50.0 [s] starting from upper left corner). Size of single side of casting is 0.5 [m]

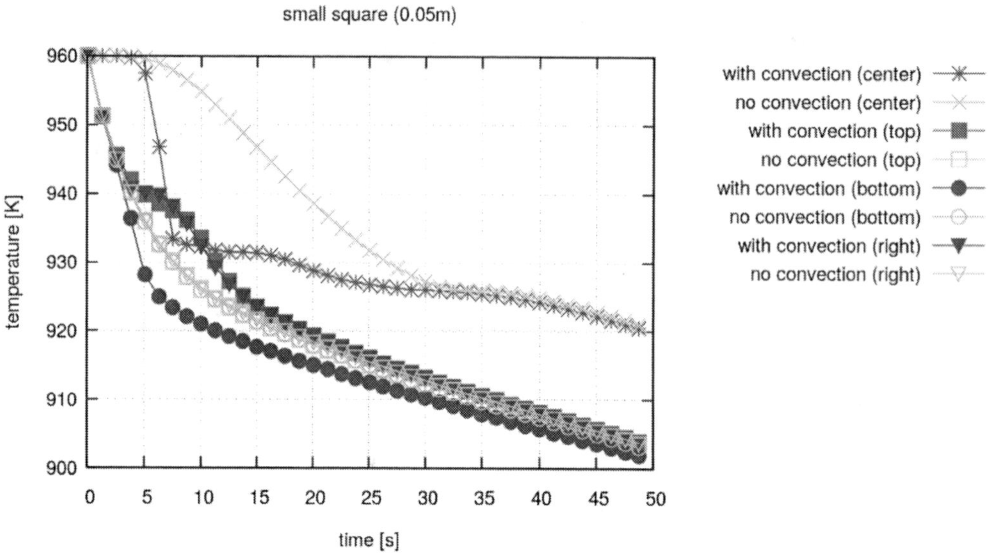

Fig. 6. Cooling curves for the small case obtained in different places of casting. As a comparison, cooling curves, for cases without convection are also plotted

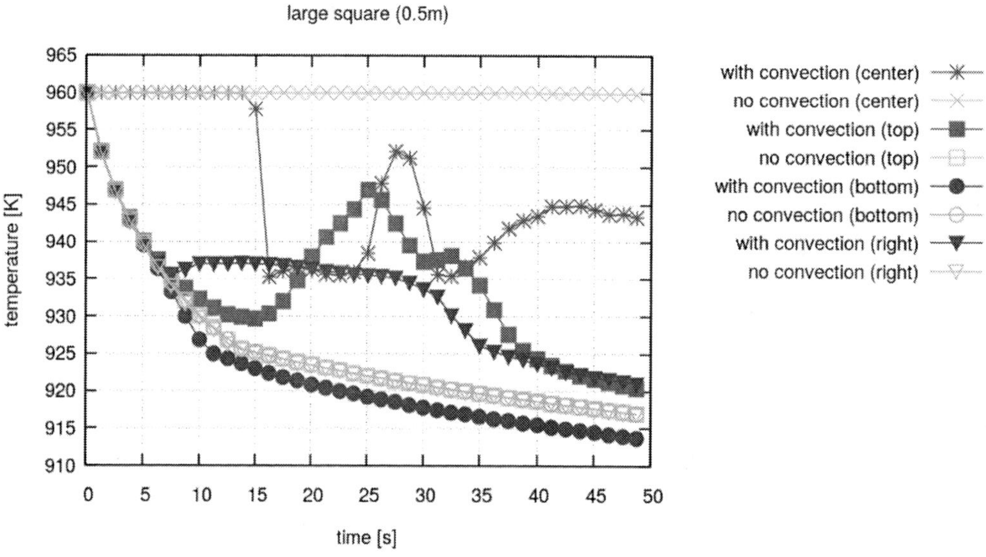

Fig. 7. Cooling curves for the large case obtained in different places of casting. As a comparison, cooling curves, for cases without convection are also plotted

Conclusion

Presented work tackles with problem of convection in solidification simulations. It presents development of model and difficulties linked with presence of momentum and continuity equations. It shows study of conditions when convection can be a significant factor. It also presents comparison with model that does not take into account this phenomena. Presented results shows that even for small casting convection can affect results in comparison with immovable liquid state.

Plans for future work include experimental comparison of results, because presented model was only checked against benchmark problems so far.

References

1. A. Bokota, S. Iskierka, *Finite element method for solving diffusion-convections problems in the presence of a moving heat point source*. Finit. Elem. Anal. Des. **17** (2), 89-99 (1994)
2. T. Skrzypczak, E. Wegrzyn-Skrzypczak, J. Winczek, *Effect of natural convection on directional solidification of pure metal*. Arch. Metal. Mater. **60** (2), 835-841 (2015)
3. D. M. Stefanescu, *Science and Engineering of Casting Solidification*. (New York, Kluwer Academic, 2002)
4. W. Feng. Q. Xu, B. Liu, *Microstructure simulation of alluminium alloy using parallel computng technique*. ISIJ International **42** (7), 702-707 (2002)
5. G. Michalski, N. Sczygiol, *Using CUDA architecture for the computer simulation of the casting solidification process*. Hong Kong : Lecture Notes in Engineering and Computer Science, 933-937 (2014)
6. E. Gawronska, N. Sczygiol, *Application of mixed time partitioning methods to raise the efficiency of solidification modelling*. 12th International Symposium on Symbolic and Numeric Algorithms (SYNASC), 99-103 (2010)
7. W. D. Bennon, F. P. Incropera, *A continuum model for momentum, heat and species transport in binary solid-liquid phase change systems - I. Model formulation*. Int. J. Heat Mass Transf. **30** (10), 2161-2170 (1987)
8. F. Brezzi, *On the existence, uniquess and approximation of saddle-point problems arising from lagrangian multipliers*. Esaim Math. Model. Numer. Anal. **8** (R2), 129-151 (1974)
9. A. N. Brooks, T. J. R. Hughes, *Streamline Upwind/Petrov-Galerkin formulations for convection dominated flows with particular emphasis on the incompressible Navier-Stokes equations*. Computer Methods in Applied Mechanics and Engineering – Special Edition on the 20th Anniversary, 199-259 (1990)
10. N. Zabaras, D. Samanta, *A stabilized volume-averaging finite element method for flow in porous media and binary alloy solidification processes*. Int. J. Numer. Methods Eng. **60** (5), 1-38 (2004)
11. R. Dyja, E. Gawronska, A. Grosser, *Numerical problems related to solving the Navier-Stokes equations in connection with the heat transfer with the use of FEM*. Procedia Eng. **177**, 78-85 (2017)
12. M. Zych, *Effect of mass matrix forms on numerical simulation results in heat conduction modeling*. J. Appl. Math. Comput. Mech. **14** (3), 149-156 (2015)
13. H. K. Kodali, B. Ganapathysubramanian, *A computational framework to investigate charge transport in heterogeneous organic photovoltaic devices*. Comput. Methods Appl. Mech. Eng. **247**, 113-129 (2012)
14. S. Balay, W. D. Gropp, L. C. McInnes, B. F. Smith, *Efficient Management of Parallelism in Object Oriented Numerical Software Libraries*. Modern Software Tools in Scientific Computing, 163-202 (1997)

Engineering calculations for complex geometric domains

Elzbieta Gawronska[1,*], *Robert* Dyja[1], *Andrzej* Grosser[1], *Jerzy* Winczek[1]

[1] Faculty of Mechanical Engineering and Computer Science, Czestochowa University of Technology, Dabrowskiego 69, 42-201 Czestochowa, Poland

Abstract. The paper aims to present the possibility of performing engineering calculations in domains of complex shapes expressed by mathematical descriptions. We focus on calculations done with the use of the Finite Element Method. An example of the compound structure is gyroid, which is a periodic structure representing an area close to the porous structure. The presented gyroid structure exhibits circular struts and a spherical core and spatially occupies a cube. The side length of the cube is defined as unit cell size, and the volume percentage of the struts inside the cube is referred to as volume fraction. The periodic cellular lattice structures are generated by our own software module. One of the uses of such structures is their employment in additive manufacturing (AM) of the so-called 3D printing (layer-by-layer AM technique), where they can contribute reducing the weight of an item and limiting the material consumption. It is important to answer the question of what effect does an element with a gyroid structure have on the thermal properties (e.g. the heat flow versus the volume fraction) compared to an element with full structure. The paper will show the results of such a comparison.

Keywords: gyroid, volume fraction, complex shapes, heat flow, numerical modeling

1 Introduction

3D printing was invented in 20th century, but its industrial use was quite limited. It has changed over the last years. Additive Manufacturing (AM) processes take information from a computer-aided design (CAD) file that is later converted to a stereolithography (STL) file. CAD systems are dedicated to the design of parts manufactured by traditional methods and are not helpful to designers who want to exploit the extensive opportunities offered by AM technology. For this process, the drawing made in CAD software is approximated by discrete elements and sliced to contain the information of each layer that is going to be printed. AM technologies build near-net shape components layer by layer using 3D model data and so they are the direct descendants of the work in 3D printing are revolutionary to many sectors of manufacturing insofar as they reduce part lead time, cost, material waste, and energy usage [1]. Furthermore, AM has the potential to enable modern designing that could not be manufactured using conventional methods and to extend the life of in-service parts through innovative repair methodologies. Applications, where legacy components are

* Corresponding author: gawronska@icis.pcz.pl

Reviewers: *Vladimír* Dekýš, *Ivan* Kuric

still necessary for operation and fabricators, no longer manufactured, for instance aging aircraft systems or older power stations, could use AM to create parts directly from a CAD file. For industrial sectors, one of the most crucial values of additive manufacturing is the ability to reduce weight while maintaining mechanical performance. The advantages are conclusive since AM can result in lower material costs, significant reductions in production time and increasing design flexibility [2]. AM is transforming approaches in medicine and making work easier for architects. For consumer products, artistic and aesthetic designs can be directly manufactured from a CAD file without concern for standard manufacturing practice. Many of the traditional designs for traditional production principles can no longer be applicable if parts with complex internal features have to be produced. AM technology is used to create elements previously difficult or impossible to be manufactured with traditional techniques [3].

One of the main advantages of AM methods is the ability to produce parts which have very complex geometry and which are made by prevailing methods in a very complicated way. For AM technology to be included in the industrial production of the real components, that high mechanical properties of produced parts are achieve. Moreover, a geometrically defined structure with complex shapes which allows weight reduction must have sufficient toughness. The final properties of manufactured parts are strongly dependent on each laser produced track and layer [4].

Gyroid structure is a complex structure which allows weight reduction and has circular and smooth struts with a spherical core. The circular and smooth struts of the unit cell continuously varying along the spherical core are presented in Fig. 1.

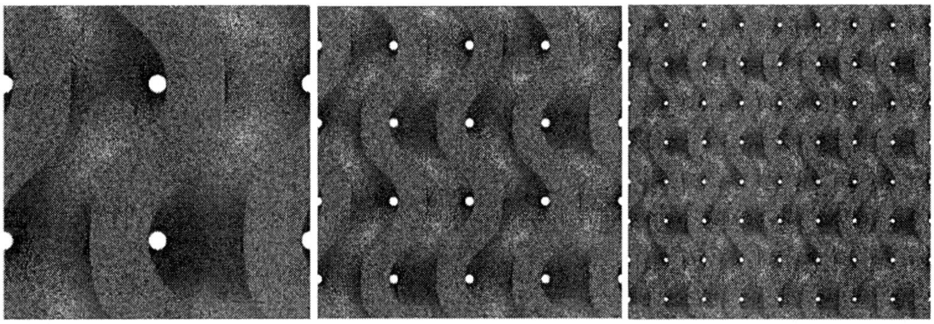

a) b) c)

Fig. 1. View through gyroid structures with a) one b) two c) four unit cells per one dimension

Additionally, the design of gyroid structures allows easy removal of the loose metal powder trapped inside of these structures after the building process of a particular part is completed. This inter-connected network of an open structure is well suited for unnecessary material removal [5]. The concept of designed cellular lattice materials is motivated by the desire to put material only where it is needed for a particular application. From a mechanical engineering viewpoint, a key advantage offered by cellular materials is high strength accompanied by a relatively low mass. These materials can provide excellent energy absorption characteristics as well good thermal and acoustic insulation properties. Designed cellular structures typically exhibit stronger structure strength per unit weight than typical foam structures [6].

In the paper, we use gyroid structure which is a homogeneous, isotropic structure, preserving the same mechanical properties in each direction. A team of researchers at MIT has designed one of the strongest lightweight materials known and has made the simulations with 3-D gyroid forms created from this matter. The results of tensile and compression tests showed that the construction could be ten times as strong as steel but much lighter. Their findings proved that the crucial aspect of performed tests has more to do with the unusual geometrical configuration than with the material itself. The researchers suggest that similar durable, lightweight materials could be made from a variety of matter by creating similar geometric features [7]. Isotropy of that structures is very useful because of there is no difference in the physical properties such as thermal expansion, heat conduction, electrical conductivity, refractive index, etc. regardless of the direction in which they are measured. In the paper, we consider heat flow. Results revealed that the type of structure and volume fraction are the main factors influencing the heat flow pace.

2 Methods

When dealing with solid 3D objects, it is often convenient to only model the object's boundaries using a mathematical representation of the surfaces. The choice of surface representation is particularly important for the computational modeling of 3D objects as each depiction has its advantages and disadvantages. These include the availability and complexity of operations that can be used to manipulate the surface (e.g. smoothing, Boolean operations) as well as the efficiency of the representation. Each formula also has an impact on how models are visualized and ultimately realized (e.g. via AM). The most common surface's depiction can be classified as one of the following forms: explicit, parametric or implicit.

The most common type of explicit surface representation is a mesh of polygons. Typically, these are triangles or quadrilaterals although others can be used. These polygons are often stored as an order list of vertex indices - the order being used to define the direction the polygon is facing (i.e., the surface normal) [8].

Unlike explicit surfaces, parametric surfaces do not store points on a surface. Instead, points on a parametric surface are expressed as a function of the parametric variables (u, v), which can be generalized to lie on the unit square [0,1]×[0,1]. Non-Uniform Rational B-Spline (NURBS) surfaces are a form of parametric surface commonly used due to their compact representation, smooth surfaces and easy of manipulation [9, 10].

Implicit surfaces are defined as an iso-surface of some function f. In 3D the surface is defined by a set of points p∈R3 satisfying the equality:

$$f(x, y, z) = 0. \tag{1}$$

As with parametric forms, implicit surfaces provide a compact representation for potentially complex surfaces. They also offer some advantages, notably their flexibility and well-defined Boolean operations. However, unlike parametric forms they provide little local shape control and manipulating them can be unintuitive. The implicit function approach for the structure design being used has been proved to be very versatile, as it allows geometries to be conceived just by pure mathematical expressions. This way, different cellular structures can be easily defined and optimized [11]. In the paper, gyroid lattice structures are mathematically defined, allowing precise control of volume fraction (depends on δ thickness) and unit cell size in the creation of an interconnected network:

$$\cos x \sin y + \cos y \sin z + \cos z \sin x = \delta, \tag{2}$$

thickness is normalized and takes values from 0 to 1. In the paper, it takes value $\delta = 0.8$.

The design of gyroid structures is based on efficiently generated and implicitly defined periodic structure and rapidly constructed finite element meshes. Finite element mesh for the full cube was generated with Gmsh software. The used version of Gmsh did not have the option of design NURBS surface, mesh for gyroid structure was generated with our own software using CGAL library (it contains procedures for describing the division of an area into finite elements). CGAL is a software project that provides easy access to efficient and reliable geometric algorithms in the form of a C++ library. CGAL is used in various areas which need geometric computation, such as geographic information systems, computer aided design, molecular biology, medical imaging, computer graphics, and robotics [12].

Gyroid lattice structures are mathematically defined, allowing precise control of volume fraction and unit cell size in the creation of an interconnected network. The volume fraction is defined as the volume percentage of the solid material in the cellular lattice structure. The gyroid surface was used to generate periodic lattice structures with the same unit cell sizes $u = 0.4$ [m] and average volume fraction of $\gamma \approx 0.52$. The designed models of the gyroid with the finite element meshes (built in the cube presented in Fig. 2) generated for every considered lattice structure are shown in Fig. 3, Fig. 4 and Fig. 5.

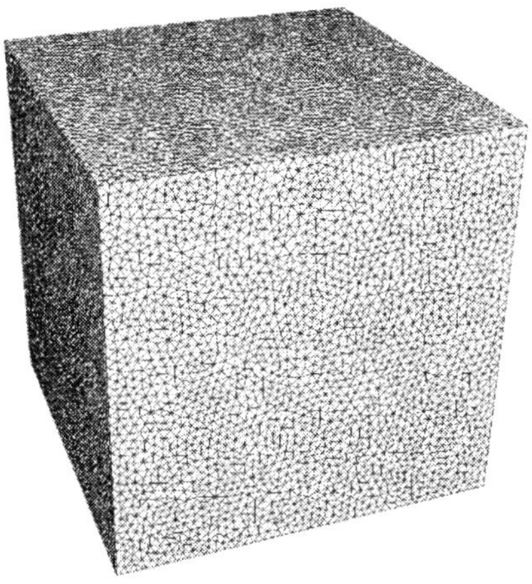

Fig. 2. Cube mesh

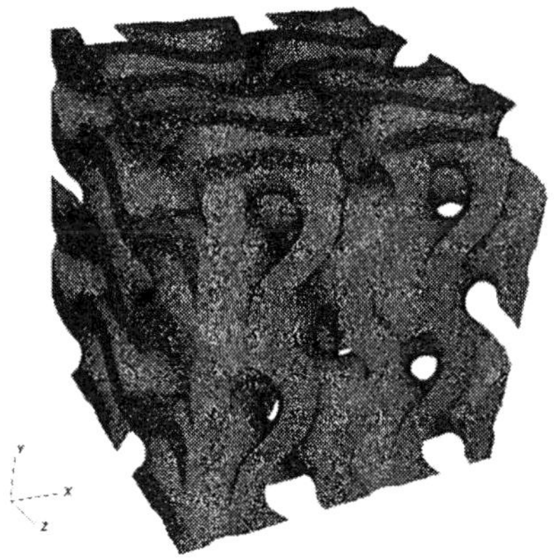

Fig. 4. Gyroid mesh (two unit cells per one dimension)

Fig. 3. Gyroid mesh (one unit cell per one dimension)

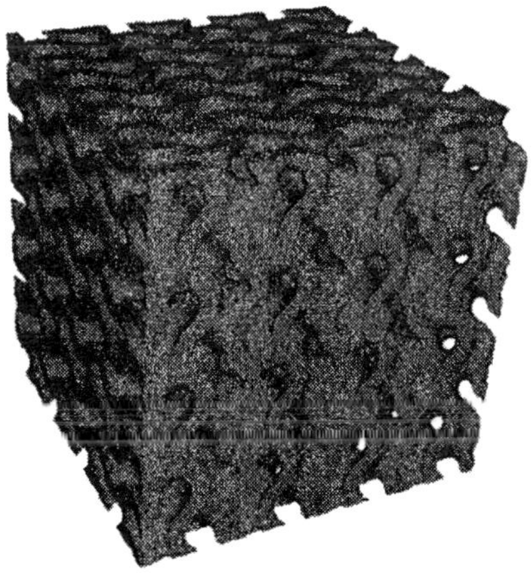

Fig. 5. Gyroid mesh (four unit cells per one dimension)

3 Numerical simulations

The article presents the results of the heat conduction simulations for a cube whose face has 0.04 [m] edge length and for the gyroid lattice structures inserted in the cube. The following physical properties were used in simulations: λ - heat transfer coefficient 260 [W/(m·K)], ρ - density 2800 [kg/m^3], c - specific heat 1000 [J/(kg·K)]. Such material properties are similar to those of aluminum alloys - material significant and often used in industry. The boundary condition is a heat flux established on one surface with a fixed value of 10000 [W/m^2] (Neumann boundary condition). Four simulations were made: one for the full cube and three for the three different areas of the gyroid. For each simulation, 500-time steps of 0.5 [s] were made. The initial temperature is 300 [K]. The temperature distributions and their values are shown in Fig. 6, Fig. 7, Fig. 8 and Fig. 9 for the cube, the gyroid with one unit cell per dimension (gyroid-1), the gyroid with two unit cells per dimension (gyroid-2) and the gyroid with four unit cells per dimension (gyroid-4), respectively.

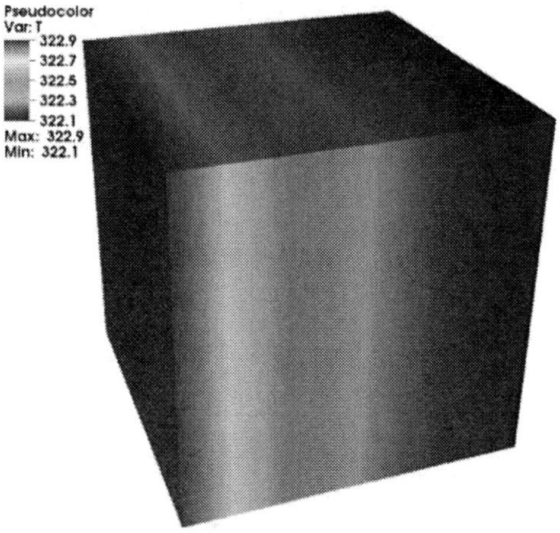

Fig. 6. The temperature distribution after 500 time steps - cube

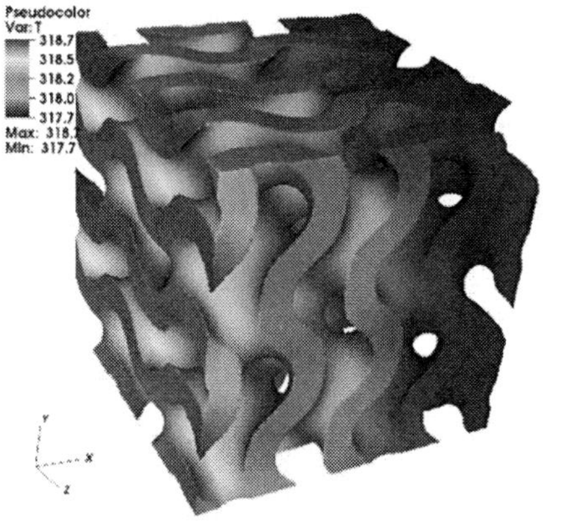

Fig. 8. The temperature distribution after 500 time steps - gyroid with two unit cells per one dimension

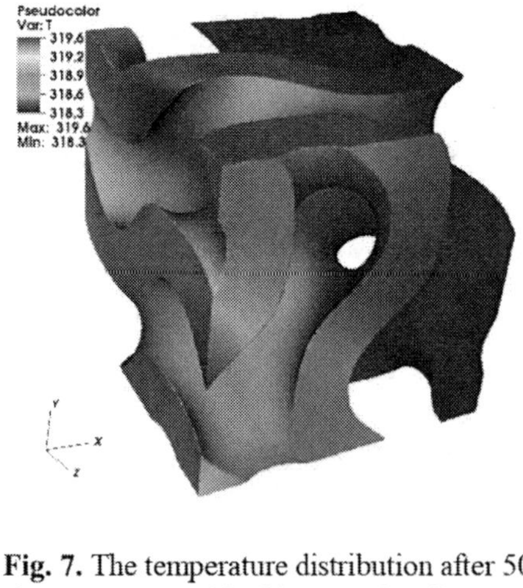

Fig. 7. The temperature distribution after 500 time steps - gyroid with one unit cell per one dimension

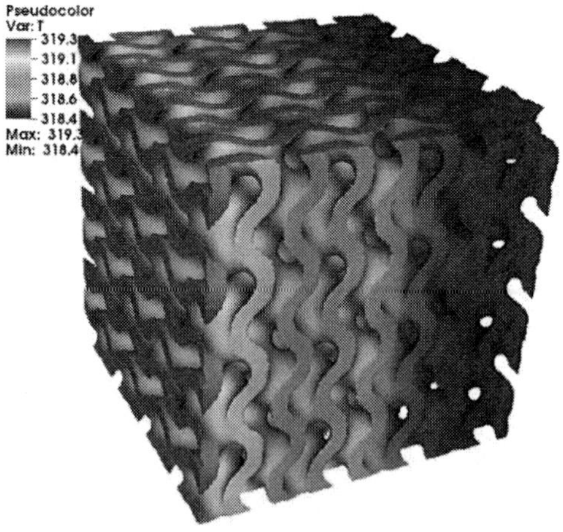

Fig. 9. The temperature distribution after 500 time steps - gyroid with four unit cells per one dimension

The gyroid meshes have much more finite elements than the cube since it was necessary to use ones which are finer to accurately recreate all details of the geometry (see Table 1).

Table 1. Characteristics of domain decomposition

Structure	Number of nodes	Number of elements
Cube	44567	241939
Gyroid-1	289543	1195995
Gyroid-2	469847	1823180
Gyroid-4	814535	3044948

Table 2 summarizes the data refers to the time that was needed to achieve the minimum temperature of 315 [K] (differences in calculated temperatures do not exceed 0.01 [K]). Of course, this temperature was at the furthest possible distance from the wall on which the boundary condition was set. The boundary condition was set only on one wall. The rest have nothing to set, which in case of heat conduction means excellent insulation. There is no air inside the gyroid structure or any heat exchange, just heat flows from the heated wall through the entire volume of the structure to the opposite wall.

Table 2. Characteristics of values important in numerical simulations

Structure	Area fraction ε	Volume fraction γ	t [s]
Cube	1	1	170.0
Gyroid-1	0.4336769	0.5215625	206.0
Gyroid-2	0.4203725	0.5223438	212.5
Gyroid-4	0.4265122	0.5228375	205.0

The presented data showcase that the gyroid structures heat up in almost the same time. That time, however, is longer than the full cube heating time. This difference can be explained by the fact that for an established boundary condition, the heat flux acting on the surface unit depends on a surface occupied by, e.g. a gyroid on the selected wall. By defining the area fraction ε as the ratio of the area held by a chosen structure a' to the area of the entire wall a:

$$\varepsilon = \frac{a'}{a}$$

for the cube a maximum value $\varepsilon_{max} = 1$ is obtained, while it is the smallest for the gyroid-2 structure. It appears that the amount of heat that flowed into the cube was the largest and thus the process of heating up was the fastest there, whereas for the structure of gyroid-2 the amount of heat was the smallest, so the structure was heating up the slowest. Using this property of gyroidal structures can be substantial for the industry in situations in which a given product should not get heated too quickly during its use.

Conclusions

The research conducted in this paper concerned engineering calculations for complex geometric domains. Gyroid structures are a good example of such domains. The use of mathematical representations in an implicit formulation has been presented to be quite straightforward and flexible for the generation of gyroid lattice structures. To recreate the gyroidal structure, this is needed much more of finer finite elements. There were conducted several numerical simulations with authorial software, and it was shown the achieved results of the heat flow through the complex shape of a structure are physically correct. The influence of area and volume fractions on heating up of chosen structure were settled - the gyroidal structures accumulate heat slower than full cubes, which can be substantial for the industry in situations in which a given product should not get heated too quickly during its use.

In future work, the realization of research for the gyroid structure with different geometry, various area and volume fractions, and different wall thickness is planned. Rapid heating of materials preceding rapid solidification induces high thermal variations resulting in residual stresses that can introduce cracks or undesired effects to the machine part. It is also an important aspect of future considerations.

References

1. H. Piili, A. Happonen, T. Väistö, V. Venkataramanan, J. Partanen, A. Salminen, *Cost Estimation of Laser Additive Manufacturing of Stainless Steel*. 15th Nordic Laser Materials Processing Conference, Nolamp, Physics Procedia **78**, 388-396 (2015)
2. C. Yan, L. Hao, A. Hussein, D. Raymont, *Evaluations of cellular lattice structures manufactured using selective laser melting*. Int. J. Mach. Tool Manu. **62**, 32-38 (2012)
3. K. V. Wong, A. Hernandez, *A Review of Additive Manufacturing*. ISRN Mechanical Engineering **2012**, 208760, (2012)
4. K. Monkova, P. Monka, *Some Aspects Influencing Production of Porous Structures with Complex Shapes of Cells*. Proceedings of 5th International Conference on Advanced Manufacturing Engineering and Technologies: NEWTECH 2017, 267-276 (2017)
5. C. Beyer, D. Figueroa, *Design and Analysis of Lattice Structures for Additive Manufacturing*, J MANUF SCI E-T ASME **138** (12), 121014 (2016)
6. H. Liang, R. D. Y. Chunze, H. Ahmed, P. Young, *Design and additive manufacturing of cellular lattice structures*. The International Conference on Advanced Research in Virtual and Rapid Prototyping (VRAP), Leiria, Innovative Developments in Virtual and Physical Prototyping, 249-254 (2011)
7. D. L. Chandler, *3-d-graphene-strongest-lightest-materials*. [Online] http://news.mit.edu/2017/3-d-graphene-strongest-lightest-materials-0106 (2017)
8. P. Zhang, J. Toman, Y. Yu, E. Biyikli, M. Kirca, M. Chmielus, A. C. To, *Efficient Design-Optimization of Variable-Density Hexagonal Cellular Structure by Additive*. Manufacturing: Theory and Validation, J. Manuf. Sci. E. **137** (2), 021004, 01 04 (2015)
9. L. Yang, O. Harrysson, D. Cormier, H. West, H. Gong, B. Stucker, *Additive Manufacturing of Metal Cellular Structures: Design and Fabrication*. JoM **67** (3), 608-615 (2015)
10. D. Ma, F. Lin, C. Chua, *Rapid Prototyping Applications in Medicine*. Part 1: NURBS-Based Volume Modelling, INT J ADV MANUF TECH **18** (2), 103-117 (2001)
11. G. Strano, L. Hao, R. M. Everson, K. E. Evans, *A new approach to the design and optimisation of support structures in additive manufacturing*. INT J ADV MANUF TECH **66** (9), 1247-1254 (2013)
12. https://www.cgal.org/ [Online] (Accessed 23 08 2017)

Noise and temperature reduction in the contact of tribological elements during braking

Juraj Gerlici[1], *Mykola* Gorbunov[2], *Kateryna* Kravchenko[1,*], *Olga* Prosvirova[2], *Tomas* Lack[1]

[1]Faculty of Mechanical Engineering, University of Zilina, Univerzitná 8215/1, 010 26 Zilina, Slovak Republic
[2]Faculty of Transport and Logistics, Volodymyr Dahl East Ukrainian National University, Central Avenue 59a, 93400 Sewerodonetsk, Ukraine

Abstract. This paper describes the problem of rising temperature of the friction elements during braking. Negative influence of brake pads heating on the vehicle efficiency and traffic safety is well known. Methods of temperature stabilization in the contact of tribological elements during braking are proposed. Variants of brake pad application with additional elements consisting of phase transition material are considered. Their effectiveness is estimated. The effect of the contact characteristics on noise generation is analysed. The method of noise reduction is considered. The target function for the tribological processes management in the contact of the brake elements has been developed. The application of this target function allows using the braking reserve with the achievement of maximum braking effect with minimal wear and noise.

Keywords: brake friction pair, noise generation during breaking, temperature stabilization, phase transition, target function

1 Introduction

Traffic safety is the key requirement when implementing new technologies in transport. Railway transport is the safest and the most energy efficient mode of transport. However, there is a problem of its use such as considerable noise, especially of freight trains at night, which negatively affects the ecological safety of the environment. Today increased noise is one of three the most harmful environmental factors. According to the EU Commission on Rail Transport, about 10% of the EU population is exposed to the harmful effects of noise from the railway transport, which is a health hazard, as well as the cause of anxiety and discomfort [1, 2]. Therefore, if the train sound level equals 60 dBA, the percentage of population anxiety is about 10%, at 75 dBA - 40%, and at 85 dBA the anxiety rate exceeds 70% [3].

The fight against noise is associated with serious costs: from € 150 to 200 million is spent annually only to install acoustic screens along the railways in Europe. The cost of replacing metal brake pads by composite ones for freight cars can range from € 2 to 2.5

[*] Corresponding author: kkatherina@ukr.net

Reviewers: *Milan* Sága, *Alžbeta* Sapietová

billion [3].The development of new technical proposals to reduce the noise impact of the rail transport requires the identification of the main sources of noise emission.

2 Research Analysis

The noise produced by the rolling stock can be conditionally divided into three groups shown in Fig.1. The noise level and its distribution in the groups, percentagewise, largely depend on the speed of movement, the design of the track, the rolling stock and its elements. The noise of equipment (compressors, traction electromotors, etc.) prevails at speeds up to 50-60 km/h. Rolling noise (the collision processes in the "wheel-rail" system and the friction in the "pad-wheel" system) prevails in the speed range of 60-300 km/h. The aerodynamic noise formed by the air flow around the rolling stock hull, the pantograph, etc. prevails at speeds over 300 km/h [1, 2, 3].

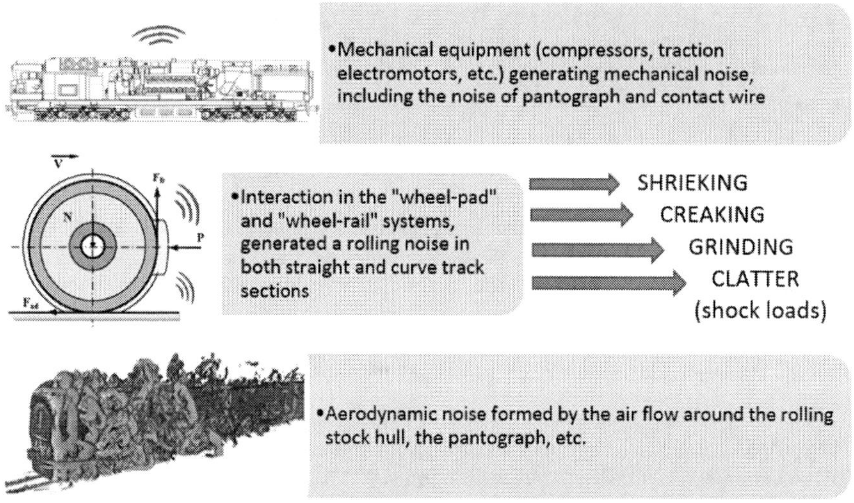

Fig. 1. Groups of noise generation by railway rolling stock operation

Despite the fact that the safety of the vehicle largely depends on the availability of an efficiently functioning braking system, its choice is the determining factor in the noise generation in the train. The results of the investigations [1-5], in which the list of train noise sources (Fig.1) and the mechanism of excitation of the main noise sources are determined, show that the reduction of the brake noise is achieved due to the stabilization of the friction coefficient (temperature) of friction bodies materials [1, 4, 5].This is due to the fact that during the braking the energy is released on friction surfaces that leads to the heating of the interacting bodies, to the structural transformations when critical temperatures are reached, to the change of the of material physical properties, and, ultimately, to the change in working conditions of friction pairs and the intensity of their wear, to the formation of defects and to an increase in noise levels. In addition, friction results in the micro-gripping of surfaces, which leads to the detachment of microparticles and the formation of squealing and creaking.

Stabilization of temperature and friction coefficient in the "wheel-pad" contact is a complex, multifactorial task, largely dependent on the conditions of interaction in the "wheel-rail" system. The pad and the rail are not interconnected constructively, however, they interact in the process of working with the same element, the wheel, and have a relationship through the loads and the physical and chemical properties of the materials. The efficiency of the "pad-wheel-rail" tribosystem, based on the properties of the materials, depends on their structure, which, in turn, varies depending on the heating conditions, that is, on the number and type of thermal cycles.

3 The target function for the tribological processes management in the contact of the brake elements

The increase in temperature in the "pad-wheel" contact and the associated adverse factors is an inevitable process. However, the use of rolling stock reserves allows temperature stabilization and prevents the formation of brake pads with critical temperatures in the tribocontact, which leads to the change in the structure of the material, and to increased wear and noise.

The reserve for the use of rolling stock is understood as follows.

In a multi-axle vehicle the maximum braking force is limited by the adhesion force of the limited, skid-prone axis, which first goes into excess slip mode, while the other wheel pairs do not reach their gripping maximums, that is, their braking capabilities, and as a result the whole vehicle capabilities are underutilized. This is due to uneven wear of wheel pairs, the difference in wheel diameters and static loads such as weight distribution, redistribution of loads from the wheel pair to the rails during braking, contact conditions, the difference of coefficients in the vertical and horizontal dynamics for each wheel pair, etc. [6, 7].The high braking force of some wheel pairs also leads to skidding, an increase in contact temperature, a decrease in safety and an increase in noise. At minimization of all unfavorable factors the reserve of braking force use by rail rolling stock will be maximum.

To estimate the braking system reserves, the target function is designed to minimize the difference between the braking force and the adhesion force of each wheel pair:

$$C_b = \sum_{i=1}^{n}\left(P_i\varphi_{i\_max} - B_i f_i\right) \rightarrow min \quad (1)$$

where P_i is the vertical load from the *i-th* wheel pair to the rails, [kN],
$\varphi_{i\ max}$ is the coefficient of adhesion of the wheel to the rail, [dimensionless value],
B_i is the force of pressing of a brake pad to a wheel, [kN],
f_i is the coefficient of pad friction, [dimensionless value].

The physical meaning of the target function is that, depending on the vertical load (P_i) and the physical and mechanical properties of the "wheel-rail" contact ($\varphi_{i\ max}$), it is necessary to adjust the force of pressing the brake pads (B_i), to stabilize the coefficient of friction in the "wheel-pad" contact (f_i) for account of the temperature regulation in this contact.

4 Innovative proposals for stabilizing the coefficient of friction (of temperature) in tribocontacts "wheel-rail", "wheel-pad"

Innovative proposals for stabilizing the coefficient of friction (of temperature) in tribocontacts "wheel-rail", "wheel-pad" and the use of reserve of the braking force by a rail vehicle.

1. An effective solution for the uniform distribution of loads from wheel pairs to rails is the use of loading devices. This solution can be used not only for braking, but also for traction [8, 9].

2. Stabilization and increase of the adhesion coefficient is achieved due to the supply of sand into the contact of the wheel with the rail. However, sand systems work inefficiently and lead to considerable wear of surfaces. This is due to the unstable supply of sand into the contact. According to Yu.I. Osenin the maximum and sufficient amount of sand in contact is 0.06 kg/m² [10]. At that, the sand should lie in one layer, the distance between the particles should be regulated and equal to three radii of the sand particles. To achieve this, it is proposed to electrify the sand before it is brought into contact. This will allow to distribute the sand in one layer with the necessary distance between the grains of sand [7, 11].

From the theoretical and experimental studies, it has been established that the angle of sanding depends on the electric charge transferred to the abrasive mixture. The determination of the charge, depending on the design of the injectors, must be carried out according to the following system of differential equations [7]:

$$m_p\ddot{y} = -E_{oy}\cdot q_p + m_p\cdot g - R_y + \frac{\varepsilon_0(\varepsilon-1)}{2}\nu\left(\dot{E}_{oy}\right)^2$$
$$m_p\ddot{r} = -E_{or}\cdot q_p - R_r + \frac{-\varepsilon_0(\varepsilon-1)}{2}\nu\left(\dot{E}_{or}\right)^2 \qquad (2)$$

where m_p, q_p are the mass and charge of the abrasive particle, E_{oy}, E_{or} is the field strength for electrostatic charging, R_y R_r is the resistance of air to particle motion, ε_o is the dielectric permeability of the medium; ε is the electric constant.

As a result of the simulation it was established that in order to achieve an effective sand distribution on the contact of the wheel with a rail of 60 mm width, the voltage applied to the charging device should be 450 V, and the electric charge of the grains should be $1{,}502\cdot10^{-11}$ C (15 pC).

As a result of the processing of the experimental data, the dependencies of friction coefficient from the temperature were obtained for different frictional states (Fig. 2, 3) [12]. The supply of electrified sand to oily rails makes it possible to increase the coefficient of friction in the contact from 0.25 to 0.4 (Fig. 3), to the rails covered with water from 0.35 to 0.5 (Fig. 2). Compared with the supply of non-electrified sand, the electrification effect is 16% for oily rails, 20% for water-covered rails.

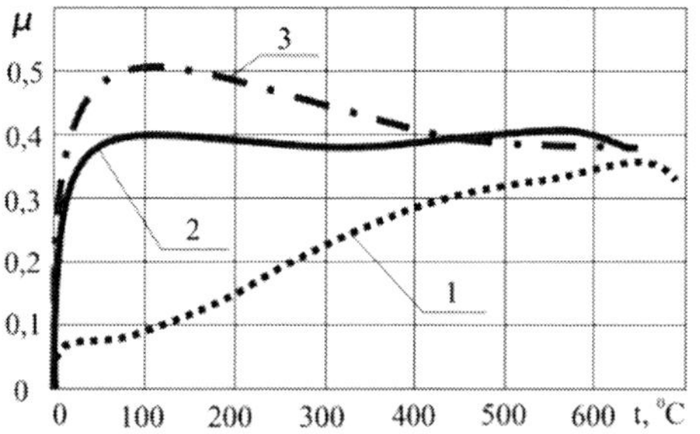

Fig. 2. The dependence of friction coefficient on temperature when supplying an electrified sand on the watered rails:

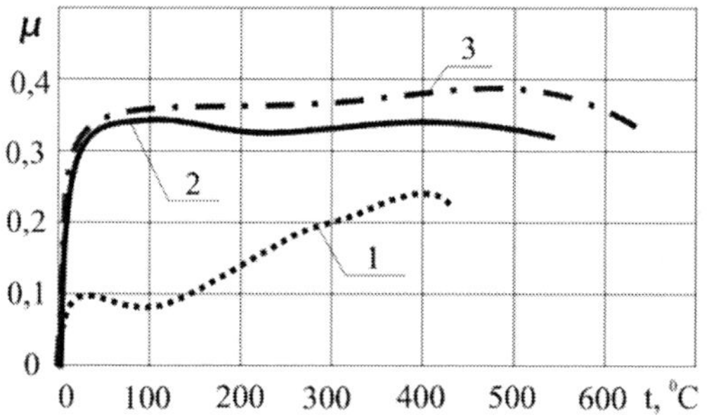

Fig. 3. The dependence of friction coefficient on temperature when supplying the electrified sand on the oiled rails:

1 – without sand; 2 – with sand; 3 – with electrified sand

3. It is advisable to stabilize the coefficient of friction by controlling the temperature in the contact of the wheel with the pad and removing heat from it.

An analysis of promising solutions for removing heat from the friction surface and stabilizing the temperature is carried out:

- forced air supply with the possibility of temperature control;
- the use of pads with porophore inserts;
- the use of pads with cooling fins;

- the use of brake linings whose outer surface is made of heat-dissipating material;
- local effect of external chemical elements (ozone, foaming agents, ionized air, etc.) on tribo-elements;
- the use of inserts from materials with thermal phase transitions in the construction of the pads [13].

One of the priority methods of creating the optimum temperature regime during braking is the use of additional materials with a phase transition of the first kind in the design of brake elements. The phase state is understood as the thermodynamic equilibrium state of the substance that differs in physical properties from other possible equilibrium states of the same substance. Within one aggregate state, the substance can be in several phase states, differing in their properties, composition and structure. The transition of a substance from one phase state to another - a phase transition - is always associated with qualitative changes in the properties of the substance (changes in the aggregate state of the substance or transitions associated with changes in the composition, structure, and properties of the substance). The first-order phase transition (thermal phase transition) is accompanied by heat absorption and is characterized by a constant temperature, changes in entropy and volume. The heat supplied to the body goes not to heating the body, but to breaking the interatomic bonds.

Structurally, the brake pad with increased energy capacity contains cavities for placing materials with thermal phase transitions of the first kind. For the reason that depending on the volume of the material and the temperature of the thermal phase transition, the absorbing capacity of the heat of the pad varies, subsequently the design provides for the presence of several types of materials with phase transitions. For example, material n1 with a thermal phase transition temperature of 150-200 ^{0}C, $n_2 - n_n$ is above 200 ^{0}C. Such solution will allow to stabilize temperatures and friction over a wide range of values.

When friction heating of the base material of the pad and reaching the temperature T_1 of the thermal phase transition of material ζ_1, a transition occurs from one phase state to another, which is accompanied by heat absorption Q_1, Q_2, ..., Q_n (Fig. 4) and temperature stabilization of the friction pair. The transition time is fixed (t_1, t_2, ..., t_n) and depends on the ζ_1 material volume. Similar processes occur with other materials of ζ_n inserts of the brake pad when reaching higher temperatures as shown in Fig. 4. When the train stops, the generation of heat from the friction pair also stops and a reverse phase transition occurs. By cooling the base material of the pad, the aggregate state returns to its original state.

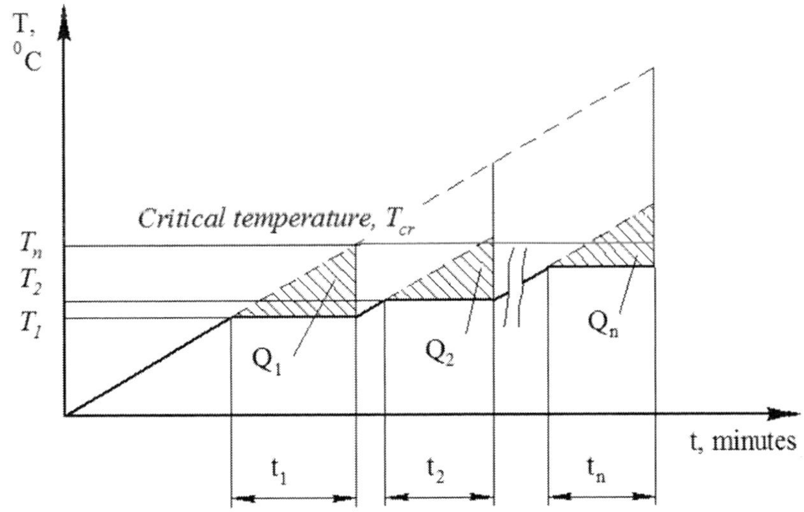

Fig. 4. Changing the brake pad temperature with increased power consumption during braking

It can be seen from the figure that the critical temperature T_{cr} of the material of the pad, which leads to deterioration of the friction parameters of the pad and to increased wear, is not achieved when using elements with a phase thermal transition, which allows to stabilize the temperature and coefficient of friction in contact, to improve the braking efficiency, to reduce wear and rolling noise level.

Stabilization of the temperature in the contact of friction pairs (wheel-rail, wheel-pad) is also provided by the supply of cooled air. When the heated friction surfaces interact with the cooled air, moisture forms on them, the so-called "self-oiling" effect. According to the studies [6, 14], the wet surface of the contacting surfaces reduces the coefficient of friction (Fig. 5), however, to ensure a regulated braking distance, the braking force is sufficient. At the same time, such solution significantly reduces the noise level.

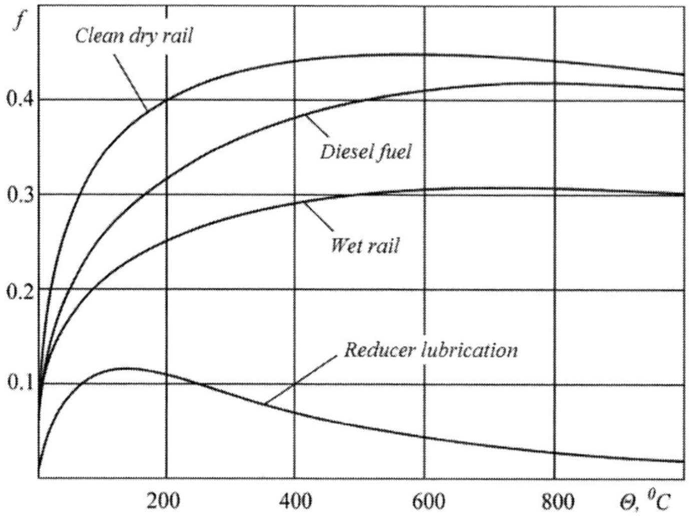

Fig. 5. Dependence of the friction coefficient on the temperature at rolling with slipping

Conclusions

Based on the analysis of promising directions for improving the brake system, innovative technical solutions and methods are proposed that consist in stabilizing the temperature and the friction coefficient of friction pairs, which ensure both an increase in braking efficiency and noise reduction. A target function has been created that makes it possible to reveal the reserves of the vehicle's braking efficiency by redistributing the forces of pressing the brake pads. The supply of electrified sand provides an increase in the stabilization of the clutch on each wheel pair.

The use of inserts from materials with thermal phase transitions in the construction of the pads provides a stable temperature (friction coefficient) in the braking process. The supply of moist air allows to stabilize the temperature and to achieve the effect of self-oiling of friction surfaces. The proposed ways of implementing the braking force reserve of the rolling stock make it possible to increase safety and to reduce the noise level from one of the noise sources - the interaction in the "pad-wheel-rail" system.

References

1. G. V. Blokland, S. Lutzenberger, *Measures on Rail Traffic Noise in Europe*. Input paper for the Interest Group on Traffic Noise. Abatement **4**, 57 (2014)

2. D. A. Kuklin, *The problem of reducing external noise of trains at the source and along the path of expansion*. dissertation abstract ... of doctor of technical Sciences: 01.04.06. Baltic state technical University VOENMECH to them. D. F. Ustinov - St. Petersburg, 52 (2016)
3. P. V. Matveev, *Calculation and reduction of rolling noise of trains*. dissertation ... of candidate of technical Sciences: 01.04.06. Baltic state technical University "VOENMECH". D. F. Ustinov - Saint Petersburg, 170 (2015)
4. A. R. Abu Bakar, *Modelling and Simulation of Disc Brake*. Contact Analysis and Squeal: Ph. D. dissertation. Liverpool (United Kingdom), 201 (2005)
5. V. I. Kolesnikov, V. P. Sergienko, S. N. Bukharov, A. P. Sychev, V. V. Sergienko, *Optimization of the dynamic characteristics of friction materials and their role in noise generation under friction*. Bulletin of the SOUTHERN SCIENTIFIC CENTER of RAS **5** (4), 3-14 (2009)
6. N. I. Gorbunov, *Improving the traction qualities of the locomotive by improving carts elastic ties*. dis. ... Ph. D.: 05.22.07 / Voroshilovgrad, 269 (1987)
7. K. A. Kravchenko, *Justification the reserves for increasing the traction qualities of the locomotive and their implementation control slide in the system of wheel with a rail*. dis. ... candidate of technical Sciences: 05.22.07. Lugansk 23 (2010)
8. M. I. Gorbunov, S. G. Grishchenko, V. A. Dogan, K. A. Kravchenko, S. V. Popov, A. I. Fesenko, Declaration patent for useful model No. 40369. B61C 15/00. *Loading device*. 18.09.2008, publ. 10.04.2009, bull. **7**. (2009)
9. N. I. Gorbunov, Y. A. Kravchenko, S. G. Grishchenko, Y. Y. Osenin. *Increased traction properties of the locomotive by improving the relations of the body with the trucks*. Collection of scientific papers of the VIII International conference "TRIBOLOGY and RELIABILITY", Saint-Petersburg, 18-29 (2008)
10. Y. I. Attending, D. M. Marchenko, I. A. Shvedchikova, *Frictional interaction of wheel with a rail*. Lugansk: Publishing house of SUDU, 227 (1997)
11. M. I. Gorbunov, O. K. Kravchenko, C. B. Popov, M. V. Kovtanets, Y. Y. Osenin, Declaration patent for useful model No. 48520. B61C 15/00. *A method of increasing adhesion of wheels with the rail*, № u200908751, Appl. 20.08.2009 g.; publ. 25.03.2010, bull. **6** (2010)
12. N. Gorbunov, A. Kostyukevich, K. Kravchenko, *Efficiency function for evaluation of the locomotive traction and adhesion qualities*. TEKA Commission of Motorization and Power Industry in Agriculture **X**, Poland, 80-86 (2010)
13. J. Gerlici, M. Gorbunov, K. Kravchenko, O. Prosvirova, T. Lack, V. Hauser, *Analysis of innovative methods for modern rolling stock brake systems efficiency increase*. SBORNÍK PŘÍSPĚVKŮ Z mezinárodní vědecké konference pro mladé vědecké pracovníky "Experimentální a výpočtové metody v inženýrství" Usti nad Labem, Czech Republic, electronic publishing (2017)
14. A. L. Golubenko, *Adhesion of wheel and rail*. 2nd edit.– Lugansk: VUGU, 476 (1999)

The role of inflow boundary layer thickness for numerical modelling of flow around bluff bodies of built-up areas

Renata Gnatowska[1,*]

[1]Czestochowa University of Technology, Faculty of Mechanical Engineering and Computer Science, 21 Armii Krajowej Av, 42-200 Czestochowa, Poland

Abstract. The paper presents the numerical results obtained with the use of the ANSYS FLUENT commercial code for analysing the flow structure around a two rectangles in line surface-mounted bluff bodies immersed in a boundary layer. The simulation of a configuration with a flow around objects has been done using both the steady and unsteady approaches. Effects of the inflow boundary layer thickness on the flow field, wall-shear stress and the grid resolution accuracy in predicting periodic vortex shedding from two tandem arrangement bodies are examined. It has been observed that the periodic vortex shedding is considerably reduced with an increase in the δ/H parameter.

Keywords: boundary layer thickness, CFD, fluid flow, rectangular bluff bodies

1 Introduction

Increased interest in the problem of bluff-body aerodynamics has been observed over the past few decades. It is governed by potential applications of the results in the design process of new buildings as well as optimization of the existing urban areas and city centers. The local wind climate influences the comfort around and between buildings, the life quality in urban areas as well as specific economic aspects of utilization of the defined zones. The flow structure of a build environment is usually studied based on simplified configurations starting from single rectangular obstacles, where a horse-shoe vortex, which extends downstream along to the obstacle side. The flow separation from the obstacle walls results in strong shear layers along which the turbulence production is high. The resulting high turbulence level increases diffusion and enhances entrainment by a shear layer of a low momentum reverse flow in the near-wake which strongly affects the local pressure gradients and increases the mixing [1]. Another often investigated configuration are cubic obstacles in a tandem arrangement, where the flow field is further complicated by mutual interference. For that case, a small separation (intermittent), lock-on and large separation regimes similar to the two-dimensional geometries are observed which are strongly affected by the influence of the ground vicinity [2]. The flow structure around surface-mounted obstacles depends on various factors. One of the most important factors is the oncoming

* Corresponding author: gnatowska@imc.pcz.pl

Reviewers: *Milan* Žmindák, *Richard* Pastirčák

boundary layer thickness δ related to the height of obstacles H. For thin boundary layers ($\delta/H<0.3$) the structure of the upstream separation is characterized by a multiple secondary recirculation upstream of a horseshoe vortex [3]. For thick boundary layers ($\delta/H>0.7$) the dynamic behavior of the pressure and velocity field is bimodal [4, 5]. Other important factors are also the oncoming flow turbulence, the non-dimensional distance ratio (S/H) and the height ratio of the two consecutive bodies. The sheared flow effect caused by the atmosphere boundary layer modifies the flow pattern on the windward side of a high-rise block. Tall buildings, defined as those which protrude above their neighbours, act as scoops to collect the wind over much of their height and deflect it to the ground level. In front of a building exposed to the wind, a vortex forms below the stagnation point. This vortex is shed on each side of the building and it can produce excessively high winds at ground level, causing discomfort or danger [6].

The studies of the wind environment around buildings are generally performed as experiments in wind tunnel. The alternative to this is the application of Computational Fluid Dynamics (CFD), which has been increasingly exploited in various ways recently. The usually applied method for computation of turbulent flows in wind engineering is the Reynolds Averaged Navier-Stokes (RANS) approach. Within this approach, the equations are averaged in time over all the turbulent scales, to directly yield a statistically steady solution of the flow variables. Another option is the use of standard turbulence models from the RANS approach in time-dependent simulations, unsteady RANS (uRANS). This approach should be applied for flows where the unsteadiness is determinable, i.e. the frequency spectrum shows a spike at a shedding frequency. This is the case of the flow around surface-mounted cubes which is analyzed in the paper.

2 Computational details

The flow around two rectangular in-line surface-mounted square cylinders immersed in a boundary layer is computed in this work. The geometry of the analysed cases are sketched in Figure 1. The calculations were carried out for configuration of two elements with different height, aligned in one line and the distance between them was $S/D=2.5$. The results presented in this work relate to a fixed ratio of object height $H_1/H_2=0.6$ and value of their "immersion" in boundary layer $H_2/\delta = 0.3\div1$. Three-dimensional steady and unsteady RANS simulations have been carried out using a commercial CFD code, FLUENT v.17.2, with the RNG version of a ε–k turbulence model. According to the literature [7-9] this model is widely used for flows in a build environment. Users of this model have found a good agreement between computations and measurements. The spatial derivatives have been discretized using a second order upwind scheme while the pressure-velocity coupling has been achieved using the SIMPLE algorithm. The unsteady calculations have been performed using time step based on an estimation through experimental identification of dominant frequency and comprises 1% of the flow oscillations period. The blockage ratio, defined as the ratio of the frontal area of the body to the computational domain cross-sectional area has been 0.7%. For computation purposes, the flow domain is divided into a number of hexahedral cells. The mesh is non-uniform in all the three directions. The grid is clustered near the object and the spacing is increased to a proper ratio of 1.2 away from the object surface. The first cell adjacent to the walls has been set with respect to the criteria required for the individual near-wall treatment. Hence, using a two-layer approach, the width of the near-wall cell has been 0.003H, which corresponds to $1 < y+ < 3$ where H is the object height. According to the literature [10] at least 10 cells per the cube root of the object volume should be used for a flow around surface-mounted obstacles. This recommendation given above has determined the initial minimum grid resolution which is 174x74x48 per cube. The grid independent solution has been obtained for 298x130x70 by systematically refining the entire mesh in each direction, increasing the number of nodes by about 50% [5, 6, 11]. The inflow boundary layer has been prescribed according to methods proposed by Richards et al. [8, 9]. The inlet flow profile have been approximated using the power law distribution *U(z)* and also distributions of the turbulent kinetic energy k and dissipation rate ε (values are compared with experimental data) [5, 6].

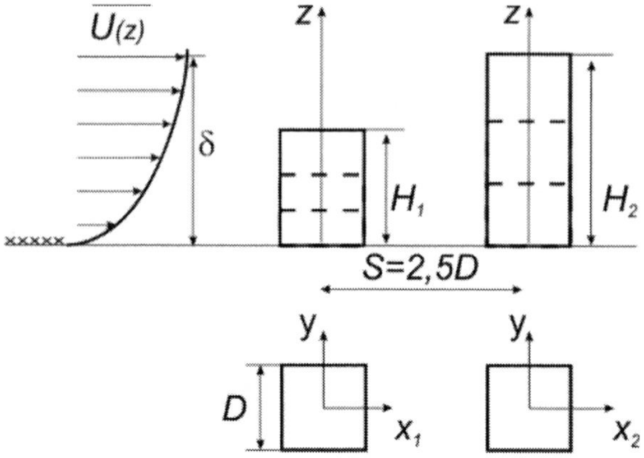

Fig. 1. The geometry of the analysed case

3 The role of inflow boundary layer thickness

The ratio of the objects height to the incoming boundary layer thickness (H/δ) has an important influence on the flow structure, and particularly, on the separation regions upstream and downstream the obstacle. As has been shown, for example, by Hunt et al. [12] and Castro and Robins [13], the shape and form of the separation region is markedly reduced by the presence of upstream near wall turbulence.

To confirm these observations, computational modelling of a flow around a multiple bluff bodies arrays immersed in a thin and thick boundary layer has been performed. The thin boundary layer is equal to the height of second cube ($H_2/\delta = 1$) while the thick boundary layer is three times higher than H_2 ($H_2/\delta = 0.3$). The flow around configuration of two rectangular tandem arrangement buildings, described by the distance $S/D = 2.5$ and the height ratio $H_1/H_2 = 0.6$, when the objects are arranged in the flow direction with the smaller one located on the upwind side is analysed. A strong downwash effect is observed for the relatively small spacing considered here. The big standing vortex which stabilizes the flow in the spacing does not allow the separated shear layers from the lateral sides of windward object to interact in between the obstacles. Additionally, it produces an excessively high velocity at the ground level what is directly correlated with the wall shear stresses distribution on the ground. It is worth mentioning that the downwash effect becomes weaker and the shear layers manage to curl in the gap [11] for larger spacing values not considered in this paper. The simulation of a configuration with a flow around objects has been done using both the steady (RANS) and unsteady (uRANS) approaches.

The wall shear stress (τ [Pa]) distribution at the symmetric line ($y/D = 0$) and at the line $y/D = 0.5$, which is located along the upstream object edge extension in the gap between objects for three flow cases is presented in Figure 2. It can be observed that the size of the cavity region is significantly reduced in the presence of thicker boundary layer. The maximum amplitude of shear stresses in the cavity zone range is also smaller. Similar conclusion has been formulated by Iaccarino et al. [14] who show the difficulty in reproducing the main characteristics of the flow. Oscillations are a result of coupled oscillations between the separated shear layers from lateral sides, which is similar to the vortex shedding process for two-dimensional obstacles. But for surface-mounted obstacles this process is modified by the shear layer along the obstacle free end (top), the oncoming flow shear gradient and horseshoe vortices. In that case using the uRANS calculations should improve the agreement between the numerical and experiment prediction of the cavity zone.

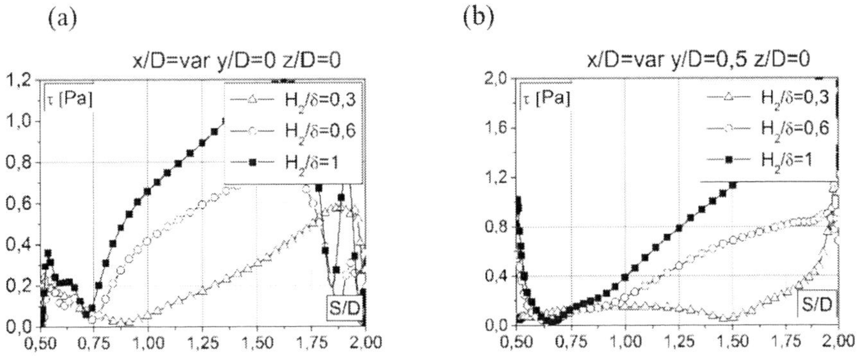

Fig. 2. Distribution of wall shear stress for steady simulation (H_2/δ=0.3; H_2/δ=0.6; H_2/δ=1.0) in the gap between objects: (a) at the line y/D=0; (b) at the line y/D=0.5

A small discrepancy between steady and unsteady results is recognized at the symmetric line (y/D=0). It is possible as the flow in the spacing is stabilized by the downwash effect. Strong differences between the results are observed at the line y/D=0.5. In this plane, the separated shear layers forming the windward obstacle's lateral sides cannot interact in the gap. The distribution of wall shear stresses of an unsteady simulation for three cases of H_2/δ is compared in Figure 3. In left column is presented the data taken along the central line y/D=0 and in right column along the line y/D=0.5. The analysis begins from the case of configuration totally immersed in a boundary layer H_2/δ=0.3 (first row). The time evolution of wall shear stresses on the ground over one period is observed and very week periodic motion in the gap between objects. The literature analysis, i.e. Castro [15], Sakamoto and Haniu [2], gives indication that a thick boundary layer inclines to suppress the periodic motion in the wake of a cube. The flow is periodic when the parameter is equal H_2/δ=1, what is also reported in Figure 3c. A similar numerical results has been also performed by Iaccarino et al. [14] and Huptas et al. [11], but analysis was performed for flow around a single cube immersed in a thin boundary layer, δ/H=0.07 and δ/H=0.1 respectively. The above results confirm the important role of the inlet boundary layer thickness on the flow structure around the obstacle.

Conclusion

A numerical analysis of the objects interaction in the bluff bodies arrays directly in the gap between bodies reveals the importance of the inlet boundary layer thickness on the flow structure, wind comfort around buildings [5, 16]. It has been shown that the intensification of the flow oscillation is significantly promotes with a decrease in the value of the relationship between the thickness of the layer and the height of the objects (δ/H). It is apparent especially in the surroundings of the second element. This indicates the need to use unsteady methods of flow modelling around objects.

Obtained from the non-stationary calculations (uRANS) averaged values of the dependent variable do not differ significantly from the stationary results (RANS). The results obtained from the uRANS methodology show better compliance with the experiment when the flow is not statistically stationary ($H/\delta>1$). In flow with a thick boundary layer ($H/\delta<1$), for non-stationary calculations, although the experimental frequency identification, do not recognize any oscillation. The cause may be a much stronger attenuation of flow through the layer in the case of numerical analysis.

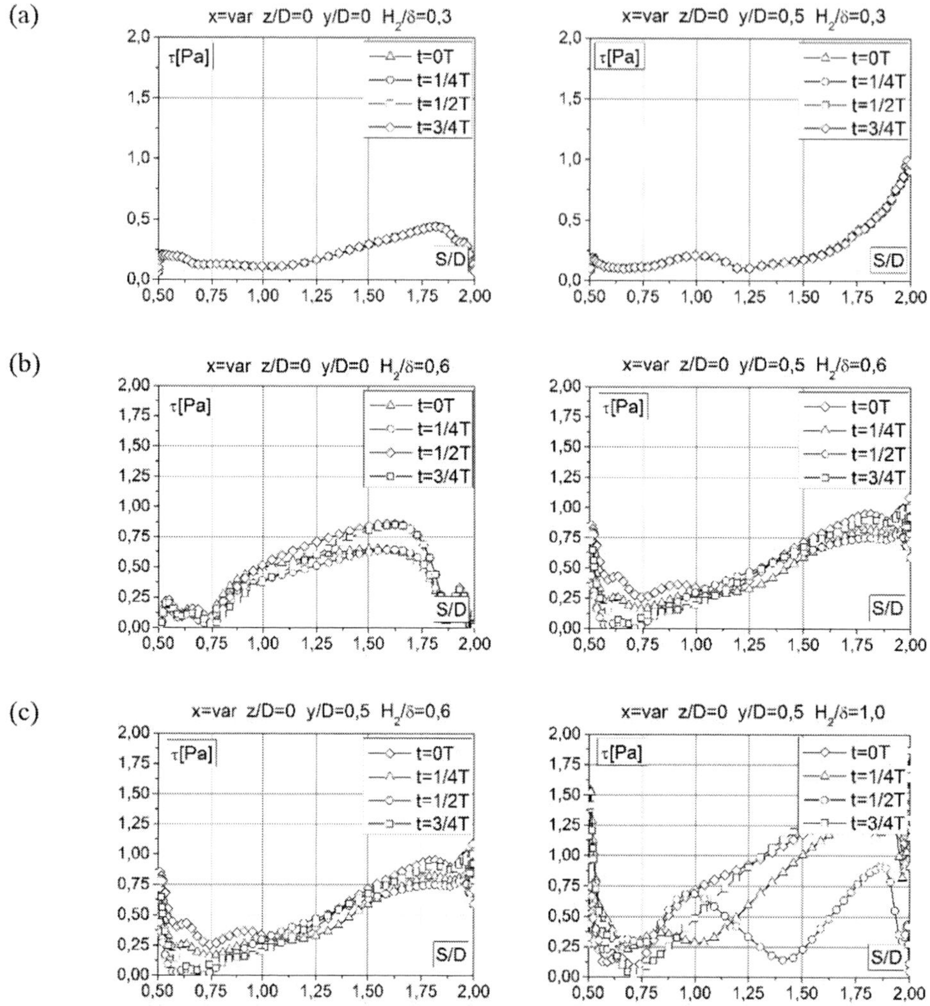

Fig. 3. Distribution of wall shear stress for unsteady simulation in the gap between objects at the line y/D=0 (left column) and at the line y/D=0.5 (right column) for (a) H_2/δ=0.3; (b) H_2/δ=0.6; (c) H_2/δ=1.0

The work was partially supported by National Science Centre, Poland (Narodowe Centrum Nauki) under grant number NCN 2017/01/X/ST8/00076.

References

1. R. Martinuzzi, C. Tropea, *The flow around surface-mounted, prismatic obstacles placed in a fully developed channel flow*. ATJFE **115**, 85-85 (1993)

2. H. Sakamoto, H. Haniu, *Aerodynamic forces acting on two square prisms placed vertically in a turbulent boundary layer*. J. Wind Eng. Ind. Aerodyn. **31** (1), 41-66 (1988)
3. E. Logan, W.H. Schofield, *Turbulent shear flow over surface mounted obstacles*. Int. J Heat Mass Transfer **112**, 376-385 (1990)
4. R.J. Martinuzzi, B. Havel, *Turbulent flow around two interfering surface-mounted cubic obstacles in tandem arrangement*. J. Fluids Eng. **122** (1), 24-31 (2000)
5. R. Gnatowska, *Aerodynamic Characteristics of Three-Dimensional Surface-Mounted Objects in Tandem Arrangement* Int. J. Turbo Jet. Eng. **28** (1), 21-29 (2011)
6. R. Gnatowska, M. Sosnowski, V. Uruba, *CFD modelling and PIV experimental validation of flow fields in urban environments*. In E3S Web of Conferences **14**, 01034 (EDP Sciences 2017)
7. A. D. Ferreira, A. C. M. Sousa, D. X. Viegas, *Prediction of building interference effects on pedestrian level comfort*. J. Wind Eng. Ind. Aerodyn. **90** (4), 305-319 (2002)
8. P. J. Richards, R. P. Hoxey, *Appropriate boundary conditions for computational wind engineering models using the k-ε turbulence model*. J. Wind Eng. Ind. Aerodyn. **46**, 145-153 (1993)
9. P. J. Richards, G. D. Mallinson, D. McMillan, Y. F. Li, *Pedestrian level wind speeds in downtown Auckland*. Wind Struct. **5** (2-3-4), 151-164 (2002)
10. J. Franke, C. Hirsch, A. G. Jensen, H. W. Krüs, M. Schatzmann, P. S. Westbury, N. G. Wright, *Recommendations on the use of CFD in predicting pedestrian wind environment*. (Final report, 2004)
11. M. Huptas, W. Elsner, *Steady and Unsteady Simulation of Flow Structure of Two Surface-mounted Square Obstacles*. Task Quarterly **12** (3), 197-207 (2008)
12. J. C. R. Hunt, C. J. Abell, J. A. Peterka, H. Woo, Kinematical studies of the flows around free or surface-mounted obstacles; applying topology to flow visualization. J. Fluid Mech. **86** (1), 179-200 (1978)
13. I. P. Castro, A. G. Robins, *The flow around a surface-mounted cube in uniform and turbulent streams*. J. Fluid Mech. **79** (2), 307-335 (1977)
14. G. Iaccarino, A. Ooi, P. A. Durbin, M. Behnia, *Reynolds averaged simulation of unsteady separated flow*. Int. J Heat Mass Transfer. **24** (2), 147-156 (2003)
15. I. P. Castro, *Measurements in shear layers separating from surface-mounted bluff bodies*. J. Wind Eng. Ind. Aerodyn. **7** (3), 253-272 (1981)
16. R. Gnatowska, *A Study of Downwash Effects on Flow and Dispersion Processes around Buildings in Tandem Arrangement*. Pol. J. Environ. Stud. **24** (4), (2015)

Design of a computational model for multi-body contact of deformable bodies in the flowing fluid

Marián Handrik[1,*], *Milan* Vaško[1]

[1]University of Žilina, Faculty of Mechanical Engineering, Department of Applied Mechanics, Univerzitná 8215/1, 010 26 Žilina, Slovak Republic

Abstract. The article deals with the design and testing of a computational model for multi-body contact of deformable bodies in the flowing fluid. The computational model will be designed to allow easy modify the impact area of deformable bodies and their shape. The computational model must allow the application of an endless cycle of bodies impact with the possibility of restarting and calculating the following time intervals for collision of bodies.

Keywords: multi-body contact, fluid structure interaction, finite element method, ADINA

1 Modelling of multi-body contact in the fluid flow

Modelling of a multi-body contact is a computationally demanding task. The computational difficulty of a model increases with a number of bodies that may come into contact. If we consider a possible contact among all bodies in the solved task, then a number of contact pairs will be determined according to the relation $P = (n-1)^2$, where n is the number of all bodies in the contact problem. When generating a mesh of finite elements, it is necessary to define a sufficiently dense mesh of finite elements on the contact bodies. The contact task has bad convergence in case of low mesh density. Mesh density must be also determined with regard to shape complexity of bodies in such a way it approximates an edge of a contact body with sufficient precision. Most of computational models of the contact implemented in software for computing by the method of finite elements divide contact bodies into a group of 'contactor' and 'target'. In such a case the contact is detected in case when nodal points of the body 'contactor' penetrate into a body of the type 'target'. On the contrary, if nodal points of the mesh of a body 'target' penetrate into a body of 'contactor', a contact is not detected. In case of a multi-body contact it is convenient to define contact pairs in such a way that each body is a 'contactor' and at the same time also a 'target'. By such a definition of contact pairs convergence of a solution of the contact task becomes rapidly improved and oscillation in nodal points is decreased [1-3].

* Corresponding author: marian.handrik@fstroj.uniza.sk

Reviewers: *Bohuš* Leitner, *Tomáš* Lack

Computational models for solving of flowing fluids require a fine mesh of elements, elements must be sufficiently small so that a flow in the boundary layer close to a body surface would be possible to solve. Both groups of computational models, contact tasks and flow, belong to a group of models with a high demand on computational means and time [4-9].

The aim of the design of a computational model is to design an efficient computational model for the analysis of a mutual contact of bodies in flowing fluid [10-11]. The designed computational model must contain following properties:

- A part of the model must be constructed in a way to enable mutual collisions of contact bodies and a shape of the area must be easily modified.
- A channel in which fluid flows and carries contact bodies must be closed, bodies proceed through the channel and re-collide.
- A computational model must be maximally efficient from the standpoint of requirements on computational means and computing time.
- A designed computational model will be created for the software package ADINA, a designed model will consider possibilities of a model design in this program.
- Contact bodies will have a circular shape.
- A shape of the channel must be designed to avoid mutual jamming and a blockage of the channel.
- A task will be solved in 2D while structural bodies will have plane state of stress.
- In flowing, a material model for viscous flow regardless a turbulence model will be considered.

In the system ADINA it is necessary to design an individual geometric model and finite element model for a structural analysis and an individual model for the flow analysis. A calculation is performed as an iterative process to search for a balance between the structural analysis and the flow analysis in every time step of the solution. A mutual interaction between models is transferred from the structural model into the model of the flow by means of a change of a position of limits if bodies in the fluid. Active forces determined from pressure on limits of bodies are transferred from the model of the flow into the structural model.

2 The geometric and FE model for the structural analysis

The designed geometric model of an enclosed channel in the structural analysis is formed by the following elements. The geometric model is shown in Figure 1:

- A structure forming an outer boundary of the channel. Geometry of the body is designed in the system ADINA via an application of the module ADINA-M which is the module for modelling of bodies by the use of a library for a graphic format parasolid. A gradual shape of the channel is designed in a way to prevent mutual jamming of contact bodies and the channel blockage.
- The structure forming an inner boundary of the channel. As in the case of the outer boundary, even the inner boundary is designed by the application of the module ADINA-M.
- Bodies in the flowing fluid. Circular bodies are divided into four segments, i.e., quarters of the circle. Bodies are designed by this method, so that it would be possible to minimize a number of elements generated on the bodies by an application of a mapping mesh by triangle elements.
- An obstacle in the centre of a widened part of the channel. Bodies stop by the obstacle and thus a mutual contact is formed. The body is formed in the module ADINA-M.
- A system of auxiliary lines enables a definition of a movement of contact bodies in the channel. The system consists of a set of lines that are meshed by the use of beam

elements. A rotation of auxiliary points moves contact bodies in the channel. A speed of rotation is defined thus the speed of movement of contact bodies is 0.1 ms^{-1} [12-13].

A diameter of contact bodies is 10 mm, the mean diameter of the circle in curves of the channel is 60 mm and the width of the channel is 11 mm.

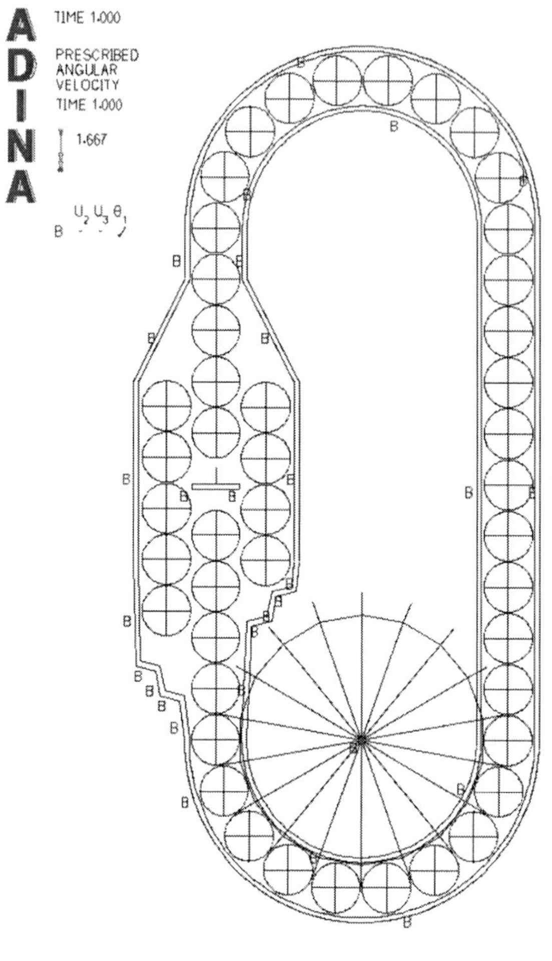

Fig. 1. Geometry of structural model and boundary condition

The program ADINA enables a definition of several contact groups, in a contact group contact bodies and contact pairs are defined. Due to less complex modelling, a design of a model, we defined following contact groups:

- The first contact group – a mutual contact of contact bodies. A geometric model contains 46 circular contact bodies. In case of a potential contact of each body with all bodies there are designed 2116 contact pairs in the model.
- The second contact group – a contact of 46 circular contact bodies and obstacles in a channel. The contact group contains 92 contact pairs. Each body is defined as a 'contactor' and a 'target' at the same time.
- The third contact group – a contact of 46 circular contact bodies and an outer boundary of the channel. There are 92 contact pairs in this contact group.
- The fourth contact group – a contact of 46 circular contact bodies with an inner boundary of the channel. There are 92 contact pairs in this group.

- The fifth contact group – a contact of 46 circular contact bodies and 18 contact bodies of auxiliary geometry. There are 828 contact pairs in this group.

In the first four contact groups we must define an occupation of contact areas within small distances and we use the value $5x10^{-2}$ mm in the model while the diameter of contact bodies is 10 mm. Occupation of a body contact is necessary for a correct function of the model of flow. Among contact bodies there must be formed a thin layer of flowing elements – at least one element. Otherwise, it would not be possible to generate a mesh of finite elements for a model of the flowing fluid, boundaries of an area in the model of flowing must not overlap or contact.

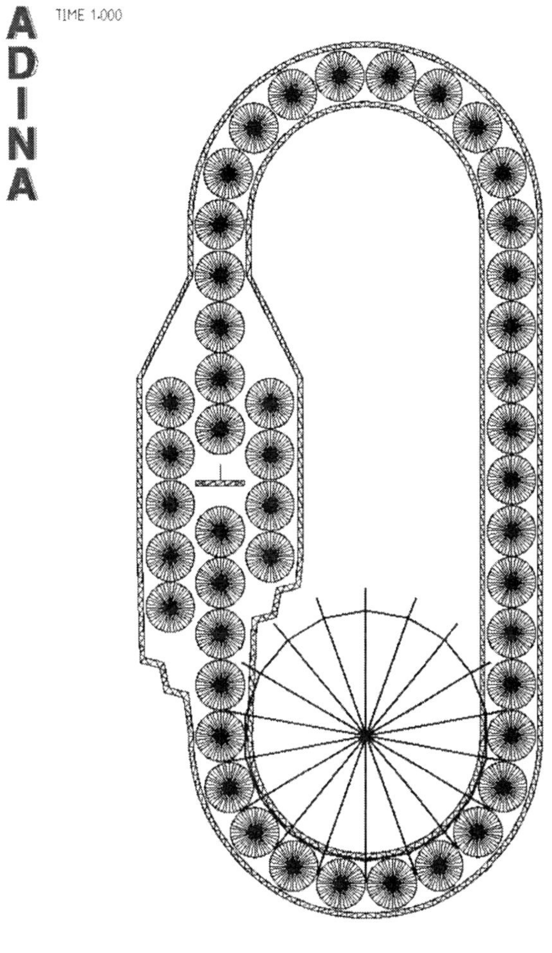

Fig. 2. Finite element mesh of structural model

In the designed structural model there is applied an elastic material model for steel, Young's modulus of elasticity $2.1x10^{11}$ Pa, Poisson number 0.3, density 7850 kg m^{-3}.

In the model following initial and boundary conditions are defined:

- A withdrawal of all stages of independence on an outer side of the boundary of the channel, they fix a position of the channel in the space.
- In the centre of auxiliary geometry, shifts in a direction of axes y and z are removed, a rotation in a direction of the axis x remains.
- A removal of shifts in two points on the obstacle in the channel, it fixes a position of the obstacle in the space.

- A load by gravitation in a direction of *z* (in negative direction).
- A definition of an angular velocity in a centre of auxiliary geometry1.6666 rads^{-1}. Contact bodies will have the speed of 0.1 ms^{-1}.
- A zero initial speed is defined for all bodies.

A calculation is solved as an implicit task by Newmark method, for plane stress state – a thickness of designed geometric model is 100 mm. Thickness of 100 mm is selected thus maximum contact forces will not exceed 100 N. Properties of the contact are selected in a way that contact forces lower than 0.1 N are not considered in the contact. Limits of a minimum value of contact forces strikingly improve convergence of a contact task and it prevents oscillation in contact detection.

Due to a reason of the maximum efficiency of a computing model we selected a mesh density to generate a minimum number of elements in which a convergence of the task is secured. In this case values of contact forces are calculated correctly, but the state of stress in the contact point is not. In a case we would intend to determine the state of stress with sufficient precision in the contact point, i tis necessary to increase density and, eventually, to change a type of elements into quadratic elements. In generating of a mesh on contact bodies we use 10 elements per a quarter of a circle. In the other parts of the model, the size of elements is set on 2 mm. Auxiliary lines are modelled via an application of one beam element per a line. The final part of the auxiliary lines is modelled with a small cross section of the beam and beam elements form an elastic element in this case. A presence of an elastic element in the model decreases maximum values of contact forces among circular contact bodies and auxiliary lines in an application of greater time span [14]. A mesh of linear triangular elements for a structural model is depicted in Figure 2.

3 The geometric and FE model for analysis of fluid flow

When designing a geometric model of flow we follow geometry designed for the structural model. For designing an outer boundary of the area boundary curves from the structural model are applied and they are supplemented by the inflow and outflow parts of the model. Boundaries of contact bodies are defined by a construction of a circle with a rotating point of 360°. Consequently, bodies in the module of ADINA-M are designed from these boundary curves and the geometric model for a boundary in the area of fluid in the model is formed by their mutual subtracting. The geometric model is depicted in Figure 3.

In the model there is applied a model of a laminar flow and following properties of the fluid (water) are defined: dynamic viscosity 0.001 kgm^{-1}s^{-1}, fluid density 1000 kg m^{-3}.

In the designed model of flow there are defined following boundary and initial conditions:

- The boundary of contact bodies is defined as „no slip FSI boundary“.
- In all the parts of the outer boundary of the area of flowing is defined as no slip boundary condition.
- In the inflow and outflow areas there is defined the velocity of fluid flow 0.1 ms^{-1}.
- In the model there is defined the initial condition with zero speed of flow in all the points of the mesh of finite elements.

To solve the issue Euler's method for solving transactional analysis of fluid flow is applied. When designing a mesh of finite elements the FCBI-C formulation of elements for flow is applied. In FCBI-C formulation of finite elements is defined one control point in the centre of gravity in every element. FCBI-C formulation enables an application of a segregated iteration solver AMG. A selection of a formulation of elements and a type of a solver is designated with a regard of minimisation of a time period needed for a solution. In meshing we used triangle elements of the size 0.2 mm and in the process of solution the automatic meshing of finite elements for flow is activated.

In solving, the constant time step $1x10^{-4}$s is applied.

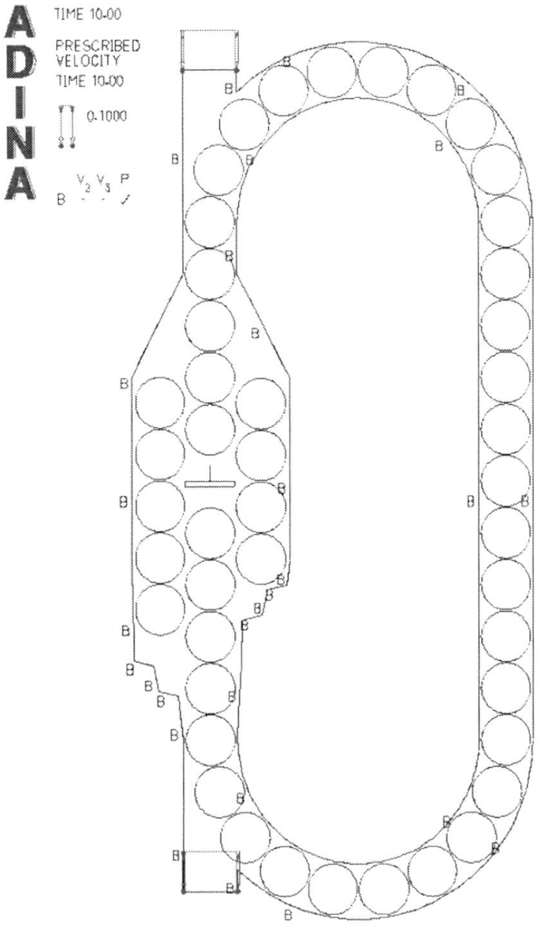

Fig. 3. Geometry of fluid model and boundary condition

4 The solution for the model of multi-body contact of bodies in the flowing fluid

When solving FSI analysis via FCBI-C elements the iteration scheme for finding the balance of the structural analysis and analysis of flow is modified in the following way. The calculation is started by a determination of a balance in the structural analysis, then there follows one outer iteration of AMG solver for computation of the model of flow, a procedure repeats and thus the preciseness of a solution of the model of flow improves. The iteration process of finding of a balance in FSI analysis is completed when the prescribed preciseness of the solution of the model of flow or the maximum number of iteration of AMG solver for flow are reached.

The designed computing model has following properties from the standpoint of a solution of a task of multi-body contact of bodies in the flowing fluid:

- The number of iterations of the Newton-Raphson method for a solution of a contact task in one time step was within the interval 3 to 25 iterations, exceptionally and more iterations were necessary.
- The number of iteration for finding the balance of a solution of interaction of a construction of a body with fluid ranged within the interval 2 to 30 outer iterations of AMG solver, in rare cases the model of flow did not converge even by 250 iterations. In the next time step the standard speed of task convergence was reached within the interval 2 – 30 iterations.
- The task was solved on the server IBM dx360 M 3. 2 x CPU Intel I7 2.4 GHz 6 cores, 96 GB RAM. In computing, parallelisation of the program ADINA of the type SMP and 12 computing nucleus were applied. Maximally, 2.5 GB RAM was used. The period of solution of one time increment of the solved task ranged within the interval 3 – 40 seconds according to a number of iterations. 180 seconds were necessary for a total meshing of the model of flow. Meshing of the model was necessary after approx. 80 time increments.

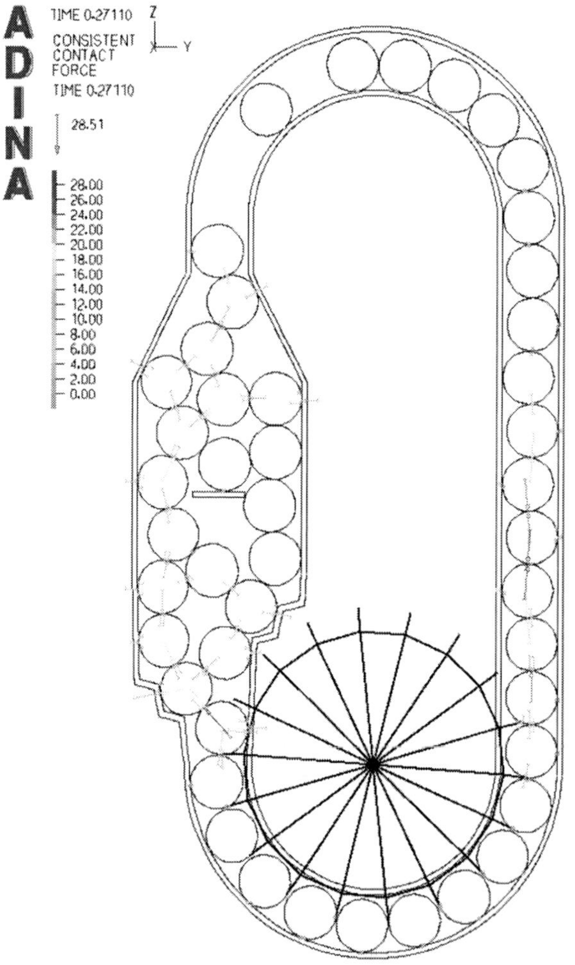

Fig. 4. Contact forces in time 0.2711 s [N]

In numerical tests of the designed computational model, correctness of a design of a gradual constriction of the channel was proved. In this spot the mutual jamming of circular contact bodies does not take place.

Presented results of contact forces are shown in the time of 0.2711 in Figure 4. The speed of fluid flow is shown in Figure 5 in the same time; the speed of flow is shown in a form of velocity vectors ms^{-1}.

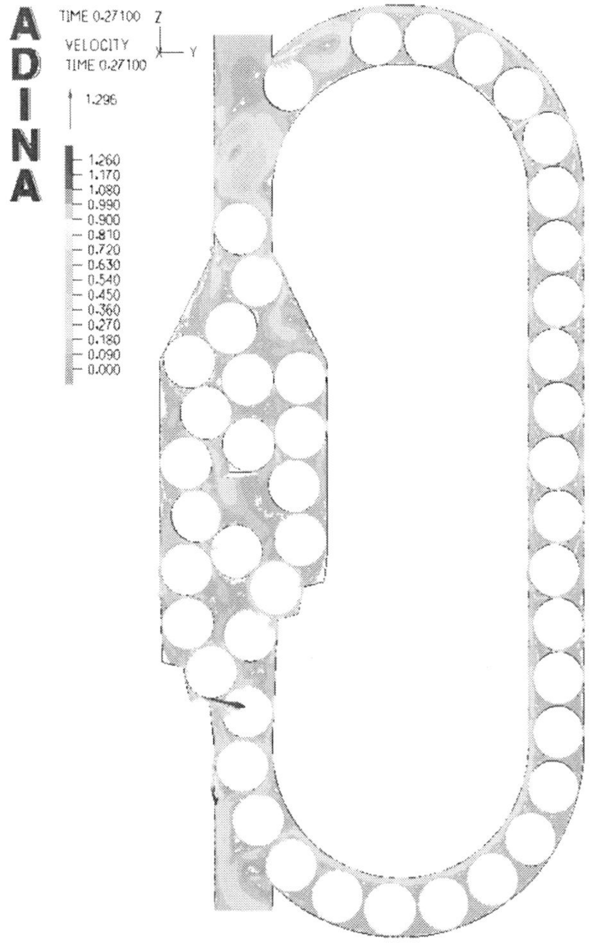

Fig. 5. Fluid flow velocity, vector results in time 0.2711 s [ms^{-1}]

Conclusion

The designed FE computational model meets all required criteria imposed on the model. It enables a mutual interaction of contact bodies together with an interaction with channel boundaries. The interaction of bodies takes place in the flowing fluid, while the speed of fluid flow is possible to modify regardless the speed of rotation of auxiliary lines formed by beam elements.

In testing calculations was proved that a gradual channel shape is correctly designed and it enables gradual ordering of bodies into a taper part of the channel. In a transition of about 50 bodies their mutual jamming did not occur.

The designed computational model enables a solution without a time limit, after completing of a computing task a restart of solution is possible and in the next solution of further time increment of a total solution.

This work is supported by: VEGA 1/0795/16 Development of effective methods for correction and optimization of coupled mechanical system.
This publication is the result of the project implementation: Development of optimal technology for analysis of limiting states of construction elements in contact ITMS: 26220220118 supported by the Operational Programme Research and development funded by the ERDF.

References

1. on-line manual, *Adina, Theory and modelling guide, Volume I: Adina*. (2016)
2. on-line manual, *Adina, Theory and modelling guide, Volume III, CFD and FSI*. (2017)
3. B. Ftorek, P. Tomasovic, B. Dorociakova, *Comparison of two methods for solving nonlinear parabolic model in porous media*. Proceedings of Aplimat 2009: 8th international conference, 259-263 (2009)
4. G. Domek, *Research on the contact area between the timing belt and the toothed pulley*. Lectures notes in engineering and computer science, 2242-2244 (2011)
5. J. Gerlici, T. Lack, *Contact geometry influence on the rail/whel surface stress distribution,* Procedia engineering **2**, 2249-2257 (2010)
6. J. Vavro Jr., J. Vavro, P. Kovacikova, R. Bezdedova, *Kinematic and dynamic analysis of the manipulator for removal of rough tyres*. Procedia Engineering **136**, 120-124 (2016)
7. M. Sofer, R. Halama, F. Fojtik, L. Vaclavek, *Rolling contact fatigue research on VSB-TUO, comparison of current test machine with it's new concept*. New methods of damage and failure analysis of structural parts, 153-158 (2010)
8. J. Zapomnel, P. Ferfecky, *A computational investigation of the disk-housing impacts of accelerating rotors supported by hydrodynamic bearings*. Journal of applied mechanics-transactions of the ASME **78**, (2011)
9. L. Jakubovičová, M. Sága, *Computational analysis of contact stress distribution in the case of mutual stewing of roller bearing rings*. Novel trends in production devices and systems **474**, 363-368 (2014)
10. M. Astorino, J.F. Gerbeau, O. Pantz, K. F. Traoré, *Fluid-structure interaction and multi-body contact: Application to aortic valves*. Computer Methods in Applied Mechanics and Engineering **198**, 3603-3612 (2009)
11. W. Schiehlen, *Research trends in multibody system dynamics*. Multibody System Dynamics **18**, 3-13 (2007)
12. M. Naď, *Modification of model properties of machining tool bodies*. Proceedings of the 7th international conference of DAAAM baltic industrial engineering, **1**, 54-59 (2010)
13. I. Malujda, D. Wilczynski, *Mechanical properties investigation of natural polymers*. Procedia engineering **136**, 263-268 (2016)
14. J. Zapoměl, V. Dekýš, P. Ferfecki, A. Sapietová, M. Sága, M. Žmindák, *Identification of material damping of a carbon composite bar and study of its effect on attenuation of its transient lateral vibrations*. Int. Journal of Applied Mechanics **7** (6), (2015)

Design of a device to simulate environmental influences on the course of brake tests

Jozef Harušinec[1,*], *Andrej* Suchánek[1], *Mária* Loulová[1], *Peter* Strážovec[1]

[1]University of Žilina, Faculty of Mechanical Engineering, Department of Transport and Handling Machines, Univerzitná 8215/1, 01026 Žilina, Slovak Republic

Abstract. When performing the technical tests of the brake system under laboratory conditions, it is necessary to create the same conditions as those in operation. The friction coefficient is dependent on a number of parameters, including airflow during ride and weather conditions. Therefore, during the take-over tests, both these conditions are obligatory to simulate. This paper describes the design of the chamber, which will allow the test stand RAILBCOT to carry out the so-called dry and wet tests with airflow as during real-life ride. It also isolates the acoustic emission and emission of braking material to the surrounding area when tested.

Keywords: railway wheel, brake block, structural analysis

1 Introduction

One of the basic requirements of European railways is reduction of the noise generated in wheel-rail contact in the service of freight railway wagons [1, 2, 3].

Since the beginning of the service it was known, that a significant reduction of freight wagons noise can be achieved by replacing iron cast brake blocks with brake blocks made from composite materials. Currently, brake blocks made from composite materials are used in service [4, 5, 8, 12, 13]. However, the value of their friction coefficient is different from the one of the iron cast brake block. If composite materials are used, there is a change in braking forces acting on wheels [7, 14, 16, 17, 18]. These types of composite brake blocks (K-block) have a higher friction coefficient than iron cast brake blocks. In service, it is necessary to adjust the vehicle braking system to the composite brake block of K-type, which is financially expensive [9, 11, 15, 19, 22]. For these reasons, it was proposed to use K-block for new vehicles. For existing vehicles, brake blocks made from composite materials with a similar friction coefficient as iron cast brake blocks (LL-blocks) have to be used to avoid the adaptation of the braking system by using a "silent" K-block.

* Corresponding author: jozef.harusinec@fstroj.uniza.sk

Reviewers: *Robert* Grega, *Ksawery* Szykiedans

2 Test stand of braking components RAILBCOT and simulator of equivalent railway traffic load SIMRAIL

The RAILBCOT test stand is a device that works on the adhesion principle [6, 10, 20, 21]. The main purpose is to analyze the change in geometry of the railway wheelset tread profiles due to wear caused by simulated operating loads under laboratory conditions. Along with the wear of the tread profile, brake block wear will be monitored for different braking modes. The purpose of the test stand is to load the wheelset with an adequate force load, simulated by a combination of engine torque, wheel load change, track gauge free channel size change, wheelset attack angle change, and variable braking mode based on the independent action of four brake units with a change of configuration, shape and the material composition of the brake blocks.

Fig. 1. 3D model of the Test stand RAILBCOT with module SIMRAIL in program CATIA

The purpose of the test consists of two processes. The first is rolling the wheel along the rotating rail. The test stand analyzes the ride characteristics of the wheelset primarily taken from the Y25 chassis with an axle load of 22.5 t. The maximum velocity is 160 km.h^{-1}. The engine has a maximum output power of 434 kW with a constant torque of 3900 Nm.

SIMRAIL (Fig. 1) is a module used to plausibly simulate the phenomena encountered during a ride of a real rail vehicle along a real track (lateral displacement of the wheelset and the change in angle of attack) and for modifying the wheel forces values. Laboratory tests must show similar results to the real Y25 bogie. The laboratory conditions will be comparable to those in real operation. Simulations will take place after several configurations of the operating mode. We will deal with the force effects in the wheelset-track contact in an arc, ride in a straight track with an empty or loaded vehicle, and also with use of the lateral displacement and the angle of attack of the wheelset.

3 Reasons for wheel-rail contact cooling

The RAILBCOT device was designed for long-term tests that show changes in geometry due to the aggressive effect of L and LL brake blocks on the wheelset tread. In such tests, it is necessary to provide sufficient cooling either for the test stand components as well as for the tested components. The cooling of the bearings, transmission and the engine is solved by the manufacturer either passively or actively (forced refrigerant circulation). There are two problems with the test pair. The first is the heating of the break block - wheel test pair and the other is the emission of dust particles (Fig. 2A) of the cast iron or the composite material of the brake blocks. In the first problem, the change of friction ratio and the possibility of damaging the wheel-brake block test pair arise due to heat. Because the wheel is driven by the rotating rail via an adhesive bond, the adhesion coefficient may be reduced due to the effect of the deposits and thus the creep between the wheel and the rail can increase (Fig. 2B, 2C).

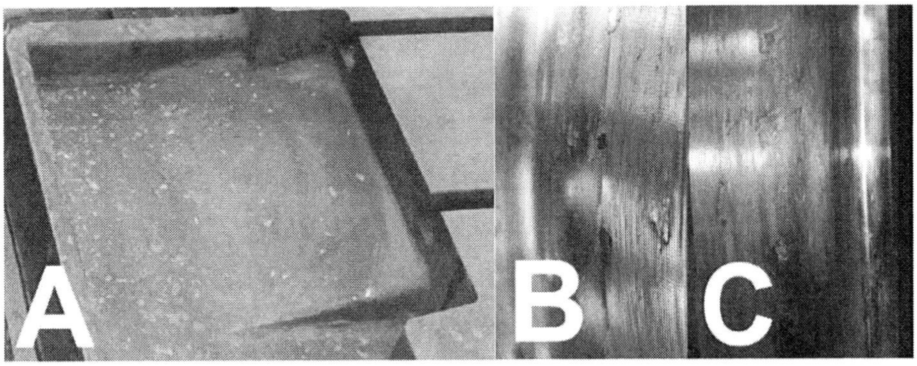

Fig. 2. Emission of the cast iron during braking with an overheated brake block (A) and cast iron sediment on the wheel tread (B, C)

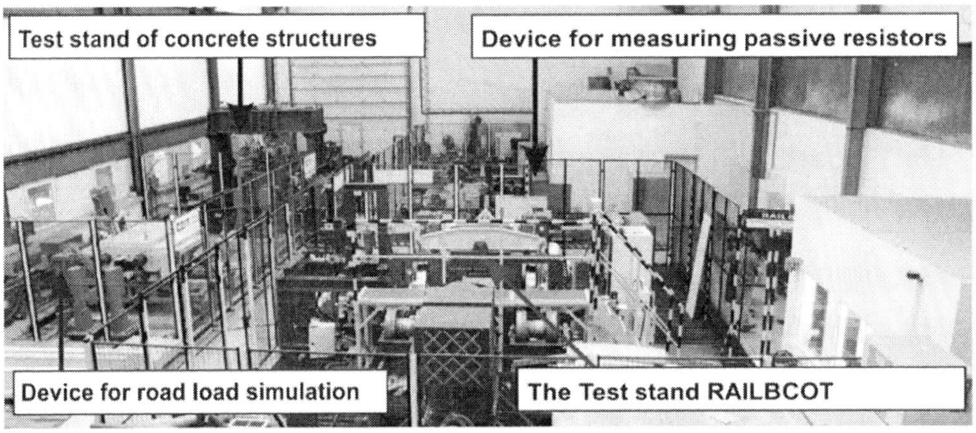

Fig. 3. Test stand RAILBCOT is installed in laboratory BI026

In the event of dust particles being emitted, electrical components may be malfunctioning or shorted. Therefore, all cabinets containing electrical circuits are equipped with an electrical enclosure of IP 65 at least. These particles also have a negative effect on the respiratory tract of the test device's operating staff. It is therefore necessary to eliminate its impact on the staff by using protective equipment (respirators).

Last but not least, it is necessary to take into account the fact that the device is located in a building, where other testing devices for research or teaching purposes are placed as well (Fig. 3). Therefore, for practical reasons, partial or total detachment of the operation area is necessary. Firstly, to protect other test locations and in secondly to protect the RAILBCOT device itself from other devices. The Department of Transport and Handling Machines also has a certificated Inertia braking state UIC. In Fig. 4 shows a view of the test stand and the air tunnel during the braking tests.

Fig. 4. Inertia braking state UIC

The air tunnel provides a standard-matching airflow around the tested components which at the same time removes dangerous particles from the test area. Two air-blowers control the air flow, which are controlled by turning the throttle. The throttle is controlled by a stepper motor via a worm gear. This configuration also served as a model for the planned RAILBCOT air tunnel.

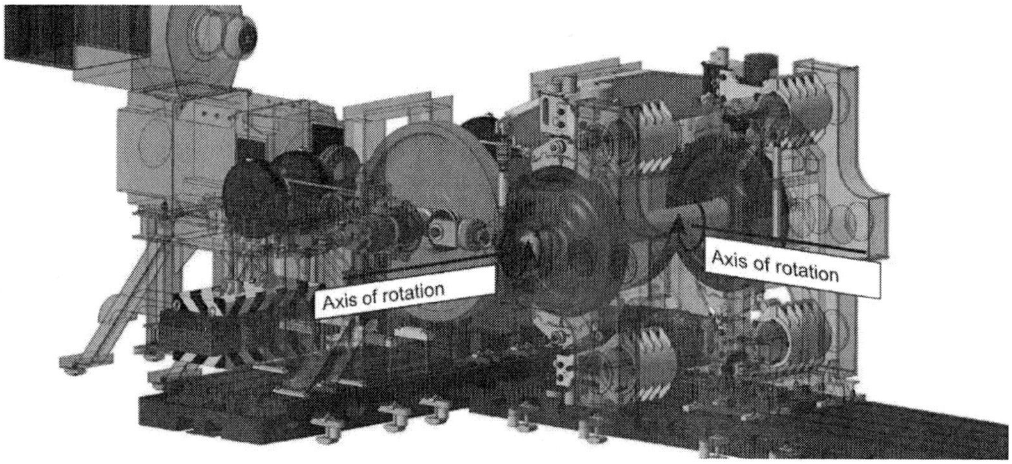

Fig. 5. Test stand of braking components RAILBCOT

The RALBCOT device is more complicated by the fact that the tested wheel in Fig. 4 performs only a rotary movement, the wheelset on Fig. 5, in addition to rotation around the wheelset axis, also performs a swinging movement around the axis of the drive motor. This increases the demands on the air tunnel design. Necessary is also a two-wheel cooling, which as well increases the demands on the fans transporting air to the wheels.

4 Air-tunnel design

The design of the pipeline is based on UIC 541-4 standard. For tests up to 80 $km.h^{-1}$, it is necessary to achieve a half speed of air above 80 $km.h^{-1}$, the required airflow rate is 40 $km.h^{-1}$. Therefore, the fan must ensure this speed. For a suitable fan size, we need to know the required flow.

For a maximum velocity of 11.11 $m.s^{-1}$ (40 $km.h^{-1}$) the following dependency will apply (Fig. 6). As is evident for constant velocity, the flow dependence is linearly dependent on channel area size. For its determination, it is necessary to take into account the dimensional possibilities of the existing structure and also to ensure a fluent flow through the chosen channel. A plausible simulation of airflow conditions requires a solution according to Fig. 7.

$$Q = S \cdot v \tag{1}$$

where:
Q – flow rate [$m^3.s^{-1}$],
S – cross-sectional area of the pipe [m^2],
v – velocity in the pipe [$m.s^{-1}$].

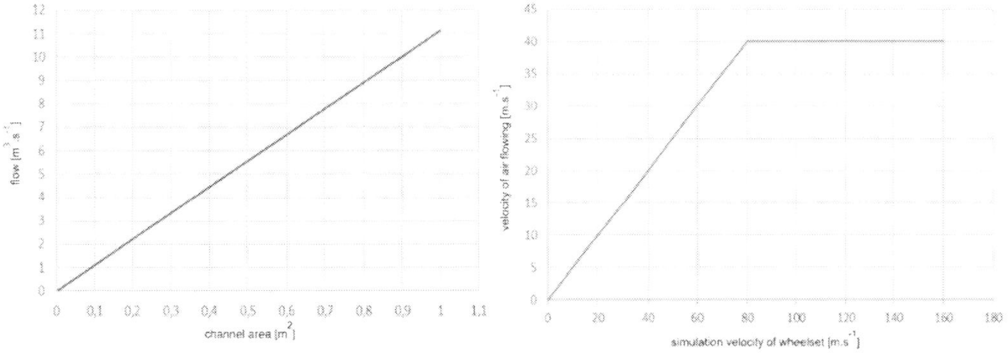

Fig. 6. Dependence between flow and channel cross-section (left) and vehicle velocity and simulated air velocity (right)

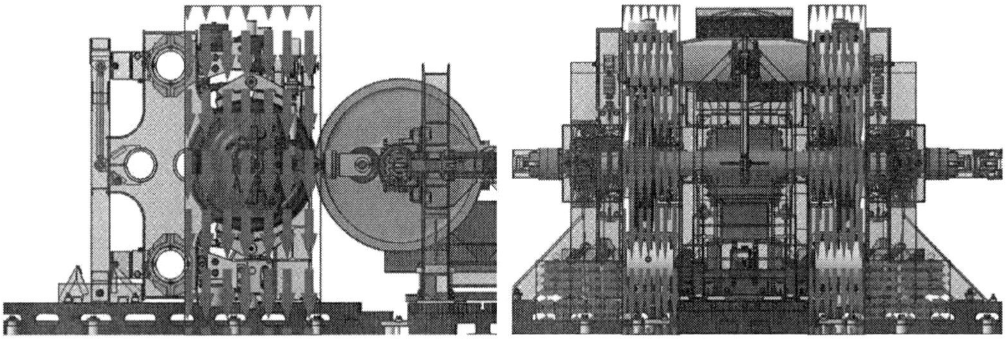

Fig. 7. Original version of the air flow travel

However, this has led to two fundamental problems. When the air flows downwards (see Fig. 7), the flow is suddenly changed by 90 degrees, which would require a suitable deflector, that can not be created for spatial reasons. Such a condition will result in shocks in the airway and thus increase of noise and increase of the load on the whole system. The second problem is the fact that the brake unit does not allow a smooth flow around the wheel and thus its better cooling. In Fig. 8, three cases are illustrated.

Case A represents a half view of the Y25 bogie and its front wheel cover by a buffer beam. The original concept of the test stand was based on simulating this type of bogie, as it is a standardized bogie for use on European tracks. Therefore, the design according to position B would not be satisfactory from the point of view of the directly circumflown area.

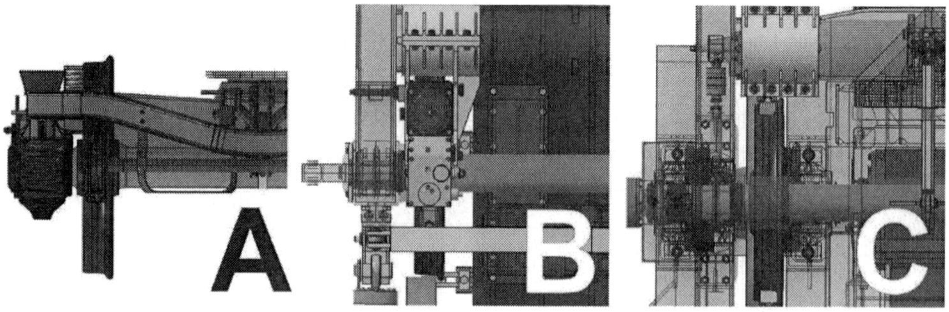

Fig. 8. Barriers in the air flow for different variations

Therefore, we chose a variant according to position C. Here the overlap is smaller and the surface of the wheel directly cooled by the air intake is similar to that of position A.

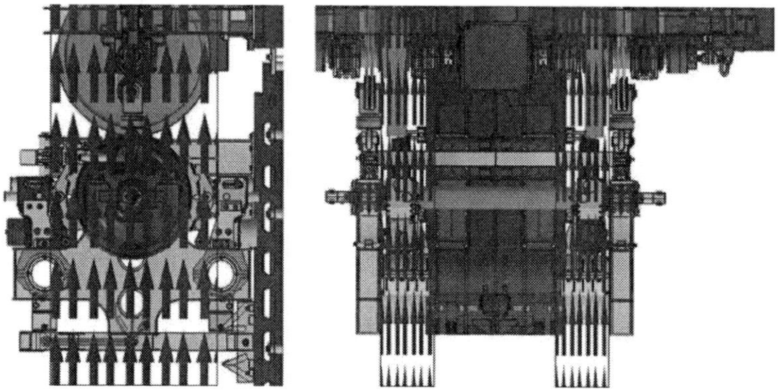

Fig. 9. Adjusted variant of wheelset cooling solution

In addition, in this variant, the air can also be used to cool the rotating rail, which in the case of the previous variant would require additional ventilation equipment (Fig. 9). The disadvantage of this solution is the fact that with only one source, at zero flow, the rail is without cooling.

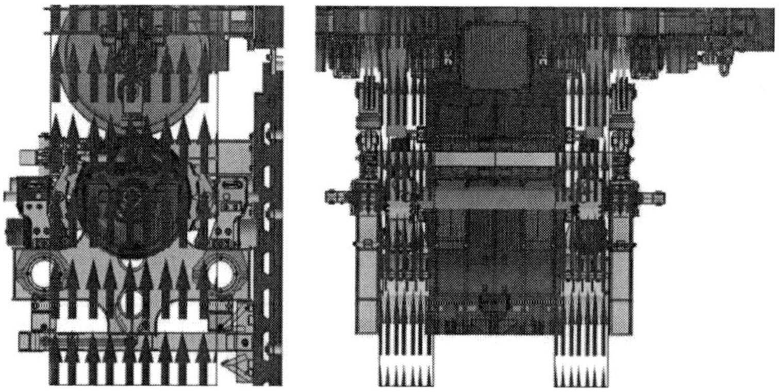

Fig. 10. Virtual air channel view

Fig. 10 shows the path of the air flowing around the wheel. This is only a basic assumption because the frame can be rotated about a 60 mrad angle around the motor's longitudinal axis. It is necessary to create this path by means of guide plates and enclosing the whole device in a package made up of steel profiles, into which boards of noise-absorbing material will be inserted.

The projected flowrate required for the fan dimensioning will be according to the dimensions of the channel from Fig. 10 determined using the equation (1) as follows.

$$Q = S \cdot v = a \cdot b \cdot v = 1.35 \cdot 0.438 \cdot 11.111 = 6.5699\, m^3.s^{-1} = 23651\, m^3.h^{-1} \qquad (2)$$

It is therefore necessary to install at the inlet two fans with a flow rate of 6.6 $m^3.s^{-1}$ or one with a flow of 13.2 $m^3.s^{-1}$.

Such a channel is, however, only virtual since its construction is demanding for manufacture and since it is necessary to maintain the possibility of relative movement between the movable frame with the wheelset and the static frame. Therefore, the channel will be designed as a set of routing plates and a protective structure will be used to ensure the insulation of the device from the surrounding area, which will also reduce the noise produced by the device.

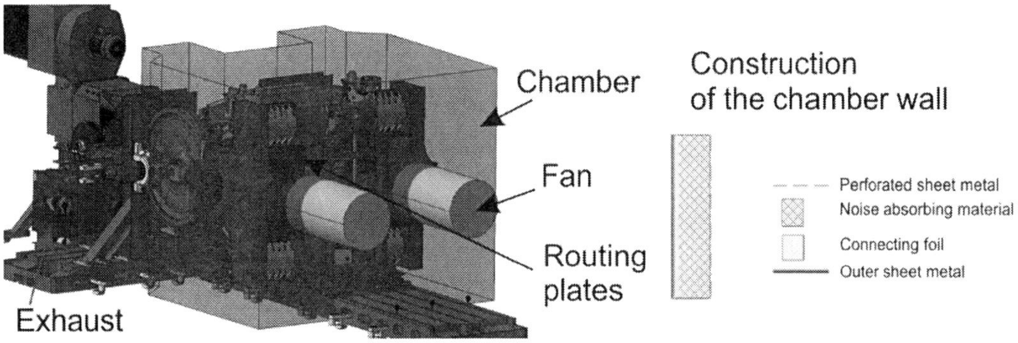

Fig. 11. Protection chamber

As can be seen in Fig. 11, the air is blown into the protection chamber by a pair of fans, where it is driven by the routing plates to the rotating axle. From there, the air is transported by the channel into the exhaust where it is cleaned in the filter and proceeds to the outer area, or it will be drained outside the laboratory. The wall of the chamber is made up of a frame structure, that is filled with metal boxes. These boxes contain noise absorbing material.

Conclusion

In our article, we have conceptually defined the basic parameters for the design of the wheel-brake block pair circumfluence. The presented problem is just an introduction to a more detailed design of the final device construction. In the next step, it will be necessary to determine the nature of the flow in the proposed channel using CFD analysis and to determine the dimensions and shapes of the exhaust. It is also necessary to choose an efficient filter at the end of the exhaust which will separate the dust particles from the air after braking.

The work was supported by the Cultural and Educational Grant Agency of the Ministry of Education of the Slovak Republic in project No. KEGA 077ŽU-4/2017: Modernization of the Vehicles and engines study program. The work was also supported by the project No. APVV-0842-11: Equivalent railway operation load simulator on the roller rig and VEGA No. 1/0927/15: Research of the use of alternative fuels and hybrid drives on traction vehicles with aim to reduce fuel consumption and air pollutants production.
Research-Educational Centre of Rail Vehicles (VVCKV)

References

1. M. Blatnický, M. Štauderová, J. Dižo, *Numerical analysis of the structure girder for vehicle axle scale calibration.* Procedia Engineering **177**, 510-515 (2017)
2. J. Dižo, *Analysis of a goods wagon running on a railway test track.* Manufacturing technology: journal for science, research and production **16** (4), 667-672 (2016)
3. J. Dižo, M. Blatnický, B. Skočilasová, *Computational modelling of the rail vehicle multibody system including flexible bodies.* Communications: scientific letters of the University of Žilina **17** (3), 31-36 (2015)
4. J. Gerlici, M. Gorbunov, K. Kravchenko, A. Kostyuchevich, O. Nozhenko, T. Lack, *Experimental rigs for wheel / rail contact research.* Manufacturing technology: journal for science, research and production **16** (5), 909-916 (2016)
5. J. Gerlici, M. Gorbunov, K. Kravchenko, R. Domin, M. Kovtanets, T. Lack, *Slipping and skidding occurrence probability decreasing by means of the friction controlling in the wheel-braking pad and wheel-rail contacts.* Manufacturing technology: journal for science, research and production **17** (2), 179-186 (2017)
6. J. Gerlici, T. Lack, *Railway wheel profile development based on the geometric characteristics shapes.* In Contact mechanics and wear of rail/wheel systems = CM2009: 8th international conference: 15th-18th September 2009, Firenze, Italy, ISBN 978-88-904370-0-7, 961-967 (2009)
7. J. Gerlici, T. Lack, *Rail geometry analysis (from the point of view of wearing in the operation).* Communications - scientific letters of the University of Žilina **5** (1), 43-51 (2003)
8. J. Gerlici, T. Lack, *Rail vehicles brake components test bench utilisation.* Applied mechanics and materials **486**, 379-386 (2014)
9. J. Gerlici, T. Lack, *Modified HHT method for vehicle vibration analysis in time domain utilisation.* In Applied mechanics and materials **486**, 396-405 (2014)
10. J. Gerlici, T. Lack, *Railway wheel and rail head profiles development based on the geometric characteristics shapes.* Wear: An international journal on the science and technology of friction, lubrication and wear **271** (1-2), 246-258 (2011)
11. J. Gerlici, T. Lack, *Contact geometry influence on the rail / wheel surface stress distribution.* Procedia Engineering **1**, 2249-2257 (2010)
12. J. Gerlici, T. Lack, *Railway wheelset and track contact.* 200 p., ISBN 80-8070-317-5, (EDIS – University of Žilina publishers, Žilina, 2004) [in Slovak]
13. J. Gerlici, T. Lack, *Transport Means Properties Analysis. Vol. I.* 214 p., ISBN 80-8070-408-2 (EDIS – University of Žilina publishers, Žilina, 2005)
14. T. Lack, J. Gerlici, *Railway wheel and rail roughness analysis.* Communications: Scientific Letters of the University of Žilina **11** (2), 41-48 (2009)
15. T. Lack, J. Gerlici, *Wheel/rail tangential contact stress evaluation by means of the modified strip method.* Communications : scientific letters of the University of Žilina **16** (3A), 33-39 (2014)
16. T. Lack, J. Gerlici, *Tangential stresses for non-elliptical contact patches computed by means of a modified FASTSIM method.* Civil-Comp Proceedings **1**, [online] (2016)
17. T. Lack, J. Gerlici, *The FASTSIM method modification in speed up the calculation of tangential contact stresses between wheel and rail.* Manufacturing technology: journal for science, research and production **13** (4), 486-492 (2013)

18. T. Lack, J. Gerlici, J. *A modified strip method to speed up the calculation of normal stress between wheel and rail.* Applied mechanics and materials **486**, 359-370 (2014)
19. T. Lack, J. Gerlici, *A modified strip method to speed up the tangential stress between wheel and rail calculation.* Applied mechanics and materials **486**, 371-378 (2014)
20. P. Pecháč, M. Sága, *Controlling of local search methods' parameters in memetic algorithms using the principles of simulated annealing.* Proc. Eng. **136**, 70-76 (2016)
21. A. Sapietová, M. Sága, P. Novák, R. Bednár, J. Dižo, *Design and application of multi-software platform for solving of mechanical multi-body system problems.* Mechatronics: Recent technological and scientific advances, 345-354 (2011)
22. M. Sága, R. Bednár, M. Vaško, *Contribution to modal and spectral interval finite element analysis.* Vibration Problems ICOVP 2011, Springer Proceedings in Physics **139**, 269-274 (2011)

Discrete element method approach to modelling VPP dampers

Pawel Chodkiewicz[1,*], *Jakub* Lengiewicz[2], *Robert* Zalewski[1]

[1]Warsaw University of Technology, Institute of Machine Design Fundamentals, Narbutta 84, 02-524 Warsaw, Poland
[2] Institute of Fundamental Technological Research of the Polish Academy of Sciences, Pawińskiego 5B, 02-106 Warsaw, Poland

Abstract. In this paper, we present a novel approach to modeling and analysis of Vacuum Packed Particle dampers (VPP dampers) with the use of Discrete Element Method (DEM). VPP dampers are composed of loose granular medium encapsulated in a hermetic envelope, with controlled pressure inside the envelope. By changing the level of underpressure inside the envelope, one can control mechanical properties of the system. The main novelty of the DEM model proposed in this paper is the method to treat special (pressure) boundary conditions at the envelope. The model has been implemented within the open-source Yade DEM software. Preliminary results are presented and discussed in the paper. The qualitative agreement with experimental results has been achieved.

Keywords: VPP, discrete element method, Yade DEM, modelling, smart structures, smart materials

1 Introduction

In the paper, we analyse dampers based on Vacuum Packed Particles (VPP). They are composed of loose granular medium encapsulated in a hermetic envelope, with controlled pressure inside the envelope, see Fig. 1. When no pressure difference is applied (atmospheric pressure inside) then the system behaves more like classical loose granular media. When the internal pressure is reduced then the system is getting stiffer and, even more important, is also gaining the ability to dissipate a significant amount of mechanical energy under the cyclic excitation [1, 2].

* Corresponding author: pawel@chodkiewicz.com.pl

Reviewers: *Tomáš* Lack, *Alžbeta* Sapietová

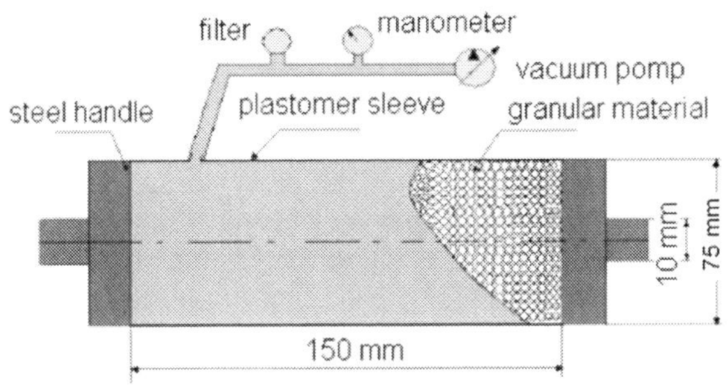

Fig. 1. VPP testing specimen

VPPs are quite novel smart structures, and are gaining increasing interest of the engineers around the world. One of the most advantageous features of VPP systems is the ability to easily change their shape when no underpressure is applied. In such an inactive state, the system can be suitably adjusted to a desired shape, and then the underpressure can be applied to ``solidify'' the system. This feature is exploited, e.g., to design robotic grippers that can carry objects of different geometries, [3, 4]. It can also be found in the medical field. Special mattresses, filled with the granular material, serve to safely transport injured patients, immobilizing their whole bodies [5]. Another medical application is the endoscope that can change its rigidity thanks to VPP [6]. VPP can also be used as dampers. For example, by properly adjusting the grain structure inside the VPP, one can filter out only selected sound frequencies [7, 8].

However, probably the most promising use of the Vacuum Packed Particles is to damp mechanical vibrations. The possibility to actively and quickly control the mechanical properties of VPP dampers makes it possible to adjust their design to various specific applications, e.g., cf. dampers with VPP core and springs [9, 10], or the engineering problem to damp beam vibrations [1, 2, 11, 12]. Vacuum Packed Particles were also used to damp vibrations in automobile industries. In [13], the authors use a damper with VPP core to eliminate vibrations of the car body. In general, dampers made of VPP can be viewed as a good replacement for magnetorheological (MR) or electrorheological (ER) devices, because VPP are more economical in production and more eco-friendly than MR and ER [14].

The VPP structures are also investigated by developing suitable mathematical models that would capture the complex mechanical response of VPP. In our earlier work [15], we tried to adapt fenomenological Gubanov model, successfully used in the past to analyze the complex properties of magnetorheological or electrorheological fluids. However, when applied to VPP, only part of mechanical response could be successfully represented by the model.

In this work, we present the approach to modelling the Vacuum Packed Particles by using Discrete Element Method (DEM). In the past, DEM has been successfully used to analyze various types of granular media, therefore, quite naturally and straightforwardly it can be adapted to modelling VPP systems. Also, such a direct approach to modelling will possibly give a tool to analyze the nature of dissipative response in VPP dampers at the microscopic (particle) level.

In Section 2, we describe the method to treat special boundary conditions at VPP envolope. In Section 3, DEM simulation details are provided and preliminary results are presented. Short conclusions are presented in Section 4.

2 Discrete element model of pressure-tight envelope

The discrete element model has been developed within the open source Yade DEM framework [16]. There are two components of Vacuum Packed Particles to be considered: the pressure-tight envelope made of foil and the granular material inside. As for the granular material, we adopted the standard frictional model for spherical particles, which is provided by Yade DEM. The DEM model of pressure-tight envelope is not standard, and its developement is presented futher in this Section. Additionally, special plugin for Yade DEM has been implemented to collect postprocessing data, necessary to further analyze the VPP time-dependent mechanical properties.

The pressure-tight envelope in VPP is made of foil that is subjected to pressure loading. Its discrete element model should capture the basic elastic mechanical response of the thin foil and should enable to properly apply the necessary pressure boundary conditions. What we propose is the special deposition of the discrete element particles building the envelope and appropriate algorithm to compute the external forces acting on the foil.

As for the geometry of the envelope, each building block is made of five spheres - one in the center (blue) and four neighbors (white), see Figs. 2 and 3. We assume further in this work that the discretization of the foil envelope is fine, i.e., the diameter of the foil particles are not larger than the thickness of the foil,

$$t_f \geq 2 \cdot r_{kf} \tag{1}$$

where t_f is the foil thickness and r_{kf} is the radius of spheres that create the envelope. This is schematically shown in the Figs. 3a-b, see also the discussion in the Remark at the end of this Section.

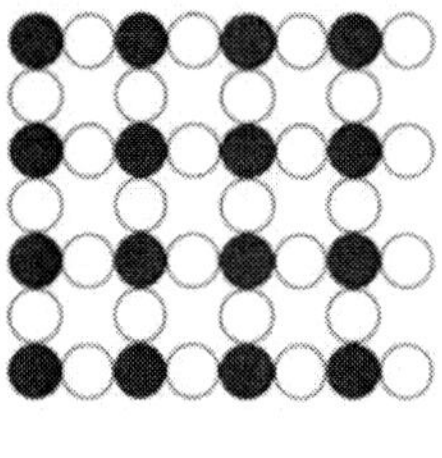

Fig. 2. Schematics of a foil envelope structure built of spheres

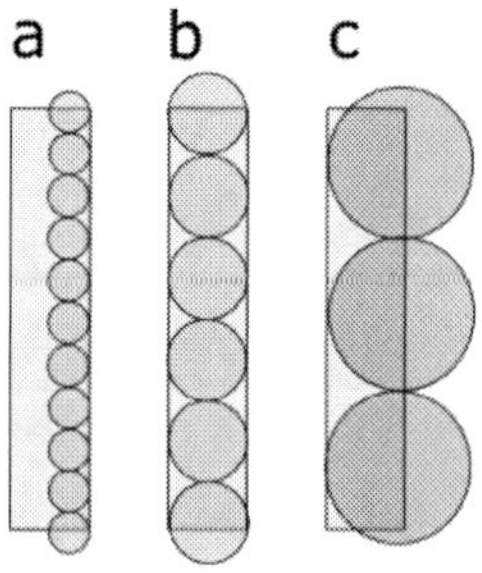

Fig. 3. Discrete element representation of a foil (through-thickness cross-section). Particle diameter can be (a) smaller than, (b) equal to and (c) larger than the thickness of the foil

If the diameters of the particles are different than the thickness of the foil, special treatment is necessary to properly represent the properties of the foil in the discrete element setting. Instead of directly applying the Young modulus of foil E_f, the effective Young modulus of the particles E_{kf} must be used to reflect the difference in the thicknesses (and also to compensate the effect of the vacancies in the discrete structure, see Fig. 2),

$$E_{kf} = 2 \cdot E_f \cdot \frac{t_f}{2 \cdot r_{kf}} \tag{2}$$

The pressure boundary conditions at the envelope, in discrete element setting, are specified by appropriate forces applied to the envelope particles. (We only apply the forces to the blue particles, which has proved to give more stable behaviour during the simulation.) The force for a given particle is computed by integrating the pressures over the actual tributary area. This must be done at every iteration because the geometry of the envelope is constantly changing. The concept of tributary area is presented in Fig. 4.

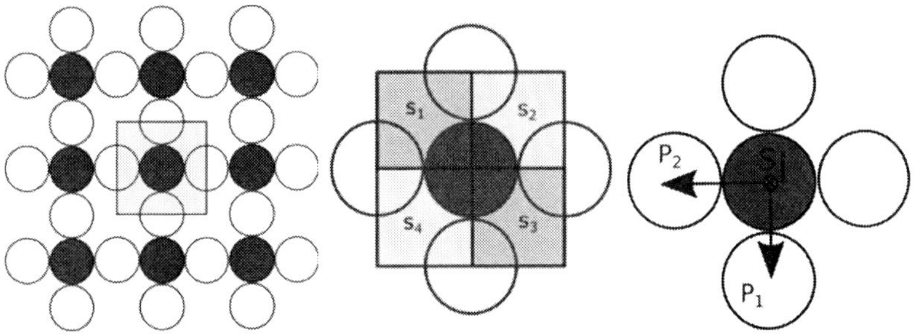

Fig. 4. The tributary area over which the pressures are integrated

The algorithm to compute the vector of external forces is the following. We compute four parts of the tributary area, s_1, s_2, s_3, s_4, as the outer product of the appropriate differences between the *current* positions of the particles,

$$\mathbf{s}_i = (\mathbf{P}_i - \mathbf{S}_j) \times (\mathbf{P}_{1+(i\%4)} - \mathbf{S}_j), \tag{3}$$

where $i = \{0, 1, 2, 3\}$ is the index of the neighbour and j is the index of the middle (blue) particle. The tributary area vector is simply the sum of the four parts

$$\mathbf{A}_j = \mathbf{s}_1 + \mathbf{s}_2 + \mathbf{s}_3 + \mathbf{s}_4 \tag{4}$$

and the vector of the pressure force $\overline{\mathbf{F}}_j$ is the constant pressure of the intensity $\boldsymbol{p}$ integrated over the tributary area, i.e.,

$$\overline{\mathbf{F}}_j = \mathbf{A}_j \cdot p. \tag{5}$$

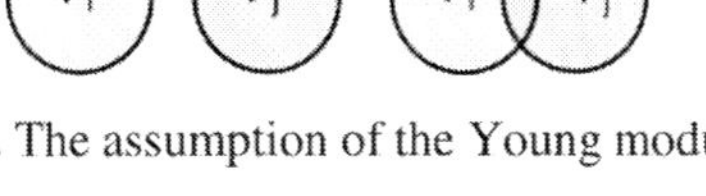

Fig. 5. The assumption of the Young modulus

Remark. In the present model, the diameter of the particles that build the foil is considered much smaller than the thickness of foil, see Fig. 3a, which assures necessary discretization quality. For example, with this assumption the expected buckling effects can be quite straightforwardly captured. Conversely, if the diameters of the particles are larger than the foil thickness then special treatment is necessary to artificially increase the compliance of the foil for in-plane compression cases. One of the simplest possible treatments is to re-scale the Young modulus of the foil for compressive part of the deformation, see Fig. 5, i.e.,

$$E_{kf} \rightarrow \begin{cases} E_{kf} \cdot \alpha & for \quad \left\| \mathbf{P}_i - \mathbf{P}_j \right\| < 2 \cdot r_{kf} \\ E_{kf} & otherwise \end{cases}, \tag{6}$$

where $\alpha \leq 1$ is a properly adjusted scaling parameter.

3 DEM simulation and results

The simulation for the VPP damper has been performed in Yade DEM environment. The simulation consists of several stages, see Fig. 6. In the first two stages, the foil envelope is built and the loose granules are placed inside and above the envelope. Next, the simulation waits until the falling granules fill out the envelope (a special membrane is applied at the top boundary of the damper to only allow the downward passage of the particles). Then, the pressure BC is imposed at the envelope and the sinusoidal kinematic excitation is applied to the top part of the setup (i.e., the upper rings of the envelope and the top boundary).

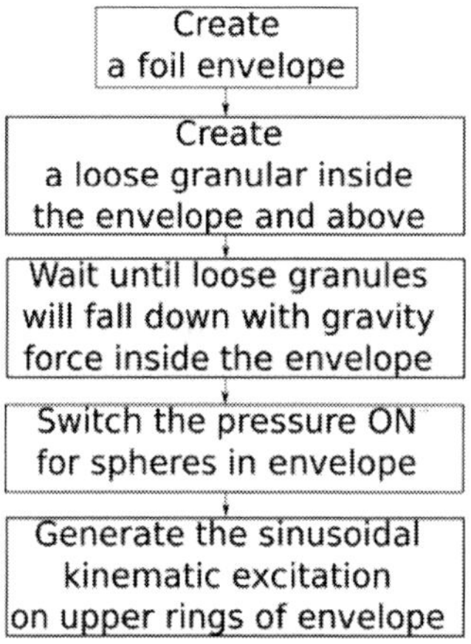

Fig. 6. The stages of the simulation

The simulation parameters are presented in Table 1, where r_{kf} is the radius of envelope particles, r_{kg} is the radius of spherical granules, s_r and s_h are the radius and the height of the damper, respectively, E_f is the Young modulus of the foil envelope, E_k is the Young modulus of the granules, ν_f is the Poisson ratio for the foil envelope and ν_k is the Poisson ratio for the granules.

Table 1. The values of the simulation parameters

Simulation parameters	r_{kf}	r_{kg}	s_r	s_h	E_f	E_k	v_f	v_k
Value	0.00025 [m]	0.001 [m]	0.01 [m]	0.05 [m]	100 [MPa]	1.0 [GPa]	0.25	0.25

The geometry of the envelope, see Fig. 7, has been simply built according to the following Python code,

```
for k in range( n_kfr ):
    for j in range(2, n_kfh ):
        if(k%2 == 0 and j%2 ==0):
            continue
        else:
            x = s_rm *cos(k* φ )
            y = s_rm *sin(k* φ )
            z = j*2* r_kf
```

where s_{rm} is the radius at which the envelope particles are placed,

$$s_{rm} = \frac{r_{kf}}{\sin(\pi / n_{kfc})} , \tag{7}$$

where n_{kfc} is a number of spheres along the circumference of the envelope,

$$n_{kfc} = \text{floor}(\frac{2 \cdot \pi \cdot s_r}{2 \cdot r_{kf}}) , \tag{8}$$

and φ is the angular distance between the centres of the neighbouring envelope spheres along the circle,

$$\varphi = \frac{2 \cdot \pi}{n_{kfc}} \tag{9}$$

The number of spheres along the height of the damper, n_{kfh}, reads

$$n_{kfh} = \text{floor}(\frac{s_h}{2 \cdot r_{kf}}) \tag{10}$$

The top three rings (white colour in Fig. 7) and one bottom ring of the envelope are fixed in order to apply appropriate displacement boundary conditions.

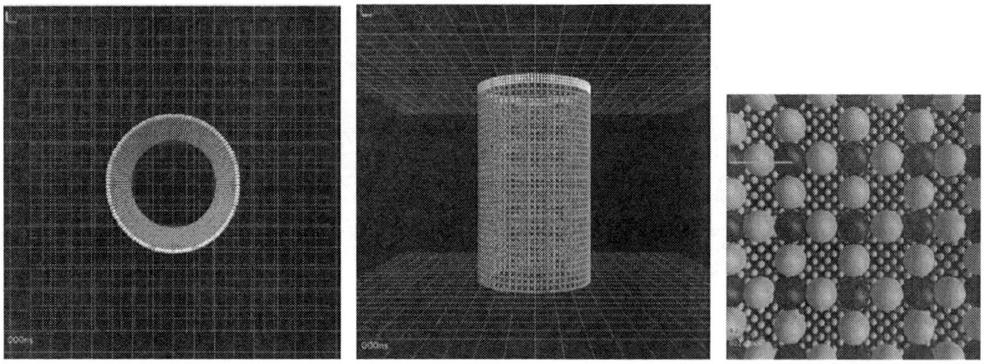

Top view Side view Zoom-in

Fig. 7. Exemplary geometry of the envelope in Yade DEM.

(a) (b) (c)

Fig. 8. The third stage of the simulation. Initially disposed granules (a) are filling out the envelope (b) until no more particles can be added (c).

For the geometry presented in Fig. 8, the whole simulation takes about 40 minutes to complete. It consists of initial four stages of the simulation, followed by 15 oscillations in the last stage. The comparison between the DEM simulation results and the earlier experimental results of [14], see Fig. 9, shows the satisfactory qualitative agreement of the force-displacement hysteresis loops.

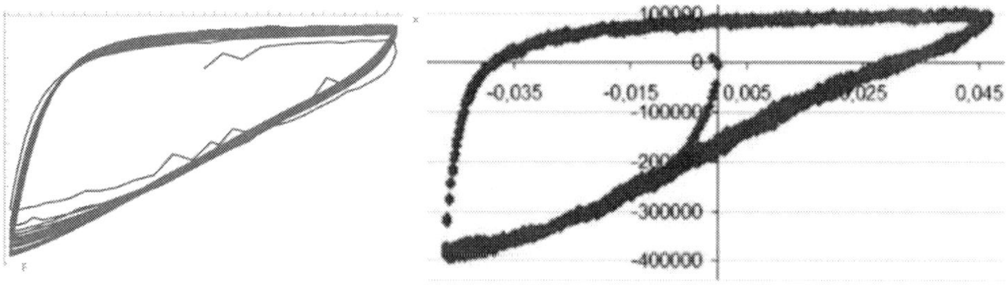

Fig. 9. Hysteresis loops. DEM simulation results (left) and the experimental results (right)

Conclusions

The preliminary results shown in this paper suggest that the DEM modelling is a promising tool to analyse VPP systems. The qualitative agreement of the hysteresis curves has been achieved between the DEM simulations and the earlier experimental results. The advantage of the DEM approach is that one can directly track the evolution of the granules, thus can possibly observe and analyse the microstructural dissipative mechanisms, characteristic to VPP systems. This is one of the very attractive research directions left for future work.

As for the VPP dampers, the next step is to calibrate the model with real experimental data. If succeeded, and the *quantitative*, not only qualitative, agreement will be achieved, this will prove the validity of the presented DEM approach.

References

1. J.M. Bajkowski, R. Zalewski, *Transient response analysis of a steel beam with vacuum packed particles*. Mechanics Research Communications **60**, 1–6 (2014)
2. R. Zalewski, T. Szmidt, *Application of Special Granular Structures for semi-active damping of lateral beam vibrations*, Engineering Structures **65**, 13–20 (2014)
3. J. Amend, N. Cheng, S. Fakhouri, B. Culley, *Soft Robotics Commercialization: Jamming Grippers from Research to Product*, Soft Robot **3**, 213–222 (2016)
4. E. Brown, N. Rodenberg, J. Amend, A. Mozeika, E. Steltz, M.R. Zakin, H. Lipson, H.M. Jaeger, *Universal robotic gripper based on the jamming of granular material*, PNAS. **107**, 18809–18814 (2010)
5. M.D. Luscombe, J.L Williams, *Comparison of a long spinal board and vacuum mattress for spinal immobilisation*, Emergency Medicine Journal **20**, 476–478 (2003)
6. A.J. Loeve, O.S. Ven, J.G. van de Vogel, P. Breedveld, J. Dankelman, *Vacuum packed particles as flexible endoscope guides with controllable rigidity*, Granular Matter **12**, 543–554 (2010)
7. R. Zalewski, M. Rutkowski, *New Controllable Sound Absorbers Made of Vacuum Packed Particles*. In: New Contributions in Information Systems and Technologies, 299–309, Springer, Cham (2015)
8. M. Rutkowski, R. Zalewski, *Semi Active Noise Damping by Vacuum Packed Particles*, Machine Dynamics Research **36** (3), 81–88 (2012)
9. R. Zalewski, P. Chodkiewicz, *Semi-active linear vacuum packed particles damper*, Journal of Theoretical and Applied Mechanics **54**, 311–316 (2016)
10. R. Zalewski, P. Chodkiewicz, M. Shillor, *Vibrations of a mass-spring system using a granular-material damper,* Applied Mathematical Modelling **40**, 8033–8047 (2016)
11. J. M. Bajkowski, B. Dyniewicz, C.I. Bajer, *Damping properties of a beam with vacuum-packed granular damper,* Journal of Sound and Vibration **341**, (2015)
12. T. Szmidt, R. Zalewski, *Inertially excited beam vibrations damped by Vacuum Packed Particles. Smart Mater.* Struct. **23**, (2014)
13. M. Makowski, R. Zalewski, *Vibration Analysis for Vehicle with Vacuum Packed Particles Suspension*, J. Theor. Appl. Mech. **53**, 109–117 (2015)
14. R. Zalewski, *Modelowanie i badania wpływu podciśnienia na właściwości mechaniczne specjalnych struktur granulowanych*, Wydawnictwo Komunikacji i Łączności (2013)
15. R. Zalewski, P. Chodkiewicz, *Gubanov model for vacuum packed particles*. Mechatronics 2013 : recent technological and scientific advances, 57–64, Springer (2014)
16. J. Kozicki, F.V. Donze, *A new open-source software developed for numerical simulations using discrete modeling methods*. Comput. Meth. Appl. Mech. Eng. **197**, 4429–4443 (2008)

Numerical analysis of stiffener for hybrid drive unite

Lenka Jakubovičová[1,*], *Milan* Sága[1], *Marián* Handrik[1]

[1]University of Žilina, Univerzitná 1, 010 26 Žilina, Slovakia

Abstract. The matter of this article is a stress-strain analysis of hybrid drive prototype unit connected directly to convention Concrete Transit Mixer Gearbox. The unite was developed with intention to do field test on existing convection machines with possibility to use existing interfaces. The hybrid drive unit consists from electric and hydrostatic motor connected through addition mechanical transmission gearbox. The question is if today standard interface is good enough or need additional support a "stiffener". Two engineering design were analysed. The first one includes using the stiffener to fixate the construction of hybrid drive unite connected to the planetary gear. The second one is without the stiffener. For strain-stress analysis, a finite element software ANSYS Workbench was used.

Keywords: finite element method/FEM, hybrid drive unite, ANSYS Workbench

1 Introduction

In order to test this new technology on existing machines was necessary solved several challenges [1, 2, 3]. One of that challenge was do the field test on existing convention solution with possibility to connect hybrid drive unit to the main gearbox where is the same interface as is today for one motor only, Fig. 1. Designers needed to be sure that the stiffness of the subassembly will be stiff enough and from this reason was proposed to use additional support part „stiffener" see Figure 4. which was further analysed. The finite element method is especially used in the ground vehicle industry [4, 5, 8, 9, 10].

The prototype hybrid drive unit for a mixing drum that is disposed on a motor vehicle which, as compared with the prior art, exhibits a further improvement to the efficiency of the drive system of the mixing drum and offering possibility eliminate unnecessary exhaust gas and noise emissions in a particular working arrears as underground and tunnels. The drive engine, in particular the internal combustion engine of the truck mixer, is capable of being shut down while the rotation of the mixing drum is maintained. The drive system will be capable of implementation inexpensively with little effort and robust and reliable in its realization.

* Corresponding author: lenka.jakubovicova@fstroj.uniza.sk

Reviewers: *Robert* Grega, *Bohuš* Leitner

Fig. 1. Concrete Transit Mixer Application standard conventional solution

System was installing on the 4 years standard transit mixer with 9 m^3 drum where the drum drive was updated from hydrostatic to hydroelectric, see Figs 2 and 3.

Fig. 2. Concrete Transit Mixer Application and prototype of hybrid drive unit

This object is achieved as a result of the fact that in addition to driving the mixing drum via a hydraulic drive train with displacement pump and hydraulic motor, the electric motor, with a relatively low output as compared with the hydraulic drive, is or can be permanently connected to the truck mixer gearbox in such a way that the mixing drum can be driven solely by the electric motor (Fig. 3).

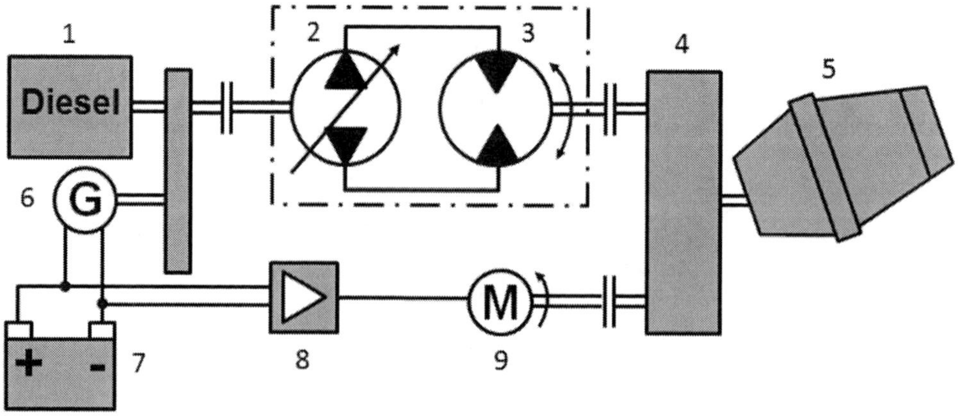

Fig. 3. The main components of a drive unit in a schematic illustration
1- Diesel engine, 2- Hydraulic pump, 3- Hydraulic motor, 4- Gearbox, 5- Drum, 6- Alternator, 7- Electric battery, 8- Inverter, 9- Electric motor

2 Numerical analysis of stiffener for hybrid drive unite

The main components of hybrid drive unite are on Figure 4.

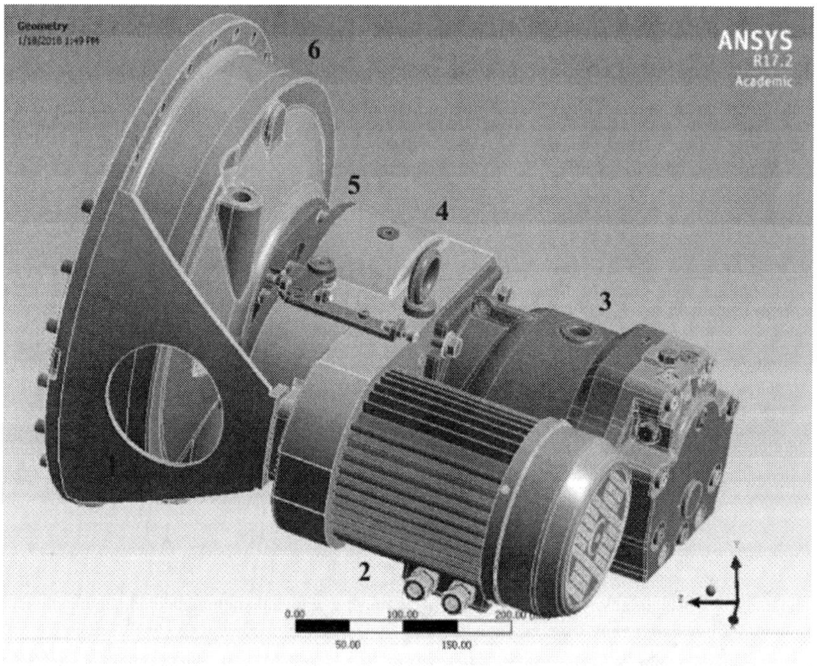

Fig. 4. The main components 1 – Stiffener, 2 – Electric motor, 3 – Hydrostatic motor, 4 – Sandwich Gearbox, 5 – Flange, 6 – Planetary gear cover

In the finite element analysis, only the main part consisted of stiffener, flange, planetary gear cover and sandwich gearbox was considered, Fig. 5. The mass of electric motor and hydrostatic motor was replaced with the resultant gravitational force acting in the centre of gravity, Fig. 5.

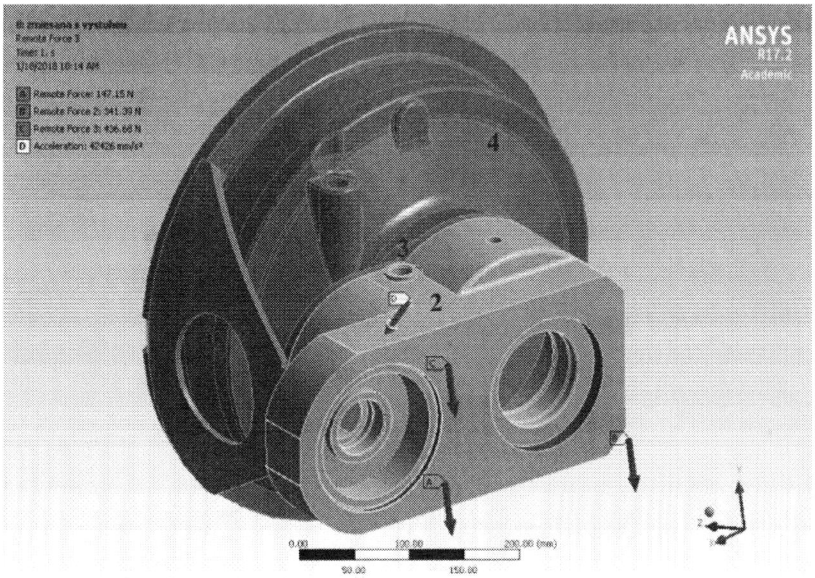

Fig. 5. The main components 1 – Stiffener, 2 – Gearbox, 3 – Flange, 4 – Planetary gear cover

At the planetary gear cover all degrees of freedom were defined - fixed support, the cover is firmly attached to the planetary gear box attached to the drum drive, Fig. 6. Through the concrete transport was considered horizontal and vertical acceleration, Fig. 5.

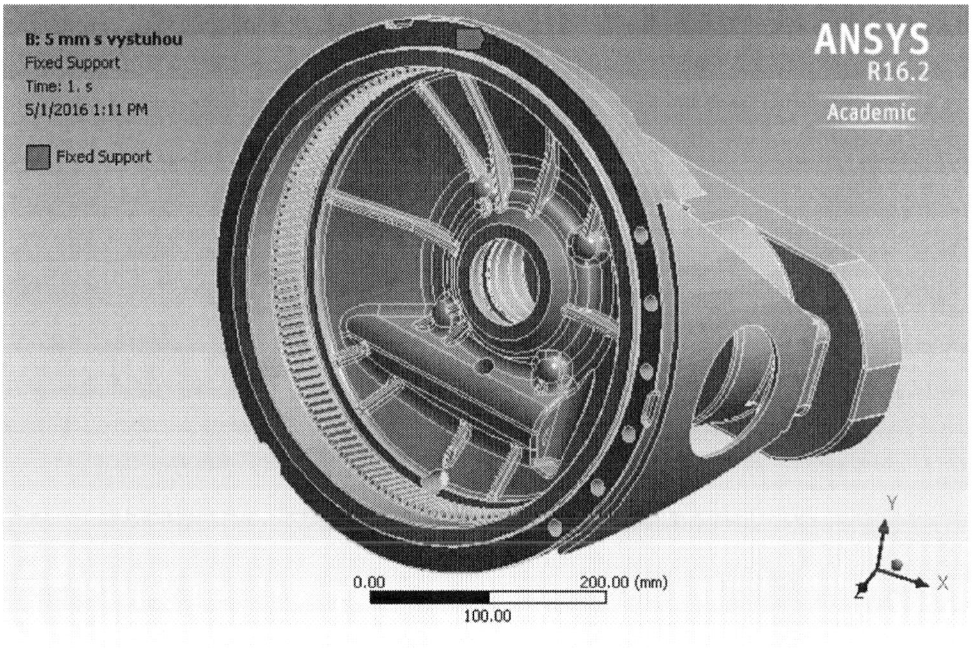

Fig. 6. Boundary condition – fixed support

Stress-strain analysis was carried out using the Ansys Workbench 16.2 software program, into which the model was imported and subsequently modified to a model suitable for finite-element analysis. The finite-element mesh was create by quadratic tetrahedron elements. Figure 7 shows flange and stiffener finite-element mesh and figure 8 shows planetary gear cover and gearbox finite-element mesh.

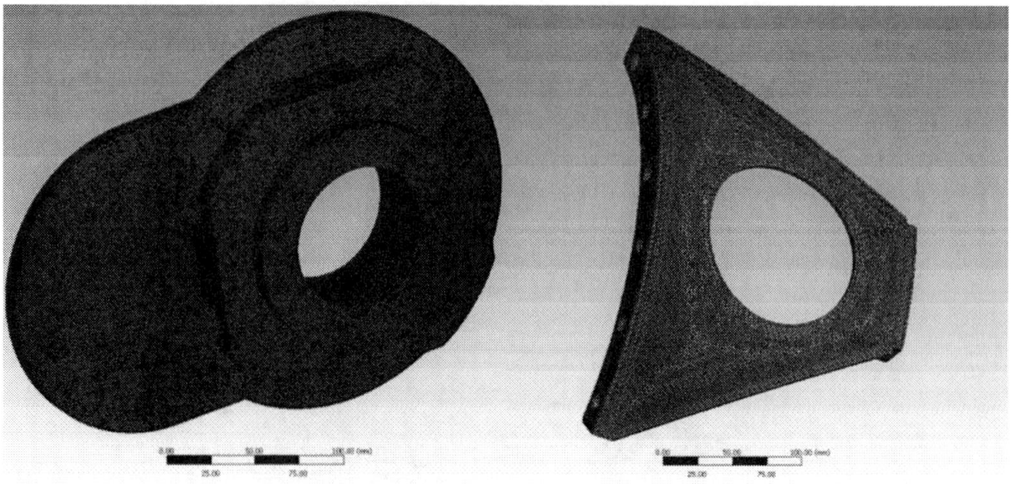

Fig. 7. Flange and stiffener finite-element mesh

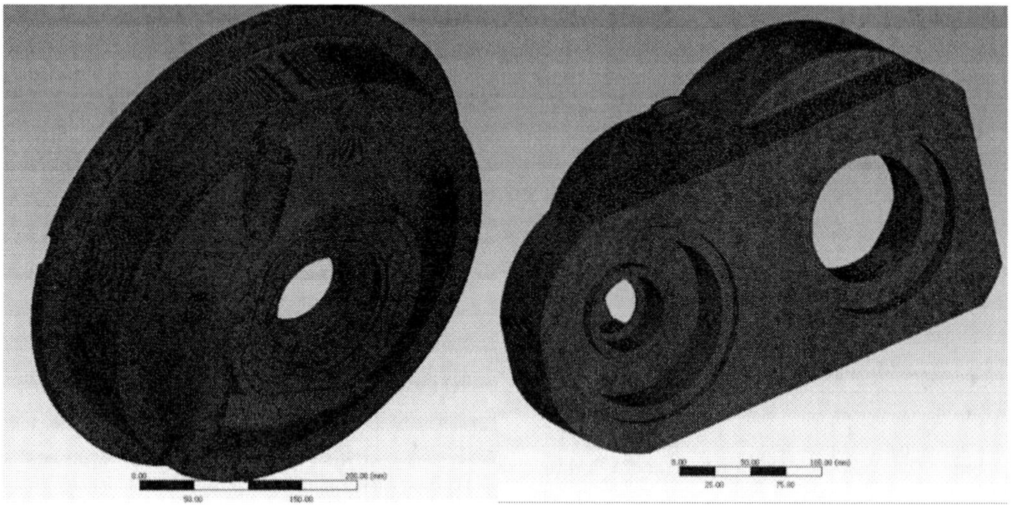

Fig. 8. Planetary gear cover and gearbox finite-element mesh

3 Results

The finite element analysis generated the maximum von Misses stress at the flange, see Figures 9 – 12. For first design with stiffener von Misses stress reached 8.3599 MPa and for second one von Misses stress reached 14.018 MPa, Table 1. Use the stiffener can increase stiffness of the subassembly by 40 percent. The maximum stress without the stiffener is enough below the permissible maximum value, and then the stiffener is no necessary to use and the prototype hybrid drive unit without the stiffener can withstand the load with the required safety level [6, 7].

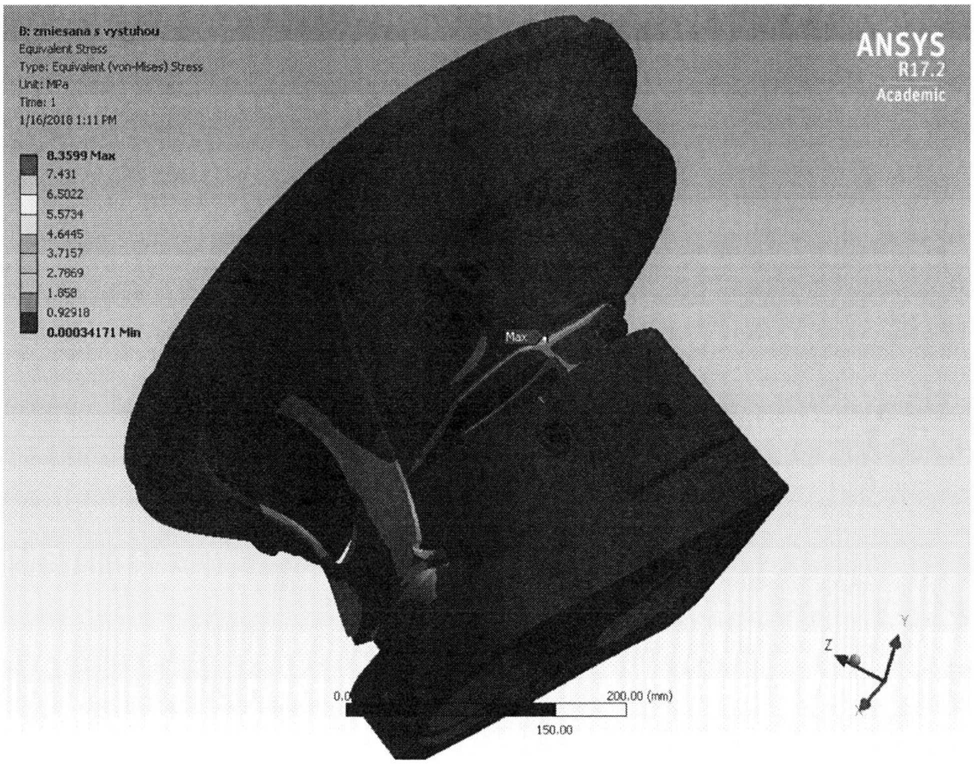

Fig. 9. Von Misses stress – block (with stiffener)

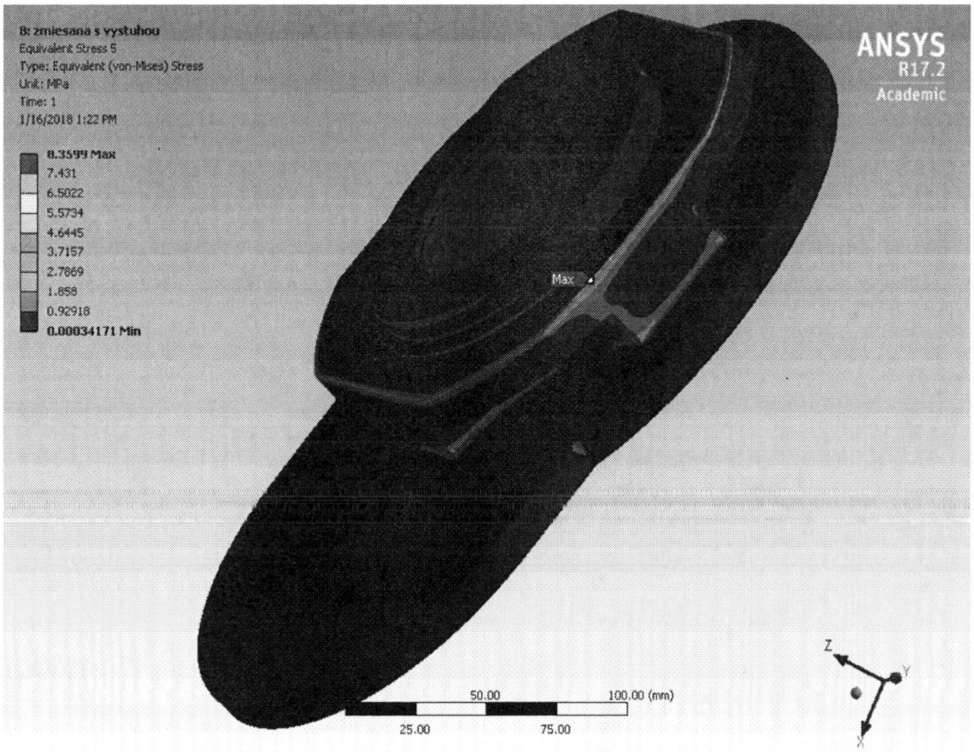

Fig. 10. Von Misses stress – flange (with stiffener)

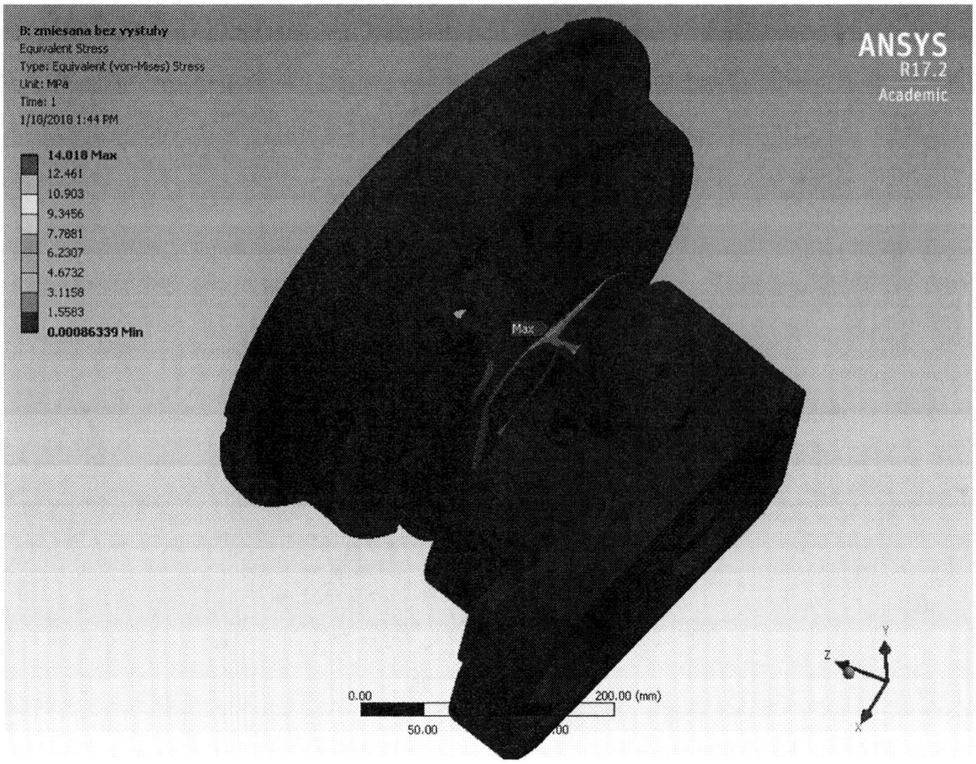

Fig. 11. Von Misses stress – block (without stiffener)

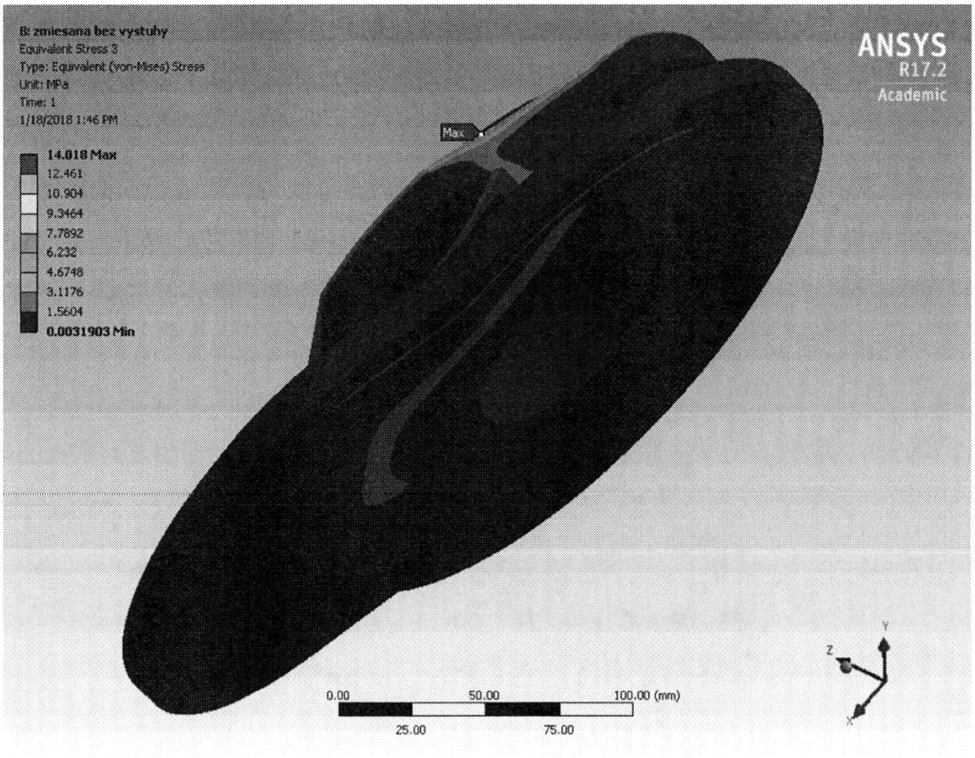

Fig. 12. Von Misses stress – flange (without stiffener)

Table 1. Equivalent (von-Misses) stress

	Equivalent Stress
With stiffener	8.3599 [MPa]
Without stiffener	14.018 [MPa]

Conclusion

The request was find out if the design proposal of the first prototype developed for field tests on existing machines is stiff enough or will be needed update stiffness or use a support part. Based on this request was developed stiffener, Fig. 5.

However at the end we can see based on result in table 1 that the design of prototype is stiff enough and is not necessary to do additional update was confirmed as well that with simple design proposal of stiffener can increase stiffness of the subassembly by 40% percent. This discovery can to the developing engineers possibility for future optimization and weight reductions if will be needed.

This work was supported by grant KEGA 015ŽU-4/2017 - Digital technologies for machine building study programs as a part of strategy "Internet of Things".

References

1. P. Zavadinka, R. Grepl, *Energy saving potential of a hydrostatic drivetrain with variable charge pump*. Proceedings of the 8th FPNI PH.D symposium on fluid power, (2014) ISBN 978-0-7918-4582-0
2. P. Zavadinka, R. Grepl, *Cam ring force simulation for variable roller pump*. Mechatronics 2013: Recent technological and scientific advances, 199-206 (2014) ISBN 978-3-319-02294-9
3. P. Zavadinka, *Simulation of vehicle transport duty cycle with using of hydrostatic units control algorithm*. Mechatronics 2011: Recent technological and scientific advances, 395-402 (2011) ISBN 978-3-642-23243-5
4. R. Grega, J. Krajňák, L. Žuľová, G. Fedorko, V. Molnár, *Failure analysis of driveshaft of truck body caused by vibrations*. Engineering Failure Analysis **79**, 208-215 (2017) ISSN 1350-6307
5. W. Piekarska, M. Kubiak, M. Žmindák, *Issues in numerical modeling of phase transformations in welded joint*. XXI Polish-Slovak Scientific conference machine modeling and simulations MMS 2016, Procedia Engineering **177**, 141-148 (2017)
6. P. Kovacikova, R. Bezdedova, J. Jr.Vavro, J. Vavro, *Comparison of numerical analysis of stress-strain states of cast iron with vermicular graphite shape and globular graphite shape*. 20th International conference machine modeling and simulations MMS 2015, Procedia Engineering **136**, 28-32 (2016)
7. W. Piekarska, M. Kubiak, M. Vaško, *Numerical estimation of the shape of weld and heat affected zone in laser-arc hybrid welded joints*. Procedia Engineering **177**, 114-120 (2017)
8. A. Vaško, M. Vaško, J. Belan, E. Tillová, *Comparison of fatigue properties of nodular cast iron at low and high frequency cyclic loading*. Procedia Engineering **177**, 576-581 (2017)

9. L. Zuľova, R. Grega, J. Krajňák, G. Fedorko, V. Molnár, *Optimization of noisiness of mechanical system by using a pneumatic tuner during a failure of piston machine*. Engineering Failure Analysis **79**, 845-851 (2017) ISSN 1350-6307

10. P. Pastorek, P. Novák, P. Kopas, M. Močilan, *Finite element analysis of bond behavior in a steel reinforced concrete structure strengthened carbon fibre reinforced polymer (cfrp) strips*. Metalurgija **56** (3-4), 405-408, ISSN 0543-5846

Experimental and computational comparative study of the specimens loaded by bending and torsion

Lenka Jakubovičová[1,*], *Milan* Vaško[1], *Milan* Sága[1], *Peter* Kopas[1]

[1]University of Žilina, Faculty of Mechanical Engineering, Department of Applied Mechanics, Univerzitná 8215/1, Žilina, Slovakia

Abstract. The article presents the theoretical and computational analysis of the fatigue tested specimens loaded by bending and torsion. The testing device allows loading by constant turn of the clamping parts. The stress and strain magnitude in test depends on the material characteristics and shape of the specimen. The calculations are realized by finite element method (FEM). The obtained results are verified with the experimental measurement applying the optical system ARAMIS.

Keywords: bending, torsion, calibration curve, Finite Element Method (FEM), optical system ARAMIS, ADINA

1 Introduction

The multiaxial fatigue equipment has been designed for the purpose of experimental measurement fatigue curves of test specimen. For evaluation of fatigue curves it is necessary to know stress and strain conditions on individual loading levels. The stress and strain magnitude in test sample is dependent on shape and material of the specimen. Exact solution of stress and strain values is difficult and FE method [1] and experimental measurement (optical system ARAMIS) were used for their specification [2].

1.1 Test equipment design

Design of experimental equipment has been based on mechanical principle. The fixed rotation is generated by excenter and linkage mechanism. A loading magnitude is possible to modify by changing of excentric magnitude. Also if we change a length of connecting crank on the experimental equipment, there will be change in a loading cycle character (proportional-nonproportional loading, tension/tension loading, etc.). Power of device is secured by two synchronic electromotors with frequency converters from 0.5 Hz to 100 Hz. Loading frequencies are identical with frequency of rotation drive. Synchronization of the electromotors is secured using electronics and allows synchronization of loading amplitudes. Synchronization of electromotors also allows setting phase shift for individual

* Corresponding author: lenka.jakubovicova@fstroj.uniza.sk

Reviewers: *Robert* Grega, *Radim* Halama

loading levels. In experimental equipment are also two force measure systems. These systems may be used to evaluate the force values during the loading process [3]. Our attention was focused on stress and strain conditions analysis by finite element method and on experimental measurement deformations by system ARAMIS in this paper.

1.2 Material characteristics

Two commercial aluminium alloys EN AW 6063.T66 (AlMgSi0,7.F25) and EN AW 2007.T3 (AlCuMgPb) were used for multiaxial fatigue testing. The chemical composition is shown in Tab. 1. Tested materials were bars with a circular section with diameter of 10 mm. The EN AW 6063.T66 material was in pressing status, while EN AW 2007.T3 in drag status. The structure of both materials has been influenced by mechanical and heat treatment process. It was examined a tensile test and hardness measurement by Vickers, and metallographic analyses [4, 5].

Table 1. Chemical composition of aluminium alloys

EN AW 6063										
Si	**Fe**	**Cu**	**Mn**	**Mg**	**Cr**	**Ni**	**Zn**	**Ti**	**Others components**	**Al**
0.2-0.6%	0.35%	0.1%	0.1%	0.45-0.9%	0.1%	-	0.1%	0.1%	Max. 0.05% each Max. 0.15% global	Remaining %
EN AW 2007										
Si	**Fe**	**Cu**	**Mn**	**Mg**	**Cr**	**Ni**	**Zn**	**Ti**	**Others components**	**Al**
0.8%	0.8%	3.3-4.6%	0.5-1.0%	0.4-1.8%	0.1%	0.2%	0.8%	0.2%	Max. 0.05% each Max. 0.15% global	Remaining %

The material characteristics of fatigue specimen were obtained experimentally from tensile testing machine by STN EN 42 0310. The Stress-Strain diagram contains two curves - true stress and stress depending up strain (Fig. 1 and 2). For FEM analyses by ADINA were useful true stresses [1, 6].

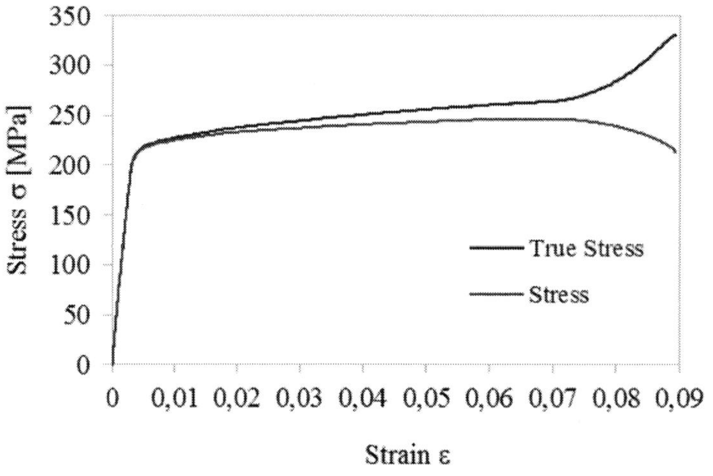

Fig. 1. Stress-Strain curve for EN AW 6063

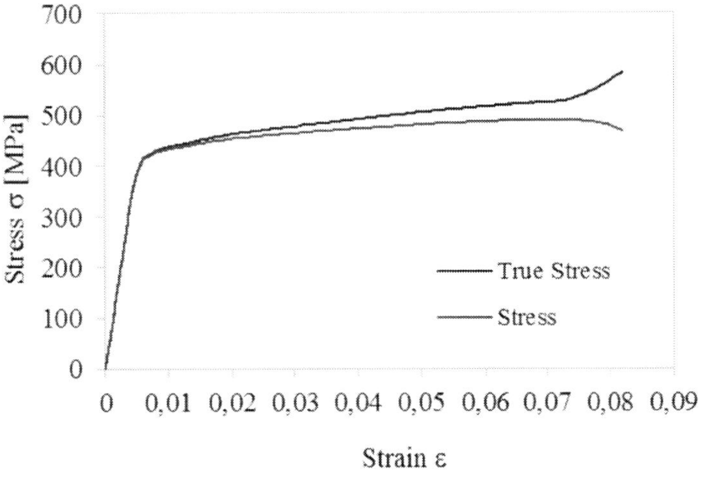

Fig. 2. Stress-Strain curves for EN AW 2007

1.3 Loading degrees setting

The loading regulation for bending and torsion were realized by excenter displacement according to body excenter. Power transfer from excenter to body excenter is assured by soft grooving what allows stepped setting whole eccentric system. This also allows exact setting of rocker arm.

Two pairs of excentric couple with fine grooving 40 under STN EN 01 4933 and setting maximum amplitude of 2 and 1 mm were prepared for both loading systems. In the case of maximal amplitude 2 mm the excenter body and excenter were made with eccentricity of 1 mm. In case of maximal amplitude 1 mm the eccentricity was 0.5 mm. Excenter has 37 teeth, so it is possible to set 37 loading levels. Setting scheme of eccentricity is presented on Fig. 3.

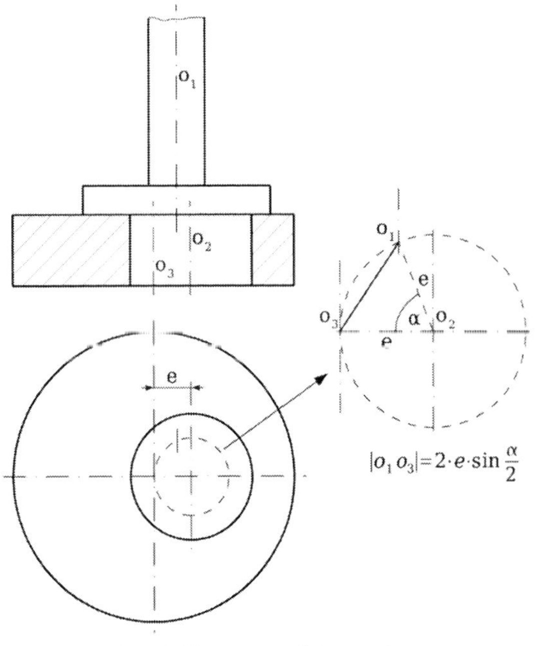

Fig. 3. The eccentricity setting

2 Stress-strain results obtained by optical system ARAMIS

ARAMIS is contactless optical system based on three-dimensional measurement to evaluate displacement at static or dynamic loading. With the software ARAMIS can be analysed, calculated and documented a deformation process where the results from these photos provide optimal interpretation of object behaviours.

ARAMIS system consists from optical camera, which caches the surface object. By the software can be recognized the structure of surface from taken digital photos and assign coordinates to showed pixels (Fig. 4). Then ARAMIS compare digital photos and calculates displacements of individual pixels and evaluate deformation. It is very important to have specimens with sprayed layer for this process. For this reason it is usual combination black and white spray colour applied on surface of specimens (Fig. 5).

Fig. 4. The camera system ARAMIS caching the surface of loading object

The measure of cyclic deformation were made under loading frequency 0.5 Hz, while camera system takes the digital photos with frequency 30 Hz (30 pictures per second) [2]. The maximum values of displacements were measured at the specimen radius, where were applied sampling mask (Fig. 5). The displacement area is displayed on each photo (Fig. 6) and time history was created at the point of maximal displacements (Fig. 7). Dominant values for cyclic loading are ε_{xx} and dominant values for cyclic torque loading are γ_{xy}.

Fig. 5. The photo analysed by system ARAMIS with "sampling mask"

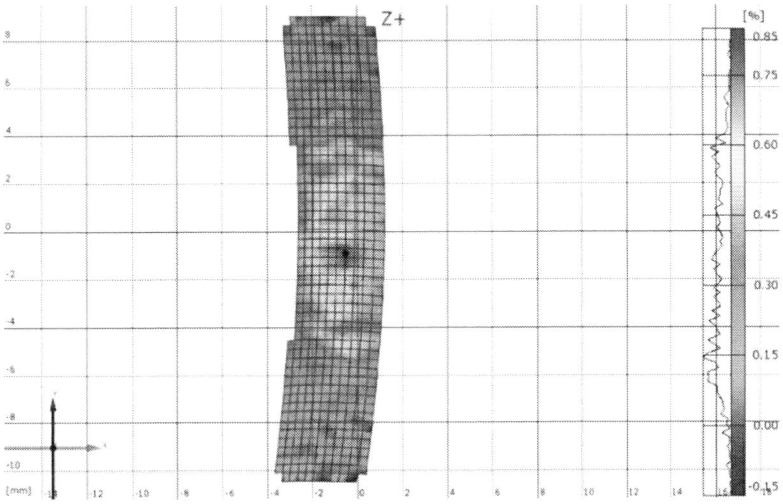

Fig. 6. Displacements at sampling mask

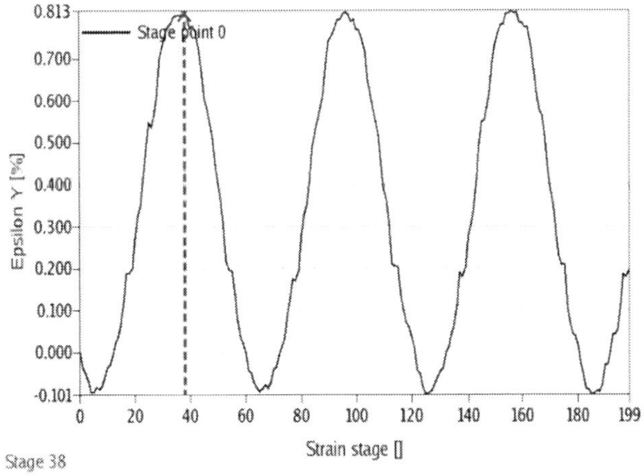

Fig. 7. The time history of displacements by each photo

3 Stress-strain FE analysis of test specimen

It is necessary to know plastic strain amplitude (Manson-Coffin curve) or stress amplitude (Wöhler curve) on each loading level for fatigue test interpretation [7]. Therefore, it is necessary to analyse stress and strain at the maximum values by FEM. The specimen model was created by FE software ADINA (Fig. 8) [6, 8].

The material was assumed as plastic-bilinear; the true stresses were obtained from real stress-strain curves on Figs. 1 and 2. The linear element of tetrahedron type was automatically generated. The "Load Plot" function was defined by excenter setting with 37 steps with eccentricity of 1 mm or 2 mm. The shell and beam elements for hammer simulation were used at the fix point.

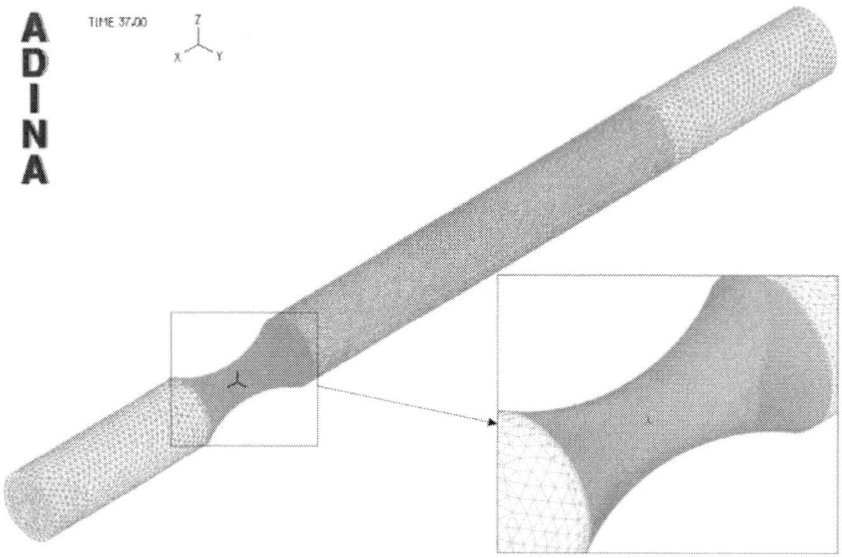

Fig. 8. FE model of specimen

The maximum stress and strain values are concentrated in specimen radius. It has been shown from computational analysis of stress conditions. There was obtained stress-strain values at the specimen radius (Fig. 9 to Fig. 12).

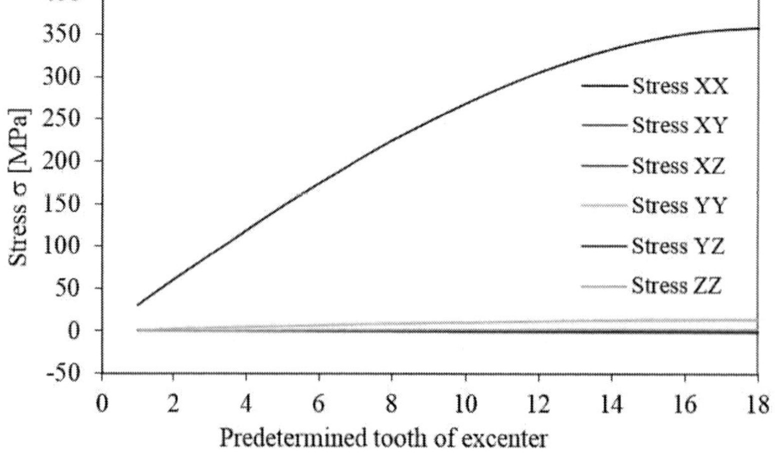

Fig. 9. Bending: Stress values for EN AW 2007 and eccentricity 1 mm

The results associate to eccentricity $e = 1$ mm. For bending we can determinate that the stress and strain values have a dominant values; it means stress σ_{xx} and strain ε_{xx}. The same phenomenon has been observed for the torsion at the stress and strain values; it means stress σ_{xy} and strain ε_{xy} [9, 10]. Obtained stress values allow determine the plastic deformation amplitude for each load step (turning step by step the gear - 37 steps = 1 turn). It was determined a deformation ε_{xx} during the bending sinusoidal cycle of loading and screw angle γ_{xy} was determined during the torsion sinusoidal cycle of loading.

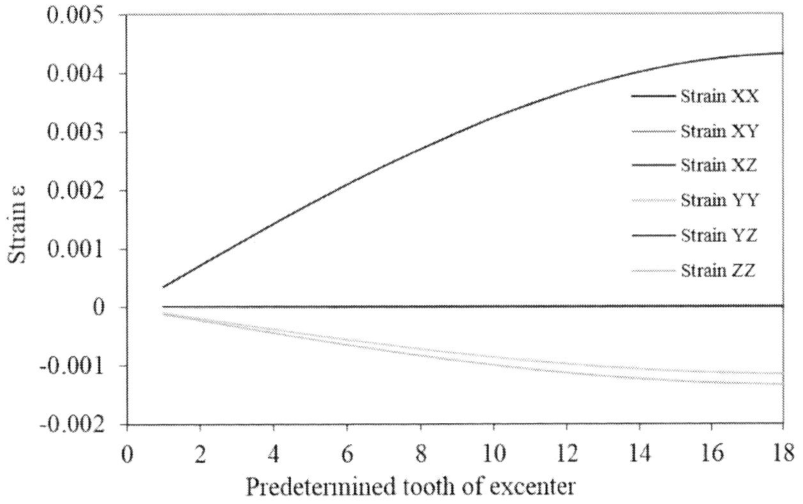

Fig. 10. Bending: Strain values for EN AW 2007 and eccentricity 1 mm

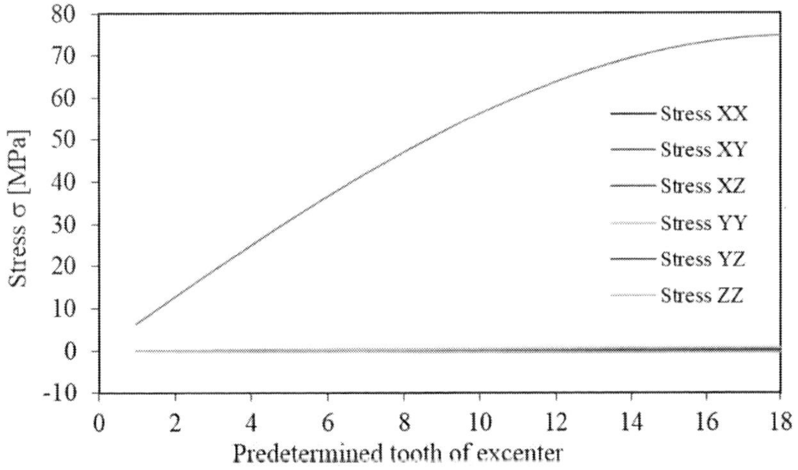

Fig. 11. Torsion: Stress values for EN AW 2007 and eccentricity 1 mm

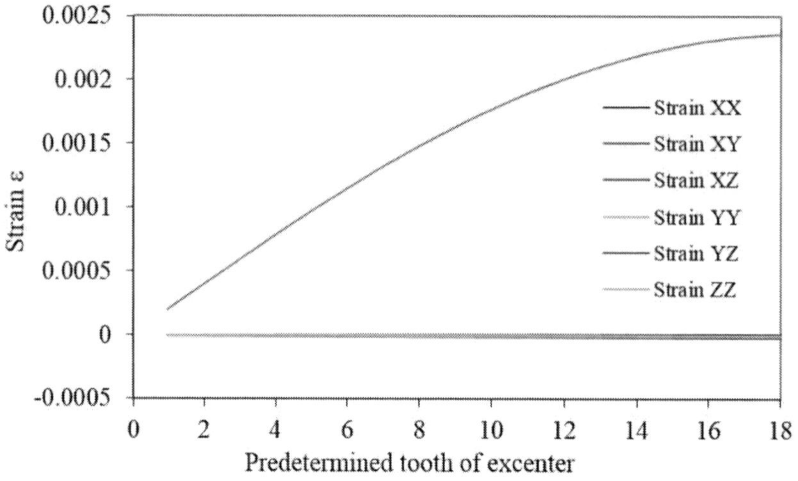

Fig. 12. Torsion: Strain values for EN AW 2007 and eccentricity 1 mm

4 Comparison of Stress-strain FE and ARAMIS analyses

The measurement by optical system ARAMIS was realized at the load level equal to eccentricity setting to uneven number of teeth. The obtained strain results from ARAMIS compared with obtained results by FE method are shown in Fig. 13 to Fig. 16. The results obtained by optical system ARAMIS at the sinusoidal cycle bending was low increasing like results obtained from FEM analyses.

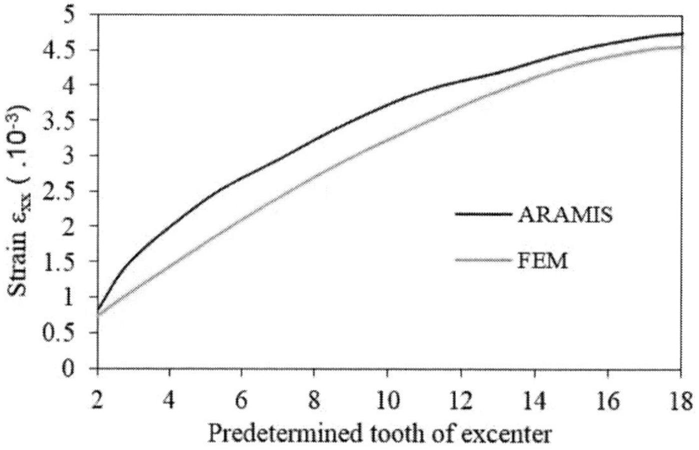

Fig. 13. The strain ε_{xx} obtained by optical system ARAMIS vs. FEM (bending, EN AW 6063, eccentricity 1 mm)

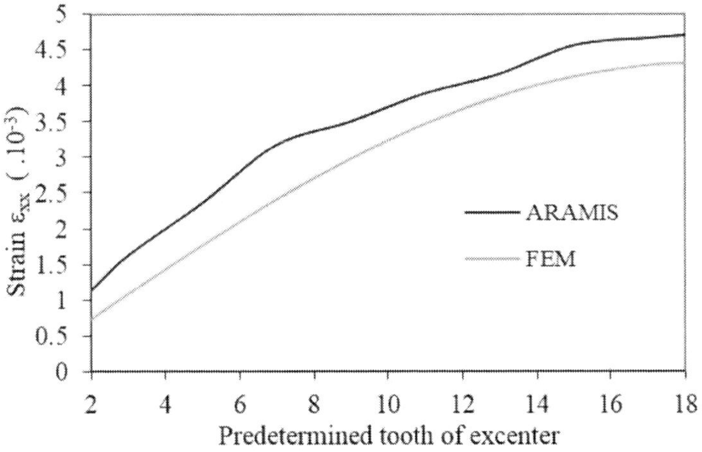

Fig. 14. The strain ε_{xx} obtained by optical system ARAMIS vs. FEM (bending, EN AW 2007, eccentricity 1 mm)

During evaluation of points displacement by ARAMIS system interpreted the measure of shearing strain value ε_{xy}, ADINA software interpreting shearing strain values γ_{xy} on the base of FE analyses [1, 11]. It can be seen that the value obtained by ARAMIS system must be for the FE analyses multiplied by two from this reason, i.e.

$$\varepsilon_{xy} = \frac{1}{2}\gamma_{xy}.$$

The obtained results from ARAMIS and from FE analysis in the case of the sinusoidal torsion are shown in Fig. 14 and Fig. 15. The strain values obtained from FEM and strain values obtained from ARAMIS system indicated relatively good agreement at all load steps for bending-torsion cycle sinusoidal loading [2, 12, 13].

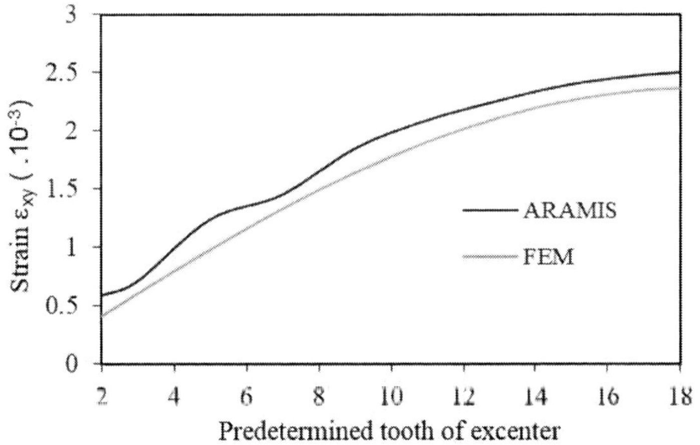

Fig. 15. The strain ε_{xy} obtained by optical system ARAMIS vs. FEM (torsion, EN AW 6063, eccentricity 1 mm)

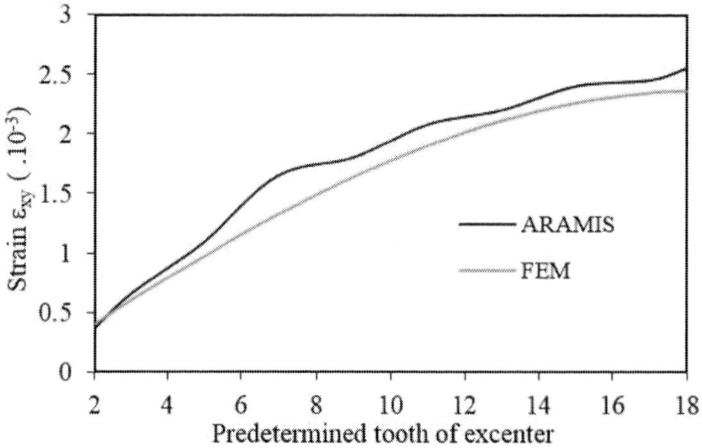

Fig. 16. The strain ε_{xy} obtained by optical system ARAMIS vs. FEM (torsion, EN AW 2007, eccentricity 1 mm)

5 Calibration curves for EN AW 6063 and EN AW 2007 materials

The measured values of deformation ε by system ARAMIS and also the commutated values of deformation ε by FEM have good correlation for aluminium alloy. In the future it will be interesting in combination of bending and torsion loading therefore it is necessary to have calibration curve for each level of loading (Fig. 17 and 18).

So if we want to set a deformation on exact value for torsion loading, we must backwards to establish a number of tooth for setting of excenter from calibration curve, see Eq. (1) to Eq. (4). Than we can provide the cycle sinusoidal loading for determine multiaxial fatigue life-time characteristic.

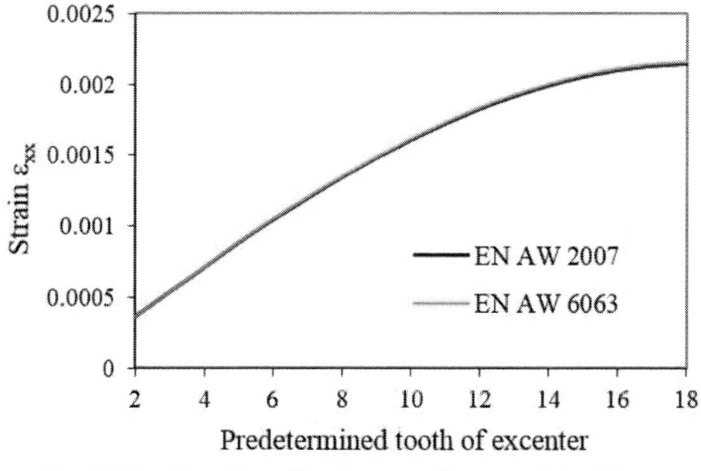

Fig. 17. Bending: The calibration curve for eccentricity 0.5 mm

Strain calibration curve (ε_{xx}) for EN AW 2007 and EN AW 6063 material and for eccentricity:

- 0.5 mm: $$n_{tooth} = \left(-0.0002\,\varepsilon_{xx}^{\;3} - 0.001\,\varepsilon_{xx}^{\;2} + 0.187\,\varepsilon_{xx} - 0.0071\right)\cdot 10^{-3}, \quad (1)$$

- 1 mm: $$n_{tooth} = \left(-0.0003\,\varepsilon_{xx}^{\;3} - 0.0017\,\varepsilon_{xx}^{\;2} + 0.372\,\varepsilon_{xx} - 0.014\right)\cdot 10^{-3}. \quad (2)$$

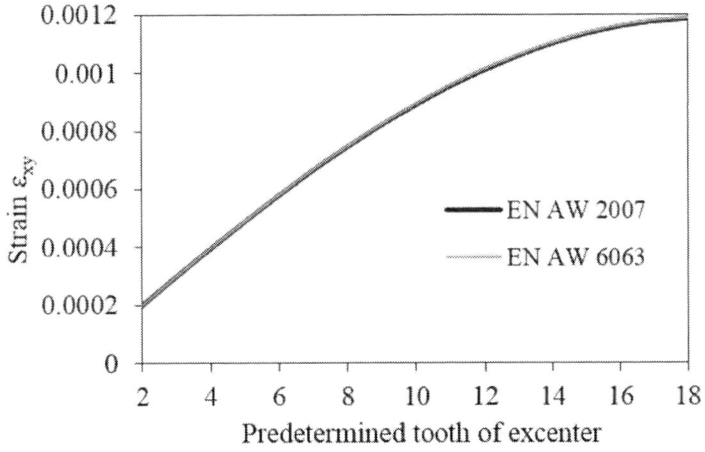

Fig. 18. Torsion: The calibration curve for eccentricity 0.5 mm

Strain calibration curve (ε_{xy}) for EN AW 2007 and EN AW 6063 material and for eccentricity:

- 0.5 mm: $$n_{tooth} = \left(-0.0001\varepsilon_{xy}{}^{3} - 0.0007\varepsilon_{xy}{}^{2} + 0.1039\varepsilon_{xy} - 0.004\right)\cdot 10^{-3}, \quad (3)$$

- 1 mm: $$n_{tooth} = \left(-0.0002\varepsilon_{xy}{}^{3} - 0.0013\varepsilon_{xy}{}^{2} + 0.2069\varepsilon_{xy} - 0.008\right)\cdot 10^{-3}. \quad (4)$$

Conclusion

Our paper deals with determination of the calibration curve for combined loading (bending and torsion) using numerical and experimental evaluating. Analyses were realized for EN AW 6063 (AlMgSi0,7.F25) and EN AW 2007 (AlCuMgPb) materials.

The numerical calculations were realized by Finite Element Method and results were verified with the experimental measurement applying the optical system ARAMIS. The results have shown an independence of the strain calibration curve on the material type.

This work was supported by the Slovak Research and Development Agency under the contract No. APVV-14-0096 and by grant KEGA 015ŽU-4/2017 - Digital technologies for machine building study programs as a part of strategy "Internet of Things".

References

1. K. J. Bathe, *Finite Element Procedures*. (Prentice Hall, New Jersey, 1982)
2. A. Sapietová, V. Dekýš, *Dynamic analysis of rotating machines in MSC.ADAMS*. Procedia Engineering **136**, 143-149 (2016)
3. M. Handrik, P. Kopas, V. Baniari, M. Vaško, M. Sága, *Analysis of stress and strain of fatigue specimens localised in the cross-sectional area of the gauge section testing on bi-axial fatigue machine loaded in the high-cycle fatigue region*. Procedia Engineering **177**, 516-519 (2017)
4. W. Piekarska, M. Kubiak, Z. Saternus, *Numerical simulation of deformations in T-joint welded by the laser beam*. Archives of Metallurgy and Materials **58** (4) 1391-1396 (2013)

5. A. Vaško, J. Belan, L. Kuchariková, E. Tillová, *Low and high frequency fatigue tests of nodular cast irons*. Metalurgija **56** (1-2), 25-28 (2017)
6. ADINA, *Theory and Modelling Guide*. [help manual] (Watertown, 2010)
7. P. Kopas, M. Blatnický, M. Sága, M. Vaško, *Identification of mechanical properties of weld joints of AlMgSi07.F25 aluminium alloy*. Metalurgija **56** (1-2), 99-102 (2017)
8. J. Kortiš, Ľ. Daniel, M. Handrik, *The Numerical Analysis of the Joint of the Steel Beam to the Timber Girder*. Procedia Engineering **91**, 160-164 (2014)
9. L. Jakubovičová, B. Ftorek, V. Baniari, A. Sapietová, T. Potoček, M. Vaško, *Engineering Design of Test Device*. Procedia Engineering **177**, 520-525 (2017)
10. J. Gerlici, T. Lack, *Contact geometry influence on the rail / wheel surface stress distribution*. Procedia Engineering **2** (1), 2249-2257 (2010)
11. R. Grega, J. Homišin, J. Krajňák, M. Urbanský, *Analysis of the impact of flexible couplings on gearbox vibrations*. Scientific Journal of Silesian University of Technology-Series Transport **91**, 43-50 (2016)
12. R. Halama, A. Markopoulos, F. Fojtík, M. Fusek, Z. Poruba, J. Famfulík, *Effect of stress amplitude on uniaxial ratcheting of aluminum alloy 2124-T851*. Materialwissenschaft und Werkstofftechnik **48** (8), 814-819 (2017)
13. R. Grega, J. Krajňák, L. Žuľová, G. Fedorko, V. Molnár, *Failure analysis of driveshaft of truck body caused by vibrations*. Engineering Failure Analysis **79**, 208-215 (2017)

Cooperation of axisymmetric connection elements under dynamic load

Andrzej Kołodziej[1*], *Marian* Dudziak[1], *Krzysztof* Talaśka[2]

[1] The President Stanisław Wojciechowski State University of Applied Sciences in Kalisz, Kalisz, 62-800 Poland
[2] Chair of Basics of Machine Design, Poznan University of Technology, Poznan, 60-965, Poland

Abstract. The article presents a method for determining the parameters that define the cooperation of the elements in the axisymmetic connection. The connection, which constitutes a shaft cooperating with a sleeve, has been tested for reaction forces in the connection during shaft rotation in the static sleeve. The shaft was characterized by deliberately modelled roundness deviations in the form of ovality, triangularity and quadrangularity. In addition, the research programme has taken into account the determination of the impact of tolerance of the outside diameter of the shaft. Determination of reaction forces has been carried out using the FEM software. The shaft has been modelled as a rigid element that rotates with a given rotational speed in the deformable sleeve. The conclusions present the impact of roundness deviation types and the tolerance value on reaction forces in the connection restraint. The method presented in the article can be used to predict the behaviour of the elements of axisymmetic connections under dynamic load, which can contribute to forecasting the durability of the connection.

Keywords: axisymmetric connections, dynamic loads, shape deviations

1 Introduction

Axisymmetric connections are some of the most basic connections in machine design. They come in two basic types: resting and movable. In the case of resting connections, their load carrying capacity is related to allowable stresses at rest. Movable connections are more vulnerable to damage, especially through wear and tear [1, 2]. This is related to friction from the contact of interacting surfaces. The key in this kind of phenomena is the state of interacting surfaces It directly affects a number of tribological phenomena, leading to accelerated wear and tear of the connection. In the case of interacting cylindrical surfaces of the shaft and sleeve (outer and inner), what is also important apart from surface roughness, are shape deviations. This applies both to shape deviations of the cylinder generatrix (we can distinguish here: conicity, barrel-shapedness and saddle-shapedness) and to roundness deviations (ovalness, triangularity, quadriangularity, multiangularity) [1, 2]. When analysing production processes of the axisymmetric connection components, from a statistical point of view, it is possible to achieve any combination of the above mentioned

* Corresponding author: a.kolodziej@pwsz.kalisz.pl

Reviewers: *Grzegorz* Domek, *Marek* Macko

shape deviations. Figure 1 shows sample results of measuring shafts with detected cross section shape deviations (multiangularity). It is possible, of course, to control the production process in order to achieve selected shape deviations with higher probability. In order to do it successfully, given the expected life cycle of the connection, methods of determining the parameters of interaction between such connections in operating conditions corresponding to dynamic loads should be developed. An example would be rotational movement of the shaft relative to the sleeve. It was decided that the main parameter defining the nature of the axisymmetric connection would be the value of the reaction force at fixed support during shaft rotation in the hub.

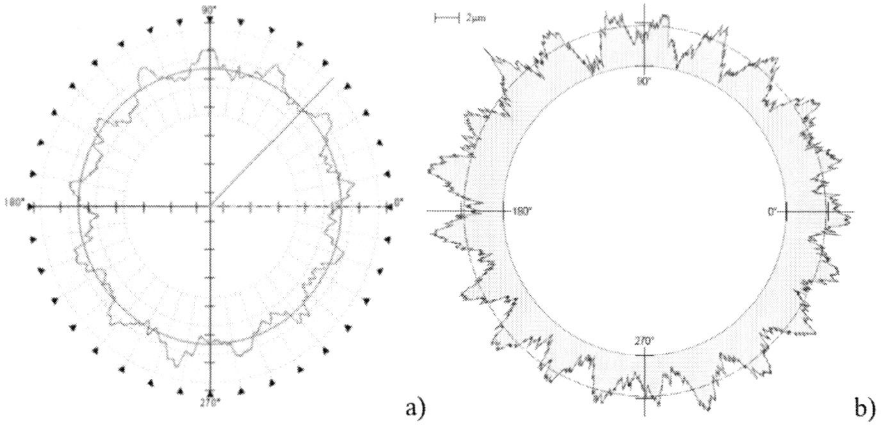

Fig. 1. a) 13-angularity chart, b) 21-angularity chart [7]

The finite element method was used for the above mentioned analyses. Model shafts characterized by saddle-shapedness and ovalness, triangularity and quadriangularity were analysed. Cross section shape deviations were made in four tolerance grades IT2, IT4, IT6 and IT9.

2 Computational model creation

Figure 2 shows the stages of computational model creation. Simplifications were used, based on symmetry in the vertical axis and by using a small surface contact area between the shaft and the sleeve. As a result, the issue was reduced to the flat problem 1 mm thick (this value of the contact between the outer shaft surface and the inner sleeve surface was assumed). The entire process of creating the computational model is analogous to the one presented in the works [3-7]. Figure 3 shows the types of studied shape deviations (diameter ø19 mm), including ovalness, triangularity and quadriangularity in four tolerance grades: IT2 (2.5 μm), IT4 (6 μm), IT6 (13 μm) and IT9 (52 μm). The shaft length was assumed as 70 mm. In the FEM model (Abaqus software), it was assumed that a fixed shaft rotates with the specified rotational speed (1 RPS, 10 RPS and 100 RPS) in a deformable sleeve. Figure 4 shows the FEM model. Material properties: Young modulus E = 210 GPa, Poisson's ratio $v = 0.3$, density equal to 7830 kg/m^3. Considering the simplification of the model, it was loaded in such a way as if the entire shaft with the assumed geometric parameters (without additional load) rotated in the sleeve. The first stage of the work with the model was to determine the minimum number of finite elements on the external surface of the sleeve. During calculations, this surface was fixed and from this place the values of reaction forces were recorded. The stages of minimising the number of finite elements are presented in Figure 4.

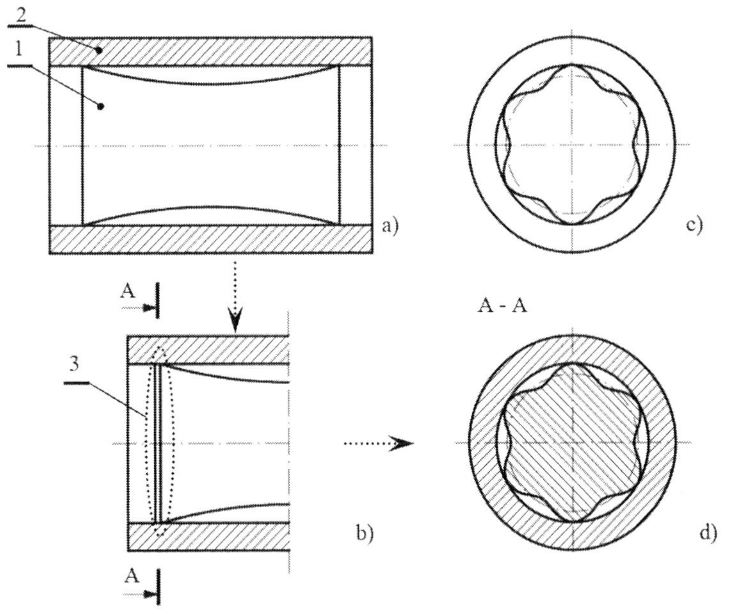

Fig. 2. Stages of computational model creation: a) real model: 1 – saddle shaft, 2 – nominally round hub, b) simplification using symmetry relative to the vertical axis: 3 – shaft-hub contact area, c) simplification to the flat problem thanks to a short length of the shaft where shaft-hub contact takes place [7]

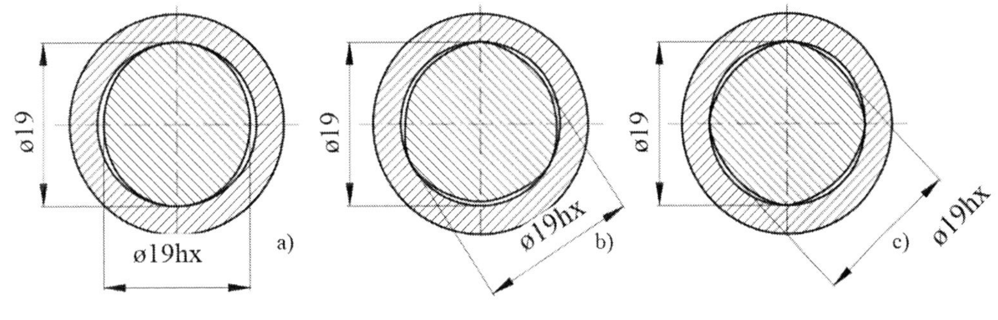

Fig. 3. Studied shape deviations: a) ovalness, b) triangularity, c) quadriangularity, x – tolerance grade, grades IT2, IT4, IT6, IT9 were analysed

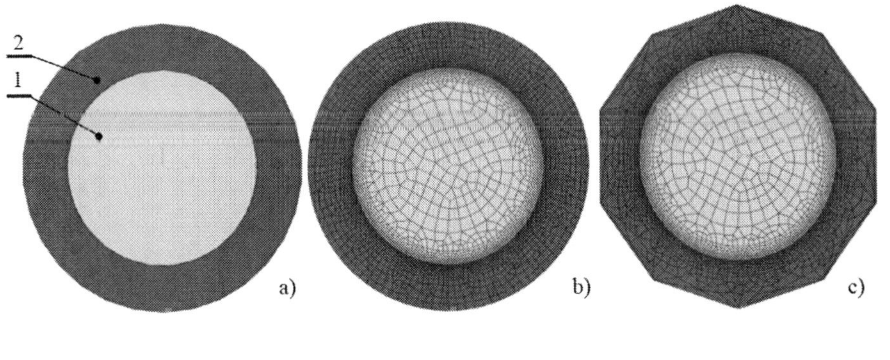

Fig. 4. FEM model: a) flat model: 1 – shaft, 2- sleeve, b) discretized model with 100 finite elements on the outer surface of the sleeve, c) discretized model with 10 finite elements on the outer surface of the sleeve

Figure 5 shows charts with the values of reaction force in the sleeve fixed support for 100 elements on the outer surface of the sleeve (fig. 5a) and with 10 elements on the outer surface of the sleeve (fig. 5b). Calculations were made for the same gravity force, these values were compared with gravity force computed analytically. It can be clearly seen that a decrease in the number of finite elements on the outer surface does not adversely affect the accuracy of the calculations. This is the basis for the simplification of the model.

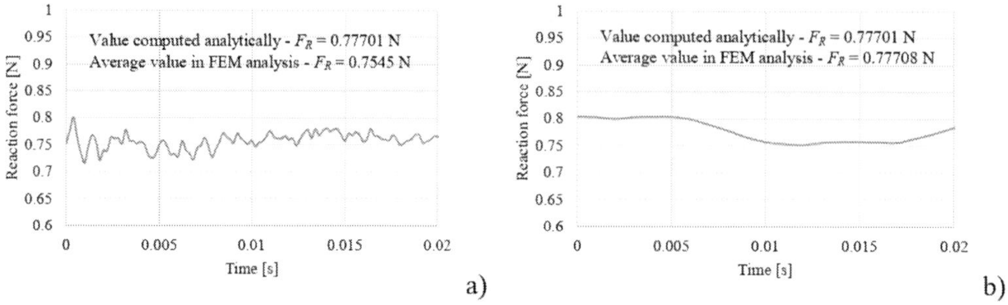

Fig. 5. Reaction force F_R: a) from the analysis performed on the model with 100 elements on the outer surface of the sleeve, b) from the analysis performed on the model with 10 elements on the outer surface of the sleeve

3 Simulation results

Figure 6 shows sample stress distributions in the sleeve in selected angular positions of the shaft (oval) relative to the sleeve. Figure 7 and 8 present the same results for a triangular and quadriangular shaft. Stress distributions are only for reference, the purpose of the charts is to indicate contact zones of the shafts with shape deviations, which is why tables with stress values are not included. Figures 9-11 present the change in the value of reaction force F_R in fixture for the oval shaft made in the 9th tolerance grade for three speeds (one rotation). A marked increase in the value of reaction force is visible. Figures 12-15 present the change in the value of reaction force F_R for the triangular shaft made in four difference tolerance grades (rotational speed 1 RPS). Figures 16 and 17 present distributions of reaction force F_R for the triangular shaft (IT9) for rotational speed 10 RPS and 100 RPS. Figures 18-21 present the change in the value of reaction force F_R for the quadriangular shaft made in four difference tolerance grades (rotational speed 1 RPS). Figures 22 and 23 present distributions of reaction force F_R for the quadriangular shaft (IT9) for rotational speed 10 RPS and 100 RPS.

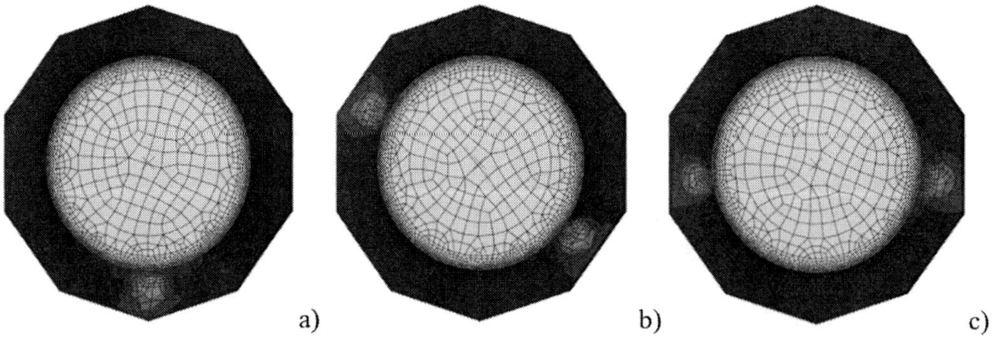

Fig. 6. Sample stress distributions for an oval shaft: a) position 0°, b) position 22.5°, c) position 45°

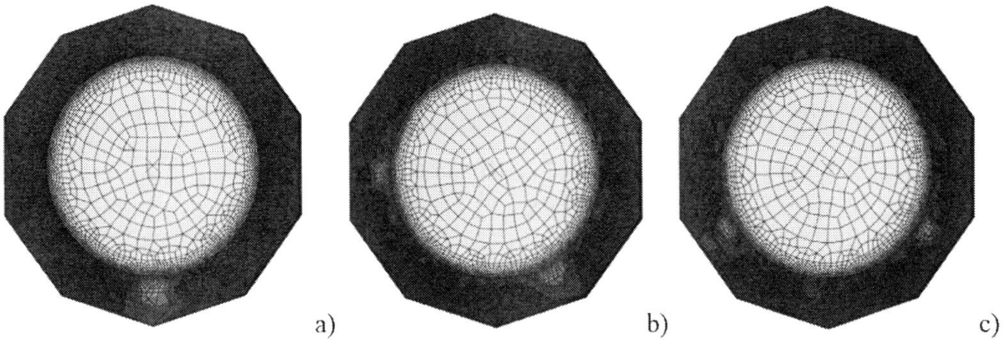

Fig. 7. Sample stress distributions for a triangular shaft: a) position 0°, b) position 30°, c) position 60°

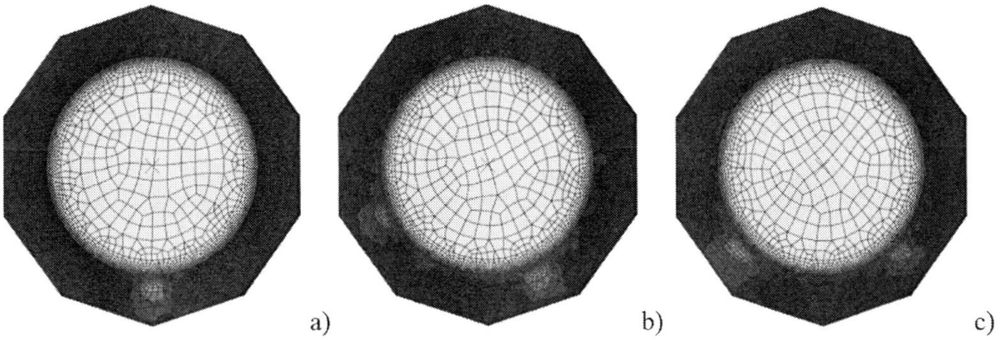

Fig. 8. Sample stress distributions for a quadriangular shaft: a) position 0°, b) position 22.5°, c) position 45°

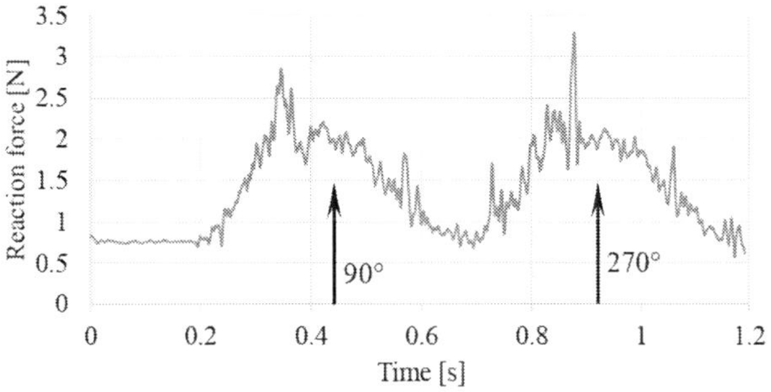

Fig. 9. Reaction force in the sleeve fixture: oval, IT9, 1 RPS

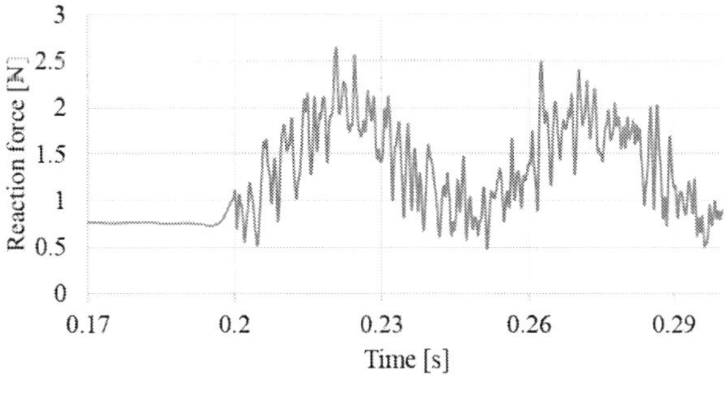

Fig. 10. Reaction force in the sleeve fixture: oval, IT9, 10 RPS

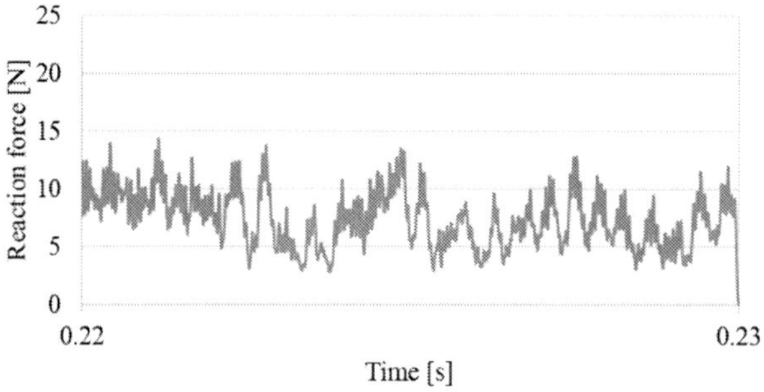

Fig. 11. Reaction force in the sleeve fixture: oval, IT9, 100 RPS

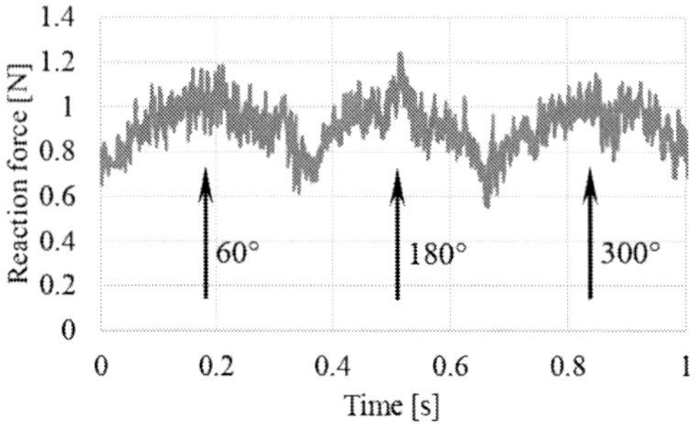

Fig. 12. Reaction force in the sleeve fixture: triangularity, IT2, 1 RPS

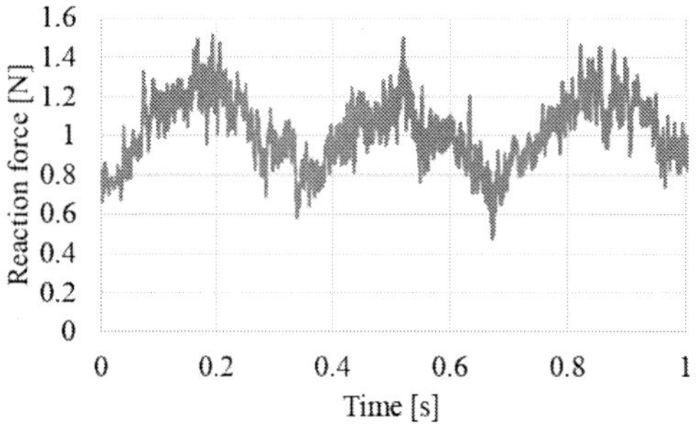

Fig. 13. Reaction force in the sleeve fixture: triangularity, IT4, 1 RPS

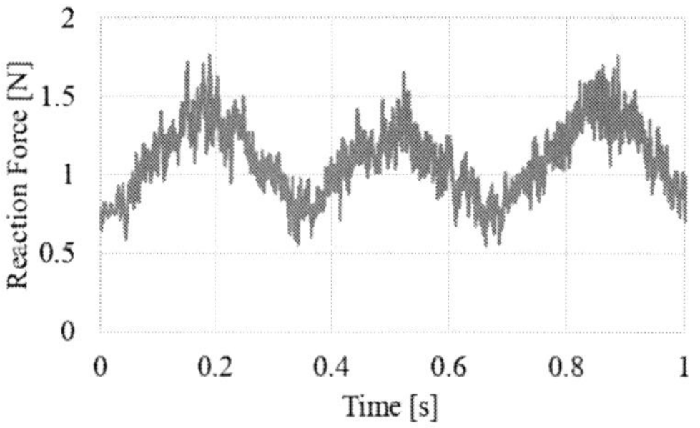

Fig. 14. Reaction force in the sleeve fixture: triangularity, IT6, 1 RPS

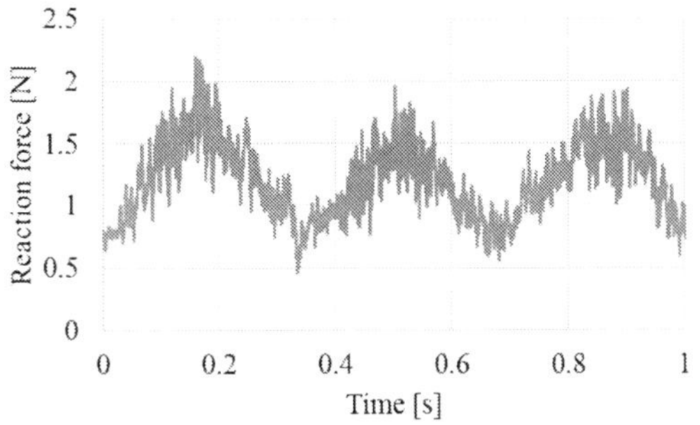

Fig. 15. Reaction force in the sleeve fixture: triangularity, IT9, 1 RPS

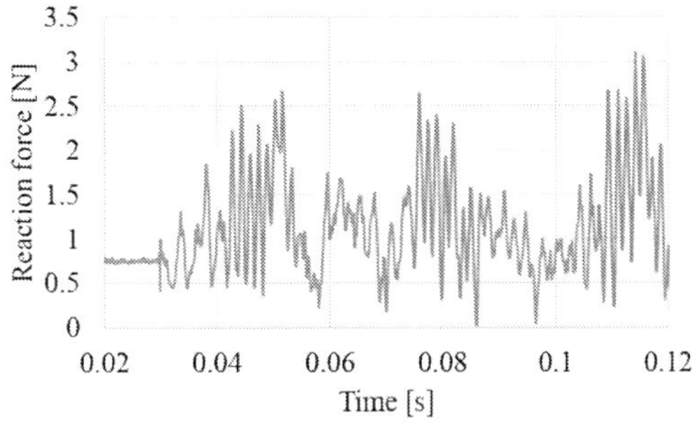

Fig. 16. Reaction force in the sleeve fixture: triangularity, IT9, 10 RPS

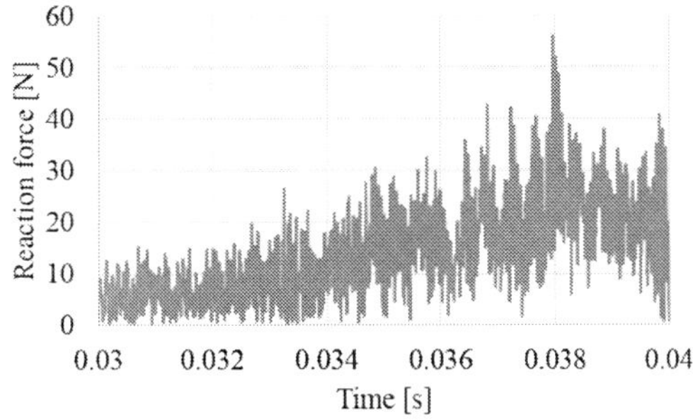

Fig. 17. Reaction force in the sleeve fixture: triangularity, IT9, 100 RPMS

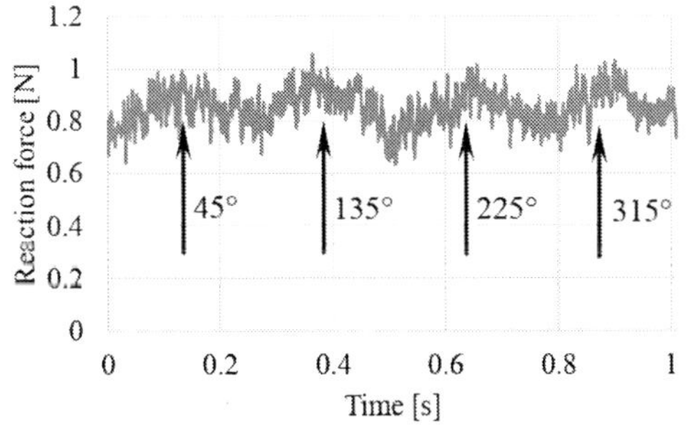

Fig. 18. Reaction force in the sleeve fixture: quadriangularity, IT2, 1 RPS

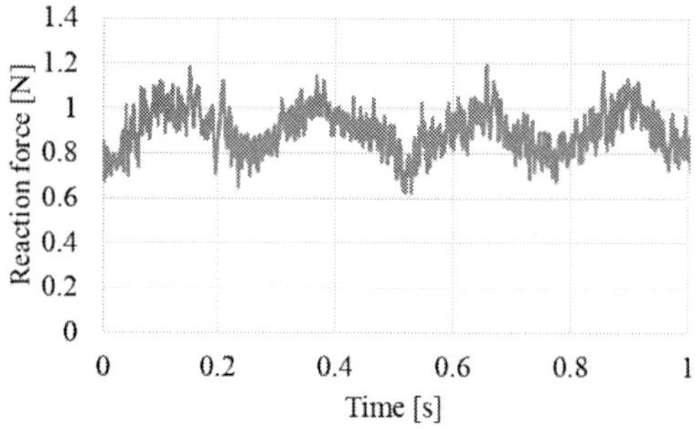

Fig. 19. Reaction force in the sleeve fixture: quadriangularity, IT4, 1 RPS

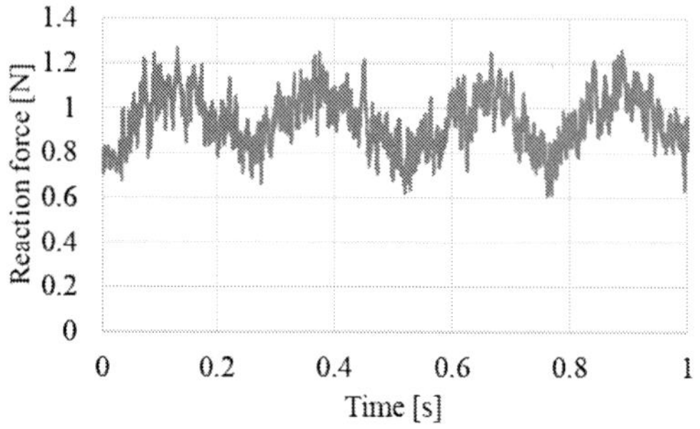

Fig. 20. Reaction force in the sleeve fixture: quadriangularity, IT6, 1 RPS

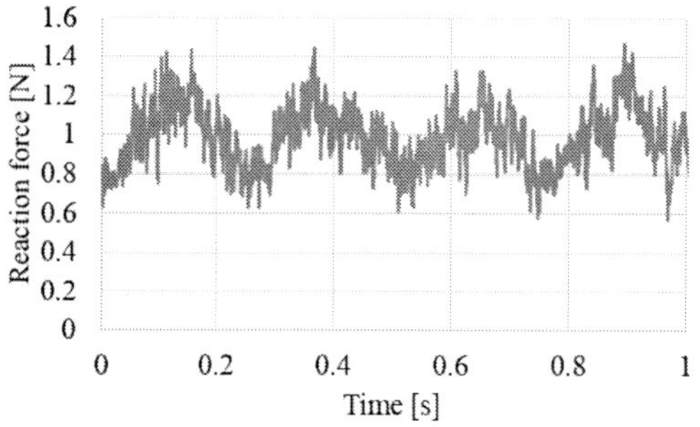

Fig. 21. Reaction force in the sleeve fixture: quadriangularity, IT9, 1 RPS

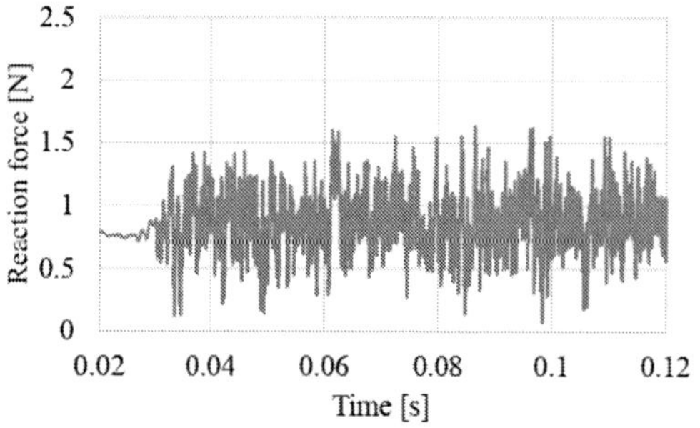

Fig. 22. Reaction force in the sleeve fixture: quadriangularity, IT9, 10 RPS

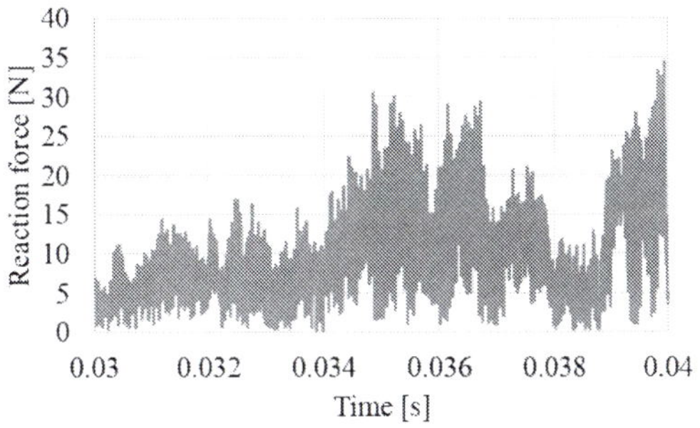

Fig. 23. Reaction force in the sleeve fixture: quadriangularity, IT9, 100 RPS

Conclusions

The method for determining the course and changes in the value of reaction forces F_R in the sleeve fixture of the axisymmetric connection with shape deviations presented in this work is the approach to the problem proposed by the authors. Analysing the course of reaction forces F_R it can be easily stated that both shape deviations and tolerance grade of the connection components have a significant impact on the value of the presented forces.

It is important that the actual cross-sectional shape of the shaft affects the nature of the changes in the value of the reaction force F_R. The identification of these forces and their analysis can contribute to better knowledge of the phenomenon of vibrations of machine systems equipped with the axisymmetric connection. The increasing rotational speed of the shaft relative to the sleeve (especially for inferior tolerance grades) unambiguously leads to a significant increase in the value of the reaction force F_R. It is quite an obvious conclusion but at the same time it confirms high usability of the presented method. The next stage of this research should be to check the impact of additional connection load on the characterization of cooperation parameters.

References

1. S. Adamczak, *Geometrical measurements of surface – shape, waviness and roughness.* (WNT, Warsaw, 2008)
2. S. Adamczak, W. Makieła, K. Stępień, *Investigating advantages and disadvantages of the analysis of a geometrical surface structure with the use of Fourier and wavelet transform*. Metrology and Measurement Systems **XVII**, 233-244 (2010)
3. M. Dudziak, G. Domek, A. Kołodziej, *Variation of static parameters of cooperation in axisymmetric connection*. Procedia Engineering **136**, 56-62 (2016)
4. M. Dudziak, G. Domek, A. Kołodziej, K. Talaśka, *Contact Problems Between the Hub and the Shaft with a Three-angular Shape of Cross-section for Different Angular Positions*. Procedia Engineering **96**, 50–58 (2014)
5. M. Dudziak, A. Kabała, A. Kołodziej, *The Topic of Contact Zone Problems in the Shaft and Hole Joint, Taking into Account Component Form Errors.* Journal of Mechanics Engineering and Automation **3**, 586-590 (2013)
6. M. Dudziak, G. Domek, A. Kołodziej, K. Talaśka, *Contact Problems Between the Hub and the Shaft with a Four-Angular Shape of Cross-Section for Different Angular Positions.* Applied Mechanics and Mechatronics II, Applied Mechanics and Materials, 54-62 (2015)
7. M. Dudziak, G. Domek, A. Kołodziej, K. Talaśka, *Multi-angularity - identification of*

parameters and compatibility conditions of the axisymmetric connection with form deviations. Procedia Engineering **177**, 431–438 (2017)

Lap weld joint modelling and simulation of welding in programme SYSWELD

Radoslav Koňár[1,*], *Miloš* Mičian[1], *Andrej* Zrak[1]

[1]University of Žilina in Žilina, Faculty of Mechanical Engineering, Univerzitná 1, 010 26 Žilina, Slovakia

Abstract. Simulations of the welding process for applications of practice using SYSWELD are presented. This paper presents simulation of welding in the repair of high-pressure gas pipeline with steel sleeve with composite filling. Material of experimental sample was steel S355. The simulations in SYSWELD divided in to two parts: the thermal simulation followed by the mechanical simulation. The results of the numerical model, which are listed in article are compared to real experiments.

Keywords: Welding simulation, SYSWELD, lap joint, steel S355

1 Introduction

Simulation tools will continue to gain importance for both scientific investigations and industrial applications. This further applies to welding technology. The present work focuses on the simulation of temperature fields, structural changes, residual stresses, distortions and hardness due to the welding of the circumferential overlap weld joint at the repair of the gas pipeline by composite sleeve.

2 Permanent repair of defects at gas pipelines with using steel sleeve

External corrosion is a major concern for pipeline operators. When areas of corrosion or other damage on operating pipelines are identified, there are significant economic and environmental incentives for performing repair without removing the pipeline from service. There are a variety of repair strategies available to pipeline operators for a given repair situation. One way is to repair by steel sleeve with composite filling[1-4].

The steel repair sleeves can be used for permanent repairing of high pressure gas pipeline defects without interrupting. With using these repair methods, we can repair defects, such as internal and external corrosion, gouges, dents, grooves, arc burns, cracks, defective girth welds, laminations and leaks [1, 2, 3].

The steel sleeve is composed of segmented steel casing, fitted on two steel distance rings, which defines the space between the sleeve and the repaired pipe (Fig. 1.). This space

* Corresponding author: radoslav.konar@fstroj.uniza.sk

Reviewers: *Josef* Soukup, *Tomasz* Domański

is filled with glass beads and epoxy (composite). When epoxide cured, provide a perfect transmission of stresses from pipeline to sleeve, while an equal stress distribution in the pipeline and sleeve [4, 5].

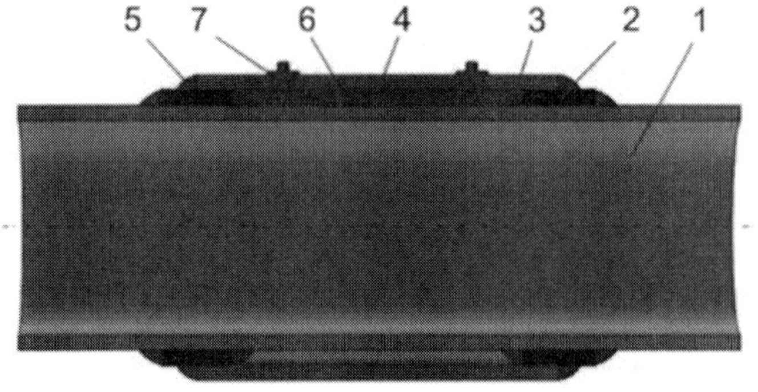

Fig. 1. Section of the gas pipeline at repair of gas pipelines with using steel sleeves with composite filling (1 – repairing gas pipelines, 2 – distance rings, 3 – circumferential fillet weld between distance ring and gas pipeline, 4 –steel segmented sleeve, 5 - circumferential fillet weld between distance ring and steel sleeve, 6 – composite filling, 7 – inspection holes) [4]

3 Architecture of simulation software Sysweld

Sysweld integrates effects linked to metallurgical transformations in the thermal, mechanical and hydrogen diffusion analysis process. Simulation is broken down into a number of successive steps due to the modular aspect of the product. Fig. 2. illustrates interaction between the various modules, with the two principle modules shown shaded. Two supplementary modules are included in the figure [1, 5, 6].

Simulation is consequently conducted in a number of successive steps (with the results of one step forming input data for the following step):

- thermal and metallurgical computation (determination of thermal cycles and metallurgical phase proportions according to space and time),
- mechanical computation (stresses and residual strains),
- computation of hydrogen diffusion, integrating the effect of temperature, stresses and traps (reversible or inreversible).

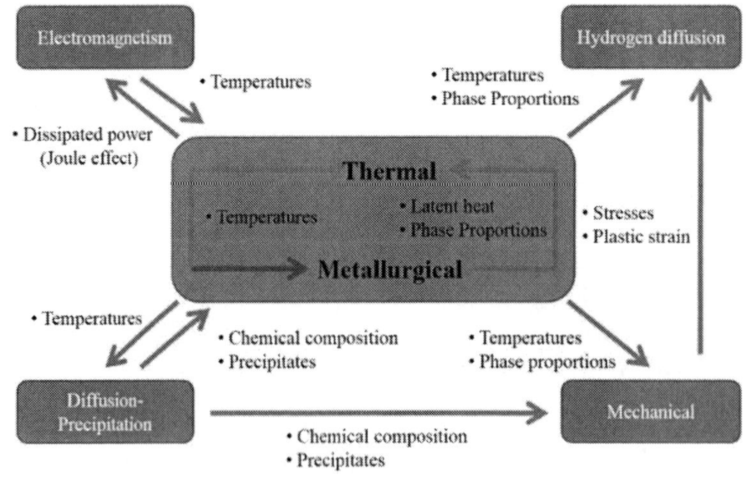

Fig. 2. General architecture of SYSWELD [4]

In principle, a finite element simulation of the welding process consists of two main parts: thermal analysis and mechanical stress analysis. In thermal analysis, the temperature field is determined as a function of time for each integration point. This temperature time-history is used as an input into the thermal stress analysis. Herein, the thermal solution can be sequentially or fully coupled with the mechanical solution of the structure. Because the rate of heat generation due to mechanical dissipation energy can be neglected in the heat transfer analysis, a sequentially coupled thermal-stress analysis is commonly applied for the simulation of a welding process in which a thermal analysis is followed by a stress analysis.

4 Experimental sample and analysis of boundary conditions

Experimental part of this article describes analysis of boundary conditions for the welding simulation at the repair of gas pipelines with steel sleeve and simulation of welding using this boundary condition for all welds.

Experimental sample (Fig. 3.) was composed of pipe and distance ring of materials L360NB in compliance with standard EN 10208-2 (WNr. 1.0582). In Tab. 1. is listed chemical composition of this material. Diameter of pipe is 323.9 mm, pipe thickness 10 mm and length 970 mm. Diameter of distance ring is 333.9 mm, ring thickness 10 mm and length 90 mm [4, 8, 10].

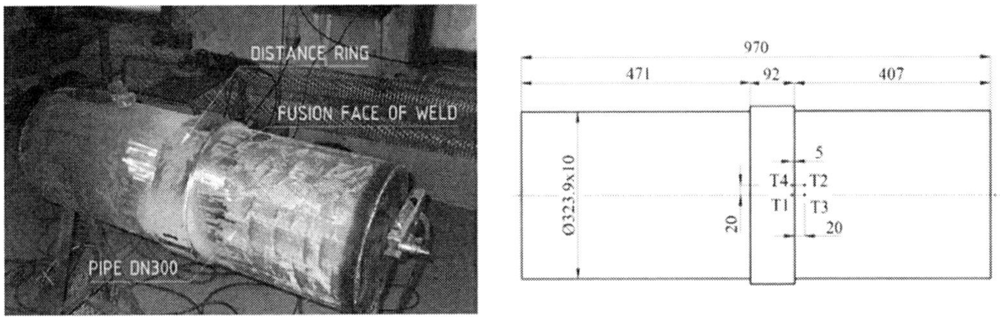

Fig. 3. Experimental sample, real sample (left), scheme with thermocouples location (right)

MMA process was used for weld joint. The filler material was coated electrode ESAB OK 48.00, which chemical composition is 0.06% C, 0.50% Si, 1.20% Mn. Experimental sample was welded by one all around weld, which contains 6 weld beads [4, 11].

Table 1. Real chemical composition of steel L360NB in weight %

C	Si	Mn	P	S	V	Nb	Ti
0.150	0.420	1.380	0.018	0.017	0.008	0.003	0.002
Al	**N**	**Cu**	**Ni**	**Cr**	**Mo**	**V+Nb+Ti**	
0.025	0.008	0.060	0.030	0.070	0.010	0.112	

During welding were measured welding parameters, welding time and thermal cycles in four points. After welding the weld was analysed. Complete analysis of weld for simulation in simulation programme SYSWELD contains:

- parameters of welding U_w, I_w and v_w,
- cross-sectional geometry of the welds (weld metal, heat affected zone),
- hardness of the weld,
- temperature cycles.

Two samples for macrostructural analysis (Fig. 4.) were taken from finished weld. Macrostructure not presented defects of weld. Digitized weld macrostructures (Fig. 4.) were created from cross-sectional parameters of welds in graphic programme AutoCAD.

Parameters are necessary for the definition geometrical parameters of Goldak heat source model penetration and width of weld [4, 5, 12].

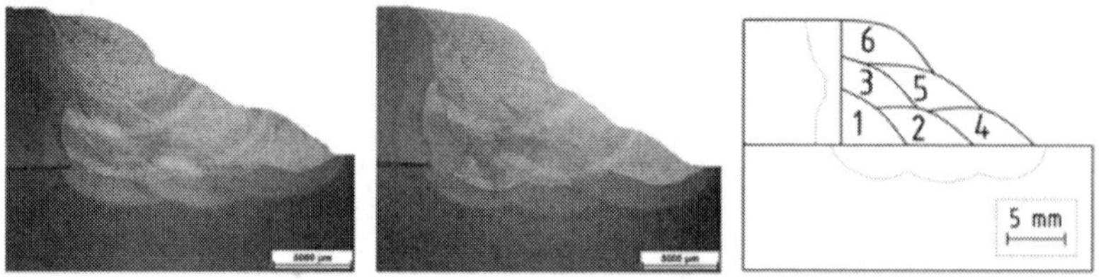

Fig. 4. Macrostructural analysis (left) with final digitized macrostructure (right)

Hardness of weld joint was measured by Vickers. The Vickers hardness profile of the weld was measured in solitary point in compliance with standard STN EN ISO 6507 (Fig. 5.). Hardness of all measured location of weld is listed in the table of the Fig. 5. Hardness of weld joint is about 230 HV, hardness of HAZ is about 200 HV and hardness of the base material is about 185 HV [4, 9, 12].

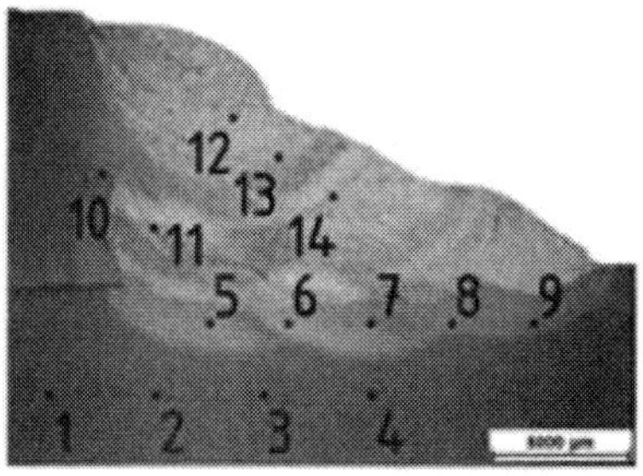

Fig. 5. Hardness of weld joint by Vickers

Table 2. Hardness of measured locations of weld

Hardness by Vickers [HV10]													
Base material				**Heat affected zone**						**Weld metal**			
1	**2**	**3**	**4**	**5**	**6**	**7**	**8**	**9**	**10**	**11**	**12**	**13**	**14**
188	184	184	185	201	204	197	197	196	189	247	232	230	232

5 FEM simulation of welding in Sysweld

For simulation process in programme SYSWELD was used 2D rotational model. Axis of rotation was y-axis. The model and experimental sample had the same geometrical dimensions - distance ring, welds. Length of experimental model for simulation was reduced to 300 mm in order to reduce computational time. Programme VisualMesh is designed for meshing 2D/3D geometrical models. In this programme was created preparation and distribution of geometrical model to FEM mesh. Meshed FEM model is shown on the Fig. 6. Distribution model has 1966 finite elements and 2885 nodes. The smallest element used in the FEM model is in the area of the weld with the dimension of 1.0 mm × 1.0 mm × 1.5 mm [4, 5, 7].

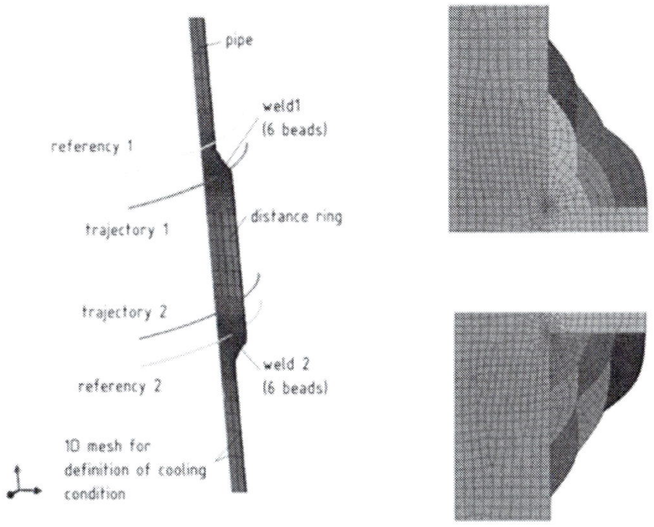

Fig. 6. 2D rotational FEM meshed model fol simulation

Material of experimental sample was steel L360NB. We could use database S355J2G3, because both of steels have the same mechanical and physical properties. Parameters of Goldak model and welding speed are in Tab. 3.

Table 3. Parameters of Goldak model

Bead	a[mm]	b [mm]	c_1 [mm]	c_2 [mm]	Q [W]	$_{sw}$ [mm.s^{-1}]
1	3.2	4.0	2.0	4.0	800	1.6
2	6.5	2.0	2.0	4.0	1250	1.6
3	4	3.0	2.0	4.0	2700	1.6
4	12	2.0	2.0	4.0	1300	1.6
5	10	3.0	2.0	4.0	1850	1.6
6	1.5	9.0	2.0	4.0	2400	1.6

S355J2G3 CCT diagram is illustrated on Fig. 7. Red curve (Fig. 7.) shown cooling rate used for simulation. Cooling rate was set to q=30 W.m^{-1}.

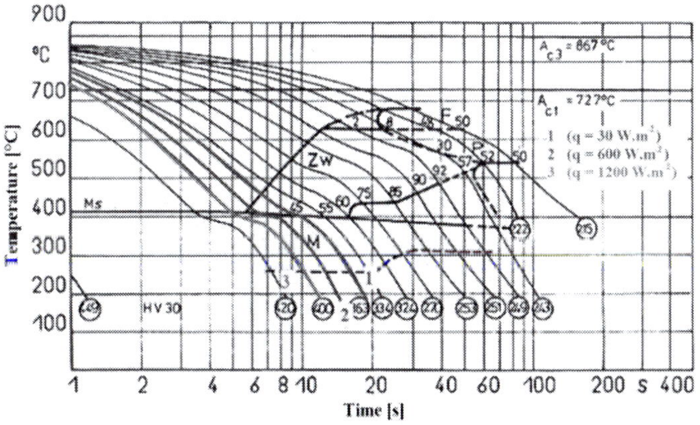

Fig. 7. CCT diagram for steel S355J2G3

Temperature fields, temperature cycles, residual stresses and hardness were simulated in the simulation for both of weld. Each weld was consists from 6 weld bead. Graphic results of temperature cycles of the simulation for finished weld are on the Fig. 8 [4, 9].

https://doi.org/10.1051/matecconf/201815702018

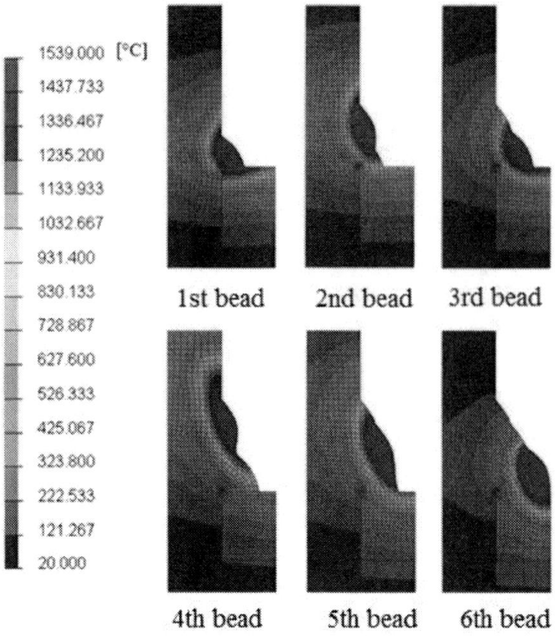

Fig. 8. Simulated temperature fields of weld

Residual stresses were only simulated, not measured. Longitudinal, transversal and reduced stresses by Von Mises were computed. The reduced stresses by Von Mises are illustrated on Fig. 9 [4, 8].

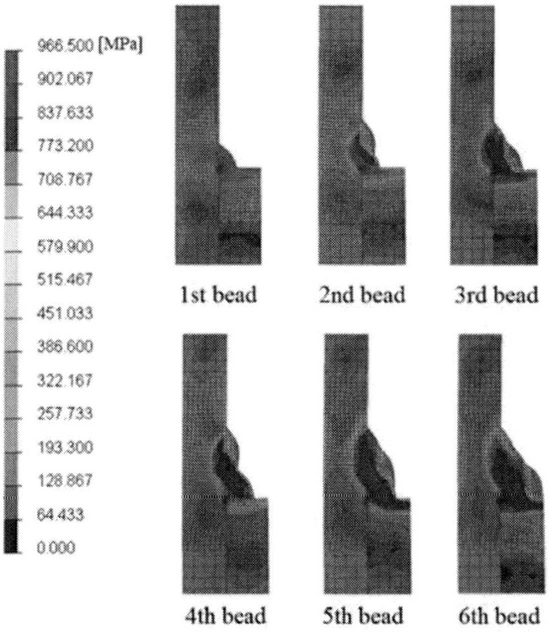

Fig. 9. Reduced simulated stresses by Von Misses theory

The model was rotational about the axis y and symmetrical about the centre of distance ring (plane XZ), therefore the results of residual stresses for weld 1 and weld 2 are same. Calculation of hardness by Vickers (Fig. 10.) depends on the proportion of structural phase. Hardness was computed for the finished weld joint, which contained two welds [4].

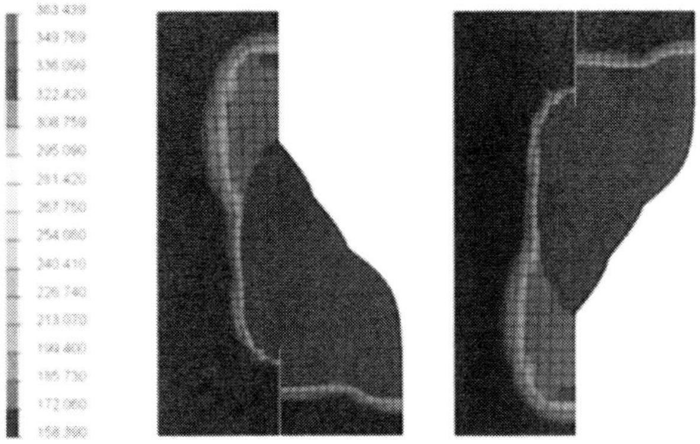

Fig. 10. Simulated hardness by Vickers [4]

Conclusion

Information about repair of gas pipeline with steel repair sleeves and welding simulation in programme SYSWELD are included in the theoretical part of this article. Experimental part describes analysis of boundary conditions and simulation of welding at the repair of gas pipelines with steel sleeves. From this information was simulated process of welding for both 6 beads welds. Results of simulation process provide information about temperature fields, temperature cycles, residual stresses and hardness of weld joint. Values of residual stress have not been verified by measurement, therefore they are only approximate.

Welding simulation has become a strong tool in technological praxis. It helps efficiently to solve complex problems in welding in a relatively short time. This article is intended as an example of welding numerical simulation capabilities in the programme SYSWELD.

This work has been supported by the Scientific Grant Agency of the Ministry of Education of the Slovak Republic, grant KEGA: 034ZU-4/2015.

References

1. A. Joseph, et al., *Evaluation of residual stresses in dissimilar weld joints*. International Journal of Pressure Vessels and Piping **82**, 700-705 (2005)
2. G. Totten, M. Howes, T. Inoue, *Handbook of Residual Stress and Deformation of Steel*. (AMS International, USA, 2002)
3. J. Moravec, *Influence of Welding Parameters on Weld Pool's Geometry in Shielding Gas Welding*. (Pollypress, Liberec, 2011)
4. R. Koňár, et al., *Finite Element Modeling and Numerical Simulation of Welding at the Repair of Gas Pipelines with Steel Sleeve*. Manufacturing technology: journal for science, research and production, **16**, 360-365 (2013)
5. R. Koňár, M. Patek, *Numerical simulation of dissimilar weld joint in SYSWELD simulation software*. Technical gazette **24**, 137-142 (2017)
6. P. Novák, J. Meško, M. Žmindák, *Finite element implementation of multi-pass fillet weld with phase changes*. Manufacturing technology : journal for science, research and production, **13**, 79-85 (2013)
7. J. Winczek, et al., *Analysis of t hermomechanical states in single-pass GMAW surfaced steel element*. AIP Conference Proceedings **1822**, 1-13 (2017)

8. M. Brezničan, P. Fabian, J. Meško, M. Drbúl, *The simulation of influence of quenching temperature on properties of bearing rings*. Manufacturing technology: journal for science, research and production **13**, 20-25 (2013)
9. J. Winczek, T. Skrzypczak, *Thermomechanical states in arc weld surfaced steel elements with heat of melted electrode material taken into account*. Archives of Metallurgy and Materials **61**, 1277-1288 (2016)
10. E. Krivoš, R. Pastirčák, R. Madaj, *Effect of technological parameters on the quality and dimensional accuracy of castings manufactured by patternless process technology*. Archives of metallurgy and materials **59**, 1069-1072 (2014)
11. P. Jankejech, P. Fabian, J. Brončck, Y. Shalapko, *Influence of tempering on mechanical properties of induction bents below 540°C*. Acta Mechanica et Automatica **10**, 81-86 (2016)
12. J. Winczek, *The analysis of stress states in steel rods surfaced by welding*. Archives of Metallurgy and Materials **58**, 1243-1252 (2013)

Numerical modal analysis for vibration-damping properties of ductile cast

Petra Kováčiková[1,*], *Ján* Vavro[1], *Ján* Vavro Jr.[1], *Andrej* Dubec[1]

[1]Department of Numerical Methods and Computational Modeling, Faculty of Industrial Technologies, Alexander Dubcek University of Trencin I. Krasku 491/30, 020 01 Púchov, Slovak Republic

Abstract. This paper deals with standard test method for the measurement of vibration-damping properties of materials, ductile cast iron - graphitic cast iron with the spherical shape of graphite (ductile cast iron). Investigation is focused on its mechanical properties, chemical composition and microstructure, which are necessary as input parameters for computational modeling and numerical analysis. The vibration or oscillating of experimental samples are closely connected with such parameters as natural shapes (eigenshapes) and natural frequencies (eigenfrequencies). Preparation of a geometric model of a homogeneous beam sample and numerical analysis was performed by finite element method in the ADINA software environment. This paper can be used for further development and growth from the aspect of application of computational tools in the area of damping and eigenfrequency properties for construction materials.

Keywords: cast iron, graphite, eigenshapes, eigenfrequencies, ADINA, finite elements methods

1 Introduction

Nowadays, increasing demands are being made on machine constructions or parts and equipment in operation. There is the increase of the performance and speed of operating construction and parts during operation, the result of which is formation of undesirable vibrations [1]. This undesirable features and critical machine vibration violation negatively impacts on their life time, reliability, and they may cause a machine damage and accident. Absorption of material vibration can be considered as its diminishing ability. Factors that affect the material damping include viscoelasticity, thermoelasticity, acoustic effects, or magnetism. Other factors include the amount of mechanical stress, load frequency, or temperature. This paper focuses on the sample of cast iron with a globular graphite shape in terms of its microstructure, chemical composition and mechanical properties. Eigenshapes and eigenfrequencies are closely related to the internal damping of the material and therefore the numerical part is focused on the finite element modal analysis [1, 2].

* Corresponding author: petra.kovacikova@fpt.tnuni.sk

Reviewers: *Vladimír* Dekýš, *Radim* Halama

2 Material characteristics

The experimental material was a cast iron with regular grained graphite, the graphite particle size was from 15 to 30 µm (Figure 1). From a microstructural point of view, the matrix consists of ferrite and lamellar perlite containing more than 2%. The dispersion of the perlite was from 0.3 to 0.8 µm (Figure 2). The microstructure was evaluated according to STN (Slovak Technical Standards) 42 0461 using the Vega 3 Tescan thermoemission electron microscope.

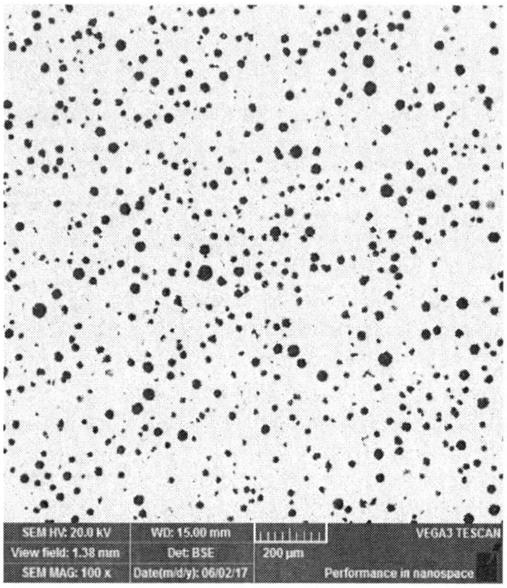

Fig. 1. Regular grained shape of ductile iron graphite

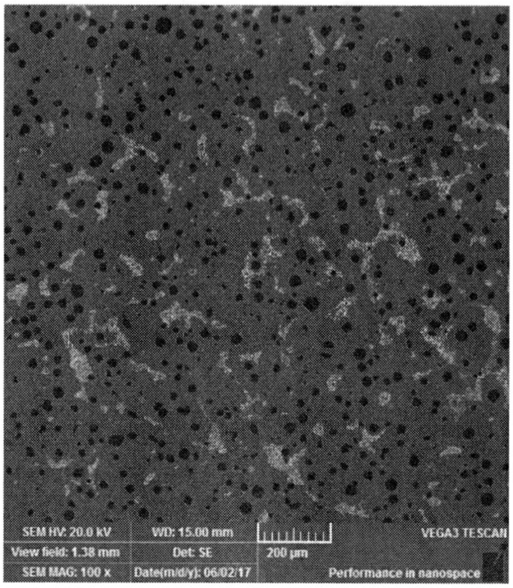

Fig. 2. Ferritic-pearlitic matrix of ductile cast iron

The amount of globular graphite in the structure is referred to nodularity, where this type of graphite can be obtained by modification. The bulb shape of the graphite does not cause a stress concentration in comparison with lamellar type of graphite (Figs 3, 4) [3-5].

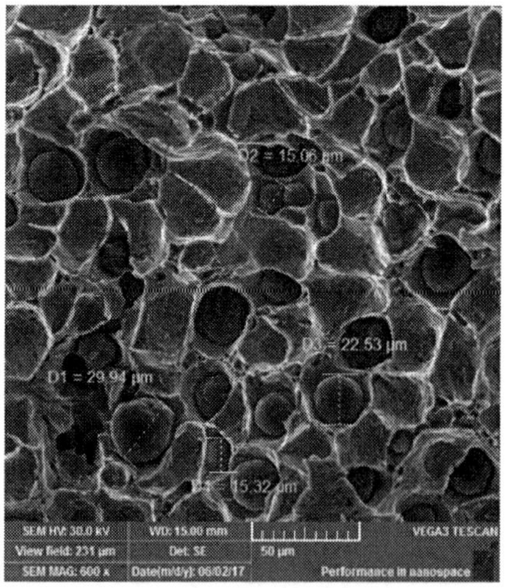

Fig. 3. Cavity micromorphology of the casting surface of ductile cast iron

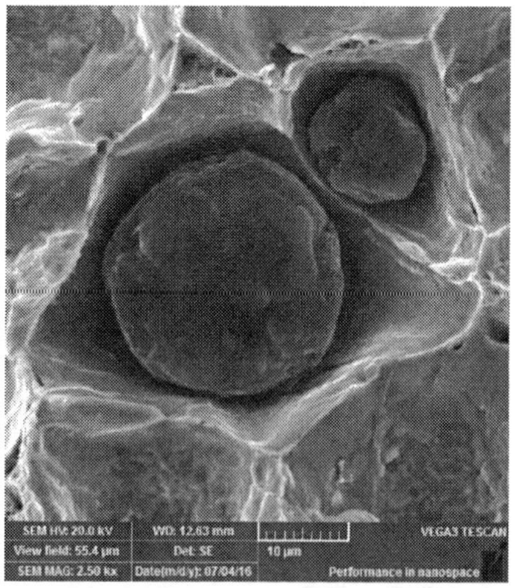

Fig. 4. Detail of globular graphite in the in the fracture area

The general specification properties for the investigated cast iron of this type can be seen in Table 1. The content of alloying elements of the ductile cast iron was measured from the whole volume of the sample by the EDX-7000 detector (Table 2).

Table 1. Mechanical properties

R_m (MPa)	$R_{p0,2}$ (MPa)	A (%)	HB	E (GPa)
380 - 900	250 - 600	2-25	120 - 350	140 - 185

Table 2. Content of alloying elements in wt. %

Si	Cu	P	Mn	Ca
2.758	0.504	0.379	0.373	0.122

3 Experimental modal analysis

Experimental modal analysis stands for the process of experimental determination of modal parameters (eigenshapes, eigenfrequencies) of linear systems. The way and the method of experimental measurement are standardised. The damping properties of the material can be used with mathematical models to design damping systems and they can be also applied for prediction of their performance. A comparison of damping capabilities is in the Figure 5.

Fig. 5. Comparison of steel and cast iron damping capabilities

The damping structure based on homogeneous layer representing the tested beam sample has good results, but in some applications the damping structure may consist of two or more layers with significantly different properties. The simplest frequency transmission requires the use of a two-channel frequency analyzer with a single channel for input or excitation measurement and a second channel for output or response measurement. The schematic image of the device for measuring of the damping properties is shown in Figure 6. As a part of the measuring device, the multi-analyzer PULSE 12 software system for vibro-acoustic measurements is used for evaluation of measurement results. When examining the dynamic effects of an excitation system, it is necessary to take into accout the linearity of the system and this fact means that the response of the system is directly proportional to the corresponding excitement [1, 2, 6, 7, 8].

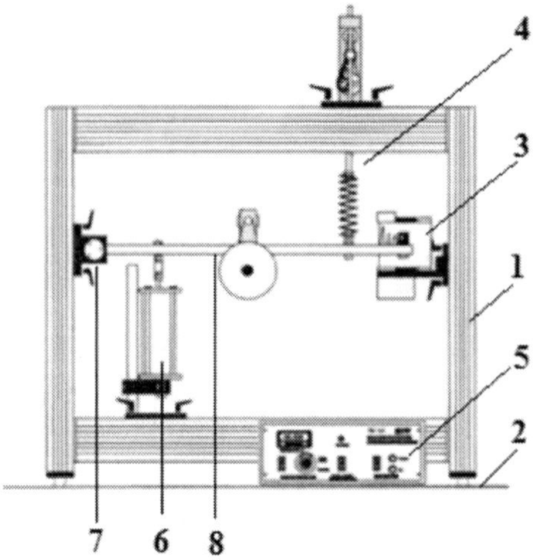

Fig. 6. Detailed description of the measuring device: 1-frame, 2- base board, 3-mechanical drum recorder, 4-suspension of free end from a helical spring, 5- control unit for excitation frequency, 6-regulation unit (attachment of springs, exciter and damper), 7- fixed mount, 8- tested beam sample

4 Results and discussions of numerical analysis

The eigenshapes (natural shapes) and eigenfrequencies (natural frequencies) represent an important factor in oscillation of structural materials. The model of sample had precisely defined dimensions and it was according to ASTM E756. Measurement of the homogeneous beam sample (Figure 7) was performed from the aspect of the strength and damping properties of the materials, while the input parameters of the investigated ductile cast iron were predefined and modal analysis (Figure 8) was performed using the finite element method in the ADINA software environment [1].

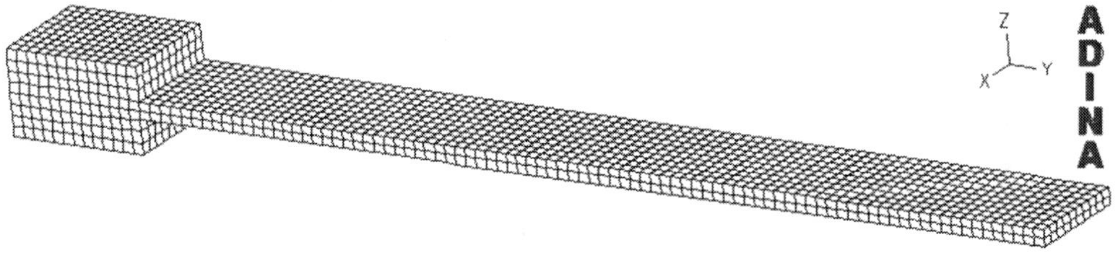

Fig. 7. Sample Model based on Finite Element Method

Table 3. Obtained frequencies for individual eigenshapes

Eigenshapes	Frequency (Hz)
1	7.76591E+01
2	4.86928E+02
3	5.11581E+02
4	1.30376E+03
5	1.36697E+03
6	2.69083E+03
7	3.04368E+03

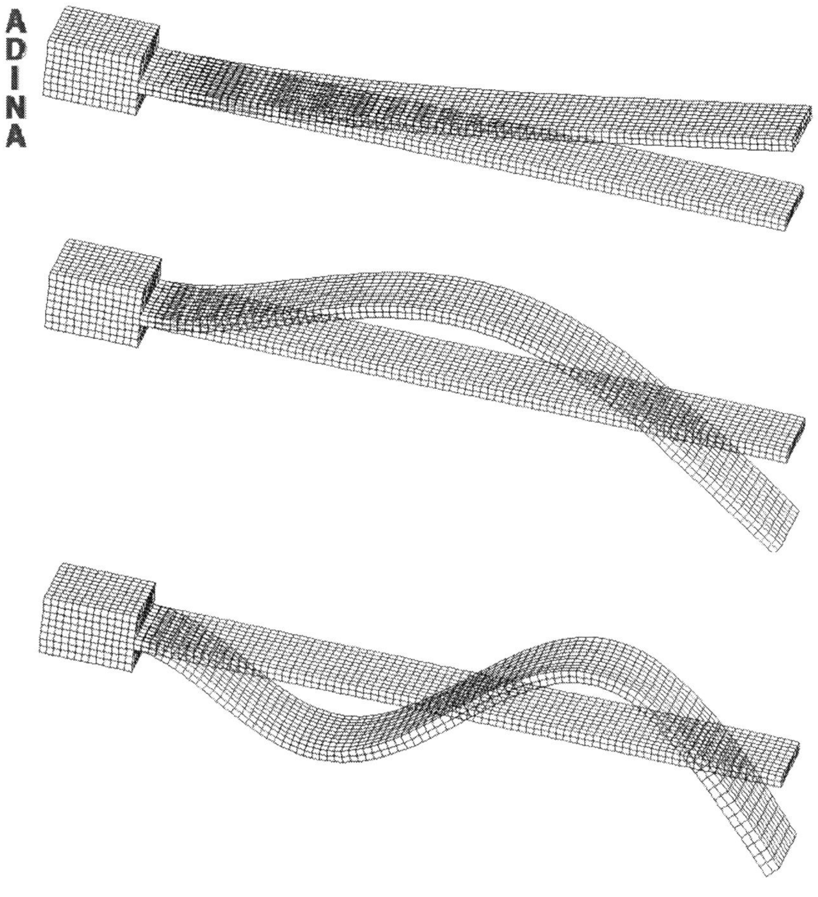

Fig. 8. Selected first eigenshapes of the homogeneous beam

Conclusion

In production of heavy duty machines and devices, which have to operate under the continuous cyclic loading, the dumping properties (including acoustics – noise and vibrations) have to be taken into account, because the increased vibrations have a negative effect on functionality of operating machines, which can be totally damaged or they can even cause any accident. Based on the verification of the results from numerical analysis and specified experimental measurement, it would be possible to determine the most suitable material to be used for high stressed parts, with are going to be made of the ductile cast iron. The mentioned facts make us look for ways to improve of limiting technical parameters and operating efficiency by selection of the right material. Modal analysis is a field of study that combines signal processing and computational interaction, mechanics theory, oscillation, acoustics, applied mathematics, and engineering prediction. Internal damping occurs in the structure of the material and can be caused by imperfection of this material. It is therefore necessary to have all important knowledge about the specific material before its usage in the operation.

This paper deal was supported by the Slovak Grant Agency KEGA 007TnUAD-4/2017, VEGA grant No. 1/0649/17, VEGA grant No. 1/0589/17 and resulted from the project "Centre for quality testing and diagnostics of materials", ITMS code 26210120046 relating to the Operational Program Research and Development funded from European Fund of Regional Development.

References

1. J. Vavro, *Numerická analýza napäťových stavov štruktúr grafitických liatin*. (Žilina, 76 p., 2013) ISBN 978-80-8075-614-7
2. A. Puškár, *Vnútorné tlmenie materiálov*. (Žilina, EDIS, 382 p., 1995) ISBN 80-7100-260-7
3. M. Handrik, L. Jakubovičová, P. Kopas, M. Sága, *Analysis of microplastic areas near graphite particles of nodular cast iron loading bellow yield stress*. Metalurgija (Metalurgy) **49** (2), 263-267, ISSN0543-5846 (2010)
4. STN 42 0461, *Hodnocení metalografické struktury litin*. (Praha, ÚNM, 16 p., 1977)
5. P. Skočovský, T. Podrábský, *Grafitické liatiny*. (Žilina, EDIS, 168 p., 2005) ISBN 80-8070-390-6
6. *Standart test method for measuring vibration-damping properties of materials*. (Code ASTM E 756-04, ASTM International, 2004)
7. D. J. Ewins, *Modal testing: Theory and Practice*. (Gilliard, Great Yarmouth, 269 p., 1988) ISBN 0-86380-017-3
8. A. Bilošová, *Experimentální modální analýza*. (VŠB - Technická univerzita Ostrava, Ostrava, 104 p., 2006)

Numerical modelling of caustics of solid state disk lasers

Marcin Kubiak*,[1]

[1]Czestochowa University of Technology, Institute of Mechanics and Machine design Foundations, Dabrowskiego 73, 42-200 Czestochowa, Poland

Abstract. This paper concerns mathematical modelling of caustics of solid state laser YAG type with the disk active medium. The heat source model is developed on the basis of interpolation algorithms using geostatistical kriging method. The real laser beam power distribution and caustics are taken into account in the model. Measurements of laser beam power distribution and diameter of the laser beam spot for different focusing are performed using UFF100 analyzer. Yb:YAG laser emitted by Trumpf laser head D70 is used in the experiment. Presented results include the comparison of experimentally determined laser power intensity distribution and caustic with results obtained by developed interpolation model.

Keywords: Yb:YAG laser, thermal load, heat source, Kriging algorithm, numerical modelling

1 Introduction

Changes in the focal length of focusing lens in the optical system of solid state laser changes not only the laser spot diameter but also the laser beam power intensity distribution. Beam divergence which is a measure for how fast the beam expands far from the beam waist is usually defined by the measurement of beam caustic – the beam radius at different positions. The measurement of laser beam power distribution is performed by using laser profilers [1-3].

In mathematical and numerical models of laser beam heat source power distribution only specific laser beam spot diameter is adopted, omitting the beam caustics. Generally, the distance from beam focusing position as a technological parameter influencing laser beam intensity distribution is usually neglected. Consequently, the laser beam intensity distribution models assumed in numerical analysis significantly differ from real Yb:YAG laser profile, obtained through experimental research [4, 5].

This paper presents interpolation algorithms allowing a precise description of Yb:YAG laser power intensity distribution and its caustic. Elaborated models take into account the real laser power distribution obtained in experimental research made using TruDisk 12002 laser. Kriging method [6-8] is used in this study in the form of point Kriging for the interpolation of laser beam heat source power distribution.

* Corresponding author: kubiak@imipkm.pcz.pl

Reviewers: *Piotr* Krawiec, *Tomáš* Lack

2 Experimental research

The light is pumped to the laser beam head by optical fibers with core diameter d_{LLK}. After the exit from the fiber, the beam is divergent by angle Θ_{LLK} to diameter d_0. The laser head is equipped with collimator lens having focal length f_c = 200 mm and focusing lens with focal length f_{og} = 400 mm. After focusing of laser light by focusing length by angle Θ_f the spot diameter (laser beam focus diameter) is equal to d_{og}. The following equations can be used to determine average diameter of the laser beam focus, Rayleigh length (Z_R) as well as the actual diameter $d(z)$ depending on beam focusing position (z).

$$\mathrm{d}_{og} = \frac{f}{f_c} d_{LLK} \tag{1}$$

$$\mathrm{Z}_{\mathrm{R}} = \frac{\pi (d_{og}/2)^2}{\lambda} \tag{2}$$

$$\mathrm{d}(z) = d_{og}\sqrt{1 + \left(\frac{z}{Z_R}\right)^2} \tag{3}$$

where λ is a wavelength of laser radiation.

The measurement of beam power distribution and caustics is performed using beam profiler Prometec UFF100 (Fig. 2). UFF100 is a beam profiler for high power lasers. It measures laser beam profile due to hollow needle. Laser beam is flowing through rotating needle and is transported to the detector. Detector signals are transformed into digital signals and processed later by computer software.

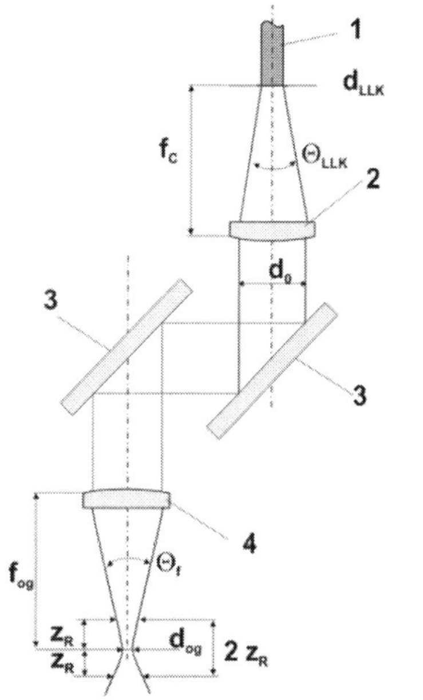

Fig. 1. Basic relations of laser beam transported through the fiber and focused by a laser head. 1 optical fiber, 2 collimator lens, 3 directional mirrors, 4 focusing lens

Fig. 2. Laser beam analyzer UFF100

The power density distribution of the laser beam in a plane perpendicular to the axis of beam propagation (assumed to be *z*-axis) for different measurement planes is analyzed. The analysis is performed at the continuous power 900W of the laser beam. The position of measurement plane changes with respect to the theoretical distance from the beam focusing position in the range of ± 10 mm. For each testing planes, the radius of the beam in two perpendicular axes (w_x and w_y) is defined from which an average radius of the laser beam spot is calculated (w) in accordance with standard PN-EN ISO 11146. Figure 3 presents experimentally obtained percentage distribution of Yb:YAG laser beam power for chosen beam focusing (z=0 and ± 5 mm). Figure 4 illustrates the beam caustic w_x, w_y and average radius w.

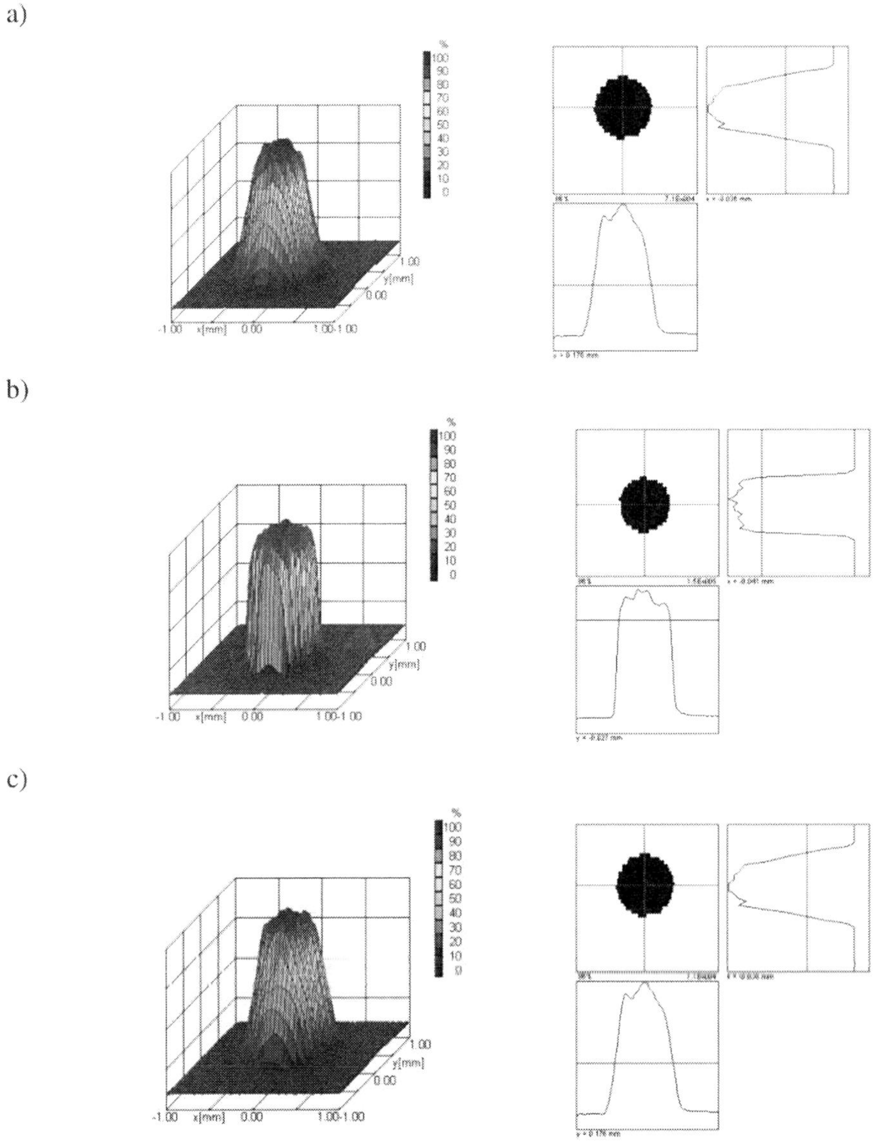

Fig. 3. Percentage laser beam power distribution at focusing a) z=5mm, b) z=0 and c) -5 mm

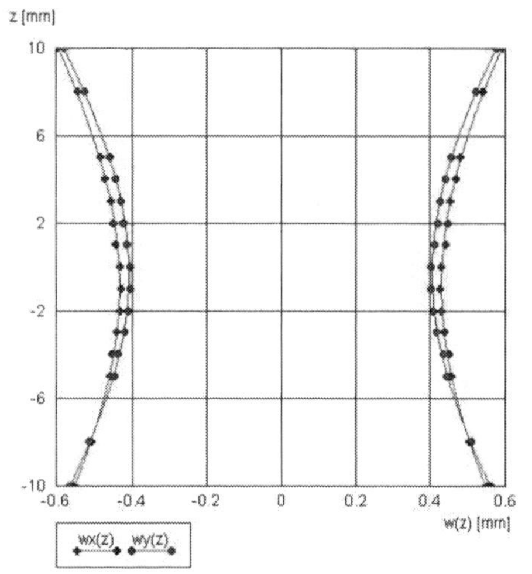

Fig. 4. Two dimensional picture of beam caustic: a) in perpendicular axes and b) average beam spot diameter

3 Interpolated laser beam power distribution based on Kriging algorithm

Kriging interpolation method of predicting the power distribution of the heat source at the point (x, y) is a linear combination of observations in basic points (the actual power distribution). The estimate is a function of the weighted average:

$$\tilde{f}(x,y)=\sum_{i=1}^{n}w_i f(x_i,y_i) \tag{4}$$

where w_i are weight coefficients assigned to particular observations, $f(x_i, y_i)$ is the real value of the function (variable) at the measured point, n is the number of sampling points that are considered in estimating of the variable within the circle of radius r_k from estimated point.

Coefficients w_i are calculated on the basic of Kriging system of equations, as follows:

$$\begin{bmatrix} 0 & \gamma(h_{12}) & \gamma(h_{13}) & \dots & \gamma(h_{1n}) & 1 \\ \gamma(h_{21}) & 0 & \gamma(h_{23}) & \dots & \gamma(h_{2n}) & 1 \\ \gamma(h_{31}) & \gamma(h_{32}) & 0 & \dots & \gamma(h_{3n}) & 1 \\ \dots & \dots & \dots & \dots & \dots & \dots \\ \gamma(h_{n1}) & \gamma(h_{n2}) & \gamma(h_{n3}) & \dots & 0 & 1 \\ 1 & 1 & 1 & \dots & 1 & 0 \end{bmatrix} \mathrm{X} \begin{bmatrix} w_1 \\ w_2 \\ w_3 \\ \dots \\ w_n \\ \lambda \end{bmatrix} = \begin{bmatrix} \gamma(d_1) \\ \gamma(d_2) \\ \gamma(d_3) \\ \dots \\ \gamma(d_n) \\ 1 \end{bmatrix} \tag{5}$$

where $\gamma(h_{ij})$ are the values of theoretical semivariogram at the distance h_{ij} between basis points in the observation set, $\gamma(d_i)$ are the values of theoretical semivariogram at the distance d_i between observed point and i-th basic point, λ is Lagrange multiplier.

Theoretical semivariogram is generally unknown, therefore values of function $\gamma(h_{ij})$ and $\gamma(d_i)$ are approximated by analytical functions. One of the main function used in the approximation is a linear function $\gamma(\mathrm{h}) = C_0 + Sh$, in which the semivariogram tends to sill

S at $h \to \infty$, while C_0 is a function discontinuity (nugget effect). Coefficients S and C_0 are estimated using empirical semivariogram and sample variance Var, described by the following relations:

$$\tilde{\gamma}(h_k) = \frac{1}{2n} \sum_{i,j=1}^{n} \left(f(x_i, y_i) - f(x_j, y_j) \right)^2$$
$$Var = \frac{1}{n-1} \sum_{i=1}^{n} \left(f(x_i, y_i) - \bar{f} \right)^2, \quad \bar{f} = \frac{1}{n} \sum_{i=1}^{n} f(x_i, y_i) \tag{6}$$

where $n = n(r_k)$ is the number of basic points in a measurement set, $\bar{f}$ is the average value of the function in a measurement set.

Finally, after solving the system of equations the heat source power distribution for interpolation grid is calculated as follows:

$$q(x, y, z) = \eta Q_S \tilde{f}(x, y) \tag{7}$$

4 Results and discussion

Figures 5-6 show the power distribution of the heat source obtained by Kriging method for interpolation grid step Δh=0.02 mm taking into account experimental data for $z = 0$ and at the distance from beam focusing position z=10mm respectively. In all figures the comparison of modelled percentage power intensity distribution with experimentally obtained distribution is illustrated in central axes of the heat source ($x = 0$ and $y = 0$).

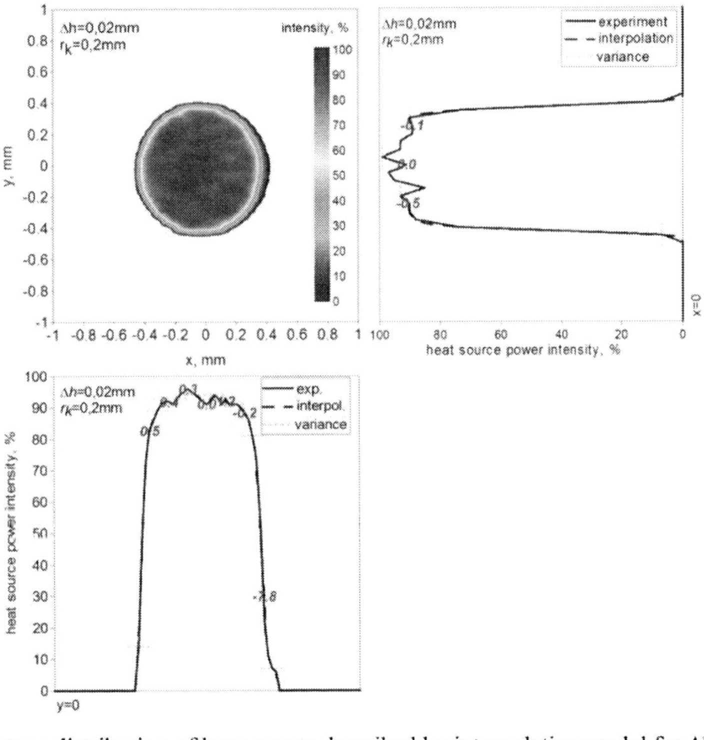

Fig. 5. Percentage distribution of laser power described by interpolation model for Δh=0.02mm. Distance from beam focusing $z = 0$

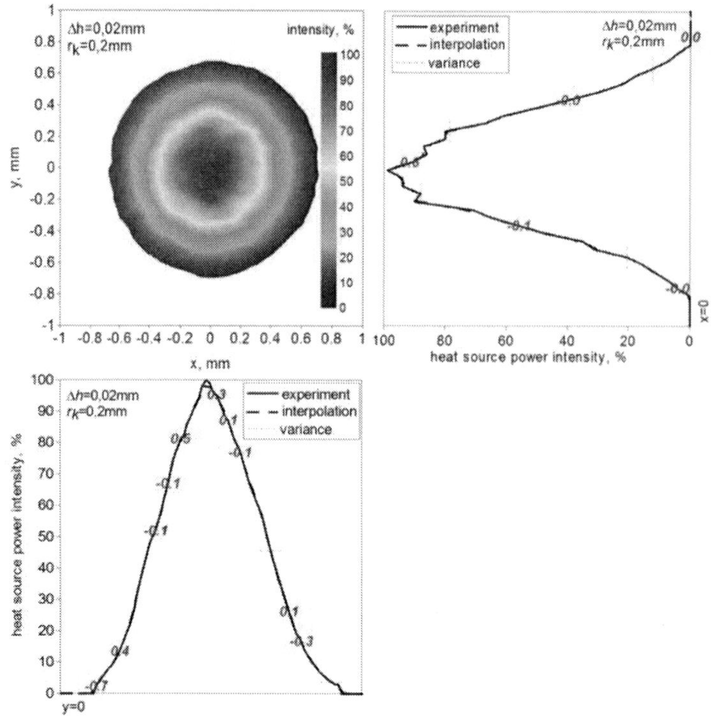

Fig. 6. Percentage distribution of laser power described by interpolation model for Δh=0.02mm. Distance from beam focusing z = 10 mm

Equations 1-3 allowed the prediction of beam caustic for f_{col} = 200 mm, f_{og} = 400 mm and different diameters of optical fibers. Calculated beam caustic is presented in Fig. 7.

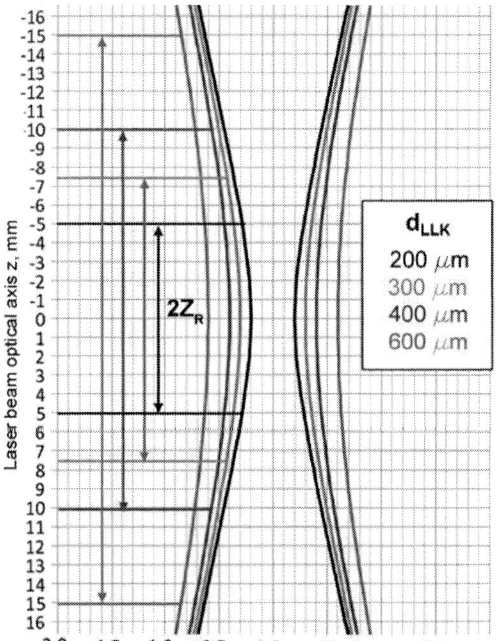

Fig. 7. Predicted beam caustic for different fibres diameters d_{LLK}

Conclusions

The increase of the distance from the beam focusing position changes the nature of intensity distribution, from top-hat to Gaussian like distribution. Therefore, classical Gauss distribution don't fit such laser beam power intensity distribution. The modelling of Yb:YAG laser power distribution by interpolation algorithms allows for a precise reproduction of the real thermal load, depending on the laser profile obtained for specified industrial laser.

Interpolation model can include technological parameter like the distance from the beam focusing position "z" for which suitable heat source distribution is selected. This model takes also into account local extremes existing in the distribution.

Developed model allows considering not only the influence of the distance from beam focusing on the laser spot diameter but also on the laser beam power intensity distribution at different distances from beam focusing position.

References

1. D. Richardson, J. Nilsson, W. Clarkson, *High power fiber lasers: current status and future perspectives*. Journal of the Optical Society of America **B.27**, 63-92 (2010)
2. T. Y. Fan, *Heat generation in Nd:YAG and Yb:YAG*. IEEE Journal of Quantum Electronics **29**, 1457-1459 (1993)
3. R. Brockmann, D. Havrilla, *Third generation of disk lasers. A new benchmark for industrial solid state lasers*. Laser Technik Journal **6** (3), 26-31 (2009)
4. X. Song, B. Li, Z. Guo, S. Wang, D. Cai, J. Wen, *Influences of pump beam distribution on thermal lensing spherical aberration in an LD end-pumped Nd:YAG laser*. Optics Communications **282**, 4779-4783 (2009)
5. C. Pfistner, R. Weber, H. P. Weber, S. Merazzi, R. Gruber, *Thermal beam distortions in end-pumped Nd:YAG, Nd: GSGG, and Nd: YLF Rods*. IEEE Journal of Quantum Electronics **30**, 1605-1614 (1994)
6. M. A. Oliver, R. Webster, *Kriging: a method of interpolation for geographical information system*. Int J Geo Inf Sys. **4** (3), 313-332 (1990)
7. S. Sakata, F. Ashida, M. Zako, *Structural optimization using Kriging approximation*. Computational Methods in Applied Mechanical Engineering **192** (7–8), 923-939 (2003)
8. M. Kubiak, W. Piekarska, Z. Saternus, T. Domański, *Numerical prediction of fusion zone and heat affected zone in hybrid Yb: YAG laser+ GMAW welding process with experimental verification*. Procedia Engineering **136**, 88-94 (2016)

Computer simulations of thermal phenomena in surface heating process using the real distribution of Yb:YAG laser power

Marcin Kubiak[1,*], *Vladimír* Dekýš[2], *Tomasz* Domański[1], *Pavol* Novák[2], *Zbigniew* Saternus[1]

[1]Czestochowa University of Technology, Institute of Mechanics and Machine design Foundations, Dabrowskiego 73, 42-200 Czestochowa, Poland
[2]University of Zilina, Department of Applied Mechanics, Univerzitna 1, 010 26 Zilina, Slovak rep.

Abstract. This work concerns mathematical and numerical modelling of temperature field during Yb:YAG laser heating of sheets made of S355 steel with the motion of liquid steel in the fusion zone taken into account. Laser power distribution and the caustics are determined on the basis of the geostatistical kriging method. Temperature field and melted material velocity field in the fusion zone are obtained from the numerical solution of continuum mechanics equations using projection method and finite volume method. Numerical algorithms are implemented into computer solver using ObjectPascal programming language. Computer simulations of Yb:YAG laser heating process are performed for different process parameters. Characteristic zones of experimentally obtained cross sections of heated elements are compared to numerically predicted fusion zone and heat affected zone.

Keywords: Yb:YAG laser, laser heating, heat affected zone, numerical modelling

1 Introduction

The heat treatment of metals is a process known for ages. Studied theoretically in terms of thermal, structural and mechanical phenomena [1-3]. Nowadays, diode pumped Yb:YAG lasers are intensively studied in terms of the use of new laser processing methods in the industry [4]. The study is performed in the field of experimental research as well as theoretical approach. The laser beam power and its distribution in heated element as well as liquid material motion in the fusion zone are major factors determining the temperature field in the laser beam processing.

Theoretical analysis of coupled physical phenomena accompanying laser-material interaction is very complex in terms of mathematical modelling and numerical solutions. Numerical models usually describe chosen phenomena, due to the computational complexity of the problem [5-8]. Consideration of liquid material motion in the model allows for the analysis of previously neglected phenomena in material melting processes and has a significant impact on the estimated temperature distribution [6,9].

* Corresponding author: kubiak@imipkm.pcz.pl

Reviewers: *Ladislav* Écsi, *Richard* Pastirčák

Moreover, changing focal length of focusing lens in the optical system determines the laser spot diameter as well as laser beam power intensity distribution [10,11]. Only a specific laser beam spot diameter is adopted in analytical models and the beam focusing as a technological parameter influencing the laser beam intensity distribution is neglected. Consequently, the laser beam intensity distribution models assumed in numerical analysis significantly differ from real Yb:YAG laser profile, obtained through experimental research.

A comprehensive theoretical model is presented in this study of thermal phenomena during Yb:YAG heating of S355 steel sheets, taking into account the real laser beam heat source intensity distribution and the motion of liquid steel in the fusion zone. In order to assess the quality of obtained results and comparison of predicted fusion zone and heat affected zone with experimental data is performed. Schematic sketch of analyses system is presented in Fig. 1.

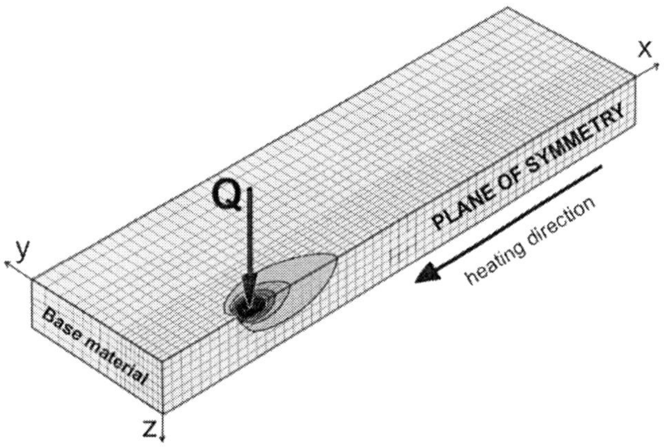

Fig. 1. Schematic sketch of laser heating simulation

2 Numerical solutions

The temperature field and liquid material velocity field are determined on the basis of the solution of mass, momentum and energy conservation equations. The motion of melted material in the fusion zone is mainly generated by natural convection according to Boussinesq model [12-15]. Mathematical model assumes fuzzy solidification front between solidus and liquidus temperatures where liquid material motion is treated as a fluid flow through porous medium with respect to Darcy's model. In theoretical considerations latent heat of fusion and evaporation are taken into account.

Initial conditions and boundary conditions complete governing equations. Momentum equation is completed by initial condition $t = 0 : \mathbf{v} = 0$ and boundary conditions implemented at the welding pool boundary determined by solidus temperature (T_S). Energy conservation equation is completed by initial condition: $t = 0 : T = T_0$ and boundary conditions, taking into account heat loss due to convection, radiation and evaporation.

Governing equations are numerically solved using projection method with finite volume method [16]. Interpolation algorithms are used for the description of heat source power distribution, allowing a precise description of Yb:YAG laser power intensity. Models elaborated on the basis of ordinary Kriging method [17] in the form of point Kriging take into account the real laser power distribution obtained in experimental research made on laser welding station equipped with TruDisk 12002 laser. Solution algorithm is presented in Fig. 2.

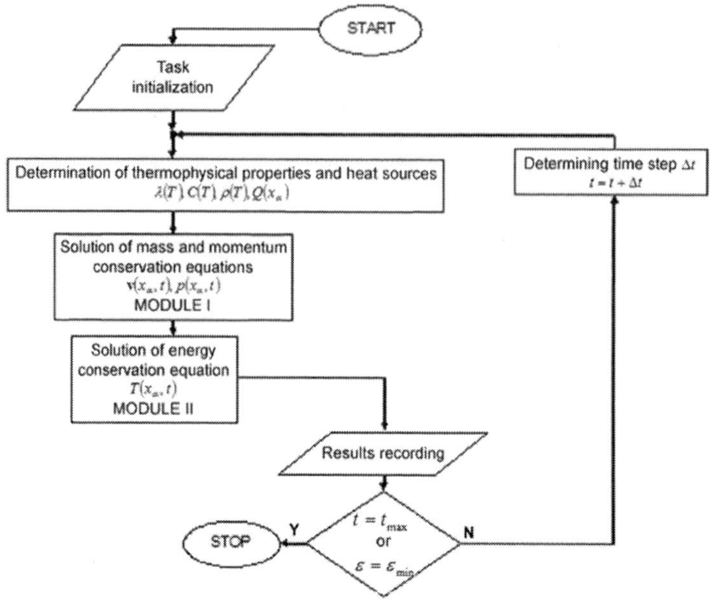

Fig. 2. Numerical solution algorithm

3 Experimental research

Trumpf D70 laser head combined with TruDisk 12002 disk laser is used in experimental research (Fig. 3a). The head is equipped with a collimator lens having a focal length f_c=200mm and a focusing lens having a focal length of f=400 mm. The fibre with a diameter $dLLK$=400μm allows to obtain double magnification of the optical system, thus laser beam diameter is d=0.8 mm. Prometec UFF100 laser beam analysis system (Fig. 3b) is used in order to analyse the density of laser power distribution and the caustic of the laser beam.

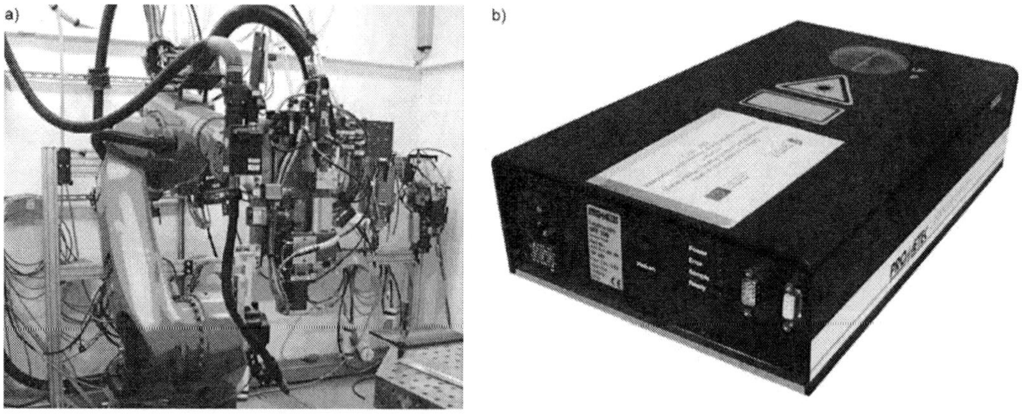

Fig. 3. Robotic laser welding station with a laser disk TruDisk 12002: a) D70 laser head mounted on the wrist of KUKA industrial robot and b) laser beam analyser UFF100

The laser heating experiment is made for S355 steel sheets with dimensions 250mm in length, 50mm in width, with a thickness of 5mm. Three heating processes are performed (Fig. 4) assuming laser head movement speed v= 3 m/min, the focusing point on the top surface of heated element (z=0) and three different laser powers: Q=900, 500 and 320W.

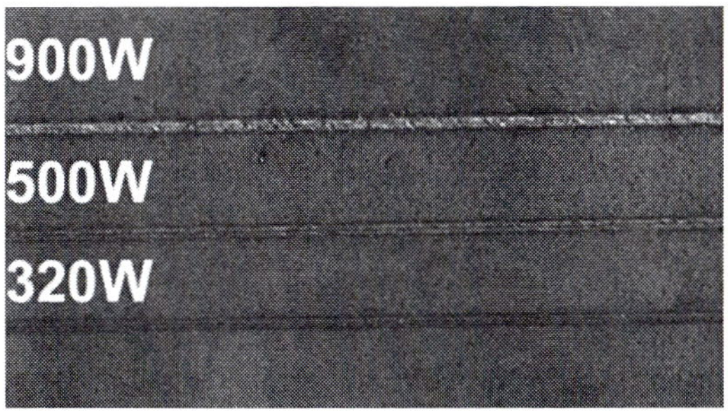

Fig. 4. Top surface of heated element for three different laser powers

4 Results and discussion

Numerical algorithms are implemented into computer solver using ObjectPascal programming language. Computer simulations of the Yb:YAG laser heating process are performed for S355 steel sheets with dimensions used in the experiment. The analysed domain is discretized by a staggered grid with the spatial step set to 0.1 mm. The numerical analysis is made using interpolated heat source model and technological parameters assumed in experimental research. Basic parameters of the laser beam used in the model are obtained using Prometec UFF100 beam analyser, a diagnostic device for measuring high power lasers profile. Fig. 5 shows heat source power distribution obtained by Kriging method for interpolation grid step Δh=0.02 mm.

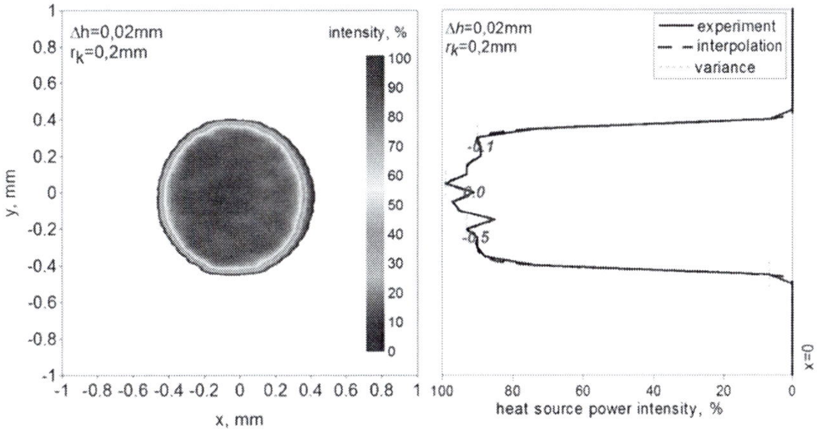

Fig. 5. Percentage distribution of laser beam power for $z = 0$

Figure 6 illustrates the temperature distribution at the top surface and in cross section (for x=3 mm) of laser heated sheet. In Figure 6b a comparison of predicted characteristic zones of heated element with the macroscopic picture of the cross section of heated element is presented. Solid line points out melted zone boundary (solidus temperature) whereas dashed line marks heat affected zone boundary (austenitization temperature, about 1000K). Figure 7 presents temperature distribution in the cross section of heated element maid for three different laser head powers in comparison with the macroscopic picture of the cross section of heated elements.

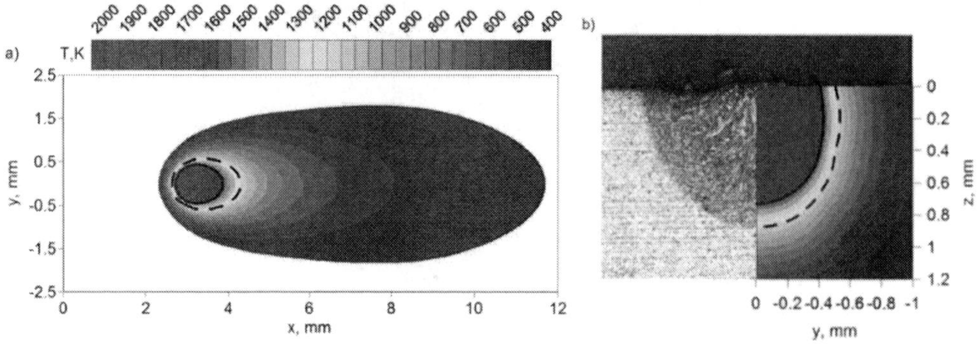

Fig. 6. Temperature distribution a) at the top surface and b) in cross section (for x=3 mm) of laser heated sheet in comparison with macroscopic picture of the real heated element

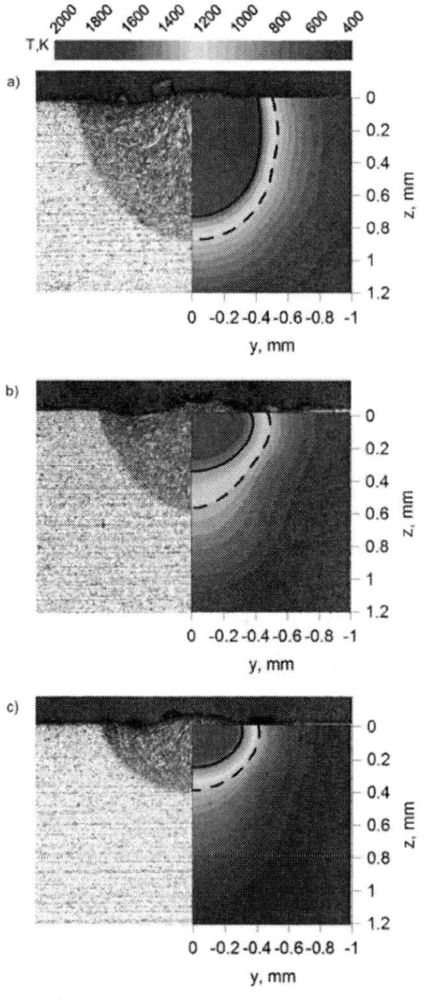

Fig. 7. Temperature distribution in the cross section (for x=3 mm) of laser heated sheets for three laser beam powers: a) 900W, b) 500W and c) 320W in comparison with macroscopic picture of the real heated elements

Conclusion

Heat affected zone in laser heated elements is narrow, up to about 1.2 mm in width for 900W laser beam power decreasing to about 0.8 for laser beam power 320W. Melted zone that appear for each experiment is also decreasing with decreasing laser beam power. Predictably, the use of lower laser beam power contributed to the reduction of melted area and heat affected zone.

Application of Kriging algorithm allowed to use experimental studies in numerical analysis and for a precise reproduction of the real thermal load. Performed comparison of the heat source power distribution model with experimental data as well as characteristic zones of heated elements corroborates the validity of performed theoretical assumptions and usability of model in practical applications.

References

1. S. H. Kang, Y. T. Im, *Three-dimensional thermo-elasto-plastic finite element modeling of quenching process of plain carbon steel in coulee with phase transformation*. J. Mater Proc Technol. **192-193**, 381-390 (2007)
2. A. Bokota, T. Domanski, *Modelling and numerical analysis of hardening phenomena of tools steel elements*. Arch Metall Mater. **54** (3), 575-587 (2009)
3. T. Domanski, A. Sapietova, M. Saga, *Application of Abaqus software for the modeling of surface progressive hardening*. Procedia Engineering **177**, 64-69 (2017)
4. D. Richardson, J. Nilsson, W. Clarkson, *High power fiber lasers: current status and future perspectives*. J. Opt Soc Am. **B.27**, 63-92 (2010)
5. A. De, T. Debroy, *Reliable calculations of heat and fluid flow during conduction mode laser welding through optimization of uncertain parameters*. Welding Journal **84** 101-112 (2005)
6. X. Jin, L. Li, Y. Zhang, *A heat transfer model for deep penetration laser welding based on an actual keyhole*. Int J Heat Mass Tran. **46**, 15–22 (2003)
7. L. Han, F.W. Liou, *Numerical investigation of the influence of laser beam mode on melt pool*. Int J Heat Mass Tran. **47**, 4385-4402 (2004)
8. J. Winczek, A. Modrzycka, E. Gawronska, *Analytical description of the temperature field induced by laser heat source with any trajectory*. Procedia Engineering **149**, 553-558 (2016)
9. M. Kubiak, W. Piekarska, *Comprehensive model of thermal phenomena and phase transformations in laser welding process*. Computers & Structures **172**, 29-39 (2016)
10. M. Kubiak, W. Piekarska, S. Stano, *Modelling of laser beam heat source based on experimental research of Yb: YAG laser power distribution*, Int J Heat Mass Tran, **83**, 679-689 (2015)
11. M. Kubiak, W. Piekarska, Z. Saternus, T. Domański, *Numerical prediction of fusion zone and heat affected zone in hybrid Yb: YAG laser+ GMAW welding process with experimental verification*. Procedia Engineering **136**, 88-94 (2016)
12. R. Gnatowska, T. Rybak, *Numerical Analysis of Heat Transfer around 2D Circular Cylinder in Pulsation Inflow*. AIP Conference Proceedings **1648**, 850125 (2015)
13. R. Dyja, E. Gawronska, A. Grosser, *Numerical problems related to solving the Navier-Stokes equations in connection with the heat transfer with the use of FEM*. Procedia Engineering **177**, 78-85 (2016)

14. T. Skrzypczak, E. Węgrzyn-Skrzypczak, *Modeling of thermal contact through gap with the use of finite element method*. J Appl Math. Comput Mech. **14** (4), 145-152 (2015)
15. T. Skrzypczak, *A finite element multi-mesh approach for heat transport between disconnected regions*. Procedia Engineering **177**, 204-209 (2017)
16. S. V. Patankar, *Numerical heat transfer and fluid flow*. (Taylor & Francis, USA, 1990)
17. M. A. Oliver, R. Webster, *Kriging: a method of interpolation for geographical information system*. Int. J Geogr Inf Sys. **4**, 313-332 (1990)

Analysis of deformation and microstructure evolution during the hot deformation of titanium alloy

Marcin Kukuryk[1,*], *Jerzy* Winczek[1]

[1]Czestochowa University of Technology, 42-201 Czestochowa, Poland

Abstract. The paper presents theoretical and experimental analysis of deformations and microstructural evolutions in the hot cogging process of Ti-6Al-4V alloy. A three–dimensional thermal – plastic coupled finite element model is employed to study the mechanical and thermal interaction between the forging anvils and the workpiece. To explore the distributions of effective strain, effective stress, mean stress and temperature of the specimens have been systematically studied. Attention has been paid to deformation, temperature, stress and strain inside the specimens and these parameters have been used to determine the evolution of the microstructure in deformed samples during hot cogging process. A comparison of theoretical with experimental results shows that the developed model may be used to accurately predict deformations and microstructural parameters.

Keywords: hot forging, titanium alloy, finite element, strain, stress, microstructure evolution

1 Introduction

Titanium alloys are well known for industrial applications such as aerospace, automotive, marine and biomedical. Ti-6Al-4V alloy is one of the most important engineering alloys, combining attractive properties such as high strength to density ratio, high strength at relatively high temperatures (up to 550 °C), good fatigue properties and excellent heat corrosion resistance [1]. The alloys are manufactured by hot plastic working methods, chiefly by open die forging or die forging. The hot workability of an alloy is normally limited by the generation of various deformation defects that eventually lead to gross failure [2]. The great non-uniformity of deformations in the forging process makes the production of internal-quality titanium forgings much more difficult than imparting the proper shape and geometry to the forgings [3]. In order to maintain a relatively small gradient of properties within the entire forging volume and, at the same time, their required level, cogging process is used in anvils of a purpose-designed shape [4].

The microstructure of titanium alloys is very sensitive to the processing parameters such as forging temperature, strain and strain rate. Especially, microstructure evolution in the hot cogging process has an influence on the mechanical properties of material because the flow

* Corresponding author: kukurykm@itm.pcz.pl

Reviewers: *Richard* Pastirčák, *Eva* Tillová

stress of the material is affected by dynamic recrystallization and grain growth [5]. The determination of temperature distribution within the deformation zone during the titanium alloy forging process is important because of the effect of temperature on the properties and structure of the material being deformed [6].

A finite element method-based model has been developed within the present work, which enables the prediction of the distribution of deformation parameters (strain, stress, temperature) and microstructure evolution during the cogging process of the Ti-6Al-4V alloy in three types of forging anvils. Here, a Johnson-Mehl-Avrami-Kolmogorov type equation describing the relationships between deformation conditions (strain, strain rate and temperature) and DRX volume in $\alpha+\beta$ phase was introduced and fitted for Ti-6Al-4V alloy [7]. The results of the experiment are compared quantitatively and qualitatively with those of simulation.

2 The material model and experimental procedures

In this study, the rigid – thermoviscoplastic finite element method is used because it is necessary to solve simultaneously the material – flow problem for a given temperature distribution and heat – transfer problem in the hot cogging process. From the variational principle, the functional, ϕ for rigid – viscoplastic material can be written as follows:

$$\phi = \int_V \bar{\sigma}\,\dot{\varepsilon}_e\,dV + \frac{1}{2}\beta\int_V \dot{\varepsilon}_V^2\,dV - \int_{S_T} T_i\,v_i\,dS \quad (1)$$

where $\bar{\sigma}$ is the effective stress, $\dot{\varepsilon}_e$ is the effective strain rate, $\dot{\varepsilon}_V$ is the volumetric strain rate, V is the volume of a deforming workpiece, T_i is the external force acting on a surface S_T and v_i velocity, β a penalty constant with a large positive value.

The flow stress (σ) is a function of strain (ε), strain rate ($\dot{\varepsilon}$) and temperature (T) (experimental results):

$$\sigma = f(\varepsilon, \dot{\varepsilon}, T)\,. \quad (2)$$

The flow curves of the model as per Eq. (2) are consistent with the experimental data in the temperature range of 1073-1273 K, the true strain range of 0.105 - 0.693 and the strain rate range of 0.1 - 1.0 s^{-1}. The flow stress of the Ti-6Al-4V alloy obtained from compression tests is input to the FEM simulation system.

The temperature distribution of the workpiece in the analysis of heat – transfer can be obtained by solving the energy balance equation, expressed by

$$\rho c_p \frac{\partial T}{\partial t} = \nabla^T (k\,\nabla T) + q \quad (3)$$

where k is the thermal conductivity coefficient, T is the temperature, ρ is the density, c_p is the specific heat, t is the time, $q=\gamma\bar{\sigma}\dot{\varepsilon}_e$ is the heat generation rate in the deformation body, γ is the fraction of mechanical energy converted into heat, and was assumed to be 0.9. The finite element method is then adapted to solve Eqs. (1) and (3) which define the thermo - mechanical state of a body in the cogging process.

The microstructure change taking place by deformation is generally called dynamic recrystallization. It is very difficult to model dynamic recrystallization concurrently during forming. The relationships describing the mechanism of dynamic recrystallization are given by

$$\varepsilon_c = 8.29 \times 10^{-4}\, d_0^{\,n_1}\, \dot{\varepsilon}^{m_1} \exp\frac{Q_1}{RT} \tag{4}$$

$$X_{DRX} = 1 - \exp\left[-\beta_d \left(\frac{\bar{\varepsilon} - \varepsilon_c}{\varepsilon_{0.5}}\right)^{k_d}\right] \tag{5}$$

$$\varepsilon_{0,5} = a_2\, d_0^{\,n_2}\, \dot{\varepsilon}^{m_2} \exp\frac{Q_2}{RT} \tag{6}$$

$$d_{DRX} = a_3\, d_0^{\,h_3}\, \bar{\varepsilon}^{n_3}\, \dot{\varepsilon}^{m_3} \exp\frac{Q_3}{RT} \tag{7}$$

$$d_{AVG} = X_{DRX}\, d_{DRX} + (1 - X_{DRX})\, d_0 \tag{8}$$

where ε_c is the critical strain at which dynamic recrystallization initiates, T the absolute temperature, X_{DRX} the volume fraction of dynamic recrystallization, $\bar{\varepsilon}$ the effective strain, $\dot{\varepsilon}$ the effective strain rate, $\varepsilon_{0.5}$ the strain for 50% recrystallization, R the gas constant, d_0 the initial grain size, d_{DRX} the dynamically recrystallized grain size, $a_{2\text{-}3}$ the material data (experimental coefficients), h_1, $n_{1\text{-}3}$, $m_{1\text{-}3}$the material data (experimental coefficients), $Q_{1\text{-}3}$ the activation energies obtained from experiment, β_d the material data (experimental coefficient); k_d the material data (experimental exponent), d_{AVG} is the average grain size.

3 Results and discussion

The studied Ti-6Al-4V alloy with the composition of 6.20 wt.% Al, 4.10 wt.% V, 0.15 wt.% Fe, 0.01 wt.% C, 0.005 wt.% H, 0.005 wt.% N, 0.14 wt.% O and the balance of Ti. The β transus temperature was measured to be approximately 1250K by thermal dilatation method. The main input temperature – dependent parameters for simulations were flow stress, thermal conductivity, and heat capacity. The flow stress was represented as a function of strain, strain rate and temperature. The variations of flow stress with temperature were generated for a range of temperatures (1073 – 1273 K with a step of 50 K) and strain rate conditions (0.05, 0.10, 0.50, and 1.0 s^{-1}) by compression tests in a MAEKAWA – Japan plastometer. The temperature of the specimens was monitored using a chromel – alumel thermocouple. The total deformation, as calculated from the specimen height change, was 0.105 - 0.693. The obtained results allowed the determination of the flow stress values, as dependent on the actual strain, strain rate and temperature, which were input to the program with the aim of evaluating the best rheological model for the material being deformed. The flow curves for lowest (0.10 s^{-1}) and highest (1.0 s^{-1}) strain rate conditions are shown in Fig. 1 for brevity. It can be clearly seen that the flow stress increases to a maximum during high temperature deformation. After attaining a peak value, the stress decreases until it reaches a steady state. This is an indication that the material softens by dynamic recrystallization.

For the titanium alloy under examination and for the anvil material X37CrMoV51, the thermal characteristics, such as density, specific heat and thermal conductivity, were assumed based on experimental data and were input as functions of temperature. The initial anvil temperature was taken as 300°C, i.e. the anvil heating temperature in industrial conditions. Some more important values of the determined coefficients occurring in formulas (4)-(8) are as follows:

$a_2 = 0.0011$, $n_1 = 0.40$, $m_1 = 0.017$, $Q_1 = 41700$, $\beta_d = 0.6903$, $a_2 = 0.0011$, $n_2 = 0.25$, $m_2 = 0.05$, $Q_2 = 24430$, $a_3 = 1600$, $h_3 = 0.70$, $n_3 = 0.80$, $Q_3 = -52235$

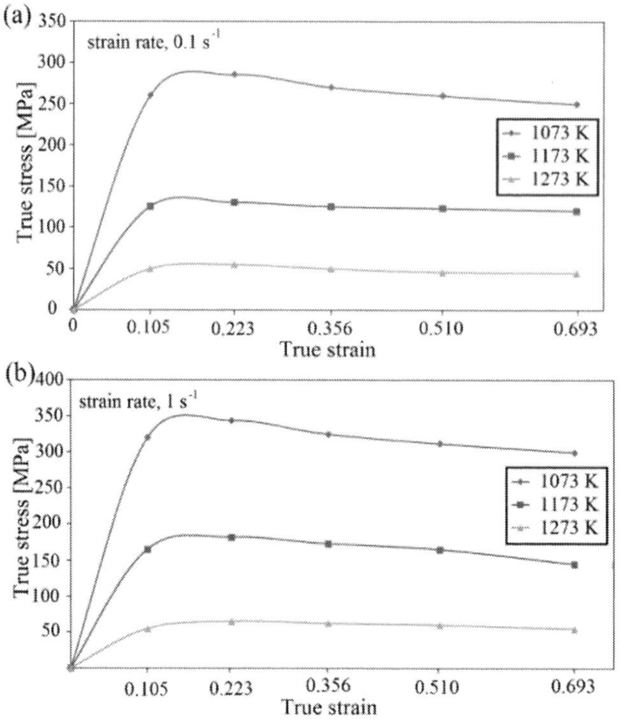

Fig. 1. Variation of flow stress for the Ti-6Al-4V alloy as a function of true strain and temperature, for different strain rate values, i.e.: (a) - $\dot{\varepsilon} = 0.1 s^{-1}$ and (b) - $\dot{\varepsilon} = 1 s^{-1}$

In numerical computations and in experimental tests, ø80 mm-diameter and 200 mm-long stock of the Ti-6Al-4V alloy was taken. The cogging process consists of many partial, interrelated reductions made in a specific order. Assuming the cyclical nature and repeatability of the phenomena, the study of the stress and strain state can be limited to the examination of one engineering pass (two consecutive reductions with the simultaneous rotation of the material by 90°), which will considerably simplify the forging process. Forging was carried out in two consecutive reductions with tilting the workpiece by an angle of 90°, while retaining a constant value of true reduction of ε_h= 0.35 and a constant relative feed of l_w= 0.75. These samples have an initial grain size (d_o) of around 50 μm before deformation. The cogging process was conducted in different shape of the anvils: with oblique working surfaces, combined (bottom anvil is V-shaped) and asymmetrical V - shaped with an angle of α_1=105° and α_2= 135°, as shown in Fig. 2. The tests carried out enabled the determination of the local values describing the strain state, the stress state, the temperature distribution, and microstructural evolution during cogging process of the Ti-6Al-4V alloy.

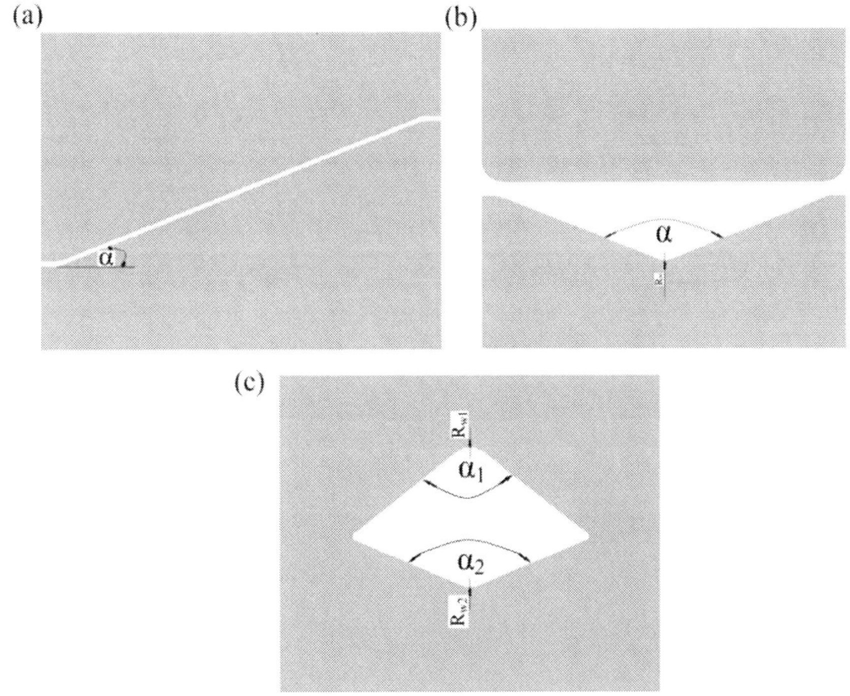

Fig. 2. Shape of anvils: (a) – with oblique working surfaces (α=30°), (b) – combined anvils (bottom anvil is V-shape) with an angle of α= 135° (R_w=0.35D_0; D_0= 80 mm), (c) – asymmetrical V – shaped with an angle of α_1=105° and α_2= 135° (R_{w1} =R_{w2} = 0.35 D_0)

In the case of the elongation of a material having a circular section on anvils with oblique working surfaces (Fig. 3a), the non-homogeneousness of deformations is comparatively large. In the presented distribution of the effective strain following the one engineering pass (ε_h = 0.70), it is possible to differentiate a number of characteristic zones. Area situated adjacently to surface of anvils is experiencing significantly small deformations ($\varepsilon_i/\varepsilon_h$ = 0.48-0.67). The largest deformation occurred in the central part of a forging, whereas the values of the effective strain ε_i significantly exceeded the value of true strain ε_h ($\varepsilon_i/\varepsilon_h$ = 1.46), the lateral zones of a forging are the area of intermediate deformations ($\varepsilon_i/\varepsilon_h$ = 0.87-1.07). This part of the material of a forging, in addition to that, is subjected to the effect of tensile stresses for the part of non-deformed external zones, and also for the part of a strongly deformed center (Fig. 3b). The process of forging was accompanied by the stable distribution of temperature in the central part of a forging resulting from the emission of the heat of plastic deformations, and the contact point of a hot metal and cool tools resulted in the formation of the significant gradients of temperature in the region adjacent to the contact surfaces (ΔT–80-180°C). Measurements with the application of a thermovision camera of the temperature of the surface of a forging following deformation demonstrated the compatibility of this temperature with the calculated one (differences ±10 °C).

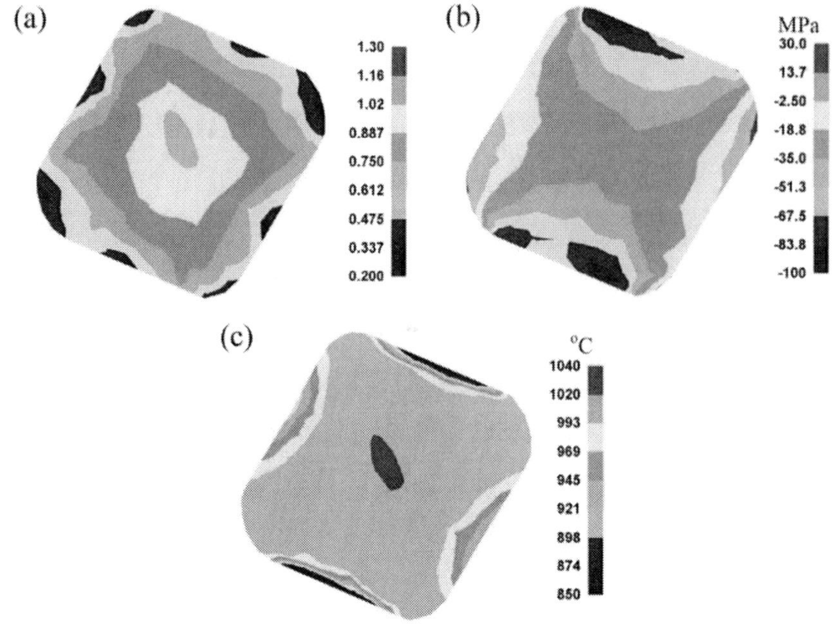

Fig. 3. Distribution of effective strain (a), mean stress (b) and temperature (c) on the cross-sectional surface of the specimens after cogging process in anvils with oblique working surfaces. True strain $\varepsilon_h = 0.70$

In Fig. 4, the distribution of the effective strain and mean stress following the cogging process of samples made of the titanium alloy (Ti-6Al-4V) on combined anvils with an angle of α= 135°, following the one engineering pass (ε_h = 0.70), were presented. The applied anvils demonstrated a similar distribution of the effective strain and mean stress in the cogging process of the titanium alloy being investigated.

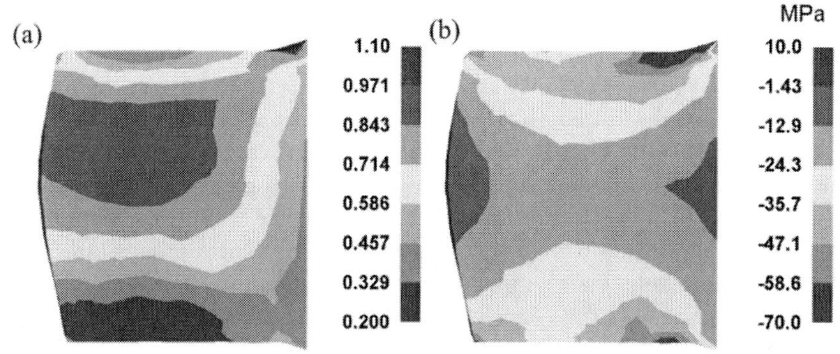

Fig. 4. Distribution of effective strain (a) and mean stress (b) in the longitudinal-section of the specimens after cogging process in combined anvils with α=135°. True strain $\varepsilon_h = 0.70$

In Fig. 5, the distribution of effective strain, mean stress and temperature on the cross-sectional surface of the specimens after cogging process in asymmetrical V – shaped anvils with an angle of α_1=105° and α_2= 135°, following the one engineering pass (ε_h = 0.70), were presented. The anvils applied in investigations demonstrated a positive effect on the value, and on the homogeneousness, of the distribution of the effective strain, and also resulted in the absence of tensile stress in the central part of the deformation basin (Fig. 5a and 5b). On the larger part of the surface of the transverse section of a sample, the effective strain significantly exceeded the value of the true strain ($\varepsilon_i / \varepsilon_h$ = 1.52 – 1.94), which had a positive effect on the re-forging of it. The lateral zones of a forging were the collections of the smaller values of the effective strain: $\varepsilon_i / \varepsilon_h = 1.12$. The presented distribution of mean stress (Fig. 5b) indicates the possibility of the appearance of tensile stress only on a very small area situated in the lateral zones of a forging. The process of forging was accompanied by the stable distribution of temperature in the central part of a forging resulting from the emission of the heat of plastic deformations, and a comparatively large surface of the contact point between the deformed metal and the cool tools resulted in the formation of the significant gradients of temperature (ΔT =150- 200 °C).

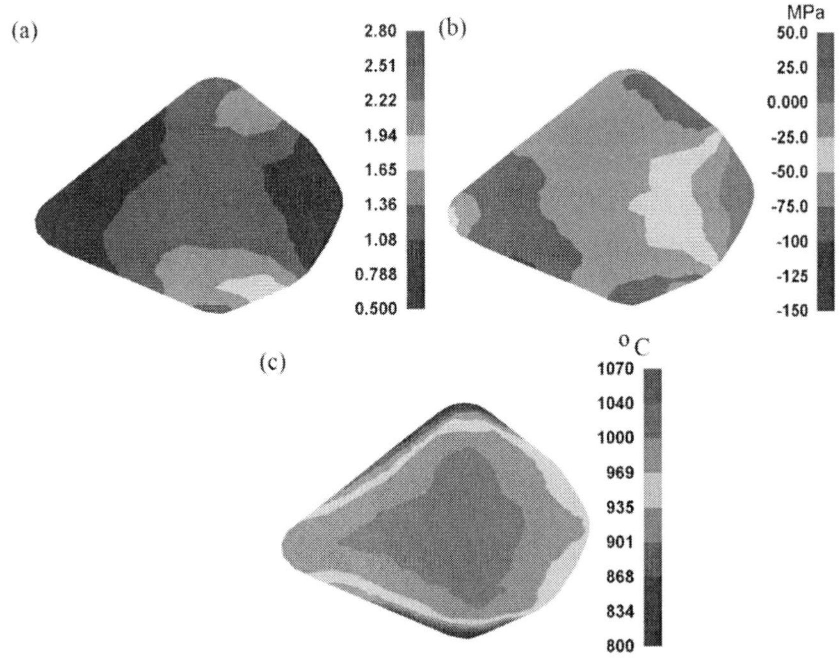

Fig. 5. Distribution of effective strain (a), mean stress (b) and temperature (c) on the cross-sectional surface of the specimens after cogging process in asymmetrical V – shaped anvils with an angle of α_1=105° and α_2= 135°. True strain ε_h = 0.70

The distribution of the dynamically recrystallized volume fraction and average grain size on the cross - sectional surface of the specimens after cogging process (after the one engineering pass) in anvils with oblique working surfaces was presented in Fig. 6. The process of deformations was progressing in an uneven manner across the sample, and that brought the fruit of the presence of zones having the different degrees of recrystallizing and different average grain size. The dynamically recrystallized volume fraction reached its maximum value in the central point of a forging, and it amounted to 61.2 % (Fig. 6a). The distribution of the average grain size on the cross - sectional surface of the specimens of was not even. The smallest grain size was obtained for the central area of a forging, which amounted to 24.1 μm (Fig. 6b). On the surface of contact between a deformed metal and the anvils, the significantly larger average grain size (44.8 μm) was obtained.

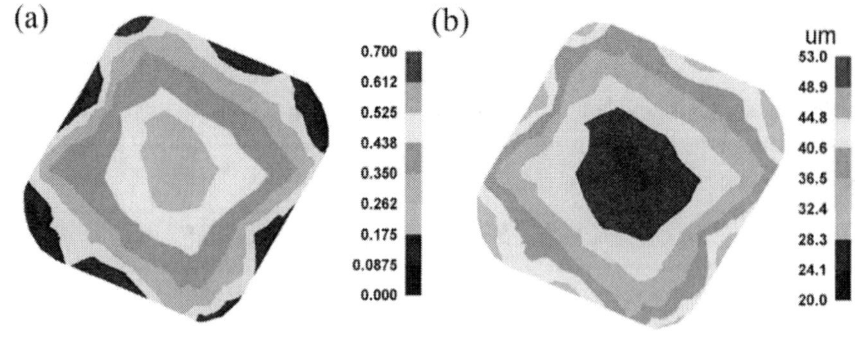

Fig. 6. Distribution of the dynamically recrystallized volume fraction (a) and average grain size (b) on the cross - sectional surface of the specimens after cogging process in anvils with oblique working surfaces. True strain $\varepsilon_h = 0.70$

For combined anvils, the dynamically recrystallized volume fraction reached its maximum value in the central point of a forging, and it amounted to 85.7 % (Fig. 7a). The smallest grain size was obtained for the central area of a forging, and it amounted to 16.3 μm (Fig. 7b). On the surface of contact between the deformed metal and the anvils, a significantly larger average grain size (41.4 - 47.7 μm) was obtained.

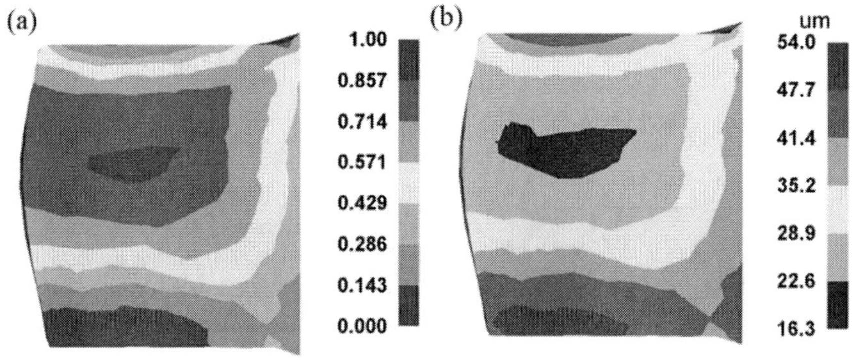

Fig. 7. Distribution of the dynamically recrystallized volume fraction (a) and average grain size (b) in the longitudinal section of the specimens after cogging process in combined anvils with α=135°. True strain $\varepsilon_h = 0.70$

The distribution of the dynamically recrystallized volume fraction, and also the distribution of the average grain size on the cross - sectional surface of the specimens after cogging process (after the one engineering pass) on asymmetrical V-shaped anvils with the angle of α_1=105° and α_2= 135° was presented in Fig. 8. The microstructure which was obtained was to a large degree recrystallized and homogeneous on a dominant part of the section of a sample. The dynamical recrystallization was commencing on the large part of the area subjected to the largest deformation situated in the axial zone of a forging, where the dynamically recrystallized volume fraction reached its maximum value. For the lateral zones of a forging, the dynamically recrystallized volume fraction reached smaller values, and it amounted to 62.5-75.0 % (Fig. 8a). The distribution of the average grain size on the cross - sectional surface of the specimens was more even in comparison with the distributions on the previously-researched anvils. The intensive re-forging of the axial zone of a forging in the asymmetrical V-shaped anvils resulted in the formation, in this area, of the most grinded structure, which amounted to 11.3 – 15.0 μm (Fig. 8b). For the lateral zone of a forging, the average grain size was larger, and amounted to 22.5 – 26.3 μm.

The conducted comparative analysis gives rise to the conclusion that the largest effect of the grinding of a grain size in the axial zone of a forging in the course of the forging of the Ti-6Al-4V titanium alloy was obtained on asymmetrical V-shaped anvils with an angle of α_1=105° and α_2= 135°, in the case of which, simultaneously with increase in true strain, there increased the effective strain, which was accompanied by a significant reduction in the grain size. Introduction, in the initial phase of cogging process on asymmetrical V-shaped anvils, will make it possible to reach even plastic processing in the volume of a forging, and also obtaining the homogeneous and finely-crystallite microstructure. The final deformation may be conducted in combined anvils, or in the ones with oblique working surfaces.

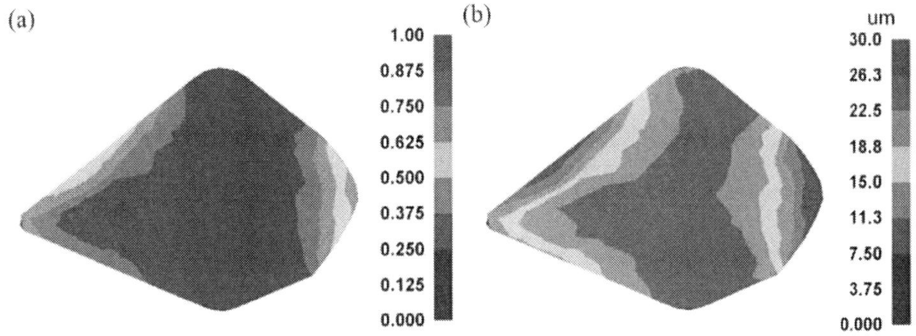

Fig. 8. Distribution of the dynamically recrystallized volume fraction (a) and average grain size (b) on the cross-sectional surface of the specimens after cogging process in asymmetrical V – shaped anvils with an angle of α_1=105° and α_2= 135°. True strain ε_h= 0.70

The experimental confirmation of the results of numerical research into the distribution of the effective strain on asymmetrical V-shaped anvils with the angle of α_1=105° and α_2= 135° was presented in Fig. 9. The state of deformations was determined with the application of the coordinate grid method. The experimental research was conducted on cylindrical samples having the following dimensions: ϕ80 x 150mm and made of the Ti-6Al-4V alloy, deformed on a hydraulic press having the pressure amounting to 2.500 kN.

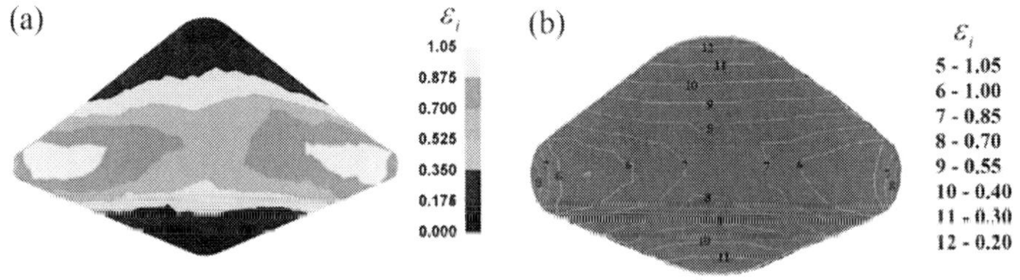

Fig. 9. Comparison of the theoretical (a) and experimental (b) distribution effective strain (ε_i) on the cross-sectional surface of the specimens after cogging process in asymmetrical V – shaped anvils with an angle of α_1=105° and α_2= 135°. Relative feed l_w= 0.75, true strain ε_h =0.35

In order to determine the constituents of the tensor of finite Cauchy deformations, the displacement field was approximated with the application of the two-parameter polynomials of degree six; in the case of that, the coefficients were determined with the application of the method of least squares, in the case of taking under consideration the condition of constant volume (with the application of the method of Lagrange multiplier). The conducted comparative analysis of the effective strain from the numerical calculations and those obtained by means of experimental method demonstrated a comparatively good compatibility of them.

The experimental measurement of temperature was performed on the surface of deformed samples made of the Ti-6Al-4V alloy with the application of a thermovision camera. The compatibility of the calculated and measured temperatures obtained for the surface layer (Fig. 10a) makes it possible to presume that the simulated distribution of temperatures on the section of a forged band are compatible with the actual ones.

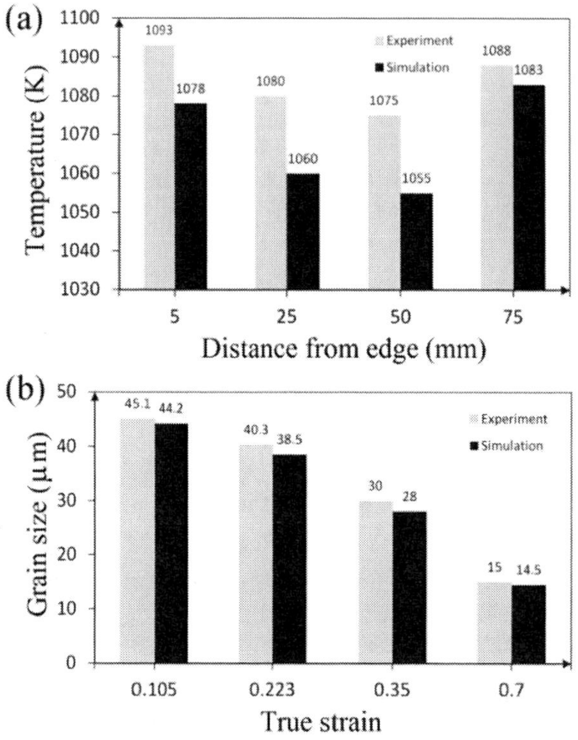

Fig. 10. Comparisons between the experimental and simulated results after cogging process in asymmetrical V –shaped anvils with an angle of α_1=105° and α_2= 135°: (a) temperature in the longitudinal section of the specimens (l_w= 0.75; ε_h = 0.70), (b) average grain sizes in the centre of the specimens deformed at different true strain

The experimental measurement of the average grain size was performed for a number of representative samples made of the Ti-6Al-4V alloy, which were re-forged on asymmetrical V-shaped anvils with an angle of α_1=105° and α_2= 135°, with the application of the following parameters of deformations: $\dot{\varepsilon}$ =0.25s^{-1}; T_o=1173K; ε_h = 0.105-0.693. On the selected samples, a number of measurements of the average grain size by means of the comparative method and in accordance with the standard: PN-EN ISO 643:2013-06 (Fig. 10b) were performed.

Conclusions

By means of the selection of the appropriate shape and geometry of the working surface of anvils, and also rational technical parameters, it is possible to exert a significant influence upon: the situation of the maximum values of the effective strain, the value of the non-homogeneousness of the distribution of deformations, and also upon reduction in the effect of tensile stress in the deformation basin. The process of forging on asymmetrical V-shaped anvils was accompanied by a small gradient of temperature in central part of the deformation basin contributing to increase in the homogeneousness of deformations, and that brought the fruit of obtaining a microstructure recrystallized to a large degree, and homogeneous. In the central part of the deformation basin, the largest reduction the average grain size was obtained (11.3-15.0 μm). In contact zones and also for the lateral surfaces of a forging, the average grain size was significantly larger, whereas it was reducing approximately two times. In the course of the cogging process on asymmetrical V-shaped anvils with an angle of α_1=105° and α_2= 135°, as a result of the intensive re-forging of the central areas of a forging, a twice smaller the average grain size was obtained in them in comparison with the average grain size was obtained in the course of forging on the remaining researched anvils.

The application in the initial phase of the cogging process on asymmetrical V-shaped anvils will make it possible to reach the large values of the effective strain in the volume of the deformed material, with the simultaneous maintenance of the high homogeneousness of the distribution of it. The final operation of the cogging process may be conducted in oblique anvils, or in the combined ones. The combination of different anvils in the process of the cogging process of the two-phase alloys of titanium provides the possibilities of summary use of their particular assets.

References

1 N. Kotkunde, A.D. Deole, A.K. Gupta, S.K. Singh, *Comparative study of constitutive modeling for Ti-6Al-4V alloy at low strain rates and elevated temperatures*. Materials and Design **55**, 999-1005 (2014)

2 T.R. Prabhu, *Simulations and experiments of the nonisothermal forging process of a Ti- 6Al-4V impeller*. Journal of Materials Engineering and Performance **25** (9), 3627-3637 (2016)

3 S. Luo, D. Zhu, L. Hua, D. Qian, S. Yan, F. Yu, *Effects of process parameters on deformation and temperature uniformity of forged Ti-6Al-4V turbine blade*. Journal of Materials Engineering and Performance **25** (11), 4824- 4836 (2016)

4 S.K. Choi, M.S. Chun, C.J. Van –Tyne, Y.H. Moon, *Optimization of open die forging of round shapes using FEM analysis*. Journal of Materials Processing Technology **172**, 88-95 (2006)

5 RG. Guan, Y.T. Je, Z.Y. Zhao, Ch.S. Lee, *Effect of microstructure on deformation behavior of Ti-6Al-4V alloy during compressing process*. Materials and Design **36**, 796-803 (2012)

6 G. Chen, Ch. Ren, X. Qin, J. Li, *Temperature dependent work hardening in Ti-6Al-4V alloy over large temperature and strain rate ranges: experiments and constitutive modeling*. Materials and Design **83**, 598-610 (2015)

7 G.Z. Quan, G.Ch. Luo, J.T. Liang, D.S. Wu, A. Mao, Q. Liu, *Modelling for the dynamic recrystallization evolution of Ti-6Al-4V alloy in two–phase temperature range and a wide strain rate range*. Computational Materials Science **97**, 136-147 (2015)

Case study of technological problems arising from the use of new material, DOMEX steel sheet, on the components of the mower discs on the example of the production company SaMASZ Sp. Z o. o.

Damian Kułak[1,*], *Jerzy* Jaroszewicz[2], *Marcin* Woronko[3]

[1]Enterprise SaMASZ Sp. Z o.o., 15-161 Białystok, ul. Grass 15, Poland
[2]Bialystok Technical University, Department of Production Management, 16-001 Kleosin, ul. Father Tarasiuk 2, Poland
[3]Graduate of Bialystok Technical University, Department of Production Management, 16-001 Kleosin, ul. Father Tarasiuk 2, Poland

Abstract. The paper describes the improvement of technological process of knife holder production used in disc mowers at SaMASZ Sp. With o.o. The improvement was aimed at eliminating cracks in the press of a 500t crank press press crush that occurred after the Boron 27 starting material was changed to Domex MC 500 high strength steel. The technology of regeneration of claws with identified operational cracks and a stand for fatigue tests was also proposed. Defectoscopy and strength testing methods have been developed to verify that the structure of the regenerated material is free from cracks or micro cracks in the material.

Keywords: knife holder, Boron 27, Domex, fatigue tests, defectoscopic examination

1 Introduction

The subject of the publication is the problem of cracking knife holders used in Perfect CUT disc mowers during pre-pressing. The problem was due to the substitution of the Boron 27 starting material on the DOMEX 500MC high strength sheet [1].

Changing the material without adjusting the stigma of the handler and cutting system on the sheet during laser cutting resulted in numerous defects in the form of cracks, micro cracks and corrugations [2].

* Corresponding author: damian.kulak@samasz.pl

Reviewers: *Richard* Pastričák, *Františka* Pešlová

2 The construction and technological characteristics of the knife holder

The knife holder is one of the parts of the rotating disc, the cutting unit of the Perfect CUT cutter [3].

Fig. 1. Perfect CUT disc strip

This feature is designed to prevent knives from blocking during operation [4]. Initially made of material Boron 27 now made of DOMEX 500MC steel. Total weight is 1.519 kg, # 5 mm sheet.

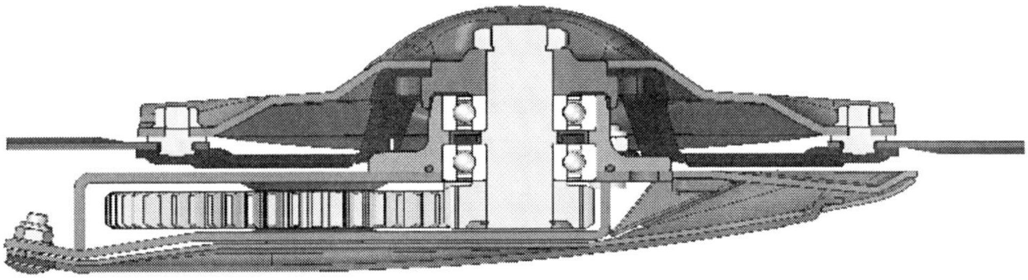

Fig. 2. Cross-section of cutting bar

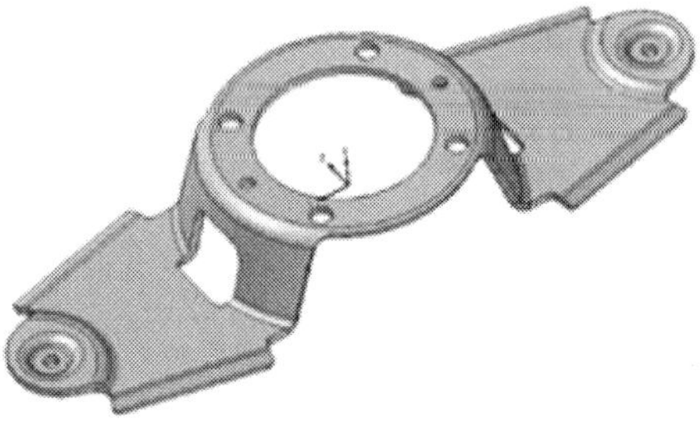

Fig. 3. Knife holder made of boron steel

2.1 Technological process of knife holder

SaMASZ Disc Mower Technology Processes:

- 005 - Cutting Sheet Sheet - LASER
- 010 - Cuts the disk holder - cracks and micro-cracks occur in the process - PRESSES
- 015 - Punch the holes of the disk holder - PRESS
- 020 - Finished (2 pages) - PRESS
- 025 - Temper and temper

Afterwards, the bracelets are subjected to a hardening process and the cracks become even more visible in the process. The heat treatment process aims to improve the mechanical properties of the material by causing significant changes in their internal structure. However, during this process, the accompanying physicochemical phenomena may affect the material so that, instead of improving, it may impair the mechanical properties of the metal materials [5]. In this case, they may cause internal stresses, which will affect the strength of the object at the point of focus and cause permanent deformation and cracks. The above disadvantages associated with the hardening process lead to cracks in the structure of the material from which they are made, which results in an increase in their costs during production. Cracks in the heat-treated parts due to internal stresses create the greatest problems in the manufacturing process of finished goods [6].

3 Analysis, identification of cracks occurring during pre-pressing, corrective actions

At the end of 2016, during the hardness testing of parts after hardening of the quality control department, he noticed cracks and micro cracks in the holders. At the beginning of the study, two parties were analyzed.

Table 1. Summary of the test batches of knife holders

Lot number	Counts pieces in the party	Material	Defective details	% Of defective details in relation to the tested lot
I	900	DOMEX 500MC	70	8 %
II	900	DOMEX 500MC	234	26 %

Then the separated holders - 304 pieces were divided into groups shown in Table 2.

Table 2. Types of defects created after pre-press

Lp.	Kind of defect	The number of items	Dealing with defects
1	Cracks are not visible	137	Conditionally accepted
2	Cracks ≤ 3 mm	21	Cracks were hardened after quenching - allowed for production
3	Cracks above> 3 mm	146	Not allowed for production

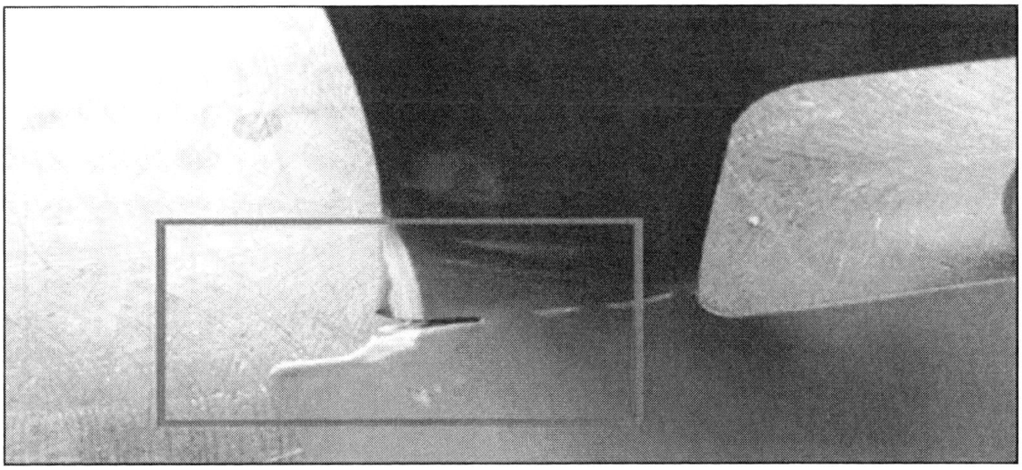

Fig. 4. Cracking occurred during the pressing process, cracking less than 3mm.

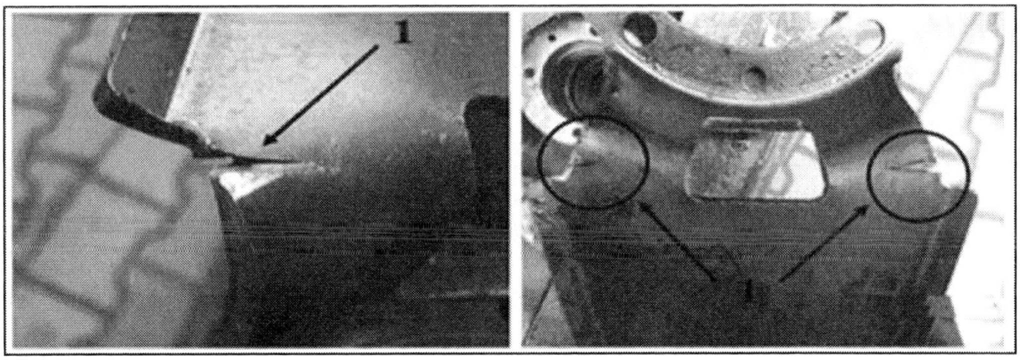

Fig. 5. Cracking occurred during the pressing process, cracking above 3mm.

3.1 Target improvements to the technology process of the holder

SaMASZ made two changes to prevent defects in parts:

A. Construction change - the handle has been redesigned, a change in the blank of the holder has been applied by increasing the radius (figure 6). This change was strategic, because after this change, the crooks do not show cracks.

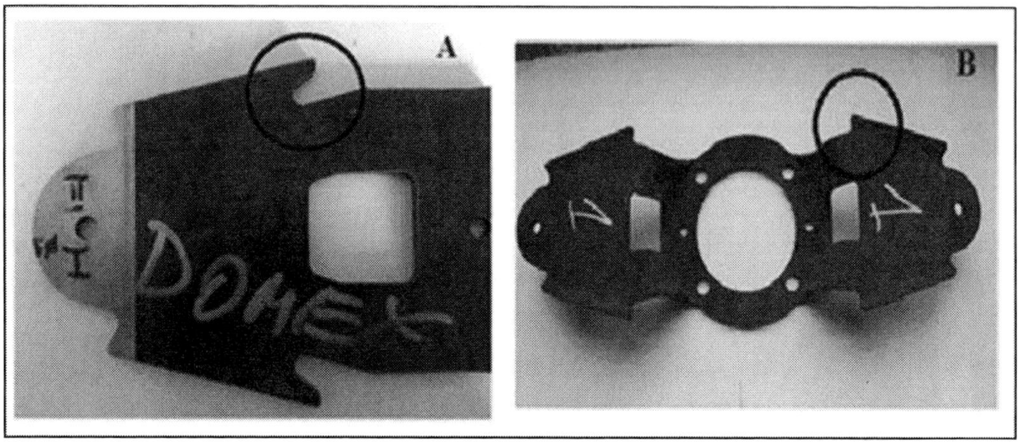

Fig. 6. A - a new type of stick holder, B - old type holder, susceptible to cracks
Both of them are made of DOMEX 500 material

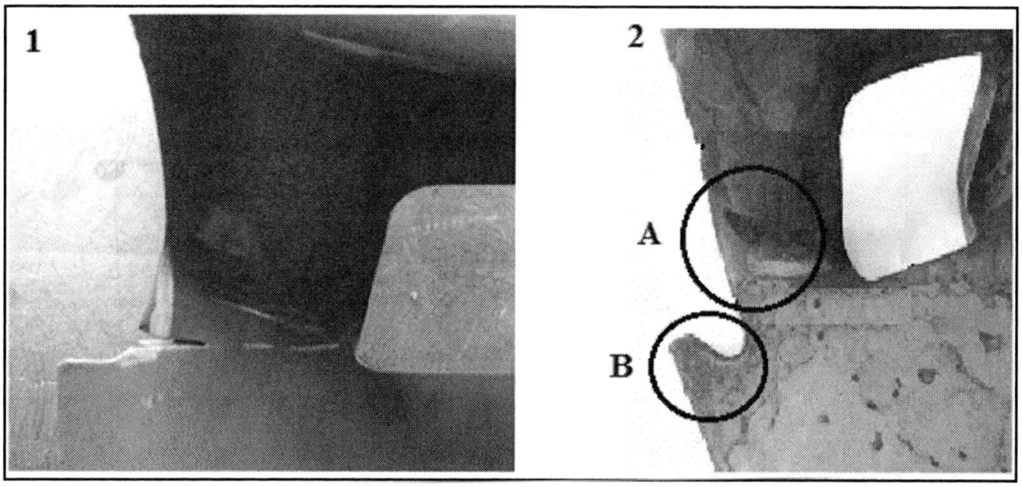

Fig. 7. 1 - old type sticker, 2 - new type sticker
A - after the structural change, there are no cracks, B - increased radius of the sting

B. Change in the organization of the technological process - the second action to avoid cracking on the products is to not postpone the crosshatch of the sheets (Figure 8) across the sheet after rolling the sheet.

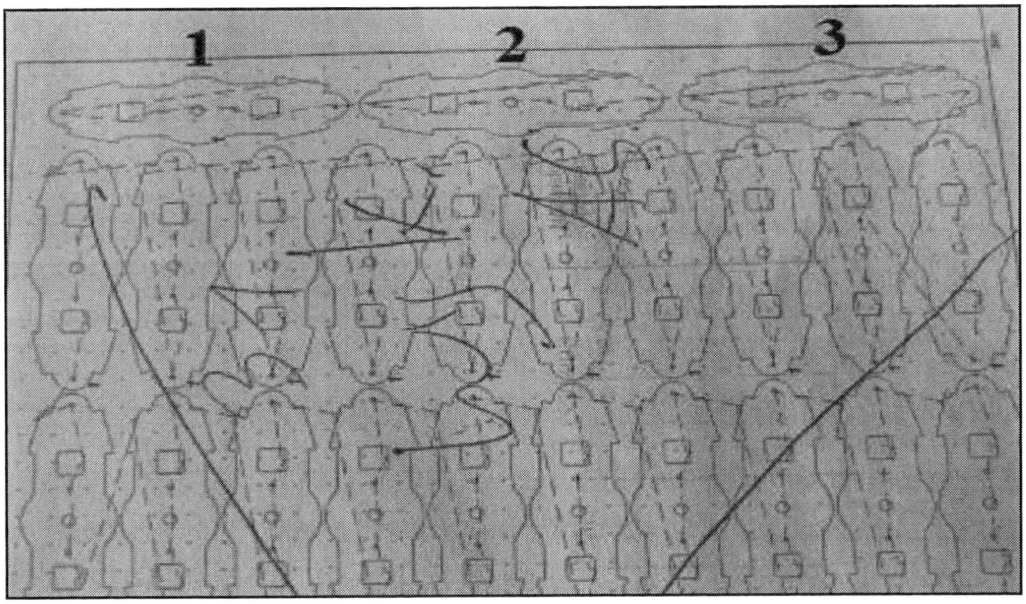

Fig. 8. Incorrect placement of parts on sheet metal - marked 1- 3

4 Endurance tests, bending simulations of the holder during blade replacement, ultrasonic defectoscopy

For the technological research of the holders we used 8 samples made in SAMASZ company. Hardened central knife holders with cracks after pressing, were prepared for checking resistance to cyclic bending- real work simulation using a specially shaped key. View of the station and the way of simulation is shown in Figure 9.

Fig. 9. The stand for research of holders resistance at the PB Institute of Innovation

Table 3. Group I - Holders No. 1 and No. 2 – Cracks not visible to 1 mm, not welded

Example number 1	Example number 2
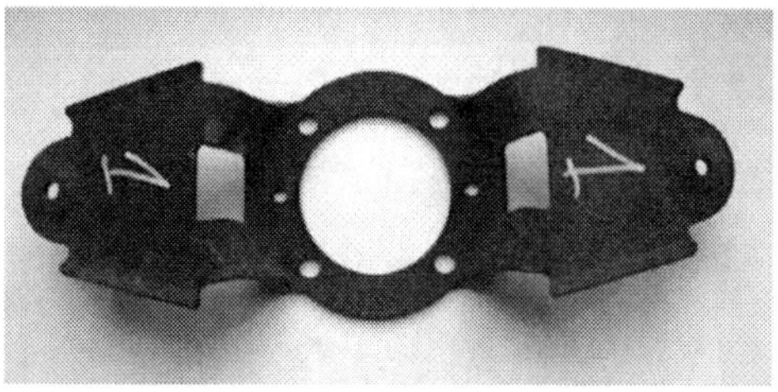	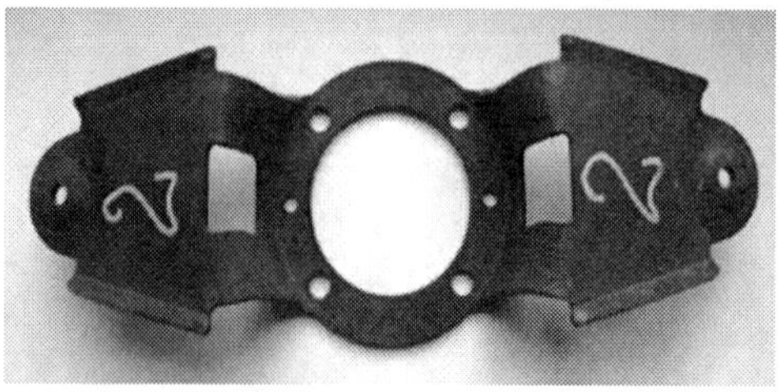

Table 4. II group - holders no 3 and no. 4 - cracks up to 3mm, welded and grinded

Example number 3	Example number 4
	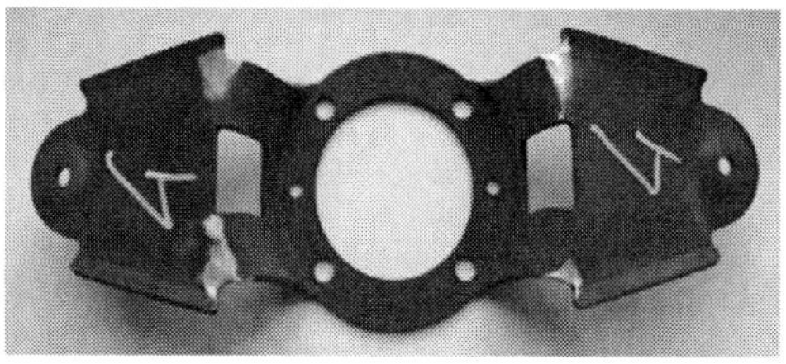

Table 5. Third group - Holders No. 5 and No. 6 - Cracks above 3 mm, welded and grinded

Example number 5 - crack 7.6 mm	Example number 6 - crack 6.3 mm

Table 6. IV group - holders No. 7 and No. 8 - no cracks

Example number 7	Example number 8
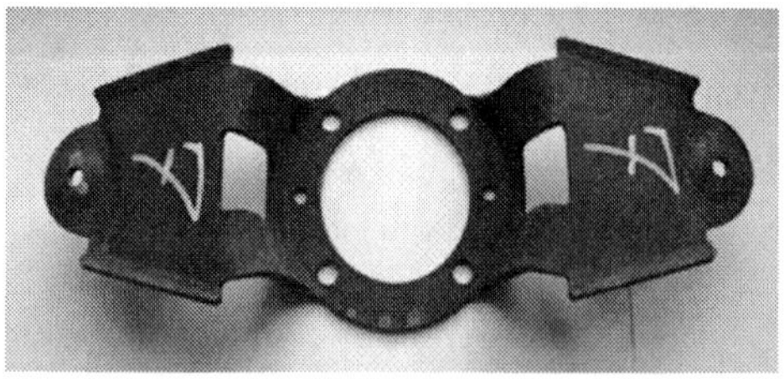	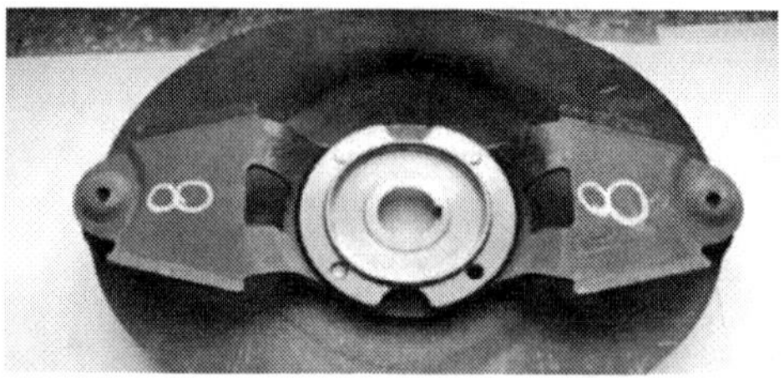

Defected holders were regenerated by welding cracked places by MAG method. The next step was to grinding welds equally with the holder. To make sure if regenerated holders are ready to use in production, We made bending tests for holders exactly with the side of insert and remove of the holders knife.

4.1 Description of research results

Test conditions:

- temperature = 23 ° C
- Frequency = 2 s
- surface condition - surface cleaned, ground
- hardness of the material (DOMEX 500HB)

Table 7. I group - holders No. 1 and No. 2 - cracks not visible to 1 mm, not welded

Sample number	Number of cycles	Conclusions
I	200	Cracks less visible to 1 mm, as at the beginning of the sample test
II	400	
III	600	

Table 8. II group - holders no 3 and no. 4 - cracks up to 3mm, covered and sanded

Sample number	Number of cycles	Conclusions
I	200	No visible cracks in material structure
II	400	
III	600	

Table 9. III group - Holders No. 5 and No. 6 - Cracks above 3 mm, covered and ground

Sample number	Number of cycles	Conclusions
I	200	No visible cracks in material structure
II	400	
III	600	

Table 10. IV group - holders No. 7 and No. 8 - no cracks

Sample number	Number of cycles	Conclusions
I	200	No visible cracks in material structure
II	400	
III	600	

4.2 DI 60 METRISON flaw detector tests

For the reliability of the study, SaMASZ Sp. Z o. o. gave a defectoscope sample without regeneration, with cracks after pressing.

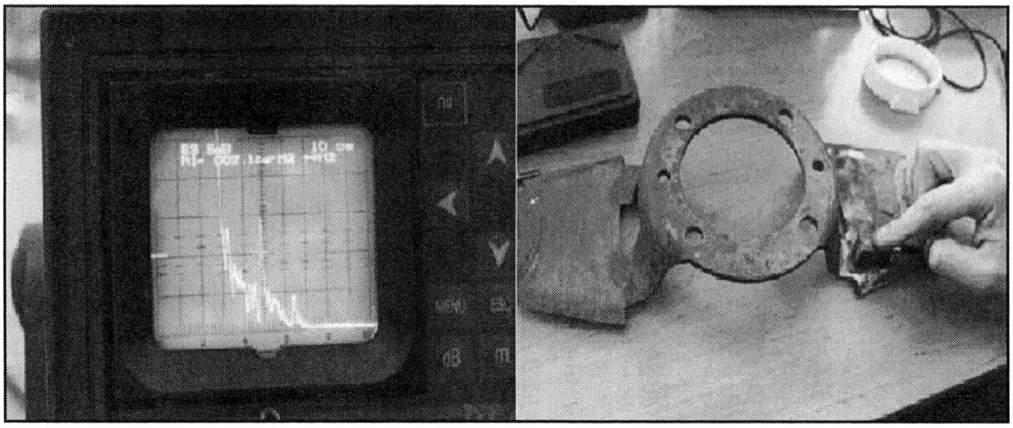

Fig. 10. Diffusion test with DI 60, holder showing cracks

In Figure 10 we see that the stick exhibits discontinuity of material. From ultrasonic flaw detection, we read a distance value of 2.1 cm. The distance of the center of the head from its origin is 1.4 cm, so the microcracks are 0.7 cm from the beginning of the head [7].

Fig. 11. Location for cracks in knife holder

Subsequently, samples subjected to knife replacement simulations were examined. Defectoscopy studies are aimed at detecting and identifying sites of discontinuity [8]. Examination of 8 samples showed no cracks and microcracks in the material structure.

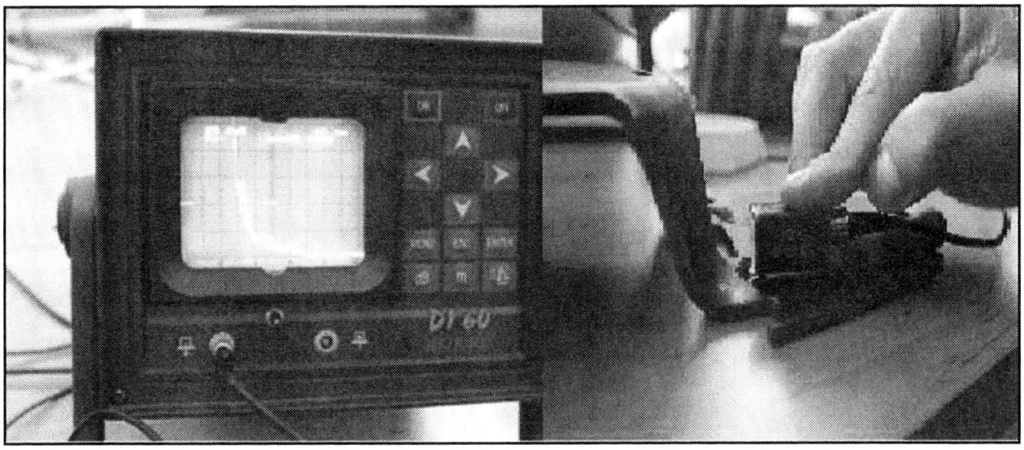

Fig. 12. Defectoscopic examination of 8 samples - no cracks

Conclusions

The analysis of endurance tests and ultrasonic flaw detection allowed us to formulate the following conclusions:

- change in the technological and organizational process used to produce the knife holder, eliminating defects in the manufacture of the work piece;
- regenerated by welding (MAG method) knife holders, can be used for further production.

So, the problem that arose from the need to convert Boron 27 steel to a new type of material (DOMEX 500MC) was resolved. Knife holders, made of new sheet metal and regenerated details, do not show any cracks, so they are safe to use.

References

1. M. Feld, *Fundamentals of technological process design of typical machine parts*. (Scientific and Technical Publishing House, Warsaw, 2003)
2. M. Kapcińska, K. Żurowski, M. Gościański, *Analysis of new generation of materials used for selected elements in machine construction working in extremely difficult conditions of erosion and tribological wear in terms of increasing their service life*. Works by PIMR, TT- **4** (2008)
3. A. Stolarski, S. Rabiczko, D. Kułak, J. Jaroszewicz, *Innovative and energy-efficient manufacturing techniques in machine construction at SaMASZ Sp. Z. o. o*. Energy in science and technology 2012, Publisher Bialystok Technical University, 148-156, Bialystok (2012)
4. J. Jaroszewicz, A. Stolarski, D. Kułak, *Innovative technological and constructional solutions in modern machines for the greenery on the example of SaMASZ Sp. Z o.o.*. Energy in science and technology 2013, Publisher Bialystok Technical University, 57-71, Bialystok (2013)
5. M. Łabęcki, M. Gościański, D. Kapcińska, Z. Pirowski, Tribological, *Strength and structural investigations of selected materials used for elements of agricultural machines working in the soil. Industrial Institute of Agricultural Machinery*. Poznań, **52**, (2007)

6. J. Napiórkowski, *Spatial impact of soil on working elements of agricultural tools,* Agricultural Engineering **17**, (2005)
7. L. Filipczyński, Z. Pawłowski, *Ultrasonic Materials Research. Materials for SITMP course.* (Ed. INCO, Warsaw, 1962)
8. A. Śliwiński, *Ultrasounds and their application.* (WNT Publishing House, Warsaw, 2001)

Fatigue life prediction of mechanical structures under stochastic loading

Bohuš Leitner[1,*], *Lucia* Figuli[1]

[1]University of Žilina, Faculty of Security Engineering, Department of technical sciences and informatics, 1. mája 32, Žilina, Slovakia

Abstract. Problems of fatigue life prediction of materials and structures are discussed in the paper. Service loading is assumed as a continuous loading process with possible discontinuous events, which are caused by various operating conditions. The damage in a material is due to a cumulative degradation process. The damaging process is then represented either by rain-flow matrices or by a fatigue damage function which is derived using some hypothesis of a fatigue failure criterion. Presented theoretical procedure enables a very effective estimation of a service life and/or reliable evaluation of residual life of any structures under various types of loading and environmental conditions. This approach creates a good basis for powerful expert systems in structural and mechanical engineering. The aim of the paper is to present briefly some results of analysis of load-bearing steel structure loads of special railway crane PKP 25/20i which was utilized in some specific ad relatively hard operating conditions. Virtual models of the structure were being used in an analysis of acting working dynamics loads influence to be able to forecast fatigue life of load-bearing of the crane jib.

Keywords: mechanical structures, stochastic loading, material degradation, fatigue damage, measurement of dynamics loads, fatigue life prediction, load-bearing structure

1 Introduction

Optimization of structure design is a decisive condition for effective production. It is exceptionally significant in mechanical and material engineering because either economy demands or waste effects could be considerable in this case. Progressive software (CAD and CAE) techniques have an indisputable advantage compared to traditional empirical-experimental methods because they are by far quicker and less expensive.

So that to accelerate calculation procedures, we must have a suitable databank of operating parameters and of their influence upon a service life. The parameters that affect a durability of structure parts are following: operating loading, structure characteristics, material properties and environmental conditions. Then we must use an appropriate method to estimate a service life and dependability for each combination of the parameters. Methods that enable these objectives are created by a combination of an empirical-

* Corresponding author: Bohus.Leitner@fbi.uniza.sk

Reviewers: *Juraj* Gerlici, *Milan* Vaško

experimental, theoretic-analytical and computer-simulation approach. The fundamental problem is to respect a complete time loading history with possible non-homogeneous events.

Usually, the simulated loading is simultaneously transformed into closed hysteresis loops using a rain-flow method, and a damage accumulation is estimated according to failure hypotheses. In the case of composite materials, various material characteristics as well as material joints must be respected. That brings much more complications in comparison with procedures used for homogeneous materials.

2 Theoretical background for operating conditions modelling

The most serious problem in modelling of damage mechanics is to analyse and simulate operating conditions. The analysis and modelling of service conditions are the input operations in the simulation steps, according to Fig.1.

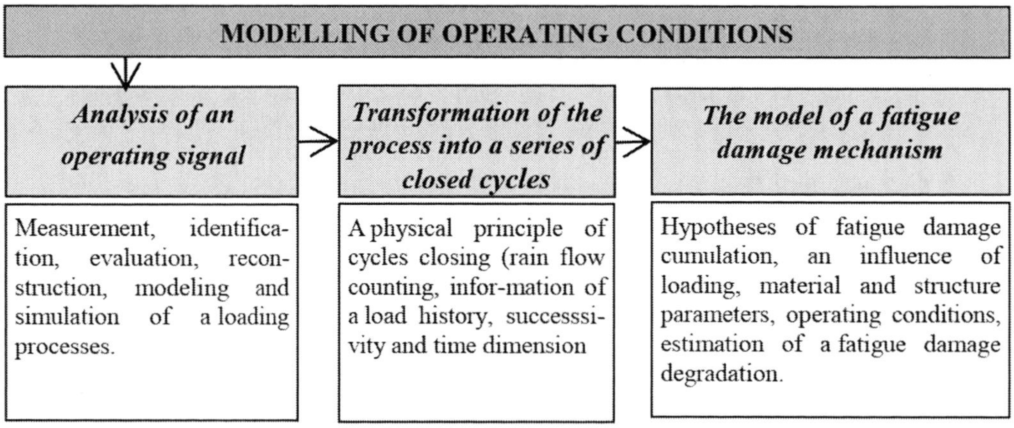

Fig. 1. Diagram of modelling procedure

The correct simulation of service conditions is the necessary condition of a successful performance. Service conditions consist of following items [1]:

- ***operating loading*** (generally, random loading with discontinuous damaging events),
- ***material characteristics*** (endurance limit, ultimate strength, S-N curve, fatigue diagrams),
- ***structure parameters*** (notch stress concentrators, surface treatment, welds, …),
- ***environmental properties*** (corrosion, friction, high/low temperatures, radiation, …).

Operational loading. The analysis and simulation of the operating loading are the most important and complicated problems in the procedure. Methods how to perform the tasks are described in [2]. The methods summary is introduced in Fig. 2.

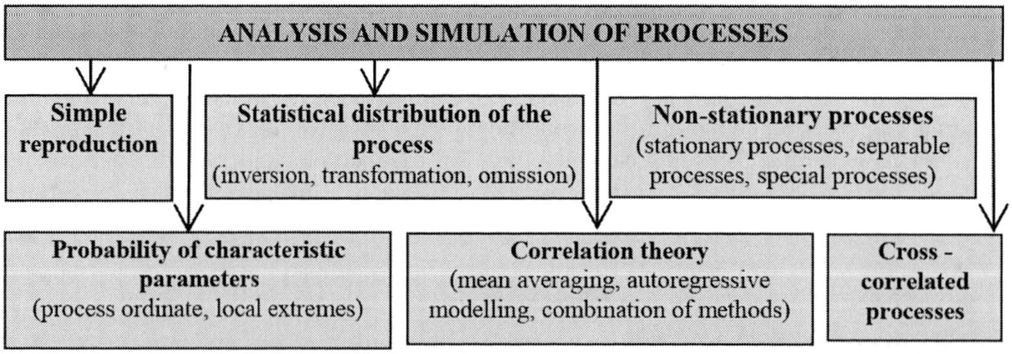

Fig. 2. Methods of a signal processing and generation

An operating loading is mostly considered as a stationary or non-stationary stochastic process. Such a process is completely described by its both probability density function (including a mean value and variance) and power spectral density (auto-correlation function). However, various discontinuous events that are mostly caused by manoeuvres and special service conditions must be respected, too. This problem significantly complicates a simulation procedure, and new non-standard methods and algorithms must be applied [3]. It is also very important to have a reliable algorithm of counting of closed hysteresis loops. Rain-flow methods mostly render possible the procedure only after a definite block of closed loops and they suppose the knowledge of the entire time history before starting to count. In such a case, it is, however, possible neither continual monitoring of a fatigue damage after each cycle nor a running estimation of a residual life during an operation. The algorithm of a simultaneous signal generation and rain-flow counting was proposed in [1]. The method enables to respect a contribution of each closed hysteresis loop in the stress/strain domain continually, corresponding to the on-line generation of a loading process.

Material characteristics. Information of material characteristics of individual structure components is necessary for estimation of a fatigue damage. The most important characteristics are e.g. elastic modulus, shear modulus, strength limit, yield stress, fatigue limit, Woehler curve, cyclic deformation curve, etc. All the parameters must be identified before a simulating procedure, in order to be included into calculation algorithms. Characteristics for each structure component could be considered separately, what enables an investigation of composite structures. It is possible to suppose the time dependence of the characteristics, too. Therefore, we can also investigate composite materials with non-stationary properties.

Structure parameters. Simulated loading must be decomposed for the structure, according to the fatigue strength theory. In order to determine a real fatigue life curve, we must know structure characteristics, e.g. stress intensity factors, notch coefficients, surface sensitivities, size factors, welding parameters, etc. These parameters are then implicitly included in the actual endurance limit, and this way, into calculation relationship, too. In the case of multi-axial loading, some hypothesis of a stress recalculation must be applied. For example, the strain energy hypothesis for a combination of bending and torsion has form [1]

$$\sigma_{eq} = \sqrt{\left(\sigma_a + \psi_\sigma . \sigma_m\right)^2 + 3.\left(\tau_a + \psi_\tau . \tau_m\right)^2} \tag{1}$$

where σ_a is the amplitude of a normal stress, σ_m is the mean value of a normal stress, τ_a is the amplitude of a shear stress, τ_m is the mean value of a shear stress, and ψ_σ, ψ_τ are corresponding sensitivities to the asymmetry of a cycle.

Environmental properties. Parameters that correspond to environmental conditions have a serious influence upon a fatigue life, too. The factors are above all: corrosion, high or low temperatures, friction, radiation, etc. They could be also implicitly included into a calculation relationship. The advantage of this approach is again in the ability to consider possible non-stationary environmental characteristics [4]. That is very convenient for composite structures, too.

3 Estimation of material degradation parameters

Two basic methods for an estimation of a material degradation can be mostly used (Fig. 3). The first one is based on a rain-flow matrices creation, and according to the other method, we must specify a fatigue damage function.

Rain-flow matrices of loading and damaging. The *rain-flow matrix of loading* (RML) is created by elements $l_{i,j}$ which represent the number of hysteresis loops that have been originated in the i-level and that have started their closing in the j-level [2].

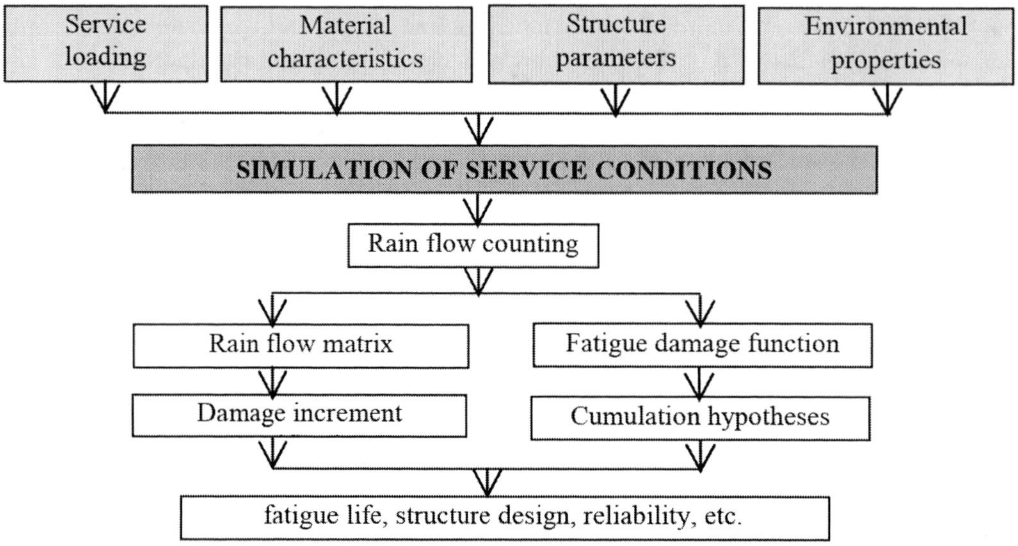

Fig. 3. The estimation procedure

In the case of non-stationary loading processes, and for non-linear hypotheses (or for cyclic unstable materials), the time moment of a given loop (or it's closing sequence) is also important, if need be an information about the foregoing hysteresis loop (or about more loops), too. Then, the rain-flow matrix must be extended by the next dimension, i.e. it must be a three-dimensional one. Each closed hysteresis loop produces some definite corresponding damaging influence. Indicating a damaging effect of the loop as $d_{i,j}$, we would obtain also a square *rain-flow matrix of damaging* (RMD) which is composed of the elements $d_{i,j}$. RMD could be three-dimensional, too. There is no problem, how to identify a damage quantity $d_{i,j}$, i.e. how to identify a damaging effect of a sole hysteresis loop. For this purpose, we must use some of hypotheses of a fatigue cumulative damage. In the case of a linear rule of fatigue damaging, we constitute two-dimensional RMD. Though, if we accept some non-linear hypothesis, the rain-flow matrix should be three-dimensional, too. Having both the RML and RMD, we can calculate the estimation of an accumulated fatigue damage D as their scalar product [2], i.e.

$$D=\sum_{i=1}^{n}\sum_{j=1}^{n} l_{i,j}.d_{i,j}\,. \tag{2}$$

Fatigue damage function. A creation of a fatigue damage function is the other possibility how to estimate degradation. Such a function is determined in the following form

$$D_n(\sigma_a)=k(\sigma_a).n^{m(\sigma_a)} \quad ; \quad k(\sigma_a)>0 \quad ; \quad m(\sigma_a)>0 \tag{3}$$

where $k_{(\sigma a)}$ and $m_{(\sigma a)}$ are definite parameters which could be identified on the basis of fatigue damage criteria. The function depends both on the number of passed closed cycles and on the character of loading, i.e. on amplitudes and mean values of the cycles. Using this method, a continuous monitoring of a fatigue damage process is possible, and the time history is respected, too. The choice of the damage criterion is the problem in this case, because the well-known linear Miner´s hypothesis is frequently not the optimal one, and other hypotheses must be adjusted to a special case, i.e. only the sole closed cycle is in each block of a macro-block, then. Frequently, the derivative function of a fatigue damage addition has the form

$$d_n(\sigma_a) = k(\sigma_a).m(\sigma_a).n^{m(\sigma_a)-1} , \tag{4}$$

that corresponds to the above mentioned damaging effect of a definite closed loop.

Summing the function, the number of cycles till the critical damage accumulation N_{kr} can be estimated according to the relationship [2]

$$\sum_{n=1}^{N_{kr}} d_n(\sigma_a) \leq 1 \leq \sum_{n=1}^{N_{kr}+1} d_n(\sigma_a) \ . \tag{5}$$

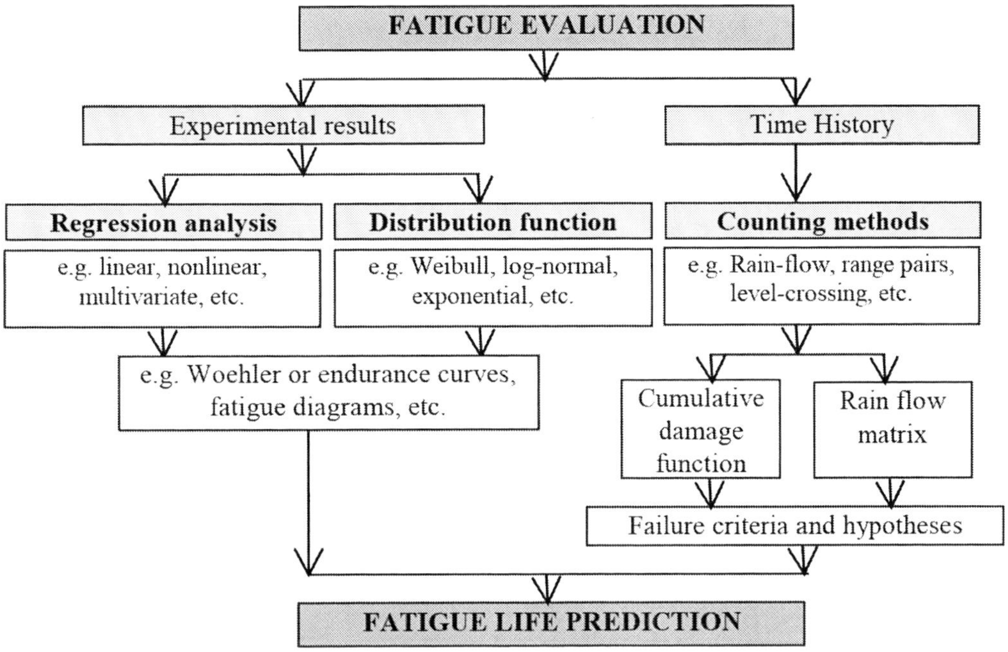

Fig. 4. The procedure of a fatigue life evaluation

4 The computational scheme of a fatigue life prediction

The procedure of a fatigue damage evaluation in order to predict a fatigue life is in Fig. 4. We can dispose either of experimental measured results or computer simulated data. Experimental data are processed using a regression analysis or a frequency distribution function is determined [5].

A selection of possible hypotheses is optional and it depends on a specific situation or on a philosophy of a fatigue damaging process. This way, a lot of tasks can be solved, e.g. an estimation of a fatigue life of a given structure under given loading, optimisation of a material choice, structure dimensioning, investigation of an environmental conditions influence, etc.

5 Case study - prediction of fatigue life of load bearing structure

A special railway crane PKP 25/20i is lifting equipment designed for exploitation in a limited space conditions which are met by the building of railway (Fig. 5). His main working purpose is to lay down or to tear off rail fields of maximal length 25 m and mass of

20 tons by a complex reconstruction or by renovation of the railway. Determination of real working conditions with a choice of typical loads was important for calculation of fatigue life of a crane bridge as a carrying steel structure. There was possible to build a theoretical load spectrum of a crane load-bearing structure on that base.

Fig. 5. A Special Railway Crane PKP 25/20i

It was necessary thoroughly analyse functions of the crane one by one and to determine how they influence the fatigue life by building of characteristic load spectrum of crane bridge. After a theoretical analysis of single working functions were following characteristic working functions of a crane:

- *tearing off rail fields* - the limiting function of a machine (high load of a crane bridge by high mass and dynamic effects by tearing off rail fields out of gravel bed),
- *laying down rail fields* - the limiting function of a crane,
- *auxiliary working functions* - manipulation of carriages, travel of a crab (type of complemented cycles, low influence on the total fatigue life of crane bridge),
- *travel of a crane* - function between basic technological workings (no important influence on fatigue damage of crane bridge),

It is obvious from the analysis of working conditions that the most exacting working conditions are the tearing off the old rail fields and that this function has the decisive influence on fatigue life from the point of view of number of cycles and loads.

5.1 Selection of measuring conditions

On the base of knowledge of fatigue life theory and analysis of working conditions was decided that an experimental measurement would be done in a real practice. The measurement was planned to realize by the renovation of a railway track. Before actual measurements there was necessary to determine critical locations of the crane bridge and to realize a theoretical analysis of working conditions of a crane. These were chosen on the base of static conditions calculations using model of FEM software package (Fig. 6).

The outputs of them were the locations for location of tensiometers on the crane bridge according to Fig.7.

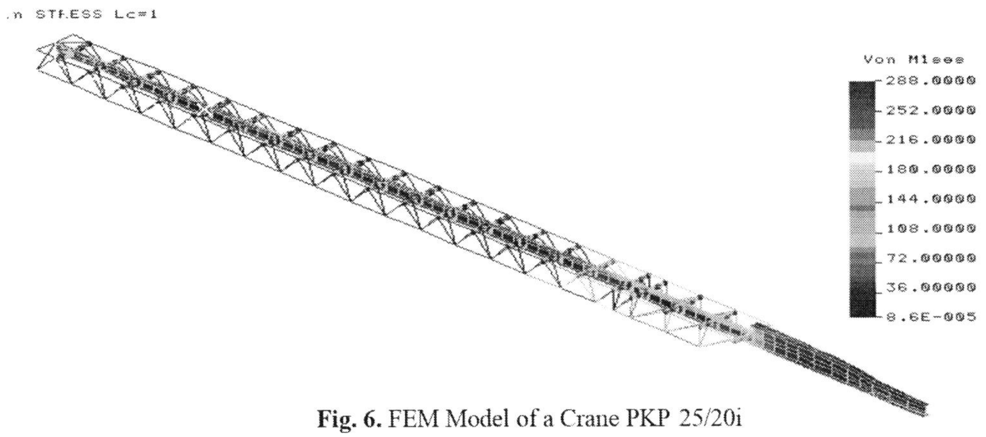

Fig. 6. FEM Model of a Crane PKP 25/20i

Analysis of measured data. By analysis of measured values of stresses was determined:
1. *Stress by tearing off rail fields has a periodical character* (Fig.8). The dynamic increase of stress was about 10-20 % of the maximal amplitude. The most expressive progress of stresses was determined during tearing off, start, stopping and lazing down rail fields. The measured value of stress amplitude was average 175 MPa.

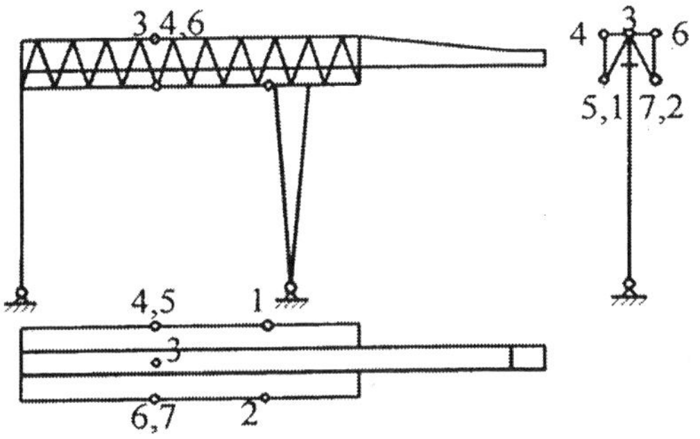

Fig. 7. Locations of tensiometers on a crane bridge

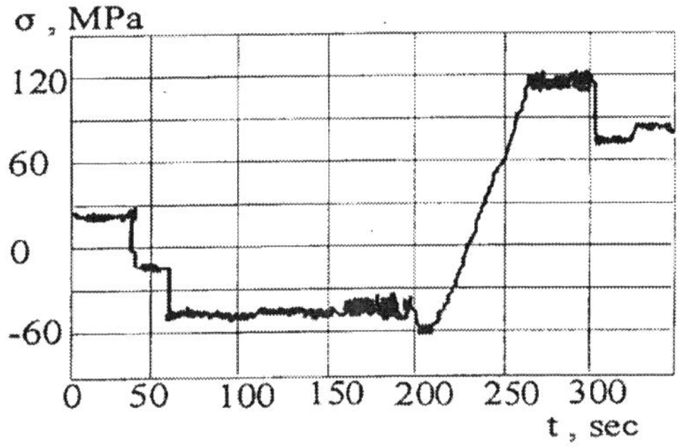

Fig. 8. Stress by tearing off, start, stopping and lazing down of rail fields

2. *The manipulation with railway carriage* (Fig.9) using turn able part of crane bridge was characterised by a non-stationary process by measured values of stress in average 20 MPa.

3. *Travel of a crane to the next rail field* was characterized by a non-stationary progress of stress increase, by which a maximal absolute value of stress increase was 34 MPa (Fig 10).

For completeness in Tab.1 are the absolute values of the stresses in selected critical cross sections of the crane structure (obtained experimental measurements

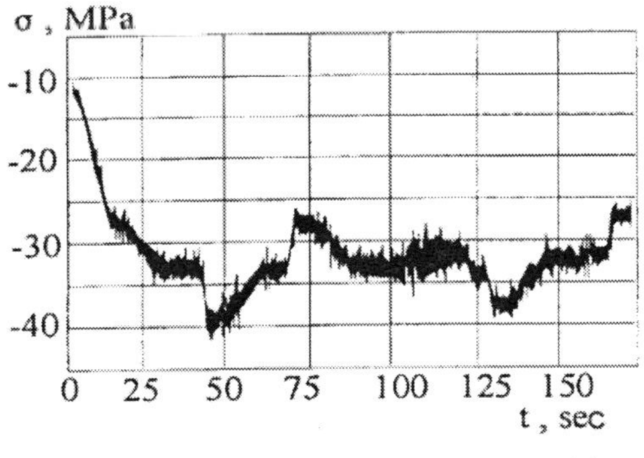

Fig. 9. Stress by manipulation using turntable part

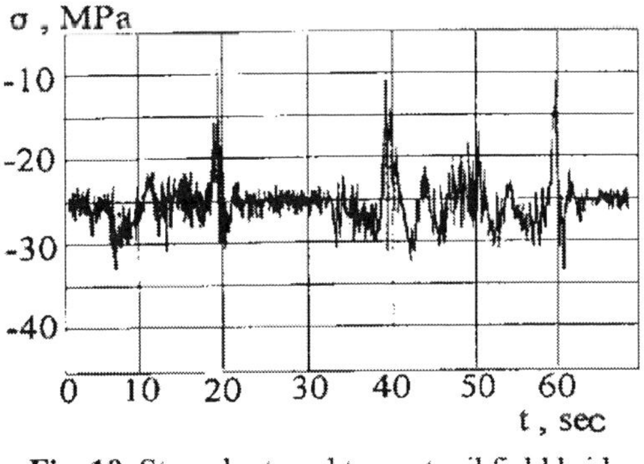

Fig. 10. Stress by travel to next rail field bridge

Table 1. Tensions in the critical points of the structure

Tensiometer number	Measurement number [MPa]				
	1	**2**	**3**	**4**	**5**
1	-85.22	-93.33	-87.27	-104.91	-95.62
2	-89.65	-89.42	-100.10	-94.76	-87.71
3	139.12	165.37	151.09	153.68	154.32
4	176.29	165.27	170.62	176.46	175.83
5	135.77	150.54	140.65	147.62	138.96
6	170.51	150.99	165.96	172.43	170.83
7	139.15	152.33	137.92	147.49	144.89

Spectrum of loads. On the base of an analysis of working conditions and results of experimentally measured increase of stresses during working of a crane was possible to construct a real spectrum of loads (Fig.7). Single working cycles introduce characteristically working conditions of a railway crane, which consist of following working cycles:

- *A - tearing off old rail fields* (σ = 175 MPa, n = 1250)
- *C - auxiliary working cycles*

  C1 - free travel of crabs (σ = 40 MPa, n = 3000 *cycles*),

  C2 - travel to the next rail field (σ = 34 MPa, n = *i cycles*),

  C3 - manipulation with of railway carriage (σ = 20 MPa, n = 650 *cycles*).

5.2 Fatigue Life Prediction of Crane Load-Bearing Structure

The general procedure of life prediction from point of view of fatigue damage was based:

- determination of the most significant working regimes and factors of typical working conditions from point of view of structure loads and their following activities [5-7],
- identification of carrying structure critical parts and determination of their working loads,
- analysis of choice material properties and estimate of effects of different types of notches, connections and nonlinearities in examined points of structure [8,9] and
- application of suitable hypothesis of fatigue damage cumulation and quantification of predicted fatigue life of selected parts of carrying structure [10, 11].

The most significant working regimes. The activities connected with technology of tearing off the old rail fields and laying down the new ones were identified as the most significant working regimes of mentioned crane and some of accessory processes as no-load run of a crab, handling of railway bogies and travelling to the following rail field too. The effect of surroundings low temperatures was chosen from surrounding conditions because they cause freezing of the rails bed and following increase of the pulling force by tearing off rail fields. The effects of surroundings conditions connected with unevenness (longitudinal or transverse) of newly laid down rail bands were not taken into account because it has not an expressive effect on crane working.

The identification of critical parts of structure. The virtual FEM model was elaborated and verified to use in the procedure of identification of supporting structure critical parts. The stresses experimentally obtained were utilized during the actualization of the FEM model in order to obtain the maximum adequacy of the model comparing to the real structure. It was determined from static and dynamic analyses realized. The critical parts are namely upper and lower beams of supporting structure (EL232, EL237, EL89 and EL93) and places near of locking connection of C and D parts (EL335) on (Fig.11).

Material properties of structure details. Supporting structure of the crane bridge is made from the low carbon steel 11 523. The Woehler curve of the used material was determined after the STN standards. Obtained Woehler curve of the load/bearing structure material and its parameters were utilized as an input into the suitable hypothesis of fatigue damage cumulation (w = 5,8, N_C = 2,5.106, σ_C = 190 MPa and R_e = 355 MPa).

Prediction of working fatigue life of a load-bearing structure. The principal hypotheses of fatigue damage cumulation were applied during prediction of working fatigue life of a crane supporting structure. It was made by means of quantification of cumulation fatigue damage process too. Three principal hypotheses of fatigue damage cumulation were applied. There were used the Palmgren-Miner (P-M) and the Corten-Dolan (C-D) hypothesis which are based on utilization of corresponding material curve and actual loading spectrum.

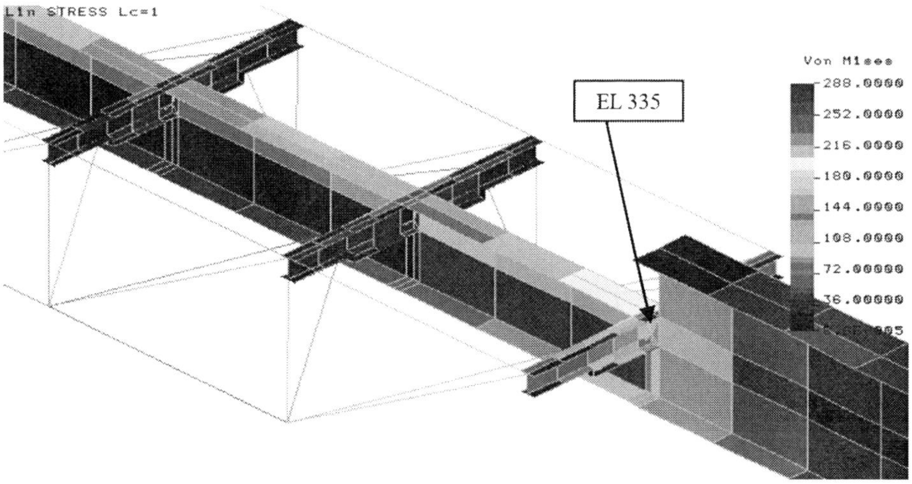

Fig. 11. Detail of *C* and *D* parts in locking connection

Rajcher's theorem defines the fatigue damage in the critical location of the structure part induced per one second and is expressed by the following equation in form [1]

$$D_s = \frac{\Gamma\cdot\left(\frac{w}{2}+1\right)\cdot\left[2\cdot\int_0^\infty f^{\frac{2}{w}}\cdot S(f)\,df\right]^{\frac{w}{2}}}{N_C\cdot\sigma_C^w} \tag{5}$$

where w is exponent of S/N curve, σ_c is fatigue limit, N_c is limit number of cycles to failure, f is frequency, $S(f)$ is spectral power density of the stress loading process, Γ is gamma function value. *Time until the next failure* (in hours) can be expressed as follows

$$T = \frac{1}{3600\cdot D_s} = \frac{N_C\cdot\sigma_C^w}{3600\cdot f_e\cdot\left(2\cdot s_\sigma^2\right)^{\frac{w}{2}}}\cdot\Gamma\cdot\left(\frac{w}{2}+1\right). \tag{6}$$

It is obvious that all the process can be realized only by means of the computer technique efficient enough. The approach in practice is that after import or calculation of the process spectral power density values $S(f)$ the process standard deviation S_σ and process effective frequency f_e of the probability density will be determined. The practical application of the presented process was realized by the program created in the *MATLAB*© environment. The worked computational program named *FATIGUE.M* was used at the fatigue life computational estimation in the selected critical points under the chosen characteristics of the crane operating conditions [6]. The Rajcher's hypothesis of fatigue damage cumulation uses Woehler curve parameters, Gamma function and power spectral density (PSD) of a structure response an acting working loads [11]. Obtained results of estimated measure of load-bearing structure fatigue life are in Tab.2.

Table 2. The estimate of fatigue life of critical elements of crane carrying structure

Fatigue life prediction based on hypothesis' [in years]	**Element 232**	**Element 89**	**Element 335**
P-M (Palmgren–Miner)	24.2	23.1	25.0
C-D (Corten – Dolan)	19.6	19.2	21.2
R (Rajcher)	20.2	19.9	20.4

Demand for this procedure was made by its user. The term of life recommended by producer was 15 years by acceptation of determined technical conditions and recommendations. It was possible to declare basing on results obtained that it is possible to utilize the crane after proposed term of technical life from the point of view of fatigue life. However, the preventive checks of critical points were recommended in each 2-yers interval.

Conclusions

It follows from obtained results from point of view of safety and reliability of the crane PKP 25/20i working that its supporting structure - bridge of a crane is designed sufficiently. But it is necessary to respect proposed technology of its utilization and maximum value of loads. However, the non-reversible deformation and structure damage can occur during certain working conditions (for example tearing off rail field from frozen bed, strong wind impact or non-qualified manipulation and control resulting in the loss of crane stability). It is necessary to state finally that the fatigue damage cumulation process and fatigue life estimation can be made with certain probability only using certain assumptions and simplification despite of having available lot of data from the areas of fatigue strength.

This work has been supported by project VEGA No. 1/0240/15.

References

1. M. Růžička, M., Hanke, M., Rost, *Dynamic Strength and Lifetime* [in Czech]. (CVUT, Prague, 1989)
2. J. Čačko, *Modelling of a Fatigue Cumulative Damage under Random Loading Conditions*. Fracture Mechanics: Applications and Challenges, Elsevier, Amsterdam 2000, 113-120 (2000)
3. A. Sorg, J. Utzinger, B. Seufert, M. Oechsner, *Fatigue life estimation of screws under multiaxial loading using a local approach*. International Journal of Fatigue **104**, 43-51
4. A. Chovanec, *Analysing and modelling off-road vehicle availability*. Transport Means: proceedings of the 16th International conference, October 25-26, 2012, Kaunas University of Technology, Lithuania, 54-57 (2012)
5. P. Kopas, L. Jakubovičová, M. Vaško, M. Handrik, *Fatigue resistance of reinforcing steel bars*. Proceedia Engineering **136**, 193-197 (2016)

6. M. Vaško, B. Leitner, M. Sága, *Computational fatigue damage prediction of the lorry frames under random excitation*. Communications **12** (4), 62-67 (2010)
7. J. Gerlici, T. Lack, *Railway wheel and rail head profiles development based on the geometric characteristics shapes*. Wear **271** (1-2), 246-258
8. J. Dižo, M. Blatnický, *Use of multibody system dynamics as a tool for rail vehicle behaviour diagnostics*. Diagnostyka **17** (2), 9-16 (2016)
9. Q. Dong, G. N. Xu, H. L. Ren, A. H. Wang, *Fatigue remaining life estimation for remanufacturing truck crane Jib structure based on random load spectrum*. Fatigue and Fracture of Engineering Materials and Structures **40** (5), 706-731
10. M. Vaško, M. Blatnický, P. Kopas, M. Sága, *Research of weld joint fatigue life of the AlMgSi07.F25 aluminium alloy under bending-torsion cyclic loading*. Metalurgija **56** (1-2), 94-98 (2017)
11. M. Sága, M. Vaško, P. Pecháč, *Chosen numerical algorithms for interval finite element analysis*. Procedia Engineering **96**, 400-409 (2014)

Numerical modelling of heat flows in the upper blast furnace of the electric arc furnace

Richard Lenhard[1,*], *Milan* Malcho[1], *Peter* Ďurčanský[1], *Katarina* Kaduchová[1]

[1]University of Zilina, Department of Power Engineering, 010 26 Zilina, Slovakia

Abstract. Contributing shows one of the alternatives to use the waste heat from an electric furnace hat with subsequent use for the need for heating water preparation and central heating. In a contribution is presents a numerical model of heat flows from electric arc furnace designed for, into a cooled furnace from which waste heat is led to the central heating system. The model was developed implementing the finite volume method in ANSYS-Fluent program code.

Keywords: heat flow, arc furnace, numerical modelling, heat recovery

1 Background

Sustainable energy supply, energy price development and legislative measures in the environmental field are factors that clearly encourage such technical solutions that are geared towards low- and medium-term waste heat saving. Waste heat as source of thermal energy is advantageous to reuse as secondary energy source preferably in the technologies that produce it. Waste heat arise in general as a consequence of fuel and energy consumption in technological processes in which they are not used because their parameters are inappropriate for the maintained technology. However, waste heat can be a source of energy for other devices in which they can wholly or partly replace fuel or energy.

One way to reduce the cost of thermal energy is to recover waste heat through heat exchangers of different designs depending on the application. Technological processes characterized by a massive potentially usable waste heat flow are characteristic for the production of ferroalloys in electric arc furnaces.

Our proposed device is part of an energy system that performs cooling of exhaust fumes and at the same time acquires unused heat from the technological process of production of ferroalloys in an arc furnace. The device is developed and designed purposefully for use in specific location. It is composed of a furnace hat, door panels and ring, as is shown in Figure 1. The hat of the furnace is designed as a pressure vessel divided into three independent segments. The segments consist of a set of plate exchangers and a tubular structure. Each of the three segments consists of sheet panel coolers and of the tubular segment. Each segment is a separate group of pressure vessels. Radiator panels are large-area plate heat exchangers. The planar walls of the plate heat exchangers are suitable for the construction of thermal systems up to a temperature of 450 °C. They are reinforced by a

* Corresponding author: richard.lenhard@fstroj.uniza.sk

Reviewers: *Tomasz* Domański, *Milan* Žmindák

system of veneers between the walls. Inside the panel there are partitions that direct the water flow across the entire panel surface. Radiator panels are installed in three groups around the perimeter of the bottom of the hat. Each of the panels disposed in a pipe section forms a separate heat transfer system, the panel and the tube segment receive radiated heat from the furnace and through the hot water in the heat exchanger transmit it to the hot water circuit.

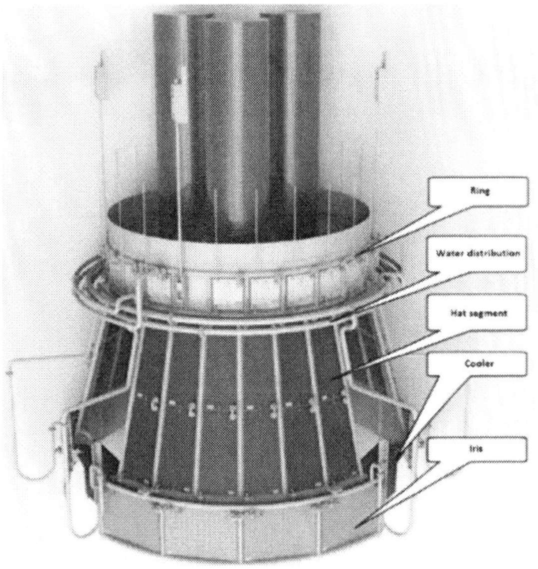

Fig. 1. Hat of furnace – 3D sketch

2 Numerical simulation of heat transfer

Furnace model was developed in order to determine theoretical performance limits of the heat recovery system and to simulate the energy flow into the modified types of heat exchangers. The basic construction nodes, hat, aperture, flue gas volume and electrodes, served as input geometry for heat flow calculation [1]. The proposed model is shown in Figure 2.

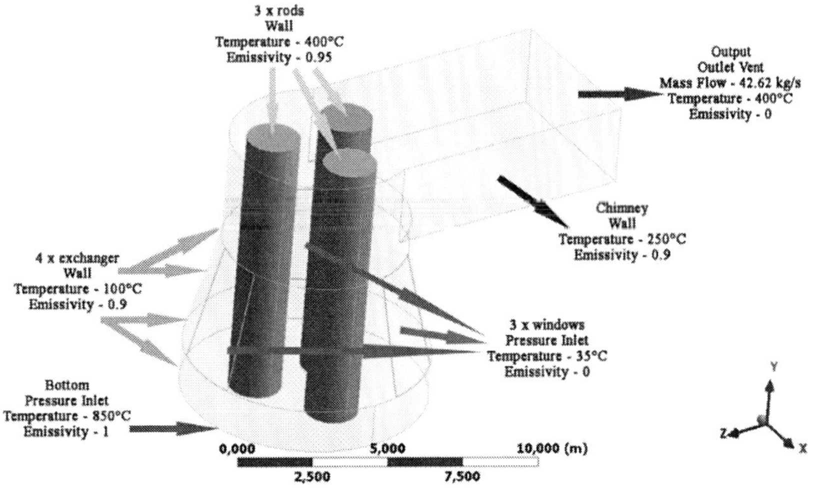

Fig. 2. Boundary conditions for the model

A side effect of the proposed equipment is to reduce the exhaust flue temperature and to improve the working environment by lowering the temperature around the furnace. The electric arc and batch in the furnace is the source of heat radiation that is out of the furnace absorbed by the hat and in the space for loading between the hat and the floor around the furnace by three vertical movable curtains with heat exchangers. [2, 3]. Thermal radiation in the furnace is not constant, has a stochastic character and changes through time and space in the furnace, depends on the type of batch and on the skills and experience of the loading operator [4, 5, 6]. The creation of the furnace three dimensional models allowed numerical simulation of the energy flow into the heat exchangers in order to find out the heat recovery system performance. In Figure 2 are the input parameters for calculation of temperatures, velocities and heat flow. [7] The geometry of the simplified furnace was created in Design Modeller. The model system consists of 4 superposed cooling panels, see Figure 3, then of 3 electrodes, see Figure 4, with openings where the electrodes are supplied, and the coating, as in Figure 5, which replaces the cover of the openings with the electrodes themselves. The other parts of the furnace, e. g. the exhaust, was also presented, to display flue gas flow, see Figure 6. The whole inner volume of the simplified furnace model was considered as fluid [8, 9].

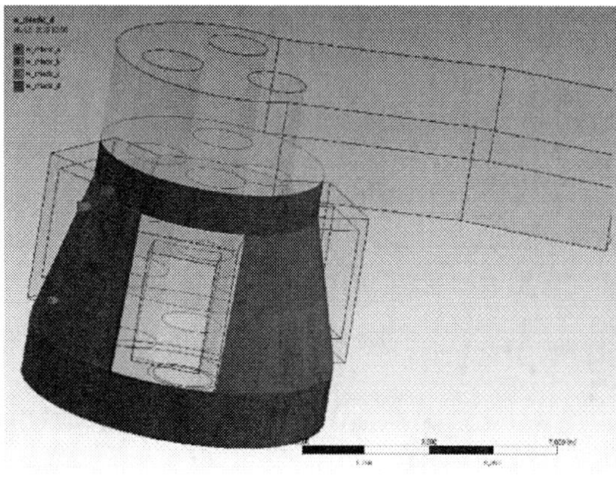

Fig. 3. Location of cooling panels

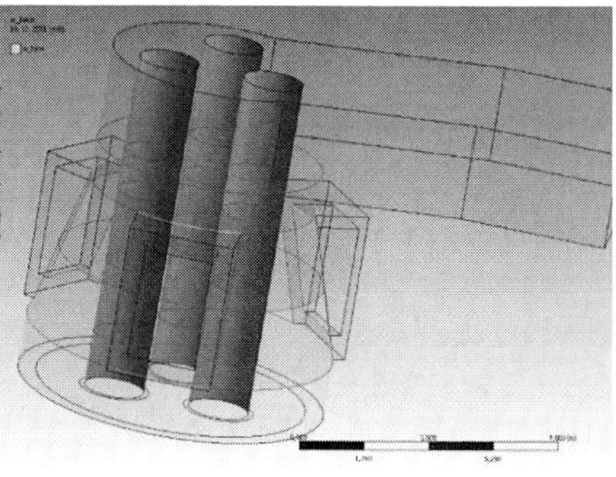

Fig. 4. Electrodes

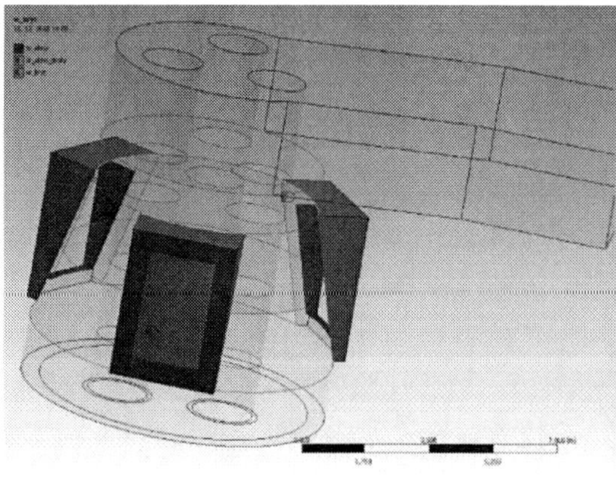

Fig. 5. Model of holes cover where the electrodes are fed

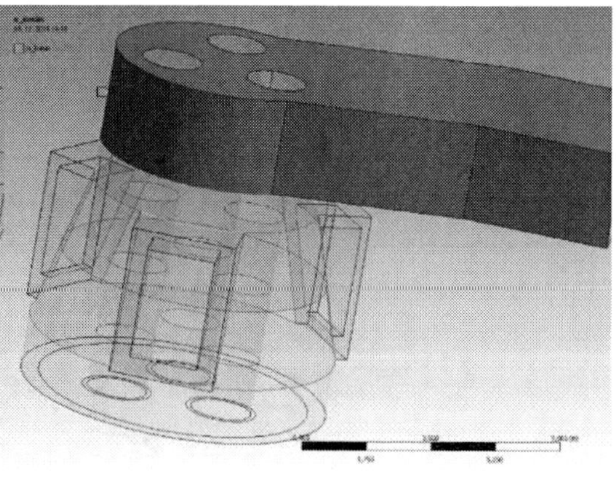

Fig. 6. Flue gas exhaust

3 Setup and numerical solution

The computing mesh for the 3D simplified model was created in the Meshing module. The network itself contains 332 564 cells. For the mesh creation a patch conforming method has been used to create tetrahedrons as shown in Figure 7. Also the higher crosslinking features Curvature and Proximity were used [10, 11].

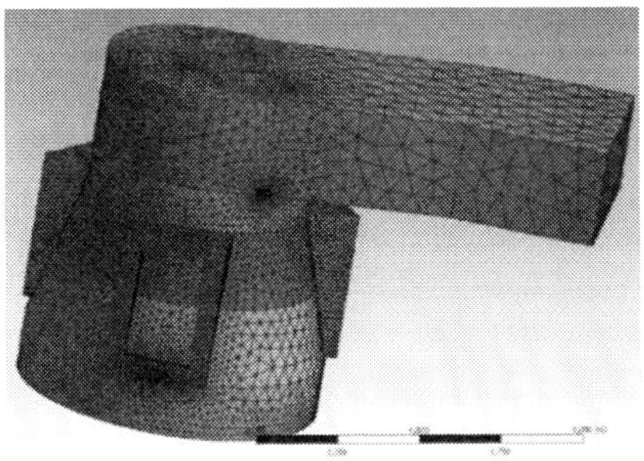

Fig. 7. Mesh before conversion

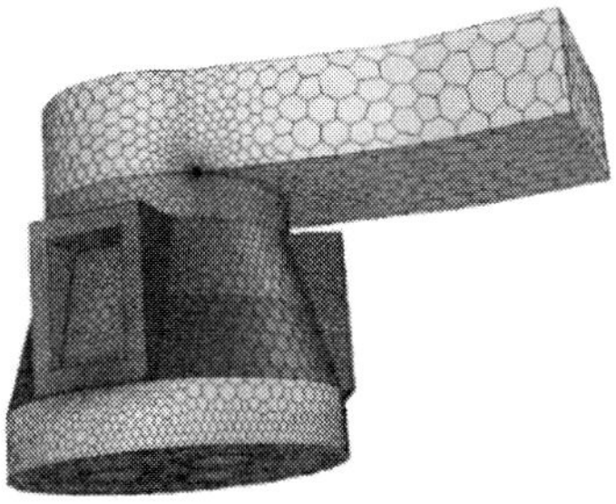

Fig. 8. Mesh converted to polyhedral mesh

To improve and accelerate numerical simulation, the network was further modified by converting to a polyhedral mesh. With this adjustment, we have achieved better job readability. The adjustment itself was done in the ANSYS Fluent software. The result is in Figure 8. The k – ε turbulence model was chosen to perform the numerical simulation, with the realizable settings. Radiation model Discrete Ordinates with a 5 to 5 resolution has been turned on [12]. This model in the radiation transmission equation addresses radiation intensity that is positional and direction dependent [13]. Radiation is counted in each quadrant or octant. The handle transport equations are computed similarly as for flow and energy solutions. The simulation itself was performed in the ANSYS Fluent program with boundary conditions set according to Table 1.

Table 1. Boundary conditions

Name of boundary	Value entry	Velocity	Temperature	Emisivity
	Condition type	kg/s	°C	
in_ground	pressure-inlet	-	1 000	1
in_window	pressure-inlet	-	35	0.5
out	outlet-vent	42.62	400	0.5
w_cool_a	wall	-	100	0.95
w_cool_b	wall	-	100	0.95
w_cool_c	wall	-	100	0.95
w_cool_d	wall	-	100	0.95
w_ground_a	wall	-	2 000	1
w_ground_b	wall	-	-	1
w_flue	wall	-	250	0.9
w_cover	wall	-	-	0.5
w_wind_wire	wall	-	-	0.5
w_wire	wall	-	400	0.95

The flue gases were considered as the working medium, the walls of the model were designed as simplified surfaces with number of layers, thickness, temperature and emissivity.

Numerical simulation took place under standard operating conditions. In the simulations, a steady transfer of heat was assumed by the radiation from the molten layer (Table 1 in_ground) of the selected parameters. To achieve the continuous rate of flue gas flow, the velocity was chosen at the outlet by specified temperature [14]. The calculation was considered as stationary.

4 Results of numerical simulation

The result of the simulation is the calculation of the radiation and convection component of the heat transfer into the four zones in the furnace hat. On the basis of entered boundary conditions, 2.5 megawatts have been obtained from the cooling panels, as is described in Table 2. The model is divided into 4 zones. The 1st zone forms the cylindrical surface on which heat exchangers are designated as lifting aperture. The 2nd zone consists of heat exchanger in the lower slanted part of the hat. The 3rd zone forms the heat exchanger in the top slanted part of the hat. The 4th zone forms the heat exchanger on the cylindrical ring at the top of the furnace hat.

Table 2. Total transferred heat power by exchangers

Name of boundary	Zone	Power (W)	Surface (m^2)	Avg. temperature (K)
w_cool_a	1	1357596.9	39.584	869.546
w_cool_b	2	619773.5	32.934	909.647
w_cool_c	3	368621.97	25.6035	784.068
w_cool_d	4	211761.59	21.426	651.34
summary		2557753.9	119.5475	

The Figure 9, Figure 10, Figure 11 and Figure 12 show the resulting local temperature fields for individual parts of the cooling panels. From the obtained field temperature results it is possible to determine locations with local overheating, or to locate areas with higher cooling intensity due to the intake of outside air through the holes. In Figure 13, Figure 14, Figure 15 and Figure 16 the temperature fields are shown, where it is possible to monitor the influence of the outdoor air temperature passing through the openings into the furnace. The average temperature achieved in furnace based on entered boundary conditions is 1 003 K. The Figures 15 and Figure 16 show velocity fields, where it is possible to track the effect of the furnace structure on the flue gas flow in the furnace section planes. It can be seen from the results that the flow is uneven in the interior, which may in some places cause greater fouling of the surfaces with carbon and thus cause local temperature increase.

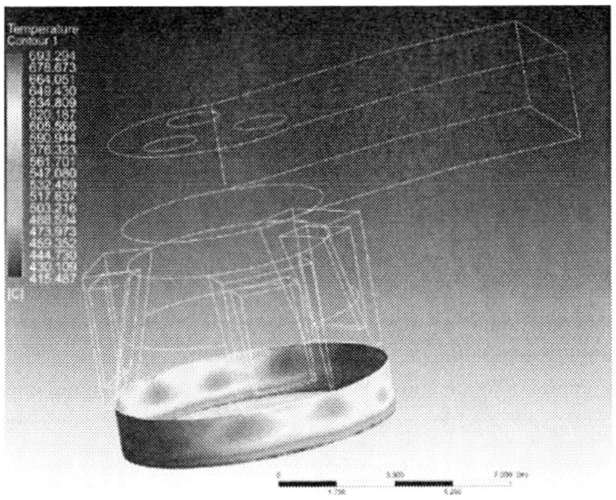

Fig. 9. Temperature fields of lower cooling panel – w_cool_a

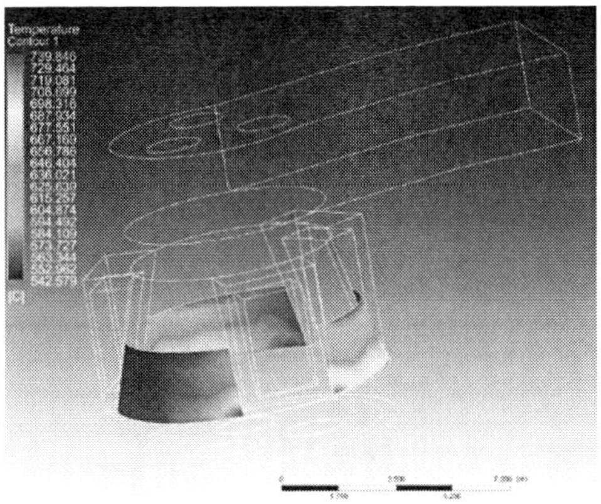

Fig. 10. Temperature fields of cooling panel – w_cool_b

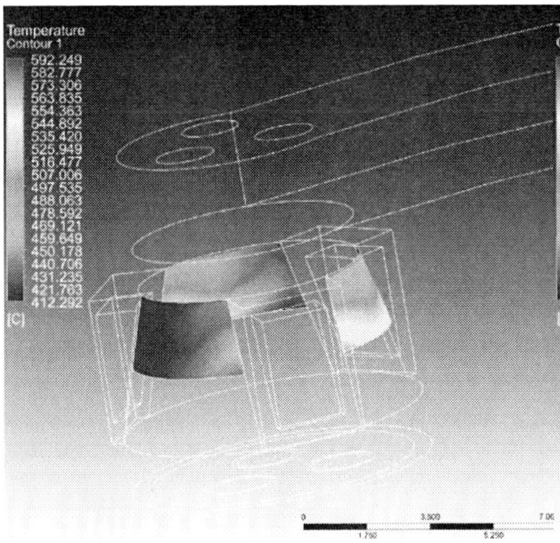

Fig. 11. Temperature fields of cooling panel – w_cool_c

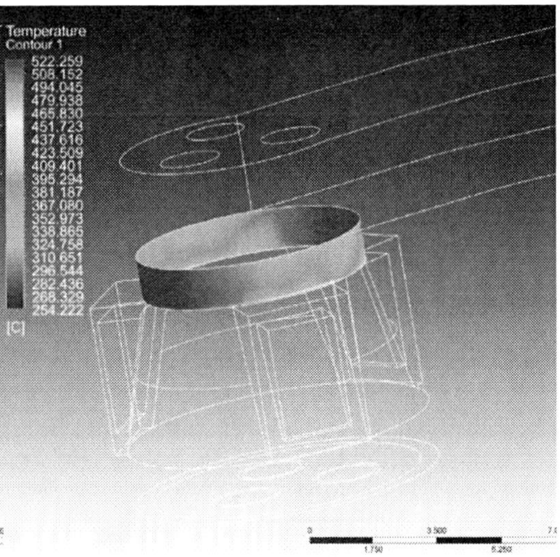

Fig. 12. Temperature fields of upper cooling panel – w_cool_d

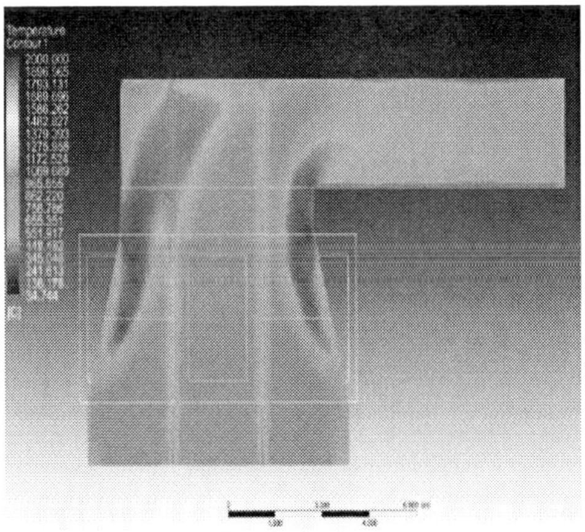

Fig. 13. Temperature fields in section area

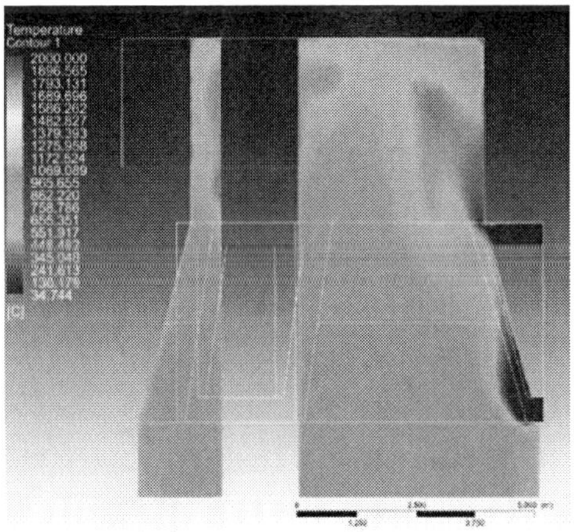

Fig. 14. Temperature fields in section area

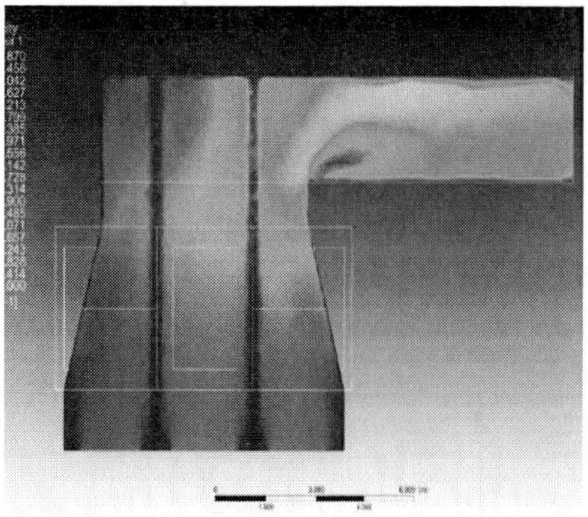

Fig. 15. Velocity fields in section area

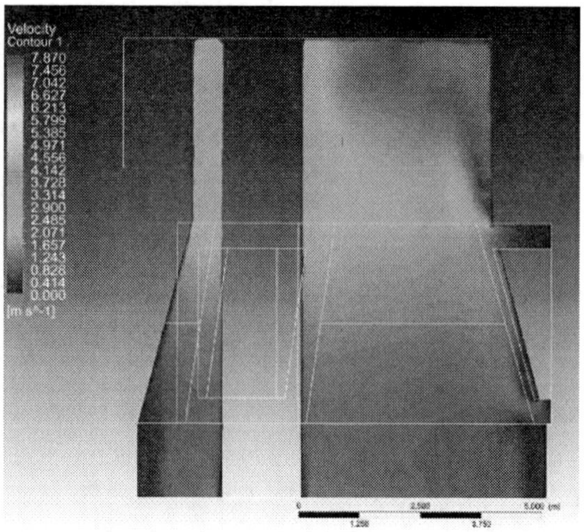

Fig. 16. Velocity fields in section area

Conclusion

The issue of recovering and using heat from electric arc furnaces is very wide and very complicated and requires some experience with this issue. Its solution should have a comprehensive approach not only to the way and form of heat generation, but also to the possibilities of its use. In our opinion, it is also necessary to propose technical solutions that accept the biggest problem of heat generation from electric arc furnaces hat - the heat exchange surface of the recuperation exchanger with a dust extraction of very specific properties, for example perfect sanding on uneven surfaces. This is also a significant restriction on the use of waste heat for the production of electricity, whether via water vapour or through the ORC cycle. Existing installations need a stable high potential heat source. Therefore, it may be necessary to deal with systems that are relatively simple and resistant to conditions in which they have to work reliably. It is possible to design significantly more physically efficient types of heat exchangers, compared to the present solution, but the functionality of such exchanger would be, as the experience has shown, already after several hours of operation question and especially their cleaning during operation is practically impossible. Great attention must be paid, in our opinion, to the quality regulation of the heat recovery equipment, which must respond flexibly to the non-stationary nature of the exchanger control resulting from the loading method of the furnace.

The research is supported from the project "APVV-15-0778 - The limits of conventional radiation and cooling through the phase change of the working fluid in the thermo-syphon loop".
This contribution is the result of the project implementation: Slovak Infrastructure for High Performance Computing, ITMS code 26230120002, supported by the Research & Development Operational Programme funded by the ERDF.

References

1. *ANSYS Help Viewer, Version: 17.2.0.* (2016)
2. P. Durcansky, P. Hrabovsky, M. Malcho, *Waste heat recovery system with heat pipe through phase change in a small biomass furnace*. SGEM (2016)
3. R. Nosek, S. Gavlas, R. Lenhard, V. Sedlak, H. M. Arvesen, *Condenser optimization of heat pipe*. Communications **16**, 62-66 (2014)

4. M. Malcho, J. Jandačka, R. Lenhard, S. Gavlas, T. Ochodek, *Numerical simulation of heat transport from the electric arc furnace to the heat recovery system*. Communications **18**, 115-121 (2016)
5. VDI heat atlas, Verein Deutscher Ingenieure, VDI-Gesellschaft Verfahrenstechnik und Chemieingenieurwesen (GVC), 2nd ed., 1585 p. (Springer, Heidelberg, 2010)
6. T. Brestovič, M. Čarnogurská, R. Pyszko, M. Kubik, *Effect of radiation on heat exchange in finned heat transfer surfaces*. XVIII Intern. scientific conference The application of experimental and numerical methods in fluid mechanics and energy, 19-26 (2012)
7. B. Skočilasová, J. Skočilas, *Determination of pressure drop coefficient by CFD simulation*. Proc. of AIP Conference **1608**, 206-210 (2014)
8. J. Kňourek, R. Matas, O. Prokeš, D. Tenkrát, *Numerical simulations of natural gas flow in pipe system with flowmeters*. EPJ Web of Conferences, **67** (2013)
9. A. Kapjor, K. Kaduchová, R. Lenhard, H. Smatanová, *Prenos tepla z orientovaných teplovýmenných plôch pri prirodzenej konvekcii*. ISBN 978-80-554-1304-4 (2017)
10. J. Černecký, J. Koniar, *The study of heat transfer in narrow gaps at low Reynold's number values*. Acta facultatis technicae **13,** 75-82 (2009)
11. E. Pivarčiová, J. Černecký, *Vizualizácia teplotných polí a prúdenia*. Aplikácia experimentálnych a numerických metód v mechanike tekutín, 194-198 (2002)
12. M. Brůna, A. Sládek, *Nové trendy v oblasti numerickej simulácie pre zlievarenstvo*. Technológ (2011)
13. Z. Trávníček, P. Dančová, J.H. Lam, V. Timchenko, J. Reizes, *Numerical and experimental studies of a channel flow with multiple circular synthetic jets*. EPJ Web of Conferences (2012)
14. W. Adamczyk, F. Szelejewski, P. Kozolub, R. Białecki, T. Kruczek, *CFD estimation of heat losses in thermal conductivity measurements*. Computer Assisted Mechanics and Engineering Sciences (2013)

Evaluation of the technical condition of medium-sized boilers

Pavel Lošák[1,*], *Tomáš* Létal[1], *Jiří* Buzík[1], *Martin* Naď[1]

[1]Brno University of Technology, Faculty of Mechanical Engineering, Institute of Process Engineering, Technická 2, 616 69 Brno, Czech Republic

Abstract. The recent trend in the steam and electricity production has been both to increase the efficiency of the facility and to keep tightening legislation concerning emission limits. The lifetime of energy equipment is greatly influenced by the operating temperature, pressure and operating characteristics. The new conditions lead the operator to more often changes of these parameters, which has negative influence to the facility in terms of service life. Precise knowledge of the facility being operated and the ability to predict the residual life of its key parts in time is therefore necessary. A new methodology for determining the residual life and evaluating problematic situations of medium size boilers was developed at Brno University of Technology. Its approaches and advantages will be presented in this paper. The methodology provides the user with approaches for the lifetime evaluation of an equipment as a whole, based on detailed knowledge of the equipment being investigated and the ongoing damage. Additionally, if the equipment is continuously evaluated, it is possible to extend the inspection interval or to achieve a significantly higher lifetime of the entire equipment, thereby reducing the economic cost. If defined criteria are met, the methodology also allows inclusion of FEM and CFD simulations for achieving higher relevance of the results.

Keywords: lifetime evaluation, boiler damage

1 Introduction

Safe operation of the equipment is one of the key requirements for technical device and must be taken into account both when designing the equipment and operating it. Inappropriate device design may lead to public safety threats as well as considerable economic losses. Together with safety, the economy of operation also plays an important role. Each part of a device has a certain lifetime, after which it is necessary to replace the part or to repair it in such a way as to avoid unplanned shutdown of the device. Appropriate design, operation and maintenance, together with appropriate and timely repairs, can extend the time the device will be able to operate safely.

With the development of science and technology, there is still a growing knowledge and refinement of the methods enabling meeting of these requirements. However, with the development of technology, there is also an increasing demand for materials. Operating

* Corresponding author: losak.p@fme.vutbr.cz

Reviewers: *Jan* Krmela, *Ireneusz* Malujda

temperatures and pressures increase as well as the need for operative changes in production that may unfavorably affect the device in many ways. It is therefore necessary to thoroughly understand the principles of damage and on the basis of these, carry out expert analyses describing the behavior of the investigated unit. Still, the current level of science and technology is not so high that it can capture the complexity of processes and take into account all the parameters that affect the behavior of the device. It is therefore important to apply appropriate simplifying assumptions leading to results without significant deviation from the reality under consideration.

There are a number of degradation mechanisms detrimental to boiler operation. Presented article focuses on the field of steam boilers and methodology for the evaluation of their technical condition. This methodology deals with a comprehensive approach contributing to safe and economically acceptable operation of the equipment. It also includes an appendix containing procedures for device diagnostics. Methodology deals with assessment of fatigue, corrosion, high temperature oxidation and erosion. In order to prevent accidents and their associated health and economic risks, it is necessary to carefully investigate and evaluate the degradation mechanisms. These mechanisms are becoming increasingly important today. Based on efforts to increase the share of renewable energy use in power generation, a significant part of electricity is produced today by solar and wind power plants. This new phenomenon leads to frequent instabilities in the power grid, which is due to probabilistic nature of these alternative sources. Suddenly, it is necessary to compensate for these instability in the power grid. These requirements may also have an impact on the operation of steam boilers, where previously almost monotonous operation transformed to operation with significant cyclical changes.

In some European Union countries, support for alternative energy sources is also provided by the government offering financial benefits to those who are willing to compensate the emerging power grid instabilities. Here the precise knowledge of damage mechanisms can help in deciding whether it is financially advantageous to continue increasing electricity production despite the accelerated damage of the equipment.

The recent trend is the tightening of emission limits and associated fines for those who are unable to meet these limits. Reducing emissions is closely linked to the need to increase efficiency. This increase in efficiency is linked to the need for higher pressures and higher temperatures in the process. Current steam parameters can be at "ultra-super critical" levels, which corresponds approximately to 600 °C and 305 bar. The next generation of ultra-super critical level II will have temperatures even higher at 700 °C. Selection of materials that are able to withstand such harsh conditions for prolonged time is seriously limited. Therefore, there is still ongoing research and development of new materials able to meet these challenges. However, the problem may occur when the increased operating conditions are applied to materials that have not been designed for these parameters at the design stage. This is the case, for example, when the operator is forced to reduce the emissions of existing plant and will therefore increase steam parameters. In such cases, there is a potential problem with damaging the oxide layers. Although oxide layer formation is common, elevated temperatures or rapid temperature changes may lead to a massive breakdown of the cohesiveness of the oxide layers. That may result in blockage of pipe throughput, turbine blade erosion, and other complications.

All these changes lead to an increased need for accurate monitoring and evaluation of the device. The presented article is devoted to the development of a methodology for the on-line and off-line assessment of damage to equipment.

2 Motivation

One of the commonly performed activities at the Institute of Process Engineering, where this methodology has been developed, is the implementation of so-called troubleshooting analyses of equipment from the energy industry. In these activities, we repeatedly encountered a critical lack of backgrounds for performing appropriate analyses to identify the causes of the problem situation, their effect on the equipment and the subsequent proposal for corrective measures. Therefore, a methodology was developed that would summarized the necessary data to perform precise simulations of the equipment operation and allowing to propose accurate and effective corrective measures in case of a failure. Moreover to allow plan changes of operating parameters in so-called "on-line" diagnostics. Such an assessment would allow the operator to manage the residual life of the plant and monitor the effects of the operating states for residual life.

The knowledge of the operating history and other documented documents is of vital importance for the evaluation of the equipment. Without these backgrounds, which is common case for many devices, troubleshooting is complicated and requires a number of simplifications. Although the results do not fully describe the reality under examination, they may still provide valuable information as they present the only available information that is relevant to decisions about corrective measures. Even in this case, the methodology can be a valuable guideline for performing the analyses, even though under simplified assumptions.

3 Evaluation methodology

The presented methodology is divided into categories that affect or are necessary to determine residual service life and troubleshooting. The methodology provides a set of requirements for monitoring, evaluating and storing data so that qualified lifetime management can be performed. Also, verification of the methodology is possible after it is used by the operator. The methodology consists of verified parts, such as calculations according to standards, for which no further verification is presumably required (although still possible) and the parts that are considered with the verification. Verification is based on comparison of the calculation results with the current state of the device according to inspections of the equipment. In the case of a disagreement between the calculated life expectancy and the actual condition of the equipment, the methodology further considers the possibility of applying corrective measures.

The methodology also focuses on establishing criteria that would make it more transparent and allow for the independent replication of finite element (FEM) and flow simulation (CFD) numeric analyses, which in certain cases presents an undisputed advantage in the device diagnostics, search for causes of damage and design of corrective measures.

3.1 Main parts

The main parts of the methodology that summarize the complex approach to device evaluation are as follows: Design calculations, Documented quality, Device monitoring, Device Diagnostics, Device status evaluation and Troubleshooting. Each chapter contains a list of documents that are relevant to that section. The more precisely will the evaluation be according to these modules, the more accurately can the life of the equipment be determined and consequently the associated activities such as revision plans optimized. In the event of unplanned damage to the equipment, a qualified assessment of the problem situation and the design of an appropriate corrective measures can be made in accordance with the proposed methodology.

3.1.1 Design calculations

Designing equipment in accordance with standards is also the basis for determining residual service life. Design is tied to the use of valid regulations and safety requirements according to the applicable regulations of the given area. An important part of the design is also the risk analysis according to the relevant regulations. Design calculations must include, in particular, the determination of the planned lifetime of the installation and also take unavoidable additional loads into account.

3.1.2 Documented quality

For the life management system, respectively for the ability to perform qualified assessments of residual lifetime or to deal with unexpected problem situations, the documents recording complete and accurate information about the activities performed are crucial. The documented quality can be divided into the areas of production, operation and exceptional cases. In the area of the manufacturing part it is necessary to keep all relevant documents related to this area which document the fabrication of the equipment and will be useful in the future in solving problematic situations. Area of operation will include the operational records of the facility, including inspections, along with information about the shutdowns and startups of the facility. Some of the exceptional cases will include data relevant to the situation, e.g. information about specific degradation of material characteristics discovered during an inspection.

3.1.3 Equipment monitoring

Equipment monitoring will be designed based on the most probable and detrimental degradation mechanisms that may occur under expected normal as well as exceptional operation conditions. Parameters that will be monitored with the appropriate degradation mechanism are as follows:

Static load:
- pressure
- temperature
- actual dimensions
- material parameters

Variable loads:
- temperature and pressure gradient
- actual dimensions
- material Parameters
- numbers of load cycles according to their character (cold, warm hot)

Additional loads due to the nature of the operation:
- media temperatures
- wall surface temperatures
- forces in anchorages
- moves

Load by operating medium - corrosion, erosion, oxidation:
- supply water composition
- fuel composition

3.1.4 Equipment diagnostic

The diagnosis of the equipment will be carried out according to the criteria given by the relevant standards. Other methods may be used to diagnose the device, when the suitability for use is demonstrated. Diagnosis can be done with commercially available software or using in-house procedures. Part of the presented methodology is a description of usable procedures for device diagnostics in a given area, which can be used for development of an evaluation software. These procedures are included in the appendix to this methodology. Diagnosis will be done on-line, with the damaging trend determined from the continuous monitoring as well as manual evaluation of the damage found during the inspections.

3.1.5 Equipment condition evaluation

The damaging trend estimated by on-line diagnostics will be compared to the original design lifetime. The evaluation of the residual life and condition of the equipment will be further extended by a lifetime correction based on the actual state of the equipment ascertained during carried out inspections. Consideration will also be given to changes in service life due to repairs and changes in the design of the equipment.

In case when the damage accumulation factor exceeds allowable limit, it is possible to modify the approach to assessing the equipment according to its current state and to set a new lifetime plan, on the basis of which the maintenance and inspection plan will be modified.

There may also be situations where the calculated cumulative damage factor is lower than the actual damage found during the inspection. In this case, it is necessary to introduce the correction factor of the damage accumulation factor until the cause of this discrepancy has been identified and corrective measures (modification of the operating conditions or assumptions for calculating the accumulation of damage) have been made.

3.1.6 Troubleshooting

If unexpected failures occur, troubleshooting will be performed. Detailed records of the production, operating and other stages can be used to identify the cause of the damage and to propose corrective measures together with a modification of the life plan. For this activity, it is advisable to use the services of expert organizations with modern computing tools.

Evaluation methods

The first stage will include visual inspection of the problem situation and the establishment of a procedure leading to appropriate corrective actions. These activities will consist of:

- material analysis
- detail analysis of operating conditions
- visual checks to detect abnormal function of parts of the device (e.g. poorly adjusted burners, inappropriate distribution of emission-reducing additives)
- checking the dimensions
- a detailed strength calculation that takes into account the actual state of the equipment and the subsequent identification of the factors that caused the damage
- detailed calculation of media flow and analysis of their impact on the problem situation
- combined strength and flow analysis and identification of problem factors

Strength calculation methods

Heat calculation, flow simulation and subsequent strength calculation will be performed for a particular part of the plant. If possible, analytical analysis will be performed. Other methods, for example, finite element method or computational fluid dynamics method can be used if analytical solution is not available.

Strength calculation by FEM

For strength calculation, it is recommended to consider the requirements of standards, for example according to ČSN EN 13445-3 [1]. The standard permits several useful methods. One method is "design by analysis". Two approaches to the solution can be used, the first is a stress categorisation route. An elastic calculation is performed, which is then evaluated by assigning the respective stresses to the categories and then comparing these stresses with the permitted stresses according to the requirements of the standard. This method yields conservative results that may overestimate expected damage in some cases. For further refinement the second method called the direct route can be used. The direct route method is the calculation performed according to the limit states. This method checks each possible structural failure separately. The design checks considered are:

- control of the structure to gross plastic deformation,
- progressive plastic deformation,
- loss of stability,
- fatigue,
- checking static equilibrium of the structure.

Further refinement of residual life estimates may be also achieved using specific non-linear analyses. These can provide accurate determination of time-dependent stress and strain results together with eventual multi-axial behaviour. These calculations help to locate dangerous areas in the structure, and consequently aid choice of sensor type and placement for continuous monitoring as well as refine revision plans. It is also possible to predict the growth rate of identified cracks and to assess their acceptability for further operation.

Flow calculations using CFDs

In some cases, where non-uniform boundary conditions cannot be determined using conventional methods due to complex flow, they may be established using CFD simulation. Useful boundary conditions from CFD include pressures, forces, heat transfer coefficients and temperatures. In case of non-uniform thermal field, thermal analysis may be also performed before final structural analysis. CFD analysis consists of the following steps:

Preparation and simplification of geometry

In this step, either 3D geometry is taken from the design solution or newly created according to the drawing documentation. If 3D geometry is assumed, it is first necessary to create an inverse volume - that is, fill the internal volume of the device with fluid in the areas where the CFD simulation will be performed, remove / suppress the volume of solid walls of the device, and then proceed to refine and simplify the geometry created. Trimming is often necessary because there are details in the design drawings that do not need to be considered in terms of flow.

The result should be "multibody part" geometry. All volumes should be coupled in one part to ensure a conformal mesh at the interfaces of the individual volumes. Basically, it's a good idea to avoid using non-conformal mesh that requires interpolation of results between meshes.

Creating a mesh

The computing mesh (grid) is a basic precondition for a correct and convergent calculation. The basic requirement is to adequately describe the areas of high gradients of each modelled variable. One of the important gradients that need to be correctly described is the speed and temperature boundary layers of the walls. Here it is necessary to take into account the used turbulent models or wall functions as well and to correctly select the size of the first cell in the wall according to the recommended dimensionless parameter y +. If we want to achieve the most accurate resolution, at least 10 cells perpendicular to the wall with the first cell of size y + <1 in the boundary layer area must be used.

Setting boundary conditions and selecting physical and numerical models
First, it is necessary to determine which physical models will be included in the calculation. For preliminary analysis, it is often advantageous to neglect the thermal expansion of fluids if there are no significant changes in the temperature in the calculation area. Likewise, it is possible in certain cases to neglect the compressibility of the fluids under small changes of pressure. Another options to consider is radiation modelling. It always depends on the temperature ratios in the computing domain and the properties of the fluid. In general, it is advantageous to perform the first calculation in a simplified form and then proceed to include more complex physical models.

One of the key aspects of flow modelling is the choice of a turbulent model when the flow is in turbulent regime.

According to the selected setting, it is necessary to proceed to determining and specifying boundary conditions. Their choice has a major impact on the results of the calculation. It should be based on measured real-time data, previous numerical simulations, or in the case of new technology from design parameters.

The choice of numerical models should be derived from the type of mesh and physical models. However, in practice, it is necessary to use higher-order discretization for all equations. For the convergence evaluation, it is always necessary to evaluate the stability of the physical quantities (e.g. velocity or temperature) in addition to so-called residuals. It is also always necessary to evaluate the total weight and energy balance.

Evaluation of results and export of data
The evaluation should always include verification of the quality and suitability of the computing mesh. It is suitable to use 2D colour contours of individual physical quantities. For these contours, the colour range showed should represent the full range that is present in the displayed area. If, for some reason, the scale is modified, it must be stated and explained in the text.

Data transfer for eventual FEM analysis is possible in several ways, depending on the software used. Ideal is to use direct interconnection within one software, which guarantees minimal distortion of the data being transferred. In the case of data transfer between different software, attention should be paid to the way of interpolation.

Conclusion

In the present article, a methodology which was designed to increase the reliability and safety of steam boiler operation was presented. The emphasis and main contribution of the methodology is to characterize the documents that are created during all phases of the steam boiler lifetime from its design to the end of its lifetime and may be required for lifetime management. The methodology is composed of the design part, the operating part, the diagnostic part and the part of the evaluation of unexpected states, where the use of modern computational methods such as FEM and CFD is promoted. Together with the implementation of these methods in the equipment evaluation, there is a need for precise specification of the use of these methods. It is necessary to adhere to certain rules and to document important steps when performing these analyses so that it is further possible to replicate these analyses by an independent person. For the purpose of replication, the results of the analyses as well as input data need to be documented. For the diagnostics of the device, whether on-line or off-line, the methodology includes an annex proposing methods for determining the proportion of individual degrading mechanisms to the overall damage of the examined unit.

The authors gratefully acknowledge financial support provided by Technology Agency of the Czech Republic within the research project No. TE02000236 "Waste-to-Energy (WtE) Competence Centre"

References

1. Český normalizační institut, *ČESKÁ TECHNICKÁ NORMA, Netopené tlakové nádoby*. ČSN EN 13445. 2015.

The use of CAD applications in the design of shredders for polymers

Marek Macko[1,*], *Krzysztof* Tyszczuk[1], *Grzegorz* Śmigielski[1], *Józef* Flizikowski[2], *Adam* Mroziński[2]

[1]Faculty of Mathematics, Physics & Technical Science, Kazimierz Wielki University in Bydgoszcz, Poland
[2]Faculty of Mechanical Engineering, University of Technology and Life Sciences in Bydgoszcz, Poland

Abstract. Computer-aided design is now the primary design tool, providing opportunities for improving project quality, allowing for in-depth analysis of the specific features of the solution. In the methodology, computer simulation applications were used. Computer simulation aims to reconstruct the course of the studied process from its mathematical model using a computer and to examine the influence of the environment (input signals) and the internal properties of the object (process parameters) on the characteristics of the object. In designing the shredder, in addition to the selection of structural features, it is also possible to perform strength and kinematic simulations, resulting in a higher quality of the target solution. The results of the simulation work presented in the paper indicate a wide range of possible applications - also in the field of food engineering and processing of polymers - eg shredder mechanisms. Particular attention has been paid to the reduction of energy consumed by the deformation of samples, eg by pre-compaction of sample before introduction to the working area of the shredder. It has been estimated that the gap between the rotating discs should be minimized. Establishing design guidelines has resulted in a number of concept designs.

Keywords: CAD, Shredder, Simulation, Energy Efficiency

1 Introduction

The grinding is an important step in the preparation of the product, including food processing and plastics technology. This process is difficult to formal explanation because of the need to take into account the characteristics of random ways of cracking and movement of elements to be crushed in the workspace machines, changing boundary conditions load, etc. There are many papers describing CAD implementations in different applications and simulations [1, 2, 3, 4, 5, 6]. In order to improve the efficiency of crushing undertaken numerous structural modifications and mechatronic implementations. They base on the use of modern systems of measurement and control in a chopper. With such implementations we can improve parameters using CAD tools as well as monitor parallel the grinding process, energy requirements as a measure of the energy required for grinding weight of 1 kg material [7, 8]. For fine grinding, the unit energy requirement is between 20

* Corresponding author: mackomar@ukw.edu.pl

Reviewers: *Milan* Sága, *Alžbeta* Sapietová

to 30 kW·h/Mg, and for very small increases up to 800 kW·h/Mg (for current designs) [9, 10]. So high values are not conducive to lowering overall expenditures for shredding in both the manufacturing and the recycling of plastics. Energy demand for fragmentation is the greater, the smaller the average grain size after grinding. It is estimated that to achieve the grain size (product of grinding) - 10 mm requires 1 kW·h/Mg (average value during grinding), while the disintegration to an average particle size of less than 1 mm need to 200 kW·h/Mg. Energy demand is related to design features shredder and implementation of mechatronic modules, e.g. the size of the gap between the knives (fixed and movable) decide on the level of demand for energy, but also affects the occurrence of dominant loads and deformations. For small size, the sample undergoes fracture gap (1 mm), and with increasing energy demand grows slot on fragmentation. For larger dimensions of the slot, pulling the material and its bending and disintegration.

The research includes implementation of CAD tools, especially in relations with geometric relations but also some simulations and mechatronic modules in laboratory/test stand in which the grinding process is employed to separate successive pieces of polymer materials as feed, by working the edge of the holes in the disks or drums [11]. Procedures of the solution of the problem as the mathematical procedure is an attempt to respond to the complexity of the creative process of inventing numerous strains of suboptimal design. The first attempts, supporting grinders, was carried out using desktops implemented the program IE_TEST-07_BIO. Power was very effective, but did not allow for mobile access to the program. There have been attempts to implement the program on e.g. mobile phones. The tests proved to be very fruitful and comfortable in co-operation for authors.

Laboratory tests shredding system implements many structural variants assemblies of shredder: drum, disc with holes cylindrical and polygonal disc with holes, selected on the basis of the matrix of possible solutions, filling the initial criteria for energy efficiency and grading curve. The experimental results expand the knowledge of the shredding and provide the material to the created and the staging database for the construction and the grinding process. These data were used in an integrated expert system in order to facilitate the selection of design shredder suitable materials. Additionally, the workspace chopper occurring complex processes of disintegration. The impact on the process of geometric features there are working elements, including shape, size, porosity shredded elements (related to their density) and a number of operating parameters that determine the behavior or degradation of intermolecular bonds, especially in the case of polymer materials. Susceptibility to grinding depends on the material properties, such as density, mechanical strength, durability, resistance to weather and environment active, thermal stability and electrical properties and thermal properties. The polymer materials are subject to physical and chemical changes that occur in the structure during processing, storage and use, which usually leads to deterioration of processability and performance. Rationalisation of factors affecting the properties of materials during their processing and use was considered in many publications [5].

Among the external factors, there are mentioned:

- climatic factors (eg. the atmospheric oxygen, water, air temperature, solar radiation, wind),
- environmental factors (corrosion, erosion, mechanical, biological),
- technical factors,
- process parameters (temperature, time, pressure),
- design features (plasticizing systems, tools and grinders),
- material (structure, shape, origin, pollution).

The largest percentage in the production of thermoplastic materials are polymers which when heated undergo plastic state and harden upon cooling while maintaining the ability to plasticize the next reheated. As a result, these materials can be processed hot and repeatedly formed. It is also possible the processing of post-consumer waste.

In the area of design shredders, it is richly represented achievements on the knife mills, including shredders with variety of geometrical relations. Requirements for shredding plastics inspire the search for new - alternative design solutions. There are known constructional solutions for special polymer materials and shredders for a wide range of design solutions and performance measured in kg/hr. Among the industrial design solutions there are well known cylindrical grinders, breakers, crushers, ball mills, vibrating rod and knife grinders with specific geometrical features, put into production (including IMPB Torun, TRYMET Pilchowo, Instalwent Bydgoszcz, Heinrich Dreher GmbH & Co. KG Hellweg Maschinenbau GmbH & Co. KG, Weima, Hosokawa Alpine and many others).

2 CAD and mechatronic implementations

Interdisciplinary, especially mechatronic aspects of our work, use advanced computer techniques especially CAD/CAE and AI, which in combination with the achievements in the field of mechanical engineering and robotics contribute to the improvement of product quality (especially in the area of materials engineering). Innovative shredding system is based on the applications and intelligent procedures that could ultimately improve the efficiency of the processes taking place in the overall loop recycling and shredding for preparing materials for further processing. It contains a high potential for dissemination of commercial, as it could be (in the future) easy to use, particularly in small and medium-sized enterprises, which dominate the regional environment, polymer processing.

Work goals can be achieved through the implementation of the proposed research methodology, which include research in several areas:

A. Development of the knowledge about the shredding and granulation of specialized systems, design, material properties (polymers, organic materials), some of the relationships between process parameters and design of the grinding system. The choice of design grinder (multi knife edge and grinders) and prepare a 3D virtual model (in SolidWorks) to strength and kinematics simulation

B. Model feedback operation on the basis of image analysis grinding device. At this stage neural network application has been used to recognize the size of the granules and analysis to design features of the grinders which are responsible for improving the condition.

To identify and assess properties of grinding products based on selected multi-edge grinding technologies in recycling, the following design solutions (from the allowable set) were assumed: which maximize (efficiency, output, degree of fineness) or minimize (power demand, unit energy consumption, energy dissipation, torque, angular, linear and rotational speed) values of selected operational characteristics.

Integrated system allows determining permissible range of variables (design properties) of the grinder in laboratory circumstances thanks to used computer aided experiment, design and operation of grinders. To allow objective assessment of the efficiency, authors used the measuring system that features recording of momentary values of torque and momentary values of rotational speed of the drive shaft (fig. 1, fig. 2). The test station was designed to allow replacement of the working assembly (knife-, disk- or beater system) as well as change of orientation (from horizontal to vertical).

Fig. 1. General view of a laboratory research stand shredding

Similar like other complex technological processes, the grinding depends on many factors that can be classified into system and design related factors (associated with grinding assembly and its equipment) and process-related factors.

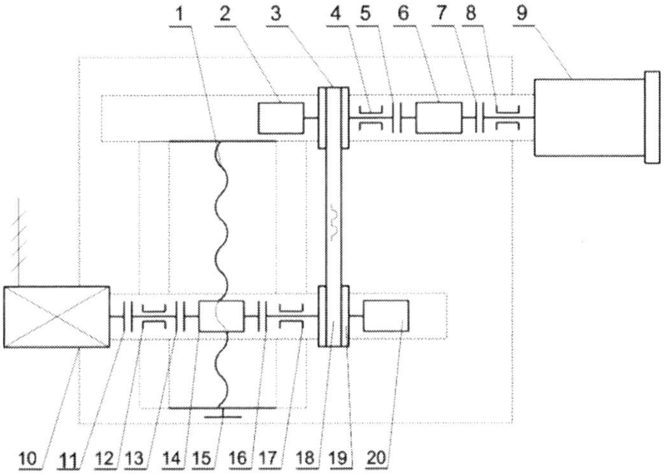

Fig. 2. Diagram of the mechanical/laboratory research stand shredding ULR-2.0/2004: 1 - screw for adjusting the distance between the motor shaft and chopper, 2 - rotary transducer - Pulse, 3 - wheel gear-toothed in strips, 4 - radial bearing, 5 - clutch splints, 6 - measuring shaft, 7 - clutch splints, 8 - radial bearing, 9 - chamber grinder, 10 - engine, 11 - clutch splints, 12 - radial bearing, 13 - clutch splints, 15 - measuring shaft, 16 - clutch splints, 17 - radial bearing, 18 - a toothed belt transmission, 19 gear wheel in strips-gear, 20 – rotary transducer

The first group of factors includes: applied grinding system (of periodical, continuous or periodically-cyclic type), number of grinding machines and their grouping, the system of connections between grinders and plasticizing equipment as well as characteristics of applied equipment (sort, type, peripheral speed, grinding components and their design properties etc.).The modular system includes a grinding expanded module to control the speed controller of management, the chamber for grinding, moisture sensors, computer chip, charge sensor and an extensive application that manages the system (Fig. 3).

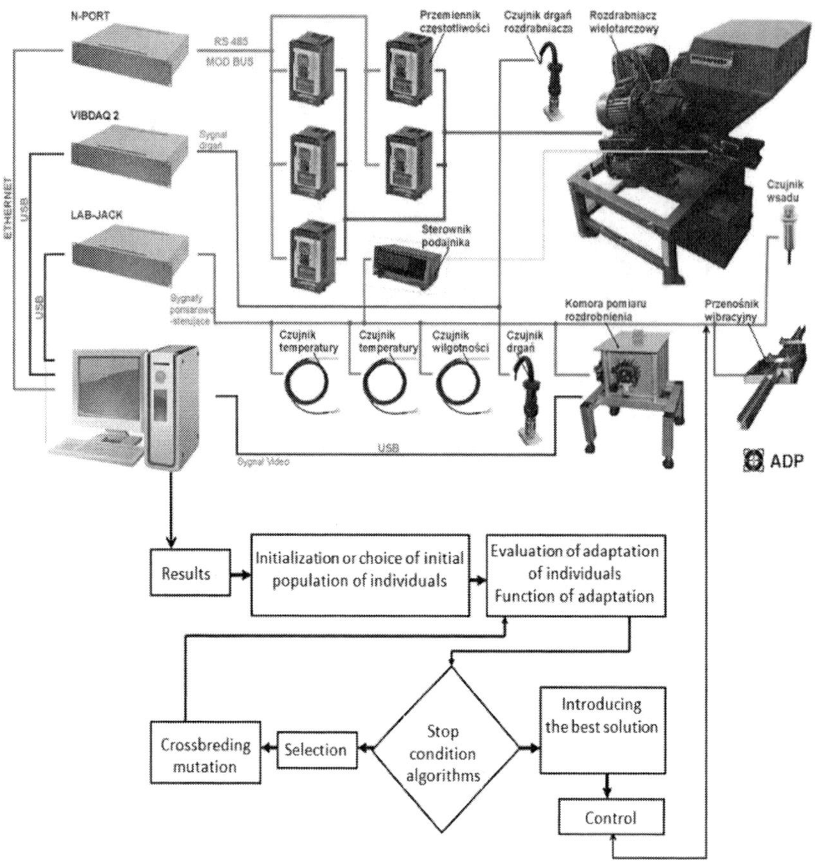

Fig. 3. Modular – mechatronic system of shredding

The computer monitor shows the dynamic characteristics of the grinding process, energy demand, the measurement of particles of particulate material (fig. 4).

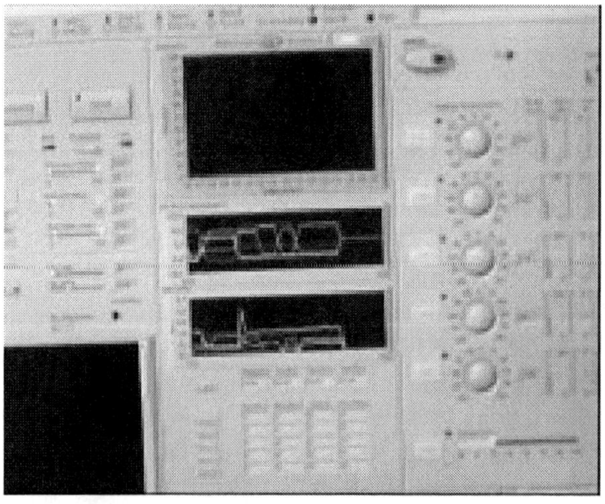

Fig. 4. The screen of research shredder based of multi-disc, RWT- 5KZ, active screen control system

3 Results

The purpose of the model is to determine the decision-making LabVIEW shares to be applied to an object. This module is based on current measurements collected from sensors and experience (stored in the knowledge module) determines decisions. They are sent to the executive module and learner. The module shall send decisions to the object, while learners take account of decisions in the learning process for the subsequent assessment of the effects of the decision. The main challenge for system designers is to create algorithms that can make the correct decisions in the case of incomplete or redundant information.

Each time you launch the experiment generates the report as a file in the .xls format. The files are placed in the directory My_documents/Mill/AG/ (fig. 5). The file names contain the date, hour and minute of the start of the experiment.

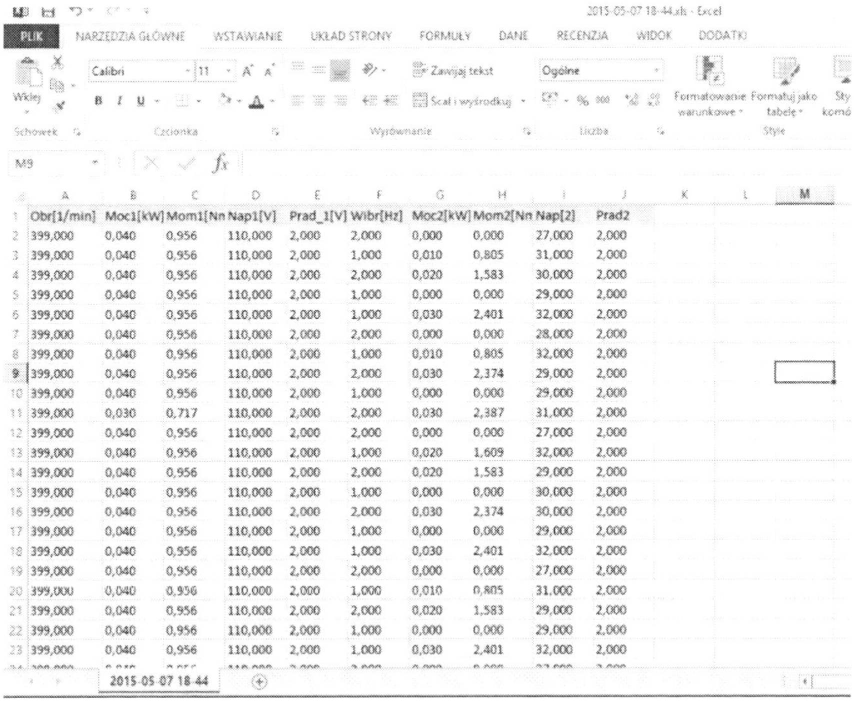

Obr[1/min]	Moc1[kW]	Mom1[Nn	Nap1[V]	Prad_1[V]	Wibr[Hz]	Moc2[kW]	Mom2[Nn	Nap[2]	Prad2
399,000	0,040	0,956	110,000	2,000	2,000	0,000	0,000	27,000	2,000
399,000	0,040	0,956	110,000	2,000	1,000	0,010	0,805	31,000	2,000
399,000	0,040	0,956	110,000	2,000	2,000	0,020	1,583	30,000	2,000
399,000	0,040	0,956	110,000	2,000	1,000	0,000	0,000	29,000	2,000
399,000	0,040	0,956	110,000	2,000	1,000	0,030	2,401	32,000	2,000
399,000	0,040	0,956	110,000	2,000	2,000	0,000	0,000	28,000	2,000
399,000	0,040	0,956	110,000	2,000	1,000	0,010	0,805	32,000	2,000
399,000	0,040	0,956	110,000	2,000	2,000	0,030	2,374	29,000	2,000
399,000	0,040	0,956	110,000	2,000	1,000	0,000	0,000	29,000	2,000
399,000	0,030	0,717	110,000	2,000	2,000	0,030	2,387	31,000	2,000
399,000	0,040	0,956	110,000	2,000	2,000	0,000	0,000	27,000	2,000
399,000	0,040	0,956	110,000	2,000	1,000	0,020	1,609	32,000	2,000
399,000	0,040	0,956	110,000	2,000	2,000	0,020	1,583	29,000	2,000
399,000	0,040	0,956	110,000	2,000	1,000	0,000	0,000	30,000	2,000
399,000	0,040	0,956	110,000	2,000	2,000	0,030	2,374	30,000	2,000
399,000	0,040	0,956	110,000	2,000	1,000	0,000	0,000	29,000	2,000
399,000	0,040	0,956	110,000	2,000	1,000	0,030	2,401	32,000	2,000
399,000	0,040	0,956	110,000	2,000	2,000	0,000	0,000	27,000	2,000
399,000	0,040	0,956	110,000	2,000	1,000	0,010	0,805	31,000	2,000
399,000	0,040	0,956	110,000	2,000	2,000	0,020	1,583	29,000	2,000
399,000	0,040	0,956	110,000	2,000	1,000	0,000	0,000	29,000	2,000
399,000	0,040	0,956	110,000	2,000	1,000	0,030	2,401	32,000	2,000

Fig. 5. Example registered in the disk file the results of measurement of the efficiency of the powertrain grinder (mill rotary vibration) [9].

The task of the executive module is to provide the control signal (designated by the decision module) to an object, for example. Settings update drives. Due to the changing conditions of grinding (change in weight of the stock, the phenomenon related to the change of humidity, e.g. unstabilized grains), it is necessary to use the adaptive formulating optimal conditions and changes actuators, set the speed of dosing, the angular velocity of each target (usually being a function of humidity, form and grain size) for the current state variable processing. The auto-monitor includes monitoring sensors: speed, acceleration, force and torque, the form and dimensions of the product; drivers, so you can make the switching characteristics of the performance; photo-optical systems, electronic counting modulation and demodulation, amplification, switching signals, comparison and logical operations control, regulation, compensation.

The genetic algorithm used in the default configuration, testing 34 combinations of settings of the drive. Amount of the combination has been adapted to the size of the tray material for grinding process speed so that one of the feed provided to carry out the experiment.

The results of experimental studies broaden the knowledge of shredding and provide the material for the staging created and the current database structure and the multi-holes grinding process. These data were used in integrated expert system to facilitate the selection of construction shredder suitable materials.

The concept of laboratory enables the experimental grinding for a wide range of factors constant:

- material properties and shape of the shredded as feed,
- dimensions of shredded,
- and variables.

Conclusions

Engineering of shredding, cognitively active monitoring industrial processing IPP is an issue the vast and emerging on the basis of fragmentary experience of technical sciences and cognitive science achievement. In designing future monitors, intelligent systems based on cognitive science, we can not forget about the use of classical control theory, which is indispensable in issues related to stability and resistance - the security processing in general.

Developed mathematical apparatus, methods of analysis and synthesis of structures regulating the loop feedback modeling algorithms and identification, theory of stability, theory of optimality - all these issues are part of the field, which is the control theory.

It is not possible to develop the concept of mechatronic systems shredding, cognitively active monitoring, apart from classical control theory. Classical control theory man, should play a leading role in the development of future advanced systems, because only in this way can be the condition of safety, that is, for control systems, regulation and compensation objectives and conditions for the industrial processing, basic. Positively verified assumptions and solved the problem of self-regulation, to obtain a product with the specified geometric dimensions in the process with minimal power requirements, rational efficiency (minimum energy consumption per unit, in a special integrated circuit grinding cereal grains.

Usefulness of the mechatronic laboratory set-up gives results for wide variety of geometric features and exploitation parameters (e.g. power consumption). Proposal of a design of shredder with maximum possible efficiency is reduced primarily to finding a design of the grinding assembly and of the machine that is specific to a given material, making it possible to obtain a desired product, and to selection of process variables ensuring minimum energy expenditure in combination with rational shredding efficiency.

Purpose of the development, which consists in designing and implementing modern computer-aided research subsystems: process efficiency, product quality, harmlessness of the product and the grinding process of high complexity has been achieved. Solved the problem of measurement, collection and analysis of the characteristics of utility, state postulated grinding. Collection, analysis and processing of data on the state indicators are based on computer support in an environment of LabVIEW. The measurement system supports the position of the grinding technology: precision (RPW-11NT), rotationally vibrating (lm-S2n2x2) sześciowalcowego (ZBPP-6W), multi-plate (RWT-7JA). Besides designating the characteristics of utility lets you explore, develop, apply and explore ideas and structures, technologies of grinding, grinding. The essence of the technical system, including the system of measuring shredding is to ensure the efficient flow of data in the shredding and adequate procedures to achieve the assumed level of knowledge of the entire technical system, in the context of specified development needs, innovation and progress.

Processing of biological materials (eg. Grain, feed, food), polymeric materials, suitable defined goals cognition takes into account the specific characteristics of the material being processed, machine design and parameters of the grinding process.

References

1. M. Sága, M. Vaško, P. Pecháč, *Chosen Numerical Algorithms for Interval Finite Element Analysis*. Procedia Engineering **96**, 400-409 (2014)
2. M. Vaško, M. Sága, *Application of Fuzzy Structural Analysis for Damage Prediction Considering Uncertain S/N Curve*. Applied Mechanics and Materials **420**, 21-29 (2013)
3. P. Pecháč, M. Sága, *Controlling of Local search Methods' Parameters in Memetic Algorithms Using the Principles of Simulated Annealing*. Procedia Engineering **136**, 70-76 (2016)
4. G. Domek, M. Dudziak, *Energy Dissipation in Timing Belts Made from Composite Materials*. Advanced Materials Research, 189-193 (2011)
5. E. Rusiński, P. Moczko, D. Pietrusiak, G. Przybyłek, *Experimental and numerical studies of jaw crusher supporting structure fatigue failure*. Strojniški vestnik - Journal of Mechanical Engineering **59** 9, (2013)
6. E. Rusiński, J. Czmochowski, D. Pietrusiak, *Selected Problems in Designing and Constructing Surface Mining Machinery*. FME Transactions **40** (4), (2012)
7. Czerniak J.M., Macko M., Ewald D.: *The CutMAG as a new hybrid method for multi-edge grinder design optimisation*. Advances in Intelligent Systems and Computing **401**, 327-337 (2016)
8. J. Flizikowski, A. Mrozinski, K. Dziadosz, *Modelling of multi discs grain grinders constructions*. CHISA 2006 - 17th Int. Congress of Chem. and Proc. Eng.
9. J. Flizikowski, A. Mroziński, A. Tomporowski, *Active monitoring as cognitive control of grinders design*. AIP Conference Proceedings 1822, 020006, (2017)
10. A. Tomporowski, J. Flizikowski, *Motion characteristics of a multi-disc grinder of biomass grain*. PRZEMYSŁ CHEMICZNY **92**, 498-503 (2013)
11. M. Macko, *Metoda doboru rozdrabniaczy wielokrawędziowych do przeróbki materiałów polimerowych*. Wyd.UKW, Bydgoszcz, (2011)

Study of the prototype mechanism of height adjustment of the bed in hospital bed

Ireneusz Malujda[1], *Dominik* Wilczyński[1,*], *Krzysztof* Talaśka[1], *Dominik* Wojtkowiak[1], *Maciej* Szulc[2]

[1]Poznan University of Technology, Chair of Basics of Machine Design, Piotrowo Str. 3, Poland
[2]MSprojekt Maciej Szulc, Strumykowa 8 Str., Opalenica, Poland

Abstract. The presented kinematic system of the mechanism is an effect of a design process taking into an account design objectives specified by the medical staff in a hospital. In order to meet these requirements, a kinematic analysis of the mechanism was performed with the objective of finding a connection between various loads of the bed deck and the minimum force needed to drive the height adjustment mechanism. The next step was the validation process, performed by building the prototype of the studied machine on which the empirical tests were completed. During testing, the drive force of the mechanism was measured for various values and distribution of the load. The obtained results were compared with the results of the simulation analysis. This made it possible to reach the final research goal, which was to determine how to place all the mechanism parts in order to achieve the lowest possible drive force and fulfil all design objectives simultaneously.

Keywords: hospital bed, height adjustment mechanism, kinematic analysis

1 Introduction

Hospital beds used by health care professionals are operated manually in most cases, which requires some physical effort [1, 4]. In the last few years there has been the development of their design dictated by the need to improve the comfort of patients and hospital workers [1, 5, 10]. It is mainly about minimising risks caused by, among others, mistakes and indisposition of the medical personnel occurring during the transport of patients in life-threatening conditions [5, 6]. Solutions include, among others, automation of bed movement by means of motorised devices, for both functional and transport purposes [2]. In addition, it concerns devices that will allow the transfer of a patient from one bed to another, or from a bed to a wheelchair [3, 4, 7], devices providing periodic, automatic change of a lying patient's position in order to prevent the emergence of bedsores [8, 9], and advanced, automated wheelchairs [11-13]. The technical development of this kind of devices and their application in the hospital environment are facing many difficulties and limitations [1].

This work presents a design solution for the hospital bed height adjustment mechanism developed within the objectives related to the requirements set for one of the hospitals in

* Corresponding author: dominik.wilczynski@put.poznan.pl

Reviewers: *Robert* Grega, *Milan* Naď

Poznan. The main objective for the mechanism design is to provide comfort and safety for the patient as well as for the medical personnel when performing diagnostic tests.

2 Design objectives

Based on the interview with the medical staff of the cardiac surgery ward, design objectives for a specialized mobile set consisting of a hospital bed and a medical equipment stand were formulated. One of the important elements of this set is the bed deck height adjustment mechanism.

The following geometric parameters and design objectives have been assumed for the bed:

- overall bed dimensions in the bed deck plane: 2090 x 1000 mm,
- maximum bed height: 1200 mm,
- adjustable bed deck height in the 500 ÷ 800 mm range,
- lifting mechanism drive started manually by staff,
- ability to introduce a probe (head of a diagnostic device) under the bed deck, in order to perform diagnostic tests without having to move (rotate) the patient,
- back angle adjustment in the 0 ÷ 70° range started manually by personnel,
- ability to measure the weight of the patient,
- ability to attach a medical equipment stand,
- large, rotating wheels with braking function,
- bumper wheels in extreme points of the design,
- handles for easy manoeuvring,
- protective railings,
- design guards.

Meeting the specified objectives and the necessary adjustment functionality required a kinematic analysis and designing the adjustment mechanism for the bed deck on its basis.

3 Design and operating principle of the bed deck height adjustment mechanism

The kinematic diagram of the bed top height adjustment mechanism is shown in Figure 1a and b.

Hospital bed deck height adjustment mechanism 12 (upper frame) consists of two systems of drive arms 4 and 10 and 18 and 22 coupled with kinematic pairs turning around the z axes, 11 and 21, operating as a scissors lifting system (fig. 1). Drive arms systems 4 and 10, and 18 and 22 connect the base frame 6 with the bed deck 12 through kinematic pairs turning around the z axes, 5 and 13 and 17 and 23, and are integrated in the extreme positions looking along the longer side of the base frame 6 and table-top 12 (along the x axis). The drive mechanism 1 is embedded at the height of the base frame 6 and constitutes an active sliding kinematic pair that sets into angular motion one of the two pairs of drive arms 4 and 10, and indirectly the second pair of the drive arms 18 and 22 by means of drive transmission beam 9 (fig. 1). Movement in the vertical direction (along the y axis) of bed deck 12 is provided by guide sleeves 14 and 19, and guide rails 15 and 20. Guide sleeves 14 and 19 are attached to the base frame 6 and guide rails 15 and 20 are attached to bed deck 12 (fig. 1).

The drive mechanism 1 can be any active sliding kinematic pair with one degree of freedom connected to the beams of the base frame 24 with a kinematic pair turning around the z axis 2. The beams of the base frame 24 constitute an element of the base frame structure 6.

In order to select geometric parameters of individual segments of the mechanism, a kinematic analysis discussed in the further part of this work was performed.

a)

b)

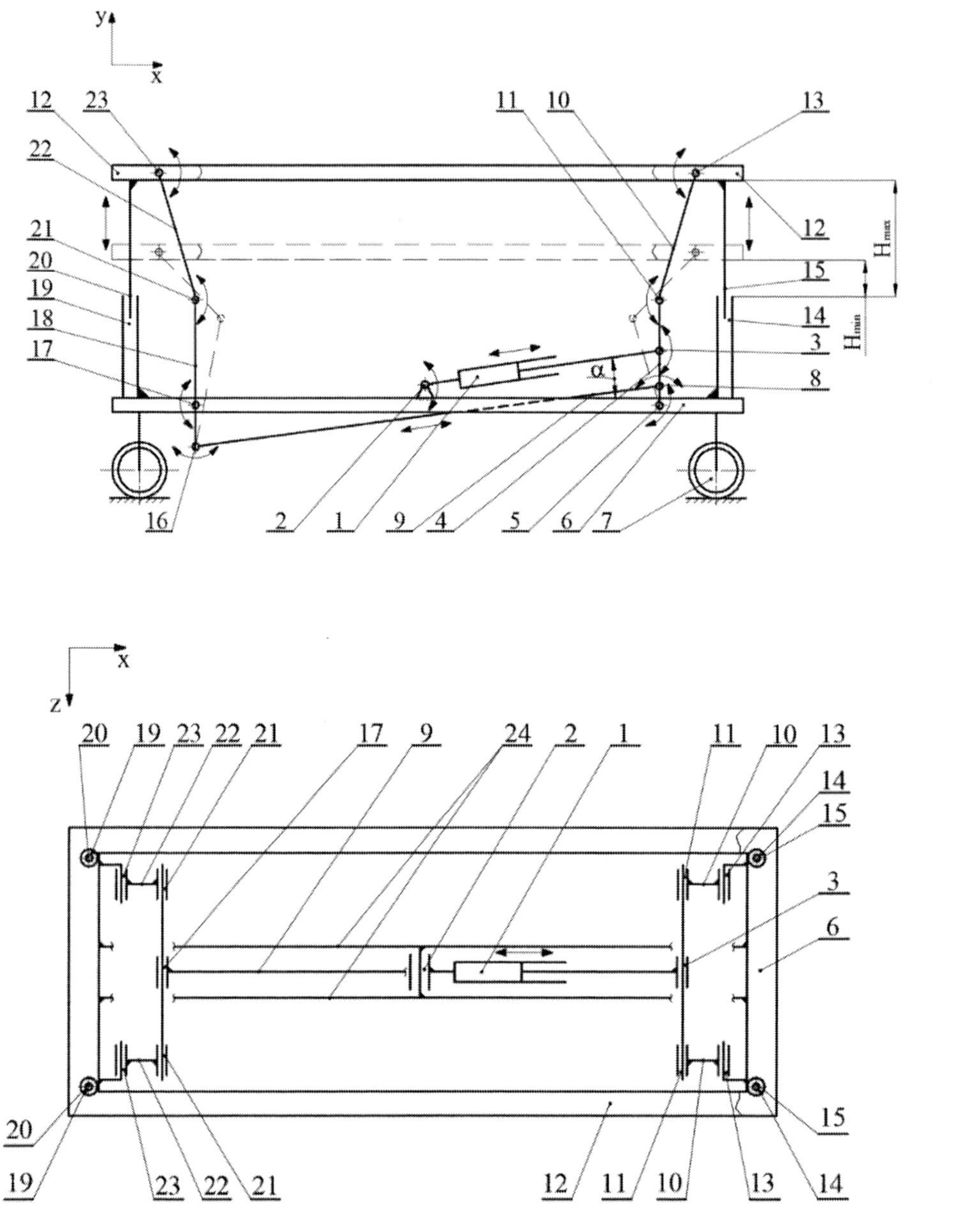

Fig. 1. Kinematic diagram of the bed top height adjustment mechanism in a hospital bed: a) view in the xy plane, b) view from the top in the xz plane, where: 1 – drive mechanism, 2 – turning kinematic pair, 3 – turning kinematic pair, 4 – drive arm, 5 – turning kinematic pair, 6 – base frame, 7 – wheel set, 8 – turning kinematic pair, 9 – drive transmission beam, 10 – drive arms, 11 turning kinematic pairs, 12 – upper frame, 13 – turning kinematic pair, 14 – guide bush, 15 – guide rail, 16 – turning kinematic pair, 17 – turning kinematic pair, 18 – drive arm, 19 – guide bush, 20 – guide rail, 21 – turning kinematic pair, 22 – drive arms, 23 – turning kinematic pair, 24 – horizontal axis, 25 – horizontal axis, 26 – base frame beams

4 Kinematic analysis of the bed deck height adjustment mechanism and its results

One of the main purposes of performing a kinematic analysis was to determine the angle α between the axis of the actuator, and the *xz* plane of the base frame (fig. 1). The value of this angle has a significant influence on the initial value of the force needed to start the bed deck height adjustment mechanism (lowest bed deck position H_{min}, looking along the y axis, fig. 1). The numerical simulation was used to specify the value of the angle α determining the lowest possible value of the forces with regards to the actuator that starts the process of lowering and raising the bed deck. It is important that the initial angle α has such a value that in the upper position of the bed deck there is enough space underneath to make it possible to manoeuvre medical diagnostic instruments. Simulations were made also at specific bed deck load values (assuming different load values depending on the patient's weight). The kinematic model (fig. 2) of the bed deck height adjustment mechanism was built using SolidWorks in order to seek the specific values of the geometric parameters: A, B, C, D, E, F, G, H, I, J, α (fig. 2). It was assumed that the drive mechanism would make use of a hydraulic actuator, whose sliding movement kicks in as a result of starting a foot-powered pump integrated with the actuator in a single body. The simulation model takes into account the actual geometric parameters of the drive, A_{min} – initial length, A_{max} – final length with a fully extended rod.

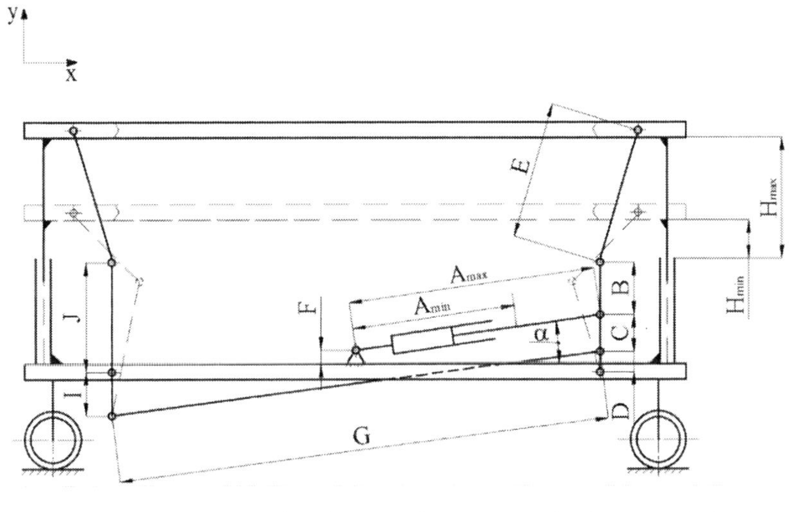

Fig. 2. Kinematic diagram of bed deck height adjustment mechanism with desired geometric parameters of individual parameters

The analysis includes the following design limitations and project assumptions:
- the distance between the ground and the bottom surface of the base frame is 200 mm (wheel diameter 150 mm + wheel stud length 50 mm),
- hydraulic actuator stroke – 300 mm, maximum strength – 10 000 N.
- bed deck height adjustment in the 500 ÷ 800 mm range, which means that the height H should be adjusted in the 300 ÷ 600 mm range.
- to prevent the mechanism from blocking, and to obtain free space under the bed deck where the patient is lying, at the maximum height of the bed, the angle between the drive arms 10 and 22 (fig. 1), and the vertical y axis should be within the 15° ÷ 30° range,
- angle α at zero actuator extension should be greater than 0° so that the actuator's body is not located below the xz plane of the structural profiles of the base frame 6 (fig. 1b).

Figure 3 presents a sample simulation result in the form of the characterization of the changes in the value of the force required to start the bed deck height adjustment mechanism. The characterization applies to changes in the value of the force needed to raise the bed deck for the angle α = 7.8° and the bed deck load 12 (fig. 1a) corresponding to the patient's weight of 86 kg.

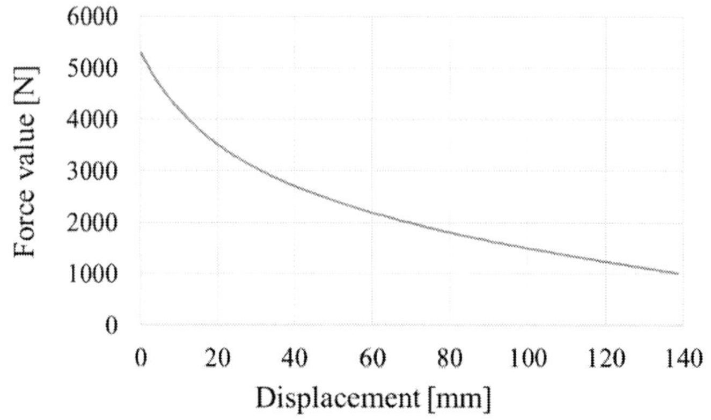

Fig. 3. Characterization of the changes in the value of the force required to start the mechanism in the function of displacement obtained on the basis of the numerical simulation α = 7.8° and m = 86 kg

The next figure shows the characterization of the changes of the force needed to start the bed deck adjustment mechanism for the angle α = 11.7° and the bed deck load 12 (fig. 1a) corresponding to the patient's weight of 86 kg.

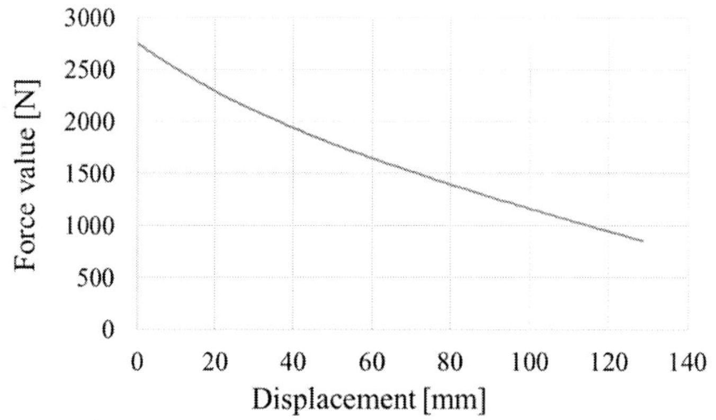

Fig. 4. Characterization of the changes in the value of the force required to start the mechanism in the function of displacement obtained on the basis of the numerical simulation α = 11.7° and m = 86 kg

As is apparent from the above characterizations, as you increase the initial value of the angle α from α = 7.8° to α = 11.7°, the initial value of the force required to start the mechanism is reduced from F = 5276 N to F = 2583 N.

In the next stage of the design process the bed prototype was built, and subsequently the verification of the simulation results was performed by comparing them with the results of measurements of the force changes obtained from the study of the bed mechanism prototype.

5 Building and testing the prototype

The kinematic analysis of the bed height adjustment mechanism formed the basis for determining the geometric parameters of the kinematic chain of the mechanism A, B, C, D, E, F, G, H, I, J, α (fig. 2), which are discussed in the previous section. This made it possible to build the bed prototype using the lifting mechanism (fig. 1 and 2) and to study its functional characteristics.

The view of the base frame with attached vertical guide rails, which in turn are connected to the upper frame is shown in Figure 5.

The build-phase bed prototype with the mounted mechanism for lifting the bed deck with the kinematic system as shown in Figure 1 and 2 is presented below (fig. 6).

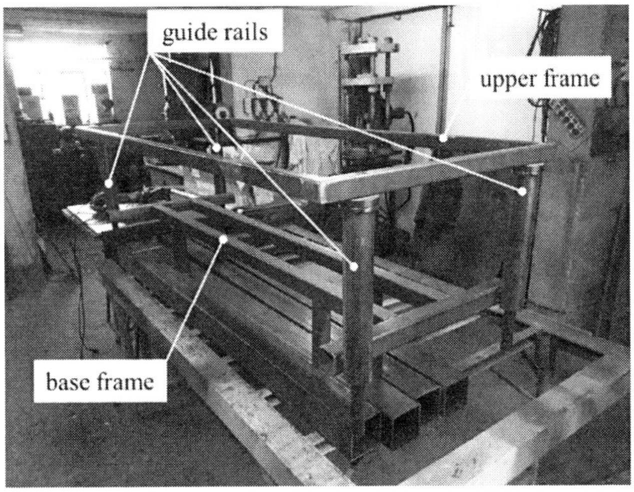

Fig. 5. Bed frame of the research and prototyping station at one of the build stages

The bed with the functional bed deck height adjustment mechanism is presented in Figure 6.

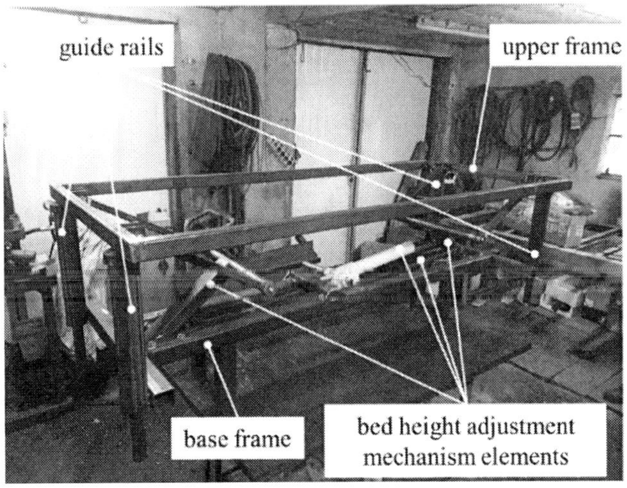

Fig. 6. Bed frame with the enclosed bed deck height adjustment mechanism

Figure 7 shows the partially elevated bed deck and partially extended piston rod of the hydraulic actuator.

Fig. 7. General view of the bed height adjustment mechanism while in operation – piston rod of the actuator partially extended, bed deck raised to height H

The drive mechanism consists of a hydraulic actuator by Bansbach, started with a foot-powered pump enclosed in a single body. The maximum axial force with which the actuator can interact with the powered mechanism is F = 10 kN and the maximum extension of the rod is s = 300 mm (fig. 8). The following figure shows the overall view of the actuator.

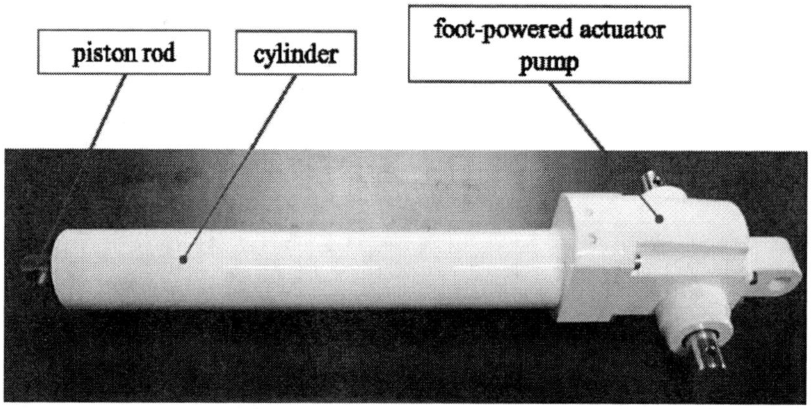

Fig. 8. Overall view of the hydraulic cylinder powering the bed deck height adjustment mechanism in a hospital bed

The mechanism was studied with regards to determining the force F necessary to start it and the results were compared to the results of the numerical analysis. Bearing in mind that the values of the force F necessary to start the mechanism obtained by means of the numerical simulation can significantly deviate from the value of the actual forces, the prototype was built using a hydraulic cylinder with the highest possible drive force. Figure 9 shows the research station consisting of the bed prototype, the force sensor and the displacement sensor.

Fig. 9. View of the bed frame when measuring the force needed to power the bed deck height adjustment mechanism

The studies of the force F needed to start the bed deck height adjustment mechanism were performed for two values of the angle α = 7.8° and 11.7°. This formed the foundation for comparing the simulation results with the actual results of prototype testing.

Figure 10 shows the characterization comparison for the changes of the force F acting on the axis of the hydraulic actuator obtained from the simulation and the experimental studies for the angle α = 7.8° and the bed deck load 12 (fig. 1a) corresponding to the load with the patient's weight m = 86 kg.

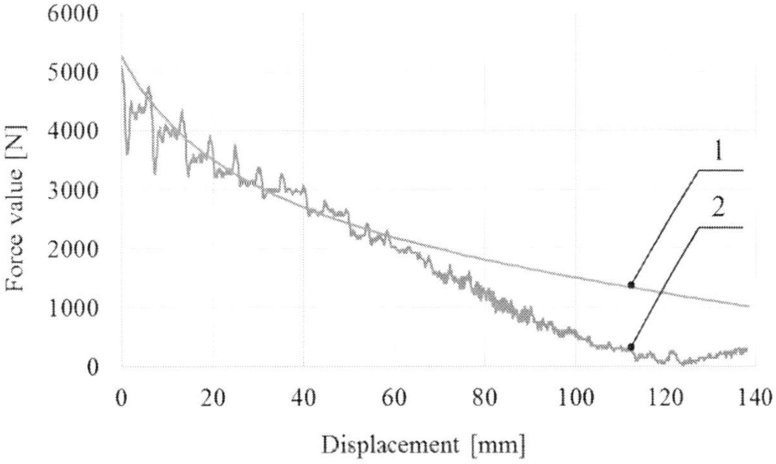

Fig. 10. Characterization comparison of the changes of the force F acting on the axis of the hydraulic actuator rod needed to power the bed deck adjustment mechanism for the angle α = 7.8° and the load corresponding to the patient's weight m = 86 kg, where: 1 – characterization obtained from simulation, 2 – characterization obtained from testing

The maximum force F obtained from the prototype tests is 5107 N whereas from the numerical analysis it is 5276 N.

Surges in changes of the force obtained from the tests are due to the surge-prone nature of the drive mechanism, which results from the foot pumping of the oil to the cylinder chamber during the piston rod extension. Pumping is done by means of the angular rotation

of the pump shaft, at which point the oil is pumped to the cylinder chamber, which in turn causes the rod to get extended and power the mechanism. After the pumping cycle movement is complete, the piston rod stops, which is visible as a decrease in the value of the force. The numerical simulation assumes a continuous interaction of the force with the mechanism, which is why its course is smooth.
As the value of the angle α increases, the force F decreases, which follows from both the numerical simulations and the actual results of the prototype tests. The value of the force obtained from the simulation falls slightly faster, which can be explained by not taking into account drag forces in the numerical model (e.g. friction in individual turning kinematic pairs).

Figure 11 presents the characterization comparison of the changes of the force (F), taken from the simulation and the actual prototype tests, for the angle α = 11.7° and the bed deck load 12 (fig. 1a) with the patient's weight m = 86 kg.
As a result of the change in the minimum value of the angle α from initial 7.8° (corresponding to the minimum bed deck height) to 11.7° (minimum bed deck height) with the bed deck load m = 86 kg, there is a 36% decline in the value of the force F from 5107 N to 3247 N. Hence, α = 11.7° is assumed as the final value of the angle α. This value is taken as the limit because along with the lifting of the upper frame (bed deck) the angle increases, and the higher the initial value, the higher the end value of the angle. This is related to a considerable angular lift of the actuator body, which can greatly reduce the space under the bed deck at its maximum height. In accordance with the objectives, effort was taken to ensure the largest possible free space under the bed deck.

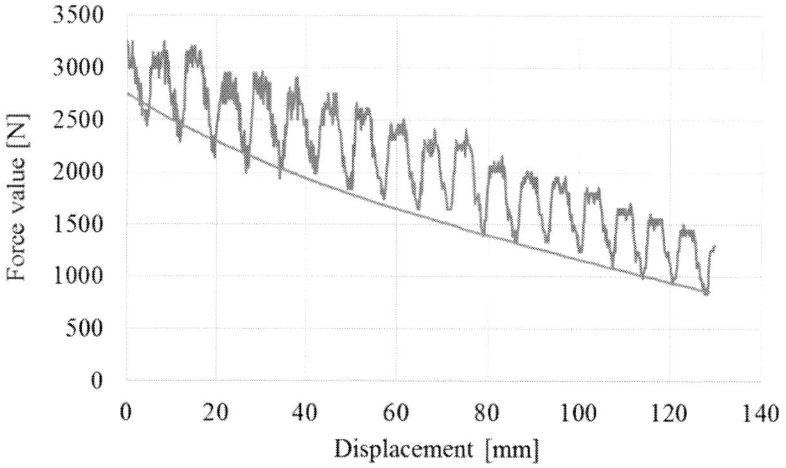

Fig. 11. Characterization comparison of the changes of the force F acting on the axis of the hydraulic actuator rod needed to power the bed deck adjustment mechanism for the angle α = 11.7° and the load corresponding to the patient's weight m = 86 kg, where: 1 – characterization obtained from simulation, 2 – characterization obtained from testing

Figure 12 presents the characterization comparison of the changes of the force (F), taken from the simulation and the actual prototype tests, for the angle α = 11.7° and the bed deck load 12 (fig. 1a) corresponding to the load with the patient's weight m = 160 kg.

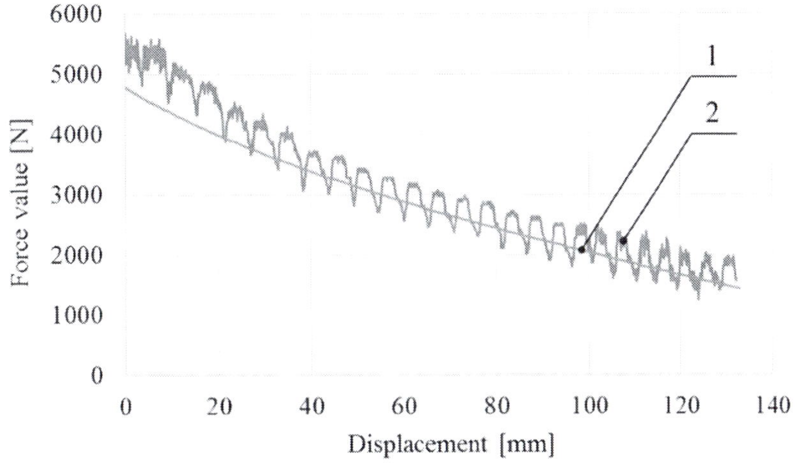

Fig. 12. Characterization comparison of the changes of the force F acting on the axis of the hydraulic actuator rod needed to power the bed deck adjustment mechanism for the angle $\alpha = 11.7°$ and the load corresponding to the patient's weight m = 160 kg, where: 1 – characterization obtained from simulation, 2 – characterization obtained from testing

For loading the upper frame of the bed prototype with the weight m = 160 kg the maximum force F is 5679 N. This result suggests it is feasible to use a hydraulic actuator with a smaller maximum force (F < 10 kN), which means it is possible to purchase a cheaper actuator.

Conclusions

This work presented the design of the hospital bed deck height adjustment mechanism formulated on the basis of the specific design objectives. In accordance with the principles of concurrent engineering, the virtual model of the bed and its height adjustment mechanism was made and a kinematic analysis was performed. The results of this analysis made it possible to select appropriate geometric parameters of the lifting mechanism, which provides the required free space under the bed deck for diagnostic tests. The mechanism also allows bed deck height adjustment within the required range.

In the next stage, after the completion of the prototype build, it is planned to test the bed prototype in operating conditions of the hospital environment.

References

1. H. H. Nguyen, T. N. Nguyen, R. Clout, H. T. Nguyen, *A novel target following solution for the electric powered hospital bed*. Proceedings of the Annual International Conference of the IEEE Engineering in Medicine and Biology Society **7319164**, 3569-3572, (EMBS 2015)
2. D. T. Johnson, P. D. Holtan, *Bed transfer system*. Patent US7533742 B2, **19** May (2009)
3. Thomas W. Hanson, R. S. Henderson, *Combination bed mover and patient transfer apparatus*. Patent US 6834402 B2, **28**, (2004)
4. P. Shih-Wei, L. Feng-Li, F. Li-Chen, *Mechanism Design and Mechatronic Control of a Multifunctional Test Bed for Bedridden Healthcare*, Mechatronics, IEEE/ASME Transactions on **15**, 234-241 (2010)

5. C. Wang, A. V. Savkin, R. Clout, H. T. Nguyen, *An intelligent robotic hospital bed for safe transportation of critical neurosurgery patients along crowded hospital corridors*. IEEE Transactions on Neural Systems and Rehabilitation Engineering **23** (5), 6879265, 744-754
6. N. Huy Hoang, N. Tuan Nghia, R. Clout, H. T. Nguyen, *An advanced control strategy of an electrical - Powered hospital bed, in Engineering in Medicine and Biology Society (EMBC)*. 36th Annual International Conference of the IEEE, (2014)
7. H. Wang, F. Kasagami, *Careful-patient mover used for patient transfer in hospital*. IEEE/ICME International Conference on Complex Medical Engineering, 20-26 (2007)
8. W. Ching-Hua, T. Ting-Chun, H. Shin-Chieh, C. Wan-Chun, C. Yen-Ming, T. Kun-Tse, Y. Chun-Went, C. Kuo-Yi, *Hospital bed with auxiliary functions of lateral positioning and transferring for immobilized patients*. The 33rd Annual Conference of the IEEE Industrial Electronics Society (IECON), 2991-2995 (2007)
9. A. Basmajian, E. Blanco, H. Asada, *The marionette bed: automated rolling and repositioning of bedridden patients*. in Proceedings of the 2002 IEEE International Conference on Robotics & Automation, 1422-1427 (2002)
10. A. Andhare, A. Onkar, P. Padole, *Design of bed for bedridden patients: Analysis and synthesis of mechanisms*. 15th National Conference on Machines and Mechanisms, NaCoMM (2011)
11. S. Mascaro, J. Spano, H. Asada, *A reconfigurable holonomic omni-directional mobile bed with unified seating (RHOMBUS) for bedridden patients*. IEEE International Conference on Robotics and Automation, 1277-1282 (1997)
12. L. Tan, S. Lu, W. Zhang, *A robotic nursing bed design and its control system*. Proceedings of the 2009 IEEE International Conference on Robotics and Biomimetics, December 19-23, (2009)
13. S. Peng, F. Lian, *Mechanism design and mechatronic control of a multifunctional test bed for bedridden healthcare*. IEEE/ASME Transactions on Mechatronics **15** (2), 234-241 (2010)

Transformation of heat into mechanical energy by means of rotating systems

Marian Mešina[1,*]

[1]Privat, D-73266 Bissingen, Mörikestraße 25, Germany

Abstract. All heat engines need two different temperatures for their work, $T_1<T_2$. The efficiency is limited by the Carnot formula from above. This article presents a new conception for heat engines. Unlike conventional heat engines, the working medium has an additional degree of freedom, the rotation around a given axis. The heat introduced and removed can not only lead to a change in the parameters of temperature, pressure, and volume, which are considered in conventional thermodynamics, but also to a change in the state of rotation. The rotational speed must also be taken into account in all phases of the cycle for all efficiency calculations. In many cases, this leads to a surprisingly different result from the results of conventional thermodynamics, that the efficiency of the cycle can exceed the Carnot limit. The efficiency values depend not only on temperatures and rotational speeds, but also on the material data. The proposed new type of heat engine makes it possible to better utilise very small temperature differences and under certain conditions, in combination with an ideal heat pump, to extract ambient heat and convert it into mechanical energy. The calculated results were presented for simple geometry and can easily be verified experimentally. In combination with an ideal heat pump, the proposed heat engine facilitates the surroundings to withdraw heat and convert it into mechanical work.

Keywords: Heat-mechanical energy conversion, efficiency, RS machine, Carnot

1 Introduction

In conventional thermodynamics, the maximum efficiency of heat engines can be fully specified by means of temperatures in the range of which the working cycle takes place. Carnot's theorem is the well-known formulation of this fact.

This article presents a whole new principle for converting heat into mechanical energy. The systems under consideration have an additional degree of freedom - the rotation around their own axis. The heat supplied can also partly be converted into the acceleration of the rotation of the system. This degree of freedom can take the energy from the surroundings and also put it back into the surroundings. Not only temperatures but also rotational states (rotational speeds) must therefore be taken into account in the description of the efficiency. We refer to the new type of heat engines proposed here as RS machines (machine operating on the basis of the rotating system).

With conventional heat engines, the thermal expansion or shrinkage of the working material (e.g. a gas) is directly converted into mechanical work. The new working principle

* Corresponding author: m-mesina@t-online.de

Reviewers: *Josef* Soukup, *Branislav* Ftorek

proposed here works differently. In the proposed concept, in the rotating systems the heat supply leads to the expansion of the working medium and thus to mass displacement in the direction of the rotational axis. This movement of the mass leads to a reduction of the inertia moment. This reduction results in the increase of the rotational frequency at a constant angular momentum and also an increase in the rotational energy. This phenomenon has long been used by ice skaters when performing pirouettes. However, the muscle work of the athletes causes a shift in mass. In an RS machine, the heat supplied is the cause of the mass shift.

The aim of the calculations carried out was to check whether this phenomenon was sufficient to increase efficiency beyond the Carnot limit. The functioning of the RS machine as described here allows the user - due to a partially selectable starting rotational frequency - to determine the amount of heat that can be converted to mechanical energy, at fixed working temperatures T_1 and T_2.

The RS machines operate cyclically, each cycle has 4 phases. In phase 1, the system is operated at the ambient temperature T_1 and accelerated at the rotational speed ω_1. In this phase, mechanical energy must be supplied to the system. In phase 2, the temperature of the working liquid is increased from T_1 to T_2. To simplify our calculations, we have assumed that the necessary heat is supplied at a constant angular momentum. The rotational speed increases to ω_2. In phase 3, the rotational energy is removed from the system at the constant temperature T_2 and the system is decelerated. In phase 4, the working liquid is cooled down to the T_1 temperature. The work medium could still work in phase 4, but we have not included this to simplify the calculations.

When the rotation in phase 1 is accelerated, only in the case of a compressible liquid, some of the energy supplied is consumed to compress the liquid. In the following calculations, we assume that when the rotation is decelerated in phase 3, the braking of the rotation leads to decompression and the decompression energy is converted into rotational energy.

2 Description of the geometry of the RS machine (simplest version)

The basic principle of an RS machine can be seen in figures 1-3. The main part of the machine is a hollow cylinder that is to rotate around its axis. The working fluid first has the T_1 temperature and is in a cylindrical shell in the rotation with the outer radius R_a and the internal radius R_{i1} (see Figure 1 and 2). The influence of gravitation on the geometry of the working fluid is not included.

In the container are at least two partition walls, which serve to ensure the rotation of the working medium together with the container. The partitions are mechanically fixed to the container. In addition, by pressure on the separating walls, the expanding liquid can change the rotational state of the system. The geometry described here has deliberately not been selected optimally! The article only intends to show that even with this simple and non-optimal design, higher efficiency values can be achieved than according to Carnot. The greatest advantage of this geometry is that the computational check can be performed very easily.

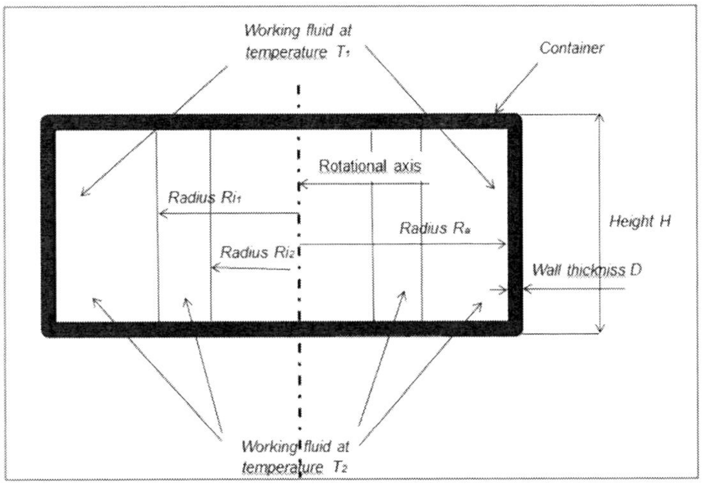

Fig. 1. RS machine, side view, perpendicular to the axis of rotation

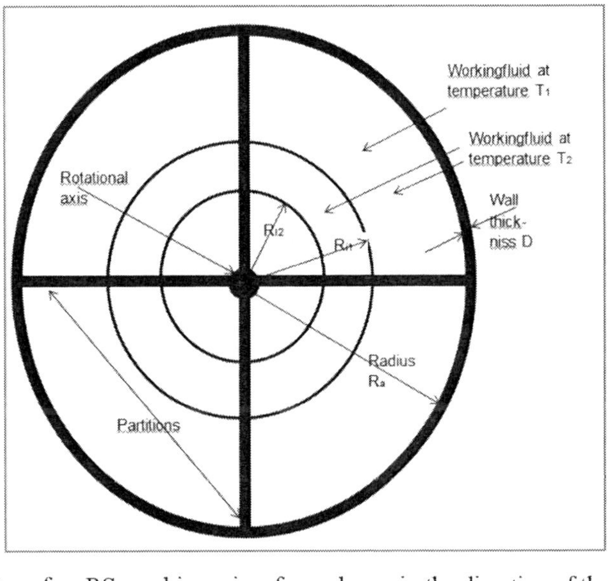

Fig. 2. Geometry of an RS machine, view from above, in the direction of the rotational axis

After heating to the temperature T_2 the liquid is in the cylinder with the outer radius Ra and in the inner radius R_{i2}. Because $T_2 > T_1$ (expansion in the direction of the axis of rotation), R_{i2} is also $< R_{i1}$. This results in the reduction of the moment of inertia. For all further considerations, we assume that with the temperature transition from T_1 to T_2 the container geometry remains virtually unchanged

3 Basic formulas for the efficiency calculations of the RS machine

Terms:

M_{flu} - Mass of the working medium,
ρ (P_1, T_1) - density of the liquid used at initial pressure P_1 and temperature T_1,
C_{vFlu} - specific heat of the liquid,
H - Container hight,
M_{Cont} - mass of the container,
ρ_{Cont} - density of the container material,
Cv_{Cont} - specific heat of the container material,
α - thermal expansion coefficient of the liquid,
P_1 - initial pressure, 100000N/m^2,
Working fluid – diethyl ether.

The geometry terms are described in the figures 1-3.

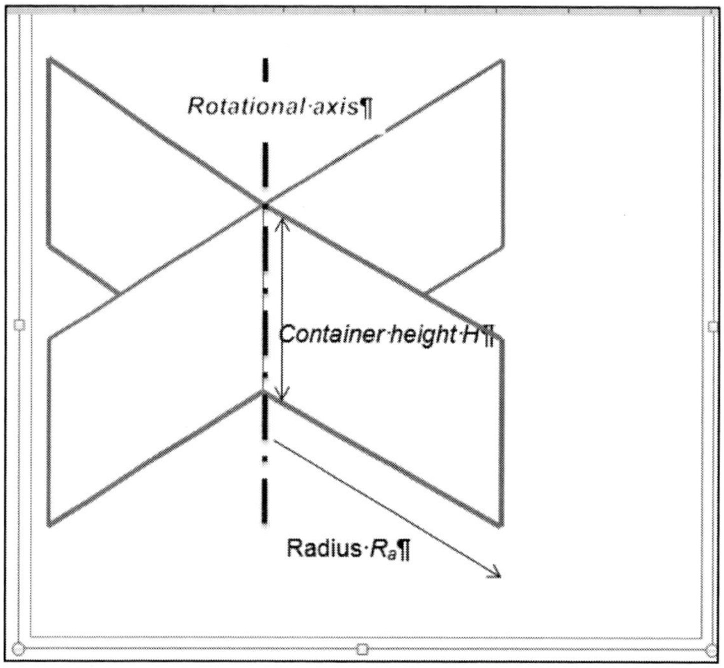

Fig. 3. Four partitions in the container of an RS machine

4 Efficiency calculation for the incompressible medium

At temperature T_1 and at pressure P_1 the following applies:

$$M_{Flu} = \pi * (Ra^2 - Ri_1^2) * H * \rho(P_1, T_1) \tag{1}$$

For incompressible media, we consider ρ as dependent on temperature only.

The M_{Flu} mass is located in the cylinder shell with the outer radius Ra and inner radius Ri_1. Mass M_{Flu} is equal to product volume (P_1, T_1) times the density (P_1, T_1). At temperature T_2 the following applies:

$$M_{Flu} = \pi * (Ra^2 - Ri_2{}^2) * H * \rho(P_2, T_2) \tag{2}$$

The M_{Flu} mass is located in the cylinder shell with the outer radius Ra and inner radius Ri_2. From the equalisation of (1) and (2) the following applies:

$$Ri_2 = \sqrt{Ra^2 - (Ra^2 - Ri_1{}^2) * \frac{\rho(P_1, T_1)}{\rho(P_2, T_2)}} \tag{3}$$

Since Ri2 $\geq$ 0 must be, the temperature T_2 cannot be arbitrarily large for a given geometry. However, this limitation can be corrected by a small change in the geometry.
For the moment of inertia of a cylinder shell with the outer radius Ra inner radius Ri, density ρ, and the height H, the following applies:

$$J(P,T) = \pi * (Ra^2 - Ri^2) * H * \rho(P,T) * \frac{(Ra^2 + Ri^2)}{2} \tag{4}$$

In which ρ is the material density of the liquid in the shell.
The formula (4) is used with various parameters for the calculations of the moments of inertia of the working fluid and the container.
We refer to J_1 as the complete moment of inertia of the whole system at the beginning of phase 2 and J_2 at the end of phase 2.

$$J_1 = J_{1Flu} + J_{Cont} \tag{5}$$

$$J_2 = J_{2Flu} + J_{Cont} \tag{6}$$

Assumption: The moment of inertia of the container J_{Cont} remains virtually unchanged in the temperature range used. The kinetic energy of rotation at the beginning of phase 2 is:

$$Ek_1 = J_1 * \frac{\omega_1^2}{2} \tag{7}$$

At the end of phase 2

$$Ek_2 = J_2 * \frac{\omega_2^2}{2} \tag{8}$$

Where ω = 2*π*f, f is the frequency of rotation. For the angular momentum L the following applies:
At the beginning of phase 2

$$L_1 = J_1 * \omega_1 \tag{9}$$

At the end of phase 2

$$L_2 = J_2 * \omega_2 \tag{10}$$

Ideally, the loss of rotational moment during phase 2 is 0. In reality, there are losses in rotational momentum (eg due to friction). These losses can be represented by factor K_{red}.

Then:

$$J_2 \cdot \omega_2 = J_1 \cdot \omega_1 * K_{red} \tag{11}$$

Where $K_{red} \leq 1$ (the loss of the angular momentum). From the equation (11), the angular velocity ω_2 at the end of phase 2 can be calculated:

$$\omega_2 = \left(\frac{J_1}{J_2}\right) * \omega_1 * K_{red} \tag{12}$$

For the kinetic energies E_{kin1} and E_{kin2} the following applies:

$$E_{k_2} = E_{k_1} * \frac{J_1}{J_2} * K_{red}^2 \tag{13}$$

When $J_1 * K^2_{red} > J_2$, then E_{k2} is also > E_{k1}.This means an increase of the rotational energy due to the heat supply.

In the course of phase 3, the system can provide the kinetic energy E_{k2} . In phase 4, the working liquid and the container are cooled from temperature T_2 to temperature T_1. In order to simplify the calculations, at the end of phase 3 the remaining temperature difference between T_2 and T_1 are deliberately not used for the production of the work! Nevertheless, the achieved efficiency values can exceed Carnot's limit.

For the obtained mechanical energy $delta_{Ek}$ the following applies:

$$delta_{Ek} = Ek_2 - Ek_1 \tag{14}$$

From equations (7-14) the following applies:

$$delta_{Ek} = \left(\left(\frac{J_1}{J_2}\right) * Kred^2 - 1)\right) * Ek_1 \tag{15}$$

For incompressible working media, we assume that the moments of inertia J_1 and J_2 are only dependent on temperatures T_1 and T_2.

Equation (15) shows that the gained energy $delta_{Ek}$ is proportional with the kinetic energy Ek_1 . As ω_1 and thereby also E_{k1} can be selected freely, the efficiency can be influenced at least in part (see figures 5-8 and tables 2-5). This is a new and the most important property of RS machine, which leads to the possibility to exceed the Carnot limit of efficiency.

5 Calculation of the efficiency η

The efficiency of the above-described working cycle η is defined as the work obtained in the cycle divided by the total amount of the heat supplied to the system.

The heat supplied must provide the following three tasks for incompressible working fluid) and four tasks for compressible liquid energy:

A1. Increase of the rotational energy by $delta_{Ek}$, (equation 15). The increase of the kinetical energy $delta_{EK}$ is the gained work in one working cycle (later the phase 2).

A2. Heating the working fluid (see Eq. 16) and the container see eq. (17) from temperature T_1 to temperature T_2.

For the required heat quantity, the following applies:

$$DQ_{Cont} = M_{Cont} * C_{V\,Cont} * (T_2 - T_1) \tag{16}$$

$$DQ_{Flu} = M_{Flu} * C_{V\,Flu} * (T_2 - T_1) \tag{17}$$

A3. As a result of the thermal expansion, some of the working liquid is displaced in the direction of the axis of rotation.
The necessary work A_{transl} serves to overcome the centrifugal force. This work can be estimated from the following formula:

$$A_{transl} \approx M_{Flu} * \frac{\alpha * (T_2 - T_1)}{1 + \alpha * (T_2 - T_1)} \omega_2^2 (Ri_1 - Ri_2) * \frac{(Ri_1 + Ri_2)}{2} \tag{18}$$

Where Ri_1 internal limit of the zone at temperature T1, Ri_2 the internal limit at temperature T_2 and $M_{Flüss}$ is the mass of the working fluid in the zone (see Fig.1 and 2). For the calculation of the efficiency η we make the pessimistic assumption that A_{transl} can only be taken into account when calculating the heat supplied to the system. In our calculation, this work does not contribute to the energy that can be supplied by the system.

A4. For compressible fluids there is a 4th task. In phase 3 for isothermal deceleration of rotation, the heat must also supply the expansion energy of the compressible medium. This task is not applicable for incompressible working media.
For the efficiency η of the RS machine with incompressible working fluid the following applies:

$$\eta = \frac{deltaE_k}{deltaE_k + DQ_{Kont} + DQ_{Flu} + A_{transl}} \tag{19}$$

Formulas (14) - (19) show that the efficiency is dependent not only on the temperatures T_1 and T_2, but also on the angular velocity ω_1 or frequency F_1 as well as the material data. By the practical application of formulas (4) - (19), it can be easily shown (s. figure 4) for many combinations of ω_1, T_1, and T_2, that the efficiency η is greater than the Carnot limit.

Because the gained energy $delta_{Ek}$ and the work A_{transl} are quadratically dependent on frequency F_1, converges with increasing frequency F_1 to a value that is less than 1, but can still be greater than Carnot's. The real liquids are compressible. Formula (19) can therefore only be used for rough estimates of the efficiency values.

Formula (19) has to be modified for calculating the efficiency.

Our calculations for compressible fluids are based on the following formula:

$$\rho(P_2, T_2) = \rho(P_1, T_1) \frac{(1 + \kappa(P_2 - P_1))}{(1 + \alpha(T_2 - T_1))} \tag{20}$$

P_1, P_2 - different pressure values,
T_1, T_2 - different temperatures,
ρ - pressure and temperature dependent density,
α - thermal expansion coefficient,
$\varkappa$ – Compressibility coefficient.

Assumption: The parameters α and ϰ are considered constant. Formula (20) can be seen as a substitute for the state equation of the liquid.

In order to calculate the location dependency of the density, we divided the total volume of the liquids into N sub-regions (cylinder shells) of the same mass. In our test examples it was found sufficient to take N = 30 and 40 radial zones, respectively. When the temperature or ω changes, only the boundaries between the subareas shift. Pressure in the radial zones is location-dependent and increases in the direction of the outer edge by the own weight of the liquid in the radial zone.
For the efficiency calculation the following applies:

$$\eta = \frac{E_{dekompr} - E_{kompr} + deltaEk}{deltaEk + DQ_{Kont} + DQ_{Flu} + A_{transl} + E_{dekompr}} \tag{21}$$

Let us call E_{kompr} the necessary energy for the compression of the liquid in the phase 1 and as $E_{dekompr}$ the releasing energy in the expansion of the liquid in phase 3. In the counter of the fraction in equation (21) is the total energy that can be taken from the system in one cycle. In the denominator is the total heat required to be supplied to the system in one cycle.

Advantages of the RS machine:

- Efficiency is much better in many situations than according to Carnot
- Numerous designs and optimisation options are possible (e.g. geometry, material selection, rotational speed)
- Use of small temperature differences
- Regeneration possibility of the used heat reservoir by the heat supply from the environment
- In principle, a new energy source (as perpetuum mobile of 2nd type)
- A technically possible use of the water-ice anomaly. When freezing water, the water can only expand in the direction of the rotational axis.

Disadvantages of the RS machine:

- Very high rotational speeds are necessary (see figures 4 and 5), but possible in the context of today's technology.
- Isothermal acceleration (phase 1) and deceleration of the rotation (phase 3) may be very slow.
- Special materials recommended (maximum thermal expansion and low compressibility)
- Other operation than with conventional heat engines - isothermal acceleration of rotation at temperature T_1 (phase 1) and isothermal deceleration of the rotation at temperature T_2 in phase 3.
- Heat supply with an angular momentum as constant as possible (phase 2)
- In order to simplify the calculations, we assume that in phase 4 the working fluid is cooled from temperature T_2 to temperature T_1 without performing work.

Further development possibilities:

- The use of the phase change (liquid gas, solid phase gas)
- Use of residual heat at the end of phase 3
- Heating the working fluid without heating the whole container
- Optimising the geometry

6 Proposal for verifying the theory

The container with the working fluid must be accelerated at temperature T_1 to the angular velocity ω_1. The necessary energy E_1 for this is measured. Then the system must be heated to temperature T_2 at a constant angular momentum. We call the heat necessary for this Q_2. After this in phase 3, the rotation is completely decelerated to $\omega = 0$ at a constant temperature T_2. The energy E_3 with drawn from this and the amount of Q_3 supplied is measured.

The heat Q_2 and Q_3 should be obtained from a heat reservoir with the temperature T_2 . For test purposes only, the required amount of heat could be generated electrically or chemically directly in the system.

The efficiency can be determined from the measured data (see equation (22)).

$$\eta = \frac{E_3 - E_1}{Q_2 + Q_3} \tag{22}$$

The internal structure of the RS machine does not have to be known for measuring the efficiency!

7 Some computed results

Table 1. Common input data for all test calculations

Working fluid:	diethyl ether
Density:	738 kg m^3 at 273 degrees K.
Compressibility ϰ	1.84E-09 /GPa
Thermal expansion coefficient α	0.00162/deg K
Specific heat of the liquid Cv:	2310 J/(Kg. degree K)
Material of container:	aluminium
Density:	2700 Kg/m^3
Specific heat Cv_{cont}	896 J/(Kg . degree K)

Other data for the calculation: Ra = 0.5; H = 0.4; D = 0.015.

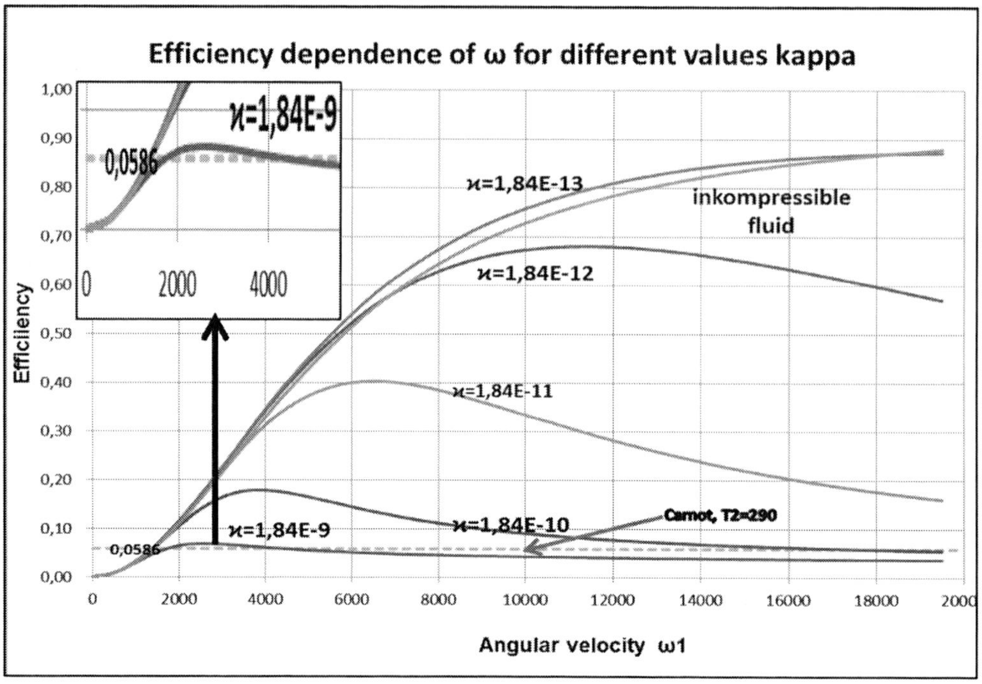

Fig. 4. Efficience dependence from ω_1 for various coefficients of compressibility and for incompressible fluid, T_2=290

Input data for calculation: Ra=0,5; H=0,4; D=0,015; T_2=290; various values of kappa.

In Fig. 4 and 5 Carnot's limits were drawn as horizontal lines (horizontal means independent from ω_1). The maximum of the efficiency curve can be clearly above the Carnot's limits (e.g. for the temperatures T2 = 275 and 290 K). For all ω_1 from interval [UGR, OGR], the calculated efficiency is greater than the Carnot's limit.

Table 2. Some important parameters of ω_1-dependence curves of efficiency, from figure 5

Temperature T_2	320	300	290	275
Eta Carnot	0,146875	0,09	0,058621	0,007273
Eta RS machine max. value	0,11720	0,08758	0,06882	0,02296
Eta max at ω1 approximately	4076	3100	2575	1275
Angular speed ω_1, UGR	no solution	no solution	1700	475
Angular speed ω_1, OGR	no solution	no solution	4400	5550

Input data for the calculation: Ra = 0.5; H = 0.4; D = 0.015; kappa=1.84E-09

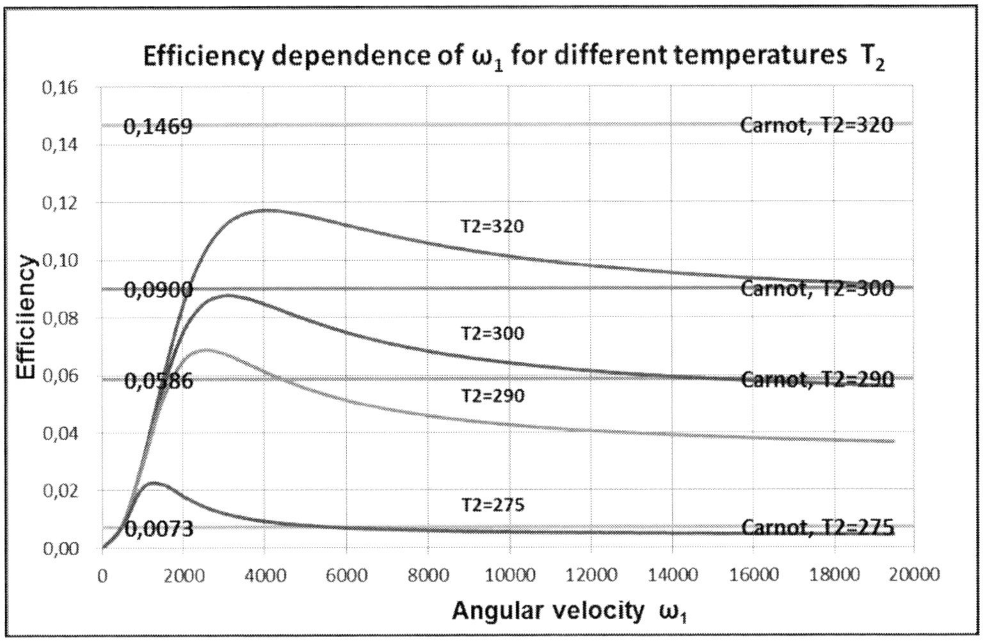

Fig. 5. Efficiency dependence of ω_1 for different temperatures T_2, comparison with Carnot's limit

Input data for the calculation: Ra = 0.5m; H = 0.4m; D = 0.015m; Kappa = 1.84E-09.

Figures 4-5 show that the efficiency of an RS machine increases as a function of the angular velocity ω_1 up to a maximum value. Reducing it after achieving maximum efficiency is thereby explained because the difference ($E_{dekompr}$ - E_{kompr}) increases in comparison to the energy growth in phase 2. For the increasing rotational frequency ω_1, the RS machine is more similar to conventional heat engines in terms of efficiency.

At the temperature T2 = 320 and T2=300 K (s. the figure 5), the Carnot limit cannot be exceeded only by the change of ω_1, at constant value of other selected parameters R_a, M_{Flu}, height H and wall thickness of the container D.

Exceeding the Carnot's limit is easier at low temperature T_2 (at low Carnot's limit) than at higher temperature T_2 (see. figure 5).

Conclusion

What makes it possible to say that even heat engines with a better efficiency than Carnot are possible?

In the patent application, the basic principle of RS machine has been presented. The efficiency of this machine can be described through formulas, which are transparent, trivial, and comprehensible. It is an attempt to simulate the working cycle of the RS machine. The simulation of the processes in nature is now a standard tool in natural science and in technology.

The most important feature of this simulation is that the efficiency of the RS machine is explicitly dependent on the rotational speed ω. It is by no means possible to eliminate the parameter ω. The calculations carried out show that the calculated efficiency can clearly exceed the Carnot's limit in many situations.

In combination with an ideal heat pump, it is possible to extract heat from the surroundings and convert it into mechanical work.

I am particularly grateful to J. Pimiskern for many helpful discussions, for many interesting references to this topic, and for his help in the design of this publication.

References

1. Patent application DE 102011104914 A1
2. H. Kuchling, *Taschenbuch der Physik*. (20th Edition, Carl Hanser Verlag Leipzig, 2011), ISBN 978-3-446-42457-9

Finite element simulation of laser cutting process of steel sheet

Jozef Meško[1], *Andrej* Zrak[1], *Rastislav* Nigrovič[1,*], *Milan* Žmindák[2]

[1]UNIZA, Faculty of Mechanical Engineering, Department of Technological Engineering, Univerzitna 8215/1, 010 26 Žilina, Slovakia
[2]UNIZA, Faculty of Mechanical Engineering, Department of Applied Mechanics, Univerzitna 8215/1, 010 26 Žilina, Slovakia

Abstract. Lasers are widely used in industry as cutting tools due to ultra flexibility of the cutting conditions, obtaining high quality end product, quick set up, non-mechanical contact between the workpiece and the tool, and small size of the heat affected zone. In the present study, laser gas assisted cutting process is examined. The laser cutting sheet solution is practically always very convenient compared to conventional technologies and brings the greatest cost savings in the manufacturing process.

Keywords: Laser cutting, FEM, Simulations of laser cutting, Finite element mesh

1 Introduction

Material cutting is still a very important manufacturing operation. A number of methods are used for this, each of which has created its area of optimal use. Each of them has some deficiency that prevents universal use. That is the reason, why there are still more and more modern procedures. The thermal cutting of the material can be included in the manufacturing process as a material preparation operation. Under this term, we mean cutting technology, working on the principles of local melting, combustion, evaporation, or possibly combining of these phenomena. In this technology, where energy is needed to initialize the process, and it is supplied by different heat sources. Laser cutting is applicable to a wide variety of construction materials: unalloyed and low alloy steel, high alloy steel and nickel alloys, non-ferrous metals and their alloys (Al, Cu), highly reactive materials and their oxygen-sensitive alloys (Mg, Ti), non-metallic materials (plastics, composites, wood, glass) [1, 2, 7].

* Corresponding author: rastislav.nigrovic@fstroj.uniza.sk

Reviewers: *Petr* Horyl, *Josef* Soukup

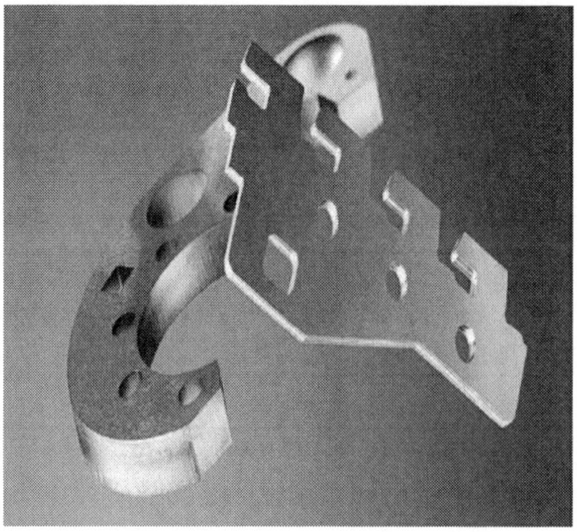

Fig. 1. A demonstration of a laser workpiece

When the laser beam impacts the material, there occurs an interaction between that material and the laser beam. The subsequent processes that take place during the material cutting and the effect on the material properties following the focused beam impact, mainly depend on the chemical composition of the material being cut and on the quality of its surface which was machined by milling in our case.

In thermal cutting of metallic materials by a laser, it is always necessary first to create a hole in the processed material, from which the cutting would continue. Making the hole is based on the laser drilling principle and has slightly different characteristics of the course than that of the actual cutting. The impacting laser beam transmits the kinetic energy of the material photons, which converts into heat that melts and partially evaporates the material being heated. The surroundings of the laser beam inflammation site (in the material being cut) contain gases that get ionized at the moment of the beam impacting the material and change to plasma. The material being cut sublimates into a gaseous state, and is blown from the material to the environment by the action of an assist gas under a relatively high pressure. Part of the material, which does not pass into a gaseous state, is blown in the liquid form by the flowing assist gas. The above-mentioned process causes formation of a pit in the material being cut, and the laser beam can penetrate continuously deeper, which results in depth melting of the material [1, 2, 7].

2 Theoretical foundations

The heat transfer condition, which includes the material characteristics of the sample for simulation, is defined by Equation.

$$\rho * c_p * \frac{6T}{6t} = \nabla * (k\nabla T) \quad (1)$$

Where:
ρ - density [kg/m^3],
cp - the specific heat capacity at the const. pressure [J/kg.K],
T - temperature [°C],
∇ - nabla operator,
k - thermal conduction [W/m.K],
t- time [s].

An introductory step for creating a model of the laser cutting process is a compact analysis of the problem and the definition of the result, especially from that point of view, so that it is clear what is required from the simulation. The output that is the desired simulation result can accurately quantify the heat in the coordinate system [x, y, z, t], where x, y, z are the coordinates defining the position of the point with respect to the axis x, y, Defines the exact time period in which the point temperature is quantified. An important step is to correctly define the total heat source, respectively. The heat input which is then dispersed in the surrounding material. Said method of defining a heat source is performed on the basis of a mathematical relationship in which the coefficient of reflectivity of the surface to which the laser beam enters, the coefficient of energy absorption of the photon flux through the material and the analytical function representing the 2D Gaussian curve are included as variables. The equation defines a point thermal source. The next step is to determine the correct networking of the model. The point heat source, which represents the power of the photons moving in the laser beam, is defined as the boundary condition in the entire domain in the simulated design. This resource is defined by Equation 2, which is specified in the Comsol Multiphysics 4.4 program as a variable, dependent on the global system x, y, z [3, 5].

$$Q(x,y,z) = Q_0 * (1 - R_c) * \frac{A_c}{\pi * 6_x * 6_y} * e^{-\left[\frac{(x-x_0)^2}{z*6_x^2} + \frac{(y-y_0)^2}{z*6_y^2}\right]} * e^{-A_c * z} \qquad (2)$$

Where:

Q_0 - total heat supplied by the laser beam [J],
R_c - coefficient of reflectivity of the cut material,
A_c - absorption coefficient,
σ_x - standard deviation of the laser in x direction [mm],
σ_y - standard deviation of the laser in y direction [mm],
x_0, (y_0) - center of beam path in x, (y) direction [mm].

3 Model FEM

3.1 Simulation of steady-state lasers with constant power

The simulation was generated as a process of heating the plate. The material properties of the board are considered to be constant - simulation does not involve a change in the material phase. The refractive index of the steel is defined by the reflection coefficient and the energy absorption coefficient by the material. The material model represents a plate whose sides except the top are considered to be heat insulated. The movement of the laser beam is characterized by x and y coordinates, and in the plane xy, z = 0 is on the surface of the plate. Parameters x0 and y0 serve to enter the shift of the center of the laser beam. The diameter of the incident laser beam in practice has a changing value, the shifting of the laser beam focus point, and therefore the change of the incident beam diameter is defined by the standard deviation parameters σ_x and σ_y.

At the beginning was created, a model of a laser, which is considered as a heat source, with a constant power that does not move on the surface of the plate. The method was designed to investigate the heat dissipation in the plate to which the laser beam enters in continuous mode for 3 seconds. The analytical feature representing the 2D Gaussian curve was then created. This function is given by a symbolic expression and has the ability to bind arguments in a repeated function solution. [4, 6]

Table 1. Variables used for simulation

Mark	Set value	Setting characteristics
x_0	0 [mm]	Center of laser beam path
y_0	0[mm]	Center of laser beam path
Q_0	3200[W]	Total laser beam power
sig_x	0,5[mm]	Standard deviation in x direction
sig_y	0,75[mm]	Standard deviation in y direction
Ac	0,50[1/cm]	Absorption coefficient
Rc	0,50	Reflection coefficient
L	100[mm]	Dimension of the square plate
Lz	10[mm]	Thickness of the board
t	3[s]	The time of the laser beam on the plate

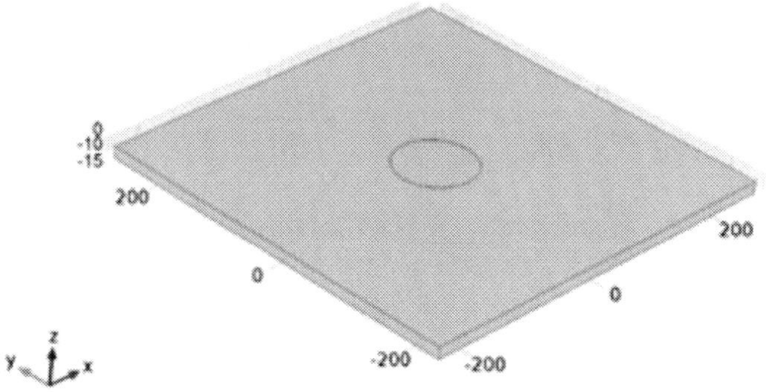

Fig. 2. Geometric model of the board

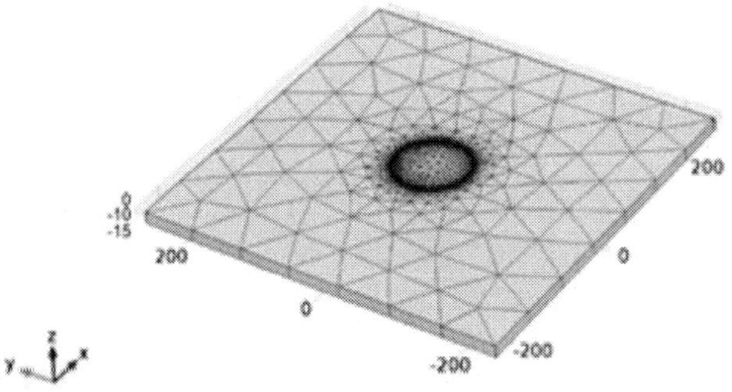

Fig. 3. Finite element mesh

4 Numerical results

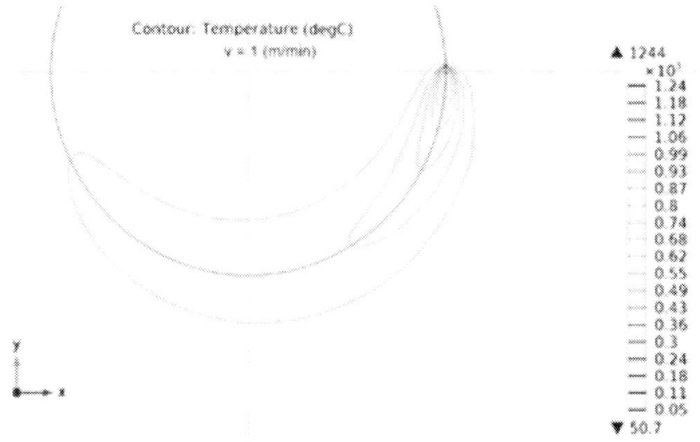

Fig. 4. Temperature fields at a cutting speed of 1 m/min

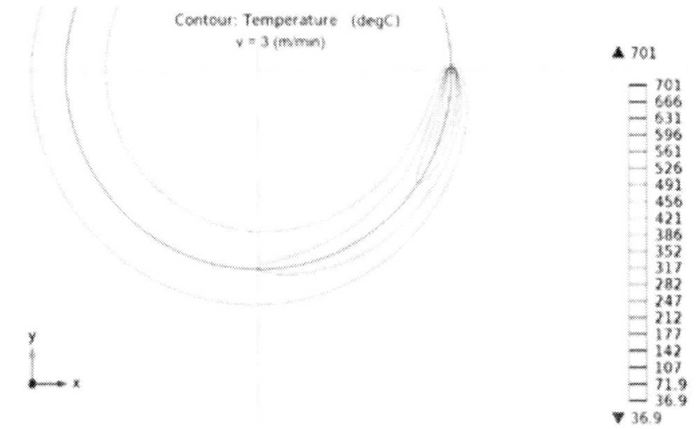

Fig. 5. Temperature fields at a cutting speed of 3 m/min

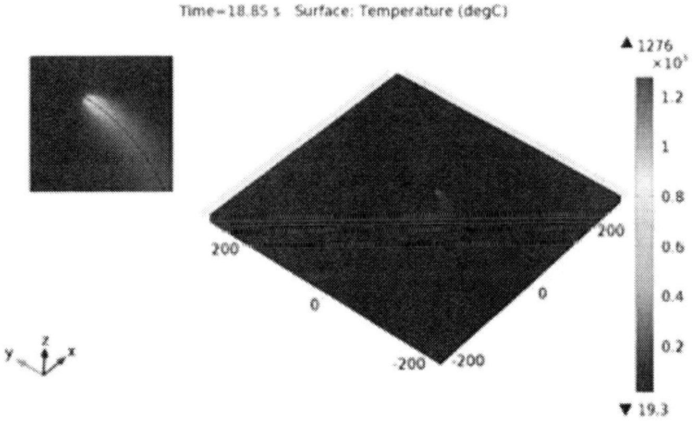

Fig. 6. Temperature field model

Conclusion

Based on the model of heat transfer in the material during laser cutting, it is possible to understand the principle of heat conduction from the cutting point to the surrounding material. This mechanism is particularly important when dealing with the problem of cutting rough metal sheets, Small parts with a complicated cutting path so that a relatively large amount of heat is introduced into the material. After is the principle of heat dissipation in the cut material known, depending on the time, it is possible to design the ideal arrangement of parts and the processing of the components so that the cutting edge of the material causing the deformation is not bounded at the defined part of the component. Based on the analysis of simulated model results, it can be stated that the temperature of the material at the point of cutting immediately at the edge of the laser beam is assumed to be approximately 1700 ° C and the maximum stress due to the thermal expansion of the material is assumed to be approximately 1300 MPa. The temperature of the material at the cutting location is relatively high, but with increasing distance from the cutting point it drops sharply, e.g. At a distance of 10 mm from the source of heating, the temperature of the material is 100 °C.

Research presented in this paper was partially financially supported through realization of projects KEGA 014ŽU-4/2016 - Responsible investigator: prof. Jozef Meško, MSc., Ph.D.

References

1. L. C. Caristan, *Laser cutting guide for manufacturing*. Dearborn, Michigan, USA: Society of Manufacturing Engineers, 452 p. ISBN 0-87263-686-0 (2004)
2. W. T. Silvfast, *Laser Fundamentals*. (Cambridge University Press, 666 p. ISBN 0-521-83345-0, 2004)
3. J. Meško, A. Zrak, *Analysis of heat transfer in laser cutting problems*. Analysis of technology in various industries : monography, Chapter **6**, Częstochowa: Oficyna Wydawnicza SMJiP, ISBN 978-83-63978-11-2 (2014)
4. P. Novák, J. Meško, M. Žmindák, *Finite element implementation of multi-pass fillet weld with phase changes*. Manufacturing technology, Journal for science, research and production **13** (1), 79-85 (2013)
5. M. Žmindák, Z. Pelagič, P. Pastorek, M. Moličan, M. Vyboštʼok, *Finite element modelling of high velocity impact on plate structures*. Procedia Engineering **136**, 162-168 (2016)
6. Z. Pelagič, M. Nágeľ, M. Žmindák, P. Pastorek, *Finite element simulation of wear in dynamic contact problems*. Dynamika tuhých a deformovatelných těles 2013, sborník přenášek z XI. mezinárodní vědecké konference, Ústí nad Labem, 9.-11. října 2013, Ústí nad Labem: FVTM UJEP, ISBN 978-80-7414-607-7, (2013)
7. H. Danielewski, R. Banak, A. Domagala, *The Experimental Analysis of Striation Pattern Created During Laser Cutting*. Transcom 2013, Žilina, Slovakia (2013)

Numerical research of biomechanical system with SMA prosthesis

Andrzej Mitura[1,*], *Rafal* Rusinek[1]

[1]Lublin University of Technology, Department of Applied Mechanics, Nadbystrzycka 36, 20-618 Lublin, Poland

Abstract. In this paper the modelling of special biomechanical mechanism and application of shape memory materials are presented. The model of the human middle ear is made as multibody system. The basic 3dof ear model of the healthy middle ear is modified to represent the damaged ear. A damaged of the ossicular chainis taken into account by adding gap in visco-elastic joints. In addition, an attempt of the ossicles chain reconstruction through prosthesis made of shape memory alloy is presented. Moreover, a new description of the hysteresis sub-loop which characterise prosthesis material is proposed. Finally, dynamic responses of healthy, damaged and reconstructed models of the middle ear are compared by quality index.

Keywords: multibody model, biomechanical system, shape memory alloy

1 Introduction

The human body is a complex object that receives information from the environment using the senses: eyesight, hearing, touch. With the development of anatomy, it was possible to know the structure of the hearing system. However, research on a living object is difficult. Therefore, the doctors and researchers strive to create models, which allow to study how a damaged or undamaged human ear works. In the literature different type of ear models are proposed. In some papers the finite element method is applied to model middle ear [1, 2]. Substitutes models, for example the lumped parametric models are created as a alternative method. In paper [3] Feng and Gan present the multibody model, where six lumped masses are connected by visco-elastic joints. This biomechanical system includes such elements as: the external ear canal terminated with tympanic membrane, three ossicles (so called: the malleus, the incus, the stapes) and the cochlea. In paper [4] the authors limited the model only to the middle ear part. This approach allows testing the transmission of the signal (sound) through the hearing ossicles.

In the first part of this work, a three degrees of freedom model of middle ear is presented. The model is tested for two cases, the healthy and damaged. The damage of the middle ear can be caused by a past illness, sudden injury and so on. As a consequence a loss of continuity of the connection between the ossicles can cause. The separation of the middle ear bone in the model is made by adding a gap in the links between: malleus-incus

* Corresponding author: a.mitura@pollub.pl

Reviewers: *Ivan* Kuric, *Alžbeta* Sapietová

or incus-stapes. The damage of this type can be cured through the middle ear prosthesis. A typical prosthesis is made of titanium and has to be cut during the operation in order to get the proper length [5]. Therefore, new material to active control length is searched. Rusinek and others propose the prostheses made of alloy with shape memory effect (in short SMA) [6], for example NITINOL. Generally, SMA is the material where the shape depends on temperature and/ or stress. In the second part of this paper a model of SMA spring used to reconstruct ossicles chain as the prosthesis is presented. The prosthesis is mounted between the tympanic membrane and the stapes. After reconstruction the model of middle ear was reduced to system with two degrees of freedom because usually the incus is removed. The dynamic responses of the considered models: the healthy ear, the damaged ear and the reconstructed ear with the help of SMA prosthesis are compared by a proposed quality index.

2 Modelling and analysis of dynamics of the biomechanical system

2.1 Dynamics of healthy middle ear model

In this section the model of the human middle ear is presented. The applied multibody system is based on proposition presented in papers [4, 6]. Three rigid body represent ossicular bones. The symbols m_M, m_I, m_S describe lumped masses of the malleus, the incus and the stapes, respectively. These ossicles are connected to each other and to the other elements of the human ear by visco-elastic joints (figure 1).

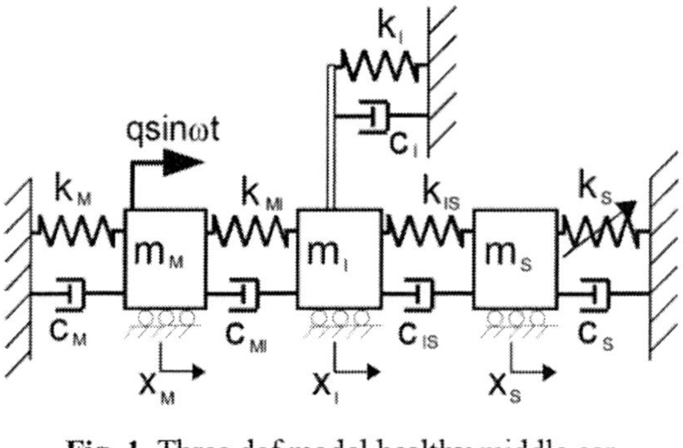

Fig. 1. Three dof model healthy middle ear

The values of system parameters have been taken from literature [4, 6] as follows : m_M=25 mg, m_I=28 mg, m_S=1.78 mg, k_M=1.1 N/mm, k_{MI}=1000 N/mm, k_I=0.4 N/mm, k_{IS}=1.2 N/mm, c_M=0.06 Ns/m, c_{MI}=0.359 Ns/m, c_I=0.055 Ns/m, c_{IS}=0.0079 Ns/m, c_S=0.00174 Ns/m. Nonlinear spring (k_s) is described by a polynomial function:

$$F_s(x_S) = k_{S0}x_S + k_{S1}x_S^2 + k_{S2}x_S^3 \quad (1)$$

where: k_{S0}= 0.825 N/mm, k_{S1}=300 N/mm^2, k_{S2}=2.25x10^{15} N/mm^3. Finally, the healthy middle ear system can be described by the following ordinary differential equations of motion:

$$m_M\ddot{x}_M + k_M x_M + c_M\dot{x}_M + k_{MI}(x_M - x_I) + c_{MI}(\dot{x}_M - \dot{x}_I) = q\sin\omega t \quad (2)$$

$$m_I\ddot{x}_I + k_I x_I + c_I\dot{x}_I - k_{MI}(x_M - x_I) - c_{MI}(\dot{x}_M - \dot{x}_I) + k_{IS}(x_I - x_S) + c_{IS}(\dot{x}_I - \dot{x}_S) = 0 \quad (3)$$

$$m_S \ddot{x}_S + k_{S0} x_S + k_{S1} x_S^2 + k_{S2} x_S^3 + c_S \dot{x}_S - k_{IS}(x_I - x_S) - c_{IS}(\dot{x}_I - \dot{x}_S) = 0 \tag{4}$$

The numerical simulations of the system described by Eqs.2-4 are made in Matlab software. Figure 2 shows the root mean square of stapes responses $RMS(x_S)$. The *RMS* values is calculated in time window from 1 to 2 second. The healthy system is excited by periodic signal, where amplitude q is changed from 0.1 N to 4 N and ω from 300 to 5000 Hz.

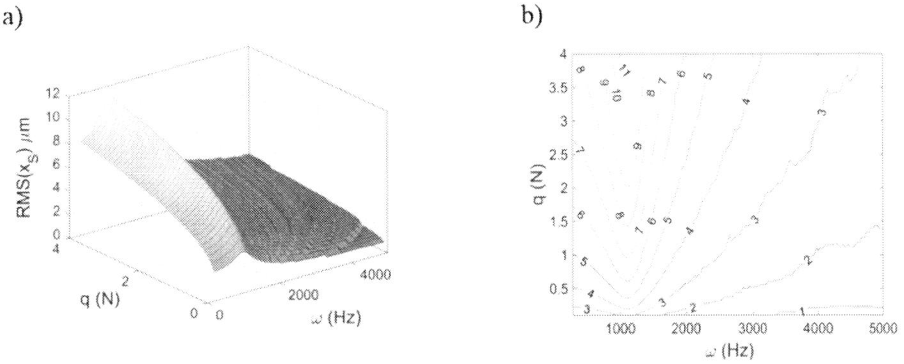

Fig. 2. The 3D characteristic of stapes responses for healthy ear (a) and its isolines interpretation (b)

In the next analyses, the characteristic presented in fig.2 is used as reference to compare the healthy system with the damaged and reconstructed by SMA prosthesis one.

2.2 Analysis of damaged middle ear models

To model the damaged middle ear the model of healthy ear (fig.1) is modified. The added modifications of the model should show how loss of connection continuity in ossicles chain influences to level of stapes vibrations. In the numerical model of the damaged system the discontinuities in two connections have been assumed. Generally, the existence of the gap in visco-elastic joint can be modelled by changing the values of the stiffnes and damping coefficients to zero with no contact [7]. Two versions of the model with defect are developed, where during simulation only one discontinuity is included, Δ_{MI} or Δ_{IS}. In these models the description of a damage has the form:

- first variant - only gap (Δ_{MI}) beetwen malleus and incus, then the coefficients k_{MI} and c_{MI} are

$$k_{MI} = \begin{cases} 0 & \text{when} \quad x_M - x_I \le \Delta_{MI} \\ 1000 N/mm & \quad\quad\quad\; x_M - x_I > \Delta_{MI} \end{cases} \tag{5}$$

$$c_{MI} = \begin{cases} 0 & \text{when} \quad x_M - x_I \le \Delta_{MI} \\ 0.359 Ns/m & \quad\quad\quad\; x_M - x_I > \Delta_{MI} \end{cases} \tag{6}$$

- second variant - only gap (Δ_{IS}) beetwen incus and stapes, then the coefficients k_{IS} and c_{IS} have form

$$k_{IS} = \begin{cases} 0 & \text{when} \quad x_I - x_S \le \Delta_{IS} \\ 1.2 N/mm & \quad\quad\quad\; x_I - x_S > \Delta_{IS} \end{cases} \tag{7}$$

$$c_{IS} = \begin{cases} 0 & \text{when} \quad x_I - x_S \le \Delta_{IS} \\ 0.0079 Ns/m & \quad\quad\quad x_I - x_S > \Delta_{IS} \end{cases} \tag{8}$$

Still the numerical simulations is made in Matlab software, where Δ_{MI}= 0.001 μm or Δ_{IS}= 0.001 μm is applied. The obtained results for the healthy system and the damaged ones (the first and the second variant) are compared using the quality index. In this paper the proposed index J_I is defined as:

$$J_I = RMS(x_S - x_{S,mean})_{damaged} / RMS(x_S - x_{S,mean})_{healthy} \tag{9}$$

To calculate the value of this index the responses of the stapes are recorded. The root mean square is calculated for the healthy ear signal and the damaged ear system. For the damaged system the stapes responses may be significantly shifted ($x_{S,mean}$ is not zero). This fact is reflected in the quality index, from displacements x_S the mean value $x_{S,mean}$ is subtracted. Generally, index J_I describes dynamic change of the stapes response. The stapes is last element of ossicles chain. Level of its vibrations has decisive influence to transmision signal to the internal ear.

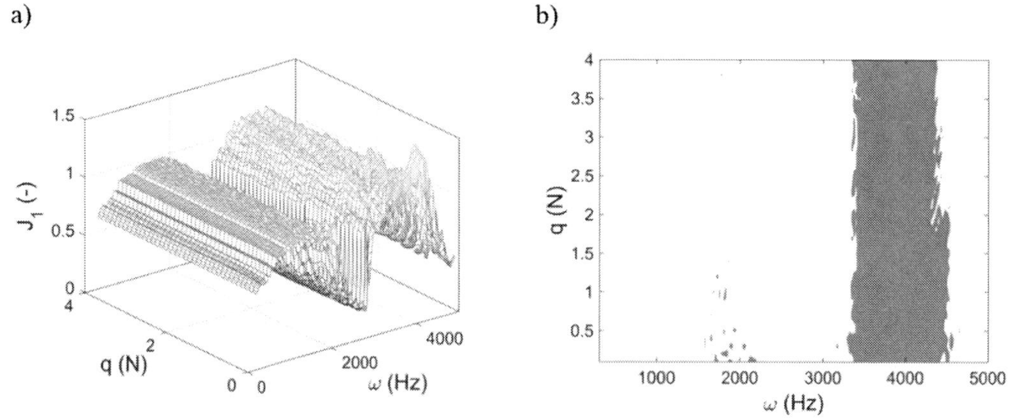

Fig. 3. The stapes responses for the damaged ear. Defect between malleus and incus, Δ_{MI}= 0.001 μm. 3D characteristic (a) and its colormap interpretation (b), white colour - J_1<0, gray colour - J_1>1

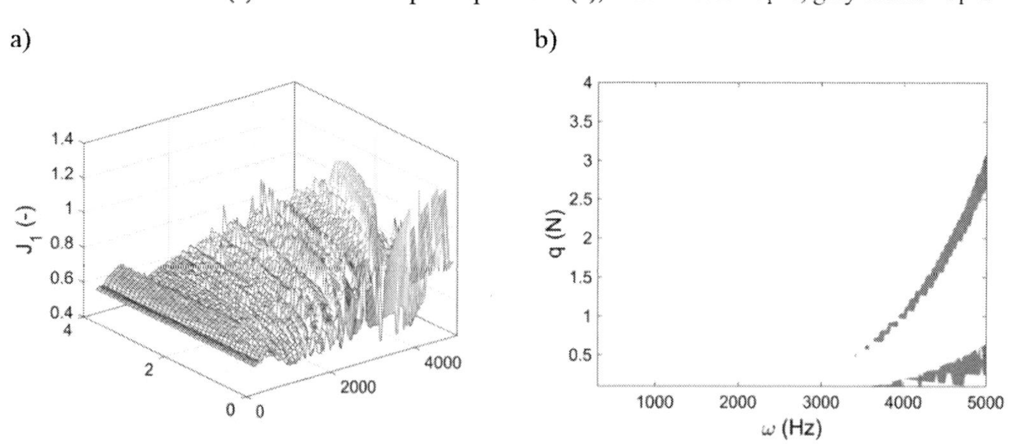

Fig. 4. The stapes responses for damaged ear. Defect between incus and stapes, Δ_{IS}= 0.001 μm. 3D characteristic (a) and its colormap interpretation (b), white color - J_I<0, gray color - J_I>1

The obtained results (Fig 3 and 4) present the influence of defects to vibrations of stapes. Very small gap causes a decrease in the vibration level of the last element of the middle ear. On the colormap in the resonance area about 1000 Hz dominates the white color. In damaged systems the vibrations were higher for higher frequencies (gray color, performance index >1).

3 Model of SMA system

3.1 Model of the SMA prosthesis

Damaged systems are reconstructed using SMA prosthesis. In this section the prosthesis is modeled as the spring made of shape memory alloy. It was assumed that prosthesis is a cylindrical rod with diameter 0.5 mm and length 4 mm. On the basic of the paper [8] the model included pseudoelastic and hysteresis effects is developed. The idealized curve of the external hysteresis loop based on points $(\Delta x_1,F_1)$, $(\Delta x_2,F_2)$, $(\Delta x_3,F_3)$, $(\Delta x_4,F_4)$ is presented in Fig.5. These points are calculated with stress and strain assumption: $\epsilon_1=1\%$, $\epsilon_3=7\%$, $\sigma_1=20$ MPa, $\sigma_3=30$ MPa. From these coordinates the cooeficients of characteristic lines are defined:

$$a_1=(F_2-F_1)/(\Delta x_2-\Delta x_1) \quad b_1=F_1-a_1\Delta x_1 \tag{10}$$

$$a_2=(F_3-F_4)/(\Delta x_3-\Delta x_4) \quad b_2=F_3-a_2\Delta x_3 \tag{11}$$

$$a_0=(F_4-F_2)/(\Delta x_4-\Delta x_2) \quad b_0=F_2-a_0\Delta x_2 \tag{12}$$

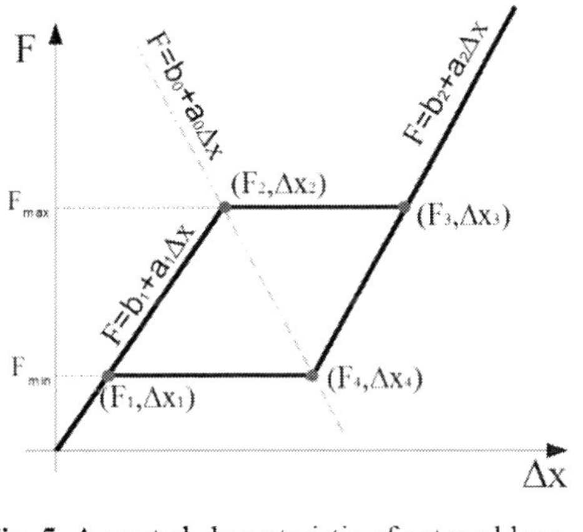

Fig. 5. Accepted characteristic of external loop

Here, a new model of sub-loops hysteresis is proposed. It does not describe the course of the process, phase deformation from martensite to austenite and vice versa. The model can only approximates the idealized shape of the curves. Nonlinearities are caused by stress, this is so-called pseudoelastic effect. Generated force by SMA element is described by

$$F = b+a\Delta x \tag{13}$$

Hovewer, the cooeficient a and b are not constant. Their values can be reinitialized. The change of coofiecitients $a(t)$ and $b(t)$ depends on following conditions:

1) if $\Delta\dot{x} \geq 0$ and $F \leq a_0\Delta x + b_0$ then $a=a_1$, $b=F-a_1\Delta x$,

2) if $\Delta\dot{x} < 0$ and $F \leq a_0\Delta x + b_0$ then $a=0$, $b=F$,

3) if $\Delta\dot{x} \geq 0$ and $F > a_0\Delta x + b_0$ then $a=0$, $b=F$,

4) if $\Delta\dot{x} < 0$ and $F > a_0\Delta x + b_0$ then $a=a_2$, $b=F-a_2\Delta x$,

5) if $F \geq F_{min}$ and $F \leq a_2\Delta x + b_2$ then $a= a_2$, $b=b_2$,

6) if $F \leq F_{max}$ and $F \geq a_1\Delta x + b_1$ then $a=a_1$, $b=b_1$.

In the proposed description of the hysteresis effect, the indicated order of checking the above conditions is important. Figure 6 and 7 present numerical results, the hysteresis loops for different load path. Coordinates of characteristics points are taken as follows: Δx_1=40 μm, F_1=3.92 N, Δx_2=60 μm, F_2=5.89 N, Δx_3=280 μm, F_3=5.89 N, Δx_4=260 μm, F_4=3.92 N.

a) b)

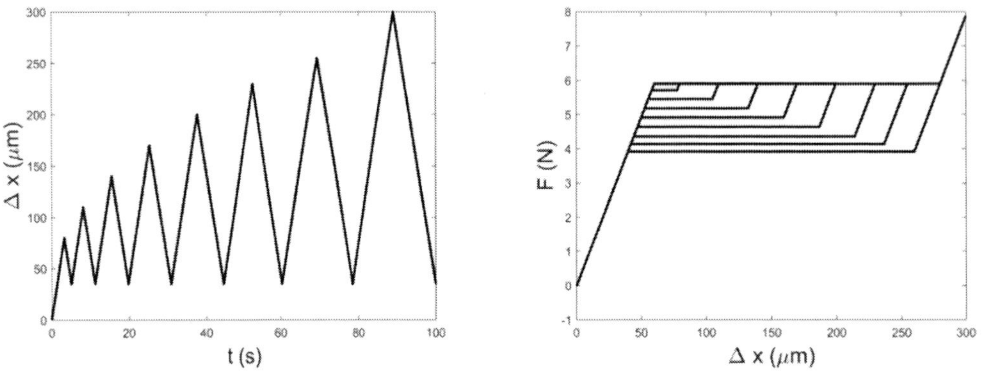

Fig. 6. First case of load path. Time series of spring deflection (a) and internal hysteresis loops (b)

a) b)

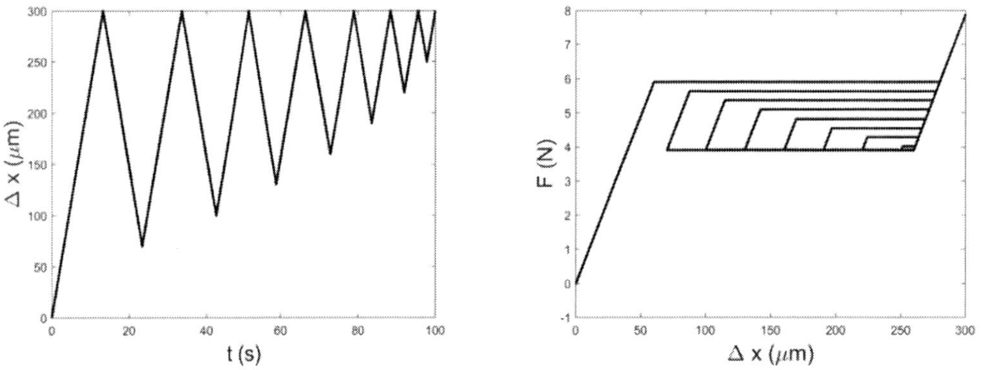

Fig. 7. Second case of load path. Time series of spring deflection (a) and internal hysteresis loops (b)

The obtained characteristics give possibility to realize the trends observed in the literature [8, 9].

3.2 Ear model with SMA prosthesis

The SMA prosthesis is usually mounted between tympanic membrane and stapes. Therefore, 2dof model of the reconstructed middle ear is proposed (figure 8). In this case the malleus and the incus are removed. The damage ossicular chain is reconstructed with prosthesis. In the proposed model the body m_p is used to replace substitute prosthesis mass. It calculated directly from the volume of the cylindrical rod is about 5 mg (density 6500 kg/m^3 [6]). The prosthesis is held in the prescribed position through the preload $F_{preload}$. The cylindrical rod made of SMA is squeezed by the surgeon between the membrane and the stapes. In the mathematical description of the reconstructed middle ear the following form of force F is included:

$$F = b+a\ (\Delta x+x_{preload}) - F_{preload} \tag{14}$$

where $\Delta x= x_P-x_S$ *and* $x_{preload}$ is initial prosthesis deformation generated by preload ($F_{preload}$=2.5 N). The total force and deflection of SMA spring are defined as $F_{SMA}=F+F_{preload}$, $\Delta_{SMA}=\Delta x+x_{preload}$, respectively. Modified hysteresis description form, the equation (14) based on (13) and it allows to easy consideration of the preload in the numerical model of SMA spring. Finally, 2dof system is described by equations:

$$m_P\ddot{x}_P + k_M x_M + c_M\dot{x}_M + F = q\sin\omega t \tag{15}$$

$$m_S\ddot{x}_S + k_{S0}x_S + k_{S1}x_S^2 + k_{S2}x_S^3 + c_S\dot{x}_S - F = 0 \tag{16}$$

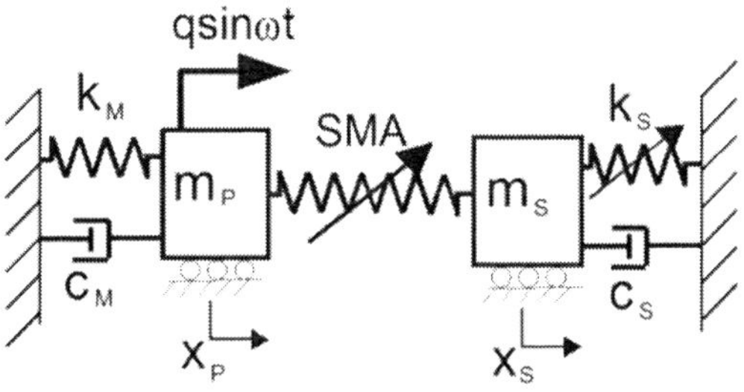

Fig. 8. Model of reconstructed middle ear with SMA prosthesis

Results of reconstructed model simulations are compared with responses of the healthy ear. A similar quality index is used as in previous section:

$$J_1=RMS(x_S-x_{S,mean})_{reconstructed}/RMS(x_S-x_{S,mean})_{healthy} \tag{17}$$

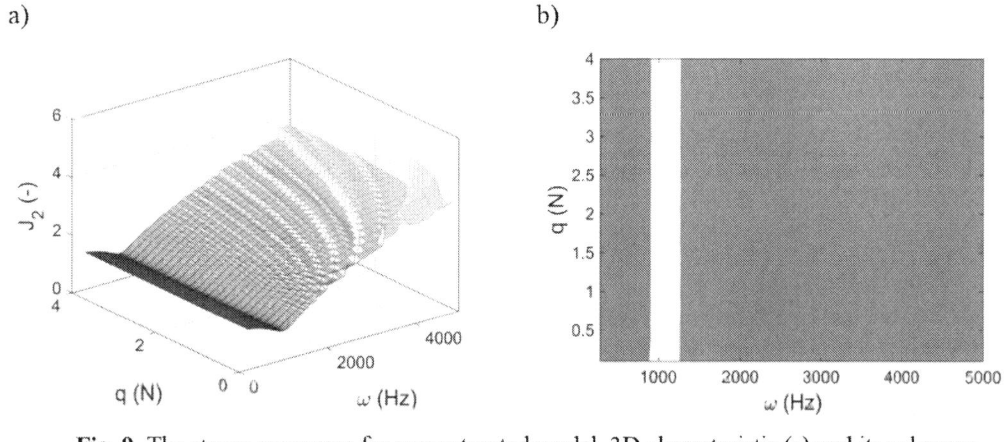

Fig. 9. The stapes responses for reconstructed model. 3D characteristic (a) and its colormap interpretation (b), white color - J_2<0, gray color - J_2>1

The obtained results show that proposed prosthesis can be used to reconstruction damaged ear. In this case for the majority of parameters the stapes vibrations are higher than in healthy system. Only near the resonance area are they smaller, white color on colormap. Minimal values of quality index J_2 is about 0.9. 3D characteristics suggests that first natural frequency of reconstructed model is shifted relative to healthy ear. It can cause the problems with normal hearing at higher frequency. The detailed impact of the prosthesis parameters on the reconstructed model dynamics will be presented in the next study. The last stage of the research, possibility of simulating the hysteresis effect by proposed model is confirmed. Figure 10, 11 and 12 show that for larger amplitudes of excitation the hysteresis can occur. Then the stress in the SMA spring must exceed assumed 30 MPa (or $F_{SMA}>5.89$ N, see section 3.1). For q=3.55 N internal sub-loops and for q=4 N only external loop can be observed.

a) b)

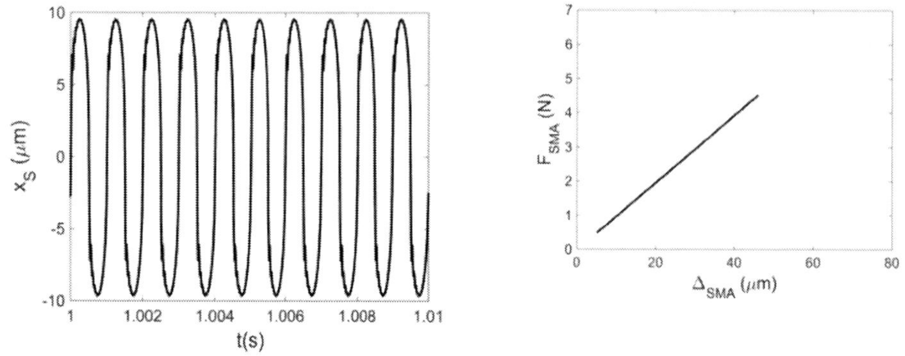

Fig. 10. Time series of stapes responses (a) and characteristic $F(\Delta_{SMA})$ (b) for reconstructed model. q=2 N, ω=1000 Hz

a) b)

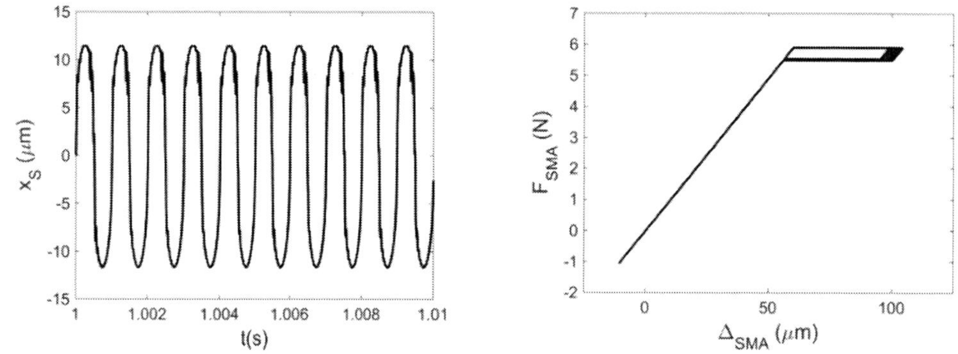

Fig. 11. Time series of stapes responses (a) and characteristic $F(\Delta_{SMA})$ (b) for reconstructed model. q=3.55 N, ω=1000 Hz

a) b)

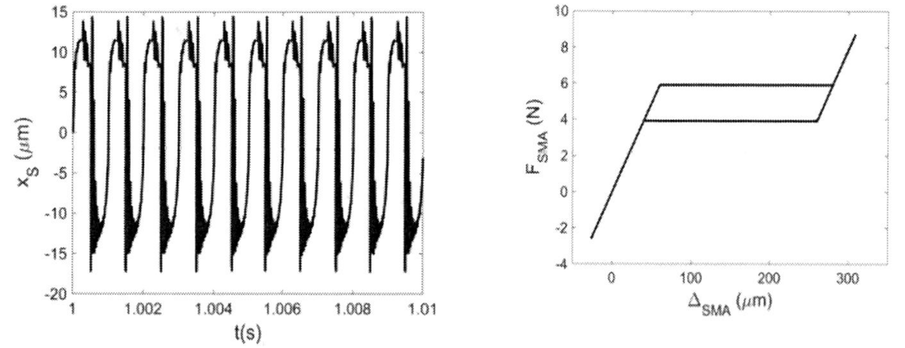

Fig. 12. Time series of stapes responses (a) and characteristic $F(\Delta_{SMA})$ (b) for reconstructed model. q=4 N, ω=1000 Hz

Conclusions

In the paper the classic models with multiple degrees of freedom (2 or 3dof) have been used in modelling the human middle ear. In comparison with previous studies two discontinuities (nonlinearities) have been introduced in the ear models. First, the discontinuity of connections between the malleus and the incus or the incus and the stapes has been tested. The numerical results for very small gap (only 0.001μm) are shown to represent ossicular chain damage. This damage causes visible decrease in stapes vibrations. From comparison both defects (gray areas in fig. 3 and 4) it can be seen that the region of undesirable signal is more increased by damage between the incus and the stapes. The reconstruction of the ossicles chain by means of SMA prosthesis gives satisfying results. This analysis will be extended in next contribution. Moreover, new model of nonlinear hysteresis effect in SMA materials is proposed. The performed simulations confirmed that proposed description is adequate to include hysteresis effect in the model of reconstructed middle ear. In the analyzed model the hysteresis effect is evident at high amplitudes of excitation. Then it is possible to obtain different hysteresis path (internal sub-loops or only external curve). In future research on reconstructed middle ear with the help of shape memory prosthesis will be developed theoretically and experimentally.

The work is financially supported under National Science Centre (Poland). Project number 2014/13/B/ST8/04047.

References

1. S.I. Chen, M.H. Lee, C.M. Yao, P.R. Chen, Y.F. Chou, T.C. Liu, Y.L. Song, C.F. Lee, *Modeling sound transmission of human middle ear and its clinical applications using finite element analysis*. Kaohsiung Journal of Medical Sciences **29**, 133-139 (2013)
2. E. Skrodzka, J. Modlawska, *Modal analysis of the human tympanic membrane of middle ear using the finite element method*. Archives of Acoustics **31**, 23-28 (2006)
3. B. Feng, R.Z. Gan, *Lumped parametric model of the human ear for sound transmission*. Biomechan Model Mechanobiol **3**, 33-47 (2004)
4. R. Rusinek, J. Warminski, M. Zadrozniak, M. Szymanski, *Nonlinear Approach to Modelling of Otosclerosis in a Human Middle Ear*. Differ Equ Dyn Syst. **21**, 45-57 (2012)
5. A. Mardassi, A. Deveze, M. Sajuan, J. Mancini, B. Parikh, A. Elbedeiwy, J. Magnan, J.P. Lavieille, *Titanium ossicular chain replacement prostheses: Prognostic factors and preliminary functional results*. European Annals of Otorhinolaryngology, Head and Neck Diseases **128**, 53-58 (2011)
6. R. Rusinek, J. Warminski, M. Szymanski, K. Kecik, K. Kozik, *Dynamics of the middle ear ossicles with an SMA prosthesis*. International Journal of Mechanical Sciences **127**, 163-175 (2017)
7. K.P. Jankowski, A. Mitura, J. Warminski, *Dynamics of a self-balancing double-pendulum system*. Journal of Computational and Nonlinear Dynamics **11**, 1-9 (2016)
8. Y. Huo, I. Muller, *Nonequilibrium thermodynamics of pseudoelasticity*. Continuum Mech Thermodyn **5**, 163-204 (1993)
9. A. Masud, M. Panahandeh, F. Aurrichio, *A finite-strain finite element model for the pseudoelastic behavior of shape memory alloys*. Comput Methods Appl Mech Engrg. **148**, 23-27 (1997)

Computer simulation of thermoforming process and its verification using a rapid tooling mould

Michał Modławski[*,1], *Tomasz* Jaruga[1]

[1]Department of Polymer Processing, Czestochowa University of Technology, Armii Krajowej 19c, 42-201 Częstochowa, Poland

Abstract. The results of computer simulation of thermoforming process made using ANSYS Polyflow software are presented in this paper. The analysis of the wall thickness distribution across an U-shaped thermoformed product manufactured using a positive mould was made. The simulation results were verified using a mould manufactured in a 3D printing process which was Fused Deposition Modelling (FDM) and a poly(ethylene terephthalate) formed sheet. It was proven that the computer simulation and a tool made with a Rapid Tooling technology can be useful for predicting the quality of a thermoformed part, particularly to find the problems with thin-walled areas.

Keywords: computer simulation, thermoforming

1 Introduction

Thermoforming technology is getting more popular recently, because of its advantages, like relatively small cost of tool, especially when compared to injection moulds. It is very common in the packaging industry, with food packaging application, because of the easy possibility to automatize the process and to reach high productivity. Another advantages is the possibility to produce large plastic parts and that's why it found its place in the automotive industry, but also the tram and train interior parts are manufactured [1,2].

Thermoplastic materials are used in this technology and one of the most important factors is the temperature of processing a plastics sheet. It should be high enough to soften the sheet and to enable its forming but also not too high because of the risk of material damage. In case of thermoforming machines it is possible to use a certain temperature distribution in order to obtain the product wall thickness as uniform as it is only possible.

A computer simulation can be made before a tool is made to predict some potential problems with the process. If the results of the simulation would indicate the problems with the tool it is necessary then to make some corrections of the tool design. The processing parameters influence on the part quality can be also predict. One of the most important parameters is the formed sheet temperature, but its distribution is also important because it impacts the wall thickness distribution of the part [3,4].

---

* Corresponding author: modlawski@ipp.pcz.pl

Reviewers: *František* Nový, *Eva* Tillová

2 Experimental

The aim of this article was to evaluate the possibility of using a Rapid Tooling mould made by Fused Deposition Modelling technology to test the correct design of a thermoformed product or to make a short run production. Moreover, the efficiency of computer simulation of product wall thickness distribution was evaluated, by comparison of the results from the thermoforming process on the machine to the result of computer simulation done using *ANSYS Polyflow* software.

2.1 Mould

The mould was designed using *Siemens NX 8.5* software. It is a positive (convex) mould with the geometrical shape corresponding to some problems occurring during this process – different radii, internal and external dimensions and a pocket - Fig. 1.

The mould was produced using Fused Deposition Modeling (FDM) 3D printing machine *golemD* from acrylonitrile-butadiene-styrene (ABS) material filament. This material was chosen because it has good temperature resistance in comparison to other polymeric materials used for 3D printing. The length and width of the mould were limited by the dimensions of the 3D printer platform.

3D printers create models layer by layer and the layers' marks can be seen on a model surface and there is a risk that they will be also visible on the product made in the thermoforming process. In the case of big models made from ABS some cracks can occur between model layers. Such gaps were eliminated from the current model by sealing the side surfaces with a polyimide tape of 0.6 mm thickness.

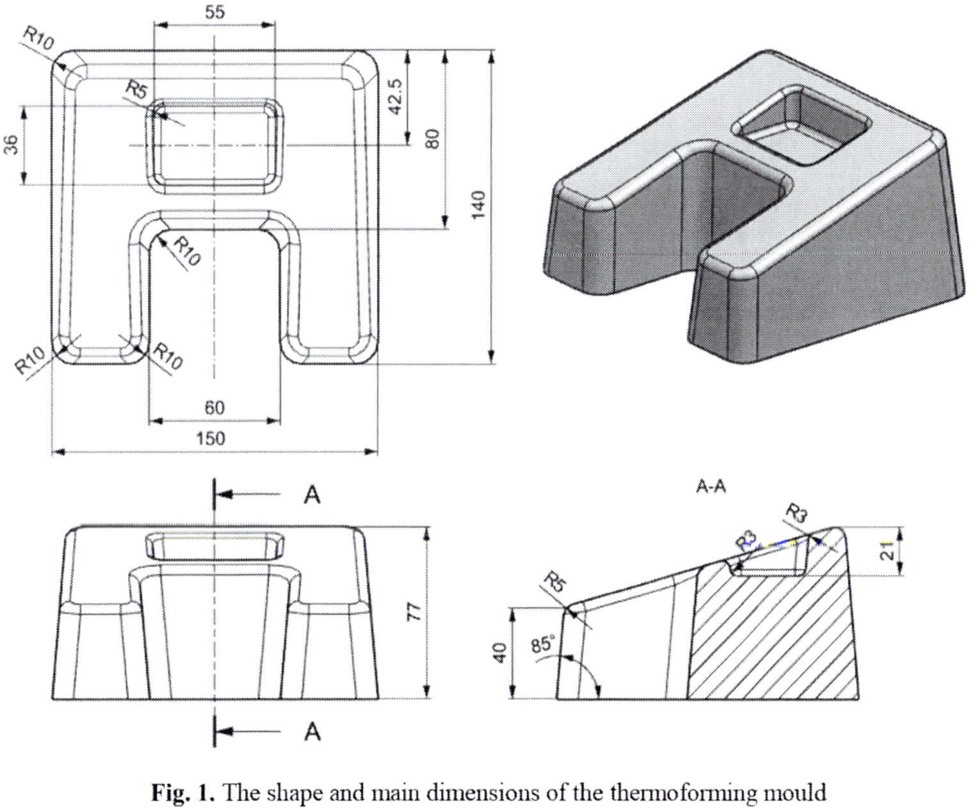

Fig. 1. The shape and main dimensions of the thermoforming mould

2.2 Thermoforming process

The test of the mould were done on the thermoforming machine *VP Minor*. The photography of the mould placed on the machine table is shown in Fig. 2.

Fig. 2. View of the mould placed on the machine table

The processed material was PET sheet of 0.35 mm thickness, made from the recycled and origin plastic in extrusion process by *Hanex* company, Poland. The outer layers of this sheet were made from the origin material, but 70-80% of the thickness was made from the recycled material [9].

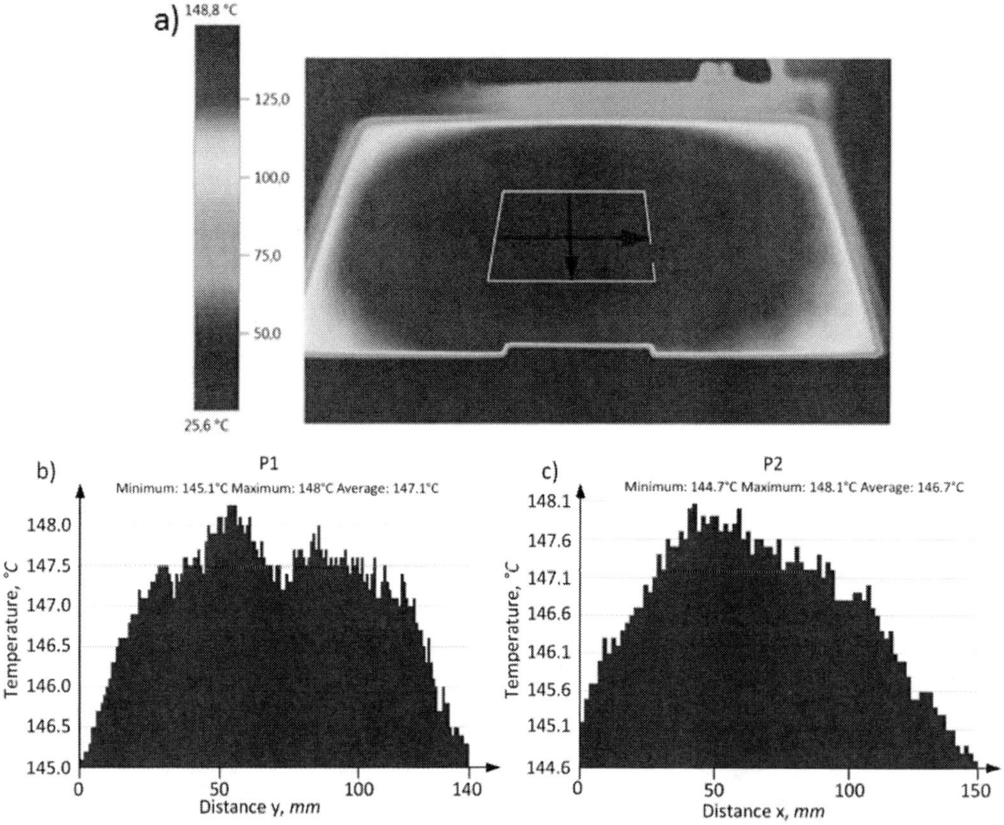

Fig. 3. PET sheet temperature distribution after 8 s of heating: a) thermovision photography of the entire sheet, b) temperature along P1line, c) temperature along P2 line

The material was warming by the machine heaters by 8 s. The distribution of temperature after the heating stage was obtained by thermal imaging camera *TESTO 890* manufactured by *Testo* company and is presented in Fig. 3. The sheet temperature above the mould was homogenous - see P1 and P2 lines and graphs in Fig. 3b and 3c. Average value of temperature on area in the white rectangle over the mould is 146.9°C.

2.3 Computer simulation of thermoforming

The computer simulation was conducted in solver *Polyflow* which is a part of *ANSYS* software. The following geometrical models were used in the simulation: the positive thermoforming mould with the table (platform) of the thermoforming machine and poly(ethylene terephthalate) formed sheet.

The mould was symmetrical to X-axis so only its half was considered in the simulation. 2D shell geometry was applied as the representation of this mould surface (Fig. 4). The model was meshed with the options: *Face Meshing* - which allows generating a mesh composed from equal squares and *Mesh Sizing*. The size of a square of one finite element was in this case equal 9 mm, but the function *Adaptive Meshing was* also applied. The maximum number of recursive subdivisions for contact for one element was equal 3.

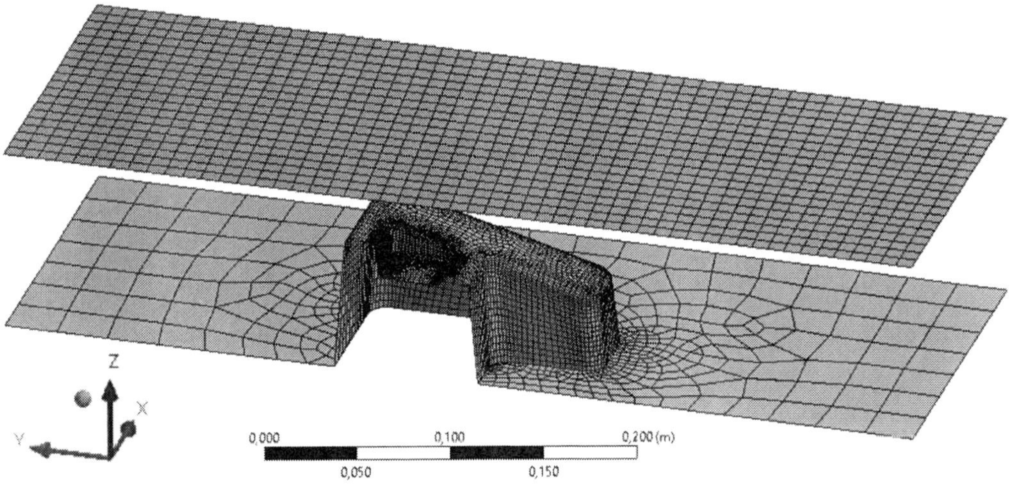

Fig. 4. Model of the plastic sheet and the mould mesh in *ANSYS Polyflow*

Thermoforming process is a time-dependent problem. The time of the total forming process considered in the simulation was equals 2 s. It included two stages:

- lifting the machine platform together with the mould, which lasts 0.5 s,
- product shaping, which starts from the time of 0.5 s and ends at 2.0 s. The value of the air underpressure was assumed to be 0,075 MPa.

The considered forming time is short therefore the changes in the temperature value were neglected.

The 0.35 mm plastic sheet is made from PET. The density of this material in the solid state is equal 1350 kg/m^3. The viscosity data of PET in the liquid state can be found in the literature [10] . Based on these data the coefficients for PET in temperature 146.9°C of Carreau model (1) were calculated.

$$\eta = 67314.3 \cdot \left[1 + (0.675\dot{\gamma})^2\right]^{-0.167} \quad (1)$$

where η - the polymer viscosity, $\dot{\gamma}$ - polymer shear rate.

A contact between the mould and the sheet was applied when defining the boundary conditions in the simulation.

3 Results and discussion

The results of the thermoformed part wall thickness distribution obtained during the experiments and from the computer simulation were compared.

3.1 The results of thermofoming on the machine

It was found out that the optimal time of the sheet heating for the machine and mould used was 8 s. The quality of the part was very good, but drilling of 1 mm diameter venting holes was necessary inside the mould pocket - in the corners and on the edges.

Fig. 4. The exemplary product manufactured with 8 s time of heating

The formed sheet thickness was measured by the thickness measurement gauge *Olympus Magna Mike 8600,* basing on Hall-effect. A steel ball of 1/16" diameter was used. The maximum thickness value which can be measured using this this ball is 2.03 mm. The measurement accuracy of this tool is in the range from ±0.025 mm to +0.04 mm and depends on the measured value. The thickness was measured along four lines, marked: A, B, C and D, presented in Fig. 5.

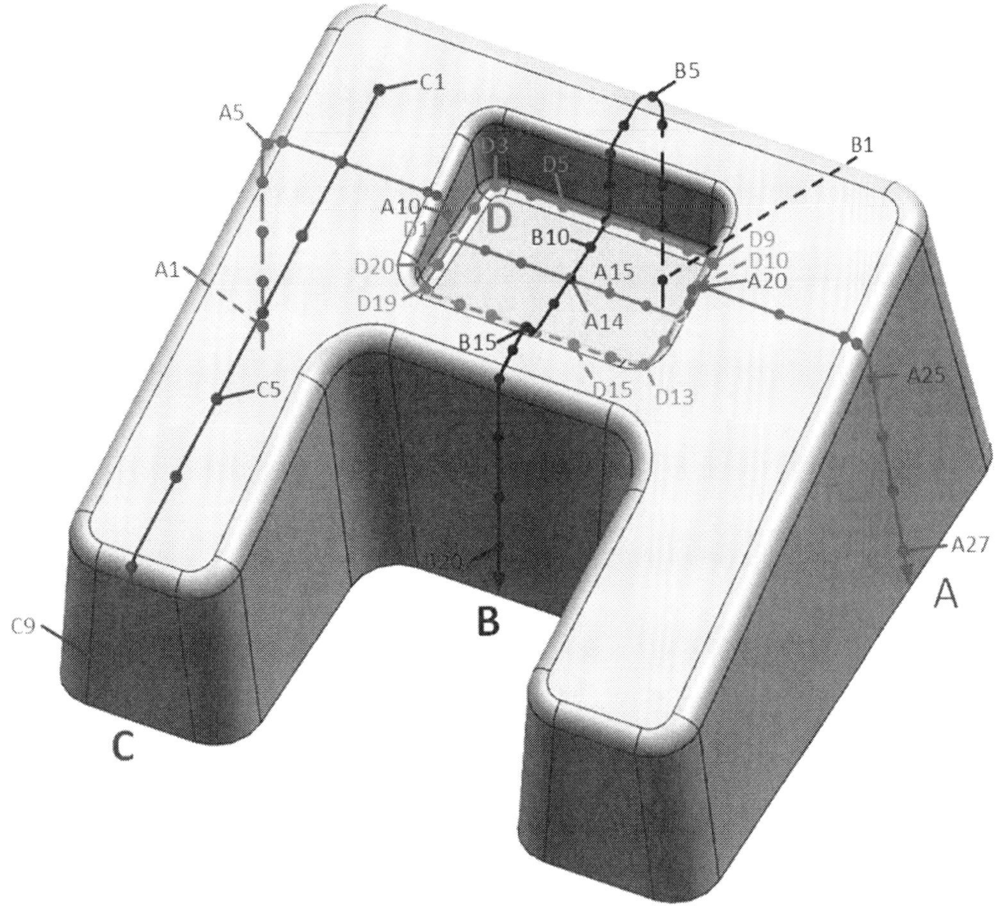

Fig. 5. The mould model with four measurement lines (A- D) and thickness measurement points marked

3.2 Computer simulation results

The result of the wall thickness distribution across the part obtained in the computer simulation is presented in Fig. 6. The distribution of thickness along A, B, C and D lines, presented in Fig. 5 was determinated by the function *Probe* from menu *Tools ANSYS CFD-Post* program. The smallest thickness value can be observed in the inside corners. It is uniform on the upper surface, because the mould touched this area of the sheet as it raised just before the vacuum forming stage.

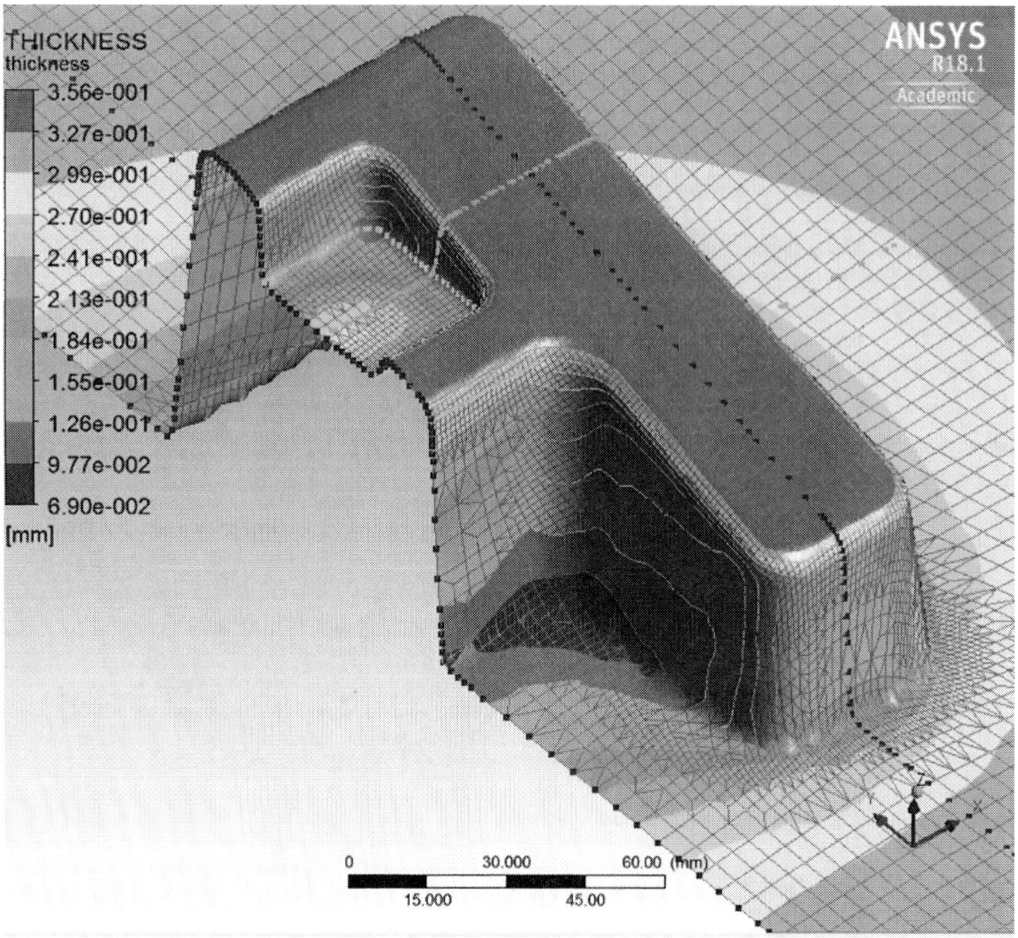

Fig. 6. Thermoformed product wall thickness distribution (with A-D lines marked)

The results of the real thermoformed product thickness (four samples of good quality were considered) are presented in Fig. 7 and compared with the case from the computer simulation. There is a good accordance between the simulation results and the real case. The thickness values from simulation are usually bigger than the real values in the higher areas of the product (further from the machine table surface) while they are smaller than the real values in the bottom areas of the product. This means that in the simulation the behaviour of the polymeric sheet is slightly different, depending on the degree of stretching.

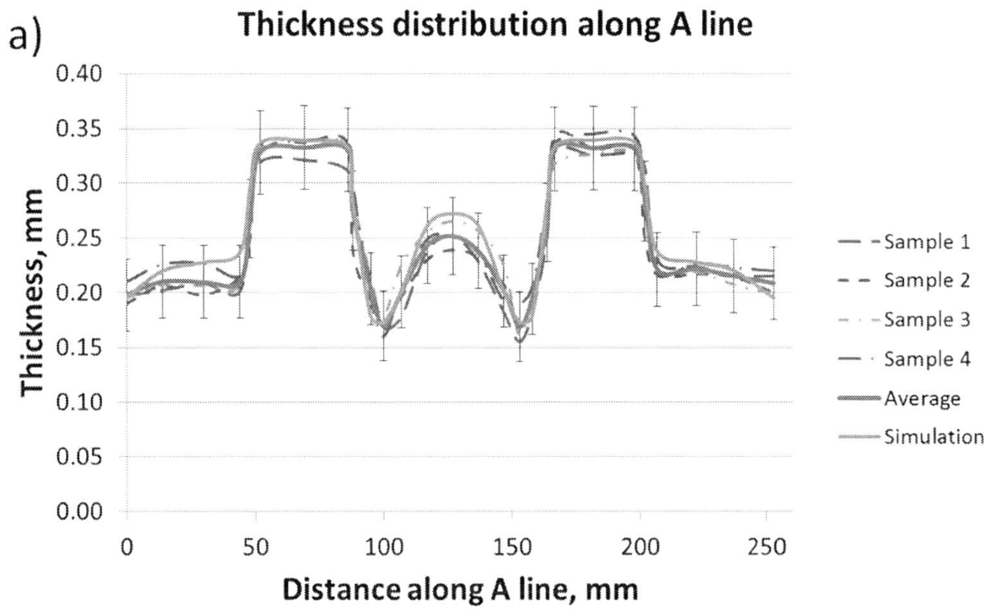
a)
Thickness distribution along A line
Thickness, mm
Distance along A line, mm
Sample 1
Sample 2
Sample 3
Sample 4
Average
Simulation

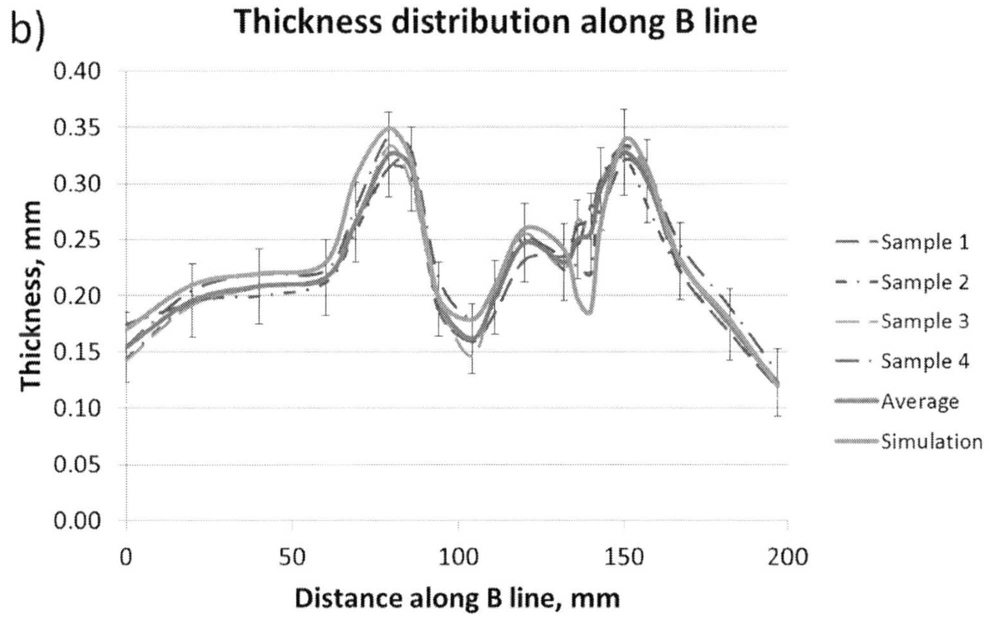
b)
Thickness distribution along B line
Thickness, mm
Distance along B line, mm
Sample 1
Sample 2
Sample 3
Sample 4
Average
Simulation

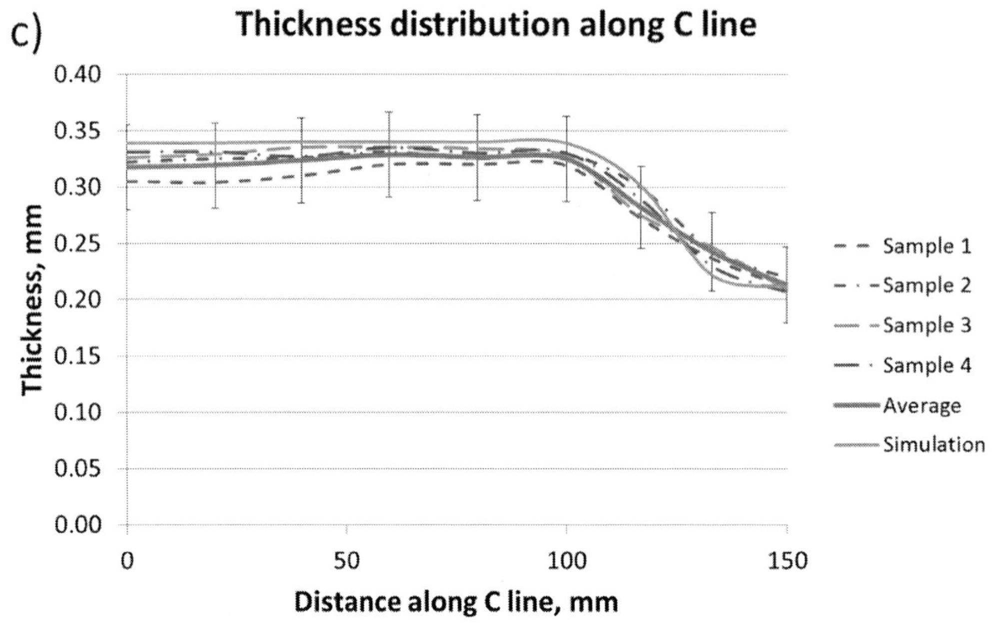

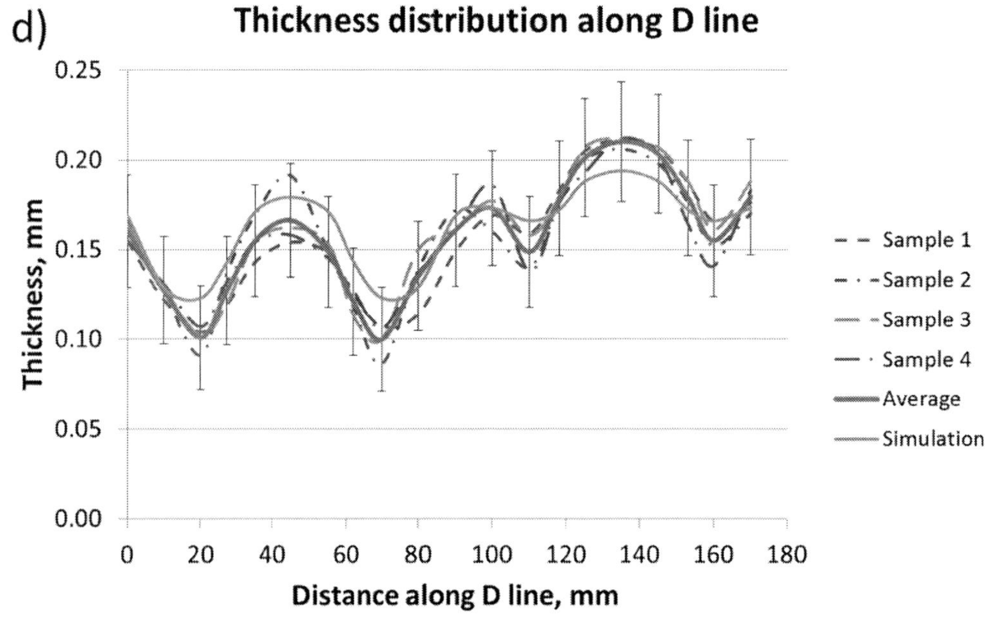

Fig. 7. Plots of wall thickness along: A-D lines

Conclusion

Computer simulation can be a useful tool to predict the quality of thermoformed parts. It was found out that the deviation of the wall thickness between the simulation and the experiment is not more than 23.4 %. Average value of this deviation is 2,5 %. The approach of using simulation in the first step of the design process can give some benefits since it can help to avoid excessively thinned areas of the product by redesigning the mould before its manufacturing.

It is possible to use Rapid Tooling moulds made by FDM technology in short run production or in trials of a newly designed tool. As it was proved on the case analysed in this paper, this kind of mould can work well and thermoformed product of a good quality can be obtained while the time of mould making is usually short.

References

1. A. Illig, *Thermoforming. A Practical Guide*. (Hanser, 2001)
2. G. L. Beall, J. L. Throne, *Hollow Plastic Parts. Design and Manufacture*. (Hanser, 2004)
3. K. Pepliński, A. Mozer, *Ansys Polyflow software use to optimize the sheet thickness distribution in thermoforming process*. Journal of Polish CIMAC **6**, 215-220 (2011)
4. D. Sykutera, K. Pepliński, *Zastosowanie program Ansys-Polyflow do wspomagania wytwarzania opakowań formowanych próżniow*. Inż. Ap. Chem. **50**, 73-74 (2011)
5. E.S. Erdogan, O. Eksi, *Prediction of Wall Thickness Distribution in Simple Thermoforming Moulds*. Journal of Mechanical Engineering **60**, 195-202 (2014)
6. M. Modławski, J. Nabiałek, T. Jaruga, *Symulacja odkształcenia wyrobu rozdmuchowego o stałej i zmiennej grubości ścianki*. Tworzywa Sztuczne w Przemyśle **34**, 94-97 (2016)
7. D. Kwiatkowski, M. Modławski, T. Jaruga, *Symulacje komputerowe grubości ścianki butelki uzyskiwanej w procesie wytłaczania z rozdmuchiwaniem*. Przetwórstwo Tworzyw **21**, 256-261 (2015)
8. ANSYS Polyflow help documents (2017)
9. R-PET film - Hanex : http://hanex.com.pl/folie (access: 07.2017)
10. G. Astarita, G. Marrucci, L. Nicolais, *Rheology, Volume 3: Applications*. (Plenum Press, 1980)

Thermal load non-uniformity estimation for superheater tube bundle damage evaluation

Martin Naď[1,*], *Zdeněk* Jegla[1], *Tomáš* Létal[1], *Pavel* Lošák[1], *Jiří* Buzík[1]

[1]Brno University of Technology, Faculty of Mechanical Engineering, Technická 2896/2, 616 69 Brno, Czech Republic

Abstract. Industrial boiler damage is a common phenomenon encountered in boiler operation which usually lasts several decades. Since boiler shutdown may be required because of localized failures, it is crucial to predict the most vulnerable parts. If damage occurs, it is necessary to perform root cause analysis and devise corrective measures (repairs, design modifications, etc.). Boiler tube bundles, such as those in superheaters, preheaters and reheaters, are the most exposed and often the most damaged boiler parts. Both short-term and long-term overheating are common causes of tube failures. In these cases, the design temperatures are exceeded, which often results in decrease of remaining creep life. Advanced models for damage evaluation require temperature history, which is available only in rare cases when it has been measured and recorded for the whole service life. However, in most cases it is necessary to estimate the temperature history from available operation history data (inlet and outlet pressures and temperatures etc.). The task may be very challenging because of the combination of complex flow behaviour in the flue gas domain and heat transfer phenomena. This paper focuses on estimating thermal load non-uniformity on superheater tubes via Computational Fluid Dynamics (CFD) simulation of flue gas flow including heat transfer within the domain consisting of a furnace and a part of the first stage of the boiler.

Keywords: boiler damage, superheater tubes, thermal load non-uniformity, Computational Fluid Dynamics (CFD)

1 Introduction

Boiler can be described as a device which produces steam (or in some cases hot water) using heat received from chemical reaction. Choice of the main boiler components depends on fuel type which is burnt and the boiler size is based on required amount of steam production for consequent use in turbine or other processes. There are three basic types of fuels which are usually burnt: gas (natural gas, methane...), liquid (various oils) and solid (dark and brown coal ...). Important distinction between various fuels is in their flue gas composition, since gas burning produces almost no ash in comparison with liquid and solid fuels that produce a lot of ash, often resulting in significant problems [1, 2].

---

* Corresponding author: nad@fme.vutbr.cz

Reviewers: *Ján* Vávro, *Juraj* Grenčík

Nowadays, the most desired features of boilers are high efficiency, reliability as well as optimal fuel and operation flexibility. Therefore, demand for boilers with higher standards, higher efficiency and possibility of burning various fuel types, but also with lower impact on the environment, is more relevant than ever.

1.1 Boiler damage types

During the boiler service life (often more than 200 000 hours), some kind of damage is basically unavoidable. Each damage type is unique with different impact, severity and influence on service life. Although various mechanisms of damage often act together, they can be divided into five basic categories based on damage root cause [3–5]:
- Temperature (Creep, Long-term and Short-term overheating)
- Corrosion (waterside, fireside)
- Flow–related (erosion, cavitation, fouling, slagging)
- Operation (change of fuel, incorrect regulation, sudden load change…)
- Construction (improper equipment design, material misuse…)

Tubes with internal pressure are crucial parts of water tube boilers as well as steam generators. They are often exposed to high temperatures and pressures both on steam side as well as flue gas side. High sensitivity of material properties to temperature at these conditions combined with imperfect design methods increase the probability of unexpected high temperature damage. The tube temperature may rise during operation time due to oxide scales growth (from inside or outside) or because of reduction of flow rate of "coolant" in the tubes. Service life of tubes is limited by effects of high temperature, stresses, aggressive environment, corrosion degradation and other. Resistance can be increased by proper material selection [6].

1.2 Temperature related damage

Damage due to high temperature is common problem of tube bundles located in boilers. Temperature may cause material overheating as well as creep or in combination with aggressive flue gasses also high temperature corrosion. Investigation of damage caused by overheating requires detail analysis of damaged components consisting of microstructure analysis and identification of other forms of material degradation caused by high temperature. It is useful to distinguish short-term and long-term overheating as they lead to different types of damage.

1.2.1 Short-term overheating

It occurs if tube material temperature rises rapidly above designed values, generally for a short time. Short-term overheating can be divided into three types [3]:
- Subcritical (material temperature under the lower critical transformation temperature – 727 °C for carbon steel)
- Intercritical (material temperature between the lower and higher critical transformation temperature)
- Upper critical (material temperature above the upper critical transformation temperature – 843 °C for carbon steel)

Metal yield strength is reduced at elevated temperatures and may be exceeded by stresses, leading to plastic deformation, often in form of bulking which leads to thinner walls and later to rupture of the tube wall. Higher temperatures and stresses may also cause rupture of the tube due to exceeding of tensile strength. In case of extremely high temperatures, short term creep may occur, during which time to damage decreases with increasing temperature. The main cause of short term overheating is lack of coolant flow, often connected with oxide scales, corrosion, fouling or slagging. Therefore, it is necessary to monitor coolant flow and the state of the pipes [3].

1.2.2 Long-term overheating

It occurs if the designed material temperature is exceeded for days, weeks, months or even longer. Long-term overheating is the most common root cause of boiler damage with 90% of cases affecting preheaters. It is result of combination of temperature, time, stress and material properties. Long-term overheating, often combined with creep, may result in tube rupture. Therefore, proper material selection and operational conditions are required to prevent this kind of damage.

1.2.3 Prevention of temperature damage

Overheating is common type of damage mainly of components, such as boiler tube bundles, located in areas with high flue gas temperature. Tube degradation due to overheating rises with rising temperature, stress and time. Improper combination of these factors may cause damage and failure. According to standards, influence of temperature is reflected in maximal allowable stress value (decreases with increasing temperature), during design of boilers and pressure vessels. The most common material in boilers is carbon steel, because of its price. However, this material is often not sufficient for higher temperatures. Therefore, tubes exposed to steam are usually made of low alloyed steel used with small amount of molybdenum (normally 0.25–1.25 % for better creep resistance) and chrome (normally 0.5–2.5 % for better creep resistance and slowing down of high temperature corrosion). [1, 3]

2 Equipment description

This article focuses on three-staged boiler with natural circulation and steam production 60 t/hr, located in a chemical plant. Several fuel types were burnt in this boiler during the service life, however it was primarily designed for natural gas, heavy fuel oil and tar mixtures.

Boiler has several crucial parts (see Fig. **1**). For various types of fuel combustion, four low emission burners are used. Combined with flue gas regulation system, they were selected to reduce NOx emissions in order to meet legislation limits. Membrane walls of combustion chamber (first stage) and of second stage work as evaporators. Two superheaters (P1 and P2) and one economiser are located in the second stage, second economiser is located in the third stage.

2.1 Boiler parameters

There are several parameters which are important during boiler damage analysis. One of them is boiler service time, which is in this case 62000 hrs (at the time of experimental data collection). Other boiler parameters can be divided into two basic categories (all parameters are for natural gas combustion):

a) Design parameters:
   - Steam production: 60 t/hr
   - Superheated steam temperature: 375 °C (+23 °C, -10 °C)
   - Superheated steam pressure: 3.82 MPa
   - Feed water temperature: 145 °C
b) Real parameters during natural gas combustion
   - Steam production: 35–95 t/hr
   - Superheated steam temperature: 340–385 °C
   - Superheated steam pressure: 3.45–3.65 MPa

Apparently, the design parameters differ from the actual operation parameters, so they should not be used for proper damage estimation. It is common that there is not enough data history with required parameters, too. Therefor it is necessary to obtain them by further calculations.

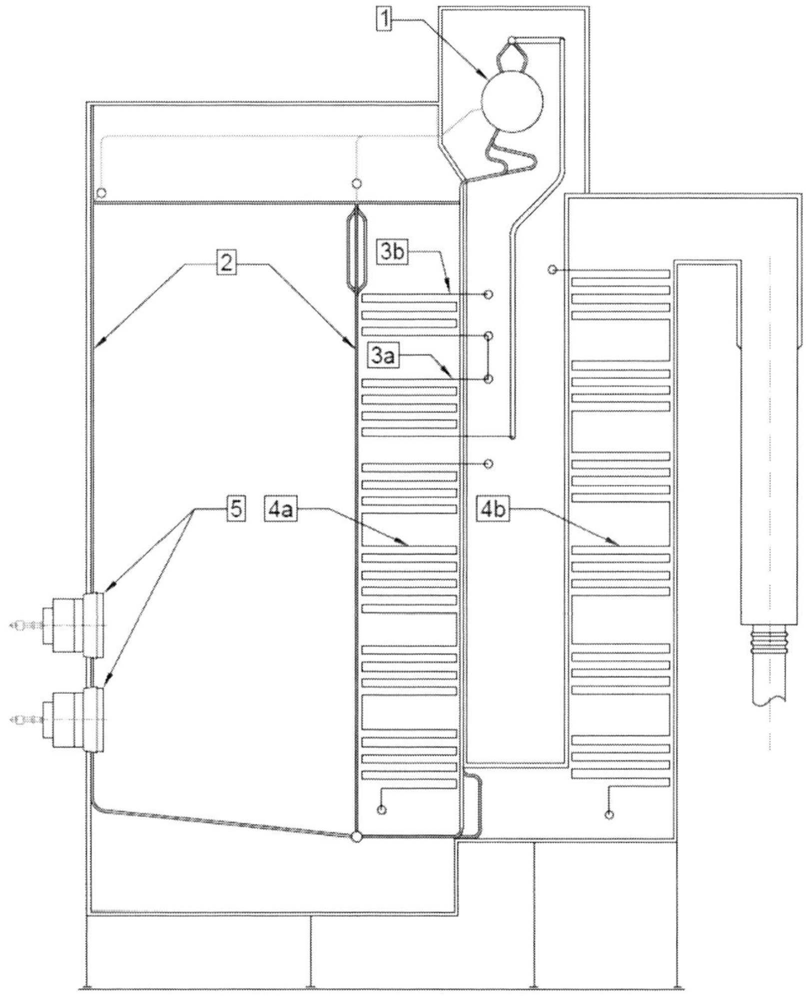

Fig. 1. Simplified boiler scheme (1 – boiler drum, 2 – evaporator/membrane walls, 3a – superheater P1, 3b – superheater P2, 4a – economizer E1, 4b – economizer E2, 5 – burners)

2.2 Superheater P2

Superheaters are generally the most damaged parts of boiler since they are located in rough areas of boiler, often exposed to high temperatures and pressures. Various types of damage may occur however temperature related are the most common.

Steam superheating in the examined boiler is realized by two horizontal superheaters (P1 and P2, see Fig. **1** pos. 3a and 3b). Both superheaters are in form of tube bundles located in second stage of boiler. This article focuses on superheater P2 (see Fig. **2**) which is the most exposed one as first one after combustion chamber. It is located at the top of the second stage (at 19.31 m elevation) and is made of 66 tubes in 3 rows (198 tubes total).

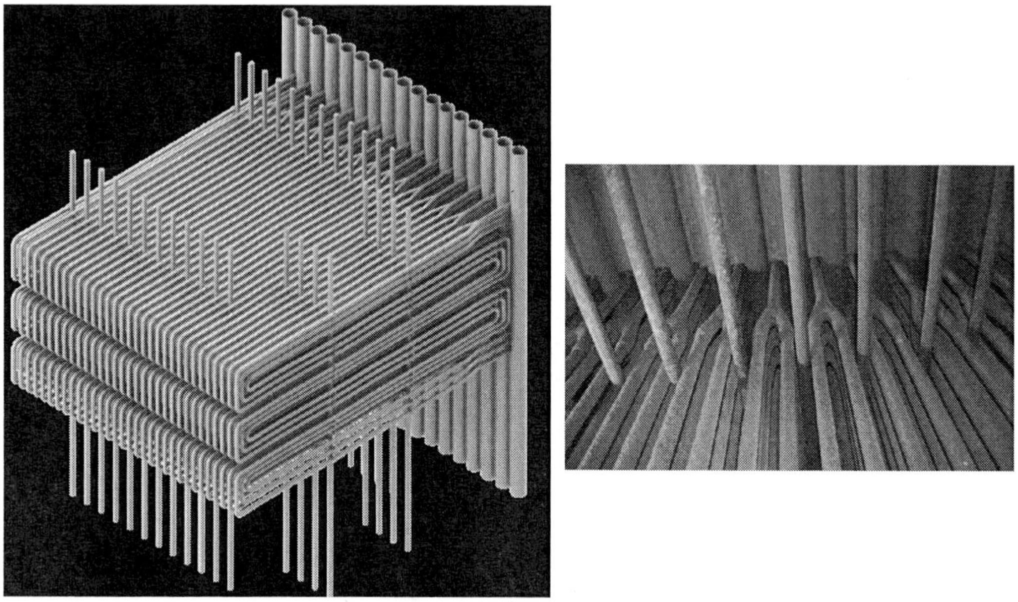

Fig. 2. Superheater a) partial 3D model (left), b) photo (right)

3 Thermal load non-uniformity simulation

To estimate temperature related damage, it is necessary to know tube surface temperatures. However, this information is not available in most cases, therefore it is necessary to estimate thermal load from key information which are measured during operation.

Combination of several approaches was used (typically combination of analytical calculations, thermal–hydraulic calculations and computational fluid dynamics (CFD) calculations for flue gas flow simulations). All simulations are performed using experimental data received during combustion of natural gas. However, several presuppositions of unknown conditions were used. One of them is optimal blowing frequency that provides clean tube surface without impurities.

3.1 Temperature field on superheater tubes

First of all, control thermal–hydraulic calculations were performed to obtain surface temperatures of the superheater P2 tubes. Commercial software was used, however, few simplifications, related to superheater geometry, had to be made since the software was able to simulate only average steam/flue gas flows through tube bundle, however for our case, it was sufficient. During surface temperature calculation, tubes were divided into ten identical subsections (294 mm each). As a basic input parameters, experimental data were used:

- Flue gas temperature after combustion chamber: 734 °C
- Flue gas velocity after combustion chamber: 17.8 m/s
- Steam input temperature: 248.2 °C
- Steam input velocity: 14.7 m/s

In order to obtain relevant results compared to experimental data, it was necessary to include the influence of oxide scales (see Fig. **2**b), which were created during operation and had influence on thermal duty. It was performed by simulating fouling using the fouling coefficient R_z = 0.0035 $m^2 \cdot K/W$. To verify the results accuracy, outlet steam temperature was compared to experimental data (369 °C). Results in form of surface temperatures (with and without influence of oxide scales) are shown on Fig. **3**.

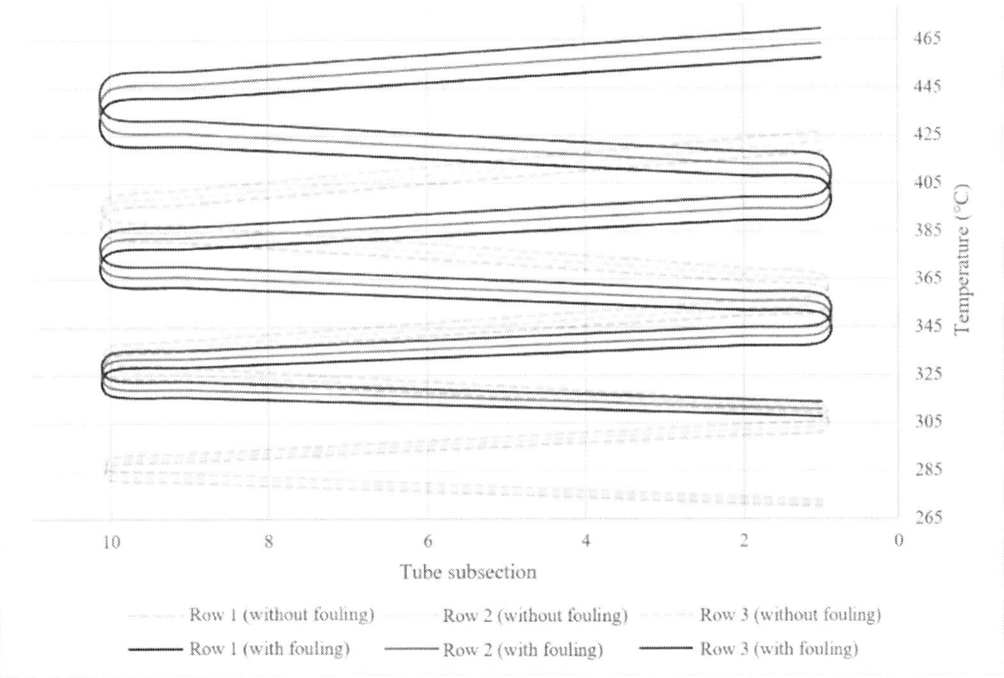

Fig. 3. Surface temperatures on tube rows of superheater P2 with and without fouling (simulation of oxide scales)

The graph shows great influence of oxide scales (simulated using fouling coefficient) on tube surface temperature. Those temperature differences are in range from 35.3 °C to 50.2 °C. The average temperature difference is 42 °C – that is high enough to have significant influence on thermal duty as well as temperature related damage resistance. Because of that, further simulations were necessary.

3.2 Flue gas boiler flow

To obtain accurate surface temperatures on all tubes, it is necessary to know thermal load first. Thus, CFD analyses were performed to simulate flue gas flow in the boiler. Combustion was not simulated, as it would presumably have just a little effect on the results in the area of interest, which is located next to the combustion chamber and encompasses the superheater.

During basic hot flue gas simulation simplifications were used:

- *Model size.* Simulation of combustion chamber and part of second stage.
- *Model geometry.* Membrane walls were replaced by smooth walls with thickness of 5 mm simulating tube wall thickness. To simulate cooling by membrane walls, wall emissivity was set up to simulate oxide scales (influencing thermal flow). Also, wall temperature, based on location in combustion chamber, was calculated using heat balance calculations.
- *Tube bundles geometry.* Superheater was simulated as a porous zone, as well as tubular grid between combustion chamber and second stage was simulated as porous jump, both to cover influence of local resistance
- *Combustion process.* Flow of hot flue gases was simulated as well as their cooling. Parameters of flue gases were used based on experimental data received. Burners were modelled as four inlets with diameter of 1 m. Inlet flue gas temperature (1569 °C) had to be calculated based on heat balance of boiler, because measured flame temperature would not properly represent the fluid gas temperature.
- *Other CFD setup.* For turbulence simulation, K-ε turbulence model with scalable wall functions was used. Radiation was simulated after stabilization of calculation and Discrete Ordinates model was used. Fluid properties of exact flue gas composition were used. Absorption coefficient was calculated by UDF, and its value varied based on position and local temperature (average value was 0.1545 1/m). Simulation was computed as steady-state and equation discretization of second order upwind scheme was used.

3.3 Temperature load non-uniformity

To evaluate heat load non-uniformity on superheater, it is necessary to know flue gas temperatures and velocities in plane directly above superheater. As can be seen on Fig. **4**, velocities in this plane are not homogenous, with average of 4.9 m/s. Those differences are caused by insufficient flow stabilization after turn in the upper part of the second stage.

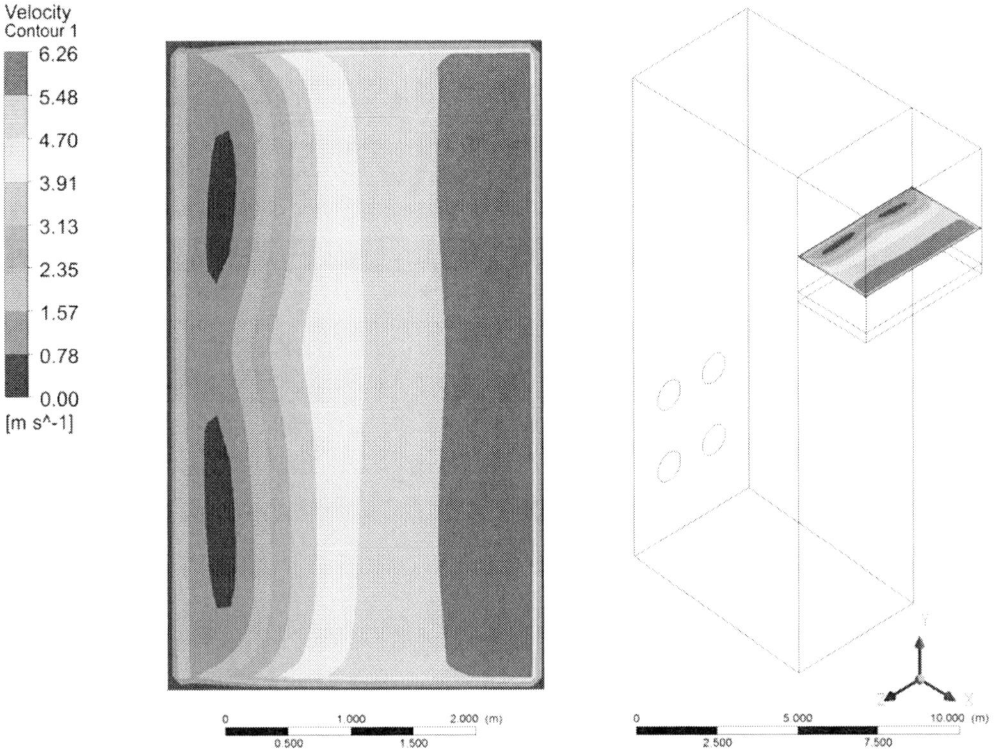

Fig. 4. Flue gas velocity above superheater P2.

Uneven flue gas flow also strongly effects the temperature distribution, as can be seen in Fig. **5**. Flue gas streams with higher velocity have shorter residence time in combustion chamber and they do not have enough time to transmit heat into evaporator (the membrane walls). On the other hand, slower flue gas streams have enough time to get cooler. Average temperature in plane above superheater is 683.1 °C.

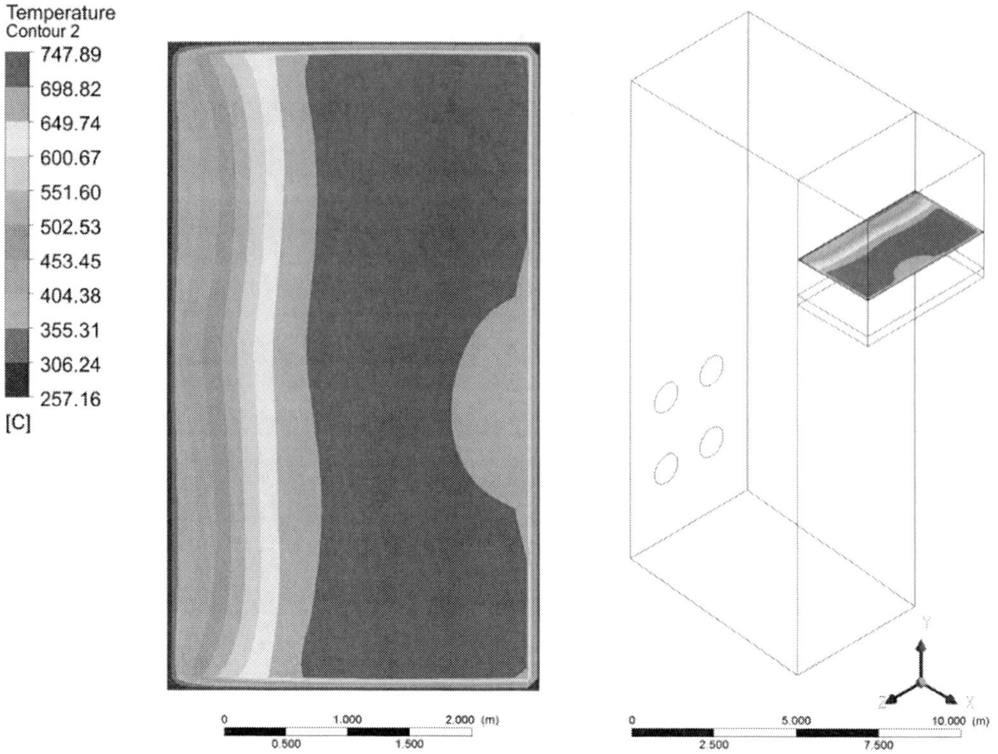

Fig. 5. Flue gas temperature above superheater P2.

Uneven flue gas velocities and temperatures pointed out problems with measuring of those parameters. Measured value is strongly dependent on location where the sensor is located.

Conclusion

Temperature related damage is a common damage type related to boilers. Superheaters are among the most influenced equipment. This article focuses on particular superheater as a part of steam boiler located in chemical plant. To consider damage importance, it is necessary to know surface temperatures of the superheater tubes. Data provided by operators often do not include those temperatures. In order to calculate them, thermal load non-uniformity has to be known. It can be calculated by various approaches. In this article, thermal-hydraulic calculations combined with CFD analysis were used. Results showed, that velocity and temperature distribution in plane directly above superheater is not homogenous, therefore, significant local variations of tube damage can be expected.

The results of this project NETME CENTRE PLUS (LO1202) were co-funded by the Ministry of Education, Youth and Sports within the support programme „National Sustainability Programme I".

References

1. K. Rayaprolu, *Boilers: a practical reference* [online]. Boca Raton, FL: CRC Press, [vid. 2017-03-20]. ISBN 978-1-4665-0054-9. http://dx.doi.org/10.1201/b12977 (2013)
2. J. L. Singer, *Combustion: Fossil Power Systems.* Windsor CT: Combustion Engineering, ISBN 978-0-9605974-0-6 (1981)

3. D. J. Flynn, R. D. PORT, NALCO CHEMICAL COMPANY, ed. *The nalco guide to boiler failure analysis*. 2nd ed. New York: McGraw-Hill, ISBN 978-0-07-174300-6 (2011)
4. X. Luo, Z. Zhang. *Leakage Failure Analysis in a Power Plant Boiler*. IERI Procedia [online] **5**, International Conference on Agricultural and Natural Resources Engineering (ICANRE 2013), 107–111, ISSN 2212-6678 (2013)
5. M. Asgaryan, N. Simms, S. M. Wu. *Prediction of the Remaining Service Life of Superheater and Reheater Tubes in Coal-Biomass Fired Power Plants*. Advanced Materials Research [online] **856**, 343–348, ISSN 1662-8985 (2014)
6. M. M. Rahman, J. Purbolaksono, J. Ahmad. *Root cause failure analysis of a division wall superheater tube of a coal-fired power station*. Engineering Failure Analysis [online] **17** (6), 1490–1494, ISSN 1350-6307, (2010)

Prediction of modal properties of circular disc with pre-stressed fields

Milan Naď [1,*], *Rastislav* Ďuriš[1], *Tibor* Nánási[1]

[1]Slovak Univeristy of Technology in Bratislava, Faculty of Materials Science and Technology in Trnava, Jána Bottu 25, 917 24 Trnava, Slovak Republic

Abstract. Many structural elements in technical devices, as well as some tools in manufacturing equipment, such as saw blade or , have a geometric shape similar to a circular disc. In many cases, these circular discs must meet the required dynamic properties. One of the techniques to achieve the required dynamic properties of circular disc is based on initiating pre-stressed fields in the disc plane. In-plane residual stresses are created by appropriate technological treatments in selected disc part, for example by rolling of annulus leading to plastic deformation which causes the change of volume distribution in this disc part. The effects of in-plane residual stresses on modal properties of circular disc are analysed. The natural frequencies modifications depending on position and width and also on the change in thickness of the roll-prestressed annulus are investigated in this paper.

Keywords: circular disc, vibration, modal properties, pre-stressed fields, finite element method

1 Introduction

There are many situations and applications in the acoustical and structural problems where the vibrating circular disc have to be modified to achieve convenient dynamical properties [1]. This is mainly concerned with the requirements for the so-called "tuning" of the modal properties of the vibrating circular disc by means of technological treatments that induce the residual stresses in the disc plane. The natural frequencies of circular disc, clamped on inner radius, are varying when the localized plastic deformation caused for example by roll-tensioning, induces residual stresses. The natural frequencies of circular disc, clamped on inner radius, are varying when the localized plastic deformation caused for example by roll-tensioning, induces residual stresses. A similar effect resulting from residual stresses can also be achieved by phase transformation during technological treatments. During the process of initiating disc in-plane stresses using the roll-tensioning process, a disc is compressed within a certain annular contact zone between two opposing rollers. The contact zone of circular disc is plastically deformed and the residual stresses are occurred in whole disc plane. Then the effects of residual stresses induced by roll-tension on modal properties (natural frequencies, mode shapes) can be analysed. The appropriate conditions for corresponding technological treatments can be predicted using the natural frequency

* Corresponding author: milan.nad@stuba.sk

Reviewers: *Milan* Sága, *Milan* Vaško

characteristics where the natural frequency values are depending on parameters causing in-plane disc stresses. The considered computational procedure for the implementation of in-plane disc residual stresses is based on idea which is similar to the formation of thermoelastic stresses.The natural frequency characteristics for various rolling positions, for various rolling depths and widths of the annulus are obtained by modal analysis using Finite Element Method (FEM). The role of residual stresses obtained by rolling can be assessed from the change in natural frequencies and modal shapes.

2 Formulation of the problem

Creating pre-stressed fields in a circular disk causes a change in its spatial properties, i.e. a change in the distribution of its mass and stiffness parameters. As a consequence of these changes, the modification of the dynamic properties of a given circular disc occurs.

2.1 Theoretical approach to modelling of a modified dynamic system

The general equation of motion for undamped system without external excitation is defined by

$$\mathbf{M}\ddot{\mathbf{u}} + \mathbf{K}\mathbf{u} = \mathbf{0}\,, \tag{1}$$

where $\mathbf{M}$ is mass matrix, $\mathbf{K}$ is stiffness matrix of the system, $\mathbf{u}$ and $\ddot{\mathbf{u}}$ are displacements and accelerations vectors, respectively.

Equation (1) can be transformed [5] using the transformation equations

$$\mathbf{u}(t) = \mathbf{\Phi}\mathbf{q}(t)\,, \qquad \mathbf{\Phi}^T\mathbf{M}\mathbf{\Phi} = \mathbf{I}\,, \qquad \mathbf{\Phi}^T\mathbf{K}\mathbf{\Phi} = \mathbf{\Lambda} \tag{2}$$

where $\mathbf{\Phi} = [\boldsymbol{\phi}_1, \boldsymbol{\phi}_2, \ldots \boldsymbol{\phi}_N]$ is matrix of modal vectors, $\mathbf{\Lambda}$ is spectral matrix..

Eigenvalue problem of the system (1) can be written in the form

$$(\mathbf{\Phi}^{\mathrm{T}}\mathbf{K}\mathbf{\Phi} - \omega^2\mathbf{\Phi}^{\mathrm{T}}\mathbf{M}\mathbf{\Phi})\mathbf{q} = (\mathbf{\Lambda} - \omega^2\mathbf{I})\mathbf{q} = \mathbf{0}\,, \tag{3}$$

where ω is natural angular frequency.

When the system described by equation (1) is modified [4], then the modification of the system has to be incorporated through mass and stiffness changes of system parameters. The mass and stiffness properties of the system are modified and equation (1) becomes

$$(\mathbf{M} + \Delta\mathbf{M})\ddot{\mathbf{u}} + (\mathbf{K} + \Delta\mathbf{K})\mathbf{u} = \mathbf{0}\,, \tag{4}$$

The modification matrices $\Delta\mathbf{M}$ and $\Delta\mathbf{K}$ characterise the mass and stiffness modifications in the spatial model. The practical modification is not carried out on matrices but on physical components or parameters of the structure.

Using equations (2), the eigenvalue problem of modified system (4) is

$$[\mathbf{\Phi}_m^T(\mathbf{K} + \Delta\mathbf{K})\mathbf{\Phi}_m - \omega_m^2\mathbf{\Phi}_m^T(\mathbf{M} + \Delta\mathbf{M})\mathbf{\Phi}_m]\mathbf{q}_m = \mathbf{0}\,, \tag{5}$$

where ω_m is the natural angular frequency of modified system.

Equation (5) provides the new natural angular frequencies (ω_m) and new modal vectors ($\boldsymbol{\phi}_m$) of the system after structural modification. Then i-th natural angular frequency is expressed

$$\omega_i = \sqrt{\frac{\boldsymbol{\phi}_i^T(\mathbf{K} + \Delta\mathbf{K}_\sigma)\boldsymbol{\phi}_i}{\boldsymbol{\phi}_i^T(\mathbf{M} + \Delta\mathbf{M})\boldsymbol{\phi}_i}}\,. \tag{6}$$

2.2 Computational model of circular disc with prestressed fields

The general shape of circular disc of outer radius r_0, inner radius r_v and thickness h_0 (Fig.1) is considered. The material of circular disc is isotropic and homogeneous. The inner radius r_v specifies a circle where the disc is clamped by flanges. To modify the modal properties of a circular disc the case with pre-stressed annulus field is considered. The geometrical shape of disc in-plane prestressed fields are defined by middle radius and width of each prestressed field, i.e. r_1 and b_1.

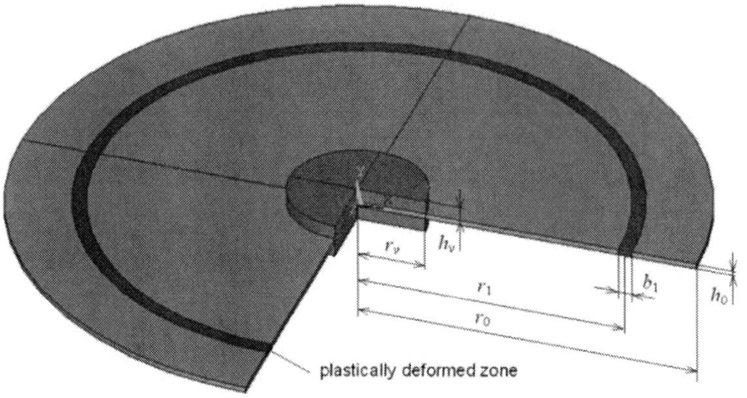

Fig. 1. The circular disc with prestressed fields

The fundamental considerations and derivation of equations of motion are based on Kirchhoff's plate theory assumptions. Using Kirchhoff's plate theory, the field of displacements in the cylindrical coordinates r, φ, z, can by written as

$$\begin{aligned} u_r(r,\varphi,z,t) &= u(r,\varphi) - z\frac{\partial w(r,\varphi,t)}{\partial r}, \\ v_\varphi(r,\varphi,z,t) &= v(r,\varphi) - \frac{z}{r}\frac{\partial w(r,\varphi,t)}{\partial \varphi}, \\ w_z(r,\varphi,z,t) &\cong w(r,\varphi), \end{aligned} \tag{7}$$

where $u(r, \varphi)$, $v(r, \varphi)$ and $w(r, \varphi)$ are displacements of point laying on neutral plane of the circular disc in coordinate directions.

Generally, the strain-displacement relations in the cylindrical coordinates for pre-stressed circular disc can be written in the form

$$\boldsymbol{\varepsilon} = \boldsymbol{\varepsilon}_0 - \widetilde{H}(r)\boldsymbol{\varepsilon}_i, \tag{8}$$

resp.

$$\begin{bmatrix} \varepsilon_r \\ \varepsilon_\varphi \\ \gamma_{r\varphi} \end{bmatrix} = \begin{bmatrix} u_{,r} \\ \dfrac{v_{,\varphi}}{r} + \dfrac{u}{r} \\ \dfrac{u_{,\varphi}}{r} + v_{,r} - \dfrac{v}{r} \end{bmatrix} - z\begin{bmatrix} w_{,rr} \\ \dfrac{w_{,\varphi\varphi}}{r^2} + \dfrac{w_{,r}}{r} \\ \dfrac{2w_{,r\varphi}}{r} - \dfrac{2w_{,\varphi}}{r^2} \end{bmatrix} - \widetilde{H}(r)\begin{bmatrix} \varepsilon_{r,i} \\ \varepsilon_{\varphi,i} \\ \gamma_{r\varphi,i} \end{bmatrix} \tag{9}$$

where $\widetilde{H}(r) = H[(r_1 - b_1/2) - r] - H[(r_1 + b_1/2) - r]$ is a modified Heaviside function describing the position of the pre-stressed zone ($H(x) = 0$ for $x < 0$; $H(x) = 1$ for $x > 0$), $\varepsilon_{r,i}$,

$\varepsilon_{\varphi,I}$, $\gamma_{r\varphi,i}$ are the initial strains inserted in pre-stressed area, $\frac{\partial(\cdot)}{\partial r} = (\cdot)_{,r}$ and $\frac{\partial(\cdot)}{\partial \varphi} = (\cdot)_{,\varphi}$ are the partial derivations.

Generally, the stress-strain relations under consideration of initial stresses and initial strains are given by

$$\boldsymbol{\sigma} = \mathbf{D}(\boldsymbol{\varepsilon}_0 - \widetilde{H}(r)\boldsymbol{\varepsilon}_i) + \boldsymbol{\sigma}_i \tag{10}$$

where $\boldsymbol{\sigma}$ and $\boldsymbol{\varepsilon}$ are stress and strain vectors, $\boldsymbol{\sigma}_i$ and $\boldsymbol{\varepsilon}_i$ are initial stress and initial strain vectors, $\mathbf{D}$ is elasticity matrix. Using the finite element formulation, the equation of motion for a free vibration of in-plane stressed disc is described by expression

$$\mathbf{M}\ddot{\mathbf{u}} + (\mathbf{K} + \Delta\mathbf{K}_\sigma)\mathbf{u} = \mathbf{0}\,, \tag{11}$$

where $\mathbf{M}$ is mass matrix, $\mathbf{K}$ is stiffness matrix, $\mathbf{K}_\sigma$ is stiffness matrix resulting from stress distribution induced by rolling, $\ddot{\mathbf{u}}$ and $\mathbf{u}$ are vector of nodal accelerations and vector of nodal displacements, respectively. We note, that the mass distribution of circular disc after rolling is not changed, but the bending stiffness is considerably changed.

Equation (10) can be transformed to modal coordinates using the transformation equations (2). After applying the these transformations, the equation of motion (11) can be used to determination of the natural angular frequencies and mode shapes of the circular disc with roll-tensioning induced residual stress distribution. We obtain the following eigenvalue problem

$$(\mathbf{K} + \mathbf{K}_\sigma - \omega_i^2\mathbf{M})\boldsymbol{\phi}_i = \mathbf{0}\,, \tag{12}$$

where $\omega_i = \sqrt{\frac{\boldsymbol{\phi}_i^T(\mathbf{K} + \mathbf{K}_\sigma)\boldsymbol{\phi}_i}{\boldsymbol{\phi}_i^T\mathbf{M}\boldsymbol{\phi}_i}}$ is i-th natural angular frequency, $\boldsymbol{\phi}_i$ is eigenvector describing i-th modal shape of the circular disc.

3 Numerical simulation and results

We consider a circular disc (Fig. 1) of the outer radius r_0 = 120 mm, flange radius is r_v = 25 mm, thickness h = 1.8 mm. The width of plastically deformed annulus is assumed as 10 mm. This width is selected arbitrarily and it is considered as a representative value for planned experimental verification of investigated phenomenon. The input data used for numerical analysis of circular disc are introduced in the Table 1.

Table 1. Input data

Young modulus	E	**[GPa]**	210.0
Poisson number	μ	**[-]**	0.3
density	ρ	**[kgm^{-3}]**	7800.0
coefficient of thermal expansion	α	**[K^{-1}]**	1.2×10^{-5}
outer radius of disc	r_0	**[mm]**	120.0
flange radius	r_v	**[mm]**	25.0
disc thickness	h_0	**[mm]**	1.8
width of plastic-afected zone	b_1	**[mm]**	10.0
depth of rolling of the plastic-afected zone	Δz	**[mm]**	{0.0; 1.0; 2.0; 3.0; 4.0}

The analysed disc is assumed to be perfectly fixed in region $r \geq r_V$. The outer edge of circular disc is free.

The change in disk stiffness after rolling, which is represented by the modified stiffness matrix $\mathbf{K}_\sigma$, must be determined from the residual stress distribution in the disc plane. To determine the residual stress distribution, the method of thermoelastic stress loading is used [2]. The thermoelastic expansion induces a stress distribution, which is analogous to the stress distribution initiated by rolling. The dependence between temperature and depth of roll-tensioning is approximately described by equation

$$\Delta T \approx \frac{\mu}{h_0 \alpha} \Delta z \,, \tag{13}$$

where μ is Poisson number, α is the coefficient of thermal expansion, h_0 is disc thickness and Δz is depth of roll-tensioning.

The matrices $\mathbf{M}$, $\mathbf{K}$ and additional matrix $\mathbf{K}_\sigma$, which follows from stress distribution arising from rolling (in this model analogy with thermoelastic expansion is used), are calculated automatically by ANSYS. The calculation processes for determination of natural angular frequencies and mode shapes are realised by ANSYS.

The distribution of radial σ_r and tangential σ_t residual stresses induced in plane of circular discs with one roll-tension annulus for various parameters is shown in Fig. 2.

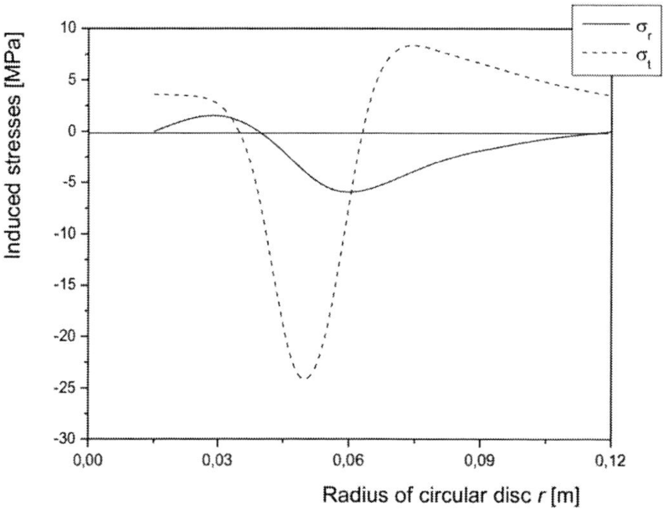

Fig. 2. Distribution of radial σ_r and tangential σ_t residual stresses in disc plane ($r_1 = 50$ mm; $\Delta z = 1$ μm; $b_1 = 10$ mm)

In Fig. 3 the natural frequency curves of the modal shapes 0/1, 0/0, 0/2, 0/3 (nodal circles/nodal lines) for different depth of rolling ($\Delta z = 1{,}0 \div 4{,}0$ μm) calculated by FEM when r_1 varies from 0,03 m to 0,11 m are shown. The natural frequencies of circular disc before roll-tensioning are marked by $r_1 = 0.0$ m. The tendency of curves for mode shapes 0/1 and 0/0 differs from curves for mode shapes 0/2 and 0/3. The natural frequencies of the mode shapes 0/2 and 0/3 increase with r_c until the maximum values near $r_c \approx 0.055$ m are reached; then they decrease. Contrary to this, the natural frequencies of the mode shapes 0/1 and 0/0 decrease with r_1 and for $r_1 \approx 0.046$ m reach the minimum; then they increase.

a)

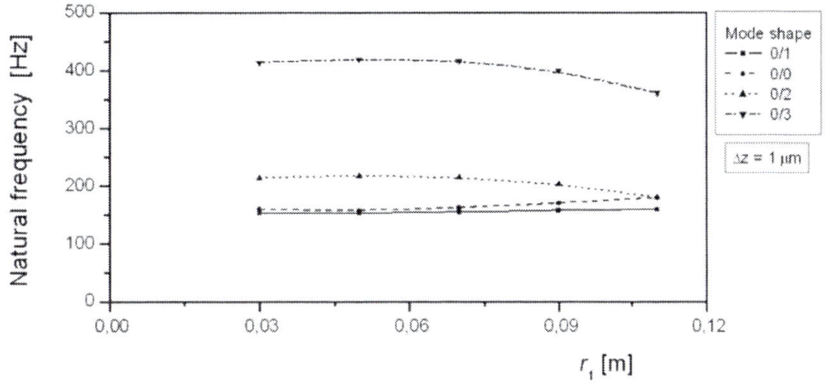

b)

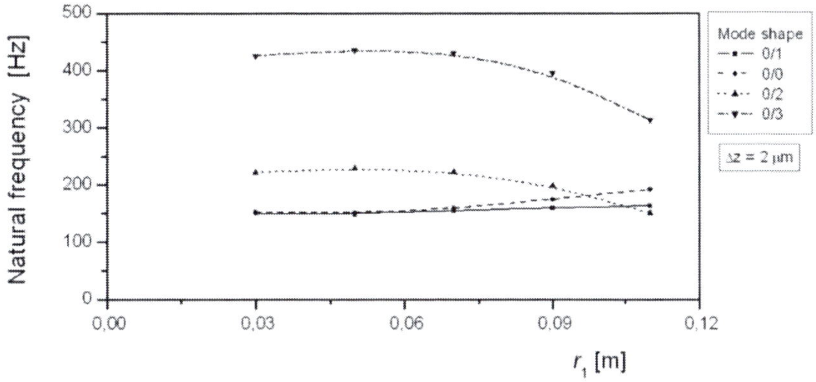

c)

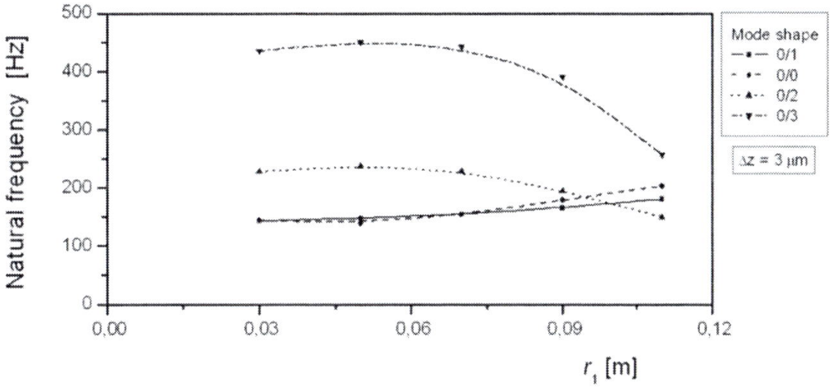

d)

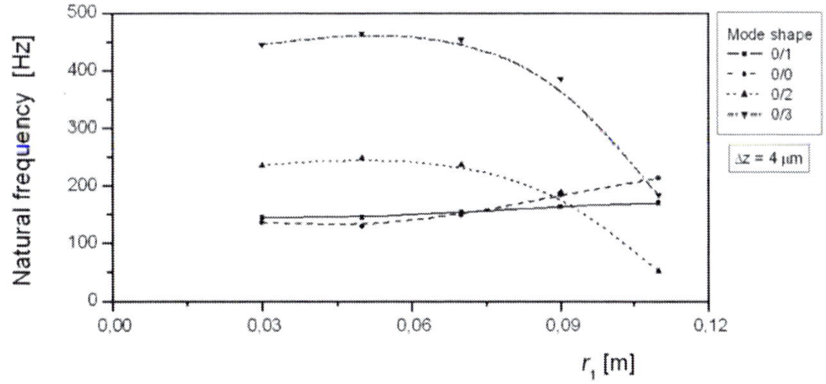

Fig. 3. Dependency of natural frequency on center radius of pre-stressed annulus r_1 for individual Δz and for first four mode shapes

In Fig. 4, the natural frequency curves of the individual modal shapes 0/1, 0/0, 0/2, 0/3 in dependency of depth of rolling are shown. The natural frequencies of the modal shapes 0/1 and 0/0 decrease with r_1 until the minimum values near $r_1 \approx 0.046$ m are reached; then they increase. Contrary to this, the natural frequencies of the modal shapes 0/2 and 0/3 increase with r_c and for $r_1 \approx 0.055$ m reach the minimum; then they increase. The effect of depth of rolling on natural frequency is evident from these graphs.

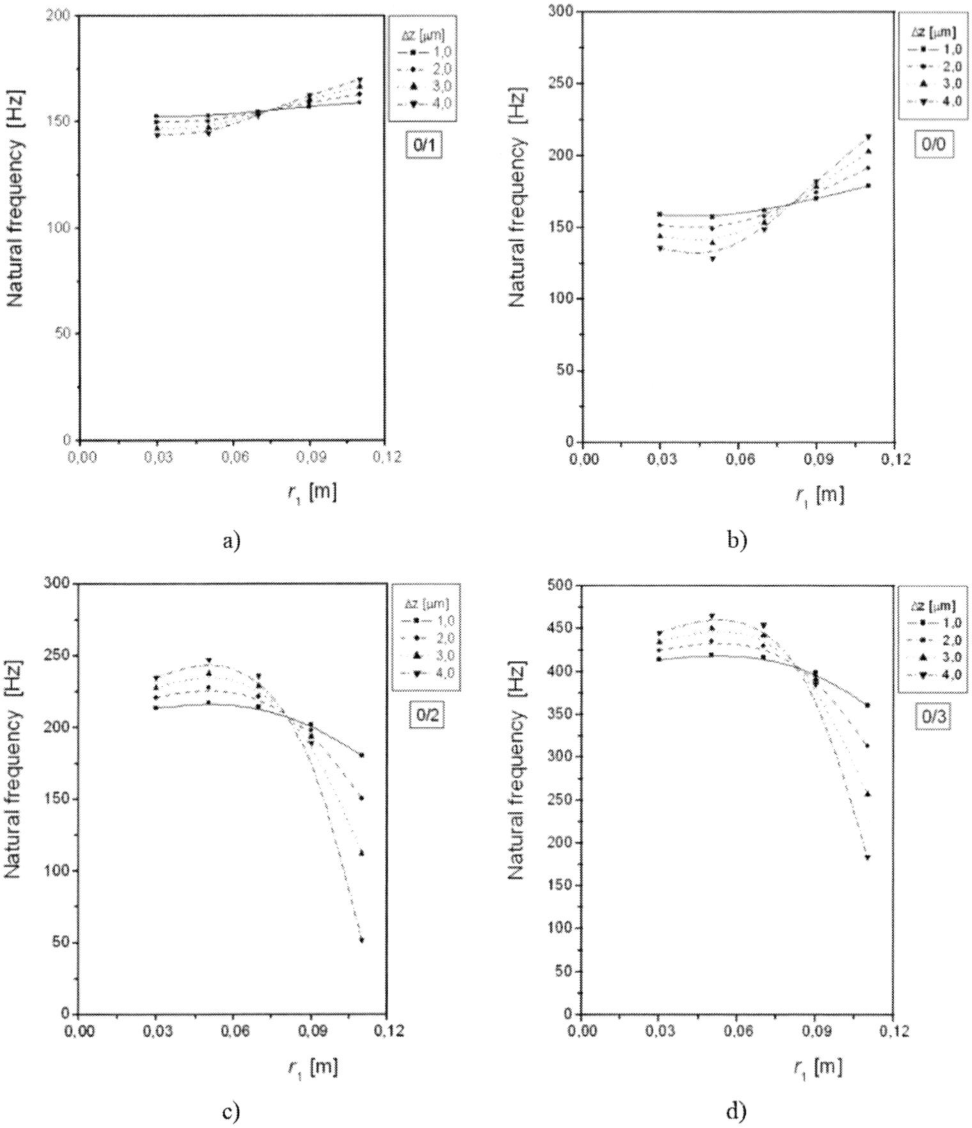

Fig. 4. Dependency of natural frequency of the first four mode shapes on center radius of pre-stressed annulus r_1 and for various Δz

Conclusions

The theoretical formulation and calculation model for analysis of dynamical properties of circular disc with residual stress distribution are presented. Finite element analysis for estimating the natural frequencies was used. For certain mean radius of pre-stressed annulus r_1, the natural frequencies of mode shapes 0/2 and 0/3 become smaller than those before roll-tensioning and the requred effect leading to growth of natural frequencies caused by pre-stresses state of circular disc cannot be achieved. Therefore, the appropriate rolling position is necessary to be determined from natural frequency characteristics calculated for various mean radius of pre-stressed annulus. This method of structural modification can be used to solve the various design problems of mechanical systems [3], [6] and it is very effective to modification of the dynamic properties of similar structural elements.

The authors wish to thank the financial support of the project VEGA 1/1010/16 and project IP MTF 1603/2017.

References

1. G. Izrael, J. Bukoveczky, L. Gulan, *Influence of Nonstandard Loads onto Life of Chosen Modules of Mobile Working Machines*. Machine Design **3**, 13-16 (2011)
2. M. Sága, R. Bednár, M. Vaško, *Contribution to modal and spectral interval finite element analysis*. Vibration Problems ICOVP 2011, Springer Proceedings in Physics **139**, 269-274 (2011)
3. M. Nad', *Modification of Modal Characteristics of Vibrating Structural Elements*. (Scientific Monographs, Köthen, 2010)
4. F. Kuratani, S. Yano, *Vibration Analysis of a Circular Disc Tensioned by Rolling Using Finite Element Method*. Archive of Applied Mechanics **70**, 279-288 (2000)
5. I. Mazurkievič, L. Gulan, G. Izrael, *Mobile Working Machines - Theory and Constructions of Basic Modules*. [in slovak] (1st ed., STU Bratislava, 2013)
6. M. Orečný, Š. Segľa, R. Huňady, *Methodology for Tuning a Semi-active Dynamic Vibration Absorber on a Passive Suspended Seat*. Applied Mechanics and Materials **81**, 63-68 (2015)

Simulation of solidification during crystallization under pressure

Richard Pastirčák[1*], *Marek* Brůna[1], *Dana* Bolibruchová[1]

Department of technological engineering, University of Žilina, Univerzitná 8215/1, 010 26 Žilina.

Abstract. The paper deals with squeeze casting technology. The influence of process parameters variation (casting temperature, mold temperature, pressure) will be observed. The experimentally obtained boundary conditions (heat flow, HTC) will be verified using the ProCast simulation software. Shape influence on the mold filling process and the temperature field under pressure will be evaluated. To evaluate the mechanical properties, a tensile test was performed. Three identical samples were cast for each parameters change. From these samples the average values of the mechanical properties were measured and calculated. The thickness of the flat test samples was 3.15, 4, 5, 6.3, 8 mm with a sample width of 10 mm. For gravity cast of the casting, the mechanical properties of the thinner parts were higher. At the pressure influenced castings, the mechanical properties were higher in the thicker parts of the casting.

Keywords: squeeze casting, heat flow, HTC, pressure, simulation

1 Introduction

The squeeze casting technology, which was used in experimental verification of the heat transfer is a relatively unusual method. This is due, in particular, to little experience with practical use combined with the low lifetime of the mold, which is caused by high thermal and mechanical stresses combined with a long cycle time. With direct squeeze casting, the cycle time can be up to about 1 minute. Such stress greatly reduces the life of the mold [1, 3, 9, 12]. The main characteristics of this method, which are distinguishing it from conventional high-pressure casting include, in particular, the slow filling of the melt into the mold cavity (roughly 0.5 $m.s^{-1}$) and considerably larger gates, than those used in the high-pressure casting. The slow filling of the mold cavity guarantees a laminar flow without turbulences. In addition, a longer filling time allows evacuation of gases from the mold cavity, which reduces the resulting porosity of the casting. Thanks to the large slit, the melt and later the casting itself are constantly under the influence of the applied pressure, while the gates solidified as last. This allows continuous melt substitution in interdendritic areas. The result of this is high integrity and minimum casting porosity. This also prevents the creation of various foundry defects. These castings shows high mechanical properties, which can be further increased by subsequent heat treatment. [2, 4, 8, 11] The pressure at

[*] Corresponding author: richard.pastircak@fstroj.uniza.sk

Reviewers: *Eva* Tillová, *Milan* Žmindák

squeeze casting is from 30 to 150 MPa, depending on the type of alloy used, the casting temperature, and more. The pressure itself prevents the formation of an air gap, which is normally formed at the gravity casting between the casting and the mold. This air gap substantially slows heat transfer and thus also the cooling rate of the casting. By eliminating the gap due to pressure, there are significant changes in the structure and also in the mechanical properties of the material. In The structure, the grains size of the primary structure is reduces and also is change the morphology of eutectic silicon and intermetallic phases, which act less harmful. The resulting structure is finer and more uniform throughout the volume. The concentration of silicon in the eutectic and the volume ratio of the primary α-phase is increased and the volume of the eutectic in the structure is reduced. The result is a change in mechanical properties, where the strength, elongation and fracture toughness are increasing. At the gravity casting achieves a better mechanical properties the parts of casting, which solidified faster, in this case is about the thinnest parts of the casting. Pressure utilization at the squeeze casting technology greatly accelerates solidification and cooling in the thicker areas of the casting. This is associated with the eliminate of an air gap formed by gravity casting. The result is a change in structure and a significant increase in mechanical properties compared to gravity casting [1, 4, 5, 6, 7, 10].

2 Experimental material and process

For the study of crystallisation under pressure, a casting with different wall thicknesses was used, which is shown in Fig. 1. The thicknesses of the individual walls were selected based on the selected numbers of the geometric series R10 with the sequence of 3.15 mm, 4 mm, 5 mm, 6.3 mm and 8 mm. The width of each wall was 22 mm with a length of 100 mm. Dimensions were chosen to evaluate the mechanical properties of individual casting walls (Rm). The casting model was created in the SolidWorks work environment, simulation calculations were carried out in the ProCast program, and the casting process was carried out in the casting laboratory at the Department of Technological Engineering at the University of Žilina in Žilina.

Fig. 1. Casting model

For experimental verification was used AlSi7Mg alloy, which is characterized by excellent foundry properties, especially good fluidity, allowing a better fluidity of the melt to all parts of the form. The chemical composition of the alloy used is shown in Table 1.

Table 1. Chemical composition of the AlSi7Mg0.3 alloy

Chemical composition of the AlSi7Mg0.3 [%]						
Fe	Si	Mn	Ti	Cu	Mg	Zn
max 0.19	6.5 – 7.5	max 0.1	max 0.25	max 0.05	0.25 – 0.45	max 0.07

The mold was made of low-carbon steel and she was composed of two parts. The lower part of the mold copied the shape of a stepped casting. The second part was a detachable upper part of the mold with the gating system. Figure 2 shows the lower mold part (left) and upper mold part (right).

Fig. 2. The lower and the upper part of the mold

Verification of filling and solidification was carried out in the simulation program and later experimentally. In the first stage was evaluated the results of the experimental melting with the results of the simulation calculation. The main evaluations for the setting of the casting process parameters was the evaluations of the fluidity of the casting at the gravity casting. The marginal conditions for the calculation were used from the experiment, for example casting temperature, mold temperature, casting method, AlSi7Mg0.3 alloy as material of the casting and low carbon steel as material of the mold. Casting time was 3.2 s. The fluidity of the material was defined as a critical proportion of the solid phase of 25%. The calculation network consisted from 199,647 elements.

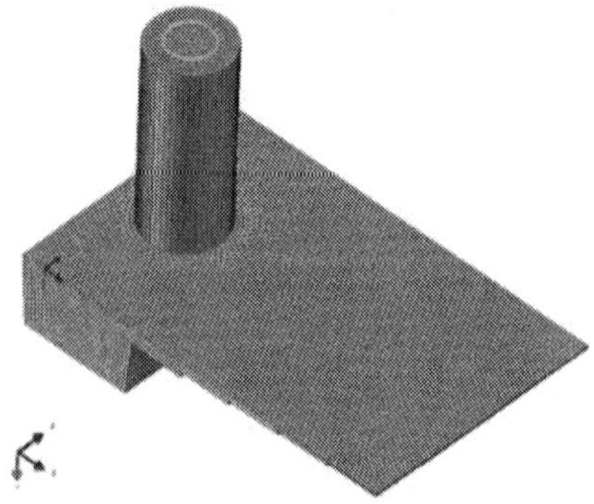

Fig. 3. Geometric design of a casting with a calculation network

The results of the simulation calculation proved, that by the selected parameters do not cause the fluidity of the casting in the full volume, as can be seen in Fig. 4 and 5, which was also confirmed experimentally.

Fig. 4. Comparison of casting after completion of the first casting and the first simulation for the first casting design at a mold temperature of 200 ° C and a casting temperature of 720 ° C

Fig. 5. Comparison of casting after completion of second casting and additional simulation for the first casting design at a mold temperature of 350 ° C and a casting temperature of 750 °C

After verifying the suitability of the proposed marginal and initial simulation calculation conditions, the geometric modification of the model was performed. At the same time, a geometric modification the depth of the inlet part of the melt was proposed in the space under the gating system. The original height of 30 mm was reduced to a half of 15 mm.

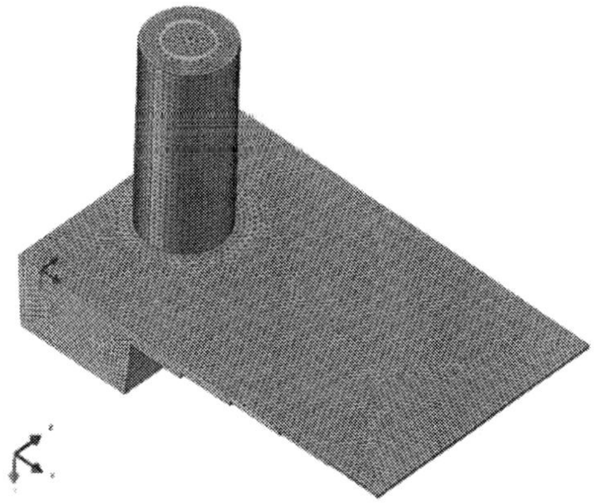

Fig. 6. Geometric design of a casting with a calculation network

The geometric model of the final casting, which was created in SolidWorks. The simulation network is shown in Fig. 6. The network consists from 158 834 elements. The simulation parameters are casting temperatures 720 and 730 ° C, mold temperature 250 ° C, casting time 3.2 s, velocity of the free surface at the mold wall 0.7 and critical proportion of solid phase 25%. As can be seen in Fig. 7, the results of the simulation calculation showed, that the modifications in the modeled casting caused to fluidity of the melt in almost the entire volume.

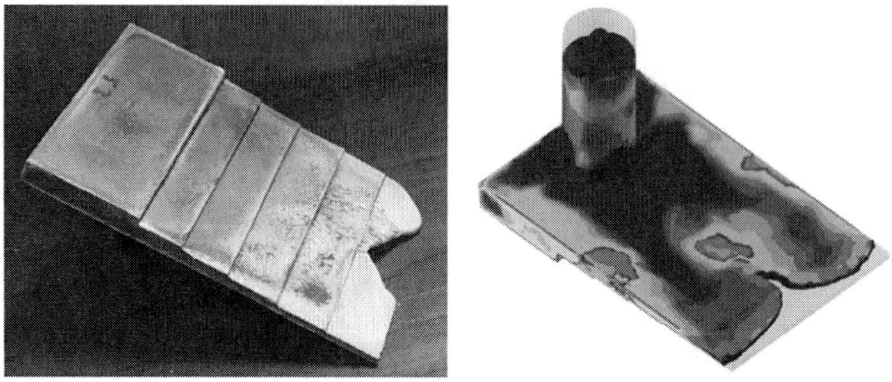

Fig. 7. The proportion of solid phase in the casting after completion of the simulation at a mold temperature of 200 ° C and a casting temperature of 720 ° C

Based on the results of the simulation calculation, a change of casting parameters was proposed to casting temperature 730 ° C and mold temperature 250 ° C. The results of the simulation calculation as well as the experimental verification confirmed the complete filling of the mold cavity during gravity casting, Fig. 8.

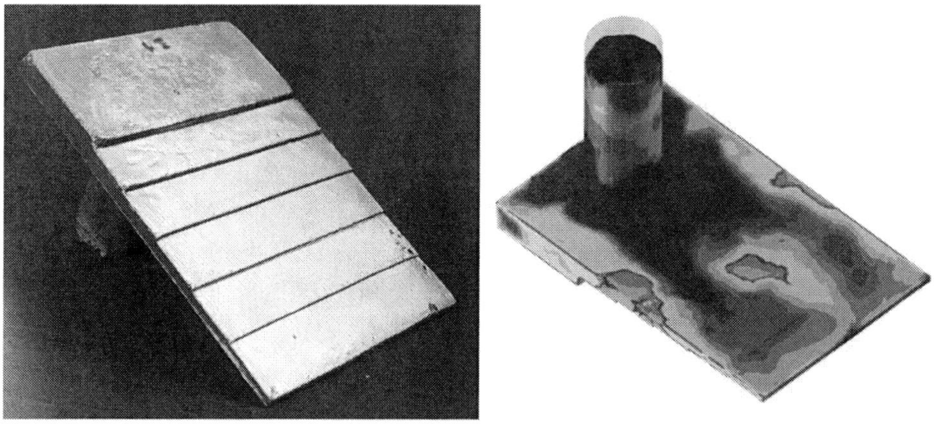

Fig. 8. Comparison of casting after completion of second casting and additional simulation for final casting design, at mold temperature 250 ° C and casting temperature 730 ° C

Based on the results of the simulation calculation, an experiment for crystallisation under pressure was proposed. Casting parameters: Casting temperature 730 ° C, mold temperature 250 ° C, pressure 30 MPa, start of pressure was shortened to 2.5 s from the start of casting.

3 Results

By defining the casting process with crystallisation under pressure, the simulation calculations and experimental casting were performed. The basic casting parameters are shown in Table 2. The time of the applied pressure was 40 seconds.

Table 2. Casting parameters of individual samples

Casting	Mold temperature [°C]	Pouring temperature [°C]	Pressure [MPa]
1.	250	730	0.1
2.	250	730	30
3.	200	735	30
4.	150	740	30

In Fig. 8 is shown the results of the simulation calculation at the individual casting conditions. The presented results represent the proportion of the solid phase at the time from begin of applied pressure to the melt. In Fig. 9-10 is shown the total solidification time of the individual samples and the casting time between the liquidus and the solidus. These times have the most significant effect on the resulting mechanical properties of castings.

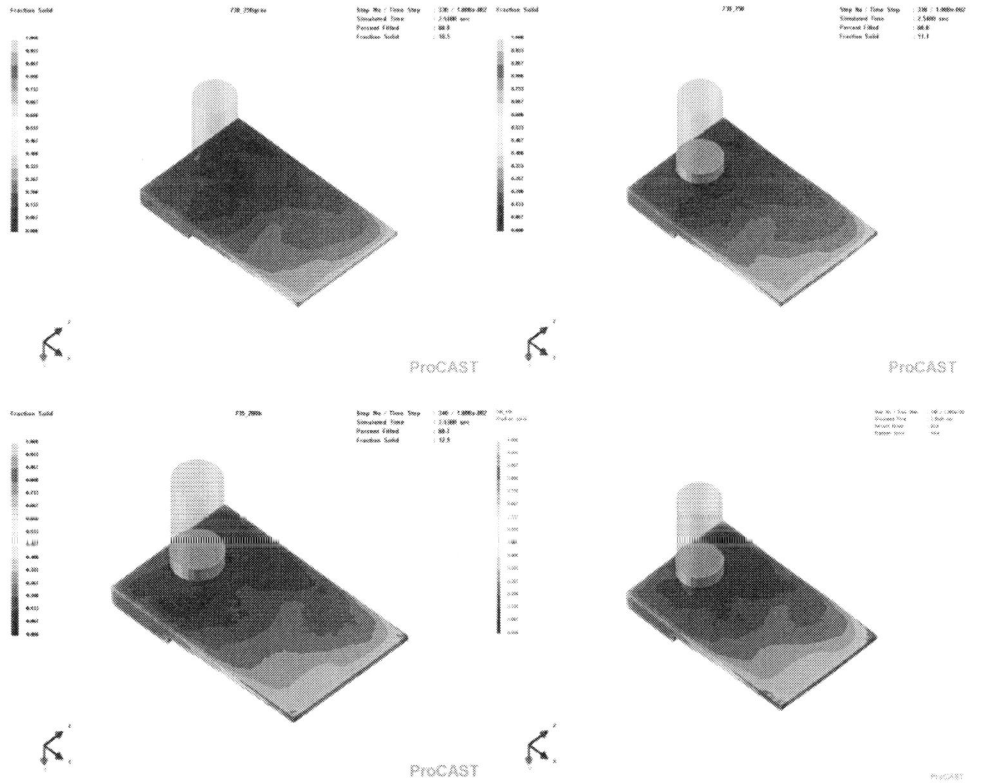

Fig. 8. The proportion of solid phase in the casting, 2.5 seconds from the start of casting

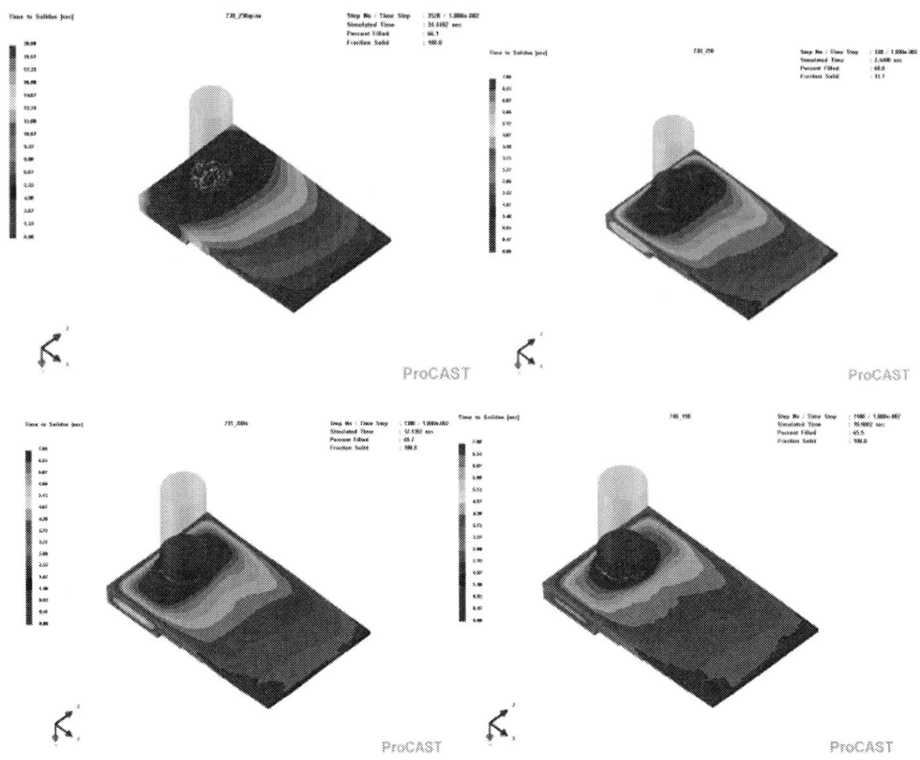

Fig. 9. Solidification time of the casting

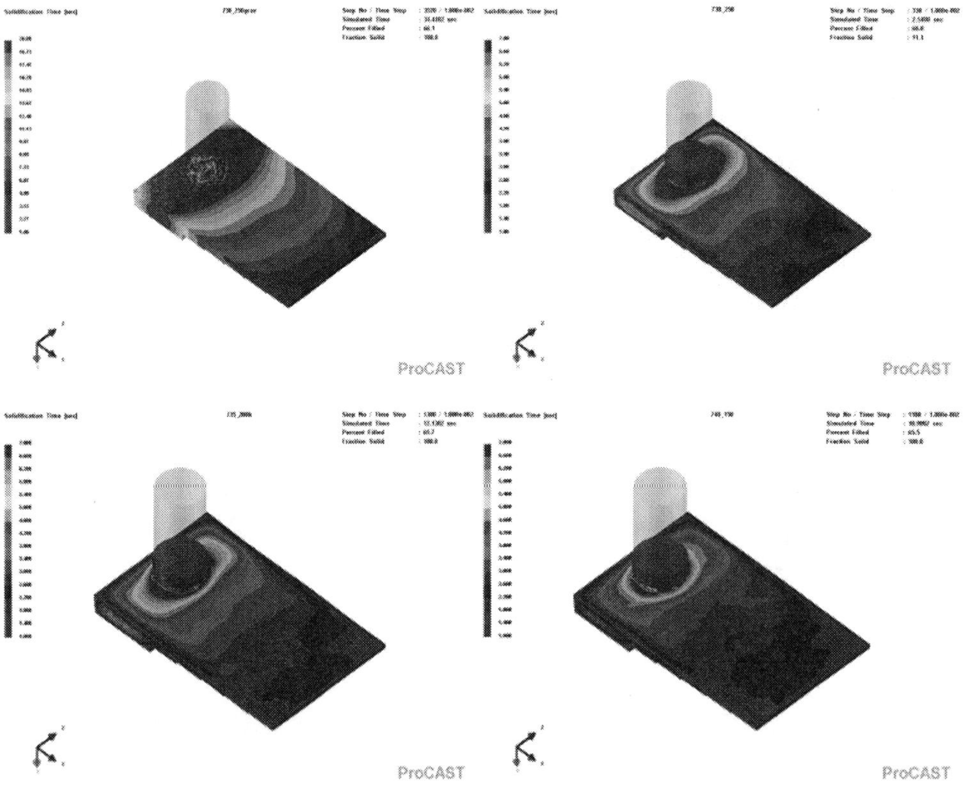

Fig. 10. Crystallization time

To evaluate the mechanical properties, a tensile test was performed. Three identical samples were cast for each parameters change. From these samples the average values of the mechanical properties were measured and calculated. The thickness of the flat test samples was 3.15, 4, 5, 6.3, 8 mm with a sample width of 10 mm. The average measured results of Rm for individual casting processes and their individual areas of different thicknesses are shown in Table 3.

Table 3. Results of the tensile strength for individual castings and wall thicknesses

Wall thickness [mm]	Casting 1 [MPa]	Casting 2 [MPa]	Casting 3 [MPa]	Casting 4 [MPa]
3.15	176	158	157	195
4	170	198	170	191
5	166	185	164	193
6.3	164	200	188	200
8	150	196	185	205

Conclusions

The first stage of the experiments was aimed on defining the marginal conditions of the simulation calculation. Casting geometry has been designed with regard to the poor fluidity of casting. In the experiment, technology of the gravity casting in to metal form was used. By defining the marginal conditions, a very good match of the results of the simulation calculation with the experimental results was achieved. These modified marginal conditions were used in the next stage of experiments, in which we focused on technology of the crystallization under pressure. The geometry of the inlet has been adapted to achieve complete fluidity of the casting. The crystallization conditions under pressure were defined from the evaluation of previous experiments. The casting temperature must be at least 730 ° C. We have adjusted the mold temperature with respect to the amount of heat contained in the casting, from 150 to 250 ° C. The start-up time of the pressure on the crystallizing melt was selected for 2.5 s, in order to quickly solidify the thinnest part of the casting. The temperature field does not change significantly for the different casting conditions within 2.5 seconds. During the total casting time of the casting it can be observed that the lower mold temperature influences on the shortening of the total solidification time of the casting. From the results of the experiments is also obvious, that the solidification time of the casting at the crystallization under pressure is greatly reduced. At the gravity cast of the casting is the solidification time of the thickest part of the casting about 16 s. At the pressure affected castings is this solidification time from 3 to 4,2 s. This effect was also confirmed in the evaluation of mechanical properties. For gravity cast of the casting, the mechanical properties of the thinner parts were higher. At the pressure influenced castings, the mechanical properties were higher in the thicker parts of the casting. The thinner parts of the casting solidify approximately as long as gravity casting. The significantly shorter solidification time of the thicker parts of the casting means, that the effect of the pressure results in the formation of a finer structure and also in the increase the tensile strength.

This work was created in the framework of the grant projekt VEGA N°1/0494/17. The authors acknowledge the grant agency for support.

References

1. R. Pastirčák, J. Ščury, M. Brůna, D. Bolibruchová, *Effect of Technological Parameters on the AlSi12 Alloy Microstructure During Crystallization Under Pressure.* Archives of foundry engineering **17** (2), 75-78 (2017)
2. R. Pastirčák, J. Ščury, *Effect of technological parameters on microstructure in alloy AlCu4Ti using squeeze casting.* American Institute of Physics Publising, AIP conference proceedings **1745**, (2016)
3. J. O. Aweda, M.B. Adeyemi, *Experimental determination of heat transfer coefficients during squeeze casting of aluminium.* Journal of Materials Processing Technology **2009** (3), 1477-1483 (2009)
4. J. H. Lee, H. S. Kim, C. W. Won, B. Cantor, *Effect of the gap distance on the cooling behavior and the microstructure of indirect squeeze cast and gravity die cast 5083 wrought Al alloy.* Materials Science & Engineering A **338** (1-2), 182-190 (2002)
5. M. Brůna, D. Bolibruchová, R. Pastirčák, *Numerical simulation of porosity for Al based alloys.* Procedia Engineering **177**, 488-495 (2017)
6. P. Fabian, G. Sládek, J. Brončok, I. Bilska, A. Szczotok, *Electrospark alloying of carbon steel with WC-Co-Al2O3 : deposition technique and coating properties.* Advanced Materials Research **874**, 101-106 (2014)
7. P. Pecháč, M. Sága, *Controlling of local search methods' parameters in memetic algorithms using the principles of simulated annealing.* Procedia Engineering **136**, 70-76 (2016)
8. D. Bolibruchová, J. Macko, M. Brůna, *Elimination of negative effect of Fe in secondary alloys AlSi6Cu4 (EN AC 45 000, A 319) by nickel.* Archives of metallurgy and materials **59** (2), 717-721 (2014)
9. P. Kopas, L. Jakubovičová, M. Vaško, M. Handrik, *Fatigue resistance of reinforcing steel bars.* Procedia Engineering **136**, 193-197 (2016)
10. M. Patek, M. Mičian, A. Sládek, D. Kadáš, *Numerical analysis of T-joint welding with different welding sequences.* Manufacturing technology : Journal for science, research and production **16** (1), 234-238 (2016)
11. D. Bolibruchová, L. Richtarech, S. M. Dobosz, K. Major-Gabrys, *Utilisation of Mould Temperature Change in Eliminating the Al5FeSI Phases in Secondary AlSi7Mg0.3 Alloy.* Archives of Metallurgy and Materials **62** (1), 357–362 (2017)
12. R. Koňar, M. Mičian, *Ultrasonic Inspection Techniques Possibilities for Centrifugal Cast Copper Alloy.* Archives of foundry engineering **17** (2), 35-38 (2017)

Heat flow and heat transfer coefficient during crystallization under the pressure

Richard Pastirčák[1,*], *Ján* Ščury[1], *Tomáš* Fecura[1]

Department of technological engineering, University of Žilina, Univerzitná 8215/1, 010 26 Žilina

Abstract. Estimation of the heat flow at the metal-mold interface is necessary for accurate simulation of the solidification processes. For the numerical simulation, a precise prediction of boundary conditions is required to determine the temperature distribution during solidification, porosity nucleation, microstructure development, and residual stresses. Determination of the heat transfer coefficients at the metal-mold interface is a critical aspect for simulation of the solidification process and the microstructure modeling of the castings. For crystallization under the pressure and for thin-walled castings, HTC evaluation is important due to the very limited freezing time.

Keywords: Squeeze casting, HTC, Heat flow, Pressure.

1 Introduction

All the thermal phenomenons in the casting-mold system are very closely related to the release and transfer of heat during solidification and cooling of the casting. The rate of heat transfer is very important to determine the solidification time and the distribution of temperature fields throughout the system. It also directly and indirectly affects the resulting microstructure, and thus the quality and properties of the future casting [1, 3, 4].

The heat flow between the casting, mold, and surrounding environment is from the initial contact of the melt with the mold up to his complete cooling a non-stationary phenomenon. In non-stationary thermal processes, the thermal content of the solid is changing. The heat content of the casting decreases and the mold rises until the heat contents and temperatures are balanced. The change rate of the thermal content of the solid, the rate of isotherm process and temperature equalization is directly proportional to his thermal conductivity λ and inversely proportional thermal-capacity cabability $c.\rho$. [1, 2, 5, 6].

The heat flow at the interface between the mold and the casting directly influences the solidification process and plays an important role in determining the solidification conditions of the casting metal. In the process of the casting under pressure, the casting quality is directly affected by the transfer of heat to the casting-mold interface, where the amount of heat transferred from the casting to the mold is changing over time. By creating an air gap at the casting-mold interface during solidification, the value of the heat transfer

* Corresponding author: richard.pastircak@fstroj.uniza.sk

Reviewers: *Františka* Pešlová, *Wieslawa Piekarska*

coefficient decreases. The change of heat transfer coefficient by the casting time is a key factor for controlling solidification. The value of this coefficient is influenced in particular by the heat-physical properties of the materials, the temperature of the mold, the temperature of the metal, the working pressure, the casting geometry, the roughness of the mold surface and the protective coating of the mold [3, 7, 8, 10].

The squeeze casting technology, which was used in experimental verification of the heat transfer is a relatively unusual method. This is due, in particular, to little experience with practical use combined with the low lifetime of the mold. The pressure value for squeeze casting is from 30 to 150 MPa. The pressure itself prevents the formation of an air gap which normally formed during gravity casting at the interface between the casting and the mold. By eliminating this air gap, there is a significant increase of the heat transfer and the heat flow from the casting into the mold. The melt and later the casting itself is constantly in contact with the walls of the metal mold, which resulting in a significant increase the rate of solidification and cooling of the casting. The result is a change of the structure and mechanical properties of the casting. In The structure, the grains size of the primary structure is reduces and also is change the morphology of eutectic silicon and intermetallic phases, which act less harmful. The resulting structure is finer and more uniform throughout the volume. The concentration of silicon in the eutectic and the volume ratio of the primary α-phase are increased and the volume of the eutectic in the structure is reduced. The result is a change in mechanical properties, where the strength, elongation and fracture toughness are increasing [2, 4, 9, 10].

2 Experimental material and process

For experimental casting, direct squeeze casting method was used. Precisely measured dose of the liquid metal was poured into the mold cavity. The mold was machined from low carbon steel. The surface of the mold cavity and the plunger has been treated with a protective paint type Terracotta. The liquid melt was pressed by using a plunger which section area was 1000 mm2. The onset of action was approximately 8 seconds by filling mold. The sample was under pressure until the cooling of the melt to 200 °C. During the experiment was casted 4 samples at various temperatures and the applied casting pressure. The casting parameters are listed in Tab. 1. The first sample was cast by gravity casting at a temperature of 710 °C. Further samples were cast under pressure by using technology squeeze casting. In both cases the metal mold was preheated to 100 °C. For measuring the temperature of the melt and mold 8 thermocouples type K were used, with a wire Do = 0.25 mm in order to respond rapidly. Position of the thermocouples is shown in Fig. 1.

Table 1. Casting parameters for individual samples

sample	pressure [MPa]	casting temperature [°C]	mold temperature [°C]
1	0.1	710 ± 5	20 ± 3
2	50	710 ± 5	100 ± 3
3	100	710 ± 5	100 ± 3
4	150	710 ± 5	100 ± 3

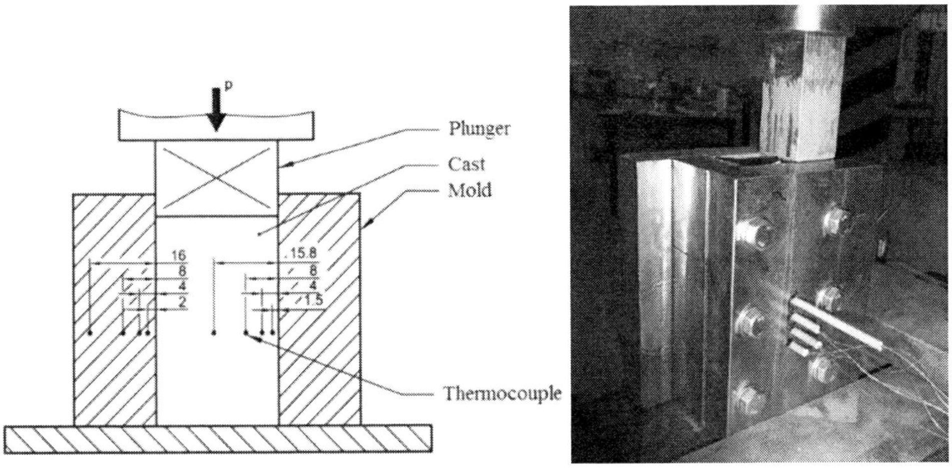

Fig. 1. Principle of thermocouples distribution

The governing heat transfer equation in one-dimensional coordinates is given by equation (1):

$$\rho c_p \frac{\partial T}{\partial t} = \frac{\partial}{\partial x}\left(\lambda \frac{\partial T}{\partial x}\right) \qquad (1)$$

The heat flux on the interface cast-mold can be calculated from the temperature gradient on the surface and the subsurface layer using the equation (2):

$$q = -\lambda \frac{\partial T}{\partial x} 5 = -\lambda \frac{T_1 - T_2}{\Delta x} \qquad (2)$$

where λ is thermal conductivity from the casting to the mold. $T_{1,2}$ are temperatures of the measured points and Δx is the distance between these points.

The average heat transfer coefficient HTC at the interphase may be expressed mathematically (3):

$$HTC = \frac{q}{(T_O - T_F)} \qquad (3)$$

where T_O is temperature casting surface, T_F is temperature mold surface.

The alloy AlCu4Ti was used for the experiment. Regarding this alloy, a significant solidification front should not occur, and also a wide freezing interval provides a sufficiently long period of time to initiate the pressure action during the time of the alloy crystallization. This alloy was therefore suitable for detecting the heat flow from the cast into the mold. The chemical composition of the AlCu4Ti alloy is shown in Table 2.

Table 2. The chemical composition of the experimental alloy AlCu4Ti

Element	**Si**	**Cu**	**Mg**	**Ni**	**Sn**	**P**	**Sb**
[at. %]	0.05	4.612	0.005	0.006	0.025	0.0005	0.00463
Element	**Ti**	**Fe**	**Mn**	**Na**	**Zn**	**Zr**	**Pb**
[at. %]	0.171	0.127	0.456	0.0001	0.008	0.001	0.002

3 Results

The measured values were processed in a tabular processor Microsoft Excel. From the measured curves of temperature in casting and mold, temperature on the surface of the mold and casting was estimated. In individual time points was by using the regression curve determined an identified temperature in the zero distance from the surface. Principle of evaluation is shown in Fig.2-3.

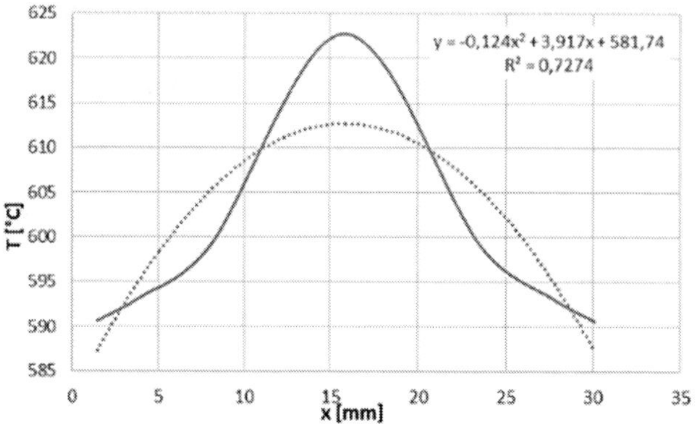

Fig. 2. The fundamentals of determining the surface temperature of the casting by regression curves

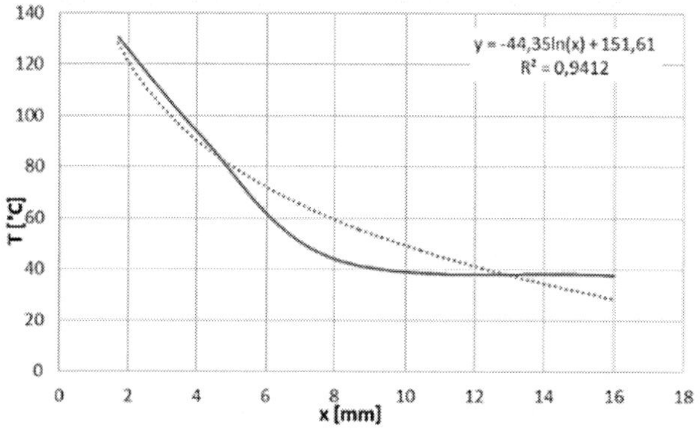

Fig. 3. The fundamentals of determining the surface temperature of the mold by regression curves

In Fig. 4 is shown the course of the temperatures at the casting without pressure.

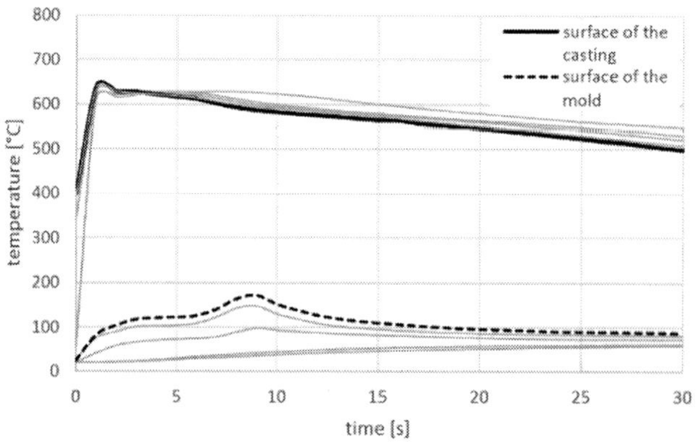

Fig. 4. Temperatures in the casting and in the mold at the gravity casting

In Fig. 5-7 is shown the course of the temperatures with using squeeze casting technology.

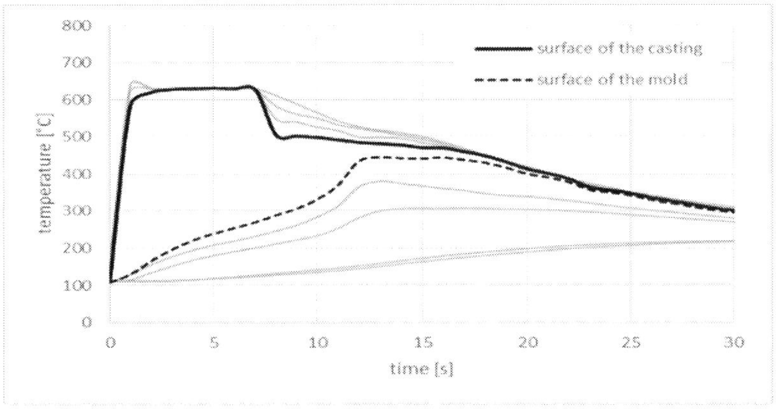

Fig. 5. The temperatures in the casting and in the mold at the operating pressure 50 MPa

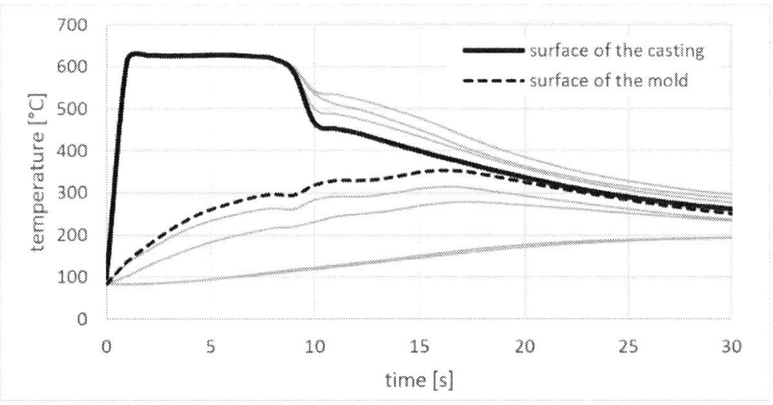

Fig. 6. The temperatures in the casting and in the mold at the operating pressure 100 MPa

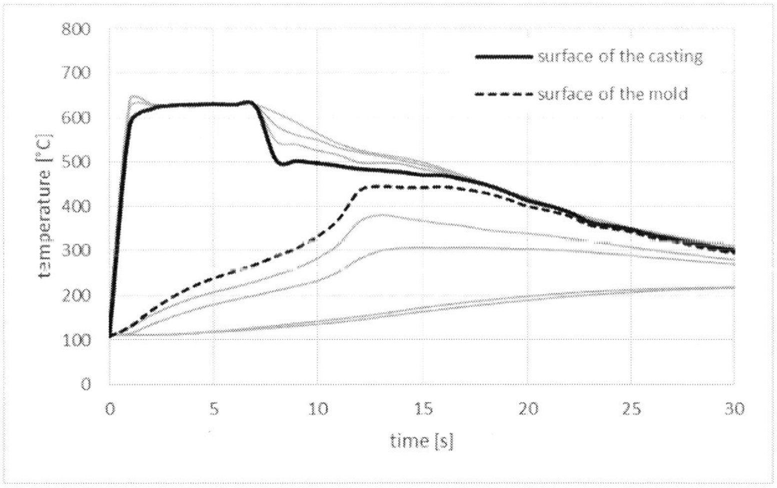

Fig. 7. The temperatures in the casting and in the mold at the operating pressure 150 MPa

3.1 Heat flux and heat transfer coefficient

The heat flux on the interface between casting and mold was determined from temperatures on the surface of a mold and temperatures 2 mm under the surface by the equation (1). The results of HTC were calculated by the equation (2). The course of heat flux and HTC for each sample is shown on the figures 8-11.

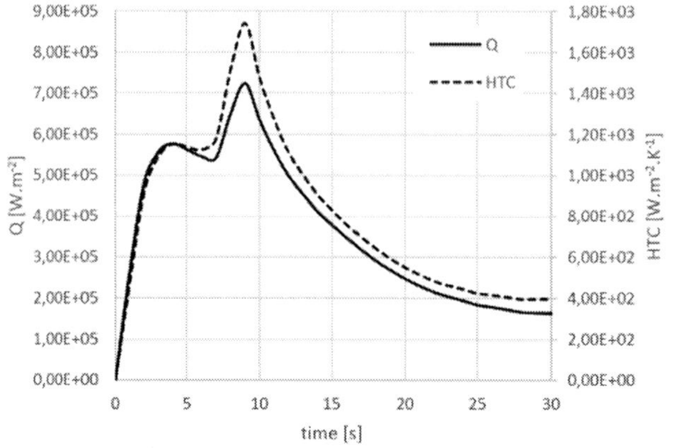

Fig. 8. Heat flux and HTC in gravity casting at pouring temperature 710 °C

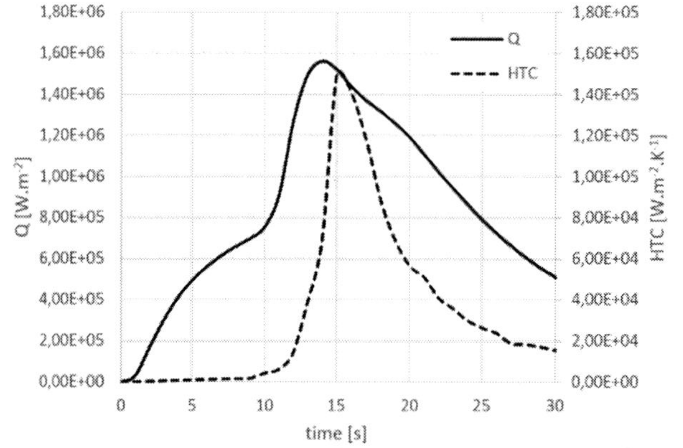

Fig. 9. Heat flux and HTC at pouring temperature 710 °C and pressure 50 MPa

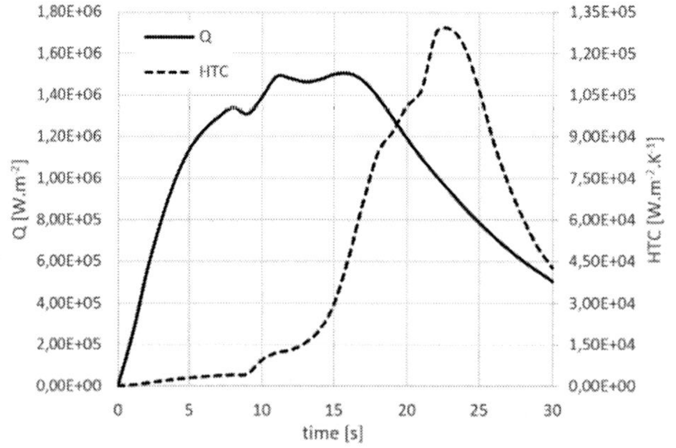

Fig. 10. Heat flux and HTC at pouring temperature 710 °C and pressure 100 MPa

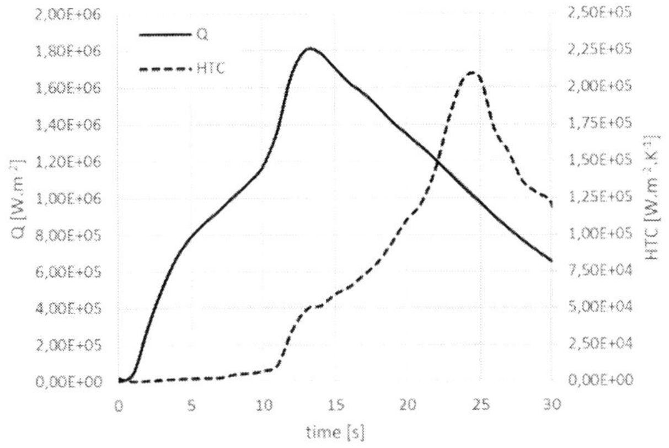

Fig. 11. Heat flux and HTC at pouring temperature 710 °C and pressure 150 MPa

From the results of the measurements it is clear, that there has been a significant increase of the heat flow and the heat transfer coefficient, which result was in faster cooling of the casting. The amount of heat Q that passes through the isothermal surface S is defined by multiplying the heat flux q and time τ

$$Q = q \cdot S \cdot \tau \tag{4}$$

The amount of heat that has been cast from the cast into the mold for 30 seconds is shown in fig. 12.

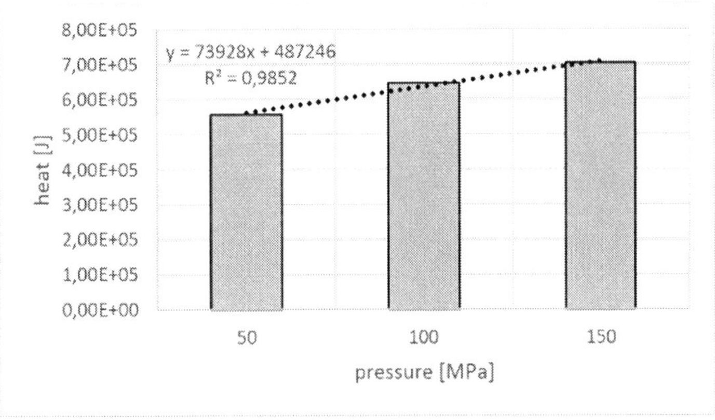

Fig. 12. Amount of heat transmitted to the casting

Conclusions

Heat flux in the gravity cast sample increased within 5 seconds after pouring. Shrinking of the casting resulted in the formation of a gap at the casting-mold interface. The formed gap caused reduced heat flux from the casting. In the unaffected sample, the heat flux at the onset of pouring increased to the value of $5.75.10^5$ $W.m^{-2}$. At time 7 seconds after pouring, heat flux through the gap started to apply. The heat flux value increased up to $7.22.10^5$ $W.m^{-2}$. Heat transfer coefficient reached the value of $1.15.10^3$ $W.m^{-2}.K^{-1}$ at the onset of pouring, only to increase to the value of $1.74.10^3$ $W.m^{-2}.K^{-1}$ when applying heat transfer through the gap. For technological reasons, pressure was applied 8 seconds after filling the mold. All samples affected by pressure experienced a significant increase in both the heat flux and heat transfer coefficient at the casting-mold interface. At the casting temperature of 710 °C, the increase in HTC was 187-fold at 50 MPa, 317-fold at 100 MPa, and 525-fold at 150 MPa. Heat transfer at the casting-mold interface significantly increases with the increasing difference in heat contents of the poured material and the mold, and also with the increasing applied pressure.

It was observed the amount of heat that has passed into the casting mold. A direct relationship between the amount of heat and the applied pressure was observed.

This work was created in the framework of the grant projekt VEGA N°1/0494/17. The authors acknowledge the grant agency for support.

References

1. A. F. Ilkhchy, M. Jabbari, P. Davami, *Effect of pressure on heat transfer coefficient of A356 aluminium alloy*. [online] [cited by 2016-03-08] ISSN 0735-1933, http://www.sciencedirect.com/science/article/pii/S0735193312000711, (2012)
2. J. O. Aweda, M. B. Adeyemi, *Experimental determination of heat transfer coefficients during squeeze casting of aluminium*. Journal of Materials Processing Technology **209** (3), 1477-1483 (2009)
3. R. Pastirčák, J. Ščury, M. Brůna, D. Bolibruchová, *Effect of Technological Parameters on the AlSi12 Alloy Microstructure During Crystallization Under Pressure*. Archives of foundry engineering **17** (2), 75-78 (2017)
4. R. Pastirčák, J. Ščury, *Effect of technological parameters on microstructure in alloy AlCu4Ti using squeeze casting*. American Institute of Physics Publising, AIP conference proceedings **1745** (2016)
5. A. Iyer, *Squeeze casting: The future*. [online]. [cited by 2017-02-11] http://issinstitute.org.au/wp-content/media/2011/05/ISS-FEL-REPORT-A-IYER-low-res.pdf (2011)
6. D. Bolibruchová, J. Macko, M. Brůna, *Elimination of negative effect of Fe in secondary alloys AlSi6Cu4 (EN AC 45 000, A 319) by nickel*. Archives of metallurgy and materials **59** (2), 717-721 (2014)
7. P. Pecháč, M. Sága, *Controlling of local search methods' parameters in memetic algorithms using the principles of simulated annealing*. Procedia Engineering **136**, 70-76 (2016)
8. D. Bolibruchová, L. Richtarech, S.M. Dobosz, K. Major-Gabrys, *Utilisation of Mould Temperature Change in Eliminating the Al5FeSI Phases in Secondary AlSi7Mg0.3 Alloy*. Archives of Metallurgy and Materials **62** (1), 357-362 (2017)
9. P. Kopas, L. Jakubovičová, M. Vaško, M. Handrik, *Fatigue resistance of reinforcing steel bars*. Procedia Engineering **136**, 193-197 (2016)
10. R. Koňar, M. Mičian, *Ultrasonic Inspection Techniques Possibilities for Centrifugal Cast Copper Alloy*. Archives of foundry engineering **17** (2), 35-38 (2017)

Analysis of stress and deformation fields of shape complex beams

Miroslav Pástor[1], *Pavol* Lengvarský[1,*]

Department of Applied Mechanics and Mechanical Engineering, Faculty of Mechanical Engineering, Technical University of Košice, Letná 9, 042 00 Košice, Slovakia

Abstract. In this paper is investigated the analysis of stress and deformation fields of shape complex beams. The shape complex beams are made from load-bearing sheet (trapezoidal sheet) circumferentially connected with strips of sheet metal, these beams are a substitute for more complex and heavier beams. The numerical analysis with static load are performed for these beams. The effect of three different types of connections between load-bearing sheet and strips of sheet metal is investigated. The first type of connection is represented by the trapezoidal sheet perfectly welded to the strips of sheet metal, the second type of connection is represented by the trapezoidal sheet welded to the strips of sheet metal only on the base sides of the trapezoidal sheet. The third one is represented by point welds. The stress and deformation fields for all types of the connections are compared and the suitable variant is chosen.

Keywords: stress, deformation, beam, contact, container

1 Introduction

Start of development of thin-walled structures (beams) can be traced back to the 19th century, due to the rapid development of the industry, it was necessary to create large-scale buildings, structures and facilities, when conventional design methodologies together with traditional materials could not be always applied. Various types of beams can be made from structural elements, from full to various hybrid beams. The use of a specific beam depends on many static, structural, manufacturing factors and requirements. The current trend is to use different types and shapes of beams and create from them a complex truss. Nowadays, the trend of optimal utilization of design and material is used for mass and cost reduction. This creates opportunities for new ideas of design structural elements. New ideas can also include trapezoidal beams, where the trapezoidal sheet is connected circumferentially with the strips of sheet metal. These beams can be used as part of roof construction, which can be used to protection of shipping containers before the different weathers [1].

The carrying frame of the containers is made from welded closed cylindrical tubes with the thickness 3 mm. The covering of container is made from panels of profiled alloy steel sheets with thickness from 1.5 mm to 2 mm [2, 3]. The alloy steel is also called Corten steel, it is characterized by high resistance to weathering. The containers (Fig. 1) are

* Corresponding author: pavol.lengvarsky@tuke.sk

Reviewers: *Radim* Halama, *Marcin* Kubiak

designed for ship transportation, they are the cheapest and fastest way to create inexpensive warehouses. They can to serve as low-cost modular skeletons for creating residential, non-residential, underground and above-ground objects. The ship containers can be use as mobile polyfunctional objects, halls offices, objects for housing or business etc. (Fig. 2). The main advantage of shipping containers is they higher resistance to weathering, fire resistance, safety and lock ability. The containers are stackable, mobile without needing to unload contents.

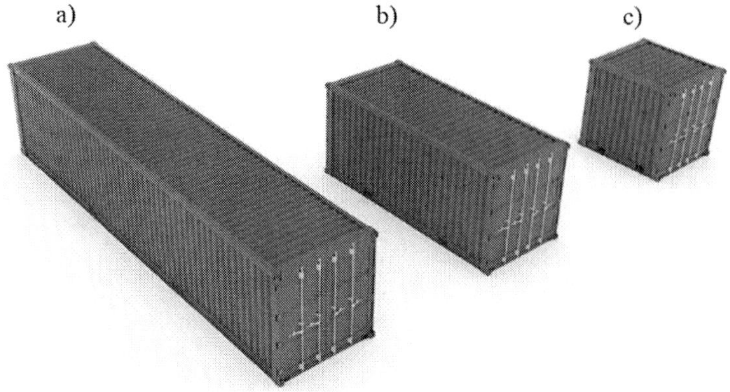

Fig. 1. Ship container a) 40´DC with length 12 m, b) 20´DC with length 6 m and c) 10´DC with length 3 m

Fig. 2. Mobile polyfunctional objects

The structural design of the shape complex beams is investigated in this paper. The beams are carrier elements of the roof structure with dimensions identical with ship container (20´DC) of length 6058 mm, width 2438 mm, height 2591 mm, mass 2200 kg and load capacity 24 800 kg [2, 3]. The design of the beams is based on the required beam length, the minimum height of the lower end and the minimum slope. The shape complex beams are made from load-bearing sheet (trapezoidal sheet) circumferentially connected with strips of sheet metal, these beams are a substitute for more complex and heavier beams. The effect of three different types of connections between load-bearing sheet and strips of sheet metal is investigated. Modern means of numerical modelling (finite element method) are used for design and optimization the beam [4-6].

2 Finite element model of shape complex beam

The shape and the basic dimensions of the designed beam are shown in Fig. 3. The beam web is made from a trapezoidal steel sheet with thickness 0.7 mm and the beam flanges are made from strips of sheet metal with thickness 5 mm. The ends of beam are closed with the same strips of the sheet metal. The reinforced strips of sheet metal are used 100 mm from the ends of the beam.

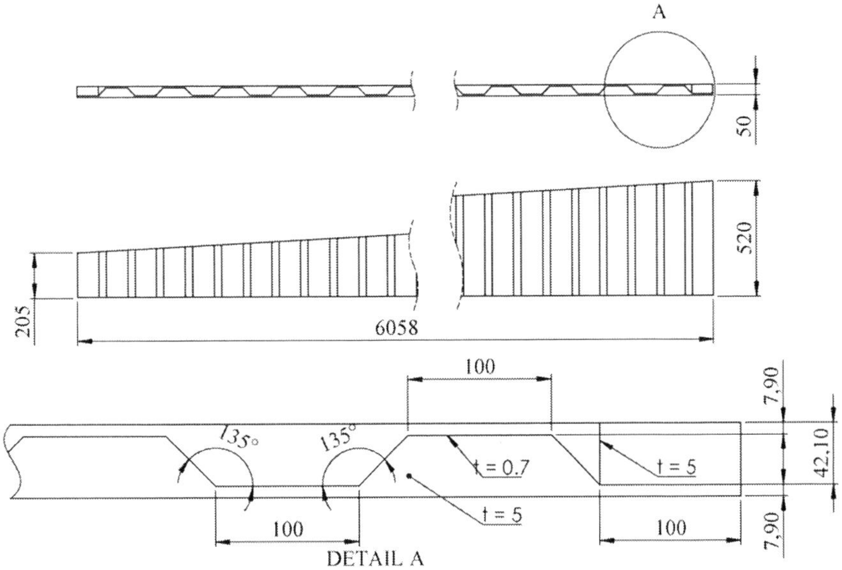

Fig. 3. Basic dimensions of designed beam

The model of beam is prepared in CAE software, where the material properties, contacts between parts, boundary conditions, loads are applied on the model. The beam is made from the steel *KOHAL 280* with the yield strength 280 MPa and the maximum tensile strength 410 MPa. The steel is suitable for non-structural presswork and for forming. The three variants of the boundary conditions are applied to connection the trapezoidal sheet with circumferential strips of sheet metal. The first type of connection is represented by the trapezoidal sheet perfectly welded to the strips of sheet metal (Fig. 4a – red dashed line). The second type of connection is represented by the trapezoidal sheet welded to the strips of sheet metal only on the base sides of the trapezoidal sheet (Fig. 4b). The third one is represented by the point welds (Fig. 4c). The connections are made only as bonded contact because the effect of the welds and stresses in the welds are not studied.

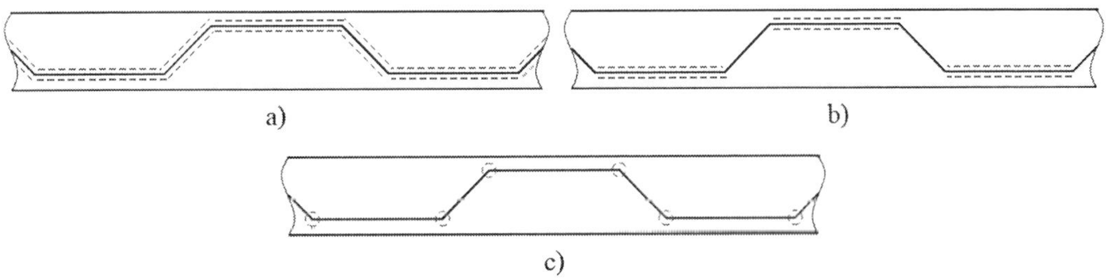

Fig. 4. Three variants of the connection of the trapezoidal sheet with strips of sheet metal

The top strip of sheet metal is loaded by the pressure with value $P = 25000$ Pa, the load is computed using the Slovak Technical Standard (STN EN 1991-1-3) [7]. The self-weight is applied too to the beam. The beam is on both ends fastened with a bolt and is supported by steel plate as part of container. The degree of freedom in the direction perpendicular to the beam on the top flange is taken every 1000 mm, this restriction represents the connection of the beam with the next beam and the support for the roofing sheet. Finally, the numerical computations are done for the finite element model consist from shell elements with maximum size of element 20 mm. The fields of equivalent stresses for the three variants of the connection of the beam web with the beam flanges are shown in Figs. 5–7. The detailed views of fields of equivalent stresses of beam for all variants are shown in Figs. 8–10.

Fig. 5. Field of equivalent stresses of beam for the first variant

Fig. 6. Field of equivalent stresses of beam for the second variant

Fig. 7. Field of equivalent stresses of beam for the third variant

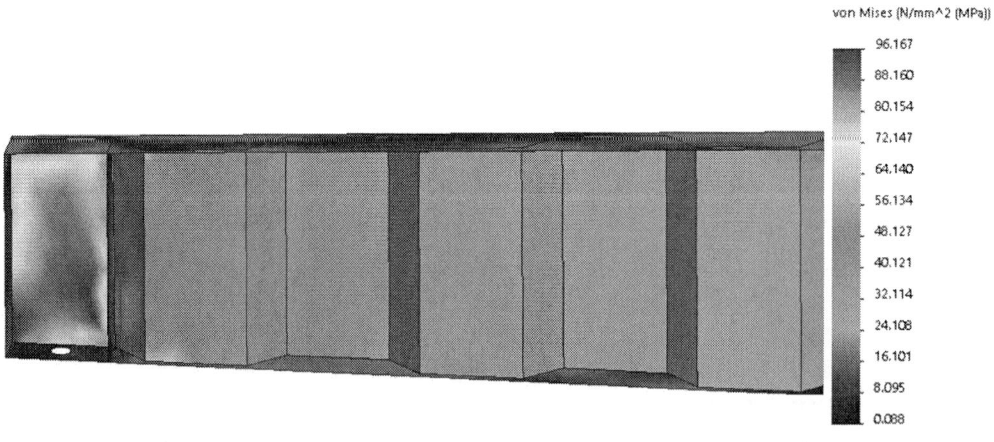

Fig. 8. Detail of field of equivalent stresses of beam for the third variant

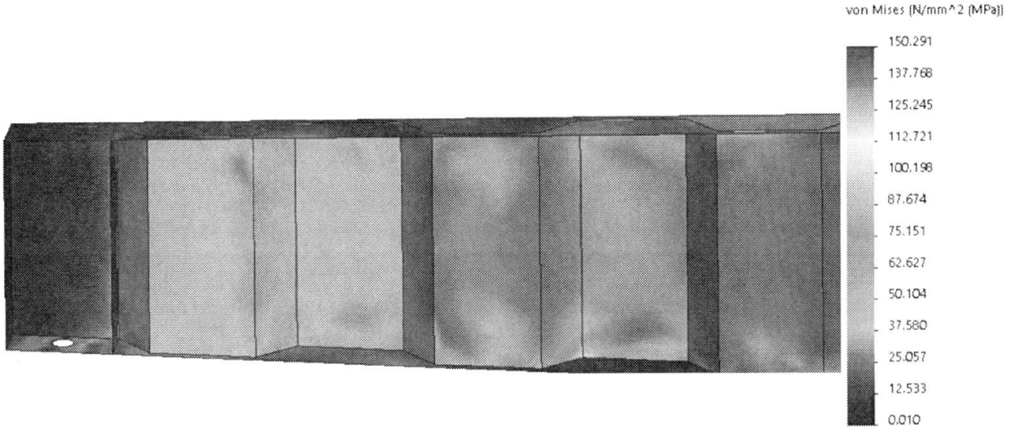

Fig. 9. Detail of field of equivalent stresses of beam for the third variant

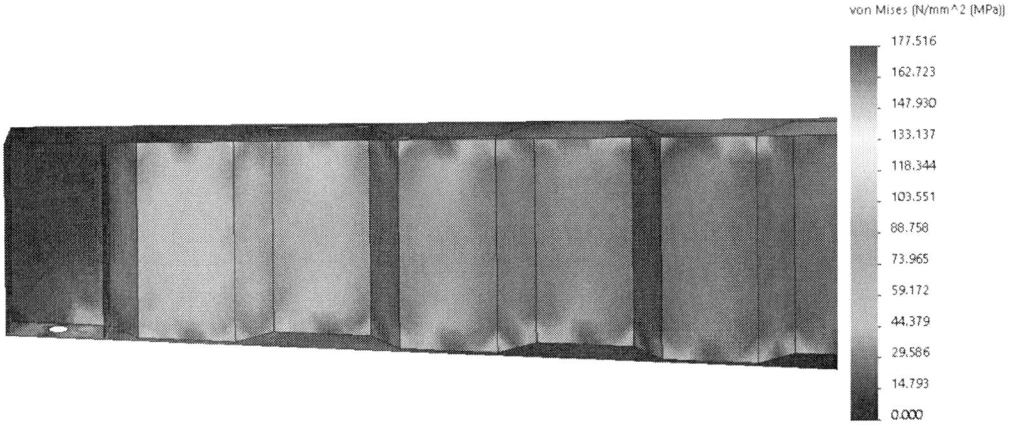

Fig. 10. Detail of field of equivalent stresses of beam for the third variant

The displacements of points of the structure in *Z* direction (perpendicular to the ground) for the first variant are shown in Fig. 11. The shape of the second and third variants after deformation is the same as the first variant, only the value of the maximum displacement is different.

Fig. 11. Displacements of points of beam for the first variant

The computed maximum equivalent stresses and displacements in the *Z* direction for the all variants are given in Table 1.

Table 1. Computed parameters of beam

Variant	Equivalent stress σ_{max} [MPa]	Displacement in *Z* direction w_{max} [mm]
1.	96.167	4.189
2.	150.291	5.656
3.	177.516	4.673

For the comparison, the numerical analysis of the beam with the trapezoidal sheet thickness equal 1 mm is performed for the all variants. The all other dimensions, parameters, boundary conditions and loads are considered to be the same. The results of computations are given in the Table 2.

Table 2. Computed parameters of beam

Variant	Equivalent stress σ_{max} [MPa]	Displacement in Z direction w_{max} [mm]
1.	87.552	3.851
2.	141.918	5.070
3.	165.505	4.458

Based on the comparison of the achieved results from Tables 1 and 2 can be said, that the all stresses are below as the Yield strength. From the point of view of the maximum equivalent stresses, the first variant is the best variant and the second variant is the worst variant. With the change of the thickness of the trapezoidal sheet from 0.7 mm to 1 mm, the investigated parameters decrease but the total weight of the beam increases from 39.926 kg to 45.508 kg, it represents 13% increase of mass.

Conclusion

The numerical computations of the three variants shape complex beam were investigated in this paper. The beam web was made from the trapezoidal sheet and the flanges were made from the strips of sheet metal. The all variants had the same dimensions, boundary conditions and loads. The connections between the beam web and flanges were varied for the variants. The first type of connection was represented by the trapezoidal sheet perfectly welded to the strips of sheet metal, the second type of connection was represented by the trapezoidal sheet welded to the strips of sheet metal only on the base sides of the trapezoidal sheet. The third one was represented by the point welds. The three variants of connection were used in the numerical simulations but in real carrier structures can be used only the first two variants. The computations were done for the beam web with the thickness 0.7 mm and 1 mm. With the change of the thickness of the beam web, the equivalent stresses decreased by 10 MPa and the mass increased by 13%. According to weld technology is better use the beam web with thickness 1 mm and the effect of the weight of the weld can be negligible.

This research was financially supported by Slovak Grand Agency through the grant number VEGA 1/0731/16 and KEGA 049TUKE-4/2017.

References

1. M. Pástor, J. Bocko, T. Kula, C.S. Roszkos, *Possibilities of Optimizing the Shape and Geometry of Castellated Beams Using Numerical and Experimental Modeling*, Am. J. Mech. Eng. **4** (7), 357-362 (2016)
2. *STN 26 9340*, ISO freight containers series 1. Specification and testing. Basic regulations.
3. *STN 29 9344 – ISO 1161*, Containers ISO series 1. Corner fitting.
4. P. Paczos, P. Wasilewicz, E. Magnucka-Blandzi, *Experimental and numerical investigation of five-layered trapezoidal beams*. Compo. Struct. **145**, 129-141 (2016)
5. J. Nagler, *Pitting corrosion investigation of cantilever beams using F.E. method*. Acta Mech. Slovaca. **16** (4), 12-19 (2012)
6. O. Tusnina, *A finite element analysis of cold-formed Z-purlins supported by sandwich panels*. Appl. Mech. Mater. **467**, 398-403 (2013)
7. *STN EN 1991-1-3, Eurocode 1*. Actions on structures, Part 1-3: General actions. Snow loads.

Numerical simulation of blanking process

Peter Pecháč[1,*], *Milan* Sága[1]

[1]University of Žilina, Faculty of Mechanical Engineering, Department of Applied Mechanics, Univerzitná 8215/1, 010 26 Žilina, Slovak Republic

Abstract. This paper presents numerical simulation of blanking process for cold-rolled steel sheet metal. The problem was modeled using axial symmetry in commercial finite element software ADINA. Data obtained by experimental measurement were used to create multi-linear plastic material model for simulation. History of blanking force vs. tool displacement was obtained.

Keywords: blanking, plasticity, axial symmetry, rupture, contact, nonlinear static analysis

1 Introduction

Blanking is the most widely used forming operation. It is used both for the preparation of semi-finished products and cutting of metal parts either for final use or as preparation for other technological operations [1, 2]. Simple analytical calculations are usually used during preparation of serial manufacturing [3, 4]. These analytical calculations are sufficient for determining basic parameters of blanking process such as maximal blanking force, however when problems occur during the manufacturing process the solution is often complicated and can be calculated only by using advanced numerical methods [5]. This paper presents a comparison of blanking process for a steel washer with and without v-ring indenter using finite element method. Use of v-ring indenter improves quality of resulting surface and also decreases the necessary blanking force.

2 Material data and model

Material data are one of the most important inputs in numerical simulation [6-8]. Precise results can be obtained only when precise material data in combination with proper material model are used [9-11]. Material data used for the presented simulations were obtained by tensile test using sheet metal samples of 1 mm thickness [12, 13]. The sheet metal samples were made from structural steel S235J.

The data obtained from tensile test were used to create multi-linear elastic-plastic material with isotropic hardening. Young's modulus E = 210000 MPa, Poisson's ratio μ = 0.3. The stress-strain data in true stress representation were used. Rupture of elements

* Corresponding author: peter.pechac@fstroj.uniza.sk

Reviewers: *Jozef* Bocko, *Tomasz* Domański

occurred at the last stress-strain point. The material model in true stress representation can be seen on Fig. 1.

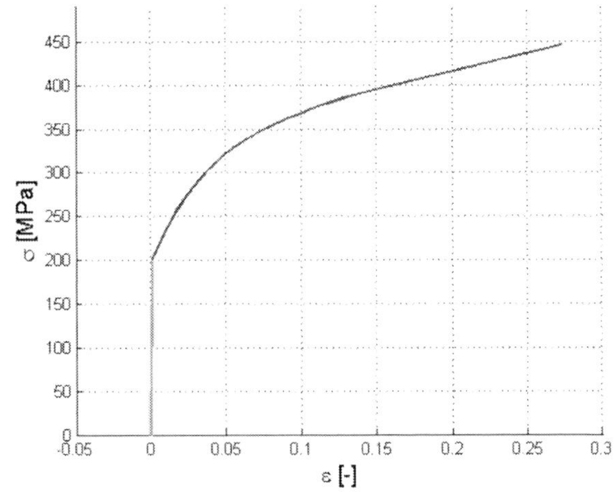

Fig. 1. Stress-strain curve for multi-linear material model in true stress representation

3 FEM model

The problem was modelled using axial symmetry in commercial FEM software ADINA. The punch, blank holders, v-ring indenter and die were modelled as rigid bodies [14]. Rigid target contact was used for contact of punch blank holders and die with sheet metal [15-17]. Friction coefficient of 0.2 was used for all contacts. The sheet metal was meshed using 2D 4 node elements, with very fine mesh in the area near punch.

In the first phase the sheet metal was pressed by blank holders using force of 62400 N. The second stage involved displacement of punch, which was prescribed by time function. The punch moved 5 mm downward and then 6 mm upward. In the model with v-ring indenter, the indention operation was performed prior to all other operations. The thickness of sheet was 2 mm and depth of dent created by v-ring indenter was 0.5 mm.

The FEM model with boundary conditions for blanking without v-ring indenter can be seen on Fig. 2:

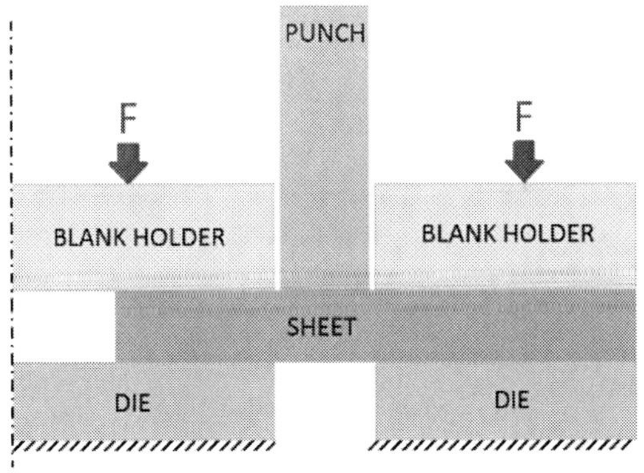

Fig. 2. FEM model without v-ring indenter

The FEM model with boundary conditions for blanking with v-ring indenter can be seen on Fig. 3.

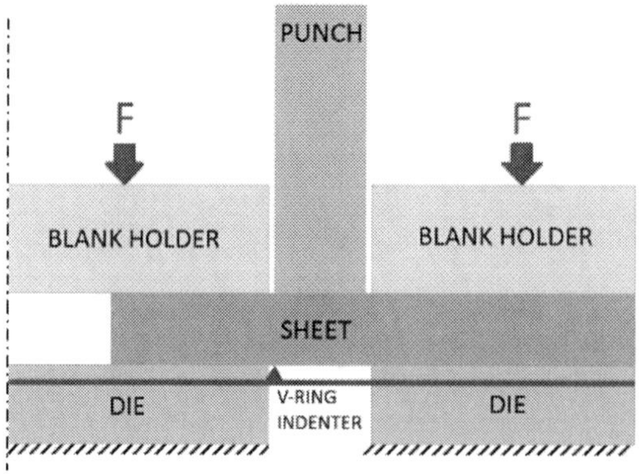

Fig. 3. FEM model with v-ring indenter

4 Results of numerical simulation

The results of FEM simulation can be seen on the following Figs. 4-6:

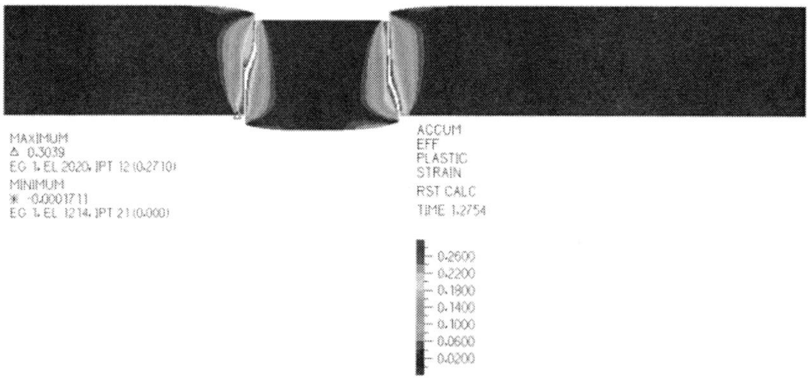

Fig. 4. Accumulated effective plastic strain – model without v-ring indenter

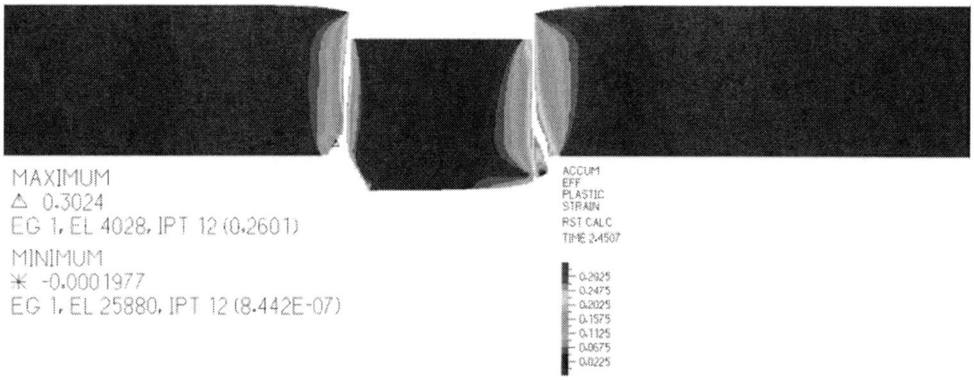

Fig. 5. Accumulated effective plastic strain – model with v-ring indenter

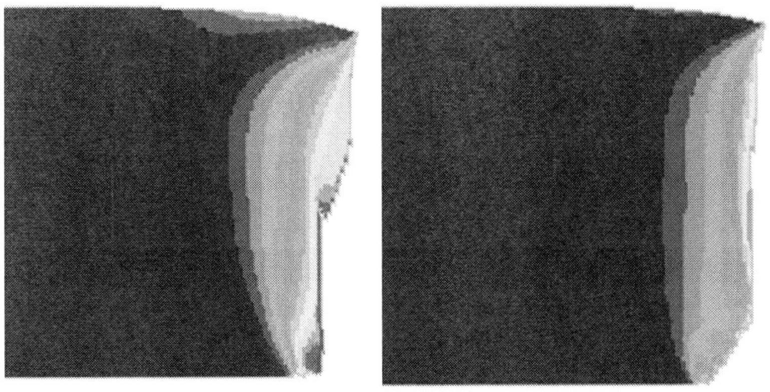

Fig. 6. Comparison of resulting surface – without v-ring indenter (left), with v-ring indenter (right)

The history of applied blanking force can be seen on Figs 7 and 8:

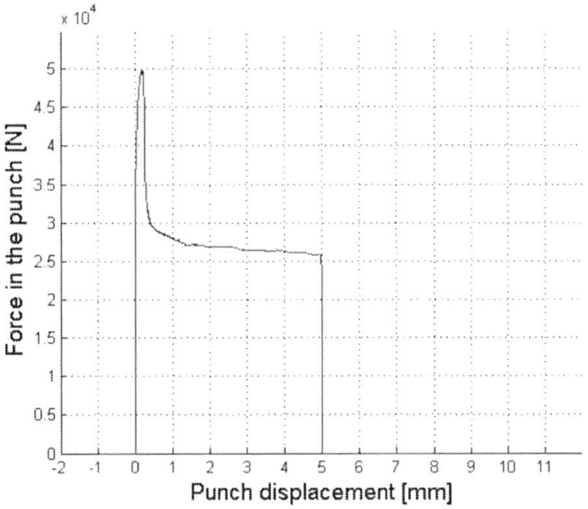

Fig. 7. Blanking force vs. punch displacement – without v-ring indenter

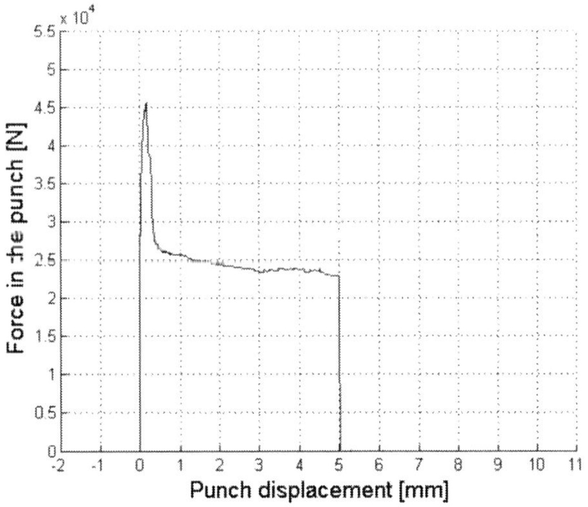

Fig. 8. Blanking force vs. punch displacement – with v-ring indenter

Conclusion

The numerical simulation confirmed that the use of v-ring indenter can benefit the blanking process. The maximum blanking force without v-ring indenter was 49921 N. The maximum blanking force with v-ring indenter was 45668 N. The use of v-ring indenter decreased the required blanking force by 8.5%. The improvement in terms of blanking force was only minor, however the use of v-ring indenter improved the quality of the resulting surface as can be seen on the Fig. 6.

The presented problem required nonlinear finite element solution which is much more complex than simple analytical solution used for calculation of blanking force. On the other hand, the results of FEM solution offer much more than information about maximal blanking force. The results of FEM simulation also include the shape and quality of resulting surface after blanking. Numerical simulation can be used to find optimal parameters for the blanking process and it is also more cost efficient than using physical prototypes.

This work has been supported by KEGA project No. 015ŽU-4/2017.

References

1. M. Blatnický, M. Štauderová, J., Dižo, *Numerical Analysis of the Structure Girder for Vehicle Axle Scale Calibration*. Procedia Engineering **177**, 510-515 (2017)
2. M. Dudziak, G. Domek, A. Kołodziej, K. Talaśka, *Contact Problems Between the Hub and the Shaft with a Four-Angular Shape of Cross-Section for Different Angular Positions*. Applied Mechanics and Materials **816**, 54-62 (2015)
3. P. Kopas, L. Jakubovičová, M. Vaško, M. Handrik, *Fatigue resistance of reinforcing steel bars*. Procedia Engineering **136**, 193-197 (2016)
4. M. Łazarska, T. Woźniak, Z. Ranachowski, A. Trafarski, G. Domek, *Analysis of acoustic emission signals at austempering of steels using neural networks*. Metals and Materials International **23** (3), 426-433 (2017)
5. J. Adamus, P. Lacki, W. Więckowski, *Numerical simulation of the fine blanking process of sheet titanium*. Archives of Metallurgy and Materials **56** (2), 431-438 (2011)
6. J. Gerlici, T. Lack, *Railway wheel and rail head profiles development based on the geometric characteristics shapes*. Wear - An International Journal on the science and technology of friction, lubrication and wear **271** (1-2), 246-258 (2011)
7. M. Macko, J. Flizikowski, Z. Szczepański, K. Tyszczuk, G. Śmigielski, A. Mroziński, J. Czerniak, A. Tomporowski, *CAD/CAE applications in mill's design and investigation*. Proceedings of the 13th International Scientific Conference on Computer Aided Engineering, Lecture Notes in Mechanical Engineering, Chapter **35**, 343-351 (2017)
8. J. Czerniak, D. Ewald, M. Macko, G. Śmigielski, K. Tyszczuk, *Approach to the Monitoring of Energy Consumption in Eco-grinder Based on ABC optimization*. BDAS 2015, Communications in Computer and Information Science **521**, 516-529 (2015)
9. M. Handrik, M. Vaško, P. Kopas, V. Mózer, *The linear and nonlinear stability loss of structures due to thermal load*. Procedia Engineering, **136**, 359-364 (2016)
10. P. Novak, V., Dekys, *Induction heating of inner rolling bearing ring in ANSYS*. Manufacturing Technology **15** (5), 881-885 (2015)

11. M. Žmindák, P. Novák, V. Dekýš, *Analysis of bond behaviour in strengthened reinforced concrete beam with carbon fiber reinforced polymer lamella*. MATEC Web of Conferences **107** (00045), (2017)
12. J. Vavro, J. Vavro Jr., P. Kováčiková, P. Kopas, M. Handrik, *Simulation and analysis of defect distribution in passenger car tire under dynamic loading*. Applied Mechanics and Materials **611**, 544 – 547 (2014)
13. J. Vavro Jr., J. Vavro, P. Kováčiková, P. Kopas, M. Handrik, R. Bezdedová, *Numerical analysis of stress states for graphitic cast iron structures*. Applied Mechanics and Materials **611**, 252-255 (2014)
14. M. Sapieta, V. Dekýš, A. Sapietová, *Thermal-stress analysis of beam loaded by 3 point bending*. Procedia Engineering **136**, 216-219 (2016)
15. T. Lack, J. Gerlici, *A modified strip method to speed up the normal stress between wheel and rail calculation*. Applied mechanics and materials **486**, 359-370 (2014)
16. M. Macko, J. Flizikowski, K. Tyszczuk, G Śmigielski, I. Rojek, A. Mroziński, J. Czerniak, A. Tomporowski, *Mechatronic implementations used for the effective monitoring and data acquisition of the shredding process*. ASME 2016 International Design Engineering Technical Conferences and Computers and Information in Engineering IDETC 2016, Charlotte, North Carolina, USA. August 21-24, (2016)
17. A. Sapietová, V. Dekýš, M. Sapieta, P. Pecháč, *Application of computational and design approaches to improve carrier stability*. Procedia Eng. **96**, 410-418 (2014)

Numerical analysis of laser welding with consideration analytical methods of determining phase transformations and mechanical properties of welded joint

Wiesława Piekarska[1], *Dorota* Goszczyńska-Króliszewska[1,*], *Zbigniew* Saternus[1], *Tomasz* Domański[1], *Marcin* Kubiak[1]

[1]Institute of Mechanics and Machine Design Foundations, Czestochowa University of Technology, Dabrowskiego 73, 42-200 Częstochowa, Poland, EU

Abstract. The numerical analysis of laser welding process with consideration analytical methods determining phase transformations and mechanical properties of welded joints are presents in this paper. The analytical CCT diagram and final structural composition of S355 steel are presented. The empirical relations presents in paper are determined by chemical compositions investigated steel and cooling rate between temperatures 800-500°C ($t_{8/5}$). Phase composition and mechanical properties each of structures of steel in weld and heat affected zone (HAZ) are determined on the basis of analytical methods. Laser welded flat is used in numerical simulations in ABAQUS. Mathematical modes of volumetric welding source are used in the calculations. Temperature fields, shape and size of melting zone for selected points in the cross-section of the joint are determined on the basis of thermal cycles obtained numerical.

Keywords: CCT diagram, analytical methods, heat affected zone, numerical modelling, laser welding

1 Introduction

The increase of usefulness of laser welding technology in industries causes an increase of technological requirements obtained to concern with obtaining good properties of welding joints. Experimental research are executed in order to understand the phenomena in welding process of laser beam. Experimental research are expensive and low-labor so the research are focused on mathematical and numerical modeling of arising phenomena [1, 2].

Numerical analysis of laser welding with consideration of analytical methods of analysis of phase transformations and mechanical properties of welded joints were presented. The analytical models concern to predicting of the structure of heat-affected zone (HAZ) on the basis of chemical composition. They can be used to a supplement to simulation results and for preliminary analysis of the material intended for welded

* Corresponding author: goszczynska@imipkm.pcz.pl

Reviewers: *Zuzana* Murčinková, *Alžbeta* Sapietová

constructions. Empirical relationships presented in paper were determined by the chemical composition of analyzed steel and cooling rate between 800 - 500°C ($t_{8/5}$). Phase composition and mechanical properties of each phases in weld and HAZ were predicted on the basis of analytical methods. Laser welded flat is used in numerical simulations in ABAQUS. Mathematical model of volumetric welding source is used in the calculations. Mechanical properties of welded joint were performed on the basis of numerically obtained phase composition and analytically obtained mechanical properties of each phases.

2 Analytical methods of determining phase transformations and mechanical properties of welded joint

Material during welding changes its thermo-physical and mechanical properties due to changing temperature field in a wide range. The most various material properties occur in the heat affected zone (HAZ). In this area a large variety of structures occurs conditioned by thermal cycles as a result of phase transformations in solid state.

Analytical models concern prediction of HAZ structure on the basis of chemical composition of steel, elaboration of simplified CCT diagrams and estimation of mechanical properties of HAZ of welded joints [3-7]. Analytical formulas are determined using the statistical analysis of results of experimental studies for a certain group of materials. The analytical relationships used in this paper concern wieldable low carbon and high strength steels, which included the S355 steel. Chemical composition of analyzed steel was shown in Table 1. Methods to the building simplify CCT diagrams for different kind of steel were by authors [3, 4, 8].

Table 1. Chemical composition of S355 steel [%]

Steel	C	Mn	Si	P	S	Al	Cr	Ni
S355	0.19	1.05	0.20	0.028	0.02	0.006	0.08	0.11

Dilatometric research on high strength steel was performed in order to verify the CCT diagram of austenite transformation obtained by using analytical relationships and to evaluate their usefulness. Dilatometric research was performed on the DIL805 Bahr Thermoanalyse GmbH dilatometer. Austenitization temperature T_A=1200°C and heating rate 100 K/s were assumed in dilatometric research as well as different cooling rates simulating thermal cycles in welding [9]. Analitycally and experimentally obtained CCT diagrams and phase volume fractions were presented in Figure 1 and 2.

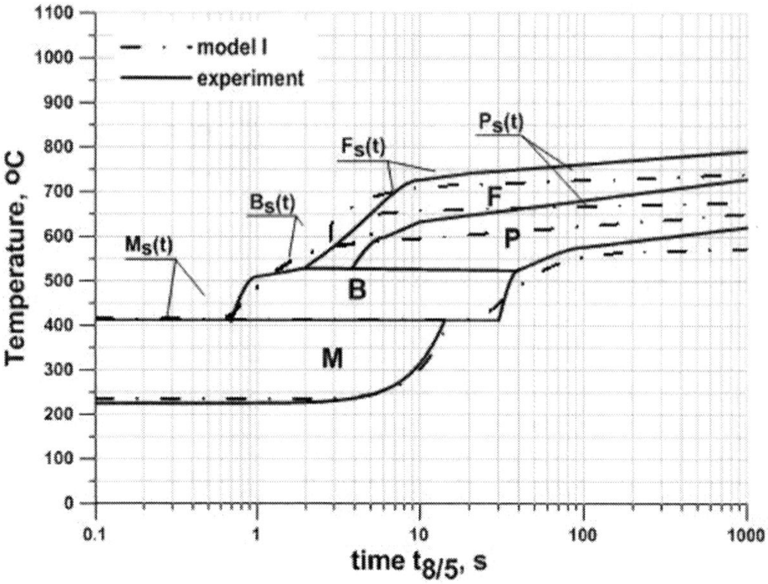

Fig. 1. Analytically and experimentally CCT diagram of S355 steel

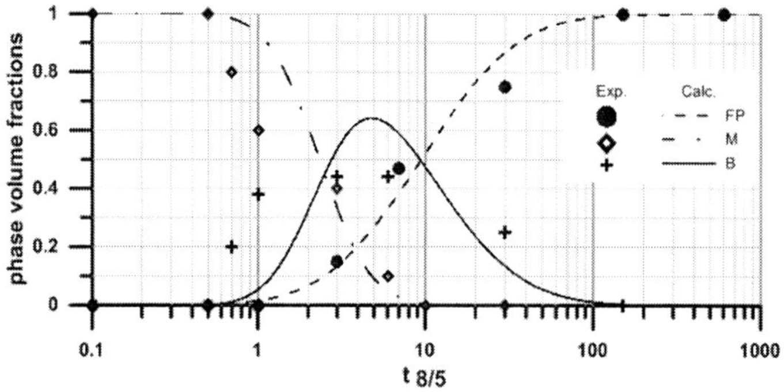

Fig. 2. Analytically and experimentally phase volume fraction of S355 steel

Mechanical properties of HAZ can be determined from the structural composition and mechanical properties of each structure. If the phase composition in the heat affected zone is known, more specifically contribution of each structure and properties of structural components W_i, there is a possibility to approximately predict properties of the entire zone [3, 4, 6].

$$W = \sum_{i=M,B,F,P} W_i \eta_i \quad (1)$$

where: W can be hardness, yield strength, tensile strength, impact strength, elongation and contraction.

Many dependencies are discussed in the literature [1, 2, 5, 6], which can determine mechanical properties such as: R_e, R_m, A_5, Z, KCU and HV. Property of each phase are defined on the base of chemical composition, whereas contributions of each phase are defined on the base of experiment or numerical simulations.

Yield strength (R_{ei}) and Tensile strength (R_{mi}) of each phase: ferrite-pearlite, martensite and bainite can be determined with a high probability by Kasatin and Seyffart equations [4]:

$$\begin{aligned} Re_{FP} &= 187 + 92C + 47Mn + 90V \\ Re_M &= 602 + 2150C + 500Mo \\ Re_B &= 500 + 460C - 120C^2 + 150V + 360Mo \end{aligned} \tag{2}$$

$$\begin{aligned} Rm_{FP} &= 297 + 1360C + 60Mn + 140V \\ Rm_M &= 798 + 3215C \\ Rm_B &= 590 + 960C + 39.7Mn + 200V \end{aligned} \tag{3}$$

Hardness *(HV_i)* of individual structural components can be defined by [3] as a function of chemical composition and velocity $v_{8/5}$ ($v_{8/5} = (800-500)/t_{8/5}$):

$$\begin{aligned} HV_{FP} &= 42 + 223C + 53Si + 30Mn + 12.6Ni + 7Cr + 19Mo + \\ &+ (10 - 19Si + 4Ni + 8Cr + 130V)\log v_{8/5} \\ HV_M &= 127 + 959C + 27Si + 11Mn + 8Ni + 16Cr + 21\log v_{8/5} \\ HV_B &= -323 + 185C + 330Si + 153Mn + 65Ni + 144C + \\ &+ 191Mo + (89 + 53C - 55Si - 22Mn - 10Ni - 20Cr - 33Mo)\log v_{8/5} \end{aligned} \tag{4}$$

Elongation (A_i) and contraction (Z_i) for structure components can be defined by [3] as a function of steel chemical composition and time *t*, where *t* is the cooling time between temperatures 800°C and 500°C:

$$\begin{aligned} A_M &= 12.267C^2 - 1.5Mn + 0.76\ln t \\ A_B &= 21.3 - 35.6C - 4.0Mn - 5.0V + 1.84\ln t \\ A_{FP} &= 36.5 - 127C + 153C^2 - 1.16Mn + 8.0V + 0.66\ln t \end{aligned} \tag{5}$$

$$\begin{aligned} Z_M &= 48.5 - 158C + 116C^2 - 0.98\ln t \\ Z_B &= 53.3 - 132C + 103C^2 - 5.1Mn - 10V + 3.4\ln t \\ Z_{FP} &= 65.4 - 88C - 82C^2 - 6.7Mn + 18V + 0.6\ln t \end{aligned} \tag{6}$$

3 Exemplary of prediction of mechanical properties in welded joint

Numerical analysis of thermal phenomena occurring during laser beam welding was carried out using the commercial Abaqus FEA program based on the finite element method [10, 11]. The laser beam butt-welding of S355 steel sheets with dimensions 150x15x4 mm was considered. Variables with temperature thermo-physical properties and properties such as: laser beam power $Q = 3.2\ kW$, welding speed $v = 1.0$ m/min, the beam radius $r_0 = 1\ mm$, depth penetration $s = 7$ mm were used in numerical calculations.

The analysis of thermal phenomena is made on the basis of the numerical solution of energy conservation equation together with Fourier law [12-15] using finite element method (FEM). The temperature field expressed in the criterion of weighted residuals method is described by the following equation:

$$\int_V \rho \frac{\partial U}{\partial t} \delta T\, dV + \int_V \frac{\partial \delta T}{\partial x_\alpha} \cdot \left(\lambda \frac{\partial T}{\partial x_\alpha} \right) dV = \int_V \delta T\, q_V dV + \int_S \delta T\, q_S dS \tag{7}$$

where: λ is a thermal conductivity [W/mK], U is the internal [J/kg], q_v is the capacity of the laser beam power [W/m^3], $T = T(x,t)$ is a temperature [K], q_s is a density of heat flux [W/m^2], T is a partial differential of T.

Equation (7) expressed in the criterion of weighted residuals method and is completed by the initial condition and boundary conditions of Dirichlet, Neumann and Newton type with the heat loss due to convection and radiation [8]. To describe distribution of power of laser beam was used Cylindrical-Power-Normal model of heat source (C-P-N) [16]:

$$Q_2(r,z) = \frac{kK_z Q_L}{\pi\left(1-e^{(K_z s)}\right)} e^{-\left(kr^2+K_z z\right)}\left(1-u(z-s)\right) \tag{8}$$

where: Q_L is a power of the heat source [W], r_0 is radius of laser beam [m], $r=\sqrt{x^2+y^2}$ is current radius [m], $K_z = 3/s$ is exponent of a power of heat [m^{-1}], $k = 3/r_0^2$ is source concentration factor [m^{-2}], and s is depth penetration of beam [m], t $u(z-s)$ is Heaviside function.

During the numerical simulation in the ABAQUS FEA program, two additional numerical procedures, HETVAL and UEXPAN, were used for the determination of volume fractions. In addition, CCT and CTA diagram of S355 steel were interpolated In UEXPAN procedure. Cross section of analyzed welding joint for selected distance in the welding line (x=75 mm) was presented in Figure 3. In addition, was marked boundary of individual zones: continuous line showed the weld zone (T_L≈1527°C), dotted line showed boundary of heat affected zone (T_g≈727°C).

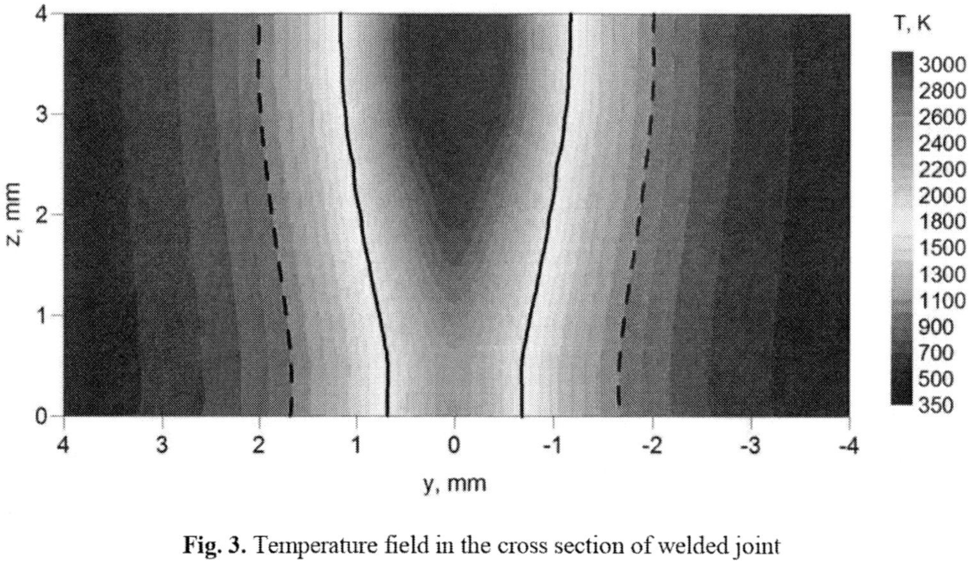

Fig. 3. Temperature field in the cross section of welded joint

Structural composition of welded joint in the cross section was determined on the basis on the numerically determined temperature distribution. Results of numerical analysis of phase composition of welded joint were presented in Figure 4.

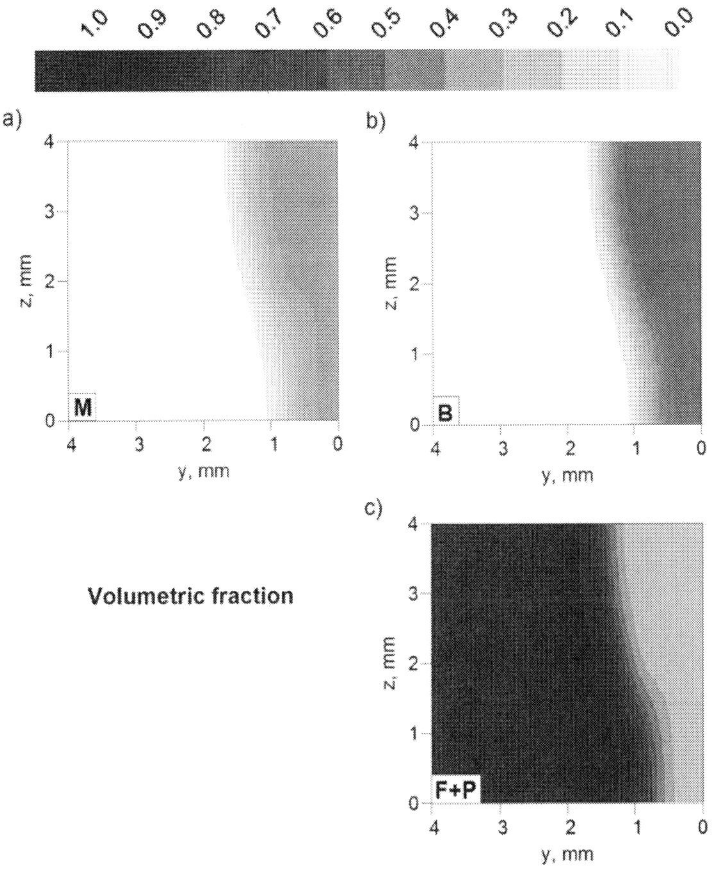

Fig. 4. Numerically estimated phase composition in weld and HAZ,
a) martensite, b) bainite, c) fearite+pearlite

Prediction of mechanical properties in weld and HAZ of welding joint was performed on the basis of relationships (1-6) and numerically obtained phase volume fraction. Distributions of mechanical properties in the cross section of analyzed steel were presented in Figures 5 to 7.

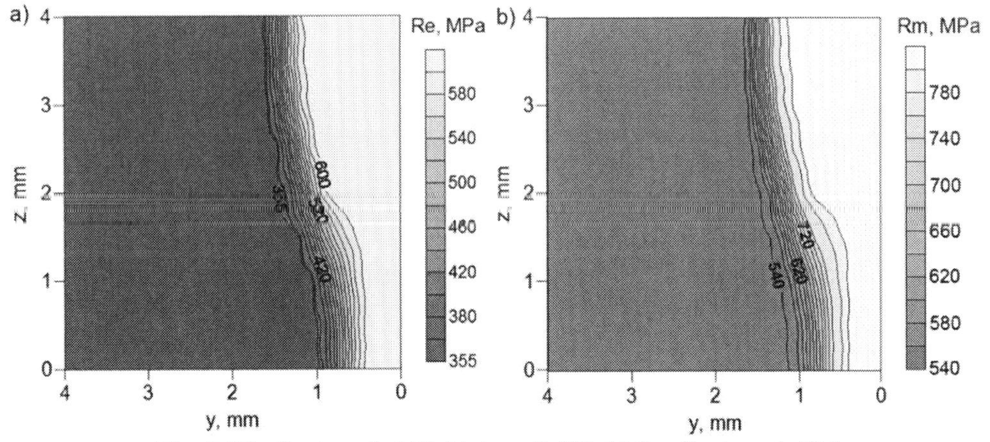

Fig. 5. Distribution of: a) Yield strength (R_e), b) Tensile strength (R_m)

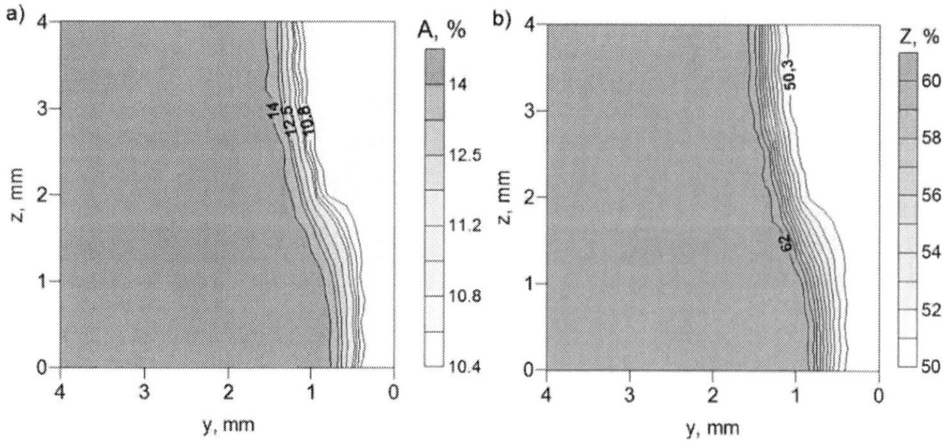

Fig. 6. Distribution of: a) elongation (A), b) contraction (Z)

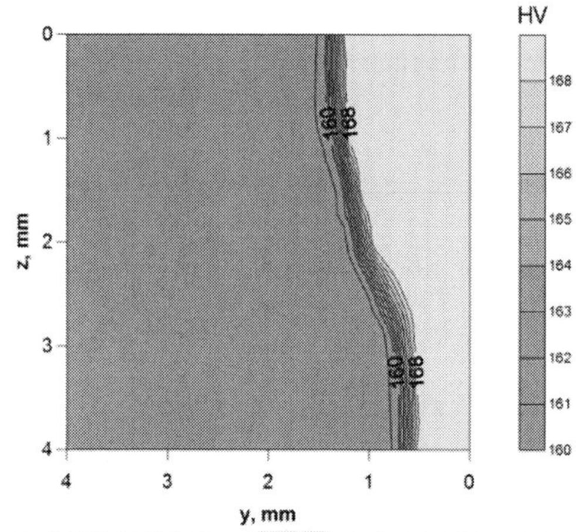

Fig. 7. Hardness distribution (HV)

Conclusions

The usefulness of analytical methods used to prediction the structure and mechanical properties in welded joint is assessed in this work. Analytical CCT diagram of S355 steels was built which verified by experiment. Phase volume fraction: ferrite, pearlite, bainite and martensite were determined. On the basis on analytical phase volume fraction performed a prediction of mechanical properties of welded joints. The numerical model build in the Abaqus FEA, which reflects laser beam welding, has been completed by additional HETVAL and UEXPAN subroutines which allow to define of phase volume fraction in the cross section of weld. Based on the numerically estimated contribution of individual phase transformations and known analytical models to determine the mechanical properties of individual structures, the R_e, R_m, A, Z and HV distributions in the cross section of the welded joint of S355 steel were predicted.

Mechanical properties obtained using analytical methods can be applied for the preliminary analysis of material properties intended for different welded constructions. They can also be used as input data in numerical analysis of stresses and deformations in welded elements, substituting expensive experimental research in this field.

References

1. W. Piekarska, M. Kubiak, Z. Saternus, *Numerical Modelling of Thermal and Structural Strain in Laser Welding Process*. Archives of Metallurgy and Materials (4), 1219-1227 (2012)
2. W. Piekarska, Z. Saternus, M. Kubiak, T. Domański, *Numerical Modelling of Stress State and Deformations in Laser Butt-Welded Sheets Made of X5CrNi18-10 Steel*. METAL 2015 Proceedings, Brno, Czech Republic, 736-741 (2015)
3. J. Mikuła, *Analityczne metody oceny spawalności stali*. Zeszyty naukowe Mechanika **85**, Politechnika Krakowska, Kraków (2001)
4. W. Piekarska, *Numerical and Experimental Study of Phase Transformations in Welding Processes*. Archives of Metallurgy and Materials (4), 2559-2568 (2015)
5. P. Seyffart, O.G. Kasatkin, *Mathematisch-statistische Beschreibung der Austenitumwandlung in der Wärmeeinflußzone*. Schweißtechnik **29**, 117-119 (1979)
6. P. Seyffart, O.G. Kasatkin, *Rascentnye modeli dla ocenki mechaniceskich svojstv metalla ZTV pri svarke nizkoegirovannych stalej*. Proc. Int. Conference mechanical Modelling and Information Technologies in Welding and Related Processes, ed. V.I. Makhnenko, Katsiveli, Crimea, E.O. Paton Welding Inst. of NAS of Ukraine, Kiev 103-106 (2002)
7. P. Seyffart, O.G. Kasatkin, *Interpolâcionnye modeli dlâ ocenki fazovogo sostava zony termičeskogo vliâniâ pri dugovoj svarke nizkolegirovannyh stalej*. Avtomat. Svarka **1**, 7-11 (1984)
8. J. Brózda, J. Pilarczyk, M. Zeman, *Spawalnicze wykresy przemian austenitu CTP-S*. (Wydawnictwo ŚLĄSK, Katowice, 1983)
9. W. Piekarska, M. Kubiak, *Theoretical investigations into heat transfer in laser-welded steel sheets*. J. Therm. Anal. Calorim. **110**, 159-166 (2012)
10. T. Domanski, A. Sapietova, M. Saga, *Application of Abaqus software for the modeling of surface progressive hardening*. Procedia Engineering **177**, 64-69 (2017)
11. A. Bokota, T. Domanski, *Modelling And Numerical Analysis Of Hardening Phenomena Of Tools Steel Elements*. Archives Of Metallurgy And Materials **54** (3), 575-587 (2009)
12. SIMULIA Dassault System, *Abaqus theory manual*. Version 6.7 (2007)
13. L. Sowa, *Effect of steel flow control devices on flow and temperature field in the tundish of continuous casting machine*. Archives of Metallurgy and Materials **60** (2A), 843-847 (2015)
14. M. Sapieta, V. Dekýš, A. Sapietová, *Thermal-stress analysis of beam loaded by 3 point bending*. Procedia Engineering **136**, 216-219 (2016)
15. L. Jakubovičová, A. Gašparec, P. Kopas, M. Sága, *Optimization of the induction heating process in order to achieve uniform surface temperature*. Procedia Engineering **136**, 125-131 (2016)
16. E. Ranatowski, *Uwagi o modelowaniu procesów cieplnych przy użyciu skoncentrowanych źródeł energii*. Przegląd Spawalnictwa **8-10**, 152-155 (2002)

Computer analysis of thermal phenomena and deformation in lap joint welded by a laser beam

Wiesława Piekarska[1], *Alžbeta* Sapietová[2], *Zbigniew* Saternus[1,*], *Marcin* Kubiak[1]

[1]Czestochowa University of Technology, Faculty of Mechanical Engineering and Computer Science, Czestochowa, Poland
[2] University of Žilina, Faculty of Mechanical Engineering, Žilina, Slovak Republic

Abstract. Laser welding technology is applied to many types of welded joints. The determination of the influence of technological parameters on the properties of resulting joint is a significant problem for technologists. Numerical estimation of the shape of the weld and its deformation is important in the initial stage of construction design. The paper concerns computer analysis of thermal and mechanical phenomena in laser welded lap joint made of austenitic steel. Three dimensional discrete model of analyzed lap joint is created in Abaqus FEA engineering software. Numerical analysis takes into account temperature dependent thermomechanical properties of austenitic steel. The movable heat source power distribution is modelled using Gaussian distribution. Computational model takes into account the gap between the joined plates. Temperature distribution in analysed joints is presented on the basis of performed numerical simulations. The shape and size of the fusion zone as well as deformation of the joint are estimated.

Keywords: Laser welding, numerical modelling, lap joint, welding deformation

1 Introduction

The application of laser technology in the production process opened new possibilities for using of welding in many industries. The main advantage of using laser welding techniques is the increase of the efficiency of the welding process and the possibility of automation and robotization of the process [1-3]. Laser beam welding technology is currently widely used in the manufacture of innovative welded constructions [4-6]. However, this welding method is related to many new, unknown issues that designers must contend. The heat distribution is completely different in laser welding in comparison to conventional, well known electric arc processes [5, 7, 8]. Other important issue is welding deformation that occurs in welded elements. In available literature the deformation depends mainly on the amount of heat input to the material during welding processes [2, 9-11]. The use of a laser heat source significantly minimizes the impact of heat load on the weldment which contributes to smaller welding deformation [7, 8]. Laser beam welding is used to perform various types of joints, including lap joints [3, 4, 6]. Lap joints are used in many industries, including the automotive industry and the aerospace industry.

* Corresponding author: saternus@imipkm.pcz.pl

Reviewers: *Františka* Pešlová, *Ján* Vavro

Recently, research centers and design offices use numerical methods to evaluate the accuracy of construction design at the initial stage of production process [12-14]. A proper mapping of the real process in a computational program and the assumption of appropriate mathematical and numerical models has a significant impact on the quality and reliability of obtained results [7, 8].

Numerical analysis of laser welding of lap joint made of austenitic steel X5CrNi18-10 is carried out in this study. Numerical calculations are performed in the Abaqus computational software on the basis of the use of finite element method (FEM). A three dimensional discrete model of the discussed system is developed with temperature dependent thermophysical properties of analyzed steel taken into account. In order to adapt the Abaqus for the simulation of welding processes the implementation of additional numerical subroutines is necessary. Subroutines are written in a FORTRAN programming language. The numerical subrotines allows to model a moveable heat sources with a specific power distribution. Appropriate contact conditions are assumed in the model at the interface between joined elements. Temperature distribution and size of the melted zone as well as stress state and deformation in the laser welded lap joint are estimated on the basis of performed numerical simulations.

2 Numerical modeling of thermal and mechanical phenomena in Abaqus program

The numerical analysis of the lap joint welding process in the Abaqus program is divided into two steps. The first step involves conducting numerical analysis of thermal phenomena. An analysis of mechanical phenomena is then performed on the basis of results obtained in the first step of calculations. Schema of the analysis is shown in Figure 1.

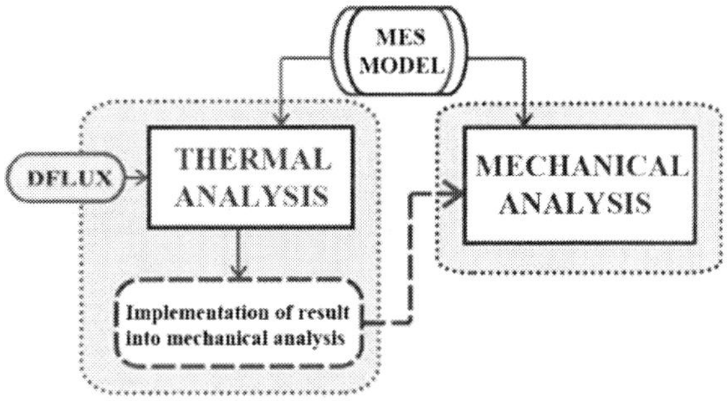

Fig. 1. Thermomechanical analysis diagram in Abaqus FEA

2.1 Analysis of thermal phenomana

The differential governing equation of temperature field in the Abaqus software is based on the energy conservation equation with Fourier's law [8, 15]. Simulations are carried out in Lagrange's coordinates, where coordinates of the center of heat source $x=v{\cdot}t$ is defined for every time increment t [s], depending on the welding speed v [m/s]. Temperature field is expressed in the criterion of weighted residuals method as follows:

$$\int_V \rho \frac{\partial U}{\partial t} \delta T \, dV + \int_V \frac{\partial \delta T}{\partial x_\alpha} \cdot \left(\lambda \frac{\partial T}{\partial x_\alpha} \right) dV = \int_V \delta T \, q_v dV + \int_S \delta T \, q_s dS \tag{1}$$

where λ is a thermal conductivity [W/m °C], $U = U(T)$ is a internal energy [J/kg], q_v is a laser beam heat source [W/m^3], $T = T(x_\alpha,t)$ is a temperature [°C], q_s is a boundary heat flux [W/m^2], δT is a variational function, ρ is a density [kg/m^3], $T = T(x_\alpha,t)$ is temperature [°C].

Equation (1) is completed by the initial condition $t=0: T=T_0$ and boundary conditions of Dirichlet and Neumann type. Newton type condition with the heat loss due to convection, radiation and evaporation [16].

During the thermal phenomena analysis of laser welding process of the lap joint, free contact between joined elements is assumed. Therefore, it is necessary to define the thermal contact between the joined elements [10, 17, 18]. The heat flow at the interface between the two bonded elements is described by the Fourier law taking account of the contact resistance r_S. Applying the fourth type of boundary condition causes continuity of temperature distribution and heat flux [16]. The scheme of the considered system is shown in Figure 2. In the discussed system, two separate systems A and B were connected to the common temperature $T_C=T_{ot}$.

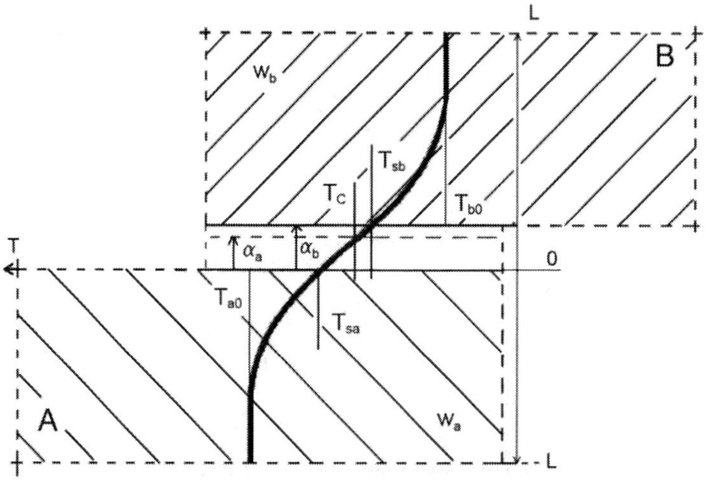

Fig. 2. Scheme of the analysis of thermal phenomena taking into account the gap between the two elements

The fourth type of boundary condition with the contact resistance r_s included can be written as [17]:

$$\lambda_a \left(\frac{\partial T_{al}}{\partial l_a} \right)_{sA} = \lambda_b \left(\frac{\partial T_{bl}}{\partial l_b} \right)_{sB} = \frac{T_{al} - T_{bl}}{r_s} \tag{2}$$

where T_{al}, T_b are respectively the temperature in the areas A and B [°C], T_{a0}, T_{b0} initial temperature in the area A and B [°C], t is time [s], α_a, α_b are coefficients of temperature equalization in areas A and B [m^2/s], λ_a, λ_b are thermal conductivity of the area A and B [W/m °C].

Fuzzy solidification front is assumed in the model [8]. Solidus and liquidus temperatures are set to T_S=1400°C and T_L=1455°C, whereas latent heat of fusion H_L=260×10^3 J/kg.

Modeling of the moveable welding source in the Abaqus is done using DFLUX subroutine. The most commonly adopted Gaussian heat source model is used in the research. The accepted model assumes a linear decrease of energy density along material penetration depth [19].

$$q_v(r,z) = \frac{Q \cdot \eta}{\pi r_o^2 s} \exp\left[\left(1 - \frac{r^2}{r_o^2}\right)\right]\left(1 - \frac{z}{s}\right) \tag{3}$$

where Q is a laser beam power [W], r_0 is a beam radius [m], $r = \sqrt{x^2 + y^2}$ is actual radius [m], s is penetration depth [m], z is actual penetration [m].

2.2 Analysis of mechanical phenomana

The mechanical analysis in elastic-plastic range is based on classic equilibrium equations, supplemented by constitutive relations [7, 20]:

$$\nabla \circ \dot{\boldsymbol{\sigma}}(x_\alpha, t) = 0, \qquad \dot{\boldsymbol{\sigma}} = \dot{\boldsymbol{\sigma}}^T \tag{4}$$

$$\dot{\boldsymbol{\sigma}} = \mathbf{D} \circ \dot{\boldsymbol{\varepsilon}}^e + \dot{\mathbf{D}} \circ \boldsymbol{\varepsilon}^e \tag{5}$$

$$\boldsymbol{\sigma}(x_\alpha, t_0) = \boldsymbol{\sigma}(x_\alpha, T_s) = 0, \qquad \boldsymbol{\varepsilon}^e(x_\alpha, t_0) = \boldsymbol{\varepsilon}^e(x_\alpha, T_s) = 0 \tag{6}$$

where $\boldsymbol{\sigma} = \boldsymbol{\sigma}(\sigma_{ij})$ is stress tensor, x_α describes location of considered point (material particle), ($\circ$) is inner exhaustive product, $\mathbf{D} = \mathbf{D}(T)$ is a tensor of temperature dependent material properties.

The total strain is defined as a sum of elastic ε^e, plastic ε^p and thermal ε^{Th} strains:

$$\varepsilon^{total} = \varepsilon^e + \varepsilon^p + \varepsilon^{Th} \tag{7}$$

Elastic strain is modelled using an isotropic Hooke's law, whereas plastic strain is calculated using plastic flow model obeying Huber-Misses plasticity condition [20].

In the case of numerical analysis of laser welding lap joint, it is necessary to define the mechanical interaction between two joined materials. As a result of mechanical interactions, the contact area of the two elements is deforming. Boundary conditions assumed in mechanical analysis are prescribed for preventing rigid body motion.

3 Discrete model

Three dimensional discrete model of analyzed lap joint is developed in Abaqus CAE software. The analyzed joint consists of two plates made of austenitic steel X5CrNi18-10 whith dimensions 50x20x1.5 mm. The model assumes that the welded plates overlap on the length 25mm. Scheme of the analyzed system is shown in Figure 3.

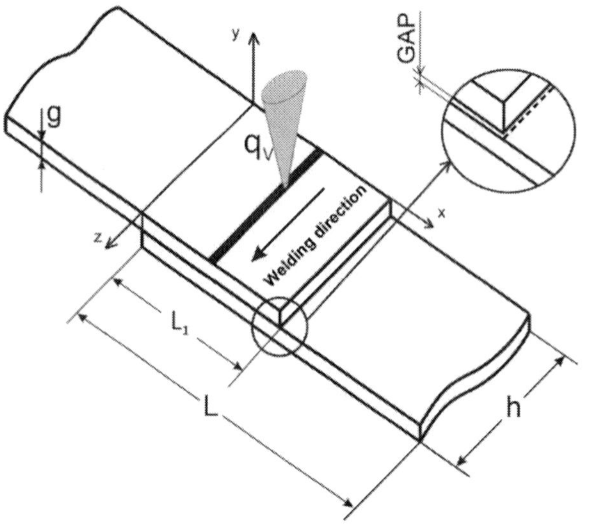

Fig. 3. Scheme of analyzed system

Perfect contact between the connected elements is assumed at the melted zone in the developed discrete model of the lap joint. Further, outside the melting zone, free contact between joined elements is assumed. The width of the melting zone at the welded plates is determined on the basis of pre-made simulation calculations. The gap between welded plates in the remaining area is 0.1 mm. The discrete model consists of 30.000 finite elements. Due to the high temperature gradient that occurs at the welding source activity zone the high density of the finite element mesh is assumed (Figure 4).

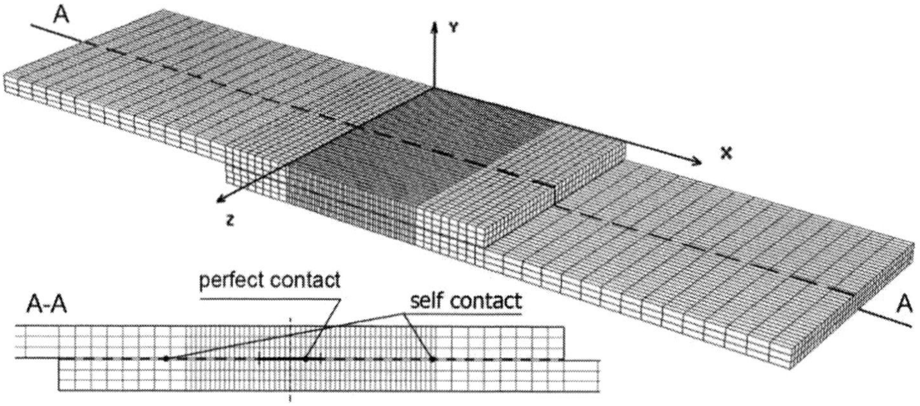

Fig. 4. Discretization of analyzed domain

Assumed thermomechanical properties of austenitic steel changing with temperature are presented in Figure 5 [8].

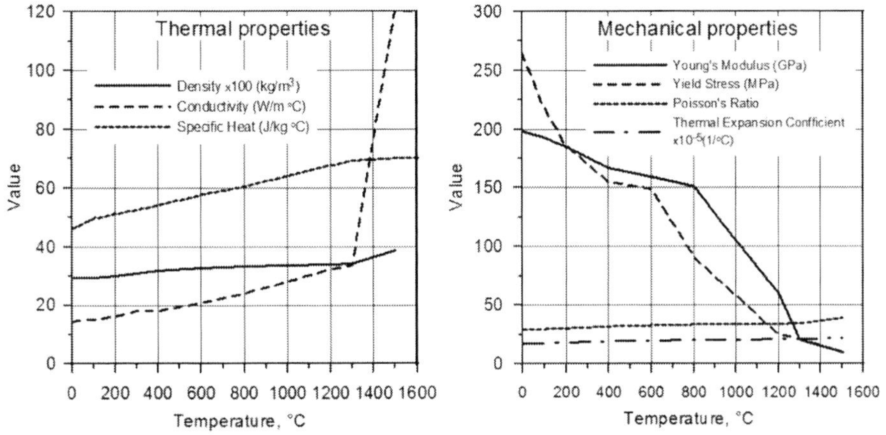

Fig. 5. Thermomechanical properties of X5CrNi18-10 stainless steel

4 Results and discussion

Numerical calculations of the laser welding process of lap joints are performed in the Abaqus simulation program. In calculations, the efficiency of the laser beam welding process is assumed on the basis of literature data as η = 75%. Process parameters assumed in calculations are summarized in Table 1.

Table 1. Process parameters used in computer simulations

Laser beam power	Welding speed	Beam radius	Penetration deep
Q=*1500 W*	v=*0.7 m/min*	r_0=*1 mm*	s=*4 mm*

In the case of the analysis of thermal phenomena with the free contact taken into account it is necessary to define the appropriate interaction condition. Thermal conductivity in the contact area of both elements must be defined The calculation program determines the conductivity as a function of the gap width [15].

Temperature field in analyzed joints are estimated on the basis of performed numerical simulations. Figure 6 shows the temperature field in welded lap joint, while Figure 7 shows distribution of temperature field in the cross-sectional area. In these figures the solid line indicates the boundaries of the melting zone.

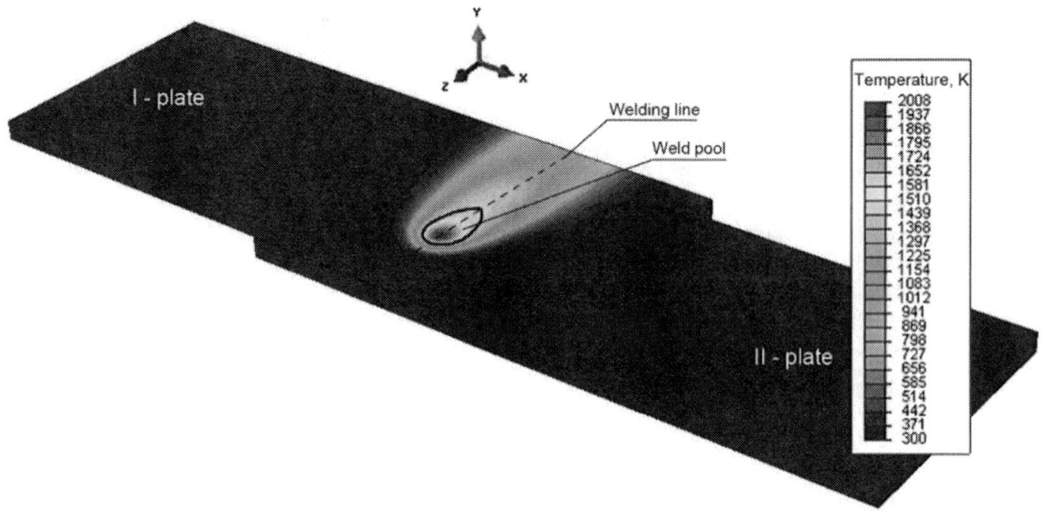

Fig. 6. Numerically estimated temperature field in laser welded lap joint

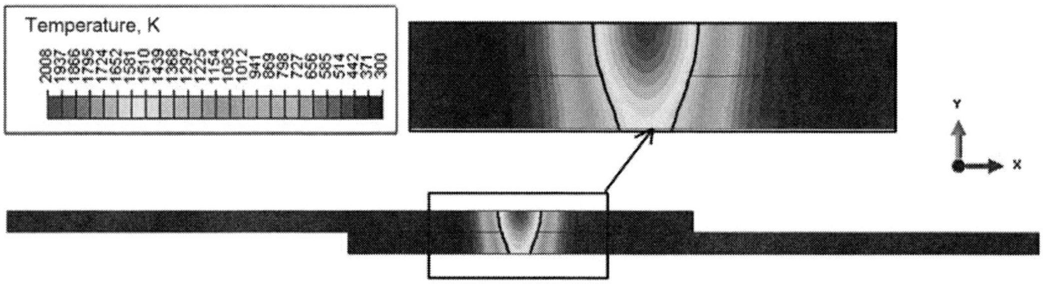

Fig. 7. Temperature field in cross section of the welded lap joint

On the basis of obtained temperature distributions, the state of stress and displacement of welded elements is determined. The developed numerical model takes into account the mechanical boundary conditions (Fig. 8) for two analyzed cases and contact conditions between welded parts.

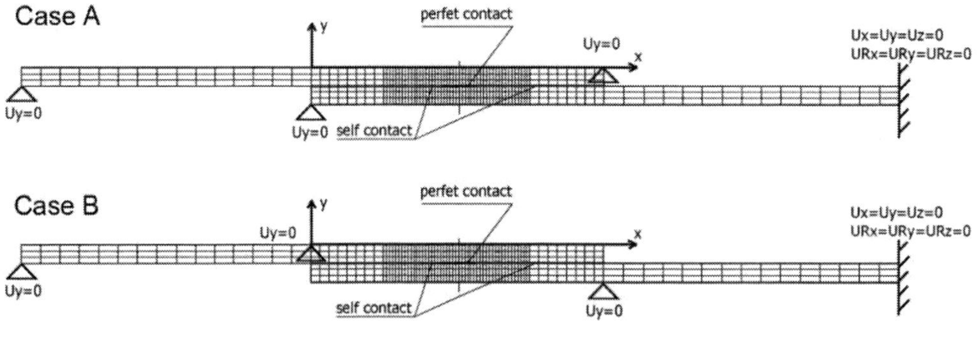

Fig. 8. Mechanical boundary conditions

Figure 9 shows the distribution of temporary and residual stresses for two different cases of support. The maximum value of residual stresses is about 170 MPa.

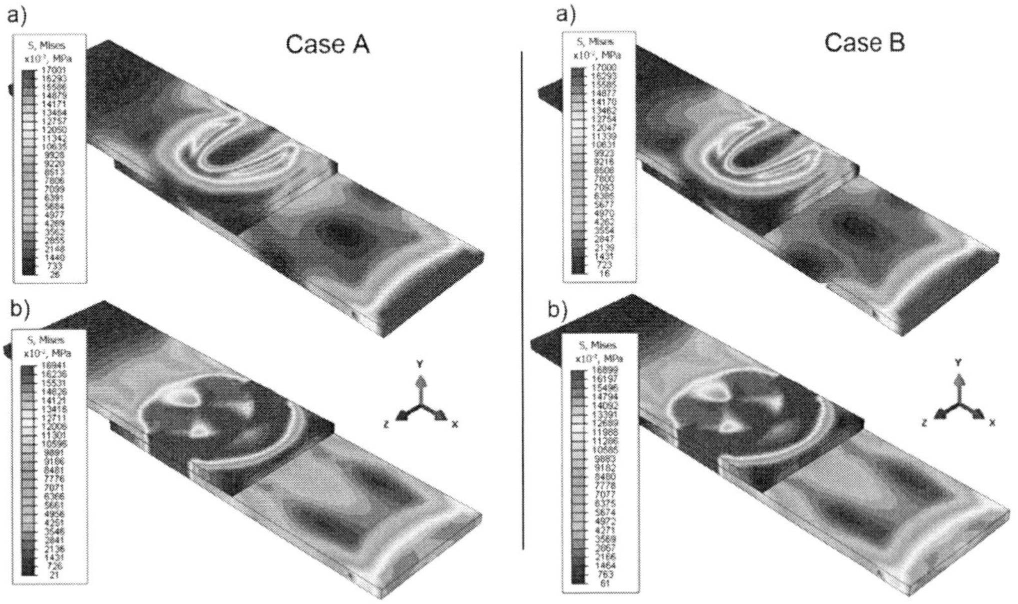

Fig. 9. Stress distribution (von Misses) for different welded times and boundary conditions, a) temporary stresses, b) residual stresses

From the analysis of results of reduced stress distributions for both cases of support it can be stated that stresses are almost identical in these cases. A completely different situation occurs in the case of displacement of the welded joint, presented in Fig. 10.

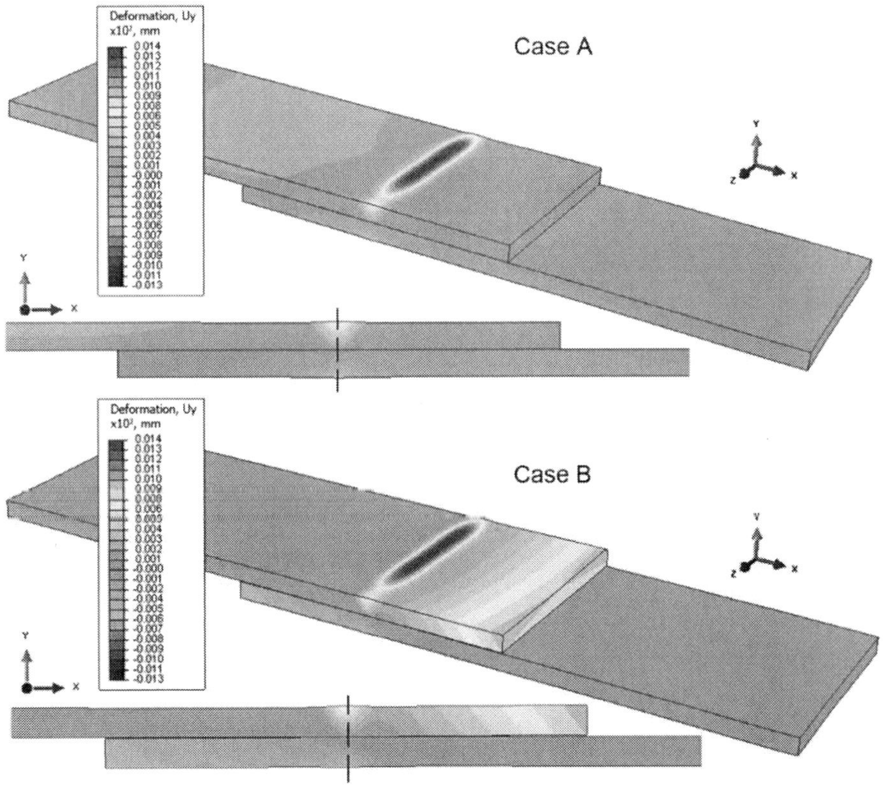

Fig. 10. Numerically estimated deflection U_Y in laser welded lap joint for two case

Assumed boundary conditions in Case A prevent free motion of the plate. In Case B, the free ends of the plate have the possibility to move which is reflected in the numerically estimated displacement presented in Fig. 10.

Conclusions

The paper presents an analysis of thermomechanical phenomena in welded laser lap joints using the Abaqus FEA software. The Abaqus has been completed by additional numerical subroutines allowing for a complete analysis of thermomechanical phenomena of the welding process. The appropriate contact conditions at the interface between joined elements are defined. The temperature distribution is determined in the lap joint. The melting zone, state of stress and welding deformation are estimated. The influence of the supporting method on stress and deformation of the joint is analyzed. In analyzed support cases the impact on the state of stress was relatively small (Fig. 9). Significant differences occurred in the fields of displacement (Fig. 10). Numerical simulations show that the developed numerical model can be a useful tool for preliminary design of laser welded lap joint.

References

1. S. Zhang, *Stresses in Laser-Beam-Welded Lap Joints Determined by Outer Surface Strains*. Welding Research, 14-18 (2002)
2. G. A. Moraitis, G. N. Labeas, *Residual stress and distortion calculation of laser beam welding for aluminum lap joints*. J Mater Process Tech. **198** (1-3), 260-269 (2008)
3. M. Hyrcza-Michalska, F. Grosman, *The Research Of Laser Welded Blanks Characteristics In Basic And Technological Drawability Tests*. 12th International Conference on Metal Forming, Cracow, Poland, Steel Research International, 780-787 (2008)
4. J. Pilarczyk, M. Banasik, S. Stano, J. Dworak, *Laser joining with wire filler material and mechanical seam tracing system*. Przegląd Spawalnictwa **12**, 9-14 (2011)
5. M. Kubiak, W. Piekarska, Z. Saternus, T. Domański, S. Stano, *Simulations And Experimental Research On Laser Butt-Welded T-Joints*. Metal 2014: 23rd International Conference On Metallurgy And Materials, 726-731 (2014)
6. J. Ma, F. Kong, W. Liu, B. Carlson, R. Kovacevic, *Study on the strength and failure modes of laser welded galvanized DP980 steel lap joints*. J Mater Process Tech. **214** (8), 1696-1709 (2014)
7. W. Piekarska, M. Kubiak, Z. Saternus, *Numerical Simulation of Deformations in T-Joint Welded by the Laser Beam*. Arch Metall Mater. **58** (4), 1391-1396 (2013)
8. M. Kubiak, W. Piekarska, Z. Saternus, S. Stano, T. Domański, *Numerical Prediction Of Deformations In Laser Welded Sheets Made Of X5CrNi18-10 Steel*. Arch Metall Mater. **60** (3), 1965-1972 (2015)
9. Y. Miyazaki, S. Furusako, *Tensile Shear Strength of Laser Welded Lap Joints*. Nippon Steel Technical Report **95**, 28-34, (2007)
10. H. Wei, Q. He, J. S. Chen, H. P. Wang, B. E. Carlson, *Coupled thermal-mechanical-contact analysis of hot cracking in laser welded lap joints*. J Laser Appl. **29** (2), (2016)
11. S. K. Choa, Y. S. Yanga, K. J. Sona, J. Y. Kimb, *Fatigue strenght in laser welding of the lap joint*. Finite Elements in Analysis Design **40**, 1059-1070 (2004)

12. A. Anca, A. Cardona, J. Risso, et al., *Finite element modeling of welding processes*. Appl Math Model **35**, 688–707 (2011)
13. A. Sapietová, V. Dekýš, M. Sapieta, P. Pecháč, *Application of computational and design approaches to improve carrier stability*. Procedia Eng. **96**, 410-418 (2014)
14. A. Sapietová, M. Sága, P. Novák, R. Bednár, J. Dižo, *Design and application of multi-software platform for solving of mechanical multi-body system problems*. Mechatronics: Recent technological and scientific advances, 345-354 (2011)
15. SIMULIA, *Abaqus FEA theory manual*. Version 6.7, Dassault System (2007)
16. L. Sowa, *Effect of steel flow control devices on flow and temperature field in the tundish of continuous casting machine*. Arch Metall Mater. **60** (2A), 843-847 (2015)
17. W. Longa, *Krzepnięcie odlewów w formach piaskowych*. (Śląsk, Katowice, 1973)
18. T. Skrzypczak, E. Węgrzyn-Skrzypczak, *Modeling of thermal contact through gap with the use of finite element method*. J. Appl. Math. Comput. Mech. **14** (4), 145-152 (2015)
19. S. A. Tsirkas, P. Papanikos, Th. Kermanidis, *Numerical simulation of the laser welding process in butt-joint specimens*. J Mater Process Tech. **134**, 59-69 (2003)
20. A. Bokota, T. Domanski, *Modelling And Numerical Analysis Of Hardening Phenomena Of Tools Steel Elements*. Arch Metall Mater. **54** (3), 575-587 (2009)

Application of analytical methods for determination of hardness distribution in welded joint made of S1100QL steel

Wiesława Piekarska[1], *Milan* Sága[2], *Dorota* Goszczyńska-Króliszewska[1,*], *Tomasz* Domański[1], *Peter* Kopas[2]

[1]Institute of Mechanics and Machine Design Foundations, Czestochowa University of Technology, Dabrowskiego 73, 42-200 Częstochowa, Poland, EU
[2]University of Žilina, Faculty of Mechanical Engineering, Department of Applied Mechanics, Univerzitná 8215/1, 010 26, Žilina, Slovak Republic

Abstract. The prediction of hardness distribution in the cross section of welded join made of S1100QL steel is performed in this study on the basis of analytical methods. Analytical CCT diagram and volume fraction of each phases of S1100QL steel as a function of cooling time $t_{8/5}$ are determined. A numerical simulation of welding process is performed in ABAQUS. Thermal cycles and temperature field in welded joints are determined. Prediction of hardness distribution in the cross section of the joint is performed on the basis of obtained cooling times $t_{8/5}$. Results of numerical simulations are compared with experimentally obtained results.

Keywords: CCT diagram, analytical methods, heat affected zone, numerical modelling, welding

1 Introduction

Analytical models presented in paper concern to building simplified CCT diagrams, prediction the structure and evaluating mechanical properties on the basis on chemical composition of steel form the group of weldable steels [1-6]. Results of analytical methods are often used for the preliminary analysis of the properties of the material, which preceded the subsequent experimental studies and the development of mathematical models. Experimental researches are very expensive so researches are focused on the mathematical modeling of arising phenomena.

Application of analytical methods to predict the structure and hardness distribution in the cross section of S1100QL steel were presented in paper. The analyzed steel belongs to a group of high-strength steels with a yield point above 900MPa [7-10]. Application of high-strength steels allows to significantly reducing the weight of construction, increasing its load capacity. Consequently the demand for this kind of steels is still growing.

From the empirical relationships, the characteristic values that were used to build CCT diagram of the S1100QL steel were determined. Numerical analysis of the butt welding of

* Corresponding author: goszczynska@imipkm.pcz.pl

Reviewers: *Grzegorz* Domek, *František* Nový

high-strength flat steel S1100QL was performed. In ABAQUS FEA built three-dimensional discrete model reflecting the geometry of considered model. To describe the power distribution of the welding source, a model of heat source with a Goldak's distribution was assumed. Temperature distributions, the shape and size of the melted zone of the weld joint were determined on the basis of numerical simulation. Mechanical properties such as: hardness (HV), yield strength (R_e), tensile strength (R_m), elongation (A) and contraction (Z) were determined on the basis of experimental determined phase compositions. The hardness distribution (HV) in the cross section of the weld joint was predicted for different distances from the heat source. The results was compared with those obtained experimentally [8, 10].

2 Analytical models of phase transformations

Analytical models created on the basis of the composition of steel were used to predict the structure composition of HAZ, further to develop simplified CCT diagrams. Equations were obtained by the use of statistical analysis of results of experimental research performed for certain material groups. As a result, models developed by different authors were given, allowing the determination of characteristic quantities of austenite transformation [1, 2, 4, 6]. Chemical composition of investigated steel was presented in Table1. Symbols of chemical elements provided by all empirical formulas means percentage of a given element, e.g. C→%C.

Table 1. Chemical composition of S1100QL steel [atomic %]

Steel	C	Si	Mn	P	S	Cr	Mo	Ni	V
S1100QL	0.18	0.50	1.6	0.017	0.003	1.45	0.50	2.39	0.12

In this paper analytical models were used to create an analytical CCT diagram with different approaches in determining the time of initiation of austenite transformation. Start and finish temperatures of each phase transformation were determined for start and finish times of phase transformations estimated in relation to the chemical composition of the analyzed steel.

Time $t_{8/5}$ and start times of diffusive transformations as: bainite t_B, ferrite t_F and perlite t_P were expressed as follows [1, 4]:

$$\begin{aligned} t_B &= -1.8 + 28.3C + 17.1Mn - 59.6Si - 20Cr + 13.2Ni + 0.1Mo - 5.3V \\ &+47.0Nb - 289.6C^2 - 3.7Mn^2 + 65.4Si^2 + 38.6Cr^2 - 7.0Ni^2 - 21.0Mo^2 \\ &-9.8CMn - 19.5MnSi + 232.1CSi \end{aligned} \tag{1}$$

$$t_F = 10^{[5.8(C+Si/29+Mn/14+Ni/67+Cr/16+Mo/6+V/425)-0.83]} \tag{2}$$

$$t_P = 10^{[5.14(C+Si/17+Mn/19+Ni/25+Cr/16+Mo/4+V/3)+0.06]} \tag{3}$$

where: t_B, t_F, t_P are start times of bainite, ferrite and pearlite transformations.

Time-dependent (t = $t_{8/5}$) start temperatures of the formation ferrite $F_s(t)$, bainite $B_s(t)$, pearlite $P_s(t)$ and the finish temperature of phase transformations $T_k(t)$ were determined by formulas [4, 5], in the following form:

$$B_s(t) = T_{B0} + \Delta T_B erf[(\ln t - \ln t_{B0})K_B] + K_{Bt} \ln t \rightarrow T_{B0} = T_{B0}(t) = M_s \tag{4}$$

$$F_s(t) = T_{F0} + \Delta T_F erf[(\ln t - \ln t_{F0})K_F] + K_{Ft} \ln t \rightarrow T_{k0} = T_{k0}(t) = M_s \tag{5}$$

$$P_s(t) = T_{P0} + \Delta T_P erf[(\ln t - \ln t_{P0})K_P] + K_{Pt} \ln t \rightarrow T_{F0} = B_s(t_{F0}) \tag{6}$$

$$T_k(t) = T_{k0} + \Delta T_k erf[(\ln t - \ln t_{k0})K_k] + K_{kt} \ln t \rightarrow T_{P0} = B_s(t_{P0}) \tag{7}$$

where: values ΔT_B, K_B, ΔT_F, K_F, ΔT_P, K_P, ΔT_k, K_k depend on the chemical composition of examined steel [4, 5].

Start and finish temperatures of martensite transformation, were described respectively [1, 2]:

$$M_s = 530 - 415C + 90C^2 - 35Mn - 30Cr - 20Ni - 15W - 10Mo \tag{8}$$

$$\begin{aligned} M_f &= 381.76 - 252.44C - 111.12Mn + 54.538Si + 114.17Cr \\ &- 23.779Ni - 57.381Mo + 215.7V + 945.4Nb + 1821.7Ti - 1746.5B \end{aligned} \tag{9}$$

The CCT diagrams of S1100QL steel by using the formulas (1-9) were given in Figure 1. For comparison the experimental CCT diagram obtained by [8] was presented in Figure 1.

Depending on cooling rates the analytical model for determining phase composition is very useful in phase transformation analysis. The volume fractions of particular phases, such as: ferrite-pearlite (η_{FP}), bainite (η_B) and martensite (η_M), as a function of time $t_{8/5}$ were described as follows [1, 2]:

$$\eta_M = 0.5[1 - erf \frac{((\ln t) - \ln \Delta t_{FP})}{\ln S_{FP}} \tag{10}$$

$$\eta_{FP} = 0.5[1 - erf \frac{((\ln t) - \ln \Delta t_M)}{\ln S_M} \tag{11}$$

$$\eta_B = 1 - \eta_M - \eta_{FP} \tag{12}$$

where: $\ln \Delta t_{FP}$, $\ln \Delta t_M$, $\ln S_{FP}$, $\ln S_M$ depend on the chemical composition of examined steel [1, 2].

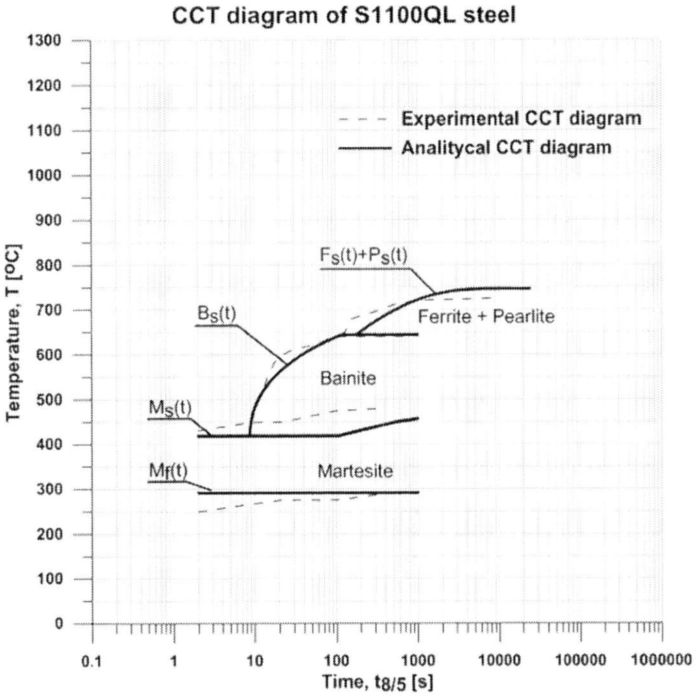

Fig. 1. Analytical and experimental [8] CCT diagram of S1100QL steel

3 Analytical methods for predicting a hardness distribution

Mechanical properties HAZ of the weld joint can be determined from the structural composition and mechanical properties of each of the structures. If the phase composition in the heat affected zone is known, more specifically contribution of each structure (ferrite-pearlite, martensite and bainite) and properties of structural components W_i , there is a possibility to approximately predict properties of the entire zone [3].

$$W = \sum_{i=M,B,F,P} W_i \eta_i \tag{13}$$

where: W can be mean hardness, yield strength, tensile strength, impact strength, elongation and contraction.

In the literature [1, 2, 5, 6] are administered many dependencies, which can be determine the mechanical properties such as R_e, R_m, A_5, Z, KCU and HV. Property of each phases: ferrite-pearlite, martensite and bainite were defined on the based on chemical composition. Whereas contributions each phases were defined based on the experiment or through used numerical simulations.

Hardness, mapping the structural heterogeneity of the HAZ, is one of the basic volume which characterized the welded joint. Hardness of HAZ is a function of phase composition and hardness for each structure of phases. HV_i (i means each phases such as: ferrite-pearlite, martensite and bainite). Hardness HV_i of individual structural components was defined by [10] as a function of chemical composition and velocity $v_{8/5}$ ($v_{8/5} = (800 - 500)/t_{8/5}$):

$$HV_{FP} = 42 + 223C + 53Si + 30Mn + 12.6Ni + 7Cr + 19Mo +$$
$$+ (10 - 19Si + 4Ni + 8Cr + 130V)\log v_{8/5}$$
$$HV_M = 127 + 959C + 27Si + 11Mn + 8Ni + 16Cr + 21\log v_{8/5} \quad (14)$$
$$HV_B = -323 + 185C + 330Si + 153Mn + 65Ni + 144C +$$
$$+ 191Mo + (89 + 53C - 55Si - 22Mn - 10Ni - 20Cr - 33Mo)\log v_{8/5}$$

Authors of papers [8, 10] performed experimental research of base material of S1100QL steel [8]. Structure of base material of S1100QL steel consisted in fine-grained martensite (grain diameter about 1-8 mm), bainite and ferrite (Fig. 2).

Fig. 2. Microscopic image of the base material of S1100QL steel [8]

Mechanical properties in base material were determined analytically on the basis of relationships in papers [1, 2, 5, 6]. Structures of base material were performed experimentally. Analytical and experimental results were presented in table 2 [10].

Table 2. Mechanical properties of base material of S1100QL steel

Results	HV	R_e [MPa]	R_m [MPa]	A [%]	Z [%]
Experimental [10]	320	1120	1430	12.5	58
Analytical	318	1139	1377	15.05	48.02

4 Field temperature, phase transformations, kinetic of transformations

The exemplary of prediction of hardness distribution in the cross section of flat butt welded by electric arc made of S100QL steel was performed in this study. Field temperature of welded joint was determined on the basis of finite element methods [11, 12, 13] in ABAQUS FEA software. The analysis of thermal phenomena is made on the basis of the solution of energy conservation equation together with Fourier law [14]. The temperature field expressed in the criterion of weighted residuals method was described by the following equation:

$$\int_V \rho \frac{\partial U}{\partial t} \delta T \, dV + \int_V \frac{\partial \delta T}{\partial x_\alpha} \cdot \left(\lambda \frac{\partial T}{\partial x_\alpha} \right) dV = \int_V \delta T \, q_V dV + \int_S \delta T \, q_S dS \quad (15)$$

where: λ thermal conductivity [W/mK], U = U(T) is internal energy, q_v is the capacity of source [W/m^3], $T = T(x_\alpha, t)$ is temperature [K], q_s is density of heat [W/m^2], and δT is a partial differential of T.

Equation (15) is completed by the initial condition and, boundary conditions of Dirichlet, Neumann and Newton type with the heat loss due to convection and radiation [15-17].

A movable welding source is implemented in Abaqus/FEA using additional numerical DFLUX subroutine [2, 14]. For the calculation of the temperature field assumed power of the heat source: Q=2200 W and welding speed: v=10 mm/s. Mathematical model of Goldak's volumetric heat source power distribution was used in the analysis [18]:

$$q_f(x,y,z) = \frac{6\sqrt{3} f_f Q}{abc_f \pi \sqrt{\pi}} \exp(-3\frac{x^2}{a^2}) \exp(-3\frac{y^2}{c_f^{\,2}}) \exp(-3\frac{z^2}{b^2}) \quad (16)$$

$$q_r(x,y,z) = \frac{6\sqrt{3} f_f Q}{abc_r \pi \sqrt{\pi}} \exp(-3\frac{x^2}{a^2}) \exp(-3\frac{y^2}{c_r^{\,2}}) \exp(-3\frac{z^2}{b^2}) \quad (17)$$

$$q(x,y,z) = q_f(x,y,z) + q_r(x,y,z) \quad (18)$$

where: a, b, c_r and c_f are dimensions of semi-ellipsoid axes, f_f and f_r are values representing energy distribution in the frond and the back of the heat source, ($f_r + f_f = 2$).

Numerical calculations of the temperature field were performed as a 3D task. Prediction of hardness distribution was performed on the basis of determined temperature distribution. Figure 3 presented results of numerical simulation in ABAQUS FEA software. Areas of weld, heat affected zone, base material, and direction of the heat source were marked.

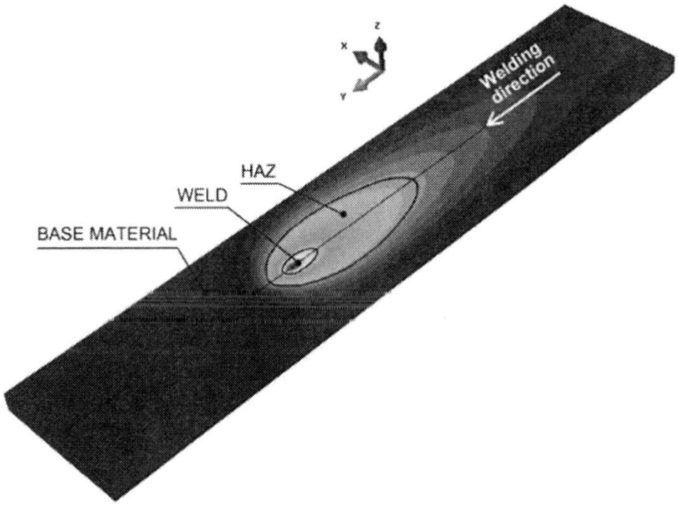

Fig. 3. Numerical model of welded flat of S1100QL steel

5 Exemplary of calculations

Analysis of hardness distribution was performed for 42 selected points of the welded joint in the cross section (Fig. 4). Prediction of hardness in weld and HAZ of welded joint were performed on the basis of equations 13 and 14 and analytically obtained phase volume fraction on the function of time $t_{8/5}$.

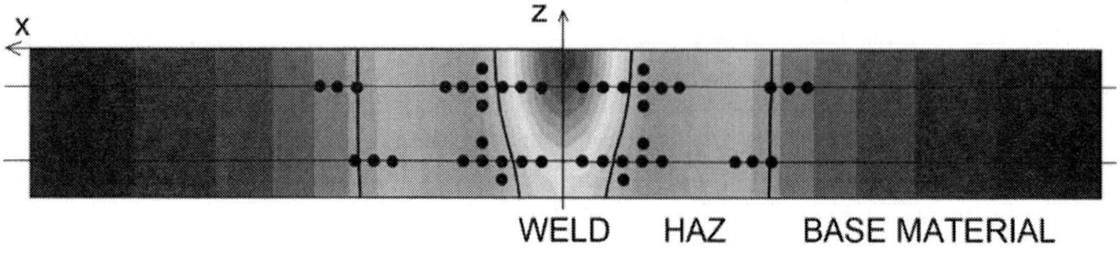

Fig. 4. Cross section of welded joints and hardness measurement points

Thermal cycles for selected point of cross section of welded joint were performed in order to determined cooling times $t_{8/5}$ (Fig. 5). Prediction of hardness distributions of cross section of welded joints was determined on the basis of analitycally obtained phase volume fraction and numerically obtained cooling times $t_{8/5}$.

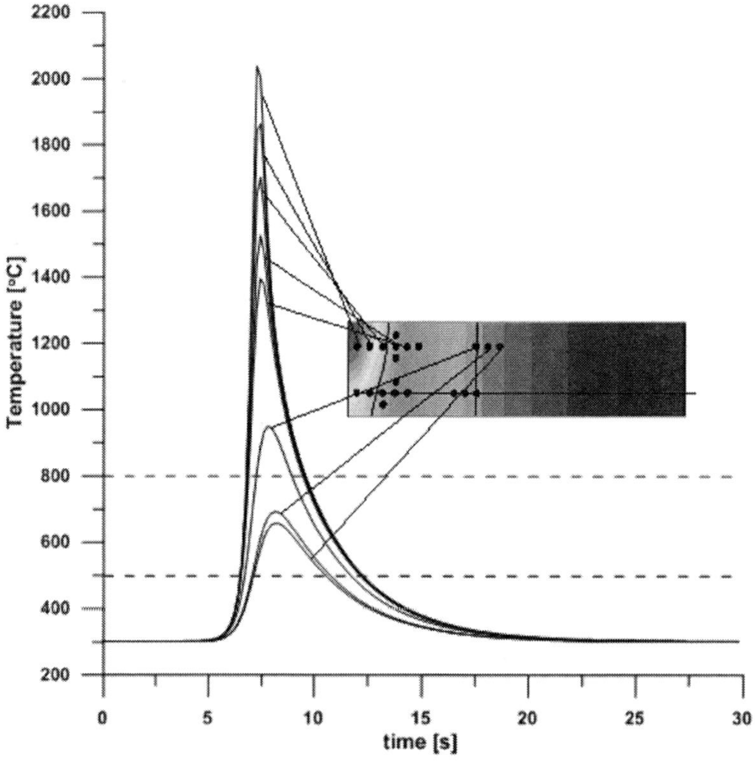

Fig. 5. Thermal cycles of measurement points (face)

Figure 6 presents results obtained analitycally of prediction of hardness distribution of selected points in cross section of welded jolint. The results of experimental research presented by the authors [8] were compared in Figure 6.

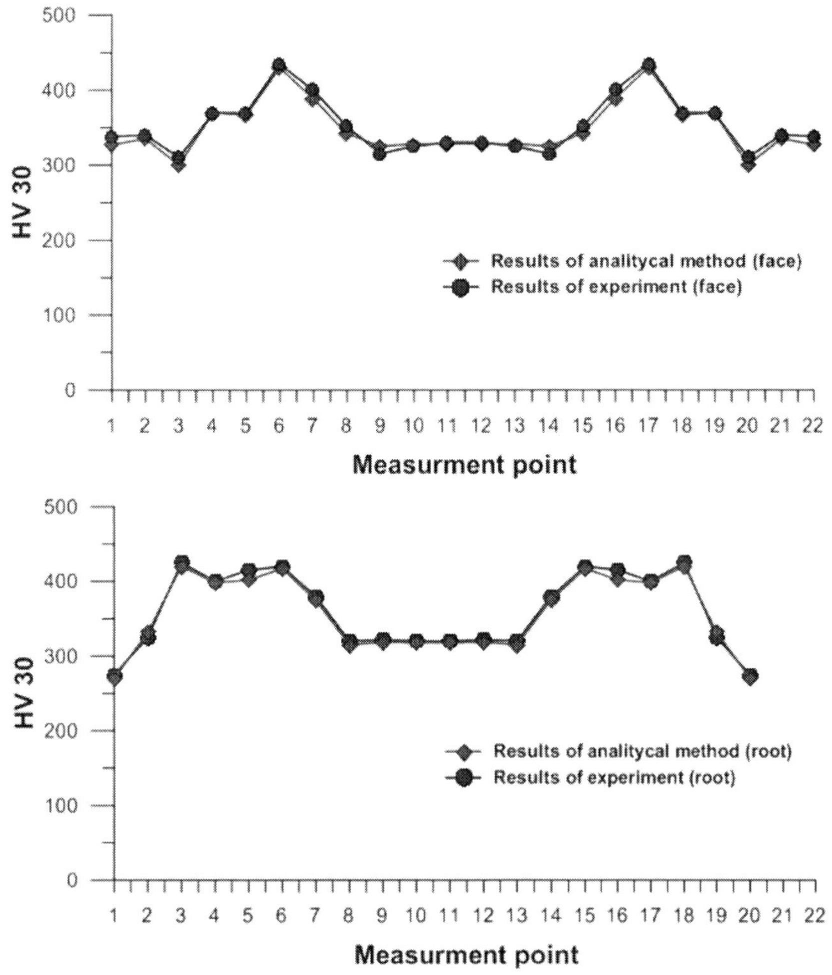

Fig. 6. Hardness distribution in the cross section of welded joint

Conclusions

The usefulness of analytical methods to predict the structure and mechanical properties in welded joint was presented in this work. Analytical CCT diagram of S1100QL steel was verified by experiment. Prediction of mechanical properties of welded joints was performed on the basis of experimentally obtained phase volume fraction. Result of hardness distribution in cross section of welding joint of S100QL steel on the basis of analytical methods were presented in paper. Results of analytical methods were compared with experimental results. Comparison of analytical and experimental results allowed to conclude that analytical methods can be applied for the preliminary analysis of material properties intended for different welded constructions. They can also be used as input data in numerical analysis of stresses and deformations in welded elements, substituting expensive experimental research in this field.

References

1. J. Mikuła, *Analityczne metody oceny spawalności stali*. Zeszyty naukowe Mechanika **85**, Politechnika Krakowska, Kraków (2001)
2. W. Piekarska, *Analiza numeryczna zjawisk termomechanicznych procesu spawania laserowego. Pole temperatury, przemiany fazowe i naprężenia*. Seria Monografie **135**, Wydawnictwo Politechniki Częstochowskiej, Częstochowa (2007)
3. H. Shen, Y. Shi, Z. Yao, J. Hu, *An analytical model for estimating deformation in laser forming*. Computational Materials Science **37**, 593-598, (2006)
4. P. Seyffart, O. G. Kasatkin, *Rascentnye modeli dla ocenki mechaniceskich svojstv metalla ZTV pri svarke nizkoegirovannych stalej*. Proc. Int. Conference mechanical Modelling and Information Technologies in Welding and Related Processes, Welding Inst. of NAS of Ukraine, Kiev 103-106 (2002)
5. P. Seyffart, O. G. Kasatkin, *Interpolâcionnye modeli dlâ ocenki fazovogo sostava zony termičeskogo vliâniâ pri dugovoj svarke nizkolegirovannyh stalej*. Avtomat. Svarka **1**, 7-11 (1984)
6. P. Seyffart, O. G. Kasatkin, *Mathematisch-statistische Beschreibung der Austenitum-wandlung in der Wärmeeinflußzone*. Schweißtechnik **29**, 117-119 (1979)
7. A. Klimpel, K. Luksa, Ł. Warcaba, T. Szorek, A. Kiełbasiński, *Zrobotyznowane spawanie GMA złączy blach ze stali S690 i S960*. Archiwum Technologii Maszyn i Automatyzacji **3**, 30, 59-68 (2010)
8. J. Nowacki, A. Sajek, P. Matkowski, *The influence of welding heat input on the microstructure of joints of S1100QL steel in one-pass welding*. Archives of Civil and Mechanical Engineering **16**, 777-783 (2016)
9. T. Ślęzak, L. Sniezek, *Fatigue Properties and Cracking of High Strength Steel S1100QL Welded Joints*. Key Engineering Materials **598**, 237-242 (2014)
10. I. Samardžić, A. Corić, M. Dunđer, *Weldability investigation of fine-grained S1100QL steel*. Metalurgija **3**, 55, 453-456 (2016)
11. T. Domanski, A. Sapietova, M. Saga, *Application of Abaqus software for the modeling of surface progressive hardening*. Procedia Engineering **177**, 64-69 (2017)
12. M. Sapieta, V. Dekýš, A. Sapietová, *Thermal-stress analysis of beam loaded by 3 point bending*. Procedia Engineering **136**, 216-219 (2016)
13. A. Sapietová, V. Dekýš, M. Dudinský, *Utilizing of sensitivity analysis in preparation of optimizing procedure*. Scientific Journal of Silesian University Of Technology, Series Transport **76**, 113-118 (2012)
14. *SIMULIA Dassault System*. Abaqus theory manual, Version 6. 7 (2007)
15. M. Kubiak, W. Piekarska, Z. Saternus, T. Domański, S. Stano, *Simulations And Experimental Research On Laser Butt-Welded T-Joints*. METAL 2014: 23RD International Conference On Metallurgy And Materials, 726-731 (2014)
16. R. Dyja, E. Gawronska, N. Sczygiol, *The effect of mechanical interactions between the casting and the mold on the conditions of heat dissipation: a numerical model*. Archiv Metall Mater. **60** (3), 1901-1909 (2015)
17. L. Sowa, A. Bokota: *Numerical simulation of the molten steel flow in the tundish of CSC machine*. Archives of Metallurgy and Materials **57** (4), 1163-1169 (2012)
18. J. A. Goldak, *Computational Welding Mechanics*. (Springer, NY, 2005)

Missing values reconstruction in sound level monitoring station by means of intelligent computing

Leszek Radziszewski[1], *Michał* Kekez[1,*], *Alžbeta* Sapietová[2]

[1]Kielce University of Technology, Faculty of Mechatronics and Mechanical Engineering, Aleja Tysiąclecia Państwa Polskiego 7, 25314 Kielce, Poland
[2]University of Žilina, Faculty of Mechanical Engineering, Univerzitná 8215/1, 01026 Žilina, Slovakia

Abstract. The aim of the paper was to reconstruct the missing data by applying the model which describes variability of sound level in the whole period from 2013 to 2016. To build the model, the computational intelligence methods, like fuzzy systems, or regression trees can be used. The latter approach was applied and we built the model with Cubist regression tree software, using equivalent sound levels recorded in 2013. For the reconstruction of sound level data in short period of time (several days), time series values and day_of_week values together should be used in the training dataset. For the reconstruction of sound level data in long period of time (several months) day_of_week values should be used in the training dataset.

Keywords: reconstruction of missing sound level data, random forest, regression trees

1 Introduction

The work presents the analysis of equivalent sound level [1] data recorded during measurements in monitoring stations located at Krakowska Street in Kielce, Poland. Krakowska street is a dual-carriageway road, which connects the south-western part of the city with the expressway towards Cracow. The monitoring station consists of sound level meter and weather station. The microphone for the acoustic pressure measurements is mounted at a distance of 4 m from the edge of the road and 4 m above ground level. Acoustic measurements were carried out [2] by using SVAN 958A, digital, four-channel, class 1, vibration and sound meter. In the research, the ½" prepolarized free field condenser microphone MIKROTECH GEFELL MK 250, which has sensitivity of 50 mV/Pa, was used together with SV 12L preamplifier. The range of frequencies was 3.5 Hz to 20 kHz, and dynamic range was 15–146 dB. The resolution of the signal RMS detector is 0.1 dB. The measurements were carried out 24 hours a day. The RMS values of the A sound level were registered in the buffer every 1 s and the results were recorded every 60 seconds. Based on the measurements that were conducted 24 hours a day, the equivalent sound levels were calculated for three time periods: day (6 a.m. to 6 p.m.), evening (6 p.m. to 10

* Corresponding author: mkekez@tu.kielce.pl

Reviewers: *Bohuš* Leitner, *Josef* Soukup

p.m.), and night (10 p.m. to 6 a.m.) [3]. The noise pollution (nuisance) due to long-term exposure to noise is very often measured by the equivalent sound level ($L_{Aeq,T}$), expressed in (dB), defined as [4]

$$L_{Aeq,T} = 10 \cdot \log\left[\frac{1}{T}\int_0^T \left(\frac{p_A(t)}{p_0}\right)^2 dt\right], \qquad (1)$$

where p_0 is the standardized reference acoustic pressure of 20 μPa. According to the ISO standard, this parameter can be determined from [5]

$$L_{eq} = 10 \log\left(\frac{1}{N}\sum_{i=1}^{i=N} 10^{0.1L_{A,i}}\right), \qquad (2)$$

where $L_{A,i}$ is the A-weighted acoustic pressure level measured in the measurement interval i. The average sound level can be determined as expected value from [6]

$$\bar{L} = \frac{1}{N}\sum_{i=1}^{i=N} L_{A,i}. \qquad (3)$$

2 Measurements results

The data collected at Krakowska Street in 2013 consists of 905 records describing equivalent sound level for a particular night (Fig. 1a), evening (Fig. 1b), or day (Fig. 1c). In the year 2013 there are missing values [7] for beginning of January, second half of January, first half of February, and most part of December.

When the missing values for a particular night, evening, and day are replaced by 0 dB, the calculated $L_{Aeq,T}$ values for the whole year are: 63.98 dB (nights), 68.13 dB (evenings), and 69.50 dB (days). When records which originally contained missing values are omitted, the equivalent sound levels are: 64.80 dB, 68.95, and 70.35 dB respectively. The correct calculation of $L_{Aeq,T}$ values requires the reconstruction of the missing data records.

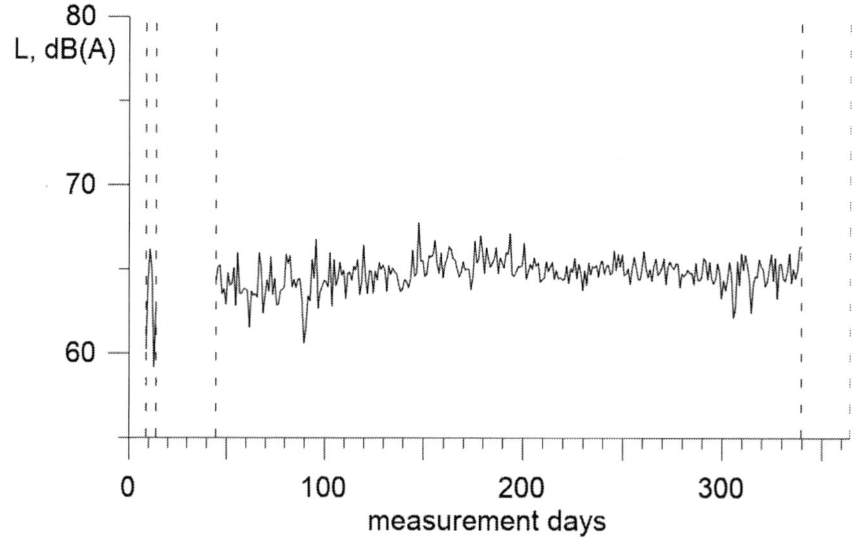

Fig. 1a. Krakowska Street, 2013, equivalent sound levels for nights (time_of_day=0)

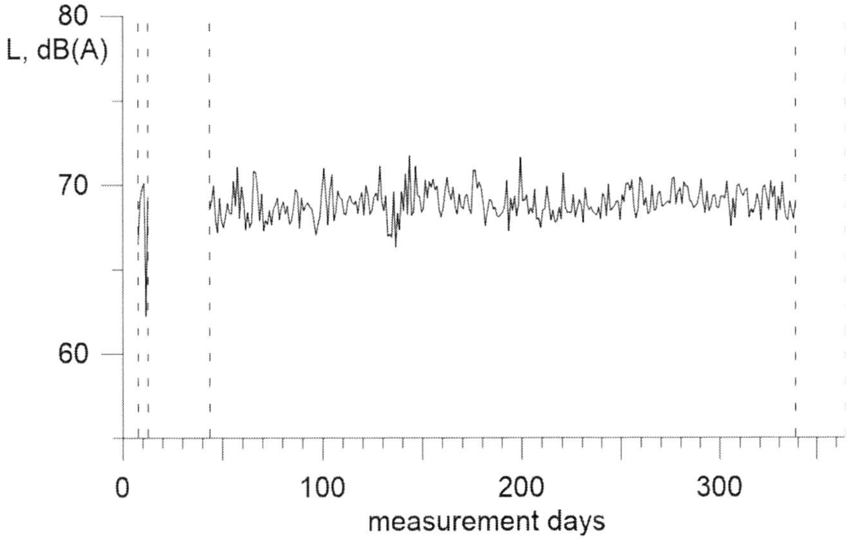

Fig. 1b. Krakowska Street, 2013, equivalent sound levels for evenings (time_of_day=1)

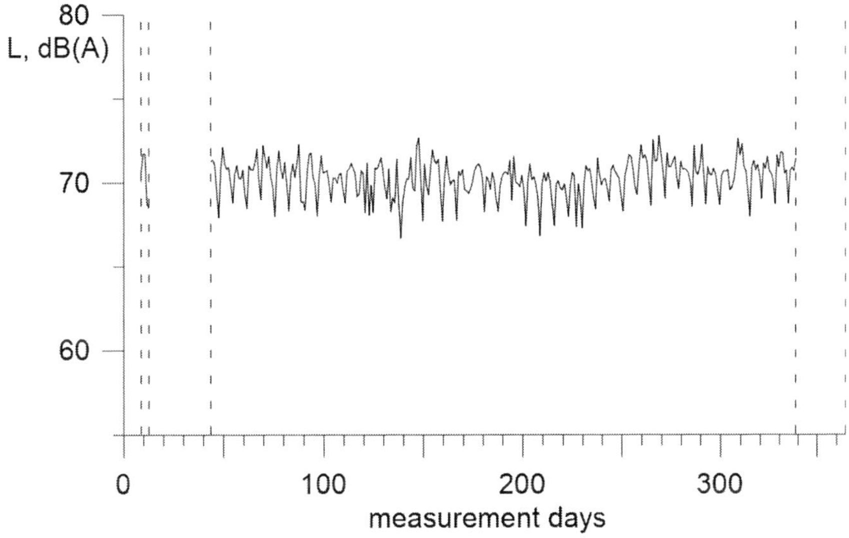

Fig. 1c. Krakowska Street, 2013, equivalent sound levels for days (time_of_day=2)

3 The proposed models for reconstruction of data for longer periods of time

The aim of this paper is to reconstruct the missing data by applying the model which describes variability of sound level in the whole year 2013. To build the model, the computational intelligence methods, like fuzzy systems [8], regression trees or time series analysis [9] can be used. The second approach was applied and we built the model with Cubist regression tree software [10], using equivalent sound levels recorded in 2013. The training data for Cubist consisted of records, each containing values of one output attribute, *dB_A* (sound level) and 2 input attributes: *day_of_the_week* (from 1 – Monday to 7 – Sunday), and *time_of_day* (0 – night, 1 – evening, 2 – day). The training dataset consisted of records collected in the first half of the year, while the test dataset contained records coming from the second half of the year.

The data was dived into 3 separate sets, one for each time of day (night, day, and evening). Cubist produced 3 rulesets (one for each dataset). Later, they were merged, and the obtained regression tree model had 5 rules:

IF day_of_the_week>6 AND time_of_day=0 THEN db_A = 63.48
IF day_of_the_week≤6 AND time_of_day=0 THEN db_A = 64.03 + 0.15 day_of_the_week
IF time_of_day=1 THEN db_A = 68.81
IF day_of_the_week>5 AND time_of_day=2 THEN db_A = 76.098 – 1.078 day_of_the_week
IF day_of_the_week≤5 AND time_of_day=2 THEN db_A = 70.74

The accuracy of the model on the data regarding the whole year 2013 is quite good (Tab. 1) in terms of mean absolute error (MAE, eq. 4) and root mean square error (RMSE, eq. 5)

$$MAE = \frac{\sum_{i=1}^{n}|\bar{y}_i - y_i|}{n}, \quad (4)$$

$$RMSE = \sqrt{\frac{\sum_{i=1}^{n}(\bar{y}_i - y_i)^2}{n}}, \quad (5)$$

where y_i denotes real *db_A* value, $\bar{y}_i$ denotes *db_A* value calculated by the model, and n is the number of records. For every day of week, and every time of day, MAE lies within the range of 0.47 dB (Saturday, day) to 0.98 dB (Sunday, night). The values of RMSE for every day of week and every time of day lies within the range of 0.59 dB (Saturday, day) to 1.28 dB (Sunday, night). The values of MAE for all days of week together are: 0.83 and 0.65 dB (nights), 0.87 and 0.61 dB (evenings), 0.58 and 0.59 dB (days), on the training and test data respectively.

Table 1. Accuracy of the Cubist regression tree model on the data from the whole year 2013

	Monday	**Wednesday**	**Friday**	**Sunday**
RMSE [dB]				
Night	1.02	1.05	0.74	1.28
Day	0.93	0.72	0.79	0.75
Evening	0.92	0.92	1.10	0.75
MAE [dB]				
Night	0.85	0.74	0.59	0.98
Day	0.67	0.54	0.60	0.59
Evening	0.77	0.71	0.81	0.62

Ten-fold cross validation (repeated several times) shows that MAE (mean absolute error) of the model for the first half of the year 2013 is 0.84–0.87 dB (nights), 0.87–0.88 dB (evenings), 0.58–0.60 dB (days). The scatter plots for prediction of sound level, for the test data (second half of the year 2013) are shown in Fig. 2.

When the missing values of sound level for a particular night, evening, and day are replaced by the values obtained by the Cubist model, the $L_{Aeq,T}$ values calculated for the whole year are: 64.74 (nights), 68.93 dB (evenings), and 70.35 (days).

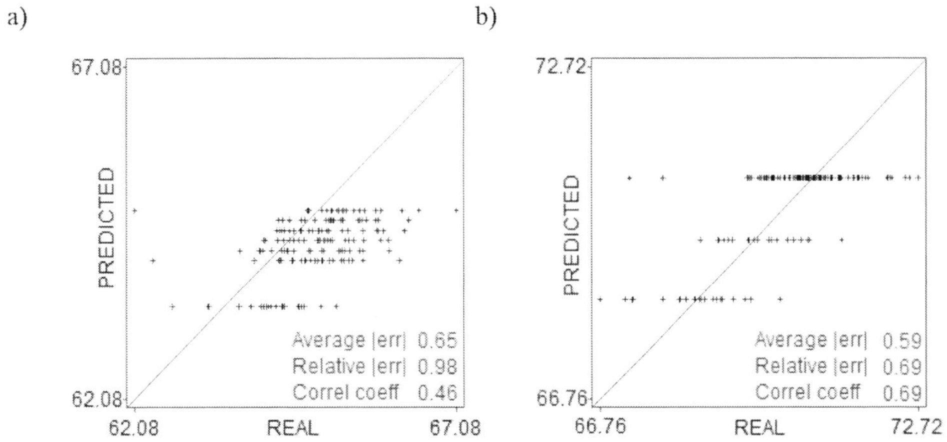

Fig. 2. Scatter plots for prediction of equivalent sound level by Cubist regression tree model in second half of the year: a) for nights (time_of_day=0), b) for days (time_of_day=2)

Later, the committee of models was built. At first, the Cubist system was run with the option „committee of 5 members". The transparency of the obtained model was low (5 regression trees in one model), and the accuracy was exactly the same as for single regression tree. Later, Cubist software was run with the option „committee of 100 members". In this case, the transparency of the models was very low (100 regression trees), and the accuracy was almost exactly the same as for the single regression tree.

Later, RandomForest algorithm from WEKA software package was used to build random forest. Because the number of features (i.e. input attributes) is only 2, the random feature selection had almost „nothing to do". The accuracy of the obtained model for evenings is better than that of the single regression tree built by Cubist. Mean average error (MAE) was: 0.84 dB and 0.65 dB for nights, 0.82 dB and 0.59 dB for evenings, 0.59 dB and 0.61 dB for days, on training and test data respectively.

Next, the neural network model (multilayer perceptron) was built by Multilayer Perceptron algorithm from WEKA software. The accuracy of the obtained model is worse than accuracy of regression tree built by Cubist. Comprehensibility of the model is very low, because of the form of the models (weights and thresholds compared to if-then rules in the previously presented models).

4 The proposed models for reconstruction of data for short periods of time

The next aim of this paper is to reconstruct the missing data in short periods of time (several consecutive days only) with high accuracy by applying the model which describes variability of sound level in the whole year 2013. In this case, the training data contains time series data *t_1, t_2,.. ,t_6*. The dataset consisted of records, each containing values of one output attribute, *dB_A* (sound level), and 8 input attributes: *day_of_the_week* (from 1 – Monday to 7 – Sunday), *time_of_day* (0 – night, 1 – evening, 2 – day), *t_1* (sound level value at the same time of day, but one day earlier), *t_2* (similarly, but two days earlier), *t_3*, *t_4*, *t_5*, and *t_6*. The training dataset consisted of records collected in the first half of the year, while the test dataset contained records registered in the second half of the year. Again, the data was dived into 3 separate sets, one for each time of day (night, day, and evening). Again, Cubist produced 3 rulesets (one for each dataset). Later, they were merged, and the obtained regression tree model had 5 rules:

IF day_week <= 6 AND time_of_day=0 THEN db_A = 31.598 + 0.343 t_1 + 0.169 t_2
IF day_week > 6 AND time_of_day=0 THEN db_A = 2.999 + 0.656 t_1 + 0.255 t_5 - 0.019 day_week + 0.023 t_2
IF time_of_day=1 THEN db_A = 56.69 + 0.176 t_1
IF day_week > 5 AND time_of_day=2 THEN db_A = 76.346 - 1.116 day_week
IF day_week <= 5 AND time_of_day=2 THEN db_A = 48.393 + 0.25 t_1 - 0.113 day_week + 0.141 t_4 - 0.068 t_5

The transparency of the model is moderate (5 complicated rules). The accuracy is the best, but the model can be applied only for short periods of missing data (up to one week). For longer periods, most input values for the model *(t_6, t_5, ..., t_1)* must be produced by the model itself, which would lead to lower accuracy. When *t_1* to *t_6* are taken from measurement data, the accuracy is 0.65 and 0.53 dB (nights), 0.79 and 0.60 dB (evenings), 0.56 and 0.56 dB (days), on training and test data respectively.

When the missing values of sound level for a particular night, evening, and day are replaced by the values obtained by this model, the $L_{Aeq,T}$ values calculated for the whole year are: 64.74 dB (nights), 68.93 dB (evenings), and 70.35 (days).

When the training dataset did not contain *day_of_the_week* and *time_of_day* input attributes (only *dB_A* and *t_1*, *t_2*, ..., *t_6* time series attributes were present), the Cubist software produced regression tree, which had much lower accuracy for nights and days than the previous model.

The comparison of results (Tab. 2) shows that the highest accuracy is obtained by the Cubist regression tree built for *day_of_week* and time series (*t_1*, …, *t_6*) training data, and this accuracy is better than that of the classifier presented in the previous section.

Table 2. Accuracy of models for equivalent sound level at Krakowska Street monitoring station

Mean Average Error, dB(A)	**Night**		**Evening**		**Day**	
	train data	test data	train data	test data	train data	test data
Cubist regression tree	0.83	0.65	0.87	0.61	0.58	0.59
Committee of 5 regression trees, Cubist software	0.83	0.65	0.87	0.61	0.58	0.59
Committee of 100 regression trees, Cubist software	0.83	0.65	0.87	0.61	0.58	0.59
Random forest, WEKA software	0.84	0.65	0.82	0.59	0.59	0.60
Neural network, WEKA software	0.93	0.61	0.87	0.62	0.63	0.61
Cubist regression tree for day_of_week and time series data	0.65	0.53	0.79	0.60	0.56	0.56
Cubist regression tree for time series data only	0.76	0.57	0.79	0.60	0.65	0.85

Conclusions

The comprehensibility of the classifier produced by Cubist regression tree for Krakowska Street monitoring station is highest among all obtained models. The comparison of results (Tab. 2) shows that for short periods of missing data, the highest accuracy is obtained by the Cubist regression tree built for *day_of_week* and time series (*t_1*, …, *t_6*) training data – 0.53 dB, 0.60 dB, 0.56 dB on test data for nights, evenings, and days respectively. For longer periods of missing data, the best accuracy is achieved by two models, obtained from training dataset which does not contain time series data, namely Cubist regression tree (for night and day equivalent sound levels – 0.83 dB and 0.58 dB respectively, on training data), and WEKA implementation of random forest (for evening equivalent sound levels – 0.82

dB on training data). The time necessary to build most of the presented models does not exceed 0.2 s.

The obtained models allow the calculation of the equivalent sound levels for the whole year: 64.74 dB (for nights), 68.93 dB (for evenings), 70.35 (for days). All the obtained results can be used for verification of sound propagation models during elaboration of city acoustic map.

References

1. W. Batko, B. Stępień, *Type A Standard Uncertainty of Long-Term Noise Indicators*. Arch. Acoust. **39** (1), 25-36 (2014)
2. A. Bąkowski, L. Radziszewski, Z. Skrobacki, *Assessment of uncertainty in urban traffic noise measurement*. Procedia Engineering **177**, 281–288 (2017)
3. K. Marciniuk, M. Szczodrak, B. Kostek, *Performance of Noise Map Service Working in Cloud Computing Environment*. Arch. Acoust. **41** (2), 297–302 (2016) DOI: 10.1515/aoa-2016-0029
4. R. Makarewicz, R. Gołębiewski, *The Uncertainty of Noise Composed of Separate Sound Events*. Arch. Acoust. **41** (1), 133–138 (2016) DOI: 10.1515/aoa-2016-0013
5. *ISO 9612 :2009, Acoustics – Determination of occupational noise exposure – Engineering method*.
6. D. I. Mihajlov, M. R. Prascevic, *Permanent and Semi-permanent Road Traffic Noise Monitoring in the City of Nis (Serbia)*. J. of Low Frequency Noise, Vibration and Active Control **34** (3), 251–268 (2015)
7. M. Spławińska, *The Problem Of Imputation Of The Missing Data From The Continuous Counts Of Road Traffic*. Archives of Civil Engineering **61** (1), 131–145 (2015) DOI:10.1515/ace-2015-0009
8. M. Kekez, L. Radziszewski, A. Sapietova, *Fuel type recognition by classifiers developed with computational intelligence methods using combustion pressure data and the crankshaft angle at which heat release reaches its maximum*. Procedia Engineering **136**, 353–358 (2016)
9. E. Solazzo, S. Galmarini, *Comparing apples with apples: Using spatially distributed time series of monitoring data for model evaluation*. Atmospheric Environment **112**, 234–245 (2015)
10. https://www.rulequest.com/cubist-info.html

Numerical estimation of temperature field in a laser welded butt joint made of dissimilar materials

Zbigniew Saternus[1,*] *Wiesława* Piekarska[1], *Marcin* Kubiak[1], *Tomasz* Domański[1], *Dorota* Goszczyńska-Króliszewska[1]

[1]Czestochowa University of Technology, Faculty of Mechanical Engineering and Computer Science, Czestochowa, Poland

Abstract. The paper concerns numerical analysis of thermal phenomena occurring in the butt welding of two different materials by a laser beam welding. The temperature distribution for the welded butt-joint is obtained on the basis of numerical simulations performed in the ABAQUS program. Numerical analysis takes into account the thermophysical properties of welded plate made of two different materials. Temperature distribution in analysed joints is obtained on the basis of numerical simulation in Abaqus/Standard solver, which allowed the determination of the geometry of laser welded butt-joint.

Keywords: Numerical modelling, laser welding, dissimilar materials, temperature field, Abaqus

1 Introduction

The main objective of optimization of welding technology is to reduce the cost of production and the mass of joined elements while maintaining optimum durability [1, 2]. One of the new trends in the welding industry is performing welded joints with dissimilar materials [2-5]. These types of welded structures are commonly used today in aviation, construction of steam power plants, chemical installations etc. Depending on the combination of materials used, for example: austenitic stainless steel – non-austenitic steel or austenitic stainless steel – copper/copper alloy, the required strength, anti-corrosion and heat properties must be obtained from performed joints [5]. Another aspect which supports the use of dissimilar joints is the economic factor.

Welding of dissimilar materials is quite complicated because of the large differences in material behavior of joined materials [2-4]. The most important parameters of the joined materials are: the melting temperature, thermal conductivity and thermal expansion coefficient. Issues in joined materials occur when more than one of these conditions are present at the same time [5].

Most often arc welding or laser beam welding are used in in this type of joints. Each of these technologies involves specific conditions of heat distribution in the analyzed joints

* Corresponding author: saternus@imipkm.pcz.pl

Reviewers: *Alžbeta* Sapietová, *Ján* Vavro *Jr.*

[3, 6, 7]. The quantity of heat impact into the welded material has a significant effect on the microstructure change and joint deformation [7, 8].

Performing experiments of dissimilar materials joining is quite expensive as it requires many experimental tests to determine the proper parameters of the welding process. Sasaki and Ikeno in the study [9] conducted experiments on laser welding of butt joints made of dissimilar materials - austenitic steel and brass. They experimentally analyzed the influence of parameters such as: laser beam focus, power change and welding speed of heat source on the quality of the resulting joints. Figure 1 shows exemplary welds for different positioning of welding heat source.

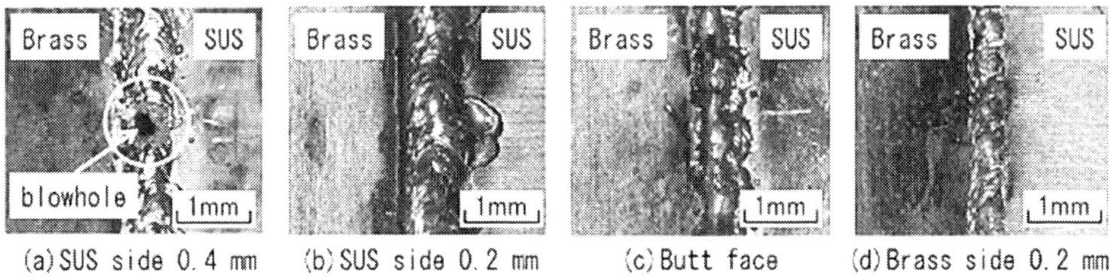

Fig. 1. View of welded beads for laser welded butt joint [9]

Due to the cost of experiments, numerical modeling becomes a useful tool for analyzing the relevant technological parameters of the process [10, 11]. Numerical modeling can greatly reduce costs of experimental research at the initial stage of construction design [6, 9, 12]. An important case of study in numerical research is a reliable representation of the real process and the appropriate adaption of mathematical and numerical models of the analyzed phenomena.

Numerical analysis of laser welded butt joint made of two dissimilar materials: austenitic steels X5CrNi18-10 and brass CuSn10 is performed in this study. The presented numerical analysis is based on experimental studies from literature [9]. Numerical simulations are performed in the Abaqus/FEA computational software. A three dimensional discrete model of the discussed system is developed with temperature dependent thermophysical properties of analyzed materials taken into account. In order to provide a moveable welding source in the model the implementation of additional numerical subroutine DFLUX is necessary. Temperature distribution and size of the melted zone in the laser welded butt joint are estimated on the basis of performed numerical simulations.

2 Modelling of thermal phenomena in Abaqus

Numerical analysis of thermal phenomena occurring in the welding process of welded butt joint made of disimilar materials is difficult to carry out. It requires taking into account the specific conditions of the laser beam welding process and taking into account the different thermophysical properties of the joined materials (Fig. 2). The numerical model of the movable welding source requires implementation of additional subroutine DFLUX in Abaqus software.

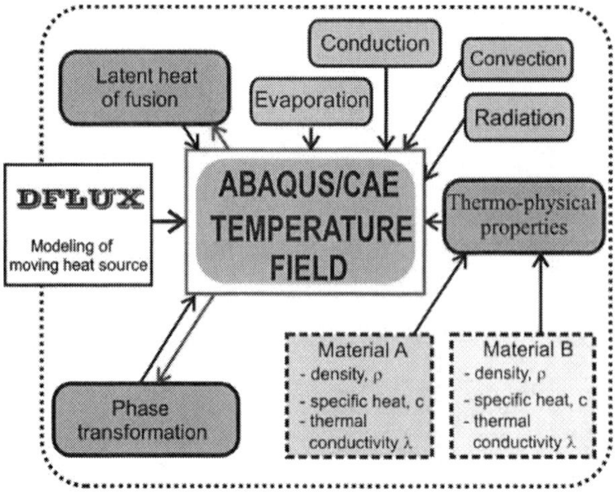

Fig. 2. Scheme of coupled thermal phenomena

The numerical analysis of heat transfer in the Abaqus software is based on the energy conservation equation and Fourier's law, expressed as follows [6, 13]:

$$\int_V \rho \frac{\partial U}{\partial t} \delta T\, dV + \int_V \frac{\partial \delta T}{\partial x_\alpha} \cdot \left(\lambda \frac{\partial T}{\partial x_\alpha} \right) dV = \int_V \delta T\, q_v dV + \int_S \delta T\, q_s dS \tag{1}$$

where λ is a thermal conductivity [W/m °C], $U = U(T)$ is a internal energy [J/kg], q_v is a laser beam heat source [W/m^3], $T = T(x_\alpha, t)$ is a temperature [°C], q_s is a boundary heat flux [W/m^2], δT is a variational function, ρ is a density [kg/m^3], $T = T(x_\alpha, t)$ is temperature [°C].

Temperature field is expressed in the criterion of weighted residuals method. Simulations are carried out in Lagrange's coordinates, where coordinates of the center of heat source is determined at each time step, depending on the welding speed [6]. Equation (1) is completed by the initial condition $t = 0 : T = T_0$ and boundary conditions of Dirichlet and Neumann and Newton type, which takes into account the heat loss to convection, radiation and evaporation [14]:

$$T|_\Gamma = \widetilde{T} \qquad q_{Sym} = -\lambda \frac{\partial T}{\partial n} = 0 \qquad q_S = -\lambda \frac{\partial T}{\partial n} = \alpha_k (T|_\Gamma - T_0) + \varepsilon \sigma (T|_\Gamma{}^4 - T_0^4) \tag{2}$$

where α_k is convective coefficient, ε is radiation (ε = 0.5), σ is Stefan-Boltzman constant and $q(r, 0)$ is the heat flux towards the top surface of welded workpiece (z = 0) in the source activity zone of radius r, T_0 = 20 °C is an ambient temperature.

Most widely used mathematical model of heat source power having Gaussian type distribution is used in numerical model [15]. The presented model assumes a linear decrease of energy density along material penetration depth (Fig. 3), expressed as follows:

$$q_v(r,z) = \frac{Q}{\pi r_o^2 s} \exp\left[\left(1 - \frac{r^2}{r_o^2}\right)\right]\left(1 - \frac{z}{s}\right) \tag{3}$$

where Q is a laser beam power [W], r_0 is a beam radius [m], $r = \sqrt{x^2 + y^2}$ is actual radius [m], s is penetration depth [m], z is actual penetration [m].

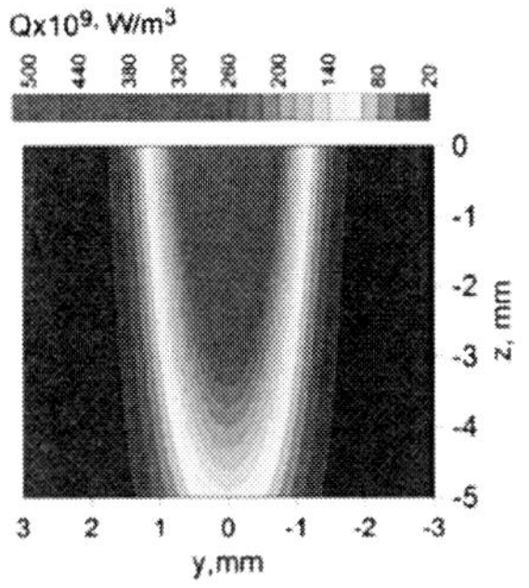

Fig. 3. Gaussian heat source power distribution along material penetration depth

The modeling of the moveable welding heat source in the Abaqus is performed through the additional DFLUX subroutine, written in the Fortran programming language.

3 Numerical model

The joint is made of two dissimilar materials: austenitic stainless steel X5CrNi18-10 and brass CuSn10. The dimensions of joined sheets are taken from the experimental study in literature [9], set as: 50 x 25x 1.5 mm. The scheme of the analyzed system is presented in Figure 4.

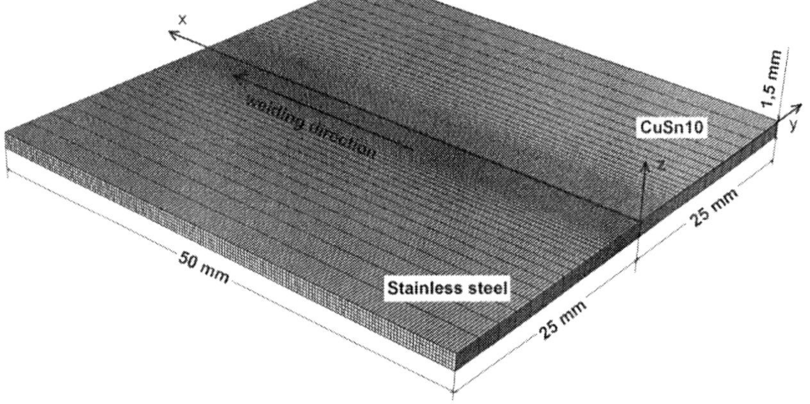

Fig. 4. Scheme of considered system

Perfect contact at the interface between joined parts is assumed in the developed numerical model of butt joints [6, 16]. The finite element mesh is used to provide good quality results with not long duration of the time simulation. The size of finite element is assumed to be quite small in welding source activity zone, increasing linearly with the distance from the welding line. The total number of elements used in the numerical model is 30500. The thermal analysis use rectangular finite element mesh, whit DC3D8 type of elements [13]. The calculations are carried out for three cases depending on the positioning center of the welding source (Fig. 5).

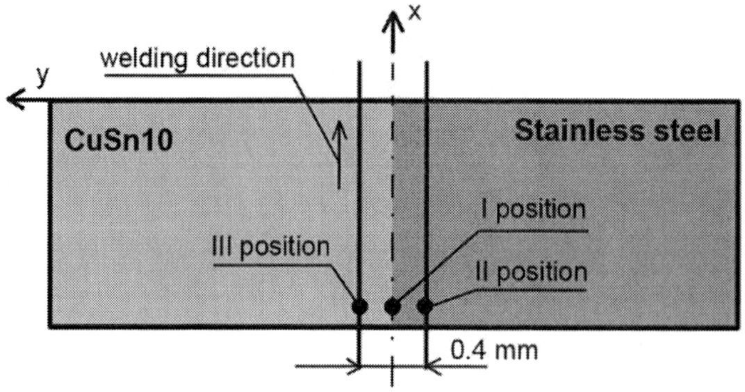

Fig. 5. Coordinates of the position of welding heat source central axis

Material model used in the analysis takes into account thermophysical properties of the welded materials changing with temperature. Figure 6 shows the assumed properties of austenitic steels, while in Figure 7 used in calculation properties of brass. Assumed in the analysis convective coefficient is set to: α_k= 100 W/m^2 °C.

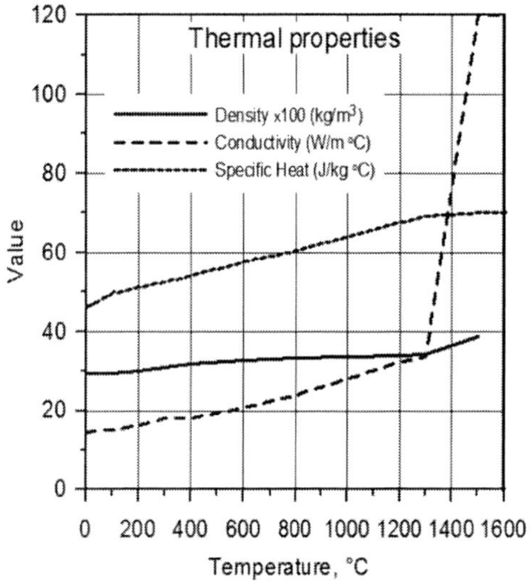

Fig. 6. Thermophysical properties of X5CrNi18-10 stainless steel

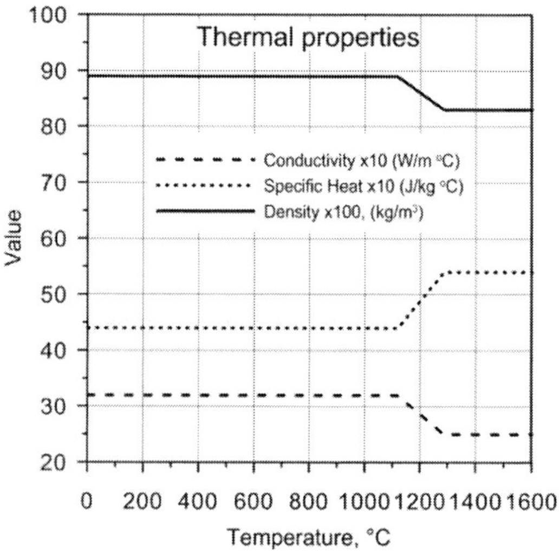

Fig. 7. Thermophysical properties of brass CuSn10

Solidus and liqidus temperatures for welded materials are:

- for austenitic stainless steel - T_S=1400°C and T_L=1455°C, whereas latent heat of fusion H_L=260×10^3 J/kg.
- for brass - T_S=1118°C and T_L=1280°C, whereas latent heat of fusion H_L=220×10^3 J/kg.

4 Results and discussion

The following numerical parameters of the heating source are assumed for numerical calculations of laser welding: beam power Q = 750 W, welding beam diameter r_o = 0.4 mm, penetration depth s = 3.5 mm and welding speed v = 300 mm / min. The welding heat source parameters are taken from the literature [9]. Numerical calculations are performed in Abaqus/Standard solver.

Temperature field and shape of melted zone are estimated in laser butt welded joints on the basis of performed numerical simulations. Figure 8 shows the temperature field in the joint for the positioning of welding source in the contact line of the joined sheets. The isoline T_L is determined at the boundary of melted zone for each material.

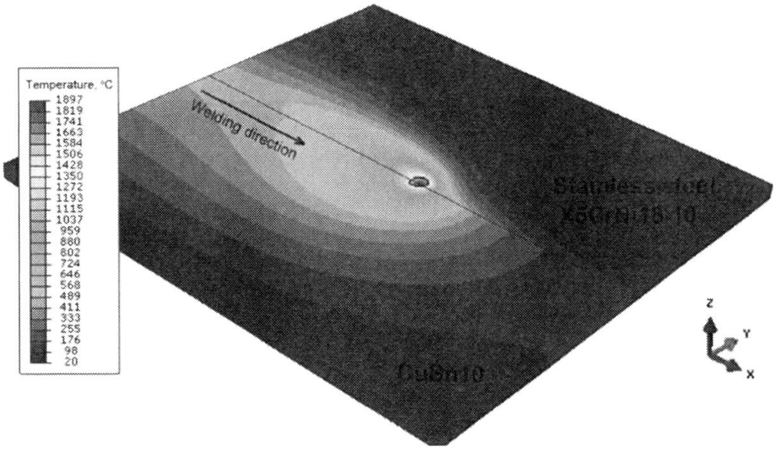

Fig. 8. Numerically estimated temperature field in laser welded butt joint

Figure 9 shows the temperature distribution in the cross section of welded joints. Presented figures indicated boundaries of melted zone for each analyzed material.

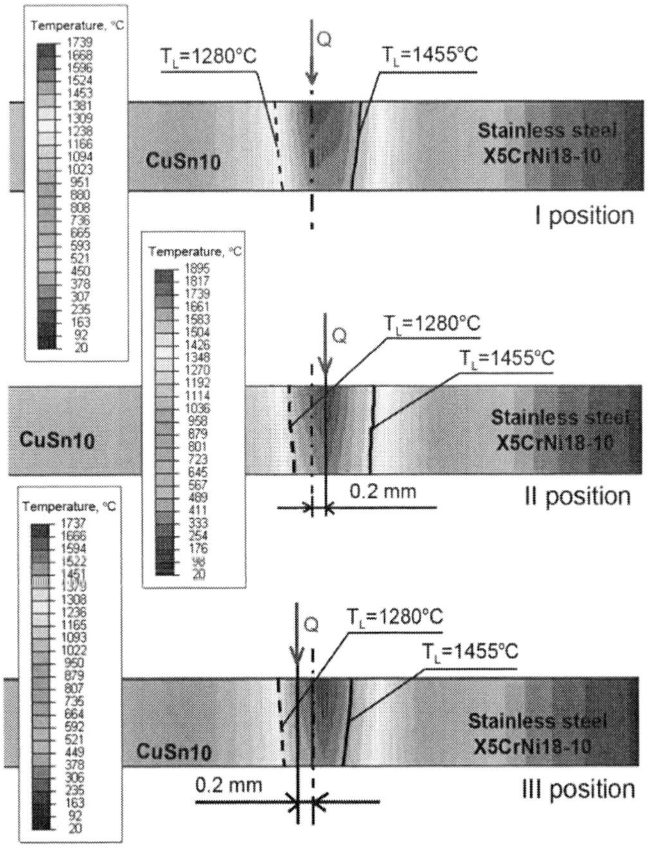

Fig. 9. Temperature field in the cross section of the joint for different positions of the heat source

From the analysis of obtained simulation results it can be seen that for the III case of heat source 0.2 mm positioning on the brass side the narrowest zone melting is achieved. This result is also confirmed by experimental research [9].

Conclusions

It is difficult to obtain good quality welded joints of dissimilar materials because of the different thermophysical properties of joined materials. For significant differences between materials it is even more difficult to perform a good welded joint, which could have a high capacity to cracking. Numerical modeling of the welding process of dissimilar materials is a useful tool for engineers to determine a proper process parameters. Numerical prediction of thermal phenomena in welded elements, mainly temperature distribution, is a significant help in the initial stage of construction design. This can also significantly reduce the cost of conducting experimental research. The Abaqus software is a universal tool for analyzing laser beam welding process.

From the analysis of laser beam welding process of austenitic stainless steel to the brass it can be seen that the position of center of the heat source has a significant influence on the melted zone. The results of numerical simulations confirm the results of experimental work carried out by the authors of [9].

References

1. K. Ciechacki, T. Szykowny, *Ocena jakości spawania różnoimiennych stali odpornych na korozje*. Inż. Ap. Chem. **49** (5), 28-30 (2010)
2. Y. Li, Sh. Hu, J. Shen, L. Liu, *Microstructures and mechanical properties of H62 brass-316L stainless steel in overlap welded joints by continuous-wave laser*. Int J Adv Manuf. Technol. **79**, 627-634 (2015)
3. N. Arivazhagan, S. Singh, S. Prakash, G. M. Reddy, *Investigation on AISI 304 austenitic stainless steel to AISI 4140 low alloy steel dissimilar joints by gas tungsten arc, electron beam and friction welding*. Materials & Design **32**, 3036-3050 (2011)
4. Ch. Shuhai, H. Jihua, X. Jun, Z. Xingke, L. Sanbao, *Influence of processing parameters on the characteristics of stainlesssteel/copper laser welding*. J Mater Process Tech. **222**, 43-51 (2015)
5. http://www.wisconsinwireworks.com/dissimilar_metals.html
6. M. Kubiak, W. Piekarska, S. Stano, Z. Saternus, *Numerical Modelling Of Thermal And Structural Phenomena In Yb:Yag Laser Butt-Welded Steel Elements*. Arch Metall Mater. **60** (2), 821-828 (2015)
7. M. Kubiak, W. Piekarska, Z. Saternus, T. Domański, *Numerical prediction of fusion zone and heat affected zone in hybrid Yb:YAG laser plus GMAW welding process with experimental verification*. Procedia Eng. **136**, 88-94 (2016)
8. T. Domański, W. Piekarska, M. Kubiak, Z. Saternus, *Determination of the final microstructure during processing carbon steel hardening*. Procedia Eng. **136**, 77-81 (2016)
9. M. Sasaki, J. Ikeno, *Laser Butt Welding of Brass and Stainless Steel*. J Adv Mech Des Syst. **5** (4), 347-357 (2011)
10. V. Dekýš, A. Sapietová, O. Števka, *Understanding of the dynamical properties of machines based on the interpretation of spectral measurements and FRF*. Experimental stress analysis, Applied Mechanics and Materials **486**, 106-112 (2014)

11. T. Domanski, A. Sapietova, M. Saga, *Application Of Abaqus Software For The Modeling Of Surface Progressive Hardening*. Procedia Eng. **177**, 64-69 (2017)
12. R. Koňár, M. Patek, *Numerical Simulation Of Dissimilar Weld Joint In Sysweld Simulation Software*. Tehnicki vjesnik **1**, 137-142 (2017)
13. SIMULIA, *Abaqus FEA theory manual*. Version 6.7, Dassault System (2007)
14. L. Sowa, A. Bokota, *Numerical simulation of the molten steel flow in the tundish of CSC machine*. Arch Metall Mater. **57** (4), 1163-1169 (2012)
15. S.A. Tsirkas, P. Papanikos, Th. Kermanidis, *Numerical simulation of the laser welding process in butt-joint specimens*. J Mater Process Tech. **134**, 59-69 (2003)
16. T. Skrzypczak, *A finite element multi-mesh approach for heat transport between disconnected regions*. Procedia Engineering **177**, 204-209 (2017)

Analysis of the influence of the pipe's wall thickness on the corrugation of the surface during inductive bending of pipes

Milan Sága[*,1], *Marián* Handrik[1], *Peter* Kopas[1], *Maroš* Tropp[2]

[1]University of Zilina, Faculty of Mechanical Engineering, Department of Applied Mechanics, Univerzitná 1, 010 26, Zilina, Slovak Republic
[2]Triple D Bending, 4707 Glenmore Trail SE, Calgary, Canada

Abstract. The paper presents a calculation model for simulation the technological process of inductive bending of pipes. In the case of an incorrect combination of geometric bending parameters (pipe diameter, bend radius and pipe wall thickness) occurs the corrugation of the inner part in the pipe bend. The proposed calculation model for the ADINA FE software allows for a relatively short time and at low cost to predict the emergence of corrugation.

Keywords: finite element analysis, induction bending, pipes, wall thickness, surface corrugation

1 Heat induction tube bending process analysis

The technological process of the hot heat induction tube bending is based on the principle of local heating of a tube and thus the change of mechanical properties of the material due to changed temperature. By means of a local increase in the temperature of the tube material, the mechanical properties are changed by plasticizing the material at lower stress values [1-3].

The first end of the tube is fixed in the bending arm, which rotates freely around the axis of rotation. The state of stress corresponding to the bend is induced in the tube by pushing the other end of the tube in the direction of the axis. The local change of mechanical properties in the point at the increased temperature allows the appearance of plastic deformations in the point of the increased temperature. The stresses in the area of elastic deformations are in other parts of the tube. Forming the boundary of the place with the local change of temperature is determined by a gradual displacement of the cold parts of the tube to the heating point, and there is a water spray on the other side of the inductor which cools the plastic deformed material. Two ways of guiding the tubes are used when bending the tubes using the induction heating: the use of guide pulleys, or the constraint at the feed point [4]. The schematic illustration of the tube bending process using the guide pulleys is shown in Figure 1.

* Corresponding author: milan.saga@fstroj.uniza.sk

Reviewers: *Jozef* Bocko, *Juraj* Gerlici

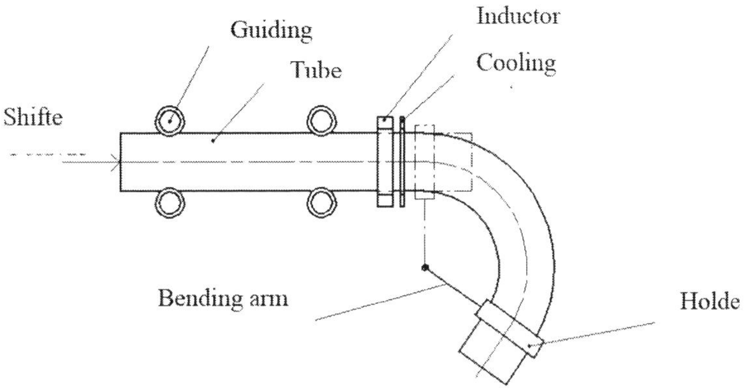

Fig. 1. The scheme of heat induction tube bending

A modern design of electrical power supplies for induction heating allows a non-contact temperature measurement at the heating point and subsequent control of the electrical quantities of the supply so as to maintain a constant temperature of the material at the heating point. A heat induction tube bending device is shown in Figure 2.

2 A simplified calculation model of heat induction tube bending

If we want to create a physically accurate model of the heat induction tube bending process, the computational model must contain the following physical fields:

- Electromagnetism – induction heating.
- Thermomechanical fields - change of mechanical properties due to temperature, plasticization of the material and contact of bodies.
- Interaction of fluid with the structure - coolant flow, heat transfer from the structure, free surface.

Such a computational analysis model becomes too complex when using the mentioned multiphysical model. The complexity of the problem lies in the large number of unknown variables that need to be solved and, at the same time, in the non-linearity of the problem solved. At present, it is possible to solve these complex physical models mainly by the finite element method (FEM). When solving the problem of the electromagnetism, fluid flow and contact, the finite element method requites a fine mesh of finite elements to achieve the convergence. The resulting models can have approximately hundreds of millions of equations and the solution is time-consuming.

In a more in-depth study of the hot heat induction tube bending process we find that it is possible to eliminate the electromagnetic fields and fluid flow from the model. The resulting model will deal with the thermomechanical analysis with the plasticity of the material:

- The purpose of the induction heating is to heat the tube at a defined constant temperature at point of inductor. In the induction heating, the heat is induced in the surface layers of the material and progressively spreads from the surface to the material. During the heating, the power of the inductor's electrical source is changed so that the temperature at the tube surface is constant.
- The purpose of the water spray is to cool the surface of the tube and to remove heat. In practice, the cooling water flow is set so that the heat removal is sufficient and the tube has cooled.

Fig. 2. Device for heat induction tube bending

There are two zones on the tube, one with the heating temperature and the other with the cooling temperature. Such a zone can be defined in the finite element method by applying appropriate limit conditions. During the bending, the tube is moved and the inductor with the water spray is not moved, which is a problem in defining the limit conditions that would also have to be moved around the surface.

The second way is to use heat transfer contact. At the inductor site, the bodies with the heating temperature are formed. Their contact with the tube as well as heat transfer in the respective contact are defined. The bodies which have coolant temperature are also formed at the point of the spray, and contact with the surface of the tube is defined [5].

3 Finite element method calculation model

If we replace the induction heating and the cooling spray with the heat transfer contact then the physical model of the heat induction tube bending is reduced to a thermomechanical analysis with heat transfer in the contact. All degrees of freedom apart from the radial movement are taken away from the formed spare bodies for inductor and spray. For all the bodies formed, the force of a small value is defined which causes that the bodies are pushed against the surface of the tube in the radial direction. The pushing the bodies against each other in the radial direction is necessary to ensure the contact lasting during the whole calculation and also when the shape of the tube is changed in the point of the plastic deformation - tube bending point, see Figure 3.

This type of task has a relatively good convergence and allows a significant reduction of the solved final equation system, in our case the final model had about 500 000 equations for the mechanical analysis and 150 000 equations for the heat transfer. An iterative solver was used in solving the thermomechanical analysis, where the structural analysis with the contact and the transfer of heat in a given time step were solved separately. Two to three iterations are needed to solve one time step.

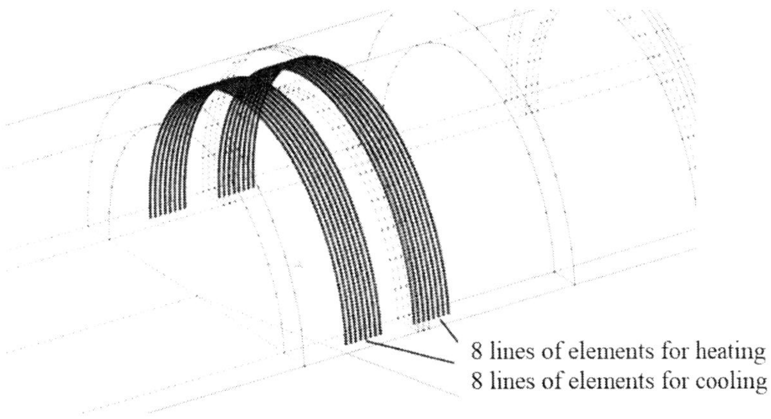

Fig. 3. Auxiliary bodies for heating and cooling the surface of the tube

On the other hand, there are separate matrices for structural analysis and temperature analysis with a smaller difference in value of elements on the main diagonal, which significantly improves the convergence of the contact problem and reduces the number of iterations required. Another simplification of the problem is the use of symmetry to create a geometric model and limit conditions. Due to the low feed rate of 2.4 m / hour, the task is solved as a quasi-static task considering large displacements and distortions. We use the temperature-dependent material bilinear model in the solution, see Figure 4 [6]. The geometric model of the tube is made of volume bodies; the need for the use of the volume bodies is given by the heat transfer from the surface of the tube in the direction of the wall thickness of the tube. If the shell elements are used, it would not be possible to determine the temperature gradient in the direction of the thickness of the tube. 5 elements for the wall thickness of the tube, 18 elements in the direction of the circumference of the tube on the half model, 10 mm in the longitudinal direction are used when generating a mesh. The finite element mesh is shown in Figure 5. The model is supplemented by the volume elements at the beginning of the tube and the beam elements that model the bending arm and its fixing to the tube [7].

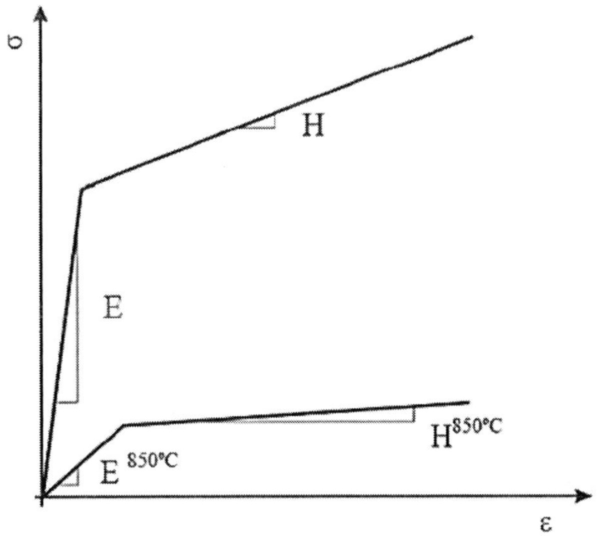

Fig. 4. Thermal-dependent bilinear material model

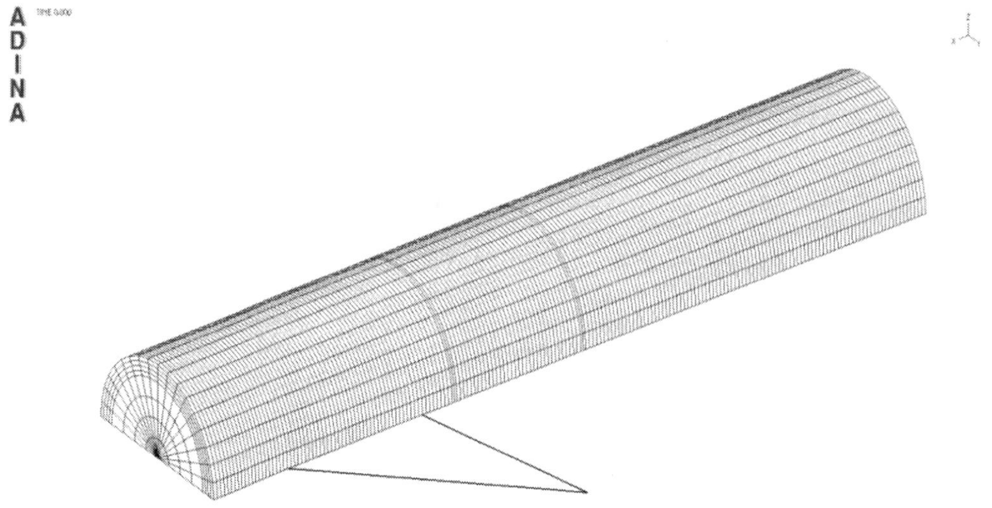

Fig. 5. Finite element mesh

4 Experimental verification of the calculation model

The proposed calculation model for simulating the heat induction hot tube bending process was experimentally verified and the results obtained by the calculation revealed a sufficient compliance with the experimental sample [8-12]. The investigated changes of forms (simulation vs. experiment) showed an error of less than 7%. The shape of the bend and change of the wall thickness of the tube were evaluated when comparing the obtained experimental results and FE model. The final shape of the tube after bending is shown in Figures 6 and 7.

Fig. 6. Final shape of the tube after bending

Fig. 7. Final shape of the tube after bending, sectional view of the tube

5 Analysis of the effect of the wall thickness of the tube on the final shape after bending

The parameters describing the geometry of the tube were observed when studying the effect of the geometrical dimensions of the tube on the final shape of the tube after bending:

- Wall thickness of the tube: 10, 15, 20, 25, 30 [mm].
- Diameter of the tube: 245 [mm].
- Bending radius: 508 [mm].
- Angle of bend: 30 [°].
- Inductor and water spray width: 30 [mm].
- Inductor and water spray distance: 30 [mm].
- Heating temperature: 900 [°].
- Cooling temperature: 20 [°].

The final shapes of the tubes after bending for the respective wall thickness are shown in Figures 8 to Figures 12. A slight waviness can be observed on the inner side of the bent tube when the wall thickness of the tube was 10 and 15 mm. An increase in the wall thickness of the tube was found in case of a higher diameter of tubes.

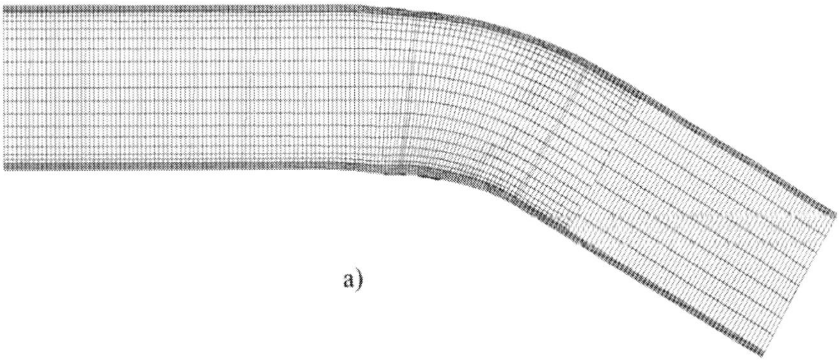

a)

Fig. 8. Calculated bend form, wall thickness 10 [mm]

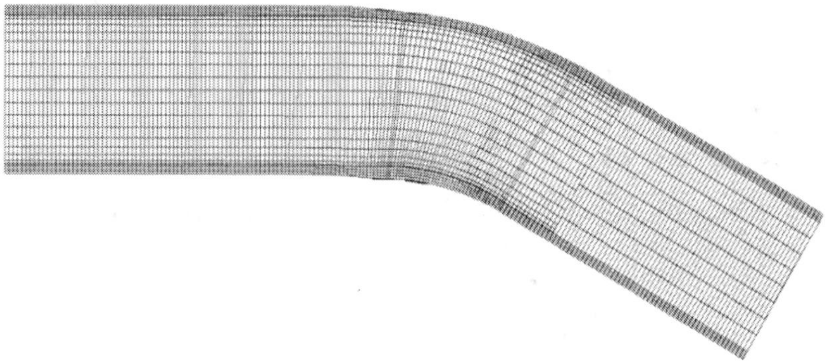

Fig. 9. Calculated bend form, wall thickness 15 [mm]

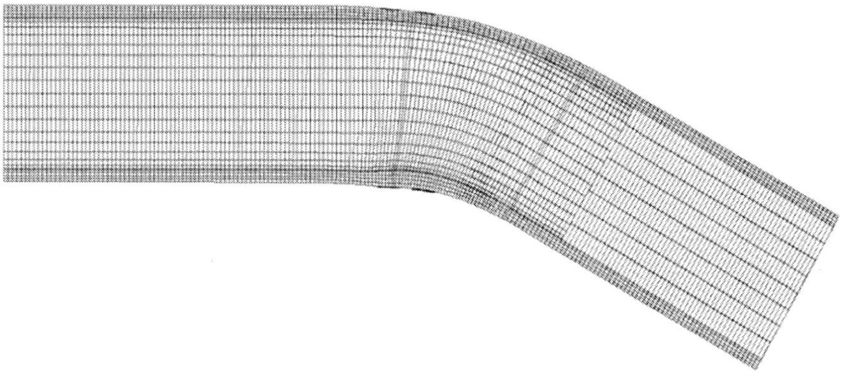

Fig. 10. Calculated bend form, wall thickness 20 [mm]

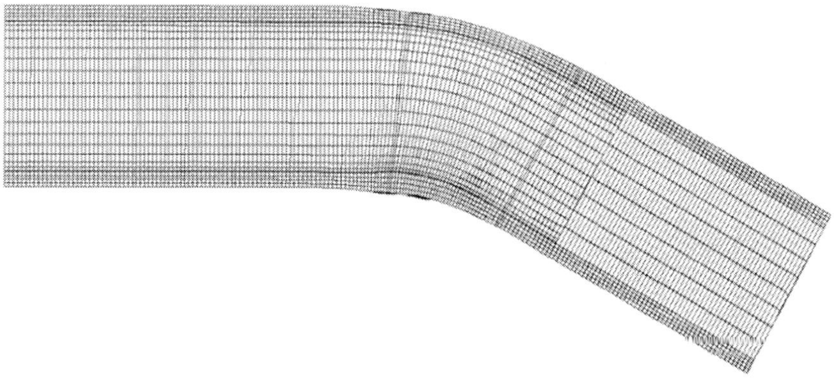

Fig. 11. Calculated bend form, wall thickness 25 [mm]

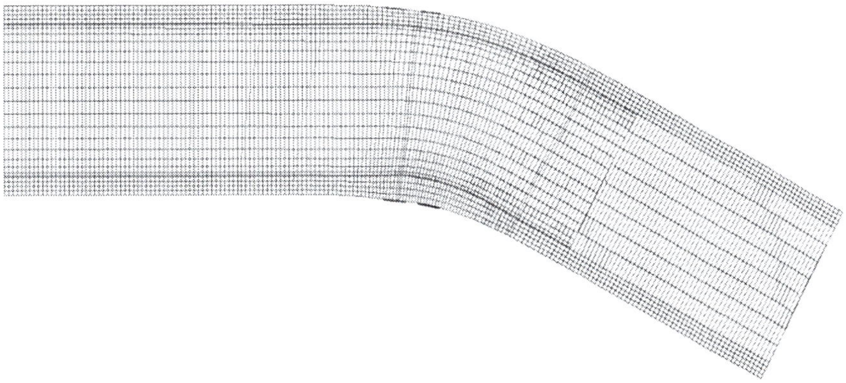

Fig. 12. Calculated bend form, wall thickness 30 [mm] is defined.

Conclusion

The created computational analysis model models the hot heat induction tube bending with the sufficient accuracy. The speed of the calculation model is acceptable for a given type of task. The calculation takes about 2 times longer than the actual bending process in the real production. The use of calculation models is effective despite the prolonged time of calculation, which leads, for example, to lower power consumption, fewer occurrences of production defects and the machine may be used for the production itself and not for the trial-and-error method experiments.

By using a computational calculation model, we can predict the shape of the tube after bending and so avoid the experiments, in which we know in advance that the final shape will not meet the required tolerances for the shape of the bend.

This work is supported by KEGA 015ŽU-4/2017 Digital technologies for machine building study programs as a part of strategy "Internet of Things".
This work was supported by the Slovak Research and Development Agency under the contract No.APVV-14-0096.

References

1. M. Handrik, P. Kopas, V. Baniari, M. Vasko, M. Saga, *Analysis of stress and strain of fatigue specimens localised in the cross-sectional area of the gauge section testing on bi-axial fatigue machine loaded in the high-cycle fatigue region*. Procedia Engineering **177**, 516-519 (2017)
2. P. Kopas, M. Saga, V. Baniari, M. Vasko, M. Handrik, *A plastic strain and stress analysis of bending and torsion fatigue specimens in the low-cycle fatigue region using the finite element methods*. Procedia Engineering **177**, 526-531 (2017)
3. M. Tropp, M. Handrik, P. Kopas, M. Saga, *Computer simulation of induction bending process*. Pipeline and Gas Journal **239**, (2012)
4. W. Piekarska, M. Kubiak, M. Vaško, *Numerical estimation of the shape of weld and heat affected zone in laser-arc hybrid welded joints*. Procedia Engineering **177**, 114-120 (2017)
5. L. Jakubovičová, B. Ftorek, V. Baniari, A. Sapietová, T. Potoček, M. Vaško, *Engineering design of a test device*. Procedia Engineering **177**, 520-525 (2017)

6. A. Vasko, M. Vasko, J. Belan, E. Tillova, *Comparison of fatigue properties of nodular cast iron at low and high frequency cyclic loading*. Procedia Engineering **177**, 576-581 (2017)
7. P. Kovacikova, R. Bezdedova, J. Vavro jr., J. Vavro, *Comparison of numerical analysis of stress-strain states of cast iron with vermicular graphite shape and globular graphite shape*. Procedia Engineering **136**, 28-32 (2016)
8. M. Nad, L. Rolnik, L. Icmancova, *Prediction of Changes in Modal Properties of the Euler-Bernoulli Beam Structures Due to the Modification of Its Spatial Properties*. International journal of structural stability and dynamics **17**, (2017)
9. R. Halama, P. Macura, L. Pecenka, F. Fojtik, M. Sofer, *Experimental Analysis of Residual Stresses in Backup Roll and FE solution*. Proceeding of 49th International Scientific on Experimental Stress Analysis, 85-90 (2011)
10. P. Zidlik, P. Ferfecki, B. Strnadel, *New computational procedures for evaluation of fracture surfaces of the drop weight tear test of the piping steel*. Proceedings of the ASME pressure vessels and piping conference Vol 1: Codes and standards, 1055-1060 (2012)
11. R. Grega, J. Krajnak, L. Zulova, G. Fedorko, V. Molnar, *Failure analysis of driveshaft of truck body caused by vibrations*. Engineering Failure Analysis **79**, 208-215 (2017)
12. M. Bruna, D. Bolibruchova, R. Pastircak, *Numerical simulation of porosity for Al based alloys*. Procedia Engineering **177**, 488-495 (2017)

Forced convection and heat transfer around a bounded cylinder

Blanka Skočilasová[1,*], *Jan* Skočilas[2], *Josef* Soukup[1]

[1]J. E. Purkyne University in Usti nad Labem, Faculty of Production Technology and Management, Pasteurova 3334/7, 400 96 Usti nad Labem, Czech Republic
[2]Czech Technical University in Prague, Faculty of Mechanical Engineering, Department of Process Engineering, Technická 4, 166 07 Praha 6, Czech Republic

Abstract. The article deals with the heat transfer solution in the fluid flow inside a pipe, the small tube is inserted into the pipe and its axis is perpendicular to the axis of the pipe. The fluid is water, the regime of the fluid flow is turbulent. The small tube is loaded by the internal source of the heat, the power of the source is constant and homogenous over the small tube surface and the cold water cools down the small tube. The analysis of the described problem helps to design the experimental model (the heat source above all) to validate the CFD results. The work included the numeric and analytic method. The model was prepared with the aim to investigate large amount of the variants of the geometric set up (the tube and pipe diameters ratio, the geometrical aspect ratio, corresponding stabilization length of the pipe for given diameter), the flow properties (variation of the velocity values, Reynolds number) and the power of the heat source (ensuring the measurable differences of the temperatures on the tube surface, large Nusselt number). The effect of the constrained space has been observed and described, compare to the free fluid flow.

Keywords: convective heat transfer, Nusselt number, flow over cylinder

1 Introduction

The most important process parameter of the heat exchanger design [1] is heat transfer surface directly linked with total heat transfer coefficient and convective heat transfer coefficient respectively. The article deals with investigation of momentum and heat transfer within specific simple.

The described problem is well known and investigated for many decades [2]. The several phenomena can be observed during inertia and heat transfer in the fluid near or behind immersed body which is overflowed by fluid. Many articles exist related to experimental [3-5] or numerical investigation [6-8] of the horizontal tube inserted into the free stream of fluid for low or high Reynolds number. In the last decade the main effort was done in investigation of the heat transfer around the heated cylinder during mixed convection, i.e. forced convection with buoyancy effect. The authors of articles related to heat transfer at surface of overflowed tube applied CFD simulation very often to investigate

* Corresponding author: blanka.skocilasova@ujep.cz

Reviewers: *Wiesława* Piekarska, *Milan* Žmindák

the problem with (assumed) unconstrained surrounding, e.g. natural convection in the fluid with immersed horizontal pipe [6-8]. Many authors preferred CFD solution of the problem by own code using standard method, e.g. SIMPLE algorithm [6] or Galerkin weighted residual finite element method [7], DNS approach [8-9] or they were using the commercial software ANSYS [10-12]. The research devoted to stream turbulence intensity and local perturbation generated behind the cylinder or generation of unsteady wake are also frequented, both experiment [4, 13-14] and numerical simulation [6-7] were performed by authors.

The similar problem with heated tube in the channel investigated Bayruk [15]. The difference of the presented article is to investigate the constrained environment effect on the heat transfer coefficient. After the suitable model will be developed, the temperature probe and the experimental stand will be build up according to [16-17]. The heated cylinder is equipped by surface thermocouples to measure the wall temperature on the cylinder perimeter, which allows to determine the local values of convective heat transfer coefficient.

2 Materials and methods

The small tube is inserted into the large pipe. The tube is positioned perpendicularly to pipe axis. Inner diameters of the pipe are $D = 60$ mm and 300 mm, outer diameter of the tube is $d = 12$ mm, the length of the pipe is $L = 4$ m, the stabilization length is 3 m, the exhaust part (behind the tube) has length 1 m. In the text of the article term "tube" represents small object inserted into the stream and term "pipe" is used for description of the channel. The physical parameters of water were used for simulation of fluid flow.

The diameters of the pipe or the ratios of the cross-section of the pipe to perpendicular surface projection of the small tube were mutually different more than five times. These geometries ensure incorporate the effect of the close environment, i.e. the effect of the pipe walls, into the final equations describing the heat transfer process under various conditions.

2.1 Numerical simulation

The models were created in the software ANSYS – Fluent 14.0. A various geometrical characteristics, flow properties and powers of heat source were applied. One diameter of the inserted tube 12 mm and two diameters of the pipe 60 and 300 mm were modeled. The model with 60 mm pipe diameter had 828 thousands of elements and with 300 mm had 1318 thousands of elements. The k-ω model with wall function SST was chosen [18]. The heat flux related to surface was set up as boundary condition to wall of the tube (small cylinder).

The simulation was performed with flow velocities 0.16, 0.53, 2.66, 5.32, 15.95 and 26.58 m.s^{-1}, in both diameters of pipe. Also 3 values of the power of the heat source were applied 26 826 W.m^{-2}, 44 711 W.m^{-2} and 89 422 W.m^{-2} which corresponds to the values of total power 60 and 100 W for small diameter of the pipe and 100, 500 and 1000 W for large diameter of the pipe. The smallest value of the power of the heat source was not used in case of pipe with 60 mm diameter, because the effect of the heat source was negligible on the temperature of the fluid and observed differences between the wall temperature and the fluid temperature were too small.

2.2 Correlations

In this part the applied relations for the heat transfer coefficient determination are described. The equations were found by the literature search [1, 18-24] and they evaluate mean value of the convective heat transfer coefficient for the transversal flow over the cylinder in the unconstrained environment. The valid range of these equations was respected.

The following equations are possible to use for the fluid overflow around the pipe (1), (2) and (3) [22], where the determining dimension is outer diameter of the tube and the determining velocity is bulk velocity of the fluid stream, the ranges of validity are as follow: equation (1a) $5 < \mathrm{Re} < 10^3$, equation (1b) $10^3 < \mathrm{Re} < 2\cdot 10^5$ and equation (1c) $2\cdot 10^5 < \mathrm{Re} < 2\cdot 10^6$

$$Nu = 0.5\,\mathrm{Re}^{0.5}\,\mathrm{Pr}^{0.38}(\mathrm{Pr}/\mathrm{Pr}_s)^{0.25}, \quad (1a)$$

$$Nu = 0.25\mathrm{Re}^{0.6}\,\mathrm{Pr}^{0.38}(\mathrm{Pr}/\mathrm{Pr}_s)^{0.25}, \quad (1b)$$

$$Nu = 0.023\mathrm{Re}^{0.8}\,\mathrm{Pr}^{0.37}(\mathrm{Pr}/\mathrm{Pr}_s)^{0.25}, \quad (1c)$$

Whitaker relation is also possible to use [23]

$$Nu = (0.4\mathrm{Re}^{1/2} + 0.06\mathrm{Re}^{2/3})\,\mathrm{Pr}^{0.4}\left(\frac{\mu}{\mu_w}\right)^{1/4}, \quad (2)$$

within the range $0.67 < \mathrm{Pr} < 380$ and $3.5 < \mathrm{Re} < 7.6\ 10^4$.

For the perpendicular overflow around the infinite cylinder [1] see (3a) for $0 < \mathrm{Re} < 4\cdot 10^3$, (3b) for $4\cdot 10^3 < \mathrm{Re} < 4\cdot 10^4$ and (3c) for $4\cdot 10^4 < \mathrm{Re} < 4\cdot 10^5$

$$Nu = 0.683\mathrm{Re}^{0.466}\mathrm{Pr}^{1/3}, \quad (3a)$$

$$Nu = 0.193\mathrm{Re}^{0.618}\mathrm{Pr}^{1/3}, \quad (3b)$$

$$Nu = 0.027\mathrm{Re}^{0.805}\mathrm{Pr}^{1/3}, \quad (3c)$$

In the literature [19] Churchill-Bernstein equation can be found

$$Nu = 0.3 + \frac{0.62\mathrm{Re}^{0.5}\,\mathrm{Pr}^{1/3}}{\left[1+\left(\frac{0.4}{\mathrm{Pr}}\right)^{2.3}\right]^{0.25}}\left[1+\left(\frac{\mathrm{Re}}{28200}\right)^{5/8}\right]^{0.8} \quad (4a)$$

for $\mathrm{Re}\cdot\mathrm{Pr} > 0.2$, and for Reynolds number in range $7\cdot 10^4 < \mathrm{Re} < 4\cdot 10^5$ it is recommended to increase the calculated value by 20%. For the estimated calculation McAdams equation can be used [19]

$$Nu = k.\mathrm{Re}^m\,\mathrm{Pr}^n, \quad (4b)$$

where the values of the coefficient k and exponents m and n depend on the Re, see [19].

3 Results and discussion

The results are divided into two sections which contain the simulation and the correlations results.

3.1 Numerical simulation

In the figure 1 there is presented the results for same geometry of the model, D = 60 mm, d =12 mm. The figures differ in the presentation of the contour temperature on the tube wall for the different velocities and power of the heat source. Only contour of small inserted object – tube is presented in the pictures.

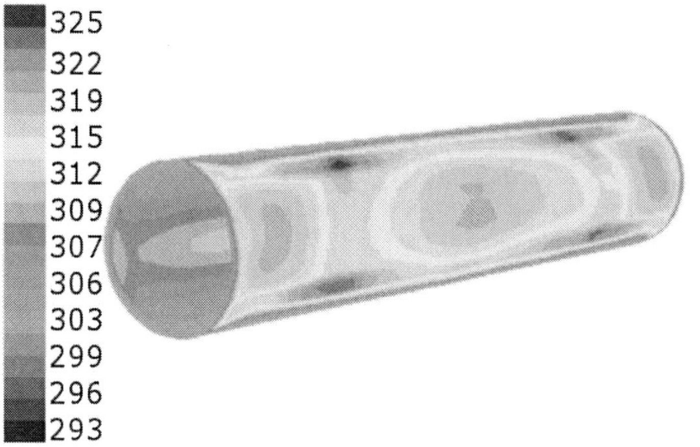

Fig. 1. Contours of static temperature on the tube wall - rear side, D = 60 mm, d = 12 mm, w = 0.16 m.s^{-1}, P = 60 W, T_{min} = 293 K, T_{max} = 325 K, T_{mean} = 309 K

Table 1. Mean value of the convective heat transfer coefficient on the tube surface, D = 60 mm, d = 12 mm, power of source P_m = 44 711 W.m^{-2}

velocity	t_s	t_s - t_f	$\overline{\alpha}$
m.s^{-2}	°C	°C	W.m^{-2}.K^{-1}
0.16	46.28	26.28	1701.12
0.53	33.09	13.09	3414.42
2.66	23.57	3.57	12528.59
5.32	21.96	1.96	22861.89
15.95	20.67	0.67	66595.67
26.58	20.38	0.38	118603.11

t_s – wall temperature, t_f – mean value of the temperature of the fluid.

The mean values of heat transfer coefficient are presented in the table 1, the profile of the local values of the heat transfer coefficient is presented in the figure 2. The symmetric profile of the coefficient presented in the figure 2 was expected, according to data from literature. The perpendicular position in the tube surface is represented by angle in the horizontal axis. The peaks of local values represent the location of the boundary layer separation from the surface. The maximum values of the coefficient appear at the both border of the tube, where the effect of the pipe wall can be observed. The rest of the variations of investigated cases were performed by the same described way.

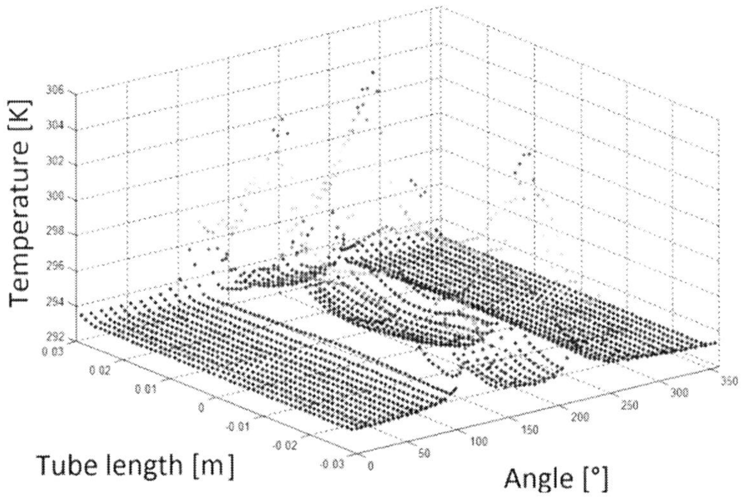

Fig. 2. Local values of the convective heat transfer coefficient for $D = 60$ mm, $d = 12$ mm, $w = 0.16$ m.s^{-1}, $P = 60$ W ($P_m = 26\,826$ W.m^{-2})

3.2 Applied correlation

The equations (1a) to (4b) were used to evaluate the heat transfer coefficient, and the computation results are summarized in the table 2 and for each velocity the corresponded equation (validity of Re range) is used.

Table 2. Mean values of heat transfer coefficient, correlations, $D = 60$ mm, $d = 12$ mm, $P_m = 44\,711$ W.m^{-2}, according to equations (1a) to (4b)

w	Re_d	α			
m.s^{-1}		W.m^{-2}K^{-1}			
		1a)	**2)**	**3a)**	**4b)**
0.16	1 916.2	2 468.87	2 135.77	2 200.53	2 110.27
		1b)	**2)**	**3b)**	**4b)**
0.53	6 347.3	5 446.89	3 988.32	4 111.96	3 826.03
		1b)	**2)**	**3b)**	**4b)**
2.66	31 856.2	13 360.80	10 396.33	11 143.54	10 071.92
		1b)	**2)**	**3c)**	**4b)**
5.32	63 712.3	20 024.04	21 585.18	18 934.6	18 171.07
		1b)	**3c)**	**4a)**	**4b)**
15.95	191 017.2	38 416.15	45 826.8	39 597.03	43 738.08
		1b)	**2)**	**3c)**	**4b)**
26.58	318 322.1	62 489.41	69 129.58	57 761.2	65 810.49

The correlations and numerical solutions were compared. The comparison served to determine the region in which it is possible to generalize the used physical conditions and derived new equation describing investigated problem.

3.3 Suggested correlation for Nusselt number

Let assume the general equation for Nusselt number in the following common form

$$Nu = \Phi.\mathrm{Re}^{0.8}\,\mathrm{Pr}^{1/3}\left(\frac{d}{D}\right)^{\Psi} \tag{5}$$

where constants Φ and Ψ have to be evaluated. The part *d/D* of equation represents the dependence both investigated geometries, ration between diameters of inserted tube and pipe. Due to the fact that there are four mutually dependent parameters, it is necessary to simplify the identification process of the parameters by the definition of several assumptions. These assumptions will make the identification process of the parameter realizable and more reliable.

For the total range of Reynolds numbers from 3000 to 320 000 we assume the value of the exponent of Reynolds number is 0.8 and is independent on the constants *Φ* and *Ψ*.

By the same method we analyzed the Prandtl number exponent. The value of this exponent oscillates around 1/3. We assume this value in the proposed equation (5).

To determine the values of constant Φ and Ψ we used the standard solver included in the common software e.g. Matlab or Excel. At the beginning we estimated initial values of the coefficients and by nonlinear regression we found out the accurate values correspond to best fit of the equation (6) and results of the simulation. The minimization of least squares was applied.

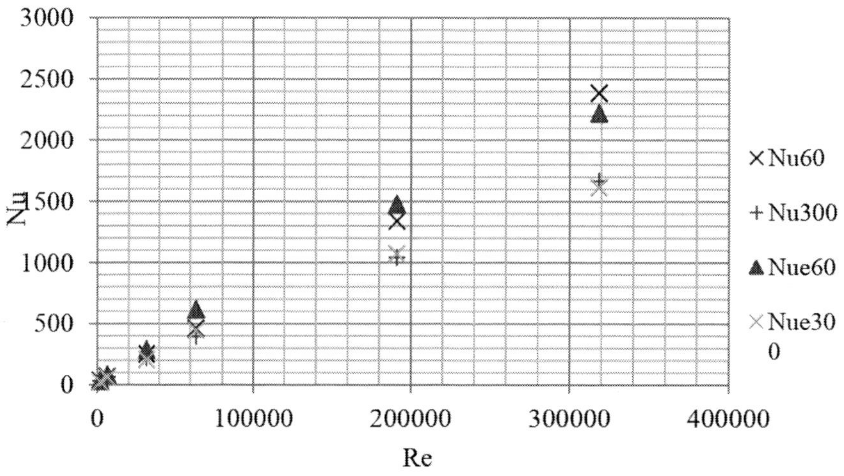

Fig. 3. Dependence of Nusselt number on the Reynolds number for $D = 60$ and $D = 300$ mm, comparison of the simulation values and result of the proposed new equation (6); Nu60, Nu300 are Nusselt numbers obtained from simulation, Nue60, Nue300 are Nusselt numbers evaluated from equation (6)

In the figure 3 the relationship between Reynolds and Nusselt number is presented. The comparison of the simulation and equation (6) with new obtained constants is shown in the figure 3.

The best fit of the trends obtained by simulation and regression analysis for constant Prandtl number and range of Reynolds number from 3 000 to 350 000 was obtained for values of constants Φ = 0.25 and Ψ = 0.2.

Final equation with optimized parameters is

$$Nu = 0.25\cdot \mathrm{Re}^{0.8}\cdot \mathrm{Pr}^{1/3}\cdot\left(\frac{d}{D}\right)^{0.2} \tag{6}$$

and it is valid for $d = 0.012$ m, $0.06 \text{ m} < D < 0.3$ m, $3\,000 < \text{Re} < 3.5 \cdot 10^5$, $\text{Pr} = 7$, which also express it limitation.

The difference of the new equation is caused by the fact, that all equation mentioned in the literature are intended for overflow in unconstrained space e.g. infinite tube length. The new equation involves the effect of the connection wall of the tube and wall of the pipe on the velocity field in this place and consequently influences the heat transfer in this area.

Conclusions

The comparison of the correlations and simulation results for evaluation of the mean valued of the convective heat transfer coefficient on the tube wall with inner source of heat placed perpendicularly to the fluid stream in the pipe was performed.

From comparison it was found that the highest differences were obtained for highest values of Reynolds number. It is obviously caused by effect of the constrained space around the tube, where the modification of the velocity field by these constrains is the most significant. It can be seen from investigation that the maximum of local values of convective heat transfer coefficient appears at these locations, i.e. near the connection of the walls, where the intensity of turbulence is highest and the biggest vortices are created. Theoretically the wall connection should slow down the fluid flow due to friction of the fluid at walls, but with respect to investigated geometry the geometric layout creates small jets at the vicinity of the wall connection, which increase the heat transfer, compared to the values at the rest of geometry.

The equation for calculation Nusselt number concerning the effect of the geometrical constrains of the investigated models was proposed with respect to similar form of the equations from literature. Constants of the equation were found by regression analysis of the results obtained by simulation.

This work was supported by SGA (Students Grant Agency) UJEP-SGS-173-05-01 Jan Evangelista Purkyně University in Ústí nad Labem, Czech Republic.

References

1. J. H. Lienhard, *A Heat Transfer Textbook*. third edition, Phlogiston Press. Cambridge, Massachusetts, USA, (2008).
2. A. Žukauskas, *Enhancement of forced convection heat transfer in viscous fluid flows*. Int. J. Heat Mass Transfer **37**, 207-212 (1994)
3. F. Dumouchel, J. C. Lecordier, P. Paranthoen, *The effective Reynolds number of a heated cylinder*. Int. J. Heat Mass Transfer **41**, 1787-1794 (1998)
4. C. K. Huang, Y. J. Cheng, Y. P. Kang, *Combined effect of grid turbulence and unsteady wake on convective heat transfer around a heated cylinder*. Int. Comm. Heat Mass Transfer **34**, 1091-1100 (2007)
5. R. Perrin, M. Braza, E. Cid, S. Cazin, P. Chassaing, C. Mockett, T. Reimann, F. Thiele, *Coherent and turbulent process analysis in the flow past a circular cylinder at high Reynolds number*. J. Fluid and Structures **24**, 1313-1325 (2008)
6. S. Bhattacharyya, A. K. Singh, *Vortex shedding and heat transfer dependence on effective Reynolds number for mixed convection around a cylinder in cross flow*. Int. J. Heat Mass Transfer **53**, 3202-3212 (2010)
7. M. M. Billah, M.M. Rahman, U. M. Sharif., N.A. Rahim, R. Saidur, M. Hasanuzzaman, *Numerical analysis of fluid flow due to mixed convection in a lid-*

driven cavity having a heated circular hollow cylinder. Int. Com. Heat Mass Trans. **38**, 1093-1103 (2011)

8. M. Boirlaud, D. Couton, F. Plourde, *Direct Numerical Simulation of the turbulent wake behind a heated cylinder*. Int. J. Heat Fluid Flow **38**, 82-93 (2012)
9. J. G. Wissink, W. Rodi, *Heat transfer from the stagnation area of a heated cylinder at Re_D = 140,000 affected by free-stream turbulence*. Int. J. Heat Mass Transfer **54**, 2535-2541 (2011)
10. G. Juncu, *A numerical study of momentum and forced convection heat transfer around two tandem circular cylinders at low Reynolds numbers. Part II: Forced convection heat transfer*. Int. J. Heat Mass Transfer **50**, 3799-3808 (2007)
11. N. Mahir, Z. Altaç, *Numerical investigation of convective heat transfer in unsteady flow past two cylinders in tandem arrangements*. Int. J. Heat Fluid Flow **29**, 1309-1318 (2008)
12. H. S. Yoon, J. B. Lee, J. H. Seo, H. S. Park, *Characteristics for flow and heat transfer around a circular cylinder near a moving wall in wide range of low Reynolds number*. Int. J. Heat and Mass Transfer **53**, 5111-5120 (2010)
13. A. Quintino, *Experimental analysis of the heat transfer coefficient enhancement for a heated cylinder in cross-flow downstream of a grid flow perturbation*. App. Thermal Eng. **35**, 55-59 (2012)
14. C. Sak, R. Liu, D. S. K. Ting, G.W. Rankin, *The role of turbulence length scale and turbulence intensity on forced convection from a heated horizontal circular cylinder*. Exper. Thermanl and Fluid Science **31**, 279-289 (2007)
15. E. Buyruk, H. Barrow, I. Owen, *The influence of adjacent tubes on Convection Heat Transfer from a Heated Tube in Cross-flow*. Fourth UK National Conference on Heat Transfer, I. Mech. E. Conference Trans., 135-139 (1995)
16. S. Sanitjai, R. J. Goldstein, *Forced convection heat transfer from a circular cylinder in crossflow to air and liquids*. Int. J. Heat and Mass Transfer **47**, 4795-4805 (2004)
17. J. W. Scholten, D. B. Murray, *Unsteady heat transfer and velocity of a cylinder in cross flow II. High freestream turbulence*. Int. J. Heat Mass Transfer **41**, 1149-1156 (1998)
18. R. Lenhard, K. Kaduchová, J. Jandačka, J., *Numerical Simulation in Indirectly heated hot Water Heater*. Adv. Material Research, 1693-1697 (2014)
19. A. Bejan, A. D. Kraus, *Heat transfer Handbook*. John Wiley and sons, Inc., Hoboken, New Jersey, (2003)
20. V. P. Isatchenko, V. A. Osepova, A.S. Sukomel, *Heat transfer (Теплопередача)*. Energoizdat, Moskva (1981)
21. A. Zhukauskas, A. Slantchiauskas, *Heat transfer at trubulent fluid flow (Теплоотдача в турбулентном потоке жидкости)*. Mintis, Vijnius, Litva (1973)
22. H. Hausen, *Heat transfer in co-current, conter-current and cross flow (Wärmeübertragung im Gegenstrom, Gleichstrom und Kreuzstrom)*. Springer Verlag, Berlin, Heidelberg, (1976)
23. M. Kaviany, *Principle of Convective Heat Transfer*. Springer-verlag, New York, Inc. Second edition, ISBN 0-387-95162-8, (2001)
24. S. O. Atayilmay, *Transient and steady-state natural convection heat transfer from a heated horizontal concrete cylinder*. Int. J. of Thermal Sci. **49**, 1933-1943 (2010)

The effect of restraints type on the generated stresses in gantry crane beam

Leszek Sowa[1,*], *Wiesława* Piekarska[1], *Tomasz* Skrzypczak[1], *Paweł* Kwiatoń[1]

[1]Czestochowa University of Technology, Institute of Mechanics and Machine Design Fundamentals, Dąbrowskiego 73, 42-201 Częstochowa, Poland

Abstract. This paper includes an analysis of the mechanical phenomena in the gantry crane beam, because the cranes are currently one of the most common devices for the transporting loads. Designing modern mechanical structures is a complex task that requires the use of appropriate tools. Such a modern tool is the numerical simulation, which uses different numerical methods. One of the best known methods is the finite element method, also used here. Simulations are limited to analysis of the strength of the gantry crane beam that was the loaded of the force load movement along its length. The numerical analysis was made to the gantry crane beam which cross-section was an I-beam and ends were fixed in different ways. As the result of numerical calculations, the stresses and displacements of the structure of gantry were obtained. The influence of the restraints type and changing the loading force position on generate the Huber-Misses stress in the gantry crane beam was estimated. The aim was to ensure that the maximum equivalent stress generated in the gantry crane beam was less than the strength of material, because then the construction is safe.

Keywords: numerical modelling, stresses analysis, computational mechanics, FEM

1 Introduction

The cranes are currently one of the most known devices for lifting and transporting horizontally heavy load from one position to another. They are applied in heavy industry, shipyards, car factories and auto workshops. The new design or improve existing mechanical constructions of cranes is a complex task that requires use of appropriate tools [1]. Such a modern tool, that make possible to analyze strength parameters of designed structures, appears to be numerical simulation which usually uses the Finite Element Method (FEM) [2-7].

Finite element analysis software offers inexpensive solutions to the problem of crane girder failure and is the primary tool for helping us identify the cause of the problem, and recommends solutions. Therefore the numerical simulation should be adopted as standard tool in failure [1]. When analyzing works on the study of these problems, it is noted that different types of finite elements are used in MES because it appears that this has an effect on the received results of numerical simulations [2, 3]. In paper [2] the finite element meshes with 4-node tetrahedral and 4-node quadrilateral shell elements were generated

* Corresponding author: sowa@imipkm.pcz.pl

Reviewers: *Josef* Soukup, *Milan* Žmindák

from the solid and shell models of FEM, respectively. After a comparison results of the finite element analyses with the conventional calculations and the measurement performances on the existing crane, it was found that the analysis using quadratic shell elements give the most realistic results. Other researchers using the solid and shell elements in FEM did not notice significant differences in the results of numerical simulations [3]. It is supposed that the reason of this situation was too great element dimension along the beam length in case using the solid element [2].

Analysis of the load motion is an important issue in the design of large gantry cranes [8]. Usually, the load movement is neglected in numerical simulations and analysis is performed for one position of load force in the center of the beam length [2-4, 6]. Numerical simulations that allow for the analysis of mechanical phenomena in subsequent positions of the load force along the whole length of the beam can be found in [5] and are presented in this paper.

The aim of this paper is to obtain information relating to the boundary conditions what should be assumed, in what number and how to position them, so that they do not cause an increase of stress in the gantry crane beam. The problem of stress concentration and localization of maximum displacement will also be discussed. Therefore, numerical simulations for different cases of beam fixing were performed to obtain the distribution of stresses and displacements in the gantry crane beam. The value of the calculated stresses is compared to the allowable stresses of the beam material. If they are less than the allowable stresses, then it is considered that the structure is safe.

2 The mathematical model of mechanical phenomena

The analysis of mechanical phenomena in elastic range is based on classic equilibrium equations, supplemented by constitutive relations and boundary conditions. The mathematical description is based on solving the following set of equations [5, 9, 10]:

– the equilibrium equations:

$$\sigma_{ij,j} = 0, \qquad \sigma_{ij} = \sigma_{ji}, \tag{1}$$

– the constitutive equations – general Hook's law:

$$\sigma_{ij} = D_{ijks}\varepsilon_{ks}, \tag{2}$$

– the Cauchy's equations:

$$\varepsilon_{ks} = \frac{1}{2}(u_{k,s} + u_{s,k}), \tag{3}$$

where σ_{ij} is stress tensor, ε_{ks} is strains tensor, u_i is displacement vector, D_{ijks} is matrix of elasticity (a tensor of material properties).

Equation (1) is completed by boundary conditions (displacements), which are assumed to ensure the static determination of considered system [5, 7, 9, 10]:

$$\sigma_{ij}n_j = p_i\big|_{\Gamma}, \quad u_i = u_i\big|_{\Gamma}, \tag{4}$$

where p_i is components of force per unit area acting on a plane (Γ) with normal n_j.

The numerical model uses FEM and is derived from the criterion of weighted residuals method [4-7]. Equation (1) is multiplied by the weighting function and integrated over the considered region. As the result of using Gauss-Green's theorem and Galerkin's method, which were described in the work [7], the following matrix equation is obtained [5, 9, 10]:

$$\mathbf{K}\cdot\mathbf{u}=\mathbf{R}\,, \tag{5}$$

where **K** is the element stiffness matrix, **u** is the vector of nodal displacement, **R** is the vector of nodal forces resulting from the boundary load.

By solving the matrix equation (5), nodal displacement is obtained in the area under investigation. Then, using the equations (3) and (2), the searched stress fields are obtained.

3 Description of the problem

To analyze the influence of beam restraint type and position of beam load force on the stress distribution in gantry crane, the static system of beam shown in Figure 1 was considered. The gantry crane beam with a length 4.8 m is restrained on two supporting column with a width of 0.6 m or as described in the variants defined below. One end of the beam (right) can not move in three directions and the second end (left) is free to shift along the *z*-axis or can not move in three directions, in according to variant restrain. Cross section of the beam is I-beam with dimensions 300x125x10 mm (Fig. 1).

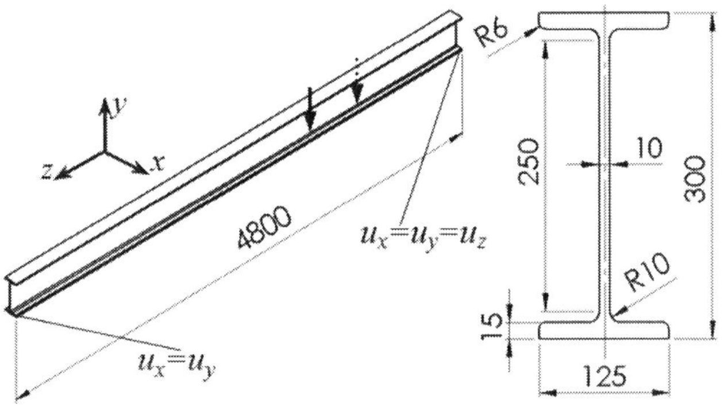

Fig. 1. Scheme of the test problem, I-beam cross section of the gantry crane beam

The structure is loaded by pressure perpendicular to the lower flange of the beam and directed downwards, i.e. opposite to the *y*-axis. The pressure value is due to the total weight of the loaded crane trolley that can move along the lower flange of the gantry crane beam. The crane trolley movement is carried out by shifting the pressure that acts on the contact surface of the trolley wheels with the beam to the next positions along the crane beam, ignoring the effects of inertia. The assumed pressure value of 13 513 513.514 Pa multiplied by the contact surface of the trolley wheels with the beam corresponds to the force load equal to 50 kN. The loads were taken into account in the elastic range and therefore the elastic constants were assumed, corresponding to the steel S235: Young's modulus (E) equal to 2.1x10^5 MPa, the Poisson's ratio (v) of 0.28 and the yield strength (R_e) of 235 MPa [2-6]. Value of the allowable stress the steel beam material (σ_{all}) amounts to 120 MPa. The value of these stresses results from the division of the yield strength by the safety factor that is assumed to be of 1.98. The safety factor should be set up in the range from 1.5 to 3 for the crane girder project in accordance with the recommendations [2, 3].

Analysis of the stresses of the gantry crane structure was performed using professional software (Star module), supplemented by an additional procedure movement of force. The gantry crane beam was discretized by a mesh 39 003 nodes, which define 19 224 finite elements. In this study, a parabolic tetrahedral solid element is used. Research carried out in order to increase the strength of the structure of the gantry crane allow for the evaluation of the stress state, indicating the critical areas and values.

4 Results and discussion

In order to simulate considered phenomena, the advanced numerical methods are used. Numerical analysis of the stresses, strains and displacements of the one-beam gantry crane structure was made using finite element method. The influence of restraints type on the generated stresses in gantry crane beam has been analyzed. To obtain stress distribution and displacement in the gantry crane beam, the numerical simulations were performed for various beam restraining cases (Figs 2-13). The difference between these cases of fixing the ends of the beam is due to the number of zero displacements and their placement on the crane beam.

The following variants of restraint were taken into consideration:

- I restraint variant: one end of the beam (left) is free to shift along the *z*-axis (displacement equal to u_x=u_y=0) and the other end can not move in three directions (u_x=u_y=u_z=0) (Figs 2-7).
- II restraint variant: left and right end of beam can not move in three directions (u_x=u_y=u_z=0) (Figs 8, 9).
- The displacement of the finite element nodes located on the contact surface of the crane beam with the support column are set to zero in the directions given in variant I or II.
- III restraint variant: displacements are set to zero at point (u_x) and on the line (u_y, u_z) laying in 0.6 m from the right end and at point (u_x) and on the line (u_y) laying in 0.6 m from the left end of beam (Figs 10).
- IV restraint variant: displacements are set to zero at point (u_x) and on the line (u_y, u_z) laying in 0.6 m from the right end and at point (u_x) and the line (u_y, u_z) laying in 0.6 m from the left end of beam (Figs 11).
- V restraint variant: displacements are set to zero at point (u_x) and on the line (u_y, u_z) laying on the right end and at point (u_x) and on the line (u_y) laying on the left end of beam (Figs 12).
- VI restraint variant: displacements are set to zero at point (u_x) and on the line (u_y, u_z) laying on the right end and at point (u_x) and on the line (u_y, u_z) laying on the left end of beam (Figs 13).

The effect of changing the loading force position on generate the equivalent stress (σ_{eq}) in the crane beam was evaluated. The Huber-Misses-Hencky stress distribution in the gantry crane beam (I-beam) for various positions of load force identified *z*-coordinate are shown in Figures 2-5. The strength condition was checked using the value of generated stresses in successive loading force positions. It sought to that the maximum value of Huber-Mises-Hencky stress induced in the gantry crane beam was less than the strength of beam material ($\sigma_{eq} < \sigma_{all}$), because then the design is safe. The strength condition has not been met in all cases of beam attachment (III and V variant). Generally, the area of increased stresses is related to the current position of the load force and shifts with it, achieving a maximum value in the middle of the beam length, but not in all cases of attachment. The displacement of the gantry crane structure is shown in Figure 6, 7 and 9. They reached a maximum value for the central position of load force (Figs 6, 9). In order to verify the value of the displacement obtained, the ratio of vertical displacement in the half of the beam to beam length should be checked. The research performed allows the evaluation of the stress state, pointing out the critical areas and values. They were made in order to increase the strength of structure of the gantry crane.

Fig. 2. Huber-Misses-Hencky stress distribution in the gantry crane beam for location of the force load z = 1.5 m, I variant

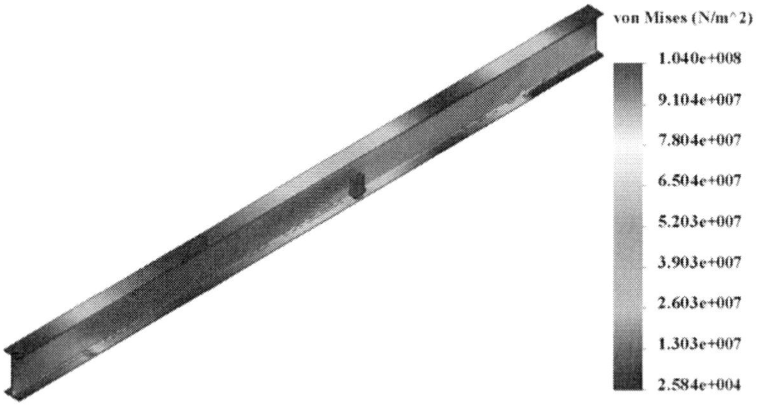

Fig. 3. Huber-Misses-Hencky stress distribution in the gantry crane beam for location of the force load z = 2 m, I variant

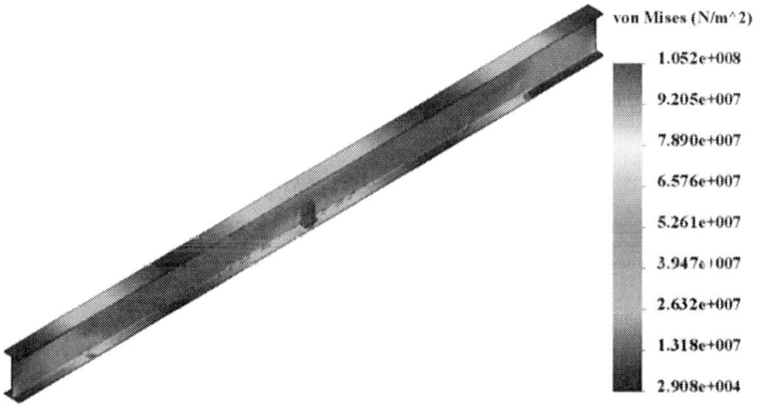

Fig. 4. Huber-Misses-Hencky stress distribution in the gantry crane beam for location of the force load z = 2.4 m, I variant

Fig. 5. Huber-Misses-Hencky stress distribution in the gantry crane beam for location of the force load z = 3.3 m, I variant

Fig. 6. Displacement field in the gantry crane beam for location of the force load z = 2.4 m, I variant

Fig. 7. Displacement field in the gantry crane beam for location of the force load z = 3.3 m, I variant

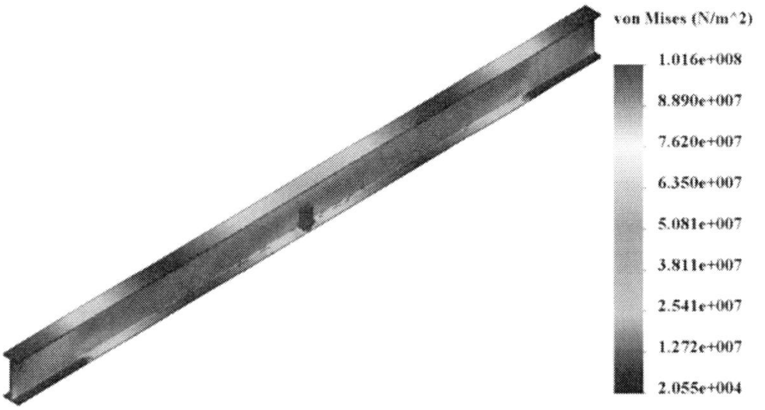

Fig. 8. Huber-Misses-Hencky stress distribution in the gantry crane beam for location of the force load z = 2.4 m, II variant

Fig. 9. Displacement field in the gantry crane beam for location of the force load z = 2.4 m, II variant

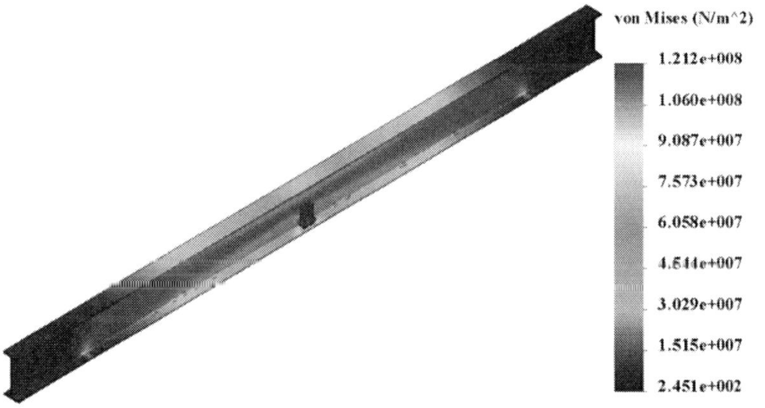

Fig. 10. Huber-Misses-Hencky stress distribution in the gantry crane beam for location of the force load z = 2.4 m, III variant

Fig. 11. Huber-Misses-Hencky stress distribution in the gantry crane beam for location of the force load z = 2.4 m, IV variant

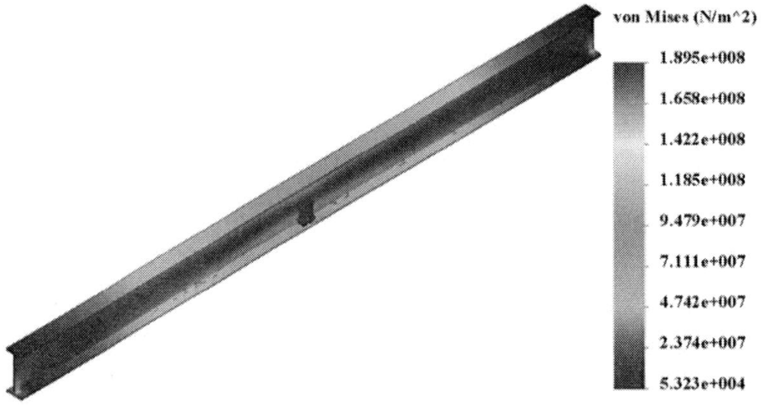

Fig. 12. Huber-Misses-Hencky stress distribution in the gantry crane beam for location of the force load z = 2.4 m, V variant

Fig. 13. Huber-Misses-Hencky stress distribution in the gantry crane beam for location of the force load z = 2.4 m, VI variant

Conclusions

This paper focuses mainly on the numerical simulations of mechanical phenomena occurring in the I-beam of gantry crane.. The influence of beam restraint type and changing the loading force position on generate the equivalent stress in the gantry crane beam was evaluated. The region of stresses increased is associated, to the current position of the load force and moves with it (Figs 2-5). These stresses reach the maximum value in the half of beam length under the load force for the restrain variant I, II, III (Figs 4, 8, 10), whereas in variants IV, V and VI (Figs 11, 12, 13) the maximal stresses occur at the points where the beam is attached. The large value of stress at this location does not significantly affect on the safety of the gantry crane structure. Equivalent stress located in the beam center exceed the allowable-values in the variants III and V while in other cases they are less than the allowable stress of the material beam. Analyzing the obtained displacement values (exemplary results in Figs 6, 7, 9), the ratio of the maximum displacement to the beam length was checked. In general, this ratio is within acceptable limits for all restrain variants [6]. The manner of restraining the beam III and especially V are not recommended because it may cause incorrect assessment of the real state of stresses in the examined crane structure. Fixing the beam by immobilizing of its contact surface with supporting column gives the most realistic results of the assessment of the crane beam strength.

References

1. P.R. Patel, V.K. Patel, *A review on structural analysis of overhead crane girder using FEA technique*. Int. J. Eng. Sci. Innov. Tech. **2** (4), 41-44 (2013)
2. B.C. Alkin, C. E. Imrak, H. Kocabas, *Solid modeling and finite element analysis of an overhead crane bridge*. Acta Polyt. **45** (3), 61-67 (2005)
3. A. Suratkar, V. Shukla, D.K.S. Zakiuddin, *Design optimization of overhead EOT crane box girder using finite element analysis*. Int. J. Eng. Res. Tech. **2** (7), 720-724 (2013)
4. C.B. Pinca, G.O. Tirian, A. Josan, G. Chete, *Quantitative and qualitative study on the state of stresses and strains of the strength structure of a crane bridge*.WSEAS Tran. Appl. Theor. Mech. **5** (4), 231-241 (2010)
5. L. Sowa, Z. Saternus, M. Kubiak, *Numerical modelling of mechanical phenomena in the gantry crane beam*. Proc. Eng. **177**, 225-232 (2017)
6. T. Haniszewski, *Strength analysis of overhead traveling crane with use of finite element method*. Tran. Prob. **9** (1), 19-26 (2014)
7. T. Skrzypczak, *A finite element multi-mesh approach for heat transport between disconnected regions*. Proc. Eng. **177**, 204-209 (2017).
8. V. Gašić, N. Zrnić, M. Milovančević, *Considerations of various moving load models in structural dynamics of large gantry cranes*. FME Trans. **41**, 311-316 (2013)
9. Z. Saternus, W. Piekarska, M. Kubiak, T. Domański, L. Sowa, *Numerical analysis of deformations in sheets made of X5CrNi18-10 steel welded by a hybrid laser-arc heat source*. Procedia. Eng. **136**, 95-100 (2016)
10. W. Piekarska, D. Goszczyńska-Króliszewska, T. Domański, A. Bokota, *Analytical and numerical model of laser welding phenomena with the initial preheating*. Proc. Eng. **177**, 149-154 (2017)

The effect of the gantry crane beam cross section on the level of generated stresses

Leszek Sowa[1,*], *Tomasz* Skrzypczak[1], *Paweł* Kwiatoń[1]

[1]Czestochowa University of Technology, Institute of Mechanics and Machine Design Fundamentals, Dąbrowskiego 73, 42-201 Częstochowa, Poland

Abstract. The gantry cranes are currently one of the most popular devices to transporting loads. They can be used in many industries. This study focuses on the numerical analysis of only the portable gantry crane. The mathematical and numerical model of mechanical phenomena in the gantry crane beam was presented in this paper. This problem was solved by using of the Finite Element Method (FEM). The analysis was made to the cross-section of gantry crane beam which was the I-beam or box beam. The numerical simulations are limited to numerical analysis of strength of the gantry crane beam which was loaded by the motion of the load force along its length. As a result of the calculations, the stresses and displacements of the structure of gantry crane were obtained. The influence of beam shape change and changing the loading force position on generate the equivalent stress in the crane beam was evaluated. The research performed, which allows the estimated of the stress state, pointing out the critical areas and values, were made in order to increase the strength of the structure of the gantry crane.

Keywords: numerical modelling, stresses analysis, gantry crane beam, FEM

1 Introduction

Manipulation of material is an essential component of any production and distribution system. The important mechanical elements of these systems are different cranes, which are used to lifting and lowering a load and moving it horizontally. They are mainly used in car factories, heavy industry and shipyards. Their construction features vary widely according to their main operational specifications, such as: the type of movement of the crane and the crane trolley, the weight and type of load, the location of the crane and the geometric features [1]. They are called overhead cranes or gantry cranes. The usual distinction drawn between the two cranes is that with gantry cranes, the entire structure is usually wheeled (often on rails) [2]. The main component of crane is girder beam which transfers the load on its other elements therefore it is tested for strength. The construction of the overhead crane with a double girder is often analyzed [2-4], while a single beam is usually used in the gantry crane [5] and analyzed in this paper. The gantry cranes with a single girder are often used in the auto workshops, they are smaller and about simple design [5, 6]. Usually, the gantry cranes are made with steel [2-5], but they can be making with selected

* Corresponding author: sowa@imipkm.pcz.pl

Reviewers: *Petr* Horyl, *Ksawery* Szykiedans

aluminium alloys because they have good mechanical properties and enhanced corrosion resistance. The main advantage of such structures is the reduced mass with retained strength parameters. Taking into account the dynamic parameters of the analyzed structures, it must be noted that those made of aluminium alloy show much less stiffness. Therefore, for larger bridge spans and larger load capacities, their use may be illegitimate because of the permitted deflection of the girder [6].

Literature review helps in a way to get the information regarding structural analysis of crane girder and it is the main source in setting new trends in design from the early stage of project. It turns out that the finite element analysis software offers inexpensive solutions to failure problem of crane girder and is an essential tool to help us in determining the cause of problem, it also recommends the solutions. Therefore the numerical simulation should be assumed as standard tool in failure [1, 2]. Usually, the solid and shell elements are used in FEM for getting best results of numerical simulations [2, 3]. After a comparison results of the finite element analyses with the conventional calculations it was found that only the analysis using quadratic shell elements give realistic results. It is believed that the reason of this situation was too large element dimension along the beam in case using the solid element [3].

The problem of moving the load is very important topic in the design of large gantry cranes. Dynamic effects in case of problems with moving loads are not often considered [7]. Usually, the load movement is neglected and numerical analysis is performed for its one position in the center of the beam length [2-4, 6] Numerical simulations that allow analyze of mechanical phenomena in subsequent positions of load force, over the entire length of beam range, can be found in [5] and will be presented in this paper.

The destination of this article is to provide information about the strength of the crane beam of different cross-section loaded with moving load force, so that it has not been damaged. The problems of stress concentration and the extreme deformations localization will also be discussed. For this purpose, the numerical simulations were performed to get the distributions of stresses, strains and displacements in the gantry crane beam. The value of the calculated stresses is compared with the permissible stresses of beam material. If they are smaller than the permissible stresses, it is supposed that the construction is safe.

2 Mathematical description

The mechanical phenomena analysis in the elastic range was made using the equilibrium equations (1), which are supplemented by constitutive and Cauchy's equations (2), and boundary conditions (3). The mathematical model is based on solving the following system of equations [5, 8-11]:

$$\sigma_{\alpha\beta,\beta} = 0, \qquad \sigma_{\alpha\beta} = \sigma_{\beta\alpha}, \tag{1}$$

$$\sigma_{\alpha\beta} = D_{\alpha\beta\gamma\kappa} \varepsilon_{\gamma\kappa}, \qquad \varepsilon_{\gamma\kappa} = \frac{1}{2}(u_{\gamma,\kappa} + u_{\kappa,\gamma}), \tag{2}$$

$$\sigma_{\alpha\beta} n_{\xi} = p_{\xi}\big|_{\Gamma}, \qquad u_{\kappa} = u_{\kappa}\big|_{\Gamma}, \tag{3}$$

where $\sigma_{\alpha\beta}$ is stress tensor, $\varepsilon_{\gamma\kappa}$ - strains tensor, u_{κ} - displacement vector, $D_{\alpha\beta\gamma\kappa}$ - tensor of material properties (Young's modulus (E) and Poisson's ratio (ν)) and p_{ξ} - components of force per unit area acting on a plane (Γ) with normal n_{ξ}.

Displacement boundary conditions are used in the discussed problems [2-6, 8-11]. The nodes located on the contact surface of the gantry crane beam with the support column on the right side of the beam can not displace in three directions (u_x=u_y=u_z=0) and the left side in two directions (u_x=u_y=0) (Fig. 1).

The numerical model uses finite element method in the weighted residuals formulation [5, 8-12]. Equation (1) is multiplied by the weighting function and integrated over the entire region. As the result of using Ostrogradsky-Gauss-Green's theorem and Galerkin's method, which were described in the work [12], the following matrix equation is obtained [5, 8-11]:

$$\mathbf{K} \cdot \mathbf{u} = \mathbf{R}, \tag{4}$$

where **K** is the stiffness matrix, **u** - sought displacement vector, **R** - vector associated with the loads and boundary conditions. The solution of the matrix equation (4) is obtained as a nodal displacement in the examined area. Then, using the equation (3) and (2) are obtained the sought fields of stresses.

3 Description of the problem

For the analysis the effect of beam shape change and the position of load force on the stress distribution in the gantry crane beam, the static system of single-span beam is presented in Figure 1. The gantry crane beam with the length 4.8 m is attached on two supporting column with a width of 0.6 m. One end of the beam (left) is free to shift along the *z*-axis (Fig. 1) and the second end can not move in three directions. Cross section of the beam is I-beam with dimensions 300x125x10 mm, box beam with dimensions 300x125x5 mm and thick-wall box beam with dimensions 300x125x10 mm.

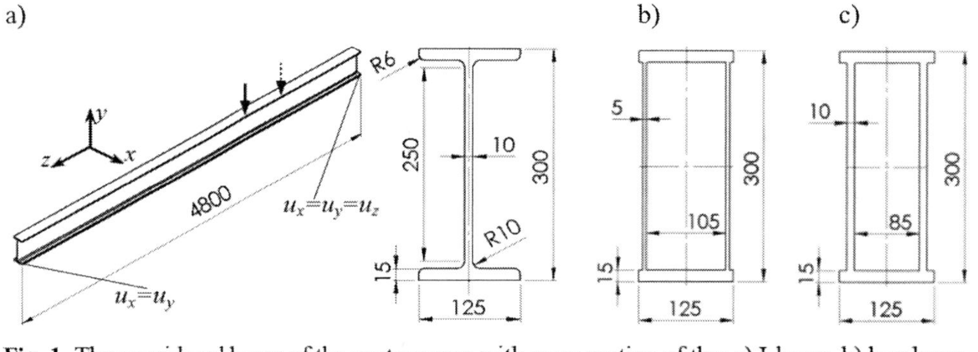

Fig. 1. The considered beam of the gantry crane with cross section of the: a) I-beam, b) box beam, c) thick-wall box beam

The structure is loaded by the pressure perpendicular to the upper flange of the beam and directed downward, i.e. contrary to the *y*-axis. The pressure value is the result of the total weight of the loaded crane trolley that can travel along the upper flange of the gantry crane beam. The movement of the crane trolley performed by moving the pressure which acts on the contact surface of the trolley wheels with the beam, to consecutive positions along the crane beam, ignoring inertia effects. The assumed pressure value of 13 513 513.514 N/m^2 multiplied by the contact surface of the trolley wheels with the beam corresponds to the load force of 50 kN. The loadings have been considered in the elastic range and therefore the elastic constants were assumed, corresponding to the steel S235: Young's modulus (E) equal to 2.1e+5 MN/m^2, the Poisson's ratio (v) of 0.28 and the yield strength (R_e) of 235 MN/m^2 [2-6, 8]. The value of the allowable stress the beam material (σ_{all}) amounts to 120 MN/m^2. The value of this stress is calculated by dividing yield strength by the safety factor is assumed that is 1.98. The safety factor should be between 1.5 and 3 for the crane girder design as recommended [2, 3].

The analysis of the stresses of the gantry crane structure was carried out using the professional program (SolidWorks) with additional force movement procedure where the move trajectory of the force was divided into the set of temporary positions. Then the calculations for the each position were performed with the use of loop to obtain the stress distribution. This aspect of presented work is a new contribution developed by the authors. The gantry crane beam has been divided into 19 224 finite elements that are defined by 39 003 mesh nodes when the beam has the cross-section of I-beam or 26 763 elements and 53 206 nodes if the beam has a cross-section of the box. In this investigation a parabolic tetrahedral solid element is used. Numerical tests carried out to increase the structure strength of the gantry crane make it possible to assess the state of stresses, indicating critical areas and values.

4 Results of numerical calculations

In this study, the numerical analysis of the strength of portable gantry crane is made. As the result of the computations performed, the stresses, strains and displacements of the gantry crane structure were obtained. The effect of beam shape changes and change of load force position on the creation of equivalent stress (σ_{eq}) in the gantry crane beam was assessed.

Fig. 2. Von Mises equivalent stress distribution in the I-beam of gantry crane for location of the load force z = 1.5 m

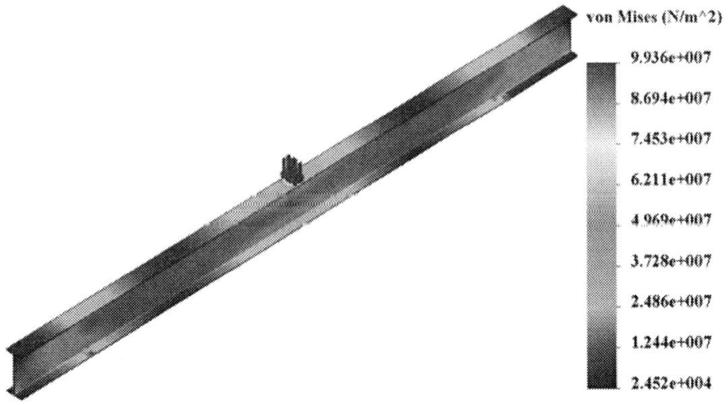

Fig. 3. Von Mises equivalent stress distribution in the I-beam of gantry crane for location of the load force z = 2.4 m

The distribution of von Mises equivalent stress in the gantry crane beam whose cross-section is the I-beam is shown in Figures 2-4 or if it is the box beam is shown in Figures 6-8 and if it is the thick-walled box beam is shown in Figure 10.

Fig. 4. Von Mises equivalent stress distribution in the I-beam of gantry crane for location of the load force z = 3.3 m

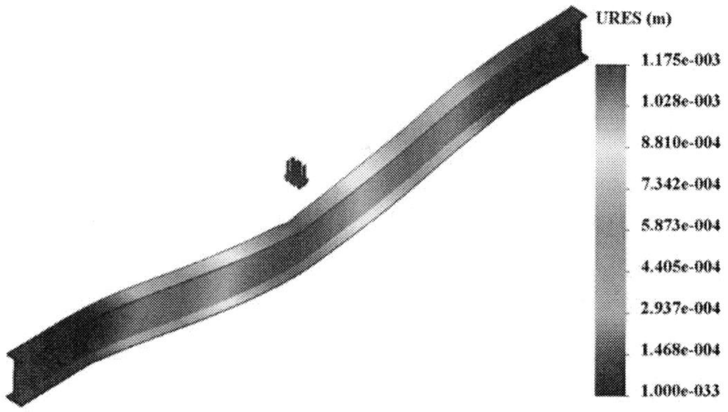

Fig. 5. Displacement field in the I-beam of gantry crane for location of the load force z = 2.4 m

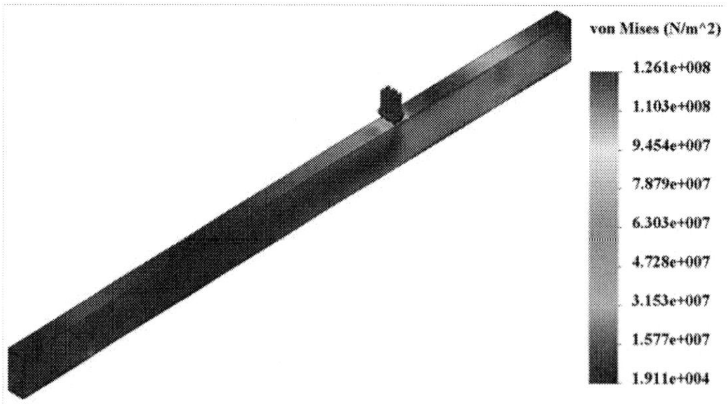

Fig. 6. Von Mises equivalent stress distribution in the box beam of gantry crane for location of the load force z = 1.5 m

The strength condition is verified by using the stresses induced in the beam in successive load force positions. It sought to that the maximum value of von Mises equivalent stress generated in the gantry crane beam was less than the strength of beam material ($\sigma_{eq} < \sigma_{all}$), because then the construction is safe. However, the strength condition was not met when the gantry crane beam had the cross-section of box because the stress peak was in place where the beam was attached (Fig. 6).

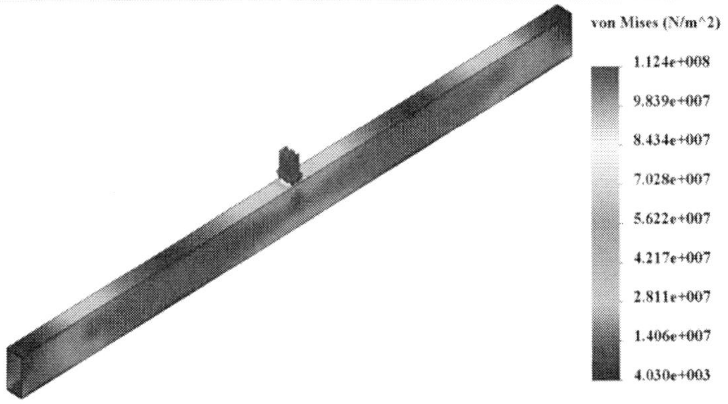

Fig. 7. Von Mises equivalent stress distribution in the box beam of gantry crane for location of the load force z = 2.4 m

Fig. 8. Von Mises equivalent stress distribution in the I-beam of gantry crane for location of the load force z = 3.3 m

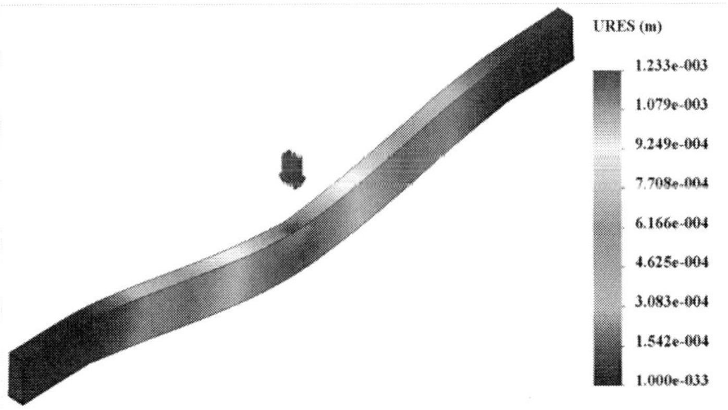

Fig. 9. Displacement field in the box beam of gantry crane for location of the load force z = 2.4 m

In general, the region of increased stress is related to the current location of the load and moves with it, resulting in a maximum value in the center of the beam length. The displacements field of the gantry crane structure is shown in Figure 5, 9 and 11, when these magnitudes reached the maximum value for the central position of load force. In order to verify the value of the displacement obtained, the ratio of vertical maximum displacement to beam length should be examined. The largest displacement value can be found in Figure 9 (box beam) and then this ratio is equal to 2.56e-4. These researches were performed to estimate the strength of the gantry crane structure depending on the cross-section used.

Fig. 10. Von Mises equivalent stress distribution in the thick-walled box beam of gantry crane for location of the load force z = 2.4 m

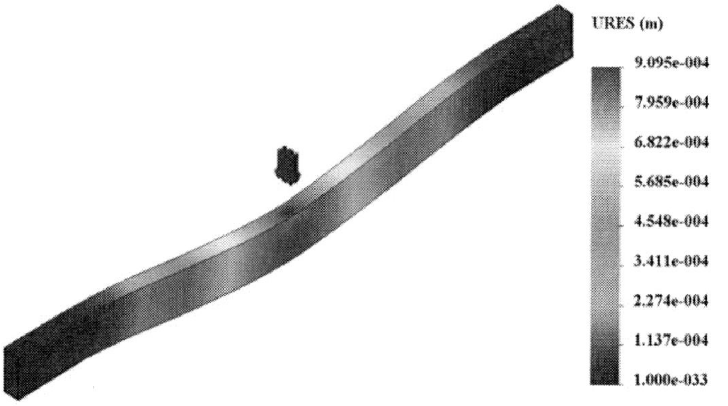

Fig. 11. Displacement field in the thick-walled box beam of gantry crane for location of the load force z = 2.4 m

Conclusions

The mathematical model and numerical simulations of the mechanical phenomena in the gantry crane beam are presented in this paper. The strength analysis of the gantry crane beam whose cross-section is the I-beam or the box beam was performed. As a result of the numerical calculations carried out, the stresses and displacements of the gantry crane structure were obtained. The influence of beam shape change and changing the loading force position on generate the von Mises equivalent stress in the crane beam was evaluated. The area of increased stresses is associated, to the current position of the load force and moves with it (Figs 2-4, 6-8). Generally these stresses reach the maximum value in the center of the beam length under the force load (Figs 3, 7, 10). The maximum von Mises equivalent stress in the middle of the gantry crane beam length whose cross-section is I-beam amounted to 99.4 MN/m^2 (Fig. 3), in the box-section beam was 112.4 MN/m^2 (Fig. 7) and in the thick-walled box beam was equal to 92.7 MN/m^2 (Fig. 10). These stresses are less than the allowable stress of the beam material. In addition, a peak of stresses is located at the points where the beam is attached (Figs 4, 6). The strength condition in this place is not met when the gantry crane beam has the cross-section of box (Fig. 6). Analyzing the distribution of displacement (exemplary results in Figs 5, 9, 11), the ratio of the maximum displacement to the beam length is checked and it is within acceptable limits [6]. It should be noted that results of numerical simulations are consistent with the results found in literature of the subject [2-5]. In these papers, the stress peak is located at the points where the beam is attached, as in this paper. The authors of paper [4] recommended reducing stresses near the end of the crane beam by improving the construction of gantry crane. The use of the I cross-section in the gantry crane beam design seems to be a better solution in terms of strength and therefore this cross section is usually used in real gantry crane structures.

References

1. P.R. Patel, V.K. Patel, *A review on structural analysis of overhead crane girder using FEA technique*. Int. J. Eng. Sci. Innov. Tech. **2** (4), 41-44 (2013).
2. A. Suratkar, V. Shukla, D.K.S. Zakiuddin, *Design optimization of overhead EOT crane box girder using finite element analysis*. Int. J. Eng. Res. Tech. **2** (7), 720-724 (2013)
3. B.C. Alkin, C. E. Imrak, H. Kocabas, *Solid modeling and finite element analysis of an overhead crane bridge*. Acta Polyt. **45** (3), 61-67 (2005)
4. C.B. Pinca, G.O. Tirian. A. Josan, G. Chete, *Quantitative and qualitative study on the state of stresses and strains of the strength structure of a crane bridge.* WSEAS Tran. Appl. Theor. Mech. **5** (4), 231-241 (2010)
5. L. Sowa, P. Kwiatoń, *Numerical analysis of stress fields generated in the gantry crane beam*. Procedia Eng. **177**, 218-224 (2017)
6. D. Gąska, T. Haniszewski, *Modelling studies on the use of aluminium alloys in lightweight load-carrying crane structures*. Transport Problems **11** (3), 13-20 (2016)
7. V. Gašić, N. Zrnić, M. Milovančević, *Considerations of various moving load models in structural dynamics of large gantry cranes*. FME Trans. **41**, 311-316 (2013)
8. W. Piekarska, M. Kubiak, Z. Saternus, S. Stano, T. Domański, *Numerical prediction of deformations in laser welded sheets made of X5CrNi18-10 steel.* Arch. Metal. Mater. **60** (3), 1965-1972 (2015)
9. T. Domański, A. Sapietova, M. Saga, *Application of Abaqus software for the modeling of surface progressive hardening*. Proc. Eng. **177**, 64-69 (2017)
10. R. Dyja, E. Gawrońska, N. Sczygiol, *The effect of mechanical interactions between the casting and the mold on the conditions of heat dissipation. A numerical model*. Arch. Metall. Mater. **60** (3), 1901-1909 (2015)
11. W. Piekarska, D. Goszczyńska-Króliszewska, T. Domański, A. Bokota, *Analytical and numerical model of laser welding phenomena with the initial preheating*. Proc. Eng. **177**, 149-154 (2017)
12. Skrzypczak T., Węgrzyn-Skrzypczak E., *Modeling of thermal contact through gap with the use of finite element method*. J. Appl. Math. Comput. Mech. **14** (4), 145-152 (2015)

Analysis of the distribution of temperature fields in the braked railway wheel

Andrej Suchánek[1,*], *Jozef* Harušinec[1], *Mária* Loulová[1], *Peter* Strážovec[1]

[1]University of Žilina, Faculty of Mechanical Engineering, Department of Transport and Handling Machines, Univerzitná 8215/1, 01026 Žilina, Slovak Republic

Abstract. The article deals with detection of reduced stress in a braked railway wheel, based on thermal transient analysis on virtual models, which influence the characteristics of the railway wheels. Structural analysis was performed by means of the ANSYS Multiphysics program system package. Thermal transient analysis deals with detection of temperature fields which are a result of braking by brake block. The applied heat flux represents the heat generated by friction of brake block. It is applied to a quarter model of the wheel to speed up the calculation. This analysis simulates two braking processes with subsequent cooling. Distribution of the equivalent stress was detected in the railway wheel cross section, at selected points. The input parameters were taken from the thermal transient analysis. These equivalent stresses result from thermal load.

Keywords: railway wheel, brake block, structural analysis

1 Introduction

The brake system of railway vehicles is an important subsystem in terms of driving safety. Investigated issue is process of non-stationary temperature fields spreading, generated by the braking railway vehicles The thermal load on the wheels has a significant rate on the influences due braking by the brakes, that lead to wear - modification and damage of the wheel tread [5, 15, 21]. In the braking process we have to take into account the fact, that the wheel is a brake block not only overheated at the point of their contact but also loaded with normal and tangential stress, which is the source and the necessary precondition for the braking process creation [7, 14]. With regard to the braking design and the braking mechanism, we include braking brakes between the adhesive brakes because the braking effect of the vehicle itself with respect to the track is realized by contacting the wheel and the rail via the contact patch [12, 18].

Researches must pay attention to studying the effects of thermal and mechanical loading wheels of railway vehicles of reasons: the operation of vehicles, protection of life and health of the traveling public, reliable transport material and minimizing the negative effects of rail traffic on the environment [10, 25].

* Corresponding author: andrej.suchanek@fstroj.uniza.sk

Reviewers: *Bohuš* Leitner, *Milan* Žmindák

Article deals with the detection of residual stress in a braked railway wheel based on thermal transient analysis. Railway wheel is loaded by heat flux, which is applied on the contact surface. This topic is discussed in more detail in mentioned literature and references [1-4, 2, 3, 4, 6, 11, 19, 22].

2 Residual stresses

In design and computational practice, a material is usually considered as homogeneous isotropic continuum [9].

The deformation of bodies and stress occurrence happens mainly due to [17, 23]:

1. Mutual power action of bodies.
2. Action of field of temperature:
   - homogenous (tension appears when thermal dilatation of body is restrained),
   - non - homogenous (tension appears even if thermal dilatation of body is free).

If the plasticity condition is not satisfied, active tensions occur in elastic area and after removing of causes of their formation it completely disappear.

If formation of elastic plastic state arises in any point of a body, then after removing of formation causes (power, distortive and thermal), some residual stresses stay in a body [13, 20, 24].

Certain residual state of stress remains almost always in structural materials due to their production technology.

Residual stresses can be:

1. Helpful – e.g. peening is introduced compressive stresses in the surface layer, leading to prolongation of lifetime.
2. Harmful – cracking, corrosion stress, reduce fatigue limit or brittle fracture resistance (especially negative effect on tensile residual stresses).

High tensile stresses, induced, for example in the area of welds cools and phase transformations, may cause rupture even without additional external forces [8].

3 ANSYS program package

The program ANSYS (Fig. 1) uses the finite element method. Modeling of the finite element method belongs to the group of numerical methods. This method develops due to the constant increase in computing power. Its core is the discretization of bodies on the files of finite elements. These elements form analogue after parts field that can be mathematically written [16].

The ANSYS program is generally nonlinear, multiphysics program including structural and thermodynamic analysis, analysis of flow continuum, analysis electrostatic and electromagnetic fields, and acoustic analysis. All these analyzes can be performed individually, but thanks ANSYS multiphysics conceived program can also be included in one common analysis. [10]. The ANSYS program allows you to not only check calculations, but also enables optimization and sensitivity analysis due to parameterized computational models, as well as the calculations of reliability.

ANSYS Mechanical product is intended to simulate the structural and thermodynamic tasks. The program includes the complete set of linear and nonlinear simulation with using linear and non-linear elements, material models, and contact non-linear algorithms.

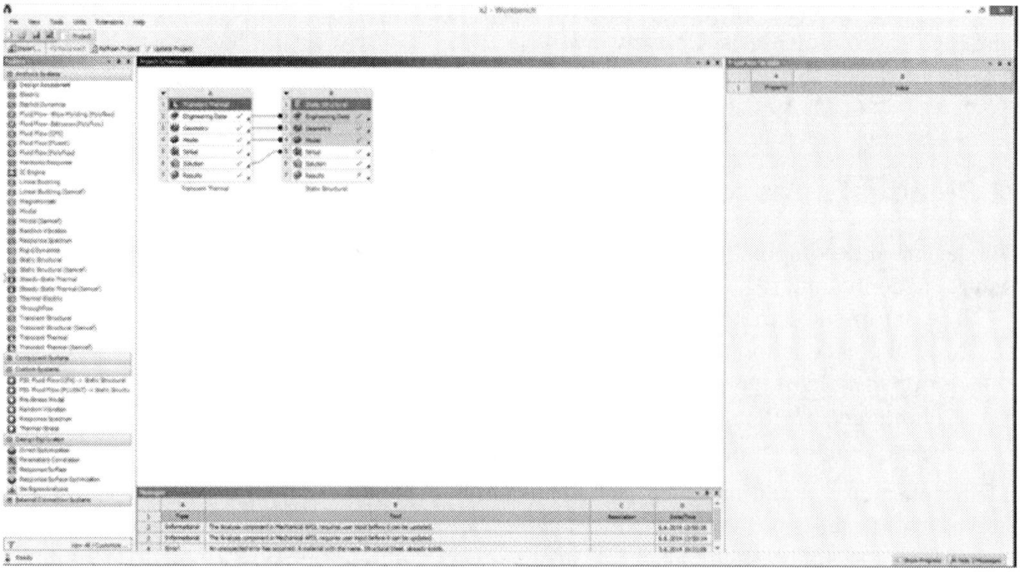

Fig. 1. ANSYS Workbench program system environment

4 Transient thermal analysis in ANSYS program

The problem simulates heating of the railway wheel tread. The railway wheel is braked by the brake block. The heat generated by the brake pad friction represents a heat flow of 32 kW (cast iron block) and 38 kW (composite block). This analysis simulates two braking for 100 seconds when the heat flow is applied to the wheel tread. Railway wheel cools for 300 seconds; after the first braking and 1000 seconds after the second braking. The value of heat flux is then zero. The composite brake block has a worse heat dissipation than cast iron and therefore heats the wheel's surface more heavily. The applied heat flux in the composite block is greater than that of the cast iron block.

A quarter model of railway wheel was created using CATIA program and imported into ANSYS program.

4.1 The definition of material properties

Railway wheel is made of steel DIN 40Mn4. The thermal properties used in the simulation are shown in Tab 1.

Table 1. Thermal properties of materials

Properties	Railway wheel	Air
Density ρ [kg.m^{-3}]	7850	1.170
Heat capacity Cp [J.kg-1.K^{-1}]	486	1100
Thermal coductivity k [W.m^{-1}.K^{-1}]	52	0.026
Emissivity – wheel tread[-]	0.20	-
Emissivity – other surfaces	0.80	
Dynamic viscosity [Pa.s]	-	1.8.10^{-5}

4.2 Definition of boundary conditions

A quarter model was used because of the acceleration calculation. The symmetry has been applied to the model.

The values of the heat flux (power) 32 kW (cast iron block) and 38 kW (composite block), which is applied to the wheel tread (Fig. 3) are shown in table 2. Dependence heat flux to time is shown in Fig. 2. Simulation did not involve wheel rotation or thermal expansion of the wheel.

Table 2. Dependence heat flux to time

Step	Time [s]	Heat flux [W]
1	0	0
1	10	32000 (38000)
2	100	32000 (38000)
3	101	0
4	400	0
5	410	32000 (38000)
6	500	32000 (38000)
7	501	0
8	1500	0

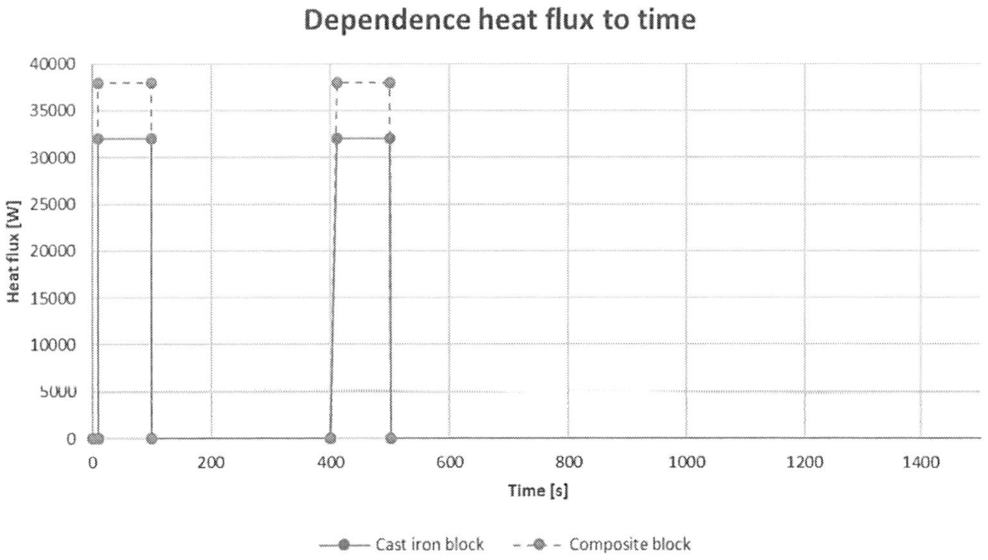

Fig. 2. Dependence of a heat flux to time

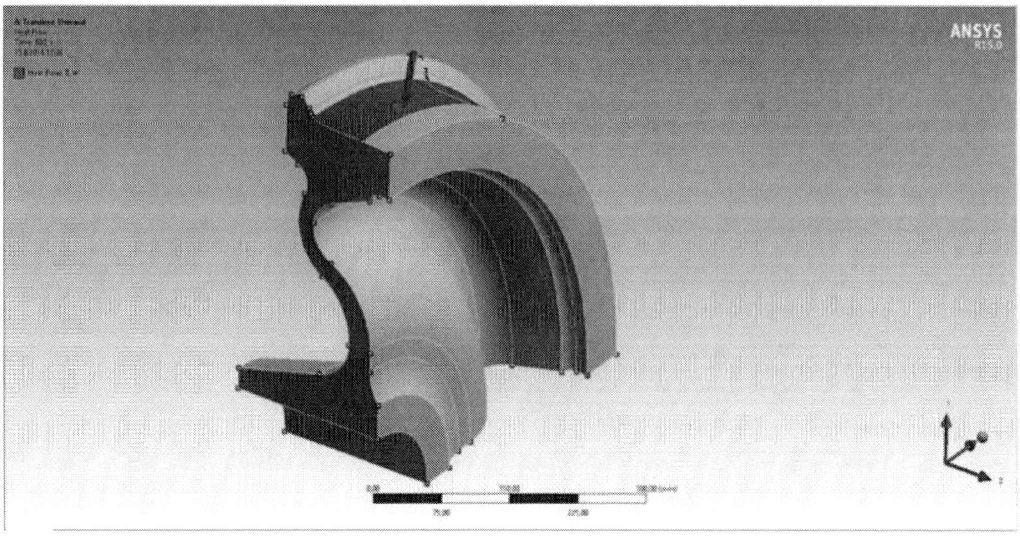

Fig. 3. Heat flux applied to the tread of the rail wheel

Reference ambient temperature was set to 22°C.
Thermal radiation was set to all areas of railway wheel.
Six degrees of freedom were taken to the railway wheel. (Fig. 4).

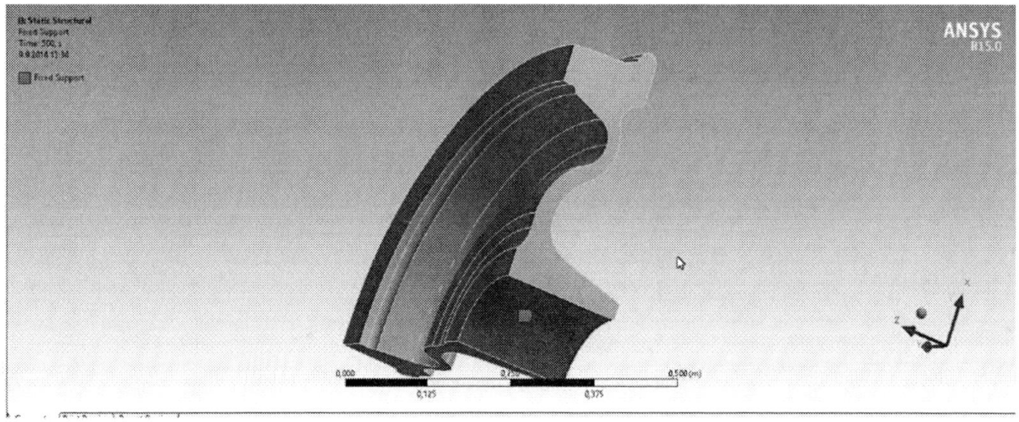

Fig. 4. Applied boundary conditions

Finite element mesh (Fig. 5) was created according to the dimensional parameters with the following parameters:

- element size: 10 mm,
- element type: SOLID 90,
- number of elements: 5075,
- number of nodal elements: 26156.

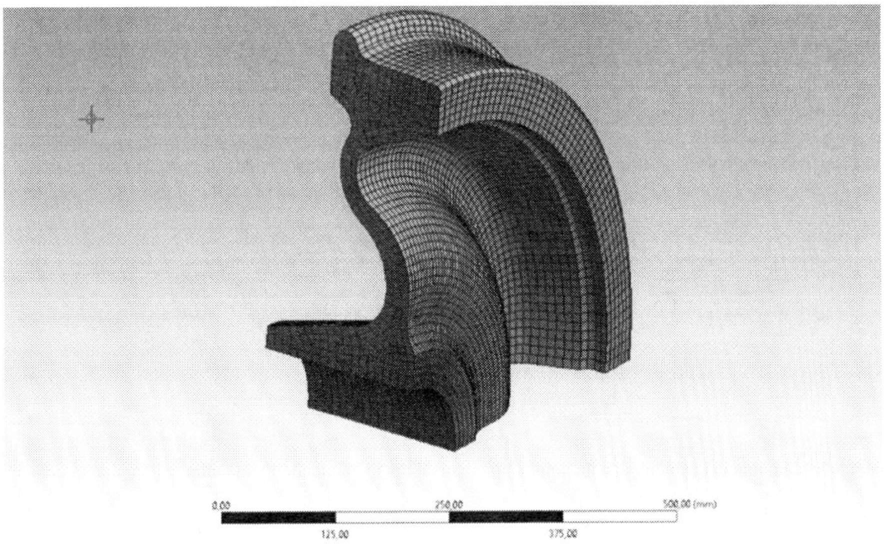

Fig. 5. Finite element model

Setting solver:

- direct solver with fixed setting step and automatic control of convergence.

Computer parameters:

- processor: Intel Core i7 3,3 GHz (6 core)
- memory (RAM): 64 GB

4.3 Results

On the basis of the analysis performed, the maximum value of the braking temperature was determined at 485 °C at 500 seconds due braking by cast iron block, which is at the end of the second braking. The temperature was 143 °C after cooling. The maximum temperature was determined at 500 seconds at 572 °C due braking by composite block. The temperature was 164 °C after cooling.

The temperatures in the cross section of the rail wheel, at selected points (Fig. 6), are shown in Fig. 7 and Fig. 8. These points were chosen so that the distribution of temperatures and stresses in the cross section of the wheel can be monitored.

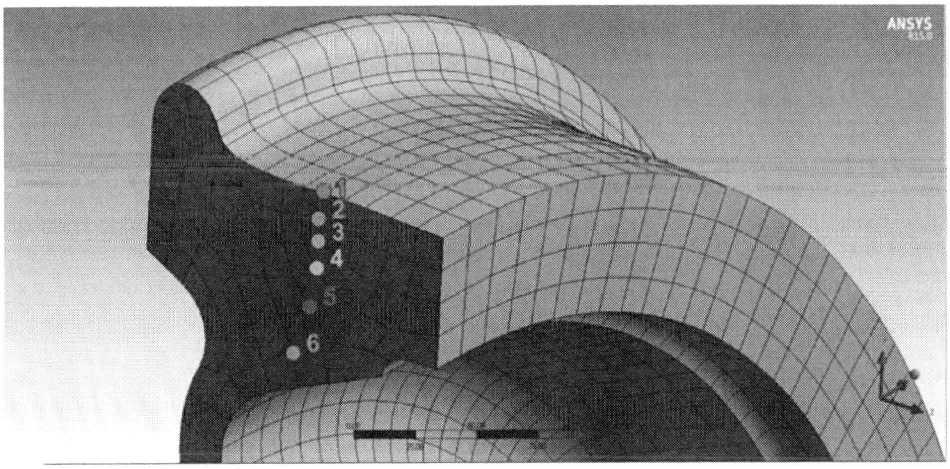

Fig. 6. The selected points, where were evaluated the temperatures and stresses

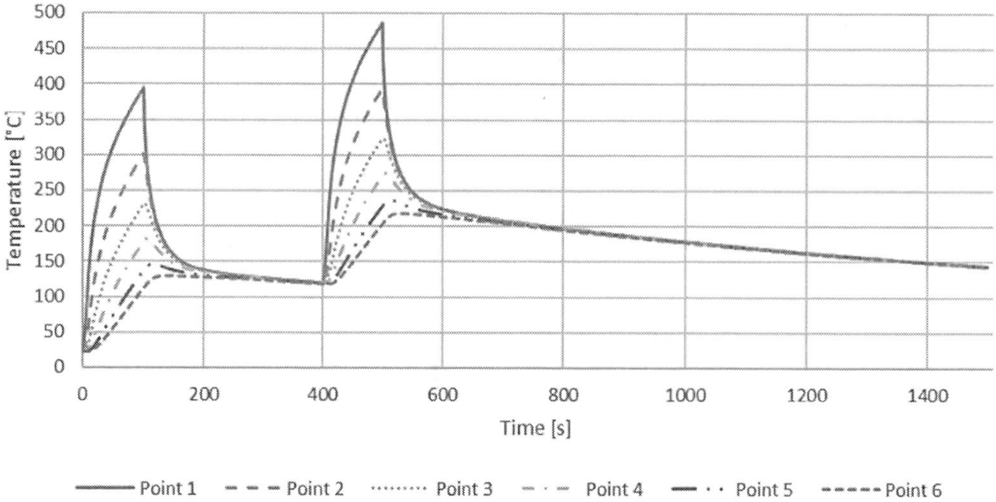

Fig. 7. Dependence temperature to time in selected points – cast iron block

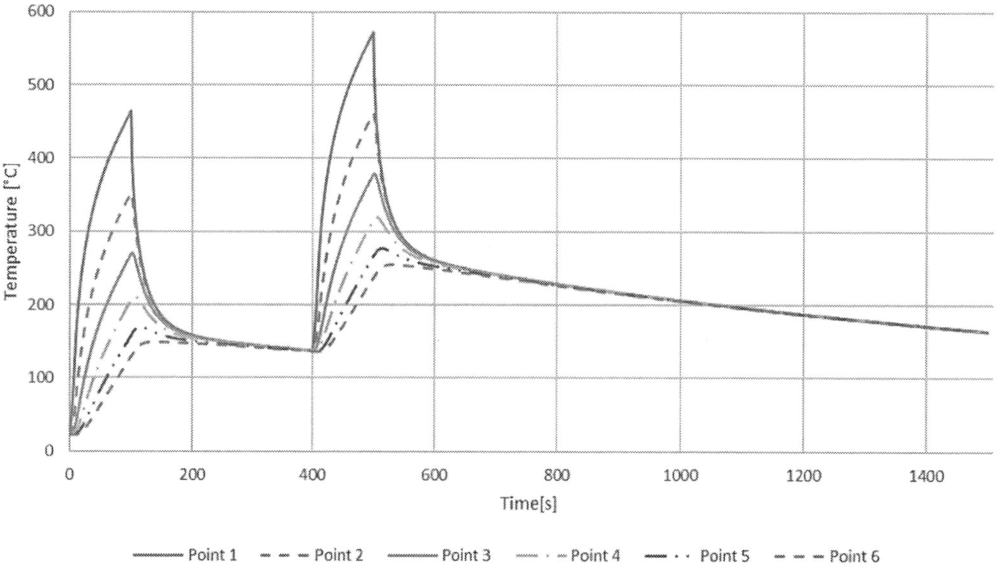

Fig. 8. Dependence temperature to time in selected points – composite block

5 Calculation of equivalent stress

On the basis of the thermal transient analysis, it was possible to detect the course of the reduced stresses in the braked railway wheel, which were caused by the thermal load. We could use the parameters from the previous thermal transient analysis thanks the ANSYS program.

Stress value is 90 MPa, determined by HMH theory on the wheel tread in time 1500 seconds. In other measured points, the value is around 24 MPa when braking by cost iron block. Stress value on the wheel tread is 197 MPa in time 1500 seconds, when braking by composite block and in other point is value 29 MPa.

The course of the reduced stresses in the cross-section of the railway wheel (Fig. 6) are shown in Fig. 9 and Fig. 10.

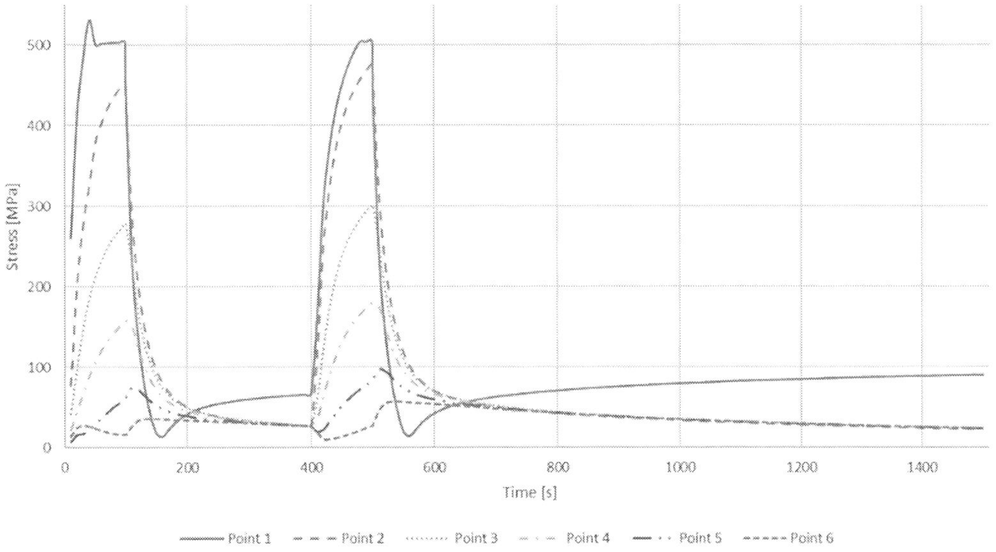

Fig. 9. Dependence stress to time in selected points – cast iron block

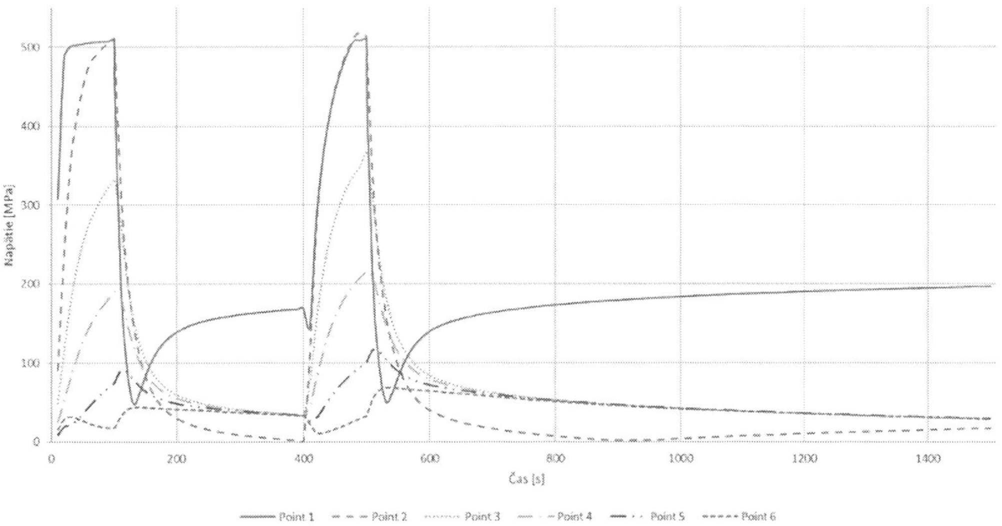

Fig. 10. Dependence stress to time in selected points – composite block

Conclusion

One possibility for identifying the impact of thermal and mechanical loading on braked railway wheel is the use of appropriate software and implementation of computer analyzes.

The article deals with the detection of structural properties braked railway wheel using a program that uses the finite element method.

On the basis of the thermal transition analysis, when the thermal fields can be determined by braking at constant heat output, a structural analysis was performed when heat inputs from the previous simulation were used as input parameters. Courses of the reduced stresses are results of simulation analyzes generated by thermal load in railway wheel. The next step is to compare the calculated values with the measured values on the test stands. It will be necessary to fine-tune the input parameters of the simulation to match the real test values. Our task was to choose the appropriate software to detect stresses based on the temperature fields. This problem was solved in ANSYS program, which proved to be appropriate for the given task type.

The work was supported by the Cultural and Educational Grant Agency of the Ministry of Education of the Slovak Republic in project No. KEGA 077ŽU-4/2017: Modernization of the Vehicles and engines study program. The work was also supported by the project No. APVV-0842-11: Equivalent railway operation load simulator on the roller rig and VEGA No. 1/0927/15: Research of the use of alternative fuels and hybrid drives on traction vehicles with aim to reduce fuel consumption and air pollutants production.
Research-Educational Centre of Rail Vehicles (VVCKV)

References

1. M. Blatnický, M. Štauderová, J. Dižo, *Numerical analysis of the structure girder for vehicle axle scale calibration*. Procedia Engineering **177**, 510-515 (2017)
2. J. Gerlici, R. Řezníček, *Temperature fields in the brake blocks when braking on an incline.* In XI. International conference "Current problems in rail vehicles". 233-244, Česká Třebová, [in Slovak] (1994)
3. J. Gerlici, V. Hlavňa, R. Řezníček, *Simulation of down hill braking with a shoe brake.* 4th mini conference on vehicle system dynamic, identification and anomalies. Technical university of Budapest (1994)
4. J. Gerlici, T. Lack, D. Kalinčák, *Laboratory simulation of braking with a shoe brake.* In: SITARZ, M.: Railway wheelsets. Monograph. ISBN 83-7335-151-5. Chapter **5**, 83-92, Silesian university of technology, Gliwice Poland (2003)
5. J. Gerlici, T. Lack, *Modified HHT method for vehicle vibration analysis in time domain utilisation*. Applied mechanics and materials **486**, 396-405 (2014)
6. J. Gerlici, T. Lack, *Contact geometry influence on the rail / wheel surface stress distribution*. Procedia Engineering, 2249-2257, Elsevier Ltd. (2010)
7. J. Gerlici, T. Lack, *Rail vehicles brake components test bench utilisation*. Applied mechanics and materials **486**, 379-386 (2014)
8. J. Gerlici, T. Lack, *Iterative method for railway wheel profile design*. Communications: Scientific Letters of the University of Žilina **11** (2), 49-56 (2009)
9. J. Gerlici, T. Lack, *Railway wheel and rail head profiles development based on the geometric characteristics shapes*. Wear: an international journal on the science and technology of friction, lubrication and wear **271** (1-2), 246-258 (2011)
10. J. Gerlici, T. Lack, J. Harušinec, *Realistic simulation of railway operation on the RAILBCOT test stand*. Applied mechanics and materials **486**, 387-395 (2014)
11. J. Gerlici, T. Lack, *Test benches computer control software tools development*. Scientific bulletin of North University of Baia Mare: Fascicle: Mechanics, Tribology. Technology of Machine Manufacturing **XVII** (C), 181-186, Baia Mare (2003)
12. L. Jakubovičová, M. Sága, *Computational analysis of contact stress distribution in the case of mutual stewing of roller bearing rings.* Novel Trends in Production Devices and Systems, Applied Mechanics and Materials **474**, 363-368 (2014)
13. L. Jakubovičová, A. Gašparec, P. Kopas, M. Sága, *Optimization of the induction heating process in order to achieve uniform surface temperature*. Procedia Engineering **136**, 125-131 (2016)

14. T. Lack, J. Gerlici, *Contact area and normal stress determination on railway wheel / rail contact*. Communications, the scientific letters of the University of Zilina **1/2005**, 38-45 (2005)
15. T. Lack, J. Gerlici, *Wheel/rail contact stress evaluation by means of the modified Strip method*. Communications: scientific letters of the University of Žilina **15** (3), 126-132 (2013)
16. T. Lack, J. Gerlici, *Wheel/rail tangential contact stress evaluation by means of the modified strip method*. Communications: scientific letters of the University of Žilina **16** (3A), 33-39 (2014)
17. T. Lack, J. Gerlici, *A modified strip method to speed up the calculation of normal stress between wheel and rail*. Applied mechanics and materials **486**, 359-370 (2014)
18. T. Lack, J. Gerlici, *The FASTSIM method modification in speed up the calculation of tangential contact stresses between wheel and rail*. Manufacturing technology: journal for science, research and production **13** (4), 486-492
19. T. Lack, J. Gerlici, *A modified strip method to speed up the tangential stress between wheel and rail calculation*. Applied mechanics and materials **486**, 371-378 (2014)
20. T. Lack, J. Gerlici, *The usage of arcs radii profile variation for the synthesis of railway wheel and rail head profiles*. Communications - scientific letters of the University of Žilina **8** (2), 57-69 (2006)
21. K. W. Man, *Contact mechanics using boundary elements*. Topics in engineering. Volume 22. (Computational mechanics publication, ISBN 1 85312 334 X, 185. p, Southampton, 1994)
22. R. Řezníček, J. Gerlici, T. Lack, *Stress analysis monoblock wheels braked by FEM*. Proceedings: ŽELSEM '93, Savings in railway", University of Zilina 1993, 155-161, Loučeň, In Slovak (1993)
23. A. Sapietová, M. Sága, P. Novák, R. Bednár, J. Dižo, *Design and application of multi-software platform for solving of mechanical multi-body system problems*. Mechatronics: Recent technological and scientific advances, 345-354 (2011)
24. A. Suchánek, J. Gerlici, J. Harušinec, T. Lack, *Analysis of temperature distribution in a braked railway wheel during braking by the brake block*. TRANSCOM 2013, section 6, 289-292. EDIS Publishing Institution of the University of Zilina, ISBN 978-80-554-0695-4 (2013)
25. Vlk, et al., *Experimental mechanics*. Chapter **6.1** Introduction into experimental determination of residual stresses. Brno university of technology, 119-123 [in Czech]

Stimulation of nerve endings via medical device

Antonin Svoboda[1,*], *Josef* Soukup[1]

[1]Department of Machine and Mechanics, Faculty of Mechanical Engineering, University J. E. Purkyne, 40001 Usti nad Labem, Czech Republic

Abstract. This solution describes solution of vibromechanism for stimulation nerve paths and nerve endings injured patients. Medical doctors stimulate nerve paths and nerve endings patients after injury of spine or after another injury where was broken nerve endings. These methods are in primary designed for tetraplegic and paraplegic patients specially for men. For this request was developed device and mechanism for vibrostimulation of nerve paths and nerve endings. In develop of construction has been paying attention to safety of mechanism. Next were taken into account the construction of other devices, especially their shortcomings in the operation on the accumulator. For mains powered vibrators 230 (110) Volt were assessed for safety and noise vibrating mechanism. The sum of all failures was proposed structure described below. The success rate of this vibration method is reported mostly between 60-80%.

Keywords: vibration, excenter, amplitude, frequency, DC engine

1 Introduction

According to available information, and clinical test have mechanisms to induce ejaculation reflex for the collection of genetic material (sperm) in affected males for several years. These methods proved to be reliable with relatively high efficiency [1]. As the most effective method of obtaining genetic material in men can be called surgical sperm collection. This method is performed under general anesthesia, which limits patient. Performance is preceded by internal examination, ECG, blood biochemistry and other examinations according to the patient's condition. Behind this method it is then electrostimulation, which again is performed under general anesthesia by electrical stimulation of the prostate. This method, however, requires a great practical experience of the doctor, as it can cause a burn or burns intestine of the patient. At least stressful method for the patient is vibrostimulation. The doctor or also informed patient stimulates via vibration nerve endings in the penis prescribed amplitude and frequency [2]. After injury of spine are medical doctors ready to help tetraplegic and paraplegic patients return them back normal or near normal life. Most man patients after injury of spine are no able to have got children and family. This status is for most men traumatic and stressful [2]. At least stressful method for the patient is vibro-stimulating.

The doctor or also informed patient vibration stimulates nerve endings in the penis specified amplitude and frequency. To induce ejaculation is after approximately eleven

* Corresponding author: antonin.svoboda.most@seznam.cz

Reviewers: *Alžbeta* Sapietová, *Ján* Vavro *Jr.*

minutes of stimulation, but it is necessary to observe the procedure three series of stimulation after three minutes and a minute break between every three minutes. Another way to evokes reflection in paraplegics and another way in tetraplegic. Paraplegic person is paralyzed legs, and then the person tetraplegic paralyzed on all four limbs. This method describes ways to return back to life without minimum patient's stresses. One of no stress method is vibro-stimulating of nerve paths [1]. Our developed mechanism is generating vibrations – amplitude and frequencies between effective ranges. After consultations injury doctor specialists, we can presuppose amplitude of the range between 1 – 3 mm and frequencies between 80 – 100 Hz.

For optimal frequency and amplitude, we can take information from doctor's injury and convalescence specialists. Finally, we can make a proposal optimal frequency and amplitude [1]. For the best solution we can mark range between 80 – 120 Hz and 1 – 4 mm [3]. To achieve the desired frequency, range between 80 – 120 Hz, it is necessary to convert frequency to RPM of DC engine speed (1).

$$f = \frac{n}{T}\ [Hz] \Rightarrow n = f \cdot T \Rightarrow n = f \cdot 60\ . \tag{1}$$

Where is f – frequency, n – DC engine speed in RPM (rotation per minute), T – time

Unbalance in rotating machines is a common source of vibration excitation [4]. We consider here a spring-mass system constrained to move in the vertical or horizontal directions and excited by a rotating machine that is unbalanced, as shown in Fig. 1, 2. The unbalance is represented by an eccentric mass m with eccentricity e that is rotating with angular velocity ω [5].

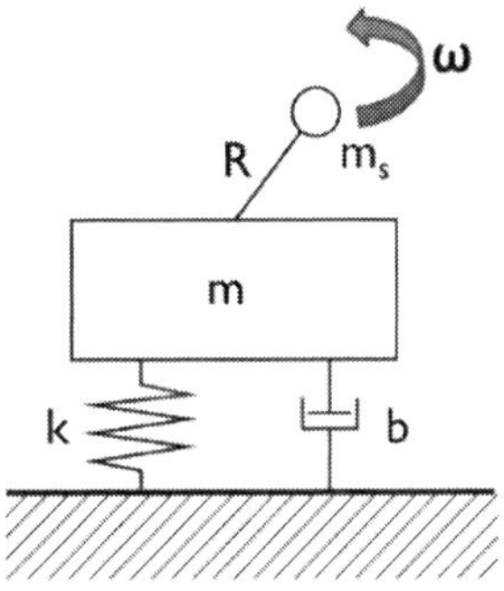

Fig. 1. Forced damped vibration system with one degree of freedom

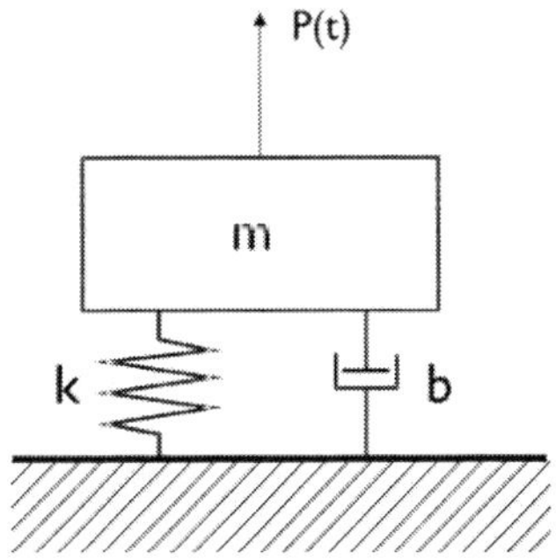

Fig. 2. Forced damped vibration system with one degree of freedom

Where is m_s – weight of particles, m – weight, R – eccentricity, ω – angular velocity, b – coefficient of shock absorber, k – stiffness of the spring

Equation of motion

$$m\ddot{x} + b\dot{x} + kx = P(t), \tag{2}$$

where is x – the displacement from the static equilibrium of position, $P(t)$ – the excitation of power

Angular speed

$$\omega^2 = \frac{k}{m}. \tag{3}$$

Harmonic oscillation is characterized by harmonic waveforms

$$y = y_0 \sin(2\pi f) = y_0 \sin \omega t \tag{4}$$

$$a = \frac{dv}{dt} = -\omega^2 y_0 \sin \omega t = a_0 \sin \omega t \tag{5}$$

$$v = \frac{dy}{dt} = y_0 \omega \cos \omega \tag{6}$$

where is ω – angular velocity, y_0 – amplitude of displacement, a_0 – amplitude of acceleration, t – time.

2 Methods

Vibration that takes place under the excitation of external forces is called forced vibration. When the excitation is oscillatory, the system is forced to vibrate at the excitation frequency. If the frequency of excitation coincides with one of the natural frequencies of the system, a condition of resonance is encountered, and dangerously large oscillations may result. The failure of major structures such as bridges, buildings, or airplane wings is an awesome possibility under resonance. Thus, the calculation of the natural frequencies of major importance in the study of vibrations [5]. The mechanism was designed in several variants. Theoretical solution can be effective via analytics method, method of final points and experimental measurement. For the best solution can be mark experimental measurement. For making these tests of experimental measurement was designed vibrogenerator with DC engine in several variants. In all variants was used the same electric engine SPEED 400 with maximum 12000 RPM and powered between 6 – 12 Vols. For medical device – vibrogenerator is effective RPM between 4800 – 7200 @ 3,32 – 4,81 Volts. Crank mechanism Fig. 3. is very difficult and expensive for production in this very small sizes. Is necessary to take into consideration more difficult assembly of mechanism device [3]. Advantages: Precisely adjustable amplitude shift crankshaft, separation of amplitude from frequency - with increasing of the engine speed (frequency) is the amplitude constant. Disadvantages: Miniature parts of the crank mechanism - higher demands on production and assembly [6].

Fig. 3. Crank mechanism

Fig. 4. Construction with the cam

A cam mechanism (Fig. 4) provides the distinct advantage the simplicity of design [7], the lower cost of the assembly apparatus, than the variants with the crank mechanism [8]. Advantages: Simplicity of design - lower assembly costs and assembly of device mechanism. Disadvantages: Complicated of cam production, noise of the device. When we would like to use a cam mechanism, it is necessary to take into account unbalance cam which causes the vibration of the whole device. In this case, unbalance works in favour of the final oscillation amplitudes, but to the detriment of engine bearings and the entire device. At the prescribed speed in the range 4800 – 7200 RPM occurs due to imbalance even more options, so-called own auto oscillations [8].

Eccentric is probably the simplest design of solution, bringing low manufacturing costs. Experimental measurements with eccentric showed higher noise of the device. Advantages: Simplicity of design. Disadvantages: Higher load of engine's bearings, noise of the device. Unbalance Fig. 5. and Fig. 6. is the simplest solution, however, the fundamental problem depending engine speed (frequency) vs. amplitude. Advantages: Simplicity of design. Disadvantages: Higher load of engine's bearings, frequency dependence of the amplitude.

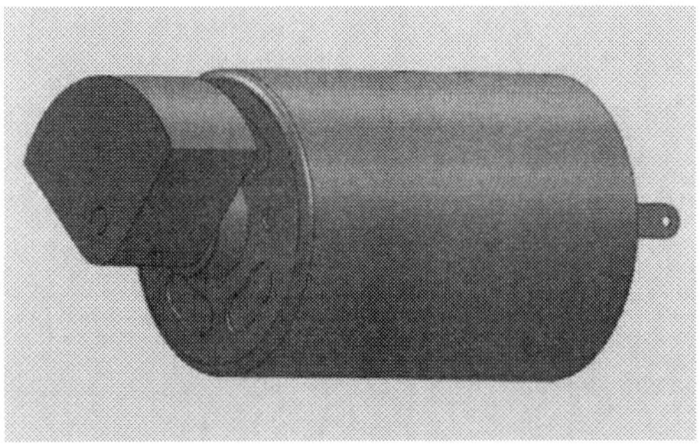

Fig. 5. Construction with the unbalance

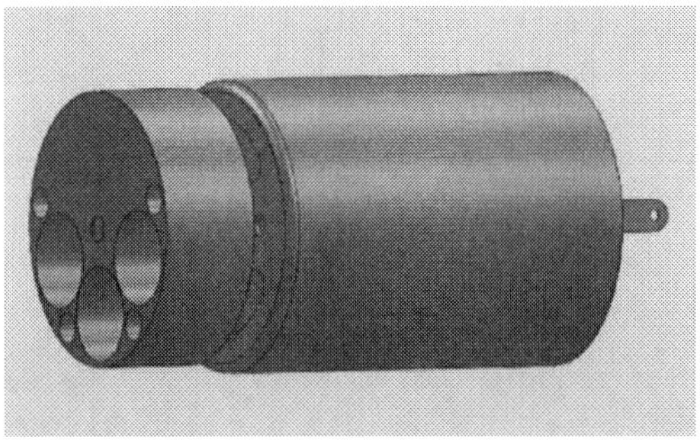

Fig. 6. Constructions with drilled unbalance holes are the most simpler from a manufacturing point of view

3 Results

With experimental measurements we found optimal values of amplitude and frequency. In the future, these results will be used for the final design of the device for generating vibrations. Unbalance method was verified in the laboratory. DC engine was powered by a stabilized source and were measured engine's RPM, the electric current (A) and frequency

(Hz). RPM of DC engine was taken via laser method of scan. The final frequency of unbalance mechanism of medical device, was taken by three axes accelerometer.

At Fig. 7, 8, 9 are shown graphical acceleration [m/s^2] (axes Y) and frequency [Hz] (axes X) what were recount to amplitude [mm] via La Grange method. Frequency of DC engine was controlled via laser RPM equation (1).

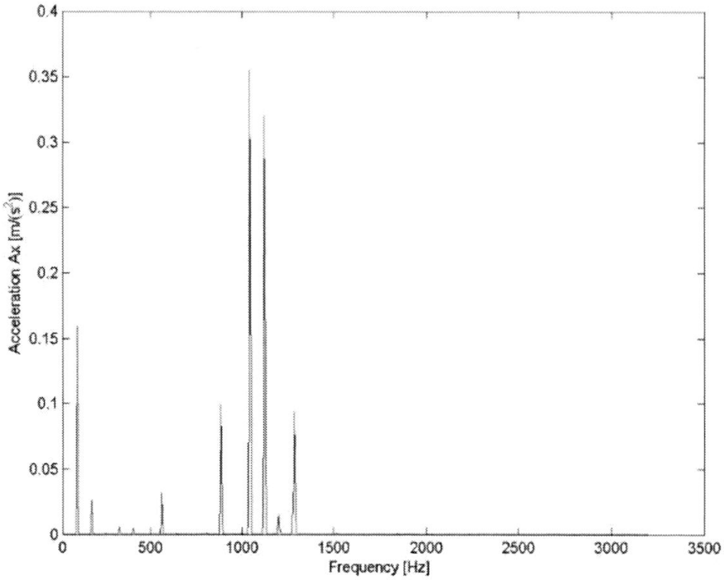

Fig. 7. Acceleration in the x direction – measurements at speed 4800 RPM

In Fig. 7 was achieve the maximum of acceleration value 0.3549 m/s^2 in frequency 1040 Hz via 4800 RPM of DC engine. These values were recalculated to amplitude and later theoretically confirmed via Lagrange numerical mathematics method.

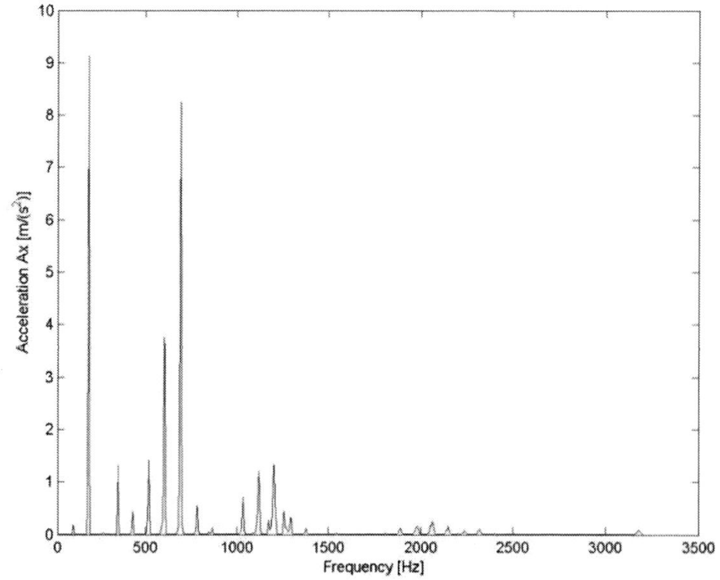

Fig. 8. Acceleration in the y direction – measurements at speed 5160 RPM

In Fig. 8 was achieve the maximum of acceleration value 9.1390 m/s^2 in frequency 172 Hz via 5160 RPM of DC engine. These values were recalculated to amplitude and later theoretically confirmed via Lagrange numerical mathematics method.

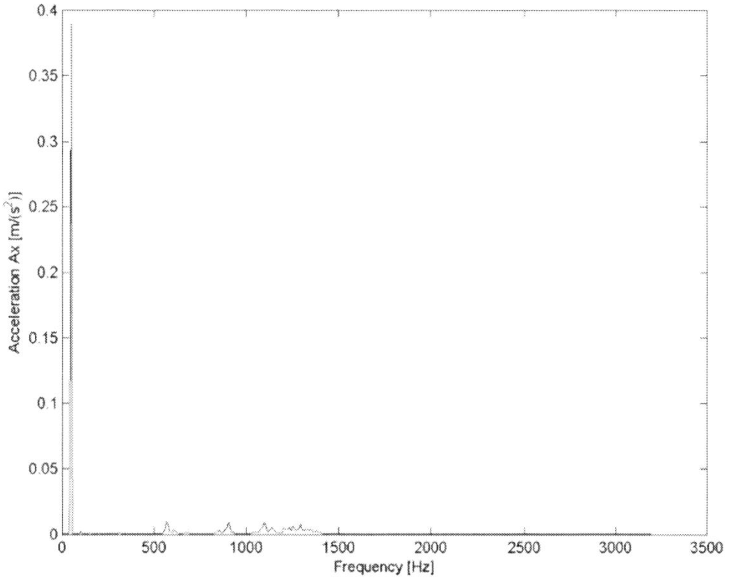

Fig. 9. Acceleration in the z direction – measurements at speed 3060 RPM

In Fig. 9 was achieving the maximum of acceleration value 0.3892 m/s^2 in frequency 52 Hz via 3060 RPM of DC engine. These values were recalculated to amplitude and later theoretically confirmed via Lagrange numerical mathematics method.

Conclusions

For measurement of vibrations was designed special stand. In all cases of tests were used the same engine SPEED 400 powered by stabilized source. RPM was taken via laser sensor and vibration was measured via accelerometer. All measurement results were recalculating from acceleration to amplitude and was used numerical method Lagrange for verifying this values. Next were processed via software MATLAB and converted to MS-EXCEL. In these methods described above was compared vibration exciter suitable mechanism design. Was selected the most appropriate method of construction equipment. This method was experimentally verified in laboratory. It will also be developed by the cheapest method of production and assembly device for vibration excitation and stimulation of nerve endings and paths. The result of this work will be quality and cheap device for stimulating man patients.

J. E. Purkyne University in Usti nad Labem, Faculty of Production Technology and Management for support of this project from SGS grant.

References

1. T. Šrámková, *Poruchy sexuality u somatických nemocných a jejich léčba*. Grada Publishing, 27-32 (2013)
2. K. Juliš, R. Brebta, a kol., *Mechanika I. – Statika a kinematika*. (SNTL Praha, 1986)

3. Z. Dejl, *Konstrukce strojů a zařízení I.* Spoje. (Montanex, Ostrava, 2000)
4. H. Benaroya, *Mechanical vibration.* (Marcel Dekker, N.Y., 2004)
5. J. R. Ellis, *Road Vehicle Dynamics.* (Ahron, USA, 1989)
6. R. Brebta, L. Půst, F. Turek, *Mechanické kmitání.* (Sobotáles, Praha, 1994)
7. A. Svoboda, *Mechanický přístroj pro stimulaci nervových drah.* Diplomová práce, Fakulta výrobních technologií a managementu UJEP v Ústí nad Labem, Ústí n. L., (2015)
8. V. Vejrosta, *Konstrukce zdravotnických přístrojů.* (Česká společnost pro zdravotnickou techniku, June, 1995)

Polyurethane foam behaviour during impact

Stanislav Špirk[1,*], *Michal* Křížek[1], *Štěpán* Jeníček[1]

[1]University of West Bohemia in Pilsen, Regional Technological Institute, Univerzitní 22, 306 14 Plzeň, Czech Republic

Abstract. This contribution is focused on the passive safety of the railway vehicles interiors. The passive safety has become a new area of the rail vehicles design. In the field of rail vehicles interior passive safety the computer simulations are most widely used. The computer simulations must be based on relevant material data. For some common materials (eg. steel, aluminium, etc.) the material data are ascertainable. The interiors of rail vehicles contain quite specific materials frequently. One of the material with more complicated mechanical properties is polyurethane foam (PUR). The aim of this paper is to present validation of material parameters for specific polyurethane foam by experiment. The material properties (for seat made from Elastoflex W 5662 covered with Holdsworth BHLH 31634) will be useful for further simulations.

Keywords: Crash, foam, PUR, dynamics, Elastoflex, Holdsworth

1 Introduction

Mechanical properties of polymer foams at ambient temperatures have been evaluated by various authors using experiments with static and shock loading. Their results are available in studies [1] and [2]. Here, foam parts are considered an important element of a vehicle interior that can protect passengers from severe deceleration by gradual and controlled absorption of energy. An optimal material should be able to absorb kinetic energy of impact and have strength below a relevant force limit.

Passenger injury is a result of the collision between a body part and the interior feature impacted. In case of collision with a padded seat, the description of the seat crash behaviour is complicated. The main aspects are connected with the soft parts of this structure. The final mechanical behaviour during impact will be the result of the response of PUR prestressed by the cover fabric. The prestress of PUR is significant but difficult to measure. The best way of material properties validation is to conduct a test which is close to the conditions during the actual passenger impact. The foam material Elastoflex W 5662 covered with Holdsworth BHLH 31634 cloth was used in experiment. The material model presented in the end of paper has the good concordance with experiment.

* Corresponding author: spirks@rti.zcu.cz

Reviewers: *Mariana* Pajtášová, *Ksawery* Szykiedans

2 Properties of foam materials

Mechanical properties of polymer foams at ambient temperatures have been evaluated by various authors using experiments with static and shock loading. Their results are available in studies [1] and [2]. Here, foam parts are considered an important element of a vehicle interior that can protect passengers from severe deceleration by gradual and controlled absorption of energy. All investigations in this field focus predominantly on the capability of controlled energy absorption. Optimal energy-absorbing materials must absorb the kinetic energy of an impact, while keeping the reaction force below a certain limit. Energy is dissipated by cell bending, buckling or fracture. Generally, the stress-strain curve shows a long horizontal section followed by a steep rise (Fig. 1). This explains the high efficiency of energy absorption achieved with foamed materials. In addition, a foamed specimen always provides lower acceleration at maximum compressive force than a solid specimen of the foam matrix material with the same volume absorbing the same amount of energy.

Foamed materials exhibit linear response within a relatively narrow interval up to approximately 5 % engineering strain. Here, the slope of the curve is given by the modulus of elasticity of the material. As the load on the foam increases, the cells begin to collapse by elastic buckling, plastic yielding or brittle crushing – depending on the mechanical properties of the specimen. The collapse continues under approximately constant stress, until the cell walls come into contact under the pressure. At this point, stress begins to rise steeply, as the material densifies. During unloading, the stress decreases non-linearly. Absorbed energy is generally represented by the area below the curve in the plot.

Upon impact, the passenger body is subjected to deceleration whose magnitude depends on the stiffness of the seat material. Consequently, the fundamental requirement is to define the dependence of true stress in a foamed material on its actual engineering strain under compression. It is usually an increasing function, whereas the unloading behaviour is defined separately through a parameter which reflects energy dissipation. Simulations were carried out using the type-45 material model (General nonlinear strainrate dependent foam with optional energy absorption) [3] (Pam-Crash software). The characteristics of PUR foam can be found in literature [4]. Figure 1 shows the static true stress-true strain plot for PUR.

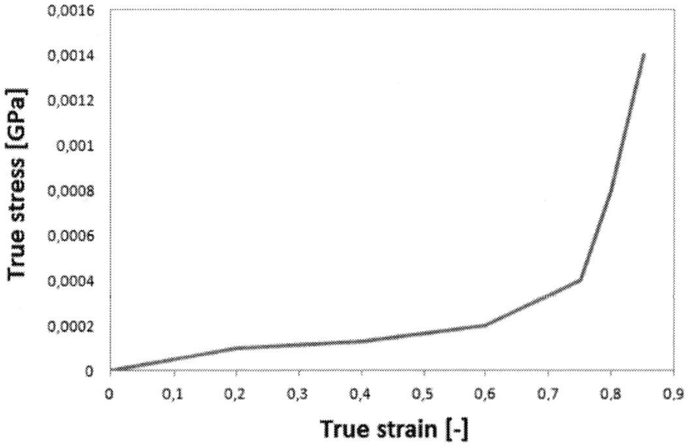

Fig. 1. Initial stress-strain dependence of the foam material (PUR)

The key question is whether the literature data offer an accurate characterization of the behaviour of the seat material. Additional factors may play a role in a real-world seat, such as foam pre-stress. Although the thickness of the cover fabric can be considered negligible, the fabric will still apply certain pre-stress to the filling. Consequently, the mechanical properties of the seat cushion never quite match the properties of the filling material alone. Clearly, a tension test cannot be carried out here. Foamed materials are typically tested in compression. However, testing a polyurethane foam seat (including the cover fabric) at higher loading rates in a conventional testing machine is complicated and difficult to carry out. Since some preliminary materials data is already known, the best course of action appears to involve a simple experiment to verify the definition of the material.

3 Experimental validation of mechanical properties of foamed material

The purpose of the simple experiment is to verify the behaviour of an actual material under conditions (loading mode, load rate, etc.) which are as close as possible to the real-world situation. In general, calculations are best validated by experiments. Various collision scenarios involve the contact between various parts of human body and the seat. Head impact is unquestionably considered the most important one. An experiment which is intended to verify the behaviour of a material should therefore follow this scenario. From the standpoint of calculation validation, a head-seat impact makes the ideal experiment. If its results are also to be used for validating a foamed material, there is a problem with a vast number of unknown and difficult-to-verify parameters. The outcome would be affected by the stiffness of the seat, by the properties of the material of the passenger head model, and by its geometry, weight and stiffness. In order to obtain relevant output for validating the data of the foamed material, the experiment should be simplified to involve an impact of a rigid ball on a foamed material specimen placed on a rigid plate. This approach is close to Multibody dummy head (seen eg. in [7]) which impact to the FE model part. In the simplest arrangement, the ball could be dropped on the seat cushion. However, this poses an unacceptable risk of the ball falling away from the seat cushion (or bouncing away), causing an injury or damaging the measuring equipment. The most basic method of guiding the movement is to use a rotating arm. The resulting pendulum would cause the ball striker to hit the same spot repeatedly. Energy losses associated with the impact would therefore be easy to evaluate. In addition, the arm would make the transmission of data to a recording equipment very straightforward. The proposed experimental arrangement is shown in Fig. 2.

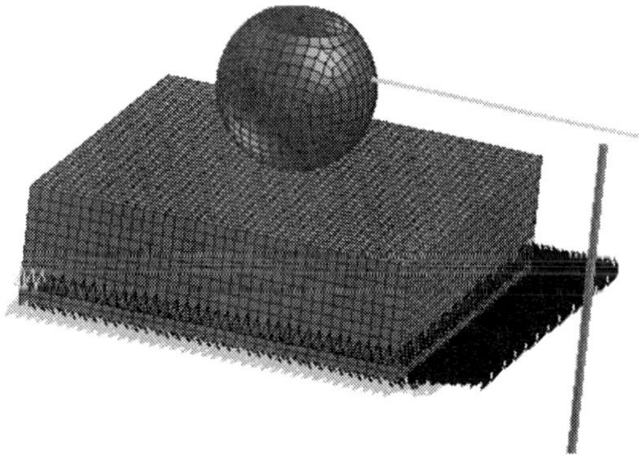

Fig. 2. Experimental set-up and simulation

The striker is a steel ball of 3 mm steel sheet filled with concrete for greater weight. The ball is provided with a threaded hole into which the threaded end of the rotating arm is screwed. The other end of the arm is attached to a hinge mounted to a fixed plate. The plate height is such that the ball comes into contact with the specimen at the point when the arm is in horizontal position. From the upper to the horizontal position, the fall can be evaluated analytically. Below that position, FEM computation should be used.

The acceleration of the ball was measured using PCB 352C33 S/N 120471 accelerometer. This is a dynamic accelerometer suitable for measuring rapidly changing accelerations. It was attached by means of a magnet. It was connected to a computer via a Highspeed USB carrier NI9234 logger. The readings were recorded by means of the Signal expres 2011 software. In order to obtain a slow-motion video sequence, a high-speed Olympus ispeed2 camera has been used. [5] Its speed was 500 fps at an 800×600 resolution (Fig 3). The size of the camera sensor means that more expensive photographic lenses are not required. C-mount lenses are easily attached to the camera and do not compromise image quality. For experiment was used the c-mount lense with low color aberration and low TV distortion, focal length 35 mm, aperture 2, mod 0.35 m, back focal length 10.14, angle of view 14° image circle 11 mm. The station was lit with two 1500 W halogen lamps. The recording from the high-speed camera was processed using the Ispeed2 Advanced PC program. The crucial task for validating material data is to verify the speed at the moment of impact of the striker. By this means, it can be confirmed whether this is a free fall case where friction of the hinge, aerodynamic drag and other aspects can be neglected. The high-speed camera can record the position of a chosen point over a period of time (where one pixel is 0.53 mm, with software 1/4 pixel approximation, accuracy average ±0.63 mm with used optics). Hence, one can evaluate the position and the speed of the striker throughout the process.

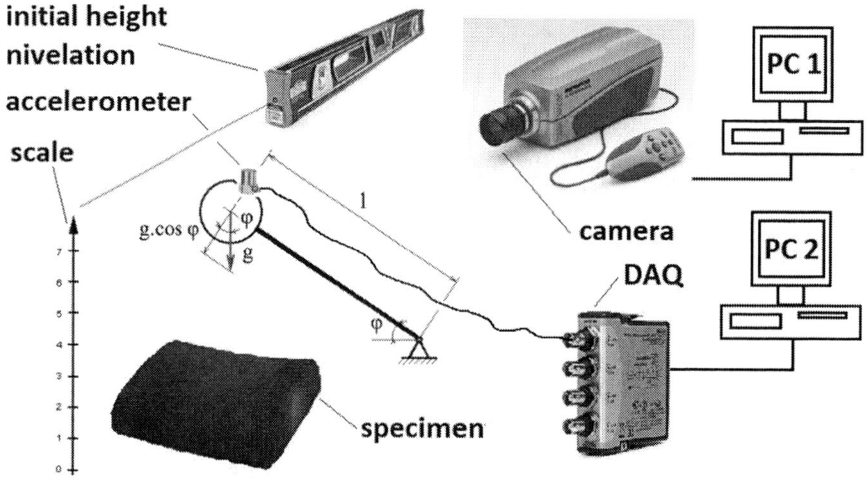

Fig. 3. Functional diagram of the test bed

If no other than the gravitational acceleration acts on the striker, its impact speed can be determined analytically. In fact, this is a special case of a pendulum. The arm, which has a length of l = 910 mm and a weight of 0.6 kg, holds a striker of m = 2.4 kg. This ball-shaped striker has a radius of 60 mm. The angle between the arm and the horizontal is denoted as φ. Several simplifying assumptions were used for solving the mathematical pendulum problem. Small displacements are not considered. Therefore, equation 1 is used for finding the displacement φ.

$$\int_0^{\varphi} \frac{g}{l} cos\varphi \; 2d\varphi = \int_0^{\omega^2} d\omega^2 \tag{1}$$

For the speed v at the time instant when $v = 0$ (striker impact), one can use the assumption that the vertical speed corresponds to the tangential speed because cos φ =1. The resulting speed v at the moment of impact of the striker is given by equation 2.

$$v = \sqrt{2 \cdot g \cdot l \cdot sin\varphi} \tag{2}$$

4 Experimental results

Comparing three values obtained with the same initial height of the striker enables to evaluate the accuracy of measurement and eliminate random effects which threaten to compromise the results. The acceleration values from the accelerometer are compared in Fig. 4.

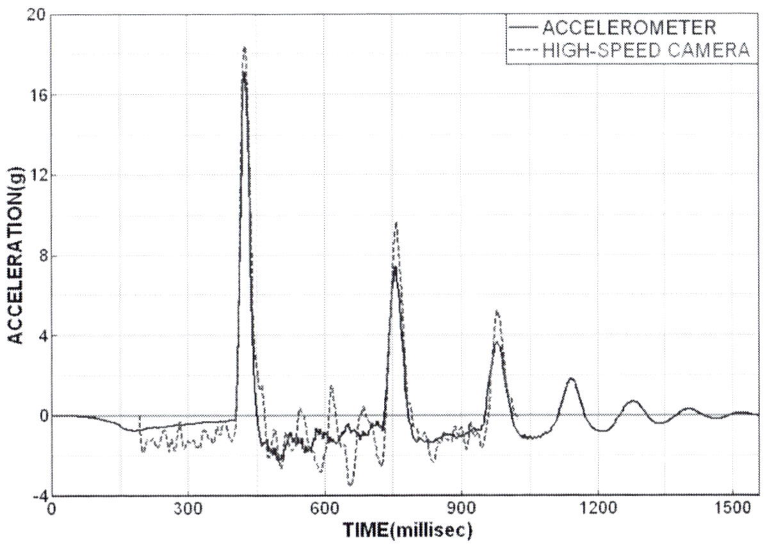

Fig. 4. Comparison of experiment results based on data from the accelerometer and high-speed camera for the initial height of 880 mm

Initial derivative calculations were carried out using the MS Excel tool. The basis here is the definition of derivative, i.e. the most straightforward method of numerical differentiation [6] expressed by the relationship (3). The Visual Viewer tool has also been useful for data filtering and other manipulations.

$$f'(x) = \lim_{h \to 0} \frac{f(x+h)-f(x)}{h} \approx \frac{f(x+h)-f(x)}{h} \tag{3}$$

The comparison between accelerometer readings and high-speed camera data is interesting. The simplest alternative is differentiation of the displacement from the high-speed camera data, and integration of the accelerometer readings. If displacement is considered as the result of horizontal and vertical displacement components, the speed profile calculated from the accelerometer readings should match the speed profile calculated from the displacements obtained from the high-speed camera data. In every frame, the high-speed camera detects the position with a certain accuracy which is related to the correct identification of the position of the point of interest. It introduces a small error in every frame which has no significant effect on the general trend. However, this error can be positive at one point and negative at the next one. Numerical solution of the second derivative of this function generates noise due to these small fluctuations. Hence, the displacement function appears very smooth but its second derivative is affected by considerable noise. Filtering of results is an effective solution here. The resulting acceleration may be affected by this filtering but both speed profiles with filtered results from the high-speed camera suggests a good correlation between both measuring methods.

5 Discussion of results

The material data is first validated at a higher speed. Figure 5 shows the results of the validation and calculation. Clearly, the behaviour of the non-validated material is rather close to the experimental findings. The deformation rate at the point of impact can be estimated with the aid of the initial striker velocity and the initial height of the part (4).

$$\dot{\epsilon}(t) = \frac{d\epsilon}{dt} = \frac{d}{dt}\left(\frac{L-L_0}{L_0}\right) = \frac{v(t)}{L_0} \tag{4}$$

Where ϵ denotes engineering strain, L_0 is the initial width of the foam, L is the instantaneous width of the foam, and v(t) is the instantaneous rate of compression of the foam at time t. Figure 5 shows results of the calculation for the initial speed of 3.9 m/s. The behaviour of the model is close to the experimental findings. By adjusting the material parameters, much better correlation can be obtained. This is clear from Fig. 5 which compares the non-validated profile and the validated profile where the latter overlaps with the experimental results at some points.

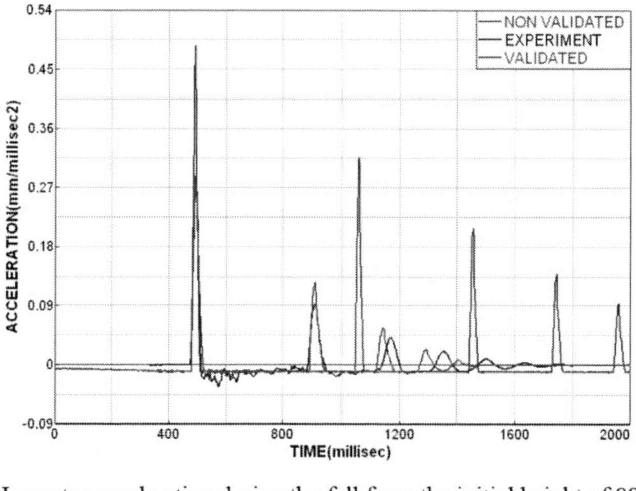

Fig. 5. Impactor acceleration during the fall from the initial height of 880 mm

Figure 6 shows results of the calculation for the initial speed of 2.8 m/s. It is obvious that at lower speeds, the simulation results are less close to the experimental findings than at higher speeds. Nevertheless, the priority is to validate the material model for the higher speed.

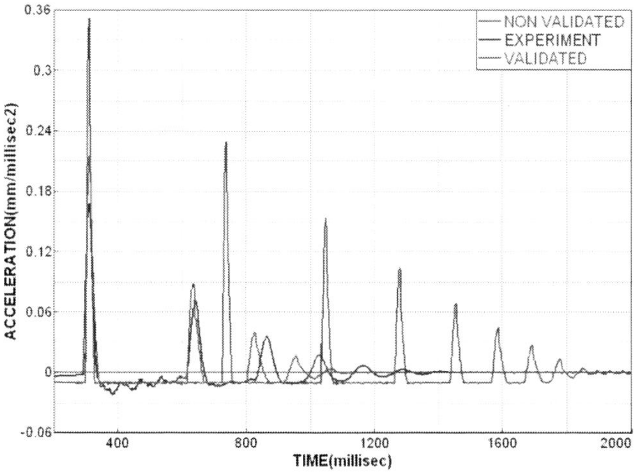

Fig. 6. Striker acceleration from the initial height of 500 mm

The most important outputs are the parameters of foam material model for foam seat padding made from Elastoflex W 5662 covered with Holdsworth BHLH 31634. The material model validated by experiments is suitable for seat padding in explicit simulations. The used stress-strain characteristics is shown in the figure 7. Young modulus is here E = 0.02 GPa, the density ρ = 4e-7 kg/mm3 and dissipation unloading factor is UNLFAC = 0.25. The example of material model definition with conventions of SW Pam-Crash is in Table 1.

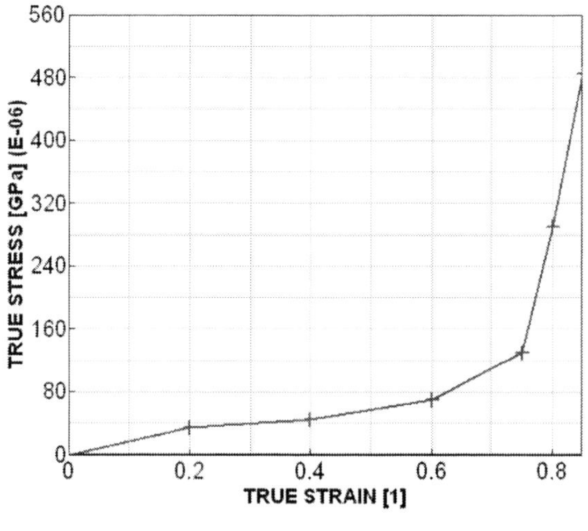

Fig. 7. The stress-strain dependence of the foam material (PUR)

Table 1. Material model parameters with conventions of SW Pam-Crash

UNIT [MM] [KG] [MS] [KELVIN]					
MATYP	**RHO**	**ISINT**	**ISHG**	**ISTRAT**	
45	4E-7	3	2	0	
QVM	**IDMPD**				
1	0				
E	**UNLFAC**	**HYDRO**	**KSI**	**SLMULT**	**Fo**
0.02	0.25	0.	0.7	0.	3.0
Q1	**Q2**	**Q3**	**ERATc1**	**ERATc2**	**ERATt1**
1.2	0.06	0.01	0	1	0
LCc1		**LCc2**		**LCt1**	

X	Y	X	Y	X	Y
0.	0.	0.	0.	0.	0.
0.2	3.5E-5	0.2	9.54E-5	0.2	3.16E-5
0.4	4.5E-5	0.4	0.000111	0.4	0.000414
0.6	7.0E-5	0.6	0.000159		
0.75	0.00013	0.75	0.000254		
0.8	0.00029	0.8	0.000446		
0.85	0.000485	0.85	0.00159		

Conclusion

The most significant result is the material model (Elastoflex W 5662 covered with Holdsworth BHLH 31634) prepared for SW Pam-Crash (see Tab 1). The material model validated by experiments is suitable for seat padding in explicit simulations. The present experiment enables us to assess whether the behaviour of a material model corresponds to the reality. In addition, the material model data can be adjusted to obtain better correlation between collision simulation and the real-world situation. To validate foamed material data, an FEM calculation was carried out, which effectively simulates the experiment. A model of a striker (a rigid body) has been created. The striker model is attached by means of a 1D element to a kinematic joint with a single degree of freedom. These bodies with weights identical to those of the real experimental parts represent the pendulum. The striker hits a cuboid created with solid eight-node elements. This cuboid represents a model of the relevant part of the seat made of a foamed material. Its height is clearly defined (by the height of the part). The horizontal size of the cuboid is increased to the point, at which further increases have no impact on the calculation. If further increases have no impact on the calculation, there is no need for simulating the entire part. The behaviour of the model is close to the experimental findings. By adjusting the material parameters, much better correlation can be obtained. This is clear from Fig. 5 which compares the non-validated profile and the validated profile where the latter overlaps with the experimental results at some points.

The present contribution was prepared under financial support of the project LO1502 'Development of the Regional Technological Institute' carried out under the auspices of the National Sustainability Programme I of the Ministry of Education of the Czech Republic aimed to support research, experimental development and innovation.

References

1. M. Avalle, G. Belingardi, R. Montanini, *Characterization of polymeric structural foams under compressive impact loading by means of energy-absorption diagram*. International Journal of Impact Engineering, 455–472, (2001)
2. M. Paulino, F. Teixeira-Dias, *On the Use of Polyurethane Foam Paddings to Improve Passive Safety in Crashworthiness Applications*. InTech, (2012)
3. Author's team of ESI Group, *Virtual Performance Solution 2010*. ESI Group, Paris, (2010)
4. Author's team of ESI Group, *Lerning Solutions PAM-CRASH TRAINING*. ESI Group, Paris, (2010)
5. Advanced Technology Group: iSPEED 2 [online]. 2013 [cit. 2013-09-6]. [WWW]: http://www.atg.cz/cz/pdf/katalogy/VT/ispeed.pdf
6. K. Kopecky: *Numerical Differentiation*. LECTURE NOTES FALL 2007

7. L. Hyncik, J. Manas, J. Spicka, S. Spirk, L. Kovar, *Development of 6 years old child virtual model by automatic scaling*. SAE Technical Papers, SAE 10th International Conference on Automotive Engineering **2**, (2014)

Computer aided structural analysis of newly developed railway bogie frame

Pavol Šťastniak[1,*], *Marián* Moravčík[2], *Peter* Baran[1], *Lukáš* Smetanka[1]

[1]University of Žilina, Faculty of Mechanical Engineering, Department of Transport and Handling Machines, Univerzitná 8215/1, 010 26 Žilina, Slovak Republic
[2]Tatravagónka Inc., Štefánikova 887/53, 058 01 Poprad, Slovak Republic

Abstract. The paper is part of the publication series, which describe the most significant and innovative research and development design solutions and computational procedures as part of European structural funds project. The object of this article is strength conditions assessment of new structure of rail bogie frame, which is characterized mainly by better dynamic properties while driving on the track, good operational properties and higher safety against derailment. For validation of the new structure design, there has been created a substitute simulation model. Results of calculations and prototype tests prove, that new structure of the bogie frame satisfies strength assessments.

Keywords: stress analysis, bogie frame, computer tools

1 Introduction

Indeed nowadays about 1.5 billion t/km are transported in Europe by lorry at distances farther than 150 km, conversely only 0.4 billion t/km (20%) are transported by train, this entails important costs for fossil fuels. In the nearby future when transportation has to be more sustainable it seems quite clear that freight railways will win the mode choice more often.

For this to happen though, it is necessary that freight railways, apart from lowering their prices, significantly improve highly increase the quality of transportation. In that sense, quality standards such as reliability, flexibility, availability, cargo security and safety, punctuality, customisation, marketability, traceability, complementary servicing and time for transport among others have to be improved by railways as well.

Hence, freight rail transport has the challenge to become excellent and to gain in reputation. There are many actions to increase quality in freight rail transport; one of them is the optimisation of the current wagon fleet to improve availability, flexibility, marketability, commercial speed, cargo security and cost. This optimisation has to respond to the actual trends of transport demand and has to be in consonance with the required and feasible infrastructure upgrades.

University of Žilina and Tatravagónka Inc. Poprad jointly respond to these requirements

* Corresponding author: pavol.stastniak@fstroj.uniza.sk

Reviewers: *Vladimír* Dekýš, *Petr* Horyl

by means of the European structural funds project with name "Development of two types of freight wagons with bogies for non-standard wheelbase or track wheelset, complying with the criteria for interoperability, environmental issues, safety and reliability", where developed, among other things, a new freight rail bogie type TVP 2009-R (Fig. 1) for "Russian" track gauge of 1520 mm.

2 Technical specification of new freight bogie

The concept is derived from the bogie type TVP 2007, which has been certified under TSI and they are currently used on the EU tracks with track gauge of 1435 mm. Designed bogie hence used:

- simplicity and reliability of friction damping by system Lenoir,
- reliability based on verified design solutions,
- real speeds increase by using bogie technology with a rigid frame,
- transition improving through curves with small radiuses for wheelset radial position ensuring.

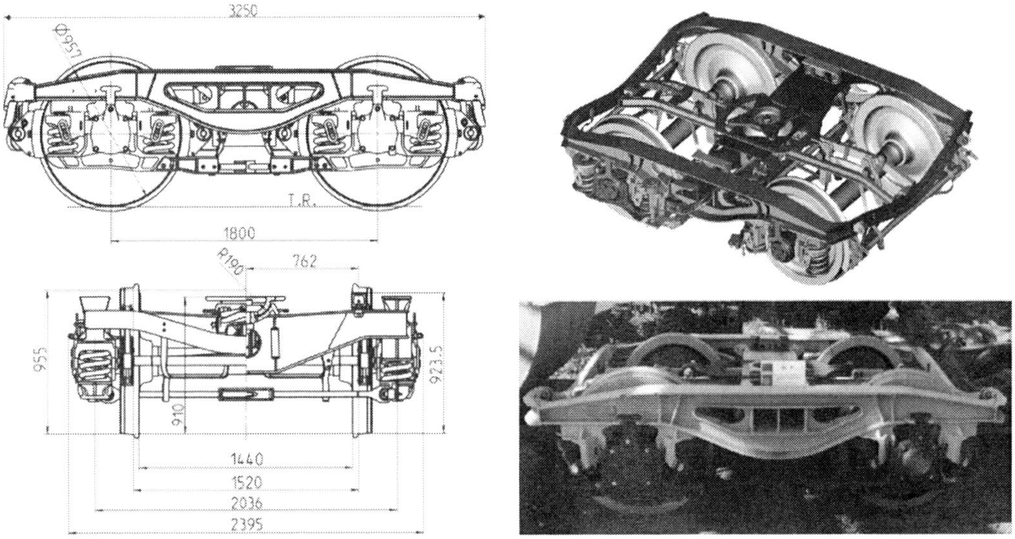

Fig. 1. The prototype of new railway bogie

The bogie type Y25 is characterized by wheelset guiding, in which is primary suspension with progressive damper as part of axlebox. The damper is always located on one side of the axlebox suspension of each wheelset. Bogies equipped with this standard configuration have a certain size of guiding forces while driving along the curve of the track and related technical consequences [1].

Table 1. Dimensional and technical parameters of the bogie

Wheelbase	1800 mm	Height of the bogie pivot above the top of rail	900 ±12 mm
Wheeldiameter	957 ±7 mm	Weight	4900 kg ±5 %
Axle pin diameter	130 mm	Maximum speed at an axle load of 25 t	120 km/h
Axle pin length	190 mm	Temperature range	+50 °C to +60 °C
Lateral clearance of all axles	±10 mm	Special equipment	ready fo Automatic system MATROSOV
Longitudinal clearance of all axles	±8 mm		composite brake blocks

In order to reduce these forces, the new design of bogie uses the connection of wheelsets by means of cross coupling. The purpose of cross coupling is radially adjustable wheelset position in curves. Therefore, the vehicle achieves better dynamic properties, higher safety against derailment and lower negative effect on the track, known as "track friendly" bogies. Basic technical and dimensional parameters of the bogie are listed in Table 1.

3 Bogie frame assessment by means of computer simulations

Every new freight bogie or wagon prototype before start in operation must be assessment and approval according to requirements of Technical Specifications for Interoperability (TSI). For assessing is authorized only notified body (NoBo). It is the organization responsible for the conformity assessment of interoperability constituents and structural subsystems as required in the TSI (in Europe e.g. Ricardo, Luxcontrol, Sconrail) and in Russian Federation only Federal Budgatery Organization "Register of Certification on the Federal Railway Transport".

As part of the documentation, which the applicant must submit, are the results of prototype tests. The aim of the validation plan is to prove that the design of the bogie frame fulfils the conditions defined in the technical specification. In addition, it shall show that the behaviour of the bogie frame, constructed according to the design, will give satisfactory service without the occurrence of defects such as catastrophic rupture, permanent deformation and fatigue cracks. It shall further demonstrate that there is no adverse influence on the associated bogie components or sub-assemblies.

The procedure for the validation of the mechanical strength of a bogie frame against the acceptance criteria shall be established on the basis of: 1) analysis (computations), 2) laboratory static tests, 3) laboratory fatigue tests, 4) track tests.

Currently, by using advanced computer technology is possible those tests at high levels predict by simulation analysis. The great advantage of the mentioned simulations is that the entire development process of rolling stock is so accelerated, leading to a reduction in overall costs. Simulations and subsequent optimization of the vehicle structure is made before production of the vehicle itself. This leads to minimizing the number of unsatisfactory results conducted on a real vehicle. This may, in such a stage of development lead to delays and increased costs. Computational models of vehicles and their components are more or less simplified compared with the actual ones. This simplification is seen when comparing the results from real tests.

The object of our calculations is the strength test of new design of freight bogie frame through FEM analysis.

3.1 Computational model

Computational model (Fig. 2) was created in program MSC.Marc. For analysis of bogie frame design by finite element method has been created mesh of shell finite elements, which lies in the planes center surfaces individual parts of analyzed design. The analysis is performed in a linear region. The distortion of the results of the analysis resulting from the introduction of the simplifications mentioned is considered negligible.

They are used as isoparametric, four-node shell elements with average size of elements 10 mm. The thickness of the shell elements was determined according to the drawing documentation. Location of forces and boundary conditions (hemispherical bogie pivot, slides, etc.) are modeled by 2-node beam elements. The springs of suspension are replaced with very soft beam elements.

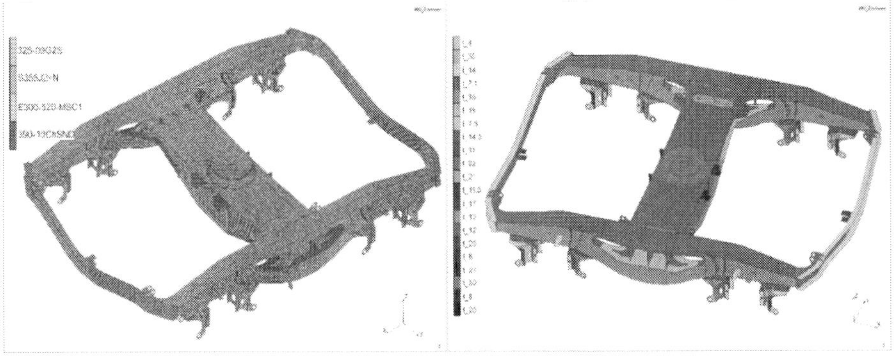

Fig. 2. Computational model – FEM mesh and elements thickness

Frame parts were designed from structural steel of 09G2S, 10ChSND, S355J2+N and their properties are shown in Table 2. Castings (axle guard) are designed of E300-520-MSC1.

Table 2. Material properties

Material	Norm	Thickness [mm]	Min. yield strength	Min. tensile strength
			Re [MPa]	Rm [MPa]
325-09G2S	GOST 19281-89	from 10 to 20	325	450
10ChSND	GOST 19281-89	up to 40	390	510
E300-520-MSC1	UIC 840-2	casting	300	520
S355J2+N	EN 10025-2	up to 16	355	470
S355J2+N	EN 10025-2	from 16 to 40	345	470

Consideration is being given to the fact, that the material is linearly, elastic and isotropic. Mechanical properties: Young modulus of elasticity $E=2.1\cdot10^5$ MPa, Poisson´s ratio $\mu=0.3$. Goodman diagrams were used according [2], curves "a1", "a2", "b", annex F.3, with conditions of application of the annex F.4, safety factor 1.1.

3.2 Boundary conditions

Boundary conditions (Fig. 3) are captured:
- in the middle of lower flangeplate of transverse beam,
- in two nodes of lower flangeplate of transverse beam (exceptional loads),
- in the middle of manganese plates of axle guide (normal service loads),
- under 8 substitute springs.

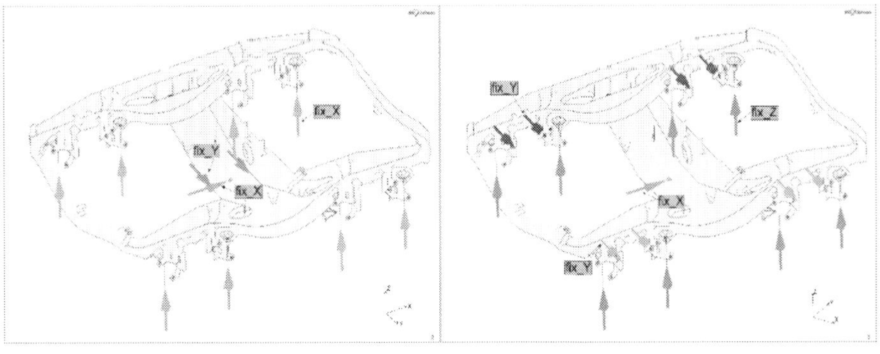

Fig. 3. Boundary conditions for exceptional loads (left) and normal service loads (right)

The first two boundary conditions are used only for the stability of calculation, because reactions in them are practically zero.

3.3 Load conditions

In reality the loads are combined in a complex manner and so it is difficult to represent them exactly in analysis. Consequently, it is generally the practice, for ease of analysis, to represent the true loads by a series of load conditions which include the above effects in a simplified form, either individualy or in combination. It is essential that the simplification ensures that the effects of the true loads are not underestimated. The load conditions required for the design and assessment of the bogie frame will be depended on the application being considered.

The source data does not indicate how the specified loads should be reacted. Therefore, the designer needs to consider load balances appropriate to the application. In particular, the reactions to the lateral loads specified as acting at the wheelset should include the bogie inertia forces to determine realistic secondary suspension forces.

Load conditions were created according [3] for axle load 25 t. These are combinations of vertical loads in hemispherical bogie pivot, in slides, lateral forces in pivot and unevenness of the track. All loads used as the basis for the bogie design should incorporate any necessary allowance for uncertainties in their values. Schematic view of loading forces are shown in Fig. 4.

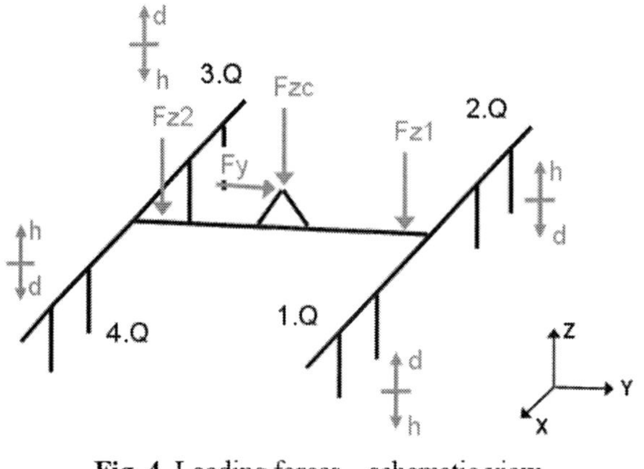

Fig. 4. Loading forces – schematic view

Loading conditions according [3] are divided into two groups (below).

A) Test under exceptional loads

Table 3. Load cases for tests under exceptional loads

Load Condition	Vertical Forces [kN]			Lateral Forces	Unevennes s of track [‰]	Brake Forces
	Sidebearer 2	Hemispherical bogie pivot	Sidebearer 1	F_y [kN]		
	$-F_{z2}$	$-F_{zc}$	$-F_{z1}$			
1		884.9				
2h		464.6	199,1		30	
2d		464.6	199,1		-30	
3		464.6	199,1	183.5		
4	199.1	464.6		-183.5		
5		530.9				F_B

According [4], annex J, counts with 1.2 times the braking forces. Longitudinal force in hemispherical bogie pivot and friction force between wheels and track are in balance during maximum braking.

B) Test under normal service loads

Table 4. Load cases for tests under normal service loads

Load Condition	Vertical Forces [kN]			Lateral Forces [kN]	Unevenness of track [‰]
	Sidebearer 2	Hemispherical bogie pivot	Sidebearer 1	F_y	
	$-F_{z2}$	$-F_{zc}$	$-F_{z1}$		
1		442.4			
2		619.4			
3		265.5			
4		495.5	123.9	98.1	
5	123.9	495.5		-98.1	
6		212.4	53.1	98.1	
7	53.1	212.4		-98.1	
4h		495.5	123.9	98.1	15
4d		495.5	123.9	98.1	-15
5h	123.9	495.5		-98.1	15
5d	123.9	495.5		-98.1	-15
6h		212.4	53.1	98.1	15
6d		212.4	53.1	98,1	-15
7h	53.1	212.4		-98.1	15
7d	53.1	212.4		-98.1	-15

3.4 Evaluation of simulation

The static strength requirements correspond to the exceptional load conditions under which the bogie shall remain fully functional. It shall be demonstrated by analysis and/or testing, that no permanent deformation, instability or fracture of the structure as a whole, or of any individual element, will occur under the exceptional design load conditions. It is important to ensure that the design loads are expressed in a form that is consistent with the method of analysis and the way in which the permissible material stress levels are defined.

Exceptional loads

Table 5. The highest values of von Mises stress and their position

Place on the bogie frame	Number of loading condition	Max. detected stress [MPa]	Allowable stress [MPa]	Safety factor [-]
Main transverse beam	1	318	325	1.02
Conection of mein transverse beam and longitudinal beam	1h	239	325	1.36
Conection of mein transverse beam and longitudinal beam	1d	238	325	1.37
Upper flangeplate of longitudinal beam	3	338	325	0.96
Upper flangeplate of longitudinal beam	4	338	325	0.96

The analysis is intended to detect the most critical places in the structure after axle load of 25 tons. We examined sizes of equivalent von Mises stress for all load conditions. The highest values, which were detected, are listed in Tab. 5. Some graphic behaviour of stresses on the undeformed bogie frame are shown in Fig. 5 and Fig. 6.

In load cases 3 and 4 was exceeded the permissible value. It is a place on the upper flangeplate of longitudinal beam (Fig. 6) with value of stress 338 MPa. Exceeding of permissible value about 4 %, but it is just in area 4 elements on upper sheet.

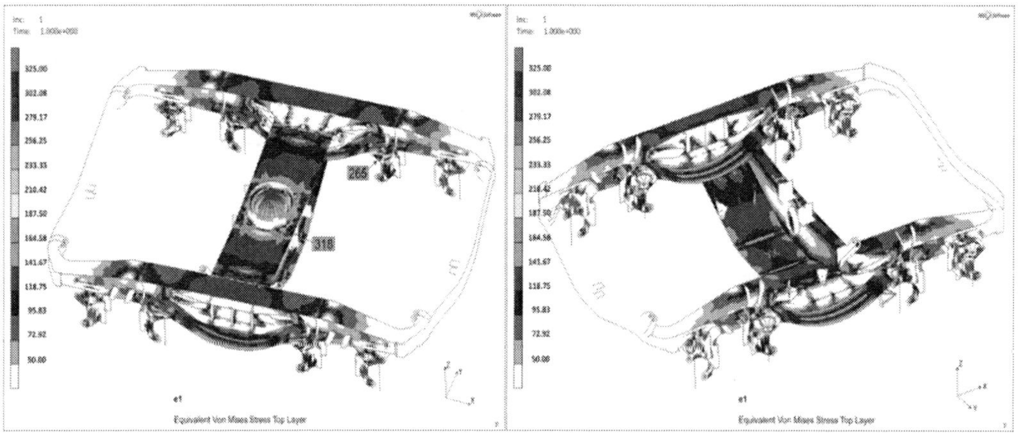

Fig. 5. Behaviour of von Mises stress for load condition no. 1

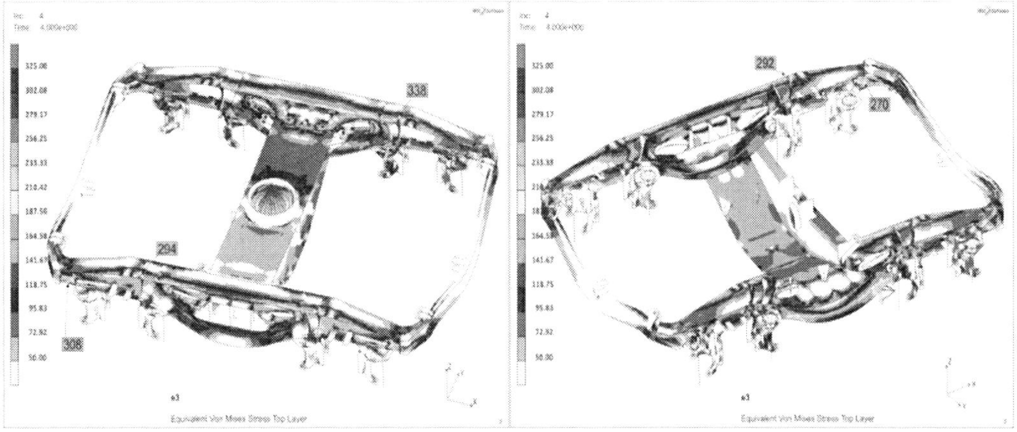

Fig. 6. Behaviour of von Mises stress for load condition no. 3

Normal service loads

For evaluation of simulation was used method according [4], annex J, point J.2. On the Fig. 7 and Fig. 8 are shown elements, which exceeded curves of Goodmans diagram (limit values for curves "a2", "a1" and "b" according [2], annex F.3, with safety factor 1.1).

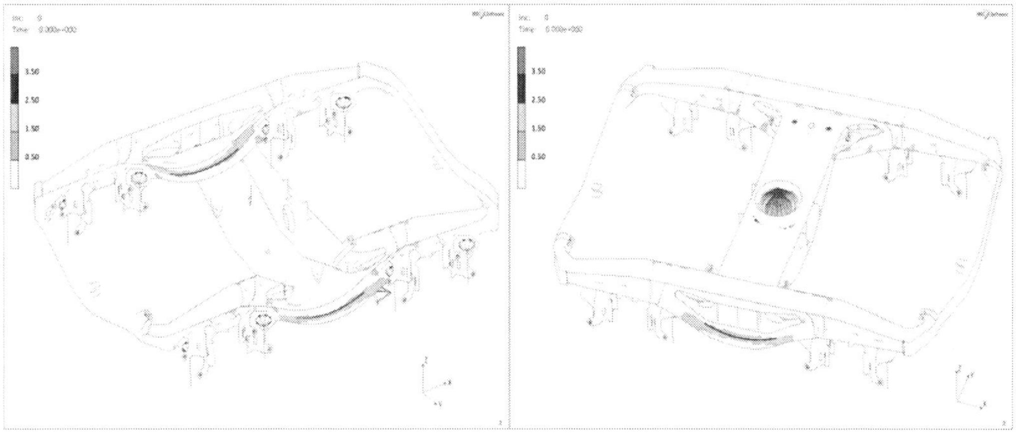

Fig. 7. Exceeded curve “b” (red color) – lower flangeplate of longitudinal beam

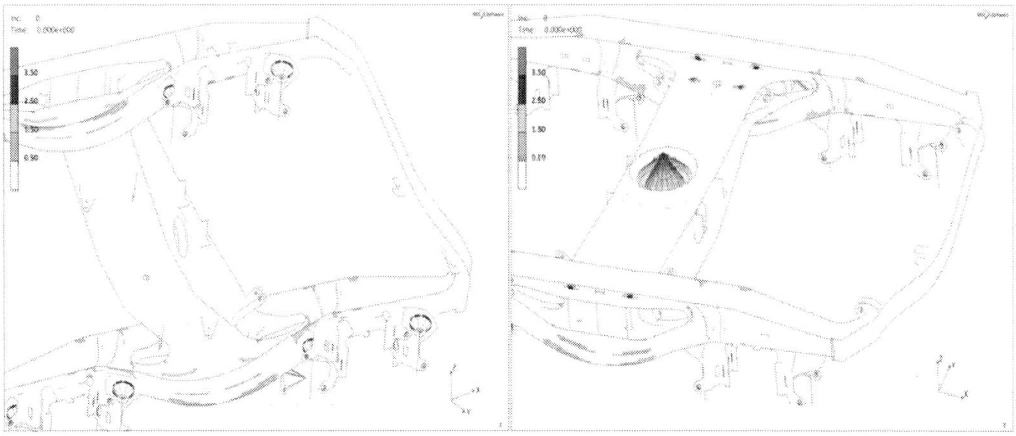

Fig. 8. Exceeded curve “a1” (blue color) –longitudinal beam web

Conclusion

Computational simulations are now an integral part of the development process of rolling stock. They allow a more detailed analysis of the behaviour of the vehicle as a whole or its individual parts. Therefore, it is possible to better optimize the design of rail vehicles and prevent potential problems in the operation, which would require increased costs. The bogies of rail vehicles are required to withstand the maximum loads consistent with their operational requirements and achieve the required service life under normal operating conditions with an adequate probability of survival.

Static and fatigue tests of bogie frame prototype (Fig. 9) have been performed in internationally accredited institution. The results of the measurements and the defectoscopic tests showed no deformations or cracks. Results of calculations and prototype tests prove, that new designed construction of the bogie frame satisfies strength conditions.

With this research and development subjects deals other publications too [5-12].

Fig. 9. Fatigue and strength tests of prototype bogie frame

This publication is the result of the project implementation Development of two types of freight wagons with bogies for non-standard wheelbase or track wheelset, complying with the criteria for interoperability, environmental issues, safety and reliability, ITMS 26220220070, supported by the Operational Programme Research and development funded by the ERDF.
The work was supported by the Cultural and Educational Grant Agency of the Ministry of Education of the Slovak Republic in project No. KEGA 077ŽU-4/2017: Modernization of the Vehicles and engines study program. The work was also supported by the project No. APVV-0842-11: Equivalent railway operation load simulator on the roller rig and VEGA Nr. 1/0927/15: Research of the use of alternative fuels and hybrid drives on traction vehicles with aim to reduce fuel consumption and air pollutants production.
Research-Educational Center of Rail Vehicles (VVCKV)

References

1. P. Čižík, *2-axle bogie according to line Y25 for freight wagons*. Dynamical problems in rail vehicles 2016: Slovak - Polish scientific workshop: Žilina December 6th and 7th, 2016, Slovak Republic: proceeding of the workshop, Žilina: University of Žilina, ISBN 978-80-554-1299-3, 146-154 (2016)

2. ERRI B12/RP17, No. 8, *Programme of Stresses to be carried out on Wagons with Steel Underframe and Body Structure*. European Rail Research Institute (1996)
3. EN 13 749: *Railway applications - Wheelsets and bogies - Method of specifying the structural requirements of bogie frames*. European Committee for Standardization (2011)
4. Commission Decision 2006/861/EC (Annex J) of 28 July 2006 and Commission decision of 23 January 2009 amending Decisions 2006/861/EC and 2006/920/EC (2009)
5. J. Gerlici, T. Lack, J. Harušinec, *Realistic simulation of railway operation on the RAILBCOT test stand*. Applied mechanics and materials **486**, 387-395 (2014)
6. J. Gerlici, T. Lack, J. Harušinec, *The Test Stand Load Modulus Implementation for the Realistic Railway Operation in the Laboratory Conditions*. Manufacturing technology: journal for science, research and production **13** (4), 444-449 (2013)
7. J. Gerlici, T. Lack, J. Harušinec, *Contact geometry influence on the rail/wheel surface stress distribution*. Procedia Engineering **2**, Fatigue 2010: 10th international fatigue congress: June 6-11, 2010 Prague, Czech Republic, (1), 2249-2257 (2010)
8. V. Hauser, et al., *Impact of three axle boxes bogie to the tram behavior when passing curved track*. Procedia Engineering **192**, 295-300 (2017)
9. M. Kassner, *Fatigue strength analysis of a welded railway vehicle structure by different methods*. International Journal of Fatigue **34** (1), 103-111 (2012)
10. M. Sága, R. Bednár, M. Vaško, *Contribution to modal and spectral interval finite element analysis*. Vibration Problems ICOVP 2011, Springer Proceedings in Physics **139**, 269-274 (2011)
11. A. Sapietová, M. Sága, P. Novák, R. Bednár, J. Dižo, *Design and application of multi-software platform for solving of mechanical multi-body system problems*. Mechatronics: Recent technological and scientific advances, 345-354 (2011)
12. M. Svoboda, J. Soukup, *Dynamic Measurement of Four-Axle Railway Wagon*. Manufacturing technology: Journal for science, research and production **13** (4), 552-558 (2013)

Modelling mechanical properties of the multilayer composite materials with the polyamide core

Krzysztof Talaśka[1,*], *Dominik* Wojtkowiak[1]

[1]Chair of Basics of Machine Design, Poznan University of Technology, Piotrowo 3, 60-965 Poznań, Poland

Abstract. Due to the wide range of application for belt conveyors, engineers look for many different combinations of mechanical properties of conveyor and transmission belts. It can be made by creating multilayer or fibre reinforced composite materials from base thermoplastic or thermosetting polymers. In order to gain high strength with proper elasticity and friction coefficient, the core of the composite conveyor belt is made of polyamide film core, which can be combined with various types of polymer fabrics, films or even rubbers. In this paper authors show the complex model of multilayer composite belt with the polyamide core, which can be used in simulation analyses. The following model was derived based on the experimental research, which consisted of tensile, compression and shearing tests. In order to achieve the most accurate model, proper simulations in ABAQUS were made and then the results were compared with empirical mechanical characteristics of a conveyor belt. The main goal of this research is to fully describe the perforation process of conveyor and transmission belts for vacuum belt conveyors. The following model will help to develop design briefs for machines used for mechanical perforation.

Keywords: multilayer polymer composites, polyamide core, conveyor belts, material modelling, Abaqus.

1 Introduction

Due to the increasing requirements for the properties of the conveyor and transmission belts, it is necessary to continuously improve them in terms of design, materials and manufacturing technology. Conveyor belts are required to have, above all, appropriate tensile strength, while maintaining adequate stiffness to ensure stability during transport and, at the same time, adequate flexibility that makes it possible to put the belt around small-diameter drive wheels. Flat belts usually have the structure of the multi-layer polymer composite (Fig. 1) and they consist of the core 1, protective gaskets 2 and two covers: bottom (return) 3 and top (load-carrying) 4 [1, 2]. Based on the analysis of flat belts available on the market three basic belt types can be distinguished [1-3]:

* Corresponding author: krzysztof.talaska@put.poznan.pl

Reviewers: *Grzegorz* Domek, *Bogdan* Posiadala

- flexible, light belts whose core is made of polyurethane (sometimes reinforced with polyester fabric), polyvinyl chloride (PVC), fabric or rubber,
- rigid, strong belts whose core is the polyamide layer,
- strong, flexible belts whose polyurethane core is reinforced with aramid fabric (Kevlar) cord.

In most belts, due to beneficial frictional properties, return and load-carrying covers are made of nitrile butadiene rubber (NBR) whose surface structure can be both smooth and rough. In order to obtain a non-flammable layer with a very smooth texture, PVC is also used as material for covers. Protective gaskets are used primarily in rigid belts, and their role is to protect the core from damage and also to improve strength properties. Most often, the role of a gasket is played by a layer of properly woven polyamide fabric, which - in the process of merging belts - is melted into the structure of adjacent layers, ensuring integration of the belt structure [1-3].

In this work, the authors focused on modelling the mechanical properties of rigid, heavy-duty belts whose core is made of a polyamide layer, the role of the gasket is played by polyamide fabric, and the return and load-carrying covers are made of nitrile rubber (NBR). Nitta conveyor belts were used in this research (Fig. 1). For an extensive analysis of belt properties, six belt types were examined, three of which have a polyamide core of the same thickness (0.5 mm), but differ in stiffness, which is affected by the thickness of the covers along with the gaskets (L – light, M – medium, XH – extra hard), whereas the remaining belts have increased strength and stiffness (TFL series) with core thickness 0.7 mm, 1 mm and 1.5 mm. The objective of these tests is to model the presented type of material, which will make it possible in further stages of the work to develop the perforation model of this kind of conveyor belt and to determine the effective geometry of the perforating tool using FEM analysis in ABAQUS environment.

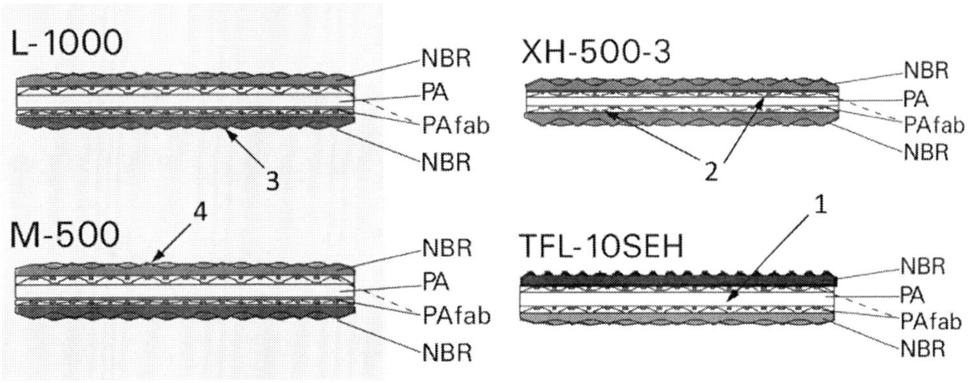

Fig. 1. Nitta PolyBelt® composite conveyor belts with the polyamide core: 1 – core, 2 – gasket, 3 – return cover, 4 – load-carrying cover, NBR – nitrile butadiene rubber, PA – polyamide, fab – fabric [3]

2 Methodology of the research

In order to determine the mechanical parameters necessary to develop the material model of composite materials with polyamide core for each of the six belts (L500, M500, XH500-4, TFL7S, TFL10S and TFL15S), quasi-static tensile testing in a longitudinal direction, compression testing and pure shear testing were performed. For all described testing, MTS Insight testing system with load capacity up to 50kN and properly adapted equipment was used. Based on the obtained results, it is possible to determine basic material constants (Young's modulus, shear modulus, Poisson's ratio, etc.), necessary to model the material in

Abaqus. In addition, the results make it possible to analyse the behaviour of the test materials, and the appropriate selection of specimens makes it possible to specify the influence of the thickness of the individual layers on the strength properties of the belt. This approach makes it possible to also apply the model to other belts with a similar structure, but with a different distribution of specific layers.

For tensile testing, a two-shouldered type B specimen compliant with ISO 527-4:1997 was used. The shape of the specimen ensures that it is not destroyed in grips or adjacent sections and it is recommended for testing fibre- or mat-reinforced plastic composites [4]. The specimen was mounted by its shoulders between the grip plates (Fig. 2a), and subsequently the bolt was strongly tightened. The surfaces of these plates are knurled to prevent the specimen from slipping during the test. Specimens were stretched at a speed of 50 mm/min, because belt producers use such test conditions in their product data sheets. For each belt type, 5 specimens were stretched and the average characteristics were determined from the obtained results.

Compression testing consisted in compressing the specimen in the form of a disc with a diameter of 10 mm using a cylindrical pin of the measuring head mounted in the grip of the MTS machine (Fig. 2b). The specimen was placed on a flat steel plate with high stiffness, which eliminated the effect of base plate deformation on the value of compressive stresses. Quasi-static compression testing was performed at a speed of 0.01 mm/sec to the value of the compression force 5000 N. Due to the very high repeatability of the results, it was limited to three attempts, which were used to determine the average characteristics of the force relative to displacement.

Fig. 2. How the specimen was mounted during testing: a) tensile, b) compression

For shear testing, the station for pure shear of belts presented in Figure 3 and designed by the authors was used. A rectangular specimen 1 with the dimensions 20 x 100 mm is sheared by two cutting edges of the moving blade 2 and stationary blade 3 that are made of fast-cutting 64 HRC steel. The moving blade is tightly fitted in the guide plate 4 whose linear motion is provided with two sets of zero backlash linear guides 5, consisting of a guide pole embedded in the base 6, a guide sleeve embedded in a guide plate and a ball bearing case. This type of guidance ensures the maintenance of appropriate, even very small clearance between the blades. Connecting the guide plate with the grip of the MTS machine 8 is done through the head plate 7 fitted with the bolts to the guide plate. The stationary blade is bolted to the movable support 9, which provides the ability to adjust the distance between the cutting edges in the range 0÷1 mm thanks to the groove guide system in the base. The second support is the fixed support 10 and both supports are bolted to the base. During the test, the belt specimen is placed on two supports and pressed to them using pressing plates (11) and bolts to ensure proper belt tension while shearing. Because of the quasi-static nature of the test, the speed of traverse for the testing machine is 0.5 mm/s. In the course of the test, the characteristics of the cutting force in the function of displacement are recorded until the specimen is cut through. In addition, the clearance value is measured using a feeler gauge. The tests presented in the were performed for the clearance value 0.05 mm. For each belt type, 5 attempts were made, from which the average characteristics were determined.

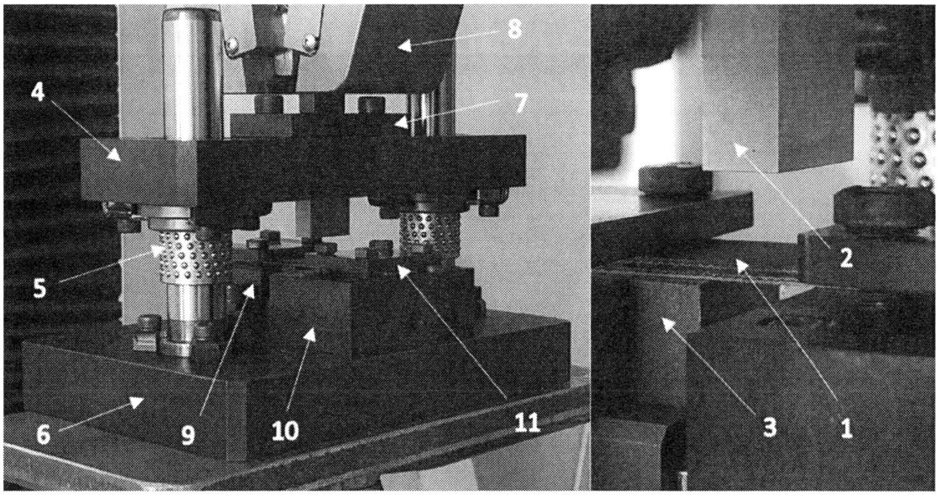

Fig. 3. The station for pure shear of belts and tapes: 1 – specimen, 2 – movable blade, 3 – stationary blade, 4 – guide plate, 5 – linear guides, 6 – base, 7 – head plate, 8 – MTS machine grip, 9 – movable support, 10 – fixed support, 11 – pressing plate

3 Results of the empirical research

In order to compare the properties of individual belt types with each other, it is necessary to know the percentage share of a given material in the composite structure. For this purpose, thickness of individual layers was measured using a microscope (Fig. 4) and share ratios δ were determined for each layer in line with the equations 1÷3. The measurement results are shown in Table 1.

$$t = t_{PA} + 2t_{PA-fab} + 2t_{NBR} \tag{1}$$

$$\delta_i = \frac{t_i}{t} \tag{2}$$

$$1 = \delta_{PA} + 2\delta_{PA-fab} + 2\delta_{NBR} \tag{3}$$

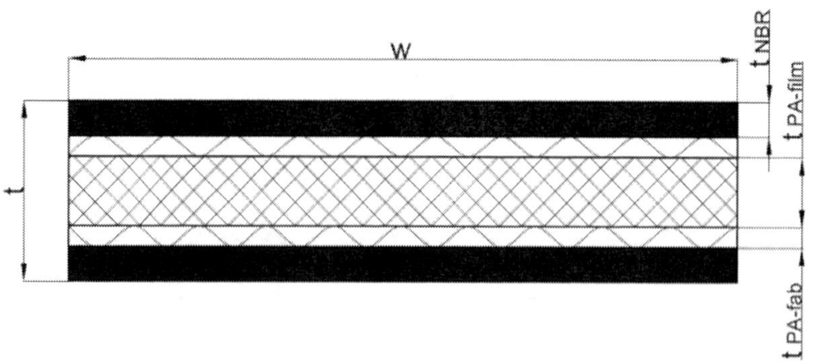

Fig. 4. Cross section of the belt with a polyamide core, together with characteristics measurements - schematic

Table 1. Geometrical parameters of the studied conveyor belts

Belt title	t [mm]	$t_{PA\text{-}film}$ [mm]	$t_{PA\text{-}fab}$ [mm]	t_{NBR} [mm]	δ_{PA} [-]	$\delta_{PA\text{-}fab}$ [-]	δ_{NBR} [-]
L500	1.55	0.5	0.3	0.225	0.3225	0.1935	0.145
M500	2.5	0.5	0.4	0.6	0.2	0.16	0.24
XH500-4	4	0.5	0.4	1.35	0.125	0.1	0.3375
TFL7S	2.4	0.7	0.4	0.45	0.292	0.167	0.1875
TFL10S	2.6	1	0.4	0.4	0.385	0.154	0.154
TFL15S	3.1	1.5	0.4	0.4	0.484	0.129	0.129

Figure 5 shows the comparison of the average characteristics of tensile force relative to displacement. A correlation can be observed that the belts whose polyamide core has the same thickness (L500, M500 and XH500-4) have very similar characteristics, whereas a further increase in core thickness causes an increase in the characteristics slope (Young's modulus) and the maximum tensile force, but the breaking point occurs much faster here than for the belts from outside the TFL series. Therefore, it can be seen that increasing core thickness has a beneficial effect on the strength of the belt while significantly increasing its stiffness.

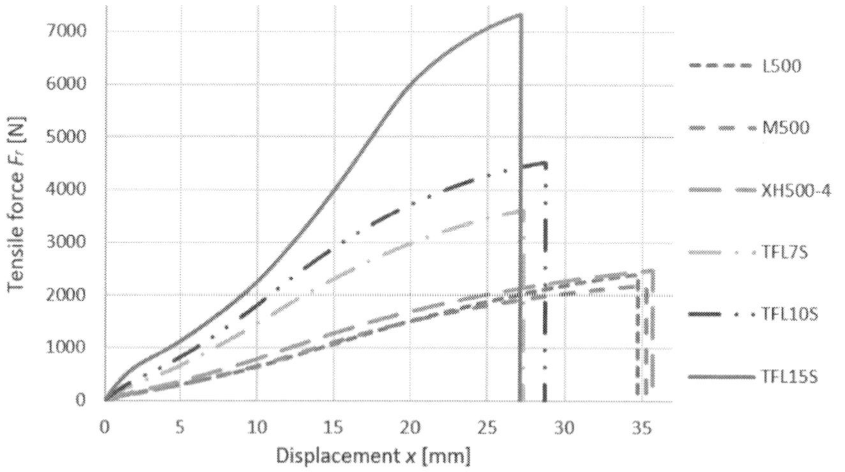

Fig. 5. Correlation of the tensile force relative to displacement $F_r(x)$ for 6 tested belt types

Figure 6 shows the comparison of the average characteristics of compression force relative to displacement. Here, the key role is played by the thickness of the nitrile rubber covers. Young's modulus for the TFL series belts and the L500 belt is very similar due to a similar share ratio of the rubber layer. The other two belts are characterized by lower Young's modulus, proportionate to their share ratio of the rubber layer.

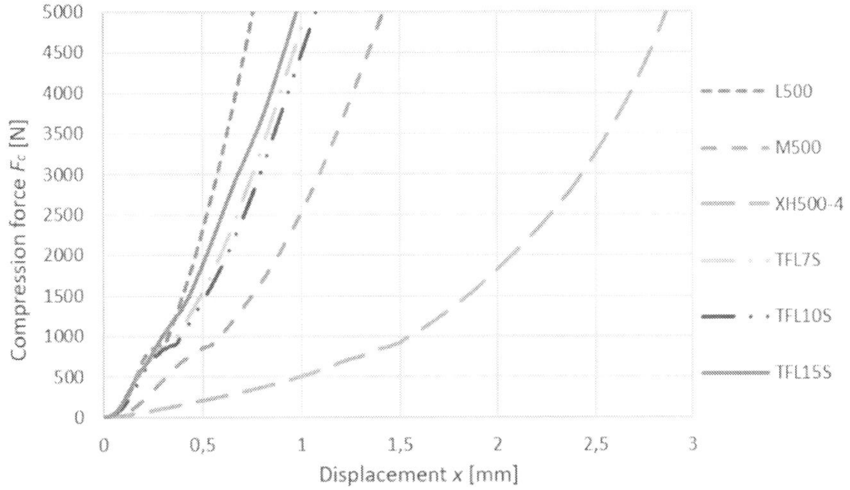

Fig. 6. Correlation of the compression force relative to displacement $F_c(x)$ for 6 tested belt types

Figure 7 shows the comparison of the average characteristics of shear force relative to displacement. Here, one can also see the effect of the correlations described for tensile and compression testing, that is the smaller the share of rubber in the belt structure, the faster the force increases, whereas the increase in polyamide code thickness leads to the increase of the shear force.

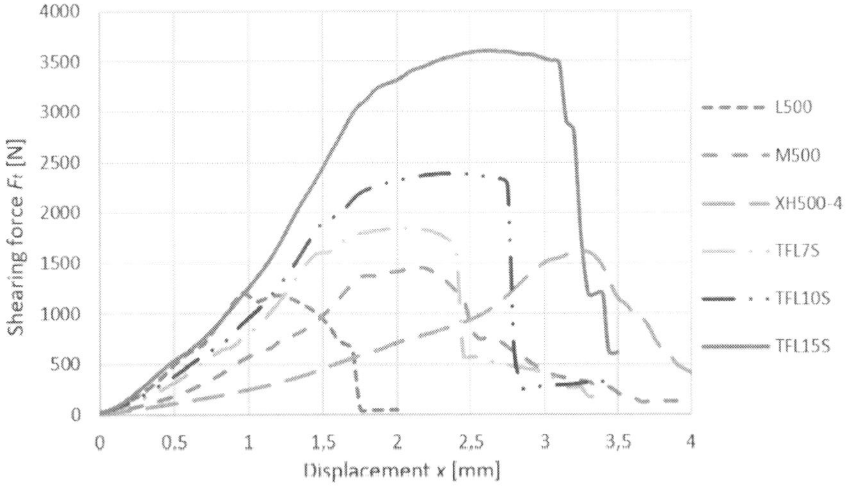

Fig. 7. Correlation of the shear force relative to displacement $F_t(x)$ for 6 tested belt types

4 Modelling mechanical properties of the multilayer material with polyamide core

In order to determine the elastic mechanical parameters of the belt, it is necessary to chart the stress characteristics in the function of deformation. In the case of the tensile test, deformation is obtained by dividing the displacement of the traverse of the testing machine by the length of the measuring section equal to 108 mm. Tensile stresses are determined from the quotient of the tensile force and the cross section of the belt, which is a rectangle with a length of 10 mm and a width equal to the thickness of the belt t. A sample chart for the tensile test for the TFL10S belt, along with the methodology for determining Young's modulus, is presented in Figure 8. Young's modulus for the compression test was determined in a similar manner (Fig. 9), with the difference being that deformation here is the ratio of displacement to the total thickness of the belt, whereas for determining stresses the circular section with a diameter of 10 mm is used. In both cases, the base model of reinforced elastic-plastic material was used, in which after achieving certain displacement there is hardening of the material – Young's modulus increases [5].

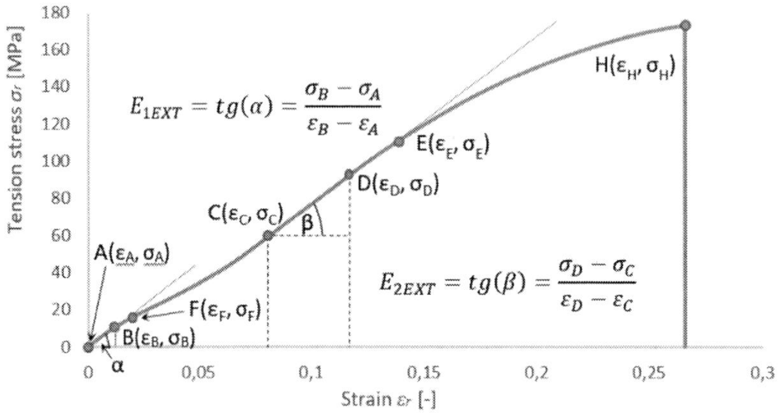

Fig. 8. Correlation $\sigma_r(\varepsilon_r)$ for the TFL10S belt, along with the methodology for determining the mechanical parameters

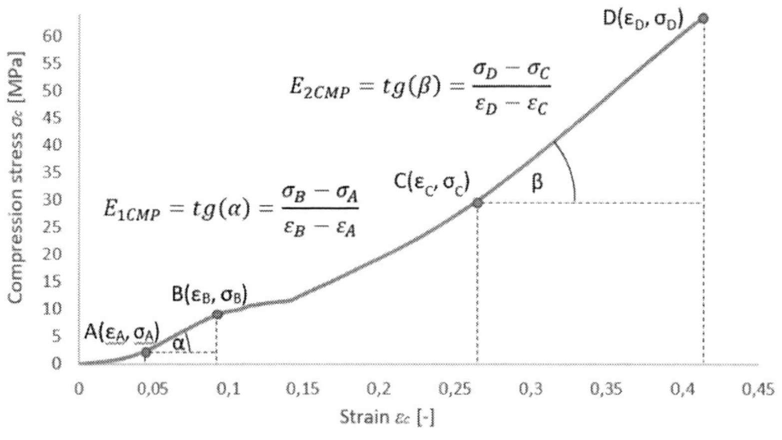

Fig. 9. Correlation $\sigma_c(\varepsilon_c)$ for the TFL10S belt, along with the methodology for determining the mechanical parameters

In the case of shear tests, it is necessary to determine the pure shear deformation angle using the correlation 4. After charting the characteristics (γ) (fig. 10), shear modulus is determined in an analogous way to Young's modulus in the case of tensile and compression testing. Knowing both moduli, it is possible to determine Poisson's ratio using the correlation 5. It should be noted, however, that this way of modelling materials is subject to significant measuring error, and is used to estimate the parameters of the belt that will be used to perform FEM analysis. The values obtained for the tested belts are presented in Table 2.

$$\gamma = arctg\left(\frac{x}{t}\right) \tag{4}$$

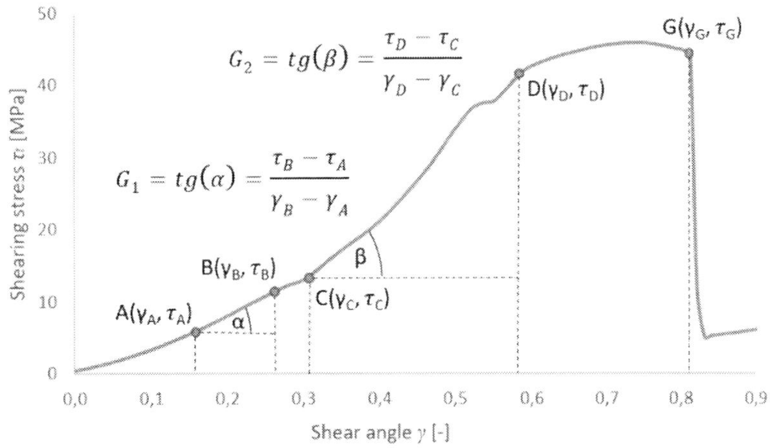

Fig. 10. Correlation $\tau_t(\gamma)$ for the TFL10S belt, along with the methodology for determining the mechanical parameters

$$\nu = \frac{E}{2G} - 1 \tag{5}$$

Table 2. Elastic mechanical parameters of the tested conveyor belts

Belt title	E_{1EXT} [MPa]	E_{2EXT} [MPa]	E_{1CMP} [MPa]	E_{2CMP} [MPa]	G_1 [MPa]	G_2 [MPa]	ν_1 [-]	ν_2 [-]
L500	537.1	618.7	116.8	223.9	41.3	92.8	0.41	0.21
M500	320.9	390.6	82	201.2	29.2	77.2	0.4	0.3
XH500-4	245.9	263.4	44.1	151.8	15.4	56.2	0.43	0.35
TFL7S	768.2	765.3	141.3	222.3	50.7	90.5	0.39	0.23
TFL10S	878.8	921.6	144.9	233.9	53.6	97.7	0.35	0.2
TFL15S	1252.6	1365.8	177.8	289	61.8	110.4	0.44	0.31

To model the plastic flow of material and damage mechanics, the Johnson-Cook (J-C) models shown in the formulas 6 and 7 [6, 7] were used. To use this model for FEM analysis, parameters *A, B, C, n, m* oraz $d_1 \div d_5$ should be determined. Because the tests were performed at room temperature and the nature of the specimen load is quasi-static, the coefficients *C* and d_5 related to speed of deformation and *m* and d_5 related to material temperature are 0. Models are therefore simplified to the form presented in the formula 8.

$$\bar{\sigma} = \left[A + B\left(\bar{\varepsilon}^{pl}\right)^n\right]\cdot\left[1 + C\ln\left(\frac{\dot{\varepsilon}}{\dot{\varepsilon}_0}\right)\right]\cdot\left[1 - \hat{\theta}^m\right] \tag{6}$$

$$\bar{\varepsilon}_f^{pl} = \left[d_1 + d_2 \cdot \exp(d_3 \frac{p}{q})\right]\cdot\left[1 + d_4 \cdot \ln\left(\frac{\dot{\varepsilon}}{\dot{\varepsilon}_0}\right)\right]\cdot\left(1 + d_5\hat{\theta}\right) \tag{7}$$

$$\bar{\sigma} = A + B\left(\bar{\varepsilon}^{pl}\right)^n, \quad \bar{\varepsilon}_f^{pl} = d_1 + d_2 \cdot \exp(d_3 \frac{p}{q}) \tag{8}$$

In order to determine the remaining Johnson-Cook parameters, 3 characteristic points on the chart of the tensile test should be designated – ultimate tensile strength (E), yield point (F) and specimen break point (H) and the point G on the chart of the shear test, where material damage takes place [7]. Subsequently, the value of engineering stresses and deformations should be converted to the actual values that take into account the change in the specimen section in the plastic range, in line with the formula 9 [8]. In the presented tests, the point F was assumed to be the point for which the difference between the actual value and the value approximated with the equation of the line CD was 1MPa. To determine the J-C parameters, the formula 10 was used, with the parameter n determined for each point after the point E and then averaged. To determine the parameters of the damage model, it is necessary to know the dimensionless pressure-deviatoric stress ratio, which - for shear - is 0, for compression -1/3, and for tensile 1/3 [9]. On this basis, the parameters of the plastic damage model were determined using the correlation 11. The determined parameters for the tested belts are presented in Table 3.

$$\varepsilon_{tr} = \ln\left(1+\varepsilon_{eng}\right), \quad \sigma_{tr} = \sigma_{eng}\left(1+\varepsilon_{tr}\right) \tag{9}$$

$$A=\sigma_F \quad n_i = \frac{\ln\left[\left(\sigma_E - A\right)/\left(\sigma_i - A\right)\right]}{\ln(\varepsilon_E / \varepsilon_i)} \quad B = \frac{\sigma_E - A}{\left(\varepsilon_E\right)^n} \tag{10}$$

$$d_1 = \varepsilon_E \quad d_2 = \gamma_G - d_1 \quad d_3 = 3\ln\left(\frac{\varepsilon_H - \varepsilon_F - d_1}{d_2}\right) \tag{11}$$

Table 3. Parameters of the Johnson-Cook models for the tested belts

Belt title	*A* [MPa]	*B* [MPa]	*n* [-]	d_1 [-]	d_2 [-]	d_3 [-]
L500	8.4	1213.2	1.37	0.117	0.541	3.91
M500	4.8	518	1.16	0.160	0.55	4.9
XH500-4	3.8	347.7	1.14	0.107	0.575	3.78
TFL7S	11.9	1066.4	1.18	0.133	0.652	6.41
TFL10S	13.6	1200.1	1.17	0.127	0.686	5.98
TFL15S	18.9	2941	1.51	0.15	0.635	7.15

On the basis of the obtained results, the model for the approximation of the elastic properties of the belts based on the thickness of their individual layers was determined, shown in Fig. 11. It was observed that in the case of tensile testing Young's modulus depends significantly on the thickness of the polyamide core – quadratic function. Modules E_1 and G_1 of the elastic range of the belt depend exponentially on the thickness of the rubber, and modules E_2 and G_2 after hardening of the material increase linearly with the increase of the polyamide layer thickness. The following model does not include the effect of the thickness of the gaskets, because the selected belts contain polyamide fabric with identical thickness (except L500) and similar ratios $\delta_{PA\text{-}fab}$. It is possible to approximate the plastic properties of the belt in a similar way or - due to insignificant differences - select the parameters of a comparable belt from Table 3.

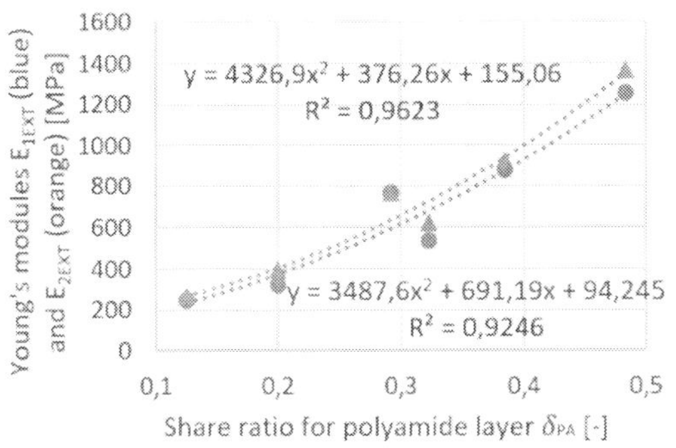

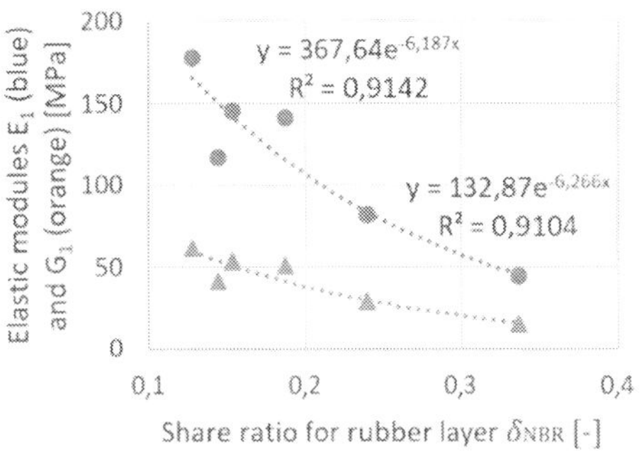

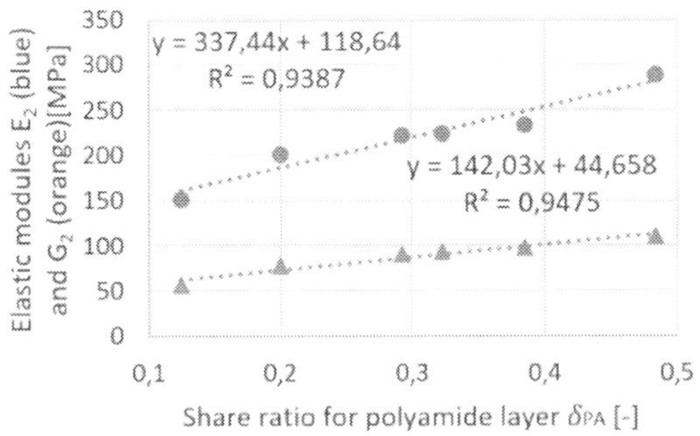

$$E_{1EXT} = 3487{,}6\delta_{PA}^{\ 2} + 691{,}2\delta_{PA} + 94{,}2$$
$$E_{2EXT} = 4326{,}9\delta_{PA}^{\ 2} + 376{,}3\delta_{PA} + 155{,}1$$
$$E_{1CMP} = 367{,}6e^{-6{,}19\delta_{NBR}}$$
$$G_1 = 132{,}9e^{-6{,}27\delta_{NBR}}$$
$$E_{2CMP} = 337{,}4\delta_{PA} + 118{,}6$$
$$G_2 = 142\delta_{PA} + 44{,}7$$

Fig. 11. Model for the approximation of the elastic mechanical properties of composite belts with polyamide core

5 FEM analyses

In order to verify the accuracy of the derived model of the mechanical belt with a polyamide core, FEM analyses were performed, aimed at recreating the tensile and compression tests using the obtained parameters for the TFL10S belt. Both analyses were performed with the Abaqus/Explicit module. The change of Young's modulus was obtained using the VUSDFLD subroutine, which caused the hardening of the material beyond a given stress value. Because of the quasi-static nature of the study, in order to reduce calculation time, the Mass Scaling feature was used during the simulation. It is justified in the case of a negligibly small impact of inertia forces on the results of the calculated values [6].

In the model of the tensile test, a two-shouldered specimen was divided into two gripping sections and a measuring section using the partition. The lower gripping section was stripped of all degrees of freedom with the boundary condition (BC) ENCASTRE. A reference point was created on the top gripping section and attached to all its surfaces, and kinematic excitation with speed 50 mm/min in the direction of the specimen axis was obtained using BC Velocity applied to this point. The results of the analysis are shown in Figure 12.

In the case of the shear test model, all the elements except the belt were modelled as rigid by attaching Rigid Body to reference points located on each of them. The support, pressing plates and the stationary blade were stripped of all degrees of freedom, whereas kinematic excitation of the moving blade at the speed of 0.5 mm/s was achieved through BC Velocity applied to the reference point to which it is rigidly attached. From this point, the reaction force present in the cutting process was also recorded. Because of the complexity of the model, the local control of the grid was used in order to thicken it in the shear plane, which made it possible to obtain more accurate results with a minimal extension of the calculatation time. Sample results of the analysis are shown in Figure 13.

Fig. 12. Results of the FEM analysis of the tensile test for the belt TFL10S

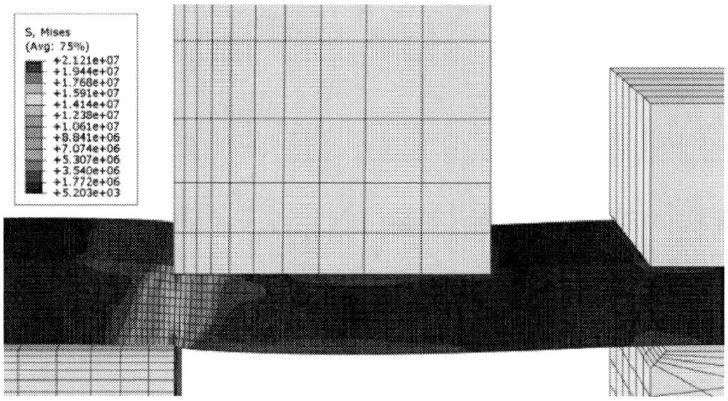

Fig. 13. Results of the FEM analysis of the shear test for the belt TFL10S

For comparison of the results of FEM analyses and experimental research, the characteristics presented in Figures 14 and 15 were charted. It can be observed that the difference between the values of the most important parameters, that is the values where damage occurs, does not exceed 15%. However, there are visible differences in the course of the characteristics, which may introduce errors when modelling the perforation process.

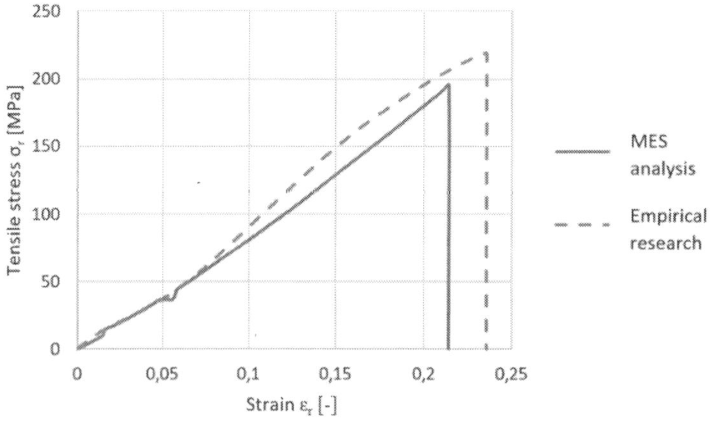

Fig. 14. Comparison of the results of empirical research and FEM analysis for the tensile test

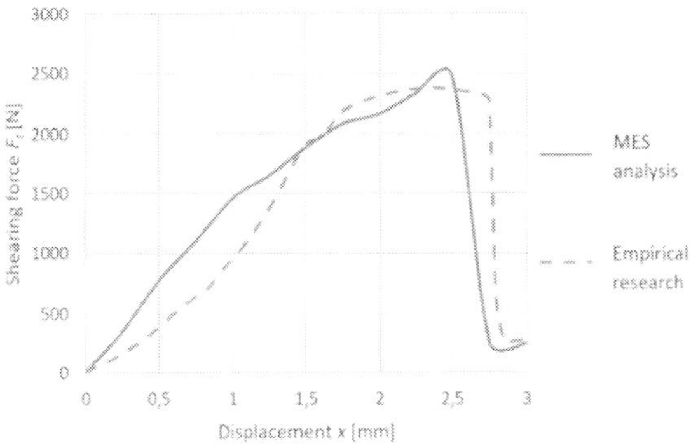

Fig. 15. Comparison of the results of empirical research and FEM analysis for the tensile test

Conclusions

Based on the results of research and FEM analysis presented in this work, it can be said that the developed model makes it possible to approximate the mechanical parameters of belts with polyamide core with 15% accuracy. Thanks to this, the model becomes useful for comparative analyses of different geometries of perforating tools, which - together with relevant empirical verification studies - will make it possible to develop effective blade geometry. This model, however, is insufficient to fully model the complex process of perforation because it does not take into account many factors, such as temperature or deformation speed. In further research, it is planned to extend the research to refine the model.

References

1. M. Hardygóra, *Taśmy przenośnikowe*. Wydawnictwo Naukowo-Techniczne, Chapter **1** and **2**, Warszawa (1999)
2. L. Gładyszewicz, *Przenośniki Taśmowe – teoria i obliczenia*. Wydawnictwo Politechniki Wrocławskiej (2003)
3. Wilhelm Herm. Müller, *Nitta Industries* – polish catalog. www.whm.pl (Access date: 06.07.2017)
4. S. Ochelski, *Metody doświadczalne mechaniki kompozytów konstrukcyjnych*. Wydawnictwo Naukowo-Techniczne, Warszawa (2004)
5. C. Yu, R. Yang et al., *Relationships between the work recovery ratio of indentation and plastic parameters for instrumented spherical indentation*. MRS Comunications **5** (1), 89-94 (2015)
6. Abaqus User's Manual, Version 6.13 2, Dassault Systémes Simulia Corp., Providence, Rhode Island, USA (2013)
7. Y. Zhang, J.C.Outerio, T. Mabrouki, *On the Selection of Johnson-cook Constitutive Model Parameters for Ti-6Al-4 V Using Three Types of Numerical Models of Orthogonal Cutting*. Procedia CIRP **3**, 112-117 (2015).
8. I. Faridmehr, M.H.Osman et al., *Correlation between Engineering Stress-Strain and True Stress-Strain Curve*. American Journal of Civil Engineering and Architecture **2** (1), 53-59 (2014)
9. S. Kut, *A simple method to determine ductile fracture strain in a tensile test of plane specimen's*. Metallurgy **49** (4), 295-299 (2010)

Selection of optimal solution heat treatment of the casting cylinder heads

Eva Tillová[1,*], *Mária* Chalupová[1], *Lenka* Kuchariková[1], *Juraj* Belan[1], *Denisa* Závodská[1]

[1]University of Žilina, Department of Materials Engineering, Univerzitná 8215/1, 010 26 Žilina, Slovak Republic

Abstract. The effect of solution treatment on mechanical properties (UTS, elongation, Brinell hardness) and microstructure (Si-morphology and Si-size) of an aluminium alloy (A356) used for casting cylinder heads was studied. The tests were carried out with specimens machined from the bulkheads of V8 engine blocks cast by the low pressure process. The samples were tested in as-cast and T6 heat treating conditions (solution heat treatment at 530°C with different time - 2, 3, 4, 5, 6, 7 hours, quenching in water at 20°C and precipitation hardened for 4 hour at 160°C). The results show that used heat treatment improves mechanical properties of the cylinder head casts. Tensile strength and hardness of specimens increase with solution treatment time. The hardness is a reflection of solution strengthening and silicon particle distribution in matrix. Solution temperature 530°C and 5 hours solution time is appropriate to obtain better morphology and distribution of Si particles in microstructure. Prolonged solution treatment (more than 5 hours) leads to a coarsening of the Si particles, while the numerical Si density decreases. As the particle density decreases, a fewer number of sites are available for crack nucleation, and hence, the fracture properties are improved. The data obtained from this study will be used to improve process control, and to help the selection of heat treatment of the casting for future products.

Keywords: Al-Si cast alloy, mechanical properties, solution treatment

1 Introduction

The cylinder head is an integral component of internal combustion engines. It conveys air and gasoline to the combustion chamber and serves as a cover for the cylinders. The main function of the cylinder head is to help the head gasket seal the cylinders properly so that they are able to build enough compression for engine operation. In the vast majority of four stroke engines, the cylinder head mounts the entire valve gear and provides the basic framework for housing the gas-exchange valves as well as the spark plugs and injectors. It also supports the different parts of cooling system [1].

The cylinder head must be strong and rigid to distribute the gas forces acting on the head as uniformly as possible through the engine block. The combustion gas, the coolant and the lubricating oil flow independently in the cylinder head and follow complex three

* Corresponding author: eva.tillova@fstroj.uniza.sk

Reviewers: *Darina* Ondrušová, *Mariana* Pajtášová

dimensional routes. Thus cylinder heads are generally produced by gravity or low-pressure die casting. In Europe, grey cast iron cylinder heads have been almost completely replaced by cast aluminium alloys during the past 20 years. Aluminium has the advantages of light weight, high thermal conductivity, and ease of production to close tolerances. Depending on the engine type, the introduction of an aluminium cylinder block results in a weight reduction of 10 to 20 kg (i.e. at least 50 %) [1-2]. An important fact is that weight reduction has a ripple effect on fuel efficiency. For example, weight reduction enables the manufacture to develop the same vehicle performance with a smaller engine, and such a smaller engine enables the use of a smaller transmission and a smaller fuel tank. With this ripple effect, it is estimated that 10 % of vehicle weight reduction results in 8-10 % of fuel economy improvement [3].

Aluminium cast alloys used in the production of such complex cast parts as engine blocks or cylinder heads must meet a combination of requirements which include low cost, excellent castability, low hot tearing tendency, good machinability, moderate strength at elevated temperatures and resistance to fatigue, i.e. high fatigue strength at elevated temperatures. As a result of the permanent increase of combustion pressures and temperatures, the potential of the common aluminium cylinder head alloys is almost fully exploited. In order to satisfy all the product requirements, optimised casting alloys and a proper control of the as-cast microstructure by the application of sophisticated casting processes are generally necessary. The applied aluminium alloys have to offer sufficient strength and hardness at room temperature for machining and assembly. Furthermore, high strength at elevated temperatures (up to 250°C) is crucial to ensure that the engine block-cylinder head assembly can withstand the combustion forces and the forces resulting from thermal expansion and contraction during service cycles without losing tightness of the cylinder head gasket. Creep strength is required in particular for the head gasket area [1, 4, 5].

To improve strength and creep properties, generally, alloying elements such as Mg and Cu are added [6]. On the other hand all alloying elements which dissolve in the α-matrix or which will reduce the amount of α-dendrites will decrease the heat conductivity. Ductility can be increased by low iron contents which on the other hand increases the price for the alloy. Further contribution to higher ductility will be given by modification of the AlSi-eutectic by means of Na or Sr [7]. Unfortunately, the addition of the modifier elements is often associated with increased porosity, possibly through altered feeding efficiency or increased risk to gas pick-up of the aluminium melt [8-9].

The best combinations of strength and ductility are offered by casting alloys with low iron content such as AlSi7Mg0.3 (A356). Therefore, in the past, most cylinder heads were made from primary aluminium alloys. But also alloys which can be produced using recycled aluminium (i.e. with a slightly increased impurity content) still provide sufficient ductility [1]. The mechanical properties of an Al-Si cast alloy for engine blocks and cylinder heads are mainly determined by its cast structure and the microstructural characteristics such as the grain size, dendrite arm spacing (DAS) [10], the size, shape and distribution of the eutectic silicon particles [6, 11-12], as well as the morphologies and amounts of intermetallic phases present [13].

Cylinder heads are usually applied in the T6 condition. Heat treatment influences the resultant mechanical properties [14-24]. The T6 heat treatment comprises three stages [19, 20]:

- Solution heat-treating at relatively high temperature is required to activate diffusion mechanisms, first, to dissolve Mg-rich phases formed during solidification and, then, to homogenize the alloying elements, such as Mg and Si, so as to achieve an elevated yield stress subsequent ageing [19]. Further, the solution heat treatment changes the morphology of eutectic Si from polyhedral or fibrous morphology in the modified alloys, to globular structure [11-12, 15-17].

- Quenching is usually carried out to room temperature to obtain a supersaturated solid solution of solute atoms and vacancies, in order to achieve an elevated strengthening subsequent ageing [19]. The most rapid quench rate gives the best mechanical properties, but it can also cause unacceptable amounts of distortion or cracking in components. Thus, parts of complex shape, often with both thin and thick sections, are commonly quenched in a medium that provides a slower cooling. This hardening medium can be hot water, an aqueous solution of polyalkylated glycol, or other fluid medium such as forced air or mist [20, 22].
- Artificial ageing consists of further heating the casting at relatively low temperatures (120-210°C) and it is during this stage that the precipitation of dissolved elements occurs [25]. These precipitates are responsible for the strengthening of the material. In AlSiMg cast alloys, the decomposition of the supersaturated solution begins with the clustering of Si atoms. This clustering leads to the formation of coherent spherical GP zones, consisting of an enrichment of Mg and Si atoms that elongate along the cube matrix direction to develop into a needle shape coherent β″ phase. With prolonged ageing, the needle shaped GP zones grow to form rods of an intermediate phase, β′, which is semicoherent with the matrix. The final stable β-Mg_2Si phase forms as an incoherent platelets on the α-Al matrix and has ordered face centered - cubic structure [19].

The present study is part of a larger research project which was conducted to evaluate the heat treatment of an AlSi7Mg0.3 cast alloy used in the manufacture of engine blocks and cylinder heads. The study was conducted on as-cast and heat treated samples that were cut from the bulkheads of cylinder heads. The experiments were carried out in cooperation with the practice. The manufacturer required selection of optimal solution heat treatment, respectively shorter dissolution annealing times.

2 Material and experimental procedure

For the experimental work were used cylinder head castings OM 661 for Daewoo (Fig. 1) that were produced from the most popular AlSi7Mg0.3 (A356) alloy. Melt was grain refined with Tiboral 6 masteralloy. Modification was achieved with AlSr 90/10 masteralloy. The chemical composition (in wt. %) of the melt prepared is shown in Table 1. Addition of Sr and Na can cause a transition of the eutectic silicon phase from coarse acicular flakes and plates (needles), which often act as crack initiators, to fine fibber's there consequently, improve the mechanical properties, especially the ductility.

Table 1. Chemical compositions of experimental material used in the present study (in wt. %)

Si	Mg	Ti	Fe	Mn	Cr
6.96	0.356	0.125	0.108	0.007	0.0005
Ni	**Zn**	**Sr**	**Ca**	**Na**	**Al**
0.0036	0.004	0.0343	0.0017	0.0005	rest.

Experimental samples (Fig. 2) for microstructure observation and mechanical properties were cut from commercial automobile cylinder head castings (supplied from the manufacturer). The castings were in the as-cast and heat treated conditions. Standard heat treatment used for this type of experimental cylinder head in the foundry is solution heat treatment for 7 hours at 530°C, quenching in water at 20°C and precipitation hardening for

4 hour at 160°C. The manufacturer required selection of optimal solution heat treatment, respectively shorter solution heat treatment times to achieve the required strength, hardness and ductility values.

Experimental castings (cylinder heads) were solution treated by different solution time (2, 3, 4, 5, 6 and 7 hours) at 530°C ± 5°C, quenched in water at 20°C ± 2°C and precipitation hardened for 4 hour at 160 °C ± 5°C. The as-cast state specimens represent the 0-hour aging time condition. Heat treatment was carried out by the manufacturer.

Fig. 1. The experimental cylinder head casting

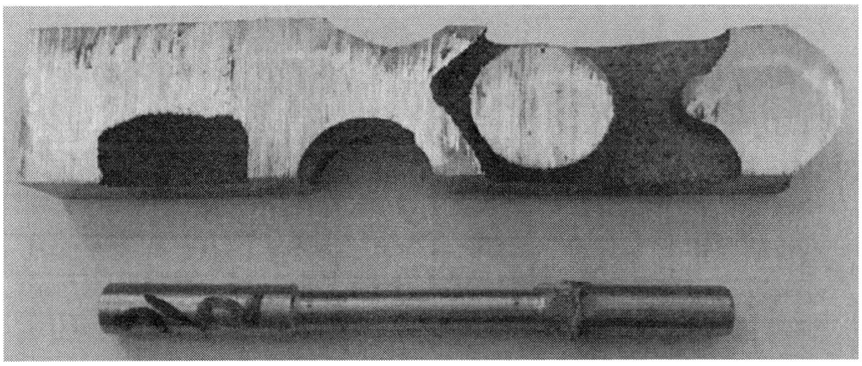

Fig. 2. Part of casting and sample for tensile test

Tensile test were carried out on a tensile machine ZDM 30 at room temperature following the STN EN ISO 6892-1 in as-cast state and after heat treatment; these specimens were of 14 mm in diameter by 120 mm in gauge length (Fig. 2). Values of ultimate tensile strength in dependence on heat treatment were determined by the average value of three test bars. Hardness measurement was performed by a Brinell hardness tester with the load of 250 kp, 5 mm diameter ball and the dwell time of 30 s according to standard STN EN ISO 6506-1. The Brinell hardness value at each state was obtained by the average of at least six measurements.

Metallographic samples were sectioned from the tensile test bars (after testing) and standard prepared for metallographic observations. A conventional grinding and polishing process, consisting of wet grinding on increasingly finer 500 and 1200 grade round SiC grinding papers (Struers TegraPol-15). The samples were then polished with 3 μm diamond pastes followed by Struers Op-S. Subsequently, etching sample preparation was carried out in these steps: a) for observation in optical microscope: standardly etched by 0.5 % HF. Thereafter the specimens were washed under running water, then ethanol, and dried; b) for observation in scanning electron microscope (SEM): standardly etched by 0.5 % HF or deep etched for 30 s in HCl solution (36 ml HCl, 100 ml distilled water) in order to reveal the three-dimensional morphology of the phases (eutectic silicon or intermetallic phases). Details of sample preparation are given in [6, 18].

The microstructures were immediately studied using an optical (Neophot 32) or scanning electron microscope VEGA LMU II. Chemical composition of phases were analysed by the energy dispersive X-ray spectroscopy (EDX analyser Brucker Quantax). Mapping of some specific areas of the polished sample surfaces was also carried out to show the distribution of alloying elements in the phases. The silicon particle characteristics (average Si particle area (μm^2)) were measured using a NIS Elements 4.0 image analyser.

3 Results and discussion

3.1 Microstructural evaluation

This type of alloy is characterized by dendrite structure (Fig. 3a) with a homogeneous distribution of eutectic silicon. The as-cast microstructure of AlSi7Mg0.3 consists of α-phase (Al-matrix), eutectic (mixture of eutectic silicon and α-matrix) and various types of Fe- and Mg- rich intermetallic phases (Fig. 3b).

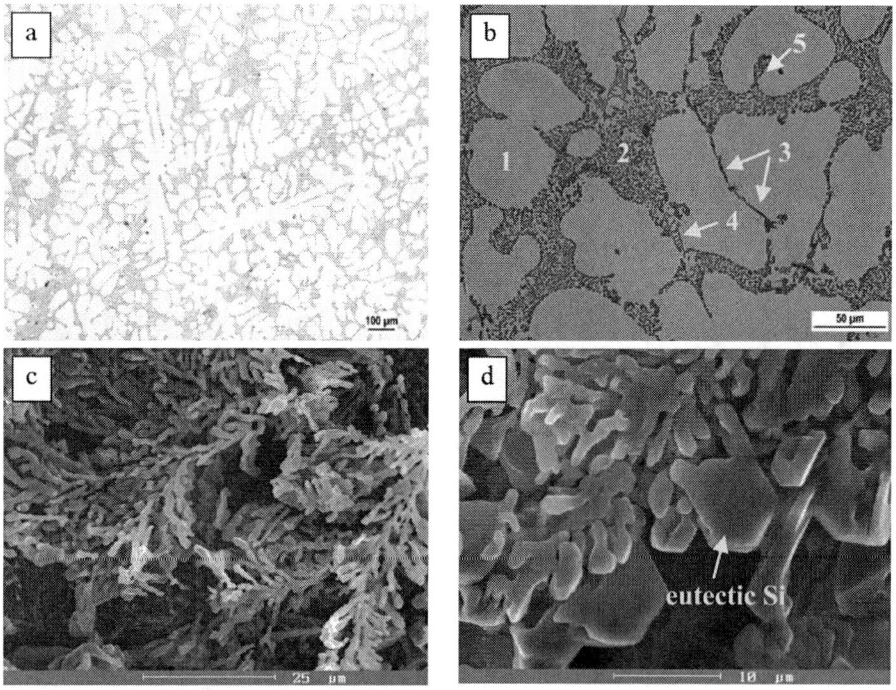

Fig. 3. Microstructure of cylinder head: a) as-cast state; b) phases (1-matrix; 2-eutectic; 3-needless Al_5FeSi; 4-compact skeleton-like $Al_{15}(FeMn)_3Si_2$; 4-Mg_2Si); c) morphology of eutectic Si (SEM); d) detail of eutectic Si-morphology (SEM)

By the addition of appropriate elements such as strontium in the liquid metal, it is possible to modify silicon morphological structure from lamellar to globular during the solidification. This modification treatment improves the mechanical properties of the Al-Si alloy by reducing the notching effect of needle shaped natural eutectic silicon [6, 9, 18]. The silicon appears as fibbers with a round cross-section (Fig. 3c). Fibbers grow as clusters from a single nucleating site. The 3-D form bars corresponds with metallographic sections, i.e. bunchy in the middle of the arrangement prevailing fine round bars and outwards, increased incidence of so called plate-like type with hexagonal shape (Fig. 3d).

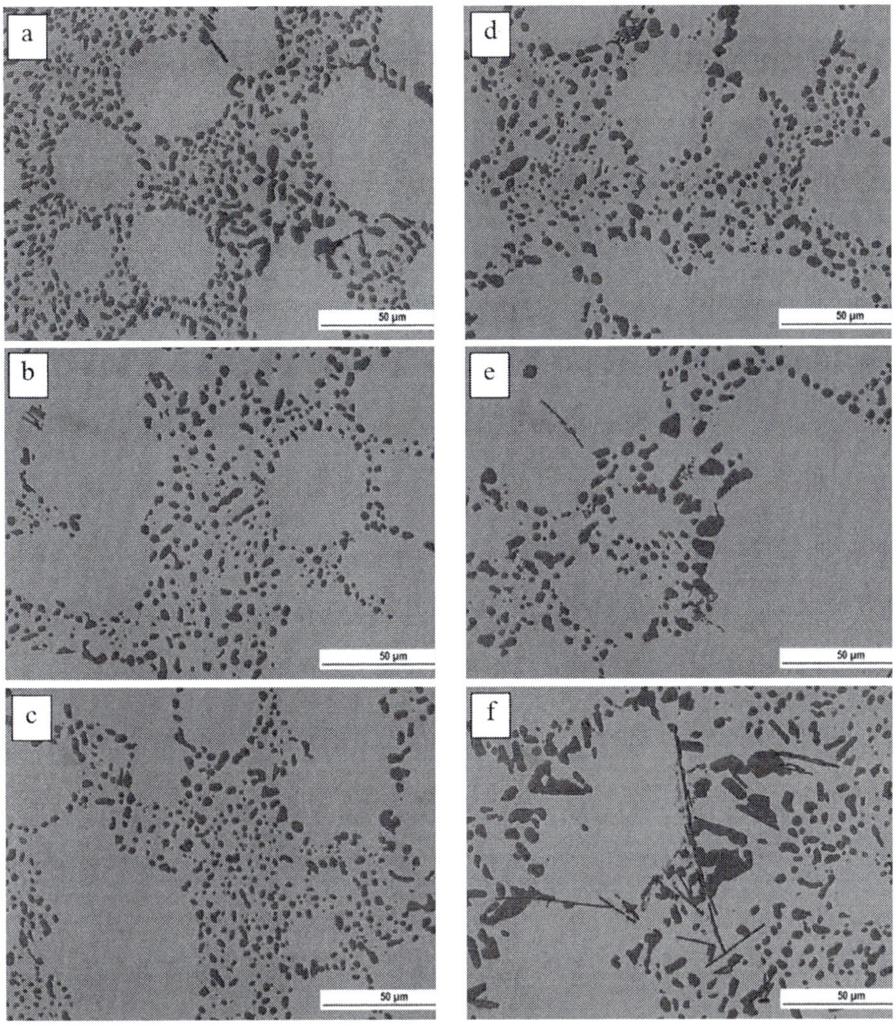

Fig. 4. Effect of solution treatment on microstructure of AlSi7Mg0.3 cast alloy, a) 2 hours; b) 3 hours; c) 4 hours; d) 5 hours; e) 6 hours; f) 7 hours, etch. 0.5 % HF

The effects of solution treatment time on microstructure and eutectic silicon morphology are shown in Fig. 4 and Fig. 5. Solution treatment causes a substantial degree of spheroidization and coarsening in modified eutectic phases. During the first 2 hours of solution treatment (Fig. 4a; Fig. 5a) Si particles gradually fragmented and rapidly spheroidized. Spheroidization process of the silicon through heat treatment takes place in two stages: fragmentation or dissolution of the eutectic fibbers/branches and spheroidization of the separated branches. The rate of spheroidization depends on the size of the fragmented segments.

Prolonged solution treatment leads to a significant coarsening of the Si particles (Fig. 4b-f, Fig. 5b-f). The former has the greatest effect on the time required to complete spheroidization and is strongly affected by the morphology of the Si particles. In Fig. 4e, Fig. 4f, Fig. 5e and Fig. 5f we can see that prolonged solution treatment time (more than 4 hours) leads to a coarsening of the Si particles, while the numerical Si density decreases. The spacing between the particles increased evidently too. It has been proved that structure containing large and elongated particles but small dislocation slip distances (within the dendrite cells) exhibited a high particle cracking rate and thus a low ductility [26]. Therefore, especially for unmodified A356 alloy, structure refinement by means of solution treatment for enough time is very important in improving the ductility.

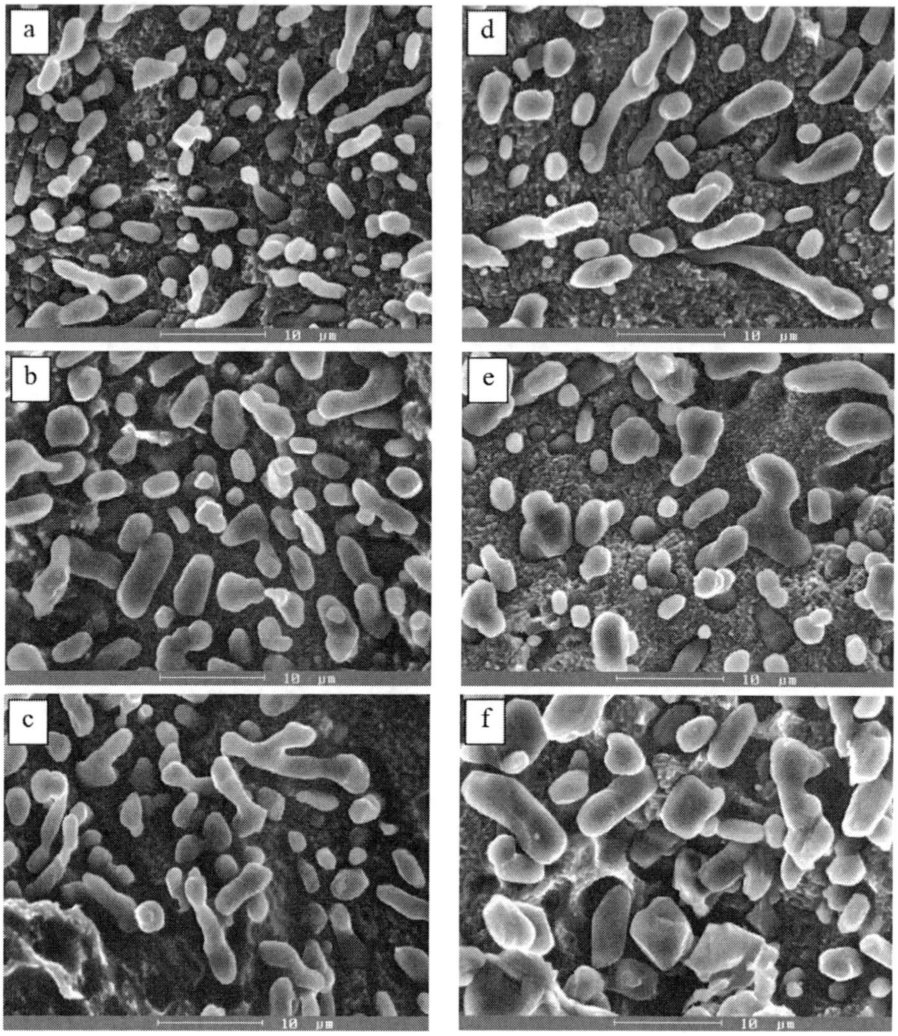

Fig. 5. Effect of solution treatment on morphology of eutectic Si (SEM),
a) 2 hours; b) 3 hours; c) 4 hours; d) 5 hours; e) 6 hours; f) 7 hours, deep etch. HCl

Fig. 6 shows the change in the average diameter of the silicon particles reflecting the particle round as a function of solution time at 530°C. Due to the presence of modifier as Sr and Na, the particles in the modified material are already fine and rounded. The average diameter of silicon particles increased rapidly within the first 2 hours at the solution temperature (from 2.14 μm^2 in as cast state to 3.9 μm^2 after 2 hours). Then they increased slowly with further increase in solution time (3.95 μm^2 after 3 hours; 4.18 μm^2 after 4 hours). After a prolonged solution treatment of 5 hours at 530°C, the average diameter of the silicon particles increased to 5.27 μm^2. These results are in agreement with [27-28].

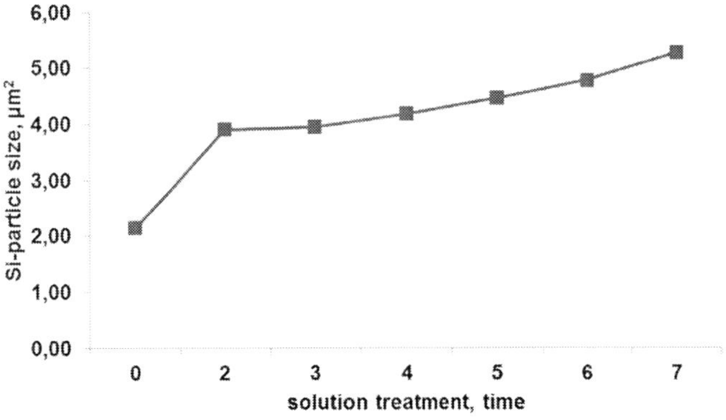

Fig. 6. Effect of solution treatment on Si-particle size

Fe-rich intermetallic phases and Mg_2Si particles are also observed in the microstructure (Fig. 1b). Fe-rich phases are just like eutectic Si completely brittle [6, 13]. It is suggests that the Fe-intermetallic phases may act as stress raisers and crack initiation sites that reduce the strength and ductility. By reason that experimental alloy contains less than 0.108 % of Fe and 0.007 Mn, compact skeleton-like $Al_{15}(FeMn)_3Si_2$ phase and very short and little Al_5FeSi needles were present. During heat treatment compact Fe-phase dissolved and fragmented to smaller skeleton particles. Long platelets Al_5FeSi phases were shortening and narrowing into small needles phases. These results are in satisfactory agreement with [11, 29].

3.2 Mechanical properties

Alloys of the A356 type (AlSi7Mg0.3) are usually heat treated in order to develop higher mechanical properties. Heat treatment of Al-Si-Mg alloy causes precipitation hardening, spheroidization of silicon particles, increased bonding between the hard second phase silicon particles and aluminium. Magnesium's role is to strengthen and harden aluminium castings. Silicon combines with magnesium is about to form the hardening phase, Mg_2Si that provides the strengthening of this alloy. According to the increase of the strength, the hardness is also improved.

The effect of experimental heat treatment on mechanical properties of the AlSi7Mg0.3 manufactured cylinder heads castings shows Fig. 7. Utility ultimate tensile strength (UTS) and hardness (HBS) of all heat treated specimens increase with solution treatment time at 530°C. Elongation (ductility) with solution treatment time decreases. The trend of increase in UTS and decrease in elongation observed in this study is in good agreement with other researcher [12, 17, 21-22, 27-28, 30-31].

The required mechanical properties used in manufacture for a typical aluminium cylinder head castings OM 661 for Daewoo includes an ultimate tensile strength of 245 MPa, Brinell hardness of 80 HBW and ductility of 6 %. These values meet more or less all experimental heat treated samples.

Ultimate tensile strength in as-cast state was measured approx. 192 MPa (Fig. 7a). With a solution time of 2 and 3 hours at 530°C, the UTS were substantially higher (approx. 236 and 250 MPa) than those obtained from the as-cast specimens, but they are still clearly lower than the values that could be achieved with a longer solution time. By increasing solution time from 2 to 4 hours, the UTS increased to 264 MPa. With further increase of the solution time up to 5 hours, the UTS increased to 268 MPa. With a prolonged solution treatment time of 6 or 7 hours at 530°C, the UTS achieved were 284.5 and 297 MPa respectively. This means that with a relatively short solution treatment time of 3 hours, more than 100 % of the maximum UTS can be achieved.

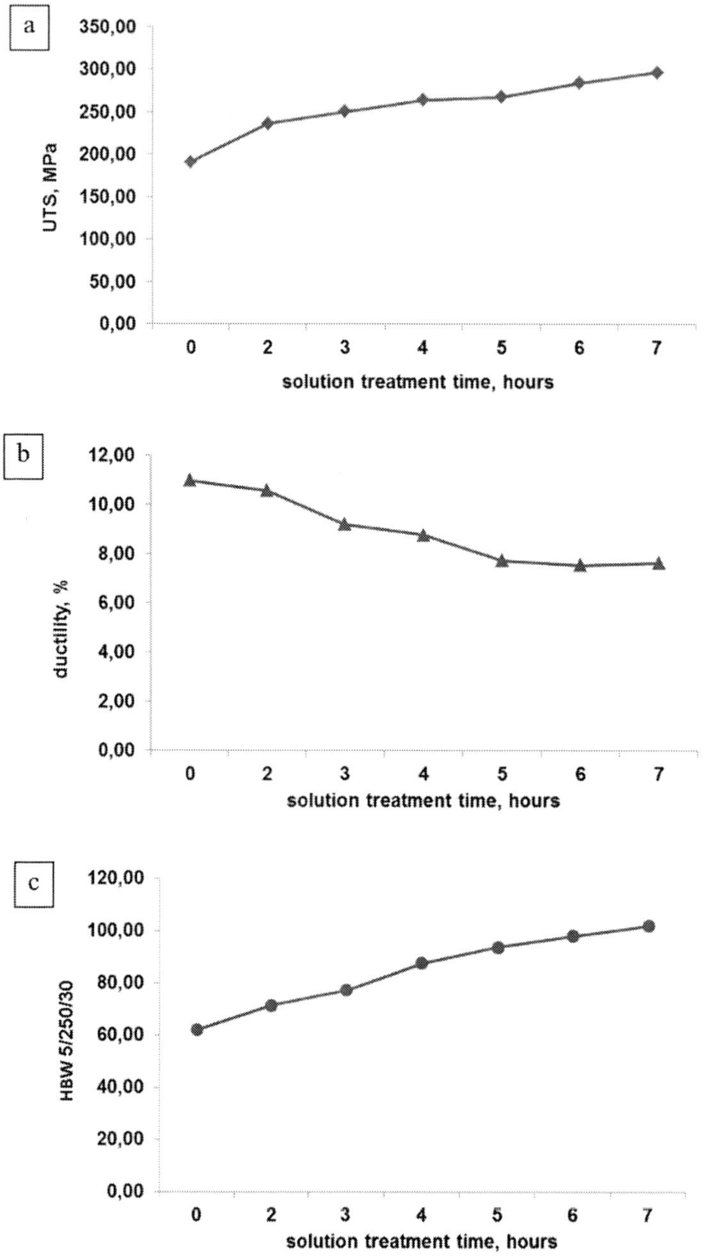

Fig. 7. Effect of solution treatment on mechanical properties:
a) UTS; b) ductility; c) Brinell hardness

The eutectic Si particles also strengthen the alloy. For the same amount of Si, smaller and more uniformly distributed particles have a stronger effect. The fragmentation of Si fibres reduces the particle size and increases the particle number and thus strengthens the alloy. The fragmentation of the eutectic Si takes place mostly at the early stage of solution treatment, and so, the strengthening effect caused by the morphology variation of Si particles occurs mostly in this time. The Si fragmentation still may be occurring, but slowly, and at the same time, Si particles are spheroidized and coarsened. In this time, the strengthening effect caused by Si fragmentation is counteracted by the weakening effects of coarsening and spheroidization, resulting in small variations in strength.

In Fig. 7b the ductility to fracture is seen to decrease with ageing time as the ultimate tensile strength increases.

Effect of solution treatment time at 530°C on hardness shows Fig. 7c. Brinell hardness in as-cast state was approx. 62 HBW 5/250/30 (Fig. 7c). With a solution time of 2 hours the HBW were higher more than 15 % (71.3 HBW). By increasing solution time from 2 to 3 or 4 hours, the HBW increased to 77.2 (grow about 24.5 %) and 87.7 HBW (41 %) respectively. With further increase of the solution time up to 5 hours, the HBW increased to 93.8 (more about 51.2 % as in as-cast state). With a prolonged solution treatment time of 7 hours at 530°C, the HBW were 102. Regularly hardness increased after experimental heat treatment. It is well-known that the Si phase is harder than Al-phase. Hence, the size, morphology and distribution of Si could affect the hardness of Al-Si alloy. Moreover, finer Si provides more obstacles to grain boundary sliding. Therefore, the treated alloys which have finer Si-particles in structure present higher hardness values.

Conclusions

The experimental heat treatment (T6 - solution heat treatment from 2 hours to 7 hours at 530°C, quenching in water at 20°C and precipitation hardening for 4 hours at 160°C) improves microstructure and mechanical properties of all samples of the cylinder head casts from AlSi7Mg0.3 (A356) alloy.

The results obtained suggest that it is possible to shorten the holding time at a temperature of solution treatment.

The mechanical properties of the cylinder head castings (OM 661 for Daewoo) are reliably achieved by the manufacturer in the following heat treatment: 4 hours at 530°C, quenching in water at 20°C and precipitation hardening for 4 hours at 160°C. In this heat treatment, castings have an ultimate tensile strength 264 MPa, ductility 8.8 % and Brinell hardness 87.7 HBW.

Solution temperature 530°C and 4 hours solution time is appropriate to obtain better Si-morphology and distribution of Si particles in microstructure of the cylinder heads casts. Solution treatment blunts the sharp edges of Si particles and promotes fracture resistance. Prolonged solution treatment (more than 5 hours) leads to a coarsening of the Si particles, while the numerical Si density decreases. As the particle density decreases, a fewer number of sites are available for crack nucleation, and hence, the fracture properties are improved.

The data obtained from this study will be used to improve process control and to help the selection of the heat treatment for future products.

The authors acknowledge the financial support of the project VEGA No 1/0533/15 and European Union - the Project ITMS: 26220220154.

References

1. EAA - European Aluminium Association. *The Aluminium Automobile Manual: Applications - Power train - Cylinder heads* (2011)
2. *ASM Handbook, vol. 15 - Casting*. (ASM International, Materials Park, Ohio, USA, 2002)

3. W. S. Miller, L. Zhuang, J. Bottema, A. J. Wittebrood, P. de Smet, A. Haszler, A. Vieregge, *Recent development in aluminium alloys for the automotive industry*. Materials Science and Engineering A **280**, 37-49 (2000)
4. R. Molina, P. Amalberto, M. Rosso, *Mechanical characterization of aluminium alloys for high temperature applications. Part1: Al-Si-Cu alloys*. Metallurgical Science and Technology **29**, 1, 5-15 (2011)
5. L. Heusler, F.J. Feikus, M.O. Otte, *Alloy and Casting Process Optimization for Engine Block Application*. AFS Transactions **01-050**, 1-9 (2001)
6. E. Tillová, M. Chalupová, *Structural analysis* (Štruktúrna analýza), (EDIS ZU, 2009)
7. COMALCO: *Modification of foundry Al-Si alloys*. Technical report, No4, (Comalco Aluminum Limited. Brisbane, Australia, 1997)
8. J. Cais, J. Svobodová, V. Weiss, *Influence of modification and heat treatment on the mechanical properties of the AlSi7Mg0.3 alloy*. Cluster - Casting - Future: international scientific conference. Rzeszów, 67-70, (2014)
9. N. Náprstková, S. Kusmierczak, J. Cais, *Modification of AlSi7Mg0.3 alloy by strontium*. Manufacturing Technology **13** (3), 373-380 (2013)
10. B. Zhang, M. Garro, C. Tagliano, *Dendrite arm spacing in aluminium alloy cylinder heads produced by gravity semi-permanent mold*. Metallurgical Science and Technology **21** (1), 1-9 (2003)
11. E. Tillová, M. Chalupová, L. Hurtalová, *Evolution of Phases in a Recycled Al-Si Cast Alloy During Solution Treatment*. Scanning Electron Microscopy, Dr. Viacheslav Kazmiruk (Ed.), InTech, (2012)
12. L. Hurtalová, J. Belan, E. Tillová, M. Chalupová, *Changes in structural characteristics of hypoeutectic Al-Si cast alloy after age hardening*. Journal Materials Science / Medžiagotyra **18** (3), 228-233 (2012)
13. J. A. Taylor, *Iron-containing intermetallic phases in Al-Si based casting alloys*. Procedia Materials Science **1**, 19-33 (2012)
14. K. L. Fan, G. Q. He, X. S. Liu, B. Liu, M. She, Y. L. Yuan, Y. Yang, Q. Lu, *Tensile and fatigue properties of gravity casting aluminum alloys for engine cylinder heads*. Materials Science and Engineering A **586**, 78-85 (2013)
15. M. A. Moustafa, F. H. Samuel, H. W. Doty, *Effect of solution heat treatment and additives on the microstructure of Al-Si (A413.1) automotive alloys*. Journal of Materials Science **38**, 4507-4522 (2003)
16. F. Paray, J. E. Gruzlesky, *Microstructure - mechanical property relationships in 356 alloy*. Cast Metals **7** (1), 29- 40 (1994)
17. E. Sjölander, S. Seifeddine, *The heat treatment of Al-Si-Cu-Mg casting alloys*. Journal of Materials Processing Technology **210**, 1249-1259 (2010)
18. E. Tillová, M. Chalupová, L. Hurtalová L., E. Ďuriníková, *Quality control of microstructure in recycled Al-Si cast alloys*. Manufacturing Technology **11**, 70-76 (2011)
19. *ASM Metals Handbook. Vol.4 - Heat treating*, (ASM International, Materials Park, OH, USA, 1991)
20. A. Manente, G. Timelli, *Optimizing the heat treatment process of cast aluminium alloys*. Recent Trends in Processing and Degradation of Aluminium Alloys, (InTech, 2011)

21. S. Shivkumar, S. Ricci, C. Keller, D. Apelian, *Effect of solution treatment parameters on tensile properties of cast aluminum alloys*. Journal of Heat Treating **8** (1), 63-70 (1990)
22. M. Tash, F. H. Samuel, F. Mucciardi, H. W. Doty, *Effect of metallurgical parameters on the hardness and microstructural characterization of as-cast and heat-treated 356 and 319 aluminum alloys*. Materials Science and Engineering A **443**, 185-201 (2007)
23. L. Hurtalová, E. Tillová, M. Chalupová, *The changes of fatigue properties in aluminium cast alloy during solution treatment*. Key Engineering Materials **592-593**, 433-436 (2014)
24. S. Capuzzi, *Development of heat treatments for automotive components die cast with secondary aluminium alloy at semi-solid state*. Metallurgia Italiana **106** (3), 3-11 (2014)
25. M. Javidani, D. Larouche, X. Grant Chen, *Dissolution of Cu/Mg bearing intermetallics in Al-Si foundry alloys*. Metallurgical and Materials Transactions A: Physical Metallurgy and Materials Science, **47** (10), 4818-4830 (2016)
26. H. C. Long, J. H. Chen, C. H. Liu, D. Z. Li, Y. Y. Li, *The negative effect of solution treatment on the age hardening of A356 alloy*. Materials Science and Engineering A **566**, 112-118 (2013)
27. D. L. Zhang, L. H. Zheng, D. H. StJohn, *Effect of a short solution treatment time on microstructure and mechanical properties of modified Al-7wt.%Si-0.3wt.%Mg alloy*. Journal of Light Metals **2**, 27-36 (2002)
28. D. Lados, D. Apelian, L. Wang, *Solution Treatment Effects on Microstructure and Mechanical Properties of Al-(1 to 13 pct)Si-Mg Cast Alloys*. Metallurgical and Materials Transactions B **42B**, 171-180 (2011)
29. L. Richtárech, D. Bolibruchová, E. Kucharčíková, *Appliciation of heat treatment for elimination of iron in secondary Al-Si alloy*. Manufacturing Technology **16** (2), 431-436 (2016)
30. D. Lados, D. Apelian, L. Wang, *Aging Effects on Heat Treatment Response and Mechanical Properties of Al-(1 to 13 pct)Si-Mg Cast Alloys*. Metallurgical and Materials Transactions B **42B**, 181-188 (2011)
31. M. Sága, P. Kopas, M. Uhríčik, *Modeling and Experimental Analysis of the Aluminium Alloy Fatigue Damage in the case of Bending - Torsion Loading*. Procedia Engineering **48**, 599-606 (2012)

Parallelization of computational algorithms for optimization problems with high time consumption

Milan Vaško[1,*], *Marián* Handrik[1], *Alžbeta* Sapietová[1], *Jana* Handriková[2]

[1]University of Žilina, Faculty of Mechanical Engineering, Department of Applied Mechanics, Univerzitná 8215/1, Žilina, Slovak Republic
[2]Armed Forces Academy of General M. R. Štefánik, Department of Informatics, Liptovský Mikuláš, Slovak Republic

Abstract. The paper presents an analysis of the use of optimization algorithms in parallel solutions and distributed computing systems. The primary goal is to use evolutionary algorithms and their implementation into parallel calculations. Parallelization of computational algorithms is suitable for the following cases - computational models with a large number of design variables or cases where the objective function evaluation is time consuming (FE analysis). As the software platform for application of distributed optimization algorithms is using MATLAB and BOINC software package.

Keywords: ADINA, MATLAB, BOINC, optimization algorithms, parallelization, distributed computing systems, evolutionary algorithms, FE analysis

1 Introduction

Solving of the optimization problem is generally based on the repeated evaluation of the objective function. Evaluation of the objective function represents FE analysis of the issue in the optimization problems in continuum mechanics [1, 2]. It is necessary to implement several types of analyses based on FEM for complex optimization problems solving (stress analysis to determine the state of stress distribution, modal–frequency analysis to determine the natural frequencies, loss of stability analysis, etc.) [3, 4].

To obtain the optimal solution is required tens to hundreds of thousands of objective function evaluations, depending on the used optimization method and the number of design variables [5]. The solution of such problems is unrealistic using conventional optimization algorithms because this solution could take several months or years.

For the purpose of more efficiently solve the optimization problem it is necessary to modify optimization algorithms. The ADINA software package will be used for evaluating the objective function value in the implementation of parallel optimization algorithms [6, 7]. It can be defined parameters and create a parametric model in the input file. This enables easy change of design variables in optimization process.

* Corresponding author: milan.vasko@fstroj.uniza.sk

Reviewers: *Radim* Halama, *Ján* Vavro *Jr.*

2 Computational systems for optimization problem solution

Using the parallel solution of linear equations in the FEM analysis is a fundamental possibility of increasing the speed of the optimization problem solution [1]. The two fundamental architectures of computational systems can be used in this case:

- Symmetric Multi Processing (SMP) technology – it means multiprocessor and multicore computers,
- Distributed Multi Processing (DMP) technology – this is a group of computers that are connected to a computer network and computers work together to solve problems.

Iterative solvers are used for solving systems of linear equations in large-scale problems. Parallel implementation of iterative solvers is effective only for a relatively small number of compute cores that are used for the problem solution. Increasing of the computing power is sometimes possible by using graphic accelerators. In the case of more required and solving tasks and provided adequate computing resources power to implement they the following calculation procedure can be used:

- Serial problem solving using the whole computing power – in this case the nodes are often only partially used.
- Parallel solutions of multiple tasks on the smaller number of computational cores – individual computing nodes are used more efficiently.

Practical tests have shown that in case of sufficient capacity of computing resources the parallel solution in terms of duration of solving across a series of tasks appears more efficient. Alternatively, increasing the speed of solving the optimization problem is the use of the above properties and modification of the optimization algorithms to the parallel implementation.

3 The optimization algorithms

Optimization methods can be divided into analytical and iterative. If in solving complex optimization problems of continuum mechanics to evaluate the objective function is using commercial FEM program, it is possible the iterative methods use only in the optimizing process. Iterative optimization algorithms can be divided into three basic groups [2]:

- the direct methods,
- the first order methods,
- the second order methods.

The group of direct methods can be divided into two subgroups – methods of random search and evolutionary algorithms. Their parallelization is simple – it is sufficient to implement a parallel evaluation of objective function, i.e. multiple FEM analyses solution work simultaneously. A subset of deterministic methods after their parallelization is transformed into the first subset or the gradient methods. Parallelization of the first order methods is based on the process of numerical differentiation parallelization [7, 8]. Based on the possibility of parallelization the iterative optimization algorithms can be divided into the following groups.

Algorithms that cannot be parallelized - this group can include algorithms where parallelization is possible only in a few steps of the algorithm (Simplex algorithm, in which is only possible to parallelize the compilation of the initial simplex).

Algorithms with the interaction - this group consists of algorithms in which needs to be completed one step of algorithm to algorithm can continued in the calculation (gradient methods). With these methods it is possible to parallelize the gradient or Hess matrix compilation while algorithm must wait for their establishment. Therefore, limiting factor for running the program is waiting for compilation the last element of the gradient or Hess matrix. Suitable computational resources for this type of algorithms are those where each computing node has the same computing power. Then, the powerful computational nodes do not have to wait for less powerful computing nodes.

Algorithms without the interaction - algorithms included in this group have to solve a high degree of tolerance to time and return the objective function solution (evolutionary algorithms). The descendants group with appropriate genetic variations is generating in the vicinity of the parent. The value of the objective function will be subsequently evaluated for this group of descendants. Member with the best objective function value becomes the new parent.

The following modification can be used for parallelization of evolutionary algorithms: Packet of descendants is created into which shall be supplemented continuously descendants generating on the base of current parents. These descendants are waiting to evaluate the objective function. The next step is parallel evaluation of descendants waiting in the packet and comparing the value of the objective function with a parent. If a descendant has a better objective function value than the current parent this child becomes the new parent.

In the packet of descendants may be found simultaneously descendants from the different parent generations but better parent can be found in the previous generation as is the current parent. Parallel evaluation of the descendant's objective function is mutually independent and sufficient condition the reasonable length of time for return the objective function. This condition allows into simultaneous evaluation include the computing resources with significantly different computing power (Fig. 1).

4 Grid computing using evolutionary algorithms

Gradient and evolutionary algorithms are significant in terms of optimization algorithms parallelization with the use of a commercial FEM program for evaluating the value of objective function. Problems of continuum mechanics cannot be considered as task with the one global extreme. More local extremes are often found in the solution area.

The term *grid computing* represents group of different computers that work together to solve complex computing tasks. Complex task is to find the optimal solution of the problem in terms of the optimization process. Partial task is to determine the value of the objective function for the vector of design variables. In evolutionary algorithms it represents determination the value of objective function for a specific descendant. There are a number of software platforms for the implementation of grid computing today.

BOINC (Berkeley Open Infrastructure for Network Computing) has been used to implement a distributed computing model. It's open source system for distributed computing. The robustness, flexibility and universality of the system is verified on a number of world famous projects in the area of grid computing (SETI, Milkyway, Einstein, etc.) [9].

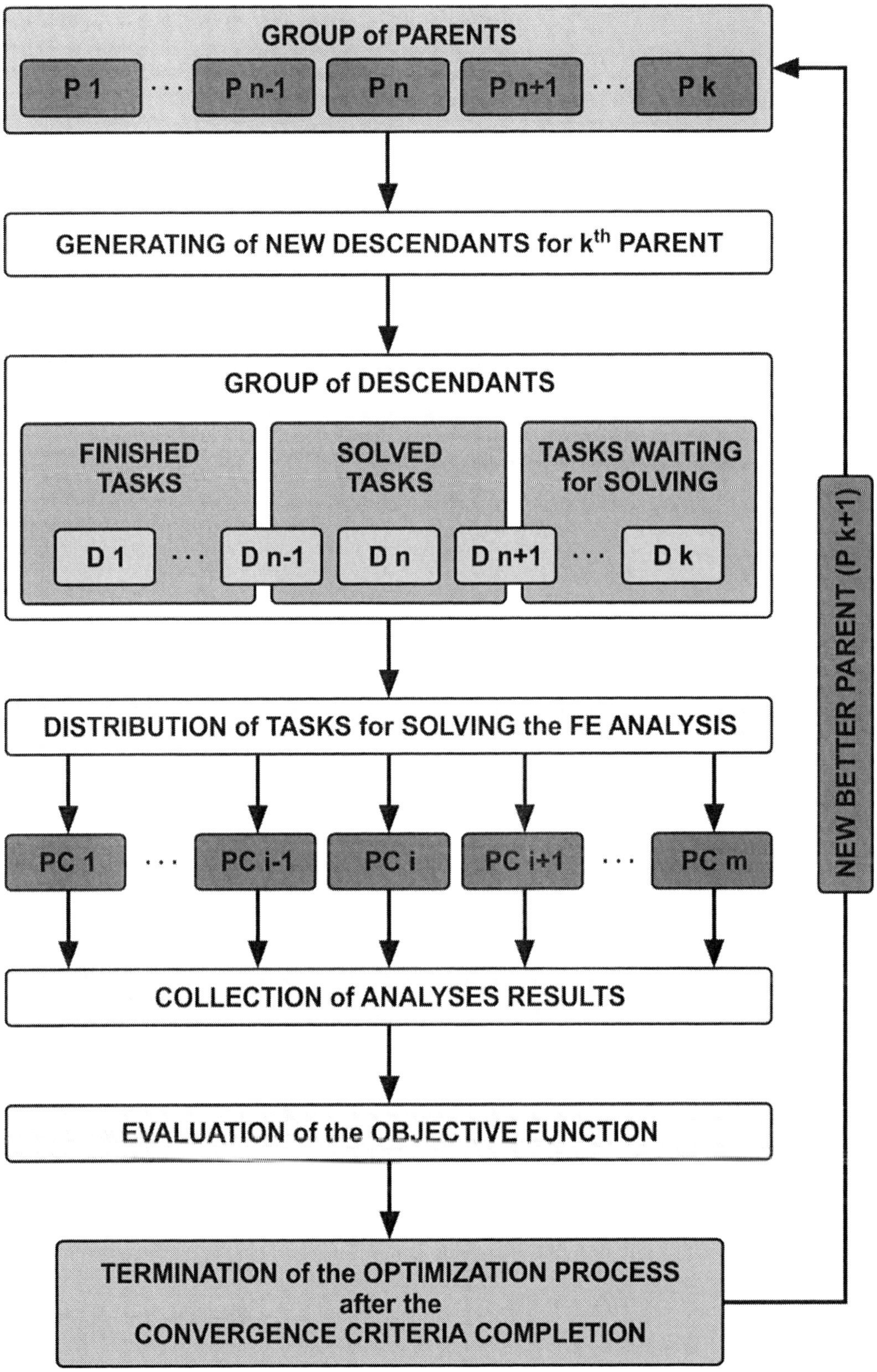

Fig. 1. Scheme of the optimization process for the grid infrastructure

Computing system consists of server and client part. The server part is designed to ensure the following functions:

- generate work units (work unit generator),
- insertion of work units into a set of tasks to send (feeder),
- identification of client's request and schedule its compliance (scheduler),
- client requirements completion – download work units and upload results (CGI scripts in Apache),
- comparison of the results for individual work units (validator),
- processing results after their validation (assimilator),
- provision of communication and reporting information about the process of the work units solution (transitioner),
- release disk space by deleting of solved work units (file deleter).

The client part is designed to ensure the following functions:

- download the main program for work unit solving,
- monitoring updates and ensure downloading a newer version,
- enable logging to the project or multiple projects,
- allocation of free system resources for their use,
- sending requests for the allocation of work units and their download on the client computer,
- send the information about status and results of solution.

A computer running Debian Linux 6.0 has been used to implement a distributed computing system. BOINC was installed directly into the system without using virtualization. The following changes were necessary perform to adjust the BOINC system for needs of optimization model implementation:

- change the general system settings,
- adjust the work units generator,
- create a client application,
- adjust the processing of the results, modifying the assimilator.

General settings change consists of changing the approach of the feeder to the prepared work units. The strategy of work units' choices was changed for insertion in the system FIFO (First In First Out). Work units, i.e. individual descendants are inserting from the oldest to newest for solving in the evolutionary strategy.

At the editing of work units generator it is based on the assumption that its task is generate new descendants with a random genetic mutation in the best parent's neighborhood. The generator monitors the number of prepared work units waiting to be processed. The best parent parameters are loaded and runs generate a new group of descendants if the number of prepared work units decreases under the specific bound.

It is monitoring before generating of descendants if in terms of the solution convergence the generating is needed. The creation of work unit consists of generating random genetic mutations, preparation of input files for the ADINA preprocessor, preparing files for the ADINA postprocesing and compression of all files to the "7zip" archive.

The advantage is that at any modification in the optimization task there is always necessary to transfer only one file. The optimization problem change is therefore realized by the change of the input files for work unit's generator on the server side. Therefore is not necessary to change the work unit's generator. The role of the client application is to ensure the following functions:

- extract the files from the work unit archive,
- run ADINA preprocessing and then solving tasks in ADINA,
- write selected results into files by running ADINA postprocesing,

- archiving initial and executive programs by "7zip" into a single file for transmission to the server.

The assimilator modification is in preparing them to execution the following steps:

- The result of the analysis solved by FEM in the program ADINA can contain multiple numeric data recorded in different files. The assimilator compiles objective function value from the partial results and then determines if the evaluated descendant becomes new parent.
- If it becomes the new parent then into the relevant files are written new parent parameters and its objective function value.
- In the evaluation of the each work unit convergence criterion is checked and in the case of its compliance gives the instruction to the termination of generating work units.
- The condition was used in the choice criteria of convergence if enough new descendants were generated without finding a new better parent.

It is necessary to make appropriate adjustments in the assimilator when is changing the form of the target function.

5 Optimization process and FE analyses

The optimization algorithm programmed in MATLAB using the BOINC system and evolutionary optimization algorithms was developed to test the proposed methodology [9-11]. The algorithm was tested on the problem of optimizing the wall thicknesses of the shell structure. The tests were performed on a simple structure consisting of a beam with an "I" cross section. The beam was loaded with continuous (distributed) load and supported at the ends.

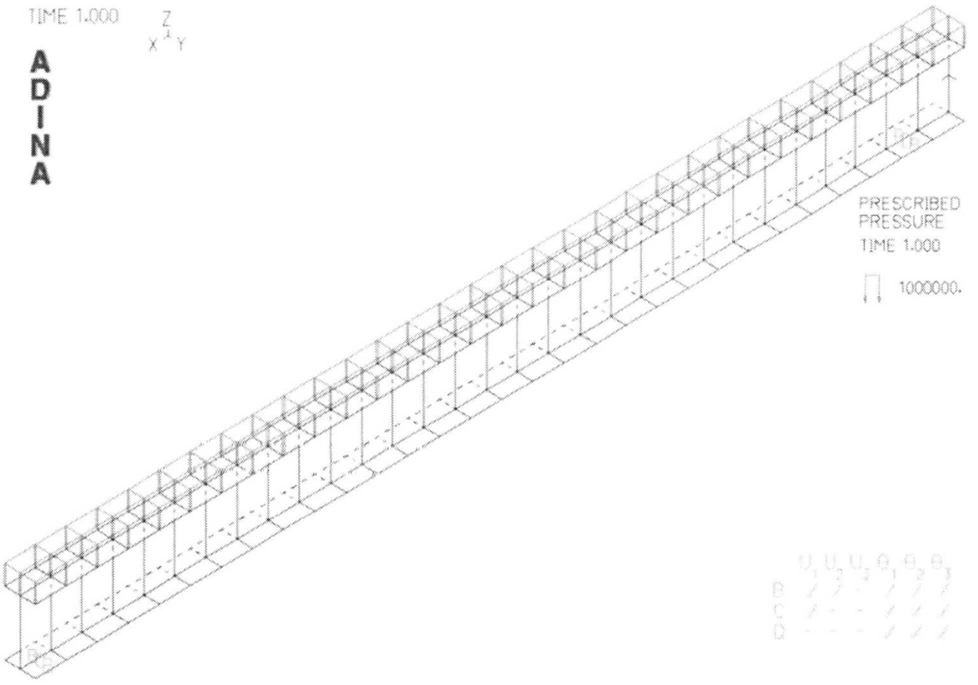

Fig. 2. Geometric model with boundary conditions

The geometric model of the beam was divided into 30 parts and each part consisted of three independent thicknesses of the beam wall. Thus, the total number of optimization variables was ninety.

An elastic material model for steel was used in the calculation. The starting thicknesses of the shell structure were 0.01 m. Created geometric model with boundary conditions is shown in Fig. 2. Tetragonal mesh of shell elements with the element size of 0.002 m was used in the finite element model. The total number of elements was 76 500 and the total number of nodes was 77 353. The target function has been choosing so that the maximum value of equivalent stresses of 100 MPa in each part of the structure was reached at optimum. A secondary condition was that the minimum thickness of the shell structure was not less than 0.001 m.

The distribution of von Mises equivalent stresses before the optimization process is shown in Fig. 3. The maximum value of the stress was 313 MPa and the minimum stress was 0.2 MPa.

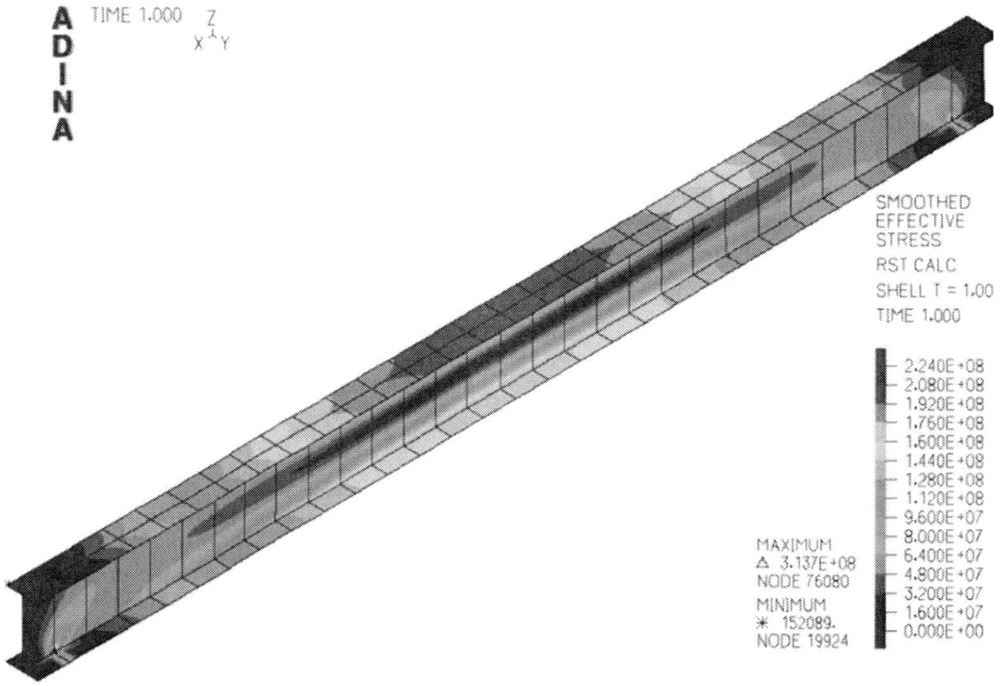

Fig. 3. Distribution of von Mises equivalent stresses before the optimization process

The resulting distribution of von Mises stress after the optimization process is shown in Fig. 4. The maximum stress reached a value of 100 MPa and minimum stress was 0.02 MPa in this case. The resulting wall thickness of the individual parts of the beam after optimization is shown in Fig. 5. The maximum wall thickness reached 0.2988 m and the minimum wall thickness was 0.001 m.

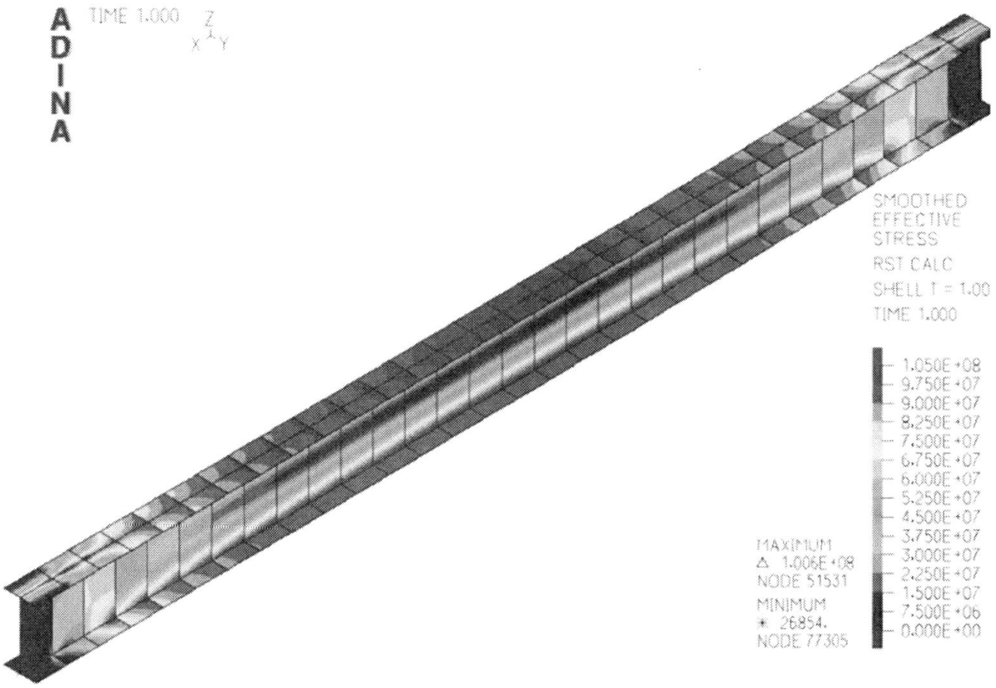

Fig. 4. Resulting distribution of von Mises stresses after the optimization process

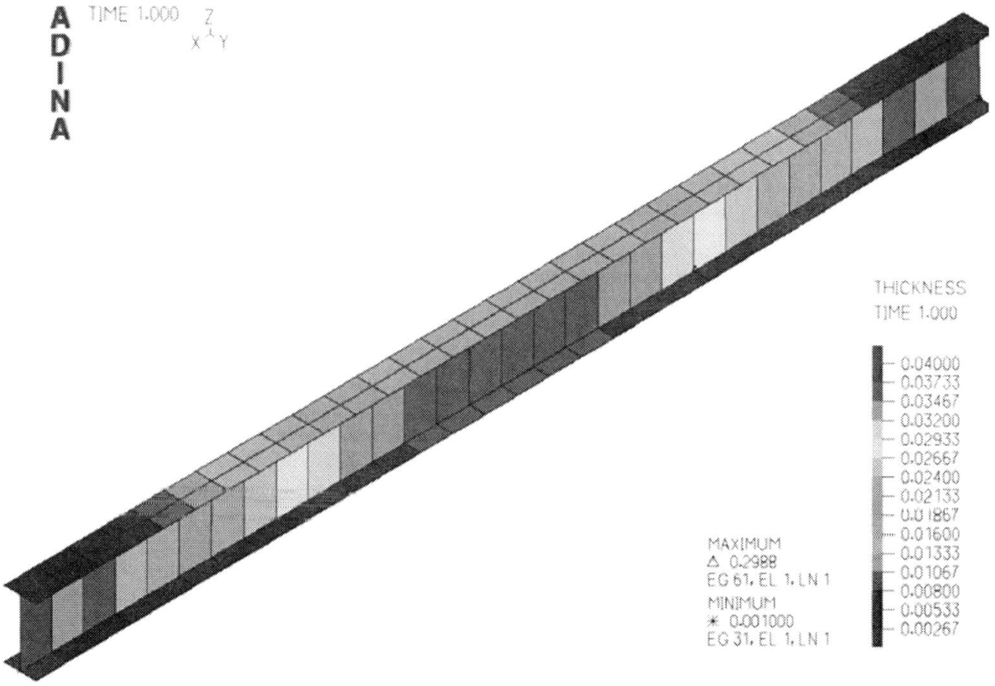

Fig. 5. Resulting wall thickness of the individual parts of the beam after optimization

Low stress values after optimization were obtained in the end segments of the beam but in places of boundary conditions the maximum stress value was reached [12]. The optimization process was conducted on a group of 26 computers. The calculation lasted 108 hours and the 69 542 iterations of the optimization task was performed.

Conclusions

Use of the BOINC in combination with MATLAB as a control system for grid computing allows very fast response of the system at the appropriate setting of its parameters (i.e. the short time between the end of the FEM analysis and processing of the results in the assimilator for the work units generator). This allows the implementation of appropriate and adapted evolutionary algorithms that require a higher level of feedback between the generated descendants and actual parents.

A small claims rate on system resources, parallel realization of the work units generation and parallel processing of the results enables involving the order of several thousand to tens of thousands of PCs in FEM analyses necessary for solving the optimization problem. Robust computing infrastructure allows the realization of optimization problems for finding the global extreme in continuum mechanics problems with hundreds of design variables in relatively short time.

This work is supported by VEGA 1/0795/16 "Development of effective methods for correction and optimization of coupled mechanical system".

References

1. K. J. Bathe, *Finite Element Procedures*. (Prentice Hall, New Jersey, 1982)
2. M. Sága, M. Vaško, N. Čuboňová, W. Piekarska, *Optimisation Algorithms in Mechanical Engineering Applications*. (Pearson Education Limited, 291 p., 2016)
3. L. Jakubovičová, M. Sága, *Computational analysis of contact stress distribution in the case of mutual stewing of roller bearing rings*. Novel Trends in Production Devices and Systems, Applied Mechanics and Materials **474**, 363-368 (2014)
4. A. Vaško, L. Hurtalová, M. Uhríčik, E. Tillová, *Fatigue of nodular cast iron at high frequency loading*. Materialwissenschaft und Werkstofftechnik **47** (5-6), 436-443 (2016)
5. P. Oršanský, B. Ftorek, P. Ďurčanský, *Mathematical model of a closed hot air engine cycle using MATLAB Simulink*. XIX. The Application of Experimental and Numerical Methods in Fluid Mechanics and Energetic 2014, AIP Conference Proceedings **1608**, 173-176 (2014)
6. ADINA, *Theory and Modelling Guide*. [help manual]. (Watertown, 2010)
7. P. Pecháč, M. Sága, *Memetic algorithm with normalized RBF ANN for approximation of objective function and sedondary RBF ANN for error mapping*. Procedia Engineering **177**, 540-547 (2017)
8. M. Vaško, M. Sága, M. Handrik, *Comparison Study of the Computational Methods for Eigenvalues IFE Analysis*. Appl. and Computational Mechanics **2** (1), 157-166 (2008)
9. *Berkeley Open Infrastructure for Network Computing (BOINC)*. [online] http://boinc.berkeley.edu/
10. *MATLAB*. [online] https://www.mathworks.com/products/matlab.html

11. L. Žuľová, R. Grega, J. Krajňák, G. Fedorko, V. Molnár, *Optimization of noisiness of mechanical system by using a pneumatic tuner during a failure of piston machine*. Engineering Failure Analysis **79**, 845-851 (2017)
12. A. Sapietová, V. Dekýš, M. Sapieta, P. Pecháč, *Application of computational and design approaches to improve carrier stability*. Procedia Engineering **96**, 410-418 (2014)

The damage analysis of the reinforced concrete beam and the prestressed reinforced concrete beam

Milan Vaško[1,*], *Marián* Handrik[1], *Matej* Rác[1], *Vladislav* Baniari[1], *Ján* Kortiš[2], *Ľuboš* Daniel[2]

[1]University of Žilina, Faculty of Mechanical Engineering, Department of Applied Mechanics, Univerzitná 8215/1, 010 26 Žilina, Slovak Republic
[2]University of Žilina, Faculty of Civil Engineering, Department of Structural Mechanics, Univerzitná 8215/1, 010 26 Žilina, Slovak Republic

Abstract. The paper deals with the finite-element analysis of damage in reinforced concrete beam and prestressed reinforced concrete beam. The steel reinforcement is modeled using non-linear rebar elements and an elastoplastic model of the reinforcement is considered. The initial prestress is defined in the ropes that are used to create the prestressed reinforced concrete beam. These ropes are also modeled using non-linear rebar elements and the elastoplastic material model. Created computational model allows the damage modeling of the reinforced and pre-stressed reinforced concrete beams under static loading.

Keywords: reinforced concrete beam, prestressed reinforced concrete beam, rebar elements, FE analysis

1 Introduction

Concrete is the most used construction material for the construction of load-bearing structures in the world. Specific physical and mechanical properties of concrete must be considered when designing structures. Concrete has a relatively high compressive strength but low ductility and tensile strength (represents only about 10 % of the compression strength). Concrete is reinforced to eliminate this deficiency. The most commonly used reinforcing material is steel and it is usually embedded passively in the concrete. In addition, a reinforcement based on glass, carbon or aramid fibers is used.

Prestressed concrete is a production technology in which a pressure stress independent of the external load is introduced into the concrete element by a mechanically tensioned reinforcement. This pressure stress totally or significantly reduces the unfavorable tensile stress caused by the external load. Bearer of prestressing is the elastic deformation of the rope (tendon). The tendon is stretched so much that its elastic deformations strain the concrete element by pressure. The combination of reinforcement and prestressing

* Corresponding author: milan.vasko@fstroj.uniza.sk

Reviewers: *Robert* Grega, *Bohuš* Leitner

guarantees a longer life and load bearing capacity, which allows the possibility of designing smaller cross sections.

2 Structural properties of concrete and reinforced concrete

Concrete is a composite material that is made of stone, cement, water, additives and impurities. Some concrete properties are improved by additives and admixtures. Concrete properties are determined by its composition, method of production, transport, compacting, treatment, protection and proper maintenance. The most important features of structural concrete are mechanical properties and durability in the environment.

Concrete behaves almost as a linear elastic material at pressures up to 30% of concrete strength $R_{m\ CON}$. When compressive stresses $(0.3\text{-}0.4)\cdot R_{m\ CON}$ are reached the microcracks occur in contact of cement putty and stone crumbs. The microcracks further propagate along the surface of the grains with increasing pressure stress up to value of $0.6\cdot R_{m\ CON}$. Microcracks are beginning to expand into the cement putty and create a coherent system of microcracks if pressure exceeds the value of $0.6\cdot R_{m\ CON}$. Concrete plasticizing and permanent irreversible deformations of the structure have occurred since that moment.

The modulus of elasticity E_{CON} is the basic deformation characteristic of concrete. Hooke's law is valid for stresses up to about 30-40 % of short-term strength and 60 % of tensile strength. However, this dependency is applied only for short-term load. Various factors affect the modulus of elasticity (water saturation, material properties of cement and stone crumbs, etc.) [1].

Reinforced concrete is a composite material in which the tension is transmitted by inserted reinforcing bars and pressure is transmitted by concrete. Coherence between the steel core and the hardened cement stone is expected. This stone with its strongly alkaline reaction creates a passive surface of the reinforcement and prevents corrosion [2].

3 Prestressing reinforcement

Prestressed concrete - thanks to a compression force in prestressed concrete cracks are found only limited or their size is small. The pressure reserve for load transfer is created by the formation of a prestress to better utilize the basic pressure properties of the concrete and eliminates the disadvantage of low tensile strength. Linear elastic material it becomes because the preload eliminates brittle tensile failure.

The prestressed beam is stiffer and consequently it has less ductility. The prestressing beams may be slimmer, with a smaller cross-sectional area and longer length due to these properties under the same requirements to deflections limit. The energy required for its breach is higher than for conventional reinforced concrete parts.

The prestressing reinforcement is an active element in prestressed concrete. Its stress is generated by mechanical tension. This is independent of the structure deflection or the external load. The required mechanical and physical properties of the concrete reinforcement are obtained by suitable chemical composition of the steel. Hot-formed rods of conventional carbon steel with a maximum carbon content of 0.22 % are used. High-strength wires, tendons and rods are used as prestressing reinforcements in construction practice. The basic requirement of the prestressing reinforcement is the largest elastic deformation [3-7].

Additional mechanical cold forming can be used to improve the strength properties of steel. The steel strength R_m = 1400-2000 MPa (ultimate stress) can be achieved by this molding. The high-strength prestressing reinforcement must also have a guaranteed tensibility (ε_{pu} up to 3.5%). The tensibility of the reinforcement is one of the most significant properties of steel. This represents the ability of reinforcement's large deformation without significant decrease in stresses [1, 8].

The properties of the concrete reinforcement and the prestressed reinforcement can be compared in Fig. 1. Yield stress R_e is basic characteristic of concrete reinforcement produced by the conventional hot-rolled steel. The reinforcement cannot be used above this value due to the subsequent large longitudinal deformations ε_s. These transformations cause large plastic deformations of the reinforcement, which cause the loss of cohesion with the concrete. Concrete reinforcements produced by cold drawing have a non-significant yield stress value therefore $\sigma_{0.2}$ is used [1, 3, 9].

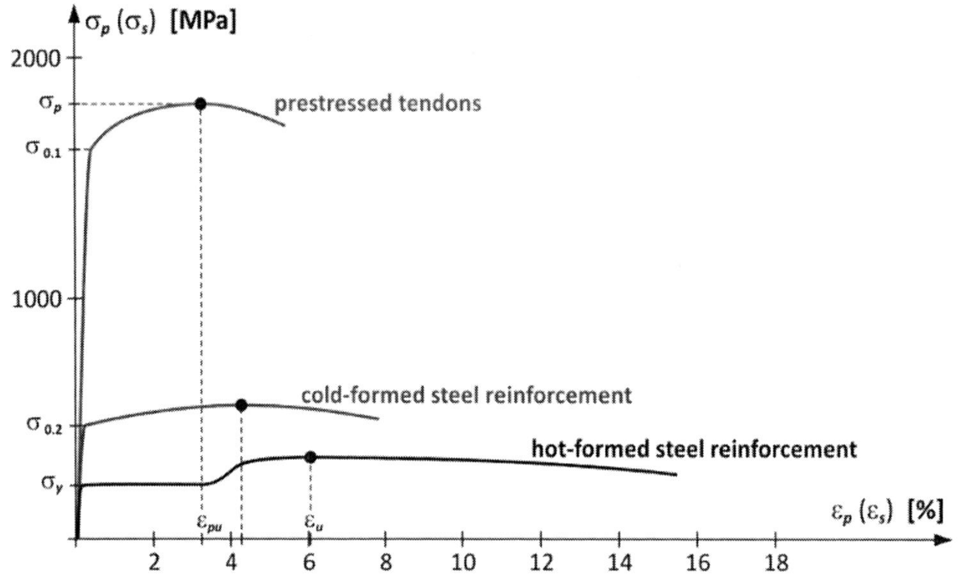

Fig. 1. Properties of the concrete reinforcement and the prestressed reinforcement

4 Numerical experiments

The basic dimensions of the tested beam are shown in Fig. 2. The total beam length is 3600 mm and the theoretical distance between the supports of the beam is 3300 mm. The pressure acting on the beam is 1.4 MPa [10].

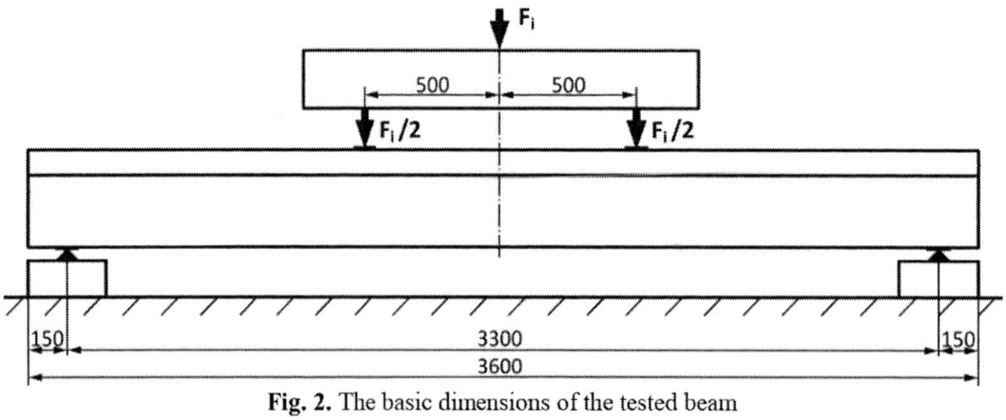

Fig. 2. The basic dimensions of the tested beam

4.1 FE model creating

The beam model was created in ADINA software [11, 12]. Parts 1 and 3 are not important for calculation, so they have a smaller mesh density. The material model of the concrete is assigned to the part 2 and it is meshed with 3D solid 8-node elements (Brick Elements) with the size of 20 mm. On the contact area of these parts is created so-called Glue Mesh. This feature allows connecting areas of the model with different types of mesh (Fig. 3).

The pressure is applied to two areas of size 60×350 mm. Boundary conditions were defined in the support locations so that the beam is statically determinate. The rotation in all 3 axes was discarded before the calculation and it was necessary to specify boundary conditions for translational movement. All three translational movements in *x*, *y* and *z* axes were removed at the point "C" and translational movement in the z-axis was removed at the point "B".

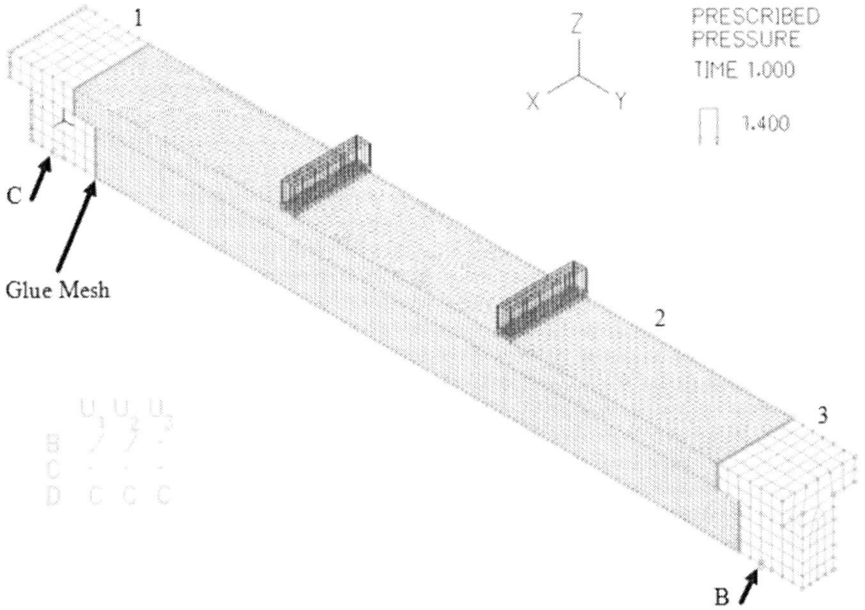

Fig. 3. Meshed beam model created in ADINA

The reinforcement was created using geometric lines that were subsequently defined as Truss-Rebars. The "Rebar element" is used to create trussed elements and limitations between beam nodes and fixed elements. The reinforcement used in the beam model is shown in Fig. 4 [12, 13].

Various material models were used in the model (for concrete, steel reinforcement and prestressed steel reinforcement). To verify the strength and fatigue properties of concrete reinforcement, an experimental test was made at the Department of Applied Mechanics (Faculty of Mechanical engineering, University of Žilina). The result of the measurement was the determination of the strength characteristics and the working diagram of the reinforcement [10, 14].

The constants for the material model of the concrete reinforcement and the prestressing reinforcement are shown in Table 1.

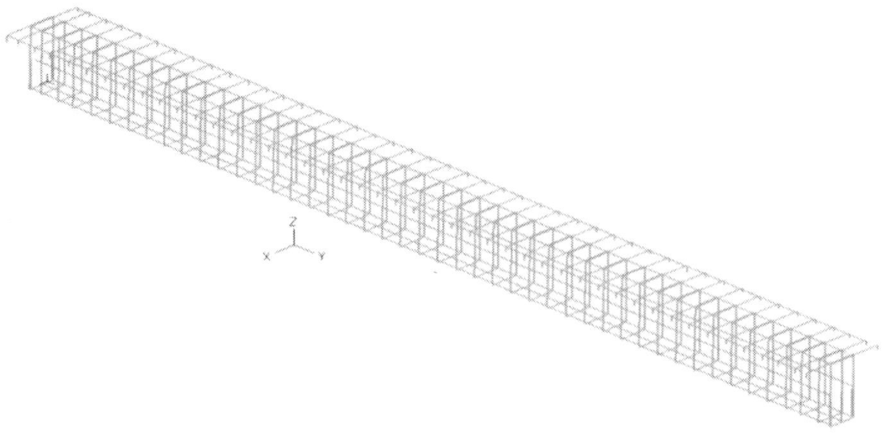

Fig. 4. Steel reinforcement in the model of reinforced concrete beam

Table 1. Constants of the concrete reinforcement and prestressing reinforcement material model

	Concrete reinforcement	**Prestressing reinforcement**
Young's Modulus [MPa]	210 000	195 000
Poissnos's Ratio [-]	0.3	0.3
Initial Yield Stress [MPa]	500	1 600
Strain Hardening Modulus [MPa]	5 000	2 500
Max. Allowable Effective Plastic Strain [-]	0.02	0

4.2 FE simulation of the reinforced concrete beam

The beam with concrete material model, which is loaded by the pressure of 1.4 MPa (i.e. 29.4 kN for each area) is shown in Fig. 5. The stress at the top of the beam is approximately 6.6 MPa and at the bottom is approximately 4 MPa. Deflection of the beam is shown in Fig. 6; the maximum deflection reached 1.2 mm.

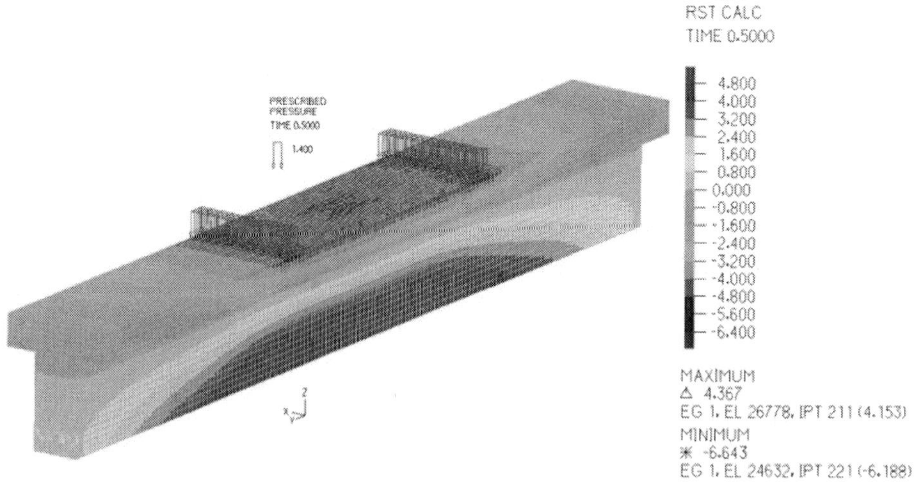

Fig. 5. FEM analysis of stresses in beam model

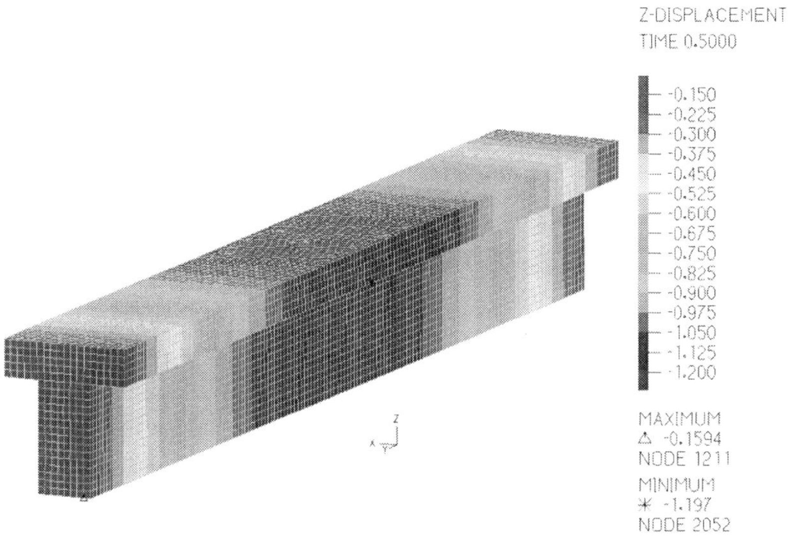

Fig. 6. Deflection of the beam in *z* axis

Deflection of the beam at the point of maximum deflection as a function of acting compressive force is shown in Fig. 7. Displacement curve had a linear course up to the value of the applied pressure of 1.12 MPa. When this pressure was exceeded, the beam began to crack and there was started a crack propagation. This caused a weakening of the beam and there was a greater and more rapid increase in the value of the beam deflection.

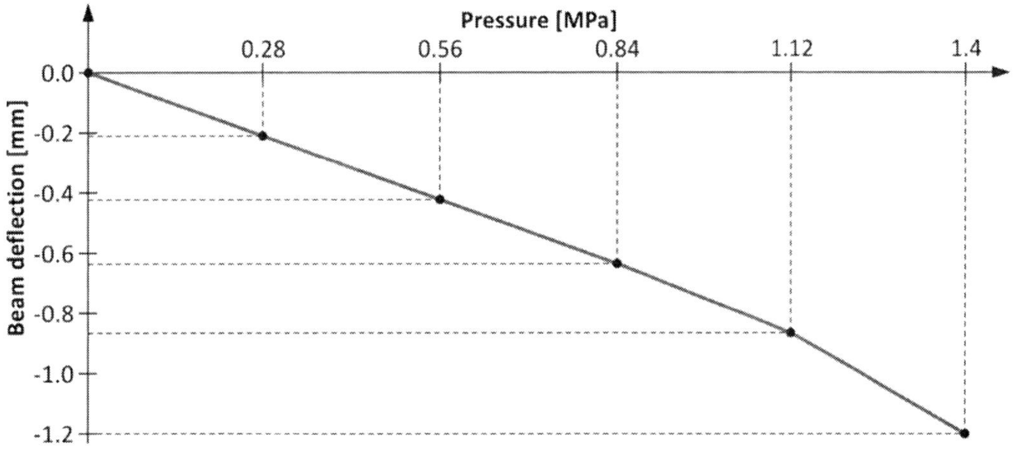

Fig. 7. Deflection of the beam at the point of maximum deflection

Similar values were reported in the FEM analysis of the beam compared to the static beam test [10, 15, 16]. Beam deflection of 1.4 mm was measured in a static test in the same place as in FEM analysis. Figs. 8 and 9 show the formation of cracks and the subsequent extinction of the elements in the FEM simulation of the static test. After the extinction of the first elements the calculation further converged, even though part of the beam was located in the plastic zone due to the reinforcement that carried the pressure load acting on the beam [17, 18].

Fig. 8. Formation of cracks and the subsequent extinction of the elements

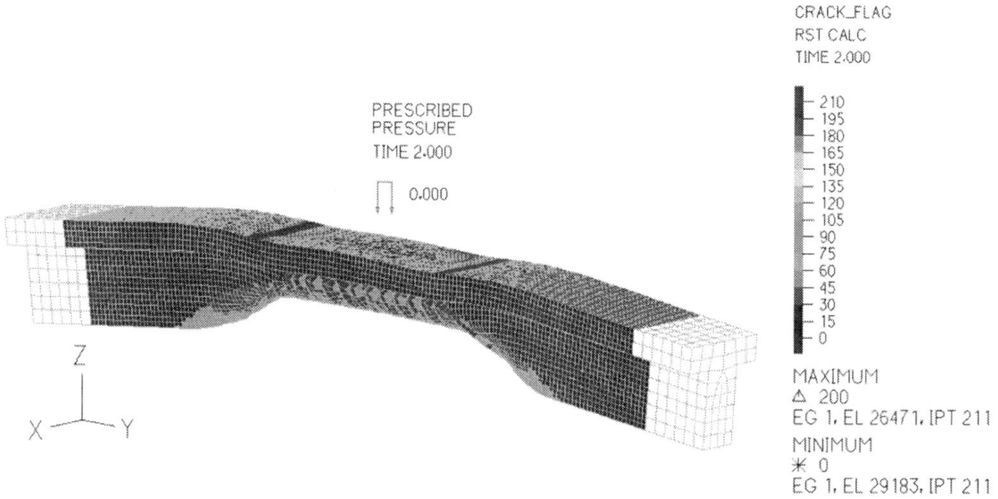

Fig. 9. Extinction of the beam elements

4.3 FE simulation of prestressed concrete beam

The prestress in the beam was created by adding two prestressed ropes. The ropes were loaded with a preload which was recalculated with an initial pressure of 1.5 MPa. The number of prestressed ropes, their position and dimensions were calculated based on the beam's bearing capacity and expected life at the request of dimensional compliance with the above tested beam (Fig. 10) [4, 19, 20].

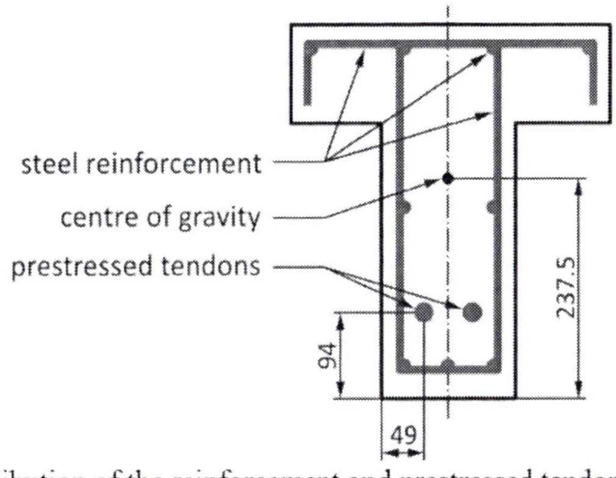

Fig. 10. Distribution of the reinforcement and prestressed tendons in the beam

The prestressed ropes cause stress at the upper part of the beam about 1.7 MPa and at the bottom -11 MPa. The prestressing of the tendons has been calculated so that no cracking of the concrete occurs in this state. The stresses in the center of the beam at the top reached 1.6 MPa and -9.4 MPa at bottom. FE simulation of the prestressed beam was done in the same way as the reinforced concrete beam test (Fig. 11). At places of the highest stress concentrations the stress values are 1.75 MPa and -11.5 MPa.

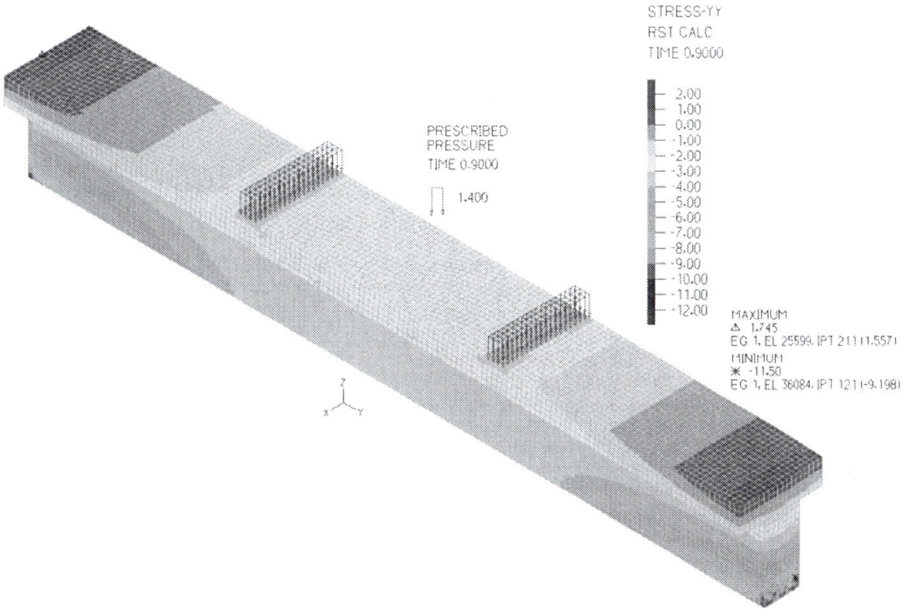

Fig. 11. FEM simulation of the prestressed beam

Further stress concentrations have arisen at locations where parts of a beam with different types of mesh are joined. The stresses at these points reached the value of 1.7 MPa and -10.7 MPa. At the center of the beam the maximum stress at the top of the beam reached -2.6 MPa and at the bottom -3.5 MPa.

FEM simulation showed that the deflection of the unloaded beam was 1.2 mm due to prestressed ropes. The deflection value was changed when an external load began to work. Upon reaching a maximum pressure value of 1.4 MPa the beam deflection reached value of 0.2 mm. The deflection of the beam in the negative direction of the z axis has not occurred due to the prestressed ropes. When the maximum pressure is applied, the beam is almost in the equilibrium position and there is no violation of the beam.

Conclusions

FEM simulation of reinforced concrete beam and prestressed beam, their mutual comparison and comparison with the experimental static test of the reinforced concrete beam was solved with in this paper. Comparison of the results showed that the prestressed beam carries a higher load and has a smaller deflection values than a reinforced concrete beam of the same dimensions. The formation of cracks is greatly reduced in the prestressed beam as opposed to the reinforced concrete beam. FEM analysis confirmed that the prestressed reinforcement in beams plays a major role.

As mentioned above, the number of prestressed tendons, their position and dimensions were calculated based on the beam's bearing capacity and expected life at the request of dimensional compliance with the reinforced concrete beam. FEM analysis also showed that by varying the number of prestressed ropes and their positions it is possible to create different variants of the prestressed beams according to their specific bearing and dimensional requirements.

This work is supported by VEGA grant No. 1/0795/16 "Development of effective methods for correction and optimization of coupled mechanical system" and work was supported by the Slovak Research and Development Agency under the contract No. APVV-0736-12.

References

1. I. Havran, *Prestressed concrete.* [in Slovak] (STU Bratislava, Bratislava, 2010)
2. L. Vítek, *Building materials.* [in Czech] (VUT Brno, Brno, 2012)
3. A. Vaško, J. Belan, *Fatigue tests of nodular cast iron at low and high frequency cyclic loading.* Materials Today-Proceedings **4** (5), 5985-5988 (2017)
4. P. Kopas, L. Jakubovičová, M. Vaško, M. Handrik, *Fatigue Resistance of Reinforcing Steel Bars.* Procedia Engineering **136**, 193–197 (2016)
5. G. Domek, A. Kołodziej, *Design of the tendon structure in timing belts.* Procedia Engineering **136**, 365-369 (2016)
6. R. Grega, J. Krajňák, L. Žuľová, G. Fedorko, V. Molnár, *Failure analysis of driveshaft of truck body caused by vibrations.* Engineering Failure Analysis **79**, 208-215 (2017)
7. M. Bara, T. Kmita, J. Korzekwa, *Microstructure and properties of composite coatings obtained on aluminium alloys.* Archives of Metallurgy and Materials **61** (3), 1107-1112 (2016)
8. W. Piekarska, M. Kubiak, Z. Saternus, *Numerical simulation of deformations in T-joint welded by the laser beam.* Archives of Metallurgy and Materials **58** (4) 1391-1396 (2013)
9. J. Stancek, V. Bulej, *Design of driving system for scissor lifting mechanism.* Acad. J. of Manufacturing Eng. **13**, 38-43 (2015)
10. A. Čavojcová, *Resistance enhancing of concrete constructions stressed by cyclic loading.* [Dissertation work, in Slovak], 130 p. (2016)
11. K. J. Bathe, *Finite Element Procedures.* (Prentice Hall, New Jersey, 1982)
12. *ADINA, Theory and Modelling Guide, Vol. III, CFD and FSI.* [help manual]. (2017)
13. L. Jakubovičová, M. Sága, *Computational analysis of contact stress distribution in the case of mutual stewing of roller bearing rings.* Novel Trends in Production Devices and Systems, Applied Mechanics and Materials **474**, 363-368 (2014)

14. Z. Stankovičová, V. Dekýš, P. Novák, *Time average synchronization in thermoelastic stress analysis*. Procedia Engineering **136**, 204-210 (2016)
15. P. Pastorek, P. Novák, P. Kopas, M. Močilan, *Finite element analysis of bond behavior in a steel reinforced concrete structure strengthened carbon fibre reinforced polymer (CFRP) strips*. Metalurgija **56** (3-4), 405-408 (2017)
16. M. Žmindák, P. Novák, V. Dekýš, *Analysis of bond behaviour in strengthened reinforced concrete beam with carbon fiber reinforced polymer lamella*. MATEC Web of Conferences **107** (UNSP 00045), 1-7 (2017)
17. M. Handrik, P. Kopas, V. Baniari, M. Vaško, M. Sága, *Analysis of stress and strain of fatigue specimens localised in the cross-sectional area of the gauge section testing on bi-axial fatigue machine loaded in the high-cycle fatigue region*. Procedia Engineering **177**, 516-519 (2017)
18. A. Sapietová, V. Dekýš, M. Sapieta, P. Pecháč, *Application of computational and design approaches to improve carrier stability*. Procedia Engineering **96**, 410-418 (2014)
19. B. Leitner, *The procedure of fatigue damage estimation for mechanical structures under service loading*. Transport Means 2014, 54-58 (2014)
20. R. Halama, P. Macura, L. Pečenka, F. Fojtík, M. Šofer, *Experimental Analysis of Residual Stresses in Backup Roll and FE solution*. Experimental stress analysis, 49th Int. Scient. Conf. on Experimental Stress Analysis, 85-90 (2011)

Analytical model of stress field in submerged arc welding butt joint with thorough penetration

Jerzy Winczek[1,*], *Ryszard* Parkitny[2], *Krzysztof* Makles[1], *Marcin* Kukuryk[1]

[1]Czestochowa University of Technology, 42-201 Czestochowa, Poland
[2]Katowice School of Technology, 40-555 Katowice, Poland

Abstract. Analytical model of temporary and residual stresses for butt welding with thorough penetration was described assuming planar section hypothesis and using integral equations of stress equilibrium of the bar and simple Hooke's law. In solution the effect of phase transformations (structure changes and structural strains) has been taken into account. Phase transformations during heating are limited by temperature values at the beginning and at the end of austenitic transformation, depending on chemical composition of steel while the progress of phase transformations during cooling is determined on the basis of TTT-welding diagram. Temperature values at the beginning and at the end of transformation are conditioned by the speed of heating. Kinetics of diffusional transformation is described basing on Johnson-Mehl-Avrami-Kolmogorov equation, while martensitic transformation, basing on Koistinen-Marburger equation. Stresses in elasto-plastic state are determined by iteration, using elastic solutions method with changeable longitudinal modulus of elasticity, conditioned by stress-strain curve. Computations of stress field have been conducted for one-side butt welded of two steel flats made from S235 steel. It has enabled a clear interpretation of influence of temperature field and phase transformation on stresses caused by welding using Submerged Arc Welding (SAW) method.

Keywords: modeling, welding, temperature field, phase transformations, stresses

1 Introduction

Modeling of thermomechanical states during welding requires the determination of temperature field. It is necessary to calculate the shares of structural elements taking into account their changes that occur as a result of phase transformations. Finally the thermal and structural strains enable the determination of temporary and residual stresses.

Modeling the temperature field in many technological processes requires the movement heat source taking into account. Such an approach is found in descriptions of temperature fields in welding processes [1-15], as well as in the machining processes [16-19]. In modelling of temperature field during welding dominate two approaches: analytical [1-6] and numerical [7-15].

Works refering to stress state analysis during welding, are mainly based on FEM (Finite Element Method) and experimental research. In most publications on FEM techniques for

* Corresponding author: winczek@imipkm.pcz.pl

Reviewers: *Ladislav* Écsi, *Janusz* Mielniczuk

welding, a homogeneous model of the material with the thermo-mechanical properties dependent on temperature is adopted. The effect of phase transformation in the solid state has been taken into consideration among others in [20-24].

2 Computational model of stress states

The analysis of the stress state was performed for steel rectangular elements whose dimension in the direction of welding dominates over the transverse dimensions. The stress state of the elements, for which considerations relating to the stress state of rods were used, is approximately uniaxial.

Considered element is freely supported and exposed to mechanical (thermal) loads which for particular x cross sections are characterized by the internal forces (sectional) $N = N(x)$, $M_y = M_y(x)$ and $M_z = M_z(x)$. Other sectional forces will be regarded as small (shear forces $T_y = T_y(x)$, $T_z = T_z(x)$), or not occuring ($M_x = M_x(x)$). The element is also subjected to the temperature field $T = T(x,y,z,t)$.

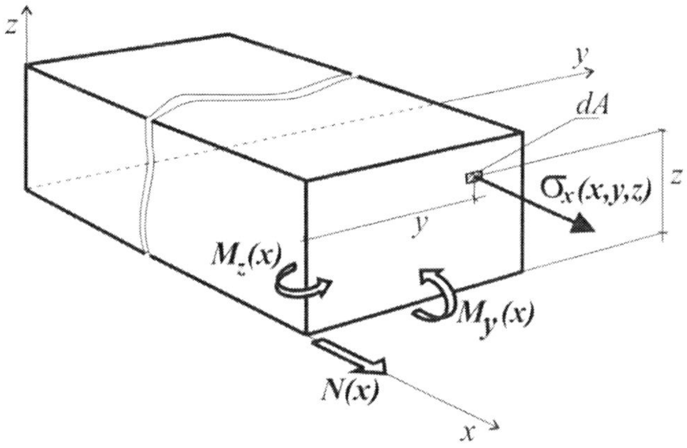

Fig. 1. Scheme of the element subjected to mechanical and thermal loads

Stress state of the element is characterized by one-dimensional stress state $\sigma_x = \sigma_x(x,y,z,t)$ (Fig. 1), which meets the requirement of the integral equations of stress equilibrium:

$$\begin{aligned} &\int_{(A)} \sigma_x dA = N \\ &\int_{(A)} \sigma_x y dA = -M_z \\ &\int_{(A)} \sigma_x z dA = -M_y \end{aligned} \tag{1}$$

The other three integral equations of equilibrium:

$$\begin{aligned} &\int_{(A)} \tau_{xy} dA = T_y \\ &\int_{(A)} \tau_{xz} y dA = T_z \\ &\int_{(A)} \left(-\tau_{xy} z + \tau_{xz} y\right) dA = M_x \end{aligned} \tag{2}$$

are omitted due to the minimal sectional forces T_y, T_z and M_x.

Cauchy relations were used and the plane cross section hypothesis was adopted in order to introduce stress formulas and element displacements. It was assumed that the cross-sections of the element are subject only to the translation and rotation and do not undergo deplanation or deformations in their plane. The assumption of cross-sectional rigidity of the element in consequence imposes the necessity of proving that linear and angular deformation in the plane are equal to zero. Poisson's ratio also adopts value zero, cross-sections do not respond to lengthening or shortening of the element longitudinal fibers. Displacements of individual cross sections are defined by three translation vectors $u_0 = u_0(x)$, $v_0 = v_0(x)$ and $w_0 = w_0(x)$ at axis x, with x, y and z directions respectively. These displacements are defined as element axial displacement and the element axis deflection in the y and z direction respectively. In the following considerations the geometrical axis of the element is closely related to the axis x. Displacement field of the points that define cross-sections of the element is described by linear dependencies, with the following restrictions:

- displacement field of the element is kinematically allowed (the strains inseparability equations are fulfilled),
- postulated element strain ε_x is small ($\varepsilon << 1$).

In the present case, in accordance with technical bendig beam theory, we do not use differential equilibrium equations but only the integral equilibrium equations and the simple Hooke's law $\sigma_x = E\varepsilon_x$, which is equivalent to accepting linear-elastic body in constitutive relations. Therefore:

$$\sigma_x = E\left(u_0' - v_0''y - w_0''z - \alpha T\right) \tag{3}$$

and the integral equilibrium conditions (1) will be written in the form:

$$\begin{aligned} &\int_{(A)} E\left(u_0' - v_0''y - w_0''z - \alpha T\right)dA = N \\ &\int_{(A)} E\left(u_0' - v_0''y - w_0''z - \alpha T\right)ydA = -M_z \\ &\int_{(A)} E\left(u_0' - v_0''y - w_0''z - \alpha T\right)zdA = -M_y \end{aligned} \tag{4}$$

Bearing in mind that element axis displacements u_0, v_0 and w_0 are solely the functions of the variable x, under integral values are symbolically saved as:

$$\begin{aligned} &A^E = \int_{(A)} EdA \quad S_y^E = \int_{(A)} EzdA \quad S_z^E = \int_{(A)} EydA \quad J_{yz}^E = \int_{(A)} EyzdA \quad J_y^E = \int_{(A)} Ez^2dA \\ &J_z^E = \int_{(A)} Ey^2dA \quad N^T = \int_{(A)} \alpha ETdA \quad M_y^T = \int_{(A)} \alpha ETzdA \quad M_z^T = \int_{(A)} \alpha ETydA \end{aligned} \tag{5}$$

for elastic modulus which is variable with respect to the coordinates (non-homogeneous element material) or temperature-dependent Young`s modulus, after solving the equations system (4), stresses defined by formula (3) will be written as:

$$\sigma_x = E\left(\left(N + N^T\right)\frac{\Delta_N}{\Delta} + \left(M_y + M_y^T\right)\frac{\Delta_{M_y}}{\Delta} + \left(-M_z + M_z^T\right)\frac{\Delta_{M_z}}{\Delta} - \alpha T\right) \tag{6}$$

where:

$$\Delta_N = J_y^E J_z^E - \left(J_{yz}^E\right)^2 + \left(S_y^E J_{yz}^E - S_z^E J_y^E\right)y + \left(S_z^E J_{yz}^E - S_y^E J_z^E\right)z \tag{7}$$

$$\Delta_{My} = S_z^E J_{yz}^E - S_y^E J_z^E + \left(S_y^E S_z^E - A^E J_{yz}^E\right)y + \left(A^E J_z^E - \left(S_z^E\right)^2\right)z \tag{8}$$

$$\Delta_{Mz} = S_y^E J_{yz}^E - S_z^E J_z^E + \left(A^E J_y^E - \left(S_y^E\right)^2\right)y + \left(S_y^E S_z^E - A^E J_{yz}^E\right)z \tag{9}$$

$$\Delta = A^E J_y^E J_z^E + 2 S_y^E S_z^E J_{yz}^E - \left(S_y^E\right)^2 J_z^E - \left(S_z^E\right)^2 J_y^E - A^E \left(J_{yz}^E\right)^2 \tag{10}$$

The stresses in elasto-plastic state are determined by iteration using the method of elastic solutions at the variable modulus of longitudinal elasticity conditioned by the stress-strain curve [25].

3 Example of computations

In the considered example it is assumed that the welded material is steel S235 with the following material properties: specific heat C_p = 670 J/(kg·K), density ρ = 7800 kg/m^3, thermal diffusivity a = 1.2 x 10^{-5} m^2/s , the temperatures of the beginning and end of austenite transformation, respectively A_1 = 996 K (723 ^{0}C) and A_3 = 1108K (835 ^{0}C), solidus temperature T_S = 1763K (1490 ^{0}C) and the liquidus temperature T_L = 1793 K (1520 ^{0}C). The model of joining two butt welded flats with length 1 m, a thickness of 0.012 m and a width of 0.1 m as shown in Fig. 2.

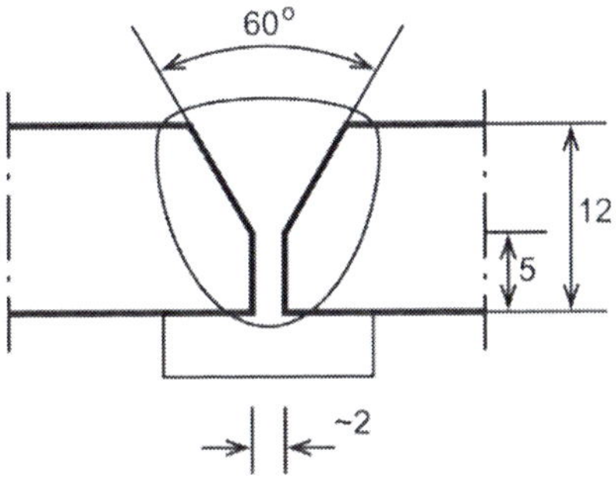

Fig. 2. Scheme of the welded joint with technology chamfering

Numerical simulations were performed with author's program created in Borland Delphi environment using Borland Pascal language. Near the control area (cross section - x = 0.5 m) calculations were performed with time step 0.05 s. Calculations with such time increase are necessary due to the high heating rates, as well as reaching maximum temperature at particular points of surfaced object at different times of the deposition process. In the anticipated areas and the heat affected zone temperature calculations were made at points in the distances of 0.5 mm, increasing the distance further from the HAZ - at the edges of the flat calculation were performed every 5 mm.

The calculations were performed assuming the speed of movement source v = 0.005 m/s, and its power P = 18000 W. Three-dimensional temporary field for butt welding with thorough penetration was described on the basis of analytical methods proposed by Geissler and Bergmann [26, 27] of an integral transformation and Green's function. The results of the temperature field calculations are presented in [28] as well as the boundaries of the heat-affected zones (fusion line, full and partial transformation zones) (Fig. 3). In the figure, the selected points of cross section were marked, for which a detailed analysis of welding thermal cycles and stresses were performed.

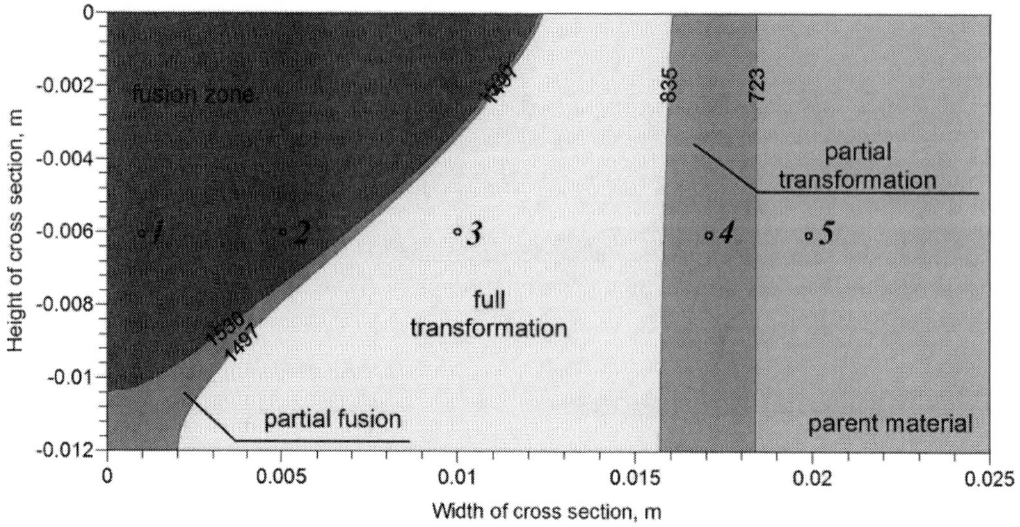

Fig. 3. Heat affected zones [25]

In calculations the effect of phase transformations (structure changes and structural strains) has been taken into account. Phase transformations during heating are limited by temperature values at the beginning and at the end of austenitic transformation, depending on chemical composition of steel while the progress of phase transformations during cooling is determined on the basis of TTT-welding diagram. Temperature values at the beginning and at the end of transformation are conditioned by the speed of heating. Kinetics of diffusional transformation is described basing on Johnson-Mehl-Avrami-Kolmogorov equation [29, 30], while martensitic transformation, basing on Koistinen-Marburger equation [31]. The calculation results of the volume fraction of the structural constituents, also thermal and structural deformations are discussed in [32]. This was the basis for the stress calculation – temporary and residual.

In strain calculations there was assumed linear expansion coefficients of particular structural elements and structural stresses (Tab. 1) determined on the basis of the dilatometric research [33, 34].

Table 1. Structural (γ) and thermal (α) expansion coefficients of phases

	α [1/°C]		γ
Austenite	$2.178 \cdot 10^{-5}$	$\gamma_{F,P,S\to A}$	$1.986 \cdot 10^{-3}$
Ferrite	$1.534 \cdot 10^{-5}$	$\gamma_{B\to A}$	$1.440 \cdot 10^{-3}$
Pearlite	$1.534 \cdot 10^{-5}$	$\gamma_{A\to F,P}$	$3.055 \cdot 10^{-3}$
Bainite	$1.171 \cdot 10^{-5}$	$\gamma_{A\to B}$	$4.0 \cdot 10^{-3}$
Martensite	$1.36 \cdot 10^{-5}$		

Tensile curves of ferrite and pearlite are assumed on the basis of works [35, 36]. In case of austenite, bainite and martensite on the basis of data from works [37, 38] tensile curves are determined according to Swift law [39].

Residual stresses distribution in the whole right half of joint cross-section has been presented in Fig. 4, while the stresses in the heat affected zone have been presented in Fig. 5. Residual stresses distribution is distinguished by high values of tensile stresses in weld and heat affected zones. In the parent material area (outside heat affected zone) compressive stress state occurs. In fusion and full transformation areas tensile stress values are determined by plastic strains. In remaining areas of parent material, stress state is determined by maintenance of stress equilibrium in the cross-section.

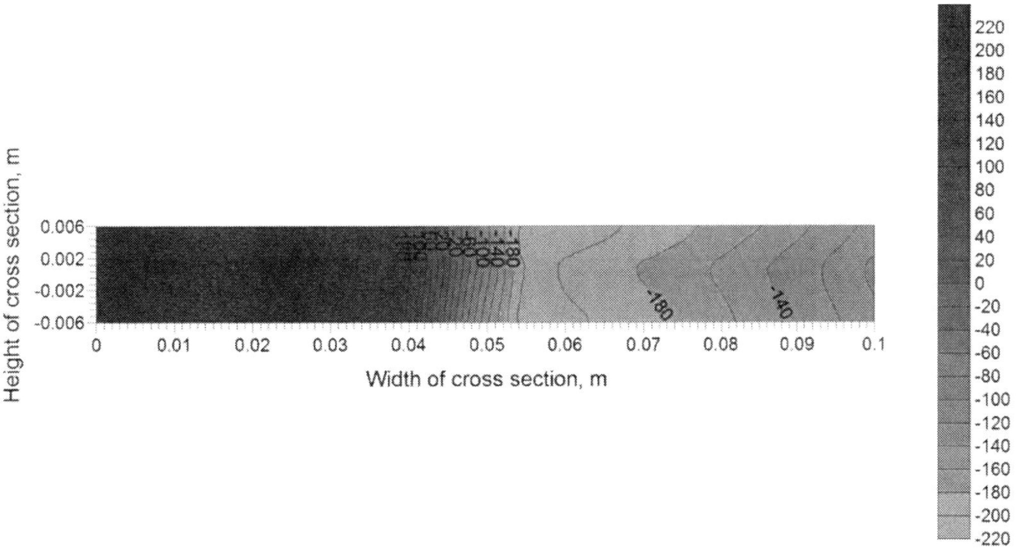

Fig. 4. Residual stress distribution in the whole right half of joint cross-section

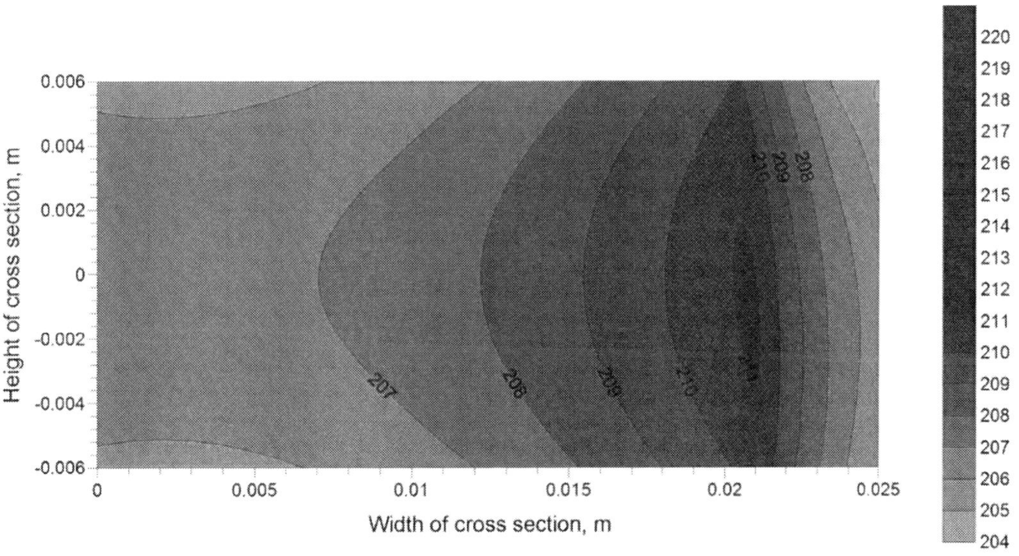

Fig. 5. Residual stress distribution in the heat affected zones

In figures 6 –11 the history of temperature and stress changes at selected points of cross section has been presented (comp. Fig. 3). At point *1* from the weld area (groove weld) (Fig. 6), stresses amount to zero as long as the point is in a state of liquid metal. After solidification during cooling tensile stresses increase. Sudden transformation of stresses into compressive is caused by phase transformation. At point *2* from fusion zone (Fig. 7) due to heating compressive stresses increase. During transformation of initial structure into austenite stresses become tensile, but then during melting they vanish. After solidification the history of stress changes is similar to that at point *1*. At point *3* (Fig. 8) from the area of full transformation due to the heating compressive stresses occur. Then, likewise at point *2*, due to transformation of initial structure into austenite, stress state changes into tensile. Next, tensile stresses, being under the influence of further temperature growth, decrease but during cooling stresses increase again. During phase transformations of overcooled austenite into ferrite and pearlite stresses plunge becoming compressive. Then during further cooling compressive stresses decrease but values of tensile stresses increase. At point *4* (Fig. 9) from the area of partial transformation during heating compressive stresses increase, but then due to austenitization transform into tensile. During transformation cooling, what causes further, much slower growth of tensile stresses. Transformations during cooling cause the decrease in values of tensile stresses and change of their sign. After phase transformations, due to further cooling, compressive stresses decrease, while tensile start to increase. At point *5* (Fig. 10) and at point *6* with coordinates (0.09, 0) - (Fig. 11) from the parent material area change of stress states results from temperature changes, changes of yield stress and from stress equilibrium in the cross-section. Compressive stresses increase during heating but decrease during cooling, becoming tensile.

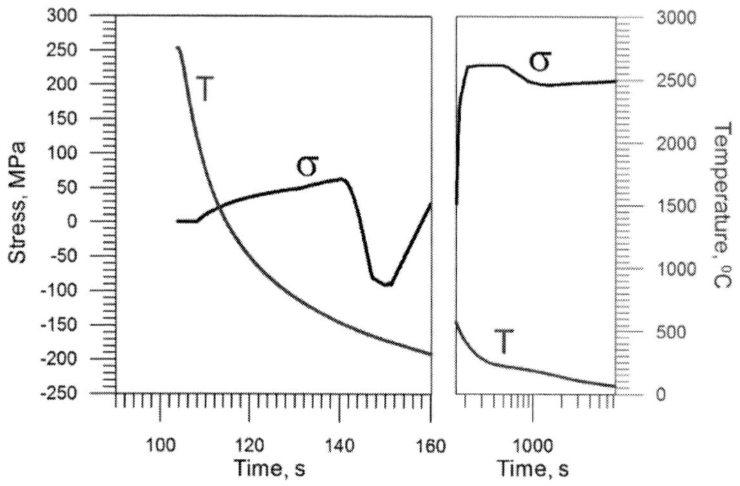

Fig. 6. Thermal cycle and stresses during welding at point 1

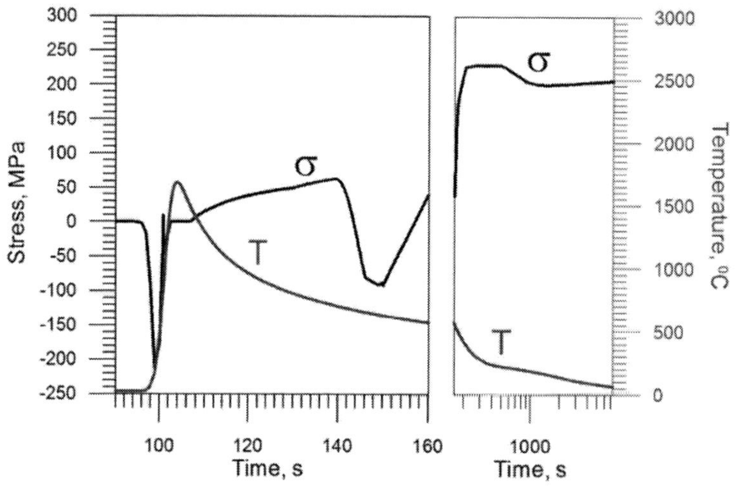

Fig. 7. Thermal cycle and stresses during welding at point 2

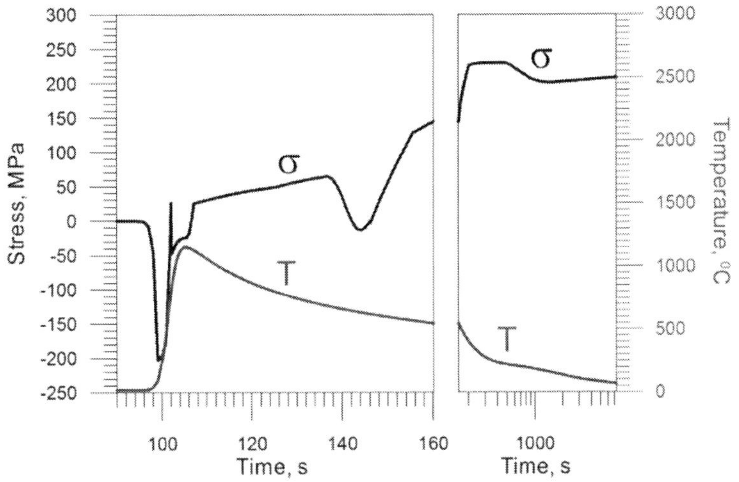

Fig. 8. Thermal cycle and stresses during welding at point 3

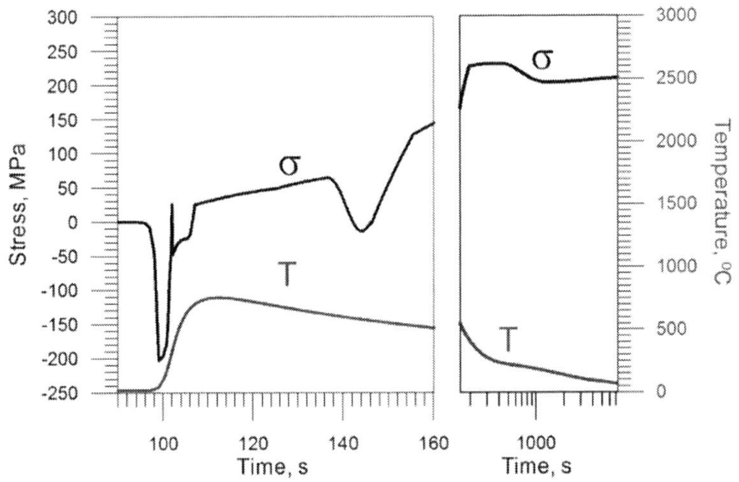

Fig. 9. Thermal cycle and stresses during welding at point 4

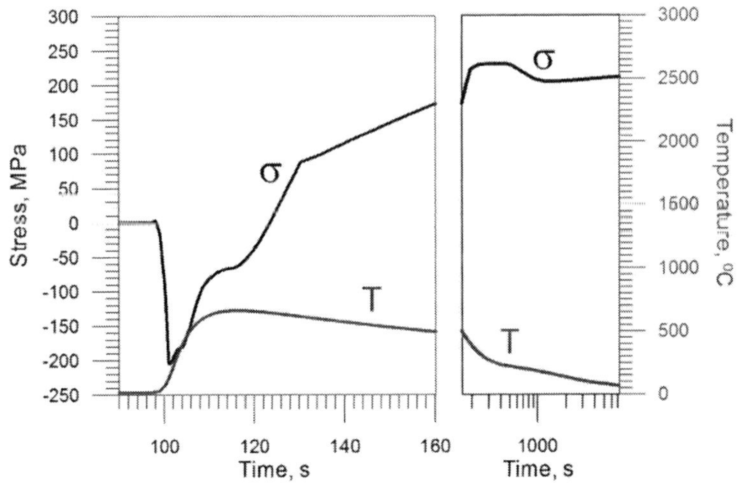

Fig. 10. Thermal cycle and stresses during welding at point 5

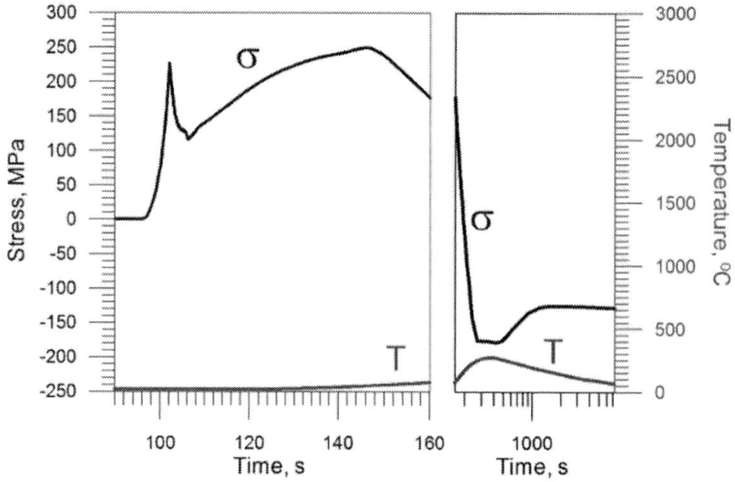

Fig. 11. Thermal cycle and stresses during welding at point 6

Conclusions

Calculated state of residual normal stresses after welding is characterised by high values of tensile stresses in weld, fusion and full transformation zones as well as by a sudden decrease of their values in partial transformation zone. Whereas in parent material areas outside heat affected zone compressive stresses occur. Correctness of such residual stress distribution was proved by experimental results presented in literature.

The obtained results of numerical simulations confirm the correctness of models and author's programs. They provide the basis for modeling of thermomechanical states during welding used in industrial practice.

References

1. R. N. S. Fassani, O. V. Trevisan, *Analytical modeling of multipass welding process with distributed heat source*. J. Braz. Soc. Mech. Sci. Eng. **25**, 302-305 (2003)
2. A. Ghosh, N. Barman; H. Chattopadhyay, S. Hloch, *A study of thermal behaviour during submerged arc welding*. Stroj. Vest. J. Mech. Eng. **59**, 333-338 (2013)
3. A. Franco, L. Romoli, A. Musacchio, *Modelling for predicting seam geometry in laser beam welding of stainless steel*. Int. J. Therm. Sci. **79**, 194-205 (2014)
4. V. D. Fachinotti, A. A. Anca, A. Cardona, *Analytical solutions of the thermal field induced by moving double-ellipsoidal and double elliptical heat sources in a semi-infinite body*. Int. J. Num. Meth. Biomech. Eng. **27**, 595-607 (2011)
5. T. Antonakakis, C. Maglioni, V. Vlachoudis, *Closed form solutions of the heat diffusion equation with Gaussian source*. Int. J. Heat Mass Transf. **62**, 314-322 (2013)
6. S. Salimi, P. Bahemmat, M. Haghpanahi, *A 3D transient analytical solution to the temperature field during dissimilar welding processes*. Int. J. Mechn. Sci. **79**, 66-74 (2014)
7. S. Joshi, C. Semetay, J. W. H. Price, H. F. Nied, *Weld-induced residual stresses in a prototype dragline cluster and comparison with design codes*. Thin Wall. Struct. **48**, 89-102 (2010)

8. J. Hildebrand, I. Starcevic, F. Werner, H. Heinemann, G. Köhler, *Numerical simulation of TIG-dressing of welded joints*. Joint Int. Conf. Comput. Decision Making Civil Building Eng., Montreal, Canada, 1487-1496 (2006)
9. M. M. Mahapatra, G. L. Datta, B. Pradhan, *Three-dimensional finite element analysis to predict the effects of shielded metal arc welding process parameters on temperature distributions and weldment zones in butt and one-sided fillet welds*. P. I. Mech. Eng. B J. Eng. **220**, 837-845 (2006)
10. S. Joshi, J. Hildebrand, A. S. Aloraier, T. Rabczuk, *Characterization of material properties and heat source parameters in welding simulation of two overlapping beads on a substrate plate*. Comp. Mater. Sci. **69**, 559-565 (2013)
11. W. Piekarska, M. Kubiak, *Three-dimensional model for numerical analysis of thermal phenomena in laser-arc hybrid welding process*. Int. J. Heat Mass Tran. **54**, 4966-4974 (2011)
12. A. Moarrefzadeh, *Numerical simulation of workpiece thermal profile in Plasma Arc Cutting (PAC) Process*. WSEAS Trans. Appl. Theor. Mech. **6** (4), 160-166 (2011)
13. Y. Chen, Y. He, H. Chen, H. Zhang, S. Che, *Effect of weave frequency and amplitude on temperature field in weaving welding process*. Int. J. Adv. Manuf. Techn. **75**, 803-813 (2014)
14. A. Sarkar, A. Datta, P. Dey, R. N. Rai, S. C. Saha *A numerical approach for modelling thermal profiles and effects of process parameters on it in submerged arc welding of AISI 1518 grade steel*. J. Therm. Eng. **1**, Special Issue 3, No 6, 505 – 516 (2015)
15. M. R. Nezamdost, M. R. Nekonie Esfahani, S. H. Hashemi, S. A. Mirbozorgi, *Investigation of temperature and residual stresses field of submerged arc welding by finite element method and experiment*. Int. J. Adv. Manuf. Technol. **87**, 615-624 (2016)
16. Z. Murčinková, K. Vasilko, *Thermo-physical aspects of chip machining*. High Temp. - High Press. **45**, 273-289 (2016)
17. O. V. Berezshnaya, E. P. Gribkov, V. D. Kuznestov, *Investigation of thermostressed state coating formation at electric contact surfacing of "Shaft" type parts*. Adv. Mater. Sci. Eng. ID 6597317, 14, http://dx.doi.org/10.1155/2016/6597317 (2016)
18. H. J. Jiang, H. L. Dai, *Effect of laser processing on three dimensional thermodynamic analysis for HSLA rectangular steel plates*. Int. J. Heat Mass Tran. **82**, 98-108 (2015)
19. S. Feng, C, Huang, J. Wang, H. Zhu, P. Yao, *An analytical model for the prediction of temperature distribution and evolution in hybrid laser-waterjet micro machining*. Precis. Eng. **47**, 33-45 (2017)
20. P. Novák; J. Meško; M. Žmindák, *Finite element implementation of multi-pass fillet weld with phase changes*. Manuf. Technol. **13** (1), 79-85 (2013)
21. T. Kik, M. Slováček, B. Wyględacz, *Numerical analysis of multipass T-joint welding and post welding heat treatment*. Weld. Technol. Rev. **88** (5), 101-106 (2016)
22. M. Patek, M. Mician, A. Sladek, D. Kadas, *Numerical analysis of T-joint welding with different welding sequences*. Manuf. Technol. **16** (1), 234-238 (2016)
23. R. Konar, M. Mician, M. Patek, D. Kadas, *Finite element modeling and numerical simulation of welding at the repair of gas pipelines with steel sleeve*. Manuf. Technol. **16** (2), 360-365 (2016)
24. R. Konar, M. Patek, M. Mician, *Experimental measurements and numerical simulation of bridge construction welding at low temperatures*. Communications: scientific letters of the University of Žilina **16** (3A), 130-135 (2014)

25. A. Bokota, R. Parkitny, *Modelling of thermal, structural and mechanical phenomena in hardening processes of steel elem*ents. A. Piela [ed.]: *Informatics in Metal Technology* (Gliwice: Silesian Univ. Technol.), Ch. **7**, 257-298 (2003)
26. E. Geissler, H. W. Bergmann, *Calculation of temperature profiles heating and quenching rates during laser processing*. Opt. Mag. **3**, 430-434 (1987)
27. E. Geissler, H. W. Bergmann, *3D Temperature Fields in Laser Transformation Hardening, Part I: Quasi-Stationary Fields*. Opt. Mag. **4**, 396-403 (1988)
28. J. Winczek, R. Parkitny, *Modelling of heat affected zone in submerged arc welding butt joint with thorough penetration*, Procedia Engineering **177C**, 241-246 (2017)
29. W. Piekarska, M. Kubiak, M. Žmindak, *Issues in numerical modeling of phase transformations in welded joints*. Procedia Engineering **177**, 141-148 (2017)
30. W. Piekarska, D. Goszczyńska-Króliszewska, *Analytical methods of predicting the structure and mechanical properties of high tensile strength steel*. Procedia Engineering **177**, 92-98 (2017)
31. T. Domański, A, Bokota, *Numerical models of hardening phenomena of tools steel base on the TTT and CCT diagrams*. Arch. Metal. Mater. **56**, 325-344 (2011)
32. J. Winczek, *The analysis of temporary temperature field and phase transformations in one-side butt-welded steel flats*, Chapter 6 in: *Joining Technologies*, M. Ishak [ed.], Intech, Rijeka, Croatia 103-130 (2016)
33. J. Winczek, A. Kulawik, *Dilatometric and hardness analysis of C45 steel tempering with different heating-up rates*. Metalurgija **51**, 9-12 (2012)
34. J. Winczek, K. Makles, M. Gucwa, R. Gnatowska, M. Hatala, *Modelling of strains during SAW surfacing taking into heat of the weld in temperature field description and phase transformations*. IOP Conf. Series: Materials Science and Engineering **225**, 012038 doi:10.1088/1757-899X/225/1/012038 (2017)
35. J. Gawąd, D. Szeliga, A. Bator, V. Pidvysockyy, M. Pietrzyk, *Interpretation of the tensile test results interpretation based on two criterion optimization*. Proc. 14. Conf. KomPlasTech, *Informatics in Metal Technology*, M. Pietrzyk et al. [ed.], Akapit, Cracow, 27-34 (2004)
36. P. M. M. Vila Real, R. Cazeli, L. Simoes da Silva, A. Santiago, P. Piloto, *The effect of residual stresses in the lateral-torsional buckling of steel I-beams at elevated temperature*. J. Construct. Steel Research **60**, 783-793 (2004)
37. M. Melander, *A Computional and Experimental Investigation of Induction and Laser Hardening*. Linkoping Studies Sci. Technol. **124**, Linkoping Univeristy (1985)
38. J. Lian, Z. Jiang, J. Liu, *Theoretical model for the tensile work hardening behaviour of dual-phase steel.* Mater. Sci. Eng. **A147,** 55-65 (1991)
39. S. K. Kim, Y. M. Kim, Y. J. Lim, N. J. Kim, *Relationship between yield ratio and the material constants of the swift equation*. Metals Materials Int. **12** (2), 131-135 (2006)

Identifying the cause of cutting tool failure by using simulation software

Andrej Zrak[1,*], *Radoslav* Koňár[1]

[1]Univeristy of Žilina, Faculty of Mechanical Engineering, Univerzitná 1, 010 26, Žilina

Abstract. The article is focused on the construction problems of cutting tools designed for manufacturing components in high numbers. Analysis of the tool with low lifetime is provided with the use of simulation software to detect the causes of the tool failure. Parameters of the simulation are close to the real conditions of the tool in production. The results from the simulation indicate that the bending moment occurs in the tool because of forces which have opposite direction on different areas. On these basis technological adjustments are made. These adjustments are also analysed by using simulation software.

Keywords: cutting tool, stress analysis, failure of cutting tool, position of energy force

1 Introduction

Forming technology is mostly applied in manufacturing parts that are made in high quantity. The reason for applying this technology is its low time consumption. The manufacturing process takes only few seconds. On the other hand, when producing low quantities, forming technology is not effective because of the high costs when manufacturing the forming tool or tools.

From the point of manufacturing and renovation costs of the forming tools, its lifetime in production environment is essential. This article describes the analyses of the tool, which was designed for progressive forming of components. The number of manufactured parts without the need to repair the tool was very low (sometimes 200 pcs. and the assumed number when designing the tool was approximately 5000 pcs.)

2 Analysis of the cutting tool

The object used for analyses is called progressive forming tool which is used for cutting components from steel S235 JR (thickness $t = 4$ mm). This tool was examined (Fig. 1) because of its low lifetime during production. The functional parts of the tool are made out of tool steel 50CrMoV1314, heat treated to 62 HRC. Fixing parts of the tool are made from steel E355. In one cutting cycle the tool cuts two identical parts for better effective use of the material.

* Corresponding author: andrejzrak@gmail.com

Reviewers: *Piotr* Krawiec, *Tadeusz* Zloto

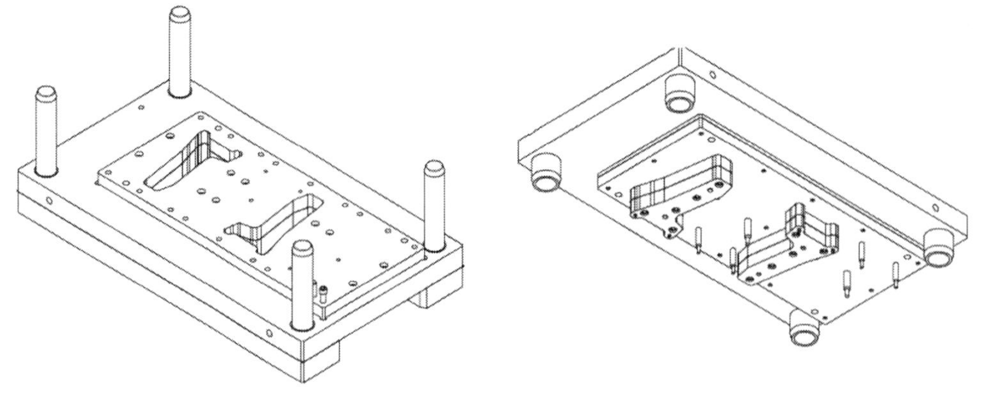

a) b)

Fig. 1. Analysed cutting tool. a) bottom part, b) upper part

Design of the tool is characterized by the different high of the punches. Half of the punches are about 4 mm higher (thickness of the steel sheet). This means that the cutting process proceeds in two phases. At first the material is cut by the left side of the punches and then by the right. Fig. 2 describes the differences of maximal cutting force when cutting by using all punches at the same time (Fmax1), and when cutting by using different high of the punches (Fmax). At different heights the cutting force is not so high and the dynamic strain on the machine and tool is lower. Cutting energy (integrated area under the cutting force curve) is for cutting with the identical height of the punches theoretically equal to the total of cutting energy for classes of punchers with the same height [8, 10, 11].

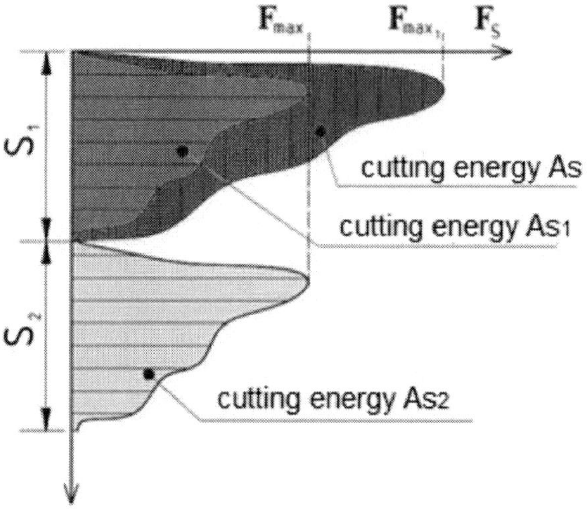

Fig. 2. Description of cutting force

Simulation of the authentic tool was carried out in Autodesk Inventor Professional software. This simulation corresponds to the realistic conditions in praxis. The force effect initiated by the press machine was located on the upper plate of the tool. Force from the press was 1300 kN for one set of punchers. This force replacement is sufficient enough, because the machine which affects the tool by force is not enough precise and stiff to ensure force effects to prevent the plate from moving in the plane direction perpendicular to the vector of force action.

Free movement in the plane which is perpendicular to the vector of force action was verified by using camera system. The magnitude of force effect was estimated on the basis of theoretical calculation of cutting force for selected tool. Measuring of realistic force effect was not provided for the lack of strain gauges containing necessary measuring scale to detect the force effect (2600 kN). Fig. 3 describes the output of simulation focused on the size of movements from each parts of the tool during cutting by the first assembly of punchers with bigger length [2, 5, 7, 9].

According to the analysis of the model it is possible to assume, that the force effect from the machine on the tool causes a bending moment. This moment is responsible for radial fatigue of the guiding elements and causes deviation of the upper part of the tool due to the lower part.

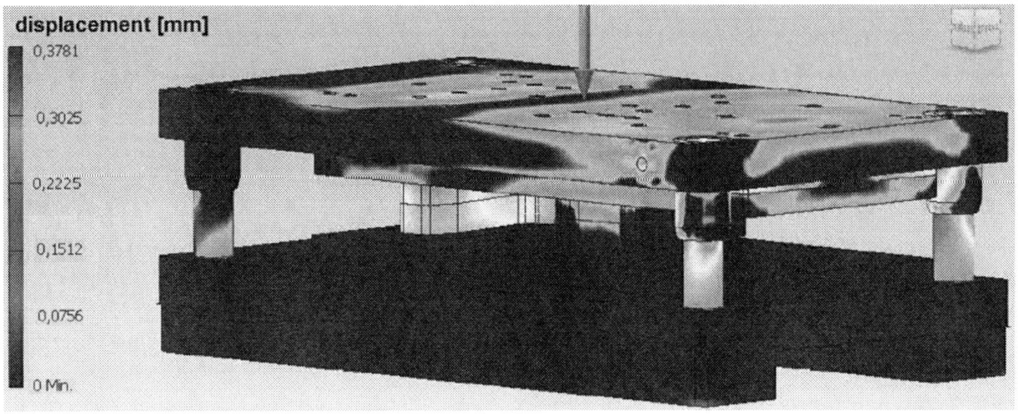

Fig. 3. The magnitude of each part movement of the tool during cutting by using the first half of the punches

Fig. 4 describes the model of cutting with displayed clearance. Clearance is a parameter related to the properties and thickness of the cut material. In this case the applied clearance has a value of 0.2 mm. If we take into considering that the functional parts of the tool move in the process of cutting 0.2 mm from each other, their misalignment is from 0.2 to 0.3 mm. The result is that the clearance is reduced to zero value on one side and doubled on the other side of the puncher. In worst case scenario the two punchers collide. This can be also observed from the result of simulation. Collision of punchers with hardness 62 HRC causes the abruption of cutting edge and the tool is then not sharp (one of the causes of tool damage from service data). Continuing wear process of the tool increases exponentially. Wear o the tool is excessive also in the case that the clearance is not optimal. By using simulation technologies there can be a chance to predict the state of zero clearance and to avoid the negative influence on the tools lifetime. Wrong clearance has also negative influence on the quality of the steel blanks. Analysis of the blanks which have been cut by using this tool had proved this assumption.

Result from the simulation (Fig. 3) represents the state during cutting by using the right set of punchers. After cutting this part the left side of the set cuts the rest. The strain of guiding elements is equal with opposite vector of orientation. From determined facts it is possible to state that the upper part of the tool during one cycle moves approximately 0.2-0.3 mm to the right side and then 0.4 – 0.6 mm to the left. This movement causes high wear of the guiding elements and increases the size of shifting in the next cycles [4, 6, 9].

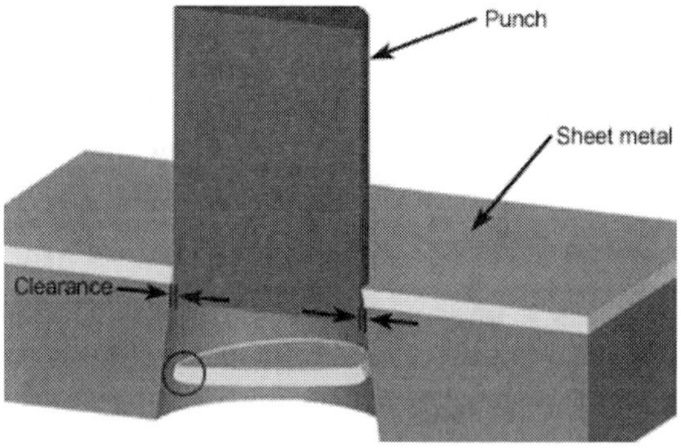

Fig. 4. Cutting scheme - clearance [1]

3 Proposition of steps leading to the increase of cutting tool lifetime

On the basis of determined facts two steps of increasing the lifetime of the cutting tool were proposed.

3.1 Cutting with the same height of the punchers

Main disadvantage, when cutting where the punchers have identical height, is that there is a need to apply bigger forces on the tool to make the cut. This force effect changes its character non-uniformly and creates shock stress of the tool. Its advantage is that the puncher does not initiate bending moment, which influences the tool because the force effect of the machine and the counter effect of the table is in one axis. Fig. 5 represents the result of simulation focused on the movement of tool parts during cutting with the same height of the punchers at two times bigger load (2600 kN) [5, 7, 9].

Fig. 5. Result from the simulation of cutting with same height of the punchers

Simulation proves the assumption that this way of cutting does not initiate force effect in the guiding elements of the tool and does not cause the misalignment of punchers from the designed motion trajectory. From detail analysis of the most strained parts, none of the areas reach the yielding strain of the material. That means the designed method is safe, preventing the tool to be damaged. Of course by using this method there is a need for a press, which can produce a force load minimally 30 000 kN for the tool.

3.2 Cutting with identical height of the punchers including angle change of the edge

Origin conception of the tool was manufactured by using the most common used method for decreasing the maximal cutting force. By using this method, it is very important to insure, that the resultant force effect is always in the middle of the force effect of the used press. In this case, from the conception of the used press, it is not possible to reach this method. That means there had to be proposed another solution [2, 5, 7].

Next way how to decrease the maximal cutting force and to secure the resultant force effect in the middle is to grind the punchers under defined angle (Fig. 6). This modification of punchers causes the cut to be executed sequentially, the force effect is not on the 4 mm path (thickness of the material) but on 8 mm path. If we take into consider that the cutting energy stays approximately the same (increased by the energy, which is consumed for the deformation of the cut material), it can be possible to state that, at tools which have symmetrical geometry, the maximal cutting force in this case is approximately half of the original. Disadvantage of the proposed method of cutting is that the blank does not have planar character, but is slightly deformed to a cylinder shape. Therefore it is necessary to include a straightening operation.

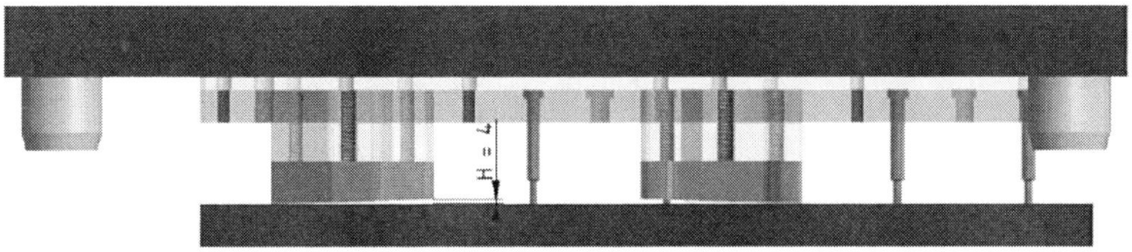

Fig. 6. Angle grinding of the punchers

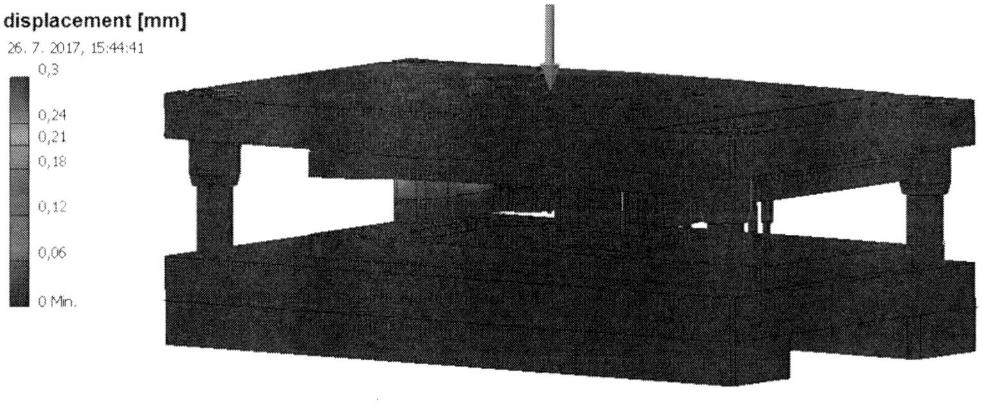

Fig. 7. Result from the simulation of punchers that were grinded under defined angle

From the analysis it is possible to state that the designed conception does not initiate movement in the radial direction on the force vector of the press(fig. 7). Here can be seen that the stated modification of the puncher induces small radial stress. This stress is caused because of the leaning cutting surface. At dimensions of this puncher the subscribed method does not negatively influence the operation of the tool and does not cause damage or abnormal misalignment.

Conclusion

The article contains analysis of a tool which is used in technical praxis to manufacture parts by using progressive cutting. The reason for analysis of this tool was that its lifetime was too short. According to the simulation the tool was designed in a way that during cutting the press machine acts on the tool by force which induces bending moment and this moment causes radial strain on the guiding elements. The tool in its origin conception would be possible to apply on a press which would contain movable parts with high stiffness and accuracy. The next option would be to change the guiding elements of the tool for stronger. However, from calculation of this option the size of guiding elements exceeded the dimensions that would be suitable for mounting to the tool. For effective fixture of this issue two changes of the design of the tool were proposed, without dramatic modification of the whole construction.

First method was to cut the openings synchronously where the radial strain of the guiding elements of the tool would be eliminated. On the other hand, the cutting force would be doubled. By using this method there is a need to use a press machine with higher force load.

Unfavourable is also the shock strain applied on the tool and machine. The second method is that the surface of the punchers would be grinned in a defined angle and would cut synchronously. Forming punchers with big dimensions can't cut on a 4 mm path (thickness of the material) but on an 8 mm path. This method decreases the maximal cutting force and does not cause radial strain of the guiding elements of the tool. The disadvantage of this method is that the blanks are deformed and therefore it is necessary to straighten them.

This article points out the importance of applying numerical simulations into designing forming tools before their manufacture. It is necessary to remark that simulations which were used to analyse the tool contained in this article are only basic analysis focused on strength and movement of the tool. However, these analyses provide important information about detecting the causes and prevention of failure or damage of the tool before putting into production.

This article was created with the support of project KEGA – 034ZU-4/2015.

References

1. *Custom PartNet*. [online] http://www.custompartnet.com (2009)
2. M. Žmindák, J. Meško, Z. Pelagić, A. Zrak, *Finite element analysis of crack growth in pipelines*. Manufacturing technology: Journal for science, research and production **14** (1), 116-122 (2014)
3. P. Novák, J. Meško, M. Žmindák, *Finite element implementation of multi-pass fillet weld with phase changes*. Manufacturing technology : journal for science, research and production **13** (1), 79-85 (2013)
4. M. Brezničan, P. Fabian, J. Meško, M. Drbúl, *The simulation of influence of quenching temperature on properties of bearing rings*. Manufacturing technology: journal for science, research and production **13** (1), 20-25 (2013)
5. R. Koňár, M. Mičian, M. Patek, D. *Kadáš, Finite element modeling and numerical simulation of welding at the repair of gas pipelines with steel sleeve*. Manufacturing Technology: journal for science, research and production **16** (2), 360-365 (2016)
6. J. Moravec, *Increase of the operating life of active parts of cold-moulding tools*. Technički vjesnik = Technical gazette **24** (1), 143-146 (2017)

7. S. Rózowicz, S. Tofil, A. Zrak, *An analysis of the microstructure, acrostructure and microhardness of NiCr-Ir joints produced by laser welding with and without preheat*. Archives of metallurgy and materials **61** (2b), 1157-1162 (2016)
8. M. Patek, M. Mičian, A. Sládek, D. Kadáš, *Numerical analysis of T-joint Welding with different welding sequences*. Manufacturing Technology: journal for science, research and production **16** (1), 234-238 (2016)
9. M. Handrik, P. Kopas, V. Baniari, M. Vaško, M. Sága, *Analysis of stress and strain of fatigue specimens localised in the cross-sectional area of the gauge section testing on bi-axial fatigue machine loaded in the high-cycle fatigue region*. Procedia Engineering **177**, 516-519 (2017)
10. P. Jankejech, P. Fabian, J. Bronček, Y. Shalapko, *Influence of tempering on mechanical properties of induction bents below 540°C*. Acta Mechanica et Automatica **10** (2), 81-86 (2016)
11. E. Krivoš, R. Pastirčák, R. Madaj, *Effect of technological parameters on the quality and dimensional accuracy of castings manufactured by patternless process technology*. Archives of metallurgy and materials **59** (3), 1069-1072 (2014)

Simulation of sound transmission through the porous material, determining the parameters of acoustic absorption and sound reduction

Peter Zvolenský[1,*], *Juraj* Grenčík[1], *Ľubomír* Kašiar[1], *Vladimír* Stuchlý[1]

[1]Department of Transport and Handling Machines, Faculty of Mechanical Engineering, University of Zilina, Slovak Republic, Univerzitná 8215/1 010 26 Zilina, Slovak Republic

Abstract. Currently, the quality of structural design of a railway coach is evaluated by so called acoustic comfort, which is characterized by achieved levels of internal noise. Therefore, acoustic parameters of car body are being developed purposely. The paper presents the results of the computer simulation of noise transmission through the wagon walls and the use of noise tests from the train running. The acoustic properties of the original and new materials in the care body are compared.

Keywords: simulation of noise transmission, noise of railway vehicles

1 Introduction

Currently in the EU, increased attention is paid to reducing rail noise, so all EU Member States are doing research in the area of noise reduction. In particular, technical solutions in the field of railway vehicle design, track structure and technological aspects of rail traffic can effectively reduce the noise emitted by the operation of a railway vehicle. Possible approaches in this area can be defined mainly in the following aspects:

- Investigation of noise sources and vibration sources and noise propagation pathways,
- Simulation and modelling of the noise fields generation and their impact on the surroundings and passengers,
- Verification of theory and computer modelling by experimental research in traffic conditions,
- Evaluation and development of noise reduction procedures and methods applied on the design of relevant railway vehicles.

The reduction of railway noise requires the research of the above-mentioned interconnected areas as components of the system. In railway traffic, we meet with two ways of noise reduction:

- Urgent acute noise abatement usually resulting from measured noise exposure situations detected mainly by state supervision, e.g. noise affecting residential areas, or passengers in the interior of a railway vehicle,

* Corresponding author: peter.zvolensky@fstroj.uniza.sk

Reviewers: *Jerzy* Jaroszewicz, *Josef* Soukup

- The conceptual task required by legislation or environmental policy of the EU, or individual states, e.g. concept of formation and international transit of "silent trains".

Often the modern rolling stock designs in terms of assessment and noise emissions are mostly at the limit of existing possibilities. Also, the approaches of the secondary solution (acoustic barriers, car body design, aerodynamics of the shape of a railway vehicle body, the use of modern materials) are relatively well worked out. Different design and interior materials also have well-defined parameters of airborne sound and sound absorption. At the same time, the use of modern materials and technologies that effectively promote recycling is of particular importance today. The rapid transfer of scientific knowledge and the results of experimental research in the field of noise emissions from rolling stock into transport practice bring new opportunities, tools and approaches to the effective improvement of the quality of the environment [1, 2, 3, 4].

2 Application of porous acoustic materials in simulation of acoustic energy transfer through the structure of a rail passenger wagon floor

Research that is associated with the acoustic properties of porous materials needs to be applied in practice. From the point of view of the objective assessment of the noise emissions generated by the rolling stock, it is necessary to identify the origin and sources of noise on a railway passenger car and also the type and pathways of spreading acoustic energy into the structure and the surrounding area. This space represents not only the transport communication and its immediate surroundings, but also the interior of a vehicle, where the passengers are exposed to the noise. In the general interest, it is necessary to address the possibilities, the ways and the new technologies for reducing the noise in the interior of the means of transport, not only in the automotive industry, which is the major leader in the research of new technologies for the production and application of acoustic materials, but also on the railway. It is an integral part of the business strategy of manufacturers and operators of the means of transport in general. By this the comfort of transport is being improved and it directly affects the final customer, that is, the passenger when choosing a suitable means of transport.

At present, numerous modern technical and technological solutions for noise reduction are being used in the development of new vehicle designs. One of the frequent trends is the research and development of new acoustic materials and the way they are used.

There is a wide range of modern thermal and sound-insulating materials on the market. These materials are used in the design of new rolling stock. In recent years, ZSSK (Railway Company Slovakia) has been modernizing its fleet and thus offering a good opportunity to improve the acoustic properties not only of the floor but also of the entire car body in the modernization of passenger railway wagon. Therefore we have also focused our research on the design the Bdghmeer railway wagon. In the design of the floor of the Bdghmeer passenger railway wagon, mineral wool is currently used as a heat-insulating and sound-insulating material. During the modernization only the old insulation layer is replaced by the new one. The research at the Department of transport and handling machines was also focused on the design of new acoustic and thermal insulation materials in the wagon floor structure.

The simulation processes of sound propagation in liquids and solids in simulation tools are based on several mathematical models. From the point of view of the requirements of the airborne sound reduction of the floor structure of a passenger railway wagon, mineral wool and newly designed STERED® material has important role. In both cases, these are porous materials with a fibrous microstructure. The acoustic energy is predominantly converted to thermal energy, while the sound propagation in the porous structures is determined by the ability of the air to propagate through the pores of the microstructure and also by vibrations that propagate through the solid structure of the fibrous material, in this case the fibres. The sound waves propagate through the fibres. An example of the flow of fluid (air) through the porous structure is shown in Figure 1.

Fig. 1. Transfer of fluid through the structure of the porous medium [5]

A scientific discipline that deals with propagation of sound in porous substances is called "porous acoustics or acoustics of porous materials". By defining appropriate parameters for a porous material, the simulation program is able to create a model of the porous structure of the material and use it to simulate the transmission of acoustic energy through this structure. From the research of the acoustic properties of porous materials it implies that the most significant effect on acoustic parameters such as sound absorption and airborne sound reduction has the so-called viscous characteristic length *Lv*. It is a parameter that determines the mean value of the macroscopic dimensions of the channels that bind the individual pores of the material structure. Simulations within the investigation of the acoustic parameters of the wall of the monitored passenger wagon were mainly focused on determining the sound reduction and sound absorption of the original solution using the mineral wool and verifying these acoustic properties using the new progressive material STERED®.

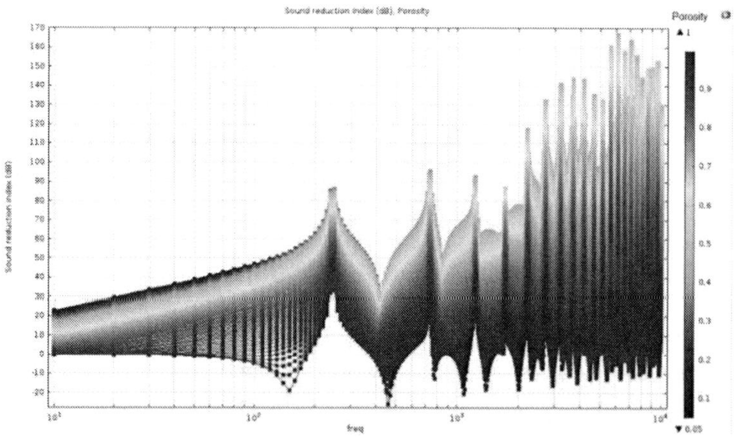

Fig. 2. Frequency dependence of airborne sound reduction index on the porosity of fibrous material

From the simulations obtained, original 3D charts have been created that declare the properties, which are damping capabilities of the investigated structure and materials used. In the figure 2 is an example of simulation for determining the sound reduction of the porous material.

3 Simulation of acoustic energy transfer through porous media

The role of simulation tools is to approximate as much as possible the actual physical actions which are constantly ongoing in nature. Simulation software works on the basis of mathematical models (Fig. 3), which were mostly derived from empirical relationships obtained on the base of sufficient number of relevant experiments.

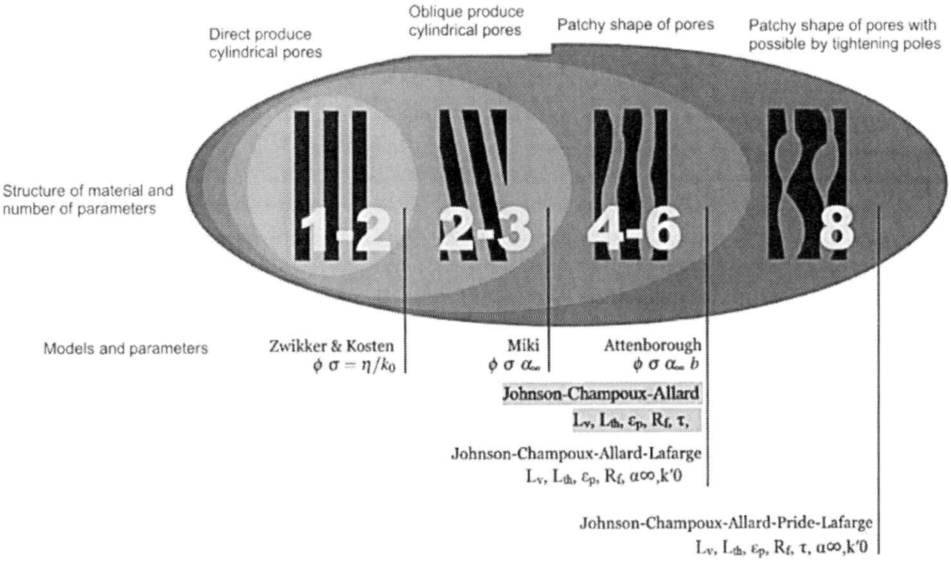

Fig. 3. The most used mathematical models and parameters necessary for the formation of a porous structure [5]

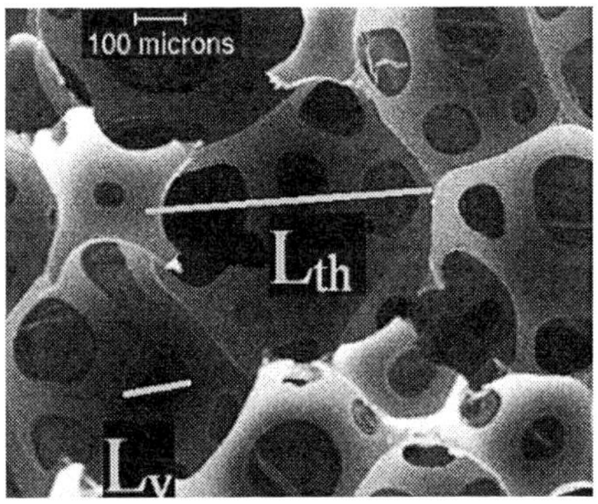

Fig. 4. Thermal and viscous characteristic pores length - from the microscopic image of polyurethane foam [6]

The simulations use different mathematical models on which the calculation and simulation of acoustic energy transfer through the porous structure runs (Fig. 4) [7].

Thermal and viscous characteristic length directly determines the texture, shape and arrangement of pores, or fibre porous structure. The equation for their calculation is as follows:

$$L_v = \sqrt{\frac{8\mu\tau}{\varepsilon_p R_f}} \qquad [m], \tag{1}$$

where:
Lv - viscous characteristic length,
μ - dynamic viscosity of air, which is in the pores,
εp - porosity,
Rf - airflow resistance.,
τ - curvature pores impact factor (Fig. 5).

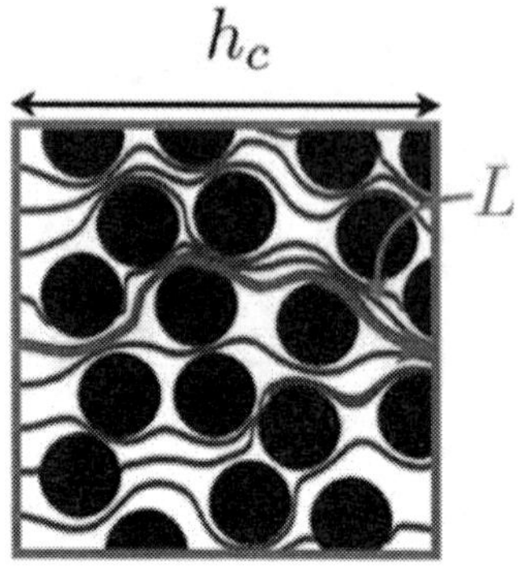

Fig. 5. Curvature pores impact factor [8]

Curvature pores impact factor can be calculated as:

$$\tau = \frac{L}{h_c} \qquad [-] \tag{2}$$

where:
L - sound pressure level,
hc - board thickness.

The empirical formula for calculating of the air flow resistance is as follows:

$$R_f = \frac{3{,}2\mu(1-\varepsilon_p)^{1{,}42}}{a^2} \qquad \left[Pa \cdot s \cdot m^{-2}\right] \tag{3}$$

where:
a - the mean fibre diameter.

Curvature pores impact factor of the fibre structure can be calculated according to the empirical formula:

$$\tau = \frac{1}{\varepsilon_p^{\frac{1}{2}}} \tag{4}$$

or it can be estimated from the detailed study of the microstructure microscopic image in the following manner. By substituting these equations into the equation 1 we obtain equation in this form:

$$L_v = \sqrt{\frac{8\mu\varepsilon_p^{\left(-\frac{1}{4}\right)}}{\varepsilon_p \dfrac{3{,}2\mu(1-\varepsilon_p)^{1{,}42}}{a^2}}} \qquad [m] \tag{5}$$

Research on acoustic properties of porous materials showed that the most significant effect on the acoustic parameters, such as sound absorption and airborne sound insulation has a viscous characteristic length *Lv*. It is a parameter which determines the mean value of dimensions macro-channels, which connect the individual pores of the material structure. Here rise the viscous losses of sound energy, when air pressure changes permeating from one pore to another through a channel that connects these pores. Parameter called thermal characteristic length Lth is derived from the viscous characteristic length. It indicates the mean macroscopic pore size of the porosity structure [9].

Mineral wool and STERED® are porous fibrous materials. Decisive influence on the acoustic properties of porous materials has just the porosity. The porosity influence of the fibrous material on the acoustic properties has been studied by using of the simulated computation by the change of the porosity in a range from 5 to 99.5% by 0.5% and the frequency from 10Hz to 10 kHz by 10 Hz.

Model presents a principle of the impedance tube (Fig. 6) and the inlet signal is defined by function in complex form. [9-11]

$$1\exp^{(-i(kx \cdot x + ky \cdot y))} \quad [Pa] \tag{6}$$

where:

x - position vector of given environment oscillator in direction of application of x-axis,
y - position vector of given environment oscillator in direction of application of y-axis,
kx - wave vector in the x-direction of sound waves propagation,
ky - wave vector in the y-direction of sound waves propagation.

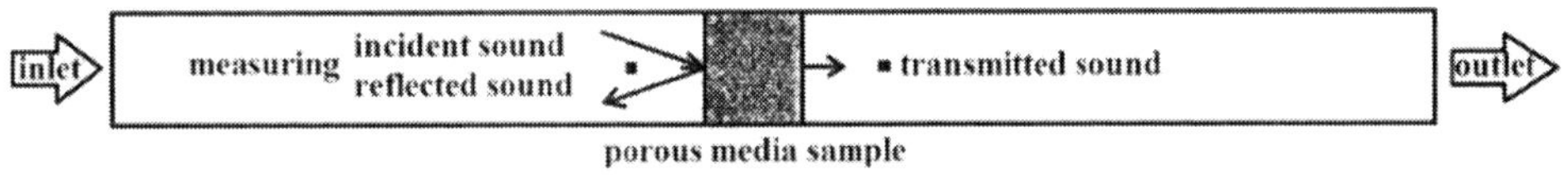

Fig. 6. Model of simulation of acoustics energy transfer through the porous media

The influence of material porosity on the sound reduction index and sound absorption coefficient is better observable on the graph, which was created by the separation of the specific frequencies (Fig. 7). The monitored characteristics of acoustic parameters at the frequency of 1 kHz are given as an example (Fig. 8).

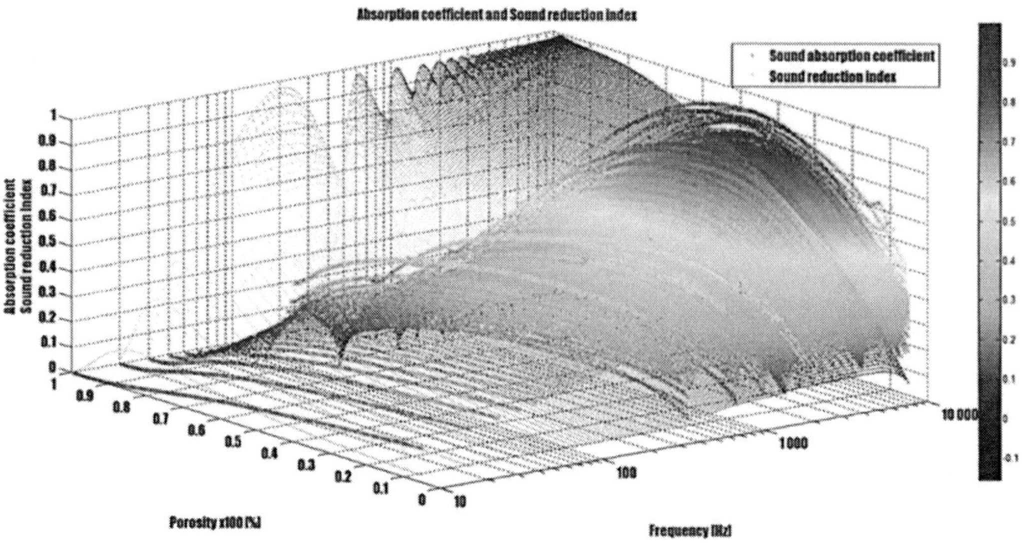

Fig. 7. Frequency dependence of sound absorption coefficient and sound reduction index on the material porosity

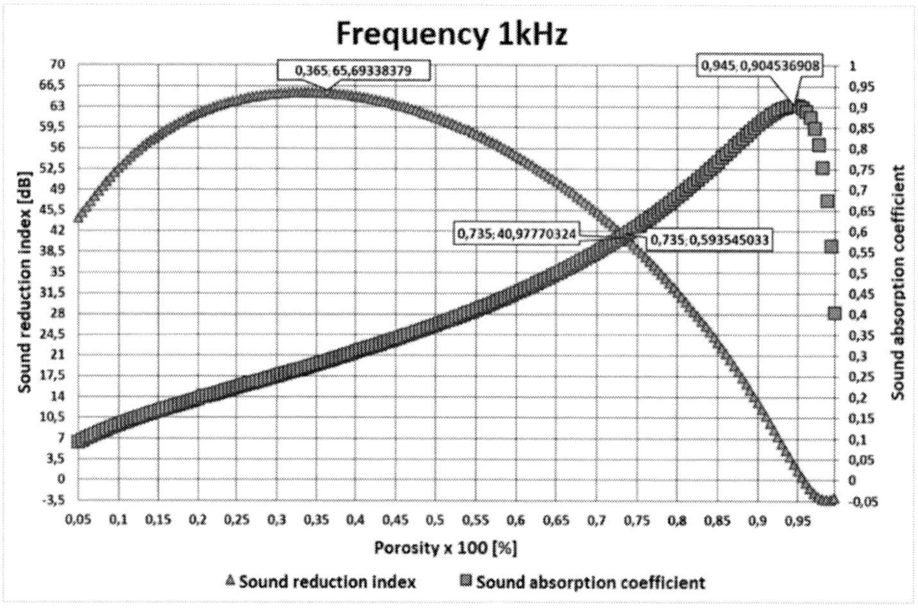

Fig. 8. The characteristic of sound reduction index and sound absorption coefficient depending on the material porosity in frequency 1 kHz

Model for investigation of the acoustic energy transfer through the porous material was created within the simulation. Related percent porosity of the material, its sound absorption and airborne sound insulation depending on the frequency of transferring acoustic energy was monitored. As an output example is presented the dependence at the frequency of 1 kHz, which is the best audible by human ear and so the simulation results of effects of noise on human body can be taken relatively objective.

The studied relationship is based on energy balance transfer of acoustic energy through the porous microstructure of fibrous material (Fig. 9).

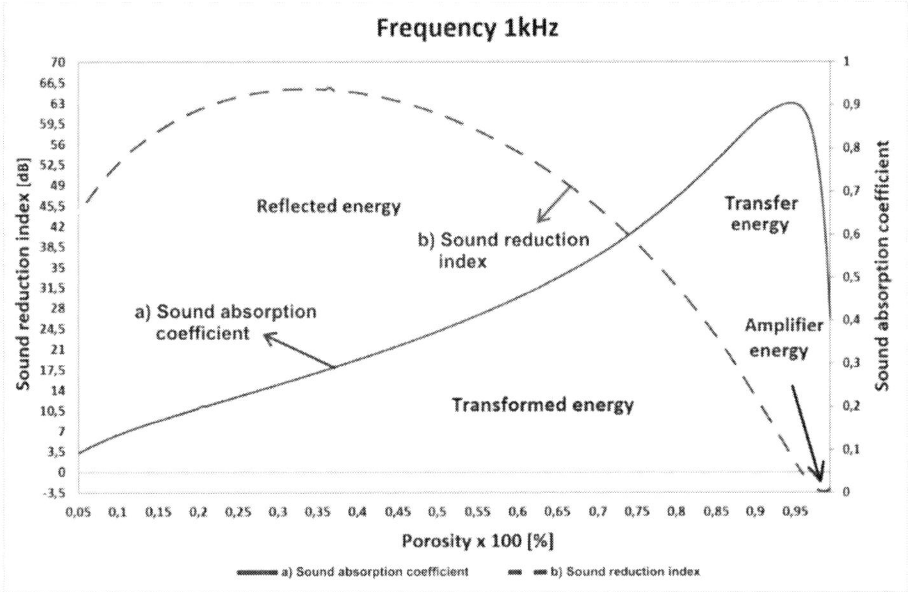

Fig. 9. Energy balance of the acoustic energy transfer through porous fibrous microstructure

The mathematical model can create a geometrical model of the fibrous material microstructure, which in principle works as a breaker of acoustic waves. The influence of the material porosity is important and therefore porosity measurement of mineral wool, material STERED® 200 and STERED® 250 (Fig. 8) was performed, by using direct porosity measurement method. This method is suitable only for very regular-shaped test samples. The measured samples had to be saturated in the water under vacuum. The volume of pores was determined on the base of the water bulk density, which was in the pores of porous materials [12, 13].

4 Measurement of internal and external noise of a passenger railway wagon

Due to the need to perform simulation calculations of acoustic energy transmission through the structure of the floor of a passenger railway wagon and verifying the suitability of using the proposed STERED® acoustic material in this design, it was necessary to measure the internal noise and external noise that is acting in the space of running gear during train run and use the data obtained as an input for simulation calculations. Measurements were carried out on the modernized railway passenger wagon series Bdghmeer no. 61 56 28-70 013-5 during the run on the selected section Prievidza-Leopoldov and Leopoldov-Bratislava track corridor. The passenger rail wagon was placed at the end of the train and locked for experimental purposes. The Bdghmeer four-axle wagon is designed for the transport of passengers and luggage and is also adapted for the transport of immobile passengers. Measurement was performed using the Pulse 3560 B measurement system from Brüel & Kjær. Three measuring microphones Brüel & Kjær, type 4188, were used for the measurements. Signal data for frequency analyses and sound pressure levels from three microphones were recorded during the measurement. The time of one measurement was 60 seconds for both measured quantities.

Measuring microphone no. 1 was placed in the exterior just above the ground and fixed to the floor frame structure to detect noise emission levels and frequency noise analysis of the source, the GP 200 bogie, during running at different speeds on the selected track section. Measuring microphone no. 2 was located in the disabled section in the rear of the wagon in the direction of travel, directly above the bogie. Data recorded by measuring microphone no. 1 and no. 2 could be compared immediately to determine the acoustic properties of the floor of the passenger wagon and the entire car body structure and at the same time they served as input data for simulation of the transmission of acoustic energy through the car body structure. The experimental results were confronted with the results of the simulations. Measuring microphone no. 3 was located in the aisle at a height of 1.2m above the floor, near the door and the lifting platform for access of disabled passengers. The measured data were informative in order to assess the noise levels at different locations of the structure of the investigated wagon.

The measurement was not intended only to obtain sufficient data for simulation calculations but also for eventual studies and identification of the dynamic track effects and running speed on the noise emitted by the rolling stock. In each of the track sections, ten measurements were carried out and, due to an objective assessment of the impact of the speed and track dynamics of the track sections on the noise in the interior of the passenger railway wagon, the data were further processed and the determined the so called mean values of the sound pressure levels recorded in relation to time and frequency during running on different selected track sections, which differed by the track structure and the train speed. By comparing the resulting frequency and sound pressure levels measured by the M2 and M3 measuring microphones on the track sections concerned, we experience the significant effect of the track condition on the interior noise in the passenger rail wagon. In the case of the resulting frequency analyses recorded by the M2 measuring microphone, the difference is almost 10 to 15 dB in the whole spectrum. Thus by measuring we have found a considerable influence of the technical condition of the track on the noise in the interior of the passenger railway wagon. The impact of the technical condition of the track on the noise in the interior of the passenger car is much higher than the effect of the running speed on the increase of the noise of aerodynamic origin.

Conclusions

As the most important outputs from the research activities in the above mentioned issues it can be summarized:

- Creating new models showing the sound reduction and absorptive properties of porous materials and determining the energy balance of the acoustic energy transfer through the porous microstructure of the material.
- Application of porous acoustics in the simulation of acoustic energy transmission through the structure of the floor of a railway passenger wagon.
- Performing acoustic analysis at the positions of individual sections representing the different shape and dimensional composition of the sandwich panel structure using simulation in the "COMSOL Multiphysics 5.0" environment.
- Original results of simulation in the form of frequency spectra of sound transition in the original and newly proposed material structure of the floor and walls of the railway passenger wagon.
- Qualitative evaluation and comparison of the results of simulation models in relation to the current state and the proposed application of new STERED® 200 material in the car body structure and the use of STERED® 250 as an alternative to a wooden grate.

The paper was supported by the Scientific Grant Agency of the Ministry of Education of the Slovak Republic and the Slovak Academy of Sciences in project no. VEGA 1/0766/15 “Research sources of noise emissions from rail transport”.

References

1. P. Zvolenský, Ľ. Kašiar, P. Volna, D. Barta, *Simulated computation of the acoustic energy transfer through the structure of porous media in application of passenger carriage body*. Procedia Engineering **187**, 100-109 (2017)
2. L. Bavlna, *Research of Internal Noise in Vehicles on Depending Floor Structure of Rail Passenger Carriage*. [Dissertation thesis], Zilina: SjF ZU, 124 p., (2016)
3. J. Gallíková, R. Poprocký, *Computer-aided analysis and the consequences of failures of the selected vehicle subsystem*. TRANSCOM 2015, University of Žilina, ISBN 978-80-554-1048-7, 58-63 (2015)

A. Pultznerová, L. Ižvolt, *Structural modifications, elements and equipment for railway noise reduction*. In: Procedia Engineering: Theoretical foundation of civil engineering. **91**, 274-279 (2014)

4. D. Johnson, J. Koplik, R. Dashen, *Theory of dynamic permeability and tortuosity in fluid-saturated porous media*. Journal of Fluid Mechanics **176**, 379-402 (1987)
5. L. Jaouen, [Online], http://apmr.matelys.com/, Available on: *APMR Acoustical Porous Material Recipes*: http://apmr.matelys.com/Parameters/Characterization/Acoustics/ThermalCharacteristicLength.html (2013)
6. T. Bourbié, O. Coussy, B. Zinszner, *Acoustics of porous media*. (Institut francais du pétrole publications, Paris, 1987)
7. R. E. Garcia, *Introduction to the Materials science of Rechergeable Batteries. Tortuosity and Porosity - Tortuosity in Porous Electrodes*. (Purdue University, Purdue, 2014)
8. Y. Champoux, J. Allard, *Dynamic tortuosity and bulk modulus in air-saturated porous media*. Journal of Applied Physics **70**, 1975-1979 (1991)
9. J. Julina, *Akustické vlastnosti vybraných stavebních materiálů*. (Univerzita Tomáša Baťu v Zlíne, Fakulta technologická, Ústav fyziky a materiálového inžinierstva, Zlín, 2008)
10. J. Oreský, S. Žiaran, *Noise and vibration in practice*. In: S. Žiaran (Ed.), Prenos vibroakustického vlnenia diskontinuitou, 63-66, (Nakladateľstvo STU Bratislava, 2011)
11. M. Matyka, J. Gołembiewski, Z. Koza, *The velocity distribution in a random porous medium*. Proceedings of the 5th International Con-ference on Porous Media and its Applications in Science and Engineering, p. 1, Kona, Hawaii: Engineering Conferences International, (2014)
12. EN ISO 3381, Brussels: CEN - European committee for standardization, (2011)

Passive safety system for small unmanned aerial vehicles

Piotr Bartkowski[1,*], *Robert* Zalewski[1]

[1]Institute of Machines Design Fundamentals, Warsaw University of Technology, Narbutta 84, 02-254 Warsaw, Poland

Abstract. In this paper a new air-bag prototype suitable for protecting valuable objects mounted on the drone is presented. The paper provides a complimentary study involving both numerical simulations and experimental study. The experimental research results are presented for typical air-bag's textile material and were used as a base for the material model calibration process. This model was used for the numerical simulations of the proposed air-bag prototype, which were carried out in the LS-Dyna environment. Based on the outcome of the study, the proposed prototype seems to be a suitable device for preventing the unmanned vehicle equipment from unexpected accidents.

Keywords: Air-bag, drone, modelling, experiments, LS-Dyna, air-bag prototype

1 Introduction

Unmanned aerial vehicles (UAVs) have been dynamically developing during recent years, especially for commercial applications. UAVs are being used not only for the simplest tasks such as film recording or capturing pictures but also are used for more complex and responsible actions such as structure inspections or even to package delivering. Teleoperated multirotors are also used by the firefighters or police. In order to perform such operations, UAVs often require the installation of fragile, sensitive and very expensive additional equipment.

Statistics show that failures of the multirotors in the air and in consequences their crash with a ground are common. The collisions are very often caused by the components failure. Despite the fact that the design process and testing methodology have been strongly developed recently [1], it is technically impossible to eliminate all the accidents. According to [2] during the flight the most dangerous stage is landing. Although there are some landing systems available for manned airplanes, such as used at major airports Instrument Landing Systems (ILS), which significantly limit the number of crashes, they are not popular for unmanned aerial vehicles because of its complexity and weight. The landing processes of UAVs are therefore generally controlled manually or supported by more or less sophisticated portable external aiding systems [3].

* Corresponding author: bartkowski_piotr@onet.pl

Reviewers: *Juraj* Gerlici, *Milan* Naď

Another important issue is related to various mishaps occurring in the air, due to uncontrolled collisions with other flying or ground objects such as electric poles, transmission and telephone lines or even birds. Such situations generally lead to serious damages of UAVs and in consequences of the equipment attached to them. Despite the fact that various systems preventing air accidents are continuously improved and developed [4] there is still need to enhance their effectiveness.

An interesting solution reducing landing threats for UAVs is proposed in the paper [5], which presents a vision-based automatic landing system that uses a dome-shaped airbag. In the proposed solution the air-filled dome enables airplanes to land softly on the specially colored copula using advanced navigation sensors. Nevertheless, such a solution is insufficient, when the object was damaged in the air. Other interesting proposition for passive safety system was described in [6], which presents a parachute mounted on the UAV.

In this paper a new concept of passive safety system for unmanned aerial vehicles will be presented. The proposed solution consists of a special parachute and airbag attached to the UAV. The parachute protects the object in case of failure at high altitudes, whereas the airbag protects the device in case of damages just above the ground, where the parachute is ineffective. In this paper the authors focus on the design process of an air-bag. In subsequent sections the full concept of this innovatory system is presented in details.

2 Material testing

The concept of the discussed airbag was created with finite element method (FEM) using LS-DYNA environment. The most important stage in the airbag modeling process is calibrating a suitable material model, which must be backed up with experiments. The air-bag is composed of a woven fabric which is rapidly inflated during a crash. For the analysis of the textile fabric it is necessary to perform a wide range of experiments. Beside the standard uniaxial tension test in various directions, it is necessary to conduct the shear test and additionally, what is highly recommended, biaxial tests. As far as the uniaxial tests is concerned, the experiment is relatively simple to carry out. Shear test and biaxial test are generally much more complex and generate more problems. Shearing tests are conducted in few different ways including bias extension test [7] or picture frame test [8, 9]. Among popular biaxial test methodologies a hemisphere test is worth mentioning, biaxial test with cross-shape sample or inflated cylinder fabric test [10].

In this paper the results of simple uniaxial tests and bias extension tests of an airbag fabric PA 6.6 are presented. Samples used in uniaxial tests (Fig. 1) were stretched on MTS testing machine, however strains were measured by additional non-contact 3D deformation measuring system (ARAMIS). During bias extension tests the strains were measured based on the testing machine yaws displacement. Fig. 2 shows the typical results of the simple uniaxial tests. In Fig. 3 the results of bias extension tests are depicted. Additionally the photos recorded by the ARAMIS system including calculated strain maps are presented in Fig. 4. More detailed description of the fabric material laboratory tests can be found in the previous papers of the authors [11, 12].

The data acquired from the experiments were used in the next stage of modeling implemented to numerical model prepared in LS-Dyna software. This topic is described in details in the next part of this paper.

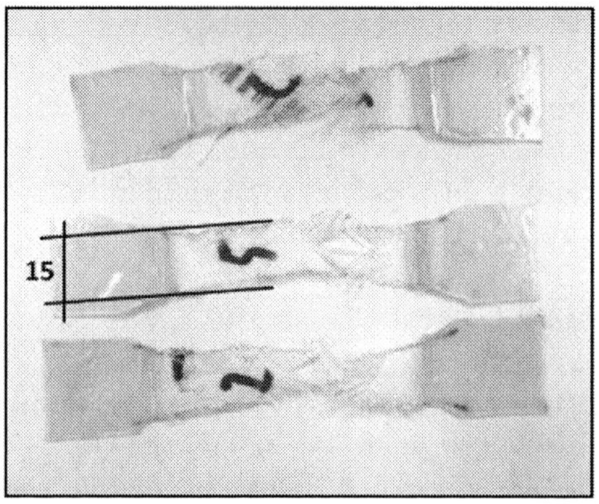

Fig. 1. Tested samples

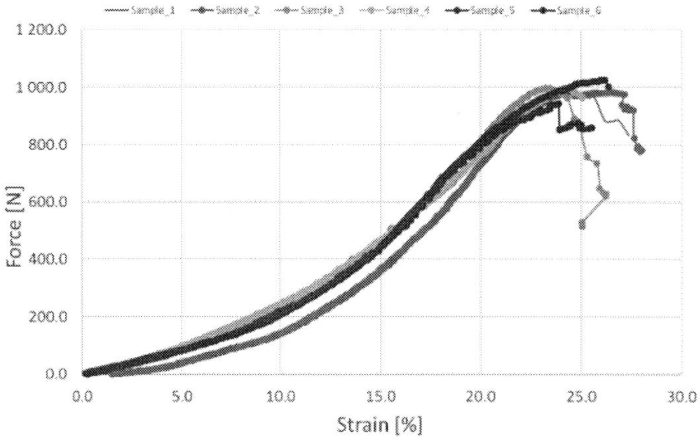

Fig. 2. Results of the uniaxial test

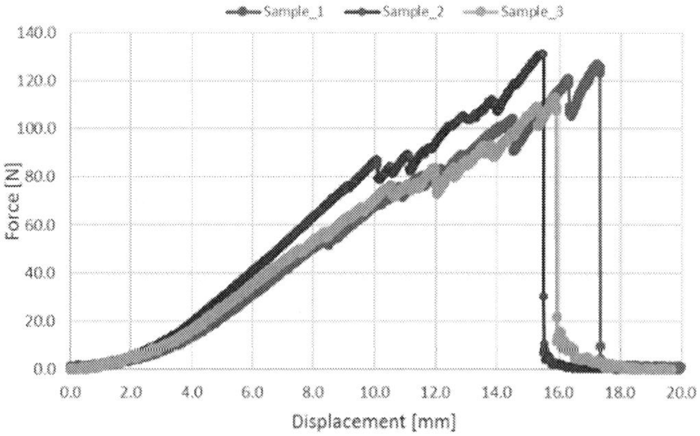

Fig. 3. Results of the bias extension test

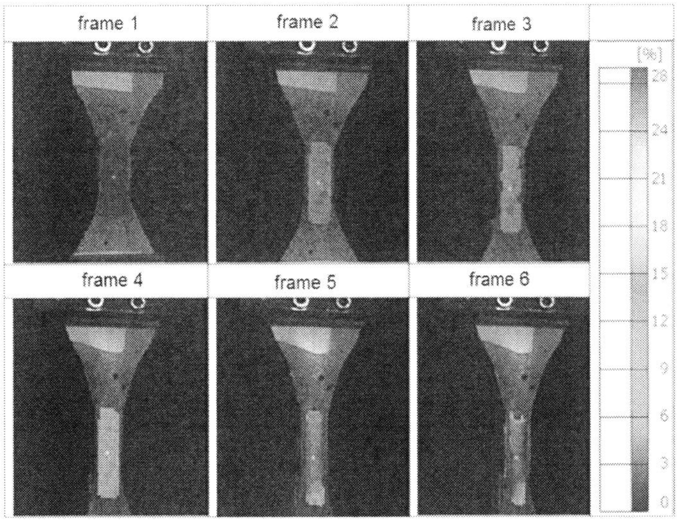

Fig. 4. Strain measured by ARAMIS system

3 Experimental research of leakage behavior

The crucial role in predicting the response of impacting object plays a gas leakage that is caused by fabric porosity or vents located on airbag surface (Fig. 5).

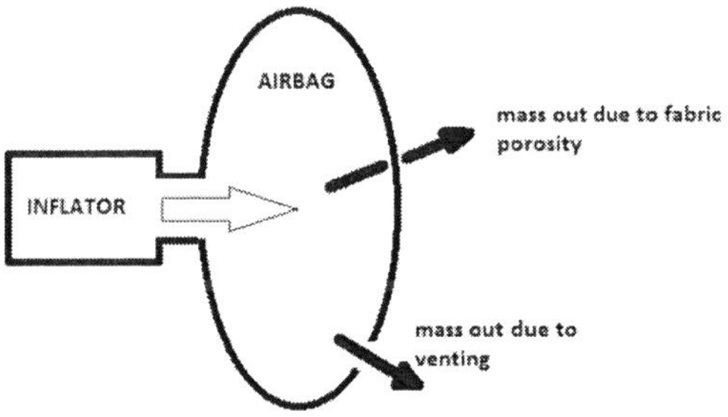

Fig. 5. Schematic illustration of the gas leakage

The modeling of leakage by vents is rather simple and will not be considered in this work. The more complicated task is modeling the leakage by material porosity. LS-DYNA provides various options for the airbag leakage modeling but the model must comply with the experiments. For this reason the airbag was tested on the drop test stand. During the experiments the internal pressure and strain of the airbag fabric were measured. The investigated airbag was filled up to predetermined pressure value and after that the intake of structure was closed. As a result the gas could only flow by fabric pores. The internal pressure was continuously measured during the outflow process. At the same time the average values of strains, calculated from the testing surfaces (Fig. 6) were recorded by the ARAMIS system. As the process of filling the air bag is rapid, the limitations of video measuring system (60 frames per second) did not allow to measure the strains during this stage. Therefore, the strains depicted in the Fig. 6 are equivalent to the maximum pressure reached in the air bag.

Fig. 7 depicts the investigated airbag mounted on the test stand with strain maps calculated by the measuring system. In Fig. 8 variations of the internal pressure in time function are shown. Three different values of predefined maximal pressures were taken into consideration in laboratory tests. The obtained results were used in the model validation process to verify numerical results with experiments. Fig. 9 depicts the recorded experimentally characteristics of pressure in function of average strain determined from the area shown in Fig. 6. Fig. 10 confirms the reproducibility of the experimental results obtained from the experiments performed at a maximum pressure in the airbag.

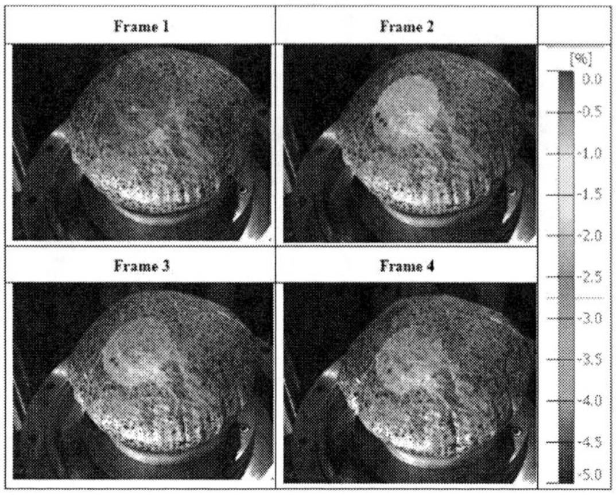

Fig. 6. Airbag strain maps

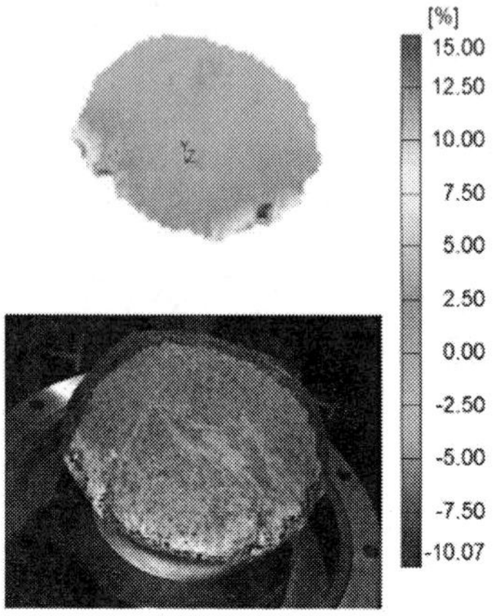

Fig. 7. Airbag with strain maps

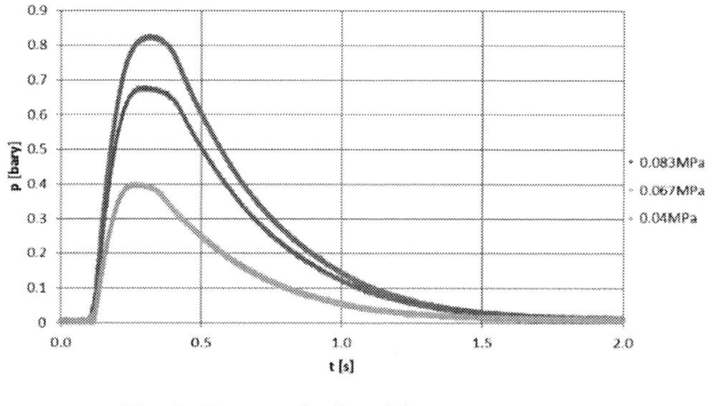

Fig. 8. Pressure in the airbag

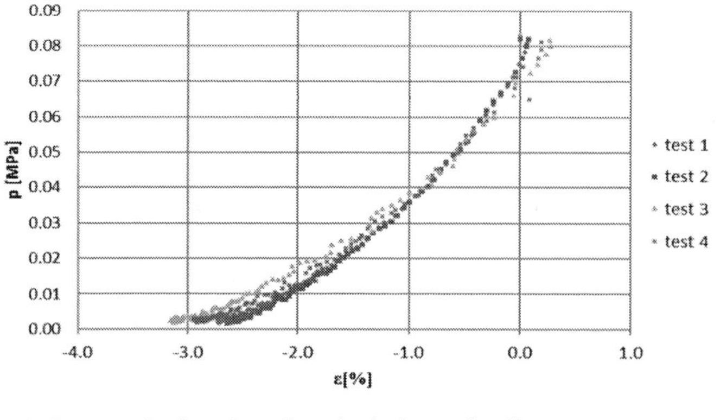

Fig. 9. Pressure in function of strain during unloading process

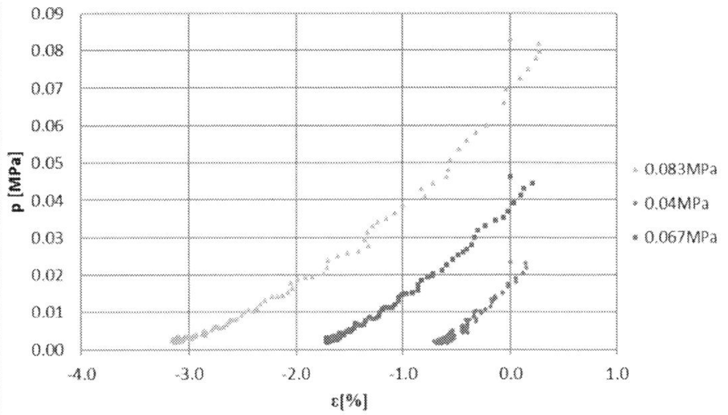

Fig. 10. Pressure in function of strain during unloading process

4 Modeling of material mechanical behavior

To model the real mechanical response of the considered fabric material in the LS-Dyna environment the material model MAT_FABRIC (#34) was used [13]. This material model has been developed for many years beginning with the simple linear constitutive law and up to the latest implementation with a nonlinear biaxial loading and unloading curves. The newest model of the material allows to simulate the porosity of the structure where the gas flows out [14].

In order to ensure that the investigated airbag material model works correctly the comparison between simulation and experimental research was performed. Based on the geometrical dimensions of the tested samples FE models were prepared. Before the experimental data were implemented to FE model it was necessary to transform it into II Piola-Kirchhoff stress tensor and Green-Lagrange strain tensor. For the tension case it can be calculated from Eq. 1 and 2. Similarly, for shear tests it can be obtained from (3) and (4). The derivation of equations (1-4) is precisely described in paper [11, 12].

$$(\tilde{T})_{22}=\frac{f}{A_0}\cdot\frac{1}{1+\varepsilon_{eng}} \tag{1}$$

$$(E)_{22}=\frac{1}{2}\cdot\left(1+\varepsilon_{eng}\right)^2-\frac{1}{2} \tag{2}$$

$$(\tilde{T})_{21}=(\tilde{T})_{12}=\frac{f\cdot(l_0-W)}{t\cdot W(lo-W+d)} \tag{3}$$

$$(E)_{12}+(E)_{21}=\left(1+\frac{d}{l_0-W}\right)^2-1 \tag{4}$$

Where T_{22}, T_{21}, T_{12} - elements of stress tensor, E_{22}, E_{21}, E_{12} - elements of strain tensor, f - force measured by testing machine, ε_{eng} - engineering strain, A_0 - sample cross section, l_o - initial length of the sample, W - initial width of the sample, t - thickness of the sample, d - yaws displacement.

When FE model was completed the analysis were performed using explicit LS-Dyna code. As a result it was possible to verify the experimental and numerical results. To ensure that the model correctly describes the stress and strain state in the material two independent tests were performed. The first one was related to the simple uniaxial tension test. In this probe the elongation of the sample corresponded to the main material direction. Forces in a strain function were determined numerically and in the final stage the results were compared with real experimental response of the material. As it is shown in Fig. 11 the model predicts precisely the behavior of the fabric.

In another experiment - bias extension test, the sample coordinate system was rotated by 45 degrees and the testing piece was again strained. In this case the observed differences between simulation and experimental results are more significant (Fig. 12). It is worth mentioning, that considering the real behavior of an airbag, the shear component in stress tensor does not play a significant role, consequently the authors assumed that model is reliable for further numerical analyses. Numerically calculated stress fields generated in loaded fabric samples in simple tensile test and bias extension test are depicted in Figs. 13 and 14 respectively.

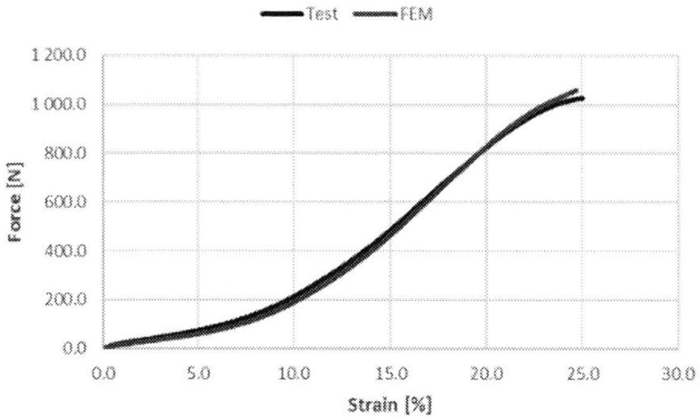

Fig. 11. Test vs experiments - simple uniaxial tension

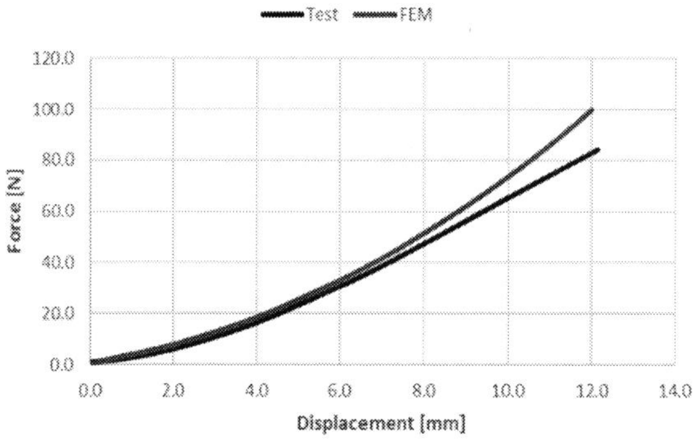

Fig. 12. Test vs experiments - bias extension test

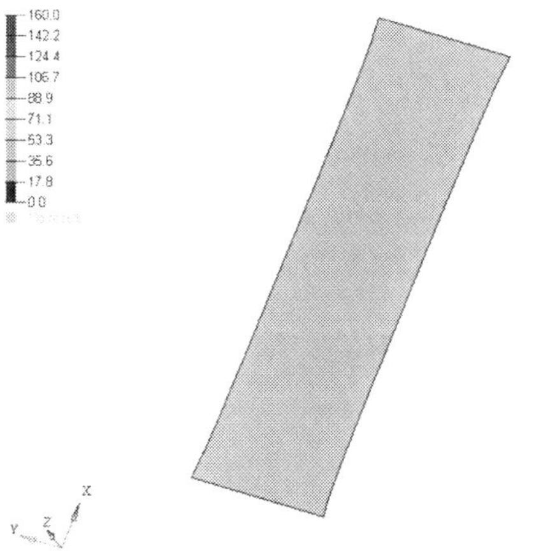

Fig. 13. Principal stress distribution in material sample - simple uniaxial tension

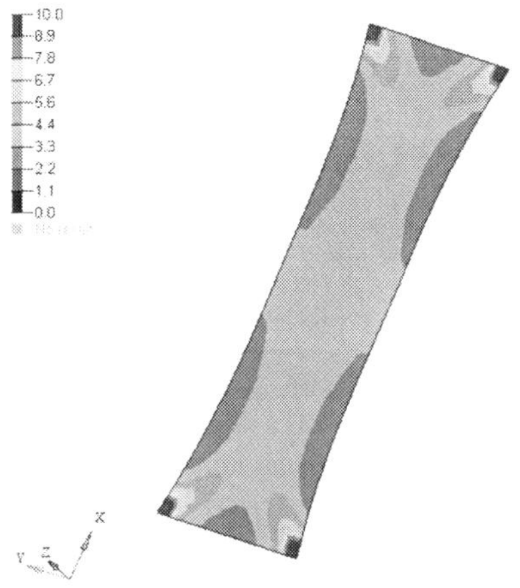

Fig. 14. Maximum shear stress distribution in material sample - bias extension test

5 Leakage modeling

Numerical evaluation of the leakage through the fabric involves computing the mass of the gas that needs to be removed from the airbag (Fig. 5). Primarly the leakage depends on three criteria: area of leakage, pressure and duration time. A more advanced model includes also blockage of the leakage area caused by the contact with an object. LS-DYNA offers few different formulations of the leakage caused by fabric porosity. In this paper Wang-Nefske's formulas with blockage of the leakage area were used. Model is described by Eq. 5.

$$\dot{m}_{out}=\sqrt{n\cdot g_c}\cdot\left[\sum_{1}^{n}(FLC(t)_n\cdot FAC(p)_n\cdot Area_n)\right]\cdot\sqrt{2\cdot p\cdot\rho}\cdot\sqrt{\frac{\gamma\cdot(Q^{\frac{2}{k}}-Q^{\frac{\gamma-1}{\gamma}})}{\gamma-1}} \qquad (5)$$

Where: g_c - gravitational conversion constant $[\frac{m}{s^2}]$, n - $1\cdot\frac{s^2}{m}$, FLC (t) - material coefficient [-], FLC (p) - material coefficient [-], Area - fabric area $[m^2]$, p - pressure [Pa], ρ - density [kg/m^3], γ - ratio of specific heats [-], Q - pressure relation [-], $\dot{m}_{out}$ - mass ratio [kg/s].

The model has two material coefficients (FLC (t), FAC (p)) that have to be identified experimentally. Those coefficients had a constant value equal 2160 and 0.0001 respectively. Measurement of the porosity coefficient can be a very difficult task, but several types of air permeability measuring instruments are available for general use in textile testing [9]. In general the idea involves controlled air flow by known fabric area at given pressure. In this paper a different approach was proposed. Experiments were performed on the real airbag. In the first step the investigated object was digitalized by reverse engineering method using Photomodeler. Afterward, the FE model was developed. The original concept of the gas leakage modeling process is presented in Fig. 15.

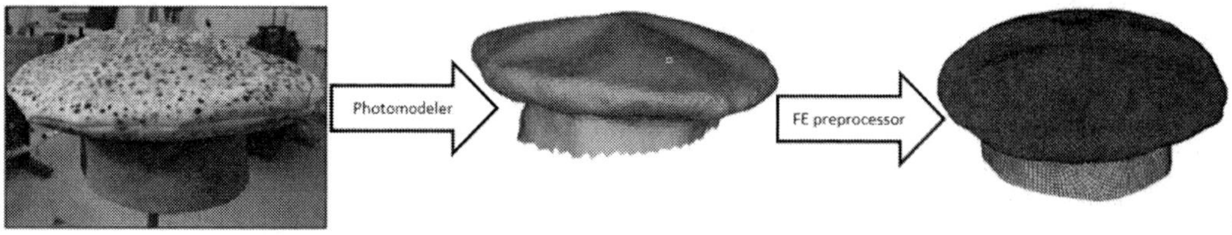

Fig. 15. Modeling approach

Fig. 16 presents FE airbag model on which the simulation process was based. The model calibration method consisted of changing iteratively the model coefficients to obtain the best compatibility between simulation and experiment. This procedure was performed for one selected initial condition (Fig. 17). When the model validation process was successfully completed, additional numerical experiments were carried out for other initial conditions that were also implemented in experimental procedure. Verification of the direct numerical data with laboratory tests results are shown in Figs. 18 and 19. As differences between numerical and experimental data are negligible, it can be assumed that the applied identification procedure is correct.

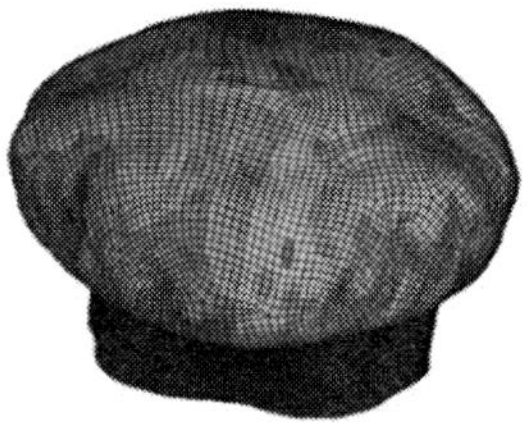

Fig. 16. FE model of airbag

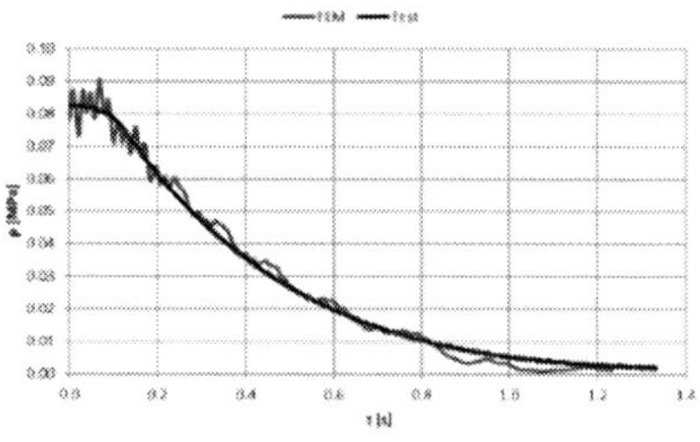

Fig. 17. Simulation vs test

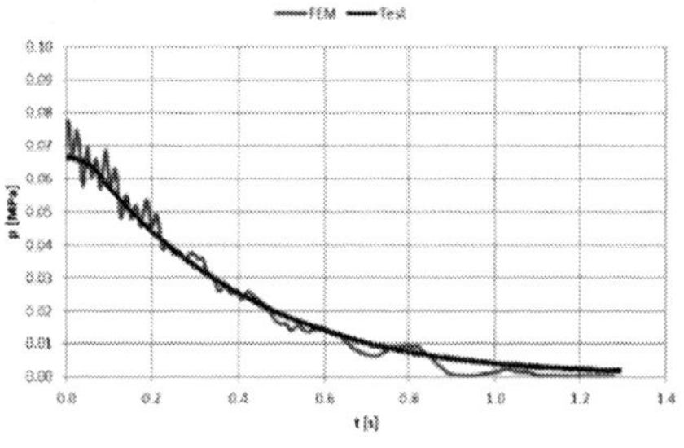

Fig. 18. Simulation vs test

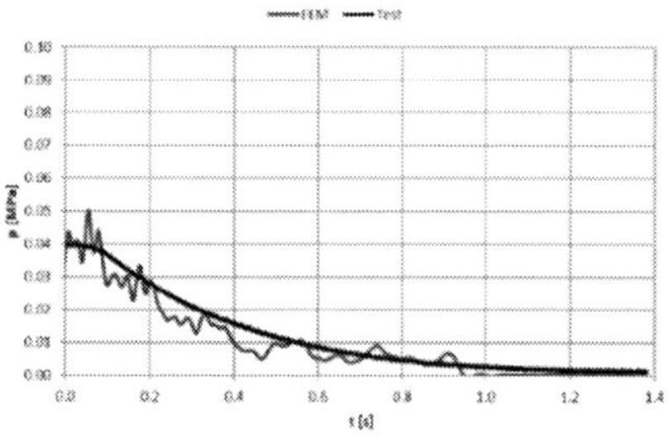

Fig. 19. Simulation vs test

6 UAV airbag concept

In the final stage of this paper the concept of UAV passive safety system will be described. The proposed system consists of two main parts, a parachute and an airbag. The purpose of applying the parachute is to protect UAVs in case of failure at high altitudes. The airbag is a suitable device when using parachute is impossible i.e. when the altitude of the object is lower than the threshold value. Fig. 20 shows the advantage of the proposed concept over the classical system including a simple parachute. Application of the additional airbag in the safety system for UAVs significantly reduces the hazardous altitudes range.

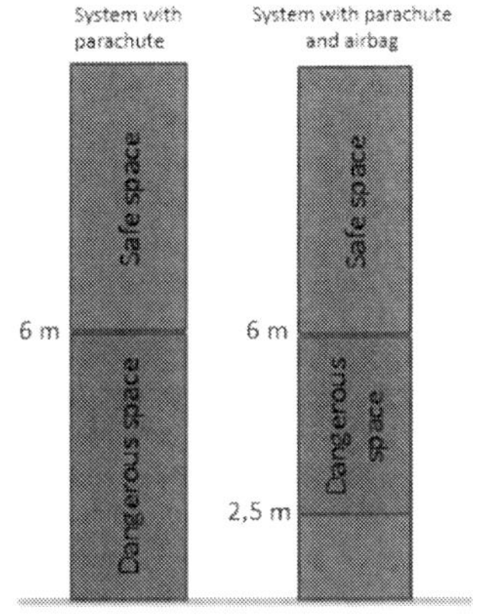

Fig. 20. Concept of airbag safety system

It is technically impossible to propose the system that protects multirotors independently on the accident location. It is assumed that the parachute is able to react when the mishap occurs on the highs over 6 meters. On the other hand it is technically impossible to consider the air-bag, which absorbs energy of the object falling from heights exceeding 6-meters because of the system mass.

Statistically most accidents occur close to the ground [2] so use of the airbag seems to be justified. The concept of the airbag was calculated with FEM applied in LS-DYNA software. It was assumed that the object crashes with the rigid obstacle at the 2.5 meters altitude, in consequences breaks down and begins to freely fall down. It was additionally assumed, that the drone is equipped with an additional device such as camera or other measuring system. The mass of the full system was estimated as 15 kg. The main objective of the proposed passive safety system is to protect the valuable thermal imaging camera transported by the unmanned vehicle. Following pictures (Figs. 21-23) illustrate several subsequent stages of the numerical simulations. Fig. 24 depicts the acceleration in time variations of the object mounted on the investigated UAV. Based on the technical specifications provided by the camera producer the possible acceleration range is limited to $294,3\frac{m}{s^2}$ only if it lasts not longer than 0.035s.

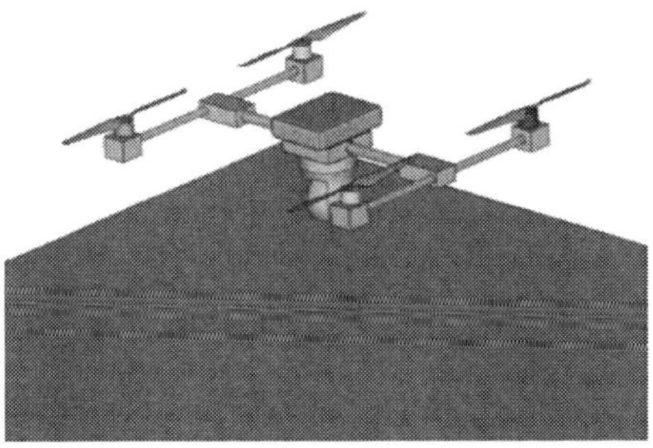

Fig. 21. Initial configuration of UAV and airbag – simulation time = 0

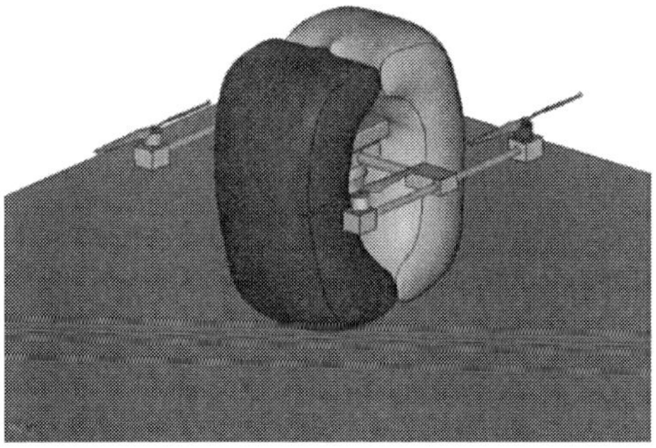

Fig. 22. Activation of the airbag – simulation time = 0.057s

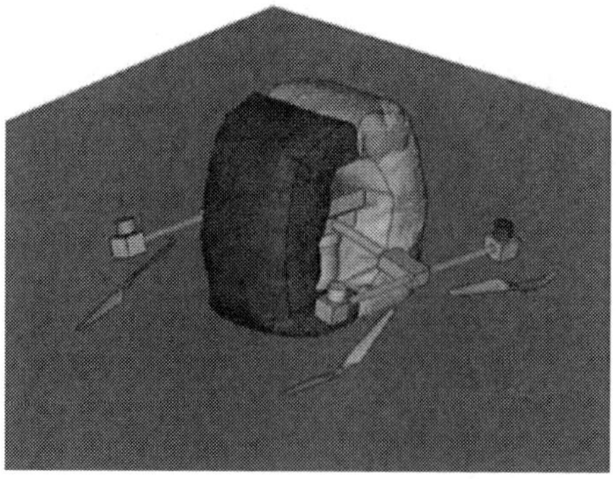

Fig. 23. Crash with the ground – simulation time = 0.12s

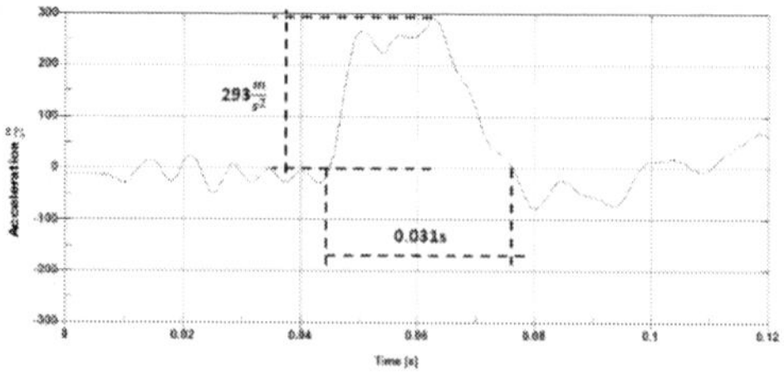

Fig. 24. Acceleration vs. time characteristics – filter SAE 180

The unfolding process of the air bag was not considered in this paper. It was assumed that this process does not have a significant influence on the air bag behavior as well as for a whole system behavior. In order to improve the transparency of Fig. 21, the airbag was hidden.

Conclusion

In this paper the new concept of the safety system for UAVs was proposed, which combines the advantages of a parachute and the special airbag provided to protect an additional device transported by the flying object. The proposed concept seems to have a great potential for future engineering applications. In the paper not only the new concept of a passive safety system was presented but also the complex methodology that can be adopted for any design process for airbags was proposed. The conducted simulations revealed a technical capability of the proposed UAVs airbag project as all design assumptions were fulfilled. The project was developed based on the models verified with the experiments, what exactly fits in the contemporary way of design. It was shown how the mechanical properties of a textile material should be recorded and modeled. Furthermore, the original approach based on the reverse engineering method for the validation of the model leakage behavior was proposed.

The models of the drone and airbag proposed in this paper were to some extent simplified, as the real dynamics of such a complex rotating systems is very complicated. Consequently, the real behavior of the investigated system could be much more problematic. To fully confirm the correctness of the proposed approach to the discussed problem conducting more complex numerical simulations is required. In the final stage, the experimental research carried out on the real object is necessary to verify the proposed numerical approach.

References

1. C. Guowei, B. M. Chen, T. H. Lee, M. Dong, *Design and implementation of a hardware-in-the-loop simulation system for small-scale UAV helicopters*. Mechatronics **19**, 1057-1066 (2009)
2. H. Sungsik, D. H. Shim, *A vision-based landing system for small unmanned aerial vehicles using an airbag*. Control Engineering Practice **18**, 812-823 (2010)

3. D. Fitzgerald, R. Walker, D. Campbell, *A Vision Based Forces Landing Site Selection System for an Autonomous UAV.* International Conference on Intelligent Sensors, Sensor Networks and Information Processing (2005)
4. M. Ptak, E. Rusinski, J. Karlinski, S. Dragan, *Evaluation of kinematics of SUV to pedestrian impact-Lower leg impactor and dummy approach.* Archives of Civil and Mechanical Engineering **12**, 68-73 (2012)
5. H. Sungsik, D. H. Shim, *A vision-based landing system for small unmanned aerial vehicles using an airbag.* Control Engineering Practice **18**, 812-823 (2010)
6. D.P. Hilliard, M. P. Hilliard, *US 6338457 B1.* US Patent (2002)
7. K. Potter, *Bias extension measurements on cross-plied unidirectional prepreg. Compos. Part A.* Appl. Sci. Manuf. **33**, 63-73 (2002)
8. G.B. McGuinness, R.A. Canavan, T.A. Nestor, C.M. O'Bradaigh, *A picture-frame intraply shearing test for sheet-forming of composite materials.* Proceedings of ASME Materials Division **69**, 1107-18 (1995)
9. G.B. McGuinness, C.M.O. Bradaigh, *Development of rheological models for forming flows and picture-frame shear testing of fabric reinforced thermoplastic sheets.* Journal of Non-Newtonian Fluid Mechanics **73**, 1-28 (1997)
10. S. Farboodmanesh, J. Chen, Z. Tao, J. Mead, H. Zhang, *Base fabrics and their interaction in coated fabrics.* Smart Textile Coatings and Laminates, A volume in Woodhead Publishing Series in Textiles, 42-94 (2010)
11. P. Bartkowski, R. Zalewski, *Experimental Research and Modeling of the Air-Bag Fabric Material.* Machine Dynamics Research **39** (1), 143-152 (2015)
12. P. Bartkowski, R. Zalewski, *Designing process of the drone's passive safety system.* New Advances in Information Systems and Technologies, Á. Rocha et al. (eds.), Advances in Intelligent Systems and Computing 445, Springer International Publishing Switzerland, DOI 10.1007/978-3-319-31307-8_74, 729-737 (2016)
13. A. Hirth, A. Haufe, L. Olovsson, *Airbag Simulation with LS-DYNA Past – Present – Future.* LS-DYNA Andwenderforum, Frankenthal (2007)
14. Schlenger, M., *A New Model for Simulation of Fabric Leakage in LS-DYNA.* LS-DYNA Forum, Bamberg, D-III-7-22 (2010)

Analysis of inertial amplification mechanism with smart spring-damper for attenuation of beam vibrations

Mateusz Barys[1,*], *Robert* Zalewski[1]

[1]Department of Automotive and Construction Machinery Engineering, Institute of Machine Design Fundamentals, Warsaw University of Technology, Warsaw, Poland

Abstract. In this paper an inertial amplification mechanism with an embedded smart spring-damper device for attenuation of longitudinal vibrations in continuous structures is analyzed. The complex systems are the extension of the already investigated inertial mechanism, here additionally equipped with the vacuum controlled spring-damper device which shows features of smart materials. This allows the semi-active control to affect different frequency vibration ranges in the real time. The fea.tures of the basic inertia amplification mechanism are preserved as a possibility to generate two neighbouring anti-resonance frequencies between resonance peaks in the low frequency range.

Keywords: inertial amplification mechanism, vacuum packed particles, damping of vibration, smart structures.

1 Introduction

This work is an attempt to combine the local resonance phenomena and a spring-damper device with vacuum packaged granular materials to mitigate rod longitudinal vibrations. The semi-active system allows to control the affected vibration frequencies in the real time by changing the level of underpressure in the spring-damper device, what leads to the different values of the stiffness and damping coeffcient of the device.

The concept of inertial amplification was presented by Flannely in 1967, in a patented device for attenuating vibrations called a "Dynamic Antiresonant Vibration Isolator" DAVI) [1]. The device is similar to a single degree of freedom (sdof) mass-spring system [2], however contains an additional mass, which is embedded between the main mass and the ground via a lever. This leaver amplifies an inertial force generated by isolation mass, and whole system works as a mechanical filter. The main mass is effectively isolated form ground vibrations where a transmissibility drops to zero for the particular frequency. The transmissibility of the standard spring-mass system for no frequency is reduced to zero. It rather approaches to zero for infinite frequencies. Furthermore, the standard spring-mass system is tuned to attenuate vibrations in a particular frequency range by the spring stiffness and its mass, where the DAVI system is tuned by the isolation mass and lever

* Corresponding author: mateuszbarys6@gmail.com

Reviewers: *Grzegorz* Domek, *Tomáš* Lack

ratio, which are more practical tuning parameters. We can increase the lever ratio to increase the inertial force from the mass. This means that using an additional small mass, amplified "effective" mass can be fairly large whereby an antiresonance frequency is obtained in the low frequency range.

The concept of inertial amplification was further developed by researches, however received less attention in the literature than local resonance phenomena. In 2002, Smith developed the inerter, which working principle is similar to Flannelly's system. The inerter is based on a generated inertial force proportional to the relative acceleration between two attachment points for vibration isolation purposes. The invention has been utilized in practical applications as a suspensions systems of Formula One cars [3].

In the next couple of years Yimaz and collaborators have presented series of papers regarding an inertial amplification phenomena focusing on periodically repeated inertial mechanisms. As an example, they investigated a periodical array of system similar to Fllanylly's DAVI to generate multiple anti-resonances and wide attenuated bands [4]. The next works concern lattice structures, where periodically repeated inertial mechanisms were a spine of vibration isolating structures [5, 6, 7].

The first attempts of utilizing the inertial amplification to attenuate vibrations in continuous structures have been presented by Frandsen and collaborators [8]. The inertial mechanism (similar to the single mechanism shown in [6]) is used to mitigate propagation of longitudinal waves through continuous 1D rod. The obtained results have shown that the system isolates vibrations at the low frequency range by generating wide and deep band gaps without applying a large isolation mass.

The system presented by Frandsen and collaborators [8] concerns a passive device for vibration attenuation. Here, a smart spring-damper is embedded to Frandsen's mechanism to create a semi-active system for mitigation of longitudinal vibrations in 1D rod. This allows a control of affected frequencies by the hybrid system in the real time. Additionally, due to damping properties of the smart spring-damper device the resonance amplitude is effectively decreased. In this paper the concept of a vibration damping prototype with variable dissipation properties, based on a core made of Vacuum Packed Particles (VPP) is taken into consideration [9], [10], or [11]. These structures are constructed of granular materials (loose grains) placed in the hermetic plastomer envelope in which a partial vacuum is generated. The difference between the atmospheric pressure surrounding the granular core and the underpressure inside the system provides friction forces between the individual grains.

These forces are all the greater the lower the internal pressure exists in the core. Bearing in mind the obvious limitations, limited by atmospheric pressure and vacuum, the vacuum parameter allows the free and smooth change of the global physical properties of VPP. In the previous papers of the authors the particular attention was paid to the analysis of the experimental results obtained for the family of granulated cores filled with granular grains. Cyclic loading of samples applying a kinematic excitation rule was considered. The studies were conducted for three different excitation frequencies and a stepwise alternating, full range of possible underpressures. Additionally, attention was paid to the basic problems encountered during the research and ways to eliminate them in the proposed solution of the damper prototype. The most important, from the point of view of potential engineering applications, are the characteristics of the cores made of special granular structures. Typical examples are depicted in Fig. 1. The graph illustrates the effect of the underpressure on the dissipative capacities of VPP.

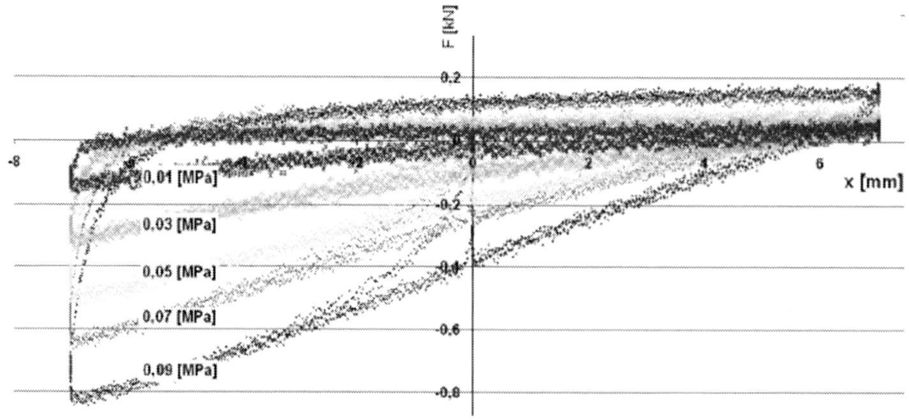

Fig. 1. Typical characteristics for VPP damper investigated under various values of underpressure

Based on the data illustrated in Fig. 1, we note that the influence of the vacuum parameter is unquestionable. Changing the vacuum parameter from $p_1 = 0.01$ MPa to $p_2 = 0.09$ MPa results in an increase in the recorded maximum force by several hundred percent. VPP proved to be an effective semi-active material for reducing vibration of beam elements [12], [13]. In the papers the authors used a granular sleeve in which the damped object was placed. The variations in internal vacuum had a significant effect on the dissipative properties of the system [14]. Compared to commercially available MR dampers and shock absorbers with variable dissipation properties, the proposed granular damper has many advantages. The main advantage is the economic aspect. In addition, the granular device appears to be less sensitive to temperature variations compared to the popular MR damper [15]. Another advantage of the proposed solution is its reduced sensitivity to pollution. In addition, in the case of granular conglomerates, the problem of tightness is automatically eliminated. The basic limitation of granular damper is its nonlinear characteristics. However, this obstacle can easily be eliminated by changing the design of the device.

2 Modelling

The attached hybrid mechanism to the elastic 1D rod by simple supports is presented in Fig. 2. Generally, the hybrid system can be split into to two separate devices i.e. the IA mechanism and a smart spring-damper. The first one consists of the bars denoted by l, which are connected by a moment free connection at the top and their inclination is controlled by an angle Θ_l. Furthermore, it is assumed that the bars are ideally rigid and massless to avoid their deformation. The isolating mass ma generates the inertial force. As it can be seen in Fig. 2, the spring-damper is attached between bars of the IA and is described by following parameters: a stiffness coefficient $k(p)$ and a damping coefficient $c(p)$, where p denotes an underpressure. All elements of the hybrid system are connected together by moment free connections. This provides that no moment is transferred through the system and allows for the elastic rod to vibrate freely. The elastic rod with Young modulus E, mass density ρ, length L has rectangular cross section with thickness t_b and width w_b.

This is an initial simplified mechanical model of the hybrid system, where the masses of the smart damper, rigid bars and bearings are omitted. Moreover, the physical construction of the spring-damper device is not investigated in this article. The attention is given on the impact of smart device in case of the control range of the affected vibration frequencies and whether it is whatsoever possible.

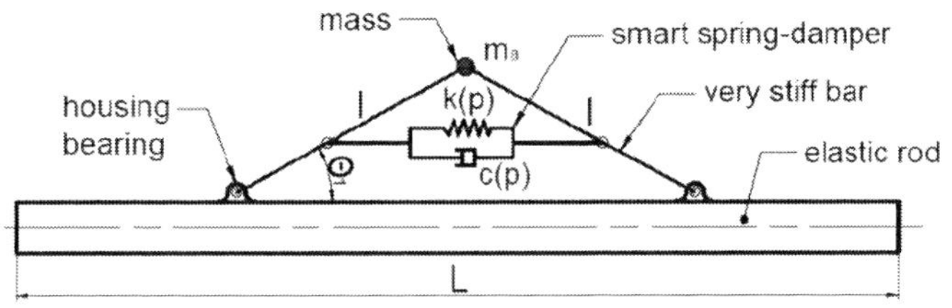

Fig. 2. The elastic rod with attached hybrid mechanism

3 Analysis

The effect of the hybrid system on the rod vibrations is discussed in this section. The parameters of the system are investigated and their impact on the hybrid mechanism effectiveness i.e., a control scope of the affected vibration frequencies is revealed.

The FE-model of the hybrid system is created in commercial FE-software Ansys. The host rod as well as bars are modeled by three-dimensional beam elements, which are based on Timoshenko beam theory. The bars are connected together and with the host beam by revolution joints to have moment free connections. The additional mass of the hybrid system is applied as a point mass at the top nodes of the attached mechanism. A spring-damper element is used to model the smart damping device, which is massless. As it can be seen in Fig. 3 the hybrid system is attached to host beam at the top and underneath. This ensures that vertical force components, which are generated by both mechanisms cancel out each other. Hence, only pure horizontal forces are transmitted to the host beam.

Fig. 3. FEM model of the system in Ansys environment

The system is excited to the vibrations by a periodical force with an amplitude F_0 = 100 [N], which is applied at the first node of the rod in the horizontal direction. The response of the system i.e., an amplitude of vibrations is examined at the second beam end. The host beam always has the same material parameters given in Tab. 1. The rigidity of the bars is obtained by setting their Young modulus $E_b = 1 \times 10^{20}$ [Pa]. Here, we only focus on the simplified system having rigid bars, of course for a physical system bars parameters will have an impact on mechanism performance in case of the vibration attenuation. This phenomena was investigated by Yilmaz and Hulbert [6], where authors show that their IA is the most effective for anti-resonance frequencies, which are looked for in the frequency range below an IA resonance frequency.

The all investigated parameters of the system are illustrated in Fig. 4.

Table 1. Beam mechanical properties.

L [m]	w_b [m]	t_b [m]	E [Pa]	ρ [kg/m^3]
0.8	*0.05*	*0.005*	2.1×10^{11}	*7850*

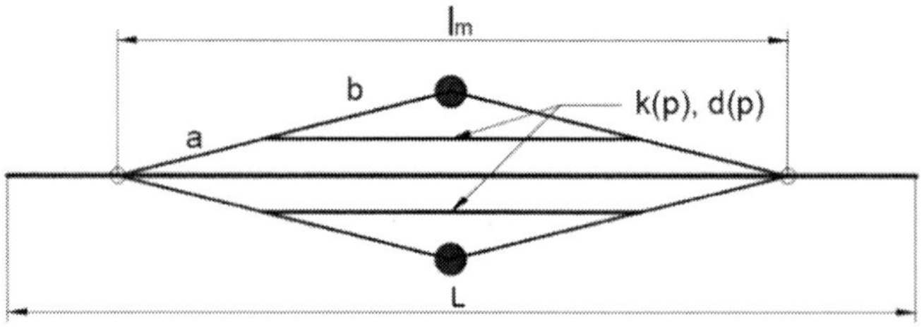

Fig. 4. System with analysed parameters

First, the effect of attaching the hybrid system to the rod is presented for mechanism parameters given in Tab. 2, where M_b is a mass of the rod and a damping ratio is determined by $\xi = c(p)/2\sqrt{m_a k(p)}$. Fig. 5 shows three FRF's: for a homogeneous rod, the rod with pure IA and the rod with the hybrid system. The excitation frequency is normalized by the first resonance frequency of the homogeneous rod and is denoted by $\overline{\omega}$.

Table 2. Mechanism parameters – case No. 1

θ_1 [rad]	m_a/M_b	l_m/L	k [m]	ξ	$\overline{h}$
$\pi/27.6$	*0.09*	*0.875*	*4 x 10⁸*	*0.28*	*0.5*

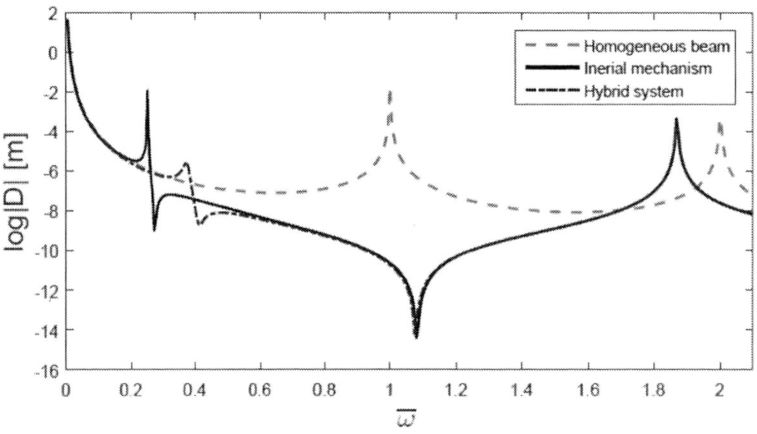

Fig. 5. Comparison of the FRF's for different systems

The black FRF in Fig. 5 shows the vibration attenuation properties of the inertial amplification mechanism i.e., it generates anti-resonance dips in the low frequency range. Furthermore, two neighboring dips are observed between two resonance peaks. This is characteristic behavior of the IA system and is described accurately in [8]. Here, we focus on the smart damper which is added to the IA mechanism construction. As it is seen in Fig. 5, for the hybrid system, the first resonance peak and anti-resonance dip are moved to the right (in dir. of the higher frequencies) in comparison to the FRF of the pure IA mechanism. This is caused by the additional spring element, which increases the stiffness of the whole system. These peak and dip for the hybrid system are no longer sharp but they are rounded and attenuated to the lower level. It means that the smart damper suppresses not only resonance vibrations but also the anti-resonance dips, which the amplitude does not approach zero any more.

What is worth emphasizing, only the first peak and the dip are affected by the smart-damper and the second order peak and dip still have the same trend as for the IA. It can be explained by a vibration form for the particular mode shape. For the first mode of vibration the attachment points of the damper device move out-of phase in horizontal direction, thus their net relative motion has maximal value and device in the most effective. For the second mode the attachments points move in-phase, thus relative motion seen by the damper device is insignificant and device does not work.

Based on Fig. 5 we can see that the smart damper embedded into IA works and can be used as an additional element to control the range of affected vibration frequencies by the system. Furthermore, it attenuates resonance peaks what is its advantage, does not seen for the pure IA mechanism. In the further sections, the particular parameters of the smart damper are analyzed and their impact on the FRF.

3.1 Hybrid system stiffness variation

The first analysed parameter is the stiffness of the hybrid system. The damping is set to zero and other parameters are given in Tab. 2. Fig. 6 shows FRF's for the pure IA as well as for the hybrid systems with different values of the stiffness coefficient.

As it can be seen in Fig. 6, increasing stiffness of the spring we move the first resonance peak and anti-resonance dip to higher frequencies. This proves that the stiffness coefficient of the spring-damper device can be a tuning parameter, and the system works with the expectations.

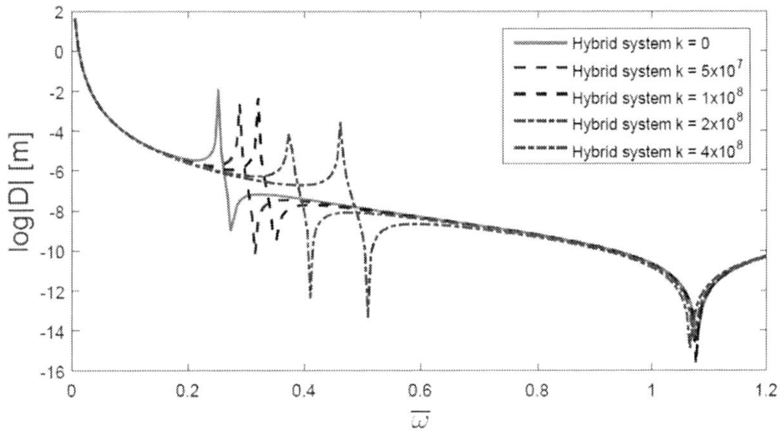

Fig. 6. Stiffness variation of the spring-damper device

3.2 Hybrid system damping variation

The damping coefficient is the next investigated parameter of the hybrid system. The FRF's are plotted in Fig. 7 for the all other parameters presented in Tab. 2.

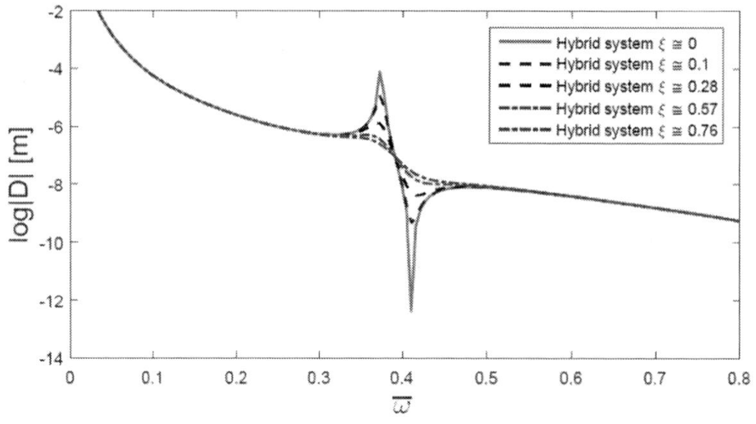

Fig. 7. Damping variation of the spring-damper device

Fig. 7 shows that the damper decrease, both resonance and anti-resonance amplitude of vibrations. The impact of the damper is the same as for Flannely's DAVI, where damping element has affected both resonance peak and anti-resonance dip [1]. Even though the damper decrease the level of the attenuation for the anti-resonance dip, still its effect is considered as a positive because the resonance peak which occur as a first along the FRF does not rise to the infinite value.

3.2 Effect of smart damper location

The last considered parameter is the height of the attachment points of the smart damper device. Fig. 8 presents FRF's for the system with a varying parameter $h = a/(a + b)$, where variables a and b are illustrated in Fig. 4. It is expected that the location of the attachment point along bar 1 can have significant impact on the affected vibration frequencies. The distance b works as a lever arm for forces generated by the spring-damper device and amplifies them for the increasing distance b.

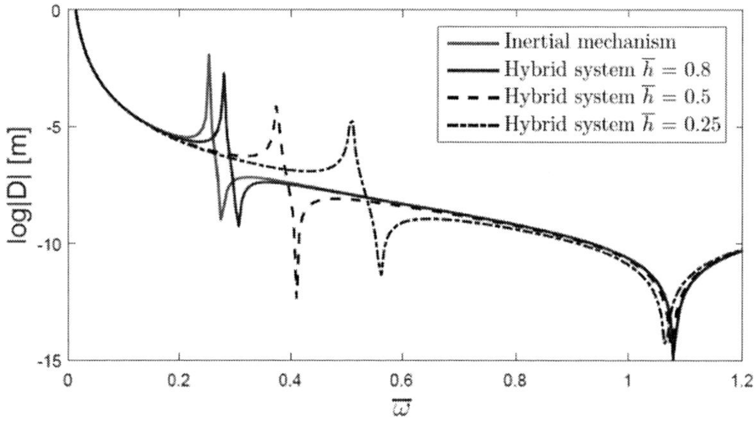

Fig. 8. Effect of the attachment points of the spring-damper device

As it is seen in Fig. 8, the obtained results are in line with above prediction. Decreasing parameter h, we can considerably increase the frequency of the first resonance peak and anti-resonance dip. Thus, in order to observe the maximal effect of the spring-mass damper, it has to be attached as close to the rod as possible. Then its stiffness can be lower, what is an advantage in the case of longitudinal vibrations of the continuous structures. Where a large stiffness is necessary to see the effect of the external device as e.g. the standard local resonator [2].

Conclusion

The hybrid system for attenuation of the longitudinal vibrations in the continuous rod has been analyzed numerically. It was shown that applying the smart spring-damper to the inertial mechanism we can control the frequency of the first resonance peak and anti-resonance dip. Furthermore the hybrid system suppresses the amplitude of the resonance and anti-resonance vibrations. Of course the all 'trademarks' of the pure IA mechanism [8] are preserved i.e., the possibility of generating two neighboring dips between resonance peaks or multiple anti-resonance dips in the FRF. Basically, the hybrid system extends the possibilities of the pure IA mechanism, develops it from the passive to the semi-active system. The spring-damper device embedded in the IA mechanism construction allows the semi-active control of the affected frequencies in the real time by the hybrid system. A physical realization of the device assumes using the under-pressure as a control factor to change stiffness and damping coefficients of the device. The device will be based on the vacuum packaged particles. The major challenge with the construction of the damper device is its stiffness, which has to be high enough to see the effect of the hybrid system in case of the control affected frequencies for longitudinal vibration. As it was shown in Fig. 8, the force generated by damper device is amplified by the construction of the system, however still the large stiffness will be necessary.

Future work assumes to realize physical model of the system and conduct experiments e.g. the modal analysis. Furthermore, the hybrid system can be applied to the continuous beam to attenuate bending vibrations and control the affected vibration frequencies.

References

1. G. Flannelly, *Dynamic antiresonant vibration isolator*. U.S. Patent 989, 958 (1967)
2. H. Frahm, *Device for damping vibration of bodies*. U.S. Patent No. 989958 (1911)
3. M. Chen, C. Papageorgiou, F. Scheibe, F.-C.Wang, M. Smith, *The missing mechanical circuit element*. IEEE Circuits and Systems Magazine **9**, 10-26 (2009)
4. C. Yilmaz, N. Kikuchi, *Analysis and design of passive low-pass filter-type vibration isolators considering stiffness and mass limitations*. Journal of Sound and Vibration **293**, 171-195 (2006)
5. C. Yilmaz, G. Hulbert, N. Kikuchi, *Photonic band gaps induced by inertial amplification in periodic media*. Physical Review B **76**, 054309 (2007)
6. C. Yilmaz, G. Hulbert, *Theory of phononic gaps induced by inertial amplification in finite structures*. Physics Letters A **374,** 3576–3584 (2010)
7. G. Acar, C. Yilmaz, *Experimental and numerical evidence for the existence of wide and deep photonic gaps induced by inertial amplification in two-dimensional solid structures*. Journal of Sound and Vibration **332**, 6389–6404 (2013)
8. N. M. M. Frandsen, O. R. Bilal, J. S. Jensen, M. I. Hussein, *Amplification of continuous structures: Large band gaps from small masses*. Applied Physics **119**, 124902 (2016)
9. R. Zalewski, P. Chodkiewicz, *Semi-active linear vacuum packed particles damper*. Journal of Theoretical and Applied Mechanics **54** (1), 311–316 (2016)

10. R. Zalewski, M. Shillor, P. Chodkiewicz, *Vibrations of a mass-spring system using a granular-material damper*. Applied Mathematical Modelling **40**, 8033–8047 (2016)
11. M. Makowski, R. Zalewski, *Vibration analysis for vehicle with vacuum packed particles suspension*. Journal of Theoretical and Applied Mechanics **53** (1), 109-117 (2015)
12. R. Zalewski, T. Szmidt, *Application of special granular structures for semi-active damping of lateral beam vibrations*. Engineering Structures **65**, 13–20 (2014)
13. T. Szmidt, R. Zalewski, *Inertially excited beam vibrations damped by vacuum packed particles*. Smart Mater. Struct. **23** (9), (2014)
14. J. Bajkowski, R. Zalewski, *Transient response analysis of a steel beam with vacuum packed particles*. Mechanics Research Communications **60**, 1–6 (2014)
15. R. Zalewski, J. Nachman, M. Shillor, J. Bajkowski, *Dynamic model for a magnetorheological damper*. Applied Mathematical Modelling **38**, 2366-2376 (2014)

AUTHOR INDEX

AUTHOR INDEX

AUTHOR INDEX

22nd International Slovak-Polls Conference on Machine Modelling and Simulations (MMS 2017)

MATEC Web of Conferences Volume 157 (2018)

Sklene Teplice, Slovakia
5 – 8 September 2017

Part 2 of 2

Editors:

M. Vasko
M. Handrik
L. Jakubovicova
P. Kopas
M. Blatnicka
V. Baniari
O. Stalmach
A. Sapietova
M. Saga

ISBN: 978-1-5108-5977-7

Printed from e-media with permission by:

Curran Associates, Inc.
57 Morehouse Lane
Red Hook, NY 12571

Some format issues inherent in the e-media version may also appear in this print version.

Printed by Curran Associates, Inc. (2018)

For additional information, please contact EDP Sciences – Web of Conferences at the address below.

EDP Sciences – Web of Conferences
17, Avenue du Hoggar
Parc d'Activité de Courtabœuf
BP 112
F-91944 Les Ulis Cedex A
France

Phone: +33 (0) 1 69 18 75 75
Fax: +33 (0) 1 69 28 84 91

contact-edps@webofconferences.org

Additional copies of this publication are available from:

Curran Associates, Inc.
57 Morehouse Lane
Red Hook, NY 12571 USA
Phone: 845-758-0400
Fax: 845-758-2633
Email: curran@proceedings.com
Web: www.proceedings.com

TABLE OF CONTENTS

PART 1

PART 2

Electromagnetic angular positioner based on DC micromotor

Maciej Bodnicki[1,*], *Jakub* Wierciak[1], *Wojciech* Credo[1], *Karol* Bagiński[1], *Leszek* Wawrzyniuk[1]

[1]Warsaw University of Technology, Faculty of Mechatronics, Institute of Micromechanics and Photonics, ul. Boboli 8, 02-525 Warszawa, Poland

Abstract. The presented works concerned launching of an angular positioner powered by an electromagnetic actuator, designed for performing angular micromovements within a range of few microradians. The principle of operation is based on balancing the electromagnetic torque of the motor with a torque that is twisting a compliant element. As electrodynamic actuators have no distinguished controlled positions, therefore in typical positioning systems desired positions are obtained applying a closed-loop position control. Usually, such systems employ also a feedback (dumping) related to velocity of the moving elements, what simplifies forming of dynamics of the system. The design of the physical model employs a DC micromotor, whose rotor is coupled with a torsional torquemeter. A feedback signal is generated by resistive strain gauges. The paper presents a mathematical model of the positioning system, results of simulation study as well as results of experimental study. The simulation study indicates that it is possible to select such design features and such type of the micoromotor that a high dynamics of positioning is ensured.

Keywords: angular positioning, DC micromotor, torque meter, simulation tests, test stand

1 Introduction

Miniature positioning drives that implement angular displacement with respect to one or more axes are used primarily in optical positioning systems. They are used for accurate, often quick positioning of mirrors or other optical elements. For systems with low dynamics, it is possible to use drives with classic rotary motors with gears characterized by high accuracy and high ratios. Both DC and stepping motors are used in such drives. An advantage of stepping motors is their ability to operate in open loop control. . SMA actuators are also used in low dynamic systems.

Often, however, the requirements for movement ranges include several dozen mrad, with the dynamics of work of, for example several hundred Hz. The use of piezoelectric elements makes it possible to achieve such parameters (these are often line actuators operating on an arm relative to the axis of rotation). Such positioners are also built as the electromagnetic actuators (VCM motors) with specially designed excitation combinations

* Corresponding author: m.bodnicki@mchtr.pw.edu.pl

Reviewers: *Piotr* Krawiec, *Ireneusz* Malujda

(permanent magnets) and windings, supplemented by suspension (often air gap suspension).

2 Idea of the positioner with torque balance

It was considered possible to build an electrodynamic positioner. The concept of mechanical structure of such a system and versions of its control has been developed. The use of an electrodynamic actuator, such as a typical, commercial DC micromotor with a coreless rotor, requires a high resolution rotor angular position transducer, far beyond the capabilities of conventional incremental converters.

For this reason, it has been proposed to use strain gauges measuring the deformation of a twisted mechanical element rigidly coupled to an rotor. The driven element (e.g. mirror) is attached to the torsion element at the location where its largest angular displacement occurs.

The ability to shape the correct torsional rigidity and high bending stiffness is guaranteed by the so called "drilled cross" elements known from the torque transducers' construction. The scheme of the proposed solution is shown in Figure 1.

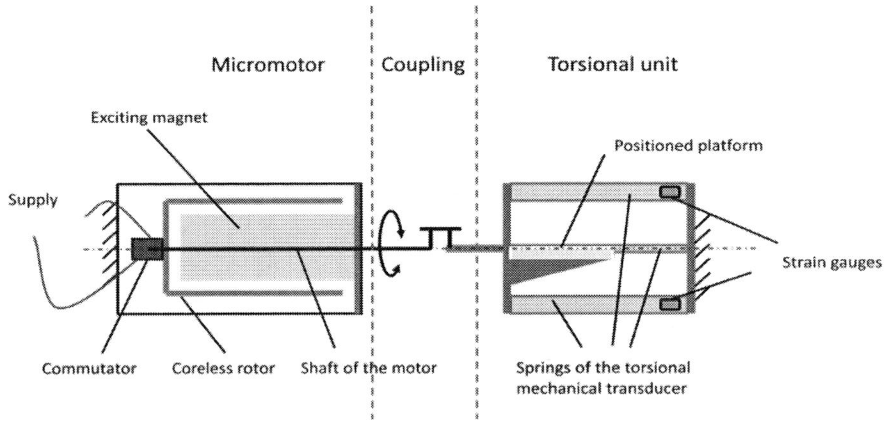

Fig. 1. Scheme of the positioner with electromagnetic activation on basis DC micromotor

According to the above assumptions, this system has the functional structure shown in Figure 2. The micromotor tracks the position by twisting the mechanical element to which the scanner table is connected. The amplified differential signal is the result of subtracting the load from the load cell transducers from the position signal. The load of the motor is the sum of the static torque (originating from twisting the element) and dynamic torque (originating from acceleration of the displaced parts). Preliminary selection of system parameters based on drive and measurement drive catalogue data leads to an estimate of the approximate native vibration frequencies of the uninhibited system, which is approximately 200 Hz.

Two types of scanner operation are possible with the proposed system:

- with feedback from the deformation signal of the torsion spring,
- without feedback, based on the linear dependence of the torque developed by the motor against the current flowing in its winding. In the latter case, it is based on the linear mechanical characteristics of the twisted element.

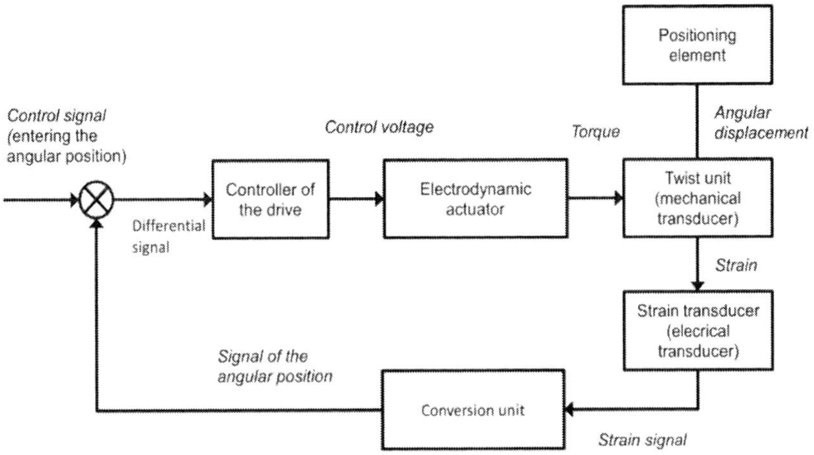

Fig. 2. Functional block diagram of the analyzed system

3 Test stand and physically experiments

For preliminary experimental research, a test stand was assembled using a DC micromotor with permanent magnet in stator and an ironless rotor in form of a cylinder, characterised by a rapid mechanical response.

The shaft of the micromotor was coupled by a bellows coupling (with significant torsion stiffness) with a flexible element. As a susceptible flexible element, a system of 4 flat springs, constituting a mechanical transducer of a stationary torque meter (SPM type, developed at the Institute), was used. The torque generated by the micromotor forces the mechanical transducer to be converted by an electric transducer (strain gauge bridge) into an electrical signal which is registered in the measurement system.

Due to the proportionality of the mechanical structure and processing circuit, the voltage at the output of the bridge is the information about the twisting of the spring system and the angular displacement of the shaft of the micro motor. There is also the possibility to control the output of the angular positioner.

The block diagram of the stand is shown in Fig. 3, its view in Fig. 4. Figure 5 shows the close-up of the spring system and its schematic view with marked places of strain gauge.

The micromotor used in experiments was characterized by a startup torque of 0.054 N·m (when the rotation speed is zero) at 12 V supply voltage. The voltage reduction resulted in a proportional reduction in the startup torque.

A SPM-100 stationary torque-meter (design in IM&P) with a measuring range of 0.1 N·m and a torsional stiffness of mechanical transducer c_{SPM} = 54 N·m/rad was used in the test stand. In the case of the applied micromotor, it was possible to achieve the maximum angle of twist of the mechanical transducer φ_{SPM} = 1 mrad (3.4'). This is not, however, the value of twisting of the modelled positioner. Due to the use of bellows coupling, the resulting rigidity value of the serial system results from the formula:

$$c = \frac{c_{SPM} \cdot c_C}{c_{SPM} + c_C} = 45{,}76 \text{ N·m/rad} \tag{1}$$

Where: c - torsional stiffness, c_{SPM} - torsion stiffness, c_C - bellows coupling torsion rigidity (300 N·m/rad).

The net maximum angle of the positioner model (with 12 V supply of the used DC micromotor) was therefore φ_{max} = 1.18 mrad (4.06').

The torque conversion processing constant is A_T = 0.381 Nm / V, with respect to the angular movement of the positioner, the processing constant has the value A_φ = 8.36 mrad/V (28.74'/V)

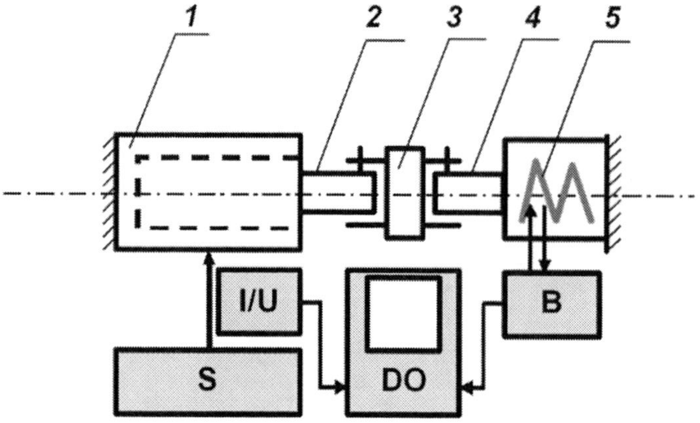

Fig. 3. Block diagram of the test stand – model of the positioner

1 – DC micromotor, 2 – shaft of the micromotor, 3 – coupling (spring bellow type), 4 – shaft of the stationary torque meter, 5 – mechanical transducer of the torque meter, S – supply, I/O – non contact, Hall – effect current/voltage transducer, D) – digitally oscilloscope, B – strain gauge bridge

Fig. 4. View of electromechanical components of the test stand (model of the positioned)

1 – DC micromotor, 2 – coupling, 3 –stationary torque meter

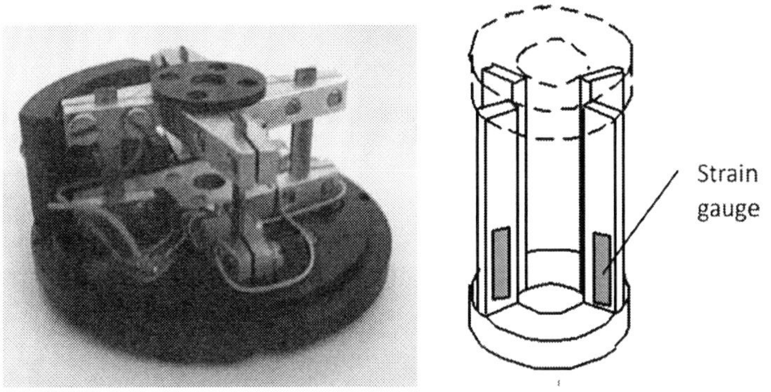

Fig. 5. System of 4 flat springs, constituting a mechanical transducer of a stationary torque meter (SPM type)

Figure 6 shows the time waveform of the current in the winding of the micromotor after stepped voltage supply with the visualisation of the time constant of the process. By properly selecting the electrical system parameters and operating mode of the power supply, it is possible to significantly shorten the time constant value. Fig. 7 shows the positioning leading to selected angular positions.

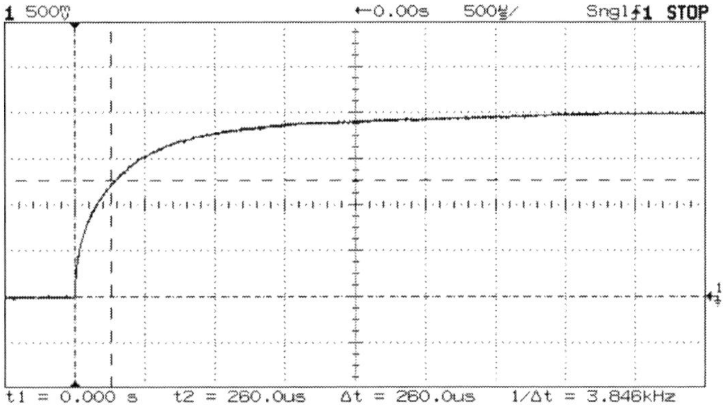

Fig. 6. Current course during step of the voltage supply (12 V; the time constant value is 280 μs)

a)

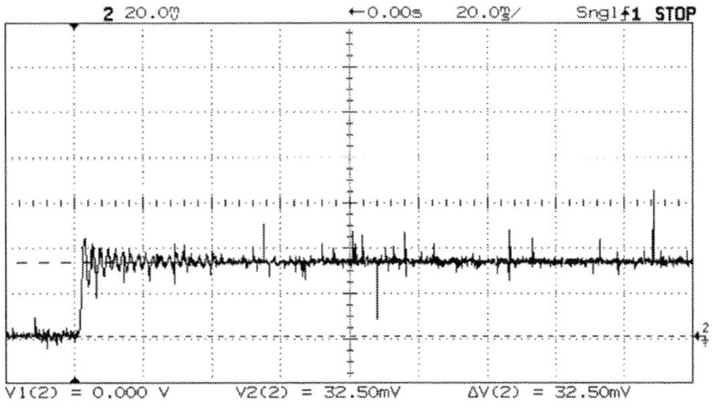

b)

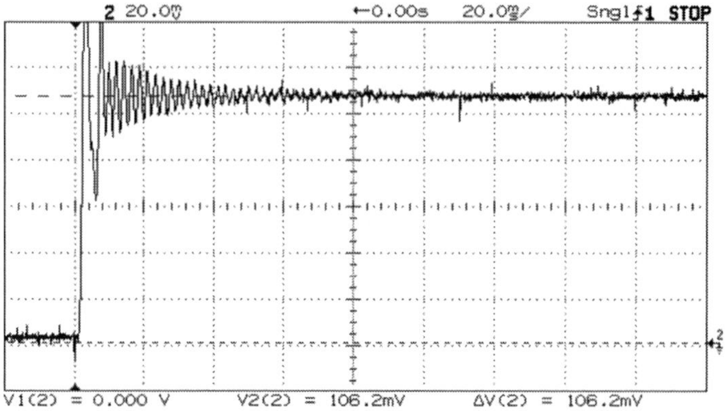

Fig. 7. Positioning

a) from position 0 to 0.295 mrad, b) from position 0 to 0,885 mrad

Fig. 8 shows cyclic switching between zero level and maximum angular deviation (1.18 mrad)

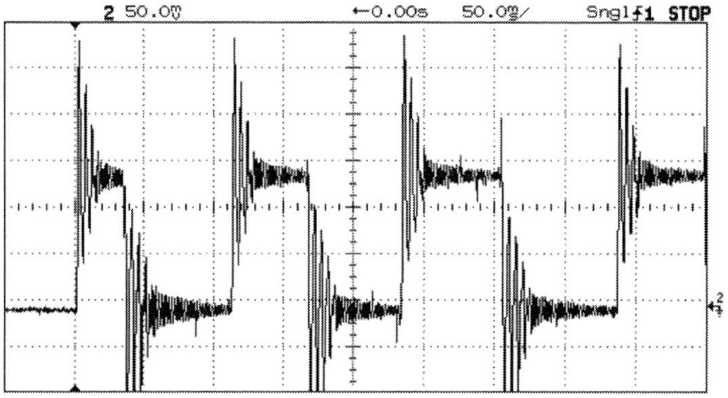

Fig. 8. Sequence of jumps between zero and max position of the positioner

4 Mathematical model

A mathematical model has been developed and includes: a classic DC motor model (which consists of two equations: balance of torque (2) and balance of voltages (3)), a torsion model in the form of a force torque linearly dependent on the angular displacement of the element and the damping moment proportional to the velocity of motion (4) and the control system model consists of equation of the strain gauge bridge channel dynamics (5) and equation of controller and the power amplifier (modeled as a PID regulator) – (6):

$$K_T i = (J_s + J_L)\frac{d^2\varphi}{dt^2} + K_D\frac{d\varphi}{dt} + (M_F \cdot sgn(\frac{d\varphi}{dt}) + M_L), \quad (2)$$

$$u = R_t i + L\frac{di}{dt} + K_E\frac{d\varphi}{dt}, \quad (3)$$

$$M_L = C_{SPM} \cdot \varphi + B_{SPM} \frac{\mathrm{d}\varphi}{\mathrm{d}t} \tag{4}$$

$$\tau \frac{\mathrm{d}U_{\mathrm{B}}}{\mathrm{d}t} + U_B = k \cdot \varphi \tag{5}$$

$$u(t) = k_p \left[\varsigma(t) + \frac{1}{T_i} \int_0^t \varsigma(t)\,d\,t + T_d \frac{d\,\varsigma(t)}{d\,t} \right] \tag{6}$$

where the symbols mean following: i - armature current, J_L - moment of inertia of positioning elements, J_s - moment of inertia of rotor of micromotor, K_D - constant of viscous damping in the motor, K_E - voltage constant, , K_T - torque constant, L - inductance of armature windings, M_F - motor friction torque, M_L - load torque, R_t - total resistance of armature circuit, u - supply voltage, φ - angular displacement of motor rotor, C_{SPM} – torsional stiffness of the spring system, B_{SPM} – constant of viscous damping in the spring system, U_B – voltage signal from strain gauges bridge (angular position signal), τ – time constant in low-pass filtering system, k – coefficient in the strain gauge bridge, k_p – gain coefficient, T^i - integral time, T_d - derivative time, ζ - the difference signal.

5 Simulation tests

Simulation models are written parallel in two software environments/languages: Matlab/Simulink (see Fig. 9) and AMIL.

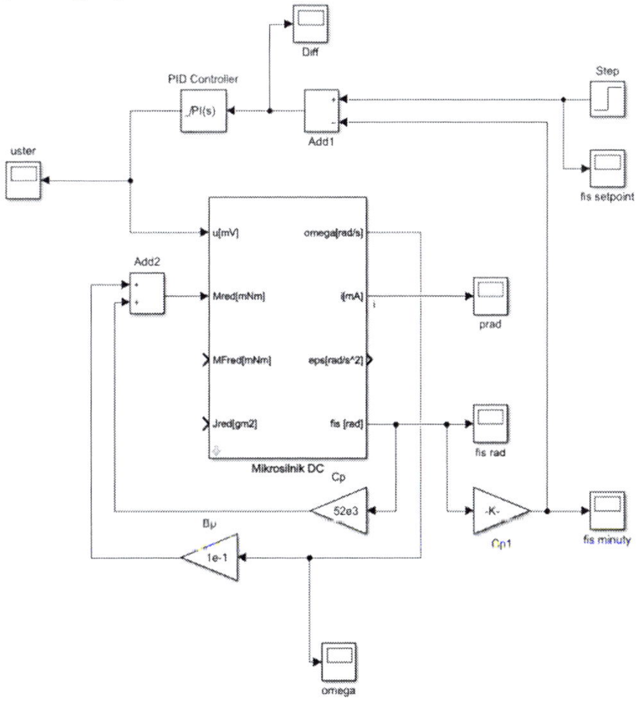

Fig. 8. Simulink model of the positioner

Sample results showing the effect of changes of dumping coefficient are shown in Fig. 9.

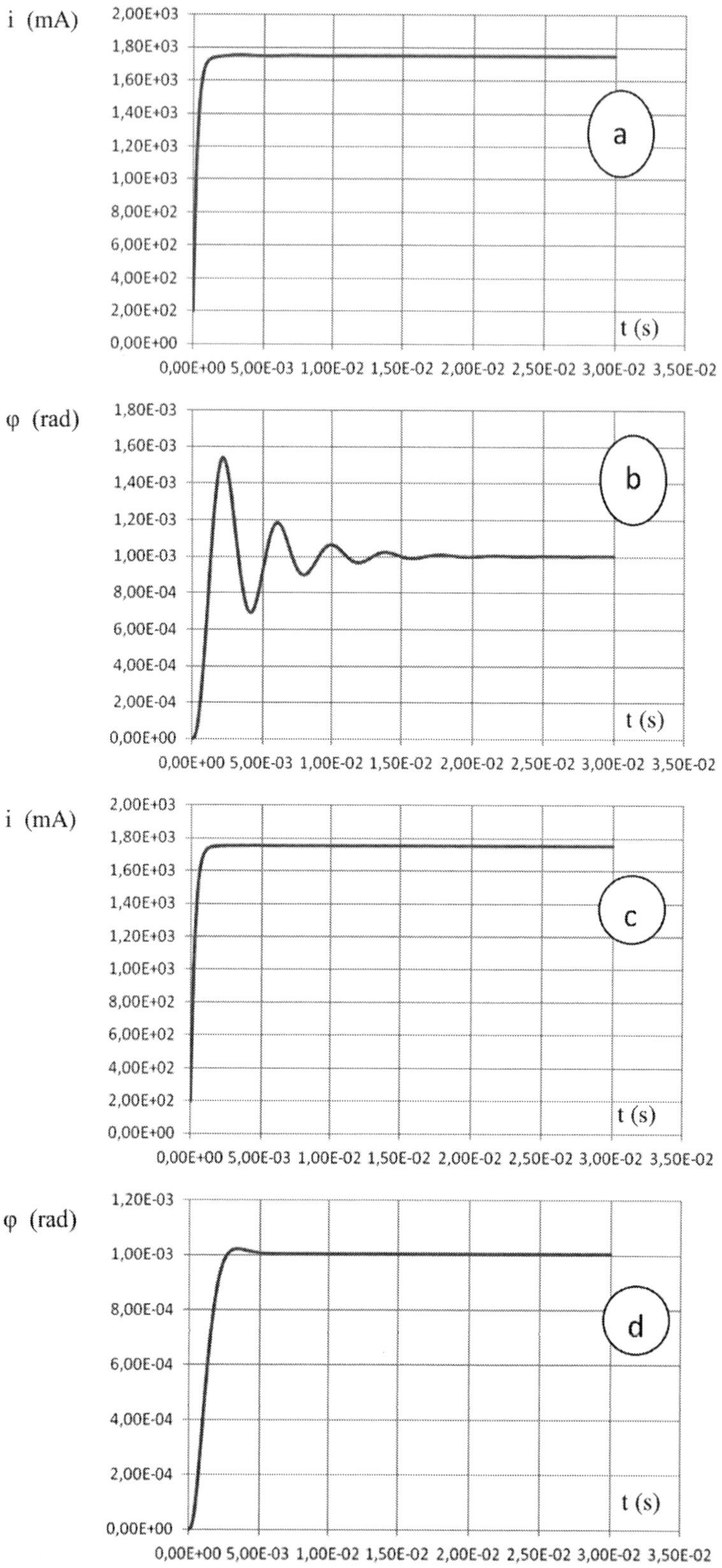

Fig. 9. Simulation of the positioning: a, c – current, b, d – angle of twist; c, d - change of dumping coefficient (multiplication by 10)

Conclusions

Design of an angular positioner with a range of single mrad has been proposed. The demonstrator (physical model) as assembled and tested in test stand -. An example of a DC motor and a strain gauge torquemeter were used. However, the parameters of the applied micromotor were not optimized. Experiments have been performed indicating the positioning capability of up to 1 mrad, with positioning time up to 40 ms.

The developed mathematical model and simulation software allow for optimization of the positioner and give the possibility e.g. of selection of the micromotor from the catalog proposals and the design of a dedicated spring system. The development of the model is anticipated by taking into account thermal phenomena.

This work has been supported by the a Grant The National Centre for Research and Development (Poland) no DOB-1-6/1/PS/2014 and partly by statutory activity of the Institute of Micromechanics and Photonics, Faculty of Mechatronics, Warsaw University of Technology.

References

1. E. Kiel, (ed), *Drive Solutions*. (Springer-Verlag Berlin Heidelberg, 2008)
2. K. Uchino, *Piezoelectric Actuators and Ultrasonic Motors*. (Springer US, 1997)
3. R. H. Bishop, (ed), *Mechatronic systems, sensors, and actuators*. (CRC Press, Boca Raton, 2008)

Assessment of a rail vehicle running with the damaged wheel on a ride comfort for passengers

Ján Dižo[1,*], *Miroslav* Blatnický[1], *Stasys* Steišūnas[2], *Blanka* Skočilasová[3]

[1]Department of Transport and Handling Machines, Faculty of Mechanical Engineering, University of Žilina, Univerzitná 8215/1, 010 26 Žilina, Slovak Republic
[2]Railway Transport Department, Faculty of Transport Engineering, Vilnius Gediminas Technical University, J. Basanavičiaus g. 28, LT-03224 Vilnius, Lithuania
[3]Department of Machines and Mechanics, Faculty of Production Technology and Management, Jan Evangelista Purkyně University, Railway Transport Department, Faculty of Transport Engineering, Vilnius Gediminas Technical University, Na Okraji 1001/7, 400 01 Ústí nad Labem, Czech Republic

Abstract. In certain conditions rail vehicles wheels can be during operation damaged. Then, the profile of wheels is no longer circular, but it is changed depending on the type and severity of defects. When such rail vehicle with the damaged wheel operates, the quality of a ride comfort for passenger is degraded. This article is focused on the assessment of ride comfort for passenger based on results obtained from dynamic analyses. Simulations and calculations were carried out in commercial multibody software. In our research we considered one type of the railway wheel untrueness – wheel-flat. This type of wheel damaging is relatively common and has such influence on the ride comfort for passenger worsening, which needs to be detected and investigated.

Keywords: rail vehicle, wheel-flat, ride comfort for passenger, computer simulation

1 Introduction

The ride comfort is one of the most important rail vehicle assessment standards. The ride comfort is given by several different adverse effects which passengers are subjected. These effects include mainly noise, air humidity, lighting, temperature, ventilation, but mainly vehicle vibrations. Negative vibrations can be caused during running due to track irregularities and also due to the wheel untrueness. One of the most frequently wheel untrueness is the wheel-flat [1]. This type of wheel defect occurs when the wheel locks and slides along the rail for the sake of malfunctioning brakes or because the overmuch braking force in relation to the practicable friction in the wheel/rail contact.

The ride comfort is one of fundamental assumptions of a rail vehicle achievement and popularity for passengers and transport operators. For this reason a big emphasis is given on the rail vehicle analysis before its operation [2, 3]. For analysis computer simulations and detailed analysis of measured experimental values are widely used.

* Corresponding author: jan.dizo@fstroj.uniza.sk

Reviewers: *Vladimír* Dekýš, *Milan* Nad'

2 Passenger ride comfort

The passenger ride comfort can be evaluated by the so-called indirect method [4-6]. For this method it is necessary to know acceleration values of a rail vehicle in analysed points. Acceleration values are filtered and weighted functions that take into account the human sensitivity to the vibration in reference directions. Such modified values are statistically evaluated subsequently and comfort indices for the floor and the standing and seated person are calculated numerically. Acceleration inputs to the calculation can be obtained by measurement [7-9] or as a result of the simulation calculation [10].

During the rail vehicle running dynamic movements of wagon body arise. These movements effect as vibrations. Passengers are subjected these negative effects during operation [11].

The ride comfort is the total sensation which the rail vehicle body movements generate in the passenger body. These rail vehicle body movements are to the whole passenger body transmitted in passenger – vehicle points [12, 13].

These sensations are classified as:

- Average sensation – due to vibrations.
- Immediate sensation – due to the immediate changing of the average sensation by the short effect.

Body accelerations are not only used in order to evaluate the passenger ride comfort but also to evaluate running properties and vehicle vibrations. Whereas comfort criterions are stricter as criterions for vehicle running properties in order to operation approval, there is to the ride comfort optimization concentrated at the rail vehicles dynamic analysis mostly, there through good running properties are achieved.

As mentioned above, rail vehicle body movements are to the human body transmitted through contact points according to the passenger position (Fig. 1) [4]:

- Standing position: floor-feet.
- Seated position: headrest-neck, armrest-arms, seat-hip, backrest-back, floor-feet.

The passenger ride comfort is valuated according to the EN 12299:2009 [4] standard based on the UIC 513. The ISO 2631 standard is also used, eventually the grade quality running W_z by Sperling. Ride comfort criterions appear from body accelerations measured namely in the vertical, lateral, eventually longitudinal direction.

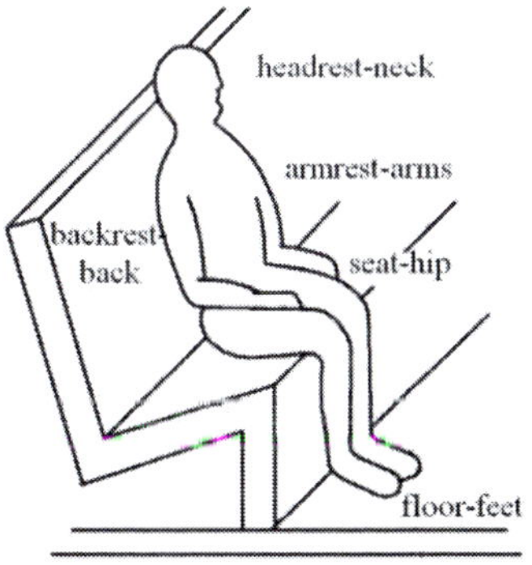

Fig. 1. Interfaces points

For the passenger ride comfort evaluation weighting filters are used which deliberate the human body sensitivity by particular of frequency. These filters are different for vertical and lateral direction. Weighting filters accordance with EN 12299:2009 standard (measurement on the floor) for the lateral direction are consistent to the ISO 2631 filters, they are different for vertical direction a bit only. Compared with the older grade quality running W_z filters for lateral direction are particularly significantly different, therefore the different criterion can reach to different ride comfort quality results in spite of the same vehicle [14].

2.1 Calculation of ride comfort indices

The ride comfort indices calculation assumes the measurement of acceleration (m/s^2) in the *x*, *y*, and *z* direction (a_x, a_y, a_z). Based on the sampling frequency f_n is determined the number of samples that are recorded in the time interval of 5 sec. At the sampling frequency of 200 Hz during 5 sec. are obtained 1000 samples. Based on the condition of samples occurrence in 5 sec. intervals shooting time is divided into 5 sec. periods. Each of the sections has clearly established the start T_1 and the end T_2 in time. For the dataset at the each time interval is performed the fast Fourier transformation (FFT). For range of frequencies 0.4 Hz - 100 Hz is performed CAW calculation [4, 6].

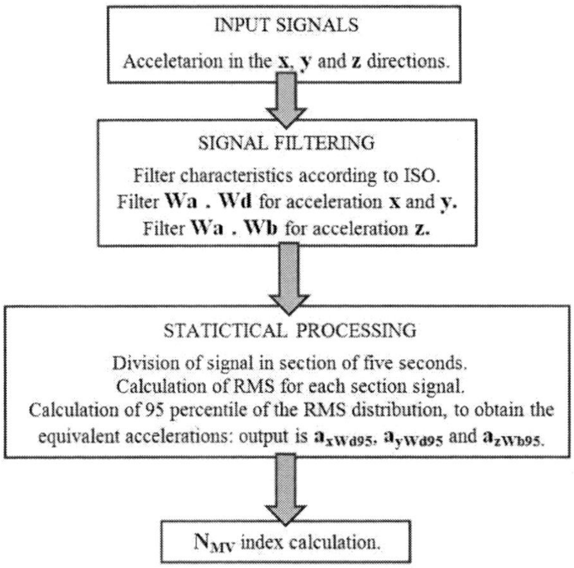

Fig. 2. Workflow for the N_{MV} index calculation

The weighting filter *w*, which takes into account the human body sensitivity at different frequencies, is applied depending on the type of evaluation – floor, standing position or seated position.

Weighting function modified acceleration values are statistically evaluated and summation functions in histograms are determined.

The ride comfort calculation procedure is shown in Fig. 2.

Result values of passenger ride comfort indices for the average comfort are calculated as follows [4]:

- Floor:

$$N_{MV}=6.\sqrt{\left(a_{XP95}^{W_d}\right)^2+\left(a_{YP95}^{W_d}\right)^2+\left(a_{ZP95}^{W_b}\right)^2} \qquad (1)$$

It can be sometimes useful depending on the application to calculate partial indices of the ride comfort:

$$N_{MV_x}=6.a_{XP95}^{W_d}, \quad (2)$$

$$N_{MV_y}=6.a_{YP95}^{W_d}, \quad (3)$$

$$N_{MV_Z}=6.a_{ZP95}^{W_b}, \quad (4)$$

- Standing position:

$$N_{VD}=3.\sqrt{16.\left(a_{XP50}^{W_d}\right)^2+4.\left(a_{YP50}^{W_d}\right)^2+\left(a_{ZP50}^{W_b}\right)^2}+5.\left(a_{XP95}^{W_d}\right)^2 \quad (5)$$

- Seating position:

$$N_{VA}=4.\left(a_{ZP95}^{W_b}\right)^2+2.\sqrt{\left(a_{YP95}^{W_d}\right)^2+\left(a_{ZP95}^{W_b}\right)^2}+4.\left(a_{XD95}^{W_c}\right)^2 \quad (6)$$

Analyses in this paper are focused on the N_{MV} index computation. To calculate this comfort index, into the calculation accelerations in longitudinal (x), lateral (y) and vertical (z) direction of each point of interest must input.

3 Assessment of the passenger ride comfort by means of simulation computations

In the development and manufacture rail vehicles process is given the care of a test and validation their static and dynamic properties. There is the main emphasis to the effective resources exploitation in a competition, therefore there are preferably used a lot of types of rail vehicle computer simulations [15, 16] as a mechanical system, its subsystems and its particular parts [17-20].

Generally there is an experimental process of data obtaining that are needed to optimization of new products designing and creating. The computer simulation is characterized by the real vehicle system replace computer model created especially for the specified conditions.

The dynamic analysis evaluates changes of rail vehicle behaviours as a mechanical system under the influence of equilibrium change.

Dynamic simulations of the mechanical system can be performed by using of a variety tools virtual reality [21-24]. For purposes of the passenger car model creation, dynamic simulation, processing of results and ride comfort for passenger assessment we have used the Simpack/Rail package.

Simpack/Rail allows modelling and analysis of rail vehicles and similar devices of every kind, from trams to high-speed trains but also from material handling systems to roller coasters.

Most parts of a rail vehicle model in Simpack are made of standard Modelling Elements, but contact between the rails and wheels and the track uses specialized Modelling Elements. Calculation of ride comfort indices requires further specialized Control Element (accelerometer) using.

As it is already mentioned above, the passenger ride comfort can be evaluated by the so-called indirect method [4, 6, 7].

3.1 Computer model of the passenger car

The ride comfort for passenger was evaluated for the passenger car with four wheelsets. All wheelsets had the same profile and there are guided by the swinging arms in bogies. Between wheelsets and bogie frame the primary suspension is mounted and the passenger car body and bogies the secondary suspension system connects. The wagon is equipped with primary and secondary suspensions. Main parts of vehicle are body of wagon, bogie frames, axle boxes and wheelsets.

The Fastsim method was used for the rail/wheel contact calculation. There is the standard method for vehicle dynamics [25-29]. During the wheel/rail contact modelling process the wheel-flat was specified. It comes to this that the nominal radius of a wheel radius varies with the wheel rotation angle. In Simpack the wheel untrueness is possible to model in three ways: radius deviation, Fourier coefficients and harmonic function (simple polygonality) [2, 30].

In our research the radius deviation is considered. In this model, radius deviation is prescribed pointwise in polar coordinates by defined input function. Its independent coordinate is the rotation angle, which must be defined in the interval [0, 2π] and the radius deviation ΔR or the actual local radius R in meters are dependent coordinates. The mean value is subtracted from the values to get the actual radius deviation. Complete data are splined to allow a continuous interpolation including the derivatives.

The wheel movement with a flat can be described follow (Fig. 3). The wheel drops for a distance from point T_1 to T_3 and it induces an impact on the top of rail in point B [31].

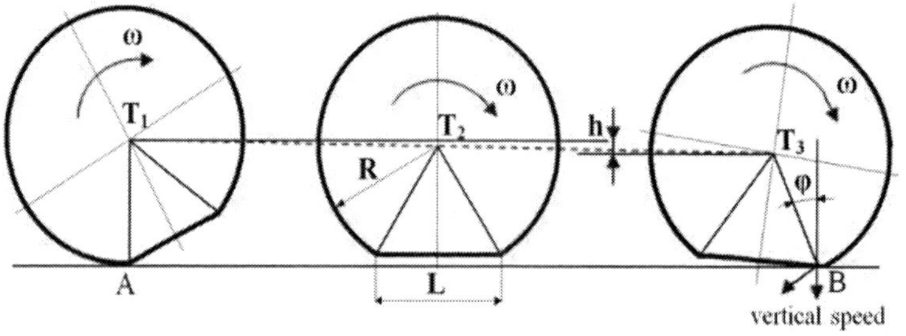

Fig. 3. Wheel/rail impact caused by the wheel-flat [25]

If we assuming that the wheel-flat length L and the wheel radius R are constant, the equation of motion can be derived (7):

$$\omega\sqrt{\frac{2.h}{g}}=2.arcsin\frac{L}{2.R}-arccos\left(1-\frac{h}{R}\right) \tag{7}$$

where ω represents the rotation speed, h is the dropping distance of the wheel, g is the gravity acceleration, L and R are the wheel-flat length and the rolling radius of the wheel respectively [31]. The impact on a rail is not only from the wheel dropping but also from the vertical speed component. This cannot be neglected. The wheel rotation speed is considerable and therefore it can induce relatively high vertical impact forces. Its vertical component is expressed as (8):

$$v_{ver.}=\omega.R.sin\varphi \tag{8}$$

where φ marks the angle shown in Fig. 2. It is calculated as [1] (4):

$$\varphi=arccos\left(1-\frac{h}{R}\right) \tag{9}$$

The passenger wagon was run on the straight track section of 5,000 m. The UIC60 rail profile and the S1002 wheel profile were used. The track model contains also measured track irregularities prescribed by discrete form.

For the assessment of the passenger ride comfort is necessary to know accelerations in individual direction *x*, *y* and *z*. Therefore specialized Control Elements (167: Accelerometer) were defined into points as Fig. 4 shows.

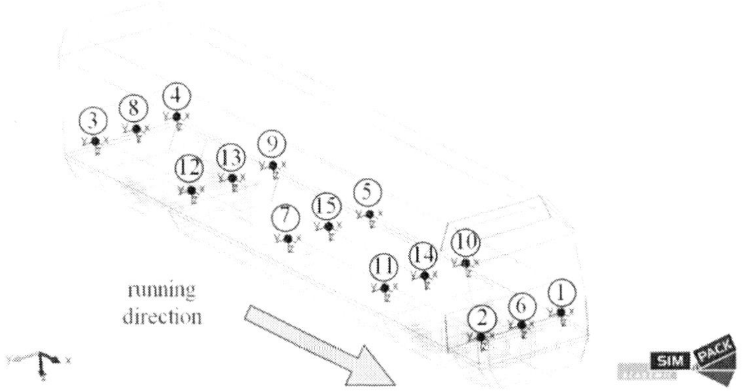

Fig. 4. Location of points for the passenger ride comfort assessment on the wagon floor

The process of the passenger ride comfort index computation at the wagon floor level consists of several steps. Outputs of procedures are frequency weighted RMS-values of acceleration $a_{XP95}^{W_d}$, $a_{YP95}^{W_d}$ and $a_{ZP95}^{W_b}$, respectively, where a is RMS-value of acceleration; X, Y and Z are directions of acceleration, P is measured position (in our case floor), number 95 represents 95-percentile, W_d is weighting curve in x and y direction at the floor level and W_b is weighting curve for z direction at the floor level.

The five-second RMS-values of the frequency weighted accelerations are calculated as:

$$a_{xj}^{W_d}(t)=\left[\frac{1}{\tau}\cdot\int_{t-\tau}^{t}\left(\ddot{x}_{W_d}^{*}(\tau)\right)^{2}d\tau\right]^{0.5} \tag{10}$$

$$a_{yj}^{W_d}(t)=\left[\frac{1}{\tau}\cdot\int_{t-\tau}^{t}\left(\ddot{y}_{W_d}^{*}(\tau)\right)^{2}d\tau\right]^{0.5} \tag{11}$$

$$a_{zj}^{W_d}(t)=\left[\frac{1}{\tau}\cdot\int_{t-\tau}^{t}\left(\ddot{z}_{W_d}^{*}(\tau)\right)^{2}d\tau\right]^{0.5} \tag{12}$$

From values of frequency weighted acceleration, the NMV indices for all points according to the equation (1) were calculated. The evaluation of the passenger ride comfort based on the estimated comfort indices and comparing with the scale in Tab. 1.

Table 1. Comfort index evaluation scale [4]

$N_{MV} < 1.5$	Very comfortable
$1.5 \leq N_{MV} < 2.5$	Comfortable
$2.5 \leq N_{MV} < 3.5$	Medium
$3.5 \leq N_{MV} < 4.5$	Uncomfortable
$N_{MV} > 4.5$	Very uncomfortable

For the assessment of negative effects of the passenger car running with the wheel-flat on the passenger ride comfort, the wagon was run at various speeds. We have chosen the speed interval from 30 km/h up to 210 km/h with the step of 30 km/h. Thus, in the aggregate there were performed seven simulations and for a comparison other seven simulations with the passenger car with a purely wheel radius (without wheel-flat). The ride comfort indices were calculated for both cases. Results of all analyses are shown in Figs. 5 - 11.

Calculated indices of the passenger ride comfort allow objectify expected subjective discomfort during travelling.

Generally, from all results we can see that the position in which the passenger experiences the least influence of acceleration on the body is near to the centre of the wagon body (centre of gravity). Areas, in which the highest values of acceleration are generated, are in the front and rear sectors of the passenger car.

Let's us now focus on how the passenger car running with the wheel-flat influences on the ride comfort at various speeds. All figures are arranged so that results from analyses with the wheel-flat are on the right side and results from analyses without the wheel-flat are on the left side (take note of Figures).

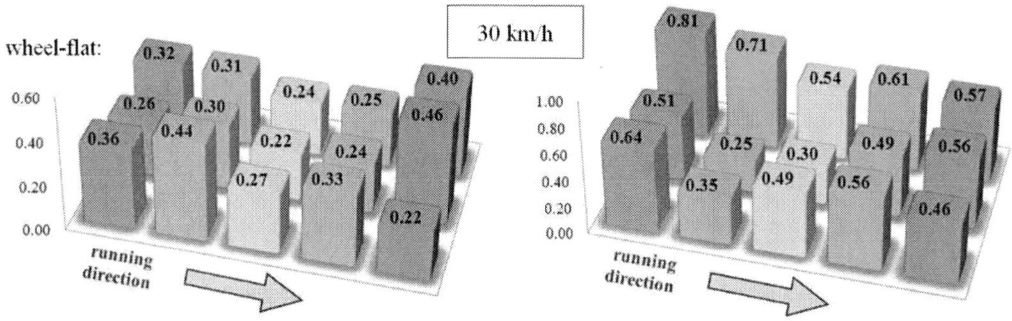

Fig. 5. The N_{MV} indices for the speed of 30 km/h with wheel-flat (left) and without wheel-flat (right)

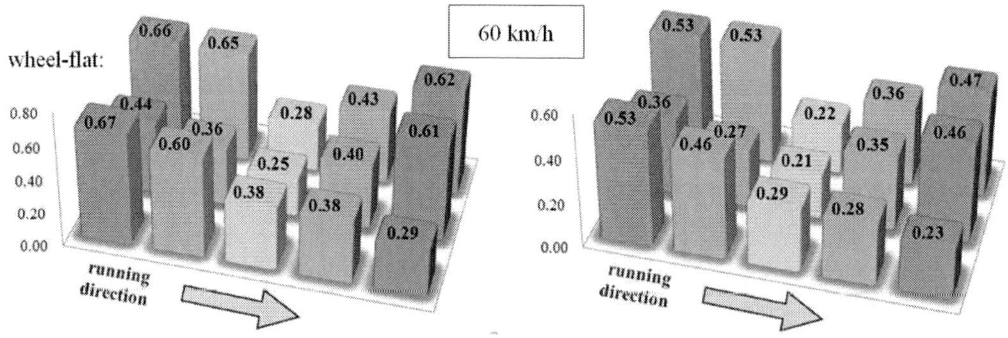

Fig. 6. The N_{MV} indices for the speed of 60 km/h with wheel-flat (left) and without wheel-flat (right)

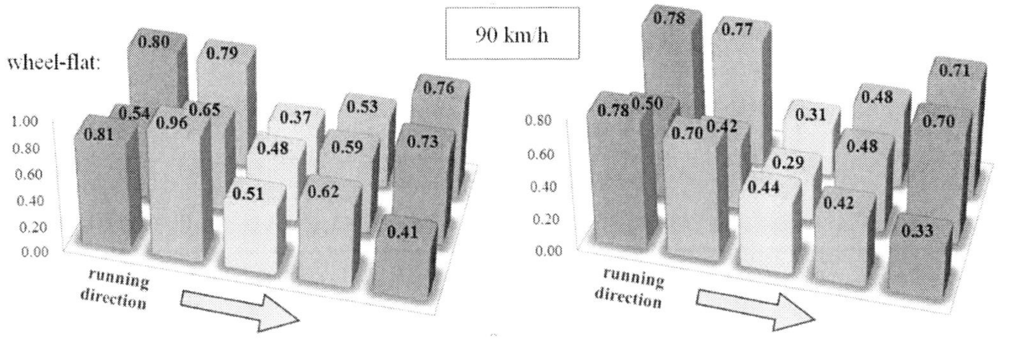

Fig. 7. The N_{MV} indices for the speed of 90 km/h with wheel-flat (left) and without wheel-flat (right)

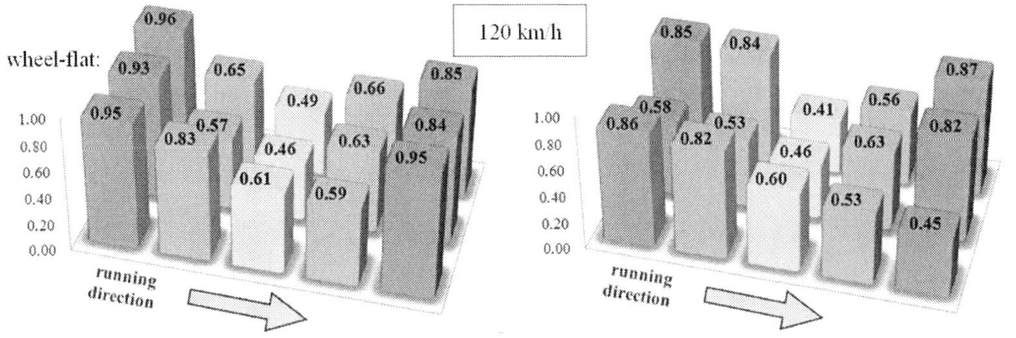

Fig. 8. The N_{MV} indices for the speed of 120 km/h with wheel-flat (left) and without wheel-flat (right)

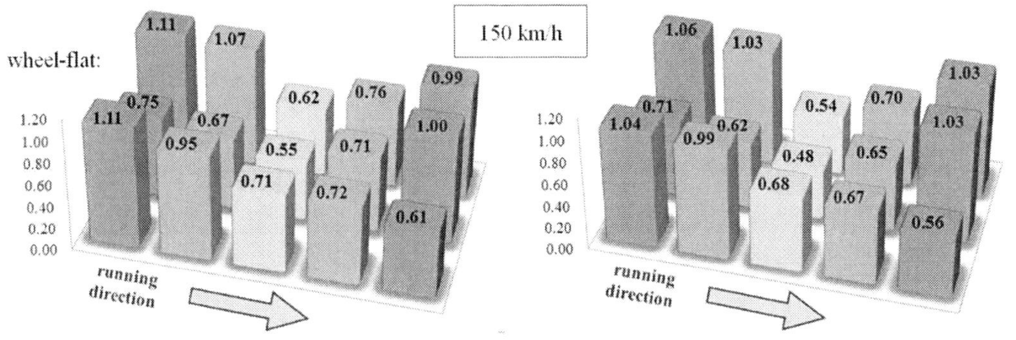

Fig. 9. The N_{MV} indices for the speed of 150 km/h with wheel-flat (left) and without wheel-flat (right)

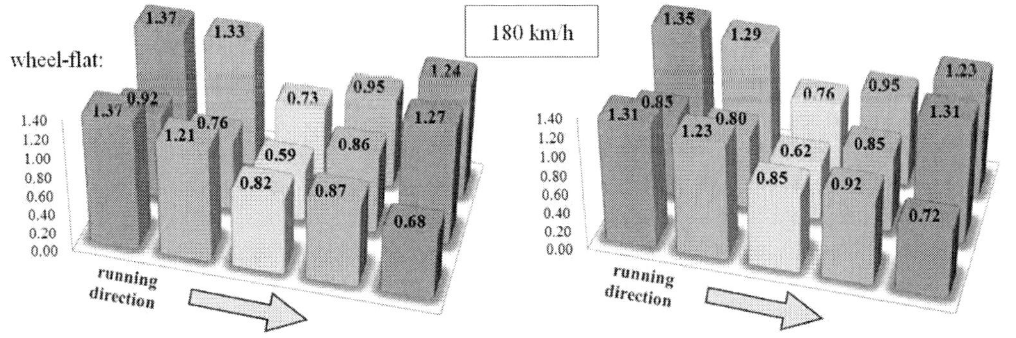

Fig. 10. The N_{MV} indices for the speed of 180 km/h with wheel-flat (left) and without wheel-flat (right)

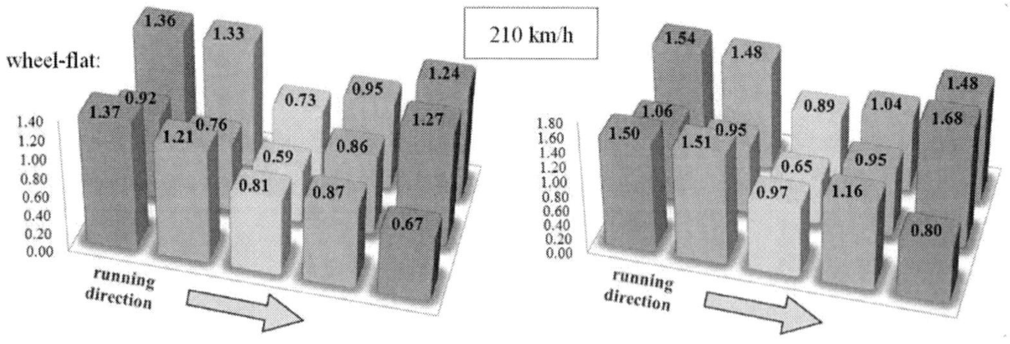

Fig. 11. The N_{MV} indices for the speed of 210 km/h with wheel-flat (left) and without wheel-flat (right)

Comparison calculated ride comfort index N_{MV} shown in Figs. 5 - 11 stepwise form 30 km/h (Fig. 5) up to 210 km/h we can see, as we are expected, that ride comfort for passenger is worse in case passenger car running with wheel-flat. But except for two investigated cases: if the wagon runs at the relatively low speed of 30 km/h and at the highest analysed speed of 210 km/h.

For these two cases we have identified that after performing the entire calculating procedure of the N_{MV} indices, that the wagon running if less comfortable in wagon without wheel-flat. This is an interesting detection primarily concerning the low speed because in published works [30, 32] there is found out such phenomenon, that exceedingly the driving a rail vehicle with damaged wheel at lower speed has more negative effects. Together herewith there is necessary to warn that in referred publications negative impacts from the track point of view were investigated but not from the passenger ride comfort point of view.

Finally, from the global view and comparison all calculated N_{MV} indices values with the scale given in EN 12299:2009 [4] (Tab. 1), the travelling in the analysed passenger car is classified for as „very comfortable" ($N_{MV} < 1.5$) except of the case, when the wagon without the wheel-flat run at the speed of 210 km/h, where one point (Point 6) in front of wagon body and three points (Points 3, 4 and 12) in the rear part of the wagon body are classified as „comfortable" (values of the N_{MV} ride comfort index exceeded 1.5).

In our future research we will focus on the more detailed investigation and comparison of the wagon running at the specific speeds and we will perform more simulations with smaller step of the running speed in order to detect, which the wagon with the damaged wheel generates most adverse effects on passengers at.

Conclusion

The passenger ride comfort is the criterion, whose assessment is the most actual today. The ride comfort is mainly from acceleration signals assessed. The EN, UIC and ISO standards states conditions, which assessment is performed. Signals are obtained by measuring accelerations at various wagon body points.

When neither the real vehicle nor measured acceleration data are available computer simulation tools allow obtaining these acceleration data. Through them we can perform computer simulations of a modelled rail vehicle on a track for various excitement values and assess vehicle properties based on prescribed parameters. Then, consistent with valid standards we can calculate indices of ride comfort for passengers based on results of dynamic model of the analysed vehicle.

This article was supported by the project VEGA 1/0927/15 "Research of the use of alternative fuels and hybrid drives on traction vehicles with aim to reduce fuel consumption and air pollutants production".
The work was supported by the Cultural and Educational Grant Agency of the Ministry of Education of the Slovak Republic in project No. KEGA 077ŽU-4/2017: Modernization of the Vehicles and engines study program.
This publication is the result of the project implementation: Modern methods of teaching control and diagnostic systems of engine vehicles ITMS 26110230107 supported by the Operational Programme Education funded by the ESF.

VÝSKUMNÁ AGENTÚRA Európska únia Štrukturálne fondy

References

1. M. Bogdevicius, R. Zygiene, G. Bureika, S. Dalydka, *An analytical mathematical method for calculation of the dynamic wheel-rail impact force caused by wheel flat*. Vehicle System Dynamics **5**, 689-705 (2016)
2. J. Dižo, S. Steišūnas, M. Blatnický, *Dynamic analysis of motion smoothness of rail vehicle with wheel-flat*. Proceedings of 22nd International conference "Current problems in rail vehicles – PRORAIL 2015" **I 22**, 83-92 (2015)
3. J. Galliková, R. Poprocký, P. Volna, *Implementation of FMEA method in maintenance of semi-trailer combination*. Diagnostyka **4**, 85-92 (2016)
4. EN 12299:2009, *Railway Applications – Ride Comfort for Passengers – Measurement and Evaluation*. European Committee for Standardization, Brussels (2009)
5. J. Gerlici, *Comfort for passenger of rail vehicle assessment*. Advanced methods in computational and experimental mechanics, 141-172 (2013)
6. Z. Ondrová, J. Gerlici, T. Lack, *Comfort for passengers of railway vehicle analysis*. Proceedings of 7th European conference of young research and science workers "TRANSCOM 2007" **7**, 205-210 (2007)
7. J. Gerlici, T. Lack, J. Harušinec, *Realistic simulation of railway operation on the RAILBCOT test stand*. Applied mechanics and materials **486**, 387-395 (2014)
8. J. Gerlici, M. Gorbunov, K. Kravchenko, A. Kostyukevich, O. Nozhenko, T. Lack, *Experimental Rigs for Wheel/Rail Contact research*. Manufacturing Technology **5**, 909-916 (2016)
9. J. Gerlici, T. Lack, *Rail vehicles brake components test bench utilisation*. Applied mechanics and materials **486**, 379-386 (2014)
10. S. Dailydka, S. Naimovičius, V. Lingaitis, *Analysis of passenger vibration comfort in a train* Transport Means – Proceedings of the International Conference, International Conference Transport Means 2012, Kaunas, Lithuania **16**, 302-305 (2012)
11. V. Hauser, O. Nozhenko, K. Kravchenko, M. Loulová, J. Gerlici, T. Lack, *Impact of three axle boxes bogie to the tram behavior when passing curved track*. Procedia Engineering **192**, 295-300 (2017)
12. T. Lack, J. Gerlici, *Vehicles Dynamical Properties Analysis from the Point of View of Comfort for Passengers*. Archives of Transport **1-2**, 91-110. (2007)
13. T. Lack, J. Gerlici, *Analysis of Vehicles Dynamic Properties from the Point of View of Passenger Comfort*. Komunikacie **2**, 10-18 (2008)

14. M. Maňurová, A. Suchánek, *The analysis of a rail vehicle with a tilting bogie*. Manufacturing Technology **5**, 1020-1027 (2016)
15. P. Šťastniak, J. Harušinec, *Computer aided simulation analysis for computation of modal analysis of the freight wagon*. Komunikacie **4**, 73-79 (2013)
16. M. Kostrzewski, R. Mełnik, *Numerical dynamics study of a rail vehicle with differential gears*. Procedia Engineering **192**, 439-444 (2017)
17. P. Hejma, M. Svoboda, J. Kampo, J. Soukup, *Analytic Analysis of a Cam Mechanism*. Procedia Engineering **177**, 3-10 (2017)
18. F. Klimenda, J. Soukup, *Modal analysis of thin aluminium plate*. Procedia Engineering **177**, 11-16 (2017)
19. P. Baran, M. Brezáni, P. Kukuča, P. Šťastniak, *Basic dynamical analysis and comparison of balancing systems of non-conventional piston machine FIK*. Procedia Engineering **192**, 34-39 (2017)
20. V. Baniari, M. Blatnická, M. Šajgalík, M. Vaško, M. Sága, *Measurement and numerical analyses of residual stress distribution near weld joint*. Procedia Engineering **192**, 22-27 (2017)
21. A. Sapietová, J. Bukovan, M. Sapieta, L. Jakubovičová, *Analysis and implementation of input load effects on an air compressor piston in MSC.ADAMS*. Procedia Engineering **177**, 554-561 (2017)
22. A. Sapietová, L. Gajdoš, V. Dekýš, M. Sapieta, *Analysis of the influence of input function contact parameters of the impact force process in the MSC.ADAMS*. Advanced mechatronics solutions **393**, 243-253 (2015)
23. R. Labuda, D. Barta, A. Kovalčík, *Effective use of the braking effect of vehicle drivetrain at deceleration*. Proceedings of XLI. International scientific conference of Czech and Slovak university departments and institutions dealing with the research of internal combustion engines "KOKA 2010" **41**, 206-211 (2010)
24. P. Kopas, L. Jakubovičová, M. Vaško, M. Handrik, *Fatigue resistance of reinforcing steel bars*. Procedia Engineering **136**, 193-197 (2016)
25. J. Gerlici, T. Lack, *Modified HHT method for vehicle vibration analysis in time domain utilisation*. Applied mechanics and materials **486**, 396-405 (2014)
26. Gerlici, T. Lack, *Contact geometry influence on the rail/wheel surface stress distribution*. Procedia Engineering **1**, 2249-2257 (2010)
27. J. Gerlici, T. Lack, *Railway wheel and rail head profiles development based on the geometric characteristics shapes*. Wear **1-2**, 246-258 (2011).
28. T. Lack, J.Gerlici, *A modified strip method to speed up the calculation of tangential stress between wheel and rail*. Applied mechanics and materials **486**, 371-378 (2014)
29. T. Lack, J. Gerlici, *A modified strip method to speed up the calculation of normal stress between wheel and rail*. Applied mechanics and materials **486**, 359-370 (2014)
30. J. Dižo, S. Steišūnas, M. Blatnický, *Simulation Analysis of the Effects of a rail Vehicle Running with Wheel Flat*. Manufacturing Technology **5**, 889-896 (2016)
31. J. Bian, G. Yuantong, M.H. Murray, *A Dynamic wheel-rail impact analysis of railway track under wheel flat by finite element analysis*. Vehicle System Dynamics **6**, 784-797 (2013)
32. M. Pawełczyk, P. Lesiak, R. Podsiadło, *Simulation study of the 4-axle wagon damaging impact on the track caused by some deformations on the wheel*. Proceedings of 22nd International conference "Current problems in rail vehicles – PRORAIL 2015" **II 22**, 115-125 (2015)

Free vibration analysis of a DST-0285 truck crane considering changes in the system load configuration

Tomasz Geisler*,[1]

[1]Institute of Mechanics and Machine Design Foundations, Faculty of Mechanical Engineering and Computer Science, Czestochowa University of Technology, ul. J. H. Dąbrowskiego 73, 42-201, Czestochowa, Poland

Abstract. A model of a truck crane, consisting of the fundamental units of the crane, is presented in this paper. The finite element method and a COSMOS/M package were applied to perform the modelling. Research into the influence of a change in the crane radius and load location on changes in the chosen free vibrations was carried out using the built models. Diagrams containing changes in the frequency of vibrations for chosen values of the crane radius and the load are presented and conclusions are drawn.

Keywords: truck crane, load, forms of vibrations, frequency

1 Introduction

The problem of modelling the dynamics of a truck crane support system subjected to a load is a complex one. To build an appropriate dynamic model of a crane, a knowledge of its basic work should be taken into account.

Crane work is associated with lifting and moving loads. This requires changes in the reciprocal locations of the fundamental systems of the crane. All the reciprocal positions of the crane units in the full range of their movements and the dynamic properties of all the units must be taken into account. This has been considered in many studies. A number of works have presented a multitude of problems related to descriptions of the influence of crane units on the dynamics of the crane system. A model that covers a full description of a load movement in relation to operational motions was presented in [1].

In the papers [2, 3] the influence of the support system, including the foundations, on the dynamics of a truck crane was analysed. The influence of changes in the mass of the load on the dynamics of the whole object was presented in [4]. The monograph [5] concerns modelling and research into the dynamics of a self-propelled truck crane.

In the paper [6] was analyzed of transverse damped vibrations in the system of changing the boom radius of a DST-0285 truck crane. Damped frequencies of vibrations and degree of amplitude decay of studied system were calculated.

* Corresponding author: geisler@imipkm.pcz.czest.pl

Reviewers: *Marek* Macko, *Krzysztof* Talaśka

The construction of crane units and research into the dynamics of a chassis frame were presented in the papers [7, 8] which were sacrificed to the crane modelling. In the papers the dynamic stability of laboratory model of the truck crane was presented [9].

In the work [10] was formulated and solved the problem of transverse damped vibrations in a crane boom of a truck-mounted crane. The influence of structural damping on the support and slides of the boom and the change of the geometry of the system to the eigenvalues of the system are presented.

Later works considered the influence of the remaining units of a truck crane, including the crane boom, on the free vibrations of the crane. Works on the verification of a discrete-continuous model and FEM model of a truck crane were also published [11]. This work is a continuation of research into the crane system [12, 13, 14] taking into account changes in the crane radius and a variable value of the load. A paper on static analysis was published [15] in the frame of research into the crane support system.

The aim of this study was to build a complete truck crane model and to obtain modal shapes and the corresponding vibration frequency, which would illustrate the dynamic behaviour of the crane system for various configurations of its fundamental units while moving cargo.

Such a parametric model allows computations on the assumed reciprocal locations of the fundamental systems of the crane to be carried out.

Changes in the crane system configuration for the reciprocal locations of the fundamental systems of the crane comprehended in the usage documentation and found in crane exploitation are also possible. Due to number limitation of the created models, computations were performed for the selected operating configuration of the system.

This paper presents an analysis for three selected forms of vibrations. Those chosen forms gave the full picture of all crane systems linked as a whole. Changes in the frequency of vibrations for the chosen forms of the crane system in dependence on changes in: an elevation angle, the length of crane boom, load value and its location, were presented as the results of research.

2 Computational model of a truck crane

Modelling the crane support system requires basic units of the crane in their complete form, and the additional units, to be taken into consideration. The proposed model of the crane support system consists of coupled main crane units: the chassis frame, the rotating frame (body) with the embedded three-stage telescopic boom and the cordage. In the modelling the weight of the equipment and the units loading the system were taken into account.

Change in the configuration of the crane system is due to its operational movements. The movements are determined in a horizontal plane of rotation of the body frame with the telescopic boom in relation to the body frame and rotation of the boom regarding the body frame in a vertical lifting plane. The crane radius depends on an angle of inclination of telescopic crane boom and its variable length.

Load of the crane lifting system is carried out by the assumed load, suspended on ropes. Load model takes into account changes in the position of the load and associated elastic properties of cordage.

The location of the assumed load is determined by the combination of the aforementioned relative positions of the crane units and the length of the crane cordage. Computations were conducted on the selected positions of the body frame with the embedded telescopic boom in relation to the body frame.

The main units of a DST-0285 crane were designed and made in „Bumar-Fablok" Factory of Building Machinery and Locomotives in Chrzanow. The models are worked out on the basis of the submitted documentation.

Construction of all main elements in the form of the main sheets, membranes, webs, ribs, beams and the other essential elements of construction was considered in the modelling crane units [7, 8, 12].

The aim of the performed modelling was to obtain the crane model making research possible with the respect to determination of frequency and free vibrations of the crane system. The basic model was built with the use of the finite element method and a COSMOS/M package [16, 17].

In the modelling, the following basic structural elements were applied to linear analysis: shell3, solid, spring, rbar, mass.

A series of necessary simplifications, resulting from a large number of elements and associated with this complexity of construction, was applied in the model.

The basic model was built parametrically. This allows to change the relative positions of crane units and to modify all geometrical dimensions of the framework and the boom, for example the thickness of metal sheets. In the basic model, only the values of parameters, characteristics of system components and their mutual positions are being changed. The number of parameters was over 600.

The load values as well as the length and the elasticity of ropes suspending the load are also introduced using the parametrical method.

The model of the analysed crane system consisting of connected chassis frame (1), body frame (2) and the telescopic crane boom (3), rope system (4) and the load (5) of a truck crane is presented in the figure (Fig. 1). Elasticity of the boom system (6) and elasticity of the support chassis system (7) were considered in the computations.

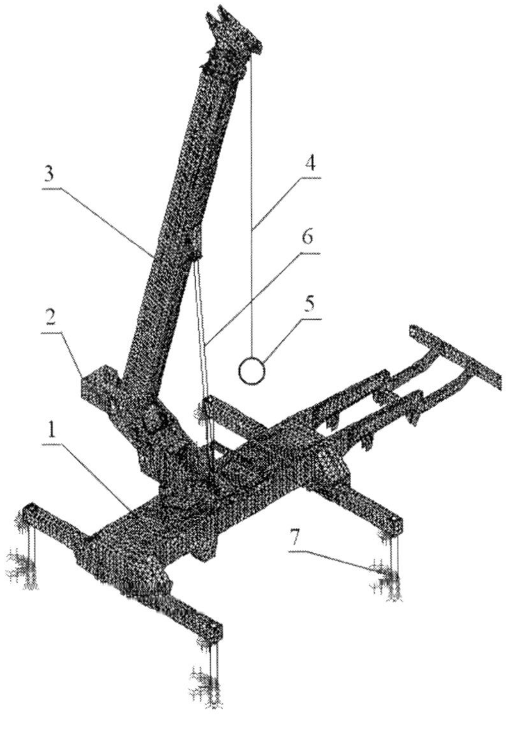

Fig. 1. The model of the truck crane system: (1) chassis frame, (2) body frame, (3) telescopic crane boom, (4) rope system, (5) load, (6) boom system, (7) support chassis system

The model was loaded with a cargo, placed as a weight suspended on a rope system. The load values of the model were given as the concentrated mass, suspended on elements modelling a rope system.

Apart from the cargo load, system was loaded with the mass elements modelling majority of elements, construction units and the equipment of the crane. According to the documentation, the mass values of the crane equipment: the operator and driver cabins, main and auxiliary winches, counterweight, engine with a gearbox, front and rear suspension, wheels, lift and telescoping actuators were considered in modelling. The mass elements modelling crane units were placed in the areas of their actual location in the structure of the crane.

Suspension of the load on a spring rope system was taken into account and the equivalent elasticity value of the rope system was accepted for the computations [14] in the carried out modelling.

The computations were conducted assuming 6 strands of rope on the sheave block and accepting the rope length from the boom head to the main winch equalled to the boom length. Parameters of the rope were assumed as WS6x36 construction.

The equivalent elasticity value of the boom supporting actuator was considered. The elasticity adopted in the computations, varied along the length of the actuator and was associated with the value of the elevation angle of the telescopic boom.

The value of the boom support stiffness varied within 4-5x10^6 [N/m]. The model took into account the constant value of elasticity of the chassis support (support actuators) [7].

3 Exemplary results of simulation research

In the study of the dynamics of whole crane it is necessary to consider not only the behaviour of the telescopic boom, but also the other crane units. The dynamic behaviour of the linked crane units is substantial.

Consideration of the crane as a complex system should also include the movement of the telescopic boom mounted on the body frame, which then is mounted on the chassis frame.

The dynamic movement of the boom forces the movement of the whole body frame and a lifting element of the chassis frame.

Forced displacements of both frames are closely connected with the boom displacement, linking the movements of the major units into movement of the entire crane support system.

The study was conducted with the assumed configurations of the reciprocal positions of the crane units and lengths of rigging in order to obtain the influence of changes of boom outreach and load positions on free vibrations of the crane support system.

In order to perform the research, models of the body frame with a telescopic boom fixed to a body frame along the rotation plane were assumed, designation "+90°" [13]. The models were built considering changes in the crane radius and load position.

Change in the crane radius encloses change in an angle of inclination and length of the boom while change in load of the crane encloses change in its value and length of its suspension. Considering constant location angle of the body frame with telescopic crane boom in relation to chassis frame, changeable parameters connected to changes in the crane radius were assumed:

- constant length of the boom lt = 10 [m], changeable lifting angles of the boom α = 0°, 25°, 50°, 75°.
- constant lifting angles of the boom α = 0°, 25°, 50°, 75°, changeable lengths of the boom lt =10, 15, 20, 24 [m] (Fig. 2).

Changeable value of load location connected to the length of the crane cordage was given for all combinations of named parameters. The length of rope, l = 2, 10, 17, 27 [m], determined as the distance of cargo from the boom head at minimal permissible value l = 0.5 [m], was assumed.

Load m was included from m_{min}=100 [kg] to m_{max} = 30000 [kg]. The values of the masses: m = 100, 1000, 5000, 10000, 20000, 30000 [kg] were accepted. Verifying computations for l = 0, 5, 12, 22 [m] and m = 0 [kg] were conducted.

Changes in the crane radius and loads assumed in modelling exceeded the values determined by real hoisting capacity which were included in (DTR) documentation and the table of hoisting capacities. However, they were necessary to determine changes in the frequency of considered vibrations.

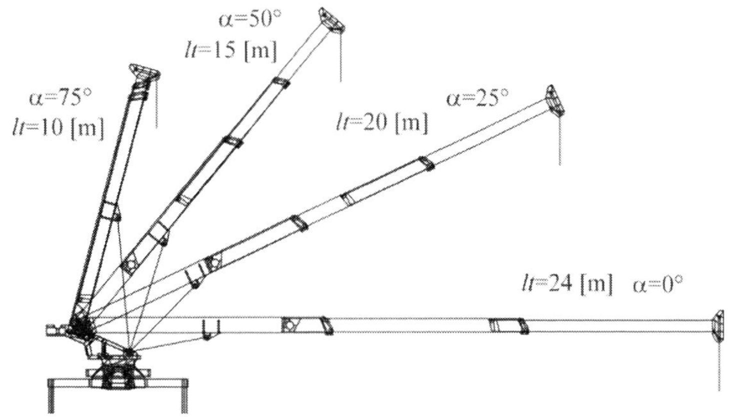

Fig. 2. Changes in the boom length and its location in relation to a body frame in lifting plane

Changes in the vibrational frequency of the chosen forms of free vibrations were presented as the results of solution to the problem of research into vibrations of the crane system. Changes in the values of the vibrational frequency for the three chosen forms of vibrations were put to the analysis. The choice of the analysed forms of vibrations resulted from the assumption that these forms have the highest influence on the dynamic behaviour of the crane support system.

For the chosen forms of vibrations the following denotations were assumed:

- form of flexural vibrations of the telescopic crane boom, denotation „G", (frequency ω_1),
- form of flexural vibrations of the boom together with frames, denotation „GR", (frequency ω_2),
- form of flexural vibrations of the boom together with frames, denotation „GP", (frequency ω_3).

„G" form of vibrations contains mainly flexural vibrations of triple telescopic crane boom in the lifting plane (Fig. 3).

The next form, „GR", contains flexural vibrations of the boom in interaction mainly with the body frame and at a lower rate with the chassis frame as the crane support system. For this form movement of the boom with the body frame produced rotation of the central part of the chassis frame around longitudinal axis (Fig. 4)

The form, denoted as „GP", also contains flexural vibrations of the boom. In this case movement of the boom is connected to considerably higher displacements of whole support part of the chassis frame. The movements of the units take place in the vertical planes. For the „GP" form, displacement amplitude of the boom elements is comparable to displacements of the chassis frame elements (Fig. 5).

To determine vibrational frequency of the analysed forms, their identification was performed on the basis of an analysis of displacements of the crane units during animation of their movements.

Figures (Figs. 3-5) present displacements of the boom and crane frames with the initial positions plotted for „G", „GR" and „GP" forms.

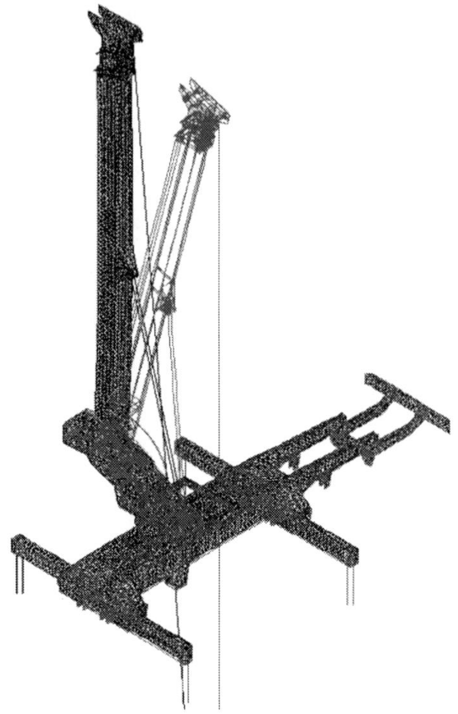

Fig. 3. Displacements of the boom for „G" form with the initial positions plotted

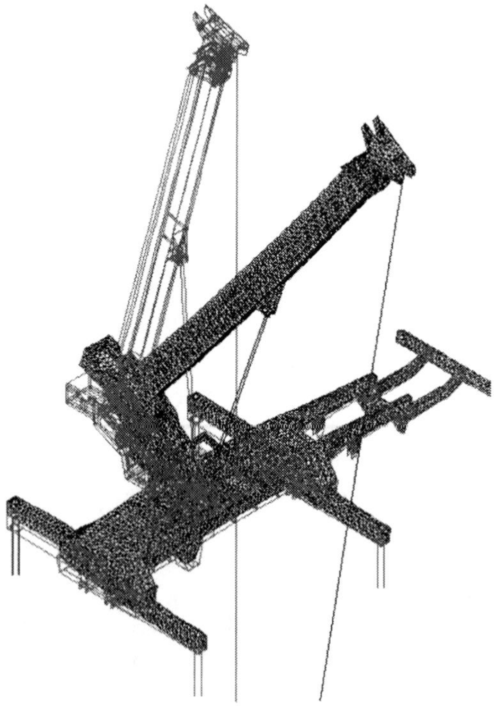

Fig. 4. Displacements of the boom and frames for „GR" form with the initial positions plotted

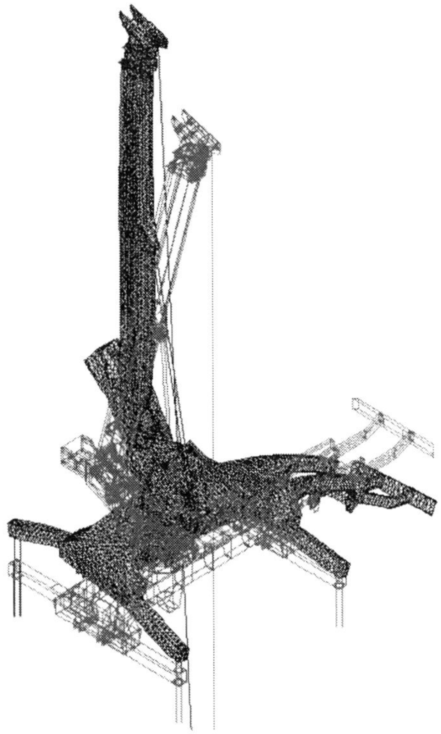

Fig. 5. Displacements of the boom and frames for „GP" form with the initial positions plotted

Computations were conducted on the chosen forms of vibrations containing the boom vibrations in the vertical lifting plane and for the displacements of body and chassis frames.

The obtained results are presented in diagrams containing courses of changes in the values of vibrational frequency in relation to chosen parameters.

Diagrams (Figs. 6-8) concern the influence of the boom lt = 10-24 [m] and lifting angle (α = 0°, 75°), for the constant value of the rope length l = 17 [m] and changeable load m = 0-30000 [kg].

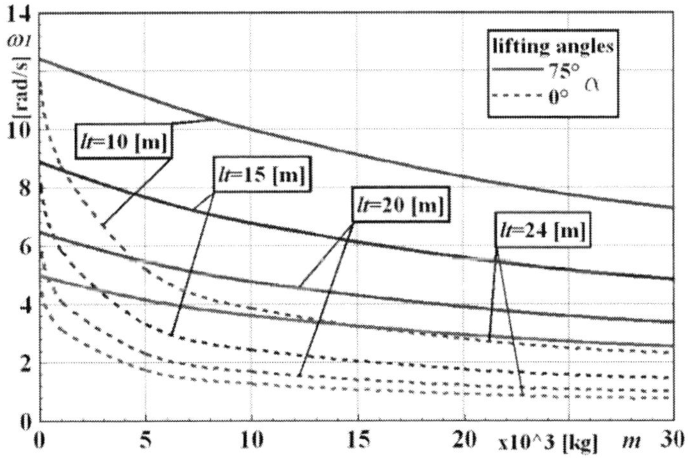

Fig. 6. Changes in vibrational frequency ω_l for „G" form

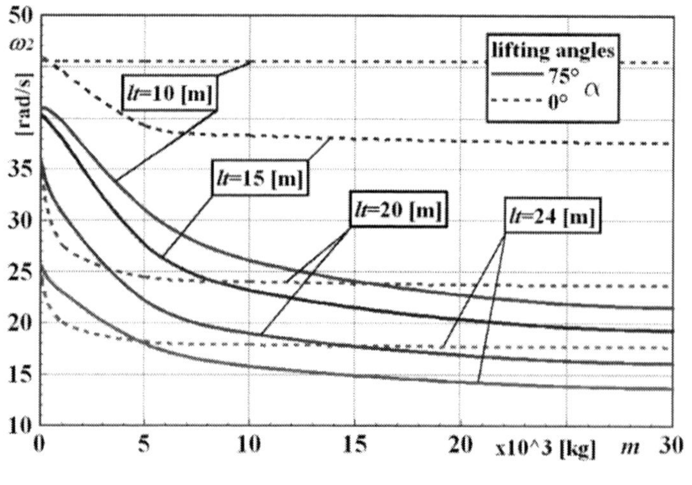

Fig. 7. Changes in vibrational frequency ω_2 for „GR" form

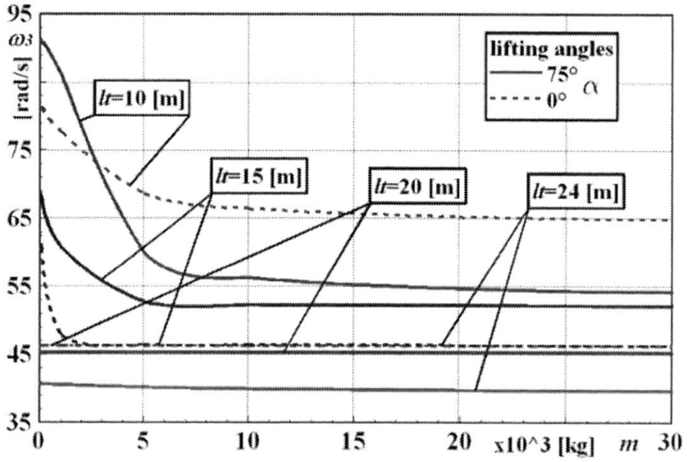

Fig. 8. Changes in vibrational frequency ω_3 for „GP" form

In the diagrams plotted for two values of the boom length (lt = 10, 24 [m]), one can observe decrease in vibrational frequencies ω_1, ω_2 and ω_3 together with an increase in the value of the load m. At constant load together with decrease in an angle of inclination of the boom α, the value of ω_1 decreases for „G" form while for „GR" form the value of ω_2 increases. This concerns every analysed length of telescopic crane boom.

For „GP" form together with decrease in the value of α angle one can observe significant change in value of ω_3. For the greater lengths of the boom this influence is lesser. The course of line is very similar for the values: lt = 20 and 24 [m] and an angle $\alpha = 0°$.

Diagrams plotted in the „XYZ" coordinate system (3D) are placed in the further part of the paper. Vertical axis „Z" contains changes in vibrational frequency and horizontal axes (X, Y) contain changes in parameters m and l, for the constant for example lt.

Diagrams present data obtained for the higher number of computational models and contain surfaces built on the lines which represent changes in frequency of the tested forms in dependence on the chosen parameters. Changes in the shape of obtained surfaces allow for detailed description of analysed changes.

Then chosen diagrams for the three analysed forms, for the constant length of the boom $lt = 10$ [m], lifting angle $\alpha = 0\text{-}75°$ and rope length $l = 17$ [m], considering changeable values of the load $m = 0\text{-}30000$ [kg] were plotted (Figs. 9-11).

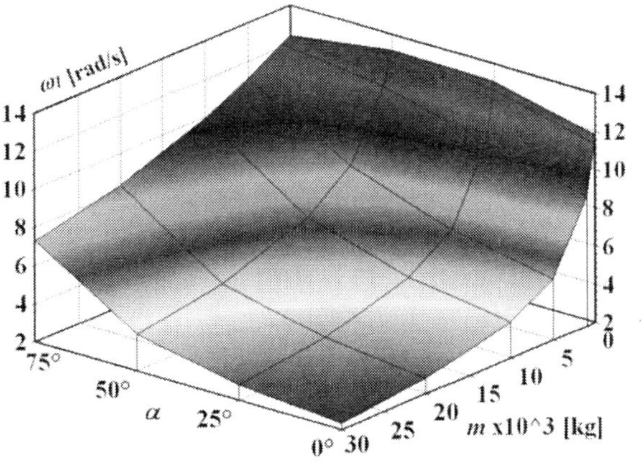

Fig. 9. Changes in vibrational frequency ω_1, for „G" form

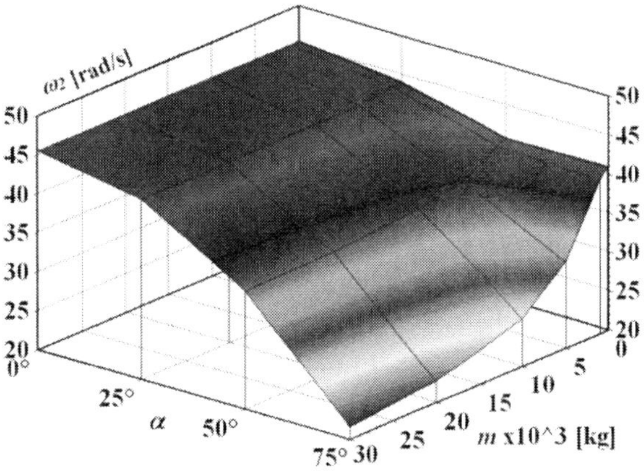

Fig. 10. Changes in vibrational frequency ω_2, for „GR" form

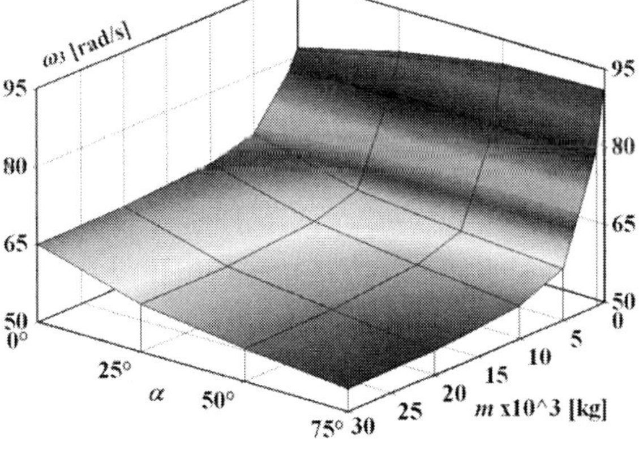

Fig. 11. Changes in vibrational frequency ω_3, for „GP" form

Increase in the value of the load m leads to decrease in the value of frequencies of every tested forms of vibrations and depends on the lifting angle of the boom α. For „G" and „GP" forms, increase in the value of the load leads to change in character of the curve dependent on the assumed value of the load.

Surfaces in the diagrams plotted for the other boom lengths have the similar course for „G" and „GR" forms. Range of changes and reduction of the value of frequency decrease together with increase in the boom length.

For „GP" form the shape of surface is changeable and depend on the assumed boom length. Range of changes in the value of frequency decreases together with an increase in the boom length and is the lowest for the boom length lt=24 [m]. Increase in the value of the load for constant length of rope and increase in the boom length leads to decrease in the values of vibrational frequency of the analysed forms. Character of decrease in the value of frequency is different and depends on the analysed lifting angle of the boom.

The next diagram (Fig. 12) was plotted for changes in frequencies of „GP" form. It contains the influence of rope length l = 2-27 [m] on the changeable load m = 0-30000 [kg] at the constant lifting angle α = 75°.

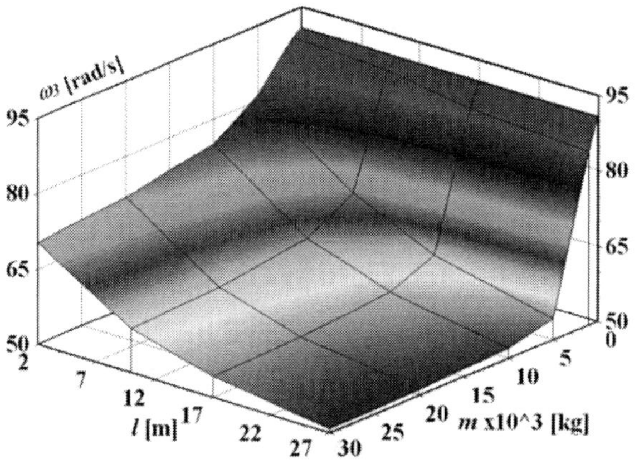

Fig. 12. Changes in vibrational frequency ω_3, for „GP" form

The influence of the load location on the range of changes (decrease) in the value of frequency is connected to the length of rope system.

For the considered forms of frequencies, changes are most of all noticeable at the boom length lt=10 m.

For the highest assumed value of the load and rope lengths, decrease in the value of frequency equals for the forms: „G" -8%, „GR" -21.7%, „GP" -27.6%, respectively. The range of changes decreases together with increase in the boom length and decrease in the angle of inclination.

An analysis of computations reveals the influence of change in the crane radius, which is a combination of lifting angle and the boom length and changes in rope length considering the changeable load, on changes in vibrational frequencies of the analysed forms.

Conclusions

The assumed aim of this work was realised through a series of computations on the worked out model to find the influence of different configurations of the crane support system on the dynamic behaviour of a truck crane during the load displacement.

Consideration of the dynamic behaviour of the crane system took into account different forms of free vibrations containing movement of the telescopic crane boom as the main operating element and of the remaining units for their different reciprocal positions.
Chosen diagrams, which enable an evaluation of the influence of the assumed parameters on the behaviour of the crane system for the chosen forms of vibrations, were included in this paper. Consideration of changes in the crane radius in the lifting plane together with changes in the load revealed their essential influence on the dynamic behaviour of the crane support system.

The choice of three analysed forms of vibrations seemed to be justified due to their visible influence on the dynamic behaviour of the truck crane system.

Limiting the analysis of crane dynamics to only vibrations of the boom present for „G" form is insufficient. It is necessary to consider the consecutive forms of vibrations. The analysed forms of vibrations („GR", „GP") are complex forms of vibrations of crane units. They provide a wider picture of the dynamic behaviour of the crane units as a whole during changes in the crane radius and the load location.

In the research conducted into a comparison of MES model and the analytical model, change in curve for different lengths of ropes was stated for the „G" form. It depends on an increase in the load value pointed out in work [11].

Consideration of the chosen forms of vibrations and the consecutive forms presented in computations is essential for further analysis of the dynamics of a DST-0285 truck crane.

Research into the chosen forms containing changes in location of the telescopic crane boom with the body frame in relation to the chassis frame in the rotation plane is also justified. This is connected with the need to take into account the influence of the changeable rigidity of the lifting part of the chassis frame on the location of the remaining units of the crane.

The built models and conducted computations can be applied in experimental research into measurements of vibrational frequencies and into modal analysis of real system of a DST-0285 truck crane.

References

1. B. Posiadała, B. Skalmierski, L. Tomski, *Motion of lifted load brought by kinematic forcing of the crane telescopic boom*. Mechanism and Machine Theory **25** (5), 547-556 (1990)
2. B. Posiadała, *Influence of crane support system on motion of the lifted load*. Mechanism and Machine Theory **32** (1), 9-20 (1997)
3. Z. Towarek, *The dynamic stability of a crane standing on soil during the rotation of the boom*. International Journal of Mechanical Sciences **40** (6), 557-574 (1998)
4. J. Strzałko, J. Grabski, *Dynamic analysis of a machine model with time-varying mass*. Acta Mechanica **112**, 173-186 (1995)
5. B. Posiadała (Ed.), *Modelling, identification of models and research into dynamics of a truck mounted crane*. (WNT, Warsaw: Scientific-Technical Publishing House, 2005) [in Polish]
6. W. Sochacki, M. Bold, *Damped vibration of the system of changing the crane boom radius*. Journal of Applied Mathematics and Computational Mechanics **14** (2), 111-122 (2015)
7. T. Geisler, B. Posiadała, *Free vibration analysis of a chassis frame of a self-propelled truck crane*. SYSTEMS – Journal of Transdisciplinary Systems Science, "Computer Aided Engineering" **7** (1), 199-205 (2002) [in Polish]

8. T. Geisler, *Research into the influence of a load on the vibrations of a chassis frame of a truck crane*. Scientific Letters of Rzeszow University of Technology, Mechanics **60**, 197-204 (2002) [in Polish]
9. W. Sochacki, *The dynamic stability of a laboratory model of a truck crane*. Thin-Walled Structures **45**, 927-930 (2007)
10. W. Sochacki, M. Bold, *Effect of Structural Damping on Vibration of a Truck-Mounted Telescopic Crane*. American Institute of Physics, AIP Conference Proceedings **1648**, 850058-1- 850058-4 (2015)
11. T. Geisler, W. Sochacki, *Modelling and research into the vibrations of a truck crane*. Scientific research of the institute of mathematics and computer science **1** (10), 49-60 (2011)
12. T. Geisler, *Research into the influence of a load on the vibrations of a crane boom of a DST-0285 truck crane*. XXI National and First International Scientific Conference on Educational Theory of Machines and Mechanisms, Theory of Machines and Mechanisms, Bielsko-Biala - Szczyrk, 197-204 (2008) [in Polish]
13. T. Geisler, *Research into the influence of a load location on the vibrations of a crane support system of a DST-0285 truck crane*. Engineering Machines Problems **33**, 15-24 (2009) [in Polish]
14. T. Geisler, *Free vibration analysis of a crane support system of a DST-0285 truck crane considering changes in the system configuration*, Acta mechanica et automatica **4** (1), 30-37 (2010) [in Polish]
15. T. Geisler, *Static analysis of a crane support system of a DST-0285 truck crane*. Mechanical Review (Przeglad Mechaniczny) **7/8**, 42-48 (2012) [in Polish]
16. E. Rusiński, *Finite Element Method*. System COSMOS/M, WKiŁ, Transport and Communication Publishers, Warsaw, (1994) [in Polish]
17. COSMOS/M *User Guide Version* 1.75. Structural Research and Analysis Corporation, Los Angeles, (1995)

Calculated estimation of railway wheels equivalent conicity influence on critical speed of railway passenger car

Juraj Gerlici[1,*], *Rostyslav* Domin[2], *Ganna* Cherniak[2], *Tomáš* Lack[1]

[1]University of Žilina, Faculty of Mechanical Engineering, Department of Transport and Handling Machines, 010 26 Žilina, Slovak Republic
[2]Volodymyr Dahl East Ukrainian National University, Department of Rail Transport, 93400 Sewerodonetsk, Ukraine

Abstract. The article presents the results of determining the equivalent conicity characterizing the geometric interaction of wheels and rails on 1520 mm track. The cases of interactions of rails and wheelsets, wheels of which have different profiles, are considered. The computer model of the motion speed of the passenger car is developed. According to the results of computer simulation, critical speeds for the bogie moving through the tracks of indented gauge are received. Based on the results of the simulation analysis, the significant dependence of the value of the equivalent conicity on the critical speed is confirmed. The necessity of considering the limit values of equivalent conicity in tests of high-speed rolling stock is emphasized.

Keywords: railway wheels, equivalent conicity, critical speed, passenger vehicle

1 Introduction

The increase of speed of passenger trains up to 160-200 km/h brings new technological level on the Ukrainian railways. Cars of high-speed trains must fully comply with international requirements, both in the level of comfort, and in terms of safety. In this regard, the task of ensuring the stability of motion of the high-speed rolling stock must be resolved as a matter of priority. According to the results of numerous researches, depending on the design and parameters of running gears the configuration of working surfaces of wheels and rails significantly affects the dynamic behavior of rail vehicles, in particular, critical speeds in respect of hunting oscillations [1, 4, 6]. The problem of optimal interaction of underframes and the track becomes of particular importance for high-speed rolling stock.

According to regulatory documents operating on the railways of the European Union, field tests on the admission of the rolling stock to operation must be carried out under certain parameters of interaction between wheels and rails, including those which reflect the conditions of the worn out surfaces in contact [5, 13]. Such an approach was implemented to test the impact of natural wear of wheels and rails on the dynamic characteristics of the

* Corresponding author: juraj.gerlici@fstroj.uniza.sk

Reviewers: *Alžbeta* Sapietová, *Josef* Soukup

rolling stock. The results of the relevant research proved that for certain units of the rolling stock critical speeds in respect of hunting oscillation are considerably sensitive to contours of the threads of wheels [3]. However, up to this day regulatory documents on running dynamic tests of rolling stocks for 1520 mm tracks do not envisage tests on the impact of the depreciation of wheels on the dynamic behavior of the rolling stock, which does not allow qualitatively assess of its running characteristics in real operating conditions [2].

2 Characteristics of equivalent conicity

According to numerous studies of oscillations and stability of motion of railway rolling stocks, geometric characteristics of the interaction of wheels and rails are the factors that determine the dynamic properties of the rolling stock [8, 11]. Therefore, in the study of conditions for safe and comfortable operation of the rolling stock intended for operation with high speeds, the impact of geometric characteristics of the interaction of wheels and rails should be estimated as a matter of priority.

Among the parameters that characterize the geometric interaction of wheelsets and rails, the so-called "equivalent conicity" takes on the generalized role. Namely this parameter provides optimal estimation of the contact of the pair "wheel-rail" on straight sections of track and in curves of large radius. According to the known definitions, the equivalent conicity is equal to tangent $tg\gamma_e$ of the cone angle γ_e of the wheelset with conical wheels, transverse movement of which has the same kinematic wavelength as this wheelset [12].

The equivalent conicity provides an opportunity to compare conditions of the contact of wheels and rails in different states based on the design and maintenance. The linearization approach is used. This approach involves the replacement of nonlinear dependence of the change of the radius of rolling circumference for linear one. The angular coefficient of linear dependence is the equivalent conicity.

It is necessary to adhere to common rules in determining the equivalent conicity in order to able to compare the results obtained for various railways. For this purpose, the International Union of Railways implemented the principles of calculating the equivalent conicity defined by UIC 519 Leaflet [12].

Methods of calculating the equivalent conicity are as follows: analytical description of profiles of wheels and rails is given; $\Delta r = f(y)$ characteristic is calculated for each lateral displacement y of the wheelset as the difference between right and left radius of rolling circumference of wheels $\Delta r = r_r - r_l$; the equivalent conicity $tg\gamma_e$ for the lateral displacement y of the wheelset is determined using linear regression for the part of $\Delta r = f(y)$ characteristic within $2y$ interval.

The equivalent conicity should be calculated with the actual wheel profile of the tested vehicle and theoretical profile of the rail of the track with corresponding gauge. The lateral displacement of the wheelset is considered in the range of $y = \pm 3$ mm.

Acceptable values of the equivalent conicity, which should not be exceeded during the test of the rolling stock on the track, depend on the speed characteristics of rolling stocks and equal to [5]: 0.5 – at $V \leq 140$ km/h; 0.4 –140 km/h < $V \leq 200$ km/h; 0.35 – 200 km/h < $V \leq$ 230 km/h; 0.3 – 230 km/h < $V \leq 250$ km/h; 0.25 – 250 km/h < $V \leq 280$ km/h; 0.15 – 280 km/h < $V \leq 350$ km/h.

Values of the equivalent conicity for options of wheels profiles used on the Ukrainian railways are calculated in accordance with the said method, in particular, with GOST 9036 and DMetI, developed at the National Metallurgical Academy (Dnipro city) [7]. The wheel profile complying with GOST 9036 is considered in two states: initial (new or turned) and worn (Fig. 1).

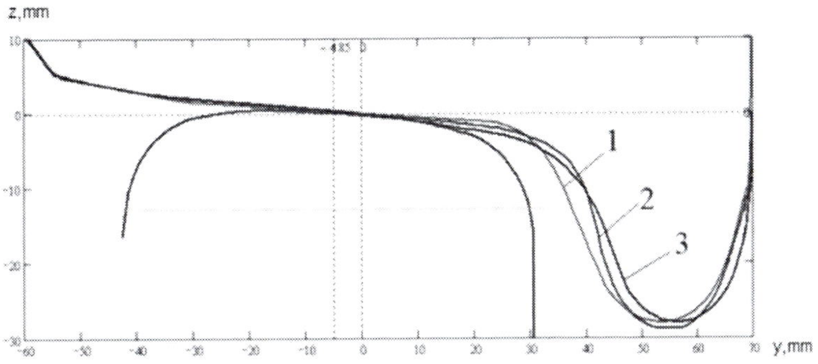

Fig. 1. Profiles of considered wheels
(1 – new profile complying with GOST; 2 – profile with wear; 3 – DMetI profile)

According to the recommendations of the Organization for Railways Cooperation, at the maintenance of track on the sections of high-speed operation minimum gauge tolerances are (+ 2, - 2), and the maximum ones are (+ 6, - 4) [9]. Thus, the gauge for 1520 mm track may be within 1516 – 1526 mm range. Equivalent conicities at the change of gauge of the track with P65 rails within the specified range in increments of 2 mm are calculated for given profiles of wheels.

The results of calculations showed that equivalent conicity $tg\gamma_e$ for wheels with new standard profile of the running surface within the gauge changes from 1516 to 1526 mm is constant and equals to 0.05. The equivalent conicity for wheels with worn running surface changes from the value of 0.247 at 1516 mm track to the value of 0.111 at 1526 mm track. When the track gauge varies from 1516 mm to 1520 mm, the value of $tg\gamma_e$ varies from 0.247 to 0.142. When the track gauge amounts to 1522 mm, $tg\gamma_e$ is equal to 0.149, and when the track gauge increases from 1524 mm to 1526 mm, the value of the equivalent conicity changes from 0.133 to 0.111.

For the wheel with the running surface complying with DMetI profile the value of the equivalent conicity gradually changes from $tg\gamma_e$ = 0.128 at 1516 mm track gauge to $tg\gamma_e$ = 0.115 at 1526 mm track gauge, which more than double the value of effective conicity in case of wheels with the standard profile of running surfaces.

Differences between the values of effective conicity for wheels with different profiles lead to a change in lengths of hunting waves L_{he} of the wheelset. Thus, in the case of wheels turned in accordance with GOST 9036 profile L_{he} is equal to 17.21 m and independent of the track gauge, and in case of worn wheels the value of L_{he} varies from 7.74 m at S = 1516 mm to 11.55 at S = 1526 mm. In case of the use of DMetI profile the length of the hunting wave of the wheelset slightly varies from 10.76 m to 11.35 m at the increase of track gauge within 1516 – 1526 mm range.

It should be noted that the reduction in the length of the hunting wave increases frequencies of horizontal oscillations of the rolling stock. Thus, at V = 160 km/h the frequency of hunting oscillations of the wheelset with GOST 9036 profile wheels at 1520 mm track is 2.60 Hz, whereas the frequency of hunting oscillations of the wheelset with worn wheels increases to 4.35 Hz.

Table 1 brings together the values of effective conicity $tg\gamma_e$ and the length of the hunting wave L_{he}, which are calculated for three configuration options of wheels running surfaces in question at different track gauges.

Table data showed that considered combinations of wheels profiles and rails differ significantly by the nature of interaction. According to the results of numerous studies, configuration of working surfaces of wheels and rails can significantly affect the dynamic behavior of rail vehicles, in particular, critical speeds in respect of hunting oscillations depending on the design and parameters of the running gears.

Table 1. The length of hunting waves at the interaction of wheelsets with wheels with different profiles and railway track

Track gauge, mm	Standard profile				DMetI profile	
	New		Worn			
	$tg\gamma_e$	L_{he}, m	$tg\gamma_e$	L_{he}, m	$tg\gamma_e$	L_{he}, m
1516	0.05	17.21	0.247	7.74	0.128	10.76
1518	0.05	17.21	0.177	9.15	0.126	10.84
1520	0.05	17.21	0.142	10.21	0.123	10.98
1522	0.05	17.21	0.149	9.97	0.12	11.11
1524	0.05	17.21	0.133	10.55	0.117	11.25
1526	0.05	17.21	0.111	11.55	0.115	11.35

3 Determination of critical speeds

Critical speeds depending on the equivalent conicity were determined for passenger car. For this purpose, the basic computer model of spatial dynamics of high-speed passenger car was used [1, 3].

The mathematical model of passenger car was made in UM package in accordance with subsystems method [10]. Structurally "Passenger car" object is an integral system consisting of individual subsystems of solid bodies interconnected by articulations and load-bearing elements. Accepted subsystems of the car are body and bogies, which in turn include wheelsets subsystems. Therefore, model of the system has three levels (Fig. 2): 1 – "Wheelsets"; 2 – "Bogie"; 3 – "Passenger car". The general model contains identical subsystems with equivalent identifiers of parameters of inertial characteristics of bodies, elastic-dissipative properties and force relations. The general system "Passenger car" composed of 19 solid bodies, 18 linear and 22 bipolar load-bearing elements.

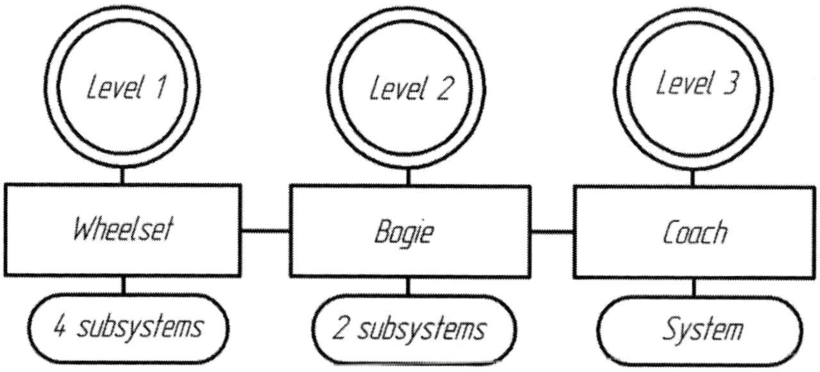

Fig. 2. Structural flow-chart of model development for car dynamics

The computer model of the rolling stock in question was supplemented with appropriate components in order to study the impact of oscillation dampers, in particular, hunting dampers, on the dynamics of the rolling stock (Fig. 3).

The simulation of running of the car through the track without destabilizations with continuous reduction of speed from maximum to the set one was carried out. The speed range of 140 – 50 m/s was considered in calculations. The rate of speed drop was 4.2 m/s/s at the specified integration time for motion equations – 30s. Time based historical graphs for lateral displacements of wheelsets were made for each calculation option. At that, the critical speed V_c was determined by the moment when self-oscillation regime stopped.

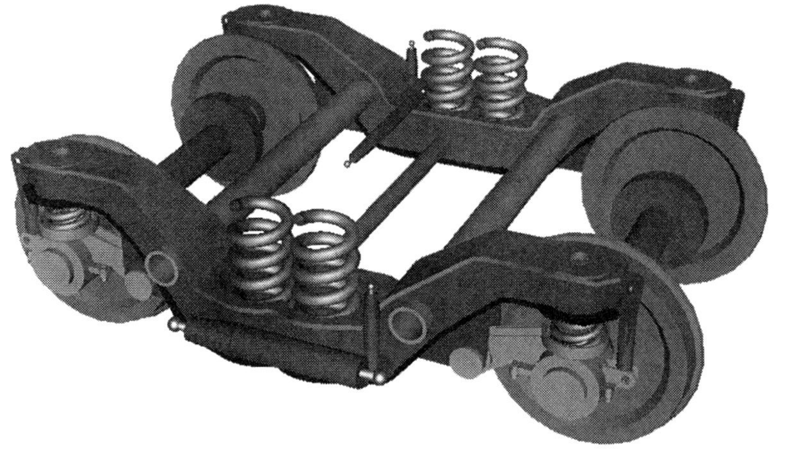

Fig. 3. Graphic representation of the model of bogie with oscillation dampers

Fig. 4 shows consolidated time based historical graphs for lateral displacements of all four wheelsets with wheels in the initial (new) state on 1520 mm track. As seen from the figure, in this case self-oscillations of wheelsets die out for 12 seconds, and at the given rate of motion speed drop this corresponds to the value of 89.4 m/s, i.e. V_c = 89,4 m/s. The results of calculations in case of new wheels turned in accordance with GOST 9036 profile, at other values of track gauge within the range of 1512 – 1526 mm, showed the insensitivity of the critical speed in respect of hunting oscillations to track gauge changes.

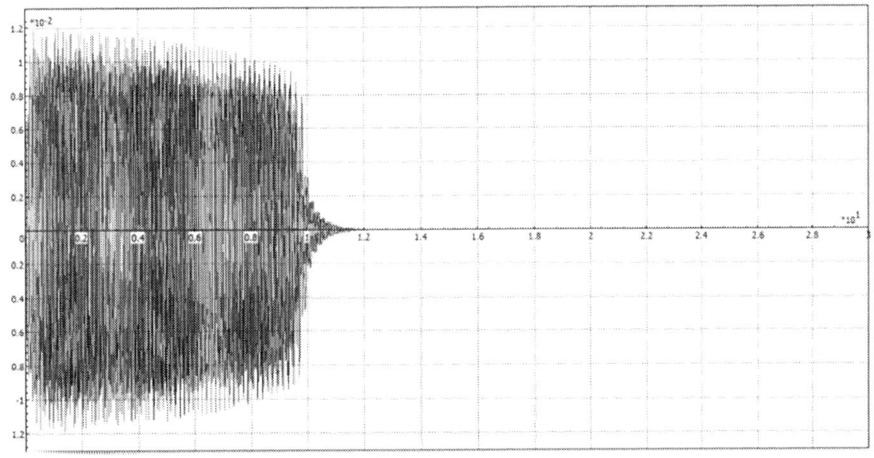

Fig. 4. Horizontal lateral displacement of wheelsets (wheel profile in accordance with GOST 9036). Horizontal axis: time [s], Vertical axis: wheelset lateral displacement [m]

Graphs for displacement of wheelsets with worn wheels at 1520 mm track are shown in Fig. 5. In this case critical speed is 48.3 m/s. The comparison of this data and data from the Fig. 4 showed that parameters of motion stability of the car when moving through the track with nominal gauge are severely affected by wheels wear. At the same time there is the following upward trend in increases in critical speeds with an increase at the track gauge: at S = 1516 mm – V_c = 5.4 m/s; at S = 1518 mm – V_c = 30.7 m/s; at S = 1520 mm – V_c = 48.3 m/s; at S = 1522 mm – V_c = 64.4 m/s; at S = 1524 mm – V_c = 77.1 m/s; at S = 1526 mm – V_c = 88.3 m/s.

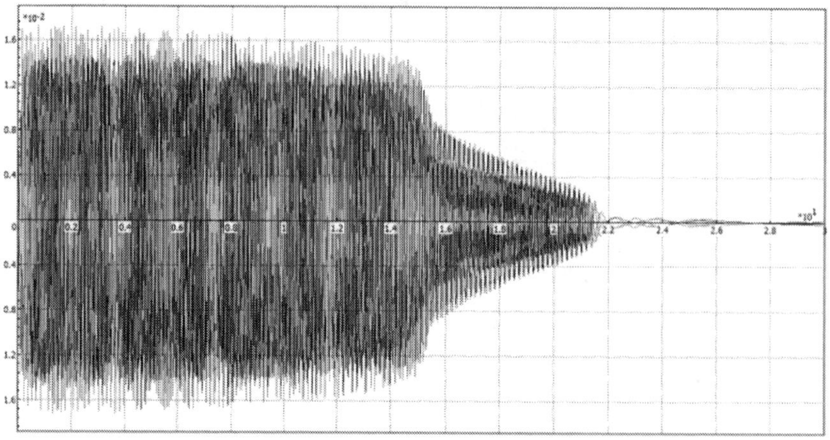

Fig. 5. Horizontal lateral displacement of wheelsets (wheels with wear). Horizontal axis: time [s], Vertical axis: wheelset lateral displacement [m]

Fig. 6 shows graphs for lateral displacements of wheelsets with DMetI profile wheels at 1520 mm track. According to the results of computer simulations in this calculation case, there is a conclusion on low sensitivity of critical speed to track gauge changes. Thus, when the track gauge is within the range of S = 1516 – 1522 mm V_c is equal to 56.0 m/s, and when the track gauge is within the range of S = 1524 – 1526 mm V_c is equal to 60.1 m/s.

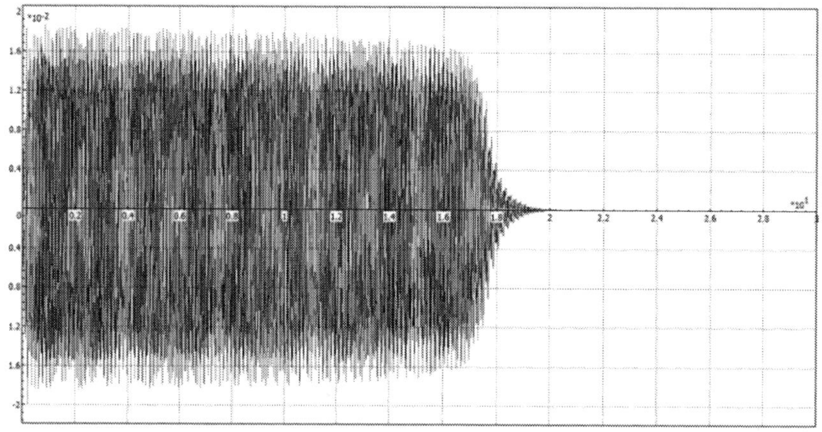

Fig. 6. Horizontal lateral displacement of wheelsets (DMetI wheel profile). Horizontal axis: time [s], Vertical axis: wheelset lateral displacement [m]

Since the problem of the loss of stability of motion is still relevant for the rolling stock with design speeds of 160 km/h and higher, hunting dampers are widely used as effective means of suppression of self-induced oscillations of railway rolling stocks. The impact of these dampers on the dynamics of high-speed car was studied in this work. According to the results of corresponding calculations, critical speeds generally increase when implementing hunting dampers. The value of the increase in critical speeds depends on the used wheel profile.

Conclusion

The value of equivalent conicity characterizing the geometric interaction of wheels and rails was calculated according to the said method for wheel profiles that are used on 1520 mm track railways. The impact of equivalent conicity on the length of the hunting wave and on critical speeds in terms of hunting self-oscillations depending on the track gauge was studied. The results of the study also confirmed the significant dependence of critical speeds on operation of hunting dampers in bogies. The value of the increase in critical speeds depends on the used wheel profile. Based on the results of studies it is recommended to include running dynamic tests in the development of regulatory documents on the admission to operation of the rolling stock for 1520 mm track with design speeds of 160 km/h and higher, taking into account the limit values of equivalent conicity.

References

1. Y. Diomin, *Stabilization of high-speed railway vehicles.* Vehicle System Dynamics Int. J. of Vehicle Mechanics and Mobility **23** (1), 107-114 (1994)
2. Y. Diomin, R. Diomin, *Procedural issues acceptance of rolling stock gauge 1435/1520 mm*. Prace naukowe Politechniki Warszawskiej. Transport **98**, 119-124 (2013)
3. Y. Diomin, L. Dlugach, M. Korotenko, O. Markova, *Self-oscillations and motion stability of rail vehicles.* [in Russian], Kyiv, Naukova dumka **160** (1984)
4. J. Elkins, A. Carter, *Testing and analysis techniques for safety assesment of rail vechicles: The State-of-the-Art.* Vehicle System Dynamics Int. J. of Vehicle Mechanics and Mobility **22**, 185-208 (1993)
5. EN 1436, *Railway applications – Testing for the acceptance of running characteristics of railway vehicles – Testing of running behaviour and stationary tests*. (2005)
6. S. Iwnicki, *Handbook of Railway Vehicle Dynamics*. (CRC Press, 535 p., 2006)
7. V. Kulieshov, A. Sladkovsky, *On the effectiveness of the work on the "wheel-rail" program in the locomotive sector of Ukraine.* [in Russian] Railway Transport of Ukraine **3**, 22-24 (2001)
8. T. Lack, J. Gerlici, *Delta r negative shape influence on the equivalent conicity assessment manner.* [in Slovak] Computational and experimental methods in applied mechanics **I**, 27-36 (2013)
9. OSJD, *Regulation on the laying and maintenance of the track on railway sections for high-speed operation (speed up to 200 km / h inclusively).* OSJD Code 788 (2011)
10. D. Pogorelov, *Simulation of Rail Vehicle Dynamics with Universal Mechanism Software.* Rail vehicle dynamics and associated problems, 13-58 (2005)
11. O. Polach, *Influence of wheel/rail contact geometry on the behaviour of a railway vehicle at stability limit.* ENOC-2005, 2203-2210 (2005)
12. UIC Code 519, *Method for determining the equivalent conicity.* International Union of Railways (2004)
13. UIC Code 518, *Testing and approval of railway vehicles from the point of view of their dynamic behaviour – Safety – Track fatigue – Ride quality.* International Union of Railways (2009)
14. A. Sapietová, M. Sága, P. Novák, R. Bednár, J. Dižo, *Design and application of multi-software platform for solving of mechanical multi-body system problems.* Mechatronics: Recent technological and scientific advances, 345-354 (2011)

The development of diagnostics methodological principles of the railway rolling stock on the basis of the analysis of dynamic vibration processes of the rail

Juraj Gerlici[1], *Olena* Nozhenko[1,*], *Ganna* Cherniak[2], *Mykola* Gorbunov[2], *Rostyslav* Domin[2], *Tomáš* Lack[1]

[1]University of Žilina, Faculty of Mechanical Engineering, Department of Transport and Handling Machines, 010 26 Žilina, Slovak Republic
[2]Volodymyr Dahl East Ukrainian National University, Department of Rail Transport, 93400 Sewerodonetsk, Ukraine

Abstract. The paper discusses the preconditions of the methodology development of diagnosis system for assessing dynamic impact of the rolling stock on the basis of processing and analysis of data obtained in operation on the results of measurement of parameters that characterize dynamic vibration processes of the mechanical system of "rolling stock – track". On the basis of usage of the processing methods of time series and stochastic processes there has been established the relationship between these dynamic processes and wheel defects, and designed experimental data processing algorithms, which in the future will be an integral part of the intellectual systems of decision making when assessing the impact level of the rolling stock on the track. The article presents some results of the experimental and theoretical research of the rail accelerations data, which have been registered during passing of the train. The computational algorithm of the specialized pre-processing of the multidimensional signal recorded by this system is described. The advantage of this algorithm is that it does not require additional information about the train speed on the section equipped with the monitoring system, about the number of wheel pairs of locomotives and train cars, and about the distances between the wheel pairs. Based on the processing results of accelerations of the rails there have been set the parameters of increase in all statistical indicators with increase in the train speed. It is noted that the level of indicators for accelerations of the rails in the vertical direction is twice as high as the corresponding parameters in the horizontal direction.

Keywords: rolling stock with the wheels defects; acceleration of the rails; the statistical signal processing techniques; the calculation algorithm

[*] Corresponding author: juraj.gerlici@fstroj.uniza.sk

Reviewers: *Grzegorz* Domek, *Alžbeta* Sapietová

1 Introduction

The modern systems of monitoring of the dynamic condition of vehicles intended to detect faults fundamentally use concepts and hypothesis, which are based on in-depth filtration methods and the time series analysis [1]. The practical implementation of such systems is carried out on the basis of evaluation of the dynamic behavior of both the rolling stock [2-6] and the track structure during passing of the train [7-9].

Currently, most of the commercial products for monitoring the railway vehicles condition are focused on the bogie system, which is due to the fact that some of its most important components are exposed to rapid change of their technical condition and have a serious impact on operability of the system "railway vehicle – railway" in whole and movement safety in particular. In such a case the key aspect is the ability of the existing monitoring technology to detect the diagnostic parameters in real time and to assess and predict the technical maintenance time [1].

The modern monitoring systems, which are installed on the track, were analyzed by Brickle B etc. [9] commissioned by the The Rail Safety and Standard Board (RSSB) UK and have been classified by the functional categories, where as a separate group there has been allocated the wheel impact load detectors (WILD), which detect the presence of a defective wheel by measuring the magnitude of the load (amount of force the wheel exerts to the rail) and comparing it to the specified threshold.

The instrumental basis of the monitoring systems can be optical sensors, accelerometers, load sensors or strain gauges. The available systems in market are GE Transportation's MATTILD, DeltaRail's Wheelchex, Teknis' WCM and Salient System's WILD [1]. However, according to authors [1, 9], these systems are not that reliable and in most cases the inspection of railway vehicles takes place in the depot before it leaves for operation. Such inspections are time-consuming and prone to human error. These techniques have been around for many years, but an increase in damaged wheelset due to higher speeds, heavier loads and modified operating conditions has led the rail stakeholders to re-evaluate these inspections strategies. A difficulty for such systems is to determine the threshold limit values of the measured magnitude which characterizes the force the wheel exerts to the rails.
And, if the threshold limit values of stresses on the rail base are standardized [10], then there arise dome difficulties concerning justification of the thresholds limit values of accelerations of the rail or the data for optical sensors under conditions of the necessity to assess different types of the rolling stock, the axle loads and the technical conditions with availability of only the output signal.

For countries with a gauge of 1520 mm, where the quality of technical maintenance and used design of the rolling stock (for example, a large unsprung weight of freight cars on the bogies of model 18-100) create additional risks of occurrence of limit-exceeding impact of the rolling stock on the track, there is an acute problem of development and implementation of a new method to monitor workloads affecting the track structure during passing of freight cars. In Ukraine, the work in this direction started in 2010. It envisages creation of a method of monitoring which is based on accelerations of elements of the track structure as magnitudes that do not require creation of specific measurement conditions (as opposed to the methods of stress freezing). This article presents the results of experimental research on impacts on the track structure of freight cars with the use of the given method of monitoring.

2 Description of the experimental monitoring system and its testing conditions

At the moment in the countries with a gauge of 1520 mm the existing regulations for assessing dynamic influence of the new and modernized rolling stock on the track are regulated by normative documents [10] and include assessment of the stress-deformed states of the rails under influence of the rolling stock, based on the recording of the stress in the cross section elements of the rail by using strain gauges (fig. 1, a). The admissibility of the stress magnitudes (edge stresses) and forces arising in the rails in the vertical and horizontal directions are estimated basing on the maximal possible values for these parameters as set out in [10].

Assessment of the dynamic impact of the rolling stock on the rails using accelerometers that underlies the developed method for monitoring of work loads is complicated by absence of threshold limit values of acceleration of the rail. These magnitudes characterize the maximum permissible value of influence of the rolling stock on the rail track that provides the permissible technical condition of the track and the level of accumulation of changes in the parameters subjected to control, for the period between two scheduled inspections during passing of the track geometry car. The proposed system is based on measuring acceleration of the rails due to the train passing the way section equipped with the sensors system [11]. It uses two-axle accelerometers with measuring acceleration range up to 700 m/s^2 in the vertical direction and up to 350 m/s^2 in the horizontal one; the error is up to 10%. It's made based on the chip ADXL278. The accelerometers are installed on the bases of the rails (fig. 1, b) symmetrically on both rail threads. The software that implements synchronization, the data collection and storage of the measured accelerations is implemented on LabView platform.

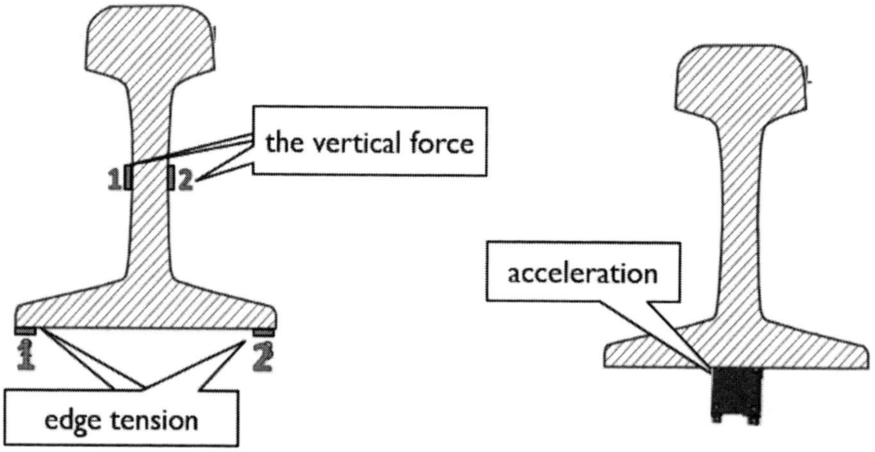

Fig. 1. Scheme of sensors location in the measuring section of the rail to assess workloads of the rolling stock on the track in accordance with [12] (a) and implemented during research conduction (b)

Comparative testing to assess the impact of the rolling stock on the rail were carried out along a straight section equipped with two measuring systems - standardized system according to DSTU 7571: 2014 (Fig. 1, a) [12] and the proposed system (Fig. 1, b.) [13]. Experimental section had rails of type R65, ferroconcrete sleepers (1840 units/km) with stone ballast. 8 cross sections of the rails (Fig. 2) were equipped with sensors in the space between the sleepers with the distance between the cross sections of about 545 mm.

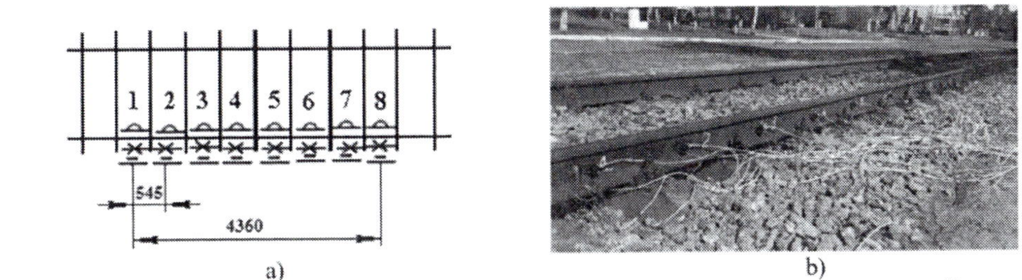

– a device for measuring of vertical force; – device for measuring the stresses in the edge of the rail flange; – a device for measuring acceleration rails

Fig. 2. Experimental section, equipped with sensors: a) scheme; b) photo

Tests involved an experimental train consisting of a locomotive of the model CHME3 and a gondola car in an empty (20.5 t) and loaded (79 t) states. The gondola car had defects on the rolling surface (1 wheelset without defects, 2 wheelset with the slid flat with depth of h_2 = 0,5 mm, 3 wheelset with the slid flat with depth of h_3 = 2,5 mm 4 wheelset with the slid flat h_4 = 1 mm). Experimental trips were carried at a constant rate, which varied in the range of $v = 10 \div 40$ km / h, with an interval of Δv = 5 km/h, 6 trips ($l = 1 \div 6$) with each speed in empty and loaded states. The sampling frequency of the accelerometers and the strain gauges was 400 Hz.

The aim of the test was to compare the magnitudes of stresses in the rails during passing of the rolling stock with accelerations of the rail for establishing relationships between the fixed magnitudes, the rolling stock parameters and test conditions (the speed, the load of rolling stock, the wheel defects) and assessment of dynamic impact of the rolling stock with defective wheel sets by statistical methods.

3 Analysis of the dynamic impact of an experimental train on the track structure

The processing procedure of the experimental data obtained during measurement by the strain gages, was performed by statistical methods in accordance with [12], and data, obtained using accelerometers, - following the procedure described in [13]. In this case, the magnitudes of edged stresses were compared (the signal example is shown in Fig. 3a) and acceleration of the rails (the signal example is shown in Fig. 3b). The initial comparison of the signals showed clear advantages of the data processing that record edged stresses, and characterize a low frequency process which does not require the special processing algorithms.

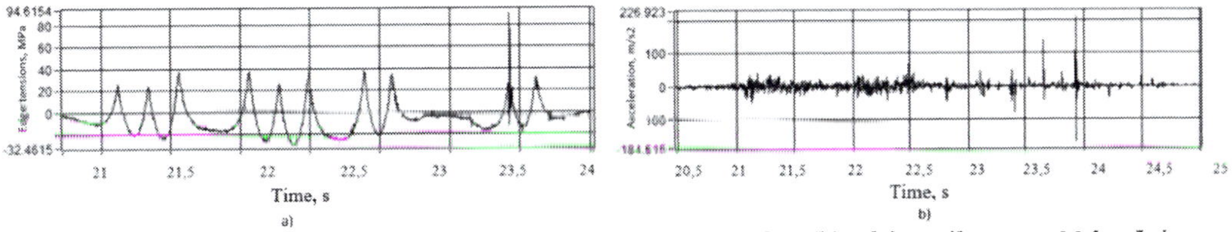

Fig. 3. Records of signals, fixing edged stresses (a) and acceleration (b) of the rails at v = 30 km/h in the vertical direction when passing of the experimental train with loaded gondola car through the third sensor

In the first stage of the signals processing there has been set a task to establish the impact of the movement speed on the magnitude of dynamic impact of the experimental train on the rail, which was assessed by the magnitude of the edge stresses of the rails according to standard procedures and magnitudes of acceleration of the rails based on the proposed methodology. For the separation of the multidimensional signal S into segments associated with individual wheels in the automatic mode, there has been developed a computational algorithm consisting of these steps:

- decomposing signal S into three parts $S = S' \cup S'' \cup S'''$, where parts S', S'', S''' correspond to measurements before the train passing, during passing and after passing of the train over the sensors. The informative part of the signal processing component is S'' for which an iterative procedure is applied using the cost function;
- filtrating signal S'' using the methodology of wavelet compression with a nonparametric noise removal using the Waveshrink function;
- building the envelopes eSi″ using Gilbert transformations for all components of the multidimensional signal;
- building reciprocal correlation functions for the envelopes eSi'' of the same type of signals and calculating the shift between signals received from different sensors on this basis;
- displacing the envelopes of signals from different sensors in a position corresponding to the signal from the first sensor and forming the summary signal (for signals of the same type), which allows to find segments associated with the individual wheels.

It should be noted that the developed algorithm does not require input information on the speed of the train on the section, equipped with the monitoring system, and the distances between the wheel pairs of the train. The magnitudes of these parameters are calculated in the process of the algorithm operation.

After envelope segments have been allocated for the all multidimensional signals, maximum values have been calculated for each segment, on the basis of which the statistical samples for processing are formed. Performed statistical processing includes grouping of the samples in the vertical Z and the horizontal Y direction for the left L and right R rails, which in the result provide selection of four main groups of samples ZR, ZL, YR, YL. In this case, each basic group contains samples comprising the maximum accelerations for all wheel sets $w = 1,...,20$ and all the way sections $m=1,...,15$. The obtained samples are checked for compliance with the normal distribution law by assessing the sample uniformity using the coefficient of variation, which by the analogy with [10] should be less than 0.4 (40%). After checking the samples on normality, statistical indicators are calculated for them, namely, the magnitudes of expectation, standard deviation, and probable maximum values $max_a ZR$, $max_{pr} ZL$, $max_{pr} YR$, $max_{pr} YL$. In more detail the algorithm of processing the acceleration of the rails is presented in [13].

Comparison of the statistical parameters of the researched magnitudes of stresses and accelerations of the rails during passing of the shunting locomotive and the loaded gondola car has showed that the speed in the range of 10 ... 40 km/h had no significant effect on the magnitudes of the statistical parameters of the rails edge stresses. In contrast to stresses, the magnitude of the rail acceleration during passing of the shunting locomotive and loaded gondola car is linearly dependent on the movement speed. The resulting dependences of the rails accelerations on the magnitude of the movement speed of the rolling stock is qualitatively consistent with the results of tests [13], where there was also observed a linear dependence of possible maximum acceleration values on the movement speed in the range of 40 ... 130 km/h during passing of a two-axle locomotive CHS8. Moreover, it was found out that several times exceeding of the permissible value of the magnitude maxσ for 3rd wheel gondola car (reaches up to 790 MPa, h_3 = 2,5 mm) for speeds of 20 - 40 km/h for individual passing, which is 240 MPa for all types of the rolling stock [12, 13]. In addition, such exceeding is observed for all the other wheels of the gondola car at individual speeds for individual passing. Considering the impact of the gondola car movement speed on the rails acceleration, there has been found increase in magnitudes of statistical characteristics for 9th (h_3 = 2,5 mm) and 10th (h_4 = 0 mm) wheel setsof the experimental train in both directions. Moreover, the range of changing of maximum possible values of edge stress of the rail max_σ and maximum possible values of the rails acceleration which arise during passing of the rolling stock max_a in a range of speeds of 10 - 40 km/h, presented in Fig. 4, clearly shows an abnormal magnitudes both for the rails acceleration and for edge stresses, which occur during passing of the gondola car wheel sets with wheels defects compared with the values of the same parameters arising during passing of the locomotive CHME3. At the same time the parameters considered in the vertical direction are illustrative, which, in comparison with the horizontal direction, more clearly show increased impact of the wheel sets of the gondola car. Assessment of the dependence max_σ on the magnitude max_a (Fig. 5) in the investigated range of speeds has demonstrated abnormal magnitudes of the rails acceleration during passing of the 6th wheel set of the locomotive (Fig. 5a) without exceeding the permissive level of stress (240 MPa), and that the level of acceleration exceeding 330 m/s^2 for 3rd and 4th wheel sets of the gondola is probably abnormal (Fig. 5b).

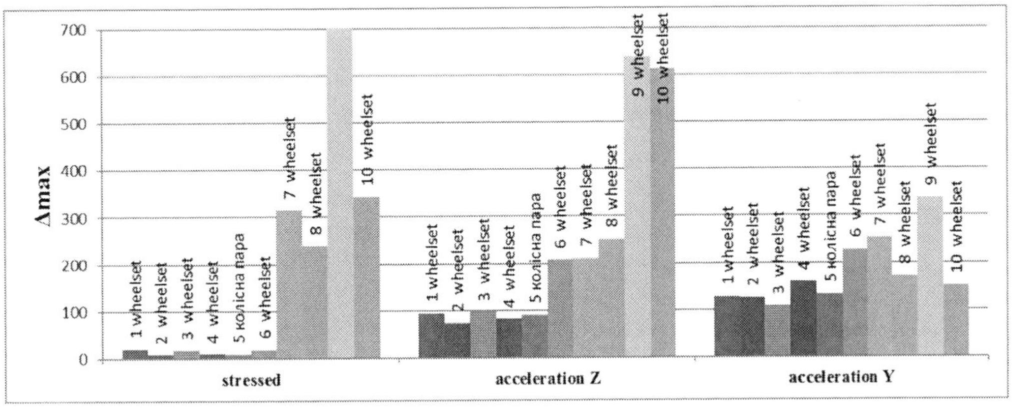

Fig. 4. The range of changes max_σ and max_a at v = 10 ÷ 40 km/h for all wheel sets of the experimental train

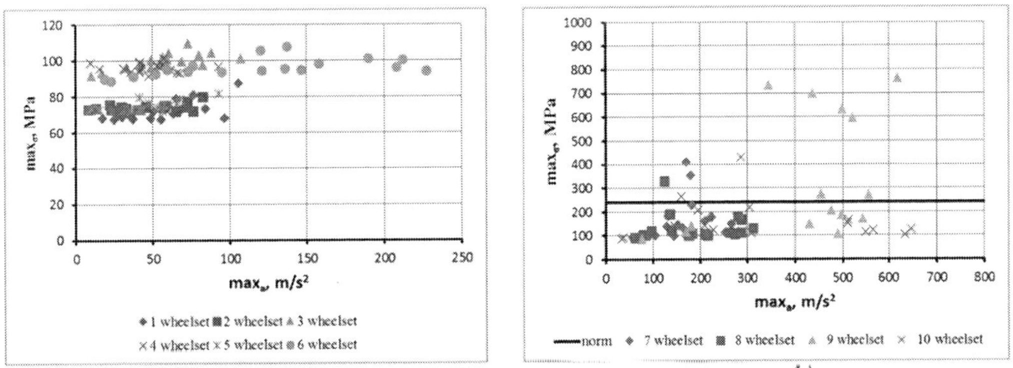

a) b)

Fig. 5. Dependence of max_σ and max_a in the vertical direction in the tested speed range for the wheel setsof the locomotive a) and the gondola car b)

Conclusion

The method for monitoring workloads of the existing rolling stock on the track structure is proposed. The developed method is based on measuring accelerations of the rails in the vertical and horizontal transverse directions. The necessary measurements are carried out on the selected track section using two-axial accelerometers installed symmetrically on the bases of the rails between the sleepers on the right and left sides of the track structure. In contrast to the method of measuring the stress and the forces, which currently being used to assess the impact of the rolling stock on the rails for the railways with a track of 1520 mm, this method does not require creation of special conditions for measurement. The total number of accelerometers installed on certain sections of track crossings with a fixed step, depends on the purpose of monitoring and characteristics of the track structure. During passing of the train on the of test section multi-dimensional signals with a number of components equal to doubled number of accelerometers are registered. The article presents the results of the field testing of the system carried out in the track section, which was equipped with accelerometers in 8 measurement cross sections, and comparison of the results with the standard method for determining the impact of the rolling stock on the rail by identifying the stress in the rails. For experimental research there has been used experimental train consisting of the shunting locomotive CHME3 and the gondola car. The train passed through the experimental section in forward and reverse directions at speeds from 10 to 40 km/h with an interval of 5 km/h.

Comparison of the statistical parameters of the stress of the rails and of acceleration during passing of the shunting locomotive showed that the speed in the range of 10 ... 40 km/h had no significant effect on statistical parameters of magnitudes of edge stress of the rails both during passing of the empty gondola car and the loaded one, the maximum possible value of which varies in the range of max_σ = 66 ... 106 MPa. Interesting information can be found in [14, 15, 16]. In contrast to stresses, the magnitude of the rails acceleration during passing of the shunting locomotive is linearly dependent on the speed. This trend is observed in both the vertical (magnitude max_aZ reaches 227 m/s^2), and the horizontal (magnitude max_aY reaches 283 m/s^2) directions.

Analysis of statistical parameters of stresses of the rails and accelerations during passing of the gondola car has showed several times exceeding of the permissible value of the maxσ for all wheels of the loaded gondola car at individual speeds.

The research was held within the framework of National Scholarship Programme of the Slovak Republic for the support of mobility of students, PhD students, university teachers, researchers and artists and on the basis of the scientific research "Development of the scientific principles of diagnostics of mechanical transport systems on the basis of the analysis of dynamic oscillation processes of their elements", funded by the Ministry of Education and Science of Ukraine.

References

1. R. Ngigi, A. Pislaru Crinela Bal, G. Fengshou, *Modern techniques for condition monitoring of railway vehicle dynamics*. J. of Physics: Conference Series **364** (2012)
2. C. Ward, P. Weston, E. Stewart, H. Li, R. Goodall, C. Roberts, T. Mei, G. Charles, R. Dixon, *Condition monitoring opportunities using vehicle-based sensors*. Proc. Inst. of Mech. Eng. Part F: J. of Rail and Rapid Transit **225** (2), 202-218 (2011)
3. T. Mei, H. Li, *Measurement of vehicle ground speed using bogie-based inertial sensors*. Proc. of the Inst. of Mech. Eng. Part F: J. of Rail and Rapid Transit **222** (2), 107-116 (2008)
4. P. Monje, G. Martinez, B. Aranguren, L. Casado, *Using bogie-mounted sensors to measure wheel rolling and sliding in railway tracks*. Proc. of the Inst. of Mech. Eng. Part F: J. of Rail and Rapid Transit (2011)
5. A. Matsumoto, Y. Sato, H. Ohno. et al., *A new measuring method of wheel–rail contact forces and related considerations*. Wear **265** (9–10), 1518-1525 (2008)

6. S. Bleakely, S. Senini, *Autonomous time frequency analysis of wagon body accelerations*. Proc. of the V Asia Pacific Industrial Engineering and Management Systems Conference Gold Coast Australia, 34.6.1-34.6.12 (2004)
7. *Bogie Condition Monitoring*. Extract from the Railway technical handbook **1**, 152-163
8. A. Mostovych, A. Cherniak, O. Nozhenko, *Application of the methods of correlation and spectral analysis for processing the results of dynamic testing of railway rolling stock*. Railway transport of Ukraine **4** (113), 20-24 (2015)
9. T. Moynihan, G. English, *Railway safety technologies*. Canada Research and traffic group (2007)
10. A. Wei, Q. Xin, W. Chung, S. Liu, H. Tam, S. Ho, *Real-time train wheel condition monitoring by fiber bragg grating sensors*. Int. J. of Distributed Sensor Networks **2012**, 1-7 (2012)
11. B. Brickle, R. Morgan, E. Smith, J. Brosseau, C. Pinney, *Identification of existing and new technologies for wheelset condition monitoring*: Report for Task T607. TTCI Ltd UK RSSB (2008)
12. DSTU 7571:2014 *Rolling railways. The norms of permissible impact on railway track 1520 mm*.
13. G. Chernyak, J. Gerlici, O. Nozhenko, R. Domin, K. Kravchenko, T. Lack, *The experimental research of the dynamic loading of the railway track*. III ročník konference pro mladé vědecké pracovníky "Experimentální a výpočtové metody v inženýrství" (2016)
14. A. Sapietová, M. Sága, P. Novák, R. Bednár, J. Dižo, *Design and application of multi-software platform for solving of mechanical multi-body system problems*. Mechatronics: Recent technological and scientific advances, 345-354 (2011)
15. M. Sága, R. Bednár, M. Vaško, *Contribution to modal and spectral interval finite element analysis*. Vibration Problems ICOVP 2011, Springer Proceedings in Physics **139**, 269-274 (2011)
16. P. Pecháč, M. Sága, *Controlling of local search methods' parameters in memetic algorithms using the principles of simulated annealing*. Proc. Eng. **136**, 70-76 (2016)

Comparation of the impact of selected technical gases on properties of a pneumatic flexible coupling

Robert Grega[1,*], *Jozef* Krajňák[1], *Daniela* Harachová[1]

[1]Technical University, Faculty of Mechanical Engineering, Košice

Abstract. An important part of the oscillating mechanical drives are flexible shaft couplings. The special use of the pneumatic flexible coupling is possibility to serve as mechanical drives tuners. Being examined in tuners mechanical drives there is also a research focus on the impact of industrial gases on the change of dynamic characteristics of pneumatic couplings. The paper investigates five different industrial gases, namely: air, helium (He-4.6), a mixture of propane and butane gas (C3H8+C4H10), argon (Ar) and nitrogen (N2) to see how these gases affect the dynamic characteristics of pneumatic flexible shaft coupling with marking 4-2/70 T-C.

Keywords: pneumatic flexible coupling, industrial gases, change of a pressure

1 Introduction

In technical operations, mechanical drives are used for a drive and coordination of technology. An inseparable part of the mechanical drives is considered to be a mechanical coupling. One of the important groups of the mechanical couplings is flexible coupling. Its significance lies in the fact that they are only ones capable of protecting the mechanical drive from a resonance [1-3]. Within the group of flexible couplings, there are so-called torsional vibration tuners. These are special flexible couplings in which the properties can be changed and the resonance can be regulated [4-7]. One of such special flexible coupling is a pneumatic flexible coupling [8-11].

The pneumatic flexible coupling contains a flexible space that can be filled with gas. By changing the pressure of a gaseous medium it can significantly influence the properties of a pneumatic flexible shaft coupling as well as mechanical properties of the whole mechanical system in which the given coupling is included [12-13].

The aim of the paper is to investigate the impact of five industrial gases on properties of a flexible pneumatic coupling. This paper presents the impact of air, helium, nitrogen, argon and mixture of propane and butane gas and follows a change of damping ability of the coupling. Damping is one of the essential dynamic characteristics of flexible couplings. Damping characteristic as a dynamic characteristic of a coupling is used mainly in the transition of the mechanical drive in resonance area. In resonance area, one of parameters

---

* Corresponding author: robert.grega@tuke.sk

Reviewers: *Milan* Vaško, *Ján* Vavro *Jr.*

that affect the efficiency of the pneumatic flexible coupling is a speed of fulfillment, respectively a speed of a change of a pressure of gaseous medium in the pneumatic flexible coupling´ s chamber [14].

2 Material and methods

The pneumatic flexible coupling which has been used for research was developed in our Department and it was marked as 4-2 / 70T-C (see Fig. 1).

Fig. 1. Pneumatic flexible coupling 4-2/70 T-C

For the purposes of our research, we used the pneumatic flexible shaft coupling marked as 4-2 / 70 T-C. To transfer a torque, it uses four flexible two-bellows pneumatic members that are evenly located around the circumference, serving as a flexible space. The distance of these flexible spaces from the centre of rotation of the coupling is 83.5 mm. A height of the flexible space in neutral position is 90 mm and diameter is 70 mm. Allowable maximum deformation of the flexible space is 25 mm and it allows the maximum twisting of the coupling of 11.5 °. These flexible spaces are anchored to steel flanges and they are joined together.

For research of damping of pneumatic flexible coupling, five industrial gases have been used. The gases are commercially available and generally used in various industries.
As first investigated gas was the air. The air is a mixture of gases (78.084 % nitrogen N2, 20.947% Oxygen O2, 0.934 % argon Ar, 0.033 % carbon dioxide CO2 and 0.002 % other gases). The air is contained in the atmosphere; it is non-flammable and easily available.

The second type of used gas is helium. On Earth, helium occurs only very rarely. The Earth's atmosphere is found in the upper layers only. Small amount can be found in the natural gas from which it is obtained by process of freezing. Rarely, helium rises from cracks in the Earth ground. Due to its extremely low density and its inert characteristic, helium is used for filling balloons and airships to replace flammable hydrogen gas. A significant disadvantage is its relatively high price. Moreover, helium atom has very small diameter and it easily diffuses through solid materials, and thus the losses occurs [15, 16]. For purposes of our measurements, we used helium marked as HE-4.6. The technical gas contains 99.996 % helium.

The third used gas is a mixture of propane and butane; in detail the percentage ratio of 50% propane and 50% butane. Normally, the propane- butane is considered to be a gas, but rather easily can be removed by cooling or pressing to liquid state. In the liquid state it occupies only 1/260 of its gaseous volume. We used the gas because of its higher density than air, despite its negative and that is its flammability.

Argon was the fourth used gas. Argon is used in welding such as pure gas or in a mixture with carbon dioxide or nitrogen. Argon gas is heavier than the air. The most important chemical characteristic of argon is its inertness, which makes it an ideal protective gas even at the temperatures necessary in metallurgy or welding arc light. Argon is obtained in an ammonia production from hydrogen and air nitrogen, where it is actually a waste. 1 m^3 of air contains 9.3 litres of argon. Argon can also be produced from the air. Oxygen binds with red hot copper, nitrogen is bonded to a metal magnesium or calcium as nitride and the remainder of the crude argon contains about 0.25% of other rare gas, which are separated from the argon by fractional distillation, absorption and desorption of the fractional carbon.

The last used gas was nitrogen. Nitrogen is non-flammable gas without colour and odor. It is slightly lighter than air. Nitrogen is so-called inert gas and its treated is very difficult with other chemical compounds. With the exception of oxygen, chlorine and fluorine, nitrogen is presented as more electronegative element, so the compound can be considered as nitrides of the elements [17].

Argon, helium and nitrogen are non-toxic, but non-supportive of life. Their different physical properties can be seen in Table 1.

Table 1. Basic physical properties of individual gases

	helium He	air O_2+N_2	propane-butane $C_3H_8+C_4H_{10}$	argon Ar	nitrogen N_2
Specific gas constant r [J kg-1 K-1]	2079	287.04	163.39	208.49	296.75
Gas density [kg.m-3] (15° C)	0.176	1.276	2.145	1.76	1.234
Molecular weight [kg kmol-1]	4.003	28.966	50.102	39.944	28.016

On Fig. 2 is shown a gas density of the individual gases. It can be seen that the greatest value is achieved by a mixture of propane and butane 2.145 $kg.m^{-3}$. Argon, nitrogen and the air achieved lower values and the lowest value of 0.176 $kg.m^{-3}$ is achieved by helium. A similar course is shown in Fig. 3 where molecular weights of the gases are visible. The maximum value 50.102 $J.kg^{-1}.K^{-1}$ also achieved a mixture of propane and butane and then sequentially followed by argon, air and nitrogen with the lowest value 4.003 J.kg is again helium.

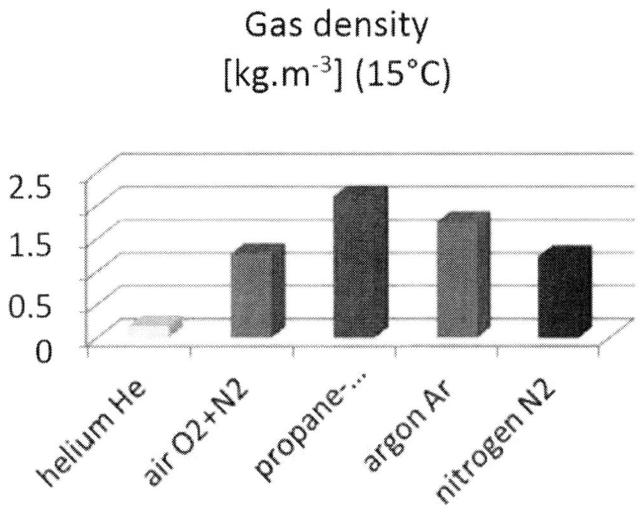

Fig. 2. The density of the gas

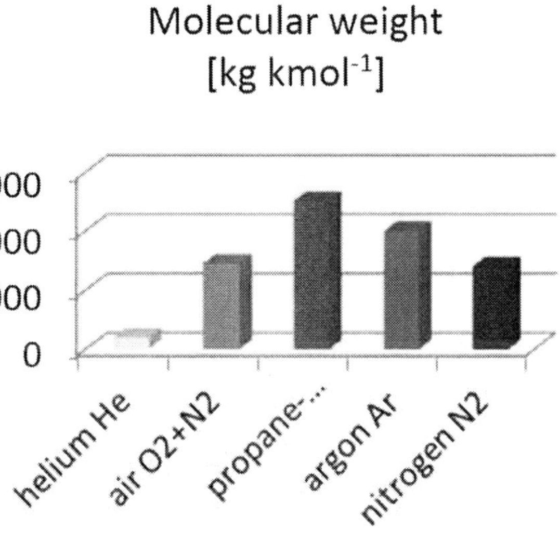

Fig. 3. The molecular weight of the gas

The change can be evident in value of a specific gas constant. Completely, the highest value is achieved by helium gas. It is a value of 2079 $J.kg^{-1}.K^{-1}$. Other gases have values of a specific gas constant much lower. Lowest value of 208.49 $J.kg^{-1}.K^{-1}$ has argon. It is value almost 10 times lower than helium. The air and nitrogen have approximately 7 times lower value than helium (the air has 287.04 $J.kg^{-1}.K^{-1}$ and nitrogen has 296.75 $J.kg^{-1}.K^{-1}$. A mixture of propane and butane has approximately 13 times lower value of specific gas constant and that is 163.39 $J.kg^{-1}.K^{-1}$ shows in Fig. 4.

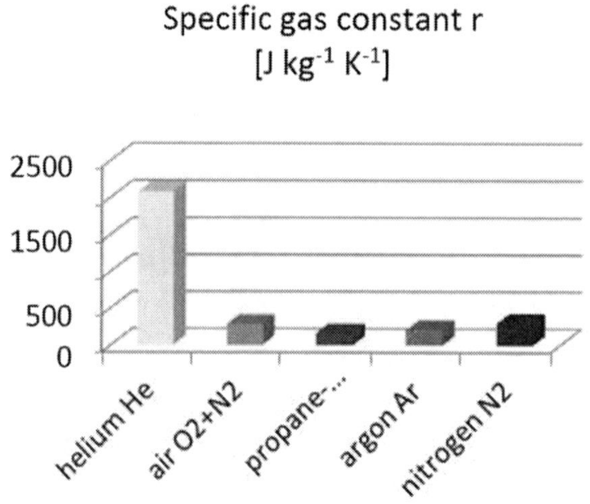

Fig. 4. The specific gas constant

3 Results and discussions

The first examined parameter was the needed time of fulfilment of the flexible coupling´s spaces. The reference time to which the other times were compared was the time of filling the spaces with the air (marked as t_{air}). Against this reference value will determine the ratio of time during fulfilment by any other gas. The interesting thing is the ratio of the time for the fulfilment of the flexible spaces to the flexible pressure limit, respectively; any pressure change in the flexible space is still the same.

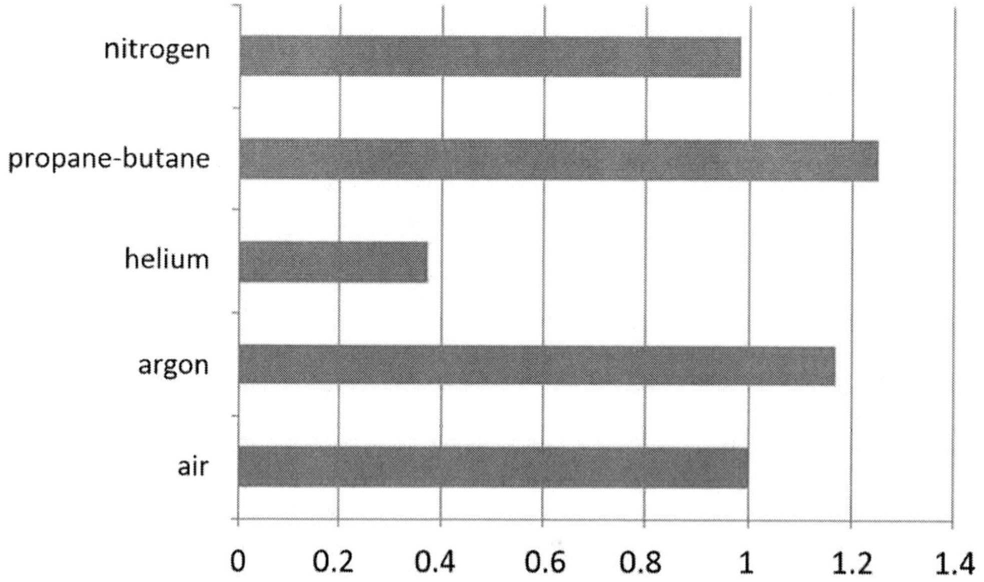

Fig. 5. Time comparison of pneumatic flexible coupling 4-2/70 T-C pressuring on required pressure using different gases

Figure 5 shows the relative values of fulfilment time of the flexible spaces for different gases. The smallest ratio and therefore the fastest fulfilment of flexible space are by helium gas. In order to pressurize the flexible space to a certain pressure, the time of fulfilment by helium is 0.372 times shorter than those ways in which we are pressurizing the flexible space by the air. This means, $t_{he} = 0.372.t_{air}$. If we use the nitrogen contained in the air, so time $t_{nitrogen} = 0.98.t_{air}$ will be almost the same as the time required to pressurize by the air. The use of argon gas that is extracted and denser than air, we have found that the time to fill the flexible space at a certain pressure is greater $t_{ar} = 1.17.t_{air}$. The longest pressurizing time took a mixture of propane and butane. This mixture is densest and it has the highest molecular weight and the lowest value of a specific gas constant and thus pressurizing the flexible space to the final pressure takes 1.25 times longer than achieve the desired pressure change by the air. It applies to $T_{Pro\text{-}Btut} = 1.25.t_{air}$.

Another parameter examine in research was a damping pneumatic coupling. The measurements were realized by means of which was determined the damping ***b***. Dynamic measurements were carried out for all five gases. The pneumatic flexible coupling was placed in the measuring device. By the method of free oscillations were determined values of damping ***b***. An example of such measurements for the investigated gas at a pressure of 500 kPa is shown in Figure 6.

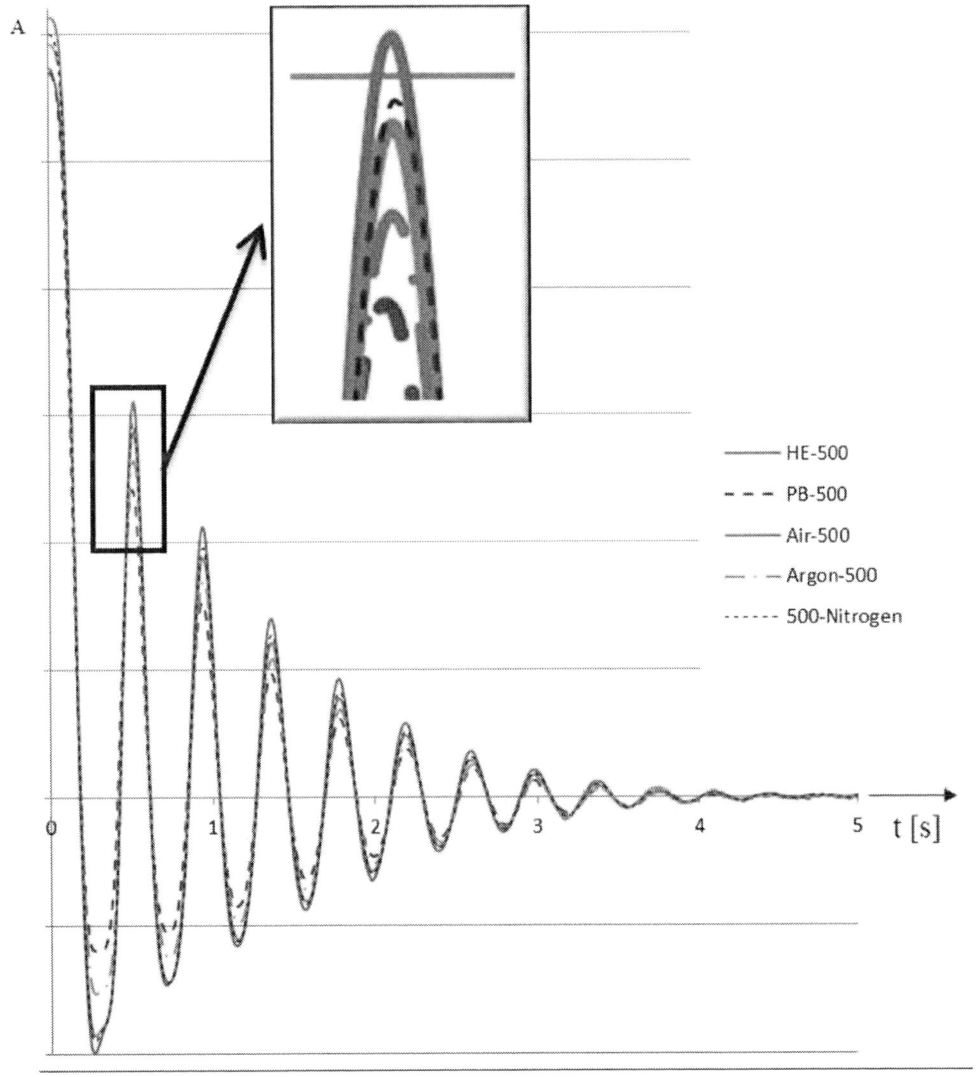

Fig. 6. Measurement at 500 kPa by using of five different gases using a method of free oscillation

The experiments were carried out on pneumatic shaft coupling which was gradually pressurized at the pressure range from 100 to 600 kPa. Subsequently, all examined gases were used in order: air, helium, propane-butane, argon and nitrogen. From the measured data were determined a damping value in the pressure range from 100 kPa to 600kPa for all investigated gases. The result damping values are shown in Figure 7.

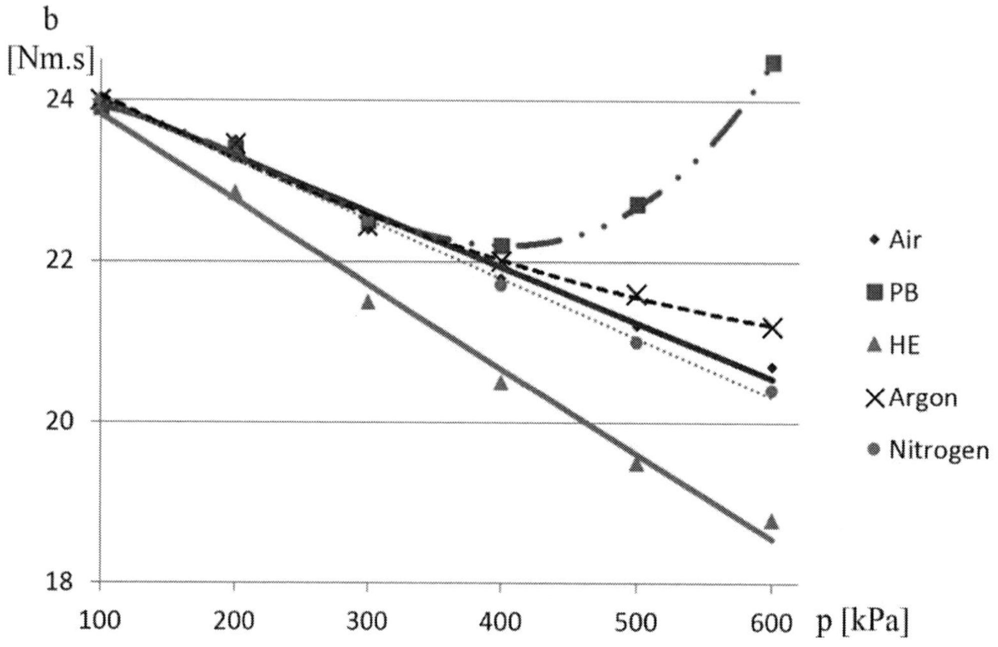

Fig. 7. Damping **b** values of pneumatic flexible coupling, in the range of pressure 100 to 600 kPa and using different gases

Fig. 7 presents the changes in damping of couplings using all investigated gases. At a pressure of 100 kPa, a type of gas is not so substantial as to all gases are approximately the same value 24Nm.s. The amount of gas in flexible coupling is low, so they do not show significant change in physical properties of gases. Gradually with increasing pressure in the coupling, properties of gases will begin to show a change of damping values of the pneumatic coupling. The flexible space filled with Helium, gradually lowers the damping while at maximum pressure 600kPa achieve the lowest value of all the gases of 18.8 Nm.s. In dependence on the pressure, the flexible space filled with the air decrease a damping value 20.7Nm.s at the minimum pressure at 600kPa. Argon and nitrogen have similar physical properties than the air and therefore the damping is similar. Gradually reduce its value depending on the pressure, whereby with a pressure of 600kPa shall nitrogen reach a value of damping 20.4Nm.s and argon 21.2Nm.s. The damping value using a mixture of propane and butane is gradually decreased up to 22.2Nm.s at pressure of 400kPa and then starts to increase to a maximum value 24.49Nm.s at a pressure of 600kPa.

Conclusions

Gaseous medium is very important part of the pneumatic flexible couplings. The type of gaseous medium and its different physical properties such as molecular weight, density gas, and specific gas constant as well as other properties can change and influence the damping of pneumatic flexible couplings. In the concrete, the paper describes pneumatic flexible shaft couplings marked as 4-2/70T-C which has been developed in our department and located in Laboratory at our department where it also performed all measurements.

The selection of gas used in pneumatic flexible coupling will have an impact on time to fulfil the pneumatic flexible couplings. The fastest pressure changes can perform the coupling filled with the gas with the lowest density and that is helium. The longest time needed to change the pressure of pneumatic flexible coupling is reached with a mixture of propane and butane that is in our case the gas with the highest density of all used gases.

Type of gas used in pneumatic flexible couplings affects its damping value. Low pressure did not reveal the change in damping values by using a variety of different gases. The differences in damping values began arise with increasing pressure. The pneumatic flexible coupling reached the lowest values during pressurization with helium. On the other hand, the highest value of damping, the pneumatic flexible coupling reached by fulfilment with a mixture of propane and butane. The pneumatic flexible couplings filled the air, argon, nitrogen achieve the damping value almost the same due to the fact that their physical properties are likely the same.

In conclusion we can state that the type of gas as well as its physical properties can greatly influence the damping of flexible pneumatic couplings and thus influence the properties of mechanical drives in which these pneumatic flexible couplings are included.

The acknowledgements VEGA 1/0198/15 Research of innovative methods for emission reduction of driving units used in transport vehicles and optimisation of active logistic elements in material flows in order to increase their technical level and reliability.

References

1. J. Kuľka, M. Mantič, M. Kopas, E. Faltinová, D. Kachman, *Heuristic optimization approach to selecting a transport connection in city public transport*. Open Engineering **7** (1), 1-5, ISSN 2391-5439 (2017)
2. J. Kuľka, E. Faltinová, M. Kopas, M. Mantič, *Diagnostics and optimisation of crane track durability in metallurgical plant*. Diagnostyka **17**, 41-46 (2016)
3. L. Jakubovičová, P. Zavadinka, J. Jakubovič, *Transport Duty Cycle Measurement of Hybrid Drive Unit for Mixing Drum*. Adv. Intell. Syst. Comput. **393**, 219-224 (2016)
4. J. Homišin, M, Urbanský, *Partial Results of Extremal Control of Mobile Mechanical System*. Diagnostyka **16** (1), 35-39 (2015)
5. L. Jakubovičová, M. Sága, M. Vaško, *Numerical study of influence of mutual slewing of roller bearing rings on the principal stresses at contact area*. Scientific Journal of Silesian University of Technology - Series Transport **84**, 83-91 (2014)
6. M. Sága, M. Vaško, P. Pecháč, *Chosen Numerical Algorithms for Interval Finite Element Analysis*. Procedia Engineering **96**, 400-409 (2014)
7. J. Zapoměl, V. Dekýš, P. Ferfecki, A. Sapietová, M. Sága, M. Žmindák, *Identification of Material Damping of a Carbon Composite Bar and Study of Its Effect on Attenuation of Its Transient Lateral Vibrations*. International Journal of Applied Mechanics **7** (6), (2015)

8. P. Czech, *Identification of Leakages in the Inlet System of an Internal Combustion Engine with the Use of Wigner-Ville Transform and RBF Neural Networks*. 12th International Conference on Transport Systems Telematics Location: Katowice Ustron, Poland, Telematics in the Transport Environment, Communications in Computer and Information Science **329**, 414-422 (2012)
9. H. Madej, P. Czech, *Discrete Wavelet Transform and Probabilistic Neural Network in IC Engine Fault Diagnosis*. Eksploatacja i Niezawodnosc - Maintenance and Reliability **4**, 47-54 (2010)
10. P. Czech, *Diagnosis of Industrial Gearboxes Condition By Vibration and Time-Frequency, Scale-Frequency, Frequency-Frequency Analysis*. METALURGIJA **51** (4), 521-524 (2012)
11. Ł. Konieczny, R. Burdzik, J. Warczek, P. Czech, G. Wojnar, J. Młyńczak, *Determination of the effect of tire stiffness on wheel accelerations by the forced vibration test method*. J. Vibroengineering **17**, 4469-4477 (2005)
12. P. Czech, G. Wojnar, R. Burdzik, J. Konieczny, Ł. Warczek, *Application of the discrete wavelet transform and probabilistic neural networks in IC engine fault diagnostics*. J. Vibroengineering **16**, (2014)
13. R. Grega, (et al.): *The chances for reduction of vibrations in mechanical system with low-emission ships combustion engines*. International Journal of Maritime Engineering. **157** (A4), 235-240 (2015)
14. J. Homišin, P. Kaššay, M. Puškár, R. Grega, J. Krajňák, M. Urbanský, M. Moravič, *Continuous tuning of ship propulsion system by means of pneumatic tuner of torsional oscillation*. International Journal of Maritime Engineering: Transactions of The Royal Institution of Naval Architects **158** (A3), A231-A238
15. M. Dehdashti-Jahromi, H. Farrokhpour, *Effect of helium nanoclusters on the spectroscopic properties of embedded SF6: Ionization, excitation and vibration*. Spectrochimica Acta Part A: Molecular and Biomolecular Spectroscopy **173**, 772-782 (2017)
16. Shanping Lu, Wenchao Dong, Dianzhong Li, Yiyi Li, *Numerical study and comparisons of gas tungsten arc properties between argon and nitrogen*. Computational Materials Science **45** (2), 327-3352 (2009)
17. J. Sunarso, S.S. Hashim, Y.S. Lin, S.M. Liu, *Membranes for helium recovery: An overview on the context, materials and future directions*. Separation and Purification Technology **176**, 335-383 (2017)

Car body and bogie connection modification for track curves passability improvement

Vladimír Hauser[1,*], *Olena S.* Nozhenko[1], *Kateryna O.* Kravchenko[1], *Mária* Loulová[1], *Juraj* Gerlici[1], *Tomáš* Lack[1]

[1]University of Žilina, Faculty of Mechanical Engineering, Department of Transport and Handling Machines, Univerzitná 1, 010 26 Žilina, Slovak Republic

Abstract. For Tram cars, it is often necessary to operate in cities on strongly curved track, which is followed by an increased effect of the vehicle on the track. Especially, this increased effect occurs in spiral transition curves situated between direct and arc sections or between two arc sections of different radius. In such case, increased guiding forces, creep in the rail - wheel contact, wear and noise generation can be observed. Exactly with the aim to reduce these undesirable effects we designed a tram bogie with steered wheelsets. This paper deals with a modification of its coupling to vehicle body in order to improve vehicle dynamics in transition curves. Proposed innovative construction of this coupling unit is registered by authors under Utility Model Nr. u201609015 and Utility Model Nr. u201703246. Description of the proposed way for a vehicle to pass through curved track with regard to bogie-body coupling and wheelset steering mechanisms with usage of multibody computing software is given in this paper.

Keywords: bogie to vehicle body coupling, track transition curves, tramcar throughput improvement

1 Introduction

The way a vehicle passes a track arc depends on many factors. The most important of these include the bogie wheelbase, the distance of pivot pins, the way the car body is mounted on the bogie, the design of the wheelset guiding in the bogie, the track gauge and the width of the track free channel with which the geometry of the wheel and rail profiles is directly related. The most favorable situation for the vehicle during a ride in a curve is the radial position, the wheelset guides should provide zero angles of attack, the wheel and rail profiles should provide one-point contact as well as lateral displacement of the wheel from the track axis such that the change of the radius of the outer and inner wheels would compensate the difference in the lengths of the outer and inner rails. Fulfilling these conditions will result in a reduction in creep velocities in wheel-rail contacts, minimizing vehicle riding resistance in curve, extending wheel and rail profile service life, reducing bogie and track load, and reducing traffic noise. It is particularly desirable to achieve such a favorable condition in an environment with frequent occurrences of small radius arcs, such

* Corresponding author: vladimir.hauser@fstroj.uniza.sk

Reviewers: *Marek* Macko, *Milan* Žmindák

as the railways of public transport vehicles, where the increased contact wear of the contact pair profiles and, in particular, the increased noise generation is significant problem [1 - 3]. The team of authors deals with the solution of this problem in the form of atypical bogie design ensuring favorable transition through track arcs [4, 5]. A key element of the proposed solution is a mechanism providing radial steering of the wheelsets while riding along a curved section of the track, where the position of wheelsets is derived from the bogie frame rotation relative to the car body.

2 The wheelset control mechanism

The purpose of the control mechanism is to define the position of the wheelsets, and it is desirable that due to the operation of the mechanism there is no excitation of the undulatory movement of the vehicle when driving on a straight track. The steering is carried out by moving of the axle boxes in the longitudinal direction by means of a steering mechanism. The input variable is the angle of rotation of the bogie towards the vehicle frame. The output variable is the angle of rotation of the wheelset towards the bogie frame, which is created by moving the axle boxes of the wheelset guiding in an opposite direction. A schematic diagram of the proposed bogie design is shown in Fig. 1a) A cross section diagram of the proposed bogie design is shown in Fig. 1b).

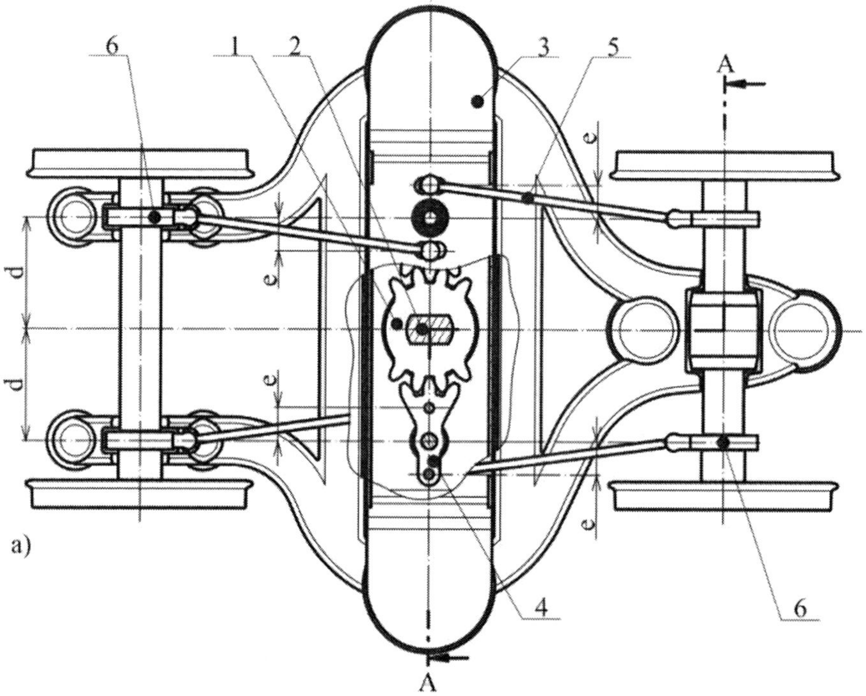

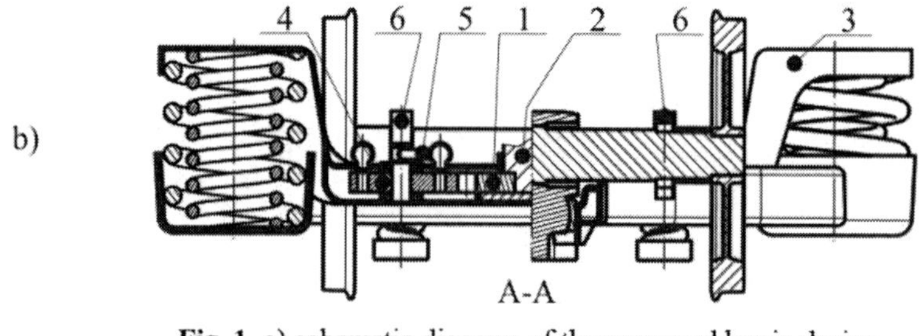

Fig. 1. a) schematic diagram of the proposed bogie design
b) cross section diagram of the proposed bogie design

The design of the wheelset steering mechanism requires a gear wheel segment (1) being mounted on the pivot coupling the car body with the bogie (2). The bogie rotation is realized around the pivot (2). Thus, the gear frame (3) has only one degree of freedom towards the car body. Levers with a gear segment are rotatably mounted on the gear frame (4). The levers (4) are coupled by means of rods (5) to the axle boxes (6). With regard to the suspension operation, spherical joints are considered to attach the rods to the levers. For a correct function of the proposed mechanism, it was necessary to determine the gear ratio between the gear segments as well as the dimensions "e" and "d".

Relationships for the calculation of these dimensions can be determined from the requirement that the vehicle has to pass through a track arc with radially rotated wheelsets, as shown in Fig. 2. The following relationships are based on the assumption that the pivot pin is located at the midpoint of the wheelbase of the bogie, shown in Fig. 2 as the dimension "b".

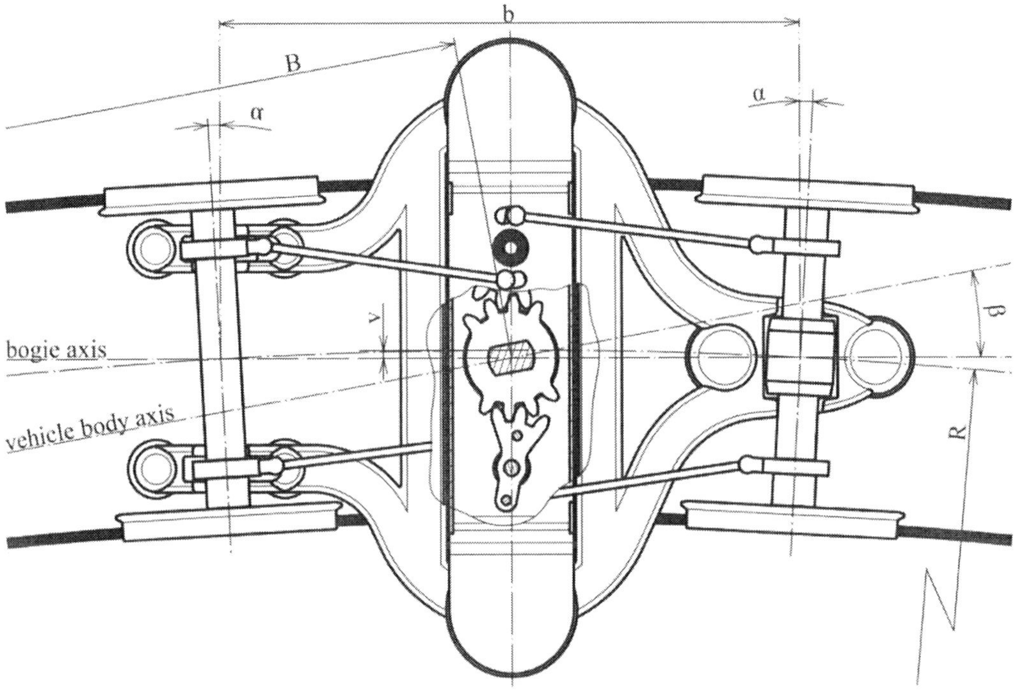

Fig. 2 Expected way of vehicle passing a track arc (ride direction to the right, front bogie shown)

For the angle of rotation of the wheelset towards the bogie frame α [rad] we can write a relationship:

$$\alpha = \arcsin \frac{b}{2R} \tag{1}$$

where variable b [mm] represents the bogie wheelbase and R [mm] the radius of the track arc. For the angle of rotation of the bogie relative to the car body, it is possible to write the relationship:

$$\beta = \arcsin \frac{B}{2(R-v)} \tag{2}$$

Where the variable B [mm] represents the distance of the pivot pins and v [mm] represents the lateral deflection of the pivot from the axis of the track in the arc. For the calculation of this value it is possible to use relation (3), or relation (4):

$$v = R - \frac{b}{2\tan\alpha} \tag{3}$$

$$v = R - \sqrt{R^2 - \left(\frac{b}{2}\right)^2} \tag{4}$$

The wheelset steering mechanism is required to rotate the wheelsets in the bogie when rotating the bogie frame towards the car body. This is done largely by longitudinally moving the axle boxes in the bogie frame. The value of this longitudinal displacement x_d [mm] can be expressed by:

$$x_d = d \cdot \sin\alpha \tag{5}$$

Where *d* [mm] is the distance of the axle box from the longitudinal axis of the bogie. It is necessary that the lever joints with the gear segment are being moved in the longitudinal direction exactly by the x_d value when the bogie is turning. Therefore, for the value of the dimension e [mm], it is possible to write the relationship:

$$e = \frac{x_d}{\sin\beta} \tag{6}$$

After fitting relations (1), (2) and (5) to (6), the value of dimension e [mm] can be modified to:

$$e = \frac{d \cdot b}{R} \cdot \frac{R - v}{B} \tag{7}$$

Since the variable v reaches values much smaller than the variable R, it can be neglected in the numerator of the second fraction. Consequently, it is possible to write the following relation for dimension e [mm]:

$$e = \frac{d \cdot b}{B} \tag{8}$$

For the considered vehicle with bogie wheelbase b = 1900 mm, the distance between the pivot pins B = 6 400 mm and the joints situated at a distance d = 350 mm from the longitudinal axis of the vehicle, the resulting value of e = 104 mm.

As the bogie design is asymmetrical, several ways of placing the bogie under the vehicle car are considered. It is necessary to find out which position will be more favorable in terms of both ride in track arcs and ride stability.

3 Examinated options of bogie mounting on the car body

There are three specific requirements for attaching the bogie to the car body:

1. Ensure wheelset steering that is derived from the pivot pin with the gear segment shown in Fig. 1 under items (1) and (2).
2. Exit the vehicle from the straight-line arc to bring the vehicle to a centered position with respect to the track axis, while the steering control derived from the pivot pin must not cause the undulatory movement of the vehicle to be excited while riding in a straight track.
3. Decrease the amplitude of the lateral force when riding through the track transition curve.

In order to meet these criteria, two ways of mounting a pivot pin were proposed. The first method considers a pivot pin firmly attached to the bottom of the car body and its rotary mounting on the frame of the bogie gear. Such a solution is fully in line with the relationships mentioned in Chapter 2, in the circular parts of the arc one can expect its correct function. However, it is necessary to check the vehicle's response to ride on the track transition curve when there may be a certain increase in angles of attack, and therefore the downgrade of the vehicle's characteristics must be examined. It is also necessary to pay attention to the undulatory movement of the vehicle after its return onto the straight track.

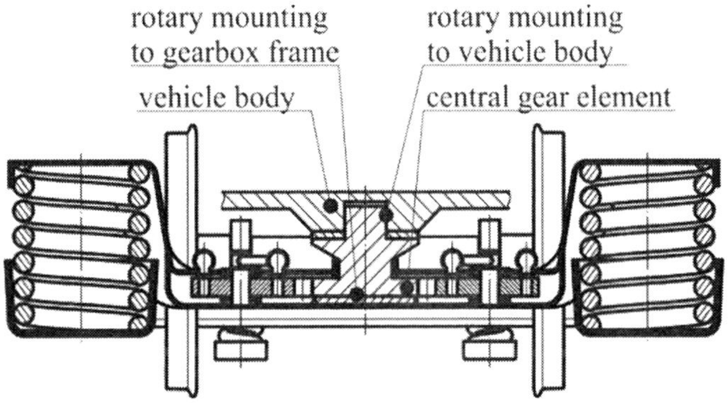

Fig. 3. Bogie with a coupling element placed in a rotary way in two places

The second method (Figure 3) considers a pivot pin attached to the bottom of the car body by means of a flexible element with a damping. At the other end, the pivot pin is also rotary mounted on the bogie gear frame. Such a solution will provide an improvement in the vehicle's properties when riding through a track transition curve when each of the vehicle bogies is in a track arc of another radius. The flexible rotation damping element of the pivot makes it possible to stabilize the vehicle on a straight track. It is necessary to pay attention to the ride through circular arcs, where the rotary mounting of the pivot pin on the bottom of the car body may cause some downgrade in the driving characteristics.

From the bogie design, its asymmetrical composition is evident in relation to the lateral plane going through the pivot pin, resulting in two possibilities of placing the bogies under the car body as shown in Fig. 4.

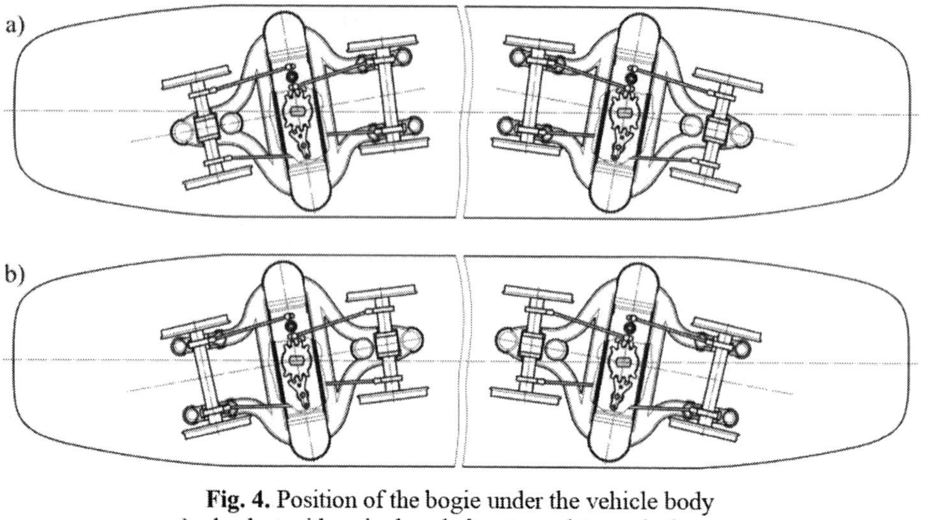

Fig. 4. Position of the bogie under the vehicle body
a) wheelset with a single axle box turned towards the stem
b) wheelset with a single axle box turned towards the vehicle centre

For both the bogie placement as well as for both the above-described pivot pin mount methods, a vehicle ride simulation analysis was performed to determine the suitability of the proposed solutions.

4 Simulation analysis of vehicle ride

Vehicle simulation analysis was performed in the Simpack 9.9 program environment using the Rail module. A vehicle model was used, the basic parameters of which are close to the T3 vehicle for the 1000 mm gauge [6], using the bogie of the described design. Creating the vehicle model itself was described by the authors in detail in [7].

It is generally known that, regarding many parameters, the ride of the vehicle in the opposite direction S-arc is very problematic [8 - 11]. The aim of the authors was to examine the response of the different variants of the location and mounting of the vehicle bogie to such situation [12 - 14]. In the simulation, a ride of the vehicle on a track according to the standard [15], which consists of straight sections, a transition curve of a clothoid shape, and two counter-arcs with a radius of 50 m was considered. The length of each section of the track and its arrangement is shown in Fig. 5. The straight section at the beginning of the track is used to stabilize the vehicle. The final straight section serves to verify the vehicle's stability against undulatory movement. The driving speed was set constantly at 10 km / h without driving and braking forces.

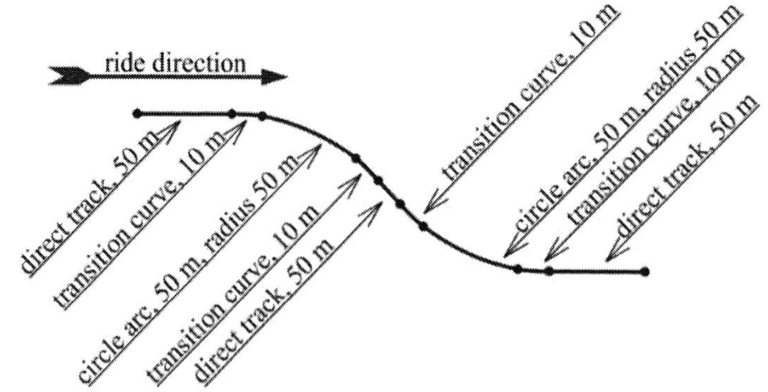

Fig. 5. Track used in simulation

During the calculation, among other variables were recorded, in particular, the value of the lateral displacement of the wheelset, the angles of attack, the force acting in the wheel - rail contact pairs and the wear number. These signals were obtained for each wheel - rail contact pair. Due to the large range of calculated data, only quantities' courses for the first bogie are given below. It is through the front bogie that the vehicle receives the initial impulse to change the ride direction. This fact causes the courses of the monitored quantities to have a less favourable character for the front bogie.

5 Evaluation of the monitored quantities' courses

The chapter presents the results of a series of simulation analyses, the individual variants being marked as follows:

Variant 1: the bogies are oriented according to Fig. 4 a), the pivot pin is firmly attached to the car body

Variant 2: the bogies are oriented according to Fig. 4 a), the pivot pin is rotatably mounted on the car body

Variant 3: the chassis are oriented according to Fig. 4b), the pivot pin is firmly attached to the car body

Variant 4: the bogies are oriented according to Fig. 4 (b), the pivot pin is rotatably mounted on the car body

The ride of the vehicle significantly describes the value of the lateral displacement of the wheelset from the track axis, which for the analysed variants is shown in Fig. 6. It can be deduced from its course that in all cases the front bogie of the vehicle passes through an arc in a close to the chord position. From the point of view of the work of the wheel profile, the basic requirement for its correct function is therefore met. Variants 3 and 4 show a similar lateral deflection on both wheelsets, while variant 4 looks more perspective in view of entering the arc. When riding on a straight track, the undulatory movement of wheelsets occurs, with the smallest amplitude of this movement being achieved in variant 1.

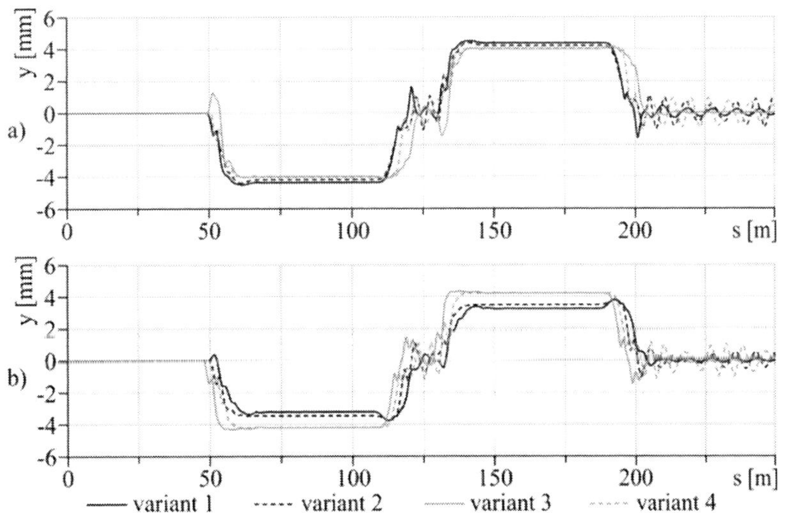

Fig. 6. Lateral displacement of the wheelset towards the track axis
a) the first wheelsets, b) the second wheelset

The rail vehicle is not only carried but also guided by its track. It is therefore necessary to know the course of the guiding forces in the wheel-rail contacts. In a favorable case, the guiding forces should act in the same direction on both wheelsets of the bogie. If their size is similar, one of the assumptions for uniform wear of the two wheelsets is created. In case of a ride in a track arc, the increase in the guiding forces should not be sudden. From Fig. 7 it can be deduced that these requirements seem to best fit the variants 3 and 4. However, it is not sufficient for the process of guiding forces to be analysed separately. It is also necessary to evaluate them in terms of safety against derailment, i.e. to analyse the course of the ratio of the guiding and vertical wheel force as shown in Fig. 8. On the basis of the calculated data, it can be concluded that all proposed solutions are safe against derailment, with variants 3 and 4 appearing more perspective, as more favorable values are achieved on the attacking wheelset.

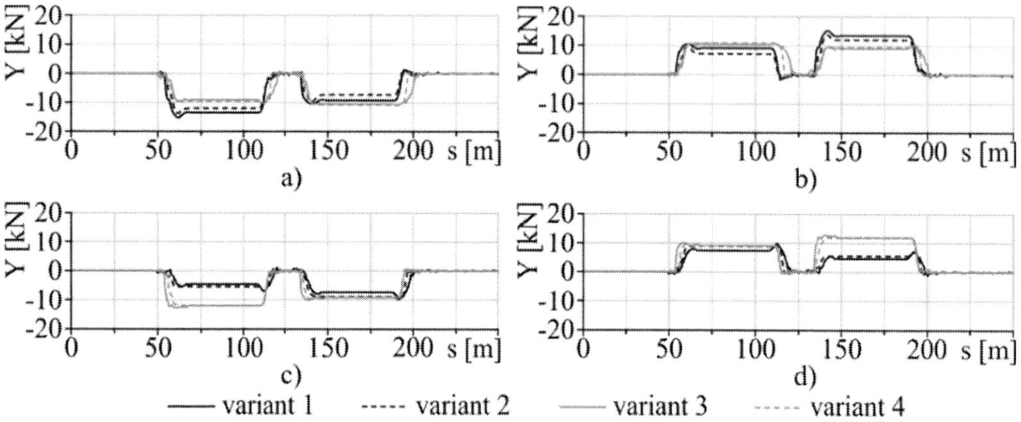

Fig. 7. Lateral forces in contact:
a) front left wheel- rail, b) front right wheel - rail,
c) rear left wheel - rail, d) rear right wheel – rail

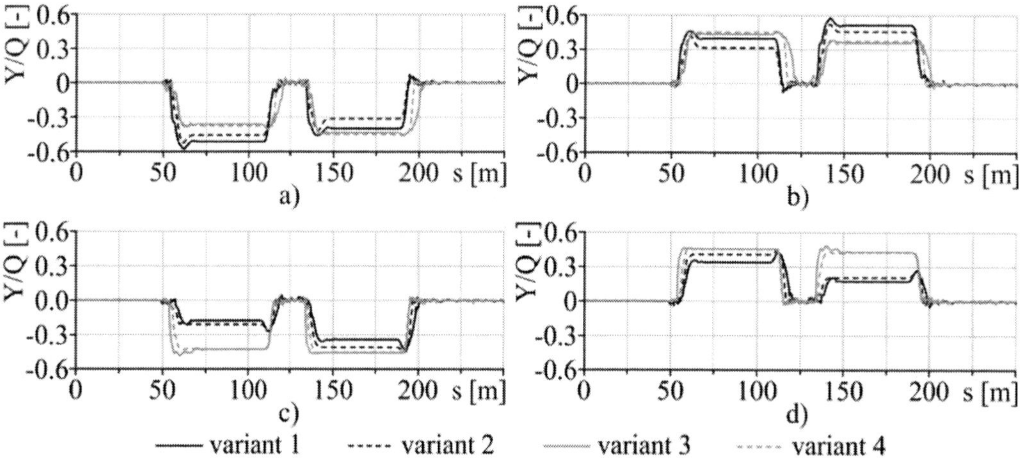

Fig. 8. Safety against derailment index:
a) front left wheel - rail, b) front right wheel - rail,
c) rear left wheel - rail, d) rear right wheel - rail

A significant indicator of the reliability of the bogie is the value of angles of attack, for the analysed variants shown in Fig. 9. In a favorable case, it should be close to zero, in reality it usually reaches values of up to 3°. In an event that non-zero angles of attack are present, the wheel-rail contact is overloaded by the transfer of more signifficant lateral forces, which often can not be transfered through the adhesive bond, significant creep occurs in the lateral and longitudinal directions. When two-point contact is achieved, the contact surfaces are often overloaded by forces circulating between the contact points of the wheel tread and the wheel flange. The resulting product of such operation is increased noise and wear of the contact pair. From Fig. 9 it can be read that although the analysed track has an arc with the smallest possible radius allowed by the standard [15] to be used, particularly small angles of attack are achieved. From this point of view, options 1 and 2 appear in perspective.

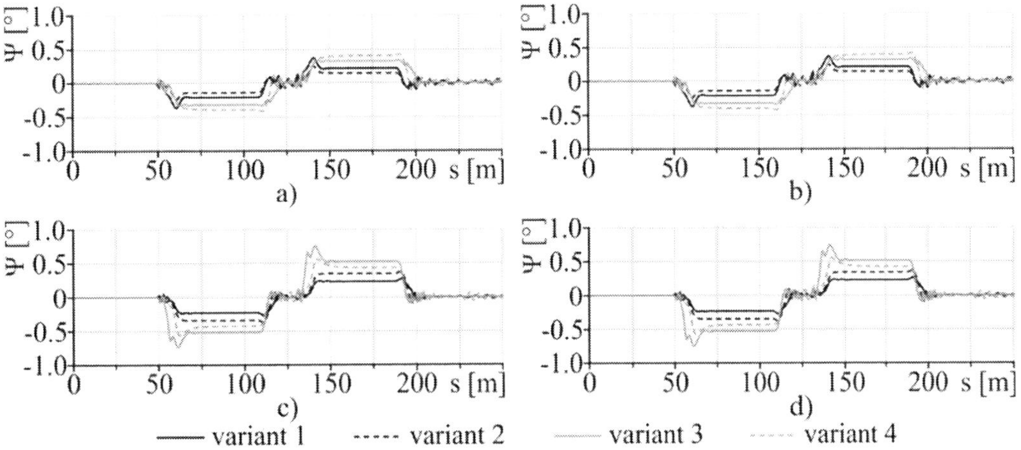

Fig. 9. Angles of attack:
a) front left wheel - rail, b) front right wheel - rail,
c) rear left wheel - rail, d) rear right wheel – rail

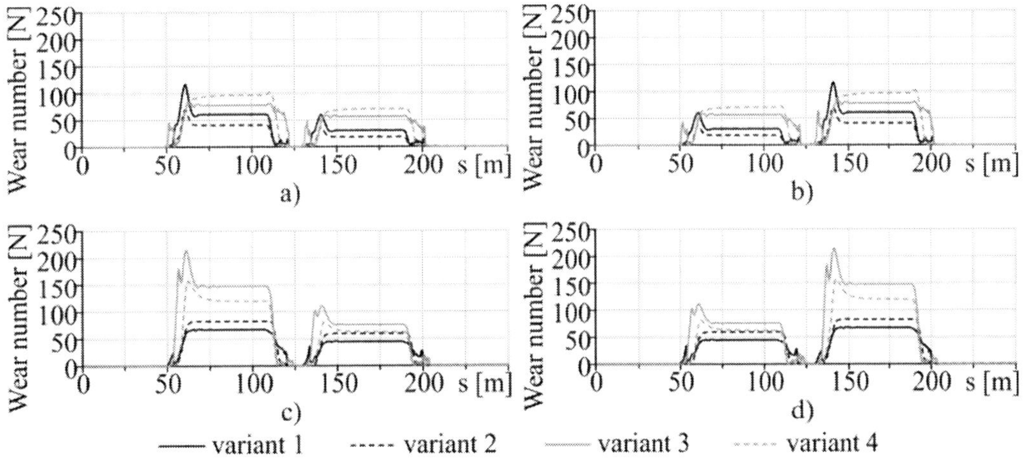

Fig. 10. Wear number in contact:
a) front left wheel - rail, b) front right wheel - rail,
c) rear left wheel - rail, d) rear right wheel - rail

Fig. 10 shows the course of the wear numbers, i.e. intensity of interaction in individual wheel-rail contacts. It can be concluded that a significant improvement is achieved in all analyzed cases compared to vehicles with no steering mechanism. From among the proposed possibilities of placing and mounting of the bogie, variant 2 seems perspective, as there are no significant amplitudes in its course and overall it reaches low values.

Conclusion

The article, by means of a simulation analysis of a rail vehicle, compares in total four variants of car body to bogie junction design differing in the way the bogie is placed and mounted. It provides a view of the problem in terms of the signal courses of deflection of the two wheelsets from the track axis, the course of the guiding forces, the safety against derailment index, the angle of attack and the wear number. Based on the results obtained, the authors consider among the analysed proposals for the given track as most perspective the second variant, i.e. the orientation of the bogies such that the wheelset with a single axle box is headed towards the stem and a pivot pit mounted by means of a pair of rotary bonds, one of which is in the junction of the car and the central element and the second in junction between the central element and the gear frame.

Further improvement in ride parameters is likely to be achieved if the junction of the car with the bogie is not located at the center of the bogie wheelbase. The authors'collective is currently engaged in optimizing the pivot pin shift value and determination of the corresponding changes in vehicle behavior.

The work was supported by the Cultural and Educational Grant Agency of the Ministry of Education of the Slovak Republic in project No. KEGA 077ŽU-4/2017: Modernization of the Vehicles and engines study program. The work was also supported by the project No. APVV-0842-11: Equivalent railway operation load simulator on the roller rig and VEGA Nr. 1/0927/15: Research of the use of alternative fuels and hybrid drives on traction vehicles with aim to reduce fuel consumption and air pollutants production.
Research-Educational Center of Rail Vehicles (VVCKV)

References

1. J. Dižo, *Analysis of a goods wagon running on a railway test track*. Manufacturing technology: journal for science, research and production **16** (4), 667-672
2. J. J. Kalker, *Three-dimensional elastic bodies in rolling contact*. 314 p. ISBN: 0-7923-0712-7 (1990)
3. L. Smetanka, P. Šťastniak, J. Dižo, *Comparison of wear laws programmed in SIMPACK*. Experimental and computational methods in engineering. ISBN 978-80-7414-698-5 (2017)
4. J. Gerlici, T. Lack, V. Hauser, M. Maňurová, O. S. Nozhenko, K. O. Kravchenko, *Carriage bogie*. Utility model no. UA 114040 U, category: B61F5/30. Kyjiv: Ukrajinskyj instytut promyslovoji vlasnosti (UKRPATENT), 9 p. (2017)
5. M. Loulová, V. Hauser, J. Gerlici, T. Lack, O. S. Nozhenko, O. P. Kravchenko, K. O. Kravchenko, *Carriage bogie*. Utility model application no. u201703246, category: B61F5/30. Kyjiv: Ukrajinskyj instytut promyslovoji vlasnosti (UKRPATENT), 9 p. (2017)
6. M. D. Ivanov, A. A. Ponomarev, B. K. M. Ieropol'skij, *Tramways T-3*. 240 p. (1977)
7. V. Hauser, O. Nozhenko, K. Kravchenko, M. Loulová, J. Gerlici, T. Lack, *Impact of three axle boxes bogie to the tram behavior when passing curved track*. Procedia Engineering. **192**, 295-300 (2017)
8. A. Chudzikiewicz, B. Sowinski, *Modelling and simulation of trams bogies with fully independently rotating wheels*. In Dynamics of vehicles on roads and tracks. 1427-1434. ISBN: 978-1-4987-7702-5 (2016)
9. A. Chudzikiewicz, *Simulation of rail vehicle dynamics in MATLAB environment*. In Vehicle system dynamics **33** (2), 107-119, ISSN 0042-3114 (2000)

10. R. Melnik, S. Koziak, *Rail vehicle suspension condition monitoring - approach and implementation*. Journal vibroengineering, Lithuania, ISSN 1392-8716
11. A. Sapietová, M. Sága, P. Novák, R. Bednár, J. Dižo, *Design and application of multi-software platform for solving of mechanical multi-body system problems*. Mechatronics: Recent technological and scientific advances, 345-354 (2011)
12. M. Blatnický, M. Štauderová, J. Dižo, *Numerical analysis of the structure girder for vehicle axle scale calibration*. Procedia Engineering **177**, 510-515 (2017)
13. J. Dižo, J. Harušinec, M. Blatnický, *Multibody system of a rail vehicle bogie with a flexible body*. Manufacturing technology **15** (5), 781-788
14. J. Dižo, M. Blatnický, B. Skočilasová, *Computational modelling of the rail vehicle multibody system including flexible bodies*. Communications: scientific letters of the University of Žilina **17** (3), 31-36 (2015)
15. TNŽ 73 63 61 *Geometrical position and layout of railways with 1000 mm gauge*. (ŽSR, Bratislava, 37 p., 2007)

Innovative wheel tread design aimed to tramcar-track interaction improvement when passing curves of a small radius

Vladimír Hauser[1,*], *Olena S.* Nozhenko[1], *Kateryna O.* Kravchenko[1], *Mária* Loulová[1], *Juraj* Gerlici[1], *Tomáš* Lack[1]

[1]University of Žilina, Faculty of Mechanical Engineering, Department of Transport and Handling Machines, Univerzitná 1, 010 26 Žilina, Slovak Republic

Abstract. Tramways in cities often ride in track curves of a small radius, which is followed by an increased effect of the vehicle on the track. In the rail - wheel contact often occurs creep in longitudinal as well as in lateral direction, which causes increase of riding resistance, wear in rail – wheel contact and noise operation. Exactly with the aim to reduce these undesirable effects we designed a tram bogie with steered wheelsets and with the double wheel tread. This inventions are registered by authors under Utility Model Nr. u201609015 and Patent Application Nr. a201701589. Description of proposed process of passing vehicle through strongly curved track is given in this paper. A comparison of creep velocities in rail - wheel contact results obtained by computer multibody simulations for a tram vehicle riding through a S-curve are presented. Purposefulness of proposed invention is investigated by constructing vehicle ride resistance dependence to the track curve radius.

Keywords: double wheel tread, strongly curved track, tramcar throughputness

1 Introduction

Special attention is taking place in the environment of public transport railways, for which higher amount of radius curves up to 17 meters being applied is specific. The outcome of such operation of vehicles is an increase in vehicle's effects on the track in the rail-wheel contact resulting in increased ride resistance, creep in the rail-wheel contact patch, speeding up the process of wear in the contact-pair as well as noise generation. At present, a variety of technical solutions for the vehicle bogie design as well as rail designs focused on decreasing of these negative effects exists [1-5]. Their use in smaller radius curves however, cannot give acceptable results and often causes complications in bogie design. The authors give a concept of creep reduction in rail-wheel contact by double wheel tread, which doesn't require complicated bogie design.

Creep in wheel-rail contact takes place in both lateral and longitudinal direction. Creep in lateral direction in small radius curves reaches higher values. It is possible to effectively minimize these values by using a bogie design, which allows steering of the wheelsets to a

* Corresponding author: vladimir.hauser@fstroj.uniza.sk

Reviewers: *Robert* Grega, *Milan* Naď

radial position [6, 7]. The authors aim at a design solution, which would remove the cause of longitudinal creep in rail-wheel contact and that would not be complicated at the same time. We suggest solving this task by using a double wheel profile and a change in rail geometry of a small radius curve [8]. The purposefulness of this solution is proven by simulation analysis of the vehicle ride dynamics.

2 Motivation

Vehicle forces affect the track through contact between wheels and rails. The effect is especially intensive when the vehicle passes through arcs of small radius. Often it is not possible to transfer it through the wheel-rail adhesion bond, and so a creep that takes place in the longitudinal and lateral direction develops. [9] Since the creep is present in the contacts and the shear friction coefficient is not zero, part of the vehicle performance serves to overcome the mentioned creep. The dependence of this power dissipation from the radius of the track curve gives an overall view on the characteristics of the vehicle ride through a track curve.

This dependence, as shown in Fig. 1, was obtained for a 10 km.h-1 speed of a vehicle similar to a T3 tram [10], with values for two types of bogies and two types of wheel profiles being presented. The first bogie has fixed wheelsets, the second bogie is equipped with a mechanism for steering the wheelsets. The E-99-00 wheel profile has lower tread conicity values, the KP-1 profile [11] has its conicity adjusted to tracks with a greater number of curves. In all cases, the power dissipation of the vehicle is divided into a component covering creep in the longitudinal and lateral directions.

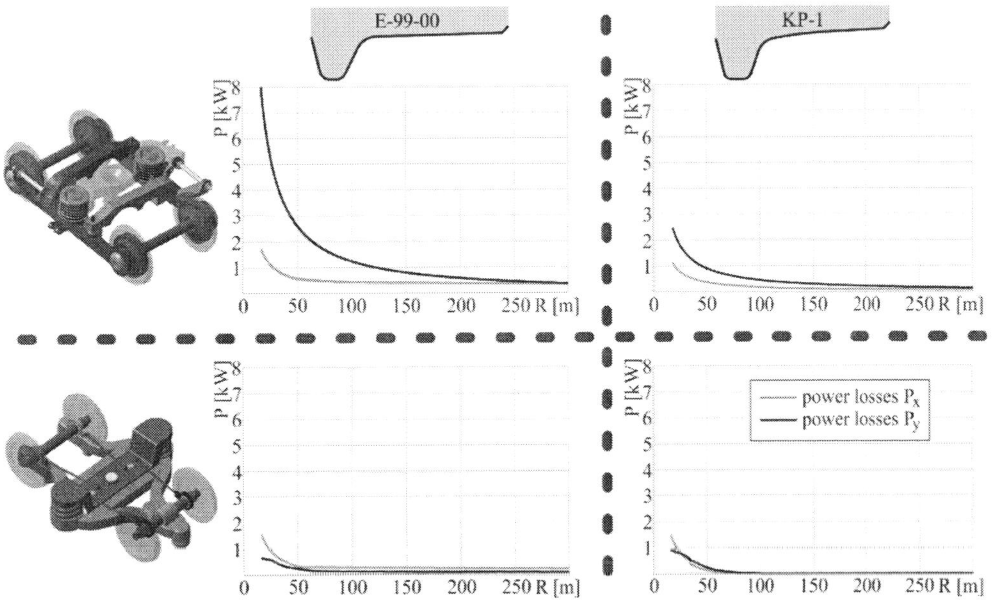

Fig. 1. Comparison of the power dissipation of two bogie types equipped with two different wheel profiles

From the comparison of the power dissipation of the vehicle shown in Fig. 1 it can be understood, that the wheelset steering mechanism can effectively minimize creep in the lateral direction particularly, even when a wheel profile not suitable for ride through track curves is used. However, even in the case of a vehicle equipped with wheelset steering mechanism with a wheel profile suitable for ride through track curves, the power dissipation in small radius curves reaches values about 2 kW.

The ultimate product of this power dissipation is the wear of the wheel-rail contact. The wear range is cumulative in relation to the number of passes of the vehicle at the given track section. It is therefore particularly desirable to minimize the power dissipation of the vehicle, thereby prolonging the operating life of the contact wheel-rail pair. We suggest to reach this state by minimizing the longitudinal creep.

3 Analysis of the wheelset ability to pass through a track curve

Curve radius, at which the wheelset is theoretically able to ride through without creep can be determined using the equation:

$$R \geq \frac{2Sr}{\Delta r}[\text{mm}] \tag{1}$$

where *2S* is contact circles distance [mm], *r* is wheel radius [mm], Δr is the value of the wheel profile delta function at the moment the clearance provided by the track channel is completely exhausted [mm].

For a tram wheelset designed for a gauge of 1000 mm, with a wheel radius of 340 mm, KP-1 profile, contact circles distance $2S$ = 1061.9 mm and value Δr 5.5 mm, the minimal value of the curve radius, that can be run theoretically without creep is 65.64 m. For example, in Bratislava the city railways vehicles commonly operate on tracks with radiuses starting from 17 m. It is therefore necessary to examine the options of increasing the Δr value in formula (1). The relation of the desired Δr value to the track radius can be expressed, based on the formula (1) as shown in Fig. 2. Increase in the Δr value by increasing the value of the ride surface conicity cannot be seen as a suitable solution. Such intervention in the wheel profile geometry would lead to a more significant adulatory movement and thus worsening the vehicle ride stability at higher velocities. Thereby a conflicting requirement on increasing the Δr value without intervening in the existing wheel profile ride surface geometry.

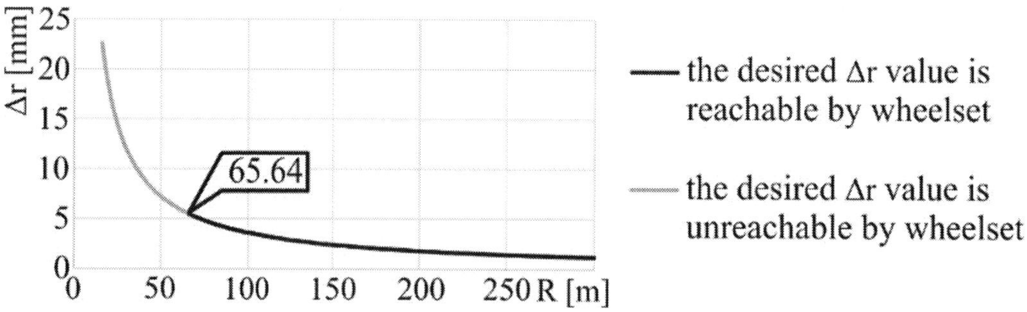

Fig. 2. Desired value Δr necessary for ride in curve without creep

Described conflicting requirement can be fulfilled by creating new point of rail-wheel contact, which will be used only for ride in strongly curved track.

4 Suggestion of a way for wheelset passing a small radius curve

We suggest solving this situation by using a wheel profile with an extended ride surface designed for a ride in small radius curves, located on the outer side of the wheel profile. A schematically picture of the wheelset with an extended profile and the principle of its ride through track curve is given in Fig. 3.

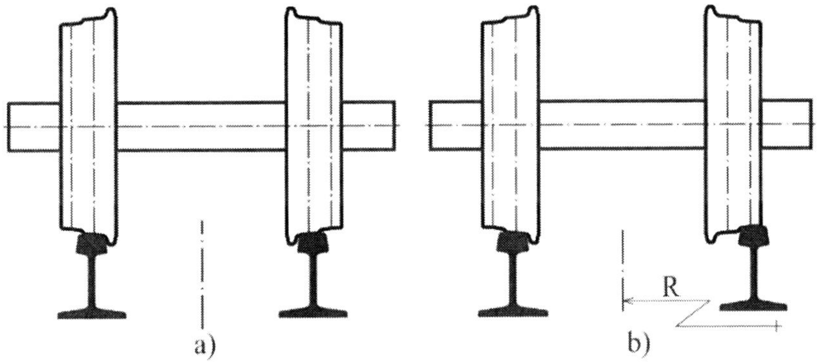

Fig. 3. Wheelset equipped with an additional tread
a) during ride on a straight track, b) during ride in track arc of small radius

The inner part of the wheel profile is identical with the former profile. Ride characteristics of the vehicle during ride on a straight track or in curved track with a large radius therefore stays unchanged. An additional tread is designed on the outer part of the profile for ride of the axle on curved track with small radius. In curved track the inner wheel therefore rides on the additional tread of a smaller radius, allowing the required shift in delta r function curve to be achieved.

For an effective use of the additional tread, it is necessary to determine its displacement from the original profile in lateral and vertical direction appropriately, defined using the dimensions y and z on Fig. 4a). For creating the geometry of the secondary tread, it is possible to use a segment of the original wheel tread.

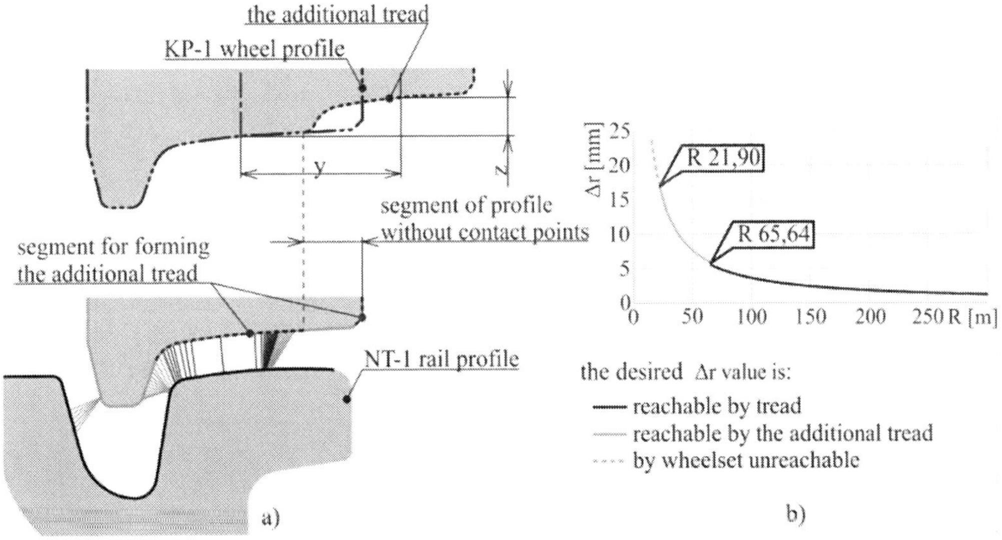

Fig. 4. a) Placement of the additional tread on the wheel profile,
b) Desired and attainable Δr value for ride of the wheelset in curve without creep

The displacement in lateral direction should be minimal, to avoid significant increase in material volume and thus the mass of the wheel rim. However, its minimal value is limited by the distribution of the given wheel-rail pair contact points. For placing the additional tread, we suggest using the outer part of the original wheel profile, where the profile doesn't have any con-tact points with the rail. A certain widening of the wheel profile is necessary nevertheless. In the considered case of profile KP-1, a lateral displacement of the additional tread from the original one y = 50 mm was chosen.

The displacement in vertical direction is dependent on the wheel profile delta r function value at the moment the clearance provided by the track channel is completely exhausted. For an effective use of the additional tread to be provided, following formula can be derived for the value z:

$$z = 2\Delta r + \frac{y.\Delta r}{2S}, \tag{2}$$

In the considered case of the original profile KP-1, the vertical displacement of the additional wheel tread value z is 11.5 mm. The required and achieved Δr value curve (as given in Fig. 2) is therefore changed in a way shown in Fig. 4b).

Based on the given theoretical consideration it is possible to expect more favourable ride in curve with a radius smaller than 65.64 m thanks to the use of an additional tread. A theoretical prerequisite for a wheelset ride without creep is created for a track curve radius of up to 21.9 m. This theoretical assumption was verified by a simulation analysis of the vehicle's ride on a track with a described geometry. The contact geometry had to be inserted into the calculation program in the desired form.

5 Creating inputs for simulation analysis

To perform the simulation analysis, it is necessary to define, among other parameters, the wheel and rail profiles in the desired shape [12-14], i.e. in the form of text files containing among other parameters, particularly the sequence of coordinate points lying on the profile treads in the desired form. During the calculation, these points are automatically interconnected by the spline curve that represents the desired profile. For commonly used wheel and rail shapes, these files are already pre-written in the SIMPACK calculation program. The wheel-rail contact of the city railway vehicles is not covered by the standards concerning the railway vehicles, therefore the profiles needed for the following analysis cannot be found among the pre-written profile files. It was therefore necessary to proceed with their creation.

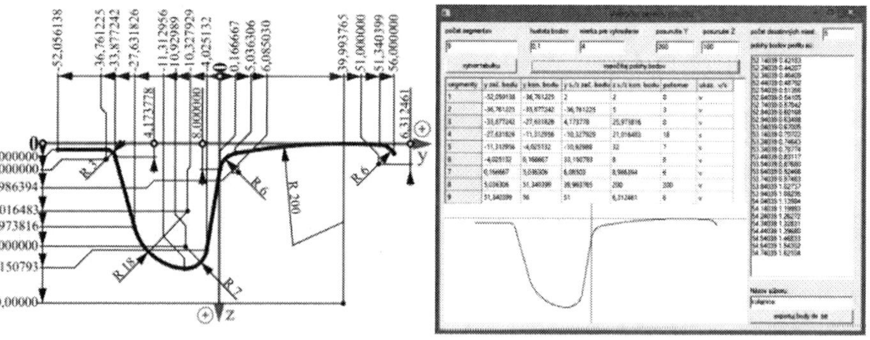

a) b)

Fig. 5. a) Drawing of the rail profile in desired shape,
b) Program dialog box for obtaining coordinate points

The required files can be added to the SIMPACK database, but SIMPACK does not have the ability to create such files directly by defining their geometry. It is therefore necessary to create the required sequence of points and define their coordinates in the appropriate file. The basic input for creating such file is the drawing of the profile in the desired shape, as shown in Fig. 5a). Based on this defined geometry, it is possible to create a set of parameter equations of the segments from which the profile is composed.

Subsequently, based on the choice of one of the coordinate points, the second coordinate of this point can be counted using these parametric equations. This procedure must be repeated cyclically in order to obtain a sufficient number of points of the wheel profile tread. Approximately 200 to 3000 points evenly positioned over the wheel profile tread, depending on the profile complexity can be considered sufficient.

Because of the complicated process that is necessary to repeat many times when defining a rail with variable geometry, the authors have created a program containing algorithm, which allows to generate a sequence of coordinate points of the profile of arbitrary density based on a simple definition of certain specific dimensions of the individual profile segments, and save them in a form suitable for direct import into the SIMPACK program afterwards. A program dialog box with given specific dimensions of the rail tread segments generating the profile coordinates and plotting the specified profile is shown in Fig. 5b).

6 Simulation analysis of a vehicle ride

For verifying the theoretical prerequisite of wheelset ride without creep using the additional tread in track curves radius up 21.9 to 65.6 m, as described in chapter 5, a series of vehicle ride dynamics simulation analysis was performed in simulation program SIMPACK with the aim of finding the dependency of the monitored quantities from the track curve radius. From among the examined quantities, the power dissipation arising from the vehicle ride resistance in track curve represents the effectiveness of the vehicle ride in curved track well. The curves of this quantity in three different situations are compared in this article.

In the first case, a vehicle without the possibility of steering the wheelsets nor the wheel profile with additional tread is considered. In the second case we consider a vehicle with a steering wheelsets possibility and without the proposed wheel profile design. The third case considers a vehicle with steering wheelsets equipped with the de-scribed new wheel profile design. The vehicle parameters basically resemble the ones of a T3 tram.

For obtaining an overall view of the vehicle behaviour it is necessary to evaluate the monitored quantities in dependence on the track curve radius. One of the ways to achieve such dependence is a simulation of vehicle ride along a track transition curve of a sufficient length. A track transition curve is a track section, where the radius of the curve is continuously changed. It can have various shapes. According to the TNŽ 73 63 61 standard [15], a transition curve in the shape of a clothoid is allowed to be used in the track. In case the clothoid connects a straight track section with a circular curve, it is possible to calculate the values of its virtual radius R [m] in the considered place with the use of the following formula:

$$R = \frac{R_2 l}{s}, \tag{3}$$

where R is track curve virtual radius [m], R_2 is radius of the circular track curve on the end of the transition curve [m], l is overall length of the transition curve [m] and s is transition curve length measured from its beginning to the considered place [m].

In case the radius changes smoothly enough, the track section on which the vehicle is located currently can be considered as a curve with a constant radius. Therefore, the distance travelled along the track represented on the horizontal axis of the considered quantity curves can be calculated with the use of formula 3 for the track curve radius in the considered place.

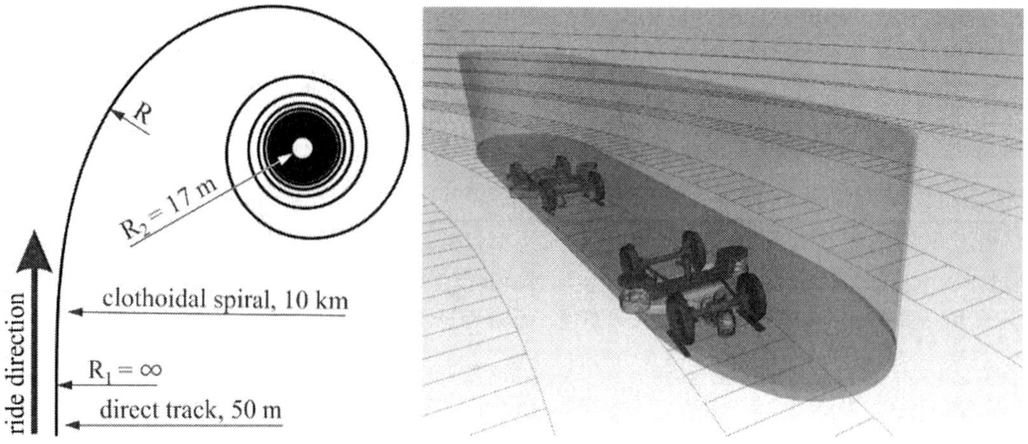

Fig. 6. Track in the shape of a transition curve

I the subsequent simulation, a ride on a track composed from a straight section designed for consolidation of the vehicle ride, a track transition curve in the shape of a clothoid with a length of 10 km and a circular arc with a radius of 17 metres. The shape of the track is shown in the Fig. 6. Comparison of the vehicle power dissipation during a ride at the speed of 10 km.h^{-1} is given in Fig. 7. The course of this quantity provides a global view for the way a vehicle rides in a track curve. Because of a big amount of the calculated values, only power dissipation curves for the curve radius interval from 17 to 100 m, hence the area, where an effective application of the proposed design can be expected are presented. Power dissipation needed to cover the creep in longitudinal direction P_x and in lateral direction P_y are given separately, but the vehicle covers both power dissipations at the same time.

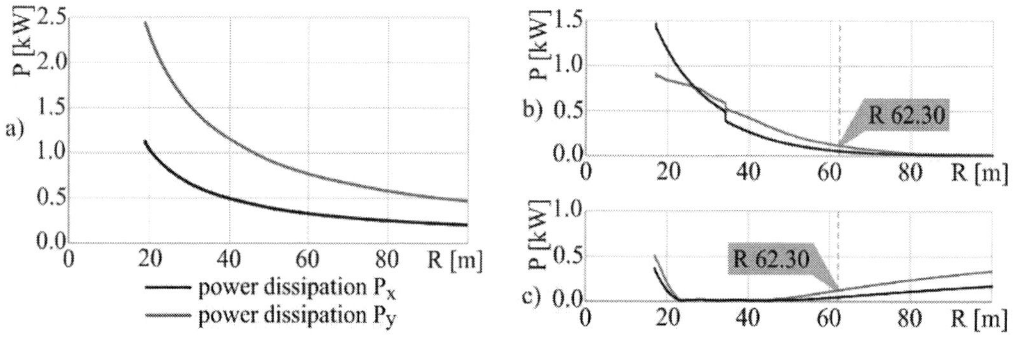

Fig. 7. Power dissipation of the vehicle
a) without the possibility to steer the wheelsets and additional tread,
b) with the possibility to steer the wheelsets and without additional tread,
c) with the possibility to steer the wheelsets and with the use of additional tread

Based on the simulation analysis of a ride in track transition curve it can be concluded, that in the case of vehicle ride in track curve with a small radius, it is possible to achieve more favorable ride parameters in a wider track curve radius interval.

From the comparison of the given simulation analysis can be evaluated, that a ride of a vehicle with use of an additional tread is more favourable in a curve with a radius smaller than 62.3 m in terms of power dissipation curves. From the given graphs it is evident, that it comes to significant reduce of negative effects in the monitored area when using the additional tread. In the track curve radius range of 23.55 to 44.88 m the power dissipation is practically eliminated. For the described way of vehicle ride in track curves with small radius to be able to be put to use, it is necessary to focus on creating an altered regime of wheel rolling during ride as well as examining vehicle response to such change.

It is generally known, that in terms of safety against derailment, the ride of the vehicle in an S-curved track with occurrence of superelevation twisting is seen to be very problematic. The goal of the authors was also to verify the vehicle's response to such situation that is further complicated by the change in the wheelset drive mode if the radius of the track exceeds the same value as the authors described in detail in Chapter 4. The track used for the calculation was defined according to TNŽ 73 63 61 standard [15], consisting of straight segments, transition segments, and two curves of opposite direction with 25 meters radius. The length of track segments and their layout is shown in Fig. 8. The rails are seated with a lateral superelevation of 33.6 mm. The track gauge is 1 000 mm.

We compare three cases: In first case the rails have a NT1 profile all along the track and the wheel profile was set to KP-1. Original bogie was used. In the other two cases the rail has a NT-1 profile on the straight section and a special shape (Fig. 8) in curve sections. In these cases, the wheel profile KP-1 with additional tread was used. Tram car was equipped with by authors proposed bogie [12]. Boogies layout facing each other with the front of the bogie frame and a facing each other with the frame rear was investigated. The part of the bogie frame with wheelset being seated in only one axlebox is considered the frame front.

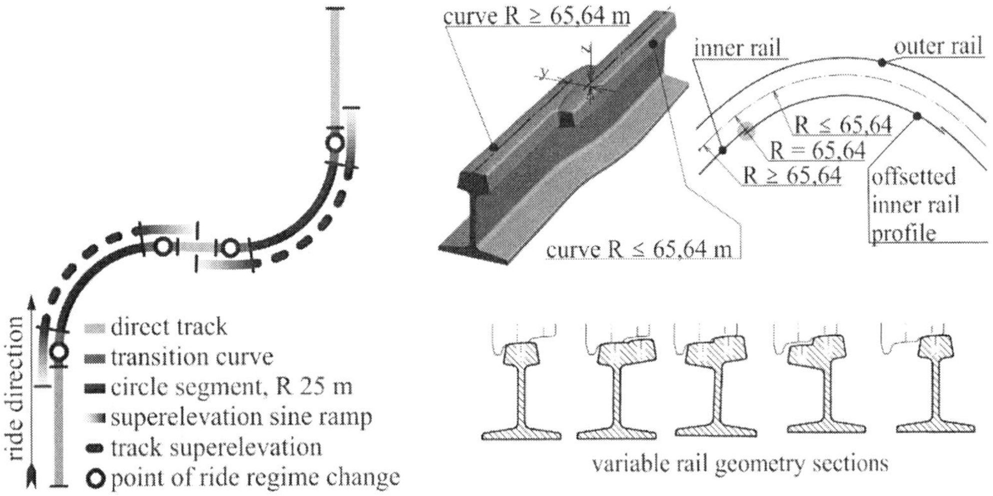

Fig. 8. S-curved track and a change point of driving mode

The tread change is performed on a wheel situated on the inner side of a track curve at the position of transition curve, where the curve radius measured from the track axis reaches a value of 65.64 m. At this point, a change in the rail profile takes place. The rail profile placed in a curve of a small radius is offset against the original profile in lateral and vertical direction for a value identical with lateral and vertical displacement of the wheel profile. The change of wheel tread is performed through vehicle passing this part of the track, as is shown in Fig. 8.

When a wheelset enters a track curve, it attacks the outer rail. In a curve with a diameter corresponding the change of the tread, the clearance provided by the track free channel is considered to be completely reduced. If the change of the tread would be performed using an instant offset of the rail profile, as shown in Fig. 3, it would result in a step change of the Δr value. The wheelset would suddenly get into a position favourable for a ride in curve with a radius of 21.9 m, although the track radius in the given place is 65,64 m, which would lead to violent attack on the rail. An instant offset of the rail profile would therefore become a source of lateral excitation, which is an unfavourable. It is therefore necessary to prevent a signifficant step change in Δr value of the wheelset at the moment of tread change. This is reached using a varying rail profile near the place of tread change, which minimizes the step change of Δr value.

The expression of the SAD (safety against derailment) as a ratio of guiding forces and wheel forces on individual wheels corresponds the guiding forces curve. Higher values represent a worse state, lower ones represent a safer ratio of horizontal and vertical forces. From the results in Fig. 9 can be seen, that all values are significantly under the generally known allowed value of 0.8. Higher values arise, when entering and coming out from the curve, but even in these cases they do not reach more than 0.5.

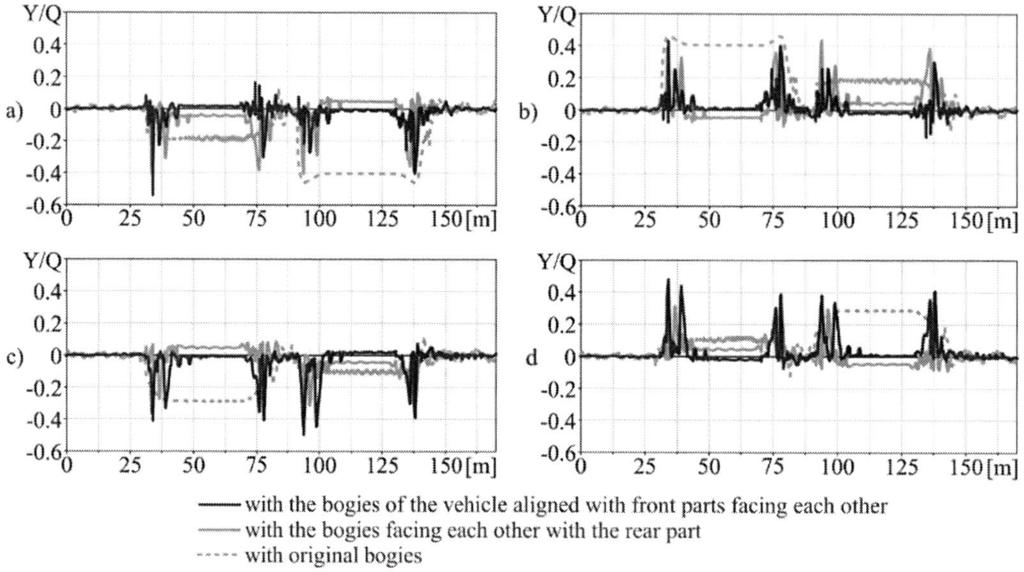

Fig. 9. Safety against derailment of the front bogie;
a) first wheelset, right wheel, b) first wheelset, left wheel,
c) second wheelset, right wheel, d) second wheelset, left wheel.

From the graphs can be concluded, that a tram with original bogies reaches worse results than bogies equipped with the steering mechanism. Bogies with the steering mechanism reach a maximal value of 0.05 and the original ones a value of 0.3. At the moment of entering and leaving the curve, values around 0.5 are reached. The wheelset radial steering mechanism as well as usage of the additional tread have a positive impact on SAD.

Conclusion

Based on the simulation analysis of a ride in track transition curve it can be concluded, that in the case of vehicle ride in track curve with a small radius, it is possible to achieve more favourable ride parameters in a wider track curve radius interval.

From the comparison of the given simulation analysis can be evaluated, that a ride of a vehicle with use of an additional tread is more favourable in a curve with a radius smaller than 62.3 m in terms of power dissipation curves. From the given graphs it is evident, that it comes to significant reduce of negative effects in the monitored area when using the additional tread. In the track curve radius range of 23.55 to 44.88 m the power dissipation is practically eliminated.

For the described way of vehicle ride in track curves with small radius to be able to be put to use, an altered regime of wheel rolling during ride as well as examining vehicle response to such change was investigated. Rail of variable rail geometry was proposed. The wheelset running regime change possibility is able to ensure a more favourable ride in track curves. With its use, decrease of the SAD index, power dissipation and wear number during vehicle ride in track curves with small radius can be reached, as shown by the performed simulation analysis of a tram vehicle model with two different types of bogie.

The work was supported by the Cultural and Educational Grant Agency of the Ministry of Education of the Slovak Republic in project No. KEGA 077ŽU-4/2017: Modernization of the Vehicles and engines study program. The work was also supported by the project No. APVV-0842-11: Equivalent railway operation load simulator on the roller rig and VEGA Nr. 1/0927/15: Research of the use of alternative fuels and hybrid drives on traction vehicles with aim to reduce fuel consumption and air pollutants production.
Research-Educational Center of Rail Vehicles (VVCKV)

References

1. J. Dižo, *Analysis of a goods wagon running on a railway test track.* In: Manufacturing technology: journal for science, research and production **16** (4), 667-672
2. M. Blatnický, M. Štauderová, J. Dižo, *Numerical analysis of the structure girder for vehicle axle scale calibration*. Procedia Engineering **177**, 510-515 (2017)
3. A. Chudzikiewicz, B. Sowinski, *Modelling and simulation of trams bogies with fully independently rotating wheels*. Dynamics of vehicles on roads and tracks, 1427-1434, ISBN 978-1-4987-7702-5, (2016)
4. A. Chudzikiewicz, *Simulation of rail vehicle dynamics in MATLAB environment*. Vehicle system dynamics **33** (2), 107-119, ISSN 0042-3114 (2000)
5. R. Melnik, S. Koziak, *Rail vehicle suspension condition monitoring - approach and implementation*. Journal vibroengineering, Lithuania, ISSN: 1392-8716
6. J. Gerlici, T. Lack, V. Hauser, M. Maňurová, O. S. Nozhenko, K. O. Kravchenko, *Carriage bogie.* Utility model no. UA 114040 U, category: B61F5/30, Kyjiv, Ukrajinskyj instytut promyslovoji vlasnosti (UKRPATENT), 9 p. (2017)
7. M. Loulová, V. Hauser, J. Gerlici, T. Lack, O. S. Nozhenko, O. P. Kravchenko, K. O. Kravchenko, *Carriage bogie*. Utility model application no. u201703246, category: B61F5/30, Kyjiv, Ukrajinskyj instytut promyslovoji vlasnosti (UKRPATENT), 9 p. (2017)
8. V. Hauser, J. Gerlici, T. Lack, M. Loulová, O. S. Nozhenko, K. O. Kravchenko, O. V. Prosvirova, *Railway bogie wheelset and curved track segment*. Patent application, number a201701589, category: 4267/3A/17, Kyjiv, Ukrajinskyj instytut promyslovoji vlasnosti (UKRPATENT), 11 p. (2017)
9. J. J. Kalker, *Three-dimensional elastic bodies in rolling contact*. ISBN 0-7923-0712-7, 314 p., 1990)
10. M. D. Ivanov, A. A. Ponomarev, B. K. M. Ieropol'skij, *Tramways T-3*. 240 p. (1977)
11. J. Zelenka, M. Kohout, *The wheel - rail interaction in the environment of tramcar operation*. [online] [cit. 2017-01-30] Available at: http://www.sdp-cr.cz/WD_FileDownload.ashx?wd_systemtypeid=34&wd_pk=W-zEzMDMsWzMwX (2012)
12. J. Dižo, J. Harušinec, M. Blatnický, *Multibody system of a rail vehicle bogie with a flexible body*. Manufacturing technology **15** (5), 781 – 788

13. J. Dižo, M. Blatnický, B. Skočilasová, *Computational modelling of the rail vehicle multibody system including flexible bodies*. Communications: scientific letters of the University of Žilina **17** (3), 31-36 (2015)
14. L. Jakubovičová, M. Sága, *Computational analysis of contact stress distribution in the case of mutual stewing of roller bearing rings*. Novel Trends in Production Devices and Systems, Applied Mechanics and Materials **474**, 363-368 (2014)
15. TNŽ 73 63 61 *Geometrical position and layout of railways with 1000 mm gauge*. (ŽSR, Bratislava, 37 p., 2007)

Universal compact lower limb turning module intended for use in orthotic robots

Mateusz Janowski[1], *Danuta* Jasińska-Choromańska[1], *Dymitr* Osiński[1], *Marcin* Zaczyk[1,*]

[1]Faculty of Mechatronics, Warsaw University of Technology, Boboli 8, 02-525, Warsaw, Poland

Abstract. In this paper, a model of an orthotic robot's lower limb rotation system is presented. The system is intended for use in typical contemporary orthotic robots such as the 'Veni-Prometheus' System for Verticalization and Aiding Motion designed at the Faculty of Mechatronics, Warsaw University of Technology. In the paper, the state of the art is briefly stated, with the relatively low number of orthotic robots allowing realization of pivoting turns highlighted. The intended two-stage pivoting turning movement is analyzed in detail and the operating conditions as well as limitations of the turning module are indicated. The conception of a turning module introduces additional degree of freedom to the existing orthotic robot designs by realizing the rotation about the lengthwise axis in the thigh link. A three-dimensional model and its analysis are shown. The proposed design ensures the necessary movement of the lower limb and the torso of an impaired person during the execution of pivoting turn while remaining compact in order to ease the introduction of the turning system to different orthotic robot designs.

Keywords: lower limb turning module, orthotic robot, exoskeleton, change of the gait direction, spinal cord injury, rehabilitation.

1 Introduction

In the paper, a concept of a turning module of the lower limb of an orthotic robot is proposed. The module is designed to be installed in the existing mechanical structures of orthotic robots, called also exoskeletons for aided rehabilitation, or to be implemented in a completely new structure. The task of the module is to enhance capabilities of the devices it is intended to work with, by the way of introducing an additional degree of freedom of the lower limb, which enables a change of direction of motion in a way resembling the natural one. At present, design solutions enabling a turn of the lower limb occur only in a few industrial and military exoskeletons.

Orthotic robots provide a possibility of motion, while the user is supported on crutches or on a walking frame (as in the case of ReWalk [1] and Ekso GT [2] robots), as well as without any additional supports (as in the case of REX [3] robot). The way of motion is considerably limited. Most of devices of this type make it possible for the handicapped to move by themselves only over a plane surface. But a few enable sitting down and rising up, as well as going up/down the stairs. Such movements as a rotation around the vertical axis

* Corresponding author: m.zaczyk@mchtr.pw.edu.pl

Reviewers: *Ivan* Kuric, *Bogdan* Posiadala

or a change of direction of motion in orthotic robots without an additional turning module, are realized by the way of raising a considerable part of the body weight using crutches or a walking frame and performing a turn on the feet of the device.

At the Institute of Micromechanics and Photonics, Faculty of Mechatronics, Warsaw University of Technology, a 'Veni-Prometheus' System for Verticalization and Aiding Motion (Fig. 1) was built, which was designed to restore lost motor abilities for the handicapped as well as to aid rehabilitation of individuals, who were not fully able [4, 5]. The device offers five functions: moving over a plane, sitting down and rising up as well as going up/down the stairs. Electric motors drive movements of flexion and extension of the hip and knee articulations. The movement of dorsal and plantar flexion at the ankle joint is not driven. Works aimed at development of a mechanical structure providing a capability of making turns are currently being carried out [6].

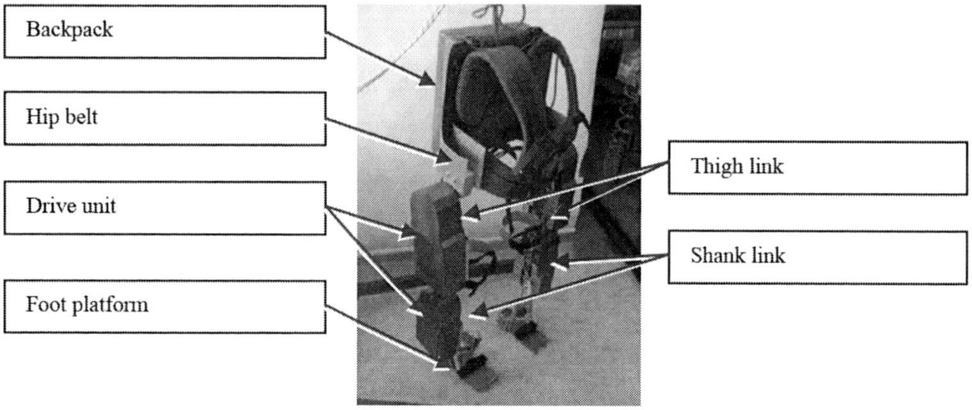

Fig. 1. 'Veni-Prometheus' System for Verticalization and Aiding Motion

2 Survey of the existing solutions

Orthotic robot equipped with a turning module is designed for individuals suffering from a serious injury of the spinal cord, such as accident victims with an injury of the lumbar region of the backbone that have completely or partially lost nervous connection between muscles and the brain from the waist down. Usually, they are not able to keep a vertical position using crutches or a walking frame due to an unsatisfactory stability of the lower limbs. A considerable part of such victims is confined to wheelchair. It is not an advantageous solution with regard to functioning of the organism, since it is limited by a sitting or lying position.

Wearable robots aiding movements of muscles, i.e. exoskeletons, which appear in civil industry as well as those created for military purposes, feature degrees of freedom that make it possible to change direction of motion. A common feature of almost all the solutions of this type is a lack of assistance for the degrees of freedom that are responsible for rotations enabling a change of direction of motion. Description of design solutions applied in outstanding exoskeletons (whose degrees of freedom of the lower limbs are not limited to the sagittal plane) are presented blow; these are NASA X1 (Fig. 2) and BLEEX (Fig. 3) exoskeletons.

The NASA X1 exoskeleton was designed for the purposes of dynamometric measurements as well as trainings for astronauts [7]. Because of micro-gravitation conditions in space bases and spacecrafts, muscles and bones of astronauts are subjected to a fast degeneration. At the initial stages, the design was also intended for individuals suffering from an injury of the spinal cord. Thus, forces resulting from the gravitation of the user's mass are transmitted through the frame of the exoskeleton to the ground. The assisted degrees of freedom are flexion and extension at the knee as well as flexion and extension of the thigh. The exoskeleton enables a turn of the leg, devoid of its drive, at the hip owing to application of a hinge located between the hip frame and the thigh section of the exoskeleton (Fig. 2).

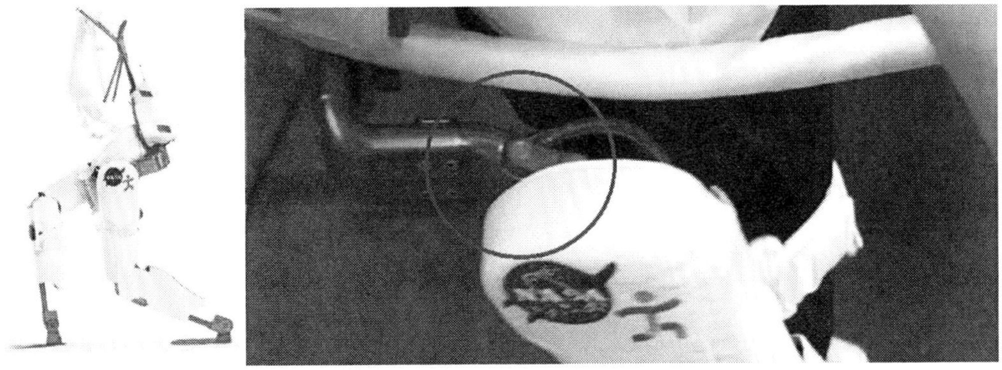

Fig. 2. NASA X1 and its turning module [7, 8]

Another example is BLEEX exoskeleton, which was designed to help carry large masses that are secured in the backpack fixed to the frame of the device [9]. Forces resulting from the gravitation of the backpack mass are almost completely transmitted by the frame of the exoskeleton, in order to reduce fatigue of the user owing to smaller loading of his joints and muscles. The assisted movements are flexion and extension at hip joints, knee joints and ankle joints as well as adduction and abduction at hip joints. It is also possible to perform movements of metatarsus flexion and hip rotation, however the movements are not powered. The movement of rotation about the vertical axis is possible both at the articulations located above the thigh connectors as well as at the middle articulation connecting both halves of the hip belt, what made it possible for the authors of the design to compare gaits performed using one of the two options (Fig. 3).

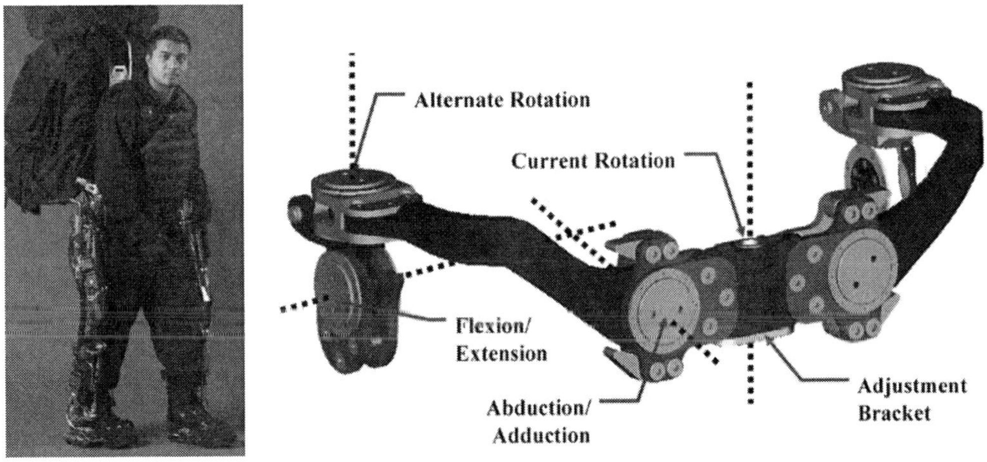

Fig. 3. Hip belt of BLEEX exoskeleton [9]

Both in the NASA X1 as well as in the BLEEX exoskeleton, rotation of the lower limb is possible owing to application of hinges enabling a motion within a transverse plane. In both designs, the rotation axis is shifted out of the outline of the human body. That results in linear displacements between the exoskeleton connector and the leg of the user while a rotation is performed, and the articulations connected with rotation of the lower limb within a transverse plane are devoid of a drive, what in turn results in the necessity of the user to be able to perform appropriate movements in order to initiate the turning by himself.

3 Conception of the lower limb turning system in the orthotic robot

In order to improve the way of locomotion provided by the existing orthotic devices that aid motion of the handicapped, one may apply an additional system making it possible to introduce a driven degree of freedom of the lower limbs, which in fact enables a turn of the lower limb at the hip with respect to the vertical axis, i.e. a hip rotation in order to ensure a possibility of changing direction of motion, i.e. turning.

It seems that the most advantageous solution to be implemented in the existing robots is a separate module of additional degree of freedom, making it possible to change direction of motion by the way of introducing a driven hip rotation. Such solution is aimed at an easy adaptation and assembly of the device in the existing orthotic robots without a necessity of introducing considerable changes of the mechanical structure.

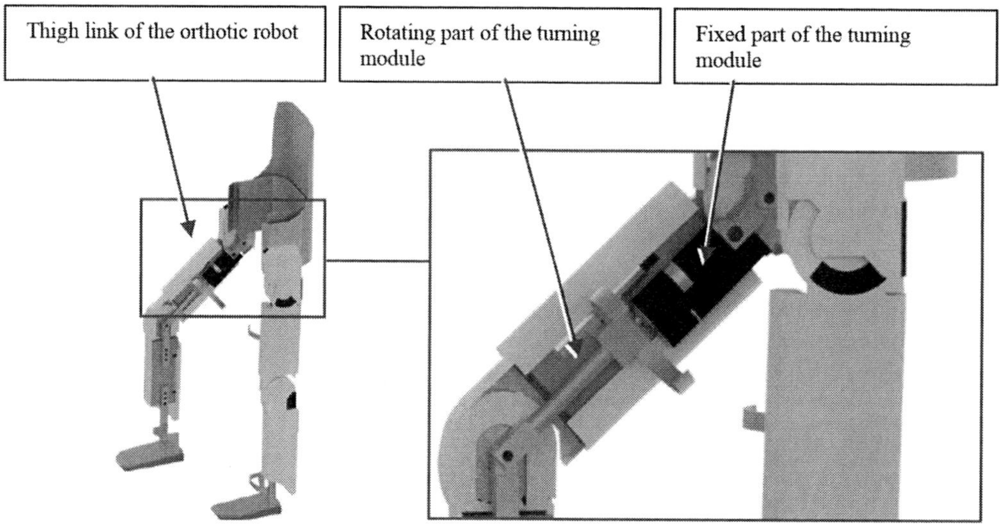

Fig. 4. Practical realization of the foot turning module in the Veni-Prometheus orthotic robot with close-up view of the applied turning module of the lower limb

While designing the turning module, one chose a solution of turning the lower limb, which consists in rotation about an axis that is parallel to the thigh within a transverse plane. It was decided to accept such solution also due to a possibility of obtaining smaller overall dimensions and masses as compared to other solutions.

4 Results

4.1 Methodology for determination of the operating conditions for a hip rotation system

The analysis covered two basic cases of loads during operation of the considered foot turning module. It was assumed that a turn of the lower limb will be realized in four stages. At the first stage, one of the lower limbs will be raised by the way of a flexion at the hip and the knee, thus it will turn outwards. At the second stage, a turn of the trunk will take place in such a way that the body, except for the leg resting upon the ground, is directed toward direction of the turn. At the third stage, the raised leg will be lowered to the ground, and at the fourth stage the leg that has been resting on the ground so far, will be raised, turned and set next to the second one.

For such scheme, one can distinguish two main cases of loads of the turning mechanism, i.e. when:

- the limb, upon which the turn is realized, is the main point of support of the exoskeleton,
- the limb is turned at the moment of being raised above the ground to a height of 50 mm, thus ensuring a collisionless turn of the foot.

The first indicated case assumes a rotation of the trunk with respect to the motionless lower limb supported on the ground. User of the orthotic robot makes use of crutches or a walking frame in order to have a support, what results in the fact that he has an additional support and can perform a rotation of the trunk using the upper groups of the muscles. Owing to such approach, the motion can be more fluent and predictable for the user, since he controls it himself, what may result in a better balance. That makes it also possible to use a smaller drive since there is no need to force and brake a motion of almost the whole body; only the lower limb is to be driven, in the second case.

The second case assumes a rotation of the lower limb, when it is raised by an appropriately small angle at the hip. It was assumed that before rotation of the limb, the shank will be positioned orthogonally with respect to the ground. Additionally, the feet of the device are always parallel to the shank, what results from the design of typical orthotic robots.

In order to calculate value of the forces and the torques occurring in the above cases, the following main assumptions were accepted:

- maximal permissible mass of the user is 110 kg,
- user's height is 1.8 m,
- Veni-Prometheus is considered as the exoskeleton the turning module is mounted to and has mass of 20 kg,
- the lower limb of the exoskeleton/orthotic robot has mass of 5 kg,
- turning module of the lower limb is fixed to one of the lower limbs and has mass of 1.5 kg,
- turning module of the lower limb is not rotated with the limb, onto which it is attached,
- time of rotating the limb while it is lifted above the ground is 0.5 s.

The considered case concerns loads of the drive while rotating a lifted lower limb, since such situation defines requirements related to the operation of the mechanism the best way. It is a conservative assumption, as it is the case with the largest loads in the turning mechanism.

4.2 Computation of the mechanism's load

4.2.1 Rotation of lifted lower limb

The first issue taken into consideration was to define the angle by which the lower limb will be lifted with respect to the trunk. In Fig. 6, this angle is denoted as α. The other denotations mean:

L – length of thigh/shin together with the foot,

R – distance between the foot and the rotation axis of the lower limb,

dx – distance between the sole and the ground,

α - angle at which the thigh is lifted with respect to the trunk.

It was assumed that the minimal distance dx should be 50 mm in order to ensure that rotation of the limb will not result in a collision, and thus the angle at which the thigh was lifted was evaluated at 26°.

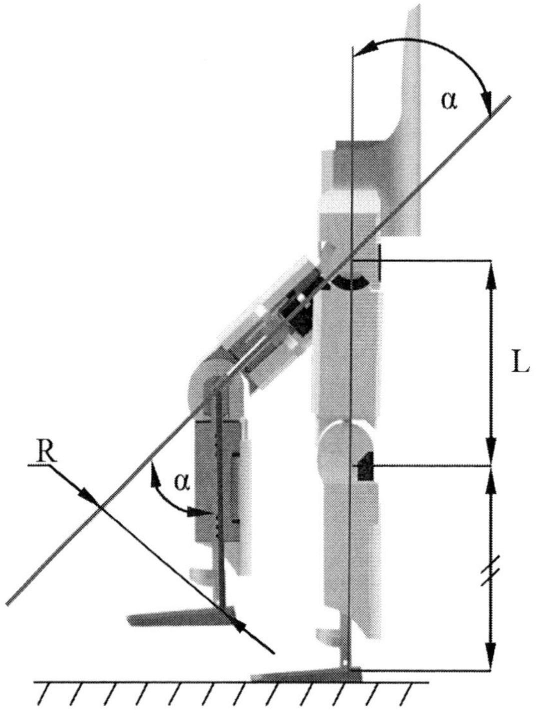

Fig. 5. Geometry of the computed arrangement

4.2.2 Consideration related to mass moments of inertia

The rotation axis of the lower limb is an axis parallel to the long thigh part of the lower limb. During a motion, the shin is inclined by an angle α=26° with respect to the rotation axis. In order to determine the mass moment of inertia of the limb while rotated around the considered axis, the following were taken into account:

- displacement between the mechanism (thigh connector) and the user's body is 150 mm,
- simplification that the user's body density is homogeneous,
- mass moments of inertia of the user's body resulting from its model,
- mass moments of inertia of the mechanism,
- increase of inertia due to shift of the axis, calculated according to the parallel axis theorem.

While computing the mass moment of inertia, a parametric model of the human body made in the MSC Adams software was employed [10]. The computations were executed for a possibly disadvantageous case of loads, when the user of the device is of the maximal weight and height. Size and mass of the turning module were over-evaluated in order to take into account additional resistance to motion, which results among other things from movements of the human body or a necessity of overcoming frictional resistance at the joints and articulations of the device.

4.2.3 Calculation of angular acceleration and maximal rotational speed

The assumed time of performing rotation of the limb at the hip is 0.3 s (shorter than the accepted requirement), and the angular way of that motion is 30°. Because of a relatively small range of operation it has been assumed that the speed profile of the considered motion is triangular, what means that within the first half of the angular way a constant acceleration is developed, whereas a constant deceleration within the second half. Thus, the angular acceleration was obtained, accepting the maximal time of motion and the required angular range of the rotation.

The considered case is related to loads of the drive while rotating a lifted lower limb, being the most representative with regard to the requirements connected with operation of the mechanism, since at the assumed way of performing a turn the biggest loads of the turning module occur.

4.2.4 Resultant maximal torque

Knowing the maximal computed values of the mass moment of inertia as well as the angular acceleration, it was possible to determine a maximal value of the torque resulting from accelerating the inert mass of the lower limb. The value was calculated on the basis of the mass moment of inertia of the member as well as a constant angular acceleration of the motion. The final value was increased by ca. 10%, taking into account a resistance to motion that has been neglected so far. Thus, the ultimate value was evaluated at 3.5 Nm.

4.2.5 Required characteristics of the drive

To drive the device, it was assumed that a drive unit will be used, which meets the following requirements:
- overall dimensions no larger than 150 x 100 x 100 mm,
- supply voltage up to DC 24 V,
- constant output torque of the drive unit exceeding 3.5 Nm,
- rotational speed of 20 rpm (2.1 rad/s) to be developed within 0.25 s.

The requirements were specified on the basis of a survey of typical commercial electric DC micro-drives available.

A minimal power of the electric drive at the output drive was determined, taking into account the maximal required torque as well as the maximal rotational speed to be developed.

The computed value is only a part of the power output of the motor, and the power of the motor driving the mechanism should be higher both due to its efficiency as well as additional resistance to motion, which was not taken into account in the above considerations, such as friction between the mechanical members and resistance of a frictional character resulting from a necessity of setting in motion body parts of the user. Finally, it was assumed that the power output of the motor should be no lower than 20 W, what is approximately 3-times higher than the value computed above. Such a big difference results also from the fact that it is advantageous to use such part of the operational characteristic of the motor, where it features high efficiency, i.e. between 15% and 50% of the stall torque.

Brushed DC motors were surveyed. Due to typical characteristics of drives, the following were assumed:

- the motor works with a gear featuring reduction ratio over 100:1 and efficiency of at least η=70%,
- the required constant torque of the motor is approximately of 50 mNm,

while formulating the requirements, typical DC electric drives and reduction gears were considered. The required values calculated above pertain to the torque and the speed of the output shaft of the turning module, i.e. the element that drives directly rotation of the limb.

4.2.6 Check of the output rotational speed

Because of the required a torque having a large value at a low rotational speed it was decided to use an electric motor coupled with a reduction gearhead. The maximal speed at the output of the motoreducer should be no lower than the assumed 20 RPM. Knowing typical designs of drives that meet the specified requirements, a turning module of the feet was designed; it is presented in Fig. 6. Electric motor is coupled with a reducer transmits the drive to a rotary member coupled with an encoder by means of a belt transmission. Assembly of the unit (without the casing) is also presented in Fig. 6.

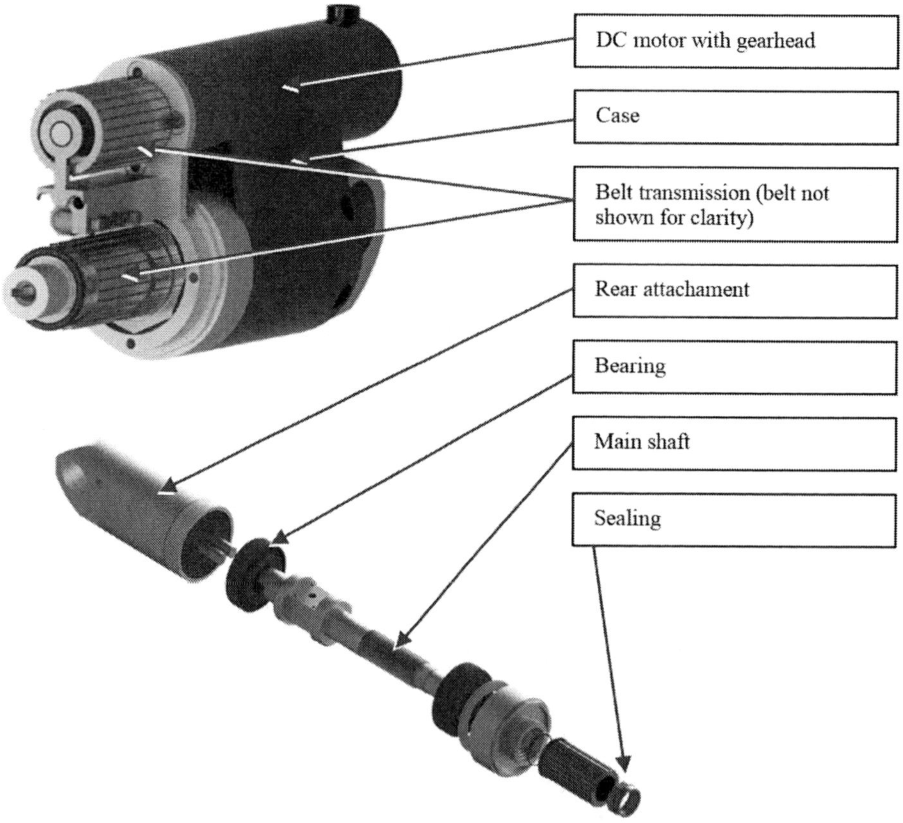

Fig. 6. Perspective and partially exploded view of the designed drive

5 Discussion

The orthotic robot taken into consideration for operating with the presented module of foot turning is the 'Veni-Prometheus' System for Verticalization and Aiding Motion. Thanks to choosing the concept of the turning module that is a separate unit attached to the frame of the exoskeleton, providing additional driven degree of freedom of the lower limb, it is also possible to adapt the turning module for other orthotic robots.

The presented concept of the device is meant to enhance turning capabilities of orthotic robots for rehabilitation purposes as well as aiding motion of individuals suffering from a spinal cord injury. The way of performing movements while changing direction of motion is strictly defined. All the available degrees of freedom, which are featured by the orthotic robot working together with the additional module, are actuated, what makes it possible to ensure a motion range which is explicitly defined and is safe for the user. It is aimed at protection against a possible injury – pulling one's tendons or muscle attachments, performing movements that are not anatomical, or performing undesired movements of the articulations, which would not be driven otherwise.

Locating the turning module in the hip connector, which overlaps the exoskeleton's axis of the rotation movement, results however in some linear displacements during a rotation, since a human limb rotates around an axis that runs through the head of the thigh bone. Such displacement results in an increase of the load of the mechanism, both due to introduction of undesired displacements as well as an increase of the mass moments of inertia (according to the parallel axis theorem), which load the limb. However, in the case of such arrangement the mass moments of inertia connected with the exoskeleton are smaller, since the motion takes place just around its axis.

Conclusions

Among the currently used design solutions of orthotic robots, a turn of the lower limb in the hip was initially considered - just as in the case of NASA X1 and BLEEX exoskeletons. It was noticed that a turn of the leg may be realized along the thigh axis within a safe range of rotation of 30°. It was verified if the new device would realize a turn of the whole limb at any position with respect to the whole robot, and it was accepted that at the assumed parameters of motion, there always took place a correct turn of the lower limb.

Computations related to the drive for the new design solution were executed on the basis of the obtained results, and it was stated that the output power generated by the motor may be over three times lower as compared with the solution employing a turn in the hip. This results in lower mass and overall dimensions of the whole device, as well as lower consumption of electric energy while activating a turn of the lower limb.

References

1. A. Esquenazi, M. Talaty, A. Packel, M. Asulino, *The ReWalk Powered Exoskeleton to Restore Ambulatory Function to Individuals with Thoracic-Level Motor-Complete Spinal Cord Injury*. Am. J. Phys. Med. Rehabil. **91**, 911-921 (2012)
2. T. Swift, *Control and Trajectory Generation of a Wearable Mobility Exoskeleton for Spinal Cord Injury Patients*. PHD Thesis, University of California, Berkeley (2011)
3. R. Irving, R. Little, *Mobility aid*. European Patent, EP 2231096 (2013)
4. D. Jasińska-Choromańska, K. Szykiedans, J. Wierciak, D. Kołodziej, M. Zaczyk, K. Bagiński, M. Bojarski, B. Kabziński, *Mechatronic system for verticalization and*

aiding the motion of the disabled. Bulletin of the Polish Academy of Sciences, Technical Sciences **61** (2), 419-431 (2013)

5. K. Bagiński, J. Wierciak, *Forming of operational characteristics of an orthotic robot by in-fluencing parameters of its drive systems*. R. Szewczyk, C. Zieliński, M. Kaliczyńska, Adv. Intel. Sys. Comput. **351** (2), 1-9 (2015)
6. D. Osiński, M. Zaczyk, D. Jasińska-Choromańska, *Conception of Turning Module for Orthotic Robot*. R. Jabłoński, T. Brezina, Adv. Intel. Sys. Comput. **393**, 147-152 (2016)
7. R. Rea, Ch. Beck, R. Rovekamp, M. Diftler, P. Neuhaus, *X1: A Robotic Exoskeleton for In-Space Countermeasures and Dynamometry*. AIAA Space 2013 Conference and Exposition (2013)
8. *X1 - Exoskeleton for Resistive Exercise and Rehabilitation*. [online] https://www.youtube.com/watch?v=XedjEBocaKo (2012)
9. A. Zoss, H. Kazerooni, A. Chu, *On the Mechanical Design of the Berkeley Lower Extremity Exoskeleton (BLEEX)*. IEEE/RSJ, 3465-3472 (2005)
10. D. Osiński, D. Jasińska-Choromańska, *Parametric Model of Human Body for Orthotic Robot Simulation Study*. Adv. Intel. Sys. Comput. (to be published)

Analysis of transverse vibration and stability issues of discrete-continuous elastic systems with nonlinearly variable parameters

Jerzy Jaroszewicz[1,*], *Łukasz* Dragun[1]

[1]Bialystok University of Technology, Faculty of Engineering Management, Ojca Tarasiuka 2, 16-001 Kleosin, Poland

Abstract. The work is devoted to methods of analysis of vibrations and stability of discrete-continuous, multi-parameter models of beams, shafts, rotors, vanes, converting to homogeneous and one-dimensional. The properties of Cauchy's influence function and the characteristic series method were used to solve the boundary problem. It has been shown that the methods are an effective tool for solving boundary problems described by ordinary fourth- and second-order differential equations with variable parameters. Particular attention should be paid to the solution of the border problem of two-parameter elastic systems with variable distribution of parameters. Universal beam-specific equations with typical support conditions including vertical support, which do not depend on beam shape and axial load type, are recorded. The shape and type of load are considered in the form of an impact function that corresponds to any change in cross-section of the support and continuous axial load, so that the functions describing the stiffness, the mass and the continuous load are complete. As a result of the solution of the boundary vibration problem of freely bent support and any change in its cross-section, loaded with any longitudinal load, arranged on the resilient substrate, strict relations between the own frequency parameters and the load parameters were derived. Using the methods, simple calculations were made, easy to use in engineering practice and conditions of use were given. Experimental studies have confirmed the high accuracy of theoretical calculations using the proposed methods and formulas.

Keywords: beams, boundary issues, transverse oscillations, intrinsic frequency stability, critical loads

1 Introduction

Components such as turbine blades, helicopter carrier blades, engine rotors, machine spindles, crane booms, drilling rigs, masts can be modeled with beams of variable distribution of parameters, the most important of which are: rigidity, mass, modulus of elasticity, cross section, substrate elasticity Longitudinal load. Sometimes discrete inclusions in the distribution of mass and in the resilient substrate should also be considered. We need to know the relevant dynamic characteristics, such as: values and

[*] Corresponding author: j.jaroszewicz@pb.edu.pl

Reviewers: *Maciej* Bodnicki, *Robert* Zalewski

forms of our own vibrations, and critical loads, in the loss of stability of the divergence and flatter type [1-3].

The exact solution of the abovementioned, in the general case, is possible in only a few specific cases. In the paper [4], the boundary of vibration of the cantilevered cone with the Bessel function is solved. The output equation of the vibration of flexible beams with a power-varying cross-section may be in the form of the Euler equation and then the general integral is expressed by the elementary functions [5-7].

The analytical and above all numerical methods used to test such beams are [8-10]. Using this method, it is quite difficult to assess the accuracy of the calculation of the required characteristics.

Among the numerical methods, finite element method and finite difference method are widely used [11, 12]. Examples of analytical methods are variation methods that, although they do not guarantee high accuracy; because of the specific form of vibrations, they give a very simple form of solution [10]. It is impossible not to mention the method of subsequent approximations. On the basis of this, Mike has derived the characteristic equation of the cantilever beam with varying cross section, but it has a very complicated form [5].

It is noted that there are no works that take into account several variable parameters at the same time, therefore the proposal of Zoryj L.M. and Jaroszewicz J. to apply for this purpose the method of characteristic series. This proposal was developed in joint work [2-7]. It must be clarified that Bernstein's previously proposed "spectral function" method was used only to analyze systems with fixed parameters without considering friction [10, 13]. At work [6] Zoryj L.M. He gave the general way of building universal determinants in the case of multi-parameter springs. The proposed method was then applied to the typical boundary problems described by the second and fourth order ordinary differential equations. The authors of this work [10,11] have used a series of characteristic methods to solve the boundary free-floating edge problem of a cantilever beam with a freely varying cross section. General formulas have been obtained which characterize the characteristic coefficients used in [5] for vibration analysis of cantilever beam of different shapes. Work [5] has also derived models that can be applied in engineering practice to calculate the basic vibration frequencies of the flexible beams and their conditions of use. The authors practically use the important property of the Cauchy function, published by L. Zoryj in 1978, which indicates that the function of influence and its subsequent derivative with respect to the parameter always form the basic arrangement of ordinary differential equation with variable parameters.

In this chapter, the boundary problem of the resilient spring flexing support beams has been formulated: flexural stiffness, continuous mass, longitudinal load, resilient substrate, which also contains discrete inclusions such as concentrated masses, resilient supports.

The replacement model of the beam in question is shown in Figure 1, on which it is marked: $f(x)$ – Rigid stiffness; $\mu(x)$ – continuous mases distribution; $k(x)$ – stiffness of the elastic substrate; $p(x)$ – longitudinal loads; $b(x)$ – external friction; N(x) – denotes distributed axial loads; M_i, b_i, c_i – discrete inclusions in continuous mass distribution, in the external friction and in the resilient substrate, applied in points x_i ; l – beam length; x, w – co-ordinates. As a result, the general form of the characteristic equation was obtained, on the basis of which it is possible to calculate the estimate of the frequencies of the own vibration and the estimates of the critical forces. Several typical examples, often found in engineering practice, were considered.

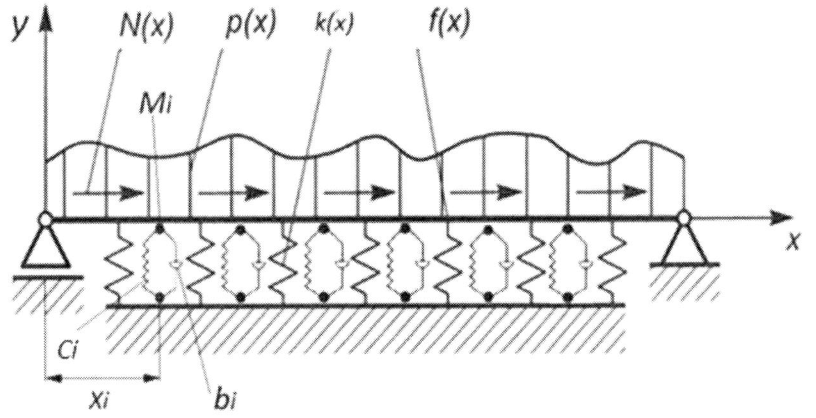

Fig. 1. The beam model

In particularly cases flexural rigidity and continuous mases may have form likes: f(x) = $EJ_0\,(1-\gamma\frac{x}{l})^n$ and M(x) = $\mu_0(1-\gamma\frac{x}{l})^{\frac{n}{2}}$, γ – coefficient of conquers; J_0- moment of inertia and mases of section x=0.

2 Definition of the problem

Vibrations of bent rods of length *l* that occur in the plane O_{yx}, can be described using the equation [4, 5]:

$$\frac{\partial^2}{\partial x^2}\left(f\frac{\partial^2 y}{\partial x^2}\right)-\frac{\partial}{\partial x}\left(N\frac{\partial y}{\partial x}\right)+\varrho F\frac{\partial^2 y}{\partial t^2}+b\frac{\partial y}{\partial t}+cy=q(x,t) \tag{1}$$

Where the signs are in accordance with Figure 1 and the previous description.

Equation (1) is considered under typical boundary conditions, two at each end of the bar and specified initial conditions (2). We will consider tasks of free vibration under typical boundary conditions, in the absence of external constraint, where q (x, t) ≡ 0. Separating variables (x, t) we come to the equation:

$$(fy'')''-(Nf')'+(\varrho F\lambda^2+b\lambda+c)y=0 \tag{2}$$

The following boundary conditions are considered:

$$y(0)=0\,,\;\; y'(0)=0\,;\quad y(l)=0\,,\;\; y'(l)=0 \tag{3}$$

$$y(0)=0\,,\;\; y'(0)=0\,;\quad y(l)=0\,,\;\; fy''|_{x=l}=0 \tag{4}$$

$$y(0)=0\,,\;\; y''(0)=0\,;\quad y(l)=0\,,\;\; y''(l)=0 \tag{5}$$

$$y(0)=0\,,\;\; y'(0)=0\,;\quad y(l)=0\,,\;\; (fy+Ny')|_{x=l}=0 \tag{6}$$

By substituting the general form of the solution for the above bounded boundary conditions (3,4,5,6) we obtain the following characteristic equations [9]:

1) For a bar with two fixed ends the problem (2), (3):

$$F(l,0) = 0 \tag{7}$$

2) For a bar with a left-right and a right freely supported issue (2), (4):

$$F'(l,0) = 0 \tag{8}$$

3) For a bar with two articulated-based issues (2), (5) clamped:

$$S'(l,0) = 0 \tag{9}$$

4) For the bracket issue (2), (6) supported cantilever:

$$R(l,0) = 0 \tag{10}$$

In the above equations, Cauchy's derivatives of their derivatives are used [10]:

$$F(x,\alpha) \equiv \phi\dot{\phi}' - \dot{\phi}\phi' \tag{11}$$

$$S'(x,\alpha) \equiv \phi\ddot{\phi}'' - \ddot{\phi}\phi'' \tag{12}$$

$$R(x,\alpha) \equiv f(x)(\phi''\dot{\phi}''' - \dot{\phi}'''\phi'') + N(x)(\phi'\dot{\phi}'' - \phi''\dot{\phi}'') \tag{13}$$

where: ϕ (x,α) – fundamental solution of the equation (2), Dots "." Denote derivatives with respect to α ', "," denote derivatives with respect to x.

We will now assume, as in the case of the string vibration task, that there is interdependence [5]. Then in place of (2) the equation [5] should be considered:

$$(fy'')'' - (Ny)' - \sum_{i=1}^{\upsilon} \alpha_i y(x_i)\varrho(x - x_i) = 0 \tag{14}$$

where

$$-\alpha_i = M_i\lambda^2 + b_i\lambda + c_i \tag{15}$$

M_i –size of concentrated mass,

b_i, c_i – coefficients of friction and stiffness of the spring at the coordinate x_i.,

In this case, the boundary conditions (3-6) correspond to the characteristic equations (7-10) corresponding to the given case, that is, the equation (14) of the functions F, S, and R obtained from (11), (12), (13) By replacement instead ϕ (x,α) solutions $Q_1(x,\alpha) = \phi_{x\alpha} + \alpha_1\phi_{1\alpha}\phi_{x1}$, determined by dependency (1.29 saved in [5]), wherein $\psi_s(x)$ is the fundamental solution of the equation:

$$(fy'')'' - (Ny')' = 0 \tag{16}$$

3 Examples of survey requirements with no quantity will come freedoms

We will examine some simple examples. Let Y_υ=1, N(x) ≡0. Then based on the formula (1.29 saved in [5]) we get:

$$Q_1(x,\alpha) = \phi_{x\alpha} + \alpha_1 \phi_{1\alpha} \phi_{x1} \tag{17}$$

where:

$$\phi(x,\alpha) = \int_\alpha^x \frac{1}{f(s)}(x-s)(s-\alpha)ds * \theta(x-\alpha) \tag{18}$$

Substitute (17) for the characteristic equation (11), which corresponds to the beam with two fixed ends, from which, after some transformations, we obtain:

$$\alpha_1 = V(l,0)\left[V(x_{1,0})\int_{x_1}^{l}\frac{(t-x_1)^2}{f(x)}dt + v(l,x)\int_0^{x_1}\frac{(t-x_1)^2}{f(t)}dt\right]^{-1} \tag{19}$$

where:

$$V(x,\alpha) \equiv \int_\alpha^x \frac{dt}{f(t)}\int_\alpha^x \frac{t^2 dt}{f(t)} - \left(\int_\alpha^x \frac{tdt}{f(t)}\right)^2 \tag{20}$$

Based on equations (8) and (9) considering (17 - 20) we find:

$$\alpha_1 = V'(l,0)\left[V(x_1,0)\frac{(l-x_1)}{f(l)} + V'(l,x_1)\int_0^{x_1}\frac{(t-x_1)^2}{f(t)}dt\right]^{-1} \tag{21}$$

and finally:

$$\alpha_1 = l^2\left[(l-x_1)^2\int_0^{x_1}\frac{t^2}{f(t)}dt + x_1^2\int_{x_1}^{l}\frac{(l-t)^2}{f(t)}dt\right]^{-1} \tag{22}$$

Using (17) and proceeding analogously to the above, we obtain the following pattern for the bracket, which carries the resiliently attached mass at the end of the equation (10) /x_1=l/:

$$\alpha_1 = \left[\int_0^l \frac{(l-s)^2}{f(s)}ds\right]^{-1} \tag{23}$$

Dependencies (19) and (21 - 23) which represent the square equations with respect to the characteristic index λ, allows you to specify frequencies / complex / extinguishing vibrations corresponding to systems with one degree of freedom for any given flexural stiffness f(x)>0 /0<x<l/ and coefficient of friction b_1> 0, satisfactory condition of continuation of vibration. We will note that in the case f(x) = 0 = EJ_0 = const., b_1= c_1 = 0, Based on these relationships it is easy to come up with patterns known from the literature for the squares of the frequencies of the corresponding vibrations, the bars in question, and also for the critical speeds of the shafts with the disk.

If instead of a concentrated mass $\tilde{\mu}$ at the point ξ there is a spring attached to the mass bar μ on spring with stiffness $\tilde{c}$ and damper with friction coefficient $\tilde{\varepsilon}$, is a parameter α_i instead (15) should be specified in the form:

$$-\widetilde{\alpha}_\iota = \widetilde{M}_\iota\lambda^2(\widetilde{\varepsilon}_\iota\lambda + \widetilde{c}_\iota)(\widetilde{M}_\iota\lambda^2 + \widetilde{\varepsilon}_\iota\lambda + \widetilde{c}_\iota)^{-1} \tag{24}$$

In the case of the presence at the point xi of both masses / rod and rod spring / add the total parameter – $\widetilde{\alpha_\iota + \alpha_\iota}$.

For example, for a support that carries a fixed mass M on its end with an elastic support with a stiffness c and a coefficient of friction b and an additional spring loaded mass $\widetilde{M}$, with stiffness of spring c with wavelength and coefficient of friction $\tilde{\varepsilon}$, based on (23) we obtain the following equation:

$$M\lambda^2 + b\lambda + c + \tilde{M}\lambda^2 \frac{\tilde{\varepsilon}\lambda+\tilde{c}}{\tilde{M}\lambda^2+\tilde{\varepsilon}\lambda+\tilde{c}} + \left(\int_0^l \frac{(l-s)^2}{f(s)} ds\right)^{-1} = 0 \tag{25}$$

Hence, it is possible to determine the frequency of a series of corresponding systems with two degrees of freedom, and in particular we come to comply with the solutions known from the literature of tasks.

As a next example we will consider the task of free vibration without a mass beam with freely supported ends bearing the masses at the axes x_i=1,2,3 And with elastic support bars at these points. General solution of the equation (14) if the longitudinal force does not work N(x) ≡0 has character (1.31 saved in [5]); equations $Q_{3s}(s)$ for s=1,4 they are referred to as the pattern accordingly (1.29 saved in [5]) under the following conditions:

$$\psi_1 = 1, \psi_2 = x, \psi_3 = \phi, \psi_4 = \phi \tag{26}$$

where:

Function ϕ (x,α) is defined by the integral (18). Assuming a parameter α equal 0 substituting (1.31 saved in [5]) to the conditions at the left end y''(0)=0, y''' (0)=0, we find out that A_3=0 i A_4=0.

Limit conditions at the right end give a linear system of homogeneous algebraic equations with respect to constants A_1,A_2. We obtain the characteristic equation in the following form:

$$\begin{vmatrix} Q''_{31} & Q''_{32} \\ Q'''_{31} & Q'''_{32} \end{vmatrix}_{\substack{x=l \\ \alpha=0}} = 0 \tag{27}$$

Note that at x_1=0 calculations are much simpler, we assume that x_1=0 and after doing some transformations from (27) we get:

$$\alpha_1\alpha_2 x_2^2 + \alpha_1\alpha_3 x_3^2 + \alpha_2\alpha_3(x_3 - x_2)^2 +$$
$$+\alpha_1\alpha_2\alpha_3\{x_2[(x_2 - x_3)\phi_{31} + x_3\phi_{32}] + x_3\phi_{21}(x_3 - x_2)\} = 0 \tag{28}$$

where:

$$\phi_{ij} = \int_{x_i}^{x_j} \frac{1}{f(s)}(x_i - s)(s - x_j)ds \tag{29}$$

Based on the above equation (28), it is possible to determine the characteristic indices, i.e., the customary frequencies of a plurality of suitable three, two, one degrees of freedom, critical shaft speed with a disk, and also with respect to (24) and exemplary (25) discrete Number of degrees of freedom / up to six [11]/. Let, for example, $\alpha_1 = \alpha_2 = \infty$, instead of (28) we obtain the following equation:

$$x_2^2 + \alpha_3\{x_2[(x_2 - x_3)\phi_{31} + x_3\phi_{32}] + x_3\phi_{21}(x_3 - x_2)\} = 0 \tag{30}$$

For x_2=l-a, x_3=l, f=EJ_0=const., $\alpha_3 = M\omega^2$ we have respectively: $\phi_{21}=1/6_{l-a}3_{f-1}$, $\phi_{31}=l_3/6_f$, $\phi_{32}=a_3/6_f$ and based on (30) we determine the critical speed of the light shaft with discs occurring at the free end:

$$\omega_{kp}^2 = \frac{3f}{a^2 l} \tag{31}$$

Based on (28), it is also possible to obtain, in specific cases, the question of the conditions under which the elastic supports are permissible to replace rigid, rigid stiffness which for non-homogeneous sections of the beam can be assumed to be large.

Let's, for example $\alpha_1 = \alpha_3 = -c$, $\alpha_2 = -M\lambda_2$. With this equation (28) for the beam with constant stiffness, we obtain the following pattern for a small beam of length l with elastic supports at the ends, which carries the mass M at the point $x_2 = \frac{1}{2} l$:

$$\omega^2 = \frac{2x}{M}\left(1 + \frac{1}{24} c l^3 / EJ\right)^{-1} \tag{32}$$

The above pattern (32) agrees with the solution of a known task [10]. Of course, at $1/24cl_3/EJ<<1$ the beam can be considered absolutely rigid.

If, for example, the stiffness of the beam section between the points of its axis x_2 i x_3 Far outstrips stiffness between points x_1 i x_2, it's in the equation (28) should be assumed $\phi_{32}=0$, $\phi_{31}=\phi_{21}$, If the beam can be rigid, then in this expression in cubes is equal to zero. In the latter case, it is also possible to consider the left end freely supported if it is directed $\alpha 1$ to infinity. In this equation (28) looks:

$$\alpha_2 x_2^2 + \alpha_3 x_3^2 = 0 \tag{33}$$

Hence it is easy to come to a range of tasks with rigid rods with masses and resilient supports.

Table 1. Results of calculations of lower $\underline{a\omega_0}^2$ and upper $\overline{a\omega_0}^2$ estimators of base frequency for square cone $C_n = 3$

No.	γ	Analytical		Numerical		ε [%]	Experiment
		$\underline{a\omega_0}^2$	$\underline{\omega_0}^2$	$\underline{a\omega_0}^2$	$\underline{a\omega_0}^2$		
1	2	3	4	5	6	7	8
1	0	12.36	12.36	12.36	12.36	0.02	–
2	0.2	13.01	13.02	13.01	13.02	0	14.02
3	0.4	13.96	13.97	13.96	13.97	0	–
4	0.6	15.46	15.49	15.46	15.49	0	14.43
5	0.8	18.37	18.46	18.37	18.46	0	–
6	1.0	28.03	28.52	29.94	31.29	7.66	32.03

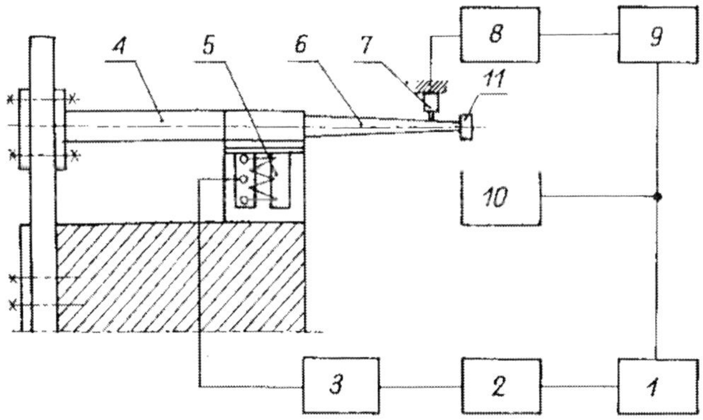

Fig. 2. Schematic diagram of the test stand by resonance method of flexural vibrations of cone with masses on end

Oscilloscope 8, oscilloscope 9 and pulse 10 are used to measure the vibration frequencies of the sample, which is realized by means of a sensor 7, oscilloscope 8, oscilloscope 9, and pulse counter 10. When equalizing the vibration force of the oscilloscope 9, When moving through the resonance, the ellipse is overturned. The LC 3 resonator circuit is necessary because of the insufficient power of the amplifier 2. The source of vibrations is electromagnet 5 supplied with variable frequency current, 1 - generator, 2 - power amplifier, 3 LC resonant system. The electromagnet 5 excites the flexural vibrations of the sample 6, rigidly fixed at the end of the support 4.

Conclusion

The proposed general expressions allow to obtain closed formulas for the second and third coefficients of a series characteristic for any function of stiffness from which it is required to be complete and have a finite value in the interval (0, l).

The presented methods (analytical and numerical) for the calculation of coefficients of a series of characteristics, based on general patterns, provide high accuracy in all cases considered (Table 1). It can be expected that equally high accuracy of calculations will be obtained for beam with focus on the end of other transverse cross sectional variations. Only cases in which the stiffness function at the free end tends to zero requires calculating the limit values of the series coefficients.

In the case of beams and shafts, even a relatively small mass at the ends of the beams, significantly lowers the base frequency. Coefficient of convergence γ this increases the frequency of beams without mass or very low mass ($\mu \leq 0{,}001$). Section coefficients n i m they show a significant effect on the baseline rate only at $\mu = 0$ and close to zero.

References

1. R. Panuszka, T. Uhl, *Wyznaczenie częstości rezonansowych ciąg-łych układów mechanicznych na przykładzie belki i płyty drgającej.* Archiwum Budowy Maszyn, Tom XXX **1-2**, 111-123 (1983)
2. J. Jaroszewicz, L. M. Zoryj, *Drgania giętne belki wspornikowej o zmiennym przekroju.* Rozprawy Inżynierskie **33** (4), 537-547 (1985)
3. J. H. Lau, *Vibration freqencies of tapered bars with end mass.* Journal of Applied Mechanics **51**, 179-181 (1984)
4. J. Jaroszewicz, L. M. Zoryj, *Izgibnyje kolebanija i dinamičeskaja ustojčivost' balok s peremennymi parametrami.* Prikladnaja mechanika, Kijev **30** (9), 75-81 (1994b)
5. J. Jaroszewicz, L. M. Zoryji, *Metody analizy drgań i statecznoœci kontynualno-dyskretnych układów mechanicznych.* Rozprawy Naukowe Politechniki Białostockiej **54**, Białystok, 126 (1997)
6. L. M. Zoryj, *Ob universalnych charakteristiceskich uravenijach v zadacach kolebanij i ustojcivosti uprugich sistem.* Mechanika Tverdogo Tiela **6**, 155-162 (1982)
7. L. M. Zoryj, *K teorii ustojcivosti sistem s raspredelonnymi parametrami.* Dokl. AN USSR **11A**, 992-995 (1986)
8. J. Jaroszewicz, K.K. Żur, Ł. Dragun, *The influence function in analysis of bending curve and reactions of elastic supports of beam with variable parameters.* Journal of Theoretical and Applied Mechanics **52** (1), 247-255 (2014)

9. J. Jaroszewicz, L. M. Zoryj, *Investigation of the effect of axial loads on the transverse vibrations of a vertical cantilever with variable parameters*. Prikladnaja Mechanika, International Applied Mechanics **36** (9), 129-137 (2000)
10. J. Jaroszewicz, Ł. Dragun, *Limitation of Cauchy function method in analysis for double estimators of free transversal vibration of cantilever tapered shafts*. Technical Sciences **16** (2), 131-142 (2013)
11. J. Jaroszewicz, L. Radziszewski, Ł. Dragun, *The Study of the Effect of Static Axial Loads on Vertically-Mounted Tapered Cantilever Transverse Vibrations Using the Cauchy Function*. Applied Mechanics and Materials **712**, 49-54 (2015)
12. W. Antoniuk, J. Jaroszewicz, L. Radziszewski, Ł. Dragun, *Theoretical stress analysis-based improvement of friction clutch disc manufacturing process*. Czasopismo Techniczne. Mechanika, Politechnika Krakowska **113** (4-M), 73-79 (2016)
13. T. Hoang, D. Duhamel, G. Foret, H.P. Yin, P. Joyezb, R. Caby, *Calculation of force distribution for a periodically supported beam subjected to moving loads*. Journal of Soundand Vibration **388**, 327-338 (2017)

The assessment of the integration methods for the rail vehicle ride dynamics solution

Tomáš Lack[1,*], *Juraj* Gerlici[1]

[1]University of Žilina, Faculty of Mechanical Engineering, Department of Transport and Handling Machines, Univerzitná 8215/1, 01026 Žilina, Slovak Republic

Abstract. The article is devoted to the analysis of the dynamic characteristics of the mechanical system composed of rigid bodies mutually connected with flexible bindings that represent springs and dampers. The analysed system represents the rail vehicle, which is distinguished by the contact link between the wheel and rail, or wheelsets and the track. Currently, analyses of similar type are carried out basically by using commercially distributed software packages, where you enter inputs (which the input forms allow) and the user may achieve results respecting the handling procedures. In accordance with the different approaches to solutions of partial solutions of the issue, the results of the simulations can differ. In the article the theoretical analysis is conducted the parameters of calculation are set out and the results are compared with the results obtained from the calculation by SIMPACK.

Keywords: integration methods, mechanical system, rail vehicle, stability, results comparison

1 Introduction

The article is devoted to the analysis of dynamic properties of multibody mechanical model representing a rail vehicle. The motivation for the creation of the article is the fact that there exists a steadily rising demand for detecting motion characteristics of transport machinery with the simulation calculation on computers [1]. Compared with the standard procedures, the transport machines used to be designed, a prototype was produced, the prototype was tested and adjusted on the basis of the results of this method, has distinct advantages in lower implementation costs [2, 3]. The price of the computer and the software comes out for sure to be cheaper than to make a new bogie, a wagon or locomotive. And it is even faster. Such considerations do not often take into account the quality of the people that such a simulation carried out.

The quality of working staff doing the computational analyses lies in education, skills with creation of relevant computation models in a particular program environment. In addition to skills when working with the running program, it is necessary to know the issue of equipment, which are the subject of simulation analyses, their standard or specific technical parameters, terms and conditions of the external action of the environment on

* Corresponding author: tomas.lack@fstroj.uniza.sk

Reviewers: *Josef* Soukup, *Milan* Žmindák

such a system and the expected response-the behavior of the device in use. The experience, which can help the staff to assess the credibility of gained results, are invaluable for verification and validation of the models [4, 5].

2 Technical support integrated in software packages

The currently, conducted simulation analysis primarily using commercially distributed software packages, where you enter inputs (which the input forms allow) and the user may achieve results respecting the handling procedures. Packages are comprehensive solutions providing the user a sophisticated analysis tool, which with its broad range often exceeds the user individual options and skills. Packages integrate a large amount of knowledge in a given area that are algorithmized into the procedures that a user uses through the programming environment. The environment is often a user-friendly and for the standard tasks programme with a high degree of autonomy will generate results. It is also possible to view the graphical tools that are also part of the environment. In view of the different approaches to achieve partial solutions, the results of the simulations can be different. In the article the theoretical analysis is conducted, parameters of calculation are set out and the results are compared with the results obtained from the calculation program SIMPACK. The similarity of the results predicts the correctness of the calculation performed by own programme code and opens the possibility of further development of the programme implementation modules for more specific calculations.

3 Mathematic model

The procedure of mathematic model creation is processed in the following text: rail vehicle equations of motion, that describe its movement in the straight track and in track arcs. The procedure was algorithmized and written into the programme language code. This programme code represents a complete software solution. By means of this programme it is possible to gain the results of the simulated ride of the rail vehicle with specified parameters of the mechanical system of the vehicle and track. Contact ellipses and the normal stress course are calculated by means of the Hertz method [6, 7, 8, 9].

For middle wheelset radius, middle contact angle value, track curvature and angle velocity of the wheelset revolution is valid:

$$r_0 = \frac{r_l + r_r}{2}; \quad \delta_0 = \frac{\delta_l + \delta_r}{2}; \quad P = \frac{\delta_0 \cdot r_0}{a}; \quad C = \pm\frac{1}{R}; \quad \Omega = \frac{v_0}{r_0};$$
$$\gamma_0 = \pm b \cdot C; \quad \dot{\gamma}_0 = \pm b \cdot C' \cdot v_0; \quad \ddot{\gamma}_0 = \pm b \cdot C'' \cdot v_0^2 \qquad (1)$$

where:
r_0...the average radius of the wheel [m],
r_l...an immediate radius of the left wheel [m],
r_r...an immediate radius of the right wheel [m],
δ_0...middle angle of tangent planes [rad],
δ_l...angle of the left wheel tangent plane [rad],
δ_r...angle of the right wheel tangent plane [rad],
a...half distance of the contact circles [m],
C...bending curvature of the track [m^{-1}] (In the case of right arc is valid the sign „+“, in the case of left arc is valid the sign: „ - “),
R...radius of a track [m],

γ_0... angle of attack, when the bogie in the track is in a chordal position [rad] (For the first (front) wheelset is valid the sign: „ - “, for pre second (rear) wheelset is valid the sign: „ + “),
Ω... angular speed of the wheelsets [rad.s^{-1}],
v_0... half distance of the contact circles [m].

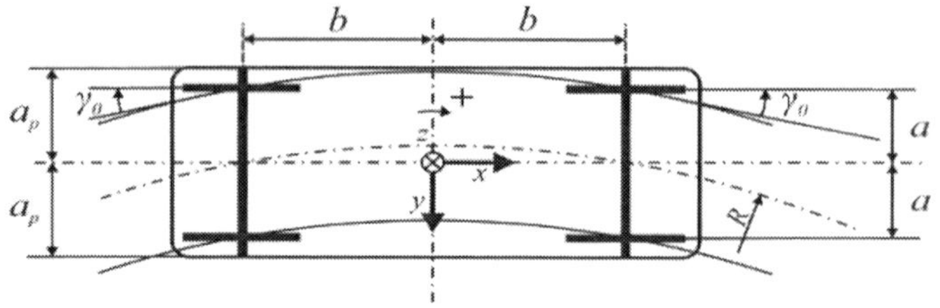

Fig. 1. Representation of the rolling stock chassis passing arc track

In the Fig. 1 there is a rail vehicle schematically depicted. It moves from the left to the right side direction in the track arc with given velocity. A vehicle approached to the track stretch of rails in arc under the angle of attack.

For the relative slip vector is valid:

$$\begin{bmatrix} s_{i,1} \\ s_{i,2} \\ s_{i,3} \end{bmatrix} = \frac{1}{v_0} \begin{bmatrix} v_0 \cdot \left(1 - \frac{r_i}{r_0} \pm a \cdot C\right) + r_i \cdot \dot{\beta} \pm a \cdot (\dot{\gamma}_0 + \dot{\gamma}) \\ \dot{y} - v_0 \cdot (\gamma_0 + \gamma) - \dot{\alpha} \cdot r_i \\ \dot{\gamma}_0 + \dot{\gamma} \pm \Omega \cdot \delta_i + v_0 \cdot C \end{bmatrix}; \quad i = l, r \tag{2}$$

where:
$s_{i,1}$... slip in the contact points of the left or the right wheel in the direction 1 (=x), 2 (= in the transverse direction of the tangential plane) and 3 (in the direction perpendicular to the tangential plane, [-], [-], [m^{-1}],
$\dot{\gamma}$ ... the angular speed of the approaching wheel of the first wheelset during the rotation about the *z*-axis [rad.s^{-1}],
γ... wheelset turning [rad],
$\dot{\alpha}$ ... the angular speed of the wheel sets in rotation around the *x*-axis [rad.s^{-1}].

For combined modulus of elasticity is valid:

$$G = \frac{2 \cdot G_W \cdot G_R}{G_W + G_R} \tag{3}$$

where:
G... combined modulus of elasticity in shear [Pa],
G_W... modulus of elasticity in wheel shear [Pa],
G_R... modulus of elasticity in rail shear [Pa].

For contact forces in contact points is valid:

$$f_{11} = (a_e \cdot b_e) \cdot G \cdot C_{11}, \quad f_{22} = (a_e \cdot b_e) \cdot G \cdot C_{22}, \quad f_{33} = (a_e \cdot b_e)^2 \cdot G \cdot C_{33}, \quad f_{23} = (a_e \cdot b_e)^{\frac{3}{2}} \cdot G \cdot C_{23} \tag{4}$$

where:
f_{11}, f_{22}... have the dimension of force [N],
f_{33}... [N.m^2],
f_{33}... in accordance to the Kalker's computational method [N.m],
a_e, b_e... are major and semi-minor axis of the ellipse in the wheel/rail contact patch [m],
C_{11}, C_{22}, C_{33}, C_{23}... Kalker's coefficients [-], [10].
For tangential forces and moments in the contact points is valid:

$$\begin{bmatrix} f_{ti,1} \\ f_{ti,2} \\ l_i \end{bmatrix} = \begin{bmatrix} f_{11} & 0 & 0 \\ 0 & f_{22} & f_{23} \\ 0 & -f_{23} & f_{33} \end{bmatrix} \cdot \begin{bmatrix} s_{i,1} \\ s_{i,2} \\ s_{i,3} \end{bmatrix}; \quad i = (l, r) \tag{5}$$

where:
$F_{ti,1}$ ... tangential force at the contact point left or right wheels, in the (lengthwise) direction of the 1 axis of the contact plane [N],
$F_{ti,2}$ ... tangential force at the contact point left or right wheels, in the (transverse) direction of the 2 axis of the contact plane [N],
l_i ... spin moment of the left or right wheel around an axis 3 perpendicular to the contact plane of the rail and wheel [N.m].
For normal forces is valid:

$$f_{nl} = \frac{1}{2 \cdot a} \cdot I_2 \cdot (\dot{\gamma}_0 + \dot{\gamma}) \cdot \Omega - \frac{1}{2 \cdot a} \cdot (r_l \cdot f_{tl,2} + r_r \cdot f_{tr,2}) + \frac{1}{2} \cdot (Q_l + Q_r) \tag{6}$$

where:
f_{nl} ... the normal force at the contact point of the left wheel [N],
I_2 ... the moment of inertia of the wheel when turning around an axis of rotation [kg.m^2],
Q_l ... the wheel force on the left wheel [N],
Q_r ... the wheel force on the right wheel [N].

$$f_{nr} = -\frac{1}{2 \cdot a} \cdot I_2 \cdot (\dot{\gamma}_0 + \dot{\gamma}) \cdot \Omega + \frac{1}{2 \cdot a} \cdot (r_l \cdot f_{tl,2} + r_r \cdot f_{tr,2}) + \frac{1}{2} \cdot (Q_l + Q_r) \tag{7}$$

where:
f_{nr} ... the normal force in the contact point of the right wheel [N].
Depiction of forces acting on a wheelset is in Fig. 2.

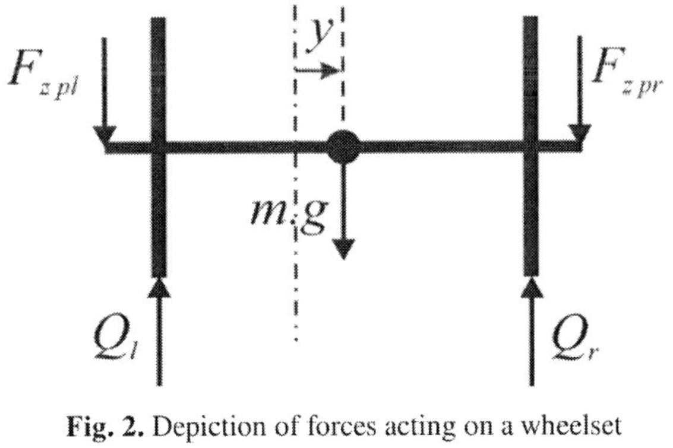

Fig. 2. Depiction of forces acting on a wheelset

For the left wheel force is valid:

$$Q_l = \frac{m \cdot g \cdot (y_r - y) + F_{z\,pl} \cdot (y_r - y + a_p) + F_{z\,pr} \cdot (y_r - y - a_p)}{y_l + y_r} \tag{8}$$

Where:
Q_l ... the wheel force on the left wheel [N],
y_r ... the coordinate of the contact point of the right wheel and rail [m],
y ... transverse displacement of the wheel [m],
$F_{z\,pl}$ ... the force in place of axle box (primary suspension) of the left wheel [N],
$F_{z\,pr}$ ... the force in place of axle box (primary suspension) of the right wheel [N],
a_p ... half distance of the bearings boxes (primary suspension on one wheelset) [m],
For right wheel force is valid:

$$Q_r = \frac{m \cdot g \cdot (y_l + y) + F_{z\,pl} \cdot (y_l + y - a_p) + F_{z\,pr} \cdot (y_l + y + a_p)}{y_l + y_r} \tag{9}$$

where:
Q_r ... the wheel force on the right wheel [N],
y_l ... the coordinate of the contact point of the left wheel and rail [m].
Contact forces and moments were computed by means of FASTSIM method.

4 Wheelset equations of motion

We will express equations of motion in the matrix shape (10)

$$\begin{bmatrix} m & 0 & 0 & 0 \\ 0 & m & 0 & 0 \\ 0 & 0 & I_2 & 0 \\ 0 & 0 & 0 & I_1 \end{bmatrix} \cdot \begin{bmatrix} \ddot{x} \\ \ddot{y} \\ \ddot{\beta} \\ \ddot{\gamma} \end{bmatrix} +$$

$$+\left\{ \frac{2}{v_0} \begin{bmatrix} 0 & 0 & r_0 \cdot f_{11} & 0 \\ 0 & f_{22} \cdot \left[1 + P + \frac{\delta_0^2}{2 \cdot a^2} \cdot (r_l^2 + r_r^2) \right] & 0 & f_{23}(1+P) \\ 0 & 0 & f_{11} \cdot \frac{r_l^2 + r_r^2}{2} & -a \cdot f_{11} \cdot \frac{r_l - r_r}{2} \\ 0 & -f_{23}(1+P) & -a \cdot f_{11} \cdot \frac{r_l - r_r}{2} & a^2 \cdot f_{11} + f_{33} \end{bmatrix} + \begin{bmatrix} 0 & 0 & 0 & 0 \\ 0 & 0 & 0 & -I_2 \cdot \frac{\delta_0 \cdot v_0}{a \cdot r_0} \\ 0 & 0 & 0 & 0 \\ 0 & I_2 \cdot \frac{\delta_0 \cdot v_0}{a \cdot r_0} & 0 & 0 \end{bmatrix} \right\} \cdot \begin{bmatrix} \dot{x} \\ \dot{y} \\ \dot{\beta} \\ \dot{\gamma} \end{bmatrix} +$$

$$+\begin{bmatrix} 0 & 0 & 0 & 2\cdot f_{22}\cdot\gamma_0 - f_{23}\cdot\left[\dfrac{\delta_l-\delta_r}{r_0}+2\cdot\left(C+\dfrac{\dot{\gamma}_0}{v_0}\right)\right] \\ 0 & (Q_l+Q_r)\cdot\dfrac{\delta_0}{a} & 0 & -2\cdot f_{22}\cdot(1+P) \\ 0 & 0 & 0 & 0 \\ 0 & 2\cdot f_{11}\cdot\dfrac{a\cdot\delta_0}{r_0} & 0 & 2\cdot f_{23} \end{bmatrix}\cdot\begin{bmatrix} x \\ y \\ \beta \\ \gamma \end{bmatrix}=$$

$$=\begin{bmatrix} \left\{2\cdot f_{22}\cdot\gamma_0 - f_{23}\cdot\left[\dfrac{(\delta_l-\delta_r)}{r_0}+2\cdot\left(C+\dfrac{\dot{\gamma}_0}{v_0}\right)\right]\right\}\cdot\gamma_0 \\ 2\cdot f_{22}\cdot\gamma_0\cdot(1+P)-f_{23}\cdot\left[\dfrac{(\delta_l-\delta_r)}{r_0}+2\cdot\left(C+\dfrac{\dot{\gamma}_0}{v_0}\right)\cdot(1+P)+\dfrac{\delta_0}{a\cdot r_0}\cdot(r_l\cdot\delta_l-r_r\cdot\delta_r)\right]+ \\ (Q_l+Q_r)\cdot\dfrac{\delta_l-\delta_r}{2}+I_2\dfrac{\delta_0\cdot v_0}{a\cdot r_0}\cdot\dot{\gamma}_0 \\ f_{11}\cdot\left\{r_l\cdot\left[1-\dfrac{r_l}{r_0}+a\cdot\left(C+\dfrac{\dot{\gamma}_0}{v_0}\right)\right]+r_r\cdot\left[1-\dfrac{r_r}{r_0}-a\cdot\left(C+\dfrac{\dot{\gamma}_0}{v_0}\right)\right]\right\} \\ -2\cdot f_{23}\cdot\gamma_0-f_{33}\cdot\left[\dfrac{\delta_l-\delta_r}{r_0}+2\cdot\left(C+\dfrac{\dot{\gamma}_0}{v_0}\right)\right]-2\cdot f_{11}\cdot a^2\cdot\left(C+\dfrac{\dot{\gamma}_0}{v_0}\right)-I_1\cdot\ddot{\gamma}_0 \end{bmatrix} \tag{10}$$

In addition to the forces acting on the wheelset, the force acts on each vehicle when operating during the transition as a result of the effects of gravity and centripetal acceleration. (11):

$$\begin{aligned} F_{yi} &= m_i\cdot\left(-g\cdot\sin\alpha_R+v_0^2\cdot C\cdot\cos\alpha_R\right) \\ F_{zi} &= m_i\cdot\left(g\cdot\cos\alpha_R+v_0^2\cdot\left|C\cdot\sin\alpha_R\right|\right) \end{aligned}\text{, where } \alpha_R = arctg\left(\frac{U}{2.a}\right), \tag{11}$$

where:
U ... cant of the outer rail in the track arc [m],
$2.a$ ... distance of contact circles [m].
The accelerations distribution is depicted in Fig. 3

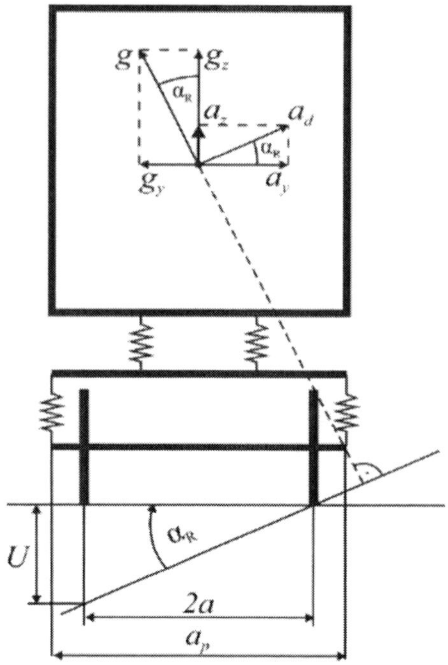

Fig. 3. Accelerations distribution

5 Vehicle equations of motion

The vehicle model consisting of two two-axis bogies (four wheelsets à four degrees of freedom, two bogie frames à six degrees of freedom) and one body of wagon (six degrees of freedom) has 34 degrees of freedom. In the following we presuppose, that the model consists of rigid bodies connected with flexible bindings. The rigid bodies are defined by mass matrix and the flexible bindings by stiffness matrix and damping matrix.

We can express the vehicle equation of motion in a matrix shape:

$$[M]\cdot\{\ddot{X}\}+[B]\cdot\{\dot{X}\}+[K]\cdot\{X\}=\{f\} \tag{12}$$

where:

[M] ... mass matrix [-],
[B] ... damping matrix [-],
[K] ... stiffness matrix [-],
$\{\ddot{X}\}$ ...vector of accelerations [-],
$\{\dot{X}\}$ ... vector of velocities [-],
$\{X\}$ ... vector of displacements [-],
$\{f\}$ ... vector of external loads [-].

A vector of the external load forms the right side of a matrix equation as well as operating force. For the calculation, it is possible to use different types of methods. In the following there are depicted all stabil methods, that were tested for equation of motion solution.

5.1 Central-difference method

Initialize $\{X_0\}$, $\{\dot{X}_0\}$ and $\{\ddot{X}_0\}$. Select time step Δt and calculate integration constants:

$$\Delta t \le \frac{T_n}{\pi}; \quad a_0 = \frac{1}{\Delta t^2}; \quad a_1 = \frac{1}{2 \cdot \Delta t}; \quad a_2 = 2 \cdot a_0; \quad a_3 = \frac{1}{a_2} \tag{13}$$

Calculate

$$\{X_{-\Delta t}\} = \{X_0\} - \Delta t \cdot \{\dot{X}_0\} + a_3 \cdot \{\ddot{X}_0\} \tag{14}$$

Form stiffness [*K*], mass [*M*], damping [*C*] matrixes and {*f*} load vector:
Form effective stiffness matrix:

$$[A] = a_0 \cdot [M] + a_1 \cdot [C] \tag{15}$$

Calculate effective force vector at a time $t + \Delta t$:

$$\{F_{t+\Delta t}\} = \{f_t\} - \left(\left[K - a_2 \cdot [M]\right]\right) \cdot \{X_t\} - \left(a_0 \cdot \left[M - a_1 \cdot [C]\right]\right) \cdot \{X_{t-\Delta t}\} \tag{16}$$

Solve for displacements at time $t + \Delta t$:

$$[A] \cdot \{X_{t+\Delta t}\} = \{F_{t+\Delta t}\} \tag{17}$$

Calculate $\{\dot{X}\}$ and $\{\ddot{X}\}$ at time t:

$$\{\dot{X}_t\} = a_1 \cdot \left(\{X_{t+\Delta t}\} - \{X_{t-\Delta t}\}\right) \tag{18}$$

$$\{\ddot{X}_t\} = a_0 \cdot \left(\{X_{t+\Delta t}\} - 2 \cdot \{X_t\} + \{X_{t-\Delta t}\}\right) \tag{19}$$

5.2 Fourth Order Runge-Kutta Method

One of the methods that can be used for such simulations with the advantage of the method is Runge-Kutta. To get results, we have used the Runge-Kutta method – in this case 4-th order.

Initialize $\{X_0\}$, $\{\dot{X}_0\}$ and $\{\ddot{X}_0\}$.

Transformed into first order matrix differential equation as follows:

$$\begin{Bmatrix} \{\ddot{X}\} \\ \{\dot{X}\} \end{Bmatrix} = \begin{bmatrix} -[M]^{-1} \cdot [C] & -[M]^{-1} \cdot [K] \\ [I] & [0] \end{bmatrix} \cdot \begin{Bmatrix} \{\dot{X}\} \\ \{X\} \end{Bmatrix} + \begin{Bmatrix} [M]^{-1} \cdot \{f\} \\ \{0\} \end{Bmatrix} \tag{20}$$

where setting:

$$\{y\}=\begin{Bmatrix}\{\dot{X}\}\\\{X\}\end{Bmatrix},\qquad \{N\}=\begin{Bmatrix}[M]^{-1}\cdot\{f\}\\\{0\}\end{Bmatrix} \tag{21}$$

then the equation becomes:

$$\{\dot{y}\}=f\left(t,\{y\},\{N\}\right) \tag{22}$$

Assuming the duration of each time step is h , $t_i=ih$, i=1,2,…then the computational formulas over the time interval [ti,ti+1] are as follow:

$$\begin{aligned}
\{K_1\}&=f\left(t_i,\{y(t_i)\},\{N\}\right)\\
\{K_2\}&=f\left(t_i+\frac{\Delta t}{2},\{y(t_i)\}+\frac{\Delta t}{2}\cdot\{K_1\},\{N_1\}\right)\\
\{K_3\}&=f\left(t_i+\frac{\Delta t}{2},\{y(t_i)\}+\frac{\Delta t}{2}\cdot\{K_2\},\{N_2\}\right)\\
\{K_4\}&=f\left(t_i+\Delta t,\{y(t_i)\}+\Delta t\cdot\{K_3\},\{N_3\}\right)
\end{aligned} \tag{23}$$

where we assume:

$$\{K_1\}=[a_1,a_2,\ldots a_n]^T,\quad \{K_2\}=[b_1,b_2,\ldots b_n]^T,\quad \{K_3\}=[c_1,c_2,\ldots c_n]^T \tag{24}$$

then:

$$\begin{aligned}
\{N_1\}&=\left\{N\left(\{y_1(t_i)\}+\frac{\Delta t}{2}\cdot a_1,\{y_2(t_i)\}+\frac{\Delta t}{2}\cdot a_2,\ldots,\{y_n(t_i)\}+\frac{\Delta t}{2}\cdot a_n\right)\right\}\\
\{N_2\}&=\left\{N\left(\{y_1(t_i)\}+\frac{\Delta t}{2}\cdot b_1,\{y_2(t_i)\}+\frac{\Delta t}{2}\cdot b_2,\ldots,\{y_n(t_i)\}+\frac{\Delta t}{2}\cdot b_n\right)\right\}\\
\{N_3\}&=\left\{N\left(\{y_1(t_i)\}+\Delta t\cdot c_1,\{y_2(t_i)\}+\Delta t\cdot c_2,\ldots,\{y_n(t_i)\}+\Delta t\cdot c_n\right)\right\}
\end{aligned} \tag{25}$$

Computational formula is:

$$\{y(t_{i+1})\}=\{y(t_i)\}+\frac{1}{6}\cdot\Delta t\cdot\left(\{K_1\}+2\cdot\{K_2\}+2\cdot\{K_3\}+\{K_4\}\right) \tag{26}$$

5.3 Newmark Beta method

Initialize $\{X_0\}$, $\{\dot{X}_0\}$ and $\{\ddot{X}_0\}$. Select time step Δt, parameter α and β, and calculate integration constants:

$$\beta\geq\frac{1}{2};\quad \alpha\geq\frac{1}{4}\cdot\left(\frac{1}{2}+\beta\right)^2 \tag{27}$$

$$a_0 = \frac{1}{\beta \cdot \Delta t^2}; \quad a_1 = \frac{\alpha}{\beta \cdot \Delta t}; \quad a_2 = \frac{1}{\beta \cdot \Delta t}; \quad a_3 = \frac{1}{2 \cdot \beta} - 1; \quad a_4 = \frac{\alpha}{\beta} - 1; \quad a_5 = \left(\frac{\alpha}{2 \cdot \beta} - 1 \right) \cdot \Delta t$$

$$a_1 = \frac{\alpha}{\beta \cdot \Delta t}; \quad a_2 = \frac{1}{\beta \cdot \Delta t}; \quad a_3 = \frac{1}{2 \cdot \beta} - 1; \quad a_4 = \frac{\alpha}{\beta} - 1; \quad a_5 = \left(\frac{\alpha}{2 \cdot \beta} - 1 \right) \cdot \Delta t \tag{28}$$

Form stiffness $[K]$, mass $[M]$, damping $[C]$ matrixes and $\{f\}$ load vector:
Form effective stiffness matrix:

$$[A] = a_0 \cdot [M] + a_1 \cdot [C] + [K] \tag{29}$$

Calculate effective force vector at a time $t + \Delta t$:

$$\begin{aligned} \{F_{t+\Delta t}\} = \{f_{t+\Delta t}\} + [M] \cdot \left(a_0 \cdot \{X_t\} + a_2 \cdot \{\dot{X}_t\} + a_3 \cdot \{\ddot{X}_t\} \right) + \\ + [C] \cdot \left(a_1 \cdot \{X_t\} + a_4 \cdot \{\dot{X}_t\} + a_5 \cdot \{\ddot{X}_t\} \right) \end{aligned} \tag{30}$$

Solve for displacements at time $t + \Delta t$:

$$[A] \cdot \{X_{t+\Delta t}\} = \{F_{t+\Delta t}\} \tag{31}$$

Calculate $\{\dot{X}\}$ and $\{\ddot{X}\}$ at time $t + \Delta t$:

$$\{\ddot{X}_{t+\Delta t}\} = a_0 \cdot (\{X_{t+\Delta t}\} - \{X_t\}) - a_2 \cdot \{\dot{X}_t\} - a_3 \cdot \{\ddot{X}_t\} \tag{32}$$

$$\{\dot{X}_{t+\Delta t}\} = a_1 \cdot (\{X_{t+\Delta t}\} - \{X_t\}) - a_4 \cdot \{\dot{X}_t\} - a_5 \cdot \{\ddot{X}_t\} \tag{33}$$

5.4 HHT method (Hilber – Hughes - Taylor)

Initialize $\{X_0\}$, $\{\dot{X}_0\}$ and $\{\ddot{X}_0\}$.
Select time step Δt, parameter α, β, α_f, α_m and γ, and calculate integration constants:

$$\beta = \frac{1}{4} \cdot (1 + \gamma)^2; \quad \alpha = \frac{1}{2} + \gamma; \qquad \alpha_f = \gamma; \quad \alpha_m = 0; \qquad \gamma = 0 \to \frac{1}{3} \tag{34}$$

where:
γ ... coeffitient of amplitude damping.

$$\begin{aligned} a_0 &= \frac{1 - \alpha_m}{\beta \cdot \Delta t^2}; & a_1 &= \frac{(1 - \alpha_f) \cdot \alpha}{\beta \cdot \Delta t}; & a_2 &= \frac{1 - \alpha_m}{\beta \cdot \Delta t}; \\ a_3 &= \frac{1 - \alpha_m}{2 \cdot \beta} - 1; & a_4 &= \frac{(1 - \alpha_f) \cdot \alpha}{\beta} - 1; & a_5 &= (1 - \alpha_f) \cdot \left(\frac{\alpha}{2 \cdot \beta} - 1 \right) \cdot \Delta t \end{aligned} \tag{35}$$

Form stiffness [*K*], mass [*M*], damping [*C*] matrixes and {*f*} load vector:
Form effective stiffness matrix:

$$[A]=a_0\cdot[M]+a_1\cdot[C]+(1-\alpha_f)[K] \tag{36}$$

Calculate effective force vector at a time $t + \Delta t$:

$$\begin{aligned}\{F_{t+\Delta t}\}&=(1-\alpha_f)\cdot\{f_{t+\Delta t}\}+\alpha_f\cdot\{f_t\}-\alpha_f\cdot[K]\cdot\{X_t\}+\\&+[M]\cdot\left(a_0\cdot\{X_t\}+a_2\cdot\{\dot{X}_t\}+a_3\cdot\{\ddot{X}_t\}\right)+\\&+[C]\cdot\left(a_1\cdot\{X_t\}+a_4\cdot\{\dot{X}_t\}+a_5\cdot\{\ddot{X}_t\}\right)\end{aligned} \tag{37}$$

Solve for displacements at time $t + \Delta t$:

$$[A]\cdot\{X_{t+\Delta t}\}=\{F_{t+\Delta t}\} \tag{38}$$

Calculate $\{\dot{X}\}$ and $\{\ddot{X}\}$ at time $t + \Delta t$:

$$\{\ddot{X}_{t+\Delta t}\}=a_0\cdot\left(\{X_{t+\Delta t}\}-\{X_t\}\right)-a_2\cdot\{\dot{X}_t\}-a_3\cdot\{\ddot{X}_t\} \tag{39}$$

$$\{\dot{X}_{t+\Delta t}\}=a_1\cdot\left(\{X_{t+\Delta t}\}-\{X_t\}\right)-a_4\cdot\{\dot{X}_t\}-a_5\cdot\{\ddot{X}_t\} \tag{40}$$

5.5 Park Stiffly stable method

Initialize $\{X_0\}$, $\{\dot{X}_0\}$ and $\{\ddot{X}_0\}$.

Select time step Δt and calculate integration constants:

$$a_0=\frac{10}{6\cdot\Delta t};\quad a_1=-\frac{15}{6\cdot\Delta t};\quad a_2=\frac{1}{\Delta t};\quad a_3=-\frac{1}{6\cdot\Delta t}. \tag{41}$$

Form stiffness [*K*], mass [*M*], damping [*C*] matrixes and {*f*} load vector:
Form effective stiffness matrix:

$$[A]=a_0^2\cdot[M]+a_0\cdot[C]+[K] \tag{42}$$

Calculate effective force vector at a time $t + \Delta t$:

$$\begin{aligned}\{F_{t+\Delta t}\}=\{f_{t+\Delta t}\}-[M]\cdot&\begin{pmatrix}a_1\cdot\{\dot{X}_t\}+a_2\cdot\{\dot{X}_{t-\Delta t}\}+a_3\cdot\{\dot{X}_{t-2\cdot\Delta t}\}\\+a_0\cdot a_1\cdot\{X_t\}+a_0\cdot a_2\cdot\{X_{t-\Delta t}\}+a_0\cdot a_3\cdot\{X_{t-2\cdot\Delta t}\}\end{pmatrix}\\&-[C]\cdot\left(a_1\cdot\{X_t\}+a_2\cdot\{X_{t-\Delta t}\}+a_3\cdot\{X_{t-2\cdot\Delta t}\}\right)\end{aligned} \tag{43}$$

Solve for displacements at time $t + \Delta t$:

$$[A]\cdot\{X_{t+\Delta t}\}=\{F_{t+\Delta t}\} \tag{44}$$

Calculate $\{\dot{X}\}$ and $\{\ddot{X}\}$ at time $t + \Delta t$:

$$\{\dot{X}_{t+\Delta t}\} = a_0 \cdot \{X_{t+\Delta t}\} + a_1 \cdot \{X_t\} + a_2 \cdot \{X_{t-\Delta t}\} + a_3 \cdot \{X_{t-2\cdot\Delta t}\} \tag{45}$$

$$\{\ddot{X}_{t+\Delta t}\} = a_0 \cdot \{\dot{X}_{t+\Delta t}\} + a_1 \cdot \{\dot{X}_t\} + a_2 \cdot \{\dot{X}_{t-\Delta t}\} + a_3 \cdot \{\dot{X}_{t-2\cdot\Delta t}\} \tag{46}$$

6 Computational model of a vehicle

The processing computational model is characterized by parameters of the vehicle, its geometric and mass parameters, parameters of flexible linkages, dampers. Others are parameters of the contact links and wheel track and the track parameters. All necessary parameters are set out deliberately to make them clearly referred to the input data, which were used to obtain the results using our own programme and a commercial package.

6.1 Vehicle parameters

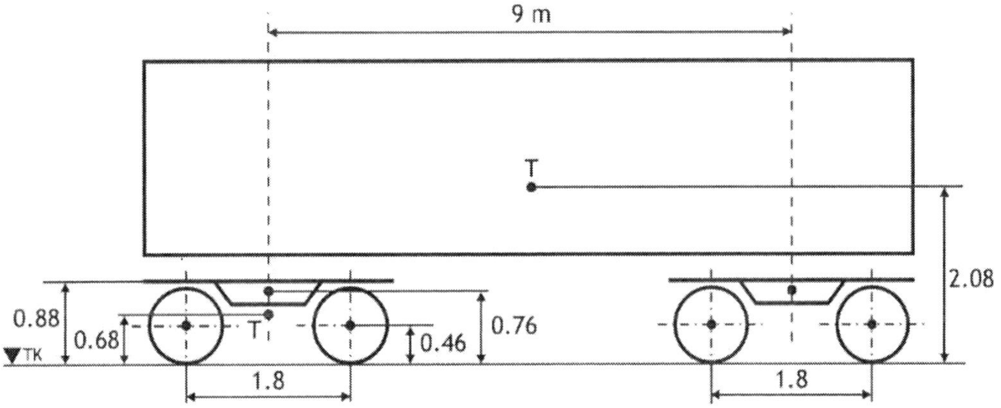

Fig. 4. Schematic drawing of the vehicle

In accordance with the specification, our model is a simple vehicle compounded from two two-wheelsets bogies and one wagon body (Fig. 4). All flexible bindings are linear, bodies are taken as rigid bodies. The data of the vehicle are introduced in the following tables from Tab. 1 to Tab. 4.

Table 1. Mass vehicle parameters

Masses				
	m **[kg]**	I_1 **[kg.m^2]**	I_2 **[kg.m^2]**	I_3 **[kg.m^2]**
Wheelset	1 375	800	100	800
Bogie	1 750	1 250	1 330	2 760
Body of wagon	81 000	100 000	2 000 000	2 000 000

Table 2. Flexible bindings parameters

Suspension	k_x **[kN.m^{-1}]**	k_y **[kN.m^{-1}]**	k_z **[kN.m^{-1}]**
Primary suspension	50 0000	1 000	2 500
Secondary suspension	0	0	1 140
Ball pin	10 000	50 000	100 000

Table 3. Dampers parameter

Damping	b_x **[kN.s.m^{-1}]**	b_y **[kN.s.m^{-1}]**	b_z **[kN.s.m^{-1}]**
Primary damping	50	150	200
Secondary damping	0	0	50
Ball pin	50	50	100

Table 4. Geometrical parameters of the vehicle and track

Parameter	**Description**	**Value**	**Unit**
Distance of the points of work of suspension and damping of the primary suspension	a_p	2	[m]
Distance of the points of work of suspension and damping of the secondary suspension		1.7	[m]
The distance of the pivots of bogies		9	[m]
Wheel base of bogies		1.8	[m]
Nominal diameter of the wheels		0.92	[m]
Height of centre of gravity of the wheelset over rail top		0.46	[m]
Height of the bogie center of gravity over rail top		0.68	[m]
Height of the body of wagon center of gravity over rail top		2.08	[m]
Height of the primary suspension over rail top		0.46	[m]
Height of the secondary suspension over rail top		0.88	[m]
Height of the pin ball over rail top		0.76	[m]
Wheel profile	S1002		
Rail head profile – measured in the track	S91700_16_sym.TWR		
Track gauge		1435	[mm]
Equivalent conicity at 3 mm		0.145	-
Coefficient of friction		0.4	-
Radius of the track arc	R	293	[m]
Track cant in the ark	U	144	[mm]
Kalker's coefficients reduced by the factor		0.67	

6.2 Geometric characteristics of a wheelset and a track

Geometric characteristics of wheelset and track can help to predict the movement of a wheelset in a track. At the same time, they depict the characteristics of reciprocal attitude of wheel tread profile and rail head profile. These characteristics represent an important input parameter for a mechanical system excitation due to the wheels of wheelset rolling in rails of track. Both functions input into the computation directly.

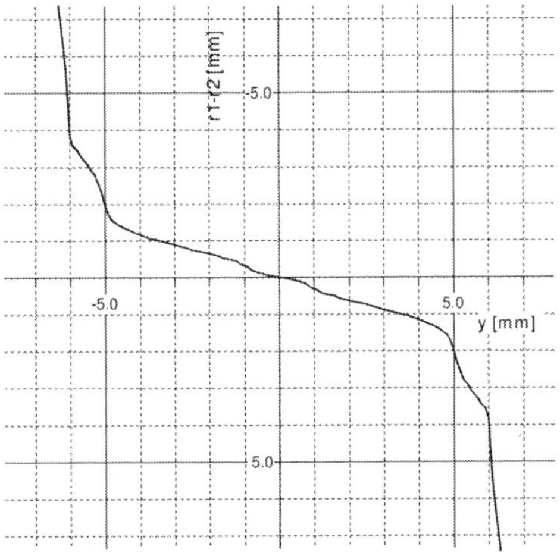

Fig. 5. View at the dependence of the Delta R function on the lateral shift of the wheel profile after rail profile.

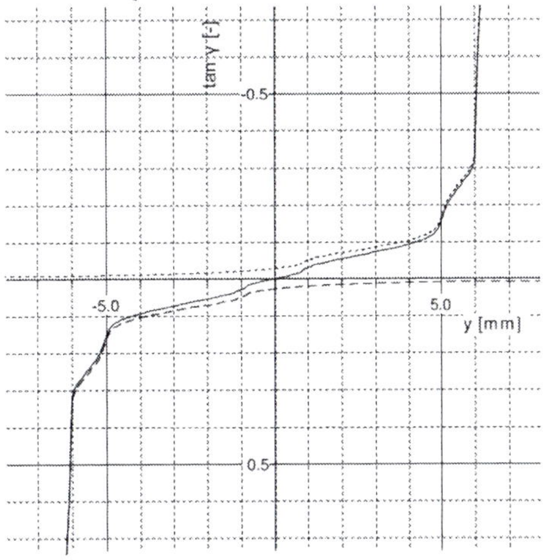

Fig. 6. View at the dependence of the Tangent Gamma function on the lateral shift of the wheel profile after rail profile.

From all geometric characteristics we are introducing Delta R and Tangent Gamma functions only. From Fig. 5 and Fig. 6 it is visible, that the most important curve changes are when the transversal movement of a wheel tread profile along a rail head profile reaches cca 5 mm.

Free track gauge channel is exhausted in this point and a wheel is via its flange climbing the rail head side. In the Tangent Gamma function curve this region is characterized by a steep change of course.

6.3 Track parameters

The model track consists of connected geometrically different track parts: straight line track sections, transition curves, superelevation ramps and from the track in arcs. It arises from Fig. 7, that length of the track is 3 km. Geometry of the track in horizontal plane is depicted by a bold black line. It has the vertical axis description on the right side.

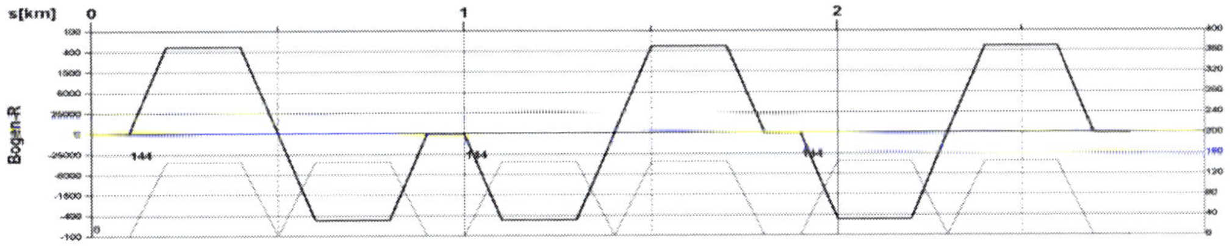

Fig. 7. Track definition

The track consists of four arcs with the radius of R 293 meters, three of them are left and three of them are right. The axis is on the left side. In the bottom part of graph there are depicted the superelevations of the outside rail 0.144 m. The vertical axis for this graph is on the right side. This track has neither climbing nor descending. For the computation transition curves and superelevation ramps of the Bloss were used.

7 Computation results

Here are compared two different physical variables for the first or third wheelset. In the following graphs we can see the course of wheelset lateral shift (displacement) in (y), slips (S) in the separate directions, guiding forces (Y), sum of guiding forces (SY), wheel forces (Q). At the same time there are evaluated and graphically depicted the forces in the contact plane of wheel and rail: normal forces (N), tangential forces (T) in special directions.

In Fig. 8 there is a view at a vehicle model in SIMPACK programme. The look of the model is demonstrative only. We utilized the basic depicting tools, that are a part of the programme environment.

Fig. 8. View at a vehicle model in SIMPACK programme

The results obtained from the computations executed by means of our own software, where are algorhitmized are the equations of motions and auxiliary relations, presented in this paper. They are compared with results obtained by means of computations performed with SIMPACK programme package.

The results gained by means our own programme are depicted with red colour. The results from SIMPACK are of black colour.

7.1 Shift and rotating of the wheelset

We may take as the same the wheelset shift in the *y*-direction obtained by both of procedures (Fig. 9).

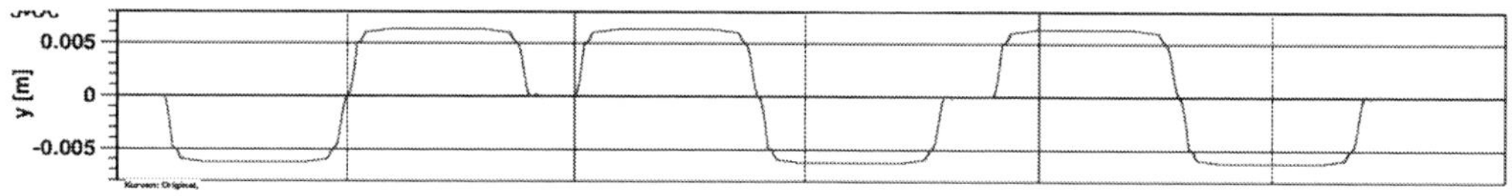

Fig. 9. Shift of the wheelset in the *y*-axis direction

We may take very similar a revolution of the wheelset around *z*-axis by both of procedures. Small differences appear at transitions from one arc to another arc (Fig. 10).

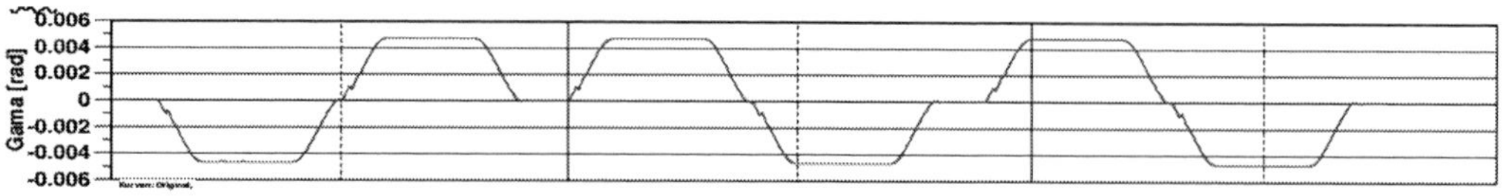

Fig. 10. Revolution of the wheelset around *z*-axis

7.2 Slips

In Fig. 11 to Fig. 16 there are depicted courses of slips in the x-direction on the right wheel of the first wheelset.

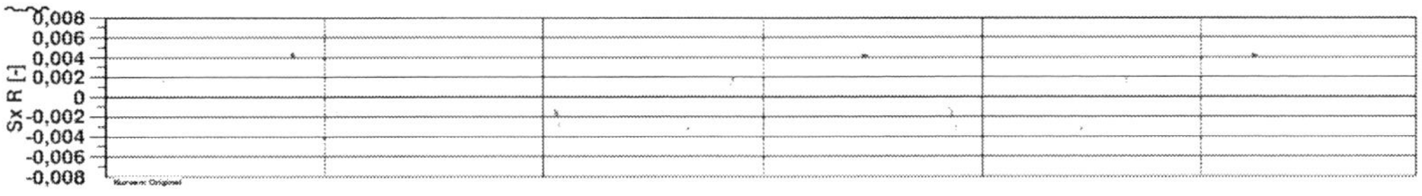

Fig. 11. Slip in the x-direction on the right wheel of the first wheelset

Slip parameters are important for tangential forces calculation. This is the reason, that the „overshootings" during the transition of curves courses that are visible at slips in the x-directions of right (Fig. 11) and left wheels (Fig. 12), appear in the tangential forces calculation (Fig. 17 and Fig. 20) too.

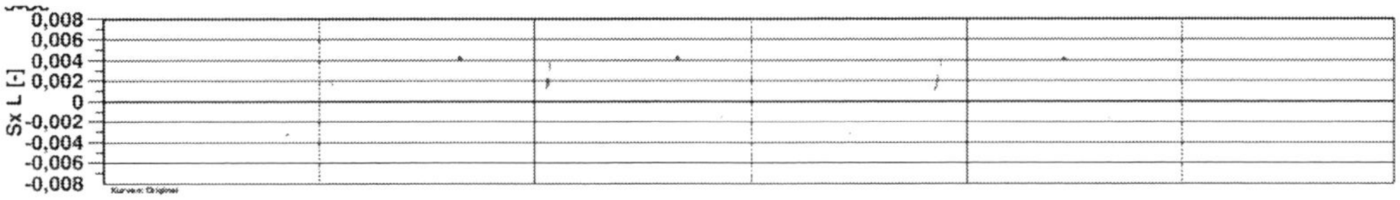

Fig. 12. Slip in the x-direction on the left wheel of the first wheelset

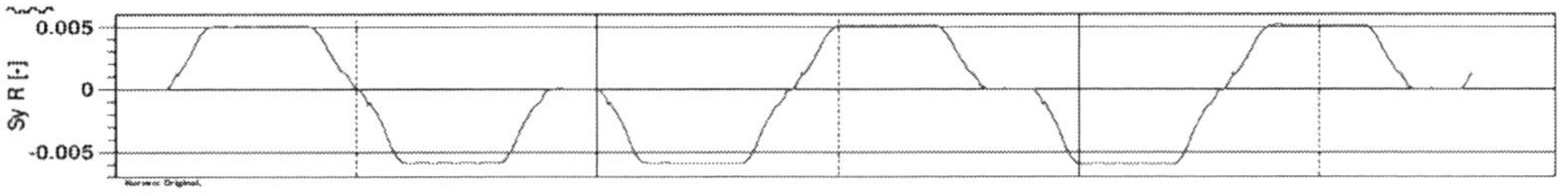

Fig. 13. Slip in the y-direction on the right wheel of the first wheelset

Slips in the transversal direction obtained by means of calculations with SIMPACK have more smooth course without the overshootings in the opposite side of slips in the longitudinal x-axis direction.

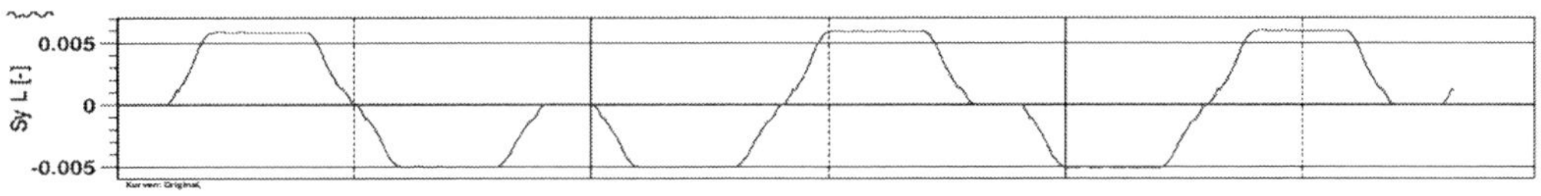

Fig. 14. Slip in the y-direction on the left wheel of the first wheelset

7.3 Normal forces

Normal forces in their results represent the force acting in the normal axis direction to the contact plane in the rail/wheel contact point. The course of their values is very similar again (Fig. 15 and Fig. 16).

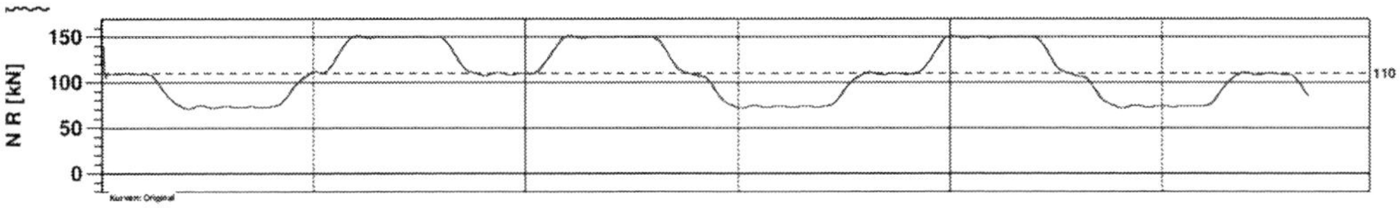

Fig. 15. Normal force on the right wheel of the first wheelset

https://doi.org/10.1051/matecconf/201815703013

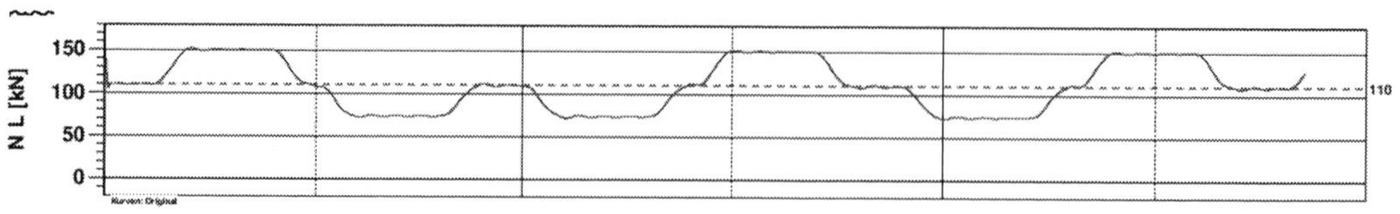

Fig. 16. Normal force on the left wheel of the first wheelset

7.4 Tangential forces

When we evaluated the tangential forces, the differences appear in the places, where vehicles come from one track geometry into the other track geometry (straight, a ramp, an arc). The course of the tangential forces in the longitudinal x-direction gained by means of SIMPACK programme is of more steep inclination of the transitions (Fig. 17 – Fig. 20). This course comes back to the stable state when a vehicle runs in a straight line track or in an arc via „overshooting". In this cases, there are short-timed detected higher values of forces. The „overshooootings" forces courses follow from the „overshooting" courses of slips.

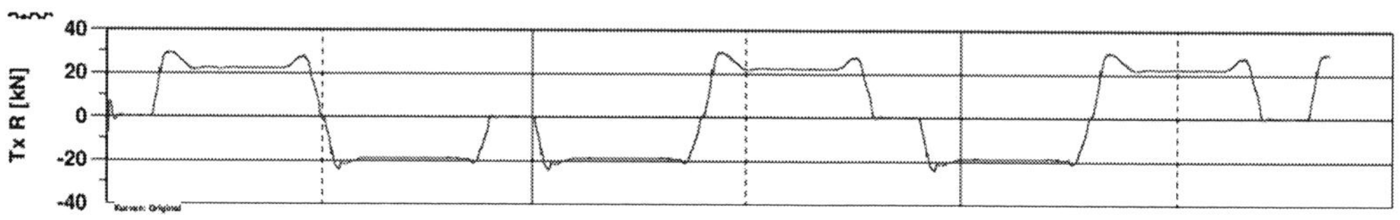

Fig. 17. Tangential force in the *x*-direction on the right wheel of the first wheelset

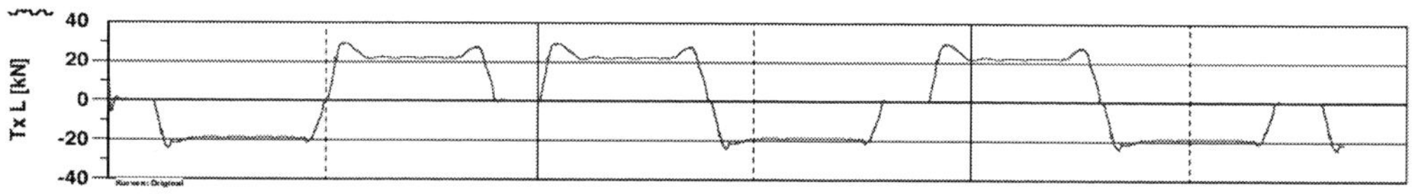

Fig. 18. Tangential force in the *x*-direction on the left wheel of the first wheelset

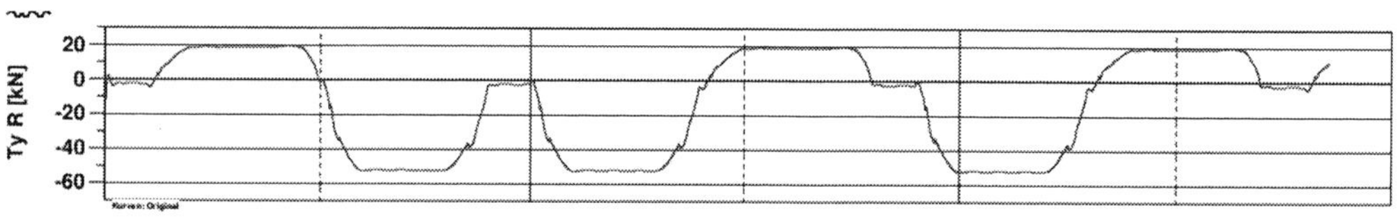

Fig. 19. Tangential force in the *y*-direction on the right wheel of the first wheelset

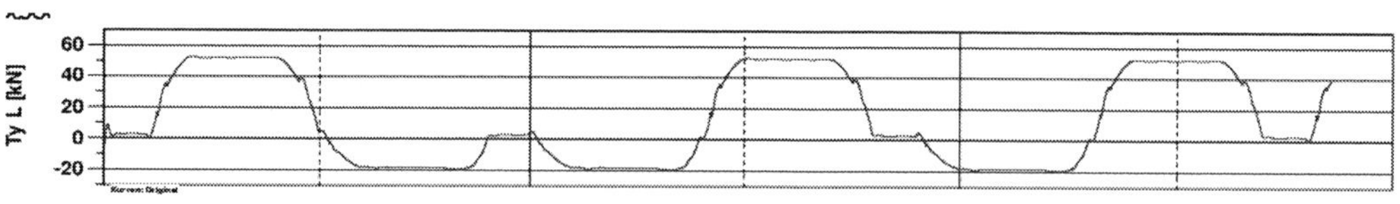

Fig. 20. Tangential force in the y-direction on the left wheel of the first wheelset

Tangential forces in the y-axis direction obtained by both of procedures are similar in the shape of their course. It relates to the slips course, they are evaluated and depicted in the Fig. 13 and Fig. 14.

7.5 Guiding forces

The guiding forces course is an important parameter for results comparison (Fig. 21 – Fig. 23). This parameter gives not only evidence of lateral force influence of the wheel to the rail, but in this case it gives the view at the result of processing of the contact parameters of the wheelset and track, contact patches and stresses and assumptions of the force acting between a vehicle and a track.

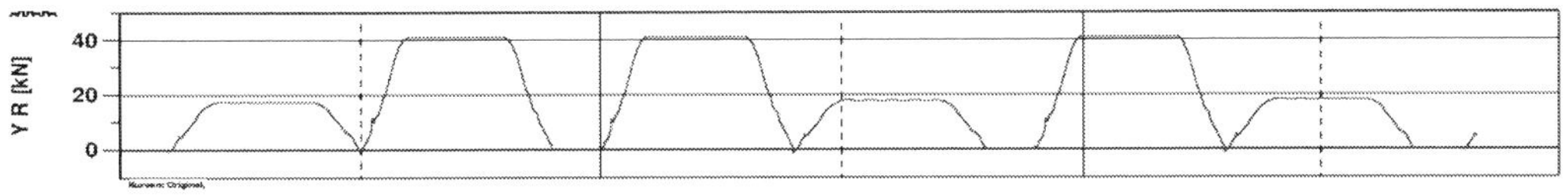

Fig. 21. Guiding force on right wheel of the first wheelset

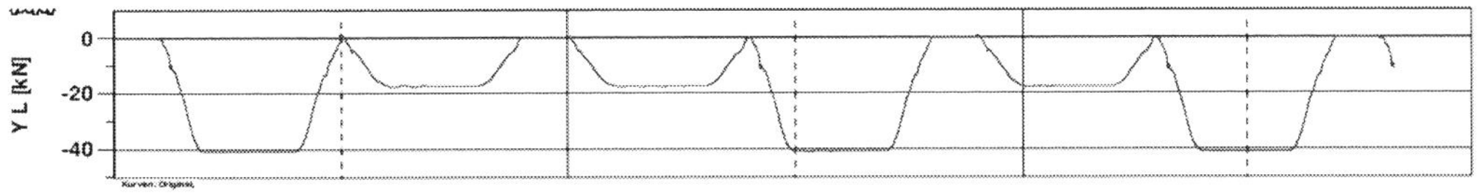

Fig. 22. Guiding force on left wheel of the first wheelset

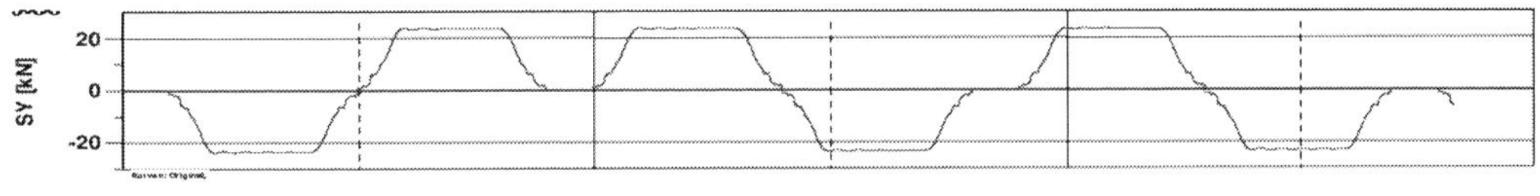

Fig. 23. Sum of the guiding forces on the first wheelset

The close course of the individual forces verified results obtained with the mathematical expression investigated the movement state of a vehicle on the basis of the results obtained from the programme, which is now widely used by reputable professionals. Or when viewed from the opposite side: the results produced by an automated computing system with a closed core and prescribed (and also some optional) calculation procedures, we would be able to get the solution to the system of equations derived in this article.

7.6 Wheel forces

The computations results of the wheel forces at the left and the right wheels are very similar to each other, see Fig. 24 and Fig. 25.

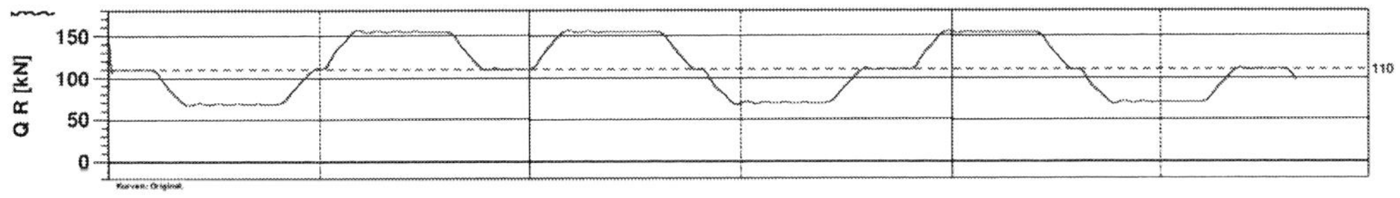

Fig. 24. Wheel force on right wheel of the first wheelset

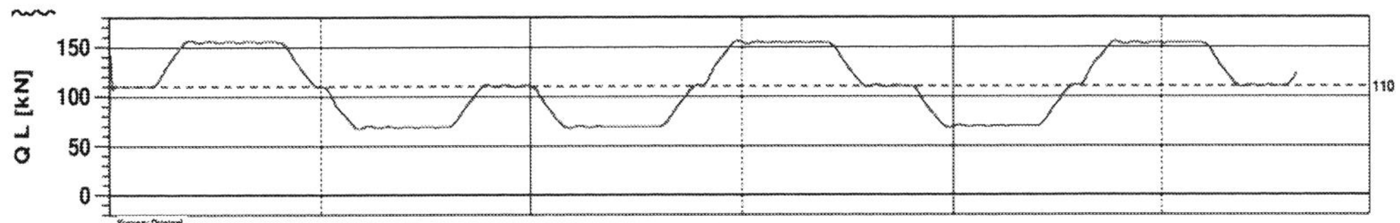

Fig. 25. Wheel force on left wheel of the first wheelset

8 Methods comparison

We tested the methods in the way, that all model bodies were moved (deflected from centered position) 0.003 m in transversal y-direction (Tab. 5). The graphical depiction of wagon body position deflection in y-axis direction is shown in Figs. 26, 27, 28, 29, 30. The difference is in computation with various damping parameters (γ). The Integration methods stability comparison is in Table 5.

Table 5. Integration methods stability comparison

Integrating method	Parameters	Stability
Central diferention		**Stabil**
Newmark	$\beta = \frac{1}{2}, \quad \alpha = \frac{1}{4} \quad alebo \quad \alpha = \frac{1}{6}$	**Stabil**
HHT	$\gamma = 0.15$	**Stabil**
Wilson Θ	Θ = 1.5	Unstabil
Houbolt		Unstabil
Park stiffly stable		**Stabil**
Runge-Kutta	4-grade	**Stabil**
Two cycles with an trapezoid rule		Unstabil

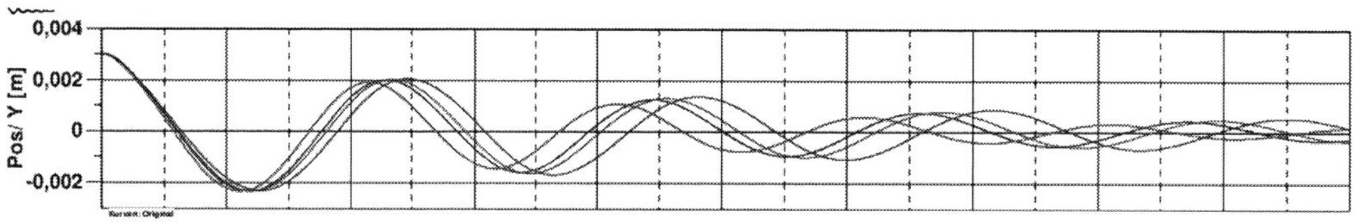

HHT ($\gamma = 0$), HHT ($\gamma = 0.15$), HHT ($\gamma = 0.2$), HHT ($\gamma = 0.3$)

Fig. 26. Wagon body position deflection in y-axis direction

$$\Lambda_{13} = \ln\frac{x_1}{x_3}; \qquad \Lambda_{35} = \ln\frac{x_3}{x_5}; \qquad \Lambda_{15} = \frac{1}{2}\ln\frac{x_1}{x_5}; \qquad \Lambda_{24} = \ln\frac{x_2}{x_4} \tag{47}$$

$$D = \frac{\Lambda}{\sqrt{(2\cdot\pi)^2 + \Lambda^2}} \tag{48}$$

Central difference, Newmark, HHT (γ = *0*), Park stiffly stabil, Runge-Kutta

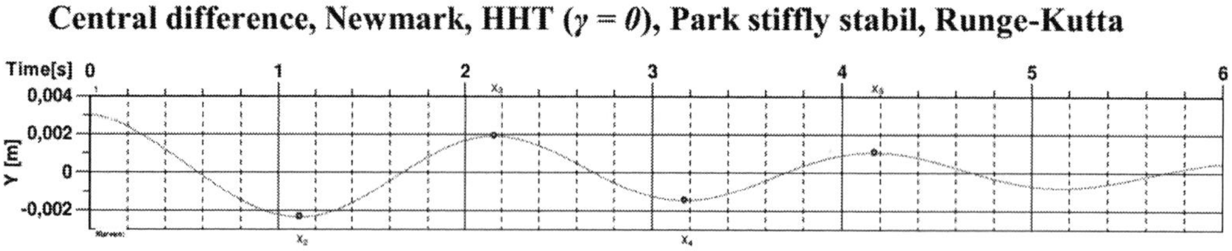

Fig. 27. Wagon body position deflection in y-axis direction calculated by HHT method when $\gamma = 0$

The parameters for computations in accordance to equations (47) and (48) when $\gamma = 0$ are in the Table 6.

Table 6. Wagon body position deflection in y-axis direction and calculated damping parameters when $\gamma = 0$

x_1	x_2	x_3	x_4	x_5	
0.003	-0.0024	0.0019	-0.0015	0.0011	
Λ_{13}	Λ_{35}	Λ_{15}	Λ_{24}	Λ_ϕ	D
0.45	0.58	0.51	0.48	0.51	0.08

HHT ($\gamma = 0.15$)

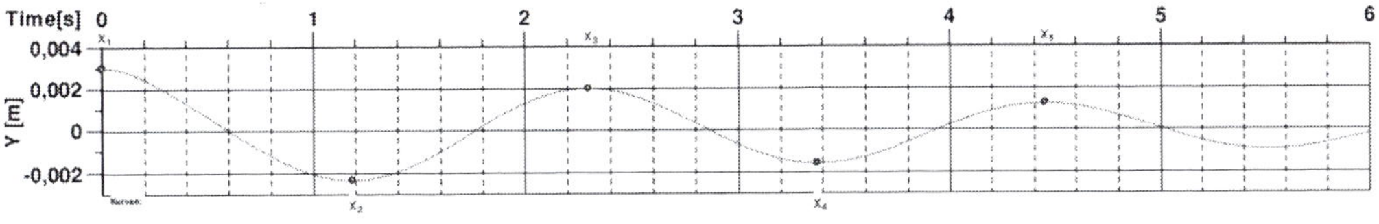

Fig. 28. Wagon body position deflection in y-axis direction calculated by HHT method when $\gamma = 0.15$

The parameters for computations in accordance to equations (47) and (48) when $\gamma = 0.15$ are in the Table 7.

Table 7. Wagon body position deflection in y-axis direction and calculated damping parameters when $\gamma = 0.15$

x_1	x_2	x_3	x_4	x_5	
0.003	-0.0023	0.002	-0.0016	0.0012	
Λ_{13}	Λ_{35}	Λ_{15}	Λ_{24}	Λ_ϕ	D
0.4	0.47	0.44	0.38	0.43	0.07

HHT ($\gamma = 0.2$)

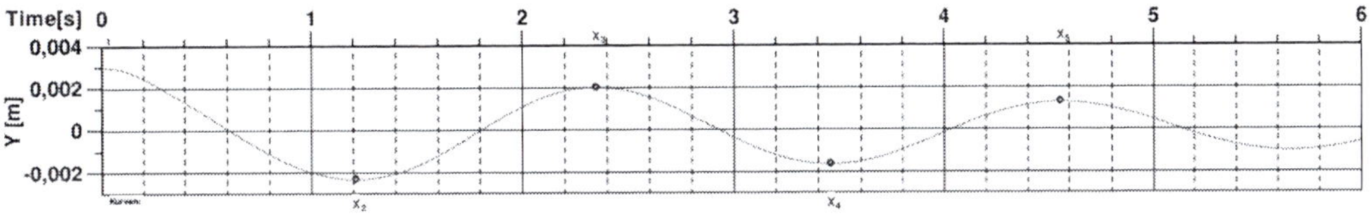

Fig. 29. Wagon body position deflection in y-axis direction calculated by HHT method when $\gamma = 0.2$

The parameters for computations in accordance to equations (47) and (48) when $\gamma = 0.2$ are in the Table 8.

Table 8. Wagon body position deflection in y-axis direction and calculated damping parameters when $\gamma = 0.2$

x_1	x_2	x_3	x_4	x_5	
0.003	-0.0023	0.002	-0.0016	0.0013	
Λ_{13}	Λ_{35}	Λ_{15}	Λ_{24}	Λ_ϕ	D
0.39	0.45	0.42	0.36	0.4	0.06

HHT ($\gamma = 0.3$)

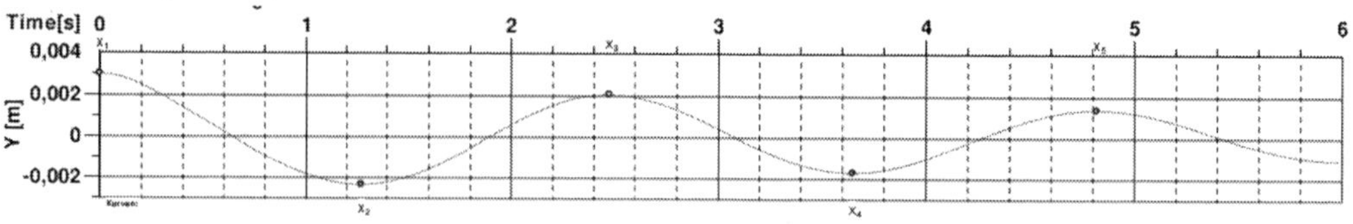

Fig. 30. Wagon body position deflection in *y*-axis direction calculated by HHT method when $\gamma = 0.3$

The parameters for computations in accordance to equations (47) and (48) when $\gamma = 0.3$ are in the Table 9.

Table 9. Wagon body position deflection in *y*-axis direction and calculated damping parameters when $\gamma = 0.3$

x_1	x_2	x_3	x_4	x_5	
0.003	-0.0023	0.002	-0.0017	0.0014	
Λ_{13}	Λ_{35}	Λ_{15}	Λ_{24}	Λ_{ϕ}	D
0.38	0.4	0.39	0.32	0.37	0.06

Conclusion

At present, the analyses of the dynamic characteristics of railway vehicles are carried out basically by using commercially distributed software packages, where you enter inputs (which the input forms allow) and the user may achieve results respecting the handling procedures. In view of the different approaches to solutions of partial solutions, the results of the simulations can differ.

In the paper the theoretical analysis of issue solution is performed, the computation parameters are specified and the results of computations performed with own programme are compared with the results gained from programme SIMPACK. The evaluated results are for the first or the third wheelset. In the figures with graphical representation of the courses: wheelset shift in the lateral direction (y), wheel-set rotation, slips (S), forces in the contact plane of rail and wheel: normal (N), tangential (T), guiding forces (Y), sum of guiding forces (SY), wheel forces (Q) there can be seen a very good coincidence at the computation with programme SIMPACK and our programme, here presented theory is processed in this. We analysed various time integration methods for judgement of the suitability, stability and numerical damping of these methods.

Finally, we can state, that the calculations met the goal, that we had set to ourselves. The results obtained by means of commercial package in general confirm the rightness of the theoretical analysis and realized algorithm in our own programme code. At the same time the theories in the investigated issue confirm the trustworthiness of the results obtained by commercial package. In the following period it can enable to us to model the flexible bindings and dampers with characteristics, that program SIMPACK does not directly enables and to use other contact theories of the rail and wheel contact with the non-elliptic contact patch.

The work was supported by the Cultural and Educational Grant Agency of the Ministry of Education of the Slovak Republic in project No. KEGA 077ŽU-4/2017: Modernization of the Vehicles and engines study program. The work was also supported by the project No. APVV-0842-11: Equivalent railway operation load simulator on the roller rig and VEGA No. 1/0927/15: Research of the use of alternative fuels and hybrid drives on traction vehicles with aim to reduce fuel consumption and air pollutants production.
Research-Educational Centre of Rail Vehicles (VVCKV)

References

1. K. Popp, W. Schiehlen, *Ground vehicle dynamics*. ISBN 978-3-540-24038-9 (Springer-Verlag Berlin Heidelberg, 357 p.)
2. R. V. Dukkipati, J. R. Amyot, *Computer-aided simulation in railway dynamics*. ISBN 0-8247-7787-5 (Marcel Dekker, New York, USA, 423 p., 1988)
3. V. Garg, R. Dukkipati, *Dynamics of railway vehicle systems. academic press Canada*. (Don Mills, Ontario, 1984)
4. M. Sága, R. Bednár, M. Vaško, *Contribution to modal and spectral interval finite element analysis*. Vibration Problems ICOVP 2011, Springer Proceedings in Physics *139*, 269-274 (2011)
5. A. Sapietová, M. Sága, P. Novák, R. Bednár, J. Dižo, *Design and application of multi-software platform for solving of mechanical multi-body system problems*. Mechatronics: Recent technological and scientific advances, 345-354 (2011)
6. S. Iwnicki, *Handbook of railway vehicles dynamics*. (CRC Press, Boca raton, Taylor&Francis Group, 2006) ISBN-13: 978-0-8493-3321-7
7. J. Gerlici, T. Lack, *Contact railway wheelset and track*. University of Žilina. ISBN 80-8070-317-5. [In Slovak] (2004)
8. T. Lack, J. Gerlici, *Rail vehicle dynamics properties investigation by means of simulation computations*. Dynamics of rigid and deformable bodies 2016. International scientific conference. Ústí nad Labem: Univerzita J. E. Purkyně, 2016. ISBN 978-80-7561-016-4, CD-ROM, 17 p. (2016)
9. P. Shackleton, S. Iwnicki, *Comparison of wheel-rail. contact codes for railway vehicle simulation*. an introduction to the Manchester Contact Benchmark and initial results, Vehicle System Dynamics **46** (1), 129-149 (2008)
10. J. J. Kalker, *Three-dimensional elastic bodies in rolling contact*. ISBN 0-7923-0712-7 (Kluwer Academic Publishers, Dordrecht/Boston/London, Netherlands, 314 p., 1990)

Y25 freight car bogie models properties analysis by means of computer simulations

Tomáš Lack[1,*], *Juraj* Gerlici[1]

[1]University of Žilina, Faculty of Mechanical Engineering, Department of Transport and Handling Machines, Univerzitná 8215/1, 01026 Žilina, Slovak Republic

Abstract. The article deals with the results of the simulation analysis of a railway wagon bogie model. We analysed four freight wagon bogie variants for its dynamics properties research. The bogie models correspond in general to the Y25 bogie concept. The models were created in SIMPACK software enhanced by the RAIL module. From the research results depicted in the graphs we found out, that the newly designed bogie variant gives the best results when compared to the other analysed versions. The newly designed model consists of a standard Y25 bogie frame with two Lenoire friction dampers. This bogie is equipped with longitudinal linkages on both sides. These linkages are completed with a radial torsion binding, torsion rod, between side bogie parts. The contact of railway wheels and rails generates active forces affecting the surface contact, affecting the size of the normal and tangential stress, wear surfaces of the wheel/rail, or directly the size of the derailment.

Keywords: chassis, Lenoir, damper, coupling, simulation

1 Introduction

Freight wagon transportation is a widespread global phenomenon, where it is necessary to transport large quantities of goods especially over long distances in an efficient, energy-saving and environmentally friendly manner [1]. The impact of goods transportation on the environment is achieved through the interaction of the railway track and wagon arising from the contact of individual wheels of the chassis with the rail [2-4].

In contact of railway wheels and rails, there are acting forces which have with their effect on the contact area, on the size of the normal and tangential stress, wear surfaces of the wheel and rail or directly on the anti-derailment stability [5-7].

The size of the forces acting on the wheel-rail contact when crossing the arc on track has a significant impact on the wheelset position in the track. If the radial position of wheelset (wheelset axis is perpendicular to the axis of the track) is at the minimal, ideally zero angle of attack, the guiding forces with minimum distortion appear [8, 9]. This has a positive impact on the value to the anti-derailment factor and on the occurrence and impact of tangential stresses in the transverse direction (y) in the wheel-rail contact [10].

The chassis construction of freight wagons has a significant impact on the size and forces of the wheel-rail contact [11-18]. One of the most common chassis for rail freight

* Corresponding author: tomas.lack@fstroj.uniza.sk

Reviewers: *Milan* Sága, *Milan* Vaško

services in Europe is the Y25-type chassis. In practice, there are several variations of this basic concept. In order to compare the results of the simulation analysis of the freight bogie railway model we have done a comparison of several construction variants right at Y25-type bogie. Various computational procedures can be used in the assessment tools [2].

2 Bogies and track models definition

Y25 basic type. Chassis Y25 (Fig. 1), in addition to other details, is characterised by wheelset guidance, where the suspension with primary progressive friction damper is part of the axle box situated always on one side of the suspension of each wheelset axle box [8-11]. The bogies in standard configuration by crossing the arc record certain sizes of guiding forces, and these imply technical consequences.

In order to reduce these values, the chassis manufacturers have changed some design parameters of guidance of Y25 bogie wheelset, respectively the bogies were modified [5].

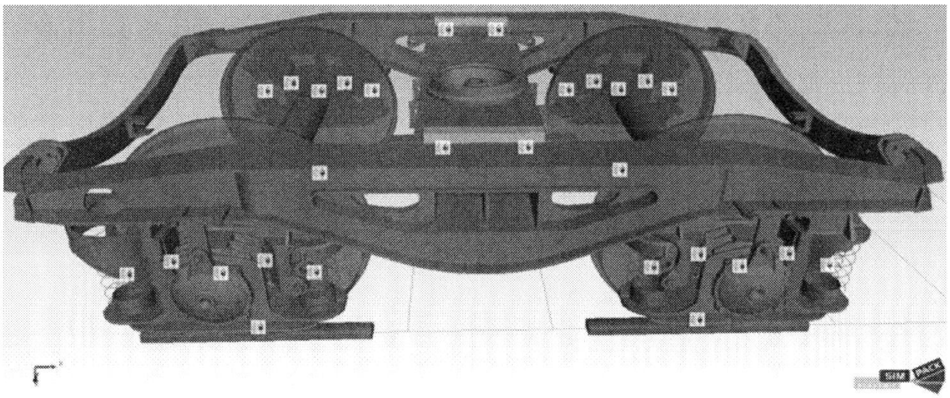

Fig. 1. Model of a standard Y25 bogie with a single Lenoir friction damper

Model of the Y25 bogie with two Lenoir dampers. One of the possible modifications is called "Release" of a wheelset in the longitudinal direction of the bogie with two Lenoir friction dampers (Fig. 2) and the connection of the front right wheel with the left rear wheel and also front left wheel with the rear right wheel using a cross coupling [5].

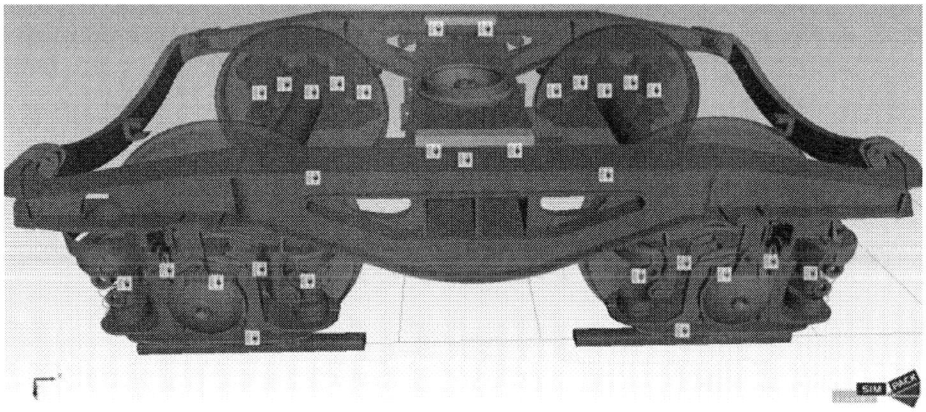

Fig. 2. Model of a Y25 bogie with two Lenoir friction damper

Model of the Y25 bogie with two Lenoir dampers. One of the possible modifications is the bogie model Z25 with two Lenoir dampers and a cross coupling (Fig. 3).

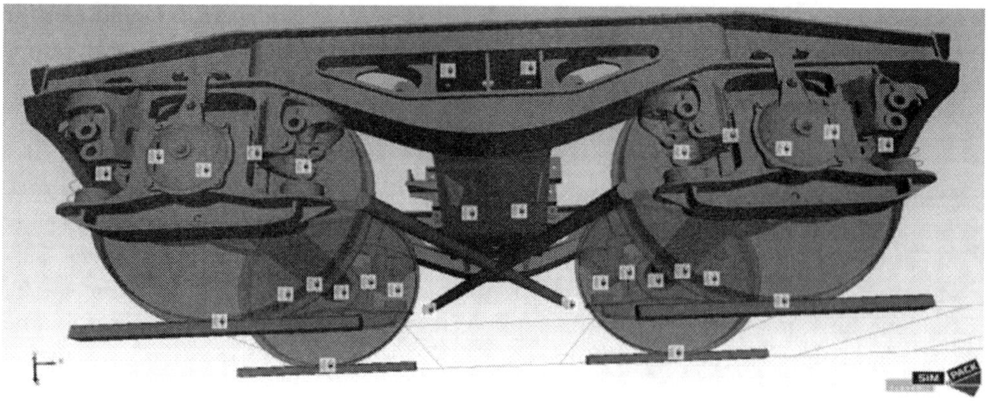

Fig. 3. Bogie Y25 linking wheelset cross-binding, view from below

Model of the Y25 bogie with two Lenoir dampers, longitudinal and transverse torsion. For modifications of a Y25 freight railway bogie we chose a modification inherent to railway vehicles of another category, designated for another purpose - the kind of vehicle, which has a bogie equipped with a longitudinal coupling between wheelsets, for illustration of real application.

The principle of application of longitudinal coupling extended by a transverse torsion coupling was implemented in the design of a new computational model of a freight bogie that can be investigated for the influence of some design changes, which we have applied to the expensive Y25 chassis (Fig. 4).

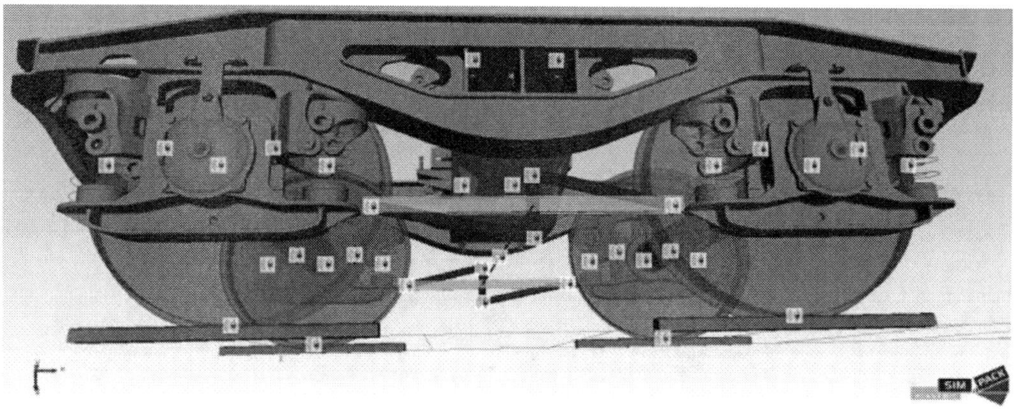

Fig. 4. Bogie Y25 with two Lenoir friction dampers linking wheelset with torsional binding – view from the bottom

Definiton of track simulation calculation. For simulations we have used standard definition of a model railway track of 1435 mm, with a rail head profile UIC60. Railway wheels are equipped with driving profile S1002.

The decisive parameters for comparative calculations were only horizontal track geometry, which consisted of direct and four interlinked arches (Fig. 5, Fig. 6).

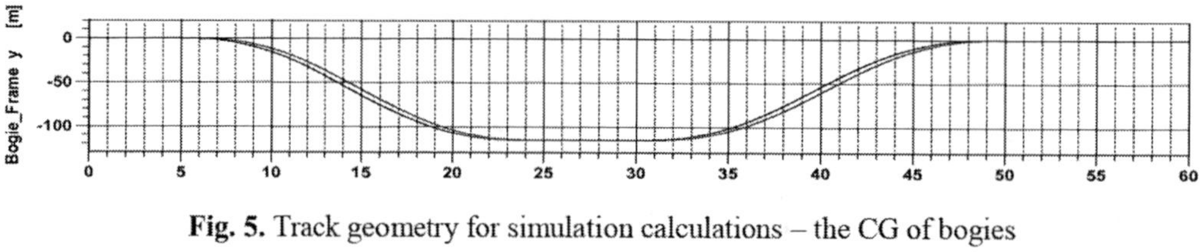

Fig. 5. Track geometry for simulation calculations – the CG of bogies

3 Comparison of the results of calculation

The following section shows the results of calculations of all the four analyzed computational models. The evaluation focused on plotting the course of the guiding forces, angle of attack, wear number, the sum of guiding forces, safety against derailment, forces in drawbars in cross bonds, drawbar forces Fx in the longitudinal direction and the view of the rotation of torsional cross bond.

3.1 Guiding forces

Guiding forces at the first bogie are on the first wheelset and the leading wheel in the original design with a single Lenoir damper, Fig.1. In track arcs, the waveform values are the highest compared to other variants. Fig. 6-9 show this in a black line. In real practice we know solutions with cross-couplings, which diagonally connect opposite axle boxes of the bogie, as seen in Fig. 3. The results of the calculations in the graph are represented by the red curve.

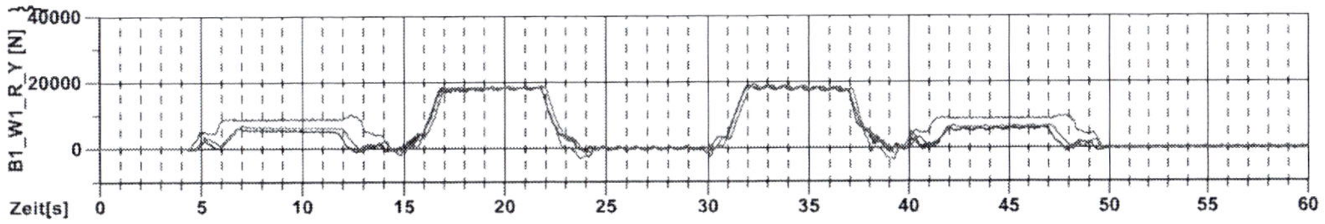

Fig. 6. Guiding force at the first bogie, the first wheelset, right wheel

In the simulation analysis, we created a model of variant without cross bond with released wheelsets with two Lenoir friction dampers which has the lowest guiding force, Fig. 2. In the graph it is shown as the blue line. This solution requires adding a stabilizing bonding of front and rear wheelset. Therefore, we have proposed an interconnection of the axle boxes by binding shown in Fig. 4. Guiding forces in this binding are shown in purple in the graph. This line closely copies variation of the forces without the cross bond. We can see that this interconnection is much more favorable than the currently used cross bond.

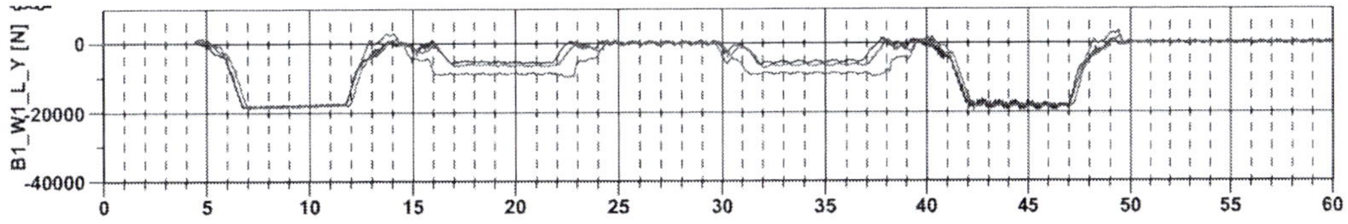

Fig. 7. Guiding force at the first bogie, the first wheelset, left wheel

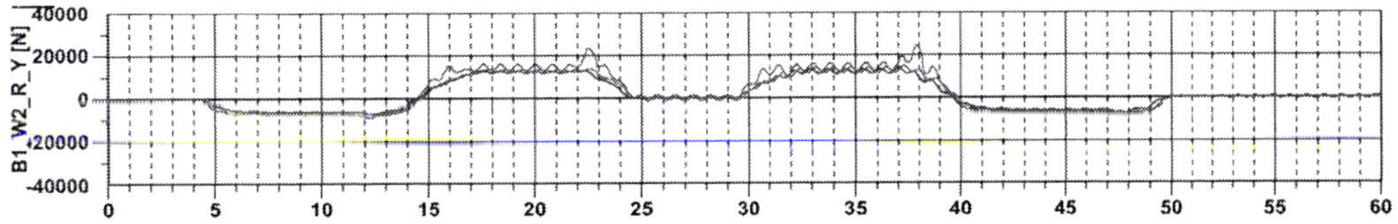

Fig. 8. Guiding force at the first bogie, the second wheelset, right wheel

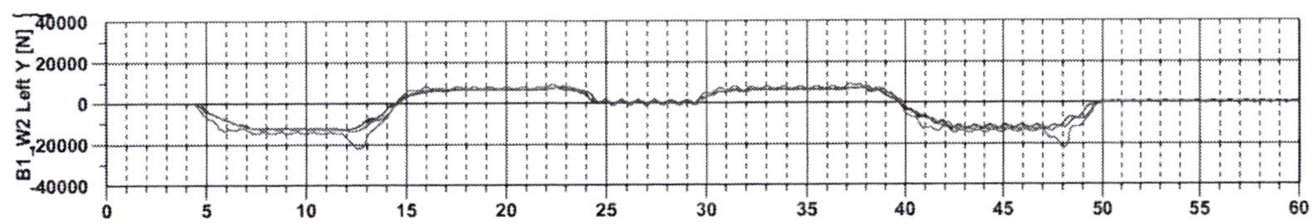

Fig. 9. Guiding force at the first bogie, the second wheelset, left wheel

3.2 Sum of guiding forces

The summaries of the guiding forces (Fig. 10 - 11) acting on the individual wheelsets are very balanced and are primarily related to the operational characteristics of the vehicle, such as geometry, speed, vehicle weight. The new solutions of all variations have less than 20 000 N.

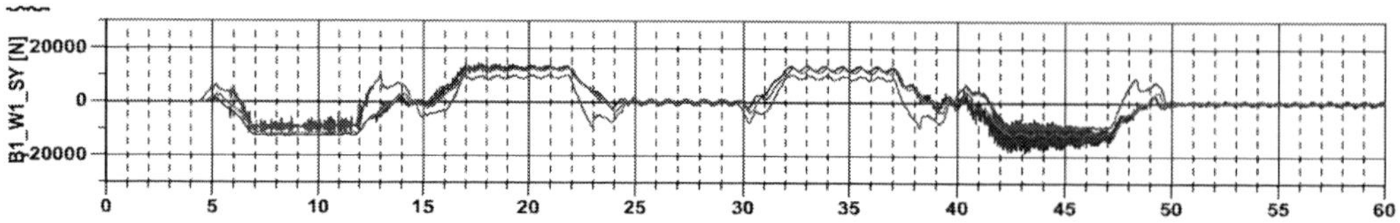

Fig. 10. Sum of guiding forces of the first wheelset first bogie

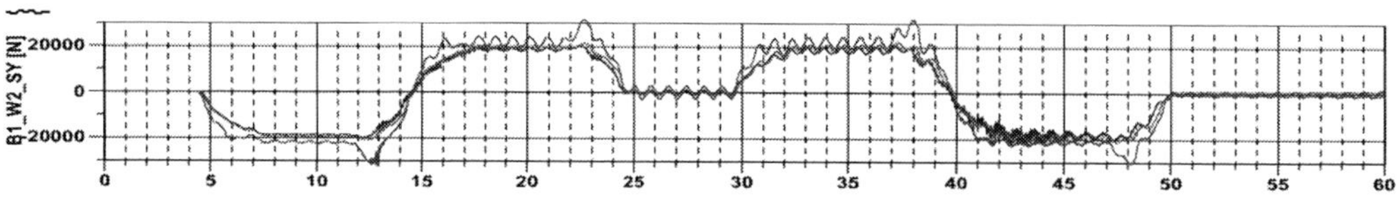

Fig. 11. Sum of guiding forces of the second wheelset first bogie

3.3 Angle of attack

When the bogie or wheelset passes the track arc, the angle of attack (Fig. 12 – 15) results from the position of the bogie or wheelset in the track when there is angular deviation of perpendicular to the axis of the wheelset (the bogie axis) and the axis of the track. In terms of evaluation of the angle of attack at the vehicle, it is the angle formed by the vehicle axis and the tangent to the arc of a track at the contact point of the loading wheel.

There are more wheelset technical solutions which should ensure passing of wheelset through the arc in the radial position to the track arc. These solutions have better or worse results.

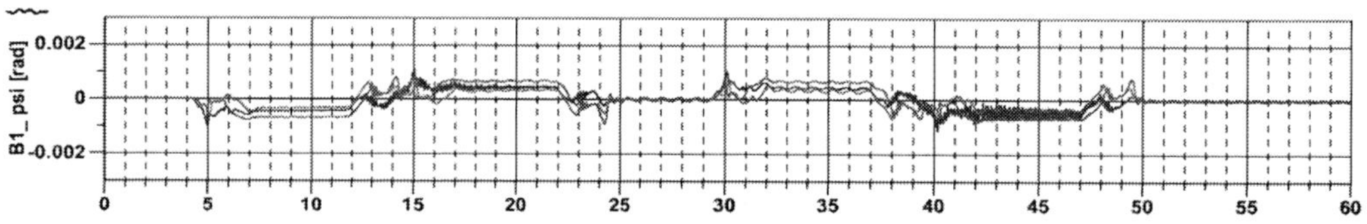

Fig. 12. Angle of attack of the first bogie

In any case, the wheelset position where the wheelset stands passing the arc of a track to a position approaching the radial position causes less guiding forces than in opposite case. This situation also affects the floating clearance in the longitudinal movement of the wheelset, which can be spent by setting the positioning.

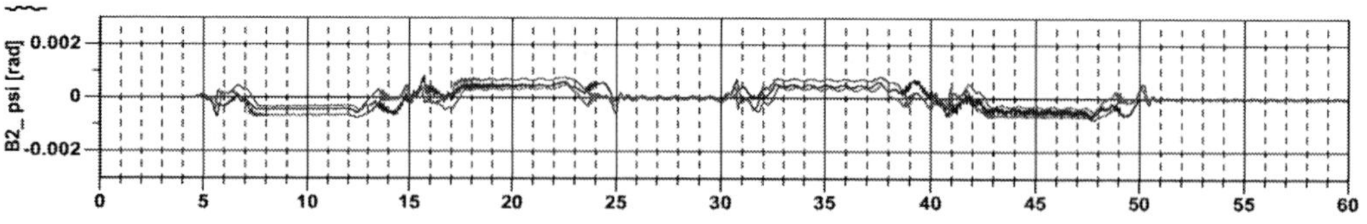

Fig. 13. Angle of attack of the second bogie

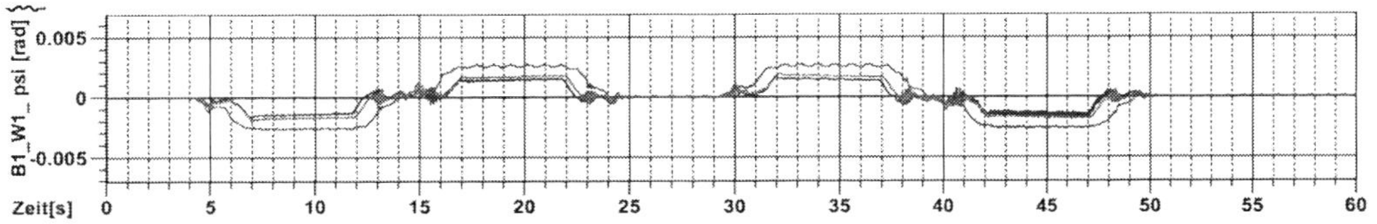

Fig. 14. Angle of attack of the first wheelset first bogie

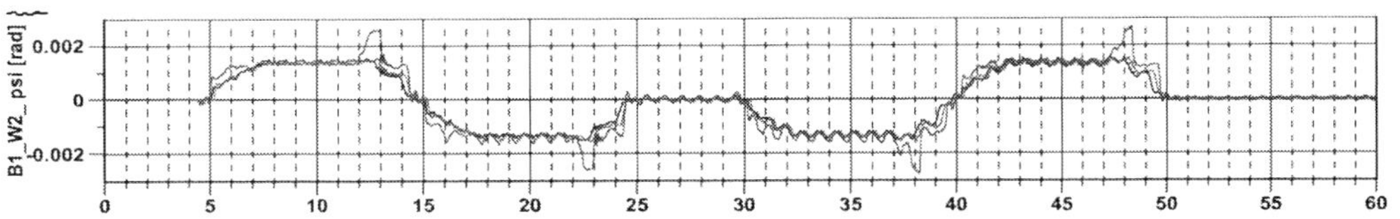

Fig. 15. Angle of attack of the second wheelset first bogie

Fig. 12 – 15 clearly show that the solution of the bogie by releasing the wheelsets with two Lenoirs with the possibility of motion transfer by torsion coupling is a suitable solution in view of minimizing the angle of attack.

3.3 Wear number

The parameter of wear number in Fig. 16 – 17 has a physical dimension of force [N]. Wear number of contact area is calculated according to the formula (1):

$$|T_x \cdot v_x| + |T_y \cdot v_y| + |M_z \cdot \varphi_z| \tag{1}$$

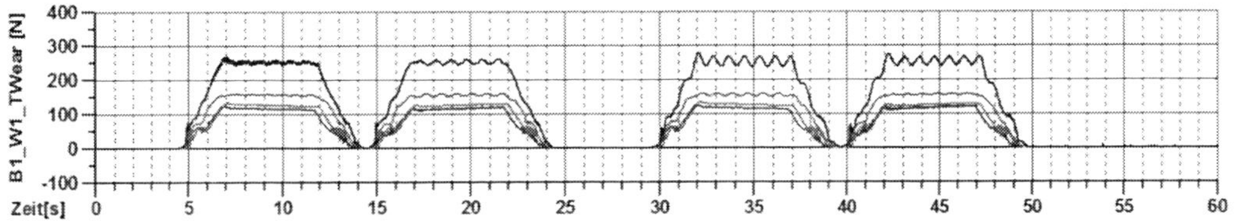

Fig. 16. Course of the wear number of the first wheelset first bogie

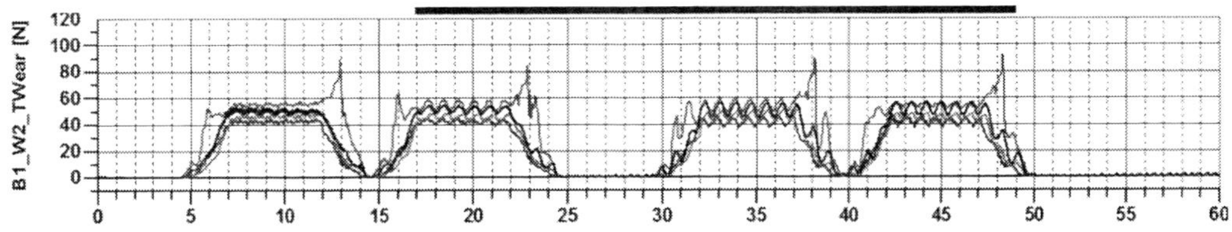

Fig. 17. Course of the wear number of the second wheelset first bogie

3.4 Safety against derailment

Expression of safety against derailment (Fig. 18 - 21) as a relative ratio equivalent to the proportion of the guiding forces and wheel forces on an individual wheels variations corresponds to its variation of the guiding forces. Higher values represents worse conditions, lower values represents a safer ratio of horizontal forces to the vertical for-ces. Although all values are significantly under the well-known lower accepted level of 0.8, in this comparison we can see the extent to which one option is more preferable than the other, from this perspective.

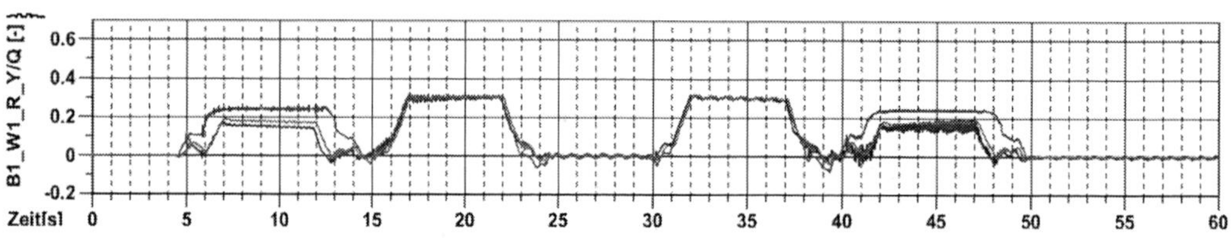

Fig. 18. Anti-derailment stability of the right wheel, first wheelset, first bogie

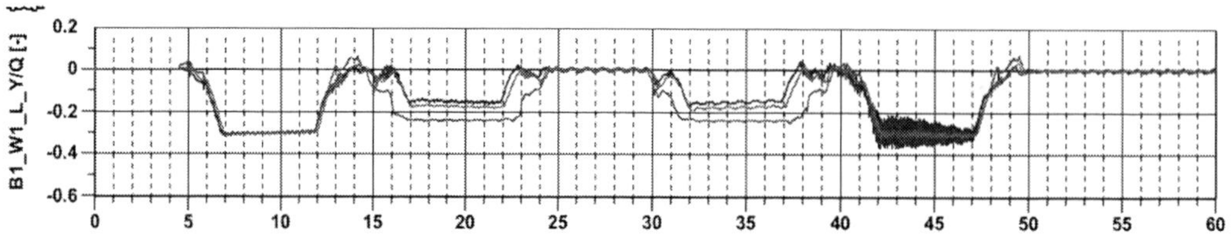

Fig. 19. Anti-derailment stability of the left wheel, first wheelset, first bogie

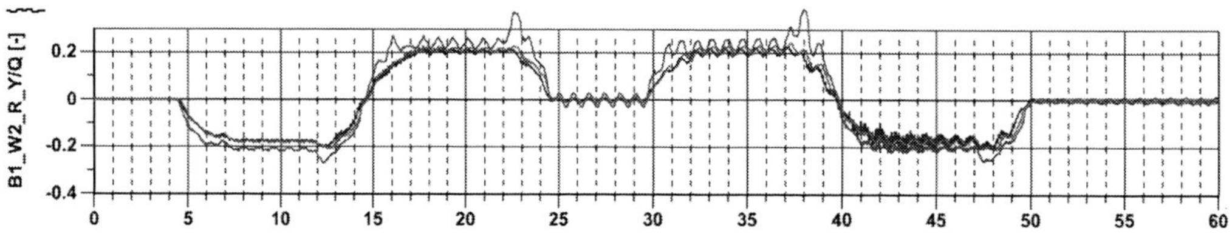

Fig. 20. Anti-derailment stability of the right wheel, second wheelset, first bogie

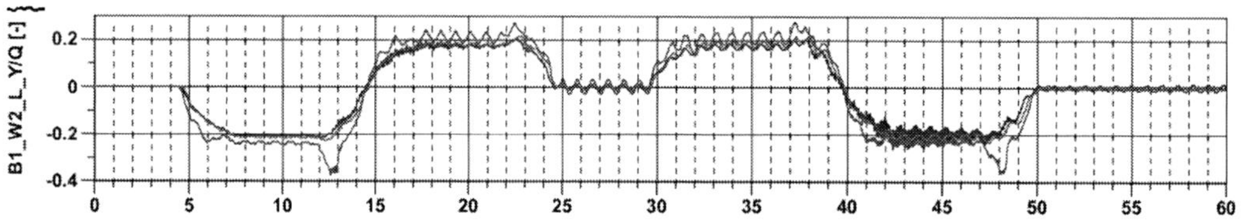

Fig. 21. Anti-derailment stability of the left wheel, second wheelset, first bogie

3.5 Forces in drawbars of cross coupling

Fig. 22 - 23 show the forces in drawbars of cross coupling, Fig. 3.

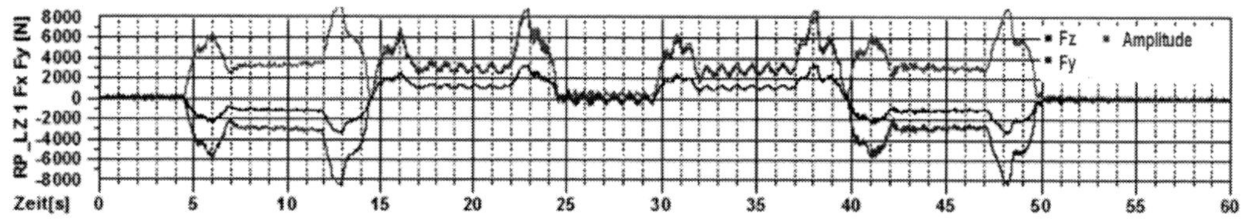

Fig. 22. First bogie, connect the right front axle box and the left rear

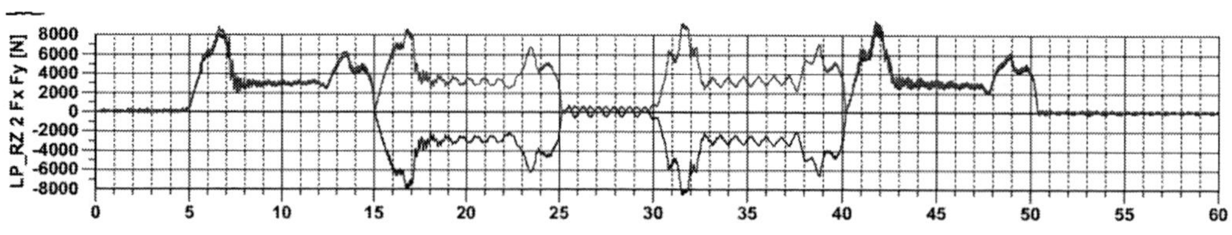

Fig. 23. First bogie, connect the left front axle box and the right rear

In this case, the drawbars are transmitting the lateral forces. Fx curve shows the force in x-direction, Fy curve shows the force in the y – direction and Amplitude shows the force in lonfitudinal drawbar direction.

3.6 Fx forces in drawbars in longitudinal cross coupling

Fig. 24 – 25 show the forces in drawbars of cross coupling pulley of the transmission of longitudinal forces on the sides of the chassis in the variant of a newly designed cross coupling. In this case, the drawbars do not transmit lateral forces. The relative longitudinal displacements of the axle boxes are transferred to the other side of the chassis through a transverse torsion bond. The forces in these drawbars in the proposed longitudinal bond between the axle boxes are more than ten times smaller compared to the detected forces in the drawbars at cross-coupling in a classical variant bogie with cross coupling.

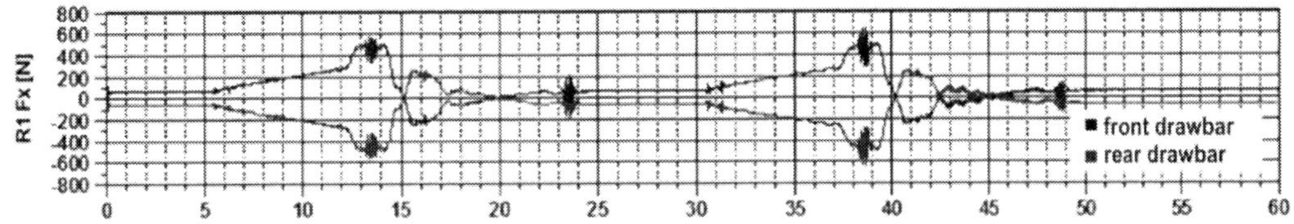

Fig. 24. Course of longitudinal forces Fx in drawbars on the right side of the first bogie

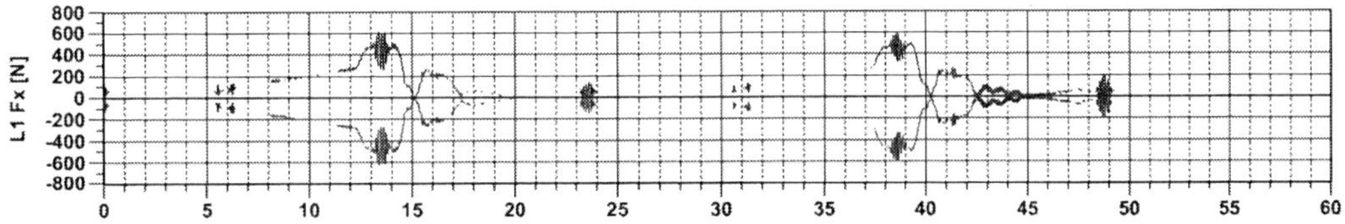

Fig. 25. Course of longitudinal forces Fx in drawbars on the left side of the first bogie

3.7 Representation of course in rotation torsion bar of cross coupling

Rotation of the torsion bar of cross bond of bogies represents a flexible rotational coupling at the sides of the bogie (Fig. 26, Fig. 27).

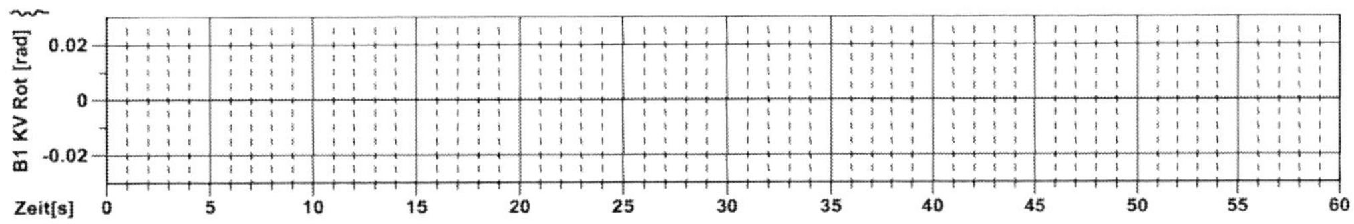

Fig. 26. Guiding force at the first bogie, the first wheelset, right wheel

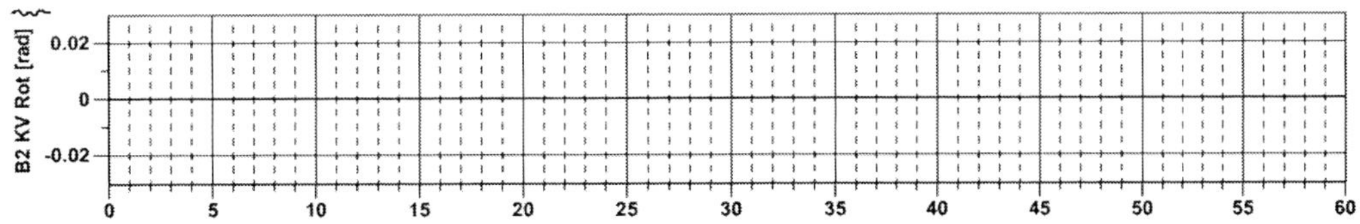

Fig. 27. Guiding force at the first bogie, the first wheelset, right wheel

Conclusions

This article deals with the results of simulation analysis of a railroad freight bogie model to determine the dynamic properties of four variants of bogie construction, which corresponds to the type Y25. The models were created in SIMPACK software extended with RAIL module. All the bogie models have the same basic parameters similar to the Y25 (frame, wheelsets, connection to the chassis). The differences are only in the changed component.

Vehicles in the simulation software traverse the track with the same geometry. It's a model combination of straight track and four consecutive arcs. To achieve more significant differences in the results, the test track is only in the horizontal plane, which means the outer rail in arcs does not have any super-elevation installed. Vehicles go through the proposed model track at a constant speed.

The first variant is the basic model concept Y25 with one Lenoir friction damper in the primary suspension of the wheelset, the opposite side suspension is equipped only with springs and friction surface is rigidly clamped to the axle box. The second option is guiding with the released both sides of the wheelset axle box, each side of the wheelset is equipped with two flexible bonds with Lenoir friction dampers. The third option is guiding with the released both sides of the wheelset axle box, each side of the wheelset is equipped with two flexible bonds with Lenoir friction dampers. Wheelsets of each bogie are connected by direct cross-coupling. The fourth option is guiding with the released both sides of the wheelset axle box, each side of the wheelset is equipped with two flexible bonds with Lenoir friction dampers. Wheelsets of each bogie (front - rear) are connected with longitudinal coupling and also with torsion cross-coupling (left - right wheelset).

The result of simulation analysis are graphs of the examined parameters, guiding forces, anti-derailment safety, longitudinal force Fx in the drawbars on the sides of the bogies (in the fourth variant) and the wear numbers resulting from the action of force at a contact point. These graphs are completed with angle of attack information of bogies and wheelsets.

From the course of the examined values shown in the graphs we can see that the newly proposed variant of bogie type Y25 with released wheelsets, two Lenoir dampers, longitudinal coupling on the sides of the bogie complemented by transverse torsion coupling between the sides has a better dynamic behaviour of the vehicle in moving on the track, lower lateral guiding forces values, anti-derailment safety with using bogies with cross-coupling. Drawbars with longitudinal coupling feature smaller forces than drawbars with cross-coupling.

The work was supported by the Cultural and Educational Grant Agency of the Ministry of Education of the Slovak Republic in project No. KEGA 077ŽU-4/2017: Modernization of the Vehicles and engines study program. The work was also supported by the project No. APVV-0842-11: Equivalent railway operation load simulator on the roller rig and VEGA No. 1/0927/15: Research of the use of alternative fuels and hybrid drives on traction vehicles with aim to reduce fuel consumption and air pollutants production. Research-Educational Centre of Rail Vehicles (VVCKV)

References

1. J. Dižo, J. Harušinec, M. Blatnický, *Multibody system of a rail vehicle bogie with a flexible body*. Manufacturing Technology: Journal for Science, Research and Production **15**, (5), 781-788 (2015)
2. M. Sága, R. Bednár, M. Vaško, *Contribution to modal and spectral interval finite element analysis*. Vibration Problems ICOVP 2011, Springer Proceedings in Physics, **139**, 269-274 (2011)

3. J. Dižo, *Evaluation of ride comfort for passengers by means of computer simulation*. In: Manufacturing Technology: Journal for Science, Research and Production **15** (1), 8-14 (2015)
4. J. Dižo, M. Blatnický, B. Skočilasová, *Computational modelling of the rail vehicle multibody system including flexible bodies*. Communications: scientific letters of the University of Žilina **17** (3), 31-36 (2015)
5. S. Iwnicki, Y. Bezin, A. Orlova, P. A. Johnsson, S. Stichel, H. Schelle, *The 'SUSTRAIL' High Speed Freight Vehicle*. Simulation of Novel Running Gear Design. Proc. of the 23rd Intern. Symposium on Dynamics of Vehicles on Roads and Tracks, 11. IAVSD. August 2013 Qingdao, China (2013)
6. T. Lack, J. Gerlici, *Modified HHT method for vehicle vibration analysis in time domain utilisation*. Applied mechanics and materials **486,** 396-405 (2014)
7. A. Sapietová, M. Sága, P. Novák, R. Bednár, J. Dižo, *Design and application of multi-software platform for solving of mechanical multi-body system problems*. Mechatronics: Recent technological and scientific advances, 345-354 (2011)
8. T. Lack, J. Gerlici, M. Maňurová, *Analysis of dynamic properties of freight bogie model 1*. In: Innovations in the concept, design, manufacturing and testing of freight II: January 2015, University of Zilina, 65-74 [In Slovak] (2014)
9. T. Lack, J. Gerlici, M. Maňurová, *Analysis of dynamic properties of freight bogie model 2*. In: Innovations in the concept, design, manufacturing and testing of freight II: January 2015, University of Zilina, 51-56 [In Slovak] (2015)
10. T. Lack, J. Gerlici, *Freight car bogie analysis by means of simulation computations*. In: PRORAIL 2015 Current problems in rail vehicles, 22nd intern. conference, Zilina, September 2015. ISBN 978-80-89276-48-6 [In Slovak] (2015)
11. T. Lack, J. Gerlici, *Railway bogie properties analysis by means of simulation computations*. Railway transport and logistics **11** (2), 54-60 [In Slovak] (2015)
12. M. Maňurová, A. Suchánek, *Determination of stiffness of triple spring built in a bogie of a rail vehicle*. In: Manufacturing Technology: Journal for science, research and production **16** (2), 390-396 (2016)
13. L. Jakubovičová, M. Sága, *Computational analysis of contact stress distribution in the case of mutual stewing of roller bearing rings*. Novel Trends in Production Devices and Systems, Applied Mechanics and Materials **474**, 363-368 (2014)
14. P. Šťastniak, J. Gerlici, T. Lack, J. Harušinec, *Computer aided simulation analysis for computation of modal analysis of the freight wagon*. TRANSCOM 2013: 10th European conference of young researchers and scientists: Žilina, June 24-26, 2013, Slovak Republic, Section 6: Machines and equipment, Applied mechanics, Žilina, University of Žilina, 297-300 (2013)
15. P. Pecháč, M. Sága, *Controlling of local search methods' parameters in memetic algorithms using the principles of simulated annealing*. Proc. Eng. **136**, 70-76, (2016)
16. F. Nangolo, J. Soukup, J., *The effect of asymmetry on vertical dynamic response of railway vehicles*. Manufacturing Technology: Journal for science, research and production **14** (3), 375-380 (2014)
17. M. Svoboda, J. Soukup, *Dynamic measurement of four-axle railway wagon*. Manufacturing Technology **13** (4), 552-558 (2013)
18. J. Skočilas, B. Skočilasová, J. Soukup, *Determination of the rheological properties of thin plate under transient vibration*. Latin American Journal of Solids and Structures. Brasil society for mechanics and engineering, 1679-7825 [online]

Contribution to numerical study of vehicle vertical stochastic vibration

Alžbeta Sapietová[1,*], *Milan* Sága[1], *Dana* Stančeková[2], *Milan* Sapieta[1]

[1]Department of Applied Mechanics, University of Žilina, Univerzitná 1, 01026, Slovakia
[2]Department of Industrial Engineering, University of Žilina, Univerzitná 1, 01026, Slovakia

Abstract. The goal of the paper is to present an application of the software MB_DYN inbuilt in MATLAB for stochastic response of the chosen vehicle computational model. The input kinematics parameters will be road irregularity with random character. The dynamic model parameters are considered as deterministic. The analysed vehicle model assumes 10 DOF. The stochastic response in time and frequency domain was solved by program MB_DYN in MATLAB using Monte Carlo method. Applying the simulation techniques the influence study of the vehicles speed and road quality in chosen points was realised.

Keywords: MATLAB, stochastic dynamics, vehicle speed, road quality

1 Introduction

The software MB_DYN has been successfully applied by authors several years ago in industrial and academic research areas. The software has been given for the dynamic analysis of multi-body mechanical systems. It comprises the computation of the symbolic equations of motion (using Matlab) and the simulation of the dynamic behaviour [1].

The dynamical analysis of a mechanical system starts with the modelling process. The real physical system is approximated by an idealized model [2, 3].

The input data for MB_DYN have to be entered in input files prepared with prompts and comments. The user has to provide only simple expressions for the description of kinematics and mass distribution with respect to arbitrary reference frames. Observation points allow the determination of position, velocity and acceleration of arbitrary points of the multibody system.

2 Vehicle dynamic model

Dynamic and computational model of the car was created in MB_DYN. The dynamic model (Fig. 1) was built-up from elements defined in the Table 1, 2, 3. The necessary geometry parameters of the vehicle are presented on Fig. 2.

* Corresponding author: alzbeta.sapietova@fstroj.uniza.sk

Reviewers: *Juraj* Gerlici, *Bohuš* Leitner

Table 1. Six rigid bodies, of mass and rotary inertia concentrated in the centre of gravity of each body

m_1 [kg]	m_2 [kg]	m_3 [kg]	m_4 [kg]	I_{x1} [kg.m^2]	I_{y1} [kg.m^2]	I_{x3} [kg.m^2]	I_{y3} [kg.m^2]
1130	42.5	25	42	504	1840	0.8	0.5

Table 2. Linear springs elements

k_1 [N/m]	k_2 [N/m]	k_3 [N/m]	k_4 [N/m]	k_5 [N/m]
200000	15500	23300	250000	180000

Table 3. Damping elements

b_2 [kgm^{2}s^{-1}]	b_3 [kgm^{2}s^{-1}]	b_4 [kgm^{2}s^{-1}]
1130	42.5	25

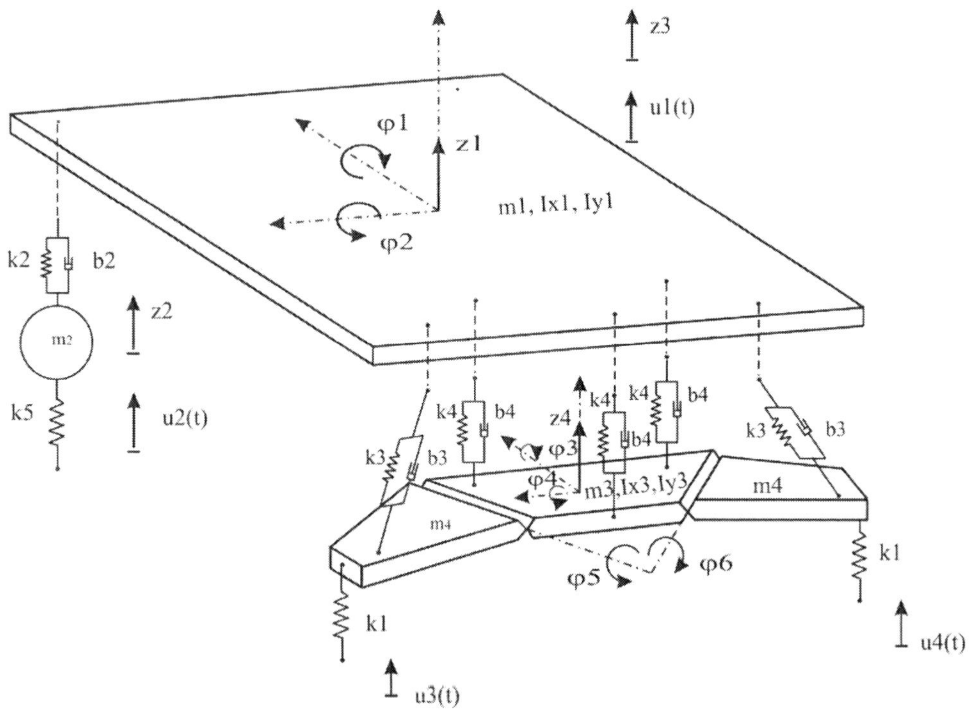

Fig. 1. Dynamic model of vehicle

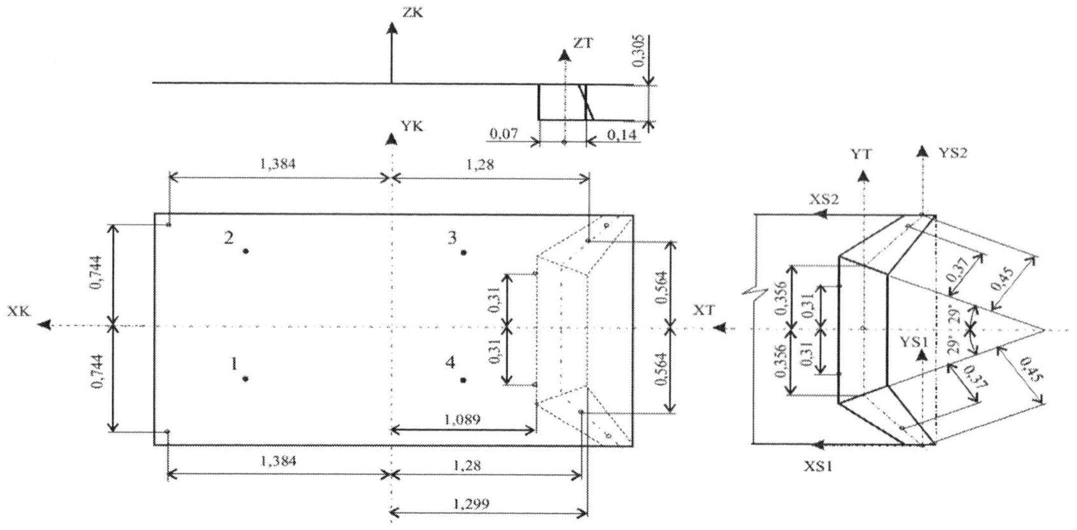

Fig. 2. Geometry of vehicle

3 Description of kinematic excitation

Generally we can found three different sorts of roadway roughness, i.e.
- local extreme unevenness,
- regular unevenness,
- irregular unevenness.

The two unevennesses mentioned first are deterministic and may be described geometrically. The irregular unevenness however are random and their probabilistic idealization involves the use of the theory of stochastic processes [4-6].

An appropriate mathematical modelling of these irregularities of the pavement surface can be based on the concept of stationary stochastic process, whose probabilistic structure can be conveniently described by the corresponding power spectral density function (PSD) [7-10]. According to the International Organization for Standardization (ISO) specifications, this power spectrum can be represented by an exponential function

$$S(\Omega) = S(\Omega_0) \cdot \left(\frac{\Omega}{\Omega_0} \right)^{-w} . \quad (1)$$

The coefficient $S(\Omega_0)$ and the exponent w are two parameters that depend on the characteristics of the roadway roughness [11-18]. Table 4 presents usual values of $S(\Omega_0)$ for different qualities of pavement, assuming $\Omega_0 = 1(\mathrm{m}^{-1})$ and $w = 2$.

The generation of an artificial discrete profile of the pavement surface can be performed based on the previous knowledge of the power spectral density function S(Ω), assuming such a profile given by a Fourier series, as follows:

$$u_{i+1} = u_i + \Delta x \cdot \sqrt{S_0} \cdot \Omega_0 \cdot w_i , \quad (2)$$

where function $w_i(t)$ is a zero mean Gaussian white noise process with the power spectral density $S = 1$ (m^2s^{-3}):

$$w(t) = \sum_{i=1}^{n} \sqrt{2 \cdot \Delta\Omega_i} \cdot \cos(\Omega_i \cdot x - \theta_i), \quad (3)$$

Parameter Ω_i is a frequency within the interval of relevant spectral content, θi is random phase angle with uniform distribution in the interval $[0, 2\pi]$, $\Delta\Omega_i$ is a frequency increment and n is the total number of harmonic waves considered. The simulation of the random response of the vehicle was realized in 8000 discrete points. These simulated road distance was 2 km.

Table 4. Usual values of $S(\Omega_0)$ for different qualities of pavement, assuming $\Omega_0 = 1$ (m^{-1}) and $w = 2$.

Quality of pavement	$S(\Omega_0)$ [cm^3]		
	lower limit	**mean value**	**upper limit**
very good	0.5	1	2
good	2	4	8
average	8	16	32
poor	32	64	128
very poor	128	256	512

4 Results

The final goal of this study was to realise the influence analyse of the vehicles speed and roadway roughness on the acceleration in chosen points of the car. The grafic presentation of the results is on Figs. 3, 4, 5, and 6. On illustrate, the power spectral density for speed 60 km/h and third degree of roadway quality is shown on figures 7 and 8. The study can attend as a guide for signification analysis in vehicle dynamics.

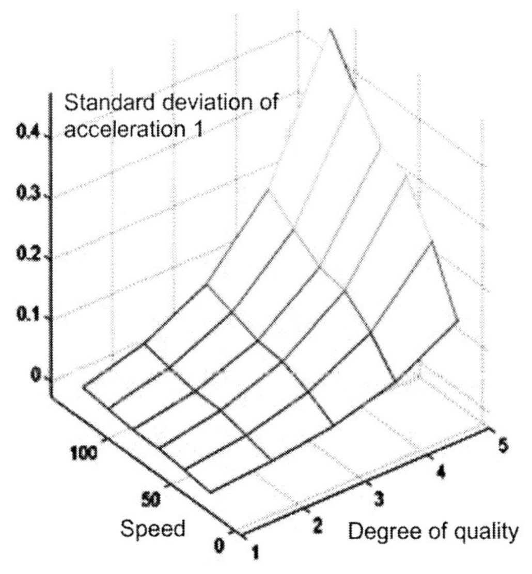

Fig. 3. Influence of the speed and roadway roughness on the acceleration

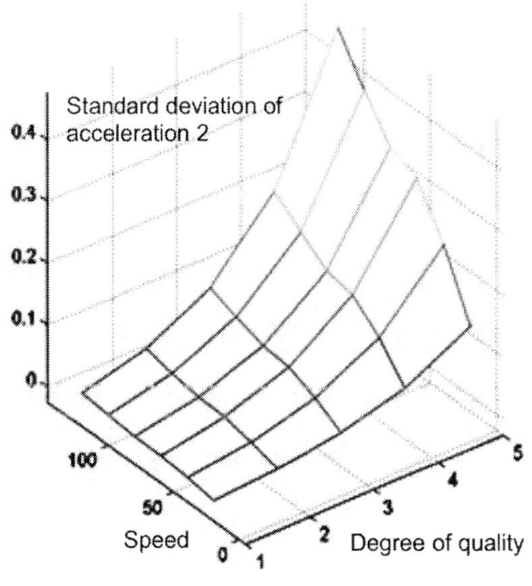

Fig. 4. Influence of the speed and roadway roughness on the acceleration

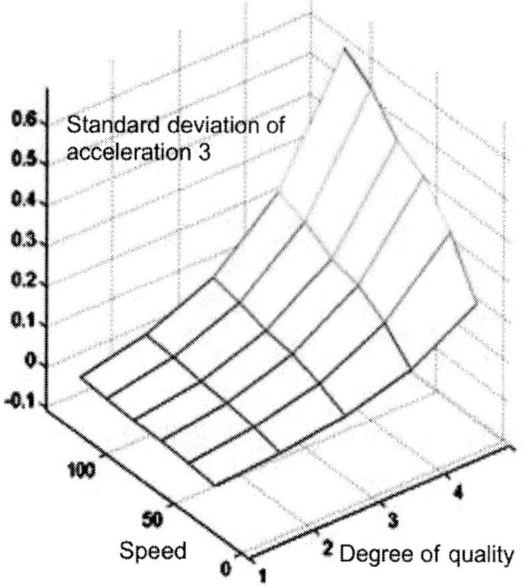

Fig. 5. Influence of the speed and roadway roughness on the acceleration

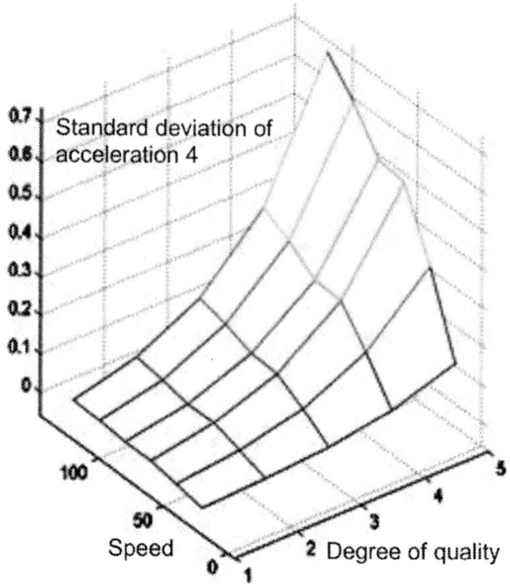

Fig. 6. Influence of the speed and roadway roughness on the acceleration

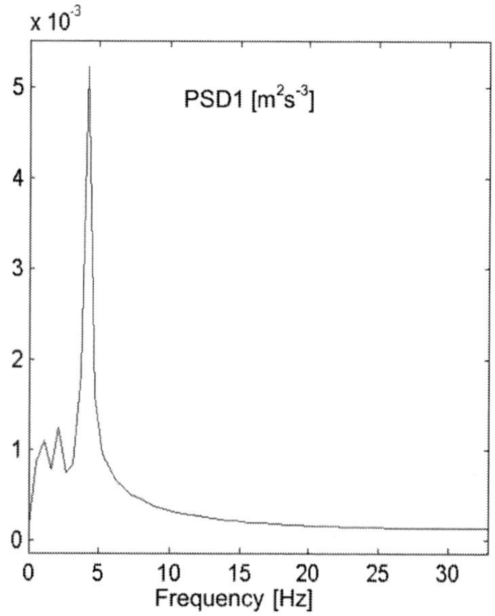

Fig. 7. Power spectral density of acceleration in point 1

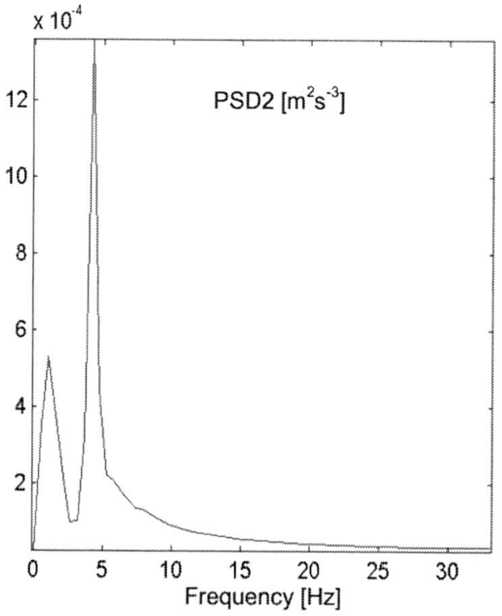

Fig. 8. Power spectral density of acceleration in point 2

Conclusion

The paper has been form on the base of the collaboration of departments of mechanics of universities in Zilina and Stuttgart. The aim was to propose the methodology of the solution of multibody dynamics systems by software packages NB_DYN. The main scientific goal was to present numerical study of the speed of movement and road quality influence on car vertical vibrations. The study can attend as a guide for signification analysis in vehicle dynamics.

The part of the results of this work was supported by grants VEGA 1/0795/16 and KEGA 017ŽU-4/2017.

References

1. W. Schiehlen, W, *Multibody Systems Handbook.* Springer-Verlag Berlin Heidelberg, **432** (1990)
2. A. Sapietová, M. Sága, P. Novák, R. Bednár, J. Dižo, *Design and application of multi-software platform for solving of mechanical multi-body system problems.* Mechatronics: Recent technological and scientific advances, 345-354 (2011)
3. J. Gerlici, T. Lack, *Railway wheel and rail head profiles development based on the geometric characteristics shapes.* Wear: an international journal on the science and technology of friction, lubrication and wear **271** (1-2), 246-258 (2011)
4. J. Gerlici, T. Lack, *Modified HHT method for vehicle vibration analysis in time domain utilisation.* Applied mechanics and materials **486**, 396-405 (2014)
5. J. Gerlici, T. Lack, J. Harušinec, *Realistic simulation of railway operation on the* RAILBCOT test stand. Applied mechanics and materials **486**, 387-395 (2014)
6. J. Ondrouch, P. Ferfecki, Z. Poruba, *Active vibration reduction of rigid rotor by kinematic excitation of bushes of journal bearings.* Metalurgija **49** (2), 107-110 (2010)
7. J. Flizikowski, M. Macko, J. Czerniak, A. Mroziński, *Implementation of genetic algorithms into development of mechatronic multi-edge's grinder design.* ASME 2011 Int. Mech. Eng. Congress and Exposition, IMECE **7**, 1227-1235 (2011)
8. M. Macko, J. Flizikowski, Z. Szczepański, K. Tyszczuk, G. Śmigielski, A. Mroziński, J. Czerniak, A. Tomporowski, *CAD/CAE applications in mill's design and investigation.* Proceedings of the 13th International Scientific Conference on Computer Aided Engineering - Editors: Rusinski, Eugeniusz, Pietrusiak, Damian (Eds.). Series Title: Lecture Notes in Mechanical Engineering, Part F **10**, 343-351 (2017)
9. T. Domanski, A. Bokota, *Numerical analysis of phenomena during progressive hardening of tool steel elements for cold work.* Procedia Engineering **177**, 70-77 (2017)
10. T. Domanski, A. Bokota, W. Piekarska, M. Kubiak, *Nmerical analysis of microstructure and stress generated by heat treated of steel elements.* Conference Proceedings, 642-647 (2015)
11. M. Łazarska, T. Woźniak, Z. Ranachowski, A. Trafarski, G. Domek, *Analysis of acoustic emission signals at austempering of steels using neural networks.* Metals and Materials International **23** (3), 426-433 (2017)
12. M. Dudziak, G. Domek, A. Kołodziej, K. Talaśka, *Contact Problems Between the Hub and the Shaft with a Four-Angular Shape of Cross-Section for Different Angular Positions.* Applied Mechanics and Materials **816**, 54-62 (2015)
13. J. Vavro Jr., J. Vavro, P. Kováčiková, R. Bezdedová, *Kinematic and dynamic analysis of planar mechanisms by means of the SolidWorks software.* Procedia Engineering **177**, 476-481 (2017)
14. J. Vavro Jr., J. Vavro, P. Kováčiková, R. Bezdedová, *Kinematic and dynamic analysis of the manipulator for removal of rough tyres.* Procedia Engineering **136**, 120-124 (2016)

15. R. Halama, R. Fajkos, P. Matusek, et al., *Contact defects initiation in railroad wheels - Experience, experiments and modelling*. 8th Conference on Contact Mechanics and Wear of Rail/Wheel Systems **271** (1-2), 174-185 (2011)
16. M. Handrik, M. Vaško, P. Kopas, V. Mózer, *The Linear and Nonlinear Stability Loss of Structures due to Thermal Load*. Procedia Engineering **136**, 359–364 (2016)
17. P. Kopas, L. Jakubovičová, M. Vaško, M. Handrik, *Fatigue Resistance of Reinforcing Steel Bars*. Procedia Engineering **136**, 193–197 (2016)
18. M. Kubiak, T. Domanski, V. Dekys, et al., *Measurement of strain during tension test of welded joint using multi-camera 3D correlation system*. Procedia Engineering **177**, 107-113 (2017)

Probability analysis of dynamical effects of axial piston hydraulic motor

Alzbeta Sapietova[1,*], *Vladimír* Dekys[1], *Milan* Sapieta[1], *Peter* Sulka[1], *Lukas* Gajdos[2], *Izabela* Rojek[3]

[1]Faculty of Mechanical Engineering, Department of Applied Mechanics, University of Zilina, Univerzitná 1, 01026, Slovakia
[2]Schaeffler Slovakia, Dr. G. Schaefflera 1, 02401 Kysucke Nove Mesto, Slovakia
[3]Kazimierz Wielki University in Bydgoszcz, Institute of Mechanics and Applied Computer Science Bydgoszcz, Kujawsko-Pomorskie, Poland

Abstract. The paper presents an analysis of impact force on stopper screw in axial piston hydraulic motor. The solution contains probabilistic description of input variables. If the output parameters of probabilistic solution are compared with arbitrary values and values acquired by analytical solution, the probability of proper operation of the device can be evaluated.

Keywords: probability, dynamic, impact force, MATLAB, hydrostatic motor

1 Introduction

We commonly encounter with procedures and approaches in which is physical physic described by a mathematical model. This description requires geomtric parameters, load values and material´s variables. The solution is the most often deterministic. In case if we want to consider the variability of the input parameters and realize philosophy of the safety coefficient in the deterministic solution, then the probabilistic approach can be used.

In case of minor variables are the analytical formulas sometimes useful because simulation methods as method Monte Carlo (MMC) are recommended for a larger amount of variables (for example, seven and more). If the correct model is intended and the appropriate solution is applied, it is also necessary to formulate a request for input data in relation to the meaning of object. If are input data in the form of probability density for particular independent variables, the issue is usually simple to sort out. Even though, if these inputs have to been estimated, problems come into being in relation to the production stage, for example: construction tolerances and the distribution of real values within these limits, the stability of the production process and the proceeding in which the production control is carried out. The same problem was solved in [1]. Indefinite input parameters can also be processed, for example, by interval arithmetic or access fuzzy, [2-4].

Equally important is the issue of load, or more precisely its determination, transfer of such data from generic previous devices, the correction of this loading for specific customer

* Corresponding author: alzbeta.sapietova@fstroj.uniza.sk

Reviewers: *Tomasz* Domański, *Juraj* Gerlici

conditions and similarly. The total allowable time for the solution and the costs of analyse are significant when deciding whether or not to use this method.

2 Axial piston hydraulic motor

The magnitude of the dynamic force acting on the stopper bolt was investigated on an axial piston hydraulic cylinder with an inclined block in Figure 1. The pressure is transmitted through the Servo Piston and the Spindle to the Valve Segment. The valve segment, when rotated, comes int contact with the Set Bolt, between which a force is formed. These dynamic effects are further absorbed by the load-bearing structure of the hydromotor. Because in the restraints of servo-binder and in the restraint between the servo-piston and the pin are relatively low friction forces so when considering the maximum pressure achieves a large angular acceleration in the valve segment. This is what they get in hydromotor has significant impact effects that may cause in the contact area creation of small plastic deformations and also in the consequence of cumulation of damage formation lead to the appearing and spreading of crevasses and cracks, [5].

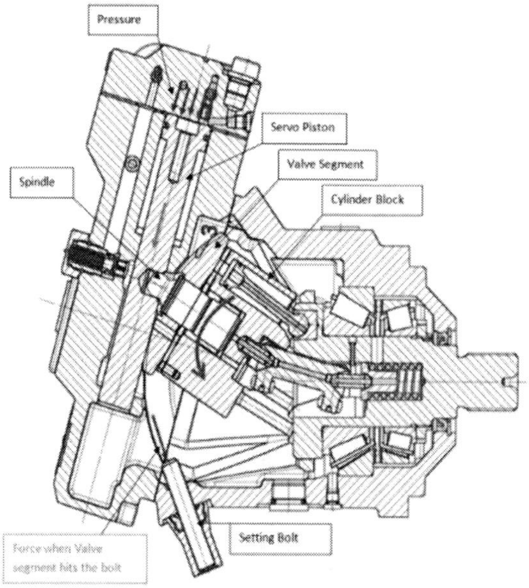

Fig. 1. Hydraulic motor scheme with axial piston with sloping block and labelling effects of load

3 Calculation of the dynamic force on the stopper crew

The solution was performed by means of reducing the amount of material and performance. Figure 2 displayed a kinematic diagram of an axial piston hydraulic engine, where: 1 – frame, 2 – servo-piston, 3 – pin, 4 – valve segments, A – impact, v2 – velocity of actuator, v_3 – velocity between pin and valve, ω_3 – angular velocity of the pin, ω_4 – angular velocity of the valce segment.

For the resulting force is possible to obtain following equation:

$$F_{RED} = \frac{F_P r_B - I_{RED}\alpha}{r_{RED}} \quad , \tag{1}$$

F_{RED} is the calculated reduced force, F_P is the pressure force acting on the servo-piston, r_B is the arm on which the force F_P acts to the rotating constraint with frame, I_{RED} is the reduced momentum of inertia, α is the angular acceleration, r_{RED} is the position of the reduced member´s center of gravity to the rotating constraint with frame.

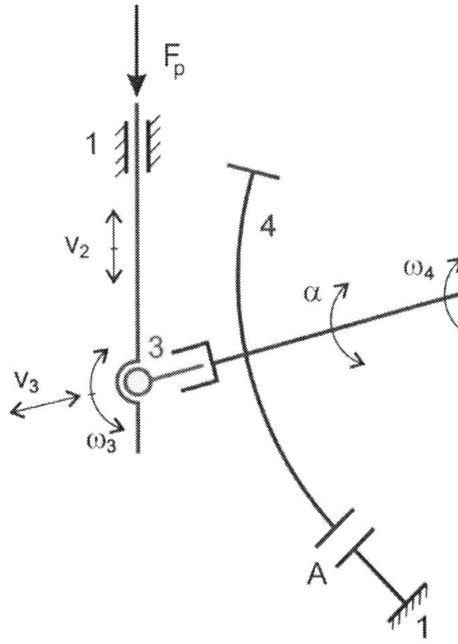

Fig. 2. Kinematic scheme of the axial piston hydraulic motor

The input variables on the right-hand side of the relationship (1) were made in this simplified model by other variables (geometric parameters, values determined on the basis of kinematic and kinetic conditions and material moments of inertia). A total of 9 independent variables, In_{k_ave}, k = 1, ... 9, were used to calculate the force F_{RED}. One of the input variables was also the angular acceleration of the member 4 of the mechanism of Figure 2. This acceleration was determined by simulation in the MSC.ADAMS program and the average value of this acceleration was estimated based on the average values of the other input variables, [6-8].

4 Probability analysis of the dynamic force acting on the stop screw

The basic calculation of the average values of the reduced force F_{RED_ave} was performed for the average values of all input variables In_{k_ave}, k = 1, ... 9. Subsequently, the task was formulated - to determine the probability P that the random variable F corresponding to the force F_{RED} will be more than 10% higher than the value F_{RED_ave}:

$$P = \text{Prob}\{F > 1.1\ F_{RED\_ave}\}. \tag{2}$$

Subsequently, for each input random variable In_k, k = 1, ... 9, the estimated interval, variable that is most likely to cover its values. To determine these intervals, an estimate of its average value of In_{k_ave} and the percentile deviation value Δ_k [%] was used so that:

$$Prob\{In_k \in (In_{k\_ave}+\delta_k,\ In_{k\_ave}+\delta_k)\} = most\ likely,\ \delta_k = 0{,}01.\ \Delta_k.In_{k\_ave}.\ k = 1, \ldots 9. \tag{3}$$

From a designer's point of view the estimate of the mean value of the average variable In_{k_ave} and its possible percentage deviation Δ_k is comprehensible. In the next step, if we want to generate values for input variables from a certain type of probability distribution for the Monte Carlo method, we need to know, for example, in the normal distribution, the mean value of the μ_k and the standard deviation σ_k for the individual input random variables. This problem can be solved by using estimations μ_{k_est} and σ_{k_est} :

$$\mu_{k\_est} = In_{k\_ave}, \quad \sigma_{k\_est} = \delta_k/w, \; k = 1, \dots 9, \; w = 1, \dots 3. \tag{4}$$

There is w a multiple of standard deviation and for example if $w = 3$, then:

$$Prob\{In_k \in (In_{k\_ave} - \delta_k, In_{k\_ave} + \delta_k)\} \approx 0.997. \tag{5}$$

The estimations of probability P by relation (1) was made for 3 cases:

The 1st case, the input random variables were simulated from normal distributions, $N(\mu_{k\_est}, \sigma^2_{k\_est})$. This case corresponds to the state when the In_k values outside the interval are also processed in the simulation, although they should be excluded, i.e.:

$$Prob\{In_k \notin (In_{k\_ave} - \delta_k, In_{k\_ave} + \delta_k)\} \neq 0. \tag{6}$$

This case corresponds to the non-control of values In_k outside the interval $(In_{k\_ave}+\delta_k, In_{k\_ave}+\delta_k)$. The 2nd case, the input random variables were simulated from truncated normal distributions, $N(\mu_{k\_est}, \sigma^2_{k\_est})$ at interval $(In_{k\_ave}+\delta_k, In_{k\_ave}+\delta_k)$. This case corresponds to the state when the In_k values outside the interval are excluded, i.e.:

$$Prob\{In_k \notin (In_{k\_ave} - \delta_k, In_{k\_ave} + \delta_k)\} = 0. \tag{7}$$

The exclusion process in 2nd case can be implemented by the production quality control system of the analysed object.

The 3rd case, the input random variables In_k were simulated from uniform distribution at interval $(In_{k\_ave} - \delta_k, In_{k\_ave} + \delta_k)$. This alternative corresponds to relationship (7).

The simulation results are in Table 1.

Table 1. The probability of exceeding an average value of reduced force F_{d_ave} by at least 10 %, point estimate P and interval estimate (P_{min}, P_{max})

			1%				2%		
Cause	w	$\hat{P}_1$	$\hat{p}_{1,min}$	$\hat{p}_{1,max}$	r_1	$\hat{P}_2$	$\hat{p}_{2,min}$	$\hat{p}_{2,max}$	r_2
1. Normal	1	0,1619	0,1608	0,1631	8,80	0,1689	0,1678	0,1700	4,56
	2	0,0241	0,0236	0,0246	1,31	0,0281	0,0276	0,0286	0,76
	3	0,0016	0,0014	0,0017	0,09	0,0021	0,0020	0,0023	0,06
2. Normal truncated	1	0,0129	0,0125	0,0132	0,70	0,0269	0,0264	0,0274	0,73
	2	0,0038	0,0036	0,0040	0,21	0,0089	0,0086	0,0091	0,24
	3	0,0004	0,0003	0,0005	0,02	0,0011	0,0010	0,0012	0,03
3. Uniform		0,0184	0,0183	0,0185	1	0,0370	0,0368	0,0372	1

An interpretation of Table 1. A total of 18 simulations were performed, combining the following factors:

- 3 different types of distribution functions of input random variables (normally, normally truncated, uniform) with only one same type of input random variables used in one simulation,
- 3 different values of parameter w (4),
- 2 different input data sets that were characterized by the parameter Δ_k as follows: in both groups, the value for the acceleration parameter was 10 %, and the other input parameters were assigned a value of 1 % (in the first group) and 2% (in the 2nd group) 3 * 3 * 2 = 18.

For each simulation, probability estimate $\hat{P}_i$, interval probability estimate $(\hat{p}_{i,min}, \hat{p}_{i,max})$ and the ratio between probability for normal, normally truncated and uniform distribution to probability estimate for a uniform distribution, for i = 1.2:

$$\hat{P}_i = \frac{n_i}{N}, \tag{8}$$

N is the total number of simulations, N=10^6, n_i is the number of simulations when the $F > 1.1\ F_{RED\_ave}$ condition is satisfied. The interval is defined as:

$$(\hat{p}_{i,\min}, \hat{p}_{i,\max}) = \left(\hat{P}_i - 2\sqrt{\frac{\hat{P}_i(1-\hat{P}_i)}{N}}, \hat{P}_i + 2\sqrt{\frac{\hat{P}_i(1-\hat{P}_i)}{N}} \right) \tag{9}$$

The value 2 corresponds to approximately 95% confidence interval in the relationship (9). In Figure 3, 4 and 5 are the histograms of the simulated F values for 2% of the input value, Figure 3 for normal distribution (w = 3, the cause1), Figure 4 for normal truncated distribution (w = 3, the cause1) and Figure 5 for uniform distribution (the cause1). The histograms for normal and normal truncated (Cause2) distributions are different, for example, the generated values of the force in the normal distribution, case 1, are in the range approximately (1300, 1900) N.

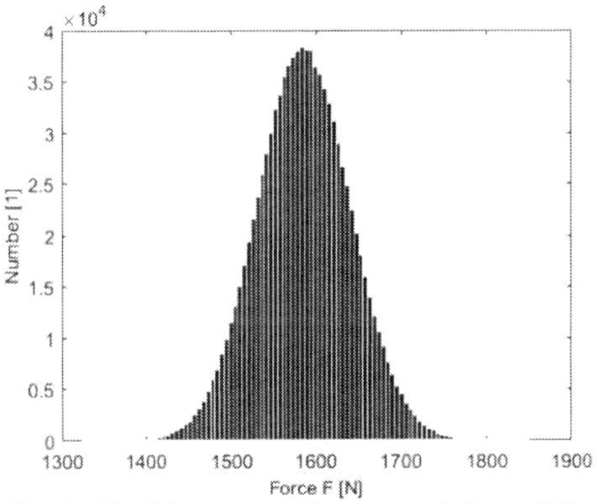

Fig. 3. The histogram of generated force F for normal distribution, the cause 1.

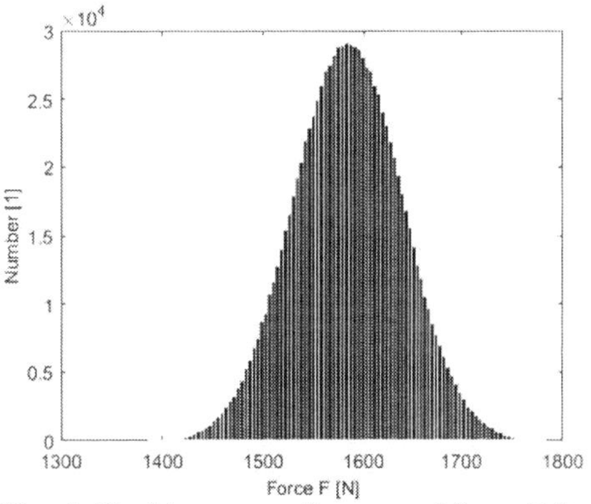

Fig. 4. The histogram of generated force F for normal truncated distribution, the cause 2.

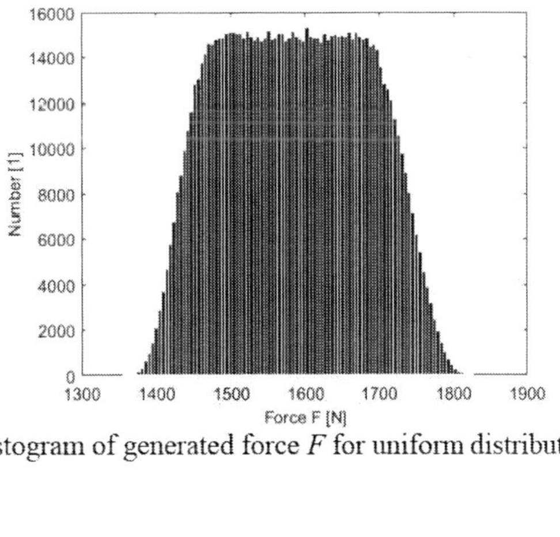

Fig. 5. The histogram of generated force F for uniform distribution, the cause 3.

Conclusions

The contribution refers to determining the probability of exceeding the limit value of the selected parameter. For these estimates, the input distribution used plays a significant role. Knowing these distributions may, however, require analysis and regular control of the stability of the production process, which may not be a simple and inexpensive matter [9-15]. If we use the initial estimate to determine the probability of a even distribution of input values, then it is a pessimistic estimate. However, this estimate respects the condition that the input parameter is or is not an element of, for example, tolerance inertial. For example, if we use truncated normal distribution as an estimate, then parameter w should be specified. However, this may imply requirements to increase the cost of a solution if a priori was considered only to discard elements that are characterized by parameters that do not fall within the allowed input value range.

This procedure can also be used to estimate the risk of a fault or a limit state when using safety factors in the design practice.

This work was supported by VEGA 1/0795/16, KEGA 017ŽU-4/2017 and by the Slovak Research and Development Agency under the contract No. APVV-0736-12.

References

1. A. Sapietová, V. Dekýš, M. Sapieta, P. Pecháč, *Application of computational and design approaches to improve carrier stability*. Procedia Eng. **96**, 410-418 (2014)
2. A. Sapietová, R. Petrech, M. Petrovič, *Analysis of the dynamical effects on housing of the axial piston hydromotor*. Novel Trends in Production Devices and Systems, Applied Mechanics and Materials **474**, 357-362 (2014)
3. M. Kekez, L. Radziszewski, *Genetic-fuzzy model of diesel engine working cycle*. B POL ACAD SCI-TECH **58**, 665-671 (2010)
4. L. Radziszewski, M. Kekez, *Application of a genetic-fuzzy system to diesel engine pressure modeling*. Int J Adv Manuf Technol. **46,** 1-9 (2010)
5. M. Kekez, L. Radziszewski, *A Genetic-Fuzzy System for Modelling of Selected Processes in Diesel Engine Fuelled by Biofuels*. Biofuel Production - Recent Developments and Prospects, 561-576 (2011)
6. Sapietova, J. Bukovan, M. Sapieta, L. Jakubovicova, *Analysis and implementation of input load effects on an air compressor piston in MSC.ADAMS*. Book Series: Procedia Engineering **177**, 554-561 (2017)
7. A. Sapietová, V. Dekýš, *Dynamic analysis of rotating machines in MSC.ADAMS*. Procedia Engineering, **136**, 143-149 (2016)
8. A. Sapietova, L. Gajdos, V. Dekys, M. Sapieta, *Analysis of the Influence of Input Function Contact Parameters of the Impact Force Process in the MSC. ADAMS*. Advances in intelligent systems and computing **393**, 243-253 (2017)
9. M. Sága, M. Vaško, P. Pecháč, *Chosen numerical algorithms for interval finite element analysis*. Procedia Engineering **96**, 400-409 (2014)
10. M. Handrik, M. Vaško, P. Kopas, V. Mózer, *The Linear and Nonlinear Stability Loss of Structures due to Thermal Load*. Procedia Engineering **136**, 359–364 (2016)
11. M. Sága, P. Kopas, M. Uhríčik, *Modeling and experimental analysis of the aluminium alloy fatigue damage in the case of bending - torsion loading*. Procedia Engineering **48**, 599-606 (2012)

12. J. Vavro Jr., J. Vavro, P. Kováčiková, R. Bezdedová, *Kinematic and dynamic analysis of planar mechanisms by means of the SolidWorks software*. Procedia Engineering **177,** 476-481 (2017)
13. J. Vavro, J. Vavro Jr., P. Kováčiková, P. Kopas, M. Handrik, *Simulation and analysis of defect distribution in passenger car tire under dynamic loading*, Applied Mechanics and Materials **611**, 544 – 547 (2014)
14. J. Vavro Jr., J. Vavro, P. Kováčiková, P. Kováčiková, R. Bezdedová, J. Híreš, *Kinematic and Dynamic Analysis and Distribution of Stress in Items of Planar Mechanisms by Means of the MSC ADAMS Software*. Manufacturing Technology **17**, 267-270 (2017)
15. P. Pastorek, P. Novak, P. Kopas, *Finite element analysis of bond behavior in a steel reinforced concrete structure strengthened carbon fibre reinforced polymer (CFRP) strips*. Metalurgija **56**, 405-408 (2017)

Wear research of railway wheelset profile by using computer simulation

Lukáš Smetanka[1,*], *Pavol* Šťastniak[1], *Jozef* Harušinec[1]

[1]University of Žilina, Faculty of Mechanical Engineering, Department of Transport and Handling Machines, Univerzitná 8215/1, 010 26 Žilina, Slovak Republic

Abstract. The wear of rails and wheels is important problem in rail traffic. The change of the shape of the wheel profile has not only a great influence on the dynamic properties of the vehicle (like stability, safety by passing curved tracks, etc.), but also affects the ride comfort of passengers and environmental insults, in extreme cases it can cause rail derailment. One of the ways to predict these undesired conditions are computer aided simulation analyzes. In this article are presented assessments of wheel profile wear by Archard wear law in Simpack simulation software, when is railway vehicle driving at different velocities.

Keywords: wear research, rail vehicle, computer tools

1 Introduction

The dynamic behavior of rail vehicles on the track is significantly affected by the interaction between the wheel and the rail. In the depth of the examined issue is an understanding of the contact between the wheels and the rails. There are problems in this more in-depth study which stand alone are broad scientific areas. One of the most important areas with significant influence on geometry change and thus the change in contact conditions is undoubtedly the wear and related geometry changes of one part of the contact pair or both, such as the corrugation of the track and its irregularity, changes in friction and slip ratio or changes in vertical and longitudinal forces of the contact pair, etc. The impact of wear certainly affects also the acoustic and vibrational expression of the car on the rail, which, by increasing the negative side, devastates the environmental environment around the tracks [1, 2].

Wear belongs to the degradation processes in which the material is discharged from the surface of one or both of the contact surfaces of rigid bodies which are in relative movement. Relative movement of the contact surfaces can directly lead to destroying of the component, or to damage of surface treatment or to surface damage, resulting in subsequent destruction of the component. According to STN 01 5050 the wear proves as removing or displacing individual particles from the worn surface by mechanical effects, or accompanied by other influences (chemical, electrical etc.). The characteristic of the relative movement between the two entities in contact defines the wear process, in general

* Corresponding author: pavol.stastniak@fstroj.uniza.sk

Reviewers: *Maciej* Bodnicki, *Janusz* Mielniczuk

it can be divided into the following areas: adhesion, abrasive, corrosive, erosive, fatigue, cavitation and vibrational wear. During the operation of the railway transport, both wheel and rail wear occur, either due to normal operation or failure. With regard to the type of wear, the wear of rolling stock is based on the character of friction and also on the basis of cyclic fatigue. Over time the process of fatigue become separate scientific area named RCF (Rolling Contact Fatigue) [3-6].

2 Archard wear law in SIMPACK software

According to several literary sources, there are three main trends in the development of the wear, patterns from the middle to the end of the twentieth century and they are: based on empirical approach, based on a contact mechanics approach and based on a reflex mechanics approach. An irreplaceable and important factor in wear is friction. The hypothesis suggests:

$$F_F = \mu \cdot F_N \tag{1}$$

where:
- F_F is frictional force,
- μ is frictional coefficient,
- F_N is applied normal force.

This frictional equation is based on two basic assumptions, first is approach that the stress distribution is independent to normal load and second that the applied load is proportional to real contact area.

We knows more wear laws which were then development and modified, but all of them are based on basic theories from authors like Reye (1860), Tabor (1939), Holm (1946) a Archard (1953). [7,8,9]

The supplement for wear determination is in SIMPACK software embedded to the module SIMPACK RAIL, it is therefore the determination of wear in the contact wheel-rail pair, where are the contact stresses solved by using the Kalker method named FASTSIM. [9,10] The developer motivated the application of supplement mainly to the issue of safety and maintenance, as is known, wheels and rails are subjected to wear and rolling contact fatigue (RCF) throughout their lifecycle, due the forces acting during frequent rail traffic. Considering that there are many different waste laws and RCF, the module has an open structure that makes it possible to implement the necessary changes to achieve the preferred type of wear. The SIMPACK software offers two different laws of wear and tear for simulation: Archard wear law and Krause/Poll wear law.

Archard wear law is for the SIMPACK software based on the following equation:

$$V = \frac{N \cdot \Delta s}{H} \cdot \begin{array}{lll} k_1 \text{..if...} & p > f \cdot H & \text{seizure} \\ k_2 \text{..if...} & p \leq f \cdot H \wedge v_s \leq v_1 & \text{mild}-1 \\ k_3 \text{..if...} & p \leq f \cdot H \wedge v_1 < v_s \leq v_2 & \text{severe} \\ k_4 \text{..if...} & p \leq f \cdot H \wedge v_2 < v_s & \text{mild}-2 \end{array} \tag{2}$$

where:
- V – wear volume [m^3],
- N – normal contact force [N],
- Δs – sliding distance [m],
- H – hardness of the softer material in contact pair [MPa],
- $k_1 - k_4$ .. wear coefficients, (seizure, mild 1, severe and mild 2) [-],

- p – contact pressure [MPa],
- f – hardness fraction H over which destruction occurs (seizure),
- v_s – sliding (relative) velocity [m/s],
- $v_{1,2}$ – boundary velocities, in which the wear coefficients are changing from mild 1 to severe and back to mild 2 [m/s].

The sliding distance is calculated from actual output step. The normal force, pressure and sliding (relative) velocity are selected from wheel-rail contact elements. The typical values for wear coefficients are used in literature like: $k_1 = 300 - 400 \cdot 10^{-4}$, $k_2 \approx k_4 = 1 - 10 \cdot 10^{-4}$ and $k_3 = 30 - 40 \cdot 10^{-4}$ (Fig. 1).

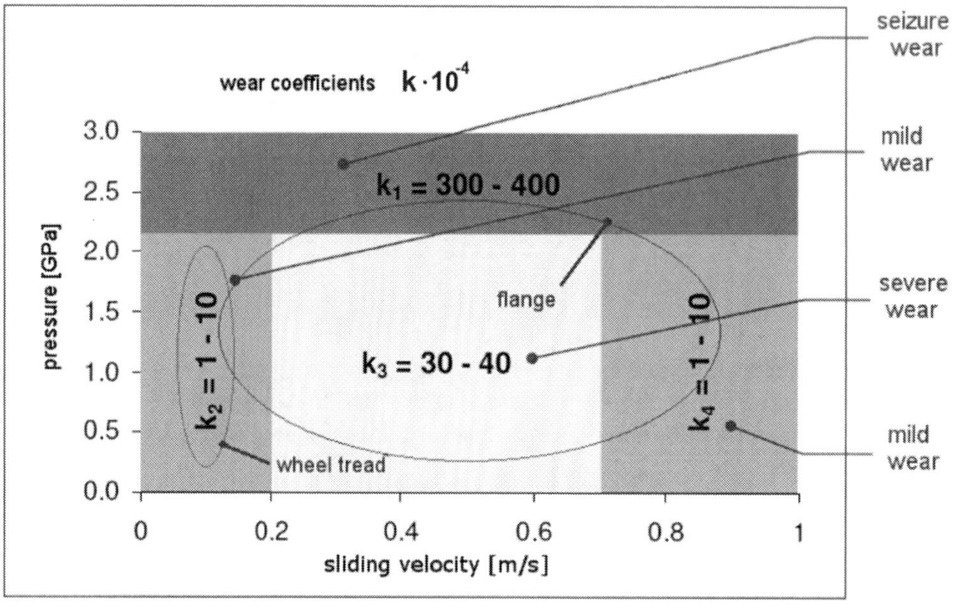

Fig. 1. Wear coefficients

3 Preparation of simulation

From the point of view of simulation of wear it was necessary to create the most accurate model of the vehicle so that the simulation showed the best possible match with the actual vehicle on the real track. The SIMPACK software has been selected as a suitable solution for this simulation, which enables us to achieve the desired behavior of the vehicle with its characteristic features. The software allows for a quick and comfortable assembly of the MBS (Multibody Simulation) model of a rail vehicle that meets the wheel-rail contact specifications.

3.1 Computational vehicle model

The computational model is characterized by vehicle parameters, geometric and weight parameters, elastic parameters of springs and dampers, and the like. Additionally, they are the parameters of the contact between wheel and rail and the track parameters. [10-14] This article presents results for one vehicle model (Fig. 2) *Model A*.

The weight parameters, elastic parameters of springs and dampers and geometrical parameters of the vehicle are described in following tables (Tab. 1, Tab. 2, Tab. 3).

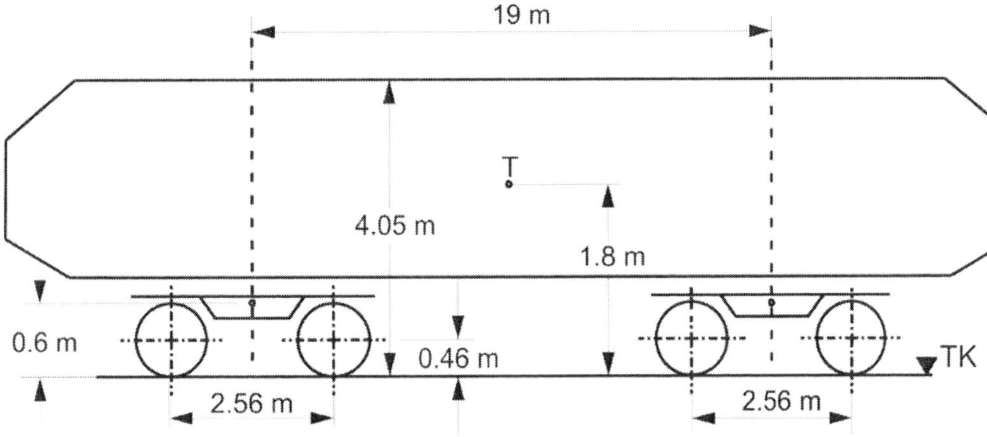

Fig. 2. Schematic sketch of the vehicle (*Model A)*

Table 1. The weight parameters (*Model A*)

	m [kg]	**I_{xx} [kg·m^2]**	**I_{yy} [kg·m^2]**	**I_{zz} [kg·m^2]**
Wheelset	1813	1120	112	1120
Bogie Frame	2615	1722	1476	3067
Vehicle Casing	32000	56 800	1 970 000	1 970 000

Table 2. Elastic parameters of springs and dampers (*Model A*)

Springing	**c_x [N·m^{-1}]**	**c_y [N·m^{-1}]**	**c_z [N·m^{-1}]**
Primary ***Springing***	31 391 000	3 884 000	1 220 000
Secondary ***Springing***	160 000	160 000	430 000
Damping	**d_x [N·s·m^{-1}]**	**d_y [N·s·m^{-1}]**	**d_z [N·s·m^{-1}]**
Primary ***Springing***	15 000	15 000	4000
Secondary ***Springing***	32 000	32 000	20 000

Table 3. The geometrical parameters of vehicle (*Model A)*

Parameter	**Description**	**Value**	**Dimension**
The transverse distance of activities of primary springing and damping		2	[m]
The transverse distance of activities of secondary springing and damping		2	[m]
The distance of rotary pin of bogies		19	[m]
The wheelbase of bogies		2.56	[m]
Nominal wheel diameter		0.92	[m]
Height of center of gravity for wheelset above TK		0.46	[m]
Height of center of gravity for bogie frame above TK		0.6	[m]
Height of center of gravity for vehicle casing above TK		1.8	[m]
Height of the primary springing above TK		0.46	[m]
Height of the secondary springing above TK		0.52	[m]
Wheel profile	S1002		
Rail gauge		1435	[mm]
Frictional coefficient	μ	0.4	[-]

3.1 Track parameters

The model track consists from the straight sections, clothoid transitions and superelevations segments in the curves of the track. Fig. 3 shows the horizontal track profile, which were by wear simulation used with aim to determine the wear of the vehicle on the track. The length of the model track is 400 km and from picture is clearly that the profile is very diverse. The track consists from 24 right and 24 left curves of track, the vehicle runs in straight track overall 366 km and 44 km in curves of track. The superelevations of the outer ribbon in the curves of track are for ech curve of another value. The model track it is not considered like ideally horizontally and vertically straight, in the simulation is considered lateral and vertical irregularities on track too. The irregularities cause the excitations of dynamics system by driving the vehicle on the track and are mostly stochastic.

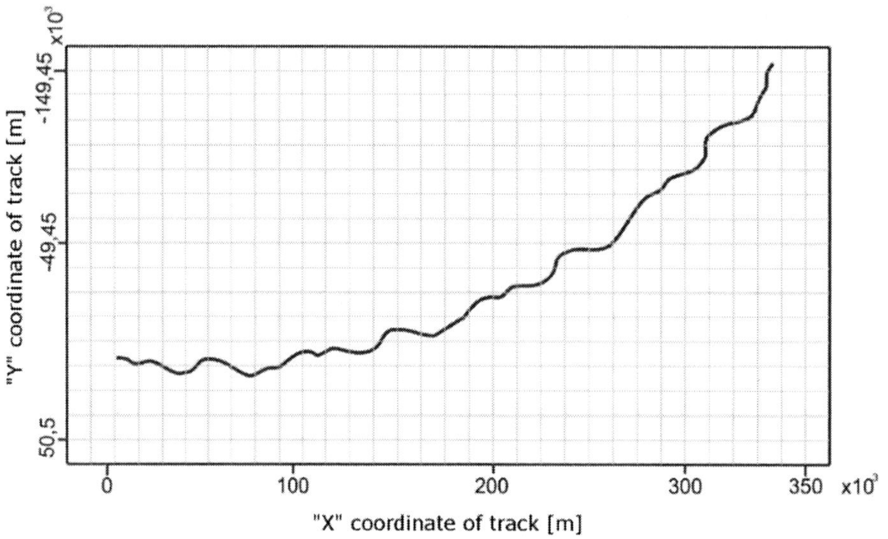

Fig. 3. Horizontal track profile

Fig. 4 shows the rail profiles which were used by wear simulations. All the analyzes are comply to condition, that the vehicle was traveling along track where the rail profile was defined by standard (UIC 60 profile) with cant of 1:40, or the track profile really measured on the track, the profile S 91700_16 with the cant of 1:20.

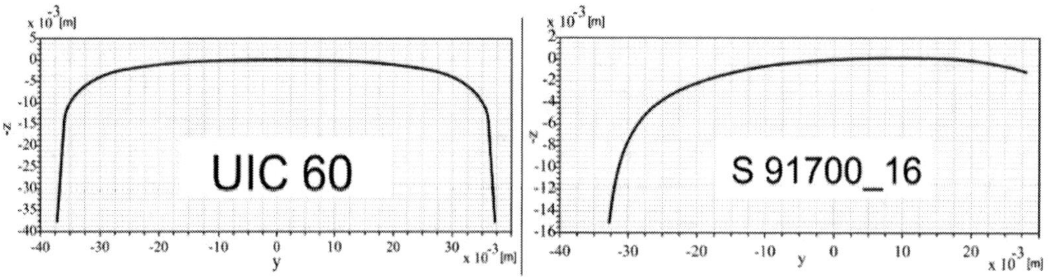

Fig. 4. The rail profiles

4 Simulations and evaluation of results

In the simulations of railway wheel wear we focused on the impact of vehicle velocity change. The simulations was performed in two variants, by driving the vehicle *Model A* on the track with rail profile UIC 60 or rail profile S 91700_16 and the vehicle velocity was different in three independent variations. The velocity was set at 15 m/s which is 54 km/h, 22.5 m/s which is 81 km/h and to 30 m/s which is 108 km/h. In the final view was performed 10 analysis (10 x 400 km = 4000 km long track) for each velocity and rail variant, together it was 60 analysis.

On following figures (Fig. 5 and Fig. 6) are displayed the curves of wear depth for two rail variants (UIC 60 and S 91700_16) and for three different velocities, 15 m/s (green line), 22.5 m/s (red line) and 30 m/s (blue line). The curves of wear depth are displayed for each wheelset of vehicle *Model A* but only for right wheels. The curves for left wheels had the similar character. For understanding the time aspect of the calculation, one simulation on 400 km long track by the 15 m/s vehicle velocity took 15 hours computer time (CPU Time).

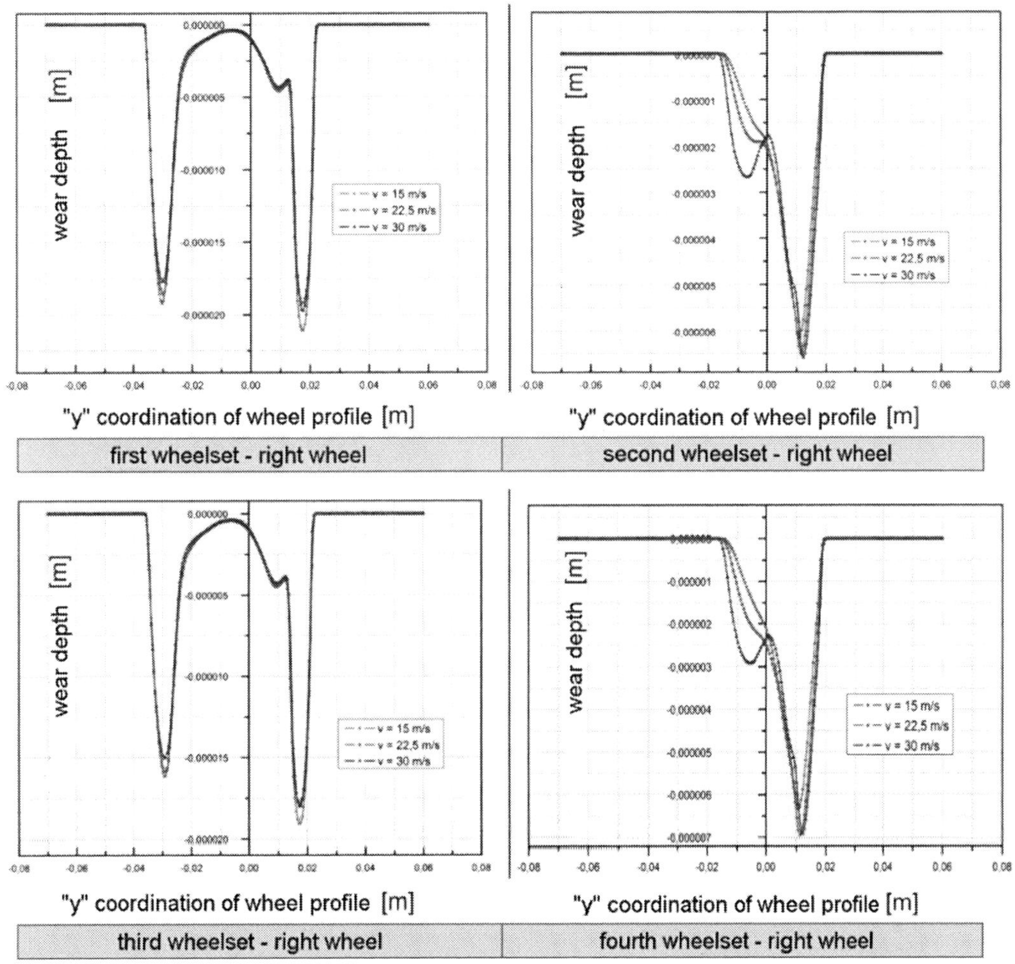

Fig. 5. The wear depth for *Model A* by driving on the *S 91700_16* rail profile

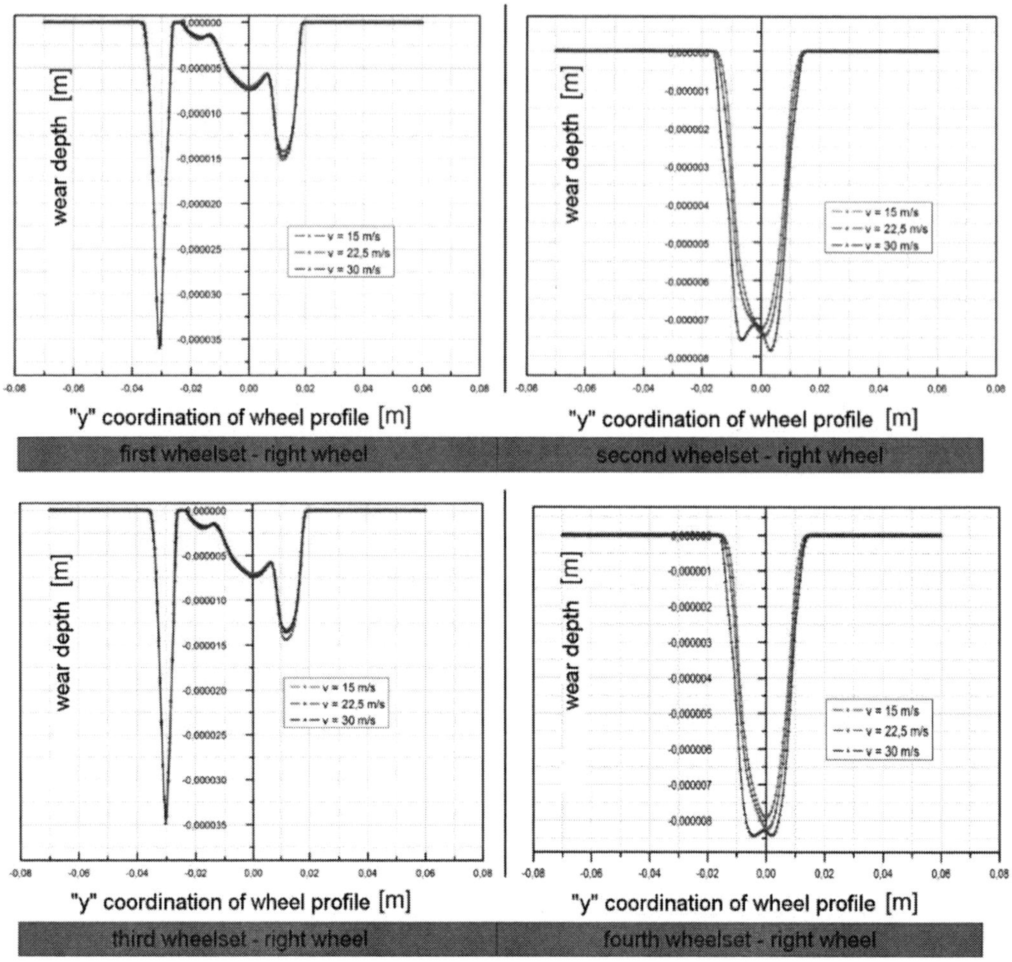

Fig. 6. The wear depth for *Model A* by driving on the *UIC 60* rail profile

The figures shown, it is possible to notice the similar character of curves on the first and third wheels, while on the second and fourth wheels are the differences in the curves more visible. In the first and third wheels, differences are visible especially in the areas of the curve peaks, while with the second and fourth wheels are curves with higher driving velocity wider. The presence of peaks in the curves of the wear depth is caused by the passage of the vehicle on the curves of track where, as is generally known, the contact surface is changing. The contact area is generally smaller and the contact stress is higher than at the given position than in the centered position of the wheel in the rail.

Since it is not entirely clear how the velocity change factor affects to the wear depth, we decided use the volume criterion (Fig. 7, Fig. 8).

 https://doi.org/10.1051/matecconf/201815703017

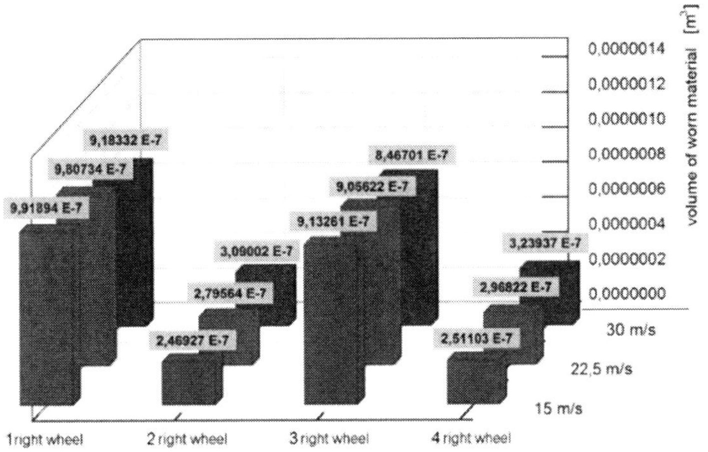

Fig. 7. The volumes of worn material for *Model* A by driving on the *S 91700_16* rail profile

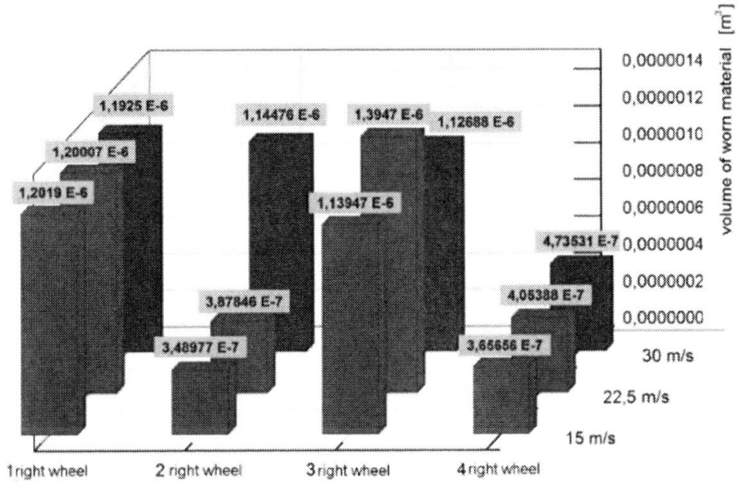

Fig. 8. The volumes of worn material for *Model* A by driving on the *UIC 60* rail profile

Conclusion

This article is aimed to determination of railway wheel wear by driving on the track. The wear simulations were focused to influence of the change of vehicle velocity on the change of wear depth for railway wheel, which were driven at two possible rails profile, UIC 60 or S 91700_16. The simulations were performed in six independent variants, where total traveled distance for one variant was 4000 km. The vehicle velocity was set to 15 m/s, 22.5 m/s and 30 m/s. On the displayed figures are compared wear depth and volumes of worn material for individual variants.

The work was supported by the Cultural and Educational Grant Agency of the Ministry of Education of the Slovak Republic in project No. KEGA 077ŽU-4/2017: Modernization of the Vehicles and engines study program. The work was also supported by the project No. APVV-0842-11: Equivalent railway operation load simulator on the roller rig and VEGA Nr. 1/0927/15: Research of the use of alternative fuels and hybrid drives on traction vehicles with aim to reduce fuel consumption and air pollutants production.
Research-Educational Center of Rail Vehicles (VVCKV)

References

1. I. I. Argatov, Y. A. Fadin, *Mathematical Modeling of the Periodic Wear Process in Elastic Contact between Two Bodies*. Journal of Friction and Wear **29** (2), 81-85 (2008)
2. C. I. Barbinta, et al., *Wheel-rail contact modelling and analysis, considering profiles types and lateral displacement.* Transport Research Arena: 5th Conference: Transport Solutions from Research to Deployment, Paris: 14.04.2014 - 17.04.2014, France: Paris, Accession Number: 01540797 (2014)
3. R. Enblom, M. Berg, *Simulation of railway wheel profile development due to wear – influence of disc braking and contact enviroment*. Wear: an international journal on the science and technology of friction, lubrication and wear **258**, 1055-1063 (2004)
4. J. Gerlici, T. Lack, *Contact of rail wheelset and track*: (Žilina, University of Žilina, EDIS, 200 p., ISBN 80-710000-317-5, 2004)
5. L. Jakubovičová, M. Sága, *Computational analysis of contact stress distribution in the case of mutual stewing of roller bearing rings*. Novel Trends in Production Devices and Systems, Applied Mechanics and Materials **474**, 363-368 (2014)
6. Z. Pelagić, M. Nágeľ, M. Žmindák, D. Riecky, *Wear simulation modeling by using the finite element method*. Manufacturing technology: Journal for science, research and production **15** (2), 191-195 (2015)
7. T. Lack, J. Gerlici, *Contact area and normal stress determination on railway wheel / rail contact.* Communications - scientific letters of the University of Žilina **7** (2), 38-45 (2005)
8. M. Svoboda, J. Soukup, *Dynamic Measurement of Four-Axle Railway Wagon*. Manufacturing technology: Journal for science, research and production **13** (4), 552-558 (2013)
9. J. Dižo, S. Steišunas, M. Blatnický, *Simulation analysis of the rail vehicle running with wheel flat.* Manufacturing technology: Journal for science, research and production **16** (5), 889-896 (2016)
10. V. Hauser, et al., *Impact of three axle boxes bogie to the tram behavior when passing curved track.* Procedia Engineering **192**, 295-300 (2017)
11. R, Kohár, S. Hrček, *Dynamics analysis of rolling bearings with an elastic cage*. Modern methods of construction design, Cham: Springer, ISBN 978-3-319-05202-1, 249-254 (2014)
12. T. Lack, J. Gerlici, *The FASTSIM method modification in speed up the calculation of tangential contact stresses between wheel and rail*. Manufacturing technology: Journal for science, research and production **13** (4), 486-492 (2013)
13. T. G. Paerce, N. D. Sherrat, *Prediction of wheel profile wear*. Wear **144**, 343-351 (1991)
14. M. Žmindák, Z. Pelagić, J. Soukup, *Analysis of fiber orientation influence to dynamic properties of composite structures*. Manufacturing technology: Journal for science, research and production **15** (3), 490-494 (2015)

Analysis and synthesis parameters influencing to the effects of impact

Peter Šulka[1,*], *Alžbeta* Sapietová[1], *Vladimír* Dekýš[1], *Milan* Sapieta[1]

[1]Department of Applied Mechanics, Faculty of mechanical engineering, University of Žilina, Univerzitná 8215/1, 010 26 Žilina, Slovak Republic

Abstract. In the solutions will be presented to the interaction of two bodies loaded with impact effects in the environment of software MSC. ADAMS. By performing a sensitivity analysis of the parameters influencing to the effects of impact and function parameters of impact will be obtained parameters for optimum setting MSC.ADAMS for acquirement feedback, as the most identical with reality.

Keywords: Analysis, impact, compression, restitution, MSC.ADAMS

1 Introduction

Analysis of a wide range of problems in technical practice uses mathematical representation of the future product in a virtual computerized form. Such modelling of the problem requires that the parameters input to the solution, and also the outputs, be consistent with reality.

Keeping this objective in mind, this paper deals with dynamic analysis of load by forces – in this case the so-called impact forces – that cause destruction of the device. If there is interaction between two bodies and the bodies do not move relative to each other, we talk about their contact. Interaction of bodies that move relative to each other is called impact. Impact is characterized by short-term significant forces occurring in the contact area. These impact forces act for a very short time – in the order of milliseconds. Despite the above, these forces cause significant changes in the velocity and thus in the momentum of the participating bodies. The change is by a final value. Therefore, each body must receive a final impulse equivalent to the change in its momentum. However, the impact forces must be very large in order to generate the final impulse in a very short time. This is then followed by the above-mentioned destruction of the material when the forces act. A better understanding of the interdependence between the impact force and the mechanism parameters allows engineers to find more efficient and innovative solutions. Therefore, we carried out analysis of the impact force in the tensioning mechanism using the IMPACT function in the MSC.ADAMS software.

* Corresponding author: peter.sulka@fstroj.uniza.sk

Reviewers: *Tomáš* Lack, *Milan* Naď

2 Theoretical background of contact forces in the MSC.ADAMS software environment

When the bodies in the virtual prototype models are in contact with each other, the CONTACT function allows making or modifying the contact in the ADAMS/View software environment. Bodies in the contact models can touch each other at points, touch points can travel along pre-described curves, or the curves of one body can be in contact with the curves upon the other body. We can also define the contact between non-deformable bodies at a standstill and in motion. We can model the contact of bodies with stiff (rigid) as well as flexible elements [1].

Solver algorithms detect the contacts of two geometric objects, determine the touch point position, calculate the position of the common normal at the touch point, and calculate the normal and slipping relative velocity of the point in contact.

In terms of dynamics, it is necessary to distinguish:

- discontinuous (pulse) contacts occurring in short periods of time – impacts, for which ADAMS/Solver creates an estimate of the contact force by modeling the course of local deformation. The damping force with the corresponding damping coefficient or restitution coefficient represents loss of energy during the collision. In the first phase, compression occurs during which kinetic energy is converted to potential and dissipative energy of the compressed material. In the second phase, the accumulated potential energy causes reverse motion accompanied by the transformation of potential energy into kinetic and again dissipative energy.
- continuous contacts represented by a model in the form of a non-linear spring-damper system in which stiffness takes into account elasticity of contacting surfaces and damping, in turn, takes into account energy dissipation, and the bodies are not separated after the collision. Calculation of contact forces is carried out separately for each point, and the resulting effect is their sum.

The force acting in the contact point is calculated either by the restitution method (POISSON model) or by the IMPACT function. Both contact force models result from the regulation of normal contact constraint penalty, where the constraint in this modelling technique is secured mathematically by applying forces according to the constraint gradient, and the magnitude of forces is a function of the constraint breach level [2, 3].

Theoretically, there is no mutual penetration of surfaces at the contact of non-deformable bodies, which can be expressed in the form of inequalities for unilateral constrain. Additional constrain conditions can be expressed also by introducing Lagrange coefficients, but the penalty regulation is advantageous due to its simplicity since no additional equations or variables are needed. The advantage is the ability to apply algorithms of active and inactive conditions for unidirectional constraint as well as for clear interpretation from the physical point of view. The disadvantage of the penalty regulation is that the user is responsible for setting a suitable penalty parameter in the form of material stiffness, a high value of which may cause difficulties in the work of solver integration algorithms. Mutual penetration of the body surfaces is monitored by the gap function g, whose positive value indicates penetration. A positive value of the normal contact force in turn means that it is trying to separate the bodies, and its value is non-zero while the surfaces are in contact [4].

The impact force model in MSC.ADAMS is implemented using the IMPACT function. An example of modelling using the IMPACT function is a ball falling towards the ground, plotted at the beginning of the investigated motion and at impact (Fig. 1). If the distance between the I and J markers reaches x_1, the impact force IMPACT turns on. This occurs when two bodies collide. As long as the distance between the I and J markers is greater than x_1; the force is zero.

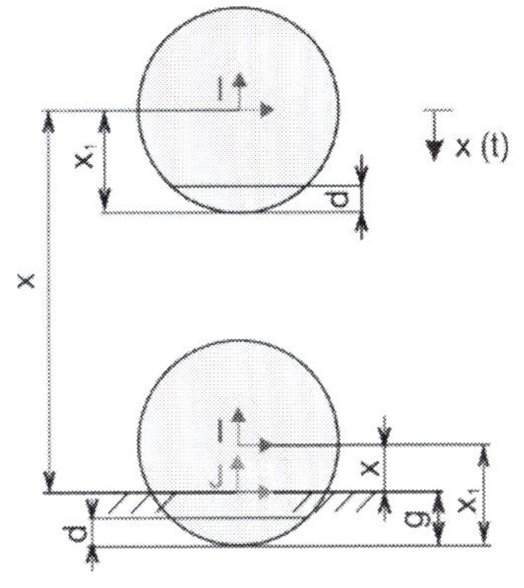

Fig. 1. Example illustrating the IMPACT function

The impact force has two components, a spring or stiffness component, and a damping or viscous component. The stiffness component is proportional to the stiffness coefficient k, that is load per unit length, i.e. it is a function of changes in the distance between the I a J markers. The stiffness component opposes the mutual penetration of the body surfaces. The damping coefficient c belonging to the damping component of the force is a function of the velocity of mutual penetration of the body surfaces. The damping opposes the direction of relative motion. The damping coefficient achieves a maximum value c_{max}, at a user-defined penetration d (Fig. 2b).

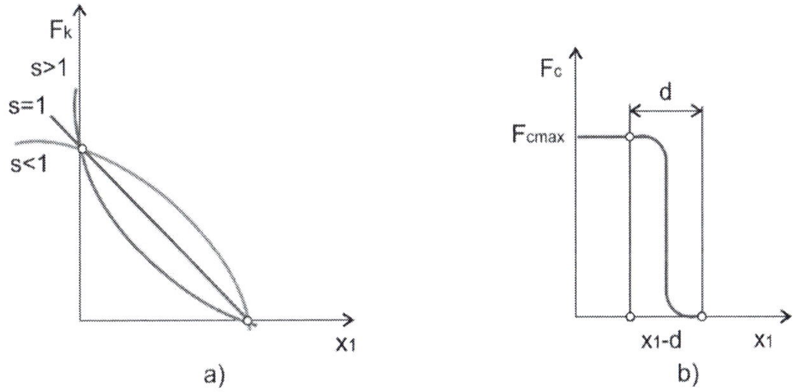

Fig. 2. a) Exponent s of the force in a non-linear spring, b) Depth d of the penetration to the maximum damping force

Figure 3 shows a local coordinate frame of the I marker that represents the first body and a local coordinate frame of the J marker that represents the second body. The impact force model consists of the force acting in the compression spring, where $F_k=kx$, and of the force acting in the damper, where $F_c = c\dot{x}$. Then the impact force magnitude is $Fimpact=F_k+F_c=kx\text{-}c\dot{x}$. Note that the IMPACT function solution in the MSC.ADAMS software environment results in a real number [5, 6].

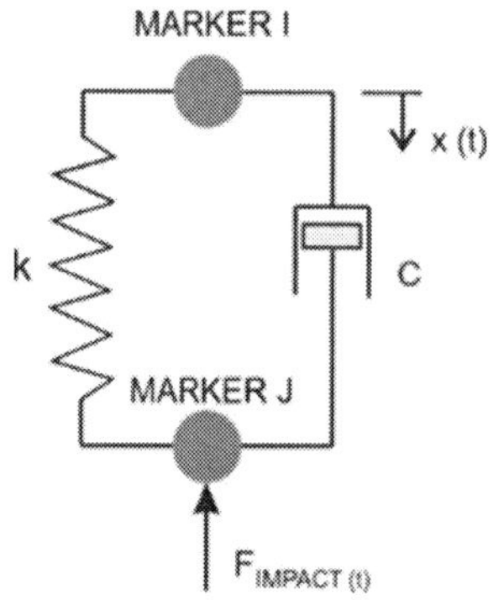

Fig. 3. Model of unilateral impact force by the IMPACT function

Given the solver problems with high stiffness values and the need to include critical material properties into the impact model, the solver algorithm works with an extended mathematical model that includes non-linear viscous damping dependent on the magnitude and velocity of mutual penetration of surfaces.
The equation defining IMPACT is:

$$IMPACT = \begin{cases} Max(0, k\,(x_1 - x)^s - STEP(x, x_1 - d, c_{max}, x_1, 0).\dot{x}) & : x < x_1 \\ 0 & : x \geq x_1 \end{cases}, \tag{1}$$

Based on the presented impact model, the following applies (Fig. 1):
$x > x_1$, no mutual surface penetration occurs and the force is zero (penetration $g = 0$)
$x < x_1$, penetration g occurs at the end closer to the *J* marker, and the force is > 0.
Also note that when $g < d$, the instantaneous damping coefficient is a cubic step function of the penetration g.

When $g > d$, the instantaneous damping coefficient is c_{max}.

The MSC.ADAMS (C ++) software solver never returns a negative force for impact. If the above expression is negative, the solver returns a force value of zero.

It must be said that in terms of theory there has been a failure so far to mathematically describe the phenomena during impact, although the impact models based on energy balance before and after the impact correlate the corresponding parameters with sufficient precision [7].

3 Setting the parameters of the pulse contact in MSC.ADAMS

In the following, we will focus on analysis and synthesis parameters for setting solutions intermittent (impulse) terminal i.e. IMPACT of MSC. ADAMS.

Fig. 4 displayed two bodies in software MSC.ADAMS, which are labeled *1* a *2*. These bodies come together and initial difference in velocity. Ordinarily they first touch at a contact point *A*. During a very short time of contact, the point A_1 on the surface of the body *1* identical to point A_2 on the surface of the body *2*. When at least one of the bodies, *1* and *2*, a surface which is smooth topologically in the contact (i.e. surface with a continuous curvature), there is a plane tangent to this surface at point *A*. Identical contact points A_1 and A_2 are situated in this contact plane. If both bodies are convex and the surfaces have continuous curvature near the contact point, then this tangent plane is tangential to both surfaces that touch at point *A*, i.e. the surfaces of the colliding bodies have a common tangent plane. Moreover, unit vector *n* specifies the direction of the normal to the tangent plane (Fig. 4). The contact force and changes in relative velocity at the contact point *A* will be resolved into components normal and tangential to the common tangent plane.

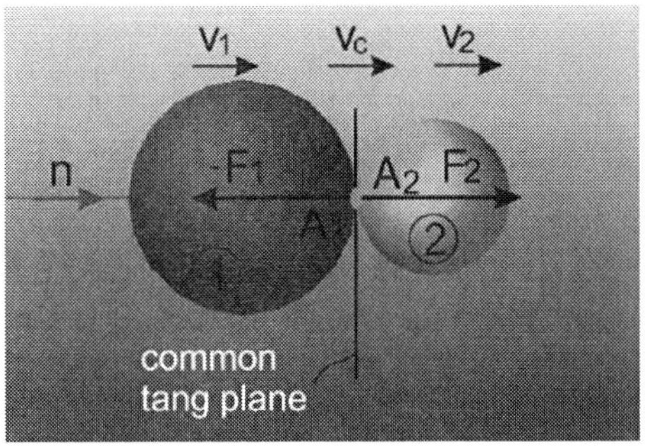

Fig. 4. Collinear impact of the two bodies

Weight of the opposing bodies *1* and *2* is $m_1=12kg$ and $m_2=10kg$, and the time-dependent velocity at the start of movement is $v_1=10m/s$ and $v_2=6m/s$ pointing in the direction parallel to *n*. In a direct collision, these bodies are not rotating when they collide, so that velocity is uniform in each body (i.e. in each point of bodies). During contact, there are equal but opposite compressive reaction forces which develop at contact points A_1 and A_2. These forces oppose interference or overlap of the contact surfaces. In the case of direct impact between collinear bodies, the relative velocity between the contact points A_1 and A_2 remains parallel to the common normal direction throughout the contact period. A reaction force develops at the contact point because of compression of the local contact region.

This force opposes relative motion during contact. In a direct collision, the reaction force acts in the normal direction, i.e. parallel to the velocities, as illustrated in (Fig. 4). If the colliding bodies are hard, the contact force is very large in comparison with any body force, consequently, in rigid body impact theory anybody or applied contact forces of finite magnitude are negligibly small in comparison with the reaction at the contact point *A*. The finite body forces are ignorable because they do not work during the vanishingly small displacements that develop during an almost instantaneous collision. This is why a body force such as gravity does not affect the changes in velocity occurring in a collision. During impact between hard bodies, the only active forces are reactions at points of contact.

3.1 Compression and restitution

In the process of action of the contact force exist two phases and the compression and restitution. After the colliding bodies first touch, the contact force *F(t)* rises as compression of bodies in contact point. We will consider with small elastic part of a total weight of which amounts to significant deformation of the contact point. If flexibility is rate-independent, the maximum indentation and maximum force occur simultaneously when the normal component of relative velocity vanishes [8, 9]. (Fig. 5) illustrates the normal force as a time function.

Fig. 6 displays the separation of the contact period into an initial phase of approach or compression and subsequent phase of restitution. During compression, kinetic energy of relative motion is transformed into internal energy of deformation by the contact force – with action of contact, force reduces the initial normal relative velocity of the colliding bodies while simultaneously an equal but opposite contact force does work that increases the internal deformation energy of the deformable particle [6, 8].

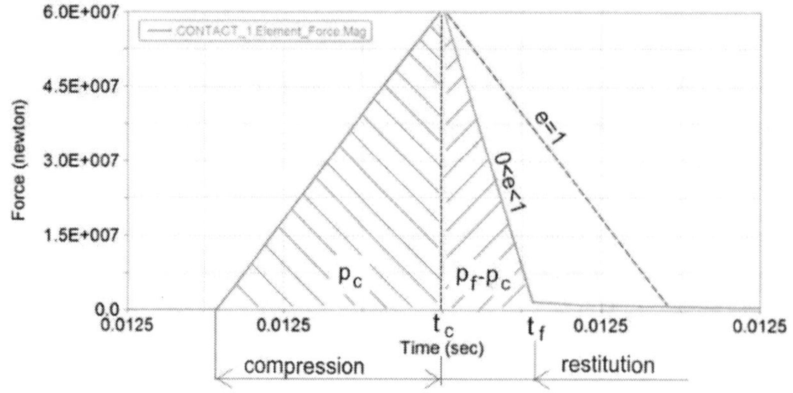

Fig. 5. Change of contact of the impact force as a function of time

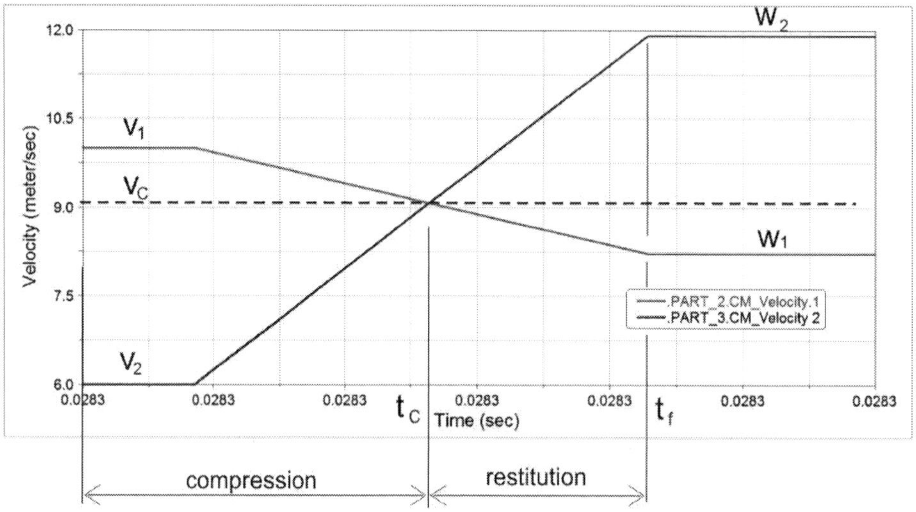

Fig. 6. Changing the normal velocities of the conflicting bodies as a function of time

The compression phase terminates and restitution begins when the normal relative velocity of the contact points vanishes. In a subsequent phase of restitution, internal elastic area drives the bodies apart. Elastic strain energy stored during compression generates the force that drives apart during restitution – the work done by this force restores part of the initial kinetic energy of relative motion. The compliance of the deforming region during restitution is smaller than that during compression [2, 9].

At any time t after incidence, the normal component of contact force *F* has and impulse *pc*, which equals the area under the curve of force, shown in (Fig. 5). During the compression of increasing monotonically pulse that slows down the body *1*, and increases the speed of the body *2* as shown in (Fig. 6). Consider the moment when changes in compression bruising to restitution as *tc*. Colliding elements have a relative speed between the contact points, which is lost at the end of compression:
v (*tc*) = 0, i.e. ends when the contact points have the same speed *vc* in the normal direction (Fig. 6). While *pf* is finite impulse in compartments (Fig. 5). In (Fig. 6) is illustrated that each contact point of the body there is a change in velocity, which is directly proportional to the normal reaction of impulse in contact point *A*. The reaction impulse $p_c = \int_0^{t_c} F(t)dt$ that causes the bodies to move at the same speed is called normal pulse compression. This impulse is characteristic that is required for analysis of the collision process.

By the direct impact finite impulse pf and the total kinetic energy absorbing directly related to the coefficient of restitution e. Restitution coefficient introduces the dissipation of (kinetic) energy due to inelastic deformation in the field of contact point. This factor has a value in the range $0 \le e \le 1$, whereas 0 means plastic deformation (i.e. without final separation, so that no part of initial kinetic energy of the normal relative motion is not restored) [5], while by the $e = 1$ attends to the perfectly elastic collision (i.e. no loss of kinetic energy of the normal relative motion).

It is needed to say, that proportion of final to initial relative velocity of a normal pulse for the restitution of the pulse at compression are directly dependent on a valid equations:

$$e = -\frac{\upsilon_f}{\upsilon_0}, \quad e = -\frac{p_f - p_c}{p_c}. \tag{2}$$

3.2 Influence of step-size to calculate the impact size

That the impact force was recorded, it is necessary to select the appropriate step size. To do this, we used sensitivity analysis during which we carry out a series of simulations to determine the impact of the integration step for calculating the contact force (Tab. 1). Further smoothing of the step-size was not performed, as the program did not converge.

Reaction forces are indefinitely large, nevertheless, they produce a finite impulse that continuously changes in relative velocity during the instant of contact.

Table 1. Step-size of iterative process

Simulation	**Contact 1**	**Contact 2**	**Contact 3**
Step-size	0.0002	0,00002	0,000002

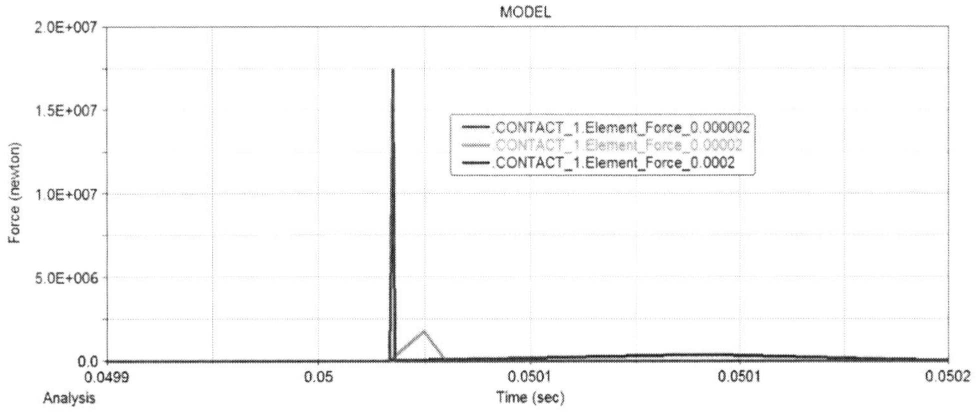

Fig. 7. Changing the step-size value regarding to the result's accuracy

From comparison of process in (Fig. 7) is illustrated, that the small integration step is justified only in the time interval in which the contact occurs. Otherwise, too small value of time step too unnecessarily slows down the simulation. Computational efficiency can be achieved by controlling the integration step simulation scripts where soften up over just before impact during collision and just after impact.

3.3 Influence of impact integrator and formulations DAE equations

In order to achieve from the simulations of virtual prototype optimal results, was carried amount of computational simulations by entering various types of integrators (GSTIFF, WSTIFF, HASTIFF) and formulations of differential and algebraic equations (DAE, I3, SI2, SI1). For comparison, we used the velocity values after the impact calculated from equation conservation of dynamics. After that is possible to analytically calculated velocities of bodies after the impact (collision) whereas for particular e have equations form

$$w_1 = \frac{m_2 v_2 (1+e) + v_1 (m_1 - e m_2)}{m_1 + m_2} \,,\; w_2 = \frac{m_2 v_2 (1+e) + v_1 (m_1 - e m_2)}{m_1 + m_2} \,. \tag{3}$$

The analysis was carried out on basis of several simulations in the choice of various types of integrators, i.e. WSTIFF, GSTIFF, and HASTIFF. Based on the sensitivity analysis was selected method SI1 in which solver algorithms monitor the deviation for the relocation, for the velocity and for the Lagrange coefficient and for the generalized momentum of dynamics and the acceleration. This means that this method is the most accurate of these methods, however, calculation takes a lot of time. Series of dynamic simulations were performed with the calculation step-size of *0.00002* with coefficient of restitution $e=1$ and the penalty coefficient of *1,5e+006*. From the (Fig. 8), it is clear that the most accurate results are observed at the choice of the integrator WSTIFF. The calculated values by (Eq. 3) are w_1= *6.3636 m/s*, w_2= *10.3636 m/s* and the values received from MSC.ADAMS are exactly the same (Fig. 8).

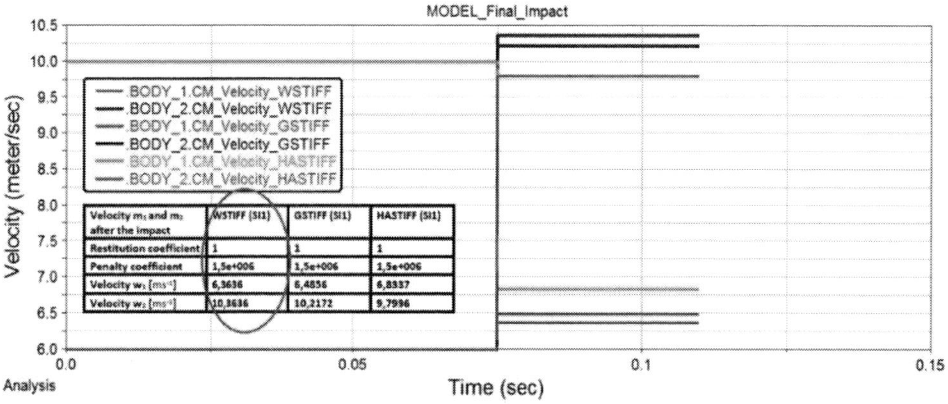

Fig. 8. The comparison of accurate particular solvers

3.4 Influence of penalty coefficient

In an environment of software, MSC.ADAMS is represented as an intermittent direct impact coefficient of restitution in addition to defining a penalty coefficient. Its task is to define an attribute stiffness between the contact materials. The value of the penalty parameter is ensured, that the penetration of one geometry to another is small. However, the large values can cause difficulties in numerical integration [1, 4, 8]. Creators of the set value 1E6 is suitable for systems modeled in kg-mm-sec. We issue from the theoretical background that by the flexible impact *(e=1)* is valid equation conservation of dynamics and equation conservation of mechanical energy. Were performed sensitivity analysis where the input variables are the coefficients of restitution and penalty coefficient (analyzed for each „*e*"), the target function was to ensure the compliance of the equation conservation of mechanical energy. The sum of the velocities of bodies by $e=1$, before the impact is equal to the amount after the impact. (Fig. 9) displays the progress of the impact force by perfectly elastic, plastic and perfectly plastic bodies. (Fig. 10) shows the velocity by significant coefficients e. By inelastic impact ($0 \leq e \leq 1$) is valid only equation conservation of momentum. During the simulations when $e = 0$ is monitored velocity in both bodies after the impact the same ($w_1 = w_2$). Fig. 11 is illustrated the process of translational kinetic energy by the same input parameters as by solving the velocity.

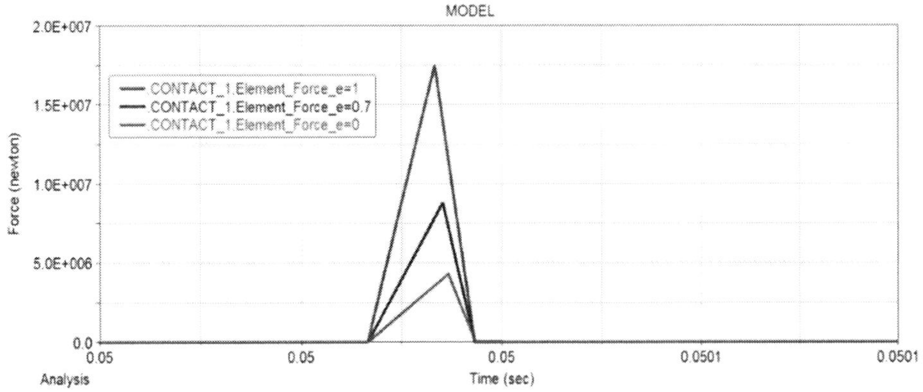

Fig. 9. Graphical representation of the contact force by different coefficients of resitution

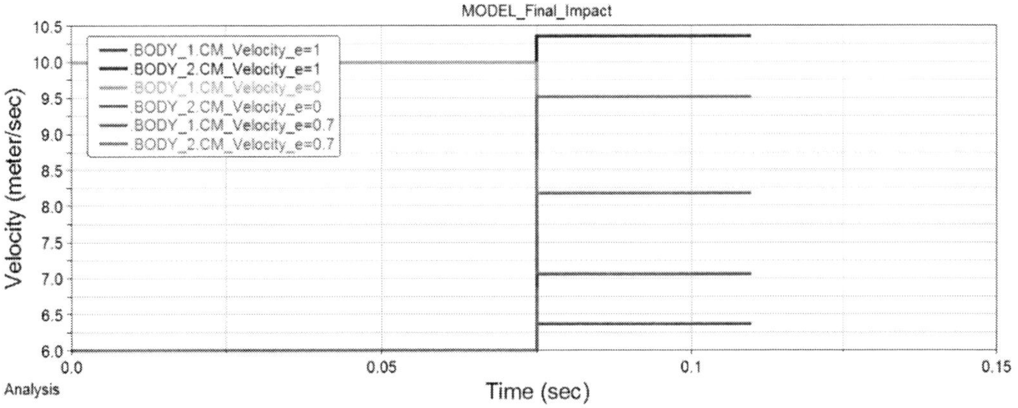

Fig. 10. Graphical representation of velocities both bodies by different coefficients of restitution

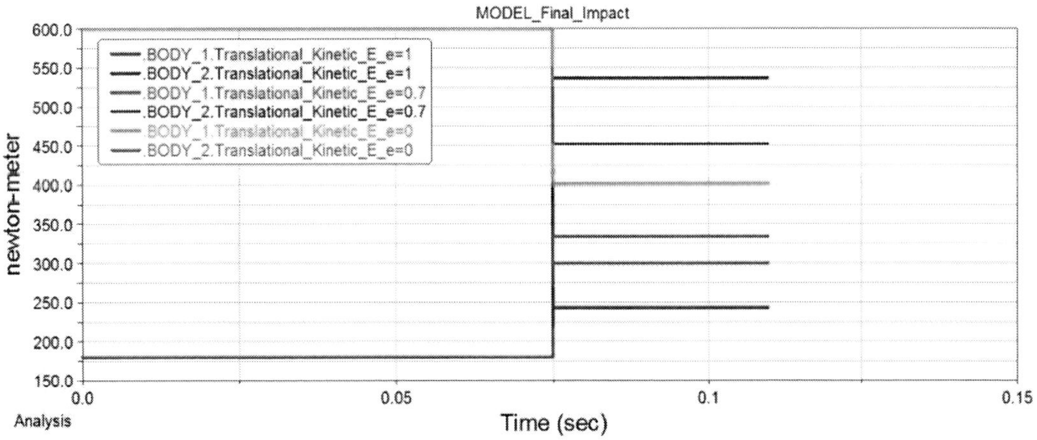

Fig. 11. Graphical representation of kinetic energy both bodies by different coefficients of restitution

From process of (Fig. 8-10) follow, that the selected parameters of solver correspond and the results are consistent with the reality. With the default parameters of solution algorithms was made as row of simulation with different coefficients of restitution and the results were verified with numerical results (Eq. 3). This approach is documented in (Table 2).

Table 2. Influence of restitution coefficient and penalty coefficient

WSTIFF, SI1	**No. 1**		**No. 2**		**No. 3**	
Coefficient of Restitution	1		0,7		0	
Penalty coefficient	1,5e+006		1,5e+007		1,5e+008	
	Analyt.	ADAMS	Analyt.	ADAMS	Analyt.	ADAMS
Velocity w_1 [ms^{-1}]	6,3636	6,3636	6,9091	7,069	8,1818	8,1819
Velocity w_2 [ms^{-1}]	10,363	10,3636	9,7091	9,5172	8,1818	8,1819
Kinetic energy before the IMPACT m_1 – EK_1	600	600	600	600	600	600
Kinetic energy before the IMPACT m_2 – EK_2	180	180	180	180	180	180
Kinetic energy after the IMPACT m_1 – EK_1^*	-	242,975	-	299,821	-	401,698
Kinetic energy after the IMPACT m_2 – EK_2^*	-	537,024	-	452,889		332,156
The fulfillment of the condition of kinetic energy before $EK=(EK_1+EK_2)$ and after $EK^*=(EK_1^*+EK_2^*)$	-	$EK=EK^*$ 780=779,99	-	$EK>EK^*$ 780>752,71	-	$EK>>EK^*$ 780>>733,84
Maximum contact force [N]	-	5,4289e+6	-	5,19e+8	-	5,7022e+6

From the comparison of the obtained contact force by the different coefficients of restitution was confirmed, that the the bodies are more plastic if the force is lower. From the (Tab. 3) and from (Fig. 8-11) is validated the theoretical background and suitability of obtained parameters of sensitivity analysis by solving the velocity, kinetic energy and impact force in progress of function IMPACT.

Conclusion

In the previous sensitivity analysis, we have seen the significant influence of input contact parameters of the impact force process, i.e., selection of suitable time steps, restitution coefficient and coefficient of penalization. An important aspect is also the correct choice of the integrator for DAE equations solution. The accuracy and precision of simulation results should always be verified by comparing the results with the solutions of equations in closed shape (if obtainable) or the monitoring of the total energy.

The part of the results of this work has been supported by VEGA grant No. 1/0795/16.

References

1. A. Sapietová, L. Gajdoš, V. Dekýš, M. Sapieta, *Analysis of the Influence of Input Function Contact Parameters of the Impact Force Process in the MSC. ADAMS,* ADVANCED MECHATRONICS SOLUTIONS Book Series: Advances in Intelligent Systems and Computing **393**, 243-253 (2016)
2. M. Handrik, P. Kopas, V. Baniari, M. Vaško, M. Sága, *Analysis of stress and strain of fatigue specimens localised in the cross-sectional area of the gauge section testing on bi-axial fatigue machine loaded in the high-cycle fatigue region*. Procedia Engineering **177**, 516-519 (2017)
3. M. Žmindák, Z. Pelagič, M. Nagel, L. Smetanka, Z. Stankovičová, *Simulation modeling of barrel bore wear under dynamic pressure load*. Communications **17** (3), 51-56 (2015)
4. A. Sapietová, M. Sapieta, B. Hyben, *Sensitivity Analysis Application for Multibody System Synthesis*. Applied Mechanics and Materials **420**, 68-73 (2013)
5. K. Johnson, *Contact Mechanics*. (Cambridge University Press, Cambridge, 2013)
6. J. Giesbers, *Contact Mechanics in MSC Adams – A technical evaluation of the contact models in multibody dynamics software MSC Adams*. [Bachelor Thesis], University of Twente, 2013
7. M. Sága, M. Žmindák, V. Dekýš, A. Sapietová, Š. Segľa, *Selected methods of analysis and synthesis of mechanical systems*. (VTS at University of Žilina, 360 p., 2009)
8. L. Jakubovičová, M. Sága, *Computational analysis of contact stress distribution in the case of mutual stewing of roller bearing rings*. Novel Trends in Production Devices and Systems, Applied Mechanics and Materials **474**, 363-368 (2014)
9. V. Dekýš, A. Sapietová, O. Števka, *Understanding of the dynamical properties of machines based on the interpretation of spectral measurements and FRF*. Experimental stress analysis, Applied Mechanics and Materials **486**, 106-112 (2014)

Kinematic and dynamic analysis and distribution of stress for four-item mechanism

Ján Vavro *Jr.*[1,*], *Ján* Vavro[1], *Petra* Kováčiková[1], *Juliána* Vršková[1]

[1]Faculty of industrial Technologies in Púchov, Alexander Dubček University of Trenčín, I. Krasku 491/30, 020 01 Púchov, Slovac Republic

Abstract. This paper presents a kinematic and dynamic analysis and distribution of the stress for four-item planar mechanism by means of the SolidWorks software. Graphic dependence of kinematic and dynamic magnitudes of some points is given in dependence on the angle of rotation of the driving item and in dependence on the time. Distribution of the stress in the items is presented in [Pa]. In relation to the kinematic and dynamic analysis and subsequent simulation of the planar as well as spatial mechanisms, it is great solution to use SolidWorks software program. The considerable advantage of this mentioned program is based on its simplicity from the aspect of modeling and moreover, it is important to point out that utilisation of the mentioned program leads to results which are precise and accurate in the case of the numerical solution of the equations in the whole magnitude referring to motion of mechanism while the given results are obtained in the graphic form.

Keywords: kinematic analysis, dynamic analysis, finite element method, planar mechanism

1 Introduction

The main objective of the kinematic and dynamic analysis of the four-item planar mechanism in the SolidWorks program was closely connected with the investigation of the position deviation tolerance, the convergence speed of equation solutions for the each one specified position of the planar mechanism. The mentioned analysis will be used for comparison of the obtained results with the results, which will be obtained by help of any other commercial software systems.

2 Kinematic and dynamic analysis of planar mechanism

The planar mechanism representative (Fig. 1) consists of four bodies and it was used as computational model. Using the kinematic analysis [4-6] and dynamic analysis and subsequent simulation [1-3], the main objective is connected with the determination and entering of the position domains, speed (velocity) domains as well as acceleration of the individual bodies in relation to the specified input values of the angular velocity for the

* Corresponding author: jan.vavro.jr@fpt.tnuni.sk

Reviewers: *Radim* Halama, *Milan* Žmindák

driving body designated as 2. The angular velocity for the body, designated as 2, is specified in this way: ω_{21}=36 [°/s], where ω_{21}=36 [°/s] is constant and it is changed in dependence on time (Fig. 2).
Specified input values can be seen in (Fig. 1).

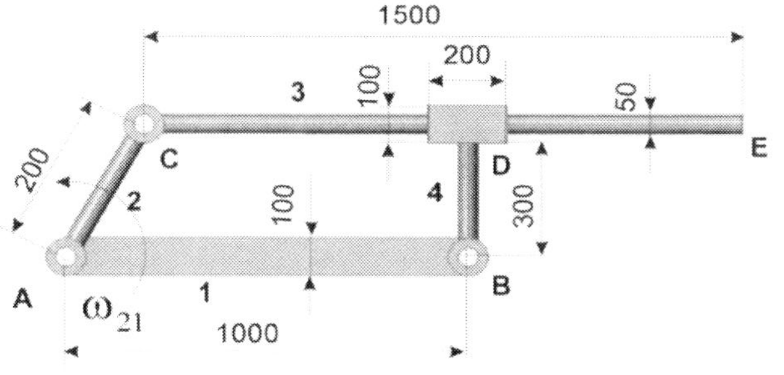

Fig. 1. Planar mechanism – computational model

Course of input value for angular velocity can be seen in Fig. 2 and angular acceleration of 2, 3, 4 bodies in dependence on time can be seen in Fig. 3.

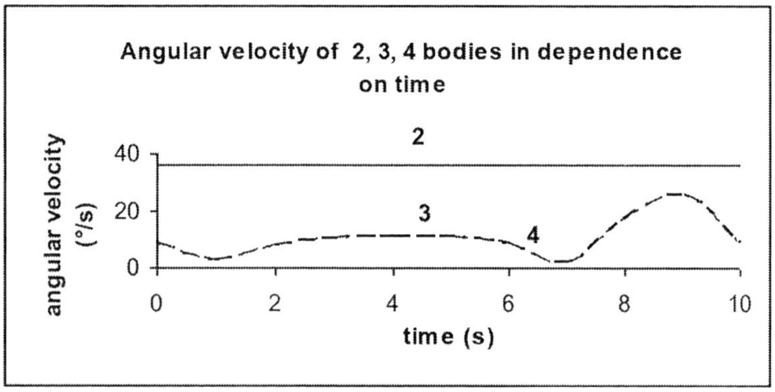

Fig. 2. Angular velocity of 2, 3, 4 bodies in dependence on time

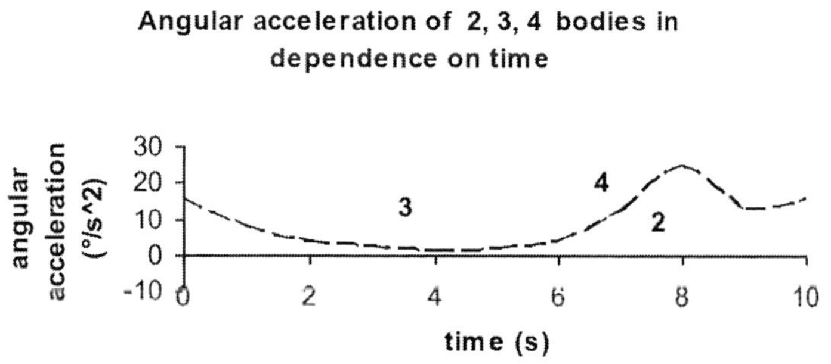

Fig. 3. Angular acceleration of 2, 3, 4 bodies in dependence on time

The simulation [12] of operation relating to planar mechanism can be seen in the Fig. 4 for time step referring to one second while the whole simulation takes place for ten seconds.

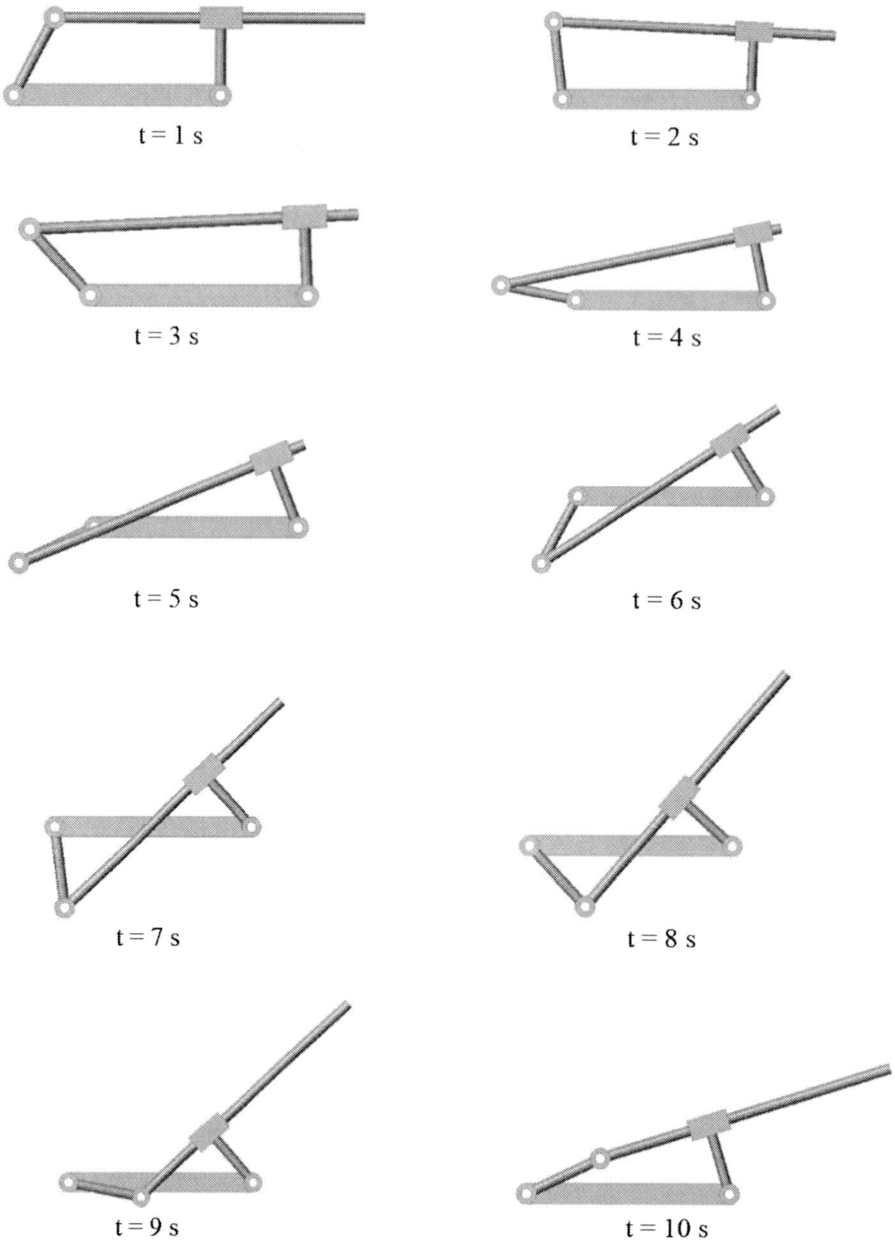

Fig. 4. Simulation of planar mechanism operation for ten positions

The whole course of the velocity and acceleration for C, D, E points of bodies can be seen in Fig. 5 and Fig. 6.

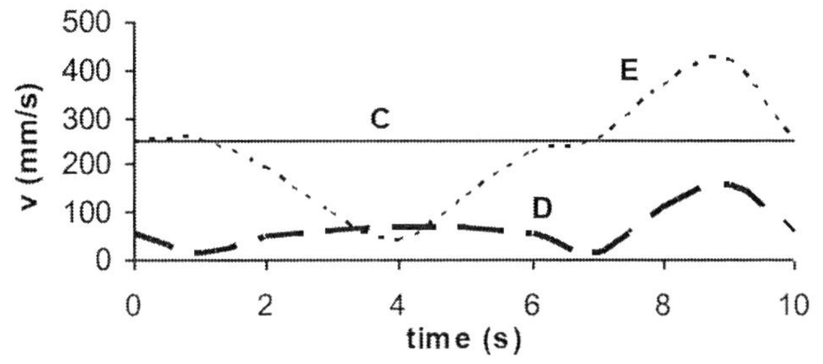

Fig. 5. Velocity in points (C, D, E) - dependent on the time

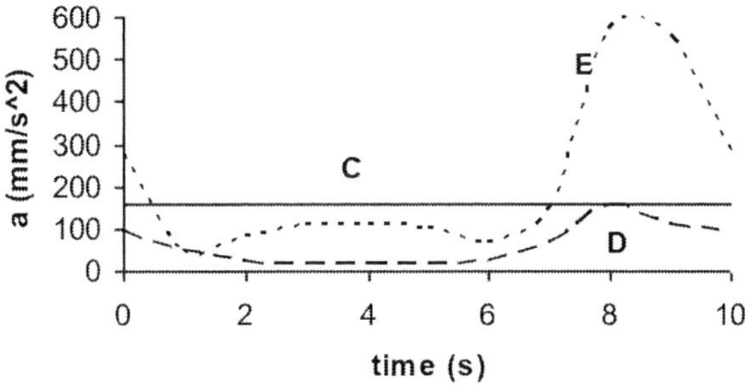

Fig. 6. Acceleration in points (C, D, E) - dependent on the time

The main objective of the dynamic analysis is connected with specification of the loading for the individual items and determination of the courses relating to mutual reactions, referring to individual kinematic connections [7], [10-11]. Fig. 7 represents the course of the reaction in C point of the body, designated as 2 and Fig. 8 represents the course of the reaction in B point of the body, designated as 4.

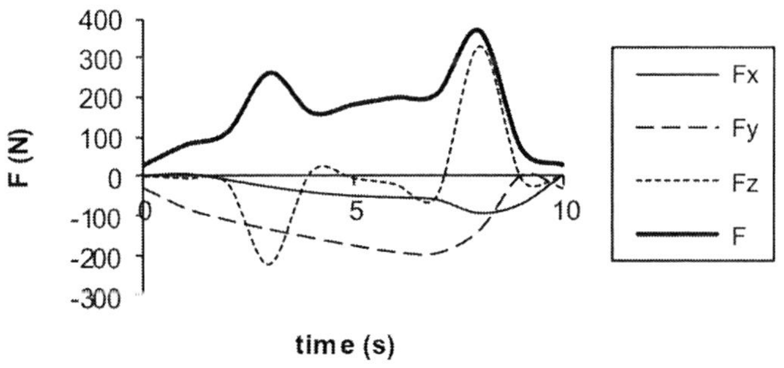

Fig. 7. Course of the reaction in C point of the body, designated as 2 - dependent on the time

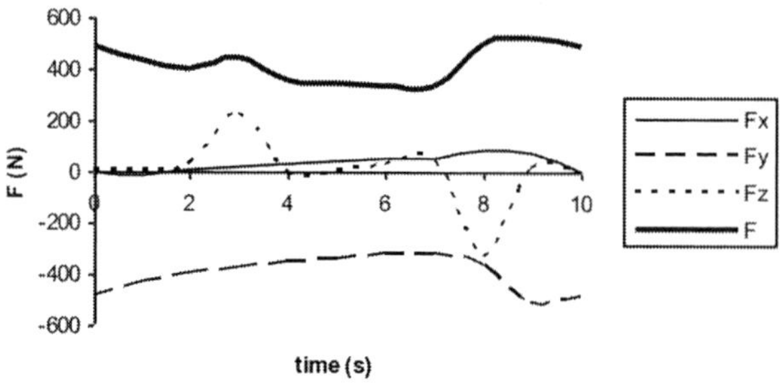

Fig. 8. Course of the reaction in B point of the body, designated as 4 - dependent on the time

3 Type of finite elements and material properties

The analysis of the planar mechanism was based on selection of the linear tetrahedral element with four nodes (see Fig. 9). The mentioned element belongs among the basic types of the volume finite elements. Including three shifts for each node, twelve deformation degrees of the freedom for the given element were defined. Considering the fact, which is mentioned above, the vector of unknown nodal shifts, can be presented by the equation (1).

$$\mathbf{u} = \begin{bmatrix} u_1 & v_1 & w_1 & u_2 & v_2 & w_2 & u_3 & v_3 & w_3 & u_4 & v_4 & w_4 \end{bmatrix}^T . \tag{1}$$

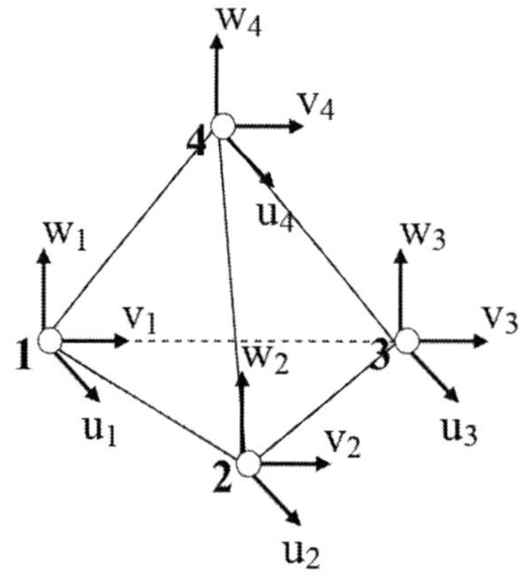

Fig. 9. Tetrahedral space element

The analysis was based on utilisation of the linear model. Relating to the analysis, the other important values were utilised:

- modulus of elasticity (Young's modulus): $E = 2.1\ e^{11}$ (Pa),
- Poisson's ratio: $\mu = 0.3$,
- density of material: $\rho = 7850$ (kg.m^{-3}).

4 Distribution of the stress in items of planar mechanism

The distribution of the stress for linked bodies [8-9], designated as 1, 2, 3, 4 can be seen in Figs. 10-17.

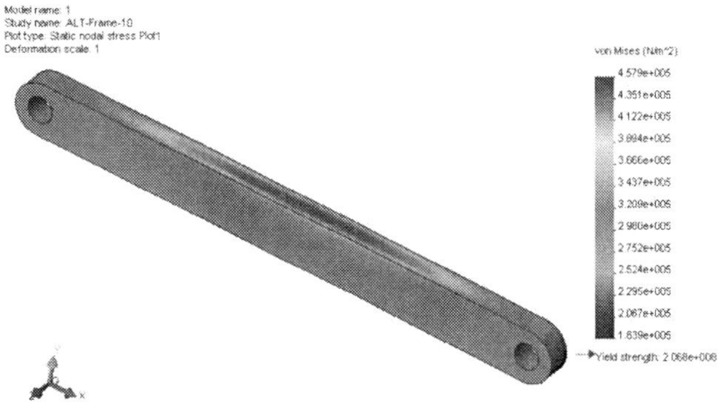

Fig. 10. Distribution of the stress for body designated as 1 in [Pa]

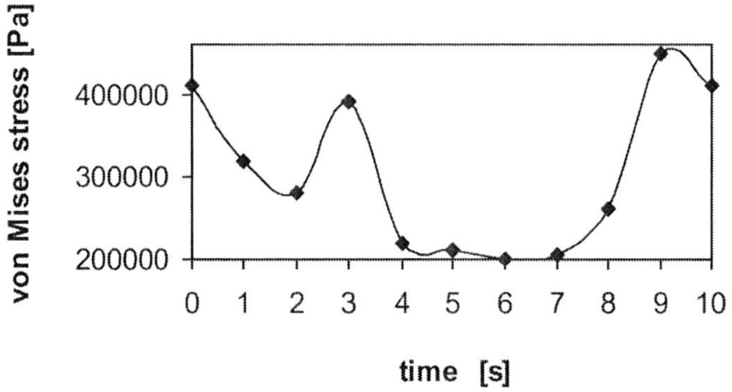

Fig. 11. Course of the stress for body, designated as 1- dependent on the time

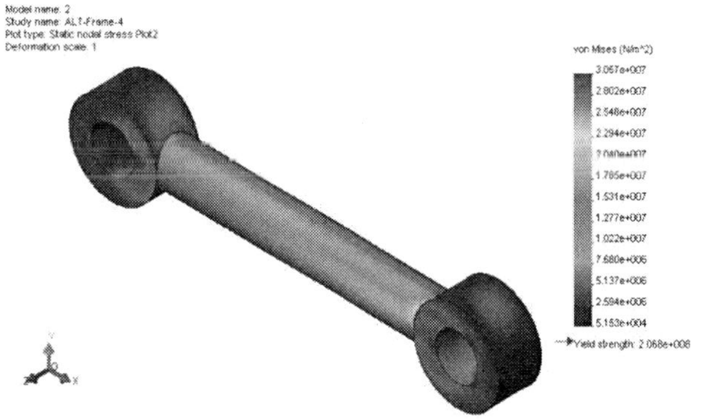

Fig. 12. Distribution of the stress for body, designated as 2 in [Pa]

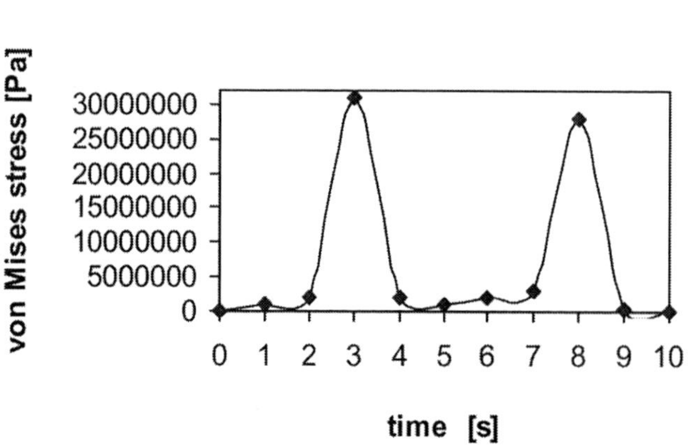

Fig. 13. Course of the stress for body, designated as 2- dependent on the time

Fig. 14. Distribution of the stress for body, designated as 3 in [Pa]

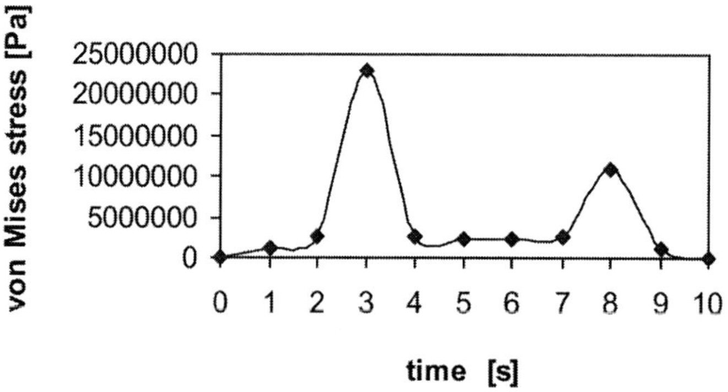

Fig. 15. Course of the stress for body, designated as 3- dependent on the time

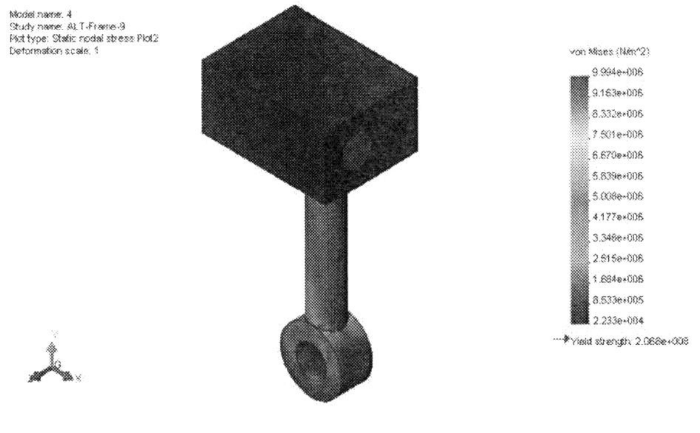

Fig. 16. Distribution of the stress for body, designated as 4 - in [Pa]

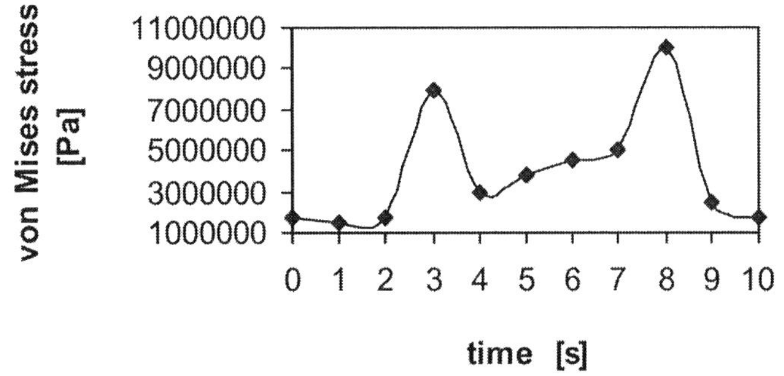

Fig. 17. Course of the stress for body, designated as 4 - dependent on the time

Conclusion

Based on the evaluation of the results, the utilisation of the Motion Program is significantly useful because it is effective way to determine all kinematic parameters of any mechanism and moreover, the loading for any point of the body system is able to be specified. The tolerance for the position deviation was also tested while the predetermined deviation was 10^{-9}. It is important to point out that from the aspect of convergence, it was not necessary to use more than five steps for each one position. On the other side, the convergence failure was connected with specification and entering of inaccurate parameters.

This work was supported by the Slovak Grant Agency VEGA 1/0649/17, VEGA 1/0589/17, KEGA 007TnUAD-4/2017, and resulted from the project "Center for quality testing and diagnostics of materials", ITMS code 26210120046 relating to the Operational Program Research and Development funded from European Fund of Regional Development.

References

1. B. Paul, *Kinematics and Dynamics of Planar Machinery*. (Prentice-Hall, New Jersey, 1979)
2. V. Brát, *Handbook of kinematics with examples*. (SNTL, Prague, 1976)
3. A. Sapietová, M. Saga, B. Hyben, M. Sapieta, *Effective methods of parameters refinement of machinery in the program MSC ADAMS*. Applied Mechanics and Materials **611**, 67-74 (2014)
4. A. Sapietová, M. Sapieta, B. Hyben, *Document Sensitivity analysis application for multibody system synthesis*. Applied Mechanics and Materials **420**, 68-73 (2013)
5. A. Sapietová, M. Sága, P. Novák, R. Bednár, J. Dižo, *Design and application of multi-software platform for solving of mechanical multi-body system problems*. Mechatronics: Recent technological and scientific advances, 345-354 (2011)
6. A. Sapietová, V. Dekýš, M. Vaško, *A numerical model of rotating machine having unbalance and the measurements of its dynamical properties*. Metalurgija (Metalurgy) **49,** 503-507 (2010)
7. L. Jakubovičová, M. Sága, *Computational analysis of contact stress distribution in the case of mutual stewing of roller bearing rings*. Novel Trends in Production Devices and Systems, Applied Mechanics and Materials **474**, 363-368 (2014)
8. F. Klimenda, J. Soukup, M. Zmindak, *Deformation of Aluminium Thin Plate*. Manufacturing Technology **16**, ISSN 1213-2489, 124-129 (2016)
9. M. Sága, R. Bednár, M. Vaško, *Contribution to modal and spectral interval finite element analysis*. Vibration Problems ICOVP 2011, Springer Proceedings in Physics **139**, 269-274 (2011)
10. M. Handrik, P. Kopas, V. Baniari, M. Vaško, M. Sága, *Analysis of stress and strain of fatigue specimens localised in the cross-sectional area of the gauge section testing on bi-axial fatigue machine loaded in the high-cycle fatigue region*. Procedia Engineering **177**, 516-519 (2017)
11. J. Zapoměl, V. Dekýš, P. Ferfecki, A. Sapietová, M. Sága, M. Žmindák, *Identification of material damping of a carbon composite bar and study of its effect on attenuation of its transient lateral vibrations*. Int. Journal of Applied Mechanics **7** (6), (2015)
12. I. Delyová, D. Hroncová, P. Frankovský, *Analysis of Simple Mechanism Using MSC Adams*. Manufacturing Technology **14**, ISSN 1213-2489, 141-145 (2014)

Kinematic and dynamic analysis and distribution of stress for six-item mechanism

Ján Vavro *Jr.*[1,*], *Ján* Vavro[1], *Petra* Kováčiková[1], *Jakub* Híreš[1]

[1]Faculty of industrial Technologies in Púchov, Alexander Dubček University of Trenčín, I. Krasku 491/30, 020 01 Púchov, Slovak Republic

Abstract. This paper presents a kinematic and dynamic analysis and distribution of the stress for six-item planar mechanism by means of the SolidWorks software. Graphic dependence of kinematic and dynamic magnitudes of some points is given in dependence on the angle of rotation of the driving item and in dependence on the time. Distribution of the stress in the items is presented in [Pa]. In relation to the kinematic and dynamic analysis and subsequent simulation of the planar as well as spatial mechanisms, it is great solution to use SolidWorks software program. The considerable advantage of this mentioned program is based on its simplicity from the aspect of modeling and moreover, it is important to point out that utilisation of the mentioned program leads to results which are precise and accurate in the case of the numerical solution of the equations in the whole magnitude referring to motion of mechanism while the given results are obtained in the graphic form.

Keywords: kinematic analysis, dynamic analysis, finite element method, planar mechanism

1 Kinematic and dynamic analysis of planar mechanism

The planar mechanism representative (Fig. 1) consists of six bodies and it was used as computational model. Using the kinematic analysis [4-6] and dynamic analysis and subsequent simulation [1-3], the main objective is connected with the determination and entering of the position domains, speed (velocity) domains as well as acceleration of the individual bodies in relation to the specified input values of the angular velocity for the driving body designated as 2. The angular velocity for the body, designated as 2, is specified in this way: $\omega_{21}=1$ [°/s] and $\alpha_{21}=0.7$ [°/s^2], where $\omega_{21}=1$ [°/s] is not constant and it is changed in dependence on time (Fig. 2). Specified input values can be seen in (Fig. 3)

[*] Corresponding author: jan.vavro.jr@fpt.tnuni.sk

Reviewers: *Vladimír* Dekýš, *Petr* Horyl

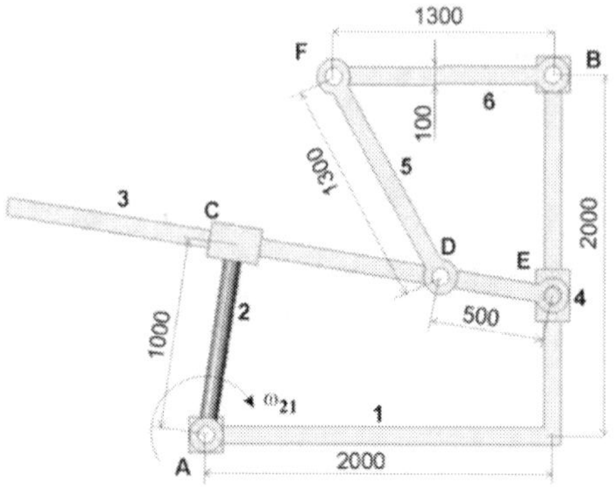

Fig. 1. Planar mechanism – computational model

Course of input value for angular velocity and angular acceleration is in Fig. 2 and Fig. 3.

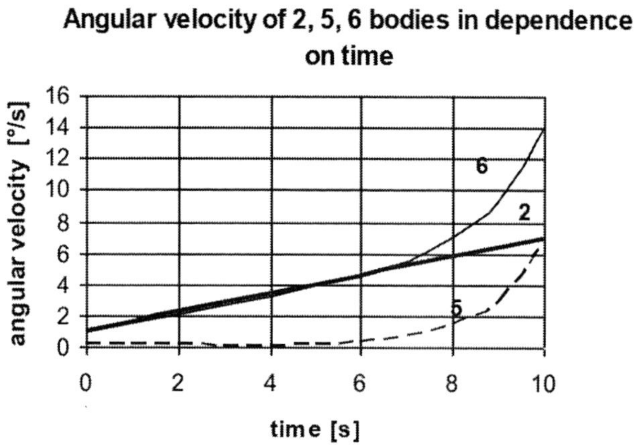

Fig. 2. Angular velocity of 2, 5, 6 bodies in dependence on time

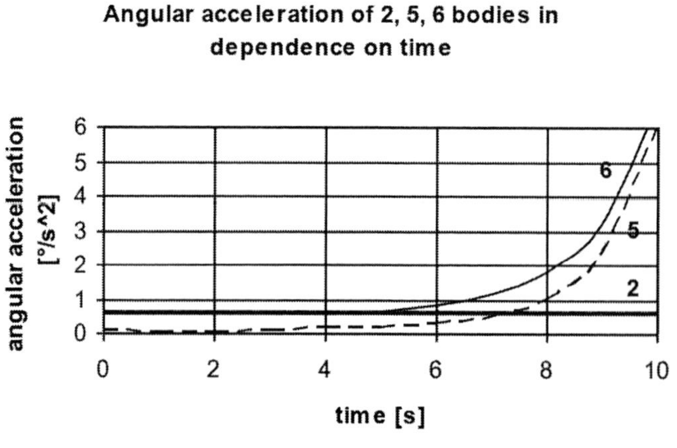

Fig. 3. Angular acceleration of 2, 5, 6 bodies in dependence on time

The simulation [12] of operation relating to planar mechanism can be seen in the Fig. 4 for time step referring to one second while the whole simulation takes place for ten seconds.

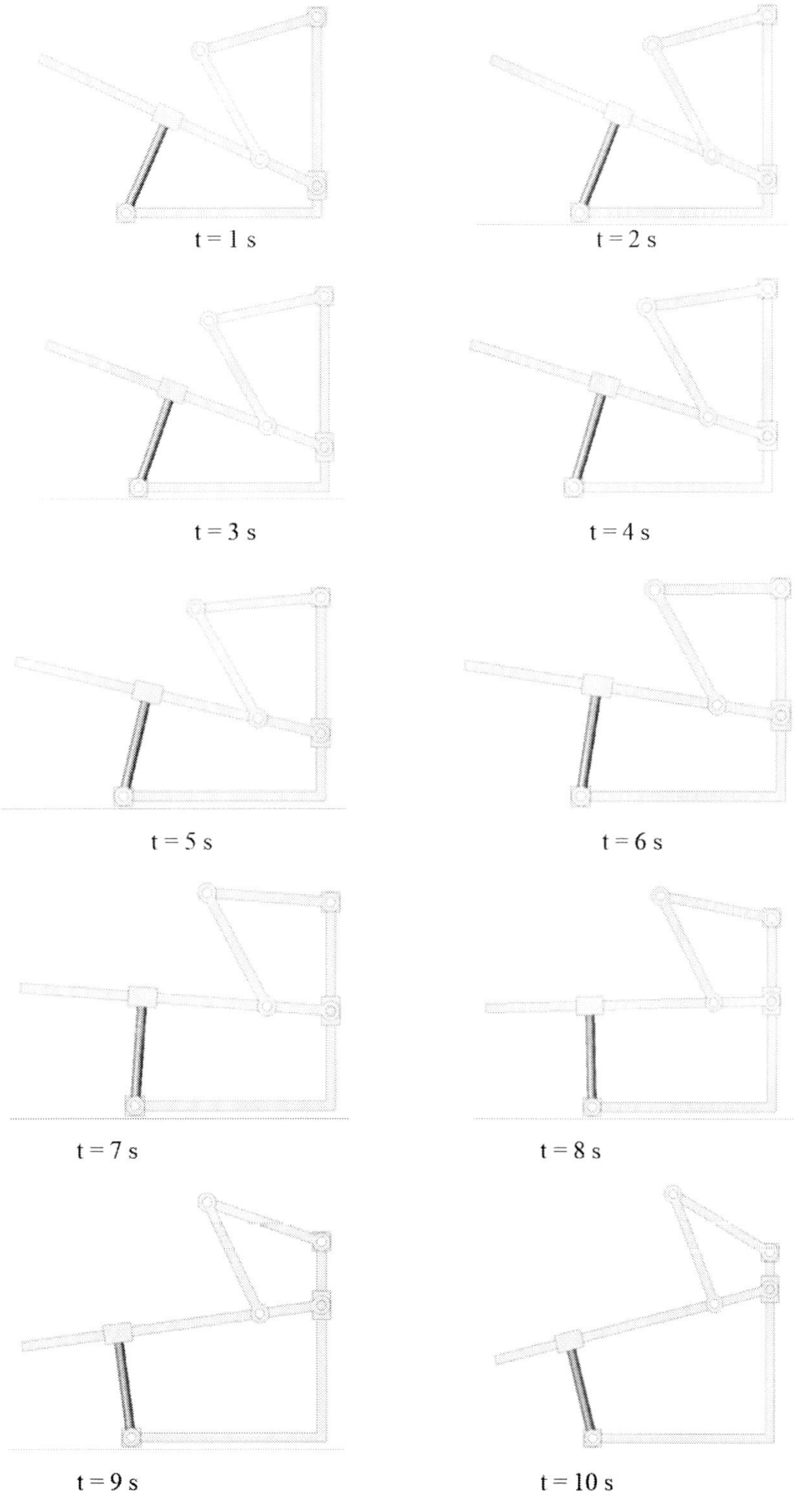

Fig. 4. Simulation of planar mechanism operation for ten positions

The whole course of the velocity and acceleration for C, D, E, F points of bodies can be seen in Fig. 5 and Fig. 6.

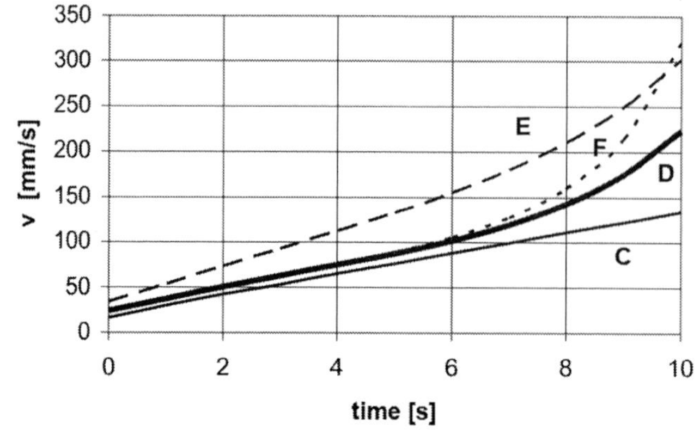

Fig. 5. Velocity in points (C, D, E, F) - dependent on the time

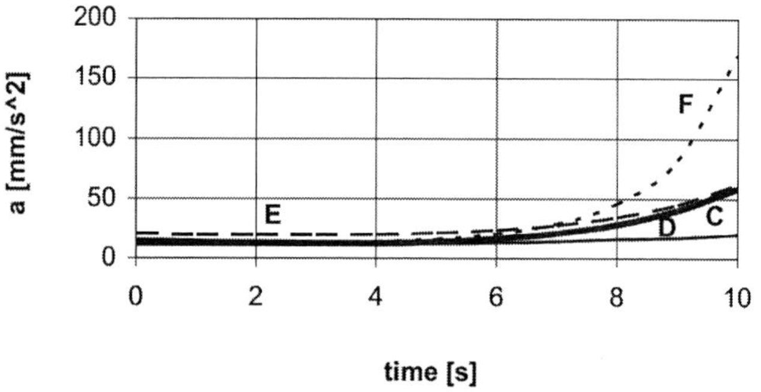

Fig. 6. Acceleration in points (C, D, E, F) - dependent on the time

The main objective of the dynamic analysis is connected with specification of the loading for the individual items and determination of the courses relating to mutual reactions, referring to individual kinematic connections [7], [10-11]. The analysis was based on utilisation of the linear model. Relating to the analysis, the other important values were utilised:

- modulus of elasticity (Young's modulus): $E = 2.1\ e^{11}$ [Pa],
- Poisson's ratio: $\mu = 0.3$,
- density of material: $\rho = 7850$ [kg.m^{-3}].

Fig. 7 represents the course of the reaction in F point of the body, designated as 6 and Fig.8 represents the course of the reaction in D point of the body, designated as 5.

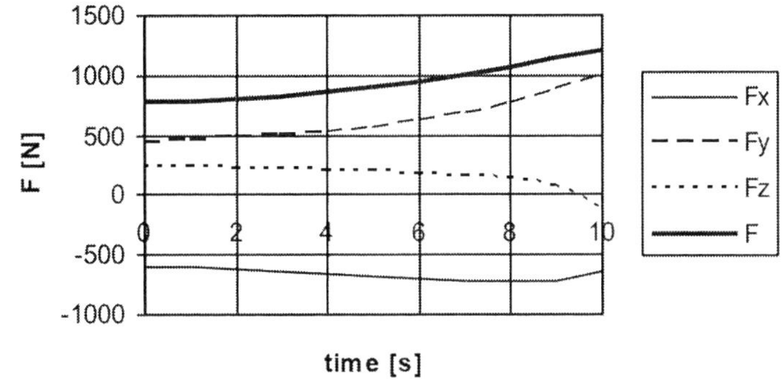

Fig. 7. Course of the reaction in F point of the body, designated as 6 - dependent on the time

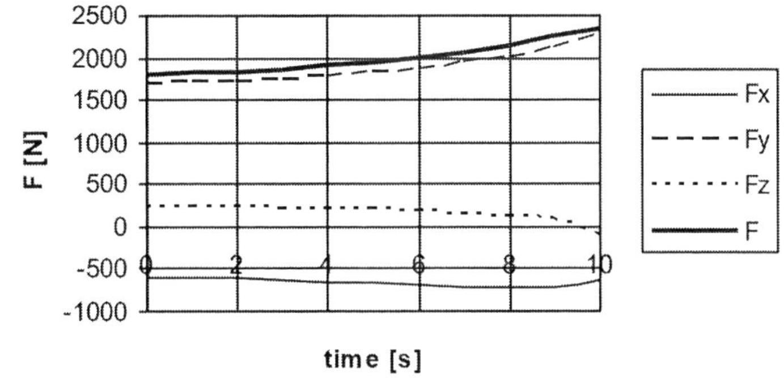

Fig. 8. Course of the reaction in D point of the body, designated as 5 - dependent on the time

2 Distribution of the Stress in Items of Planar Mechanism

The distribution of the stress for linked bodies [8-9], designated as 1, 2, 3, 5 can be seen in Figs. 9-16.

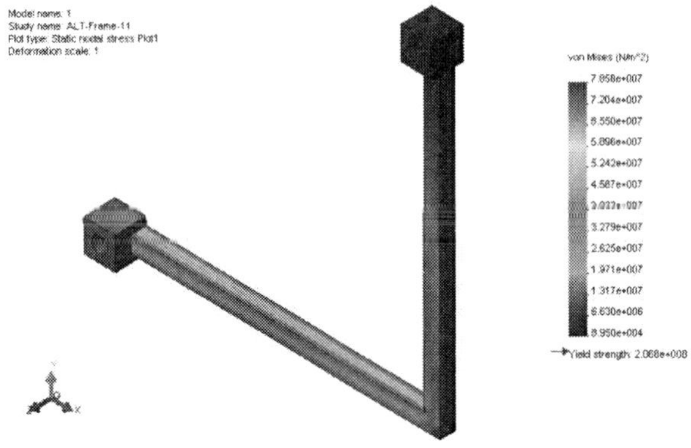

Fig. 9 Distribution of the stress for body designated as 1 in [Pa]

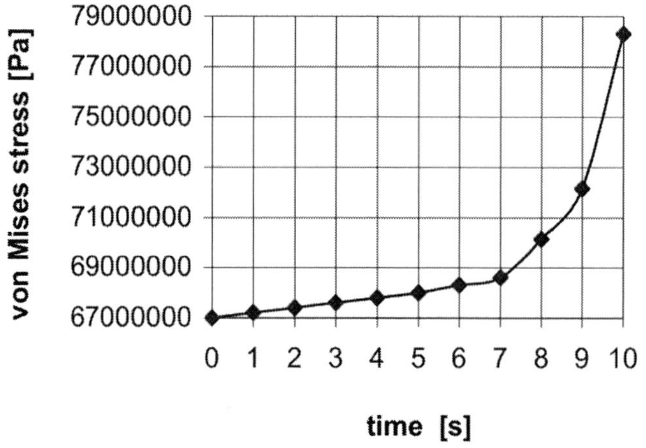

Fig. 10 Course of the stress for body, designated as 1 - dependent on the time

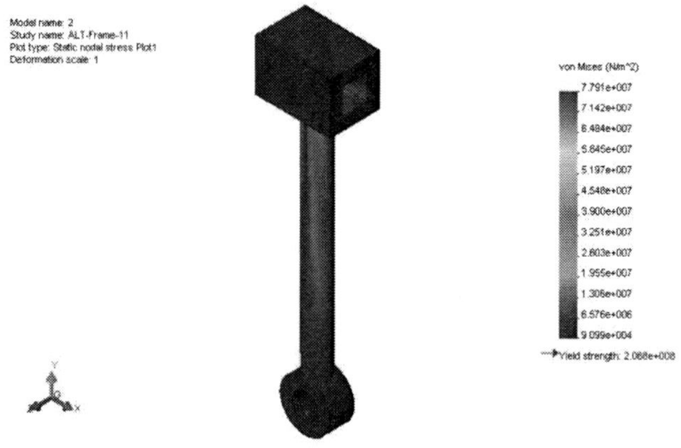

Fig. 11 Distribution of the stress for body, designated as 2 in [Pa]

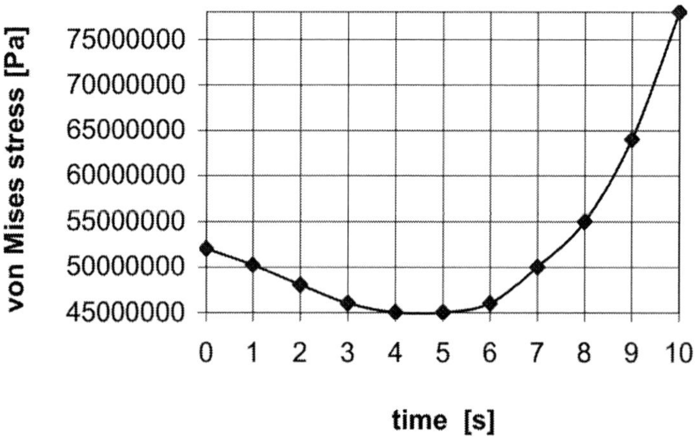

Fig. 12. Course of the stress for body, designated as 2 - dependent on the time

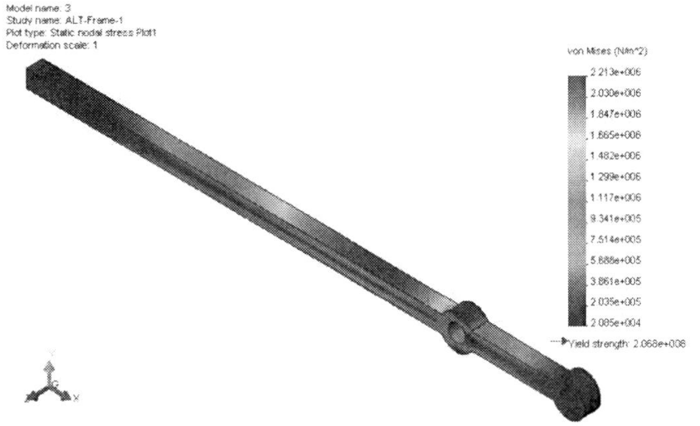

Fig. 13. Distribution of the stress for body, designated as 3 in [Pa]

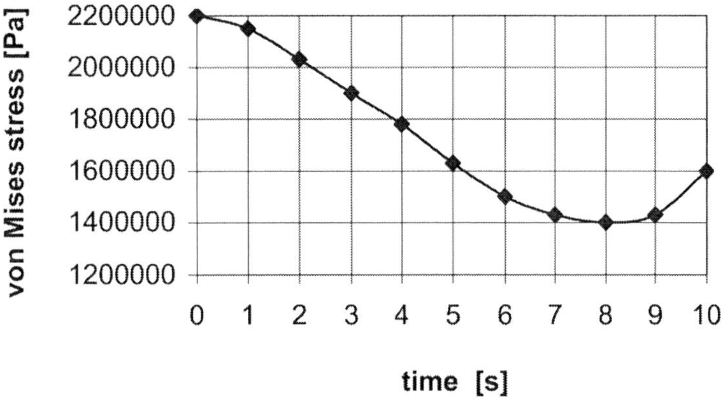

Fig. 14. Distribution of the stress for body, designated as 3 in [Pa]

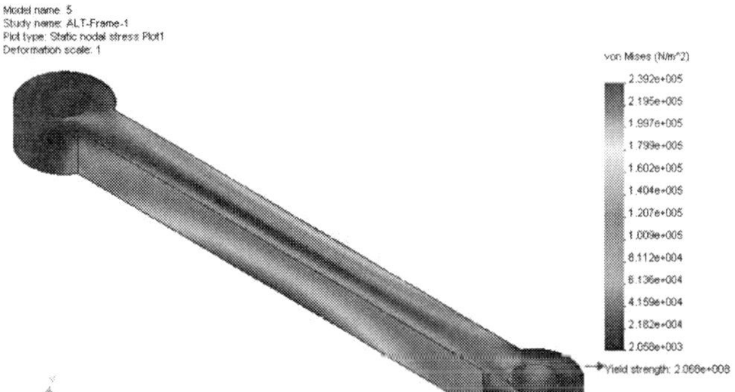

Fig. 15. Course of the stress for body, designated as 5 - dependent on the time

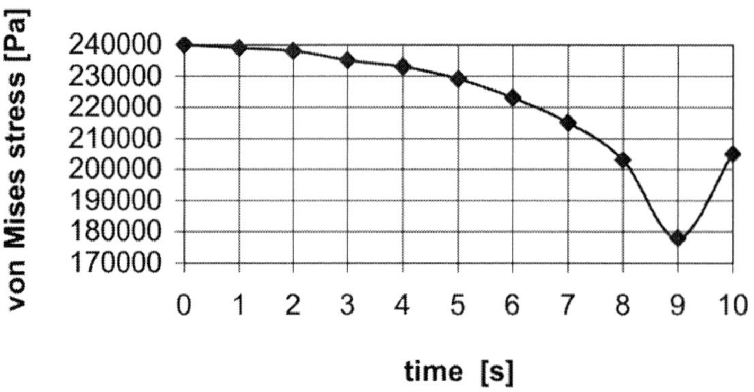

Fig. 16. Course of the stress for body, designated as 4 - dependent on the time

Conclusion

Based on the evaluation of the results, the utilisation of the Motion Program is significantly useful because it is effective way to determine all kinematic parameters of any mechanism and moreover, the loading for any point of the body system is able to be specified. The tolerance for the position deviation was also tested while the predetermined deviation was 10^{-9}. It is important to point out that from the aspect of convergence, it was not necessary to use more than five steps for each one position. On the other side, the convergence failure was connected with specification and entering of inaccurate parameters.

This work was supported by the Slovak Grant Agency VEGA 1/0649/17, VEGA 1/0589/17, KEGA 007TnUAD-4/2017, and resulted from the project "Center for quality testing and diagnostics of materials", ITMS code 26210120046 relating to the Operational Program Research and Development funded from European Fund of Regional Development.

References

1. B. Paul, *Kinematics and Dynamics of Planar Machinery*. (Prentice-Hall, New Jersey, 1979)
2. V. Brát, *Handbook of kinematics with examples*. (SNTL, Prague, 1976)
3. A. Sapietová, M. Saga, B. Hyben, M. Sapieta, *Effective methods of parameters refinement of machinery in the program MSC ADAMS,* Applied Mechanics and Materials **611**, 67-74 (2014)
4. A. Sapietová, M. Sapieta, B. Hyben, *Document Sensitivity analysis application for multibody system synthesis*. Applied Mechanics and Materials **420,** 68-73 (2013)
5. M. Sága, M. Vaško, P. Pecháč, *Chosen numerical algorithms for interval finite element analysis*. Procedia Engineering **96**, 400-409 (2014)
6. A. Sapietová, V. Dekýš, M. Vaško, *A numerical model of rotating machine having unbalance and the measurements of its dynamical properties*. Metalurgija (Metalurgy) **49**, 503-507 (2010)
7. A. Sapietová, M. Sága, P. Novák, R. Bednár, J. Dižo, *Design and application of multi-software platform for solving of mechanical multi-body system problems*. Mechatronics: Recent technological and scientific advances, 345-354 (2011)

8. F. Klimenda, J. Soukup, M. Zmindak, *Deformation of Aluminium Thin Plate*. Manufacturing Technology **16**, ISSN 1213-2489, 124-129 (2016)
9. A. Sapietová, V. Dekýš, M. Sapieta, P. Pecháč, *Application of computational and design approaches to improve carrier stability*. Procedia Engineering **96**, 410-418 (2014)
10. M. Handrik, P. Kopas, V. Baniari, M. Vaško, M. Sága, *Analysis of stress and strain of fatigue specimens localised in the cross-sectional area of the gauge section testing on bi-axial fatigue machine loaded in the high-cycle fatigue region*. Procedia Engineering **177**, 516-519 (2017)
11. J. Zapoměl, V. Dekýš, P. Ferfecki, A. Sapietová, M. Sága, M. Žmindák, *Identification of material damping of a carbon composite bar and study of its effect on attenuation of its transient lateral vibrations*. Int. Journal of Applied Mechanics **7** (6), (2015)
12. I. Delyová, D. Hroncová, P. Frankovský, *Analysis of Simple Mechanism Using MSC Adams*. Manufacturing Technology **14**, ISSN 1213-2489, 141-145 (2014)

Timing belt in power transmission and conveying system

Grzegorz Domek*[1], *Piotr* Krawiec[2], *Michał* Wilczyński[1].

[1]Kzimierz Wielki University in Bydgoszcz (UKW), Faculty of Mathematics, Physics and Technical Sciences, The Institute of Technology, Chodkiewicza 30, 85-061Bydgoszcz, Poland
[2]Poznan University of Technology, Faculty of Machines and Transport, Chair Machine Design Piotrowo 4, 60-138 Poznań, Poland

Abstract. This paper presents the problem of phenomena occurring at the contact of a timing belt and a pulley. Depending on a belt size range these phenomena differ significantly. There is no indication as to what solutions are optimal for drive belts. The analysis of the coupling process and performance tests have shown that the drive belt should have a cord of very good mechanical properties and its raceway side should be made from the material of a low friction coefficient against the pulley material. A flat belt in power transmission and conveying systems cooperates with several elements consisting of timing pulleys, tensioners or guiding rails. In gear with timing belts they depend strongly on characteristics of the process as well as the type of friction. In recent constructions, producers of timing belts are very much concerned about achieving as much slippery surface as possible. The work describes the problem of friction on different surfaces as well as its influence on gear lifetime. Research results confirm that on many surfaces bigger coefficient of friction is expected.

Keywords: Timing belts, power transmission belts

1 Introduction

A belt drive operation can be represented as a wheel running over a race. This can be compared to the movement of a vehicle wheel on the road, but the road, unlike the belt, is non-deformable. Meshing of a belt and pulleys causes wear to the mating surfaces, mainly due to the rolling friction and sliding friction. To improve the belt life, designers consider improvement of phenomena occurring between a belt and a pulley. In order to reduce the rolling friction, the diameters of pulleys and the hardness of a belt are increased, among others by making them of metal. The sliding friction in a belt transmission gear is associated with compensation of the driving strand elongation on pulleys. To minimize the effects of this phenomenon the elongation-resistant cord is used such as made of steel, polyamide or carbon fiber, thus many belts are named "Carbon" [4, 9].

* Corresponding author: gdomek@ukw.edu.pl

Reviewers: *Michał* Śledziński, *Milan* Vaško

2 Coupling between a timing belt and pulleys

The phenomena occurring in pulleys complicate in the case of timing belt transmission gears. Wear of belts is additionally caused by the process of belt teeth coupling and decoupling with pulley teeth. The attempts to shape teeth of a belt and groves of pulleys so that the coupling is of rolling nature have been unsuccessful so far. Type schedules of rounded and involute teeth (HTD - high torque driver, STD- super torque driver) have not ensured the coupling (meshing) process similar to that in cylindrical transmission gear. They only facilitated sliding of belt teeth into the groves of toothed pulleys [1, 7]. Comparing the process of a belt tooth coupling with a pulley tooth to the motion of a pendulum of variable length (Fig. 1) one can present the equation of vibrations associated with this process as follows:

$$\frac{d}{dt}[m(l_0+\Delta l)^2\dot{\varphi}] + mg(l_0+\Delta l)\varphi = 0, \tag{1}$$

or:

$$m(l_0+\Delta l)^2\ddot{\varphi} + 2m(l_0+\Delta l)\Delta l\dot{\varphi} + mg(l_0+\Delta l)\varphi = 0. \tag{2}$$

In the above equation m- means the belt tooth weight, l_0 - the distance between the belt tooth center and the central point of meshing "C" (Figure 1), Δl - the value of the belt or wheel pitch (depending on whether the moving part is a pulley or a belt) [8].

Fig. 1. Comparison of the process of a belt tooth coupling with a pulley tooth to a pendulum, c - moving pivot point of a tooth, k- moving pulley.

Due to the flexibility of a cord and gradual transfer of load by successive teeth onto the pulley, the pulley and belt teeth coupling process is of "collision" nature. In the belt drive with too little tensioned belt the interference of these teeth occurs and the belt breaks. The teeth of the loaded belt hit the tips of the pulley teeth and then they slip over their sides until full meshing. This phenomenon causes rapid wear of belt teeth and, after long time, also of pulley teeth [6,10]. In order to reduce effects of this phenomenon and increase the life of the transmission gear, the friction coefficient of the belt-pulley materials pair must be reduced. Another primary actions include reduction of the strand elongation. Changes of teeth arrangement over the width have been aimed at reduction of noise associated with coupling and improvement of belt guiding on pulleys. From among these solutions least design problems were caused by belts of herringbone teeth (Fig. 2).

Fig. 2. Herringbone toothed belts in a passenger lift

3 Timing belt slip over pulleys

The above-mentioned activities are not identical with measures undertaken for other belt drives. However, the following question must be answered: what is the status of a toothed belt meshing with a pulley and what status are we going to achieve? In most toothed belt drives coupling is of combined form-fitting and friction nature while in structures we most often expect form-fitting [2,3]. The nature of the meshing between a belt and a pulley in this case is dependent on the drive design and belt load. In the standard design of most toothed pulleys, the groove is larger than the belt tooth and thus there is a lateral clearance and often a tip clearance between them. If there is no peripheral force causing strand (cord) elongation exceeding the value of the lateral clearance, then it can be assumed that there is only friction coupling in the drive (Fig.3).

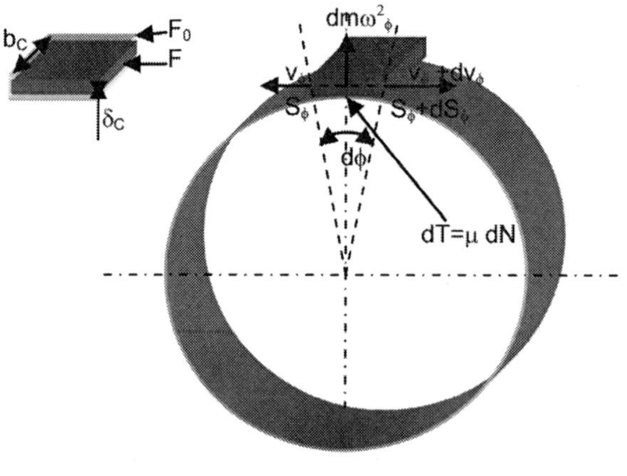

Fig. 3. Contact of an elementary segment of a belt with a pulley

If the belt strand (cord) was non-tensile as in Euler equation, then only friction coupling would occur. The coupling quality depends on the belt on pulley friction factor μ_z and internal belt friction factor μ_w. The coupling of the belt drive can be considered as follows:

$$\frac{dS_1}{dS_2} = e^{\beta\mu}, \quad (3)$$

where: $$\mu = \mu_z + \mu_w \quad (4)$$

while β is the contact angle between the belt and the pulley in radians.

In fact, even a high quality cord is tensile and after exceeding the limit value of the peripheral force the belt starts to cooperate with the wheel by form-fitting coupling. As the load is increasing and thus the belt is elongating, the form-fitting coupling prevails in the toothed drive while the share of friction coupling is negligible. The quality of the coupling is also influenced by the number of teeth at the pulley contact arc [4,5]. Depending on the belt and cord material, the optimum number of teeth varies from 12 to 16. This is the number of teeth, which shall transmit the entire peripheral force from the belt onto the pulley. Successive teeth do not participate in the coupling. Reduced number of teeth participating in the coupling (meshing) process can lead to overloading of the teeth.

4 The design of the raceway side of a belt

Belt designers have unsuccessfully been trying to reach the consensus on the desired design features of the raceway side of a belt for years. Polyurethane belts are most often manufactured without additional coating on the raceway side, while rubber belts are covered with additional fabrics. There are many different raceway side coating types available on the market, characterized by different friction coefficients against metal. In the aforementioned cases there is a huge difference between the friction coefficient values of belt material against pulley material. It varies from 0.3-0.4 in the case of belts without fabric coating (Fig. 4) to 0.008 - 0,01 for "slippery" fabric over steel (Fig. 5,6).

Fig. 4. Polyurethane belt of HTD type without additional coating on the raceway side

Fig. 5. The rubber HTD belt coated with modern polyester fabric

Fig. 6. The Synchrochain belt coated with modern polyamide fabric

The results of performance tests have clearly confirmed that a driving timing belt should be provided with such a raceway so that its friction coefficient against metal is as small as possible, it should have rigid teeth and elongation-resistant cord [11]. The coupling nature in the belt drive will then be close to the form-fitting one and the friction losses will be reduced. However care should be taken to increased rigidity of the belt which can prevent the belt from operating on small diameter pulleys.

Conclusions

The latest designs of toothed belts aim at reduction of the friction coefficient. Their names include the word "chain". Technical parameters of those belts match the parameters of chains and thus belts are used interchangeably with chains in many applications. However, they have lost an important feature of the belts - i.e. silent-running and currently their use instead of chains reduces the emission of noise to a small extent only. The results of tests confirmed the need to reduce the friction of the driving toothed belt. In applications where the belt is used for control or conveying purposes, these properties must be considered individually.

References

1. H. Dressing, F. Holzweissig, *Dynamics of Machinery, Theory and Applications*. Springer Verlag, Berlin Heidelberg (2010)
2. M. Dudziak, G. Domek, *Gear with timing belts in mechatronic drives*. Machine Dynamics Problems **28** (3), 83-88 (2004)
3. M. Dudziak, G. Domek, *Analysis of Timing Belt's Shape in Meshing with Timing Pulley*. Machine Dynamics Problems **30** (2), 32-39 (2006)
4. G. Domek, M. Dudziak, *Algorithm for choosing the gears with timing belts*. Machine Dynamics Problems **30** (3), 65-71 (2006)
5. G. Domek, A. Kołodziej, *The surface conditions of pulleys in use*. Machine Dynamics Problems **30** (3), 72-78 (2006)
6. M. Dudziak, G. Domek, *Analisis of antitorque in gear with timing belts*. Materials Engineering **11** (1), 121-124 (2004)
7. G. Domek, P. Krawiec, *Methods of Designing of Timing Belts Pulley*. University Reviev, Vol **1**, No.3 /2007, Aleksander Dubcek University of Trencin, Izhevsk State Technical University, 15-20 (2007)
8. T. Nagel, Zahnriemengetriebe 2010- *neueste Entwicklungen und Trends*. Tagungsband zur 14. Internationalen Fachtagung Zahnriemengetriebe, Dresden, 5-8 (2010)
9. R. Perneder, *Handbuch Zahnriementechnik*. Springer Verlag, Berlin (2009)
10. M. Sága, P. Kopas, M. Uhríčik, *Modeling and experimental analysis of the aluminium alloy fatigue damage in the case of bending - torsion loading*. Procedia Engineering **48**, 599-606 (2012)
11. J. Vollbarth, *Anforderungen an Zugtraeger fur Zahnriemen und deren Eigenschaften im Vergleich*, Tagungsband zur 14. Internationalen Fachtagung Zahnriemengetriebe, 57-63 (2010)

Use of risk assessment methods in maintenance for more reliable rolling stock operation

Juraj Grenčík[1,*], *Roman* Poprocký[1], *Jana* Galliková[1], *Peter* Volna[1]

[1]Department of Transport and Handling Machines, Faculty of Mechanical Engineering, University of Žilina, Univerzitná 8215/1, 010 26 Žilina, Slovak Republic

Abstract. Operation of the rolling stock is associated with a number of risks. The consequences of failures are sometimes very serious - safety or the environment, sometimes only economic. A significant factor affecting the safe operation of railway vehicles is maintenance. The paper presents methods of risks assessment and possibilities of their reduction in design, operation and maintenance of railway vehicles. Special attention is given to the new "silent brakes" and safety related issues for freight wagons.

Keywords: risk assessment, railway safety, reliability

1 Introduction

It is well known fact that European railways are the safest mode of land transport. The safety level has improved at impressive pace over the past decade and the railway industry can be proud of its achievements, mostly achieved through technical advances. However, although extremely rare, catastrophic, multi-fatality have a heavy impact on the confidence of passengers, customers, public founders and investors. As well as the human cost, every accident, whether they result in injuries or not, represents a significant business cost in a highly competitive environment. Catastrophic accidents have the potential to close otherwise viable businesses and reduce services altogether.

Security issues are also addressed by institutions at the top level, for example European Union Agency for Railways (ERA). Monitoring safety performance is a priority task of the Agency in its mission to promote a harmonised approach to railway safety in Europe. A harmonised Safety Management System (SMS) is the foundation for managing and controlling risks, and building trust among railway undertakings and infrastructure managers in the European Union [1].

Common Safety Indicators (CSIs) are used by National Safety Authorities to gather information from railway undertakings and infrastructure managers, which combined with other relevant data, makes a comparative analysis possible, and serves as basis for policy recommendations at EU level. CSIs are a common set of rail safety data, gathered to facilitate the assessment of achievement of Common Safety Targets (CSTs) and monitor the development of safety in Member States [2].

* Corresponding author: juraj.grencik@fstroj.uniza.sk

Reviewers: *Marcin* Kubiak, *Ivan* Kuric

Common safety indicators are divided into the following groups:

- Indicators relating to accidents,
- indicators relating to dangerous goods,
- indicators relating to suicides and attempted suicides,
- indicators related to precursors of accidents,
- indicators to calculate the economic impact of accidents,
- indicators related to technical safety of infrastructure and its implementation.

Common Safety Targets (CST) are the qualitative measures for assessing the railway safety risk in the Member States of the Union. Rail transport is the only mode of transport for which these objectives have been prescribed by European legislation. The common safety targets are EU-wide maximum risk and the National Reference Values (NRVs) are the maximum risk values for each Member State. The level of risk is measured in terms of the number of killed and severely injured persons per train kilometre. There are categories of risk for passengers, employees, crossings users, unauthorized persons on railway and others, as well as society as a whole [3].

Around 2 000 significant accidents occur each year on the railways of the EU Member States. Collisions and derailments represent a mere ten per cent of them. Accidents to persons caused by rolling stock in motion and level-crossing accidents constitute the majority of significant accidents, excluding suicides. The number of significant accidents per accident type in the period 2012–2014 is presented in the Fig. 1. The number of significant accidents increased by 5 % in 2014 year-over-year in EU-28 Member States. This is the first Year over Year (YoY) increase in ten years [4].

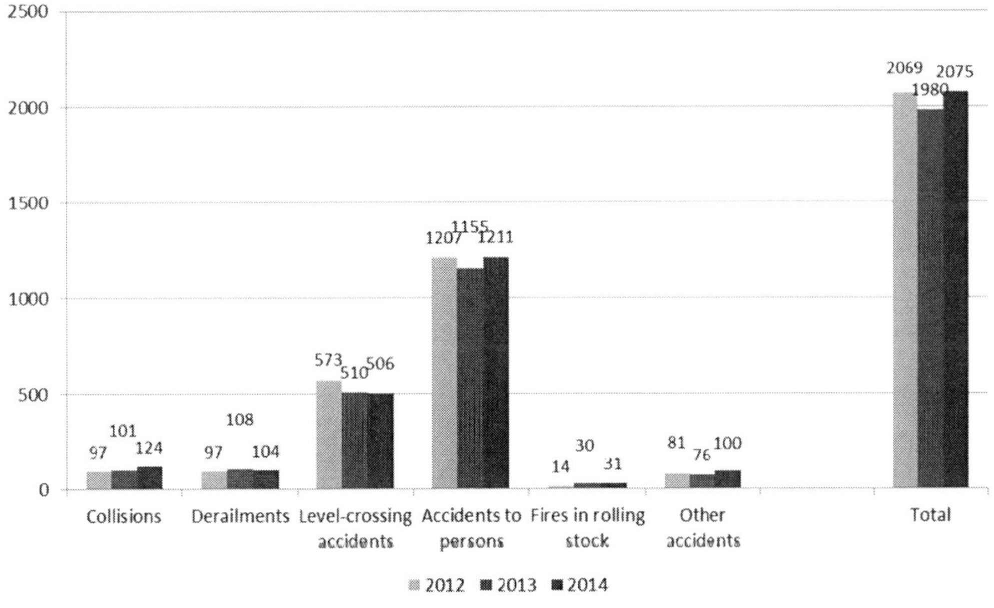

Fig. 1. Significant accidents per type of accidents [4]

2 Risks in general

Every process, system, or human activity is affected by risks that can have both a positive and a negative impact. In relation to maintenance activities, the occurrence of each failure is a negative phenomenon that can be expressed in terms of the magnitude of the risk depending on the probability of the occurrence of the disorder being encountered and the consequences. Hazards such as sources of potential damage must be analysed in relation to all activities of the organization. These sources can be, for example:

- Technical equipment, technological processes, chemicals and materials,
- safety requirements and health protection at work,
- political activities,
- environment and its impacts,
- legal and commercial relations,
- economic aspects in the country of operation,
- market situation,
- the level and method of management,
- logistics, etc.

In general, an accident prevents the build-up of faults in a device or a deviation from its normal operation. This phase may take minutes, hours, sometimes even years. Defects or deviations from normal activity do not yet cause an accident but not create appropriate conditions for it. Operators usually do not notice this phase if they are not performing the prescribed work or do not have information about functioning of the object. That's why they do not feel threatened. In the next phase a sudden event arises, which will significantly change the situation. If operators try to restore the normal operation of the technological process and do not have complete information, they only exacerbate the development of the accident. In the last phase there is a sudden event that represents the impulse, after which the technical system ceases to be subordinated to a man and there is a negative phenomenon.

The source of risks is sometimes nature, sometimes human being, sometimes technology or technological processes. Any undesirable event may have a link to a certain loss associated with the risk object.

The risk factors in connection with reliability, safety and maintenance in transportation systems are analysed in a complex manner e.g. by Dhillon in [5, 6].

The relationship of the risk object to adverse events allows the risk to be divided into:

- Individually,
- technical,
- ecological,
- social,
- economical,
- other risks.

Each type of risk has characteristic sources and factors (Table 1.).

Table 1. Objects, resources and undesirable consequences of the different types of risks

Type of risk	Object of risk	Source of risk	Undesirable consequences
individual	person	live conditions	illness, trauma, disability, death
technical	technical systems and objects	technical incompetence, disruption of the operation of technical systems and objects	crash, explosion, fire, catastrophe, destruction
environmental	ecological systems	anthropogenic interference with the natural environment, unusual situations technogenous	environmental catastrophes, natural disasters
social	social groups	unusual situation, reduced quality of life	group trauma, illness, death of people, increase in mortality
economical	material resources	reducing the safety of production or the natural environment	increased security costs, damage caused due to lack of protection

3 Risk management

Risk management is the systematic application of management policies, procedures and established practice. It is a complex of coordinated management and control activities with respect to risk. This process is dependent on the experience, knowledge, imagination, creativity and ability of the team (individuals) performing these activities. Applying these procedures, without involving teamwork and competent staff, cannot provide proper and thorough risk analysis results. An important step is to choose the appropriate risk assessment method. The risk assessment based on a risk analysis assesses the severity of the estimated magnitude of the risk and assesses the need to reduce it.

Risk management is a structured sequence of logical steps (Fig. 2.), where the first step is a risk analysis that examines the potential negative consequences that may result from failures in the performance of technical systems, deviations in technological processes, or errors by service personnel. This also means that negative impacts on humans and the surrounding environment even in the normal operation of the technical system can be investigated. Subsequently, it is necessary to identify the likelihood and extent of the consequences of a negative event resulting from a given work or other activity of equipment or a system. Based on hazard identification, the magnitude of the risk is estimated [7, 8].

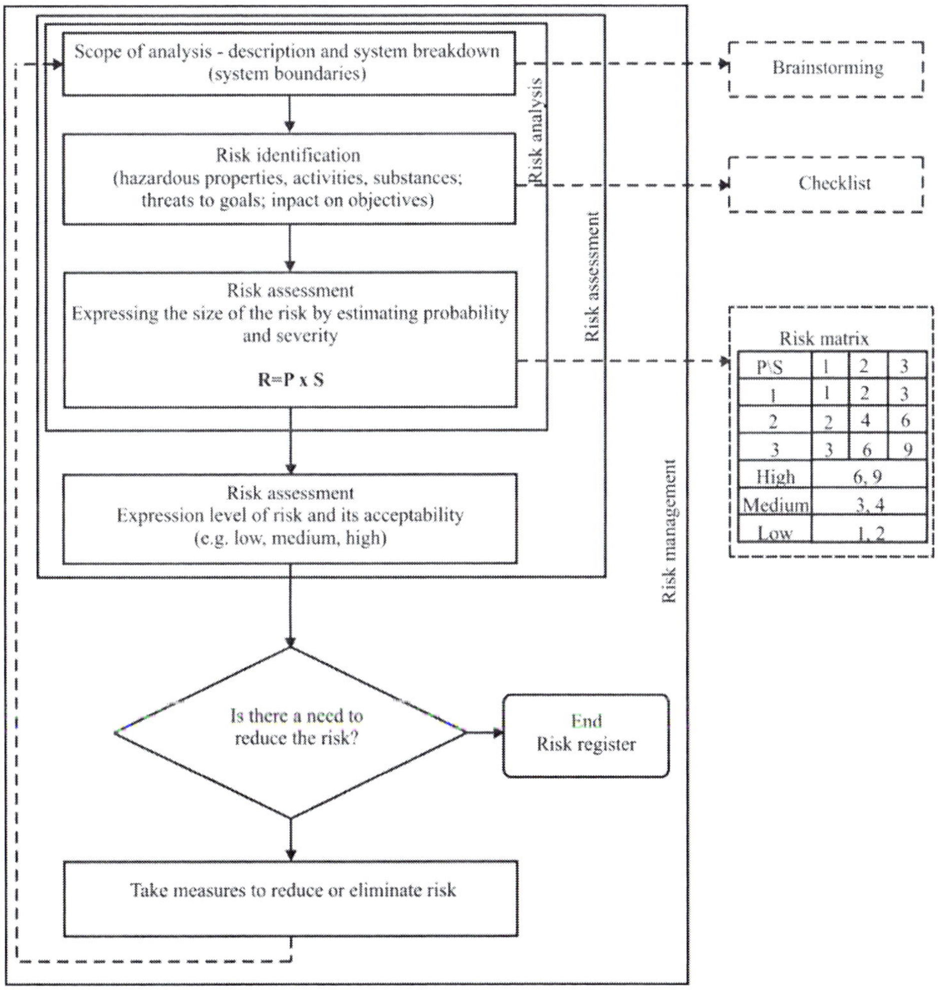

Fig. 2. Risk management chart flow (adapted from [8])

The main task of risk identification is to identify and accurately describe all possible risks that may arise in the given system, based on system information, results of expertise and experience from similar systems. This is an important stage of the analysis, because if the risks are not identified at this stage, then they are not analysed. It is very important to anticipate possible human errors and failures and to consider the relationship "human-system-environment". There are a number of formal risk identification methods.

Risk evaluation is the process of determining risk parameters and their relationship. The risk associated with the type of threat is a function of the severity of the damage (consequence) that may cause and the likelihood of this damage occurring.

The initial information and the results of the risk assessment (risk assessment) must also be documented. In principle, the risk analysis process can be completed at the stage of its identification. After the risk identification phase, the risk assessment stage follows. The final step is to develop risk-mitigation recommendations (risk management) if the level of risk is higher than the admissible level. The risk assessment process is closely linked to the risk analysis. It is a process aimed at determining the magnitude (extent) of the risk of the hazard analysed for the protection of human health, material values, the environment and other situations associated with the emergence of danger. Risk assessment can be defined as a systematic process of evaluating and interpreting real system information that identifies the threat, the consequences of the threat, and quantifying or qualitatively assessing the magnitude of the risk and deciding whether or not it is acceptable.

In assessing the risk, it is first necessary to obtain an overall risk estimate that we can derive from a combination of the likelihood of occurrence of the risk and the severity of its consequences. By combining them we get a matrix of risk. Consequently, the risk can be expressed, even though it is actually invisible and intangible [9].

Risk assessment is the step in which the identified risks must be compared on the basis of criteria with an acceptable level of risk to eliminate the hazard with an unacceptable level of risk. This step serves as a basis for developing recommendations and risk mitigation measures. Acceptable risk criteria as well as risk assessment results can be expressed qualitatively, quantitatively or semi-quantitatively.

In the process of risk management, the so called principle of ALARP ("as low as reasonably practicable") is usually applied, the fundament of which is to reduce the risk to the level that is reasonably acceptable. This principle works with risks located in an acceptable area (Fig. 3.).

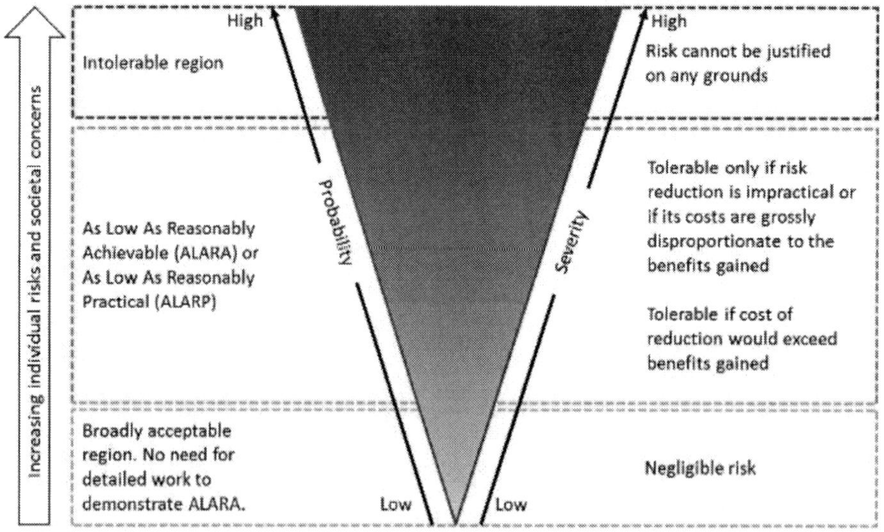

Fig. 3. ALARP principle [8]

When risk management measures are adopted, residual risk must be documented, regularly monitored and reviewed in cooperation with stakeholders. It is necessary to take into account the effectiveness and efficiency of the measures. Bad management decision made on the basis of ignorance or lack of experience may to increase the risk or create a different type of threat by adopting inappropriate measures.

3.1 Safety criticality – ERA approach

EU member states have until now developed their safety rules and standards mainly on national lines, based on national technical and operational concepts. Simultaneously, differences in principles, approach and culture have made it difficult to break through the technical barriers and establish international transport operations.

Directive 91/440/EEC, Directive 95/18/EC on the licensing of railway undertakings and Directive 2001/14/EC on the allocation of railway infrastructure capacity and the levying of charges for the use of railway infrastructure and safety certification provide the first steps towards regulation of the European rail transport market by opening the market for international rail freight services. However, the provisions on safety have proved to be insufficient and differences between safety requirements remain, which affect the optimum functioning of rail transport in the Community. It is of particular importance to harmonise the content of safety rules, safety certification of railway undertakings, the tasks and roles of the safety authorities and the investigation of accidents [10].

Important role in railway operation safety was recognised in the area of maintenance. Therefore, ERA (European Railway Agency) is responsible for certification of ECM (Entities in Charge of Maintenance) with the aim of enhancing maintenance system, in the first stage, of freight wagons, focused on risk reduction by properly managed and executed maintenance works. Each freight wagon has to have its ECM registered who is responsible for its maintenance.

In 2016, the 4th railway package1 introduced the term "safety critical components" [1]. EU railway legislation does not contain though any clear definition of which component can be characterised as safety critical. In 2016, ERA conducted several informal and formal consultations to define the state of play in the area. In general terms, the only output of the consultation conducted was that currently there is no list defining which components of the railway system are safety critical [11].

A harmonised list could vary among the different railway systems in the EU member states, utterly considering potential diverse factors present in the different member states, such as the environmental conditions including the geographical scope, the safety objectives, the km or the operational hours, the operational processes, the maintenance context, the time (lifecycle) and the design of each different railway system. Hence, depending on the situation, a harmonised list for all EU member states could be either non-complete or too exhaustive, which could unavoidably result in non-sustainable increase of cost in design, use and maintenance of the technical systems.

Safety criticality approach is closely related not only with the systems but also with the human's performance. Multi-disciplinary teams should be assigned in order to identify and classify significant Safety Critical Events by taking into account not only the probable failures but also the operational and maintenance procedures which contribute to the limitation of the failures.

4 Selected safety critical components on railway vehicle

Undoubtedly, wheelsets and brake are safety critical components on a railway vehicle. However, their failures have different potential consequences which is discussed below.

4.1 Serial systems

Typical subsystems on rolling stock without possibility of redundancy (back-up) are running gears, in general consisting of wheelsets (2 wheels pressed on axle), bearings, suspensions and wheelsets guidance. In particular wheels and axles are exposed to high dynamic loads exposed to very high number of loading cycles – typical for fatigue loads. Any serious damage on wheels or axle will result in serious accident.

Serial system (system, where failure on a single subsystem causes failure of the whole system) is schematically shown in (Fig. 4.).

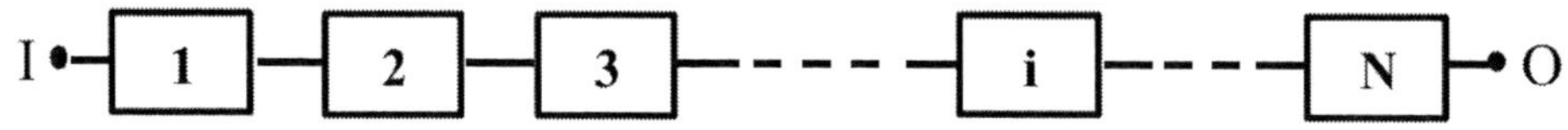

Fig. 4. Reliability block diagram of a serial system

Reliability of the serial system $R_S(t)$ consisting of N components (subsystems)can be calculated by multiplication of reliabilities of the individual components $R_i(t)$ [13]:

$$R_S(t) = R_1(t) \cdot R_2(t) \cdot .... \cdot R_i(t) \cdot .... \cdot R_N(t) = \prod_{i=1}^{i=N} R_i(t) \quad (1)$$

In case the components have exponential reliability distribution with failure rate λ_i, the reliability of the serial system can be calculated by multiplication of reliabilities of the individual components (subsystems):

$$R_S(t) = \exp(-\lambda_S \cdot t) = \prod_{i=1}^{i=N} \exp(-\lambda_i \cdot t) = \exp\left[-\sum_{i=1}^{i=N} \lambda_i \cdot t\right] \quad (2)$$

According to [2] for technical systems where a functional failure has credible direct potential for a catastrophic consequence, the associated risk does not have to be reduced further if the rate of that failure is less than or equal to $\lambda = 10^{-9}$ per operating hour.

Typical component with such catastrophic failure consequences is a wheelset. Just to have an idea what this requirement means let us consider a typical freight train composed of 25 four-axle wagons, which is with 100 wheelsets. With "rough" estimation of 2500 hours operation per year we get 0.99975003125 reliability of the train related to running gears, or 0.00024996875 failure probability.

For estimation of real failure rate of broken wheelsets in Europe (EU28) we use the statistics on European railways during the years 2012-14 (Fig. 5) where there are about 70 failures on average per year. From the statistics of freight transport in EU28 [4] we get about 800 million train kilometres per year which gives about 8000 freight trains operated 2500 hours at average velocity of 40 km/h within one year in the EU28. Using the assumed composition of trains (100 wheelsets) we get failure rate $\lambda = 10^{-7.456}$ per operating hour. Comparing with the required value of $\lambda = 10^{-9}$ there is still space for improvements (more than 10 times) resulting in about 5 broken wheelsets on freight trains in Europe per year.

Sufficiently low enough? The estimated numbers are "rough" and highly simplified, but prove a realistic value of 10^{-9} requirements for failure rate of safety critical components.

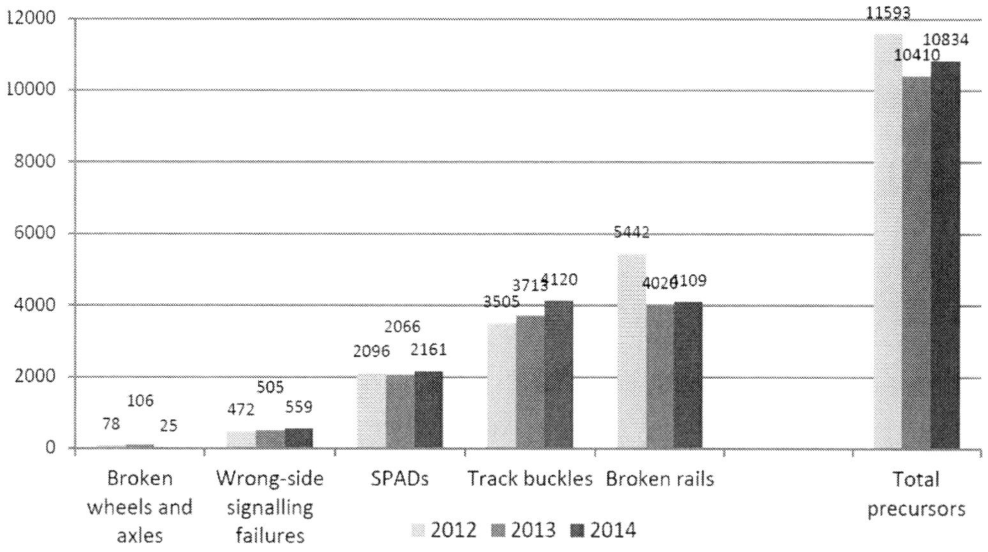

Fig. 5. Accident precursors (EU28. 2012- 2014)

4.2 Parallel (redundant) systems

It may seem strange, but a subsystem on a train that has multiple redundancies is a brake system. Many railway people say that brake system is the most important from the reliability point of view. This is only partly true – stop safely train is very important, but failure of the individual brake on a wagon in a long train creates almost no problem as there are other functioning brakes in the remaining wagons of the train.

Parallel system where failure on a single subsystem is backed-up by other subsystems is schematically shown in the Fig. 6.

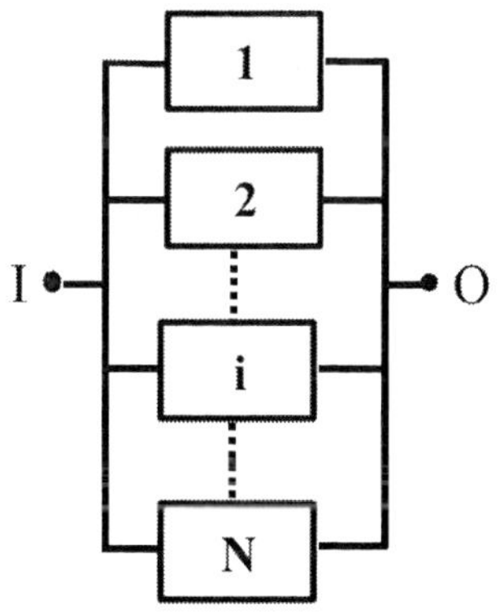

Fig. 6. Reliability block diagram of a parallel system

Reliability of the parallel system $R_S(t)$ consisting of N components (subsystems)can be calculated by multiplication of reliabilities of the individual components $R_i(t)$:

$$R_S(t) = 1 - \prod_{i=1}^{i=N} [1 - R_i(t)] \tag{3}$$

Failure probability $F_S(t)$ of the whole system is:

$$F_S(t) = 1 - R_S(t) = \prod_{i=1}^{i=N} [1 - R_i(t)] = \prod_{i=1}^{i=N} F_i(t) \quad (4)$$

To be more accurate, the brakes are not fully backed up. In fact, in 25 wagons freight train when on brake does not work, the braking effect is reduced by 4%, but still sufficiently high for effective braking. For emergency stopping the train usually 80% of active brake could be sufficient, which on 25 wagons train represent 20 wagons with functioning brake.

For this calculation of reliability, it is more appropriate to use case of components in a k-out-of-n configuration. In this case, the reliability of the system with such configuration can be evaluated using the binomial distribution:

$$R_S(k, n, R) = \sum_{r=k}^{n} \binom{n}{r} R^r (1-R)^{n-r} \quad (5)$$

where n is the total number of units (subsystems), k is the minimum number of units required for system success, R is the reliability of each unit.

So when we take data recorded from freight wagons maintained by one maintenance workshop in Slovakia (Fig. 7.), where there were 280 brake failures on 346 wagons during 5-year period. Count represents Pareto analysis of number of failures during 5-year period. Brake failure modes are coded "3XX" and one can see that among the top 6 most frequent failures 4 of them are brake failures (on X-axis). From the data we get 0.162 failures on each wagon per year on average, which is $6.474 \cdot 10^{-5}$ failure rate per operating hour; much higher (about 10^4) than required for safety critical components.

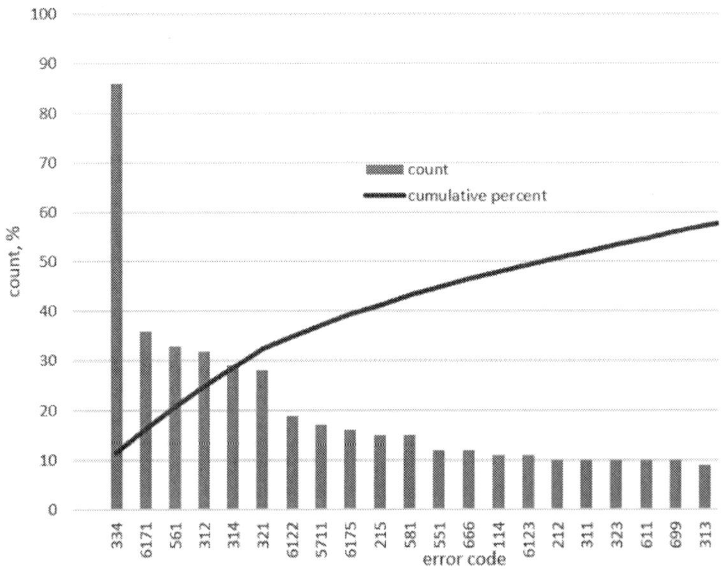

Fig. 7. Pareto analysis of failures on monitored freight vehicles

But if we consider 20 wagons with functioning brake out of 25 sufficient for fail-safe braking of a train, substituting to (5) we receive $R_S = 0.84$, which is acceptable value. In Fig. 8 is a diagram showing increasing reliability with decreasing number of functioning brakes needed for fail-safe (emergency) stop of a freight train.

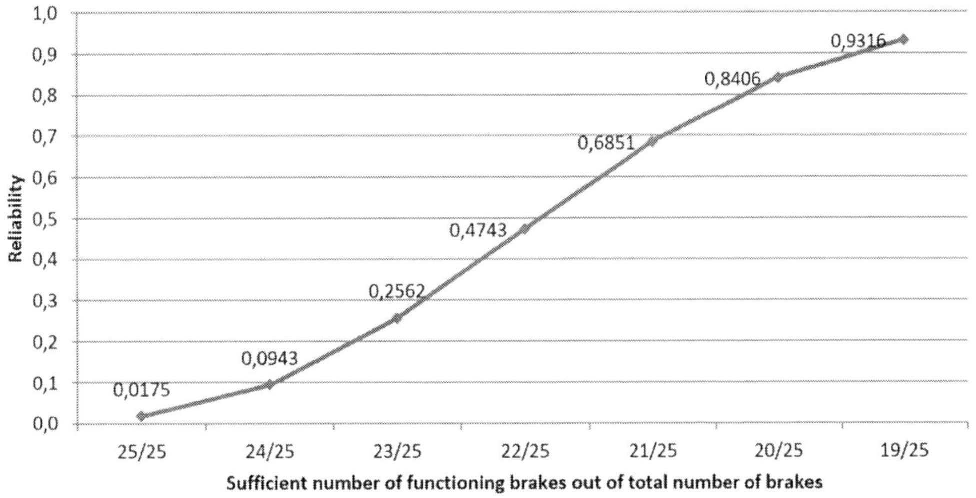

Fig. 8. Reliability of brake on train considering k-out-of-n configuration

Conclusion - play with risks

A "risky combination" is envisaged on European railways. We have discussed situation with accidents where there are some 70 broken wheels and axles annually in EU28 countries. Not too high but should be less if target failure rate of safety critical components will be achieved. However, there are projects with replacement of standard cast iron brake shoes with composite brake shoes. The reason is that they are silent during braking, but have lot of side effects literally threatening railway safety as they damage wheel running surface and in worst case they cause cracks of wheels (Fig. 9). Wheels are typical and perhaps the most important safety critical components used on railway vehicles. There are strict rules to be kept in operation and maintenance to avoid unacceptable damage and to prevent accidents, which usually end in serious consequences. So it is a risky play to change operational conditions caused by new material of brake shoes on a proven design of railway wheelset. Or railway operators must be very cautious with introducing new type of brakes.

Fig. 9. Damaged wheel [14]

The paper was supported by the Scientific Grant Agency of the Ministry of Education of the Slovak Republic and the Slovak Academy of Sciences in project no. VEGA 1/0766/15 "Research sources of noise emissions from rail transport".

References

1. Directive (EU) 2016/798 of the European Parliament and of the Council of 11 May 2016 *On Railway Safety* (2016)
2. Commission Regulation (EC) No 352/2009 of 24 April 2009 On the adoption of a common safety method on risk evaluation and assessment (2009)
3. D. Jovovic, *Collection of examples of risk assessments and of some possible tools supporting the CSM Regulation*. European Railway Agency. Valenciennes: European Railway Agency (2009)
4. Railway safety performance – biennial report, European Union Agency for Railways, ISBN 978-92-9205-049-8 (2016)
5. B. S. Dhillon, *Human Reliability, Error and Human Factors in Engineering Maintenance*. USA: CRCpress. ISBN 978-4398-0383-7 (2009)
6. B. Dhillon, *Transportation Systems Reliability and Safety*. (CRC Press, Boca Raton, ISBN 978-1-4398-4640-7, 2011)
7. H. Pačaiová, J. Sinay, J. Glatz, *Safety and risks of technical systems*. (Vienala, Košice, ISBN 978-80-553-0180-8-60-30-10, 2009)
8. H. Pačaiová, Š. Markuliak, A. Nagyová, *The importance of risk in management systems*. (BEKI Design, s.r.o., Košice, ISBN 978-80-553-2618-4, 2016)
9. P. Baybutt, *Calibration of risk matrices for process safety*. Journal of Loss Prevention in the Process Industries **38**, 163-168 (2015)
10. P. Zvolenský, V. Stuchlý, J. Grenčík, R. Poprocký, *Evolution of maintenance systems of passenger and freight wagons from the ECM certification point of view*. Communications: scientific letters of the University of Žilina **16** (3A), 38-45 (2014)
11. Directive 2001/14/EC On the allocation of railway infrastructure capacity and the levying of charges for the use of railway infrastructure and safety certification (2001)
12. Commission Regulation (EU) No 445/2011 on a system of certification of entities in charge of maintenance for freight wagons and amending Regulation (EC) No 653/2007 (2011)
13. V. Stuchlý, R. Poprocký, *Údržba strojov a zariadení*. (Žilinská univerzita, 359 p., ISBN 978-80-554-0845-3, 2013)
14. Letter ANSF protocol 003081/2017 pf 21/03/2017, Update of the Safety Alert related the broken wheel (Manufacturer RAFIL) (2017)

Design of stirling engine operating at low temperature difference

Josef Sedlák[1], *Adam* Glváč[1], *Andrej* Czán[2,*]

[1]Department of Machining Technology, Institute of Manufacturing Technology, Faculty of Mechanical Engineering, Brno University of Technology, Technická 2896/2, Brno 616 69, Czech Republic
[2]Faculty of Mechanical Engineering, University of Zilina, Univerzitna 8215/1, 010 26 Zilina, Slovakia

Abstract. There are many sources of free energy available in the form of heat that is often simply wasted. The aim of this paper is to design and build a low temperature differential Stirling engine that would be powered exclusively from heat sources such as waste hot water or focused solar rays. A prototype is limited to a low temperature differential modification because of a choice of ABSplus plastic as a construction material for its key parts. The paper is divided into two parts. The first part covers a brief history of Stirling engine and its applications nowadays. Moreover, it describes basic principles of its operation that are supplemented by thermodynamic relations. Furthermore, an analysis of applied Fused Deposition Modelling has been done since the parts with more complex geometry had been manufactured using this additive technology. The second (experimental) part covers 4 essential steps of a rapid prototyping method - Computer Aided Design of the 3D model of Stirling engine using parametric modeller Autodesk Inventor, production of its components using 3D printer uPrint, assembly and final testing. Special attention was devoted to last two steps of the process since the surfaces of the printed parts were sandpapered and sprayed. Parts, where an ABS plus plastic would have impeded the correct function, had been manufactured from aluminium and brass by cutting operations. Remaining parts had been bought in a hardware store as it would be uneconomical and unreasonable to manufacture them. Last two chapters of the paper describe final testing, mention the problems that appeared during its production and propose new approaches that could be used in the future to improve the project.

Keywords: Stirling engine, Stirling thermodynamic cycle, additive method, rapid prototyping, fused deposition modelling, 3D print

1 Introduction

A Stirling engine was officially presented to the public on 27th September 1816, when twenty-six-years-old Scottish farmer Robert Stirling (1790-1878) patented his discovery under the number 4081. In that era this marvelous device represented an alternative to a steam engine, which was preferred type of driving during and long after the Industrial

* Corresponding author: andrej.czan@fstroj.uniza.sk

Reviewers: *Ivan* Kuric, *Richard* Pastirčák

Revolution. The fact that the machine had been patented eight years before the Sadi Carnot's monograph "Reflections on the Motive Power of Fire and on Machines Fitted to Develop that Power" was published confirms Robert's mastery of thermos-dynamical principles long before any comprehensive heat-engine theory existed [1]. The original patent actually specifies two separate discoveries. The first part and the principle number one was so-called "economizer"- a device nowadays known as a regenerator [2, 3]. This was definitely a very significant invention and has been used until nowadays not only in the Stirling engines, but also in many other devices. The original description of the regenerator speaks about a space filled with "successive layers of plates of the thinnest iron in use, pierced with holes....kept at a distance of two or three times of their own thickness from one another" [2]. The purpose of the regenerator is to act as a temporary heat storage element, so that as hot and cold gas is passed back and forth between the regenerator, heat is removed from the gas and then put back into it. This reduces the amount of heat needed to be transferred by the heat exchangers and increases dramatically the overall efficiency of the engine. The second part of the patent was the engine itself, which was presented as "A Closed-Cycle Air Engine", using the economizer to "diminish the consumption of fuel" [2]. Devices that were based on the principle of Stirling's patent, were being manufactured in a great variety of shapes and sizes until the turn of the centuries. They broke through thanks to their simplicity, safety, a silent operation and a possibility to be powered by any type of the fuel. Compared with the steam engines, they were environmentally-friendly and their efficiency was significantly higher, which was the main reason of their fast grow [2, 3]. In the course of the 19th century and at the beginning of the 20th century, Stirling engines were manufactured for a variety of applications like pumping water for livestock, on the railways or to supply the water to municipalities [2]. Back then its development was quite limited by the metallurgical possibilities of the era. For this reason, as well as due to its higher weight it has been slowly replaced by recently developed and fast-growing internal combustion and an electric motor [2, 3, 4]. However, nowadays it is experiencing kind of renaissance thanks to its characteristics, which are irreplaceable by any other type of drive. At present the manufacturers can use modern materials to build the engine, therefore it can work under higher pressures and higher temperature differences, which makes it far more effective and powerful than it used to be. The most prominent of European companies is company Solo Stirling GmbH, a German company which is engaged in the production of cogeneration units. The cogeneration unit made by this concern can achieve a power of 7.5 kW at the pressure of 13 MPa with the efficiency of 24%. The device is filled with helium with a temperature rising up to 650 ° C in operation. A service interval is 5,000-8,000 hours and normally it is necessary to change only an oil and fill the cylinders with helium [3]. The leader in a development and production of Stirling engines is Swedish company cKockums AB, which has been focused to the development and production of high-performance power units for submarines. These have been powered by flawless engines with an output of 75 kW. Use of the Stirling engine in the submarine's power unit is appropriate especially due to its silent operation and absence of vibrations (submarines become more difficult to be spotted). In addition, the unit does not require any air for its operation and is surrounded by water from all sides, which makes the cooling very simple and effective [3].

2 Parts and modifications of Stirling engine

The Stirling engine consists of 4 main parts that are indispensable for its operation. In the Fig. 1 there is a simple diagram of the engine developed by Robert Stirling. In this case, the regenerator and the cooler is separated from the engine body, however it does not have to be always like that. The function of heat exchangers is to supply heat and create two poles with two different temperatures [2, 5]. A displacer moves a gas between a hot side and a cold side of the engine evenly which warms it up and cools it down periodically [2, 5]. This ensures a cyclic expansion and a compression of the gas. The periodical expansion and compression of the gas move a power piston, that transmits work to a driven mechanism of the machine [2, 5]. The most sophisticated part of the Stirling engine is the regenerator. It is usually made of a conductive material and works as a temporary storage of the heat. The regenerator can quickly take the heat from the warm gas that flows through it and releases it back right at the moment when it starts to flow in the opposite direction. The result is a significant increase of a device efficiency. John Ericsson stressed the regenerator importance, by his equation: "...We will show practically that bundles of wire are capable of exerting more force than shiploads of coal..." (Said in 1855) [2].

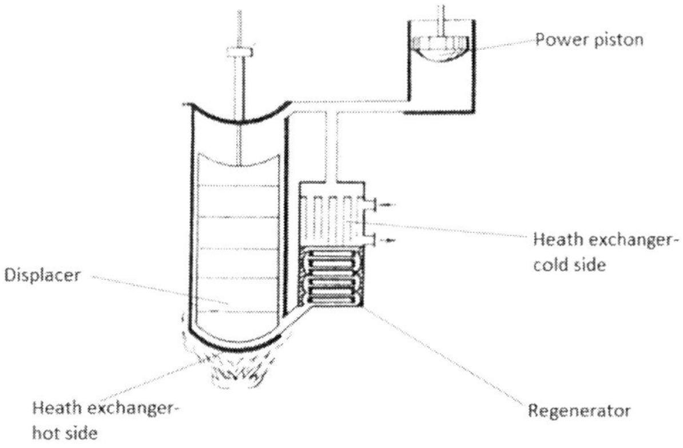

Fig. 1. Drawing of Engine Developed by Stirling Brothers in 1840 [1]

There are 3 different modifications of the engine and are usually marked by letters of Greek alphabet α, β and γ [1, 2, 3, 5]. They all are based on the same principle, which is shown in the Fig. 1. They only differ from each other by an arrangement and shape of the individual components.

The α configuration has two pistons, which are stored in two separate cylinders. These are connected to the heater, regenerator and cooler. From a mechanical point of view this is a simple configuration that stands out for its high ratio of performance to the cylinders volume. Its disadvantage is that both pistons must be sealed against a leakage of gas, which is a difficult task particularly in the case of the piston that works on the hot part of the engine [1, 2, 3, 5]. For the configuration β it is typical that the displacer and the power piston share only one cylinder. Therefore, it is a quite compact design, which stands out thanks to its small dead volume (the volume of gas that does not take part in the cycle). From the mechanical point of view, it is a very simple device, since the displacer and power piston are both connected to the same point at a crankshaft. The only complication is that the displacer's rod, which connects it to the shaft, passes through the power piston. This produces higher demands on the sealing of the gas in the cylinder [1, 2, 3, 5]. The γ and the β configuration are very similar. The only difference is that the displacer and the power piston of the modification γ are stored in two separate cylinders. This minimalizes demands on the sealing of the gas. The modification is only suitable for very small engines, which power is in hundreds of Watts [1, 2, 3, 5]. However, in comparison with others, this configuration can be built quite easily from cheap materials using exclusively simple hand tools. If the engine is designed ingeniously, it is possible to run it even when the temperature difference between two sides of the heat exchanger is almost imperceptible. Because of the use of the ABS plastic as the construction material, this is one of the main requirements for this project.

3 Thermal efficiency of engine with configuration γ which uses Stirling cycle for its operation

A special characteristic of the modification γ is that the cylinder that stores the power piston is often located at the cold side of the engine. Therefore, there is always a small amount of the unheated gas in the cylinder, even after the stage of isochoric heating. To simplify a calculation, it is often necessary to make an assumption that the temperature of the unheated gas equals the temperature of the cold side of the heat exchanger [5].

A total amount of the working gas in the cylinder can be calculated using the following equation:

$$V_{total} = V_{hot} + V_{cold} \qquad [m3] \tag{1}$$

Where: $V_{total}\,[m^3]$ - is the total volume of the working gas,
$V_{hot}\,[m^3]$ - is the volume of the heated working gas,
$V_{cold}\,[m^3]$ - is the volume of the unheated working gas.

A total amount of the substance can be calculated using the following equation:

$$n_{total} = n_{hot} + n_{cold} \qquad [mol] \tag{2}$$

Where: $\eta_{total}\,[mol]$ - is the total amount of the working gas,
$\eta_{hot}\,[mol]$ - is the amount of the heated working gas,
$\eta_{cold}\,[mol]$ - is the amount of the unheated working gas.

The ideal gas equation for the working gas in the point 3 of a p-V diagram, which is shown in the Fig. 2, has the following form:

$$n_{total} = \frac{pV_{hot}}{RT_{hot}} + \frac{p(V_{total} - V_{hot})}{RT_{cold}} \qquad [mol] \tag{3}$$

Where: p [Pa] - is the pressure of the working gas,
$R\,[J.K^{-1}.mol^{-1}]$ - is a gas constant,
T_{hot} [K] - is the temperature of the heated working gas,
T_{cold} [K] - is the temperature of the heated working gas.

From the p-V diagram shown in Fig. 2, it is obvious that the work done during the expansion of the isothermal expansion is equal to the surface under the curve W_e between the points 3 and 4. Because there are no losses to consider, the work done for each part of the cycle is the same as heat added to the gas:

$$W_e = Q_e = \int_3^4 p dv \qquad [J] \tag{4}$$

Where: W_e [J] - is the work done during the isothermal expansion,
Q_e [J] - is heat added to the gas.

By expressing the pressure p used in the equation number (3), and by its substitution in equation (4), the following relation can be obtained:

$$Q_{expansion} = W_e = n_{total} RT_{cold} \int_3^4 \frac{dv}{V_{total} - V_{hot}\left(\frac{T_{hot} - T_{cold}}{T_{hot}}\right)} \qquad [J] \tag{5}$$

The compression ratio of the engine is the quotient of the maximum and the minimum volume of the system in the cycle. In this case:

$$\varepsilon = \frac{V_4}{V_3} \quad [-] \tag{6}$$

Where: ε [-] - is the compression ratio,
V_4 [m^3] - is the volume of the working gas in the point 3,
V_3 [m^3] - is the volume of the working gas in the point 4.

By integrating of the expression (5) and using the equation for the compression ratio (6), it is possible to derive an equation to calculate heat added to the system during the isothermal expansion:

$$Q_e = W_e = n_{total} R T_{cold} \ln \frac{\varepsilon - \frac{T_{hot} - T_{cold}}{T_{hot}}}{1 - \frac{T_{hot} - T_{cold}}{T_{hot}}} \quad [J] \tag{7}$$

Furthermore, it is necessary to derive the equation which would define the work that performs the gas in the stage of the isothermal compression. This action takes place between points 1 and 2 in the Fig. 2. At this stage, there is no heated gas in the system. Therefore the temperature equals to T_{cold}. The gas equation for this stage has the following form:

$$pV_{total} = n_{total} R T_{cold} \quad [Pa.m3] \tag{8}$$

The work that is necessary for the compression of the working gas during the isothermal compression is equal to the surface under the curve 1 and 2 in the p-V diagram in the Fig. 2. Substituting the pressure p by (8):

$$W_{compresion} = \int_1^2 p dv = n_{total} R T_{cold} \int_1^2 \frac{dv}{V_{total}} \quad [J] \tag{9}$$

The total net work obtained in one cycle W_{net} is the sum of the work done during the compression and the following expansion of the gas:

$$W_{net} = W_e + W_c \quad [J] \tag{10}$$

Where: W_{net} - is the total net work obtained in once cycle.

Substituting (7) and (9) to (10):

$$W_{net} = n_{total} R T_{cold} \ln \frac{\varepsilon - \frac{T_{hot} - T_{cold}}{T_{hot}}}{1 - \frac{T_{hot} - T_{cold}}{T_{hot}}} - n_{total} R T_{cold} \ln \varepsilon \quad [J] \tag{11}$$

The total cycle efficiency is defined as the ratio between the total net work output and the total heat input. Substituting W_{net} by (7) and Q_e by (11) the following equation is obtained:

$$\eta_{t,\gamma} = \frac{W_{net}}{Q_e} = 1 - \frac{n_{total} R T_{cold} \ln \dfrac{\varepsilon - \dfrac{T_{hot} - T_{cold}}{T_{hot}}}{1 - \dfrac{T_{hot} - T_{cold}}{T_{hot}}} - n_{total} R T_{cold} \ln \varepsilon}{\ln\left(\varepsilon - \dfrac{T_h - T_c}{T_h}\right) - \ln\left(1 - \dfrac{T_h - T_c}{T_h}\right)} \quad [-] \qquad (12)$$

Where: $\eta_{t,\gamma}$ - is the thermal efficiency of the modification γ.
Or more simplified (12):

$$\eta_{t,\gamma} = \frac{W_{net}}{Q_e} = 1 - \frac{\ln \varepsilon}{\ln\left(\varepsilon - \dfrac{T_h - T_c}{T_h}\right) - \ln\left(1 - \dfrac{T_h - T_c}{T_h}\right)} \quad [-] \qquad (13)$$

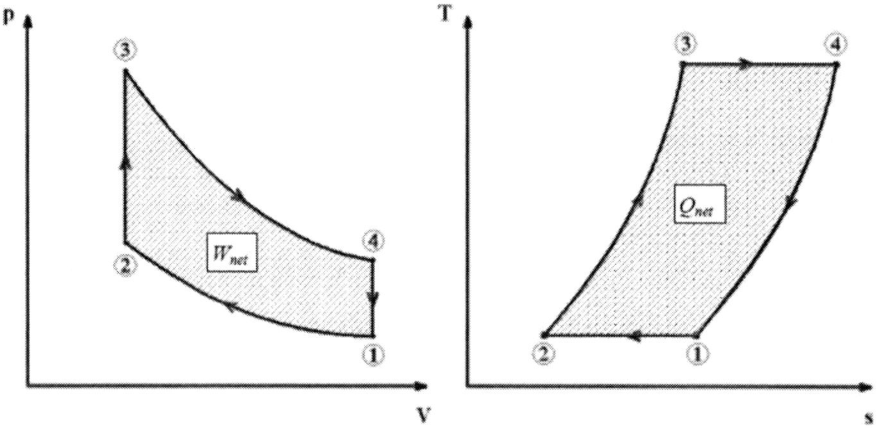

Fig. 2. p-V and T-s Diagram for Ideal Stirling Cycle during Isothermal Compression [2]

Fig. 3 and Fig. 4 have been created using the equation (13). Even though this relation is valid only for the ideal cycle, it expresses two essential facts about the γ modification of the engine:

- When the compression ratio rises, the thermal efficiency of the engine decreases.
- When the temperature difference rises, the efficiency increases logarithmically.

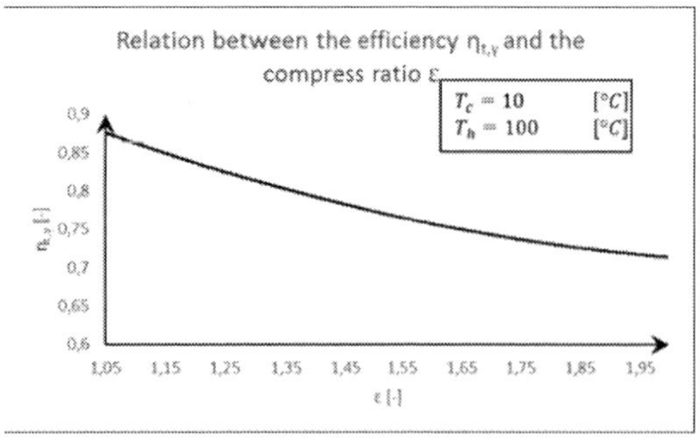

Fig. 3. Relationship between Efficiency and Compress Ratio [3]

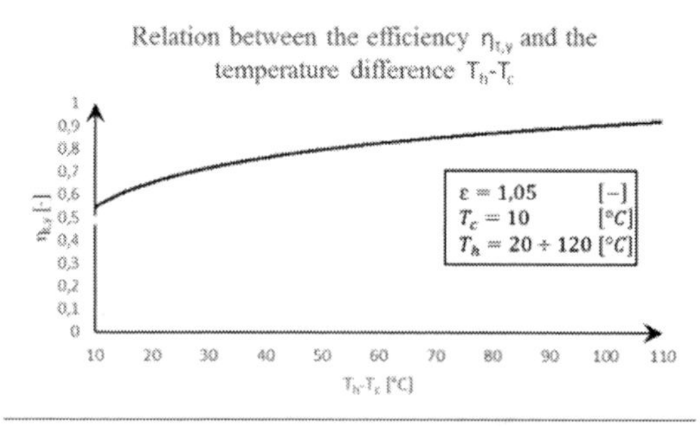

Fig. 4. Relationship between Efficiency and Temperature Difference [3]

4 Design of Stirling engine

The aim of this project is to make a model of the engine that uses the principle of the Stirling cycle for its operation. All parts with the complex shape are manufactured using the additive technology Fused Deposition Modeling (FDM). Remaining parts are either produced using hand tools or simple industrial machines, or bought in the hardware store. These are mainly screws, sealing and wires, which would be uneconomical to manufacture, or the sheets of metals coming to a direct contact with augmented heat during the engines operation. Because the ABS plus plastic has low resistance against high temperatures, the whole concept is designed to work at the temperature difference lower than 50 °C.

The idea of the whole design is to make the compact engine that would be extremely easy to assemble and disassemble and would be built from the parts that can be easily replaced in a case of optimization. It is intended that the engine would serve for training purposes, therefore there is an effort to create a structure that would withstand rather incautious treatment. For this reason, it is not expected to achieve significant performance or high revolutions. The goal is to make the engine strong enough to be able to stay in the operation as long as it is possible at extremely low temperature difference. Digital visualization of the model is shown in the Fig. 5 and Fig. 6.

The engine has been composed of 11 parts produced by the technology FDM. Low thermal conductivity of the ABS plus plastic is beneficial for the parts like the displacer or cylinder. For the rest of the printed parts, its thermal conductivity does not affect their proper function.

In cases, where material properties of the ABS plus plastic impede the proper function of the parts are designed to be as simple as possible, so they could be manufactured using hand-tools or simple industrial machines. Those are the aluminum plates forming sides of the heat exchanger and have to be made out of materials with a high thermal conductivity. It is decided to use a 2.5 mm thick aluminum sheet that is shaped by shearing. The work piston and the cylinder are machined using a lathe. It is not possible to find more adequate technology, because the power piston has to be stored in the cylinder with high precession so it can move from end to end with minimal friction.

The remaining parts that formed the engine are standardized and available in the hardware store. These are a set of M3 size screws of various lengths, nuts, washers, sealing, wires, bearings and a shaft.

Fig. 5. Digital Visualization of Stirling Engine [6]

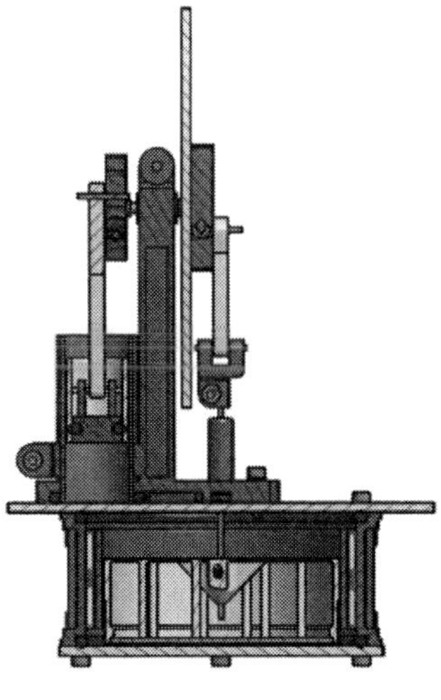

Fig. 6. Cut through Model [6]

5 Manufacturing of model

The digital model is created in a parametric modeler Autodesk Inventor. All parts are saved in a stereolithography (*.stl) format afterwards. The *.stl format is the standard format for a majority of rapid prototyping technologies. A volume of parts saved in the *.stl format is a compound of an unorganized mesh of triangular shaped surfaces [7, 8, 9, 10]. The position of each of these triangles is given by the coordinates X, Y, and Z of each of its vertices and a normal vector to its surface. The direction of the vector determines the orientation of the triangle. It defines the side of the triangle that would be oriented outwards of parts body [7, 8, 9]. The next essential finding is that a geometry of the *.stl part is always an approximation of the shape created by the parametric modeler. How close the shape of the *.stl part would get to the shape of the parametric model is determined by the density of the triangular mesh [9, 11]. When the model is converted to the *.stl format, various types of defects can be generated in the mesh. The most common are flipped triangles, boundary edges, bad edges, holes, shells, intersecting or overlapping triangles [7, 9, 12]. The overwhelming majority of defects that are originated during the export of the parts to the *.stl format, are successfully eliminated using the software MiniMagics. Few overlapping triangles are not eliminated in a case of the parts with the most complex geometry. However, this type of defects does not affect the final result of a manufacturing process, therefore it is possible to ignore it [10, 11, 12].

After all the critical defects are eliminated, all the parts are opened one by one in the software called CatalystEx. CatalystEx is the software that communicates with all kinds of 3D printers from series Dimension made by the leading manufacturer of 3D printers - Stratasys [13, 14]. The software is able quickly and easily open 3D drawings of parts, prepare them for printing and send a print command to the printer to create the part. The software is used to scale the parts and to place them on the desired place of a pad subsequently. Correct orientation of each part in a build chamber is chosen afterwards. Finally, the support material, which prevents a collapse of the model, is added (this process is highly automated and works wonderfully).

When all these adjustments are done, estimated print time is calculated and the print command is sent to the printer. The printing process consists of two steps. All parts except the stand of the engine are printed in the first step. Printing of these 10 parts take 11 hours and 28 minutes. 101.79 cm^3 of the building material and 45.65 cm^3 of the support material is used. The Fig. 7 shows parts in the middle of the printing process and the Fig. 8 shows the parts when the printing is completed. The black colored support material is still visible attached to the model.

Fig. 7. Parts in Middle of Process

Fig. 8. Parts after Successful Printing

In the second step the stand of the engine is printed. As it is the most complex part from the whole project, it take 4 hours and 42 minutes to print it. 38.7 cm^3 of the building material and 14.42 cm^3 of the support material are used.

When the printing of the parts has been completed, all parts are separated from the pad and are immersed into the ultrasonic cleaner VGT1990QT from the manufacturer GT SONIC with the goal to dissolve the support material. The active substance is prepared by mixing WaterWorks Soluble Concentrate P400SC from Stratasys and water. After the parts have been removed from the cleaners tub, smoothness of surfaces is satisfactory. Therefore only the outer side of the displacer and the inner side of the cylinder have to be finished afterwards. These surfaces are sanded to achieve a smooth touch in the contact area. Sandpapers of roughness P400 and P600 are used for the job. After the surfaces have been sanded, an acrylic spray paint is applied to cover minor bumps. Sanded surfaces are satisfactory and the displacer could move along the cylinder without rubbing its inner side. All the holes presented at the model have been purposely undersized when the model has been designed. Therefore, the last part of this stage is to adjust them to the wanted size. Finally, all the parts are assembled. The Fig. 9 shows the engine after the assembly.

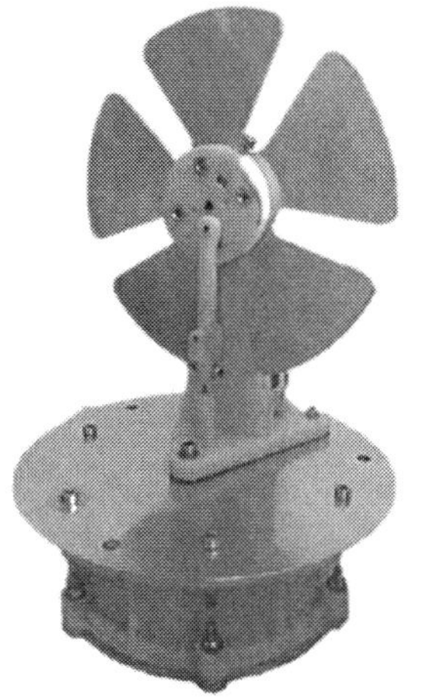

Fig. 9. Fully Assembled Engine

6 Discussion

Once the model was assembled, tests were performed. The warm side of the heat exchanger was placed on the cup filled with hot water. To ensure the sufficient temperature difference between the poles of the motor, ice was laid on the cold side. In this state, the engine was left for two minutes to allow the gas to get warm. After that, the engine was started giving the light spin to the flywheel. The flywheel made a couple of turns, but then it started to slow down gradually until a standstill. The engine while in work was too loud, which was due to the friction between the parts of the engine. The model was subsequently dismantled. The reason for the increased friction could be identified. The wire that connects the fork with the displacer was damaged. The wire was straightened and lightly sanded. The airtightness of the adhesive bond between the two parts of the displacer was tested as well. After dipping the displacer under water to a depth of 20 cm no leakage of the gas was observed. The airtightness was therefore evaluated as sufficient. Subsequently, the model was re-assembled and ready for the second test.

During the second test the engine was significantly quieter, but it failed to run smoothly. The reason for the failure was a high mass of the aluminum flywheel. It has been replaced by gluing 3 compact discs together, which together weigh 45 grams. Compared to the original aluminum flywheel it was 40% lighter.

In the third test the flywheel was spun before and after heating. It could be observed that the warm engine makes more spins as the cold engine. This difference was not observed when the aluminum flywheel was used. It could therefore be postulated that the weight of the flywheel has the significant influence on engine performance and that the weight of 45 grams is more appropriate than the weight of 75 grams.

After the third failed test the overall airtightness of the engine was tested. The test was performed by raising the power piston to its upper position and its free falling. The piston fell to its lowest position at a steady rate. This process was repeated 10 times and the average time of a descent of 0.98 seconds was measured. Expected time providing good sealing were 4 seconds. It was therefore clear that the engine was not sufficiently sealed. For further measurements the engine was dismounted, the displacer was removed and the hole in the stand has been blocked. The airtightness test was carried out again. In the second measurement average time of the descent was 1.32 s. An additional significant leakage of the gas took place in the area between the power piston and the cylinder. It was found that abrasive wearing of the outer wall of the working piston took place and has reduced the size of the piston to such an extent that there was now a gap large enough for air to leak. The reason for abrasive wearing was the ill centered piston in the cylinder.

Conclusion

Even when several modifications were done, the steady operation of the engine was not achieved. Main factors that impede the proper function of the device are:

- As the power piston was not placed in the cylinder correctly, it rubbed against the inner side of the cylinder. Therefore the contact surfaces suffered a slight damage by an abrasion.
- The cylinders were not sealed sufficiently and the gas leaked from the engine excessively.
- The original aluminum flywheel was too heavy to operate.

Nevertheless, the model is great for further editing and enhancement, which should lead to eliminate deficiencies of the first prototype. The more precise assembly should eliminate first two factors. To reduce the weight of the flywheel it is possible to choose the material with the lower density (e.g. the ABS plastic), or by implementing a number of lightening holes into the shape of the aluminum flywheel. Ones the engine reaches the steady operation, it will be suitable for experiments that would test different types of regenerators and its influence to engines performance.

Despite the fact that the steady operation was not achieved, the model would serve for educational purposes to demonstrate the mechanism of the Stirling engine and to familiarize the audience with the principle of its operation.

This article was supported and co-financed from a specific research FSI-S-16-3717 called "Research in Field of Modern Production Technologies for Specific Applications".

References

1. J. Hromádko, *Speciální spalovací motory*. (Česká zemědělská univerzita, Praha, 1. vyd., 151 p., ISBN 978-80-213-2168-7, 2011)

2. C. Lloyd, *A low temperature differential Stirling engine for power generation. Cantebury*. Master's Thesis. University of Canterbury. Department of Electrical and Computer, available at: http://hdl.handle.net/10092/2916 (2009)
3. J. Blaha, *Stirlingův motor*. 80 p., Vysoké učení technické v Brně. Fakulta strojního inženýrství. Ústav automobilního a dopravního inženýrství. Vedoucí diplomové práce doc. Ing. Zdeněk KAPLAN, CSc. (2008)
4. Kongtragool, Bancha a Somchai WONGWISES, *A review of solar- powered Stirling engines and low temperature differential Stirling engines*. Renewable and Sustainable Energy Reviews [online] **7** (2), 131-154 [seen 22.3.2015]. DOI: 10.1016/s1364-0321(02)00053-9, ISSN 1364-0321, Available at: http://linkinghub.elsevier.com/retrieve/pii/S1364032102000539 (2003)
5. D.G. Thombare, S.K. Verma, *Technological development in the Stirling cycle engines*. Renewable and Sustainable Energy Reviews [online] **12** (1), 1-38 [seen 2015-03-22] DOI: 10.1016/j.rser.2006.07.001. ISSN 13640321. Available at: http://linkinghub.elsevier.com/retrieve/pii/S1364032106000906 (2008)
6. Autodesk Inc. ©2015, *3D CAD Sotware: Inventor 3D CAD, Autodesk: 3D Design, Engineering & Entertainment Software*. [online] [seen 2015-04-22] Available at: http://www.autodesk.com/products/inventor/overview (2015)
7. C. Chua, K. Leong, C. Lim, *Rapid prototyping: principles and applications*. (New Jersey: Word Scientific, ©2010, p. XXV, 512 p. 8. MCAE SYSTEMS, spol. s.r.o., Kuřim, ČR. Dimension uPrint (2010)
8. J. Sedlak, D. Rican, M. Piska, *Study of Materials Produced by Powder Metallurgy Using Classical and Modern Additive Laser Technology*. Procedia Engineering **2015** (1), 1232-1241, ISSN 1877-7058 (2015)
9. J. Sedlak, M. Ptackova, J. Nejedly, M. Madaj, J. Dvoracek, J. Zouhar, M. Piska *Material Analysis of Titanium Alloy Produced by Direct Metal Laser Sintering*. International Journal of Metal Casting **7** (2), 43-50, ISSN 1939-5981 (2013)
10. J. Novak-Marcincin, M. Janak, L. Novakova-Marcincinova, *Increasing of Product Duality Produced by Rapid Prototyping Technology*. Manufacturing Technology **12** (12), 71-75, ISSN 1213-2489 (2012)
11. J. Sedlak, M. Slany, Z. Fiala, A. Jaros, *Production Method of Implant Prototype of Knee-Joint Femoral Component*. Manufacturing Technology **15** (2), 195-204, ISSN 1213-2489 (2015)
12. Ch. Ch. Kuo, L. Ch. Liu, W. F. Teng, H. Y. Chang, F. M. Chien, S. J. Liao, W. F. Kuo, Ch. M. Chen, *Preparation of starch/acrylonitrile-butadiene-styrene copolymers (ABS) biomass alloys and their feasible evaluation for 3D printing applications*. Composites Part B: Engineering **86**, 36-39. DOI: 10.1016/j.compositesb.2015.10.005. ISSN 1359-8368. Available at: http://linkinghub.elsevier.com/retrieve/pii/S1359836815005983 (2016)
13. uPrint SE. Stratasys Ltd. [online] [seen 2015-05-17] Available at: http://www.stratasys.com/3d-printers/idea-series/uprint-se (2015)
14. FDM Technology. Stratasys Ltd. [online] [seen 2015-05-17] Available at: http://www.stratasys.com/3d-printers/technologies/fdm-technology (2015)

Improving the quality of cutting flat glass

Martin Svoboda[1,*], *Josef* Soukup[1], *Milan* Sapieta[2]

[1]Faculty of Pruduction Technology and Management Jan Evangelista Purkyně University in Ústí nad Labem, Department of Machines and Mechanics, 400 96, Pausterova 1, Ústí n/L, Czech Republic
[2]Faculty of Mechanical Engineering, Univerzity of Zilina. Univerzitna 1, 010 26 Zilina. Slovak Republic

Abstract. The article deals with improving the quality of the cut flat glass. The aim of the study was to determine mechanism of origin temporary and permanent stress in the glass. This affects the quality of the cut flat glass. Of the commonly measured process parameters were adjusted following an evaluation, better cooling of flat glass between the furnace and the point of cutting into a prescribed format glass. The results are applicable for flat glass production lines.

Keywords: flat glass, float system, measurement, stress

1 Introduction

The research deals with the problem of cutting quality of flat glass. There is described a draft general procedure for obtaining the best adjustment of the cooling equipment for a given thickness, width and shade of the glass, which is manufactured on a production line of flat glass - float system.

The most of the world's production of float glass produced by the process of "float" (float glass). Other methods of production (drawing, rolling) either gradually or completely disappear or are intended for smaller production of specialty flat glass.

The typical properties of glass are that during cooling from high temperatures crystallises after reaching the liquidus temperature immediately throughout its mass. Another cooling it still retains the character of the liquid. For this reason this material is called a super cooled liquid [1-3,8 ,9].

2 Production of flat glass by float system

Float process is a special way of forming the glass where the glass stream entering the chamber with the molten tin. On the tin was uniformly flowing and obtains a uniform thickness and a smooth surface (Fig. 1). At present all flat glass production methods are also used for the manufacture of insulating glass. They are formed from two or more pieces of glass with an air cavity or a cavity filled with an inert gas (ie. insulating glazing or triple glass).

* Corresponding author: svoboda@fvtm.ujep.cz

Reviewers: *Andrej* Czán, *Juraj* Gerlici

Modern companies producing the flat glass on one line only. The problem with this production is a lingering residual stress. Enduring value temporary stress has a negative impact on the quality of the cut and the refractive of the glass.

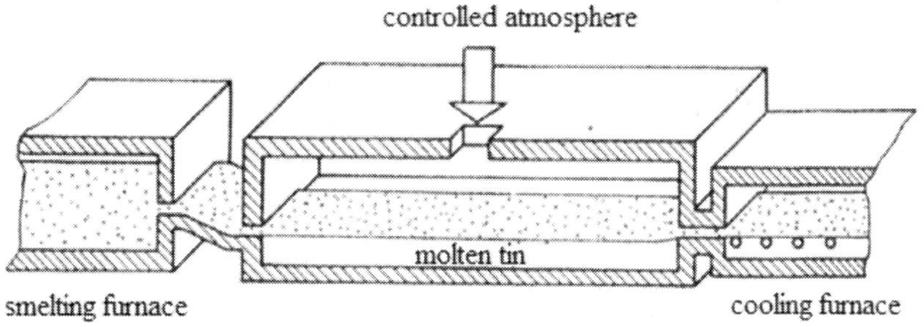

Fig. 1. The principle of manufacturing flat glass by the float system [4]

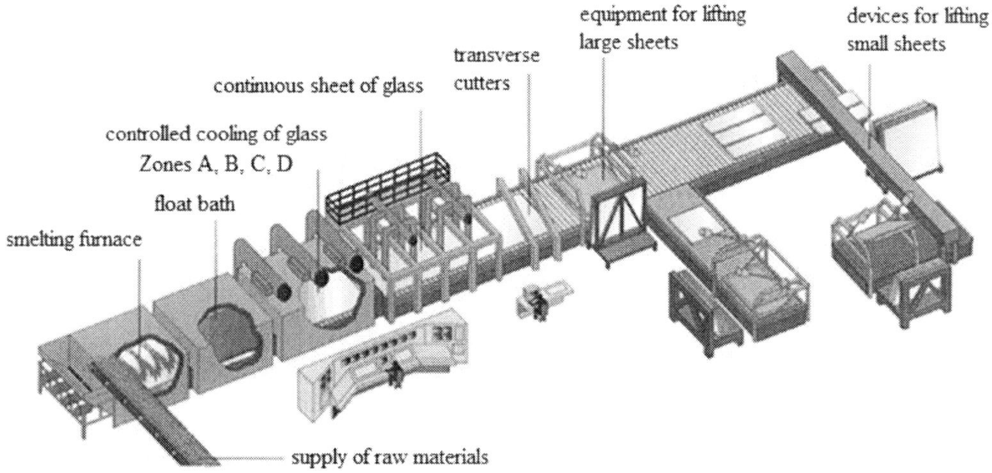

Fig. 2. Scheme flat glass production line [5]

Figure 2 illustrates the production of flat glass "float system". The cooling chamber has four cooling zone A, B, C, D in a direction of flow. In these zones is adjusted gradually cooling the glass sheets by means of air nozzles.

With the first phase of cutting glass (making an incision in the glass) is also linked to the creation a large field of residual stresses in the material being cut. These stresses in the material can be easily observed in polarized light. Residual stress is a result of plastic deformation caused by the cutting tool. The brittle material is important because residual stress to produce uniform and crack propagation at the contact point of the cutting tool and the material to be cut. Field of residual stress also significantly affects the growth of cracks due to follow the voltage is applied during the actual breaking glass [4-6].

For a good a quarry of glass has a decisive influence its structure. This depends on its composition and on the method of manufacture. Glass body deforms elastically until the moment when comes to an abrupt a quarry of glass [7-9]. The quarry is brittle with no apparent plastic deformation. Strength of glass is largely influenced by the quality of the surface (surface defects), the type of specimen and loading conditions.

3 Measurement of parameters flat glass production

By the measurements we have found that the tension is constant in the passage of the glass under unchanging production conditions. It has also been shown that changes in temperature during the day do not greatly affect the magnitude of stress and that the unbalanced setting of cooling across the glass pane greatly influences the shape of the stress curve.

Measurement of the stability of the stress in the glass was a preparatory work and is carried out to check the stress and temperature stability under the constant other conditions of production (especially the setting of the cooling device). The waist stress was measured continuously in front of the longitudinal notch device. The measured values were displayed on the display and saved in *.xls format.

From the measured values the temperature and stress dependence in the glass was processed depending on the width of the glass passage (Fig. 3). The camera begins to measure the passport of the glass at a point where a beam of polarized light strikes the edge of the glass, then the next value of the temperature and magnitude of the transient stress is always measured after 10 cm.

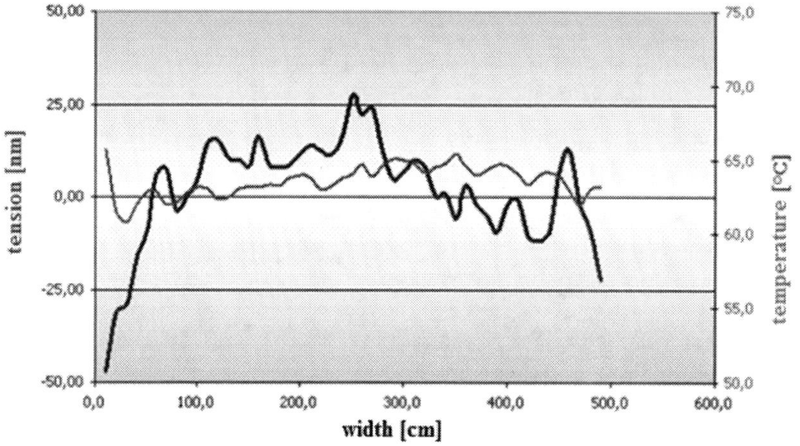

Fig. 3. Tension in the glass waist - glass thickness 4 mm

The flat glass was measured five times immediately in succession (for verify the stability of the stresses). For all measured values, the average value (both internal and internal stress) and the standard deviation were determined in each waist width of the glass. When measuring the temperature across the waist of the sheet glass, the standard deviation of the measurement ± 0.38°C was determined. For stresses, the standard deviation was ± 2.01 nm/cm.

The change of the pressure and temperature of the glass waist was also evaluated, depending on the change in the temperature in the production hall (again under the constant other physical and production conditions). The sheet of glass moves from the end of the cooling device to the breaking point for 2 to 5 minutes (according to the width of the glass waist). Fluctuations in air temperature in the refrigerating device, which would significantly affect the stresses in the passage of the glass, have also not been demonstrated. It can be stated that the temperature and stress behavior of the glass during the day did not correspond to changes in the day's room temperature. Air humidity was not measured.

It was important to determine the stress values depending on the width of the waist (Fig. 4). This dependence results in considerable unevenness in the width of the stresses across the width (Fig. 5.). This is caused by the uneven cooling of the glass over the waist width due to the setting of the cooling nozzles.

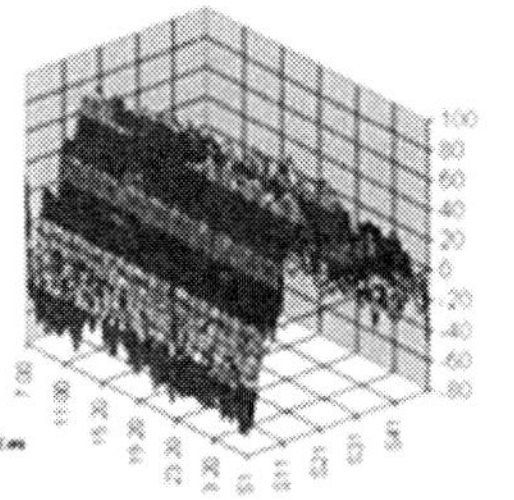

a) the stresses in the glass (24 hours/3D

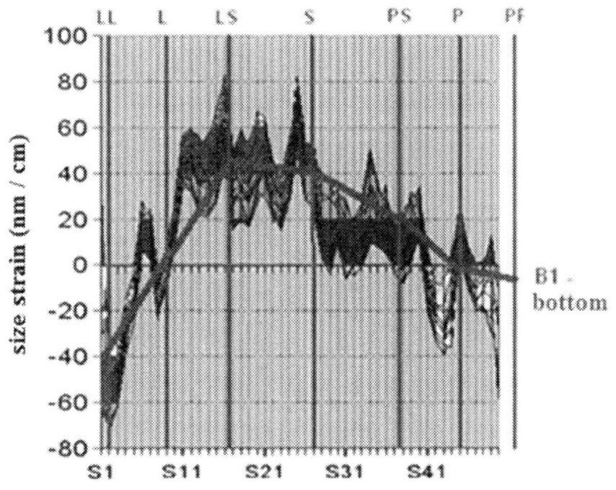

width of the waist of the glass
(S1 to S11 = 1000 mm)

b) the strain on the width of the waist of the glass

Fig. 4. Dependence of tension at the waist of the glass over 24 hours

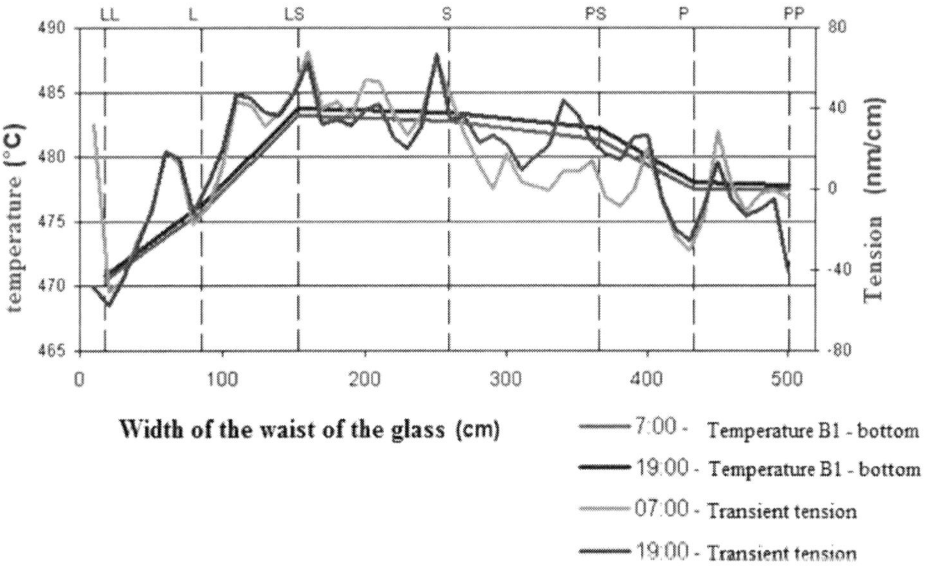

Fig. 5. Tension and temperature of the waist glass zone B - Bottom of the glass

4 Solutions and discussions

In the cooling zones A, B, C and D of the production line, seven temperature sensors are located for the cooling temperature control. Their deployment across the waist of the glass can be seen in Tab. 1.

Until now, when changing the thickness of the glass passport, the temperatures in the cooling device have been set according to the settings previously made. When adjusting the cooling device, the last acceptable setting was always found, but not the optimal or at least the setting that exhibited the best stresses or the smallest percentage of glass cut glass produced.

Table 1. Location of glass temperature sensors

Sensor designation	**L**	**ML**	**CL**	**C**	**CR**	**MR**	**R**
Distance from waist center [cm]	242	174	106	0	106	174	242

Choosing the most suitable chiller setting from a given dataset was done only for a width of 475 cm (the most common waist width). For this breadth, over a one-and-a-half year period, most of the days when stress measurements were made (both during production and residual) and quality control were also carried out (Tab. 2).
On the basis of the measurements, the best specified cooling parameters have been set. The choice was made mainly according to the shape of the stress curves, which is measured continuously during production (behind the cooling device). The reason for the selection was that individual cut flat glass formats were cut and subsequently fractured with the still persisting transient stress in the glass. The choice was made by gradually eliminating individual settings of the furnace, which were considered the worst of the gradually decreasing amount of stress curves.

Table 2. Faulty boards due to a bad cut

Date	**Glass width (cm)**	**Number of samples checked**	**Type of defect**				**Number of defective products**	**Faulty products [%]**
			glued shard	**Mussels**	**Bad quarry**	**Other defect**		
27.10.2008	475	56	0	1	0	0	1	1.79
29.10.2008	475	88	1	4	0	1	6	6.82
30.10.2008	475	78	0	3	0	0	3	3.85
26.11.2008	475	86	0	2	0	0	2	2.33
27.11.2008	475	125	0	2	0	0	2	1.60
29.11.2008	475	60	0	0	1	0	1	1.67
30.11.2008	475	60	0	0	0	0	0	0.00
1.12.2008	475	102	1	1	0	0	2	1.96
2.12.2008	475	110	0	2	0	0	2	1.82

Figure 6 shows the residual stress curve of the day when no defective fractures were found during the production of the glass waist to the predetermined sheet formats. The right part of the chart shows lower stresses in the glass pane (highlighted with red lines).

Proposal for setting a cooling device for the thickness of 3.15 mm, width of 475 cm in the Tab. 3. It was obtained by mirroring the right half that had a lower stress around the central heat sensor.

On the basis of the measurements made, it can be stated that the stresses are constant during constant production conditions in the waist of the flat glass.
Looking at the transverse distribution of stresses in the waist of the glass, it was important to find out that the cross section of the glass has a disproportionate temperature distribution

at the bottom of the cooling device. The shape of the transient stress curve showed reflection in the heat exchanger settings under the glass pane in zones A and B.

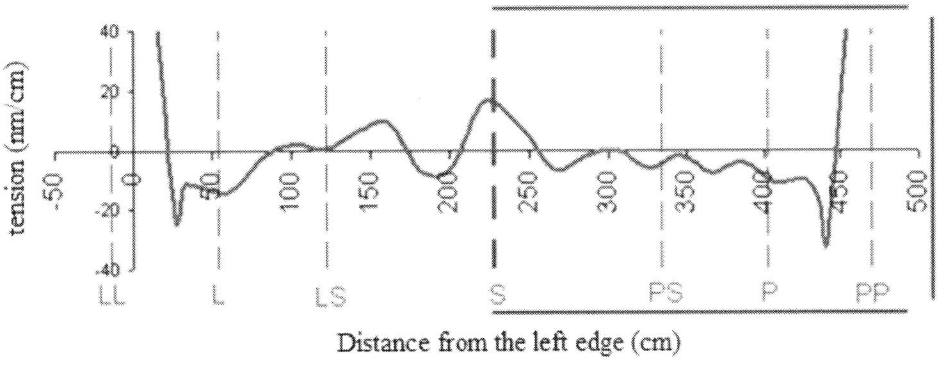

Fig. 6. Tension course during the day (24 hrs) - average depending on the width of the passport of 30 January 2008

Table 3. Design for setting the cooling in the cooling device (Glass thickness 3,15 mm, width 475 cm glass)

Top of the waist of the glass							
Cooling zones	**LL**	**L**	**LS**	**S**	**PS**	**P**	**PP**
A	514	528	528	529	528	528	514
B	443	455	456	457	456	455	443
C	390	393	390	385	390	393	390
D	263	280	275	270	275	280	263
Bottom of the waist of the glass							
	LL	**L**	**LS**	**S**	**PS**	**P**	**PP**
A	505	534	539	539	539	534	505
B	449	470	475	473	475	470	449
C	388	419	426	423	426	419	388
D	290	329	342	332	342	329	290

It has been demonstrated that the same temperature setting in the individual zones of the refrigeration device (even at a few weeks apart) has shown a constant value.
For the most suitable setting of the cooling device, it should be set for each thickness, width and color of glass produced, and include the metrological conditions.

- The following recommendations were made from the above measurements and analyzes:
- create an internal database for individual shades, thickness and width of the flat glass. This database will store data on the amount of tension created (measured on-line camera before the cut and measured residual stress on the chilled glass sheet), setting the glass waist cooling and the quantity of defective products. It will also keep records of temperature and humidity in the production hall and the outdoor environment. For this reason, it would be necessary to purchase a meteorological station (for outdoor measurements) that will allow connection to the intranet. Perform installation of temperature, pressure and humidity sensors into the air intake into the cooling furnace

- the creation of a database (history) of the values measured by the meteorological station and the air intake to the cooling furnace together with the values set in the cooling for each thickness, shade and width of the waist of the glass. This will allow quick adjustment of the cooling values.
- Refine cooling, set the bottom (and top) part of the cooling device to the desired temperature at the given point of the glass pane. This applies mainly to zones A and B. Consider the use of autoregulation at the bottom of the cooling furnace.

Conclusions

The paper describes the problems encountered in the production of sheet glass by the "float" process. Data collection options have been reported, which are normally measured and stored on the production line. Based on the evaluation of the measured data, a procedure was developed for setting the cooling part of the line for individual thicknesses, shades and width of the waist of the glass. Additionally, recommendations were made to create a value database for quick cooling settings to provide lower stresses in the glass, depending on the thickness, width and color of the glass. This recommendation includes expanding the collection of meteorological data in the hall and intake of air to the cooling device, creation of a cooling furnace database including all measured parameters and the possibilities of working with them.

This article was supported by Student Grant Competition of J. E. Purkyně University in Ústí nad Labem.

References

1. J. Dluhoš, *Materiály a technologie – plasty a vybrané nekovové materiály*. (Ostrava, Ostravská Univerzita, ISBN 80-7042-052-9, 1993)
2. A. Smrček, F. Oldřich, *Sklářské suroviny*. (Praha, Informatorium, ISBN 80-85427-47-8, 1994)
3. F. Schill, *Chlazení skla*. (Praha, Informatorium, ISBN 80-85427-34-6, 1993)
4. J. Belda, *Sklářské a keramické stroje*. (Liberec, Fakulta strojní, ISBN 80-7083-149-9, 1994)
5. http://www.autoskloczech.cz/autoskla.htm [cit. 2.11.2008]
6. L. Dušek, P. Němec, D. Veselý, *Teorie řezu skla – výzkumná zpráva*. (Pardubice, Univerzita Pardubice, Fakulta chemicko-technologická, 2003)
7. J. Menčík, *Pevnost a lom skla a keramiky*. (Praha, SNTL - Nakladatelství technické literatury, ISBN 80-03-00205-2, 1990)
8. J. Dižo, M. Blatnický, B. Skočilasová, *Computational modelling of the rail vehicle multibody system including flexible bodies*. Communications: scientific letters of the University of Zilina, **17** (3), 31-36, ISSN 1335-4205, (2015)
9. M. Sapieta, V. Dekys, O. Stevka, *Localization Lüders band in infrared spectrum*. Applied mechanics and materials **420**, 194-201 (2013)

Method of controlling innovative articulation for articulated vehicle

Mateusz Szumilas[1,*], *Sergiusz* Łuczak[1], *Maciej* Bodnicki[1], *Marcin* Stożek[2], *Tomasz* Załuski[2]

[1]Warsaw University of Technology, Faculty of Mechatronics, ul. Boboli 8, 02-525 Warszawa, Poland
[2]EC Engineering Sp. z o.o., ul. Jasnogórska 9, 31-358 Kraków, Poland

Abstract. Operation of an articulated vehicle is dependent on an appropriate damping action taking place in its rotary articulation. In order to analyse an impact of the control of the articulation on the motion of the vehicle a model of the vehicle with a controllable hydraulic damping system has been developed. A 90 degree turn and lane change manoeuvres were simulated using LabVIEW software. Modification of the damping parameters of the articulation, according to the velocity and articulation angle of the vehicle, proved to have a significant impact on the vehicle motion stability. Moreover, the sensor layer necessary for the control algorithm as well as the diagnostic system is described.

Keywords: articulated vehicle, simulation, hydraulic damping, LabVIEW, MEMS

1 Introduction

Articulated urban buses play an important role in the organisation of public transport. They provide high passenger capacity needed in big and constantly growing cities. Such buses operate in a high traffic and on routes comprising narrow roads and sharp turns. In addition, these vehicles have to develop high speeds when they enter the arterial roads or expressways. The use of a rotary articulation allows articulated buses to take sharper turns than conventional buses having equal overall length. At the same time, articulated vehicles are under risk of developing uncontrolled pivoting movements of rear and front sections with respect to one another, especially when the rear axle is driven, which is a popular arrangement for low floor vehicles. Under such conditions of operation the articulation has to provide low damping when driving at low speed and, on the contrary, increase the damping when the bus is heading straight or is changing lane at a high speed. Stiffness of the articulation can be controlled with a proper setup of hydraulic dampers combined with electro-hydraulic valves.

The paper presents an implementation and verification of a control algorithm proposed for the articulation with hydraulic damping. The verification is based on the analysis of the results of simulation of the articulated vehicle performing a 90 degree turn and lane change

* Corresponding author: m.szumilas@mchtr.pw.edu.pl

Reviewers: *Bogdan* Posiadala, *Krzysztof* Talaśka

manoeuvres. Moreover, a sensor layer employed by the control algorithm is described. The proposed sensor layer has been enhanced so that it can be also used for diagnostic purposes.

2 Method

In order to analyse impact of the articulation control algorithm on the motion of the articulated vehicle a planar model has been developed. The model includes a hydraulic damping system used in the articulation. Similar planar models developed for articulated vehicles, especially for tractor and trailer configurations, can be found in [1-3].

The following assumptions were made for the model:

- the vehicle is reconstructed by its single track model;
- there are no roll and pitch motions of the vehicle, only rotation around the yaw axis;
- normal forces acting on the tyres are computed under the static load conditions;
- aerodynamics of the vehicle is not considered.

2.1 The vehicle model

The articulated bus is modelled as two rigid bodies including front and rear unit. The front unit has a steerable axle and an unpowered middle axle. The rear unit has one powered axle. These two units are connected by a rotational joint. The model is described in the three coordinate systems:

- the inertial coordinate system fixed to the ground (X_zY_z);
- the front unit coordinate system ($X_1C_1Y_1$), with its origin at the centre of gravity of this unit (C_1) and a yaw velocity ω_1;
- the rear unit coordinate system ($X_2C_2Y_2$), with its origin at the centre of gravity of this unit (C_2) and a yaw velocity ω_2.

The geometrical and inertial parameters of the single track model of the vehicle are presented in Fig. 1. The m_1, m_2 are masses and the J_1, J_2 are mass moments of inertia about the vertical central axes of the front and rear units respectively.

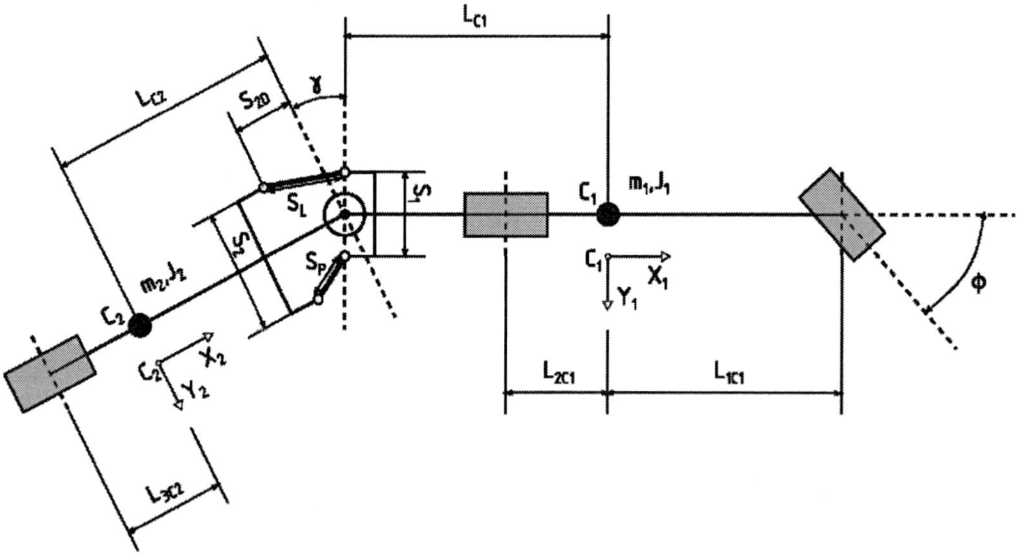

Fig. 1. The geometrical and inertial parameters of the single track model of the articulated bus

The dimensions and forces acting in the articulation are presented in Fig. 2. The relationship between angle β and articulation angle γ is:

$$\beta_L = \beta_0 + \gamma \ , \ \beta_P = \beta_0 - \gamma \tag{1}$$

where:

$$\beta_0 = 90° - \arctan\left(\frac{\left(\frac{S_2}{2}\right)}{S_{2D}}\right) \tag{2}$$

The lengths of the hydraulic dampers are:

$$S_{L/P} = \sqrt{\left(\frac{S_1}{2}\right)^2 + \left(\frac{S_2}{2}\right)^2 + {S_{2D}}^2 - 2\left(\frac{S_1}{2}\right)\left(\sqrt{\left(\frac{S_2}{2}\right)^2 + {S_{2D}}^2}\right)\cos(\beta_{L/P})} \tag{3}$$

The individual angles in the articulated joint can be expressed as:

$$\alpha_{2L/P} = \arcsin\left(\frac{\frac{S_1}{2}\sin\beta_{L/P}}{S_{L/P}}\right) + \arctan\frac{S_{2D}}{\left(\frac{S_2}{2}\right)} \tag{4}$$

$$\alpha_{1L} = 180° - \gamma - \alpha_{2L} \ , \ \alpha_{1P} = 180° + \gamma - \alpha_{2P} \tag{5}$$

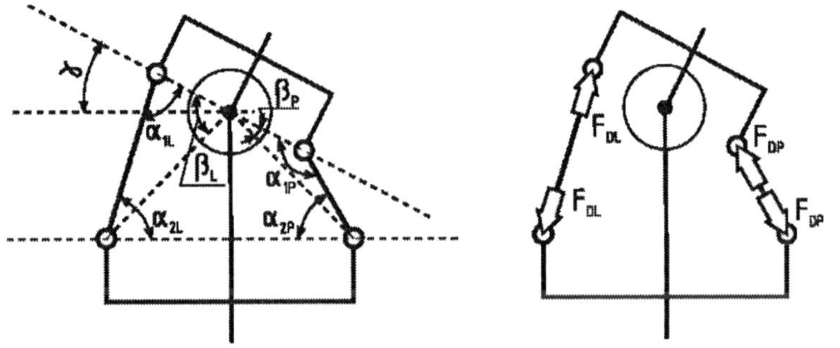

Fig. 2. The dimensions and forces acting in the articulated joint

The speed of articulation is given by equation:

$$\dot{\gamma} = \omega_1 - \omega_2 \tag{6}$$

The accelerations of the centres of gravity of the vehicle's front and rear units in their respective coordinate systems (Fig. 3) are as follows:

$$\begin{matrix} a_{1X} = \dot{V}_{1X} - \omega_1 V_{C1Y} & a_{1Y} = \dot{V}_{C1Y} + \omega_1 V_{1X} \\ a_{2X} = \dot{V}_{2X} - \omega_2 V_{C2Y} & a_{2Y} = \dot{V}_{C2Y} + \omega_2 V_{2X} \end{matrix} \tag{7}$$

The slip angles of the tyres are:

$$\delta_{k1} = \arctan\frac{V_{k1Y}}{V_{1X}} - \phi \ \wedge \ V_{k1Y} = V_{C1Y} + \omega_1 L_{1C1} \tag{8}$$

$$\delta_{k2} = \arctan\frac{V_{k2Y}}{V_{1X}} \ \wedge \ V_{k2Y} = V_{C1Y} - \omega_1 L_{2C1} \tag{9}$$

$$\delta_{k3} = \arctan\frac{V_{k3Y}}{V_{2X}} \ \wedge \ V_{k3Y} = V_{C2Y} - \omega_2 L_{3C2} \tag{10}$$

The forces acting in the articulated joint (Fig. 2, Fig. 4) are as follows:

$$\begin{aligned} a_{1X} &= \dot{V}_{1X} - \omega_1 V_{C1Y} \qquad & a_{1Y} &= \dot{V}_{C1Y} + \omega_1 V_{1X} \\ a_{2X} &= \dot{V}_{2X} - \omega_2 V_{C2Y} \qquad & a_{2Y} &= \dot{V}_{C2Y} + \omega_2 V_{2X} \end{aligned} \tag{11}$$

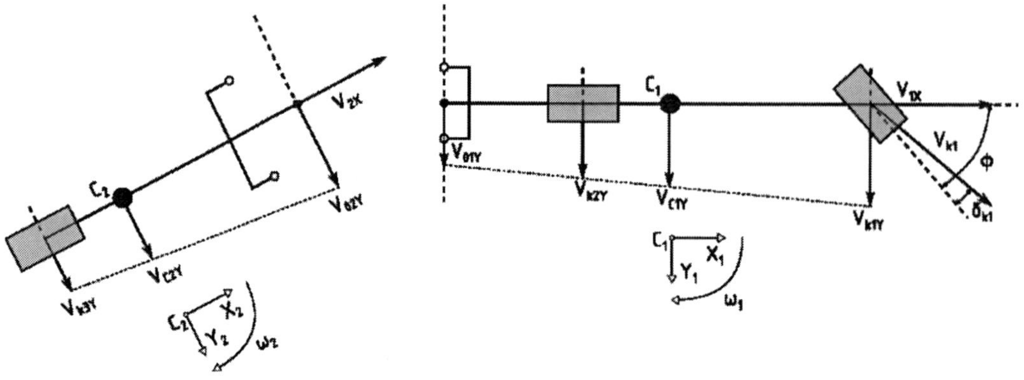

Fig. 3. The kinematics of the articulated vehicle

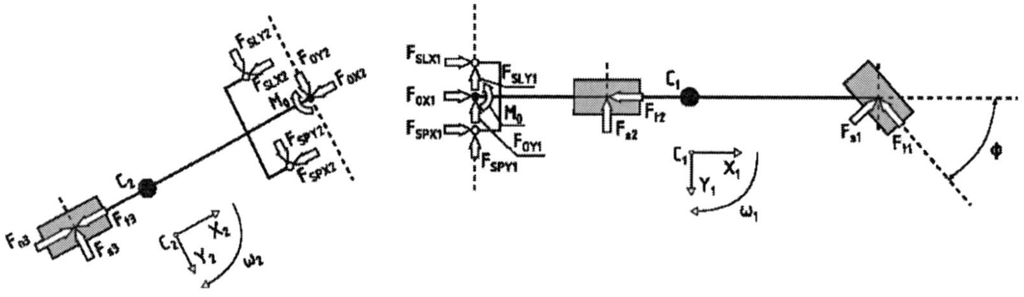

Fig. 4. The dynamics of the articulated vehicle

The forces acting in the articulated joint (Fig. 2, Fig. 4) are as follows:

$$\begin{aligned} F_{SLX1} &= F_{DL}\sin\alpha_{1L} \qquad & F_{SLY1} &= F_{DL}\cos\alpha_{1L} \\ F_{SPX1} &= F_{DP}\sin\alpha_{1P} \qquad & F_{SPY1} &= -F_{DP}\cos\alpha_{1P} \end{aligned} \tag{12}$$

$$\begin{aligned} F_{SLX2} &= F_{DL}\sin\alpha_{2L} \qquad & F_{SLY2} &= -F_{DL}\cos\alpha_{2L} \\ F_{SPX2} &= F_{DP}\sin\alpha_{2P} \qquad & F_{SPY2} &= F_{DP}\cos\alpha_{2P} \end{aligned} \tag{13}$$

The forces acting on the steerable axle in the front unit coordinate system:

$$F_{1X1} = -F_{t1}\cos\phi + F_{s1}\sin\phi \qquad F_{1Y1} = -F_{t1}\sin\phi - F_{s1}\cos\phi \tag{14}$$

The final equations representing the planar motion of the articulated vehicle:

$$\begin{aligned} &(m_1 + m_2)(\dot{V}_{1X} - \omega_1 V_{C1Y}) + m_2(\omega_1^2 L_{C1} + \omega_2^2 L_{C2}\cos\gamma - \dot{\omega}_2 L_{C2}\sin\gamma) = \\ &= \cos\gamma(F_{n3} - F_{t3} - F_{SLX2} - F_{SPX2}) - \sin\gamma(F_{s3} - F_{SPY2} - F_{SLY2}) + \\ &+ F_{SLX1} + F_{SPX1} - F_{t2} + F_{1X1} \end{aligned} \tag{15}$$

$$\begin{aligned} &(m_1 + m_2)(\dot{V}_{C1Y} + \omega_1 V_{1X}) - m_2(\dot{\omega}_1 L_{C1} + \omega_2^2 L_{C2}\sin\gamma + \dot{\omega}_2 L_{C2}\cos\gamma) = \\ &= -\sin\gamma(F_{n3} - F_{t3} - F_{SLX2} - F_{SPX2}) - \cos\gamma(F_{s3} - F_{SPY2} - F_{SLY2}) - \\ &-F_{SLY1} - F_{SPY1} - F_{s2} + F_{1Y1} \end{aligned} \tag{16}$$

$$J_1\dot{\omega}_1 = F_{1Y1}L_{1C1} + F_{s2}L_{2C1} - M_O + \frac{S_1}{2}(F_{SLX1} - F_{SPX1}) + \\ +L_{C1}[F_{1Y1} - F_{s2} - m_1(\dot{V}_{C1Y} + \omega_1 V_{1X})] \tag{17}$$

$$J_2\dot{\omega}_2 = F_{s3}L_{3C2} + M_O + \frac{S_2}{2}(F_{SPX2} - F_{SLX2}) - S_{2D}(F_{SLY2} + F_{SPY2}) + \\ +L_{C2}\{F_{s3} + m_2[sin\gamma(\dot{V}_{1X} - \omega_1 V_{C1Y} + \omega_1^2 L_{C1}) + \\ +cos\gamma(\dot{V}_{C1Y} + \omega_1 V_{1X} - \dot{\omega}_1 L_{C1}) - \dot{\omega}_2 L_{C2}]\} \tag{18}$$

2.2 The tyre model

The traction forces acting on each wheel of the vehicle are calculated assuming constant rolling resistance coefficient (19). The equation introduces nonlinearity for low longitudinal velocities of the wheel with accordance to:

$$F_{tn} = F_{kzn}C_{rr}\tanh\left(\frac{|V_{kXn}|}{V_{thr}}\right) \tag{19}$$

where: F_{tn} and F_{kzn} are the longitudinal (traction) and normal forces, respectively, acting on the wheel which belongs to the n-th axle, C_{rr} is the rolling resistance coefficient, V_{kXn} is the X component of the velocity of the wheel axle (i.e. in the rolling direction), V_{thr} is the threshold velocity.

The side forces acting on the tyres (these forces are responsible for adjusting the driving direction of the vehicle) are calculated according to the Dugoff model [4]. The longitudinal slip is omitted in this model. The side slip angle of the n-th axle is given with:

$$\alpha_n = \tan\frac{V_{kYn}}{V_{kXn}} \tag{20}$$

The equations of the Dugoff model are as follows:

$$\lambda = \frac{\mu F_{kzn}\left(1-\epsilon V_{kXn}\sqrt{s_n^2+\tan^2\alpha_n}\right)(1-s_n)}{2\sqrt{(K_{sn}s_n)^2+(K_n\tan\alpha_n)^2}} \tag{21}$$

$$F_{sn} = \frac{K_n tan\alpha_n}{1-s_n}f(\lambda) \quad \wedge \quad f(\lambda) = \begin{cases}\lambda(2-\lambda) & \text{if } \lambda < 1 \\ 1 & \text{if } \lambda \geq 1\end{cases} \tag{22}$$

where: F_{sn} is the lateral force acting on the wheel which belongs to the n-th axle, V_{kYn} is the Y component of the velocity of the wheel axle (i.e. perpendicular to the rolling direction), K_n is the cornering stiffness of the tyre, K_{sn} is the longitudinal stiffness of the tyre, ε is the friction reduction coefficient, μ is the nominal tyre-ground friction coefficient, s_n is the longitudinal slip ratio.

2.3 The damper model

The hydraulic damper (Fig. 5) is an important part of the discussed articulation. The hydraulic setup of the articulation consists of two cylinders, each paired with an electrically-driven proportional valve.

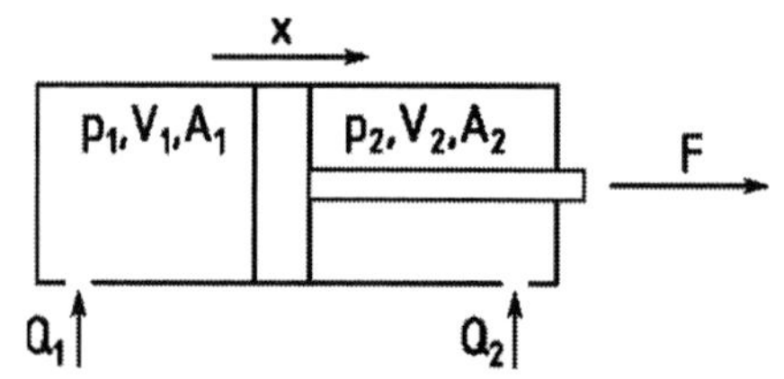

Fig. 5. The model of the hydraulic damper

The piston motion is modelled with the following equations [5-6]:

$$\begin{cases} \frac{dp_1}{dt} = \frac{K_{bm}}{V_1}\left(Q_1 - A_1\frac{dx}{dt}\right) \wedge V_1 = V_{0/1} + A_1 x \\ \frac{dp_2}{dt} = \frac{K_{bm}}{V_2}\left(Q_2 + A_2\frac{dx}{dt}\right) \wedge V_2 = V_{0/2} - A_2 x \end{cases} \tag{23}$$

$$F = p_2 A_2 - p_1 A_1 \tag{24}$$

where: x is the piston position, F is the force acting on the piston, K_{bm} is the bulk modulus of the hydraulic fluid, p_n is the pressure in the n-th chamber, Q_n is the volumetric flow to the n-th chamber, $V_{0/n}$ is the initial volume of the n-th chamber, V_n is the volume of the n-th chamber, A_n is the cross-sectional area of the piston in the n-th chamber.

The fluid flow through the proportional valve is as follows:

$$Q_Z = A_{orif} C_d \sqrt{\frac{2|\Delta p|}{\rho}}\, \mathrm{sgn}(\Delta p) \tag{25}$$

$$Q_1 = -Q_2 = Q_Z \quad \wedge \quad \Delta p = p_2 - p_1 \tag{26}$$

where: Δp is the pressure drop across the valve, ρ is the hydraulic fluid density, A_{orif} is the valve opening area, C_d is the discharge coefficient.

The opening area of the proportional valve is adjusted by the valve itself and it depends on the pressure drop across the valve and on the set point value. The opening area is modelled according to the equations:

$$A_{orif}(\Delta p) = \begin{cases} A_{orif\_min} & \text{if } \Delta p \le p_{set} \\ A_{orif\_min} + k(\Delta p - p_{set}) & \text{if } p_{set} < \Delta p \le p_{set} + p_{reg} \\ A_{orif\_max} & \text{if } p_{set} + p_{reg} < \Delta p \end{cases} \tag{27}$$

$$k = \frac{A_{orif\_max} - A_{orif\_min}}{p_{reg}} \tag{28}$$

where: A_{orif_min} is the smallest opening area, A_{orif_max} is the largest opening area, p_{set} is the valve set point, p_{reg} is the pressure regulation range.

2.4 The control algorithm

The control algorithm of the articulation mechanism is presented in Fig. 6. Taking as inputs information about articulation angle, vehicle speed, angular position of the steering wheel and vehicle emergency state (i.e., occurrence of collision, slip, excessive articulation angle, rapid deceleration, joint malfunction), the algorithm outputs the required damping level.

The damping level control is achieved by setting the pressure drop across the proportional valves of the dampers as follows:

- no damping: $\Delta p = 5 \cdot 10^5$ Pa
- 1st order damping: $\Delta p = 60 \cdot 10^5$ Pa
- 2nd order damping: $\Delta p = 120 \cdot 10^5$ Pa
- dampers locked: $\Delta p = 220 \cdot 10^5$ Pa

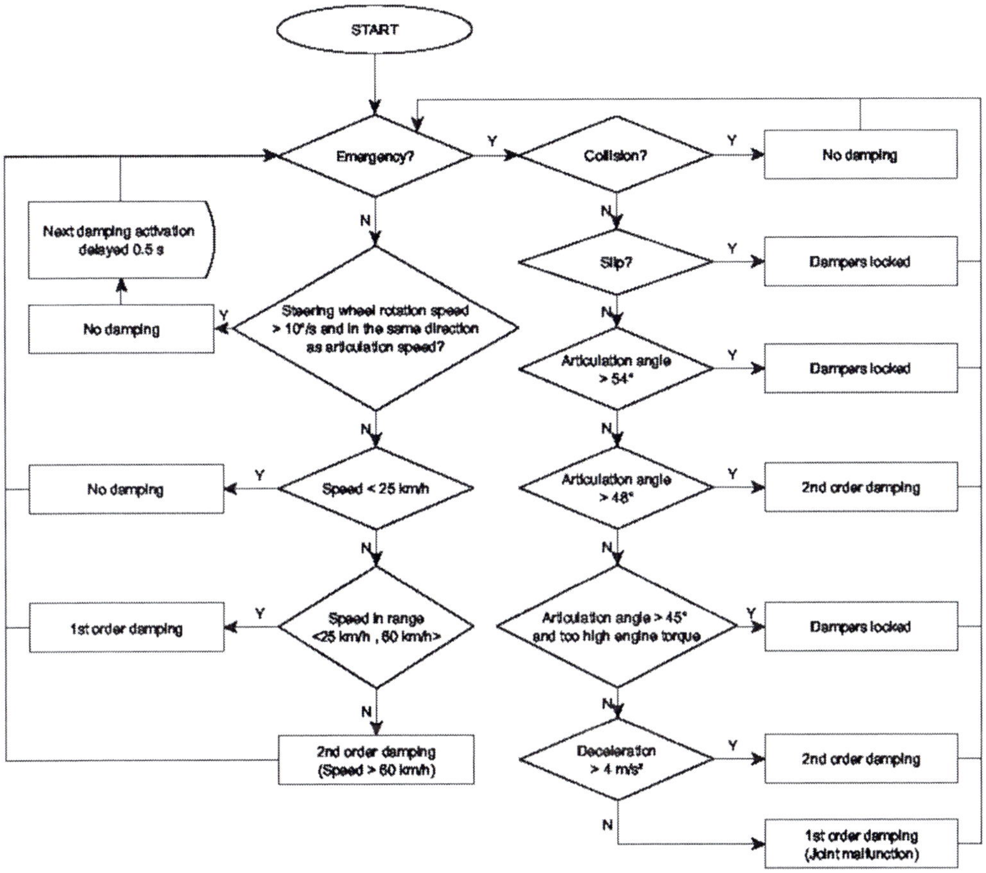

Fig. 6. The control algorithm

2.5 Simulation

The vehicle model was implemented in the LabVIEW environment. The simulation was performed with a constant step (0.2 ms) and the equations were solved using Runge-Kutta fourth-order method. Parameters of the vehicle model are presented in Table 1.

The simulated manoeuvres are: 90 degree turn (at 15 m and 30 m turning radius) and lane change (lane width 2.5 m, manoeuvre time 4 s). Each manoeuvre is simulated for empty and loaded vehicle.

Table 1. The parameters of the vehicle model

Parameter	Value
	Front unit empty/loaded
m_1	10,170 / 16,740 kg
J_1	82,761 / 127,017 kg·m^2
L1C1	2.875 / 2.525 m
L2C1	2.255 / 2.605 m
LC1	3.846 / 4.196 m
F_{z1}	48.2 / 86.1 kN
F_{z2}	33.2 / 66.5 kN
K_1	143 / 189 kN/rad
K_2	126 / 115 kN/rad
	Damper and Valve
K_{bm}	1.5 GPa
$V_{0/1}$, $V_{0/2}$	0.00136 m^3
A_1, A_2	0.00341 m^2
C_d	0.611
ρ	844 kg/m^3 (Tellus DO10)
A_{orif_min}	10^{-12} m^2
A_{orif_max}	10^{-4} m^2

Parameter	Value
	Rear unit empty/loaded
m_2	6,690 / 12,120 kg
J_2	40,335 / 77,742 kg·m^2
L3C2	-1.488 / -0.519 m
LC2	6.402 / 5.433 m
F_{z3}	84.0 / 130.5 kN
K_3	132 / 172 kN/rad
	Tyres
C_{rr}	0.008
V_{thr}	0.004 m/s
ε	0.0115 m/s
μ	0.9
s_n	0
	Articulation
S1	0.781 m
S2	1.857 m
S2D	0.648 m

3 Results

A summary of conditions, under which the manoeuvres were simulated is presented in Table 2.

Table 2. Simulated manoeuvres.

Manoeuvre	**Initial vehicle speed (+/- 1%)**				
	10 km/h	**25 km/h**	**45 km/h**	**60 km/h**	**80 km/h**
90° turn 15 m radius	Alg = D0, D1, D2, Dmax	Alg, D0, D1, D2, Dmax	X	X	X
90° turn 30 m radius	X	Alg, D0, D1, D2, Dmax	Alg = D1, D0, D2	X	X
lane change 2.5 m / 4 s	X	X	X	Alg = D1, D0, D2 = Dmax	D0, Alg=D1= D2=Dmax

X - not simulated; Alg - damping set according to the algorithm output; D0 - no damping enforced; D1 - 1st level damping enforced; D2 - 2nd level damping enforced; Dmax - dampers lock enforced; (x = y = ...) - similar results achieved with configurations x, y, etc.

During simulation of the manoeuvres, movement direction of the vehicle was changed by setting the steer angle of the front axle. The simplified steering input followed trapezoidal and sinusoidal profiles for the 90 degree turn and lane change, respectively. All manoeuvres indicated in the Table 2 were successfully accomplished, i.e., the vehicle followed the required path and did not lose stability when its front axle returned to the straight-ahead position.

As a measure of the quality of the applied control a mean absolute slip angle of the tyres was calculated for each manoeuvre (Fig. 7). As it can be seen, in most cases the control algorithm chooses the damping level which results in the lowest, or nearly the lowest, slip of the vehicle.

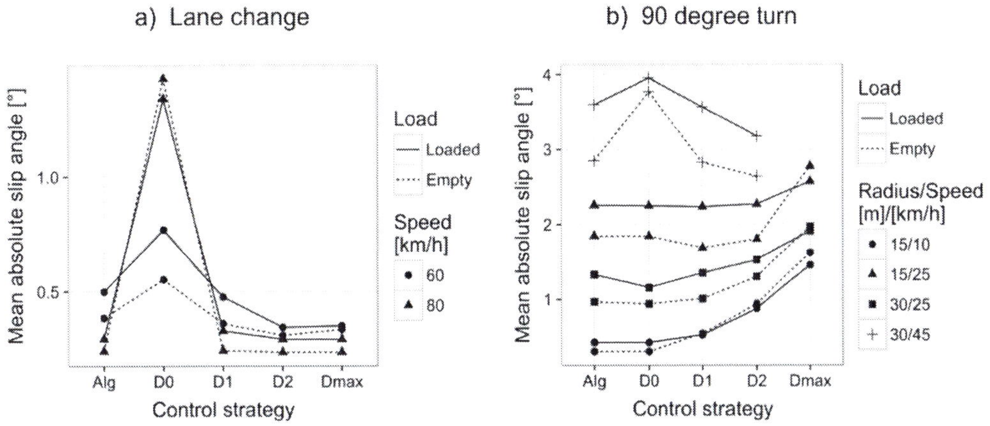

Fig. 7. The mean absolute slip angle for manoeuvres simulated with various control strategies; used abbreviations: according to Table 2 footnotes

Simulation of the lane change with no damping enforced showed a potentially dangerous situation. Such manoeuvre, performed at 60 km/h, was followed by self-limiting oscillations of the front and rear unit with respect to one another. However, after changing the speed to 80 km/h, the amplitude of these oscillations increased in time, what made the vehicle movement unstable. Loss of the articulation damping may result from a failure of the proportional valve and may lead to an accident, such as discussed in the accident report presented in [7].

4 Sensor layer

The control algorithm for the articulation requires application of only four sensors: sensor of angular position of the steering wheel of the vehicle (embedded in the steering column), sensor of angular position between the two units of the vehicle and two sensors indicating pressure in hydraulic dampers (embedded in the dampers structure). So, as far as the algorithm is concerned, only one additional sensor must be installed.

However, more sensors are required for diagnostic purposes. It is highly probable that some members of the articulation will wear out or get damaged within the assumed life of this device (of 7 years at least). These members are: the rolling bearing, metalastiks (2 pcs.), rubber bumping blocks limiting the angular range of motion of the articulation (2 pcs.) and mountings of the hydraulic dampers (2 pcs.). Except for the rolling bearing, all the other members require application of some proximity sensors, detecting an excessive range of relative linear motion within their vicinity (in the case of metalastiks, displacements in two perpendicular axes).

Let us briefly analyze each kind of the required sensors, starting with the angular position of the articulation. A typical approach is to apply an absolute angle encoder or an incremental angle encoder coupled with an additional sensor indicating some reference mutual angular position (e.g. when both units of the vehicle are aligned). However, such solution is protected by some relevant patent claims. Therefore, other propositions were put forward. One of them is application of two sensors, one in each unit of the vehicle, indicating an absolute angular position of the unit. For this purpose, a magnetometer, a gyroscope (especially MEMS sensor) or even a whole Inertial Measurement Unit (IMU) could be used (there are available automotive versions of IMU fabricated in microtechnology, e.g. for motorcycle ABS/TCS systems [8]). Another idea is to measure the mutual angular position indirectly. Then, measurements of a linear displacement of an additional member, its strain or an interaction force can be employed.

Because of the significance of the considered sensor, it would be advantageous to add another redundant sensor of angular position of the articulation for diagnostic purposes. This sensor could be employed also by the control unit in the case of a failure of the primary sensor. The sensor can be also reduced to a detector, indicating only occasionally a selected position of the units (e.g. while aligned), and only then comparing its output signal with its counterpart generated by the main sensor of angular position of the articulation.

With regard to the rolling bearing, a standard diagnostics requires temperature and vibration measurements and monitoring of the generated noise as well as lubrication conditions [9]. However, operation of the considered rolling bearing will be untypical due to its low rotational speed and rather an intermittent cycle of operation. It will be rather its radial plays (i.e. small linear displacements) that will be indicators of wear. It is difficult to tell how effective will vibration measurements be in this case. However, application of accelerometers enables monitoring both vibration as well as calculation of linear displacements (e.g. by integrating the output signal). The second option is quite easy to be realized, especially at a specified frequency (e.g. one of the natural frequencies of the articulation), as the measurement range and frequency are strictly related [10]; in such a case, by transforming the equation of motion we obtain the following formula [11]:

$$x = \frac{a}{\omega^2} = a \cdot \text{const} \tag{29}$$

where: x - displacement, a - acceleration, ω - angular frequency.

So, according to (29), it is very easy to determine the maximal linear displacement on the basis of accelerometer indication, provided it is at a fixed and known frequency. However, to determine a displacement between two different points assigned to different structural members (e.g. a pivot and a bearing), always a pair of accelerometers is necessary in order to calculate a difference of the respective linear displacements. Moreover, there are other complications connected with application of accelerometers, like e.g. a necessity of their appropriate alignment [12] or calibration of their output signals [13]. Yet, these issues are typical and can be easily dealt with.

As far as the wear or damage of the other aforementioned members of the articulation (metalastiks, bumping blocks, mountings of the dampers) are concerned, the same kind of measurement/sensors can be applied. However, in the case of metalastiks dual-axis accelerometers must be employed, whereas in other cases uni-axial accelerometers would be satisfactory.

To summarize, it seems an advantageous solution to apply MEMS sensors, especially accelerometers in the sensor layer of the articulation. These sensors are cheap and robust, have their own self-test functionality and some models are integrated with temperature sensors. Thus, vibration, linear displacement and temperature measurements can be realized using one sensor (or one pair of sensors). Some additional effects can be obtained using appropriate spatial arrangement of MEMS accelerometers, as described e.g. in [14,15] - in this way, e.g. a tilt angles of the vehicle units can be determined or accuracy of measurements can be increased, if need be. Tilt measurements can be employed to increase safety of the vehicle (e.g., by comparing rolls of the front and the rear unit).

Conclusions

The simulation of motion of the articulated vehicle showed that the proposed control algorithm ensures stability of the motion under the tested conditions. A possibility of losing the damping in the articulation as a result of a hardware failure (i.e., failure of the proportional valve) is indicated as an important risk source for the discussed control system, according to the simulation results.

It was managed to develop quite simple sensor layer to be employed by the control algorithm and diagnostic system. As proven by the presented analysis, application of MEMS sensors seems to be an advantageous solution, especially for diagnostic purposes, due to their low-cost, high reliability, self-test function and robustness.

This work has been supported by The National Centre for Research and Development, within a project, "Elaboration of innovative inter-cars articulations having a unique lighter design capable of absorbing crash energy in order to ensure passenger safety in public transport".

References

1. K. Kim, H. Guan, B. Wang, R. Guo, F. Liang, *Active steering control strategy for articulated vehicles*, Front. Inf. Technol. Electron. Eng. **17**, 576-586 (2016)
2. S. T. Oreh, R. Kazemi, S. Azadi, *A new desired articulation angle for directional control of articulated vehicles*, Proc. Inst. Mech. Eng. Part K J. Multi-body Dyn. **226**, 298-314 (2012)
3. R. Rusev, R. Ivanov, G. Staneva, G. Kadikyanov, *A Study of the Dynamic Parameters Influence Over the Behavior of the Two-Section Articulated Vehicle During the Lane Change Manoeuvre*, Transp. Probl. **11**, 29-40 (2016)
4. H. Dugoff, P. Fancher, L. Segel, *An Analysis of Tire Traction Properties and Their Influence on Vehicle Dynamic Performance*, SAE Trans. **79** (1970)
5. O.A. Bauchau, H. Liu, *On the Modeling of Hydraulic Components in Rotorcraft Systems*, Journal of the American Helicopter Society **51**, 175-184 (2006)
6. B. Titurus, J. L. du Bois, N. Lieven, *Modeling and testing of a semiactive hydraulic damper in periodic working regimes*, AIAA Journal **50**, 844-854 (2012)
7. G. Rosvall, H. Elinder, *Accident involving a Mercedes Benz articulated bus, registration no STK 595, on Glömstavägen in Stockholm/Huddinge, AB County, Sweden, on the 2nd of April 2002*, The Swedish Accident Investigation Board, Report RO 2003:01e, ISSN 1400-5751 (2003)
8. S. Łuczak, *Tilt Measurements in BMW Motorcycles*, in R. Jabłoński, R. Szewczyk (Eds.), Recent Global Research and Education: Technological Challenges. (Springer International Publishing, Cham, Switzerland, 2017)
9. *SKF Poradnik Obsługi Technicznej Łożysk*, (SKF Group, 2010)
10. M. Looney, *An Introduction to MEMS Vibration Monitoring*, Analog Dialogue **48-06**, (Analog Devices, USA, 2014)
11. R. Heaton, S. Hewitt, *The use of Operating Deflection Shapes (ODS) to model the vibration of sanders and polishers*. (Health&Safety Laboratory, Buxton, UK, 2006)
12. S. Łuczak, *Fast Alignment Procedure for MEMS Accelerometers*, in R. Jabłoński, T. Brezina (Eds.), Advanced Mechatronics Solutions. (Springer International Publishing, Cham, Switzerland, 2016)

13. S. Łuczak, *Experimental Studies of Hysteresis in MEMS Accelerometers: A Commentary*. IEEE Sensors J. **15**, 3492-3499 (2015)
14. S. Łuczak, *Guidelines for Tilt Measurements Realized by MEMS Accelerometers*. Int. J. Precis. Eng. Manuf. **15**, 2012-2019 (2014)
15. S. Łuczak, R. Grepl, M. Bodnicki, *Selection of MEMS Accelerometers for Tilt Measurements*, J. Sensors **2017**, Article ID 9796146 (2017)

A concept of smart multiaxial impact damper made of vacuum packed particles

Piotr Bartkowski [1,*], *Robert* Zalewski[1]

[1]Institute of Machines Design Fundamentals, Warsaw University of Technology, Narbutta 84, 02-254 Warsaw, Poland

Abstract. The conceptual design of smart crash energy absorber made of Vacuum Packed Particles (VPP) is presented in this work. This structure is composed of granular material inside a plastomer sleeve. The absorber can change the flexural stiffness and as a result the amount of dissipated energy depending on the impact conditions. The special properties of the granular structure create the possibility of various applications as different working modes can be used. When the pressure inside is equal or higher than atmospheric pressure the system behaves similarly to a liquid, otherwise it has properties similar to viscoplastic solid. The physical properties of the structure may be rapidly changed by adjusting the pressure inside the system. Due to these properties Vacuum Packed Particles is considered to be a smart material.

Keywords: Adaptive Structure, Crashworthiness, Smart Materials

1 Introduction

The topic of controllable materials and structures, so-called "intelligent materials" or "smart structures," has been thoroughly studied over the last years, focusing on mechanical or electrical characteristics which are easy to change during the operation of the system. However, the greatest progress has been made over the last two decades. Materials such as electrorheological (ER) and magnetorheological (MR) fluids, piezoelectrics, electrostrictive ceramics or sensors based on fiber optics are relatively well understood and widely commercialized [1, 2, 3]. On the other hand, materials such as shape memory polymers, magnetostrictive materials, conductive polymers, or micro-machined electromechanical systems are not well understood yet and are in early stages of commercialization.

This paper deals with a new type of smart structures, clearly belonging to the second group, which are systems which mechanical characteristics such as shape, stiffness, or damping, are easy to change and control. It is a contribution to the rapidly growing field of smart materials and it offers a simple and easy way to modify or upgrade many mechanical or structural elements for better control of their mechanical properties in a relatively economical and environmentally friendly manner.

The term Vacuum Packed Particles (VPP) refers to a structure that is composed of a plastomer granular material with granules having a diameter of about 1–5mm. They are

* Corresponding author: bartkowski_piotr@onet.pl

Reviewers: *Mariana* Pajtášová, *Jan* Krmela

typically semi-finished products used in the production of packaging of liquid beverages or plastic cutlery. The loose granules are enclosed in a hermetic, flexible sleeve and when air is pumped out of it, by a suitable valve mounted on the body surface, an underpressure is generated causing controllable changes in the mechanical and rheological properties of the system. A VPP typically behaves as a plastic material at the atmospheric pressure and as a solid when the underpressure is greater. They feature typical properties of a solid, such as tensile or compressive and shear strengths, damping capacity, internal "intergranular" friction and the ability to absorb energy under impact. These are all non-linear functions of the internal underpressure. It is of importance to note that all the physical and mechanical or rheological changes induced by the underpressure are fully reversible upon restoring atmospheric pressure. Indeed, such a material can almost instantaneously, only depending on the rate of the vacuum pump, be solidified and plasticized by supplying and removing air. The viscosity in the plastic state can be increased by a factor of ten or more by applying an appropriate underpressure.

Vacuum Packed Particles has been proven to be of use in the real applications, which was confirmed by several works. The most popular application build of VPP is the vacuum mattress (vacmat), which is used for immobilization of patients. Scientist proposed the VPP for building the medical pillow for stabilization of some parts of the human body [4]. The Authors of work [5] described the concept of endoscope made of VPP, which is able to change the stiffness. Besides the medical applications, Vacuum Packed Particles are applied in robotic industry, for example paper [6] showed the possibility of using VPP in manipulators and robotic gripper. Despite many useful applications, until now VPP were not used for crashworthiness applications.

The aim of this work is to present a concept of a new adaptive crash energy absorber made of the Vacuum Packed Particles. As it is possible to change the properties of this structure in the real time it is possible to classify the Vacuum Packed Particles as a smart material. The main purpose of this device is to predict the velocity of the striking object and adjust appropriate characteristic to minimalize the value of load transferred to the object. Some papers have been published on the adaptive control of structure crash characteristic [7, 8, 9], but what distinguishes the structure proposed in this article is the material and the way of controlling it. The absorber is made of granular material closed in plastomer sleeve, which was in detail described in [10, 11]. This structure allows to regulate the mechanical properties of the material by changing the vacuum value inside the sleeve.

2 Smart structures for energy absorption

The possibility to control and change the mechanical properties according to environment conditions is desirable in crash zones in cars and other vehicles. The characteristic of the crash zone in modern cars can be treated with some approximation as a linear strengthening. Due to this during crash with low speed contact forces are determined by crash boxes and as a result the acceleration has a smaller value than in a crash with high velocity. In the real situation the length of the crash zone is limited and imposed by the design space, therefore it would seem that a characteristic without strengthening but with the equal force throughout the whole length would be more appropriate. In this case the amount of the crash energy dissipated would be higher with the same peak acceleration. Unfortunately, the big disadvantage of such a characteristic is the fact that even in a crash with low speed the injuries of the passengers would be significant. Therefore it seems that the most favorable would be to regulate the force-displacement characteristic according to crash conditions. This is a solution presented in this paper, which focuses on the conceptual design of crash energy absorber which is able to change the stiffness characteristic and as a result the amount of dissipated energy in the real time. The mechanical properties of the device can be changed by regulating the underpressure inside the plastomer sleeve. The assumed mechanical characteristics is presented in Fig. 1.

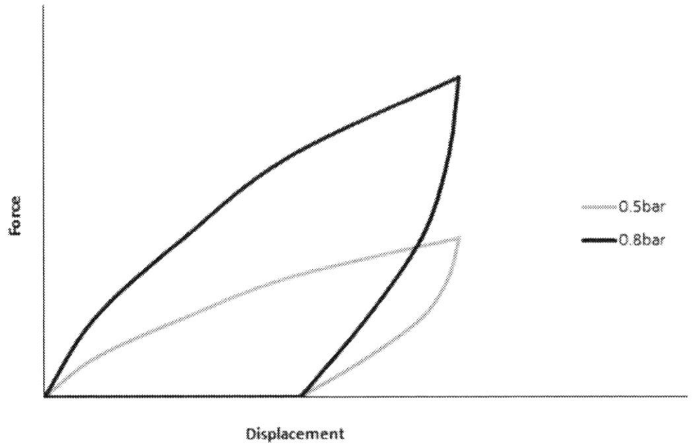

Fig. 1. Characteristics of force in funcion of displacement for the proposed absorber

The chart in Fig. 1 shows the force/displacement characteristics for two different initial values of underpressure inside the plastomer envelope. By changing the pressure parameters it is possible to regulate the amount of dissipated energy, which is equal to the area under the curve force/displacement. A damper having such a characteristics would incorporate both advantages, high crash energy absorption capability in case of the crash with high velocity and relatively low deacceleration transferred to the protected structure when the velocity is low.

3 The concept of the absorber

In this part of the paper the detail description of absorber design is presented (Fig. 2). The absorber consists of two aluminum plates (2, 3) connected by the tube (1) that is attached to plate (3) pivotally (5, 5a, 5b). The space between the plates is filled with granular material PS (Polistyren) (6) surrounded by rubber sleeve (4). Joint transfers the shear forces between the plates, and the moments are transmitted through the granular material. With the ability to change the interaction forces between the granules by changing the vacuum, it is possible adjust the flexural stiffness of the absorber, controlling the response of the system to impact forces. After the impact, the granulate deforms in a pseudo-plastic way. After the partial vacuum elimination, the forces between the granular material are relatively low, so the elasticity of the rubber sleeve enables the damper to return to equilibrium. It is also possible to apply additional spring element to increase the stiffness of the system.

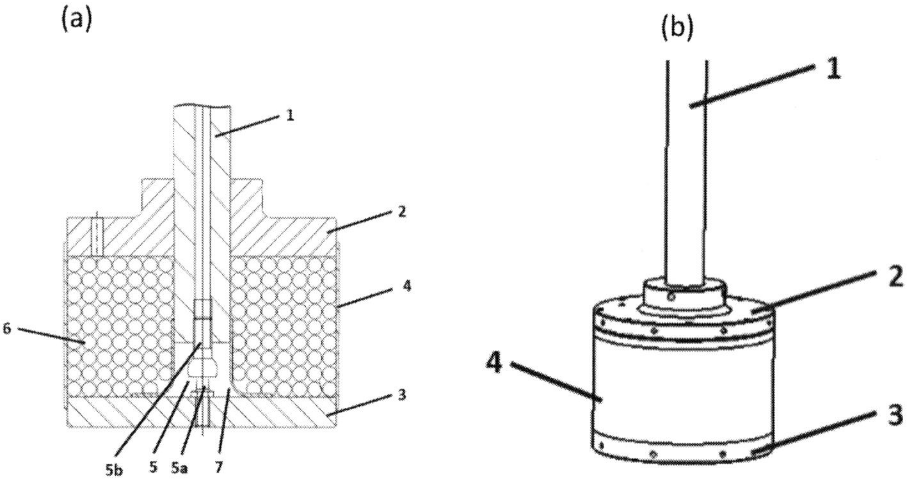

Fig. 2. The absorber design (a) cross section and (b) general view

Based on the conceptual design the prototype of the absorber was built (Fig. 3). In the preliminary configuration of the structure the plate were made with aluminum alloy and the PVC sleeve was used. This object was created to perform the preliminary experimental test allowing for verification of the concept. In this paper only the mechanical part of the device is described but the real object is additionally equipped with sensors, valves and vacuum pump with retention tank. When the velocity sensor registers the impact velocity, it changes the structure properties, by pulling out the air from the sleeve, to minimize the loads transmitted to the object while ensuring proper crash energy absorption capability.

Fig. 3. The absorber prototype

4 Preliminary experimental research

In order to verify the concept of the absorber the experimental research on the prototype was performed in the Institute of Machines Design Fundamentals of Warsaw University of Technology. As a preliminary experimental work, a with quasi static tests were performed on a dedicated test stand (Fig. 4). This stand consists of base (4), drive (3), displacement sensor (1) and force sensor (2), which allows to measure the force displacement characteristic of the tested object. During the experiments the vacuum inside the container was changed and measured by the pressure sensor (6). All the data were registered by a measuring card.

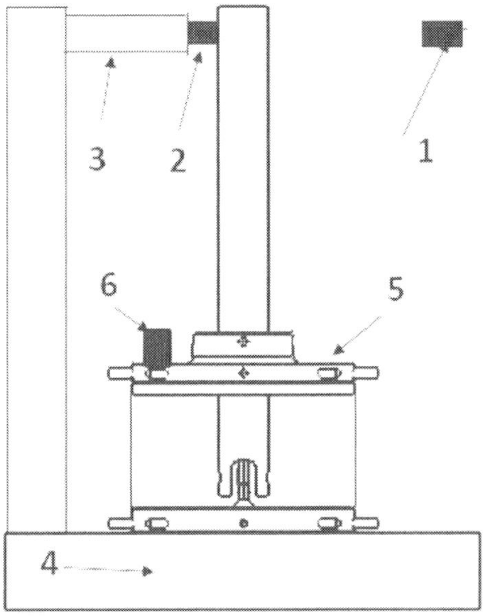

Fig. 4. Dedicated test stand

Experiments were performed for 7 underpressure values starting with 0.2M Pa and ending with 0.8 MPa. A test for each value was repeated three times and the result were similar what is illustrated in Fig. 5.

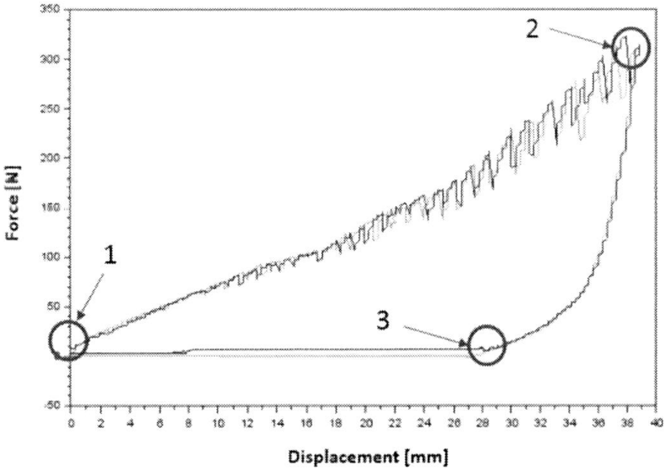

Fig. 5. The absorber characteristics

The curve 0-1 illustrated in Fig. 6 describes the loading process of the absorber. The curve 1-2 illustrates the behavior of the absorber during unloading. In point 2, the absorber is no longer subjected to any external force. In this case, the granulate was deformed in a plastic way, the absorber does not return to the equilibrium position. After reducing the vacuum to 0 (atmospheric pressure) the interaction between the granules are relatively small, therefore, by choosing the proper spring elements which transmit moments in parallel with the granular material, the absorber is able to return to the equilibrium state. As a result, it is possible to reuse the absorber, which is very important from the utility point of view.

The experiments were repeated for different underpressure conditions in order to check the possibility of changing the mechanical properties. The characteristics prove that it is possible to regulate the absorber properties by changing the value of vacuum inside the container. In Fig. 6 and Fig. 7 the characteristic for selected underpressure values are presented.

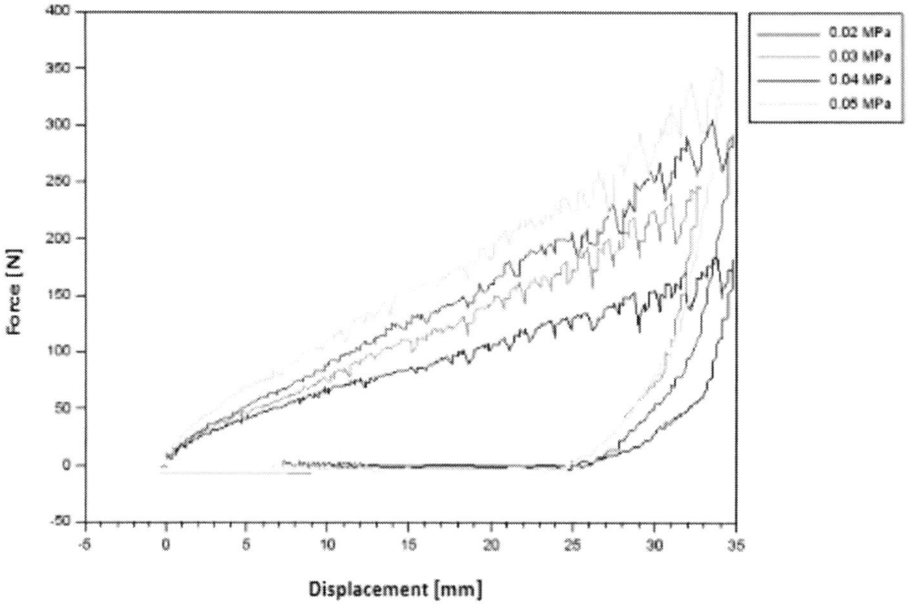

Fig. 6. Characteristic for the pressure 0.02-0.05MPa

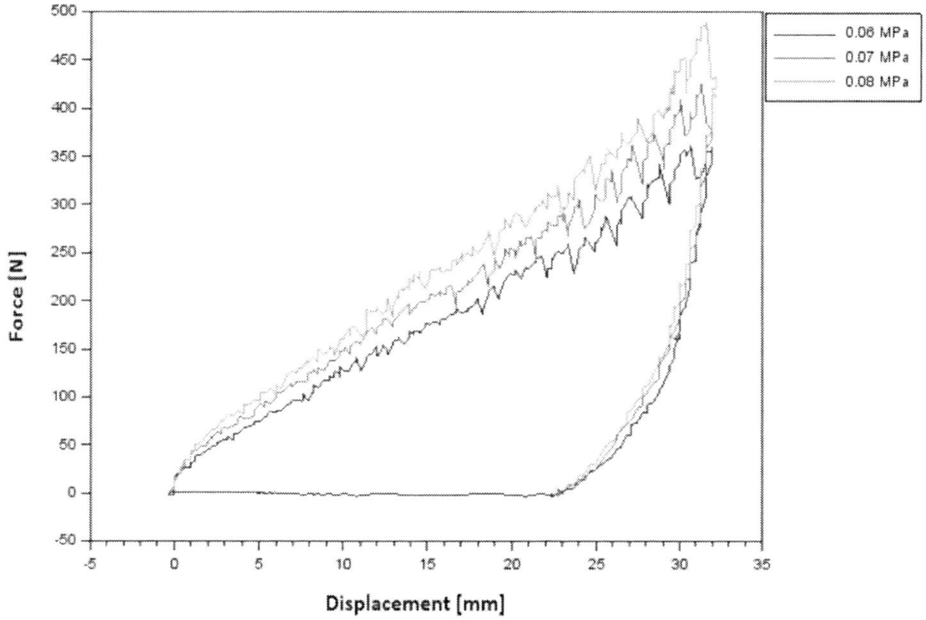

Fig. 7. Characteristic for pressure 0.06-0.08MPa

In Fig. 8 it can be seen that the force oscillates around some value. At first, the force smoothly increases and when some value is achieved it suddenly decreases, which is probably caused by the particles relocation.

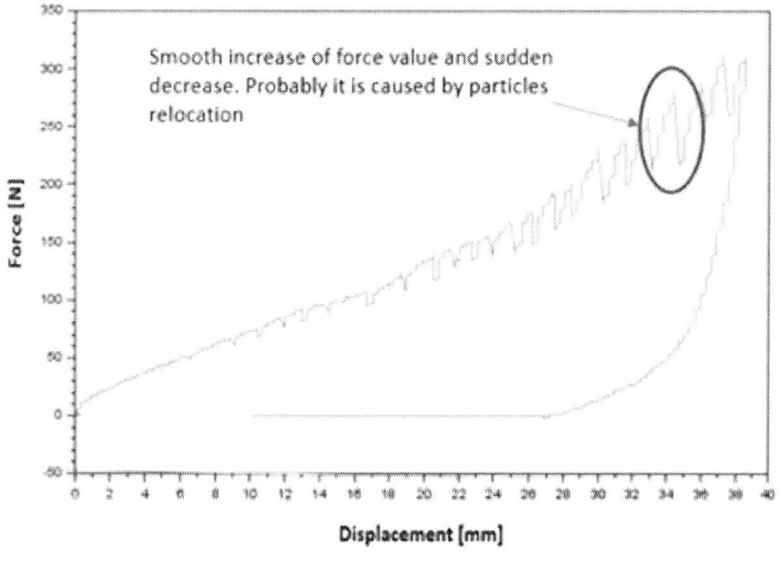

Fig. 8. The effect of the particles relocation

Conclusions

In this work the new concept of the impact energy absorber made of Vacuum Packed Particles was presented. The authors showed the basic assumption of the device and present the prototype built for performing the experiments. In the final stage of the work the preliminary results of quasi static test were presented. The experiments showed that all design assumption were met. It is possible to control and regulate the stiffness characteristic of the device and as a result the amount of the absorbed energy. Taking into consideration this special ability, the Vacuum Packed Particles absorber will be able to be implemented in various uses. The research showed that the device has a potential for commercialization and is worth further investigating. Vacuum Packed Particles behave similar to viscoplastic material so strain rate has the influence on their behavior. As the quasi-static test is not suitable to check all the properties of the absorber, the author is planning to perform the dynamic tests in the future. Additionally, the patent application of this object was filed in Polish Patent Office.

This work was carried out within the project no. 2016/23/N/ST8/02056 founded by National Science Centre, Poland.

References

1. P. Bartkowski, R. Zalewski, *Designing process of the drone's passive safety system*. New Advances in Information Systems and Technologies, Advances in Intelligent Systems and Computing **445**, Springer Int. Publishing Switzerland, 729-737 (2016)
2. E. Brown, N. Rodenberg, J. Amend, et al. , *Universal robotic gripper based on the jamming of granular material*. Proceedings of The National Academy of Sciences of the United States of America **107**, 18809–18814 (2010)
3. S.R. Gorodkin, Kordonskii, Protasevich, *Estimate as to the possibility of utilizing magnetorheological suspensions in compression equipment*. Magnetohydrodynamics, **24** (4), 511-514 (1989)
4. C. Graczykowski, *Mathematical models and numerical methods for the simulation of adaptive inflatable structures for impact absorption*. Comput Struct (2015)
5. J. Holnicki-Szulc, L. Knap, *Adaptive Crashworthiness Concept*. International Journal of Impact Engineering (2004)
6. A.J. Loeve, O.S. Van de Ven, J.G. Vogel, P. Breedveld, J. Dankelmann , *Vacuum packed particles as flexible endoscope guides with controllable rigidity*. Granul Matter. **12** (6), 543-54 (2010)
7. M.D. Luscombe, J.L. Williams, *Comparison of a long spinal board and vacuum mattress for spinal immobilization*. Emergency Medicine Journal **20** (1), 476-478 (2003)
8. M. Ostrowski, J. Holnicki-Szulc, and P. Griskevicius, *Adaptive Crashworthiness of Front-End Structure of Motor Vehicles*. SAE Technical Paper (2007)
9. P. Skalski, R. Zalewski, *Viscoplastic properties of an MR fluid in a damper*, J. Theor. Appl. Mech. **52** (4), 1061-1070 (2014)
10. P. Wang, Z. Wang, *Determination of the flow stress of a magnetorheological fluid under three-dimensional stress states by using a combination of extrusion test and FEM simulation*. Journal of Magnetism and Magnetic Materials **419**, 255-266 (2016)
11. R. Zalewski, T. Szmidt, *Application of Special Granular Structures for semi-active damping of lateral beam vibrations*. Engineering Structures (2014)
12. R. Zalewski, M. Shillor, P. Chodkiewicz, *Vibrations of a mass-spring system using a granular-material damper*. Applied Mathematical Modelling **40,** 8033–8047 (2016)

Analysis of the porous bone-implant fixation – experimental verification of numerical model

Maciej Berdychowski[1,*], *Janusz* Mielniczuk[2],

[1]Poznan University of Technology, ul. Piotrowo 3, 61-666 Poznań, Poland
[2]The Rail Vehicles Institute "TABOR", ul.Warszawska 181, 61-055 Poznań, Poland

Abstract. In this paper a new type of endoprothesis THRA with needle-palisade fixation system has been described. A numerical model of such fixation has been created. The developed model was subjected to experimental verification. During the test, the prototype of the needle palisade structure was inserted into the cancellous bone. The test results were compared with numerical simulation.

Keywords: bone, implant, numerical model, bone-implant connection

1 Endoprothesis

The hip joint is one of the most exposed to overload-degenerative changes elements of the osteoarticular human system. Its function is to transfer loads both static and dynamic generated by body mass, muscle strength acting on the joint and the accelerations occurring during movement. The most common types of hip injuries are:
- fracture of femoral neck,
- congenital hip/birth defect (eg. hip dysplasia),
- avascular necrosis of the femoral head,
- postinflammatory and posttraumatic changes,
- osteoporosis,
- cancer.

Hip injuries cause severe pain, disabling movement and normal functioning in everyday life. Currently, the only method of treatment of overload and degenerative lesions of the hip is surgical treatment, which involves replacement of damaged joint surfaces with implant.

Depending on the severity of the disease, more or less invasive treatments are used. For more invasive treatment the endoprothesis with long stem is used (Fig. 1a). It is used when the disease is very advanced or during revision surgery. For less invasive treatment the BHR (Birmingham Hip Resurfacing) method is used (Fig. 1b). In this case the endoprothesis with short stem can be used. This method allows for the exchange of diseased areas of the acetabulum and femoral head surfaces. The bone loss is minimal, and there is also minimal risk of dislocationand loosen of implant.

* Corresponding author: maciej.berdychowski@put.poznan.pl

Reviewers: *Andrej* Czán, *Jerzy* Jaroszewicz

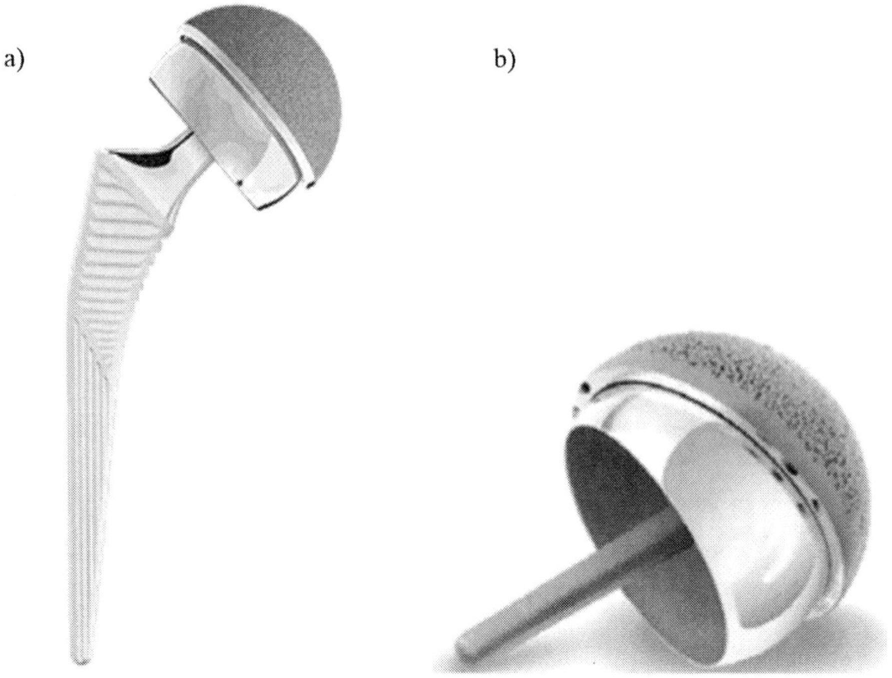

Fig. 1. Hip endoprosthesis: a) endoprosthesis with long stem [1], b) endoprosthesis with short stem so-called BHR (Birmingham Hip Resurfacing) [2]

There are different types of implants and different are methods of mounting it into bone (with cement or without) but there no matter what we do, always after some time, the new surgery and new implant is needed. It is caused by non-physiological load distribution in the bone – implant connection that causes an improper bone reconstruction. There is bone weakness, loosening of the implant and even bone fracture.

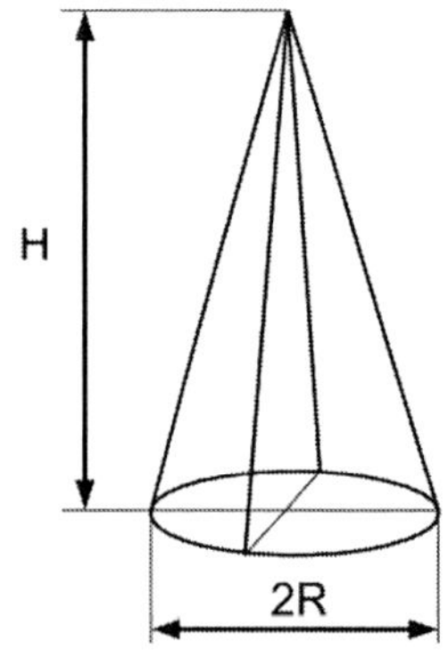

Fig. 2. A sketch illustrating geometric assumptions of modeled needles

The solution of this problem could be the application of a new type of bone-implant fixation. Instead of using a single stem, one can use many small stems similar to pins [3, 4, 5, 6]. This type of solution proposed P. Rogala [7], but he didn't determine a shape of the needle. In the patent are only guidelines and geometric assumptions between high of the needle and a radius of the circle described on the polygon in the base of the needle (Fig.2.). The finally geometry of the needles was determined during analyzes and research, which allowed us to design a prototype of the implant presented in Fig. 3. The implant consists of two parts: a cap with a projection of pins on the inside and a acetabulum with a structure of pins on the outside. Fig. 3c.

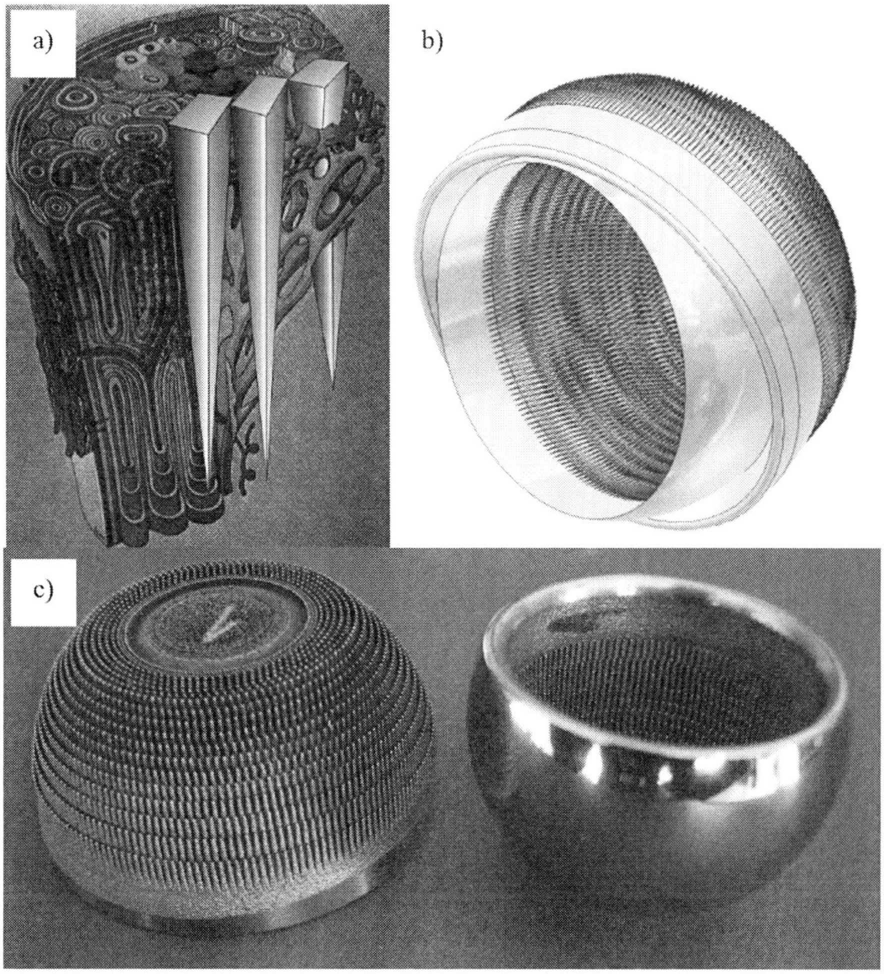

Fig. 3. The idea of the new type of endoprothesis: (a) needles penetrate into cancellous bone [8]; (b) model of the endoprothesis with needle palisade fixation system based on the European Patent EP072418 B1; (c) prototype made with SLM technology from titanium alloy.

2 Numerical analysis

The aim of the study is to build a numerical model of porous biomaterial and its application in modeling bone-implant fixation by numerical representation of the process of insertion of needles into the cancellous bone.

The geometric model was built from two parts: a bone model and a model of needles which will be inserted into the bone (Fig. 4). The needles are arranged concentrically on a base of about 8mm radius.Because of the amount of finite element needed to generate the calculations, and due to the complexity of the model, some simplifications were made to the actual object used in the tests. The base has been omitted in modeling, replacing it with appropriate bindings. The second modeled object was the bone in which the pyramid palisade was pressed. It was modeled as a cylindrical object with hollow square-shaped tubules. The arrangement of the tubules corresponded to the placement of the needles in the first object. During the simulation, the needle system is lowered and the pyramid are inserted into the bone model. In the model there was defined contact between the surfaces of the needles and the channels.

The prepared geometric model was discretized with finite elements. There were used a eight nodal hexagonal elements. The bone has been modeled as a deformable material. Titanium needles were modeled as a non-deformable body by giving the "rigid body" constrain.In the described model, material was used as isotropic, elastic-plastic with values E=0,2 GPa and ν=0,03. Calculations were made in the Abaqus program.

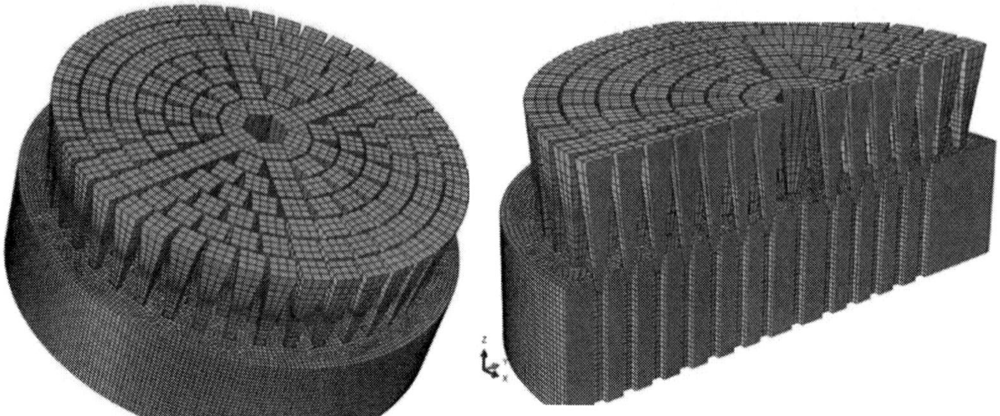

Fig. 4. Numerical model of the bone – implant connection

The figures presented below (Fig. 5, Fig. 6) show the results of the calculations. Base on the results, significant deformations of the bone material in the area between pins were observed. Hollowing of needles, of the order of 50% of the length, already causes distortion far beyond the yield point. This leads to permanent destruction of the trabecular structure of the bone, the breaking and thickening of the bone material, which consequently wedges between the pins.

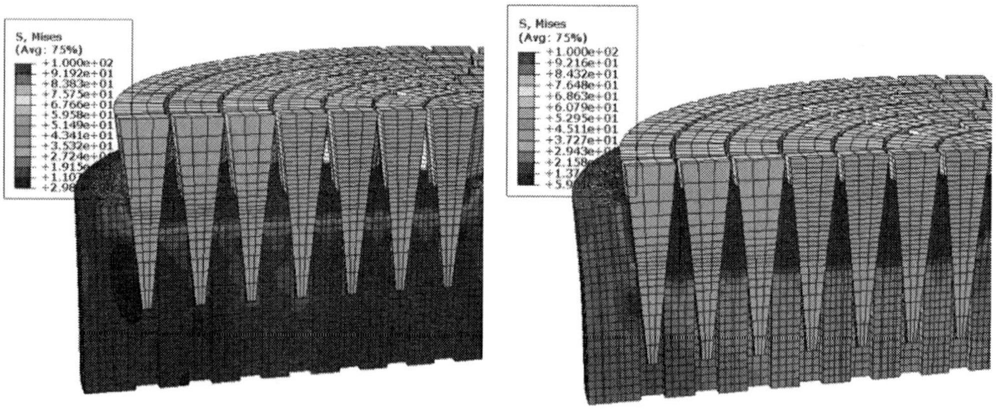

Fig. 5. Changing the stress distribution during the insertion of needles into the cancellous bone

The results of these analyzes confirm the studies. During the extraction of the samples, the bone fragments jammed between the pins. By analyzing the results, it can be seen that not only the material in the space between the needles, but also the lower one, is affected by deformation. The crushed and compacted bone forms with the specimen a type of punch that presses on the underlying layers of material causing it to deform. With large forces can lead to the destruction of the material theoretically not taking part in bone implant connection.

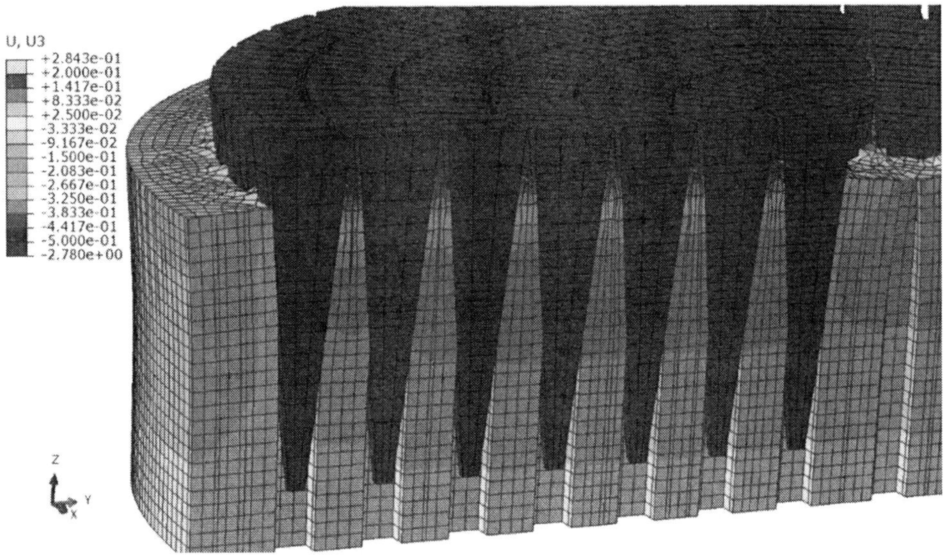

Fig. 6. Distribution of displacement in the vertical direction

3 Experimental verification of numerical model

The study aimed to find the relation between the force and the value of depth of the insertion of needle for different types of pyramids forming needle - palisade fixation system.

These characteristics will be used also to verify the numerical model of mentioned connection. It is also important to observe the changes and deformations that occur in the bones when making a connection.

The samples were designed with a cylindrical base on which a pyramid system of pins was placed. The base diameter was 16mm and its height was 6mm. Samples were made using rapid prototyping technology. With available methods, the SLM (Selective Laser Melting) were chosen. The titanium alloy powder Ti6Al4V has been used. This material has been successfully used in prosthetics for many years.

Samples were pressed in pork bone imported from a slaughterhouse. The bone was used in the research within several hours of slaughter and was stored under refrigeration. Preliminary bone processing, ie removal of soft tissues, cartilage and cutting to 1/3 of the length, took place in the slaughterhouse. Before testing, the bones were laid in a box and covered with gypsum. This solution provides a cheap and stable mounting with the desired high stiffness.

Fig. 7. Preparation the bone for tests - milling the bone head

After the gypsum dried, the bones were milled to obtain a flat surface (Fig.7). The tests were done on a machine TIRA test 2450. The insertion process took place at a speed of about 2mm / min.

Fig. 8. The bone and the samples after tests

The studies mentioned above showed, that it is not possible to completely press the needle structure into the cancellous bone. In extreme cases, when the strength reached high values there was a cracking of the bone and its breaking. Nevertheless, there has never been a process of total depression. In situations when the needle injection into the bone was interrupted earlier, and then an attempt was made to extract the needles from the bone, in many cases the bone fragments was pulled out (Fig. 8). This situation always appeared when the needles were pulled down over 50-60% of their height.

With this in mind, a series of studies were conducted to determine the maximum needle depth that would not break the bone fragments during the sample extraction. The studies let to collect a data (the force as a function of displacement) when the needles are inserted into the cancellous bone (Fig. 9).

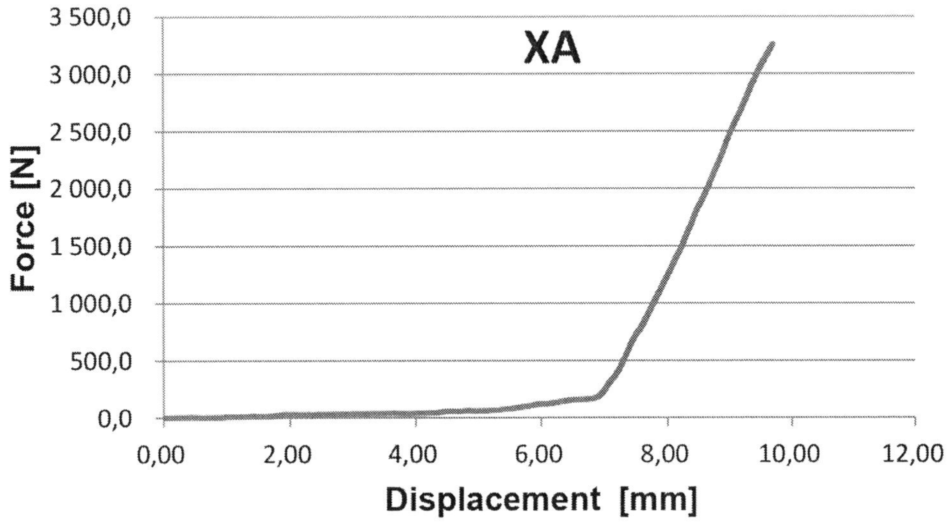

Fig. 9. Graph of dependence of the force on the needle displacement into cancellous bone. Sample with the needles in the form of pyramids with square base and dimensions: base area 1mm2 , ratio A/H=1:5

The average value of the force required to insert the needles into a given depth is shown in the Table 1. It was combined with the force value obtained by numerical simulations. The difference between those values is less than 10%.

Table 1. Summary of research results with numerical simulation results

Shape of the base	**Needles' high A/H**	**displacement [mm]**	**Force**		
			Tests [N]	**Simulation [N]**	**Difference [%]**
Square	1:5	2.1	2645	2902	8.8
Square	1:9	3.7	3143	2951	6.1
Triangle	1:5	2.0	2690	2938	8.4
Triangle	1:9	3.5	3890	3626	6.8

Conclusions

The aim of the study was to verify the numerical model of implant - bone connection. The numerical model will be used to study the process of insertion of needles into the cancellous bone. The results of the study will be necessary to formulate guidelines for the design process of implant - bone connection. The results of the research presented in this paper indicate the correctness of the constructed numerical model. However, the effect on the convergence of the experimental and simulation results is largely determined by the material (bone), the accuracy of the needle geometry mapping in the SLM process and the numerical simplicity and assumptions adopted in the numerical model.

References

1. www.tourmedica.pl
2. www.carolina.pl
3. P. Rogala, R. Uklejewski, M. Winiecki, J. Mielniczuk, A. Auguściński, *The concept and geometrical model of modern endoprosthesis concerning minimal invasive total hip resurfacing arthroplasty (THRA)*. Machine Dynamics Problems **30** (4), 89-96 (2006)
4. P. Rogala, R. Uklejewski, J. Mielniczuk, M. Winiecki, *Prototyp małoinwazyjnej endoprotezy powierzchniowej stawu biodrowego wytworzony metodą SLM*. TS Raport **48**, 6-7 (2008)
5. P. Rogala, R. Uklejewski, *Needle-palisade fixation for total resurfacing arthroplasties of hip and other joints – Biomechanical principles*. Journal of Biomechanics **39** (1), 513 (2006)
6. R. Uklejewski, P. Rogala, M. Winiecki, J. Mielniczuk, A. Auguściński, M. Berdychowski, *Modern trends in bioengineering design of low-invasive joint endoprostheses*. Inżynieria biomateriałów (Engineering of Biomaterials) **77-80**, 32-33 (2008)
7. P. Rogala, *Nowy model i rozwiązanie biomechaniczne endoprotezy stawu biodrowego*, XXXIII Zjazd naukowy Polskiego Towarzystwa Ortopedycznego i Traumatologicznego, Kraków, 21-23-09, (2000)
8. J. Mielniczuk, P. Rogala, R. Uklejewski, M. Winiecki, G. Jokś, A. Auguściński, M. Berdychowski, *Modelling of the needle-palisade fixation system for the total hip resurfacing arthroplasty endoprosthesis*. Trans VŠB-TU Ostrava, Metallurgical Series **51** (1), 160-166 (2008)

The contribution to the modal analysis using an infrared camera

Vladimír Dekys[1,*], *Zuzana* Stankovicova[1], *Pavol* Novak[1], *Ondrej* Stalmach[1]

[1]Faculty of Mechanical Engineering, Department of Applied Mechanics, University of Zilina, Univerzitna 1, 01026, Slovakia

Abstract. The paper deals with modal analysis using an infrared camera. The test objects were excited by the modal exciter with narrowband noise and the response was registered as a frame sequence by the high speed infrared camera FLIR SC7500. The resonant frequencies and the modal shapes were determined from the infrared spectrum recordings. Lock-in technology has also been used. The experimental results were compared with calculated natural frequencies and modal shapes.

Keywords: infrared camera, vibration, modal analysis, resonance, modal shape

1 Introduction

When an object is excited, for example by acting on forces in the deformation object, part of the energy supplied is also radiated in the infrared frequency range by the detectable infrared camera. This phenomenon is used, for example, in detecting Luders bands [1], non-destructive material testing [2, 3] or detection of plastic areas around stress concentrators. The temperature of the object is then determined after the use of calibration relationships, the relationship between the level of infrared radiation build up and the black body temperature. If this excitation causes only elastic deformations in the test, and if the load frequency is high enough to meet adiabatic conditions, then the relationship between temperature change and the change of trace of the stress tensor $\Delta(tr\boldsymbol{\sigma})$ can be expressed according to the relationship [4, 5]:

$$\Delta T = -\frac{\alpha}{\rho C_p} T_0 \, \Delta(tr\,\boldsymbol{\sigma})$$

$\Delta(tr\boldsymbol{\sigma})$ is the response of the object to the stimulus forces, α is the linear coefficient of thermal expansion; C_p is the heat capacity at the constant pressure, ρ is density of the analysed object, T_0 is starting temperature. Adiabatic conditions mean a sufficiently high object load frequency for steel greater than 2 Hz, for aluminium alloys greater than 20 Hz.

Changing the temperature is often very small, for a steel change of the stress by 1 MPa will cause a change of about 1 mK, [6]. Such a change is, however, at the infrared camera noise level, and its detection may require a specific procedure to increase the signal/noise

* Corresponding author: vladimir.dekys@fstroj.uniza.sk

Reviewers: *Ivan* Kuric, *Grzegorz* Domek

ratio, e.g. using lock-in method [3, 7] to determination of change of temperature. In the case of determination of temperatures it is important to know the emissivity of the tested objects [8].

2 Experiment

If, for example, an impulse hammer is used as a stimulus, then we are able to excite the object within a certain frequency range, depending on the used hammer tip, respectively from the duration of the generated pulse. This can also be achieved by excitement of white noise in the same frequency range, although finding the acquired dependencies may require a greater number of averages. Upon exciting the above mentioned noise, the object vibrates at resonant frequencies, corresponding to its natural frequencies. Excited shapes associated with individual frequencies determine the resulting vibration of the object. Subsequently, the vibration of an object causes changes in its deformations, and there is also a change in the radiated energy detected by the infrared camera.

The frequency response function FRF is generally defined, for example, as the ratio of the output response spectrum of the system to the input stimulus spectrum. In our case, the output signal is measured by the IR camera and the input spectrum was considered to be constant in the frequency range of exciting white noise. Experimental determination of resonant frequencies is based, for example, on the determination of peaks at frequencies in the amplitude spectrum. The corresponding shapes to their natural frequencies (resonance frequencies) are determined by the imaginary spectrum of the frequency response function. The imaginary spectrum was preferred when processing the camera outputs because this spectrum corresponds to the deformations and thus to the changes in the stress field and consequently to the changes in infrared spectral radiations that are registered by the camera.

Two experiments were performed. In the first case, a rectangular S235J steel sheet with dimensions of 160x60 mm with a thickness of 1mm was used. The plate was excited by the TIRA200 modal exciter in Figure 1. The plate was screwed directly onto the moving core of the modal exciter. The input signal was narrowband white noise in the range of 10-500 Hz in Figure 2. The radiation of vibration sheet was registered in mid wavelength (MWIR) using the FLIR SC7500 at native frame rate of 383 Hz. The record was made in ResearchIR Max and then imported and processed in MATLAB.

Fig. 1. The test specimen excited by TIRA200, cause 1, the plate was screwed directly onto the moving core of the modal exciter

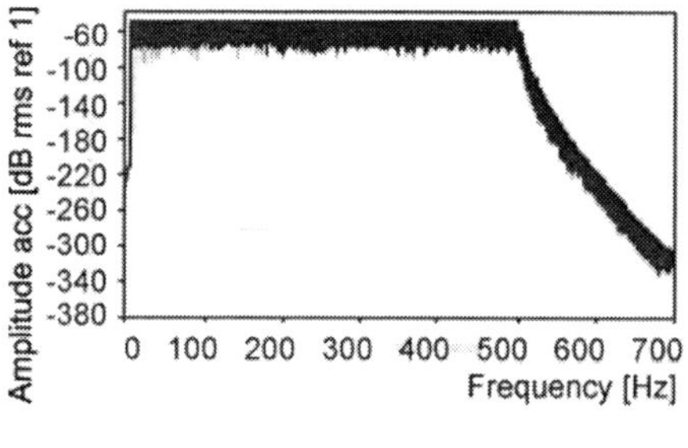

Fig. 2. The spectrum of input signal was narrowband white noise in the spectral range of 10-500Hz

The sequence of frames was processed by the lock-in method. This method is a specific form of Fourier Transform. In the case of the classical transformation the minimum frequency of the Fourier series is determined based on the number of samples and sampling frequency. For the lock-in method this frequency is equal to the load frequency. Subsequently, if the response to the input frequency f_0 of the narrowband white noise is to be determined then this frequency is the lowest frequency of the Fourier series for lock-in, $f_0 = f_{lockin}$.

It is assumed that the system is stimulated by the load force with the frequency f_0 and the response of the system $s(t)$ sufficiently describes the harmonic function with frequency $f_{lockin} = f_0$, amplitude A and phase shift φ

$$s(t)=A\ \sin(2\pi f_{lockin}t+\ \varphi)=a\ \cos(2\pi f_{lockin}t)+b\ \sin(2\pi f_{lockin}t), \tag{1}$$

$$a=A\ \sin(\varphi),\ \ b=A\ \cos(\varphi). \tag{2}$$

Let the amplitude of this response $s(t)$ is very small and the signal $S(t)$ being measured also contains a much higher noise $N(t)$, according to the relationship

$$S(t) = s(t) + N(t). \tag{3}$$

If $S(t)$ is expressed by Fourier's series than

$$\begin{aligned} S(t) &= \frac{A_0}{2} + \sum_{k=1}^{+\infty} A_k \sin(2\pi k f_{lockin}\, t + \varphi_k) = \\ &= A_1 \sin(2\pi f_{lockin}\, t + \varphi_1) + \left[\frac{A_0}{2} + \sum_{k=2}^{+\infty} A_k \sin(2\pi k f_{lockin}\, t + \varphi_k)\right] = \\ &= s(t) + N(t). \end{aligned} \tag{4}$$

If $A = A_1$ and $j = j_1$, than the coefficients a, b in (2) can be estimated by using the relationships of Discrete Fourier transformation

$$a = \sum_{k=0}^{N-1} S_t \cos(2\pi k f_{lockin}\, \Delta t), \quad b = B_k \sin(2\pi k f_{lockin}\, \Delta t) \tag{5}$$

where S_k is the k-th frame recorded by infrared camera (the frames are numbered 0, ... N-1), Δt is the period of frame frequency of camera. By using a, b, the amplitude A, phase φ, the real and the imaginary part of s were determined for f_{lockin}.

The region of interest (ROI) was selected in the image. For the selected f_{lockin}, an imaginary part of s was determined for this ROI by the lock-in method. Subsequently, the variance of the values obtained in this ROI was determined, that is, a pair of values (variance, f_{lockin}). The procedure was repeated for different frequencies f_{lockin} and the resulting data set is shown in Figure 3.

In Figure 4 is the narrowband response spectrum obtained by processing the vibration velocity measurement by Laser Doppler Vibrometer LDV100.The numerical values of the resonant frequencies determined by IR camera and vibrometer measurements as well as the results of the natural frequencies determined by the ANSYS modal analysis for the given object are presented in Table 1.

A comparison of the resonant and the natural frequencies has been demonstrated for frequencies No. 1, 3 and 5. Frequencies No. 2 and 4 were not detected by an infrared camera because a small value of $\Delta(tr\sigma)$ was due to the presence of torsion. Frequencies No. 2 and 4 were not detected by the vibrometer because this device is primarily sensitive to the vibration in a plane perpendicular to the direction of the laser beam and may therefore be less sensitive to the torsion of the sample.

Determination of extreme variance values for the imaginary part versus different f_{lockin} was used to determine resonance frequencies. The extreme variance value is given by the excitation of the corresponding oscillation shape, which is reflected in the response of the object. By this procedure, it was not possible to detect other than bending modal shapes, because torsion modal shapes did not cause a sufficient change in the principal stresses (Fig. 5 – 18).

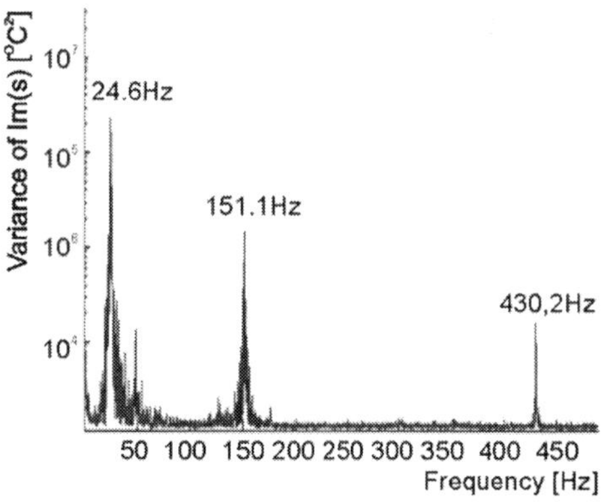

Fig. 3. The spectrum of variance from IR camera

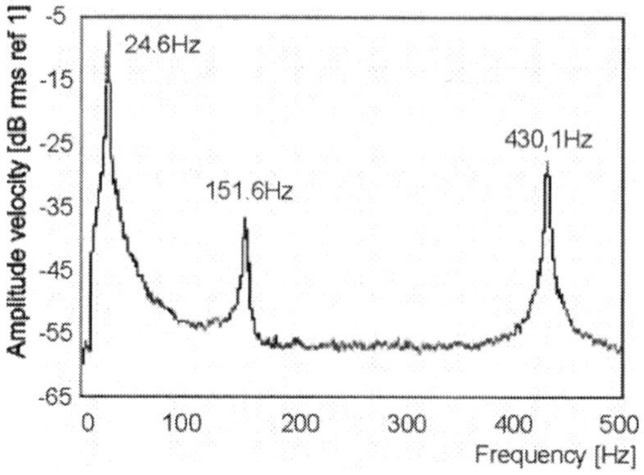

Fig. 4. The spectrum of velocity of vibration from laser beam vibrometer

Table 1. The comparison of resonant frequencies determined by IR camera, laser beam vibrometer and FEA ANSYS

	Frequency [Hz]				
Frequency No.	1.	2.	3.	4.	5.
Infrared camera	24.6	-	151.1	-	430.2
LD vibrometer	24.6	-	151.5	-	430.1
Modal analysis, FEA	24.3	125.2	152.2	399.9	429.1

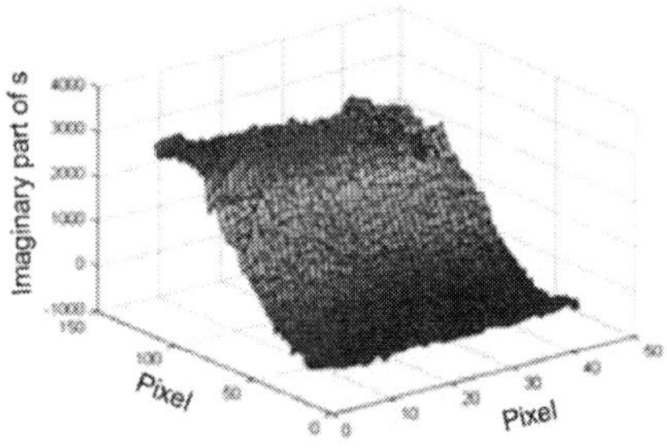

Fig. 5. Deformed bending shape at resonant frequency 24.6 Hz, IR camera, No. 1, axonometric view

Fig. 6. Modal bending shape at natural frequency 24.3 Hz, FEA ANSYS, No. 1, axonometric view

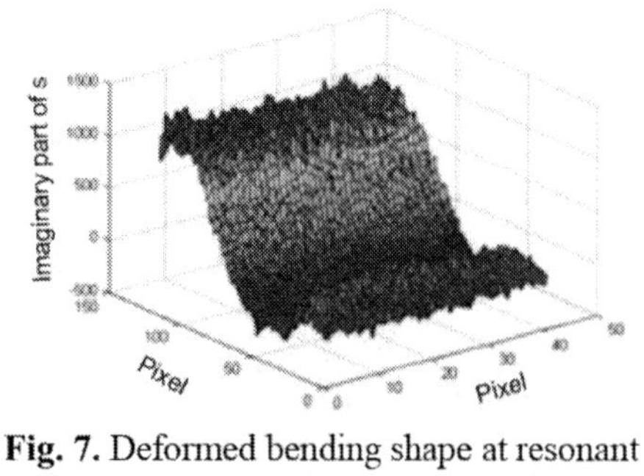

Fig. 7. Deformed bending shape at resonant frequency 151.1 Hz, IR camera, No. 3, axonometric view

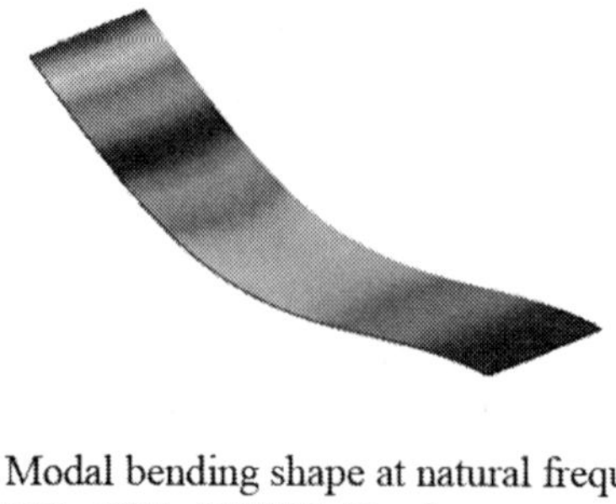

Fig. 8. Modal bending shape at natural frequency 152.2 Hz, FEA ANSYS, No. 3, axonometric view

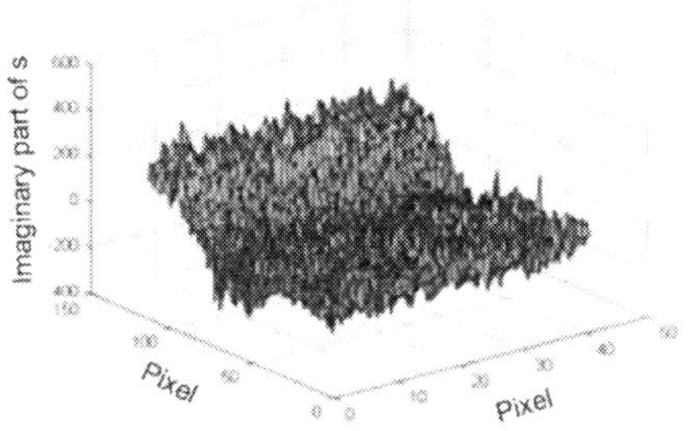

Fig. 9. Deformed bending shape at resonant frequency 430.2 Hz, IR camera, No. 5, axonometric view

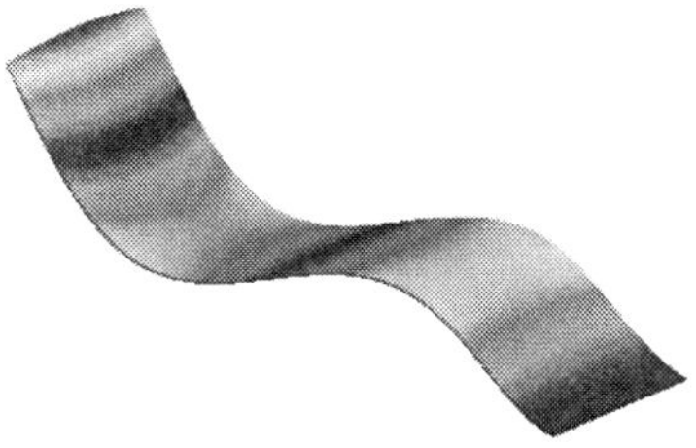

Fig. 10. Modal bending shape at natural frequency 429.1 Hz, FEA ANSYS, No. 5, axonometric view

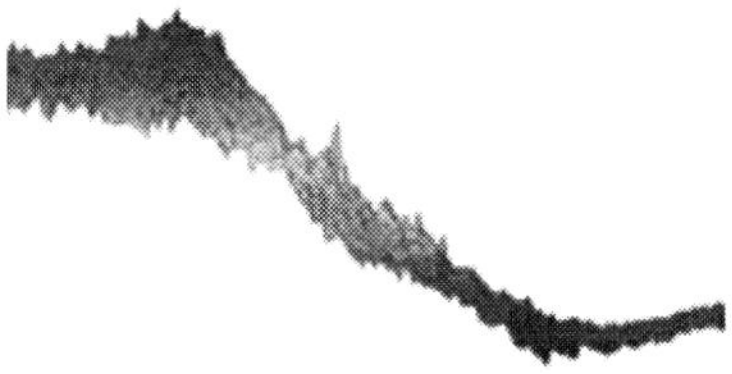

Fig. 11. Deformed bending shape at resonant frequency 24.6 Hz, IR camera, No. 1, a side view

Fig. 12. Modal bending shape at natural frequency 24.3 Hz, FEA ANSYS, No. 1, a side view

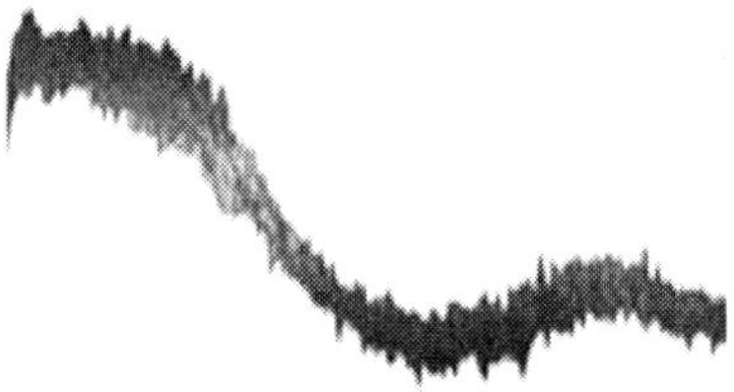

Fig. 13. Deformed bending shape at resonant frequency 151.1 Hz, IR camera, No. 3, a side view

Fig. 14. Modal bending shape at natural frequency 152.2 Hz, FEA ANSYS, No. 3, a side view

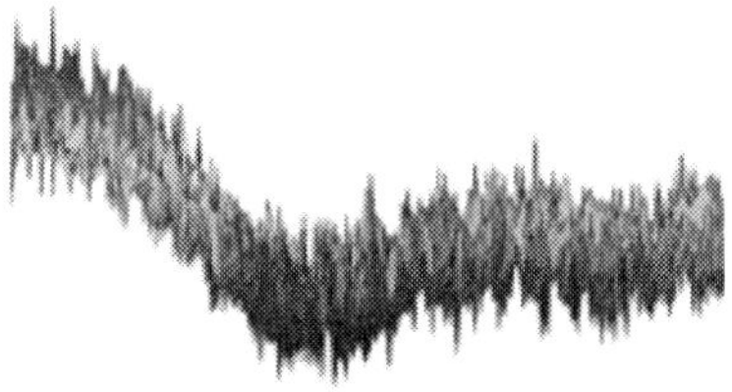

Fig. 15. Deformed bending shape at resonant frequency 430.2 Hz, IR camera, No. 5, a side view

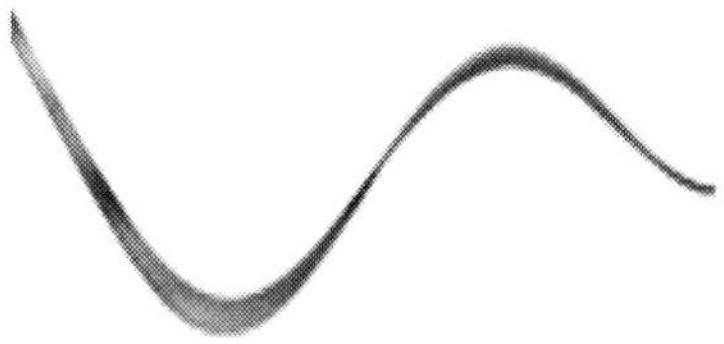

Fig. 16. Modal bending shape at natural frequency 429.1 Hz, FEA ANSYS, No. 5, a side view

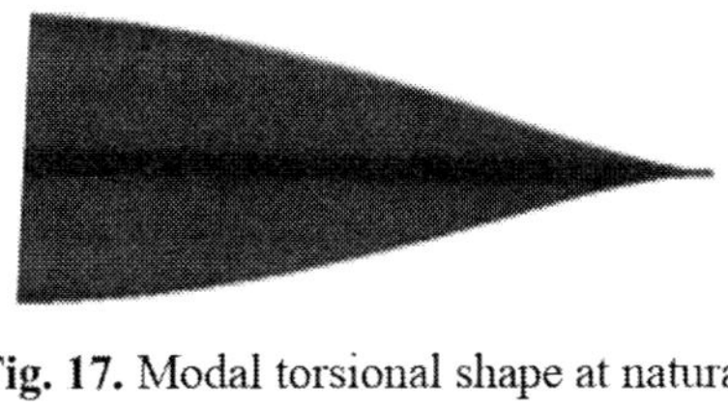

Fig. 17. Modal torsional shape at natural frequency 125.2 Hz, FEA ANSYS, No. 2, a side view, the shape not detected by the IR camera

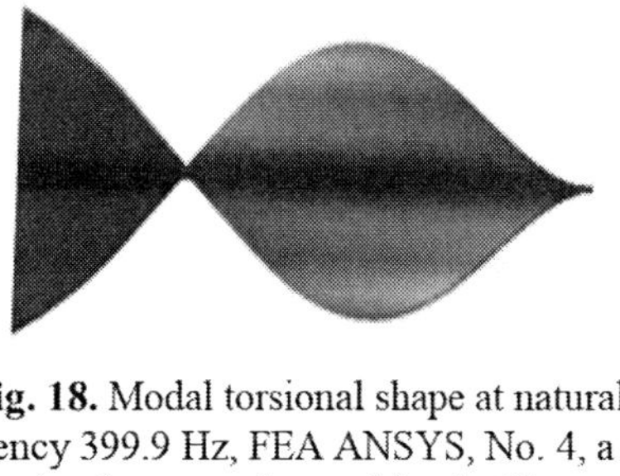

Fig. 18. Modal torsional shape at natural frequency 399.9 Hz, FEA ANSYS, No. 4, a side view, the shape not detected by the IR camera

Experiment No. 2 measured the vibration of the plastic propeller of the table fan. The bandwidth excitation spectrum is again narrowband in the range of 10-192 Hz, measured by an infrared camera and a laser vibrometer. ANSYS was used to calculate the natural frequency of the fan. Larger discrepancies were recorded between the specified first resonant frequency: the IR camera (43.8 Hz), the laser vibrometer (42.1 Hz - 42.9 Hz), and the frequency specified in the FEA ANSYS (45.7 Hz). In Figure 19 and 20 the modal shape for these frequencies are shown. Although the IR camera output is not as brilliant as the FEA, it is possible to observe the bend of propeller at this frequency as determined by the experiment and the FEA. Frequency deviations and a less clear reconstruction of the oscillation shape attribute the blade curvature and the need for a better identification of the FEA material model.

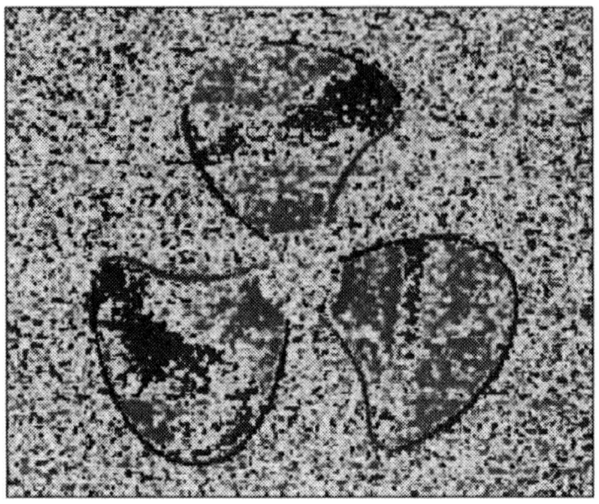

Fig. 17. Deformed bending shape at resonant frequency 43.8 Hz, IR camera, a front view

Fig. 18. Modal bending shape at natural frequency 44.7 Hz, FEA ANSYS, a front view

Conclusion

In the paper was presented an approach to determine resonant frequencies and corresponding natural frequencies by means of radiation detection in the mid infrared band (MWIR). Resonant frequencies were determined using lock-in technology. The variance of imaginary parts for individual pixels in the region is interesting. The peak values of this variance were used for resonant frequency detection. We consider the match between IR camera, laser vibrometer and ANSYS is sufficient.

Corresponding modal shapes were obtained for these frequencies. In the measurements, the objects were excited by narrow-band white noise.

In this contribution, smoothed or approximate modal shapes were not used, as we would like to analyse the properties of excitation signal on the quality of the response in future.

This contribution is related to dynamic analysis work [9, 10] at the Faculty of Mechanical Engineering, Department of Applied Mechanics, University of Zilina and [11, 14].

This work was supported by the Slovak Research and Development Agency under the contract No. APVV-0736-12 and by VEGA 1/0795/16, KEGA 017ŽU-4/2017.

References

1. V. Dekys, P. Kopas, M. Sapieta, O. Stevka, *A detection of deformation mechanisms using infrared thermography and acoustic emission*. Applied Mechanics and Materials **474**, 315-320 (2014)

2. Z. Stankovicova, V. Dekys, F. Novy, P. Novak, *Nondestructive testing of metal parts by using infrared camera*. Procedia Engineering **177**, 562-567 (2017)
3. A. Mendioroz, A. Castelo, R. Celorrio, A. Salazar, *Defect Characterization from Lock-in Vibrothermography Data*. Int J Thermophys **36**, 1208–1216 (2015)
4. X. P. V. Maldague, *Theory and Practice of Infrared Technology for Nondestructive Testing*. (Wiley, New York 2001)
5. M. Sapieta, V. Dekys, A. Sapietova, *Thermal-stress analysis of beam loaded by 3 point bending*. Procedia Engineering **136**, 216-219 (2016)
6. P. Bremond, *New developments in ThermoElastic Stress Analysis by Infrared Thermography*. (IV conferencia Panamericana de END. Buenos Aires, 2007)
7. O. Breitenstein, *Lock-in thermography*. (Springer, New York,: Springer. 2010)
8. M. Honner, P. Honnerova, *Survey of emissivity measurement by radiometric methods*. Applied Optics **54**, 669-683 (2015)
9. M. Zmindak, Z. Pelagic, *Modeling of shock wave resistance in composite solids* Procedia Engineering **96**, 517-526 (2014)
10. A. Sapietova, R. Petrech, M. Petrovic, *Analysis of the dynamical effects on housing of the axial piston hydromotor*. Applied Mechanics and Materials **474**, 357-362 (2014)
11. J. Kortis, L. Daniel, M. Skarupa, *Experimental modal test of the laboratory model of steel truss structure*. Civil and Enviromental Engineering **12** (2), 116-121 (2016)
12. J. Bocko, V. Nohajova, T. Harcarik, *Symmetries of differential equations describing beams and plates on elastic foundations*. Procedia Engineering **48**, 40-45 (2012)
13. F. Klimenda, J. Soukup, *Modal analysis of thin aluminium plate.* Procedia Engineering **177**, 11-16 (2017)
14. M. Nad, L. Rolnik, L. Cimancova, Conference: *Prediction of Changes in Modal Properties of the Euler-Bernoulli Beam Structures Due to the Modification of Its Spatial Properties.* International Journal of spatial Properties **17** (1740014), (2017)

Examples of modern methods of measuring deformations

Tomasz Domański[1,*]

[1]Institute of Mechanics and Machine Design Foundations, Czestochowa University of Technology, Dabrowskiego 73, 42-200 Częstochowa, Poland, EU

Abstract. The work presents examples of possibilities created by multi-unit 3D correlation system. Experience is the non-contact measurement of displacement and strains of slender compression bars or tensile samples. Experimental research show that as a result of axial compression of slender elements with the length substantially exceeding the cross-sectional dimensions, element destruction occurs at loads less than the respective compressive strength, resulting in deformation of the object. We present the results of numerical analysis, compared with exercise of carrying on a test bench.

Keywords: image correlation system, strains, deformations

1 Introduction

Today's facilities are characterized by high functional diversity, which, while maintaining high quality of workmanship, is very difficult. Products are increasingly demanding, and consequently products have increasingly complex 3D geometry. Measurement tools should combine the speed of collection of measurement points with high accuracy. The use of a non-invasive measurement method makes it possible to detect defects much faster without the need for specialist preparation of test specimens. The advantage of using optical measurement methods of deformation, strains or stress distribution is the ability to identify changes in the surface of the examined material at the microscale level, allowing for early identification of the process before its dynamic development. The digital image correlation method uses images taken at the same time by several optical cameras with high sensitivity to deformation and vibration of the object being observed. This method of measurement is increasingly used to determine the components of stresses, deformations or displacements in laboratory conditions, and to identify defects in machine components under the influence of static loads or dynamic variables over time [1]. Measurement methods allow for easier adaptation to the measurement of parts of machine parts in their natural industrial environment under real operating conditions.

* Corresponding author: domanski@imipkm.pcz.pl

Reviewers: *Vladimir* Dekýš, *Grzegorz* Domek

2 Image correlation system 3D

The Q-400 system used with the ISTRA 4D software is a multifunctional non-contact tool for measuring the deformation of a tested object in both two- and three-dimensional coordinate systems Fig. 1.

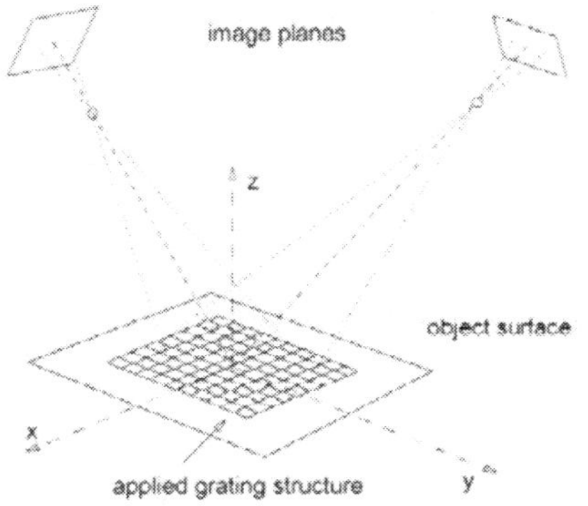

Fig. 1. Schematic diagram [2]

Dimensions and positions of the two points in the state before and after the deformation are considered. The measurement method consists in correlating digital images recorded with two or more cameras. The surface of the object is covered with a layer of white and black paint. Measurement is done by tracking the surface covered by the affected surface, subjected to a load.

3 Examples of research into mechanical phenomena

Two tests were performed to verify system performance. The first was the determination of the deformation in tensile test specimens made of the base material and with a welded joint, while the other consisted in determining the displacement of the compression slim bar.

3.1 Tensile test of samples with welded joint

The tensile test is one of the basic tests for determining the mechanical properties of materials.Universal strength machine Zwick&Roel Z100 is used with extensometer Multisens in all performed tension tests [3]. The accuracy of the strength machine is up to 1N in force and 1μm in displacement Fig. 2. Samples for the tension test were made from welded joints according to norm PN-EN ISO 6892-1.

Fig. 2. Tension test: universal strength machine Zwick&Roel Z100, with the Dantec system

Butt welding is performed for sheets made of S355 steel. Welding process proceed without a gap in the shielding gas 82% Ar + 18% CO_2, the gas flow is 18 l/min, wire speed 6 m/min and welding speed set to v=1 m/min. The distance between heat sources d=2mm. Laser beam is the leading heat source in the tandem.

Optical cameras used to record strains in the tension test are equipped with 50 mm lenses and have maximum resolution 2048x2048 pixels each. This allowed determining the full size of analyzed sample in the working area of universal testing machine. System of three cameras is used in the experiment. Cameras are mounted on the beam which is supported by two fully adjustable tripods. Strain fields are measured for the entire tension cycle. Trigger mechanism is created in Istra4D software for the measurement. Pictures are made for every time increment Δt=0.4 s.

Three tension tests were performed for the base material as well as hybrid welded samples. Results of performed tests are presented in F→ΔL diagrams (Fig. 3). There has been a slight decrease in yield strength and tensile strength in tensed welded joints simultaneously with a decrease of elongation. This can be caused by non heterogeneity of structure in the area of heat influence on the material.

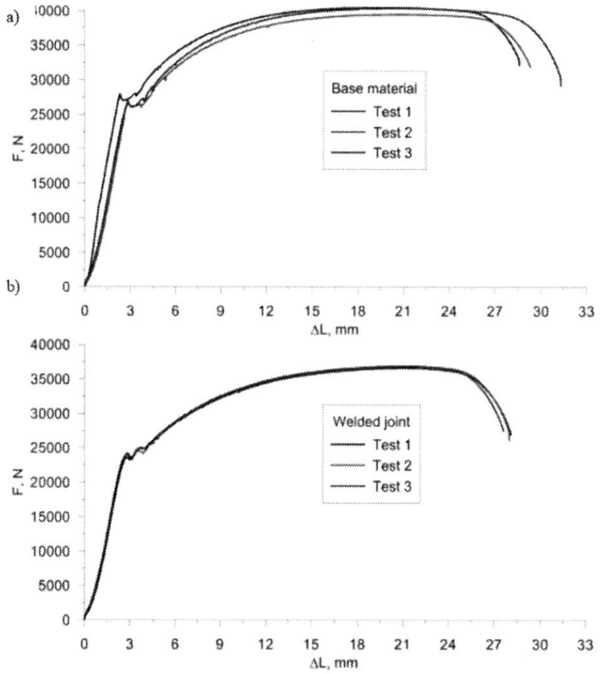

Fig. 3. F→*ΔL* diagrams: a) tensed base material, b) tensed welded joint

Figure 4 presents comparison of strain fields in tensed base material and welded joint at the various times the formation of the neck. It can be observed that the weld and HAZ has a significantly lower strain compared to the base material, which is directly related to the nature of the microstructure in these areas (see. Fig. 4).

Distribution of strain ε_y along the length of tensed sample (55 mm in both direction from the central point of the sample) is illustrated in Fig. 5. Strain distribution is shown in the central axis y for the base material and welded joint at different times. Visible decrease of strain is present in the joint where HAZ is present [4,5]. The slight increase of strain in the weld can be because of the inaccurate selection of an additional material in GMAW method. It can be observed that for the base material higher values of strain ε_y are present in comparison to welded joint.

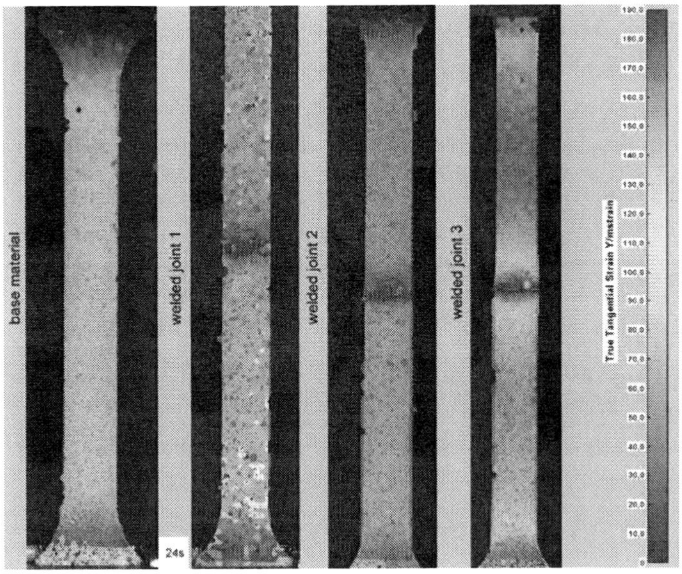

Fig. 4. Comparison of longitudinal strains for time t-24 s

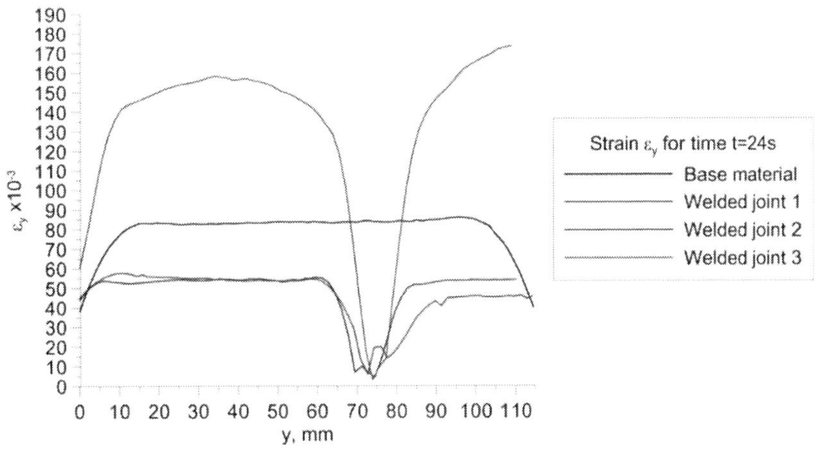

Fig. 5. Distribution of strain ε_y in the central axis *y*

3.2 Compression test of slender bar

In the conducted experiment, two cases were considered. The tests performed differed in the way the bar was loaded, as shown in the figure below (Fig 6.).

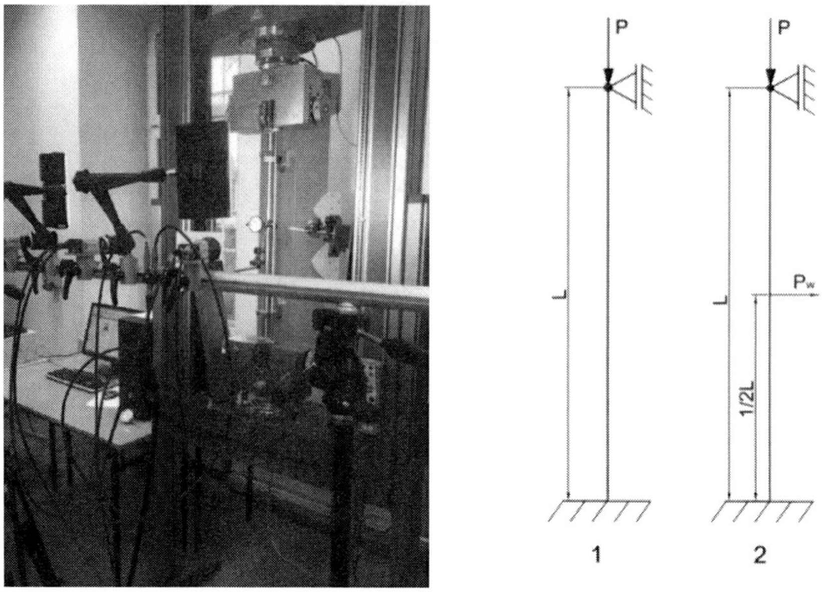

Fig. 6. View of the test stand, bar fixings for the cases considered

Test object in the form of a bar of total length L=708.9 [mm] and circular cross-section with diameter D=6 [mm], made of stainless steel AISI 304, it is fixed at the base and joint at the place of axial force P (case no.1 and no.2). In addition, for case no. 2, the tested element was additionally loaded by a horizontal force that forces the buckling direction P_w, applied at half the total length of the element. Strain of the element due to loss of stability was recorded independently by three cameras in the working area, defined as the distance 170 [mm] from bar anchorage [6,7].

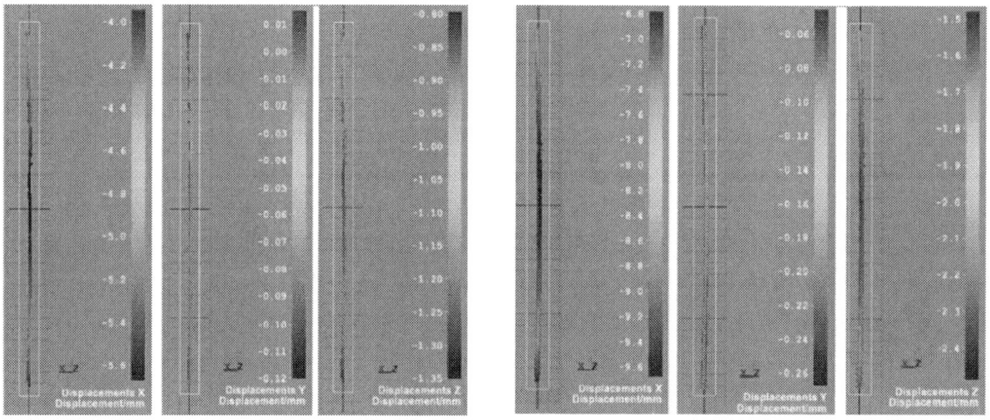

Fig. 7. Results for option 1 and 2

Conclusions

The use of multi-camera 3D correlation system allowed the analysis of deformation for the selected material points throughout the measuring sample. Use of the system allowed the analysis of deformation and strain distributions in the weld and heat affected zone during tension test. It can be observed that lower strains occur in heat affected zone in comparison to the base material which in reference to pictures of microstructure confirm partial hardening of this zone. Visible increase of strain ε_y in the weld may be due to the inaccurate

selection of an additional material in GMAW method. Obtained results allow for a thorough analysis of mechanical properties separate zones of the joint (weld, HAZ) and do not limit the tensile test to determine the global strength and plastic properties of welded material.
In the second example, the use of the Dantec Dynamics measuring system has allowed analysis of displacements and deformations in all directions while attempting to ax the rod. The action of the additional P_w force applied horizontally at half the length of the rod results in larger deformations of the element in the direction of the force. Obtained experimental results may be very helpful in verifying mathematical and numerical models of thermomechanical phenomena occurring in welding processes.

References

1. K. Patorski, *Interferometria laserowa*. Oficyna Wydawnicza Politechniki Warszawskiej, 214-261 (2005)
2. C. Herbst, K. Splitthof, *Basic of 3D Digital Image Correlation*. (Ulm, Dantec Dynamics GmbH)
3. T. C. Chu, W. F. Ranson, M. A Sutton, W. H. Peters, *Application of digital-image-correlation techniques to experimental mechanics*. Experimental Mechanics **25** (3), 232-244 (1985)
4. T. Skrzypczak, *A finite element multi-mesh approach for heat transport between disconnected regions*. Procedia Engineering **177**, 204-209 (2017)
5. M. Kubiak, W. Piekarska, *Comprehensive model of thermal phenomena and phase transformations in laser welding process*. Computers & Structures **172**, 29-39 (2016)
6. M. R. Gower, R. M. Shaw, *Towards a planar cruciform specimen for biaxial characterization of polymer matrix composites*. Applied Mechanics and Materials **24-25**, 115-120 (2010)
7. L. Sowa, Z. Saternus, M. Kubiak, *Numerical modelling of mechanical phenomena in the gantry crane beam*. Procedia Engineering **177**, 225-232 (2017)

Experimental research of vehicle traction properties for reconstruction of traffic accidents

Marian Dudziak[1] , *Andrzej* Lewandowski[2], *Michał* Śledziński*[3]

[1]The President Stanisław Wojciechowski State University of Applied Sciences in Kalisz, Nowy Świat Street 4, 62-800 Kalisz, Poland
[2]Institute of Forensic Research, Cracow, ul. Westerplatte 9, 31-033 Kraków, Poland
[3]Chair of Basics of Machine Design, Poznan University of Technology, Poznan, 60-965, Poland

Abstract. In order to broaden the database of motor vehicle traction properties in unusual conditions, the research team has performed experimental studies: on wet and snow-covered surfaces. Tests of vehicles equipped with winter tyres with non-skid snow chains have been performed on snow-covered surfaces. It has been shown that on snowy surface chains affect vehicle traction properties, mostly during acceleration. They increase the rate of acceleration up to 50% compared to a vehicle with winter tyres without chains. The results of the performed research can be the basis for the full reconstruction of road accidents under these conditions. Knowledge of traction properties of cars in difficult and unusual conditions is an important cognitive factor and serves to improve road safety.

Keywords: traffic accidents, reconstruction, snow chains, *Traction Control System*, the coefficient of adhesion

1 Introduction

Road incidents often take place under difficult conditions arising from the state of the surface (eg. wet, snowy). Reconstruction of a traffic accident [1] requires knowledge about the behaviour of a vehicle in such conditions. The most effective method of obtaining data about the vehicle's traction properties is experimental research [2] in real traffic conditions, which is what drives the authors' interest in this issue.

In snowy driving conditions, traction characteristics of a vehicle depend directly on the shape-frictional force couple between car tyres and the surface. The value of this couple is a function of the road conditions, the type of tyres and weather conditions [3]. The driver, assisted by the *traction control system*, among others, has a significant influence on the optimal use of the tyre adhesion to the road surface.

The purpose of the research was to broaden the database of vehicle traction properties on a snowy road surface [4]. The behaviour of a vehicle equipped with winter tyres and non-skid snow chains has been studied during acceleration and deceleration of the car. This allowed a comparative analysis of traction characteristics of a car equipped with winter

* Corresponding author: michal.sledzinski@put.poznan.pl

Reviewers: *Maciej* Bodnicki, *Marek* Macko

tyres and a car additionally equipped with anti-skid chains on a snow-covered surface. The influence of the traction control system (TCS) on car acceleration has also been studied.

2 Methodology

Road tests, conducted in two phases, consisted in accelerating the test car from a standing start to a speed of about 50 – 60 km/h (phase one), and then in heavy braking to a complete stop (phase two). The tests have been conducted in four cycles: with chains on the front axle and without chains, with the traction control system turned on and turned off.

The Ford Mondeo car has been used for testing. Road conditions were characterised by the surface covered with a layer of beaten snow at a temperature of -4^{0}C. In this research, accelerometers by Analog Devices fitted to the bodywork of the car have been used. During testing, acceleration on measurement sections has been recorded in the acceleration phase and deceleration phase [5, 6, 7]. Figure 1 shows the test car during the measurement and the wheel with the non-skid chain.

a) b)

Fig. 1. The car during testing – a), the wheel with the non-skid chain – b)

The recorded acceleration runs have been processed by means of a 6th order low-pass Butterworth filter, which made it possible to determine speed and distance variability in the function of time. Based on the ECE Regulation 13, mean fully developed deceleration (MFDD), has been determined according to [8]:

$$MFDD = \frac{v_b^2 - v_e^2}{S_e - S_b}\left[\frac{m}{s^2}\right], \tag{1}$$

where:
v_{max} – maximum speed of the car,
v_b – speed equal to 0.8 of the maximum speed,
v_e – speed equal to 0.1 of the maximum speed,
S_b – distance travelled between speeds v_{max} and v_b,
Se – distance travelled between speeds v_{max} and v_e.

To determine the average acceleration of the car a_{avg}, the following formula has been used

$$a_{avg} = \frac{v_b^2 - v_e^2}{S_b' - S_e'}\left[\frac{m}{s^2}\right], \tag{2}$$

where:
S_b' - distance travelled during acceleration to speed v_b,
S_e' - distance travelled during acceleration to speed v_e.

3 Experimental research results

3.1 Road testing results - the car with chains

On the selected section of the road, 5 road tests with the TCS turned on and 6 tests with the TCS turned off have been conducted with a car equipped with winter tyres with chains. Table 1 shows the results of average acceleration during the acceleration phase and the braking phase.

Table 1. Results of the tests with chains

Test No.	Max speed	Acceleration a_{avg}	Braking MFDD	Use of TCS
-	m/s	m/s^2	m/s^2	
CH1	12.6	1.6	-2.5	-
CH2	12.6	1.8	-2.6	-
CH3	14.4	1.3	-2.5	-
CH4	15.2	1.5	-2.6	-
CH5	15.3	1.3	-2.7	-
Ch6	15.9	1.2	-2.2	-
CH7	16.6	1.3	-2.0	TCS
CH8	16.4	1.2	-2.8	TCS
CH9	15.3	1.2	-2.7	TCS
CH10	16.7	1.3	-2.7	TCS
CH11	16.9	1.3	-2.5	TCS

Designation: CH1÷CH11 applies to the tests with chains, TCS - Traction Control System.

In the six conducted acceleration tests of the car with chains and TCS turned off, the obtained average acceleration values were in the 1.2 to 1.8 m/s^2 range. The average acceleration in this group is 1.44 m/s^2 with the standard deviation of 0.22 m/s^2. The range of acceleration variation in individual tests has been influenced by environmental factors and the driving style. In the test shown in Figure 2, marked as CH2, the highest average acceleration in the car acceleration phase has been obtained, ie. 1,8 m/s^2.

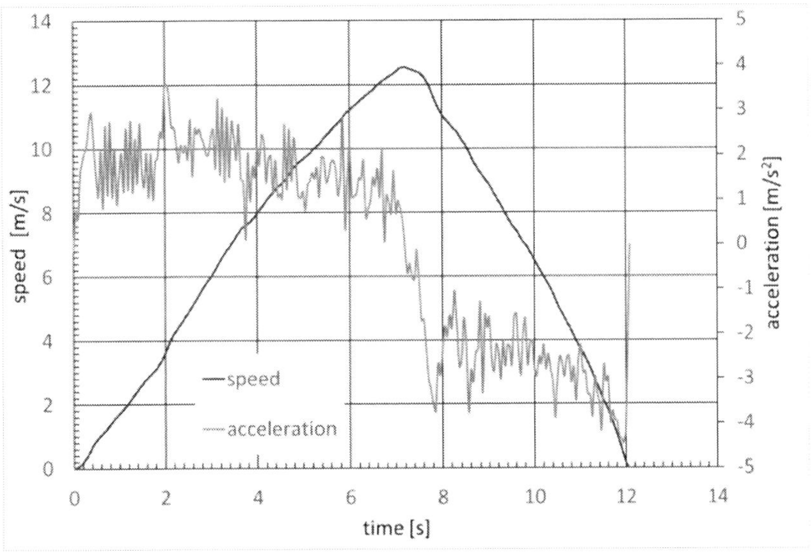

Fig. 2. The course of acceleration and speed of the car with chains and TCS turned off (CH2)

In the test CH6 the lowest acceleration value has been obtained. Figure 3 shows the differences between the speed runs in the function of time for tests CH2 and CH6. The graph shows that in the starting phase (for about 1 second), in both tests the vehicle accelerated with a similar intensity. In the test CH6, after that time, there was a reduction in acceleration, visible as a reduction of the car speed increase, which was probably related to a temporary increase in drive wheel slip.

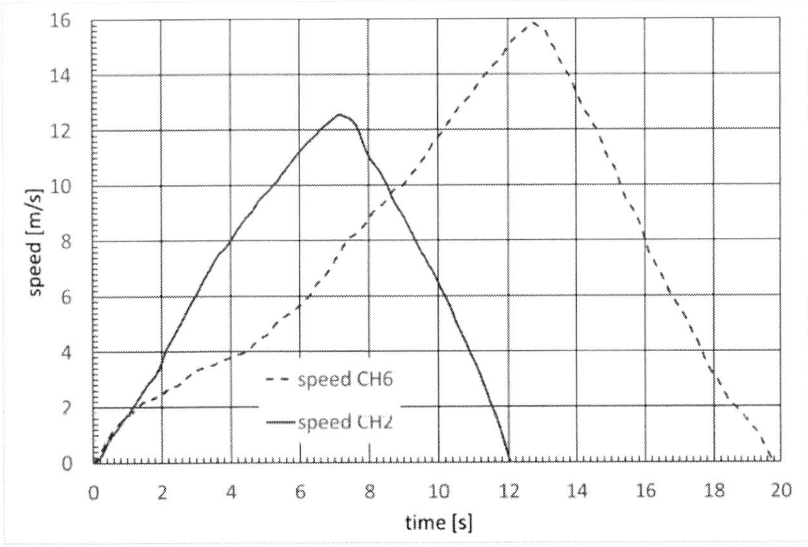

Fig. 3. Comparison of speed changes in tests CH2 and CH6 – the car with chains and TCS turned off

Figure 4 shows the run of the test CH7 with chains and TCS turned on. The diagram shows significantly smaller changes of acceleration in the test car acceleration phase relative to the tests with TCS turned off. In the five tests, the obtained average acceleration values were in the narrow range between 1.2 ÷ 1.3 m/s^2, which is clearly the effect of having the support system. The average acceleration in this group was 1.26 m/s^2 with the standard deviation of 0.05 m/s^2. Table 1 shows that maximum acceleration values have been lowered, which should be associated with the characteristics of the traction control system.

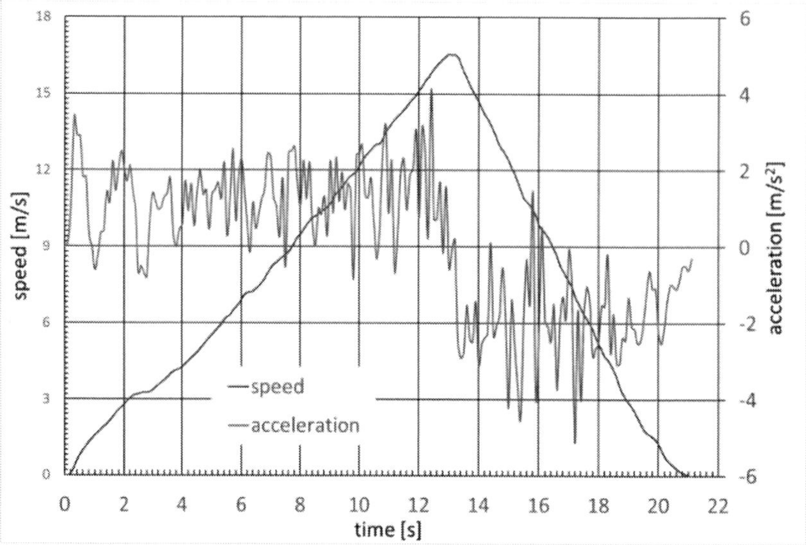

Fig. 4. The course of acceleration and speed of the car with chains and with TCS turned on (test CH7)

During the braking process of the car equipped with chains the obtained brake delays were in the 2.0 ÷ 2.8 m/s^2 range. The average brake delay in this group was 2.52 m/s^2 with the standard deviation of 0.22 m/s^2.

3.2 Road testing results - the car without chains

6 road tests have also been conducted with the car equipped with winter tyres, in two cycles (as in section 3.1), with TCS turned on and turned off. Table 2 shows the results of average accelerations during acceleration and braking.

In the six acceleration tests on the car with chains and TCS turned off, the obtained average acceleration values were in the 0.8 ÷ 1.1 m/s^2 range. The average acceleration in this group was 0.96 m/s^2 with the standard deviation of 0.10 m/s^2. The scale of variability in individual tests has been influenced by environmental factors [9] and the driving style.

Table 2. Test results – the car without chains

Test No	Max speed	Acceleration a_{avg}	Braking MFDD	Use of TCS
-	m/s	m/s^2	m/s^2	
WCH1	14.9	0.9	-2.2	-
WCH2	14.1	0.8	-2.3	-
WCH3	14.6	0.9	-2.8	-
WCH4	15.4	1.1	-2.8	-
WCH5	15.3	1.0	-3.0	-
WCH6	15.3	1.1	-2.4	-
WCH7	15.9	1.1	-2.8	TCS
WCH8	16.3	1.1	-2.7	TCS
WCH9	15.9	1.1	-2.7	TCS
WCH10	15.3	1.0	-2.6	TCS
WCH11	15.7	1.1	-2.6	TCS
WCH12	15.5	1.1	-2.6	TCS

Designation: WCH1 ÷ WCH12 applies to the tests without chains.

In the test shown in Figure 5, marked as WCH4, the highest average acceleration in the car acceleration phase has been obtained, ie. 1.1 m/s^2.

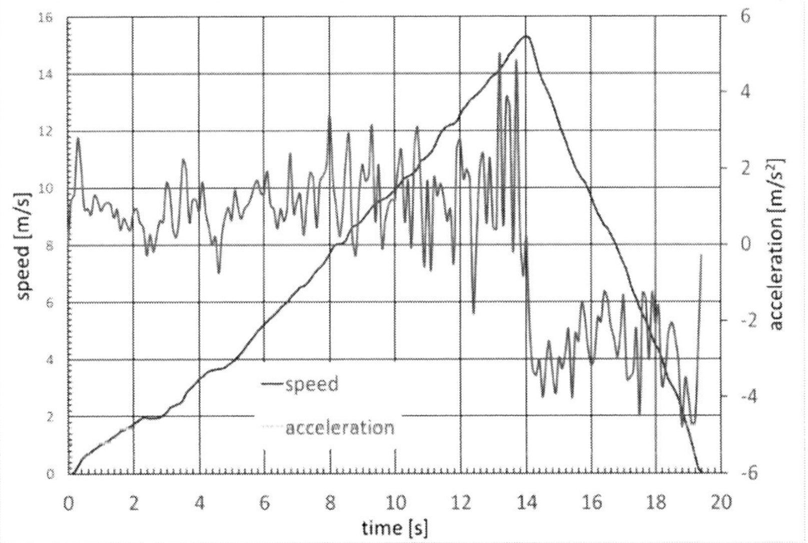

Fig. 5. The course of acceleration and speed of the car without chains and with TCS turned off (test WCH4)

Figure 6 shows the course of the test WCH8 without chains and with TCS turned on. In the six conducted tests the obtained average acceleration values were in the narrow range 1.0 ÷ 1.1 m/s^2. The average acceleration in this group was 1.10 m/s^2 with the standard deviation of 0.05 m/s^2.

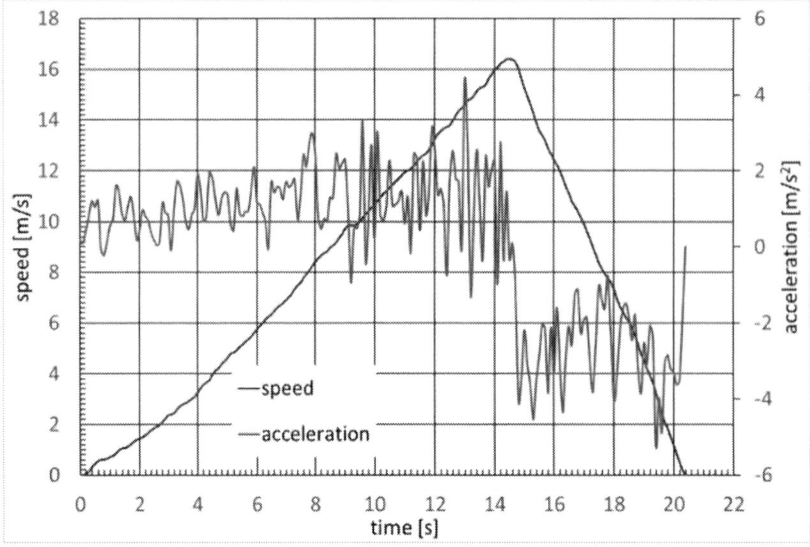

Fig. 6. The course of acceleration and speed of the car without chains and with TCS turned on (WCH8)

During the braking process of the car without chains the obtained brake delay was in the $2.2 \div 3.0$ m/s^2 range. The average brake delay in this group is 2.64 m/s^2 with the standard deviation of 0.22 m/s^2.

Conclusions

The conducted road tests have showed that the use of chains increases the ability of a car to accelerate by approximately 50% compared with driving without chains. Taking into account the relatively low values of the coefficient of adhesion on snowy and icy surfaces, it is a significant factor that increases traffic safety.

Based on the analysis of the research results, it has been found that the operation of the traction control system reduces the acceleration of a car equipped with non-skid chains only to a small extent. In case of movement of a car without chains, the TCS has favourable influence and increases acceleration values comparing to the car where TCS is turned off. Based on the research results, it has also been found that, contrary to expectations, during the braking phase the range of brake delay values for the car equipped with chains has been smaller than for one without chains. We conclude that this situation is related to the properties of ABS, which responds to local changes of adhesion of the wheels to the surface during shape-frictional contact of the chain with the surface. This issue requires further research.

The obtained results of road tests in snowy conditions expand the range of knowledge on traction properties and on the behaviour of a vehicle equipped with non-skid chains and on driving without chains, as well as contribute important insights about the impact of the traction control system. These data are important in expanding the databases used in the reconstruction of the road incidents and in project work. They can also be used in the practice of driver training and to assist their response process in actual emergency situations, which may contribute to an increase in the level of safety for vehicles in winter conditions.

References

1. Praca zbiorowa, *Wypadki drogowe. Vademecum biegłego sądowego*. Wydanie II, Institute of Forensic Research Publishers Kraków (2010)
2. F. Navin, M. Macnabb, C. Nicolletti, *Vehicle traction experiments on snow and ice*. SAE Technical Paper Series, 960652, Detroit, Michigan, USA (1996)
3. Praca zbiorowa pod redakcją M. Dudziaka, *Proces hamowania samochodu a bezpieczeństwo w ruchu drogowym*. Wydawnictwo Instytutu Technologii Eksploatacji w Radomiu, Poznań-Radom (2002)
4. D.P. Martin, G.F. Schaefer, *Tire-Road Friction in Winter Conditions for Accident Reconstruction*. SAE Technical Paper Series, 960657, (1996)
5. A. Lewandowski, J. Warszczyński, M. Dudziak, K. J. Waluś, *Analysis of the passenger car breaking process on an inclined road under winter conditions*. Machine Modeling and Simulations, ed. Marian Dudziak et all., Rokosowo, Politechnika Poznańska, Katedra Podstaw Konstrukcji Maszyn, 147-152 (2012)
6. K. J. Waluś, *Comparing the intensity of the acceleration of a passenger car eguipped with summer and winter tires in sub-zero road surface temperatures*. Technolog **4**, 262-264 (2013)
7. K. Kędziora, K. J. Waluś, *Experimental and simulation research on car acceleration and braking on snow-covered roads*. Activities of Transport Telematics, 13th International Conference on Transport Systems Telematics, TST 2013 selected papers, Katowice-Ustroń, Poland, October 23-26, 2013, ed. Jerzy Mikulski, Springer Berlin Heidelberg, Communications in Computer and Information Science **395**, 433-440 (2013)
8. Regulation No. 13 (UNECE), *Uniform provisions concerning the approval of vehicles of categories M, N and O with regard to braking*.
9. M. Dudziak, A. Lewandowski, J. Warszczyński, M. Anioła, K., Kędziora, K. J. Waluś, *Wpływ cech nawierzchni i otoczenia na parametry ruchu pojazdu w warunkach zimowych (badania pilotażowe)*. Materiały XX Francusko-Polskiego Seminarium Mechaniki VII Konferencji Międzynarodowej „Tarcie 2012", Warszawa, Politechnika Warszawska, 24-29 (2012)

Influence of geometrical parameters of convergent sleeve on the value of limit stress

Jan Górecki[1,*], *Ireneusz* Malujda[1] , *Krzysztof* Talaśka[1], *Dominik* Wilczyński[1], *Dominik* Wojtkowiak[1]

[1]Chair of Basic Machines Design, Poznan University of Technology, ul. Piotrowo 3, Poznan, Poland

Abstract. This paper presents the results of research on improving the effectiveness of the agglomeration process. Improving effectiveness was obtained as a result of the application of the convergent sleeve. The sleeve is mounted before the multi-holes die in the dry ice agglomeration machine. The empirical part of the paper presents the results of research on which FEM model was based. The numerical part of research presents the FEM model of the agglomeration process. The FEM model with a known uncertainty level was used to determine the influence of geometrical parameters of the sleeve on the limit value of the agglomeration forces. The model will be one of the starting points for the design and construction of the machine for the compaction and granulation of dry ice.

Keywords: dry ice, convergent sleeve, compaction, densification

1 Introduction

The by-products of manufacturing processes can be reused, for example as raw material. One material that undergoes the recovery process is carbon dioxide, which is a waste, for example, in the production process of ammonia compounds [1]. This process has the highest percentage share compared with other sources of CO_2 emissions (fig. 1). This gas can be compressed, resulting in its condensation, and in this form it is stored for further use. However, due to the large amounts of recycled waste material, plants very often have a problem with its full management. Therefore, liquid carbon dioxide is supplied to interested recipients [1-3].

Liquefied carbon dioxide, subjected to decompression, undergoes crystallization [4, 5]. The resulting product has temperature of -78.5 °C and sublimes under the surrounding conditions [1, 6-8]. Due to its special properties, it has been commonly called dry ice. Both maintaining low temperature and sublimation in standard conditions make the material widely used in the food industry, e.g. in the process of water carbonation, cleaning and disinfecting surfaces or as an input to refrigerated containers [4-6, 8-14].

* Corresponding author: jan.gorecki@put.poznan.pl

Reviewers: *Marek* Macko, *Wiesława* Piekarska

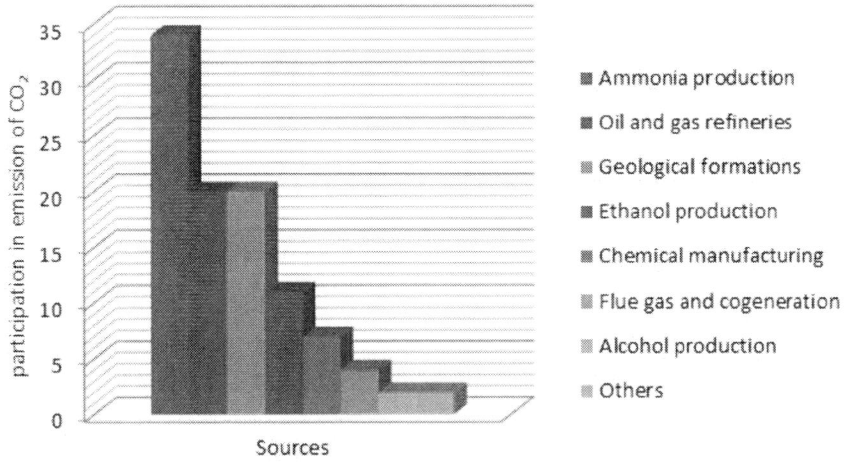

Fig. 1. Percentage breakdown of major sources of CO_2 emissions [12]

Agglomeration of crystallized carbon dioxide is performed in order to reduce its volume and extend the time of sublimation. Once compacted, it stays in the solid phase for longer, which affects the efficiency of its use for cooling purposes [1, 6, 15, 16].

The design of the piston working unit of the dry ice compaction machine is very often based on the systems with hydraulic or crank-piston drives (fig. 2) [17]. The review of the available literature does not provide sufficient information that would make it possible to perform a detailed analysis of the process of dry ice agglomeration. It is necessary for the assessment of such a process and in order to propose new technical solutions directed at its efficiency.

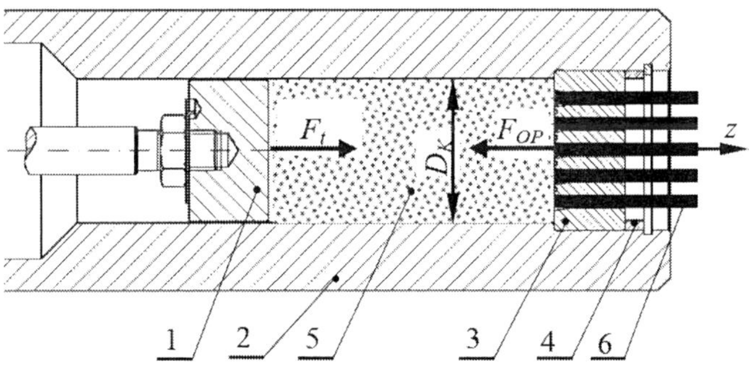

Fig. 2. Pressure agglomeration unit in piston technology. 1 – piston, 2 – cylindrical sleeve, 3 – multi-holes die, 4 – spacer sleeve, 5 – crystallized carbon dioxide, 6 – compacted dry ice

The limit force on the compacting piston necessary for the execution of the agglomeration process is the value of the drag force F_{OP} pushing the material through the multi-holes die (fig. 2, 3).

The value of F_{OP} changes in the function of the geometric parameters of the die. One of the properties that affect the value of this force is the surface area S_{PP}. In order to reduce the indicated area, it was proposed in the patent application P.419432 that the working unit is equipped with the convergent sleeve. The element is mounted on the input surface of the die, through which the agglomerated material is injected (fig. 4). The position of the element in the example working unit is presented in fig. 4

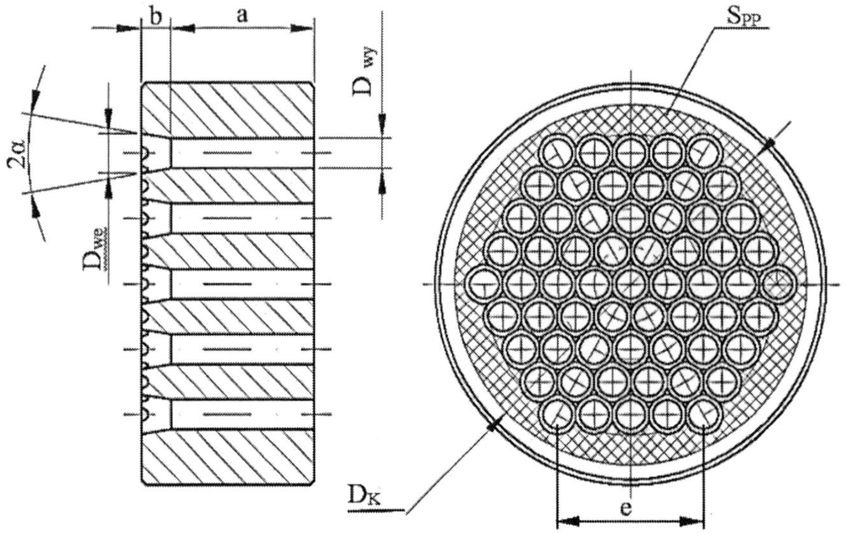

Fig. 3. Multi-holes die, D_{we} – input diameter of the forming channel, D_{wy} – output diameter of the forming channel, a – length of the cylindrical segment, b – length of the conical segment , 2α – convergence angle of the conical segment, D_K – working chamber diameter, S_{PP} – resistance surface area

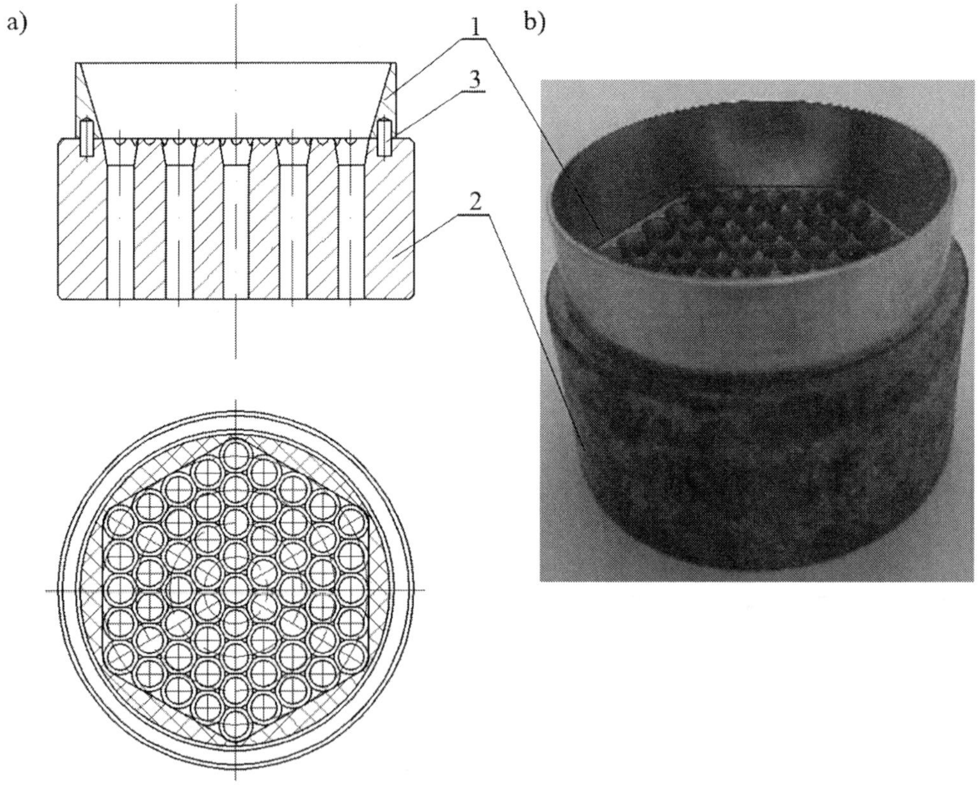

Fig. 4. Multi-holes die system with the convergent sleeve: a) system, b) stress condensing sleeve 1 – stress condensing sleeve, 2 – multi-holes die, 3 – dowel pin

Due to the non-circular symmetric shape of the sleeve, resulting from the transition from a round section to a hexagonal section, it was necessary to determine the relative positions of elements relative to each other using 2 dowel pins (3) (fig. 4).

The system assembled in such a way, with determined mutual positioning of the elements, can be mounted in the channel compacting crystallized carbon dioxide in crank-piston technology (fig. 5).

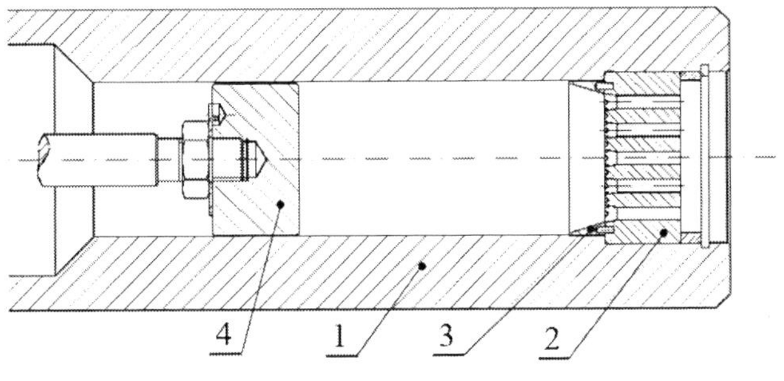

Fig. 5. The working unit of the agglomerating machine in piston technology: 1 – cylindrical sleeve, 2 – multi-holes die, 3 – convergent sleeve 4 – piston [18].

FEM analysis will be performed in this work in order to determine the impact of the application of the convergence sleeve to the value of the resistance force F_{OP}. The FEM model was based on the results of the empirical research. The model formulated in this way will be used to compare the effect of geometric parameters of the sleeve on the value of limit stress in the dry ice agglomeration process for sleeves of different heights h (fig. 6).

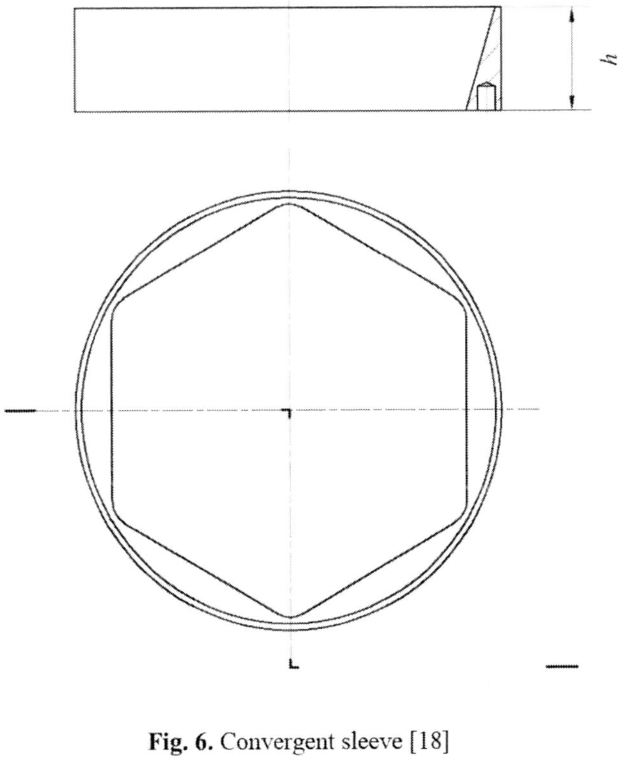

Fig. 6. Convergent sleeve [18]

2 Empirical research

In order to determine the force on the piston F_t, necessary to carry out the process of agglomeration of crystallized carbon dioxide, the study has been performed in accordance with the methodology set out in the literature [1, 9, 16]. For this purpose, the MTS machine Insight 50 kN has been used, recording values of the force and the displacement of the measuring head with 10 Hz frequency.

A special compaction head has been used to perform research (fig. 7). In order to reduce the measurement error resulting from the possibility of non-axial mounting of the compaction head, the clamps of the MTS machine have been equipped with the system for retaining perpendicularity of the compaction force relative to the specimen cross section (5).

Before the study, the compaction chamber (1) was filled with crushed dry ice. Then the stamp was inserted (3) into the compaction chamber (2). In the final stage, the head was mounted in the clamps of the MTS machine. After tarring the machine, the experiment started, during which the piston (3) was moved down at a constant speed. Once the value of the force F_t became equal to the value of the force F_{OP}, the material was pressed through the multi-holes die (4).

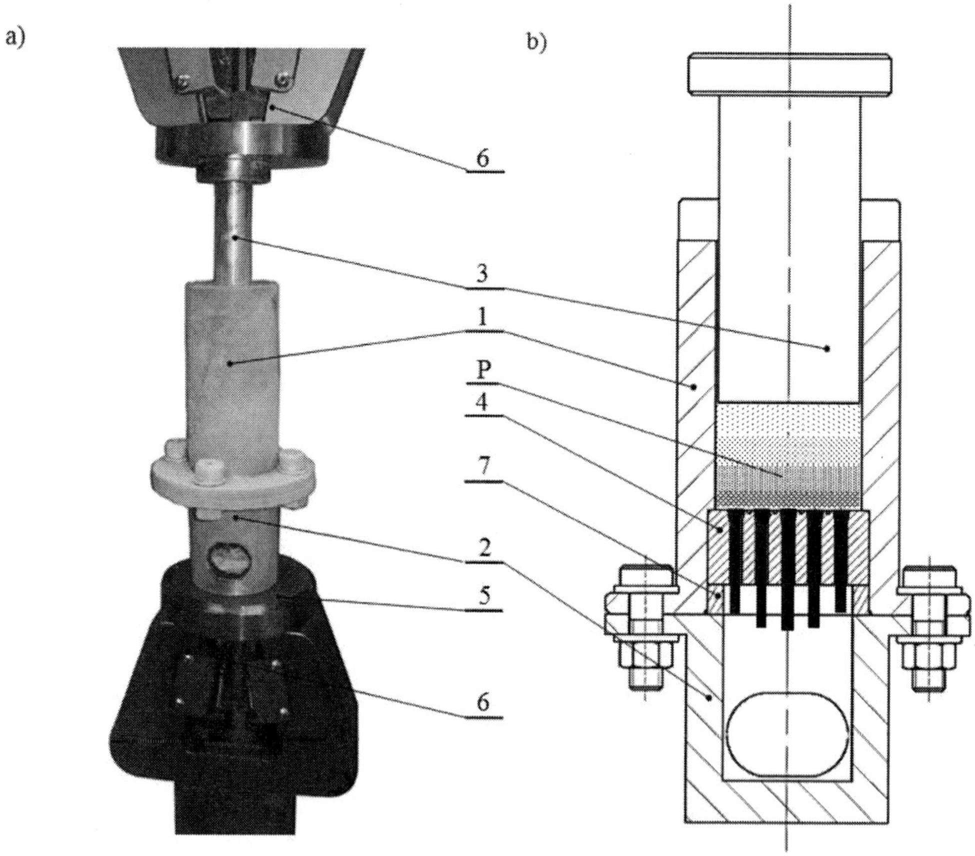

Fig. 7. Measurement system a) clamping system of the MTS machine with the measurement head and the alignment system, b) cross section of the measurement head: 1 – cylindrical sleeve, 2 – head base, 3 – piston, 4 – multi-holes die, 5 – perpendicularity retention system, 6 – MTS machine clamps, 7 – spacer sleeve, P – agglomerate. [9]

The research was conducted in 10 repetitions, which were averaged and presented in the form of compaction characterization describing the change of the force F_T in the function of displacement z (fig. 8).

The average value of the force F_{OP} has been determined at 24.1 kN. It should be noted that this value is lower than the one determined in the studies reported in the literature for the identical multi-holes die [15].

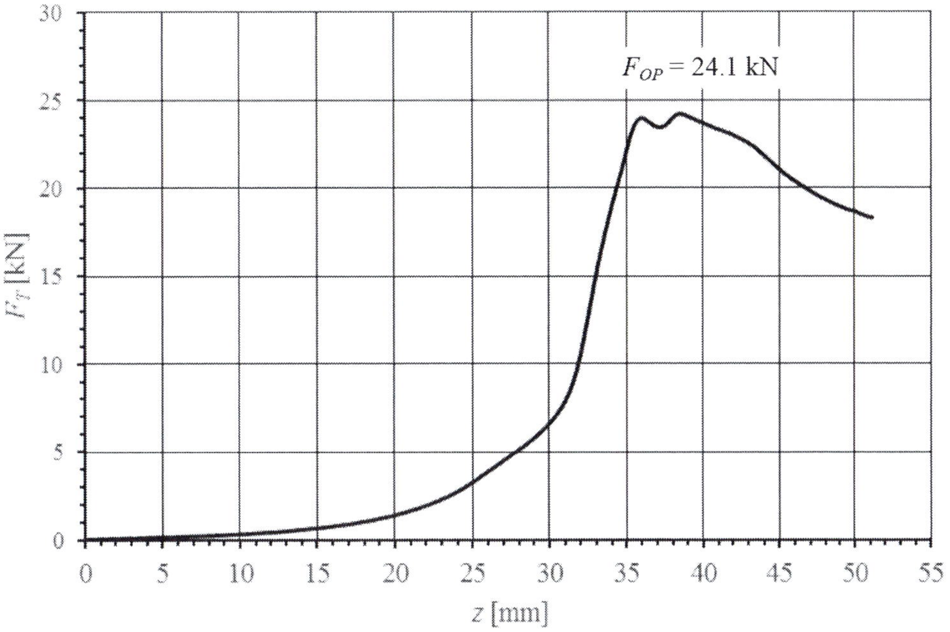

Fig. 8. Compaction characterization using the system of multi-holes die and convergent sleeve.

3 FEM Model

Figure 9 shows sample results of the analysis of pumping pre-compacted crystallized carbon dioxide. In the FEM model, piston 1 and die 2 have been modelled as rigid. The basis for this simplification is the fact that the material they are made of is many times stiffer than the material being pressed. The process of agglomeration can be divided into two phases. The first one relates to the initial compaction of the material to a specific pressure. Only in this form can it be pressed through the multi-holes die in order to shape the final geometric form. The pressed material has been modelled as elastic-plastic with a predefined hydrostatic pressure of the material arising from initial compaction. This pressure has been established on the basis of previous studies [5, 6] and its value has been set to 14 MPa. Due to insufficient information about the mechanical parameters of the pre-compacted carbon dioxide, which is to be pressed through the die, their values have been chosen so that the base convergent sleeve (h = 7.5 mm) can achieve a similar value of the force F_{OP}. The model calibrated in this way has been used only to determine the change trend of the force F_{OP}. In order to model the process of pressing the material through the die (the process of material delamination), the meshfree finite elements description method – SPH (Smoothed Particle Hydrodynamics) – has been used. More information about this method can be found in [20]. Kinematic parameters of the process have been established with the same values as in [15]. Figure 10 shows the change of drag force F_{OP} for the working unit, with the convergent sleeve height h = 7.5 mm. The chart applies only to the phase of pressing the pre-compacted material. Therefore, the characteristic change of force F_{OP} corresponding to the initial compaction of the material is not visible in the chart. Figure 11 shows the change of drag force F_{OP} for the working unit, with the convergent sleeve of height h = 5 mm. Figure 12 shows the change of drag force F_{OP} for the working unit, with the convergent sleeve of height h = 10 mm.

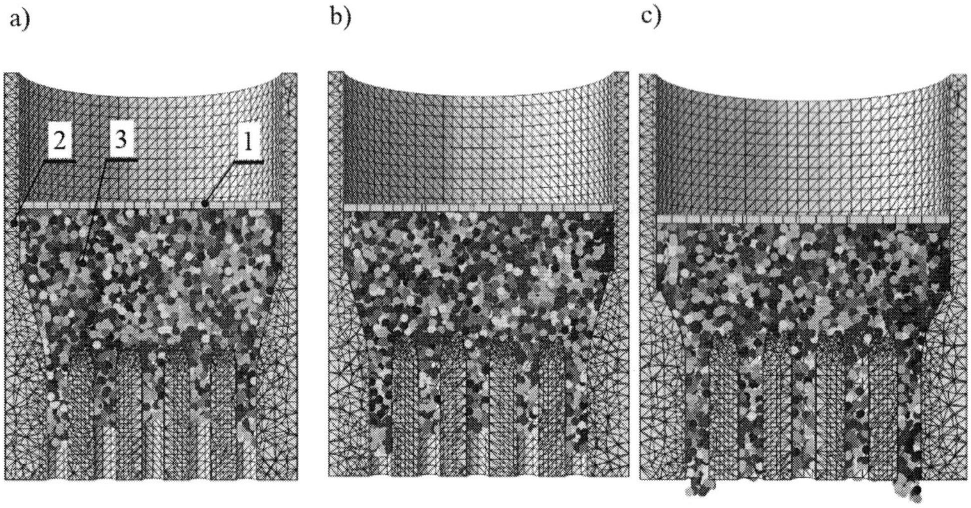

Fig. 9. FEM model of pressing pre-compacted dry ice through the multi-holes die: 1 – piston, 2 – multi-holes die, 3 – pressed material a) convergent sleeve height h – 10 mm, b) convergent sleeve height h – 7.5 mm, c) convergent sleeve height h – 5 mm

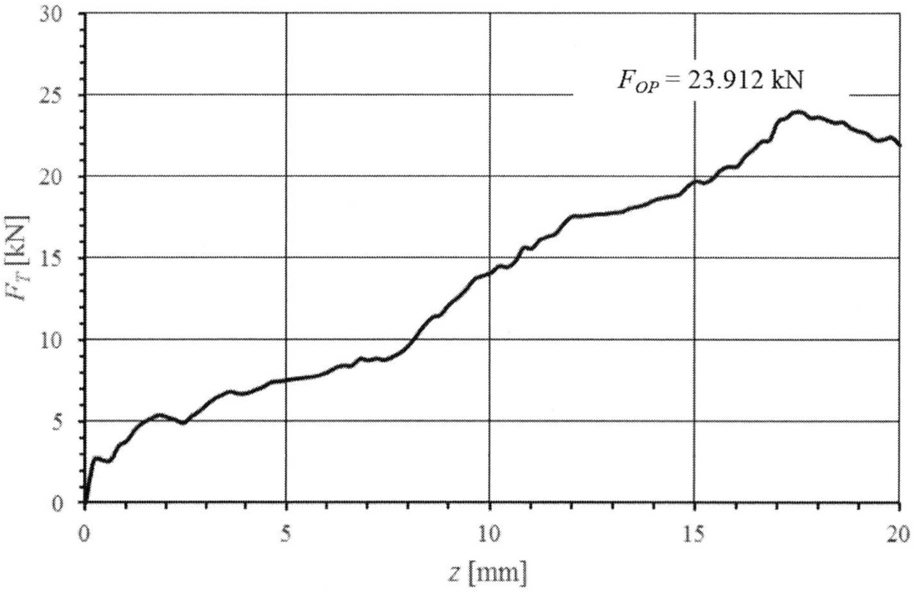

Fig. 10. Change of drag force F_{OP} for the working unit with the convergent sleeve height h = 7.5 mm

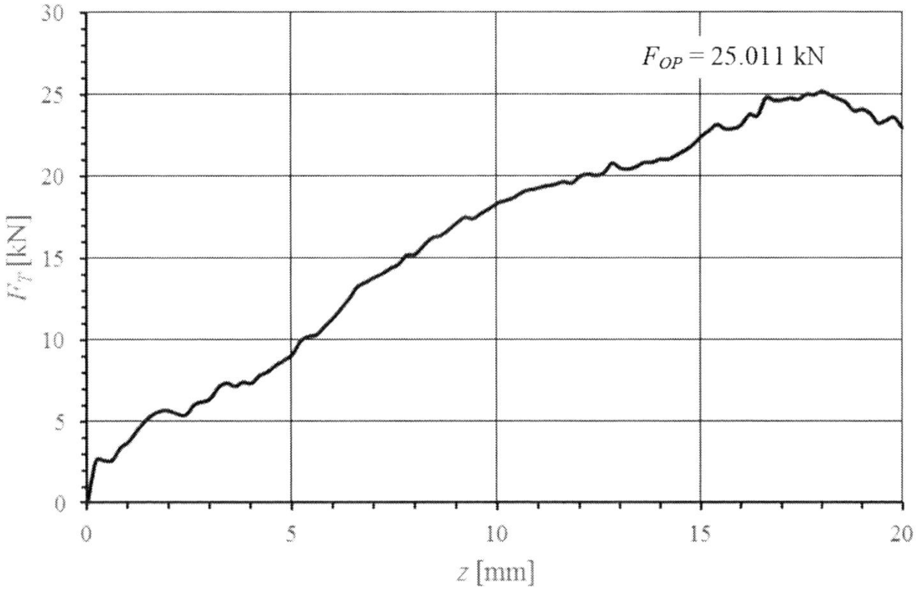

Fig. 11. Change of drag force F_{OP} for the working unit with the convergent sleeve height h = 5 mm

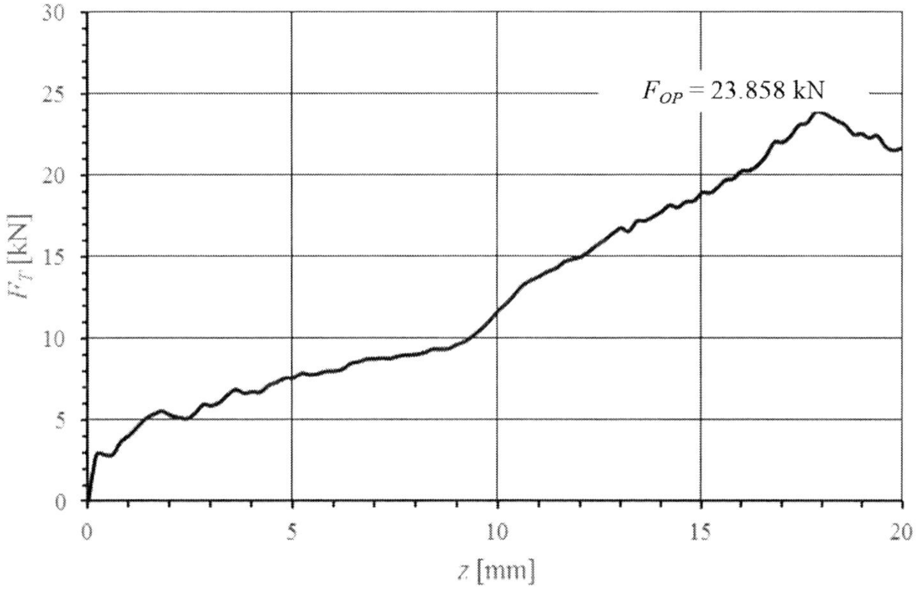

Fig. 12. Change of drag force F_{OP} for the working unit with the convergent sleeve height h = 10 mm

4 Influence of the sleeve height *h* on the value of the force F_{OP}

Based on the results of the numerical analysis, the initial characterization that describes the change in drag force F_{OP} in the function of the convergent sleeve height h has been formulated (fig. 13). The chart shows that the characterization in the range h equal between 5 and 7.5 mm has a negative gradient value. After exceeding the 7.5 mm value it stabilizes and oscillates in the error range of the MES analysis in relation to the empirical study.

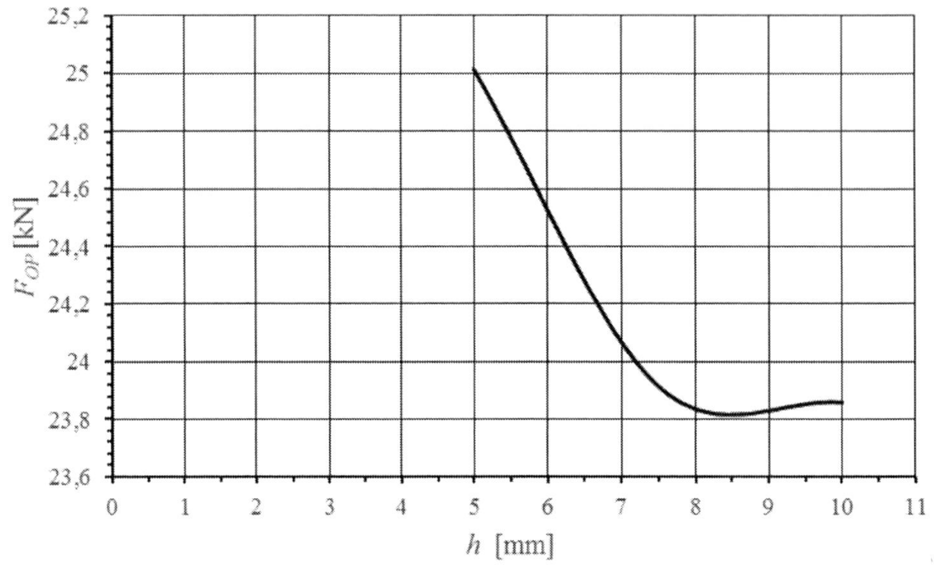

Fig. 13. Initial characterization of the changes to the drag force F_{OP} in the function of the sleeve height h

Conclusions

On the basis of the performed empirical and numerical research, it was found that:
- reduction of the surface area S_{PP} (fig. 3) has an impact on the value of drag force F_{OP},
- the largest impact on the value of drag force F_{OP} is observed in the sleeve height h range from 5 to 7.5 mm.

Deepening the current research can have a significant impact on the formulation of the assumptions for the machines for pressure agglomeration of crystallized CO_2.

References

1. L. Chen, X. Zhang, *A review study of solid-gas sublimation flow refrigeration: Form basic mechanism to applications*. International Journal of Refrigeration **40**, 61-83 (2014)
2. J. Vansant, *Carbon dioxide emission and merchant market in the European union*. In Aresta M. (eds) *Carbon Dioxide Recovery and Utilization*. Springer Science & Business Media. (2013)

3. Y. Liu, H. Maruyama, S. Matsusaka, *Agglomeration process of dry ice particles produced by expanding liquid carbon dioxide*. Advanced Powder Technology **21**, 652-657 (2010)
4. Y. Liu, G. Calvert, C. Hare, M. Ghadiri, S. Matusaka, *Size measurement of dry ice particles produced form liquid carbon dioxide*. Journal of Aerosol Science **48**, 1-9 (2012)
5. Y. Liu, D Hirama D., Matusaka S., *Particle removal process during application of impinging dry ice jet*. Powder Technology **2017**, 607-613 (2012)
6. V. Masa, P. Kuba, *Efficient use of compressed air for dry ice blasting*. Journal of Cleaner Production **111**, 76-84 (2016)
7. A. Mazzoldi, T. Hill, J. Colls, *CO_2 transportation for carbon capture and storage: Sublimation of carbon dioxide from a dry ice bank*. International Journal of Greenhouse Gas Control **2**, 210-218 (2008)
8. E. Uhlmann, M. Kretzschmar, F. Elbing, V Mihotovic, *Deburring with CO_2 Snow Blasting*. In: J. Aurich, D. Dornfeld (eds) *Burrs - Analysis, Control and Removal*. Springer, Berlin, Heidelberg
9. S. Dong, B. Song, B. Hansz, et al., *Modeling of dry ice blasting and its application in thermal spray*. Material Research Innovations **16**, 61-66 (2012)
10. M. Li, W. Liu, X. Qing, et al., *Feasibility study of a new approach to removal of paint coatings in remanufacturing*. Journal of Materials Processing Technology **234**, 102-112 (2016)
11. C. Otto, S. Zahn, F. Rost, et al., *Physical Methods of cleaning And Disinfection of Surfaces*. Food Engineering Review **3**, 171-188 (2011)
12. G. Spur, E. Uhlmann, F. Elbing, *Dry-ice blasting for cleaning: process, optimization and application*. Wear **233–235**, 402–411 (1999)
13. A. Witte, M. Bobal, R. David, et al., *Investigation of the potential of dry ice blasting for cleaning and disinfection in the food production environment*. LWT - Food Science and Technology **75**, 735-741 (2017)
14. H. Yamaguchi, X. Niu, K. Sekimoto, P. Neksa, *Investigation of dry ice blockage in an ultra-low temperature cascade refrigeration system using CO_2 as a working fluid*. International Journal of refrigeration **34**, 466-475 (2011)
15. J. Górecki, I. Malujda, K. Talaśka, *Research on Densification of solid carbon dioxide*. Journal of Mechanical and Transportation engineering **65**, 5-12 (2013)
16. J. Górecki, I. Malujda, K. Talaśka, *Influence of the Value of Limit densification Stress on the Quality of Pellets During the Agglomeration Process of CO_2*. Procedia Engineering **136**, 269-274 (2016)
17 Z. Drzymała, *Basics of Compaction Engineering of Material Compaction* (in polish), PWN, (1988)
18. Patent application No.: P.419432, *Piston assembly designed for dry ice snow compression*. (2016)
19. J. Górecki, I. Malujda, K. Talaśka, M. Kukla, P. Tarkowski, *Investigation of internal friction of agglomerated dry ice*. Procedia Engineering **136**, 275-279 (2016)
20. *Abaqus Documentation*. Dassault Systemes, Paris, 2017

Experimental investigation of properties of GFRP foam cored sandwich joints

Jana Gulanová[1,*], *Matúš* Margetin[1], *Papa-Birame* Gning[2], *Andrej* Chríbik[1]

[1]Faculty of Mechanical Engineering, Slovak University of Technology, Bratislava, Slovakia
[2]Institute of Automotive and Transport Engineering, Université Bourgogne Franche-Comté, Nevers, France

Abstract. Presented paper focuses on intermediate outcomes of the bilateral project between the Faculty of Mechanical Engineering, STU in Bratislava, Slovakia and the Institute of Automotive and Transport Engineering, UBFC in Nevers, France. Thus, mechanical properties and behaviour of glass fibre reinforced polymer composite foam cored sandwich panels joints were researched to estimate the best possible joint technology. Three-point bending and four-point bending were used to evaluate flexure strength and static force-flexure curve. Based on such testing, one specific joint composition was chosen to be broadly investigated. Finally, chosen type of joint panels was tested under four-point cyclic loading to obtain its fatigue properties.

Keywords: bending, fatigue, sandwich.

Nomenclature

F_{max}	-	maximal force during loading cycle
F_u	-	ultimate force that specimen type can survive
$\%_{Fu}$	-	ratio between F_{max} and F_u
F_m	-	mean value of force loading cycle
F_{nf}	-	force corresponding to given number of cycle to failure
δ_m	-	mean value of flexure cycle
N_f	-	number of cycle to failure
K	-	fatigue strength coefficient
B	-	fatigue strength exponent
RL	-	regress line
UPI	-	upper predictive interval
LPI	-	lower predictive interval
UCI	-	upper confidence interval
LCI	-	lower confidence interval
R	-	loading cycle ratio of symmetry

* Corresponding author: jana.gulanova@stuba.sk

Reviewers: *Juraj* Gerlici, *František* Nový

1 Introduction and experiment pre-processing

Presented project was partially focused on an optimal selection of sandwich composition and possible joining of sandwich panels to build specific assemblies within automotive industry. Mechanical validation is complicated and any case of loading analysis needs to be performed as a nonlinear task. In [2,4] authors focus on specific properties of different sandwich structure or joints comparison. But generally, there is a lack of such research dealing with selection and comparison of different permanent joints used within sandwich structures. Such a joint might be useful enabling series production of sandwich beams, which would be possible to be connected to build complex geometry and structures. Possible application here might be a carriage body of trucks or a floor of any kind of vehicle. Similar structures are known to be used in space industry, but here it is without any published research outcomes.

1.1 Technology – fabrication process

Sandwich material panel is made of top layers called skin which is sandwiching a core. For the presented project, skin was made of glass fibre reinforced epoxy composite consisting of 8 layers of prepreg cloths. There was used a special PVC foam, which was glued to skins by using epoxy adhesive film. Layer composition is shown in Fig. 1.

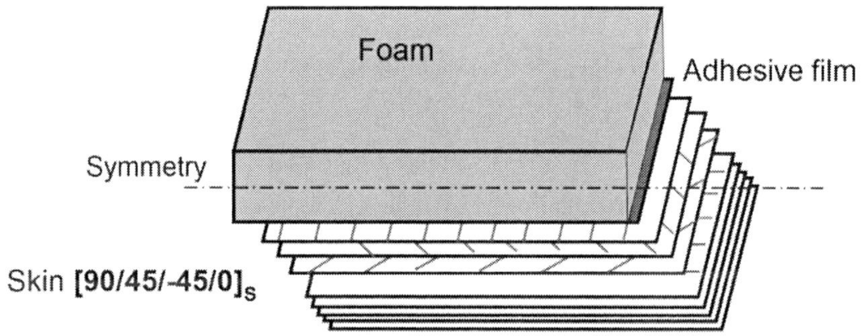

Fig. 1. Layer composition used for the monolith specimen

Fabrication process might be different for real components and structures, there is described panel fabrication process in following sentences. Cloth of prepreg is cut into specific rectangles. Without any glue, eight cloths are placed adhering specific position on each other. Then there is applied a thin film of epoxy glue and skin is stuck together with foam. Prepared sandwich is placed in vacuum bag to an oven and tempered to final form.
Fabrication process varies with different joint composition. In Fig. 2, there are shown various joints, which were tested under static loading. For further research, there was chosen aluminium insert of 50mm width as joining element.

Fig. 2. Various joint structures used for static testing – left column monolith, centre column slope and right column step

1.2 Testing procedure

In following subsections, there will be described results of static testing and fatigues testing. Firstly, specimen was tested using static three-point bending and four-point bending to find out the best joining composition. Later, there was done fatigue testing using four-point bending. Fatigue tests were performed on monolith structure and 50 mm width aluminium insert as well. Finally, results of fatigue behaviour are described.

1.2.1 Quasi-static

For the purpose of mechanical verification, there were investigated normal stress, displacement and stiffness of specimens fabricated using three kinds of joint and compared to monolith specimens within the same parameters. Such a study is shown in Fig. 3. There are 6 different benches of specimens, M30 as step of size 30mm, M60 as step of 60mmsize, B20 as slope of 20°, B40 as slope of 40°, I25 as aluminium insert of 25mm size and finally I50 as aluminium insert of 50mm size. In Fig. 3, there is shown scattering in comparison to monolith specimen testing. Static mechanical properties are not described more broadly within presented paper, since the main goal was to focus on fatigue behaviour of chosen joint structure.

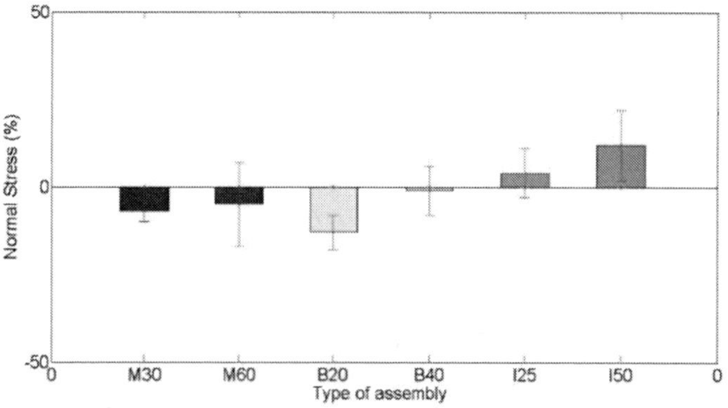

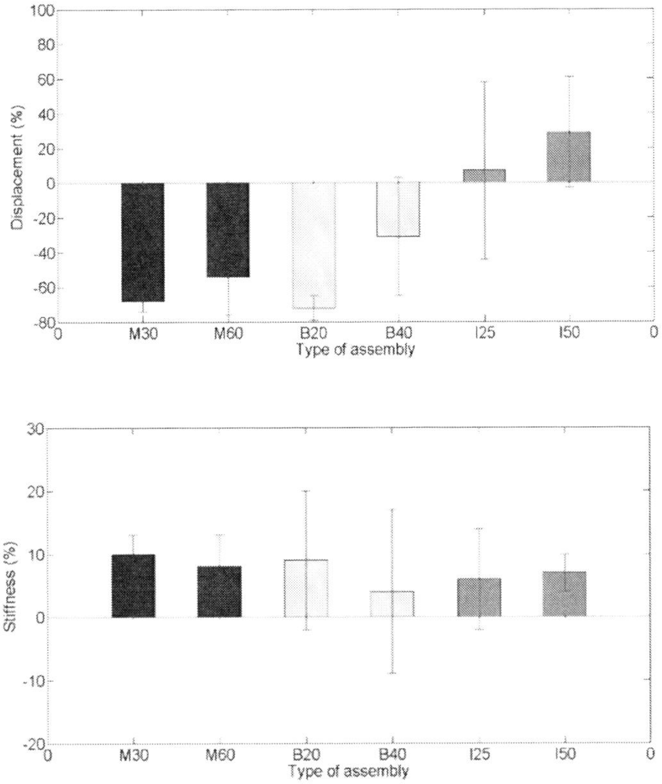

Fig. 3. Scattering in comparison to monolith specimen testing for 6 various joint structures

1.2.2 Fatigue

In order to investigate fatigue behaviour of composite sandwich structure, an experiment was conducted in which experimental specimens were tested at different levels of four-point bending loading. Experiment was carried out in the Strength and Elasticity Laboratory of the Faculty of Mechanical Engineering STU, using experimental stand Inova EDYZ testing system with maximal capacity of 200 kN. Based on results from static testing, two sets of experimental specimens were manufactured, the monolith type specimen and the insert type specimen, as can be seen in Fig. 4.

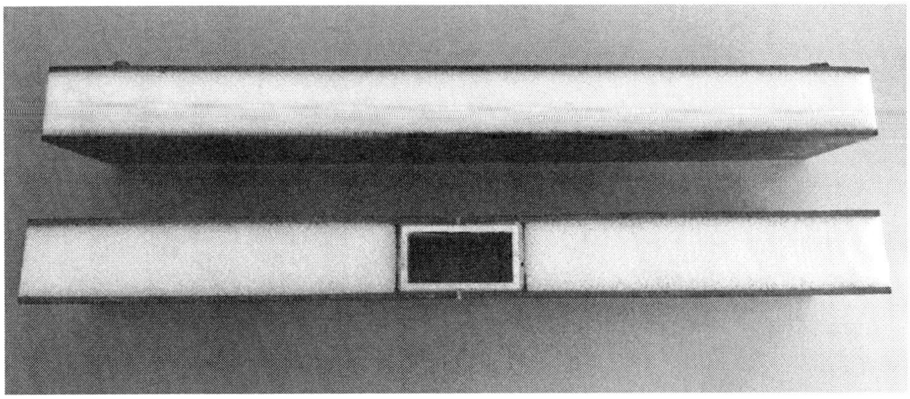

Fig. 4. Testing specimens: monolith – top; insert - bottom

The deflection of tested specimens has been measured using induction sensor placed in servo hydraulic valve and the loading force has been sensed using load cell mounted in crossbar of testing stand. To achieve four-point bending loading condition, specimens have been placed in mounting device shown in Fig. 5.

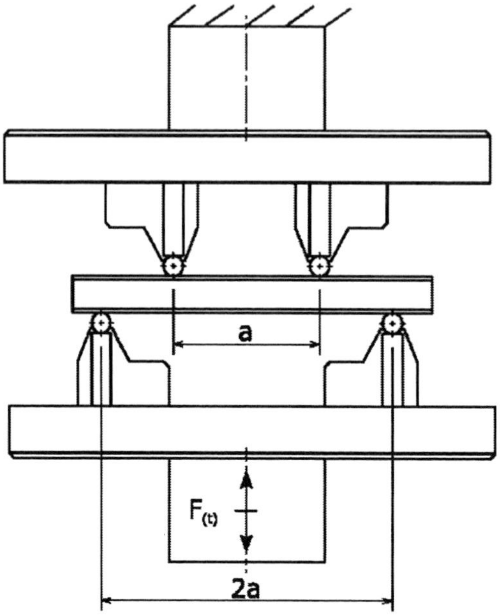

Fig. 5. Setup of a fatigue testing four-point bending

Cyclic testing of both type of specimens has been executed under force controlled loading mode. According to [6], a pulsating loading cycle has been chosen witch loading ratio $R = 0{,}1$. The main characteristics of loading cycle are shown in Fig. 6(a).

Experimental specimens were loaded in the force control mode. Failure condition of the experimental specimens was defined by the moment of the so-called "technical initiation of a fatigue crack" (which, in case of our composite sandwich structure, can be considered as a crack in the length of 3-5 mm). The number of cycles prior to the initiation of the fatigue crack was determined on the basis of a continuous measurement of the deformation response to the loading regime of the test specimen F_a = const. Completion of the test was defined either by the significant increase of the deformation in reference to the mean value or by the achievement of the life time of 10^6 cycles. The loading frequency has been set up to 5 Hz, thermo camera (Fig. 6(b)) has been used to ensure that specimen temperature has not increase more than 2°C as a result of inner heating under chosen loading condition.

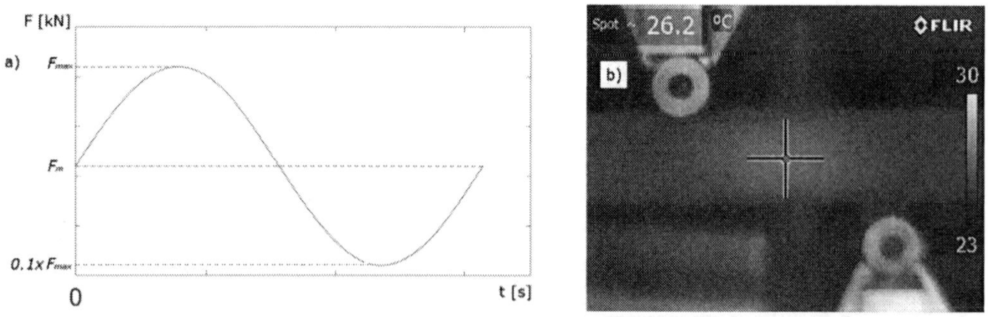

Fig. 6. (a) loading cycle description; (b) thermo camera measurement

2 Results and discussion

Based on quasi-static, there was only one composition chosen to be tested regarding fatigue behaviour.

2.1 Fatigue behaviour/properties

2.1.1 Fatigue curve

Fatigue life time curves in a form of an exponential function were used to describe cyclic behavior of experimentally tested specimens. The exponential form has been chosen due to the fact that it is the most widely used form of fatigue curve [7, 8] which describes the fatigue behavior of various materials quite accurately. The final fatigue life time curve can then be defined as a relation between maximal force in one loading cycle F_{nf} and a number of cycle repetitions until failure N_f. The curve is described by equation (1).

$$F_{nf} = K\left(2N_f\right)^b \tag{1}$$

The K and b can be called fatigue strength coefficient and fatigue strength exponent respectively. They represent material characteristics of tested specimens and have to be obtained statistically from measured data. The N_f is commonly multiplied by two in this type of function, as K then represents theoretical static ultimate force. The measured data are in table 1. The parameters K and b for regress line as well as for 95% upper and lower confidence and predictive intervals are in Table 2.

Table 1. Experimental data

MON.	**%$_{Fu}$** [%]	**F$_{nf}$** [kN]	**Nf** [cycles]	**INSERT**	**%$_{Fu}$** [%]	**F$_{nf}$** [kN]	**Nf** [cycles]
MO-09-85	85	4.21	32584	I50-02-85	85	4.59	5078
MO-03-75	75	3.71	390638	I50-01-80	80	4.32	15500
MO-08-75	75	3.71	1000000*	I50-03-75	75	4.05	195376
MO-02-70	70	3.47	462935	I50-07-75	75	4.05	544046
MO-07-68	68	3.37	1000000*	I50-04-70	70	3.78	769041
MO-06-65	65	3.22	1116473*	I50-05-65	65	3.51	1000000*
MO-05-58	58	2.87	677425	I50-06-58	58	3.13	1000000*
MO-04-58	58	2.87	1000000*	-	-	-	-
MO-01-40	40	1.98	1599973*	-	-	-	-

The fatigue life time curves for monolith and insert specimens can be seen in Fig 7. Instead of maximum force per cycle, the fraction of ultimate force in percentage (formula (2)) has been used for Y axis. In this form, the fatigue curves can be easily compared to each other.

$$\%_{uf} = \frac{F_{nf}}{F_u}100 \tag{2}$$

Table 2. Parameters of fatigue curves

MONOLITH			INSERT		
-	**K** [kN]	**b** [-]	-	**K** [kN]	**b** [-]
RL	7.27	-0.0025	RL	8.86	-0.0032
UPI	7.41	-0.0025	UPI	9.26	-0.0032
LPI	7.13	-0.0025	LPI	8.48	-0.0032
UCI	7.79	-0.0025	UCI	12.6	-0.0039
LCI	6.77	-0.0026	LCI	6.95	-0.0027

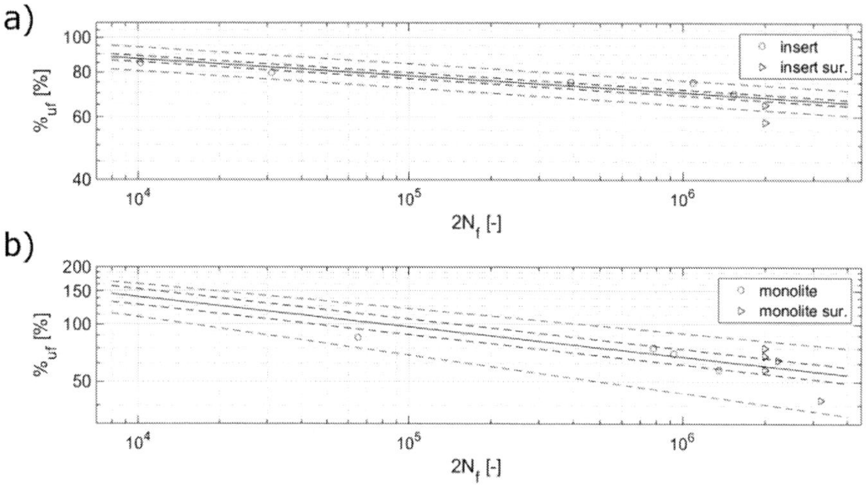

Fig. 7. (a) Insert fatigue curve; (b) monolith fatigue curve

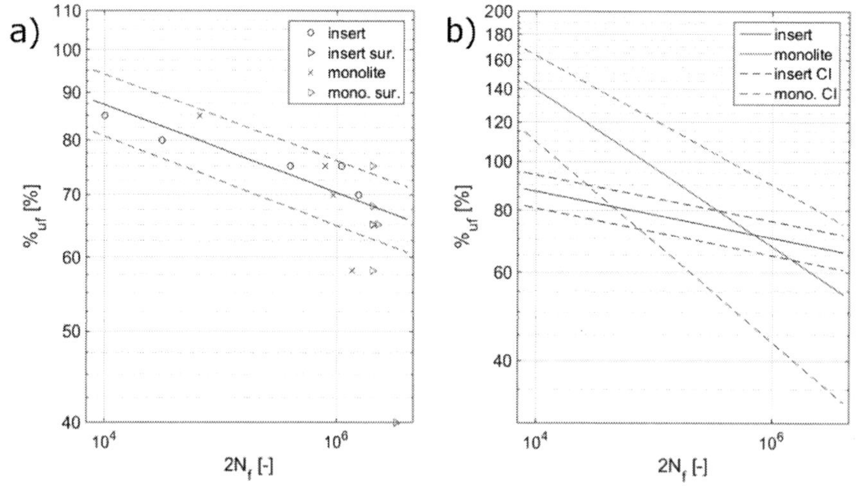

Fig. 8. (a) Insert curve with monolith points; (b) comparison of insert and monolith fatigue curve

During the experiment, we have found out that some of monolith specimens tend to fail after relatively small number of cycles (after tens or hundreds cycles) even under low loading levels. These fails were not caused by fatigue damage accumulation, but rather by manufacturing defects of adhesive film between foam and shell. As a result of this, the fatigue life curve for monolith specimens has much wider confidence interval and curve has significantly different slope than fatigue curve of insert specimens as can be seen in Fig. 8(b).

2.1.2 Cyclic ratcheting

Alongside fatigue life time curves, other useful information can be obtained from fatigue testing. Cyclic ratcheting describes behaviour of materials under non-symmetrical loading condition. Due to the fact that in real operation the loading cycles are rarely symmetrical, it is very useful to have a knowledge of how sensitive material is under non-symmetrical loading conditions. This sensitivity can be described by cyclic ratcheting curves which show increase of static (mean) flexure during life time. The ratcheting curves for monolith and insert specimens are shown in Fig. 9.

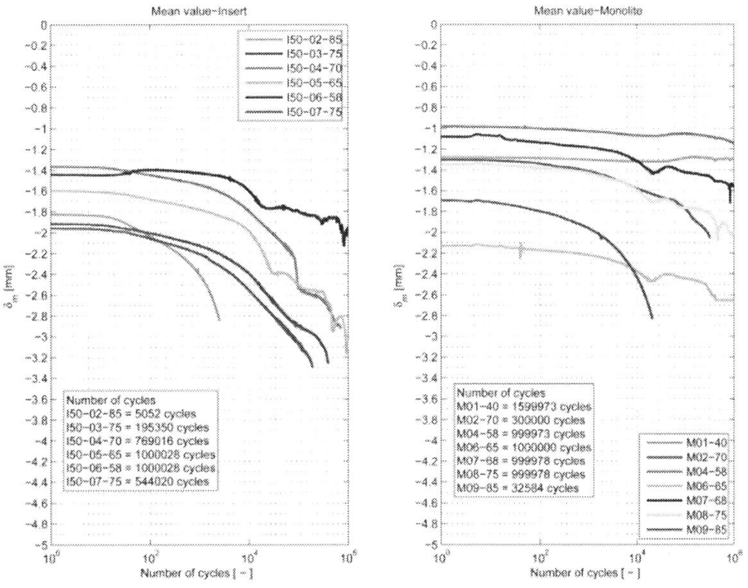

Fig. 9. Insert (left) and monolith (right) ratcheting curves

As can be seen in Fig. 9, the mean value of flexure is relatively stable during the first several cycles, and then starts to increase slowly, up to rapid increase in the last stage of life time. The small fluctuations in curves are caused by the error of a displacement sensor.

During the testing, it was observed that static flexure of tested specimens has disappeared after specimens had been unloaded. However, after several cycles of follow-up loading, it has come back into the state before unloading. Because of this, we can assume that the mean (static) flexure is caused by damage accumulation rather than a slow response of material to the variable loading condition.

2.1.3 Failure modes

After investigation of tested specimens, two kinds of failure modes have been observed in both types of specimens. Both of these crack types occurred in region between upper and lower support.

In the first case, crack initiated in the boundary layer between skin and foam. Crack started to grow in the place under upper support where maximal combination of bending moment and transverse force exists. Crack then grew parallel to the skin shell in skin/foam boundary region. When crack size caused loss of flexure stiffness, foam failed in static crack under 45° angle. This type of crack is shown in Fig. 10(a), the red arrow pointing to crack initiation region. The driven force of this crack is combination of normal stress caused by bending moment and shear stresses caused by transverse force and hertzian pressure from support.

In the second case, crack initiated in the middle of specimens' cross section in occurrence of maximal shear stress caused by transverse force Fig. 11. Crack then grew in direction of maximal shear stress planes (0° or 90° to specimens' cross section Fig. 10(c) and Fig. 10(b) respectively). When growing crack reached significant size, specimens lost flexure stiffness which caused foam fail in static crack under 45° angle. This type of crack is shown in Fig. 10(b,c), the red arrow pointing to crack initiation region. The driven force in this cracking mode is shear stress caused by transverse force.

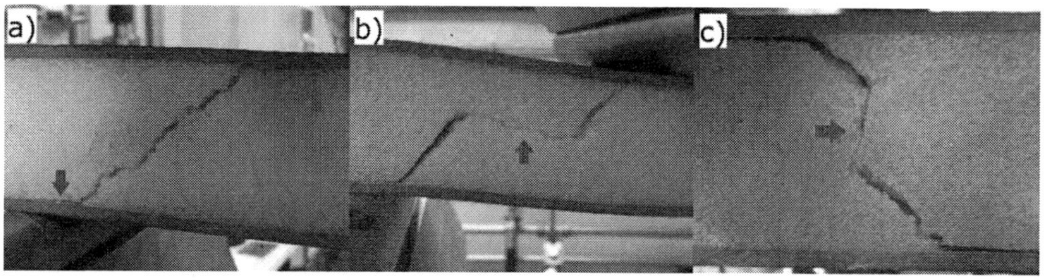

Fig. 10. Failure cracks: (a) skin/foam boundary region; (b) middle of cross section – parallel; (c) middle of cross section - perpendicular

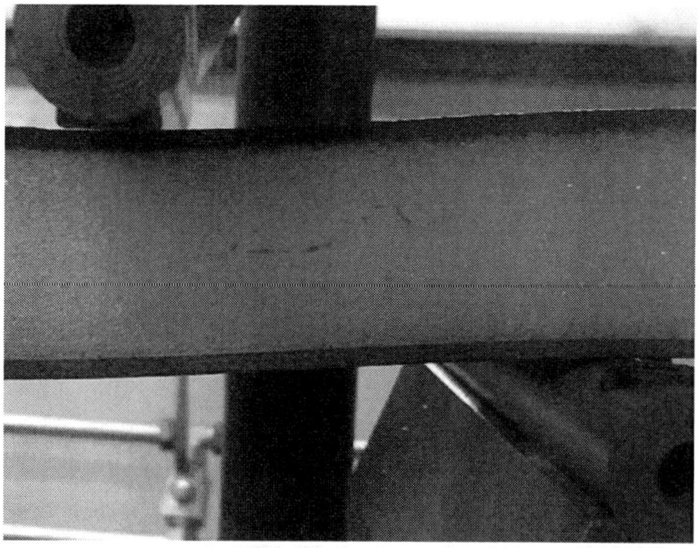

Fig. 11. Fatigue crack initiation and first stages of crack growth

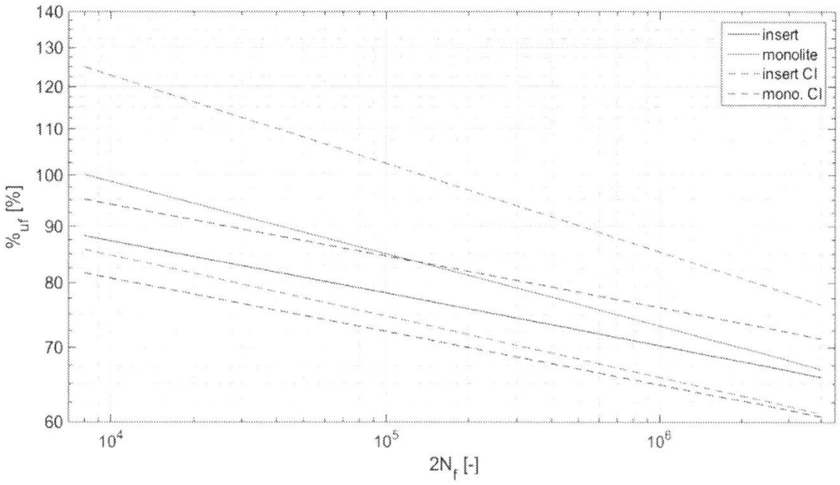

Fig. 12. Corrected comparison of fatigue curves for monolith and insert specimen

Conclusion

The investigation of sandwich composite structure under cyclic loading has been presented in this paper. Two types of sandwich specimens have been tested under four-point bending loading mode. Based on observations discussed in previous chapters, the following conclusions can be made:

- The driven force of fatigue damaging and fatigue crack initiation is the shear stress acting in foam part of a sandwich structure. Whether it is in the middle of cross section or in the boundary region between foam and foam/skin adhesive, fatigue crack initiates and grows in shear stress plane orientation till it reaches some critical length. Then the specimen cracks under 45° angle to previous crack orientation.
- Cyclic ratcheting after several loading cycles has been observed for both specimen types and under various loading levels. Static (or mean) flexure slowly disappears after unloading, but after several followed up loading cycles comes back to levels before unloading.
- When comparing data points for both types of specimens in $\%_{uf}$-N_f coordinate system, it seems that both specimens types fatigue behavior could be described by the same fatigue curve (Fig. 8a)). The different slopes of fatigue curves can be ascribed to inconsistency of monolith specimens data. By excision of problematic monolith data point the corrected fatigue curve is much similar to insert curve. However this hypothesis should be prove on bigger data sets, because after excision of problematic monolith data points, the data aren't representative anymore.

This work was supported by the Slovak Research and Development Agency under Contracts-No. APVV-0015-12and was also supported by the Scientific Grant Agency under the Contract No. VEGA 1/0301/17.
This contribution has been elaborated under the bilateral project APVV SK-FR-2015-0016.

References

1. G. Belingardi, P. Martella, L. Peroni, *Fatigue analysis of honeycomb-composite sandwich beams*. Composites Part A: Applied Science and Manufacturing **38** (4), 1183-1191, ISSN 1359-835X, (2007)
2. E. Bozhevolnaya, A. Lyckegaard, *Structurally graded core inserts in sandwich panels*. Composite Structures **68** (1), 23-29, ISSN 0263-8223, (2005)
3. M. Burman, D. Zenkert, *Fatigue of foam core sandwich beams - 1: undamaged specimens*. International Journal of Fatigue **19** (7), 551-561, ISSN 0142-1123, (1997)
4. A. C. Manalo, T. Aravinthan, W. Karunasena, M. M. Islam, *Flexural behaviour of structural fibre composite sandwich beams in flatwise and edgewise positions*. Composite Structures **92** (4), 984-995, ISSN 0263-8223, (2010)
5. H. Mathieson, A. Fam, *High cycle fatigue under reversed bending of sandwich panels with GFRP skins and polyurethane foam core*. Composite Structures **113**, 31-39 (2014)
6. *Plastiques renforcés de fibres – Détermination des propriétés de fatigue en conditions de chargement cycliques*. Standard ISO 13003:2003 (F)
7. J. A. Bannantine, J. J. Comer, J. L., Handrock, *Fundamentals of Metal Fatigue Analysis*. (Prentice-Hall, New Jersey, 1990)
8. M. Klesnil, P. Lukas, *Fatigue of Metallic Materials*. (Elsevier Science, ISBN: 9780444987235, 1992)

A new accelerated technique for validation of cyclic plasticity models

Radim Halama[1,2,*], *Petr* Gál[2,3], *Zbyněk* Paška[2], *Josef* Sedlák[2]

[1]IT4Innovations, VŠB – Technical University of Ostrava, 17. listopadu 2172/15, 708 33 Ostrava-Poruba, Czech Republic
[2]Faculty of Mechanical Engineering, VŠB – Technical University of Ostrava, 17. listopadu 2172/15, 708 33 Ostrava-Poruba, Czech Republic
[3]Integrity and Technical Engineering Division, Újv Řež a. s., Hlavní 130, 250 68 Husinec - Řež, Czech Republic

Abstract. This contribution presents a new methodology of fatigue tests evaluation based on digital image correlation application. The full-field strain analysis brings the possibility of 3D strain measurement evaluation on a curved part of specimen standardly used under axial-torsional loading. Resulting strain response curves can subsequently serve for evaluation of cyclic plasticity model predictions. The crucial idea is the way of FE model creation to can evaluate simulation in the same points of the structure as in the DIC measurement. The new methodology saves amount of material and leads to a shorter experimental time.

Keywords: digital image correlation, ratcheting, cyclic plasticity, fatigue testing

1 Introduction

Accurate cyclic plasticity modelling based on phenomenological approach requires numerous experimental results. Extensive low-cycle fatigue testing of metallic materials brings knowledge about suitability of individual approach [1]. Transient stress/strain behaviour of metals have been identified for various effects of cyclic plasticity, considering cyclic hardening/softening mechanisms, for instance in cyclic creep (or so called ratcheting) and mean stress relaxation [2]. Additional hardening due to non-proportional loading is also a complicated factor. It is not surprising, that researchers have been looking for new testing procedures to obtain more information for calibration of cyclic plasticity models in order to accelerate the whole process [3]. Recently, specimen's geometry is usually designed by means of finite element analyses support [4-5].

Cyclic plasticity models available in commercial software are usually not enough accurate for non-proportional loading cases [6]. On the other hand, the proportional loading cases, the Neuber like methods based on elastic solution are more efficient making possible the propabilistic approach [7].

This contribution follows previous research presented elsewhere [8]. The new approach of fatigue test evaluation is based on digital image correlation method application and

* Corresponding author: radim.halama@vsb.cz

Reviewers: *Grzegorz* Domek, *Milan* Sága

brings a lot of new possibilities in mechanical testing of materials. All described cases deals with ratcheting under proportional loading, but the technique proposed here might be applied also for various cases of non-proportional loading to study stress-strain behaviour of materials more effectively.

2 Experiments

2.1 Description

All tests were realized on axial-torsional testing machine LabControl 100kN/1000Nm at the VŠB –TU Ostrava [8]. In this study, uniaxial cases results are presented only. Digital image correlation (DIC) measurements were done via the Mercury RT® system (2x2.3Mpx@40Hz), provided by Sobriety s.r.o. company. An optical contrast coating was prepared on each specimen for DIC method application. The frequency of triangular loading was 0.1Hz during DIC measurement in each test. Tubular specimens made of R7T wheel steel were made according to scheme in the Fig.1. The testing part had outer diameter 12.5mm and inner diameter 10mm respectively.

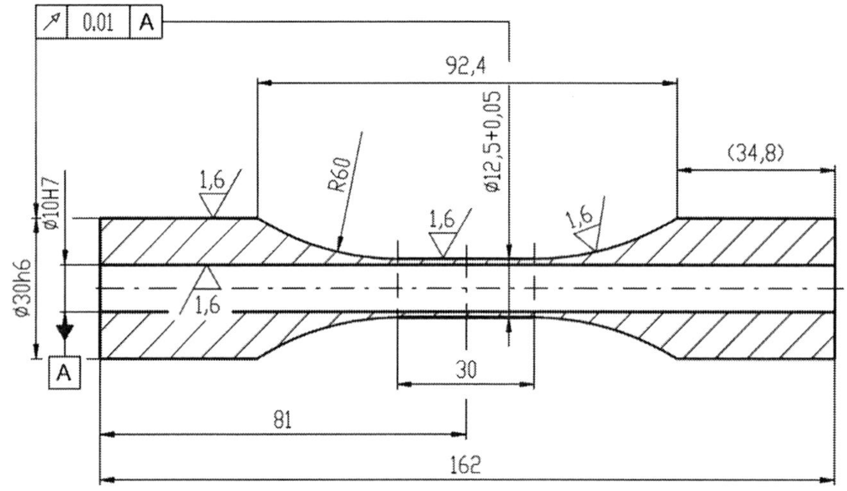

Fig. 1. A scheme of used specimen

2.2 Results of strain controlled test

A sequential test was performed under strain control to obtain the cyclic stress-strain curve of the R7T material. The monotonic stress-strain curve together with evaluated hysteresis loops is shown in the Fig. 2. A cyclic softening was observed for small strain amplitudes. A slight cyclic hardening for higher strain amplitudes is also visible.

2.3 Results of stress controlled test

The next realized experiment was intended to study uniaxial ratcheting of R7T steel [8]. A harmonic loading with the stress amplitude of 450MPa and mean stress of 50MPa was applied (the loading frequency was increased to 0.5Hz after 200 cycles). The test was done until fracture. The material reveals a ratcheting behaviour with steady state after stabilization of strain response, see Fig. 3, where DIC measurement results are also depicted.

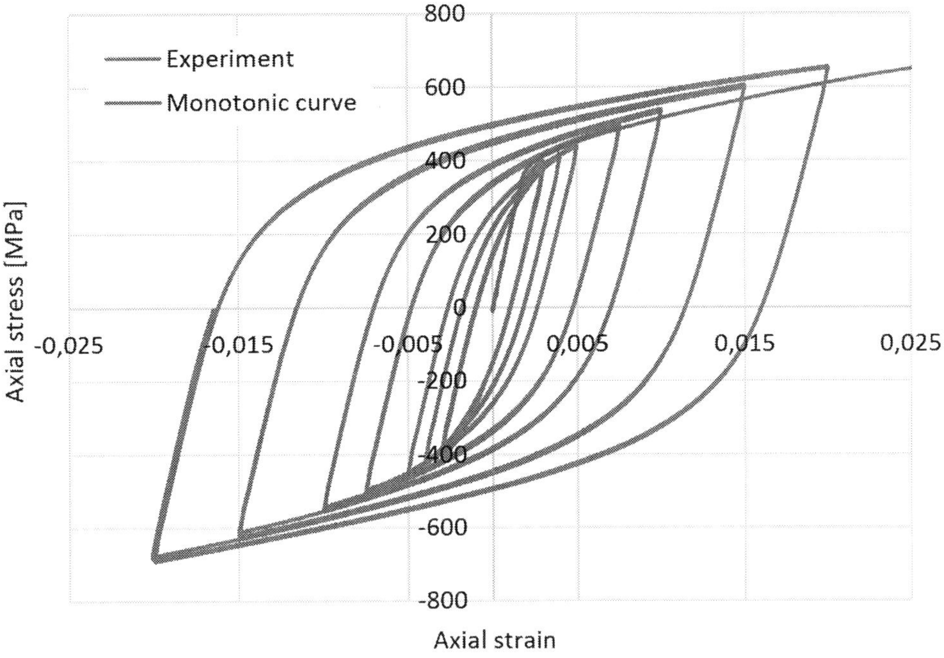

Fig. 2. Uniaxial hysteresis loops of investigated material gained from the sequential uniaxial test

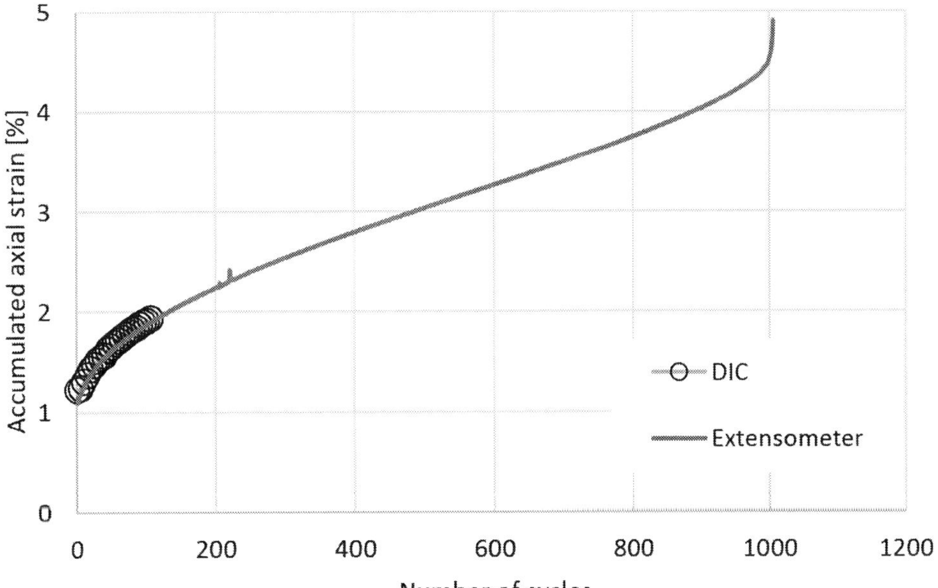

Fig. 3. Accumulation of axial strain in the uniaxial ratcheting test

Strain history was evaluated from DIC measurement in selected points as shown in the Fig. 4. The point A lies on the testing part of the specimen. The other points correspond to the curved part. The larger cross-section the less axial strain accumulation. The same behaviour was observed in realized measurement as will be presented in the next chapter.

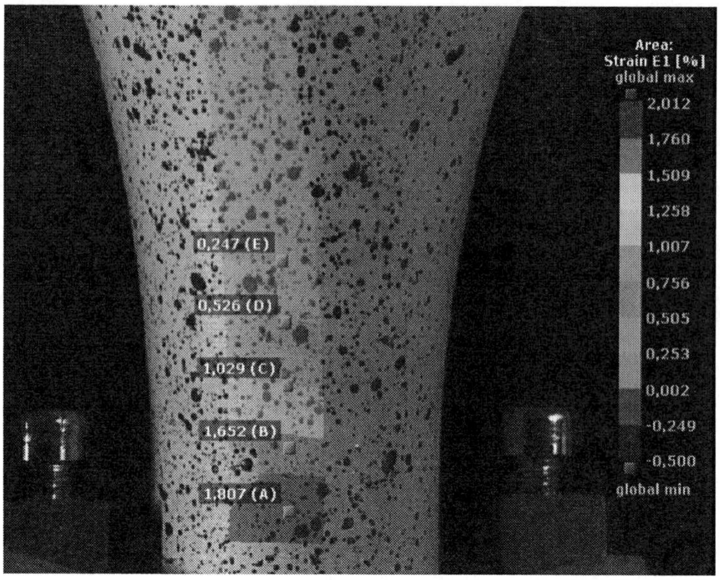

Fig. 4. First principal strain contours and points selected for evaluation of DIC measurement

Position of selected DIC points can be identified based on idea of approximation by circle with known radius (radius $R = 60$ mm), see Fig. 5. Coordinates of points were gained using pixel coordinates and a scale factor calculated from ratio between diameter of testing part in mm and pixels. The approximation function is

$$y = y_0 + \sqrt{R^2 - (x - x_0)^2}\ , \tag{1}$$

where x_0 and y_0 defines the centre of the approximation circle. The most important parameter for FE model creation is x_0. Its resulting value is $x_0 = 1.75$ mm.

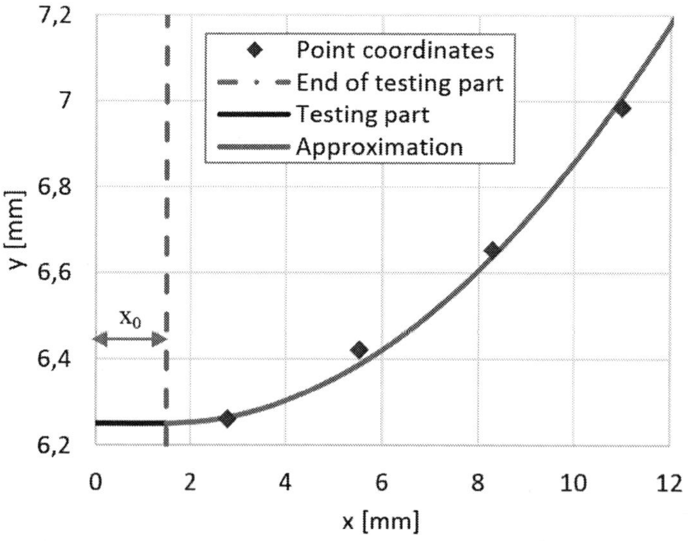

Fig. 5. Approximation of points evaluated from DIC measurement

3 Finite element modelling

3.1 FE model

Commercial FE software ANSYS was used for the simulation of uniaxial ratcheting test. For simplicity, the axisymmetric model corresponding to the half of specimen was created (Fig. 6). The model consists of 204 elements (4 node structural 2D element, PLANE182). The model was prepared by APDL macro (Ansys Parametric Design Language) to can parametrically specify position of nodes accordingly the DIC measurement evaluation (points A-E). Boundary conditions shown in the Fig. 6 were applied. An overload in first loading cycle was also set as in actual experiment [8].

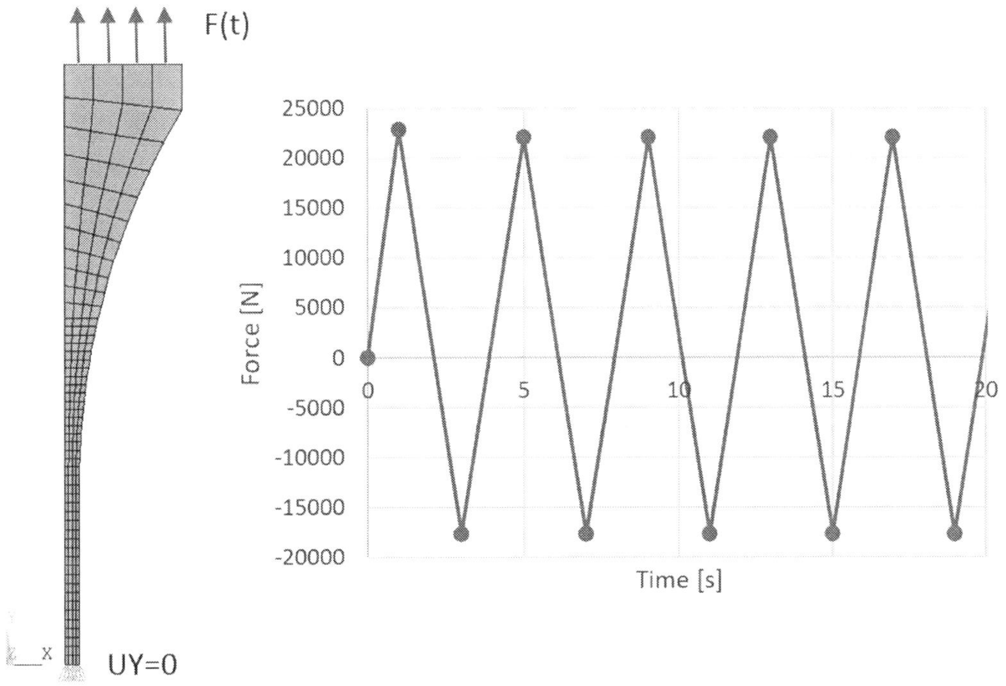

Fig. 6. FE model considered in the simulation

3.2 Material model description

A classic cyclic plasticity model with pure kinematic hardening was employed. The rate-independent cyclic plasticity theory considered in the ratcheting analysis consists of von Mises yield criterion

$$f = \sqrt{\frac{3}{2}(\mathbf{s}-\mathbf{a}):(\mathbf{s}-\mathbf{a})} - \sigma_Y = 0 \tag{2}$$

the associative flow rule

$$d\boldsymbol{\varepsilon}_p = d\lambda \frac{\partial f}{\partial \boldsymbol{\sigma}} \tag{3}$$

and the Chaboche kinematic hardening rule [9]

$$\boldsymbol{a} = \sum_{i=1}^{M} \boldsymbol{a}_i \,, \; d\boldsymbol{a}_i = \frac{2}{3} C_i d\boldsymbol{\varepsilon}_p - \gamma_i \boldsymbol{a}_i dp \,, \tag{4}$$

where $\boldsymbol{s}$ is the deviatoric part of stress tensor $\boldsymbol{\sigma}$, $\boldsymbol{a}$ is the deviatoric part of back-stress $\boldsymbol{\alpha}$, σ_Y is the initial size of the yield surface, $d\lambda$ is the plastic multiplier, which in associative plasticity corresponds to the equivalent plastic strain increment dp, and $d\varepsilon_p$ is the plastic strain tensor. The symbol ":" means the inner product between two second order tensors

$$\mathbf{x} : \mathbf{y} = x_{ij} y_{ij} \,. \tag{5}$$

Yield stress σ_Y and material parameters C_i and γ_i of i-th back stress part $\boldsymbol{a}_i$ were identified from the largest hysteresis loop shown in the Fig. 2. The parameter γ_3 influencing ratcheting rate was set to 0.65 to predict ratcheting rate in testing part of the specimen correctly. All material parameters used in the FE simulation is stated in the Table 1. These elastic constants were also used in the FEA: Young modulus $E = 202000$ MPa, Poisson's ratio $\nu = 0.3$.

Table 1. Parameters of cyclic plasticity model

Model	**Parameters**
Chaboche	$\sigma_Y = 200MPa$ $C_{1-3} = 243100{,}29370{,}8479\ MPa$ $\gamma_{1-3} = 1423,\ 213,\ 0.65$

3.3 Results

In total, 250 cycles was simulated. Contours of axial stress at the time corresponding to the first tension peak are shown in the Fig. 7, where evaluated nodes are also numbered.

The trend of ratcheting rate is predicted well for all cases. The best prediction in initial cycles was obtained for point C (node 132). For points A and B less ratcheting rate was obtained in initial cycles than in the experiment. In remaining points a strong over-prediction of strain accumulation was obtained by the Chaboche model.

Complete stress-strain history was evaluated in the five nodes. It was checked that the same value of fatigue asymmetry factor $R = -0.8$ was obtained in the evaluated surface nodes. An example of stress-strain prediction is visible in the Fig. 8. Very slight change of stress amplitude occurs in initial cycles only.

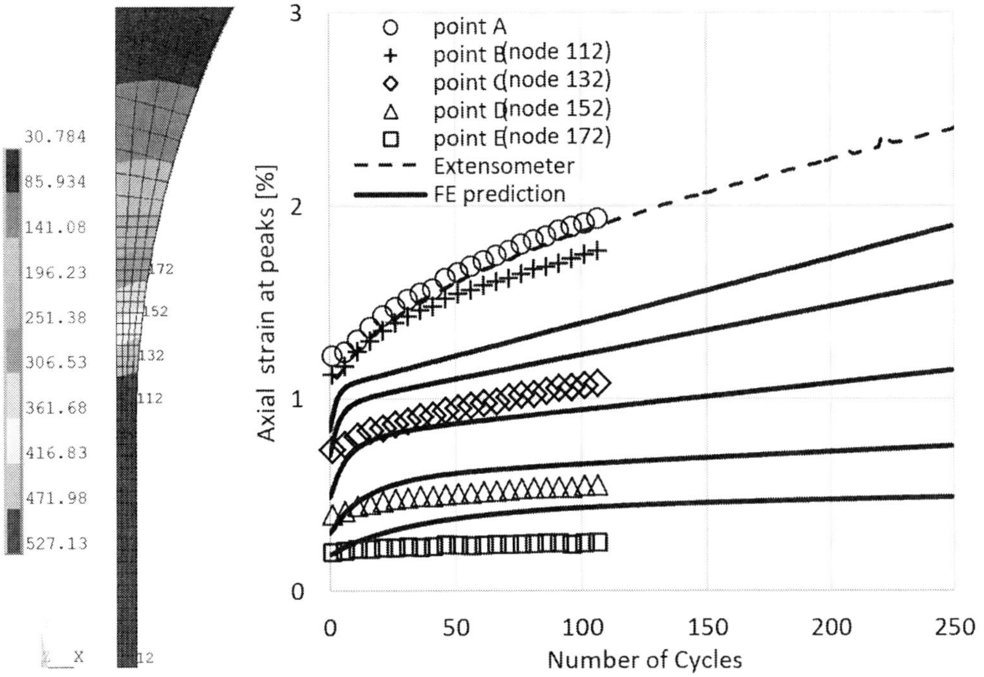

Fig. 7. Axial stress contours and resulting prediction of strain accumulation in comparison with values measured by DICM

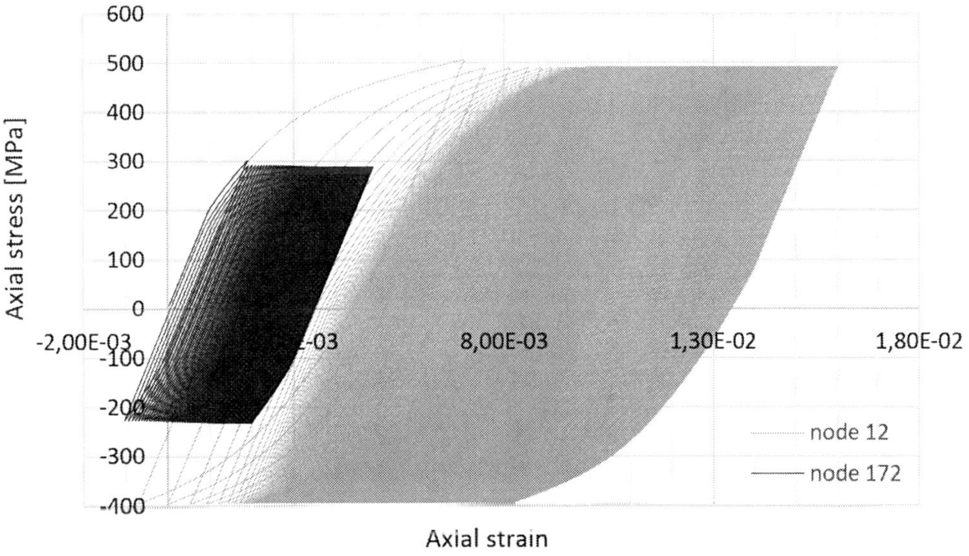

Fig. 8. Predicted hysteresis loops for node 12 and node 172

Conclusion

A new approach for low-cycle fatigue tests evaluation has been presented in this paper. Modern optical methods of full-field strain analysis make possible to get more information about stress-strain behaviour of the investigated material subjected to cyclic load. The proposed technique was validated using digital image correlation method. A possibility of

FE model creation was also described. The technique can help researches to save time and number of specimens. Good correlation between FE prediction and experiments was obtained in steady state. Better results of ratcheting prediction can be obtained for instance by application of a new constitutive model including memory surface based on the AbdelKarim-Ohno kinematic hardening rule [10]. The numerical study will be presented in a future paper.

The digital image correlation technique application brings new possibility of transient material behaviour investigation not only for the case of ratcheting, but also in research of mean stress relaxation, cyclic hardening/softening, additional hardening due to non-proportional loading or in creep testing (optically screenable cases only).

This work was supported by The Ministry of Education, Youth and Sports from the National Programme of Sustainability (NPU II) project „IT4Innovations excellence in science - LQ1602", by the Czech Science Foundation (GA15-18274S) and by the Specific Research (SP2017/136). The support is gratefully acknowledged.

References

1. R. Halama, J. Sedlák and M. Šofer, *Phenomenological Modelling of Cyclic Plasticity*, Chapter in Numerical Modelling, Peep Miidla (Ed.), InTech, 329-354 (2012) [online] http://www.intechopen.com/books/numerical-modelling/phenomenological-modelling-of-cyclic-plasticity
2. Y. Jiang, J. Zhang, *Benchmark experiments and characteristic cyclic plastic deformation behavior*. Int. J. Plast. **24**, 1481–1515 (2008)
3. R. Halama, M. Fusek, Z. Poruba, *Influence of Mean Stress and Stress Amplitude on Uniaxial and Biaxial Ratcheting of ST52 Steel and Its Prediction by the AbdelKarim–Ohno Model*. Int. J. Fatigue **91** (2), 313-321 (2016)
4. M. Handrik, P. Kopas, V. Baniari, M. Vasko, M. Saga, *Analysis of stress and strain of fatigue specimens localised in the cross-sectional area of the gauge section testing on bi-axial fatigue machine loaded in the high-cycle fatigue region*. Proc. Eng. **177**, 516-519 (2017)
5. P. Kopas, M. Saga, V. Baniari, M. Vasko, M. Handrik, *A plastic strain and stress analysis of bending and torsion fatigue specimens in the low-cycle fatigue region using the finite element methods*. Proc. Eng. **177**, 526-531 (2017)
6. R. Halama, J. Sedlák, M. Fusek, Z. Poruba, *Uniaxial and Biaxial Ratcheting of ST52 Steel under Variable Amplitude Loading – Experiments and Modeling*. Proc. Eng. **101**, 185-193
7. M. Krejsa, *Probabilistic failure analysis of steel structures exposed to fatigue*. Key Eng. Mat. **577-578**, 101-104 (2014)
8. R. Halama, L. Lipinová, R. Páleník, *Measurement of Cyclic Creep by Digital Image Correlation Method*. In Proceedings of 55th annual conference on Experimental stress analysis 2017, Hotel Atrium, Nový Smokovec, Slovakia, May 30th - June 1st, 2017, 300-306 (2017)
9. S. Bari, T. Hassan, *Anatomy of Coupled Constitutive Models for Ratchetting Simulations*, Int. J. of Plast. **16**, 381-409 (2000)
10. R. Halama, M. Fusek, M. Šofer, Z. Poruba, P. Matušek, R. Fajkoš, *Ratcheting Behavior of Class C Wheel Steel and Its Prediction by Modified AbdelKarim-Ohno Model*. In Proceedings of the 10th International Conference on Contact Mechanics CM2015, Colorado Springs, Colorado, USA, August 30 – September 3, 2015

Use of the BOINC system for distributed data collection in the "Internet of Things"

Marián Handrik[*,1], *Jana* Handriková[2], *Lenka* Jakubovičová[1]

[1]University of Žilina, Faculty of mechanical engineering, 010 26 Žilina, Slovak Republic
[2]Armed forces academy of general M. R. Štefánik, Department of informatics, 031 06 Liptovský Mikuláš, Slovak Republic

Abstract. The BOINC software package allows the creation of massive computational systems for distributed computing. The system is designed to use heterogeneous computing devices that are differing in hardware architecture and operating systems. The variability and versatility of the BOINC system allows its use to create a distributed system for measurement and data collection. This is done through the standard HTTP and HTTPS web communication protocol that is used on the Internet in connected devices, so called "Internet of things".

Keywords: data acquisition, BOINC, internet, http, https, image recognition

1 Introduction

In the field of distributed computing systems, BOINC is used to manage the computations in heterogeneous distributed computing systems. This system can also be used for ARM processor architecture. ARM processors are used, for example, in tablets, mobile phones, or single-rack minicomputer of Raspberry PI.

The single-board Raspberry PI computer is the credit-card sized computer. Raspberry PI 3 model B has a 1.2 GHz 64-bit quad-core ARM Cortex-A53 processor, 1 GB RAM, with the connector for 6 megapixel camera. The graphic output is possible via HDMI and composite video. Internet connection is possible via 100 Mbit Ethernet, WIFI and Bluetooth. Raspberry PI 3 model B has 4 x USB to connect other devices, such as keyboard, mouse, memory media USB keys - disks, cameras, measuring devices. Thanks to USB, this device can be also connected to the 4G mobile networks. Raspberry PI 3 Model B includes even 40pin General-purpose input/output (GPIO) that is a generic pin on an integrated circuit or computer board whose behaviour-including whether it is an input or output pin—is controllable by the user at run time. The GPIO interface allows the connection of specific measuring devices designed for the purpose of certain measurement.

The official Raspberry PI operating system is based on the Linux Debian operating system, containing almost all packages of this system. The price of Raspberry PI 3 B model, including the camera, does not exceed 70€, therefore it is used for bulk data collection or measurement. Raspberry PI Model B is shown in Figure 1.

* Corresponding author: marian.handrik@fstroj.uniza.sk

Reviewers: *Robert* Grega, *Milan* Naď

The BOINC system enables simple system management for data collection via Raspberry PI. In the following chapter, the methods of gathering image data are described by image recognition methods in the field of machine vision through the BOINC system [1-4].

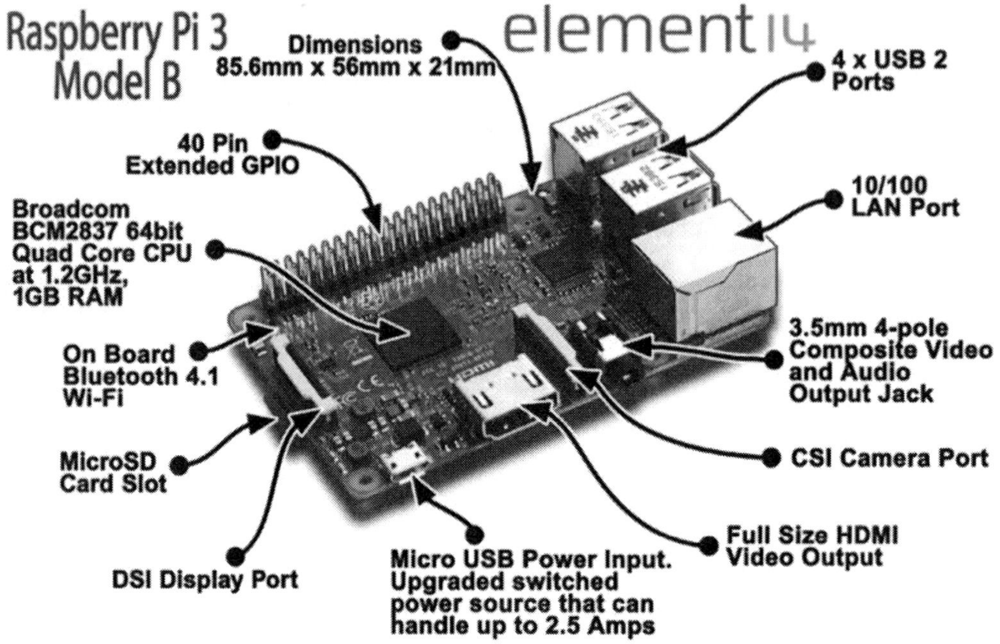

Fig. 1. Raspberry PI model B

2 Image recognition and computer vision

Image recognition systems can be designed for various specific tasks:

- For 3D reconstruction, the crucial step is to determine the relationships between the corresponding points across the image reliably. With the help of these relationships, it is possible to reconstruct the frames of cameras that recorded images as well as 3D spatial points corresponding to these points based on multi-view geometry and robust object detection methods.
- Object recognition can be divided into a specific object recognition and general object recognition of a certain type. A brief overview of object recognition methods is given in [5, 6]. The objective of category recognition is to classify an object into one of several predefined categories. The main task is to recognize the object from different angles, scales, lighting conditions or against cluttered backgrounds, etc. The aim of the object classification is to specify the object category or the group of objects to which the object belongs.
- The aim of the systems that choose the object based on the content is to return a number of most liked objects from a large database. Therefore, there are a huge number of local descriptors describing these objects as there are a huge number of objects. Special index structures, such as k-d trees, are used to search in these systems.
- The aim of simultaneous localization and mapping is to reconstruct the surrounding environment and locate the object within it in real time. These systems are based on various sensors and captured image by cameras. They are used in the automated navigation of vehicles and aircrafts, when creating extended reality.

The recognition process consists of two basic steps - the object detection and object or category recognition. The object is described (represented) by a descriptor, certain values that specifically describe the surroundings of a particular point in the image for later recognition. Descriptors can be used to find the match between objects in a database and an input image (recognizing a predetermined object), or serve to achieve a robust image representation that is the model of the given category (determining the group of objects to which the object belongs).

The procedure of recognizing a particular object can be demonstrated on pedestrian or vehicle recognition. When detecting the object from video sequence, the recording quality is as important as software for subsequent processing. In general, pedestrian and vehicle detection according to [6, 7] follow these steps:

- detection of a candidate region on which the object is likely to be,
- description of this region by using a descriptor,
- classification of this region, or verification whether a given candidate region contains the object of interest or not,
- post-processing.

The detection of candidate regions can be done, for example with the help recognizing patterns in the frame of a video sequence. In the case of a video sequence, the background subtraction method can be used - moving objects are separated from the background by calculating the difference between the current frame and the reference background.

The descriptor consists of features organized into the structure. It is expected that the descriptor is able to describe the object from different camera viewpoints and in different poses. Features are calculated for individual pixels (points of interest, key points) or for local image regions. Both types of descriptors can describe an object, in this case a pedestrian or a vehicle, as a whole or as a set of its individual parts (the parts do not have to correspond with the individual anatomical parts of the human body or vehicle components). Descriptors describing the parts of an object can be created independently in different processes.

The most common way how to create a descriptor is to link the local features into a multidimensional vector. The features used to create a descriptor can describe the shape of the object (e.g. edge detectors), object appearance (e.g. Local Binary Feature and its modification), object movement (histogram of optical flows).

The combination of multiple detection methods can increase system performance, improve its ability to detect an object. For example, in [8], the accuracy of vehicle detection at night is increased by the fusion of different features.

After gaining a descriptor from the candidate regions, the detention whether the candidate area contains an object or not follows.

Classification can use a generative or discriminative approach. Generative models of appearance are trying to provide us with as precise description as possible, they detect the data from the object category by creating a pattern - a template, with which the candidate regions are compared and if there is sufficient similarity to the pattern, they are classified as an object of a certain category. Discriminative models detect the object with the help of binary classification and robust supervised methods and algorithms, such as Support Vector Machine or cascaded AdaBoost classifier.

The classifier training process can be run incrementally - after processing each of the following tasks, the classifier is modified and is capable of classifying the object, and this corresponds to online learning tasks. If the classifier training process takes place over a set of training tasks before classifying an unknown object, it is a non-incremental learning that corresponds to offline learning tasks.

The training process can be run on a set of training tasks, which belong to a predetermined class (learning with a teacher). The goal is to detect the rule based on which an unknown object belongs to a particular class. In case of learning without a teacher, training tasks do not include information about certain classification. The goal is to divide objects with a specified combination of features into groups so that the similarity of combinations of object features assigned to one group was as large as possible, and vice versa, so that the difference between combinations of object features from different groups was greatest.

Image and video processing, which we can consider as the time sequence of individual image frames puts high demands on memory and the computing resources used. Therefore, it would be advisable to consider using parallel processing options, either using cloud or distributed systems.

Image processing methods, machine learning as well as other algorithms can be found in OpenCV free software package. Vehicle license plate recognition is available, for example, in free OpenALPR software package. There are also many other available image and video editing software packages on the Internet.

3 BOINC system

The BOINC (Berkeley Open Infrastructure for Network Computing) software platform is used for volunteer computing or grid computing creation.

BOINC is designed to support applications that have large computation requirements, large storage requirements, or both. The main requirement of the application is that it will be divisible into a large number (thousands or millions) of jobs that can be done independently. This task group also includes data collection by analyzing the image from the camera. BOINC is free software distributed under the Lesser General Public License (LGPL) version 3 or higher.

The main purpose of BOINC development is to support the public resource computing paradigm, to support the creation of many projects, to encourage a large community of computer owners to participate in one or more projects [9].

Other BOINC development goals are [9]:

Reduce the barriers of entry to public-resource computing. BOINC allows the scientists with moderate computer skills to create and operate a large public-resource computing project with about a week of initial work and an hour of maintenance per week. The server for a BOINC based project can consist of a single machine configured with common open-source software (Linux, Apache, PHP, MySQL, Python).

Share resources among autonomous projects. BOINC-based projects are autonomous; projects are not centrally authorized or registered. Each project operates its own servers and stands completely on its own. PC owners can easily participate in multiple projects and can determine how the resources will be divided among these projects. If most users are involved in multiple projects, then overall resource utilization is improved: if one project is closed for repairs, then other projects temporarily inherit its computing power. The CPU might work for one project while the network is transferring files for another.

Support diverse applications. BOINC accommodates a wide range of applications; providing flexible and scalable mechanism for distributing data and its scheduling algorithm intelligently connects, matches requirements with available resources. Existing applications in common, widely used languages can run as BOINC applications with little or no modifications. An application can consist of several files. New versions of applications can be deployed without the involvement of volunteers, centrally from BOINC servers.

Reward participants. Projects, which encourage the participation of volunteers, must provide them with some form of reward. The reward is provided on the numeric measure of the volunteer's contribution to the overall computation of the task. BOINC provides a credit accounting system that reflects the usage of the volunteer's computing resources in all projects. BOINC also makes it easy to add visualization graphics to volunteer applications that can be used as screensaver graphics. For example, the PlanetQest project allows a volunteer to name a planet found with the help of his or her computer.

The basic features of BOINC [10]:

- Project autonomy. Projects are independent; each one realizes computation and operates its own servers and databases. There is no central directory or approval process.
- Volunteer flexibility. Volunteers can participate in multiple projects. They control which projects they participate in, and how their resources are divided among these projects.
- Flexible application framework. Existing applications in common programming languages C, C++, Fortran can run as BOINC applications with little or no modification. An application can consist of several files.
- Security. BOINC protects against several types of attacks. For example, a digital signature based on public-key encryption protects against the distribution of viruses.
- Source code availability. BOINC code is distributed freely.
- Server performance and scalability. The BOINC server software is extremely efficient, so that a mid-range server can manage millions of jobs per day. The server architecture is highly scalable, making it easy to increase server capacity or availability by adding several computers.
- Support for large data. BOINC supports applications that produce or consume large amounts of data, or that use large amounts of memory. Data distribution and collection can be spread across many servers. Users can specify limits on disk usage and network bandwidth. Work is dispatched only to computers able to handle such data.
- Multiple participant platforms. The BOINC core client is available for most common platforms (Mac OS X, Windows, Linux and other Unix systems). The client can use multiple CPUs as well as graphics cards for calculations.
- Open, extensible software architecture. BOINC provides a documented interface to many of its key components, making it possible for third-party developers to create software and websites that extend BOINC.
- Volunteer community support. BOINC provides web-based tools, such as user profiles and private messaging that encourage volunteers to form an online community.
- Quick and easy integration of a new computer into the task solving process. Connecting a computer on Linux operating systems can be accomplished by running the appropriate script over the Internet, and the device does not need to be manually configured. In the case of manual configuration, the process is short and does not last more than one minute for each new computer.

The BOINC architecture is depicted in Figure 2.

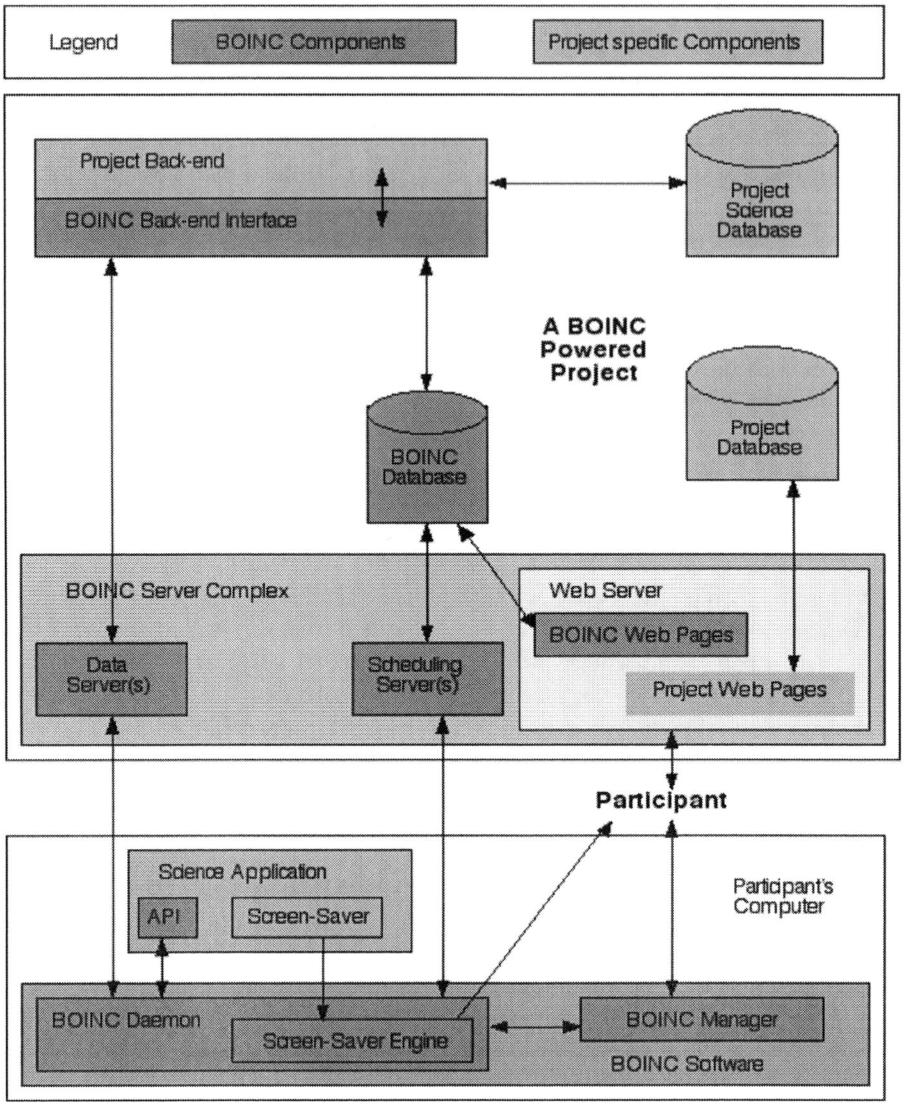

Fig. 2. BOINC architecture

Conclusion

The BOINC system is designed as a massive computational system in order to manage computing in high demand. BOINC can also be used when data collection from a large number of regions is needed. In the example presented in the text above, thanks to the single-board Raspberry PI 3 model B, it is possible to analyse an image from camera in a real time by using software Openalpr that is able to detect the vehicle license number from camera streams.

The presented examples, BOINC and Openalpr software, can be used, for example for automatic detection of the vehicle entry to toll roads, and still millions of devices for collecting data can be integrated into the system. The robustness of the BOINC system from the view of managing a large number of tasks is practically tested via large projects such as SETI@home, LHC@home, in which millions of computers around the world are involved in computing.

This work is supported by: KEGA 015ŽU-4/2017 Digital technologies for machine building study programs as a part of strategy "Internet of Things" and VEGA 1/0795/16 Development of effective methods for correction and optimization of coupled mechanical system.

References

1. P. Pecháč, M. Sága, *Memetic algorithm with normalized RBF ANN for approximation of objective function and sedondary RBF ANN for error mapping,* Procedia engineering **177**, 540 – 547, ISSN 1877-7058, (2017)
2. M. Sága, R. Bednár, M. Vaško, *Contribution to modal and spectral interval finite element analysis,* Springer proceedings in physics **139**, 269-274, ISSN 0930-8989, (2011)
3. P. Orsansky, B. Ftorek, P. Durcansky, *Mathematical model of a closed hot air engine cycle using MATLAB Simulink,* AIP conference proceedings **1608**, 173-176, ISSN 0094-243X, (2014)
4. J. Judova, M. Pajtasova, D. Ondrusova, *Effects on the bee colonies vitality in Slovakia.* International multidisciplinary scientific geoconference-SGEM **3**, 274-254, ISSN 1314-2704 (2016)
5. J. Valiska, S. Merchevsky, R. Kokoska, *Object tracking by color – based particle filter techniques in video sequences,* IEEE Radioelektronika 2014, ISBN 978-1-4799-3715-8, (2014)
6. C. M. Sukanya, G. Roopa, P. Vince, *A survey on object recognition methods.* IJCSET **6**, 48-52, ISSN 2231-0711, (2016)
7. M. Al-Smadi, K. Abdulrahim, R. A. Salam, *Traffic surveillance a rewiew of vision based vehicle detection, recognition and tracking*, International journal of applied engineering research **6**, 713-726, ISSN 0973-4562, (2016)
8. S. Sivaraman, M. M. Trivedi, *A review of recent developments in vision-based vehicle detection*, IEEE Intelligent vehicles symposium IV, 23–26 june 2013, Gold coast, Australia
9. K. Hulin, Ch. Long, G. Feng, C. Jiajie, C. Leanne, Y. Hong, *Combining region-of-interest extraction and image enhancement for nighttime vehicle detection,* IEEE Intelligent systems **31**, 57-61, (2016)
10. D. P. Anderson, Boinc: A system for public-resource computing and storage, https://boinc.berkeley.edu/grid_paper_04.pdf
11. Boinc web site, https://boinc.berkeley.edu

Studying of cutting conditions when hard turning and their impact on integrity of shaped-complex surfaces

Jozef Holubjak[1,*], *Jozef* Pilc[1], *Tatiana* Czanova[1], *Pavol* Martikan[1], *Dusan* Mital[2], *Jozef* Mrazik[1]

[1]Faculty of Mechanical Engineering, University of Zilina, Univerzitna 8215/1, 010 26 Zilina, Slovakia
[2]Faculty of Manufacturing Technologies Technical University of Košice with a seat in Presov, Bayerova 1, 080 01 Presov, Slovakia

Abstract. This article deals with impact of cutting conditions when hard turning of shaped-complex surfaces which are concentration origin of cracks especially in the area of notches. These areas significantly reduce the lifetime and functionality of surface by degradation of surface integrity where are the significant number of destruction cracks. Actual experiments are focused on detection of impact of each individual cutting condition on the generation of residual stress and its impact in each surface and subsurface layer of material. Results and evaluations explain what way is necessary to design and apply cutting conditions when hard turning of shaped-complex surfaces.

Keywords: hard turning, cutting conditions, residual stress, surface integrity

1 Stress states of materials

Stress conditions are ones of demonstrations of used machining technologies. After manufacturing process, they remain in parts and constructions and they operate continuously even without load. Their impact significantly affects the functionality of machined surfaces. [1, 2]

If the outer forces or moments are affecting on the system, then the stress is occurred, which are called as inserted. Stresses, which are in the system without affecting of outer forces and moments, are called inner. Equilibrium of inner stresses act the change of dimensions (deformations) of system if it breaches its integrity. There will be the conversion of inner forces and moments. [2, 3, 4]

Permanent residual stresses have the largest share on the functionality of part, and they cannot be detected by conventional methods. [2, 5, 6]

Obviously, to realize the benefits of understanding the residual stresses in parts and structures, tools are needed to measure them. Several techniques are available, with varying

* Corresponding author: jozef.holubiak@fstroj.uniza.sk

Reviewers: *Tomasz* Domański, *Ivan* Kuric

degrees of sophistication. Some of them are rather limited in their application, but one stands out as having widespread applications and being readily available.

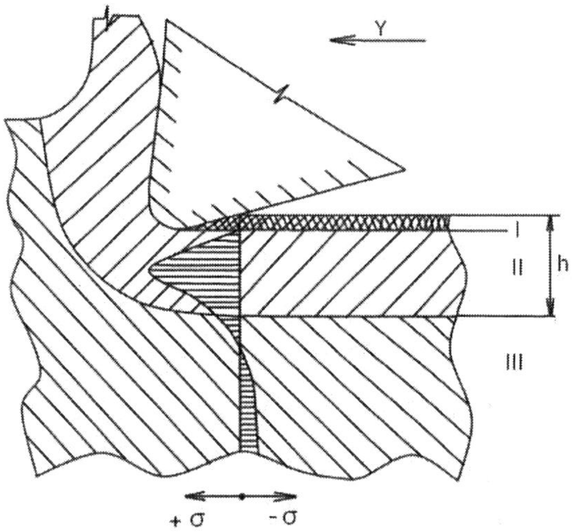

Fig. 1. Illustration of formation of residual stresses in surface layer under the influence of plastic deformations [2]

Macroscopic stresses, which extend over distances that are large relative to the grain size of the material, are of general interest in design and failure analysis. Macroscopic stresses are tensor quantities, with magnitudes varying with direction at a single point in a component. The macroscopic stress for a given location and direction is determined by measuring the strain in that direction at a single point. When macroscopic stresses are determined in at least three known directions, and a condition of plane stress is assumed, the three stresses can be combined using Mohr's circle for stress to determine the maximum and minimum residual stresses, the maximum shear stress, and their orientation relative to a reference direction. Macroscopic stresses strain many crystals periodically in the surface. This periodical distortion of the crystal lattice shifts the angular position of the diffraction peak selected for residual stress measurement.

In practise, residual stress is stress which affects the entire volume of part or the majority of itself, i.e. macroscopic character (Fig. 1). This includes the stress in infinitely thin or large area. It is important, that violation of compactness of part causes the change in macro-geometry. Methods of machining, casting, forming, etc. can cause these changes. For full classification, it should be noted that residual stresses are called sometimes as technological stresses, because they arise from the action of technological processes during the producing of parts. Direction of residual stress (tension or compression) depends on the kind of deformation. [7, 8]

For measuring of residual stresses, the iXRD X-ray diffractometer by PROTO (Fig. 2). X-ray diffraction is portable non-destructive method that can quanti-atively measure residual stress in crystalline and semi- crystalline materials. High speed detector technology enables measurements to be perfor-ed easily on metals and ceramics. iXRD uses the coherent domains of materials (grain of structure) as tensiometer, which reacts on actually condition of stress in material. Residual stresses enlarge or narrow the distances in atomic lattice (d). [9, 10]

This system measures strain and convert to stress. Distances in atomic lattice (d-spacings) are calculated using Bragg's Law λ=2.d.sinα. If a monochromatic X-ray beam impinges upon a sample with an ordered spacing (d), constructive interference will occur at an angle α. Changes in strain and thus the d-spacing translate into changes in the diffraction angle α measured by the x-ray detectors. The diffraction pattern is in the shape of a cone for polycrystalline materials. The shape of the diffraction peaks can also be related to the dislocation density and coherent domain size. [11, 12, 13]

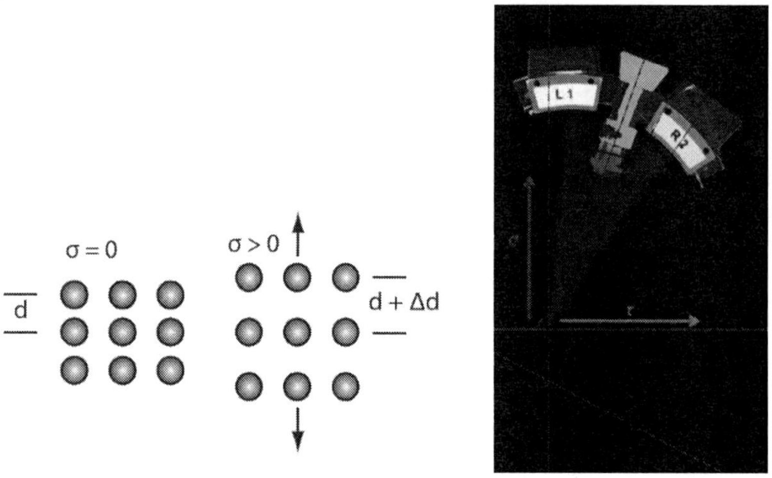

Fig. 2. Schematic presentation of influence of stress on the distance of the atomic lattice and illustration of the diffraction cone captured by detectors

2 Experiments conditions

Experimental studies were made on a high carbon rolled steel bar (60mm diameter and 500mm long) purchased from the market. The chemical analysis of the steel conducted on a direct reading spectrometer determined its chemical composition as: 1.14% C, 0.46% Mn, 0.16% Si, 0.11% S and 0.04% P (microstructure in the Fig. 3). Round slices cut from the steel bar were shaped as shafts (59mm x150 mm) by subsequent machining. The thickness of samples subjected to inhomogeneous plastic deformation were approximately 30mm whereas those subjected to thermal and phase transformation were approximately 15mm. All these samples were made free from residual stresses by annealing them in a muffle furnace as follows: heating rate, 165 $^{o}C.h^{-1}$; soaking time, 1h; soaking temperature, 850 ^{o}C; cooling rate, 30 $^{o}C.h^{-1}$.

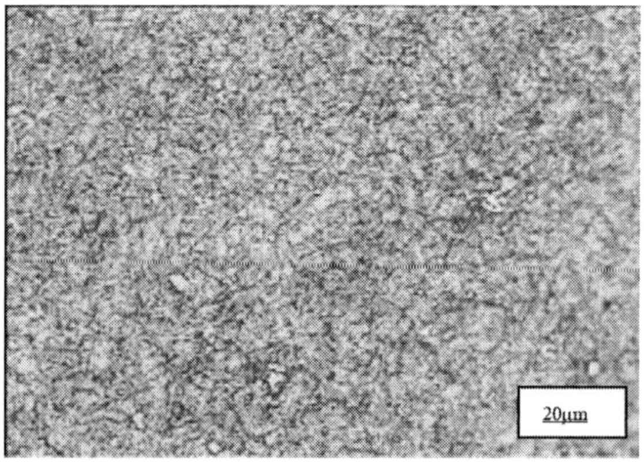

Fig. 3. Microstructure result of specimen, etching Murakami

Residual stress measurements were made on five machined samples, three samples having thermal residual stresses and two samples having thermal and phase transformation stresses (Fig. 4). The specimens were carefully prepared and made free from scale and dirt.

Clamping of the samples to the work holder was done with chuck tightening to avoid their sliding during experiment. Residual stresses were measured in the area located at the middle, e.g., at l/2 length in all the samples. The residual stresses were measured on the X-ray diffractometer using the strain flex X-ray analyzer. From theory of elasticity the relationship between residual stress (σ) and strain (ε) on the specimen surface under plane stress is given by The Bragg equation, λ=2d sin θ, relating incident X-ray wave length (λ), lattice interplanar spacing (d) and diffraction angle (θ) gives the following relationship.

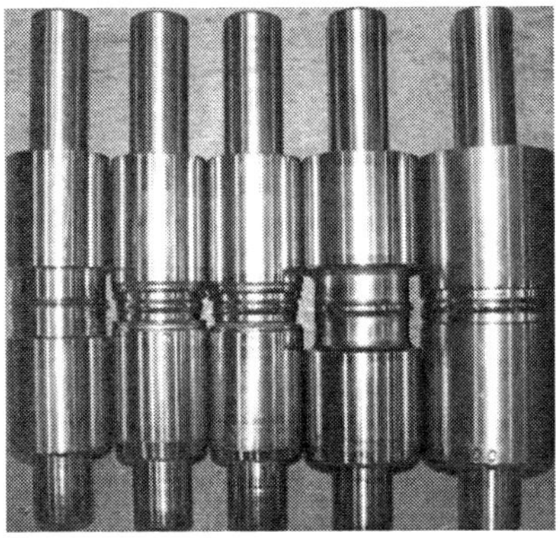

Fig. 4. Thorns on rolling bearing rings

Determination of the Magnitude and Direction of the Maximum Residual Stress Produced by Machining. The direction of maximum residual stress, that is, most tensile or least compressive, is assumed to occur in the cutting or grinding direction during most machining operations. This is frequently the case, but the maximum stress often occurs at significant angles to the cutting direction. Furthermore, the residual stress distributions produced by many cutting operations, such as turning, may be highly eccentric, producing a highly tensile maximum stress and a highly compressive minimum stress.

The residual stress field at a point, assuming a condition of plane stress, can be described by the minimum and maximum normal principal residual stresses, the maximum shear stress, and the orientation of the maximum stress relative to some reference direction. The minimum stress is always perpendicular to the maximum.

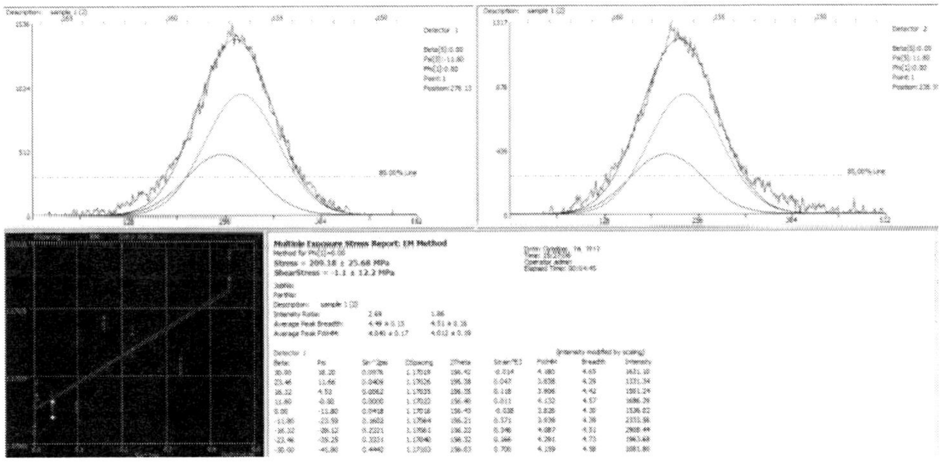

Fig. 5. Residual stress analysis of calculations with WINXRD 2.0

To investigate the minimum and maximum normal residual stresses and their orientation produced by turning of samples, x-ray diffraction residual stress measurements were performed in the longitudinal, 45°, and circumferential directions at the surface and at subsurface layers to a nominal depth of 0.1 mm, exposing the subsurface depths by electropolishing complete cylindrical shells around the cylinder. The cylinder was nominally 59 mm in diameter and uniformly turned along a length of several inches. The irradiated area was limited to a nominal height of 1 mm around the circumference by 2.5 mm along the length. Measurements were conducted using a Cr Kα (420) two-angle technique, separating the $K\alpha_1$, peak from the doublet using a Cauchy peak profile Fig. 5.

The measurements performed independently in the three directions were combined using Mohr's circle for stress at each depth to calculate the minimum and maximum normal residual stresses and their orientation defined by the angle φ, which was taken to be a positive angle counterclockwise from the longitudinal axis of the cylinder. Figure 6 illustrates the results, showing the maximum and minimum principal residual stress profiles and their orientation relative to the longitudinal direction. The maximum stresses are tensile at the surface, in excess of 140 MPa, dropping rapidly into compression at a nominal depth of 0.005 mm. The maximum stress returns into tension at depths exceeding 0.025 mm and remains in slight tension to the maximum depth of 0.1 mm examined. The minimum residual stress is in compression in excess of -480 MPa at the turned surface and diminishes rapidly in magnitude with depth to less than -138 MPa at a depth of 0.013mm. The minimum stress remains slightly compressive and crosses into tension only at the maximum depth examined. The orientation of the maximum stresses is almost exact in the circumferential direction (90° from the longitudinal) for the first two depths examined. For depths of 0.013 mm to the maximum depth of 0.1 mm, the maximum stress is within approximately 10° of the longitudinal direction.

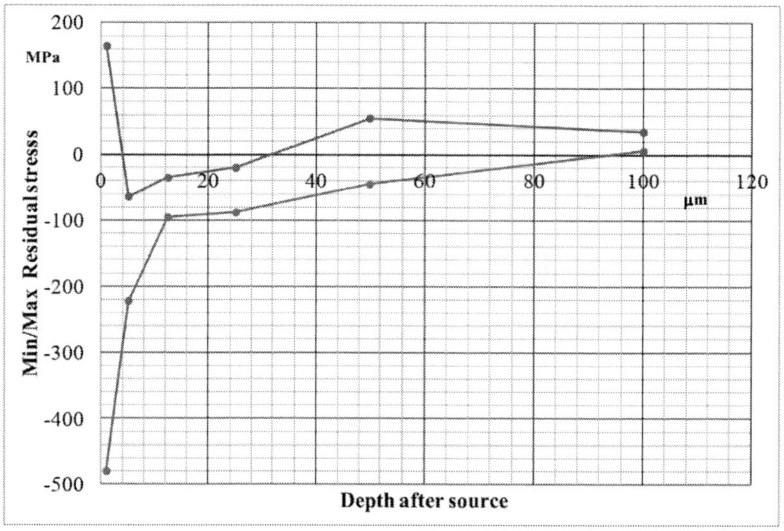

Fig. 6. Minimum and maximum principal residual stress profiles and their orientation relative to the longitudinal direction in a turned sample

2 Experiments results

Stresses were measured in the axial and radial direction components. After preparing the parts began to emit X-ray apparatus and x-rayed previously selected point in the part. The device measured the stress to a depth of 12μm in the range of angles around 123 ° -171 °. Then the computer depicted by graphs, which calculated the residual values for shear and residual stresses. These were processed into tables and these graphs are created.

Turning Roughing Operation when using a cutting speed of v_c = 100m.min^{-1} were measured axially compressed nature of the residual stress, whose value has hovered around-360MPa, radial residual stress in the application of the same cutting speed showed a value-175MPa pressure also character. With increasing cutting speed there was a reduction of residual stresses in the axial and radial direction. The value of v_c = 150m.min^{-1} to reduce tension in the axial direction compared with v_c = 100m.min^{-1} by 40%, in the axial direction decreased by 30%. When using v_c = 200m.min^{-1} decreased tension in the axial direction when applied to the values of v_c = 100m.min^{-1} by 50%, in the radial direction was changed to tensile stresses of pressure and change of 250% Fig. 7.

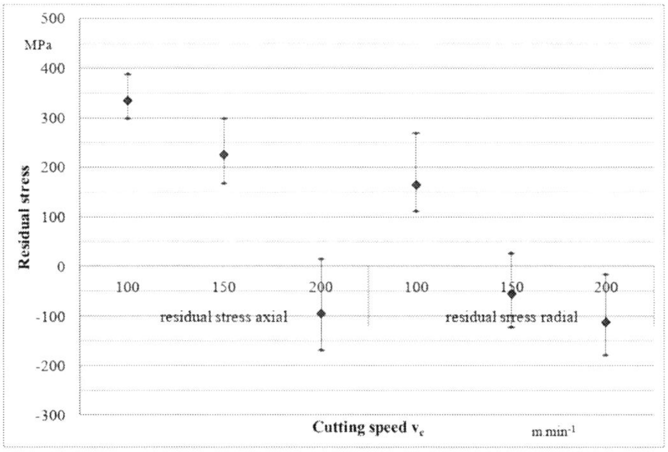

Fig. 7. Graph of residual stresses in turning roughing

Turning operation completion when finishing in all cases the measured residual stress nature of the pressure. When using a cutting speed of v_c = 100m.min^{-1} were measured axial tension, whose value corresponds to the stress -850MPa, radial residual stress when applying the same cutting speed showed a value -580MPa. The value v_c = 150m.min^{-1} is the tension in the axial direction increased compared with v_c = 100m.min^{-1} by 2%, in the radial direction increased by 20%. When using v_c = 200m.min^{-1} increased tension in the axial direction when applied to the values of v_c = 100m.min^{-1} by 250%, in the radial direction of the residual stress increased by 25% Fig. 8.

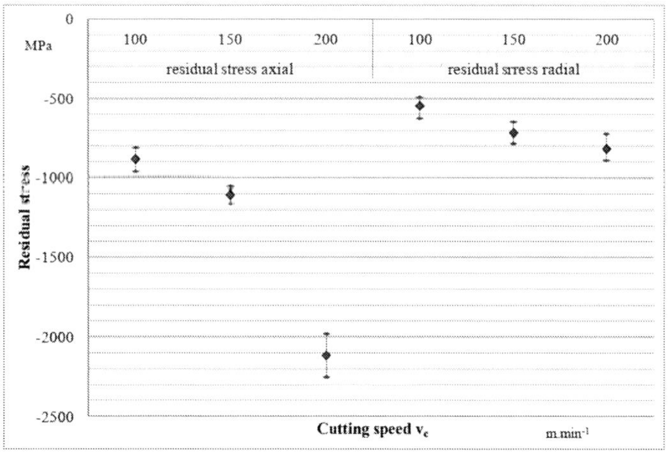

Fig. 8. Graph of residual stresses in turning finishing

That component is used as rolling tool that operates its surface on material to a ductile strength. In places where there are tensile residual stress, have a negative effect on the functional area components. They have a great impact on the spread of cracks in components. Compressive stress is appropriate because the distance between the atoms themselves are very small, they tend to associate and act against cracks in the workpiece. The value of compressive stress should not reach high values, for example 2GPa when in the body are trends the crevices and cracks. The optimal value of the residual stress varies in the range 500-700MPa. Residual stresses in a given case should not exceed 1000MPa, the most extreme value for the steel is 2550MPa. In this experiment, measurement of residual stresses in some places the size 2GPa evaluated. Such a large residual stress was caused by previous use of the rolling thorn. For this thorn in the process of rolling worked great forces that have a large impact on the results of measurements of residual stresses in the experiment Fig. 9.

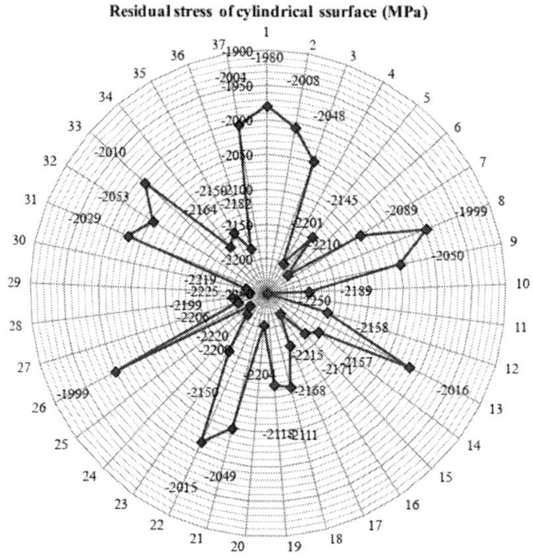

Fig. 9. Graph of residual stresses in polar coordinations when turning finishing with cutting speed $v_c = 200 m.min^{-1}$

The results appear to indicate that stresses within approximately 0.013 mm of the sample surface are dominated by machining, which resulted in a maximum stress direction essentially parallel to the cutting action. At greater depths, the stress distribution may be governed not by the machining as much as by stresses that may have been present due to forging or heat treatment.

Conclusion

Distributions of residual stresses on the surfaces and along the depth for the welded steel samples have been presented. Stress distribution for the sample with the cylinder sample is characterised by compressive stresses on the surface and by tensile stresses at the thorns hardness steel. Residual stress distributions for samples with circular surface are more complicated.

Conclusion about reversing of compressive residual stress on the surface of the weld seam to tensile stress in the depth made by analysis of equilibrium equations have been confirmed experimentally.

X-ray diffractometry has great potential for the measurement of residual stresses of forming tools. From measured data, we can improve and optimize not only the machining operations and other treatment of this tools but operations that are performed by these tool. Because it is one of the non-destructive methods, there is the chance, even after the measurement, the part is able to resume into the manufacturing process. This is a great asset especially in renovation of tools.

This article was funded by the University of Žilina project APVV 15-0405 – "Complex use of X-ray diffractometry for identification and quantification of functional properties of dynamically loaded structural elements from important technical materials" and project of University of Zilina: OPVav-2009/2.2/04-SORO number (26220220101) – "Intelligent system for nondestructive technologies on evaluation for the functional properties of components of X-ray diffraction" and KEGA project 025ŽU-4/2017 Integration of innovative detectional-visaulization multi-disciplinary technologies as on-line tool for learning of progressive CNC technologies.

References

1. K. Tiitto, *Testing Shot Peening Stresses in the Field*. The Shot Peener **4**, ISSN: 1069-2010 (1991)
2. P. S. Prevéy, *X-ray diffraction residual stress techniques*. Metals Handbook **10**. Metals Park: American Society for Metals, 380-392, ISSN: 0096-7416 (1986)
3. T. S. Wojtas, A.S., *Future of Residual Stress Testing in Shot Peened Parts during Processing*. SURFAIR IX - 9th International Conference on Surface Treatments in Aeronautic and Aerospace Industry, Cannes - France 3-5 June '92 ISBN 0375502777 (1992)
4. A. Czan, M. Sajgalik, A. Martikan, V. Kuzdak, *Identification of residual stress and structure properties in surface and subsurface layers of austenitic steel designed for nuclear energetics*. In. EAN 2015 – 53rd Conference on Experimental Stress Analysis, 36-40, ISBN 978-800105734-6 (2015)
5. A. Czan, E. Tillova, J. Semcer, J. Pilc, *Surface and subsurface residual stresses after machining and their analysis by X-ray diffraction*. Comunications **15** (2), 69-76, ISSN 1335-4205 (2013)
6. S. Tiitto, *On the influence of microstructure on magnetization transitions in steel*. Acta Polytechnica Scandinavica, Applied physics series **119**, Helsinki (1977)
7. P. Jacob, S. Marrone, *Non-destructive Evaluation of Residual Stress Depth-profiles by Barkhausen Noise Analysis and their Validation by XRD Method Combined with Electrochemical (destructive) Surface Removal*. 4th International Conference on Barkhausen Noise and Micromagnetic Testing July 3 – 4, Brescia – Italy, ISBN 951-98400-4-4 (2003)
8. M. E. Hilley, *Residual Stress Measurement by XRay Diffraction*. SAE J784a, Society of Automotive Engineers, Warrendale, PA, 21-24 (1971)
9. B. D. Cullity, *Elements of X-Ray Diffraction, Addison Wesley*. (2nd ed., 470 p., 1978)
10. A. Gorog, D. Stancekova, M. Kapustova, I. Gorogova, I. Miturska, *Influence of Selected Cutting Conditions on Cutting Forces In Advances In Science And Technology*. Research Journal **11** (1), 179-185, ISSN 2299-8624
11. M. Handrik, L. Jakubovičová, P. Kopas, M. Sága, *Analysis of microplastic areas near graphite particles of nodular cast iron loading bellow yield stress*. Metalurgija (Metalurgy) **49** (2), 263-267, ISSN 0543-5846 (2010)

12. R. Cep, L. Ocenasova, J. Novakova, L. Perkovska, A. Czan, D. Stancekova, *Intereupted Machining Tests of Ceramic Cutting Tools*. TMT 2009 Proc., Hammamet, **13** (1), 733-736, ISSN 1840-4944 (2009)
13. A. Czan, M. Sajgalik, A. Martikan, J. Mrazik, *Observation of dynamic processes in cutting zone when machining nickel alloys*. Communications – Scientific Letters of the University of Zilina **16** (3A), 161-168, ISSN 1335-4205 (2014)

Influence of structure sensitising of the AISi 316Ti austenitic stainless steel on the ultra-high cycle fatigue properties

Michal Jambor[*1,2], *František* Nový[1], *Otakar* Bokůvka[1], *Libor* Trško[2], *Monika* Oravcová[2]

[1]University of Zilina, Faculty of Mechanical Engineering, Department of Material Engineering, Univerzitná 1, 010 26, Zilina, Slovak Republic
[2]Research Centre of the University of Žilina, Univerzitná 1, 010 26, Zilina, Slovak Republic

Abstract. Austenitic stainless steels are the wide-spread materials, used mainly in the power industry. In that kind of engineering application, structural parts of rotating elements reach during their lifetime very high numbers of loading cycles, exceeding 10^7 numbers of cycles. With regard to this fact, the data of ultra-high cycle fatigue properties are needed to be used in the qualified design. Increasing demands on the efficiency cause the increase of the operating temperature, and exposition of these materials to the elevated temperatures can cause some important structural changes, which result in the sensitising of the structure. In this study authors present their own experimental results about fatigue properties of AISI 316Ti austenitic stainless steel after sensitising, in the ultra-high cycle region ($N_f = 10^6 \sim N_f = 3\times10^9$ cycles). Fatigue tests were carried out using ultrasonic fatigue testing device with frequency $f = 20$ kHz at the coefficient of cycle asymmetry $R = -1$, and temperature $T = 20\pm5$°C. In the ultra-high cycle region was observed the continuous decrease of the fatigue properties of the AISI 316Ti, and there was recorded the negative effect of the sensitising on the ultra-high cycle fatigue properties of the AISI 316Ti.

Keywords: AISI 316Ti, sensitisation, ultra-high cycle fatigue

1 Introduction

Austenitic stainless steels are the important category of the technical materials, which have found the wide range of the applications. These steels are highly resistant to many of corrosion environments, and they are successfully used in the chemical and power industry. Excellent corrosion properties of these steels are given by the chemical composition, where the high chromium content enables to create the stable passive layer which prevents degradation by the corrosion processes. In special cases where these steels are exposed to increased temperatures for some time (after welding or in the application in the power industry), significant changes of the microstructure can occur, which affect and decrease the ability to create the protective passive layer. This phenomenon is called sensitisation [1, 2]. Sensitisation of austenitic stainless steels is caused by the extensive precipitation of the

[*] Corresponding author: michal.jambor@fstroj.uniza.sk

Reviewers: *Jan* Krmela, *Františka* Pešlová

chromium-rich phases on the grain boundaries, mainly carbides $Cr_{23}C_6$ and the σ phase, which consume chromium dissolved in the matrix and locally decrease its concentration under the critical value, 11.6%, which is considered as a lowest concentration to provide creation of the stable passive layer [3-5]. Precipitation of these phases occurs preferentially in the grain boundaries regions, which can result in the formation of the continuous network of the carbides on the grain boundaries, which leads to corrosion resistance decrease and the formation of the intercrystalline corrosion and the failure even at very low loadings [6]. In austenitic stainless steels containing molybdenum (as a 316 grade) these carbides also contain Mo, so these areas are also depleted in Mo [6]. Susceptibility of the austenitic stainless steels to sensitisation strongly depends on the chemical composition. Decreasing of the C content and increasing amount of Cr, Mo, and N decrease the susceptibility to sensitisation of austenitic stainless steels [6].

To increase the resistance to sensitisation, special grades of austenitic stainless steels were developed. Improving resistance to the sensitisation could be obtained by the decreasing the carbon content to the lowest possible level. Grades AISI 304L, 316L and 321 has the maximum carbon content 0.03%. The second way how to improve resistance to sensitisation is the alloying steels by the small addition of the strong carbide former, as a Ti or Nb (Grades 316Ti and 347). Recently was introduced another option of reducing susceptibility to sensitisation, using grain boundary engineering [2, 7]. Stabilized steels exhibit a higher level of resistance to sensitisation than steels with lowering carbon content, while niobium is more effective stabilizer than titanium. Despite this fact, even in these more resistant grades, sensitisation can occur, especially in low-temperature exposures, what was documented in the work [1], where austenitic stainless steel AISI 321 (austenitic stainless steel stabilized with the Ti) sensitised after exposures at temperatures around 600°C.

Sensitisation caused degradation of the austenitic stainless steels properties, increase the susceptibility to intercrystalline corrosion, stress corrosion cracking and also degradation of mechanical properties. Sensitisation affects the fatigue properties indirectly, as the premature fatigue crack initiation occurs on the corrosion pits, which can form when the sensitised steel has decreased the corrosion properties [8]. The direct effect is the increasing the fatigue crack propagation by the intercrystalline manner, which is an important factor in the low cycle fatigue. Other aspects of the direct effect of the sensitisation on the fatigue properties, especially in the ultra-high cycle area, where the fatigue crack initiation and the propagation of short cracks plays a dominant role in the fatigue process, are still not well examined. Ultra-high cycle fatigue is the region beyond the conventional fatigue limit ($N_f = 10^7$), and during cyclic loading in this region, different mechanisms of the failure may take places, which makes necessity to experimental evaluation of the effect of the microstructural changes on the fatigue properties in this area.

Authors in this study present their own experimental results of the influence of sensitisation on the ultra-high cycle fatigue properties of the AISI 316Ti austenitic stainless steel.

2 Experimental procedures

2.1 Material

As the experimental material in this study was used steel AISI 316Ti in the form of bars with the diameter of 15 mm, which represents the austenitic stainless steel with improved susceptibility to the sensitisation by the addition of Ti stabilizer. The chemical composition of the tested material is shown in Table 1, and it was obtained by the emission spectrometry device ICP (JPY 385). Evaluation of the basic mechanical properties was carried out by the standard tensile tests. These were conducted on the ZWICK Z050 testing machine at an ambient temperature 20±5°C and the initial strain rate was $\varepsilon_m = 10^{-3}\ s^{-1}$. Specimens used for the tensile test fulfilled the requirements of the EN 10002-1 standard for tensile tests, five specimens were used and the average values of the mechanical properties were than calculate. Mechanical properties are shown in Table 2. The microstructure of the tested material was examined using optical microscopy and the scanning electron microscopy, and the samples were prepared by the standard procedure for the preparation of metallographic samples and etched by the solution of 30 ml HCl, 10 ml HNO_3 and 30 ml glycerol.

Table 1. Chemical composition of experimental material

Cr	Ni	Mo	Mn	Ti	C	Si	P	S	Fe
17.55	12.96	2.54	1.63	0.37	0.058	0.81	0.033	0.037	Bal.

Table 2. Mechanical properties of experimental material

YS [MPa]	UTS [MPa]	Elongation[%]	Reduction in area [%]	HV10
251	773	54	48	213

Microstructure observation (Fig. 1) reveals austenitic polyedric structure, with a high density of the annealing and deformation twins. In the structure were also present different kinds of secondary particles, their nature was examined using EDX analysis, and they were identified as a Cr, Ti, and Mo-rich particles, with the average size up to the 5 μm. The structure also contains considerably large Ti-rich particles, which were identified as carbides and nitrides of titanium, with the average size 10-20 μm. A small amount of delta ferrite, in a form of elongated particles, was present and the structure also contains a small amount of Si-rich inclusion. No σ phase was observed in the structure of the as-delivered state.

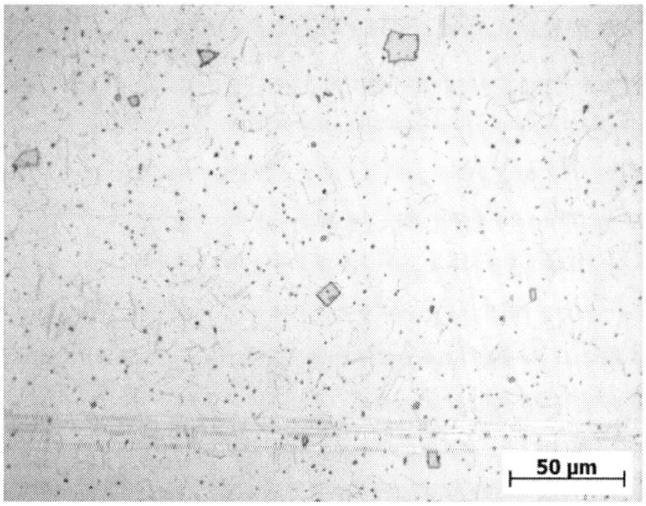

Fig. 1. Microstructure of AISI 316Ti in as-delivered state

Specimens for the fatigue tests were machined from the bars, according the drawing on the Fig. 2. Dimension of the specimens were calculated to fulfill the resonance conditions of the ultrasonic fatigue testing machines. After machining, the surface of the specimens was polished, to prevent any affecting of testing results by the specimen´s surface state.

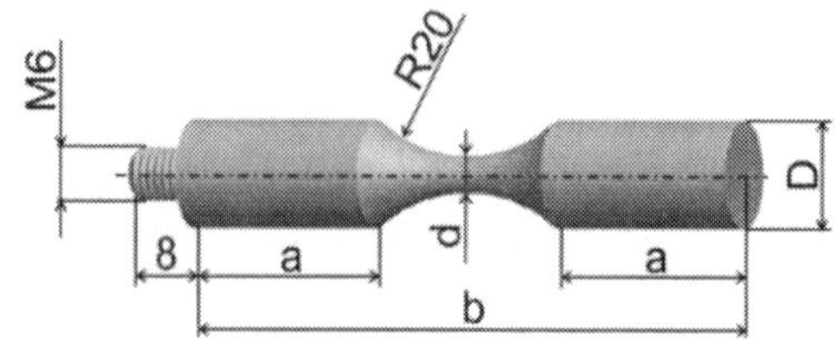

Fig. 2. Drawing of the fatigue test specimen

2.2 Sensitisation treatment and evaluation

Part of samples for the fatigue test was heat treated to provide sensitising of the structure. Sensitising heat treatment has been carried out in resistance furnace in the air atmosphere. Heat treatment consists of the heating to the 600 °C, holding at this temperature for 100 hours followed by the slow cooling on the air. Conditions of sensitisation were purposely selected, and this treatment represents the standard operating temperature of these steels in the power industry. Based on the published studies, precipitation of carbides on the grain boundaries can occur at this temperature, and also diffusion rates of chromium are considerably low, so the chromium content in the depleted zones cannot be restored by the diffusion from the grains interiors. Another important fact is that at 600 °C precipitation of Cr-rich carbides is more favourable than Ti-rich carbides [1].

For the determination of the degree of sensitisation, the ASTM standard 262 practice A has been used. This method is based on the electrolytic etching of the specimens, prepared by the standard method for the preparation of the metallographic samples, in the 10% water solution of oxalic acid at the constant potential 10V to reveal the chromium depletion zones. Recommended practice for evaluation of sensitisation degree based on the ASTM standard 262 practice A characterize three degrees of sensitisation. Step – represent no sensitised structure, where no attack on the grain boundaries is observed. Dual – structure contains a small amount of pits on the grain boundaries, without pits surrounding whole grains. Ditch – strong attack on the grain boundaries, pits surround at least one whole grain.

2.3 Fatigue tests

Determination of fatigue properties in the ultra-high cycle area was carried out on the ultrasonic fatigue testing device KAUP, which schematic diagram is shown on the Fig.3. KAUP ultrasonic fatigue testing device loads specimen with the symmetrical tension-compression loading, at the resonant frequency 20 kHz. The detailed description of this device and the principles of the ultrasonic fatigue testing are more detailed described in the references [9-11]. During whole tests, constant temperature 20±5°C was maintained. Due to a high level of internal friction, typical for austenitic stainless steels, additional cooling of the specimens was needed. Some authors in their study of ultrasonic fatigue of austenitic stainless steels [12] used a stream of cold air to cooling specimens, but this was not to be effective enough in our case, thus water cooling was used. Specimens were submerged in the water with additives, to effectively lead away heat generated during high-frequency loading. Tests were carried out in the range $10^6 < N_f < 3\times10^9$ number of cycles. Two series of specimens were used, the initial state and after sensitisation at 600°C/100h. Fatigue tests results were shown in the S-N diagrams, and the regression of the curves was done by the Basquin equation (1), in the form:

$$\sigma_a = \sigma'_f (N_f)^b \tag{1}$$

where b is exponent of fatigue live curve and σ'_f is the coefficient of fatigue strength obtained by extrapolation of stress amplitude σ_a on the first loading cycle.

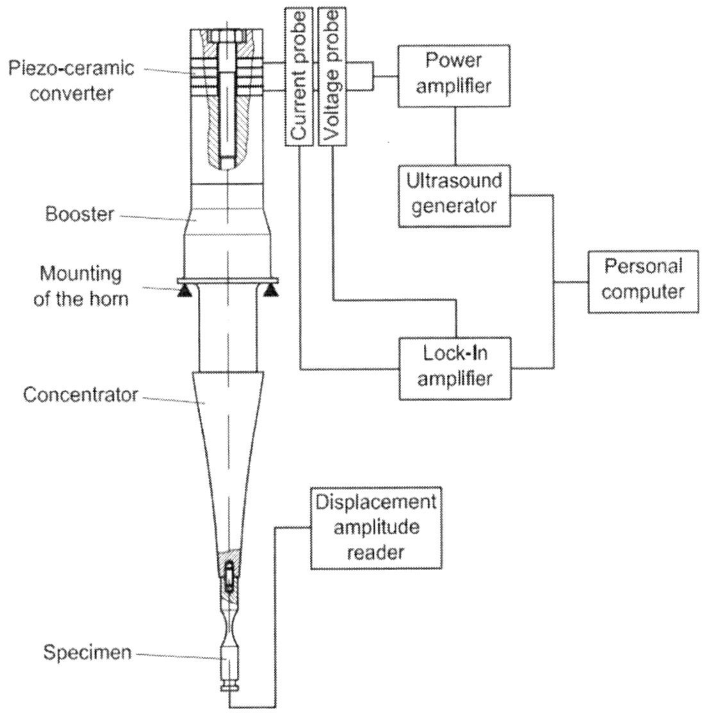

Fig. 3. Ultrasonic fatigue testing device KAUP

3 Results

3.1 Sensitisation evaluation

Sensitisation degree was evaluated according to ASTM 262-A standard. This test reveals, that applied heat treatment (600°C/100h) result in sensitisation of austenitic stainless steel AISI 316Ti. The degree of sensitisation was evaluated as a ditch, because the careful observation taken on the multiple places shows a high degree of intergranular attack and individual grains were wholly surrounded by the pits (Fig. 4).

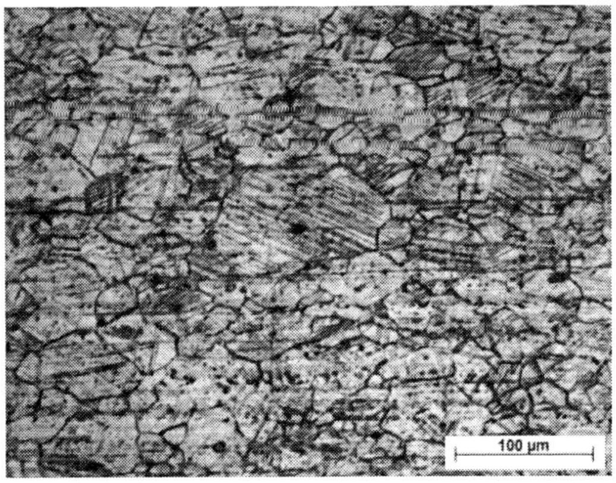

Fig. 4. Microstructure of sensitised AISI 316Ti electrolytically etched by oxalic acid

3.2 Fatigue tests

S-N curves were constructed from the fatigue tests results. Curves were approximated by Basquin equation, using the least square method [13]. Regression curves coefficients are shown in Table 3. In both testes states was recorded decrease of fatigue strength with increasing number of loading cycles, even after conventional fatigue limit (10^7 loading cycles). Figure 5 shows a comparison of the S-N curves for initial state and a state after sensitising treatment. In both curves was not recorded the presence of any plateau, which suggests no changing of preferential mechanisms of fatigue crack initiation.

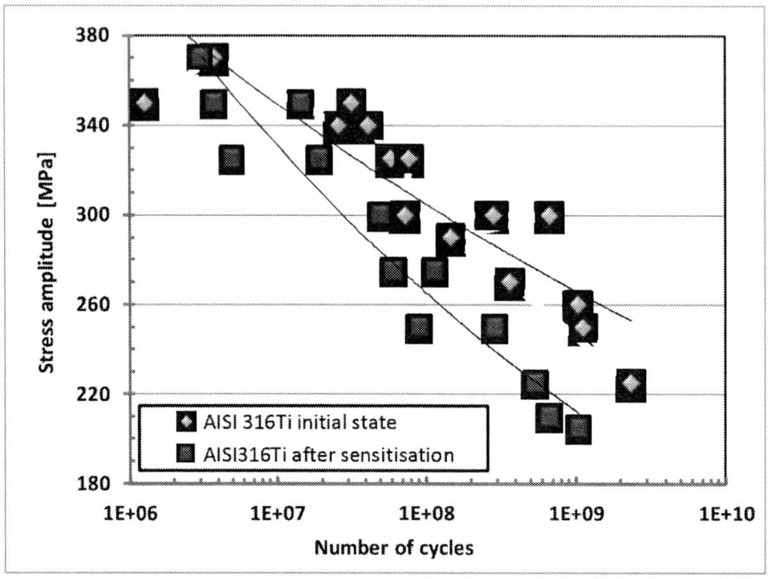

Fig. 5. S-N curves of AISI 316Ti in initial state and after sensitisation

Using regression curves coefficients; fatigue strength for different numbers of loading cycles was calculated for both tested states. Results are shown in Table 3, as it can be seen, the decrease of fatigue strength for the sensitised state is more significant at higher numbers of loading cycles. For the 10^9 number of cycles exhibits sensitised state 20% decrease of fatigue strength in comparison with the initial state.

Table 3. Regression curve coefficients and comparison of fatigue strength for different numbers of loading cycles

	σ'_f [MPa]	b	Fatigue strength for $N = 10^7$ [MPa]	Fatigue strength for $N = 10^8$ [MPa]	Fatigue strength for $N = 10^9$ [MPa]
Initial state	910.58	-0.059	351.8	307.1	268.1
Sensitised	1564.7	-0.096	333	266.9	214

3.3 Fractography analysis

Fractography analysis was carried on the broken specimens, with the aim to describe the fatigue process and to find any differences which can be related to the sensitisation treatment. Every broken specimen was carefully examined, to rule out any affecting of initiation stage by possible cavitation. Despite observations made by other authors [14], no cavitation was recorded on the specimens, even they were submerged in the water to prevent self-heating of the testing specimens during fatigue tests. Two different types of initiation of fatigue crack were recorded. In both, the crack forms in the surface and subsequently propagate along slip plane, usually at 45°degrees to the loading direction and when the stress intensity increase, the crack will start to propagate perpendicular to the loading axis. At high loading amplitudes, the first stage of the fatigue crack growth is quite small, in just a few μm, because the higher stress level allows to transition to the other slip planes, which are perpendicular to the loading axis. This type of initiation was observed for both states, in similar values of the loading stress (Fig. 6). The second type of crack initiation occurs at the lower loading stress levels. In this type of crack initiation, fatigue crack initiates also on the surface, but due to lower stress intensity, the length of crack propagation along the initial slip plane is much longer, and it may extend across several grain diameters before the increase of the stress intensity allows to promote slip in other systems then primary slip system [15], see Fig. 6. The occurrence of these two types of fatigue initiations was recorded for the both tested states so the change of initiation mechanism cannot be related to the decrease of fatigue strength in sensitised specimens. Occurrence of the second type of fatigue crack initiation was not observed in every specimen broken at lower loading amplitude, and this is caused by the nature of this mechanism, where very important role plays the crystallographic orientation of the grain where the initiation starts.

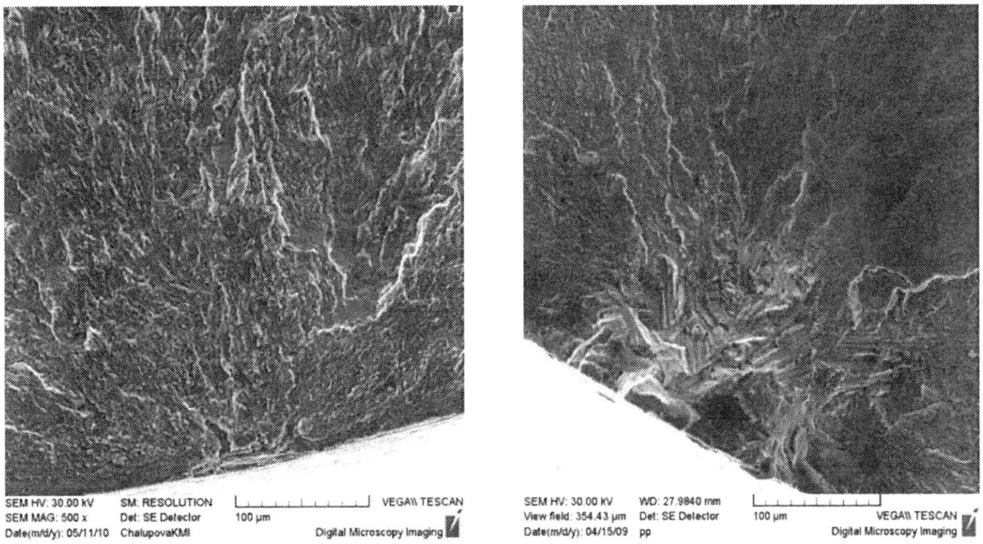

Fig. 6. Initiation of fatigue cracks. Left figure shows sensitised specimen broken after $N_f = 3.8\times10^7$ cycles at $\sigma_a = 350$ MPa. Righ picture shows crystallographic initiation of fatigue crack, sensitised specimen, $N_f = 2.9\times10^8$, $\sigma_a = 250$ MPa

Observation of the crack surfaces shows that after initiation, fatigue cracks propagate mostly by transcrystalline manner (Fig. 7), and on the cracks surfaces were observed ductile striations. In all specimens, a few regions of intercrystalline fatigue crack propagation were recorded (Fig. 7). In the initial state specimens, intercrystalline crack propagation regions did not exceed 1% of the total crack surface area. Sample after sensitising treatment exhibits the higher content of the intercrystalline crack propagation area which however did not exceed 5% of the total crack surface area.

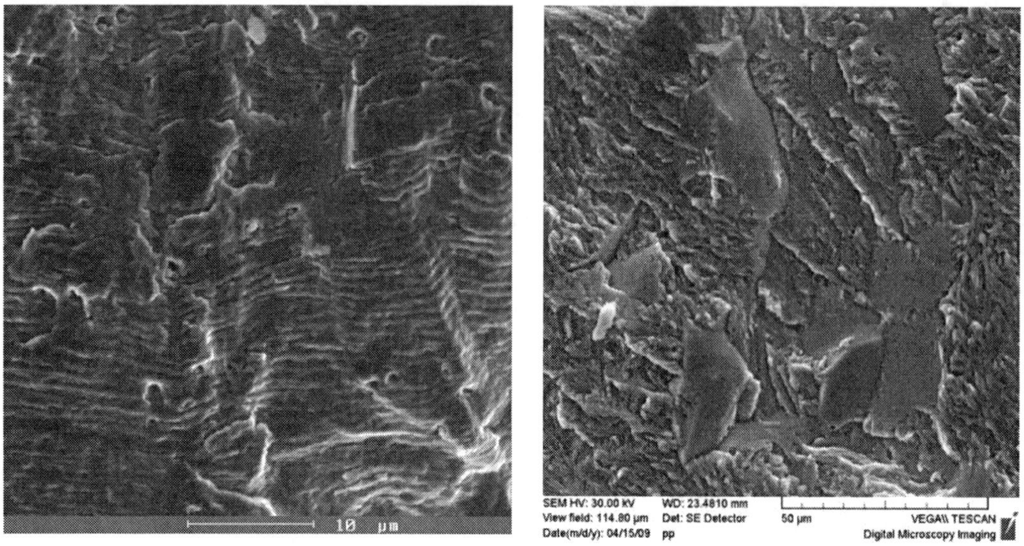

Fig. 7. Fatigue crack propagation. Left figure: transcrystalline fatigue crack propagation, clearly visible striations, sensitised specimen, $N_f = 3.8\times10^7$, $\sigma_a = 350$ MPa. Right figure: intercrystalline fatigue crack propagation, the same specimen

4 Discussions

4.1 Sensitisation

At first, the sensitisation treatment will be discus. Most previous studies were dedicated to sensitisation at high-temperature exposures. In real applications, the sensitisation of austenitic stainless steel can have two causes. The main and the most studied case is the sensitisation of welded joint, where the heat affected zone is subjected to temperature regimes ideal for the precipitation of chromium-rich carbides at the grain boundaries. Heat input cause that all carbon will solute in the matrix and during subsequent cooling there is enough time to precipitation of carbides along the grain boundaries but not sufficient time to chromium diffusion from grain interior to the depleted zone in the grain boundary areas. This can be suppressed by special heat treatment after welding, like cooling with the controlled cooling rate or by the soaking time at elevated temperatures in order to realize desensitisation. For improving resistance to sensitisation after welding, low carbon and stabilized grades of AISI 316 steel were introduced. The second case of sensitisation occurs when these steels are subjected to intermediate (550-650°C) temperatures for long exposures. In the real applications, this could be easily achieved, while these steels are standard using in this temperatures range. According to literature, at these temperatures, chromium-rich carbides precipitate in favor of titanium carbides [1]. During sufficient long exposures at high temperature, desensitisation can occur, which means, that the chromium depleted zones disappeared because there was enough time for diffusion chromium atoms to the depleted zones at grain boundaries from the grain interiors. For the austenitic stainless steels with improved resistance to sensitisation (ex 316L) was fully desensitisation observed after 2h of exposure to 950°C and after 8h at 900°C. As it's seen, the rate of desensitisation process is fully driven by the temperature and by the diffusion rates of chromium [16]. However, at a lower temperature, as in the study used 600°C, based on the observation made in the study [16], no desensitisation occurs in 316Ti grade after 100h at 600°C. For these reasons, temperature regime 600°C/100h was chosen for the sensitisation treatment in these study, and predictably, this treatment results in fully sensitisation of the structure of austenitic stainless steel AISI 316Ti.

4.2 Fatigue properties

Fatigue test result shows two important conclusions. At first, the fatigue strength of AISI 316Ti decrease with increasing number of loading cycles, what is in strong contrast with the work [12] where the constant fatigue limit in the range 10^7-10^9 of loading cycles was recorded and no decrease of fatigue strength was observed. However, results of other authors [17] made on the similar material (316L) show similar result as in our study. The second important conclusion is that sensitisation negatively affects the fatigue properties in ultra-high cycle area, and this influence is increasing with increasing numbers of loading cycles. Data displayed in Table 3 shows that this effect is stronger with increasing numbers of loading cycles. For 10^7 of loading cycles this decrease is about 5%, for 10^8 of cycles it is about 13% and for 10^9 of loading cycles is this decrease more than 20%. The decrease of fatigue strength by 20% is significant and this value must be considered in the design of construction operating under similar conditions. The reason of this decrease is unclear. Fractography analysis did not show any differences in the fatigue crack initiation, although two different mechanisms were recorded but it occurrence was similar for both tested states. Small differences were observed in the fatigue crack propagation, when the crack surface of broken sensitised specimens contains the high amount of intercrystalline fatigue crack propagation areas. In high cycle fatigue and especially ultra-high cycle fatigue, fatigue crack propagation takes only a small number of cycles and the most loading cycles are related to the fatigue crack initiation. For 10^9 cycles fatigue crack propagation takes less than 0.01% fraction of all loading cycles, thus small changes in the fatigue crack propagation in sensitised specimens cannot be the reason for the fatigue strength decrease in the ultra-high cycle area. This decrease must be connected with the fatigue crack initiation stage. Fatigue strength is strongly related with the grain size [18] and long exposures to elevated temperatures could result in the grain boundary migration, but temperature 600°C is not sufficient for the grain growth, what was verified by the grain size observation and measurement, where no signs of boundary migration were recorded. In the similar study [19], made on the AISI 316L stainless steel, sensitised at similar conditions, no effect of sensitisation on the fatigue properties in the ultra-high cycle area was recorded. The main differences between these grades are that 316Ti contains 0.37% of Ti but also, the carbon content is more than double compared with 316L. One theory should be considered in relation to these two studies. Precipitation of carbides on the grain boundaries and formation of the chromium depleted zones has another significant consequence. Chromium in austenitic stainless steels is the substituent in the solid solution. Chromium is the main element responsible for corrosion properties, but also, it strengthens the matrix by the solid solution strengthening mechanism and chromium content affect the strain induced martensitic transformation. Lower chromium content in the depleted zone thus will cause a local decrease of mechanical properties. Fatigue crack initiation and the propagation of the short fatigue cracks determine resulting fatigue properties in the ultra-high cycle region and formation of zones, with a local decrease of mechanical properties can result in a preferential slip in this areas and this can cause the acceleration of fatigue crack initiation process and the decrease of fatigue strength. This theory requires another precise experiment, focused on the determining the effect of the chromium content on the local mechanical properties.

Conclusions

Based on the experiments carried out in this study, the following conclusion could be stated:

- Heat treatment 600°C/100h result in the sensitisation of austenitic stainless steel AISI 316Ti.
- Fatigue test of AISI 316Ti in the ultra-high cycle area show continuously decrease of fatigue strength with increasing number of cycles.
- There was recorded the negative effect of the sensitisation on the fatigue properties of the AISI 316Ti in the ultra-high cycle region. This effect was stronger at lower loading amplitudes and higher numbers of loading cycles, where the decrease of fatigue strength by 20% was recorded in comparing with the initial state, while at higher loading amplitudes and lower numbers of loading cycles was the difference in fatigue strength almost zero. Nature of this decrease is unclear and it must be subjected to further investigation.

The research was supported by Scientific Grant Agency of Ministry of Education, Science and sport of Slovak Republic and Slovak Academy of Sciences, grants VEGA No. 1/0123/15 and No. 1/0951/17 and project "Research Centre of the University of Žilina, 2nd phase, ITMS 26220220183.

References

1. A. S. Lima, A. M. Nascimento, *Sensitization evaluation of the austenitic stainless steel AISI 304L, 316L, 321 and 347.* Journal of Materials Science **40**, 139-144 (2005)
2. E. Kocsisová, M. Dománková, I. Slatkovský, M. Sahul, *Study of the sensitization on the grain boundary in austenitic stainless steel AISI 316.* Research papers Faculty of Materials Science and Technology in Trnava **22,** 131-136 (2014)
3. T. Sourmail, *Review: Precipitation In Creep-Resistant Austenitic Stainless Steels.* Materials Science and Technology **17**, 1-14 (2001)
4. E. G. Na, *Evaluation of sensitization and corrosive damages of the weldment for SUS 316stainlesssteel.* Journ. of Mech. Science and Technology **27**, 2715-2719 (2013)
5. N. Parvathavarthini, R. K. Dayal, *Time-temperature sensitization diagrams and critical cooling rates of different nitrogen containing austenitic stainless steels.* Journal of Nuclear Materials **399,** 62-67 (2010)
6. H. Shaikh, N. Sivaibharasi, B. Sasi, T. Anita, R. Amirthalingan, B. P. C. Rao, T. Jayakumar, H. S. Khatak, B. Raj, *Use of eddy current testing method in detection and evaluation of sensitisation and intergranular corrosion in austenitic stainless steels.* Corrosion Science **48**, 1462-1482 (2006)
7. N. Parvathavarthini, S. Mulki, R. K. Dayal, I. Samajdar, K. V. Mani, B. Raj, *Sensitization control in AISI 316L(N) austenitic stainless steel: Defining the role of the nature of grain boundary.* Corrosion Science **51,** 2144-2150 (2009)
8. F. Nový, V. Zatkalíková, O. Bokůvka, K. Miková, *Gigacycle Fatigue Endurance of Marine Grade Stainless Steels with Corrosion Pits.* Periodica Polytechnica Transportation Engineering **41/2**, 99-103 (2013)
9. L. Trško, M. Guagliano, O. Bokůvka, F. Nový, M. Jambor, Z. Florková, *Influence of Severe Shot Peening on the Surface State and Ultra-High-Cycle Fatigue Behavior of an AW 7075 Aluminum Alloy.* Journal of Materials Engineering and Performance **26** (6), 2784-2797 (2017)

10. L. Trško, O. Bokůvka, F. Nový, M. Guagliano, *Effect of severe shot peening on ultra-high cycle fatigue of low-alloy steel.* Materials & Design **57**, 103-113 (2014)
11. M. Sága, P. Kopas, M. Uhričík, *Modeling and Experimental Analysis of the Aluminium Alloy Fatigue Damage in the case of Bending - Torsion Loading.* Modelling of Mechanical and Mechatronics Systems, Procedia Engineering **48**, 599-606 (2012)
12. C. Műller-Bollenhagen, M. Zimmermann, H. J. Christ, *Very high cycle fatige behaviour of austenitic stainless steel and the effect of strain-induced martensite.* International Journal of Fatigue **32**, 936-942 (2010)
13. J. Kohout, S. Věchet, *A new function for fatigue curves characterization and its multiple merits.* International Journal of Fatigue **23**, 175-183 (2001)
14. K. Takahashi, T. Ogawa, *Evaluation of Giga-cycle Fatigue Properties of Austenitic Stainless Steels Using Ultrasonic Fatigue Test.* Journal of Solid Mechanics and Materials Engineering **72/723A**, 1731-1736 (2006)
15. O. K. Chopra, *Mechanism of Fatigue Crack Initiaion in Austenitic Stainless Steels in Light Water Reactor Environments.* Transactions of the 17th International Conference on Structural Mechanics in Reactor Technology, 1-8 (2003)
16. P. Atanda, A. Fatudimu, O. Olowole, *Sensitisation study of normalized 316L stainless steel.* Journal of Minerals & Materials Characterization & Engineering **9** (1), 13-23 (2010)
17. T. Naoe, Z. Xiong, M. Futakawa, *Gigacycle fatigue behavior of austenitic stainless steels used for mercury target vessels.* Journal of Nuclear Materials **468**, 331-338 (2016)
18. M. Akita et al., *Effect of sensitization on corrosion fatigue behavior of type 304 stainless steel annealed in nitrogen gas.* Material Science & Engineering A **640**, 33-41 (2015)
19. J. Lago, M. Jambor, F. Nový, O. Bokůvka, L. Trško, *Giga-cycle fatigue of AISI 316L after Sensitising of structure.* Procedia Engineering **192,** 528-532 (2017)

Determination of the influence of adherent surface on the adhesive bond strength

Tomas Kalina[1,*], *Frantisek* Sedlacek[1], *Jan* Krystek[2]

[1]RTI – Regional Technological Institute, Faculty of Mechanical Engineering, University of West Bohemia, Univerzitní 8, 306 14 Pilsen, Czech Republic.
[2]European Centre of Excellence, NTIS – New Technologies for Information Society, Faculty of Applied Sciences, University of West Bohemia, Technicka 8, 30100 Pilsen, Czech Republic

Abstract. A type of surface and a surface treatment of the adherends has a great influence on strength and durability of bonded joints. This paper deals with an experimental testing of influence of adherent surface on bonded joints strength. Several combinations of materials, surface treatments and two types of adhesives has been designed and manufactured. Those properties were tested on the lap-joints. Samples were made according to the standard ASTM D5868. The materials used were steel S355, aluminium alloy EN AW 7075 T6, carbon fibre laminate and glass fibre laminate. The surface treatments used were: Sandblasted, Mechanically roughened, Anodised, Cataphoretically painted and No-surface treatment. The experimental tests of test specimens were done on the Zwick/Roell Z050 machine. Suitable combinations of material, adhesives and surface treatments were selected based on testing.

Keywords: Adherend Surface, Surface treatment, Adhesive bond, lap-joint, ASTM D5868

1 Introduction

Composite materials are becoming more and more used materials not only in aerospace and automotive industries but also in other industrial sectors where a high strength and stiffness at low weight is required.

In aerospace and automotive industries the fiber reinforcement composite materials are most frequently used. For joining of those composite materials with metals the adhesive bond is most frequently the best.

Within our research we are designing the metodology of predicting the strength of bonded composite material with metals when designing their structure. With the use of experimental testing and numerical simulations. In practice this methodology should be simply applicable to the majority of real components.

The overall quality of the adhesive bond is influenced by many factors. To the most significant ones belongs: the material of adhesive components [1, 2], the surface treatment of adhesive areas [1, 2], the quality of degreasing of adhesive areas, the type of

* Corresponding author: tkalina@rti.zcu.cz

Reviewers: *Jan* Krmela, *Janusz* Mielniczuk

adhesive [1, 2], the thickness of adhesive layer [1, 2], the technology of adhesion, etc. And just the material of adhesive components and the surface treatment of adhesive areas are the most important. Therefore this article is focused on this two factors. But there exists an inexhaustible quantity of combinations of adhesive materials, surface treatments and adhesives. And for this reason it was necessary to focus on selected narrower part of this set.

Every adhesive has different mechanical properties and different adhesion to different materials and to their surface treatments. For us to be able to design a suitable combination of „material-adhesive-surface treatment" we have to consider the influence of cohesion and adhesion to a particular material. For effective and objective design we create the databases of mechanical properties of adhesives, materials and surface treatments.

The goal of this article is to determine the adhesive properties of adhesive for the most commonly bonded materials in the automotive industry. The evaluation will be done not only from technical but also from economical point of view. Based on the results it will be possible, according to general assignment (adhesive areas, required load capacity, etc.) to design a suitable adhesive and surface treatment, or vice versa.

2 Test specimens

2.1 Materials and adhesives

Two epoxy structural adhesives were chosen: the adhesive Scotch-Weld DP490 and Scotch-Weld 7260 (producer 3M). These adhesives are often used in automotive industry.

Two most commonly used composite materials and two most commonly used metals were chosen. Specifically it is fiberglass, carbon fiber laminate, aluminium alloy EN AW 7075 T6 and steel (S355 J0, according to EN10025-2:2004). For these four chosen materials multiple surface treatments were designed. The effect of the surface treatment of individual materials on the adhesive properties of the glued joint was monitored.

It was considered to bond the fiberglass with aluminium alloy using the Scotch-Weld DP490 adhesive and adhesion of carbon fiber laminate with steel using the Scotch-Weld 7260.

2.2 Dimensions of the test samples

Samples for shear adhesion testing were designed according to ASTM D5868 [3]. This is a Lap-joint. The dimensions of the samples and the layers of adhesive are shown in the Fig. 1.

Name	Parameter	Dimension [mm]
Adhesive thickness	A	0.75
Specimen width	B	25
Specimen length	L	100
Overlap	O	25
Material thickness	T	depending on the type of material

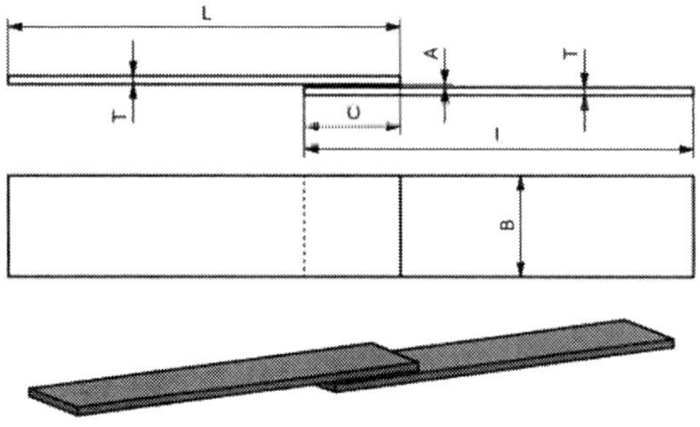

Fig. 1. Sample dimensions for adhesion detection according to the ASTM D5868

2.3 Preparation tool for samples adhesion

To ensure high accuracy were the base material cut very precisely using water jet cutter machine. To achieve high bonding accuracy a product was designed preparation tool.

The construction of preparation tool is a modular design with an adjustable main plate. Therefore the preparation tool can be set for any thickness of a material and adhesive. Moreover, it is possible to set two overlap sizes of 12.5 mm and 25mm (according to ASTM D5868 [3]). The position of the base material is determined by two auxiliary plates. The basic structure of the preparation is described in the Fig. 2.

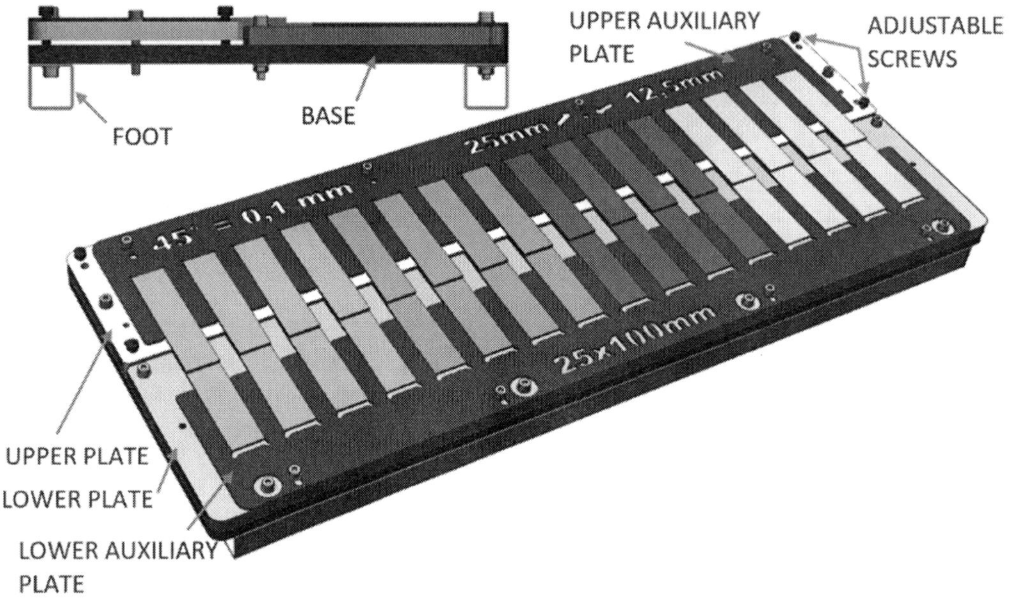

Fig. 2. Preparation tool for lap-joint samples´ adhesion

2.4 Surface treatments of the samples

Fiberglass and carbon laminate can be used without any surface treatment or it is possible to mechanically roughen the surface. It is also possible to use various chemical treatments of adhesive surfaces by means of so called primers etc. but those products are bound to specific types of adhesives and materials used. For fiberglass and carbon laminate, only two variants have been proposed: no surface treatment and the surface mechanically roughened with P80 sandpaper.

Aluminium and steel alloys are more capable of surface preparation. For the steel and aluminium alloy, the following surface treatments have been proposed: without surface treatment (surface only degreased), abrasive blasted surface, anodized surface, cataphoretic painted surface with subsequent roughening with sanding foam.

The selected materials, surface treatments, the number and size of samples are given in Table 1.

Table 1. Proposed samples

Material	Surface treatment	Code	Qty. of samples [-]	Dimension of sample [mm]
glassfiber	without	GO	4	**T1.8** - 25x100
	mechanically roughened**	GZ	4	**T1.8** - 25x100
carbon fiber laminate smooth side	without	KHO	4	**T2.5** - 25x100
	mechanically roughened**	KHZ	4	**T2.5** - 25x100
carbon fiber laminate wrinkled side	without	KDO	4	**T2.5** - 25x100
	mechanically roughened**	KDZ	4	**T2.5** - 25x100
aluminium alloy EN AW 7075	without	HO	4	**T2** - 25x100
	abrasive blasted	HP	4	**T2** - 25x100
	anodized	HE	4	**T2** - 25x100
	cataphoretic painted*	HK	4	**T2** - 25x100
steel S355	without	OO	4	**T1.5** - 25x100
	mechanically roughened**	OZ	4	**T1.5** - 25x100
	abrasive blasted	OP	4	**T1.5** - 25x100
	cataphoretic painted*	OK	4	**T1.5** - 25x100

* with subsequent roughening with sanding foam
**with P80 sandpaper

3 Experimental testing

Experimental testing of samples was performed on the Zwick / Roell Z050 machine. It is a static materials testing machines with a maximum pulling force of 50 kN, equipped with extensometers and several types of jaws.

Lap-joint samples were clamped directly into the jaws of the machine. The difference between the bottom and the top of the sample was eliminated by moving the upper jaw axes relative to the lower jaw. Samples were clamped into jaws, always 25 mm in length on each side. The distance between the jaws was set to 125mm. The distance between the arms of the strain gauges was set to 80 mm.

Lap-joint samples were loaded by pulling. The testing happened quasi-statically. The speed of jaws´s shift was set to 13mm/min according to ASTM D5868 [26]. In this measurement, the loading force [N] and the deformation [mm] of the samples were subtracted using extensometers.

Fig. 3. Static materials testing machines - Zwick/Roell Z050

4 Evaluation of tests

There is evaluated mainly maximum force to failure of the joint recalculated to average shear strength. Respectivelly, comparison of average shear strength for different surfaces. The deformation is relevant here, and is significantly affected by the properties of the bonded material.

Fiberglass samples

Samples were made of pre-preg ("pre-impregnated" composite fibers) HexPly® M49/32%/600S8/GE-300. The composition of textile weave is of 8H satin type. The nominal area weight of this material is 882g/m^2. Tensile strength $X_T = 550\ MPa$. The laminate layout is formed by alternating 0 and 45 °, so that the resulting layout was symmetrical.

The geometric parameters of the base material and final samples can be seen in Fig. 1 and Table 1.

The samples from fiberglass were glued using adhesive Scotch-Weld DP490. Measured values from experimental testing can be seen in Fig. 4. Resulting average shear strength and its comparison for fiberglass can be seen in Table 2.

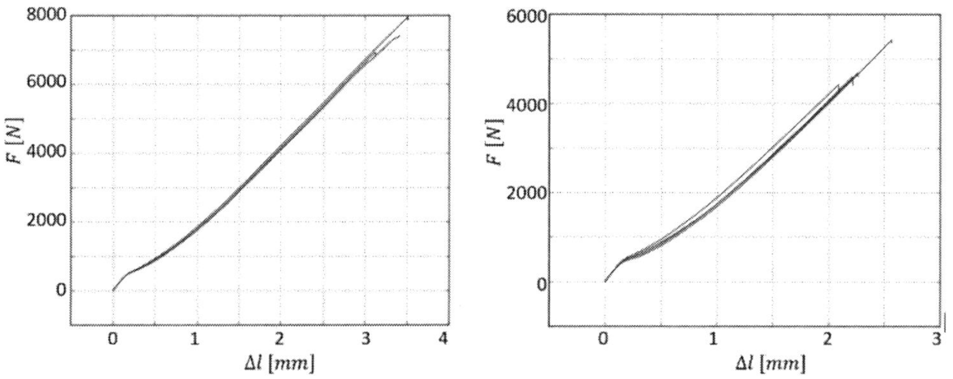

Fig. 4. Measured values from experimental testing for fiberglass
a) GO – without surface treatment, b) GZ – mechanically roughened

Table 2. Results and comparison for fiberglass

Material	**Surface treatment**	**Code**	**Average load**	**Average shear strength**
			F_A [N]	**τSA [MPa]**
glassfiber	without	GO	7 448	**11.9**
	mechanically roughened**	GZ	4 794	**7.7**

* with subsequent roughening with sanding foam
**with P80 sandpaper

With all the samples there occured a adhesion crack of an adhesive bond. Test specimen details after the test can be seen in Fig. 5.

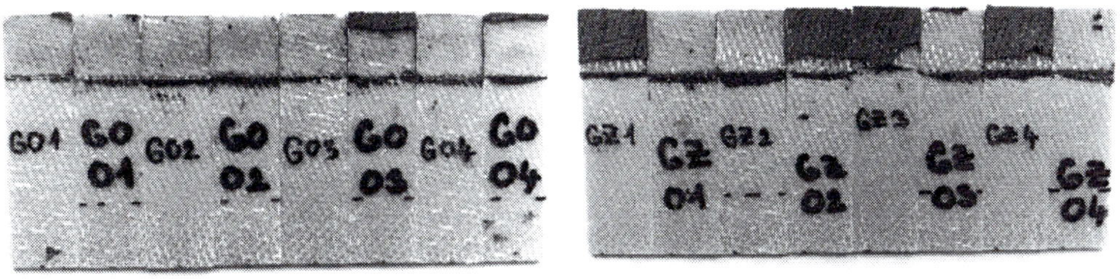

Fig. 5. Test specimen details from fiberglass after the test

Carbon fiber laminate samples

Samples were made of MTM57/CF3202VQ pre-preg. The composition of textile weave is of Twill 2x2 type. The nominal area weight of this material is 245g/m^2. Tensile strength $X_T = 709\ MPa$. The laminate layout is formed by alternating 0 and 45 °, so that the resulting layout was symmetrical.

The geometric parameters of the base material and final samples can be seen in Fig. 1 and Table 1.

The specimens from carbon fiber laminate were glued using adhesive Scotch-Weld 7260. With these samples both sides (the smooth one as well as the wrinkled one due to stripping foil) were tested. Measured values from experimental testing can be seen in Fig. 6. Resulting average shear strength and its comparison for carbon fiber laminate can be seen in Table 3.

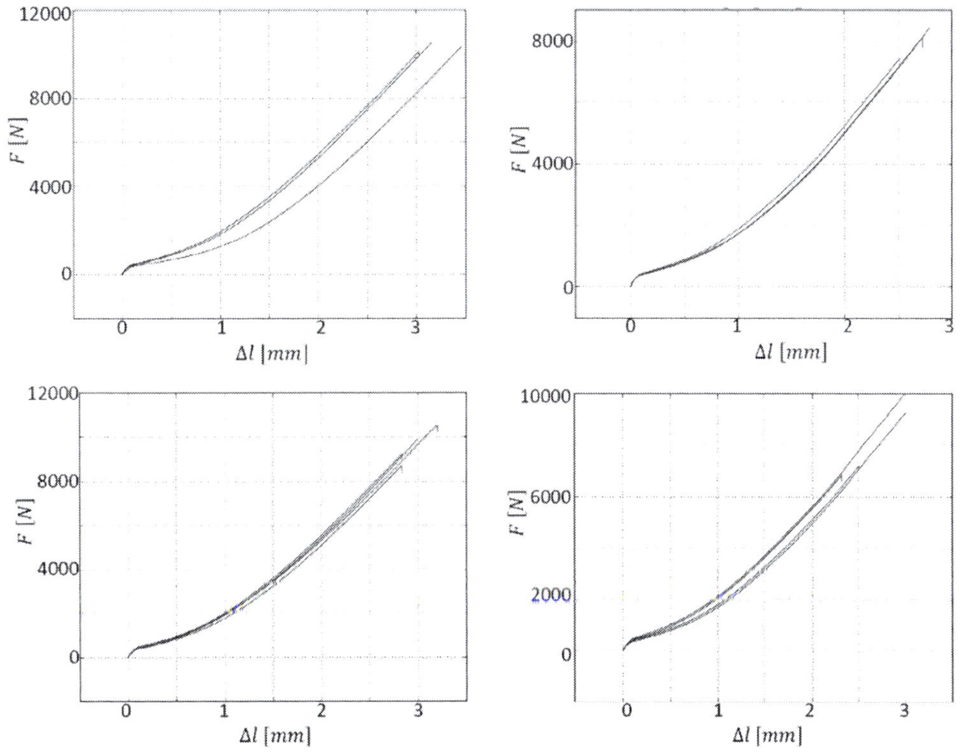

Fig. 6. Measured values from experimental testing for carbon fiber laminate
a) KHO – smooth side without surface treatment, b) KHZ – smooth side mechanically roughened, c) KDO – wrinkled side without surface treatment, d) KDZ – wrinkled side mechanically roughened

Table 3. Results and comparison for carbon fiber laminate

Material	Surface treatment	Code	Average load F_A [N]	Average shear strength τSA [MPa]
carbon fiber laminate smooth side	without	KHO	10 343	**16.5**
	mechanically roughened**	KHZ	8 013	**12.8**
carbon fiber laminate wrinkled side	without	KDO	9 602	**15.4**
	mechanically roughened**	KDZ	8 328	**13.3**

**with P80 sandpaper

With all the samples there occured a adhesion crack of an adhesive bond. Test specimen details after the test can be seen in Fig. 7. The rest of the glue with the teared fibers from carbon fiber laminate can be seen in Fig. 8.

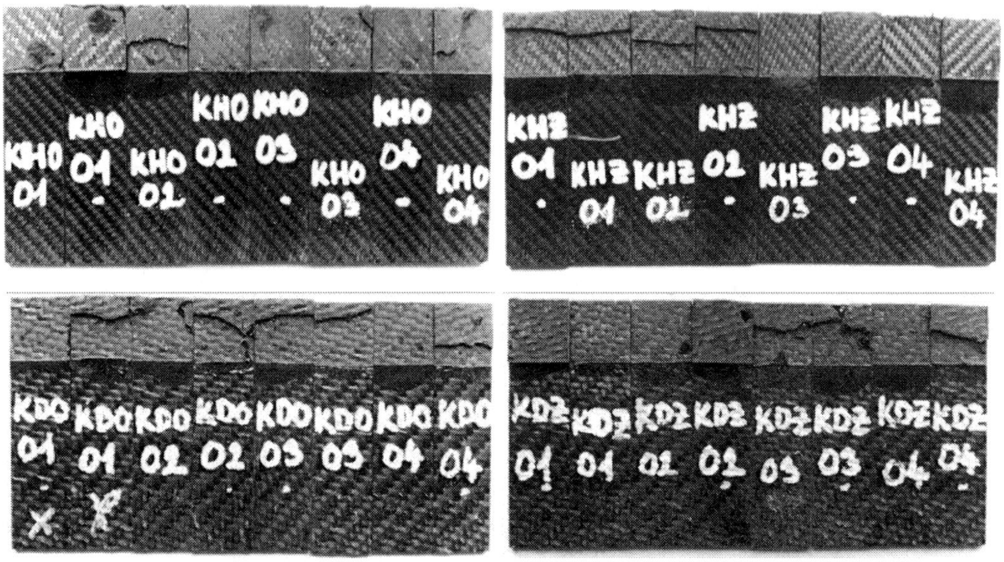

Fig. 7. Test specimen details from carbon fiber laminate after the test

Fig. 8. The rest of the glue with the teared fibers from carbon fiber laminate

Aluminium alloy samples
Specimens are made of sheet-metal aluminium alloy EN AW 7075 T6. Tensile strength is 530 MPa. The geometric parameters of the base material and final samples can be seen in Fig. 1 and Table 1.

The specimens from aluminium alloy were glued using adhesive Scotch-WeldDP490. Measured values from experimental testing can be seen in Fig. 9 and Fig. 10. Resulting average shear strength and its comparison for aluminium alloy can be seen in Table 4.

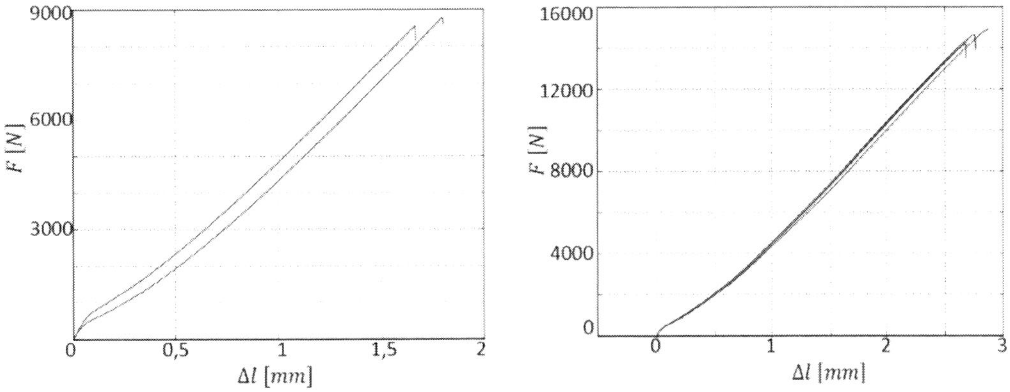

Fig. 9. Measured values from experimental testing for aluminium alloy
a) HO - without surface treatment, b) HP – abrasive blasted

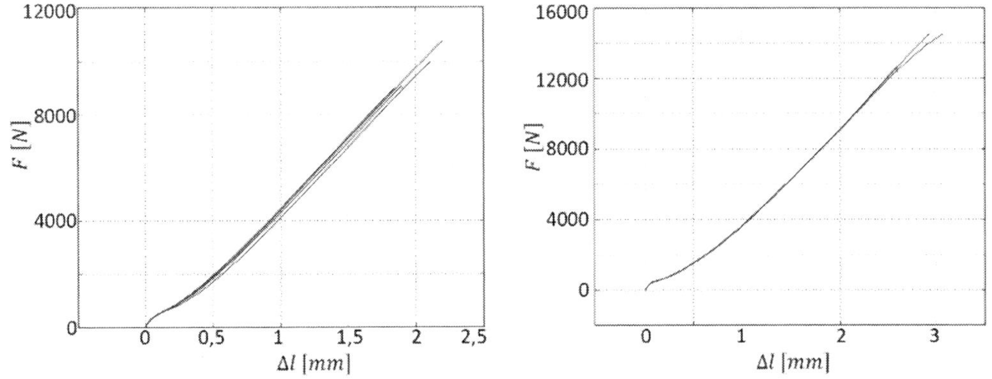

Fig. 10. Measured values from experimental testing for aluminium alloy
c) HE – anodized, d) HK – cataphoretical painted

Table 4. Results and comparison for aluminium alloy

Material	Surface treatment	Code	Average load	Average shear strength
			F_A [N]	τSA [MPa]
aluminium alloy EN AW 7075	without	HO	8 673	**13.9**
	abrasive blasted	HP	14 532	**23.3**
	anodized	HE	9 686	**15.5**
	cataphoretic painted*	HK	13 879	**22.2**

* with subsequent roughening with sanding foam

With all the samples there occured a adhesion crack of an adhesive bond. Test specimen details after the test can be seen in Fig. 11. The rest of the glue with the peeled anodized ply from aluminium alloy can be seen in Fig. 12.

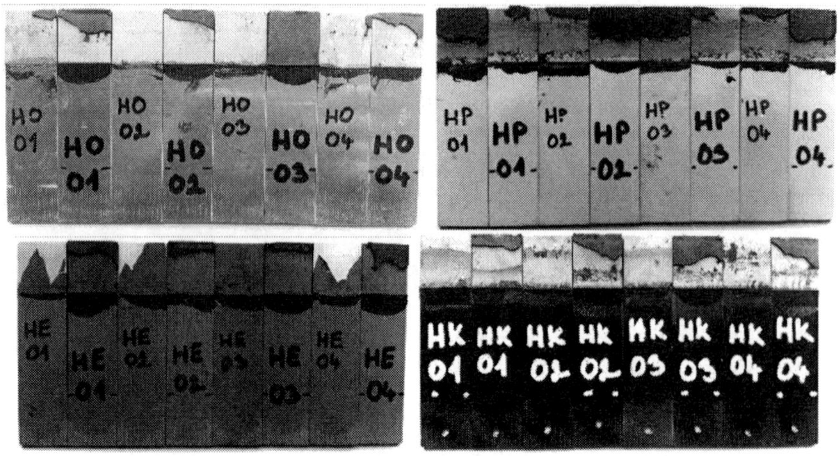

Fig. 11. Test specimen details from aluminium alloy after the test

Fig. 12. The rest of the glue with the peeled anodized ply from aluminium alloy

Steel samples

Specimens are made of steel S355 J0. Tensile strength is 550 MPa. The geometric parameters of the base material and final samples can be seen in Fig. 1. and Table 1.

The specimens from steel were glued using adhesive Scotch-Weld 7260. Measured values from experimental testing can be seen in Fig. 13. Resulting average shear strength and its comparison for steel can be seen in Table 5.

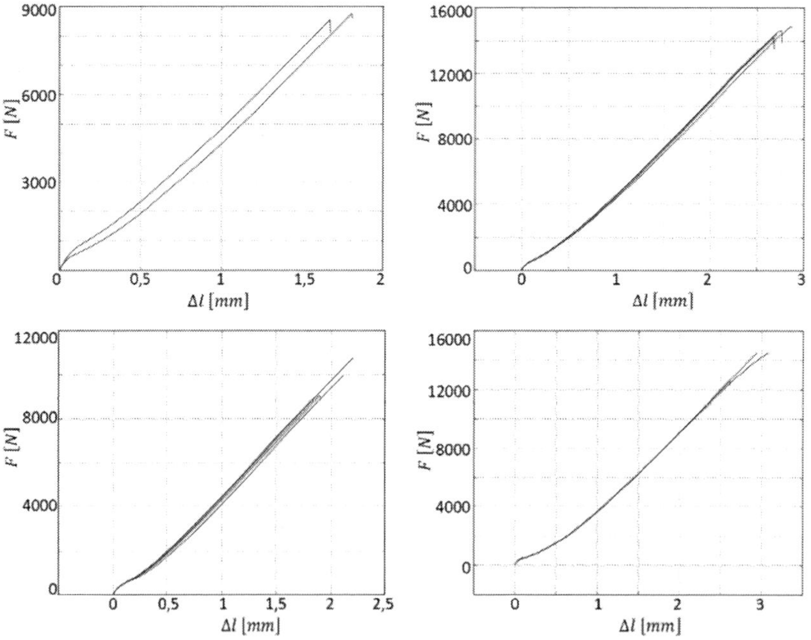

Fig. 13. Measured values from experimental testing for steel
a) OO – without surface treatment, b) OZ – mechanically roughened, c) OP – abrasive blasted, d) OK – cataphoretical painted

Table 5. Results and comparison for steel

Material	Surface treatment	Code	Average load	Average shear strength
			F_A [N]	τSA [MPa]
steel S355	without	OO	6 357	**10.2**
	mechanically roughened**	OZ	7 771	**12.4**
	abrasive blasted	OP	7 853	**12.6**
	cataphoretic painted*	OK	6 453	**10.3**

* with subsequent roughening with sanding foam
**with P80 sandpaper

With all the samples there occured a adhesion crack of an adhesive bond. Test specimen details after the test can be seen in Fig. 14.

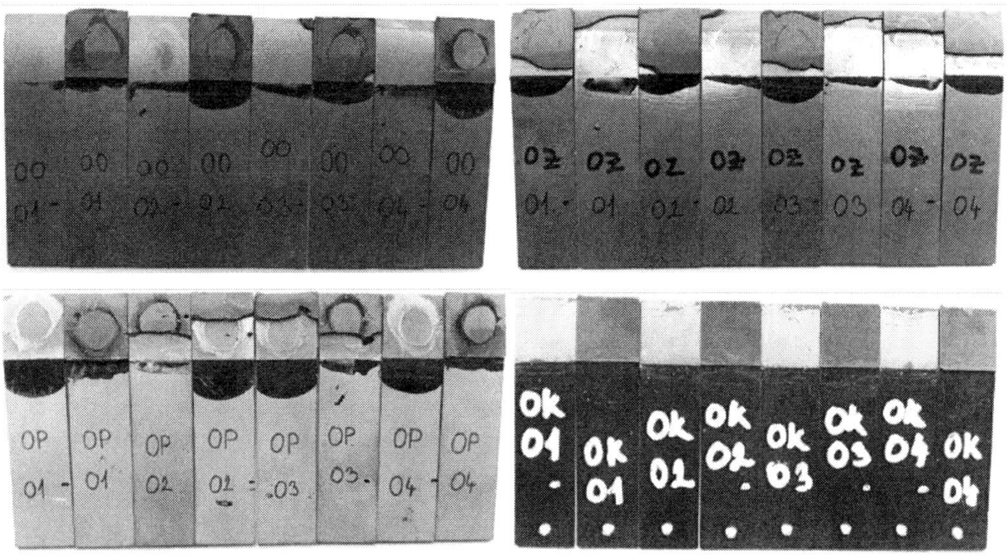

Fig. 14. Test specimen details from steel after the test

5 Summaries

There is evaluated mainly maximum force to failure of the joint recalculated to average shear strength. Respectively, comparison of average shear strength for different surfaces. The deformation is relevant here, and is significantly affected by the properties of the bonded material.

No purely cohesive damage occurred in any of the tests. This implies that the cohesive is very strong and that it is likely that the glue adhesion to the glued material will probably be the weak point of the glued joint using this adhesive.

Resulting average shear strength and its comparison for all tested material and for all tested surface treatments can be seen in Table 6.

In general, we would expect that by roughening the surface will increase adhesive strength. However, this may not always be true. Based on several studies (eg [4, 5, 6], etc.), it appears that for different materials there is a differently high value of optimum roughness to achieve the maximum strength of the bonded joint. When this limit is exceeded, the strength of the glued joint is again decreased. For example, here [5] is an optimum roughness for bonded steel samples ranging from Ra of 0.37 to 0.48 μm obtained by grain sanding paper 400. Or Budhe [4] provides an optimum surface roughness for the bonding of aluminum alloy specimens of Ra = 1.68 ± 0.14 μm.

Based on the above, we can justify why in our tests reach roughened samples of the composite materials of less strength than samples without mechanical roughening. For the roughening of samples from composite materials, the relatively thick P80 paper was used here. With this rough sanding paper, the matrix was significantly distorted to the fibers and this allowed the material to be easily pulled out of the material. We can see that the samples without surface treatment did not demage the matrix. However, samples without surface treatment achieve higher average shear strength.

Therefore, it can not be concluded from these results that the roughened surface of the laminate leads to a decrease in strength. To find the optimum roughness of the material, a large number of samples with a finely graded roughness needs to be produced. We should probably find the roughness of the surface of the laminate at which the strength of the glued joint would be higher than that of the cleaned samples. On the basis of the optimum roughness of other materials, it can be assumed that the laminate should achieve less roughness than that obtained on the current samples. Opinions about the optimum roughness of the laminate are different in the literature [1, 2, 4, 5, 6] but are most often recommended for P150-P400 grain papers.

In Table 6 is a comparison of the strength of the glued joint for various aluminum alloy surface treatments. The measured values show that the highest shear strength achieved sandblasted samples. This is a low-cost surface treatment, but the surface of the material is not protected from environmental influences. If it is desired to protect the bonded material against the influence of the surrounding area, it is more appropriate to use a cataphoretic painted material. It is also a relatively low-cost surface treatment, which also protects the part against corrosion and other environmental influences. The next one is an anodized sample that achieves a 30% lower strength than a cataphoretic painted samples but can be chosen for aesthetic reasons, for example. The lowest strength is achieved by the sample of only degreased aluminum without surface treatment.

The last of the tested materials is steel. Average shear strength of surface treatment of steel is lower than for samples of aluminium alloy. The highest strength is achieved by abrasive blastedand mechanically abraded samples. Almost identical strength reaches samples without surface treatment and cataphoretic painted samples.

The above results show that the cataphoretic painting has less adhesion to steel than to the aluminum alloys. For both metals, the paint layer was detached (i.e., adhesion of the adhesive to the lacquer is higher than the adhesion of the paint to the material.) However, the thickness of the steel sample varnish requires significantly less force than the aluminum alloy sample.

Table 6. Overall results of adhesion average shear strength

Material	Surface treatment	Code	Average load F_A [N]	Average shear strength τSA [MPa]
glassfiber	without	GO	7 448	**11.9**
	mechanically roughened**	GZ	4 794	**7.7**
carbon fiber laminate smooth side	without	KHO	10 343	**16.5**
	mechanically roughened**	KHZ	8 013	**12.8**
carbon fiber laminate wrinkled side	without	KDO	9 602	**15.4**
	mechanically roughened**	KDZ	8 328	**13.3**
aluminium alloy EN AW 7075	without	HO	8 673	**13.9**
	abrasive blasted	HP	14 532	**23.3**
	anodized	HE	9 686	**15.5**
	cataphoretic painted*	HK	13 879	**22.2**
steel S355	without	OO	6 357	**10.2**
	mechanically roughened**	OZ	7 771	**12.4**
	abrasive blasted	OP	7 853	**12.6**
	cataphoretic painted*	OK	6 453	**10.3**

* with subsequent roughening with sanding foam
**with P80 sandpaper

Conclusion

Experimental testing of the adhesive properties of the most used adhesives, materials and surface treatments was carried out. Two epoxy structural adhesives were chosen: the adhesive Scotch-Weld DP490 and Scotch-Weld 7260 (producer 3M). The materials used were steel S355 J0, aluminium alloy EN AW 7075 T6, carbon fibre laminate and glass fibre laminate.

For fiberglass and carbon fiber laminate, two surface treatments have been proposed: no surface treatment and the surface mechanically roughened with P80 sandpaper. Greatest adhesion strength of both bonded composite materials were achieved with samples without surface treatment. This adhesion strength achieves values: $\tau_{SAGO} = 11.9\ MPa$ for glassfiber, $\tau_{SAKHO} = 16.5\ MPa$ for smooth side of carbon fiber laminate and $\tau_{SAKDO} = 15.4\ MPa$ for wrinkled side of carbon fiber laminate.

For the steel and aluminum alloy, the following surface treatments have been proposed: without surface treatment (surface only degreased), abrasive blasted surface, anodized surface, cataphoretic painted surface with subsequent roughening with sanding foam. Greatest adhesion strength of adhesive bonds were achieved with abrasived blasted ($\tau_{SAHP} = 23.3\ MPa$) and cataphoretic painted ($\tau_{SAHK} = 22.2\ MPa$) samples from aluminium alloy EN AW 7075 T6. Greatest adhesion strength of adhesive bonds were achieved with abrasived blasted ($\tau_{SAOP} = 12.6$ MPa) and mechanically roughened ($\tau_{SAOZ} = 12.4$ MPa) samples from steel S355J0. Average shear strength of surface treatment of steel is lower than for samples of aluminium alloy. Cataphoretic painting can be done on both aluminium alloy and steel. But as can be seen from the results, the adhesion of the lacquer to the steel is considerably lower.

The result is an overview of the acquired values of shear strength for particular combinations of material and surface treatment.

Based on these values, we can choose the appropriate surface treatment to provide the transfer of required force. Or, on the contrary, on the basis of the selected surface treatment, we can calculate the minimum required area of the glued joint. However, it should be emphasized that this is a simplified calculation with a equally stress assumption because in

 https://doi.org/10.1051/matecconf/201815705012

reality the stress in the glued joint is not constant over its entire surface. It is therefore necessary to count with sufficient safety.

The article has been prepared under project LO1502 'Development of the Regional Technological Institute' under the auspices of the National Sustainability Programme I of the Ministry of Education of the Czech Republic aimed at supporting research, experimental development and innovation.

References

1. Collective of authors, *Loctite - Worlwide Design Handbook*. Mainz: Erasmusdruck GmbH. 452. ISBN 0-9645590-0-5 (1998)
2. Grundlagen, Verarbeitung, Anwendungen, *Faserverbundwerkstoffe Handbuch*. Waldenbuch: RG Faserverbundwerkstoffe GmbH. **237**. (2009)
3. ASTM International. *Standard Test Method for Lap Shear Adhesion for Fiber Reinforced Plastic (FRP) Bonding*. Designation: ASTM D5868-01
4. S. Budhe, A. Ghumatkar, N. Birajdar, M. D. Banea, *Effect of surface roughness using different adherend materials on the adhesive bond strength*. Applied Adhesion Science 2015, DOI: 10.1186/s40563-015-0050-4. (2015)
5. M. Muller, *Vliv drsnosti ocelového povrchu na pevnost lepeného spoje*. MM Průmyslové spektrum, Article code: 030908. (2003)
6. S. W. Lee, D. Lee, *Static and Dynamic Torque Characteristic of Composite Cocured Single Lap Joint*. Journal of Composite Materials **31**, 2188-2201, (1997)

Low-cycle fatigue behaviour of laser welded high-strength steel DOMEX 700 MC

Peter Kopas[1,*], *Milan* Sága[1], *František* Nový[2], *Bohuš* Leitner[3]

[1]University of Zilina, Faculty of Mechanical Engineering, Department of Applied Mechanics, Univerzitná 1, 010 26, Zilina, Slovak Republic
[2]University of Zilina, Faculty of Mechanical Engineering, Department of Material Engineering, Univerzitná 1, 010 26, Zilina, Slovak Republic
[3]University of Zilina, Faculty of Mechanical Engineering, Department of Technical Sciences and Informatics, Univerzitná 1, 010 26, Zilina, Slovak Republic

Abstract. The article presents the results of research on low cycle fatigue strength of laser welded joints vs. non-welded material of high-strength steel DOMEX 700 MC. The tests were performed under load controlled using the total strain amplitude ε_{ac}. The operating principle of the special electro-mechanic fatigue testing equipment with a suitable clamping system was working on 35 Hz frequency. Fatigue life analysis was conducted based on the Manson-Coffin-Basquin equation, which made it possible to determine fatigue parameters. Studies have shown differences in the fatigue life of original specimens and laser welded joints analysed, where laser welded joints showed lower fatigue resistance. In this article a numerical analysis of stresses generated in bending fatigue specimens has been performed employing the commercially available FEM-program ADINA.

Keywords: low-cycle fatigue, laser, welding, high-strength steel

1 Introduction

Static or quasistatic loading is rarely observed in modern engineering practice, making it essential to the engineer to implications of repeated loads, fluctuating loads and rapidly applied loads. By far, the majority of engineering design projects involves machine parts subjected to fluctuating of cyclic loads. Such loading includes fluctuating or cyclic stresses that often result in failure by fatigue. The most difficult aspect of fatigue is to detect the progressive changes in material properties that occur during cyclic stressing and the failure may therefore occur with no apparent warning. Also, periods of rest, with the fatigue stress removed, do not lead to any measurable healing or recovery from the effects of the prior cyclic stressing. Thus the damage done during the fatigue process is cumulative and generally unrecoverable [1]. Fatigue failure is an extremely complex physical process which is governed by a great number of parameters related to, for example, local geometry and material properties of the structural region surrounding the crack growth path. It is commonly recognized that it is impossible for a physical model to account for all fatigue

[*] Corresponding author: peter.kopas@fstroj.uniza.sk

Reviewers: *Mariana* Pajtášová, *Ján* Vavro

influencing parameters, thus a lot of approximate models have been conceived for practical fatigue assessments. In every stadium of fatigue cumulative damage dominates a definite mechanism controlled by more or less known and verified rules [2, 3]. There exists a stage of micro-plastic process in the total volume of material with a following stage of fatigue crack nucleation and stage of their growing with a more or less detailed zoning. Despite of this research no results have been achieved, which could be considered as successful ones.

Fatigue failure investigations over the years have led to the observation that the fatigue process actually embraces two domains of cyclic stressing or straining that are significantly different in character, and in each of which failure is probably produced by different physical mechanisms. One domain of cyclic loading is that for which plastic strain occurs during each cycle. This domain is associated with high loads and short lives or low-cycle fatigue. The other domain of cyclic loading is that for which the strain cycles are largely confined to elastic range. This domain is associated with lower loads and long lives, or high cycle fatigue. Low-cycle fatigue is typically associated with cycle lives from one to about 10^4 or 10^5 cycles, and high-cycle fatigue for lives greater than about 10^4 or 10^5 cycles [1, 4]. In recent years, it has been recognized that the fatigue failure process involves three phases. A crack initiation phase occurs first, followed by crack propagation phase and finally, when the crack reaches a critical size, the final phase of unstable rapid crack growth to fracture completes the failure process. The modelling of these phases has been under intense scrutiny, but the models have not yet been developed in a coordinated way to provide widely accepted engineering design tools. The basic premise of the local stress-strain approach is that the local fatigue response of the material at critical point, that is, the site of crack initiation is analogous to the fatigue response of a small, smooth specimen subjected to the same cyclic strains and stresses. The cyclic stress-strain response of the critical material may be determined from the characterizing smooth specimen through appropriate laboratory testing [5, 6].

To properly perform such laboratory tests, the local cyclic stress-strain history at the critical point in the structure must be determined, either by analytical or experimental means. Thus valid stress analysis procedures, finite element modelling or experimental strain measurements are necessary, and the ability to properly account for plastic behaviour must be included. In performing smooth specimen tests of this type, it must be recognized that the phenomena of cyclic hardening, cyclic softening and cycle-dependent stress relaxation, as well as sequential loading effects and residual stress effects that may be experienced by the specimen as it accumulates fatigue damage are presumed to be the same as at the critical point of the structure member being simulated. Since, including all these factors in a test is inconvenient, inaccurate and expensive, the use of finite element method has become a powerful tool to calculate the cyclic stress-strain response of any structure or mechanical component. The finite element method is especially used in the ground vehicle industry where discontinuities of the geometry such as notches and holes produce difficulties to calculate the local cyclic stresses and strains, which are essential to predict the fatigue life of any structure or component [7, 8].

The strength analysis of welded structures does not deviate much from that for other types of structures. Various failure mechanisms have to be avoided through appropriate design, choice of material, and structural dimensions. Design criteria such as yielding, buckling, creep, corrosion, and fatigue must be carefully checked for specific loading conditions and environments. It is, however, the fact that welded joints are particularly vulnerable to fatigue damage when subjected to repetitive loading. Fatigue cracks may initiate and grow in the vicinity of the welds during service life even if the dynamic stresses are modest and well below the yield limit. The problem becomes very pronounced if the structure is optimized by the choice of high strength steel. The very reason for this choice is to allow for higher stresses and reduced dimensions, taking benefits of the high strength material with respect to the yield criterion [9]. However, the fatigue strength of a welded joint is not primarily governed by the strength of the base material of the joining members; the governing parameters are mainly the global and local geometry of the joint. Hence, the yield stress is increased, but the fatigue strength does not improve significantly. As already stated, the fatigue behaviour of welded joints is random by nature. Very few load-bearing details exhibit such large scatter in fatigue life as welded joints. This is true even in controlled laboratory conditions. As a consequence, it becomes an important issue to take scatter into consideration, both for the fatigue process and for the final life. Furthermore, the in-services stresses may often be characterized as stochastic processes [10, 11]. However, a welded joint has some peculiar features that make some of the subjects and parameters play a much more important role than others [10, 12]. Let us start with an S-N curve for a welded joint and compare it with other specimens. Fig. 1 shows the S-N curves based on tests with one smooth plate, one plate with a bore hole, and one plate with a fillet welded transverse attachment.

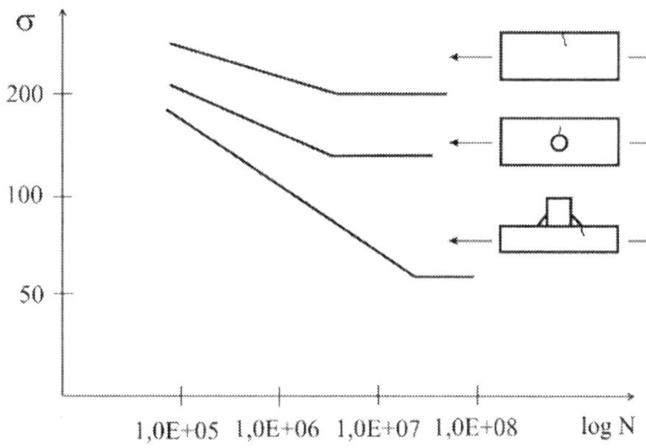

Fig. 1. Fatigue life curves for various details [11]

This type of fillet welded joint is often referred to as non-load-carrying due to the fact that the load is carried straight through the base plate. It is also noted that the applied force is transverse to the welding direction [3, 10]. In the plate, the fatigue cracks can appear anywhere in the longitudinal edges of the plate, whereas they will appear at the inner edges of the bore hole transverse to the applied loading in the specimen with the hole. The latter phenomenon is due to the stress concentration at the edge of the hole. In the fillet welded joint the cracks will appear at the weld toe and grow through the base plate in a direction perpendicular to the applied principal stresses.

The difference between the S-N curve for the specimen with the hole and the S-N curve for the welded joint is rather surprising considering the fact that the bore hole creates a stress concentration factor close to 3 at the edge of the hole. The relatively short fatigue life of welded detail is explained, in the main, by three factors:

- severe notch effect due to the attachment and the weld filler metal,
- presence of non-metallic intrusions or micro-flaws along the fusion line,
- presence of large tensile residual stresses.

2 Test equipment design

The design of experimental equipment is based on a mechanical principle. The constant rotation is generated by excenter and linkage mechanism. By changing of eccentric magnitude it is possible to change a loading magnitude. Also if we change the length of connecting crank on the experimental equipment, there will be a change in the loading cycle character (proportional-non-proportional loading, bending/torsion loading, etc.) [13, 14]. Power of the device is secured by two synchronic electro motors with frequency converters from 0.5 Hz to 100 Hz. Loading frequencies are identical with the frequency of rotation drive. Synchronization of the electro motors is secured by electronics and allows synchronization of the loading amplitude. The synchronization of electro motors also allows setting the phase shift for individual loading levels. There are also two force measurement systems included in the experimental equipment. These systems may be used for measurement of force values during the loading process. For evaluation of fatigue curves it is necessary to know stress and strain conditions on individual loading levels [12, 15, 16].

3 Experimental material and strain-life data results

Twenty smooth specimens were tested under strain controlled conditions in order to identify the strain-life behaviour of the base and laser welded experimental material. After machining, the specimen surfaces were mechanically polished. The experiments were carried out in an electro mechanic fatigue test machine, developed on University of Žilina.

For evaluation of fatigue curves it needs to know stress and strain conditions on individual loading levels. A sinusoidal waveform was used as command signal. The fatigue tests were conducted with constant strain amplitudes, at room temperature, in air. The specimens were cyclic loaded under strain control with symmetrical proportional bending loading, with a nominal strain ratio, $R_\varepsilon = -1$. The computational fatigue tests were performed under cyclic loading with the zero mean value. Frequency of each analysis was equal to 35 Hz.

Domex cold forming steels are thermo-mechanically rolled in modern plants where the heating, rolling and cooling processes are carefully controlled. The chemical analysis, consisting of low levels of carbon and manganese has precise addition of grain refiners such as niobium, titanium or vanadium. This together with a clean structure, makes Domex Steels the most competitive alternative for cold formed and welded products. The typical mechanical and chemical composition of our experimental material - high strength streel is shown in Table 1 to Table 3. The material used in this research was delivered in the form of a cylindrical shape with a diameter 10 mm. The length of cylindrical bars was 150 mm.

Table 1. Chemical composition of high strength steel DOMEX 700 MC (weight %)

C	Si	Mn	Ni	P	S	Cr	Mo
0.06	0.04	1.18	0.04	0.013	0.002	0.03	0.1
V	**N**	**Ti**	**Cu**	**Al**	**Nb**	B	C_{ekv}
0.01	0.004	0.1	0.01	0.039	0.06	0.0001	0.41

Table 2. Mechanical properties of high strength steel DOMEX 700 MC

Ultimate tensile strength	849 MPa
Tensile yield strength	790 MPa
Elongation at fracture	17 %

Table 3. Mechanical properties of laser welded high strength steel DOMEX 700 MC

Ultimate tensile strength	810 MPa
Tensile yield strength	622 MPa
Elongation at fracture	7 %

From experimentally measured values of number of cycles to failure was created uniaxial fatigue curve, which is shown in Fig. 2. For fatigue test interpretation on each loading levels it is necessary to know the plastic strain amplitude (Manson-Coffin curve) or stress amplitude (Wöhler curve) applied on each cycle loading level. For that it is necessary to analyze the stress and strain maximum values by FEM [11-13].

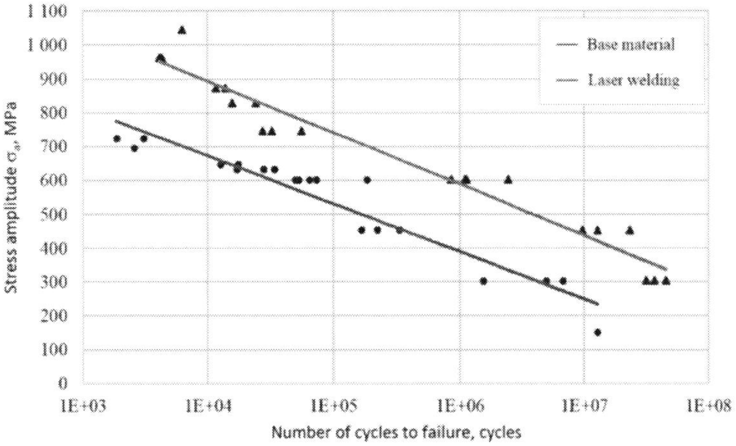

Fig. 2. S-N curves for high strength steel DOMEX 700 MC

The specimen model was created by the finite-element program ADINA. The material was assumed plastic-bilinear; the true stresses were obtained from a real stress-strain graph. The tetrahedron linear element type was automatically generated. The "Load Plot" function was defined by excenter setting with excentricity of 1 mm. At the fix point shell and beam elements were used for hammer simulating. From the computational analysis can be seen that the area with the greatest concentration of stresses or the place with the higher deformation was localized in the middle of the rod radius, see Fig 3.

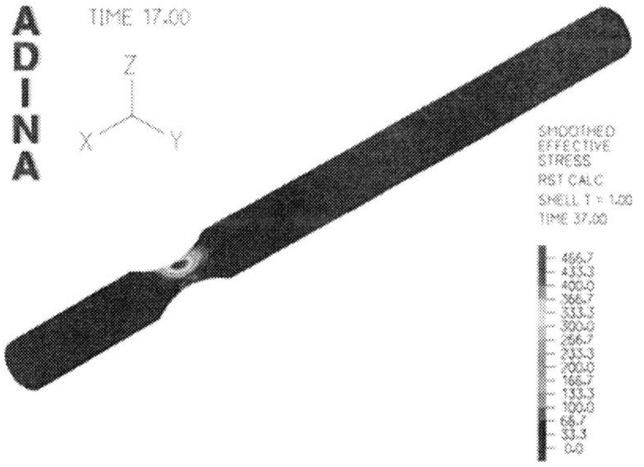

Fig. 3. Result of FEM analysis

Conclusion

In this study, the low-cycle fatigue behaviour of high-strength steel Domex 700 MC smooth bars from the base and laser welding joint condition was experimentally evaluated. Tests were performed under strain controlled conditions bending loading on machined bar specimens by means of a simple loading apparatus that allowed to reach high strain levels. Generally we can say that the results are in good agreement with the results published by other authors [6, 15]. On the basis of the tests performed, the following conclusions were formulated: The fatigue resistance of both experimental materials increases with decreasing stress amplitude continuously in the cycles of number region. From the comparison of the low cycle fatigue data of the base material and the laser welded joint stat able, that in the case of identical total and plastic strain amplitudes, the welded joint can be tolerate smaller number of cycles until the failure, then the base material. The ratio between the cycle to failure of base material and the welded joint s is decreasing with the increasing of the total and plastic strain amplitudes. Residual stress is an important factor influencing the structural behaviour in all instability failures as well as in fatigue crack initiation and propagation when cyclic service stresses are superposed onto the residual stresses. The presence of residual stresses in engineering components and structures can significantly affect the fatigue behaviour during external cyclic loading. The effect of residual stresses may either be beneficial or detrimental, depending on magnitude, sign and distribution of the stresses with respect to the load-induced stresses. Residual stresses in tension are detrimental and are often in the magnitude of the materials yield strength. The tensile residual stresses will reduce the fatigue life of the structure by increasing the growth of the fatigue crack, while compressive residual stresses will decrease fatigue crack growth rate. The existence of tensile residual stresses in a surface layer accelerates crack initiation reducing fatigue life due to the increase of local mean stress.

This work was supported by the Slovak Research and Development Agency under the contract No. APVV-14-0096.

References

1. O. Bokůvka, G. Nicoletto, L. Kunz, P. Palček, M. Chalupová, *Low & high frequency fatigue testing*. (CETRA and Univerzity of Žilina, 2002)
2. G. Rosenberg, Ľ. Juhar, *Fatigue resistance of dual phase steels in presence of microstructural inhomogeneities.* Manufacturing Technology **12** (13), 217-221 (2012)
3. F. Trebuňa, M. Buršák, *Medzné stavy, lomy*. (Grafotlač, Prešov 2002)
4. P.J.E. Forsyth, *The Physical Basis of Metal Fatigue*. (Blackie and Son, London, 1969)
5. P. Kovacikova, R. Bezdedova, J. Vavro Jr., J. Vavro, *Comparison of numerical analysis of stress-strain states of cast iron with vermicular graphite shape and globular graphite shape*. 20 th International conference machine modeling and simulations MMS 2015, Procedia Engineering **136**, 28-32 (2016)
6. J. Lago, M. Guagliano, F. Nový, O. Bokůvka, *Influence of laser shock peening surface treatment on fatigue endurance of welded joints from S355 structural steel*. Manufacturing technology, **16** (1), 154-159 (2013)
7. R. Grega, J. Krajnak, L. Zul'ova, G. Fedorko, V. Molnar, *Failure analysis of driveshaft of truck body caused by vibrations*. Engineering Failure Analysis **79**, 208-215 (2017)
8. L. Jakubovičová, B. Ftorek, V. Baniari, A. Sapietová, T. Potoček, M. Vaško, *Engineering design of a test device*, XXI Polish-Slovak Scientific conference machine modeling and simulations MMS 2016, Procedia Engineering **177**, 520-525 (2017)
9. W. Piekarska, M. Kubiak, M. Žmindák, *Issues in numerical modeling of phase transformations in welded joint.* XXI Polish-Slovak Scientific conference machine modeling and simulations MMS 2016, Procedia Engineering **177**, 141-148 (2017)
10. S. J. Maddox, 1991. *Fatigue strength of welded structures*. (Second Edition, Woodhead Publishing, UK, 1991)

11. H. J. Gough, H. V. Pollard, 1935. *The strength of metals under combined alternating stresses*. Proc. Inst. Mech. Engrs **131**, 93-103 (1935)
12. P. Kopas, M. Blatnický, M. Sága, M. Vaško, *Identification of Mechanical Properties of Weld Joints of AlMgSi07.F25 Aluminium Alloy*. Metalurgija **56** (1-2), 99-102 (2017)
13. J. Kortis, L. Daniel, M. Handrik, *The Numerical Analysis of the Joint of the Steel Beam to the Timber Girder*. XXIII R-S-P Seminar, Theoretical Foundation of Civil Engineering TFOCE, Procedia Engineering **91**, 160-164 (2014)
14. M. Kubiak, T. Domanski, V. Dekys, A. Sapietova, *Measurement of strain during tension test of welded joint using multi-camera 3D correlation system*. XXI Polish-Slovak Scientific conference machine modeling and simulations MMS 2016, Procedia Engineering **177**, 107-113 (2017)
15. L. Susmel, H. Askes, *Modified Wöhler Curve Method and multiaxial fatigue assessment of thin welded joints*. International Journal of Fatigue **43**, 30-42 (2012)
16. J. Xiong, P. Guo, Y. Cai, B. Strnadel, J. Brumek, Y. He, H. Gu, *Structural, magnetic and nanomechanical properties in Ni-doped AlN films*. Journal of Alloys and Compounds Volume **606**, 55-60 (2014)

Tensile and bending test of carbon/epoxy and carbon/geopolymer composites after temperature conditioning

Jan Krystek[1,*], *Vladislav* Laš[1], *Vilém* Pompe[1], *Pavlína* Hájková[2]

[1] European Centre of Excellence, NTIS – New Technologies for Information Society, University of West Bohemia, Univerzitní 8, 301 00 Plzeň, Czech Republic
[2] Unipetrol Centre for Research and Education a.s., Revoluční 1521/84, 400 01 Ústí nad Labem, Czech Republic

Abstract. Composites with epoxy matrix cannot be used in high temperature, while geopolymer matrix excel in high temperature resistance. First, prismatic specimens were subjected to conditioning temperature. Second, the tensile and bending test were performed at room temperature. This paper present comparison of mechanical properties of carbon/epoxy and carbon/geopolymer composites. Numerical simulations of tensile and bendidng tests were performed in finite element system Abaqus.

Keywords: bending test; composite; geopolymer; numerical simulation; tensile test.

1 Introduction

Currently, composite materials are applied in many industrial areas. Usually, composites are made with carbon or glass fibers and epoxy matrix. These composites have very good mechanical properties, for example low weight, high strength and high stiffness, but they cannot be used in high temperature environments. Mechanical properties of these composites significantly degrade with increasing temperature. This shortcoming can be removed using a geopolymer matrix. Geopolymer matrix is an inorganic polymer material. Preparation of this material is based on aluminosilicate alkali activation. Geopolymers excel in many properties. Generally, the papers present high temperature resistance [1 – 5] or resistance against acids and organic solvent agents. Any paper, which present specific mechanical properties of carbon/geopolymer composites, was not found.

This paper presents analysis of mechanical properties of carbon/epoxy and carbon/geopolymer composites. Composites with geopolymer matrix were subjected to high temperature. Two types of geopolymer matrix were used. Tensile and bending test were performed for identification of mechanical properties. Numerical simulations of these tests were performed in finite element system Abaqus.

* Corresponding author: krystek@kme.zcu.cz

Reviewers: *Justín* Murín, *Milan* Žmindák

2 Materials and specimens

The composites plates were made from 10 layers of plain weave carbon fabric (Table 1). The prismatic specimens were cut using diamond blade. Epoxy matrix L285 and hardener 285 MGS was used for specimens with label CE1. Two types of geopolymer matrix were used in this work. Geopolymer matrix FC4 consists of potassium water glass, potassium hydroxide (KOH), silica fume, constituent with high content of metakaolinite, and boric acid. Molar ratios of the components are presented in Table 2. Geopolymer matrix B3P1 consists of potassium water glass, constituent with high content of metakaolinite, and ingredients with calcium. Constituent with metakaolonite Mefisto L_{05} (produced by České lupkové zavody, a.s.) was used for both types of geopolymer matrix.

Table 1. Properties of carbon fabric

Material of fibers	Binding	Area weight	Thickness	Density
Toray 3K 200 tex	plain	200 [g/m^2]	0.32 [mm]	1 760 [kg/m^3]

Table 2. Molar ratios

	alkali activator modulus (SiO_2/M_2O)	Si:Al	M:Al	Ca:Al	H_2O:Al
FC4	1.08	17.07	4.35	-	24.62
B3P1	1.56	1.80	1.00	0.12	6.01

The specimens with geopolymer matrix were subjected to conditioning temperature of 23 °C, 200 °C, 400 °C or 600 °C, specimens with epoxy matrix only of 23 °C. Temperature resistance of the used epoxy matrix is lower than 200 °C. Afterwards, the tensile or bending test were performed using universal testing machine *Zwick/Roell Z050* at room temperature. Designation of specimens is presented in Fig. 1.

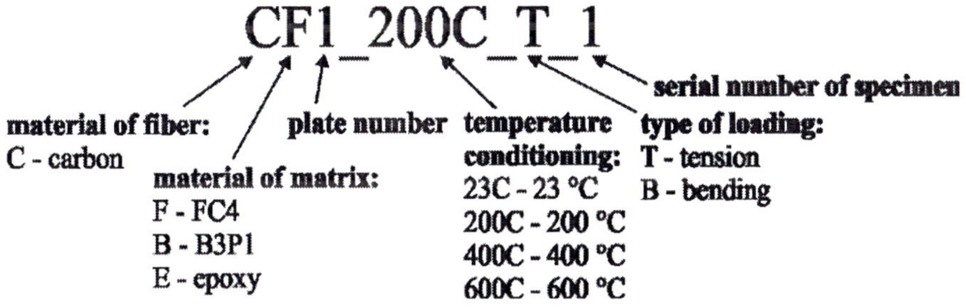

Fig. 1. Designation of specimens

The geometric properties of all specimens with epoxy matrix were: width $W_e = 25$ mm, thickness $H_e = 2.4$ mm, total length $L_e = 180$ mm. The geometric properties of specimens with geopolymer matrix were: width $W_g = 25 \div 26$ mm, thickness $H_g = 2.9 \div 3.6$ mm, total length $L_g = 150$ mm. The exact dimensions for each specimen are given in [5].

3 Tensile test

The force–displacement (F–Δl) dependencies were obtained from tensile test complying with ASTM D 3039. The specimen size was modified according to possibilities resulting from the plate size. An initial grip distance was $l_j = 100$ mm. An extensometer was used for measuring the elongation. Gage length was $l_e = 60$ mm. The load velocity (crosshead displacement) was $v_T = 2$ mm/min.

The stress-strain dependencies were calculated using

$$\sigma = \frac{F}{W \cdot H}, \qquad \varepsilon = \frac{\Delta l}{l_e}, \tag{1}$$

where F is loading force, W is width of specimen, H is thickness of specimen, Δl is elongation and l_e is initial gage length. The effective modulus was identified according to the standard *ASTM D 3039* on interval of strain $\varepsilon \in \langle 0.1\%, 0.3\% \rangle$. The stress–strain dependencies of CB4 and CF1 specimens for all temperature conditioning are shown in Fig. 2 and Fig. 3, for CE1 specimens at 23 °C in Fig. 4. The values of maximum stress and effective modulus are presented in Table 3 and Table 4.

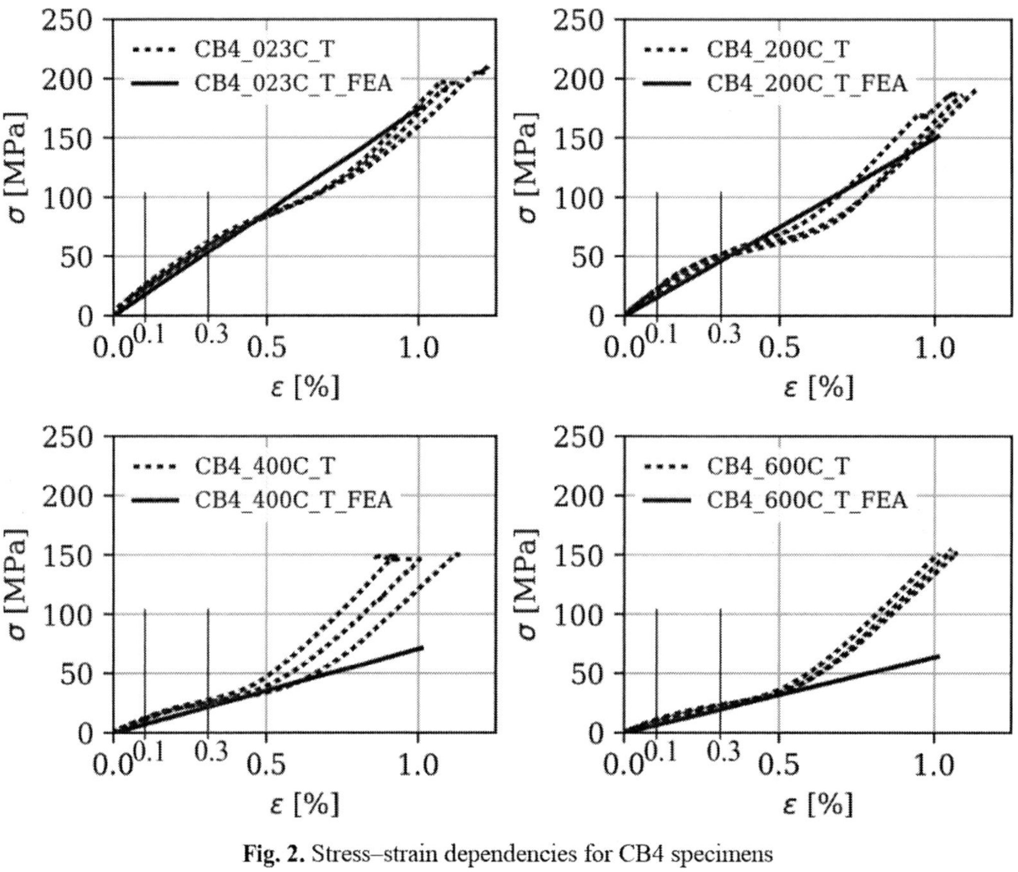

Fig. 2. Stress–strain dependencies for CB4 specimens

Table 3. Maximum stress, effective modulus and averaged modulus for CE1 specimens

specimen	σ_{max} [MPa]	E [GPa]	$\bar{E}$ [GPa]
CE1_23C_T_1	537.9	53.5	
CE1_23C_T_2	581.6	54.7	53.67
CE1_23C_T_3	510.2	52.8	

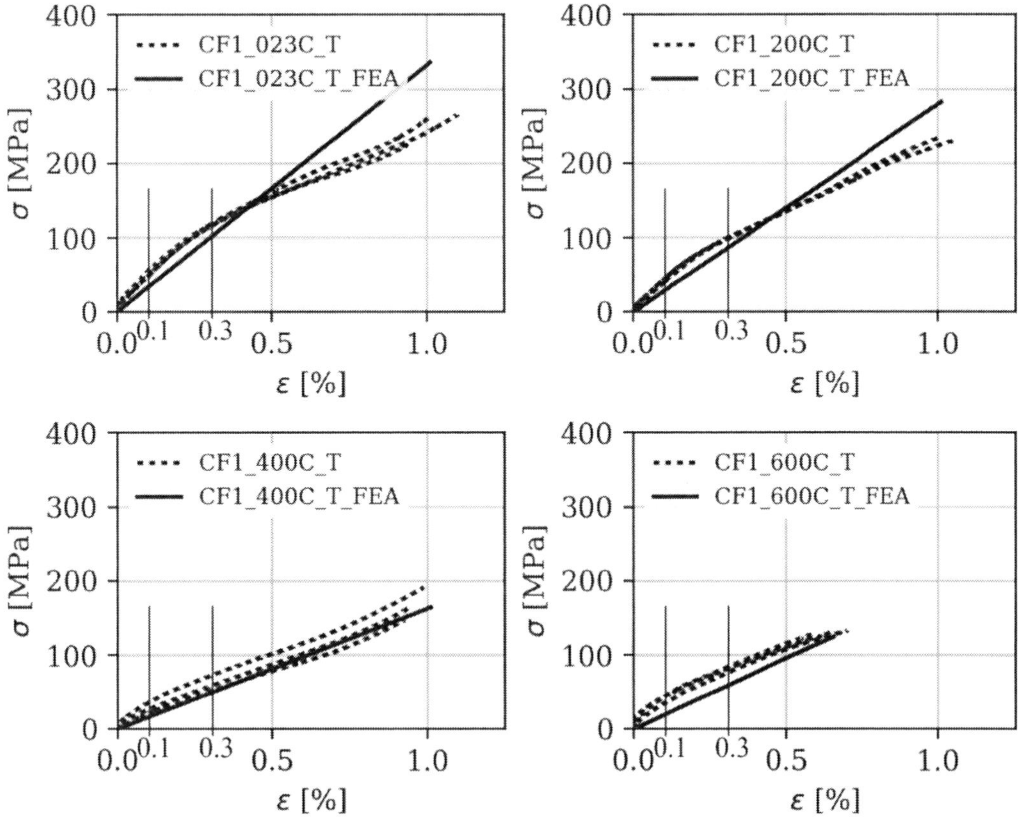

Fig. 3. Stress–strain dependencies for CF1 specimens

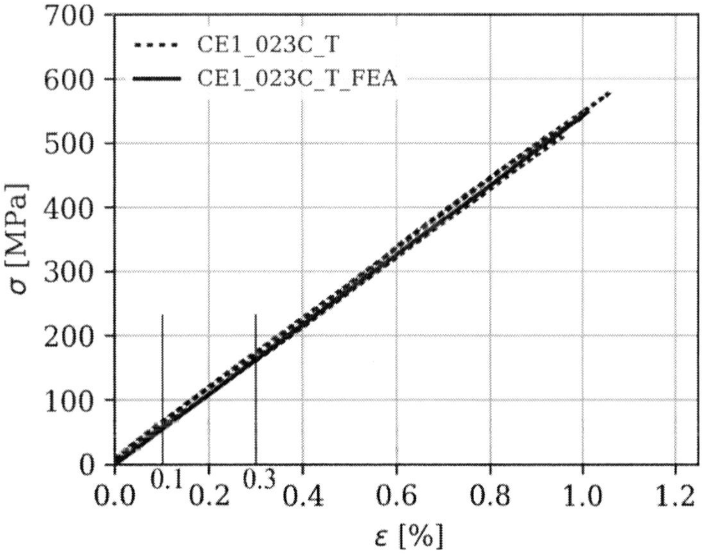

Fig. 4. Stress–strain dependencies for CE1 specimens

The stress–strain dependencies are linear for specimens with epoxy matrix and they are nonlinear for specimens with geopolymer matrix. The effective modulus in tension for CB4 specimens with B3P1 geopolymer matrix is approximately half that of the CF1 specimen with FC4 geopolymer matrix. The effective modulus in tension of CE1 specimen with epoxy matrix is larger by 60% that of CF1 specimen for 23 °C temperature conditioning.

Table 4. Maximum stress, effective modulus and averaged modulus CB4 and CF1 specimens

specimen	σ_{max} [MPa]	E [GPa]	$\overline{E}$ [GPa]	specimen	σ_{max} [MPa]	E [GPa]	$\overline{E}$ [GPa]
CB4_23C_T_1	211.2	17.3		CF1_23C_T_1	260.1	31.8	
CB4_23C_T_2	198.6	18.1	17.50	CF1_23C_T_2	225.1	33.8	33.27
CB4_23C_T_3	197.4	17.1		CF1_23C_T_3	265.5	34.2	
CB4_200C_T_1	190.6	15.4		CF1_200C_T_1	235.5	27.0	
CB4_200C_T_2	186.6	14.6	15.00	CF1_200C_T_2	229.5	29.5	28.03
CB4_200C_T_3	189.4	15.0		CF1_200C_T_3	216.7	27.6	
CB4_400C_T_1	149.4	6.6		CF1_400C_T_1	190.9	17.9	
CB4_400C_T_2	151.7	7.5	7.13	CF1_400C_T_2	146.6	15.3	16.30
CB4_400C_T_3	150.9	7.3		CF1_400C_T_3	165.5	15.7	
CB4_600C_T_1	151.4	6.4		CF1_600C_T_1	132.6	20.0	
CB4_600C_T_2	154.3	6.1	6.40	CF1_600C_T_2	127.2	19.2	19.20
CB4_600C_T_3	151.1	6.7		CF1_600C_T_3	130.2	18.4	

3.1 Numerical simulation of the tensile test

The finite element system *Abaqus* was used for the numerical simulation of the tensile test. Quadratic hexahedral elements with 20 nodes were used in a parametrically created model (Fig. 5). The loading was controlled by the displacement of the upper crosshead. One numerical model was created for each group of specimens (material of matrix + temperature conditioning). This model had averaged geometric parameters. Isotropic material model with average effective modulus (Table 2) was used in numerical analysis. The stress–strain dependencies obtained using finite element analysis (FEA) are presented in Fig. 2 – Fig. 4.

Fig. 5. Finite element model – tension (left) and bending (right)

4 Bending test

The force–displacement (F–Δl) dependencies were obtained from 3-point bending test complying with ASTM D 7264. The support span was $l_s = 80$ mm. An extensometer was used for measuring the displacement of loading nose (deflection of specimen). The load velocity (crosshead displacement) was $v_B = 2$ mm/min.

4.1 Numerical simulation of the bending test

The finite element system Abaqus was used for the numerical simulation of the bending test. Quadratic hexahedral elements with 20 nodes were used in a parametrically created model One numerical model was created for each group of specimens (material of matrix + temperature conditioning). This model had averaged geometric parameters. Isotropic material model with average effective modulus (Table 2) was used in numerical analysis. The force–displacement dependencies obtained using finite element model are presented in Fig. 6 - Fig. 8.

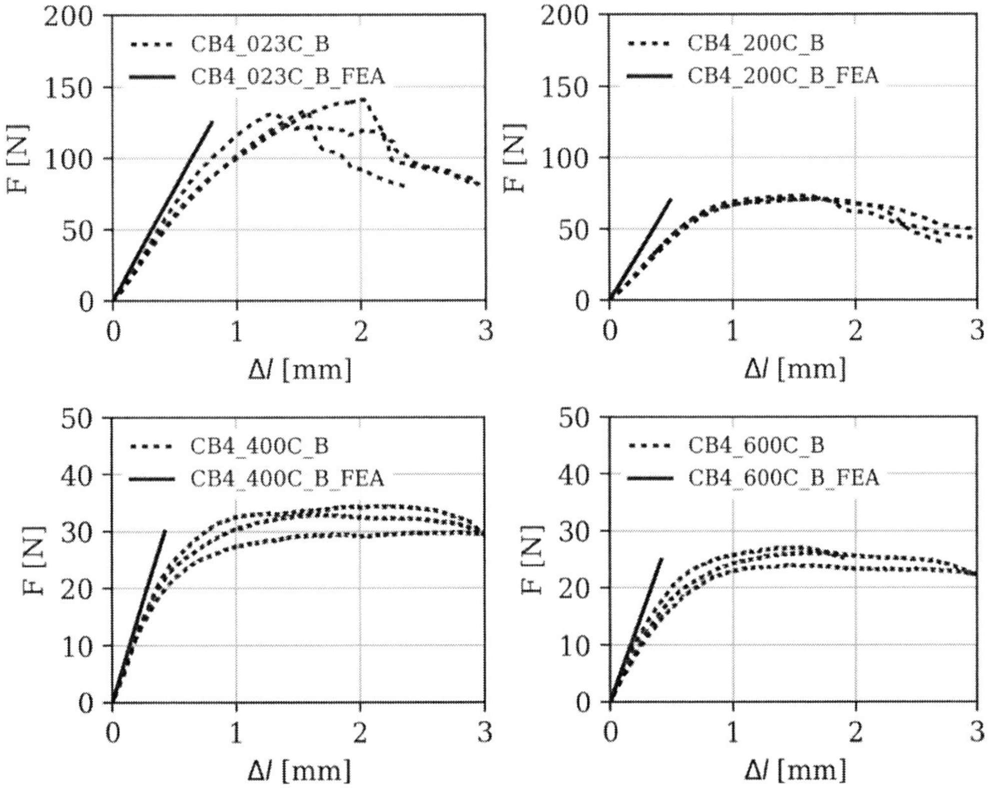

Fig. 6. Force–displacement dependencies for CB4 specimens

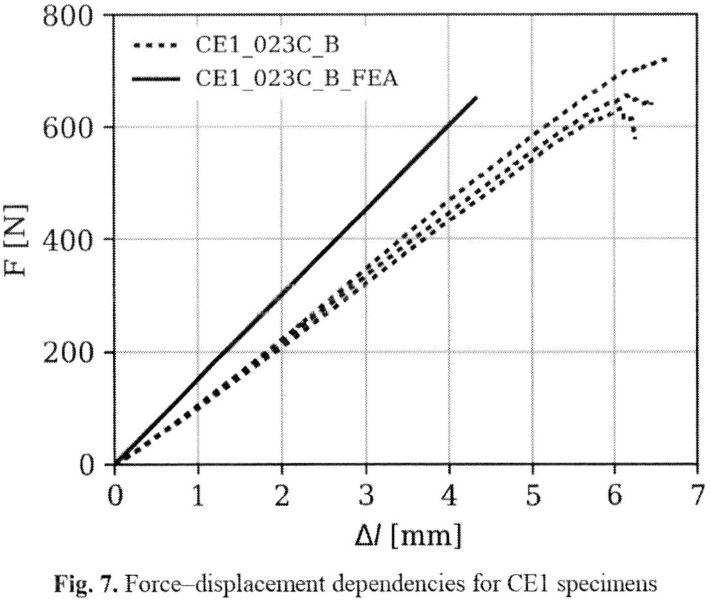

Fig. 7. Force–displacement dependencies for CE1 specimens

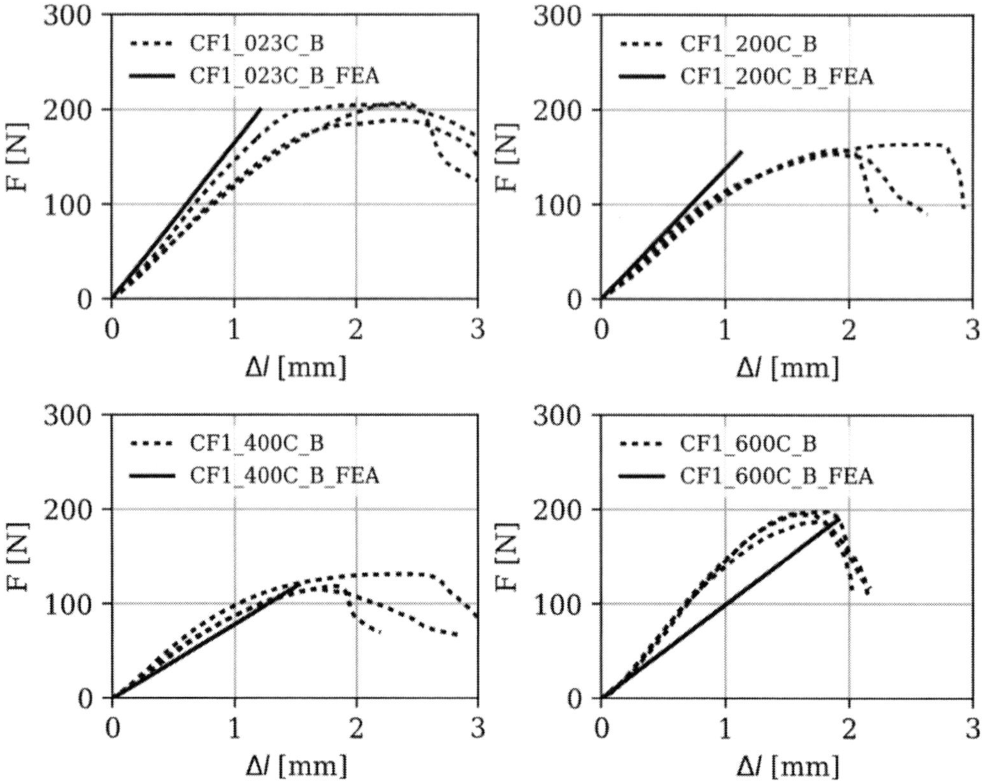

Fig. 8. Force–displacement dependencies for CF1 specimens.

Conclusion

The composite specimens were composed of carbon fibers and geopolymer or epoxy matrix. Two types of geopolymer matrix were used. The specimens were subjected to conditioning temperature at 23 °C, 200 °C, 400 °C or 600 °C. Afterwards the tensile or bending tests were performed at room temperature. The numerical simulation of tensile and bending tests were performed in finite element system Abaqus.

The stress-strain dependencies were calculated for tensile test. The stress-strain depedencies are linear for specimens with epoxy matrix and they are nonlinear for specimens with geopolymer matrix. The effective modulus in tension for CB4 specimens with B3P1 geopolymer matrix is approximately half that of the CF1 specimen with FC4 geopolymer matrix. The effective modulus in tension of CE1 specimen with epoxy matrix is larger by 60% that of CF1 specimen for 23 °C temperature conditioning.

In case of bending test, the force-displacement dependencies were obtained from 3-point bending test. The effective modulus in tension was used in numerical simulation of bending tests. These numerical models showed higher stiffness than the experiment.

The multilayer composites show different value of the effective elasticity modulus for tension and bending. Therefore, in the 3D solid finite element model, the effective elasticity for bending has to be used in the 3 point bending model.

This publication was supported by the project LO1506 of the Czech Ministry of Education, Youth and Sports. The publication is a result of the project Development of the UniCRE Centre (project code LO1606) which was financially supported by the Ministry of Education, Youth and Sports of the Czech Republic under the National Programme for Sustainability I. The authors would like to thank Robert Zemčík for helpful comments and proofreading of the manuscript.

References

1. J. Mills-Brown, K. Potter, S. Foster, T. Batho, *Thermal and tensile properties of polysialate composites*. Ceramics International **39**, 8917-8924 (2013)
2. D. Pernica, P. Reis, J. Ferreira, P. Lauda, *Effect of test condition on bending strength of geopolymer-reinforced composite*. J Mater Sci **45**, 744-749 (2010)
3. S. Samal, B. Marvalová, I. Petrikova, K. Vallons, S. Lomov, H. Rahier, *Impact and post impact behavior of fabric reinforced geopolymer composite*. Construction and Building Materials **127**, 11-124 (2016)
4. I. Petríková, B. Marvalová, A. Hrouda, L. Paur, *Properties of Composite with Geopolymer Matrix Reinforced by Basalt Fabric*. Experimental Stress Analysis 2017 Conference Proceeding CD-ROM [Full Text], ISBN 978-80-553-3166-9, 543-548 (2017)
5. J. Krystek, V. Laš, V. Pompe, P. Hájková, *Influence of Temperature on Selected Mechanical Properties of Geopolymer Composites*. Experimental Stress Analysis 2017 Conference Proceeding CD-ROM [Full Text], ISBN 978-80-553-3166-9, 382-387 (2017)

Development of methods for performing the maximum voluntary contraction (MVC) test

Mateusz Kukla[1*], *Bartosz* Wieczorek[1], *Łukasz* Warguła[1]

[1]Poznań University of Technology, Chair of Basics of Machine Design, 3 Piotrowo St., 60-965 Poznań, Poland

Abstract. The aim of the study is to determine the most effective isometric exercises required in order to perform MVC (Maximum Voluntary Contraction) tests under given static muscle contraction conditions. To this end, a group of four muscles was defined and on this basis, a series of exercises was selected. During electromyography studies, the electrical activity of the muscles was recorded. The article presents the results obtained for all exercises in the whole group of subjects. The data obtained provide a basis for evaluating and selecting the method of loading the analysed muscles during MVC tests. This is necessary in order to perform a normalization of raw (unprocessed) EMG signal, thus enabling comparison of results between individual subjects. Analyses concerning this kind of diagnosis of the electrical activity of muscles are widely applied in research related to rehabilitation and medicine, sports science, but also ergonomics.

Keywords: surface electromyography (EMG), maximum voluntary contraction (MVC) test, normalization

1 Introduction

Kinesiological electromiography, also called surface electromiography, is an experimental technique for examining the electrical activity of muscles. It involves detecting differences in the potentials of electromagnetic fields generated by the body during muscle contraction. The electrical voltage value determined in this way depends, among other things, on the load that is being overcome. This method of analysis allows to define the activity of selected muscle groups over time during performance of specific exercises. The character of the measurement, however, does not allow to determine the force generated by a given muscle or a number of muscle fibres that become activated. Also, it is impossible to make a straightforward comparison between the values measured in individual measurement sites or for individual persons. This is due to constantly changing physico-chemical conditions in the human body. Despite these limitations surface electromiography is frequently used in research related to rehabilitation and medicine. The methods described are also applied in the engineering practice. Due to their non-invasive character they are in common use in studies within the field of ergonomics, biomechanics or design of rehabilitation equipment

* Corresponding author: E-mail address: mateusz.kukla@put.poznan.pl , Tel.: +48 61 224 45 14
Reviewers: *Gregorz* Domek, *Marcin* Kubiak

such as wheelchairs [1, 2]. They are also widely used for determining muscle strength during performance of various physical activities [3, 4].

The works under this project pertain to the biomechanics of manual wheelchair propulsion. Their outcome will be a basis for the design of new types of manual and hybrid wheelchair drive systems. To achieve this objective a number of secondary tasks were defined, including tasks related to the analysis of muscular activity during wheelchair propulsion. Depending on what users they are intended for, individual types of wheelchair drive systems differ with the source of driving energy and with design. The surface electromiography methods enable comparative analysis of the physical effort required to propel different types of wheelchairs.

2 Methods

The study was conducted based on the SENIAM project [5] methodology and information provided in the "The ABC of EMG" manual [6]. Signals were recorded using Noraxon Myotrace 400 apparatus and analysed with MyoResearch XP MT400 4CH Clinical Edition software. The research was conducted on a group of nine subjects characterized in Table 1.

The research methodology involved an analysis of the upper extremity muscles participating in propelling the wheelchair by means of pushrims. Measurements were made on the following muscles: deltoid muscle – anterior and posterior fibres, triceps brachii and extensor carpi radialis longus. After determining the electrode placement sites (Figure 1), the skin surface was prepared by degreasing and hair removal. The purpose of the preparation was to improve the transfer and electrical measurements conditions, thus improving the impedance conditions. In the next step 20 mm diameter round electrodes were attached in the area of the centres of the bellies of selected muscles at the distance of approx. 20 mm from each other. An example of the layout of electrodes on one of the subjects is represented in Figure 1. Next the impedance test was performed and if the result was positive (a value below 15 kΩ was obtained) the research proper was started.

Table 1. Study subjects' characteristics

No.	Sex	BMI [kg/m^2]	Height [cm]
I	F	19.8	162
II	M	26.6	169
III	M	22.3	182
IV	F	20.4	173
V	M	26.4	173
VI	F	20.7	166
VII	M	21.7	173
VIII	F	22.4	169
IX	M	35.0	169

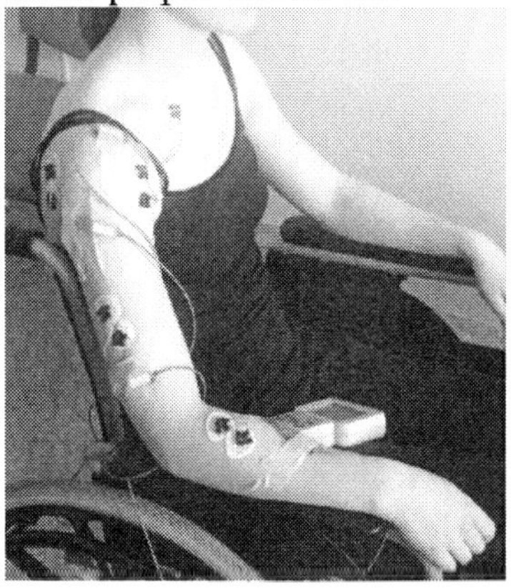

Fig. 1. An example of electrode layout

The study's target group is the disabled. For this reason, a number of important conditions must be taken into consideration when planning future studies of a similar type. To normalize the results popular weight training exercises were chosen and modified in such a way that they could be performed by a subject seated in a wheelchair. The advantage of this choice is the simplicity of the technique of individual exercises and availability of a wide body of practical materials referring to them. Where possible, exercises involving both arms at the same time were chosen. The aim was to attempt to minimize the risk of compensation of the loaded extremity with muscles in other parts of the body, for instance, through shoulder balancing or incorrect position of the back. The measurement taking cycle included the following subsequent stages: zeroing of the measurement at the starting position, start of measurement, start of movement – light load on the patient's muscles during performance of the exercise, end of movement – full load on the patient's muscles at the exercise's terminal position, retaining the state of full load over 3 to 5 seconds, release of the load. Next the cycle of the exercise was repeated 3 to 5 times including the intervals, and the measurement was ended. The tests were conducted under the static contraction conditions, with manual loading of individual groups of muscles. Intermediate elements, for instance an element simulating a barbell, with possibly the lowest weight were used at the process. The list of exercises is shown in Table 2. Schematic representation of individual exercises is shown in Figure 2.

Table 2. The list and numbering of individual exercises performed by the subjects

No.	Muscle	Type of exercise
1.1	**Deltoid muscle anterior (A) and posterior (B)**	External rotation of arm
1.2		Internal rotation of arm
1.3		Barbell chin raise
1.4		Barbell raise in front of the head
1.5		Barbell raise behind the head
1.6		Barbell front raise
1.7		Lateral arm raise
2.1	**Triceps brachii (C)**	Winding under load
2.2		Wrist bend with barbell, overhand grip
2.3		Wrist bend with barbell, underhand grip
2.4		Wrist rotation
3.1	**Extensor carpi radialis longus (D)**	French press with both hands
3.2		Forearm extension
3.3		Barbell forearm extension, overhand grip
3.4		One arm French press

The obtained EMG traces were smoothed, and in the next step their RMS (Root Mean Square) values were determined. To normalize the results processed in this way it was necessary to determine a reference value. An arithmetic mean ($\bar{x}$) of the amplitude of the highest signal portion with 1000 ms fixed duration was defined as this value's estimator. From the values determined by this method the highest one was selected and it was subsequently used for the EMG signal normalization. The standard deviation of the arithmetic mean ($\sigma_{\bar{x}}$) was adopted as the searched value error. For more detailed analysis of the recorded EMG traces, apart from the mean value, the median (Me) was also determined as a parameter less sensitive to extreme values of the chosen portion of the amplitude. Normalization of the amplitude performed in the above-described manner does not change the nature of the EMG signal trace. It changes the Y-axis scaling representing the signal as a percentage of the maximum measured value. The MVC test is one of the most commonly used methods of the EMG signal normalization [7-10]. It forms a basis for the comparison and evaluation of individual study results. In this manner a total of 198 mean amplitude values were determined (22 amplitude values from individual exercises for the analysed muscles for each of the 9 patients).

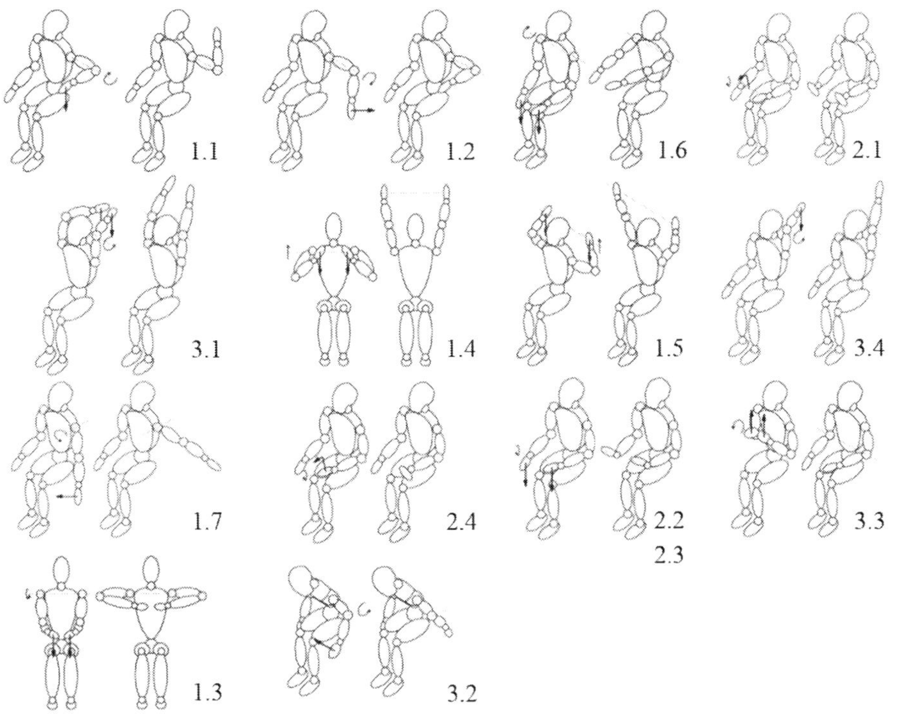

Fig. 2. Schematic representation of exercises performed by the subjects

Results

The research results were processed using the earlier-described methodology. The examples of the EMG traces measured and windows used are shown in Figure 3. Due to a large body of data collected only values for selected exercises and subjects are presented in a tabular form as Table 3. A graphic representation of the values determined is shown in Figure 4.

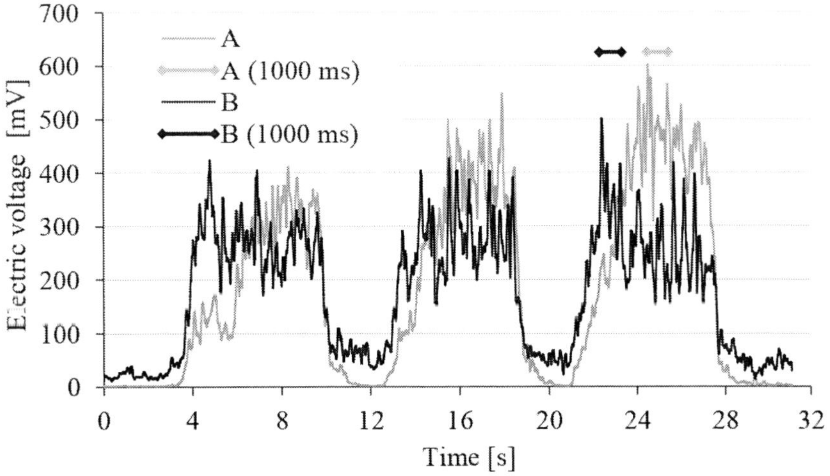

Fig. 3. Examples of measurement results for patient IV with selected 1000 ms signal windows marked, exercise: Lateral arm rise (1.7), deltoid muscle – anterior part (A) and posterior part (B)

Table 3. Compilation of values determined for selected exercises and subjects

Exercise		1.2		1.3		1.4		1.7	
No.	**Muscle**	**A**	**B**	**A**	**B**	**A**	**B**	**A**	**B**
I	$\bar{x}$	23.6	293.0	16.2	219.7	49.5	142.1	60.7	145.0
	$\sigma_{\bar{x}}$	4.7	53.7	4.1	31.4	6.5	41.4	11.2	20.2
	Me	24.4	292.9	15.9	216.6	48.7	126.9	59.1	144.1
III	$\bar{x}$	141.1	473.8	265.4	290.5	273.5	38.2	365.6	270.4
	$\sigma_{\bar{x}}$	17.1	121.6	45.5	39.5	39.2	3.7	41.5	73.7
	Me	143.5	459.8	265.6	285.5	272.0	38.1	365.5	283.5
VIII	$\bar{x}$	206.5	94.7	94.1	61.8	183.1	34.6	428.5	30.7
	$\sigma_{\bar{x}}$	44.0	18.8	16.1	12.5	16.3	5.2	67.0	5.7
	Me	197.3	97.9	96.9	60.7	184.1	34.6	435.1	30.9
IX	$\bar{x}$	109.9	24.6	130.6	47.1	210.4	16.2	246.8	25.4
	$\sigma_{\bar{x}}$	15.8	4.9	16.0	6.8	20.7	2.2	39.3	4.7
	Me	111.4	23.4	126.2	46.3	216.1	16.5	249.6	24.8

Exercise		2.1	2.2	2.4	3.1	3.2
No.	**Muscle**	**C**	**C**	**C**	**D**	**D**
I	$\bar{x}$	272.7	270.4	99.5	25.1	34.8
	$\sigma_{\bar{x}}$	29.7	27.1	9.9	3.6	4.3
	Me	273.6	267.9	99.5	24.3	34.1
III	$\bar{x}$	307.0	333.0	96.1	236.9	260.5
	$\sigma_{\bar{x}}$	68.2	39.9	13.0	46.9	46.2
	Me	293.1	327.6	96.2	224.6	245.5
VIII	$\bar{x}$	171.8	213.8	44.3	157.7	274.0
	$\sigma_{\bar{x}}$	40.2	20.3	7.4	23.8	45.7
	Me	164.6	214.9	43.6	157.5	270.7
IX	$\bar{x}$	66.5	80.4	36.1	14.3	20.0
	$\sigma_{\bar{x}}$	6.8	42.4	10.6	2.1	3.8
	Me	65.3	64.6	36.4	14.1	19.7

Analysis of the results obtained reveals big differences between mean amplitudes both for individual exercises and for individual study subjects. The standard deviation ranged from 10 to 30% of the mean value. These high values result both from the nature of the measurements and from the method used to determine the mean amplitude involving use of a time window of constant width of 1000 ms. Additionally, a number of considerable deviations from a typical range of standard deviation values were observed. These included for instance: $\sigma_{\bar{x}}$ = 49.8% (patient VI, exercise 1.2, muscle A), $\sigma_{\bar{x}}$ = 48.9% (patient V, exercise 1.4, muscle B) and $\sigma_{\bar{x}}$ = 65.5% (patient V, exercise 2.3, muscle C). The median and mean amplitude were similar for a great majority of the subjects.

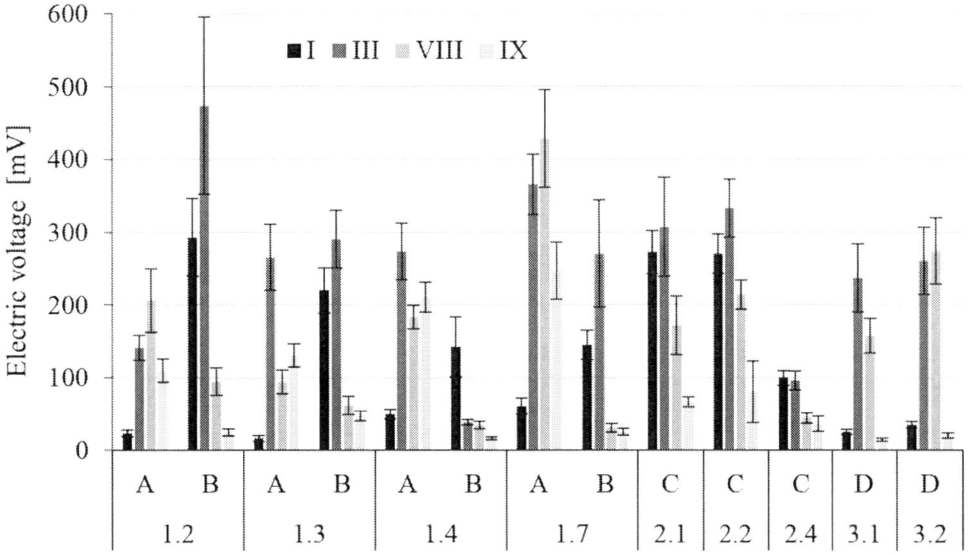

Fig. 4. Graphic representation of values obtained for selected exercises and subjects

Analysis of the data obtained revealed that big differences between these values occurred in cases of big measurement errors, that is, when considerable standard deviations were obtained. If the contamination of the EMG signal by potential artifacts was avoided or eliminated in a proper way, this can indicate non-static load application by the study subjects. Measurement results obtained in this manner cannot be used in further analyses. Thus, it is necessary to eliminate incorrectly determined measurement points or to repeat the test allowing to determine them again.

There are a number of limitations regarding the body mass index. This index cannot be used for people suffering from disease and children. Also attempts of classifications regarding sportsmen and women (for instance athletes and culturists) based on the BMI can lead to erroneous interpretations. No study subject can be classified as belonging to one of the abovementioned groups. Having this in mind, a relationship between the BMI value and the determined amplitudes can be observed. Usually, the greater the body mass index the lower the values of the calculated parameters. Human body is a relatively good conductor of electrical current. However, the conductance individual types of body tissues may be different and depends also on the thickness of a given layer. It is probably due to a greater thickness of the adipose tissue layer that the amplitudes obtained tend to be lower for persons with higher BMI values.

The analysis of all the results obtained revealed that exercises 2.1 and 2.2, chosen for the extensor carpi radialis longus, allow to obtain the most similar values of normalization amplitudes for all study subjects. Therefore, it can be stated that they are among the most effective resistance exercises necessary to conduct MVC tests under the given conditions of static contraction of muscles. In turn, the greatest spread of results was noted for exercise 2.3. From the technical point of view this exercise is performed correctly when the forearm is rotated in a way enabling grasping the intermediate element with underhand grip. In the process the belly of the investigated muscle moves under the electrodes. As a result small values, additionally characterized by a large spread, were obtained. Thus it can be concluded that this particular exercise is not useful for carrying out the normalization test. The greatest spread of the results for an individual muscle was observed during the tests regarding the triceps brachii. It was also observed that a large proportion of patients had problems with correct performance of exercises 3.1-3.4. It was concluded that the selected group of exercises for triceps brachii (D) should be significantly modified or new activities should be proposed.

Conclusions

Under this research project a number of MVC tests were performed for defined groups of muscles and exercises on a group of nine persons. For this purpose adequate procedures were developed, pertaining both to the preparation of patients for the tests and to the performance of the measurements. Suitable data processing methods as well as algorithms for determining the searched values were prepared. This formed a basis for an analysis of the parameters obtained, which allowed the authors to specify a group of the most effective resistance exercises required for the performance of the MVC tests under the set conditions of static muscle contraction. A group of exercises that should be rejected or modified was also indicated.

The study results and the experience gained from the investigation will allow the researchers in the future to increase the effectiveness of the works related to determining the activity of selected muscle groups during the performance of specific movements and exercises. This, in turn, will enable quality assessment of the biomechanics of propelling of manual wheelchairs of different types. The data obtained in the study form a basis for determining muscular activity occurring while propelling a wheelchair. The results will be compared and evaluated and will further serve as necessary input data for the design of innovative wheelchair drive systems to be used in vehicles intended for persons with physical disabilities.

The study was financed from the means of the National Centre for Research and Development under LIDER VII programme, research project no. LIDER/7/0025/L-7/15/NCBR/2016.

References

1. R. A. Cooper, D. P. VanSickle, R. N. Robertson, M. L. Boninger, G. J. Ensminger, *A Method for Analyzing Center of Pressure During Manual Wheelchair Propulsion*. IEEE Transactions on Rehabilitation Engineering **3** (3), 289-298 (1995)
2. M. L. Boninger, A. M. Koontz, S. A. Sisto, T. A. Dyson-Hudson, M. Chang, R. Price, R. A. Cooper, *Pushrim biomechanics and injury prevention in spinal cord injury: Recommendations based on CULP-SCI investigations*. Journal of Rehabilitation Research & Development **42** (3), 9-20 (2005)
3. E. Świtoński, A. Głowacka, *Określanie sił mięśniowych podczas chodu na podstawie sygnałów sEMG, Modelownie inżynierskie*. t. **16** (47), 185-189 (2013)
4. D. Winter, H. Yack, *EMG Profiles During Normal Human Walking: Stride-to-Stride and Inter-Subject Variability*. Electroencephalography and Clinical Neurophysiology **67**, 402-411 (1987)
5. Online at: http://www.seniam.org/
6. P. Konrad, *„ABC EMG" Praktyczne wprowadzenie do elektromiografii kinezjologicznej*. (Wyd. Technomex Sp. z o. o., ISBN 83-920818-1-1, Gliwice 2007)
7. B. Cestelein, B. Cagnie, T. Parlevliet, L. Danneels, A. Cools, *Optimal Normalization Tests for Muscle Activation of the Levator Scapulae, Pectoralis Minor, and Rhomboid Major: An Electromyography Study Using Maximum Voluntary Isometric Contractions*. Archives of Physical Medicine and Rehabilitation **96**, (10), 1820-1827 (2015)

8. H. Nieminen, E.-P. Taklal, E. Viikari-Juntura, *Normalization of electromyogram in the neck-shoulder region*. European Applied Journal of Physiology and Occupational Physiology **67** (3), 199-207 (1993)
9. R. A. Ekstrom, G. L. Soderberg, R. A. Donatelli, *Normalization procedures using maximum voluntary isometric contractions for the serratus anterior and trapezius muscles during surface EMG analysis*. Journal of Electromyography and Kinesiology **15** (4), 418-428 (2004)
10. M. Halaki, K. Ginn, *Normalization of EMG Signals: To Normalize or Not to Normalize and What to Normalize to?*, In: Computational Intelligence in Electromyography Analysis - A Perspective on Current Applications and Future Challenges, Chapter **7**. (Publisher: Rijeka, InTech, ISBN 978-953-51-0805-4, 2012)

Utility of an unitary-shredding method to evaluate the conditions and selection of constructional features during grinding

Marek Macko[1,*], *Krzysztof* Tyszczuk[1], *Grzegorz* Śmigielski[1], *Adam* Mroziński[2]

[1]Faculty of Mathematics, Physics & Technical Science, Kazimierz Wielki University in Bydgoszcz, Poland
[2] Faulty of Mechanical Engineering, UTP University of Science and Technology, Bydgoszcz, Poland

Abstract. In order to evaluate and improve the efficiency of the process of grinding, various investigations are conducted, based on the relevant research methodology. One of them is the method in which the crushed sample is subjected to single stroke loads. On this basis, the influence of the geometric features of the chipper system and the dynamic process on the efficiency of the grinding is determined. Charpy hammers instrument were used to perform these modifications so that the momentary force of the resistance could be recorded with varying sample alignment, blade geometry changes and others. In addition, it was proposed to use a super fast camera (up to 1200 fps) to record the deformation of the sample and its destruction, in order to interpretation the burdens there. Under such idealized conditions, a range of variables has been identified that significantly affect the reduction of energy demand during grinding.

Keywords: shredding, constructional features, breaking up, energy efficiency

1 Introduction

Charpy's hammer is a device that has been proposed to be used to evaluate the suitability of the selected workpiece construction. Basically, this method is used to measure impact strength of materials. By measure of impact, it is possible to mean the ratio of the work involved to the destruction of the sample to the area of the cross-section. There are many possibilities to estimate behaviour of materials, especially dynamic mechanical properties [1-7]. By testing with the hammer, the energy required for the impact fracture (fracture) of the specimen supported by the ends on the two supports is determined by the central pendulum impacting it. Breaking work K is the potential energy difference of the hammer in the starting position of K_{max} and the end of the K_{min}. The device consists of a base and two columns in which a pendulum with a beater is mounted. The pillars are supported by the pillars on which the sample is laid. This should be done in such a way that the notch is on the same side as the locking supports. When releasing the hook the pendulum with the beater drops from the highest position determined by the angle α, thus changing the initial potential K_{max} to kinetic. In the impact test, a swing hammer is broken in one hit, the

* Corresponding author: mackomar@ukw.edu.pl

Reviewers: *Milan* Sága, *Ján* Vavro *Jr.*

notched sample in the middle. This sample is supported by both ends. The energy used to break the sample is expressed in J - a measure of the impact strength of the material.

Polymer materials, compared to other materials, have a very high durability and resistance to atmospheric conditions, resulting in a long period of biodegradation in the environment [8-10]. They have a weak or medium resistance to mechanical factors depending on their purpose. As regards the preparation of plastics at the stage of processing as well as the development of plastics after the service life, the necessary operation is the comminution [11-14]. Because of problems with detailed estimates of energy demand and high energy consumption, research is being conducted to improve this. One of the test methods is single comminution. Thanks to these tests, one can estimate the energy requirement for a single sample crushing while selecting the appropriate parameters (eg knife shape, knife stroke speed).

2 Own research

It was proposed to use impact testing equipment; Charpy hammer (Fig. 1) to assess the impact of changes in geometric features of the chopper operating system. By varying the angle of the blade and the variable materials from which the samples were made, the effects of these changes on the efficiency of the process were assessed.

a) b)

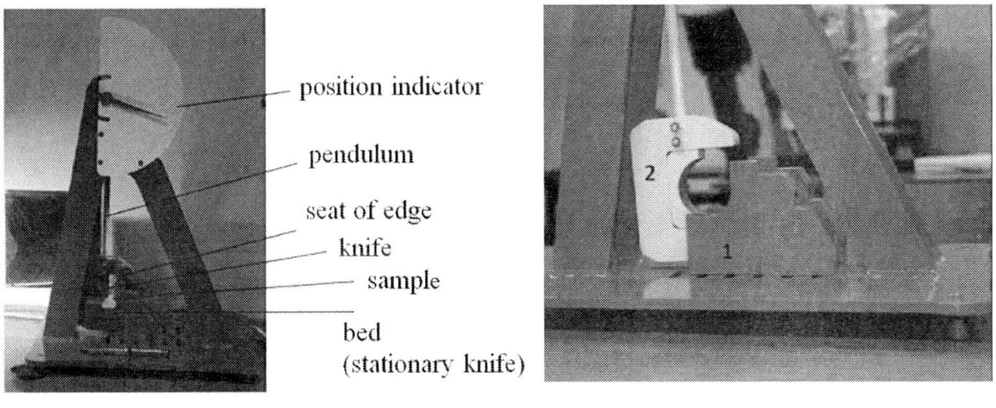

Fig. 1. View of the modified Charpy hammer - a) and working space; 1 - support, 2 - pendulum of modified shape.

In addition, temporary load values were recorded during the process (Fig. 2), and cameras were recorded for fast-change processes at an accuracy of 1200 frames per second (fps) to isolate the key features of the process [15, 16]. Laboratory tests were performed according to predefined parameters and precision test equipment. The instrumentation used was a four-channel Spider8 system, which was compatible with the computer, whose results could be recorded using a graph in the CATMAN program or LabVIEW application. As a measuring device, the S9 10 kN force gauge was used.

a) b)

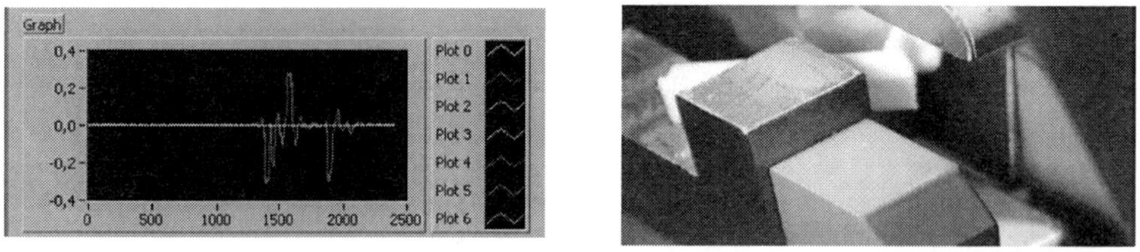

Fig. 2. Example of changes in momentary loads during a single crushing test - a, view of the sample crumbling space – b.

Blades of various geometric shapes (blade angles: 90°, 80° and 70°) were used in the modified test system. In addition, the pendulum swing angle was observed. Polypropylene samples were tested with 1% filler (blowing agent).

2.1 Destruction of polypropylene samples - 1% filler - blade angle - 70°

During the test a breakthrough was observed at the point of contact of the sample with the blade. As the blade moves and intensifies the impact surface of the attack plane, after the initial compression of the sample, the fragment of contact plane of the attack plane with the material is torn off (Fig. 3). It has been observed that the deflection of the specimen extends radially from the blade towards the outer edges of the sample. Circles marked by the segments indicate where there are additional shear stresses.

a) b) c)

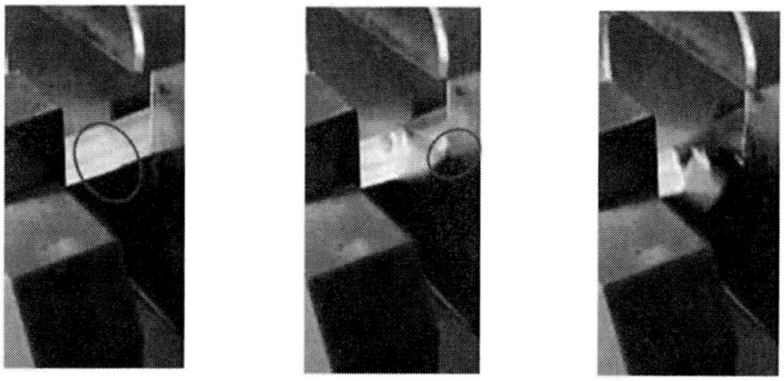

Fig. 3. Destruction of PP sample 1% at a blade angle of 70°: a) Blade stroke and crack initiation; b) material rupture associated with rupture of part of the sample to which the racial plane part was active; c) disintegration of the sample fragment.

2.2 Destruction of polypropylene samples - 1% filler - blade angle - 80°

A polypropylene sample with a filler content of 1% (at the moment of impact) is locally compressed and deformed at the contact area (pictured in red circle) at the time of impact. The sample area is also compressed against the yoke on the blade side. The entire length of the sample is bent and at the same time there is a fragile fracture perpendicular to the edge of the sample. The crack is approximately 1 mm from the point of impact of the blade in the attack plane. There have also been observed cases of spreading microcracks from the blade

At an angle of about 82°. In each of these cases, the sample portion was cut between the base (yoke) and the knife application plane.

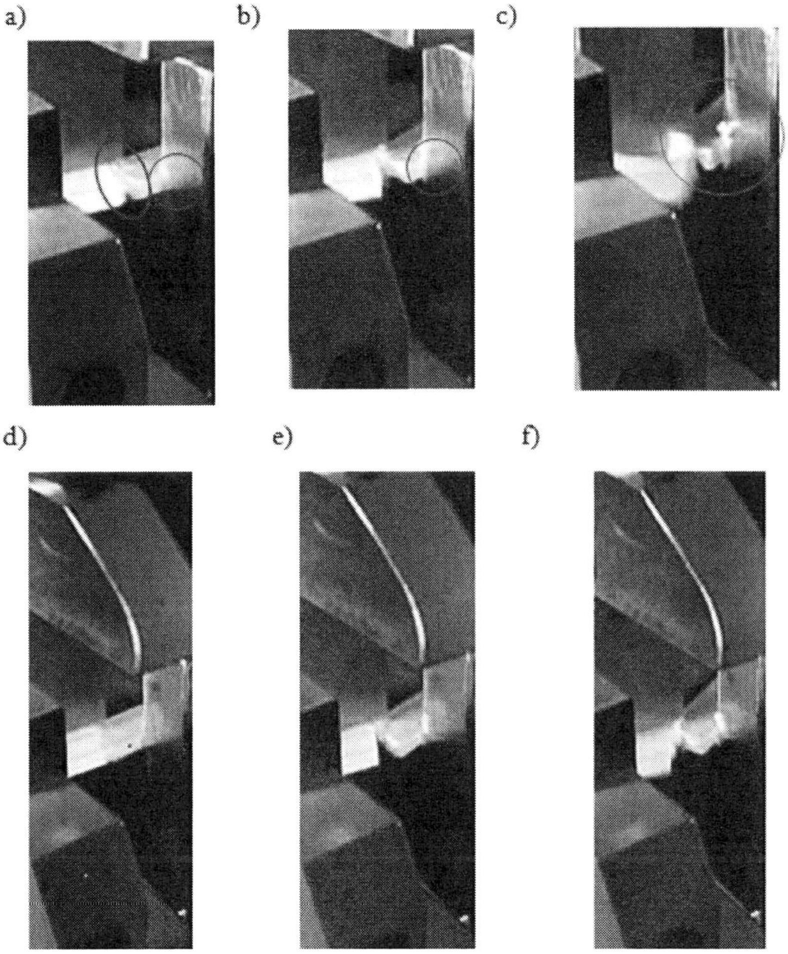

Fig. 4. Destruction of the sample PP1% at a blade angle of 80°: a, d) Blade stroke and breakthrough; b, e) further compression of the sample (deformation within the solid knife); c, f) Secondary cut off of the beam part.

2.3 Destruction of polypropylene samples - 1% filler - blade angle - 90°

In this case, the fracture propagates between the blade and the edge of the support, but no complete breakthrough occurs. Part of the sample is torn off, and its residues abut against the attack plane and are cut off from the rest of the material (Fig. 5). The figure also shows a semicircular element that disintegrated at the moment of impact. For the blade angle 90°, the impact of the entire surface of the attack plane is affected. Material decontamination occurs in this case in an uncontrolled manner.

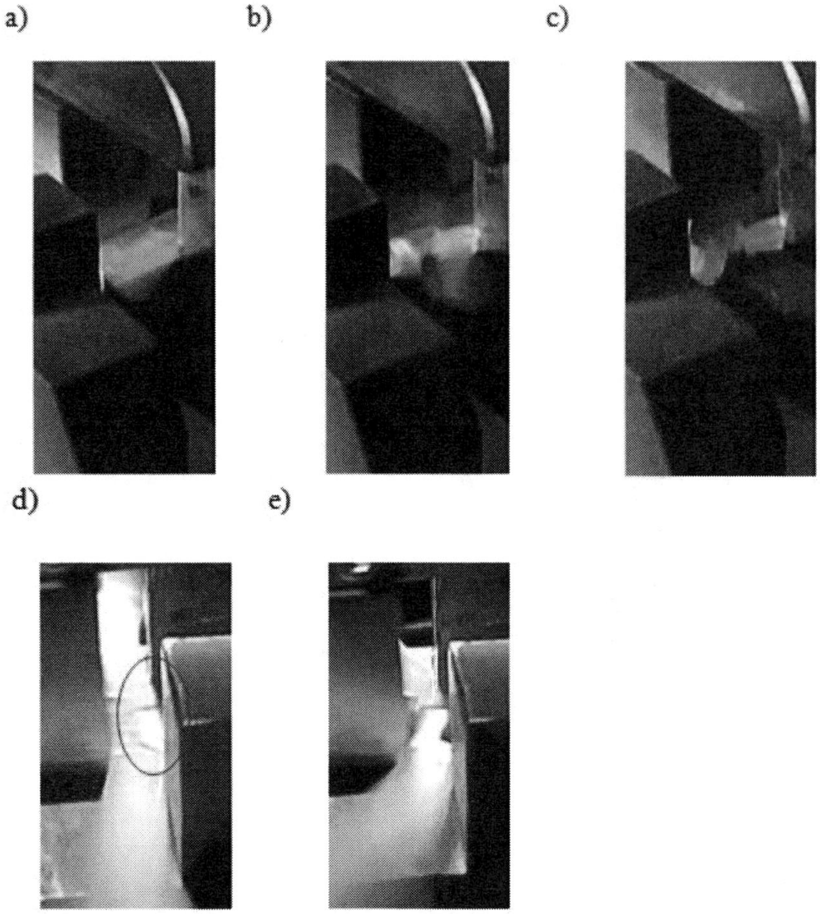

Fig. 5. Destruction of the PP1% sample at a 90° blade angle: a) blade blast and cracks; b) material rupture associated with rupture of part of specimen to which part of the rake surface has been acting; c) ripped sample element; d, e) view in a plane perpendicular to the plane of pendulum movement, (e - notice able detachment of a part of the material in contact with the attack plane).

Conclusions

In the study, a breakthrough could be observed at the moment the blades were brought closer to the sample. As the knife moves and significantly increases the impact surface of the attack plane, after the sample has been compressed, a portion of the rake face contact with the material is torn apart. Research has shown that the specimen bends radially from the blade towards the outer edges of the sample. A polypropylene sample with a filler content of 1% (tensile test strip) at the time of impact is subjected to local compression and deformation at the point of contact. A sample of the yoke on the side on which the blade works is also squeezed. The whole surface is bent. At the same time, the brittle fracture is perpendicular to the edge of the sample. Cracks appear approximately 1 mm from the area where the blade hits the blade in the attack plane. There are also cases of spreading microcracks from the blade at an angle of about 82°. Each time in this type of situation, a sample of the sample is cut between the base (yoke) and the plane of application of the knife. In this case, the cracks come between the blade and the edge of the support. However, there is no complete breakthrough of the sample. Its part is torn off, and the residue is based on the plane of attack. It comes to cut from the rest of the material. One of the semicircular elements is disintegrating at the moment of impact. For the angle of the 90° blade, the entire surface of the attack plane can be observed. Material dehiscence is uncontrolled for this test.

These studies confirm the complexity of the processes that affect the grinding of polymeric materials. A decisive factor in the de Cohesion process and the amount of energy demand is the angle of the knife blade. If the sample is made of polypropylene with 1% filler content, the energy demand varies, depending on what is the angle of the blade. This is confirmed by observations of the course of the phenomenon, and more precisely the optimally oriented stresses that appeared at a blade angle of 70°. For a blade of 80°, the fragmentation has changed. In addition to the original breakthrough, there were also observed associated stresses which resulted in further cracks in the plane of application. A piece of sample material was located between the breakthroughs. In the case of a knife whose blade reaches the 90° angle, it does not only break the middle part of the sample (without breakthrough). In addition, it had its rotation and slip between the blades. Under the conditions of energy tests, no breakthrough occurred at the horizontal alignment of the sample. In other cases, it had optimum disintegration. After analysis the results of the research it is possible to draw conclusions about the nature of the structure. Of course, for any material, these will be different conclusions, because the parameters will be different. It is important to capture the sample in detail, since only then can you choose (looking through the prism of minimizing energy demand) material variant. In theory supplemented by practice, we notice that Charpy's hammer is uncomplicated in design, but it is worth considering its modifications - as in the case of this study. It can make it possible to get smaller deviations in individual measurements, which translates into their accuracy.

References

1. Bauer, D. Wustenberg, *Fracture Behavior of Polypropylene under Dynamic Cutting and Shearing*. Actions in Granulators Chemical Engineering & Technology **25** (11), 1047–1051 (2002)
2. J. Flizikowski, *Rozdrabnianie tworzyw sztucznych*. WMN – ATR Bydgoszcz, (1998)
3. H. Kalman, M. Hubert, E. Grant, Y. Petukhov, M. Haim, *Fatigue behavior of impact comminution and attrition units*. Powder Technology **146**, 1– 9 (2004)
4. M. Sága, P. Kopas, M. Uhríčik, *Modeling and experimental analysis of the aluminium alloy fatigue damage in the case of bending - torsion loading*. Procedia Engineering **48**, 599-606 (2012)
5. M. Sága, R. Bednár, M. Vaško, *Contribution to modal and spectral interval finite element analysis*. Vibration Problems ICOVP 2011, Springer Proceedings in Physics **139**, 269-274 (2011)
6. M. H. Pahl, *Zerkleinerungstechnik, Praxiswissen Verfahrenstechnik*. Mechanische Verfahrenstechnik: Auflage, Leipzig, (1997)
7. F. K. Wittel, *Single particle fragmentation in ultrasound assisted impact comminution*. Granular Matter, Springer-Verlag (2010)
8. T. Rydzkowski, *Properties of recycled polymer mixtures obtained in the screw-disc extrusion process*. Polimery 2011 **56** (2), 135-139 (2011)
9. T. Klepka, *Construction of axial-symmetric polymeric extrudates of complex forms*. Polimery 2008 **53** (5), 390-395 (2008)

10. K. Pepliński, P. Czyżewski, D. Górecki, D. Sykutera, M. Bieliński, *Selected geometrical features and strength indicators of elements manufactured by fused deposition modeling technology*. Polimery 2017 **62** (3), 198-207 (2017)
11. J. Flizikowski, T. Topoliński, M. Opielak, T. Tomporowski, A. Mroziński, *Research and analysis of operating characteristics of energetic biomass micronizer*. Eksploatacja i Niezawodność – Maintenance and Reliability **17** (1), 19-26 (2015)
12. A. Tomporowski, *Filling model for the working multi-disc biomass grain grinding unit*. Archive of Mechanical Engineering **59** (2), 155-174 (2012)
13. M. Macko, *Metoda doboru rozdrabniaczy wielokrawędziowych do przeróbki materiałów polimerowych*. Wyd. UKW, Bydgoszcz, (2011)
14. M. Macko, *Economic-energetic analysis of multi-edge comminution of polymer recyclates*. Przemysł Chemiczny **92** (8), 1499-1502 (2013)
15. G. Śmigielski, W. Toczek, R. Dygdała, K. Stefański, *Metrological analysis of precision of the system of delivering a water capsule for explosive production of water aerosol*. Metrology and Measurement Systems **23** (1), 47-58 (2016)
16. M. Sasso, M. Fardmoshiri, E. Mancini, M. Rossi, L. Cortese, *High speed imaging for material parameters calibration at high strain rate*. Eur. Phys. J. Spec. Top. **225** (2), 225 - 295 (2016)

Experimental casting of forging ingots from model material

Ján Moravec[1], *Peter* Kopas[2,*], *Lenka* Jakubovičová[2], *Bohuš* Leitner[3]

[1]University of Zilina, Faculty of Mechanical Engineering, Department of Technological Engineering, Univerzitná 8215/1, 010 26, Zilina, Slovak Republic
[2]University of Zilina, Faculty of Security Engineering, Department of Technical Sciences and Informatics, Univerzitná 8215/1, 010 26, Zilina, Slovak Republic
[3]University of Zilina, Faculty of Security Engineering, Department of Technical Sciences and Informatics, Univerzitná 8215/1, 010 26, Zilina, Slovak Republic

Abstract. The paper describes the process of casting ingots from the model material into a special mold made for these tests. The material was chosen stearin, which proved to be suitable for this type of laboratory test. During the solidification process of the ingot model under laboratory conditions, it was observed how gradually the layer formed on the contour of the casting. Gradual cooling of the ingot resulted in a decrease in the volume of the liquid phase in his body. The fog is readily observable by the naked eye and this is manifested by the formation of a gap between the ingot mold wall and the ingot body. A silicone oil has been used as a separating melt separating layer and the wall of the ingot that has reliably fulfilled this task. Casting was done in two ways, with a standing and lagging mold. The process of filling the cavity itself was to create conditions for the linear flow of the melt. Observation of the ingot after its solidification confirmed the fact that the filling of the cavity proceeded under such conditions in terms of the melt flow rate.

Keywords: ingot, mold, stearin, solidification, casting

1 Introduction

In technological practice we constantly search for methods and ways that can ensure comparably usable parameters for a real process, obtained in experiments [1-4]. The following text focuses on one of the several ways enabling shrinkage formation in the body of the ingot, and also on the study of the shrinkage process using modelling materials. For quite a long time experimental work has been done using easily melt able materials [5-8].

1.1 Ingot casting – practical aspects

In terms of the theory of solidification and crystallisation, the most preferred way is casting of dimensionally smallest ingots. This is used when the aim is to achieve optimal homogeneity – both chemical and structural. Practical manufacture of steel ingots is a

[*] Corresponding author: peter.kopas@fstroj.uniza.sk

Reviewers: *Andrej* Czán, *Ján* Vavro *Jr.*

challenging technical and organisational process and a problem depending on several factors, such as: basic type of cast steel (killed and rimmed steel); required shape of the ingot for further processing; chemical composition of the material; size of the production unit; cycle of ingot mould set preparation; method of ingot casting (bottom, top); number of ingot types and production capacity of the forge. Methods of experimental research into the solidification process are divided into direct and indirect.

Direct methods:
- Discharge of liquid metal from the ingot mould and measuring the thickness of the remaining solidified melt.
- Monitoring the solidification process by means of a bar immersed in the liquid steel at a certain angle (bar test).
- Dipping a mandrel into the steel.

Indirect methods:
- Measurement of the steel temperature using thermocouples.
- Addition of trace elements or radioisotopes to the liquid steel.
- Disturbance of steel crystallisation by mechanical effects (shock, vibration).
- Modelling using low-melting compounds.

Of great importance for good formability (malleability, ductility) is the size and arrangement of crystals and segregates in ingots, which are influenced by the number of crystal nuclei and rate (velocity) of crystallisation, as well as by the volume change in the melt's transition into a solid state [9]. The most commonly used ingot for forges is an ingot with a wider top end. In case of small ingots, solidification occurs simultaneously in the whole cross-section of the part of the ingot in the ingot mould. Although dendritic segregation can be detected in these ingots, but the chemical composition is uniform all over the cross-section. Hardly any differences are manifested in the content of individual elements between the surface and the centre. There are only a few differences between the top and bottom parts of the ingot. A large ingot of fully killed steel features various regions corresponding to the course of crystallisation.

1.2 Faults of steel ingots

Main shrinkage - in case of ingots with lost tops, the main shrinkage must remain in the ingot top. When it appears in the ingot body, it is the result of an under cast or too-little ingot top. Shrinkage porosities are caused by insufficient refilling of liquid steel. Total porosity is the result of insufficient deoxidation. Distinctive dendritic segregation occurs due to a too high crystallisation rate at a small number of crystallisation nuclei. Longitudinal cracks on the walls are the result of stress at too fast a hot casting, or at premature removal of the ingot from the ingot mould and its rapid cooling. Transverse and diagonal cracks are caused by obstacles in the shrinkage of the ingot – either by the ingot mould or by unsuitable casting temperature and rate.

2 Experimental work

The experimental material we used was stearin. It is a mixture of free fatty acids, in particular palmitic and stearic acids. The melting point is indicated as 55–59 °C. We used a steel ingot mould and cast experimental rollers. These were monitored with sodium thiosulphate. When applying stearin, differences became apparent in the quality of the cast samples. The bodies of samples after solidification in the ingot mould were relatively suitable for subsequent observations. Figure 1 shows the samples after completion of the experiment.

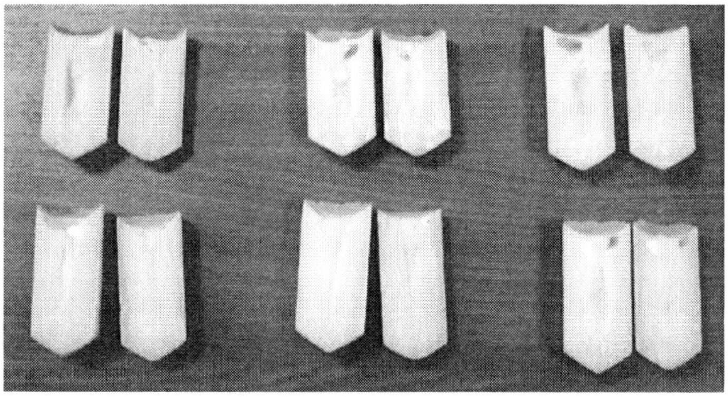

Fig. 1. Shrinkages and cavities in the samples

2.1 Model equipment

There was manufactured a model ingot mould. We used this ingot mould to carry out experimental work. Sensor wires were sealed using a sealing paste. To record the temperature we used three sensors at different heights. Values were read from each sensor at pre-defined time intervals. The values were then processed in a chart. We used ALMENO 2890-9 Ahlbonn measuring station. A bad casting was produced at a high melting temperature, but the result was better at a lower temperature. Optimum temperature proved to be 58 °C, when the cavity filling was best. Casting conditions turned much worse when overheating the melt.

2.2 Filling the ingot mould cavity

We used silicon oil as a separating layer to separate the melt from the ingot mould wall. Silicon oil reliably accomplished the task. Figures 2 and Figure 3 thoroughly illustrate the whole course of the process and provide a good overview of all monitored parameters.

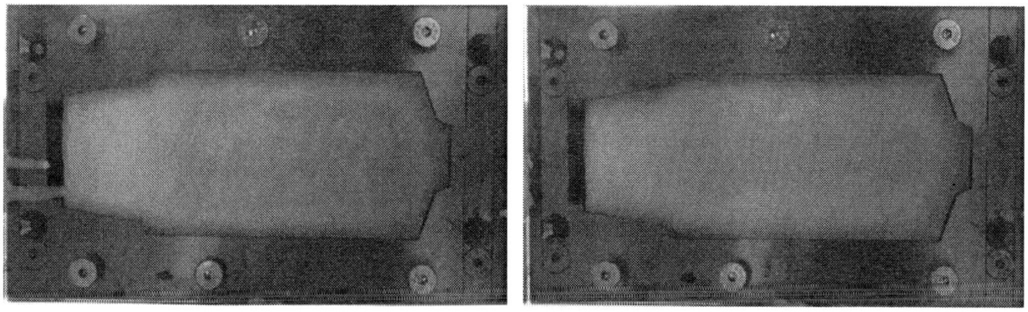

Fig. 2. Solidification of the model ingot in the longitudinal axis
left – after 20 minutes; right – after 40 minutes

https://doi.org/10.1051/matecconf/201815705017

Fig. 3. Shrinkages and cavities in the samples

We carried out two methods of filling the ingot mould cavity: First Case (Fig. 4) status I, and Second Case (Fig. 4) status II. The filling port (inlet), which is not used in either method, is plugged with an auxiliary plug during filling the ingot mould cavity. In the actual cavity filling process we attempted to create the conditions for a linear flow of the melt. Observation of the ingot after it solidified confirmed the fact that the cavity filling was carried out under such conditions in terms of the melt flow velocity. The total time of the cavity filling was 42s.

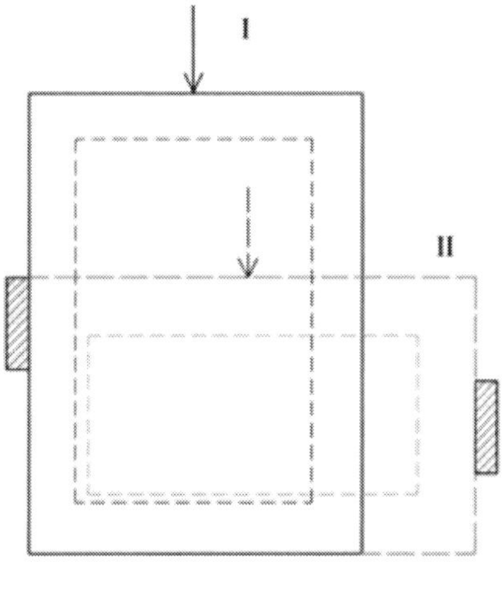

Fig. 4. Two positions for casting ingot

2.3 The course of experimental work

In the course of the model ingot solidification process in laboratory conditions we observed how a layer was gradually forming along the contour of the ingot casting. Cooling of the ingot was gradual, which resulted in reducing the volume of liquid phase in the ingot body, which was observed during the experiment. Shrinkage is readily observable with the naked eye, and this is manifested by formation of a gap between the ingot mould wall and the ingot body. Based on the temperature readings of the thermocouples we plotted a graph – dependance of temperature on time during the ingot cooling. The temperature of the ingot decreased rapidly during the first five minutes, then it stabilised at the same value (Fig. 5).

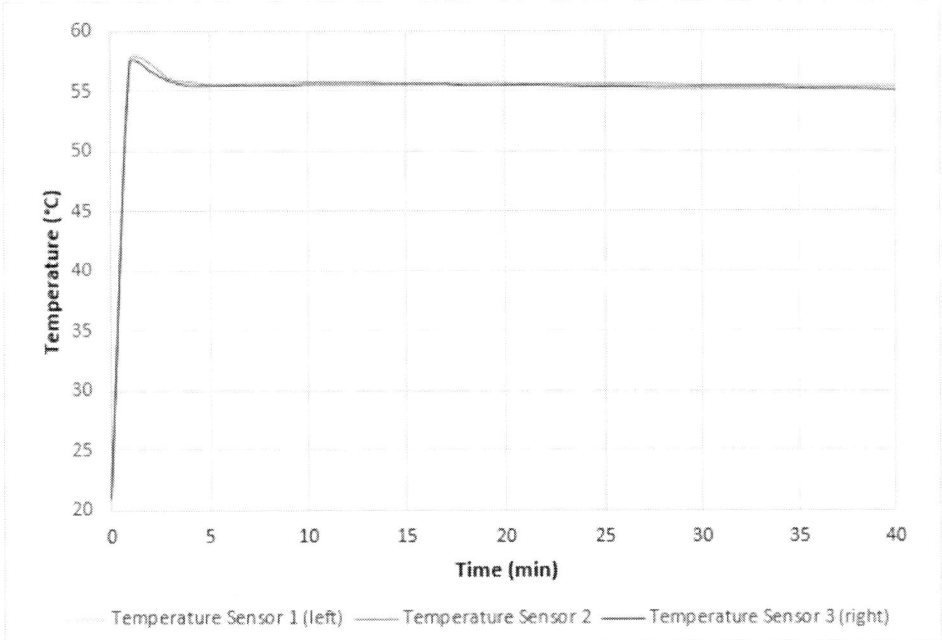

Fig. 5. Graphs showing temperatures over the sample solidification

2.4 Solidification

Regarding solidification it is desirable to continue to determine the effect of chemical composition of steels (high carbon and high alloyed steels) on the liquidus and solidus temperatures. It is also necessary to deal with the determination of the coefficients of heat transfer between the ingot and the mould, and between the ingot and the pad, both for conventional casting as well as for continuous casting. The above-mentioned data are required for a very close approximation of numerical procedures to real-life conditions during cooling of steel ingots. Special attention should be paid to the study of solidification of very heavy ingots of non-alloyed as well as alloyed steels, because other properties of forgings are largely influenced by solidification and primary crystallisation of ingots. The solidification process is directly associated with the development of macro-segregations, segregates and sedimentation cone. Heavy forgings are the basis for the development of mechanical engineering and chemistry. It is also equally important to study solidification of the ingot surface layer in order to comprehensively improve surface quality of raw ingots and prevent cracks resulting from heat stress [10, 11].

2.5 Steel shrinkage during cooling and solidification

Shrinkage in ingots is formed due to steel shrinking during its cooling and solidification. The size of shrinkage can be calculated using the equation by [12]:

$$V_S = \alpha_V + \alpha_{\Delta t}(t_1 - t_S) - \frac{1}{2}\beta(t_S - t_2) \tag{1}$$

where V_S is the volume of shrinkage; α_V - the coefficient of volumetric shrinkage during solidification (dimensionless); β - the coefficient of volumetric shrinkage in the solid state; t_1 - the mean temperature of liquid steel at the beginning of solidification; t_2 - the mean temperature of solidified steel at the end of solidification; t_s – temperature of solidification; $\alpha_{\Delta t}$ - the coefficient of volumetric shrinkage in the liquid state.

Shrinkage size when casting steel into an ingot mould and if it is $\alpha_{\Delta t} = 0.9.10^{-4}K^{-1}$, $\alpha_V = 0.0034$, $\beta = 0.64.10^{-4}K^{-1}$, $t_1 = 1525\,°C$, $t_2 = 1275\,°C$, $t_S = 1500\,°C$ can be written as :

$$V_S = 0.034 + 0.9.10^{-4}(1525 - 1500) - \frac{1}{2}.0.64.10^{-4}(1500 - 1275)$$
$$V_S = 0.034 + 2.25.10^{-3} - 7.2.10^{-3}$$
$$V_S = 0.029$$

The shrinkage volume is 2.9 % of the ingot volume. The equation does not include metal solidification during casting, which of particular importance in ingot casting. The shrinkage size is governed by the relation by [13]:

$$V_S = \alpha_{4t}(t_1 - t_S) + \alpha_V\eta - \beta(t_1 - t_2)\varphi - \beta(t_S - t_2)\eta \quad (2)$$

where: φ – is the coefficient of deformation of solidified crust due to ferro-static pressure, t_1 – is the ingot surface temperature at the end of casting, t_2 – is the temperature of the ingot at the end of solidification.

The above-mentioned relationship not only includes steel solidification during casting, but it also notes the solidified crust deformation. Differences in the shrinkage volumes can be explained in case of ingots of different sizes by the solidified crust deformation. However, accurate representation of the third member of the equation is difficult. According to [14] the equation was modified by taking into account the proportion of steel solidified during casting, leading to the following equation:

$$V_S = \alpha_V + \alpha_{\Delta t}(t_1 - t_S) - 0.5\beta(t_S - t_2)\frac{V_{i+h} - \vartheta_1}{V_{i+h}} \quad (3)$$

2.6 Vertical solidification of the ingot

Solidification of the ingot from the bottom in the vertical direction affects the central uniformity of the steel is shown on Fig.6. A theoretical calculation of the course of vertical solidification of a cylindrical ingot according to equation (4) [15]:

$$h = \frac{r}{2} ln \frac{r}{r - 2k\sqrt{\tau}} \quad (4)$$

Derivation of equation (4) is based on the following simplifying assumption:
- the ingot pad is a flat plate,
- the heat removed in vertical and horizontal directions is calculated according to a parabolic relationship, $\xi = k\sqrt{\tau}$; $Q = \frac{k}{n}\sqrt{(\tau)}\pi r^2$ where n is an empirical coefficient (m^3J^{-1}),
- the volume of solidified metal for the time period $d\tau$ is directly proportional to the amount of heat removed for the same time period:

$$\frac{dV}{d\tau} = n\frac{dQ}{d\tau} \quad (5)$$

$$dV = (\pi r^2 - 2\pi r\xi)dH \quad (6)$$

$$\frac{dV}{d\tau} = (\pi r^2 - 2\pi r\xi)\frac{dH}{d\tau} \quad (7)$$

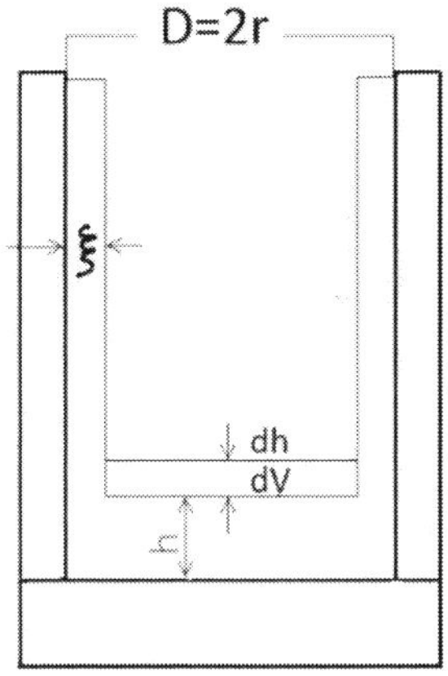

Fig. 6. Scheme of solidifying ingot with designated individual parameters used in the equation

The resulting equation is as follows:

$$\frac{dQ}{d\tau} = \frac{1}{2}\frac{k}{\sqrt{\tau}}\frac{\pi r^2}{n} \tag{8}$$

$$\frac{dh}{d\tau} = \frac{kr^2}{2(r^2\sqrt{(\tau)} - 2rk\tau)} \tag{9}$$

After integration and determination of the integration constant from the condition (*t = 0* is *h = 0*), the resulting equation will be:

$$h = \frac{r}{2}\ln\frac{r}{r - 2k\sqrt{\tau}} \tag{10}$$

Difference between the predefined course of vertical solidification according to the probe measurements and the theoretical calculation using equation (10) is determined by sedimentation of equiaxic crystals, which is not taken into account in equation (10).

Conclusion

Some authors [11] utilised model experiments on ingots with sodium thiosulphate, cast into ingot moulds made of Plexiglas, to observe the effect of the ingot mould conicity and slenderness on the course of vertical solidification. An increase in conicity for 45 t forging ingots very significantly improved homogeneity in the axial part, as well as a reduction in the H/D ratio, because vertical solidification significantly slowed. When we look at the curve depicting the vertical solidification process with respect to time, we can conclude that the smaller the curve slope, the better ingot homogeneity in the axial part. Acceleration of vertical solidification in the later stages of the ingot solidification can be explained by the horizontal solidification fronts getting closer to the ingot axis. Efficient and long-term heating of the ingot top (by electric arc or induction) may affect the central part of the ingot by slowing down vertical solidification. Input power must be regulated so that, at the time of the ingot solidification, too intense heating does not result in the creation of a wide two-phase zone of solidification in the ingot axis. This paper described the process of ingot solidification in a model ingot mould when placing the given mould in two positions. The graph show the course of temperatures recorded by three sensors during 40 minutes. At the stearin casting temperature of 58 °C we can observe that inner shrinkage is smaller. Shrinkage could be completely eliminated using a raiser. Material to forming (forging) processes is supplied in the form of castings, i.e. this is what forming inherently affects. Therefore, it is necessary to deal with this phenomenon that, as soon as at its initial input (first entry), affects the resulting quality of shaped parts.

Research presented in this paper was partially financially supported through realization of projects KEGA 015ŽU-4/2017 and of the project implementation: „Centrum for industrial research and development in the field of light metals and composites“ (ITMS 26220220154) supported by the Operational Programme Research and development funded by the ERDF. The authors gratefully acknowledge this support.

References

1. W. Piekarska, M. Kubiak, M. Vasko, *Numerical estimation of the shape of weld and heat affected zone in laser-arc hybrid welded joints*. Procedia Engineering **177**, 114-120, (2017)
2. L. Jakubovicova, B. Ftorek, V. Baniari, A. Sapietova, T. Potocek, M. Vasko, *Engineering design of a test device*. Procedia Engineering **177**, 520-525, (2017)
3. A. Vasko, M. Vasko, J. Belan, E. Tillova, *Comparison of fatigue properties of nodular cast iron at low and high frequency cyclic loading*. Procedia Engineering **177**, 576-581, (2017)
4. J. Zapoměl, V. Dekýš, P. Ferfecki, A. Sapietová, M. Sága, M. Žmindák, *Identification of material damping of a carbon composite bar and study of its effect on attenuation of its transient lateral vibrations*. International journal of applied mechanics **7** (6), (2015)
5. L. Šmrha, *Crystalization and Solidification of Steel Ingots*. (Praha, STNL, 1983)
6. L. Šmrha, J. Chvojka, *Foundry Letters*. **63** (9), (1991)
7. L. Šmrha, *Foundry*. **10**, 245-249 (1962)
8. W. James, R. Middleton, *British Foundryman*. January **58**, 36-46 (1987)
9. J. Přibil, *Controlled solidification of steel castings*. (Praha, STNL, 1986)
10. N. Chvorinov, *Giesserei*. **10**, 177-186 (1975)
11. L. Kiesel, K. Smuin, W. Pabst, *Freiberger Forschungshefte*. **6** (B137), 79-92 (1967)
12. N. Chvorinov, *Giesserei*. **11**, 201-208 (1975)
13. J. Nechenedzi, *Litejnoje proizvodstvo*. Moskva, **3**, 14-19 (1982)
14. B. B. Guljajev, *Zatverdevanije i neodrodnost stali*. (Metallurgizdat, Moskva, 1986)
15. L. Kiesel, K. Smuin, W. Pabst, *Freiberger Forschungshefte*. **5** (B137), 81-93 (1985)

Analysis of the material configurations and influence on the dynamic response

Zuzana Murčinková[1,*], *Michal* Halapi[1]

[1]Faculty of Manufacturing Technologies, Department of Designing and Monitoring of Technical Systems, Bayerova 1, 080 01 Prešov, Slovak Republic

Abstract. The paper presents the analysis of the material configuration of composites based on the measurement of their dynamic response. The article presents the measurement scheme, the design of the measuring stand together with the analysis of the results. Moreover, it analyses the FFT spectrums of layered long fibre composite, short fibre composites of different fibres materials and homogeneous materials as steel and aluminium alloy.

Keywords: natural frequency, damping time, fibre, sample, measuring stand

1 Introduction

Each mechanical system (e.g. production machine-tool) under a particular type of dynamic force provides the dynamic response characterized by specific behaviour depending on various factors. The objective of the machine-tool designer is to maximize the dynamic stiffness between tool centre point-workpiece and tool centre point-drive motors while keeping the machine mass light for high speed positioning and high-productivity machining [1]. A broad variety of materials can be found in modern machine tool structures ranging from steel and cast iron to fibre reinforced composite materials [2]. The (short, long) fibre composite materials provide wide options of various material configurations regarding fibre volume fraction, fibre length, fibre orientation in bulk, laminate layer orientation etc. The fibre direction relative to the structure internal force flows has to be considered carefully during structural design [3]. The paper is contribution to research of dynamic characteristics (natural frequencies, material damping) of material configuration in case of low velocity impact intending for utilization in machine-tool applications mainly by experimental, also by numerical approaches.

2 Measuring setup

Fig. 1 illustrates the measuring chain. The measuring equipment is Polytec IVS 400. The main advantage is contactless and wear-free measurement what eliminates the influence of environmental conditions regardless some servo-mechanisms or noise protection to perform the measuring. Polytec is precise and reliable vibrometer based on a phenomenon of

* Corresponding author: zuzana.murcinkova@tuke.sk

Reviewers: *Vladimír* Dekýš, *Alžbeta* Sapietová

Doppler Shift in laser beam to measure e.g. the vibration velocity, amplitude of a moving object etc. The light beam is reflected by moving surface and thus the frequency of light is shifted proportionally to its velocity. The laser-Doppler vibrometry is significant by the independence of measured data on reflected light intensity. Hence, the laser vibrometer is suitable for surfaces of low reflectivity.

If a wave is reflected by a moving object and detected by a measurement system, the measured frequency shift f_D (Doppler frequency) of the wave can be described as:

$$f_D = \frac{2.v}{\lambda} \tag{1}$$

where v is the velocity of measured object and λ is the wavelength of the emitted wave (of laser light beam). To be able to determine the velocity of an object, the (Doppler-) frequency shift has to be measured at a known wavelength. This is done using a laser interferometer. The optical interference, requiring two coherent light beams, with their respective light intensities I_1 and I_2, to overlap. The resulting intensity is not just the sum of the single intensities, but is modulated according to the formula:

$$I(\mathbf{r}) = I_1 + I_2 + 2\sqrt{I_1 I_2}\, cos(2\mathbf{k}''.\mathbf{r} + 2\Delta\phi) \tag{2}$$

with a so-called "interference" term. This interference term relates to the path length difference between both beams, **r** is position vector, k is wave number related to λ:

$$k = \frac{2\pi}{\lambda} \tag{3}$$

and thus **k** is wave vector, vector size is the same as size of wave number k and vector direction is the same as direction of wave propagation. Superpositioning of two waves, the wave vectors $\mathbf{k}_1$ and $\mathbf{k}_2$ are needful to be known (θ is angle between vectors), then after the decomposition to individual rectangular components where:

$$\mathbf{k}' = \frac{\mathbf{k}_1 + \mathbf{k}_2}{2} \qquad \mathbf{k}'' = \frac{\mathbf{k}_1 - \mathbf{k}_2}{2} \tag{4}$$

Then size of vector $\mathbf{k}''$ is:

$$|\mathbf{k}''| = \frac{2\pi}{\lambda} sin\frac{\theta}{2} \tag{5}$$

Physical description of Polytec equipment is according to [4, 5].

Some technical data: operating temperature +5 ÷ +40°C; velocity ranges: ±20, 100, 500 m.s^{-1}; frequency range: 0 ÷ 22 kHz (digital output), weight: 2.6 kg. More in [4].

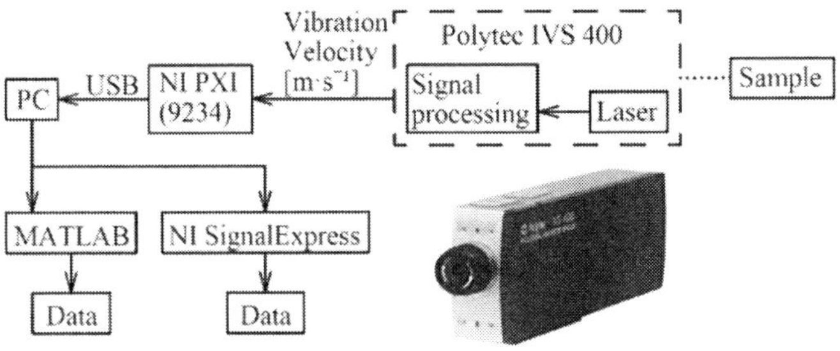

Fig. 1. Measuring chain with Polytec IVS 400

The samples were fixed on one side and excited by impulse force generated by steel ball of specific weight. The ball was rolling at inclined track of specific length.

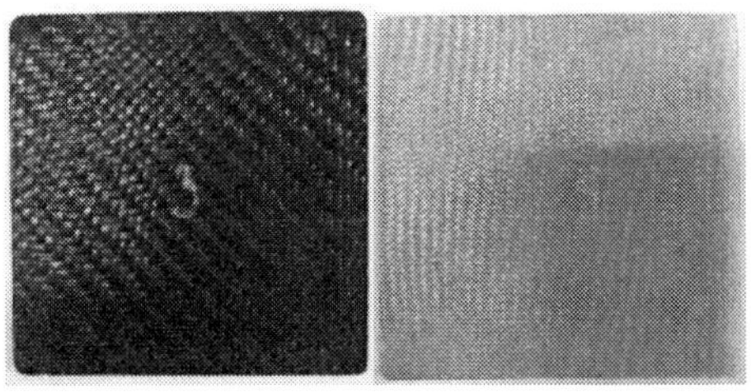

Fig. 2. Laminate with carbon (C#X#) and glass fibres (IGXXX)

The measurement was performed on six square laminate samples, size 115x115x0,9 mm, each of which is composed of three layers of carbon or glass fiber twill fabrics in matrix (Fig. 2). The carbon and glass fibre laminate samples are numbered as 1, 2, 3 and 4, 5, 6, respectively. The orientation of individual layers is different (Table 1) to be able to analyse the influence of that orientation. The carbon fabric is twill weave labelled as 200g/m^2, twill 2/2 and glass fabric 280g/m^2, twill 2/2, respectively. The three layers are in mixture of epoxy resin (EPIKOTETM Resin MGS® LR 285) and fixative (EPIKURETM Couring Agent MGS® LH286) of mutual ratio 10:4.

Table 1. Sample Layup description

Number of sample	1 (4)			2 (5)			3 (6)		
Layer	1	2	3	1	2	3	1	2	3
Fabric orientation	#	#	#	X	X	X	#	X	#

Note: # - fiber orientation parallel to the edges of the sample,
X - diagonal fiber orientation (45°)

The samples are under the low velocity impact in range 1.5-2.5 m/s. There is not one general definition with clear distinction between low and high velocity impact. The many researchers recognize the low velocity impact in case of velocity up to 10 m/s. Other researchers insisted that the upper limit of low-velocity impact vary from 1 to 10 m/s depending on the material properties of target and the mass and stiffness of impactor. More in [6].

The proposed design of the measuring stand (Fig. 3) defines a position-adjustable measuring stand. Priority has been proposed for the research (testing, measurement) of behavior of composite materials – natural frequencies, shapes and damping factors. When designing the stand, consideration was given not only to the variety of materials but also to the shape and size of the samples to be measured - flat samples up to about 700 mm height and width and rotating shapes up to a diameter of 125mm, it is possible to measure essentially any shape of the sample if it is possible to clamp it in the clamp of measuring stand. Stand is of rigid construction but with the ability to adjust the various positions needed for measurement purposes. Fig. 3 shows the snapshot of impact of ball into sample.

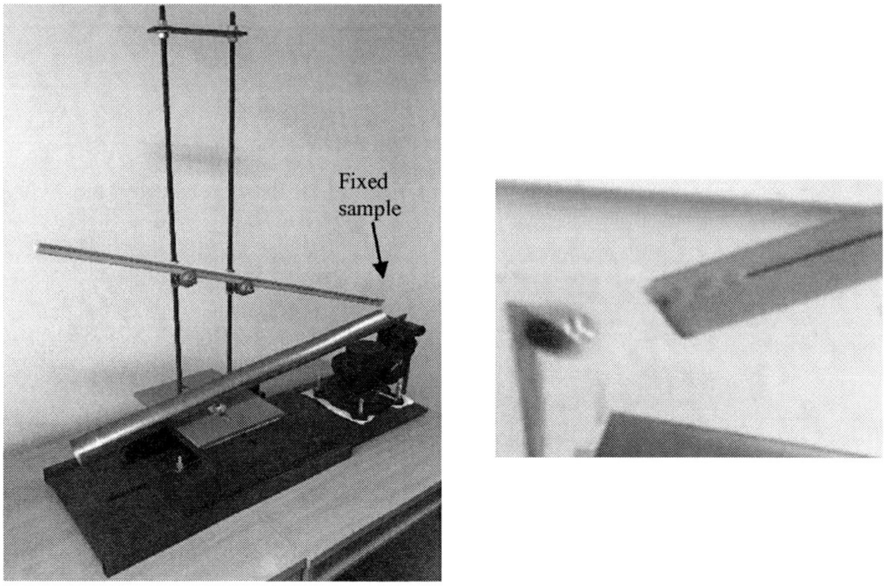

Fig. 3. Measuring stand and detail of impact

3 Experimental results

The graph in Fig. 4 evaluates the measured first two natural frequencies of samples based on the FFT (Fast Fourier Transformation) analysis of the measured vibration velocity values. Samples 1 (4), 2 (5), 3 (6) are of same carbon (glass) fiber, same matrix material, weight, number of layers, dimensions. The difference is only in material configuration. In this case it is different orientations of laminate layers and thus the different orientation of material properties. The impact forces are of four values: 332 N, 680 N, 2917 N and 5978 N. The applied values of impact forces has not important influence for evaluation of natural frequencies of individual samples. The differences are changing in very small range. It is caused by both the material configuration and the measuring error. We can stated that material configuration has not significant influence on natural frequency.

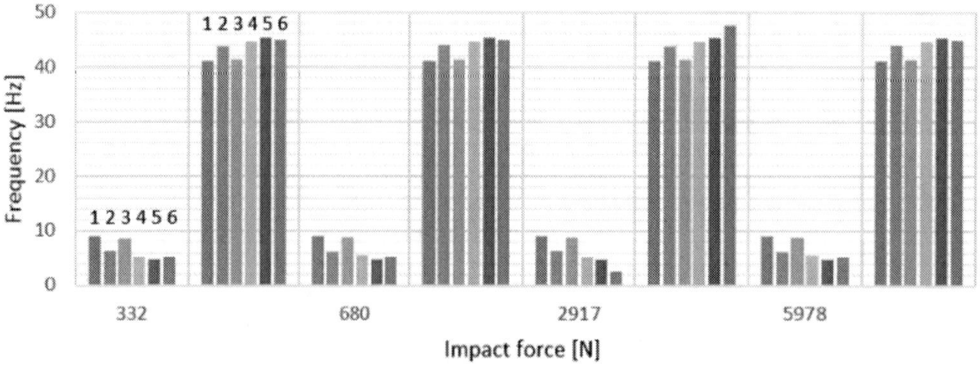

Fig. 4. The natural frequencies for 6 samples according to Table

The samples of different material configuration are of different material damping time (Fig. 5). Naturally, the damping time of carbon laminate fibers is shorter comparing to glass fibers while the weight is lower. The longest damping time is for samples numbered 3 (2 s) and 6 (2,3 s) that have confiuration #X#. The shortest damping time is for samples numbered 2 (1,3 s) and 5 (1,5 s) that have configuration XXX. We can make statement that the material configuration influencing more damping time as natural frequency.

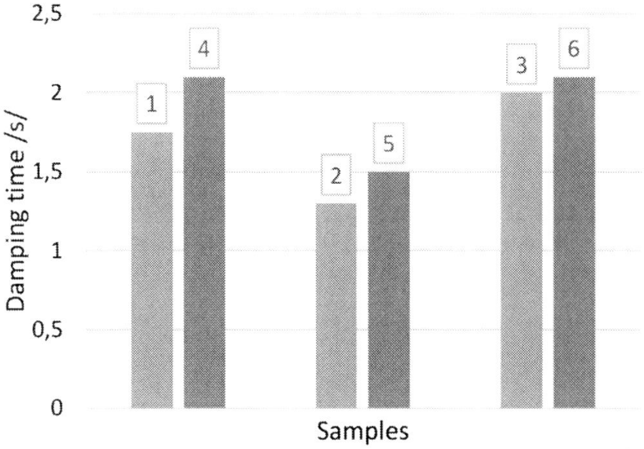

Fig. 5. Damping time for individual samples and same impact force

The measurements shows the different damping time of individual layers (Fig. 6). The circles in Fig. 6 marks the jumps (of various size) that are caused by change of wave phase. In generally, the reason of the phase change is interface layer-layer improving the damping.

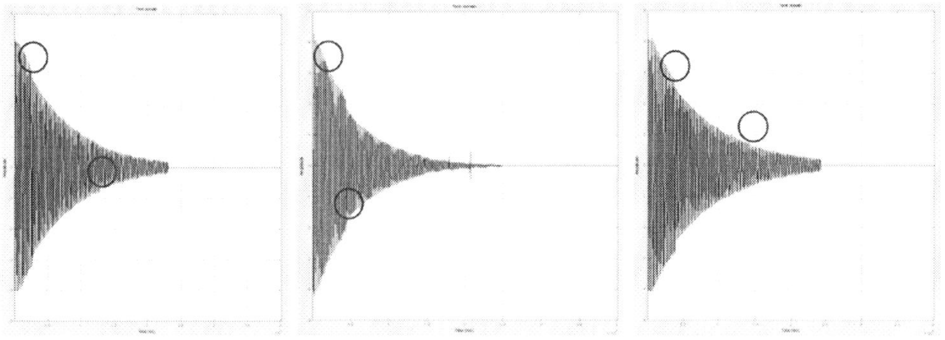

Fig. 6. Vibration velocity behaviour; dependence of vibration velocity amplitude on time for glass fiber laminate samples

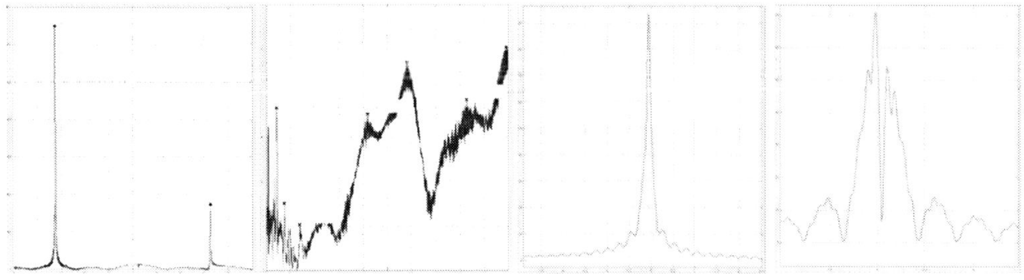

a) b) c) d)

Fig. 7 FFT spectrum of layered long fibres (laminate), (a), short fibre composite (b), steel (c) and aluminium alloy (d); frequency vs. magnitude

Evaluating and observing the results, we can make statement that dynamic response of homogenous and non-homogeneous (composite) materials is different (Fig. 7). Natural frequency of layered laminate and steel is characterized by one sharp peak (Fig. 7a, 7c) comparing to several peaks of short fibre composite (Fig. 7b) and aluminium alloy (Fig. 7d) that form the wide frequency range without significant frequencies. Such materials is appropriate for dynamic performance no matter how homogeneous is material. Homogeneity is only the idealization.

The experimental results were confirmed by numerical simulation in Abaqus/CAE software that involve simulation of laminates (similar to Ansys Composite PrepPost, Creo Simulate, Autodesk Simulation Composite Design) based on principle of Classic Laminate Theory. The boundary conditions of numerical model responded to real conditions. The material properties of individual laminate layer were defined according to provided data of manufacturer of samples. The twill was idealized and one twill layer was replaced by two layers of half thickness and orientation of fibres was defined perpendicular to each other. This way one layer was modelled. (More in [7]) The maximum difference between the experimental and numerical results is 0.2 Hz.

Conclusions

The material configuration influence the final dynamic mechanical material properties. Our experiment were made on continuous fibre reinforced carbon and glass laminates as available materials with possibility to alternate material configurations. The material damping and natural frequencies were influenced by material configuration. In our experiment the configuration of XXX (45°angle) provide the best material damping, lower first natural frequency (higher second natural frequency). These results are outcomes for other research of material configuration of composites and other construction materials regarding their various FFT spectrums.

Authors thank for supporting this research by grant VEGA 1/0910/17 of Agency of Ministry of education of Slovak Republic.

References

1. M. Law, Y. Altintas, A.S. Phani., *Rapid evaluation and optimization of machine tools with position-dependent stability*. Int. J. Mach. Tools Manuf. **68**, 81-90 (2013)
2. H. Ch. Möhring et al., *Materials in machine tool structures*. CIRP Ann. - Manuf. Technol. **64** (2), 725-748 (2015)
3. J. E. Ashton, *Implications of the Behavior of Advanced Composite Materials*. J. Franklin Inst. **290** (6), (1970)
4. www.polytec.com
5. F. Trebuňa et al., *Experimentálne metódy mechaniky – laserová vibrometria a ESPI*. (TU Košice, 2015)
6. Z. Shen, *Characterisation of Low Velocity Impact Response in Composite Laminates*. (University of Hertfordshire , PhD Thesis, 2014)
7. Z. Murčinková, M. Halapi, *Micro-mechanical response of continuous and discontinuous fiber composites with support of computational software*. Int. J Innovative Sci., Eng. Technol. **3** (1), 300-304 (2016)

Influence of microscopic casting defects on fatigue endurance of ductile cast iron

František Nový[1,*], *Peter* Kopas[2], *Otakar* Bokůvka[1], *Michal* Jambor[1,3], *Libor* Trško[3]

[1]University of Zilina, Faculty of Mechanical Engineering, Department of Material Engineering, Univerzitná 1, 010 26, Zilina, Slovak Republic
[2]University of Zilina, Faculty of Mechanical Engineering, Department of Applied Mechanics, Univerzitná 1, 010 26, Zilina, Slovak Republic
[3]Research Centre of the University of Žilina, Univerzitná 1, 010 26, Zilina, Slovak Republic

Abstract. In this work, there are published results about fatigue endurance of ductile cast iron obtained at high-frequency sinusoidal cyclic push-pull loading in the ultra-high cycle fatigue region. The main attention was focused on the fatigue lifetime data scatter caused by the influence of microscopic casting defects (microshrinkages, microbubbles, microcracks, non-metallic inclusions and local clusters of big graphitic nodules).

Keywords: ultra-high cycle fatigue, microscopic casting defects, microshrinkages, microbubbles, microcracks

1 Introduction

Graphite cast irons are the important group of cast structural materials, with the wide range of application in the industry. The special important is the group of cast irons with the spheroidal shape of graphite particles (LGG), which are frequently used in the automotive industry and in the heavy earthmoving machinery. Their wide range of applications is due to their economist, good technological properties and their ability to damp vibrations. In the most applications, the components made from the cast iron are exposed to the time-varying loading during their operating. Fatigue properties of LGG, despite intensive research, are not yet fully described. It is well known, that fatigue tests of this category of materials show a significant scattering of the measured data. The magnitude of this scattering is most often justified by the presence of structural heterogeneities and the casting defects, which significantly affect the fatigue properties [1-8].

Since requirements for the reliability and quality of castings has a steadily rising tendency, improving and the developing of the new methods and devices for the quality control and its testing are necessary. Every product made by the casting technology contains some defects or heterogeneities, which assessment and tolerance should depend on the purpose of the casted parts. The admissibility of the individual defects and their size is usually given in the acceptance protocol. In real practice, the wide range of casting defects

* Corresponding author: frantisek.novy@fstroj.uniza.sk

Reviewers: *Zuzana* Murčinková, *Bohuš* Leitner

appears in the casted parts. Among the most difficult to detect are cavities, micro-cavities, inclusions, structural heterogeneities and the different grains size, and its detection is required especially for parts with guaranteed mechanical and fatigue properties.

Authors in this study present their own experimental results of the influence of small (microscopic) casting defects on the ultra-high cycle fatigue properties of LGG.

2 Experimental material and procedures

2.1 Material

Experiments were carried out on the specimens from the two melts of LGG, whose chemical composition and the mechanical properties are shown in the Tab.1. The first melt (designated as A) represents the pearlitic-ferritic LGG, with the 4% volume fraction of ferrite in the structure. The second melt, (designated as B) is the ferritic-pearlitic LGG with the 25% volume fraction of the perlite in the structure. Both melts were tested in the as-cast state, without any following heat treatment (Fig. 1 and Fig. 2).

Table 1. Chemical composition and the mechanical properties of used LGGs

melt	C	Mn	Si	P	S	Cu	Mo	HB 5/2450/10	UTS [MPa]
	weight %								
A	3.55	0.53	3.01	0.030	0.033	0.49	0.025	224	702
B	3.38	0.08	2.93	0.037	0.012	0.02	0.037	144	480

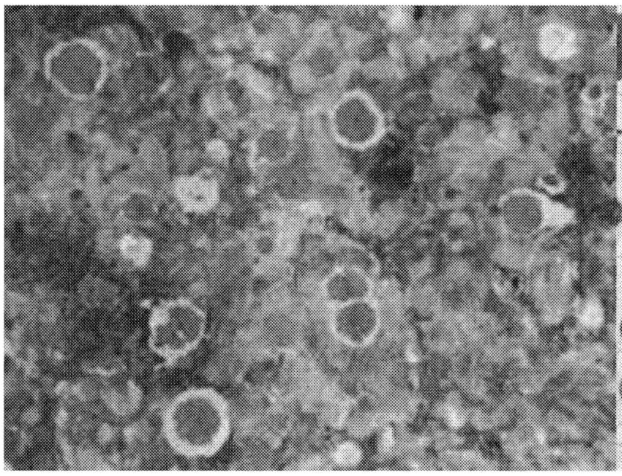

Fig. 1. Structure of the perlitic-feritic LGG (A), etch. 3% Nital, mag. 100 x

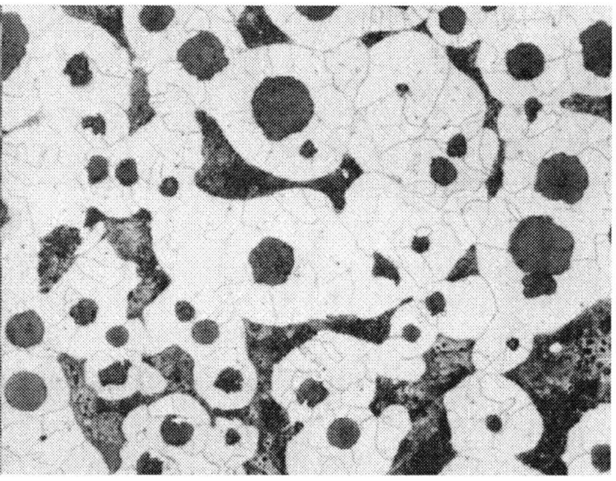

Fig. 2. Structure of the feritic-perlitic LGG (B), etch. 3% Nital, mag. 100 x

2.2 Casting defects causes of their formation

One fifth of the overall scraps of castings in the foundries is due to cavities. The castings with these defects are mostly unrepairable and in the production of a casting of the dynamically loaded parts with such errors should be discarded. These defects are caused by the presence of gases in the molten metals and the form and by the volume changes of the casting during the solidification of metal.

According to their size, they can be divided into macroscopic cavities (bubbles, shrinkages, shrinkage porosity, slags, cracks), and microscopic cavities (microbubbles, microshrinkages, microcracks).

There are two main types of porosity namely gas porosity and shrinkage porosity. In case of gas porosity, the shape of the bubbles (Fig. 3) is very diverse and they could be present in the spherical to ellipsoidal shape but also they could have irregular, elongated shape. Bubbles can be divided by origin to endogenous and exogenous. Endogenous bubbles forms from the gases dissolved in the molten metal when the pressure of the dissolved gasses exceeds the resistance against them. Endogenous closed bubbles have the metallic shiny, unoxidized surface of the walls, or covered with just a thin oxide layer. Exogenous bubbles are formed during casting if the gas pressure in the casting area exceeds the pressure of the solidifying surface of the molten metal and the gases penetrate to the casting. In such a case they can be also classified as a flooded gas. These defects have many common characteristics and often it is not possible to clearly distinguish these two types of casting defects. The susceptibility to bubble formation increases with the increasing wall thickness of castings and decreases with the increasing metallostatic pressure. These conclusions are based on the theory of heterogeneous nucleation of bubbles in castings and proved by the practical experiences. When assessing the susceptibility of casting to the formation of hydrogen, oxygen and nitrogen bubbles, the high content of carbon and silicon and their impact on the solubility limits of oxygen, hydrogen, and nitrogen in the cast iron must be taken into account. Carbon and silicon increase the solubility limits of oxygen, but at the same time reduce the solubility of hydrogen and nitrogen in the cast iron. Increasing carbon and silicon content restrict its own solubility in the cast iron. All types of bubbles have a common the underlying cause. Temperature drop during the casting and cooling of the metal decreases the solubility limits of gases in the metal and when this limit value of the solubility depending on the temperature, pressure, and composition is exceeded, the gases are excreted which result in the bubbles formation. This takes place by the mechanism of heterogeneous nucleation. Hydrogen and nitrogen bubbles are formed during solidification of the cast iron due to supersaturation of the melt by dissolved hydrogen and nitrogen. Oxygen (except the absorbed atmospheric oxygen and the water vapor) does not cause the bubble formation directly but through reaction with the carbon in molten or solid metal, which result in the formation of carbon monoxide. Bubbles made by the carbon monoxide are mostly concentrated in the near surface regions and if they are not directly connected with the free surface, their surface does exhibit neither decarburization nor oxidization. The main causes of the bubbles formation in cast iron castings are: using of inappropriate batch (large volumes of many times re-melted waste, use of corroded, wet and not blasted waste and the general use of the wet batch), unnecessary prolongation of melting process and insufficient deaeration of molds and cores, combined with their high moisture level. After pouring the molten metal into the mold cavity of sand form, a large number of gases and vapors are produced. Their amount and nature depend on the type of used molding mixtures, the weight of casting etc. In the raw molds, the moisture contained in the molding mixture is immediately converted to vapor what results in the generating considerable pressure of gases on the mold-metal interface in both directions (towards the mold and towards the metal). At this stage, two factors determine the gas flow direction, namely the permeability of the molding mixture and the crystallization process of the casting. At this moment, the metallic crust on the casting surface is of great significance, as it is impermeable. Compressed gases try to escape from their origins through the least resistance. If the permeability of the molding mixture is sufficient, the gases penetrate throughout the grains of forming material to the environment. Otherwise, the partial pressure of the gases will increase at the mold-metal interface and therefore there is the danger that the gases will penetrate into the crystallizing casting and form the slags. The formation of slags and sand blow of mold or cores is unambiguous and is related to releasing of water vapor or the gases from the organic substances contained in the raw molding mixture or in the binder of synthetic resin mixtures [9].

Shrinkage is the most common type of casting defect. There are four types of shrinkage that can occur in metal castings: cavity, sponge, filamentary and dendritic. Cavity shrinkage occurs when two different sources of molten material are joined to create a common front while solidification is already taking place. A lack of additional feed material to fill in the accumulating gaps can further exacerbate the cavity shrinkage problem. Sponge shrinkage arises in the thicker mid-section of the casting and causes a thin lattice texture similar to filament or dendrites develop. Filamentary shrinkage results in a network of continuous cracks of various dimensions and densities, usually under a thick section of the material. It can be difficult to detect and the fracture lines tend to be interconnected. Dendritic shrinkage or dendritic fractures are narrow, randomly distributed lines or cavities that are often unconnected. They are typically thinner and less dense than filamentary cracks.

Internal closed shrinkages have an irregular shape, influenced by the different rates of heat removal from the mold wall and edges, cores, the inlet systems etc. At their nonoxidized surfaces are dendrites of varying size. Usually, they are found at the heat center or under the solidified part of casting in the lower parts of casting with the interrupted supply of molten metal while this part is not still solidified. They are formed due to reducing the metal volume upon its solidification.

The shrinkage porosity (Fig. 4) can be characterized as small shrinkages which are formed in the heat center of the casting. They form in the final stage of solidification due to bulk volume shrinking in the case of imperfectly directed metal solidification. In the casting, generally, local cluster or chains of shrinkage porosity are formed along the entire length or height of the casting in its thermal axis. The clusters of shrinkage pores are most often formed in the thicker walls or their connections, or in the other local thermal centers of the casting.

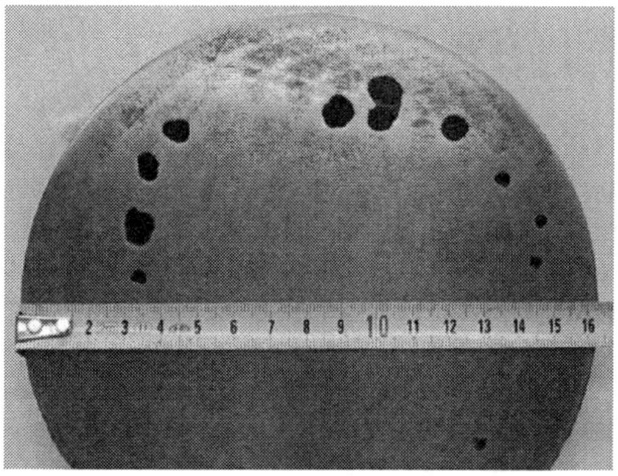

Fig. 3. Bubbles

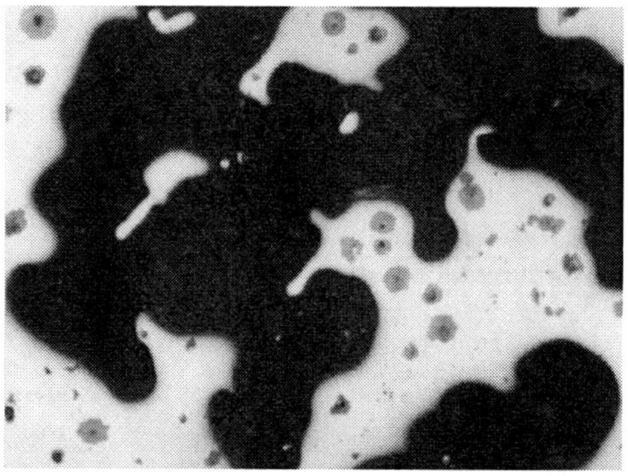

Fig. 4. Shrinkage porosity, magnification 50x

The principal causes of the microscopic bubbles formation (microbubbles, micro shrinkages, and microcracks) are similar as in the macro defects formation (bubbles, shrinkages, and hot-cracks). Similarly, they differ in both, the morphology and in the physical nature of their origin.

Microbubbles, small spotted cavities, are formed during the solidification of the alloy, just after the previous dendrite solidification, which are the most favorable places for their heterogeneous nucleation. They form due to supersaturation of the molten metal with the dissolved gases, with the same mechanism as the endogenous bubbles, but with the difference that their growth stage is limited as a consequence of significantly lower concentrations of the dissolved gases in the molten metal than in the case of large bubbles. Microbubbles are sometimes occurring together with the macroscopic bubbles and they often form in the combination with microshrinkages. The microshrinkages (Fig. 5) have predominantly crystalline character (Fig. 6) and they are formed during the solidification in the dendrites contact areas of grains of solidifying alloy. They are most often formed at places with limited molten metal supply and in alloys with long solidification interval and large shrinkage coefficient. During solidification, the most of additives, alloying elements and impurities are displaced to the remaining molten metal in the interdendritic spaces, which result in decreasing of solidus temperature and caused the formation of brittle phases and eutectics, which subsequently serves as a micro cracks sources. The microshrinkages surface is formed by a rounded end of the higher order dendrites, which growth has stopped on the closed surface of microshrinkages due to the stopped supply of the molten metal.

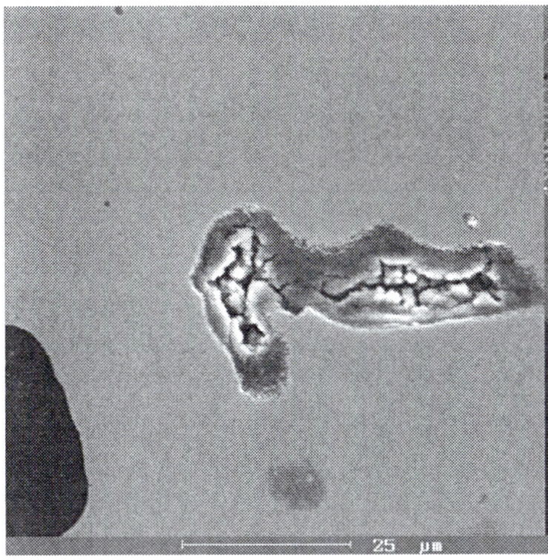

Fig. 5. Detail of microshrinkage

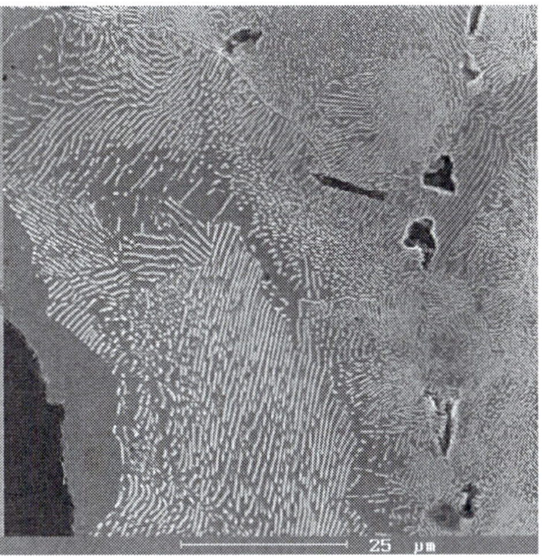

Fig. 6. Intercrystalline microshrinkages

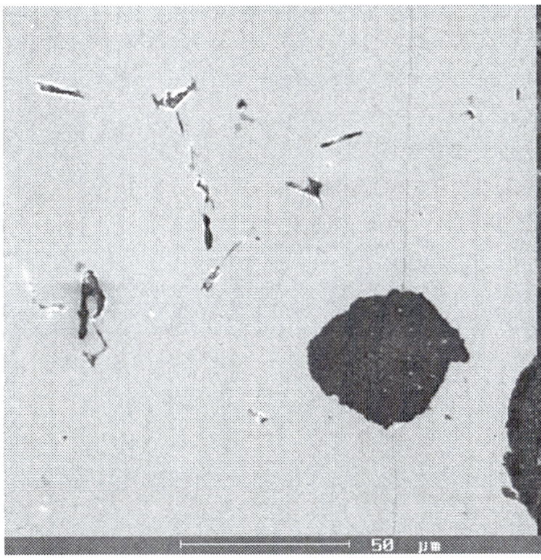

Fig. 7. Microshrinkages and microcracks

Fig. 8. Microshrinkages and microcracks on the fracture surface of the broken specimen

A sequence of defects the microbubble and the microshrinkage very often end with a formation of microcracks at high or lower temperatures. Microcracks start to form preferentially at the dendrites or grain boundaries, but also in the dendrite and grain interiors. The formation of microcracks along the dendrites and grains boundaries is characteristics for the high-temperature crack formation. The microcracks formation in the dendrite and grain interiors is more frequently at lower temperature levels and usually is associated with the phase transformations. This division of microcracks is just approximate, and origin of individual microcracks must be examined carefully. Microcracks very often coexist together with the microshrinkages, and they continuously pass into each other which make very difficult to distinguish them (Fig. 7 and Fig. 8). In the cast iron castings, which are much more heterogeneous than steel castings, the microscopic cavities are usually located at the peripheral regions of eutectic cells, most often at the point of contact of the several cells.

2.3 Fatigue tests

The fatigue tests were carried out on the ultrasonic fatigue testing system KAUP-ŽU, which is more detailed described in the works [10, 11] and the specimens were loaded by the high-frequency tension-compression sinusoidal loading ($f \approx 20$ kHz, $R = -1$, $T = 20\pm10$°C, specimens were cooled by submerging in the water with anticorrosion inhibitor). The surface of the specimen's middle section was ground and polished to eliminate any possible influence of the surface state on the fatigue properties.

3 Results

Theory of permissible defects and the linear fracture mechanics, which are used in the fatigue life prediction of parts containing casting macro defects, are well described. But these are applicable only to the low cycle fatigue ($N < 10^6$ cycles). The effect of the microscopic casting defects on the fatigue properties is usually examined in the high cycle region ($N \approx 10^8$ cycles) and the very high cycle region ($N > 10^9$ cycles). This reflects the fact that unlike in the low cycle fatigue, where the fatigue behaviour of the materials is generally controlled by the fatigue crack propagation process (initiation stage is neglected), the fatigue properties in the very high cycle region are determined by the initiation stage of the fatigue process. Due to a limited scope of the paper, the following part will deal only with the examination of microscopic cavities effect on the fatigue life of the LGG in the very high cycle region.

The result obtained from the experimental measurement of the fatigue life of the ferritic-pearlitic and pearlitic-ferritic LGGs in the very high cycle region depicted as a function $\sigma_a = f(N)$ in the Fig. 9 shows a high degree of scattering. Scattering of the fatigue life values increases with the decreasing amplitude of cyclic loading. Such a large scattering in the fatigue life of LGG in the very high cycle region are provable caused by the microscopic casting defect such as microbubbles, microshrinkages, microcracks, inclusions, a local cluster of graphite particles, imperfectly spheroidal graphite, structural heterogeneity and different grain size.

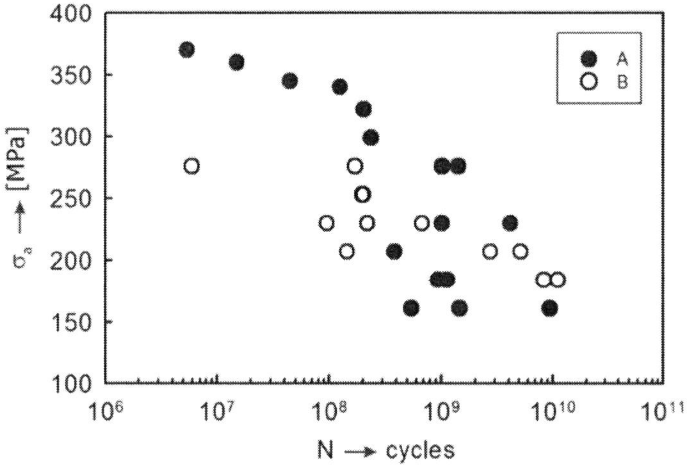

Fig. 9. Fatigue life curve, $\sigma_a = f(N)$

Above listed casting microdefects have a significant influence on the initiation stage of the fatigue process. The time after fatigue crack initiation required to the propagation of fatigue crack to such a degree that fracture occurs in the high cycle region is an only negligible fraction of the total fatigue life of the testing specimen.

Fractographic analysis of the LGG specimens broken during the fatigue tests reveals, that similarly, as in the low cycle fatigue, even in the fatigue tests in the very high cycle region there are two principal factor, affecting the fatigue life (in the material quality point of view) namely: location and size of casting microdefects. These effects were more obvious in fatigue tests in the very high cycle region, due to prolonged initiation stage, when compared with the low cycle fatigue testing. The micro defects connected with the surface of the testing specimens (Fig. 10) are from the initiation of fatigue crack point of view, the second most dangerous, immediately after the macroscopic defects, in which the initiation stage is shortened to almost zero. Microdefects in the near subsurface layers (Fig. 11) are also very detrimental, as they significantly accelerate the fatigue crack initiation process. Microscopic casting defects, which are dispersed in the volume of the testing specimens rarely serves as the fatigue crack initiators. In case of the fatigue crack initiation on the isolated microscopic defect in the bulk material, in the most cases, this initiation of the fatigue crack occurs on the larger cluster of defects, while very important role plays the notch effect and the stress fields surrounding the defect. The assessment of the notch effect of microscopic casting defects is generally very complex process and brings with it many dangers, on which the fatigue life estimations can easily fail. This is mainly related to the existence of very complex stress fields around the microscopic casting defects.

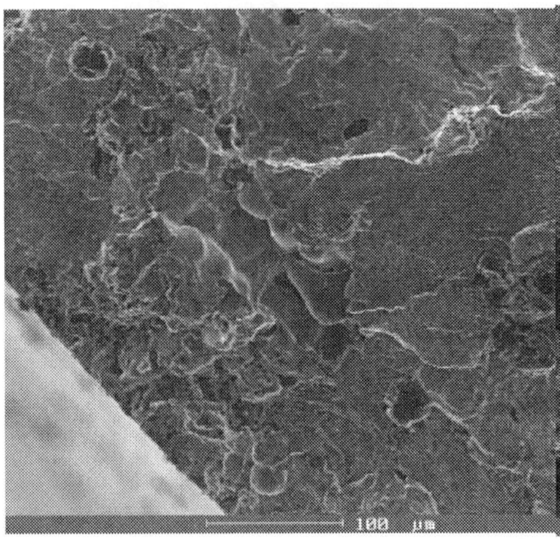

Fig. 10. Iniciation of the fatigue crack from the surface casting defect

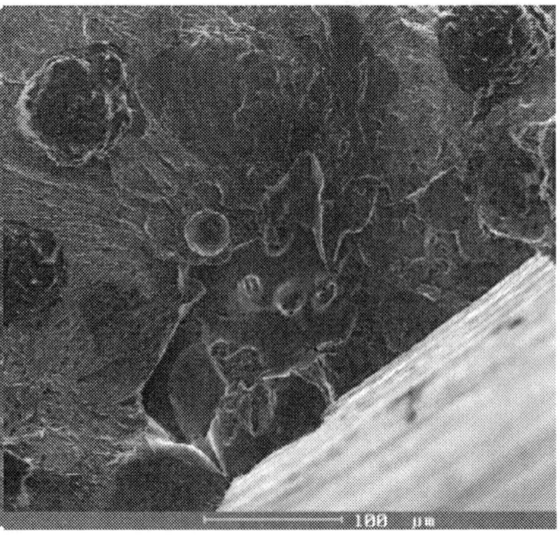

Fig. 11. Initiation of the fatigue crack from the subsurface casting defect

In many cases can occurs that the part of the fracture surfaces with a local occurrence of a cluster of microscopic casting defects do not exhibit any signs of fatigue failure (Fig. 12). In these cases, is common that the crack tip missed the microscopic defects and the area containing microdefects broke suddenly due to local overloading. As the primary initiation sites for LGG without any macroscopic casting defects are considered microcavities and the agglomerates of graphite particles which come to the surface.

Interaction of propagating fatigue crack with the microscopic defect is much more complicated and usually requires an individual approach on the case-by-case basis. There is no way to automatically assume that if there is a casting microdefect near the tip of the fatigue crack, this crack will propagate towards it. In many cases could be observed, that propagating fatigue crack in the in close proximity miss the defect, or the crack pass just through its edge and continuous in the propagation perpendicular to the load direction without deviation or branching (Fig. 13), which would be often mistakenly expected from it. On the other hand, there are also opposite cases, when the crack is unexpectedly deflected from its original direction of propagation, due to at first sight, insignificant defect, which seems to be in the sufficient distance from the former direction of crack propagation.

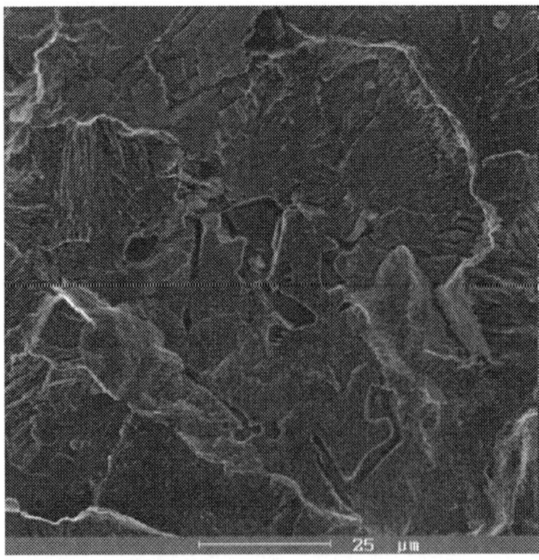

Fig. 12. Cluster of microcavities on the fracture surface

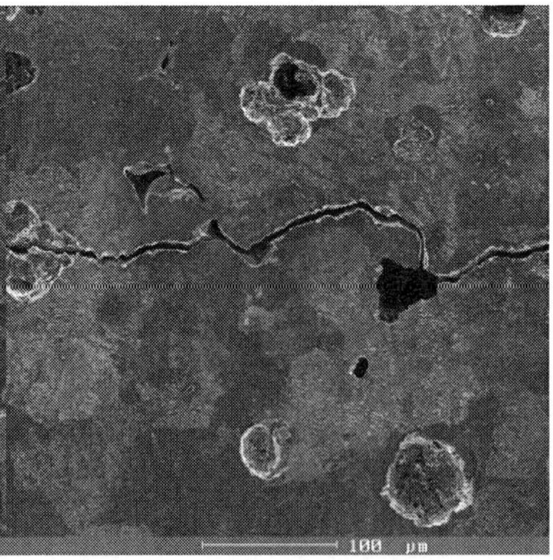

Fig. 13. Fatigue crack propagation (surface of the specimen for the fatigue crack propagation rate evaluation)

Conclusions

In this paper, authors present fatigue tests results of the pearlitic-ferritic and ferritic-pearlitic LGG in the high cycle ($N \approx 10^8$ cycles) and the very high cycle region ($N \approx 10^{10}$ cycles) obtained by the high-frequency loading. For both examined LGGs, the dependence $\sigma_a = f(N)$ has a decreasing tendency in the investigated interval. With the increasing number of loading cycles substantially increases the scattering of the fatigue life values, which is mainly caused by the extension of the initiation stage of the fatigue process. The length of the initiation stage depends on the presence of microscopic casting defects, with the most important factors were their position and size.

The increase of fatigue life of LGG casting can be successfully achieved mainly by improvements in the casting technology, which can prevent the formation of casting defect as cavities and microcavities. In particular, the following improvement should be taken into account:

- Provide a suitable composition of the molding mixture and its moisture.
- Provide sufficient permeability for molds and cores (reduce gasification level of the molten metal as are microbubbles).
- Ensure adequate collapsibility of the form.
- If it is feasible, do not make large differences in the thickness of the castings parts and the sharp transitions between them.
- Optimize the molten metal supply.
- Ensure a directed solidification of the casting by the use of suitable placed risers and chills.
- Choose an appropriate temperature of casting and ensure sufficient molten metal supply thought risers.

The research was supported by Scientific Grant Agency of Ministry of Education, Science and sport of Slovak Republic and Slovak Academy of Sciences, grants VEGA No. 1/0951/17 and No. 1/0123/15 and project "Research Centre of the University of Žilina, 2nd phase", ITMS 26220220183.

References

1. R. Konečná, M. Kokavec, G. Nicoletto, *Surface conditions and the fatigue behavior of nodular cast iron*. Procedia Engineering **10**, 2538-2543 (2011)
2. A. Vaško, L. Trško, R. Konečná, *Fatigue behavior of synthetic nodular cast iron*. Metalurgija **54** (1), 19-22 (2015)
3. A. Vaško, M. Vaško, J. Belan, E. Tillová, *Comparison of fatigue properties of nodular cast iron at low and high frequency cyclic loading*. Procedia Engineering **177**, 576-581 (2017)
4. A. Vaško, J. Belan, L. Kuchariková, E. Tillová, *Low and high frequency fatigue tests of nodular cast irons*. Metalurgija **56**, (1 2), 25 28 (2017)
5. R. Konečná, G. Nicoletto, L. Bubenko, S. Fintová, *A comparative study of the behavior of two heat-treated nodular cast irons*. Engineering Fracture Mechanics **108**, 251-262 (2013)
6. Y. Nadot, J. Mendez, N. Ranganathan, *Influence of casting defects on the fatigue limit of nodular cast iron*. Internation Journal of Fatigue **26**, (3), 311-319 (2004)
7. L. Collini, A. Pirondi, R. Bianchi, M. Cova, P. P. Milella, *Influence of casting defects on fatigue crack initiation and fatigue limit of ductile cast iron*. Procedia Engineering **10**, 2898-2903 (2011)

8. N. Shiraki, T. Watanabe, T. Kanno, *Relationship between Fatigue Limit and Defect Size in Spheroidal Graphite Cast Iron with Different Graphite Spheroidization Ratios and Microstructures.* Material Transactions **56**, (12), 2010-2016 (2015)
9. T. Elbel, F. Havlíček, P. Jelínek, P. Levíček, J. Rous, K. Stránský. *Defects of castings produced from iron alloys*. (MATECS, Brno, 1992)
10. O. Bokůvka, G. Nicoletto, M. Guagliano, L. Kunz, P. Palček, F. Nový, M. Chalupová, *Fatigue of materials at low and high frequency loading*. (EDIS ŽU, Žilina, 2015)
11. F. Nový, M. Činčala, P. Kopas, O. Bokůvka, *Mechanisms of high-strength structural materials fatigue failure in the ultra-wide life region*. Materials Science and Engineering A **462**, 189-192 (2007)

Vibration of the steel frame of the primary sand classifier

Josef Soukup[1,*], *Blanka* Skočilasová[1], *Jan* Skočilas[2]

[1]Faculty of Production Technology and Management, J. E. Purkyne University in Usti nad Labem. Pasteurova 3334/7, 400 96 Usti nad Labem. Czech Republic
[2]Czech Technical University in Prague, Faculty of Mechanical Engineering, Department of Process Engineering, Technická 4, 166 07 Praha 6, Czech Republic

Abstract. Article deals with the presentation of measurement that have been carried out on the steel frame of the primary sand classifier. The aim of the investigation was to identify the sources of the abnormal vibration of the frame. The vibration significantly reduces the effective performance of the classifier. The frame anchorage has been stripped out from concrete footing by these vibrations. The measurement of dynamic properties of the frame were performed. The displacements of the particular trusses have been experimentally investigated by accelerometers. The tensile test were performed in the places of expected maximum amplitude of the vibration. Based on the experiment results and new design calculation of the steel frame (truss construction), the new construction design was developed and improved by frame reinforcing. The new designed construction was investigated by vibration measurements again. The acceleration and total amplitude of particular trusses were recorded. The steel construction of the supporting frame has frequencies in range from 13.07 to 13.10 Hz. The significant enhancement of the noise and vibration of the main bearing have been observed during the classifier operation. The system of precautions was proposed to improve operation of the classifier and after realization the classifier run with no limitation.

Keywords: vibration, frame, acceleration, steel structures

1 Introduction

The reduction of the maximum amplitude of the vibration of the rigid construction and related noise is an optimization process which can be solved by several methods, e.g. experimental or numerical [1]. The improvement of the measuring methods or developing new methods of measurement allows to obtain more detailed or high accuracy information about investigated systems [2 - 4]. For the numerical simulation the actual common methods are used for investigation of the dynamic system, e.g. SEM or FEM [5]. The nowadays dynamic system observation has been based on classical approach of identification of the modal parameters of the system [6]. By increasing the computing capabilities, the new algorithms of numerical methods were developed, e.g. OKID [7],

* Corresponding author: josef.soukup@ujep.cz

Reviewers: *Robert* Grega, *Ján* Vavro *Jr.*

MLK [8] or FDD [9]. But usually the customer needs the analysis and improvement of the system in very short period. In such cases there is no time for numerical study and its validation. Therefore this study is pure experimental and arises from preliminary analysis [10]. From measurements the new design construction is proposed and validated its positive effect on the noise and vibration reduction.

The technological line for exploited gravel sand treatment consists of two vibrating classifiers, a coarse (with sand washer) and fine one; the screens are fitted with the respective conveying routes (belt conveyors). The gravel sand is exploited from water using a chain-bucket dredger, conveyed by the belt conveyor to the technical facility of the sand pit where the above-mentioned classifiers and aggregate yards are located.

The excavated gravel sand is conveyed by the 800 mm wide and 40 m long belt conveyor to a METSO CVB 2661 III spraying coarse vibrating classifier; the sand then travels from the classifier to a two-shaft knife washer. Big rocks are conveyed from the classifier to the rock crusher, from where the crushed rocks return to the coarse classifier, non-crushed rocks are sorted and stored for other purposes. The sand is conveyed from the coarse classifier by a steep 650 mm wide and 27 mm long conveyor to another METSO CVB 2050 III spraying vibrating classifier (not subject of the design) that classifies the sand to the individual fractions, storing them in the sand year by fraction.

In operation, the coarse classifier makes high vibrations, which are transmitted by the steel structure to the foundations and further to its surroundings. Vibrations are caused by the crusher itself; high structural deviations reducing durability of the classifier bearings.

The facility operator requested the measurement of the steel structure and casing and mechanism vibrations of the coarse classifier.

The steel structure is anchored to the reinforced-concrete foundation slab. Drawing documentation of the reinforced-concrete slab was not provided. According to the investor and findings noted in the submitted technical documentation, the reinforced-concrete slab is 1,000 mm thick, anchoring axes of supporting posts are spaced in the longitudinal direction at 4,800 mm and 3,900 mm, the total span between the end posts is 8,700 mm. Anchoring axes of the posts in the transversal direction are spaced at 3,230 mm, the post base is 400 mm wide, total anchoring width is 3,630 mm (total foundation slab width in transversal direction is 4,200 mm). The steel structure scheme is shown in Figure 1.

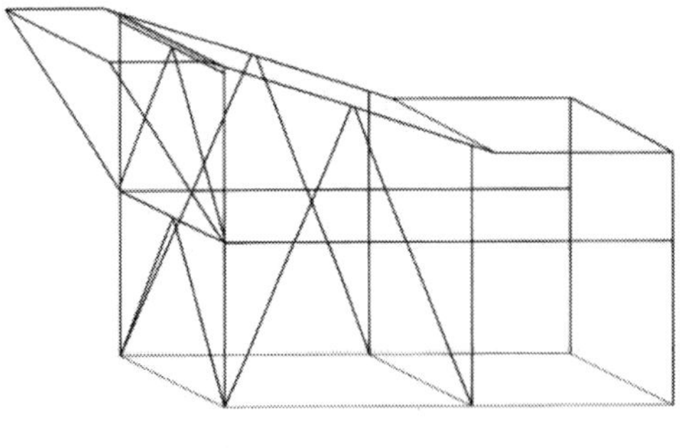

Fig. 1. Classifier steel structure scheme

Fig. 2. View of the classifier

The classifier supporting structure consists of 6 posts with transversal and longitudinal spokes. The classifier is placed on longitudinal spokes. A gallery is attached to the consoles around the classifier (on spokes). The walkway floor is made from floor grids; the walkways and gallery are fitted with railing. Posts have foundation slabs, anchored to the foundation using pasted-in HILTI M30 anchors (effective anchoring depth is 270 mm). The posts are made from HEB 240 profile (classifier installation); the main longitudinal spokes from profile HEB 240, HEA 160, U 240, U 200. Post height in longitudinal direction (3 posts) varies. Post 1 (the longest one) is 6,574 mm long, post 2 (the middle one) is 5,018 mm long, post 3 is 4,780 mm long.

The main longitudinal spoke is a welded assembly of 3 parts, welded from HEA 240 profile. The oblique part (according to placement on posts) is 6.147 mm long, 2 pieces (one from each side) made from HEA 240 profile with the length of 1,938 mm are welded to it (free bearing on the tallest post) and 2,990 mm in case of the shortest post. Welding is missing in the support area (in the centre of the post); the posts are missing in area of the classifier supports. The steel structure is made of S235 galvanised steel.

Empty coarse classifier placed on the steel structure weighs 12,640 kg, material inside the classifier weighs 1,500 to 2,000 kg (depends on the filling of the classifier).

2 Methodology

Regarding the fact that during the first assessment, the steel structure displayed design errors, the requirements for steel structure completion were not met, particularly the requirements for completing nodal points, completing transversal spatial reinforcements, spatial dimensioning (too narrow structure), insufficient dimensions of foundation slam, etc.

Sub-base in the area of the classifier steel structure contains an approx. 1 m layer of floury soil with gravel sand underneath with the thickness of as much as 12 m and the underground water level reaches as high as to the floury soil (almost to the surface in the areal of the borrow pit). As the steel structure vibrations are transmitted to the surroundings, it was not possible to directly measure the steel structure deviations.

The measurement methodology has been processed, dividing the measurements into 3 stages. In stage 1, an indicative measurement of acceleration on one web was conducted, based on which the acceleration measurement methodology was drafted. This indicative measurement provided the initial information about the steel structure vibrations, in particular, about the frequency and dominating direction of oscillation. It was determined that the steel structure oscillation frequency is 13 Hz in constant operation of the classifier, the main oscillation direction is along the transversal axis. The frequency of the frame resonance was identified during the classifier deceleration and fixed at 10 Hz. Regarding the fact that the classier driving motor speed cannot be adjusted, it is impossible to state with certainty that the frequency of 13 Hz is the frequency of resonance.

Fig. 3. Location of the 3-axis acceleration sensor

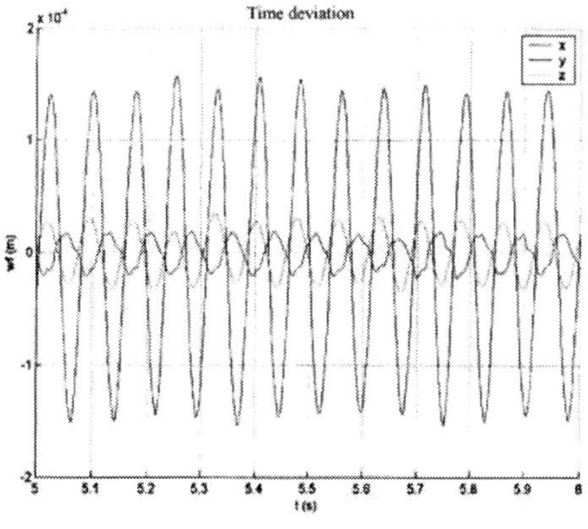

Fig. 4. Time progression of deviation

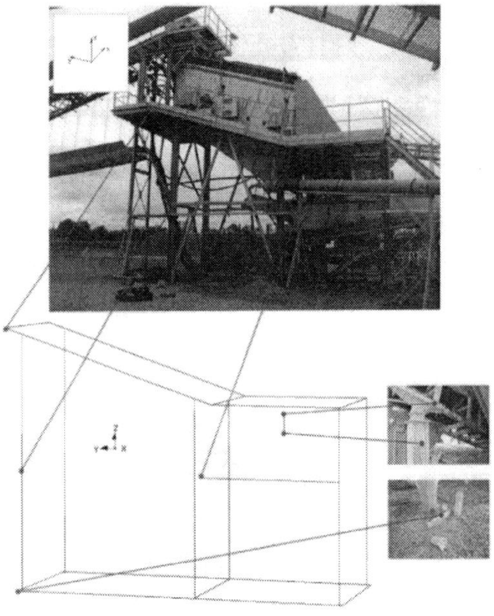

Fig. 5. Acceleration (shift) measurement model

Based on the above-mentioned measurement, a more detailed measurement was subsequently conducted, using Brül&Kjaer three-axis (4506 type, 2 pcs) and single-axis (4507 B004 type, 2 pcs) accelerometers with a PULSE (6/1) analyser, type 3560 C.

A reference accelerometer was installed on the rear post to measure acceleration along the axis. Another 3 accelerometers were relocated during the classifier steel structure measurement as required. Effective deviation values (RMS) were recorded, including their spectrum and wave shape. Maximum deviations occurred at the frequency of 13 Hz.

Another step was strain measurement. The measurement was conducted using HBM strain gauges, type 1-LY11-10/20, with an active length of 10 mm, resistance of 120 Ω, application of Z70 bonding agent.

This measurement was used to determine the steel structure oscillation frequency of 13.1 Hz (FFT analysis which indicates whether it is the dominating frequency), maximum stress range 17.92 MPa. In terms of strength, the structure is compliant, which is normal in such situations.

These measurements indicate that the steel structure is a little rigid in the horizontal plane (*x*, *z* axes) and it was recommended to reinforce the structure.
Regarding the fact that it was impossible to prepare a brand new design of the steel structure for operating and economic reasons, we did the new (static and dynamic) calculation and proposed a design adjustment.

The steel structure supplier performed internal reinforcement of the structure by inserting diagonal reinforcements in the structure (Figure 6). However, the reinforcement was performed inconsistently with the recommendation, conditions for lattice work were not met, nodal points were not created, etc. This introduced torsional and bending stress to the structure.

3 Experimental examination of reinforced steel structure

Acceleration values at selected points were measured on the reinforced steel structure. Deviations were then determined by integration (the direct determination of deviations was impossible to perform due to vibrations in the surroundings). For purposes of acceleration measurement, points between the steel structure nodes were selected so that shifts can be determined in the next step if it turns out to be meaningful (for eventual further specification of results). Acceleration in the selected points was measured in dependence on movement of the classifier (identical time axis). Measurement was conducted using 7 single-axis acceleration sensors, of which one was always installed on the classifier. The other 6 sensors were installed in various points on the steel structure (see the respective figures of the layout in the text to follow). There were 6 measurements conducted (according to layout of sensors), each measurement for 3 conditions of classifier operation

a) classifier start-up (marked x-1),

b) operation with material (marked x-2),
c) machine deceleration without material (marked x-3),

where x is the sequence number of measurement (from 1 to 6). The individual sensors are numbered from 0 to 6. Measurement time was fixed at 20 seconds. Measurement was conducted with the use of:

- Type HBM B12/200acceleration sensors, serial No. 51587, 073810245, 073810249, 073810242, 52918, 110410320, 104810051. All sensors have been calibrated
- measuring AD card NI 6225, serial No. 121B821 (HA 4198431) – calibrated
- standard laptop.

Figures 6 and 7 show positioning of acceleration sensors in 2 measurement cases (cases 1 and 6)

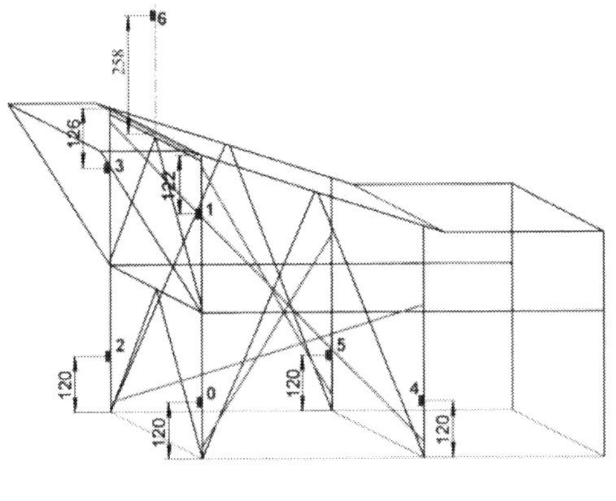

Fig.6. Reinforced steel structure scheme Location of acceleration sensors – measurement 1

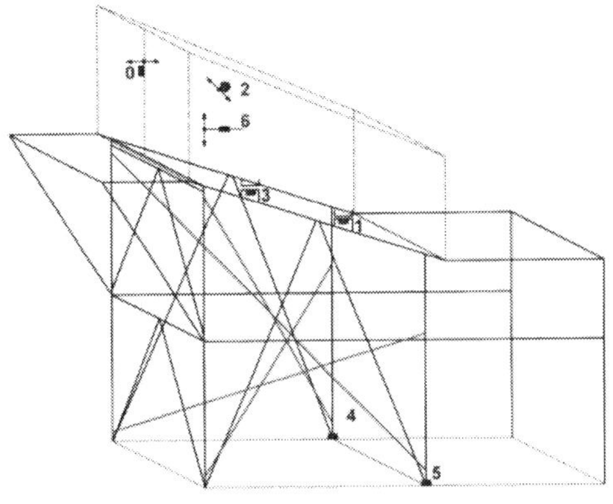

Fig. 7. Reinforced steel structure scheme Location of acceleration sensors — measurement 6

Loading forces of the steel structure from the classifier were determined based on maximum and minimum values of acceleration

$$F = m.a \tag{1}$$

where F – force [N], m – mass [kg], a – acceleration [ms^{-1}].

The forces were determined in both the horizontal direction (x, y) and in vertical the direction (z), where the vertical acceleration values were measured (measurements 4, 5 and 6). The forces were determined based on the amount of the raw material in the classifier, which was determined from the classifier output. Weight during operation with material is based on the line output (1,000 $t.hour^{-1}$ i.e. about 0.3 $t.s^{-1}$). The gravel sand holding time in the classifier is about 5-6 seconds.

The graphical acceleration presentation (Figures 8-10) only includes a part of the measured values, however, they always include the critical section. They show the acceleration trend for sensors 0 and 1 according to Figure 6. This is acceleration along the x axis on the same web in different heights from the foundation (the distance from the foundation and upper web end is shown in Figure 6).

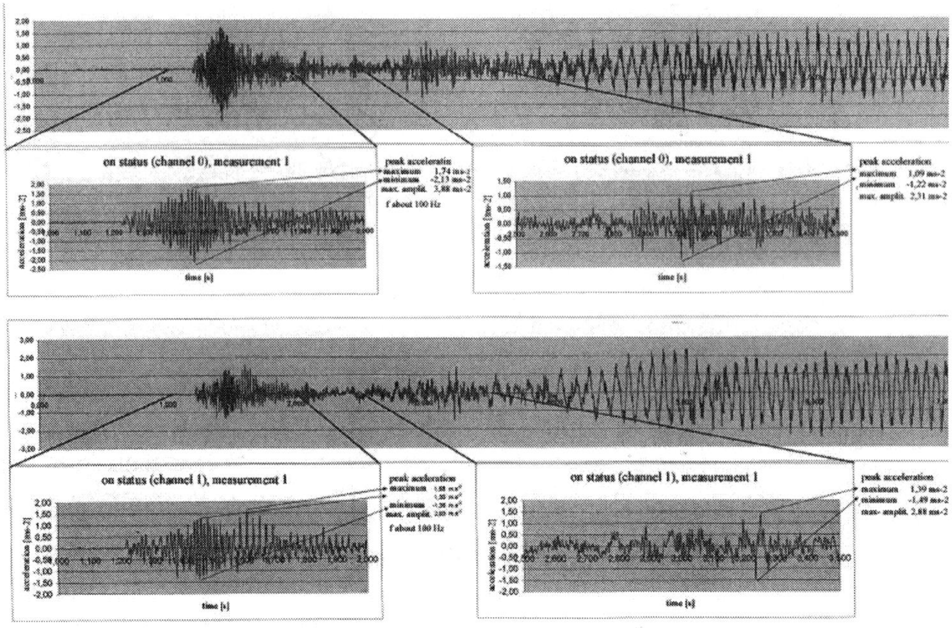

Fig. 8. Measurement 1 – classifier start-up, sensor 0 (up) and 1 (down)

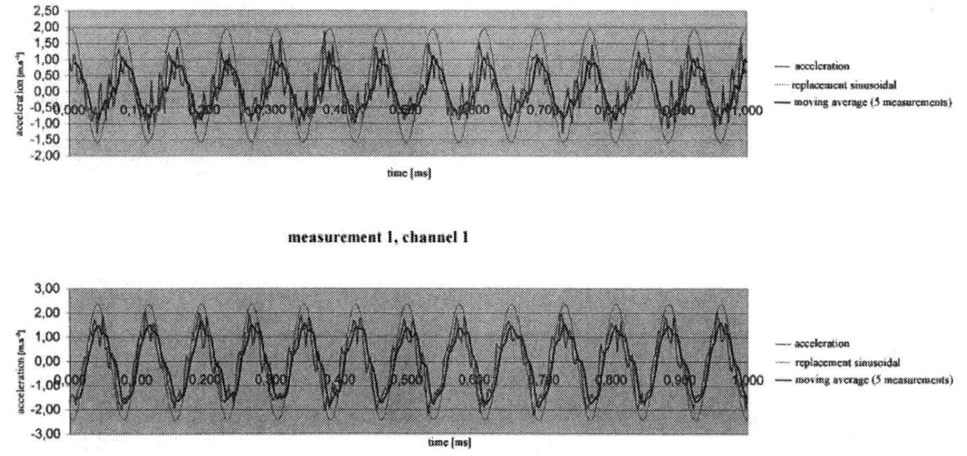

Fig. 9. Measurement 1 — classifier operation with material, sensor 0 (up) and 1 (down)

The above-mentioned figures indicate that during the classifier start-up (Figure 8), increasing speed results in a sharp increase in acceleration; acceleration decreases after passing the frequency of resonance, and increases again afterwards. Similar behaviour can be observed in all sensors. Classifier operation with material (i.e. after screens are loaded) is evidently smoother. The acceleration trend essentially copies the classifier shaft speed (Figure 9).

A big vibration of the steel structure is evident during the classifier deceleration (Figure 10) before stopping. Even after the classifier stops (about 2,800 seconds), damping occurs fast, whereas acceleration again increases in more than 10 seconds. The steel structure stops vibrating in about 18 seconds.

Maximum and minimum values of acceleration, oscillation, amplitude, mean value and acting forces were evaluated (only on the classifier). Table 1 shows the characteristic values of acceleration for measurements 1 and 6 for all sensors.

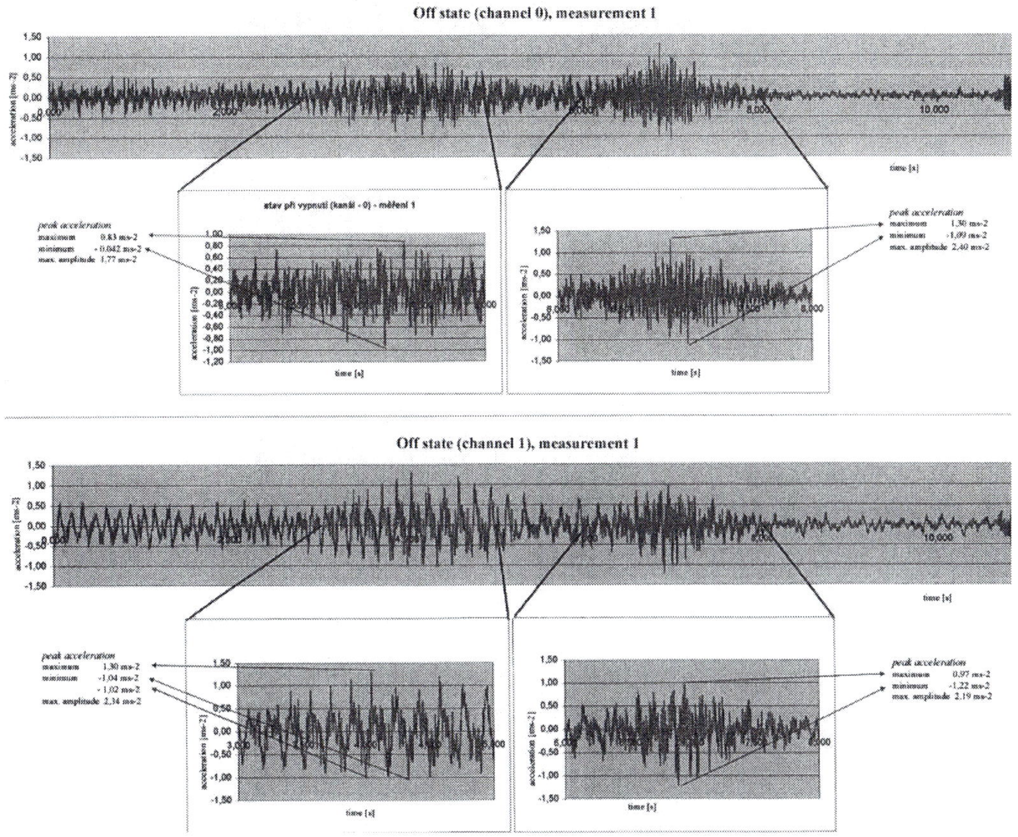

Fig. 10. Measurement 1, classifier deceleration Sensors 0 (up) and 1 (down)

Table 1. Acceleration values of the individual sensors in measurements 1 and 6

Acceleration	Unit	Sensor						
		0	1	2	3	4	5	6
Measurement 1								
Classifier start-up (without material) *interval 1.0-2.0 seconds*								
a_{max}	m.s^{-2}	1.74	1.58	1.17	0.82	1.93	2.57	3.11
a_{min}	m.s^{-2}	2.13	1.36	1.19	-0.85	-1.25	-2.44	-3.67
Classifier operation with material interval 0.0-1.0 second								
a_{max}	m.s^{-2}	1.85	2.09	0.32	2.28	1.90	2.11	29.94
a_{min}	m.s^{-2}	-1.16	-2.05	-3.27	-1.71	-2.14	-1.79	-30.68
Classifier deceleration interval 3.0-5.0 seconds								
a_{max}	m.s^{-2}	0.83	1.30	1.44	1.49	1.40	3.30	*
a_{min}	m.s^{-2}	-1.09	-1.04	-1.50	-1.75	-1.70	3.49	
Measurement 6								
Classifier start-up (without material) interval 0.0-2.0 seconds								
a_{max}	m.s^{-2}	3.28	0.79	1.84	4.52	0.20	1.23	1.40
a_{min}	m.s^{-2}	-3.08	-0.75	-1.74	-6.64	-0.17	-0.87	-1.20
Classifier operation with material interval 0.0-1.0 second								
a_{max}	m.s^{-2}	47.11	0.94	3.01	1.46	0.51	1.33	42.12
a_{min}	m.s^{-2}	-42.91	-1.25	-3.62	-1.99	-0.31	-0.97	-41.03

Table 1. (continued)

Acceleration	Unit	Sensor						
		0	1	2	3	4	5	6
Classifier deceleration		interval 4.0-5.0 seconds						
a_{max}	$m.s^{-2}$	10.15	0.96	0.91	0.79	0.21	0.67	17.25
a_{min}	$m.s^{-2}$	-9.61	-0.88	-1.61	-0.71	-0.23	-0.65	-12.96

a_{max} – maximum acceleration value
a_{min} – minimum acceleration value

The acceleration trend on sensors 2 (transversal acceleration of classifier casing) and 3 (vertical acceleration in classifier placement on supporting steel structure under silent-blocks) is shown in Figures 11-13.

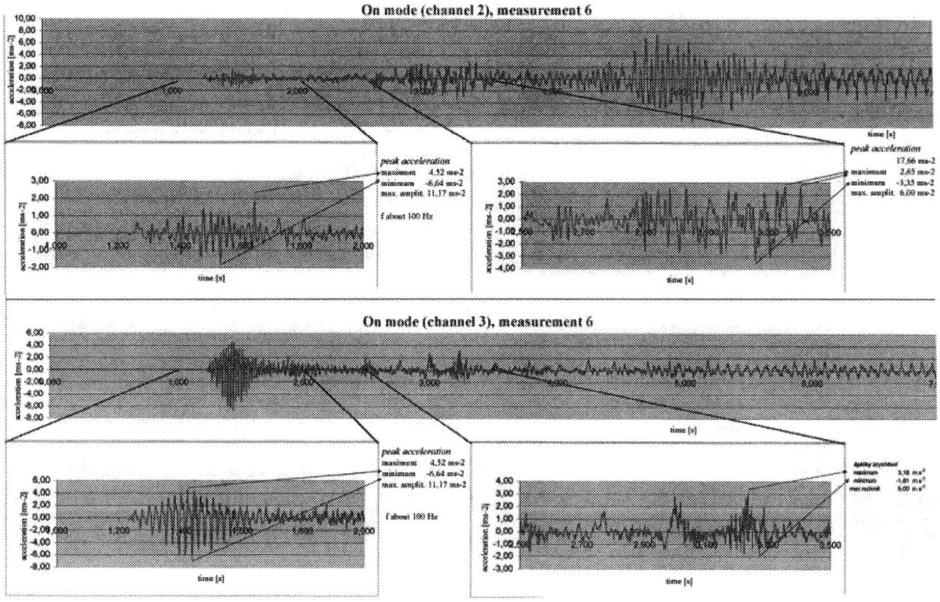

Fig. 11. Measurement 6 – classifier start-up, sensor 2 (up) and 3 (down)

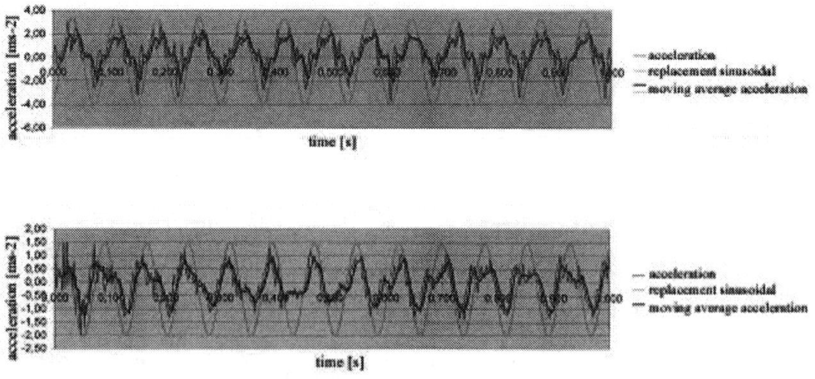

Fig. 12. Measurement 6 – with material, sensor 2 (up) and 3 (down)

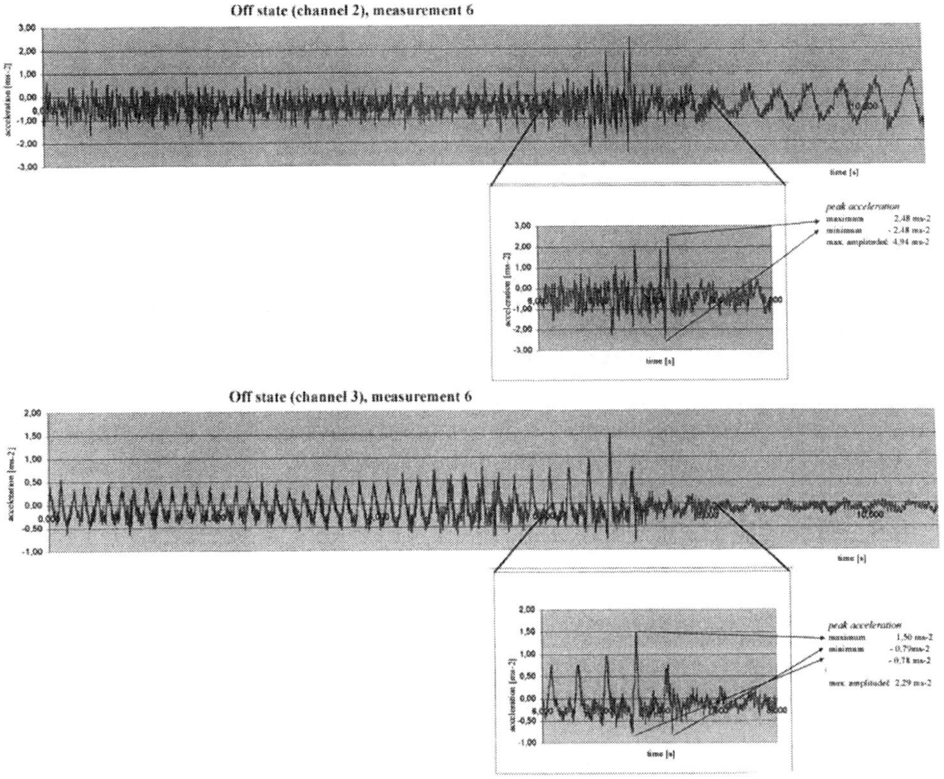

Fig. 13. Measurement 6 – classifier deceleration, sensor 2 (up) and 3 (down)

Conclusion

The measurement results indicate that the frequency of the entire structure oscillation ranges from 13.07 to 13.1 Hz in all measurement cases, which demonstrates that all measurements are commensurable.

Although the steel structure of the CVB 2661 III classifier was made stronger (reinforced), which definitely reinforced the entire structure, we have reservations about its structural design, which was also confirmed by the measurements. The structure rigidity in the horizontal plane (along both axes) is low, mainly in the transversal direction. The structure is too narrow with respect to the acting load and height.

Due to the low rigidity of the structure, big deviations occur on its upper side, which are amounting to± 12.5 mm in transversal direction on the longest web. This results in big acceleration and thus in big forces produced by the classifier operation, acting on the steel structure. Measurements on the bases of the steel structure indicate that the steel structure is affected by torsional forces; vertical forces in bases against the anchorage are too high (see the new stress analysis).

New static calculation of the entire reinforced steel structure was conducted, also considering dynamic forces from acceleration as the input values of forces from the classifier in addition to the dead weight.

This new calculation indicates that the steel structure is compliant in terms of strength; however, the structure deflection is not compliant. This is the reason why it was recommended to refit all internal reinforcements of the steel structure according to the original proposal (create nodal points) and design the steel structure reinforcement from the outside. This will require the foundation slab to be expanded for purposes of brace anchoring.

Insufficient rigidity of the steel structure and thus high values of vibrations result in the limited durability of the main shaft bearings of the coarse gravel sand classifier.

References

1. V. Wowk, *Machinery vibration: measurement and analysis*. Boston: McGraw-Hill, 358 p., ISBN 0-07-071936-5, (1991)
2. S. P. Hyo, P. Keunhyoung, K. Yousok, W. Ch. Se, *Deformation Monitoring of a Building Structure Using a Motion Capture System*. IEEE/ASME Transactions on Mechatronics **20** (5), 2276-2284 (2014)
3. T. U. Yılmaz, U. O. Özugurel, K. Bulut, M. N. Inci, *Vibration amplitude analysis with a single frame using a structured light pattern of a four-core optical fibre*. Optics Communications **249** (4–6), 515-522 (2005)
4. D. Zhu, J. Guo, Ch. Cho, Y. Wang, K. Lee, *Wireless Mobile Sensor Network for the System Identification of a Space Frame Bridge*. IEEE/ASME Transactions on Mechatronics **17** (3), 499-507 (2012)
5. S. L. Zuo, F. M. Li, Ch. Zhang, *Numerical and experimental investigations on the vibration band-gap properties of periodic rigid frame structures*. Acta Mechanica **227** (6), 1653-1669 (2016)
6. G. Fraraccioa, A. Bruggerb, R. Bettic, *Experimental studies on damage detection in frame structures using vibration measurements*. Shock and Vibration **17**, 697-721 (2010)
7. J. N. Juang, M. Phan, L. G. Horta, R.W. Longman, *Identification of Observer/Kalman FilterMarkov Parameters: Theory and Experiments*. Journal of Guidance, Control, and Dynamics **2**, 320–329 (1993)
8. R. Brincker, L. Zhang, P. Andersen, *Modal identification of output-only systems using frequency domain decomposition*. Smart Material and Structure **10**, 441–445 (2001)
9. M. De Angelis, H. Lus, R. Betti, R. W. Longman, *Extracting Physical Parameters of Mechanical Models From Identified State Space Representations*. Journal of Applied Mechanics **69**, 617–625 (2002)
10. B. Skočilasová, J. Volek, J. Soukup, *Report on Steel Structure Experimental Examination Results*. Ústí n. L., (2009)

Fatigue limit estimation using IR camera

Zuzana Stankovičová[1,*], *Vladimír* Dekýš[1], *Milan* Uhríčik[1], *Pavol* Novák[1], *Bohumír* Strnadel[2]

[1]University of Žilina, Faculty of Mechanical Engineering, Department of Applied Mechanics, Slovak Republic
[2]Center of Advanced Innovation Technologies, VŠB-Technical University Ostrava,Czech Republic

Abstract. Determination of fatigue limit by traditional method is associated with testing of large number of specimens. This is caused by more scatter in test results and therefore a higher amount of tests are needed to generate a statistically reliable trend. The paper presents rapid and reliable estimation of fatigue limit using infrared thermography. By analysing the temperature of the material surface during cyclic loading it is possible to evaluate the dynamic behaviour of a structure and determine the fatigue limit. The method provides reliable results using a very little number of specimens in a very short time compared with traditional method.

Keywords: Fatigue limit, Wöhler diagram, infrared camera, thermography

1 Introduction

Repeated mechanical, thermal or mechanical-thermal stresses of the material result in a gradual accumulation of damage. This process is known as material fatigue. Material failure due to fatigue can occur only in the presence of microcraks or defects in material. If a static load occurs on a sample, local plasticization and redistribution of stress to the surrounding material does not create a critical condition and material breaks only at higher stresses according to the strength condition. In the case of cyclic loading, if the material comes into the state of local microplasticity, a microcracking occurs. Repeated application of the same loads results in crack propagation up to the fracture [1].

The fatigue limit σ_C is used as a material resistance against such repeated loading. This quantity is theoretically defined as the minimum stress level below which material is not damaged. It is an experimentally determined value that indicates the stress or plastic strain level at which the material can withstand an infinite number of cycles [2].

1.1 A traditional approach to fatigue assessment

By classic approach, the fatigue limit is determined by experimentally compiling the fatigue curve. It is necessary to perform a series of fatigue test at different stress levels into the failure. The numbers of cycles, in which samples are broken, are observed. The fatigue curve is constructed, usually in the coordinate σ-N (stress level – number of cycles to

* Corresponding author: zuzana.stankovicova@fstroj.uniza.sk

Reviewers: *Wiesława* Piekarska, *Ján* Vavro

fracture). This task is time-consuming, especially in the case of high-cycle fatigue, where there is more scatter in the results. Therefore more testing is needed to create a statistically reliable trend. The fatigue test for high-cycle fatigue is usually performed in 10^7 cycles and for non-ferrous metals up to 5.10^8 cycles.

The area of high cycle fatigue is characterized by the Wöhler diagram. This area can be characterized by Wöhler's relationship in the alternating cycle:

$$\sigma_{max} = \sigma_f'(N_f)^b \ , \qquad (1)$$

where σ_f' is coefficint of the fatigue strength obtained by extrapolating the lifetime curve to first load cycle and b is exponent of fatigue lifespan and its value for metals is moving in the ranfe of -0.05 to -0.12. These parameters are determined experimentally. In the past two decades, several articles have been published [3-5] related to the use of IR thermography in fatigue tests. Based on the energy emitted from the material and captured by the IR camera during a fatigue test it is possible to determine the fatigue limit and even the whole fatigue curve with a smaller number of samples than the classic fatigue test.

1.2 Using infrared thermography for fatigue tests

In one of the first works [6] the authors compared the traditional method for the construction of a fatigue curve with temperature measurements on the outer surface 57 samples, using the same loading frequency for the duration of the test (more than 250 hours of recording). A thorough examination of the recorded data highlighted the important aspects that were then subject to further investigation [4]. If the applied stress was below the fatigue limit, the temperature increase in the warmest place was practically zero. If the value of stress reaches the fatigue limit, the temperature increases during the first part of the test (phase 1), then it remains constant (phase 2) until the short time before the failure, followed by a rapid increase of the temperature until the fracture (phase 3). Fig. 1 shows the given phases during the test.

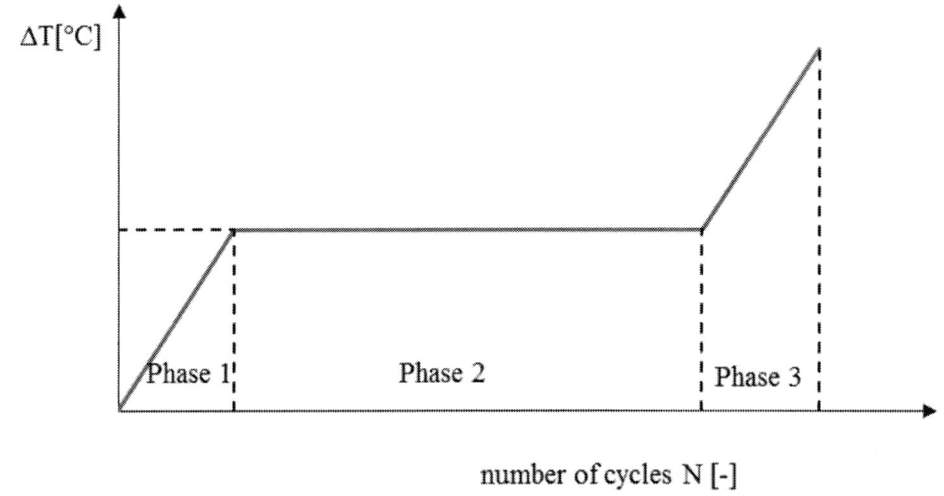

Fig. 1. Phases during the fatigue test [4]

Analysis of fatigue damage using infrared thermography can be done by evaluating dissipative energy, where heat production correlates with energy dissipation in the material. Luong [5] developed a method for the rapid evaluation of fatigue limit based on the detection of dissipated energy from the material loaded with cyclic loading.

Based on Luong's approach, Bremond and Potet [7] developed the "D-mode" technique. This is digital imaging, which is implemented in the commercially available Altair LI software (module is enable to detect heat generation due to local plastic deformation at various levels of fatigue amplitude). There is minimal energy dissipation for loading under fatigue limit. Under loading greater the fatigue limit a microplastic deformation occurs resulting in a rise in temperature. The test is carried out by gradually increasing the stress level. The dependence of the stress level on the dissipative energy in the form of temperature $T_{D,i}$ is graphically constructed. The voltage at which the temperature increases sharply corresponds to the fatigue limit value obtained from the classical fatigue test [8].

2 Fatigue limit determination using IR thermography

Determination of the fatigue limit using infrared thermography was used for cold-rolled material S235J. In the experiment flat bars were used. The shape of the sample and its dimensions are in Fig. 2.

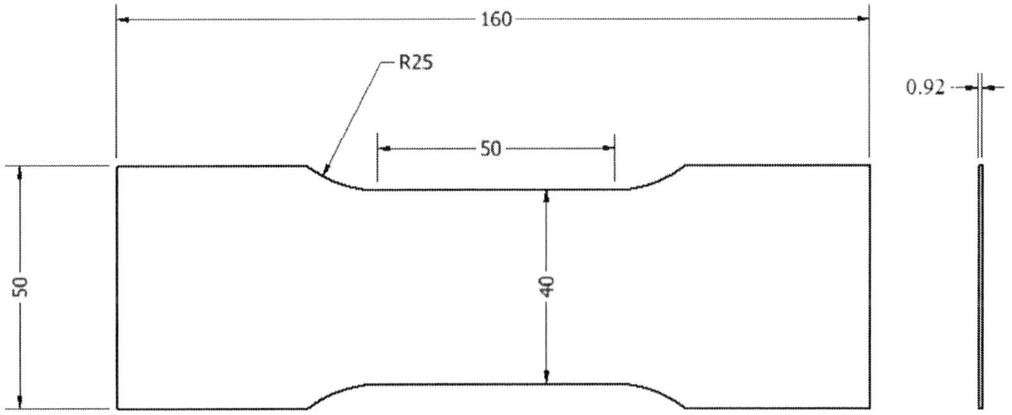

Fig. 2. Sample used in experiment

The test was carried out using a ZWICK Roel Vibrophores Amsler 150 HFP 5100 Resonant Pulsator. The samples were loaded sequentially, with a static and dynamic force value being entered, which are essentially the mean value and the amplitude of the load cycle. Since a thick flat sheet was used for the measurement, an asymmetry $R = 0$ was chosen to avoid sample buckling. Table 1 shows the amplitude values of the force and the corresponding stress at three measurements. Some parts of the table are not filled because they have been skipped in these modes, for example due to the properties of the test device.

Table 1. Amplitude values

	Amplitude value of the force[N] / **Amplitude value of the stress [MPa]**												
1.	2720	2880	3040	3200	3360	3520	3680	3760	3840	3920	4000	4160	4480
	74	78	82	87	91	95	100	102	104	106	108	113	121
2.	2720	2880	3040	3200	3360	3520	3680	3760	3840	3920	4000	-	-
	74	78	82	87	91	95	100	102	104	106	108	-	-
3.	2720	2880	3040	3200	3360	3520	-	-	3840	-	4000	4160	4480
	74	78	82	87	91	95	-	-	104	-	108	113	121

FLIR SC7500 IR camera was used to measure temperature changes. The measurement was used frame frequency $f_{frame} = 383$ Hz. A white emission spray was applied to the samples to reduce the reflection from the surroundings (the spray is for LWIR, in our case the spray was designed to homogenize the surface to be measured). The sample was shielded with polystyrene blocks because there was a change of weather during the day, which could affect the measurement. The evaluation was performed in the Altair LI software in the D-module dissipation. The processing was selected for 10 seconds, representing 3830 frames. At each stress level, the measurement was started only after the 10 000 cycles that the material had needed to stabilize the temperature. In the thermograms, the area was selected (the same in each measurement) and processed. This processing was carried out for all stress levels on three samples. Subsequently, a graph of the dissipative energy dependence in the form of temperature from the stress level was plotted - Fig. 3 because the change of this energy during the test can be expressed by changing the temperature of the sample.

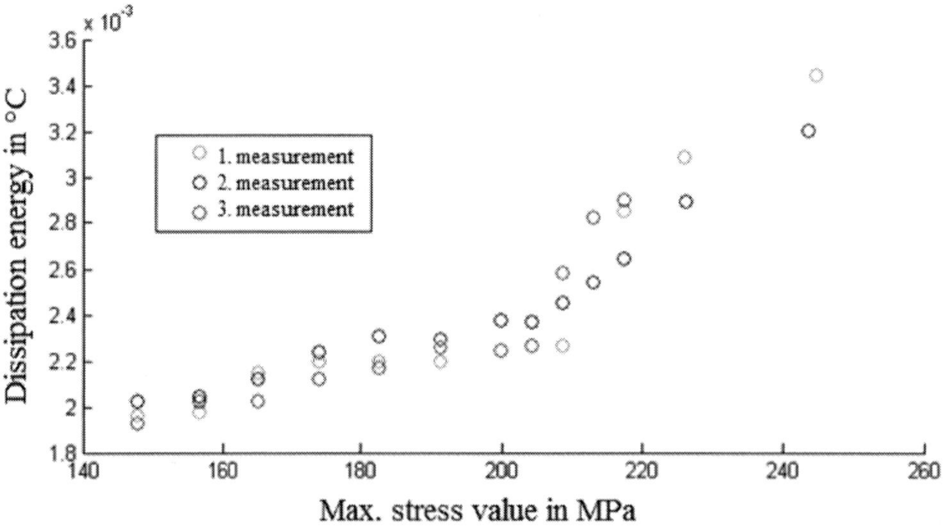

Fig. 3. Results of measurements

On further processing we decided to average the results of the measured data for each stress level. The resulting graph is in Fig. 4, where there is also a visible break within 200-210 MPa. The points are the order numbers of each load level.

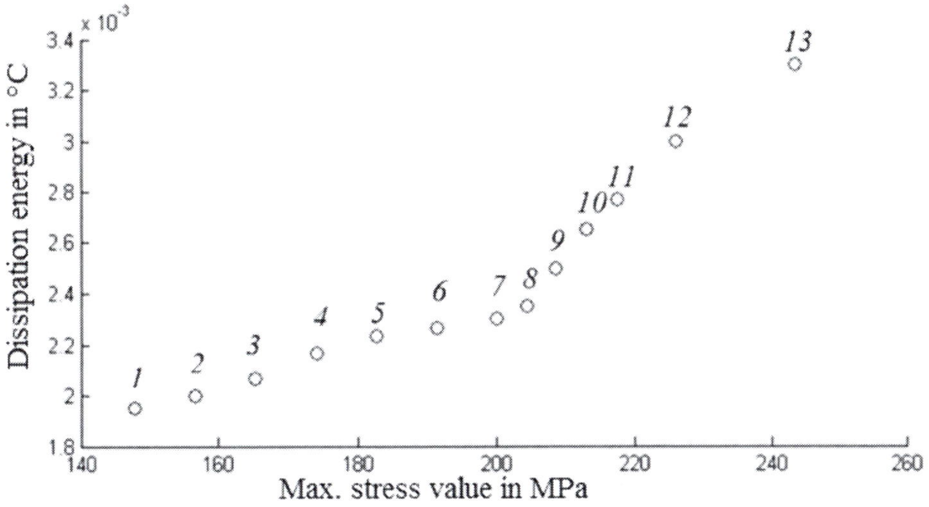

Fig. 4. Results of measurements

The method of slopes was used to evaluate the fatigue limit. For each two points, the slope was calculated from their coordinates. Then a graph of the changes of slopes was plotted, on which an increase in 7th ratio is visible. It is difference between the 7th and 8th slopes, which are points 7, 8 and 9. In the figure 5a is the resulting graph, where the red dot highlights a significant break that indicates a fatigue limit 204.35 MPa. Estimation was corrected using 2 linear regression lines. Each of two groups of points, by which were determined the lines, included red dot. In Fig. 5b is the results of the approximation, the intersection of lines corresponds the stress value of 201.54 MPa. This corresponds to the level of dissipative energy expressed as a temperature 0.0024°C.

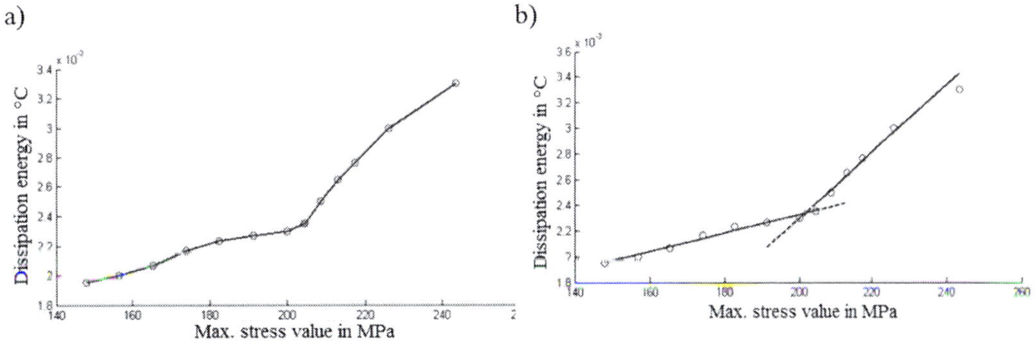

Fig. 5. Results of measurements

Value of fatigue limit from the experimental measurement from literature for steel sheet and material S235J equals to 195 MPa [9]. The difference between the values is close to 4 percent.

Conclusion

Paper deals with the application of infrared thermography in fatigue tests. The experiment was performed on a material S235J. For a quick determination of a fatigue limit the evaluation of dissipation energy was used. The measured energy values were processed in the D-module of the Altair LI software in the form of temperature. In total, three measurements were performed in order to average the results. The result of fatigue limit was compared with the experimental measurement of the material from the literature. The difference between the values is close to 4 percent.

This contribution is related to dynamic analysis work [9, 10] at the Faculty of Mechanical Engineering, Department of Applied Mechanics, University of Žilina [11-14] and partners [15, 16].

This work is supported by the Slovak Research and Development Agency under the contract No. APVV-0736-12 and by KEGA 015ŽU-4/2017 Digital technologies for machine building study programs as a part of strategy "Internet of Things" and VEGA 1/0983/15.

References

1. G. Fargione et al., *Rapid determination of the fatigue curve by the thermographic method.* International journal of fatigue **24**, 11-19 (2002)
2. A. Puškár, R. Pohl, *Dynamická pevnosť a únavová životnosť.* (Bratislava, Alfa, 1987)
3. F. Risitano, et al, *Determination of the fatigue limit by semi static tests.* Cassino, Italy, 13-15 Giungo, ISBN 978-88-9594-36-6 (2011)
4. G. Rosa, A. Risitano, *Thermographic methodology for rapid determination of the fatigue limit of materials and mechanical components.* Internationl journal of fatigue **22**, 65-73 (1999)
5. M. P. Luong, *Nondestructive damage evaluation of reinforced concrete structure using infrared thermography.* Society of photo-optical instrumentation engineers, Newport Beach, CA, USA, 98-107 (1995)
6. G. Curte et al, *Analisi tramite infrarosso termico della "temperatura limite" in prove di fatica.* [in Italian] 14th AIAS Italian National Conference, Catania, Italy, 211–20 (1986)
7. P. Bremod, P. Potet, *Lock-in thermography: a tool to analyze and locate thermomechanical mechanisms in materials and structures.* Proceedings of SPIE 4360, **560**, (2001)
8. A. Akai et al., *Relationship between dissipated energy and fatigue limit for austenitic stainless steel.* 15th International conference on experimental mechanics, Portugal, (2001)
9. www.totalmateria.com
10. F. Novy, M. Cincala, P. Kopas, O. Bokuvka, *Mechanisms of high-strength structural materials fatigue failure in ultra-wide life region.* Materials Science and Engineering A-Structural Materials Properties Microstructure and Processing **462** (1-2), 189-192 (2007)
11. G. Nicoletto, O. Bokuvka, L. Collini, P. Kopas, *Fatigue resistance in a very high-cycle regime.* Transaction of FAMENA **29** (1), 9-16 (2005)

12. M. Zmindak, P. Pastorek, *Finite element analysis of cohesion between reinforced concrete beam and polymer lamella reinforced by carbon fibers*. Procedia engineering **177**, 582-589 (2017)
13. P. Kopas, L. Jakubovičová, M. Vaško, M. Handrik, *Fatigue Resistance of Reinforcing Steel Bars*. Procedia Engineering **136**, 193–197 (2016)
14. M. Handrik, P. Kopas, V. Baniari, M. Vaško, M. Sága, *Analysis of stress and strain of fatigue specimens localised in the cross-sectional area of the gauge section testing on bi-axial fatigue machine loaded in the high-cycle fatigue region*. Procedia Engineering **177**, 516-519 (2017)
15. B. Strnadel, P. Ferfecki, P. Židlík, *Statistical characteristics of fracture surfaces in X70 steel DWTT samples*. Engineering Fracture Mechanics **112**, 1-13 (2013)
16. J. Xiong, P. Guo, Y. Cai, B. Strnadel, J. Brumek, Y. He, H. Gu, *Structural, magnetic and nanomechanical properties in Ni-doped AlN films*. Journal of Alloys and Compounds Volume **606**, 55-60 (2014)

The experimental measurement of the tyre casing defects for the freight vehicles at the dynamic loading

Ján Vavro[1,*], *Ján* Vavro *Jr.*[1], *Petra* Kováčiková[1], *Jakub* Híreš[1]

[1]Faculty of Industrial Technologies in Púchov, Alexander Dubček University of Trenčín, I. Krasku 491/30, 020 01 Púchov, Slovac Republic

Abstract. The paper deals with the experimental measurement of the tyre casing defects for the freight vehicles at the dynamic loading. The given investigation was mainly focused on cracks and defects in tyre casing, which is commonly used for freight vehicles. The mentioned measurement or investigation was performed under the strictly specified conditions. The given measurement of crack propagation in car tyre casing was based on changes of speed, changes of loading as well as changes of service time interval. It is important to point out that the pressure in the tyre was predetermined to be the constant value. The experimental procedures were carried out by help of special drum testing machine with designation: IGTT 25 kN – C and with the 1707 mm drum diameter. The mentioned measurement was performed for the area of tyre casing shoulder and it was in the site between the first and the second buffer lining. Using the ITT- 1 program relating to non-destructive testing analyser, the CCD camera for testing procedure of tyre casing was used for the camera images which were taken from the internal side of the tyre casing.

Keywords: measurement, holograph, dynamic loading, non-destructive method

1 Introduction

The effort to improve the road-traffic safety leads to systematic enhancement of tyres. All producers endeavour to manufacture such tyres which would be reliable, safety, easily mountable as well as they would have long-term lifetime which is connected with wear-resistance. Moreover, the given tyres should be available from the aspect of cost and they should not have any negative influence on living environment. Enormous attention is also paid to occurrence of various material defects and failures, such as separations, air pockets, bulge and so on. Mutual adhesion between particular materials of which the tyre is made is also seriously investigated [1, 2]. Detection of defects, failures, separations in a tyre casing can be done by several methods. Destructive as well as non-destructive methods can be used for investigation and they help us to find defects occurring in new, re-treaded as well as old and worn-down tyres. Real disadvantage in relation to destructive methods is closely connected with the fact that the given tyre casings which are tested by destructive method

* Corresponding author: jan.vavro@fpt.tnuni.sk

Reviewers: *Jozef* Bocko, *Vladimír* Dekýš

are not able to be used again and again for testing, because destructive testing means destruction or rupture of tested object and in our case it is the tyre casing. It means that destructive methods are difficult from these aspects:

- from the economic aspect, because destructive testing means usage of high number of tested tyres,
- from the aspect of preservation of specific homogeneity for sample.

Therefore, there is the tendency to use non-destructive methods of testing in relation to tyre casings and the main contributions and advantages are:

- analysis of tyre casing without its damage or rupture,
- possibility to control the product and its following utilization for other purposes,
- the given method can be used for the process of tyre development as well as during the changes relating to manufacturing, during the control of quality of tyre casings and moreover, the mentioned method of testing can be also used for fatigue tests, speed tests and lifetime tests
- non-destructive analysis can be done again and again in relation to the same testing sample or product
- according to the type of used non-destructive method, it is possible to detect product defects in the course of a few minutes.

2 Structure of the tyre casing

It is necessary to have knowledge relating to structure of the tyre casing because only this is the correct way for specification of areas where the separations are initiated and moreover, it is also much easier to specify the distribution of the mentioned separations. The structure of tyre casing [3] consists of:

1) Tread – it is important part of tyre casing, because it is in the close or direct contact with the road surface. It is made of blend which has good adhesive properties and high wear resistance;
2) Buffer linings – it absorbs circumferential stresses as well as side forces and it absorbs also impacts which occur during the contact with road. It consists of individual layers of rubberized cord while the given layers are laid in a criss-cross formation;
3) Cord carcass of casing – it is the basic and supporting part which comprise of one or several linings from rubberized cord and they are laid around the bead plies;
4) Filling linings – they are the shaped rubber profiles and they are used for better and smoother mutual junction of individual construction parts relating to tyre casing;
5) Bead bundle – it reduces deformation in the area where there is the end of cord carcass and protection of bead;
6) Sidewall – it protect side part against damage and weather conditions. It is made of blend which is resistant to flex cracking and cracking in common;
7) Inner skim rubber – it is rubber lining which can be found on the internal side of tyre casing. It is used for protection of cord carcass and in the case of tubeless tyre casings, it avoid diffusion of air into the cord carcass of casing;
8) Bead reinforcement – it can be fabric as well as steel;
9) Bead plies – they consist of steel wires or strips which have high strength. They provide smooth and safe anchoring of cord carcass linings as well as attachment of tyre casing to wheel rim.

The Fig. 1 shows the given parts mentioned hereinbefore.

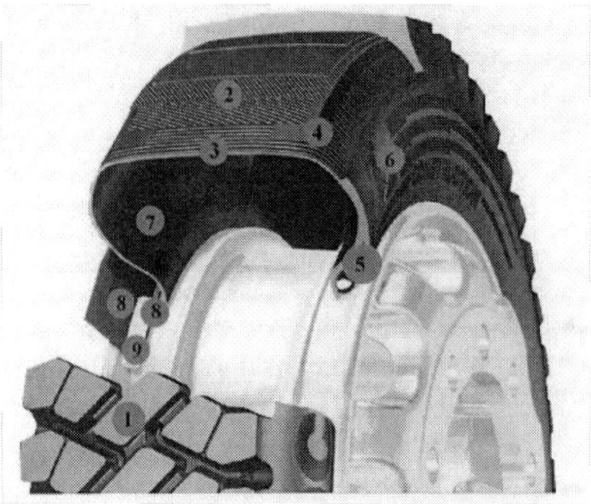

Fig. 1. Structure of tyre casing: 1) Tread; 2) Buffer linings; 3) Cord carcass; 4) Filling linings; 5) Bead bundle; 6) Sidewall; 7) Inner skim rubber; 8) Bead reinforcement; 9) Bead plies

3 Descriptions of tyre defects

The defects which have serious influence on the road-traffic safety and reliability can be classified, as follows:

- defects which occur during manufacturing (manufacturing defects)
- defects which occur during tyre operating (operating defects)

Especially, air pocket in the tyre carcass can be included into manufacturing defects because it can be source of later separation and it means that the lifetime as well as reliability of given tyre decrease. Some other manufacturing defects are closely concerned with irregular or asymmetric construction of tyre casing and these defects cause lower wear resistance. Advanced producers of tyres use the special register or list relating to defects and on the base of this register or list, the final products are classified into the specific quality grades or classes.

Operating defects involve:

a) defects which are caused during mounting on vehicle and they can occur after non-professional service or after wrong operating of service devise.
b) defects which are caused by own incorrect operating due to the overloading of vehicle, incorrect tyre pressure or even the wrong driving technique.

4 The process of measurement by help of non-destructive analyser

The distribution of separations was investigated by help of special drum testing device IGTT 25 kN – C where the diameter of drum was 1707 mm (Fig.2) and by help of non-destructive analyser see (Fig. 3).

Fig. 2. Drum testing device

Fig. 3. Non-destructive analyser

Non-destructive analyser enables us to recognize tyre structure defects quickly and easily – it is connected with closed separations (Fig. 4), the extension of which we will observe by the dynamic test. Typical arrangement of the tyre control with help of the simple cycle for tyre impurity finding is in the Fig. 5.

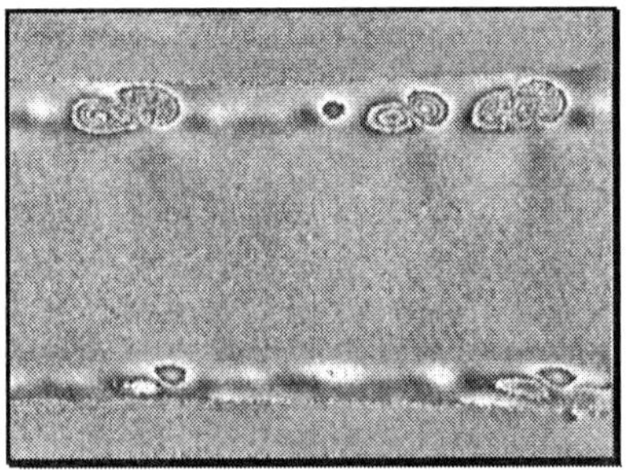

Fig. 4. Defect displayed on monitor

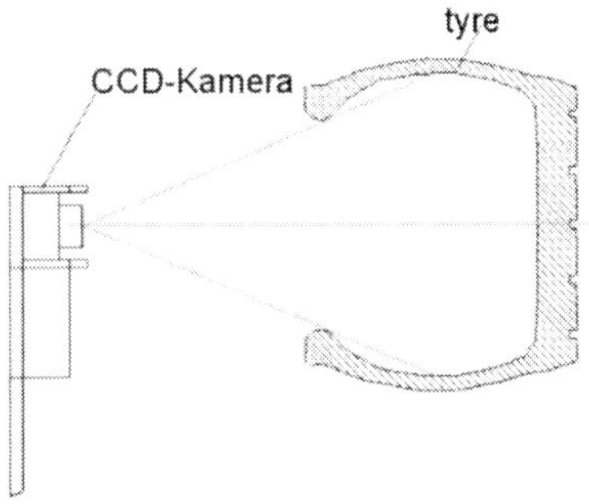

Fig. 5. Testing assembly by the simple cycle

For observation of extension of the tyre separations, the specific tyre was selected and as it can be seen in Fig. 6, the separations in the shoulder area as well as crown area were found by the help of the ITT-1 testing analyser. After 20 hours of the dynamic test, there were only small changes in separations extension, therefore we introduce the display only after 200 hours of the dynamic test (Fig. 7) [4–6].

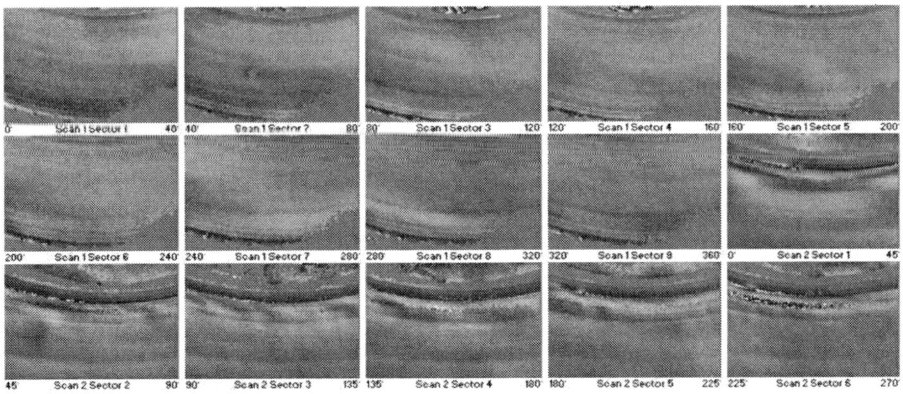

Fig. 6. First measurement before dynamic test

https://doi.org/10.1051/matecconf/201815705022

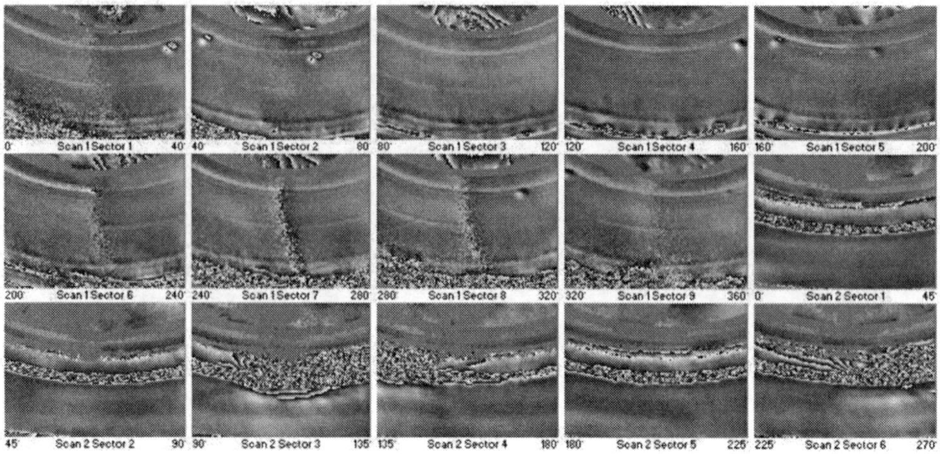

Fig. 7. 12th measurement after 200 hours of the dynamic test

5 The results of measurement for separations in the tyre casing which were obtained with the help of laser interferometry

The Figs. 8 - 21 reveal that there is the occurrence of given separations and cracks in various places in tyre. It can be seen that distribution of the cracks or separations is random or stochastic and it is influenced by many specific factors therefore the best way for evaluation is with help of statistic methods of probability for analysis of variance.

5.1 The basic types of separations in the site of shoulder and crown area of tyre casing

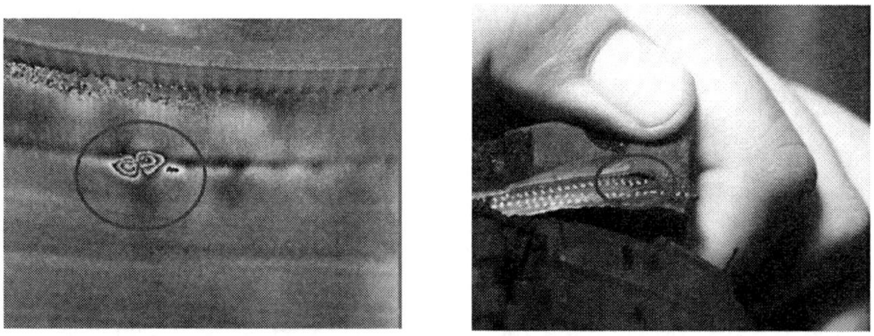

Fig. 8. The local separation in the site where the second buffer lining ends

Fig. 9. The longitudinal continuous separation in the shoulder between the first and second buffer lining

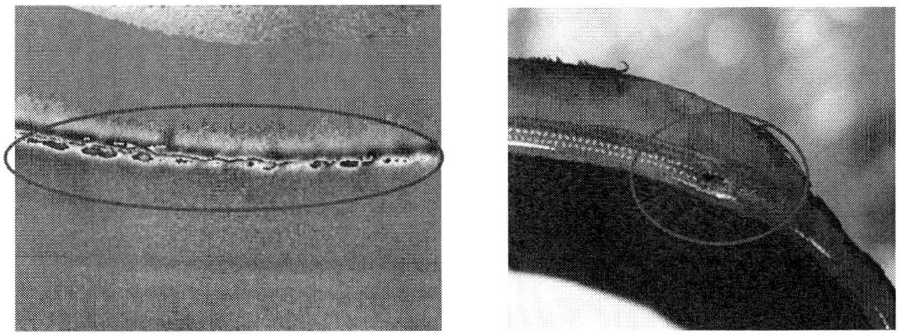

Fig. 10. The longitudinal chain separation in the shoulder between the first and second buffer lining

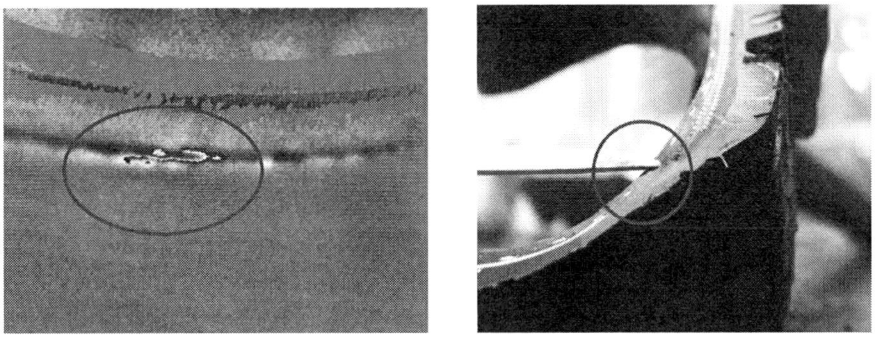

Fig. 11. The local separation in the shoulder where the first buffer lining ends

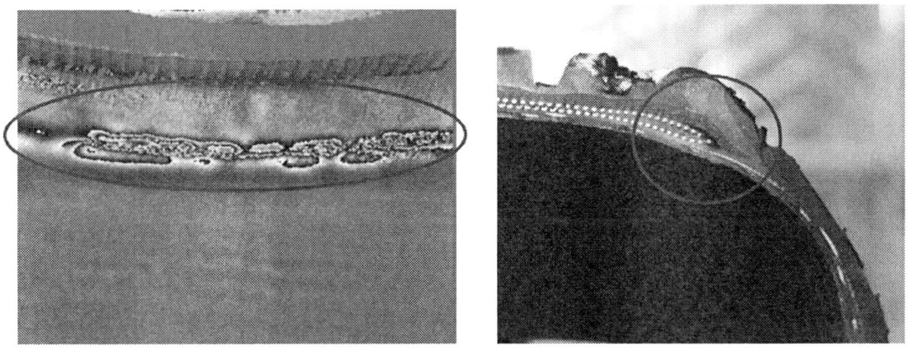

Fig. 12. The longitudinal separation in the shoulder where the first and second buffer linings ends and its expanding to the crown

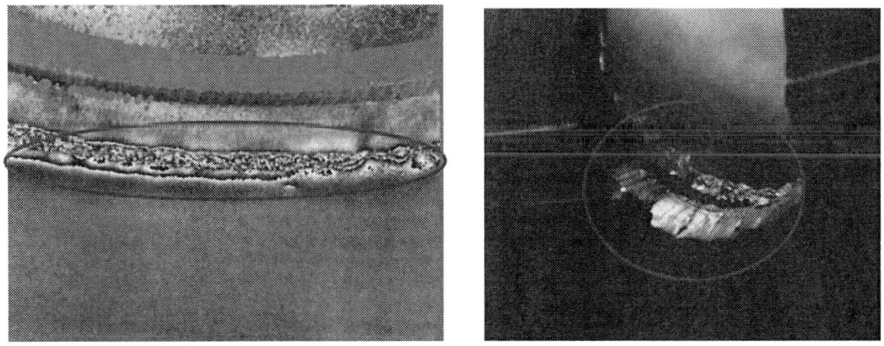

Fig. 13. The longitudinal separation in the shoulder and its expansion to the side directions

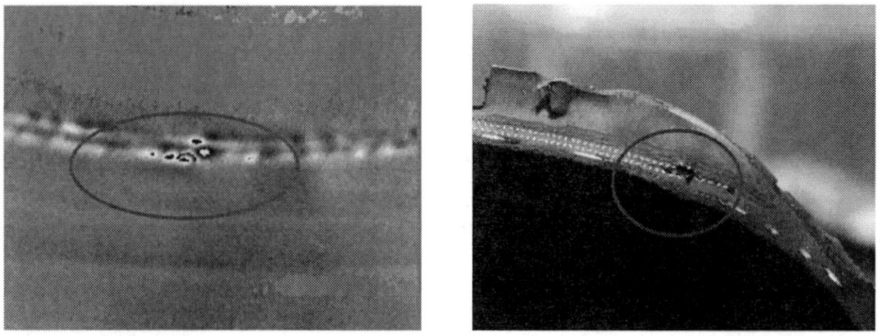

Fig. 14. The local separations in the shoulder in the site where the second buffer lining ends

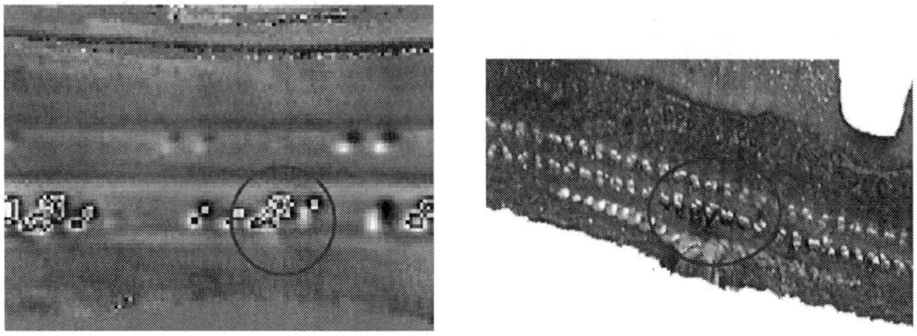

Fig. 15 The local separations in the crown between the first buffer lining and the cord carcass

5.2 The basic types of separations in the sidewall area and bead area of tyre casing

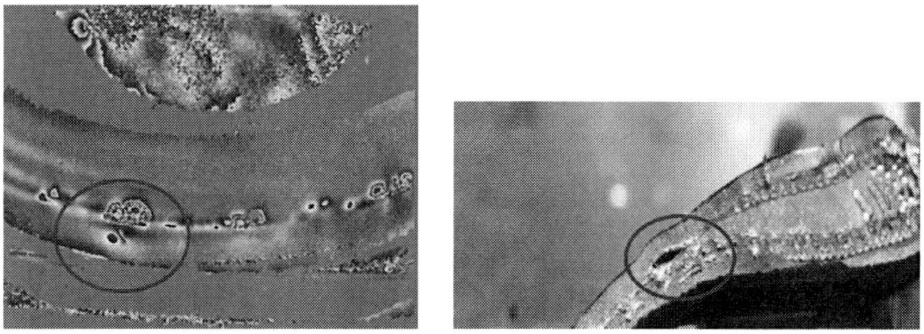

Fig. 16. The local separation in the sidewall between the end of carcass lining and bead bundle

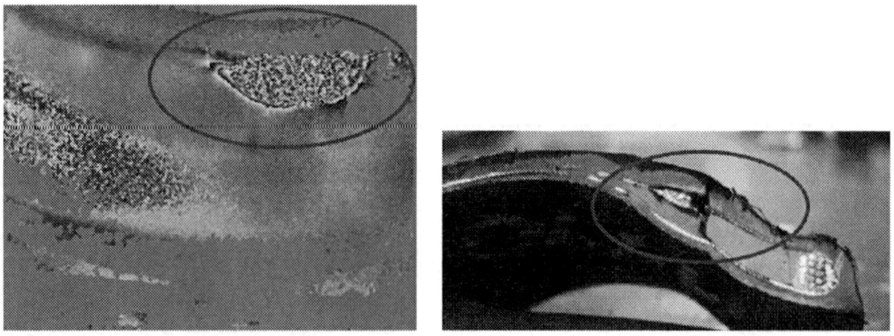

Fig. 17. The local separation in the sidewall - burst bead rubber and bead bundle, separation is between the end of carcass lining and bead bundle

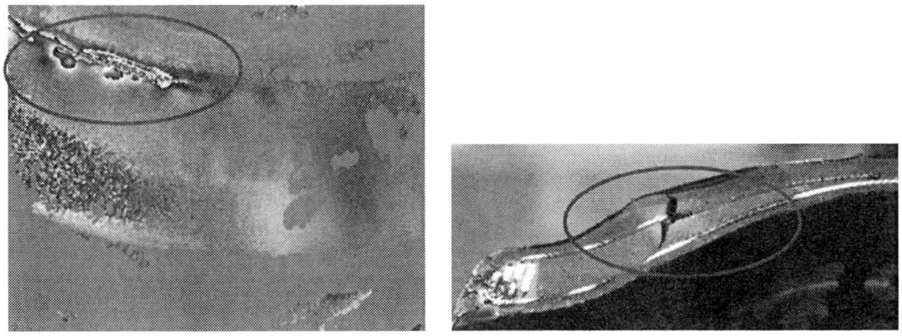

Fig. 18. The local separation of the tyre casing - burst bead rubber and bead bundle

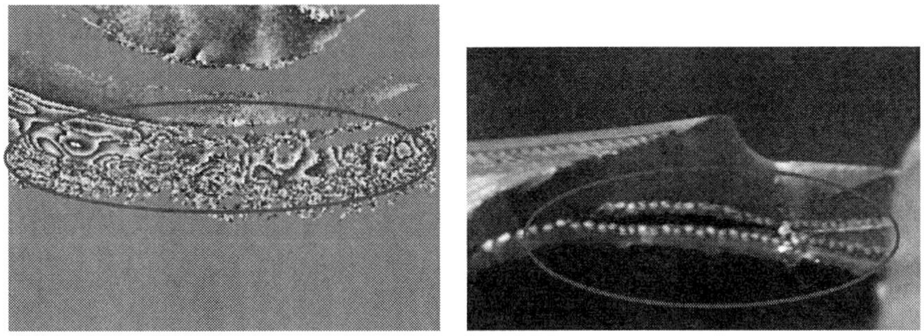

Fig. 19. The longitudinal separation in the sidewall of the tyre casing – it is in the site where the carcass lining is bent

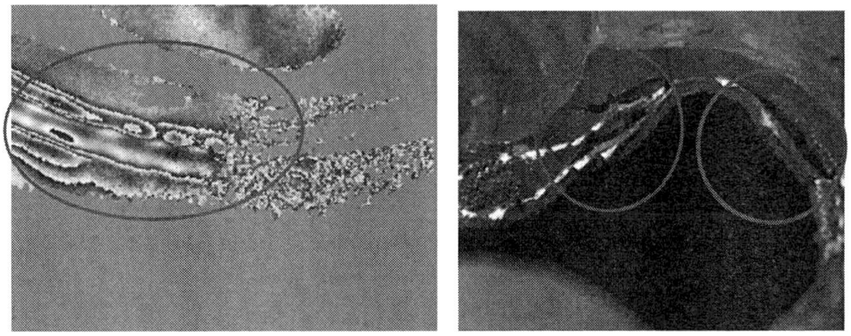

Fig. 20. The longitudinal separation in the shoulder and sidewall of the tyre casing between the bent carcass lining and filling linings of bead

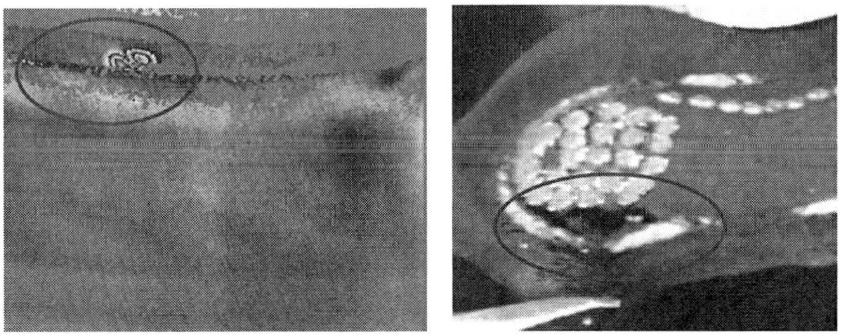

Fig. 21. The local separation in the bead area between bead plies and bent - burst bead rubber and bead bundle carcass lining

Conclusion

The individual scans of Figs. 6 and 7 reveal that there is the distribution of separations and their following mutual connection. On the base of scans in Fig. 7, we could conclude that there is the occurrence of separations around the whole periphery. The detailed and much closer observation of scan 1 (Fig.7) gives us information that separations in the shoulder can be even seen in outer or external part of tyre. The separations in the crown of tyre casing were not changed after 16000 km of dynamic testing. After this non-destructive testing of separations with help of ITT-1, the given tested tyre casing was tested in a destructive way. It was cut up, and it was inspected to confirm the occurrence of separations which were seen in the scans relating to non-destructive testing. There was the real confirmation of occurrence of separations. The distribution of separations and their mutual connection is shown in individual figures hereinafter (see Figs. from 8 to 21). The Figs. from 8 to 15 show the basic types of separations in the site of shoulder and crown area of tyre casing. The Figs. from 16 to 21 reflect the basic types of separations in the sidewall area and the bead area of tyre casing. For comparison, the images which are on the left side in the individual figures were obtained by non-destructive analysis and the images which are on the right side were obtained after slitting of the tire casing in the given site of separation.

This work was supported by the Slovak Grant Agency VEGA 1/0649/17, VEGA 1/0589/17, KEGA 007TnUAD-4/2017, and resulted from the project "Center for quality testing and diagnostics of materials", ITMS code 26210120046 relating to the Operational Program Research and Development funded from European Fund of Regional Development.

References

1. I. Kováč, J. Krmela, D. Bakošová, *Parametrizing of Material Input for Modal Analyses of FEA Tire Models*. Metallurgical Journal LXIV **7**, 73-78 (2011)
2. D. Ondrušová, M. Pajtášová, *Rubber components and their influence on rubber properties and enviromental aspects of production*. 1st Ed. Spolok Slovákov v Poľsku, Krakow, ISBN 978-83-7490-385-1, 1-166 (2011)
3. J. Marcín, P. Zítek, *Rubber productions I – Tyres*. (SNTL, Prague, 1985)
4. J. Vavro, J. Vavro jr., P. Kováčiková, R. Bezdedová, *Experimental Measurement of the Crack Propagation in the Passenger Car Tyre Casing at Dynamic Loading*. Metallurgical Journal LXVIII **5**, 109-111 (2015)
5. J. Krmela, V. Krmelová, L. Beneš, *Experiment of tire-crown for computational modeling of tire*. Scientific papers of the University of Pardubice: Series **B 19**, 81-88 (2014)
6. J. Krmela, *Systems Approach to the Computational Modeling of Tyres*. Monograph, ISBN 978-80-7399-365-8, 1-102 (2008)

Measurement system for special surface mapping using miniature displacement sensors

Martyna Zowade[1], *Maciej* Bodnicki[1,*]

[1]Warsaw University of Technology, Faculty of Mechatronics, Institute of Micromechanics and Photonics, ul. Boboli 8, 02-525 Warszawa, Poland

Abstract. The aim of the work was to design a special system for measurements of elements with repetitive geometry or assemblies with repeating components, set in a linear patterns. The main focus was based on developing a computer program for signal analysis from variable number of miniature displacement sensors. It was set that the response for displacement of measuring tip from each sensor was a 0-5 V voltage signal with possibility of using different type of sensors. Requirements were determined based on projected measurement method. A special design of sensor was made for testing the computer program. If the characteristics of the sensor is known, it is possible to compute the type A evaluation of uncertainty. The results are presented in XY chart on computer screen. The program allows the user to choose any number of the sensors and determine the distance between them. Also, the possibility of calibration of sensors' set was provided. The test were conducted on a prototype handle for sensors, made on a 3D printer.

Keywords: linear displacement sensor, matrix of sensors, surface measurement,

1 Introduction

Geometric dimensioning is a crucial stage of a production process and obtaining shape and position deviation is no less important than taking the measurements of dimensions. Measurement of shape and position deviation requires some advanced equipment. This, in turn, needs a test method and auxiliary elements, e.g. custom made sensors. An accurate measurement system is needed in the case of precision machines, or when engineering tolerance is low. However, a single one-off method of production in small companies poses an obstacle as maintaining low unit cost of production is fundamental and their access to costly measurement equipment is limited.

Using standard, general purpose sensors might solve the problem. It could be done either by mounting them to a dedicated handle or, taking an entirely different approach, by designing a low- cost custom-made measurement equipment.

In the Institute of Micromechanics and Photonics (IM&P), Warsaw University of Technology a research process led to the design of Hall effect based linear displacement/position sensors, useful for analysed devices [1].

The structure of the system is shown symbolically in Figure 1.

* Corresponding author: m.bodnicki@mchtr.pw.edu.pl

Reviewers: *Ivan* Kuric, *Alžbeta* Sapietová

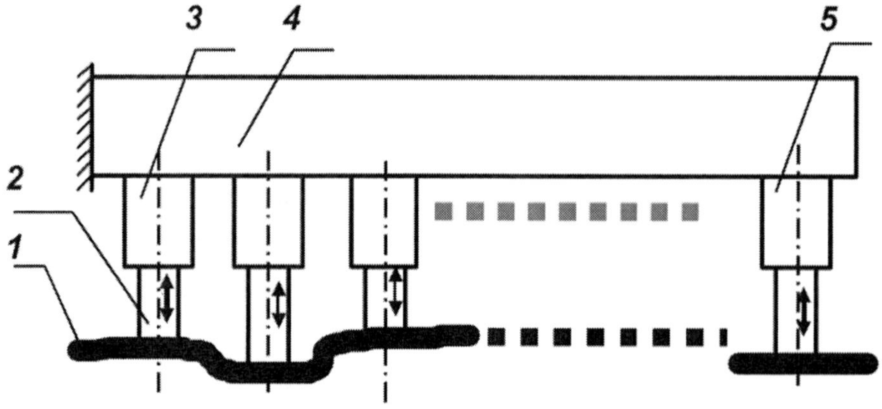

Fig. 1. Symbolic presentation of the idea of the measurement system with set of linear position/displacemnt transducers used for analysis of shape of surface,
1 – analysed object, 2 – measurement tip of the transducer no 1, 3 – body of the transducer no 1, 4 – handle, 5 – body of the transducer no N.

Potential applications include measuring the heights of repetitive elements e.g. such as guitar neck.

2 Mechanical design

Figure 2 shows the designed handle for mounting the transducer assembly. Method of communication between a computer workstation and sensors was necessary when the software was tested. The configuration was tried using a prototype of a linear displacement Hall effect sensor (see [1]) and results were satisfactory. However, the sensor design needs readjustments, e.g. creating an electronic circuit that is mechanically integrated with the sensor.

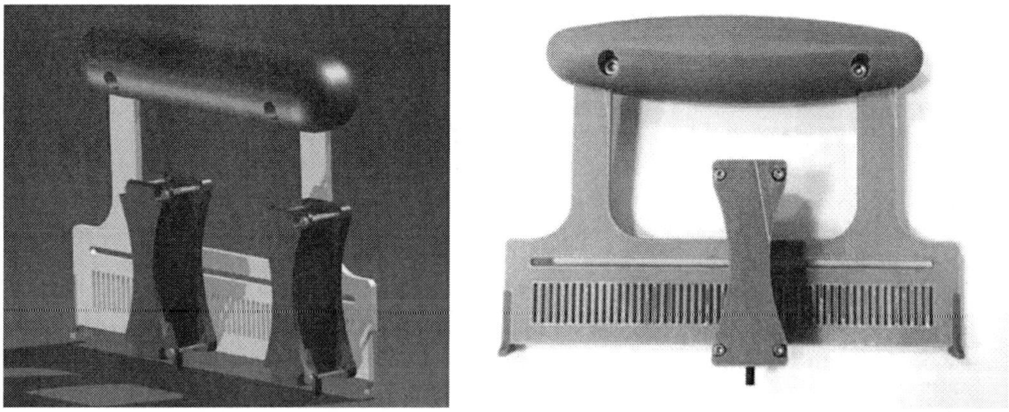

Fig. 2. Visualisation and real view of the handle with sensors

This solution concerns the manner of positioning the handle in relation to the surface. The device is manually brought close to the measured element on two flat supports. In this design, the handle is made of austenitic steel and the tips have a proper tolerance of flatness.

In order to prevent errors during the research, it was checked/controlled for a possible motion of material due to applied load – see Fig. 3.

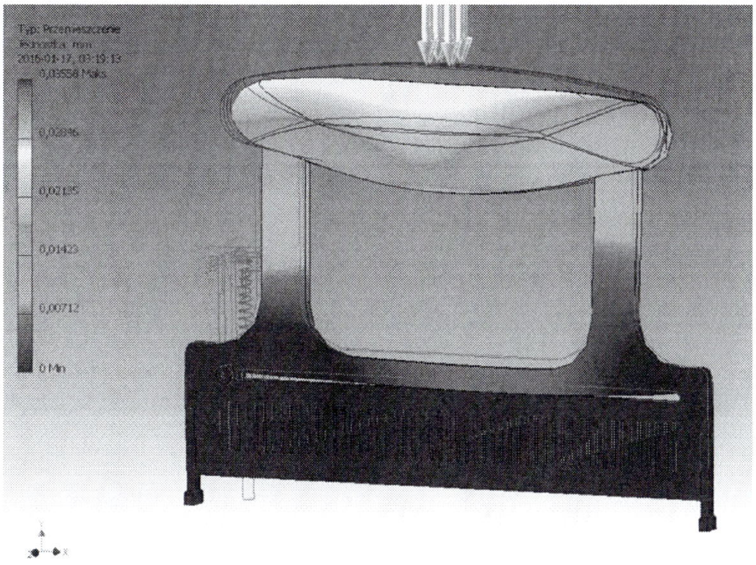

Fig. 3. MES analysis of the handle

3 Calibration of the sensor system

Since reliable measurements/diagnostics require information on the position of sensor's measuring tips, maintaining calibration of sensor position is a necessary condition for properly operating stand. This can be done twofold – mechanically or with a use of dedicated software.

The first method is to design a technique of mounting sensors to a handle so that their position is fixed. This, however, requires reduced/limited engineering tolerance in the production of parts and is not possible to achieve in FDM 3D printing which is a technology employed to create displacement sensor for a matrix.

The alternative involves hardware calibration and taking prefabricated specimens of flatness as a reference. Here, calibration would be done in an early stage of measurement/dimensioning or when the position of sensors is adjusted. Although this method does not require highly precise mounting of sensors, it is nevertheless essential that the shape of a handle does not impede placing it in proximity/close to the specimen of flatness.

4 Software and control system

Figure 4 depicts main elements of the hardware-software system. It comprises mentioned handle with a set of sensors mounted to it. Signals generated by sensors are captured by data acquisition system through controller area network and it is later processed by software. Data transmission is executed by a software-controlled microcontroller. Measurements results are displayed on a user operating panel.

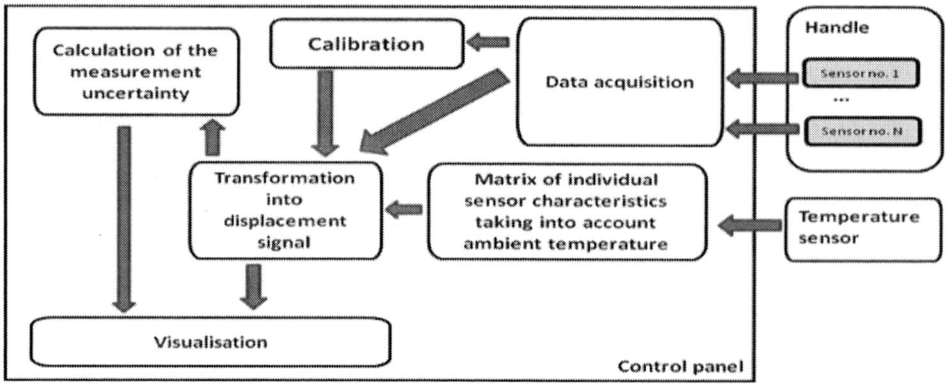

Fig. 4. Structure of the hardware-software system

The software that reads results from sensors through analogue ports was written and installed on a microcontroller. Arduino Platform was chosen for the task as it is user friendly and offers good communication with software.

In the first stage, types of variables are defined and their values are initially set equal to 0. Then, at launching the program, setup loop is run and serial communication (with a selected speed of transmission) begins. When the program is looped, it reads data from analogue ports of Arduino Leonardo board and starts transmitting it with a 10 ms time lag so that LabVIEW program can be run and the synchronisation is not disrupted. Function keys are placed on the left-hand side of operating panel.

User can configure the equipment by choosing a number, characteristics and spacing of sensors along the handle. These settings are saved in the program and can be easily read and applied again if needed. Operating panel and visualisation of the uncertainty are depicted in Figure 5.

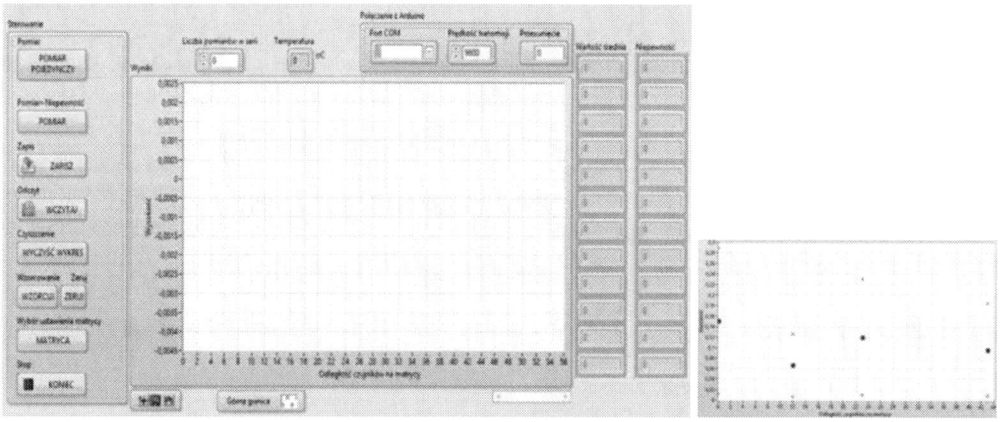

Fig. 5. The control panel and example of the uncertainty analysis for set of sensors

User can choose from available actions:

- one-off measurement,
- full measurement, including Type-A estimate of standard uncertainty (in this case the number of samples must be indicated),
- saving the output data (results of measurement) into a file,
- loading previous measurements,
- clearing a graph before new measurement,
- calibration of the handle,
- sensor settings: selecting number, type and their spacing,

- turning the program off.

The output from sensors depends on the voltage signal on analog input pin of Arduino controller chip and its value ranges from 0 to 1023, in conformity with resolution of the device. In order to obtain a value of the motion of measurement tip these two values need to be properly correlated.

A special test stand was prepared with that purpose, where sensors were tested, as presented in the paper [2]. Settings are saved (with names corresponding to particular sensors) and stored on a computer hard drive.

The handle should be recalibrated if any adjustment of sensor position on the handle was done and before every new measurement. User chooses a number, spacing and characteristics of sensors and every change of their position is stored in the program. It is also important to set proper values in control panel of Arduino controller. Once a calibration is requested by the user, a single measurement is taken and processed in a subprogram. The outcome expresses the length of a strip where sensor tips are in zero point. This is later subtracted from every measurement taken after calibration.

A display where processed data from sensors is plotted as a graph is a central point of operator panel. The graph changes depending on spacing of sensors (as chosen by user) and results of measurement. It is also embedded on a panel at the bottom of the program. There is calibration data, results of single measurement and results of a series of measurements displayed, together with three graphs that accompany it: of results of measurement, upper and lower bounds of uncertainty for particular sensors. Since sensor characteristics are sensitive to ambient temperature, a specific code was implemented which enables user to choose particular values in the course of data processing. This means that individual characteristics of sensor can be updated accordingly to ambient conditions. Sensor is mounted to the handle in two places: by a push-fit into positioning slots and by driving a screw into the handle. If spacing of slots is done properly, this manner of mounting allows using casing of various dimensions and shapes.

Conclusions

Detailed properties of a test system were formulated. A review of solutions for particular functions was done, concerning both the handle for sensors and the computer program. A specific design was chosen and, subsequently, an equipment was built and a program was written and launched. A documented and built system based on a set of sensors of motion is of the aim. All the assumptions concerning the design and functionality of the finally version were met.

A prototype of a handle carrying a set of linear displacement transducers was created. The measurement system was tested for correct operation of the computer program. A demonstration model of the handle was completed and a user manual with measuring algorithm prepared.

Some additional assumptions are met by the computer program, eg. any number of sensors connected to the computer can be supported. This makes the program useful for further research and development of the system. There is a built-in algorithm in the program that sets characteristics of sensor accordingly to ambient temperature. It is done automatically in the course of processing the measurement signal into corresponding displacement. A function that prevents the program from looping in erroneous signal - one that does not match the characteristics of a particular sensor - was also implemented. Thanks to the selected system of storage of measurement data in memory of the computer, it is possible to save information regarding date of measurement, user's id, etc., while using small amount of memory.

It was observed that in case of using prefabricated mounting elements a shrinkage of the material results in sensors protruding from mounting surface. This can be overcome and prevented by either a specific technology of 3D printing that does not cause material shrinkage or a material with better characteristics/properties, eg. PETG. This material has low contractility and, like ABS (used in the research), it is also magnetically neutral.

System was analysed/studied not only for its potential application in a research on guitar necks but also for testing performance of production of parts, with certain tolerance applied.

Moreover, system could be useful in research on wear of material in elements with regular structure, e.g. toothed bars that are used in sliding gates to transform circular to linear motion.

This work has been supported by the statutory activity of the Institute of Micromechanics and Photonics, Faculty of Mechatronics, Warsaw University of Technology.

References

1. M. Bodnicki, P. Pakuła, M. Zowade, *Miniature Displacement Sensor*. Advanced Mechatronics Solutions. (Springer International Publishing, Cham, Switzerland, 2016)
2. M. Bodnicki, P. Pakuła, M. Zowade, *Measurement of small displacements realised by miniature Hall sensors system*. French-Polish Seminar in Mechatronics, 18.-19.05.2015, Besancon, France (poster session, **23**)

Experimental testing of PLA biodegradable thermoplastic in the frame of 3D printing FDM technology

Juraj Beniak[1], *Peter* Križan[1], *Miloš* Matúš[1], *Michal* Šajgalík[2,*]

[1]Faculty of Mechanical Engineering, Slovak University of Technology in Bratislava, Nám. Slobody 17, 812 31 Bratislava, Slovak Republic
[2]Faculty of Mechanical Engineering, University of Zilina, Univerzitna 8215/1, 010 26 Zilina, Slovakia

Abstract. In the present time there are many different plastic materials and composite materials suitable for 3D printing by deposition of semi-melted material. The proper selection of correct material with suitable material properties is dependent on the situation how the produced 3D model should be used. If we need to take into account just the visual look of used material or also the mechanical properties as strength is important for loaded models for final use. The aim of this paper is to publish outputs of experimental testing for 3D models from selected materials with regards to mechanical properties of produced testing parts. Produced 3D models are from PLA biodegradable thermoplastic. Models are prepared on Fused Deposition Modelling (FDM) 3D printer. Testing is based on prepared full factors experiment with four factors on its two levels. Measured values are Tensile strength of PLA testing 3D models. In the same time there are gathered information regarding the 3D printing process and compared to measured tensile strength values for each sent of testing parts. All the measured data are statistically evaluated also by Analysis of Variance (ANOVA method).

Keywords: 3D printing, rapid prototyping, fused deposition modelling, PLA thermoplastic, fused filament fabrication

1 Introduction

There are many research articles and measurements focused on conventional material which are used for 3D model creation with using Rapid Prototyping technologies. Different rapid prototyping technologies using different materials in different consistence, with different conditions for processing and also with different properties of final model. The consistence of processed material can be as a powder, liquid material or solid material. The most wide-spread FDM (Fused Deposition Modelling) technology use solid material in form of wire. All this used material for FDM technology are based on plastic materials. The first used and also now most wide-spread material with professional FDM devices is ABS (Acrylonitrile butadiene styrene) material. In the present time we can also see many others

* Corresponding author: michal.sajgalik@fstroj.uniza.sk

Reviewers: *Grzegorz* Domek, *Ladislav* Écsi

plastic materials as PolyCarbonate (PC), PLA (Polylactic acid) but also many composite materials as Laywood what is composite of polymer and wood particles. All of this materials have some advantages and disadvantages. They have different processing requirements as for example required temperature but also very important material properties as tensile strength, flexural strength or strength in compression. Selection of material for produced model depends mostly from available device type. Because not all FDM devices are able to process all available materials. The other reason how to choose the model material, when the device can process more types of materials is the final use of produced model and the material properties.

As we can see in Table 1, there is really big difference between of some material types. Some of materials have high tensile strength, others have high elongation ratio or high deformation temperature.

Table 1. Material properties used for FDM technology

Material	**ABS plus**	**PC-ABS**	**PC-ISO**	**PC**	**Nylon 12**	**PLA**
Tensile strength (MPa)	37	41	57	68	48	65
Elongation (%)	3	6	4,3	4,8	30	6
Flexural Strength (MPa)	53	68	90	104	69	80
Deformation temperature (°C)	96	110	133	138	82	65

So in case when we need the final 3D model just for design presentation, we can use any of usable material. The only decision criterion is processing difficulties. But when we need the model as a part for final use, we have to care also about material properties, especially when the part will be loaded with some force or pressure. Because of this, we have to know what are the material properties of produced model. These properties are different from properties produced by conventional technologies, for example by plastic extrusion or injection moulding. The reason why these material properties are different is the way of part producing.

2 Fused deposition modelling process

The main principle is based on semi-melting of plastic material in heated nozzle and the semi-melted material is deposited fibre next to fibre in horizontal plane as one layer and then the same way layer on layer (Fig. 1). So whole model is divided to final number of layers and is built from bottom to the top.

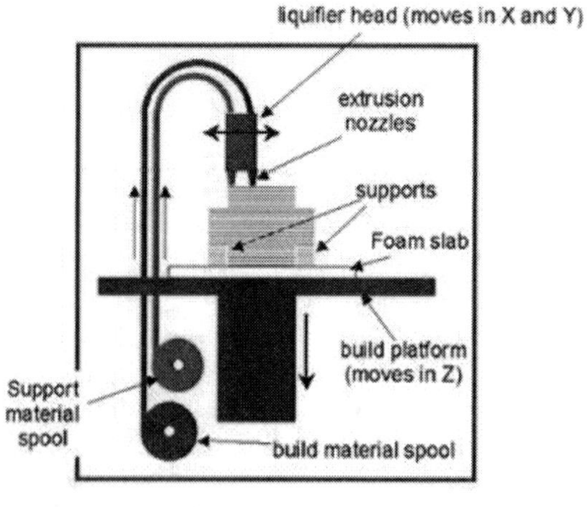

Fig. 1. Scheme of Fused Development Modeling process [1]

Fig. 2. Microscopic view on cross-section of built 3D model

With reference to this production process, the volume of produced part is not homogenous and is not for 100% filled in with material. As we can see on Fig. 2, the volume is created with fibre mesh. Also if we set the maximum filling of model, it is not the same as model produced by conventional technology. The material properties mentioned in Tab. 1 are for materials produced by conventional technology.

If we need to know the real properties of FDM part, we have to make experiments focused on measurements of required parameters. There are already few research articles which provides such information mostly about ABS plastic [2]. But there are no available information about PLA plastic and parts produced on small FDM 3D printers.

PLA plastic material is lately really preferred in FDM technology for their easy application and lower demands in processing. This material is not so predisposed to deformation by influence of cooling process as ABS plastic. There is also not necessary to use heating chamber and heating building platform. The advantage of PLA plastic is also its environmental friendliness. This material is natural polymer, produced from corn, potatoes or sugar-beet. This natural material transmit the light but is not transparent. Its mechanical properties are lower as in case of ABS material. For example the elasticity is worse, has lower melting temperature, what can be also disadvantage because the model can be deformed at lower temperature. The final models have better design look.

3 Measurement of tensile strength of PLA models

Before we start with own measurements, we had to prepare the plan of experiment or design of experiment (DOE). Because we want to know how is the tensile strength affected by different parameters, first of all we chose four important parameters. The chosen parameters are important from the point of device settings, what could affect the final tensile strength. The list of chosen factors is mentioned in Table 2.

Table 2. Selected factors and their levels

Factor	Level 1	Level 2
A – Model filling (%)	90	50
B – Filling shape	Perimeter lines	Honeycomb
C – Layer thickness (mm)	0,125	0,25
D – Model orientation	0 deg	45 deg

Depending on selected factors and their levels we prepared full factors experiment (complete experiment plan). This plan consists from all possible combinations of all factor levels. It is the simplest and the most comprehensive plan of experiment. Allows to estimate all parameters of regression model and easy find out influence and weight of most important factors and their interactions to measured parameters [3]. If we have in our case $k = 4$ factors and measurement will be realized on $h = 2$ levels and with accepted $q = 3$ repetitions, the total number of measurement will be $N_c = q \cdot h^k = 3 \cdot 2^4 = 48$ repetitions. The design of experiment is shown in Table 3.

Table 3. Design of full factor experiment

exp.	A (x_1)	B (x_2)	C (x_3)	D (x_4)
1	1	1	1	1
2	2	1	1	1
3	1	2	1	1
4	2	2	1	1
5	1	1	2	1
6	2	1	2	1
7	1	2	2	1
8	2	2	2	1
9	1	1	1	2
10	2	1	1	2
11	1	2	1	2
12	2	2	1	2
13	1	1	2	2
14	2	1	2	2
15	1	2	2	2
16	2	2	2	2

The measured values of tensile strength are presented in Fig. 3. The measurement was realized on universal tensile machine Inspekt Desk 5 kN. The testing specimen and the experiment process was prepared and realized by ISO 527-1. Each experiment, each combination, presented in Tab. 3 was repeated 5 times, to ensure the possibility of statistical evaluation and maximize the precision of reached values. On Fig. 3 are visible the average values for each experiment [5].

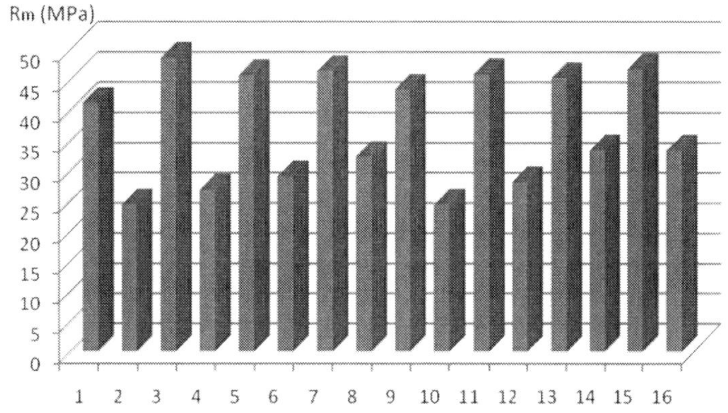

Fig. 3. Graphical illustration of measured tensile strength of PLA samples

The measured data have been evaluated by modern statistical methods. We used the ANOVA method as the base and for verification we used Students criterion and linear regression model. For processing of measured data we can use also regression analysis. This is suitable for exact specification of coefficients, which presents the weight of investigated factors. For this we selected empirical model of experiment:

$$y=\varphi(x,\beta)+s \quad (1)$$

Where $\boldsymbol{x}$ is vector of selected factors, $\boldsymbol{\beta}$ is vector of unknown parameters and $\boldsymbol{s}$ is vector of errors. Its parameters are estimated from empirical data by regression analysis methods. The model (1) can be replaced by power law series [4]:

$$y = \beta_0 + \sum_{i=1}^{k} \beta_i x_i + \sum\sum \beta_{ij} x_i x_j + \ldots + \beta_{12\ldots k} x_1 x_2 \ldots x_k \quad (2)$$

where βi parameter is estimated from empirical data and where β_{12} to $\beta_{12\ldots k}$ present correspondent interactions between two to k factors. For simplicity we take linear regression:

$$R_m = b_0 + b_1 x_1 + b_2 x_2 + b_3 x_3 + b_4 x_4 + b_{1,2} x_1 x_2 + b_{1,3} x_1 x_3 + b_{1,4} x_1 x_4 + b_{2,3} x_2 x_3 + b_{2,4} x_2 x_4 + b_{3,4} x_3 x_4 + b_{1,2,3} x_1 x_2 x_3 + b_{1,2,4} x_1 x_2 x_4 + b_{2,3,4} x_2 x_3 x_4 + b_{1,2,3,4} x_1 x_2 x_3 x_4 \quad (3)$$

In formula (3) are b_0, b_1, b_2, b_3 ..., point estimation β_0, β_1, β_2, β_3 , ...

Verification of each coefficient is made independently. For this verification can be used Student criterion. When using the full factors experiment or repeated measurements, the determining intervals are the same for all coefficients.

Conclusions

When we see the measured values in graphical form (Fig. 3), we can notice some regularity. Every unpaired values (1st, 3th, ...) is much higher than the paired values (2nd, 4th, ...). If we compare this measured values with plan of experiment (Tab. 3) we realize that this significant change is caused by factor A, what is quantity of produced model filing. This is logical if we realize that higher filling means more plastic fibres in the cross-section. This statement is visible in Fig. 4. where the factor A have the biggest gap between level 1 and level 2. So the factor A is the most significant factor from all of mentioned factors. The others two significant factors are also factor C and factor B, represented by layer thickness and filling shape. They are not so valuable as factor A but we can still see the noticeable difference.

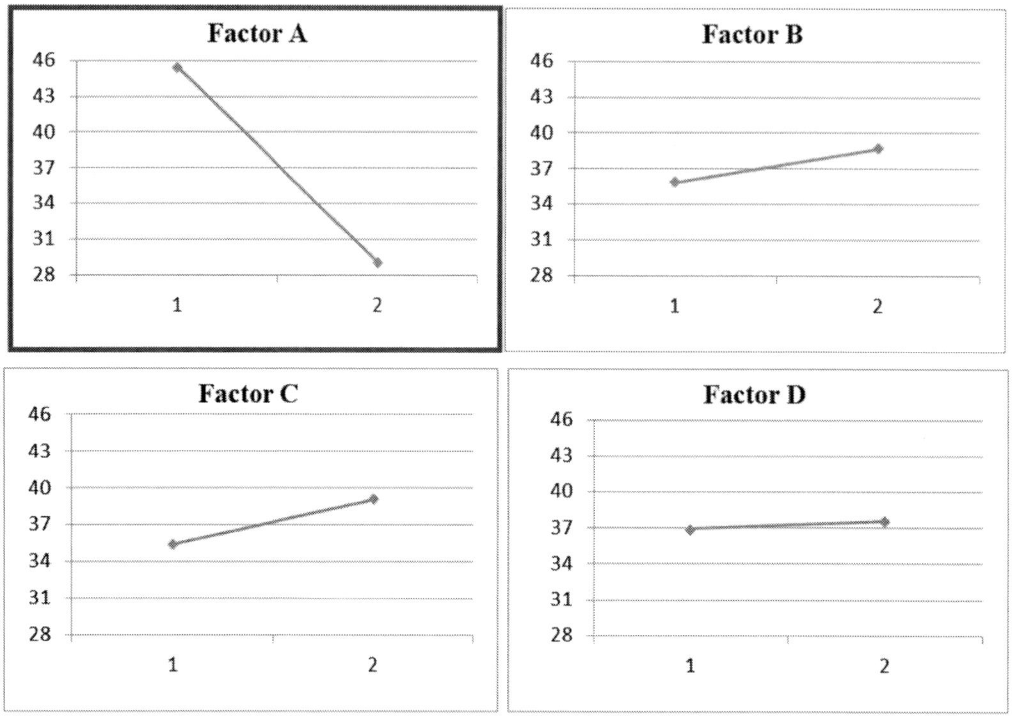

Fig. 4. Significancy overview of individual factors in monitored levels

Only the factor D is not significant, what was confirmed also with ANOVA statistical method. From the measured values we can also recognize the maximum value which have been reached in experiment number 3. The maximum value is 48,63 MPa. This value is 75,8% from value mentioned in Tab. 1 (65 MPa). This is real confirmation of our expectation stated previously.

The research presented in this paper is an outcome of the project No. APVV-0857-12 "Tools durability research of progressive compacting machine design and development of adaptive control for compaction process" funded by the Slovak Research and Development Agency.

References

1. A. Husseini, *3D printing and the construction industry; technical placement*. NBS, http://www.thenbs.com/topics/constructionproducts/articles/3d-printing-and-the-construction-industry.asp#figure-2 [online] (2014)
2. J. Lipina, P. Kopec, V. Krys, *Implementation of Polycarbonate Material Mechanical Properties of Rapid Prototyping into System Creo, Laboratory Verification of the Results*. INES 2014, IEEE 18th International Conference on Intelligent Engineering Systems, Tihany, Hungary, ISBN 978-1-4799-4615-0 (2014)
3. E. Jarošová, *Navrhování experimentů*. (Česká společnost pro jakost, 1997)
4. R. Grofik, J. Kubíček, J. Hrubý, J. Dufek, B. Kába, R. Zeipelt, *Štatistika*. (Edícia Ekonomika poľnohospodárstva, Príroda, 1987)

5. J. Beniak, *The factors affecting the production process of models by FDM device*. Habilitation thesis, Slovak University of Technology, ISBN 978-80-227-4300-6 (2014)

Buckling analysis of graphene nanosheets by the finite element method

Jozef Bocko[1], *Pavol* Lengvarský[1,*]

[1]Department of Applied Mechanics and Mechanical Engineering, Faculty of Mechanical Engineering, Technical University of Košice, Letná 9, 042 00 Košice, Slovakia

Abstract. The paper is devoted to the problems related to buckling analysis of graphene sheets without and with vacancies in the structure under different boundary conditions. The analysis was performed by the classical numerical treatment – the finite element method (FEM). The graphene sheets were modelled by beam elements. Interatomic relations between carbon atoms in the structure were represented by the beams connecting individual atoms. The behaviour of the beam as structural element was based on the properties that were established from relations of molecular mechanics. The vacancies in single layer graphene sheets (SLGSs) were created by elimination of randomly chosen atoms and corresponding beam elements connected to the atoms in question. The computations were accomplished for different percentage of atom vacancies and the results represent an obvious fact that the critical buckling force decreases for increased percentage of vacancies in the structure. The numerical results are represented in form of graphs.

Keywords: buckling, graphene sheet, vacancy, beam, finite element method

1 Introduction

A carbon is a chemical element with symbol *C*, it is a structural element of many structures. The atoms of carbon can bond together in different ways as graphite, diamond, graphene, etc. Graphene is a thin two-dimensional layer of carbon atoms. It has remarkable mechanical properties and many potential applications of these properties in the future. Among the remarkable mechanical properties belong high Young's modulus, shear modulus, strength, etc. These properties are high only for graphene sheets without vacancies or various crystallographic defects and therefore many researchers study effect of these defects on the mechanical properties.

Researchers started study the elastic properties of graphene sheets as Sakhaee-Pour [1]. He investigated the elastic characteristics (Young's modulus, shear modulus, Poisson's ratio) of defect-free single-layered graphene sheet. The graphene sheets were modelled as a beam finite element model. Georgantzinos et al. [2] computed the elastic material properties of graphene using spring finite element model and they concluded that the numerical results of the Young's modulus, shear modulus and Poisson's ratio for the graphene structures depend on their chilarity, width, length and number of layers. Ansari et

* Corresponding author: pavol.lengvarsky@tuke.sk

Reviewers: *Milan* Naď, *Milan* Sága

al. [3] investigated two types of structural defects (Stone–Wales and single vacancy) on the mechanical properties of single-layered graphene sheets. The authors compared effects of both types of defects on material properties and concluded that vacancy defects cause more reduction the material properties as the Stone–Wales defects.

In the paper [4], the buckling characteristic of the defect-free graphene sheet while considering the influences of chirality, boundary conditions and geometrical parameters were explored. The buckling analysis of a defective annular graphene sheet was investigated by Fadaee [5]. The graphene sheet was modelled as a shell and the defect was modelled as an eccentric hole. It was observed that the eccentric defect has significant effect on the critical load.

In this study, the buckling analysis of graphene sheets is investigated. The graphene sheet is modelled using the finite element method. The interatomic interactions are modelled using beam elements and carbon atoms are modelled as nodes. On the graphene sheets are applied boundary conditions and the buckling analysis is performed. The analyses are performed for the sheets with and without defects. The critical buckling force is decreased for the increased number of defects in the structure.

2 Modelling of graphene sheet

The graphene sheets can be regarded in molecular mechanics as large molecules consisting of carbon atoms and the atomic nuclei as material points. The interatomic interactions between atoms are shown in Fig. 1.

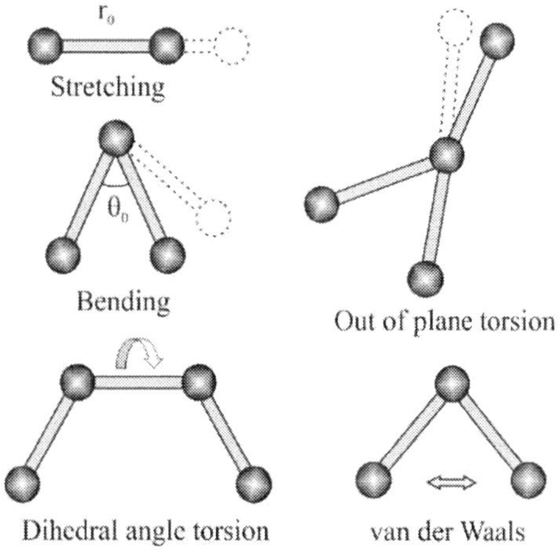

Fig. 1. Interatomic interactions in molecular mechanics

These interatomic interactions between carbon atoms can be expressed by relation

$$U_{total} = \sum U_r + \sum U_q + \sum U_f + \sum U_w + \sum U_{vdw}, \tag{1}$$

where $U_r, U_q, U_f, U_w, U_{vdw}$ are a bond stretching, a bond angle bending, a dihedral angle torsion, an improper (out of plane) torsion and a non-bonded van der Waals interaction, respectively. These interatomic interactions can be modelled and represented in structural mechanics using beam elements. The material properties of the beam elements are computed in terms of the covalent bond stiffness to simulate the interatomic interactions. The properties are obtained from the comparison of the potential energies of the molecular and structural mechanics. Li and Chou [6] defined relations between structural and molecular parameters as

$$k_r = \frac{EA}{L}, k_q = \frac{EI}{L}, k_t = \frac{GJ}{L}, \tag{2}$$

where L is the beam length, E is the Young's modulus, G is the shear modulus, I is the moment of inertia, J is the polar moment of inertia, respectively. The constants k_r, k_q, k_t are force constants (stiffness) of bond stretching, bond bending, and torsional resistance, respectively. If the interatomic interactions will be modelled by cylindrical beam elements (Fig. 2), we can obtain the input data for the finite element model. The carbon atoms will be modelled as the nodes of the beam elements. The all data are collected and shown in Table 1 [6, 7].

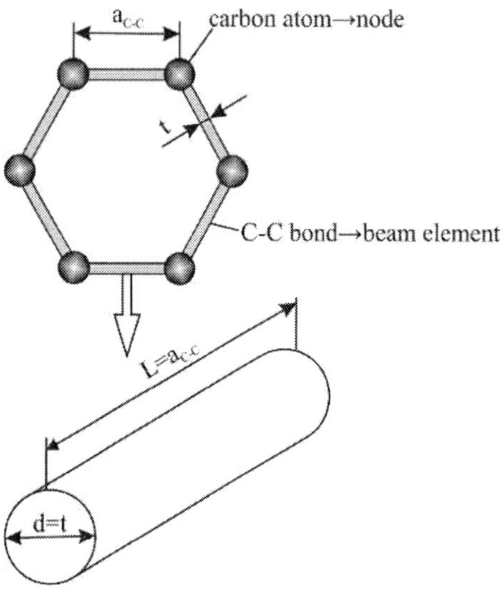

Fig. 2. Hexagon of modelled graphene sheet

Table 1. Basic and input parameters for graphene sheets

Parameter	Symbol	Value
Force constant	k_r	$6.52\text{x}10^{-7}$ Nnm^{-1}
Force constant	k_θ	$8.76\text{x}10^{-10}$ Nnmrad^{-2}
Force constant	k_τ	$2.78\text{x}10^{-10}$ Nnmrad^{-2}
Beam/C-C bond length	$L = a_{C\text{-}C}$	0.1421 nm
Diameter	d = t	0.1466 nm
Young's modulus	E	5.488 TPa
Shear modulus	G	0.871 TPa

3 Buckling of graphene sheet

The buckling of graphene sheets is investigated in this chapter. At the first, the graphene sheets without defect are modelled and the buckling analysis is performed. The buckling analysis is performed for the two types of sheets with different types of internal structure (armchair and zigzag types) and lengths (from 2.5 nm to 50 nm). The structure and dimensions of modelled graphene sheets are shown in Table 2. The chirality of the graphene sheet defines the size and types of edges of the graphene sheets.

Table 2. Structure and dimensions of modelled graphene sheets

Type of edge	Chirality	Width (nm)
Armchair	(6,6)	2.5578
	(8,8)	3.2683
	(10,10)	4.1209
Zigzag	(10,0)	2.4612
	(13,0)	3.1996
	(17,0)	4.1841

For these analyses two different boundary conditions are applied, the first type is represented as sheet fixed on one end and the second type is represented as sheet simply supported on two ends. In semi-logarithmic graph in Fig.4 are shown critical buckling forces for all modelled graphene sheets with the cantilever boundary conditions. The critical buckling forces for all modelled graphene sheet with the simply supported boundary conditions are shown in semi-logarithmic graph in Fig. 5.

From the figures is obvious that graphene sheets of armchair structure have a little bit higher buckling forces than the other structures.

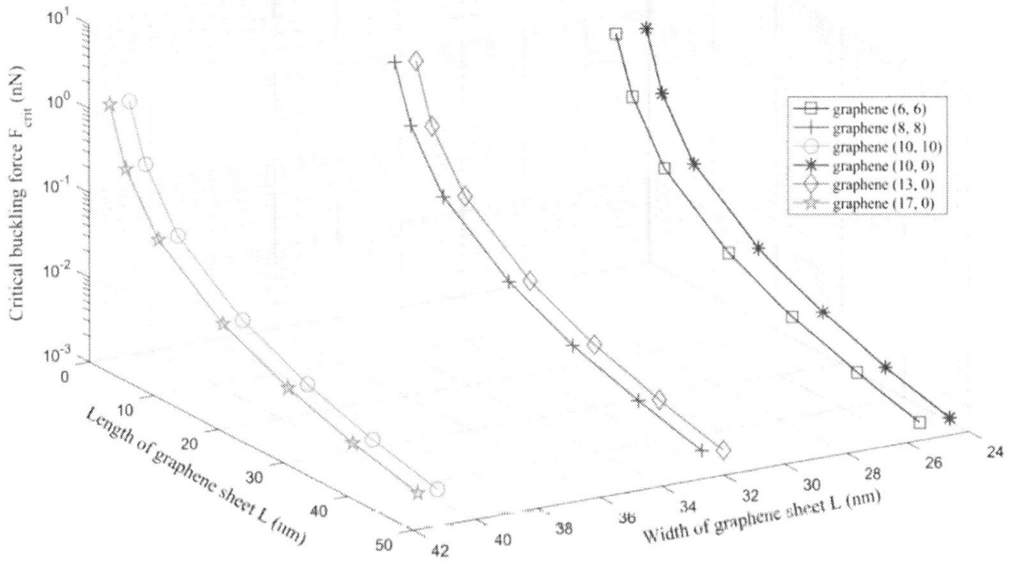

Fig. 4. Critical buckling force of graphene sheets (cantilever sheets)

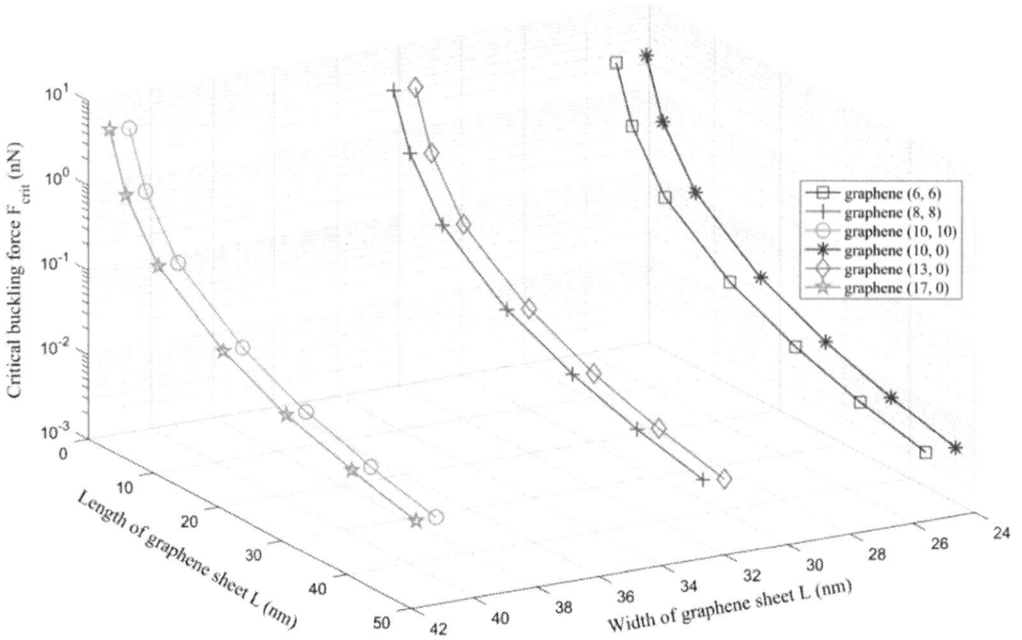

Fig. 5. Critical buckling force of graphene sheets (simply supported sheets)

The defected graphene sheets are modelled by removing random carbon atoms and beam elements connected to these atoms (nodes). From the sheet are removed 1%, 3% and 5% of carbon atoms from the total amount of atoms of ideal structure (Fig. 6). The effect of amount of defects on the critical buckling force is investigated. Dependencies of structural defects on buckling forces for different boundary conditions are given in Tables 3 (cantilever sheet) and 4 (simply supported sheet).

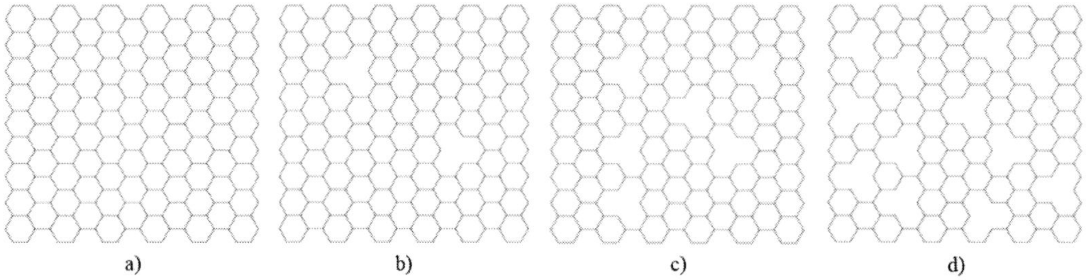

Fig. 6. Modelled graphene sheets with a) 0%, b) 1%, c) 3% and d) 5% of defects

Table 3. Influence of structural defects on buckling force (cantilever sheet)

Graphene sheet			Decrease of F_{crit} (%)		
Type of edge	**Chirality**	**Width (nm)**	**Defect 1%**	**Defect 3%**	**Defect 5%**
Armchair	(6,6)	2.5578	3.9011	11.0704	18.6456
	(8,8)	3.2683	3.8570	10.8901	18.0937
	(10,10)	4.1209	4.1043	11.3089	18.4064
Zigzag	(10,0)	2.4612	4.2806	11.6256	18.7916
	(13,0)	3.1996	4.0093	11.2019	18.6756
	(17,0)	4.1841	3.9077	11.2548	18.3841

Table 4. Influence of structural defects on buckling force (simply supported sheet)

Graphene sheet			Decrease of F_{crit} (%)		
Type of edge	**Chirality**	**Width (nm)**	**Defect 1%**	**Defect 3%**	**Defect 5%**
Armchair	(6,6)	2.5578	4.1168	11.5566	18.8878
	(8,8)	3.2683	4.0328	11.3523	18.2771
	(10,10)	4.1209	3.9452	11.7698	18.9616
Zigzag	(10,0)	2.4612	4.6785	11.7845	19.7319
	(13,0)	3.1996	3.9811	11.6983	19.0965
	(17,0)	4.1841	4.0355	11.7920	19.4753

Conclusion

The buckling analysis of graphene sheets without and with defects was described in this paper. The graphene sheets with different types of chirality, sizes and boundary conditions were modelled by the finite element method. The interatomic interactions were modelled by beam elements and carbon atoms were modelled as joints of these beams. The input properties for the finite element model were obtained from connection between the molecular and structural mechanics. The computations were accomplished for the different percentage of atom vacancies (0%, 1%, 3%, 5%) and the results represents an obvious fact that the critical buckling force decreases for increased percentage of vacancies in the structure. For the given percentage of errors, we have got buckling forces by 4, 11, 19 percent, respectively, smaller than those for ideal structures. The results were represented in form graphs and tables. The critical buckling force of the graphene sheets with armchair structure is a little bit higher than the buckling forces of other ones.

This research was supported by a grant from the Slovak Grant Agency VEGA No. 1/0731/16 - Development of Modern Numerical and Experimental Methods of Mechanical System Analysis.

References

1. A. Sakhaee-Pour, *Elastic properties of single-layered graphene sheet*. Solid State Commun. **149**, 1-2 (2009)
2. S.K. Georgantzinos, G.I. Giannopoulos, N.K. Anifantis, *Numerical investigation of elastic mechanical properties of graphene structures*. Mater. & Design **31**, 10 (2010)
3. R. Ansari, S. Ajori, B. Motevalli, *Mechanical properties of defective single-layered graphene sheets via molecular dynamics simulation*. Superlattices Microstruct. **51**, 2 (2012)
4. A. Sakhaee-Pour, *Elastic buckling of single-layered graphene sheet*. Comp. Mater. Sci. **45**, 2 (2009)
5. M. Fadaee, *Buckling analysis of a defective annular graphene sheet in elastic medium*. Appl. Math. Model. **40**, 3 (2016)
6. Ch. Li, T. Chou, *A structural mechanics approach for the analysis of carbon nanotubes*. Int. J. Solid. Struct. **40**, (2003)
7. J. Bocko, P. Lengvarský, *Buckling of single-walled carbon nanotubes with and without defects*. J. Mech. Sci. Technol. **31**, 4 (2017)

An alternative J_2 material model with isotropic hardening for coupled thermal-structural finite-strain elastoplastic analyses

Ladislav Écsi[1,*], *Pavel* Élesztos[1], *Roland* Jančo[1]

[1]Institute of Applied Mechanics and Mechatronics, Faculty of Mechanical Engineering, Slovak University of Technology in Bratislava, Slovakia

Abstract. In this paper an alternative J_2 material model with isotropic hardening for finite-strain elastoplastic analyses is presented. The model is based on a new non-linear continuum mechanical theory of finite deformations of elastoplastic media which allows us to describe the plastic flow in terms of various instances of the yield surface and corresponding stress measures in the initial and current configurations of the body. The approach also allows us to develop thermodynamically consistent material models in every respect. Consequently, the models not only do comply with the principles of material modelling, but also use constitutive equations, evolution equations and even 'normality rules' during return mapping which can be expressed in terms of power conjugate stress and strain measures or their objective rates. Therefore, such models and the results of the analyses employing them no longer depend on the description and the particularities of the material model formulation. Here we briefly present an improved version of our former material model capable of modelling ductile-to brittle failure mode transition and demonstrate the model in a numerical example using a fully coupled thermal-structural analysis.

Keywords: finite-strain formulation, thermodynamically consistent formulation, thermal-structural finite element analysis, strong coupling, generalized J_2 plasticity, isotropic hardening, material damping, ductile-to-brittle failure mode transition

1 Introduction

Modelling materials within the framework of finite-strain thermoelastoplasticity represents a challenging task in computational mechanics. While plastic behaviour of structural materials within the framework of small-strain thermoelastoplasticity is now well understood, due to the fact, that small-strain flow plasticity theories work well and their results are in agreement with experiments, the same cannot be said for finite-strain flow plasticity theories [1]. Although innumerable material models for finite-strain elastoplasticity have by now been proposed [2-10], such models in general lack universality, as their analysis results depend on the description used in the model and the particularities of the model formulation. The modelling method might simply need some

* Corresponding author: ladislav.ecsi@stuba.sk

Reviewers: *Milan* Sága, *Ján* Vávro *Jr.*

developments in the non-linear continuum theory of finite deformations of elastoplastic media in order that the related theories could be considered to be complete.

Contemporary flow plasticity theories in finite-strain phenomenological thermoelastoplasticity are either based on an additive decomposition of a strain rate tensor into an elastic part, a plastic part and a thermal part, or the multiplicative decomposition of a deformation gradient into an elastic part, a plastic part and a thermal part. The first type of theories are considered to be ad hoc extensions of infinitesimal flow plasticity theories into the area of finite deformations of elastic media to cover large displacements, but small strains of the deforming body. The related material models use an additive decomposition of a strain rate tensor into an elastic part, a plastic part and a thermal part and are based on a hypoelastic stress-strain relationship while utilizing the nonlinear continuum mechanical theory of elastic media to describe the plastic behaviour of the material [2, 11-16].

The second type of flow plasticity theories are now generally accepted as "proper theories", utilizing the theory of single-crystal plasticity, to describe the micromechanics of irreversible deformations in the material. The related material models use a multiplicative split of a deformation gradient into an elastic part, a plastic part and the thermal part, the classical flow plasticity models from small-strain elastoplasticity while utilizing the nonlinear continuum mechanical theory of elastic media to describe the plastic behaviour of the material [2, 3, 17-23].

Our ongoing research however has shown, that both types of the aforementioned theories are just variants of our modified nonlinear continuum theory of finite deformations of elastoplastic media, using an additive decomposition of the displacement field into an elastic part, a plastic part and a thermal part, which describes the plastic flow in terms of various instances of a yield surface and stress measures in the initial or current configuration of the body. Moreover, the theory allows for the generalization of the contemporary flow plasticity theories and the development of alternative material models which are thermodynamically consistent in every respect. As a result, the analysis results of the models no longer depend on the particularities of the model formulation. We will show that the contemporary strain rate tensor additive decomposition based theories are in fact finite-strain theories, but they are constrained when the material model in them is defined in terms of a Cauchy stress tensor based yield surface in the current configuration of the body. That is why they appear as if they had mixed finite-strain small-strain formulations. The contemporary deformation gradient multiplicative split based theories on the other hand are rather incomplete and not quite consistent with the theory of nonlinear continuum mechanics, because they neglect the displacement fields in the definition of the deformation gradient and their parts. Moreover, the deformation gradient cannot have a Lagrangian form unless the Lagrangian displacement field has an additive decomposition. These are a few major implications of our nonlinear continuum theory of finite deformations of elastoplastic media.

We will not present the non-linear continuum theory of elastoplastic media herein, nor will we explore any of its part. Our aim in this paper is to present an alternative J_2 material model with isotropic hardening, which is the only model at present that meets the requirement of thermodynamic consistency. Here we briefly present a modified version of our former material model capable of imitating ductile-to brittle failure mode transition and demonstrate the model in a numerical example using a fully coupled thermal-structural analysis.

2 Theory

We will use Lagrangian description to describe the motion of a material particle of a deformable body. Though a single form is sufficient to define a thermodynamically consistent material model, we will provide all forms of the material model in terms of different stress and strain measures or their objective rates respectively, as similar forms cannot be found anywhere in the contemporary scientific literature.

2.1 Kinematics of the deformation

In order to describe the kinematics of deformation of an elastoplastic media, we assume that the material/Lagrangian displacement field can additively be decomposed into an elastic part, a plastic part and a thermal part $\mathbf{u} = \mathbf{u}^{el} + \mathbf{u}^{pl} + \mathbf{u}^{th}$, where

$$\mathbf{u}^{th} = \left[\int_{X_{ref}}^{X} \alpha_X \cdot \left[T(\mathbf{X},t) - T_{ref}\right] \cdot dX, \quad \int_{Y_{ref}}^{Y} \alpha_Y \cdot \left[T(\mathbf{X},t) - T_{ref}\right] \cdot dY, \quad \int_{Z_{ref}}^{Z} \alpha_Z \cdot \left[T(\mathbf{X},t) - T_{ref}\right] \cdot dZ \right]^T.$$

Here $\alpha_X, \alpha_Y, \alpha_Z$ denote the coefficients of thermal expansion defined in the initial configuration of the body and $T(\mathbf{X},t), T_{ref}$ are the material temperature field and the reference temperature respectively. In this case neither the Green strain tensor $\mathbf{E} = 1/2 \cdot \left(\mathbf{F}^T \cdot \mathbf{F} - \mathbf{I}\right)$, nor the Almansi strain tensor $\mathbf{e} = 1/2 \cdot \left(\mathbf{I} - \mathbf{F}^{-T} \cdot \mathbf{F}^{-1}\right)$ has a decomposition into an elastic part, a plastic part and a thermal part, but an additive decomposition exists when one evaluates the objective time derivatives of the tensors. The material $\dot{\mathbf{E}}$ and the spatial $\mathbf{d} = \mathcal{L}_v(\mathbf{e})$ strain-rate tensors then take the following forms

$$\dot{\mathbf{E}} = \frac{1}{2} \cdot \left(\dot{\mathbf{F}}^T \cdot \mathbf{F} + \mathbf{F}^T \cdot \dot{\mathbf{F}}\right) = \dot{\mathbf{E}}^{el} + \dot{\mathbf{E}}^{pl} + \dot{\mathbf{E}}^{th}, \tag{1}$$

where

$$\dot{\mathbf{E}}^{el} = \frac{1}{2} \cdot \left[\left(\frac{\partial \dot{\mathbf{u}}^{el}}{\partial \mathbf{X}} \right)^T \cdot \mathbf{F} + \mathbf{F}^T \cdot \frac{\partial \dot{\mathbf{u}}^{el}}{\partial \mathbf{X}} \right], \tag{2}$$

$$\dot{\mathbf{E}}^{pl} = \frac{\dot{\lambda}}{2} \cdot \left[\left(\frac{\partial^P \Psi}{\partial \mathbf{P}} \right)^T \cdot \mathbf{F} + \mathbf{F}^T \cdot \frac{\partial^P \Psi}{\partial \mathbf{P}} \right], \tag{3}$$

$$\dot{\mathbf{E}}^{th} = \frac{1}{2} \left[\begin{pmatrix} \alpha_X & 0 & 0 \\ 0 & \alpha_Y & 0 \\ 0 & 0 & \alpha_Z \end{pmatrix} \cdot \mathbf{F} + \mathbf{F}^T \cdot \begin{pmatrix} \alpha_X & 0 & 0 \\ 0 & \alpha_Y & 0 \\ 0 & 0 & \alpha_Z \end{pmatrix} \right] \dot{T}, \quad \frac{\partial \dot{\mathbf{u}}^{th}}{\partial \mathbf{X}} = \begin{pmatrix} \alpha_X & 0 & 0 \\ 0 & \alpha_Y & 0 \\ 0 & 0 & \alpha_Z \end{pmatrix} \dot{T}, \tag{4}$$

$$\frac{\partial \dot{\mathbf{u}}^{pl}}{\partial \mathbf{X}} = \dot{\lambda} \cdot \frac{\partial^P \Psi}{\partial \mathbf{P}}, \quad \text{and} \quad \frac{\partial^P \Psi}{\partial \mathbf{P}} \neq \left(\frac{\partial^P \Psi}{\partial \mathbf{P}} \right)^T, \tag{5}$$

$$\mathbf{d} = \mathbf{d}^{el} + \mathbf{d}^{pl} + \mathbf{d}^{th}, \quad \mathbf{d}^{el} = \mathbf{F}^{-T} \cdot \dot{\mathbf{E}}^{el} \cdot \mathbf{F}^{-1}, \quad \mathbf{d}^{pl} = \mathbf{F}^{-T} \cdot \dot{\mathbf{E}}^{pl} \cdot \mathbf{F}^{-1}, \quad \mathbf{d}^{th} = \mathbf{F}^{-T} \cdot \dot{\mathbf{E}}^{th} \cdot \mathbf{F}^{-1}. \tag{6}$$

Here $\mathbf{X}$ denotes the position vector of a material point and $\mathbf{x} = \mathbf{X} + \mathbf{u}$ is the position vector of the corresponding spatial point. The deformation gradient $\mathbf{F} = \mathbf{I} + \partial \mathbf{u} / \partial \mathbf{X} =$ $= \mathbf{I} + \partial \mathbf{u}^{el} / \partial \mathbf{X} + \partial \mathbf{u}^{pl} / \partial \mathbf{X} + \partial \mathbf{u}^{th} / \partial \mathbf{X}$ can then be expressed either as a function of the material displacement field $\mathbf{u}$ alone, or as a function of its elastic $\mathbf{u}^{el}$, plastic $\mathbf{u}^{pl}$ and thermal $\mathbf{u}^{th}$ parts. The symbols $\dot{\mathbf{E}}^{el}, \dot{\mathbf{E}}^{pl}, \dot{\mathbf{E}}^{th} / \mathbf{d}^{el}, \mathbf{d}^{pl}, \mathbf{d}^{th}$ denote the elastic, the plastic and the thermal material/spatial strain rate tensors, in which the plastic flow is defined by Eqn. $(5)_1$ as a product of a plastic multiplier $\dot{\lambda}$ and an appropriate yield surface normal $\partial^P \Psi / \partial \mathbf{P}$ in terms of the 1st Piola-Kirchhoff stress tensor $\mathbf{P}$. Please note that we have also simplified the calculation of the material thermal strain rate tensor (Eqn. $(4)_1$) in which we neglected the off-diagonal elements of the material gradient of the thermal velocity field (Eqn. $(4)_2$) as their contribution is small unless the temperature gradient is high within the element. Here the symbol $\mathcal{L}_e(\bullet)$ denotes the Lie derivative operator $\mathcal{L}_e(\bullet) = \mathbf{F}^{-T} \cdot \left[\partial\left(\mathbf{F}^T \cdot (\bullet) \cdot \mathbf{F}\right) / \partial t \right] \cdot \mathbf{F}^{-1}$ of a spatial strain tensor. Please note, that the elastic, the plastic and the thermal strain rate tensors have similar forms as the strain rate tensor itself. Furthermore, it can be shown, that the plastic flow defined by Eqn. $(5)_1$ is not constrained, resulting in Eqns. (3) and $(6)_3$ respectively, as the only non-degenerated forms of the material and spatial plastic strain rate tensors.

2.2 The constitutive equation of the material

Proper formulation of a material model for finite-strain thermoelastoplasticity allows for the definition of a constitutive equation of the material in terms of various stress and strain measures or their objective rates in both the body's initial and current configurations. As a result, the constitutive equation of a material cannot be unique, but has to have various forms. These however have to comply with the principles of material modelling, particularly meet the requirements of material objectivity and moreover be thermodynamically consistent in order that they defined the same material. Furthermore, because the additive decompositions defined by Eqns. (1), $(6)_1$ exist in rate forms only, the constitutive equation too must have rate forms. In fact, Eqns. (7)-(10) define a true hypoelastic based thermoelastoplastic material model, which does not have a form in terms of finite strain measures.

In this research we have modified our former material model capable of imitating ductile-to-brittle failure mode transition in a ductile material at high strain rates [24]. In agreement with the above, the rate form of the constitutive equation of the material can take any of the following forms:

$$\dot{\mathbf{S}} = {}^{mat}\mathbf{C}^{el} : \left(\dot{\mathbf{E}} - \dot{\mathbf{E}}^{th} - x \cdot \dot{\mathbf{E}}^{pl}\right) + {}^{mat}\mathbf{C}^{vis} : \left[\ddot{\mathbf{E}} - (1-x) \cdot \ddot{\mathbf{E}}^{pl}\right], \tag{7}$$

$$\mathcal{L}_P(\mathbf{P}) = \mathbf{F} \cdot \dot{\mathbf{S}} = \mathbf{F} \cdot \left\{ {}^{mat}\mathbf{C}^{el} : \left(\dot{\mathbf{E}} - \dot{\mathbf{E}}^{th} - x \cdot \dot{\mathbf{E}}^{pl}\right) + {}^{mat}\mathbf{C}^{vis} : \left[\ddot{\mathbf{E}} - (1-x) \cdot \ddot{\mathbf{E}}^{pl}\right] \right\}, \tag{8}$$

$$\mathcal{L}_O(\boldsymbol{\tau}) = \mathbf{F} \cdot \dot{\mathbf{S}} \cdot \mathbf{F}^T = J \cdot {}^{spat}\mathbf{C}^{el} : \left(\mathbf{d} - \mathbf{d}^{th} - x \cdot \mathbf{d}^{pl}\right) + J \cdot {}^{spat}\mathbf{C}^{vis} : \left[\mathcal{L}_e(\mathbf{d}) - (1-x) \cdot \mathcal{L}_e\left(\mathbf{d}^{pl}\right)\right], \tag{9}$$

$$\mathcal{L}_T(\boldsymbol{\sigma}) = J^{-1} \cdot \mathbf{F} \cdot \dot{\mathbf{S}} \cdot \mathbf{F}^t = {}^{spat}\mathbf{C}^{el} : \left(\mathbf{d} - \mathbf{d}^{th} - x \cdot \mathbf{d}^{pl}\right) + {}^{spat}\mathbf{C}^{vis} : \left[\mathcal{L}_e(\mathbf{d}) - (1-x) \cdot \mathcal{L}_e\left(\mathbf{d}^{pl}\right)\right], \tag{10}$$

where $${}^{mat}\mathbf{C}^{el} = 2 \cdot G \cdot \mathbf{I} + \lambda^{el} \cdot \mathbf{1} \otimes \mathbf{1}, \quad {}^{mat}\mathbf{C}^{vis} = 2 \cdot G^{vis} \cdot \mathbf{I} + \lambda^{vis} \cdot \mathbf{1} \otimes \mathbf{1}, \quad G = \frac{E}{2 \cdot (1+\nu)}, \tag{11}$$

$$\lambda^{el} = \frac{\nu \cdot E}{(1+\nu) \cdot (1 - 2 \cdot \nu)}, \quad G^{vis} = \frac{E^{vis}}{2 \cdot \left(1 + \nu^{vis}\right)}, \quad \lambda^{vis} = \frac{\nu^{vis} \cdot E^{vis}}{\left(1+\nu^{vis}\right) \cdot \left(1 - 2 \cdot \nu^{vis}\right)}, \tag{12}$$

$${}^{spat}\mathbf{C}^{el}_{ijkl} = J^{-1} \cdot F_{im} \cdot F_{jn} \cdot F_{ko} \cdot F_{lp} \cdot {}^{mat}\mathbf{C}^{el}_{mnop}, \tag{13}$$

$${}^{spat}\mathbf{C}^{vis}_{ijkl} = J^{-1} \cdot F_{im} \cdot F_{jn} \cdot F_{ko} \cdot F_{lp} \cdot {}^{mat}\mathbf{C}^{vis}_{mnop}. \tag{14}$$

In Eqns. (7)-(14) the symbols $\mathbf{S},\mathbf{P},\boldsymbol{\tau},\boldsymbol{\sigma},\dot{\mathbf{S}},\mathcal{L}_P(\mathbf{P}),\mathcal{L}_O(\boldsymbol{\tau}),\mathcal{L}_T(\boldsymbol{\sigma})$ denote the 2nd Piola-Kirchhoff stress tensor, the 1st Piola-Kirchhoff stress tensor, the Kirchhoff stress tensor, the Cauchy stress tensor and their objective rates respectively. They are the time derivative of the 2nd Piola-Kirchhoff stress tensor $\dot{\mathbf{S}}$, the Lie derivative of the 1st Piola-Kirchhoff stress tensor $\mathcal{L}_P(\mathbf{P})$, defined in terms of the Lie derivative operator of a mixed spatial-material stress tensor $\mathcal{L}_P(\bullet)=\mathbf{F}\cdot\left[\partial\left(\mathbf{F}^{-1}\cdot(\bullet)\right)/\partial t\right]$, the Oldroyd rate of the Kirchhoff stress $\mathcal{L}_O(\boldsymbol{\tau})$ defined in terms of the Lie derivative operator of a spatial stress tensor $\mathcal{L}_O(\bullet)=\mathbf{F}\cdot\left[\partial\left(\mathbf{F}^{-1}\cdot(\bullet)\cdot\mathbf{F}^{-T}\right)/\partial t\right]\cdot\mathbf{F}^{T}$ and the Truesdell rate of the Cauchy stress $\mathcal{L}_T(\boldsymbol{\sigma})$ defined in terms of the Truesdell derivative operator of a spatial stress tensor $\mathcal{L}_T(\bullet)=J^{-1}\cdot\mathbf{F}\cdot\left[\partial\left(J\cdot\mathbf{F}^{-1}\cdot(\bullet)\cdot\mathbf{F}^{-T}\right)/\partial t\right]\cdot\mathbf{F}^{T}$. The latter operator actually carries out Lie differentiation, but with rearranged terms of its final form. Here the fourth order material elasticity tensor ${}^{mat}\mathbf{C}^{el}$ and the fourth order material viscosity tensor ${}^{mat}\mathbf{C}^{vis}$ have similar forms as the fourth order elasticity tensor of the St.-Venant Kirchhoff material [25], using two independent material parameters E,ν and E^{vis},ν^{vis} respectively. The fourth order spatial elasticity and viscosity tensors ${}^{spat}\mathbf{C}^{el}$, ${}^{spat}\mathbf{C}^{vis}$ the can be determined in accordance with Eqns. (13) and (14), where $J=\det(\mathbf{F})$ is the Jacobian of the deformation. The variable x denotes the ratio of ductile and total damage increment [24]. Please also note that the objective rates $\dot{\mathbf{S}},\mathcal{L}_P(\mathbf{P}),\mathcal{L}_O(\boldsymbol{\tau}),\mathcal{L}_T(\boldsymbol{\sigma})$ transform in the same way from one form to another as do the stress tensors $\mathbf{S},\mathbf{P},\boldsymbol{\tau},\boldsymbol{\sigma}$, which ensure that the formulation is thermodynamically consistent.

2.3 On the thermodynamic consistency of the formulation

It is essential that the formulation of the material model be thermodynamically consistent, as it ensures that the analysis results of the model is independent of the description and the particularities of the model formulation. The thermodynamic consistency of the formulation is then guaranteed by the appropriate transformations mentioned in the above and by the following equations ensuring the equivalence of the rate of change of the internal elastic deformation energy Eqn. (15), the internal thermal deformation energy Eqn. (16) and the internal plastic deformation energy Eqn. (17)

$$\dot{\mathbf{E}}^{el}:\mathbf{S}\cdot dV_0=\frac{\partial\dot{\mathbf{u}}^{el}}{\partial\mathbf{X}}:\mathbf{P}\cdot dV_0=\mathbf{d}^{el}:\boldsymbol{\tau}\cdot dV_0=\mathbf{d}^{el}:\boldsymbol{\sigma}\cdot dv, \tag{15}$$

$$\dot{\mathbf{E}}^{th}:\mathbf{S}\cdot dV_0=\frac{\partial\dot{\mathbf{u}}^{th}}{\partial\mathbf{X}}:\mathbf{P}\cdot dV_0=\mathbf{d}^{th}:\boldsymbol{\tau}\cdot dV_0=\mathbf{d}^{th}:\boldsymbol{\sigma}\cdot dv, \tag{16}$$

$$\dot{\mathbf{E}}^{pl}:\mathbf{S}\cdot dV_0=\frac{\partial\dot{\mathbf{u}}^{pl}}{\partial\mathbf{X}}:\mathbf{P}\cdot dV_0=\mathbf{d}^{pl}:\boldsymbol{\tau}\cdot dV_0=\mathbf{d}^{pl}:\boldsymbol{\sigma}\cdot dv, \tag{17}$$

which eventually ensure the equivalence of the rate of change of the overall internal deformation energy [26].

2.4 Mathematical modelling of the plastic flow

Similarly as in the case of the rate forms of the constitutive equation of the material, proper formulation of a finite-strain flow plasticity theory allows for the description of the plastic flow in terms of various instances of a yield surface and corresponding stress measures in either the initial or current configuration of the body. Let the instances of the yield surface be defined as ${}^{S}\Psi = {}^{S}\Psi(\mathbf{S},\mathbf{q})$, ${}^{P}\Psi = {}^{P}\Psi(\mathbf{P},\mathbf{q})$, ${}^{\tau}\Psi = {}^{\tau}\Psi(\boldsymbol{\tau},\mathbf{q})$, ${}^{\sigma}\Psi = {}^{\sigma}\Psi(\boldsymbol{\sigma},\mathbf{q})$ in terms of the stress measures $\mathbf{S}, \mathbf{P}, \boldsymbol{\tau}, \boldsymbol{\sigma}$ and a vector of internal variables $\mathbf{q}$. After introducing similar kinematic equations as the ones defined by Eqns. (1)-(6), they serve a basis for the first nonlinear continuum mechanical theory of finite deformations of elastoplastic media. Moreover, since they define the same admissible stress space and the same plastic flow respectively, the instances of the yield surface cannot be independent of each other. In fact ${}^{S}\Psi$, ${}^{P}\Psi$, ${}^{\tau}\Psi$, ${}^{\sigma}\Psi$ are related by the following formulas

$$\frac{\partial\,{}^{P}\Psi}{\partial\mathbf{P}}\cdot\mathbf{F}^{-1} = \frac{\partial\,{}^{\sigma}\Psi}{\partial\boldsymbol{\sigma}}, \quad \frac{\partial\,{}^{P}\Psi}{\partial\mathbf{P}}\cdot\mathbf{F}^{-1} = \frac{\partial\,{}^{\tau}\Psi}{\partial\boldsymbol{\tau}}, \quad \mathbf{F}^{T}\cdot\frac{\partial\,{}^{P}\Psi}{\partial\mathbf{P}} = \frac{\partial\,{}^{S}\Psi}{\partial\mathbf{S}}. \tag{18}$$

Furthermore, one of the yield surfaces ${}^{S}\Psi$, ${}^{P}\Psi$, ${}^{\tau}\Psi$, ${}^{\sigma}\Psi$ has to be chosen as a reference yield surface to define the material model. It can be shown, that when ${}^{\sigma}\Psi$ or ${}^{\tau}\Psi$ is chosen as a reference yield surface in the current configuration of the body, we recover the contemporary flow plasticity theories. Moreover, please note that Eqns. $(18)_1$ and $(18)_2$, represent constraint equations, making the contemporary additive decomposition based theories appear as if they had mixed finite-strain-small-strain formulations.

Crucial part in finite-strain material modelling is thermodynamic consistency of the plastic flow. It ensures, that plastic deformations in the material are independent of the description and the particularities of the model formulation. The thermodynamic consistency of the plastic flow can then be expressed as follows

$$\dot{\mathbf{E}}^{pl} : \dot{\mathbf{S}}\cdot dV_0 = \frac{\partial\dot{\mathbf{u}}^{pl}}{\partial\mathbf{X}} : \mathcal{L}_P(\mathbf{P})\cdot dV_0 = \mathbf{d}^{pl} : \mathcal{L}_O(\boldsymbol{\tau})\cdot dV_0 = \mathbf{d}^{pl} : \mathcal{L}_T(\boldsymbol{\sigma})\cdot dv, \tag{19}$$

where dV_0 is an infinitesimal volume element in the body's initial configuration and $dv = J\cdot dV_0$ is its spatial counterpart. In terms of the above one can prove, that Eqn. (19) has an equivalent form which can be expressed as follows

$$\frac{\partial\,{}^{S}\Psi}{\partial\mathbf{S}} : \dot{\mathbf{S}} = \frac{\partial\,{}^{P}\Psi}{\partial\mathbf{P}} : \mathcal{L}_P(\mathbf{P}) = \frac{\partial\,{}^{\tau}\Psi}{\partial\boldsymbol{\tau}} : \mathcal{L}_O(\boldsymbol{\tau}) = J\cdot\frac{\partial\,{}^{\sigma}\Psi}{\partial\boldsymbol{\sigma}} : \mathcal{L}_T(\boldsymbol{\sigma}), \tag{20}$$

which is known as the 'normality rule' and which defines a rate form of a thermodynamically consistent return mapping procedure. The result is of fundamental importance in computational mechanics as it states, how the plastic multiplier ought to be calculated during return mapping when finite-strain elastoplastic or thermoelastoplastic analysis is carried out.

2.5 The reference yield surface

It has been shown in the above, the choice of the reference yield surface governs the material model. As a result, alternative material models can be developed. In our research we have generalized the J_2 flow plasticity theory with isotropic hardening, where we used the ${}^{P}\Psi = {}^{P}\Psi(\mathbf{P},\mathbf{q})$, Eqn. $(21)_1$ yield surface, as the reference yield surface, to define our material model. Please note that the ${}^{P}J_{2}(\mathbf{P}) = \mathbf{P} : \mathbf{P}$ invariant no longer bases on the deviatoric part of the 1st Piola-Kirchhoff stress tensor $\mathbf{P}$. This is due to the fact, that the 1st Piola-Kirchhoff stress tensor transforms under the change of the observer as $\mathbf{P}^{*} = \mathbf{Q}_{R} \cdot \mathbf{P}$, and ${}^{P}J_{2}(\mathbf{P})$ is the only invariant, which is not affected by the change, i.e. ${}^{P}J_{2}(\mathbf{P}) = {}^{P}J_{2}(\mathbf{P}^{*})$, where $\mathbf{Q}_{R}$ is an arbitrary rotating tensor expressing the relative rotation of the coordinate systems of the observer with respect to the reference coordinate system. The resulting yield surface is then no longer a cylinder, but a sphere.

$$ {}^{P}\Psi = {}^{P}\sigma_{eq} - {}^{P}\sigma_{y} \leq 0, \quad \text{where } {}^{P}\sigma_{eq} = {}^{P}\sigma_{eq}(\mathbf{P}) = \sqrt{{}^{P}J_{2}(\mathbf{P})} = \sqrt{\mathbf{P} : \mathbf{P}}, \tag{21} $$

$$ {}^{P}\sigma_{y} = F_{UT11} \cdot \sqrt{r^{2} - \left[a \cdot e^{pl} - \text{center}\right]^{2}}, \; r = \sigma_{y} + Q, \; \text{center} = \sqrt{r^{2} - \sigma_{y}^{2}} \text{ and } a = \frac{\text{center} + r}{b}, \tag{22} $$

$$ \dot{e}^{pl} = \dot{e}^{pl}\left(\dot{\mathbf{F}}^{pl}\right) = \sqrt{\dot{\mathbf{F}}^{pl} : \dot{\mathbf{F}}^{pl}} = \dot{\lambda}, \; e^{pl} = \int_{0}^{t} \dot{e}^{pl} \cdot dt, \; \mathbf{F}^{pl} = \mathbf{I} + \frac{\partial \mathbf{u}^{pl}}{\partial \mathbf{X}}, \; \dot{\mathbf{F}}^{pl} = \frac{\partial \dot{\mathbf{u}}^{pl}}{\partial \mathbf{X}} = \dot{\lambda} \cdot \frac{\partial {}^{P}\Psi}{\partial \mathbf{P}}. \tag{23} $$

The actual yield stress ${}^{P}\sigma_{y}$, which is a 1st Piola-Kirchhoff stress measure, determines the radius of the yield surface and is defined by Eqn.(22)₁. It is the only nonzero component of a stress tensor $\mathbf{P}_{UT}$ (i.e. ${}^{P}\sigma_{y} = [\mathbf{P}_{UT}]_{11}$) coming from an uniaxial tensile test of the modelled material, where the operator $[(\bullet)]_{11}$ extracts the element in the first row and the first column of a 2nd order tensor $(\bullet)$, written as a 3x3 square matrix. The corresponding deformation gradient and the Jacobian of deformation are denoted as $\mathbf{F}_{UT}, J_{UT}$, where $F_{UT11} = [\mathbf{F}_{UT}]_{11}$ and $J_{UT} = \det(\mathbf{F}_{UT})$. Please also note, that the only nonzero element of the corresponding 2nd Piola-Kirchhoff stress tensor $\mathbf{S}_{UT}$, coming from the tensile test of the material is $[\mathbf{S}_{UT}]_{11} = {}^{S}\sigma_{y} = \sqrt{r^{2} - \left[a \cdot e^{pl} - \text{center}\right]^{2}}$. The equation defines an arc of a circle using three material parameters, the constant yield stress of the material σ_{y}, the maximum stress Q by which the material can harden and the maximum accumulated strain value $b = e^{pl}_{max}$, at which the material loses its integrity, i.e. ${}^{S}\sigma_{y} = 0$. The relationship between the corresponding stress measures then can be written in tensor form as $\mathbf{P}_{UT} = \mathbf{F}_{UT} \cdot \mathbf{S}_{UT}$, in which the parameters σ_{y}, Q are 2nd Piola-Kirchhoff stress measures and $e^{pl} \subset \langle 0, b \rangle$. One may note here, that we have also changed the definition of the accumulated plastic strain rate $\dot{e}^{pl}$ (Eqn. (23)₁), in the definition of which $\mathbf{F}^{pl}$ denotes the deformation gradient of pure plastic deformations at a particle of the material, whose time derivative is assumed to be in the form of Eqn.(5)₁. Other changes in the definitions of the accumulated plastic strain rate $\dot{e}^{pl}$ and the equivalent stress ${}^{P}\sigma_{eq}$ have been needed in order to meet the requirements of thermodynamic consistency in booth a one-dimensional (1D) stress state and a three-dimensional (3D) stress state respectively.

2.6 Calculation of the plastic multiplier

The calculation of the plastic multiplier is a crucial step in finite-strain elastoplastic stress analyses as it determines the value of the stress rate tensor Eqn. (7)-(10), and the plastic part of the strain rate tensors $\dot{\mathbf{E}}^{pl}, \mathbf{d}^{pl}$ during return mapping. Moreover, the return mapping procedure has to be thermodynamically consistent, i.e. it has to comply with Eqn. (20). The condition has not yet been met in any formulation in finite-strain computational plasticity. The thermodynamically consistent return mapping procedure then utilizes the objective differentiation of the yield surface ${}^{P}\Psi$ and it can be expressed as follows

$$\frac{\partial^{P}\Psi}{\partial\mathbf{P}}:\mathcal{L}_{\mathrm{P}}(\mathbf{P})-\left[\mathcal{L}_{\mathrm{P}}(\mathbf{P}_{UT})\right]_{11}=0, \tag{24}$$

where $\mathcal{L}_{\mathrm{P}}(\mathbf{P})$ is then replaced by the rate form of the constitutive equation of the material Eqn. (8), and the second term of Eqn. (24) by the form $\left[\mathcal{L}_{\mathrm{P}}(\mathbf{P}_{UT})\right]_{11}=$ $=F_{UT11}\cdot\left\{\left[-a\cdot\left(a\cdot e^{pl}-\text{center}\right)\right]/\sqrt{r^{2}-\left[a\cdot e^{pl}-\text{center}\right]^{2}}\right\}\cdot\dot{e}^{pl}$. Please also note, that the first term of Eqn. (24) can be replaced by any other term of Eqn. (20), because of the thermodynamic consistency of the formulation.

2.7 The ratio of ductile and total damage increment

The idea of the ratio of ductile and total damage increment x was first introduced by Écsi and Élesztős in order to take into account the internal damping of the material properly during plastic deformations. The ratio allows for the proportional redistribution of the plastic flow between the spring and the damper of a 1D frictional device representing the rheological model of the material [24]. The ratio is determined in an elastic predictor phase during return mapping and its value is then kept constant. Since the return mapping procedure in our material model is carried out in the 1st Piola-Kirchhoff stress space, we had to modify the definition of the ratio as follows

$$x=\frac{\left\langle\mathbf{N}:\mathbf{F}\cdot\left[{}^{mat}\mathbf{C}^{el}:\left(\dot{\mathbf{E}}-\dot{\mathbf{E}}^{th}\right)\right]\right\rangle}{\left\langle\mathbf{N}:\mathbf{F}\cdot\left[{}^{mat}\mathbf{C}^{el}:\left(\dot{\mathbf{E}}-\dot{\mathbf{E}}^{th}\right)\right]\right\rangle+\left\langle\mathbf{N}:\mathbf{F}\cdot\left({}^{mat}\mathbf{C}^{vis}:\ddot{\mathbf{E}}\right)\right\rangle},\qquad x\in\langle 0,1\rangle, \tag{25}$$

where
$$\frac{\partial^{P}\Psi}{\partial\mathbf{P}}=\mathbf{N},\quad \mathbf{N}=\frac{\mathbf{P}}{\sqrt{\mathbf{P}:\mathbf{P}}}=\frac{\mathbf{P}}{\|\mathbf{P}\|}, \tag{26}$$

and
$$\langle y\rangle=\frac{y+|y|}{2}\geq 0, \tag{27}$$

denotes the McCauly's brackets, which return zero if $y<0$ and where $\mathcal{L}_{P}(\mathbf{P})=\mathbf{F}\cdot\dot{\mathbf{S}}$. Please also note that all terms of the right-hand-side of Eqn. $(25)_1$ are objective stress rates, so that the value of x is not affected by the change of the observer.

2.8 The heat equation

In order to describe the conservation of heat energy at a particle of the body, we have modified our former heat equation [27]. The heat equation can imitate elastic heating and dissipation induced heating and its material form takes the following form

$$\rho_0 \cdot c_p \cdot \dot{T} + \dot{\mathbf{S}} : \mathbf{E}^{thp} = -\nabla_0 \bullet \mathbf{Q} + R, \quad R = 0.8 \cdot \left(\mathbf{S}^{el} : x \cdot \dot{\mathbf{E}}^{pl} + \mathbf{S}^{vis} : \dot{\mathbf{E}} \right), \tag{28}$$

where
$$\mathbf{E}^{thp} = diag\left[\alpha_X \quad \alpha_Y \quad \alpha_Z\right] \cdot T + \frac{1}{2} diag\left[\alpha_X^2 \quad \alpha_Y^2 \quad \alpha_Z^2\right] \cdot T^2. \tag{29}$$

Here the symbols $\rho_0, c_p, T, \dot{\mathbf{S}}, \mathbf{E}^{thp}, \mathbf{Q}, R$ denote the material density, the specific heat at constant pressure, the absolute temperature, the rate of change of the 2nd Piola-Kirchhoff stress tensor, (Eqn. (7)), a specific Green thermal strain tensor originating from thermal expansion and expressing the thermal strain with respect to the absolute zero temperature, the material heat flux vector and the heat generation rate per unit volume respectively. The elastic heating in the equation is defined by the term $\dot{\mathbf{S}} : \mathbf{E}^{thp}$ and the dissipation induced heating by the heat generation rate per unit volume, where we assumed that 80% of the dissipated mechanical energy changes into heat Eqn. $(28)_2$. Here $\mathbf{S}^{el}$ is the elastic part of the stress tensor and $\mathbf{S}^{vis}$ is its viscous part.

3 Numerical experiment

In our numerical experiment a cross-shaped specimen in biaxial tension was studied. The specimen geometry has been proposed by Müller [28] and it was fabricated from an AlMgSi05 alloy. Table 1 outlines the material properties of the AlMgSi05 alloy specimen used in the finite element analysis.

Table 1. Material properties of the AlMgSi05 alloy specimen

E [Pa]	$6.89 \cdot 10^{10}$
E^{vis} [Pa·s]	$6.89 \cdot 10^{7}$
$\nu = \nu^{vis}$ [-]	0.33
σ_y [Pa]	$100.0 \cdot 10^{6}$
Q [Pa]	$50.0 \cdot 10^{6}$
b [-]	1.0
ρ_0 [kg·m^{-3}]	2700.0
c [J·kg^{-1}·°K^{-1}]	895.0
$k_{XX} = k_{YY} = k_{ZZ}$ [W·m^{-1}·°K^{-1}]	218.0
$\alpha_X = \alpha_Y = \alpha_Z = \alpha$ [°K^{-1}]	$23.4 \cdot 10^{-6}$
h [W·m^{-2}·°K^{-1}]	10.0
ψ [-]	1.0
σ_{EMS} [W·m^{-2} °K^{-4}]	$11.341 \cdot 10^{-9}$

In order to assess the value of the axial component of the deformation gradient coming from the tensile stress of the material F_{UT11}, we solved the one-dimensional (1D) rate form of the constitutive equation of the material (Eqn. (7)) for the unknown component of the derivative of the axial elastic displacement field with respect to the axial material coordinate $\partial u_x^{el} / \partial X$. The rate form of the constitutive equation of this specific 1D stress analysis, after neglecting the internal damping and the thermal strain in the material, can be expressed in the following finite-strain form

$$^{S}\dot{\sigma}_{y} = \dot{S}_{11} = E \cdot \left[\frac{\partial \dot{u}_x^{el}}{\partial X} \cdot \left(1 + \frac{\partial u_x^{el}}{\partial X} + \frac{\partial u_x^{pl}}{\partial X} \right) \right], \tag{30}$$

where $^{S}\dot{\sigma}_{y} = \dot{S}_{11}$ is the axial component of the 2nd Piola-Kirchoff stress rate tensor coming from the tensile test of the material and E is the Young's modulus of the material. Furthermore considering that the accumulated plastic strain rate Eqn. $(23)_1$ in this 1D stress state is $\dot{e}^{pl} = \partial \dot{u}_x^{pl} / \partial X = \dot{\lambda}$, and that its integral is $e^{pl} = \partial u_x^{pl} / \partial X$ (Eqn. $(23)_2$), one can find F_{UT11} as a function of the accumulated plastic strain e^{pl} only in the form

$$F_{UT11} = 1 + \frac{\partial u_x^{el}}{\partial X} + \frac{\partial u_x^{pl}}{\partial X} = 1 + \left[-\left(1 + e^{pl}\right) + \sqrt{\left(1 + e^{pl}\right)^2 + 2 \cdot \frac{^{S}\sigma_{y}}{E}} \right] + e^{pl}, \tag{31}$$

where $^{S}\sigma_{y} = {}^{S}\sigma_{y}\left(e^{pl}\right)$ see also Eqn. (22).

In the numerical study 1/8 of the specimen body was modelled employing three planes of symmetry. The specimen was loaded at its four ends using $v = 0.84667\ mm/s$ prescribed velocity. Convective and radiation heat transfer was considered through all surfaces, applying 273.15 °K environmental temperature and radiation source temperature respectively. The bodies were initially at rest with 273.15 °K initial temperature. The analysis was run as transient-dynamic analysis applying 0.005 s time step size.

4 Numerical results

Figure 1 shows a few selected results at the end of the finite element analysis, at time 6 s. These are the absolute temperature distribution, the 1st principal stress distribution in terms of the Cauchy's stress measure, the accumulated plastic strain distribution over the current volume of the body and the temperature change time-history at the centre of the cross-shaped specimen respectively.

As can be seen in the figure, the results are very realistic and moreover the temperature change time-history is in agreement with similar temperature change time-histories coming from biaxial tension tests of a cross shaped specimen of an identical material [29]. Moreover, the presented theory is of great importance even from the material testing point of view of ductile materials, as it shows that the contemporary tensile tests are simply not sufficient for finite-strain material property determination without the inclusion of the deformation gradient characterizing the deformation of the tested material during its uniaxial tensile testing. The presented theory thus also serve a basis for improved material testing and it has widespread applications in industry using various manufacturing processes during which finite deformations take place in the processed material [30].

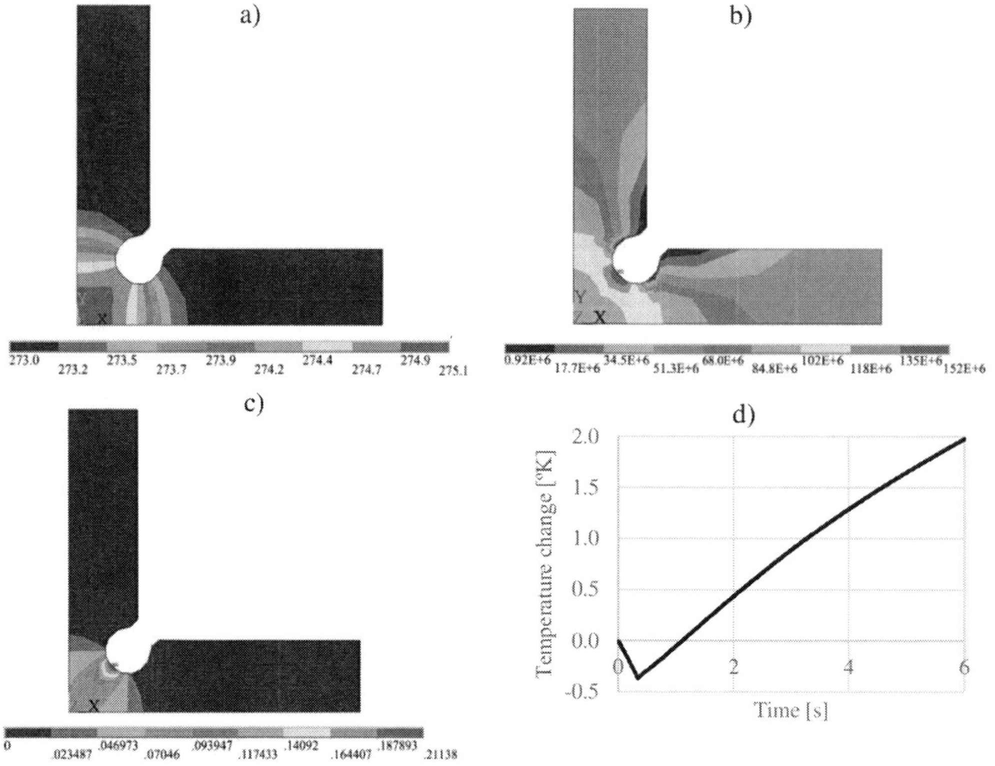

Fig. 1. Selected results from the analysis: **a.** Absolute temperature distribution [°K], **b.** The 1st principal stress distribution in terms of the Cauchy's stress measure [Pa], **c.** Accumulated plastic strain distribution [-], **d.** Temperature change time-history at the centre of the specimen

Conclusions

In this paper an alternative J_2 material model with isotropic hardening for finite-strain elastoplastic analyses was presented. The model is based on a new non-linear continuum mechanical theory of finite deformations of elastoplastic media which allows for the description of the plastic flow in terms of various instances of the yield surface and corresponding stress measures in the body's initial and current configurations. The resulting formulation is thermodynamically consistent, thus the results of the analyses, employing the model, no longer depend on the description and the particularities of the material model formulation. In this study the model behaviour was demonstrated in a numerical example using biaxial tension of a cross-shaped specimen and a fully coupled thermal-structural analysis, the results of which are in agreement with available experiments. The presented theory not only significantly improves upon contemporary thermoelastoplastic analyses, but it also serves as a basis for improved material testing.

This publication is the result of the project implementation: Research of friction stir welding (FSW) application as an alternative to melting welding methods no. 26240220031 supported by the Research & Development Operational Programme funded by the ERDF. Funding from the VEGA grant 1/0740/16 resources is also greatly appreciated.

References

1. G. A. Maugin, *The thermomechanics of plasticity and fracture*. (Cambridge University Press, Cambridge, 1992)
2. J. C. Simo, T. J. R. Hughes, *Computational inelasticity*. (Springer, New Yoerk, 1998)
3. E. A. de Souza Neto, D. Perić, D. R. J. Owen, *Computational methods for plasticity, Theory and applications*. (John Wiley & Sons Ltd., Singapore, 2008)
4. S. Nemat-Nasser, *Plasticity, A treatise on finite deformation of heterogenous inelastic Materials*. (Cambridge University Press, Cambridge, 2004)
5. A. Bertran, *Elascity and plasticity of large deformations, An introduction*. 3rd ed. (Spinger, Berlin, Heidelberg, 2013)
6. D. R. J. Owen, E. Hinton, *Finite elements in plasticity, Theory and practice*. (Pineridge Press, Swansea, 1980)
7. T. Belytschko, W. K. Liu, B. Moran, *Nonlinear finite elements for continua and structures*. (John Wiley & Sons LTD, Chichester, 2000)
8. M. A. Crisfield, *Non-linear finite element analysis of solids and structures, Essentials*, Vol. 1 (John Wiley & Sons Ltd., Chichester, 2000)
9. M. A. Crisfield, *Non-linear finite element analysis of solids and structures, Advanced topics*, **Vol. 2.** (John Wiley & Sons Ltd., Chichester, 2001)
10. K. J. Bathe, *Finite element procedures*. (Prentice-Hall INC., Englewood Cliffs, New Jersey, 1995)
11. C. Truesdell, R.A. Toupin, *The classical field theories, Handbook der physic*, III/1 (Springer-Verlag, Berlin, 1960)
12. C. Truesdell, W. Noll, *The Nonlinear Field Theories, Handbook der Physic*, III/3 (Springer-Verlag, Berlin, 1965)
13. M. E. Gurtin, *An introduction to continuum mechanics*. (Academic Press, Orlando, 1981)
14. J. E. Marsden, T. J. R. Hughes, *Mathematical foundations of elasticity*. (Dover, New York, 1994)
15. R. W. Ogden, *Non-linear elastic deformations*. (Ellis Horwood Ltd., West Sussex, 1984)
16. P. Cairlet, *Mathematical Elasticity. Three-dimenasional elasticity*, Vol. 1 (North-Holland Publishing Company, Amsterdam, 1988)
17. R. J. Asaro, *Micromechanics of crystals and polycrystals, Advances in Applied Mechanics*, Vol. 23 (Academic Press, New York, 1983)
18. G. I. Taylor, *Analysis of plastic strain in a cubic crystal, in Stephen Timoshenko 60th aniversary volume, ed., J.M. Lessels*. (Macmillan, New York, 1938)
19. R. Hill, *Generalized constitutive relations for incremental deformation of metal crystals by multislip*. J Mech Phys Solids **14**, 95-102 (1966)
20. R. Hill, J.R. Rice, *Constitutive analysis of elastic-plastic crystals at arbitrary strain*. J. Mech Phys Solids. **20**, 401-413 (1972)
21. R.J. Asaro, J.R. Rice, *Strain localization in ductile single crystals*. J Mech Phys Solids. **25**, 309-338 (1977)
22. R. J. Asaro, *Geometrical effects in the inhomogeneous deformation of ductile single crystals*. Acta Metall. **27,** 445-453 (1979)

23. C. Sansour, J. Bocko, *On the numerical implications of multiplicative inelasticity with an anisotropic elastic constitutive law.* Int. J. Num. Meth. Engrg. **5** (14), 2131-2160 (2003)
24. L. Écsi, P. Élesztős, *Moving toward a more realistic material model of a ductile material with failure mode transition.* Mat.-wiss. u. Werkstofftech. **5** (43), 379-387 (2012)
25. J. Bonet, R. D. Wood, *Nonlinear continuum mechanics for finite element analysis.* 2nd ed. (Cambridge University Press, Cambridge 2008)
26. G. A. Holzapfel, *Nonlinear solid mechanics, A continuum approach for engineering.* (John Wiley & Sons LTD., Chichester, 2001)
27. L. Écsi, P. Élesztős, *An improved heat equation to model ductile-to-brittle failure mode transition at high strain rates using fully coupled thermal-structural finite element analysis.* Defect Diffus Forum. **354**, 1-23 (2014)
28. W. Müller, *Beitrag zur Charakterisierung von Blechwerkstoffen unter merachsiger Beamspruchung.* (Springer Verlag, Berlin-Heidelberg, 1996)
29. H. Franke, *Lexikon der Physik, Franckh'sche Verlagshandlung.* (W. Keller & Co, Stuttgart, 1959)
30. G. Izrael, J. Bukoveczky, L. Gulan, P. Filípek, *Modeling and operating verification in the design process of mobile working machines.* Journal of Engineering **3**, 195-198 (2011)

Analysis of torque cam mechanism

Iryna Hren[1,*], *Petr* Hejma[1], *Štefan* Michna[1], *Martin* Svoboda[1], *Josef* Soukup[1]

[1]Faculty of Pruduction Technology and Management Jan Evangelista Purkyně University in Ústí nad Labem, Department of Machines and Mechanics, 400 96, Pausterova 1, Ústí n/L, Czech Republic

Abstract: The article deals with the process of torque camshaft at a constant speed. The method of release was determined by the normal force acting between the cam and the tappet. Subsequently, was expressed torque. A prerequisite for the calculations are a constant friction coefficient between the components of the mechanism and constant speed of the camshaft. Cam used for the purposes of this study generated polynomial lifting dependence. Part of this work is the calculation methodology lifting addiction. Course of torque is the conclusion presented on the values of the corresponding measuring station, which is the result of previous work on the subject.

Key words: cam, polynomial stroke, flat cam, hoist, torque.

1 Introduction

Cam mechanisms are an integral part of many machines. The cams are used as a generator of well-defined path in the so-called "hard" automation, also they are part of the fitness machines, which provide the required course of torque. By default, the cam mechanisms used for opening the valve of the internal combustion engine. This is also the most common use. From the above it is clear that cam mechanisms can have many configurations. This work deals with the arrangement of the cam mechanism, which is flat tappet is pressed against the surface of the cam disk using coiled torsion spring. An important parameter is the stroke of the cam dependence, which in this case assigned to the particular rotational angle position of the cam follower of the cam base circle. Cam base circle is the smallest circle which is circumscribed point of the cam profile and cam profile is the area that is in contact with the plunger. There are many approaches to the formation of lifting dependence. Due to the planned automation of calculation was chosen for this work approach, which was presented in 1952 Dudlym [1,2]. This is the definition of lifting dependence using algebraic polynomial (1).

$$s_n = \sum_{i=0}^{n} \alpha_i \cdot \phi^i \quad (1)$$

where α_i is the coefficients of the polynomial, φ [rad] the rotation angle of the cam and n is an odd integer that is greater than 1. Yet unknown polynomial coefficients are calculated using boundary conditions at prominent points of the cam profile. This is about

* Corresponding author: hren@fvtm.ujep.cz

Reviewers: *Zuzana* Murčinková, *Ján* Vavro

the points A and B on fig. 1, where the cam is shown in three positions. The first position is marked with index 1, point B (active connection point on the base circle of the cam profile) lies on the y-axis and is valid $\varphi = 0$. Index 2 is marked with the general position of the cam. Index 3 is marked with the position in which it reaches the tappet lift z $[\mathrm{m}]$, point A lies on the y axis and $\varphi = \varphi_2$. The angle φ_2 is therefore the minimum angle needed to reach the stroke z. from the definition to the n-th degree polynomial is $n + 1$ conditions are needed. To uniquely define the n-th degree polynomial that you need $n + 1$ terms.

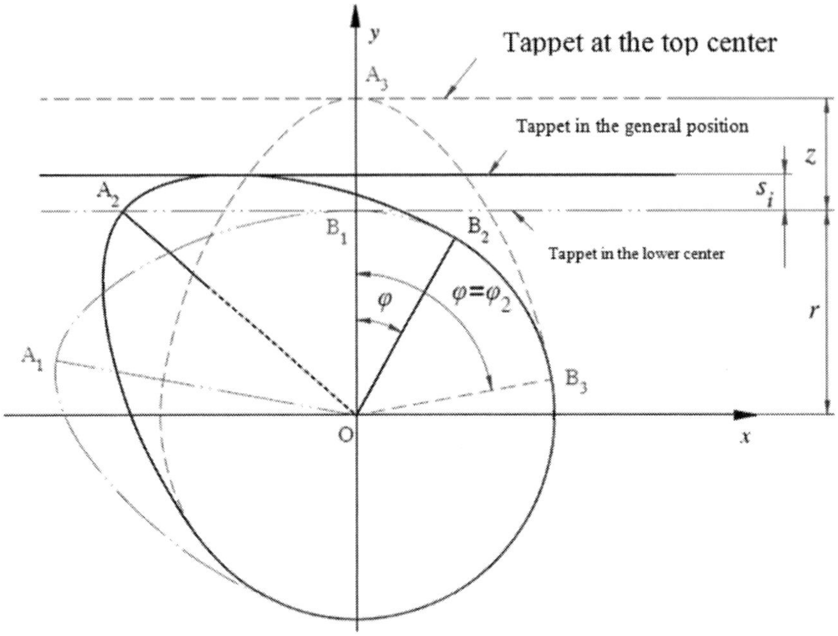

Fig. 1. The position of the cam

For the angle of rotation φ = 0 (index 1 in Fig. 1) we introduce boundary conditions given by the relation (2).

$$\frac{\mathrm{d}^u s_n(0)}{\mathrm{d}\phi^u} = 0 \tag{2}$$

where u is the order of the derivative that $u \in \mathbf{N}_0 \wedge u \in \langle 0; (n-1)/2 \rangle$. Of these conditions implies that the coefficients of the polynomial α_0 to $\alpha_{\frac{n-1}{2}}$ are equal to 0. The form of the polynomial is then reduces to equation (3).

$$s_n = \sum_{i=\frac{n+1}{2}}^{n} \alpha_i \cdot \phi^i \tag{3}$$

The angle of rotation $\varphi = \varphi_2$ match the following boundary conditions.

$$s_n(\phi_2) = z \tag{4}$$

$$\frac{\mathrm{d}^v s_n(\varphi_2)}{\mathrm{d}\varphi^v} = 0 \tag{5}$$

where v is again the order of the derivative, for which the $v \in \mathbf{N} \wedge v \in \langle 1; (n-1)/2 \rangle$. To simplify the math, we will introduce the following variable.

$$p = (n+1)/2 \tag{6}$$

Substituting the boundary conditions (4), (5) into equation (3) form a system p of equations p unknowns that has a matrix notation in the following form.

$$\mathbf{A} \cdot \mathbf{x} = \mathbf{b} \tag{7}$$

where A is a square matrix of constants with the size p, x is a column vector of unknown coefficients of the polynomial with p rows, and b is the column vector of the right sides of dimension p. Index e denote the serial number of the line and f let's label the columns. The members of the matrix A are defined as constants programmer formula (8), where $e, f \in \mathbf{N} \wedge e, f \in \langle 1; (n+1)/2 \rangle$.

$$a_{ef} = \varphi_2^{p+f-e} \prod_{k=1}^{e-1} (p+f-k) \tag{8}$$

For $e = 1$ is the first member of the right sides of the vector is given by (9).

$$b_1 = z \tag{9}$$

For $e > 1$ the remaining members of the vector b given by equation (10).

$$b_e = 0 \tag{10}$$

Members of the unknown vector is then defined by equation (11).

$$x_e = \alpha_{p+e-1} \tag{11}$$

Matrix equation (7) then has a solution (12).

$$\mathbf{x} = \mathbf{A}^{-1}\mathbf{b} \tag{12}$$

The proposed solution (12) using the inverse matrix computation is not from the perspective of computer time efficient, but clearly leads to the calculation of the unknown. The aim of this work is not solving the optimization equation system, but analysis of torque.

2 Torque

The following sections will be based methodology to calculate the torque[3-5]. For passive resistances are included in the mathematical model of the frictional forces between the cam and the tappet, between the tappet and the further conduit. First will be determined normal force between the cam and the tappet, subsequently will be expressed by the torque required to drive the mechanism. The calculation will be made by the method of release. A prerequisite is the constant angular velocity of the cam [6-8]. This paper deals only with the stroke, because during the course of work on the mechanism of the greatest strength. Stroke phase is thus decisive for the choice of a suitable propellant. Figure 2 is released tappet (linkages replaced bond strengths).

The cam rotates clockwise. The arrows indicated the forces acting on the plunger. f_k [1], ($k = \{1, 2, 3\}$) the friction coefficient between the individual components and N_k [N]

($k = \{1, 2, 3\}$) the normal force. F_p [N] the spring force which prevents the plunger rebound from the cam profile during the stroke or the return of the plunger does not change direction, and is dependent on the stroke s_n [m], F_s the inertia force of the plunger and materials related to it during the return stroke and changes direction. F_g [N] the weight lifters and materials related to it, during the return stroke and does not change direction or size. Forces N_1 [N] and $N_1 \cdot f_1$ when lifting or returning the plunger does not change direction. a_m, b_m, d_m [m] are the lengthwise dimensions of the mechanism. The spring force is given by equation (13).

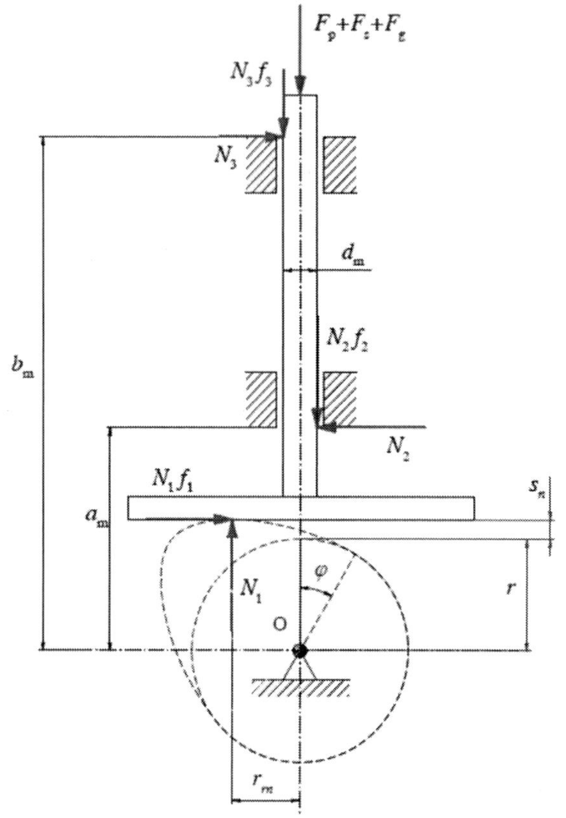

Fig. 2. Relaxed tappet

$$F_p = \frac{G \cdot d_p^4}{8n_p \cdot D_p^3}(s_n + y_1) \tag{13}$$

where G [Pa] is the shear modulus, d_p [m] the diameter of the wire from which the spring is made, D_p [m] the pitch diameter of the spring, n_p [1] is the number of active coils of the spring and y_1 [m] a longitudinal spring bias. The weight of the plunger and materials related to it is given by equation (14).

$$F_g = m_z \cdot g \tag{14}$$

where m_z [kg] is the mass of the plunger and materials related to it , and g [m/s^2]. Inertia force of the tappet and the masses attached to it is given by (15).

$$F_s = m_z \cdot a_n \tag{15}$$

where a_n [m/s^2] acceleration tappet which is given as the second derivative of the polynomial stroke curve (3), i.e. equation (16).

$$a_n = \omega_k^2 \sum_{i=\frac{n+1}{2}}^{n} i(i-1)\alpha_i \cdot \phi^{i-2} \tag{16}$$

where ω_k [rad/s] is a constant angular velocity of rotation of the camshaft. The aforementioned configuration of the cam mechanism, we can say that the place planar motion. The body of the plane has three degrees of freedom. To show unambiguously describe equilibrium tappet is thus sufficient force equation in two axes and one torque equation. Force balance at the horizontal axis (an imaginary x-axis) can be described by the following equation.

$$N_1 \cdot f_1 - N_2 + N_3 = 0 \tag{17}$$

By considering the d'Alembert's principle, we can build a relationship describing the balance of the tappet in the vertical axis (imaginary axis y).

$$-F_p - F_s - F_g + N_1 - N_2 \cdot f_2 - N_3 \cdot f_3 = 0 \tag{18}$$

The balance of moments of force acting on the plunger is related to the rotational axis of the cam and is given by equation (19).

$$-N_1 \cdot f_1(s_n + r) - N_1 \cdot r_{rn} - N_2 \cdot f_2 \frac{d_m}{2} + N_2 \cdot a_m - N_3 \cdot b_m + N_3 \cdot f_3 \frac{d_m}{2} = 0 \tag{19}$$

Given the relative size of individual members can be ignored $N_2 \cdot f_2(d_m/2)$ and $N_3 \cdot f_3(d_m/2)$. Equation (19) is then reduced to form (20).

$$-N_1 \cdot f_1(s_n + r) - N_1 \cdot r_{rn} + N_2 \cdot a_m - N_3 \cdot b_m = 0 \tag{20}$$

We solve the three equations in three unknowns (17), (18) and (20). Expressing N_3 from the equation (17) and substituting into the remaining two equations (18) and (20) we obtain the system of two equations (21) and (22) with two unknowns.

$$-F_p - F_s - F_g + N_1 - N_2 \cdot f_2 - (N_2 - N_1 \cdot f_1)f_3 = 0 \tag{21}$$

$$-N_1 \cdot f_1(s_n + r) - N_1 \cdot r_{rn} + N_2 \cdot a_m - (N_2 - N_1 \cdot f_1)b_m = 0 \tag{22}$$

By multiplying the brackets, we obtain a set of equations (23) and (24).

$$-F_p - F_s - F_g - N_2 \cdot f_3 + N_1 \cdot f_1 \cdot f_3 - N_2 \cdot f_2 + N_1 = 0 \tag{23}$$

$$-N_2 \cdot b_m + N_1 \cdot f_1 \cdot b_m + N_2 \cdot a_m - N_1 \cdot f_1(s_n + r) - N_1 \cdot r_{rn} = 0 \tag{24}$$

Expressing N_2 from the equation (23), subsequent substitution into equation (24) and adjustments we obtain equation (25) describing the axial force N_1 between the cam and the tappet.

$$N_1 = \frac{(F_p + F_s + F_g)\cdot(b_m - a_m)}{f_1 \cdot f_3 \cdot b_m + b_m - f_1 \cdot b_m(f_3 + f_2) - f_1 \cdot f_3 \cdot a_m - a_m + f_1(s_n + r)(f_3 + f_2) + r_{rn}(f_3 + f_2)} \tag{25}$$

The size of arm of force r_{1n} varies depending on the rotation angle of the cam. Figure 3 are shown the geometrical relations. According to Newton's law of action and reaction from acting on the cam follower force N_1 and the cam follower on the same reverse force.

$$w_n^2 = (r + s_n)^2 + r_{rn}^2 \quad \rightarrow \quad r_{rn} = \sqrt{w_n^2 - (r + s_n)^2} \tag{26}$$

where r [m] is the radius of the basic circle and w_n [m] the cam profile equations in polar coordinates, that was derived in [2]. The size of arm strength r_{1n} is therefore expressed by the equation (27).

$$r_{rn} = \sqrt{\left((s_n + r)\cdot\cos^{-1}\left(\operatorname{arctg}\left(\frac{\mathrm{d}s_n}{\mathrm{d}\phi}\frac{1}{s_n + r}\right)\right)\right)^2 - (r + s_n)^2} \;. \tag{27}$$

Torque balance to the axis of cam rotation is given by equation (28).

$$-M + N_1 \cdot r_{rn} + N_1 \cdot f_1(s_n + r) = 0 \,. \tag{28}$$

Torque which is needed to bring the cam shaft to rotate the cam constant angular speed ω_k, is given by equation (29)

$$M = N_1\left(r_{rn} + f_1(s_n + r)\right). \tag{29}$$

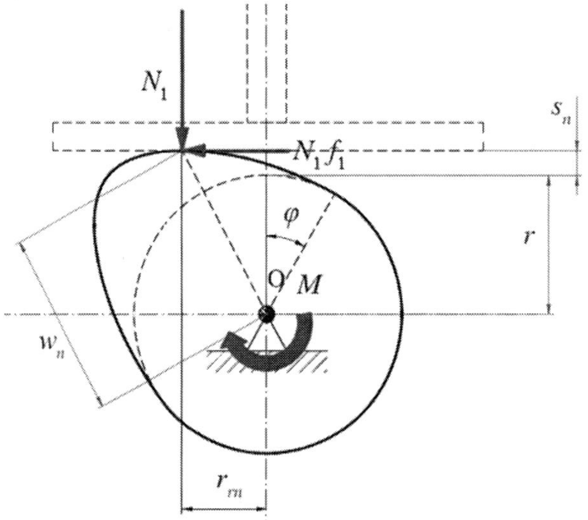

Fig. 3. Release of the cam

3 Torque specific mechanism

The following table lists the specific parameters of the cam mechanism. These are the parameters of the measuring stations, a draft of which has been the subject of previous work on this issue.

Table 1. Values of specific mechanism

Parameter	Unit	Description
$n = 9$	[1]	lifting dependence degree of the polynomial (odd integer)
$\phi_2 = \pi/2$	[rad]	the angle of connection of the cam profile on the base circle
$z = 0.01$	[m]	stroke
$r = 0.05$	[m]	the radius of the base circle
$\omega_k = 149.2257$	[rad/s]	angular velocity (equivalent to 1425 revolutions per minute)

$y_1 = 0.01$	$[\mathrm{m}]$	linear spring preload
$G = 7.9 \cdot 10^{10}$	$[\mathrm{Pa}]$	shear modulus of the material of the spring
$d_\mathrm{p} = 0.005$	$[\mathrm{m}]$	diameter of the spring wire
$D_\mathrm{p} = 0.025$	$[\mathrm{m}]$	mean spring diameter
$n_\mathrm{p} = 12.5$	$[1]$	number of active coils of the spring
$f_1 = 0.1$	$[1]$	the coefficient of friction between the cam and the tappet
$f_2 = 0.1$	$[1]$	the coefficient of friction between the tappet and leadership 1
$f_3 = 0.1$	$[1]$	the coefficient of friction between the tappet a leadership 2
$a_\mathrm{m} = 0.2$	$[\mathrm{m}]$	Length of fig. 2
$b_\mathrm{m} = 0.7$	$[\mathrm{m}]$	Length of fig. 2
$c_\mathrm{m} = 0.04$	$[\mathrm{m}]$	Length of fig. 2
$m_\mathrm{z} = 0.6$	$[\mathrm{kg}]$	weight lifters and materials related to it

The following figure shows the torque waveform of the stroke for the parameters given in tab. 1. The graph 4 is constructed based on the above theory.

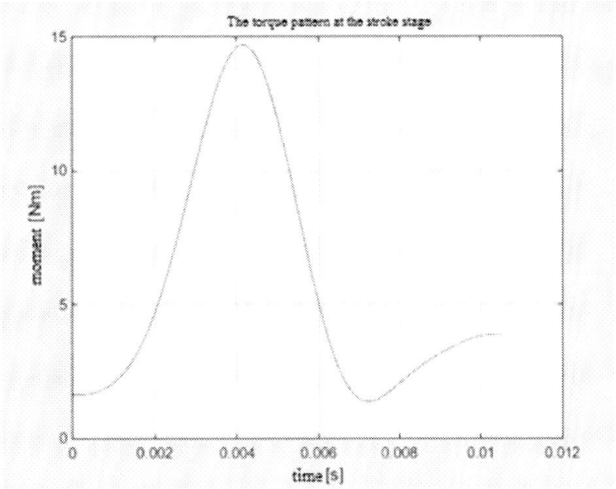

Fig. 4. Torque curve

Conclusion

Was developed a method of creating stroke dependency. Stroke dependence is generated based on the desired stroke of polynomial degree and angle of connection of the cam profile on the base circle. The degree of the polynomial is expressed by an odd natural number. The method of release was expressed by the normal force between the cam and the tappet. Subsequently, it was determined the course of torque. What is given greater degree polynomial, it is a requirement for minimum theoretical base circle radius larger (the result of previous work). Where certain basic circle radius is sufficiently large so that it is possible without changing the radius increasing degree polynomial (in a certain range), the higher degree polynomial increases as the maximum torque value.

This contribution was created in support of project SG 48202 15 0004 01/2016 at UJEP.

References

1. H. A. Rothbart, *Cam design handbook*. (The McGRAW-HILL Companies. Inc., 587 p., 2004)
2. P. Hejma, M. Svoboda, J. Kampo, J. Soukup, *Analytic analysis of a cam mechanism*. Procedia Engineering **177**, 3-10 (2017)
3. R. G. Budynas, *Shigley's Mechanical Engineering Design*. (8th ed. in SI Units, The McGraw-Hill Companies. Inc., 2008)
4. S. Borkowski, *Toyotarity. Innovations. Improvement.* (Polski Instytut Jakości sp.z o.o., 190 p., 2016)
5. M. Apetaur, F. Klimenda, *Mechanika výrobních strojů. Manipulátory a průmyslové roboty*. (1. vydání, 95 p., 2014)
6. R. Křižík, K. Weigner, J. Svoboda, *Stavba a provoz strojů III Mechanizmy*. (SNTL - Nakladatelství technické literatury, 1983)
7. P. Hejma, F. Klimenda, J. Kampo, *Návrh profilu vačky*. (Dynamika tuhých a deformovatelných těles, 2015)
8. I. Blanco, V. Diaz, J., L., P. *Cam design*. (Universidat Carlos III, Madrid, 2005)

Material property determination of the lining layers of a versatile helmet

Radek Kottner[1,*], *Richard* Hynek[1], *Tomáš* Mandys[1], *Jan* Bartošek[1]

[1]NTIS - New Technologies for the Information Society, Faculty of Applied Sciences, University of West Bohemia, Technicka 8, 301 00 Pilsen, Czech Republic

Abstract. This paper deals with material property identification of a helmet lining consisting of an outer layer of an expanded polystyrene (EPS) and inner layer of an open-closed cell foam (OCCF). A combined numerical simulation and experimental testing was used for the material property identification. Compression and drop tests were performed. The ABAQUS finite element commercial code was used for numerical simulations in which the OOCF was modelled as a rate dependent viscoelastic material, while the EPS as a crushable foam. The reaction force time histories coming from the numerical simulation and the experiment have been used as a criterion for material parameter determination. After the identification of the material properties, numerical drop-tests were used to study the behaviour of a plate and a conical composite OOCF and EPS liners to decide which of them suits more for the helmet.

Keywords: helmet liner, foam, expanded polystyrene, drop test, identification

1 Introduction

Common property of composites and foams is their low specific weight. A utilization of these two materials in suitable combination allows the creation of lightweight impact energy absorbers where composite layers advantageously distribute impact forces into foam layers. Expanded polystyrene (EPS) is a crushable foam which, in case of compression, absorbs energy during the plastic deformation stage when the stress reaches a roughly constant plateau value (plateau region in force compression diagram) over a large percentage of total strain (typically 60-70%)[1]. Open-closed cell foam (OCCF) is modern energy absorbing material having different viscoelastic mechanical response according to strain rate (produced e.g. by D3O®).

The aim of this work is the identification of parameters of material models which would sufficiently characterize the foams for the thickness determination of a versatile sport helmet lining using the finite element method (FEM) model. The versatile sport helmet having a removable chin guard and a peak is shown in Fig. 1. The helmet should be suitable for cycling, inline skating, skateboarding, horse riding, paragliding and other similar sports.

---

* Corresponding author: kottner@kme.zcu.cz

Reviewers: *Tomasz* Domański, *Ladislav* Écsi

It has outer carbon fibre reinforced plastic (CFRP) shell with thickness of 2 mm (more detail description of the composite shell and its modelling can be found in [1]). The helmet lining consists of an outer layer of the EPS and inner layer of the OCCF.

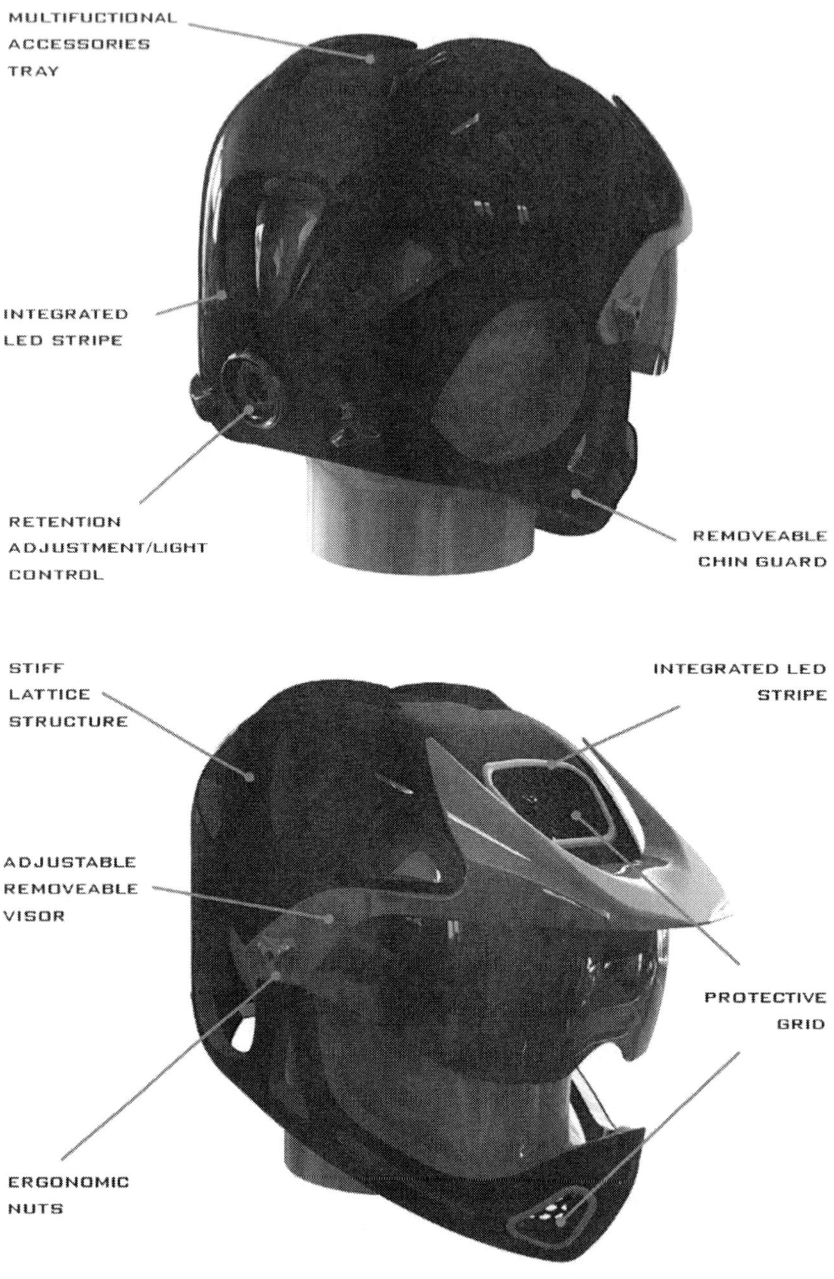

Fig. 1. Versatile sport helmet

2 Experiments

Compression and drop tests were performed using a Zwick/Roell Z050 universal testing machine and a drop tower designed by the authors [3], respectively. The temperature was 23 ± 1 °C, the atmospheric moisture was 50 ± 6 %. Rectangular samples of the foams measured 20×20 mm and had a height (thickness) of 10 or 20 mm. Densities of the EPS and OCCF were 100 kg/m^3 and 500 kg/m^3, respectively.

Two identical loading cycles were applied on the samples in the compression test using flat clamping grips. The samples were loaded and unloaded with velocity of 10 mm/s. Value of nominal strain 0.35 was prescribed as the upper limit. After reaching the upper limit of nominal strain, the testing machine maintained the strain for one minute. Zero strain was maintained also for one minute between the loading cycles.

The total weight of impactor were 246 g or 350 g in dependence of the used impactor head (spherical or flat head). The impactor was equipped by KISTLER 9712B load cell that enabled to record the time-force response (contact force) between the impactor head and a tested sample. The radius of the spherical head of the impactor was 15 mm (detailed description of the impactor geometry can be found in [3]), the flat head of the impactor had a square contact surface with side length of 30 mm. Reached impact velocity v varied from 1 m/s to 5.5 m/s. The time-displacement response (impactor displacement during the impact) was investigated using the Micro-Epsilon optoNCDT laser. Data were supplemented with images from the high-speed camera OLYMPUS i-SPEED 2 (frame rate 2000 fps).

3 Numerical simulations

The tests were simulated in the FEM system ABAQUS. The crushable foam material model was used for the EPS tests simulations. In this model, the contribution of the mean stress on the yield function is realised through a material parameter known as a shape factor [4, 5]. It defines the aspect ratio of the elliptical stress. The yield function Y is given by

$$Y = \sqrt{q^2 + \alpha^2 p^2} - B = 0, \qquad (1)$$

where q is the von Mises equivalent stress and p is the mean stress. B defines the size of the yield ellipse and is given by

$$B = \alpha p_c = \sigma_c \sqrt{1 + (\alpha/3)^2}, \qquad (2)$$

where p_c is the yield strength in hydrostatic compression. σ_c is the absolute value of the yield strength in uniaxial compression.

$$\alpha = \frac{3k}{\sqrt{9-k^2}}, \qquad (3)$$

is the shape factor of the yield surface and k is the ratio of initial yield stress in uniaxial compression and initial yield stress in hydrostatic compression.

The Prony series were used to describe viscoelastic behaviour of the OCCF. The relaxation function was assumed in the normalized series expansion [6]

$$g_R(t) = \frac{G_R(t)}{G_0} = 1 - \sum_{i=1}^{N} g_i\left(1 - e^{-t/\tau_i}\right), \qquad (4)$$

where G_R is the stress relaxation shear modulus and G_0 is the instantaneous shear modulus. N, g_i, and τ_i, $i = 1, 2, \ldots, N$, are material parameters. Since long-term Young's modulus E_{OCCF} was identified, the instantaneous Young's and shear moduli can be determined from

$$G_{\infty} = G_0(1 - \textstyle\sum_{i=1}^{N} g_i) \tag{5}$$

and

$$E_{\text{OCCF}} = 2G_{\infty}(1 + \nu_{\text{OCCF}}), \tag{6}$$

where ν_{OCCF} is the Poisson's ratio of the OCCF.

The finite-strain forms of the models were considered. Parameters of the material models were identified based on comparison of data from the tests and the numerical simulations using an optimization algorithm in optiSLang software. Since the thickness of the helmet lining can be determined using numerical simulations of helmet drop tests according to standards ČSN EN 1078 (Fig. 2) and ČSN EN 1385, the emphasis was placed on the fitting of the drop tests. Moreover, the OCCF parameters were identified separately for four different impact velocities of the flat impactor.

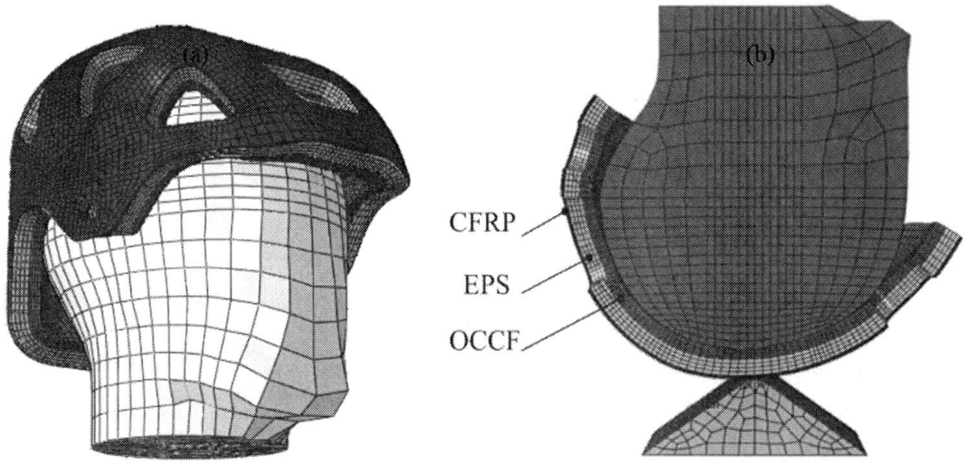

Fig. 2. FEM model of helmet (a), numerical simulation of one of helmet drop test according to ČSN EN 1078 (b)

Performed simulations were supplemented by the numerical drop test of a sample representing one cell of helmet foam liners. Two types of samples having different shapes of the interaction surface (see **Fig. 3**) were tested using a virtual flat impactor.

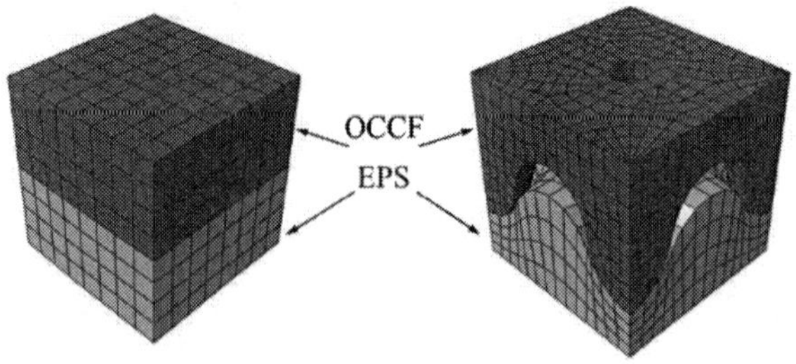

Fig. 3. "Plate" (left) and "conical" (right) design of liners

4 Results and discussion

Different mechanical behaviour of the EPS and the OCCF is obvious from the compression test (Fig. 4(a)). The EPS absorbed a great amount of energy in first loading only. Size of the hysteresis loop in case of the second EPS loading was nearly negligible, which is in agreement with [7]. Size of the first loading hysteresis loop in case of the OCCF compression test was lesser, nevertheless, OCCF behaviour was almost identical in both loading cycles.

Average maximum forces measured in the drop test using the flat impactor are shown in Fig. 4(b). It is obvious that the value of the maximum force increased with the impact velocity more significantly in case of the OCCF.

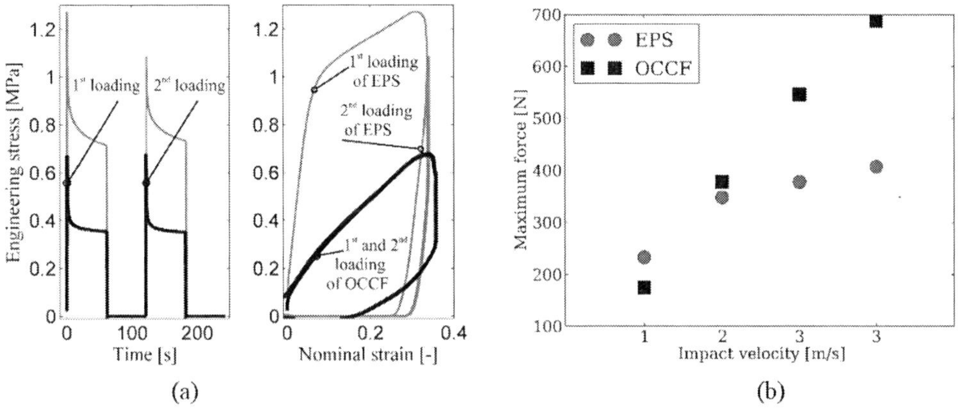

Fig. 4. Typical dependences of engineering stress on time and nominal strain in compression test (a), maximum forces measured in drop test using flat impactor (b)

Identified parameters of the material models are listed in Tables 1-2. The EPS parameters are: Poisson's ratio ν_{EPS}, Young's modulus E_{EPS}, compression yield stress ratio k, initial yield stress S_0, yield stress S_1 corresponding to uniaxial plastic strain $\varepsilon_{\mathrm{p1}}$, and yield stress S_2 corresponding to uniaxial plastic strain $\varepsilon_{\mathrm{p2}}$. The OCCF parameters are: Poisson's ratio ν_{OCCF}, Young's modulus E_{OCCF}, normalized model parameters g_1, g_2 and g_3, and relaxation times τ_1, τ_2 and τ_3. It is obvious that the OCCF parameters differ based on the impact velocity (energy). Therefore, the experimental values were approximated as energy dependent functions in the numerical simulations. The comparison of the force response in case of the OCCF drop test using the spherical impactor for the impact velocity $v = 4.5$ m/s is shown in Fig. 5.

Table 1. Identified parameters of EPS

Param.	ν_{EPS}	E_{EPS}	k	S_0	S_1	$\varepsilon_{\mathrm{p1}}$	S_2	$\varepsilon_{\mathrm{p2}}$
Units	-	MPa	-	MPa	MPa	-	MPa	-
Value	0.07	37.7	1.14	0.53	0.85	0.11	1.2	0.32

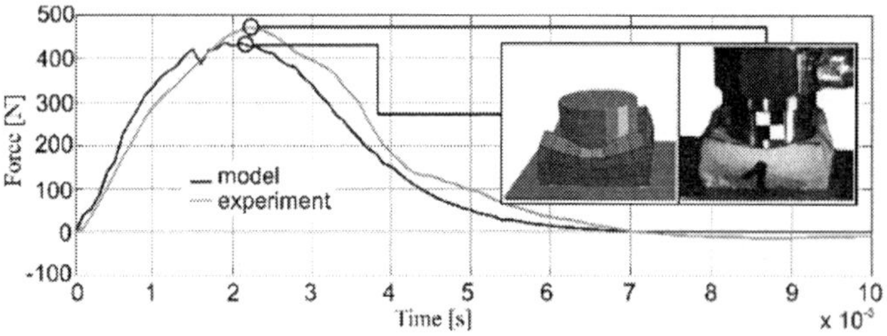

Fig. 5. OCCF drop test using spherical impactor for velocity of 4.5 m/s

Table 2. Identified parameters of OCCF

Param.	**Units**	***v* [m/s]**			
		1	**2**	**3**	**4**
ν_{OCCF}	-	0.38			
E_{OCCF}	MPa	1.99	0.69	0.62	0.55
g_1	-	0.67	0.54	0.56	0.65
τ_1	μs	103	48	43	11
g_2	-	0.23	0.43	0.41	0.30
τ_2	μs	242	113	101	25
g_3	-	0.046	0.023	0.023	0.043
τ_3	μs	1693	791	708	204

The numerical drop test shown that a significant weight save can be achieved using the "conical" design of the liner. Fig. 6 shows the comparison of the force impact response in case of the "plate" and the "conical" design of the liner. It is obvious that differences between maximum forces are not significant, however, the difference of weight is 20 %. Moreover, the "conical" design has lower stiffness in tangential direction, which is important for the brain protection [8].

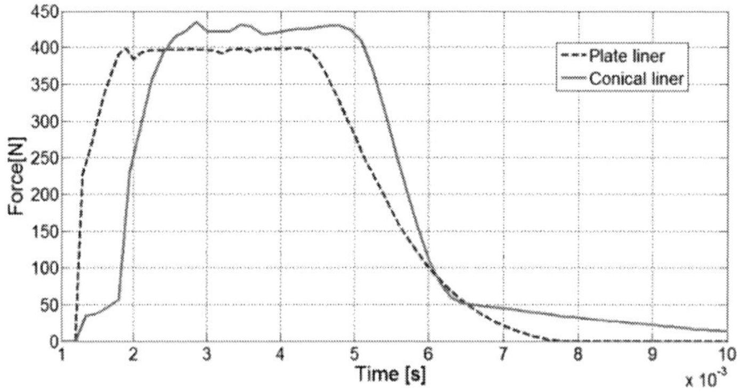

Fig. 6. Comparison of the plate and conical liners design in numerical drop test for velocity of 4 m/s

Numerical simulations of helmet drop tests according to standards ČSN EN 1078 and ČSN EN 1385 showed that the thickness of the EPS layer must be at least 10 mm and the thickness of the OCCF layer must be at least 7 mm (for analysed type of helmet).

Conclusions

The crushable foam material model and the Prony series were used to describe mechanical behaviour of the EPS and the OCCF, respectively. The parameters of the material models were identified, with the emphasis on the fitting of the drop tests. The "conical" design of the helmet liner allows weight savings of up to 20 % and it is also more suitable because of its lower tangential stiffness. Based on the results of the numerical simulations, the thickness of the versatile helmet lining must be at least 17 mm.

This publication was supported by the project LO1506 of the Czech Ministry of Education, Youth and Sports.

References

1. R. Hynek, *Versatile Sport Helmet.* (University of West Bohemia, Pilsen, 2014)
2. L. Cui, S. Kiernan, M.D. Gilchrist, *Designing the energy absorption capacity of functionally graded foam materials*. Materials Science and Engineering A **507**, 215-225 (2009)
3. R. Kottner, T. Mandys, J. Barotšek, J. Kocáb, *Impact force response of foams and rubbers*. Proc. of the 54th Conf. on Experimental Stress Analysis, Srní, 1–7 (2016)
4. W. Tanwongwan, J. Carmai, *Finite element modelling of titanium foam behaviour for dental application*. Proc. Of the World Congress on Engineering 2011, Vol. III, London, 1–6 (2011)
5. V.S. Deshpande, N.A. Fleck, *Isotropic constitutive models for metallic foams*. Journal of the Mechanics and Physics of Solids **48**, 1253-1268 (2000)
6. J. Bergström, *Mechanics of Solid Polymers*. (Elsevier, San Diego, 2015)
7. F.M. Shuaeib, A.M.S. Hamouda, S.V. Wong, R.S. Radin Umar, M.M.H. Megat Ahmed, *A new motorcycle helmet liner material: The finite element simulation and design of experiment optimization*. Materials and Design **28**, 182-195 (2007)
8. V.G. Ivancevic, *New mechanics of traumatic brain injury*. Cogn. Neurodyn **3**, 281-293 (2009)

Calculation of stress in FGM beams

Justín Murín[1,*], *Juraj* Hrabovský[1], *Vladimír* Kutiš[1]

[1]Department of Applied Mechanics and Mechatronics, Institute of Automotive Mechatronics, Faculty of Electrical Engineering and Information Technology, Slovak University of Technology in Bratislava, Ilkovičova 3, 812 19 Bratislava, Slovakia

Abstract. Content of the paper is oriented to calculation of elastic normal stress in the Functionally Graded Beams (FGM). Spatial variation of material properties is considered in the lateral, transversal and longitudinal direction of the straight beam. The displacements and internal forces are calculated using our new FGM finite beam element. Heterogeneous material properties are homogenized by extended mixture rules, laminate theory and reference volume element (RVE). Obtained results by our approach are evaluated and compared with the ones obtained by the 3D solid finite elements.

Keywords: FGM beams, Spatially varying material properties, Normal and shear stress calculation

1 Introduction

Many papers deal with static analysis of the FGM single 2D beams with transverse variation of material properties only [1-5]. Less attention is paid to longitudinal and lateral variation of material properties. However, the authors did not find papers which consequently deal with all the longitudinal, transversal and lateral variation of material properties by single beams or beam structures built of such FGM.

The presented contribution is the continuation of our previous work dealing with the derivation of a 3D FGM beam finite element with longitudinal varying effective material properties, which is suitable for analysis of beam structures made of spatially varying FGM. Effect of the axial and shear forces and Winkler elastic foundation is included, as well. Homogenization of the spatial varying material properties in the real FGM beam and the calculation of effective parameters are done by the multilayer or direct integration method (MLM or DIM) [6]. If only transversal and lateral variations of material properties are considered in the real FGM beam, longitudinally constant effective material properties arise from the homogenization. This method can also be used in the homogenization of multilayer beams with discontinuous variation of material properties in transversal and lateral direction. Numerical experiments are performed to calculate the elastostatic response of chosen spatial FGM beam structures with rectangular cross-sections with symmetrically lateral, transversal and longitudinal variations of material properties. Expressions for normal stress calculation in the cross-section caused by the axial force as well by the

* Corresponding author: justin.murin@stuba.sk

Reviewers: *Robert* Zalewski, *Milan* Žmindák

bending moments are established. For torsional loading only the twist angle and internal torsional moment is evaluated. The solution results are discussed and compared to those obtained by means of very fine 3D – solid finite element mesh.

2 3D FGM beam finite element with spatially varying material properties

Let us consider a 3D straight finite beam element (Timoshenko beam theory and Saint-Venant torsion theory) of doubly symmetric cross-section – Fig. 1. The nodal degrees of freedom at node i are: the displacements u_i, v_i, w_i in the local axis direction x, y, z and the cross-sectional area rotations – $\varphi_{x,i}, \varphi_{y,i}, \varphi_{z,i}$. The degrees of freedom at the node j are denoted in a similar manner. The internal forces at node i are: the axial force N_i, the transversal forces $R_{y,i}$ and $R_{z,i}$, the bending moments $M_{y,i}$ and $M_{z,i}$, and the torsion moment $M_{x,i}$. The first derivative with respect to x of the relevant variable is denoted with an apostrophe "'".

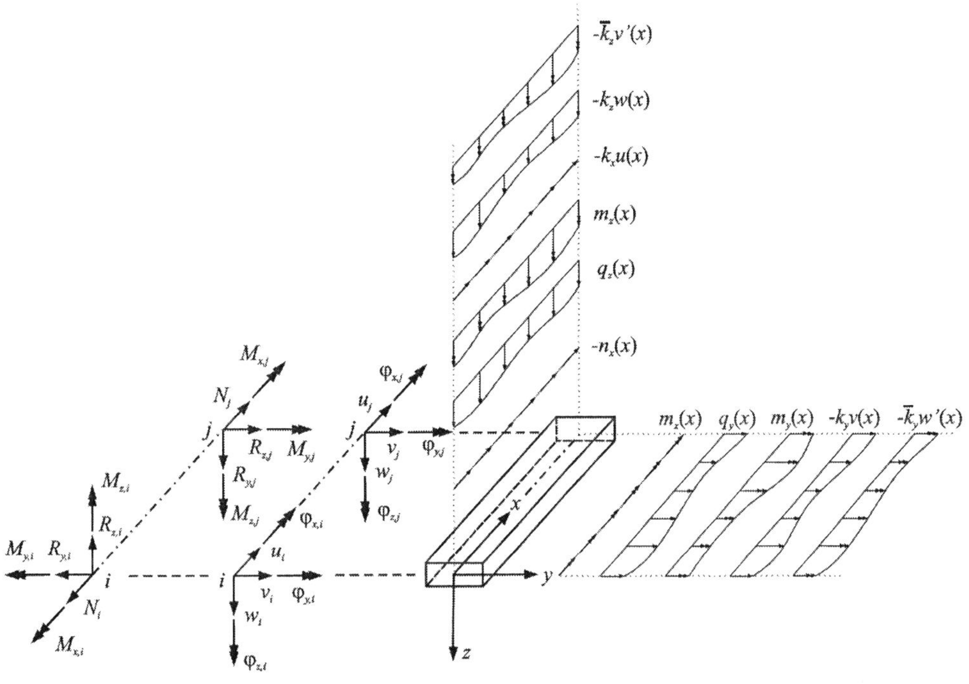

Fig. 1. The local internal variables and loads

Furthermore, $n_x = n_x(x)$ is the axial force distribution, $q_z - q_z(x)$ and $q_y - q_y(x)$ are the transversal and lateral force distributions, $m_x = m_x(x), m_y = m_y(x)$ and $m_z = m_z(x)$ are the distributed moments, A is the cross-sectional area, I_y and I_z are the second area moments, $I_p = I_y + I_z$ is the polar area moment, $k_x = k_x(x), k_y = k_y(x), k_z = k_z(x), \bar{k}_y = \bar{k}_y(x), \bar{k}_z = \bar{k}_z(x)$ are the elastic foundation modules (the torsional elastic foundation is not considered). The effective homogenized and longitudinally varying stiffness reads: $EA = E_L^{NH}(x)A$ is the axial stiffness ($E_L^{NH}(x) \equiv E_L^{NH}$ is the effective elasticity modulus for axial loading), $EI_y = E_L^{M_y H}(x)I_y$ is the flexural stiffness about the y-axis ($E_L^{M_y H}(x) \equiv E_L^{M_y H}$ is the effective elasticity modulus for bending about axis y), $EI_z = E_L^{M_zH}(x)I_z$ is the flexural stiffness about the axis z, ($E_L^{M_zH}(x) \equiv E_L^{M_zH}$ is the effective elasticity modulus for bending about axis z), $G\bar{A}_y = G_{Ly}^H(x)k_y^{sm}A$ is the reduced shear stiffness in y – direction ($G_{Ly}^H(x) \equiv G_{Ly}^H$ is the effective shear modulus and k_y^{sm} is the average shear correction factor in y – direction), $G\bar{A}_z = G_{Lz}^H(x)k_z^{sm}A$ is the reduced shear stiffness in z – direction ($G_{Lz}^H(x) = G_{Lz}^H$ is the effective shear modulus and k_y^{sm} is the average shear correction factor in z – direction), $G_L^{M_xH}(x)I_T$ is the effective torsional stiffness ($G_L^{M_xH}(x) = G_L^{M_xH}$ is the torsional elasticity modulus and I_T is the torsion constant).

Mixture rules are one of the methods for micromechanical modeling of heterogeneous materials. Extended mixture rules [7] are based on the assumption that the constituents volume fractions (formally only denoted here as fibres – f and matrix – m) continuously vary as polynomial functions, $v_f(x,y,z)$ and $v_m(x,y,z)$. The condition $v_f(x,y,z) + v_m(x,y,z) = 1$ has to be fulfilled. The appropriated material property distribution in the real FGM beam (Fig. 2a.) then reads

$$p(x,y,z) = v_f(x,y,z)p_f(x,y,z) + v_m(x,y,z)p_m(x,y,z) \tag{1}$$

Here, $p_f(x,y,z)$ and $p_m(x,y,z)$ are the spatial distributions of material properties of the FGM constituents. The extended mixture rule (1) can be analogically used for FGM material made of more than two constituents. The assumption of a polynomial variation of the constituent's volume fractions and material properties enables an easier establishing of the main appropriated field equations and allows the modeling of many common realizable variations.

In literature and in practical applications, mostly the one directional variation of the FGM properties is considered. There, an exponential law for transversal variation of the constituents volume fractions is often presented, e.g. in [8, 9, 10] and in references therein.

For the FGM beams the transversal variation (continuously or discontinuously, symmetrically or asymmetrically) has been mainly considered. The homogenization of such material properties variation is relatively simple. If the material properties vary only with respect to the longitudinal direction, the homogenization is frequently not needed since there are new FGM beam and link finite elements established that consider such variations in a very accurate and effective way, e.g. in [6, 11]. A more complicated case is, if the material properties vary in three directions - namely in transversal, lateral and longitudinal direction of the FGM real beam and the torsion is included as well.

In this contribution, the homogenization techniques for spatially varying (continuously or discontinuously and symmetrically in transversal and lateral direction, and continuously in longitudinal direction) material properties of FGM beams of selected doubly-symmetric cross-sections are considered. The expressions are proposed for the derivation of effective elasticity modules for axial loading, the transversal and lateral bending, the shear modules for transversal and lateral shear and for uniform torsion by the extended mixture rules (EMR) and the multilayer method (MLM).

Let us consider a two nodded 3D straight beam element with double symmetric cross-sectional area A (Fig. 2). The composite material of this beam arises from mixing two components. The continuous polynomial spatial variation of the elasticity moduli and mass density can be caused by continuous polynomial spatial variation of both the volume fraction ($v_f(x,y,z)$ and $v_m(x,y,z)$) and material properties of the FGM constituents ($p_f(x,y,z)$ and $p_m(x,y,z)$).

In our case the elasticity modulus $E(x,y,z)$, the Poisson ratio for the real beam have been calculated by expression (1). The FGM shear modulus can be calculated by expression:

$$G(x,y,z)=\frac{E(x,y,z)}{2(1+\nu(x,y,z))}. \tag{2}$$

If the constituents Poisson's ratio are approximately of the same value and the constituent volume fractions variation is not strong, then the FGM shear modulus can by calculated using a simplification [12]:

$$G(x,y,z)=\frac{E(x,y,z)}{\xi}, \tag{3}$$

where ξ is an average value of the function $\xi(x,y,z)=2(1+\nu(x,y,z))$

$$\xi=\frac{1}{L}\int_0^L\left(\frac{1}{A}\int_{(A)}\xi(x,y,z)dA\right)dx. \tag{4}$$

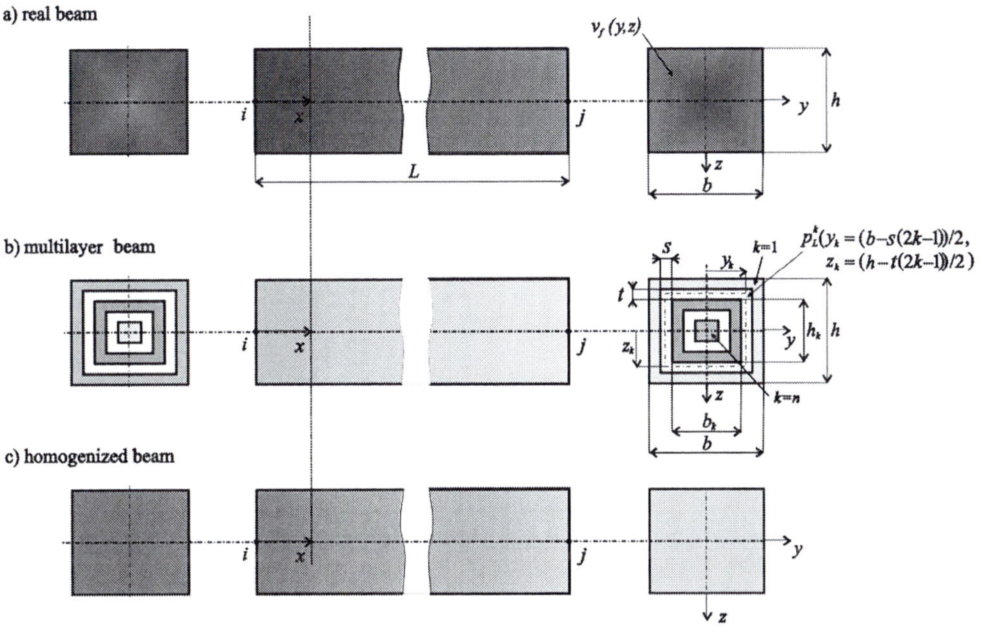

Fig. 2. FGM beam with rectangular cross-section

Homogenization of the spatially varying material properties (the reference volume is the volume of the whole beam) are done in two steps. In the first step, the real beam (Fig. 2a) is transformed into a multilayer beam (Fig. 2b). Material properties of the layers are calculated with the EMR [13]. We assume that each layer cross-section at position x has constant material properties. They are calculated as an average value from their values at the boundaries of the respective layer cross-section. Polynomial variation of these properties appears in the longitudinal direction of the layer. Sufficient accuracy of the proposed substitution of the continuous transversal and lateral variation of material properties by the layer-wise constant distribution of material properties is reached if the division to layers is fine enough. In the second step, the effective longitudinal material properties of the homogenized beam are derived using the MLM [14]. These homogenized material properties are constant through the beam's height and depth but they vary continuously along the longitudinal beam axis. Accordingly, the beam finite element equation is established for the homogenized beam (Fig. 2c) in order to calculate the primary beam unknowns – the displacements and internal forces and critical buckling force in our case. The stress has to be calculated on the real beam and it is the main goal of this paper that will be presented in the next chapter.

3 Normal stress calculation in the FGM rectangular cross-sectional area

Stress calculation depends of the cross-sectional area form. In the next we focus on the calculation of normal stress distribution on rectangular cross section. Calculation of the shear and torsional stress in the FGM beams is more complicated in our approach. We deal with this task very active and the results will be published in our future publication.
The rectangular cross-section at position x with doubly symmetric variation of material properties is considered - Fig. 3. At the centroid acts the normal force $N(x)$, the bending moments $M_y(x)$ and $M_z(x)$. $A = bh$ is the cross-sectional area, $I_z = bh^3/12$ and $I_y = b^3h/12$ are the quadratic moments of the rectangular cross-section.

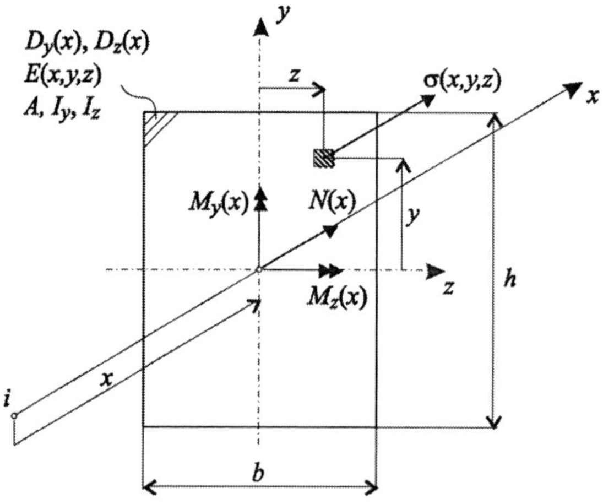

Fig. 3. To normal stress calculation

The resultant normal stress at point (x,y,z) is the sum:

$$\sigma(x,y,z) = \sigma^N(x,y,z) + \sigma^{M_z}(x,y,z) + \sigma^{M_y}(x,y,z) \tag{5}$$

There, $\sigma^N(x,y,z) = \varepsilon^N(x)E(x,y,z) = \dfrac{\sigma^N(x)}{E_L^{NH}(x)}E(x,y,z)$ is the normal stress caused by normal force and $\sigma^N(x) = \dfrac{N(x)}{A}$ is the effective normal stress in homogenized cross-section while constant effective normal strain $\varepsilon^N(x) = \dfrac{\sigma^N(x)}{E_L^{NH}(x)}$ over the whole cross-sectional area is assumed. Here $E_L^{NH}(x)$ is the effective elasticity modulus for tension-compression [7].

If the effective bending stiffness about the axis z is $D_z(x) = I_z E_L^{M_z}(x)$, then the distribution of the effective bending strain is $\varepsilon^{M_z}(x,y) = \dfrac{M_z(x)}{D_z(x)}y$ and the effective bending stress is $\sigma^{M_z}(x,y) = \varepsilon^{M_z}(x,y)E_L^{M_z}(x)$. The real bending stress at position (x,y,z) is

$\sigma^{M_z}(x,y,z)=\varepsilon^{M_z}(x,y)E(x,y,z)=\dfrac{E(x,y,z)}{E_L^{M_z}(x)}\sigma^{M_z}(x,y)$. Here, $E_L^{M_z}(x)$ is the effective elasticity modulus for bending about z-axis [7].

Analogically, if the effective bending stiffness about the axis y is $D_y(x)=I_y E_L^{M_y}(x)$, then the distribution of the effective bending strain is $\varepsilon^{M_y}(x,z)=\dfrac{M_y(x)}{D_y(x)}z$ and the effective bending stress is $\sigma^{M_y}(x,z)=\varepsilon^{M_y}(x,z)E_L^{M_y}(x)$. The real bending stress at position (x,y,z) is $\sigma^{M_y}(x,y,z)=\varepsilon^{M_y}(x,z)E(x,y,z)=\dfrac{E(x,y,z)}{E_L^{M_y}(x)}\sigma^{M_y}(x,z)$. Here, $E_L^{M_y}(x)$ is the effective elasticity modulus for bending about z-axis [7]. It is not simply obtaining functional distribution of the normal stress. For discontinuous variation of the material properties in transversal and lateral direction, the stress has to be calculated at discrete points of cross-section.

4 Numerical experiments

In the chapter, the results of numerical investigations concerning the normal stress calculation in the FGM beams with spatially varying material properties are presented.

4.1 Normal stress calculation in the FGM cantilever beam

The cantilever FGM beam is considered (as shown in Fig. 4), which is loaded at its free end by forces $F_y = F_z = 10\,\text{N}$ and $F_x = -5\,\text{kN}$ and by torsion moment $M_x = 10\,\text{Nm}$. Its rectangular cross-section is constant with height h = 0.005 m and width b = 0.01 m. The length of the beam is L = 0.1 m. The local coordinates system is denoted by the axis x, y, z.

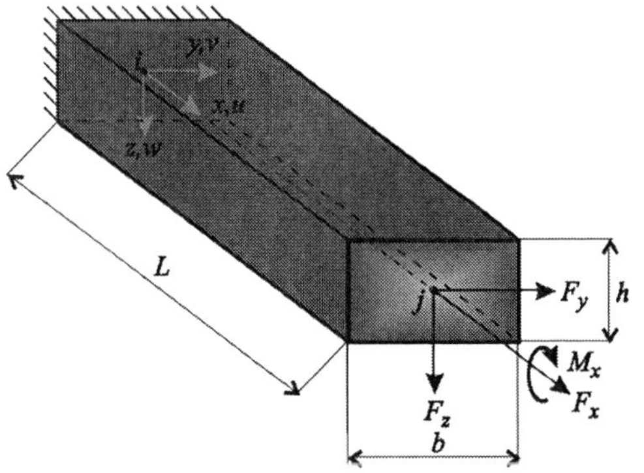

Fig. 4. Cantilever FGM beam with spatial variation of material properties

Material of the beam consists of two components: Aluminum Al6061-TO – denoted with index m and Titanium carbide TiC –denoted with index f. The material properties of the components are assumed to be constant and their values are: Aluminum Al6061-TO – the elasticity modulus $E_m = 69.0\,\text{GPa}$, the Poisson's ratio $\nu_m = 0.33$; Titanium carbide TiC – the elasticity modulus $E_f = 480.0\,\text{GPa}$, the Poisson's ratio $\nu_f = 0.20$.

The TiC volume fraction varies in the y and z direction linearly and symmetrically according to the x-y and x-z planes: $\lfloor v_f(y,z) \rfloor_{\substack{y=0\\z=0}} = 0, \lfloor v_f(y,z) \rfloor_{\substack{y=\pm h/2\\z=\pm b/2}} = 1$ - the core of the beam is made from pure Al6061-TO and linearly varies to the edges that are made from pure TiC. Constant effective material properties are considered in the local x – direction of the beam. The average shear correction factors in y – direction $k_y^{sm} = 5/6$ and in z – direction $k_z^{sm} = 5/6$ have been considered (constant Poisson ratio has been assumed in the example).

Using the EMR and MLM, the homogenized elasticity modules (in [GPa]) for: axial loading - $E_L^{NH} = 342.8$ MPa, bending about axis y and z - $E_L^{M_yH} = E_L^{M_zH} = 397.5$ MPa, shear - $G_{Ly}^H = E_{Lz}^H = 138.9$ MPa, and torsion - $G_L^{M_xH} = 162.6$ MPa have been calculated [15].

The FGM beam, clamped at the node i, has been studied by the elastic-static and buckling analysis. All the calculations were done with our 3D FGM beam finite element (NFE) which we have implemented into the code MATHEMATICA [16]. Additionally, the effect of axial force was considered. It has to be pointed out that the entire structure is discretized using only one our finite element. The critical buckling force is $N_{Ki}^{II} = 10.28$ kN. In the elasto-static analysis the axial force (tension and compression) $N^{II} \equiv N$ have been chosen as a part of the critical buckling force N_{Ki}^{II}. The same problem has been solved using a very fine mesh – 28889 of SOLID186 elements of the FEM program ANSYS [17]. The results of ANSYS as well as the results of the NFE are presented in Table 1. The average relative difference Δ [%] between displacements calculated by our method and the ANSYS solution has been evaluated.

Table 1. Displacements and rotations at the free beam end

Displacements [mm], [rad]	**NFE $F_x = 0$ N**	**ANSYS $F_x = 0$ N**	Δ **[%]**	**NFE F_x=-5kN**	**ANSYS F_x=-5kN**	Δ **[%]**
u_j	0	0	---	-0.029230	-0.029385	0.53
v_j	0.020349	0.020271	0.38	0.022281	0.022173	0.49
w_j	0.080879	0.080645	0.29	0.158012	0.157130	0.56
φ_{xj}	0.021533	0.021844	1.42	0.021534	0.021185	1.47
φ_{yj}	-0.001211	-0.001206	0.43	-0.000346	0.000344	0.75
φ_{zj}	0.000303	0.000302	0.03	0.002417	0.002410	0.26

As is shown in Table 1, the accuracy of our results is excellent compared to the ANSYS solution. The bending moments about the y and z – axis, the torsional moment, the axial, transversal and shear forces are calculated. To save space, the distribution of the bending moment about y-axis is shown in Fig. 5 only. As expected, the axial force F_x influence distribution of the bending moment significantly. The high efficiency of our method (NFE) is obvious since our results are evaluated using only one finite beam element compared to the large number of 41924 elements used in the continuum mesh (ANSYS).

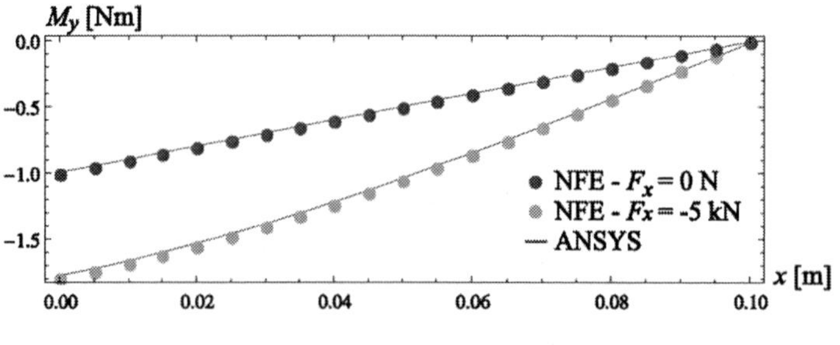

Fig. 5. Bending moment about the y - axis

Due to stress calculation, the internal forces at the clamped end $(x=0)$ are evaluated and they are: $M_x=10$ Nm; $M_y=1.78$ Nm; $M_z=1.12$; $N_x=-5$ kN; $Q_y=Q_z=10$ N. According the expressions in chapter 2, the normal stress caused by axial force are calculated. Because of discontinuous variation of elasticity modulus in real beam, the cross section is discretized on layers with constant elasticity modulus - **Fig. 6**.

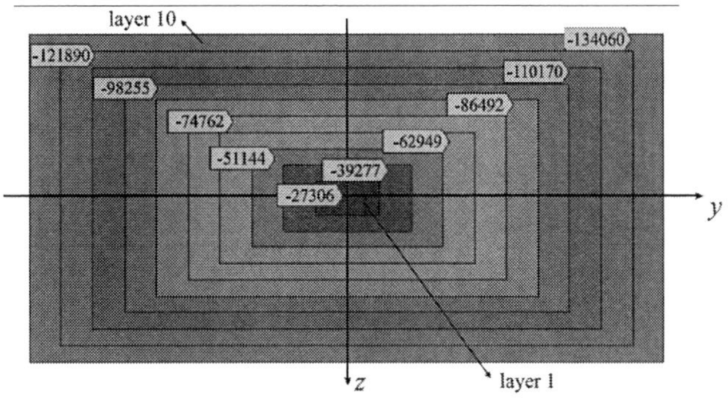

Fig. 6. Compression normal stress distribution caused by the axial force

The maximal axial force stress are in the layer number 10 and minimal in layer 1 - see Table 2. Accuracy of the calculation approach depends of the number of the layers. As shown in Table 2, for then layers is the difference between our results (NFE) and the ones obtained by very fine mesh of 3D-ssolid FE (ANSYS) very small. By very fine division to layers a continuous distribution of the stress can be obtained.

Table 2. Axial force normal stress at the clamped cross-section

Layer	σ [kPa] NFE	σ [kPa] ANSYS	Δ [%]
1	-26175	-27306	4.14
2	-38189	-39277	2.70
3	-50203	-51144	1.84
4	-62216	-62949	1.16
5	-74230	-74762	0.71
6	-86244	-86492	0.29
7	-98258	-98255	0.01
8	-110272	-110170	0.09
9	-122285	-121890	0.32
10	-134299	-134060	0.18

Distribution of the bending normal stress is shown in Fig. 7 and Table 3.

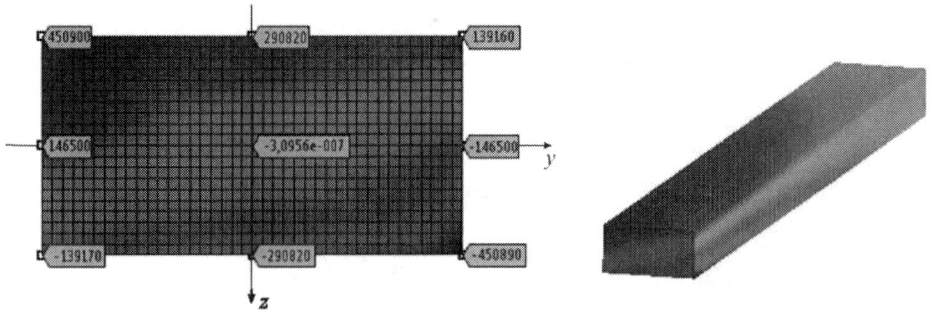

Fig. 7. Resultant biaxial bending normal stress calculated by ANSYS

Table 3. Comparing of the biaxial bending normal stress at chosen position in clamped cross-section

Position	σ [kPa] NFE	σ [kPa] ANSYS	Δ [%]
$y = 0, z = 0$	0	0	0
$y = 0, z = -h/2$	290594	290820	0.08
$y = 0, z = h/2$	-290594	-290820	0.08
$y = -b/2, z = 0$	145297	146500	0.83
$y = b/2, z = 0$	-145297	-146500	0.83
$y = -b/2, z = -h/2$	435891	450900	3.44
$y = b/2, z = h/2$	-435891	-450900	3.44
$y = b/2, z = -h/2$	145297	139170	4.22
$y = -b/2, z = h/2$	-145297	-139170	4.22

3.2 Normal stress calculation in the FGM cantilever beam on elastic foundation

The cantilever FGM beam on varying Winkler foundation is considered (as shown in Fig. 8). Its rectangular cross-section is constant with height $h = 0.005$ m and width $b = 0.01$ m. The length of the beam is $L = 0.1$ m.

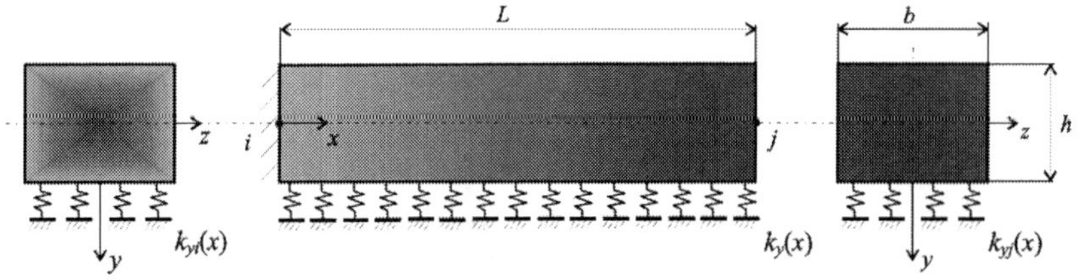

Fig. 8. Cantilever beam on elastic foundation with spatially varying material properties

The beam is made of a mixture of two components: Aluminum Al6061-TO and Titanium Carbide TiC, their constant constituent's material properties are given in Case I. The Aluminum volume fraction, in this case, varies linearly and symmetrically according to the x–y and x–z planes: At node i is $\left[v_{fi}(y,z)\right]_{\substack{y=0\\z=0}} = 1, \left[v_{fi}(y,z)\right]_{\substack{y=\pm h/2\\z=\pm b/2}} = 0$ and then vary continuous linearly in the longitudinal direction to the constant value at node j ($v_{fi} = 1$).

Using EMR and MLM with n = 20 layers the effective elastic modulus for axial loading E_L^{NH} , for bending about axis y - $E_L^{M_yH}$ and about axis z - $E_L^{M_zH}$, shear modules G_{Ly}^H and G_{Lz}^H and torsional shear modulus $G_L^{M_xH}$.

$$E_L^{NH}(x) = 342.109 - 3731.095x \text{ GPa;}$$

$$E_L^{M_yH}(x) = E_L^{M_zH}(x) = 396.429 - 3274.293x \text{ GPa;}$$

$$G_{Ly}^H(x) = G_{Lz}^H(x) = 138.581 - 1129.418x \text{ GPa;}$$

$$G_L^{M_xH}(x) = 162.233 - 1362.936x \text{ GPa.}$$

The FGM cantilever beam has been studied by elastic-static analysis. All the calculations were done with our 3D FGM beam finite element (NFE). Additionally, the effect of axial force was considered. It has to be pointed out that the entire structure is discretized using only one herein proposed finite element. The cantilever FGM beam resting on varying vertical Winkler elastic foundation $k_y = 5000 - 30000x + 600000x^2$ kN/m2 is loaded by forces $F_y = F_z = 50\,\text{N}$ and $F_x = -2\,\text{kN}$ at node j (Fig. 9). The average shear correction factors in y' – direction $k_y^{sm} = 5/6$ and in z' – direction $k_z^{sm} = 5/6$ are used [12]. The displacements at node j are evaluated using the only one new FGM beam finite element (NFE). The same problem is solved using a very fine mesh – 23015 of SOLID186 elements of the FEM program ANSYS [18 34]. The results of ANSYS as well as the results of the NFE are presented in Table 4. The average relative difference Δ[%] between displacements calculated by our method and the ANSYS solution is evaluated.

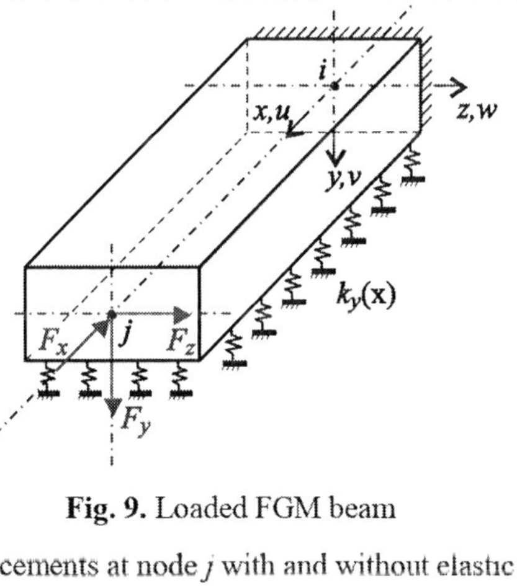

Fig. 9. Loaded FGM beam

Table 4. Displacements at node j with and without elastic foundation

Displacements at node j [mm], [rad]	NFE with foundation	ANSYS with foundation	NFE without foundation	ANSYS without foundation
u_j	-0.02445	-0.02507	-0.02445	0.02507
v_j	0.24641	0.24933	0.74414	0.75348
w_j	0.14452	0.14791	0.14452	0.14791
φ_{yj}	-0.00247	-0.00256	-0.00247	-0.00256

Displacements at node *j* [mm], [rad]	NFE with foundation	ANSYS with foundation	NFE without foundation	ANSYS without foundation
u_j	-0.02445	-0.02507	-0.02445	0.02507
v_j	0.24641	0.24933	0.74414	0.75348
w_j	0.14452	0.14791	0.14452	0.14791
φ_{zj}	0.00503	0.00527	0.1304	0.01321

The comparison of the vertical beam deflection curve with and without elastic foundation is shown in Fig. 10.

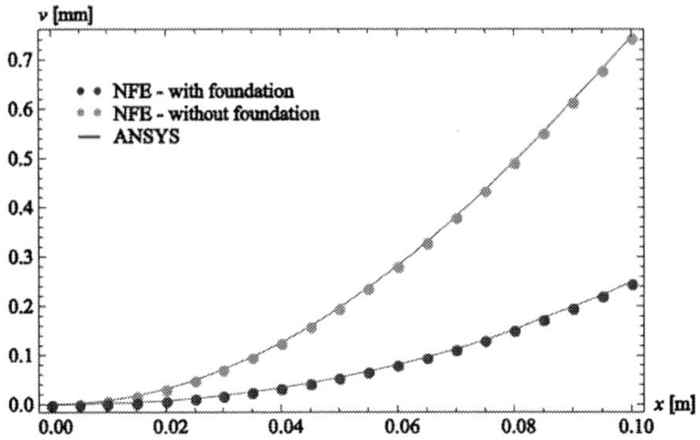

Fig. 10. Vertical beam deflection curve with and without elastic foundation

The bending moments about the *y* and *z* – axis for case without elastic foundation are shown in Fig. 11 and Fig. 12, respectively. The Fig. 13 and Fig. 14 show the transversal force in *y* and *z* – axis. The comparison of the bending moments $M_y(x=0)$, $M_z(x=0)$ and transversal forces $R_y(x=L)$, $R_{yz}(x=L)$ for the case F_x = -2 kN calculated by our approach and by ANSYS are compared in Table 5.

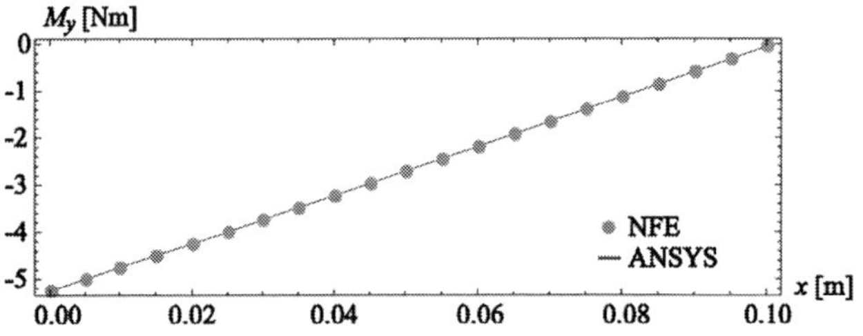

Fig. 11. Bending moment about the y – axis (without elastic foundation)

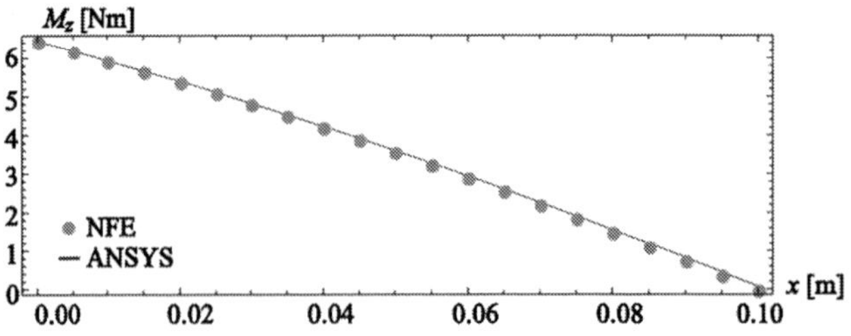

Fig. 12. Bending moment about the *z* – axis (without elastic foundation)

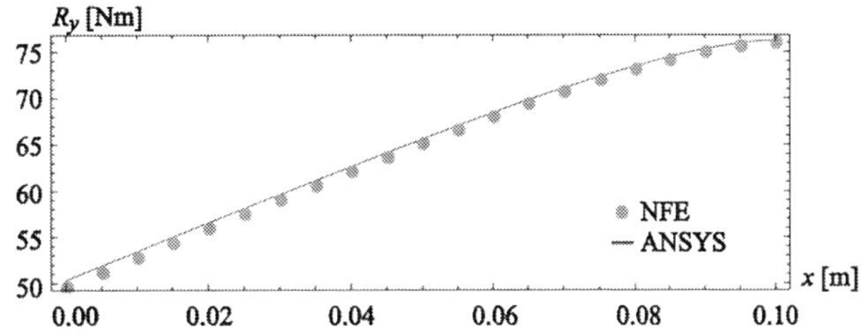

Fig. 13. Transversal force in y – axis (without elastic foundation)

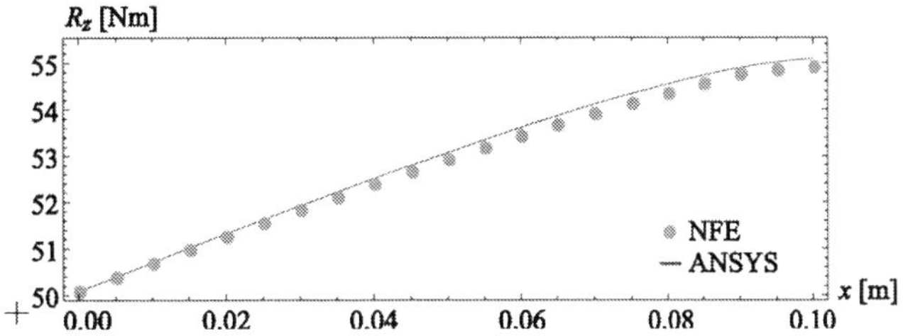

Fig. 14. Transversal force in z – axis (without elastic foundation)

Table 5. Bending moments and transversal forces

Internal variables	NFE F_x = -2 kN	ANSYS Fx = -2 kN	Δ [%]
$M_y(x=0)$ [Nm]	-5.2182	-5.2416	0.45
$M_z(x=0)$ [Nm]	6.4691	6.4404	0.45
$R_y(x=L)$ [N]	54.9472	55.0873	0.25
$R_z(x=L)$ [N]	76.2088	76.2251	0.02

In Table 6 the comparison of the resultant normal stress, caused by axial, transversal and lateral forces, in the clamped cross-section ($x = 0$) calculated by our approach (Fig. 15) and by ANSYS is shown.

Table 6. Comparison of the normal stress at chosen position in clamped cross-section

position	σ [kPa] NFE	σ [kPa] ANSYS
$y = h/2$, $z = b/2$	-320.40	-324.21
$y = -h/2$, $z = -b/2$	208.20	209.35
$y = h/2$, $z = b/2$	-245.20	-249.32
$y = 0$, $z = -b/2$	19.80	21.56
$y = h/2$, $z = 0$	-245.20	-249.32
$y = -h/2$, $z = 0$	131.50	136.22

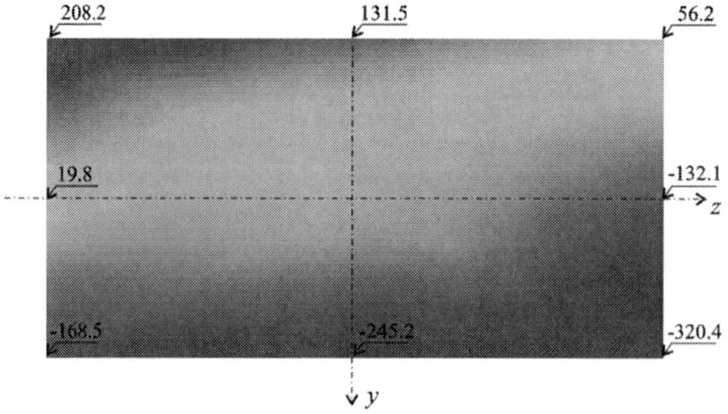

Fig. 15. Resultant normal stress $\sigma(x, y, z)$ [MPa] at position $x = 0$ calculated by our approach

In Fig. 16, the resultant normal stress, caused by axial, transversal and lateral forces (marked in blue flags), in the clamped cross-section is shown that was calculated by ANSYS (using a very fine mesh – 23015 of SOLID186). As can be seen, a very good agreement of both solution method has been obtained at all marked points excluding the corners. As is well known, the solutions with 3D solid finite elements produce in the sharp corners incorrect stress first of all by the very fine meshes. In the nearby points of the sharp corners is the match of the results very good (see the details in Fig. 16).

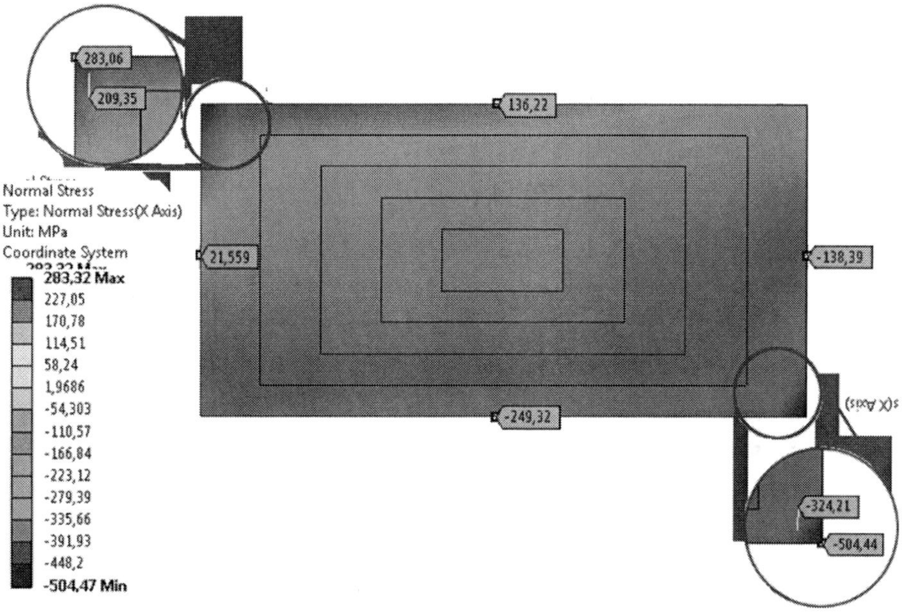

Fig. 16. Normal stress $\sigma(x, y, z)$ [MPa] at position $x = 0$ calculated by ANSYS (SOLID186)

Conclusions

In this contribution, our new 3D beam finite element for elasto-static and buckling analysis of the FGM single beam and beam structures has been used for calculation of the deformation and internal forces and moments in FGM beams with spatial variation of material properties. Effect of varying Winkler elastic foundations and shear force deformation effect is taken into account. The effect of axial force has been taken into account for the flexural loading. For an elastic-static analysis, the expressions for distribution of normal stress on the beam cross-section is established and applied. The obtained results have been studied and compared with results obtained using very fine continuum and beam meshes by the FEM program ANSYS. An excellent agreement of our solution results is obtained, which confirms respectable accuracy and effectiveness of our approach.

This work was financially supported by grants of Scientific Grant Agency of the Ministry of Education of Slovak Republic and the Slovak Academy of Sciences and VEGA No. 1/0453/15.

References

1. V. Kahya, M. Turan, *Finite element model for vibration and buckling of functionally graded beams based on the first-order shear deformation theory*. Composites Part B **109**, 108–115 (2017)
2. N. Tan-Tien, T.T. Pham, J. Lee, *Lateral buckling analysis of thin walled functionally graded open-section beams*. Compos. Struct. **160**, 952–963 (2017)
3. M. Simsek, *Bi-directional functionally graded materials (BDFGMs) for free and forced vibration of Timoshenko beams with various boundary conditions*. Compos. Struct. **133**, 968–978 (2015)
4. T. Kocaturk, M. Simsek, S.D. Akbas, Large displacement static analysis of a cantilever Timoshenko beam composed of functionally graded material. Sci Eng Compos Mater **18**, 21–34 (2011)
5. M. Meradjah, A. Kaci, M.S.A. Houari, A. Tounsi, S.R. Mahmoud, *A new higher order shear and normal theory of fictionaly graded beams*. Steel Compos Struct **18**, 793-809 (2015)
6. V. Kutis, J. Murin, R. Belak, J. Paulech, *Beam element with spatial variation of material properties for multiphysics analysis of functionally graded materials*. Comput. Struct. **89**, 1192 – 1205 (2011)
7. J. Murin, V. Kutis, *Improved mixture rules for composite (FGMs) sandwich beam finite element*. In Computational Plasticity IX. Fundamentals and Applications. Barcelona, Spain, 647-650 (2007)
8. A.E. Alshorbagy, M.A. Eltaher, F.F. Mahmoud, *Free vibration of a functionally graded beam by finite element method*. Appl. Math. Modell. **35**, 412 – 425 (2010)
9. M. Simsek, *Vibration analysis of a functionally graded beam under a moving mass by using different beam theories*. Compos. Struct. **92**, 904-917 (2010)
10. T. Rout, *On the dynamic stability of functionally graded material under parametric excitation*. PhD thesis. National Institute of Technology Rourkela, India (2012)
11. J. Murin, S. Kugler, M. Aminbaghai, J. Hrabovsky, V. Kutis, J. Paulech, *Homogenization of material properties of the FGM beam and shells finite elements*. In. 11th World Congress on Computational mechanics (WCCM XI), Barcelona, Spain (2014)
12. J. Murin, M. Aminbaghai, J. Hrabovsky, V. Kutis, S. Kugler, *Modal analysis of the FGM beams with effect of the shear correction function*. Composites Part B **45**, 1575–1582 (2013)
13. J. Murin, M. Aminbaghai, J. Hrabovsky, V. Kutis, J. Paulech, S. Kugler, *A new FGM beam finite element for modal analysis*. In. 11th World Congress on Computational mechanics (WCCM XI), Barcelona, Spain (2014)

14. J. Murin, S. Kugler, M. Aminbaghai, J. Hrabovsky, V. Kutis, J. Paulech, *Homogenization of material properties of the FGM beam and shell finite elements*. In. 11th World Congress on Computational mechanics (WCCM XI), Barcelona, Spain (2014)
15. J. Murin, M. Aminbaghai, J. Hrabovsky, V. Kutis, J. Paulech, *3D beam finite element for elastic-static analysis of the FGM structures*. In 8th International Congress of Croatian Society of Mechanics (8th ICCSM), Opatija, Croatia (2015)
16. S. Wolfram MATHEMATICA 5, Wolfram research, Inc., 2003
17. ANSYS Swanson Analysis System, Inc., 201 Johnson Road, Houston, PA 15342/1300, USA

Numerical modelling of sandwich panels with a non-continuous soft core

Jolanta Pozorska[1,*]

[1]Institute of Mathematics, Czestochowa University of Technology, Armii Krajowej 21, 42-201 Częstochowa, Poland

Abstract. The paper presents the problem of static analysis of sandwich structures with a non-continuous soft core. In the numerical 3D FE models, the core is divided into separated parts. The contact between these parts has the form of unilateral constraints. The model also allows for local debonding of the facing and local imperfections of sandwich panel geometry. Particular attention is paid to the problem of local instability of the facing that is compressed during bending. The phenomenon of progressive damage and the influence of non-continuity of the core on the structural behavior of the sandwich panel is also discussed.

Keywords: sandwich panels, soft core, multi-layered structures, finite element method, unilateral contact, local instability

1 Introduction

Sandwich structures are commonly used in various branches of industry. The subject of the paper are sandwich panels used in civil engineering. They are built from three layers: two metal facings and a flexible core. A typical core is made of polyurethane foam, polystyrene or mineral wool. The core made of polyurethane foam is a continuous structure, while the core made of polystyrene or mineral wool is discontinuous. In the latter case, (mineral wool) the core structure is very interesting because it consists of small blocks which have the shape of elongated cuboids. These blocks are called lamellae. The lamellae have the same thickness as the thickness of the core in the panel. Lamellae are displaced parallel to each other but are not connected together. The core parts are only glued to the facings. This paper is aimed at analyzing the behavior of sandwich panels with a discontinuous core.

The rapid development of technology has forced the development of appropriate numerical models allowing for an appropriate analysis of laminated composites, including sandwich structures. An overview of numerical models has been presented, among others in [1-3]. The behavior of layered structures is closely related to certain phenomena. Many of these phenomena concern the appropriate failure mechanisms determining the scope of application of the structure and the safety of its use. Various failure mechanisms are analyzed and explained in numerous research papers. Among the important issues it is necessary to mention local instability (wrinkling of facing) [4, 5], shear of the core [6, 7],

* Corresponding author: jolanta.pozorska@im.pcz.pl

Reviewers: *Milan* Žmindák, *Bohuš* Leitner

global loss of stability [8], debonding, which is often associated with the occurrence of concentrated dynamic loads [9] and indentation caused by the crushing of the soft core under the acting force [10, 11]. Of course, the analysis of these phenomena requires the use of appropriate models.

Against the background of the rich literature concerning various issues related to sandwich panels, surprisingly few papers concern the discontinuous core and the core made of mineral wool. On the one hand, there are considerable difficulties in finding complete information about the properties of mineral wool used in sandwich panels. This is because most of these cores are covered by patents. On the other hand, this topic is unpopular, perhaps due to strong ties to the technology of the production of sandwich panels. One of the few papers devoted to the structural behavior of sandwich panels with mineral wool concentrates on wrinkling phenomenon [12]. In the paper [13], the results of laboratory tests and numerical simulations of panels with a hybrid core were presented. The influence of different proportions of mineral wool and polyurethane foam on the results of the four-point bending tests were investigated. It should be clearly noted that a number of research projects devoted to panels with mineral wool core dealt with issues of fire resistance [14].

This paper presents the extension of the existing numerical model developed for laminated panels with a homogeneous core [15]. Structures with a discontinuous core are very interesting because of their complex structural behavior, impossible for providing an accurate analytical description. The numerical model was developed taking into account bending, shear, support capacity, local stability and debonding effects. The model is suitable for prediction of structure behavior with various boundary conditions.

2 Problem formulation

The layered structure shown in Fig. 1 is considered. Both facings are made of zinc-coated steel. In the discussed analyses, only cases of flat or micro-profiled facings are considered, although the presented model also allows for the analysis of sandwich panels with deep-profiled facings. The core is made of mineral wool. Please, note that the core material is orthotropic. The mineral wool is manufactured as a board with mainly horizontally arranged fibers. The board is cut into narrow and long pieces (lamellae). The lamellae are rotated 90 degrees in the sandwich structure, which provides higher stiffness of the core. The contact surfaces between lamellae are not glued. Therefore, the core is non-continuous with longitudinal and transversal, whole depth discontinuities (Fig. 1).

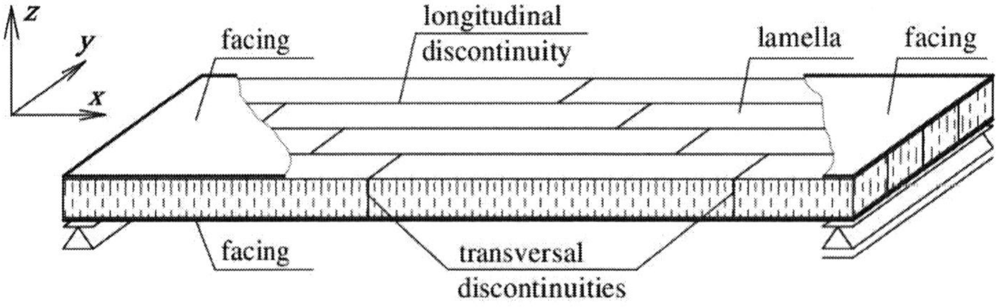

Fig. 1. Structure of a sandwich panel with mineral wool core

The aim of the paper is to analyze the influence of the core discontinuities on the sandwich panel behavior. It is important to explain the mechanism of damage and to explain the influence of material and geometric parameters on its initiation. The simplest static scheme of a single-span plate simply supported on two opposite (shorter) edges is considered. The plate is loaded uniformly on one surface. In all examples, suction was applied to the bottom facing of the sandwich panel. Such a simple static scheme was assumed because of the ease of interpretation of the results, although there are no obstacles to apply more complex boundary conditions.

3 Numerical model

The numerical model of sandwich structure was created based on previously carried out experimental and theoretical results. The model was prepared in the ABAQUS system environment. Material of steel facings was assumed as elastic - ideal plastic with: modulus of elasticity $E_F = 210$ GPa, Poisson ratio $\nu_F = 0.3$ and yield stress $f_y = 280$ MPa. Facings were modelled using S4R elements (four node, doubly curved, thin or thick shell, reduced integration, hourglass control and finite membrane strain). The core of the panel was modelled using eight node linear brick elements C3D8R. The behavior of the core material was defined using transversely isotropic elasticity. The following parameters of the core material were introduced: $E_1 = E_2 = 3$ MPa, $E_3 = 14$ MPa, $G_{12} = 1.43$ MPa, $\nu_{12} = \nu_{21} = 0.05$, $G_{13} = G_{23} = 6.36$ MPa, $\nu_{13} = \nu_{23} = 0.03$, $\nu_{31} = \nu_{32} = 0.14$. There are only five independent constants because of the relations $\nu_{31} / E_3 = \nu_{13} / E_1$, $\nu_{32} / E_3 = \nu_{23} / E_2$ and $G_{12} = E_1 / 2(1+\nu_{12})$. The total depth of the plate is 0.12, which consists of two facings, each 0.0005 m thick and the core (core with interfaces) of 0.119 m thick.

To enable initiation and propagation of the damage of the sandwich panel, between the facings and the core, a layer of interface was introduced. The 0.0005 m thick interface was modeled using COH3D8 8-node, 3D cohesive elements. Interactions between all parts were assumed as a TIE type, which makes equal displacements of nodes. The following elasticity uncoupled law for cohesive material of the interface was used:

$$\begin{bmatrix} t_n \\ t_s \\ t_t \end{bmatrix} = \begin{bmatrix} K_{nn} & 0 & 0 \\ 0 & K_{ss} & 0 \\ 0 & 0 & K_{tt} \end{bmatrix} \begin{bmatrix} \varepsilon_n \\ \varepsilon_s \\ \varepsilon_t \end{bmatrix} \tag{1}$$

where t_n is normal traction (stress) and t_s, t_t are shear tractions. Corresponding nominal strains are defined as $\varepsilon_n = \delta_n/T_0$, $\varepsilon_s = \delta_s/T_0$, $\varepsilon_t = \delta_t/T_0$ using separation δ and constitutive thickness of cohesive element T_0. The damage initiation criterion has the form of a quadratic nominal stress function. The damage is initiated when the function reaches a value of one:

$$\left\{\frac{\langle t_n \rangle}{t_n^0}\right\}^2 + \left\{\frac{t_s}{t_s^0}\right\}^2 + \left\{\frac{t_t}{t_t^0}\right\}^2 = 1 \tag{2}$$

where the notation of Macaulay brackets is used:

$$\langle t_n \rangle = \begin{cases} 0, & t_n < 0 \ (compression) \\ t_n, & t_n \geq 0 \quad (tension) \end{cases} \tag{3}$$

The following parameters of the interface were used initially: $K_{nn} = 12000$ kPa, $K_{ss} = K_{tt} = 6000$ kPa, $t_n^0 = 140$ kPa, $t_s^0 = t_t^0 = 100$ kPa.

The mineral wool lamellae are glued to the facings, but there are no connections between the lamellae. The contact between the lamellae was modelled as the surface contact of 3D deformable bodies. Two types of contact in normal direction were used (Fig. 2): "hard" and an exponential pressure-overclosure relationship. The model of "hard" contact is described by the contact pressure p between two surfaces at a point, as a function of "overclosure" h:

$$\begin{cases} p=0 \;\; for \;\; h<0 \quad (open), \\ h=0 \;\; for \;\; p>0 \;\; (closed). \end{cases} \tag{4}$$

In the case of exponential definition of the contact, clearance c_0 and pressure p_0 must be defined. To model tangential behavior of contacting bodies, the classical isotropic Coulomb friction model with the friction coefficient μ was applied:

$$\tau_{crit} = \mu p \tag{5}$$

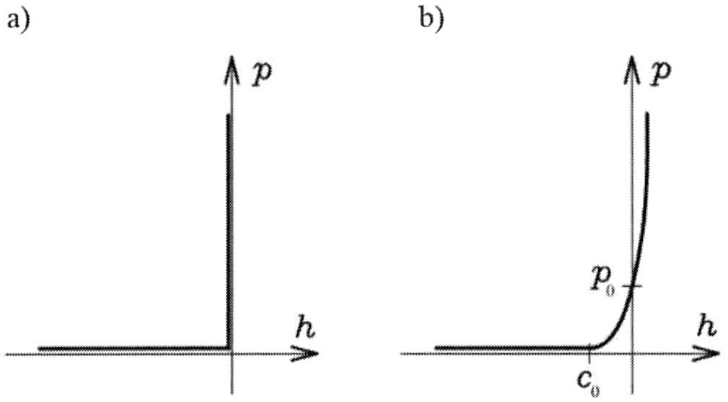

Fig. 2. Types of a contact definition in a normal direction: a) "hard" contact, b) exponential pressure-overclosure relationship

The numerical analyses were carried out for the following geometry: panel length $L = 4.80$ m, panel width $B = 0.50$ m and depth $D = 0.12$ m. The panel is based on two rigid supports with a width of 0.10 m, therefore, the span of the structure is 4.70 m. Both supports can rotate around the y-axis and, in addition, one of the supports can move in the x-direction.

The core consists of four rows of lamellae, each 0.125 m width and 0.119 m depth. The lamellae in the rows are mutually shifted 0.30 m along the x-axis. The original length of the lamellae is equal to 1.2 m, however a couple of them are cut due to the geometry of the panel. The arrangement of lamellae is shown in Fig. 3.

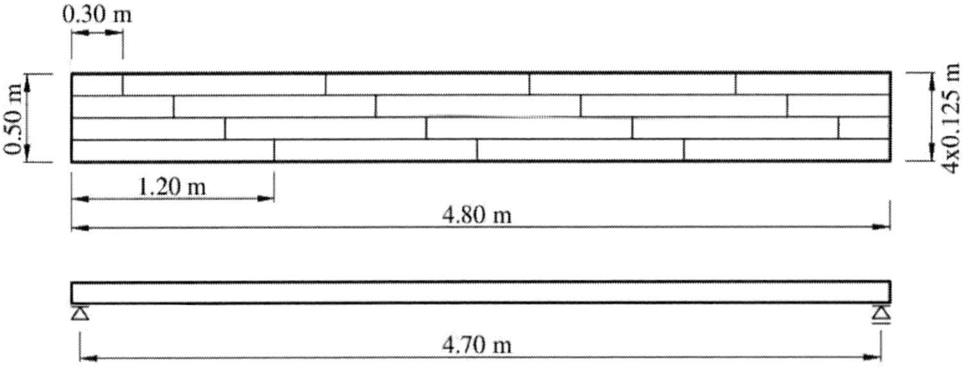

Fig. 3. The arrangement of lamellae in the sandwich panel

4 Results and discussion

In order to analyze the impact of the model parameters on the response of the structure, 6 models with different parameters of the interface and the contact between lamellae were considered. In each model, it was assumed that the facings are flat. A summary of these parameters is given in Tab. 1. The last column of this table contains one of the final results: the value of extreme tensile stress in the lower facing (in x direction). This value corresponds to the moment when the numerical solution loses convergence. Loss of convergence of the solution is due to interface damage and/or large deformations of the thin facing. The large deformation is hard to identify with the typical wrinkling of the facings, characteristic for panels with a continuous core, because in the case of a non-continuous core the local deformations of the facing are forced by vertical movement of the lamellae to each other. Two typical sandwich panel damages are shown in Figs. 4 and 5.

In the case of the Tep17_5 model (Fig. 4), the interface has been damaged near the support. When the interface has a lower strength (Fig. 5), the damage appears closer to the center of the plate. In every case, damage is initiated at the point of a contact between the two lamellae.

Table 1. Parameters and results of the numerical models

Model	Parameters of the contact between lamellae			Parameters of the interface		Extreme stress in the lower facing
	Normal behavior		Tangential behavior			
	p_0 [kPa]	c_0 [m]	Friction coefficient μ	t_n^0 [kPa]	$t_s^0 = t_t^0$ [kPa]	σ[MPa]
Tep17_1	0	0	0.10	140	100	177.3
Tep17_2	10	0.01	0.10	140	100	178.5
Tep17_3	10	0.01	0.50	140	100	184.5
Tep17_5	1	0.001	0.50	140	100	177.5
Tep17_6	1	0.001	0.50	70	50	105.2

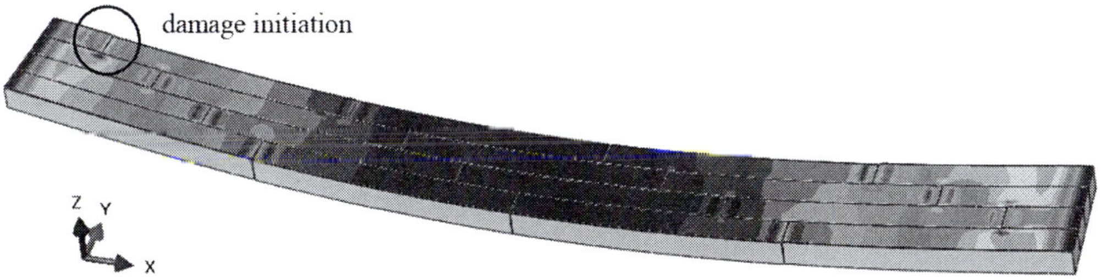

Fig. 4. Deformations that cause damage of the sandwich panel – model Tep17_5

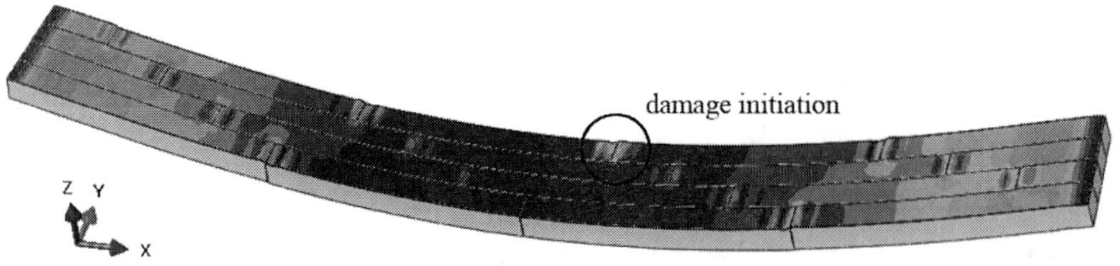

Fig. 5. Deformations that cause damage of the sandwich panel – model Tep17_6

The results shown in Tab. 1 show that the normal contact parameters have little effect on the final solution. Much more important is the coefficient of friction between the lamellae, although even a fivefold change in this factor has increased the load that causes damage by less than 4%. The strength of the interface layer is decisive. This may seem obvious, although it should be noted that in the case of a continuous and homogeneous core, the interface parameters are not as significant (assuming its reasonable strength). This was shown, among others in [16]. In the case of the core made of lamellae, however, the situation is somewhat different. The mutual movement of the lamellae causes strong deformation of the facings (and the interface), which leads to the failure of the interface and progressive damage. At low interface strength, there is damage near the support (this can be attributed to the shear effect). With higher interface strength, the damage occurs in the zone closer to the center due to the interaction of shear stresses and stresses caused by local instability of the facing. Damage always occurs at the surface of contact of lamellae.

To illustrate the typical phenomena occurring at the lamellae contact, Fig. 6 shows COPEN (contact opening at surface node) and CPRESS (contact pressure at surface node). The drawings clearly show that the opening between the lamellae occurs in the mid-height of the core.

a)

b)

Fig. 6. Contact between lamellae: a) contact opening at surface node [m], b) contact pressure at surface node [kPa]

It is worth explaining one additional aspect. Models with defined contact are characterized by the dependence of results on mesh size. In the presented models, a mesh size of 0.025 m (5 spatial elements per core thickness) was used. Analyses based on a mesh size 0.06 m led to a very rapid loss of solution convergence. Several attempts on a denser mesh showed no significant difference in relation to the 0.025 m mesh size.

After identification of the failure mechanism, additional analyses were conducted to determine the impact of initial imperfections. For the Tep17_5 model, the first 10 buckling modes were determined. As the initial imperfection, the sum of the first and the second buckling mode multiplied by 0.00001 (Tep17_5_imp), 0.01 (Tep17_5a_imp) and 0.1 (Tep17_5c_imp) was assumed, respectively. As a result, in the Tep17_5c model extreme geometric imperfections reached locally 1.6 mm. Parameters of these models and the solution results are shown in Tab. 2. The presented results indicate that small imperfections are not significant. This is because the system of lamellae is a sufficient disturbance of the model needed to locate the failure.

Table 2. Parameters of the numerical models with initially defined imperfections

Model	**Parameters of the contact between lamellae**			**Parameters of the interface**		**Extreme stress in the lower facing**
	Normal behavior		**Tangential behavior**			
	p_0 [kPa]	c_0 [m]	Friction coefficient μ	t_n^0 [kPa]	$t_s^0 = t_t^0$ [kPa]	σ[MPa]
Tep17_5	1	0.001	0.50	140	100	177.5
Tep17_5_imp	1	0.001	0.50	140	140	177.5
Tep17_5a_imp	1	0.001	0.50	140	140	177.5
Tep17_5c_imp	1	0.001	0.50	140	140	146.3

Finally, the effect of micro-profiling of the facings and discontinuity of the core was also analyzed. Model Tep17_6 was used as a basis. In Tep17_4 model, the flat facings were replaced by micro-profiled elements (depth of profiling 0.5 mm). In Tep17_7 model, the discontinuous core (lamellae) was replaced by a continuous core having the same parameters as the lamella. The results are presented in Tab. 3.

Table 3. Parameters and results of the numerical models

Model	**Parameters of the contact between lamellae**			**Parameters of the interface**		**Extreme stress in the lower facing**
	Normal behavior		**Tangential behavior**			
	p_0 [kPa]	c_0 [m]	Friction coefficient μ	t_n^0 [kPa]	$t_s^0 = t_t^0$ [kPa]	σ[MPa]
Tep17_6	1	0.001	0.50	70	50	105.2
Tep17_4	1	0.001	0.50	70	50	106.3
Tep17_7	1	0.001	0.50	70	50	121.1

From Tab. 3, it follows that micro-profiling (Tep17_4) did not have much impact, although in my opinion this conclusion should be combined with the attention that this refers to the case of low strength of the interface. In the case of higher interface strength, the greater importance of micro-profiling of the facings (particularly the compressed facing) should be expected. Replacing the discontinuous core by the continuous element (Tep17_7) resulted in an increase in the sandwich panel capacity by about 15%. The research conducted by manufacturing companies shows that this difference may be greater. The influence on this may have several parameters related to the technology of the production of sandwich panels.

Conclusions

The analyses presented in the paper indicate a specific mechanism of failure of sandwich panels with a discontinuous core. Due to the lack of core continuity, as a result of load increment, there is a mutual displacement of the lamellae. The induced deformations result in damage of the connection between the core and the facing. In the numerical model, this corresponds to the failure of the interface layer. Depending on the interface parameters, damage occurs either in the support zone or in the center zone of the panel, whereby damage is always initiated near the contact surface of the lamellae.

Unlike in the case of panels with a continuous core, the interface parameters significantly affect the capacity of the structure. Of course, the stiffness of the core also has an influence on the load capacity of the panel, but the analysis of this problem has been omitted in this paper. In addition, unlike the structure with a continuous core, it turned out that micro-profiling of the facings and minor imperfections are much less important. These are interesting observations, although their generalization requires further numerical and experimental studies.

Discontinuity of the core reduces the capacity of the sandwich panel. Unfortunately, due to technological reasons, there is no way to avoid these discontinuities in the case of mineral wool core. For this reason, there are further interesting research areas, such as the optimization of an arrangement. Based on the review of the literature, it should be stated unequivocally that the problem of analysis of sandwich panels with discontinuous core has unfortunately not been thoroughly investigated.

References

1. A. K. Noor, W. S. Burton, C. W. Bert, *Computational models for sandwich panels and shells*. Applied Mechanics Reviews **49** (3), 155-199 (1996)
2. J. Hohe, L. Librescu, *Advances in the structural modelling of elastic sandwich panels*. Mechanics of Advances Materials & Structures **11**, 395-424 (2004)
3. I. Kreja, *A literature review on computational models for laminated composite and sandwich panels*. Central European Journal of Engineering **1** (1), 59-80 (2011)
4. E. E. Gdoutos, I. M. Daniel, K. A. Wang, *Compression face wrinkling of composite sandwich structures*. Mechanics of Materials **35** (3-6), 511-522 (2003)
5. V. Koissin, A. Shipsha, V. Skvortsov, *Effect of physical nonlinearity on local buckling in sandwich beams*, Journal of Sandwich Structures and Materials **12**, 477-494 (2010)
6. C. A. Steeves, N. A. Fleck, *Collapse mechanisms of sandwich beams with composite faces and a foam core, loaded in three-point bending, Part II: experimental investigation and numerical modelling*. International Journal of Mechanical Sciences, **46**, 585-608 (2004)

7. M. Chuda-Kowalska, M. Malendowski, *Sensitivity analysis of behavior of sandwich plate with PU foam core with respect to boundary conditions and material model*. Advances in Mechanics: Theoretical, Computational and Interdisciplinary Issues, CRC Press, 125-128 (2016)
8. J. B. Dafedar, Y. M. Desai, A. A. Mufti, *Stability of sandwich plates by mixed, higher-order analytical formulation*. International Journal of Solids & Structures **40**, 4501-4517 (2003)
9. L. R. Xu, A. J. Rosakis, *Impact failure characteristics in sandwich structures, Part I: Basic failure mode selection*. International Journal of Solids and Structures **39**, 4215-4235 (2002)
10. E. Lolive, J-M. Berthelot, *Non-linear behaviour of foam cores and sandwich materials, Part 2: indentation and three-point bending*. Journal of Sandwich Structures and Materials **4** (4), 297-352 (2002)
11. J. Pozorska, Z. Pozorski, *Analysis of the failure mechanism of the sandwich panel at the support*. Procedia Engineering **177**, 168-174 (2017)
12. L. Heselius, *Wrinkling analysis of flat faced sandwich panels with respect to bonding strength and core properties with mineral wool as core material*. Technischen Universität Darmstadt (2005)
13. R. Studziński, Z. Pozorski, *Experimental and numerical analysis of sandwich panels with hybrid core*. Journal of Sandwich Structures and Materials, DOI: 10.1177/1099636216646789 (2016)
14. A. W. Giunta d'Albani et al., *Mass loss and flammability of insulation materials used in sandwich panels during the pre-flashover phase of fire*. Fire and Materials, DOI: 10.1002/fam.2418 (2017)
15. J. Pozorska, Z. Pozorski, *The influence of the core orthotropy on the wrinkling of sandwich panels*. Journal of Applied Mathematics and Computational Mechanics, **14** (4), 133-138 (2015)
16. Z. Pozorski, *Sandwich panels in civil engineering – theory, testing and design* (Wydawnictwo Politechniki Poznańskiej, Poznań, 2016)

Numerical modelling of the rigid polyurethane foam microstructure

Zbigniew Pozorski[1,*]

[1]Institute of Structural Engineering, Poznan University of Technology, Piotrowo 5, 60-965 Poznań, Poland

Abstract. The paper presents a numerical model of 2-D microstructure of rigid polyurethane foam. The selection of geometric and material parameters is presented. For a particular structure, its behavior has been studied for typical cases of external loads (or forced displacements). The characteristic phenomena have been identified and described. A parametric analysis was performed due to the dimension of the cross-section of the struts which form the cell edges. An analysis of the impact of support and loading conditions on the behavior of the cell structure was performed.

Keywords: polyurethane foam, microstructure, honeycomb structure, finite element method

1 Introduction

Rigid cell structures of low density are very common in nature and have been produced by humans for many years. The relative density (the density of the cellular material divided by that of solid from which the cell walls are made) of typical rigid foam structures is in the range 0.025-0.05 and the Young modulus of the foam is about 1-5 MPa. Many believe that the importance of this class of structures is much greater than, for example, fiber composites. Despite this, the mechanics of rigid foam structures is often neglected.

The problem of deformation of cellular structures has been repeatedly taken up in the literature. The starting point is usually the study of two-dimensional cellular solids [1]. Describing phenomena taking place in these structures, we can distinguish in principle elastic behavior, elastic buckling and formation of plastic hinges. One of the fundamental determinants of cell structure and its behavior is the relative density of the foam [2]. Searching for relationships between the relative density factor and material properties was widely discussed in the literature [3, 4], while the analysis of spatial structures is far more complex than 2-dimensional systems. The principal mechanism of linear-elastic deformation (cell wall bending) was correctly identified by Menges & Knipschild [5]. Much later it turned out that the density of the foam is not sufficient for a reliable characterization of the foam structure. The effect the manufacturing process on the foam microstructure and the properties of the matrix material is important but has received limited attention in the literature [6].

* Corresponding author: zbigniew.pozorski@put.poznan.pl

Reviewers: *Krzysztof* Talaśka, *Zuzana* Murčinková

This paper presents a relatively simple model of a rigid cellular foam structure. The elementary problem of the influence of the structure stiffness and density on the macroscopic behavior is discussed. For typical schemes of external loads, the phenomena leading to the mechanism of cells damage are observed. Analyzing these issues (including numerical issues) is essential for further, in-depth studies on phenomena occurring in foam cores of sandwich panels.

2 Problem formulation

A two-dimensional, honeycomb structure consisting of regular hexes is considered (Fig. 1). Such structure has been repeatedly described [1, 2, 7] and recommended for preliminary analysis of the behavior of foam cores. Interestingly, it was mainly used to find the relationship between foam density and foam mechanical parameters. This structure corresponds roughly to the actual foam structure (Fig. 2), although of course the actual cells are of varying size and shape. Another issue is the presence of shells in closed cell structures (Fig. 2). Despite these shells, many authors emphasize that their effect on the strength of the cell structure is small because the shells rapidly lose stability. In addition, the edges of the cells are much stiffer than the shells are.

The structure consisting of $60\times60 = 3600$ open-cell hexagons was analyzed in the paper. This is a typical honeycomb structure. The length of the edge of the hex is 0.1 mm, so the whole structure is 10.48×9.05 mm. It is worth noting that this system is not symmetrical due to the small displacement of neighboring cells and the even number of cells. This was introduced consciously to avoid the symmetry of the numerical problem.

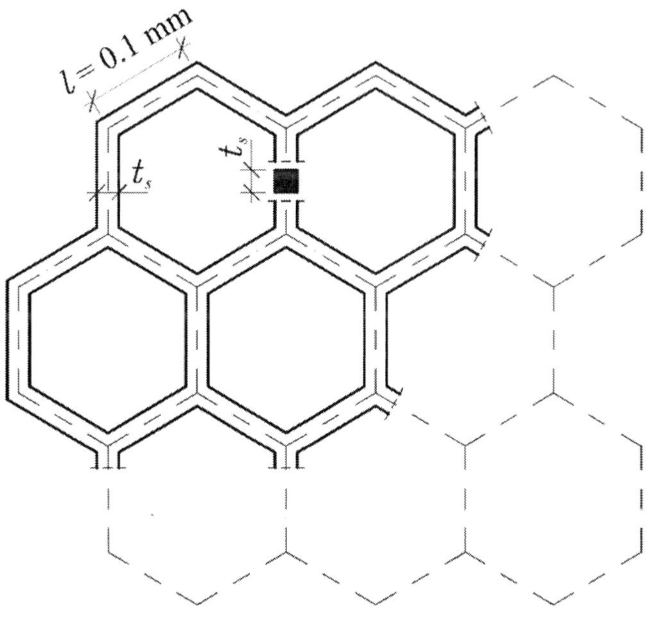

Fig. 1. A periodic foam microstructure made of hexagon elements

The aim of the study is to analyze the behavior of the system under the influence of unidirectional compression, tension and shear. For this purpose, 3 groups of models were created in the Abaqus system environment. The appropriate boundary conditions were assumed. In addition, the impact of structural rigidity on the cell deformations and stress values was investigated.

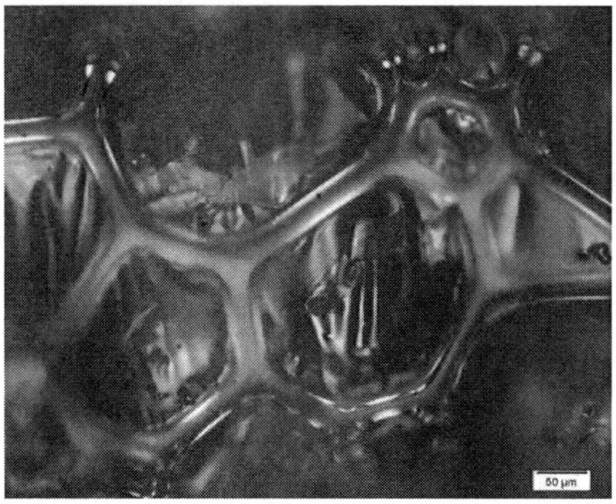

Fig. 2. Actual structure of rigid polyurethane foam

3 Numerical models

The structure shown in Fig. 1 (60×60 cells) has been modeled using two-dimensional B23 beam elements (two-node element, cubic shape functions). Each hexagon edge consists of 3 finite elements. All connections of elements are rigid, but of course the nodes can move and rotate. Based on the available literature, it is assumed that the structure is made of the elastic material characterized by E_s = 1200 MPa and ν_s = 0.3. Of course, Poisson's ratio of the material does not affect the deformations observed macroscopically. Although only the elastic material was considered in the first models, a simple definition of plasticity was finally adopted. It was assumed that at the stress of 127 MPa the plastic strain is 0.0, and at the stress 140 MPa plastic strain is 0.1.

One of the challenges was determining the dimensions of the cross-section of the beam elements from which the structure is composed. From the macroscopic observations (stress-strain relation), the Plateau effect is initiated at a stress level of 90-100 kPa, which corresponds to a strain of 2-2.5%. Based on this information, and additionally taking into account that some deformations result from the shortening of the elements due to compression, a square cross-section of 0.015×0.015 mm was assumed. These dimensions are of a similar order as the values observed in reality (cf. Fig. 2). Finally, the structure consists of beams with a square cross-section of dimension t_s = 0.015 mm, which form regular hexes connected to a system of 60×60 cells. Analyses were also carried out for different cross-sections, namely for square sections of dimension 0.0125 mm and 0.010 mm.

Due to boundary conditions, three basic cases were distinguished (compression, tension, shear). In certain areas, the appropriate displacement was forced, which quite well reflect the conditions of real experiments. On the bottom surface, displacements in both directions have been blocked (x, y). The upper surface had the following boundary conditions:

a) compression: blocked horizontal displacement, forced vertical displacement downward,
b) tension: blocked horizontal displacement, forced vertical displacement upward,
c) shear: blocked vertical displacement, forced horizontal displacement to the right.

In models 6, 7 and 8 (with a note: c – compression, t – tension, s – shear), these conditions were applicable to all nodes located on the respective surfaces of the model. Models 6, 7 and 8 correspond to the structures made of the beam elements with the cross-section dimension 0.015 mm, 0.0125 mm and 0.010 mm, respectively.

In order to analyze the other effects, 3 other models (9c_a, 9c_b, 9c_c) were also considered, wherein the starting point was model 6c. In the 9c_a model, instead of limiting the horizontal displacement (x-direction) of all the nodes on the two external surfaces, such a restriction was introduced only at one central point on the surfaces. As a result, the other nodes were free to move horizontally. In the 9c_b model, the horizontal displacement restriction was applied to all nodes located on two faces (as in the model 6c), but the vertical constraint was only applied to 6 points in the center of the top surface. Model 9c_c was different from model 6c in such a way that compression was forced in another direction (in x instead of y-direction). The displacements of the nodes on the left surface were blocked, and horizontal displacements (leftward) of all the extreme nodes on the right surface were forced.

4 Results

For each considered model, a numerical calculation was carried out until a loss of convergence of the solution. The loss of convergence occurred with a specified displacement. In the case of compression and tension, these were displacements in the y-direction (u_2), whereas in the case of shear the displacements were in the x-direction (u_1). The characteristic results, corresponding to the moment when the solution convergence was lost, are shown in Tab. 1.

Analyzing the results presented in Tab. 1, it should first be noted that the numerical problems lost the convergence at similar values of the compressive normal stress (of the order of 140 MPa). The stress level is equivalent to the yield stress of the material, but a very similar result was obtained for a material defined as ideal elastic. This means that the loss of convergence is rather related to the magnitude of the deformation of the elements. This is illustrated in Fig. 3, where the deformations of the entire specimens during compression (model 6c), shear (6s) and tensile (6t) are presented. The symbol S11 denotes normal stress in the beam element. It is worth noting that the extreme deformations occur in farthest cells located at the bottom and top of the specimens.

Table 1. Characteristic results of numerical simulations

Model		Extreme displacement [mm]	Extreme normal stresses [MPa]	Absolute value of RF2 [N]	
Name	t_s [mm]			Extreme value	Average value
6c	0.0150	-0.6871 (u_2)	136.8 / -140.1	0.003102	0.0015
6s		3.214 (u_1)	148.5 / -220.7	0.04546	-
6t		2.578 (u_2)	149.6 / -274.1	0.07486	0.030
7c	0.0125	-0.7921 (u_2)	135.7 / -140.4	0.00170	0.00090
7s		2.440 (u_1)	136.9 / -160.0	0.01535	-
7t		2.340 (u_2)	144.2 / -274.1	0.0496	0.018
8c	0.0100	-0.7790 (u_2)	136.2 / -136.9	0.00080	0.00036
8s		3.931 (u_1)	140.0 / -254.6	0.025	-
8t		2.154 (u_2)	140.0 / -244.7	0.03015	0.010

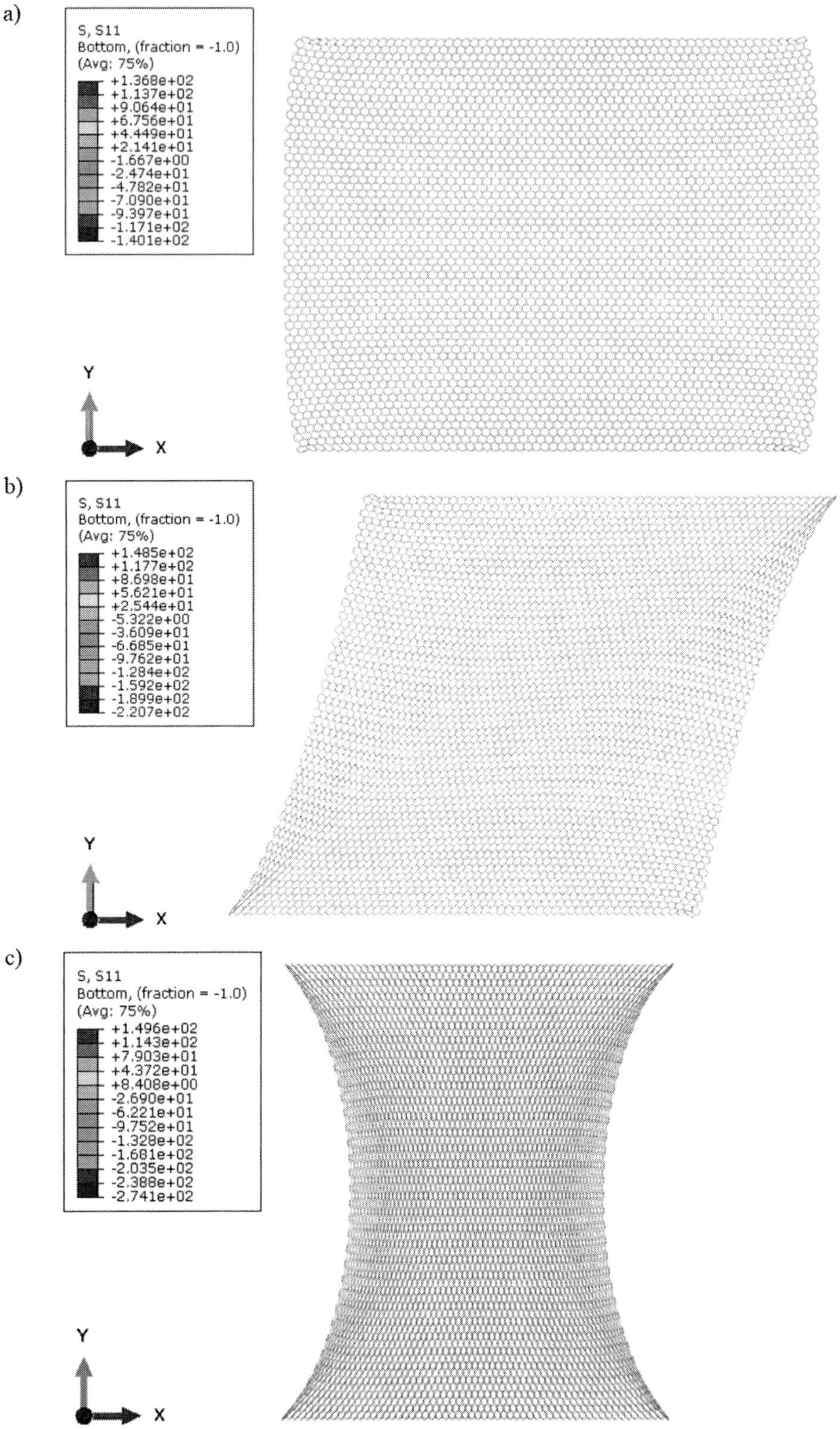

Fig. 3. Deformations of the cellular structure: a) compression (model 6c), b) shear (model 6s), c) tension (model 6t); the presented values (S11) denote normal stresses in cell elements

The extreme values of support reactions RF2 (Tab. 1) are very different and obviously decrease considerably as the beam section size t_s decreases. The mean values of the RF2 (reaction in y-direction) are much more interesting. These values are practically equivalent to the mean value of normal force in vertical pillars of the cellular structure. In the case of model 6c, the normal force in the pillar is 0.0015 N, which at the cross-section area of 2.25×10^{-4} mm^2 means the normal stress of 6.67 MPa. This means that the bending of the structural elements is decisive for the level of normal stress and deformation of the system. It is easy to see this by comparing the strain in the pillar caused by the axial compression:

$$\varepsilon_s = \frac{\sigma_s}{E_s} = \frac{6.67}{1200} = 0.0055\,[-] = 0.55\% \tag{1}$$

with the deformation of the structure observed macroscopically:

$$\varepsilon = \frac{\Delta L}{L} = \frac{0.6871}{9.05} = 0.0759\,[-] = 7.59\%\,. \tag{2}$$

Assuming a certain thickness of the structure, e.g. $b_s = 0.1$ mm, we can also calculate the macroscopically observed stress level and (effective) modulus of elasticity:

$$\sigma = \frac{N}{A} = \frac{0.0015}{0.1732\times0.1} = 0.0866\text{ MPa} = 86.6\text{ kPa}\,, \tag{3}$$

$$E = \frac{\sigma}{\varepsilon} = \frac{0.0866}{0.0759} = 1.141\text{ MPa}\,. \tag{4}$$

If the density of the material from which the structure is made is assumed in accordance with literature [2, 3] as $\rho_s = 1200$ kg/m^3, the above values are obtained for a foam structure with a density of 30.96 kg/m^3. When we reduce the cross-sectional dimension of the beam component of the structure t_s (models 7c i 8c), the density, structural rigidity and effective E-modulus decrease.

These results provoke some reflections. First of all, the rigidity of the structural elements is too low. Certainly the E_s modulus can be taken at the limit of maximum values known from the literature (1600 MPa). Moreover, the cross-sectional dimension of the beam element can be slightly increased to obtain a specimen density of 40 kg/m^3, which is consistent with typical values for structural applications. It is much more difficult to achieve the correct (geometrically and physically) nonlinear behavior of the structure and to obtain well-known mechanisms of destruction.

Analyzing deformations of a structure (including cellular) it is difficult not to refer to the issue of Poisson's ratio. In the case of cellular structure, a question immediately arises on how to determine and interpret the ratio. This is not the value set for one cell. Rather it is a value averaged over a certain area. Looking at the deformations of the entire structure in model 6c, it can be seen, however, that it is difficult to define the area for which we should determine Poisson's ratio. Boundary conditions (limitation of displacements u_1) significantly disturb deformations of the structure. If you rely on the ratio of the maximum expansion of the specimen to its shortening, then we will get the value $\nu = 0.985$. To test how boundary conditions affect deformations of the system, in model 9c_a the compression was induced, but the horizontal displacements were limited only in one point of the lower and upper external surface (Fig. 4). The system deforms more regularly, however, measurements of the displacements of the entire specimen give $\nu = 0.834$.

The regular hexagonal cell systems have different properties if we apply load (or forced displacement) in another direction. The anisotropy of the structure highly influences the structural behavior of mechanical systems [8]. In the 9c_c model, the boundary conditions have been changed. These are analogous to those in model 6c, except that the limitation of displacements u_1 and u_2 refers to the left surface (instead of the lower one), while on the right-hand side, displacements u_2 are blocked and the compression is induced horizontally. The deformed system was slightly less oval than in the case of the model 6c, but the Poisson ratio determined using the macroscopic extreme displacements of the specimen was equal to $\nu = 0.802$. It is worth noting that all the obtained Poisson's ratios are higher than 0.5, which is greater than the theoretical limit for natural materials. Moreover, it is much higher than the values given in the literature for cellular structures.

a)
b)

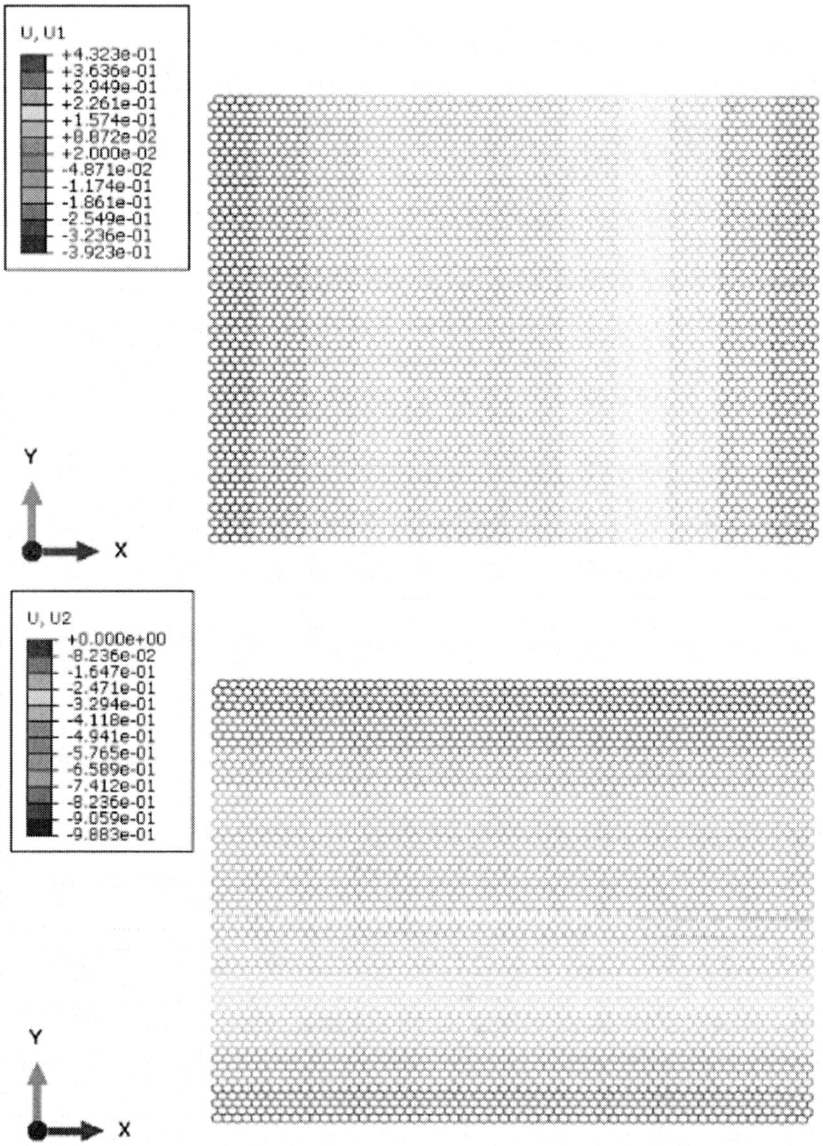

Fig. 4. Displacements of the nodes of the cell structure subjected to vertical compression (model 9c_a): a) horizontal displacements u_1 [mm], b) vertical displacements u_2 [mm]

An interesting issue is also the impact of concentrated load on deformations of cell structures. Model 9c_b was developed based on model 6c, except that the vertical displacement (compression) was imposed not on the entire upper surface, but only on the six nodes located near the center of the upper surface. Despite the considerable rigidity of the system, surprisingly rapid vertical displacements were achieved at the place of extortion. These displacements occurred despite the fact that the deformations of individual cells were not too large (Fig. 5). In addition, the concentrated extortion was very quickly redistributed to a large area of the structure.

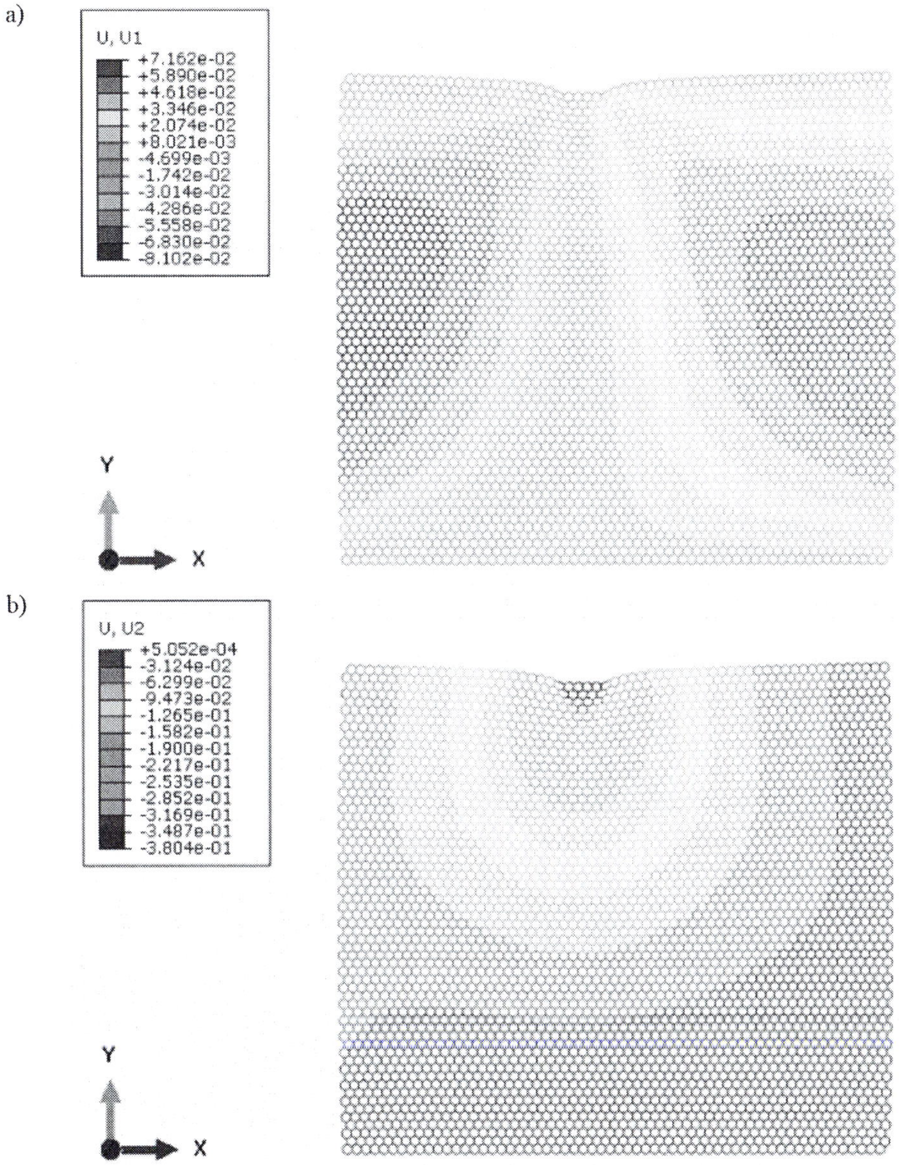

Fig. 5. Displacements of the nodes of the cell structure subjected to concentrated compression (model 9c_b): a) horizontal displacements u_1 [mm], b) vertical displacements u_2 [mm]

Conclusions

The paper presents a simple 2-dimensional cellular structure. This structure was modeled using beam elements. Linear elasticity and plasticity have been assumed for the material from which the structure is constructed. The issues of the influence of stiffness of structural elements and boundary conditions (support and load) on structural behavior were discussed. The problem of Poisson's ratio determination for the structure was discussed.

The results of the conducted analyses revealed surprisingly large local deformations of cells. These cells are degraded, leading to a loss of convergence of the numerical solution. Based on the deformations achieved for different structures, it can also be stated that the stiffness of the elements from which the structure is made is too low. For rigid foams used in civil engineering (e.g. used as cores in sandwich panels), a higher stiffness of beam elements should be assumed. On the other hand, the structural elements should lose stability much earlier, which is in some contradiction to the earlier conclusion. The probable solution to this problem is to use a model with a different geometry, different edge geometry, or a spatial model. This latter solution is appropriate after a slightly deeper understanding of the problems encountered in 2-D structures.

The intriguing issue is the determination of the Poisson ratio for the foam structures. The results obtained for the honeycomb structure are very high and do not confirm the observations made for rigid polyurethane foams. The redistribution of local loads also deserves attention. This issue is important from an engineering point of view [9]. The results presented in the paper indicate rapid redistribution of forces and surprisingly small deformations of individual cells.

References

1. M. F. Ashby, *The mechanical properties of cellular solids*. Metallurgical Transactions **14A** (9), 1755-1769 (1983)
2. L. J. Gibson, M. F. Ashby, *Cellular solids: structures and properties*. (Cambridge University Press, 1997)
3. L. J. Gibson, M. F. Ashby, *The mechanics of three-dimensional cellular materials*. Proceedings of the Royal Society of London. Series A, Mathematical and Physical Sciences **382** (1782), 43-59 (1982)
4. D. Weaire, M. A. Fortes, *Stress and strain in liquid and solid foams*. Advances in Physics **43** (6), 685-738 (1994)
5. G. Menges, F. Knipschild, *Estimation of mechanical properties for rigid polyurethane foams*. Polymer Engineering & Science, **15** (8), 623-627 (1975)
6. P. Yue, J. J. Feng, C. A. Bertelo, H. H. Hu, *An arbitrary Lagrangian–Eulerian method for simulating bubble growth in polymer foaming*. Journal of Computational Physics **226**, 2229-2249 (2007)
7. W. Szymczyk, D. Miedzińska, *Numerical analysis of an open cell foam structure with the use of models based on 2D finite elements*. Journal of KONES Powertrain and Transport **16** (2), 441-448 (2009)
8. J. Pozorska, Z. Pozorski, *The influence of the core orthotropy on the wrinkling of sandwich panels*. Journal of Applied Mathematics and Computational Mechanics **14** (4), 133-138 (2015)
9. Z. Pozorski, J. Pozorska, *Stress redistribution at the support of transversely loaded sandwich panel*. Advances in Mechanics: Theoretical, Computational and Interdisciplinary Issues. (CRC Press, 485-488, 2016)

The fractography analysis of IN718 alloy after three-point flexure fatigue test

Juraj Belan[1,*], *Alan* Vaško[1], *Lenka* Kuchariková[1], *Eva* Tillová[1]

[1]University of Žilina, Faculty of Mechanical Engineering, Department of Materials Science, Univerzitná 8215/1, 01026 Žilina, Slovakia

Abstract. In this study, the high cycle fatigue (HCF) properties of IN718 superalloy with given chemical composition were investigated at three-point flexure fatigue test at room temperature. INCONEL alloy 718 is nickel-chromium-iron hardenable alloy and due to its unique combination of mechanical properties (high-strength; corrosion-resistant and so on) used for production of heat resistant parts of aero jet engine mostly. Mechanical properties of this alloy are strongly dependent on microstructure and on presence of structural features such are principal strengthening phase gamma double prime, gamma prime and due to its morphology less desired delta phases. The mentioned phases precipitate at various temperature ranges and Nb content as well. The three-point flexure fatigue test was performed on ZWICK/ROELL Amsler 150 HFP 5100 test equipment with approximate loading frequency f=150 Hz. The S – N (Stress – Number of cycles) curve was obtained after testing. With the help of scanning electron microscope (SEM), fractography analyses were performed to disclose the fracture features of specimens in different life ranges. The brief comparison of three-point flexure and push-pull fatigue loading modes and its influence on fatigue life is discussed as well.

Keywords: IN 718 alloy, low frequency fatigue loading, three-point flexure, fatigue crack initiation, fatigue crack propagation, S–N curve, SEM observing, fractography analysis

1 Introduction

Ni-base superalloys such as IN718 derive their strength from the solid solutions of alloying elements, and to even greater extent from precipitates within the solid solution matrix. Ni-base superalloys are widely used for structural applications at high temperature, especially where creep and fatigue are the important deformation modes. Among the Ni-base superalloys, Inconel 718 (IN718) is predominantly used in high-temperature applications due to its highly satisfactory price/overall performance ratio, good corrosion resistance and mechanical properties with excellent weldability. The major applications of this alloy are turbine discs in gas turbine engines. However, its mechanical properties are degraded at temperatures above 650°C [1].

* Corresponding author: juraj.belan@fstroj.uniza.sk

Reviewers: *Richard* Pastirčák, *Marta* Kianicová

The microstructures of Ni-base superalloys are composed of a face-centred cubic matrix γ (Ni stacked on Ni plus Fe), with large additions of alloying elements that give solution hardening, corrosion resistance and a low stacking fault energy. The strengthening γ′ (~1/3 μm) and γ″ (~1/20 μm) precipitates are randomly interspersed throughout the γ matrix and at grain boundaries; it is these that make the lattice planes highly resistant to slip band formation at high temperature [2].

While microstructurally simple, superalloys are chemically complex. In addition to Al, they have Ti (substitutes for Al in the γ′, Fig. 1a, increases the fraction of γ′, and decreases the fault energy in the γ′), Cr (primarily in γ, confers oxidation resistance), Mo (in both γ and γ′, may increase or decrease the relative mismatch depending on other elements). Some residual carbon may be present in alloys used for discs (relatively low temperature) as a processing impurity while in other alloys it may be intentionally added to form carbides which reduce creep by inhibiting grain boundary sliding. Other elements may be added to provide solid solution strengthening, (e.g. Ta) to control the volume fraction of γ′, to inhibit grain boundary sliding by migrating to boundaries (e.g. odd size atomic species such as Zr), to improve castability (Hf) and to control the shapes of carbides (Hf). The volume fraction of γ′ in these alloys ranges from 25% to over 50%, with more modern alloys having a higher volume fraction. The effect of various elements on superalloys has been widely studied and much detailed information may be found in various references [3–5].

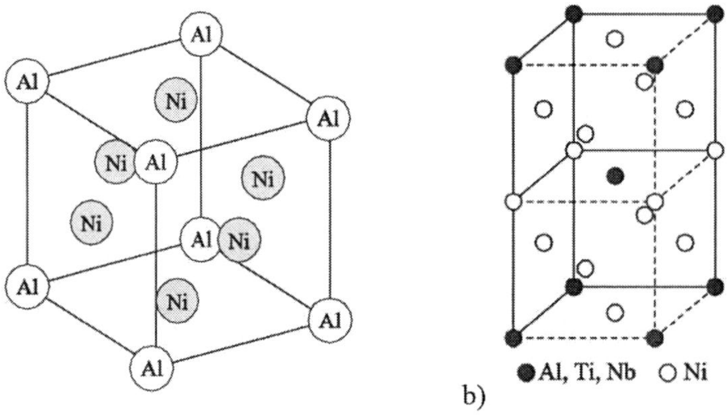

a) b)

Fig. 1. (a) Crystal structure of Ni_3Al γ′ . In the idealized structure there are three complete atoms at the centres of the cube faces and one complete atom at the cube corners. The FCC-like crystallographic structure is designated as $L1_2$; (b) Crystal structure of γ″. This crystal structure is denoted DO_{22} and may be viewed as stacking two FCC-like ordered unit cells to produce the complete unit cell [7].

Over the years, as the function of various elements has become better understood, compositions have been simplified. Superalloy processing depends on the intended application. The major forms in which they are used are:

a) single crystals (for jet engine turbine blades, with [001] axis aligned along the blade axis);
b) directionally solidified (DS) which are polycrystalline but with all crystals aligned longitudinally in the [001] direction (for land-based turbines);
c) billets formed from forged or rolled ingots (disks);
d) powder metallurgy products (HIP or HIP + Forged, used in jet engine disks).

Similar considerations apply to the Ni–Fe base superalloys such as In 718 which, in addition to Fe, also contains Nb to form coherent precipitates with a disk-like structure [6] in an FCC γ matrix. These precipitates have a DO_{22} crystal structure, and are based on the

Ni_3Nb ordered compound shown in Fig. 1b, and are denoted by γ'' . As with the γ' Ni-base alloys, alloying elements such as Al and Ti may be added and also occupy the Nb sites. The γ'' precipitates have a disc-like shape with the small dimension aligned along the three [001] directions. In addition, there are some γ' precipitates present on which the particles γ'' have been shown to precipitate. These alloys are primarily used for disks in jet engines as well as components in the power generation industry. Components from these alloys are generally produced by traditional ingot metallurgy followed by mechanical forming to obtain a uniform fine grain size (ASTM 6-7).

The aim of this study is to provide information about three point flexure fatigue test at room temperature made at low frequency with sinusoidal loading applied on simple blocky shape specimens. The frequency of fatigue test varied from 143 up to 151 Hz. Employed static force of fatigue test was set on value F = -15kN. The dynamic loading as itself changes from 6.31 kN up to 12.8 kN. The number of cycles 2.10^7 (according to standard for this kind of non-ferrous alloys is number of cycles up to 10^7 considered as fatigue lifetime limit) was considered as fatigue life limit. Next S-N curve were plotted and also scanning electron microscope (SEM) fracture surface analysis was done with focus to describe micro-mechanism of initiation and fatigue crack propagation (FCP) in Inconel 718 alloy. Received data was compared with results of other authors who made fatigue tests at the same experimental material but at different fatigue loading.

2 Experimental material and methods

The wrought Ni-Fe alloy 718 was used as an experimental material. The chemical composition of our material was (in wt. %): C – 0.026; Si – 0.09; Mn – 0.07; P – 0.008; S <0.001; Al – 0.57; B – 0.004; Bi (ppm) <0.1; Co – 0.14; Cr – 19.31; Cu – 0.03; Mo – 2,99; Nb – 5.30; Ni – 53.32; Pb (ppm) – 0.1; Se (ppm) < 3; Ta <0.01; Ti – 0.96; Ni + Co – 53.46; Nb + Ta – 5,31; and Fe – balance.

The experimental material was heat treated, according to the supplier's BIBUS Ltd. (CZ) material sheet, 980 °C / 1 hrs. AC + heating at 720 °C / 8 hrs. followed FC (50 °C per hour) to temperature 620 °C holding time 8 hrs. and air cooled. Achieved mechanical properties of material with grain size ASTM 12 are in Table 1.

Table 1. Mechanical properties of IN 718 as received from BIBUS Ltd. after applied heat-treatment and for grain size ASTM 12.

	Temperature [°C]	
	20	649
Yeld strength, Rp 0.2 [MPa]	1 213	986
Maximum strength, Rm [MPa]	1 549	1 123
Elongation, A [%]	21.3	22.6
Reduction area, Z [%]	33.3	68.0
Hardness, HB 10/3000	429	-
Creep rupture, $\sigma_{T/649}$ [MPa]	-	689
Rupture life [hrs.]	-	26.8
Creep elongation, A_{creep} [%]	-	45.7

The experimental material of mentioned chemical composition was cut into simple blocky samples with dimensions (b x h x l) 11mm x 10mm x 56mm where side of sample under fatigue loading was polished to avoid unwanted fatigue crack initiation from surface roughness. The three point flexure fatigue tests, Fig. 2, has been carried out on testing machine ZWICK/ROELL Amsler 150 HFP 5100 at room temperature with static pre-load F_{static} = -15 kN and dynamic force $F_{dynamic}$ varying from 6.31 kN up to 12.8 kN. The value of 2.10^7 numbers of cycles was set as reference and when specimen has reached this value without break, so called runout, that bending stress was considered as fatigue lifetime limit. The frequency of fatigue tests was approximately f = 150 Hz. The S-N curve was plotted from measured values, which gives relation between maximum bending stress σ_{omax} and number of cycles N_f.

For calculation of maximum bending stress formula (1) was used:

$$\sigma_{omax} = \frac{3 \times F \times L}{2 \times b \times h^2} \text{ [MPa]}, \quad (1)$$

where σ_{omax} is maximum bending stress [MPa], F – is dynamic load [N], L – distance of supports [mm], b – specimens width [mm], and h – is specimen high [mm].

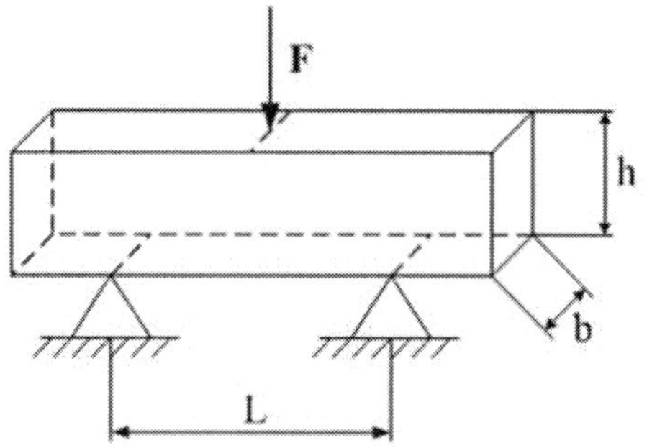

Fig. 2. A schematic drawing of setting three point flexure fatigue test

The fractography analysis of broken specimens was also done. For fractography analysis scanning electron microscope TESCAN Vega II LMU was used. All fractography analysis was done in order to describe micro-mechanisms of fatigue crack initiation, fatigue crack propagation (FCP) and final static fracture of specimens.

3 Results and discussion

A typical microstructure of wrought alloy 718 is on Fig.3. The microstructure consist of light grey blocks of carbides and fine lenticular and lamellar particles of delta phase (Ni_3Nb) distributed in the FCC matrix, with grain size approximately 10 μm and a few deformation twins. In some cases, especially after long temperature dwell coarse γ' phase and more delta phases are also visible.

For low frequency fatigue testing were used 10 specimens with simple blocky shape as seen at Figure 2. Figure 4 shows the S-N curve of IN718 obtained from the three point flexure fatigue tests at room temperature with frequency 150 Hz under the load ratio of R = 0.11. Obtained results were approximated with equation (2) what is a Basquin formula for S-N presentation and approximation.

$$\sigma_a = 2591 \times N_f^{-0.0811} \quad (2)$$

where σ_f' = 2591 is a coefficient of fatigue strength and -0.0811 = b is lifetime curve exponent.

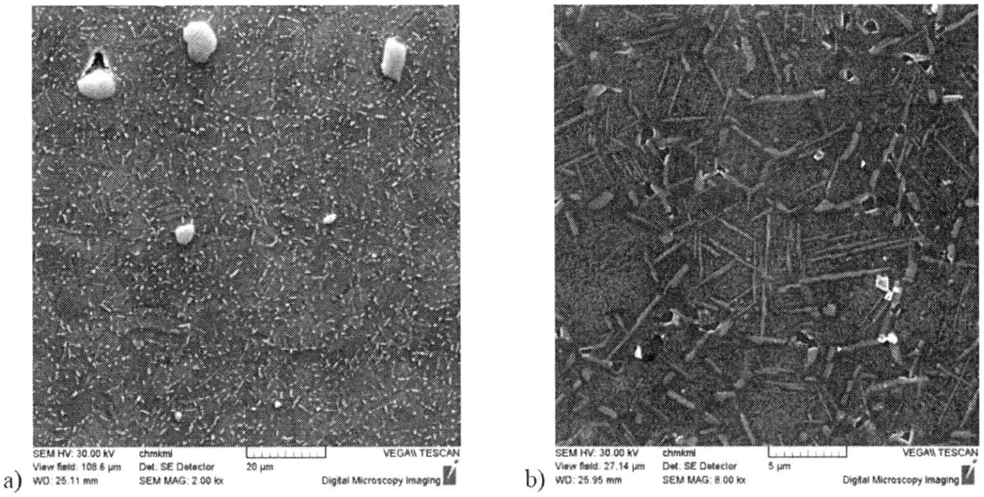

Fig. 3. Micrographs of wrought Inconel alloy 718, with presence of a) delta Ni_3Nb phase and primary carbides NbC, and b) coarse γ' phase with more delta presence, SEM, etch. Kallings.

From measured S-N curve is clearly seen that the fatigue life increases with decreasing stress amplitude and the S-N curve appears to continuously decline as the life extends. The limit number of cycles 2.10^7 (20 000 078) was reach at maximum bending stress σ_{omax} = 740 MPa.

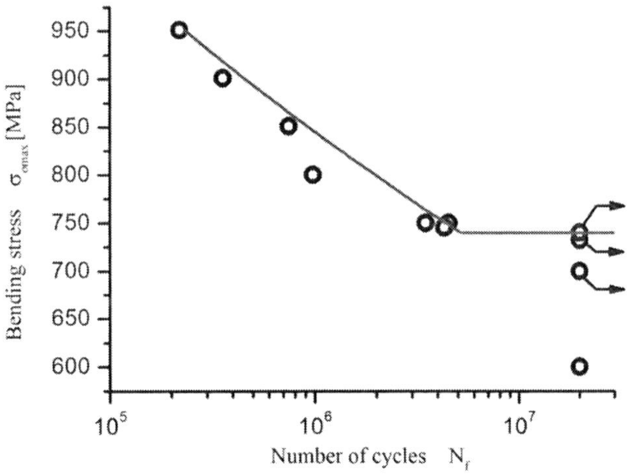

Fig. 4. The S-N curve for three point flexure fatigue test of Inconel alloy 718 with runouts after reaching 2.10^7 numbers of cycles.

According fatigue lifetime stress obtained at different testing method – ultrasonic push-pull loading with R = -1 [8 – 10]; is this result about one time higher (push-pull fatigue lifetime about 386 MPa compared to three point flexure fatigue lifetime 740 MPa at the same number of cycles). Of course, these values may be compared only roughly because of different frequency of test and different mechanism of loading.

Since Ni-base and Fe–Ni base alloys are used in jet engine discs, where both fatigue crack propagation and low cycle fatigue (LCF) are important in design. During fatigue in a strain controlled test, the stress responds in accordance to the applied strain levels and the influence of the microstructure. When an annealed specimen undergoes strain control fatigue, the stress generally increases with increasing cycling for a certain number of cycles. Depending on the deformation mode and its internal stress state, the response stress will either stabilize or reach a maximum value followed by softening. The response stress is also dependent upon the test temperature, strain rate and the applied plastic strain amplitude since these variables strongly influence the deformation substructure [7]. It is understood that the variation of frequency affects the fracture mode of this material. At higher frequencies transgranular failure occurs, as fatigue dominates, while at lower frequencies creep/oxidation exert greater influence. Shahinian and Sadananda [11] describe their observation that the fatigue crack growth rate (FCGR) increases with increasing stress ratio. Furthermore, the mode of fracture is transgranular in vacuum and, at relatively high frequency of loading, also in air, while it is predominantly intergranular in an oxidising environment at intermediate frequencies. Different fatigue crack growth mechanisms in IN718 alloy have been proposed by various researchers to describe different effects listed above. Andersson and Persson [12] carried out microscopic examination of fatigue crack surfaces in Inconel 718 using scanning electron microscope at room temperature, and observed two different kinds of fracture surfaces: locally flat fracture surfaces, and striations. It was confirmed by Andersson and Persson that both mechanisms are likely to be active simultaneously, but their effects are not additives. It was also concluded that crack growth was dominated by the crystal slip-based mechanism. Even so, striations could be found only in a few grains.

The occurrence of ductile striations on fracture surfaces indicates the presence of the transgranular (ductile) fracture mechanism of FCG at room temperature. The spacing between striations (i.e., crack growth rate) increased with the stress intensity factor range. Andersson and Persson's room temperature experiments proved the dominance of transgranular crack growth mechanism for all cases examined under various ΔK and maximum stress values. Nevertheless, different crack growth mechanisms were identified in this analysis, such as microcracking followed by subsequent linking, and pure shearing. Grain orientation exerted a strong influence on transgranular crack propagation and the crack growth rate da/dN, due to the interaction between different mechanisms active simultaneously. Andersson and Persson state that rough surfaces created by striations and crack meandering induce premature crack closure. Crack growth was predominantly found to occur either by growth on slips planes in a shearing mode, or by linking of microcracks ahead of the main crack, which did not require detectable crack opening.

All this fact about fatigue crack initiation, stable fatigue crack propagation, pretty close correspond to result received after three point flexure fatigue testing at room temperature which has been done at our experiment, as report Fig. 5.

Fig. 5a and 5b are showing initiation sites of fatigue crack at high and lower cyclic loading. The free surface at lower frequency of loading and so called „fish-eye“ at higher frequency of loading are common as initiation sites. In this case, no evidence of „fish-eye“ or initiation from undersurface oxides was observed. Figures 5c and 5d represent the stable area of fatigue crack propagation characterised by transcrystalline ductile mechanism with striation presence. However, at same cases, especially at different mode of loading (e. g. push-pull loading) is this second area of fatigue fracture characterised with mix of fatigue crack growth mechanism – starts as transcrystalline through small intercrystalline growth and finally transcrystalline again. As shows Fig. 5c, intercrystalline cleavage with facets and river morphology is presented in a few places of fracture surface.

The distance of striation shows fatigue crack growth a tone cycle loading. At higher values of loading is occurred also secondary fatigue crack perpendicular to major fatigue crack, Fig. 5e. The last Fig. 5f shows a final stage of fatigue test, static fracture, which represents the third stage. Mechanism of fracture is strictly ductile transcrystalline with dimple morphology where crushed primary carbides NbC can be observed.

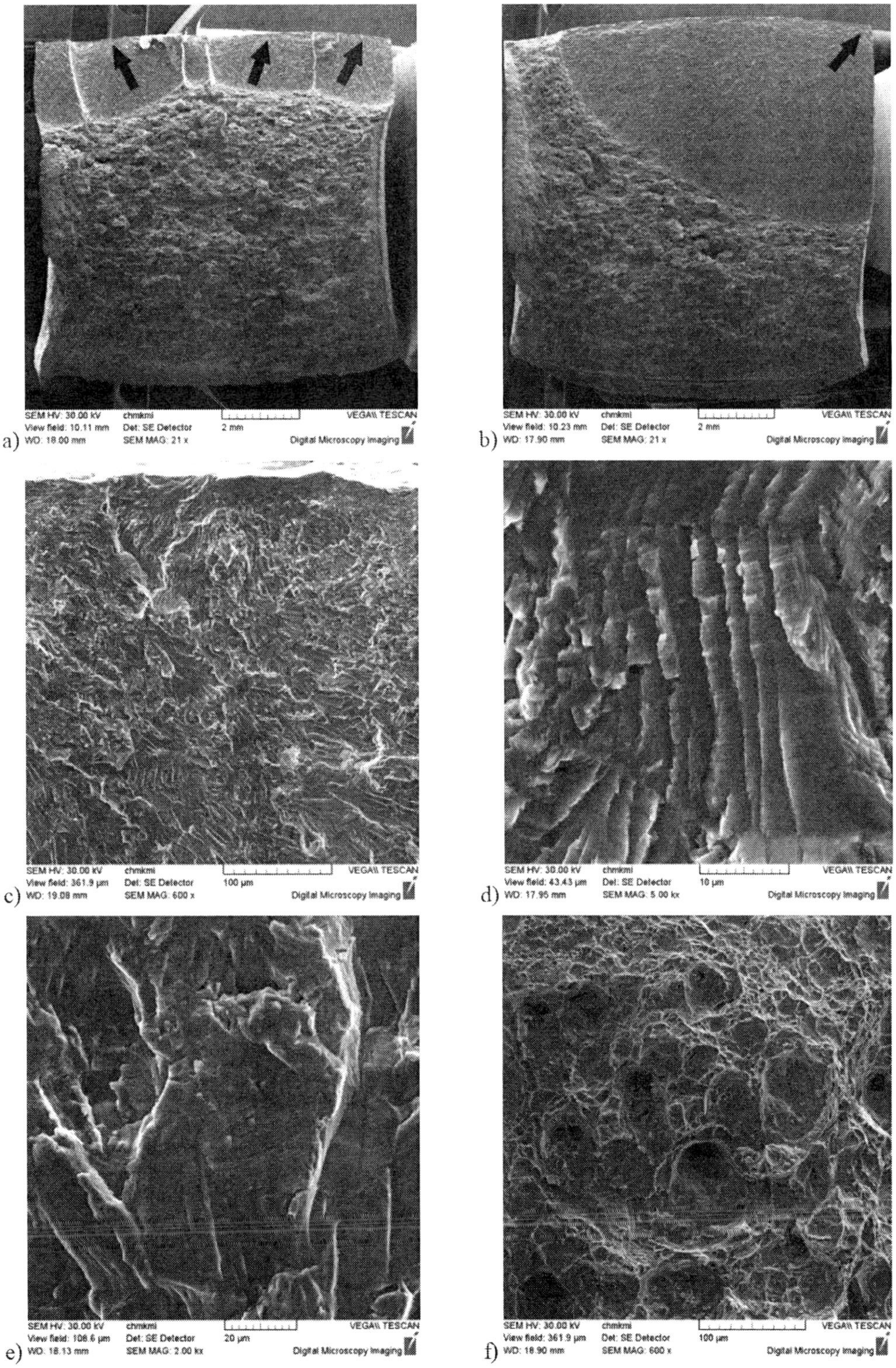

Fig. 5. Fractography surfaces of Inconel 718 after three point flexure fatigue test, SEM: a) initiation site at higher cyclic loading, note three initiation points; b) initiation site at lower cyclic loading, only single one, compared to fig.a; c-d) stable fatigue crack propagation characterised by striation; e) secondary fatigue crack occurrence at higher level of loading; f) static part of fatigue fracture.

Conclusions

In this article is given brief information about fractography and fracture mechanism of Inconel alloy 718 under three point flexure bending mode at room temperature. From measured S-N curve is clearly seen that the fatigue life increases with decreasing stress amplitude and the S-N curve appears to continuously decline as the life extends. The limit number of cycles 2.10^7 (20 000 078) was reach at maximum bending stress σ_{omax} = 740MPa. Obtained results were compared with experimental results of other authors whose provide their tests at different load condition. Our results show pretty good correlation with their measurements.

The free surface at lower frequency of loading and so called „fish-eye" at higher frequency of loading are common as initiation sites. In this case, no evidence of „fish-eye" or initiation from undersurface oxides was not observed. The FCG continue with transcrystalline ductile mechanism with striation presence. At higher values of loading are secondary fatigue crack perpendicular to major fatigue crack also occurred. Mechanism of fracture is strictly ductile transcrystalline with dimple morphology where crushed primary carbides NbC can be observed.

This work has been supported by Scientific Grant Agency of Ministry of Education of Slovak Republic and Slovak Academy of Sciences, No. 1/0533/15, No. 049ŽU-4/2017 and project EU ITMS 26220220154.

References

1. D. G. Leo Prakash, M. J. Walsh, D. Maclachlan, A. M. Korsunsky, *Crack growth micro-mechanisms in the IN718 alloy under the combined influence of fatigue, creep and oxidation.* International Journal of Fatigue **31**, 1966-1977 (2009)
2. D. Boyd-Lee, Fatigue *Crack Growth Resistant Microstructures in Polycrystalline Ni-base Superalloys for Aeroengines.* International Journal of Fatigue, **21** (4), 393-405 (1999)
3. J. F. Radavich, *The Physical Metallurgy of Cast and Wrought Alloy* 718. Superalloy 718 – Metallurgy and Applications, Ed.: E., A., Loria, The Minerals, Metals & Materials Society, 229 – 240 (1989)
4. C. T. Sims, W. C. Hagel, *The superalloys*. New York: John Wiley & Sons; 1972
5. J. K. Tien, T. Caufield, *Superalloys, supercomposites and superceramics*. New York: Academic Press; 1989
6. D. F. Paulonis, J. M. Oblak, D. S. Duvall, *Precipitation in nickel-base alloy 718.* Trans ASM **62**, 611-622 (1969)
7. Pineau, S. D. Antolovich, *High temperature fatigue of nickel – base superalloys.* Engineering Failure Analysis **16**, 2668-2697 (2009)
8. L. Trško, M. Guagliano, O. Bokůvka, F. Nový, M. Jambor, Z. Florková, *Influence of Severe Shot Peening on the Surface State and Ultra-High-Cycle Fatigue Behavior of an AW 7075 Aluminum Alloy*. Journ. Of Mat. Eng. and Perf. **26** (6), 2784-2797 (2017)
9. J. Belan, *High frequency fatigue test of in 718 alloy – microstructure and fractography evaluation.* Metalurgija **54** (1), 59-62 (2015)
10. M. Vaško, M. Blatnický, P. Kopas, M. Sága, *Research of weld joint fatigue life of the AlMgSi07.F25 aluminium alloy under bending-torsion cyclic loading*. Metalurgija **56** (1-2), 94-98 (2017)
11. P. Shahinian, K. Sadananda, *Effects of Stress Ratio and Hold-Time on Fatigue Crack Growth in Alloy 718.* J Eng Mater Technol **101**, 224-230 (1979)
12. H. Andersson, C. Persson, *In-situ SEM study of fatigue crack growth behaviour in IN718.* Int J Fatigue **26** (3), 211-219 (2004)

Study of selected thermoplastics using dynamic mechanical analysis

Marcel Kohutiar[*,1], *Mariana* Pajtášová[1], *Róbert* Janík[1], *Iveta* Papučová[1], *Jana* Pagáčová[1], *Beáta* Pecušová[1], *Ivan* Labaj[1]

Department of Materials Technologies and Environment, Faculty of Industrial Technologies in Puchov, Alexander Dubcek University of Trencin, I. Krasku 491/30, 020 01 Puchov, Slovak Republic

Abstract The aim of given paper is to study selected polymers using dynamic mechanical analysis method (DMA). DMA is one of the most useful techniques for the study of the viscoelastic behaviour of thermoplastic polymers. In relation to DMA, an oscillatory stress and strain is applied to the material at specific frequencies and temperatures and based on this mentioned fact hereinbefore, the resulting changes after the loading in the material are measured. This technique allows detecting the melting temperature and the glass transition temperature of the thermoplastic materials. Furthermore, some spectroscopy techniques, such as energy dispersive X-ray spectroscopy (EDX) and infrared spectroscopy (IR), were also used for the investigation of the thermoplastics. The thermoplastics used for examination, namely polyethylene, polystyrene, polypropylene and polyethylene terephthalate, were gained from the waste of the packaging.

Keywords: Dynamic mechanical analysis, infrared spectroscopy, dispersive X-ray spectroscopy, thermoplastics, packaging

1 Introduction

Dynamic mechanical analysis (DMA) is a method commonly used to measure the viscoelastic behaviour of plastic materials [1]. Plastics are never purely liquid or purely solid, because they exhibit both viscous and elastic characteristics when undergoing deformation. Oscillatory stress and strain is applied to the material at appropriate frequencies and temperatures [2]. Based on this mentioned fact, the resulting changes after the loading in the material are measured. The use of DMA technique along with differential scanning calorimetry (DSC) allows detecting the melting temperature and the glass transition temperature of the thermoplastic materials [3].

Infrared spectroscopy is widely used for characterization of various materials, such as plastics, ceramics or chemicals. Infrared spectrum helps us to identify or characterize an unknown material, and also analyze contaminants or chemical differences in observed material [4]. Infrared spectroscopy can be also used for analyzing additives in plastic compounds. Along with the energy dispersive X-ray spectroscopy (EDX), which provides

* Corresponding author: marcel.kohutiar@gmail.com

Reviewers: *František* Nový, *Eva* Tillová

elemental and chemical analysis of the sample, this method has become more important for the identification of potential harmful substances occurring in packaging [5, 6]. Materials for packaging can not contain toxic substances and have to be tough and flexible enough to protect the food from damage and to keep its freshness [7]. Nowadays, the most commonly used thermoplastics in the manufacture of food packaging are polyethylene, polystyrene, polypropylene and polyethylene terephthalate [8, 9].

2 Materials and methods

For the experiment, two types of low density polyethylene (LDPE), polypropylene (PP) and polyethylene terephthalate (PET) were selected as the examined materials. The thermoplastics used for examination were gained from the waste of the packaging.

Dynamic mechanical analysis was performed by the Q800 DMA from TA Instruments. The dimensions of each sample for DMA were 45 x 10 mm. The testing temperature was from -70 °C to 100 °C for LDPE, from -70 °C to 140 °C for PP, from -70 °C to 160 °C for PET at a heating rate of 3°C per minute and frequency was 10 Hz.

The DSC measurements described in this paper were performed using Mettler Toledo, TGA/DSC 2 instrument. The weight of the samples was about from 10 to 20 mg and they were heated from 0 °C to 350 °C at a heating rate of 10 °C min^{-1}.

In Fourier transform infrared spectroscopy (FTIR), the spectra were measured by ATR technique using a Nicolet iS50 FT-IR spectrometer with a resolution of 4 cm^{-1} in the range of 4000 - 400 cm^{-1}. The elemental and chemical analysis of the samples, were carried out using energy-dispersive EDX-7000 X-ray fluorescence spectrometer. All measurements were carried out at the CEDITEK (Center for quality testing and diagnostics of materials) workplace at Faculty of Industrial Technologies in Puchov.

3 Results and discussions

3.1 Dynamic mechanical analysis (DMA)

Fig. 1. shows the scan temperature data at frequency of 10 Hz for transparent and red LDPE. There is a decrease in storage modulus at approximately – 46 °C for red and at approximately – 30 °C for transparent sample which can be attributed to the β-relaxation or glass transition temperature (T_g). The loss modulus of red LDPE shows one well-defined transition and two small transitions while at transparent LDPE, all three transitions are well evident. The first peak of red sample at -29 °C may be attributed to the β-relaxation. The second and the third small peaks at approximately 25 °C and 49 °C may correspond with the α-relaxation, but it can also correspond with the melting of the small crystallites.

The possible β-relaxation of transparent sample occurs at approximately 0 °C. The second and the third peaks associated with the α-relaxation and the melting of small crystallites are at approximately 28 °C and at approximately 55 °C. There is increase in tan δ for both samples with peak of red LDPE at approximately 56 °C associated with α-relaxation.

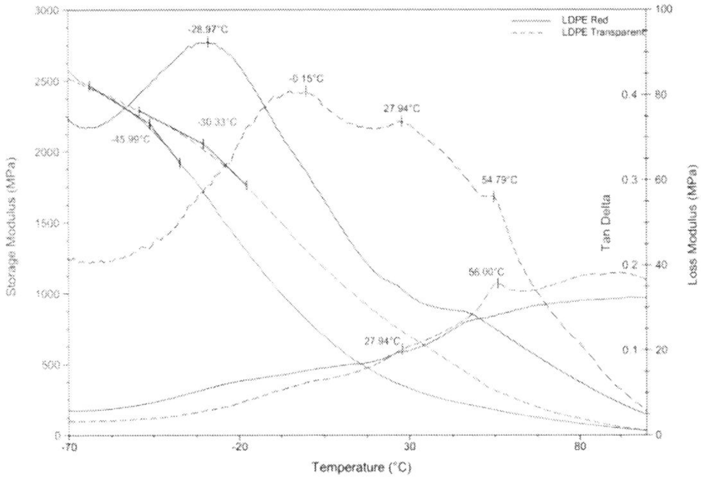

Fig. 1. Dynamic mechanical analysis of LDPE samples

The scan temperature data at frequency of 10 Hz for transparent and yellow PP are shown in Fig. 2. There is a decrease in storage modulus at approximately – 18 °C for yellow and at approximately – 7 °C for transparent PP which corresponds with glass transition temperature of materials. The loss modulus shows two temperature peaks for both samples at approximately - 41 °C and 4 °C for transparent sample and at approximately – 42 °C and -5 °C for yellow sample. These peaks are associated with the α- and β-transitions of polypropylene. The tan δ also shown two peaks associated with these transitions. The peaks of transparent PP are at approximately - 42 °C and 10 °C for transparent sample and at approximately - 43 °C and 2 °C for yellow sample.

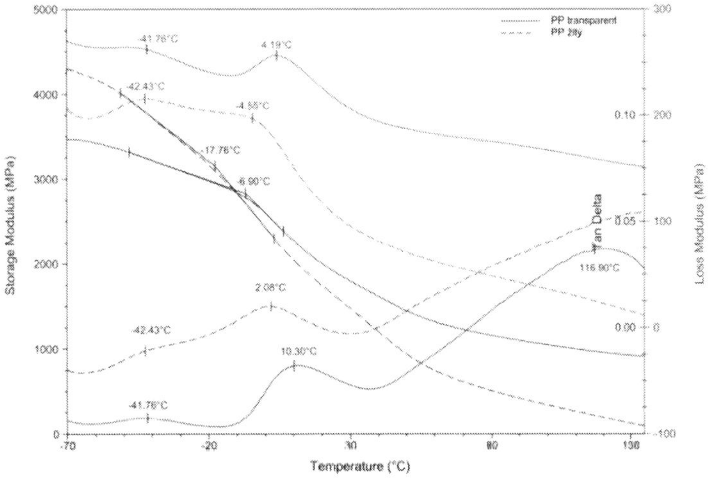

Fig. 2. Dynamic mechanical analysis of PP samples

The scan temperature data of pink and green PET at frequency of 10 Hz are shown in Fig. 3. The storage modulus of both samples shows well-defined peaks which correspond with glass transition temperature of PET. The peak for pink sample is determined at approximately 73 °C, while for green sample, it is at approximately 94 °C. The glass transition temperature can be also determined from loss modulus and from tan δ. From loss modulus, the T_g for pink sample is at approximately 93 °C and for green sample, it is approximately 103 °C, whereas from tan δ, the T_g for pink sample is at approximately 105 °C and for green sample, it is at approximately 117 °C.

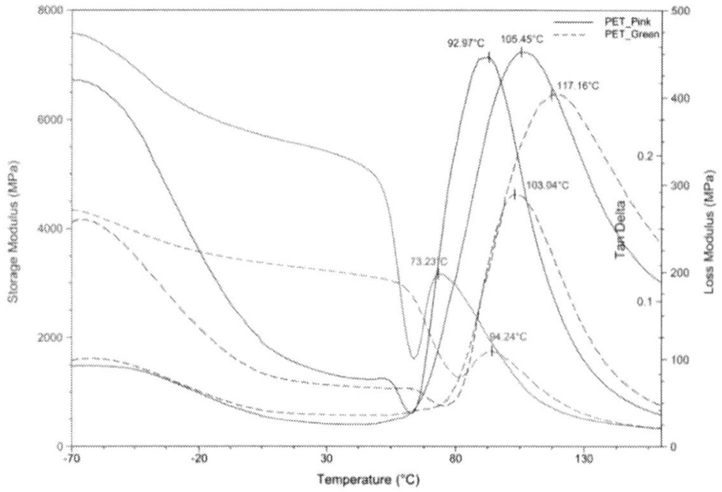

Fig. 3. Dynamic mechanical analysis of PET samples

3.2 Differential scanning calorimetry (DSC)

Differential scanning calorimetry was used for detecting the glass transition temperature and melting point of investigated materials. Fig. 4. presents DSC measurements of red and transparent LDPE samples. As can be seen from the graph, the melting peak of red LDPE is less sharp and wider in comparison with transparent LDPE. The melting point of red sample is at approximately 116 °C. The second peak is at approximately 164 °C. This peak is most likely result of the residual reactants from LDPE synthesis or contamination before analysis. Furthermore, the melting point of transparent LDPE is at approximately 121 °C.

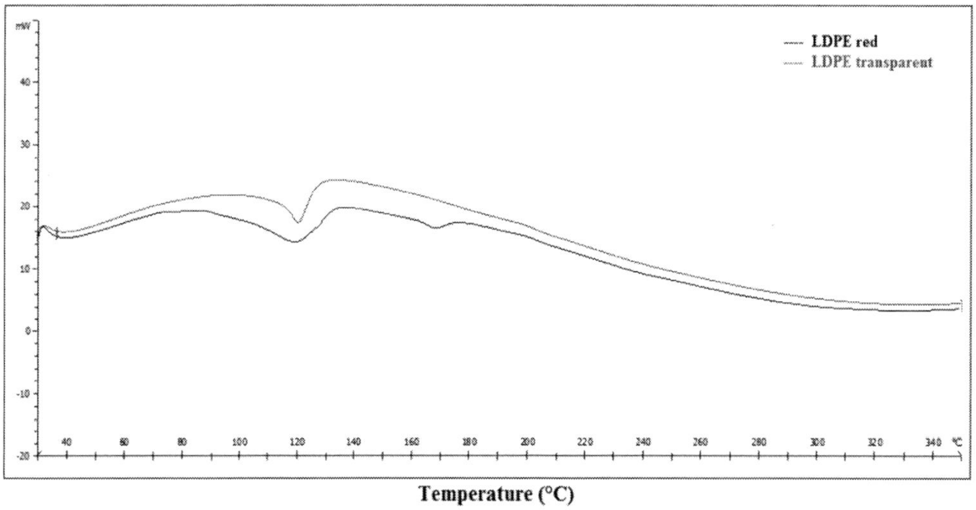

Fig. 4. DSC measurement of LDPE samples

DSC measurements of yellow and transparent PP samples are presented in Fig. 5. The melting peak of yellow PP is sharper and narrower in comparison with the transparent one. The melting point of yellow sample is at approximately 173°C while the melting point of transparent sample is at approximately 171°C.

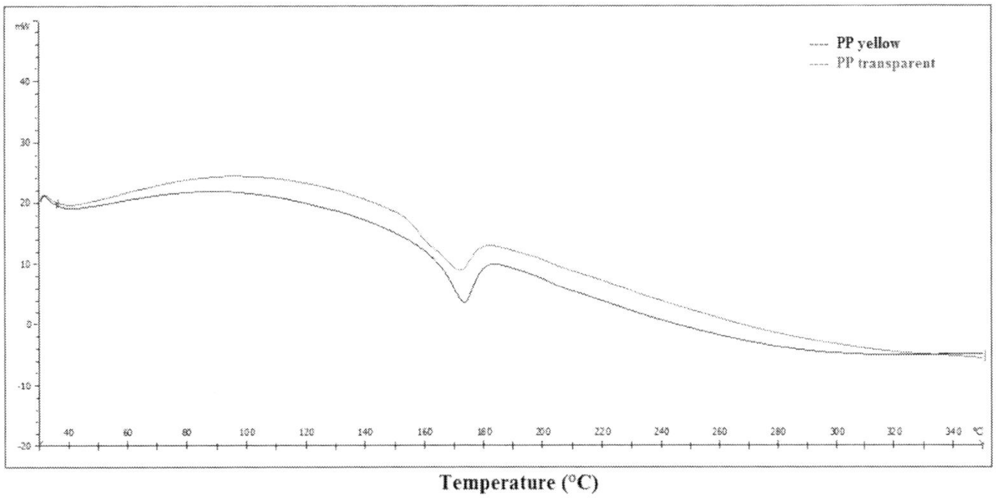

Fig. 5. DSC measurement of PP samples

Fig. 6. shows DSC measurements of pink and green PET samples. In contrast to the samples described above, the glass transition temperature is shown in the graph. The glass transition peak of pink and green sample is at approximately 104 °C and at approximately 98 °C, respectively. The melting point of pink sample is at approximately 151°C while the melting point of green sample is at approximately 152°C.

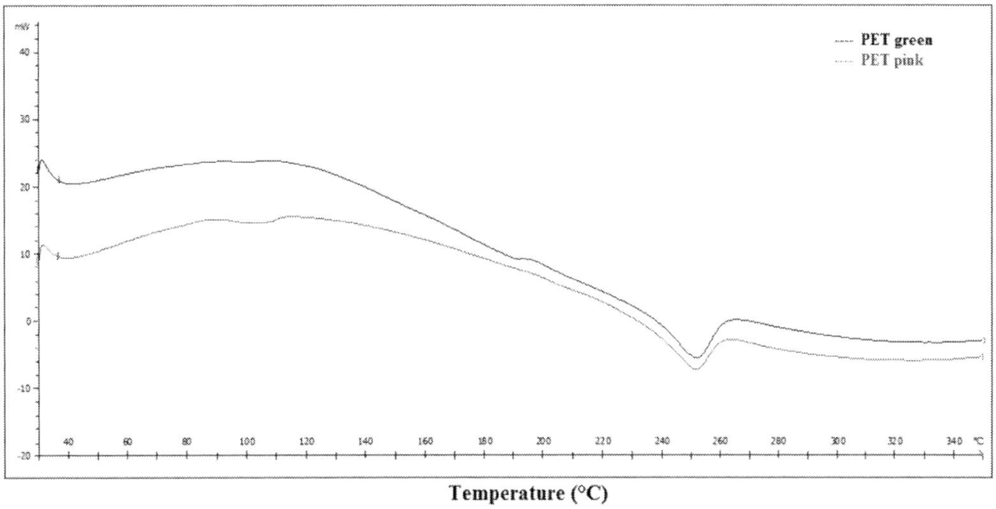

Fig. 6. DSC measurement of PET samples

3.3 Fourier transform infrared spectroscopy (FTIR)

The spectra of transparent and red LDPE samples obtained in absorbance using ATR technique is shown in Fig. 7. As can be seen from data, LDPE samples are very similar, because there are only differences in peak intensity. Characteristic peaks at around 3000 - 2800 cm^{-1} correspond with methylene group symmetric stretching $\nu_s(CH_2)$ and asymmetric stretching $\nu_{as}(CH_2)$. Peak at around 1500 – 1400 cm^{-1} can be associated with CH_2 due to sccisoring. The rocking of methylene group is presented by peaks between 750 and 700 cm^{-1}.

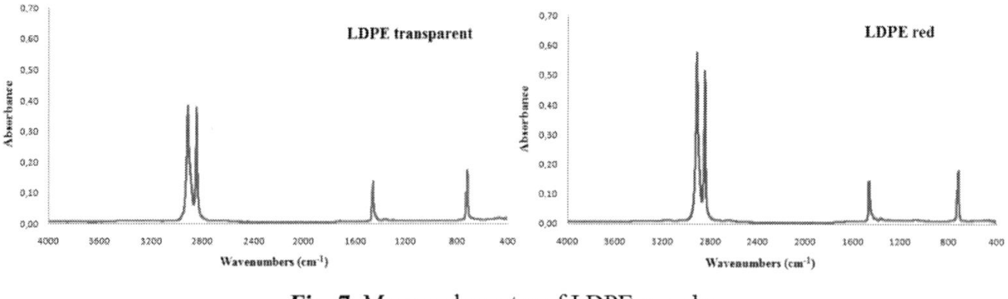

Fig. 7. Measured spectra of LDPE samples

Fig. 8. presents measured spectra of transparent and yellow PP samples obtained in absorbance using ATR technique. From the FTIR spectra scan be seen, that there is similarity between two samples. They are different in 800 – 400 cm^{-1} IR spectra region. At around 3000 - 2800 cm^{-1}, polypropylene spectral areas present absorption bands which correspond with asymmetric and symmetric stretching vibration of methylene and methyl groups. Different peaks have been also observed in the 1500 - 1350 cm^{-1} spectral area, where absorption bands can be associated with scissoring vibration of the methylene group (at around 1470 cm^{-1}) and to symmetric deformation of the methylene group (at around 1380 cm^{-1}).

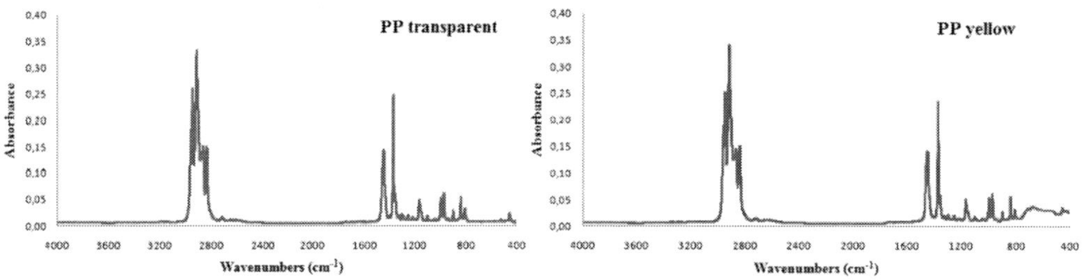

Fig. 8. Measured spectra of PP samples

Measured spectra of pink and green PET samples obtained in absorbance using ATR technique are presented in Fig. 8. There is no significant difference between the two PET samples compared with the FTIR spectra scan, because there are differences only in spectra intensity. The weak peak at around 2950 cm^{-1} represents C-H asymmetric stretching. Spectra of PET also confirm a carbonyl group along with aromatic ring appear as the second strongest peak at around 1720 cm^{-1}. Two peaks at around 1410 cm^{-1} and 1340 cm^{-1} represent the deformation of C-H alkane and peak at around 1510 cm^{-1} belongs to aromatic C-C stretching. The asymmetric C-C-O stretching involving the carbon in aromatic ring appears at around 1230 cm^{-1} as a strong peak. The strongest peak at around 720 cm^{-1} is associated with aromatic C-H wagging. The peak at approximately 870 cm^{-1} corresponds with aromatic C-H bending. The O-C-C asymmetric stretching is at around 1105 cm^{-1}.

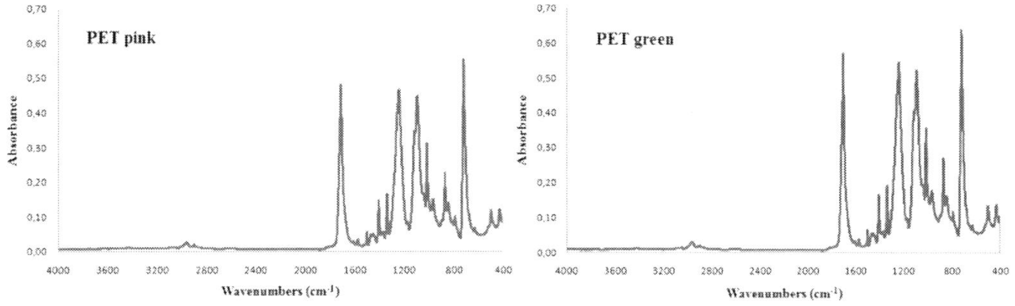

Fig. 9. Measured spectra of PET samples

3.4 Energy-dispersive X-ray fluorescence spectroscopy (EDX)

Thermoplastics used for food packaging contain low concentrations of heavy metals. Additives presented in quantity of products make heavy metals quantification very difficult. Antioxidants can contain Ni, while thermal stabilizers can contain Ni, Pb and Sb. The polymerization process is often carried out with an antimony (Sb), germanium (Ge), titanium (Ti), cobalt (Co), magnesium (Mg) or zinc (Zn) based catalyst [10].

EDX analysis of LDPE, PP and PET samples revealed percentage representation of individual elements in polymer matrix. As can be seen in Table 1., red sample contains increased amount of titanium and little amount of silicon and calcium, compared with transparent one. The presence of these elements in red sample can be due to used colorants or UV stabilizers.

Table 1. EDX analysis of LDPE samples

LDPE red		LDPE transparent	
Element	**Content (%)**	**Element**	**Contents (weight %)**
Ti	0.498	S	0.008
Si	0.093	P	0.007
Ca	0.018	Fe	0.001
S	0.004	K	0.001
K	0.003	Cr	0.001
Fe	0.001	Cu	0.001

According to Table 2, yellow PP sample also contain increased amount of titanium and calcium, compared with transparent LDPE and PP. As mentioned above, residue of titanium can be from polymerization process, where it was used in the form of catalyst.

The percentage representation of elements in PET samples is shown in Table 3. From the table, it can be clearly seen that both samples contain small amount of sulphur. Pink sample also contains increased amount of antimony which was most likely used in the form of catalyst during polycondensation of PET.

Table 2. EDX analysis of PP samples

PP yellow		PP transparent	
Element	**Contents (weight %)**	**Element**	**Contents (weight %)**
Ti	0.390	P	0.008
Ca	0.127	S	0.007
P	0.018	K	0.002
S	0.013	Ca	0.002
K	0.003	Fe	0.001
Cr	0.002	Cr	0.001

Table 3. EDX analysis of PET samples

PET pink		PET green	
Element	**Contents (weight %)**	**Element**	**Contents (weight %)**
Sb	0.017	S	0.019
S	0.016	Ca	0.004
P	0.011	K	0.002
Ca	0.005	Co	0.001
K	0.001	Cu	0.001
Cu	0.001	Fe	0.001

Conclusions

The measurements were carried out for six samples of thermoplastics, namely LDPE (red and transparent), PP (yellow and transparent) and PET (pink and green). Using DMA and DSC analysis, all the samples were measured and from the resulting graphs, the glass transition temperature and the melting point of each polymer was determined. There are several results of glass transition temperature, because it can be derived from the loss modulus, the storage modulus and the tan δ. It is because each of these results has its own physical essence.

Using the ATR-FTIR spectrometry, the spectra of the individual pairs of polymers were compared. Based on the results obtained in this work, it can be concluded that the ATR - FTIR spectrometry is useful tool for examination of the chemical structure of thermoplastics.

Furthermore, the EDX analysis revealed the presence of some elements and heavy metals in tested polymers.

KEGA 007TnUAD-4/2017, VEGA grant No. 1/0649/17, VEGA grant No. 1/0589/17, project "Center for quality testing and diagnostics of materials", ITMS code 26210120046 relating to the Operational Program Research and Development, EF of Reg. Development.

References

1. K. P. Menard, *Dynamic mechanical analysis: A practical introduction*. (CRC Press, Boca Raton, 1999) ISBN 0849386888
2. J. Babíková, *Využití metod DMA pro studium polymerních materiálů*. Bachelor's thesis, Tomáš Baťa University in Zlín, Faculty of Technology (2011)
3. I. Škrombachová, *Dynamic mechanical properties of polymer blends*. Bachelor's thesis. TnU AD, FPT (2011)
4. Da-Wen Sun, *Infrared Spectroscopy for Food Quality Analysis and Control.* (Elsevier, 2009) ISBN 978-0-12-374136-3 [online] Available at: <http://app.knovel.com/hotlink/toc/id:kpISFQAC0I/infrared-spectroscopy/infrared-spectroscopy>
5. K.L. Yam, *Wiley encyclopedia of packaging technology*. (John Wiley & Sons, 3rd Edition, 2009) ISBN 978-0-470-08704-6 [online] Available at: <http://app.knovel.com/hotlink/toc/id:kpWEPTE002/wiley-encyclopedia-packaging/wiley encyclopedia-packaging>
6. G. L. Robertson, *Foodpackaging : Principles and practice*. (Taylor & Francis, 3rd Edition, 2013) ISBN 978-1-4398-6242-1 [online] Available at:

<http://app.knovel.com/hotlink/toc/id:kpFPPPE00I/food-packaging-principles/food-packaging-principles>

7. S. Allahvaisi, *Polypropylene : Polypropylene in the industry of food packaging*. InTech (2012) ISBN 978-953-51-0636-4 [online] Available at: <https://www.intechopen.com/books/polypropylene>
8. M. Bělašková, *Characterization of wrappaging films based on polyethylene and polypropylene*. Bachelor´s thesis, Brno University of Technology, Faculty of Chemistry (2015)
9. T. Liptáková, *Polymérne konštrukčné materály*. (Vydavateľstvo Žilinskej univerzity, 2012) ISBN 978-80-554-0505-6
10. M. E. Conti, *Heavy metals in food packaging*. (University of Rome, Italy, 2009)

Porosity formation and fatigue properties of AlSiCu cast alloy

Lenka Kucharikvá[1,*], *Eva* Tillová[1], *Milan* Uhríčik[1], *Juraj* Belan[1]

[1]University of Žilina, Department of Materials Engineering, 01026 Žilina, Slovakia

Abstract. The need for aluminium alloys having a good toughness, high strength, adequate damage tolerance capability, good fatigue resistance and good corrosion resistance for use in the industries applications of aerospace, automotive and even commercial products led to study of the properties and structure of the AlSi9Cu3 cast alloy. The most important metallurgical parameters affecting the aluminium alloy's resistance to fatigue load are the amount, types and size of casting defects. Therefore quantitative analysis was used for determination of casting defects size by optical microscopy on metallographic sections and an analysis of the size distribution by extreme value statistics. The software NIS Elements was used for quantitative analysis of casting defects and results show that porosity have a great influence on to the fatigue properties of AlSi9Cu3 cast alloy.

Keywords: fatigue properties of aluminium alloys, quantitative analysis, porosity, extreme value statistic

1 Introduction

The most common aluminium alloys for automotive industry can be found in the groups of AlSi, AlSiMg and AlSiCu. These groups of aluminium alloys are the most versatile materials, comprising 85-90 % of the total aluminium cast parts produced by the automotive industry due to their highest strength-to-weight ratio, good thermal conductivity, excellent fluidity, hot tear resistance and feeding characteristics that allow casting intricate shapes such as engine blocks, cylinder head-sore chassis components [1-5]. These materials are important for their Si, Mg and Cu content, because Si considerably increase castability, tensile properties and hardness while Cu or Mg addition improve mechanical properties thanks heat treatment possibilities. The final mechanical properties of aluminium castings depends not only on chemical compositions (Si, Cu, Mg and Fe content) but on the solidification time, heat treatment, micro-voids, porosity, dendrite arm spacing, and distribution, shape and size of the silicon particles, second phases (especially Fe–rich phases) [6]. Bangyikhan [7] also pointed to the fact that: cast aluminium alloys are, at the best, only achieving and using one per cent of their potential mechanical properties, particularly in terms of fatigue life. The research shows that a great problem of lower mechanical and fatigue properties is the presence of oxides films, porosity and Fe-rich

* Corresponding author: lenka.kucharikova@gmail.com

Reviewers: *Darina* Ondrušová, *Milan* Sága

intermetallic phases in the castings [7,8]. Eady and Smith [9] also showed that pores could act as stress concentrators, initiation sites for fatigue cracks, and reduce tensile properties, particularly elongation.

Nowadays are known the new classification of defects and imperfections and is based on three-level approach [10, 11]:

I. Morphology (location) of defects (internal, external, geometrical);
II. Metallurgical origin of defects (gas-related defects, solidification shrinkage, shrinkage filling-related defects, undesired phases, thermal contraction defects, metal-die interaction defects, etc.);
III. Specific type of defects (the same metallurgical phenomenon may generate various defects at handling, finishing and machining operations).

In the past were developed three main point which should be used at producing casting for improving the casting quality [7, 12]: 1) control of the liquid metal prior to casting, 2) control of the pouring of liquid metal into the mould and any metal/mould reaction, 3) control of casting microstructure and defects during solidification. The problem is that in the casting are still presence of porosity which is not desirable for casting type of engine blocks, cylinder head-sore chassis components. These castings have a very complicated geometry which is essential to the correct operation of the engine. The consequences of fatigue damage are therefore not desirable. Fracture occurs after continuous accumulation of damage due to repeated mechanical, thermal or mechanic-thermal loading prior to failure. In the case of fatigue failure there is no macroscopic deformation of the solid. The whole process occurs in a micro volume at the crack tip or at a local site on the surface during the creation of a fatigue crack nucleus [13, 14].

The material group Al-Si-Cu are also used for components which work under condition of dynamic loading and that is why it is necessary to study the microstructure and properties of such alloys. Based on the knowledge above, a study was carried out to investigate the influence of porosity on the fatigue behaviour of secondary aluminium cast alloy of the Al-Si group (AlSi9Cu3) used extensively in automotive applications, which was casted into a metallic moulds at gravity die casting. The quantitative assessment was used in order to determine the nature of the porosity in secondary AlSi9Cu3 aluminium alloy. Also was used the analysis of casting defects according to Murakami's statistical method using Largest Extreme Value Distribution theory (LEVD) for prediction of largest defect size on the fracture surface. Statistical analysis of casting defects present in the structure has been shown to be an useful tool in the procedure of the fatigue life prediction.

2 Experimental material and methods

The experimental material was secondary cast AlSi9Cu3 alloy (prepared from recycled aluminium scrap) delivered from Confal a.s. company. The experimental material was delivered in form of ingots (Fig. 1a) and from these ingots were gravity die casted samples for experimental work into the metallic moulds (chill casting Fig. 1b). The experimental material was not modified or grain refined. The samples for mechanical and fatigue properties were made from the casted bars (Fig. 1b) with turning and milling operations. The dimension of samples were according to standards and regulations for mechanical and fatigue properties measurements. The chemical composition of the experimental material was checked out using arc spark spectroscopy, and it is shown in Table 1.

Table 1. The chemical composition of AlSi9Cu3 cast alloy according arc spark spectroscopy, wt. [%]

Si	Cu	Mn	Zn	Mg	Fe	Ni	Ti	Sn	Pb	Al
9.4	2.4	0.24	1.0	0.28	0.9	0.05	0.04	0.03	0.09	remainder

Mechanical properties were measured according to the following standards: STN EN ISO 6892-1:2010 and STN EN ISO 6506-1 [15, 16]. Hardness measurement for secondary aluminium alloy was performed using a Brinell hardness (HBW) tester with a load of 62.5 kp (1 kp = 9.81 N), 2.5 mm diameter ball and a dwell time of 15 s. Tensile strength (UTS) was measured using ZDM 30 testing machine. The evaluated UTS and HBW reflect average values of at least six separate specimens.

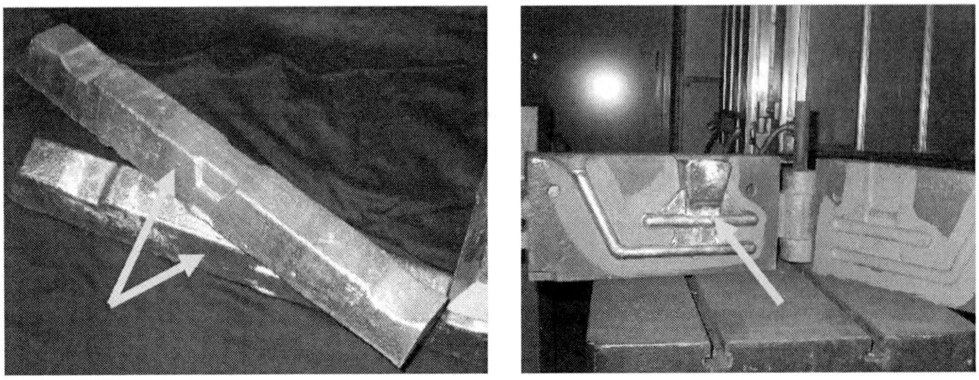

a) b)

Fig. 1. Experimental material
a) ingots delivered from Confal a.s.; b) gravity die casted samples for experimental work

Fatigue properties of chill casted specimens were carried out using Vibrophores Amsler 50-250 HFP 5100 testing machine with symmetrical push-pull load, and at a room temperature of 20 ± 5°C (Fig. 2). The dimensions of specimens for fatigue tests were ø18 x 150 mm. The stress amplitude was from 150 to 50 MPa and specified number of cycles for test was 2 x 10^7.

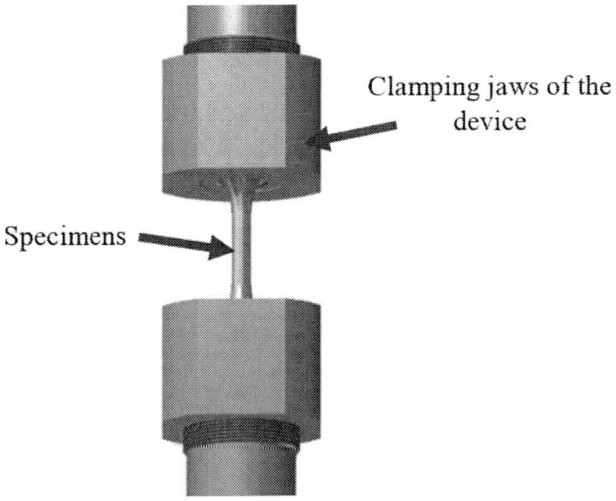

Fig. 2. The clamping scheme of fatigue samples for symmetrical push-pull load

The formation of macro-porosity in experimental material was analysed with using a MOTIC SM 2-168 microscope with the Moticam 1000 camera. The magnification was 7.5 times.

Light microscopy was used to analyse the microstructure and porosity of the material after fatigue tests. The main point was quantitative analysis of porosity in the specimens, which was carried out using a NEOPHOT 32 light microscope equipped with a computer running NIS Elements 4.2 image analyser software [17]. The pore size (area) and amount (surface fraction) were detected on longitudinal and transverse sections of the specimens. The resulting value of the measured parameters reflect average values of at least twenty-one separate measurements on each used specimens for fatigue tests. The specimens were prepared by standard metallographic procedures (wet ground on SiC papers, DP polished with 3 μm diamond pastes followed by Struers Op-S and etched for study underan optical microscope by standard etcher Dix-Keller and 0.5 % HF).

The results of quantitative analysis with using NIS Elements 4.2 analyser software (pore size) were completely used for Murakami's method on all fatigue experimental specimens. Murakami's method was applied to the characterization of the pore size population according to the largest extreme value distribution (LEVD) [18, 19]. The theory is based on the evaluation of the size of the largest defects on the controlled area marked - S_0 (field of view). The defect size was characterized in terms of the square root of its area (marked - $area^{1/2}$). The measurement was done on **n** number of the controlled areas. On each metallographic specimen at least 21 measurements were done ($n = 21$). The number of the controlled areas has to be sufficient for reasonable statistical description, but is dependent on specimen size and on chosen magnification. For this study a magnification of 100 x was used. According to the return period, marked – T ($T = S/S_0$), it is possible to predict defect sizes on areas larger than the controlled area [18, 19].

3 Experimental results and discussions

3.1 Mechanical properties

The results of as-cast AlSi9Cu3 mechanical properties are: the tensile strength 211 MPa, the Brinell hardness 98 HBW 2.5/62.5/15 and show that secondary material (experimental material) have suffice properties as the some primary material used for casting of automotive products.

3.2 Fatigue properties

Fatigue properties were measured on samples without heat treatment, modification or grain refining, i.e. as-cast state. The results of fatigue properties show Fig. 3.

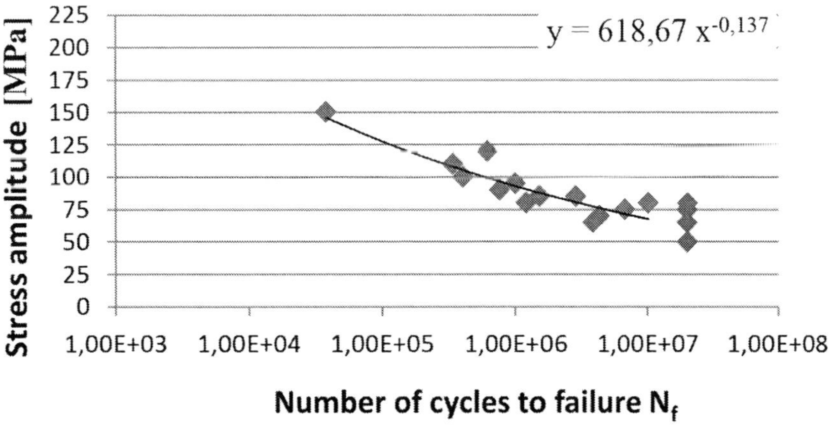

Fig. 3. Results of fatigue properties

The blue marks on the figure 3 represent specimens which were broken by load effect before reaching the specified number of cycles for fatigue test 2 x 10^7 and red are samples which were not broken before reaching the specified number of cycles for fatigue test 2 x 10^7 (the specimens that have reached these numbers of cycles without breaking are defined as run-out). The range of fatigue lifetime for 2 x 10^7 cycles was from 50 to 80 MPa → average value of fatigue lifetime for 2 x 10^7 cycles was about 67,5 MPa in experimental material (Fig. 3). The predicted value of fatigue lifetime for 10^7 cycles for material cast into the metallic mould was 68 MPa according to Basquin`s equation obtained from the diagram (Fig. 3). The specimens under stress amplitudes 70, 75 and 80 MPa were ones broken but ones not broken and reach the 2 x 10^7 cycles. There are assumptions that specimens which were broken have greater amount and size of defect as the specimens which were not broken under the same stress amplitude. Therefore were used quantitative and statistical methods of porosity evaluation in experimental material.

3.3 Porosity formation

The macro analysis of porosity formation in experimental material (Fig. 4a) confirmed existence of small amount of porosity in experimental material. The light microscopy analysis of porosity in experimental material after fatigue test showed presence of porosity, too (Fig. 4b). The analysis of macro-porosity and micro-porosity showed presence especially shrinkage porosity in experimental material (Fig. 4b).

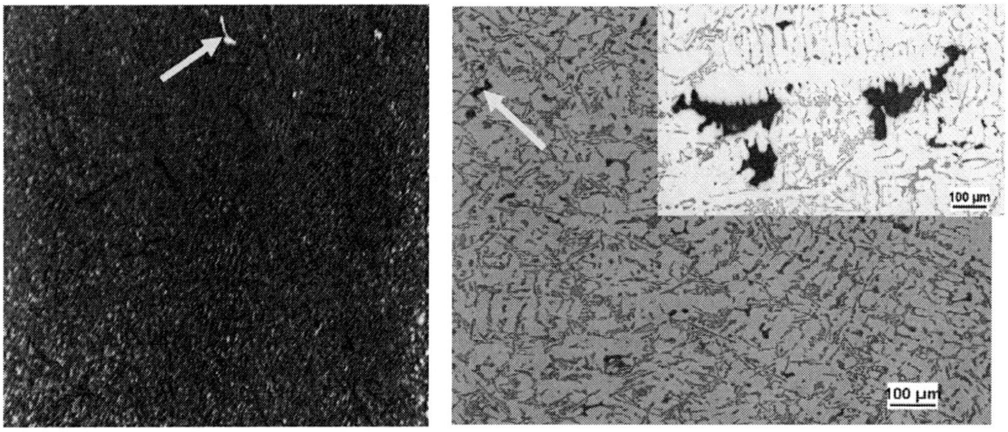

a) b)

Fig. 4. Porosity formation in experimental material

a) Macro analysis, 7.5x ; b), c) Light microscopy analysis, 100x

3.4 Quantitative analysis

The measurements of quantitative analysis were focused on size and amount of porosity in the specimens (Fig. 5). The evaluations were done on three samples marked A, B, and C. The specimen A was not broken under fatigue loading and B, C were broken under fatigue loading. The results of quantitative analysis are in Table 2. The measurements showed that the size of pores is different in longitudinal and transverse section. For prediction of largest porosity transverse section are especially used. Specimens A have smaller amount and size of porosity and therefore it is possibilities that this specimens will have better properties as the other (B, C) which have higher amount and size of porosity. The presence porosity near the surface of experimental specimens was clearly visible in both section (longitudinal and transverse) (Fig. 5c, d). The quantitative method also confirmed smaller amount of porosity in specimen A near the surface in comparison to specimens B and C.

Table 2. The results of quantitative analysis of porosity in experimental material

Specimens	Average pore size in transverse section	Average pore size in longitudinal section	Amount of porosity (surface fraction)
A	880.47 μm²	1711.80 μm²	0.04 %
B	3689.1 μm²	6023.83 μm²	0.7 %
C	7208.26 μm²	3393.54 μm²	1.2 %

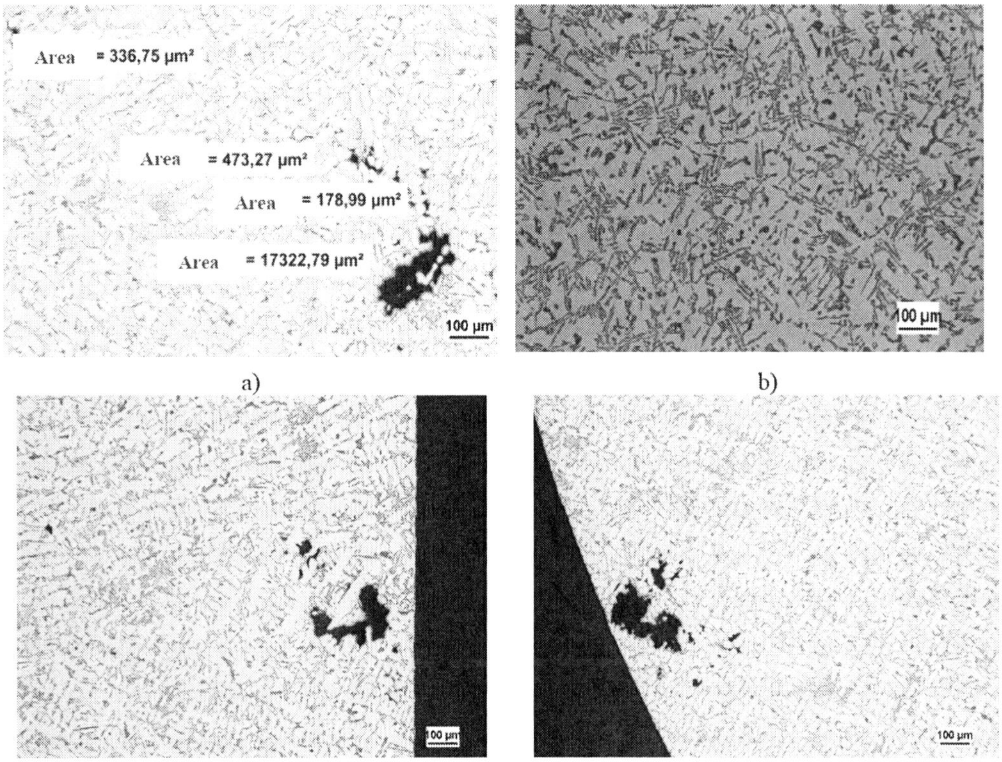

a) b)

c) d)

Fig. 5. The quantitative analysis of porosity
a) pore size; b) amount of porosity; c) longitudinal section; d) transverse section

3.5 Largest extreme value distribution of porosity

Murakami's method was used for prediction defect size in all fatigue specimens. For this study results of the largest defect size for three samples (marked A, B and C) tested at the same stress amplitude of 80 MPa were used. The results are shown in Fig. 6 and Table 3. One of these samples was not broken after 2 x 10^7 (marked A - the specimens that have reached these numbers of cycles without breaking are defined as run-out), the two others (marked B and C) were broken before reaching the specified number of cycles for fatigue test 2 x 10^7 (Table 3). The largest defects size, measured on experimental samples, is presented as a Gumbel plot. The data were extrapolated for the values corresponding to the largest defects which could occur in area of S = 50 mm² marked on the **T** axis. The ratio between the projected area and the controlled area is $T = S/S_0 = 26.82$ in this study. The results of predicted defect size are shown in Table 3.

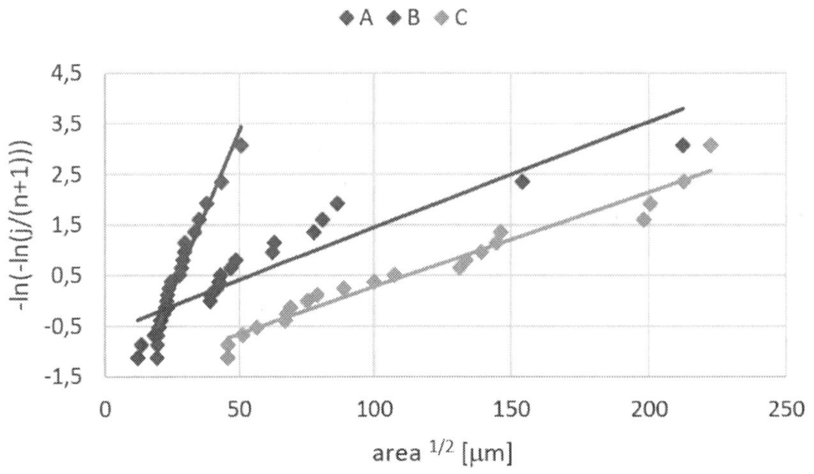

Fig. 6. The results of LEVD theory

Table 3. Results of quantitative method, LEVD theory and fatigue properties

Specimens	Measured maximum defect size at quantitative analysis [μm²]	Predicted largest defect size with using LEVD [μm²]	The number of cycles reached after fatigue test
A	2578.7	3338.69	20000086
B	23716.45	34993.9	10185036
C	49515.58	88026.1	1210151

The predicted defect size are higher in comparison with measured maximum defect size on fracture surface of specimens, but there is not possibilities to measure pores which are on fatigue fracture. The specimens were tested under the same stress amplitude but have other properties (Table 3). The specimen A had smallest defect size (predicted and measured) compared to other specimens and reach the highest number of cycles and was not broken. Specimen B had middle size of porosity and reach higher number of cycles in comparison to sample C, but was broken (did not reach 2 x 10^7 cycles). Specimen C had highest size of porosity comparison to A and B specimens and reach the smallest number of cycles up to failure of specimens under stress amplitude. These results shows that porosity have great influence to fatigue properties of materials.

Conclusions

The present study confirmed that secondary material have comparable mechanical properties as the some primary material. The gravity die casting into the metallic mould caused formation of porosity which have influence on the fatigue properties of experimental material.

Quantitative methods show that samples whose were not broken at the same stress amplitude have smaller amount and size of pores comparing to specimens which were broken.

The Murakami's method confirmed the presence of largest pores in the material which were broken, and using this method seems to be useful for prediction of the largest or limit pore size and how its size change (increase or on the other hand decrease – like in this study) fatigue lifetime. But of course it has some limitation. Calculated pore size is a little bit higher that can be observed in real materials and it validate the fact: the greater amount of pores the lower properties (UTS, hardness and fatigue properties as well) compared to materials with smaller amount of pores.

This work has been supported by The Scientific Grant Agency of the Ministry of Education of the Slovak Republic N° 1/0533/15, N° 049ŽU – 4/ 2017 and EU project ITMS 26220220154.

References

1. H.N. Girisha, K.V. Sharma, *Influence Of Ageing Heat Treatment And Magnesium On Wear And Corrosion Characteristics Of Aluminium Copper Alloy*. International Journal of Engineering Research & Technology **2** (8), 1702-1706 (2013)
2. S.K. Shaha, F. Czerwinski, W. Kasprzak, J. Friedman, D.L. Chen, *Ageing characteristics and high-temperature tensile properties of Al–Si–Cu–Mg alloys with micro-additions of Cr, Ti, V and Zr*. Materials Science & Engineering **A652**, 353-364 (2016)
3. H.R. Ammar, A.M. Samuel, F.H. Samuel, *Porosity and the fatigue behavior of hypoeutectic and hypereutectic aluminum–silicon casting alloys*. International Journal of Fatigue **30**, 1024-1035 (2008)
4. P. Pucher, H. Böttcher, H. Kaufmann, H. Antrekowitsch, P.J. Uggowitzer, *Mechanical properties and casting characteristics of the secondary aluminium alloy AlSi9Cu3(Fe) (A226)*. The Minerals, Metals and Materials Society **2**, 237-244 (2011)
5. L. Palyga, M. Stachowicz, K. Granat, *Influence of High-presure-die-casting second stage parameter on structure of AlSi9Cu3(Fe) alloy*. Manufacturing technology **16** (2), 410-416 (2016)
6. C. Lee, J. Youn, Y. Lee, Y. Kim, *Effect of eutectic Si particles on the defect susceptibility of tensile properties to microporosity variation in Al-xSi binary alloys*. Materials Science & Engineering **A 678,** 227-234 (2016)
7. K. Bangyikhan, *Effects of oxide film, Fe-rich phase, porosity and their interactions on tensile properties of cast Al-Si-Mg alloys*. PhD. Thesis, 2015.
8. G. Nicoletto, R. Konečna, P. Baicchi, V. Majerova, *Casting porosity and long-life fatigue strength of a cast Al-alloy*. In: Procs. Int. Conf. MSMF 5, Brno, Czech Republic; 2007.
9. J.A. Eady, D.M. Smith, *The effect of porosity on the tensile properties of aluminium alloy castings*. Mat. Foum **9** (4), 217-223 (1986)
10. E. Fiorese, F. Bonollo, G. Timelli, L. Arnberg, E. Gariboldi, *New classification of defects and imperfections for aluminium alloy castings*. International Journal of Metalcasting **9** (1), 55-66 (2015)
11. F. Bonolo, G. Timlli, E. Fiorese, E. Gariboldi, P Parona, L. Arnberg, *New Quality and design standards fo aluminium alloys cast products - Database on Defects*, (2013)
12. D.P.K. Singh, D.J. Mitchell, *An Improved Technique for Minimising Turbulence and Particle Stimulation During Molten Metal Delivery into Furnaces,* Technical Report RDD-01-06, Ford Motor Company (2001)
13. P. Palček, M. Chalupová, G. Nicoletto, O. Bokůvka, *Prediction of machine element durability, Education Aid for multimedia lectures*. CETRA 177 (2003)
14. O. Bokůvka, M. Jambor, F. Nový, L. Trško, J. Lago, *Influence of long-term heating on th fatigue properties of low-alloyed cast steel STN 42 2707*, Production Engineering archives **12** (3), 30-33 (2016)

15. STN EN ISO 6892-1:2010, Metallic materials. Tensile testing. Part 1: Method of test at ambient temperature.
16. STN EN ISO 6506-1, Metallic materials. Brinell hardnes measurements. Part 1: Method of test at ambient temperature.
17. P. Skočovský, P. Benko, A. Vaško, *Quantitative appreciation of structure and fracture surface of casting materials.* Proccedings of Advanced manufacturing and repair technologies in vehicle industry, Zielona Góra - Łagów, Poland, 51-55 (2000)
18. Y. Murakami, *Inclusion Rating by Statistics of Extreme Values and Its Application to Fatigue Strength Prediction and Quality Control of Materials.* Journal of Research of the National Institute of Standards and Technology **99** (4), 345-351 (1994)
19. S. Fintová, R. Konečná, G. Nicoletto, *Statistical description of largest pore size in modified Al-Si alloys.* Materials Engineering **16** (3), 24-28 (2009)

Study of new rubber to steel adhesive systems based on Co(II) and Cu(II) sulphides coats

Ivan Labaj[1,*], *Darina* Ondrušová[1], *Andrej* Dubec[1], *Mariana* Pajtášová[1], *Marcel* Kohutiar[1], *Beata* Pecušová[1]

[1]Alexander Dubček University of Trenčín, Faculty of Industrial Technologies in Púchov, I. Krasku 941/30, 020 01 Púchov, Slovak Republic

Abstract. The presented paper deals with the preparation of new rubber to steel adhesive systems using the steel surface treatment with deposition of adhesive coats based on Co(II) and Cu(II) sulphides. Efficiency of new prepared adhesive systems containing Co(II) and Cu(II) sulphides has been compared with the efficiency of double layer adhesive system commonly used in industry. The chemical composition of prepared adhesive systems was determined using the EDX analysis. Scanning Electron Microscopy (SEM) was used for study of topography and microstructure of prepared rubber to steel adhesive systems (Co(II), Cu(II) sulphide, double layer adhesive system). For determination of adhesion strength between rubber blends and metal pieces with various adhesive systems deposited on these pieces, the test according to ASTM D429 standard relating to Rubber to metal adhesion, method A was used. For all test samples, the same type of rubber blend and the same curing conditions have been used.

Keywords: Rubber blend, adhesive systems, cobalt sulphide, copper sulphide, double adhesive system

1 Introduction

In many applications of rubber blends, the reinforcing materials is used for achievement of the required properties. Adhesive systems have important function in the interface between two main materials, such as rubber blend and steel reinforcement material. There are many types of adhesive systems (e.g. adhesive systems mixed into rubber blends, spray application adhesive systems and two components adhesive systems), the effectivity of which is influenced by many factors. Effectivity of adhesive systems can be defined by strength of joint, which is expressed as force (N) on the surface of the joint (mm^2). The other factors, that affect effectivity of adhesive systems are: composition of rubber blend, type of reinforcing material and surface treatment of reinforcing material. The gradual development of reinforcing materials leads to development and innovation of the adhesive systems. Adhesive systems have application in many sectors of industry, where there is a requirement to ensure bonding of rubber blends and reinforcing materials, such as steel, fibers or textile. The automobile industry represent one of the largest consumers of

* Corresponding author: ivan.labaj@fpt.tnuni.sk

Reviewers: *Andrej* Czán, *Eva* Tillová

adhesive systems, because these systems are used for tyres, silenblocks and sealings. The life time as well as safety of final product can be ensured the hight strength of joint.

2 Experimental

2.1 Materials and methods

2.1.1 Preparation of Co(II) and Cu(II) sulphides coats

Co(II) and Cu(II) sulphides coats have been prepared using the adsorption procedure involving treatment of the working electrode representing the tested metal piece in a solution containing heptahydrate of cobalt sulphate (high purity, purchased from Kemcore), hydroxylamine sulphate, and ammonia (pH 12) for Co(II) sulphide coat and pentahydrate of copper sulphate, hydroxylamine and ammonia(pH 12) for Cu(II) sulphide coat, respectively. After preparation of coats, the reduction to sulphides in the solution of sodium sulphide was carried out and then the samples were washed with distilled water [1, 2].

2.1.2 Preparation of samples for adhesion test

Steel test samples have been prepared according to ASTM D 429 method A (ASTM D429 - Rubber to Metal Adhesion Test Equipment). For preparation of metal parts of test samples, the common type of steel was used. The test samples preparation procedure is in Table 1.

Table 1. Sample number and following preparation procedure

Sample number	Test samples preparation procedure
Sample 1	sample without surface treatment, only degreasing in KOH
Sample 2	surfaces of sample treated only with sandblasting
Sample 3	surfaces of sample treated only with phosphating
Sample 4	sandblasting with following application of commonly used double adhesive system (this process is commonly used in industry)
Sample 5	phosphating with following application of commonly used double adhesive system (this process is commonly used in industry)
Sample 6	sandblasting with following preparation of Co(II) sulphide coat
Sample 7	sandblasting with following preparation of Cu(II) sulphide coat
Sample 8	phosphating with following preparation of Co(II) sulphide coat
Sample 9	phosphating with following preparation of Cu(II) sulphide coat
Sample 10	degreasing in KOH solution, etching in HCl, with following preparation of Co(II) sulphide coat
Sample 11	degreasing in KOH solution, etching in HCl, with following preparation of Cu(II) sulphide coat

All of the mentioned steel test samples (Fig. 1) were used for the preparation of samples for adhesion test, where these steel samples were combined with the industrially used rubber blend of the same composition. Samples for adhesion test were cured in the curing press at the temperature of 160 °C during 6 minutes. After the curing procedure, the samples were conditioned for 16 hours. After the condition, all samples were tested with HOUNSFIELD H20K-W tensile testing machine.

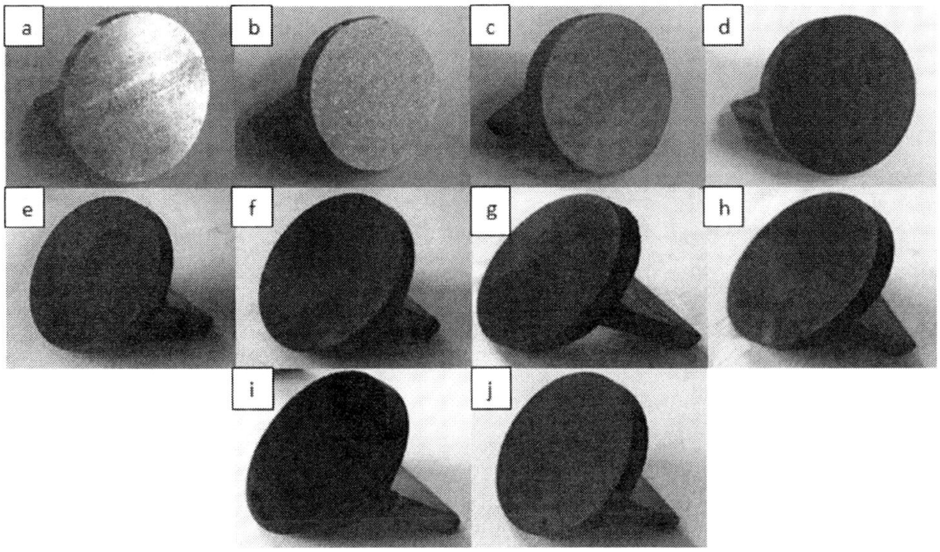

Fig. 1. Steel test sample after preparation procedure : a) Sample 1, b) Sample 2, c) Sample 3, d) Sample 4 and sample 5 – the same image of surface after preparation procedure, e) Sample 6, f) Sample 7, g) Sample 8, h) Sample 9, i) Sample 10, j) Sample 11

2.2 Testing and analysis

EDX analysis

The chemical composition of prepared Co(II) and Cu(II) sulphides coats and chemical composition of basic steel material was evaluated using Shimadzu EDX-7000 Energy Dispersive X-ray Fluorescence Spectrometer.

Microstructural SEM analysis

Microstructure and thickness of Co(II) and Cu(II) coats were studied using TESCAN VEGA 3 scanning electron microscope. The microanalysis was focused on the evaluation of topography and microrelief of the sulphide layers using detector operating in the secondary electrons mode in a high vacuum.

Adhesion test

Adhesion test of prepared test samples was performed using the HOUNSFIELD H20K-W tensile testing machine. Adhesive strength was calculated using the equation [3]:

$$R_A = \frac{F_A}{A_A} \tag{1}$$

R_A = adhesion strength ($N.mm^{-2}$)
F_A = maximum force (N)
A_A = surface of sample (mm^2)

3 Results and discussion

In results and discussion, there are summarized the results of : EDX analysis, tensile test of samples with adhesive strength, study of topography and microrelief of prepared sulphides coats and double adhesive system.

3.1 Energy Dispersive X-Ray Analysis (EDX)

Prepared Co(II) and Cu(II) sulphides coats and basic material were subjected to the EDX analysis for demonstration of chemical composition. In Table 2., there are the chemical composition of basic material (steel) and the chemical compositions of the prepared sulphides coats. The common type of steel was used for preparation of metal parts of test samples (basic material).

Table 2. Chemical composition of Co(II) and Cu(II) sulphide coat and basic material

Elements / Sample	Fe	Mn	Cr	Cu	Co	S
Basic material	98.301	0.914	0.182	0.245	-	-
Basic material + Cu(II) sulphide coat	50.192	0.462	0.201	33.254	-	15.686
Basic material + Co(II) sulphide coat	79.541	0.696	0.304	0.723	3.253	15.332

In both cases, sulphide coats were too thin, and therefore EDX analysis machine measured also the basic material together with coats. Thickness of sulphides coats was measured with scanning electron microscope.

3.2 Microstructural SEM analysis

In the Figs. 2-7. there are SEM images of topography and microrelief of double adhesive system, Co(II) and Cu(II) sulphides coats.

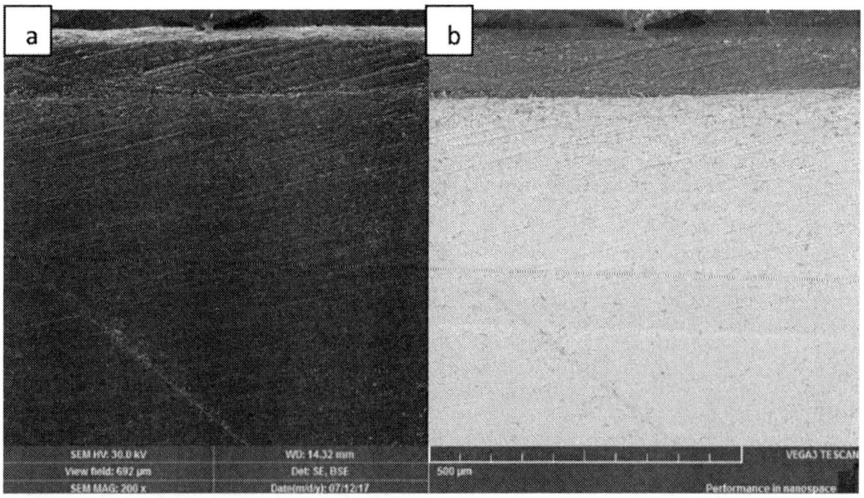

Fig. 2. SEM images of double adhesive system a) SE detection, b) BSE detection

In figure 2, it can be seen the structure of double adhesive system. In the figure 2b, it can be seen the interface between double adhesive system and basic material. It is interesting that in the figure 2b (BSE detection of double adhesive system, which is compound of two layers), the interface between two layers in double adhesive system is not visible. From this finding it possible to assume that the layers in double adhesive systems have similar composition. Lighter area in the figure 2a (SE detection) represents only the surface of sample, because the sample was not placed perpendicularly in the holder.

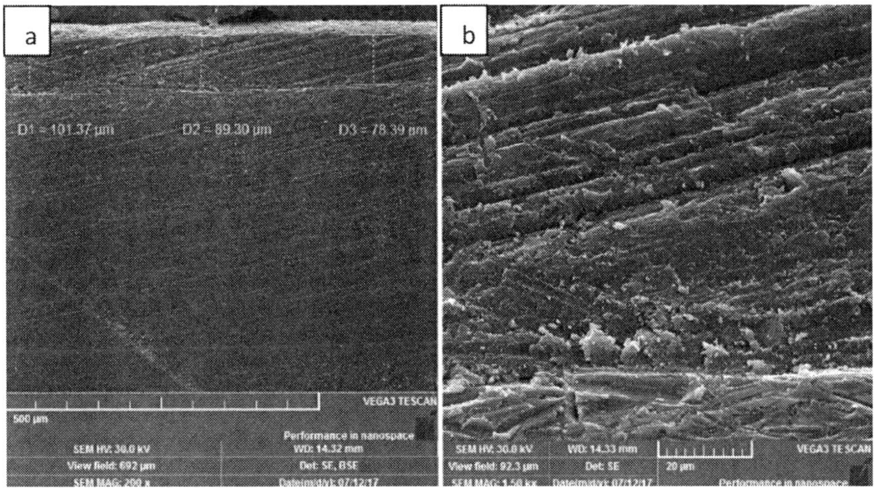

Fig. 3. SEM images of double adhesive system a) thickness measurement, b) detail of structure of double adhesive system

From the figure 3, it is visible that thickness of double adhesive system was uniform in the interval from 78 to 102 µm. Even the detailed view of double adhesive system did not reveal the interface between two individual layers of double adhesive system.

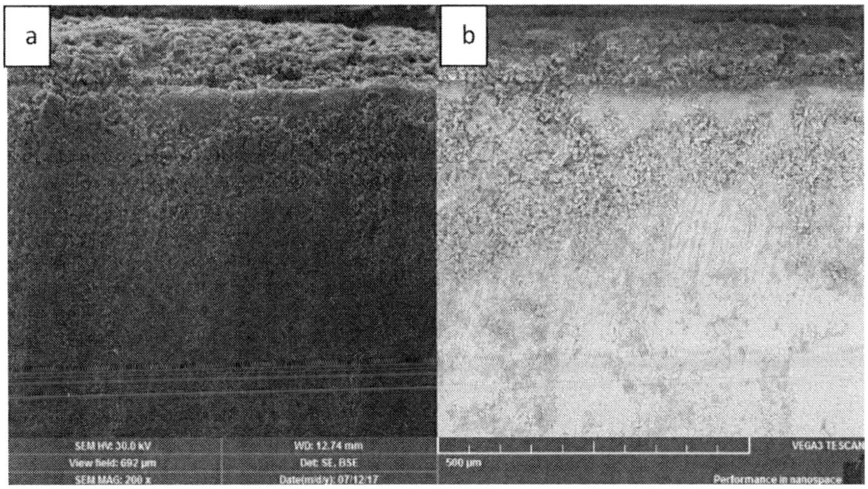

Fig. 4. SEM images of Co(II) sulphide coat a) SE detection, b) BSE detection

In figure 4, the structure of Co(II) sulphide coat can be seen. In figure 4b, the interface between Co(II) sulphide coat and basic material can be seen. In figure 4a, the pores, which are in coat of Co(II) sulphide, are visible.

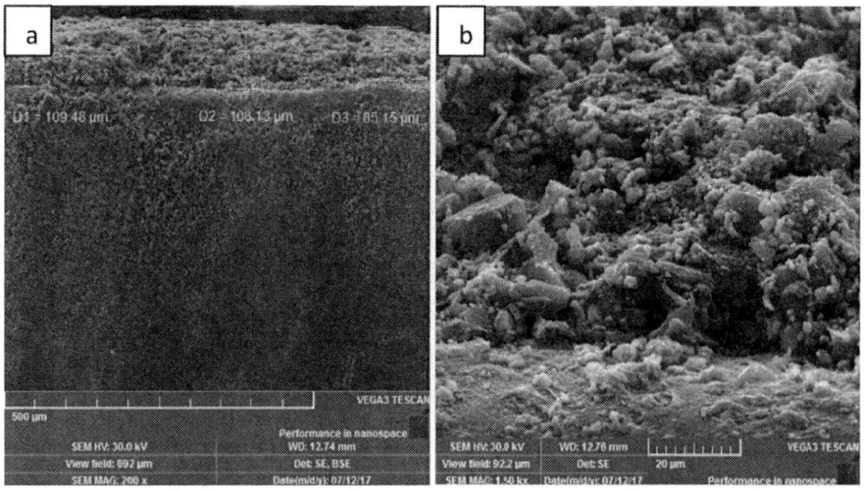

Fig. 5. SEM images of Co(II) sulphide coat a) thickness measurement, b) detail of structure of Co(II) sulphide coat

Prepared Co(II) sulphide coat is uniform and thickness of coat was in interval from 85 to 110 μm. In figure 5b, crystals in the coat can be seen. Detailed view (Fig. 5b) was compared with the images of cobalt sulphides from scientific paper [4]. From comparison of images, it is possible to consider that crystals in figure 5b are Co(II) sulphide crystals. In detailed view of Co(II) sulphide coat (Fig. 5b), there are the pores visible in coat (dark areas), and these pores can be considered as beneficial feature for adhesion strength of this coat.

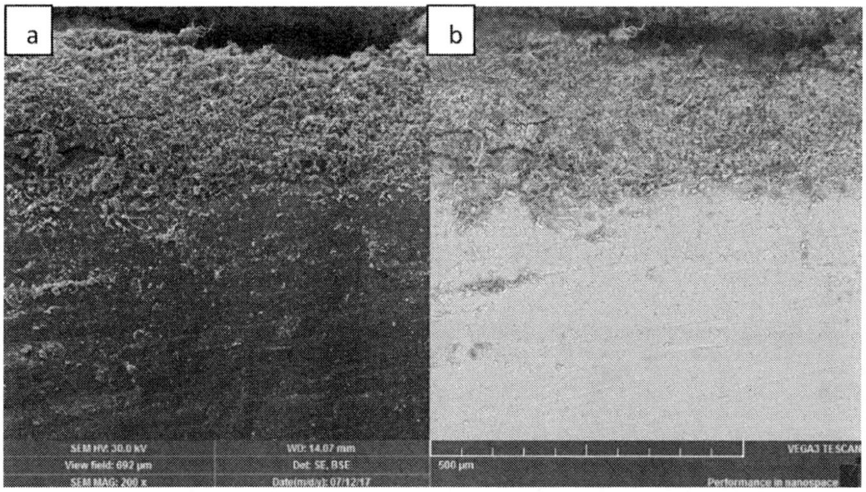

Fig. 6. SEM images of Cu(II) sulphide coat a) SE detection, b) BSE detection

In figure 6, the structure of prepared Cu(II) sulphide coat can be seen. In figure 6b, there are the visible cracks in the coat, which can lead to brittleness of prepared coat. In figure 6a, there is also the visible crack in interface area between Cu(II) sulphide coat and basic material.

Fig. 7. SEM images of Cu(II) sulphide coat a) thickness measurement, b) detail of structure of Cu(II) sulphide coat

Prepared Cu(II) sulphide coat had greater thickness in comparison with double adhesive system and Co(II) sulphide coat. The figure 7a represents measurement of Cu(II) sulphide coat. Thickness of prepared coat was in interval from 195 to 207 μm. The detailed view (Fig. 7b) revealed the more visible pores in Cu(II) sulphide coat in comparison with Co(II) sulphide coat. This phenomenon can be beneficial feature for adhesion strength of this coat.

3.3 Adhesion test

In Table 3, there are records of the results (maximum force, adhesion strength) of adhesion test of prepared samples. These results are also graphically depicted in Fig. 8 and Fig. 9. During the test of adhesive strength, two cases of breakage may occur. There can be either breakage in interface between rubber blend and basic metal part or breakage in rubber blend (it was seen in the case of sample 4 and 5) and it means that adhesive strength is higher that strength of rubber blend.

Table 3. Values of maximum force and adhesive strength of samples

Sample number	Maximum force [N]	Adhesion strength [N.mm^2]
Sample 1	268.67	0.214
Sample 2	915.33	0.728
Sample 3	1590.00	1.265
Sample 4	9286.00	7.390
Sample 5	6553.33	5.215
Sample 6	2023.33	0.178
Sample 7	2107.33	1.677
Sample 8	1648.67	1.312
Sample 9	1580.67	1.258
Sample 10	776.00	0.618
Sample 11	2091.33	1.664

According to the obtained results of adhesion test, it can be concluded that maximum force and thus maximum adhesive strength were seen for the sample 4, the preparation of which was based on sandblasting with following application of commonly used double adhesive system.

The second highest value of maximum force and adhesive strength was represented by the sample 5, which was prepared by phosphating with following application of commonly used double adhesive system. These two samples present the commonly used process in industry. In contrast, the minimal values of measured characteristics were seen for Sample 1, which was only degreased in KOH. Adhesive strength of Sample 1 had only mechanical character. Mechanical character of adhesive strength is based on mechanical interlocking. From results of adhesion test, it can be concluded, that surface treatment can influence the value of adhesive strength and it was confirmed by the results of adhesive strength for sample 2 and 3. Sample 2 and 3 had higher values of adhesive strength than Sample 1. Increasing of adhesive strength may be caused by greater amount of pores and cavities after surface treatment, because there is the increase in amount of mechanical interlocking. Influence of surface treatment (sandblasting, phosphating) was reflected in Samples with new adhesive systems based on Co(II) and Cu(II) sulphides. In comparison with the other samples with Cu(II) sulphide coat, the highest value of adhesive strength was reached by sample 7. The sample 6 had the highest value of adhesive strength in comparison with the other samples with Co(II) sulphide coat. Before preparation of new adhesive system, surfaces of samples 6 and 7 were treated with sandblasting. The mutual comparison of the samples with new adhesive systems revealed the lowest values of adhesive strength in relation to sample 9 (phosphating with following preparation of Cu(II) sulphide coat)as well as sample 10, which is based on Co(II) sulphide coat (degreasing in KOH solution, etching in HCl, with following preparation of Co(II) sulphide coat).

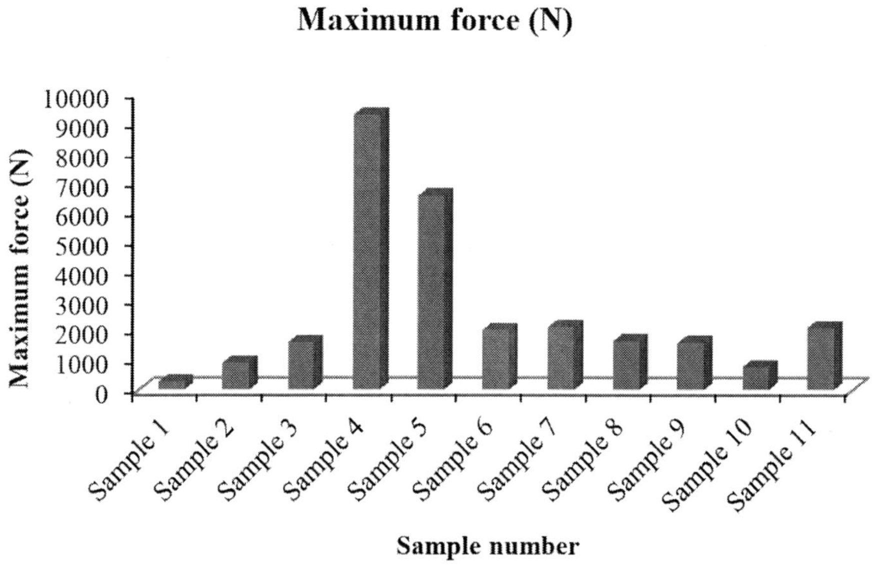

Fig. 8. Graphical representation of values of maximum force [N]

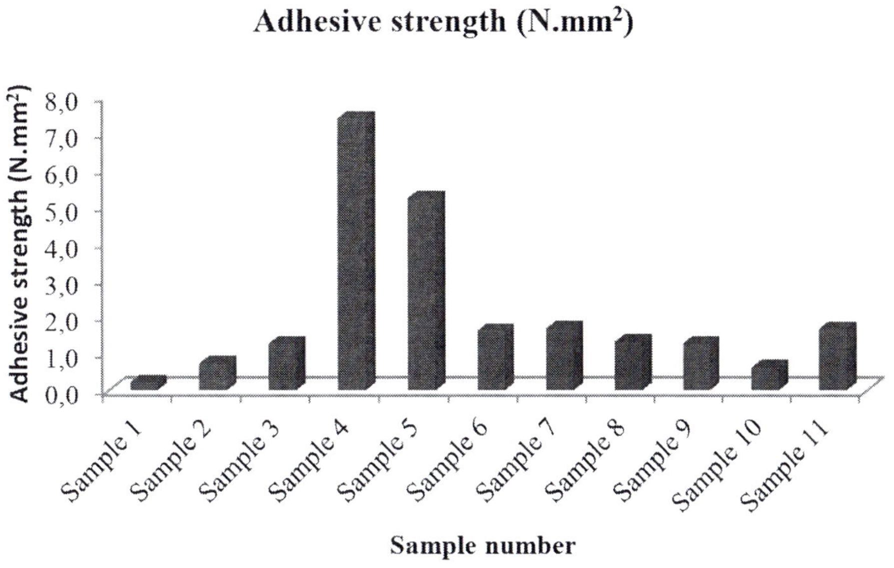

Fig. 9. Graphical representation of adhesion strength (N.mm^2)

Conclusion

Results of EXD analysis have confirmed the presence of Co, Cu and S elements contained in Co(II) sulphide coat and Cu(II) sulphide coat. SEM microstructural analysis was used for measurement of thickness of all adhesive systems and structure of adhesive systems. According to the obtained results of adhesion test, it can be concluded that maximum force and thus maximum adhesive strength were revealed in the case of the Sample 4, which was prepared by sandblasting with following application of commonly used double adhesive system. The second highest value of maximum force and adhesive strength was seen for sample 5, which was prepared by phosphating with following application of commonly used double adhesive system. From the Samples with new adhesive systems, Sample 7 with adhesive system based on Cu(II) sulphide coat exhibited the highest value of adhesive strength. New adhesive systems based on Co(II) and Cu(II) sulphide coats reached lower values than double adhesive system, which is commonly used in industry. Based on the obtained results and following analysis, it can be summarised that the surface treatment had influence on adhesive strength. Surface treatment had also influence on adhesive strength of all adhesive systems.

VEGA No. 1/0589/17, VEGA No. 1/0649/17, KEGA 007TnUAD-4/2017, project "Center forqualitytesting and diagnostics of materials", ITMS code 26210120046, Operational Program Research and Development, EF of Reg. Development

References

1. G. Valiuliene, A. Želiene, V. Jasulaitiene, I. Movzginskiene, *Composition of the Cobalt Sulfide Coating after Its Electrochemical Reduction*. Russian Journal of Applied Chemistry **76** (1), 71-75 (2003)
2. N. Petrašauskienė, R. Stokienė, S. Žalenkienė, V. Janickis, *Formation of cobalt sulfide layers on polyamide 6 by sorption–diffusion method using solutions of dodecathionic acid, $H_2S_{12}O_6$*. Chemija **26** (2), 93-97 (2015)

3. J. Škývara, *Hodnocení kvality spoje kov-pryž*. (Univerzita Tomáše Bati ve Zlíně, Fakulta technologická, 106, 2016)
4. H. Emadi, M. Salavati-Niasari, F. Davar, *Synthesis and characterization of cobalt sulfide nanocrystals in the presence of thioglycolic acid via a simple hydrothermal method*. Polyhedron **31**, 438-442 (2012)
5. ASTM D429 - *Rubber to Metal Adhesion Test Equipment. Test Resources*. [online] http://www.testresources.net/applications/standards/astm/astm-d429rubber-to-metal-adhesion-tests
6. I. Labaj, D. Ondrušová, A. Dubec, M. Pajtášová, M. Kohutiar, *Preparation and study of new rubber to steel adhesive systems*. Priemyselná toxikológia 2017, 37. Medzinárodné vedecké sympózium, Bratislava, Slovenská technická univerzita, 126-131 (2017)

Optical attributes influence on mechanical properties of laser cutted PMMA

Jozef Meško[1], *Rastislav* Nigrovič[1,*]

[1]UNIZA, Faculty of Mechanical Engineering, Department of Technological Engineering, Univerzitna 8215/1, 010 26 Žilina, Slovakia

Abstract. The article describes and characterizes the technology and principle of laser cutting PMMA material. Experimental samples with thickness t = 3 mm of PMMA were made with a constant optimized technological parameters of laser cutting. In the experimental part are compared values of hardness of cutted surface, tensile strength of different colour variations and impact on selected mechanical properties by change of technological procedures of selected material processing - tempering of plastics.

Keywords: PMMA, Laser cutting, Mechanical properties, Optical attributes of PMMA

1 Introduction

The principle of the laser cutting of metallic or non-metallic materials is based on the action of a focused laser beam on the material being cut. When cutting materials in technical practice, the surface is impacted by a focused circular beam of a diameter 0.1 to 0.4 mm, depending on the device construction and the thickness of the material to be cut. The laser beam of the above parameters, impacting the material being cut, will cause its rapid heating up. The material being cut is heated in milliseconds up to the melting or to the evaporation temperature.

When the laser beam hits the material being cut, there occurs an interaction between the material being cut and the laser beam. The subsequent processes, which take place during the cutting of the material, and the effect on the material properties following the focused beam impact, mainly depend on the chemical composition of the material being cut and on the quality of its surface, as well.

In thermal cutting of metallic or non-metallic materials by a laser, it is always necessary first to create a hole in the processed material, from which the cutting would continue. Making the hole is based on the laser drilling principle and has slightly different characteristics of the course than that of the actual cutting. The impacting laser beam transmits the kinetic energy of the material photons, which is converted into heat that melts and partially evaporates the material being heated.

---

* Corresponding author: rastislav.nigrovic@fstroj.uniza.sk

Reviewers: *Andrej* Czán, *Ján* Vavro

The surroundings of the laser beam inflammation site (in the material being cut) contain gases that become ionized instantaneously as the beam hits the material and change to plasma.

The material being cut sublimates into a gaseous state, and is blown away from the material to the environment by the action of an assist gas under a relatively high pressure. The part of the material, which is not converted into a gaseous state, is blown by the flowing assist gas in the liquid form. The above-mentioned process causes formation of a pit in the material being cut, and the laser beam can penetrate continuously deeper, which results in depth melting of the material (Fig. 1).

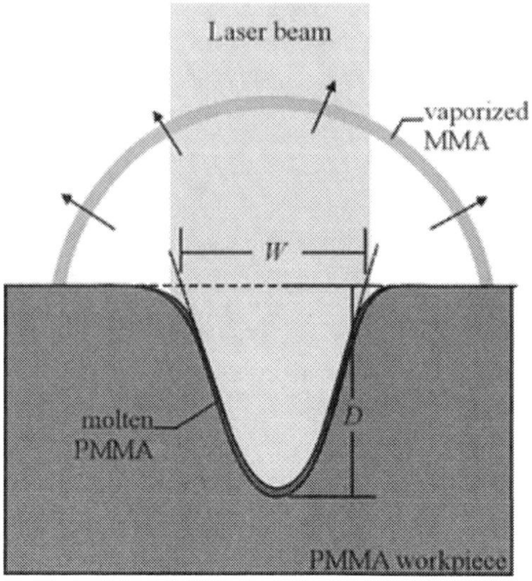

Fig. 1. The principle of PMMA laser cutting

CO_2 lasers are particularly well suited to the cutting of PMMA. Applications of thermoplastic PMMA have grown considerably in many cases, for example, lenses, light pipes, outdoor signs, light covers, bathroom fittings, skylights, meter covers, baths and toys (Fig. 2) [1-6].

Fig. 2. A demonstration of a laser workpiece from PMMA

2 Experimental conditions

2.1 Experimental material

General Poly(methacrylates) are polymers of the esters of methacrylic acids. The most commonly used among them is poly(methyl metacrylate) (PMMA). Poly(methyl metacrylate) or poly(methyl 2 – methylpropenoate) is the polymer of methyl methacrylate, with chemical formula C5H8O2. I tis a clear colourless polymer available on the market in both pellet and sheet form under the names Plexiglas, Acrylite, Perspex, Plazcryl, Acylplast, Altuglas, Lucite etc. I tis commonly called acrylic glass or simply acrylic. [5]

Table 1. Selected properties of PMMA

Density	1,15-1,19 [g.cm^{-3}]
Water absorption	0,3 – 2 [%]
Hardness, Rockwell M	63 – 97
Tensile Strength, Ultimate	47-49 [MPa]
Tensile Modulus	2.2 – 3.8 [MPa]
Specific Heat Capacity	1.46 – 1.47 J/g. °C]
Thermal Conductivity	0.19 – 0.24 [W/m.K]
Maximum Service Temperature , Air	41 – 103 [°C]
Melting Point	130 [°C]
Vicat Softening Point	47 – 117 [°C]

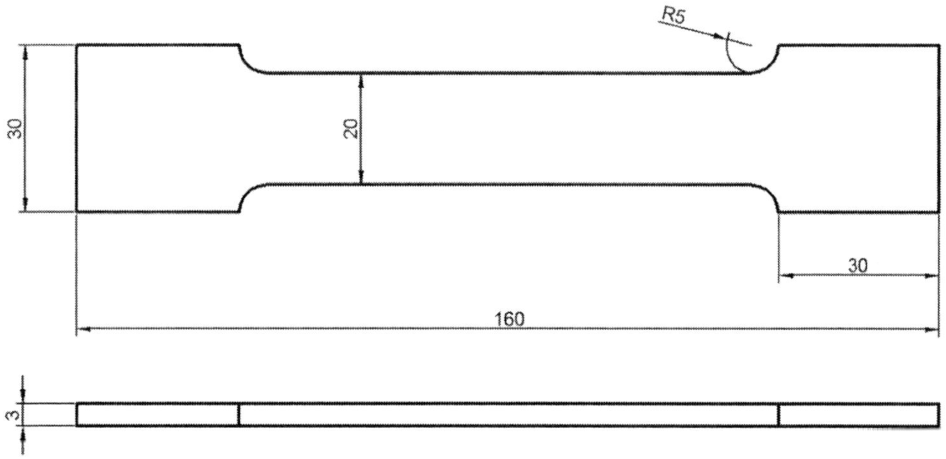

Fig. 3. Sketch of comparative samples for tensile test

2.2 Experimental machine

Laser cutting machine – XL 1200

With the laser cutting and engraving machine XL-1200 for acrylics eurolaser offers the possibility to process extremely broad materials on a comparatively small system. The processing area is 2270 mm x 1230 mm. With the optional camera recognition system which is also available for other systems the production flow can be automated, leading to an increased economy of the laser processing [7].

Cutting parameters :

- power – 100 [W]
- cutting speed – 13 [mm.s^{-1}]

All experimental samples were tempered for 5 hours at 80 °C before to mechanical tests, resulting in improved mechanical properties, due to higher confidence in measurement results and to avoid high brittleness of the measured samples.

3 Experimental part

The experiment should demonstrated the interaction between the optical properties of PMMA (Table 1.) material and mechanical properties after laser cutting. The material was made with constant conditions, but with other colour variation and light transmission (Fig. 6.). The laser for cutting of material uses a concentrated photon beam. The optical properties of the cut material greatly affecting the quality of the cut and the properties of the cut material. For the experiment were selected four colour variations of PMMA, with different light absorption (Fig. 4).

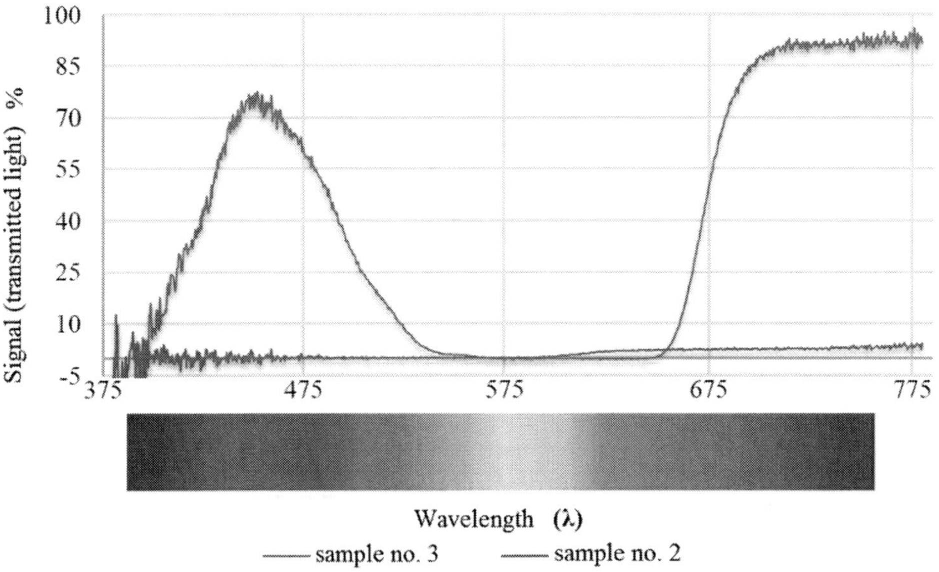

Fig. 4. Graphical representation of visible light transmission

For the experimental measurements were selected four types of samples:

- Samples no. 1 - black colour - 100% light absorption,
- Samples no. 2. - red colour - light absorption (Fig. 4),
- Samples no. 3. - blue colour - light absorption (Fig. 4),
- Samples no. 4. - white colour - 100% light absorption.

Each of these samples (Fig. 3.) was subjected to a tensile strength measurement (Fig. 7.) and measurement of hardness. For the optimization of the measurements, 10 samples of each colour variation were selected.

3.1 Measure of tensile strength

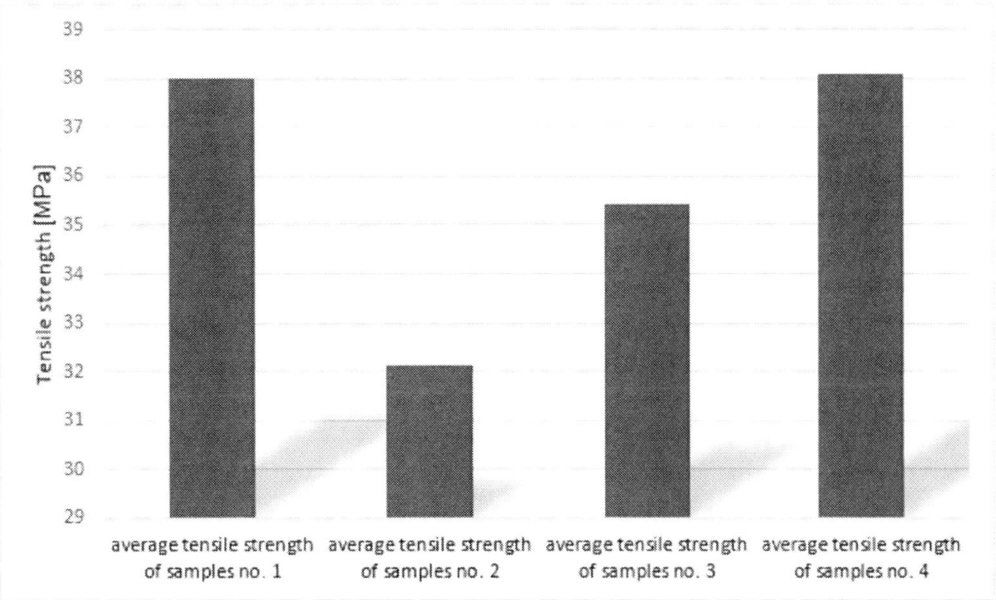

Fig. 5. The average values of tensile strength

From the measured values, it can be stated that samples with light transmission at a certain wavelength have lower tensile strengths. This fact may be attributed to the optical properties of PMMA when changing colour. Samples with 100% light absorption (no. 1 and no. 4) have an average tensile strength of 38.01 MPa, respectively. 38.09 MPa. Samples no. 2 and 3 achieve a tensile strength of 32.13 MPa, respectively. 35.44 MPa (Fig. 5).

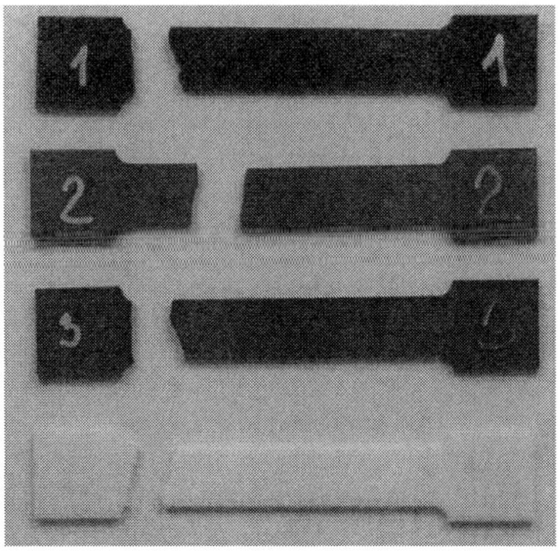

Fig. 6. Image of the used samples

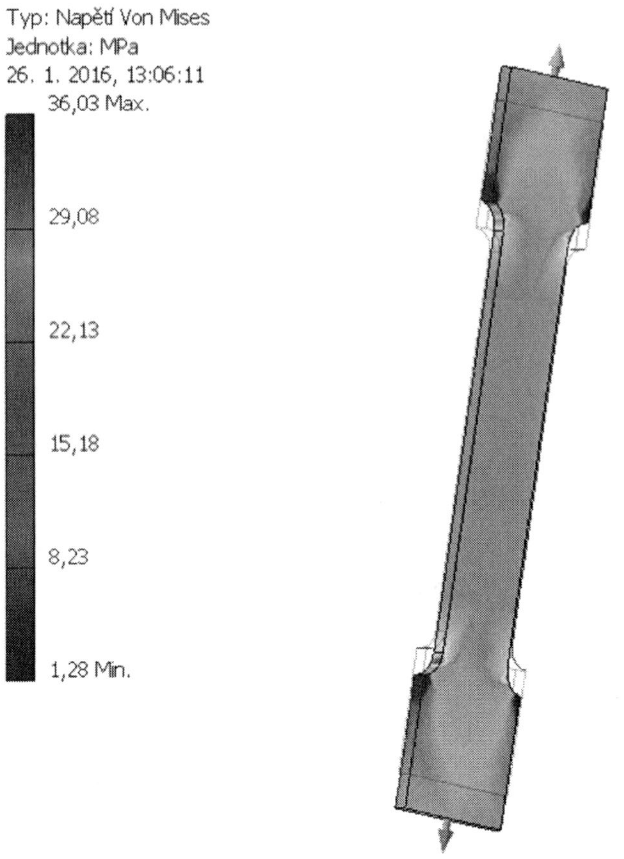

Fig. 7. Simulation tests of tensile strength

3.2 Measure of hardness

From the measured values, it can be conclude that samples with light transmission at a certain wavelength have higher hardness. Due to the higher hardness it is possible to verify lower tensile strength. This fact may be attributed to the optical properties of PMMA when changing colour. Samples with 100% light absorption (no. 1 and no. 4) have an average HB hardness of 27.6 HB, respectively 29.5 HB, which almost does not differ from the hardness of the base material. Samples no. 2 and 3 achieve a HB hardness of 32.7 HB, respectively 31.0 HB, which represents a relatively large increase over the hardness of the base material (Fig. 8).

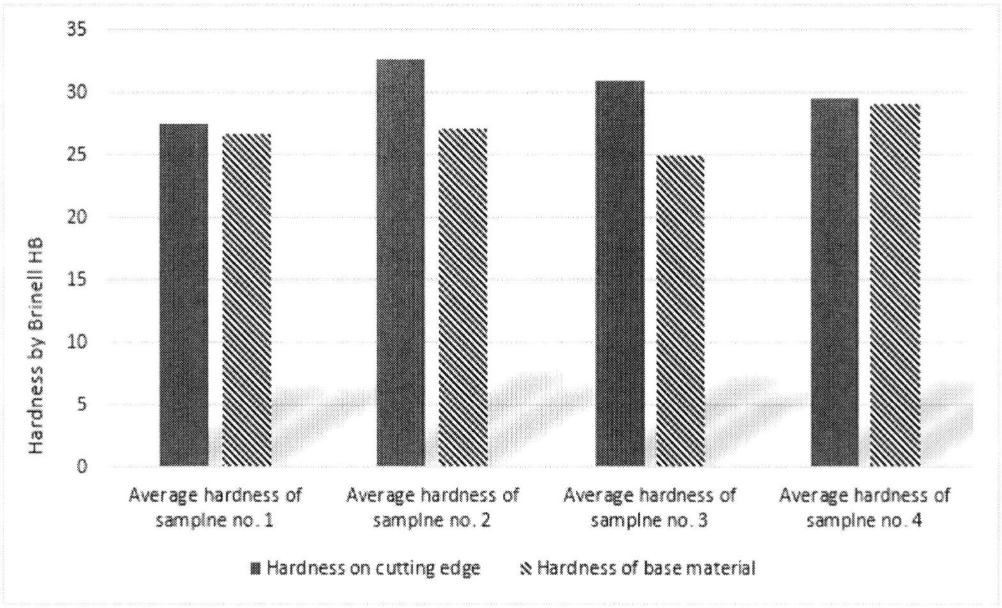

Fig. 8. The average values of hardness HB

Conclusion

The experiment unambiguously confirmed that changes in the optical properties of the same chemical composition material - PMMA, affect the mechanical properties (tensile strength and hardness) of the material after laser cutting.

In this experimental measurement were used the constant parameters of the laser cutting for each samples variant (while maintaining the dimensions and material of the samples). From the experiment it suggests that the laser cutting parameters for PMMA must be optimized for each colour option for maintaining the required mechanical properties. The concreate measurement results are presented in the experimental section of the article.

Research presented in this paper was partially financially supported through realization of projects KEGA 014ŽU-4/2016 - Responsible investigator: prof. Jozef Meško, MSc., Ph.D.

References

1. L. C. Caristan, *Laser cutting guide for manufacturing*. (Dearborn, Michigan, USA: Society of Manufacturing Engineers, 452 p., 2004)
2. W. T. Silvfast, *Laser Fundamentals*. (Cambridge University Press, 666 p., 2004)
3. J. Meško, A. Zrak, *Analysis of heat transfer in laser cutting problems*. Analysis of technology in various industries, Monography, Chapter **6**, Częstochowa: Oficyna Wydawnicza SMJiP, (2014)
4. H. Danielewski, R. Banak, A. Domagala, *The Experimental Analysis of Striation Pattern Created During Laser Cutting*. Transcom 2013, Žilina, Slovakia (2013)
5. D. W. Van Krevelen, *Properties of polymers*. ISBN-9780080548197 (2009)
6. Y. Huang, S. Liu, W. Yang, Ch. Yu, *Surface roughness analysis and improvement of PMMA-based microfluidic chip chambers by CO2 laser cutting*. Applied surface science **256**, 1675-1678

7. *EUROLASER* [online] http://www.eurolaser.com/cz/products/laser-systems-for-acrylic/xl-1200-acrylic/ (2015)
8. J. Meško, A. Zrak, K. Mulczyk, S. Tofil, *Microstructure analysis of welded joints after laser welding*. Manufacturing technology **14** (3), 355-359 (2014)

Effect of bentonite modified by silane on rubber blends properties

Zuzana Mičicová*,1, *Slavomíra* Božeková[1], *Mariana* Pajtášová[1], *Darina* Ondrušová[1]

[1]Alexander Dubček University of Trenčín, Faculty of Industrial Technologies in Púchov, I. Krasku 941/30, 020 01 Púchov, Slovak Republic

Abstract. The presented paper deals with the preparation of bentonite modified by silane and its application into polymer matrix. Natural bentonite was modified with 3-(Trimethoxysilyl) propyl-methacrylate in two different solutions. These two solutions had the same composition (ethanol, water and 3-(Trimethoxysilyl) propyl-methacrylate) but in one solution, pH was modified. These bentonites modified by silane were characterized by Fourier transform infrared spectroscopy. The modified bentonites were mixed into rubber matrix as partial replacement of commonly used filler – carbon black of the N339 type. The polymer blends were investigated from the aspect of the rheological and curing characteristics (minimum torque M_L, maximum torque M_H, optimum time of cure $t_{(c90)}$, processing safety of blend $t_{s,}$). Moreover, the quality of the silanization reactions was investigated. Measurements were done using PRPA 2000.

Keywords: bentonite, silanization, fillers, curing characteristics, silane

1 Introduction

Surface modification is a good way to improve the dispersive properties and increase the surface activity of filler. The most of particulate fillers are inorganic and their surfaces have poor compatibility with hydrocarbon polymers. Among other effects, this can result in processing problems, such as high viscosity, poor dispersion and poor mechanical properties [1, 2].

In order to obtain a positive effect of the surface treatment, it is necessary to select the suitable modifier for the treatment, with respect to the filler and the matrix. Organosilanes elong among the most frequently used surface modifiers [1, 3, 4].

In the material research of the polymeric materials, the current trend includes research of polymers in combination with clay minerals. The economic as well as environmental aspects are the reasons for the mentioned investigation of clay minerals with polymers [5, 6]. Clay fillers, in contrast to carbon black, have lower compatibility with the rubber matrix, and for this reason, the silane-binding agents are used to improve the interaction of the clay filler with the rubber chains [1, 7, 8].

---

* Corresponding author: zuzana.micic@gmail.com

Reviewers: *Andrej* Czán, *Eva* Tillová

Surface modification of clay fillers has attracted much attention because the obtained products exhibit properties suitable for many applications in material science and environmental engineering Modification is generally considered to be an attractive and effective method because due to this method, the clay particles (e.g. kaolin and montmorillonite) are more compatible with rubber matrix [9].

2 Experimental

2.1 Preparation of modified bentonite

The natural bentonite was used as initial material to be modified bentonite and this initial material was extracted from the locality called Jelsovy Potok. The chemical composition of natural bentonite is given in Tab. 1 while content of montmorillonite was 80 %. Before the surface modification, natural bentonite was dried in dryer at the temperature of 80 ± 5 °C to get constant weight. Then, this silane type was chosen to modify the bentonite surface: 3-(Trimethoxysilyl) propyl-methacrylate (TSPMA).

Table 1. Composition of natural bentonite in Weight %

Oxide	SiO_2	Al_2O_3	Fe_2O_3	TiO_2	CaO	MgO	K_2O	Na_2O
Weight %	59	20	0.3	0.2	2.2	4	0.7	0.4

The modification was carried out for two different solutions. The first solution consisted of ethanol-water mixture (70:30) with subsequent addition of silane (3 wt parts of modifying agent per 100 wt parts of filler) [9, 10]. The second solution had the same composition but its pH was treated with acetic acid. A bentonite sample was then added to both solutions and the two mixtures were stirred at 60 °C and 450 rpm for 2 hours. The resulting suspensions were filtered under reduced pressure and the prepared samples were dried in dryer at the temperature of 80 ± 5 °C. For further experiments, two prepared modified bentonites with a particle size < 25 μm were used.

2.2 Preparation of rubber blends

The five tread blends were prepared by two-step mixing in laboratory mixer of Brabender type with chamber volume of 80 cm^3 and mixing process involved 60 revolutions per minute. The proper order of the steps in relation to additives and cure system was performed as it was predetermined. The specification of tread blends is given in Table 2.

Table 2. Specification and ratio of fillers

blends	**Filler**	**ratio of fillers**
R	-	-
S	carbon black	1
SB	carbon black + unmodified bentonite	3:1
SBM	carbon black + modified bentonite by silane agent MPSTA	3:1
SBM +pH	carbon black + modified bentonite by silane agent MPSTA +pH	3:1

2.3 Characterization of the modified bentonite

The detailed characteristics of unmodified and modified bentonites were obtained by measurements based on spectral (FTIR) analysis. The infrared absorption spectra of investigated unmodified and modified bentonites were obtained with Tensor 27 FTIR spectrometer in the 4000-400 cm^{-1} spectral range with the resolution of 4 cm^{-1}. The KBr disk technique was used.

2.4 Characterization of rubber blends

The processing capacity of rubber blend can be predicted using their curing characteristics. The rheological properties and curing characteristics were investigated by help of PRPA 2000 at the temperature of 150 °C during 30 min. The quality of the silanization reactions was investigated by PRPA 2000 at temperature of 100 °C, frequency of 100 CPM and at strain in the range of (1-450) %.

3 Results and discussion

3.1 Modified bentonite

IR spectroscopy is a well-known method to identify the functional groups and to make quantitative analysis. The IR spectra can commonly provide useful information on the surface modification of the bentonite. The FTIR spectra of the unmodified and modified bentonites are shown in Figure 1 and 2. The main peaks from the observed FTIR spectra of the unmodified and modified bentonites are summarized in Table 3.

Table 3. Selected stretching vibrations (cm^{-1}) of the modified bentonite

Type of vibrations	SB	SBM	SBM +pH
ν (Al-OH)	3627	3629	3628
ν (H-O-H)	3454	3457	3452
ν_{as} (CH_3)	-	2958	2960
ν_{as} (CH_2)	-	2931	2924
ν_s (CH_3)	-	2895	2854
ν (C-O)	-	-	1126
ν (Si-O)	1056	1038	1035

In the case of the all samples, the band at 3627-3629 cm^{-1} corresponds to the stretching vibration of the hydroxyl groups are bonded with aluminium atoms of the clay mineral. The broad band at 3452-3457 cm^{-1} is assigned to hydroxyl stretching vibrations of the H–O–H groups, of the free and interlayer water molecules on montmorillonite clay, and the band at 1035-1056 cm^{-1} is associated with the stretching vibrations of the Si–O groups. The intensity of the absorption band decreased with surfactant modification, and it gives evidence about a lower hydrophilicity in the interlayer region. For SBM and SBM +pH samples, the bands at 2854-2960 cm^{-1} are assigned to the C–H stretching vibrations of aliphatic CH_2 and CH_3 groups, indicating the presence of the organic component (silane) [11, 12].

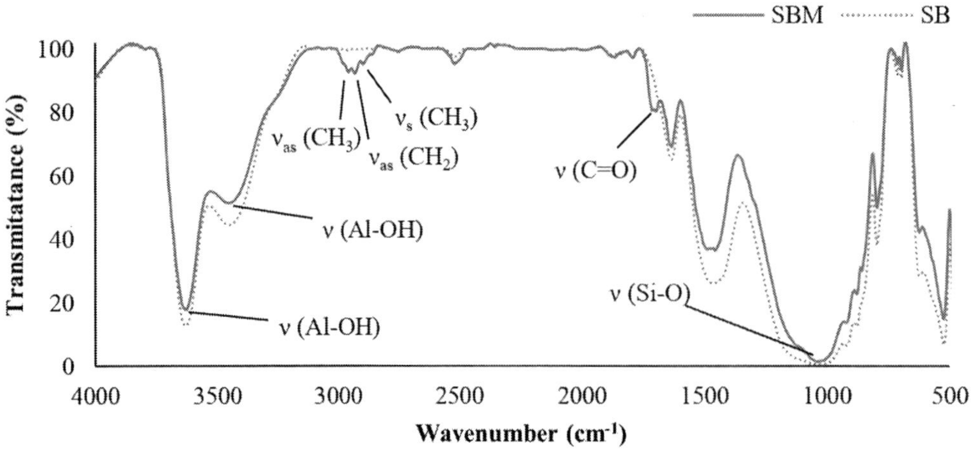

Fig. 1. IR spectra of sample SBM and SB

In case SBM +pH sample, the band at 1126 cm^{-1} is assigned to stretching vibrations of the C-O groups, indicating the presence of the organic component (silane) [11, 12].

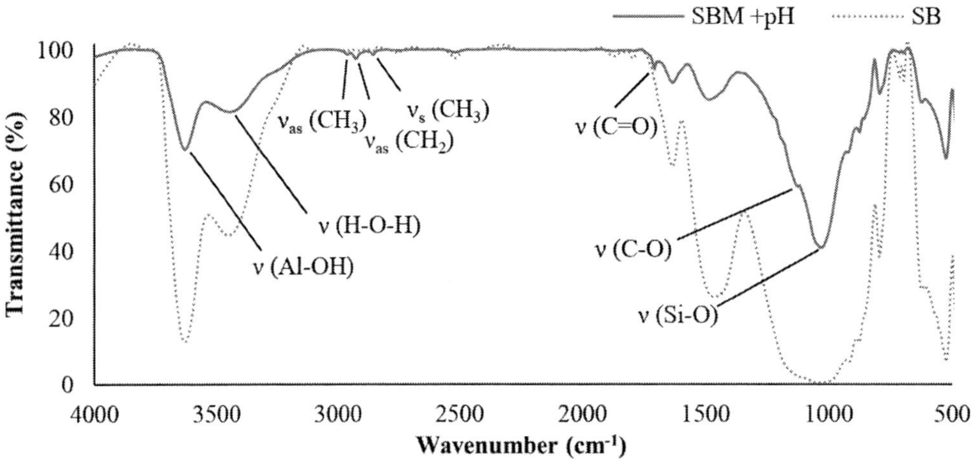

Fig. 2. IR spectra of sample SBM +pH and SB

3.2 Cure behaviour of the rubber blends

The effect of the modified filler on the course of sulphur curing process of tread blends was evaluated on the basis of curing characteristics. Based on the evaluated rheometric results of curing curves, the values of rheometric properties and curing characteristics are shown in Table 4. Graphical presentation of the curing curves is shown in Figure 3.

Table 4. Rheological and curing characteristics of rubber blends

blends	M_L (dN.m)	M_H (dN.m)	t_s (min)	$t_{c(90)}$ (min)
R	1.69	8.78	2.30	6.95
SB	1.47	6.88	3.55	12.43
SBM	1.36	6.94	3.64	11.78
SBM +pH	1.48	6.86	4.56	14.23

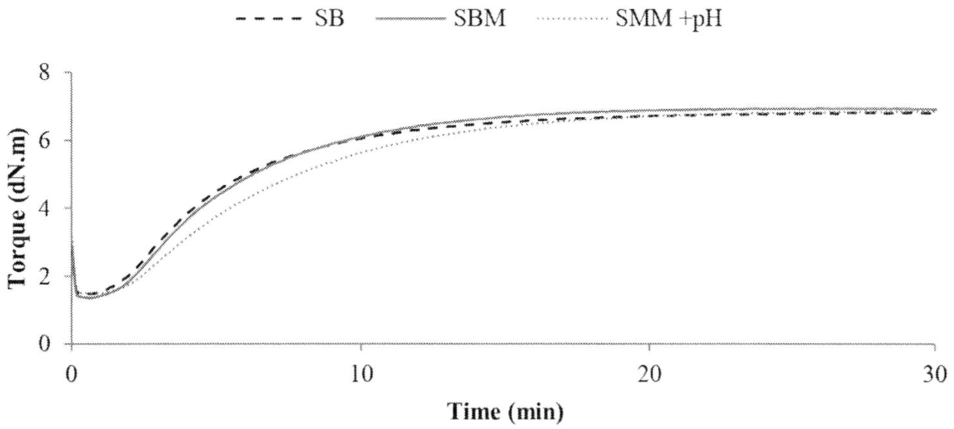

Fig. 3. Cure records of SB, SBM SBM +pH and R blends

The values of minimum torque and maximum torque for tread blend are lower in comparison with the reference blend values. The decrease of values indicates lower viscosity as well as lower stiffness of the tread blends at the beginning and end of the cure. From Table 4, it can be seen that the values of processing safety of tread blends are higher in comparison with the reference blend values. It can be assumed that the content of fillers in the rubber blends has a positive effect on processing safety, and it may affect the processing properties of the blends. In the case the optimum time, blends exhibited higher values in comparison with reference blend values [5].

3.3 The quality of the silanization reactions

The test was divided into two parts where the first part of the filler-filler interaction was determined at low strain (1%) and by this way, the size of the elastic moment (G´) was recorded. In the second part, there was the higher strain (450%) and using this strain, the size of the torque moment (S´) was determined (cohesion polymer network). The lower the filler-filler interaction, the better the dispersion of the filler used. This results in a higher efficiency of the silanization reaction. The torque value at high deformation angles indicates whether the effect of temperature in the silanization reaction could cause undesired curing of blend. The torque value should be as low as possible, otherwise there will be an increase in the degradation of the silane [13, 14]. The results are shown in Table 5 and depicted graphically in Figure 4.

Table 5. Values of G′ and S′ for determination of the silanization reaction quality

Blends	**G′ (1%)**	**S′ (450%)**
SBM	230.34	9.93
SBM +pH	247.76	11.15

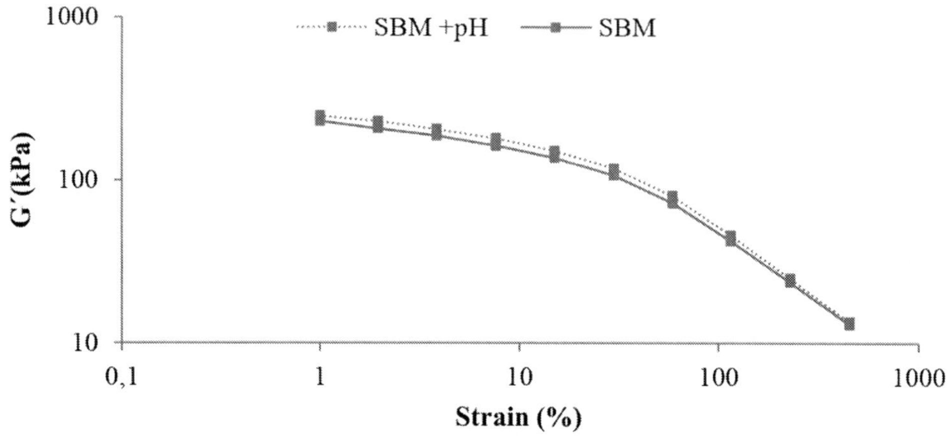

Fig. 4. The dependence of G′ on strain for SBM and SBM +pH blends

From the measured results, the lowest value of the elastic modulus, and thus the lowest efficiency of the silanization reaction, showed a SBM blend and it indicates that the filler is well dispersed in the polymer matrix.

Conclusions

This study was based on modification of bentonite which was used as partial replacement of the high reinforcing filler – carbon black of N339 type. The polymer blends with the modified bentonites were used as a partial carbon black replacement and they were investigated from the aspect of rheological properties and curing characteristics. In addition, the quality of silanization reaction was investigated. The determination of the quality of silanization reaction shows that there is the lowest value of the elastic modulus - G (1%) and thus there is the highest efficiency of the silanization reaction in relation to blend containing filler based on modified bentonite (SBM) and it leads to facts that the filler is well dispersed in the polymer matrix.

This work was supported by the Slovak Grant Agency KEGA 007TnUAD-4/207, VEGA 1/0589/17, VEGA 1/0649/17 and resulted from the project "Center for quality testing and diagnostics of materials", ITMS code 26210120046 relating to the Operational Program Research and Development funded from European Fund of Regional Development.

References

1. Z. Mičicová, S. Božeková, M. Pajtášová, D. Ondrušová, *Povrchová modifikácia bentonitu (3-aminopropyl)trietoxysilánom a jeho vplyv na reologické vlastnosti v polymérnej matrici*. Hutnícké listy **5**, 95-99 (2016)

2. P. Djomgoue, D. Njopwouo, *FT-IR Spectroscopy Applied for Surface Clays Characterization*. Journal of Surface Engineered Materials and Advanced Technology **3**, 275-282 (2013)
3. A.M. Motawie, M.M. Madany, A.Z. El-Dakrory H.M Osman, E.A. Ismail, M.M. Badr, D.A. El-Komy, D.E. Abulyazied, *Physico-chemical characteristics of nano-organo bentonite prepared using different organo-modifiers*. Egyptian Journal of Petroleum **23**, 331-338 (2014)
4. D.S. Kim, V. Dhand, K.Y. Rhee, S.J. Park, *Study on the Effect of Silanization and Improvement in the Tensile Behavior of Graphene-Chitosan-Composite*. Polymers **7**, 527-551 (2015)
5. M. Pajtášová, Z. Mičicová, D. Ondrušová, B. Pecušová, A. Feriancová, L. Raník, S. Domčeková, *Study of properties of fillers based on natural bentonite and their effect on the rubber compounds*. Procedia Engineering **177**, 470-475 (2017)
6. S. Palaty, R. Joseph, *Synergism of Xanthate/Dithiocarbamate Accelerator in Carbon Black Filled NR Compounds*. Iranian Polymer Journal **13**, 85-91 (2004)
7. S. Khaje, M. Jamshidi, *The Effect of Aging and Silanization on the Mechanical Properties of Fumed Silica-based Dental Com*posite. Journal of Dental Biomaterials **2**, 124-132 (2015)
8. A. Khoshniyat, A. Hashemi, A. Sharif, J. Aalaie, C. Duobis, *Effect of surface modification of bentonite nanoclay with polymers on its stability in an electrolyte solution*. Polymer Science Series B **54**, 61-72 (2012)
9. S. Božeková, D. Ondrušová, Z. Mičicová, M. Pajtášová, R. Moravčík, L. Buňová, *Effect of modified alternative filler on the properties of rubber compounds*. Procedia Engineering **136**, 245-250 (2016)
10. S. Božeková, Z. Mičicová, D. Ondrušová, M. Pajtášová, *Vplyv modifikovaného alternatívneho plniva silánom Bis3-(triethoxysilyl)-tetrasulphid na vlastnosti gumárenských zmesí*. Hutnické listy **5**, 48-52 (2016)
11. M.E. Parolo, G.R. Pettinari, T.B. Musso, M.P. Sánchez-Izquierdo, L.G. Fernández, *Characterization of organo-modified bentonite sorbents: The effect ofmodification conditions on adsorption performance*. Applied Surface Science **320**, 356-363 (2014)
12. C.H. Champos, B.F. Urbano, B.L. Rivas, *Synthesis and characterization of organic–inorganic hybrid composites from poly(acrylic acid)-[3-(trimethoxysilyl)propyl methacrylate]-Al_2O_3*. Composites, Part B, **57**, 1-7 (2014)
13. O. Wilczynski, *Vlivy odporu pneumatiky vs. Životní prostředí*. Plasty a kaučuk **46**, 68-73 (2009)
14. D. Ondrušová, S. Božeková, M. Pajtášová, Z. Mičicová, I. Labaj, *Modifikácia klinoptilotitu, jeho aplikácia do gumárenských zmesí a štúdium priebehu silanizačnej reakcie*. Hutnické listy **5**, 53-57 (2016)

Modification of alternative additives and their effect on the rubber properties

Darina Ondrušová[1,*], *Slavomíra* Božeková[1], *Lenka* Buňová[1], *Mariana* Pajtášová[1], *Ivan* Labaj[1], *Andrej* Dubec[1], *Juliána* Vršková[1]

[1]Alexander Dubček University of Trenčín, Faculty of Industrial Technologies in Púchov, I. Krasku 941/30, 020 01 Púchov, Slovakia; darina.ondrusova@fpt.tnuni.sk

Abstract. The present paper deals with a targeted modification of two kinds of alternative additives - waste from glass production and natural mineral filler and explores their effect on the properties of polymeric materials. In the function of first alternative filler was used sludge from weighing the ingredients of glass batch in the glass production. The second used was natural aluminosilicate material based on zeolite (clinoptilolite). These alternative fillers have been modified in order to increase its efficiency, using the silanes: 3-aminopropyl-triethoxysilane, bis(triethoxysilyl)propyl-tetrasulfide and 3-(triethoxysilyl)propyl-methacrylate. In the case of alternative filler based on zeolite the influence of silanization conditions on the filler efficiency have been also studied. Prepared modified fillers were mixed into rubber compounds as partial replacement of commonly used filler – carbon black. The influence of prepared fillers on rheology and curing characteristics of rubber compounds and also on physical and mechanical properties of vulcanizates has been studied. Obtained results of measured characteristics of polymeric systems containing prepared alternative fillers were compared with the results obtained in the case of reference rubber compound with a commertially used filler – carbon black.

Keywords: rubber, alternative filler, zeolite, glass waste, silanization

1 Introduction

The application of inorganic materials into organic polymers is one of the usual ways to improve up of physical and mechanical properties of polymers specifically hardness, tensile strength, modulus [1]. Composite materials show generally better properties than pure and homogeneous materials. The positive changes have been observed at very low concentration of inorganic mineral component in polymers [1, 2]. In the rubber industry the fillers based on SiO_2, calcium carbonate or glass are frequently used and also mineral fillers based on clay minerals such as kaolin [3-5]. Alternative filler used in present experiments comes from the weighing process of raw materials in glass production, where creates the waste in a sludge form.

* Corresponding author: darina.ondrusova@fpt.tnuni.sk

Reviewers: *František* Nový, *Eva* Tillová

The chemical composition of this alternative filler is similar to the chemical composition of kaolin, which is commonly used in rubber industry. Chemical composition of kaolin contains oxides like SiO_2, Al_2O_3, CaO and other trace elements, while the total content of SiO_2 in kaolin is 73,45% SiO_2 [6].SiO_2 and CaO also are a major components of this studied waste from glass production, therefore it has been applied into polymer materials in the function of filler.

The other kind of additivum used was natural mineral material - zeolite. Natural zeolites form a group of aluminosilicatenanoporous materials with the special porous structure allowing lots of practical applications as an ion-exchange, adsorbent, material for reversible hydratation and dehydratation and also as ecological nanofiller in polymeric materials [3]. The best known type of zeolite is clinoptilolite. Its structure (Fig.1) is based on treedimensional frameworks composed of tetrahedral SiO_4 and AlO_4 units, which are connected by shared oxygen atoms [4,5].

Fig. 1. 3D Crystalline Grid of Clinoptilolite

Surface modification of inactive mineral fillers (for example zeolites, clay minerals), has attracted much attention because the obtained products exhibit properties suitable for many materials and environmental applications [7]. Modification is generally considered to be an attractive and effective method to render inorganic mineral filler particles more compatible with polymeric matrix [8-10]. As the compatibility between fillers and polymers is low, a reduction of the polarity differences is required. This can be done by silane coupling agents such as (3-Aminopropyl)triethoxysilane, which is capable of acting with the filler surface and the polymer, and is commonly applied in silica-filled rubber [11, 12].

The present paper deals with the modification of two kinds of alternative additives based on waste from glass production and on zeolite clinoptilolite by silanization process and application of prepared modified forms into polymeric materials in the function of fillers.

2 Experimental

2.1 Preparation and modification of alternative additives

2.1.1 Preparation of alternative fillers based on the glass waste

The unmodified alternative filler - sludge was dried at the temperature of 105 ± 5 °C to constant weight. For experiments the alternative filler with a particle size < 25 μm was used.

The unmodified sample of alternative filler was analysed using X-ray fluorescence spectrometer S8 TIGER for the analysis of light elements. Results of chemical analysis of sludge sample studied are given in Table 1.

Table 1. Chemical composition of unmodified alternative filler - sludge

Oxide	SiO_2	CaO	Al_2O_3	BaO	Na_2O	K_2O	Sb_2O_3	Fe_2O_3	TiO_2	MgO	Cr_2O_3	Er_2O_3	Cl	SO_3
Mass (%)	73.09	15.14	3.95	2.68	1.32	1.24	0.93	0.67	0.19	0.16	0.1	0.1	0.07	0.06

The results of chemical analysis have confirmed that main components of studied sludge are SiO_2 and CaO, what is a positive precondition for its further application in the function of filler. Subsequently the sample of alternative filler was modified with three following silanes:
- (3-Aminopropyl)triethoxysilane (A),
- bis3-(Triethoxysilyl)propyl-tetrasulfide (E),
- 3-(Trimethoxysilyl)propyl-methacrylate (M).

The modification was carried out using 3 weight parts of modifying agent per 100 weight parts of filler according to the standard procedures [7]. To ensure homogeneous cover of alternative filler surface, the modification was performed in a mixture of water and ethanol under the same conditions using three different silane agents. Appropriate amounts of alternative filler, modifying agent and solvent were stirred for 2 hours and obtained modified forms of fillers after filtration were dried (Fig. 2). For further experiments three prepared modified alternative fillers with a particle size < 25 μm were used.

Fig. 2. Preparation of modified alternative fillers based on glass waste

2.1.2 Preparation of alternative fillers based on clinoptilolite

The clinoptilolite with a particle size < 7 μm was used for the experiments. Two series of the silanized samples of clinoptilolite were prepared by hydrolysis of (3-Aminopropyl)triethoxysilane (APTES) in the mixture of ethanol / water (90:10). The clinoptilolite with a particle size < 7 μm was finally added into the mixture. The first series of three samples of silanisedclinoptilolite were stirred at 250 rpm in a water bath with a temperature of 50 ° C. The second set of samples was stirred at 250 rpm at room temperature. Stirring times of samples were 1, 2 and 3 hours. The separated modified samples of clinoptilolite were dried in dryer for 60 minutes at temperature 110 ° C.

Six samples of the modified zeolite fillers were prepared by silanization under various conditions (Table 2).

Table 2. Labeling of prepared silanized forms of clinoptilolite

Samples	Characteristics
2 Z1H	cliloptilolitesilanized 1 hour at 50 ° C
3 Z2H	cliloptilolitesilanized 2 hours at 50 ° C
4 Z3H	cliloptilolitesilanized 3 hours at 50 ° C
5 N1H	cliloptilolitesilanized 1 hour at room temperature
6 N2H	cliloptilolitesilanized 2 hours at room temperature
7 N3H	cliloptilolitesilanized 3 hours at room temperature

2.2 Preparation of rubber compounds

2.2.1 Rubber compounds with alternative fillers based on the glass waste

The four model rubber compounds were prepared by two-step mixing in laboratory mixer Brabender in the standard way [13]. Composition of tread rubber compounds were different in type of filler. The carbon black was replaced by unmodified and modified alternative filler in a ratio of 2:1. Labelingof samples and used types of fillers are shown in the Table 3.

Table 3. Labeling of samples and dosage of fillers based on glass waste

Samples	Filler	Ratio of fillers
BK 2:1	Carbon black + unmodified alternative filler	2:1
BKA 2:1	Carbon black + modified alternative filler by silane agent A	2:1
BKE 2:1	Carbon black + modified alternative filler by silane agent E	2:1
BKM 2:1	Carbon black + modified alternative filler by silane agent M	2:1

Temperature 145 °C at the 50 rotations per second was the conditions relating to the first mixing step. In the second mixing step, curing agent (sulfur) and accelerator were added to the rubber compounds. Temperature 100°C at the 50 rotations per second was the condition relating to the second mixing step. After mixing process, rubber compounds were left for 24 hours.

2.2.1 Rubber compounds with alternative fillers based on clinoptilolite

The eight model rubber compounds were prepared by two-step mixing in laboratory mixer Plastograph-Brabender in the standard way [13, 14]. The first step of mixing was carried out at the temperature of 130 °C and a rate of pinions rotation was 50 rpm. The activator of vulcanization (ZnO) together with the fillers - carbon black, natural clinoptilolite (PZ) or silanized forms of clinoptilolite (as a replacement forcarbon black) and the natural rubber were used. In the second step which was carried out at the temperature 110 °C sulphur in a function of vulcanization agent and accelerator of vulcanization were added [13].After mixing process, rubber compounds were left for 24 hours.The fillers content in prepared model rubber compounds with the labeling of samples are indicated in Table 4.

Table 4. Fillers content in prepared model rubber compounds (in phr) with the labeling of samples

Sample/Filler	Carbon black	PZ	2 Z1H	3 Z2H	4 Z3H	5 N1H	6 N2H	7 N3H
Š	70	-	-	-	-	-	-	-
1PZ	25	45	-	-	-	-	-	-
2 Z1H	25	-	45	-	-	-	-	-
3 Z2H	25	-	-	45		-	-	-
4 Z3H	25	-	-	-	45	-	-	-
5 N1H	25	-	-	-	-	45	-	-
6 N2H	25	-	-	-	-	-	45	-
7 N3H	25	-	-	-	-	-	-	45

*phr = parts per hundred rubber

2.3 Study of Rheology and Curing characteristics

The rheological properties and curing characteristic (M_L, M_H, t_{s2}, t_{90}, R_v) of rubber compounds prepared were measured using Vulcameter Monsanto 100 at the temperature of 150 °C during 60 min, according to specifications in STN [15]. Obtained values of rheological and curing characteristics of rubber compounds containing modified and unmodified fillers were compared to the values of standard rubber compound (Š) filled with carbon black and all obtained values were also compared between themselves.

2.4 Study of Physical and Mechanical properties

The physical and mechanical properties of vulcanizates prepared were determined by help of tearing machine INSTRON according to STN [16]. The tensile strength and elongation at break were evaluated properties. Hardness of vulcanizate was measured by hardness tester IRHD according to STN [17]. The measured values of physical and mechanical properties of vulcanizates were compared with the values of standard rubber compound (Š) with carbon black filler and were compared between themselves too.

2.5 Microstructural SEM analysis

Microstructure of fillers and vulcanizates was studied using Scanning Electron Microscop TESCAN VEGA 3. The microanalysis was focused on the evaluation of topography and microrelief of filler particles using detector in a high vacuum, which is operating in the secondary electrons mode. The filler particles were fixed on adhesive carbon discs that provide conductivity between the sample and the sample stage. For optimal microstructure evaluation the surface of filler particles was metal plated using conductive layer based on the elements Au and Pd. This layer is also formed under high vacuum and under an argon atmosphere. Thanks to such sample preparation conditions, it was possible to apply the maximum incandescent cathode voltage 30 kV, which allows to get greater sharpness images from the area as well as a short distance of analyzed area from the lens.

3 Results and discussion

3.1 Effect of additives on Rheology and Curing characteristics

3.1.1 Rubber compounds with alternative fillers based on glass waste

The results from the study of rheological properties and curing characteristic (M_L, M_H, t_{s2}, t_{90}, R_v) are shown in Figs. 3 – 5. The minimum torque values M_L (Fig. 3) of all rubber compounds prepared were comparable.

In contrast, the maximum torque M_L values of rubber compounds with modified alternative fillers are higher than value of rubber compound with unmodified alternative filler (Fig. 4). The higher values of maximum torque show higher stiffness at the end of the vulcanization and they also show higher viscosity.

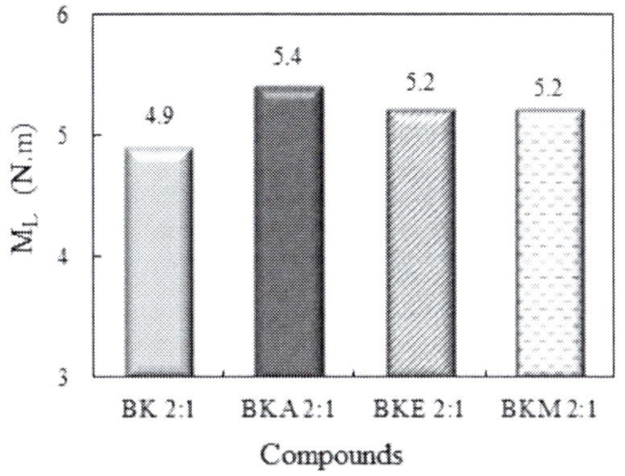

Fig. 3. Values of minimum torque M_L of rubber compounds

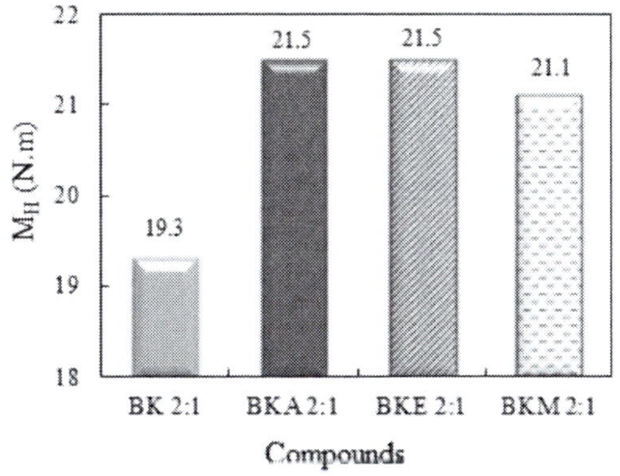

Fig. 4. Values of maximum torque M_H of rubber compounds

Modification of filler did not have significant influence on the start time of vulcanization. On the other hand, from Fig.5, it can be seen that the blends samples containing modified alternative fillers showed a shortening the optimum vulcanization time compared to the blend sample BK 2:1 filled with unmodified filler.

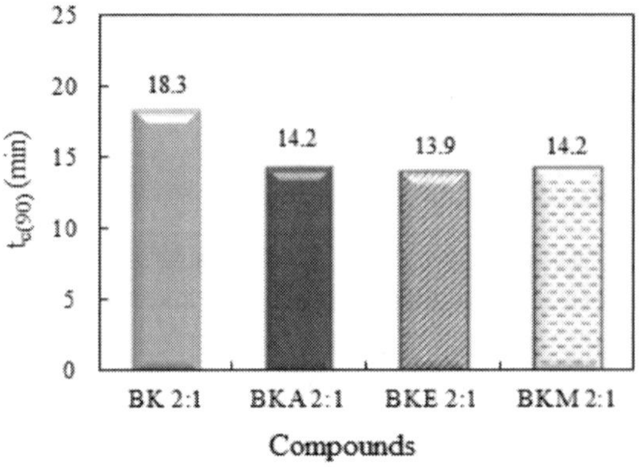

Fig. 5. Values of optimum time of vulcanization t_{90} of rubber compounds

This is confirmed in a positive impact of alternative filler modifications with silanes. This can also have a positive impact on the overall shortening of the production process.Shortening the optimum vulcanization time was reflected also in the increase in cure rate coefficient R_v (Fig. 6) in the cases of blends samples BKA 2:1, BKE 2:1 and BKM 2:1, containing alternative fillers modified with silanes in comparison with the rubber compound sample BK 2:1 filled with unmodified alternative filler .

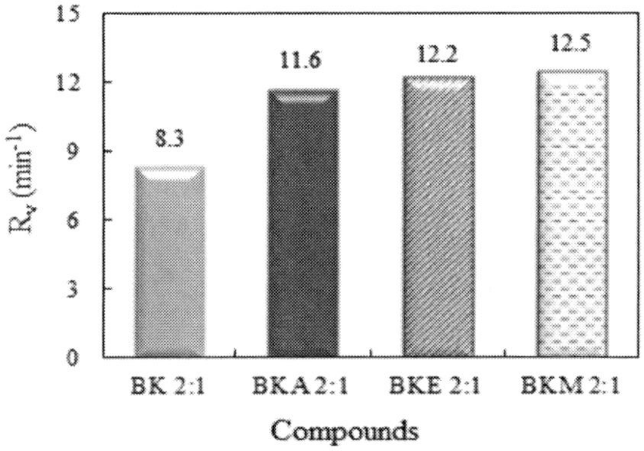

Fig. 6. Values of cure rate coefficient R_v of rubber compounds

3.1.2 Rubber compounds with alternative fillers based on clinoptilolite

The obtained values of rheological and curing characteristics(M_L, M_H, t_{s2}, t_{c90}, R_v) are shown in Figs. 7 – 9. From the values of scorch of time t_{s2} (Fig. 7) it is observed that the highest t_{S2}value showed sample 1PZ. A decrease of scorch of time values in the cases of samples 2-7,in comparison with a blend containing an unmodified zeolite filler, is due to modification of silanisation.

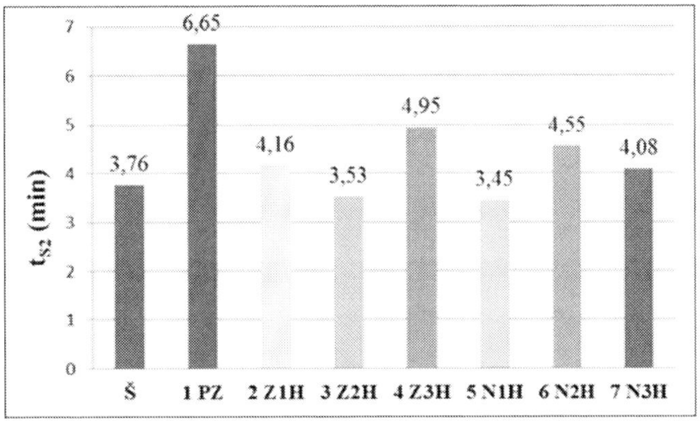

Fig. 7. Values of Scorch of time t_{s2} of rubber compounds

The rubber compounds 2Z1H, 4Z3H, 6N2H and 7N3H that were silanized under various conditions, showed a slight increase in scorch of time values compared to standard blend. That might be good in terms of the optimal course of rubber compounds processing. All rubber compounds containing modified silanized forms of filler based on clinoptilolite showed the scorch of time values comparable to standard rubber blend.

The same trend of a positive influence of silanization process has also been observed in the case of the obtained values of optimum of vulcanization (Fig. 8). Although, different conditions of silanisation of filler did not have significant influence on reology and curing characteristics, the positive influence of silanization of clinoptilolite filler is considerable.

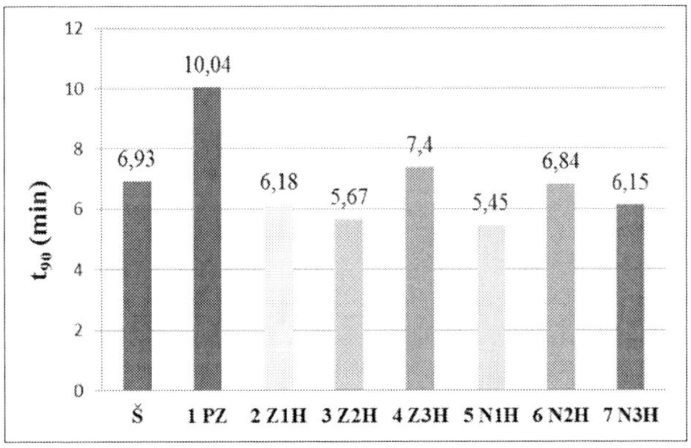

Fig. 8. Values of optimum time of vulcanization t_{90} of rubber compounds

From the measured values of coefficient of cure rate (Fig. 9) it can be concluded, that the modification of fillers had a positive effect on the cure rate. Rubber blends 2-7 containing modified fillers by silanization under various conditions show values of Rv higher than mixture with standard filler carbon black. The lowest Rv value showed the mixture containing unmodified clinoptilolite 1PZ. Thus, the use of silane APTES showed a positive effect on the activity of all components in the rubber blends containing a modified zeolite filler.

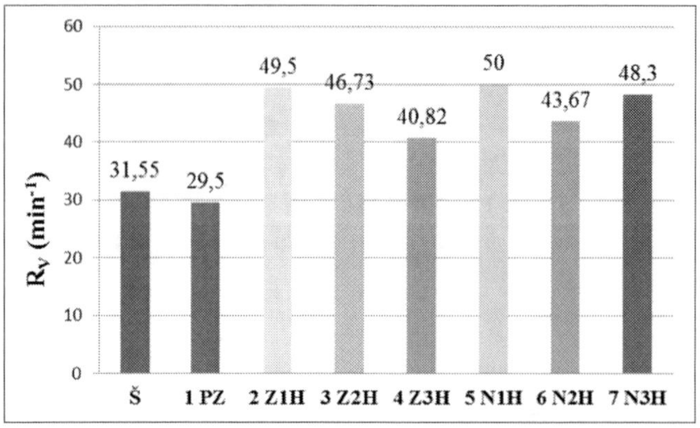

Fig. 9. Values of cure rate coefficient R_v of rubber compounds

3.2 Effect of additives on physical and mechanical properties of vulcanizates

3.2.1 Rubber compounds with alternative fillers based on glass waste

The selected physical and mechanical properties (tensile strength, elongation at breakand hardness) were tested and evaluated. Resulting graphic images of these properties are shown in Figs. 10 - 11. From Fig. 10 it can be seen, that samples BK 2:1 and BKM 2:1 show the lower values of tensile strength. On the other side, samples of vulcanizates BKA 2:1 and BKE 2:1 show higher tensile strength values in comparison with vulcanizate sample BK 2:1 which was filled with unmodified alternative filler. Modification of alternative filler showed improvement in relation to evaluation of tensile strength. The highest value of tensile strength showed the BKA 2:1 sample which was filled with alternative filler modified with silane (3-Aminopropyl)triethoxysilane.

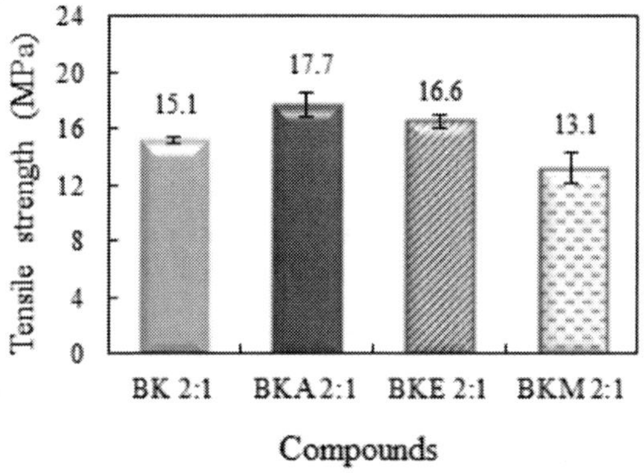

Fig. 10. Values of tensile strength of vulcanizates

Fig. 11 shows that blends samples BKA 2:1, BKE 2:1 and BKM 2:1, containing alternative fillers modified with silanes show an increase in hardness value in comparison with the rubber compound sample BK 2:1 filled with unmodified alternative filler.

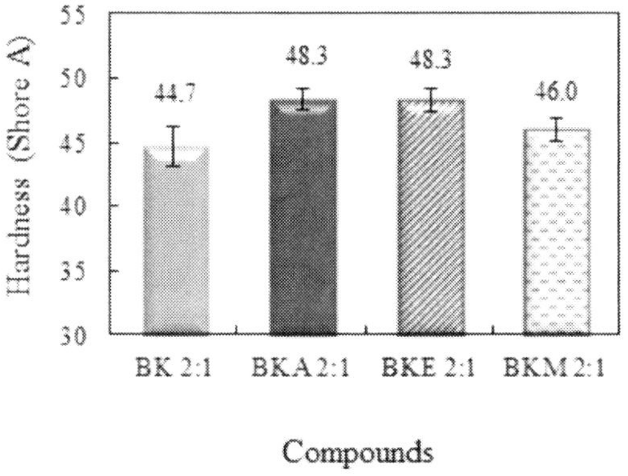

Fig. 11. Values of hardness of vulcanizates

The increase of hardness is a manifestation of improving the interaction between the filler and the polymer matrix due to increased activity of the modified filler surface.

3.2.2 Rubber compounds with alternative fillers based on clinoptilolite

The selected physical and mechanical properties (Tensile strength, Elongation at break and Hardness) were tested and evaluated. The results from the measurements of physical and mechanical properties of prepared model vulcanizates are given in Table 5. From the obtained values it can be concluded, that modification of filler based on clinoptilolite showed improvement of tensile strength values. The greatest value of tensile strength achieved vulcanizate 2Z1H with silanized filler that showed a slightly higher value of tensile strength than standard vulcanizate Š with carbon black. It can be concluded that the modified filler based on clinoptilolite 2Z1H and rubber matrix formed a stronger crosslinks network during vulcanization process than in standard vulcanizate Š. The lowest tensile strength value showed vulcanizate 1PZ. It can be concluded that the silanization of clinoptilolite filler under various conditions had a positive effect on the results of tensile strength.

Table 5. Fillers content in prepared model rubber compounds (in phr) with the labeling of samples

Vulcanizate	Tensile strength (MPa)	Sd	Tensibility (%)	Sd	Hardness (IRHD)	Sd
Š	19.17	± 0.95	180.5	± 17.08	79.4	± 1.82
1PZ	17.00	± 0.93	301.4	± 12.20	61.8	± 2.17
2 Z1H	19.35	± 0.61	224.4	± 4.56	64.4	± 2.41
3 Z2H	17.78	± 0.87	212.6	± 7.99	64.8	± 0.84
4 Z3H	17.46	± 0.66	231.8	± 4.38	65.4	± 1.14
5 N1H	17.10	± 0.75	205.2	± 10.89	66.0	± 1.22
6 N2H	17.75	± 0.75	220.0	± 3.39	64.0	± 2.00
7 N3H	17.53	± 1.05	209.4	± 3.13	66.8	± 0.84

The obtained values of elongation at break indicate, that higher values showed vulcanizate 1PZ with the addition of unmodified natural clinoptilolite.

Also vulcanizates with modified fillers 2-7 showed higher values of elongation at break compared to the vulcanizate with the standard filler. These results indicated that the vulcanizates containing the zeolite fillers silanized under the various conditions show a higher elasticity in comparison to standard vulcanizate, which can be utilized in applications where high elasticity of products is required.Based on the measured values of hardness it can be found that the hardness values decreased substantially by applying a zeolite fillers. It is clear that zeolite fillers slightly softens the rubber compounds. Vulcanizates containing silanized forms of clinoptilolite achieved higher hardness values compared to vulcanizate 1PZ with a natural clinoptilolite. From the above it can be concluded that the silanization had a positive effect on the increase of hardness values.

3.3 Microstructural SEM analysis

Microstructure of fillers and vulcanizates was studied using Scanning Electron Microscopy (SEM). For the determination of glass waste filler particles size (Fig. 12) were realized several measurements, which were based on applications 2D elements in an image from SEM. The measurements were carried out at 1000 times magnification of filler particles. The main purpose of the measurements was to determine the changes in particle size of the filler before and after modification using silanes.

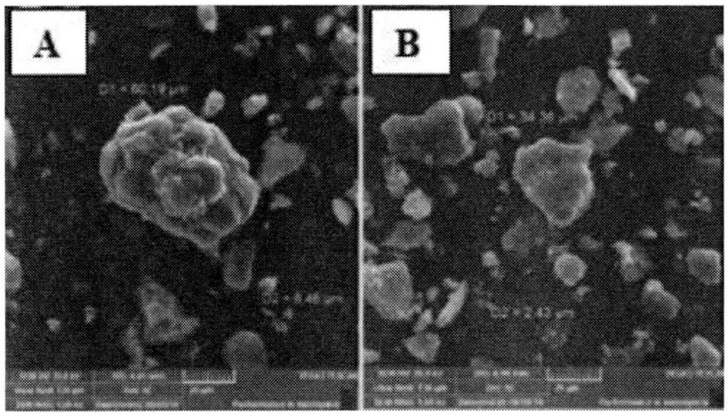

Fig. 12. Alternative filler (glass waste) particles size at 1000 x magnification
A – unmodified filler; B – filler after modification using (3-Aminopropyl)triethoxysilane

From Fig. 13 it can be seen, that the surface modification of the alternative filler using silane (3-Aminopropyl)triethoxysilane caused a reduction in the size of filler particles and thus an increase in the specific surface of alternative filler. The increase in specific surface contributed to increase in activity of filler particles and considerably improves the interaction between filler and polymer matrix (Fig. 13).

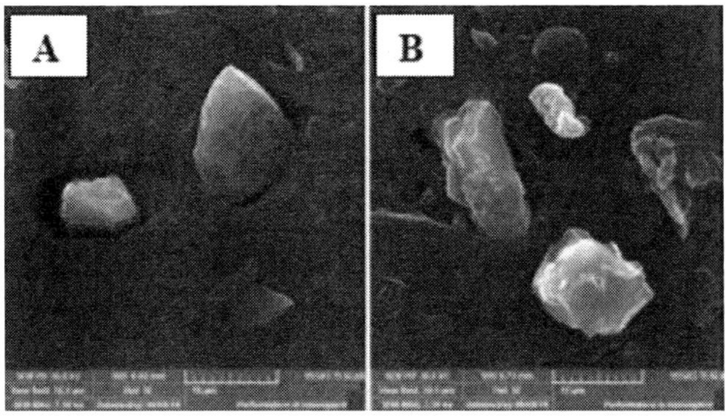

Fig. 13. Microstructure of vulcanizates with alternative filler (glass waste) 3500 x magnification A – with unmodified filler; B – with filler after modification using (3-Aminopropyl)triethoxysilane

An improvement of interaction between polymer matrix and alternative filler (after modification using (3-Aminopropyl)triethoxysilane was manifested in images of microstructure of vulcanizates (Fig. 13). Thanks to detector that works in the SE mode it can be clearly seen and evaluated the incorporation of filler particles in the rubber matrix. Evaluation of the samples was carried out on the sample quarry at 3500 times magnification. The same trend of improvement of interaction between polymer matrix and modified alternative additivum was found in the case of alternative filler based on clinoptilolite (Fig. 14).

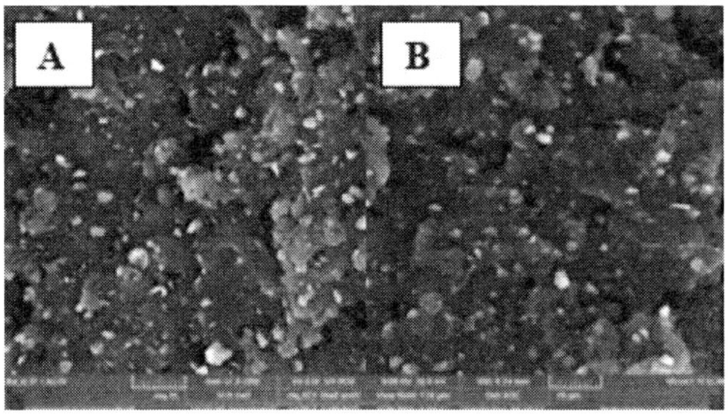

Fig. 14. Microstructure of vulcanizates with alternative filler (clinoptilolite)1000 x magnification A – with unmodified filler; B – with filler after modification using (3-Aminopropyl)triethoxysilane

Conclusion

According to the obtained results from the study of prepared modified alternative additives effect on rheology, curing characteristics of rubber compounds, physical and mechanical properties of vulcanizates and microstructural analysis of alternative filler and vulcanizates it was found an improvement of properties due to surface modification of alternative filler using silanes. In the case of alternative filler based on glass waste the best achieved properties showed sample BKA 2:1 which was filled with carbon black and alternative filler modified using (3-Aminopropyl)triethoxysilane in the ratio of fillers 2:1.

In the case of mineral alternative filler based on clinoptilolite the positive influence of silanization of clinoptilolite filler is considerable and the use of silane APTES showed a positive effect on the activity of all components in the rubber blends with the addition of modified zeolite filler.

This work was supported by the projects: VEGA No. 1/0589/17, VEGA No. 1/0649/17, KEGA 008TnUAD-4/2017 and resulted from the project ''Center for quality testing and diagnostics of materials'', ITMS code 26210120046 relating to the Operational Program Research and Development funded from European Fund of Regional Development

References

1. E. P. Giannelis, *Polymer layered silicate nanocomposites*. Adv. Mater. **8**, 29-35 (1996)
2. E. P. Giannelis, *Polymer-layered silicate nanocomposites: synthesis, properties and applications*. Appl. Organomet. Chem. **12**, 675-680 (1998)
3. L. Domka, A. Malicka, N. Stachowiak, *Production and structural investigation of polyethylene composites with modified kaolin*. Acta physica polonica **A114**, 413-420 (2008)
4. G. Wypych, *Handbook of fillers*. (ChemTec publishing, Toronto, 1999)
5. J. C. Dai, J. T. Haung, *Surface modification of clays and clay-rubber composite*. Appl. Clay Sci. **15**, 51-65 (1999)
6. G. Kakali , T. Perraki, S.Tsivilis, E. Badogiannis, *Thermal treatment of kaolin: The effect of mineralogy on the pozzolanic activity*. Appl. Clay Sci. **20**, 73–80 (2001)
7. H. He, Q. Tao, P. Zhu, W. Shen, S. Yang, *Silylation of clay mineral surfaces*. Appl. Clay Sci. **71**, 15-20 (2013)
8. S. Letaief, I. K. Tonle, T, Diaco, C. Detellier, *Nanohybrid materials from interlayer functionalization of kaolinite. Application to the electrochemical preconcentration of cyanide*. Appl. Clay Sci. **42**, 95-101 (2008)
9. E. Moretti, L. Storaro, G. Chessa, A. Talon, E. Callone, K. J. Mueller, F. Enrichi, M. Lenarda, *Stepwise dansyl grafting on the kaolinite interlayer surface*. J of Coll. and Interf.Sci. **375**, 112-117 (2012)
10. Y. Yuan, H. Chen, J. Lin, Y. Ji, *Surface modification of calcined kaolin with toluene diisocyanate based on high energy ball milling*. Appl. Surf. Sci. **284**, 214-221 (2013)
11. G. Wypych, *Handbook of Fillers*. (ChemTec Publishing, Toronto, 2010)
12. S. Mihara, *Reactive Processing of Silica-Reinforced Tire Rubber*. Ph.D thesis, University of Twente, Enschede, the Netherlands, ISBN: 978-90-365-2839-9 (2009)
13. D. Ondrušová, M. Pajtášová, *Rubber Components and their Influence on Rubber Properties and Environmental Aspects of Production*. Towarzystwo Słowaków w Polsce, 166 p., ISBN 978-83-7490-385-1 (2011)
14. J. Vavro Jr., J. Vavro, P. Kováčiková, P. Kopas, M. Handrik, *Simulation and analysis of defect distribution in passenger car tire under dynamic loading*. Applied mechanics and materials. **611**, 544-547 (2014)
15. STN 62 1416, *Determination of vulcanization characteristics on vulcameter*.
16. STN ISO 37, *Determination of tensile stress-strain properties*.
17. ISO 48, *Determination of hardness*.

Temperature dependent measurement of internal damping of austenitic stainless steels

Monika Oravcová[1,*], *Peter* Palček[1], *Mária* Chalupová[1], *Milan* Uhríčik[1]

[1]Department of Materials Engineering, Faculty of Mechanical Engineering, University of Žilina, Univerzitná 8215/1, 010 26 Žilina, Slovakia

Abstract. This article is aimed on the analysis of the internal damping changes of austenitic stainless steels AISI 304, AISI 316L and AISI 316Ti depending from temperature. In experimental measurements only resonance method was used which is based on continuous excitation of oscillations of the specimens and the whole apparatus vibrates at the frequency near to the resonance. Microplastic processes and dissipation of energy within the metals are evaluated and investigated by internal damping measurements. Damping capacity of materials is closely tied to the presence of defects including second phase particles and voids. By measuring the energy dissipation in the material, we can determine the elastic characteristics, Youngs modulus, the level of stress relaxation and many other.

Keywords: austenitic stainless steels; internal damping; plastic deformation, deformation-induced martensite

1 Introduction

Austenitic stainless steels are a class of alloys with a face-centered-cubic lattice structure of austenite which maintains from room temperature (and below) to the melting point. When 18% chromium and 8% nickel are added, the crystal structure of austenite remains stable over all temperatures. They are the most widely used grade of stainless steel. These steels have a very good mechanical and technological properties finding widespread not only in industrial, chemical and food applications, but also in medical field. They are corrosion resistant, paramagnetic in the annealed condition, however they can become slightly magnetic after plastic deformation, e.g. after cold working [1, 2].

During shaping the component into the desired geometry or shape, steels often undergo a thermomechanical processing which changes the material properties. A material's microchemistry may be changed during thermomechanical processing and its functional properties are changed.

Plastic deformation changes the properties of steels by increasing the dislocation density which affects mainly the mechanical properties and helps in diffusion acceleration, phase transformations, nucleation, etc. During the cold working process, a martensitic phase

* Corresponding author: monika.oravcova@fstroj.uniza.sk

Reviewers: *Františka* Pešlová, *Marta* Kianicová

called deformation-induced martensite is evoked. The amount of deformation-induced martensite is increasing with severity of deformation.

Cold working and deformation-induced martensite influences the damping properties of austenitic stainless steels. Several damping mechanisms – movement of crystal defects are present in the metallic materials. The damping capacity of the structural material used for various applications is an important property from the fatigue point of view. The internal damping can be defined as the capacity of a material to damp mechanical vibrations. The capacity change of material to damp mechanical vibrations as function of time or temperature corresponds to a structural, microstructural and substructural change [3-5].

2 Experimental procedures

2.1 Experimental material

For the measurement of internal damping commercially available austenitic stainless steels, namely AISI 304, AISI 316 with low carbon content and AISI 316 stabilized by Ti, were used as an experimental material. The real chemical composition of the experimental material is listed in the Table 1.

The experimental material was submitted to metallographic analysis before the measurement of internal damping. Microstructures of the austenitic stainless steels were analysed in the initial state and are presented on the Fig. 1 – Fig. 3. Metallographic analysis of longitudinal cross-section of the specimen in initial state after chemical etching in HNO_3+HF+glycerine solution showed that the structure is non-homogeneous and have polyedric austenite grains with different grain sizes and occurrence of annealing twins and deformation martensite after plastic deformation as a result from the final treatment. In the case of AISI 316Ti titanium carbides TiC where present in the structure.

Table 1. Chemical composition of austenitic stainless steels [in weight %]

Element	C	Cr	Ni	Mo	Mn	Si	S	P	Ti
AISI 304	0.05	19.59	7.84	0.49	1.57	0.29	0.03	0.05	-
AISI 316L	0.03	17.41	9.91	2.38	1.34	0.51	0.03	0.07	-
AISI 316Ti	0.03	17.83	10.62	2.36	1.54	0.44	0.03	0.06	0.14

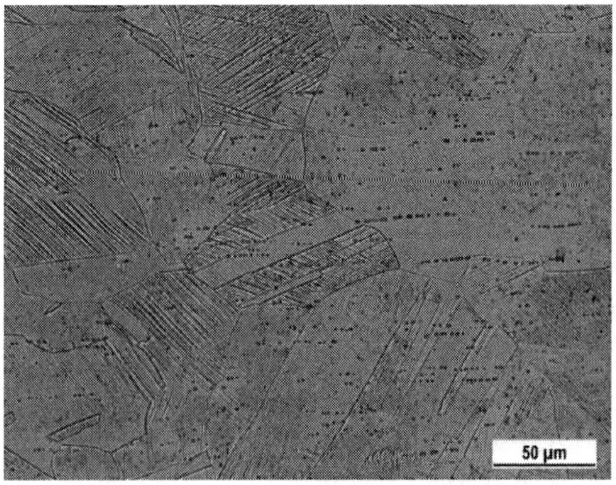

Fig. 1. Microstructure of AISI 304 in initial state

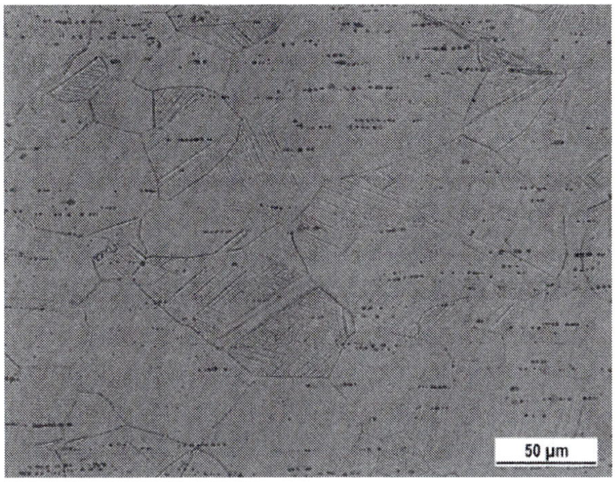

Fig. 2. Microstructure of AISI 316L in initial state

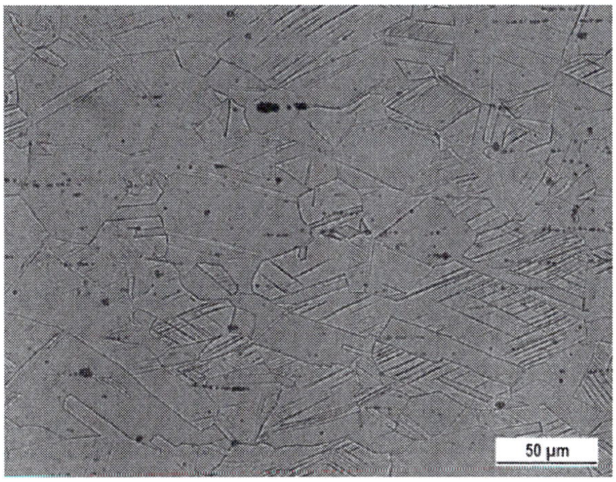

Fig. 3. Microstructure of AISI 316Ti in initial state

2.2 Experimental equipment

The experimental equipment consists of mechanical and electronical part and is shown on the Fig. 4. The mechanical part (Fig. 5a) consists of three basic parts: the transducer, which serves like the source and at the same time like detector of ultrasonic waves. Then the rod, which is made of titanium and has a cylindrical shape with a diameter of 12 mm. Its function is to isolate the heat from the test specimen, which is heated in a furnace and the transducer then maintains at a constant room temperature. And the last part is the testing specimen (Fig. 5b). Dimensions and shape of the test specimen are designed to fulfill the resonance condition, that means the natural frequency must be approximately the same like the frequency of the test equipment ±10 Hz.

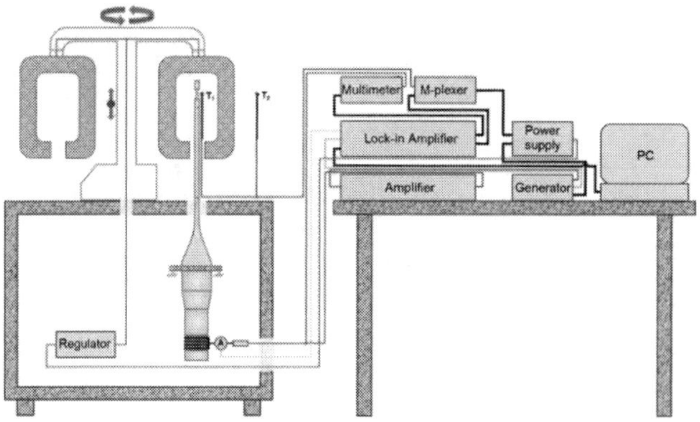

Fig. 4. Ultrasonic resonance device for internal damping measurement

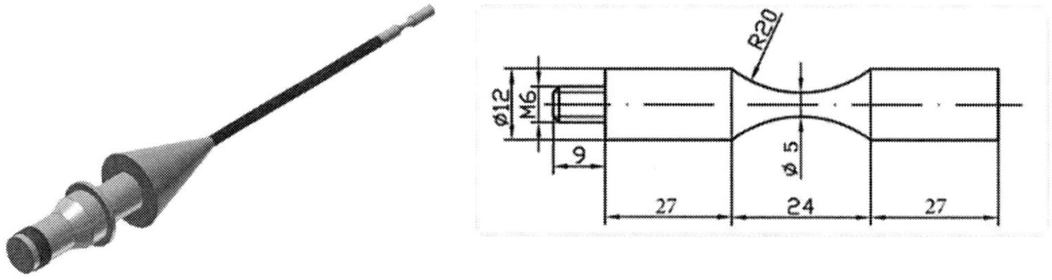

(a) (b)

Fig. 5. Mechanical part of the device (a); dimensions of the specimen (in mm) (b)

3 Results and discussion

The Fig. 6 to the Fig. 11, documents the results of the internal damping measurement of austenitic stainless steel AISI 304, AISI 316L and AISI 316Ti in intial state and after plastic deformation depending on the temperature.

In the Fig. 6, Fig. 8 and Fig. 10 are presented results of internal damping in initial state after first measurement, where the peak of internal damping during the heating is in the temperature range of 80 – 100 °C. Also these graphs contains the second measurement of internal damping, on the same samples, where not such a high peak was observed as in the first measurement. This fact that in the second measurement is no significant peak is related to the movement of disslocations and energy dissipation which has taken place during the first measurement.

All the testing samples were heated from room temperature upto 400 °C and followed by reverse cooling. Further heating above 400 °C was not necessary because all the processes has taken place in the temperature range of 25-100 °C. After this interval there is a steep decline of internal damping and after 200 °C the values of internal damping change only a little.

The highest value of internal damping, the local maximum, was reached in the temperature range 80-100 °C where deformation-induced martensite transforms to austenite [6]. The highest local maximum was observed within the AISI 316Ti testing sample. In the backward measurement, cooling, the value of internal damping is lower in comparison with starting value and there is no local maximum reached.

In the Fig. 7, Fig. 9 and Fig. 11 are presented results of internal damping after plastic deformation. The graphs also presents the first and second measurement of internal damping showing that there was an increase of damping values again in the range of 80 – 100 °C. This increase indicates that microplastic deformation which causes irreversible movemet increases the disslocation density in small volumes of the steel and causes strengthening which leads to creation of deformation martensite. This local maximum of internal damping is a result of deformation-induced martensite transformation to austenite [7].

By heating the steel, a reversible process begins, in which the elastic deformation is removed and then the stress in the deformed grains is reduced and relaxaction of disslocations begins in the disslocation accumulated grains. The deformation-induced martensite transformation into austenite is accompanied by migration and reduction of vacancies an creation of additional lattice imperfections [8]. The reverse process is then accompanied by the formation of annealing twins where the accumulated excessive energy is reduced [9]. The plastic deformation also changes magnetic properties because deformation-induced martensite has slightly feromagnetic properties [10].

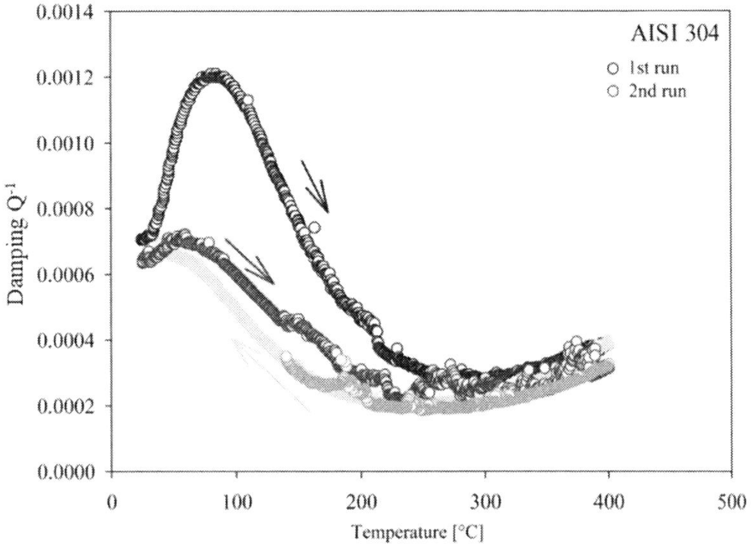

Fig. 6. Results of internal damping measurement of AISI 304 in initial state

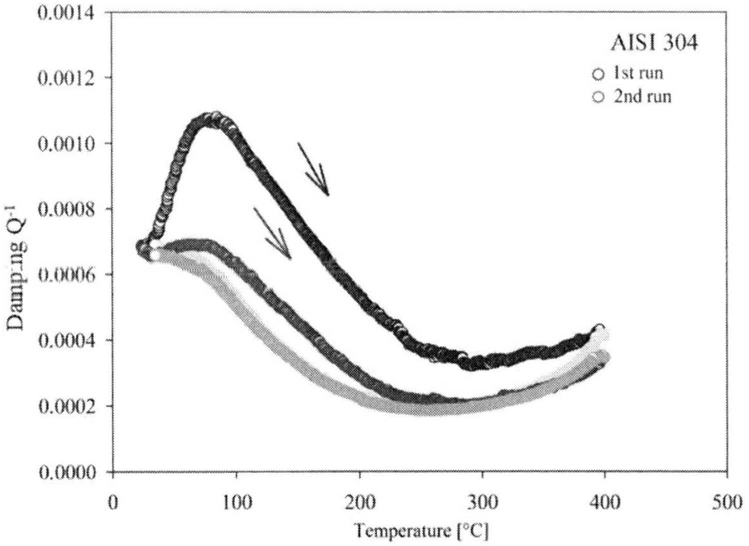

Fig. 7. Results of internal damping measurement of the deformed AISI 304

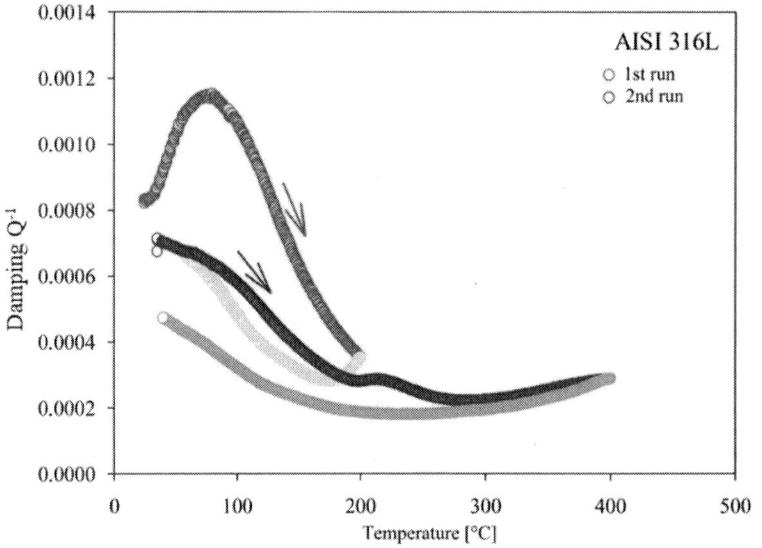

Fig. 8. Results of internal damping measurement of AISI 316L in initial state

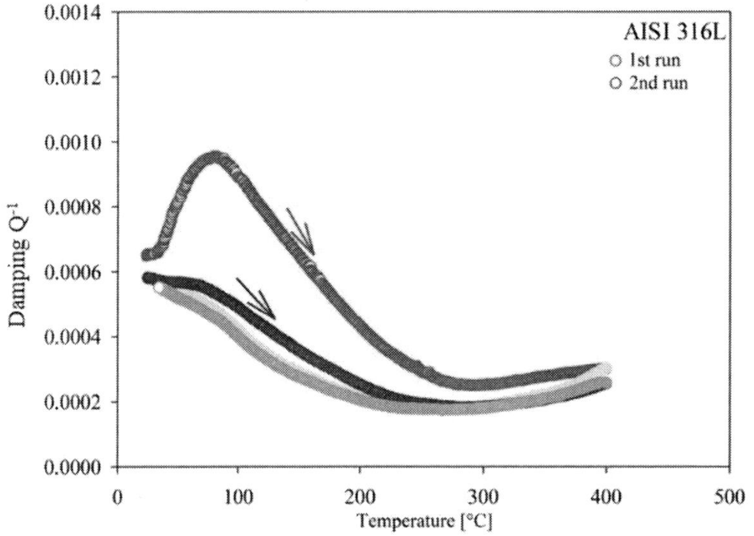

Fig. 9. Results of internal damping measurement of the deformed AISI 316L

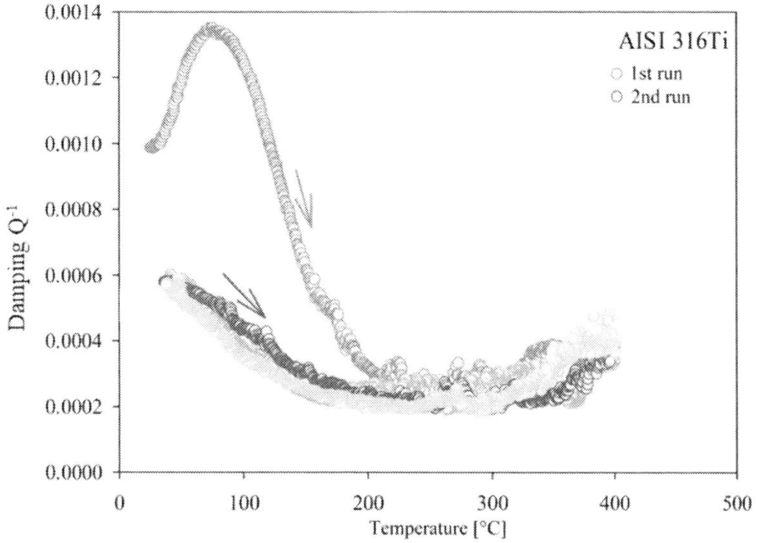

Fig. 10. Results of internal damping measurement of AISI 316Ti in initial state

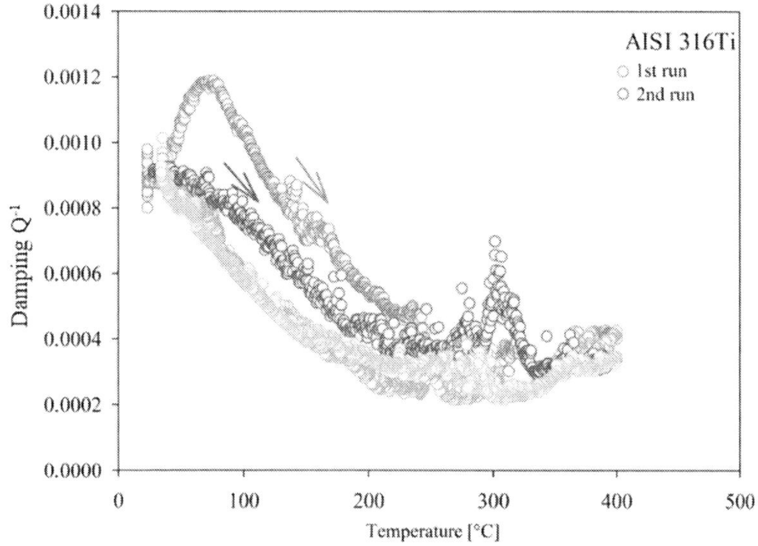

Fig. 11. Results of internal damping measurement of the deformed AISI 316Ti

Conclusion

Based on experimental results from internal damping measurements of austenitic stainless steels AISI 304, AISI 316L and AISI 316Ti can be stated:

- The peak of internal damping during the heating was in the temperature range of 80-100 °C for all the experimental material.
- We can conclude that all relaxation process has taken place during the first measurement.
- No significant local maximum was observed during second measurement nor during the cooling phase.

- An increase of internal damping values in the case of plastically deformed testing samples was observed again in the range of 80-100 °C.
- The noticed peak during the measurement in austenitic stainless steels are based on the fact if the material is plastically deformed during cold working or not.
- Deformation-induced martensite initiates the steel's strengthening and the dislocation density is higher at the grain boundary than in its bulk.
- The reversed process was accompanied by the formation of annealing twins where the accumulated excessive energy was reduced.

This work has been supported by Scientific Grant Agency of Ministry of Education of Slovak Republic and Slovak Academy of Sciences, No. 1/0683/15.

References

1. M. F. McGuire, *Stainless steels for design engineers.* ASM International (2008)
2. V. Čihal, *Korozivzdorné oceli a slitiny*. Academia Praha (1999)
3. T. F. A. Santos, M. S. Andrade, *Internal Friction on AISI 304 Stainless Steels with Low Tensile Deformations at Temperatures between -50 and 20 °C.* Advances in Materials Science and Engineering **2010**, 1-8 (2010)
4. R. Singh, B. Ravikumar, A. Kumar, P. K. Dey, I. Chattoraj, *The effect of cold working on sensitization and intergranular corrosion behaviour of AISI 304 stainless steel.* Springer Publishing : Metallurgical and Materials Transactions A **34**, 2441-2447 (2003)
5. J. Talonen, H. Hänninen, *Damping properties of austenitic stainless steels containing strain-induced martensite.* Metallurgical and Materials Transactions A **35A**, 2401-2406 (2004)
6. N. Solomon, I. Solomon, *Deformation induced martensite in AISI 316 stainless steel.* Revista de Metalurgia **46**, 121-128 (2010)
7. P. Hedström, *Deformation and Martensitic Phase Transformation in Stainless Steels.* Doctoral thesis. Luleå University of Technology (2007)
8. E. Dryzek, M. Sarnek, M. Wróbel, *Reverse transformation of deformation-induced martensite in austenitic stainless steel studied by positron annihilation.* Springer US **49**, 8449-8458 (2014)
9. Y. Jin, M. Bernacki, G. S. Rohrer, A. D. Rollett, B. Lin, N. Bozzolo, *Formation of annealing twins during recrystallization and grain growth in 304L austenitic stainless steel.* Materials Science Forum **753**, 113-116 (2013)
10. T. Oršulová, P. Palček, J. Kúdelčík, *Effect of plastic deformation on the magnetic properties of selected austenitic stainless steels.* Production Engineering Archives **14**, 15-18 (2017)

Use of waste materials in rubber matrix

Mariana Pajtášová[1,*], *Zuzana* Mičicová[1], *Darina* Ondrušová[1], *Slavomíra* Božeková[1], *Róbert* Janík[1], *Beáta* Pecušová[1], *Lukáš* Raník[1]

[1]Alexander Dubček University of Trenčín, Faculty of Industrial Technologies in Púchov, I. Krasku 941/30, 020 01 Púchov, Slovak Republic

Abstract. The presented paper deals with the use of waste materials as ecological fillers into rubber matrix. Waste materials were used as partial replacement of the commercial filler – carbon black, designated as N339. These prepared rubber compounds were characterized on the basis of the rheology and vulcanization characteristics – minimum torque (M_L), maximum torque (M_H), optimum time of vulcanization ($t_{(c90)}$), processing safety of compound (t_s), rate coefficient of vulcanization (R_v). In the case of the prepared vulcanizates, physical-mechanical properties (tensile strength, tensibility and hardness) and dynamic-mechanical properties (storage modulus, loss modulus, loss angle tan δ) were investigated. Using the dependency of loss angle on temperature, the selected properties for tyre tread vulcanizates were evaluated, including traction on snow and ice, traction on the wet surface and rolling resistance.

Keywords: waste materials, filler, physical-mechanical properties, dynamic-mechanical properties

1 Introduction

The rubber compounds are composed of polymers, fillers, oils, and numerous chemicals used for crosslinking or "vulcanization" of the compounds [1].

The fillers provide a large degree of strengthening of the rubber network, resulting in a substantial increase in stiffness, tensile strength, and resistance to abrasion [2].

The preparation of coffee leads to formation of relatively bulky waste in the form of coffee grounds and the most of these residues remain unutilized, being discharged to the environment, where they cause severe contamination and subsequent environmental pollution problems due to the toxic nature [3, 4].

In this study, we added the waste material – coffee extract into the rubber compounds in order to replace the conventional filler in an amount of 3 phr with different granularity. We studied the rheological properties and vulcanization characteristics for non-vulcanized rubber compounds as well as physical-mechanical properties and dynamic-mechanical properties of resultant vulcanizates. The measured results were compared with values of standard.

[*] Corresponding author: mariana.pajtasova.fpt@gmail.com

Reviewers: *Grzegorz* Domek, *František* Nový

2 Experimental

2.1 Material

The coffee extract was dried in dryer at temperature of 70 °C for 24 h to obtain the constant weight and then it was ground and filtered through a sieve with mesh size of 0.040, 0.090 and 0.200 μm.

2.2 Preparation of rubber compounds

The rubber compounds were prepared by two-step mixing process in laboratory mixer of Brabender type with chamber volume of 70 cm^3 and 50 rpm according to Slovak technical standards – STN 62 1425 [5]. The composition of fillers in compounds is shown in Table 1.

Table 1. The composition of fillers in compounds

compounds	filler	(phr)	filler	(phr)	(μm)
S	carbon black	87	coffee extract	-	-
LC	carbon black	84	coffee extract	3	0.040
MC	carbon black	84	coffee extract	3	0.090
BC	carbon black	84	coffee extract	3	0.200

The rheological properties and vulcanization characteristics were determined using a MONSANTO 100 vulcameter. The samples (8 g) of the compounds were tested at the vulcanization temperature of 150 °C during 60 min and then, the rubber compounds were moulded at 150 °C according to optimum time of vulcanization ($t_{c(90)}$).

2.3 Preparation of vulcanizates

Prepared vulcanizate test specimens in the shape of the double-sided blades were mechanically cut out from the vulcanized slabs. Five samples were prepared for the each vulcanizate. Tensile properties were measured using INSTRON universal testing machine. Hardness of prepared vulcanizates was measured by IRHD hardness tester with hardness ranging from 30 to 85 IRHD (International Rubber Hardness Degree). Dynamic-mechanical properties of the prepared vulcanizates were measured using DMA Q 800 (TA Instruments) – dynamic mechanical analyser at the temperature range between -70 and 70 °C, at frequency of 1 Hz and at heating rate of 3 °C/min.

3 Results and discussion

3.1 Rheology and vulcanization characteristics of rubber compounds

The rheology and vulcanization characteristics – minimum torque (M_L), maximum torque (M_H), optimum time of vulcanization ($t_{(c90)}$), processing safety of compound (t_s), rate coefficient of vulcanization (R_v) of prepared rubber compounds are shown in graphical form in Figure 1-5.

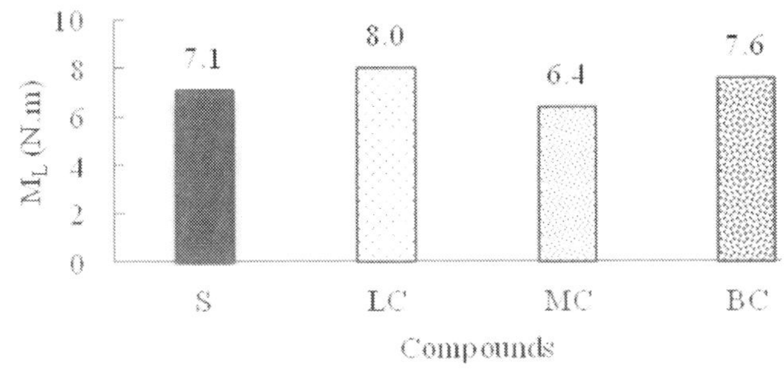

Fig. 1. Minimum torque of rubber compounds

The values of minimum torque for LC and BC rubber compounds are higher in comparison with standard compound. The increase of values of the minimum torque shows the higher stiffness as well as viscosity of the compounds at the beginning of vulcanization and it is probably due to good compatibility of rubber matrix with filler [6]. The MC rubber compound has lower value of minimum torque in comparison with standard compound.

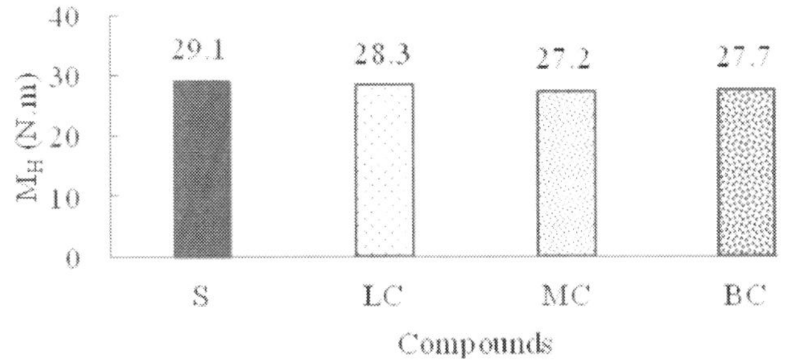

Fig. 2. Maximum torque of rubber compound

In the case of the maximum torque, all prepared rubber compounds exhibit lower values in comparison with standard compound.

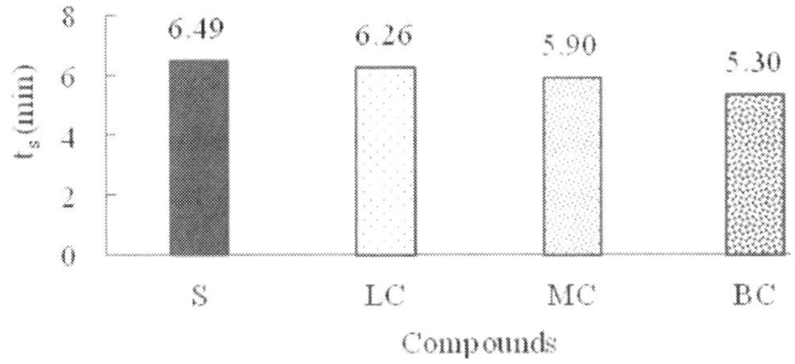

Fig. 3. Processing safety of compound

The resultant values of processing safety of compound for prepared rubber compounds are lower in comparison with standard compound.

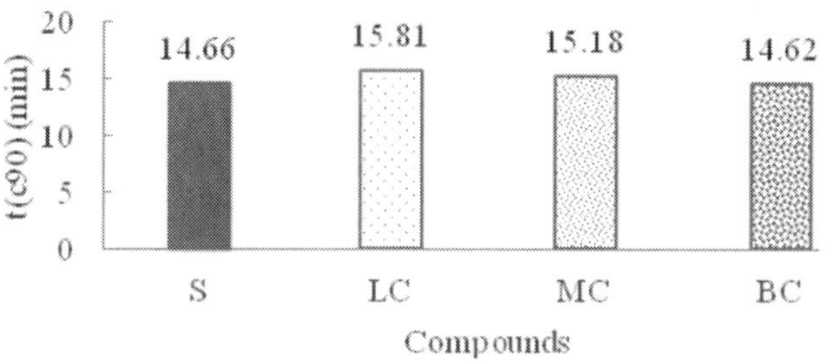

Fig. 4. Optimum time of vulcanization

From Figure 5, it can be seen, that BC rubber compounds has comparable optimum time of vulcanization to standard compound. The LC and BC rubber compounds exhibit higher values in comparison with standard compound. The extension of the optimum vulcanization time can be helpful for application in thick-walled rubber product, because the main purpose is to make vulcanization process slower [7].

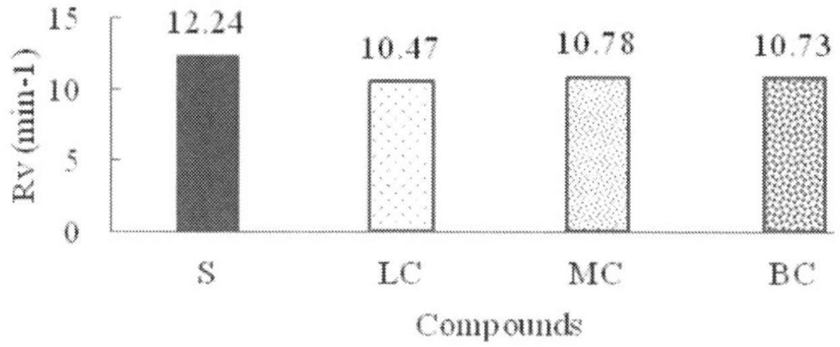

Fig. 5. Processing safety of compound

In the case of the processing safety of compound, all prepared rubber compounds exhibit lower values in comparison with standard compound.

3.2 Physical-mechanical properties of rubber vulcanizates

Selected physical and mechanical properties (tensile strength, tensibility and hardness) were tested and evaluated for standard vulcanizate and vulcanizates containing coffee extract with different grain size. Measured results of selected physical and mechanical properties are shown in graphical form in Figure 6, 7 and 8.

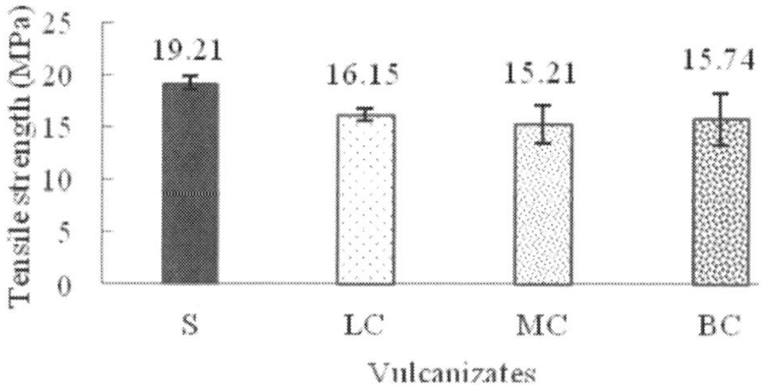

Fig. 6. Tensile strength of vulcanizates

From Figure 6, it can be seen, that all prepared vulcanizates show lower measured values of tensile strength in comparison with standard vulcanizate.

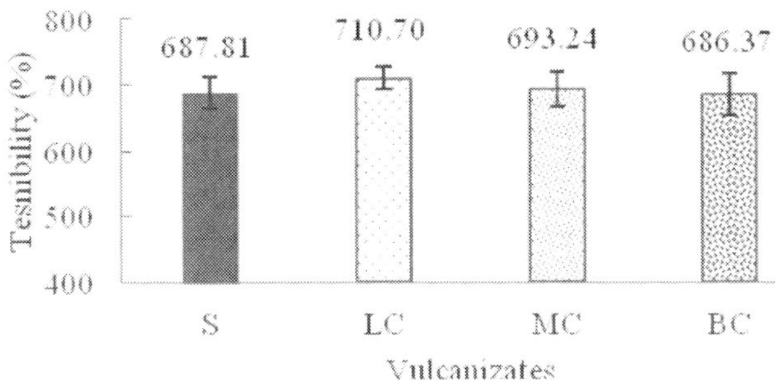

Fig. 7. Tensibility of vulcanizates

In the case of tensibility, LC and MC vulcanizate exhibit higher values in comparison with standard vulcanizate. On the other side, BC vulcanizate has comparable tensibility to standard vulcanizate.

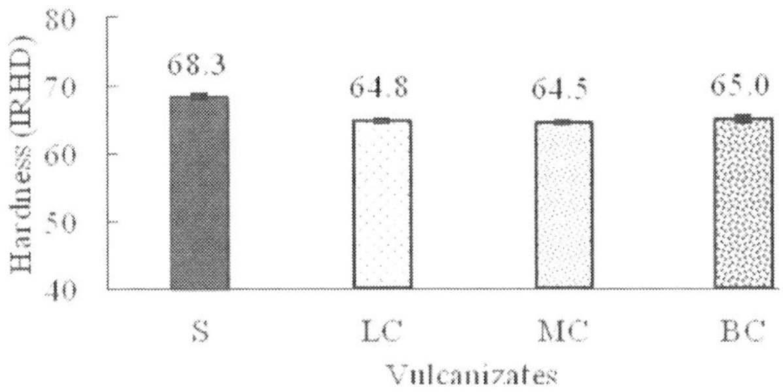

Fig. 8. Hardness of vulcanizates

The measured values of hardness for prepared rubber vulcanizates are lower in comparison with standard vulcanizate. The lower values of hardness indicate good incorporation of the filler into the compound and good compatibility with the rubber matrix [8].

3.3 Dynamic-mechanical properties of rubber vulcanizates

The viscoelastic behaviour of the prepared rubber vulcanizates was evaluated on the basis of the results of the dynamic-mechanical analysis measurements in the temperature range from -70 ° C to 70 ° C. The dependence of the dynamic-mechanical properties – the storage modulus (E') and the loss modulus (E"), the loss angle (tan δ) on the temperature is graphically depicted in Fig. 9, 10 and 11.

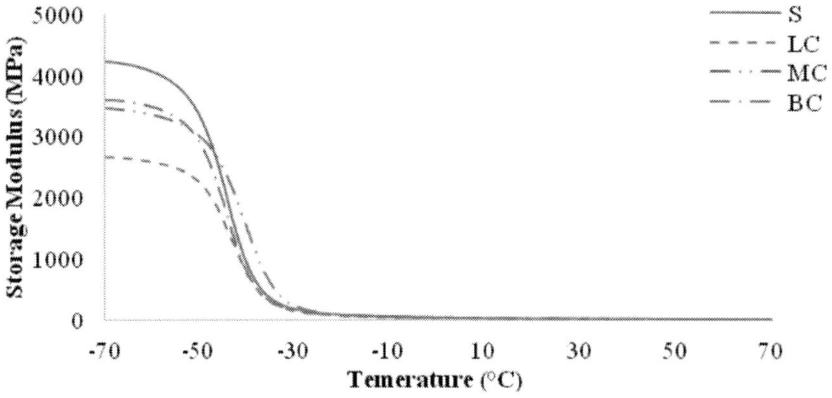

Fig. 9. Storage modulus of vulcanizates

As it is shown in Figure 9, the vulcanizates exhibit a decrease of the storage modulus in comparison with the standard vulcanizate in the following order: LC < MC < BC < S. These differences of the storage modulus (E') are probably related to the different degree of interaction and dispersion between the waste material and the polymer matrix, where the lower storage modulus of the investigated vulcanizates exhibits better interaction at the interfacial interface and the filler dispersion [9, 10].

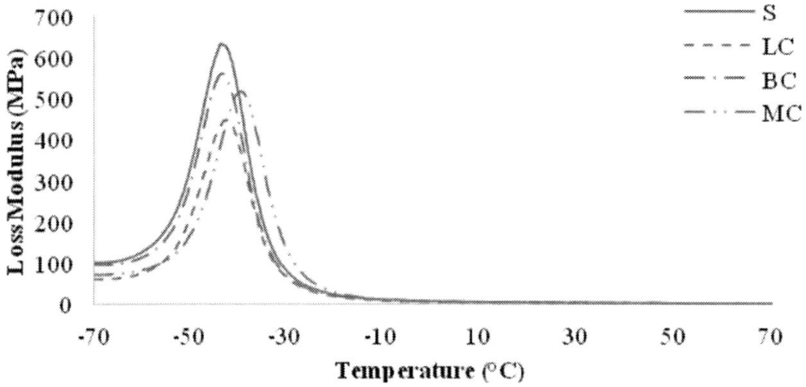

Fig. 10. Loss modulus of vulcanizates

The same course is observed for the loss modulus (E").

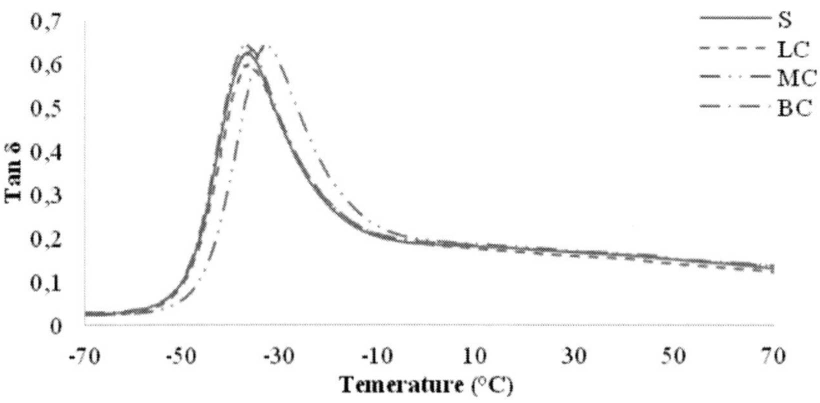

Fig. 11. Loss angle of vulcanizates

The glass transition temperatures of the prepared rubber vulcanizates are determined from the maximum peaks of the loss angle (tan δ). Their values are shown in Table 2.

Table 2. Tg and Tan δ results for vulcanizates

Vulcanizates	Tg (°C)	tan δ_{MAX}	tan δ (-25 °C)	tan δ (0 °C)	tan δ (60 °C)
S	-36.35	0.625	0.353	0.184	0.139
LC	-35.92	0.595	0.360	0.183	0.130
MC	-32.89	0.630	0.356	0.185	0.140
BC	-36.52	0.631	0.455	0.190	0.139

The stiffening effect of the filler can be determined from the max. peak of tg δ. If the max. peak of tan δ is lower, the stiffening of the resulting material is the better [11]. Values of max. peak of tan δ decrease for vulcanizates in the following order: LC < S < MC < BC. The best stiffening can be assumed for vulcanized LC.

Based on the dependency of loss angle (tan δ) on temperature, the selected properties for tyre tread vulcanizates were evaluated, including traction on snow and ice, traction on the wet surface and rolling resistance [12].

Tan δ for -25 °C represents the traction on snow and ice. The BC vulcanizate exhibits best traction on snow and ice in comparison with standard vulcanizate. Tan δ for 0 °C represents traction on the wet surface. The BC vulcanizate exhibits best traction on the wet surface in comparison with standard vulcanizate. Tan δ for 60 °C represents the rolling resistance. The LC vulcanizate exhibits the best rolling resistance in comparison with standard vulcanizate.

Conclusion

The main aim of the paper was to investigate effect of the waste materials – coffee extract in rubber matrix. According to the results of rheological properties and vulcanization characteristics, physical-mechanical properties and dynamic-mechanical properties, we can conclude that carbon black can be replaced partially by fillers based on investigated waste materials. The best achieved values in properties were observed in the case of LC vulcanizate.

This work was supported by the Slovak Grant Agency KEGA 007TnUAD-4/207, VEGA 1/0589/17, VEGA 1/0649/17 and resulted from the project "Center for quality testing and diagnostics of materials", ITMS code 26210120046 relating to the Operational Program Research and Development funded from European Fund of Regional Development.

References

1. L.N. Carli, O. Bianchi, R.S. Mauler, J.S. Crespo, *Crosslinking kinetics of SBR composites containing vulcanized ground scraps as filler*. Polym. Bull. **67**, 1621-1631 (2011)
2. Z. Mičicová, M. Pajtášová, S. Domčeková, D. Ondrušová, L. Raník, T. Liptáková, *Inorganic materials and their use in polymeric materials*. Procedia Engineering **136**, 239-244 (2016)
3. M. Haile, *Integrated volarization of spent coffee grounds to biofuels*. Biofuel Research Journal **2**, 65-69 (2014)

4. M. C. Echeverria, M. Nuti, *Valorisation of the Residues of Coffee Agro-industry: Perspectives and Limitations*. The Open Waste Management Journal **10**, 13-22 (2017)
5. STN 62 1425: *Raw rubber. A preparation of rubber compounds*. 1984 [in Slovak]
6. D. Ondrušová, S. Domčeková, M. Pajtášová, A. Dubec, Z. Mičicová, B. Pecušová, *Alternative filler based on the waste from glass production and its effect on the rubber properties*. Procedia Engineering **177**, 462-469 (2017)
7. A. V. Chapman, A. J. Tinker, *Vulcanization of Blends – Crosslink Distribution and its Effect on Properties*. KGK-Kautschuk Gummi Kunstsoffe **56**, 533-544 (2003)
8. D. Ondrušová, M. Pajtášová, *Rubber components and their influence on rubber properties and environmental aspects of production*. (Spolok Slovákov v Poľsku, Krakow, 166 p., 2011) ISBN 978-83-7490-358-1
9. M. Bhattacharya, A. K. Bhowmick, *Synergy in carbon black-filled natural rubber nanocomposites. Part I: Mechanical, dynamic mechanical properties, and morphology*. Journal of Materials Science **45**, 6126-6138 (2010)
10. A. Das, R. Jurk, K. Werner Stöckelhuber, G. Heinrich, *Effect of Vulcanization Ingredients on the Intercalation-Exfoliation Process of Layered Silicate in an Acrylonitrile Butadiene Rubber Matrix*. Macromolecular Materials and Engineering **293**, 479-490 (2008)
11. A. Gantowski, P. Palutkiewicz, E. Boceąga, *Numerical analysis of stress state during single point bending in DMTA examination*. Journal of Achievements in Materials and Manufacturing Engineering **28**, 47-50 (2008)
12. S. Božeková, Z. Mičicová, D. Ondrušová, M. Pajtášová, *Vplyv modifikovaného alternatívneho plniva silánom Bis3-(triethoxysilyl)-tetrasulphid na vlastnosti gumárenských zmesí*. Hutnícké listy **5**, 48-52 (2016)

Study of clay minerals effect on curing characteristics of polymer blends and physical-mechanical properties of prepared vulcanizates

Beáta Pecušová[1,*], *Mariana* Pajtášová[1], *Zuzana* Mičicová[1], *Darina* Ondrušová[1], *Andrea* Feriancová[1], *Marcel* Kohutiar[1], *Ivan* Labaj[1]

[1]Department of Materials Technologies and Environment, Faculty of Industrial Technologies in Puchov, Alexander Dubcek University of Trencin I. Krasku 491/30, 020 01 Púchov, Slovakia

Abstract. The given paper deals with the study of the properties of clay minerals, namely montmorillonite and moreover, it is focused on effect of these clay minerals on the curing characteristics of the polymer blends and the physical-mechanical properties of prepared vulcanizates. Montmorillonite is a major clay mineral which has a wide application in many industrial branches. It belongs to the group of dioctahedral smectite minerals with structural type in the ratio of 2:1. Characteristics of prepared modified and organomodified clay minerals are based on sulphur vulcanisation accelerators which are used for the preparation of real polymer blend where they represent a partial replacement of the common carbon black filler and then, the effect on the curing characteristics of polymer blends as well as physical-mechanical properties of the prepared vulcanizates are investigated. The results exhibit that the clay-based filler (modified and organomodified clay minerals) can be used as a partial replacement while the quality of the prepared blends is preserved.

Keywords: clay minerals, montmorillonite, fillers, physical-mechanical properties, curing characteristics

1 Introduction

Clay minerals are hydrated silicates (hydro-silicates) with a layered structure. Montmorillonite (MMT) is the typical and important representative of the mentioned clay minerals. Montmorillonite belongs to the group of dioctahedral smectite minerals with the expanding structure. Particular interest in montmorillonite results from the fact that it has the ability to expand and absorb large organic molecules through its ion exchange reactions as well as other types of layered silicates. In addition to increasing the size of the interspaces, the character of the filler surface is changed from hydrophilic to hydrophobic (organophilic) [1]. The three-layered structure of MMT is the basic structural unit, which consists of two continuous tetrahedral networks and one octahedral network is placed between two tetrahedral networks. In relation to the structure, the significant substitution can be observed and it means that there is strong negative charge of the layer and this

* Corresponding author: beata.pecusova@fpt.tnuni.sk

Reviewers: *František* Nový, *Eva* Tillová

charge is compensated by the presence of exchangeable cations (Na^+, K^+, Ca^{2+}, Mg^{2+}) in the interlayer [2]. Due to their properties, the given clays and clay minerals are applied in many industries – petroleum and cement industry. Moreover, they are also used for the production of animal feedstuffs, foodstuff, paper, rubber, plastics, synthetic leather, external and internal coverings, pencils, pastel colours, porcelain, heavy clay and fine ceramic products. In tread rubber blends, the clay minerals lead to tire adhesion improvement to wet road and better adhesion of rubber to steel cords. Clays and clay minerals are commonly used as a partial replacement of the carbon black representing the filler into rubber blends in order to reduce the prices of products as well as to make these products to be eco-friendly [3, 4, 5].

2 Material and methods

2.1 Preparation of modified and organomodified clay minerals

Natural bentonite was crushed in a mortar and then, it was sieved out through a sieve, where the size of hole was 0.040 mm in diameter. Natural bentonite can be mainly found in the calcium form of MMT or in the sodium-calcium form of MMT and therefore, the calcium form of MMT was prepared from the calcium chloride solution. Undesirable salts were eliminated by help distilled water until the reaction with the silver nitrate solution exhibited negative presence of chloride anions. Using the calcium form of MMT and the mentioned procedures hereinbefore, further modified form of Co^{2+} ions was prepared from the solution of cobalt chloride. For the preparation the of organomodified clay mineral forms, organic blends based on dithiophosphate (V) and benzothiazole (B) were used as accelerators of sulphur curing in the rubber industry and these organic blends we prepared by saturation of the accelerator in gas state into the interlayer of modified clay minerals. All testing steps and procedures were carried out at the workplace of Faculty of Industrial Technologies in Puchov.

2.2 Preparation of tread rubber blends

The tread rubber blends were prepared by a two-step stirring of two types of the rubber while these two types of the rubber were subsequently stirred with the other ingredients and the process of stirring was carried out by help of Brabender Plastograph – the laboratory stirring machine where the volume of chamber was 70 cm^3 and a stirrer speed was 50 rev/min. It is important to point out that the given stirring process was performed according to standards [6]. After the stirring process, the prepared blends were investigated from the aspect of the curing characteristics, using the Monsanto 100 Vulcameter. The given testing process was based on standardized procedures at temperature of 150 °C and the recording of time for curing curves was 60 minutes according to standards [7]. The physical-mechanical properties of vulcanizates were measured by help of Instron (tearing machine) at a laboratory temperature of 25 °C and the deformation velocity was 50 mm/min, while it was in accordance with predetermined standards [8]. The hardness of the vulcanizates was measured by the IRHD hardness tester at temperature 23 ± 2 °C and 6 mm test samples thickness [9]. All testing steps and procedures were carried out at the workplace of Faculty of Industrial Technologies in Puchov. Designation of tread rubber blends is shown in (Table 1).

Table 1. Designation of tread rubber blends

Designation of tread rubber blends		
designation	**filler**	**phr**
LB R	carbon black - N 339	87
LB 1	MMT Ca	0.5
LB 2	MMT Ca	1
LB 3	MMT Ca	1.5
LB 4	MMT Ca-V	0.5
LB 5	MMT Ca-V	1
LB 6	MMT Ca-V	1.5
LB 7	MMT Ca-B	0.5
LB 8	MMT Ca-B	1
LB 9	MMT Ca-B	1.5
LB 10	MMT Co	0.5
LB 11	MMT Co	1
LB 12	MMT Co	1.5
LB 13	MMT Co-V	0.5
LB 14	MMT Co-V	1
LB 15	MMT Co-V	1.5
LB 16	MMT Co-B	0.5
LB 17	MMT Co-B	1
LB18	MMT Co-B	1.5

*phr – per hundred rubber, filler amount in blend

3 Results and discussion

3.1 Curing characteristics of tread rubber blends

Curing characteristics of tread rubber blends are investigated on the basis of rheological records of curing curves relating to given blends. The evaluation of the curing characteristics was carried out for: minimum torque (M_L), the maximum torque (M_H), safety of rubber blends (t_s), optimal time of vulcanization (t_{C90}) and velocity coefficient of vulcanization (R_v). The measured values of the M_L (Fig. 1) for the LB 1 - LB 6 tread rubber blends are almost comparable to the reference tread rubber blend (R), except for the LB 5 blend containing the MMT Ca-V filler in amount of 1 phr, because in this case, the value is higher. Values of the M_L for LB 7 - LB 9 rubber blends with MMT Ca-B filler were higher, compared to R. In LB 10 - LB 18 blends, the M_L values were almost comparable to R, except for LB 12 containing MMT Co filler and LB 15 containing MMT Co-V filler where the amount of given filler was 1.5 phr for both cases. Containing the MMT Co-B filler, LB 16 blend (filler in the amount of 0.5 phr) and LB 18 blend (filler in the amount of 1.5 phr) indicate the higher stiffness of the sample at the beginning of curing process as well as higher viscosity in comparison with R.

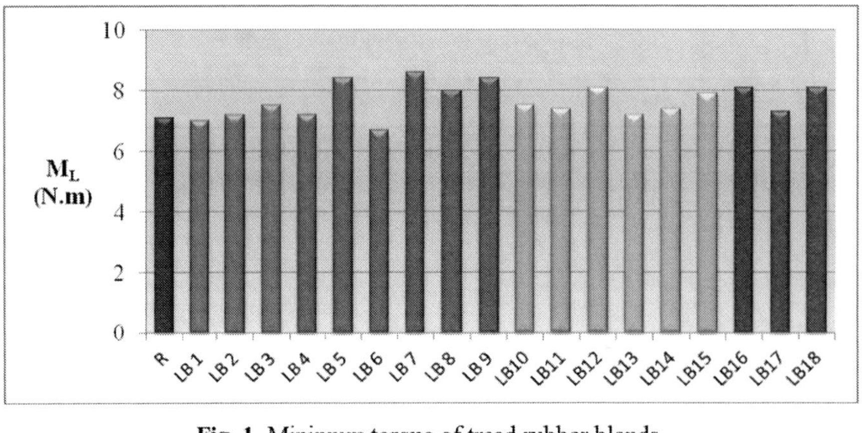

Fig. 1. Minimum torque of tread rubber blends

The values of the M_H (Fig. 2) for almost all tread rubber blends are comparable to R value, except for LB 7, LB 9, LB 16, LB 17 blends, where M_H values are higher. Lower M_H values were measured for LB 1 - LB 3, LB 6, LB 10, LB 13, LB 14 blends and it indicates a lower stiffness and the viscosity of the blends at the end of the curing process.

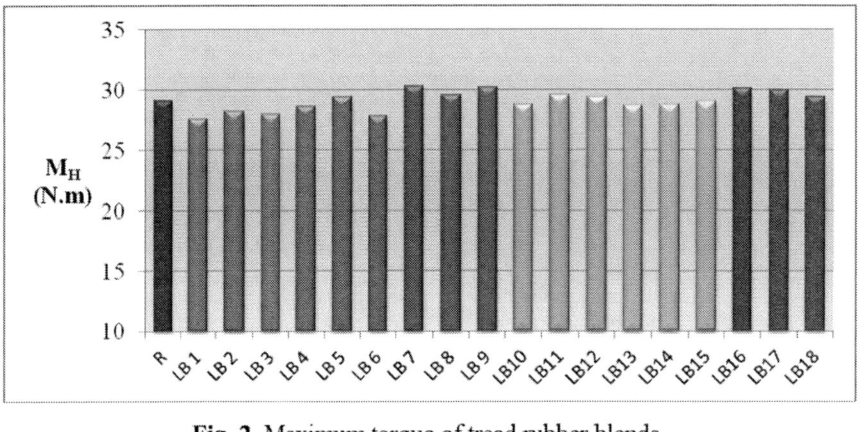

Fig. 2. Maximum torque of tread rubber blends

The t_s values (Fig. 3) were higher in the case of LB 1 tread rubber blend containing the MMT Ca filler in the amount of 0.5 phr and LB 2 with a content of MMT Ca filler of 1 phr, compared to the reference tread rubber blend. This mentioned phenomenon can positively influence the processing properties of the blends and the physical-mechanical properties of the resultant vulcanizates. For the other blends, the t_s values were lower in comparison with R.

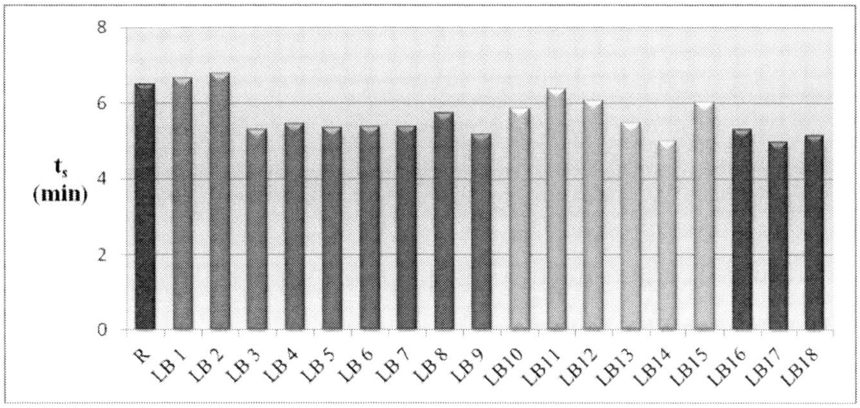

Fig. 3. Safety of tread rubber blends

The t_{c90} values (Fig. 4) are lower for the LB 4, LB 6, LB 7, LB 9, LB 10, LB 13, LB 14, LB 16 - LB 18 rubber blends in comparison with R and it has a beneficial effect on the curing process. Values of LB 3 containing MMT Ca, LB 5 containing MMT Ca-V and LB 8 containing the MMT Ca-B filler are comparable to the t_{c90} values of R.

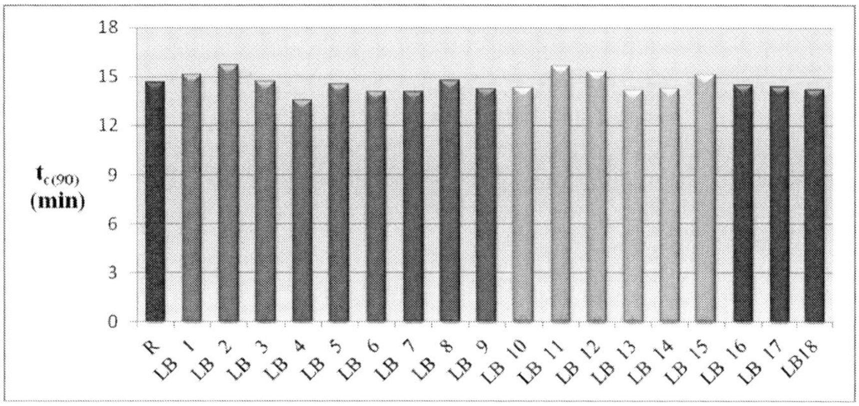

Fig. 4. Optimal time of tread rubber blends curing process

The results of velocity coefficient of curing process (R_v) are shown graphically in Fig. 5. The LB 4 tread rubber blend containing the MMT Ca-V filler in the amount of 0.5 phr has a higher value of R_v of curing process in comparison with R standing for reference blend and it leads to the higher the activity of the components of the rubber blend. LB 6 containing MMT Ca-V filler, LB 7 containing MMT Ca-B, LB 10 containing MMT Co and LB 13 containing MMT Co-V are comparable to R. In relation to the other investigated blends, they show lower values resulting in lower activity of the components.

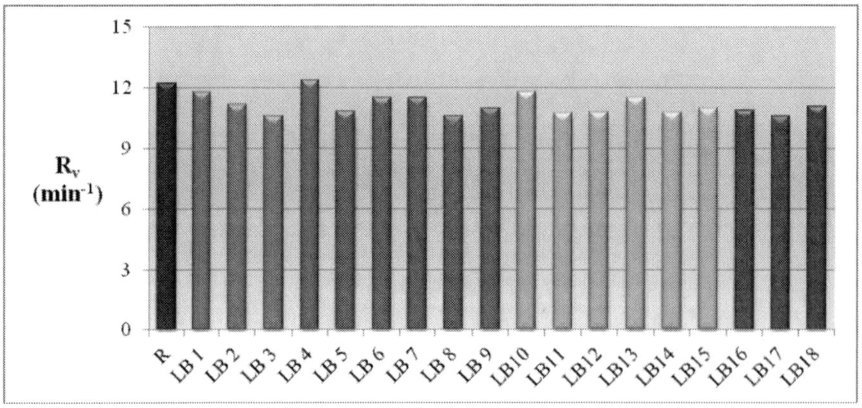

Fig. 5. Velocity coefficient of tread rubber blends curing

3.2 Physical-mechanical properties of tread rubber blends

Prepared tread vulcanizates in the form of test samples were used to determine the resulting physical-mechanical properties. The results for the reference vulcanizate (R) and for the tread vulcanizates (LB 1 - LB 18) are graphically depicted (Fig. 6 - 8). For LB 2 tread rubber vulcanizate containing MMT Ca, LB 5 vulcanizate containing MMT Ca-V, LB 8 vulcanizate containing MMT Ca-B in the amount of 1 phr, the tensile strength values were comparable to the reference vulcanizate. The other vulcanizates exhibit lower strength values, compared with reference vulcanizate. This is probably related to weaker interaction and bond between clay mineral filler particles and rubber matrix.

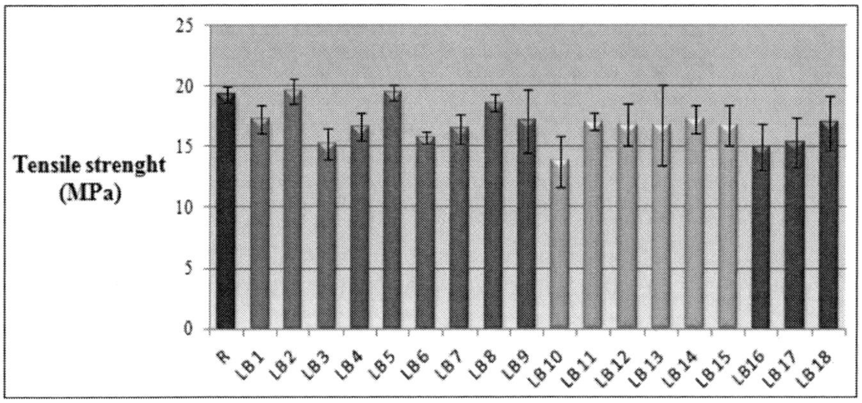

Fig. 6. Tensile strength of tread vulcanizates

In the case of the evaluation of the resulting tensibility (Fig. 7), it is observed that almost all of the prepared tread vulcanizates exhibit lower values in comparison with the reference vulcanizate, except for LB 1, LB 2, LB 5, LB 8 LB 13 and LB 14 blends, because these mentioned blends exhibit higher values of resulting tensibility. The LB 2 vulcanizate containing MMT Ca filler, LB 5 containing MMT Ca-V filler, LB 8 containing MMT Ca-B filler in the amount of 1 phr have the highest values of the resulting tensibility and it leads to indication of a higher elasticity.

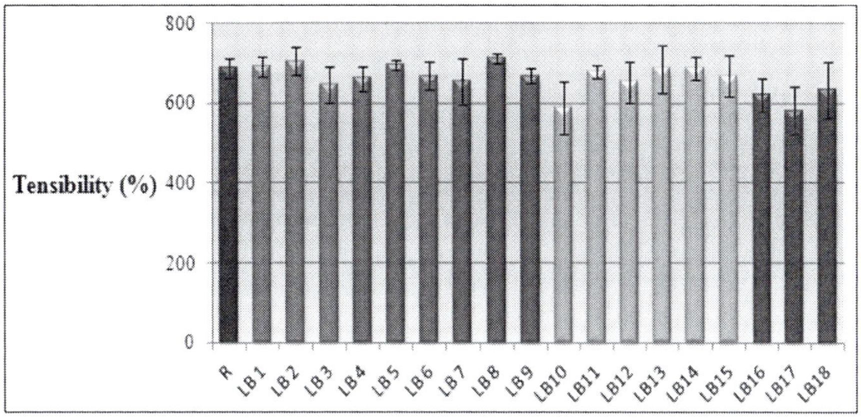

Fig. 7. Tensibility of tread vulcanizates

In relation to the hardness values of the resultant tread vulcanizates (Fig. 8), the given values are comparable to the reference vulcanizates value. Compared with the reference vulcanizate, the lowest hardness values were measured in the case of LB 6 tread vulcanizate containing MMT Ca-V filler in the amount of 1.5 phr and LB 7 containing MMT Ca-B filler in the amount of 0.5 phr and it is connected with a good incorporation of the filler into the blends and its good compatibility with the rubber matrix.

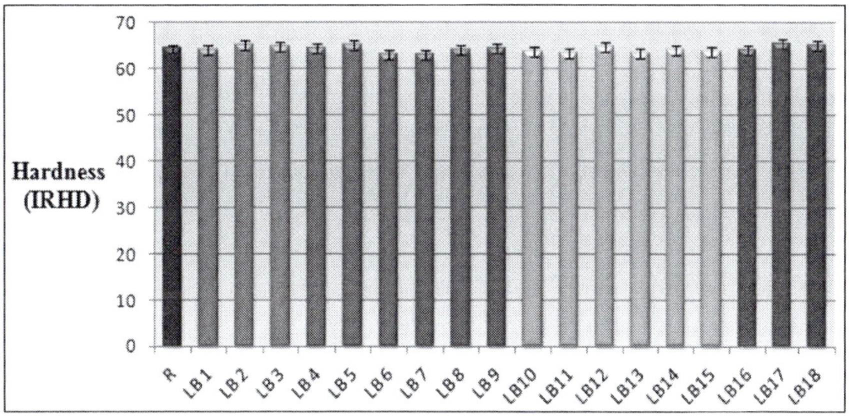

Fig. 8 Hardness of tread vulcanizates

Conclusion

The paper deals with the study of the effect of prepared modified and organomodified forms of MMT clay mineral filler on the properties of tread rubber blends. The investigation process involved the preparation of 19 tread rubber blends, while the filler in 18 blends was based on combination of specified amount of carbon black with clay. The reference sample represented the commonly used tread rubber blend in rubber industry, where the N 339 carbon black is filler. The given reference blend was used for comparison of specific curing characteristics leading to evaluation of the blend function. The evaluation of the curing characteristics was carried out for: minimum torque (M_L), the maximum torque (M_H), safety of rubber blends (t_s), optimal time of vulcanization (t_{C90}) and velocity coefficient of vulcanization (R_v).

In the form of double-sided blades, the resulting vulcanizates containing the filler based on combination of carbon black with clay were evaluated from the aspect of physical-mechanical properties (tensile strength, tensibility, hardness) and it was in accordance with valid standards. During the evaluation, the prepared samples were compared with the reference sample in order to specify the further application of the modified and organomodified clay mineral, especially MMT, in rubber industry. The obtained values indicate that fillers containing organomodified clay minerals with an intercalated accelerator on the basis of dithiophosphate (V) have higher influence on acceleration of the curing process in comparison with the fillers containing organomodified clay minerals with an intercalated accelerator on the basis of benzothiazole (B).

Acknowledgments: KEGA 007TnUAD-4/2017, VEGA grant No. 1/0649/17, VEGA grant No. 1/0589/17, project "Center for quality testing and diagnostics of materials", ITMS code 26210120046 relating to the Operational Program Research and Development, EF of Reg. Development.

References

1. V. Šucha, *Íly v geologických procesoch*. (Univerzita Komenského v Bratislave, Bratislava, 159 p., ISBN 80-223-1547- 8, 2001)
2. B. Čičel, I. Novák, I. Horváth, *Mineralógia a kryštalochémia ílov*. (Veda, Bratislava, Vydavateľstvo SAV, Slovenská akadémia vied, 1981)
3. M. Badík, et al., *Správa o stave životného prostredia žilinského kraja*. (Slovenská agentúra životného prostredia Banská Bystrica, Žilina, 6 p., 1999)
4. J. Paliesková, M. Pajtášová, A. Feriancová, D. Ondrušová, K. Holcová, J. Vavro, *Thermal properties of fillers based on organoclays in the polymeric materials*. Journal of thermal analysis and calorimetry, **119** (2), 939-943 (2015)
5. D. Ondrušová, M. Pajtášová, *Rubber components and their influence on rubber properties and environmental aspects of production*. (Spolok Slovákov v Poľsku, Krakow, 166 p., ISBN 978-83-7490-385-1, 2011)
6. STN 62 1425 *Príprava behúňových gumárenských zmesí*.
7. STN 62 1416 *Stanovenie vulkanizačných charakteristík na vulkametri*.
8. STN 62 1436 *Stanovenie ťahových vlastností*.
9. STN 62 1433 *Stanovenie tvrdosti vtláčaním hrotu tvrdomera IRHD*.

Structural changes of selected materials for glass industry after thermal loading

Františka Pešlová[*,1], *Mariana* Janeková[1], *Stanislav* Krum[2]

[1]University of Alexander Dubček in Trenčín, Faculty of Industrial Technologies, I. Krasku 1809/34, 020 01 Púchov, Slovakia
[2]Czech Technical University in Prague, Department of Materials Engineering, Karlovo nam. 13, 121 32 Prague 2, Czech Republic

Abstract. The presented paper is focused on study and investigation of structural changes in relation to the selected materials which are exposed to the thermal loading. The given work is connected with closer investigation of such material as Dominial ZF 2 which is commonly used for production of steel moulds and components in the glass industry. Moreover, the given work is also devoted to the investigation of materials which are commercially called Toolox 33 and Toolox 44 because these materials were proposed by firm to be used as substitutive materials of Dominial ZF 2. Both of the mentioned and investigated materials have not been used in glass industry yet. Experimental procedures were proposed on the basis of the fact that the investigated materials are exposed to thermal as well as mechanical loading and they are in direct interaction with hot molten glass. After simulated thermal loading, the investigated materials were exposed to the static tensile test and by this way, it was possible to obtain samples with fracture surfaces and these fracture surfaces were evaluated in terms of characteristic features relating to rupture. On the basis of the obtained results and knowledge from experimental procedures, Toolox 33 and Toolox 44 are not recommended to be used for production of steel moulds and components for glass products.

Keywords: thermal loading, glass moulds, fracture, material fatigue, Cr-Ni steel, stress-strain states, degradation

1 Introduction

Machine production of glass products consists of pouring the molten glass into metallic molds, where it is cooled and solidifies in order to obtain the desired glass shape. During the operation, undesired processes may occur in metallic material. This is reflected in the overall life-time of the glass molds and consequently in the quality of finished glass products.

In glass manufacturing plants, the high-alloyed chromium-nickel austenitic steel with high declared heat resistance, high and low temperature corrosion resistance under the designation DOMINIAL ZF2 is predominantly used. From an economic point of view, the

* Corresponding author: frantiskap@seznam.cz

Reviewers: *Eva* Tillová, *Ján* Vavro

company is considering replacing this material with a cheaper steel grade. Toolox 33 and Toolox 44 martensitic steels with similar declared material properties as DOMINIAL ZF2 were selected as potential replacements.

The purpose of this work was to investigate the behavior of materials subjected to critical heat load until the moment of the first crack initiation [1 - 4]. Simulation of heat load cycles was performed using a custom made device. Based on the formation of the fracture surface and the character of the crack propagation, the resistance against the failure by cyclic temperature changes of all the studied materials was compared.

For purpose of the simulated high temperature testing, test specimens were produced in a way that allowed observation of the effects of these changes after the cyclic loading. The samples were rapidly heated to app. 800 °C and then cooled down (in app. 5 seconds). During the heat load breaks, a visual check between the individual cycles was conducted to identify the initiation of the failure. The fracture surfaces prepared by the finishing failure using a static tensile test machine were compared fractographically. Evaluation was carried out on all materials after the identical number of cycles (50 cycles).

2 Characteristics of the evaluated materials

Toolox 33 may be described as a quenched and subsequently tempered tool steel. It is made with low residual stress, resulting in very good dimensional stability. This material does not require further heat treatment and has excellent workability. It is weldable and can be used with a polished surface. The steel is one of the low alloyed materials (it can be tempered at approx. 600 °C). The standard does not recommend this steel to be exposed to temperature higher than 590 °C, where the declared properties are not guaranteed anymore [5].

Table 1. Chemical composition of the selected materials

Material	Chemical composition [wgt.%]									
	C	Si	Mn	Cr	Ni	Mo	P	S	Al	V
Dominial ZF 2 – as delivered	0,12	2	0,9	25	20	*	*	*	*	*
Dominial ZF 2 Chemical analysis– CTU	*	1.31	1.06	28.61	15.36	*	*	0.10	0.26	*
Toolox 33 - as delivered	0.25	0.60	0.90	1.20	0.70	0.40	0.010	0.004	*	0.125
Toolox 33 – Chemical analysis– CTU	*	0.92	0.84	1.47	*	0.71	*	*	0.15	0.17
Toolox 44 – as delivered	0.31	0.60	0.90	1.35	0.70	0.80	0.010	0.004		0.145
Toolox 44 Chemical analysis– CTU	*	1.01	0.79	1.44	*	*	*	0.24	0.19	0.20

Toolox 44 may be considered identical to Toolox 33 when based on several properties. It is also a quenched and subsequently tempered tool steel with excellent dimensional stability, workability and excellent polishing capability. The working conditions are similar, which means that it should not be exposed to temperatures higher than 590 °C for a longer time periods. Unlike Toolox 33, it has a different chemical composition (Table 1) and is shows better toughness. These materials should be homogeneous with high metallurgical purity and good thermal conductivity. They are used for machining, welding and various surface treatments [6].

3 Evaluation of microstructure changes

By the evaluating the micropurity of the specifically delivered steels (Toolox 33 and Toolox 44), relatively large amounts of carbonitrides and complexes based on oxycarbonitrides and relatively high content of sulfide inclusions were found in the structure. In cases of all the investigated materials, (incuding Dominial ZF2) the local occurrence of oxide phases was observed.

The microscopic examination of the delivered materials showed that both proposed steels had a structure consisting of lower bainite, respectively of sorbite with local occurrence of martensitic needles and residual austenite (Fig. 1, Fig. 4). The microstructure of the Dominial ZF2 steel is austenitic with the occurrence of fine carbides along the grain boundaries (Fig. 7).

Samples were made from all of these materials for a simulated cyclic heat load test which was followed by a finishing failure using the instrument used for static tensile testing. The heat load cycling of individual materials took place until the first crack initiation occurred. The cyclical heating of Toolox 33 was finished after app. 80 cycles, Toolox 44 after about 50 cycles, and DOMINIAL ZF2 after about 90 cycles or more (when the first initiation of the failure occurred). Attention was paid to the failure and fracture surfaces of all materials after the same number of cycles (50 cycles) (Fig. 3, 6 and 9).

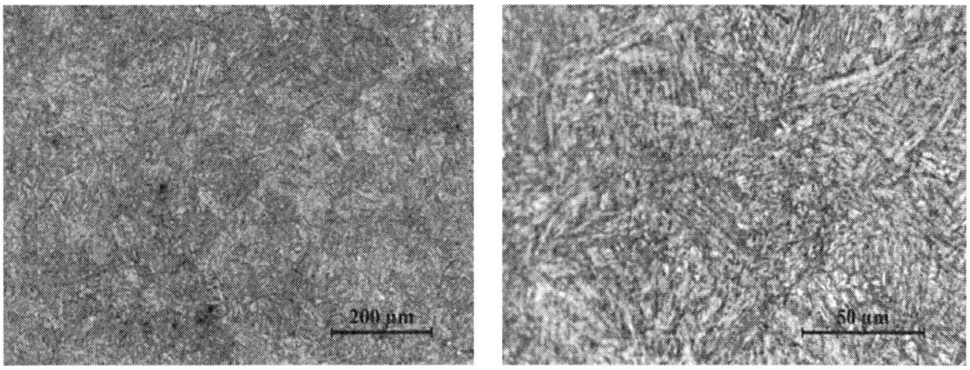

Overview Detail

Fig. 1. Microstructure of the as delivered Toolox 33 material – uncycled

On metallographically prepared sections (Fig. 2) of the area under the fracture surface (after the heat load cycles), structural changes that characterized the fatigue process around the crack were observed. In this area, there occurred strain-strass state, which subsequently resulted in a failure. The fracture surface shows the defined locations in the structure, which are typical for the coexistence of individual phases with different ductile and strength characteristics (Fig. 3). These areas support the local concentration of microdeformation processes leading at different sites to cohesion failure (the red arrow indicates the site of initiation and the occurrence of secondary cracks).

 https://doi.org/10.1051/matecconf/201815707011

Fig. 2. Microstructure in the proximity of the failure – Toolox 33 – after 50 cycles

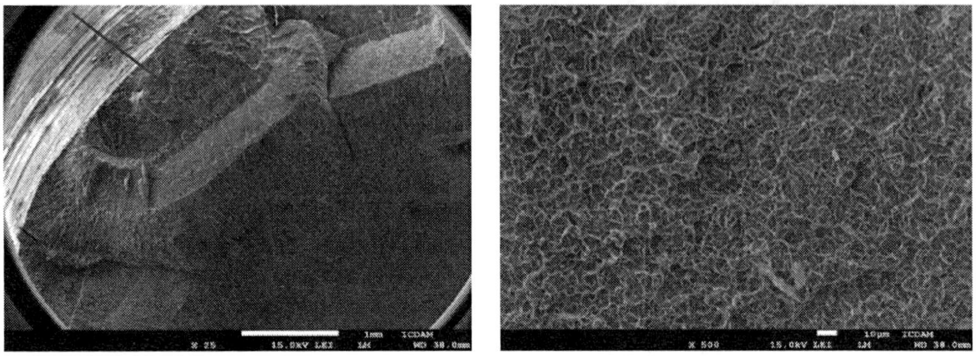

Overview Detail

Fig. 3. Fracture surface of Toolox 33 after 50 cycles

Toolox 44 has a similar microstructure (Fig.4) as Toolox 33 including the occurrence of inclusions formed due to chemical heterogeneity of the material.

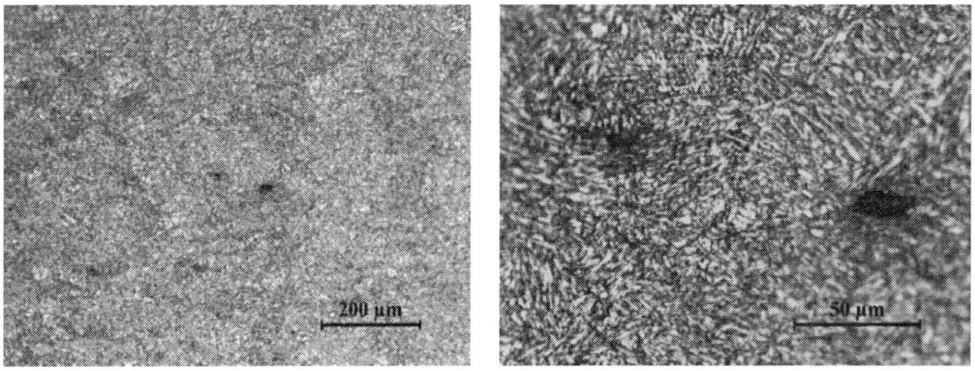

Overview Detail

Fig. 4. Microstructure of the as delivered Toolox 44 material – uncycled

Even in this case, there is a structurally changed zone after the heat load cycles, where exist high stress - strain states (Fig. 5) as in the previous material. The scale of this area will be dependent on character of the as delivered steel.

Fig. 5. Microstructure under the fracture surface - Toolox 44 – after 50 cycles

In the given case, it is a failure of the mixed nature, where ductile and fragile locations alternate in the fracture surface depending on the character of phases present in the structure (Fig. 6). The arrow point at the initiation sites of the failures.

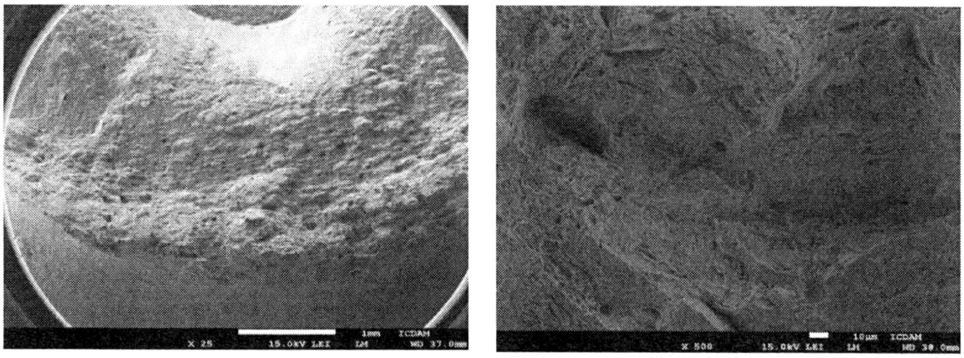

Overview Detail

Fig. 6. Fracture surface - Toolox 44, after 50 cycles

Dominial ZF2 steel is heterogeneous depending on size of austenitic grains (Fig. 7), which react to changes of deformation effects (after 50 heat load cycles) in different ways. There occurs a texture, which cease to exist in proximity of the fracture surface (Fig. 8).

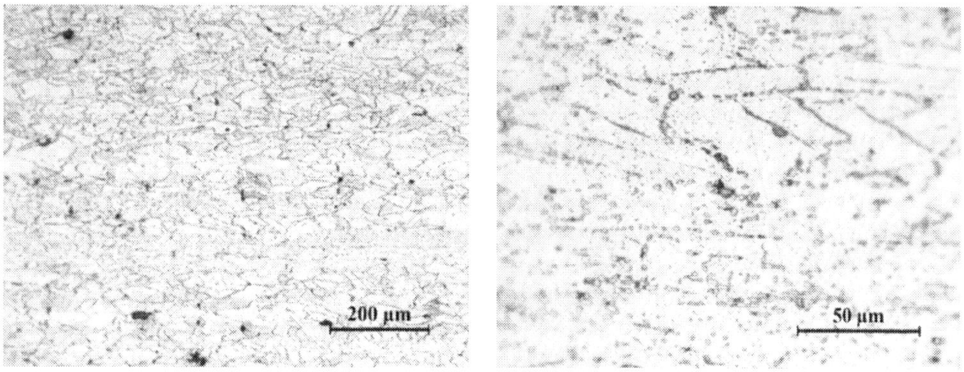

Overview Detail

Fig. 7. Microstructure of the base material - Dominial ZF – uncycled

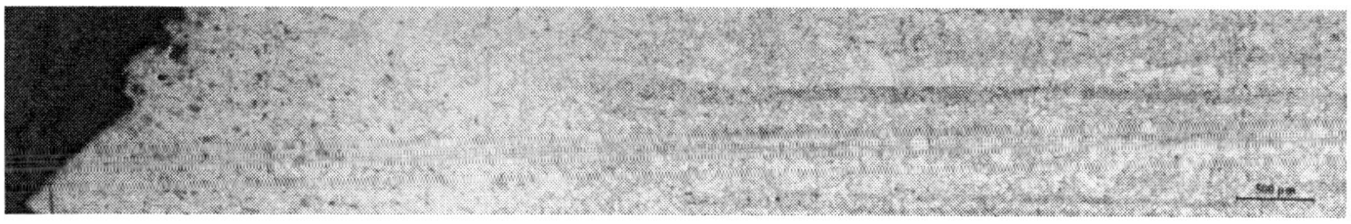

Fig. 8. Microstructure in proximity of the fracture – Dominial ZF 2 – after 50 cycles

Here, it is possible to mark the failure of the Dominial ZF 2 steel as ductile with the different size of the dimples on the edge of the sample and in its core (Fig. 9).

Overview Detail

Fig. 9. Fracture surface - Dominial ZF 2 – after 50 cycles

As can be seen on the figures above the evaluated materials are typical by their heterogeneous microstructure, with which the stress – strain states manifesting in the final character of the fracture surfaces corresponded. Due to the fact that the microscopic and fractographic examination of the structures reveal different mechanisms of the failure (ductile, mixed and fragile), it is possible to attribute a strong influence of occurrence of intermediary phases to the crack initiation occurrences in the evaluated steels.

Conclusions

Based on the morphological changes of the fracture surfaces, it is possible to state that both the temperature fatigue and the creep occur during the process. In operating conditions, this is a thermo-mechanical fatigue leading to a failure of metallic material. That is why the initial structure, micro-purity, distribution and morphology of the intermetallic phases are important. This will be reflected in the fact that each micro-area will have different thermal expansion and conductivity leading to uneven strain states in the microstructure. The number of heat load cycles that the material will withstand without failing characterizes the resistance of the given material to the failure. Simulation has shown that the DOMINIAL ZF2 has always sustained the highest number of cycles until failing.

Based on these findings, it can be stated that the evaluated steels, which are supposed to replace the materials currently used, will be very sensitive to temperature changes. There is a risk of uncontrollable processes in materials initiated by thermal-mechanical load. This will further lead to a greater reduction of the service life of components made of this material. That is the reason why Toolox 33 and 44 steels are not recommended for the production of glass molds.

This work was supported by the Slovak Grant Agency VEGA 1/0649/17, VEGA 1/0589/17, KEGA 007TnUAD-4/2017, and resulted from the project "Center for quality testing and diagnostics of materials", ITMS code 26210120046 relating to the Operational Program Research and Development funded from European Fund of Regional Development.
This work was also supported by the Ministry of Education, Youth and Sport of the Czech Republic, program NPU1, project No. LO1207.

References

1. J. Koutský, F. Jandoš, V. Karel, *Lomy ocelových částí*. (Praha SNTL, 1976)

2. I. F. C. Smith, *The behaviour of short fatigue cracks*. Eds. K. J. Miller, E. R. de los Rios, 1-77, ISBN 981-02-4180-1, (MEP, London, 1986)
3. J. Polák, *Cyklická plasticita a nízkocyklová únavová odolnosť kovových materiálov*. (ČSAV Praha, 136 p., 1986)
4. D. Hull, *Fractography*. (University Press, Cambridge, 378 p., 1999)
5. *Data Sheet 605en Toolox 33*. http://www.damatech.com [online, cit. 2017-07-30] (2015)
6. *Product sheet: Toolox® 44, 450 HBW (~45 HRC) with ESR properties*. https://eng.stahlnetz.de [online, cit. 2017-07-30]

Triaxial analysis of residual stress in surface layers after high feed machining using X-ray diffractometer

Michal Sajgalik[1,*], *Tatiana* Czanova[1], *Lucia* Zauskova[1], *Mario* Drbul[1], *Marek* Sadilek[2], *Jan* Valicek[2]

[1]Department of Machining and Manufacturing Technology, University of Zilina, Univerzitna 1, 010 26 Zilina
[2]Department of Machining, Assembly and Engineering Metrology, Technical University of Ostrava, 17. Listopadu 15/2172, 708 33 Ostrava - Poruba

Abstract. Continuously increasing of productivity is main cause of finding of new ways and methods of machining. High feed machining (HFM) is one of methods of high-productivity machining developed to achieve higher metal removal rates and decrease machining time. Each machining or treatment operation introduces residual stress into the material and its surface layers. Roughing methods of machining generate tensile residual stress obviously. Tensile character of residual stress is not suitable for functional properties of produced parts. In some cases, finishing methods of machining can improve residual stress to suitable character. New detection method of residual stress by applying of x-ray diffraction allows to measure residual stress as triaxial tensor and distribution of residual stress. When we know analyse the residual stress thoroughly, we can design right finishing method and so improve the character of residual stress. This article is focused on analysing given measuring method of residual stress in triaxial direction and next possibilities of their eventual improvement.

Keywords: x-ray diffractometry, residual stress, stress tensor

1 Introduction

Given that there is currently increasing serial production of components, machining technologies are also being developed. One of them is high-feed milling technology, which is characterized by material removal at a low depth of cut and high feed per tooth. During high-feed milling, stresses are formed in the material. These stresses, also called residual stresses, exist in the material even when the material is not already loaded by external forces. The common cause of these stresses is the manufacturing process (machining technologies, welding, forming, heat treatment, etc.). [1, 2]

Residual stresses are formed in the surface layers only if there is a plastic deformation in the surface layer. Whether the stress will be tensile or compressive depends on how the

* Corresponding author: michal.sajgalik@fstroj.uniza.sk

Reviewers: *Ivan* Kuric, *Tomáš* Lack

deformation has been occurred. It can be caused by the following factors: thermal phase transformation, thermal and plastic deformation, and mechanical deformation (shown in Fig. 1). In the case of thermal phase transformation, the residual stress is caused by a volume change (Fig. 1(a)). If the phase change causes a decrease in volume, the surface layer will want to contract, but the underlying bulk material will resist. In the result the surface layer will be under tension. In this case, the tensile stress is produced. If the phase transformation causes an increase in volume, the residual stress will be compressive. This is the case with the conventional heat treatment of steel. In the model of thermal and plastic deformation, the unit event heat will cause expansion of the surface layer. The expansion is relieved (whilst the heating is maintained) by plastic flow, which is restricted only to the surface layer. After the heat is removed, the surface layer will contract, resulting in a tensile residual stress (Fig. 1(b)). In the model of mechanical deformation, the residual stress is compressive after the surface layer is compressed by some form of mechanical action (Fig. 1(c)). There are no heating effects. Processes consisting of only chemical unit events produce zero residual stress [1-3].

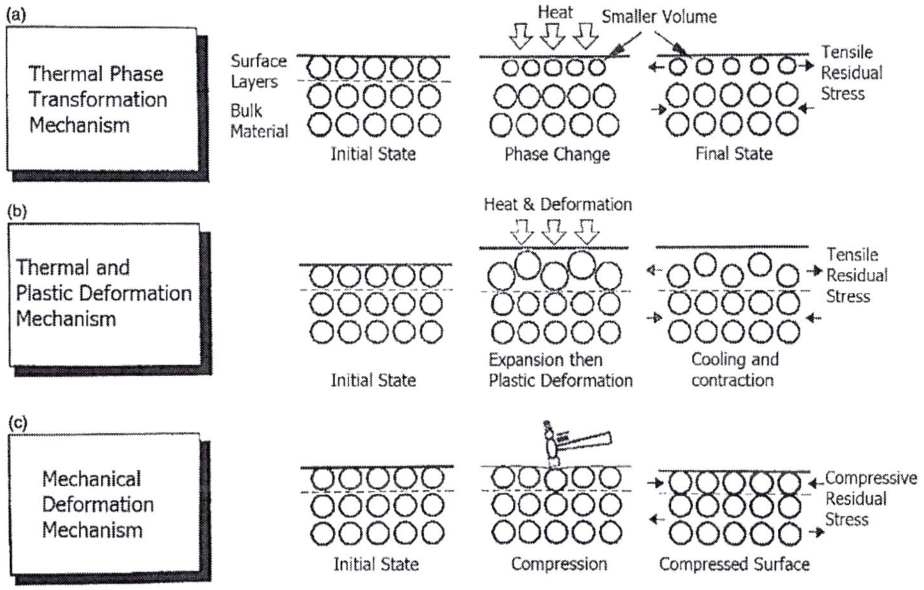

Fig. 1. The residual stress formation models [3]

Many detection methods are used to detect the residual stress in material. One of them is the non-destructive method of X-ray diffractometry. The principle is based on the influence of residual stress in the crystallographic lattice and distance of its planes (d), which is measured. As long as the stress is zero, the distance of crystallographic planes only depends on the material properties. The planes distances vary by the influence of the residual stress. Any change in the crystallographic planes distance d results in the corresponding diffraction angle. The incident X-ray beam penetrates the surface layers of the material, where depending on the crystallographic planes distance it will diffract to the diffraction cone recorded on the detectors [4-7].

The triaxial residual stress measurement is characteristic by measuring at various angles φ, when it is possible to gain a stress tensor containing normal and shear stress components acting at the measurement point. When the measurements for three directions in the point are recorded, (typically $\varphi = 0°, 45°, 90°$), the full stress tensor is calculated, comprising the normal stress for each of the measured directions, and the shear stresses also [2, 5, 8, 9].

The stress tensor in material can be written:

$$\begin{bmatrix} \sigma_{11} & \tau_{12} & \tau_{13} \\ \tau_{21} & \sigma_{22} & \tau_{23} \\ \tau_{31} & \tau_{32} & \sigma_{33} \end{bmatrix} = \begin{bmatrix} \sigma_{xx} & \tau_{xy} & \tau_{xz} \\ \tau_{yx} & \sigma_{yy} & \tau_{yz} \\ \tau_{zx} & \tau_{zy} & \sigma_{zz} \end{bmatrix} \quad (1)$$

where σ is normal stress component and τ is the shear stress component acting in the place of measurement (as shown in Fig. 2) [2, 10].

When calculating the stress tensor, it is generally assumed that the stress normal to the specimen surface (σ_{33}) is zero. There can be no stress normal to a free surface in equilibrium. However, a stress can be maintained in this surface-normal direction at some depth below the surface. Hence for some deeply penetrating measurements (tens of microns), it is possible that a non-zero value of stress could be present. In this case, calculation of the stress tensor requires knowledge of the stress-free lattice parameter d_0 for the examined material [5, 6, 11].

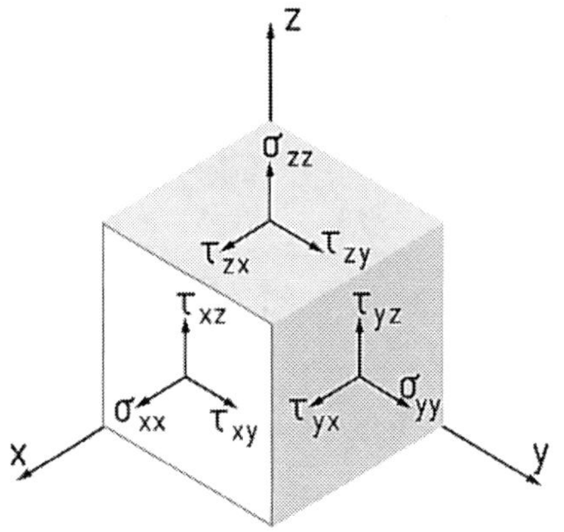

Fig. 2. Normal and shear residual stress components acting in the volume of material

In most practical applications, such as surfaces created by machining, grinding or quenching, the crystallographic lattice plane distance changes due to plastic deformation or heat treatment, which prevents the independent detection of the stress-free plane distance with sufficient precision. Extensive data collection and dependence on knowledge of d_0 limits the method of determining a complete stress tensor primarily for research purposes [8, 12, 13].

2 Experimental measurements

Samples made of X37CrMoV5-1 (19 552) steel hardened and tempered to hardness of 45 HRC with dimensions 40x40x340 mm, were used as experimental material. The samples were milled with a Φ25 mm 2-insert holder and JDMW09T320ZDSR-FTVP30RT high-feed inserts. When selecting cutting parameters, a constant cut speed v_c = 200 m.min^{-1} was selected. A cut depth of a_p = 0,5 mm was chosen, with a feed per tooth of f_z = 0,5 mm; 1 mm; 1,5 mm; 2 mm.

Following the high-feed milling operations on the vertical 3-axis CNC milling machine and subsequent cleaning and degreasing of milled surfaces, residual stresses were measured on the samples by triaxial analysis on a x-ray diffractometer (Fig. 3). Triaxial detection of residual stresses consisted of measurements at three φ angles (0°, 45°, 90°). The stress tensor with corresponding error has been then created. The software of measuring device generates stress tensor with the condition that the value of σ_{33} is zero. For the measurement,

 https://doi.org/10.1051/matecconf/201815707012

a Cr Kα lamp and collimator of 1 mm diameter were selected. Residual stresses in the samples were measured in the milling head onset area, center of the milled area, and milling head exit area.

Fig. 3. Sample on the measuring table of x-ray diffractometer

Stress tensor for the parameter of $f_z = 0,5$ mm:

Stress tensor / Error

Milling head onset area

$$\begin{bmatrix} -41,71 & -175,42 & -31,69 \\ & 276,74 & -27,40 \\ & & 0,00 \end{bmatrix} \qquad \begin{bmatrix} 5,10 & 7,35 & 1,88 \\ & 5,10 & 1,88 \\ & & 2,10 \end{bmatrix} \tag{2}$$

Center of the milled area

$$\begin{bmatrix} 166,35 & -111,93 & -25,29 \\ & 293,21 & -32,74 \\ & & 0,00 \end{bmatrix} \qquad \begin{bmatrix} 0,27 & 0,39 & 0,10 \\ & 0,27 & 0,10 \\ & & 0,11 \end{bmatrix} \tag{3}$$

Milling head exit area

$$\begin{bmatrix} 363,24 & -109,73 & -13,92 \\ & 300,41 & -14,49 \\ & & 0,00 \end{bmatrix} \qquad \begin{bmatrix} 7,36 & 10,60 & 2,71 \\ & 7,36 & 2,71 \\ & & 3,03 \end{bmatrix} \tag{4}$$

Stress tensor for the parameter of $f_z = 1$ mm:

Stress tensor / Error

Milling head onset area

$$\begin{bmatrix} 177,00 & -211,23 & -19,76 \\ & 321,13 & -28,25 \\ & & 0,00 \end{bmatrix} \qquad \begin{bmatrix} 6,56 & 9,44 & 2,41 \\ & 6,56 & 2,41 \\ & & 2,70 \end{bmatrix} \tag{5}$$

Center of the milled area

$$\begin{bmatrix} 32,75 & -74,03 & -5,17 \\ & 303,52 & -22,00 \\ & & 0,00 \end{bmatrix} \qquad \begin{bmatrix} 12,99 & 18,71 & 4,78 \\ & 12,99 & 4,78 \\ & & 5,35 \end{bmatrix} \tag{6}$$

Milling head exit area

$$\begin{bmatrix} 138,48 & -191,48 & 11,25 \\ & 307,50 & -18,31 \\ & & 0,00 \end{bmatrix} \qquad \begin{bmatrix} 0,65 & 0,93 & 0,24 \\ & 0,65 & 0,24 \\ & & 0,27 \end{bmatrix} \tag{7}$$

Stress tensor for the parameter of f_z = 1,5 mm:

$$\begin{array}{lcc} & \text{Stress tensor} & \text{Error} \\ \text{Milling head onset area} & \begin{bmatrix} -23{,}09 & -143{,}74 & -4{,}00 \\ & 467{,}00 & -10{,}32 \\ & & 0{,}00 \end{bmatrix} & \begin{bmatrix} 2{,}59 & 3{,}73 & 0{,}95 \\ & 2{,}59 & 0{,}95 \\ & & 1{,}07 \end{bmatrix} \end{array} \tag{8}$$

$$\begin{array}{lcc} \text{Center of the milled area} & \begin{bmatrix} 146{,}22 & -161{,}01 & -4{,}87 \\ & 469{,}29 & -29{,}85 \\ & & 0{,}00 \end{bmatrix} & \begin{bmatrix} 8{,}78 & 12{,}64 & 3{,}23 \\ & 8{,}78 & 3{,}23 \\ & & 3{,}61 \end{bmatrix} \end{array} \tag{9}$$

$$\begin{array}{lcc} \text{Milling head exit area} & \begin{bmatrix} -135{,}60 & -5{,}84 & 19{,}09 \\ & 511{,}29 & -31{,}82 \\ & & 0{,}00 \end{bmatrix} & \begin{bmatrix} 38{,}14 & 54{,}93 & 14{,}03 \\ & 38{,}14 & 14{,}03 \\ & & 15{,}69 \end{bmatrix} \end{array} \tag{10}$$

Stress tensor for the parameter of f_z = 2 mm:

$$\begin{array}{lcc} & \text{Stress tensor} & \text{Error} \\ \text{Milling head onset area} & \begin{bmatrix} 102{,}21 & -323{,}61 & -12{,}96 \\ & 391{,}16 & -14{,}43 \\ & & 0{,}00 \end{bmatrix} & \begin{bmatrix} 7{,}43 & 10{,}71 & 2{,}73 \\ & 7{,}43 & 2{,}73 \\ & & 3{,}06 \end{bmatrix} \end{array} \tag{11}$$

$$\begin{array}{lcc} \text{Center of the milled area} & \begin{bmatrix} -178{,}84 & -115{,}58 & -16{,}32 \\ & 385{,}42 & -3{,}25 \\ & & 0{,}00 \end{bmatrix} & \begin{bmatrix} 7{,}51 & 10{,}82 & 2{,}76 \\ & 7{,}51 & 2{,}76 \\ & & 3{,}09 \end{bmatrix} \end{array} \tag{12}$$

$$\begin{array}{lcc} \text{Milling head exit area} & \begin{bmatrix} 188{,}63 & -72{,}74 & -4{,}13 \\ & 375{,}36 & -34{,}86 \\ & & 0{,}00 \end{bmatrix} & \begin{bmatrix} 1{,}96 & 2{,}82 & 0{,}72 \\ & 1{,}96 & 0{,}72 \\ & & 0{,}81 \end{bmatrix} \end{array} \tag{13}$$

From the stress tensors, the graphical dependence of the normal residual stress components oriented in the direction of the milling (σ_{yy}) and the changing feed per tooth was made.

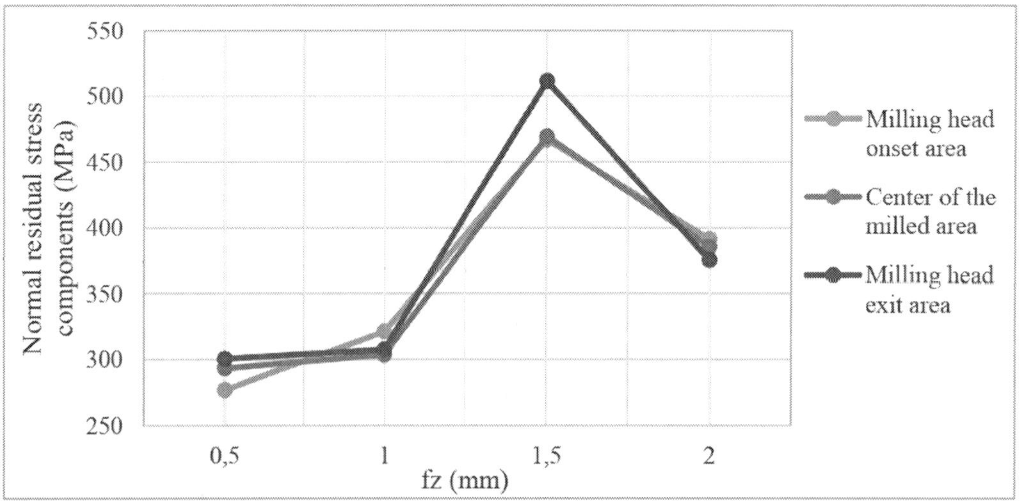

Fig. 4. Dependence of the normal residual stress components oriented in the direction of the milling and the changing feed per tooth

From the graph in Fig. 4 we can see, that the stresses in the measured places showed tensile character. Furthermore, we can see that the normal residual stress components σ_{yy} also grow with the increasing feed per tooth. The difference is only for the selected f_z = 2 mm, where the stresses decreased. The decrease could be due to the fact that the f_z = 2 mm was too high for the machined material and chosen inserts. The lowest values were measured at f_z = 0,5 mm, namely 276,74 MPa in the milling head onset area, 293,21 MPa in the center of the milled area, and 300,41 MPa in the milling head exit area. Then a sample machined with f_z = 1 mm was measured. Stress of 321,13 MPa was measured in the milling head onset area, 303,52 MPa in the center of the milled area, and 307,50 MPa in the milling head exit area. The highest values were measured at f_z = 1,5 mm, namely 467,00 MPa in the milling head onset area, 469,29 MPa in the center of the milled area, and 511,29 MPa in the milling head exit area. For f_z = 2 mm, a value of 391,16 MPa was measured in the milling head onset area, 385,42 MPa in the center of the milled area, and 375,36 MPa in the milling head exit area.

Conclusion

The selected experimental material caused machining problems and the quality of the machined surface has been lower. Improvement of the quality and functional properties of the machined surface, including the residual stresses, could be realized by the use of inserts with different geometry, or by the choice of more suitable cutting parameters. Subject of an article is the method of triaxial measurement of residual stresses. Triaxial analysis of residual stress can be used in closer examination of the impact of cutting tool on the residual stress formation in the material. Method offers many possible applications in order to verificate and extent the theoretical knowledge in this topic.

This article was funded by the University of Žilina project APVV 15-0405 – "Complex use of X-ray diffractometry for identification and quantification of functional properties of dynamically loaded structural elements from important technical materials" and project of University of Zilina: OPVav-2009/2.2/04-SORO number (26220220101) – "Intelligent system for nondestructive technologies on evaluation for the functional properties of components of X-ray diffraction" and KEGA project 025ŽU-4/2017 Integration of innovative detectional-visaulization multi-disciplinary technologies as on-line tool for learning of progressive CNC technologies.

References

1. B. Griffiths, *Manufacturing Surface Technology: Surface Integrity & Functional Performance*. (Penton Press, 256 p., 2001)
2. W. Grzesik, *Advanced Machining Processes of Metallic Materials: Theory, Modelling and Applications*. (Elsevier, 446 p., 2008)
3. A. Czan, E. Tillova, J. Semcer, J. Pilc, *Surface and subsurface residual stresses after machining and their analysis by x-ray diffraction*. Communications – Scientific Letters of the University of Zilina, 69-76 (2013)
4. N. Ganev, I. Kraus, *X-ray diffraction measurement of residual stresses*. Material Structure **9** (2), (2002)
5. A. Czan, M. Sajgalik, J. Holubjak, M. Piesova, T. Czanova, *Simulation of residual stresses in surface and subsurface layer after machining*. Communications – Scientific letters of the University of Zilina, 95-100 (2016)

6. L. Jakubovicova, M. Saga, *Computational analysis of contact stress distribution in the case of mutual stewing of roller bearing rings*. Novel Trends in Production Devices and Systems, Applied Mechanics and Materials **474**, 363-368 (2014)
7. M. Harnicarova, J. Valicek, R. Cep, et. al., *Surface integrity analysis in laser beam cutting process*. Metal 2013: 22nd International conference on metallurgy and materials, 1057-1062 (2013)
8. P. S. Prevey, P.W. Mason, D.J. Hornbach, J.P. Molkenthin, *Effect of prior machining deformation on the development of tensile residual stresses in weld-fabricated nuclear components*. Journal of Materials Engineering and Performance **5** (1), 51-56 (1996)
9. R. Cep, L. Ocenasova, J. Novakova, L. Perkovska, A. Czan, D. Stancekova, *Intereupted Machining Tests of Ceramic Cutting Tools*. TMT 2009 Proc., Hammamet, **13** (1), 733-736, ISSN 1840-4944 (2009)
10. A. Czan, M. Sajgalik, A. Martikan, J. Mrazik, *Observation of dynamic processes in cutting zone when machining nickel alloys*. Communications – Scientific Letters of the University of Zilina **16** (3A), 161-168 (2014)
11. D. Mital, J. Zajac, Z. Mitalova, Monitoring influence of selected parameters on surface quality after polymer powder coating. MM Science Journal **2016**, 1077-1081 (2016)
12. Gorog, A., Stancekova, D., Kapustova, M. Gorogova, I., Miturska, *Influence of Selected Cutting Conditions on Cutting Forces*. Advances In Science And Technology-Research Journal **11** (1), 179-185, ISSN 2299-8624
13. Czán, A., Martikáň, A., Holubják, J., Struhárňansky, J.: *Identification of stress and structure properties in surface and subsurface layers of nuclea reactor austenitic steel*. Manufacturing Technology **14** (3), 2776-281, ISSN 1213-2489 (2014)

The influence of the structure on the fatigue properties of aluminium alloys for the casting

Milan Uhríčik[1,*], *Peter* Palček[1], *Mária* Chalupová[1], *Martin* Frkáň[1]

[1]University of Žilina, Faculty of Mechanical Engineering, Department of Materials Engineering, Univerzitná 8215/1, 01026 Žilina, Slovakia

Abstract. The article will be focused on monitoring the influence of the structure on the fatigue properties of Al-Mg cast alloy. As an experimental material were used aluminium alloys EN AC 51200 and EN AC 51500, supplied in a cast state without a heat treatment, which were produced by the continuous casting method. These alloys were selected on the basis of the chemical composition, where the content of most alloying elements is comparable. Fatigue properties of aluminium alloys were tested by three-point bending cyclic loading. The fracture surface of the testing sample was examined using scanning electron microscopy (SEM), where samples were observed on various stages of the fatigue process, their characteristics and differences of fracture surfaces.

Keywords: fracture surface, Al-Mg cast alloy, structure

1 Introduction

Aluminium alloys with magnesium as the major alloying element constitute a group of heat treatable alloys with medium strength, high ductility, excellent corrosion resistance and weldability [1-3]. Wrought Al–Mg alloys are used as structural materials in marine, automotive, aircraft and cryogenic applications while the cast forms are used mainly for their corrosion resistance in dairy, food handling and chemical processing applications. Unlike the heat treatable alloys, these materials derive their strength primarily from solid solution strengthening by Mg, which has a substantial solid solubility in aluminium. However, in order to obtain strength levels approaching the regime of the precipitation hardening alloys, high Mg levels are required. Such high levels of Mg pose processing challenges and can increase the susceptibility of the alloys to stress corrosion cracking [3-6]. Fatigue failures in metallic structures are a well-known technical problem. In a specimen subjected to a cyclic load, a fatigue crack nucleus can be initiated on a microscopically small scale, followed by crack grows to a macroscopic size, and finally to specimen failure in the last cycle of the fatigue life. Understanding of the fatigue mechanism is essential for considering various technical conditions which affect fatigue life and fatigue crack growth, such as the material surface quality, residual stress, and

* Corresponding author: milan.uhricik@fstroj.uniza.sk

Reviewers: *Richard* Pastirčák, *Marta* Kianicová

environmental influence. This knowledge is essential for the analysis of fatigue properties of an engineering structure [7-9].

2 Material characteristics

As an experimental material were used aluminium cast alloys of type Al-Mg (EN AC-51200 and EN AC-51500), supplied in a cast state without a heat treatment in the form of plates (Fig. 1). They were produced by the continuous casting method. Test samples for determining the fatigue properties were made by the cutting operation from these plates, which were cut from casts (Fig. 2). The chemical composition of both alloys was examined by spark emission spectrometer SPECTROMAXx and results are shown in the Table 1. These alloys were selected based on this chemical composition, where the content of most of alloying elements is comparable, only a concentration of magnesium is considerably different.

Fig. 1. The plate cut from a casting

Fig. 2. The sample for a fatigue test

Table 1. The chemical composition of alloys, in wt. %.

	Si	Fe	Cu	Mn	Mg	Cr	Pb	Al
EN AC 51200	1.3	0.102	0.003	0.416	10.32	0.006	0.027	balance
EN AC 51500	1.53	0.11	0.004	0.529	6.28	0.008	0.031	balance

These materials have excellent corrosion resistance, very good workability, polishability and good weldability. The advantage of these alloys is that they are used without a heat treatment, thereby saving the costs of production of parts. Products from these alloys are mainly produced by high pressure casting method.

3 Experimental methodology

3.1 Metallographic analysis

Preparation of samples, which need to be given a very good attention because of its quality depends on the quality of the analysis of structures, consisted of the following operations:

Cutting – Specimens for metallographic analysis were taken from supplied plates, using a precision automatic saw MTH MIKRON 3000. Samples were cutting without damage by deformation or without heat affection.

Preparation – For the sample preparation was used a hot mounting equipment STRUERS CITOPRESS and mixtures Dentacryl and Bakelite. Pressing of samples took about 3 minutes at temperature 180 °C and pressure 25 MPa.

Grinding – After mounting of samples, they were grinded by using a semiautomatic equipment TEGRASYSTEM. For this operation were used two abrasive discs with a various grain size. The rough grinding was done by an abrasive disc with a grain size 500 and for the fine grinding was used an abrasive disc with a gain size 1200. Both grinding was carried out for 1 minute. During this operation, discs were wetted with water to prevent the heat affected of samples.

Polishing – It was also performed on a semiautomatic machine TEGRASYSTEM and it was carried out in three steps. In the first step of polishing, it was used a polishing disc MDMol with a suspension DiaProMol, in the second step was used a disc MDNap with a suspension DiaProNap and in the last step was used a disc MDChem with a paste OP-S.

Etching – Samples were etched with 0.5% hydrofluoric acid solution for 10 seconds.

The thus prepared samples were used for metallographic analysis, using a light metallographic microscope NEOPHOT 32.

3.2 Fatigue tests

Fatigue tests were carried out on the test equipment ZWICK/ROELL AMSLER 150 HFP 5100 (Fig. 3) according to norm STN 42 0363. This equipment is using the resonance principle with constant or variable amplitude and mean load. Smooth rectangular specimens without notches (10x18x40 mm) (Fig. 2) were loaded by cyclic three-point bending loading with mean stress σ_m = -144 MPa and frequency 120 Hz.

Fig. 3. The test equipment ZWICK/ROELL AMSLER 150 HFP 5100

3.3 Fractography

Fractographical assessment of fracture surfaces was realized on samples from each material after the fatigue test. On samples were observed various stages of fatigue process, their characteristics and differences of fracture surfaces between samples loading in high and low amplitude of stress. It has also been observed an area of final fracture. Fracture surfaces were evaluated by scanning electron microscopy TESCAN VEGA LMU II.

4 Results and discussion

4.1 Microstructural analysis and phase analysis

Considering that alloys were casted under the same technological conditions, it can be said that the structure affects only the chemical composition of alloys. Microstructure of materials is expressively dendritic (Fig. 4). The highest percentage representation of dendrites has α-phase (matrix) that is a solid solution of aluminium together with other additive elements. In the structure, there is also an intermetallic phase Mg_2Si. This phase has a different morphology in alloys. In the case of EN AC 51200 alloy, it has a shape of skeleton formation and is distributed in interdendritic areas. There is also β-phase, which is formed in interdendritic areas and has a stoichiometric basis Al_3Mg_2. In alloy EN AC 51500, intermetallic phase Mg_2Si has a finer morphology and does not have a typical shape of skeleton formation.

Because elements such as iron and manganese are also in alloys, can be observed a small amount of Fe-rich phase in structure of both materials, probably Al_6 (FeMn). β-phase is not present in this alloy, because of the smaller content of magnesium.

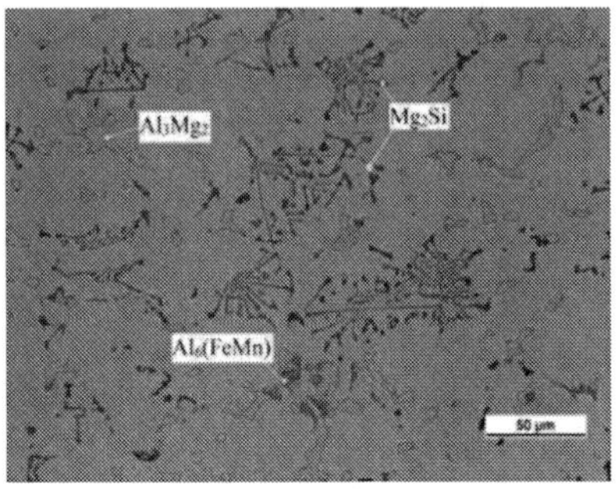

a) EN AC 51200

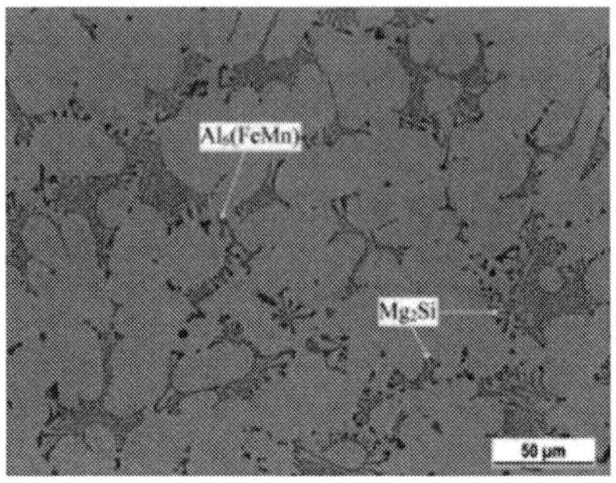

b) EN AC 51500

Fig. 4. Microstructures of aluminium alloys

For identification and confirmation of structural components was used square and line analysis. Results are shown in Fig. 5 – Fig. 8.

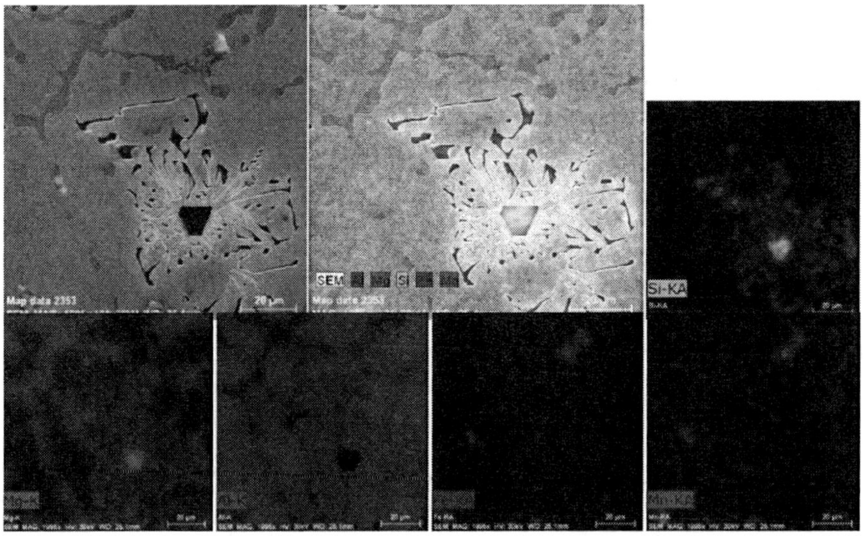

Fig. 5. Identification of structural components of aluminium alloy EN AC 51200 – a square analysis

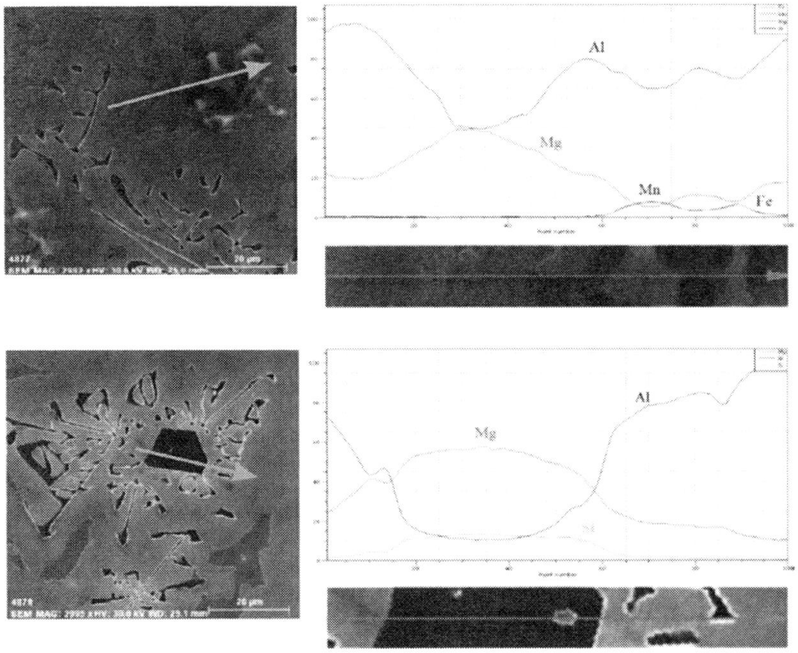

a) Al_6 (FeMn) and Al_3Mg_2

b) Mg_2Si

Fig. 6. Identification of structural components of aluminium alloy EN AC 51200 – a line analysis

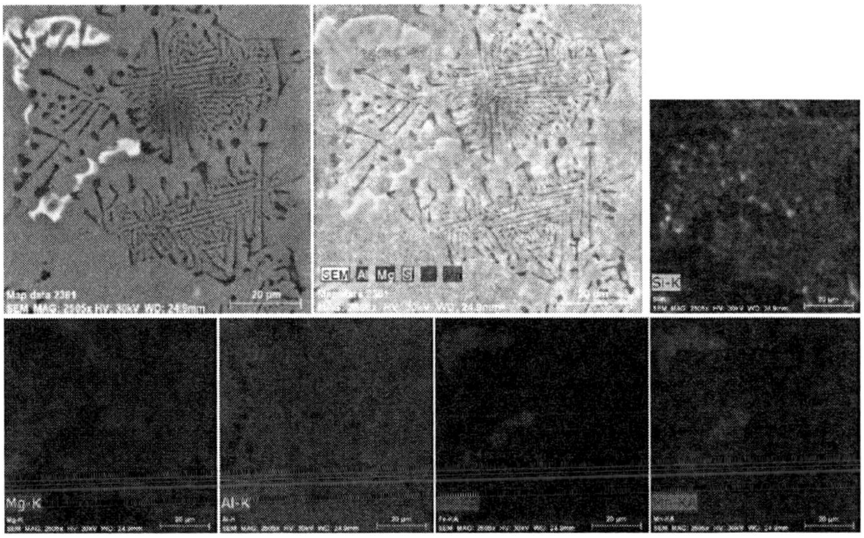

Fig. 7. Identification of structural components of aluminium alloy EN AC 51500 – a square analysis

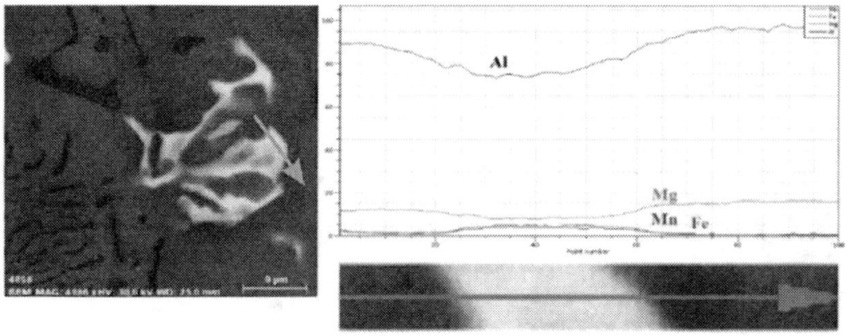

a) Al_6 (FeMn)

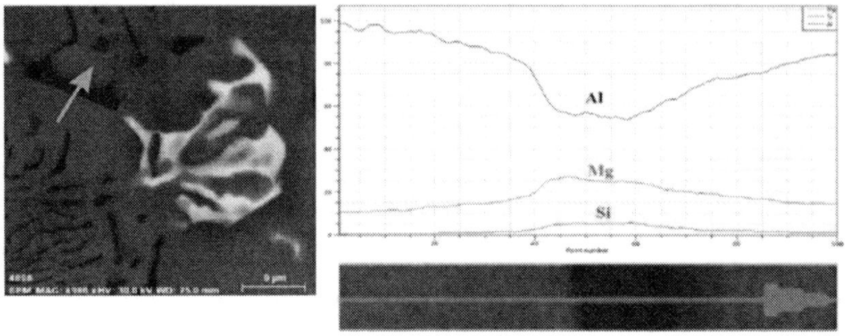

b) Mg_2Si

Fig. 8. Identification of structural components of aluminium alloy EN AC 51500 – a line analysis

It was also made a percentage assessment of shares of individual phases (Fig. 9 and Fig. 10). Results are shown in the Table 2. It can be concluded that the phase Al_6(FeMn) is comparable in both alloys. Mg_2Si phase is about 4% higher in alloy EN AC 51500. This phase can improve the hardness and tensile strength, but can also aggravate the fracture toughness and the ductility, depending on the morphology and the amount. Only the alloy EN AC 51200 contained the β-phase, in an amount of 8%. This phase is a brittle and reduced ductility at temperatures below 350°C.

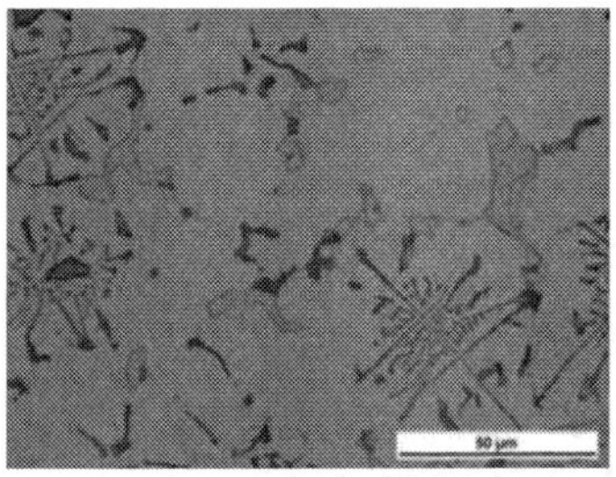

a) a share of Al_6 (FeMn)

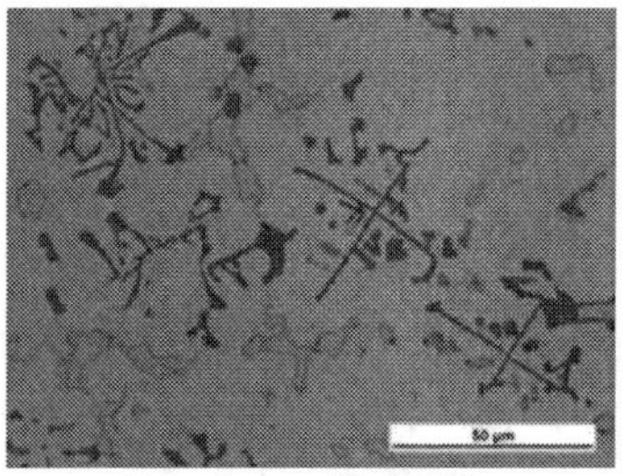

b) a share of Mg_2Si

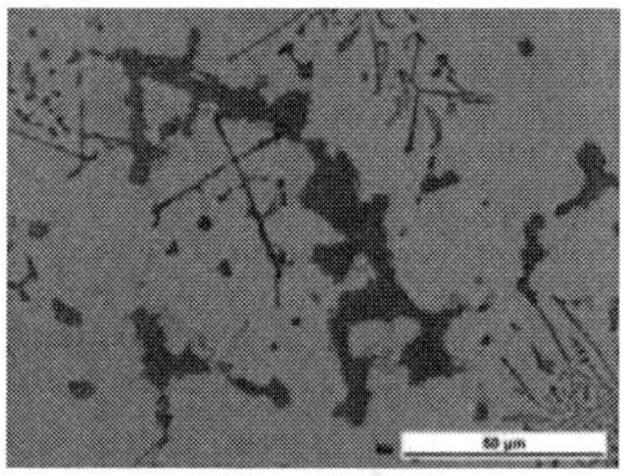

c) a share of Al_3Mg_2

Fig. 9. A measuring of percentage assessment of phases in the alloy EN AC 51200

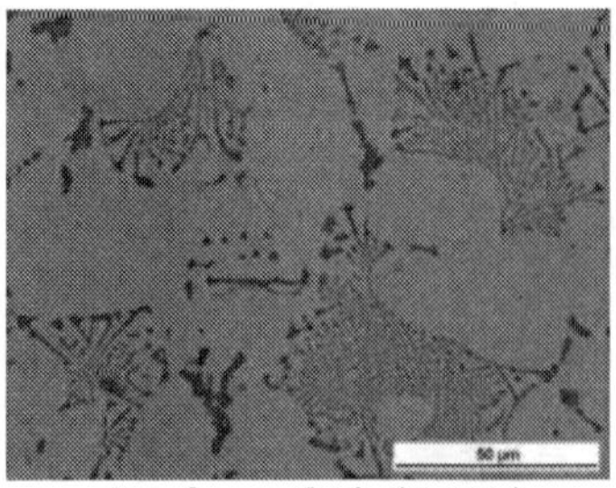

a) a share of Al_6 (FeMn)

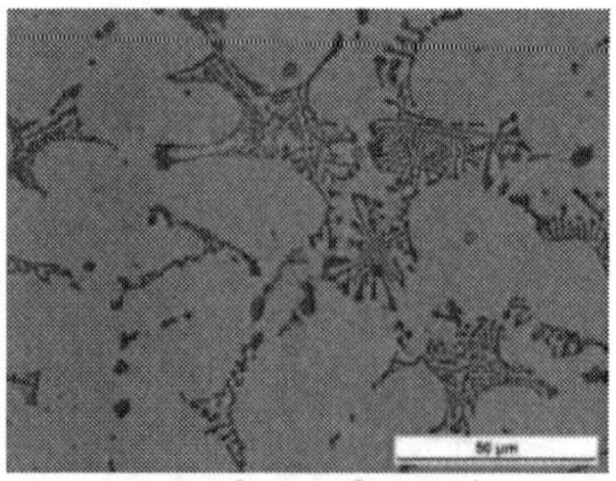

b) a share of Mg_2Si

Fig. 10. A measuring of percentage assessment of phases in the alloy EN AC 51500

Table 2. A percentage assessment of phases in alloys [%].

	Al_3Mg_2	Mg_2Si	$Al_6(FeMn)$	α-phase
EN AC 51200	8	7	0.7	balance
EN AC 51500	0	11	1.2	balance

4.2 Fatigue lifetime

Alloys EN AC 51200 and EN AC 51500 were loaded by cyclic three-point bending loading with the same parameters, only stress amplitude was changing. From results of fatigue tests were constructed graphs σ_a-N in semi-logarithmic coordinates, for every alloy (Fig.11 and Fig. 12). By comparing of graphs can be stated that alloy EN AC 51500 has better fatigue properties than alloy EN AC 51200 in analyzed area of the number of cycles.

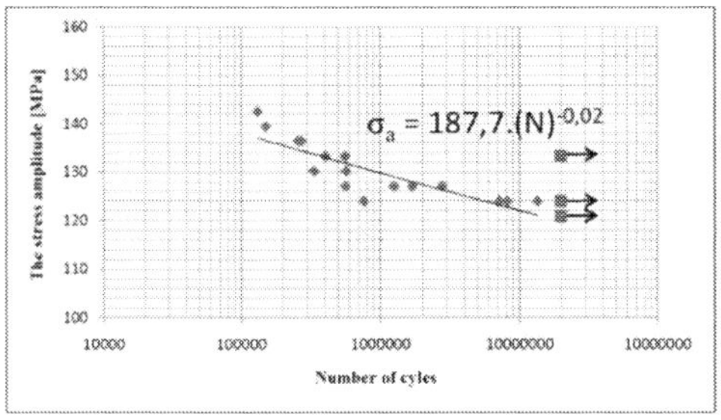

Fig. 11. Results of fatigue tests for alloy EN AC 51200

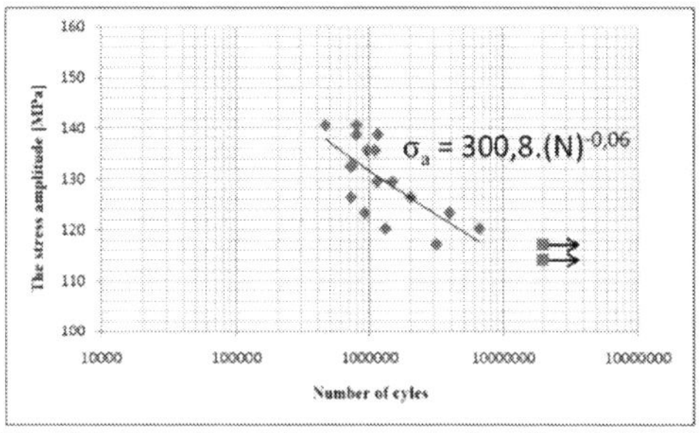

Fig. 12. Results of fatigue tests for alloy EN AC 51500

4.3 Fatigue lifetime

Fracture surfaces of test specimens were evaluated at first in term of macrofractographic. From this analysis it is possible to determine initiation (nucleation) of fatigue crack, surfaces of individual areas of fracture, namely the fatigue fracture area and the area of static final failure.

A general view of the fracture area for the alloy EN AC 51500 after 1314148 cycles at a stress of 120 MPa it documented in Fig. 13a, where can be seen various stages of fatigue process. Detail of initiation site is shown in Fig. 13b. Transgranular fatigue failure of α-phase (matrix) and there were observed typical signs of fatigue – striations and also an interfacial and a cleavage mechanism and failure of Mg_2Si phase, this all is shown in Fig. 13c. Fig. 13d displays the area of static final failure, where can be seen plastic´s transformed ridges of α-phase, together with a transgranular ductile fracture of α-phase with dimple morphology and also is visible an interfacial failure of Mg_2Si phase.

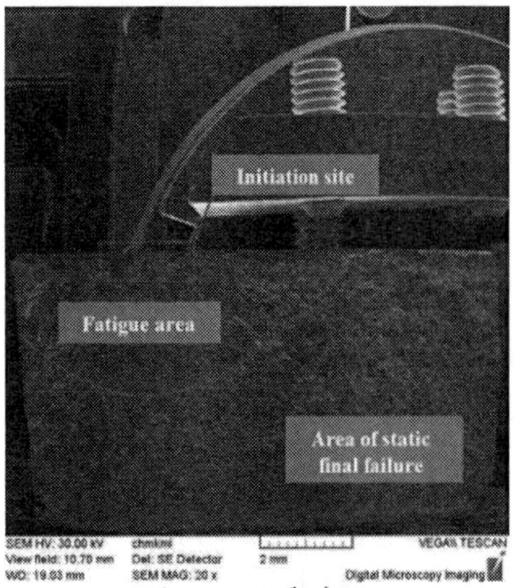

a) a general view

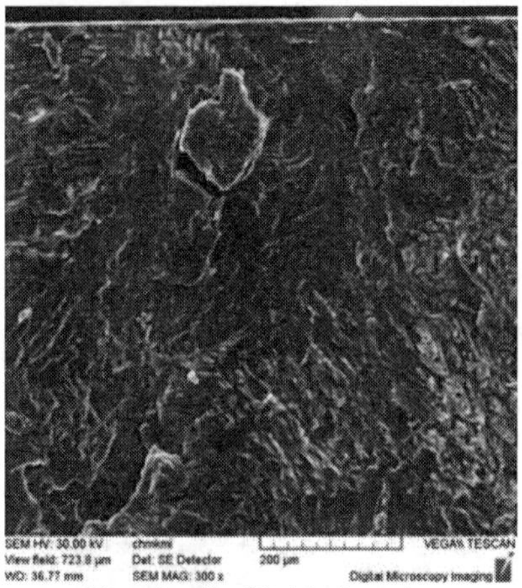
b) a detail of initiation site

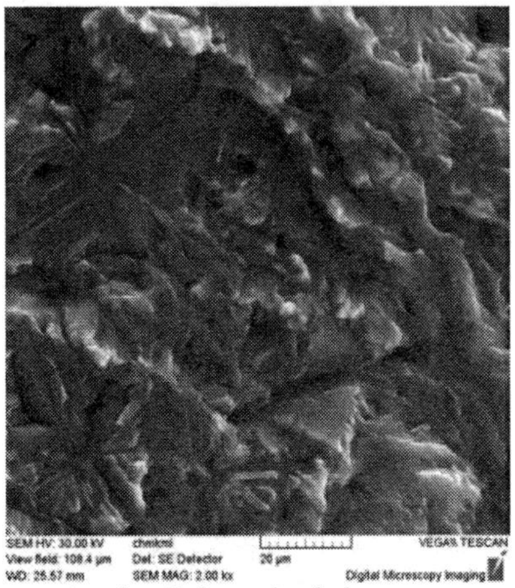
c) a detail from the fatigue area

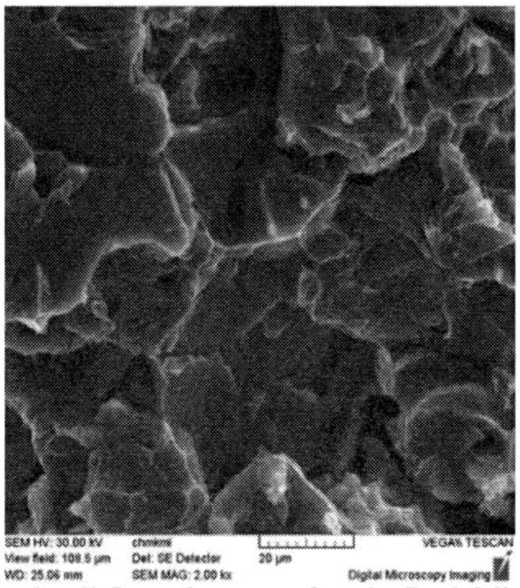
d) a detail from the area of static final failure

Fig. 13. Fracture surface of aluminium alloy 51500, σ_a = 120 MPa, N_f = 1314148 cycles

A general view of the fracture area for the alloy EN AC 51500 after 799229 cycles at a stress of 138 MPa is shown in Fig. 14a. Here can be seen that when higher amplitude of stress was applicate, the fatigue crack was spread from several initiation sites on the specimen surface and there was also a reduction of the fatigue area. Detail of one initiation site is shown in Fig. 14b. Transgranular fatigue failure of α-phase (matrix) and there were observed typical signs of fatigue – striations a it is shown in Fig 14c. Fig. 14d displays the area of static final failure, where can be also seen plastic´s transformed ridges of α-phase, together with transgranular ductile fracture of α-phase with dimple morphology and also is visible an interfacial failure of Mg_2Si phase.

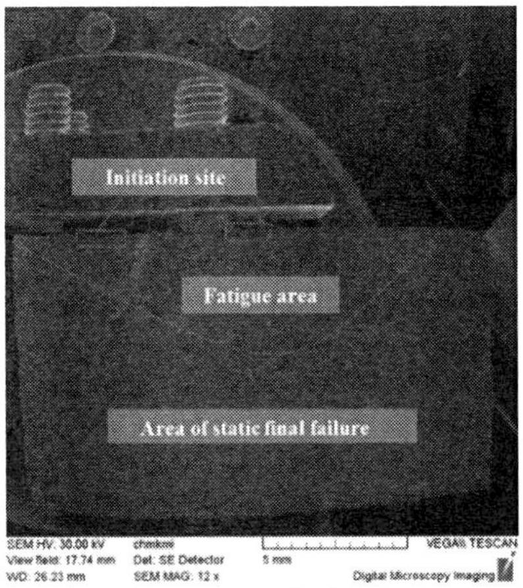

a) a general view

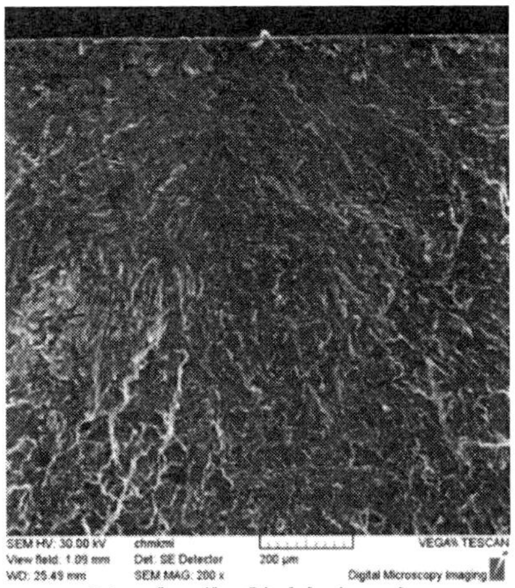

b) a detail of initiation site

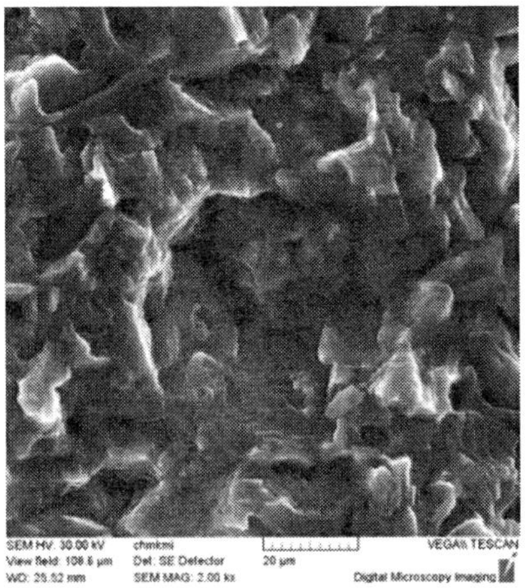

c) a detail from the fatigue area

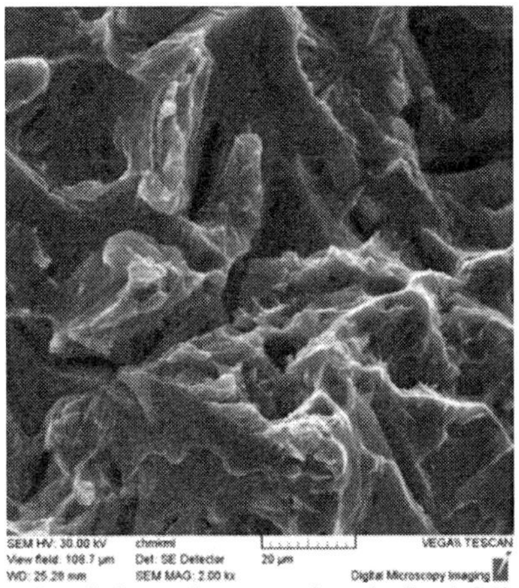

d) a detail from the area of static final failure

Fig. 14. Fracture surface of aluminium alloy 51500, σ_a = 138 MPa, N_f = 799229 cycles

A general view of the fracture area for the alloy EN AC 51200 after 761899 cycles at a stress of 124 MPa is documented in Fig. 15a, where can be seen various stages of fatigue process. Detail of initiation site is shown in Fig. 15b. In Fig 15c can be seen an interfacial failure of intermetallic phases (Al_3Mg_2, Al_6(FeMn) and Mg_2Si) and also a transgranular fatigue failure of α-phase (matrix). Fig. 15d displays the area of static final failure, where can be seen interfacially broken intermetallic phases and plastic´s transformed ridges of α-phase, together with transgranular ductile fracture of α-phase with dimple morphology.

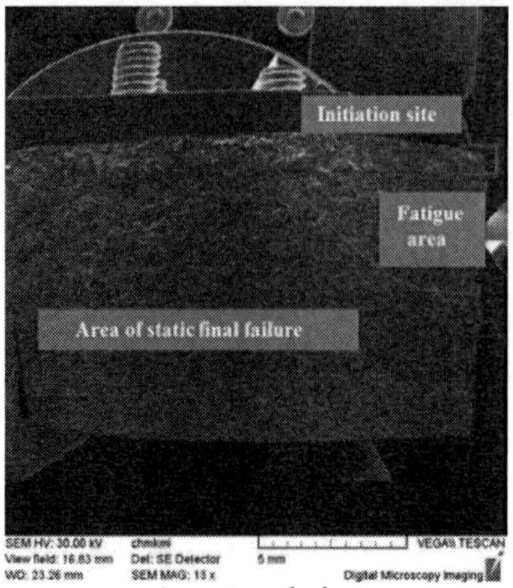

a) a general view

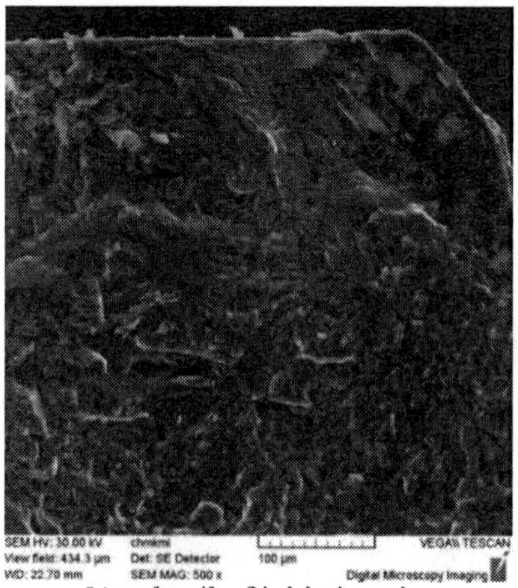

b) a detail of initiation site

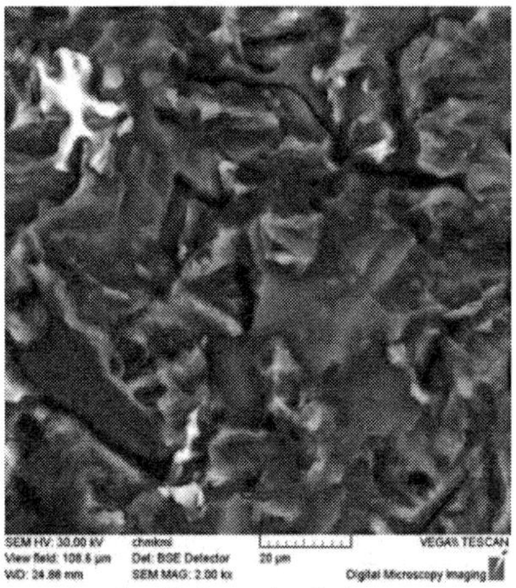

c) a detail from the fatigue area

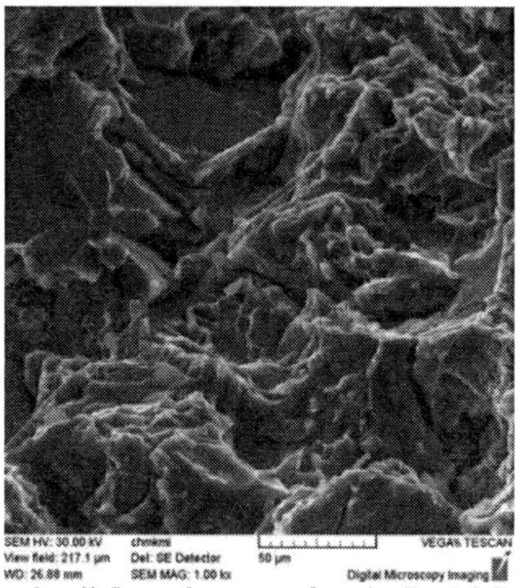

d) a detail from the area of static final failure

Fig. 15. Fracture surface of aluminium alloy 51200, σ_a = 124 MPa, N_f = 761899 cycles

A general view of the fracture area for the alloy EN AC 51200 after 150530 cycles at a stress of 139 MPa is shown in Fig. 16a. Here can be seen that when higher amplitude of stress was applicate, the fatigue crack was spread from several initiation sites on the specimen surface and there was also a reduction of the fatigue area and a smaller number of cycles to fracture. Detail of one initiation site is shown in Fig. 16b. In Fig 16c can be seen a transgranular fatigue failure of α-phase (matrix) and also an interfacial failure of intermetallic phases (Al_3Mg_2 and Al_6(FeMn)). Fig. 16d displays the area of static final failure, where can be seen plastic´s transformed ridges of α-phase, together with a transgranular ductile fracture of α-phase with dimple morphology and also is visible an interfacial failure of Mg_2Si and Al_3Mg_2 phases.

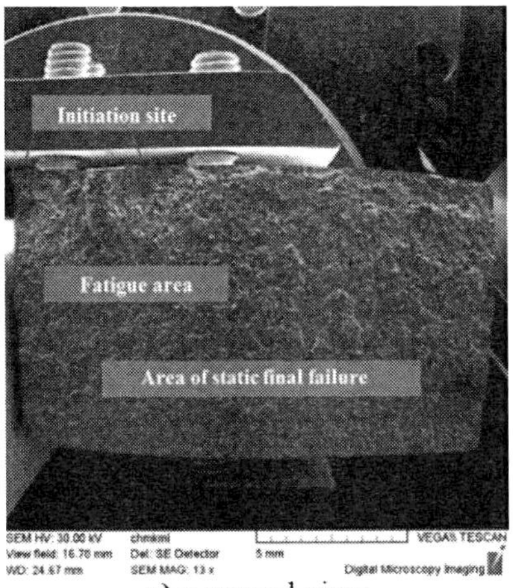

a) a general view

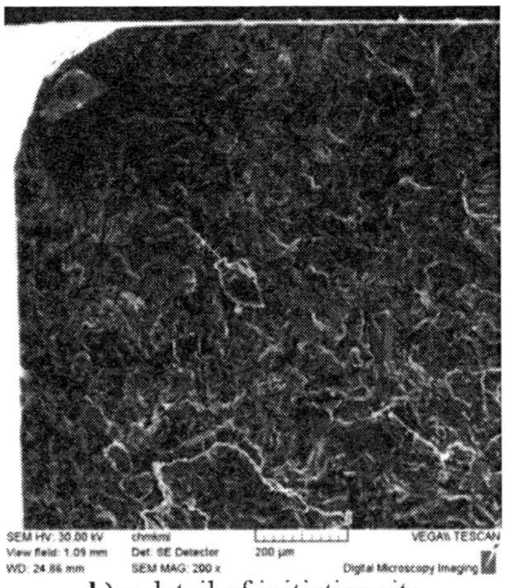

b) a detail of initiation site

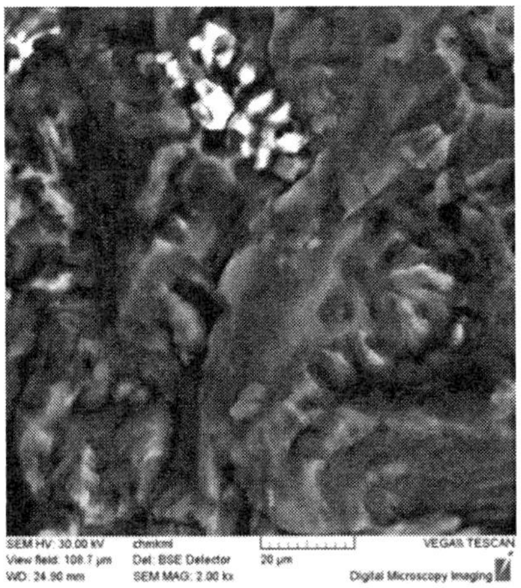

c) a detail from the fatigue area

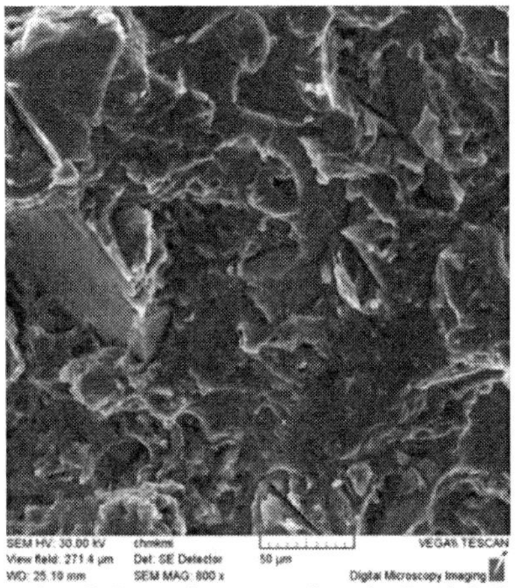

d) a detail from the area of static final failure

Fig. 16. Fracture surface of aluminium alloy 51200, σ_a = 139 MPa, N_f = 150530 cycles

Conclusion

On the basis of performed experimental fatigue tests can be drawn following conclusions:

- Square and line analysis showed, what elements and phases are found in selected alloys.
- Alloys have a clearly dendritic structure consisting mainly α-phase. In the structure of both alloys are intermetallic $Al_6(FeMn)$ and Mg_2Si phases. In alloy EN AC 51200, there is also an intermetallic Al_3Mg_2 phase.
- Alloys are different in terms of morphology and amount of intermetallic Mg_2Si phase. The alloy EN AC 51200 has this phase in a shape of skeleton formation in the amount of 7%. In the case of alloy EN AC 51500 is a finer morphology of this phase and the percentage is 11%.
- Alloys are comparable in terms of morphology and the amount of $Al_6(FeMn)$ phase.
- From fatigue tests, it can conclude that alloy EN AC 51500 has better fatigue properties than alloy EN AC 51200, in the analyzed areas of number of cycles under the same conditions of loading.
- Fractographic analysis did not reveal significant changes of micromechanisms of fracture surfaces after fatigue tests. Fatigue area is characterized by transgranular fatigue failure of α-phase (matrix). Intermetallic phases in this area were violated

interfacially, only Mg_2Si phase had also the cleavage mechanism and failure somewhere. The area of static final failure is typical of a transgranular ductile fracture of α-phase with dimple morphology with plastic´s transformed ridges. When comparing both of alloys, it can be stated that the only difference is that on the fracture surface of alloy EN AC 51500 is not an interfacially broken Al_3Mg_2 phase, because this alloy does not contain it.

- In terms of stress amplitude can be said that in case of a higher amplitude of cyclic loading there was a multiple initiation of fatigue crack.

From the analysis of the influence of the structure on the fatigue properties, it can be stated that the alloy EN AC 51500 has better fatigue properties due to a smaller amount of intermetallic phases, which are finer.

This study has been supported by the Scientific Grant Agency of the Ministry of Education of the Slovak Republic VEGA 01/0683/15 and VEGA 01/0533/15.

References

1. I. J. Polmear, *Light alloys-metallurgy of the light metals*. London (1995)
2. L. Hurtalová, E. Tillová, *Elimination of the negative effect of FE-rich intermetallic phases in secondary (recycled) aluminium cast alloy*. Manufacturing Technology **13** (1), 44-50 (2013)
3. *ASM metals handbook, properties and selection: nonferrous alloys and special-purpose materials*. Ohio, USA (1990)
4. S. Lathabai, P. G. Lloyd. Acta Materialia **50** (17), 4275-4292 (2002)
5. V. Zatkalíková, L. Markovičová, J. Belan, T. Liptáková, *Variability of local corrosion attack morphology of AISI 316Ti stainless steel in aggressive chloride environment*. Manufacturing Technology **14** (3), 493-497 (2014)
6. B. Hussey, J. Wilson, *Light alloys*. London (1998)
7. J. Belan, L. Hurtalová, A. Vaško, E. Tillová, *Metallography evaluation of IN 718 after applied heat treatment.* Manufacturing Technology **14** (3), 262-267 (2014)
8. F. Nový, V. Zatkalíková, O. Bokůvka, K. Miková, *Gigacycle fatigue endurance of marine grade stainless steels with corrosion pits*. Periodica Polytechna: Transportation Engineering **41** (2), 99-103 (2013)
9. C. Bathias, *Fatigue in the very high cycle regime*. Vienna, Austria (2001)

Study of the fatigue behaviour of synthetic nodular cast irons at low and high frequency cyclic loading

Alan Vaško[1,*], *Juraj* Belan[1], *Eva* Tillová[1]

[1]University of Žilina, Faculty of Mechanical Engineering, Department of Materials Engineering, Univerzitná 8215/1, 010 26 Žilina, Slovakia

Abstract. The paper presents the results of low and high frequency fatigue tests carried out on nodular cast iron. The specimens of synthetic nodular cast irons from three different melts were studied in the high cycle fatigue region (from 10^5 to 10^8 cycles) using fatigue experimental equipments for low and high frequency cyclic loading. Low frequency fatigue tests were carried out at frequency $f \approx 120$ Hz using the fatigue experimental machine Zwick/Roell Amsler 150HFP 5100. High frequency fatigue tests were carried out at frequency $f \approx 20$ kHz using the ultrasonic fatigue testing device KAUP-ZU. Both of them were carried out at sinusoidal cyclic push-pull loading (stress ratio $R = -1$) at ambient temperature ($T \approx 20$ °C). The relationship $\sigma_a = f(N)$ and fatigue strengths were determined experimentally; mechanical properties, microstructures and fracture surfaces were investigated.

Keywords: nodular cast iron, fatigue test, low frequency fatigue, high frequency fatigue, fatigue fracture

1 Introduction

The majority of service interruptions and cracks at the present time is caused by fatigue processes, degradation of materials properties in long time service and influence of external environment. The most significant of them is the failure caused by mechanical fatigue. The importance of fatigue results especially from the economical point of view (costs for development of new materials, testing equipments, methods of testing of prototypes and final products etc.) and safety of people whose life also depends on working reliability of devices.

Various internal and external factors significantly influence the fatigue characteristics of materials. The structural state of material and its chemical composition, grain size, shape and distribution of non-metallic inclusions, value of previous deformation etc. belong to the inside (internal) factors. The frequency of loading and load ratio, value of mean stress, state of stress, environment agresivity and temperature, history of loading, thickness and size of body, existence of apriory cracks, notches and other geometrical discontinuities, properties of surface layer etc. belong to outside (external) factors [1].

* Corresponding author: alan.vasko@fstroj.uniza.sk

Reviewers: *Darina* Ondrušová, *Františka* Pešlová

Although fatigue has been studied intensively over many years and many scientific papers and books were published, further study is necessary because the knowledge base is partly obsolete and new materials and treatments are continuously being developed [2]. Therefore the paper deals with the study of the fatigue behaviour of synthetic nodular cast irons at low and high frequency cyclic loading.

Nodular cast iron is a group of cast structural materials with a wide application in engineering practice (especially in the automotive industry). It combines high tensile strength and plasticity with high fatigue strength.

Nodular cast iron can be produced according to the classical or synthetic casting procedure (Fig. 1) which is more economical [3]. Classical melts use pig iron as a basic charging raw material. Nowadays, the production of nodular cast iron has been from an economic point of view orientated to synthetic melts where a part of more expensive pig iron in a metal charge is substituted for cheaper steel scrap. The transition from the traditional use of pig iron to synthetic nodular cast iron prepared from steel scrap requires the regulation of chemical composition of melt. Steel scrap has low content of silicon therefore increasing of content of silicon to eutectic composition ($S_C \approx 1$) is reached by using of ferrosilicon (FeSi) or metallurgical silicon carbide (SiC) additive. Addition of metallurgical SiC to a nodular cast iron melt shows specific advantages in comparison with other Si and C carriers what has been described as a special pre-inoculating effect [4-7].

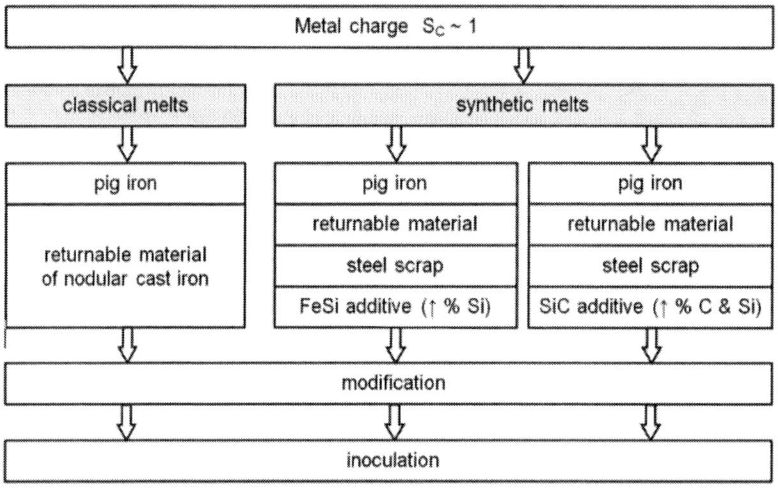

Fig. 1. Scheme of the production of nodular cast iron

2 Experimental material and methods

Three different melts of nodular cast iron were chosen for experiments and the specimens from them were used for metallographic analysis, mechanical tests, fatigue tests and microfractographic analysis. Similar chemical composition of the melts was achieved by different charge composition (Table 1). The basic charge of individual melts consisted of pig iron, steel scrap and some additives for the regulation of chemical composition. The content of these additives was chosen to achieve similar resultant chemical composition of the melts (eutectic degree $S_C \approx 1.2$).

The basic charge of the melt 1 was formed by 40 % of pig iron and 60 % of steel scrap with the additives of carburizer and metallurgical silicon carbide SiC90. The basic charge of the melt 2 was formed only by steel scrap (it did not contain any pig iron) with the same additives for the regulation of chemical composition (carburizer and silicon carbide SiC90).

The basic charge of the melt 3 was also formed by 40 % of pig iron and 60 % of steel scrap but there were used different additives for the regulation of chemical composition (carburizer and ferrosilicon FeSi75). Modifier FeSiMg7 and inoculant FeSi75 were used in the same amount for all the melts.

Table 1. Charge composition of experimental melts

Melt number	Pig iron (%)	Steel scrap (%)	Additives	Modifier & inoculant
1	40	60	carburizer + SiC90	FeSiMg7 + FeSi75
2	0	100	carburizer + SiC90	
3	40	60	carburizer + FeSi75	

Experimental bars of diameter 32 mm and length 350 mm were cast from all the melts and consequently specimens for metallographic analysis, tensile test, impact bending test, hardness test and fatigue tests were made.

The metallographic analysis of specimens from experimental melts was made by the light metallographic microscope Neophot 32. The specimens for metallographic analysis were taken out from the cast bars and prepared by usual metallographic procedure. The microstructure of specimens was evaluated according to STN EN ISO 945 and by automatical image analysis [8, 9]. The image analysis system NIS Elements, interfaced with the light microscope, was used for the evaluation of shape factor, count of graphitic nodules per unit of area and content of ferrite in the matrix. The final value is the average of at least 10 measurements, each at different place of the specimen.

The tensile test was carried out according to STN EN ISO 6892-1 by means of the testing equipment ZDM 30 with a loading range F = 0 to 50 kN. The impact bending test was carried out according to STN EN ISO 148-1 by means of the Charpy hammer PSW 300 with a nominal energy of 300 J. The Brinell hardness test was carried out according to STN EN ISO 6506-1 by means of the testing equipment CV-3000 LDB with a hardmetal ball of diameter D = 10 mm forced into specimens under the load F = 29 430 N (3000 kp) [10, 11].

The fatigue tests were carried out according to STN 42 0362 at low and high frequency sinusoidal cyclic push-pull loading (stress ratio R = -1) at ambient temperature (T = 20 ± 5 °C). Low frequency fatigue tests were carried out at frequency $f \approx$ 120 Hz using the fatigue experimental machine Zwick/Roell Amsler 150HFP 5100 (Fig. 2). High frequency fatigue tests were carried out at frequency $f \approx$ 20 kHz using the ultrasonic fatigue testing device KAUP-ZU (Fig. 3) [12, 13].

Test specimens of circular cross-section were used for fatigue tests. Shape and parameters of the specimens for low frequency cyclic loading are shown in Fig. 2c and for high frequency cyclic loading in Fig. 3c. For both fatigue tests, ten specimens from each melt were used to obtain the relationship between the applied amplitude of cyclic stress and the number of cycles to failure (Wöhler curve) and to determine the fatigue strength [14].

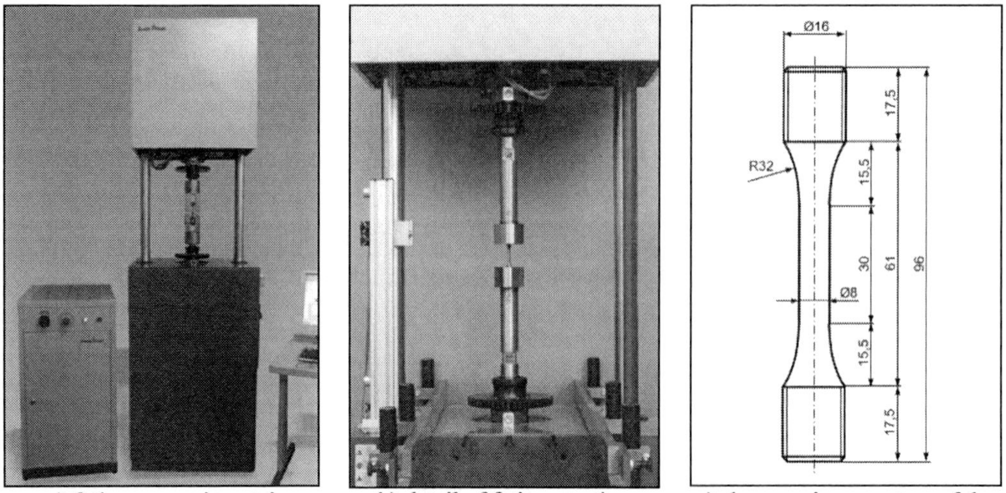

a) fatigue experimental machine Zwick/Roell Amsler 150HFP 5100 b) detail of fatigue testing c) shape and parameters of the test specimens

Fig. 2. Experimental device and specimen for low frequency fatigue test

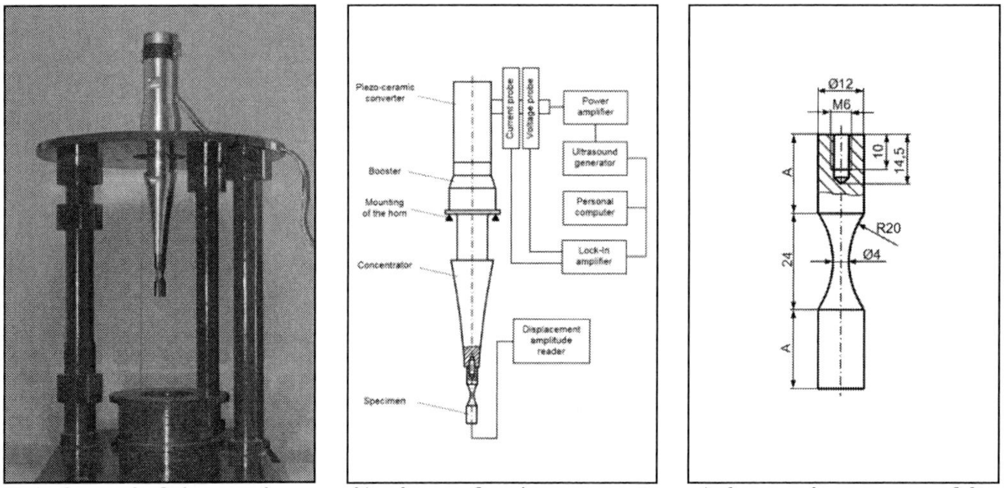

a) ultrasonic fatigue testing device KAUP-ZU b) scheme of testing apparatus c) shape and parameters of the test specimens

Fig. 3. Experimental device and specimen for high frequency fatigue test

The microfractographic analysis of fracture surfaces is an inseparable part of every work in the field of fatigue. Fracture surfaces of the specimens fractured by fatigue tests were investigated by the scanning electron microscope VEGA II LMU [15].

3 Experimental results and discussion

3.1 Metallographic analysis

Microstructure of the specimens from experimental melts is shown in Fig. 4. All the specimens are ferrite-pearlitic nodular cast irons with similar content of ferrite and pearlite in the matrix, similar shape, size and count of graphitic nodules (Table 2). Differences in the microstructure of individual specimens are caused by different ratio of pig iron and steel scrap in the charge and by different kind of additive for the regulation of the chemical composition (SiC or FeSi).

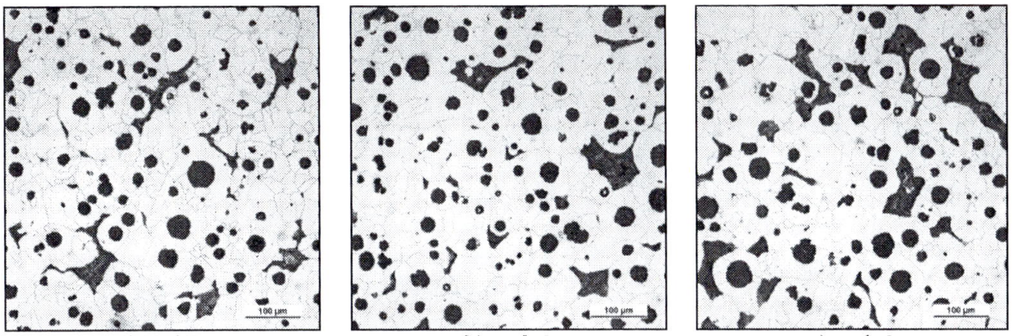

a) melt 1 b) melt 2 c) melt 3

Fig. 4. Microstructure of the specimens from cast bars, etched by 1% Nital

The results of evaluation of the microstructure according to STN EN ISO 945 and by image analysis (shape factor, count of graphitic nodules and content of ferrite) are presented in Table 2. The microstructural parameters evaluated in the specimens from the melts with SiC additive are higher than in the specimens from the melt with FeSi additive. The highest content of ferrite (approximately 78%) was reached in the melt 2, formed only by steel scrap with SiC additive and in the melt 1, formed by 40 % of pig iron and 60 % of steel scrap with SiC additive. Graphite occurs only in a perfectly-nodular and imperfectly-nodular shape in all the specimens. The size of graphite is from 30 to 120 μm, but in all the specimens the size within 30-60 μm predominates. The highest ratio of perfectly-nodular graphite (80 %), the highest shape factor (0.83) and the highest count of graphitic nodules per unit of area (almost 200 mm^{-2}) were reached in the melt 1, formed by 40 % of pig iron and 60 % of steel scrap with SiC additive.

Table 2. Results of evaluation of the microstructure

Melt number	STN EN ISO 945	Shape factor *	Count of graphitic nodules (mm^{-2})	Content of ferrite (%)
1	80%VI5/6 + 20%V6	0.83	199.8	74.0
2	70%VI5/6 + 30%V6	0.82	179.8	78.0
3	70%VI5/6 + 30%V6	0.79	151.0	65.2

* $S = 4\pi A/P^2$ (A – area of an graphitic nodule, P – perimeter of an graphitic nodule)

3.2 Mechanical tests

Mechanical properties (tensile strength R_m, elongation A, absorbed energy $K0$ and Brinell hardness HBW) of the specimens from experimental melts are summarised in Table 3.

Table 3. Mechanical properties of the specimens

Melt number	R_m (MPa)	A (%)	K0 (J)	HBW 10/3000
1	539.0	4.0	30.6	192.3
2	515.7	3.7	17.2	182.3
3	462.6	2.7	24.0	181.3

The specimens from the melts with SiC additive have better mechanical properties than the specimens from the melt with FeSi additive. It has connection with the microstructure of the specimens, especially with the character of matrix (content of ferrite and pearlite) and also with the shape, size and count of graphitic nodules. The best mechanical properties were reached in the melt 1, formed by 40% of pig iron and 60% of steel scrap with SiC additive, which has the highest ratio of perfectly-nodular graphite, the highest shape factor, the highest count of graphitic nodules and the smallest size of graphite.

3.3 Fatigue tests

For fatigue tests, ten specimens from each melt were used to obtain Wöhler fatigue curves $\sigma_a = f(N)$ and determine fatigue strength σ_c for $N = 10^7$ cycles (in the case of low frequency cyclic loading) or σ_c for $N = 10^8$ cycles (in the case of high frequency cyclic loading).

The results of fatigue tests (relationship between stress amplitude σ_a and number of cycles to failure N_f) obtained at low frequency cyclic loading ($f \approx 120$ Hz) are shown in Fig. 5a. The number of cycles to failure increases with a decreasing stress amplitude.

The values of fatigue strength σ_c determined for $N = 10^7$ cycles in comparison with tensile strength R_m are given in Table 4. The fatigue strength in analysed specimens of nodular cast iron increases with an increasing tensile strength. The fatigue strength in the specimens from the melts with SiC additive is higher than in the specimens from the melt with FeSi additive. The highest fatigue strength (255 MPa) was reached in the melt 1, formed by 40% of pig iron and 60% of steel scrap with SiC additive, which also has the best mechanical properties.

The results of fatigue tests obtained at high frequency cyclic loading ($f \approx 20$ kHz) are shown in Fig. 5b. The number of cycles to failure also increases with decreasing stress amplitude.

The values of fatigue strength σ_c determined for $N = 10^8$ cycles in comparison with tensile strength R_m are given in Table 4. At higher values of tensile strength there was observed an increase in fatigue strength. The highest fatigue strength (218 MPa) was reached in the melt 1, formed by 40% of pig iron and 60% of steel scrap with SiC additive, similarly to low frequency cyclic loading.

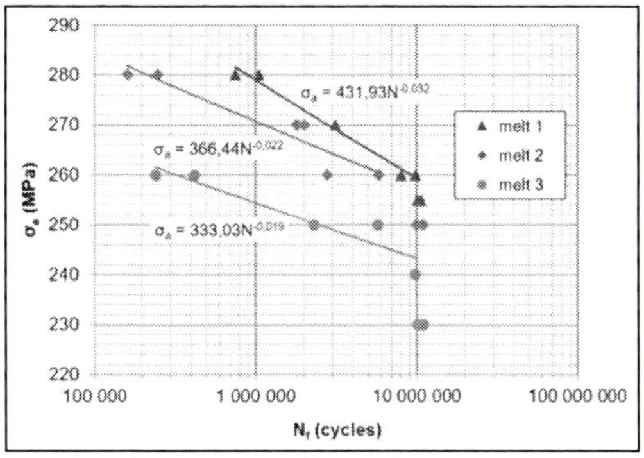

a) low frequency cyclic loading ($f \approx 120$ Hz)

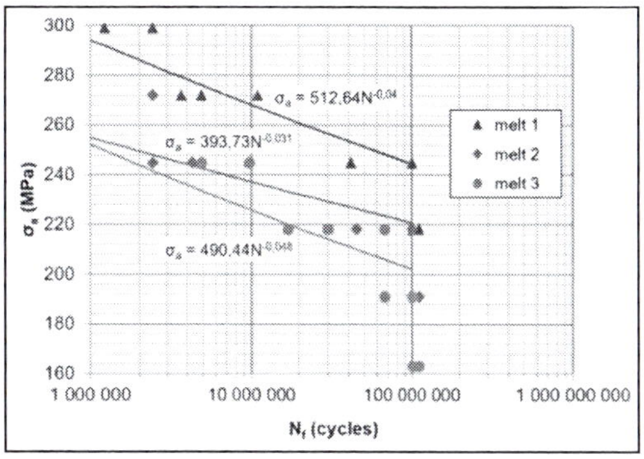

b) high frequency cyclic loading ($f \approx 20$ kHz)

Fig. 5. Wöhler curves $\sigma_a = f(N)$

Table 4. Comparison of tensile strength R_m and fatigue strength σ_c for low frequency cyclic loading (LFCL) and high frequency cyclic loading (HFCL)

Loading/ frequency		**LFCL** $f \approx 120$ **Hz**	**HFCL** $f \approx 20$ **kHz**	
Melt number	R_m **(MPa)**	σ_c **(MPa) for** $N = 10^7$ **cycles**	σ_c **(MPa) for** $N = 10^8$ **cycles**	$\sigma_{c\,10^7}$ **(MPa)** (calculated)
1	539.0	255	218	268
2	515.7	250	191	237
3	462.6	230	163	226

The values of fatigue strength σ_c for $N = 10^7$ cycles obtained at low frequency cyclic loading are in a good agreement with the values of fatigue strength σ_c for $N = 10^8$ cycles obtained at high frequency cyclic loading as well as the values of timing fatigue strength $\sigma_{c\,10^7}$ calculated for $N = 10^7$ cycles.

3.4 Microfractographic analysis

The fracture surfaces of analysed specimens after fatigue failure do not show any remarkable differences; they are characteristic of mixed mode of fracture. Examples of failure micromechanisms of the specimens at low and high frequency cyclic loading are documented in Fig. 6 and 7.

Fig. 6 shows the fracture surface of the specimen after low frequency cyclic loading with stress amplitude σ_a = 270 MPa and number of cycles to failure $N_f = 3.2 \times 10^6$. The fatigue fracture was initiated by casting defect. The fatigue fracture is characteristic of intercrystalline fatigue failure of ferrite around graphitic nodules and transcrystalline fatigue failure of ferrite and pearlite in the rest of the area (Fig. 6a). The final rupture is characteristic of transcrystalline ductile failure of ferrite with dimple morphology (Fig. 6b) and transcrystalline cleavage of ferrite and pearlite with river drawing on facets (Fig. 6c).

Fig. 7 shows the fracture surface of the specimen after high frequency cyclic loading with stress amplitude σ_a = 272 MPa and number of cycles to failure $N_f = 2.4 \times 10^6$. The fatigue failure of the specimen after high frequency cyclic loading has a mixed character of fracture, similarly to low frequency cyclic loading. The fatigue fracture is characteristic of intercrystalline fatigue failure of ferrite around graphitic nodules and transcrystalline fatigue failure of ferrite and pearlite in the rest of the area (Fig. 7a). The final rupture is characteristic of transcrystalline ductile failure of ferrite (Fig. 7b) and transcrystalline cleavage of ferrite and pearlite (Fig. 7c).

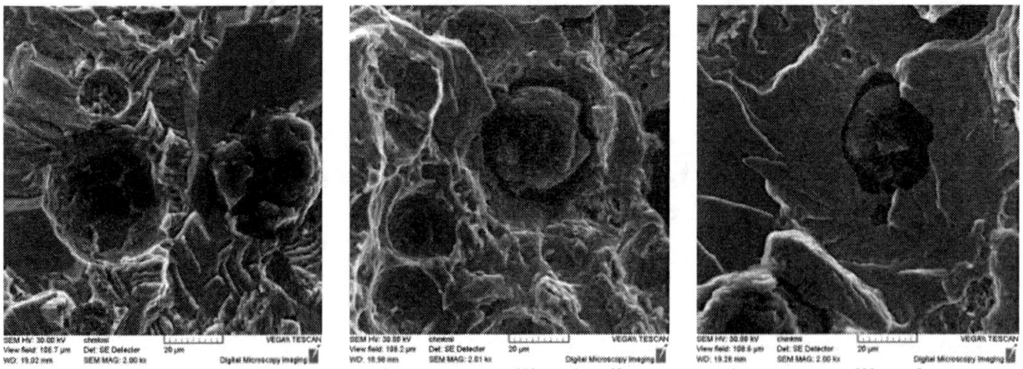

a) inter- and transcrystalline fatigue failure

b) transcrystalline ductile failure

c) transcrystalline cleavage

Fig. 6. Failure micromechanisms of the specimen at low frequency cyclic loading (melt 1, $f \approx 120$ Hz, $\sigma_a = 270$ MPa, $N_f = 3.2 \times 10^6$ cycles), SEM

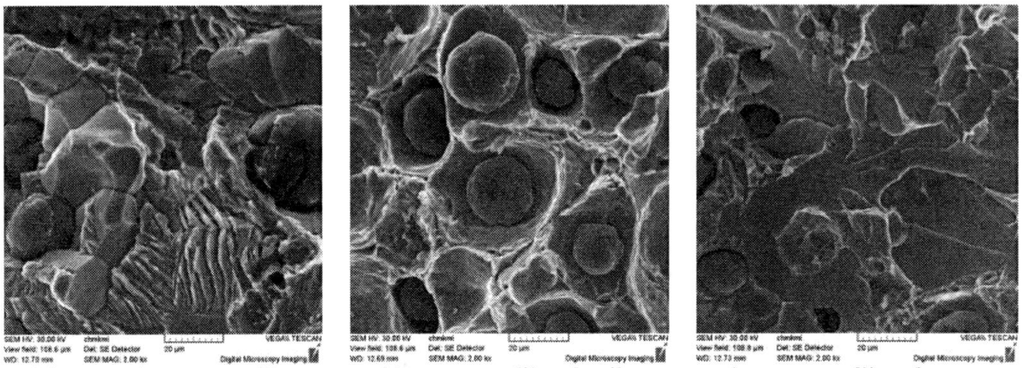

a) inter- and transcrystalline fatigue failure

b) transcrystalline ductile failure

c) transcrystalline cleavage

Fig. 7. Failure micromechanisms of the specimen at high frequency cyclic loading (melt 2, $f \approx 20$ kHz, $\sigma_a = 272$ MPa, $N_f = 2.4 \times 10^6$ cycles), SEM

Conclusions

The experimental results can be summarized to the following points:

- The substitution of a part of pig iron for steel scrap in the charge together with SiC additive positively influences the microstructure of nodular cast iron and consequently the mechanical and fatigue properties are improved;
- The highest microstructural parameters (ratio of perfectly-nodular graphite, shape factor, count of graphitic nodules and content of ferrite) were reached in the specimen from the melt 1, formed by 40 % of pig iron and 60 % of steel scrap with SiC additive;
- The best mechanical properties (tensile strength, elongation, absorbed energy and Brinell hardness) as well as fatigue properties (fatigue strength for low and high frequency cyclic loading) were also reached in the specimens from the melt 1, formed by 40 % of pig iron and 60 % of steel scrap with SiC additive;
- The values of fatigue strength obtained at high frequency cyclic loading are in a good agreement with the results obtained at low frequency cyclic loading. Moreover, application of high frequency cyclic loading is characteristic with the significant time, energy and work saving;

- No significant differences were observed by the comparison of fracture surfaces of the specimens from the analysed melts. The fatigue failure has a mixed character of fracture in all the specimens; the intercrystalline fatigue failure predominates near graphitic nodules and the transcrystalline fatigue failure predominates in the rest of the area.

This work has been supported by the Scientific Grant Agency of Ministry of Education, Science, Research and Sport of Slovak Republic, grant project VEGA No. 1/0533/15 and by the Culture and Educational Grant Agency of Ministry of Education, Science, Research and Sport of Slovak Republic, grant project KEGA No. 049ŽU-4/2017.

References

1. L. Várkoly, J. Zuidema, B. Várkolyová, M. Chalupová, *Fatigue failure of materials – theory and solved examples*. (EDIS, Žilina, 1998)
2. O. Bokůvka, G. Nicoletto, M. Guagliano, L. Kunz, P. Palček, F. Nový, M. Chalupová, *Fatigue of materials at low and high frequency loading*. (EDIS, Žilina, 2014)
3. R. Konečná, M. Kokavec, G. Nicoletto, *Surface conditions and the fatigue behavior of nodular cast iron*. Procedia Engineering **10**, 2538-2543 (2011)
4. T. Benecke, S. Venkateswaran, W.D. Schubert, B. Lux, *Investigation of the influence of silicon carbide in the production of ductile cast iron*. Foundryman **87**, 355-360 (1994)
5. K. W. Copi, Y. S. Lerner, N. S. Laukhin, *SiC vs. 75% FeSi: Comparing pre-inoculation effects*. Modern Casting **93**, 29-31 (2003)
6. W. H. Zhang, J. Ding, F. R. Nie, *SiC inoculation pretreatment of cast iron*. Foundry (Zhuzao) **58**, 279-281 (2009)
7. A. Vaško, *Influence of SiC additive on microstructure and mechanical properties of nodular cast iron*. Materials science (Medžiagotyra) **14**, 311-314 (2008)
8. P. Skočovský, A.Vaško, *Quantitative evaluation of structure of cast irons*. (EDIS, Žilina, 2007, in Slovak)
9. R. Konečná, S. Fintová, *Methods of structure study*. (EDIS, Žilina, 2014, in Slovak)
10. A. Vaško, P. Skočovský, *Properties and using of metal materials*. (EDIS, Žilina, 2014, in Slovak)
11. P. Kopas, M. Vaško, M. Handrik, *Computational modeling of the microplasticization state in the nodular cast iron*. Applied Mechanics and Materials **474**, 285-290 (2014)
12. L. Trško, O. Bokůvka, F. Nový, M. Guagliano, *Effect of severe shot peening on ultra-high-cycle fatigue of a low-alloy steel*. Materials and Design **57**, 103-113 (2014)
13. R. Ulewicz, P. Tomski, *The effect of high-frequencies loading on the fatigue cracking of nodular cast iron*. Metalurgija **56**, 33-36 (2017)
14. S. Věchet, J. Kohout, O. Bokůvka, *Fatigue properties of nodular cast iron*. (EDIS, Žilina, 2001, in Czech)
15. P. Palček, M. Chalupová, *Fractography and influence of the structure on character of the fracture*. Proceedings of Summer school of fatigue of materials. (EDIS, Žilina, 2006, in Slovak)

Smart materials activation analysis on example of nickel and titanium alloys

Bartosz Wieczorek[1,*], *Mateusz* Kukla[1], *Łukasz* Warguła[1], *Konrad* J. Waluś[1], *Klaudia* Wiesner[1]

[1]Poznan University of Technology, Chair of Basic of Machine Design, Piotrowo Street 3, 60-965 Poznan, Poland

Abstract. This paper is focused on research concerning activation time of elements made of Ni-Ti alloy (55/45% vol.) The activation time is a period of time required for alloy to reach it's austenitic transformation (Af) temperature. For examined wire it reached values up to 60 °C. Heating of NiTi wire was conducted by retaining heat. In this paper the influence of wire length and electric current power on heating time is presented. This research allows to determine the correlation between the increase of temperature and time. For given electric current values. This data is useful for effective design of SMA actuators..

Keywords: NiTinol, shape memory materials, activation, short circuit, austenitic transformation

1 Indrotuction

The need for miniaturisation of actuators (effectors) used in mechatronic systems calls for innovative engineering materials. Examples of such materials are Shape Memory Alloys (SMA) [1]. These materials remember and recover the programmed shape under certain external stimuli, such as a change of temperature or magnetic field. One of the most widely known SMA's is nitinol - a material developed by Bühler in 1962 [2] and patented, together with Wiley, in 1965 [3]. Nickel and titanium alloys feature so-called mechanical memory. The shape memory effect in nickel and titanium alloys can be attributed to their phase structure which at higher temperatures is made up of β-phase with the B2 lattice. This structure is transformed into transitional R-phase with rhombohedral crystal lattice at the temperature above the martensite start temperature Ms, followed by transformation to martensitic structure with monoclinically distorted B19 lattice The reverse transformation takes place without R-phase transition i.e. martensite returns directly to β-phase with B3 lattice. The austenite finish temperature (Af) depends on the chemical composition of alloy and can vary over the range of -150°C to +150°C [1].

An example application of SMA materials in mechatronic systems are nitinol pull-type springs. Springs of this type are tensioned under external force. When heated up to the

* Corresponding author: bartosz.wieczorek@put.poznan.pl

Reviewers: *Eva* Tillová, *Marta* Kianicová

temperature Af the spring returns to the programmed shape, as a result of reverse transformation of martensite to austenite.

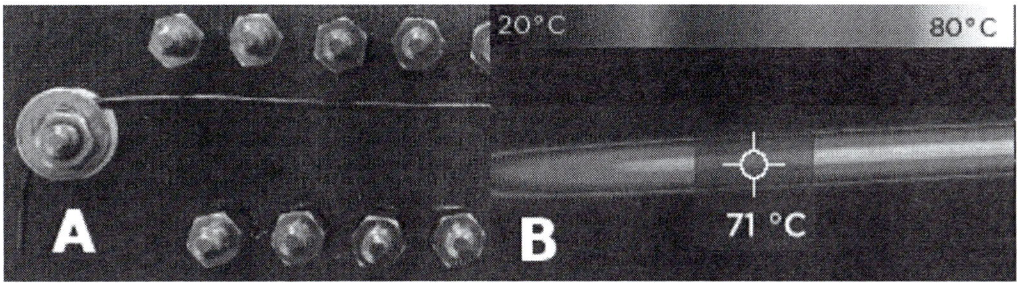

Fig. 1. View of the test specimen mounted in the test stand (A) and thermal image of 100 mm long Ni-Ti wire during heating (B)

The austenite finish temperature (Af) of the analysed alloy was ca. 60°C [4]. The easiest way to obtain such temperature is by using the test specimen as a resistor and thus increase the temperature of the wire by passing current. Uneven distribution of temperature in the initial phases of heating is a disadvantage of this method (Fig. 1).

2 Test procedure

The objective of these experiments was to investigate the effect of the duration of flow of electric current through the Ni-Ti alloy wire on its final temperature. The effect of the current on the time needed to reach the Af temperature was also investigated. The test specimens were prepared from 1 mm Ni-Ti wires in two different lengths, supplied from a regulator providing controlled current and voltage values (Table 1). For comparison 33 mm long controls were made of the same wire.

Table 1. Current and voltage values for the respective specimens

Specimen material: 1 mm wire of nickel-titanium alloy (55% Ni and 45% Ti), Af temperature of 60°C							
L = 100 [mm]				**L = 66 [mm]**			
I [A]	U [V]	P [W]	ΔP [W]	I [A]	U [V]	P [W]	ΔP [W]
5.5	1.2	6.60	0.67	5.5	1.0	5.50	0.65
6.0	1.3	7.80	0.73	6.0	1.1	6.60	0.71
6.5	1.5	9.75	0.80	6.5	1.2	7.80	0.77
7.0	1.6	11.20	0.86	7.0	1.4	9.80	0.84
7.5	1.7	12.75	0.92	7.5	1.5	11.25	0.90
8.0	1.8	14.40	0.98	8.0	1.6	12.80	0.96
8.5	1.9	16.15	1.04	8.5	1.7	14.45	1.02
9.0	2.0	18.00	1.10	9.0	1.8	16.20	1.08
9.5	2.1	19.95	1.16	9.5	2.0	19.00	1.15

Where: L – length of tested wire, I – current, U – voltage, P – power, ΔP – power measurement uncertainty.

The test stand was based on textolite plate acting as an insulator and used for mounting wires of the specified lengths. Power was supplied by Voltcraft PS 1440 benchtop power supply unit, connected to the test specimen by cables. The last element of the test stand was the Reveal PRO thermal imaging camera from Seek thermal whose role was to record the variation of temperature of the tested wire (Fig. 2).

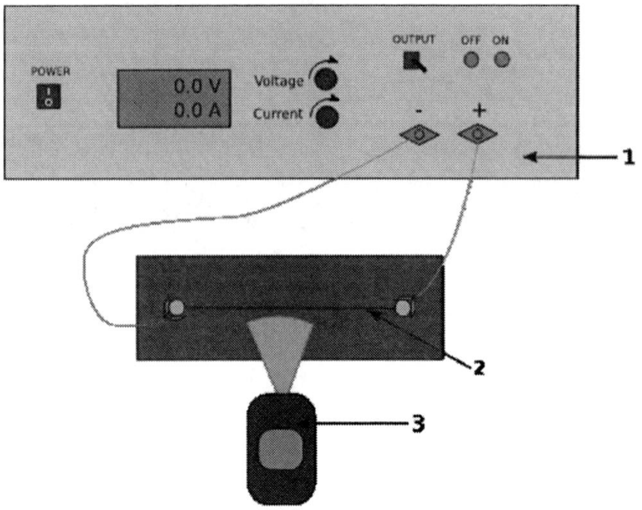

Fig. 2. Test stand setup: 1 - Voltcraft PS 1440 benchtop power supply, 2 - specimen of NiTi wire, 3 - Reveal PRO thermal imaging camera

According to the test procedure, specimens of different lengths were mounted on the textolite base plate. Then the electricity was applied at the specified current and voltage levels. For each power level the heating process was repeated five times. The time needed to increase the wire temperature from 24°C to ca. 70°C was the measured parameter.

3 Results of experiments

The research program comprised series of tests performed on Ni-Ti wires of different lengths, namely 33 mm, 66 mm and 100 mm. All the test specimens were made of straight sections of 1 mm wire. Examples of time-temperature curves are presented in Fig. 3.

The experimental results were used to determine the time needed to increase the temperature of wire made of Ni-Ti alloy up to the activation temperature of ca. 70°C. The analysis of results started with determining of effect of the wire length on the time needed to increase the specimen temperature to Af (Fig. 3). The results allow us to conclude that the length of wire has a significant effect on the length of time needed to reach the activation temperature. The values recorded at 6,6 W power input were: ca. 45 sec. for 100 mm length, ca. 22 sec. for 66 mm length and ca. 35 sec. for 33 mm length. With the shortest specimens (33 mm) the whole test set-up acted as a radiator transferring heat to the environment. As a result, it took longer to heat up the 33 mm long specimen than the 66 mm long one. Consequently, the heating time depends on the ambient temperature and the volume of parts to which the Ni-Ti wire is connected.

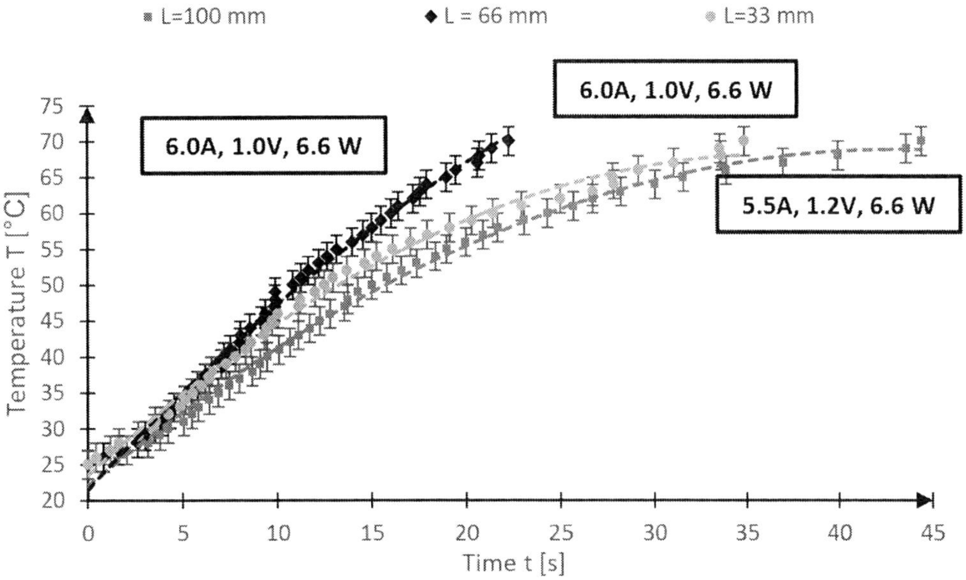

Fig. 3. Time-temperature curves for 100 mm and 66 mm long specimens

Fig. 4 presents the time-temperature curves obtained in heating 100 mm long specimens of Ni-Ti alloy up to the activation temperature at different power input levels. Heating the specimens with 6.5 A and 1.2 V current (9.75 W) produced time-temperature curves having a curved (bulged) shape. Higher values of current make the curve more linear, allowing determination of the trendline slope. The trendline slope depends on the level of power applied to the Ni-Ti wire. For input powers above 10 W the curve can be treated as linear. There is over 98% match between the trend line established for such power inputs and the experimentally determined curve. This approximation allows us to assume possibility of modelling the behaviour of nitinol wire specimens as a function of their length and power input level.

The test results were used to draw the curve representing the effect of power input on the time needed to heat up the specimen to the activation temperature. The obtained curve was subjected to linearization following determination of slope as a function of time. The slope depends on the values of applied current and voltage. Thus, by controlling these parameters we can control the behaviour of the specimen made of NI-Ti alloy. Over 99% fit is established between the experimental values and the values obtained with the function of slope variation vs. input power. The graph in Fig. 5 presents the variation of the trendline slope as a function of input power.

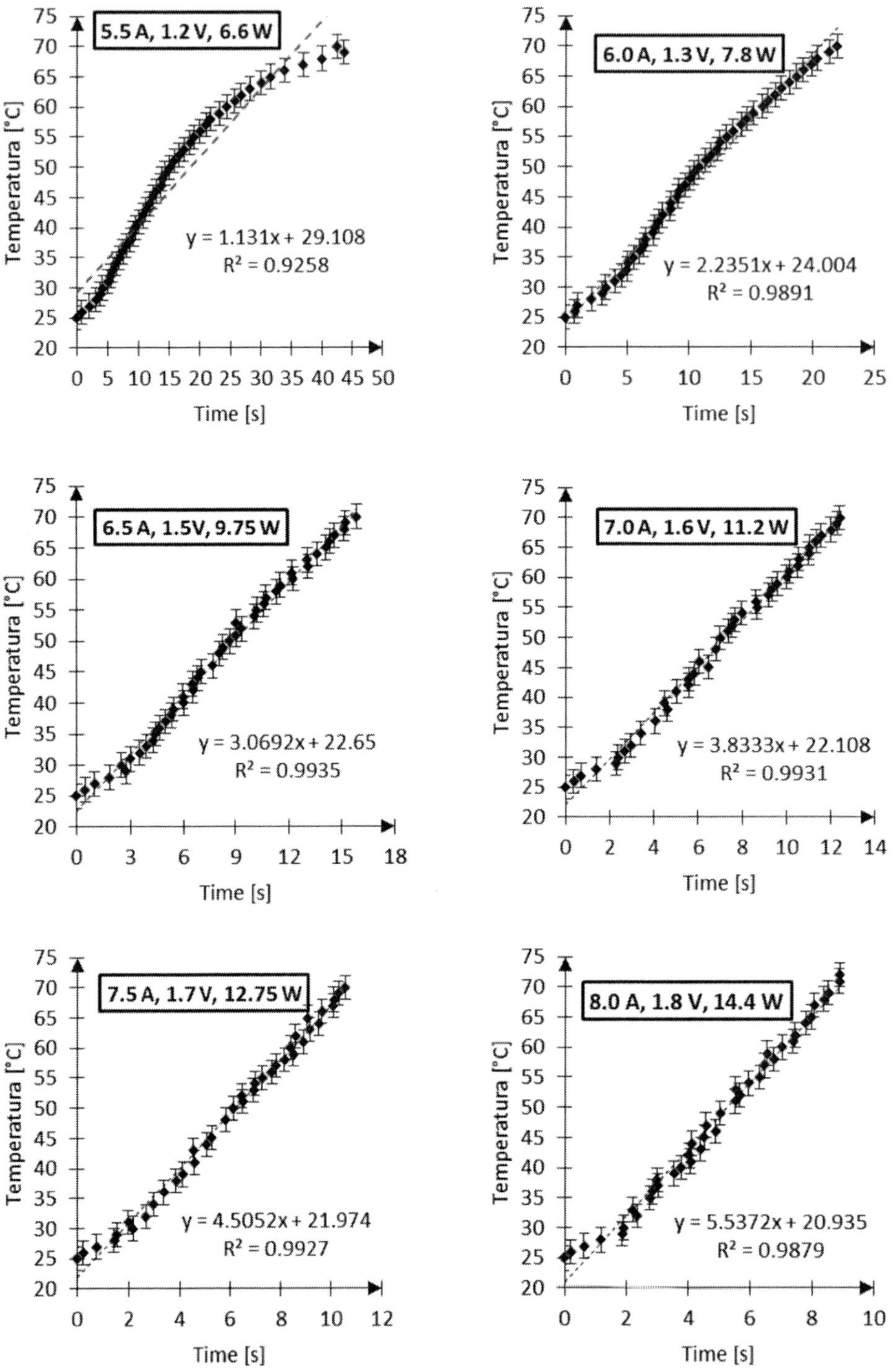

Fig. 4. Time-temperature curves obtained for 100 mm long specimens of Ni-Ti alloy depending on the heating time at different power input levels

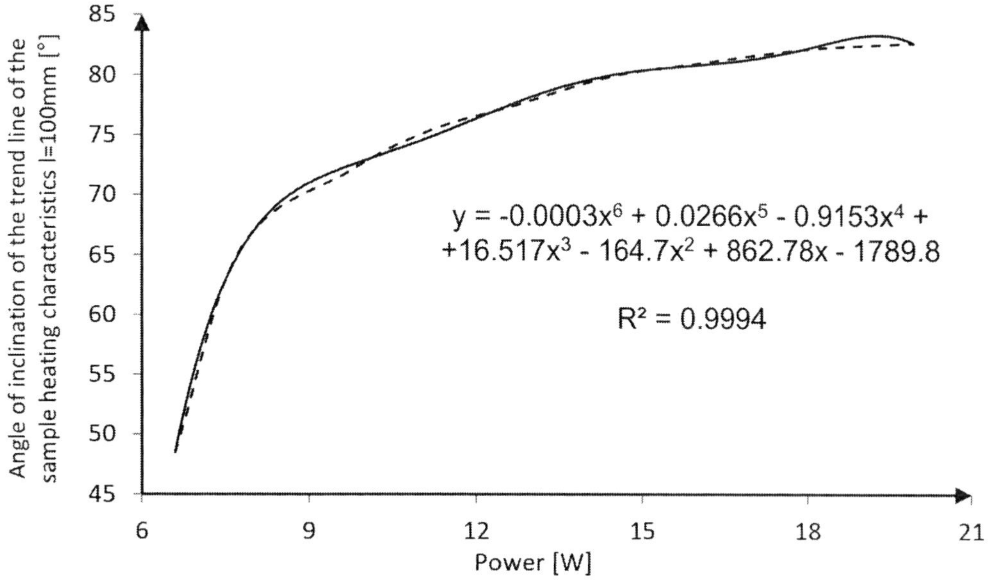

Fig. 5. Variation of the trendline slope as a function of input power

The experiment included also evaluation of the effect of low input power values (not exceeding 9 W) on the degree of bulging (curvilinearity) of the time-temperature curve. Such bulging of curve for low input power levels results from ineffective heating of the specimen with excessive loss of heat in the final stage of the heating process. This was caused by too small current which did not made up for the heat losses determined by the test stand elements and by the ambient temperature of ca. 23°C. That the curved shape (bulging) of the time-temperature curve can be attributed to insufficient current can be confirmed by comparing 100 mm and 66 mm long samples (Fig. 6). The time-temperature curves (Fig.6) show that with the decreasing volume of specimen smaller currents become more effective.

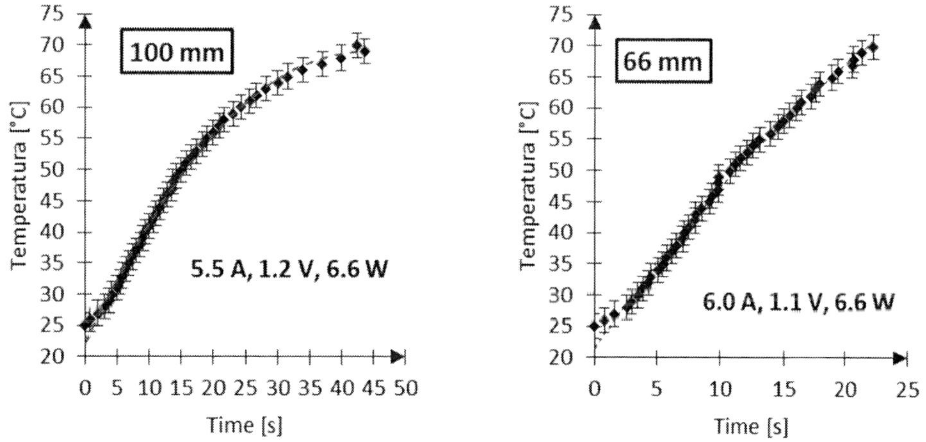

Fig. 6. Time-temperature curves for Ni-Ti alloy specimens heated by passing current of 6.6 W

Conclusions

The present experimental research has demonstrated that the heating time strongly depends on the volume of the specimen and its mounting fixtures, as well as on the ambient temperature. The shape of specimen is yet another factor influencing the heating time and this is associated with the mechanism of heat dissipation to environment [5].

This preliminary research can be used as the first step on the way to develop a mathematical model representing the relationship between the temperature of Ni-Ti alloy wire and the heating time and also the input power. The objective of this research was to find the effective current and voltage enabling linearisation of the obtained curves. This approach enables building a mathematical model based on the slope coefficient of the trend line, representing the relationship between the temperature and the heating time for a given power input and the length of specimen.

The results can be useful in designing control systems of mechatronic systems equipped with SMA actuators. The curves obtained in this research enable selection of times required to achieve the specified temperature of activation (Af) at specific power input levels. The collected data enable estimating the values of constants used in already existing models representing the process of heating resistance wires by passing current [6].

References

1. W. Kucharczyk, A. Mazurkiewicz, W. Żurowski, *Nowoczesne materiały konstrukcyjne. Wybrane zagadnienia*. Politechnika Radomska (2011)
2. A. Ćwikła, *Medical applications of intelligent materials*. Scientific Bulletin of Chelm Section of Technical Sciences, **1** (2008)
3. G. Markiewicz, *Materiały inteligentne – zastosowanie w systemach aktywnej redukcji hałasu i drgań*. Bezpieczeństwo pracy, **12** (2015)
4. S. Bohua, *Smart materials and structures*. Lecture at Swiss Federal Institute of Technology Zurich (ETH), Zurich (2015)
5. K. Biereg, *Porównanie wybranych równań konstytutywnych stopów z pamięcią kształtu*. Modelowanie inżynierskie **32**, p. 38, (2006)
6. A. Bellini, M. Colli, E. Dragoni, *Mechatronic Design of a Shape Memory Alloy Actuator for Automotive Tumble Flaps: A Case Study*. ieee transactions on industrial electronics **56** (7), 2644-2649 (2009)

Experimental research on biomass cutting process

Dominik Wilczyński[1,*], *Krzysztof* Talaśka[1], *Ireneusz* Malujda[1], *Piotr* Jankowiak[1]

[1]Poznan University of Technology, Chair of Basics of Machine Design, Piotrowo Str. 3, Poland

Abstract. This paper contains the results of the biomass cutting process in which wheat straw was used. Tests were made on the prototype test stand which made it possible to control the parameters of the cutting process and based on that define its energy consumption. During the research the following parameters were measured: rotational speed of electric motor, which drives the rotor with a set of cutting knives, electric current supply and torque on the shaft. Tests were made for the variable moisture content of the biomass and with knives with three different back-rake angles 5°, 15° and 30°. Using these measured parameters it was possible to calculate the power and the cutting force necessary to disintegrate biomass. The main goal was to find out if there is any influence of moisture content and the back-rake angle of the cutting knife on the dynamic parameters of the process - power and cutting force. The presented results can be very useful in a design process for a working assembly of biomass disintegrating machines which are directed to the energetic efficiency.

Keywords: biomass, biomass cutting process, humidity of biomass, energetic efficiency of cutting process

1 Introduction

Biomass is the third largest source of energy in the world, following coal and oil [1]. Undensified biomass is characterized by low density on the order of 0.1 ÷ 0.2 kg/l = 100 ÷ 200 kg/m^3 [1]. According to other sources, its density is in the 100 ÷ 200 kg/m^3 range for straw and grass, and 150 ÷ 200 kg/m^3 for wood biomass in the form of chips and sawdust [1-3]. This is a serious limitation for the use of this material as a source of energy due to storage and transport difficulties [1]. The densification process makes it possible to obtain the agglomerate of density equal to approximately 1.2 kg/l = 1200 kg/m^3, which affects the chemical structure of biomass that loses some of the water under pressure [4, 5]. The essence of this process has been broadly discussed in many works, including [6-10]. Densification promotes the hydrolysis of hemicelluloses and lignin to hydrocarbon compounds. This is needed in the production process of methane [1]. Factors affecting the strength and durability of the agglomerate are a type of material (materials) being densified, moisture content, particle size and distribution, process temperature, additional elements in the form of an adhesive or other additives, and densification process parameters (temperature, size of the press die, press die speed, etc.) [11, 12]. As was mentioned

* Corresponding author: dominik.wilczynski@put.poznan.pl

Reviewers: *Bogdan* Posiadala, *Michał* Śledziński

previously, particle size has an impact on the quality of the agglomerate. The process of biomass material fragmentation and its course are therefore very important for its densification [4, 13].

In this work research into the biomass cutting process as preceding biomass agglomeration into briquette form is presented. The straw material being fragmented was Triticale. A station for biomass cutting of our own proprietary design was used for research. The station design as well as research methodology and process are discussed. In following sections sample results of the experiment with conclusions is presented.

2 Research station and methodology

A research station consisting of two main systems was used in the experimental tests (Fig. 1). The first one is the system of cutters 2 used for cutting biomass material. Straw designated for cutting is transported to the cutter blades using the system of rollers 3 which are used to pull in and press biomass material. Actual cutting takes place in the cutter blades and counter cutter area. The cutter system 2 is driven by a three-phase electric motor 4 with power P = 2.2 kW and maximum speed n = 2850 RPM. Motor crank pin 4 is connected directly to the shaft of the cutting system 2 using the dog clutch. To control the motor 4, the inverter 9 Schneider Electric type MX pro 4V2.2 was used.

Fig. 1. Research station for the biomass material cutting process: 1 – shredded biomass material, 2 – system of cutters, 3 – electric motor driving rotational movement of the system of cutters, 4 – system of rollers, 5 – reduction drive of the rollers system, 6 – electric motor driving the rollers system, 7 – moisture analyzer, 8 – inverter of the electric motor of the rollers system, 9 – inverter of the electric motor of the cutters system, 10 – PC recording performance of the electric motor of the cutters system, 11 – PC recording performance of the electric motor of the rollers system

The system of rollers 3 is powered by the drive system in the form of a three-phase electric motor 6 with power P = 3 kW and maximum speed n = 1430 RPM, and a two-stage cylindrical gear reducer 5 with total gear ratio i_c = 14.5. Gearbox 5 is connected to the input shaft of the rollers system 3 using the dog clutch. The inverter Schneider Electric type 4V4.0 was used to control engine rotational speed 6. Registration of performance parameters of the engines 4 and 6 was made using Matrix software installed on the computers 10 and 11, where each was communicated with the inverter, respectively, 8 and 9. Matrix is a Schneider Electric software dedicated to support the company's inverters. This made it possible to register data on torque changes M [Nm] on the electric motor

 https://doi.org/10.1051/matecconf/201815707016

shafts 4 and 6, as well as amperage changes I [A] and voltage changes U [V] of these engines. The cutting module 2 has the ability to mount the blades with different back-rake angles of the cutting edge α (fig. 2), and the blade angle β (fig. 2). Angles α and β were successively α = 5°, α = 15° and α = 30° and β = 90°, 45° and 30°.

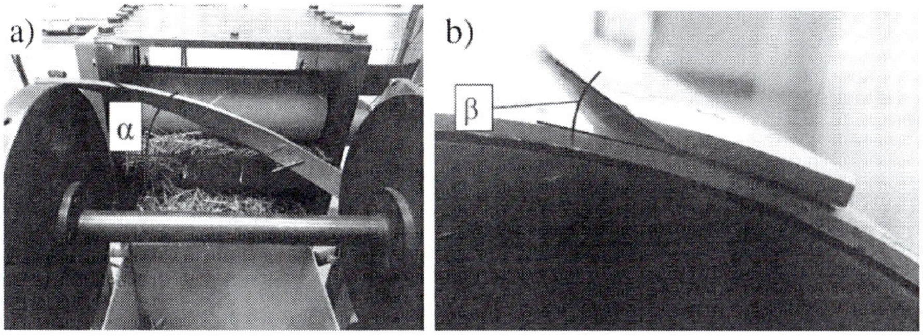

Fig. 2. View illustrating: a) back-rake angle α, the cutting edge of the knife, b) blade angle β

The conducted experimental studies made it possible to determine the change of the following performance parameters of the electric motors 4 and 6 depending on moisture of shredded straw, and the back-rake angle of the knife blade α and the angle of its blade β. Straw moisture content was determined using the moisture analyzer 7 (fig. 1 and 3).

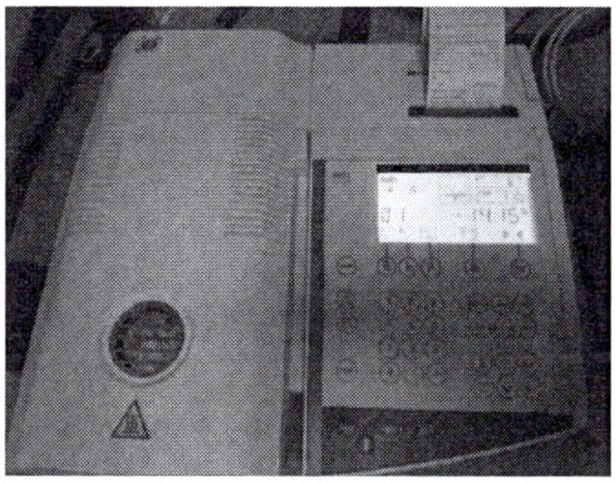

Fig. 3. Moisture analyzer used for measuring straw moisture content before shredding

For testing purposes, a straw portion with initial average moisture 16.94% was weighed out (initial moisture is understood as straw with moisture content as delivered directly from the field), intended for shredding and for seasoning for the purpose of studying the cutting process of straw with increased water content. The weight of the portion was 36.5 g ± 1 g (weight tolerance is due to losses during specimen transfer from and to the seasoning container) The prepared portions were stored in special containers so that the straw with initial moisture content did not dry up and so that some specimens were moisturised.

In order to moisturise it, the straw was seasoned for 648 hours (fig. 4) in containers with increased humidity content in the air. This made it possible for the straw to achieve average moisture content equal to 28.75%.

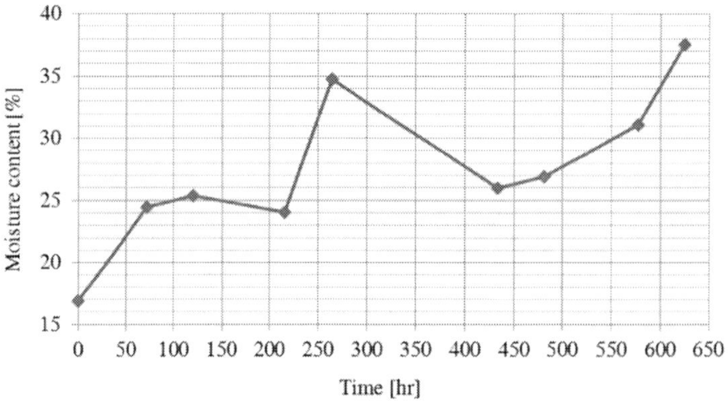

Fig. 4. Change of straw moisture content during seasoning

Straw specimens prepared in such a way were fed (fig. 5) into the system of rollers, where they were pulled, pressed and passed on to the system of cutters. For each specified angle value α and blade angle β, both during shredding the straw with moisture 16.94% and 28.75%, twenty specimens were shredded. In total, 180 specimens were shredded. The system of cutters has the ability to set the distance between the cutting edge of the knives and the edge of the counter cutter. In the current research, this distance was 0.2 mm and the feeler gauge was used for this purpose.

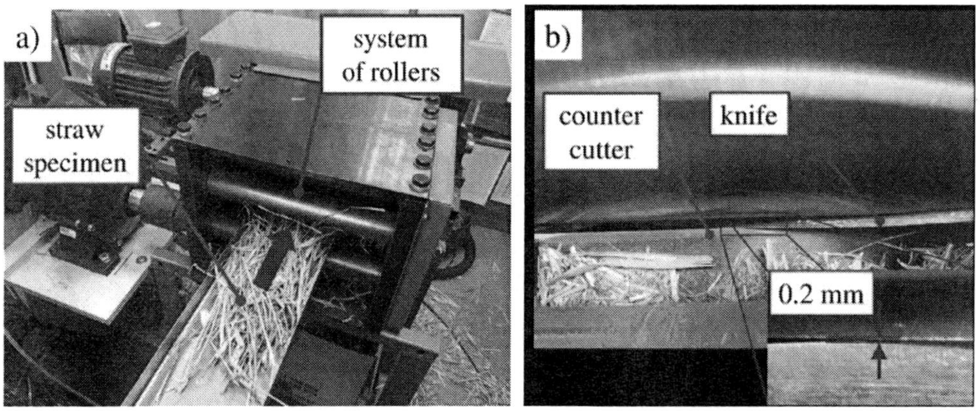

Fig. 5. General view: a) prepared straw specimen which is passed on to the system of rollers, b) cutting knives and counter cutter and the gap between them

On Figure 6 shredded straw is presented.

Fig. 6. General view of shredded straw

3 Research results and their analysis

On Figure 7 sample characterizations illustrating the changes of torque, current and voltage of the electric motor driving the system of cutters are presented.

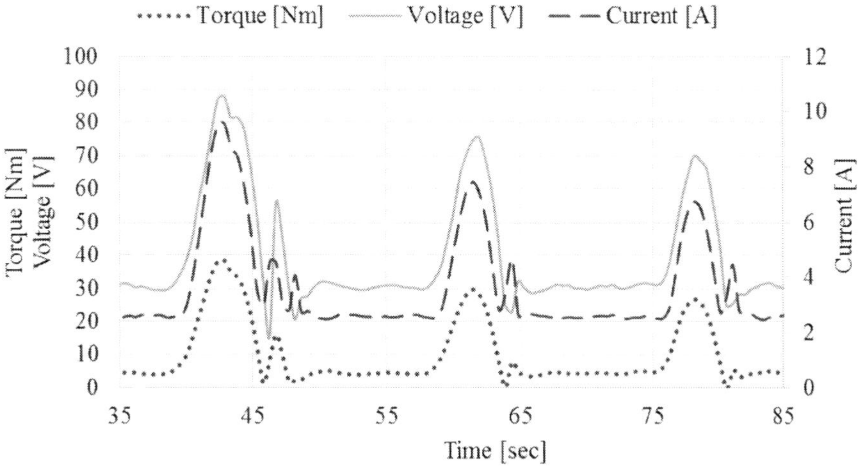

Fig. 7. Sample change characterizations of torque, voltage and current of the motor driving the system of cutters, for the process of shredding the straw with moisture content 16.94%, back-rake angle $\alpha = 15^{\circ}$ and blade angle $\beta = 90^{\circ}$

On Figure 8 sample characterizations illustrating the changes of torque, current and voltage of the electric motor driving the system of rollers are presented.

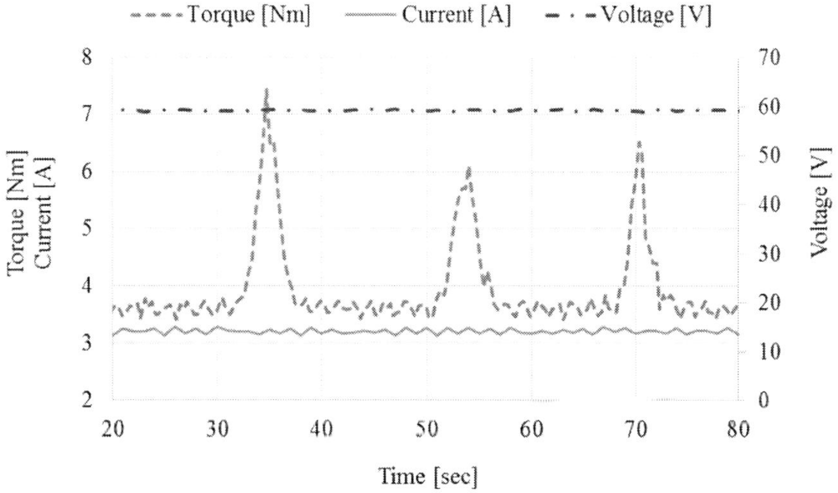

Fig. 8. Sample change characterizations of engine torque, voltage and current of the motor driving the system of rollers, for the process of shredding the straw with moisture content 16.94%, back-rake angle $\alpha = 15^{\circ}$ and blade angle $\beta = 90^{\circ}$

By dividing the recorded values of the torque on the motor shaft of the cutting system by the value of the radius of the cylindrical surface to which the knives were attached, the values of the cutting force needed to shred a given straw specimen were obtained. In this way, the characterizations shown in the following figures were obtained.

On Figure 9 the changes of the cutting force depending on the value of the back-rake angle of the cutting knives α and blade angle β, for shredding straw with medium moisture content 16.94% are presented.

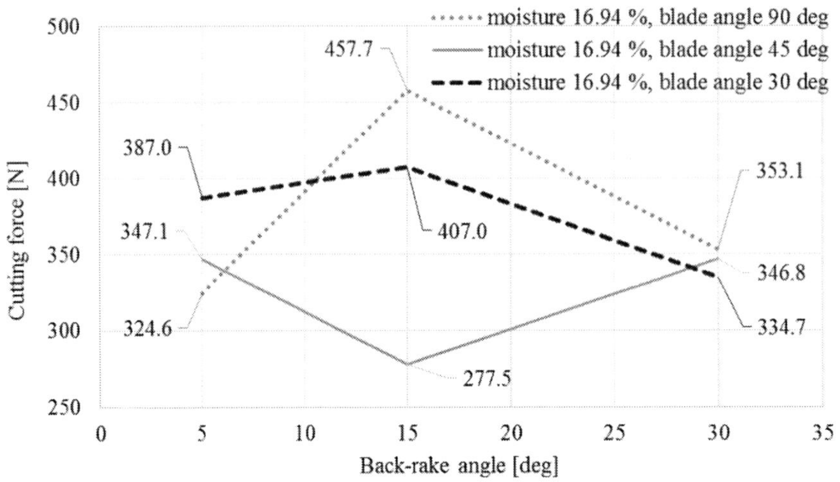

Fig. 9. Characterizations of the cutting force depending on the back-rake angle α and blade angle β, for the shredding process of straw with medium moisture 16.94%

According to Figure 9, no clear trend in the change of the cutting force was observed. The smallest dispersion of the force is visible for the back-rake angle α = 30° and all three blade angles β. The largest dispersion is visible for the back-rake angle 15°.
The smallest force value was recorded for the angle α = 15° and the blade angle β = 45°. This is also confirmed in the case of shedding the straw with average moisture content 28.75% (fig. 10). On Figure 10 change characterizations for the values of the cutting force, for shredding the straw with average moisture content 28.75%, depending on the changes in the value of the back-rake angle α and blade angle β are presented.

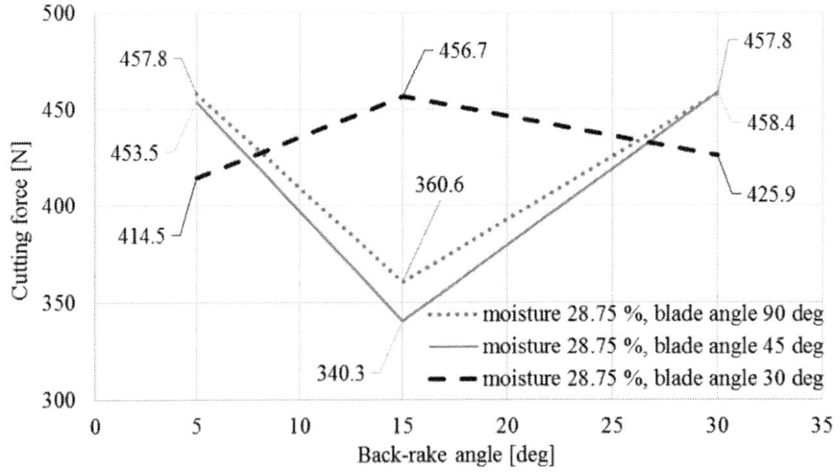

Fig. 10. Change characterizations of the cutting force value depending on the back-rake angle α and blade angle β for the shredding process of straw with medium moisture 28.75 %

According to the characterizations in Figure 10, it can be observed that the value of the cutting force increased with the increase in straw moisture content. For the back-rake angle α = 15° and the blade angle β = 45° and 90°, the cutting force is the smallest. It can be said that this is a kind of confirmation of the trend of declining force for α = 15° and the blade angle β = 45°, which is visible when shredding straw with moisture content 16.94% (fig. 9).

Conclusions

Based on the obtained research results, the following conclusions can be reached:

- as straw moisture content increases, there is an increase in the force needed to shred the straw,
- the lowest values of the cutting force was registered for the back-rake angle α = 15° and the blade angle β = 45° in both straw moisture cases; the humidity of straw; for confirmation, more research needs to be performed by shredding more specimens for each specified angle α and angle β,
- to increase measurement accuracy, additional torque sensors encased in the pins of the motors driving individual systems should be used,
- the smallest dispersion of the cutting force was registered for the angle α = 15° and the blade angle β = 30°,
- in further research, the impact of the gap between the knife blade and the edge of the counter cutter should be taken into account,
- further studies should be performed for the straw with average moisture content of approximately 17% and lower, since shredding the straw with higher moisture content requires greater cutting force, which negatively affects the increase in energy intensity of this process.

References

1. J. S. Tumuluru, C. T. Wright, J. R. Hess, K. L. Kenney, *A review of biomass densification system to develop uniform feedstock commodities for bioenergy application*. Biofuels, Bioproducts and Biorefining **5**, 683-707 (2011)
2. S. Sokhansanj, J. Fenton, *Cost benefit of biomass supply and pre-processing: BIOCAP (Canada) research integration program synthesis paper*. (online), available at: http://www.biocap.ca/rif/reprt/Sokhansanj_S.pdf (June 22, 2011)
3. P. Mitchell, J. Kiel, B. Livingston, G. Dupont-Roc, *Torrefied biomass: A foresighting study into the business case for pellets from torrefied biomass as a new solid fuel*. All Energy, University of Aberdeen, ECN, Doosan Babcock, and ITI Energy (2007)
4. C. A. N. Xavier, V. Moset, R. Wahid, H. B. Moeller, *The efficiency of shredded and briquetted wheat straw in anaerobic co-digestion with dairy cattle manure*. Biosystems Engineering **139**, 16-24 (2015)
5. P. D. Grover, S. K. Mishra, *Biomass briquetting: Technology and practices*. Field Document No. **46**, Bangkok, (1996)
6. C. K. W. Ndiema, P. N. Manga, C. R. Ruttoh, *Influence of die pressure on relaxation characteristics of briquetted biomass*. Energy Conversion and Management **43**, 2157-2161 (2002)
7. P. K. Adapa, J. Bucko, L. G. Tabil, G. Schoenau, S. Sokhansanj, *Pelleting Characteristics of Fractionated Suncure and Dehydrated Alfalfa Grinds*. ASAE/CSAE North-Central Intersectional Meeting, Saskatoon, Saskatchewan, Canada, (September 27-28, 2002)

8. P. K. Adapa, G. J. Schoenau, L. G. Tabil, S. Sokhansanj, B. Crerar, *Pelleting of Fractionated Alfalfa Products*. ASAE Annual International Meeting, Las Vegas, Nevada, USA, (July 27-30, 2003)
9. Y. Li, H. Liu, *High pressure densification of wood residues to form an upgraded fuel*. Biomass Bioenergy **19**, 177-186 (2000)
10. S. Mani, L. G. Tabil, S. Sokhansanj, *Specific energy requirement for compacting corn stover*. Bioresource Technology **97**, 1420-1426 (2006)
11. K. Ishii, T. Furuichi, *Influence of moisture content, particle size and forming temperature on productivity and quality of rice straw pellets*. Waste Management, **34** (12), 2621-2626 (2014)
12. V. S. P. Bitra, A. R. Womac, C. Igathinathane, P. I. Miu, Y. T. Yang, D. R. Smith, N. Chevanan, S. Sokhansanj, *Direct measures of mechanical energy for knife mill size reduction of switchgrass, wheat straw, and corn stover*. Bioresource Technology **100**, 6578-6585 (2009)
13. V. S. P. Bitra, A. R. Womac, N. Chevanan, P. I. Miu, C. Igathinathane, S. Sokhansanj, D. R. Smith, *Direct mechanical energy measures of hammer mill comminution of switchgrass, wheat straw, and corn stover and analysis of their particle size distributions*. Powder Technology **193**, 32-45 (2009)

Numerical computation of the damping and stiffness coefficients of the classical and magnetorheological squeeze film damper

Petr Ferfecki[1,2,*], *Jaroslav* Zapoměl[2,3], *Michal* Šofer[2], *František* Pochylý[4], *Simona* Fialová[4]

[1]IT4Innovations National Supercomputing Center, VŠB-Technical University of Ostrava, 17. listopadu 15/2127, 708 33 Ostrava-Poruba, Czech Republic
[2]Department of Applied Mechanics, Faculty of Mechanical Engineering, VŠB-Technical University of Ostrava, 17. listopadu 15/2127, 708 33 Ostrava-Poruba, Czech Republic
[3]Department of Dynamics and Vibration, Institute of Thermomechanics, Czech Academy of Sciences, Dolejškova 1402/5, 182 00 Praha 8, Czech Republic
[4]Victor Kaplan Department of Fluid Engineering, Faculty of Mechanical Engineering, Brno University of Technology, Antonínská 548/1, 601 90 Brno, Czech Republic

Abstract. Technological solution, frequently used to suppress vibrations in rotating machines, consists in adding damping devices between the rotor and its frame. This is enabled by dampers working on the principle of a squeezing thin classical or magnetorheological fluid film. The Navier-Stokes equations, Reynolds equation, and modified Navier-Stokes equations are used to determine the pressure distribution in the thin fluid film. The damping and stiffness coefficients are computed by the developed procedure presented in this paper. The proposed computational approach is based on the perturbation of the synchronous circular whirling motion. The carried-out computational simulations show that the investigated mathematical models of the squeeze film damper and magnetorheological squeeze film damper allowed computation of the damping and stiffness coefficients. It has been found that the stiffness coefficients computed by the proposed mathematical models may be different.

Keywords: Squeeze film damper, Magnetorheological fluid, Hydraulic forces, Navier-Stokes equations, Reynolds equation

1 Introduction

Unbalance of rotating machines is the main source of their lateral vibrations. Technological solution, frequently used to suppress vibrations in rotating machines, consists in adding the damping devices between the rotor and its frame.

The work of the damping devices is based on different, often mutually coupled, physical, mechanical, electric, electromagnetic, piezoelectric, hydraulic, or magneto-rheological principles. A simple device to reduce attenuation of vibration is an elastomeric ring [1] that supports the outer ring of the bearing. The vibration reduction and suppression

* Corresponding author: petr.ferfecki@vsb.cz

Reviewers: *Vladimír* Dekýš, *Zuzana* Murčinková

of the journal bearing instability by the actively controlled journal bearing consists of a movable bushing with the use of piezoactuators as reported in [2]. In paper [3], an electromagnetic damper for the control of the lateral vibration of aero-engines is presented.

On the other hand, squeeze film dampers (SFDs) are based on the mutually coupled hydraulic and mechanical physical principle. SFDs [4] have been used for many years to reduce the amplitude vibration and force transmitted to the foundation in the rotating machinery. The traditional SFDs [4] need to be designed for one free vibration mode shape, which should be eliminated in a rotor layout. Therefore, controllable SFDs working in active [5, 6] or semi active concepts [7-11] have been developed. The concept of semi active working mode is represented by SFDs lubricated by electrorheological [7, 8] or magnetorheological [9-11] fluids.

Due to the physical properties of magnetorheological oils, the magnetorheological SFDs are highly nonlinear damping elements. In order to determine the higher harmonic components in the transient or steady-state response of the rotor system [12-15], it is appropriate to use the frequency spectrums processed by Fourier transform.

The main parts of SFD (Fig. 1a) include two concentric rings with a thin layer of the lubricating film in between. The outer ring is firmly coupled with the damper body while the inner ring is connected with the rotor journal and damper housing by the rolling element bearing and squirrel cage spring, respectively. The spring enables the ring to vibrate in the radial direction but prevents its rotation together with the shaft.

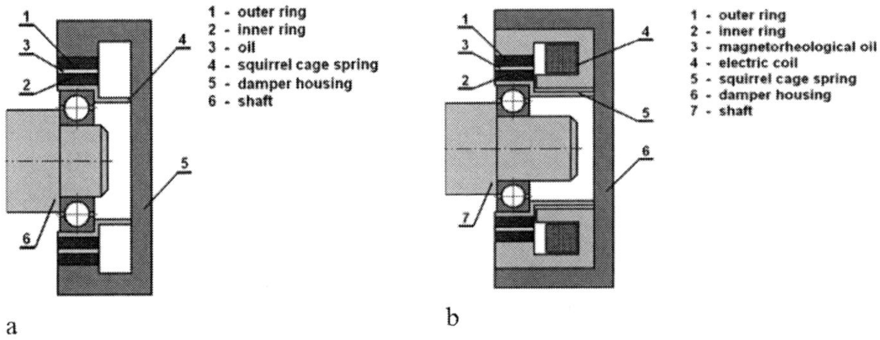

Fig. 1. The scheme of the traditional SFD (a) and magnetorheological SFD (b)

A new concept of the semi-active damping device is represented by SFDs lubricated by magnetorheological oils. The damper body is equipped with an electric coil generating magnetic flux passing through the film of magnetorheological liquid (Fig. 1b). As resistance against its flow depends on magnetic induction, the change of the applied current can be used to control the damping force.

In this paper, the computational procedure for determining the damping and stiffness coefficients of the classical and magnetorheological SFD have been developed and tested with the computational model of the damper presented in [11]. The pressure distribution in SFD is computed with mathematical models of the damper, based on the Navier-Stokes equations (NSE), modified NSE, Reynolds equation (RE), and RE modified for the damper with the short bearing approximation (SBA). A new mathematical model of the magnetorheological SFD with SBA based on the application of a bilinear material has been used to compute the pressure distribution. The carried-out computational simulations show that the developed mathematical models of the traditional SFD and magnetorheological SFD allowed computation of the damping and stiffness coefficients. It has been found that the stiffness coefficients computed by the proposed mathematical models may be different.

2 Pressure distribution in the classical and magnetorheological SFDs

The governing equations for the pressure distribution in the lubricating layer of SFD include NSE and the continuity equation [16], which are set up with the following assumptions: (i) the lubricant behaves as the Newtonian liquid; (ii) the flow in the fluid film is incompressible, laminar, and isothermal

$$\frac{\partial u}{\partial t}+u\frac{\partial u}{\partial x}+v\frac{\partial u}{\partial y}+w\frac{\partial u}{\partial z}=f_{\mathrm{x}}-\frac{1}{\rho}\frac{\partial p}{\partial x}+\frac{\eta}{\rho}\left(\frac{\partial^2 u}{\partial x^2}+\frac{\partial^2 u}{\partial y^2}+\frac{\partial^2 u}{\partial z^2}\right), \tag{1}$$

$$\frac{\partial v}{\partial t}+u\frac{\partial v}{\partial x}+v\frac{\partial v}{\partial y}+w\frac{\partial v}{\partial z}=f_{\mathrm{y}}-\frac{1}{\rho}\frac{\partial p}{\partial y}+\frac{\eta}{\rho}\left(\frac{\partial^2 v}{\partial x^2}+\frac{\partial^2 v}{\partial y^2}+\frac{\partial^2 v}{\partial z^2}\right), \tag{2}$$

$$\frac{\partial w}{\partial t}+u\frac{\partial w}{\partial x}+v\frac{\partial w}{\partial y}+w\frac{\partial w}{\partial z}=f_{\mathrm{z}}-\frac{1}{\rho}\frac{\partial p}{\partial z}+\frac{\eta}{\rho}\left(\frac{\partial^2 w}{\partial x^2}+\frac{\partial^2 w}{\partial y^2}+\frac{\partial^2 w}{\partial z^2}\right). \tag{3}$$

Where x, y, z are the Cartesian coordinates, which define the directions of the fixed frame of reference, u, v, w are the x, y, z components of the velocity, f_{x}, f_{y}, f_{z} are the components of the body force, p is the pressure, t is the time, and ρ and η is the density and dynamic viscosity of the lubricant, respectively.

The equation of continuity represents the conservation of mass, and in the case of the incompressible fluid it can be written in the following form

$$\frac{\partial u}{\partial x}+\frac{\partial v}{\partial y}+\frac{\partial w}{\partial z}=0. \tag{4}$$

RE [16] for the generation of the fluid film pressure can be derived from NSE (1-3) and the continuity equation (4) when taking into account the assumptions of the hydrodynamic lubrication [16, 17]: (i) the lubricant is the Newtonian liquid; (ii) the flow in the fluid film is incompressible, laminar, and isothermal; (iii) the inertial forces in the fluid film are neglected; (iv) the fluid pressure is constant in the radial direction; (v) the inner and outer surface of the damper ring is absolutely rigid and smooth; (vi) the width of the damper gap is small relative to the damper ring radius. Under these assumptions, the pressure distribution in the thin viscous fluid film of SFD is governed by the following RE

$$\frac{\partial}{\partial X}\left(h^3\frac{\partial p}{\partial X}\right)+\frac{\partial}{\partial Z}\left(h^3\frac{\partial p}{\partial Z}\right)=12\eta\frac{\partial h}{\partial t}. \tag{5}$$

Where X, Z denotes the local coordinates describing positions in the lubricating film in the circumferential (X) and axial (Z) direction (Fig. 2).

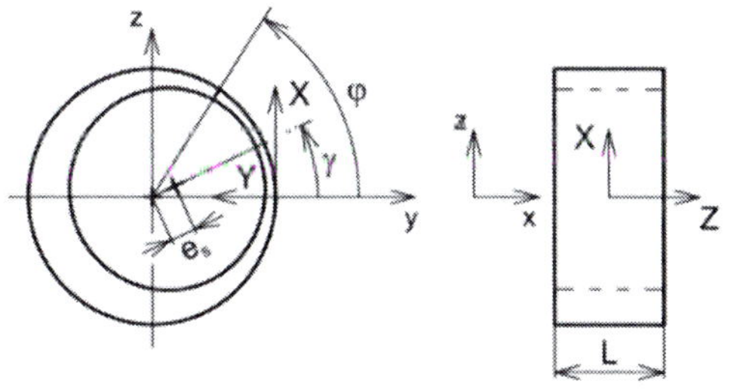

Fig. 2. The damper (xyz) and fluid film (XYZ) coordinate systems

The thickness of the fluid film at a location given by the circumferential coordinate is defined by the relation

$$h=c_0-e_{\mathrm{S}}\cos(\varphi-\gamma), \tag{6}$$

where h denotes the thickness of the fluid film, c_0 is the width of the gap between the inner and outer ring of the damper, e_S is the eccentricity, φ is the circumferential coordinate, and γ is the position angle of the line of the centres (Fig. 2).

It is assumed that the damper is symmetric relative to the plane perpendicular to the journal axis, the damper length to the radii of both rings is small, and the design arrangement of the damper makes it possible to be considered as SBA [17]. For such dampers, the prevailing flow in the lubricating layer occurs in the axial direction and RE (5) reduces to

$$\frac{\partial}{\partial Z}\left(h^3\frac{\partial p}{\partial Z}\right) = 12\eta\frac{\partial h}{\partial t}. \qquad (7)$$

By the double integration of the equation (7), the pressure profile in the axial direction of the damper is obtained in the form

$$p = \frac{6\eta}{h^3}\left(Z^2 - \frac{L^2}{4}\right)\frac{\partial h}{\partial t} + p_A. \qquad (8)$$

The boundary condition expressing that the pressure at the ends of the damper is equal to

$$p = p_A, Z = \pm\frac{L}{2}. \qquad (9)$$

L is the length of the damper, and p_A is the pressure in the ambient space.

The developed mathematical model of the magnetorheological SFD is based on the assumptions of the hydrodynamic lubrication theory [17] except those for the lubricant. The pressure distribution in the full fluid film is governed by RE adapted for the bilinear material [18]

$$\frac{\partial}{\partial Z}\left(h^3\frac{\partial p}{\partial Z}\right) = 12\eta_C\frac{\partial h}{\partial t}, \qquad (10)$$

$$\frac{\partial}{\partial Z}\left[h^3\frac{\partial p}{\partial Z} + 3h^2\tau_y + 8\frac{\tau_C^3}{\left(\frac{\partial p}{\partial Z}\right)^2} - 12\frac{\tau_y\tau_C^2}{\left(\frac{\partial p}{\partial Z}\right)^2} - 8\frac{\eta}{\eta_C}\frac{\tau_C^3}{\left(\frac{\partial p}{\partial Z}\right)^2}\right] = 12\eta\frac{\partial h}{\partial t}. \qquad (11)$$

Z_C is the axial coordinate, which defines the border between an unsheared region called the core and the region where the shear stress exceeds the τ_y yielding shear stress, τ_C is the shear stress at the core border, and η_C and η are the dynamic viscosities of the lubricant in and outside the core area, respectively. RE (10) is referred to the extent of axial coordinates for which it holds $0 \leq Z \leq Z_C$, and Eq. (11) describes the pressure distribution in the region where it holds $\partial h/\partial t < 0$ and $0 < Z_C < Z$.

The relation for the axial coordinate Z_C, which defines the border between these two regions, is as follows

$$Z_C = -\frac{\tau_C h^2}{6\eta_C\frac{\partial h}{\partial t}}. \qquad (12)$$

The solution of RE adapted to the bilinear material (10, 11) is obtained for the boundary conditions expressing that the pressure at the fluid film ends is equal to the pressure in the ambient space (9).

The yielding shear stress of the magnetorheological fluid, needed for solving the pressure gradient in the region outside the core (11) depends on magnetic induction. Based on the measurements, dependence of the yielding shear stress on magnetic induction can be approximated by a power function [19]

$$\tau_y = k_y B^{n_y}. \tag{13}$$

B is the magnetic induction, and k_y, n_y are the proportional and exponential material constants of the magnetorheological fluid, respectively.

In the most simple design case, the inner and outer rings of the damper can be considered as a divided core of an electromagnet. Then magnetic induction in the fluid film can be expressed

$$B = k_B \mu_0 \mu_r \frac{I}{h}, \tag{14}$$

where I is the electric current, μ_0 is the vacuum permeability, μ_r is the relative permeability of the magnetorheological fluid, and k_B is the product of the number of the coil turns and the damping element magnetic efficiency introduced as a ratio of magnetic flux passing through the lubricating layer with respect to the total flux generated by the electric coil. The magnetic efficiency can be determined by computational simulations using a magnetostatic problem utilizing a finite element method [20].

3 Determination of the damping force components

The components of the damping force are obtained by integration of the pressure distribution around the circumference and along the length of the damper

$$f_y = 2R \int_0^{\frac{L}{2}} \int_{\varphi_1}^{\varphi_2} p_d \cos(\varphi) d\varphi dZ, \quad f_z = 2R \int_0^{\frac{L}{2}} \int_{\varphi_1}^{\varphi_2} p_d \sin(\varphi) d\varphi dZ. \tag{15}$$

f_y, f_z are the y and z components of the damping force, R is the inner ring radius, p_d is the pressure distribution, and φ_1, φ_2 are the circumferential integration limits. The pressure distribution and circumferential integration limits are set according to the cavitation model.

The damping forces for SFD with SBA and for both the non-cavitated and cavitated operational regimes can be expressed in the analytical form, the force components expressions of which are given in [4].

The mutual interaction between the magnetorheological damper rings is accomplished by the hydraulic (damping) and magnetic forces. The hydraulic force is produced due to squeezing the magnetorheological fluid film and pushes the rings from each other. On the contrary, the magnetic force attracts. It is induced by the magnetic flux generated in the electric coils. In the paper [21] it was shown that the magnetic attractive force in the magnetorheological damper is much smaller than the hydraulic damping force, and therefore it is neglected in this study.

4 Computational procedure for determining the damping and stiffness coefficients of dampers

The influence of dampers on the performance of the rotating system dynamics has been studied with the mathematical modelling and experimental identification techniques [22].

The damping and stiffness coefficients of SFD and the magnetorheological SFD are obtained under the assumption of a circular centred orbit. The displacement and velocity components for synchronous circular whirling of the inner damper ring are defined by the relations

$$y_I = e\cos(\omega t), \; z_I = e\sin(\omega t), \; v_I = -e\omega\sin(\omega t), \; w_I = e\omega\cos(\omega t), \tag{16}$$

where y_I, z_I are the y and z components of the displacement of the inner damper ring, v_I, w_I are the y and z components of the velocity of the inner damper ring, ω is the whirl frequency, and e is the radius of the orbit.

The perturbed motion is composed by synchronous circular whirling and the superimposed harmonic motion with small amplitude as well as frequency which is higher than the whirl frequency. The components of the perturbed displacements of the inner damper ring are given by the relations

$$y_{\mathrm{I,p}} = \left[e + e_{\mathrm{p}}\cos(\omega_{\mathrm{p}}t)\right]\cos(\omega t),\quad z_{\mathrm{I,p}} = \left[e + e_{\mathrm{p}}\sin(\omega_{\mathrm{p}}t)\right]\sin(\omega t), \tag{17}$$

where y_{Ip}, z_{Ip} are the perturbed displacements of the inner damper ring in the y and z direction, ω_{p} is the perturbed whirl frequency, and e_{p} is the amplitude of the perturbed motion.

The damping and stiffness coefficients may be obtained on the basis of the computed or measured increments of the damping force

$$\Delta\mathbf{f} = \mathbf{f}_{\mathrm{p}} - \mathbf{f},\quad \Delta\mathbf{f} = \begin{bmatrix}\Delta f_{\mathrm{y}}\\ \Delta f_{\mathrm{z}}\end{bmatrix}, \tag{18}$$

where $\mathbf{f}_{\mathrm{p}}$, $\mathbf{f}$ is the vector of the perturbed and unperturbed damping forces in the stationary reference frame, $\Delta\mathbf{f}$ is the increment of the damping force, and Δf_{y}, Δf_{z} are the y and z components of the damping force increment.

Taking into account that the increment of the damping force $\Delta\mathbf{f}$ in the stationary reference frame acting on the inner damper ring can be expressed by the increment of displacement $\Delta\mathbf{q}$ and corresponding increment of the velocity $\Delta\dot{\mathbf{q}}$ of the inner damper ring

$$\Delta\mathbf{f} = \mathbf{B}\,\Delta\dot{\mathbf{q}} + \mathbf{K}\,\Delta\mathbf{q}, \tag{19}$$

$$\mathbf{B} = \begin{bmatrix}b_{\mathrm{y,y}} & b_{\mathrm{y,z}}\\ b_{\mathrm{z,y}} & b_{\mathrm{z,z}}\end{bmatrix},\qquad \mathbf{K} = \begin{bmatrix}k_{\mathrm{y,y}} & k_{\mathrm{y,z}}\\ k_{\mathrm{z,y}} & k_{\mathrm{z,z}}\end{bmatrix},\qquad \Delta\dot{\mathbf{q}} = \begin{bmatrix}\Delta\dot{q}_{\mathrm{y}}\\ \Delta\dot{q}_{\mathrm{z}}\end{bmatrix},\qquad \Delta\mathbf{q} = \begin{bmatrix}\Delta q_{\mathrm{y}}\\ \Delta q_{\mathrm{z}}\end{bmatrix}.$$

Δq_{y}, Δq_{z}, $\Delta\dot{q}_{\mathrm{y}}$, $\Delta\dot{q}_{\mathrm{z}}$ are the y and z components of the increment of displacement and the increment of velocity respectively, $\mathbf{B}$, $\mathbf{K}$ are the matrixes of the damping and stiffness coefficients in the stationary reference frame. In the damping $b_{\mathrm{i,j}}$ and stiffness $k_{\mathrm{i,j}}$ matrix, all diagonal terms are called direct coefficients and off-diagonal terms are called cross-coupled terms.

The increment of displacement $\Delta\mathbf{q}$ and corresponding increment of the velocity $\Delta\dot{\mathbf{q}}$ of the right-hand side of the equation (19) is obtained by computations for predefined kinematic parameters of the unperturbed and perturbed motion. Therefore, the unknown damping and stiffness coefficients are solved from linear equations in the following form

$$\begin{bmatrix}b_{\mathrm{y,y}}\\ b_{\mathrm{y,z}}\\ k_{\mathrm{y,y}}\\ k_{\mathrm{y,z}}\end{bmatrix} = \mathbf{A}^{+}\begin{bmatrix}\Delta f_{\mathrm{y,1}}\\ \vdots\\ \Delta f_{\mathrm{y,n}}\end{bmatrix},\quad \begin{bmatrix}b_{\mathrm{z,y}}\\ b_{\mathrm{z,z}}\\ k_{\mathrm{z,y}}\\ k_{\mathrm{z,z}}\end{bmatrix} = \mathbf{A}^{+}\begin{bmatrix}\Delta f_{\mathrm{z,1}}\\ \vdots\\ \Delta f_{\mathrm{z,n}}\end{bmatrix},\quad \mathbf{A} = \begin{bmatrix}\Delta\dot{q}_{\mathrm{y,1}} & \Delta\dot{q}_{\mathrm{z,1}} & \Delta q_{\mathrm{y,1}} & \Delta q_{\mathrm{z,1}}\\ \vdots & \vdots & \vdots & \vdots\\ \Delta\dot{q}_{\mathrm{y,n}} & \Delta\dot{q}_{\mathrm{z,n}} & \Delta q_{\mathrm{y,n}} & \Delta q_{\mathrm{z,n}}\end{bmatrix}. \tag{20}$$

The damping and stiffness coefficients in the stationary reference frame are a periodical function of time and due to the increase in both accuracy of their computation and the numerical stability of the matrix $\mathbf{A}$, the coefficients must be set during one period at a number of the time points ($n > 20$). The Moore-Penrose pseudoinverse ($\mathbf{A}^{+}$) is applied for solving the linear equations (20).

5 Computational simulations

Applicability of the developed methodology is demonstrated on the magnetorheological SFD, presented in [11]. Its photography is drawn in Fig. 3.

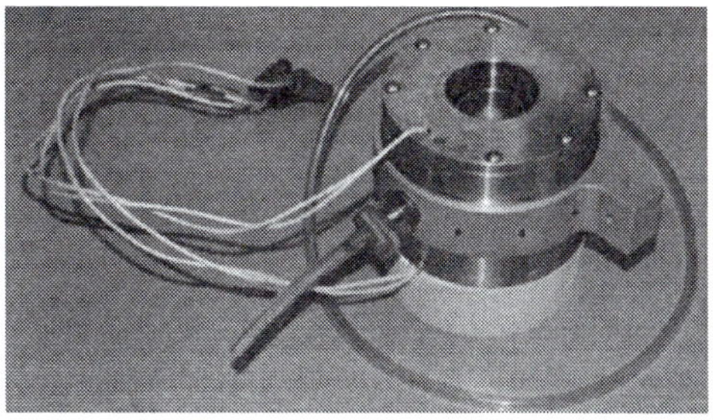

Fig. 3. The image of the investigated magnetorheological SFD

The technological parameters of the investigated magnetorheological SFD are as follows: the radius of the outer ring is 76 mm, length of the damper is 44 mm, width of the gap between the inner and outer ring of the damper is 1 mm, oil dynamic viscosity not affected by a magnetic field is 0.1 Pa·s, oil dynamic viscosity in the core is 500 Pa·s, magnetorheological oil density is 970 kg·m^{-3}, oil relative permeability is 5, damper design parameter is 60, and magnetorheological oil proportional and exponential material constants are 15 000 Pa·T^{-2} and 2, respectively.

The parameters of synchronous circular whirling of the inner damper ring are defined as follows: the whirl frequency is 1800 rpm, radius of orbit is 0.05 mm, perturbed whirl frequency is 18 000 rpm, and amplitude of the perturbed motion is 0.05 μm.

The task was to study the damping and stiffness coefficients (i) for the traditional SFD (the magnetorheological damper coils are not connected to current supply) and (ii) for the magnetorheological SFD.

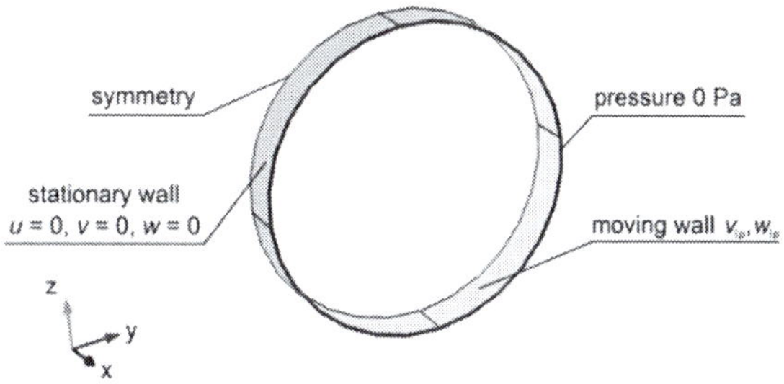

Fig. 4. Schematic representation of the SFD geometry with applied boundary conditions; v_{Ip}, w_{Ip} are the y and z components of the perturbed velocity

In this work, the commercial COMSOL software was used for solving the forces acting in SFD. The orbit radius is equal to 5% of the width of the gap between the inner and outer ring of the damper. Due to the small radius of circular whirling, the changes in the domain boundaries are negligible, the shape of the fluid domain is almost the same and therefore only one geometry model in COMSOL is used. The three-dimensional geometry of the damper filled with oil is simplified to the thin cylinder film (Fig. 4). The boundary conditions were defined in the form of ambient pressure at the outlet, symmetry at the half of the damper length, stationary outer ring, and movable inner damper ring (Fig. 4). The fluid is assumed to be incompressible, gravity forces are not taken into account, and the Reynolds number is equal to 1.8, and therefore the viscous model is set to laminar. NSE (1-3) and the continuity equation (4) are numerically solved using a time-varying solver.

The pressure distribution on the surface of the lubricating film for the non-cavitated (a) and cavitated (b) operational regime of the traditional SFD is depicted in Fig. 5. The location of the inner ring is offset in the horizontal (y) direction by 0.05 mm. The pressure distribution is almost constant in the radial direction, which clearly shows that the basic assumption of RE is satisfied.

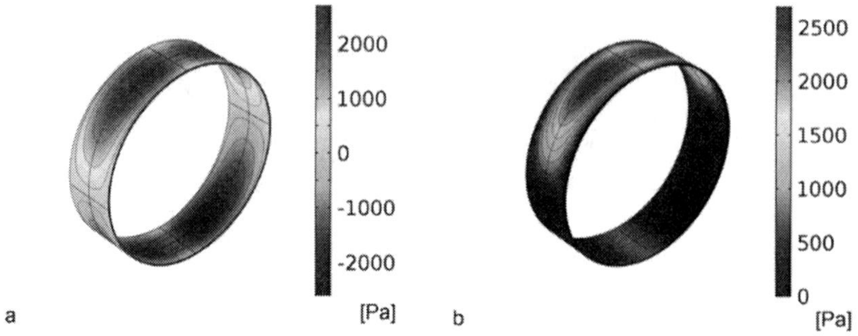

Fig. 5. Pressure distribution in the non-cavitated (a) and cavitated (b) operational regime of SFD

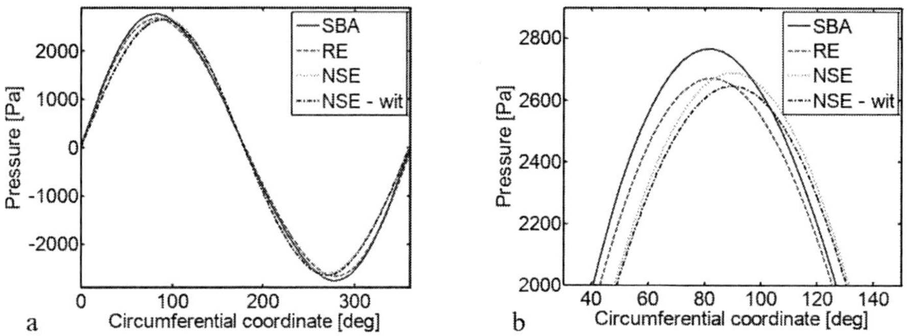

Fig. 6. Pressure distribution in the damper middle plane (a) and the detail of the maximum pressure distribution (b); NSE-wit (Navier-Stokes equations - without inertial terms)

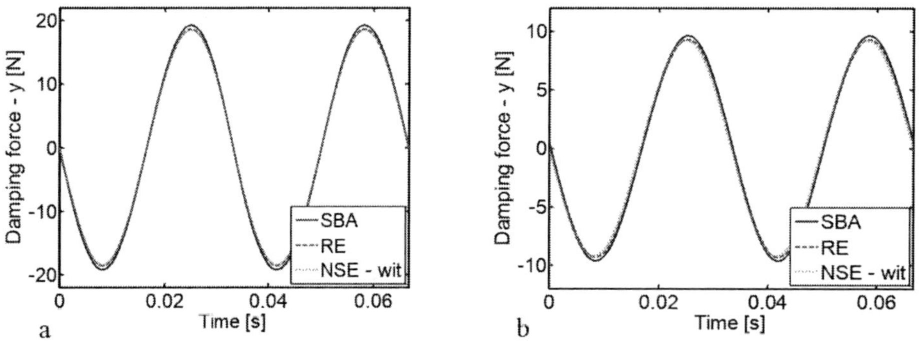

Fig. 7. Time history of the damping force in the horizontal direction in the non-cavitated (a) and cavitated (b) operational regime of SFD

The pressure distribution in the damper middle plane for the non-cavitated operational regime computed by means of several mathematical models of SFD is drawn in Fig. 6. The small differences in the amplitude and phase shift of the pressure distribution are determined between both mathematical models based on NSE. These differences are caused by negligence of the inertial effects in the NSE-wit mathematical model of the damper. The pressure distribution computed using the mathematical models (SBA, RE, and NSE-wit) of the damper neglecting the inertial forces is almost identical.

The time history of the damping force in the horizontal direction is depicted in Fig. 7. It is evident that the damping force for the non-cavitated operational regime is higher than the force related to the cavitated operational regime of SFD. The mathematical model of the damper with the assumptions of SBA arrives at the small increase in the damping force. And in this case, the flow in the damper is only in the axial direction. On the contrary, the mathematical model with NSE-wit arrives at a small reduction of the amplitude of the damping force and flow in the damper is three-dimensional.

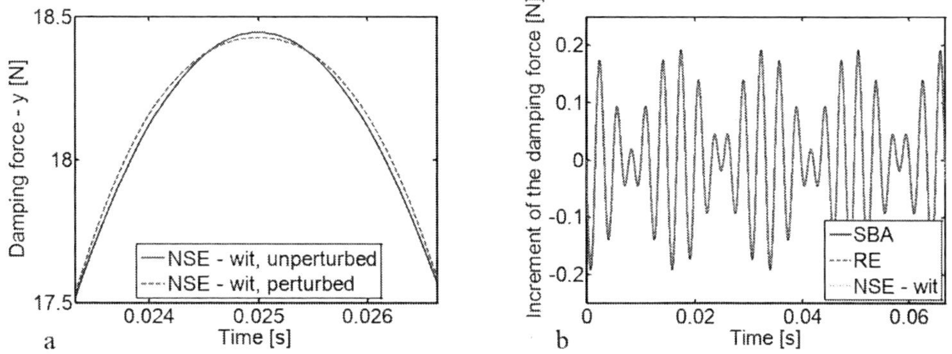

Fig. 8. Detail of the time history of the damping force (a) and time history of the increment of the damping force in the horizontal direction (b)

In order to compute the damping and stiffness coefficients, the perturbed motion is considered and the unperturbed and perturbed damping force is plotted in Fig. 8a. The corresponding increment of the damping force is depicted in Fig. 8b. SFD is operating in the non-cavitated regime. Fig. 8b shows that the increment of the damping force for the three mathematical models (SBA, RE, NSE-wit) of the damper is almost the same.

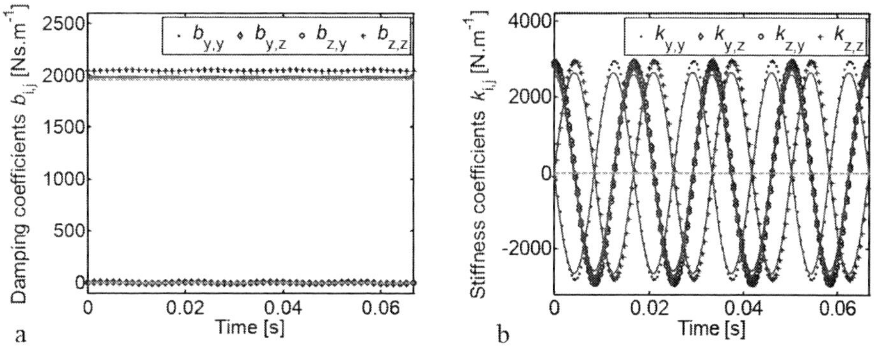

Fig. 9. Time history of the damping (a) and stiffness (b) coefficients in the non-cavitated regime of SFD; SBA (blue symbols), RE (continuous red line), NSE-wit (dashed green line)

The time-varying damping and stiffness coefficients in the stationary reference frame of SFD and the three mathematical models (SBA, RE, NSE-wit) of the damper are depicted in Figs. 9 and 10 (the non-cavitated regime and cavitated operational regime, respectively). The magnitude of the direct damping coefficients for the non-cavitated regime is greater than the one for the cavitated operational regime. The direct damping coefficients are almost equal for the three mathematical models (SBA, RE, NSE-wit). The cross-coupled coefficients of the damping are the same size and its value of magnitude is small for the mathematical models based on SBA and RE. For the NSE-wit model, the cross-coupled coefficients are equal to zero. The results show that all stiffness coefficients are equal to zero only for NSE-wit mathematical model of the damper. This applies to the both SFD operating regimes. For the non-cavitated and cavitated operational regimes of SFD (Fig. 9, Fig. 10), the values of the direct stiffness coefficients are different in comparison with the damping coefficients.

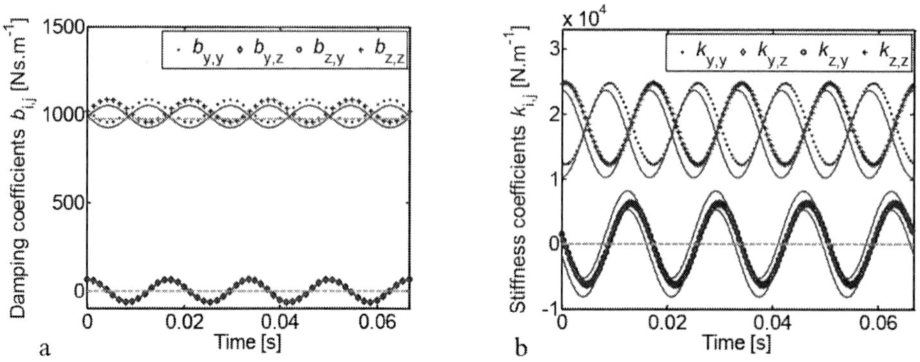

Fig. 10. Time history of the damping (a) and stiffness (b) coefficients in the cavitated operational regime of SFD; SBA (blue symbols), RE (continuous red line), NSE-wit (dashed green line)

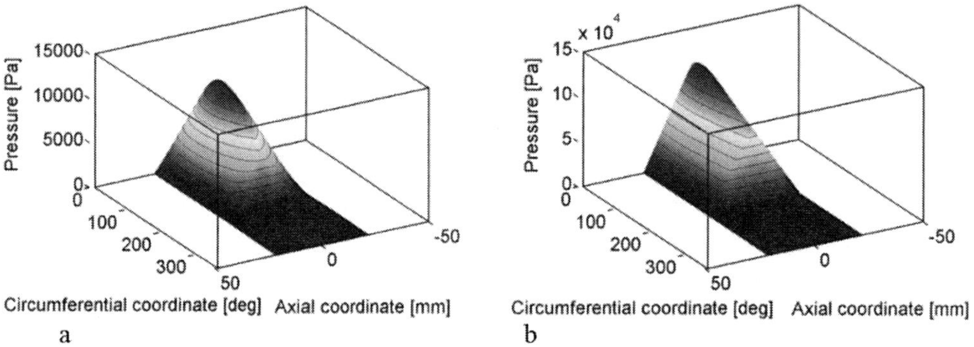

Fig. 11. Pressure distribution in the cavitated operational regime of the magnetorheological SFD, $I = 0.25$ A (a), $I = 1.0$ A (b)

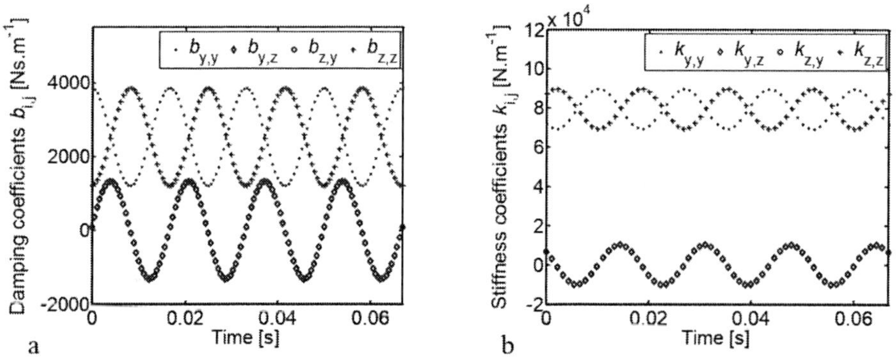

Fig. 12. Time history of the damping (a) and stiffness (b) coefficients in the stationary reference frame of the magnetorheological SFD, $I = 0.25$ A

The mathematical model of the magnetorheological SFD with the assumptions of the SBA and based on the application of a bilinear material (10, 11) has been used. The location of the inner ring is offset in the horizontal (y) direction by 0.05 mm. The pressure distribution in the area of the oil film for two magnitudes of the applied current is drawn in Fig. 11. The results show that a rising electric current increases the magnitude of the pressure.

The corresponding time-varying damping and stiffness coefficients in the stationary reference frame of the magnetorheological SFD are depicted in Fig. 12 and Fig. 13. The magnetorheological damper is operating in the cavitated regime. The higher the value of the current supplied in the coils the larger the magnitude of the damping and stiffness coefficients.

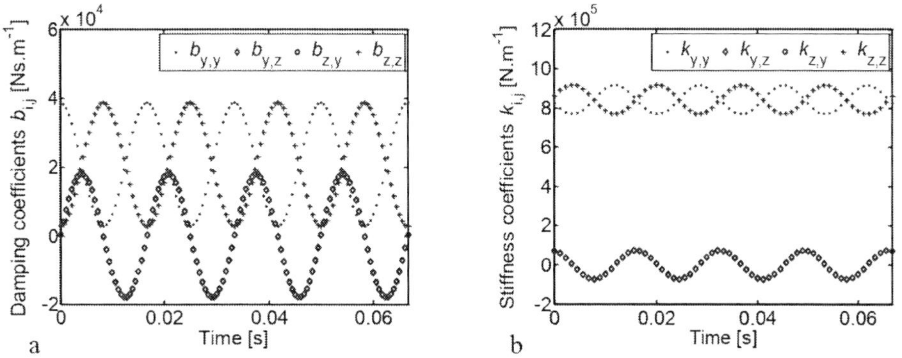

Fig. 13. Time history of the damping (a) and stiffness (b) coefficients in the stationary reference frame of the magnetorheological SFD, I = 1.0 A

Conclusion

The development of a novel computational procedure for computing the damping and stiffness coefficients of the traditional SFD and magnetorheological SFD, learning more about the effect of the inertial forces in SFD on the pressure distribution and damping forces, and the knowledge about the behaviour of the damping and stiffness coefficients are the principal contributions of this article.

Mathematical models of the damper based on NSE, modified NSE, RE, and RE modified for the damper with SBA were implemented for computation of the pressure distribution and the damping forces. It was found that the stiffness coefficients computed by the proposed mathematical models in SFD may be different.

The mathematical model of the magnetorheological SFD with SBA, where the magnetorheological fluid is represented by the bilinear material, has been used to achieve a more accurate description and increased computational stability. The simulation results show that with the rising electric current the magnitude of the damping and stiffness coefficients of the magnetorheological SFD increases.

This work has been supported by The Ministry of Education, Youth and Sports from the National Programme of Sustainability (NPU II) project „IT4Innovations excellence in science - LQ1602“, by the Czech Science Foundation 15-06621S, and by the Specific Research SP2017/136.

References

1. C. C. Thomazi, M. B. dos Santos, F. P. L. Neto, *Numerical analysis of a rotor-bearing system supported by elastomeric dampers*. Rev. Cient. **3**, 79-95 (2016)
2. J. Tůma, J. Šimek, J. Škuta, J. Los, *Active vibrations control of journal bearings with the use of piezoactuators*. Mech. Syst. Signal Process. **36**, 618-629 (2013)
3. A. Tonoli, N. Amati, A. Bonfitto, M. Silvagni, B. Staples, E. Karpenko, *Design of electromagnetic dampers for aero-engine applications*. J. Eng. Gas Turbines Power **132**, 112501 (2010)
4. J. M. Vance, *Rotordynamics of turbomachinery*. John Willey & Sons (1988)

5. A. El-Shafei, *Active control algorithms for the control of rotor vibrations using hybrid squeeze film dampers*. J. Eng. Gas Turbines Power **124**, 598-607 (2002)
6. H. R. Heidari, P. Safarpour, *Design and modeling of a novel active squeeze film damper*. Mech. Mach. Theory **105**, 235-243 (2016)
7. A. Bouzidane, M. Thomas, *An electrorheological hydrostatic journal bearing for controlling rotor vibration*. Comput. Struct. **86**, 463-472 (2008)
8. Y. Sun, M. Thomas, *Control of torsional rotor vibrations using an electrorheological fluid dynamic absorber*. J. Vib. Control **17**, 1253-1264 (2011)
9. C. Carmignani, P. Forte, E. Rustighi, *Design of a novel magneto-rheological squeeze-film damper*. Smart Mater. Struct. **15**, 164-170 (2006)
10. G. J. Wang, N. Feng, G. Meng, E. J. Hahn, *Vibration control of a rotor by squeeze film damper with magnetorheological fluid*. J. Intell. Mater. Syst. **17**, 353-357 (2006)
11. J. Zapoměl, P. Ferfecki, P. Forte, *A computational investigation of the transient response of an unbalanced rigid rotor flexibly supported and damped by short magnetorheological squeeze film dampers*. Smart. Mater. Struct. **21**, 105011 (2012)
12. A. Sapietová, V. Dekýš, *Dynamic analysis of rotating machines in MSC.ADAMS*. Procedia Eng. **136**, 143-149 (2016)
13. V. Dekýš, P. Novák, O. Dvouletý, L. Radziszewski, *The contribution to the analysis of the run up test*. Procedia Eng. **136**, 181-186 (2016)
14. V. Dekýš, A. Sapietová, O. Števka, *Understanding of the dynamical properties of machines based on the interpretation of spectral measurements and FRF*. Appl. Mech. Mater. **486**, 106-112 (2014)
15. V. Dekýš, P. Kalman, P. Hanák, P. Novák, Z. Stankovičová, *Determination of vibration sources by using STFT*. Procedia Eng. **177**, 496-501 (2017)
16. A. Z. Szeri, *Tribology: Friction, lubrication, and wear*. Hemisphere Publishing Corporation (1980)
17. Y. Hori, *Hydrodynamic lubrication*. Springer-Verlag (2006)
18. J. Zapoměl, P. Ferfecki, J. Kozánek, *Modelling of magnetorheological squeeze film dampers for vibration suppression of rigid rotors*. Int. J. Mech. Sci. **127**, 191-197 (2017)
19. O. Marinică, D. Susan-Resiga, F. Bălănean, D. Vizman, V. Socoliuc, L. Vékás, *Nano-micro composite magnetic fluids: Magnetic and magnetorheological evaluation for rotating seal and vibration damper applications*. J. Magn. Magn. Mater. **406**, 134-143 (2016)
20. P. Ferfecki, J. Zapoměl, J. Kozánek, *Analysis of the vibration attenuation of rotors supported by magnetorheological squeeze film dampers as a multiphysical finite element problem*. Adv. Eng. Softw. **104**, 1-11 (2017)
21. P. Ferfecki, J. Zapoměl, M. Marek, *Multi-physical analysis of the forces in the flexible rotor supported by the magnetorheological squeeze film dampers*. Adv. Mech. Design II. Mech. Mach. Sci. **44**, 89-95 (2017)
22. R. Tiwari, A. W. Lees, M. I. Friswell, *Identification of dynamic bearing parameters: A review*. Shock Vib. Dig. **36**, 99-124 (2004)

Parametric oscillations of the mechanical systems

Branislav Ftorek[1,*], *Pavol* Oršanský[1], *Helena* Šamajová[1]

[1]Department of Applied Mathematics, Faculty of Mechanical Engineering, University of Žilina, Univerzitná 1, 01026 Žilina, Slovakia

Abstract. This paper deals with an approximative solution in the analytical form of some second order linear differential equation with time depend parameters. The analytic solution is compared with the numerical solution obtained by Runge-Kutta method and the conditions of the convergence of this analytical method are discussed.

Keywords: oscillations, asymptotic solution, Mathieu equation

1 Introduction

Time-periodic modulation of system parameters may lead to resonances, instabilities and some pattern formation. The simplest system exhibiting parametric resonances is the harmonic oscillator with a frequency that varies periodically with time. The equation of motion of a simple harmonic oscillator whose frequency is modulated by a time-dependent zero-average perturbation $\xi(t)$ is

$$\ddot{x} + \omega_0^2[1 + \xi(t)]x = 0, \tag{1}$$

where x is the displacement and ω_0 is the natural frequency of the oscillator. This equation of motion can be solved with a friction term. Unlike a forced harmonic oscillator, the equation (1) has a fixed point at $x = 0$ regardless of the type of modulation. When the modulation is absent, the fixed point is neutrally stable. That is, near $x = 0$ the displacement neither grows without bound nor decays.

However, in the modulation case the fixed point may be unstable for certain temporal modulations. It is interesting that even linear oscillator with a modulated frequency is a "nonlinear problem" because the modulation occurs in a system parameter and is thus multiplicative (see [1, 2]). An example of unstable mode of oscillation system with its phase portrait can be seen in Figure 1.

* Corresponding author: branislav.ftorek@fstroj.uniza.sk

Reviewers: *Radim* Halama, *Jerzy* Winczek

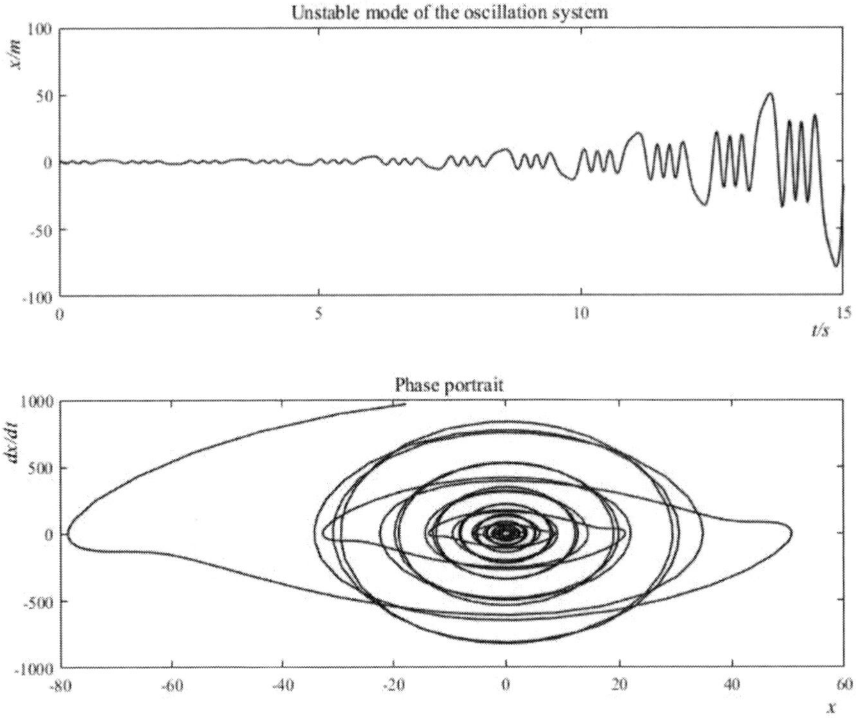

Fig. 1. Unstable mode of oscillation system with its phase portrait:
$\ddot{x} + \omega_0^2[1 + \sin Bt] = 0$ where $\omega_0 = 20, B = \frac{\omega_0}{4}$.

In most general form, the equation of motion (1) is called Hill's equation

$$\ddot{x} + p(t)\dot{x} + q(t)x = 0, \tag{2}$$

in which functions $p(t)$ and $q(t)$ are continuous and have a common period τ. If the modulation is specified to be cosinusoidal, the equation (2) is called the Mathieu equation

$$\ddot{x} + [\omega_0^2 - \varepsilon\cos t]x = 0. \tag{3}$$

The solution of the Mathieu equation or the solution of some nonlinear equation may be find by using general asymptotic methods for sufficiently small parameter ε, $0 < \varepsilon \ll 1$. Many interesting physical phenomena are described by general differential equations with nonlinear or time dependent terms

$$\ddot{x} + \omega_0^2 x + \varepsilon f(t, x, \dot{x}) = 0, \tag{4}$$

where $f(t, x, \dot{x})$ is a real function and ε is a small parameter. If $\varepsilon = 0$ then the solution of the previous equation has a form

$$x(t) = A\sin(\omega_0 t + \varphi), \tag{5}$$

where the amplitude A and the phase φ are constants. We suppose the perturbation solution for $\varepsilon \neq 0$ in the form of an asymptotic series [4]

$$x(t) = A(t)\sin(\omega_0 t + \varphi(t)) + \varepsilon\phi_1(t) + \varepsilon^2\phi_2(t) + \cdots + \varepsilon^n\phi_n(t) + \cdots, \tag{6}$$

where $A(t), \varphi(t), \phi_1(t), \ldots, \phi_n(t), \ldots$ are unknown functions. Another asymptotic method is well known as a WKB-method, or Wentzel–Kramers–Brillouin method (see [5]) for the equation with a large parameter λ, that is, $\lambda \to \infty$

$$\ddot{x} + f(x, \lambda)x = 0. \tag{7}$$

2 Asymptotic solution

Now we consider a second order differential equation with a friction term and with a fast time periodic modulation, that is, the parameter B is sufficiently large

$$\ddot{x} + 2\gamma\dot{x} + \omega_0^2[1 + \lambda \sin Bt]x = 0, \tag{8}$$

$$x(0) = 1, \tag{9}$$

where $\lambda > 0, \gamma > 0, \omega_0 > 0$, are parameters of the oscillation system. Now we will use the basic idea from the previous section and from the paper [6]. If $\lambda = 0$ (with regard to the initial condition (9)), the solution has an exact periodic form with damping

$$x_0(t) = e^{-\gamma t} \cos \omega t, \tag{10}$$

where $\omega^2 = \omega_0^2 - \gamma^2 > 0$. For $\lambda \neq 0$ we assume the solution of (8) as a harmonic modulation of the solution (10)

$$x(t) = e^{-\gamma t} \cos(\Omega t)[1 + \varepsilon \sin Bt], \tag{11}$$

where Ω, ε are unknown constant parameters (not functions) and the parameter ε is sufficiently small, that means, $0 < \varepsilon \ll 1$. From (11) we have

$$= e^{-\gamma t}[-\gamma \cos \Omega t\,(1 + \varepsilon \sin Bt) - \Omega \sin \Omega t (1 + \varepsilon \sin Bt) + \varepsilon B \cos \Omega t \cos \mathrm{B}t], \tag{12}$$

$$\begin{gathered} \ddot{x} = e^{-\gamma t}[\gamma^2 \cos \Omega t\,(1 + \varepsilon \sin Bt) + 2\gamma\Omega \sin \Omega t (1 + \varepsilon \sin Bt) \\ -2\gamma\varepsilon B \cos \Omega t \cos \mathrm{B}t - \Omega^2 \cos \Omega t (1 + \varepsilon \sin Bt) \\ -2\Omega\varepsilon B \sin \Omega t \cos \mathrm{B}t - \varepsilon B^2 \cos \Omega t \sin \mathrm{B}t], \end{gathered} \tag{13}$$

Substituting the previous relations in (8) and with regard to

$$\sin^2 Bt = \frac{1 - \cos 2Bt}{2}, \tag{14}$$

we get

$$\begin{gathered} e^{-\gamma t}\{\cos \Omega t\,[-\gamma^2 + \omega_0^2 - \Omega^2 + \frac{\varepsilon\lambda\omega_0^2}{2} \\ + \sin \mathrm{B}t\,(\varepsilon\gamma^2 - \varepsilon\Omega^2 - \varepsilon B^2 - 2\varepsilon\gamma^2 + \omega_0^2\lambda + \varepsilon\omega_0^2) \\ -\frac{\varepsilon\lambda\omega_0^2}{2} \cos 2Bt] + \sin \Omega t\,(-2\varepsilon\Omega B \cos \mathrm{B}t)\} = 0. \end{gathered} \tag{15}$$

It is obvious that the constant term $\left(-\gamma^2 + \omega_0^2 - \Omega^2 + \frac{\varepsilon\lambda\omega_0^2}{2}\right)$ must be equal to zero, that is

$$\Omega = \sqrt{\omega_0^2 - \gamma^2 + \frac{\varepsilon\lambda\omega_0^2}{2}}. \tag{16}$$

Now we consider the $sinBt$ term equals to zero

$$\varepsilon\gamma^2 - \varepsilon\Omega^2 - \varepsilon B^2 - 2\varepsilon\gamma^2 + \omega_0^2\lambda + \varepsilon\omega_0^2 = 0. \tag{17}$$

According to (16), from the previous condition we have for the second unknown parameter ε the quadratic equation

$$\frac{-\lambda\omega_0^2}{2}\varepsilon^2 - \varepsilon B^2 + \omega_0^2\lambda = 0. \tag{18}$$

The root of this equation ($\varepsilon > 0$) has the following form

$$\varepsilon = \frac{-B^2}{\lambda\omega_0^2} + \sqrt{\frac{B^4}{\lambda^2\omega_0^4} + 2} \tag{19}$$

The relation (19) represents a condition of applicability of the main approximative method. It is desirable that the parameter ε is sufficiently small, that is, $\varepsilon \to 0^+$ if $B \to \infty$. From (15) we have a residual term

$$R = e^{-\gamma t}\left[\cos\Omega t\left(-\frac{\varepsilon\lambda\omega_0^2}{2}\cos 2Bt\right) + \sin\Omega t(-2\varepsilon\Omega B\cos Bt)\right] \tag{20}$$

and we suppose that the residuum R is sufficiently small ($R \sim 0$). The validity of this assumption is verified by numerical experiments. We obtained numerical results by the MATLAB environment. The comparison between analytical solutions and numerical solutions for the large scale of parameters are shown in the graphs, where numerical results represent black dot-dashes lines and analytical results the blue line (see Figures 2-5).

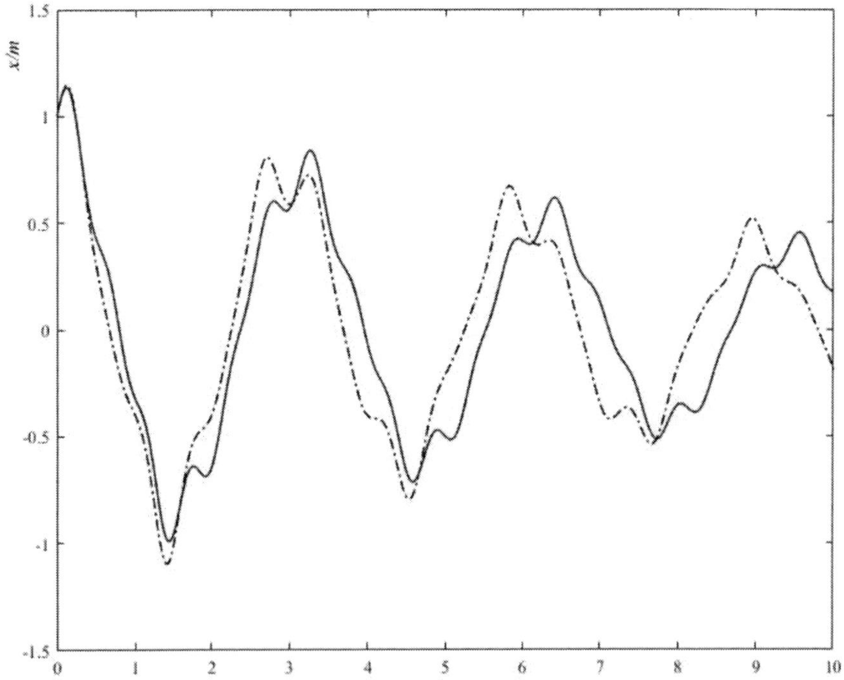

Fig. 2. Comparison between numerical (black) and analytical (blue) solution for parameters: $\gamma = 0.1, \omega_0 = \sqrt{2}, \lambda = 10, B = 10, \varepsilon = 0.196$.

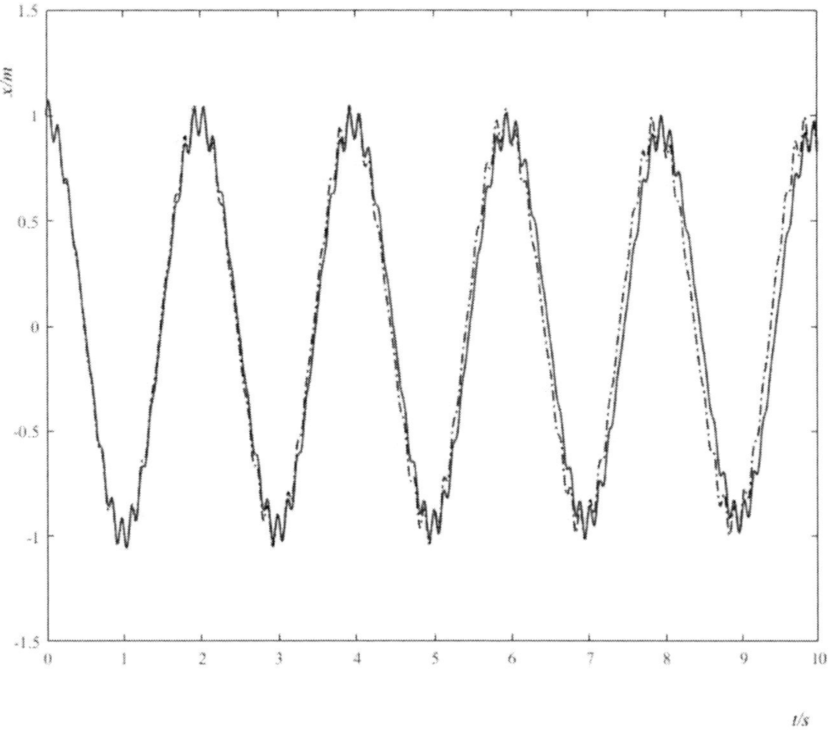

Fig. 3. Comparison between numerical (black) and analytical (blue) solution for parameters: $\gamma = 0.1, \omega_0 = \sqrt{2}, \lambda = 100, B = 50, \varepsilon = 0.08.$

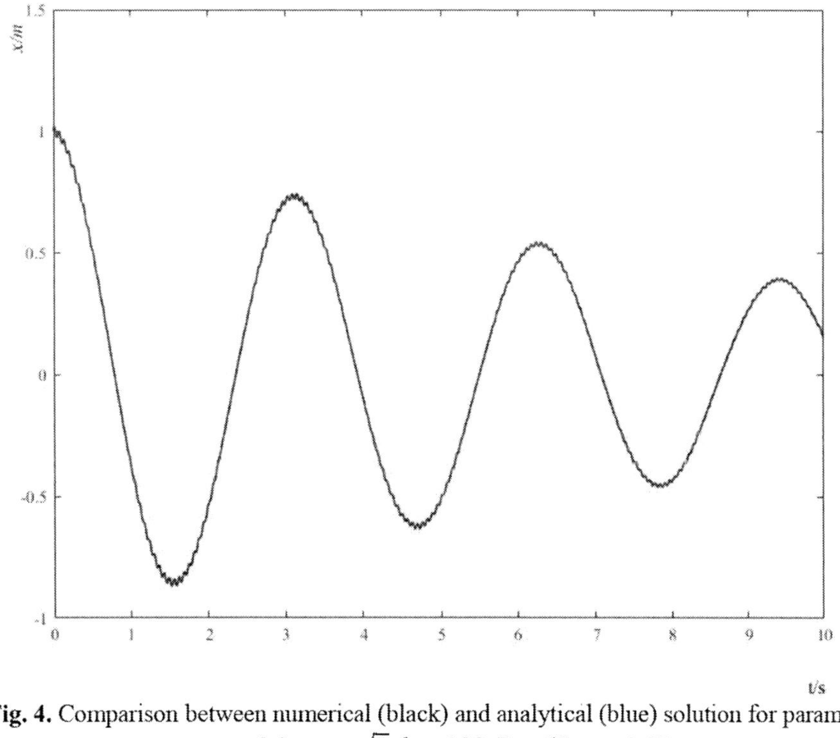

Fig. 4. Comparison between numerical (black) and analytical (blue) solution for parameters: $\gamma = 0.1, \omega_0 = \sqrt{2}, \lambda = 100, B = 50, \varepsilon = 0.02.$

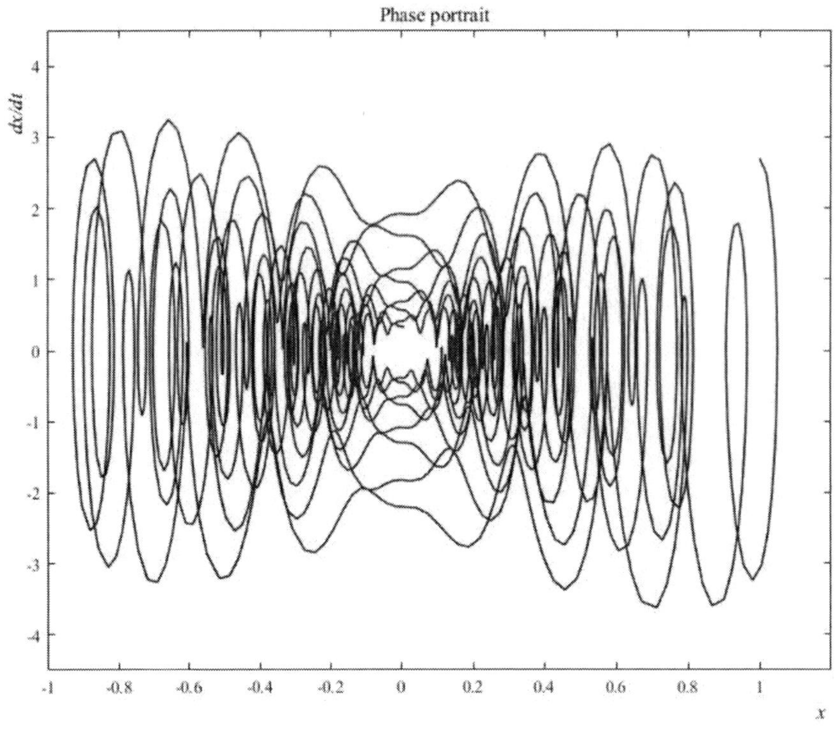

Fig. 5. Stable mode of oscillation system with its phase portrait.

Conclusion

The equation of motion (8) of a damped harmonic oscillator whose frequency is modulated by a time-dependent zero-average perturbation was investigated. An approximative solution

$$x(t) = e^{-\gamma t}\cos(\Omega t)[1 + \varepsilon \sin Bt],$$
$$\Omega = \sqrt{\omega_0^2 - \gamma^2 + \frac{\varepsilon \lambda \omega_0^2}{2}}, \qquad (21)$$
$$\varepsilon = \frac{-B^2}{\lambda \omega_0^2} + \sqrt{\frac{B^4}{\lambda^2 \omega_0^4} + 2},$$

for a fast time periodic modulation (if B is sufficiently large then the parameter ε is sufficiently small, that is, $\lim_{B\to\infty} \varepsilon = 0$) was obtained. This obtained asymptotic solution was compared with the numerical solution gained by the Runge-Kutta method.

This work was supported by KEGA under the Grant No. 015ŽU-4/2017 Digital technologies for machine building study programs as a part of strategy "Internet of Things".

References

1. R. Kawai, K. Lindenberg, C. Van den Broeck: *Parametrically modulated oscillator dimer: an analytic solution*. Physica A **312**, 119-140, (2002)
2. S. H. Strogatz: *Nonlinear dynamics and chaos*. Westview Press, Cambridge, (1994)

3. J. Zapoměl, V. Dekýš, P. Ferfecki, A. Sapietová, M. Sága, M. Žmindák: *Identification of Material Damping of a Carbon Composite Bar and Study of Its Effect on Attenuation of Its Transient Lateral Vibrations*. International Journal of Applied Mechanics **7**, 1-18, (2015)
4. R. A. Struble: *Nonlinear differential equations*. McGraw-Hill book company, New York, (1962)
5. J. D. Murray: *Asymptotic analysis*. Applied mathematical sciences, Springer-Verlag, New-York, (1984)
6. E. Wiszt, E. Wisztová: *An approximative solution of some second order differential equation with variable coefficients*. Studies of the University of Žilina, Math. Series **21**, 45-52 (2007)

Model calculations of posterior reliability indicators for the proposal of the maintenance system

Jana Galliková[1,*], *Vladimír* Stuchlý[1], *Roman* Poprocký[1], *Peter* Volna[1]

[1]University of Žilina, Faculty of Mechanical Engineering, Department of Transport and Handling machines, Univerzitná 1, 010 26 Žilina

Abstract. Designing the content and scale of maintenance of machines and equipment by a priori and posterior reliability methods in considered crucial to reducing the cost of the machine's life cycle, maintaining high operational readiness and reducing the consequences of failures. In the presented paper, attention is paid to the analysis of the calculation methods of posterior reliability for calculation indicators of reliability and to the use of the specified Weibull model for reliability calculations. The obtained results are further developed for models of optimal process calculations to perform scheduled maintenance interventions. Calculations of the other RAMS (reliability, availability, maintainability and safety) indicators that are critical to the design of an optimal engineering design with regard to maintenance and which do not receive sufficient attention in technical practice are also assessed.

Keywords: maintenance, reliability methods, Weibull model for reliability

1 Introduction

Analysis of individual failures is the specified failure of functional units and subsequently the reliability of the most distorted parts of electrical multiple unit series 671 is evaluated using the Weibull probability of distribution, which is a general probability distribution for determining the reliability of the mechanical components of the track-side vehicle. Based on these facts, it is then possible to optimize the vehicle maintenance system and to propose new approaches to care and secure reliability and especially passenger comfort [1, 2].

2 Description of the monitored object

The electric multiple unit 671 allows multiple control of the train set, which can consist of a maximum of three vehicles. In Fig. 1 we can see EMU 671 photo (driveway) running in Fig.1. Made to ensure a comfortable long-distance ride of 1,435 mm. The power supply is also provided by a voltage system of 3kV DC as well as 25kV / 50Hz AC.

* Corresponding author: jana.gallikova@fstroj.uniza.sk

Reviewers: *Ivan* Kuric, *Bohuš* Leitner

Fig. 1. Electric multiple unit series 671 (EMU 671)

2.1 Overview of HDV 671 maintenance

Technical capability of EMU in operation is maintained by maintenance. Periodic maintenance is the foundation of maintenance systems and consists of regularly alternating maintenance stages, the main feature of which is the prevention of failure.

They are:

- In operation servicing,
- minor inspection,
- major inspection,
- main repair,
- regular technical inspections.

Periodic maintenance includes the activities:

- Defectoscopy,
- diagnostic measurements and prophylactic controls,
- repair of the wheel drive profile (turning, turning)
- inspections and pressure tests of pressure vessels,
- preparation of the vehicle for winter / summer operation,
- revision of electrical equipment,
- vehicle weighing - checking and adjusting wheel pressure on the track,
- transformer oil exchange, of operating fluids.

Non-periodic maintenance includes the following maintenance steps:

- Scheduled repairs, includes:
    - checking and setting of vehicle technical parameters (e.g. test site, dry or water resistance),
    - checking the lateral clearance when pivoting the chassis to EMU cabinet frame,
    - preservation and decontamination of EMU,

  - exchange of two-wheelers,
  - exchange of hoops and full wheels,
- unplanned repair performed in periodic maintenance,
- unplanned repair performed separately, i.e. beyond the periodic maintenance.

In addition to periodic and non-periodic maintenance, construction changes (reconstruction, modernization, operational modifications) are performed on locomotives.

For locomotive delivery to the operating servicing and to a minor inspection, the kilometric norms with the permitted margin according to the ZSSK regulations apply. The upper limit is mandatory and must not be exceeded [3, 4, 5].

2.2 Methodology of following reliability of EMU 671

The most reliable results of the reliability monitoring are the results obtained by statistical methods if the primary data are recorded directly for the purpose of establishing reliability indicators by statistical methods. However, it is not excluded to use as a source of information also data recorded for another purpose: details of repairs performed, defects found during regular maintenance interventions, etc.

We will create a form that can also be used as a maintenance record. For this reason, we used the correction book form as the basis for the proposal. The information contained in the repair book is recorded in the form. The following may be considered as the most important information's:

- the number of the propelling vehicle,
- date of decommissioning of the vehicle,
- description of the fault,
- the date of return to service.

For the subsequent evaluation of the initial information, the necessary data can be directly recorded via MS Excel.

3 Engine failures

In statistical evaluation, it is always necessary to process a large amount of information. Based on available data from ZSSK Slovakia, we have developed a list of individual types of repairs on the individual components of the vehicle on the EMU 671.

As a decision-making tool, we used Pareto's analysis, which showed that as the first failure we will analyse the occurrence of repair of trolley failures (Fig. 2). High repairs occur also in parts such as electrical equipment, collector, compressor, brake [6, 7, 8].

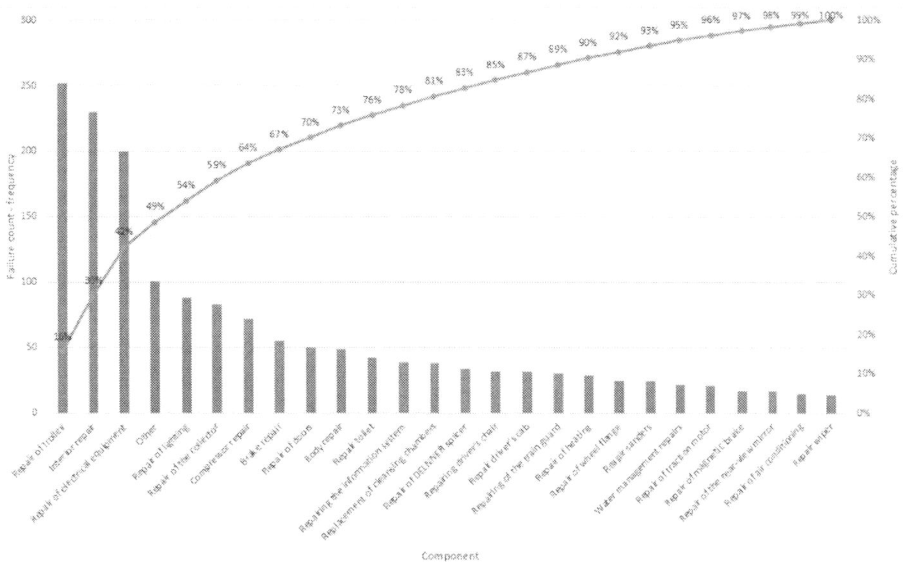

Fig. 2. Pareto's analysis of failures at EMU 671

4 Statistical evaluation of reliability

4.1 Regression curve

The obtained and classified data was subjected to a regression analysis. Table 1 shows a part of summary input and calculated values.

Table 1. Analysis of trolley failures – preparation of calculation

Ti	Component	MEDIAN RANK	1 / (1 - MR)	ln [ln (1 / (1 - MR))]	ln (T_i)
4	1	0,002773376	1,00278109	-5,886301782	1,38629436
6	2	0,006735341	1,00678101	-4,99700972	1,79175947
9	3	0,010697306	1,01081298	-4,53239071	2,19722458
9	4	0,014659271	1,01487736	-4,215307508	2,19722458
19	5	0,018621236	1,01897457	-3,974068946	2,94443898
20	6	0,022583201	1,02310499	-3,779149642	2,99573227
29	7	0,026545166	1,02726903	-3,615485838	3,36729583
32	8	0,030507132	1,0314671	-3,474343717	3,4657359
36	9	0,034469097	1,03569963	-3,350204786	3,58351894
37	10	0,038431062	1,03996704	-3,239358726	3,61091791
38	11	0,042393027	1,04426976	-3,139190651	3,63758616
46	12	0,046354992	1,04860823	-3,047788261	3,8286414
55	13	0,050316957	1,0529829	-2,963710696	4,00733319
56	14	0,054278922	1,05739422	-2,885845266	4,02535169
58	15	0,058240887	1,06184266	-2,813314783	4,06044301
60	16	0,062202853	1,06632869	-2,745415456	4,09434456

The first column (Ti) of Table 1 contains the days when the train driver malfunctioned. The second column (Component) contains the number of failures. In the third column (MEDIAN RANK - MR) for calculating the share estimation, the formula is used:

$$p = ((i - 0{,}3) \div (n + 0{,}4)) \tag{1}$$

where i is a sequence of failures and n is the total number of observed failuress, with the number 1 being from the "Component" column. In the fourth column, the formula is used:

$$1 \div (1 - MR) \tag{2}$$

In the fifth column for calculating the value, the double natural logarithm is used, then the formula:

$$ln[ln(1 \div (1 - MR))] \tag{3}$$

In the last column, the following formula is used to calculate the value:

$$ln(Ti) \tag{4}$$

We plot the values from the fifth and sixth columns in the graph (Fig. 3), where the orange curve represents the real course and the straight line represents the linear course of the *ln (ln (1 / 1- (MR)) dependence of ln (Ti)*.

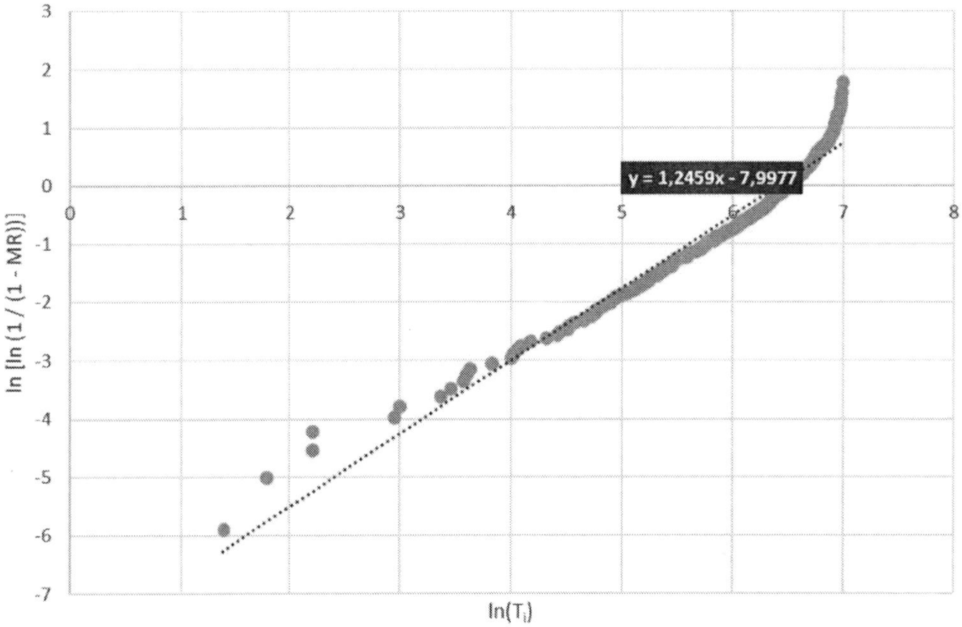

Fig. 3 Line fit pilot for trolley

After plotting dependence, the line equation, whose general shape is [8]:

$$y = mx + b \tag{5}$$

4.2 Weibull's model

In order to statistically model reliability in engineering, we will use the Weibull's model of probability. Weibull's distribution is used in cases where reliability depends on age, hours worked, or operating cycles. In terms of reliability, Weibull's distribution is used to determine reliability indicators that are important information needed to predict, evaluate and compare the life of products, evaluate construction and technology changes, compare alternative designs or technologies, compare the life of products of different technologies or different manufacturers, creating a warranty policy in a proactive approach to inventory management or repairs planning [9].

Weibull's distribution can be transformed to the form of the line equation:

$$F(x) = 1 - e^{-\left(\frac{x}{\alpha}\right)^{\beta}} \tag{6}$$

where: F(x) – cumulative distribution function,
α – scale parameter of the function (characteristic life),
β – shape parameter of the distribution.

$$1 - F(x) = e^{-\left(\frac{x}{\alpha}\right)^{\beta}} \tag{7}$$

$$\ln(1 - F(x)) = \left[-\left(\frac{x}{\alpha}\right)\right]^{\beta} \tag{8}$$

$$ln\left[ln\left(\frac{1}{1 - F(x)}\right)\right] = \beta\, ln\left(\frac{x}{\alpha}\right) \tag{9}$$

$$ln\left[ln\left(\frac{1}{1 - F(x)}\right)\right] = \beta\, ln\, x - \beta\, ln\, \alpha \tag{10}$$

By comparing this formula with a simple equation we can see that the left side of the equation corresponds to y, lnx corresponding to x, β corresponds to m and βlnα corresponds to b. Therefore, if we want to perform a linear regression, we need to know the parameter estimation. Estimating the Weibull parameter β comes directly from the straight line. Estimate for parameter α must be calculated [10-12]:

$$\alpha = e^{-\left(\frac{b}{\beta}\right)} \tag{11}$$

From the line equation: y = 1.2459-7.9977 we get β = 1.2459 where we can calculate the parameter:

$$\alpha = e^{-\left(\frac{7,9977}{1,2459}\right)} = 613.4185 \tag{12}$$

I choose number 0,9 as the value of reliability R(x), while the probability of failure Q(x) we can calculate:

$$Q(x) = 1 - R(x) \tag{13}$$

For calculation of the cycle (km) we use the formula:

$$Cycle(days) = \alpha \cdot (-\ln R(x))^{\frac{1}{\beta}} \tag{14}$$

$$101 days = 613{,}4185 \cdot (-\ln 0{,}9)^{\frac{1}{1{,}2459}} \tag{15}$$

The resulting value means the probability of failure the trolley 10% trolley occurs every 101 days.

Based on the Weibull model, we processed a calculation whose graphical output can be seen in the figure. There is a reliability function *R(x)*, the probability of failure *Q(x)*, failure rate *λ(x)*, and the probability density function *f(x)* (Fig. 4).

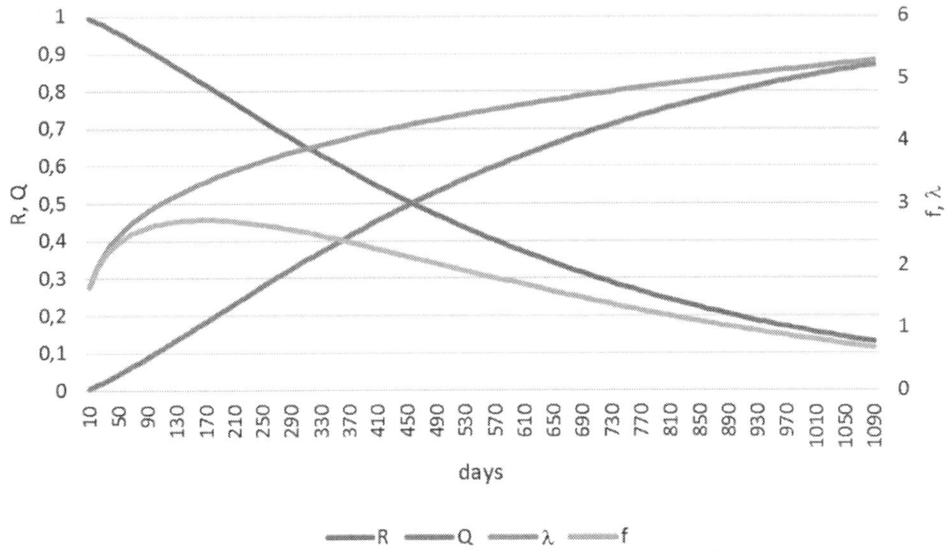

Fig. 4 Reliability indicators for Trolley

4.3 Use the Analysis ToolPak to perform complex data analysis in Excel

If you need to develop complex statistical or engineering analyses, you can save steps and time by using the Analysis ToolPak. You provide the data and parameters for each analysis, and the tool uses the appropriate statistical or engineering macro functions to calculate and display the results in an output table. Some tools generate charts in addition to output tables.

The data analysis functions can be used on only one worksheet at a time. When you perform data analysis on grouped worksheets, results will appear on the first worksheet and empty formatted tables will appear on the remaining worksheets. To perform data analysis on the remainder of the worksheets, recalculate the analysis tool for each worksheet.

The Regression analysis tool performs linear regression analysis by using the "least squares" method to fit a line through a set of observations. You can analyse how a single dependent variable is affected by the values of one or more independent variables. For example, you can analyse how an athlete's performance is affected by such factors as age, height, and weight. You can apportion shares in the performance measure to each of these three factors, based on a set of performance data, and then use the results to predict the performance of a new, untested athlete. The Regression tool uses the worksheet function **LINEST** (Fig. 5).

m = β	b
1,2459355	-7,99772
0,0157505	0,09515

α
613,41846

R	Ti
Reliability	Days
0,01	2090
0,1	1198
0,5	457
0,9	101
0,99	15

SUMMARY OUTPUT

Regression Statistics	
Multiple R	0,980603
R Square	0,961583
Adjusted R Square	0,961429
Standard Error	0,248038
Observations	252

ANOVA

	df	*SS*	*MS*
Regression	1	384,9813071	384,9813071
Residual	250	15,38067201	0,061522688
Total	251	400,3619791	

	Coefficients	*Standard Error*	*t Stat*
Intercept	-7,997719	0,095152689	-84,0514233
ln (Ti)	1,245935	0,015750465	79,10467857

Beta (or Shape Parameter) = 1,25
Alpha (or Characteristic Life) = 613,42

Fig. 5 Using the function LINEST in MS Excel

The LINEST function calculates the statistics for a line by using the "least squares" method to calculate a straight line that best fits your data, and then returns an array that describes the line. You can also combine LINEST with other functions to calculate the statistics for other types of models that are linear in the unknown parameters, including polynomial, logarithmic, exponential, and power series. Because this function returns an array of values, it must be entered as an array formula.

4.4 Interpreting the results LINEST

The Weibull shape parameter, called β, indicates whether the failure rate is increasing, constant or decreasing. And β <1.0 indicates that the product has a decreasing failure rate. This scenario is typical of "infant mortality" and indicates that the product is failing during its "burn-in" period. And β=1.0 indicates a constant failure rate. Frequently, components that have survived burn-in will subsequently exhibit a constant failure rate. And β >1.0 indicates an increasing failure rate. This is typical of products that are wearing out. Such is the case with the spring housings-both Trolley and Electrical equipment have β values much higher than 1.0 equipment, i.e., they wear out.

The Weibull characteristic life, called, is a measure of the scale, or spread, in the distribution of data. It so happens that α equals the number of days at which 63.2 percent of the product has failed. In other words, for a Weibull distribution (R = 0.368), regardless of the value of β. For example, with Trolley, about 37 percent of the Trolleys should survive at least 613.4185 days.

Conclusion

In the paper, we tried to explain the basic use of Weibull's distribution in the EMU 671 failure analysis. In the first part we processed the basic description and defects of EMU 671. In the second part we analysed failures that occurred during the operation on the critical parts of unit.

An important parameter in failure analysis was the number of days until a failure occurred. Depending on the number of days, preventative maintenance may be performed by checking the condition or replacing the individual components in which failures occur, thus preventing the occurrence of the consequences of the failure.

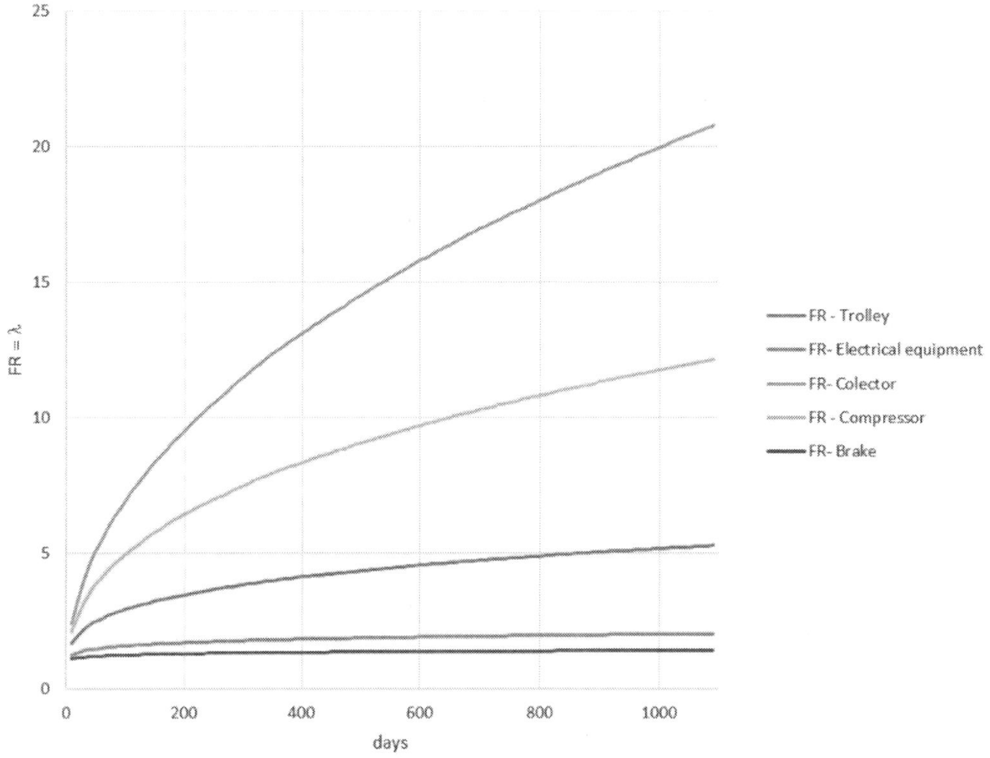

Fig. 6 Graph of failures rate

The graph (Fig. 6) shows failures rate that we can express on preventive maintenance of individual components. We see that it is not appropriate to define a planned preventive maintenance system in the case of the graphical course of the failure rate of the electrical equipment and brake because the increase of the fault intensity value is almost constant in the case of the trolley and compressor elements and it is advisable to consider planned preventive maintenance [13, 14].

This article was supported by the Scientific Grant Agency of the Ministry of Education of the Slovak Republic and the Slovak Academy of Sciences in project no. VEGA 1/0766/15 "Research sources of noise emissions from rail transport and ways of their effective reducing".

References

1. J. Galliková, R. Poprocký, *Maintenance according to the technical state with use of the enterprise asset management systems.* Zeszyty naukowe Instytutu Pojazdów: Mechanika, ekologia, bezpieczeństwo, mechatronika **3** (103), (2015)
2. J. Galliková, R. Poprocký, P. Volna, *Implementation of FMEA method in maintenance of semi-trailer combination.* Diagnostyka **17** (4), (2016)
3. J. Grenčík, et all, *Maintenance Management. Synergy of Theory and Practice.* (1. vyd., Košice: Beki Design, 629 p., ISBN 978-80-89522-03-3, 2013)
4. F. Ruman, J. Grenčík, *Proposal of new maintenance scheme of air brake system on semi-triler combination.* Diagnostyka **16** (2), 11-19 (2015)
5. M. Klinko, J. Grenčík, *Tilting body vehicles on Slovak railways - potential for use and parameters to be considered.* Communications : Scientific Letters of the University of Žilina **10** (3), 45-49 (2008)
6. H. Pačaiová, Š. Markuliak, A. Nagyová, *The importance of risk in management systems.* (Košice, BEKI Design, s.r.o., 2016)
7. L. Kašiar, P. Zvolenský, D. Barta, L. Bavlna, P. Drozdziel, *Diagnostics of electric motor of locomotive series 757.* Diagnostyka **17** (3), 95-101 (2016)
8. H. Pacaiova, J. Grencik, G. Dravecky, *Maintenance audit as a means for asset management improvement.* Euromaintenance: conference proceedings: 30 May - 1 June, 2016: Athens, Greece, Athens: EFNMS, ISBN 978-618-82601-0-8, 463-469 (2016)
9. H. Pačaiová, J. Sinay, J. Glatz, *Safety and risks of technical systems.* (ISBN 978-80-553-0180-8-60-30-10, Košice, Vienala Košice, 2009)
10. V. Stuchlý, P. Zvolenský, *History and Presence of Maintenance Education at the Faculty of Engineering of the University of Žilina.* Řízení & údržba průmyslového podniku - Slovenský špeciál **6**, (2016)
11. H. Pačajová, J. Grenčík, G. Dravecký, *Maintenance audit as a means for asset management improvement.* Euromaintenance: conference proceedings: ISBN 978-618-82601-0-8., 463-469 (2016)
12. J. Dižo, M. Blatnický, *Use of multibody system dynamics as a tool for rail vehicle behaviour diagnostics.* Diagnostyka, science journal. Polish Society of Technical Diagnostics **17** (2), 9-16 (2016)
13. M. Rakyta, et all, *Proactive approach to smart maintenance and logistics as a auxiliary and service processes in a company.* Journal of applied engineering science **14** (4), 433-442 (2016)
14. M. Gregor, et all, *Model of intelligent maintenance systems.* Annals of DAAAM: proceedings of the 26th DAAAM international symposium on intelligent manufacturing and automation, [Online], 1097-1101 (2016)

Stability and vibrations control of a stepped beam using piezoelectric actuation

Krzysztof Kuliński[1], *Jacek* Przybylski[1,*]

[1]Częstochowa University of Technology, Institute of Mechanics and Machine Design Fundamentals, Częstochowa, Dabrowskiego 69 street, Poland, EU

Abstract. The objects of this studies are the stability and transversal vibrations of the system composed of three segments, where in the centre part of the system two piezoelectric patches are perfectly bonded to the top and bottom surface of the host beam. The system is kinematically loaded as a result of prescribed displacement of one or both end supports. For the analysis purposes three different beam end supports have been taken into consideration, which prevent longitudinal displacements i.e. clamped-clamped, clamped-pinned and pinned-pinned. This type of beam loading not only affect its natural vibration frequencies but also the system's stability. By introducing the electric field to the piezo patches, depending on its vector direction, in-plane stretching or compressive residual force may be induced. Presented results show that piezo actuation can significantly modify both the critical buckling force and the vibration frequency.

Keywords: stability, vibrations, stepped beam, piezo actuation

1 Introduction

Rapid technological and research development have a crucial effect on mechanical properties of newly produced materials, allowing thereby application of beams and columns with higher slenderness at comparable level of buckling capacity. From the mechanical point of view, in case of slender systems the examination of only stability criterion is insufficient, thus additional element failure due to the vibrations should have been studied. It is well known that elements exposed to significant vibrations should be protected from the resonance phenomenon. There are variety of ways to modify fundamental systems stability and vibrations in order to prevent the resonance such as bonding another material to the host beam/column, changing the type of support or supports, adding another supporting point etc. In this paper additional material bonding method to a host beam is presented, applying piezoelectric patches, which are also called in literature as "smart materials". These materials allow not only change the beam/column structural parameters as axial and bending stiffness, which result in increased buckling capacity, but may be also used as passive modification of system vibrations frequency and furtherly the buckling capacity via the piezo actuation. Piezo actuation is the reverse piezoelectric effect, where electric field is applied to the material in order to change its dimensions. According to

* Corresponding author: jacek.pr@imipkm.pcz.pl

Reviewers: *Vladimír* Dekýš, *Tomáš* Lack

above additional pretension axial force can be generated in a beam/column, affecting its stability and vibrations.

Passive and active buckling enhancement and vibrations control were the subject of interest by many researchers. Different analytical models concerning extension, bending and shear deformations induced via strain actuators bonded and embedded in one-dimensional structures, specific characteristics and implications of piezo actuators practical use were discussed in [1]. Experimental investigation of active buckling control and theoretical discussion of the pin-ended composite carbon-epoxy column were performed by Thompson and Loughan [2]. It was showed that imperfections and geometrical nature significantly affect the system's critical buckling load. Ultimate load was significantly lower than the critical buckling load determined for the ideal column. Via the active control the out of plane deflection could be completely removed, restoring the column behaviour under compression. Despite the fact that the ultimate load was increased, its value was still lower than the critical buckling load determined for the ideal column. Different experimental studies in order to obtain optimum thickness of piezoelectric element for cantilever column were presented in [3]. Faria proposed another way of enhancing the buckling capacity of composite columns [4]. Author presented that stress stiffening effect have significant effect when both column ends have longitudinal displacements fixed. On the basis of above statement in-plane axial residual force is generated, which counteract the effects of a compressive force resulting from prescribed support displacement. Thus, fundamental beam buckling resistance may be considerably altered. Broader literature overview and mathematical models concerning buckling of structural members may be found in [5].

Two significant papers concerning the effect of piezo actuation on static and dynamic behaviour of the beams were elaborated by Faria [4] and Zehetner with Irschik [6]. Faria [4] demonstrated that determination of buckling in composite beams composed of a host beam and bonded piezoelectric patches with the use of pin-force model should have been used with extreme caution. Despite that pin-force model may be used to correctly predict the stress stiffening effect. Zehetner and Irschik [6] had also announced that assumptions taken by other authors concerning inextensible beam axis, the absence of longitudinal displacement under applied load and piezoelectric axial actuation were inadmissible and led to incorrect conclusions. Static and dynamic behaviour of multilayer beams with piezoelectric bending actuators were extensively discussed in [7]. Free vibration eigenvalue problem of thin composite plates was the subject of interest of Hernandes et al. [8], who formulated the problem with finite element method on the basis of Reissner–Mindlin theory including non-linear strain–displacement relations. It was shown that the stress stiffening effect depends on several factors i.e. material properties, placement of piezo actuators, boundary conditions and model geometry. Moreover, via the numerical analysis the effect of in-plane stress on free vibrations even in unconstrained plates was revealed. Vibrations control in composite plates were also discussed in [9-11]. Piezoelectric actuation influence on the flexural vibrations of sandwich beam was discussed in [12]. It was stated that the arrangement of piezoelectric material on the host beam have significant influence on the dynamics characteristics of studied system. Additionally, the relationship between mode shape distribution of the electric potential in the piezoelectric layer in the longitudinal direction and beam curvature, which depends on adopted boundary conditions was presented.

The main purpose of this work is to investigate the influence of piezoelectric actuation in stepped beam on buckling critical load and linear frequency. System's inhomogeneity results from a pair piezoelectric patches, perfectly bonded to the host beam on its top and bottom surface. Three different beam ends that support, eliminating longitudinal displacements are taken into considerations. In the system load is applied as a prescribed displacement of one of the beam end support. This type of loading can be also represented with temperature field applied to the beam structure, which with both ends fixed in longitudinal direction result in axial internal force occurrence. In order to enhance buckling and alter the linear vibrations in-plane axial compressive or tensile piezo force is induced, depending on the electric field vector.

2 Problem formulation

The analysed scheme composed of a host beam with both ends pinned and two piezoelectric patches mounted on beam's top and bottom surface is presented in Fig. 1.

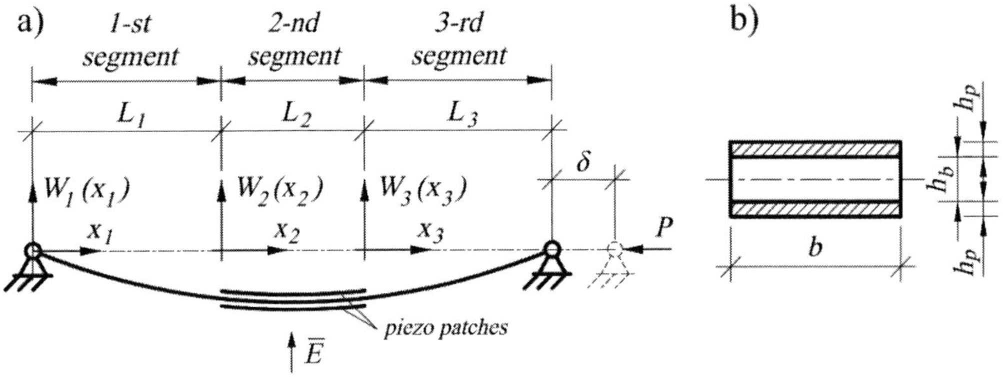

Fig. 1. Scheme of the pinned-pinned beam with two piezoelectric patches symmetrically bonded in the middle segment (a), cross section of the analysed beam (b).

In the study three different beam end supports preventing longitudinal displacements are taken into considerations: both ends clamped, both ends pinned and clamped-pinned. Presented in Fig. 1b cross-section of the beam have following dimensions i.e. exactly the same width of the host beam and piezo patches equal 20 mm, beam height equal 3 mm and each piezo layer equal 0.5 mm, respectively. It is assumed that piezo elements are perfectly bonded to the beam at the entire length, and adhesive layer thickness have insignificant influence on obtained results. Piezosegment located in the system divides beam symmetrically i.e. first and third segment have same length value. Beam is made of homogeneous, elastic isotropic aluminium material, whereas for the piezo patches homogeneous, elastic, transversely isotropic P41 material is assumed. On the basis of manufacturer [13] piezo elements Young's modulus is E_p = 83.33 GPa, density ρ = 7450 kg/m^3, whereas beam Young's modulus is assumed as 70 GPa and density ρ = 2700 kg/m^3, respectively. In the system presented in Fig. 1a, load is applied via the prescribed one of the supports displacement and before the load application the beam is assumed to be rectilinear. Electric field applied to the piezo patches is exactly the same for the top and bottom one. Thus in-plane axial compressive or tensile piezo force can be induced. On the basis of derivation of residual force equation along the stepped beam with n-pairs of piezoelectric actuators presented in [14] for three segmented system one obtains:

$$F_r = F\left[1+\eta\left(\frac{L}{L_2}-1\right)\right]^{-1} \tag{1}$$

where: F = -2$be_{31}V$ is the piezoelectric force, which depends on piezoceramic patches width b, e_{31} constant describing piezo material and the applied voltage V, η denotes the relation of the piezosegment axial stiffness to that of the beam, L is the overall beam length and L_2 is the length of piezosegment.

The non-dimensional governing equation for i-th segment beam with the residual force term for stability and vibrations problem can be expressed as:

- stability problem

$$\frac{\partial^4 w_i(\xi_i,\tau)}{\partial \xi_i^{\ 4}} \pm \varphi_i p^2 \frac{\partial^2 w_i(\xi_i,\tau)}{\partial \xi_i^{\ 2}} = 0, \quad p^2 = \left(f_r^2 + p_u^2\right), \quad \text{for } \ i = 1,2,3 \qquad (2_1)$$

- vibrations frequency

$$\frac{\partial^4 w_i(\xi_i,\tau)}{\partial \xi_i^{\ 4}} \pm \varphi_i f_r^2 \frac{\partial^2 w_i(\xi_i,\tau)}{\partial \xi_i^{\ 2}} + \mu_i \omega^2 \frac{\partial^2 w_i(\xi_i,\tau)}{\partial \tau^2} = 0, \quad \text{for } \ i = 1,2,3 \qquad (2_2)$$

where the dimensionless parameters are defined through substitutions:

$$w_i(\xi_i,\tau) = \frac{W_i(x_i,t)}{L}, \quad \xi_i = \frac{x_i}{L}, \quad \tau = \Omega t, \quad l_i = \frac{L_i}{L}, \quad \varphi_i = (1 + r_m)^{-\frac{1}{2}\left(j^{2i}+1\right)},$$
$$r_m = \frac{E_p I_p}{E_b I_b}, \quad j = \sqrt{-1}, \quad f_r^2 = \frac{F_r L^2}{E_b I_b}, \quad p_u^2 = \frac{PL^2}{E_b J_b}, \quad \mu_i = \left(\frac{\alpha_1 + (\eta - 1)\alpha_2}{\alpha_1(1 + r_m)}\right)^{\frac{1}{2}\left(j^{2i}+1\right)}, \qquad (3)$$
$$\alpha_1 = \frac{E_p}{E_b}, \quad \alpha_2 = \frac{\rho_p}{\rho_b}, \quad \eta = \frac{E_b A_b + E_p A_p}{E_b A_b}, \quad \omega^2 = \Omega^2 L^4 \frac{\rho_b A_b}{E_b I_b}$$

For the substitutions following notation is introduced: w_i denotes the transversal displacement, t – time, E_pJ_p, E_bJ_b – piezo element and beam bending stiffness, respectively, ρ_p, ρ_b – densities of the actuators and beam, respectively, A_p, A_b – the cross section area of piezopatches and beam, respectively, ω - natural vibration frequency.

Non-dimensional boundary conditions necessary to obtain the solution of stability and transversal vibrations problem for all analysed schemes are expressed as:

- clamped-clamped system

$$w_1(\xi_1,\tau)\big|_{\xi_1=0} = w_1^I(\xi_1,\tau)\big|_{\xi_1=0} = w_3(\xi_3,\tau)\big|_{\xi_3=0} = w_3^I(\xi_3,\tau)\big|_{\xi_1=0} = 0 \qquad (4_1)$$

- clamped-pinned

$$w_1(\xi_1,\tau)\big|_{\xi_1=0} = w_1^I(\xi_1,\tau)\big|_{\xi_1=0} = w_3(\xi_3,\tau)\big|_{\xi_3=0} = w_3^{II}(\xi_3,\tau)\big|_{\xi_1=0} = 0 \qquad (4_2)$$

- pinned-pinned system

$$w_1(\xi_1,\tau)\big|_{\xi_1=0} = w_1^{II}(\xi_1,\tau)\big|_{\xi_1=0} = w_3(\xi_3,\tau)\big|_{\xi_3=0} = w_3^{II}(\xi_3,\tau)\big|_{\xi_1=0} = 0 \qquad (4_3)$$

where: I and II are the Roman numerals denoting the order of the derivative with respect to the space variable ξ.

Continuity boundary conditions which describe the equality of transverse displacements, slopes, bending moments and shear forces between segments are:

$$w_i(\xi_i,\tau)\Big|_{\xi_i=li} = w_{i+1}(\xi_{i+1},\tau)\Big|_{\xi_{i+1}=0} \tag{4_4}$$

$$w_i^I(\xi_i,\tau)\Big|_{\xi_i=li} = w_{i+1}^I(\xi_{i+1},\tau)\Big|_{\xi_{i+1}=0} \tag{4_5}$$

$$(1+r_m)^{\frac{1}{2}\left(j^{2i}+1\right)} w_i^{Rn}(\xi_i,\tau)\Big|_{\xi_i=l_i} = (1+r_m)^{\frac{1}{2}\left(j^{2(i+1)}+1\right)} w_{i+1}^{Rn}(\xi_{i+1},\tau)\Big|_{\xi_{i+1}=0}, \tag{4_6}$$

$$i=1,2;\ Rn = II, III$$

3 Approximate solutions

3.1 Stability problem

By separating space and time variable with the following formula:

$$w_i(\xi_i,\tau) = w_i(\xi_i)\cos\tau \tag{5}$$

which is introduced to the equations (2_1) one obtains:

$$\frac{\partial^4 w_i(\xi_i)}{\partial \xi_i^{\ 4}} \pm \varphi_i p^2 \frac{\partial^2 w_i(\xi_i)}{\partial \xi_i^{\ 2}} = 0, \quad p^2 = \left(\pm f_r^2 + p_u^2\right) \quad \text{for } i=1,2,3 \tag{6}$$

The general solution of equation (6), can be demonstrated as follows:

$$w_i(\xi_i) = A_i\cos(p\xi_i) + B_i\sin(p\xi_i) + C_i(\xi_i) + D_i, \quad p^2 = f_r^2 + p_u^2, \quad i=1,2,3 \tag{7}$$

By substituting equation (7) into the adequate set of boundary conditions (4_1 – 4_6) one obtains a system of twelve homogeneous linear equations with respect to unknown integration constants A_i, B_i, C_i, D_i where i=1, 2, 3. Equating to zero the determinant of matrix coefficients of that system one derives a transcendental equation from which the set of successive critical loads can be numerically determined. Each of the buckling modes can be obtained after introduction of a particular eigenvalue into the normalized eigenfunction expressed by equation (7).

3.2 Vibrations frequency

Introducing into the governing equation of motion (2_2) the separation of time and space variable formula (5), one obtains:

$$\frac{\partial^4 w_i(\xi_i)}{\partial \xi_i^{\ 4}} \pm \varphi_i p^2 \frac{\partial^2 w_i(\xi_i)}{\partial \xi_i^{\ 2}} - \mu_i\omega^2 w_i(\xi_i) = 0, \qquad p^2 = \left(\pm f_r^2 + p_u^2\right) \quad \text{for } i=1,2,3 \tag{8}$$

The general solution takes the form:

$$w_i(\xi_i) = A_i\cosh(\gamma_1\xi_i) + B_i\sinh(\gamma_1\xi_i) + C_i\cosh(\gamma_2\xi_i) + D_i\sinh(\gamma_2\xi_i), \quad i=1,2,3 \tag{9}$$

where coefficients γ_1 and γ_2 are expressed as:

- in case of compressive force $p^2 > 0$:

$$\gamma_1 = \sqrt{\frac{-\varphi_i p^2 - \sqrt{\varphi_i p^4 + 4\mu_i \omega^2}}{2}}; \quad \gamma_2 = \sqrt{\frac{-\varphi_i p^2 + \sqrt{\varphi_i p^4 + 4\mu_i \omega^2}}{2}} \tag{10}$$

- in case of stretching force $p^2 < 0$:

$$\gamma_1 = \sqrt{\frac{\varphi_i p^2 - \sqrt{\varphi_i p^4 + 4\mu_i \omega^2}}{2}}; \quad \gamma_2 = \sqrt{\frac{\varphi_i p^2 + \sqrt{\varphi_i p^4 + 4\mu_i \omega^2}}{2}} \tag{11}$$

Once again substituting equation (9) into the adequate set of boundary conditions (4_1 – 4_6) twelve homogeneous linear equations with respect to unknown integration constants A_i, B_i, C_i, D_i where i=1, 2, 3 are obtained. Equating the matrix of the coefficients standing within the unknown integration constants to zero, the solution describing successive natural frequencies is determined.

4 Numerical results and discussion

In this chapter particular attention has been dedicated to the buckling enhancement and vibration control depending on the piezosegment length and the electric field vector direction. The piezosegment length in the range of $l_2 \in (0; 1)$ has been investigated. The influence of piezosegment placement and other physical and structural parameters on buckling and natural vibration frequency is presented in [15,16]. Knowing that both ends of the beam are constrained against longitudinal displacement, axial tensile or compressive force can be introduced to the system via piezo patches. Thus piezoelectric force parameter f have been assumed to be equal the pinned-pinned beam critical buckling load parameter:

$$f = \sqrt{\frac{P_{cr} L^2}{E_b J_b}} = \pi \tag{12}$$

It is worth noting that the residual force (1) can be equal to piezoelectric force only in case, if the piezoceramic actuators covers the entire beam length and axial stiffness of the host beam and actuators is exactly the same, in other cases $f_r < f$.

However, all numerical results are going to be presented in non-dimensional form, an analysis of practical model should also be demonstrated. Beam with length of L = 400 [mm] is assumed in order to verify necessary voltage to generate the piezoelectric force. On the basis of the P-41 material and properties provided by the manufacturer [13], a piezoelectric constant of value d_{31} = $6.65 \cdot 10^{-11}$ [C/N] has been adopted. The critical buckling load for the pinned-pinned beam on the basis of equation (9) is equal to 194.31 [N], hence an individual piezoceramic actuator should generate a force equal to 97.16 [N]. On the basis of above, necessary value of the applied electric field should reach the value of 582.95 [V], which is far beyond the piezoceramics depolarisation point equal 1000 [V]. Hence, assumed piezoelectric force can be safely treated as a tool for enhancing buckling and alter the natural vibration frequency.

4.1 Buckling enhancement

In Figs. 2a-2d enhancement ranges of critical buckling force for the clamped-clamped, clamped-pinned and pinned-pinned beam have been presented. In Figs. 2a-2c with the solid line type critical buckling load in the function of piezosegment length without piezo actuation $f = 0$ is presented. With the dashed line tensile piezoelectric force influence on buckling is presented, whereas with dash-dot line compressive piezo force, respectively. In the Fig. 2d percentage enhancement of buckling load in the function of piezosegment length is shown.

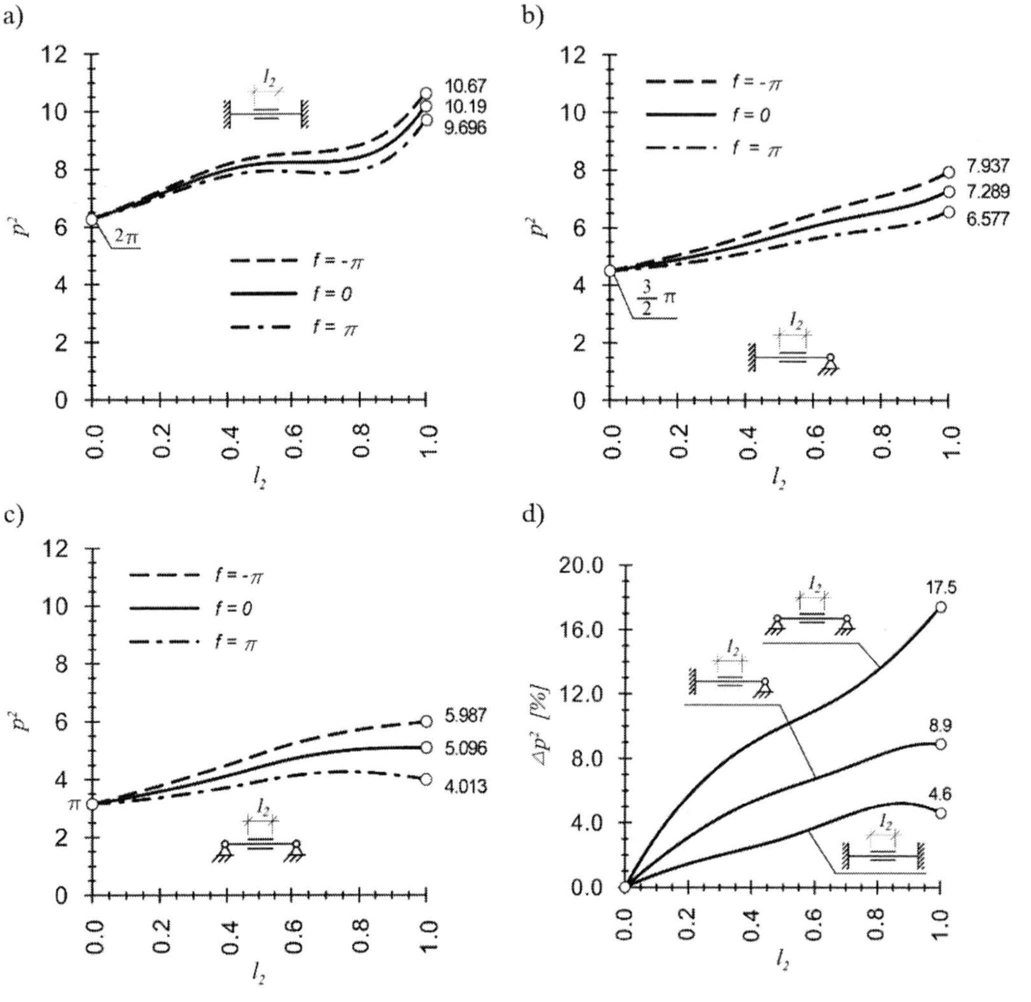

Fig. 2. Modification range of the critical buckling load for different piezosegment length, (a) clamped-clamped beam, (b) clamped-pinned beam, (c) pinned-pinned beam, (d) percentage range of critical buckling load enhancement via axial tensile residual force for different beam ends supports

Comparing results presented in Figs. 2a-2c one can state that regardless of beam ends supports, for systems without piezo actuation, critical buckling load increases together with increased length of the piezosegment. In all discussed systems, the tensile piezoelectric force result in buckling load increase, whereas compressive piezo force acts in opposite way. It is worth noting that for the beam with both ends clamped, the highest percentage influence of piezoelectric actuation (see Fig. 2d) is obtained in case of piezosegment length $l_2 = 0.88$ and is equal 5.23%. Further increase of piezosegment length in discussed beam results in decreased effectiveness. For the beam where supporting ends are assumed pinned-pinned (Fig. 2b) and clamped-pinned (Fig. 2c) together with increased piezosegment length, buckling load is increased. It should be noted that the lower rotation stiffness provided by the adopted supports in the system, the stronger effect of the induced axial residual force and wider range of possible buckling force modification is obtained (see Fig. 2d)

4.2 Vibrations control

In Figs. 3a-3c non-dimensional natural vibrations frequency is shown for the system with following ends supports: clamped-clamped, clamped-pinned, pinned-pinned, respectively. Once again with solid line system without piezo actuation is presented, dashed line denotes the compressive piezo force, whereas dash-dot line corresponds to the system with tensile axial piezo force induced. In Figs. 4a-4b percentage modification of natural system frequency for compressive and tensile piezoelectric force is shown for all analysed systems.

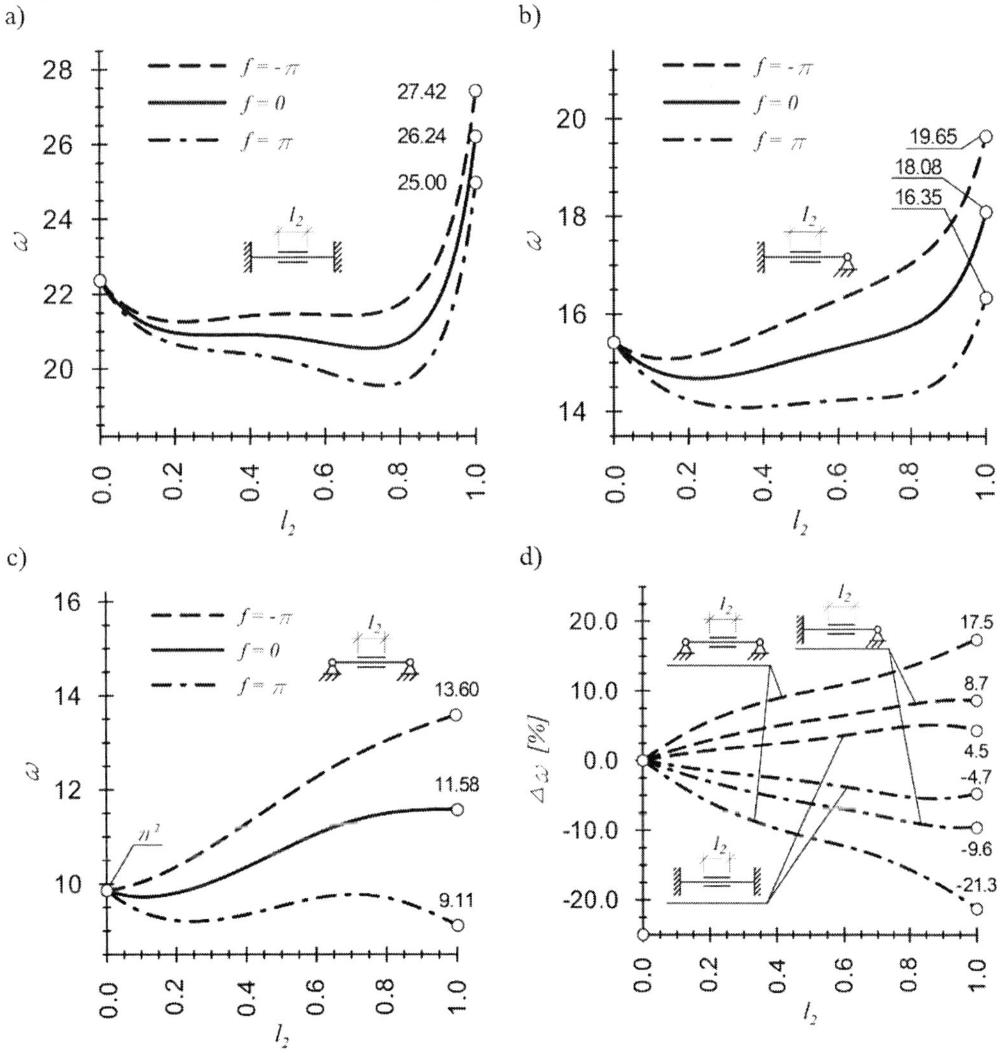

Fig. 3. The first natural frequency in the function of piezosegment length and different values of piezoelectric force induced, (a) clamped-clamped beam, (b) clamped-pinned beam, (c) pinned-pinned beam, (d) percentage range of natural system's frequency modification via piezo actuation for different beam ends supports

Considering numerical results presented in Fig. 3a-3c one can state that axial tensile piezoelectric force increases the first natural vibration frequency of each system, whereas compressive piezo force diminishes that frequency, respectively. It should be noted that natural frequency curves changes irregularly regardless of system supporting. The reason of that behaviour are: the relation of piezosegment axial stiffness to the host beam stiffness, piezo patches bending stiffness to the beam stiffness and mass distribution per unit length. For the clamped-clamped (Fig. 3a) and clamped-pinned (Fig. 3b) system at small values of piezosegment length, decrease of first natural frequency can be observed, furtherly it starts to increase. Independently of the piezoelectric force value for the clamped-clamped beam (Fig 3a) rapid increase of the first natural frequency can be observed. Similar behaviour may be observed in the clamped-pinned beam (Fig. 3b) in case of piezo patches length greater than 0.85 entire beam length. On the basis of presented curves in Fig. 3d, again one can state that the widest range of possible first natural frequency modification with the axial tensile and compressive residual force is obtained for the pinned-pinned beam, whereas the narrowest range for clamped-clamped system is obtained, respectively.

Conclusion

In this study the problem of buckling enhancement and alteration of first natural frequency via the piezo actuation for the system with various type of ends supports preventing longitudinal displacement have been discussed. The priority parameters concerned the support conditions, geometrical parameters and the piezoelectric actuation.

Regardless of the beam ends supporting induce of the axial tensile residual force counteracts the external loading, significantly enhancing the critical buckling force. Moreover, same piezoelectric force results in increased first natural vibration frequency in any analysed system in regard to that system without piezo actuation. Compressive force acts in the opposite manner. Additionally, one can be observed that the lower system's supports rotational resistance the wider range of possible system's buckling and vibrations control is obtained. It is worth noting that the range of possible critical load or natural vibration frequency control depends strongly on piezo patches geometry, relation of piezo axial and bending stiffness to the host beam and maximum allowable voltage avoiding the risk of piezo elements depolarisation. Presented numerical results concern only one of many possible configurations of the system, thus additional research and experimental tests should be provided continuing this study.

References

1. E. F. Crawley, E. H. Anderson, *Detailed models of piezoelectric actuation of beams*. Journal of Intelligent Material Systems and Structures **1**, 4-25 (2016)
2. S. Thompson, J. Loughan, *The active buckling control of some composite column strips using piezoceramic actuators*. Composite Structures **32**, 59-67 (1995)
3. M. J. Cunningham, D. F. L. Jenkins, W. W. Clegg, M. M. Bakush, *Active vibration control and actuation of a small cantilever for applications in scanning probe instruments*. Sensors and Actuators A: Physical **50**, 147-150 (1995)
4. A. R. de Faria, *On buckling enhancement of laminated beams with piezoelectric actuators via stress stiffening*. Composite Structures **65**, 187-192 (2004)
5. C. M. Wang, C. Y. Wang, J. N. Reddy, *Exact solutions for buckling of structural members*. (Computational mechanics and applied analysis, CRC Press, 2005)

6. C. Zehetner, H. Irschik, *On the static and dynamic stability of beams with an axial piezoelectric actuation*. Smart Structures and Systems **4** (1), 67-84 (2008)
7. R. G. Ballas, *Piezoelectric multilayer beam bending actuators: Static and dynamic behaviour and aspects of sensor integration*. (Springer, 2007)
8. J. A. Hernandes, S. F. M. de Almeida, A. Nabarrete, *Stiffening effect on the free vibrations behaviour of composite plates with PZC actuators*. Composite Structures **49** (1), 55-63 (2000)
9. K. Narwal, D. Chhabra, *Analysis of simple supported plate for active vibration control with piezoelectric sensors and actuators*. Journal of Mechanical and Civil Engineering **1** (1), 26-39 (2012)
10. R. C. Batra, X. Q. Liang, *The vibration of a rectangular laminated elastic plate with embedded piezoelectric sensors and actuators*. Computers & Structures **63** (2), 203-216 (1997)
11. R. C. Batra, X. Q. Liang, J. S. Yang, *Shape control of vibrating simply supported rectangular plates*, AIAAJ Journal **34** (1), 116-122 (1996)
12. Q. Wang, S. T. Quek, *Flexural vibration analysis of sandwich beam coupled with piezoelectric actuator*. Smart Materials and Structures **9** (1), 103-109 (2000)
13. *Piezoelectric material properties given by manufacturer*. www.annon-piezo.com/pztmaterials.asp (date accessed: 20 July 2017)
14. J. Przybylski, *Static and dynamic behaviour of slender mechanical systems with integrated piezoceramic elements*. (Wydawnictwo Politechniki Częstochowskiej, Częstochowa, 2012) [in Polish]
15. K. Kuliński, J. Przybylski, *Lateral displacements of sandwich beams laying on elastic foundation*. Modelowanie Inżynierskie **42**, 33-40 (2015) [in Polish]
16. K. Kuliński, J. Przybylski, *Buckling of stepped beams resting on an elastic foundation*. Journal of Applied Mathematics and Computational Mechanics **14** (4), 115-126 (2015)

Novel algorithm for tilt measurements using MEMS accelerometers

Sergiusz Łuczak[1,*]

[1]Warsaw University of Technology, Faculty of Mechatronics, ul. A. Boboli 8, 02-525 Warszawa, Poland

Abstract. A novel algorithm for determining pitch and roll under static or quasi-static conditions is presented. It is based on application of a triaxial micromachined accelerometer for the necessary measurements. In order to obtain a possibly high accuracy, a simple data fusion is applied. Depending on the nominal value of pitch/roll, appropriate mathematical formula is selected. Additionally, the algorithm makes it possible to check if the operating conditions are dynamic, as indications of accelerometers are then not reliable. Implementation of the algorithm makes it possible to determine pitch and roll with accuracy of ca. 0.2° over 360° or better.

Keywords: tilt, MEMS, accelerometer, algorithm, data fusion

1 Introduction

Micromachined (or Micro Electromechanical Systems - MEMS) accelerometers have been applied over the last years in numerous applications and various devices [1]. Some of the devices are typical and known for a long time (e.g. automotive airbags [2]), yet other are very novel and appeared only recently (e.g. new generation of motorcycle traction control systems [3]).

Except for their crucial advantages (low cost, high reliability, compatibility with electronics, high resistance to mechanical shocks, low power consumption), their miniature size is of a big importance, especially in the case of mobile microrobots, whose control systems in many cases require information on the robot attitude. Such information is related to tilt measurement - one of the most typical application of low-g accelerometers [4].

However, it must be stated that the most limiting shortcoming of micromachined accelerometers in tilt measurements are operation only under static or quasi-static conditions as well as relatively low accuracy, as compared to their conventional counterparts. With regard to the first issue, an interesting solution is a simultaneous application of a MEMS gyroscope (or even an Inertial Measurement Unit - IMU, integrating an accelerometer, a gyroscope and optionally a magnetometer) [5]; then any operating conditions are permissible. With regard to the second issue, there is a lot of solutions of increasing the aforementioned accuracy. One of them is to apply a sensor fusion (using output signals generated by few different sensors) or a data fusion (using

* Corresponding author: s.luczak@mchtr.pw.edu.pl

Reviewers: *Milan* Žmindák, *Maciej* Bodnicki

output signals generated by one sensor only). The presented novel algorithm for tilt measurements employs the later concept: 3 output signals of a standard triaxial MEMS accelerometer are used in a smart way in order to increase accuracy of tilt measurements as well as to simplify some mathematical transformations executed by means of a microprocessor system.

The algorithm addresses also such issues as: calibration of MEMS accelerometers and verification of the operation conditions. Besides, problems pertaining to physical alignment of MEMS accelerometers or anisotropy of triaxial MEMS accelerometers are briefly discussed.

2 Mathematical formulas

Geometrical relations between components (g_x, g_y, g_z) of the gravitational acceleration (g) and tilt angles (α - pitch, β - auxiliary angle, γ - roll, φ - axial tilt [6]) are illustrated in Fig. 1. (Static or quasi-static operating conditions exclude existence of any other constant accelerations.)

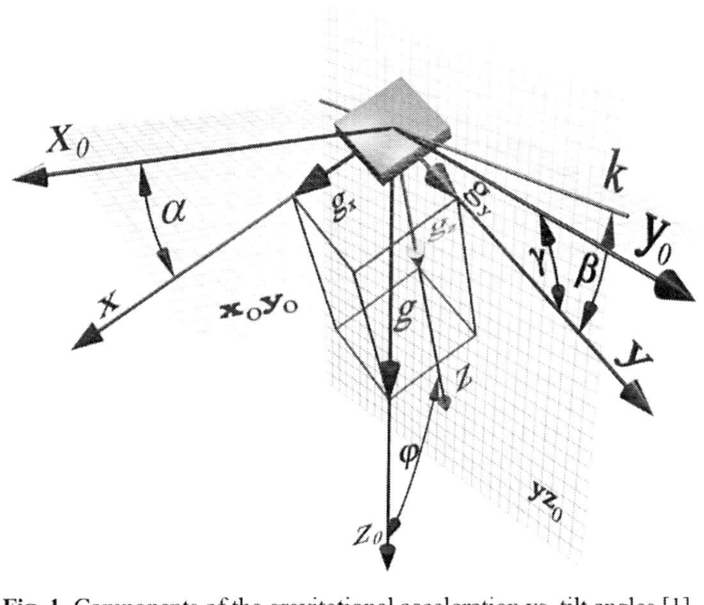

Fig. 1. Components of the gravitational acceleration vs. tilt angles [1].

There are few equivalent ways of calculating the tilt angles (basically pitch and roll) on the basis of the components of the gravitational acceleration, as discussed e.g. in [1,7]. Three basic mathematical relations between the aforementioned quantities are: arc sine, arc cosine and arc tangent. Even though application of any of these equivalent formulas yields the same nominal value, each formula is characterised by different sensitivity [7].

In order to simplify the succeeding formulas, let us introduce two additional quantities, being geometric sums of the respective pairs of Cartesian components [8]:

$$g_{xz} = \sqrt{g_x^2 + g_z^2} \quad (1)$$

$$g_{yz} = \sqrt{g_y^2 + g_z^2} \quad (2)$$

The proposed algorithm is based on application of arc cosine and arc tangent formulas; so, for each component tilt angle, two formulas (denoted by respective subscripts) will be employed [1,7]:

$$\alpha_1 = \arctan\frac{g_x}{g_{yz}} \tag{3}$$

$$\beta_1 = \arctan\frac{g_y}{g_{xz}} \tag{4}$$

$$\alpha_2 = \arccos\frac{g_{yz}}{g} \tag{5}$$

$$\beta_2 = \arccos\frac{g_{xz}}{g} \tag{6}$$

In some cases it may turn out that the auxiliary angle β is more convenient for further computations. However, if the roll angle γ must be determined, it can be calculated directly [9]:

$$\gamma_1 = \arctan\frac{g_y}{g_z} \tag{7}$$

or on the basis of the angle β [10]:

$$\gamma_2 = \arcsin\frac{g_y}{g\cos\alpha} = \arcsin\frac{\sin\beta}{\cos\alpha} \tag{8}$$

In most of measurement cases, arc tangent formulas ensure the highest sensitivity of tilt measurements [7] (thus, the highest accuracy). However, tangent is a discontinuous function at inputs of approximately ±90°. Moreover, within the vicinity of these inputs, values of the function are very high (e.g. ⟨-89.4, 89.4⟩ for the angular domains ⟨-91°, -89°⟩ or ⟨89°, 91°⟩), as presented in Fig. 4. These two features are very disadvantageous, as far as microprocessor systems are concerned.

However, arc cosine function not only features continuity, but also sensitivity similar to arc tangent function [7] within the vicinity of tilt angles of ±90°. At the same time its inputs are very small (of few thousandths of V for a typical scale factor of MEMS accelerometers).

Thus, the basic idea of the proposed algorithm is to use both functions in turns. In most of the cases, arc tangent function is to be employed, except for the vicinity of tilt angles of ±90°, where arc cosine function is to be used. Then, a basic shortcoming of the latter function with regard to tilt measurements, i.e. sensitivity of zero value at tilt angles of ca. 0° [8] does not matter whatsoever.

3 Operating conditions

It is well know, that indications of a triaxial accelerometer operating under static conditions must theoretically meet the following equation [11]:

$$\sqrt{g_x^2 + g_y^2 + g_z^2} = g\,. \tag{9}$$

In the case of qausi-static conditions, output signals of the accelerometer should be first processed by means of a low-pass filter (with the cutoff frequency set e.g. to 1 Hz or lower). Then, Eq. (9) is valid; yet, in such case a decrease of accuracy of the output signals due to amplitude attenuation [12] must be taken into account.

Because of the existing random errors (especially accelerometer inherent noise), Eq. (9) should be replaced with the following set of inequalities:

$$g - \Delta g < \sqrt{g_x^2 + g_y^2 + g_z^2} < g + \Delta g\ , \tag{10}$$

where Δg is the permissible margin of measurement uncertainty.

At this point it must be emphasized that even though Eq. (10) is satisfied, the accelerometer may still operate under dynamic conditions. It results from the fact that there exists an infinite group of vectors of external acceleration, disturbing tilt measurement, which added to the vector of gravitational acceleration (senses of these two kinds of vectors are opposite with respect to one another) yield a vector with an absolute value meeting conditions of Eq. (9).

The group of vectors (a_1, a_2, a_3, ...) is illustrated in Fig. 2, where the blue circle represents all the possible positions of their tips, and the red circle represents all the possible positions of the tip of vectors (b_1, b_2, b_3, ...), which are resultants of the gravitational acceleration and vectors (a_1, a_2, a_3, ...).

Let us consider the simplest case, when external constant acceleration a_2 acts in vertical direction, has the absolute value of $2g$ and the sense opposite with respect to the gravitational acceleration g. The accelerometer in this case will sense a resultant acceleration vector b_2, which has the absolute value of 1g, vertical direction, yet opposite sense with respect to the gravitational acceleration g. Then, the error of measuring tilt will have the maximal possible value of 180°. The other two acceleration vectors (a_1, a_3) yield resultant vectors (b_1, b_3) having also the absolute value of 1g, yet completely inconsistent with the attitude of the accelerometer.

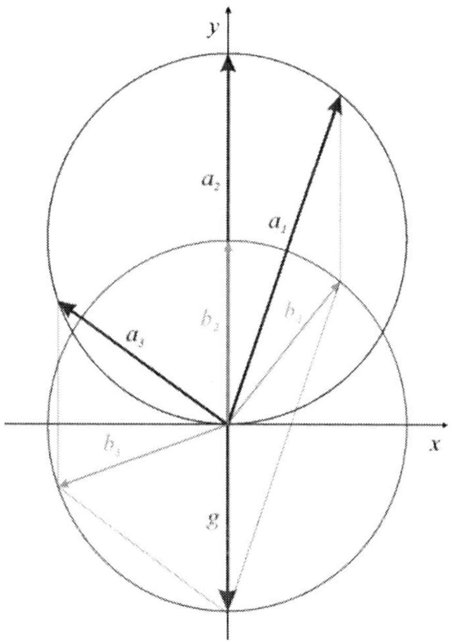

Fig. 2. Undetectable dynamic conditions

If no additional source of information about constant accelerations acting upon the accelerometer is available (or an information about the attitude of the accelerometer), generally, it is not possible to state if tilt measurements are correct. An exception may be a case, when it is known that the only constant external acceleration may act in a strictly defined direction. Then, on the basis of the measured Cartesian components of the resultant acceleration, it possible to determine both the tilt angles as well as the external acceleration, as proposed e.g. in [13].

If it is difficult to state whether the operation conditions are static, a gyroscope can be used. MEMS gyroscopes are sometimes even integrated with MEMS accelerometers (constituting an IMU), so it is a reasonable option to use both kinds of sensors. Even though gyroscopes do not directly respond to linear acceleration, one of their shortcomings is a cross coupling error caused by linear acceleration [14]. So, if it is expected that possible constant disturbing accelerations may be of a certain value, that can be distinguished in the gyroscope noise and drift, it is an interesting and inexpensive option of verifying the operation conditions of the tilt sensor. Moreover, some constant accelerations (e.g. centripetal) are interconnected with a constant or variable rotational speed, which is a standard input of a gyroscope. So, such accelerations can be easily detected.

On the other hand, if it is certain that no constant accelerations occur, the presented considerations pertaining to the operating conditions may be ignored.

4 Calibration of MEMS accelerometers

With regard to the output signal, there are two kinds of MEMS accelerometers: digital and analogue. A course of an output signal of an analogue accelerometer while tilted over a full angle (i.e. when the sensed gravitational acceleration changes within the range ⟨-1g, 1g⟩) is presented in Fig. 3, where a is the offset and b is the scale factor. Course of a digital output signal is similar, yet the offset is approximately of $a = 0$.

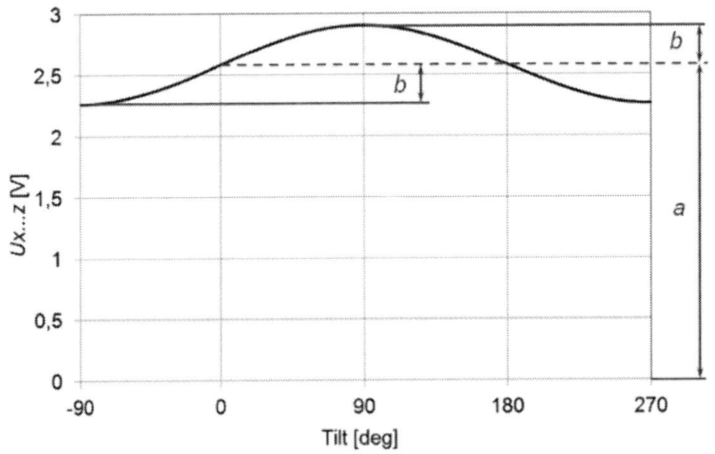

Fig. 3. Offset and scale factor of a typical MEMS accelerometer while tilted

If a decrease of measurement accuracy is acceptable, one may assume average catalogue values for the offset and the scale factor assigned to each sensitive axis. However, if a possibly high accuracy is striven for, values of offsets and the scale factors must be determined for each sensitive axis by the way of an experimental calibration. The most accurate method of calibration is rotation of each sensitive axis of the accelerometer around a horizontal axis (the calibrated sensitive axis should be perpendicular to the rotation axis with accuracy no lower than ca. 1°). In the case of triaxial accelerometers, two sensitive axes can be calibrated at the same time (i.e. x- and z-axis while rotated around y_0-axis, or y- and z-axis while rotated around x_0-axis – see Fig. 1), so two calibration procedures must be performed. Result of such calibration is illustrated in Fig. 3, and can be expressed by the following formula [15]:

$$U_{x...y} = a_{x...y} + b_{x...y}\sin\psi + c_{x...y} \quad (11)$$

$$U_z = a_z + b_z\cos\psi + c_z \quad (12)$$

where: $U_{x..z}$ is the output voltage assigned to particular sensitive axis, ψ is in this case either α (pitch) or $\beta = \gamma$ (roll), and $c_{x..z}$ is the angular phase shift, which can be regarded as an misalignment with respect to the test rig (difference between particular phase shifts represent nonorthogonality between respective sensitive axes).

It should be mentioned that calibration can be also realized by setting the accelerometer in 6 characteristic positions, where each sensitive axis once senses acceleration of -1*g* and then of 1*g* [16], or even implementing an idea of autocalibration, when the accelerometer is to be oriented in few unknown arbitrary positions, and its parameters are determined according to various algorithms, proposed e.g. in [17]-[19].

Values of the offsets $a_{x..z}$, scale factors $b_{x..z}$, and phase shifts $c_{x..z}$ should be estimated employing e.g. the least squares method [16]. Appropriate software for statistical analysis (e.g. Statgraphics or Origin) makes it not only possible to determine values of these parameters, but also to evaluate uncertainties of determining them, and thus to estimate appropriate prediction or confidence intervals of the measured output voltages or the component tilt angles. Such intervals may be used for determining the permissible margin of measurement uncertainty Δg specified in Eq. (10).

After the calibration process, the output signals of the accelerometer may be regarded relative component accelerations of gravity, calculated as follows,

$$\frac{g_{x...z}}{g} = \frac{U_{x...z} - a_{x...z}}{b_{x...z}} = m_{x...z} \quad (13)$$

Uncertainty of the relative component accelerations can be determined on the basis of a maximal width of the aforementioned prediction or confidence interval $\Delta U_{x..z}$ of the measured output voltages $U_{x..z}$, according to the following formula,

$$\Delta m_{x..z} = \frac{\Delta U_{x..z} - a_{x..z}}{b_{x..z}} \approx \Delta m \quad (14)$$

So, Eq. (10) can be transformed into:

$$1 - \Delta m < \sqrt{m_x^2 + m_y^2 + m_z^2} < 1 + \Delta m \quad (15)$$

Finally, tilt angles can be calculated according to the following formulas:

$$m_{xz} = \sqrt{m_x^2 + m_z^2} \quad (16)$$

$$m_{yz} = \sqrt{m_y^2 + m_z^2} \quad (17)$$

$$\alpha_1 = \arctan\frac{m_x}{m_{yz}} \quad (18)$$

$$\beta_1 = \arctan\frac{m_y}{m_{xz}} \quad (19)$$

$$\alpha_2 = \arccos m_{yz} \quad (20)$$

$$\beta_2 = \arccos m_{xz} \quad (21)$$

However, because there are two possible solutions for Eq. (20)-(21), negative or positive value of the tilt angle can be determined on the basis of the sign of the third relative component acceleration, i.e. m_x in the case of pitch, and m_y in the case of roll. The relative components m_x and m_y represent a sinus function of respective tilt angle [1,7], so they have a course like in Fig. 3, yet with $a = 0$ and $b = 1$. Thus, for a tilt angle of ca. -90° their value is negative, whereas positive for a tilt angle of ca. 90°. Therefore, for the indicated range of application of Eq. (20)-(21), the following is true:

$$m_x > 0 \Rightarrow \alpha_3 = \alpha_2 \quad (22)$$

$$m_x < 0 \Rightarrow \alpha_3 = -\alpha_2 \quad (23)$$

$$m_y > 0 \Rightarrow \beta_3 = \beta_2 \quad (24)$$

$$m_y < 0 \Rightarrow \beta_3 = -\beta_2 \quad (25)$$

So, in this way we have two options for calculating each of the tilt angles, what is the essential idea of the proposed algorithm: we use either arc tangent function or arc cosine function.

5 The Algorithm

The idea of the algorithm is illustrated in Fig. 4. It presents courses of two functions: cosine and tangent (Eq. (3)-(6) are inverse functions of these relationships). As can be seen, the range of the cosine function is ⟨-1, 1⟩, whereas ⟨-∞, +∞⟩ for the tangent function. For obvious reasons, not only infinite numbers cannot be handled by digital systems. It is also very disadvantageous, when numbers assume once small values (cosine function) and once large values (tangent function) due to their digitalization - especially, when we strive for high accuracy and a possibly low bit number of a dedicated microprocessor system (e.g. 8-bit).

So, the graphs prove that it is very inconvenient to use arc tangent function for calculation of tilt for angles of approximately -90° or 90°. If limited ranges of application of the tangent function (and its derivatives (3) and (4)) were accepted, compare to application of cosine function only (and its derivatives (5) and (6)), the range of the tangent function could be considerably reduced, e.g. to approximately ⟨-4, 4⟩ for angles ⟨-180°, -105°⟩ ∪ ⟨-75°, 75°⟩ ∪ ⟨105°, 180°⟩ or approximately ⟨-8, 8⟩ for angles ⟨-180°, -98°⟩ ∪ ⟨-82°, 82°⟩ ∪ ⟨98°, 180°⟩. The above reasoning is illustrated in Fig. 4, where a considerable difference between outputs of tangent and cosine function is evident for angles in the proximity of angles of ±90°.

Let us call the number determining the range of application of the tangent function a critical value *CV* that can be handled by the dedicated microprocessor system. The corresponding tilt angle a critical angle *CA*, calculated as follows:

$$\mathrm{CA} = \mathrm{arctanCV}\,. \quad (26)$$

Generally, the range of outputs of the tangent function would be then reduced to ⟨-*CV*, *CV*⟩, and the corresponding angular domain to ⟨-180°, (-180°+*CA*)⟩ ∪ ⟨-*CA*, *CA*⟩ ∪ ⟨(180°-*CA*), 180°⟩. So, the proposed idea can be expressed for pitch angle as:

$$\alpha_2 \leq \mathrm{CA} \Rightarrow \alpha_4 = \alpha_1 \quad (27)$$

$$\alpha_2 > CA \Rightarrow \alpha_4 = \alpha_3 \quad (28)$$

and for roll angle as:

$$\beta_2 \leq CA \Rightarrow \beta_4 = \beta_1 \tag{29}$$

$$\beta_2 > CA \Rightarrow \beta_4 = \beta_3 \tag{30}$$

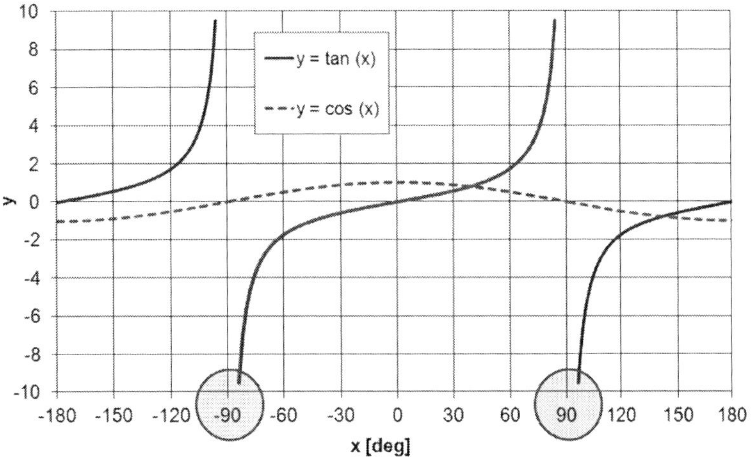

Fig. 4. Angular regions of data fusion

A problem similar to application of Eq. (20) or (21) is related to measurements over the full angular range of ⟨-180°, 180°⟩ while using Eq. (18) or (19). It must be distinguished whether the calculated tilt angle is contained within a sub-range of ⟨-90°, 90°⟩ or the remaining sub-range of ⟨-180°, 90°) ∪ (90°, 180°⟩. This time, the sign of the m_z relative component acceleration is decisive. If it has a positive value (see the red part of the cos(x) course in Fig. 4), the tilt angle is contained within the first sub-range, whereas its negative value indicates the second sub-range (see the blue part of the cos(x) course in Fig. 4).

So, formulas valid for the pitch angle are the following [20]:

$$m_z > 0 \Rightarrow \alpha_5 \approx \alpha_4 \tag{31}$$

If Eq. (22) is true, then:

$$m_z < 0 \Rightarrow \alpha_5 \approx 180° - \alpha_4 \tag{32}$$

otherwise Eq. (23) is true, so:

$$m_z < 0 \Rightarrow \alpha_5 \approx -180° - \alpha_4 \tag{33}$$

Analogical formulas are valid for the roll angle:

$$m_z > 0 \Rightarrow \beta_5 \approx \beta_4 \tag{34}$$

If Eq. (24) is true, then:

$$m_z < 0 \Rightarrow \beta_5 \approx 180° - \beta_4 \tag{35}$$

otherwise Eq. (25) is true, so:

$$m_z < 0 \Rightarrow \beta_5 \approx -180° - \beta_4 \tag{36}$$

To summarise, the presented algorithm consists of the following steps:

1. Assume the maximal value of parameter *CV*
2. Compute value of the critical angle *CA* according to (26)
3. Calibrate the accelerometer in order to estimate: a_x, a_y, a_z, b_x, b_y, b_z, ΔU_x, ΔU_y, ΔU_z
4. Compute the permissible margin of the relative acceleration Δm according to (14)
5. Set an unknown orientation of the accelerometer to be determined
6. Record the output signals of the accelerometer: U_x, U_y, U_z
7. Compute the relative component accelerations m_x, m_y, m_z according to (13)
8. Verify condition (15); if not satisfied get back to step 6
9. Compute the resultant accelerations m_{xz}, m_{yz} according to (16)-(17)
10. Determine value of pitch and roll according to (20) and (21)
11. If condition (28) is satisfied, proceed to step 16
12. Determine value of pitch according to (18)
13. Condition (27) is satisfied
14. Determine the final value of pitch according to (31)-(33)
15. Proceed to step 17
16. Determine the final value of pitch according to (22) or (23)
17. Final value of pitch has been determined
18. If condition (30) is satisfied, proceed to step 23
19. Determine value of roll according to (19)
20. Condition (29) is satisfied
21. Determine the final value of roll according to (34)-(36)
22. Proceed to step 24
23. Determine the final value of roll according to (24) or (25)
24. Final value of roll has been determined

6 Experimental verification

In order to evaluate accuracy of tilt measurements according to the proposed algorithm, few experiments were performed, using a computer-controlled test rig minutely described in [21]. Low-g triaxial MEMS accelerometers by Analog Devices Inc. were tested.

The error was calculated as an absolute value of a difference between the real tilt angle (applied by means of the test rig) and the estimated tilt angle (calculated on the basis of the average value related to output signals of the tested accelerometer, using the calibration parameters defined in Eq. (11) and (12), which had been determined beforehand). It was accepted that the arc cosine formula was used within the range of ⟨-105°, -75°⟩ ∪ ⟨75°, 105°⟩, and the arc tangent formula within the remaining range.

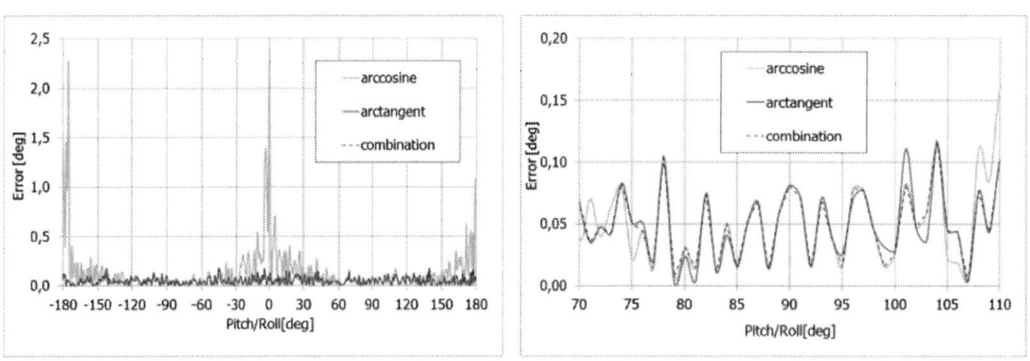

Fig. 5. Error of deterring component tilt angle

The obtained results prove that the presented algorithm makes it possible to obtain errors typical for application of arc tangent formula (in other words, almost at constant measurement sensitivity - in contrary to application of arc cosine formula - see Fig. 5a), yet without using this function within the angular regions where its outputs are very large. Fig. 5b proves that application of arc cosine formula within the range of ⟨75°, 105°⟩, does not result in increasing the measurement error. The presented course is of a random character.

The experimental studies have proven correctness of the proposed algorithm. A relatively high accuracy has been obtained: better than 0.2° over the full angular range, so similar to the value reported in [20], when an algorithm was used, which employed only arctangent formula.

Conclusions

The proposed algorithm makes it possible to determine pitch and roll within the full angular range with accuracy of ca. 0.2 degrees arc. It makes it possible to employ a simple microprocessor system for the necessary computations.

Using accurate low-range MEMS accelerometers, called MEMS inclinometers [22], [23], the accuracy of tilt measurements may be considerably increased, at least to 0.1 degrees arc [23]. Yet, since the inclinometers are single- or dual-axis accelerometers, in order to implement the proposed algorithm, at least two inclinometers must be applied, without neglecting the problem of aligning their sensitive axes.

Even higher accuracies are reported while applying differential MEMS accelerometers [24]. More detailed information regarding application of MEMS accelerometers for tilt measurements (e.g. selecting their types, configurations, spatial arrangements and imperfections) can be found in [1]. It must not be neglected that alignment of the sensitive axes of MEMS accelerometers is a problem important especially in tilt measurements, as proved in [25], [26].

References

1. S. Łuczak, R. Grepl, M. Bodnicki, *Selection of MEMS Accelerometers for Tilt Measurements*. J. Sensors **2017**, Article ID 9796146 (2017)
2. M. Jrad, M.I. Younis, F. Najar, *Modeling and Design of an Electrically Actuated Resonant Switch*. MATEC Web of Conf. **1**, Article No. 04001 (2012)
3. S. Łuczak, *Tilt Measurements in BMW Motorcycles*, in R. Jabłoński, R. Szewczyk (Eds.), *Recent Global Research and Education: Technological Challenges*. (Springer International Publishing, Cham, Switzerland, 2017)
4. J. S. Wilson, *Sensor Technology Handbook*. (Newnes, Burlington, 2005)
5. L. Beran, P. Chmelar, L. Rejfe, *An Increase in Estimation Accuracy Position Determination of Inertial Measurement Units*. MATEC Web of Conf. **75**, Article No. 05001 (2016)
6. S. Łuczak, *Specific Measurements of Tilt with MEMS Accelerometers*, in R. Jabłoński, T. Brezina (Eds.), *Mechatronics. Recent Technological and Scientific Advances*. (Springer-Verlag, Berlin Heidelberg, 2012)
7. S. Łuczak, *Guidelines for Tilt Measurements Realized by MEMS Accelerometers*. Int. J. Precis. Eng. Manuf. **15**, 2012-2019 (2014)
8. S. Łuczak, W. Oleksiuk, M. Bodnicki, *Sensing Tilt With MEMS Accelerometers*. IEEE Sensors J. **6**, 1669-1675 (2006)

9. D. Jurman, M. Jankovec, R. Kamnik, M. Topič, *Calibration and data fusion solution for the miniature attitude and heading reference system*. Sens. Actuators A: Phys. **138**, 411-420 (2007)
10. S. Popowski, *Determination of Pitch and Roll Angles in Low Cost Land Navigation Systems*. J. Aeronautica Integra **1**, 93-97 (2008) (in Polish)
11. J.C. Lötters, J. Schipper, P.H. Veltink, W. Olthuis, P. Bergveld, *Procedure for in-use calibration of triaxial accelerometers in medical applications*, Sens. Actuators A: Phys. **68**, 221-228 (1998)
12. A. Albarbar, A. Badri, J. K. Sinha, A. Starr, *Performance evaluation of MEMS accelerometers*. Measurement **42**, 790-795 (2009)
13. R. Dao, *Inclination Sensing of Moving Vehicle*. Application Note n/r 3/12/03, MEMSIC Inc. (2003)
14. S. Liu, R. Zhu, *Micromachined Fluid Inertial Sensors*, Sens. **17**, 367 (2017)
15. S. Łuczak, *Experimental Studies of Hysteresis in MEMS Accelerometers: A Commentary*. IEEE Sensors J. **15**, 3492-3499 (2015)
16. S. Stančin, S. Tomažič, *Time- and Computation-Efficient Calibration of MEMS 3D Accelerometers and Gyroscopes*. Sens. **14**, 14885-14915 (2014)
17. I. Frosio, F. Pedersini, N. A. Borghese, *Autocalibration of Triaxial MEMS Accelerometers With Automatic Sensor Model Selection*. IEEE Sensors J. **12**, 2100-2108 (2012)
18. M. Sipos, P. Paces, J. Rohac, P. Novacek, *Analyses of triaxial accelerometer calibration algorithms*. IEEE Sensors J. **12**, 1157-1165 (2012)
19. W. Yang, B. Fang, Y. Y. Tang, J. Qian, X. Qin, W. Yao, *A robust inclinometer system with accurate calibration of tilt and azimuth angles*. IEEE Sensors J. **13**, 2313-2321 (2013)
20. S. Łuczak, *Advanced Algorithm for Measuring Tilt with MEMS Accelerometers*, in R. Jabłoński, M. Turkowski, R. Szewczyk (Eds.), *Recent Advances in Mechatronics*. (Springer-Verlag, Berlin Heidelberg, 2007)
21. S. Łuczak, *Dual-Axis Test Rig for MEMS Tilt Sensors*. Metrol. Meas. Syst. **21**, 351-362 (2014)
22. *TS1000T Colybris Acceleration Inclinometer*. Preliminary Datasheet, Safran Colibrys SA
23. *The SCA103T differential inclinometer series*, Datasheet, Murata Electronics Oy
24. S. Zhao, Y. Li, E. Zhang, P. Huang, H. Wei, *Differential amplified high-resolution tilt angle measurement system*. Rev. Scientific Instrum. **85**, 096104 (2014)
25. S. Łuczak, *Fast Alignment Procedure for MEMS Accelerometers*. R. Jabłoński, T. Brezina (Eds.), *Advanced Mechatronics Solutions*. (Springer International Publishing, Cham, Switzerland, 2016)
26. S. Łuczak, *Effects of Misalignments of MEMS Accelerometers in Tilt Measurements*. T. Brezina, R. Jabloński (Eds.), *Mechatronics 2013. Recent Technological and Scientific Advances*. (Springer International Publishing, Cham, Switzerland, 2014)

Euler-Lagrange optimization of electric drives with DTM method

Pavol Oršanský[1,*], *Branislav* Ftorek[1], *Helena* Šamajová[1], *Ján* Vittek[2]

[1]Department of Applied Mathematics, Faculty of Mechanical Engineering, University of Žilina, Univerzitná 1, 01026 Žilina, Slovakia
[2]Department of Electrical Engineering, Faculty of Electrical Engineering, University of Žilina, Univerzitná 1, 01026 Žilina, Slovakia

Abstract. Euler-Lagrange optimization method is exploited here to develop energy saving position control strategy valid for a.c. drives. The optimization principle respects copper losses minimization only. Boundary value problem for Euler-Lagrange equations is solved by Differential transformation method (DTM).

Keywords: DTM-method, Euler-Lagrange equations, boundary value problem

1 Introduction

The consumption of electric motors represents nearly one half (46%) of global electricity consumption. Therefore, even small decrease of energy demands of electric drives, can result in significant energy savings with subsequent contribution to the protection of the environment. The main contribution of this paper is a mathematical analysis of the energy optimal speed and position control of the drives with a permanent magnet synchronous motor (PMSM), taking into account stator copper losses. In [1, 2] the optimal mathematical trajectory is modified to correspond to the triangular and trapezoidal profiles. This procedure is necessary for the technical realization of the near-energy optimal control to any controlled drive with PMSM. The mathematical theory of optimization methods may be found in [3].

If we consider some functional and the fundamental problem: To determine the control function $\boldsymbol{u}(t)$ for which

$$I_0 = \int_{t_A}^{t_B} L_0(\mathbf{x}, \mathbf{u}, t)dt \to \min.$$

The problem of plastic deformation considered due the minimization of total energy in [4].

* Corresponding author: pavol.orsansky@fstroj.uniza.sk
Reviewers: *Tomáš* Lack, *Jerzy* Winczek

The function L_0 describes a minimization problem with the constraints (equations of motion for some physical problem, e.g. Newton's eq. of motion $F = m\mathbf{a}$)

$$\frac{d\mathbf{x}}{dt} = f(\mathbf{x}, \mathbf{u}, t)$$

and the complete cost functional is defined by the relation

$$\mathfrak{I} = L_0(\mathbf{x}, \mathbf{u}, t) + \lambda^T \left[\frac{d\mathbf{x}}{dt} - f(\mathbf{x}, \mathbf{u}, t)\right], \tag{1}$$

where λ^T is a vector of Lagrange multipliers. The solution of the minimizing problem requires to find a solution of Euler-Lagrange equations with some boundary conditions

$$\frac{d\mathfrak{I}}{d\mathbf{x}} = \frac{d}{dt}\left(\frac{\partial\mathfrak{I}}{\partial\dot{\mathbf{x}}}\right), \qquad \frac{d\mathfrak{I}}{d\mathbf{u}} = \frac{d}{dt}\left(\frac{\partial\mathfrak{I}}{\partial\dot{\mathbf{u}}}\right). \tag{2}$$

By means of Euler-Lagrange equations we can find the optimal control laws for electrical drive system.

Following differential equations describe dynamics of the rotor position ϕ, speed ω, electrical motor torque γ_e and load torque γ_l

$$\begin{aligned} \frac{d\phi}{dt} &= \omega, \\ J_r\frac{d\omega}{dt} &= \gamma_e - \gamma_l, \\ \frac{d\gamma_e}{dt} &= -\frac{1}{T_S}\gamma_e - \frac{k_t^2}{L_S}\omega + \frac{k_t}{L_S}u_q, \end{aligned} \tag{3}$$

where J_r is the reduced moment of an inertia to the shaft PMSM, L_S and R_S are stator phase inductance and resistance respectively, $T_S = \frac{R_S}{L_S}$ is a stator time constant. The electrical torque of the motor can be expressed as

$$\gamma_e = k_t i_q \tag{4}$$

where $k_T = c\psi_{PM}$ and $c = (3p)/2$, p is a number of pol-pairs, ψ_{PM} is a linkage flux of the permanent magnet and $i_p = p(t)$ is a torque component of the stator current.

An energy consumption of the control system depends on the minimization of copper losses and is given as (see [5])

$$I_0 = R_S \int_{t_A}^{t_B} i_q^2 dt \to \min.$$

With regard to (4), we have the relation

$$I_0 = \frac{R_S}{k_t^2}\int_0^{T_m} \gamma_e^2 dt \to \min.$$

where T_m is a manoeuvre time. If we assume that the load torque is a time dependent function $\gamma_l = f(t)$, then the complete cost functional (1) has a form (see also (3))

$$\mathfrak{I} = \frac{R_S}{k_t^2}\gamma_e^2 + \lambda_1(J_r\dot{\omega} - \gamma_e + \gamma_l) + \lambda_2(\dot{\phi} - \omega) + \lambda_3\left(\dot{\gamma}_l - \dot{f}(t)\right).$$

The corresponding Euler-Lagrange equations (2) for the position control are

$$\begin{aligned} 2\gamma_e \frac{R_S}{k_t^2} - \lambda_1 &= 0, \\ -\lambda_2 &= J_r \dot{\lambda}_1, \\ 0 &= \dot{\lambda}_2, \\ \lambda_1 &= \dot{\lambda}_3. \end{aligned} \tag{5}$$

According to (5), we have $\gamma_e = \frac{k_t^2}{2R_S}$ and $\gamma_l = f(t)$ (the assumption). Finally, from Euler-Lagrange equations with constrains (3) follows a boundary value problem (BVP)

$$\begin{aligned} \dot{\lambda}_1 &= \frac{1}{J_R}\lambda_2, \\ \dot{\lambda}_2 &= 0, \\ \dot{\omega} &= \frac{1}{J_R}\left[\frac{k_t^2}{2R_S}\lambda_1 - f(t)\right], \\ \dot{\phi} &= \omega. \end{aligned} \tag{6}$$

With conditions

$$\begin{aligned} \omega(0) = \omega(T_M) &= 0, \\ \phi(0) &= 0, \\ \phi(T_M) &= \phi_{dem}, \end{aligned} \tag{7}$$

where the manoeuvre time T_M and the demanded position ϕ_{dem} are given.

2 DTM – Differential transformation method

We give a brief overview of Differential transformation method that proposes a technique, which does not evaluate the derivatives symbolically, the DTM of the n-th derivative of a function $y(t)$ is defined by

$$Y(n) = \frac{1}{n!}\left[\frac{d^n y(t)}{dt^n}\right]_{t=t_0}$$

where $y(t)$ is an original and $Y(n)$ is a transformed function.
An inverse transformation of $Y(n)$ leads to

$$y(t) = \sum_{n=0}^{\infty} Y(n)(t - t_0)^n \tag{8}$$

and if we consider $t_0 = 0$ then the representation of the function reduces to

$$y(t) = \sum_{n=0}^{\infty} Y(n)t^n.$$

The properties of the DTM with their proofs are published in [6, 7], the used ones of them are introduced below.

Let functions G, G_i, for $i = 1,2, \dots, N$ be differential transformations of functions g, g_i, for $i = 1,2, \dots, N$, then the following correspondences hold

$$g(t) = g_1(t) \pm g_2(t), \qquad G(n) = G_1(n) \pm G_2(n),$$

$$g(t) = \alpha g_1(t), \qquad G(n) = \alpha G_1(n),$$

$$g(t) = g_1(t) \cdot g_2(t), \qquad G(n) = \sum_{l=0}^{\infty} G_1(n) \cdot G_2(n-l),$$

$$g(t) = t^k, \qquad G(n) = \delta_{n,k}, \; t_0 = 0,$$

$$g(t) = \frac{d^k}{dt^k} g_1(t) \qquad G(n) = \frac{(n+k)!}{n!} G_1(n+k),$$

$$g(t) = \prod_{i=1}^{N} g_i(t),$$

$$G(n) = \sum_{s_1=0}^{n} \sum_{s_2=0}^{n} \dots \sum_{s_N=0}^{n-s_1-\dots-s_{N-1}} G_1(s_1) \cdot \dots \cdot G_N(n - s_1 - \dots - s_N),$$

where $\delta_{n,k}$ is the Kronecker delta symbol, $n, l, k, N \in \mathbb{Z}_0^+$.

3 The BVP solution using DTM method

The standard using of the DTM transform is given for initial value problems and the solution has a form (8). In the next, we derive an approximate solution of the BVP problem (6), (7) by using some properties of DTM method. So, we have

$$\lambda_1 \to U_1(n), \lambda_2 \to U_2(n), \omega \to U_3(n), \phi \to U_4(n), f(t) \to F(n)$$

$$\begin{gathered} (n+1)U_1(n+1) = -\frac{1}{J_r} U_2(n), \\ (n+1)U_2(n+1) = 0, \\ (n+1)U_3(n+1) = \frac{1}{J_r}\left[\frac{k_t^2}{2R_S} U_2(n) - F(n)\right], \\ (n+1)U_4(n+1) = U_3(n). \end{gathered} \tag{9}$$

In praxis, the manoeuvre time $T_m \leq 1s$ and we consider the approximate solution in the form, i.e. the polynomials of the third degree

$$\begin{gathered} \lambda_1 = U_1(0) + U_1(1)t + U_1(2)t^2 + U_1(3)t^3, \\ \lambda_2 = U_2(0) + U_2(1)t + U_2(2)t^2 + U_2(3)t^3, \\ \omega = U_3(0) + U_3(1)t + U_3(2)t^2 + U_3(3)t^3, \\ \phi = U_4(0) + U_4(1)t + U_4(2)t^2 + U_4(3)t^3. \end{gathered}$$

Technically, from (9) we have 12 linear equations ($n = 0,1,2$) for 16 unknown coefficients $U_i(k)$, $i = 1,2,3,4$ and $i = 0,1,2,3$. The other four linear equations we get from the boundary conditions (7), i.e.

$$\begin{gathered} U_3(0) = 0, \\ U_3(1)T_m + U_3(2)T_m^2 + U_3(3)T_m^3 = 0, \\ U_4(0) = 0, \\ U_4(1)T_m + U_4(2)T_m^2 + U_4(3)T_m^3 = \phi_{dem}. \end{gathered}$$

4 Numerical results

The parameters of the numerical calculation are: $R_S = 1.3\Omega$, $k_T = 0.975$, the moment of inertia $J_r = 0.005kg \cdot m^2$, the prescribed manoeuvre time and the demanded position are $T_m = 0.5s$, $\phi_{dem} = 2\pi$. We assume two cases of the load torque γ_l
constant

$$\gamma_l = 2,$$

exponential

$$\gamma_l = 2 + e^{2t}.$$

The corresponding energy optimal solutions for these particular problems are for $\gamma_l = 2$,

$$\omega(t) = 150.7964t - 301.5929t^2,$$
$$\phi(t) = 75.3982t^2 - 100.531t^3.$$

and for $\gamma_l = 2 + e^{2t}$

$$\omega(t) = 84.1298 - 101.5929t^2 - 133.3333t^3,$$
$$\phi(t) = 42.0649t^2 - 33.8643t^3.$$

Results for energy optimal control of the drive speed and the drive position in the case of the constant load torque $\gamma_l = 2$, are shown in Figure 1 (respectively for the exponential load torque $\gamma_l = 2 + e^{2t}$ in Figure 2)

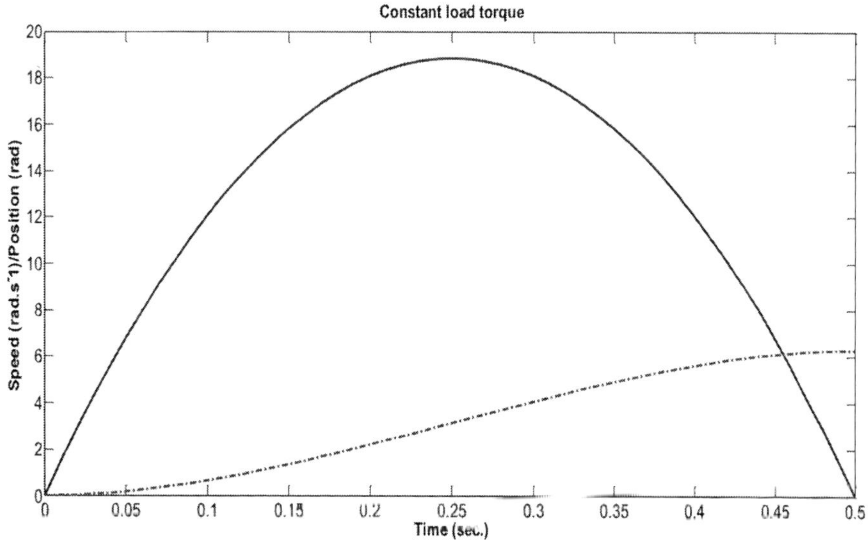

Fig. 1. Energy optimal control of the drive speed and the drive position in the case of the constant load torque $\gamma_l = 2$ (speed $\omega(t)$ – blue line, position $\phi(t)$ dash-dot line).

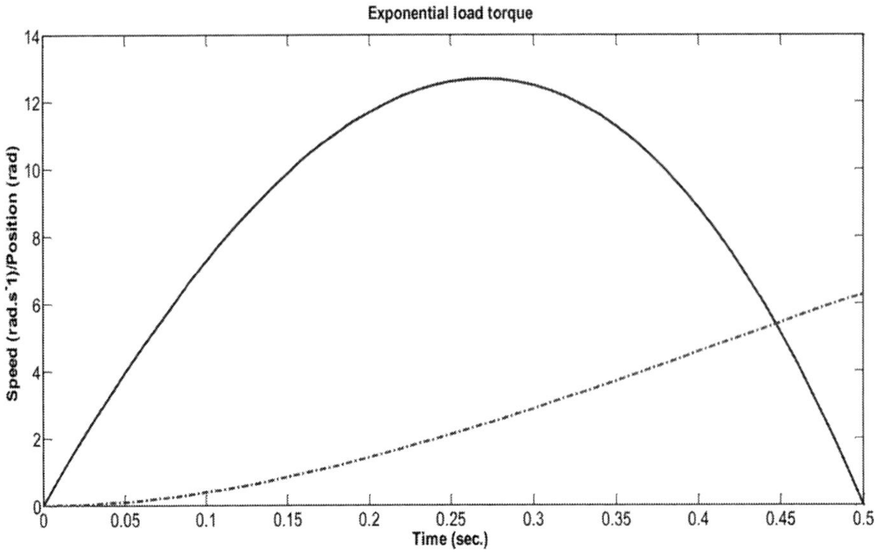

Fig. 2. Energy optimal control of the drive speed and the drive position in the case of the exponential load torque $\gamma_l = 2 + e^{2t}$ (speed $\omega(t)$ - blue line, position $\phi(t)$ - green dash-dot line).

Conclusion

In this paper the Differential transformation method is used to solve linear system of differential equations derived for optimal speed and position control of the drives with a permanent magnet synchronous motor (PMSM), taking into account stator copper losses. This problem needs to solve Euler-Lagrange equations with some constraints and boundary value conditions. The different types of load torques are considered, namely the constant and exponential load torques. The approximate solutions of BVP are given in the form of series with required accuracy, some numerical examples are provided.

The authors gratefully acknowledge the Scientific Grant Agency (KEGA) of the Ministry of Education of Slovak Republic for supporting this work under the Grant No. 015ŽU-4/2017 Digital technologies for machine building study programs as a part of strategy "Internet of Things"

References

1. J. Vitek, B. Ftorek: *Energy efficient speed and position control of electric drives with PMSM*, Communications. Scientific letters of the University of Žilina **16**, 64-71 (2014)
2. J. Vitek, B. Ftorek, P. Butko, T. Fedor: *Energy optimal control of PMSM Drive for time-varying load torque*, Applied mechanics and materials **710**, 67-75 (2015)
3. H. Tolle: *Optimization Methods*, Springer-Verlag, New-York, (1975)
4. J. Zapoměl, V. Dekýš, P. Ferfecki, A. Sapietová, M. Sága, M. Žmindák: *Identification of Material Damping of a Carbon Composite Bar and Study of Its Effect on Attenuation of Its Transient Lateral Vibrations*. International Journal of Applied Mechanics **7**, 1-18 (2015)
5. I. Bivol, C. Vasilache: *The application of Euler-Lagrange method of optimization for electromechanical motion control*. The Annals of Dunarea De Jos University of Galati **III**, 5-11 (2000)

6. A. Arikoglu, I. Özkol: *Solution of difference equations by using differential transform method*. Appl. Math. Comput. **174**, 1216-1228 (2006)
7. Y. Khan, Z. Svoboda, Z. Šmarda: *Solving certain classes of Lane-n type equations using differential transformation method*. Advances in difference equations **2012:174**, 1-11 (2012)

Transmuted Weibull distribution and its applications

Ivana Pobočíková[1], *Zuzana* Sedliačková[1], *Mária* Michalková[1,*]

[1]Department of Applied Mathematics, Faculty of Mechanical Engineering, University of Žilina, Univerzitná 1, 010 26 Žilina, Slovakia

Abstract. In this paper we study new distribution called transmuted Weibull distribution. Some properties of this distribution are described. The usefulness of the distribution for modelling data is illustrated using real data set.

Keywords: transmuted Weibull distribution, properties, maximum likelihood estimation, application; Akaike's information criterion, Bayesian information criterion, coefficient of determination, root mean square error

1 Introduction

The Weibull distribution was originally introduced by Swedish physicist Waloddi Weibull [1]. He applied it on modelling the distribution of the yield strength of materials. The Weibull distribution is popular and widely used, with many applications in areas such as engineering (reliability, failure analysis, lifetime analysis ...), material science, quality control, physics, medicine, meteorology, hydrology and others [2].

However, there are cases when standard Weibull distribution fails to model data suitably enough. This is where it is necessary to apply generalized distribution because of its flexibility and better fit of the data than standard Weibull distribution. The importance of such generalization has been proved in recent years on various problems and many standard distributions have been generalized. Some recent generalizations which are based on the Weibull distribution, are discussed in [3].

An interesting idea of generalization where the distribution is derived using the quadratic rank transmutation map, was studied first by Shaw and Buckley in [4]. Recently, many transmuted distributions have been proposed in literature. See, for example, transmuted Gumbel distribution [5], transmuted Rayleigh distribution [6], transmuted modified Weibull distribution [7], transmuted inverse Weibull distribution [8], transmuted Pareto distribution [9] and others.

Aryal and Tsokos [10] used the quadratic rank transmutation map to develop new generalization of the Weibull distribution, called the transmuted Weibull distribution. They studied mathematical properties of this distribution and the maximum likelihood estimates of the parameters.

* Corresponding author: maria.michalkova@fstroj.uniza.sk

Reviewers: *Vladimír* Dekýš, *Urszula* Siedlecka

In this paper, we focus on demonstrating the usefulness of the transmuted Weibull distribution for modelling lifetime, illustrated on three real data sets. All three data sets are modelled with the transmuted Weibull distribution, the 2-parameter and 3-parameter Weibull distribution. For comparison of the distributions the following criteria are used: the Akaike's information criterion, the corrected Akaike's information criterion, the Bayesian information criterion, the coefficient of determination and the root mean square error.

The parameters of the probability distributions are estimated using the maximum likelihood method. The calculations are performed using statistical software STATISTICA and software MATLAB.

The rest of the paper is organized as follows: In Section 2 we introduce the transmuted Weibull distribution. In Section 3 we define basic statistical properties including moments, quantiles, reliability and hazard rate functions. The maximum likelihood estimates of the transmuted Weibull parameters are presented in Section 4. Finally, in Section 5 we provide the real data applications of the transmuted Weibull distribution.

2 Transmuted Weibull distribution

A random variable X is said to have the 3-parameter Weibull distribution (WD) $W(a, b, c)$ with parameters $a > 0$, $b > 0$, $c \geq 0$ if its cumulative distribution function (CDF) for $x \geq c$ is given by

$$G(x) = 1 - \exp\left(-\left(\frac{x-c}{b}\right)^{a}\right) \tag{1}$$

and probability density function (PDF) is given by

$$g(x) = \frac{a}{b^{a}}(x-c)^{a-1} \exp\left(-\left(\frac{x-c}{b}\right)^{a}\right) \tag{2}$$

where a is the dimensionless shape parameter, b is the scale parameter and c is the location parameter.

Setting c=0, (1) and (2) transform into CDF and PDF, respectively, of the 2-parameter Weibull distribution $W(a, b)$.

The transmuted distribution can be obtained by adding real number λ, $|\lambda| \leq 1$, into CDF that provides more flexibility in the form of new distribution [7]. λ denotes the transmuting parameter.

A random variable X is said to have the transmuted distribution if its CDF $F(x)$ and PDF $f(x)$ are given by

$$F(x) = (1+\lambda)G(x) - \lambda G(x)^{2} \tag{3}$$

$$f(x) = g(x)[(1+\lambda) - 2\lambda G(x)] \tag{4}$$

where $|\lambda| \leq 1$, $G(x)$ and $g(x)$ are CDF and PDF of the base distribution, respectively. Observe that for $\lambda = 0$ we have the base distribution of the random variable X.

Let us consider the 2-parameter Weibull distribution. Adding (1) into (3) we have CDF of the transmuted Weibull distribution (TWD) $W(a, b, \lambda)$ with parameters $a > 0$, $b > 0$ and $|\lambda| \leq 1$

$$F(x) = \left[1 - \exp\left(-\left(\frac{x}{b}\right)^a\right)\right]\left[1 + \lambda \exp\left(-\left(\frac{x}{b}\right)^a\right)\right] \tag{5}$$

Similarly, combining (2), (3) and (4) we have PDF of the TWD

$$f(x) = \frac{a}{b^a} x^{a-1} \exp\left(-\left(\frac{x}{b}\right)^a\right)\left[1 - \lambda + 2\lambda \exp\left(-\left(\frac{x}{b}\right)^a\right)\right] \tag{6}$$

3 Statistical properties

In this section we define statistical properties including moments, quantiles, reliability and hazard rate functions. The origin and other aspects of this distribution can be found in [10].

3.1 Moments and quantiles

The r-th moment, $r = 1, 2, \ldots$, of the TWD is defined as

$$E(X^r) = b^r \Gamma\left(\frac{r}{a} + 1\right)\left[1 - \lambda + \lambda\, 2^{-r/a}\right] \tag{7}$$

Consequently, the mean *E(X)* and the variance *Var(X)* are given by

$$E(X) = b\Gamma\left(\frac{1}{a} + 1\right)\left[1 - \lambda + \lambda\, 2^{-1/a}\right] \tag{8}$$

$$Var(X) = b^2 \left\{\Gamma\left(\frac{2}{a} + 1\right)\left[1 - \lambda + \lambda\, 2^{-2/a}\right] - \Gamma^2\left(\frac{1}{a} + 1\right)\left[1 - \lambda + \lambda\, 2^{-1/a}\right]^2\right\} \tag{9}$$

respectively, where $\Gamma(p)$ is gamma function defined by $\Gamma(p) = \int_0^\infty x^{p-1} e^{-x} dx, p > 0$.
The α-th quantile, $\alpha \in (0, 1)$, of the TWD is given by

$$x_\alpha = b\left[-\ln\left(1 - \left(\frac{1 + \lambda - \sqrt{(1+\lambda)^2 - 4\lambda\alpha}}{2\lambda}\right)\right)\right]^{1/a} \tag{10}$$

3.2 Reliability analysis

The TWD can be useful in characterization of the lifetime data analysis. The reliability function if the TWD is given by

$$R(x) = \exp\left(-\left(\frac{x}{b}\right)^a\right)\left[1 - \lambda + \lambda \exp\left(-\left(\frac{x}{b}\right)^a\right)\right] \tag{11}$$

This function can be interpreted as probability that an item does not fail prior to some time x.

Another important characteristics is the hazard rate function that can be interpreted as conditional probability of failure when the item has already survived to time x.

The hazard rate function of the TWD is given by

$$H(x) = \frac{a}{b^a} x^{a-1} \left[\frac{1 - \lambda + 2\lambda \exp\left(-\left(\frac{x}{b}\right)^a\right)}{1 - \lambda + \lambda \exp\left(-\left(\frac{x}{b}\right)^a\right)} \right] \tag{12}$$

3.3 Random number generation

The random numbers from the TWD can be generated using the inverse function of CDF (5)

$$X = b\left[-\ln\left(1 - \left(\frac{1 + \lambda - \sqrt{(1+\lambda)^2 - 4\lambda U}}{2\lambda}\right)\right)\right]^{1/a} \tag{13}$$

where U has uniform distribution over the interval $(0,1)$.

4 Maximum likelihood estimates

As a method for estimation of parameters of the TWD we choose the maximum likelihood method (MLM).

Let $X_1, X_2, \dots, X_n$ be a random sample of size n from the TWD with PDF (6) and let $x_1, x_2, \dots, x_n$ be a realization of the random sample. The likelihood function $L = L(x_1, x_2, \dots, x_n; a, b, \lambda)$ of this random sample is given by

$$L = \left(\frac{a}{b^a}\right)^n \exp\left(-\frac{1}{b^a}\sum_{i=1}^{n} x_i^a\right) \prod_{i=1}^{n} x_i^{a-1}\left[1 - \lambda + 2\lambda \exp\left(-\left(\frac{x_i}{b}\right)^a\right)\right] \tag{14}$$

and the log-likelihood function $\ln L = \ln L(x_1, x_2, \dots, x_n; a, b, \lambda)$ of this random sample is given by

$$\ln L = n \ln\frac{a}{b^a} - \frac{1}{b^a}\sum_{i=1}^{n} x_i^a + (a-1)\sum_{i=1}^{n} \ln x_i + \sum_{i=1}^{n} \ln\left[1 - \lambda + 2\lambda \exp\left(-\left(\frac{x_i}{b}\right)^a\right)\right] \tag{15}$$

Differentiating (15) with respect to a, b and λ, respectively, and equating each derivative to zero we obtain the equations

$$\frac{n}{a} + \sum_{i=1}^{n} \ln\left(\frac{x_i}{b}\right)\left[1 - \left(\frac{x_i}{b}\right)^a\right] - 2\lambda \sum_{i=1}^{n} \frac{\left(\frac{x_i}{b}\right)^a \ln\left(\frac{x_i}{b}\right) \exp\left(-\left(\frac{x_i}{b}\right)^a\right)}{1 - \lambda + 2\lambda \exp\left(-\left(\frac{x_i}{b}\right)^a\right)} = 0 \tag{16}$$

$$\frac{a}{b}\sum_{i=1}^{n}\left[\left(\frac{x_i}{b}\right)^a - 1\right] - 2\lambda\,\frac{a}{b}\sum_{i=1}^{n} \frac{\left(\frac{x_i}{b}\right)^a \exp\left(-\left(\frac{x_i}{b}\right)^a\right)}{1 - \lambda + 2\lambda \exp\left(-\left(\frac{x_i}{b}\right)^a\right)} = 0 \tag{17}$$

$$\sum_{i=1}^{n} \frac{2\exp\left(-\left(\frac{x_i}{b}\right)^a\right) - 1}{1 - \lambda + 2\lambda\exp\left(-\left(\frac{x_i}{b}\right)^a\right)} = 0 \tag{18}$$

The MLM estimates $\hat{a}$, $\hat{b}$, $\hat{\lambda}$ of the parameters a, b, λ are obtained iteratively from the nonlinear system of equations (16) – (18).

5 Application

In this section we present the analysis of the real data sets using the TWD and compare it with the WD. In order to compare the distributions we consider the criteria:

- the Akaike's information criterion (*AIC*) – calculated using [11]

$$AIC = -2\ln L(\hat{\theta}) + 2m \tag{19}$$

where $L(\hat{\theta}) = L(x_1, x_2, \dots, x_n; \hat{\theta})$ is the maximized value of the likelihood function for the estimated model, $\hat{\theta}$ is the MLM estimate of the parameter θ, m is number of parameters to be estimated, n is number of observed data;

- the corrected Akaike's information criterion (*AICC*) – calculated using [12]

$$AICC = AIC + \frac{2m(m+1)}{n-m+1} \tag{20}$$

- the Bayesian information criterion (*BIC*) – calculated using [13]

$$BIC = -2\ln L(\hat{\theta}) + m\ln(n); \tag{21}$$

- the coefficient of determination (R^2) – calculated using

$$R^2 = \frac{\sum_{i=1}^{n}\left(\hat{F}(x_i) - \bar{F}\right)^2}{\sum_{i=1}^{n}\left(\hat{F}(x_i) - \bar{F}\right)^2 + \sum_{i=1}^{n}\left(F_n(x_i) - \hat{F}(x_i)\right)^2}, \tag{22}$$

where $\hat{F}(x)$ is estimated cumulative distribution function and

$$\bar{F} = \frac{1}{n}\sum_{i=1}^{n} \hat{F}(x_i) \tag{23}$$

Function $F_n(x)$ is the empirical distribution function, defined as follows

$$F_n(x) = \frac{1}{n}\sum_{i=1}^{n} I\left(x_{(i)} \le x\right) \tag{24}$$

where $I\left(x_{(i)} \le x\right) = 1$ if $x_{(i)} \le x$ and 0 otherwise, $x_{(1)}, x_{(2)}, \dots, x_{(n)}$ are observations in ascending order, i.e. $x_{(1)} \le x_{(2)} \le \dots \le x_{(n)}$;

- the root mean square error (*RMSE*) – calculated using

$$RMSE = \left[\frac{1}{n}\sum_{i=1}^{n}\left(F_n(x_i) - \hat{F}(x_i)\right)^2\right]^{1/2} \tag{25}$$

The distribution better fitting the data corresponds to smaller values of *AIC*, *BIC*, *AICC* and *RMSE* and larger value of R^2.

Data set 1

The first data set represents lifetimes of Kevlar 49/epoxy strands subjected to constant sustained pressure at 90% stress level until the strand failure. For previous studies with the data see [14 – 17]. The data are as follows:

0.0251, 0.0886, 0.0891, 0.2501, 0.3113, 0.3451, 0.4763, 0.5650, 0.5671, 0.6566, 0.6748, 0.6751, 0.6753, 0.7696, 0.8375, 0.8391, 0.8425, 0.8645, 0.8851, 0.9113, 0.9120, 0.9836, 1.0483, 1.0596, 1.0773, 1.1733, 1.2570, 1.2766, 1.2985, 1.3211, 1.3503, 1.3551, 1.4595, 1.4880, 1.5728, 1.5733, 1.7083, 1.7263, 1.7460, 1.7630, 1.7746, 1.8275, 1.8375, 1.8503, 1.8808, 1.8878, 1.8881, 1.9316, 1.9558, 2.0048, 2.0408, 2.0903, 2.1093, 2.1330, 2.2100, 2.2460, 2.2878, 2.3203, 2.3470, 2.3513, 2.4951, 2.5260, 2.9911, 3.0256, 3.2678, 3.4045, 3.4846, 3.7433, 3.7455, 3.9143, 4.8073, 5.4005, 5.4435, 5.5295, 6.5541, 9.0960.

Table 1 gives the descriptive statistics for the data set. We can see that the empirical distribution is skewed to the right. Table 2 presents the MLM estimates of the parameters together with the log-likelihood ln *L*, *AIC*, *AICC* and *BIC* values. Results are rounded to four decimal places.

Table 1. Descriptive statistics for data set 1

Min	Max	Mean	Variance	Lower Quantile	Median	Upper Quantile	Skewness	Kurtosis
0.0251	9.0960	1.9592	2.4774	0.8982	1.7362	2.3041	2.0196	5.6004

Table 2. MLM estimates of the parameters and model selection criteria for data set 1

Distribution	Parameter estimates	Statistical test					
		ln *L*	*AIC*	*AICC*	*BIC*	R^2	*RMSE*
$W(a,b)$	$\hat{a}$ =1.3256 $\hat{b}$ =2.1328	-122.5247	249.0494	249.2094	253.7108	0.9753	0.0423
$W(a,b,c)$	$\hat{a}$ =1.3169 $\hat{b}$ =2.1228 $\hat{c}$ =0.0058	-122.5141	251.0282	251.3525	258.0204	0.9750	0.0425
$W(a,b,\lambda)$	$\hat{a}$ =1.0509 $\hat{b}$ =1.4419 $\hat{\lambda}$ = -0.7955	-121.4300	248.8600	249.1843	255.8522	0.9838	0.0351

In Table 2 we see that values of *AIC*, *AICC* and *RMSE* are smallest ones and value of ln *L* and R^2 are highest ones for TWD, compared to values for 2-parameter and 3-parameter WDs. Hence, we can conclude that the TWD provides better fit to the data than the other two distributions.

Data set 2
The second data set represents survival times (in days) of 72 guinea pigs infected with virulent tubercle bacilli. The data were observed and reported in [18]. For previous studies with the data see [19]. The data are as follows:

10, 33, 44, 56, 59, 72, 74, 77, 92, 93, 96, 100, 100, 102, 105, 107, 107, 108, 108, 108, 109, 112, 113, 115, 116, 120, 121, 122, 122, 124, 130, 134, 136, 139, 144, 146, 153, 159, 160, 163, 163, 168, 171, 172, 176, 183, 195, 196, 197, 202, 213, 215, 216, 222, 230, 231, 240, 245, 251, 253, 254, 254, 278, 293, 327, 342, 347, 361, 402, 432, 458, 555.

Table 3 gives the descriptive statistics for data set 2. The value of skewness indicates that the empirical distribution is skewed to the right. Table 4 presents the MLM estimates of the parameters and criteria for comparison of TWD with 2-parameter and 3-parameter WDs.

Table 3. Descriptive statistics for data set 2

Min	Max	Mean	Variance	Lower Quantile	Median	Upper Quantile	Skewness	Kurtosis
10	555	176.8333	10705.0986	108	149.5	226	1.3700	2.2221

Table 4. MLM estimates for the parameters and model selection criteria for data set 2

Distribution	Parameter estimates	Statistical test					
		ln L	AIC	$AICC$	BIC	R^2	$RMSE$
$W(a,b)$	$\hat{a}$ =1.8253 $\hat{b}$ =199.6177	-427.3700	858.7400	858.9090	863.2933	0.9694	0.0477
$W(a,b,c)$	$\hat{a}$ =1.7610 $\hat{b}$ =193.3730 $\hat{c}$ =5.0001	-427.1846	860.3692	860.7121	867.1992	0.9711	0.0463
$W(a,b,\lambda)$	$\hat{a}$ =1.3534 $\hat{b}$ =139.6319 $\hat{\lambda}$ = -0.9506	-425.6935	857.3870	857.7299	864.2170	0.9814	0.0382

Values in Table 4 indicate that the TWD provides better fit to the data than the 2-parameter and 3-parameter WDs.

Data set 3
The third data set represents age of 155 patients suffering from the breast cancer, recorded from June to November 2014, who entered the Breast Tumours Detection Unit, Benha Hospital University, Egypt. The data were observed and reported in [20]. The data are as follows:

12, 17, 18, 18, 20, 24, 24, 25, 26, 28, 28, 28, 29, 30, 30, 30, 30, 31, 31, 31, 32, 32, 32, 32, 33, 34, 34, 35, 35, 35, 35, 35, 35, 35, 36, 36, 36, 36, 36, 36, 38, 38, 38, 38, 38, 38, 38, 38, 38, 38, 38, 38, 38, 39, 39, 40, 40, 40, 40, 40, 40, 40, 40, 40, 40, 40, 40, 40, 40, 40, 40, 40, 41, 42, 42, 42, 42, 42, 42, 42, 42, 43, 43, 44, 44, 45, 45, 45, 45, 45, 45, 45, 46, 46, 47, 48, 48, 48, 48, 48, 48, 48, 49, 49, 50, 50, 50, 50, 50, 50, 50, 50, 50, 50, 50, 50, 50, 50, 50, 50, 50, 50, 50, 50, 50, 52, 52, 52, 53, 54, 54, 55, 56, 56, 58, 58, 58, 59, 60, 60, 60, 60, 60, 60, 60, 63, 65, 65, 66, 69, 80, 90, 90.

Table 5 gives the descriptive statistics for data set 3. Table 6 presents the MLM estimates of the parameters and the comparison of TWD with the 2-parameter and 3-parameter WDs.

Table 5. Descriptive statistics for data set 3

Min	Max	Mean	Variance	Lower Quantile	Median	Upper Quantile	Skewness	Kurtosis
12	90	43.6516	146.0597	36	42	50	0.6915	2.2738

Table 6. MLM estimates and model selection criteria for data set 3

Distribution	Parameter estimates	Statistical test					
		ln *L*	*AIC*	*AICC*	*BIC*	R^2	*RMSE*
$W(a,b)$	$\hat{a}$ =3.6871 $\hat{b}$ =48.1100	-610.2967	1224.5934	1224.6713	1230.6802	0.9640	0.0491
$W(a,b,c)$	$\hat{a}$ =2.9515 $\hat{b}$ =38.3366 $\hat{c}$ =9.2768	-607.4815	1220.9629	1221.1189	1230.0932	0.9656	0.0489
$W(a,b,\lambda)$	$\hat{a}$ =4.0312 $\hat{b}$ =55.0570 $\hat{\lambda}$ =0.8380	-605.5600	1217.1200	1217.2768	1226.2502	0.9733	0.0438

Values in Table 6 indicate that the TWD provides better fit to data than the 2-parameter and 3-parameter WDs.

Figures 1, 2 and 3 illustrate the histograms and the fitted PDFs of the TWD, 2-parameter WD and 3-parameter WD for data set 1, 2 and 3, respectively.

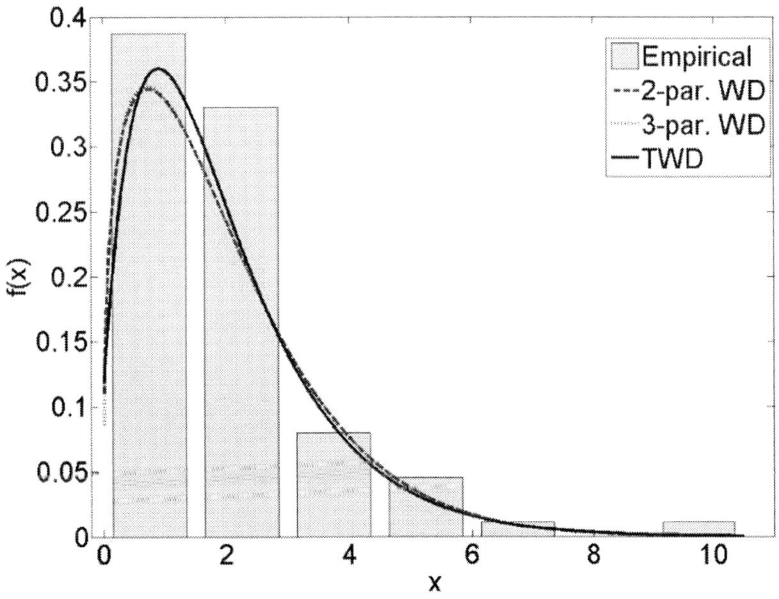

Fig. 1. Histogram of data set 1 and the fitted PDFs

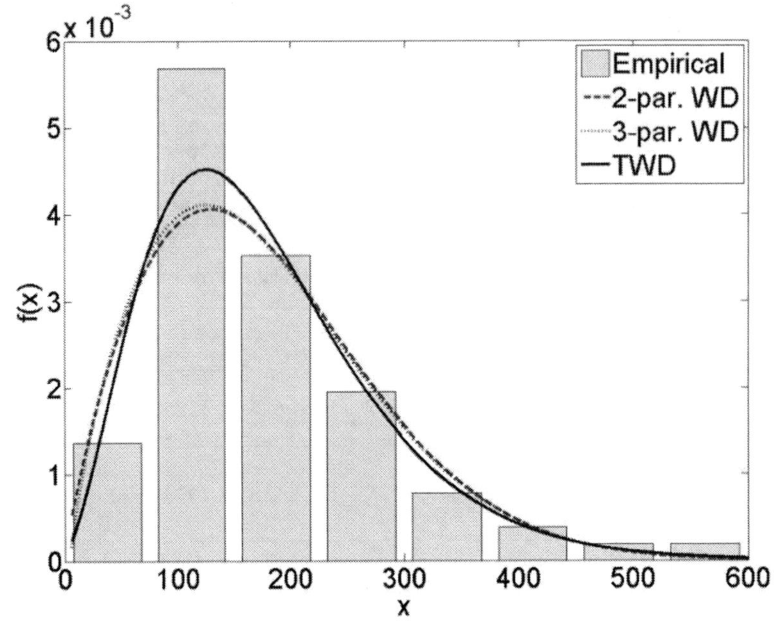

Fig. 2. Histogram of data set 2 and the fitted PDFs

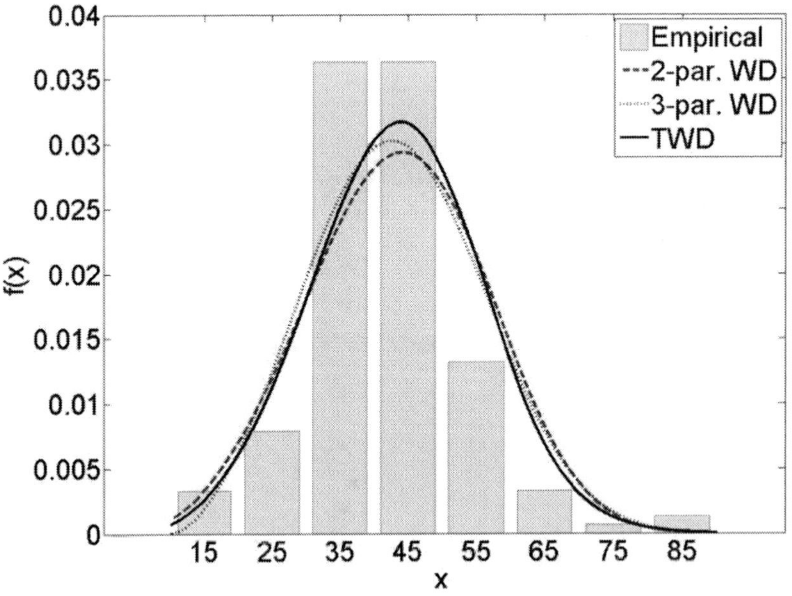

Fig. 3. Histogram of data set 3 and the fitted PDFs

Conclusion

In this paper we studied modification of the Weibull distribution, called the transmuted Weibull distribution. The usefulness of this distribution for modelling lifetime was illustrated using three real data sets. We compared the performance of the transmuted Weibull distribution with performances of the 2-parameter and the 3-parameter Weibull distributions. Comparison of distributions was based on criteria as the Akaike's information criterion, the corrected Akaike's information criterion, the Bayesian information criterion, the coefficient of determination and the root mean square error. We demonstrated that the transmuted Weibull distribution is more flexible and it better models lifetime data of animate (breast tumour patients, infected guinea pigs) and inanimate (Kevlar/epoxy strands) objects than the 2-parameter or the 3-parameter Weibull distribution.

This paper was supported by the Slovak Grant Agency VEGA through the project No. 1/0812/17.

References

1. W. Weibull, *A statistical distribution function of wide applicability*. Journal of Applied Mechanics **18** (3), 293-297 (1951)
2. H. Rinne, *The Weibull distribution: a handbook*. (Chapman & Hall, CRC Press, 2008)
3. H. Pham, C. D. Lai, *On recent generalizations of the Weibull distributions*. IEEE Transactions on Reliability **56** (3), 454-458 (2007)
4. W T. Shaw, I. R. Buckley, *The alchemy of probability distributions: beyond Gram-Charlier expansions and a skew kurtostic-normal distribution from a rank transmutation map*. Research report (2007)
5. G. R. Aryal, Ch. P. Tsokos, *On the transmuted extreme value distribution with application*. Nonlinear Alalysis: Theory, Methods and Applications **71**, 1401-1407 (2009)
6. F. Merovci, *Transmuted Rayleigh Distribution*. Austrian Journal of Statistics **42** (1), 21-31 (2013)
7. M. S. Khan, R. King, *Transmuted Modified Weibull Distribution: A Generalization of the Modified Weibull Probability Distribution*. European Journal of Pure and Apllied Mathematics **6**, 66-88 (2013)
8. M. S. Khan, R. King, R. Hudson, *Characterization of the transmuted inverse Weibull Distribution*. ANZIAM **55**, 197-217, (2013)
9. F. Merovci, L. Puka, *Transmuted Pareto Distribution*. ProbStat Forum **7**, 1-11 (2014)
10. G. R. Aryal, Ch. P. Tsokos, *Transmuted Weibull Distribution: A Generalization of the Weibull Probability Distribution*. European Journal of Pure and Apllied Mathematics **4** (2), 89-102 (2011)
11. H. Akaike, *A new look at the statistical model identification*, Automatic Control, IEEE Transactions **19** (6), 716-723 (1974)
12. C. M. Hurvich, C. L. Tsai, *Regression and time series model selection in small samples*. Biometrika **76** (2), 297-307 (1989)
13. G. Schwarz, *Estimating the dimension of a model*, Annals of Statistics **6** (2), 461-464, (1978)
14. R. E. Barlow, R. H. Toland, T. Freeman, *A Bayesian analysis of stress rupture life of Kevlar 49/epoxy spherical pressure vessels*. Procedings conference on applications of statistics, Marcel dekker, New York (1984)

15. D. F. Andrews, A. M. Herzberg, *Data: A Collection Of Problem from Many Fields for the Student and Researcher Worker*. (Springer Series in Statistics, New York, 1985)
16. I. B. Abdul-Moniem, *Transmuted Burr Type III Distribution*. Journal of Statistics: Advances in Theory and Applications **14** (1), 37-47 (2015)
17. E. A. Owoloko, P. E. Oguntunde, A. O. Adejumo, *Performance rating of the transmuted exponential distribution: an analytical approach*. Springer Plus **4**, 818 (2015)
18. T. Bjerkedal, *Acquisition of resistance in guinea pigs infected with different doses of virulent tubercile bacilli*. American Journal of Hygiene **72**, 100-148 (1960)
19. E. A. Owoloko, P. E. Oguntunde, A. O. Adejumo, *A Comparative Analysis on the Performance of the Convoluted Exponencial Distribution and the Exponential Distribution in terms of Flexibility*. Journal of Mathematics and Statistics **12** (1) (2016)
20. M. M. Mansour, E. M. Abd Elrazik, M. S. Hamed, S. M. Mohamed, *A New Additive Weibull Distribution: Based on a New Method for Adding Parameter to a Family Distributions*. International Journal of Applied Mathematical Sciences **8** (1), 01-54 (2015)

Heat conduction in a composite sphere - the effect of fractional derivative order on temperature distribution

Urszula Siedlecka[1,*], *Stanisław* Kukla[1]

[1]Institute of Mathematics, Czestochowa University of Technology, Częstochowa, Poland

Abstract. The aim of the contribution is an analysis of time-fractional heat conduction in a sphere with an inner heat source. The object of the consideration is a solid sphere with a spherical layer. The heat conduction in the solid sphere and spherical layer is governed by fractional heat conduction equation with a Caputo time-derivative. Mathematical (classical) or physical formulations of the Robin boundary condition and the perfect contact of the solid sphere and spherical layer is assumed. The boundary condition and the heat flux continuity condition at the interface are expressed by the Riemann-Liouville derivative. An exact solution of the problem under mathematical conditions is determined. A solution of the problem under physical boundary and continuity conditions using the Laplace transform method has been obtained. The inverse of the Laplace transform by using the Talbot method are numerically determined. Numerical results show the effect of the order of the Caputo and the Riemann-Liouville derivatives on the temperature distribution in the sphere.

Keywords: fractional heat conduction, heating source, Robin boundary condition

1 Introduction

The fundamental of the classical heat transfer theory is the Fourier law which leads to the parabolic partial differential equation of the heat conduction [1]. A consequence of the Fourier's law is unrealistic speed of heat flow in the medium. This inconvenience can be avoided by a generalization of the Fourier law which leads to a fractional heat conduction equation [2]. The heat conduction governed by the fractional differential equation is the subject of papers [3-9]. Applications of fractional order calculus are presented in books [10-12] and papers [13-15].

If the heat transfer in a bounded medium is considered then the heat equation is complemented by boundary conditions. The Dirichlet, Neumann and Robin boundary conditions are often used in describing the heat transfer between the body and the surroundings. In the classical heat theory, the Neumann and Robin boundary conditions include the normal derivative at the boundary of the considered region. Introducing the time-fractional derivative in the Neumann and Robin boundary conditions, the physical formulation of these conditions is obtained [16].

* Corresponding author: urszula.siedlecka@im.pcz.pl

Reviewers: *Ján* Vavro *Jr.*, *Branislav* Ftorek

A solution of the linear fractional differential equation under classical boundary conditions can be determined in an analytical form. To solve the fractional equation under physical boundary conditions, the Laplace technique can be applied. This approach leads to a solution of the problem in the Laplace domain. The temperature distribution in the time domain by using an algorithm for numerical inversion of the Laplace transform can be obtained. The methods of numerical inversion of the Laplace transforms used in classical analysis can be also applied to Laplace transforms obtained by solving the problems with fractional derivatives. Selected methods of numerical inversion of the Laplace transforms are presented in papers [17-20].

In this paper, we present the solution of the fractional heat conduction problem in a sphere consisting of an inner solid sphere and a spherical layer. The mathematical and physical formulation of the Robin boundary conditions is considered. The perfect thermal contact of the inner sphere and the spherical layer is assumed. The effect of the fractional order on the temperature distribution in the sphere has been numerically investigated.

2 Formulation of the problem

We consider the time-fractional radial heat conduction problem in a sphere. The two regions of the sphere are distinguished: $0 \le r \le r_1$ - a solid inner sphere and $r_1 \le r \le b$ - a spherical layer, where r is the radial coordinate. The heat transfer in the regions is governed by the fractional heat conduction equation [3]:

$$\frac{1}{r^2}\frac{\partial}{\partial r}\left(r^2\frac{\partial T_i}{\partial r}\right)+\frac{1}{\lambda_i}g_i\left(r,t\right)=\frac{1}{a_i}\frac{\partial^{\alpha_i}T_i}{\partial t^{\alpha_i}}, \qquad i=1,2 \tag{1}$$

where $g_i\left(r,t\right)$ is the volumetric rate of heat generation, a_i is the thermal diffusivity, λ_i is the thermal conductivity and α_i denotes the fractional order of the Caputo derivative with respect to time t. The Caputo derivative is defined by [21]

$${}_0^C D_t^\alpha f\left(t\right)=\frac{d^\alpha}{dt^\alpha}f\left(t\right)=\begin{cases}\dfrac{1}{\Gamma\left(m-\alpha\right)}\displaystyle\int_0^t\frac{f^{(m)}\left(\tau\right)}{\left(t-\tau\right)^{\alpha+1-m}}d\tau, & m-1<\alpha\le m\\ f^{(m)}\left(t\right), & \alpha=m\in N\end{cases} \tag{2}$$

where Γ denotes the gamma function. The boundary condition and the continuity condition at interface are assumed in a form with the Riemann-Liouville fractional derivative $D_{RL}^{1-\beta}$ which is defined by [21]

$$D_{RL}^{\beta}f\left(t\right)=\frac{d}{dt}\left[\frac{1}{\Gamma\left(1-\beta\right)}\int_0^t\frac{f\left(\tau\right)}{\left(t-\tau\right)^{\beta}}d\tau\right], \qquad 0<\beta\le 1 \tag{3}$$

On the outer surface of the sphere, the Robin boundary condition [16] is assumed

$$\lambda_2 D_{RL}^{1-\beta_2}\frac{\partial T_2}{\partial r}\left(b,t\right)=a_\infty\left(T_\infty\left(t\right)-T_2\left(b,t\right)\right) \tag{4}$$

where a_∞ is the outer heat transfer coefficient and T_∞ is the ambient temperature. The perfect thermal contact at the interface between the inner sphere and the spherical layer is described by conditions:

$$T_1(r_1,t) = T_2(r_1,t) \tag{5}$$

$$\lambda_1 D_{RL}^{1-\beta_1} \frac{\partial T_1}{\partial r}(r_1,t) = \lambda_2 D_{RL}^{1-\beta_2} \frac{\partial T_2}{\partial r}(r_1,t) \tag{6}$$

and the initial condition is

$$T(r,0) = F_i(r) \ . \tag{7}$$

The conditions (4) and (6) for $\beta_1 = \alpha_1$ and $\beta_2 = \alpha_2$ are called the physical conditions [16], if $\beta_1 = \beta_2 = 1$, there are so called mathematical conditions. In the second case, the D_{RL}^0 in equations (4) and (6) means an identity operator and can be omitted. We further consider the case of $\alpha_i \in (0,1]$.

3 Solution to the problem

In order to transform the heat conduction equation (1) into the fractional equation with a constant coefficient, we introduce new functions $U_i(r,t)$ given by the following relationship

$$U_i(r,t) = r\left(T_i(r,t) - T_\infty(t)\right), \qquad i = 1,2 \tag{8}$$

Taking into account equation (8) in the initial-boundary problem (1) and (4-7) , we obtain formulation of the problem for the function $U_i(r,t)$ in the form

$$a_i \frac{\partial^2 U_i(r,t)}{\partial r^2} + r g_i^*(r,t) = \frac{\partial^{\alpha_i} U_i(r,t)}{\partial t^{\alpha_i}}, \qquad i = 1,2 \tag{9}$$

$$\lambda_2 D_{RL}^{1-\beta_2}\left(\frac{\partial U_2}{\partial r}(b,t) - \frac{1}{b} U_2(b,t)\right) = -a_\infty U_2(b,t) \tag{10}$$

$$U_1(r_1,t) = U_2(r_1,t) \tag{11}$$

$$\lambda_1 D_{RL}^{1-\beta_1}\left(\frac{\partial U_1(r_1,t)}{\partial r} - \frac{1}{r_1} U_1(r_1,t)\right) = \lambda_2 D_{RL}^{1-\beta_2}\left(\frac{\partial U_2(r_1,t)}{\partial r} - \frac{1}{r_1} U_2(r_1,t)\right) \tag{12}$$

$$U_i(r,0) = r\left(F_i(r) - T_\infty(0)\right), \qquad i = 1,2 \tag{13}$$

Moreover, the conditions (10-12) are complemented by a condition for $r = 0$, which is obtained using equation (8)

$$U_1(0,t)=0 \tag{14}$$

The functions $g_i^*(r,t)$ in equation (9) are given by the formula

$$g_i^*(r,t)=\frac{a_i}{\lambda_i}g_i(r,t)-\frac{d^{\alpha_i}T_\infty(t)}{dt^{\alpha_i}} \tag{15}$$

The solutions of the initial-boundary problem (9-14) for mathematical and physical formulations will be presented below.

3.1 Mathematical formulation of boundary and continuity conditions

The heat conduction problem (9-14) under mathematical conditions for $\alpha_1=\alpha_2=\alpha$ can be solved analytically. We search for the solution to this problem in the form of the series of orthogonal functions $\Psi_{i,k}(r)$:

$$U_i(r,t)=\sum_{k=1}^{\infty}\Lambda_k(t)\Psi_{i,k}(r),\qquad i=1,2 \tag{16}$$

In the first step, we find the functions $\Psi_{i,k}(r)$ by solving the following eigenvalue problem

$$\frac{d^2\Psi_{i,k}(r)}{dr^2}+\frac{\gamma_k^2}{a_i}\Psi_{i,k}(r)=0 \qquad i=1,2 \tag{17}$$

$$\Psi_{1,k}(0)=0 \tag{18}$$

$$\Psi_{1,k}(r_1)=\Psi_{2,k}(r_1) \tag{19}$$

$$\lambda_1\left(\frac{d\Psi_1(r_1)}{dr}-\frac{1}{r_1}\Psi_1(r_1)\right)=\lambda_2\left(\frac{d\Psi_2(r_1)}{dr}-\frac{1}{r_1}\Psi_2(r_1)\right) \tag{20}$$

$$\frac{d\Psi_2(b)}{dr}=\left(\frac{1}{b}-\frac{a_\infty}{\lambda_2}\right)\Psi_2(b) \tag{21}$$

The functions $\Psi_{i,k}(r)$ are

$$\Psi_{1,k}(r)=B_{1,k}\sin\mu_{1,k}r \tag{22}$$

$$\Psi_{2,k}(r)=A_{2,k}\cos\mu_{2,k}(r-r_1)+B_{2,k}\sin\mu_{2,k}(r-r_1) \tag{23}$$

where $\mu_{i,k}=\gamma_k/\sqrt{a_i}$ and γ_k are roots of the equation

$$Q_1\lambda_2\mu_1\sin\mu_1 r_1+Q_2 Q_3=0 \tag{24}$$

where

$$Q_1 = \left(\frac{a_\infty}{\lambda_2} - \frac{1}{b}\right)\cos\mu_2(b - r_1) - \mu_2 \sin\mu_2(b - r_1),$$

$$Q_2 = \left(\frac{a_\infty}{\lambda_2} - \frac{1}{b}\right)\sin\mu_2(b - r_1) + \mu_2 \cos\mu_2(b - r_1)$$

and

$$Q_3 = \lambda_1\mu_1\cos\mu_1 r_1 + \frac{\lambda_2 - \lambda_1}{r_1}\sin\mu_1 r_1 .$$

These functions fulfil the orthogonality condition in the form

$$\frac{\lambda_1}{a_1}\int_0^{r_1}\Psi_{1,k}(r)\Psi_{1,k'}(r)dr + \frac{\lambda_2}{a_2}\int_{r_1}^{b}\Psi_{2,k}(r)\Psi_{2,k'}(r)dr = \begin{cases} 0 & \text{for } k' \neq k \\ N_k & \text{for } k' = k \end{cases} \tag{25}$$

The coefficients $B_{1,k}$, $A_{2,k}$, $B_{2,k}$, occurring in equations (22-23), are determined by using conditions (19-21). Assuming $B_{1,k} = 1$, we obtain $A_{2,k} = \sin\mu_{1,k} r_1$ and $B_{2,k} = Q_3 / \lambda_2\mu_{2,k}$.

The function $\Lambda_k(t)$ satisfies a fractional equation which is obtained substituting equation (16) into equation (9) and using orthogonality condition (25). The initial condition for the function $\Lambda_k(t)$ is obtained in a similar way by using (16) and (25) in condition (13). The fractional differential equation and initial condition have the form

$$\frac{d^\alpha \Lambda_k(t)}{d t^\alpha} + \gamma_k^2 \Lambda_k(t) = \frac{1}{N_k^r}\left(\int_0^{r_1} r g_1^*(r,t)\Psi_{1,k}(r)dr + \int_{r_1}^{b} r g_2^*(r,t)\Psi_{2,k}(r)dr\right) \tag{26}$$

$$\Lambda_k(0) = \frac{1}{N_k^r}\left(\frac{\lambda_1}{a_1}\int_0^{r_1} r\left(F_1(r) - T_\infty(0)\right)\Psi_{1,k}(r)dr + \frac{\lambda_2}{a_2}\int_{r_1}^{b} r\left(F_2(r) - T_\infty(0)\right)\Psi_{2,k}(r)dr\right) \tag{27}$$

where $N_k^r = \frac{\lambda_1}{a_1}\int_0^{r_1}\left[\Psi_{1,k}(r)\right]^2 dr + \frac{\lambda_2}{a_2}\int_{r_1}^{b}\left[\Psi_{2,k}(r)\right]^2 dr$. A solution of the problem (26-27) is given by [22]

$$\begin{aligned}\Lambda_k(t) = \frac{1}{N_k^r}\Bigg(&\int_0^{r_1} r\Psi_{1,k}(r)\int_0^{t}(t-\tau)^{\alpha-1} g_1^*(r,\tau)E_{\alpha,\alpha}\left(-\gamma_k^2(t-\tau)^\alpha\right)d\tau dr \\ &+ \int_{r_1}^{b} r\Psi_{2,k}(r)\int_0^{t}(t-\tau)^{\alpha-1} g_2^*(r,\tau)E_{\alpha,\alpha}\left(-\gamma_k^2(t-\tau)^\alpha\right)d\tau dr\Bigg) \\ &+ \frac{1}{N_k^r}E_{\alpha,1}\left(-\gamma_k^2 t^\alpha\right)\left(\frac{\lambda_1}{a_1}\int_0^{r_1} r\left(F_1(r) - T_\infty(0)\right)\Psi_{1,k}(r)dr + \frac{\lambda_2}{a_2}\int_{r_1}^{b} r\left(F_2(r) - T_\infty(0)\right)\Psi_{2,k}(r)dr\right)\end{aligned} \tag{28}$$

where $E_\alpha(z)$ is the Mittag-Leffler function defined by

$$E_\alpha(z)=\sum_{k=0}^{\infty}\frac{z^k}{\Gamma(\alpha k+1)} \tag{29}$$

Finally, taking into account equation (8) and (16), we obtain the temperature distribution in the sphere under mathematical formulation of the boundary and continuity conditions in the form

$$T_i(r,t)=T_\infty(t)+\frac{1}{r}\sum_{k=1}^{\infty}\Lambda_k(t)\Psi_{i,k}(r),\qquad i=1,2 \tag{30}$$

where functions $\Lambda_k(t)$ are given by (28) and $\Psi_{i,k}(r)$ are defined by (22) and (23). If $F_i(r)=T_{init}=const$, $T_\infty(t)=T_\infty=const$ and $g_1(r,t)=G=const$, $g_2(r,t)=0$, then the functions $\Lambda_k(t)$ are given by

$$\begin{aligned}\Lambda_k(t)=&\frac{G}{N_k^r\mu_{1,k}^2\gamma_k^2}\frac{a_1}{\lambda_1}\left(\sin\mu_{1,k}r_1-\mu_{1,k}r_1\cos\mu_{1,k}r_1\right)\left(1-E_\alpha\left(-\gamma_k^2t^\alpha\right)\right)\\&+\frac{T_{init}-T_\infty}{N_k^r}E_\alpha\left(-\gamma_k^2t^\alpha\right)\left\{\frac{\lambda_1}{\mu_{1,k}^2a_1}\left(\sin\mu_{1,k}r_1-\mu_{1,k}r_1\cos\mu_{1,k}r_1\right)+\frac{\lambda_2}{a_2}\frac{1}{\mu_{2,k}^2}\cdot\right.\\&\left[A_{2,k}\left(b\mu_{2,k}\sin\mu_{2,k}(b-r_1)+\cos\mu_{2,k}(b-r_1)-1\right)\right.\\&\left.\left.+B_{2,k}\left(-b\mu_{2,k}\cos\mu_{2,k}(b-r_1)+\sin\mu_{2,k}(b-r_1)+r_1\mu_{2,k}\right)\right]\right\}\end{aligned} \tag{31}$$

3.2 Physical formulation of boundary and continuity conditions

A solution of the heat conduction problem (9-14) under physical boundary and continuity conditions ($\beta_1=\alpha_1$, $\beta_2=\alpha_2$ in equations (10) and (12)) will be obtained by using the Laplace technique. The Laplace transform is defined as

$$\bar{f}(s)=\int_0^{\infty}f(t)e^{-st}dt \tag{32}$$

where $f(t)$ for $t>0$ is a given function of the exponential type and s is a complex parameter. After applying the Laplace transformation to the equations (9-12) and (14), and using the properties of the Laplace transform, we obtain

$$\frac{d^2\bar{U}_i}{dr^2}-\frac{s^{\alpha_i}}{a_i}\bar{U}_i=\bar{h}_i(r,s) \tag{33}$$

$$\bar{U}_1(0,s)=0 \tag{34}$$

$$\bar{U}_1(r_1,s)=\bar{U}_2(r_1,s) \tag{35}$$

$$\lambda_1s^{1-\beta_1}\left(\frac{d\bar{U}_1(r_1,s)}{dr}-\frac{1}{r_1}\bar{U}_1(r_1,s)\right)=\lambda_2s^{1-\beta_2}\left(\frac{d\bar{U}_2(r_1,s)}{dr}-\frac{1}{r_1}\bar{U}_2(r_1,s)\right) \tag{36}$$

$$\frac{d\bar{U}_2(b,s)}{dr}+\left(\frac{a_\infty}{\lambda_2 s^{1-\beta_2}}-\frac{1}{b}\right)\bar{U}_2(b,s)=0 \tag{37}$$

where

$$\bar{h}_i(r,s)=r\frac{s^{\alpha_i-1}}{a_i}\left(s\bar{T}_\infty(s)-F_i(r)\right)-\frac{r}{\lambda_i}\bar{g}_i(r,s).$$

The solution of the equation (33) for $i=1$, using the condition (34), can be written in the form

$$\bar{U}_1(r,s)=B_1\sinh S_1 r+\frac{1}{S_1}\int_0^r \bar{h}_1(u,s)\sinh S_1(r-u)\,du \tag{38}$$

and the general solution of equation (33) for $i=2$, is as follows

$$\bar{U}_2(r,s)=A_2\cosh S_2(r-r_1)+B_2\sinh S_2(r-r_1)+\frac{1}{S_2}\int_{r_1}^r \bar{h}_2(u,s)\sinh S_2(r-u)\,du \tag{39}$$

where $S_i=\frac{s^{\alpha_i/2}}{\sqrt{a_i}}$. Using conditions (35-37), a system of linear equations with respect to unknown constants B_1, A_2 and B_2 is obtained

$$B_1\sinh S_1 r_1-A_2=P_1 \tag{40}$$

$$B_1 s^{1-\beta_1}\lambda_1\left(S_1\cosh S_1 r_1-\frac{1}{r_1}\sinh S_1 r_1\right)+A_2\frac{s^{1-\beta_2}\lambda_2}{r_1}-B_2 s^{1-\beta_2}\lambda_2 S_2=P_2 \tag{41}$$

$$\begin{aligned}&A_2\frac{1}{s^{\beta_2}}\left(a_\infty b s^{\beta_2}S_2\cosh S_2(b-r_1)-sS_2\lambda_2\cosh S_2(b-r_1)+bsS_2^2\lambda_2\sinh S_2(b-r_1)\right)\\&+B_2\frac{1}{s^{\beta_2}}\left(bsS_2^2\lambda_2\cosh S_2(b-r_1)+a_\infty b s^{\beta_2}S_2\sinh S_2(b-r_1)-sS_2\lambda_2\sinh S_2(b-r_1)\right)=P_3\end{aligned} \tag{42}$$

where

$$P_1=-\frac{1}{S_1}\int_0^{r_1}\bar{h}_1(u,s)\sinh S_1(r_1-u)\,du\,,$$

$$P_2=-\frac{s^{1-\beta_1}\lambda_1}{S_1 r_1}\left(S_1 r_1\int_0^{r_1}\bar{h}_1(u,s)\cosh S_1(r_1-u)\,du-\int_0^{r_1}\bar{h}_1(u,s)\sinh S_1(r_1-u)\,du\right),$$

$$P_3=-\frac{1}{s^{\beta_2}}\left(s\lambda_2 S_2 b\int_{r_1}^{b}\bar{h}_2(u,s)\cosh S_2(b-u)\,du+\left(a_\infty b s^{\beta_2}-s\lambda_2\right)\int_{r_1}^{b}\bar{h}_2(u,s)\sinh S_2(b-u)\,du\right).$$

Substituting the determined constants B_1, A_2 and B_2 into equations (38-39), we get the complete solution of the problem in the Laplace domain. The temperature distribution in the sphere is given by

$$T_i(r,t) = T_\infty(t) + \frac{1}{r} L^{-1}\left[\bar{U}_i(r,s)\right], \quad i = 1,2 \tag{43}$$

The inverse of the Laplace transform of the functions $\bar{U}_i(r,s), i = 1,2$ is calculated numerically. The fixed Talbot algorithm to numerical inversion of the Laplace transform has been used, [18-19]. Applying this algorithm, the approximate values of the function $U_i(r,t) = L^{-1}\left[\bar{U}_i(r,s)\right]$ are determined using the formula

$$U_i(r,t) = \frac{p}{M}\left\{\frac{1}{2}\bar{U}_i(r,p)\exp(pt) + \sum_{k=1}^{M-1}\mathrm{Re}\left[\exp(ts(\theta_k))\bar{U}_i(r,s(\theta_k))(1 + j\sigma(\theta_k))\right]\right\} \tag{44}$$

where $s(\theta) = p\theta(\mathrm{ctg}\theta + j)$, $\sigma(\theta) = \theta + (\theta\mathrm{ctg}\theta - 1)\mathrm{ctg}\theta$, $p = 2M/(5t)$, $j = \sqrt{-1}$, and $\theta_k = k\pi/M$, M is a number of precision decimal digits.

4 Results of numerical calculations

The effect of the fractional order of time-derivatives occurring in the heat conduction equation and boundary conditions on the temperature distribution in the sphere has been numerically investigated. For the heat conduction model with the mathematical formulation of the boundary and continuity conditions, the results obtained by using the numerical inversion of the Laplace transform has been compared with the exact solution. The calculation has been performed for the following geometrical and thermal data: outer radius of the sphere is $b = 1.0\,[m]$, the radius of the inner sphere is $\hat{r}_1 = r_1/b = 0.6$, the thermal diffusivities are $a_1 = 3.352 \cdot 10^{-6}\,\left[m^2/s^\alpha\right]$, $a_2 = 1.474 \cdot 10^{-5}\,\left[m^2/s^\alpha\right]$, the thermal conductivities are $\lambda_1 = 16\left[W/(m \cdot K)\right]$, $\lambda_2 = 54\left[W/(m \cdot K)\right]$, the outer heat transfer coefficient is $\alpha_\infty = 500\,\left[W/(m^2 \cdot K)\right]$, the ambient temperature is $T_\infty = 25\,\left[{}^\circ C\right]$ and the initial temperature is assumed as $T_{init} = 25\,\left[{}^\circ C\right]$.

Table 1. The non-dimensional temperature $\hat{T}(\hat{r},\hat{t})$ for $\hat{t} = tb^2/a_2 = 1.0$, calculated by using the exact solution and numerical inversion of the Laplace transform (NILT)

$\hat{r}$	$\alpha = 0.7$		$\alpha = 0.8$		$\alpha = 0.9$		$\alpha = 1.0$	
	Exact	**NILT**	**Exact**	**NILT**	**Exact**	**NILT**	**Exact**	**NILT**
0	1.44421	1.44404	2.25696	2.25679	3.73501	3.73484	4.77135	4.77118
0.2	1.44176	1.44175	2.20899	2.20898	3.53158	3.53158	4.44462	4.44462
0.4	1.40995	1.40995	1.98699	1.98699	2.87100	2.87100	3.46198	3.46198
0.6	1.13143	1.13143	1.32024	1.32023	1.61769	1.61768	1.81594	1.81594
0.8	1.03183	1.03182	1.11964	1.11962	1.27291	1.27289	1.37645	1.37643
1.0	1.00653	1.00652	1.03246	1.03245	1.08066	1.08065	1.11348	1.11347

The non-dimensional temperature $\hat{T} = T/T_{init}$ in the selected points of the sphere for the mathematical boundary and continuity conditions ($\beta_1 = \beta_2 = 1.0$, $\alpha_1 = \alpha_2 = \alpha$) for different orders of the Caputo derivative α are presented in Table 1. The volumetric rate of heat generation in the inner sphere is $g_1 = 20000[W/m^3]$ and in the spherical layer $g_2 = 0$. The results presented in the Table 1 fulfil the condition: $|Exact - NILT|/Exact < 3.6 \cdot 10^{-5}$. The small differences of the Exact and NILT results allow for the use of the algorithm of numerical inversion of the Laplace transform to the heat conduction problem under physical formulation of boundary and continuity conditions.

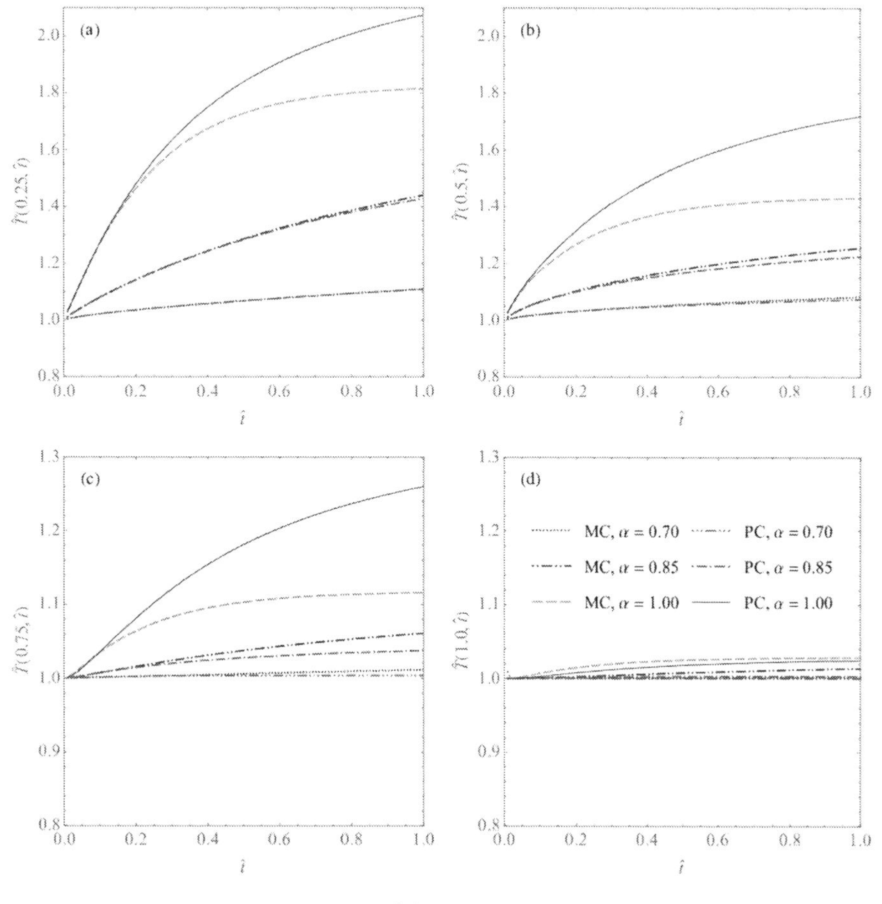

Fig. 1. The non-dimensional temperature $\hat{T}(\hat{r},\hat{t})$ as a function of time $\hat{t}$ for various values of fractional derivatives α and β. (a) $\hat{r} = 0.25$; (b) $\hat{r} = 0.5$; (c) $\hat{r} = 0.75$; (d) $\hat{r} = 1.0$

The time-histories temperatures in selected points of the sphere are presented in Fig. 1. The calculations were performed for different orders of the derivative occurring in the heat equation. In the problem under mathematical conditions (MC) one was assumed $\alpha_1 = \alpha_2 = \alpha = 0.7; 0.85; 1.0$ and $\beta_1 = \beta_2 = 1.0$ and for the problem under physical conditions (PC) the calculations were performed for $\alpha_1 = \alpha = 0.7; 0.85; 1.0$, $\alpha_2 = \beta_2 = 0.9$ and $\beta_1 = \alpha_1$. The volumetric rate of heat generation was assumed : $g_1 = 5000[W/m^3]$ and $g_2 = 0$. As was expected, the differences between the temperatures obtained for various orders of α 's decrease if the distance from the source increases. The significant effect on the temperature distribution in the sphere has order of the derivatives in the heat conduction model.

Conclusions

The problem of the fractional heat conduction in a sphere with inner heat source under mathematical and physical Robin boundary and continuity conditions by using the Laplace transform technique has been solved. It was noted a high agreement of the results obtained on the basis of the exact solution and the results computed by using numerical inversion of the Laplace transform for the heat conduction under mathematical conditions. The effect of the order of the Caputo derivative occurring in the heat conduction equation on the temperature distribution in the sphere has been numerically investigated. It was stated that the temperatures in the sphere obtained for models with the same fractional order α with mathematical and physical Robin conditions differ slightly. The significant differences in the temperatures have been observed for different orders of the fractional derivative in the heat conduction equation.

References

1. M.N. Özişik, *Heat conduction*. (Wiley, New York, 1993)
2. Y. Povstenko, *Fractional heat conduction in a semi-infinite composite body*. Communications in Applied and Industrial Mathematics **6** (1), e-482 (2014)
3. Y. Povstenko, *Fractional heat conduction in an infinite medium with a spherical inclusion*. Entropy **15**, 4122-4133 (2013)
4. Y. Povstenko, J. Klekot, *The fundamental solutions to the central symmetric time-fractional heat conduction equation with heat absorption*. Journal of Applied Mathematics and Computational Mechanics **16** (2), 101-112 (2017)
5. S. Blasiak, *Time-fractional heat transfer equations in modeling of the non-contacting face seals*. International Journal of Heat and Mass Transfer **100**, 79-88 (2016)
6. T.H. Ning, X.Y. Jiang, *Analytical solution for the time-fractional heat conduction equation in spherical coordinate system by the method of variable separation*. Acta Mechanica Sinica **27** (6), 994-1000 (2011)
7. S. Kukla, U. Siedlecka, *Laplace transform solution of the problem of time-fractional heat conduction in a two-layered slab*. Journal of Applied Mathematics and Computational Mechanics **14** (4), 105-113 (2015)
8. S. Kukla, U. Siedlecka, *An analytical solution to the problem of time-fractional heat conduction in a composite sphere*. Bulletin of the Polish Academy of Sciences – Technical Sciences **65** (2), 179-186 (2017)
9. U. Siedlecka, *Radial heat conduction in a multilayered sphere*. Journal of Applied Mathematics and Computational Mechanics **13** (4), 109-116 (2014)
10. T. M. Atanacković, S. Pilipović, B. Stanković, D. Zorica, *Fractional Calculus with Applications in Mechanics*. (John Wiley & Sons, New York, 2014)
11. A. A. Kilbas, H. M. Srivastava, J. J. Trujillo, Theory and applications of fractional differential equations. (Elsevier, Amsterdam, 2006)
12. J.S. Leszczyński, *An Introduction to Fractional Mechanics*. The Publishing Office of Czestochowa University of Technology, Czestochowa (2011)

13. M. Dalir, M. Bashour, *Applications of fractional calculus*. Applied Mathematical Sciences **4** (21), 1021-1032 (2010)
14. A. Dzieliński, D. Sierociuk, G. Sarwas, *Some applications of fractional order calculus*. Bulletin of the Polish Academy of Sciences – Technical Sciences **58** (4), 583-592 (2010)
15. W. E. Raslan, *Application of fractional order theory of thermoelasticity to a 1D problem for a spherical shell*. Journal of Theoretical and Applied Mechanics **54** (1), 295-304 (2016)
16. Y. Povstenko, *Time-fractional heat conduction in an infinite medium with a spherical hole under Robin boundary condition*. Fractional Calculus & Applied Analysis **16** (2), 354-369 (2013)
17. K. L. Kuhlman, *Review of inverse Laplace transform algorithms for Laplace-space numerical approaches*. Numerical Algorithms **63** (2), 339-355 (2013)
18. J. Abate, P. P. Valkó, *Multi-precision Laplace transform inversion*. International Journal for Numerical Methods in Engineering **60**, 979-993 (2004)
19. B. Dingfelder, J. A. C. Weideman, *An improved Talbot method for numerical Laplace transform inversion*. Numerical Algorithms **68**, 167-183 (2015)
20. H. Sheng, Y. Li, Y. Chen, *Application of numerical inverse Laplace transform algorithms in fractional calculus*. Journal of the Franklin Institute **384**, 315-330 (2011)
21. I. Podlubny, *Fractional differential equations*. (Academic Press, San Diego, 1999)
22. K. Diethelm, *The analysis of fractional differential equations*. (Springer-Verlag Berlin Heidelberg, 2010)

Numerical modeling of the contact between three-dimensional regions with the use of FEM

Tomasz Skrzypczak[1,*], *Ewa* Węgrzyn-Skrzypczak[2], *Leszek* Sowa[1]

[1]Czestochowa University of Technology, Faculty of Mechanical Engineering and Computer Science, Institute of Mechanics and Machine Design Fundamentals, Dąbrowskiego 73, 42-201 Częstochowa, Poland
[2]Czestochowa University of Technology, Faculty of Mechanical Engineering and Computer Science, Institute of Mathematics, Armii Krajowej 21, 42-201 Częstochowa, Poland

Abstract. The numerical approach based on the finite element method (FEM) for modeling of mechanical interaction between three-dimensional objects is presented in the paper. The model of contact is based on the assumption that the nodes of the region which is the source of contact cannot overlap with the nodes of the region being the target. The procedure of the detection of collision between surfaces of the source and the target is discussed in details. The behaviour of surfaces being in contact depends on their rigidity and is numerically modeled in the case of perfectly rigid source and deformable target. Each modeled object has an independent mesh of finite elements. These meshes can be freely moved relative to each other. Example of calculation using original program written in C++ is presented and discussed.

Keywords: computational mechanics, contact problem, finite element method, multi-mesh approach

1 Introduction

Mechanical contact between working elements is very common in engineering practice. Various kinds of machine parts work together going into the mechanical and also thermal interactions. Mechanical contact is the result of direct impacts of their external boundaries. In general such interactions occur between the deformable bodies although often for the sake of simplicity is assumed perfect rigidity of one or more bodies.

The first work focused on the classical contact mechanics was written by Heinrich Hertz in 1881 [1]. Hertz investigated the contact problem between two elastic bodies with curved surfaces. Nearly one hundred years later Johnson, Kendall, and Roberts found a similar solution for the contact with adhesion [2]. Their model of contact is known as JKR model of elastic contact. Different approach called DMT model was proposed by Derjaguin and co-workers in 1975 [3]. Bowden and Tabor investigated the importance of surface roughness in contact process [4]. Archard in 1957 was the first who found that the area of contact is approximately proportional to the normal force [5]. Nowadays advanced numerical methods such as finite element method (FEM) and the development of computer

* Corresponding author: t.skrzypczak@imipkm.pcz.pl

Reviewers: *Marek* Macko, *Branislav* Ftorek

technology makes possible numerical modeling of complex problems [6-8]. This work aims to present a method of detecting collisions between objects surfaces and matching them during elastic deformation.

2 Mathematical model

Two elastic objects are considered (Fig. 1). The first one called the target is fixed and not deformable, while the second called the source is mobile and deformable. In the initial phase, the objects are separated, then the movement of the source causes contact and deformation.

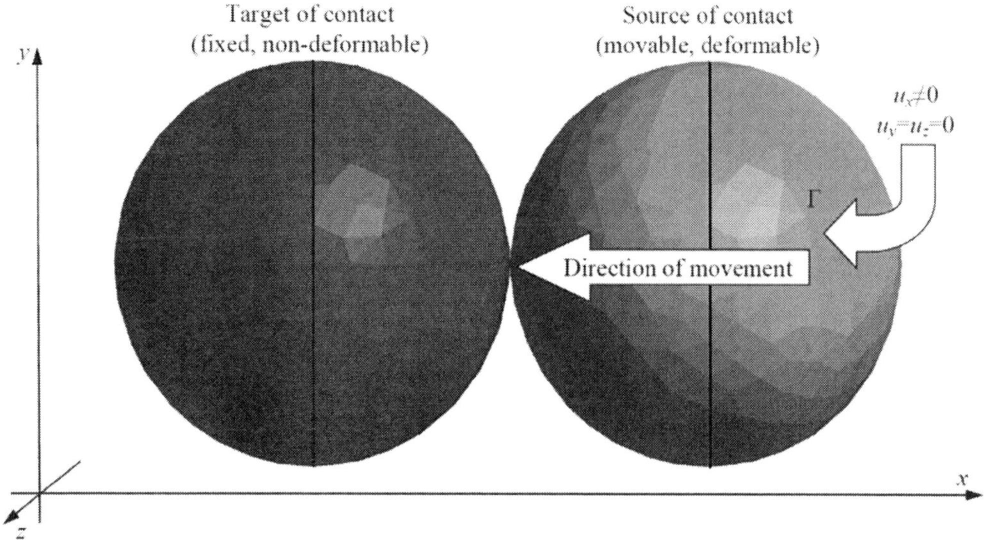

Fig. 1. Objects in contact

Mathematical model of deformations during the mechanical contact is based on the equilibrium equations of linear elasticity. Because the target of the contact is non-deformable and fixed below equations is only written for the source of the contact:

$$\begin{aligned} \frac{\partial \sigma_x}{\partial x} + \frac{\partial \tau_{xy}}{\partial y} + \frac{\partial \tau_{xz}}{\partial z} = 0 \\ \frac{\partial \tau_{xy}}{\partial x} + \frac{\partial \sigma_y}{\partial y} + \frac{\partial \tau_{yz}}{\partial z} = 0 \\ \frac{\partial \tau_{xz}}{\partial x} + \frac{\partial \tau_{yz}}{\partial y} + \frac{\partial \sigma_z}{\partial z} = 0 \end{aligned} \tag{1}$$

where: σ_x, σ_y, σ_z, τ_{xy}, τ_{xz}, τ_{yz} are the components of the symmetrical stress tensor ($\tau_{xy} = \tau_{yx}$, $\tau_{xz} = \tau_{zx}$, $\tau_{yz} = \tau_{zy}$) and x, y, z - Cartesian coordinates.

Above equilibrium equations contain six unknowns and must be converted in order to be the functions of displacements u_x, u_y, u_z because the number of equations must be the same as the number of unknowns. The components of stress tensor can be written in the following form:

$$\begin{gathered}
\sigma_x = \frac{E}{1+\nu}\left[\varepsilon_x + \frac{\nu}{1-2\nu}\left(\varepsilon_x + \varepsilon_y + \varepsilon_z\right)\right] \\
\sigma_y = \frac{E}{1+\nu}\left[\varepsilon_y + \frac{\nu}{1-2\nu}\left(\varepsilon_x + \varepsilon_y + \varepsilon_z\right)\right] \\
\sigma_z = \frac{E}{1+\nu}\left[\varepsilon_z + \frac{\nu}{1-2\nu}\left(\varepsilon_x + \varepsilon_y + \varepsilon_z\right)\right] \\
\tau_{xy} = \frac{E}{2(1+\nu)}\gamma_{xy} \\
\tau_{yz} = \frac{E}{2(1+\nu)}\gamma_{yz} \\
\tau_{xz} = \frac{E}{2(1+\nu)}\gamma_{xz}
\end{gathered} \tag{2}$$

where: E - Young's modulus, ν - Poisson's ratio, ε_x, ε_y, ε_z, γ_{xy}, γ_{xz}, γ_{yz} - the components of symmetrical strain tensor defined as follows:

$$\begin{gathered}
\varepsilon_x = \frac{\partial u_x}{\partial x}, \quad \gamma_{xy} = \frac{\partial u_x}{\partial y} + \frac{\partial u_y}{\partial x} \\
\varepsilon_y = \frac{\partial u_y}{\partial y}, \quad \gamma_{yz} = \frac{\partial u_y}{\partial z} + \frac{\partial u_z}{\partial y} \\
\varepsilon_z = \frac{\partial u_z}{\partial z}, \quad \gamma_{xz} = \frac{\partial u_x}{\partial z} + \frac{\partial u_z}{\partial x}
\end{gathered} \tag{3}$$

Equations (2), (3) make possible to convert equilibrium equations (1) into the functions of unknown displacements u_x, u_y, u_z:

$$\begin{gathered}
f_1\frac{\partial^2 u_x}{\partial x^2} + f_3\frac{\partial^2 u_x}{\partial y^2} + f_3\frac{\partial^2 u_x}{\partial z^2} + f_2\frac{\partial^2 u_y}{\partial x\partial y} + f_3\frac{\partial^2 u_y}{\partial y\partial x} + f_2\frac{\partial^2 u_z}{\partial x\partial z} + f_3\frac{\partial^2 u_z}{\partial z\partial x} = 0 \\
f_2\frac{\partial^2 u_x}{\partial y\partial x} + f_3\frac{\partial^2 u_x}{\partial x\partial y} + f_3\frac{\partial^2 u_y}{\partial x^2} + f_1\frac{\partial^2 u_y}{\partial y^2} + f_3\frac{\partial^2 u_y}{\partial z^2} + f_2\frac{\partial^2 u_z}{\partial y\partial z} + f_3\frac{\partial^2 u_z}{\partial z\partial y} = 0 \\
f_2\frac{\partial^2 u_x}{\partial z\partial x} + f_3\frac{\partial^2 u_x}{\partial x\partial z} + f_2\frac{\partial^2 u_y}{\partial z\partial y} + f_3\frac{\partial^2 u_y}{\partial y\partial z} + f_3\frac{\partial^2 u_z}{\partial x^2} + f_3\frac{\partial^2 u_z}{\partial y^2} + f_1\frac{\partial^2 u_z}{\partial z^2} = 0
\end{gathered} \tag{4}$$

where physical parameters f_1, f_2, f_3 are defined in the following way:

$$\begin{gathered}
f_1 = \frac{E(1-\nu)}{(1+\nu)(1-2\nu)} \\
f_2 = \frac{E\nu}{(1+\nu)(1-2\nu)} \\
f_3 = \frac{E}{2(1+\nu)}
\end{gathered} \tag{5}$$

Presented mathematical model of elastic deformations is supplemented by the appropriate boundary conditions:

$$u_x\big|_\Gamma \neq 0, \quad u_y\big|_\Gamma = u_z\big|_\Gamma = 0 \tag{6}$$

3 Numerical model

The numerical model is based on the finite element method [9] and is derived from the criterion of the method of weighted residuals [10]. Equations (4) are multiplied by the weighting function w and integrated over the volume Ω of the deformable object. Weak form of equations (4) is shown below:

$$\int_{\Omega}\left(f_1\frac{\partial w}{\partial x}\frac{\partial u_x}{\partial x}+f_3\frac{\partial w}{\partial y}\frac{\partial u_x}{\partial y}+f_3\frac{\partial w}{\partial z}\frac{\partial u_x}{\partial z}\right)d\Omega+\int_{\Omega}\left(f_2\frac{\partial w}{\partial x}\frac{\partial u_y}{\partial y}+f_3\frac{\partial w}{\partial y}\frac{\partial u_y}{\partial x}\right)d\Omega+$$
$$+\int_{\Omega}\left(f_2\frac{\partial w}{\partial x}\frac{\partial u_z}{\partial z}+f_3\frac{\partial w}{\partial z}\frac{\partial u_z}{\partial x}\right)d\Omega=\oint_{\Gamma}wp_x d\Gamma \qquad (7)$$

$$\int_{\Omega}\left(f_2\frac{\partial w}{\partial y}\frac{\partial u_x}{\partial x}+f_3\frac{\partial w}{\partial x}\frac{\partial u_x}{\partial y}\right)d\Omega+\int_{\Omega}\left(f_3\frac{\partial w}{\partial x}\frac{\partial u_y}{\partial x}+f_1\frac{\partial w}{\partial y}\frac{\partial u_y}{\partial y}+f_3\frac{\partial w}{\partial z}\frac{\partial u_y}{\partial z}\right)d\Omega+$$
$$+\int_{\Omega}\left(f_2\frac{\partial w}{\partial y}\frac{\partial u_z}{\partial z}+f_3\frac{\partial w}{\partial z}\frac{\partial u_z}{\partial y}\right)d\Omega=\oint_{\Gamma}wp_y d\Gamma \qquad (8)$$

$$\int_{\Omega}\left(f_2\frac{\partial w}{\partial z}\frac{\partial u_x}{\partial x}+f_3\frac{\partial w}{\partial x}\frac{\partial u_x}{\partial z}\right)d\Omega+\int_{\Omega}\left(f_2\frac{\partial w}{\partial z}\frac{\partial u_y}{\partial y}+f_3\frac{\partial w}{\partial y}\frac{\partial u_y}{\partial z}\right)d\Omega+$$
$$+\int_{\Omega}\left(f_3\frac{\partial w}{\partial x}\frac{\partial u_z}{\partial x}+f_3\frac{\partial w}{\partial y}\frac{\partial u_z}{\partial y}+f_1\frac{\partial w}{\partial z}\frac{\partial u_z}{\partial z}\right)d\Omega=\oint_{\Gamma}wp_z d\Gamma \qquad (9)$$

The weighting function w depends on the method used in the process of spatial discretization. Presented model is based on the Galerkin method where w is the same as the shape functions of the finite element used during discretization of considered volume. Final form of the global FEM equations is presented below:

$$\mathbf{K}\cdot\mathbf{u}=\mathbf{B}, \qquad (10)$$

where **K** is the global stiffness matrix, **B** - vector where known boundary displacements are stored, **u** - vector of unknown nodal displacements.

3.1 Collision detection

The collision occurs when the surface nodes of the moving object enter the interior of the stationary object (Fig. 2). Surface nodes of the source are then located within the boundary of finite elements of the target. The mesh may contain millions of elements, especially when it consists of tetrahedrons, so it is important to mark only those elements of the target that may collide with the source before starting of calculation process. Analogous selection also takes place in the elements of the source. This pre-selection of the surfaces of potential contact allows for huge savings in computational time during computer simulation. Collision is detected when any surface node of the source is inside the boundary of finite element of the target.

Fig. 3 shows the node N that is inside the tetrahedral finite element $ABCD$ due to a collision. Collision detection is very simple in this case. At the beginning it is necessary to determine the volume V of the tetrahedron E. Next, the volumes V_1, V_2, V_3, V_4 of tetrahedrons *ABCN*, *BDCN*, *ACDN*, *ABDN* constructed from the successive walls of element *E* and node *N* are calculated.

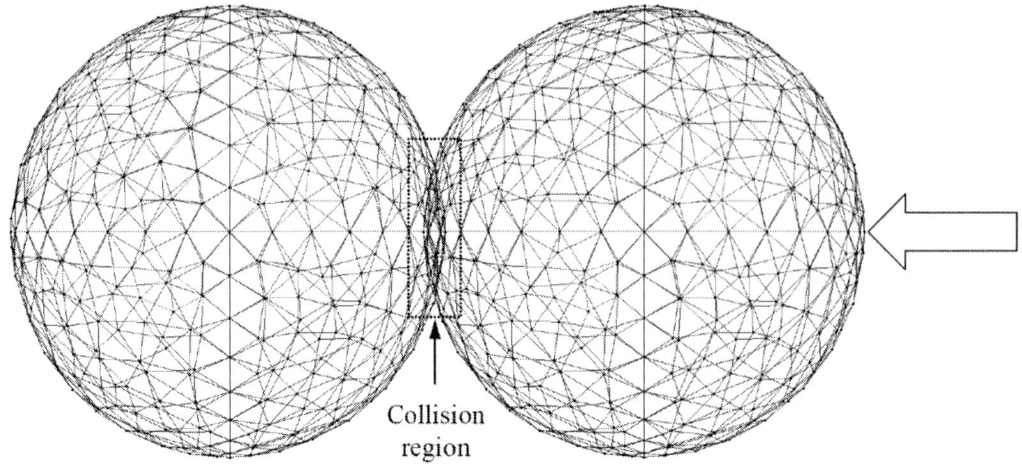

Fig. 2. Collision of objects

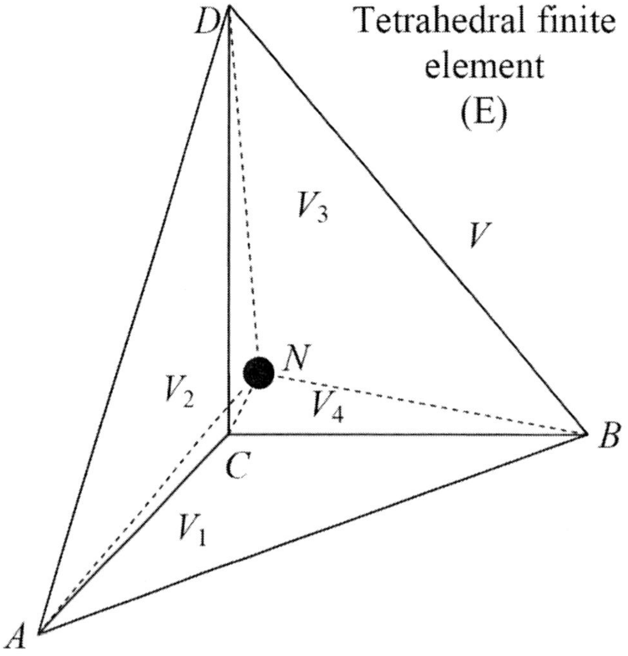

Fig. 3. A node inside a tetrahedral finite element during collision

If *N* is inside *E*, the following condition must be fulfilled:

$$V_1 + V_2 + V_3 + V_4 = V \tag{11}$$

Equation (11) is the collision criterion. Each calculation step requires searching the set of contact nodes of the source and checking this criterion for all selected boundary elements of the target. Nodes that meet condition (11) are marked as "collision nodes" because they must be moved from the inside of the element to its surface.

3.2 Object deformation

The surfaces of the objects in contact can not penetrate each other, so all the collision nodes must be translated to the surface of the target. The translation vector of each node is calculated perpendicular to the contact surface. In fact, the nodes are not directly moved, but the calculated translations modify vector **B**. Then the displacements of each node are obtained after solving the set of equations (10). Modified shape of the collision region in the source is shown in Fig. 4.

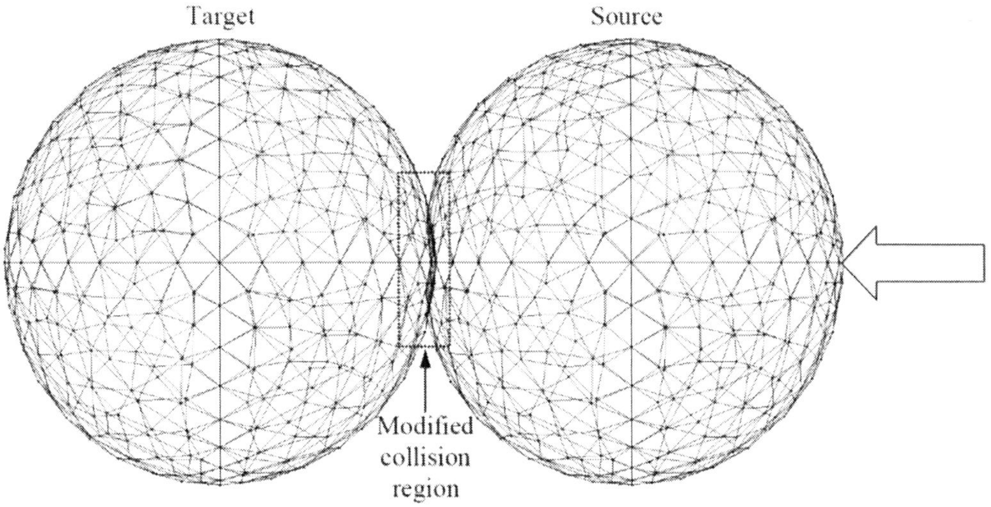

Fig. 4. Modification of the surface of the source after collision

The deformation of the source during contact progresses gradually step by step. It is important that the displacement of the source at each time step is less than the thickness of the surface layer of the elements in the mesh of the target. The algorithm during each time step can be expressed as follows:

1. Translation of the source
2. Detection of the collision nodes
3. Calculation of the boundary conditions in the collision region
4. Calculation of the nodal displacements of the source
5. Modification of the nodal coordinates
6. Return to point 1

4 Example of calculation

An original computer program using Visual C++ 2008 has been written. Finite element meshes were prepared with the use of mesh generator GMSH 2.9. A system containing two balls of the same diameters D=0.1 m is analyzed. The initial configuration of spatially discretized volumes is presented in Fig. 5. Each ball has an independent mesh built of tetrahedrons of average size h_e=0.003 m. First mesh contains 69747 elements with 12575 nodes, the second one has 67826 elements with 12251 nodes. Ball 1 is divided into finite elements despite the assumption of non-deformability because the algorithm for collision detection requires spatial discretization of each object in the system. Ball 2 is made of rubber with Young modulus E=0.05 GPa and Poisson's ratio ν=0.49. Ball 2 moves along the x axis at a speed V_x=1 mm/s. Time step Δt=1 s, total time of the analysis t_{total}=25 s. On the right half of the surface of the ball 2, displacement boundary conditions are defined u_y=u_z=0 m. Displacement u_x=1 mm is constant and cumulative due to modification of nodal coordinates at each step of analysis. This means that after 25 seconds the nodes on the right half of the ball 2 move along the x axis by 25 mm.

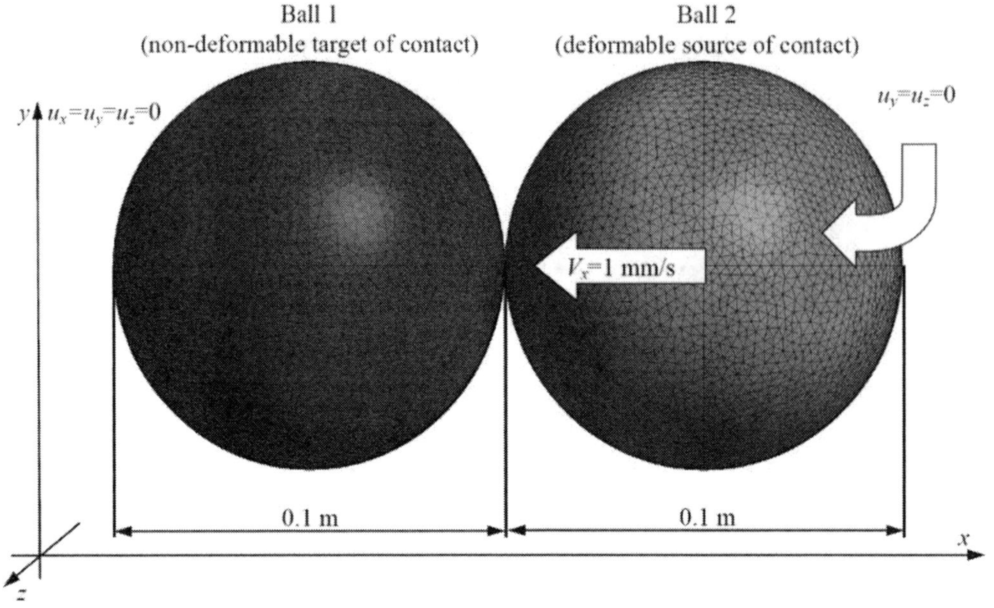

Fig. 5. Initial configuration of the system

The behaviour of balls in contact is presented in Fig. 6a-d. Right half of balls 2 remains unchanged due to assigned boundary conditions while the left one is subjected to large deformations.

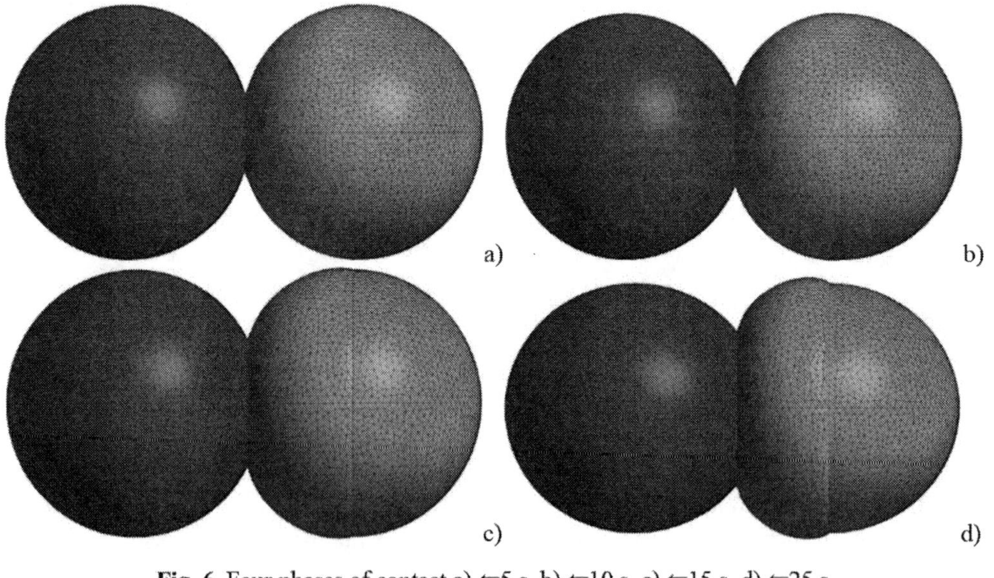

Fig. 6. Four phases of contact a) t=5 s, b) t=10 s, c) t=15 s, d) t=25 s

Fig 7a-d show the stages of deformation of ball 2. It is clearly visible how the mesh of ball 2 fits the shape of ball 1. The gradual increase in displacement allows for accurate modeling of even large deformations.

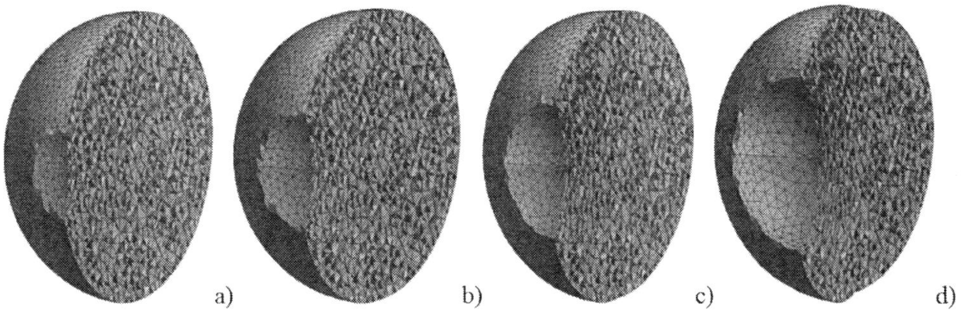

Fig. 7. Cross section through ball 2 in the *xy* plane during successive contact phases a) t=5 s, b) t=10 s, c) t=15 s, d) t=25 s

Conclusions

The numerical method of contact modeling in three dimensional space was presented. Described method of collision detection and modification of finite element meshes during the contact is straightforward and efficient. Presented approach may be useful in simulation of various processes occurring in the systems containing multiple elements. For example, the solidification of the metal in the mold is a process where two regions are in mechanical and also thermal contact. Further work will be focused on the expansion of the numerical model.

References

1. H. Hertz, *Über die berührung fester elastischer Körper*, Journal für die reine und angewandte Mathematik **92**, 156-171 (1881)
2. K. L. Johnson, K. Kendall, A. D. Roberts, *Surface energy and the contact of elastic solids*, Proc. R. Soc. Lond. A. **324**, 301-313 (1971)
3. B.V. Derjaguin, V.M. Muller, Y.P. Toporov, *Effect of contact deformations on the adhesion of particles*, J. Colloid Interface Sci. **53** (2), 314-326 (1975)
4. F.P. Bowden, D. Tabor, *The area of contact between stationary and between moving surfaces*, Proc. R. Soc. Lond. A. Mathematical and Physical Sciences **169** (938), 391-413 (1939)
5. J.F. Archard, *Elastic deformation and the laws of friction*, Proc. R. Soc. Lond. A. Mathematical and Physical Sciences **243** (1233), 190-205 (1957)
6. L. Sowa, Z. Saternus, M. Kubiak, *Numerical modelling of mechanical phenomena in the gantry crane beam* Procedia Eng. **177**, 225-232 (2017)
7. A. Bokota, T. Domański, *Modelling and numerical analysis of hardening phenomena of tools steel elements*, Arch Metall Mater. **54** (3), 575-587 (2009)
8. W. Piekarska, M. Kubiak, Z. Saternus, S. Stano, T. Domański, *Numerical prediction of deformations in laser welded sheets made of X5CrNi18-10 steel*, Arch Metall Mater. **60** (3), 1965-1972 (2015)
9. O.C. Zienkiewicz, *Metoda elementów skończonych*. (Arkady, Warszawa, 1972)
10. H. Grandin, *Fundamentals of the Finite Element Method*. (Waveland Press, Long Grove, 1991)

Numerical modeling of heat flow between three-dimensional regions in contact problems

Tomasz Skrzypczak[1,*], *Ewa* Węgrzyn-Skrzypczak[2]

[1]Czestochowa University of Technology, Faculty of Mechanical Engineering and Computer Science, Institute of Mechanics and Machine Design Fundamentals, Dąbrowskiego 73, 42-201 Częstochowa, Poland
[2]Czestochowa University of Technology, Faculty of Mechanical Engineering and Computer Science, Institute of Mathematics, Armii Krajowej 21, 42-201 Częstochowa, Poland

Abstract. The method of numerical modeling of heat transfer between three-dimensional objects being in contact is described in the paper. Presented approach is based on the finite element method (FEM) with independent spatial discretization of considered regions. The gap between external surfaces of the interacting objects has variable width and is filled with gas or liquid. The medium in the gap introduces thermal resistance into heat transfer process. The mathematical model of considered problem is based on the equation of heat diffusion supplemented by the appropriate initial and boundary conditions. The deformations of the regions resulting from the thermally dependent changes of their volumes are also included in the model. The results of numerical simulations are presented and discussed.

Keywords: computational mechanics, contact problem, heat transport, finite element method, multi-mesh approach

1 Introduction

Machine parts during the work come into thermal and mechanical contact. Mechanical contact is the result of the interactions of their external boundaries. These interactions may be caused by motion or thermal deformations. The ideal thermal contact occurs when the width of the gap equals zero. If the gap width is greater than zero the contact with thermal resistance is observed. The mathematical model of the heat transfer process is based on the heat conductivity equation with appropriate boundary and initial conditions. A fourth type of boundary condition plays very important role in the mathematical model of such process. One can distinguish two kinds of such mathematical condition describing the ideal contact or the contact with the presence of the gap of variable width, filled with gaseous or liquid medium.

Mathematical descriptions of the non-ideal thermal contact was described for the first time in [1]. Analytical solutions of the problem were presented in [2]. Numerical solution of heat transport using finite difference method (FDM) was shown in [3]. Nowadays many scientists from different branches of science investigate the process of the heat transport

* Corresponding author: t.skrzypczak@imipkm.pcz.pl

Reviewers: *Krzysztof* Talaśka, *Milan* Žmindák

through the gap. For example the mathematical description of the considered phenomenon is required in the numerical analysis of the solidification in the mold [4, 5]. Currently many advanced numerical methods such as the FEM are used to model complex mechanical phenomena [6-9]. Numerical investigations based on the FEM in the case of air gap formation process during solidification of pure metal was discussed in [10].

Presented mathematical model also contains the description of thermal deformations of considered regions which are needed to calculation of the local width of the gap.

2 Mathematical model

Fig.1 presents two objects Ω_1 and Ω_2 in contact. Boundaries Γ_1, Γ_2 lie close together forming a gap of width h. The width of the gap depends on the coordinates x, y, z and also may change according to time. At the boundaries Γ_5, Γ_6 temperature is defined while Γ_3, Γ_4 are fixed. The process of heat transfer in Ω_1 and Ω_2 is transient and the objects are subjected to the thermal deformations.

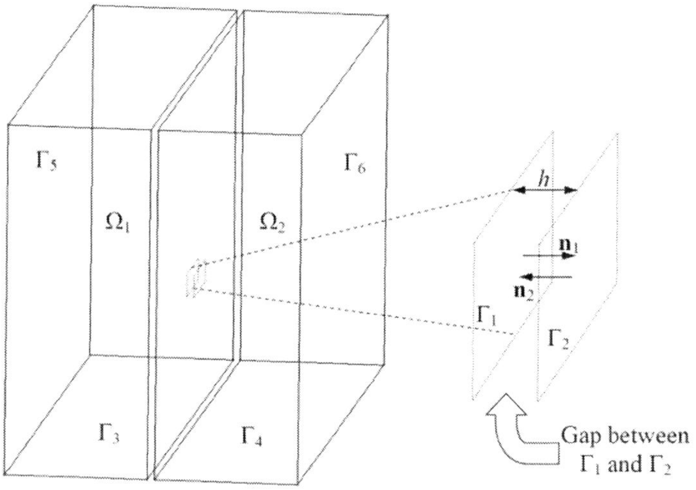

Fig. 1. Scheme of the problem

The governing equations of presented model are as follows:

- the heat conduction equation describing transient heat transfer in the three-dimensional volume

$$\frac{\partial}{\partial x}\left(\lambda_{(i)}\frac{\partial T^{(i)}}{\partial x}\right)+\frac{\partial}{\partial y}\left(\lambda_{(i)}\frac{\partial T^{(i)}}{\partial y}\right)+\frac{\partial}{\partial z}\left(\lambda_{(i)}\frac{\partial T^{(i)}}{\partial z}\right)=(c\rho)_{(i)}\frac{\partial T^{(i)}}{\partial t} \tag{1}$$

- the equilibrium equations

$$\begin{aligned}\frac{\partial \sigma_x^{(i)}}{\partial x}+\frac{\partial \tau_{xy}^{(i)}}{\partial y}+\frac{\partial \tau_{xz}^{(i)}}{\partial z}=0\\ \frac{\partial \tau_{xy}^{(i)}}{\partial x}+\frac{\partial \sigma_y^{(i)}}{\partial y}+\frac{\partial \tau_{yz}^{(i)}}{\partial z}=0\\ \frac{\partial \tau_{xz}^{(i)}}{\partial x}+\frac{\partial \tau_{yz}^{(i)}}{\partial y}+\frac{\partial \sigma_z^{(i)}}{\partial z}=0\end{aligned} \tag{2}$$

where λ is the coefficient of thermal conductivity, $c\rho$ - volumetric heat capacity, T - temperature, σ_x, σ_y, σ_z, τ_{xy}, τ_{xz}, τ_{yz} - the components of symmetrical stress tensor, t – time, x, y, z – Cartesian coordinates, i - index denoting volume 1 or 2.

Equation (1) is supplemented by the following boundary and initial conditions:

- heat flux at the contact boundaries

$$-\lambda_1 \frac{\partial T^{(1)}}{\partial n_1} = \frac{\lambda_g}{h}\left(T_b^{(1)} - T_b^{(2)}\right) = \lambda_2 \frac{\partial T^{(2)}}{\partial n_2} \tag{3}$$

- known temperature at the boundaries Γ_5, Γ_6

$$T^{(1)}\Big|_{\Gamma_5} = T^{(1)}(x,y,z), \quad T^{(2)}\Big|_{\Gamma_6} = T^{(2)}(x,y,z) \tag{4}$$

- initial temperature in the volumes 1 and 2

$$T^{(1)}(x,y,z,0) = T_0^{(1)}, \quad T^{(2)}(x,y,z,0) = T_0^{(2)} \tag{5}$$

where λ_g denotes thermal conductivity of the medium filling the gap, h is the local width of the gap, n_1, n_2 are the local directions of the vectors normal to the boundaries Γ_1, Γ_2, $T_b^{(1)}$, $T_b^{(2)}$ - the local temperatures of the volumes Ω_1, Ω_2 at the contact boundaries, $T_0^{(1)}$, $T_0^{(2)}$ - initial temperature of the volumes Ω_1, Ω_2.

There are 6 unknowns in the equations (2), so they need to be transformed into displacement-dependent functions. This is done by using the following relations:

- relations between the stress and strain tensor components

$$\begin{aligned}
\sigma_x^{(i)} &= \frac{E_{(i)}}{1+\nu_{(i)}}\left[\varepsilon_x^{(i)} + \frac{\nu_{(i)}}{1-2\nu_{(i)}}\left(\varepsilon_x^{(i)} + \varepsilon_y^{(i)} + \varepsilon_z^{(i)}\right) - \frac{1+\nu_{(i)}}{1-2\nu_{(i)}}\alpha_{(i)}\Delta T^{(i)}\right] \\
\sigma_y^{(i)} &= \frac{E_{(i)}}{1+\nu_{(i)}}\left[\varepsilon_y^{(i)} + \frac{\nu_{(i)}}{1-2\nu_{(i)}}\left(\varepsilon_x^{(i)} + \varepsilon_y^{(i)} + \varepsilon_z^{(i)}\right) - \frac{1+\nu_{(i)}}{1-2\nu_{(i)}}\alpha_{(i)}\Delta T^{(i)}\right] \\
\sigma_z^{(i)} &= \frac{E_{(i)}}{1+\nu_{(i)}}\left[\varepsilon_z^{(i)} + \frac{\nu_{(i)}}{1-2\nu_{(i)}}\left(\varepsilon_x^{(i)} + \varepsilon_y^{(i)} + \varepsilon_z^{(i)}\right) - \frac{1+\nu_{(i)}}{1-2\nu_{(i)}}\alpha_{(i)}\Delta T^{(i)}\right] \\
\tau_{xy}^{(i)} &= \frac{E_{(i)}}{2\left(1+\nu_{(i)}\right)}\gamma_{xy}^{(i)} \\
\tau_{yz}^{(i)} &= \frac{E}{2\left(1+\nu_{(i)}\right)}\gamma_{yz}^{(i)} \\
\tau_{xz}^{(i)} &= \frac{E}{2\left(1+\nu_{(i)}\right)}\gamma_{xz}^{(i)}
\end{aligned} \tag{6}$$

- relations between the strain tensor components and the displacements

$$\begin{aligned} \varepsilon_x^{(i)} &= \frac{\partial u_x^{(i)}}{\partial x}, & \gamma_{xy}^{(i)} &= \frac{\partial u_x^{(i)}}{\partial y} + \frac{\partial u_y^{(i)}}{\partial x} \\ \varepsilon_y^{(i)} &= \frac{\partial u_y^{(i)}}{\partial y}, & \gamma_{yz}^{(i)} &= \frac{\partial u_y^{(i)}}{\partial z} + \frac{\partial u_z^{(i)}}{\partial y} \\ \varepsilon_z^{(i)} &= \frac{\partial u_z^{(i)}}{\partial z}, & \gamma_{xz}^{(i)} &= \frac{\partial u_x^{(i)}}{\partial z} + \frac{\partial u_z^{(i)}}{\partial x} \end{aligned} \tag{7}$$

where: ε_x, ε_y, ε_z, γ_{xy}, γ_{xz}, γ_{yz} are the components of symmetrical strain tensor, u_x, u_y, u_z - the components of displacement vector, E - Young's modulus, ν - Poisson's ratio, α - the linear thermal expansion coefficient, ΔT - the difference between the reference and the current temperatures $\Delta T = T - T_{ref}$.

Using relations (7) in equations (6) and then inserting the stress components into equilibrium equations (2), the following equations are obtained:

$$\begin{aligned} & f_1 \frac{\partial^2 u_x^{(i)}}{\partial x^2} + f_2\left(\frac{\partial^2 u_y^{(i)}}{\partial x \partial y} + \frac{\partial^2 u_z^{(i)}}{\partial x \partial z}\right) + f_3\left(\frac{\partial^2 u_x^{(i)}}{\partial y^2} + \frac{\partial^2 u_x^{(i)}}{\partial z^2} + \frac{\partial^2 u_y^{(i)}}{\partial y \partial x} + \frac{\partial^2 u_z^{(i)}}{\partial z \partial x}\right) + f_4 \alpha_{(i)} \Delta T^{(i)} = 0 \\ & f_1 \frac{\partial^2 u_y^{(i)}}{\partial y^2} + f_2\left(\frac{\partial^2 u_x^{(i)}}{\partial y \partial x} + \frac{\partial^2 u_z^{(i)}}{\partial y \partial z}\right) + f_3\left(\frac{\partial^2 u_x^{(i)}}{\partial x \partial y} + \frac{\partial^2 u_y^{(i)}}{\partial x^2} + \frac{\partial^2 u_y^{(i)}}{\partial z^2} + \frac{\partial^2 u_z^{(i)}}{\partial z \partial y}\right) + f_4 \alpha_{(i)} \Delta T^{(i)} = 0 \\ & f_1 \frac{\partial^2 u_z^{(i)}}{\partial z^2} + f_2\left(\frac{\partial^2 u_x^{(i)}}{\partial z \partial x} + \frac{\partial^2 u_y^{(i)}}{\partial z \partial y}\right) + f_3\left(\frac{\partial^2 u_x^{(i)}}{\partial x \partial z} + \frac{\partial^2 u_y^{(i)}}{\partial y \partial z} + \frac{\partial^2 u_z^{(i)}}{\partial x^2} + \frac{\partial^2 u_z^{(i)}}{\partial y^2}\right) + f_4 \alpha_{(i)} \Delta T^{(i)} = 0 \end{aligned} \tag{8}$$

where f_1, f_2, f_3, f_4 are defined in the following way:

$$\begin{aligned} f_1 &= \frac{E_{(i)}\left(1-\nu_{(i)}\right)}{\left(1+\nu_{(i)}\right)\left(1-2\nu_{(i)}\right)}, & f_2 &= \frac{E_{(i)}\nu_{(i)}}{\left(1+\nu_{(i)}\right)\left(1-2\nu_{(i)}\right)} \\ f_3 &= \frac{E_{(i)}}{2\left(1+\nu_{(i)}\right)}, & f_4 &= \frac{E_{(i)}\left(1+\nu_{(i)}\right)}{\left(1+\nu_{(i)}\right)\left(1-2\nu_{(i)}\right)} \end{aligned} \tag{9}$$

Equations (8) are supplemented by the following boundary conditions:

$$u_x^{(1)}\Big|_{\Gamma_3} = u_y^{(1)}\Big|_{\Gamma_3} = u_z^{(1)}\Big|_{\Gamma_3} = 0, \quad u_x^{(2)}\Big|_{\Gamma_4} = u_y^{(2)}\Big|_{\Gamma_4} = u_z^{(2)}\Big|_{\Gamma_4} = 0 \tag{10}$$

3 Numerical model

The numerical model of presented problem is based on the finite element method [11] and derived from the criterion of the method of weighted residuals [12]. Equations (1) and (8) are multiplied by the weighting function w and integrated over the volume Ω_i . Equations (1) and (8) written in the weak form are shown below:

$$\int_{\Omega_i} \lambda_{(i)}\left(\frac{\partial w}{\partial x}\frac{\partial T^{(i)}}{\partial x} + \frac{\partial w}{\partial y}\frac{\partial T^{(i)}}{\partial y} + \frac{\partial w}{\partial z}\frac{\partial T^{(i)}}{\partial z}\right) d\Omega_i + \int_{\Omega_i} (c\rho)_{(i)} w \frac{\partial T^{(i)}}{\partial t} d\Omega_i = -\oint_{\Gamma_i} w q d\Gamma_i \tag{11}$$

$$\int_{\Omega_i}\left(f_1\frac{\partial w}{\partial x}\frac{\partial u_x^{(i)}}{\partial x}+f_3\frac{\partial w}{\partial y}\frac{\partial u_x^{(i)}}{\partial y}+f_3\frac{\partial w}{\partial z}\frac{\partial u_x^{(i)}}{\partial z}\right)d\Omega_i+\int_{\Omega_i}\left(f_2\frac{\partial w}{\partial x}\frac{\partial u_y^{(i)}}{\partial y}+f_3\frac{\partial w}{\partial y}\frac{\partial u_y^{(i)}}{\partial x}\right)d\Omega_i+$$
$$+\int_{\Omega_i}\left(f_2\frac{\partial w}{\partial x}\frac{\partial u_z^{(i)}}{\partial z}+f_3\frac{\partial w}{\partial z}\frac{\partial u_z^{(i)}}{\partial x}\right)d\Omega_i=\oint_{\Gamma_i}f_4\alpha_{(i)}\frac{\partial w}{\partial x}\Delta T^{(i)}d\Gamma_i \quad (12)$$

$$\int_{\Omega_i}\left(f_2\frac{\partial w}{\partial y}\frac{\partial u_x^{(i)}}{\partial x}+f_3\frac{\partial w}{\partial x}\frac{\partial u_x^{(i)}}{\partial y}\right)d\Omega_i+\int_{\Omega_i}\left(f_3\frac{\partial w}{\partial x}\frac{\partial u_y^{(i)}}{\partial x}+f_1\frac{\partial w}{\partial y}\frac{\partial u_y^{(i)}}{\partial y}+f_3\frac{\partial w}{\partial z}\frac{\partial u_y^{(i)}}{\partial z}\right)d\Omega_i+$$
$$+\int_{\Omega_i}\left(f_2\frac{\partial w}{\partial y}\frac{\partial u_z^{(i)}}{\partial z}+f_3\frac{\partial w}{\partial z}\frac{\partial u_z^{(i)}}{\partial y}\right)d\Omega_i=\oint_{\Gamma_i}f_4\alpha_{(i)}\frac{\partial w}{\partial y}\Delta T^{(i)}d\Gamma_i \quad (13)$$

$$\int_{\Omega_i}\left(f_2\frac{\partial w}{\partial z}\frac{\partial u_x^{(i)}}{\partial x}+f_3\frac{\partial w}{\partial x}\frac{\partial u_x^{(i)}}{\partial z}\right)d\Omega_i+\int_{\Omega_i}\left(f_2\frac{\partial w}{\partial z}\frac{\partial u_y^{(i)}}{\partial y}+f_3\frac{\partial w}{\partial y}\frac{\partial u_y^{(i)}}{\partial z}\right)d\Omega_i+$$
$$+\int_{\Omega_i}\left(f_3\frac{\partial w}{\partial x}\frac{\partial u_z^{(i)}}{\partial x}+f_3\frac{\partial w}{\partial y}\frac{\partial u_z^{(i)}}{\partial y}+f_1\frac{\partial w}{\partial z}\frac{\partial u_z^{(i)}}{\partial z}\right)d\Omega_i=\oint_{\Gamma_i}f_4\alpha_{(i)}\frac{\partial w}{\partial z}\Delta T^{(i)}d\Gamma_i \quad (14)$$

According to the Galerkin method the weighting function w is the same as the shape functions of the finite elements used during meshing operation. Time discretization of equation (11) is based on the implicit Euler method. Final form of the global FEM equations is presented below:

$$\left(\mathbf{K}_T+\frac{1}{\Delta t}\mathbf{M}_T\right)\mathbf{T}^{f+1}=\mathbf{B}_T+\frac{1}{\Delta t}\mathbf{M}_T\mathbf{T}^f \quad (15)$$

$$\mathbf{K}_D\mathbf{u}=\mathbf{B}_D \quad (16)$$

where $\mathbf{K}_T$ is the global thermal conductivity matrix, $\mathbf{M}_T$ - the global thermal capacity matrix, $\mathbf{B}_T$ - global vector associated with the thermal boundary conditions, $\mathbf{T}$ - vector of unknown nodal temperatures, f - time level, $\mathbf{K}_D$ - the global stiffness matrix, $\mathbf{B}_D$ - vector where known boundary displacements are stored, $\mathbf{u}$ - vector of unknown nodal displacements.

The solution of equation (15) is obtained at every step through an iterative process, independently in each region. Calculated temperatures are then used to find thermal deformations of each object. It is crucial to properly incorporate the condition (3) to the heat transport solution process [13]. During the deformations of Ω_1 and Ω_2 relative positions of objects and the width of the gap between them are changing. Because the global set of equations is built and solved independently for each region the presented approach is based on the disconnected finite element meshes.

4 Example of calculation

Two cuboids of the same dimensions 0.02x0.1x0.1 m are considered (Fig. 2). Object on the left is made of steel and its initial temperature T_0=600 K. Object on the right is initially colder, its temperature T_0=300 K and the material from which its made is copper. Both cuboids are fixed at their bottoms, $u_x = u_y = u_z = 0$ m. Between them is a gap of initial width h=0.1 mm and thermal conductivity λ_g=0.5 J·s^{-1}·m^{-1}·K^{-1}.

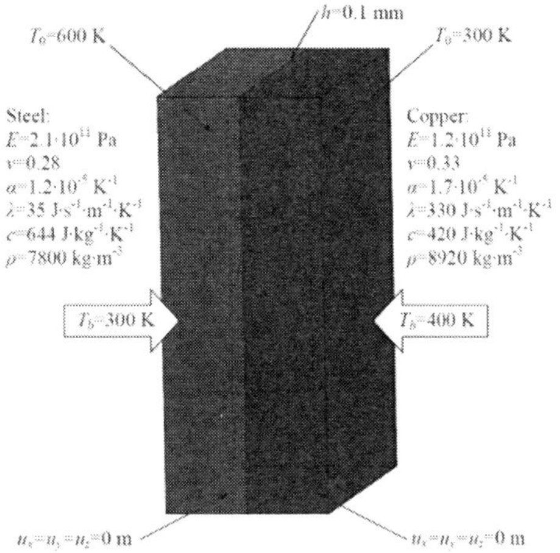

Fig. 2. Material properties, boundary and initial conditions used in calculations

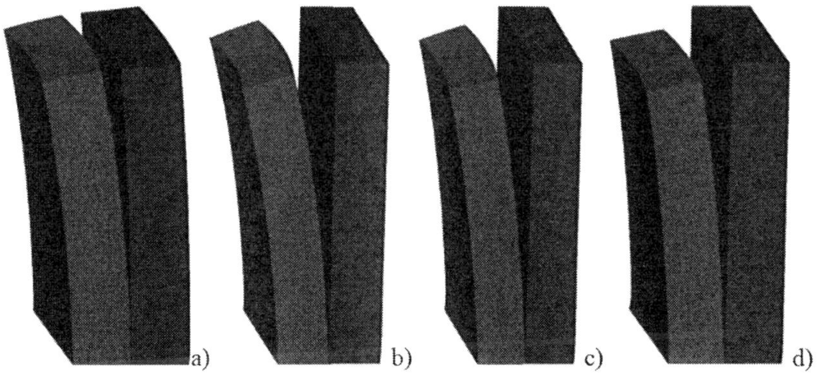

Fig. 3. Thermal deformations of the cuboids (*def_factor*=20) a) t=1 s, b) t=5 s, c) t=10 s, d) t=17 s

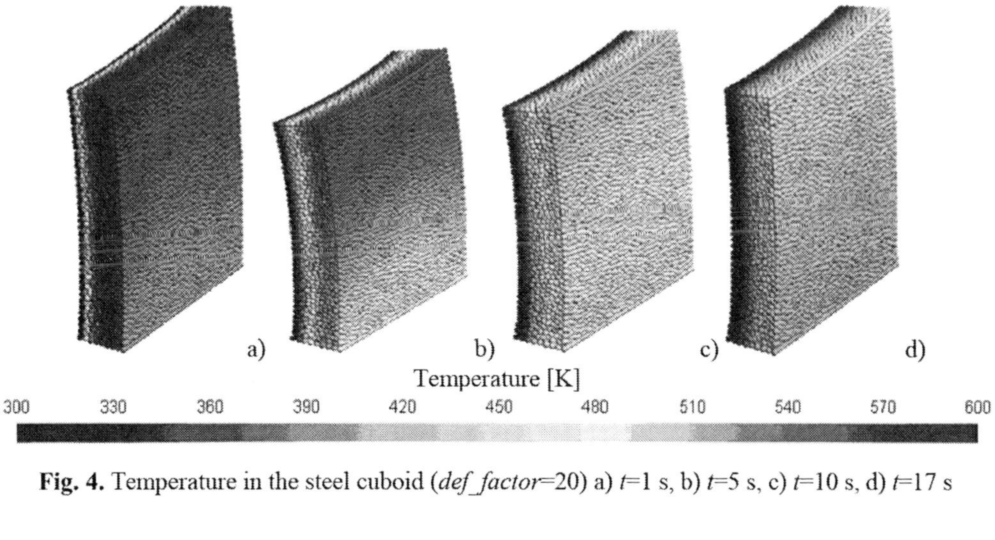

Fig. 4. Temperature in the steel cuboid (*def_factor*=20) a) t=1 s, b) t=5 s, c) t=10 s, d) t=17 s

According to the boundary and initial conditions heat transfer between the objects and their thermal deformations are simulated with the use of an original computer program written in C++. Finite element meshes are made with the use of mesh generator GMSH 2.9. Left cuboid is divided into 49498 tetrahedrons with 11468 nodes, the right one has 50315 tetrahedral elements and 11602 nodes. Time step used in the calculation process $\Delta t = 0.1$ s, total time of the analysis $t_{total} = 17$ s.

Fig. 3a-d show finite element meshes deformed due to temperature after 1, 5, 10, 17 s. The gap is widest in the upper part of the cuboids, it reaches the highest value 0f 0.7 mm after 5 s (Fig3.b). Fig. 4a-d show instantaneous temperature distributions in the steel cuboid. The impact of the local gap width on the temperature distribution on the wall in contact with the copper cuboid is evident. In both figures the deformations are exaggerated (*def_factor* = 20).

Conclusions

The numerical method of heat transfer during the contact of two three-dimensional objects taking into account thermal deformations was presented. The method using the independent spatial discretization for each volume saves a lot of operational memory because of sequential solving of the sets of equations. Obtained results proves significant impact of the variable width of the gap on the global heat transfer. Presented approach may be useful in computer simulations of various processes such as the solidification of the molten metal in the permanent mold. Further work will be focused on the expansion of the numerical model by the mechanical interaction between the objects.

References

1. M.G. Cooper, B.B. Mikic, M.M. Yovanovich, *Thermal contact conductance*. Int. J. Heat Mass Transf. **12** (3), 279-300 (1969)
2. W. Longa, *Solidification of the castings in sand molds*. (Slask, Katowice, 1973)
3. G.E. Schneider, A.B. Strong, M.M Yovanovich, *Transient thermal response of two bodies communicating through a small circular contact area*. Int. J. Heat Mass Transf. **20** (4), 301-308 (1977)
4. E. Majchrzak, J. Mendakiewicz, A. Piasecka-Belkhayat, *Algorithm of the mould thermal parameters identification in the system casting-mould-environment*. J. Mater. Process. Technol. **164**, 1544-1549 (2005)
5. R. Dyja, E. Gawrońska, N. Sczygiol, *The effect of mechanical interactions between the casting and the mold on the conditions of heat dissipation: a numerical model*. Arch. Metall. Mater. **60** (3A), 1901-1909 (2015)
6. L. Sowa, P. Kwiaton, *Numerical analysis of stress fields generated in the gantry crane beam*. Procedia Eng. **177**, 218-224 (2017)
7. L. Sowa, *Effect of steel flow control devices on flow and temperature field in the tundish of continuous casting machine*. Arch. Metall. Mater. **60** (2A), 843-847 (2015)
8. T. Domański, A. Sapietova, M. Saga, *Application of Abaqus software for the modeling of surface progressive hardening*. Procedia Eng. **177**, 64-69 (2017)
9. M. Kubiak, W. Piekarska, S. Stano, Z. Saternus, *Numerical modelling of thermal and structural phenomena in Yb:YAG laser butt-welded steel elements*. Arch. Metall. Mater. **60** (2), 821-828 (2015)

10. J. Matlak, D. Słota, *Solution of the pure metals solidification problem by involving the material shrinkage and the air-gap between material and mold*. Arch. Foundry Eng. **15** (Spec. Iss. 3), 47-52 (2015)
11. O.C. Zienkiewicz, *Finite element method*. (Arkady, Warsaw, 1972)
12. H. Grandin, *Fundamentals of the Finite Element Method*. (Waveland Press, Long Grove, 1991)
13. T. Skrzypczak, E. Węgrzyn-Skrzypczak, *Modeling of thermal contact through gap with the use of finite element method*. J. Appl. Math. Comput. Mech. **14** (4), 145–152 (2015)

Numerical solution of Rayleigh-Lamb frequency equation for real, imaginary and complex wavenumbers

Michal Šofer[1,*], *Petr* Ferfecki[1,2], *Pavel* Šofer[3]

[1]Department of Applied Mechanics, Faculty of Mechanical Engineering, VŠB-Technical University of Ostrava, 17. listopadu 15/2127, 708 33 Ostrava-Poruba, Czech Republic
[2]IT4Innovations National Supercomputing Center, VŠB-Technical University of Ostrava, Studentská 6231/1B, 708 33 Ostrava-Poruba, Czech Republic
[3]Department of Control Systems and Instrumentation, Faculty of Mechanical Engineering, VŠB-Technical University of Ostrava, 17. listopadu 15/2127, 708 33 Ostrava-Poruba, Czech Republic

Abstract. Guided waves, especially Lamb waves or shear-horizontal waves, are widely used types of waves for ultrasonic inspection of large structures. Well known property of guided waves is their dispersive character, which means that the propagation velocity of the particular wave mode is not only a function of physical properties of the material, in which the wave propagates or the wave´s frequency, but also depends on the geometry of the structure in itself. Dispersion curves provide us the information related to the dependency between the wavenumber and the frequency of the particular mode and can be obtained by a numerical solution of Rayleigh-Lamb frequency equation. A solution of Rayleigh-Lamb frequency equation forms for a given frequency and plate thickness a set of a finite number of real and pure imaginary wavenumbers and an infinite number of complex wavenumbers. Proposed paper presents a complete procedure of how to obtain all three kinds of wavenumbers for a given geometry and frequency interval. The main emphasis is placed on the effectiveness of the procedures, which are used for finding the roots of dispersion equation for all three kinds of wavenumbers.

Keywords: Rayleigh-Lamb equation, Dispersion equation, Lamb wave

1 Introduction

Ultrasonic testing is, besides eddy current testing method, one of most commonly used non-destructive techniques in the case of inspection of plate-like structures [1]. Guided waves, especially Lamb waves, are ideal solution for inspection of large structures such as metal sheets or composite structures [2]. The ability of the effective crack detection by use of various modes of Lamb wave has been demonstrated in several publications [2-4]. The structure of the particular mode of guided Lamb wave in itself is relatively complex compared to the longitudinal or transversal bulk waves. This fact appears to be an advantage, since it is possible to select a particular mode shape in order to increase the

* Corresponding author: michal.sofer@vsb.cz

Reviewers: *Jerzy* Jaroszewicz, *Branislav* Ftorek

sensitivity of such mode to specific types of flaws [3]. This advantage, is however, directly related with increased complexity of the results interpretation. In most cases, one cannot perform further analyses without the numerical predictions, since the more modes are present in the structure, the more information about the flaw is available with rapid increase of the interpretation difficultness at the same time.

The main feature of guided Lamb waves is their dispersive character, defined by dispersion curves, which describe us the dependency between the wavenumber and the frequency of particular wave mode for given geometry. The dispersion curves in case of single or multi-layered media can be obtained numerically and, as well as in particular frequency range, also experimentally [3-6]. Various methods of numerical approaches for dispersion curves computations has been published so far [3,7]. The main scope of the article is to present fast and robust procedure for calculation a set of real, imaginary and complex wavenumbers, which are the roots of dispersion equation, for wide range of frequency thickness product.

2 Theory of Lamb Waves

The governing equation will be derived under plane strain condition with the assumption of traction-free surfaces (Fig. 1).

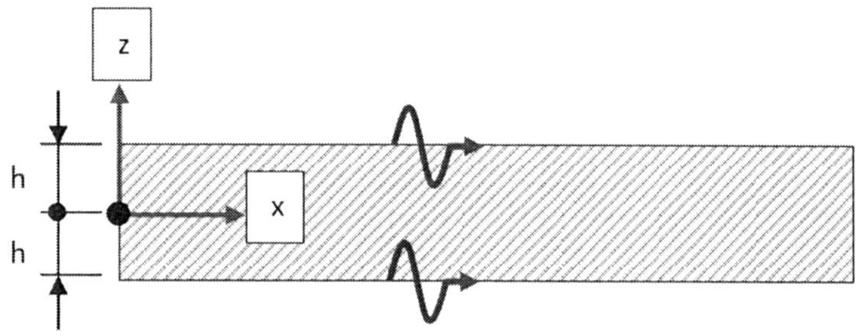

Fig. 1. Geometry of the free plate

The unknown displacement vector **u** will be defined with the use of the Helmholtz decomposition theorem [8,9]:

$$\boldsymbol{u} = \nabla\phi + \nabla \times \boldsymbol{\psi} \tag{1}$$

with

$$\nabla\boldsymbol{\psi} = 0, \tag{2}$$

where x and z are spatial coordinates, $\phi = \phi(x,z)$ and $\boldsymbol{\psi} = (0, -\psi(x,z), 0)$ is potential and vector function respectively and ∇ is differential operator. The equation of motion can be expressed in terms of displacements, resulting in following form:

$$\mu\, \nabla^2 \mathbf{u} + (\lambda + \mu)\, \nabla\, (\mathrm{div}\, \mathbf{u}) = \rho\, \ddot{\mathbf{u}}, \tag{3}$$

where λ and μ denotes Lame constants, ρ is density of the material and $\ddot{\mathbf{u}}$ denotes second time t derivate of displacement vector **u**. Substitution of Eq. 1 into Eq. 3 will result in two independent governing equations for longitudinal and transversal waves:

$$\nabla^2 \phi - \frac{1}{c_L^2}\, \ddot{\phi} = 0, \tag{4}$$

$$\nabla^2 \psi - \frac{1}{c_T^2} \ddot{\psi} = 0, \tag{5}$$

where c_L is the speed of longitudinal (dilatational) waves, c_T is the speed of transversal (shear) waves. The solutions of Eqs. 4 and 5 can be assumed in following forms:

$$\phi = \Lambda(z)\, \mathrm{e}^{\mathrm{i}(kx-\omega t)}, \psi = \Omega(z)\, \mathrm{e}^{\mathrm{i}(kx-\omega t)}, \tag{6}$$

where k represents wavenumber, ω denotes angular frequency, functions $\Lambda(z)$ and $\Omega(z)$ represent standing waves in z direction. Functions ϕ and ψ (Eq. 6) have to be further substituted into Eq. 4 and 5 in order to obtain final solutions for $\Lambda(z)$ and $\Omega(z)$:

$$\Lambda(z) = \theta_1 \sin(K_L z) + \theta_2 \cos(K_L z), \tag{7}$$

$$\Omega(z) = \delta_1 \sin(K_T z) + \delta_2 \cos(K_T z), \tag{8}$$

where

$$K_L^2 = k_L^2 - k^2 \text{ and } K_T^2 = k_T^2 - k^2. \tag{9}$$

θ_1, θ_2, δ_1, δ_2 denotes constants dependent of boundary conditions, K_L and K_T represents complex components along z axis, $k_L = \omega / c_L$ and $k_T = \omega / c_T$ represents wavenumbers of longitudinal and transversal wave and k represents wavenumber along direction of wave propagation. The displacements and stresses can be with help of Lame equations, Eq. 1 and Eqs. 6-9 expressed in following form:

$$\begin{aligned} u_x &= \left[\frac{\mathrm{d}\Omega}{\mathrm{d}z} + \mathrm{i}k\Lambda\right] \mathrm{e}^{\mathrm{i}(kx-\omega t)}, \\ v_y &= 0, \\ w_z &= \left[\frac{\mathrm{d}\Lambda}{\mathrm{d}z} + \mathrm{i}k\Omega\right] \mathrm{e}^{\mathrm{i}(kx-\omega t)}, \end{aligned} \tag{10}$$

$$\begin{aligned} \sigma_{xx} &= \left[\lambda\left(\frac{\mathrm{d}^2\Lambda}{\mathrm{d}z^2} - \Lambda k^2\right) - 2\mu(k^2\Omega - \mathrm{i}k\frac{\mathrm{d}\Omega}{\mathrm{d}z})\right] \mathrm{e}^{\mathrm{i}(kx-\omega t)}, \\ \sigma_{xz} &= \mu\left(k^2\Omega + \frac{\mathrm{d}^2\Omega}{\mathrm{d}z^2} + 2\mathrm{i}k\frac{\mathrm{d}\Lambda}{\mathrm{d}z}\right) \mathrm{e}^{\mathrm{i}(kx-\omega t)}, \\ \sigma_{zz} &= \left[\lambda\left(\frac{\mathrm{d}^2\Lambda}{\mathrm{d}z^2} - \Lambda k^2\right) + 2\mu(\frac{\mathrm{d}^2\Lambda}{\mathrm{d}z^2} - \mathrm{i}k\frac{\mathrm{d}\Omega}{\mathrm{d}z})\right] e^{\mathrm{i}(kx-\omega t)}. \end{aligned} \tag{11}$$

Finally, Eqs. 10 and 11 can be split into two sets of modes according to odd or even functions about $z = 0$ axis [3].

The dispersion equation, or in other words Rayleigh-Lamb frequency equation, for both modes can be determined by applying the traction-free boundary condition $\sigma_{zz} = \sigma_{xz} = 0$ for $z = \pm h$ (See Fig. 1), which results in:

$$\frac{\tan(K_T h)}{\tan(K_L h)} - \left[\frac{4k^2 K_T K_L}{\left(K_T^2 - K_L^2\right)^2}\right]^{\pm 1} = 0, \tag{12}$$

where +1 applies for modes of symmetric Lamb wave and -1 applies for modes of antisymmetric Lamb wave.

3 Solution of Rayleigh-Lamb Frequency Equation

As mentioned in chapter 1, the solution of Eq. 12 for given frequency will produce a set of finite number of propagating modes with real wavenumber, finite number of non-propagating modes with imaginary wavenumber and an infinite number of inhomogeneous modes with complex wavenumber [1]. In the following chapters are discussed individual approaches of solving the Rayleigh-Lamb frequency equation for individual types of wavenumbers. It has to be noted, that the entire code has been written in Matlab® software.

3.1 Numerical solution for real wavenumbers

For solving the real wavenumbers, it is preferable to rewrite Eq. 12 in following form [3]:

$$\frac{\tan(K_T h)}{K_T} + \frac{4k^2 K_L \tan(K_L h)}{\left(K_T^2 - K_L^2\right)^2} = 0, \tag{13}$$

for symmetric modes and

$$K_T \tan(K_T h) + \frac{\left(K_T^2 - K_L^2\right)^2 \tan(K_L h)}{4k^2 K_L} = 0 \tag{14}$$

for antisymmetric modes. The procedure of finding the real roots of the dispersion equation is the same for both modes (Eqs. 13 and 14). The robustness of the procedure is based on the vectorization, which is incorporated in Matlab® software. The wavenumber vector, through which were computed the values of the left-hand side of dispersion equations ranged from zero to k_{max}, where:

$$k_{max} = 1.2\omega_{max} c_T^{-1}, \tag{15}$$

where ω_{max} is maximum angular frequency within the computed frequency interval. The incremental step of wavenumber interval is equal to:

$$\Delta k_{step} = \frac{k_{max}}{1\cdot 10^3 \div 2\cdot 10^5} \ [\mathrm{m}^{-1}], \tag{16}$$

the larger incremental step is suitable in case of lower values of frequency·thickness product (approximately up to 3 MHz·mm), which delivers sufficient resolution for finding the roots. For higher values of frequency·thickness product, it is necessary to increase the resolution of the wavenumber vector for proper root localization. The roots of the dispersion equation for given frequency are located at the zero crossings when the left-hand side changes its functional value from positive to negative sign. – see Fig 2.

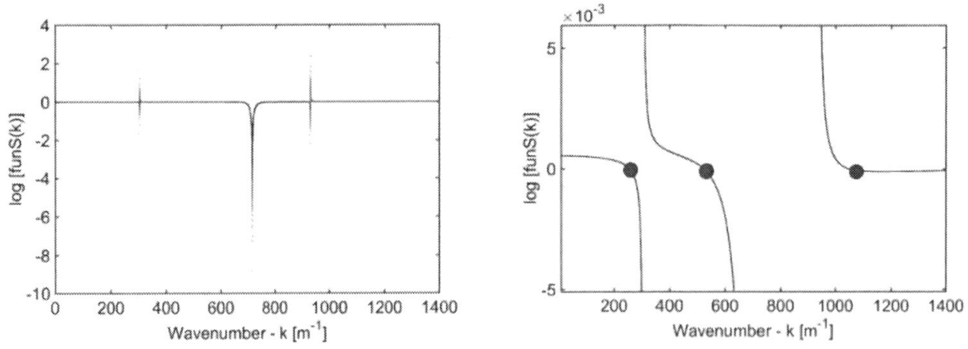

Fig. 2. Functional values of the left-hand side of dispersion relation for given wavenumber interval (left graph) and the detail of roots (right graph) for aluminium plate of c_L = 6300 m/sec, c_T = 3100 m/sec and frequency·thickness = 4 MHz·mm

Resulting dispersion curves in terms of phase velocity as a function of frequency and wavenumber as a function of frequency respectively are displayed in Fig. 3. It has to be noted, that the dispersion curves were constructed for aluminium plate of c_L = 6300 m/sec, c_T = 3100 m/sec and the thickness of the plate was equal to 8 mm. The frequency interval has been divided into 200 steps with a solving time equal to 5.5 seconds (CPU i7 3.4 GHz, 32 GB RAM). For proper estimation of the cut-off frequencies, the number of frequency interval division has to be refined.

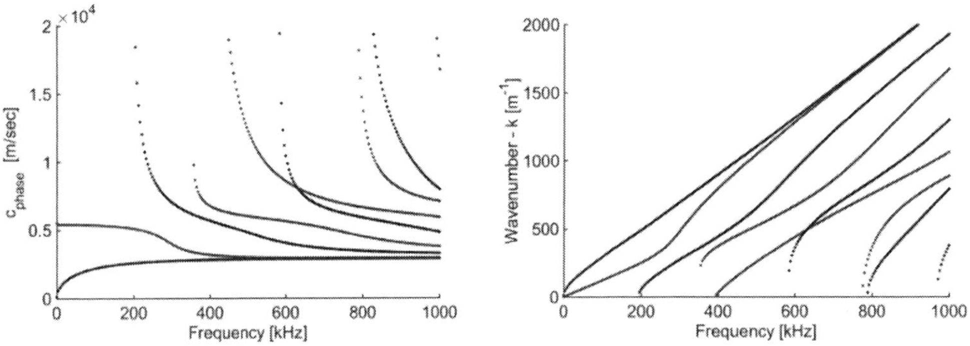

Fig. 3. Dispersion curves in terms of c_{phase} as a function of frequency (left) and wavenumber as a function of frequency (right). The symmetric Lamb modes are represented by red dots, the antisymmetric modes by blue dots

3.2 Numerical solution for imaginary wavenumbers

The pure imaginary wavenumbers, which represent evanescent waves, are supposed to be in following form:

$$k = \mathrm{i}k_{im}. \tag{17}$$

Eqs. 13 and 14 can be therefore rewritten as follows:

$$\frac{\tan(K_{T\,im}\,h)}{K_{T\,im}} - \frac{4k_{im}^2\,K_{L\,im}\tan(K_{L\,im}\,h)}{\left(K_{T\,im}^2+K_{L\,im}^2\right)^2} = 0, \tag{18}$$

for symmetric modes and

$$K_{T\,im}\tan(K_{T\,im}h)-\frac{\left(K_{T\,im}^2+K_{L\,im}^2\right)^2\tan(K_{L\,im}h)}{4k_{im}^2K_{L\,im}}=0\,, \tag{19}$$

for antisymmetric modes, where:

$$K_{L\,im}^2=k_L^2+k_{im}^2 \text{ and } K_{T\,im}^2=k_T^2+k_{im}^2. \tag{20}$$

The entire procedure, including solution time requirements of finding the roots of Eqs. 18 and 19 is essentially the same. Figure 4 displays the pure imaginary wavenumbers for aluminium plate of c_L = 6300 m/sec, c_T = 3100 m/sec and the thickness equal to 8 mm. The symmetric Lamb modes are represented by red dots, the antisymmetric modes by blue dots.

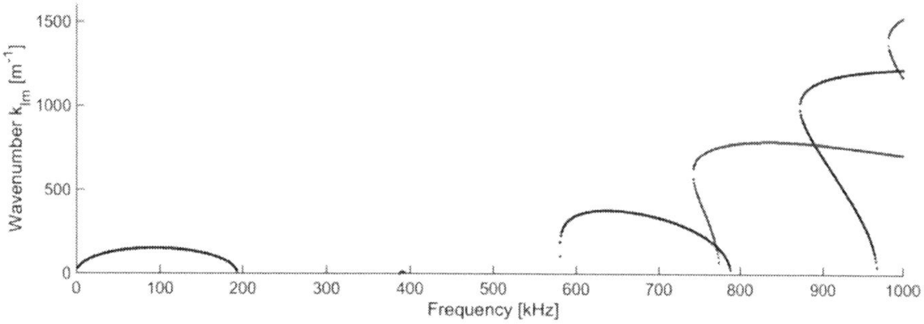

Fig. 4. Dispersion curves in terms of k_{im} as a function of frequency. The symmetric Lamb modes are represented by red dots, the antisymmetric modes by blue dots

3.3 Numerical solution for complex wavenumbers

The complex wavenumbers represent inhomogeneous modes, whose amplitude decreases exponentially with distance. Let´s suppose the complex wavenumber in following form:

$$k_{complex}=k_{real}+\mathrm{i}k_{im}. \tag{21}$$

When substituting Eq. 21 into the exponential representation of vertical displacement as a solution of the 1D wave equation: $u(x,t)=A\mathrm{e}^{\mathrm{i}(kx-\omega t)}$, we'll have:

$$A\mathrm{e}^{\mathrm{i}(k_{complex}x-\omega t)}=A\mathrm{e}^{-k_{im}x}\mathrm{e}^{\mathrm{i}(k_{real}x-\omega t)}, \tag{22}$$

which expresses us the wave travelling from left to right with exponentially decreasing amplitude A according to increasing x-coordinate. Eqs. 9 have to be therefore modified in following way:

$$\begin{aligned}K_L^2&=k_L^2-k_{real}^2+k_{im}^2-2\mathrm{i}k_{real}k_{im},\\K_T^2&=k_T^2-k_{real}^2+k_{im}^2-2\mathrm{i}k_{real}k_{im}.\end{aligned} \tag{23}$$

With respect to the resulting form of the complex components defined by Eq. 23, the solutions for complex wavenumbers, given by Eq. 13 for symmetric modes and Eq. 14 for antisymmetric modes cannot be solved independently for real and imaginary part of the complex wavenumber. Therefore, for a given frequency·thickness product, the left-hand sides of dispersion equations (Eq. 13 and 14) need to be evaluated with respect of two variables – k_{real} and k_{im}.

With Matlab's incorporated *Meshgrid* function, which is also supported by, above mentioned, vectorization property, it is possible to relatively quickly and efficiently obtain the zero crossings for the real and imaginary parts of left-hand side of the dispersion relations for symmetric and antisymmetric modes. Figure 5 displays us the evaluated real part of left-hand side of dispersion equation for symmetric modes for particular value of frequency·thickness product.

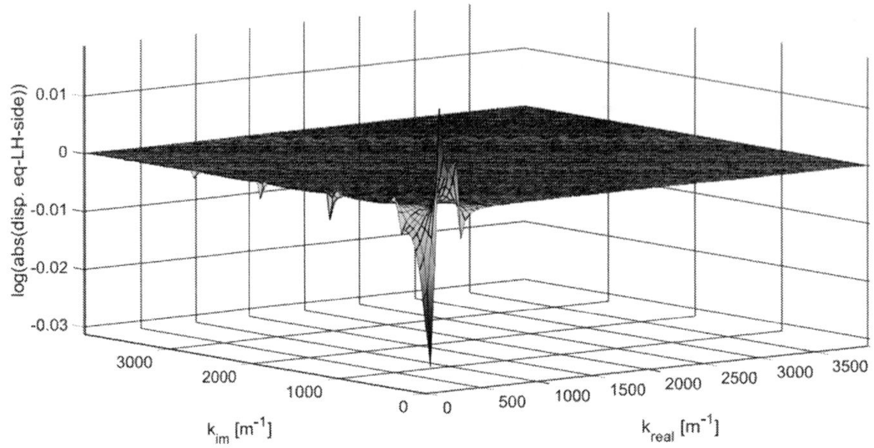

Fig. 5. Real part of left-hand side of dispersion equation (symmetric mode) in case of aluminium plate of c_L = 6300 m/sec, c_T = 3100 m/sec, thickness of 8 mm and frequency equal to 0.2 MHz

The resulting real and imaginary parts of complex wavenumbers are given by the intersections between the solution of the real and imaginary parts of left-hand side of the dispersion relation for particular mode and frequency·thickness product (See Fig. 6). As mentioned before, the solution of the dispersion equation will give an infinite set of complex wavenumbers.

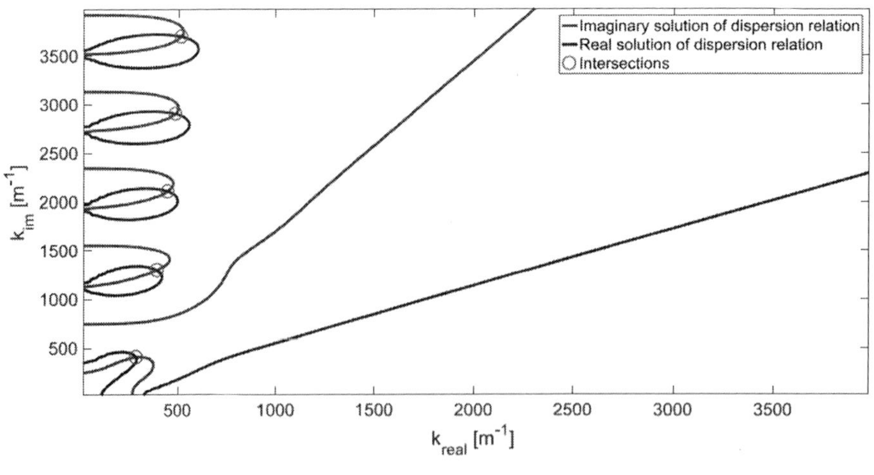

Fig. 6. Real and imaginary parts of complex wavenumbers in case of aluminium plate of c_L = 6300 m/sec, c_T = 3100 m/sec, thickness of 8 mm and frequency equal to 0.2 MHz

In our case, we will present the results for aluminium plate of c_L = 6300 m/sec, c_T = 3100 m/sec, thickness of 8 mm and the frequency interval was $(0; 1\rangle$ MHz, divided into 100 steps with a solving time equal to 9.2 seconds (CPU i7 3.4 GHz, 32 GB RAM). The ranges for real and imaginary parts of complex wavenumbers were within

$\langle 20; 2500 \rangle\ m^{-1}$. The incremental step of wavenumber vector, for both the real and imaginary part, was equal to $max(k_{im/real})\cdot 10^{-3}\ m^{-1}$. Fig. 7 displays the real and the corresponding imaginary parts of complex wavenumbers for above-mentioned input parameters.

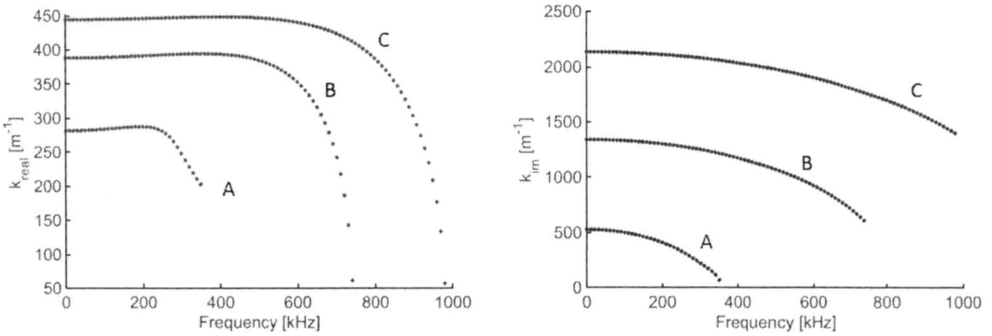

Fig. 7. Real (left) and imaginary (right) parts of complex wavenumbers in case of aluminium plate of $c_L = 6300$ m/sec, $c_T = 3100$ m/sec, thickness of 8 mm and frequency interval of $(0; 1\rangle$ MHz. Capital letters denotes corresponding real and imaginary parts

Conclusion

Dispersion curves are an indispensable tool in case of wave scattering problems [1, 9, 10] or problematics of wave propagation. The main aim of this paper was to present an overall overview of the procedures of finding the real, imaginary and complex wavenumbers for given material parameters and frequency·thickness interval. The procedures of finding the individual types of wavenumbers have been written in Matlab software with incorporation of functions, which support the vectorised operations. Thanks to this fact, there was no extra need to put emphasis on the proper selection of the wavenumber intervals in order to decrease the number of steps, related to the interval range, with the aim of reducing the computational time. The entire set consisting from all three types of wavenumbers can be for given material parameters and frequency range with reasonable division obtained within 20 seconds. Currently, authors are working on development of more time-efficient algorithm in order to apply the results with cooperation of Lamb wave structural health monitoring for various experiments [11-16] and also for further application in case of composite structures [17].

This work was supported by Specific Research (SP2017/106 and SP2017/136) and by The Ministry of Education, Youth and Sports from the National Programme of Sustainability II (LQ1602). The support is acknowledged.

References

1. B. Poddar, V. Giurgiutiu, *Scattering of Lamb waves from a discontinuity: An improved analytical approach*. W. Motion **65**, 79-91 (2016)
2. B. Poddar, V. Giurgiutiu, *Complex modes expansion with vector projection using power flow to simulate Lamb waves scattering from horizontal cracks and disbonds*. J. Ac. Soc. of America **140**, 2123-2133 (2016)
3. J. L. Rose, *Ultrasonic guided waves in solid media*. (Cambridge University Press, 2014)

4. E. Moreno, P. Acevedo, *Thickness measurement in composite materials using Lamb waves*. Ultrasonics **35**, 581-586 (1998)
5. M. Šofer, P. Ferfecki, P. Šofer, *Experimental construction of Lamb wave dispersion curves in plates*. Conference paper, EAN 2017, 55th International Conference on Experimental Stress Analysis (2017)
6. Z. Lašová, R. Zemčík, Determination of group velocities of Lamb waves in *unidirectional carbon-epoxy plate*. Conference paper, EAN 2017 - 55th International Conference on Experimental Stress Analysis (2017)
7. P. Hora, O. Červená, *Determination of Lamb wave dispersion curves by means of Fourier transform*. App. Mech. And Com. Mech. **6**, 5-16 (2012)
8. M. H. Sadd, *Wave motion and vibration in continuous media*. Univ. Of RI, Dept. Of Mech. Engr., Kingston (1990)
9. J. D. Achenbach, *Wave propagation in elastic solids*. New York: North Holland (1984)
10. M. Šofer, P. Ferfecki, P., J. Neugebauer, *Effect of the shape of geometric discontinuity on nature of Rayleigh wave back reflection*. Conference paper, EAN 2016, 54th International Conference on Experimental Stress Analysis (2016)
11. R. Halama, M. Fusek, Z. Poruba, *Influence of mean stress and stress amplitude on uniaxial and biaxial ratcheting of ST52 steel and its prediction by the AbdelKarim–Ohno model*. Int. J. of Fat. **91**, 313-321 (2016)
12. M. Handrik, P. Kopas, V. Baniari, M. Vasko, M. Saga, *Analysis of stress and strain of fatigue specimens localised in the cross-sectional area of the gauge section testing on bi-axial fatigue machine loaded in the high-cycle fatigue region*. Procedia Engineering **177,** 516-519 (2017)
13. P. Kopas, M. Saga, V. Baniari, M. Vasko, M. Handrik, *A plastic strain and stress analysis of bending and torsion fatigue specimens in the low-cycle fatigue region using the finite element methods*. Procedia Engineering **177,** 526-531 (2017)
14. P. Kopas, M. Blatnický, M. Sága, M. Vaško, *Identification of mechanical properties of weld joints of AlMgSi07.F25 aluminium alloy*, Metalurgija **56**, 99 -102 (2017)
15. P. Kopas, L. Jakubovičová, M. Vaško, M. Handrik, *Fatigue resistance of reinforcing steel bars*, Procedia Engineering **136**, 193-197 (2016)
16. V. Baniari, M. Blatnicka, M. Sajgalik, M. Vasko, M. Saga, *Measurement and numerical analyses of residual stress distribution near weld joint*, Procedia Engineering **192**, 22-27 (2017)
17. R. Kocich, L. Kunčická, A. Macháčková, M. Šofer, *Improvement of mechanical and electrical properties of rotary swaged Al-Cu clad composites*. Materials & Design **123**, 137-146 (2017)

Application of the vibration method for damage identification of a beam with a variable cross-sectional area

Izabela Zamorska[1*], *Dawid* Cekus[1], *Mateusz* Miara[1]

[1]Czestochowa University of Technology, Faculty of Mechanical Engineering and Computer Science, Czestochowa, Poland

Abstract. The subject of the paper is an application of the non-destructive vibration method for identifying the location of two cracks occurring in a beam. The vibration method is based on knowledge of a certain number of vibration frequencies of an undamaged element and the knowledge of the same number of vibration frequencies of an element with a defect. The analyzed beam, with a variable cross-sectional area, has been described according to the Bernoulli-Euler theory. To determine the values of free vibration frequencies the analytical solution, with the help of the Green's function method, has been used.

Keywords: Bernoulli-Euler beam, crack, Green's function, identification, non-destructive method, vibration

1 Introduction

The beams, whose geometry and/or material properties change along their length, are important for instance in the design of aircraft, robot arms and tall buildings, where they are used both to reduce weight or volume, and to increase strength and stability. However, in these types of objects even minor damage can cause structural failure and consequently even catastrophe. Therefore, it is very important to continuously control these elements and to detect damage as early as possible. To detect this type of damage, usually non-destructive methods (a visual method [1], a penetration method [2], a magnetic particles method [3], a radiographic method [4], an ultrasonic method [5] and a vibration method [6-9]) are used. If cracks appear, then the system parameters such as the rigidity, the vibration frequency or damping are subjected to change. Selection of the method is due to a number of factors, including the type of material, the size of the test element, and the type of internal or external damage.

In this paper, the vibration method has been used for identifying the location of symmetric cracks in the cantilever beam with a variable cross-sectional area. The vibration method is based on knowledge of a certain number of vibration frequencies of an undamaged element and the knowledge of the same number of vibration frequencies of an element with a defect (the changes in the structure cause the changes of vibration

* Corresponding author: izabela.zamorska@im.pcz.pl

Reviewers: *Robert* Zalewski, *Milan* Sága

frequencies). On this basis, the normalization factors are determined, and then three-dimensional graphs are created to plot contour lines to identify the crack. The number of searched parameters of one or more defects depends on the number of known frequencies. In the case of three-dimensional graphs (knowledge of the first three vibration frequencies), it is possible to identify two parameters of one damage (e.g. position and depth of gap [9]) or, as will be shown in this paper, one parameter (position of defects) of two cracks. The Green's function method [9-11, 13] was used to determine the vibration frequencies of the analyzed beam described according to the Bernoulli-Euler theory.

2 Formulation and solution of the free vibration problem of the beam with a variable cross-sectional area

In the paper, the cantilever beam with a varying cross sectional-area along the length and with two symmetrical cracks (Figure 1) has been analyzed.

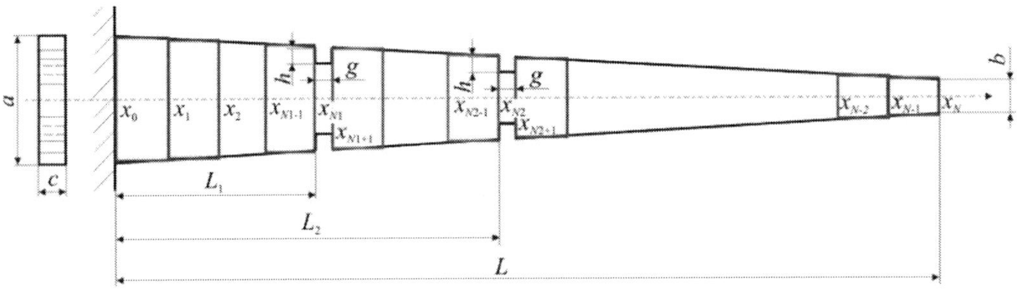

Fig. 1. A scheme of the system under study

Because the identification damage process using the frequency method requires the knowledge of vibration frequencies of an undamaged element and the same number of vibration frequencies of an element with a defect, the analytical formulation and solution of the problem have been carried out.

Each N beam's segment (length $\Delta_i = x_i - x_{i-1}$, where $i=1,\dots,N$, $x_0=0$, $x_N=L$) have the same physical properties and geometrical parameters.

The governing differential equation of motion of i-th system's segment is:

$$\overline{\Lambda}_i\left[Z_i(x,t)\right] = -\overline{S}_{i-1}(t)\delta(x-x_{i-1}) + \overline{S}_i(t)\delta(x-x_i) + \overline{M}_{i-1}(t)\delta'(x-x_{i-1}) - \overline{M}_i(t)\delta'(x-x_i), \quad (1)$$

where: $\overline{\Lambda}_i = EI_i \frac{\partial^4}{\partial x^4} + \rho A_i \frac{\partial^2}{\partial t^2}$ is the linear differential operator, $Z_i(x,t)$ is the transverse displacement, EI_i is the flexural rigidity and ρA_i is the mass per unit length. Moreover, functions $\overline{S}_i$ and $\overline{M}_i$ are the shear force and the bending moment acting on the right end of i-th segment, $\delta(\ )$ and $\delta'(\ )$ are the Dirac's delta and its derivative (the doublet function [12]). The functions Z_1 and Z_N satisfy homogeneous boundary conditions:

$$\overline{\mathbf{B}}_0\left[Z_1(x,t)\right]\Big|_{x=0} = 0, \quad \overline{\mathbf{B}}_1\left[Z_N(x,t)\right]\Big|_{x=L} = 0 \quad (2)$$

and the continuity conditions at the dividing points x_i:

$$Z_i(x,t)\Big|_{x=x_i} = Z_{i+1}(x,t)\Big|_{x=x_i}, \quad Z_{i,x}(x,t)\Big|_{x=x_i} = Z_{i+1,x}(x,t)\Big|_{x=x_i}, \quad i=1,\dots,N-1. \quad (3)$$

Assuming free vibration of the beam, we can separate variables: $Z_i(x,t)=z_i(x)e^{j\omega t}$, $\bar{S}_i(t)=S_i e^{j\omega t}$, $\bar{M}_i(t)=M_i e^{j\omega t}$, where $j^2=-1$ and ω is the free vibration frequency. Next, introducing parameters: $\Omega_i^4=\frac{\rho A_i}{EI_i}\omega^2$, $A_i=A(x_i^*)$, $I_i=I(x_i^*)$ for $x_i^*=\frac{x_i-x_{i-1}}{2}$, $s_i=\frac{S_i}{EI_i}$, $m_i=\frac{M_i}{EI_i}$ and operator $\Lambda_i=\frac{d^4}{dx^4}-\Omega_i^4$, the equation (1) and conditions (2), (3) may be written as follows:

$$\Lambda_i[z_i(x)]=-s_{i-1}\delta(x-x_{i-1})+s_i\delta(x-x_i)+m_{i-1}\delta'(x-x_{i-1})-m_i\delta'(x-x_i), \tag{4}$$

$$\mathbf{B}_0[z_1(x)]\big|_{x=0}=0, \quad \mathbf{B}_1[z_N(x)]\big|_{x=L}=0, \tag{5}$$

$$z_i(x)\big|_{x=x_i}=z_{i+1}(x)\big|_{x=x_i}, \quad z_{i,x}(x)\big|_{x=x_i}=z_{i+1,x}(x)\big|_{x=x_i}, \quad i=1,\dots,N-1. \tag{6}$$

The solution of the problem (4-6) can be obtained with the help of the Green's function method [9-11, 13].

The Green's functions of the operators $\mathbf{\Lambda}_i$ satisfy the non-homogeneous equations $\mathbf{\Lambda}_i[G_i]=\delta(x-\xi)$ and the same boundary and continuous conditions as the transverse displacement functions z_i. Functions G_i can be written as the sum:

$$G_i(x,\xi)=\sum_{k=1}^{4} C_{ik}y_k(x)+\left(2\Omega_i^3\right)^{-1}\left[y_4(x-\xi)-y_2(y-\xi)\right], \tag{7}$$

where: $y_1(x)=\cos(\Omega x)$, $y_2(x)=\sin(\Omega x)$, $y_3(x)=\cosh(\Omega x)$, $y_4(x)=\sinh(\Omega x)$ constitute a system of fundamental solutions to the homogeneous equation $\mathbf{\Lambda}_i[G_i]=0$ and coefficients C_{ik} are obtained from the boundary conditions corresponding to the i-th segment of the beam.

For a clamped-free beam function G_1 satisfies conditions corresponding with a clamped left end and free a right end. In that case we can present coefficients C_{1k} as follows:

$$C_{11}=(W_1)^{-1}\left[\alpha_{21}\gamma_1(x_1,\xi)-\beta_{21}\gamma_2(x_1,\xi)\right],$$

$$C_{12}=(W_1)^{-1}\left[\alpha_{11}\gamma_2(x_1,\xi)+\beta_{11}\gamma_1(x_1,\xi)\right],$$

$$C_{13}=-a_1(0)C_{11}-b_2(0)C_{12},$$

$$C_{14}=b_1(0)C_{11}-a_2(0)C_{12}. \tag{8}$$

The functions $G_2,\dots, G_N$ satisfy conditions corresponding with free-free ends. In that case constants C_{ik} ($i=2,\dots, N$) are:

$$C_{i1}=(W_2)^{-1}\left[\alpha_{22}\gamma_1(x_i,\xi)-\beta_{22}\gamma_2(x_i,\xi)\right],$$

$$C_{i2}=(W_2)^{-1}\left[\alpha_{12}\gamma_2(x_i,\xi)+\beta_{12}\gamma_1(x_i,\xi)\right],$$

$$C_{i3}=a_1(x_{i-1})C_{i1}+b_2(x_{i-1})C_{i2},$$

$$C_{i4} = -b_1(x_{i-1})C_{11} + a_2(x_{i-1})C_{i2}\,. \tag{9}$$

The designations used in the formulas are as follows:

$$a_s(x) = y_1(x)y_3(x) - (-1)^s y_2(x)y_4(x),$$

$$b_s(x) = y_2(x)y_3(x) - (-1)^s y_1(x)y_4(x),$$

$$\alpha_{ps} = \alpha_{ps}(x_{i-1}, x_i) = a_s(x_i) + (-1)^p a_s(x_{i-1}),$$

$$\beta_{ps} = \beta_{ps}(x_{i-1}, x_i) = b_s(x_i) + (-1)^p b_s(x_{i-1}),$$

$$W_s = \alpha_{1s}\alpha_{2s} + \beta_{1s}\beta_{2s}, \qquad p, s = 1,2,$$

$$\gamma_1(x,\xi) = \frac{1}{2\Omega_i^3}\left\{\left[y_2(x-\xi) + y_4(x-\xi)\right]y_3(x) - \left[y_1(x-\xi) + y_3(x-\xi)\right]y_4(x)\right\},$$

$$\gamma_2(x,\xi) = \frac{1}{2\Omega_i^3}\left\{\left[y_1(x-\xi) + y_3(x-\xi)\right]y_3(x) - \left[y_2(x-\xi) + y_4(x-\xi)\right]y_4(x)\right\}. \tag{10}$$

If the Green's function is known, we can determine the solution of (4) in the form [9-11, 13]:

$$z_i(x) = -\mu_{i-1}s_{i-1}G_i(x, x_{i-1}) + s_i G_i(x, x_i) + \mu_{i-1}m_{i-1}G_{i,\xi}(x, x_{i-1}) - m_i G_{i,\xi}(x, x_i), \tag{11}$$

where: $\mu_{i-1} = \dfrac{EI_{i-1}}{EI_i}$ are flexural rigidity distribution coefficients and $s_0 = m_0 = s_N = m_N = 0$. Substituting solution (11) into continuity conditions (6) and introducing functions:

$$P_i(x,\xi) = \mu_{i-1}G_i(x,\xi), \quad Q_i(x,\xi) = G_i(x,\xi) + \mu_i G_{i+1}(x,\xi), \quad R_i(x,\xi) = G_{i+1}(x,\xi), \tag{12}$$

we obtain a linear set of $2N$-2 equations with unknown parameters s_i, m_i (i=2,..., N-1). In a matrix form, it is given as follows:

$$\mathbf{A}(\omega)\mathbf{X} = \mathbf{0}, \tag{13}$$

where a non-zero elements of a main matrix **A** are given below:

$$a_{1,1} = Q_1(x_1, x_1), \qquad a_{1,2} = -R_1(x_1, x_2),$$

$$a_{2,1} = -Q_{1,x}(x_1, x_1), \quad a_{2,2} = R_{1,x}(x_1, x_2),$$

$$a_{1,N} = -Q_{1,\xi}(x_1, x_1), \quad a_{1,N+1} = R_{1,\xi}(x_1, x_2),$$

$$a_{2,N} = -Q_{1,\xi x}(x_1, x_1), \quad a_{2,N+1} = R_{1,\xi x}(x_1, x_2),$$

$$a_{2i-1,i-1} = -P_i(x_i, x_{i-1}),\ a_{2i-1,i} = Q_i(x_i, x_i), \qquad a_{2i-1,i+1} = -R_i(x_i, x_{i+1}), \qquad i=2,\dots,N\text{-}1$$

$$a_{2i,i-1} = P_{i,x}(x_i, x_{i-1}), \quad a_{2i,i} = -Q_{i,x}(x_i, x_i), \quad a_{2i,i+1} = R_{i,x}(x_i, x_{i+1}),$$

$$a_{2i-1,N+i-2} = P_{i,\xi}(x_i, x_{i-1}), \; a_{2i-1,N+i-1} = -Q_{i,\xi}(x_i, x_i), \quad a_{2i-1,N+i} = R_{i,\xi}(x_i, x_{i+1}),$$

$$a_{2i,N+i-2} = P_{i,\xi x}(x_i, x_{i-1}), \; a_{2i,N+i-1} = -Q_{i,\xi x}(x_i, x_i), \quad a_{2i,N+i} = R_{i,\xi x}(x_i, x_{i+1}),$$

$$a_{2N-3,N-2} = -P_{N-1}(x_{N-1}, x_{N-2}), \quad a_{2N-3,N-1} = Q_{N-1}(x_{N-1}, x_{N-1}),$$

$$a_{2N-2,N-2} = P_{N-1,x}(x_{N-1}, x_{N-2}), \quad a_{2N-2,N-1} = -Q_{N-1,x}(x_{N-1}, x_{N-1}),$$

$$a_{2N-3,2N-3} = P_{N-1,\xi}(x_{N-1}, x_{N-2}), \; a_{2N-3,2N-2} = -Q_{N-1,\xi}(x_{N-1}, x_{N-1}),$$

$$a_{2N-2,N-3} = P_{n-1,\xi x}(x_{N-1}, x_{N-2}), \; a_{2N-2,2N-2} = -Q_{N-1,\xi x}(x_{N-1}, x_{N-1}). \tag{14}$$

The frequency equation det**A**(ω)=0 is solved numerically with respect to the frequency ω. The mode shapes corresponding to the multi-stepped beam's frequency are in the form (11), where unknown vector $\mathbf{X}=[s_1 \ldots s_{N-1}\; m_1 \ldots m_{N-1}]^T$ is determined from (13), assuming the value of one of the coefficients s_i, m_i (for example $m_{N-1}=1$).

3 Sample numerical results of identification of the cracks' position

Using the presented model, the algorithm and computer program that enable the determination of the free vibration frequencies of the cantilever beam with a variable cross-sectional area have been worked out.

The exemplary results of numerical calculations illustrate the identification of distance (L_1, L_2) of cracks from the attachment. The following values of parameters have been accepted: a = 75 mm, b = 20 mm, c = 5 mm, g = 1 mm, h = 10 mm and L = 500 mm.

On the basis of the determined first three natural frequencies, the normalization process has been performed. In this case, the normalization is defined as the ratio of the frequency of the damaged beam (ω_u) to the frequency of the beam without defect (ω_n):

$$\lambda = \frac{\omega_u}{\omega_n} \tag{15}$$

Based on the normalization factors, the three-dimensional graphs are created. In Figures 2-4, the changes of the values of the first three normalized vibration frequencies depending on the location of two cracks and sample contour lines corresponding to them (used later in the identification process) are shown. On the basis of three-dimensional graphs, the individual contour lines for each examined frequency having the same parameters of the damage must be plotted (Figs. 5-7). As an example, in Figure 8, the contour lines for the cracks located at the beginning of the beam (first) and close to the free end of the beam (second) were plotted.

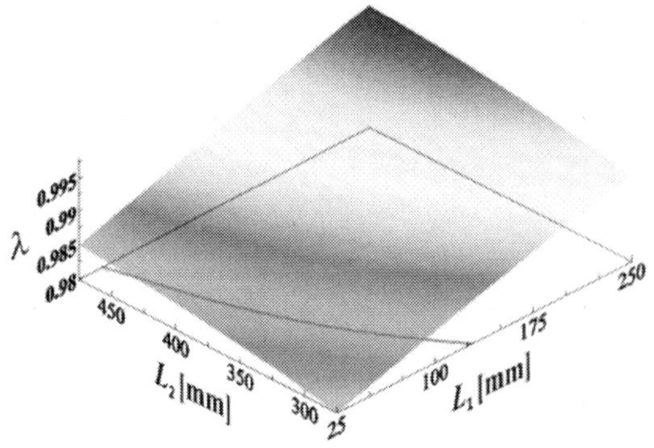

Fig. 2. The first normalized vibration frequency depending on the location of two cracks

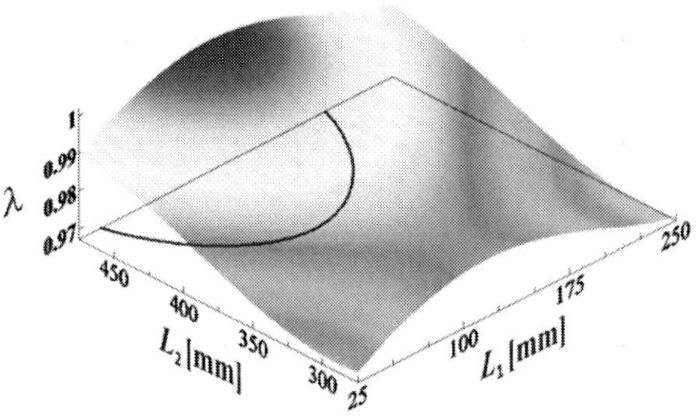

Fig. 3. The second normalized vibration frequency depending on the location of two cracks

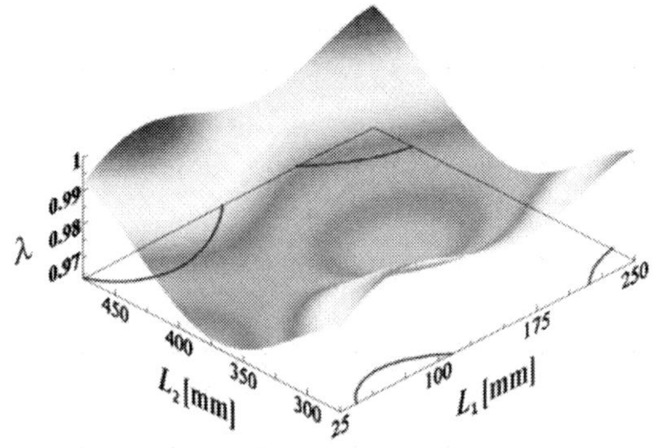

Fig. 4. The third normalized vibration frequency depending on the location of two cracks

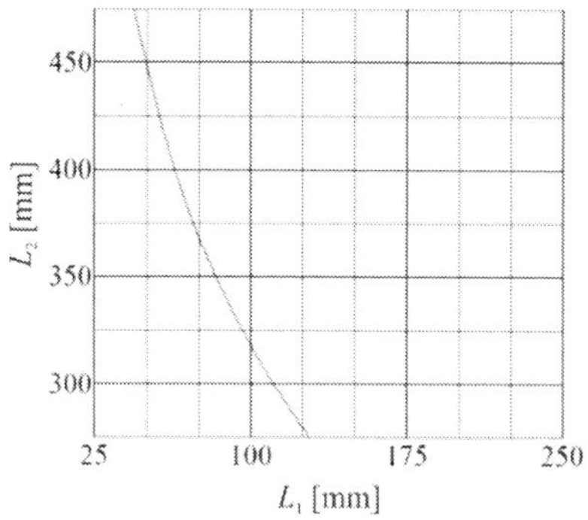

Fig. 5. Contour line for the first normalized vibration frequency

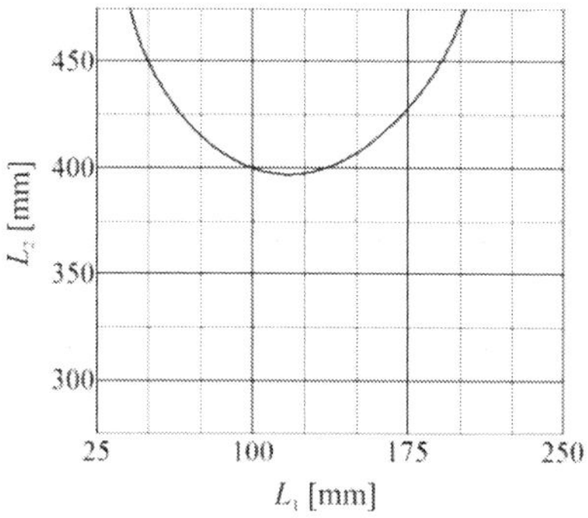

Fig. 6. Contour lines for the second normalized vibration frequency

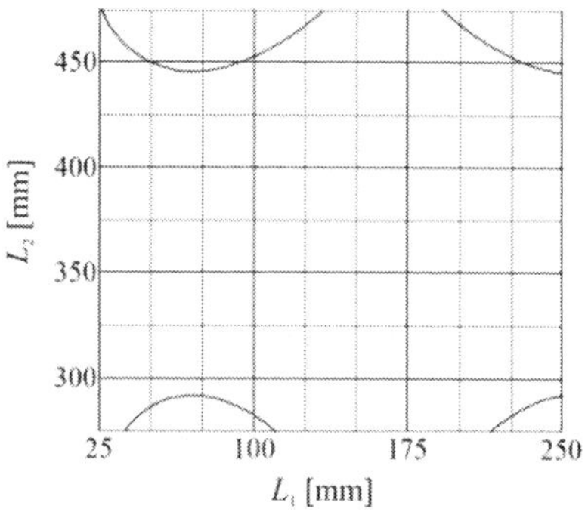

Fig. 7. Contour lines for the third normalized vibration frequency

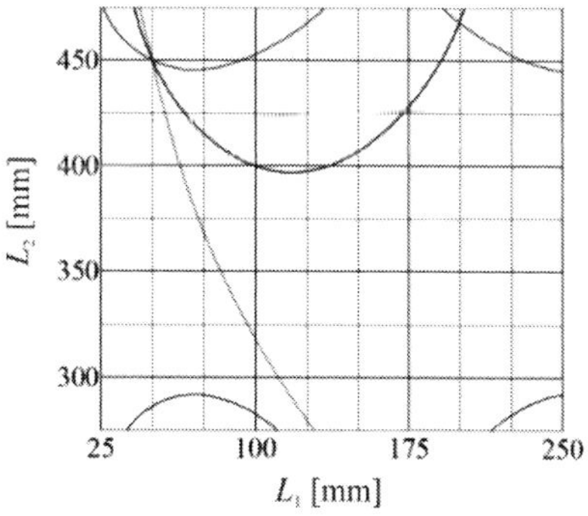

Fig. 8. The identified position of cracks

As a result of the intersection of the contour lines (Figure 8), the point defining the places where the cracks are located was determined. The distances of the cracks from the attachment obtained in the identification process (L_{1d}=49.7 mm and L_{2d}=450.2 mm) practically coincide with the values (L_{1a}=50 mm, L_{2a}=450 mm) assumed during the numerical simulations. To verify the accuracy of the discussed identification process, the cases specified in Table 1 were additionally executed. The relative error defines the difference between the assumed positions of cracks (L_{1a}, L_{2a}) and the values calculated in the identification process (L_{1d}, L_{2d}):

$$\Delta L = \left(\frac{|L_{1a} - L_{1d}|}{2L_{1d}} + \frac{|L_{2a} - L_{2d}|}{2L_{2d}} \right) \cdot 100\% \tag{16}$$

Table 1. Positions of cracks obtained with the help of the non-destructive vibration method

Assumed positions of cracks		Determined positions of cracks		The relative error ΔL [%]
L_{1a} [mm]	L_{2a} [mm]	L_{1d} [mm]	L_{2d} [mm]	
50	300	49.5	299.8	0.54
125	300	123	302	1.14
225	300	224.7	299.5	0.15
50	375	45	375.2	5.58
125	375	132	376.5	2.85
225	375	222.6	373	0.81
50	450	49.7	450.2	0.32
125	450	125.5	450.8	0.29
225	450	225.8	450.5	0.23

Based on the results compiled in Table 1, it can be seen that the relative error in most cases does not exceed 3%. The largest value of the relative error is for damages occurring at points L_{1a}=50 mm and L_{2a}=375 mm. The smallest relative error (0.15%) is the case where the cracks are closest to each other in relation to all considered instances. Moreover, for L_2=375 mm or L_2=450 mm and with an increase L_1 the error decreases.

Conclusions

In the present work, the identification of the location of two symmetrical cracks occurring in the cantilever beam with a variable cross-sectional area has been done. For identification, the non-destructive vibration method has been applied. This method requires knowledge of the frequencies of the system without and with the damage. The solution of the free vibration problem of the beam described according to the Bernoulli-Euler theory has been obtained with the help of Green's functions.

In the computational models, a rigid restraint has been used. However, this type of boundary condition is impossible to achieve in real objects. Therefore, mathematical models should be experimentally verified. In the case when the system response diverges from the theoretical results, then an identification of the model must be performed [14].

Although in this study only exemplary numerical research has been carried out, this method can be successfully used for the identification of damage of real objects. Identifying structure damage at an early stage has a great influence on the repair cost and has a significant impact on safety.

References

1. T. Yamaguchi, S. Hashimoto, *Fast crack detection method for large-size concrete surface images using percolation-based image processing*. Machine Vision and Applications **21** (5), 797–809 (2010).
2. H. Miyazaki, Y. Nomura, H. Sugai, M. Iijima, S. Inasawa, H. Kamiya, *Liquid penetration as a simple detection method for structural differences in particulate films prepared from slurries*. Powder Technology **303**, 59–67 (2016)
3. X. Li, H. Jiang, G. Yin, *Detection of surface crack defects on ferrite magnetic tile*, Journal of Sound and Vibration **375** (4), 200–216 (2016)
4. N. Boaretto, T.M. Centeno, *Automated detection of welding defects in pipelines from radiographic images DWDI*. NDT & E International, http://dx.doi.org/10.1016/j.ndteint.2016.11.003
5. S. Mezil, N. Chigarev, V. Tournat, V. Gusev, *Evaluation of crack parameters by a nonlinear frequency-mixing laser ultrasonics method*. Ultrasonics **69**, 225–235 (2016)
6. G.M. Owolabi, A.S.J. Swamidas, R. *Seshadri, Crack detection in beams using changes in frequencies and amplitudes of frequency response functions*. Journal of Sound and Vibration **265**, 1-22 (2003)
7. P. Yamuna, K. Sambasivarao, *Vibration Analysis of Beam With Varying Crack Location*. International Journal of Engineering Research and General Science **2** (6), ISSN 2091-2730, (2014)
8. K. Sokół, *Linear and nonlinear vibrations of a column with an internal crack*. J. Eng. Mech **140** (5), 04014021, http://dx.doi.org/10.1061/(ASCE)EM.1943-7889.0000719 (2014)
9. D. Cekus, M. Miara, I. Zamorska, *Damage identification of a beam with a variable cross-sectional area*. Journal of Mathematics and Computational Mechanics **15** (4), 23-32 (2016)
10. S. Kukla, I. Zamojska, *Frequency analysis of axially loaded stepped beams by Green's function method*. Journal of Sound and Vibration **300**, 1034-1041 (2007)
11. S. Kukla, *Dynamiczne funkcje Greena w analizie drgań własnych ciągłych i dyskretno-ciągłych układów mechanicznych*. seria Monografie **64**, Wydawnictwo Politechniki Częstochowskiej (1999)
12. J. Mikusiński, R. Sikorski, *Elementarna teoria dystrybucji*. PWN, Warszawa (1964)
13. J. Lee, L.A. Bergman, *The vibration of stepped beams and rectangular plates by an elemental dynamic flexibility method*. Journal of Sound and Vibration **171**, 617-640 (1994)
14. D. Cekus, P. Waryś: *Identification of parameters of discrete-continuous models*. AIP Conf. Proc. 1648 , 850055 (2015)

Load of the slipper-swash plate kinematic pair of an axial piston pump

Tadeusz Złoto[1,*], *Konrad* Kowalski[1]
[1]Czestochowa University of Technology,Institute of Mechanical Technologies, Armii Krajowej 21, Czestochowa,Poland

Abstract. The paper presents problems related to the twisting moment of the slipper. The load of the slipper and the piston has been presented and the complex formula of twisting moment of the slipper has been established. Achieved results has been presented graphically. The conducted research has indicated that the value of the twisting moment relays on both the exploitation and geometrical parameters.

Keywords: load, slipper-swash plate, axial piston pump, hydrostatic relief, twisting moment

1 Introduction

Axial piston pumps have many applications in different industrial hydraulic systems and can be used at the highest pressures and powers. Axial piston pumps are expensive, but provide exceptional performance and high efficiencies. The reliability and durability of an axial piston pump depends significantly on the condition of hydrostatic supports. Hydrostatic slipper of an axial piston pump is of relatively small size and transfers large loads in tough conditions. The swash plate is inclined at a certain angle with respect to the axis of piston symmetry. Because the piston load is located off the axis, the presence of friction and inertia forces can turn the slipper over.

A satisfactory scientific mathematical model of the whole process of designing hydrostatic slippers has been missed so far. Pump manufacturers must therefore rely on laborious and time-consuming empirical studies [1, 2]. The increasing demand for new types of pumps and for improving their performance necessitates developing universal and reliable constructions of high efficiency.

2 The piston load model

During normal operation of a piston-cylinder unit in a swash-plate axial piston pump, the hydrostatic slipper is loaded by the radial force coming from a discharge pressure (fig. 1).

* Corresponding author: zlotot@o2.pl

Reviewers: *Zuzana* Murčinková, *Jaroslav* Zapoměl

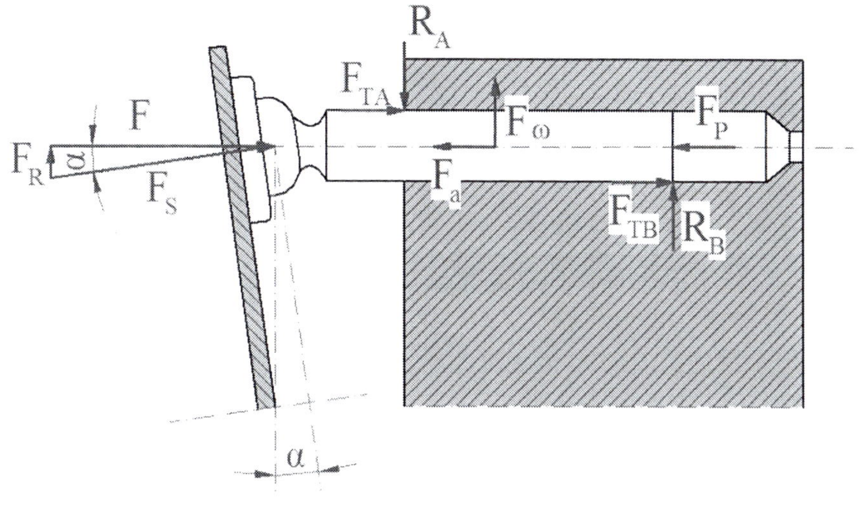

Fig. 1. Force distribution in a piston-cylinder unit

The radial force F_R results in inclined position of the piston and therefore results in reactions R_A and R_B of large value [3]. As an effect, friction force F_T occurs. This friction force was obtained from [4, 5]:

$$F_T = \mu(R_A + R_B) \tag{1}$$

where μ - friction coefficient between the piston and the cylinder.

In calculation model there have been assumed the following forces: force F_p coming from the discharge pressure, dynamic force F_a that occurs as a result of velocity changes of a reciprocating piston unit, centrifugal inertia force F_ω coming from rotation of the piston unit round the rotor axis, spring force F_{spr} pushing the piston unit to the swash plate, force F_s pressing the slipper against the swash plate, friction forces between the piston and the μR_A and μR_B and friction force F_{TS} between the slipper and the swash plate. Reactions R_A and R_B were assumed as passive forces.

3 Hydrostatic relief of the slipper

The slipper performs the task of transferring all the axial forces acting upon the piston system, and it is desirable to minimize energy losses and wear-out, and to maximize reliability within the full range of the piston pump reliability [6, 7]. The slipper surface can have various shapes [8]. Regardless of the shape, there should be a central neck connecting the slipper with the piston pressure chamber.

Standard hydrostatic slipper (Fig. 2) has an internal and external radius. The geometrical parameters of the hydrostatic slipper should be selected in such a way that the load carrying force resulting from the hydrostatic pressure field takes over 90÷95% of the normal load [8] and the hydrodynamic force takes over the other 5÷10% of the load. In the derivation of the formula for the pressure change on the slipper sealing surface it is assumed that the mating surfaces are ideally smooth, the flow is laminar and isothermal. The pressure in a typical hydrostatic slipper can be described by the following formula:

$$p = p_o \frac{\ln \frac{r_2}{r}}{\ln \frac{r_2}{r_1}} \tag{2}$$

where:
p, p_0 - the pressure along the slipper radius and in the slipper chamber, respectively,
r_2, r, r_1 - the external, current and internal radius of the slipper, respectively.
The repulsive force of a typical hydrostatic slipper is [9]:

$$F_o = p_o \frac{\pi}{2} \cdot \frac{r_2^2 - r_1^2}{\ln \frac{r_2}{r_1}} \tag{3}$$

The slipper construction affects power losses. The total power loss consists of the power lost due to leakage and friction loss. The gap height h_s (Fig. 2) is of the greatest importance. With the increase in h_s the mechanical losses decrease, the volumetric losses, however, dramatically increase. Mechanical losses can be reduced by reducing the surfaces surrounding the gap. Another way to reduce mechanical losses is to use slippers with additional bearing surfaces [10], that enable significant hydrodynamic effects on a flat surface.

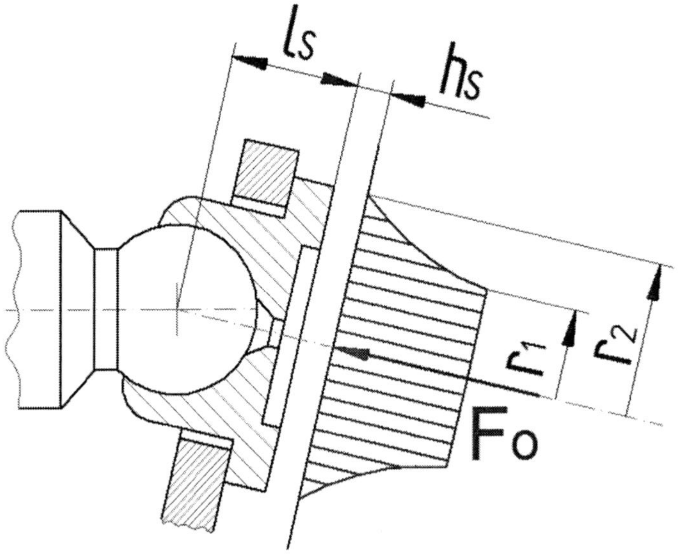

Fig. 2. Hydrostatic relief of the slipper

It is difficult to design an axial piston pump of variable yield, in which it is necessary to transfer variable loads, resulting from the changes in the inclination angle a of the swash plate.

4 Simulation of a slipper's load

During operation the slipper is under the effect of twisting moment, which is trying to turn the slipper over.

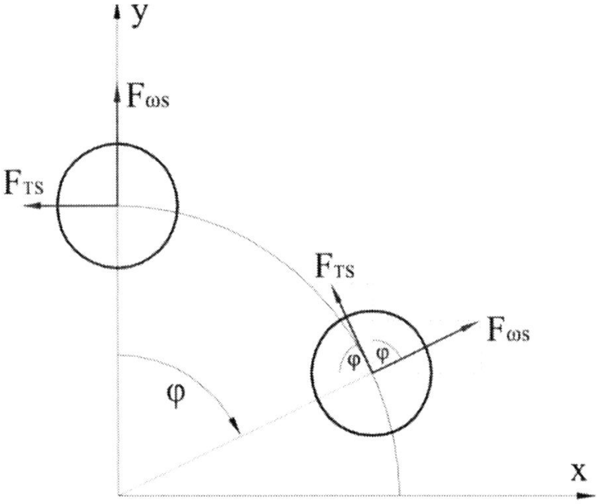

Fig. 3. Forces acting on the slipper

Fig 3 presents friction force F_{TS} between the slipper and the swash plate and the centrifugal inertia force $F_{\omega s}$ acting upon the slipper. The friction force is obtained by multiplication of pressing force (difference between the normal force F_s and the relief force F_o of the slipper) and the friction coefficient μ_s:

$$F_{TS} = \mu_s \left[\frac{0.25\pi d^2 p_1 + \mu_T (R_A + R_B) + F_{spr} + m_C \omega^2 r_p \tan\alpha \cos\varphi}{\cos\alpha} - C_p \frac{\pi p_1}{2} \left(\frac{r_2^2 - r_1^2}{\ln \frac{r_2}{r_1}} \right) \right] \quad (4)$$

where C_p - the pressure drop between the cylinder and the slipper chamber.
Centrifugal inertia force can be obtained from:

$$F_{\omega s} = m_s \left(\frac{\omega \cos\alpha}{\cos^2\varphi + \cos^2\alpha \cdot \sin^2\varphi} \right)^2 r_p \sqrt{1 + \tan^2\alpha \cdot \cos^2\varphi} \quad (5)$$

The resultant force acting upon the slipper was obtained as a geometrical sum of the projections of the component forces in the set of coordinates xy:

$$F_W = \sqrt{F_{WX}^2 + F_{WY}^2} \quad (6)$$

After substituting and bringing inertia force onto the plane on which the slipper and the swash plate mate:

$$F_W = \sqrt{\left(-F_{TS}\cos\varphi + F_{\omega s}\frac{l_s}{l_{sc}}\sin\varphi \right)^2 + \left(F_{TS}\sin\varphi + F_{\omega s}\frac{l_s}{l_{sc}}\cos\varphi \right)^2} \quad (7)$$

where l_s - the distance between the coupling centre and the mating surface, l_{sc} - the distance between the gravity centre and the mating surface.
The slipper's twisting moment is as follows:

$$M_s = l_s \cdot F_W \quad (8)$$

Applying formula 8 an influence of discharge pressure, friction coefficient between the slipper and the swash plate, mass of the slipper and the angular velocity, on value of the twisting moment M_s of the slipper were analysed. The following initial data were assumed in the computations:

- discharge pressure $p_t = p_I = 32$ MPa,
- friction coefficient $\mu_s = 0.003$ [11],
- friction coefficient $\mu_T = 0.1$ [6],
- piston diameter $d = 0.015$ m,
- mass of the slipper $m_s = 0.025$ kg,
- mass of the slipper and the piston $m_c = 0.07$ kg,
- spring force $F_{spr} = 75$ N,
- angular velocity $\omega = 150$ rad/s,
- the pressure drop of 10% between the cylinder and the slipper operating $\rightarrow C_p = 0.9$ [8],
- the value of reaction R_A and R_B depends on the cylinder block rotation angle φ [12].

The achieved results are shown in Fig. 4. The conducted analyses has shown that the angular velocity ω of the cylinder block has the highest influence on value of the twisting moment M_s of the slipper. With the increase of the angular velocity, the twisting moment increases. Significant impact on twisting moment has the slipper mass. If the slipper mass is higher, the twisting moment is higher too. Furthermore, friction coefficient between the slipper and the swash plate and the discharge pressure has smaller influence on the twisting moment of the slipper.

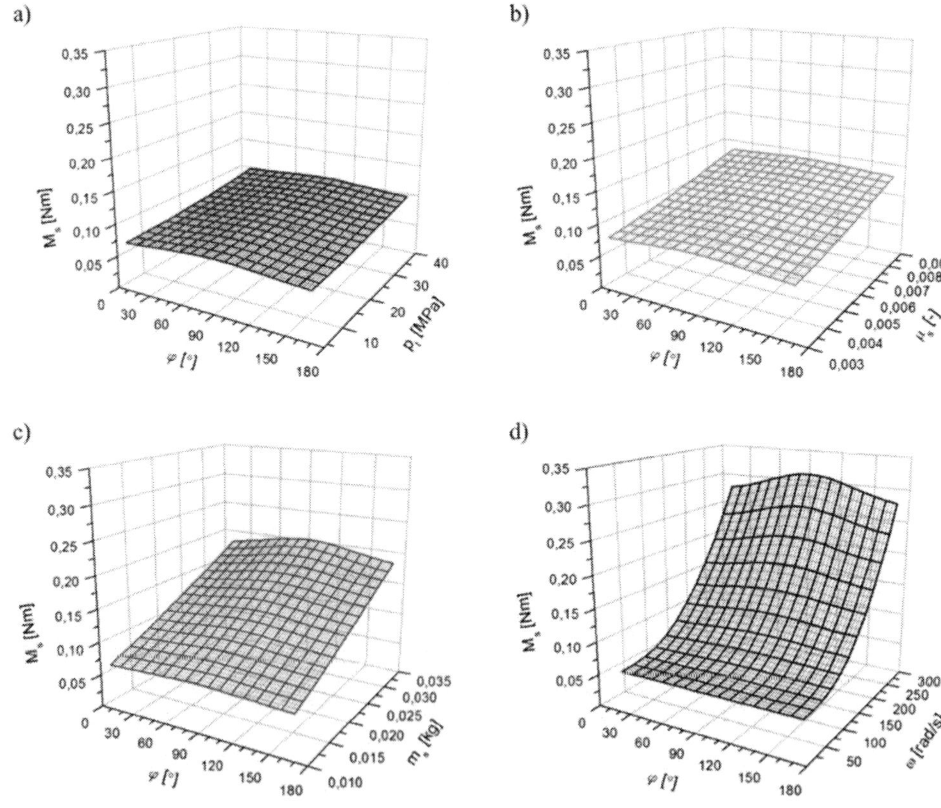

Fig. 4. Twisting moment M_s of the slipper in the pressure zone depending on: a) discharge pressure, b) friction coefficient between the slipper and the swash plate, c) mass of the slipper, d) angular velocity of the cylinder block

Twisting moment M_s has negative influence on the operation of axial piston pumps, and deteriorates the mating conditions of the slipper and the swash plate. The conducted research has shown that the twisting moment tends to turn the slipper over.
Fig. 5 presents an example of damages on the slipper surface because the oil film between the slipper and the swash plate has disappeared. The pump was turn on periodically. The reason of this destruction is the wrong installation of the pump, with leak drain channel on the wrong side of the pump. Each time the pump was turn on, especially after longer period of being switched off, the pump worked dry, and the twisting moment turned over all slippers, pressing them onto the swash plate.

Fig. 5. Hydrostatic slippers of an axial piston pump destroyed due to metal contact with a swash plate surface

Conclusions

The hydrostatic slipper of pistons machines transfers large loads and is loaded by the twisting moment of significant value. The angular speed of the cylinder block has the highest influence on the value of the twisting moment of the slipper. Apart from this, significant influence on the value of the twisting moment have friction coefficient between the slipper and the swash-plate, mass of the slipper and the discharge pressure. Another design parameters like mass of a piston unit and spring force, as well as exploitation parameters like the inclination angle of a swash plate and friction force between piston and the cylinder have small influence on the value of the twisting moment of the slipper.

References

1. E. Brzuchowski, H. Kertyńska, D. Kraszewski, *The results of model research and analysis of the influence of friction on the load state of hydrostatic bearing elements.* (In Polish : *Wyniki badań modelowych oraz analiza wpływu tarcia na stan obciążenia elementów łożysk hydrostatycznych*). Przeglad Mechaniczny **21**, (1974)
2. K.T. Renius, *Experimental investigations on sliding shoes of axial piston machines* (In German : *Experymentelle Untersuchungen an Gleitschuhen von Axialkolbenmaschinen*). Ölhydraulik und Pneumatik **3**, (1973)
3. A. Osiecki, *Hydrostatic machines drive* (In Polish : *Hydrostatyczny napęd maszyn*). WNT, Warszawa (1998)
4. Ch. Kögl, *Adjustable hydrostatic displacement units in the speed and torque control circuit on the mains with adapted supply pressure* (In German : *Verstellbare hydrostatische Verdrängereinheiten im Drehzahl- und Drehmomentregelkreis am Netz mit angepaßtem Versorgungsdruck*). Dissertation RWTH, Aachen (1995)
5. S. Stryczek, *Hydrostatic drive* (In Polish : *Napęd hydrostatyczny*). WNT, Warszawa, **1**, (1995)
6. W. Balas, *Hydrostatic bearings in axial piston pumps* (In Polish : *Łożyska hydrostatyczne w osiowych pompach tłokowych*). Przegląd Mechaniczny **11**, (1976)
7. A. Osiecki, Osiecki L., *Development works on a new construction of axial piston pumps* (In Polish : *Prace rozwojowe nad nową konstrukcją pomp wielotłoczkowych osiowych*). Przegląd Mechaniczny **4**, (1998)
8. J. Ivantysyn, M. Ivantysynova, *Hydrostatic Pumps and Motors*. Akademia Books International, New Delhi (2001)
9. Z. Szydelski, J. Olechowicz, *Elements of hydraulic and pneumatic drive and control* (In Polish : *Elementy napędu i sterowania hydraulicznego i pneumatycznego*). WNT, Warszawa (1986)
10. H. Lachner, *Hydrostatic bearings in axial piston machines* (In Polish : *Hydrostatische Lagerungen in Axialkolbenmasinen*). Ölhydraulik und Pneumatik **8**, (1974)
11. J. Niegoda, *Study of the possibility of using pistons with non-articulated hydrostatic support in axial multi-piston pumps and motors* (In Polish : *Badanie możliwości zastosowania tłoków z bezprzegubowym podparciem hydrostatycznym w pompach i silnikach wielotłoczkowych osiowych*). Dysertacja, Gdańsk (1977)
12. T. Złoto, *Modeling of hydrostatic relief and analysis of flow phenomena in the axial displacement of the axial piston pump* (In Polish : *Modelowanie odciążenia hydrostatycznego i analiza zjawisk przepływowych w szczelinie rozrządu tarczowego pompy wielotłoczkowej osiowej*). Seria monografie 133, Częstochowa (2007)

AUTHOR INDEX

AUTHOR INDEX

AUTHOR INDEX